MW01090238

McDougal Littell

ALGEBRA 1

TEACHER'S EDITION

Ron Larson
Laurie Boswell
Timothy D. Kanold
Lee Stiff

McDougal Littell
A HOUGHTON MIFFLIN COMPANY
Evanston, Illinois • Boston • Dallas

CONTENTS

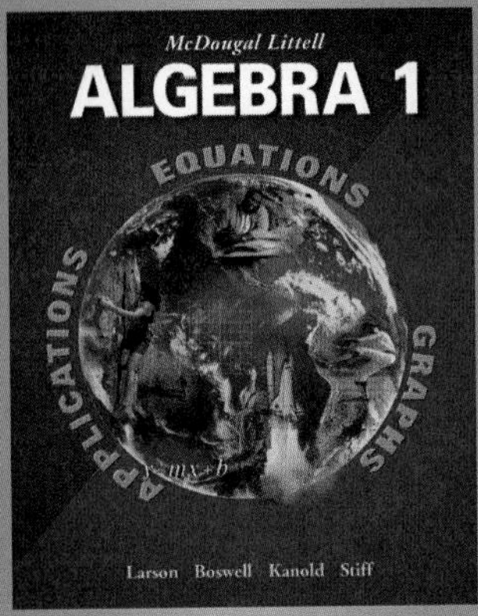

About the Cover

Algebra 1 brings math to life with many real-life applications. The cover illustrates some of the applications used in this book. Examples of mathematics in baseball, space exploration, mountain climbing, and graphic design are shown on pages 8, 66, 158, 175, 184, 304, 318, 501, and 695. Circling the globe are three key aspects of Algebra 1—the *equations, graphs*, and *applications* that you will use in this course will help you understand how mathematics relates to the world. As you explore the applications presented in the book, try to make connections between mathematics and the world around you!

Contributing Authors and Reviewers

The authors wish to thank the following individuals for their contributions.

Universal Access (*pp. T44–T47*) — Catherine Barkett, Vice President, Director of Training, Calabash Professional Learning Systems, Sacramento, CA

English Learners (*pp. T48–T49*) — Olga Bautista, Vice Principal, Will C. Wood Middle School, Sacramento, CA

Reading, Writing, Notetaking; CRISS (*pp. T53–T55*) — Joan Smathers, National CRISS Trainer, Former Language Arts Supervisor, Brevard County, FL; Marie Pettet, Reading and Language Arts Curriculum Specialist, Former K–12 Language Arts Coordinator, Fulton County, GA

ISBN: 0-618-25019-0 3456789–DWO–07 06 05 04

Internet Web Site: http://www.classzone.com

About the Authors

Ron Larson is a professor of mathematics at Penn State University at Erie. He is the author of a broad range of mathematics textbooks for middle school, high school, and college students. He is one of the pioneers in the use of multimedia and the Internet to enhance the learning of mathematics. Dr. Larson is a member of the National Council of Teachers of Mathematics and is a frequent speaker at NCTM and other national and regional mathematics meetings.

Laurie Boswell is a mathematics teacher at Profile Junior-Senior High School in Bethlehem, New Hampshire. She is active in NCTM and local mathematics organizations. A recipient of the 1986 Presidential Award for Excellence in Mathematics Teaching, she is also the 1992 Tandy Technology Scholar and the 1991 recipient of the Richard Balomenos Mathematics Education Service Award presented by the New Hampshire Association of Teachers of Mathematics.

Timothy D. Kanold is Director of Mathematics and a teacher at Adlai E. Stevenson High School in Lincolnshire, IL. In 1995 he received the Award of Excellence from the Illinois State Board of Education for outstanding contributions to education. He served on NCTM's Professional Standards for Teaching Mathematics Commission. A 1986 recipient of the Presidential Award for Excellence in Mathematics Teaching, he served as president of the Council of Presidential Awardees of Mathematics.

Lee Stiff is a professor of mathematics education in the College of Education and Psychology of North Carolina State University at Raleigh and has taught mathematics at the high school and middle school levels. He served on the NCTM Board of Directors and was elected President of NCTM for the years 2000–2002. He is the 1992 recipient of the W. W. Rankin Award for Excellence in Mathematics Education presented by the North Carolina Council of Teachers of Mathematics.

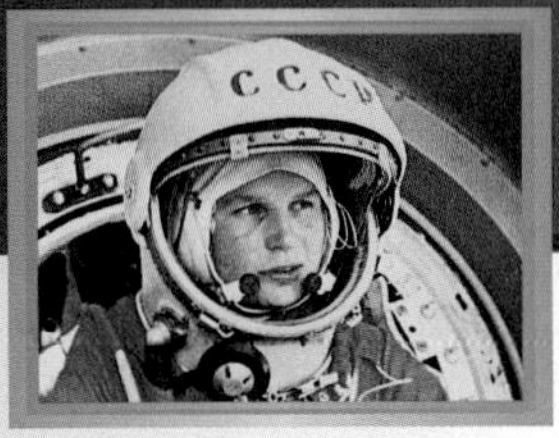

REVIEWERS

Kay S. Cassidy
Mathematics Facilitator
Cactus High School
Glendale, AZ

José Castro
Mathematics Teacher
Liberty High School
Bakersfield, CA

Tammy Parker Davis
Mathematics Teacher
Valdosta High School
Valdosta, GA

Edna Horton Flores
Mathematics Department Chair
Kenwood Academy
Chicago, IL

Doyt M. Jones
Mathematics Department Head
John Bartram High School
Philadelphia, PA

Warren Zarrell
Mathematics Department Chair
James Monroe High School
North Hills, CA

CALIFORNIA TEACHER PANEL

Courteney Dawe
Mathematics Teacher
Placerita Junior High School
Valencia, CA

Dave Dempster
Mathematics Teacher
Temecula Valley High School
Temecula, CA

Pauline Embree
Mathematics Department Chair
Rancho San Joaquin Middle School
Irvine, CA

Tom Griffith
Mathematics Teacher
Scripps Ranch High School
San Diego, CA

Diego Gutierrez
Mathematics Teacher
Crawford High School
San Diego, CA

Roger Hitchcock
Mathematics Teacher
Buchanan High School
Clovis, CA

Joseph Jacobs
Mathematics Teacher
Luther Burbank High School
Sacramento, CA

Louise McComas
Mathematics Teacher
Fremont High School
Sunnyvale, CA

Viola Okoro
Mathematics Teacher
Laguna Creek High School
Elk Grove, CA

Jon Simon
Mathematics Department Chair
Casa Grande High School
Petaluma, CA

CLEVELAND TEACHER PANEL

Laura Anfang
University Supervisor, Teacher of Secondary Math Methods
John Carroll University
University Heights, OH

Patricia Benedict
Mathematics Department Liaison
Cleveland Heights High School
Cleveland Heights, OH

Carol Caroff
Mathematics Department Chair/Teacher
Solon High School
Solon, OH

Fred Dillon
Mathematics Teacher
Strongsville High School
Strongsville, OH

Bill Hunt
Past President, Ohio Council of Teachers of Mathematics
Strongsville, OH

Dr. Margie Raub Hunt
Executive Director, Ohio Council of Teachers of Mathematics
Strongsville, OH

Robert Jones
Mathematics Supervisor (K–12)
Cleveland City School District
Cleveland, OH

Andrea Kopco
Mathematics Teacher
Midpark High School
Cleveland, OH

Sandy Sikorski
Mathematics Teacher
Berea High School
Berea, OH

Gil Stevens
Mathematics Teacher
Brunswick High School
Brunswick, OH

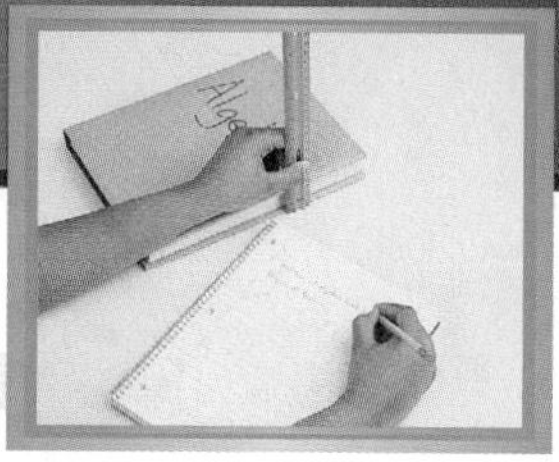

KANSAS TEACHER PANEL

Rosemary Arb
Mathematics Department Chair
North Kansas City High School
Kansas City, MO

Jerry Belshe
Mathematics Teacher
Indian Woods Middle School
Overland Park, KS

Carol Edwards
Mathematics Teacher
Belton High School
Belton, MO

Robert Franks
Mathematics Teacher
Park Hill High School
Kansas City, MO

Cynthia Hardy
Mathematics Teacher
Oxford Middle School
Overland Park, KS

Julie Knittle
Secondary Mathematics Resource Specialist
Shawnee Mission District Office
Shawnee Mission, KS

Glenda Morrison
Mathematics Teacher
Central High School
Kansas City, MO

Mitch Shea
Mathematics Teacher
Leavenworth High School
Leavenworth, KS

Ginny Taylor
Mathematics Department Chair
Santa Fe Trail Junior High School
Olathe, KS

STUDENT REVIEW PANEL

Luke Aguirre
Lincoln Middle School
Illinois

R-Jay Albano
James Lick High School
California

Fred Anthony III
Longfellow Junior High School
Michigan

Curtis Antoine
Springfield Central High School
Massachusetts

Yuri Awanohara
Tilden Middle School
Maryland

Gill Beck
Milton High School
Georgia

Kevin Berghamer
John W. North High School
California

Jared W. Billings
Meadow Park Middle School
Oregon

James Bottigliero
Jonathan Law High School
Connecticut

Lindsay Chapman
Strongsville High School
Ohio

Alison Davis
Hinsdale Middle School
Illinois

Emily Dougherty
Huron High School
Michigan

Roxanne Ilano
Gahr High School
California

Brad Larsen
Thomas Jefferson Junior High School
Utah

Wendy Mathis
Carlsbad High School
New Mexico

Agnes Mazurek
Streamwood High School
Illinois

Calvin Medious
Jim Hill High School
Mississippi

Megan Ann Merrifield
Douglas County Comprehensive High School
Georgia

Julie Nguyen
Lincoln High School
California

Omar Ramos
University High School
New Jersey

Natalie Rock
Les Bois Junior High School
Idaho

Elizabeth Sabol
Park Forest Middle School
Pennsylvania

Divino S. San Pedro, Jr.
Laguna Creek High School
California

Kristen Sands
Taunton High School
Massachusetts

Chris Sarabia
Buchanan High School
California

Jonathan Tin
Woodinville High School
Washington

Virginia Thao Tran
Fountain Valley High School
California

Jenny Waxler
Parkway North High School
Missouri

Amy E. Williams
Brookhaven High School
Ohio

Stephanie Williams
Cholla High Magnet School
Arizona

***ALGEBRA 1* CONTENT CLUSTERS**

The Larson series emphasizes the major topics of the high school curriculum through a series of content clusters.

The following list shows how *Algebra 1* covers the broad concepts of algebra.

- **Operations with Real Numbers and Powers**
  Ch. 1, Ch. 2, Ch. 8
- **Expressions, Equations, and Inequalities**
  Ch. 3, Ch. 4, Ch. 5, Ch. 6
- **Linear Functions**
  Ch. 4, Ch. 5, Ch. 6, Ch. 7
- **Polynomial and Nonlinear Functions**
  Ch. 8, Ch. 9, Ch. 10, Ch. 11
- **Rational Functions and Radicals**
  Ch. 9, Ch. 11, Ch. 12

CHAPTER 1

Connections to Algebra

APPLICATION HIGHLIGHTS

MATH & HISTORY

INTERNET CONNECTIONS

ASSESSMENT

▶ **READING, WRITING, AND NOTETAKING**

CHAPTER 2

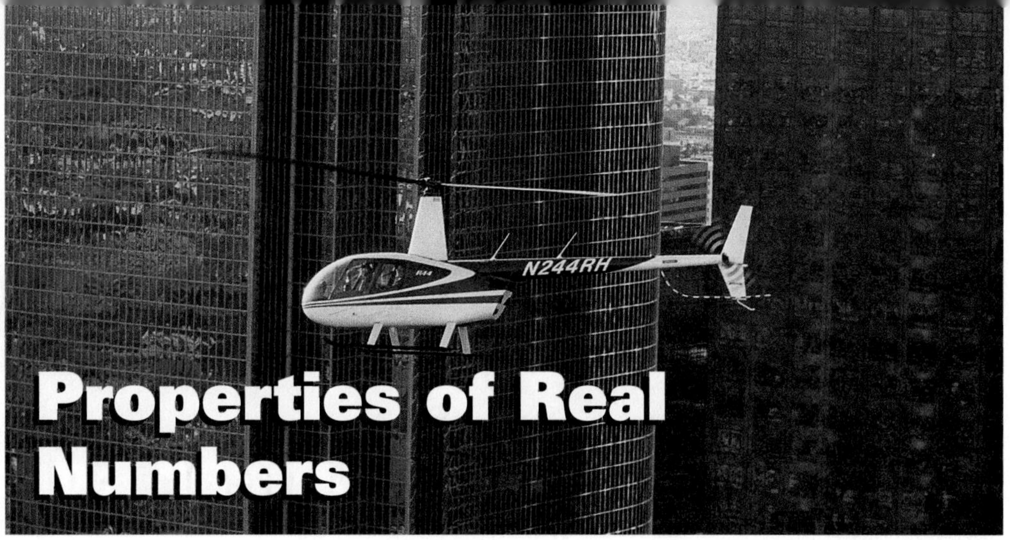

Properties of Real Numbers

APPLICATION HIGHLIGHTS

MATH & HISTORY

INTERNET CONNECTIONS

www.mcdougallittell.com

ASSESSMENT

▶ READING, WRITING, AND NOTETAKING

CHAPTER 3

Solving Linear Equations

APPLICATION HIGHLIGHTS

MATH & HISTORY

INTERNET CONNECTIONS

www.mcdougallittell.com

PROJECT

ASSESSMENT

▶ **READING, WRITING, AND NOTETAKING**

CHAPTER 4

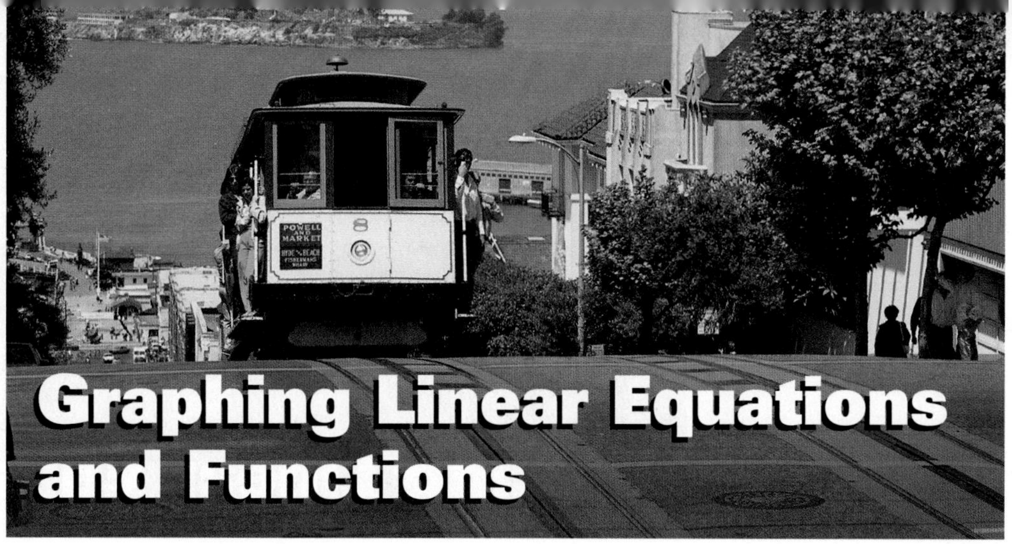

Graphing Linear Equations and Functions

CHAPTER

5

Writing Linear Equations

APPLICATION HIGHLIGHTS

MATH & HISTORY

INTERNET CONNECTIONS

www.mcdougallittell.com

ASSESSMENT

▶ **READING, WRITING, AND NOTETAKING**

CHAPTER 6

Solving and Graphing Linear Inequalities

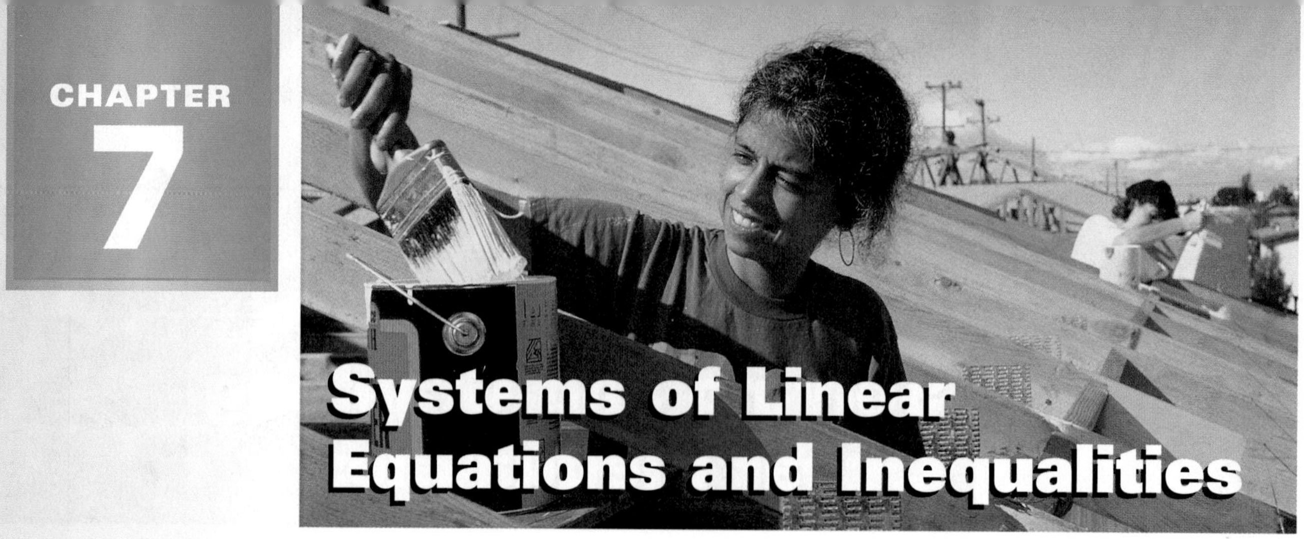

APPLICATION HIGHLIGHTS

Housing *395, 422*
Coastal Population *402*
Museum Admissions *407*
Softball Sizes *409*
Weight of Gold *413*
Steamboat Speed *415*
Career Decisions *420*
Statue Heights *428*
Theater Lighting *434*
Food Budget *436*

MATH & HISTORY

Systems of Linear Equations *417*

INTERNET CONNECTIONS

www.mcdougallittell.com

Application Links
395, 407, 413, 415, 417, 419, 422, 423

Student Help
399, 404, 409, 412, 423, 430, 433

Career Links
400, 430, 434, 436

Extra Challenge
403, 410, 416, 424, 431, 437

ASSESSMENT

Skill Review, *396*
Quizzes, *417, 438*
Test Preparation Questions, *403, 410, 416, 424, 431, 437*
Chapter Summary, Review, and Test, *439*
Chapter Standardized Test, *444*

▶ **READING, WRITING, AND NOTETAKING**

Reading *396, 432*
Writing *410, 416*
Notetaking *396, 399, 405, 411, 418, 426, 439*
Vocabulary *396, 401, 414, 429, 435, 440*

CHAPTER 8

Exponents and Exponential Functions

APPLICATION HIGHLIGHTS

Bicycling *447, 480*
Irrigation *452*
Windmills *454*
Shipwrecks *460*
Stock Exchange *465*
Olympic Rowing *468*
Amazon River *472*
Louisiana Purchase *474*
Bird Eggs *481*
Purchasing Power *484*

MATH & HISTORY

History of Microscopes *491*

INTERNET CONNECTIONS

www.mcdougallittell.com

Application Links
447, 460, 474, 481, 491

Student Help
451, 456, 462, 464, 471, 474, 492

Career Links
454, 465, 489

Extra Challenge
455, 468, 475, 482, 490

ASSESSMENT

Skill Review, *448*
Quizzes, *469, 491*
Test Preparation Questions, *455, 461, 468, 475, 482, 490*
Chapter Summary, Review, and Test, *493*
Chapter Standardized Test, *498*

▶ **READING, WRITING, AND NOTETAKING**

Reading *448, 450, 463*
Writing *454, 461, 468, 474*
Notetaking *448, 450, 456, 463, 477, 484, 487, 493*
Vocabulary *448, 453, 459, 466, 473, 480, 488, 494*

CHAPTER

9

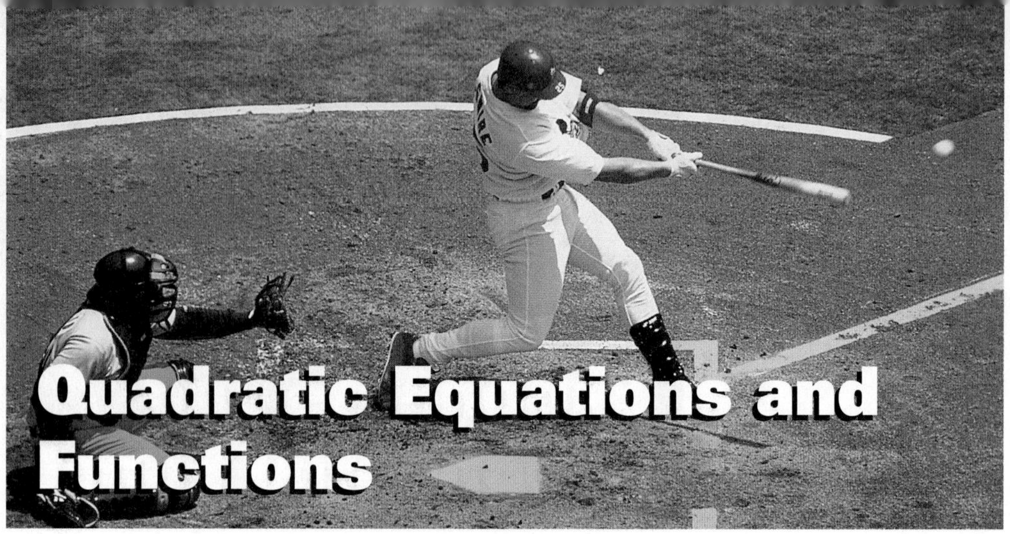

Quadratic Equations and Functions

APPLICATION HIGHLIGHTS

MATH & HISTORY

INTERNET CONNECTIONS

www.mcdougallittell.com

PROJECT

ASSESSMENT

READING, WRITING, AND NOTETAKING

CHAPTER 10

Polynomials and Factoring

APPLICATION HIGHLIGHTS

Arecibo Observatory *573, 601*
Energy Use *581*
Videocassette Sales *588*
Punnett Squares *592*
Landscape Design *606*
Cliff Diving *613*
Gymnastics *615*
Block and Tackle *621*
Pole-Vaulting *623*
Gravity *630*

MATH & HISTORY

Plant Genetics *596*

INTERNET CONNECTIONS

www.mcdougallittell.com

Application Links
573, 588, 596, 608, 630

Student Help
577, 583, 585, 594, 601, 605, 615, 618, 620, 628

Career Links
581, 594, 606

Extra Challenge
582, 589, 609, 616, 624, 631

ASSESSMENT

Skill Review, *574*
Quizzes, *596, 617, 632*
Test Preparation Questions, *582, 589, 595, 602, 609, 616, 624, 631*
Chapter Summary, Review, and Test, *633*
Chapter Standardized Test, *638*

▶ **READING, WRITING, AND NOTETAKING**

Reading *574, 576, 585, 626*
Writing *602, 616, 630*
Notetaking *574, 590, 597, 599, 604, 619, 628, 633*
Vocabulary *574, 579, 587, 593, 600, 607, 622, 629, 634*

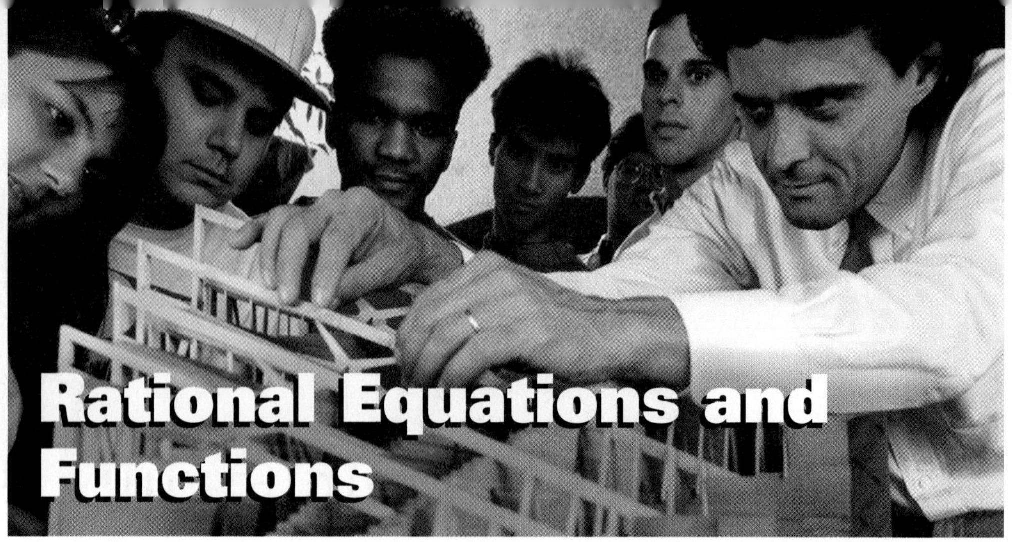

CHAPTER 11

Rational Equations and Functions

APPLICATION HIGHLIGHTS

MATH & HISTORY

INTERNET CONNECTIONS

www.mcdougallittell.com

ASSESSMENT

▶ **READING, WRITING, AND NOTETAKING**

CHAPTER 12

Radicals and Connections to Geometry

APPLICATION HIGHLIGHTS

Amusement Park Rides *707, 726*
Dinosaurs *711*
Medicine *713*
Pole-Vaulting *720*
Plane De-icing *724*
Penguins *735*
Staircases *742*
Soccer *746*
Parasailing *754*
Lawyers *760*

MATH & HISTORY

History of Pythagorean Theorem *744*

INTERNET CONNECTIONS

www.mcdougallittell.com
Application Links *707, 726, 735, 744*
Student Help *710, 715, 717, 731*
Career Links *711, 713, 749, 760*
Extra Challenge *721, 736, 743, 750, 757, 763*

PROJECT

Investigating the Golden Ratio *774*

ASSESSMENT

Skill Review, *708*
Quizzes, *728, 744, 764*
Test Preparation Questions, *714, 721, 727, 736, 743, 750, 757, 763*
Chapter Summary, Review, and Test, *765*
Chapter Standardized Test, *770*
Cumulative Practice, Chapters 1–12, *772*
Project, Chapters 10–12, *774*

▶ **READING, WRITING, AND NOTETAKING**

Reading *708*
Writing *714, 742, 743, 757*
Notetaking *708, 722, 730, 732, 738, 739, 745, 747, 752, 758, 765*
Vocabulary *712, 719, 725, 734, 741, 748, 755, 761*

Student Resources

Who Uses Mathematics in Real Life?

Here are some careers that use the mathematics you will study in Algebra 1.

FOCUS ON CAREERS

REAL LIFE

SURVEYOR *p. 231*
Surveying land is the responsibility of surveyors, survey technicians, and mapping scientists. They measure distances, angles between points, and elevations.

ZOOLOGIST *p. 236*
Zoologists are biologists who study animals in natural and controlled surroundings.

LANDSCAPE DESIGNER *p. 606*
Landscape designers plan and map out the appearance of outdoor spaces like parks, gardens, golf courses, and other recreational areas.

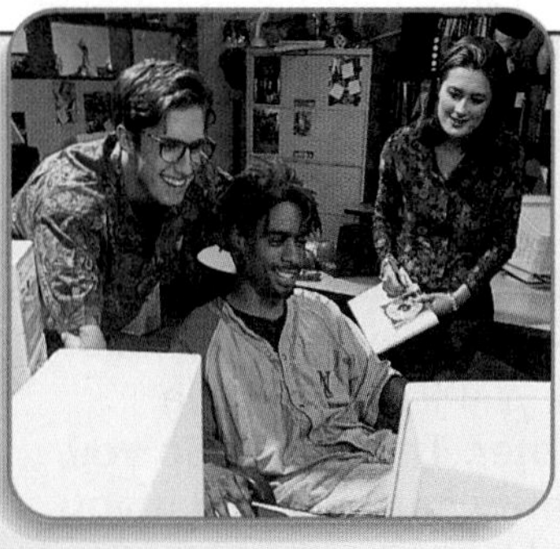

WEBMASTER *p. 400*
Webmasters build Web sites for clients. They often update their technical skills to meet the demands for first-class Web sites.

NURSE *p. 252*
Registered nurses are often responsible for measuring vital signs and giving correct doses of medication.

STUDENT HELP

YOUR BUILT-IN LEARNING SUPPORT

Your textbook contains *many special elements to help you learn. It provides several study helps that may be new to you. For example, every chapter begins with a Study Guide.*

Chapter Preview The Study Guide starts with a short description of what you will be learning.

Key Vocabulary This list highlights important new terms that will be introduced in the chapter as well as reviewing terms that you already know.

Skill Review These exercises review key skills that you'll apply in the chapter. They will help you identify any topics that you need to review.

Study Strategy The study strategies suggest ideas to help you better understand the math you are learning as well as help you prepare for tests.

CHAPTER 1

Study Guide

PREVIEW

What's the chapter about?

Chapter 1 is an **introduction to algebra**. In Chapter 1 you'll learn to

- write and evaluate expressions.
- check solutions to equations and inequalities and use mental math.
- use verbal and algebraic models to represent real-life situations.
- organize data and represent functions.

KEY VOCABULARY

▶ New	• exponent, p. 9	• inequality, p. 26
• variable expression, p. 3	• base of a power, p. 9	• function, p. 46
• unit analysis, p. 5	• order of operations, p. 16	• input-output table, p. 46
• verbal model, p. 5	• equation, p. 24	• domain, p. 47
• power, p. 9	• solution of an equation, p. 24	• range, p. 47

PREPARE

Are you ready for the chapter?

SKILL REVIEW Do these exercises to review key skills that you'll apply in this chapter. See the given reference page if there is something you don't understand.

STUDENT HELP

Study Tip
"Student Help" boxes throughout the chapter give you study tips and tell you where to look for extra help in this book and on the Internet.

Write the percent as a decimal and as a fraction or a mixed number in lowest terms. (Skills Review, pp. 784–785)

1. 60% **2.** 33% **3.** 0.08% **4.** 150%

Complete the statement using > or <. (Skills Review, pp. 779–780)

5. 899 _?_ 901 **6.** 64.1 _?_ 64.03 **7.** 2050 _?_ 2005 **8.** 0.099 _?_ 0.01

Find the area and perimeter of the figure. (Skills Review, pp. 790–791)

9. **10.** **11.**

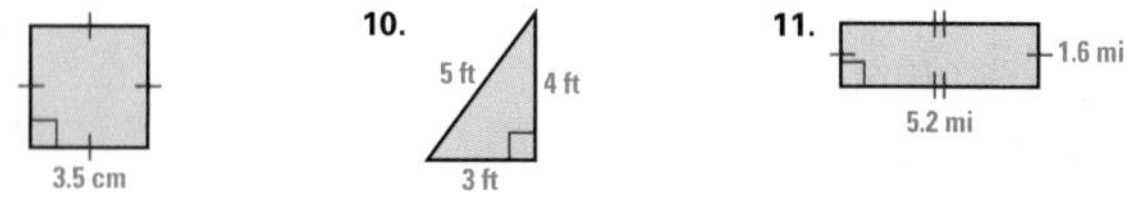

STUDY STRATEGY

Here's a study strategy!

Keeping a Math Notebook

- Keep a notebook of math notes about each chapter, separate from your homework exercises. This will help you remember new concepts and skills.
- Review your notes each day before you start your next homework assignment.

2 Chapter 1

Also, in every lesson you will find a variety of Student Help notes.

STUDENT HELP

In the Book

Study Tip The study tips will help you avoid common errors.

Skills Review Here you can find where to review skills you've studied in earlier math classes.

Look Back Here are references to material in earlier lessons that may help you understand the lesson.

Extra Practice Your book contains more exercises to practice the skills you are learning.

Homework Help Here you can find suggestions about which Examples may help you solve Exercises.

On the Internet

Homework Help: *Extra Examples* These are places where you can find additional examples on the Web site.

Homework Help: *Problem Solving Help* Here you can find additional suggestions for solving an exercise.

Keystroke Help These provide the exact keystroke sequences for many different kinds of calculators.

THE LEFT-TO-RIGHT RULE Operations that have the same priority, such as multiplication and division *or* addition and subtraction, are performed using the *left-to-right rule*, as shown in Example 2.

EXAMPLE 2 ***Using the Left-to-Right Rule***

a. $24 - 8 - 6 = (24 - 8) - 6$ Work from left to right.

$= 16 - 6$

$= 10$

b. $15 \cdot 2 \div 6 = (15 \cdot 2) \div 6$ Work from left to right.

$= 30 \div 6$

$= 5$

c. $16 + 4 \div 2 - 3 = 16 + (4 \div 2) - 3$ Divide first.

$= 16 + 2 - 3$

$= (16 + 2) - 3$ Work from left to right.

$= 18 - 3$

$= 15$

STUDENT HELP

Study Tip
In part (c) of Example 2, you do not perform the operations from left to right because division has a higher priority than addition and subtraction.

ORDER OF OPERATIONS

1. First do operations that occur within grouping symbols.
2. Then evaluate powers.
3. Then do multiplications and divisions from left to right.
4. Finally, do additions and subtractions from left to right.

A fraction bar can act as a grouping symbol: $(1 + 2) \div (4 - 1) = \frac{1+2}{4-1}$

EXAMPLE 3 ***Using a Fraction Bar***

$\frac{7 \cdot 4}{8 + 7^2 - 1} = \frac{7 \cdot 4}{8 + 49 - 1}$ Evaluate power.

$= \frac{28}{8 + 49 - 1}$ Simplify the numerator.

$= \frac{28}{57 - 1}$ Work from left to right.

$= \frac{28}{56}$ Subtract.

$= \frac{1}{2}$ Simplify.

STUDENT HELP

HOMEWORK HELP
Visit our Web site www.mcdougallittell.com for extra examples.

Skills Review
For help with writing fractions in lowest terms, see pp. 781–783.

PRACTICE AND APPLICATIONS

STUDENT HELP

Extra Practice
to help you master skills is on p. 800.

USING GRAPHS TO FIND INTERCEPTS **Use the graph to find the *x*-intercept and the *y*-intercept of the line.**

14. 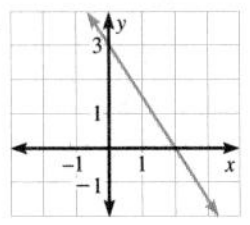

15. 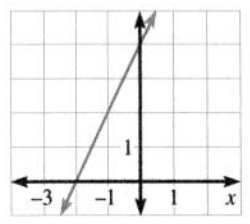

16. 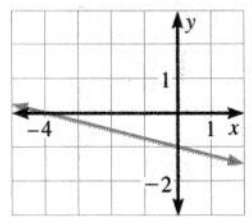

FINDING *X*-INTERCEPTS **Find the *x*-intercept of the graph of the equation.**

17. $x + 3y = 5$ **18.** $x - 2y = 6$ **19.** $2x + 2y = -10$

20. $3x + 4y = 12$ **21.** $5x - y = 45$ **22.** $-x + 3y = 27$

23. $-7x - 3y = 42$ **24.** $2x + 6y = -24$ **25.** $-12x - 20y = 60$

FINDING *Y*-INTERCEPTS **Find the *y*-intercept of the graph of the equation.**

26. $y = -2x + 5$ **27.** $y = 3x - 4$ **28.** $y = 8x + 27$

29. $y = 7x - 15$ **30.** $4x - 5y = -35$ **31.** $6x - 9y = 72$

32. $3x + 12y = -84$ **33.** $-x + 1.7y = 5.1$ **34.** $2x - 6y = -18$

USING INTERCEPTS **Graph the line that has the given intercepts.**

35. x-intercept: -2, y-intercept: 5 **36.** x-intercept: 4, y-intercept: 6 **37.** x-intercept: -7, y-intercept: -3

38. x-intercept: -3, y-intercept: -7 **39.** x-intercept: -12, y-intercept: -8 **40.** x-intercept: -7, y-intercept: 15

STUDENT HELP

HOMEWORK HELP
Example 1: Exs. 17–34
Example 2: Exs. 35–55
Example 3: Exs. 44–55
Example 4: Exs. 60–63

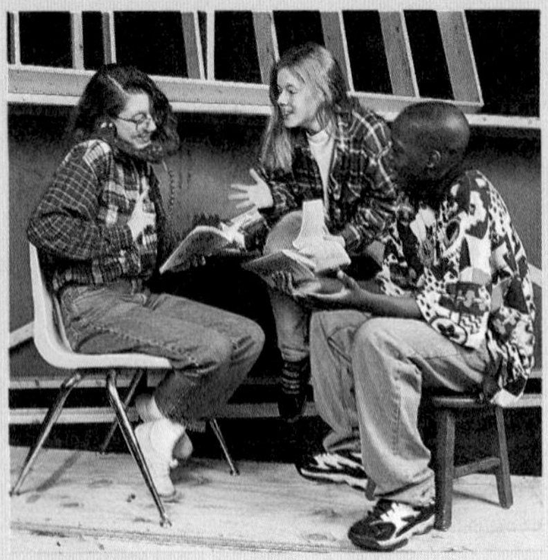

Plot a course to success with McDougal Littell Algebra 1!

HOW... can I make algebra understandable to all my students?

WHERE... can my students go for help with their homework?

WHAT... resources are available to help me succeed as a teacher?

STUDENT HELP

Study Tip
When you know two ways to solve a problem, it's a good idea to use one method to get a solution and the other method to check the solution. This is done in Examples 1 and 2.

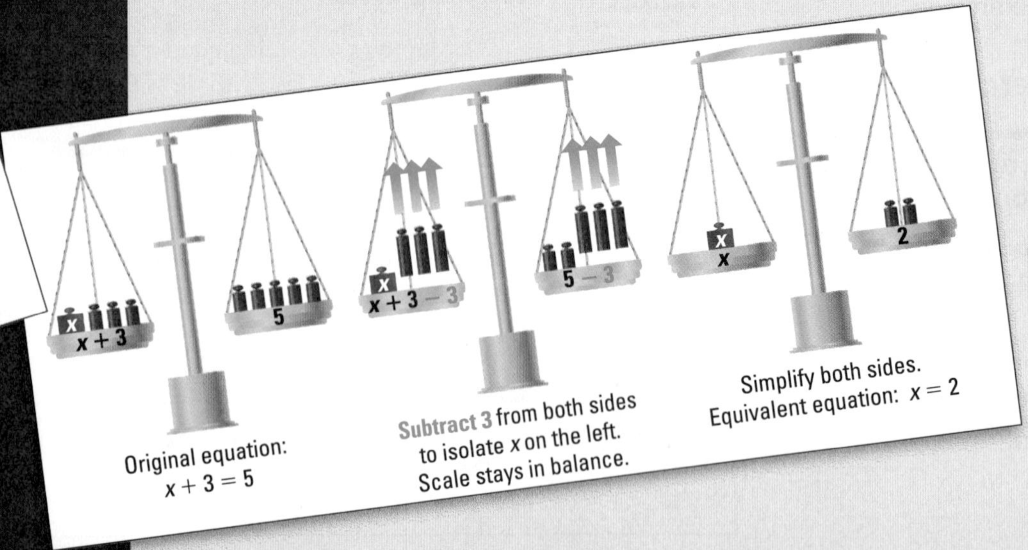

x
x + 3
5
x
x + 3 − 3
5 − 3
x
x
2
Original equation:
x + 3 = 5
Subtract 3 from both sides
to isolate x on the left.
Scale stays in balance.
Simplify both sides.
Equivalent equation: x = 2

FIND THE ANSWERS you're looking for in McDougal Littell Algebra 1!

How?

Important concepts are made understandable to all students through instructional diagrams and graphics, interactive activities, and numerous examples throughout the text.

Height of books	Distance between ruler and books	Rise	Run	Slope
7 cm	12 cm	7	12	$\frac{7}{12}$
11 cm	5.5 cm	?	?	?

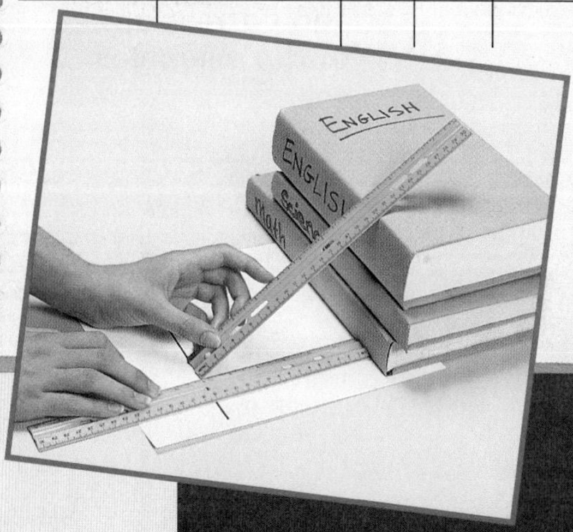

Where?

Students receive support when they are learning on their own through Student Help notes throughout the book, including homework exercises correlated to the examples and Homework Help on the Internet.

What?

A variety of resources help you adapt the program to your teaching styles and to the needs of your students.

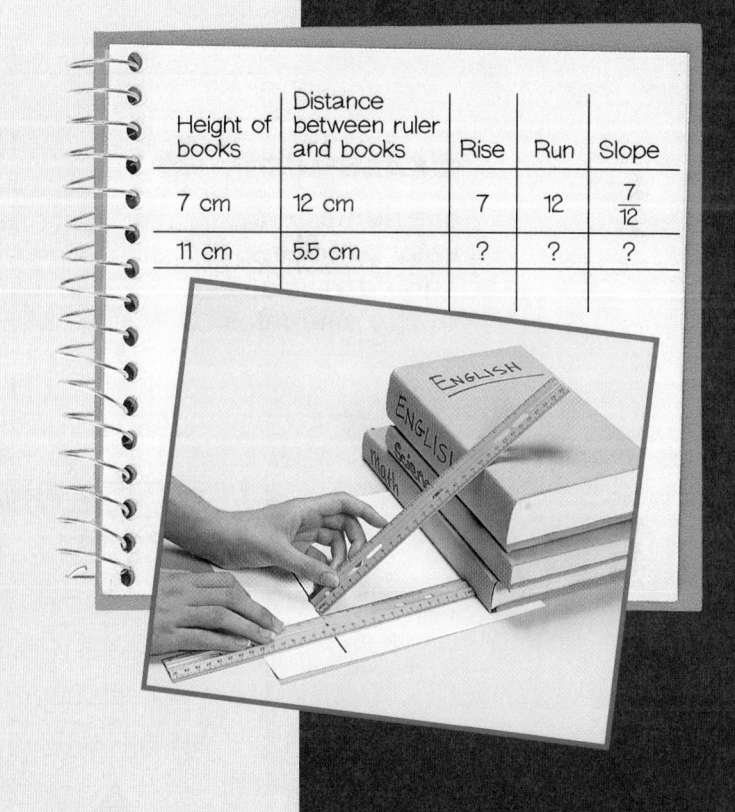

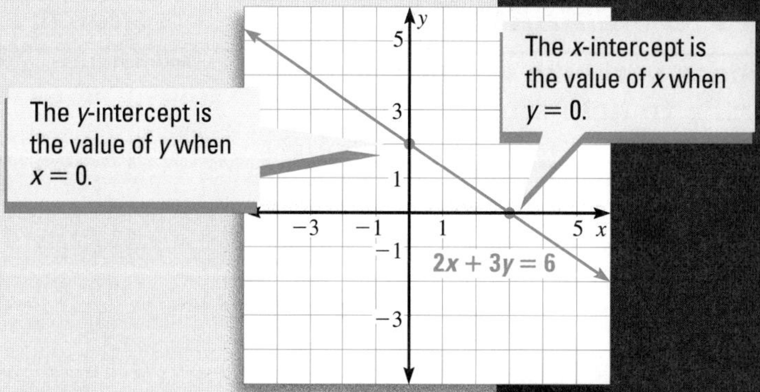

The y-intercept is the value of y when x = 0.
The x-intercept is the value of x when y = 0.
2x + 3y = 6
y
x
5
3
1
−1
−3
−3
−1
1
5

How... can I make algebra understandable to all my students?

McDougal Littell Algebra 1 provides clear, visual explanations that every student can follow.

Examples
are numerous, easy to follow, and correspond to the exercises.

Concept Summary Boxes
present ideas verbally, visually, and algebraically to help students master key concepts.

EXAMPLE 4 *Slope of a Vertical Line is Undefined*

Find the slope of the line passing through (2, 4) and (2, 1).

SOLUTION Let $(x_1, y_1) = (2, 1)$ and $(x_2, y_2) = (2, 4)$.

$m = \frac{y_2 - y_1}{x_2 - x_1}$ ← Rise: Difference of y-values; ← Run: Difference of x-values

$= \frac{4 - 1}{2 - 2}$ Substitute values.

$= \frac{3}{0}$ Division by 0 is undefined.

▶ Because division by 0 is undefined, the expression $\frac{3}{0}$ has no meaning. The slope of a vertical line is undefined.

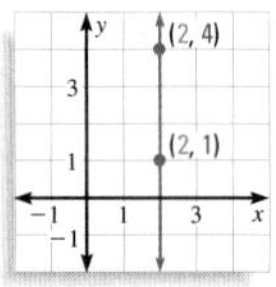

Undefined slope: line is vertical.

CONCEPT SUMMARY CLASSIFICATION OF LINES BY SLOPE

A line with positive slope *rises* from left to right.

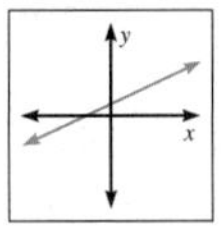

$m > 0$

A line with negative slope *falls* from left to right.

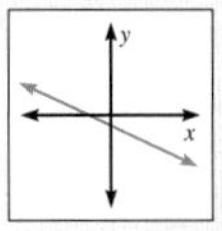

$m < 0$

A line with zero slope is *horizontal*.

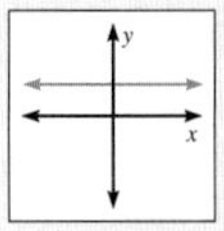

$m = 0$

A line with undefined slope is *vertical*.

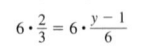

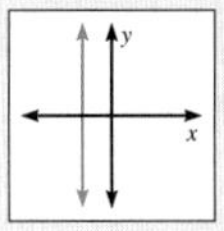

m is undefined.

EXAMPLE 5 *Given the Slope, Find a y-Coordinate*

Find the value of y so that the line passing through the points $(-2, 1)$ and $(4, y)$ has a slope of $\frac{2}{3}$.

SOLUTION Let $(x_1, y_1) = (-2, 1)$ and $(x_2, y_2) = (4, y)$.

$m = \frac{y_2 - y_1}{x_2 - x_1}$ Write formula for slope.

$\frac{2}{3} = \frac{y - 1}{4 - (-2)}$ Substitute values for m, x_1, y_1, x_2, and y_2.

$\frac{2}{3} = \frac{y - 1}{6}$ Simplify.

$6 \cdot \frac{2}{3} = 6 \cdot \frac{y - 1}{6}$ Multiply each side by 6.

$4 = y - 1$ Simplify.

$5 = y$ Add 1 to each side.

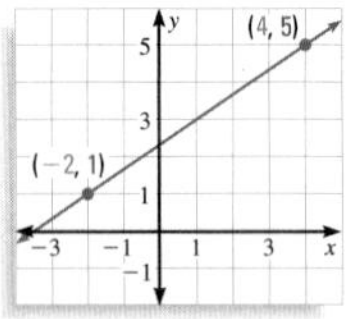

STUDENT HELP
HOMEWORK HELP
Visit our Web site www.mcdougallittell.com for extra examples.

Guide

s the chapter about?

4 is about **graphing linear equations**. In Chapter 4, you'll learn

o graph linear equations.

ways to graph linear equations quickly.

to tell whether an equation or a graph represents a function.

Y VOCABULARY

Review	▶ New	• y-intercept, p. 218
ariable expression, p. 3	• coordinate plane, p. 203	• slope, p. 226
solution of an equation, p. 24	• scatter plot, p. 204	• rate of change, p. 229
function, p. 46	• graph of a linear equation, p. 210	• slope-intercept form, p. 241
• linear equation, p. 133	• x-intercept, p. 218	• function notation, p. 257

Are you ready for the chapter?

SKILL REVIEW Do these exercises to review key skills that you'll apply in this chapter. See the given reference page if there is something you don't understand.

Rewrite as a decimal and as a fraction in lowest terms. (Skills Review, p. 784)

1. 50% **2.** 75% **3.** 1% **4.** 20%

Use the function $y = 5x + 70$, where $x \geq 0$. (Review Examples 1 and 2, pp. 46–47)

5. For several inputs x, use the function to calculate an output y.

6. Represent the data with a line graph.

7. Describe the domain and range of the function.

Evaluate the expression for the given values of the variables.
(Review Example 3, p. 109)

8. $\frac{x - y}{2}$ when $x = -3$ and $y = -1$ **9.** $\frac{x + 2y}{x}$ when $x = 6$ and $y = 3$

Here's a study
strategy!

Getting Your Questions Answered

Each day after you finish your math homework,
write a list of questions about things you don't
understand. Ask your teacher or another
student to answer your questions and write
the explanations in your notebook.

hapter 4

$(-2, 1)$ $(4, 5)$

90 2000
(proj.)
•Northeast

ata at

nd the

t of the graph of

$= 2x - 3$ 2; −3

$3x + 4y = 16$ $-\frac{3}{4}$; 4

$12x + 4y - 2 = 0$
$-3; \frac{1}{2}$

24. $y = x + 5$

27. $y = 3x + 7$

30. $y = 4x + 9$

33. $y = 2$

cept form. Then graph
$y = -\frac{4}{5}x + 3$
$-\frac{3}{2}$ **36.** $4x + 5y = 15$

39. $x + y = 0$ $y = -x$

42. $2x - 3y - 6 = 0$
See margin.
45. $2x + 3y - 4 = \frac{1}{5}x + 5$
$y = -\frac{1}{3}x + 3$

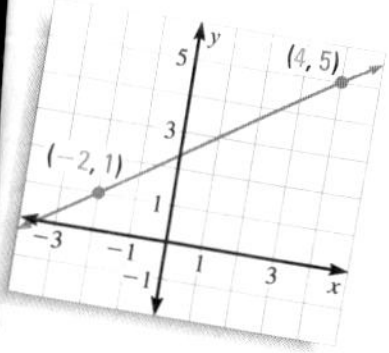

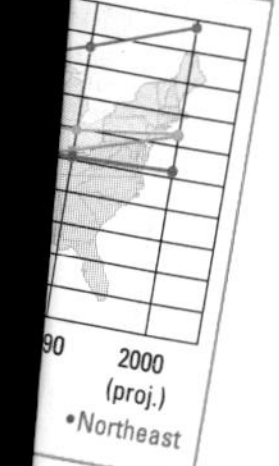

PRACTICE SETS

- **include exercises of three difficulty levels—Levels A, B, and C—which are labeled in the Teacher's Edition.**
- **Also available: Practice Worksheets (Levels A, B, and C)**

CHALLENGE PROBLEMS

(not pictured) are included in every practice set, with extra challenge online at www.mcdougallittell.com

WHERE... can my students go for help with their homework?

Student Help features throughout program provide students with a va[riety] of tips and homework support.

EXTRA EXAMPLES

show where students can find additional support on the Internet at www.mcdougallittell.com

EXAMPLE 5 **Given the Slope, Find a y-Coor[dinate]**

Find the value of y so that the line passing through t[he] ... $(4, y)$ has a slope of $\frac{2}{3}$.

SOLUTION Let $(x_1, y_1) = (-2, 1)$ and $(x_2, y_2) = (4,$ ...

$m = \frac{y_2 - y_1}{x_2 - x_1}$ Write formula for slope.

$\frac{2}{3} = \frac{y - 1}{4 - (-2)}$ Substitute values for m, x_1, y_1, x_2, and y_2.

$\frac{2}{3} = \frac{y - 1}{6}$ Simplify.

$6 \cdot \frac{2}{3} = 6 \cdot \frac{y - 1}{6}$ Multiply each side by 6.

$4 = y - 1$ Simplify.

$5 = y$ Add 1 to each side.

STUDENT HELP
HOMEWORK HELP
Visit our Web site www.mcdougallittell.com for extra examples.

PROBLEM SOLVING

provides online homework help and suggestions for solving exercises.

STUDENT HELP
HOMEWORK HELP
Visit our Web site www.mcdougallittell.com for help with problem solving in Exs. 62–64.

POPULATION RATES **In Exercises 62–65, use the line graph**

62. During which decade did the population of the South increase the most? Estimate the rate of change for this decade in people per year.

63. During which decade did the population of the Midwest increase the most? Estimate the rate of change for this decade in people per year.

64. INTERPRETING A GRAPH The line graph for the Northeast appears to be horizontal between 1970 and 1980. Write a sentence about what this means in terms of the real-life situation.

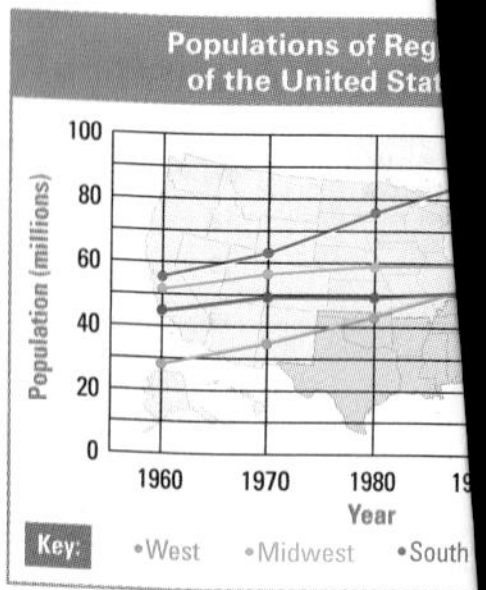

DATA UPDATE of U.S. Bureau of the Census ... www.mcdougallittell.com

65. RESEARCH Find the population of your state in 1980 and in 1990. F... annual rate of change of this population.

KEYSTROKE HELP

guides students to the exact keystroke sequences for many different kinds of calculators (not pictured).

Skills Review

guides students to review skills they've studied in earlier math classes to help them succeed with current concepts.

EXAMPLE 3 ***A Line with a Negative Slope Falls***

Find the slope of the line passing through (0, 0) and (3, −3).

SOLUTION

Let $(x_1, y_1) = (0, 0)$ and $(x_2, y_2) = (3, -3)$

$m = \frac{y_2 - y_1}{x_2 - x_1}$ ← Rise: Difference of *y*-values ← Run: Difference of *x*-values

$= \frac{-3 - 0}{3 - 0}$ Substitute values.

$= \frac{-3}{3}$ Simplify.

$= -1$ Slope is negative.

STUDENT HELP
Skills Review
For help with simplifying fractions, see p. 780.

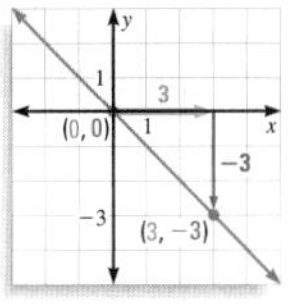

Negative slope: line falls from left to right.

Look Back

contains references to material in earlier lessons that may help students understand the lesson.

EXAMPLE 1 ***A Line with a Positive Slope Rises***

Find the slope of the line passing through (−2, 2) and (3, 4).

SOLUTION

Let $(x_1, y_1) = (-2, 2)$ and $(x_2, y_2) = (3, 4)$.

$m = \frac{y_2 - y_1}{x_2 - x_1}$ ← Rise: Difference of *y*-values ← Run: Difference of *x*-values

$= \frac{4 - 2}{3 - (-2)}$ Substitute values.

$= \frac{2}{3 + 2}$ Simplify.

$= \frac{2}{5}$ Slope is positive.

STUDENT HELP
Look Back
For help with evaluating expressions, see p. 79.

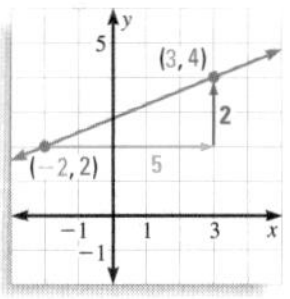

Positive slope: line rises from left to right.

EXAMPLE 2 ***A Line with a Zero Slope is Horizontal***

Find the slope of the line passing through (−1, 2) and (3, 2).

Extra Practice

is provided in the book for every section, in addition to the large number of exercises found with each lesson.

Numerous Examples

in every lesson are correlated to the exercises and support students with their homework.

PRACTICE AND APPLICATIONS

STUDENT HELP
Extra Practice
to help you master skills is on p. 800.

DESCRIBING SLOPE **Plot the points and draw a line through them. Without calculating, state whether the slope of the line is *positive, negative, zero,* or *undefined.* Explain your reasoning.**

12. (6, 9), (4, 3) **13.** (7, 4), (−1, 8) **14.** (5, 10), (5, −4) **15.** (1, 1), (4, −3)

16. (−2, 5), (3, 5) **17.** (0, 0), (−5, 3) **18.** (1, 3), (−2, 1) **19.** (2, −2), (2, −6)

GRAPH AND CALCULATE **Plot the points and find the slope of the line passing through the points.**

20. (4, 5), (2, 3) **21.** (1, 5), (5, 2) **22.** (2, 3), (−3, 0)

23. (0, −6), (8, 0) **24.** (0, 6), (8, 0) **25.** (2, 4), (4, −4)

26. (−6, −1), (−6, 4) **27.** (0, −10), (−4, 0) **28.** (1, −2), (−2, 2)

29. (3, 6), (3, 0) **30.** (−6, 2), (4, −2) **31.** (−1, −1), (−3, −6)

32. $\left(0, \frac{1}{2}\right)$, (0, 0) **33.** (2, 2), (−3, 5) **34.** (4, 1), (6, 1)

GRAPHICAL REASONING **Find the slope of each line.**

STUDENT HELP
HOMEWORK HELP
Example 1: Exs. 12–37
Example 2: Exs. 12–37
Example 3: Exs. 12–37
Example 4: Exs. 12–37
Example 5: Exs. 38–46
Example 6: Exs. 56, 57

35.

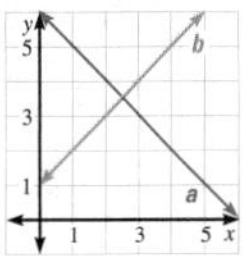

36.

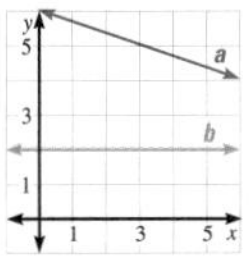

37.

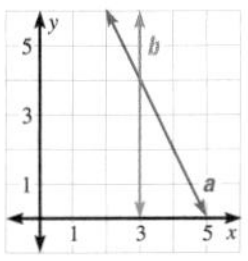

WHAT... resources are available to help me succeed as a teacher?

***The McDougal Littell resources** are conveniently organized and include a variety of materials to help you adapt the program to your teaching style and to the specific needs of your students!*

TEACHER'S RESOURCE PACKAGE

- Chapter Resource Books (one for each chapter, organized by lesson)
- Basic Skills Workbook: Diagnosis and Remediation (TE)
- Practice Workbook with Examples (TE)
- Standardized Test Practice Workbook (TE)
- Warm-Up Transparencies and Daily Homework Quiz
- Worked-Out Solution Key

Finally! Resources organized the way you teach... one lesson at a time!

McDougal Littell's Chapter Resource Books *allow you to easily carry the resources you have for a chapter in one manageable book. The materials are organized by lesson so that you can see everything you have available for a specific section.*

Chapter Resource Books include:

- Tips for New Teachers
- Parent Guide for Student Success
- Prerequisite Skills Review
- Strategies for Reading Mathematics
- Lesson Plans
- Lesson Plans for Block Scheduling
- Activity Support Masters
- Graphing Calculator Activities with Keystrokes
- Practice (Levels A, B, and C)
- Reteaching with Practice
- Quick Catch-Up for Absent Students
- Cooperative Learning Activities
- Interdisciplinary Applications
- Real-Life Applications: When Will I Ever Use This?
- Math and History Applications
- Quizzes
- Chapter Review Games and Activities
- Chapter Tests (Levels A, B, and C)
- SAT/ACT Chapter Test
- Alternative Assessment with Rubric and Math Journal
- Project with Rubric
- Cumulative Review
- Resource Book Answers

WHAT... materials are available for me to use while I teach?

A number of transparency packages give you many easy-to-use options for reviewing homework, starting class, and teaching a lesson.

Answer Transparencies for Checking Homework

Warm-Up Transparencies and Daily Homework Quiz

- Warm-Up Exercises
- Daily Homework Quizzes

Extra Example Transparencies with Standardized Test Practice

- Extra Examples
- Checkpoint Exercises
- Standardized Test Practice Questions

Notetaking Guide Transparencies

Students take notes in their own Notetaking Guide Workbook as teachers present a lesson using transparencies.

- Promote notetaking skills
- Reinforce key concepts

Starting Points: Alternative Lesson Opener Transparency Package

There are many ways to introduce a lesson — and we provide alternative ideas for teachers on ready-to-use transparencies.

- Application Lesson Openers
- Graphing Calculator Lesson Openers
- Activity Lesson Openers
- Visual Approach Lesson Openers

4.3 Lesson Opener
For use with pages 218–224

Activity

SET UP: Work with your class.

1 Create a human coordinate plane with memb... Each square on the diag...

4.4 Lesson Opener
For use with pages 226–233

Visual Approach

In Exercises 1–3, use the following information.

The slanted side of the triangle shown represents the escalator at the Wilshire/Vermont Metro Rail Station in Los Angeles. You can find the slope of this side by writing the following ratio:

$$\frac{\text{length of vertical side}}{\text{length of horizontal side}}$$

76 feet

132 feet

1 What is the length of the vertical side shown in the diagram?

2 What is the length of the horizontal side shown in the diagram?

3 What is the ratio of the length of the vertical side to the length of the horizontal side? This is the slope of the escalator.

Find the length of the vertical side and the length of the horizontal side shown in each triangle. Then find the slope of the slanted side of the triangle.

4 44 feet 191 feet

5 3 feet 5 feet

See answers in Chapter Resource Book

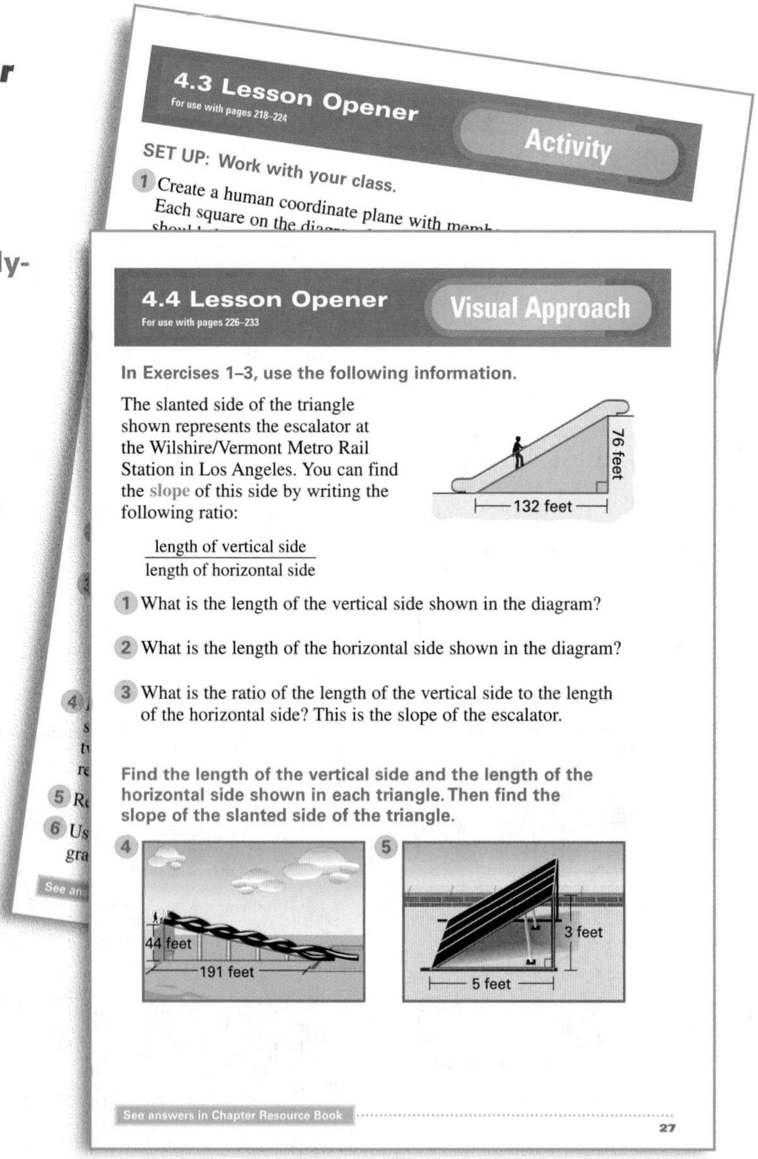

WORKBOOKS

- **Basic Skills Workbook: Diagnosis and Remediation**
- **Practice Workbook with Examples**
- **Standardized Test Practice Workbook**

SPANISH RESOURCES

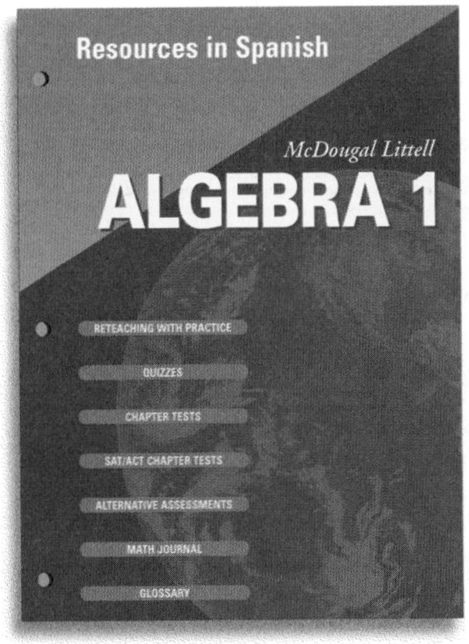

Resources in Spanish Include:

- **Reteaching with Practice**
- **Quizzes**
- **Chapter Tests**
- **SAT/ACT Chapter Tests**
- **Alternative Assessments with Rubrics**
- **Math Journal**
- **Glossary**

Exercises in Spanish

English-Spanish Chapter Reviews and Tests

OTHER RESOURCES

Data Analysis Sourcebook

- **Lessons**
- **Activities**
- **Exercises**

Explorations and Projects Book

Algebra Tile Investigations

Student Algebra Tile Kit

Teacher Algebra Tile Kit

How... can I incorporate technology into my classroom?

McDougal Littell technology resources help you and your students meaningfully use technology to enhance lessons and build understanding.

Technology for PLANNING AND TEACHING

EasyPlanner Plus Online

- Create customized lesson plans.
- Adjust to schedule changes.
- Adapt the program to local and state objectives.

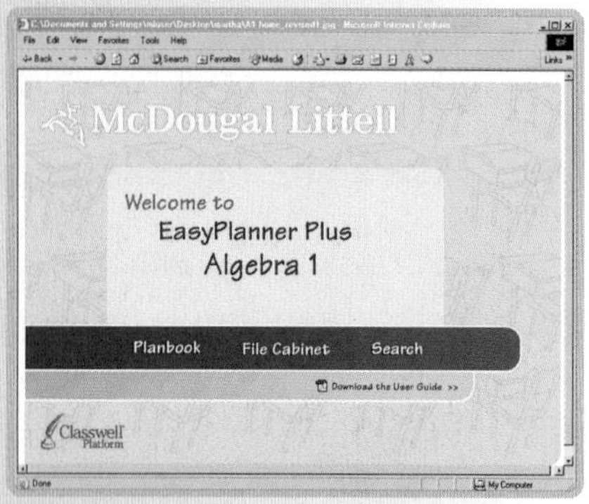

EasyPlanner CD-ROM

This handy tool provides all your teaching resources on one CD-ROM.

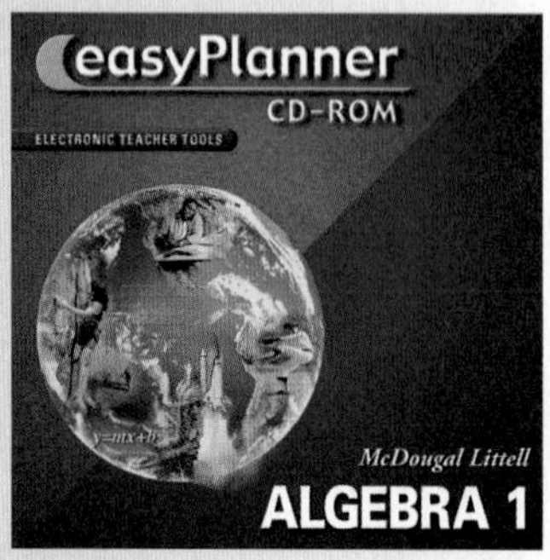

Power Presentations CD-ROM

These electronic lesson presentations help you introduce concepts through colorful diagrams and animated instructional techniques.

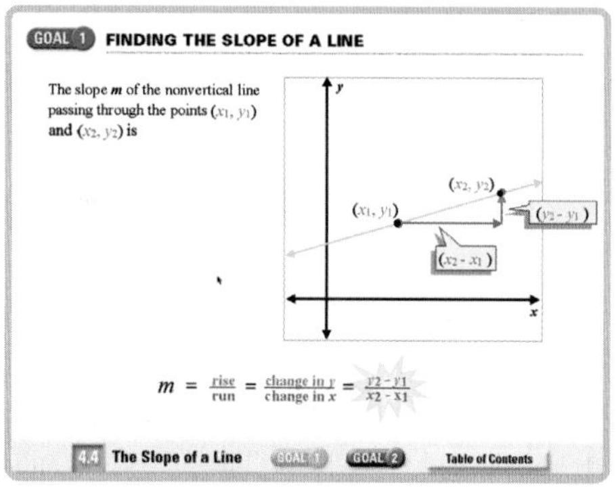

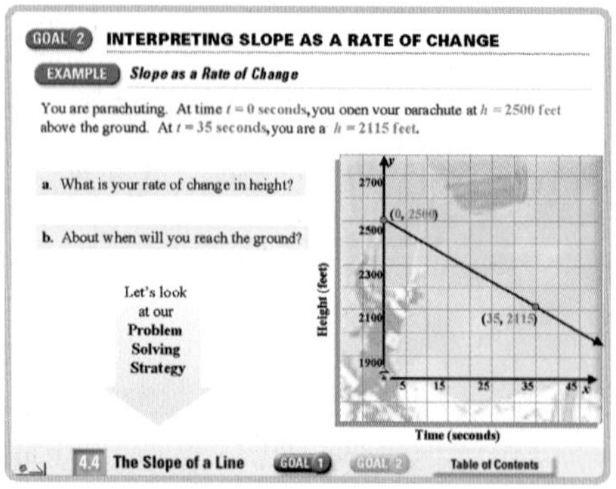

Technology for STUDENT SUPPORT

Personal Student Tutor CD-ROM

This tutorial program is correlated to the McDougal Littell series. It provides:

- animated examples
- student hints
- exercises that are automatically graded
- student progress reports

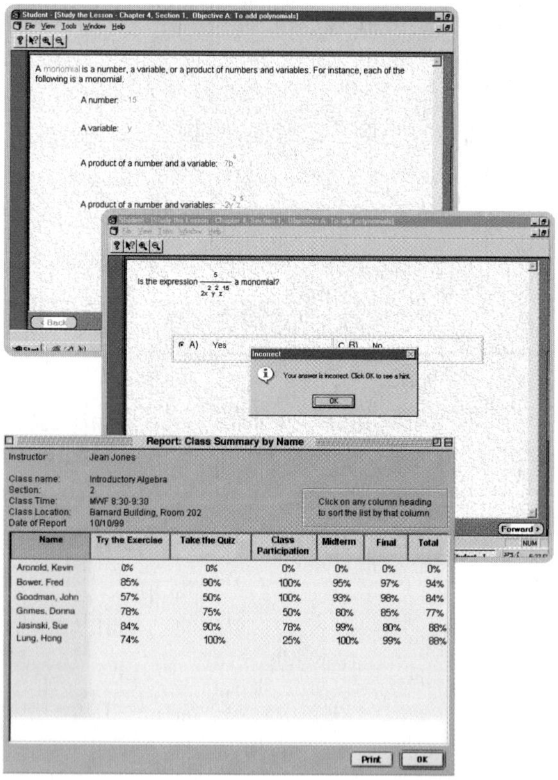

Technology for
REVIEW AND ASSESSMENT

ClassZone
ClassZone is the companion Web site to McDougal Littell Algebra 1 that includes Student Help, career links, and more. To access ClassZone, go to www.classzone.com.

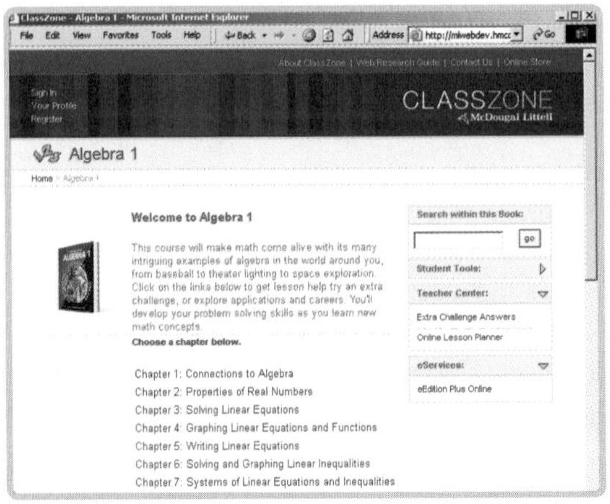

eEdition Plus Online
This online version of the book contains all the textbook materials plus interactive applications.

Chapter Audio Summaries CD
Audio CDs provide lesson-by-lesson summaries of the McDougal Littell series. Also available in Spanish and Haitian Creole.

Test and Practice Generator
- Develop tests and practice sheets.
- Instantly create answer keys.

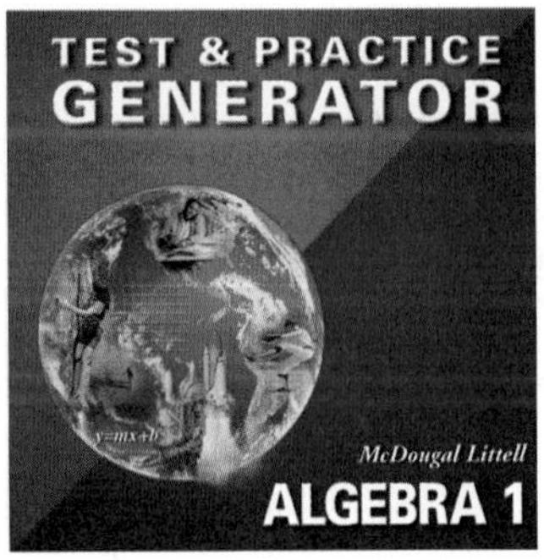

Instant Replay: Video Review Games
- Video Review Games provide a fun, interactive way to help students review the chapter – as well as keep previously learned skills sharp.
- Real-life Motivators provide a great start to each chapter and help students see how the concepts they are about to learn are applied to real situations.

PACING THE COURSE

COURSE PACING CHART

The Pacing Chart below shows the number of days allotted for each chapter. The Regular Schedule requires 160 days. The Block Schedule requires 80 days. These time frames include days for review and assessment: 2 days per chapter for the Regular Schedule and 1 day per chapter for the Block Schedule. Semester and trimester divisions are indicated by red and blue rules, respectively. Recommended pacing for Algebra 1 over two years is also provided.

					Trimester		Semester		Trimester			
Chapter	1	2	3	4	5	6	7	8	9	10	11	12
Regular Schedule	13	13	13	15	13	13	11	12	14	15	14	14
Block Schedule	6.5	6.5	6.5	7.5	6.5	6.5	5.5	6	7	7.5	7	7
Two-Year Pacing	26	26	26	30	26	26	22	24	28	30	28	28
							End of year one					

Assignments are provided with each lesson for a basic course, an average course, an advanced course, and a block-scheduled course. Each of the four courses covers all twelve chapters.

BASIC COURSE

The basic course is intended for students who enter with below-average mathematical and problem-solving skills. Assignments include:

- substantial work with the skills and concepts presented in the lesson
- straightforward applications of these skills and concepts
- test preparation and mixed review exercises

AVERAGE COURSE

The average course is intended for students who enter with typical mathematical and problem-solving skills. Assignments include:

- substantial work with the skills and concepts presented in the lesson
- application of these skills and concepts
- test preparation and mixed review exercises

ADVANCED COURSE

The advanced course is intended for students who enter with above-average mathematical and problem-solving skills. Assignments include:

- substantial work with the skills and concepts presented in the lesson
- more complex applications and challenge exercises
- test preparation and mixed review exercises

BLOCK-SCHEDULED COURSE

The block-scheduled course is intended for schools that use a block schedule. The exercises assigned are comparable to the exercises for the average course.

TWO-YEAR PACING

The two-year pacing allows more time for pre-course review using the prerequisite skills material. The schedule also provides more opportunities to use activities or reteaching and practice materials.

PACING EACH CHAPTER

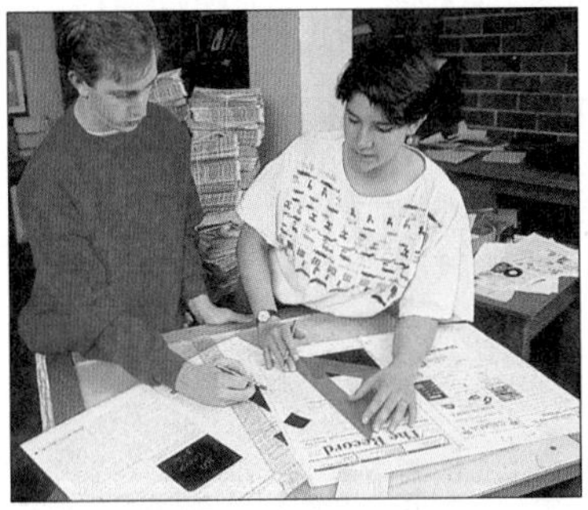

The Pacing Chart for each chapter is located on the interleaved pages preceding the chapter. Part of the Pacing Chart for Chapter 4 is shown here. The regular schedule chart provides pacing for the basic, average, and advanced courses. The block schedule chart provides pacing for the block-scheduled course.

REGULAR SCHEDULE

Day 11

4.7

STARTING OPTIONS
- Homework Check
- Warm-Up or Daily Quiz

TEACHING OPTIONS
- Motivating the Lesson
- Les. Opener (Calculator)
- Graphing Calc. Activity
- Examples 1–5
- Closure Question
- Guided Practice Exs.

APPLY/HOMEWORK
- See Assignment Guide.
- See the CRB: Practice, Reteach, Apply, Extend

ASSESSMENT OPTIONS
- Checkpoint Exercises
- Daily Quiz (4.7)
- Stand. Test Practice

Each day provides a number of options for the lesson, including starting, teaching, applying, and assessing.

The Chapter Resource Book includes a number of follow-up options for the lesson. These options include three levels of practice, reteaching examples with practice, inter-disciplinary and real-life applications, and challenge exercises.

A black dot indicates that the option can be found in the Student Edition, a green dot in the Chapter Resource Book, and a red dot in the Teacher's Edition.

BLOCK SCHEDULE

Day 3

4.5 & 4.6

DAY 5 START OPTIONS
- Homework Check
- W-Up 4.5 or D. Quiz 4.4

TEACHING 4.5 OPTIONS
- Motivating the Lesson
- Les. Opener (Activity)
- Examples 1–4
- Closure Question
- Guided Practice Exs.

BEGINNING 4.6 OPTIONS
- Warm-Up (Les. 4.6)
- Motivating the Lesson
- Concept Act. & Wksht.
- Les. Opener (Appl.)
- Graphing Calc. Activity
- Examples 1–3
- Guided Pract. Exs. 1–10

APPLY/HOMEWORK
- See Assignment Guide.
- See the CRB: Practice, Reteach, Apply, Extend

ASSESSMENT OPTIONS
- Checkpoint Exercises
- Daily Quiz (Les. 4.5)

ASSIGNMENT GUIDE

An Assignment Guide for each lesson is provided at point-of-use at the beginning of the exercise set.

Assignments are provided for basic, average, advanced, and block-scheduled courses.

ASSIGNMENT GUIDE

BASIC
Day 1: pp. 244–245 Exs. 14–50 even
Day 2: pp. 245–247 Exs. 56–59, 62–65, 69, 70, 81–89, Quiz 2 Exs. 1–19

AVERAGE
Day 1: pp. 244–245 Exs. 14–50 even
Day 2: pp. 245–247 Exs. 56–65, 70, 81–89, Quiz 2 Exs. 1–19

ADVANCED
Day 1: pp. 244–245 Exs. 14–50 even
Day 2: pp. 245–247 Exs. 62–77, 81–89, Quiz 2 Exs. 1–19

BLOCK SCHEDULE
pp. 244–245 Exs. 14–50 even (with 4.5),
pp. 245–247 Exs. 56–65, 69, 70, 78–88 even,
Quiz 2 Exs. 1–19 (with 4.7)

PLANNING FOR THE NEXT COURSES:

What books come next in the McDougal Littell series?

The same solid approach and features that are found in McDougal Littell Algebra 1 to help you and your students succeed are continued in the Geometry and Algebra 2 books.

McDougal Littell GEOMETRY

1 Basics of Geometry
2 Reasoning and Proof
3 Perpendicular and Parallel Lines
4 Congruent Triangles
5 Properties of Triangles
6 Quadrilaterals
7 Transformations
8 Similarity
9 Right Triangles and Trigonometry
10 Circles
11 Area of Polygons and Circles
12 Surface Area and Volume

McDougal Littell ALGEBRA 2

1 Equations and Inequalities
2 Linear Equations and Functions
3 Systems of Linear Equations and Inequalities
4 Matrices and Determinants
5 Quadratic Functions
6 Polynomials and Polynomial Functions
7 Powers, Roots, and Radicals
8 Exponential and Logarithmic Functions
9 Rational Equations and Functions
10 Quadratic Relations and Conic Sections
11 Sequences and Series
12 Probability and Statistics
13 Trigonometric Ratios and Functions
14 Trigonometric Graphs, Identities, and Equations

Professional *Articles*

The following pages include information about the *Algebra 1* program and explain how it reflects current research in mathematics instruction.

A Program *you can* trust

The right math, presented in the right way.

McDougal Littell's *Larson: Algebra 1* is a program you can count on to teach the mathematical concepts and methods that your students need to know in order to meet high curriculum standards and succeed on high-stakes tests. Including the appropriate content, however, is not enough. The math must be presented in a way that students can understand and that will motivate them to learn. The distinguished author team of thc Larson high school series and the thorough, research-based planning and development process ensure that this program both includes the right math and teaches the math in the right way, so that students gain conceptual understanding and achieve success on important assessments.

DISTINGUISHED AUTHOR TEAM

The experienced, expert author team of Ron Larson, Laurie Boswell, Timothy Kanold, and Lee Stiff brings a wealth of mathematical expertise, writing talent, and classroom teaching and curriculum planning experience from middle school through college to the creation of this program. Lead author Ron Larson has been writing highly-respected and widely-used textbooks for more than 20 years, and each new book benefits from the comments of the many teachers and students who used earlier editions.

STANDARDS-BASED INSTRUCTION AND ASSESSMENT

State and National Standards In planning the outlines and writing the textbook, the authors paid careful attention to state curriculum standards and state assessment objectives from all states to make sure that the important mathematical concepts and skills were included and given appropriate emphasis. The authors also made sure that the outline and course content fully addressed the standards of national organizations such as the National Council of Teachers of Mathematics (NCTM) and the National Assessment of Educational Progress (NAEP). A correlation to the NAEP objectives appears on page T58 and a correlation to the NCTM Standards can be found on page T59.

SUPPORTING BEST PRACTICES FROM RESEARCH

Educational and Cognitive Research Recent research studies have confirmed strategies for increasing student achievement. These strategies, which reflect the best practices of successful teachers, can help all students learn more effectively. The authors kept this research in mind as they planned and wrote *Algebra 1*, so that the content, organization, and instructional strategies in the program would make it easy for you to implement best-practice instruction in your classroom.

...validated by research

COMPREHENSIVE RESEARCH, REVIEW, AND FIELD-TESTING

Several years prior to the publication of this book, the authors and teams of editors, consultants, graphic designers, professional researchers, and experts in content, instruction, and technology began gathering and analyzing the extensive data on which the plans for this program are based. Data were gathered in a variety of ways, starting with school visits and concluding with field testing.

- **Classroom Visits** Discussions and classroom observations took place in schools throughout the country to determine the key needs of teachers and students, the obstacles that they face in achieving their goals, and the types of materials that can help them achieve success.

- **Nationwide Research Surveys** Comprehensive mail surveys on high school math curriculum needs, instructional practices, student achievement levels, and teacher preferences regarding instructional materials were conducted early in the development process to guide the planning of the program.

- **Teacher Panels** Panels of expert teachers from different areas of the country participated in the development of the program by identifying instructional and curriculum needs, reviewing prototype outlines and sample materials, and providing suggestions for teaching-support publications.

- **Student Review Panels** Students from a wide range of schools nationwide reviewed selected lessons from the student edition in prepublication format for overall clarity and appeal. They also commented on the usefulness of specific features such as worked-out examples and study strategies.

- **Focus Tests** Focus tests in which teachers discussed their instructional goals and evaluated sample student and teacher materials were held in different areas of the country. The teachers participating in the focus groups were chosen to represent the wide range of types of schools, philosophies of instruction, and teacher characteristics (like number of years of teaching) in the teaching population. The feedback from these diverse groups of teachers was used to revise and refine the project plans prior to writing.

- **Teacher Reviewers** The Reviewers listed in the textbooks read manuscript or proof for all the chapters in the student edition in detail with regard to clarity, accuracy, and appropriateness for classroom use and made suggestions for revision when appropriate. Their feedback guided the revision of this material.

- **Field-Testing** Selected chapters from each course were taught in the classroom to test how successful students were in learning from the textbook. Some of the aspects tested were student achievement as measured by pre- and post-test scores, students' ease of learning from the material, and students' interest in the material.

Letter from the Authors

Dear Colleagues,

While writing this Algebra 1, Geometry, Algebra 2 series, we kept two goals in mind. First, we wanted to help your students learn the concepts and skills they need to be successful in mathematics. Second, we wanted to make your work easier by providing a comprehensive selection of teaching resources.

With these goals in mind, we reviewed many state curriculum and assessment guidelines, as well as those of the National Council of Teachers of Mathematics. This will ensure that students have the opportunity to learn the mathematics required to succeed in this course and on important assessments. We also used our extensive teaching experience to plan each course so that students are well prepared for succeeding courses.

Our philosophy is to write books that are mathematically correct, student friendly, and pedagogically sound. Our books also reflect the valuable input from the many teachers and students nationwide who have used earlier editions or pilot materials in the classroom. We believe that a balance of teaching approaches is effective. Thus, we offer clear, straightforward instruction in concepts and skills, combined with engaging student activities, motivating real-world applications, and useful problem solving and communication strategies.

As an author team that has written together for more than 14 years, we are committed to setting high mathematics standards for all students. We are committed to creating textbooks that have the support that students need, whether this is differentiated instruction for underachieving students, helpful study tips and worthwhile practice for most students, or thoughtful challenges for advanced students and other interested students.

We wish you the best in your pursuit of a quality mathematics education for every student in your classroom.

Ron Larson

Tim Kanold

Laurie Boswell

Lee Stiff

The Bigger Picture *of Mathematics*

by Dr. Ron Larson,
Lead Author of the Larson Mathematics Series

Ron Larson

What is Mathematics? You would think that this is an easy question to answer. But, it isn't.

Was mathematics discovered or was it invented? Is mathematics a logical system that exists devoid of applications or is mathematics a problem solving language whose very essence is solving real-life problems? Is correct calculation an important part of mathematics or is it only important to be able to "set the problem up" correctly? Is mathematics a system of rules or is mathematics a way of thinking?

As mathematics teachers, it is important for us to realize that mathematics is all of these things. We need to be careful to pass this understanding on to our students so that they will not have a narrow answer to the question "What is Mathematics?" We need to show them the bigger picture of mathematics.

MATHEMATICS AS CALCULATION

Most people equate mathematics with calculation. While this view of mathematics is too limited, it is still true that calculation plays a critical role in mathematics.

I worry when I hear educators dismissing calculation as "something best left to calculators" or even worse, as "drill and kill." I prefer to think about other phrases that describe the importance of calculation, such as "practice makes perfect."

When I drive across a bridge, I am thankful that the teacher of the civil engineer who designed the bridge believed in the importance of correct calculation.

The point is that part of our job as mathematics teachers is to help prepare skilled professionals who can fly planes, prescribe medicines, and do any of thousands of other technically difficult jobs.

But, there may be a stronger reason for teaching students to calculate correctly. It is simply the way that students learn best. They learn specific examples first. Generalization comes later.

After a career in mathematics, it is clear to me that we derive important insights into a mathematical theory when we see how it relates to actual calculations.

Continued

The Bigger Picture of *Mathematics*

Mathematics must be approached carefully, with the confidence that comes with knowing its rules.

MATHEMATICS AS A FIELD OF STUDY

Mathematics, of course, is not simply a collection of techniques for calculating sums, products, slopes, and so on. Mathematics has content. It has a vocabulary of defined and undefined terms. It has a collection of axioms and theorems.

The vocabulary and rules of mathematics comprise much of the curriculum we teach from grade school through high school.

Like it or not, the vocabulary of mathematics is formal. It has to be in order to provide for clear communication. It is important that students know the difference between an expression and an equation. We *evaluate* an expression. We *solve* an equation.

The rules of mathematics are also formal—they are rigid and unforgiving. We don't do students favors when we allow them to approach mathematics in a sloppy manner. Mathematics must be approached carefully, with the confidence that comes from knowing its rules.

MATHEMATICS AS A MODELING LANGUAGE

It is difficult to know which of the many faces of mathematics was the first to show itself on our planet. Surely, one of the first was the use of symbols and calculations to model and solve real-life problems.

To me, the perennial debate about the virtues of applied versus pure mathematics isn't productive. From a historical point of view, it seems clear that both must be taught—hand-in-hand—from grade school through high school.

To be good at mathematics, students must understand the various types of models we use: linear, quadratic, cubic, radical, rational, exponential, logarithmic, and trigonometric. They must learn to translate real-life situations to mathematical models, obtain mathematical solutions, and then translate those solutions back into the context of the real-life application.

MATHEMATICS AS LOGICAL THOUGHT

Perhaps the most difficult task we face as mathematics teachers is to teach our students to think logically.

The real world is filled with logical fallacy —advertisers and vendors and politicians and newscasters—who frequently use illogical arguments. As mathematics teachers, we can improve our nation and our world by teaching careful logical thought to our students.

In light of all these things, I hope that this text helps your students get a view of the bigger picture of mathematics.

The Bigger Picture in Algebra 1

The following chart shows how the various ways of looking at the mathematics described in the article on pages T41–T42 are developed across the *Larson: Algebra 1* book.

Because these strands of thought are integrated throughout the book, the lesson references here are selected examples in each category, not a comprehensive listing.

Mathematics as Calculation	
Performing calculations	**Evaluating expressions (1.3), operations with real numbers (2.2, 2.3), solving multi-step equations (3.3), finding slope (4.4), statistical measures (6.7), solving systems by combination (7.3), factoring polynomials (10.6)**

Mathematics as a Field of Study	
Using mathematical vocabulary	**Definitions of solution of an equation (1.4), rate and unit rate (3.8), intercepts (4.3), direct and indirect variation (4.5, 11.3), standard form of linear and quadratic equations (5.6, 9.1), polynomials (10.8)**
Knowing mathematical principles	**Order of operations (1.3), distributive property (2.6), equivalent inequalities (6.1), properties of exponents (8.1)**

Mathematics as a Modeling Language	
Using mathematical models	**Real number line (2.1), linear equations (4.2), predicting with linear models (5.7), linear systems (7.1), exponential growth and decay (8.5, 8.6), quadratic functions (9.3)**
Translating real-life situations to mathematical models, obtaining solutions, and then translating solutions back into real-life contexts	**Distance, rate, time (1.5), water pressure (1.7), income and expenses (2.2), membership fees (3.4), burning calories (4.2), phone charges (5.1), advertising trends (5.7), web site hits (7.1), octane mixtures (7.4), population growth (8.5), depreciation (8.6), shot put (9.4), pendulum period (9.8), genetic traits (10.3), circle graphs (11.2), batting average (11.8), nozzle pressure (12.3)**

Mathematics as Logical Thought	
Thinking logically	**Counterexample (2.1), equivalent equations (3.1), inductive and deductive reasoning (Ch. 3 Extension), logical equivalence (10.4), converse (12.5), conjecture and indirect proof (12.8)**

Providing *Universal Access*

DIVERSE STUDENTS

With careful planning, teachers can help all students reach a level of mathematical competence needed to continue their education in mathematics.

In most classrooms, students present a variety of achievement levels, skills, and needs. The goal for all students is the same: We want them to develop sufficient computational, procedural, and problem solving skills to provide a solid foundation for further study in mathematics. However, all students do not arrive at these competencies at the same time or in the same way. In this article we suggest research-based strategies teachers can use to modify curriculum and instruction for special needs students. The basic instructional plan in *Algebra 1* is designed for students who are achieving at near grade level; but throughout each chapter we include specific suggestions for students who are achieving above and below grade level, and for students who are not fluent in English (see also the article titled "Adapting Curriculum and Instruction for English Learners").

Student Groups Teachers may find it helpful to view students as members of four basic groups, as shown on the facing page. (English learners can be found in all four groups.) Teachers do not need to place students in these groups; the categories are suggested so teachers can plan ahead to meet different needs of students. Note the use of the term underachievers. This term is not synonymous with special education. It may include some special education pupils but includes many more students whose low achievement levels are the result of inadequate prior schooling or attendance, high mobility rates, or a host of other reasons that have nothing to do with their abilities or disabilities. The term underachievers is not meant to be a negative term. On the contrary, we believe that students achieving below grade level can be successful in mathematics given carefully designed instruction.

SETTING THE RIGHT TONE

There are three key strategies recommended for teachers as they adapt any program to students' needs:

- Use frequent assessment as a way to determine what each student does or does not know, and use that assessment as the basis for planning.
- Plan modifications of curriculum and instruction ahead of time so that you are ready to differentiate as the need arises.
- Use a variety of grouping strategies to facilitate learning. A combination of whole class instruction and temporary groupings of students, with groups organized around students' needs, will facilitate management of the variety of achievement levels and learning needs in the classroom.

Assessment, planning, and flexible grouping are essential to ensuring that your students have the optimal chance for success. In addition to these three key strategies, general guidelines for establishing a classroom designed to meet students' needs are:

1. Establish an atmosphere where students feel comfortable asking questions and are rewarded for asking about things they don't understand.
2. Maintain the same goals for all students. Allow additional time and practice for students who need it, and provide challenging alternatives for those who are ready to move more quickly.
3. Clearly identify the skill, concept, or standards you are working on and measure progress toward those ends.
4. Have students show their work. It is much easier for teachers to understand where a student gets confused if they have evidence of the student's thought process.

5. Try small modifications in curriculum and instruction before more drastic ones.

6. Don't persist with a strategy that is not working. Try something else.

7. Encourage effort and persistence, and celebrate successes with your students.

VARYING CURRICULUM AND INSTRUCTION

1. Time Most students whose achievement is below grade level will need more time. Students who are not fluent in English will need more time. The contents of this book might be offered over a two-year period, or two periods a day. Perhaps the day can be extended through study hall, regular homework assignments, tutoring, or Saturday, summer, or "off track" catch-up sessions. Advanced students might "test out" of portions of the book and complete the material in half a year, or they may compact two courses into one.

2. Presentation Instructing in a variety of ways and taking a single concept and explaining it verbally and visually with concrete and abstract examples provide students multiple opportunities for understanding. Factoring, for example, is a key concept for success in algebra. It can be introduced and practiced in a variety of ways: with tree diagrams, with repeated division by hand or by calculator, and with oral problems or games that require students to factor mentally. If students don't understand factoring when presented one way, they can be retaught using a different approach.

3. Task parameters Multi-step problems can be especially difficult for students. These types of problems are just combinations of simpler problems and can be broken down into those simpler steps, with additional help and practice at each step. Confusing elements can be minimized and extraneous material can be eliminated. For advanced students, simpler problems can be eliminated and more challenging ones (as suggested in each chapter) may be substituted.

4. Methods of assessment Students learning English may be able to demonstrate on paper what they cannot yet verbalize. Students with physical challenges may be unable to draw a graph but may be able to select the right graph from a series of options or verbally describe the graph so that someone else can draw it. Allow students to demonstrate their knowledge in a variety of ways while helping all students to master the skills and knowledge necessary to exhibit their understanding in standard ways.

Larson: Algebra 1 is organized so that much of the differentiation for special needs students is built into the design of the program. Note that simpler concepts are introduced before more complex ones. Ample practice is provided. Challenge

FOUR BASIC STUDENT GROUPS

Advanced Group

Advanced students have already completed some of the grade-level material. They make rapid progress and become bored with repetition. They may or may not have been formally identified as gifted or talented in the area of mathematics.

Suggested Plan

1. Assess what these students already know.
2. Allow these students to "test out" of chapters or assignments.
3. Substitute challenging assignments for easier ones.
4. Modify instruction so that it is more complex or more in-depth.

Grade Level Group

Students achieving at grade level may have minor, occasional difficulties but they can be assisted to maintain their progress with extra practice and individual or group assistance on an ad hoc basis.

Suggested Plan

1. Assess what these students already know.
2. Progress through *Algebra 1* at the recommended pace and sequence.
3. On an ad hoc basis, review or provide additional practice as needed.

Underachieving Group

These learners are not achieving at expected grade level but can, with a carefully designed program that provides targeted assistance. Systematic differentiation such as preteaching, reteaching, and additional instructional time should be planned for these students, as suggested in each chapter.

Suggested Plan

1. Assess what these students already know.
2. Provide additional scaffolding and the instructional variations suggested in this book.
3. Focus on the key concepts and present material systematically.
4. Vary the kinds of instruction so that students have several opportunities to understand.
5. Provide additional practice homework

Intensive Needs Group

Intensive needs students are those whose performance is two or more standard deviations below the mean on standardized measures. These students will probably already be eligible for special education services. This is a very small percentage of the general population.

Suggested Plan

1. Assess what these students already know.
2. Determine if these students have an IEP.
3. Refer students for special education testing or child study team discussion; enlist the help of specialists.
4. Carefully consider each student's most appropriate placement in mathematics.
5. Use the specific suggestions for underachieving learners.

Continued

exercises are included throughout the pupil text. The program provides students with models for conceptual understanding of the mathematical reasoning behind each key concept. Mathematical reasoning is stressed throughout each chapter. Vocabulary words, examples, and Guided Practice exercises are standard features of each chapter. Each lesson includes Mixed Review exercises so that students recall and use skills and understandings from previous chapters. These features were designed to help you meet the needs of the students in your class.

FOR STUDENTS WHO HAVE TROUBLE PAYING ATTENTION

Some students in your classroom may be formally identified as having Attention Deficit Hyperactivity Disorder (ADHD) or Attention Deficit Disorder (ADD). Others may exhibit the same learning challenges but may not be formally identified. Whether formally identified or not, students who have trouble paying attention generally share the following characteristics:

- Trouble paying attention is not just occasional. It occurs most or all of the time, across content areas, and is inappropriate for the age of the child.
- Forgetfulness, memory problems, losing things, disorganization
- Restlessness, fidgeting
- Socially inappropriate behavior such as excessive talking, interrupting others, and difficulty waiting their turn

These students may be very bright and capable in mathematics but have a hard time staying focused for long periods of time. They need to be taught strategies for organizing their work and keeping track of where they are. In general, students with attention problems need to be helped to develop coping strategies. The teacher should approach the student in a problem solving mode: "Let's find ways to help you concentrate and organize your work," rather than using one of the following strategies in a punitive way.

1. Present the work in smaller chunks over smaller time periods and then gradually increase expectations. If students have trouble completing long tests, for example, break the material into smaller quizzes and increase the length of the quizzes as the year progresses.
2. Use cumulative review and practice. Have students periodically review what they learned in previous chapters and provide additional practice if they have forgotten.
3. Make it more obvious what the student should focus on. For example, use the test generator to put only four problems on each page; use a large font; or use an index card or piece of cardboard with a hole cut out of the middle to place on the page so that the student can focus on one problem at a time. A pencil, finger, highlighter, or sticky paper can also be used by the student to keep track of which problem he or she is working on.
4. Have students race against the clock. For some students, racing against the clock to see how many problems can be completed accurately within a five-minute time period is more motivating than doing the same number of problems at their leisure. The time period can be extended gradually.
5. Help students develop simple strategies for bringing work to and from class. A two-pocket folder, where homework goes home in the left pocket and comes back in the right, is a simple way to keep track of assignments.
6. Allow movement and schedule breaks.
7. Minimize distractions by seating students that are easily distracted near the teacher and away from hallway noise. Tables with several students at a table are more distracting than rows of desks. When students are to work quietly, offer headphones to block out noise. Headphones can be set to play quiet music, "white noise," or can be used just as earplugs to help block out noise.
8. Graphic organizers such as Venn diagrams, tree diagrams, lists, outlines, tables, and charts can all provide structures for organizing and remembering information. Mental images, choral responses, or even hand signals can help students remember. Highlighters can be used to make sure that decimal points are lined up. Graph paper is excellent for keeping homework problems neat, even when a graph is not required.
9. Keep instructions simple and clear, especially at the beginning of the year. Establish routines (e.g., the week's homework is always due on Thursday; assignments are written in a specific place on the board; the last ten minutes of class is used to make sure everyone understands what homework is expected and how to do it). Students who know the routine find it easier to work independently.

FOR STUDENTS WHO HAVE TROUBLE UNDERSTANDING THE CONCEPTS

Success in mathematics, as in music, sports, or other areas, comes for most students only with hard work and persistent effort. Concepts may seem difficult at first, but with repeated teaching and practice virtually all students can master the mathematics they need to graduate from high school, accessa variety of jobs, and lay the foundation for further study in mathematics or a related field.

Several strategies can help students make steady progress in mathematics. These include:

1. Focus on key mathematical concepts.
2. Review key concepts and skills from earlier lessons, chapters, or years.
3. Preteach key concepts and vocabulary.
4. Anticipate problem areas.
5. Provide scaffolding (guided practice) for students who need extra help.
6. Think out loud to show hidden steps.
7. Provide a sample problem to which students can return when they get stuck.
8. Break problems into simpler components.
9. Explicitly teach students a variety of problem solving strategies and help them select one that fits the situation.
10. Present concepts in a variety of ways: visually, verbally, concretely, abstractly, etc.
11. Encourage students to draw a picture or use a visual aid such as a number line, graph, or diagram.
12. Provide sufficient practice.

Finally, good teachers are perpetual students themselves. They are always looking for ways to deepen their understanding of mathematics and for good ways to explain and teach mathematics to others.

FOR ADVANCED STUDENTS

Occasionally students can demonstrate mastery of all the mathematics expected to be learned in a given grade level. Repeating previously learned material for a year is deadly to these students. It can make them dislike mathematics. For these students, moving them up a grade level for math is a simple and cost-effective solution.

Most advanced students, however, are advanced in some areas but not in others. They tend to learn quickly and need more instructional material, as well as more difficult material. The student edition, teacher's edition, and ancillaries for this program provide challenging exercises that can be used when students have demonstrated competence in a particular area. These challenge exercises should be substituted for the easier exercises in a homework assignment or lesson. When they have the time and interest, all students should be encouraged to work the challenge exercises.

General strategies for differentiating the curriculum for advanced learners include:

1. Vary the pacing. Allow advanced students some flexibility in how they progress through the course. Students who can demonstrate mastery of the objectives for a given lesson or chapter can be working on challenge exercises. Advanced students may become fascinated with a particular aspect of mathematics and want to spend more time on it.
2. Differentiate in terms of depth. Encourage advanced students to delve more in depth into mathematics. Looking at the details and the patterns; studying the language of the discipline; and looking at trends, themes, properties, theorems, proofs, and unanswered questions can enrich the curriculum for advanced students.
3. Differentiate in terms of complexity. Advanced students may be ready to connect ideas across disciplines in ways characteristic of older students or adults. Encourage them to investigate relationships between mathematics and art, history, science, and music, and to look at the development of mathematics over time.

USING GROUPING TO BENEFIT ALL STUDENTS

Grouping advanced learners together for investigations of challenge problems can provide you with time to work more closely with a group of students who need help in a particular area. Alternatively, while students who need more help are working on additional reinforcement activities or practice, you can work with a group of advanced students on a challenge project. Groups can be organized and revised daily, weekly, or by lesson according to how proficient students are with the concepts and skills targeted for that day, week, or lesson. At times you may have only one student who is ready for a challenge problem; at other times the whole class may be ready. Flexible grouping is the key to ensuring that students do not become "tracked." Asking advanced students to report to the whole class on their progress on challenge problems can provide the opportunity for the whole class to engage in more abstract and theoretical thinking.

Adapting Curriculum and Instruction *for English Learners*

English learners come to the classroom with all the variety of English speakers in regard to mathematics achievement. They may be at, behind, or ahead of grade expectations in mathematics. They may be gifted or eligible for special education services. They may have been born in the United States, or they may have arrived in this country very recently. They may speak one or more languages, and they may be literate in one or more languages other than English. They may be nearly fluent in English or have beginning or intermediate levels of understanding and production. They have in common one characteristic: They are all learning English.

With careful planning, teachers can maximize success for English learners in the mathematics classroom. Assessing each student's competencies in mathematics and English will form a basis for program planning.

GETTING TO KNOW YOUR STUDENTS

Before school starts, check the cumulative folder on each student in your class to determine which ones are learning English. See if there is recent testing. Two types of testing are most useful: mathematics achievement and reading achievement levels. If no recent test information is available, you may instead administer a pre-course math test and ask English learners to write a dictated paragraph in English to assess their reading and writing skills. Use the chart on the facing page as a guide to understanding student assessment data.

SUGGESTIONS FOR MATHEMATICS TEACHERS OF ENGLISH LEARNERS

1. Allocate additional time for mathematics. Many students will be translating from English to their primary language and back again. The meaning of many words will not be immediately clear. When you ask questions, allow extra time for students to respond. Reading mathematics textbooks and understanding what is asked for in a word problem will be slower.

2. Use student's background knowledge. Some English learners will have developed substantial background in mathematics; others will have very little. Find out what students know and then build on that knowledge.

3. Reduce the amount and sophistication of the English language used. This may be done by reordering the lessons in each chapter to begin with key vocabulary, followed by problems with a minimum of written English, followed by at least one word problem each day. Choose word problems that don't rely on assumed background knowledge. Monitor and simplify the speech you use. Speak more slowly, avoid idioms and slang, be precise and concise, and use short sentences and simple vocabulary. Using hand gestures and pictures as well as words aids communication.

4. Introduce one concept per day. Keeping the focus of each day simple will aid students in understanding the point of the lesson. Focus on key concepts, and use mathematics instructional time well.

5. Use a variety of different methods for getting a point across. Presenting concepts verbally and visually, with concrete examples and in abstract mathematical symbols, and using pictures, graphs, diagrams, and charts will enhance the chance that students will understand at least one of the presentations. As you introduce a new word, rule, or property, write it down.

6. Provide opportunities for English learners to interact with their English-speaking peers. Students who are learning a language need to hear native speakers using the language, and they need opportunities to use their new mathematics vocabulary in their speech and in their writing.

7. Provide opportunities for English learners to discuss their understandings with each other, confirm the homework assignments, or ask questions of each other in whatever language they may have in common.

8. Allow English learners to demonstrate what they know in a variety of ways. When students first learn a language, they are usually shy about speaking. They generally understand spoken language before they can produce it. Students who have recently arrived from another country with good prior schooling may be able to read in English but not speak it. Allow students to point, nod, gesture, draw a picture, or work math problems without words as they learn English.

9. Extend mathematics instructional time through homework, an extra class period, summer school, or tutoring. Many of the language-related suggestions for English learners in this series can be carried out in collaboration with the language arts teacher.

10. Keep on hand picture dictionaries, foreign language dictionaries, multi-language math glossaries, and drawing materials.

SPECIFIC SUGGESTIONS

In many of the lessons in the Teacher's Edition you will find suggestions to help you modify curriculum and instruction so that the content is accessible to English learners. In these *English Learners* notes in the side margins, we will provide you suggestions designed to (1) teach vocabulary commonly used in mathematics and (2) explain mathematical concepts in a variety of ways. Some of the vocabulary study in this book may be review for your students but for those students who need more systematic study, using the suggestions in your classroom will ensure that students have refreshed their understanding of basic terms prior to statewide testing that generally occurs toward the end of each school year. We recommend that if you have English learners in your classroom, you skim all the suggestions for English learners in a chapter so that you may use them as you need them.

ENGLISH LEARNERS

	Low Mathematics Achievement: Who is this student?	Low Mathematics Achievement: What to do?	High Mathematics Achievement: Who is this student?	High Mathematics Achievement: What to do?
Low Reading Achievement	Student may be new to the class, school, or country. Student may have had inadequate schooling. Student may have moved often. Student may be unmotivated or have test anxiety. Low reading achievement may be depressing mathematics scores. Student may have gaps and holes in knowledge. Student may need special education assistance.	Examine cumulative folder for other testing, notes, etc. Delay any testing for a week or two. Help student feel comfortable in the class during that period of time. Administer mathematics achievement test and reading test, preferably in an individual setting. Plan to assess this student at weekly intervals and closely monitor classroom work to determine if progress is being made. Look at the English Learners suggestions in each chapter.	Student has had good prior mathematics instruction. Mathematics is an area where this student can excel. Math achievement level may actually be higher than scores indicate. (Limited English reading skills affect mathematics achievement as well.) Word problems will be especially difficult.	Mathematics instruction should proceed at normal or near normal pace. Student should be involved in a systematic English language development program and intensive reading program outside of mathematics class. Spend part of each class period on mathematics vocabulary study. Provide a bilingual dictionary or math glossary and a grade-level mathematics text in the home language for home use. Look at the English Learners suggestions in each chapter for those that are most useful.
High Reading Achievement	Student may have been designated as an English learner because oral skills lag behind reading skills. Student may not test well in mathematics. Most students can make rapid progress in mathematics; a few may have learning difficulties that require the help of a specialist.	Assess mathematics achievement in a variety of ways. Concentrate on developing oral fluency. Focus on vocabulary specific to mathematics. Use a student's reading ability to improve his or her math scores.	May be a student who is ready for re-designation as a fluent English speaker. May need extra study in academic vocabulary, i.e., the specialized vocabulary of mathematics. Given systematic instruction, this student should be able to achieve at or above grade level.	Scan all of the suggestions for English learners in each chapter and progress through the ones the student needs as quickly as possible. Monitor carefully to make sure this student continues to progress at a reasonable pace.

Differentiating *Instruction*

No Child Left Behind

The new federal law known as No Child Left Behind (NCLB) highlights the need to ensure that all students, whether they are struggling, average, or advanced learners, have opportunities to make continuing progress in developing the skills they need to become successful adults.

This law charges the states with the responsibility of establishing statewide accountability systems based on challenging standards and annual statewide progress objectives.

McDougal Littell's *Larson: Algebra 1* series supports the goals of NCLB. The program is based on state curriculum standards and assessment objectives (see page T38). The series provides you with helpful materials for diagnosing how well students understand the material and for providing remediation. It also emphasizes important test-taking skills and problem-solving strategies.

ONGOING DIAGNOSIS

Intervention begins with diagnosis.

Materials to diagnose student understanding are provided before, during, and following each chapter and lesson.

- The **Basic Skills Workbook: Diagnosis and Remediation** helps diagnose how well students have mastered key prerequisite skills for the course.
- Each **Chapter Resource Book** contains a **Prerequisite Skills Review** for the chapter.
- **Key Vocabulary** for each chapter, both new and review, is listed on the **Study Guide** page to prepare for the upcoming chapter.
- There is a **Skill Review** for each chapter on the **Study Guide** page.
- **Warm-Up Exercises** and **Daily Homework Quizzes** on transparencies for the beginning of each lesson provide two mechanisms for reviewing pre-lesson skills.
- **Checkpoint Exercises** and **Closure Questions** in the Teacher's Edition help you monitor how well students are grasping the skills and concepts as you present each lesson.
- **Vocabulary Check**, **Concept Check**, and **Skill Check** items in the Student Edition help you determine if the students are ready for the homework exercises.
- **Homework Check** in the Teacher's Edition identifies exercises from the homework assignment that you can use to determine whether students have mastered the key skills and concepts.
- You can also use the **Test and Practice Generator** to create online practice sheets and see a report of the results to monitor individual progress.

...in the diverse classroom

DIFFERENTIATED INSTRUCTION AND PRACTICE

Being able to differentiate instruction and practice can help you reach all students. The *Larson: Algebra 1* program is designed to help you in this effort. You can use **EasyPlanner Plus Online** and **EasyPlanner CD-ROM** to preview and select the resources as you develop your lesson plans.

DIFFERENTIATED INSTRUCTION

- **Hands-On Activities** in the Student Edition and **Alternative Lesson Opener** transparencies
- **Extra Examples** and **Meeting Individual Needs** in the Teacher's Edition
- **Notetaking Guide Workbook** and transparencies
- **Chapter Audio Summaries CDs** in English, Spanish, and Haitian Creole
- **Power Presentations** provide electronic lesson presentations
- **eEdition Plus Online** offers the Student Edition in an online format

DIFFERENTIATED PRACTICE

- **Leveled exercises** labeled in the Teacher's Edition; leveled homework assignments; leveled practice worksheets
- **Challenge exercises** in the Student Edition; **Daily Puzzlers** in the Teacher's Edition; and **Challenge: Skills and Applications** in the Chapter Resource Books
- **Chapter Projects with Rubrics** in the Chapter Resource Books and an **Explorations and Projects Book**
- **Exercises in Spanish** with exercise sets for each lesson translated into Spanish
- **Resources in Spanish** and an **English-Spanish Glossary**

PROBLEM SOLVING STRATEGIES

In order for students to demonstrate mastery on state and national tests, they must be able to read and interpret word problems and apply appropriate strategies to solve them. *Algebra 1* incorporates problem solving throughout the textbook to help students learn to apply computational skills in context.

- **Problem Solving Plan** A four-step problem solving plan—in which students read, plan, solve, and look back—is introduced in the *McDougal Littell Middle School Math* series and continued in the high school series.
- **Problem Solving Strategies** and **Modeling** are reviewed and expanded in all chapters of the high school books.
- **Word Problems at All Levels** Word problems appear at all three difficulty levels in the exercises.
- **Multi-Step Problems** Problem Solving exercises in the Test Preparation section help students prepare for multi-step problems on state and national tests.
- **Test Practice** Test-practice exercises are at the end of each lesson and each chapter.

Strong problem solving skills help students become successful adults.

Continued

Differentiating *Instruction* *Continued*

ASSESSMENT

Students need practice with the various types of questions on standardized tests.

Students need practice with the various types of questions on standardized tests. The *Larson: Algebra 1* program provides diagnostic, formative, and summative assessment resources for measuring student progress on an ongoing basis.

- The **Student Edition** has test-practice questions at the end of every exercise set, including multiple choice, multi-step problems, and quantitative analysis items. Included also are quizzes, reviews, traditional and standardized chapter tests, and cumulative tests.
- The **Teacher's Edition** has a **Quiz** for every lesson, available on transparency.
- The **Standardized Test Practice Workbook** contains test practice for each lesson.
- The **Chapter Resource Books** have leveled chapter tests, alternative assessment, and cumulative review.
- The **Test and Practice Generator** can customize quizzes and tests. **Online quizzes** and **State Test Practice** are available at www.classzone.com.

BUILDING TEST-TAKING SKILLS

It is important for students to build strong test-taking skills in order to be successful on annual high-stakes assessments required by the NCLB Act. *Larson: Algebra 1* provides instruction and practice with test-taking skills throughout the program.

- A **Test Preparation** section is in every exercise set. These sections include multiple choice, multi-step, quantitative comparison, and writing questions.
- A **Chapter Standardized Test** appears at the end of each chapter, including a Test-Taking Strategy.
- The **Chapter Resource Book** for each chapter includes an **SAT/ACT Chapter Test.**
- The **Standardized Test Practice Workbook** has further practice to prepare students for standardized tests.

RETEACHING AND REMEDIATION

Students sometimes need reteaching in order to understand concepts better or additional practice in order to master key skills. The *Larson: Algebra 1* program provides a variety of resources to help students achieve success. Some of these resources can also be used by absent students to help them catch up.

- **Student Edition** includes **Skills Review** notes, **Chapter Reviews** with more vocabulary and skill practice, **Cumulative Practice** at the end of every third chapter, a **Skills Review Handbook** with reteaching and practice for pre-book skills, and **Extra Practice** for every lesson.
- **Teacher's Edition** includes **Extra Examples** and **Common Error** notes that can be used to help clarify understanding.
- **Technology Resources** include interactive and engaging material for reteaching and practice.
- **Chapter Resource Books** include **Prerequisite Skills Review**, **Practice** masters (Levels A, B, C), **Challenge: Skills and Applications** masters, **Reteaching with Practice** masters with worked-out examples, **Quick Catch-Up for Absent Students**, and **Chapter Review Games and Activities**.
- **Practice Workbook with Examples** includes reteaching and practice pages for each lesson.

Reading, Writing, Notetaking
Vital Skills for Today and the Future

Vital Skills Today more than ever, students need strong skills in reading, writing, and notetaking in mathematics in order to understand course content, be successful on important state and national assessments, and develop the ability to become independent learners. Acquiring these skills in high school will build an important foundation for more advanced courses and for adult life. McDougal Littell's *Larson: Algebra 1* provides many opportunities in the textbook and in the teacher's materials to help students develop their reading, writing, and notetaking skills.

Recent Research Recent brain research and classroom research in reading and writing have provided new insights into learning and also confirmed the value of well-known practices of successful teachers. Although the focus of this research is often on reading in language arts and social studies, many of the strategies also help those reading mathematical material. Two important aspects of reading addressed by research are vocabulary development and reading comprehension. *Larson: Algebra 1* offers substantial learning support in these core areas.

VOCABULARY DEVELOPMENT

The textbook provides strong support to students in learning, practicing, and reviewing vocabulary. On the Study Guide page at the start of each chapter, key review words are listed along with new words from the chapter. Throughout a chapter, new vocabulary in a lesson is emphasized by boldface type with yellow highlighting. In the Teacher's Edition, Focus on Vocabulary notes suggest ways teachers can help students read and learn new vocabulary words and give questions teachers can ask to diagnose students' vocabulary knowledge. *See pp. 130, 139–140; TE pp. 110, 140, 229, 355.*

> **FOCUS ON VOCABULARY**
> What is the difference between the reciprocal of a number and the opposite of a number? **The product of a number and its reciprocal is 1; whereas the sum of a number and its opposite is zero.**

In the Exercises, the Guided Practice exercises (which help students get ready for homework) include vocabulary as well as concept and skill practice. The Chapter Reviews provide a list of key vocabulary. In addition, there is a complete Glossary that includes examples and diagrams at the back of the book. *See pp. 237, 264, 817.*

READING COMPREHENSION

The teacher's materials give specific suggestions for helping your students read their textbook. ***See, in particular, the Chapter Resource Books, pp. 7 and 8.***

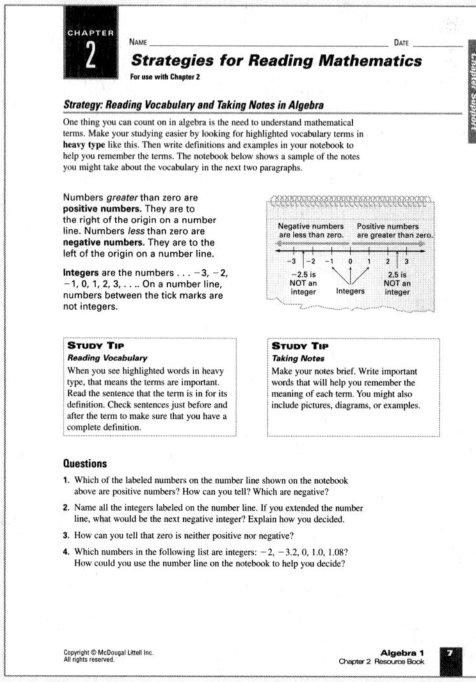

Reading, Writing, Notetaking

Vital Skills for Today and the Future Continued

Establishing a Context A useful comprehension strategy supported by both brain and classroom research is connecting new learning to prior knowledge. This strategy is incorporated throughout *Algebra 1* in the What you should learn/Why you should learn it lists at the beginnings of lessons; in the chapter-opening applications that review prerequisite skills; and in the Study Guide pages at the start of each chapter. In the Teacher's Edition, Motivating the Lesson notes in some lessons help teachers give students a real-life context to help them anticipate the math facts covered in the lesson. ***See pp. 234, 271, 272; TE pp. 167, 301, 369.***

Facilitating Understanding In order to create a student-friendly book, the authors kept these principles in mind as they wrote: Students can learn new concepts more easily when they are presented in short sentences that use simple syntax and are accompanied by appropriate tables, charts, graphs, and diagrams.

Clear definitions that enable students to determine easily whether a particular mathematical object fits the definition or not are essential for comprehension. Students need special help in understanding the symbols of mathematics and knowing how to use them in writing algebraic expressions. ***See pp. 9, 140, 234, 477.***

Student Help boxes in the side margins of the student book, such as Skills Review, Look Back, and Homework Help, show students where to look for help in many situations. In addition, Study Tip boxes help students avoid common errors or point out subtleties in the vocabulary or concepts in a lesson. ***See pp. 64, 66, 109, 230, 232, 251, 335, 577.***

Reflecting on Learning An effective strategy for increasing both reading comprehension and thinking skills is reflection on what has been read or learned. The Chapter Summary page in each chapter reviews what students learned in the chapter, and why, and suggests how the topic of the chapter fits into the bigger picture of algebra. Throughout the book, students explain their reasoning in Critical Thinking and Writing exercises and in the Project features. In the Teacher's Edition, Closure Questions, Activity Assessments, and Concluding the Projects all suggest questions to ask to test understanding and promote classroom discussion. ***See pp. 263, 372, 373, 393, 571; TE pp. 198, 229, 483.***

Using Graphic Organizers Graphic organizers such as charts, Venn diagrams, or concept maps can be especially helpful for classifying mathematical objects such as types of numbers, operations, or geometric figures. Verbal models are shown throughout the text to help students translate a real-life situation into an algebraic model. ***See pp. 32, 33, 132, 156, 228, 346, 348, 397, 765, 777, 785.***

WRITING OPPORTUNITIES

In order to become good writers, students need frequent opportunities to practice their writing skills. These opportunities occur throughout the textbook in the Critical Thinking and Writing exercises, Developing Concepts Activities, Math and History notes, and Projects. ***See pp. 39, 137, 207, 245, 307, 392–393.***

EFFECTIVE NOTETAKING

Taking effective notes is an important reading, learning, and review strategy, and yet math teachers throughout the country report that some students enter high school with few if any notetaking skills. Thus, the authors identified the goal of helping students develop their notetaking skills as an important objective of the program, and they have incorporated many notetaking aids into the program. See, especially, the Study Strategy notes on Study Guide and Chapter Summary pages as well as Concept Summary and Property boxes in many lessons and Examples in the Chapter Reviews. ***See pp. 2, 17, 62, 65, 73, 202, 213, 264–266, 309, 396, 477, 642, 708.***

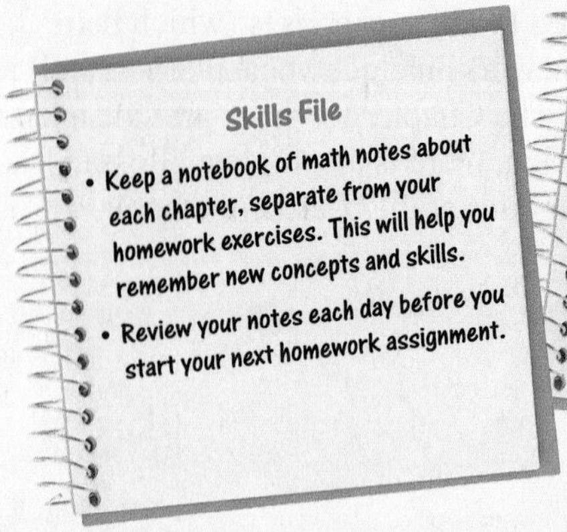
Skills File
• Keep a notebook of math notes about each chapter, separate from your homework exercises. This will help you remember new concepts and skills.
• Review your notes each day before you start your next homework assignment.

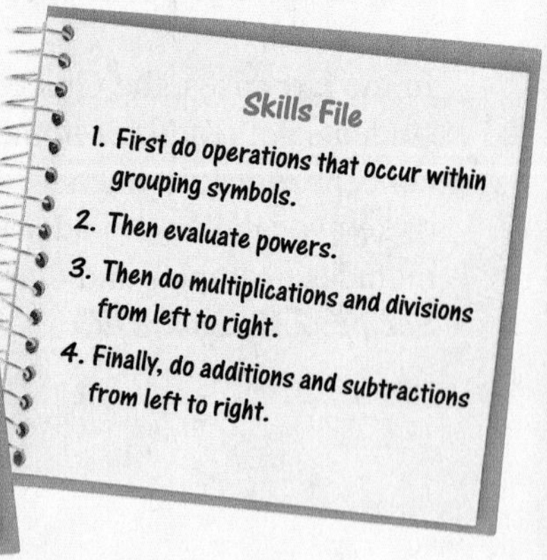
Skills File
1. First do operations that occur within grouping symbols.
2. Then evaluate powers.
3. Then do multiplications and divisions from left to right.
4. Finally, do additions and subtractions from left to right.

CRISS

CREATING INDEPENDENCE THROUGH STUDENT-OWNED STRATEGIES

Project *CRISS* was founded to help develop thoughtful, independent readers and learners through instruction based on strategies arising from scientifically-based cognitive and social learning research of the past 25 years. The chart below lists the key CRISS principles, together with examples from McDougal Littell's *Larson: Algebra 1* that support them.

Background Knowledge Background knowledge is a powerful determinant of reading comprehension. Look for introductory material that both builds upon and provides background knowledge.	Study Guide, 2, 62, 130, 202, 396, 448, 502, 574, 642, 708; Skills Review, 4, 22, 26, 31, 62, 130, 202, 272, 332, 396, 448, 502, 574, 642, 708; Look Back, 73, 76, 80, 97, 110; Think and Discuss, 1, 61, 129, 271, 331, 395, 573, 641, 707
Active Involvement Good readers are actively involved with the text and in their learning. Look for methods and activities that show students how to be active and that provide ways for them to be active.	Keeping a Math Notebook, 2, 53; Chapter Review, 54–56, 190–192, 440–442; Worked-out Examples, 3–4, 9–10, 16–17
Discussion Students need many opportunities to talk with one another about their reading and about what they are learning. Look for activities that open doors for discussion, in pairs, in groups, and with the whole class.	Developing Concepts, 16, 23, 25, 31; Think and Discuss, 1, 61, 129, 271, 331, 395, 573, 641, 707
Metacognition Good readers are metacognitive. They are goal-directed, and they know how to interact with print to construct meaning. Look for opportunities to help students become more aware of and to discuss the how and why of their learning.	Skills Review, 4, 17, 22, 26, 31, 62, 130, 202, 272, 332, 396, 448, 502, 574, 642, 708; Homework Help, 6–7, 10, 12, 17, 19, 27, 29; Look Back, 73, 76, 80, 97, 110; Keystroke Help, 11, 74, 209, 367, 492; Key Vocabulary, 2, 62, 130, 202, 272, 332, 396, 448, 502, 574, 642, 708; Vocabulary Check, 6, 12, 19, 27, 35; Study Tip, 2, 21, 33; Test Preparation, 8, 14, 30, 194, 388, 468; Error Analysis, 27, 673, 679
Writing Students need multiple opportunities to write about what they are learning. Look for activities that occur naturally throughout the textbook and activities in teacher's materials that are correlated to the textbook.	Keeping a Math Notebook, 2, 53; Think and Discuss, 1, 61, 129, 271, 331, 395, 573, 641, 707; Multi-Step Problems, 14, 21; Writing, 14, 20–21, 30, 45, 50, 70, 84, 90, 99, 106, 232, 245, 254, 260, 284, 307, 373, 377, 389, 397, 410, 416, 689, 757
Organizing for Learning Good readers know a variety of ways to organize information for learning. Look for suggestions for use of tables, charts, graphic organizers, and other devices that help students organize and interpret their learning.	Keeping a Math Notebook, 2, 53; Problem Solving Strategies, 5, 33–34, 141, 162, 258, 275; Study Strategies, 53, 62, 121
Explanation and Modeling Students become strategic when teachers model processes. Look for modeling that explains the why of a method or strategy.	Worked-out Examples, 3–4, 9–10, 16–17, 405–406, 643–644; Problem Solving Strategies, 5, 33–34, 141, 162, 258, 275; Using Technology, 15, 18, 92, 209, 299, 382, 462
Teaching for Understanding Students come to understand by doing a variety of activities. Look for a rich variety and choice of activities, exercises, and projects.	Challenge, 8, 14, 21, 30; Think and Discuss, 1, 61, 129, 271, 331, 395, 573, 641, 707; Developing Concepts, 16, 23, 25, 31; Multi-Step Problems, 14, 21; Writing, 14, 20–21, 30, 45, 50, 70, 84, 90, 99, 106, 232, 245, 254, 260, 284, 307, 373, 377, 389, 397, 410, 416, 689, 696, 757; Critical Thinking, 13, 21, 49, 76, 91, 106, 127, 143, 150–151, 158, 165, 170, 185, 455, 460, 462, 523, 538, 540, 546, 630, 698, 714, 726

Further reading: A helpful reference is Teaching Reading in Mathematics, by Mary Lee Barton and Clare Heidema, published by Mid-continent Research for Education and Learning (MCREL), Aurora, Colorado.

Research-Based *Solutions*

Research supports a variety of learning strategies.

The *Larson: Algebra 1* program reflects current research in education. The Student Edition (**SE**), Teacher's Edition (**TE**), Chapter Resource Books (**CRB**), and other ancillary materials provide opportunities for teachers and students to experience a number of different learning strategies both in school and at home.

One group of instructional strategies used in this program are the instructional strategies presented in *Classroom Instruction that Works**, a publication from the Association for Supervision and Curriculum Development. The nine strategies discussed in that publication are summarized below, along with some specific instances of the strategy's use in the *Larson: Algebra 1* program.

1. Identifying Similarities and Differences

This strategy includes comparing and classifying, and suggests representing comparisons in graphic or symbolic form.

Algebra 1 examples include: visually comparing mathematical ideas through the use of diagrams, tables, and graphs, and organizing and classifying information through the use of Concept Summary Boxes.

See, for example, SE204, SE234, SE256, SE463, CRB4 pp. 7, 20

2. Summarizing and Notetaking

This strategy includes deciding when to delete, substitute, or keep information when writing a summary and using a variety of notetaking formats — e.g., outlines, webbing, or a combination technique — and suggests encouraging students to use notes as a study guide for tests.

Algebra 1 examples include: Concept Summary and Property boxes, Guided Practice, Chapter Study Guide, Chapter Summary (SE); Focus on Vocabulary (TE); Study Strategy (CRB)

See, for example, SE250, SE256, SE259, SE396, SE561, TE236, CRB4 p. 3

3. Reinforcing Effort and Providing Recognition

This strategy includes making the connection between effort and achievement clear to students and providing recognition for attainment of specific goals to stimulate motivation.

Algebra 1 examples include: Guided Practice, Quiz, Cumulative Practice (SE); Checkpoint Exercises, Closure Question, Homework Check (TE); Parent Guide for Student Success, Warm-Up Exercises and Daily Homework Quiz (CRB)

See, for example, SE296, SE352, SE390, TE227, TE229, TE230, CRB4 pp. 3–4, 52

4. Homework and Practice

This strategy includes making the purpose of homework assignments clear to students and focusing practice assignments on specific elements of a complex skill.

Algebra 1 examples include: Practice and Applications, Homework Help boxes, Mixed Review, Chapter Review, Extra Practice (SE); Chapter Pacing Guides (TE); Prerequisite Skills Review, Practice A, B, or C, Reteaching with Practice (CRB)

See, for example, SE303–305, SE324, SE801, TE330C–330D, CRB4 pp. 5, 87, 89

5. Nonlinguistic Representations

This strategy includes creating nonlinguistic representations — including creating graphic organizers, making physical models, generating mental pictures, drawing pictures and pictographs, and engaging in kinesthetic activity — to help students understand content in a whole new way.

Algebra 1 examples include: Chapter Study Guide, In-Class Activity (SE); Multiple Representations, Managing the Activity (TE); Alternative Lesson Openers (CRB)

See, for example, SE153, SE240, SE708, TE153, TE228, CRB4 p. 3

6. Cooperative Learning

This strategy includes a description of the five defining elements of cooperative learning — positive interdependence, face-to-face interaction, individual and group accountability, interpersonal and group skills, and group processing — and gives suggestions for grouping techniques.

Algebra 1 examples include: In-Class Activity (SE); Managing the Activity, Managing the Project (TE); Cooperative Learning Activity, Activity Lesson Opener (CRB)

See, for example, SE153, SE476, TE299, TE525, TE570, CRB4 pp. 27, 39, 61

7. Setting Objectives and Providing Feedback

This strategy includes using instructional goals to narrow what students focus on and suggests providing feedback that is specific to a criterion and encouraging students to personalize their teacher's goals and to provide some of their own feedback.

Algebra 1 examples include: Chapter Study Guide, What You Should Learn lists in each lesson, Mixed Review exercises, Chapter Summary (SE); Chapter Goals, Common Error, Planning the Activity, Project Goals, Grading the Project rubrics (TE); Quiz, Daily Homework Quiz, Alternative Assessment Rubric, Project: Teacher's Notes (CRB)

See, for example, SE202, SE210, SE239, SE263, TE394, TE483, TE570–571, CRB4 pp. 49, 52, 129, 131

8. Generating and Testing Hypotheses

This strategy includes using a variety of structured tasks to guide students through generating and testing hypotheses, using induction or deduction, and suggests asking students to clearly explain their hypotheses and conclusions to help deepen their understanding.

Algebra 1 examples include: In-Class Activity, Why You Should Learn It lists in each lesson, Projects, Math and History (SE); Closing the Activity, Concluding the Project, Mathematical Reasoning (TE); Real-Life Application (CRB)

See, for example, SE397, SE411, SE417, SE662, TE71, TE392, CRB4 p. 106

9. Cues, Questions, and Advance Organizers

This strategy includes asking questions or giving explicit cues before a learning experience to provide students with a preview of what they are about to experience; using verbal and graphic advance organizers, or having students skim information before reading as an advance organizer.

Algebra 1 examples include: Chapter opener Applications, Study Guide/Preview, pre-reading lesson elements such as What You Should Learn lists and Example heads, questions in In-Class Activities, problem solving plan (SE); Motivating the Lesson (TE); Alternative Lesson Opener, Warm-Up Exercises (CRB)

See, for example, SE34, SE61, SE62, SE234–236, SE333, TE341, TE585, CRB4 pp. 12, 39, 67, 97

*Marzano, Robert J., Debra J. Pickering, and Jane E. Pollock, *Classroom Instruction that Works: Research-Based Strategies for Increasing Student Achievement* and its accompanying handbook. Alexandria, Virginia: Association for Supervision and Curriculum Development, 2001

NAEP
National Assessment of Educational Progress

*The **NAEP** is used to assess student understanding of math across the nation. The chart below lists the topics assessed by the **NAEP**, and lessons and features from* McDougal Littell's *Larson: Algebra 1 that address them.*

Number Properties and Operations	
1) Number sense	1.4, 1,5, 2.1, Act. 2.2, 2.2, Act. 2.3, 2.3, 3.6, Act. 8.1, 8.1 through 8.4, 9.1, 9.2, 11.1, Skills Rev. Handbook
2) Estimation	Act. 1.5, 3.6, 4.7, 4.8, 5.3, 8.5, Tech Act. 9.4, 11.2, Appendix 3
3) Number operations	1.2, 1.3, Ch. 2, 3.6, Act. 8.1, 8.1 through 8.5, 9.1, 9.2, 11.1, 11.2, 11.3, 11.4, 11.5, 11.6, Skills Rev. Handbook
4) Ratios and proportional reasoning	2.2, 9.1 through 9.6, Ch. 11, 12.1, 12.3, 12.4, Act. 12.7, 12.7, Skills Rev. Handbook, Appendix 2
5) Properties of number and operations	1.2, 1.3, Ch. 2, 3.1 through 3.4, Ch. 3 Extension, 8.1 through 8.6, 9.1, 11.1, 12.8, Skills Rev. Handbook

Measurement	
1) Measuring physical attributes	1.1, 1.2, 2.1, 3.2, 3.3, 3.4, 3.5, 3.6, 3.7, 3.8, Chs. 1–3 Project, 4.2, 4.3, 4.4, 4.5, 4.6, 4.7, 4.8, Ch. 5, 6.1, 6.2, 6.5, Ch. 7, 8.1, 8.2, 8.3, 8.4, 8.5, 8.6, Ch. 9, 11.1, 12.1, 12.3, 12.4, Skills Rev. Handbook, Appendix 3
2) Systems of measurement	1.1, 2.5, 3.3, 3.4, 3.7, 3.8, 5.1 (Ex. 3), 11.1, Act. 12.7, Skills Rev. Handbook

Geometry	
1) Dimension and shape	1.1, 1.3, 3.7 (Exs. 40–44), 6.3, 7.5, 12.5, Act. 12.7
2) Transformation of shapes and preservation of properties	Act. 12.5, Act. 12.7, 12.7
3) Relationships between geometric figures	2.6, 3.2, 5.2, 5.3, 6.3, 7.5, 8.1 (Exs. 75, 76), 11.4, 12.5, 12.6, 12.7
4) Position and direction	12.6, 12.8
5) Mathematical reasoning	3.2, 3.5, 6.3, 7.5, Act. 12.5, 12.6, Act. 12.7, 12.8

Data Analysis and Probability	
1) Data representation	1.6, 2.4, 3.5, Act. 3.6, 3.8, Ch. 4, Ch. 5, 6.2, 6.6, 6.7, Ch. 7, 8.2, 8.3, 8.5, 8.6, Ch. 9, Chs. 7–9 Project, Ch. 10, 11.3 through 11.5, 11.8, 12.1, 12.2, 12.3, 12.4, Skills Rev. Handbook, Appendix 1
2) Characteristics of data sets	1.6, 2.4, 5.4, 6.6, 6.7, Tech Act 8.6, Chs. 7–9 Project
3) Experiments and samples	2.8, Act. 3.6, Chs. 4–6 Project, Act. 8.6, Chs. 7–9 Project, Appendix 1
4) Probability	2.8, 3.6 (Ex. 51), 8.1, 8.3, Act. 8.6, Math and History p. 596, 11.4, 11.5 (Exs. 46, 47), Skills Rev. Handbook, Appendix 2

Algebra	
1) Patterns, relations, and functions	Ch. 1 opener, Act 1.4, 1.7, 2.3, 2.6, 2.7, 3.7, Ch. 3 Extension, Ch. 4, Ch. 5, 6.1 through 6.5, Ch. 7, Act. 8.1, 8.1, 8.2, 8.4, Act. 8.6, 8.6, 9.2 through 9.8, 10.3 through 10.8, 11.3, 11.6 (Ex. 35), 11.8, 12.1, Act 12.4, 12.4
2) Algebraic representations	1.1, Act. 1.2, 1.5, 1.7, 2.5, Act. 2.6, 2.6, 2.7, Ch. 3, Ch. 4, Ch. 5, 6.1 through 6.5, 7.1, Ch. 7, 8.1 through 8.6, Ch. 9, Ch. 10, Ch. 11, 12.1 through 12.7
3) Variables, expressions, and operations	1.1, Act. 1.2, 1.5, 1.7, 2.2, 2.3, 2.5 through 2.7, Ch. 3, Ch. 4, Ch. 5, 6.1 through 6.5, Ch. 7, 8.1 through 8.3, 8.5, 8.6, Ch. 9, Ch. 10, Ch. 11, 12.1 through 12.7
4) Equations and inequalities	1.1, 1.4, 1.5, 1.7, Ch. 3, Ch. 4, Ch. 5, 6.1 through 6.5, Ch. 7, 8.1 through 8.3, 8.5, 8.6, Ch. 9, Ch. 10, 11.1 through 11.3, 11.8, 12.1 through 12.7

NCTM
National Council of Teachers of Mathematics

The chart below lists the lessons and other features in the textbook that address the **NCTM** *Standards.*

Content Standards	
1) Number and Operations Understand numbers, ways of representing numbers, relationships among numbers, and number systems; understand meanings of operations and how they relate to one another; compute fluently and make reasonable estimates.	2.1 through 2.3, 2.5 through 2.7, Act. 3.1, 3.1, 3.2, Act. 3.4, 3.4, 3.6, Tech. Act. 3.6, 3.8, Ch. 4, 5.1, 5.2, 5.3, 5.5, Ch. 6, Ch. 7, Ch. 8, Ch. 9, Ch. 10, Ch. 11, 12.1, 12.2, 12.3, 12.4, 12.5, 12.6, 12.7, Appendix 3
2) Algebra Understand patterns, relations, and functions; represent and analyze mathematical situations and structures using algebraic symbols; use mathematical models to represent and understand quantitative relationships; analyze change in various contexts.	Ch. 1, 2.1, Act. 2.2, 2.2, 2.4, Tech. Act. 2.4, Act. 2.6, 2.6, 2.7, Ch. 3, 4.8, Ch. 5, 6.1 through 6.5, Ch. 7, 8.1 through 8.3, Act. 8.5, 8.5, Act. 8.6, 8.6, Tech. Act. 8.6, Ch. 9, Ch. 10, Ch. 11, 12.1, 12.1 through 12.5
3) Geometry Analyze characteristics and properties of two- and three-dimensional geometric shapes and develop mathematical arguments about geometric relationships; specify locations and describe spatial relationships using coordinate geometry and other representational systems; apply transformations and use symmetry to analyze mathematical situations; use visualization, spatial reasoning, and geometric modeling to solve problems.	3.2, 3.7, Ch. 4, 5.2 through 5.5, Act. 5.7, 5.7, Act. 6.1, 6.1, 6.3, 6.5, Tech. Act. 6.5, Act. 7.1, 7.1, Tech. Act. 7.1, 7.4, 7.6, 8.2, Tech. Act. 8.2, Act. 8.5, 8.5, Act. 8.6, 8.6, Tech. Act. 8.6, Act. 9.3, 9.3, Tech. Act. 9.3, 9.4, Tech. Act. 9.4, 9.7, Act. 10.1, 10.1, Tech. Act. 10.1, Act. 10.6, 10.6, 11.3, Tech. Act. 11.3, 11.4, 11.8, Tech. Act. 11.8, 12.1, Tech. Act. 12.1, 12.6 through 12.8
4) Measurement Understand measurable attributes of objects and the units, systems, and processes of measurement; apply appropriate techniques, tools, and formulas to determine measurements.	3.2, 3.7, 5.1 through 5.5, 12.5 through 12.7
5) Data Analysis and Probability Formulate questions that can be addressed with data and collect, organize, and display relevant data to answer them; select and use appropriate statistical methods to analyze data; develop and evaluate inferences and predictions that are based on data; understand and apply basic concepts of probability.	2.8, 3.5, 4.1, Tech. Act. 4.1, 6.6, 6.7, Tech. Act. 6.7, 8.3, 9.8, 11.4, Appendix 1, Appendix 2

Process Standards	
6) Problem Solving Build new mathematical knowledge through problem solving; solve problems that arise in mathematics and in other contexts; apply and adapt a variety of appropriate strategies to solve problems; monitor and reflect on the process of mathematical problem solving.	Ch. 1, 2.1, Act. 2.2, 2.2, Act. 2.3, 2.3, 2.5, 3.1 through 3.6, 3.8, 4.1, Tech. Act. 4.1, 4.3, Act. 4.4, 4.4, 4.5 through 4.8, Ch. 5, 6.2, 6.3, 6.5, Tech. Act. 6.5, Ch. 7, Act. 8.1, 8.1, 8.4, 9.2 through 9.8, 10.1 through 10.3, 11.1 through 11.4, 12.1 through 12.3, Act. 12.5, 12.5, Act. 12.7, 12.7, Appendix 1
7) Reasoning and Proof Recognize reasoning and proof as fundamental aspects of mathematics; make and investigate mathematical conjectures; develop and evaluate mathematical arguments and proofs; select and use various types of reasoning and methods of proof.	Act. 1.4, 1.4, 2.1 (Example 9; Exs. 54–57), 2.5 (Activity), 2.6 (Exs. 92, 93), 3.3 (Ex. 44), 3.4 (Exs. 42–44), Ch. 3 Extension, 4.6 (Activity), 4.7 (Ex. 51), 6.7 (Example 3), 7.3 (Exs. 2–7), 7.4, Act 7.5, 8.1 (Exs. 84, 85), 9.3 (Exs. 77–80), 10.4 (Activity), 11.5 (Exs. 42, 43), Tech. Act. 12.1, 12.5, 12.8
8) Communication Organize and consolidate their mathematical thinking through communication; communicate their mathematical thinking coherently and clearly to peers, teachers, and others; analyze and evaluate the mathematical thinking and strategies of others; use the language of mathematical ideas precisely.	3.1 through 3.4, 3.8, Ch. 3 Extension, 5.1, 5.3, 5.5, Act. 5.6, 5.6, 6.2, 6.3, 6.5, Tech. Act. 6.5, Ch. 7, Act. 8.1, 8.1, 8.4 through 8.6, 10.1 through 10.3, Act. 10.7, 10.7, 10.8, 11.1 through 11.5, 12.1 through 12.3, Act. 12.5, 12.5, Act. 12.7, 12.7, 12.8
9) Connections Recognize and use connections among mathematical ideas; understand how mathematical ideas interconnect and build on one another to produce a coherent whole; recognize and apply mathematics in contexts outside of mathematics.	Ch. 1, 2.1 through 2.3, 2.5, Act. 3.1, 3.1 through 3.4, 3.6, Tech. Act. 3.6, 3.8, Ch. 4, Ch. 5, 6.2, 6.3, 6.5, Tech. Act. 6.5, 6.7, Tech. Act. 6.7, Ch. 7, Act. 8.1, 8.1, 8.4, 9.2 through 9.8, 10.1 through 10.3, 11.1 through 11.3, 11.5, 12.1, Tech. Act. 12.1 through 12.3, Act. 12.5, 12.5, Act. 12.7, 12.7
10) Representation Create and use representations to organize, record, and communicate mathematical ideas; select, apply, and translate among mathematical representations to solve problems; use representations to model and interpret physical, social, and mathematical phenomena.	1.1, 1.2, Tech. Act. 1.2, 1.5 through 1.7, 2.1, Act. 2.2, 2.2, 2.4, Tech. Act. 2.4, 3.1 through 3.5, Tech. Act. 3.6, 3.7, 3.8, Ch. 4, Ch. 5, 6.1 through 6.3, 6.5 through 6.7, Ch. 7, Act. 8.1, 8.1, 8.4 through 8.6, 9.2 through 9.8, 10.1 through 10.3, Act. 10.5, 10.5, Act. 10.6, 10.6, 11.1 through 11.5, Act. 11.7, 11.7, 11.8, Tech. Act. 11.8, 12.1 through 12.3, Act. 12.5, 12.5, Act. 12.7, 12.7, Appendix 1

McDougal Littell

ALGEBRA 1

STUDENT EDITION

CHAPTER 1

PLANNING THE CHAPTER

Connections to Algebra

LESSON	GOALS	NCTM	ITED	SAT9	Terra-Nova	Local
1.1 *pp. 3–8*	GOAL 1 Evaluate a variable expression. GOAL 2 Write a variable expression that models a real-life situation.	2, 6, 9, 10	RQWN, RQF	2, 3	16, 18, 52	
1.2 *pp. 9–15*	GOAL 1 Evaluate expressions containing exponents. GOAL 2 Use exponents in real-life problems. TECHNOLOGY ACTIVITY: 1.2 *Create a table on a graphing calculator to evaluate an expression for different values of a variable.*	2, 6, 9, 10	MCE, APE	2, 3, 4, 34	10, 14, 16, 17, 52	
1.3 *pp. 16–22*	GOAL 1 Use the order of operations to evaluate algebraic expressions. GOAL 2 Use a calculator to evaluate real-life expressions.	2, 6, 9	RQWN, RQF	2	16, 51, 52	
1.4 *pp. 23–30*	CONCEPT ACTIVITY: 1.4 *Investigate how to use algebra to describe a pattern.* GOAL 1 Check solutions and solve equations using mental math. GOAL 2 Check solutions of inequalities in a real-life problem.	2, 6, 9	RQWN, RQF	2, 3, 6	10, 16, 17, 52	
1.5 *pp. 31–39*	CONCEPT ACTIVITY: 1.5 *Investigate how problem solving strategies help to find solutions to problems.* GOAL 1 Translate verbal phrases into algebraic expressions. GOAL 2 Use a verbal model to write an algebraic equation or inequality to solve a real-life problem.	2, 6, 9, 10	RQWN, RQF	3	16, 17, 18	
1.6 *pp. 40–45*	GOAL 1 Use tables of organized data. GOAL 2 Use graphs to organize real-life data.	2, 5, 6, 8, 9, 10	MIT, IIT	10, 19, 20, 21	15, 16	
1.7 *pp. 46–52*	GOAL 1 Identify a function and make an input-output table for a function. GOAL 2 Write an equation for a real-life function.	2, 6, 9, 10	MIT, ITT	20, 21	16	

RESOURCES

Chapter Resource Booklets

Chapter Support

Tips for New Teachers	p. 1
Parent Guide for Student Success	p. 3
Prerequisite Skills Review	p. 5
Strategies for Reading Mathematics	p. 7

Lesson Support

Lesson Support	1.1	1.2	1.3	1.4	1.5	1.6	1.7
Lesson Plans (regular and block)	p. 9	p. 23	p. 37	p. 53	p. 66	p. 80	p. 93
Warm-Up Exercises and Daily Quiz	p. 11	p. 25	p. 39	p. 55	p. 68	p. 82	p. 95
Activity Support Masters				p. 56			
Lesson Openers	p. 12	p. 26	p. 40	p. 57	p. 69	p. 83	p. 96
Graphing Calculator Activities & Keystrokes	p. 13	p. 27	p. 41				p. 97
Practice (3 levels)	p. 15	p. 29	p. 44	p. 58	p. 70	p. 84	p. 98
Reteaching with Practice	p. 18	p. 32	p. 47	p. 61	p. 73	p. 87	p. 101
Quick Catch-Up for Absent Students	p. 20	p. 34	p. 49	p. 63	p. 75	p. 89	p. 103
Cooperative Learning Activities						p. 90	p. 104
Interdisciplinary Applications		p. 35		p. 64		p. 91	p. 105
Real-Life Applications	p. 21		p. 50		p. 76		
Math & History Applications					p. 77		
Challenge: Skills and Applications	p. 22	p. 36	p. 51	p. 65	p. 78	p. 92	p. 106

Review and Assessment

Quizzes	pp. 52, 79
Chapter Review Games and Activities	p. 107
Chapter Test (3 levels)	pp. 108–113
SAT/ACT Chapter Test	p. 114
Alternative Assessment with Math Journal	p. 115
Project with Rubric	p. 117
Cumulative Review	p. 119
Resource Book Answers	p. A1

Transparencies

Transparencies	1.1	1.2	1.3	1.4	1.5	1.6	1.7
Warm-Up Exercises and Daily Quiz	p. 2	p. 3	p. 4	p. 5	p. 6	p. 7	p. 8
Alternative Lesson Opener Transparencies	p. 1	p. 2	p. 3	p. 4	p. 5	p. 6	p. 7
Examples/Standardized Test Practice	✓	✓	✓	✓	✓	✓	✓
Answer Transparencies	✓	✓	✓	✓	✓	✓	✓

Technology

- EasyPlanner CD
- EasyPlanner Plus Online
- Classzone.com
- Personal Student Tutor CD
- Test and Practice Generator CD
- Instant Replay: Video Review Games
- Electronic Lesson Presentations
- Chapter Audio Summaries CD

Additional Resources

- Basic Skills Workbook
- Explorations and Projects
- Worked-Out Solution Key
- Data Analysis Sourcebook
- Spanish Resources
- Standardized Test Practice Workbook
- Practice Workbook with Examples
- Notetaking Guide

CHAPTER 1

PACING THE CHAPTER

Resource Key
- ● STUDENT EDITION
- ● CHAPTER 1 RESOURCE BOOK
- ● TEACHER'S EDITION

REGULAR SCHEDULE

Day 1

1.1

STARTING OPTIONS
- Prereq. Skills Review
- Strategies for Reading
- Warm-Up

TEACHING OPTIONS
- Motivating the Lesson
- Les. Opener (Activity)
- Graphing Calc. Activity
- Examples 1–5
- Closure Question
- Guided Practice Exs.

APPLY/HOMEWORK
- See Assignment Guide.
- See the CRB: Practice, Reteach, Apply, Extend

ASSESSMENT OPTIONS
- Checkpoint Exercises
- Daily Quiz (1.1)
- Stand. Test Practice

Day 2

1.2

STARTING OPTIONS
- Homework Check
- Warm-Up or Daily Quiz

TEACHING OPTIONS
- Motivating the Lesson
- Les. Opener (Calculator)
- Graphing Calc. Activity
- Examples 1–4
- Guided Practice Exs. 1–15

APPLY/HOMEWORK
- See Assignment Guide.
- See the CRB: Practice, Reteach, Apply, Extend

ASSESSMENT OPTIONS
- Checkpoint Exercises, p. 10

Day 3

1.2 *(cont.)*

STARTING OPTIONS
- Homework Check

TEACHING OPTIONS
- Examples 5–6
- Technology Activity
- Closure Question
- Guided Practice Ex. 16

APPLY/HOMEWORK
- See Assignment Guide.
- See the CRB: Practice, Reteach, Apply, Extend

ASSESSMENT OPTIONS
- Checkpoint Exercises, p. 11
- Daily Quiz (1.2)
- Stand. Test Practice

Day 4

1.3

STARTING OPTIONS
- Homework Check
- Warm-Up or Daily Quiz

TEACHING OPTIONS
- Motivating the Lesson
- Les. Opener (Application)
- Graphing Calc. Activity
- Examples 1–3
- Guided Practice Exs. 1–11

APPLY/HOMEWORK
- See Assignment Guide.
- See the CRB: Practice, Reteach, Apply, Extend

ASSESSMENT OPTIONS
- Checkpoint Exercises, p. 17

Day 5

1.3 *(cont.)*

STARTING OPTIONS
- Homework Check

TEACHING OPTIONS
- Examples 4–5
- Closure Question
- Guided Practice Ex. 12

APPLY/HOMEWORK
- See Assignment Guide.
- See the CRB: Practice, Reteach, Apply, Extend

ASSESSMENT OPTIONS
- Checkpoint Exercises, p. 18
- Daily Quiz (1.3)
- Stand. Test Practice
- Quiz (1.1–1.3)

Day 6

1.4

STARTING OPTIONS
- Homework Check
- Warm-Up or Daily Quiz

TEACHING OPTIONS
- Motivating the Lesson
- Concept Act. & Wksht.
- Les. Opener (Application)
- Examples 1–3
- Guided Practice Exs. 1–9

APPLY/HOMEWORK
- See Assignment Guide.
- See the CRB: Practice, Reteach, Apply, Extend

ASSESSMENT OPTIONS
- Checkpoint Exercises, p. 25

Day 9

1.5 *(cont.)*

STARTING OPTIONS
- Homework Check

TEACHING OPTIONS
- Examples 2–3
- Closure Question
- Guided Practice Exs. 9–12

APPLY/HOMEWORK
- See Assignment Guide.
- See the CRB: Practice, Reteach, Apply, Extend

ASSESSMENT OPTIONS
- Checkpoint Exercises, pp. 33–34
- Daily Quiz (1.5)
- Stand. Test Practice
- Quiz (1.4–1.5)

Day 10

1.6

STARTING OPTIONS
- Homework Check
- Warm-Up or Daily Quiz

TEACHING OPTIONS
- Motivating the Lesson
- Les. Opener (Application)
- Examples 1–3
- Closure Question
- Guided Practice Exs.

APPLY/HOMEWORK
- See Assignment Guide.
- See the CRB: Practice, Reteach, Apply, Extend

ASSESSMENT OPTIONS
- Checkpoint Exercises
- Daily Quiz (1.6)
- Stand. Test Practice

Day 11

1.7

STARTING OPTIONS
- Homework Check
- Warm-Up or Daily Quiz

TEACHING OPTIONS
- Les. Opener (Visual)
- Graphing Calc. Activity
- Examples 1–3
- Closure Question
- Guided Practice Exs.

APPLY/HOMEWORK
- See Assignment Guide.
- See the CRB: Practice, Reteach, Apply, Extend

ASSESSMENT OPTIONS
- Checkpoint Exercises
- Daily Quiz (1.7)
- Stand. Test Practice
- Quiz (1.6–1.7)

Day 12

Review

DAY 12 START OPTIONS
- Homework Check

REVIEWING OPTIONS
- Chapter 1 Summary
- Chapter 1 Review
- Chapter Review Games and Activities

APPLY/HOMEWORK
- Chapter 1 Test (practice)
- Ch. Standardized Test (practice)

Day 13

Assess

DAY 13 START OPTIONS
- Homework Check

ASSESSMENT OPTIONS
- Chapter 1 Test
- SAT/ACT Ch. 1 Test
- Alternative Assessment

APPLY/HOMEWORK
- Skill Review, p. 62

Day 7

1.4 *(cont.)*

Starting Options
- Homework Check

Teaching Options
- Examples 4–5
- Closure Question
- Guided Practice Exs. 10–13

Apply/Homework
- See Assignment Guide.
- See the CRB: Practice, Reteach, Apply, Extend

Assessment Options
- Checkpoint Exercises, p. 26
- Daily Quiz (1.4)
- Stand. Test Practice

Day 8

1.5

Starting Options
- Homework Check
- Warm-Up or Daily Quiz

Teaching Options
- Motivating the Lesson
- Concept Activity
- Les. Opener (Application)
- Example 1
- Guided Practice Exs. 1–8

Apply/Homework
- See Assignment Guide.
- See the CRB: Practice, Reteach, Apply, Extend

Assessment Options
- Checkpoint Exercises, p. 33

BLOCK SCHEDULE

Day 1

1.1 & 1.2

Day 1 Start Options
- Prereq. Skills Review
- Strategies for Reading
- Warm-Up (Les. 1.1)

Teaching 1.1 Options
- Motivating the Lesson
- Les. Opener (Activity)
- Graphing Calc. Activity
- Examples 1–5
- Closure Question
- Guided Practice Exs.

Beginning 1.2 Options
- Warm-Up (Les. 1.2)
- Motivating the Lesson
- Les. Opener (Calc.)
- Graphing Calc. Activity
- Examples 1–4
- Guided Prac. Exs. 1–15

Apply/Homework
- See Assignment Guide.
- See the CRB: Practice, Reteach, Apply, Extend

Assessment Options
- Checkpoint Exercises
- Daily Quiz (Les. 1.1)

Day 2

1.2 & 1.3

Day 2 Start Options
- Homework Check
- Daily Quiz (Les. 1.1)

Finishing 1.2 Options
- Examples 5–6
- Technology Activity
- Closure Question
- Guided Practice Ex. 16

Beginning 1.3 Options
- Warm-Up (Les. 1.3)
- Motivating the Lesson
- Les. Opener (Appl.)
- Graphing Calc. Activity
- Examples 1–3
- Guided Practice Exs. 1–11

Apply/Homework
- See Assignment Guide.
- See the CRB: Practice, Reteach, Apply, Extend

Assessment Options
- Checkpoint Exercises
- Daily Quiz (Les. 1.2)
- Stand. Test Prac. (1.2)

Day 3

1.3 & 1.4

Day 3 Start Options
- Homework Check
- Daily Quiz (Les. 1.2)

Finishing 1.3 Options
- Examples 4–5
- Closure Question
- Guided Practice Ex. 12

Beginning 1.4 Options
- Warm-Up (Les. 1.4)
- Motivating the Lesson
- Concept Act. & Wksht.
- Les. Opener (Appl.)
- Examples 1–3
- Guided Practice Exs. 1–9

Apply/Homework
- See Assignment Guide.
- See the CRB: Practice, Reteach, Apply, Extend

Assessment Options
- Checkpoint Exercises
- Daily Quiz (Les. 1.3)
- Stand. Test Prac. (1.3)
- Quiz (1.1–1.3)

Day 4

1.4 & 1.5

Day 4 Start Options
- Homework Check
- Daily Quiz (Les. 1.3)

Finishing 1.4 Options
- Examples 4–5
- Closure Question
- Guided Practice Exs. 10–13

Beginning 1.5 Options
- Warm-Up (Les. 1.5)
- Motivating the Lesson
- Concept Activity
- Les. Opener (Appl.)
- Example 1
- Guided Practice Exs. 1–8

Apply/Homework
- See Assignment Guide.
- See the CRB: Practice, Reteach, Apply, Extend

Assessment Options
- Checkpoint Exercises
- Daily Quiz (Les. 1.4)
- Stand. Test Prac. (1.4)

Day 5

1.5 & 1.6

Day 5 Start Options
- Homework Check
- Daily Quiz (Les. 1.4)

Finishing 1.5 Options
- Examples 2–3
- Closure Question
- Guided Practice Exs. 9–12

Teaching 1.6 Options
- Warm-Up (Les. 1.6)
- Motivating the Lesson
- Les. Opener (Appl.)
- Examples 1–3
- Closure Question
- Guided Practice Exs.

Apply/Homework
- See Assignment Guide.
- See the CRB: Practice, Reteach, Apply, Extend

Assessment Options
- Checkpoint Exercises
- Daily Quiz (Les. 1.5, 1.6)
- Stand. Test Practice
- Quiz (1.4–1.5)

Day 6

1.7 & Review

Day 6 Start Options
- Homework Check
- W-Up 1.7 or D. Quiz 1.6

Teaching 1.7 Options
- Les. Opener (Visual)
- Graphing Calc. Activity
- Examples 1–3
- Closure Question
- Guided Practice Exs.

Reviewing Options
- Chapter 1 Summary
- Chapter 1 Review
- Chapter Review Games and Activities

Apply/Homework
- See Assignment Guide.
- See the CRB: Practice, Reteach, Apply, Extend
- Chapter 1 Test (practice)
- Ch. Standardized Test

Assessment Options
- Checkpoint Exercises
- Daily Quiz (Les. 1.7)
- Stand. Test Prac. (1.7)
- Quiz (1.6–1.7)

Day 7

Assess & 2.1

(Day 7 = Ch. 2 Day 1)

Assessment Options
- Chapter 1 Test
- SAT/ACT Ch. 1 Test
- Alternative Assessment

Ch. 2 Start Options
- Skill Review, p. 62
- Prereq. Skills Review
- Strategies for Reading

Teaching 2.1 Options
- Warm-Up (Les. 2.1)
- Motivating the Lesson
- Les. Opener (Visual)
- Graphing Calc. Activity
- Examples 1–9
- Closure Question
- Guided Practice Exs.

Apply/Homework
- See Assignment Guide.
- See the CRB: Practice, Reteach, Apply, Extend

Assessment Options
- Checkpoint Exercises
- Daily Quiz (Les. 2.1)
- Stand. Test Practice

CHAPTER 1 MEETING INDIVIDUAL NEEDS

BEFORE THE CHAPTER

The *Chapter 1 Resource Book* has the following materials to distribute and use before the chapter:

- **Parent Guide for Student Success**
- **Prerequisite Skills Review (pictured below)**
- **Strategies for Reading Mathematics**

PREREQUISITE SKILLS *Pages 5–6*

CHAPTER 1

NAME ______________________ DATE ________

Prerequisite Skills Review

For use before Chapter 1

Chapter Support

EXAMPLE 1 ***Writing Percents as Fractions and Decimals***

Write each percent as a decimal and as a fraction or a mixed number in lowest terms.

a. 70% **b.** 136%

SOLUTION

a. $70\% = \frac{70}{100} = \frac{7}{10} = 0.7$

b. $136\% = \frac{136}{100} = \frac{34}{25} = 1\frac{9}{25} = 1.36\%$

Exercises for Example 1

Write each percent as a decimal and as a fraction or mixed number in lowest terms.

1. 55%	**2.** 12%	**3.** 0.02%
4. 105%	**5.** 41%	**6.** 0.05%

EXAMPLE 2 ***Working with Inequalities***

Complete the statement using > or <.

a. 331 __?__ 301 **b.** 103.01 __?__ 103.1

SOLUTION

a. First, plot the numbers on a number line. Because 331 is to the right of 301, it follows that 331 > 301.

301 331

b. First, plot the numbers on a number line. Because 103.01 is to the left of 103.1, it follows that 103.01 > 103.1.

103.00 103.01 103.02 103.03 103.04 103.05 103.06 103.07 103.08 103.09 103.10

Exercises for Example 2

Complete the statement using > or <.

7. 501 __?__ 510	**8.** 98.02 __?__ 98.021	**9.** 7895 __?__ 7885
10. 0.85 __?__ 0.085	**11.** 4041 __?__ 4401	**12.** 0.0022 __?__ 0.02

PREREQUISITE SKILLS REVIEW These two pages support the Study Guide on page 2. They help students prepare for Chapter 1 by providing worked-out examples and practice for the following skills needed in the chapter:

- **Write percents in various forms.**
- **Compare decimals.**
- **Find areas and perimeters.**

DURING EACH LESSON

The *Chapter 1 Resource Book* has the following alternatives for introducing the lesson:

- **Lesson Openers (pictured below)**
- **Graphing Calculator Activities with Keystrokes**

LESSON OPENER *Page 12*

LESSON 1.1

NAME ______________________ DATE ________

Available as a transparency

Activity Lesson Opener

For use with pages 3–8

Lesson 1.1

Set Up: Work in a group. Be sure at least one classmate in your group is not the same age as you.

Here is a fun puzzle for you to solve. When you are finished, you will try to explain why it works!

Start with your current age.

PUZZLE: Start with your current age.

STEP 1 Add 5 to the age.

STEP 2 Multiply the result of Step 1 by 2.

STEP 3 Subtract twice the age you started with from the result of Step 2.

STEP 4 Subtract 10 from the result of Step 3.

1. Compare your answer with the answers of other members of your group. What do you notice?
2. Choose an age of an adult. Be sure the age you choose is different from the ages chosen by other members of your group. Repeat the puzzle and Question 1 for this age.
3. Choose an age of someone younger than you. Be sure the age you choose is different from the ages chosen by other members of your group. Repeat the puzzle and Question 1 for this age.
4. What conclusion can you make about this puzzle? Explain why you think the puzzle works this way.
5. Work with your group to make up a puzzle of your own. Use different numbers to check your puzzle. Then explain why your puzzle works the way it does.

ACTIVITY LESSON OPENER This Lesson Opener provides an alternative way to start Lesson 1.1 in the form of an activity. In this activity, students work with a number puzzle using their ages as they learn about universal properties of algebra.

FOLLOWING EACH LESSON

The *Chapter 1 Resource Book* has a variety of materials to follow-up each lesson. They include the following:

- **Practice (3 levels) (pictured below)**
- **Reteaching with Practice**
- **Quick Catch-Up for Absent Students**
- **Interdisciplinary Applications**
- **Real-Life Applications**

PRACTICE *Pages 70–72*

LESSON 1.5

NAME ______________________ DATE ________

Practice B

For use with pages 32–39

Lesson 1.5

Write the verbal phrase as an algebraic expression. Use x for the variable in your expression.

1. Four more than a number
2. Six less than a number
3. Difference of 12 and a number
4. The sum of a number and two
5. Five times a given number
6. One third of a given number
7. A number divided by eight
8. Nine more than twice a given number
9. Two less than a number, divided by three
10. Three more than the product of ten and a number
11. Five times the sum of a number and one
12. The sum of a number and five, divided by two

Write the verbal sentence as an equation or an inequality.

13. Seven more than a number x is ten.
14. The sum of a number y and six is 13.
15. Eight more than a number y is greater than or equal to ten.
16. The difference of a number a and two is eight.
17. Six less than a number z is less than 21.
18. Thirteen minus a number b is two.
19. The product of 11 and a number x is 22.
20. Fourteen is less than seven times a number x.
21. Four times a number b plus one is 17.
22. The quotient of a number t and three is nine.
23. A number a divided by two is greater than nine.
24. Three less than the product of six and a number a is nine.

In Exercises 25–28, which equation correctly models the situation?

25. ***Model Planes*** Your model plane collection consists of 15 models. Each plane is either a propeller plane or a jet. There are seven fewer propeller planes than jets. Let x be the number of jets.
    a. $x - (x - 7) = 15$ b. $x + 7 = 15$
26. ***Bake Sale*** You want to make six dozen cookies for a bake sale. If you follow the recipe, one batch makes two dozen cookies. Let b be the number of batches you need to bake.
    a. $2b = 6$ b. $\frac{b}{6} = 2$
27. ***Height*** You are 65 inches tall. You are 18 inches taller than your younger sister. Let h be your sister's height in inches.
    a. $h - 18 = 65$ b. $h + 18 = 65$
28. ***Music*** An eighth note is played twice as fast as a quarter note. Eight eighth notes can be played in one measure of music. Let q be the number of quarter notes played in one measure of music.
    a. $2q = 8$ b. $\frac{q}{2} = 8$

Recycling **In Exercises 29–31, use the following information.**
A recycling center pays \$.03 apiece for aluminum cans and \$.05 apiece for certain glass bottles. You were paid \$6 for recycling cans and bottles.

29. Write a verbal model that relates the income from cans, income from bottles, and the amount of money you were paid.
30. Assign labels and write an algebraic model based on your verbal model.
31. If you recycled 100 cans, use mental math to solve the equation. Check your answer.

PRACTICE Each lesson has three levels of practice: basic (A), average (B), and advanced (C). This Practice B involves more work with real-life modeling than Practice A, but does not have as much work with verbal models as Practice C.

TECHNOLOGY RESOURCE
Teachers can use the Time-Saving Test and Practice Generator to create additional practice worksheets for Lesson 1.5 and for the other lessons in Chapter 1.

ASSESSING THE CHAPTER

The *Chapter 1 Resource Book* has the following review and assessment materials:

- **Quizzes**
- **Chapter Review Games and Activities**
- **Chapter Test (3 levels) (pictured below)**
- **SAT/ACT Chapter Test**
- **Alternative Assessment with Rubric and Math Journal**
- **Project with Rubric**
- **Cumulative Review**

CHAPTER TEST *Pages 108–113*

CHAPTER 1

Review and Assess

NAME ______________________ DATE ________

Chapter Test B

For use after Chapter 1

Evaluate the expression for the given value of the variable.

1. $4.25q$ when $q = 6.2$
2. $\frac{16.8}{x}$ when $x = 2$
3. If you are driving at a constant speed of 65 miles per hour, how long will it take you to travel 260 miles?
4. The area of a rectangle is the product of its base and height. Find the area of the rectangle below.

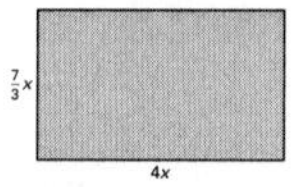

Evaluate the expression for the given value of the variable.

5. 3^n when $n = 4$
6. $(6x)^4$ when $x = 2$
7. $8 + 5a^2$ when $a = 7$
8. $64 - \frac{32}{b}$ when $b = 4$

Evaluate the expression for the given values of the variables.

9. y^2 when $x = 5$ and $y = 9$
10. $(a - b)^4$ when $a = 10$ and $b = 6$
11. A storage container is 22 inches long, 22 inches wide, and 24 inches deep. The volume of the container is the product of its length, width, and height. What is the volume of the container?

Evaluate the expression.

12. $49 \div 7 + 3 \cdot 6$
13. $4[(29 - 12) + 10]$
14. $[44 \div (10 - 8)^2] \div 7$
15. $\frac{1}{2} \cdot 18 - 3^2$

Check whether the given number is a solution of the equation.

16. $16x + 2 = 29 - 3x$; 2
17. $10x - 4 \le 20$; 5
18. If you invested \$250 in a bank account for 2 years and received \$7.50 in interest, what was the annual interest rate for the account?

Answers
1. ________
2. ________
3. ________
4. ________
5. ________
6. ________
7. ________
8. ________
9. ________
10. ________
11. ________
12. ________
13. ________
14. ________
15. ________
16. ________
17. ________
18. ________

CHAPTER TEST There are three versions of this two-page Chapter Test, one each for basic (A), average (B), and advanced (C) students. Level B involves more work with evaluating expressions when the value of the variable is given, but does not have as much work with writing verbal sentences as equations or inequalities as Level C.

TECHNOLOGY RESOURCE
Teachers can use the Time-Saving Test and Practice Generator to create customized review and assessment materials for Chapter 1.

CHAPTER OVERVIEW

CHAPTER GOALS

In this chapter, students will learn how to write and evaluate expressions using exponents and the order of operations, check solutions to equations and inequalities, use verbal and algebraic models to represent real-life situations, and use tables and graphs to organize data and to represent functions. They will translate and calculate expressions that model real-life situations and write equations for real-life functions.

APPLICATION NOTE

An important consideration for scuba divers is the effect of increased pressure at deeper depths. While a diver's body will not feel the effect of added pressure, the body's air spaces will. Air is very compressible under added amounts of pressure. Mathematically, the volume of air space is inversely proportional to the amount of pressure at a given depth. Thus, if the pressure on the diver is doubled from 1 atmosphere (of pressure) to 2 atmospheres (this happens at a depth of 33 feet), the volume of air space in his or her lungs is decreased by one-half. To avoid this, the original air space in the body must be maintained while breathing. This is done by complicated regulators in the scuba equipment.

Decompression sickness, commonly known as "the bends," happens when greater amounts of pressure on the diver cause an excessive amount of nitrogen to dissolve in the body without being expelled. Symptoms can include difficulty in breathing, weakness, dizziness, shock, or numbness. Decompression sickness is avoided by paying attention to complicated dive tables, which are determined by the depth and the length of previous dives.

Additional information about scuba diving is available at **www.mcdougallittell.com.**

CONNECTIONS TO ALGEBRA

How does water pressure affect a scuba diver?

CHAPTER

1

APPLICATION: Scuba Diving

On the surface of Earth **only the atmosphere is pressing on you. When you dive under the water, the weight of the water also presses on you. The volume of air in your lungs gets squeezed by these forces.**

To design equipment for a dive, people need to know what the variables are, such as the depth of a dive and the dive time. You'll learn to write mathematical models for such real-life problems in Chapter 1.

Think & Discuss

The table shows that for every 10 feet a diver descends, the weight of the water on the diver increases by about 4.45 pounds per square inch.

Depth (feet)	Additional Pressure (pounds per square inch)
10	4.45
20	4.45×2
30	4.45×3
40	4.45×4

1. You are diving at a depth of 40 feet. What is the pressure of the water on you?
   17.8 pounds per square inch
2. Use the pattern in the table to predict the pressure of the water on a diver at 100 feet.
   44.5 pounds per square inch

Learn More About It

You will learn more about diving and water pressure in Example 3 on page 48.

APPLICATION LINK Visit www.mcdougallittell.com for more information about scuba diving.

ADDITIONAL RESOURCES

Another way to begin the chapter is to show the video clip of a real-life motivator for Chapter 1 on the *Instant Replay: Video Review Games.*

PROJECTS

A project covering Chapters 1–3 appears on pages 198–199 of the Student Edition. An additional project for Chapter 1 is available in the *Chapter 1 Resource Book,* p. 117.

TECHNOLOGY

Software

- *Electronic Teaching Tools*
- *Online Lesson Planner*
- *Personal Student Tutor*
- *Test and Practice Generator*
- *Electronic Lesson Presentations* (Lesson 1.5)

Video

- *Instant Replay: Video Review Games*

Internet Connections
www.mcdougallittell.com

- **Application Links**
  1, 7, 13, 29, 39
- **Data Update**
  37
- **Student Help**
  4, 10, 15, 17, 41, 51
- **Career Links**
  26, 34, 42, 48
- **Extra Challenge**
  8, 14, 21, 30, 38, 45, 51

PREPARE

DIAGNOSTIC TOOLS

The **Skill Review** exercises can help you diagnose whether students have the following skills needed in Chapter 1:

- Write percents as decimals.
- Write fractions or mixed numbers in lowest terms.
- Complete statements with > or <.
- Find area and perimeter.

The following resources are available for students who need additional help with these skills:

- Prerequisite Skills Review (*Chapter 1 Resource Book,* p. 5; *Warm-Up Transparencies,* p. 1)
- *Personal Student Tutor*

ADDITIONAL RESOURCES

The following resources are provided to help you prepare for the upcoming chapter and customize review materials:

- ***Chapter 1 Resource Book***
  Tips for New Teachers (p. 1)
  Parent Guide (p. 3)
  Lesson Plans (every lesson)
  Lesson Plans for Block Scheduling (every lesson)
- ***Electronic Teaching Tools***
- ***Online Lesson Planner***
- ***Test and Practice Generator***

ENGLISH LEARNERS

Mention that many words that are part of mathematical language have other meanings in English. Point out the additional meanings of common algebra words, such as *inequality* and *solution.* Then draw out the relationship between the mathematical and non-mathematical meanings to help students understand and remember algebra terms.

CHAPTER 1

Study Guide

PREVIEW

What's the chapter about?

Chapter 1 is an **introduction to algebra**. In Chapter 1 you'll learn to

- write and evaluate expressions.
- check solutions to equations and inequalities and use mental math.
- use verbal and algebraic models to represent real-life situations.
- organize data and represent functions.

KEY VOCABULARY

▶ New

- variable expression, p. 3
- unit analysis, p. 5
- verbal model, p. 5
- power, p. 9
- exponent, p. 9
- base of a power, p. 9
- order of operations, p. 16
- equation, p. 24
- solution of an equation, p. 24
- inequality, p. 26
- function, p. 46
- input-output table, p. 46
- domain, p. 47
- range, p. 47

PREPARE

Are you ready for the chapter?

SKILL REVIEW Do these exercises to review key skills that you'll apply in this chapter. See the given **reference page** if there is something you don't understand.

STUDENT HELP

Study Tip
"Student Help" boxes throughout the chapter give you study tips and tell you where to look for extra help in this book and on the Internet.

Write the percent as a decimal and as a fraction or a mixed number in lowest terms. (Skills Review, pp. 784–785)

1. 60% $0.6, \frac{3}{5}$ **2.** 33% $0.33, \frac{33}{100}$ **3.** 0.08% $0.0008, \frac{1}{1250}$ **4.** 150% $1.5, 1\frac{1}{2}$

Complete the statement using > or <. (Skills Review, pp. 779–780)

5. 899 _?_ 901 < **6.** 64.1 _?_ 64.03 > **7.** 2050 _?_ 2005 > **8.** 0.099 _?_ 0.01 >

Find the area and perimeter of the figure. (Skills Review, pp. 790–791)

9.

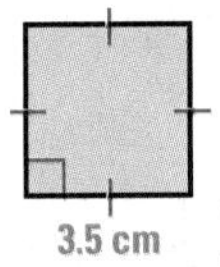

$12.25\ cm^2$, 14 cm

10.

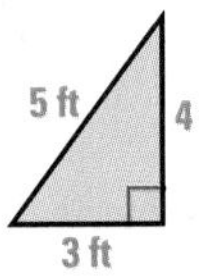

$6\ ft^2$, 12 ft

11.

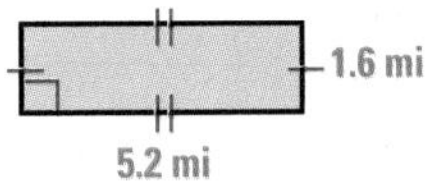

$8.32\ mi^2$, 13.6 mi

STUDY STRATEGY

Here's a study strategy!

Keeping a Math Notebook

- Keep a notebook of math notes about each chapter, separate from your homework exercises. This will help you remember new concepts and skills.
- Review your notes each day before you start your next homework assignment.

1.1

Variables in Algebra

What you should learn

GOAL 1 Evaluate a variable expression.

GOAL 2 Write a variable expression that models a **real-life** situation, as in the hiking problem in **Example 5**.

Why you should learn it

▼ To express **real-life** number relationships, such as pitching speed in **Exs. 48 and 49**.

GOAL 1 EVALUATING VARIABLE EXPRESSIONS

A **variable** is a letter that is used to represent one or more numbers. The numbers are the **values** of the variable. A **variable expression** is a collection of numbers, variables, and operations. Here are some examples.

VARIABLE EXPRESSION			MEANING	OPERATION
$8y$	$8 \cdot y$	$8(y)$	8 times y	Multiplication
$\frac{16}{b}$	$16 \div b$		16 divided by b	Division
$4 + s$			4 plus s	Addition
$9 - x$			9 minus x	Subtraction

The expression $8y$ is usually not written as $8 \times y$ because of possible confusion with the variable x. Replacing each variable in an expression by a number is called **evaluating the expression**. The resulting number is the value of the expression.

Write the variable expression. → **Substitute values for variables.** → **Simplify the numerical expression.**

EXAMPLE 1 *Evaluating a Variable Expression*

Evaluate the expression when $y = 2$.

a. $8y$ **b.** $\frac{10}{y}$ **c.** $y + 3$ **d.** $14 - y$

SOLUTION

a. $8y = 8(2)$ Substitute 2 for y.

$= 16$ Simplify.

b. $\frac{10}{y} = \frac{10}{2}$ Substitute 2 for y.

$= 5$ Simplify.

c. $y + 3 = 2 + 3$ Substitute 2 for y.

$= 5$ Simplify.

d. $14 - y = 14 - 2$ Substitute 2 for y.

$= 12$ Simplify.

1 PLAN

PACING

Basic: 1 day
Average: 1 day
Advanced: 1 day
Block Schedule: 0.5 block with 1.2

***LESSON OPENER* ACTIVITY**

An alternative way to approach Lesson 1.1 is to use the Activity Lesson Opener:

- Blackline Master (*Chapter 1 Resource Book,* p. 12)
- Transparency (p. 1)

MEETING INDIVIDUAL NEEDS

- ***Chapter 1 Resource Book***
  Prerequisite Skills Review (p. 5)
  Practice Level A (p. 15)
  Practice Level B (p. 16)
  Practice Level C (p. 17)
  Reteaching with Practice (p. 18)
  Absent Student Catch-Up (p. 20)
  Challenge (p. 22)
- ***Resources in Spanish***
- ***Personal Student Tutor***

NEW-TEACHER SUPPORT

See the Tips for New Teachers on pp. 1–2 of the *Chapter 1 Resource Book* for additional notes about Lesson 1.1.

WARM-UP EXERCISES

Transparency Available

Evaluate the expression.

1. 120(0.05)(3.5) 21

2. $\frac{3}{4} \cdot \frac{1}{3}$ $\frac{1}{4}$

3. $\frac{1}{3} + \frac{1}{3}$ $\frac{2}{3}$

4. $\frac{100}{0.25}$ 400

5. Find the perimeter of the square. 28.4 cm

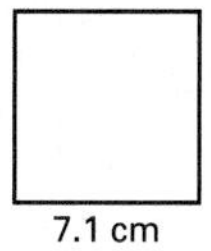

2 TEACH

MOTIVATING THE LESSON
Tell students that the average speed of a race-car in the *Indianapolis 500* race is 500 miles divided by the number of hours the race takes. For example, in 1997, the average speed of the winning car was 500 miles ÷ 3.4 hours, or about 147 miles per hour. This is an application of evaluating a variable expression, the focus of today's lesson.

EXTRA EXAMPLE 1
Evaluate the expression when $y = 3$.

a. $4y$ **12**   **b.** $\frac{12}{y}$ **4**

c. $y + 8$ **11**   **d.** $18 - y$ **15**

EXTRA EXAMPLE 2
Find the average speed, $\frac{d}{t}$, of a car that traveled 89 miles in 2 hours. **44.5 mi/h**

EXTRA EXAMPLE 3
Find the perimeter, $a + b + c$, of the triangle. The dimensions are in centimeters. **5 cm**

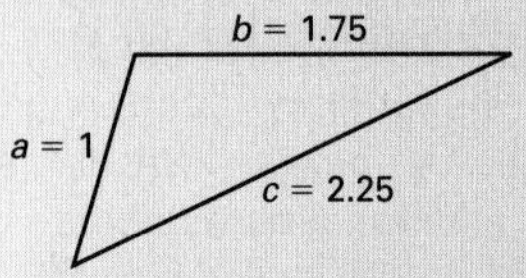

CHECKPOINT EXERCISES

For use after Examples 1–2:

1. Find the average speed of a car that travels 225 miles in 4.5 hours. **50 mi/h**

For use after Example 3:

2. Find the perimeter of the triangle. The dimensions are in inches. **22.7 in.**

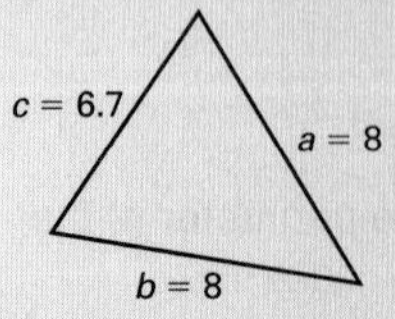

REAL LIFE — Travel

EXAMPLE 2 *Evaluating a Real-Life Expression*

Average speed is given by the following formula.

$$\text{Average speed} = \frac{\text{Distance}}{\text{Time}} = \frac{d}{t}$$

Find the average speed (in miles per hour) of a car that traveled 180 miles from Boise, Idaho, to the Minidoka National Wildlife Refuge in 3 hours.

SOLUTION

Average speed	$= \frac{d}{t}$	Write expression.
	$= \frac{180}{3}$	Substitute 180 for *d* and 3 for *t*.
	$= 60$	Simplify.

▶ The average speed was 60 miles per hour.

EXAMPLE 3 *Evaluating a Geometric Expression*

GEOMETRY CONNECTION The perimeter of a triangle is equal to the sum of the lengths of its sides: $a + b + c$.

Find the perimeter of the triangle. The dimensions are in feet.

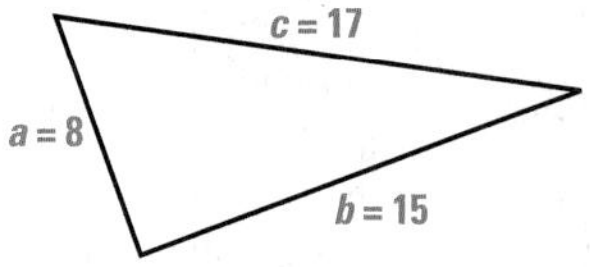

SOLUTION

Perimeter	$= a + b + c$	Write expression.
	$= 8 + 15 + 17$	Substitute values.
	$= 40$	Simplify.

▶ The triangle has a perimeter of 40 feet.

REAL LIFE — Interest

EXAMPLE 4 *Evaluating Simple Interest*

The simple interest earned by money P (called the *principal*) at an annual interest rate r for t years is given by Prt. You deposit \$650 at a rate of 8% per year. How much simple interest will you earn after one half of a year?

SOLUTION

Simple interest	$= Prt$	Write expression.
	$= (650)(0.08)(0.5)$	Substitute 650 for *P*, 0.08 for *r*, and 0.5 for *t*.
	$= 26$	Simplify.

▶ After one half of a year, you will have \$26 of simple interest.

STUDENT HELP
Skills Review
For help with writing percents as decimals, see pp. 784–785.

FOCUS ON APPLICATIONS

REAL LIFE **COASTAL REDWOODS,** found in California and Oregon, are the tallest trees on Earth. Only 4% of original redwood forests remains. The tallest living redwood is 370 feet high.

GOAL 2 MODELING A REAL-LIFE SITUATION

Writing the units of each variable in a real-life problem helps you determine the units for the answer. This is called **unit analysis** and it is often used in problem solving in science. When the same units of measure occur in the numerator and the denominator of an expression, you can cancel the units.

In real-life problems, you may need to translate words into mathematics. One way to do this is to use the **verbal model** shown in Example 5.

EXAMPLE 5 *Finding Time*

HIKING AMONG REDWOODS You plan to go hiking in the Jedediah Smith Redwoods State Park in California. You estimate you'll hike at a rate of 2 miles per hour on a steep trail. How long will it take you to hike from Howland Hill Road along the Boy Scout Tree Trail and back?

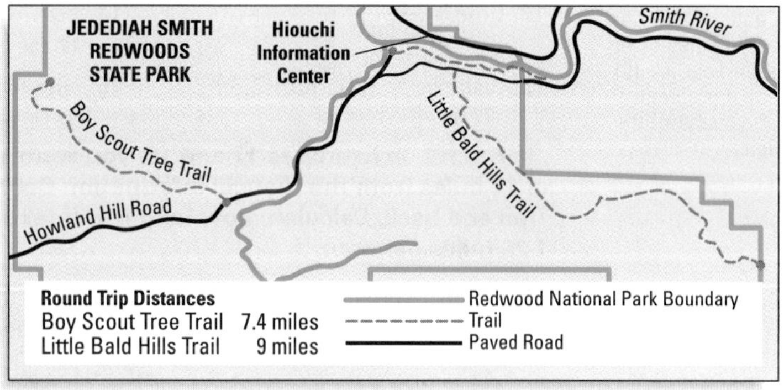

▶ Source: Redwood National and State Parks

SOLUTION

The map shows that the round-trip distance is about 7.4 miles.

PROBLEM SOLVING STRATEGY

VERBAL MODEL $\text{Time} = \dfrac{\text{Distance}}{\text{Rate}}$

LABELS

Time = t (hours)

Distance = **7.4** (miles)

Rate = **2** (miles per hour)

ALGEBRAIC MODEL

$t = \dfrac{7.4}{2}$ **Write algebraic model.**

$= 3.7$ **Simplify.**

▶ It should take you about 3.7 hours to hike the Boy Scout Tree Trail.

UNIT ANALYSIS Use unit analysis to check that *hours* are the units of the solution.

$$\text{Time} = \frac{\text{Distance}}{\text{Rate}} = \frac{\text{mi}}{\text{mi/h}} = \cancel{\text{mi}} \cdot \frac{\text{h}}{\cancel{\text{mi}}} = \text{h}$$

EXTRA EXAMPLE 4
Find the simple interest, *Prt,* you will earn in two years on a deposit of $870 at a rate of 5% per year. **$87**

EXTRA EXAMPLE 5
You plan to ride your bicycle to school and back at an average rate of 6 miles per hour for a distance of 10.5 miles. How long will it take you? **1.75 h**

CHECKPOINT EXERCISES

For use after Example 4:

1. Find the simple interest you will earn after one half of a year with a deposit of $1250 at a rate of 6% per year. **$37.50**

For use after Example 5:

2. Your parents drive to the grocery store 9 miles away at 12:30 P.M. and return at 2:00 P.M. If the grocery shopping took 45 minutes, what was the average speed of their car? **24 mi/h**

FOCUS ON VOCABULARY
Ask students to explain the difference between a variable and a variable expression.

CLOSURE QUESTION
How do you evaluate an expression that contains variables? **You substitute the values of the variable in the expression and simplify the resulting numerical expression.**

DAILY PUZZLER
For positive numbers can you tell which expression $x + x + x + x$ or $x \cdot x \cdot x \cdot x$ is greater? If so, which. If not why not? **No; if you evaluate the expressions for certain numbers such as $\frac{1}{2}$ or 1, $x + x + x + x$ is greater. If you evaluate for other numbers such as 2 or 10, then $x \cdot x \cdot x \cdot x$ is greater.**

3 APPLY

ASSIGNMENT GUIDE

BASIC
Day 1: pp. 6–8 Exs. 19–32, 35–38, 45–47, 50–68 even

AVERAGE
Day 1: pp. 6–8 Exs. 19–39, 45–47, 50–68 even

ADVANCED
Day 1: pp. 6–8 Exs. 20–30 even, 40–49, 50–68 even

BLOCK SCHEDULE WITH 1.2
pp. 6–8 Exs. 19–39, 45–47, 50–68 even

EXERCISE LEVELS

Level A: *Easier*
19–30, 45–47, 58–69

Level B: *More Difficult*
31–38, 40–42, 52–55,

Level C: *Most Difficult*
39, 43, 44, 48–51, 56, 57

✓ HOMEWORK CHECK

To quickly check student understanding of key concepts, go over the following exercises: Exs. 22, 26, 29, 32, 36. See also the Daily Homework Quiz:

- Blackline Master (*Chapter 1 Resource Book,* p. 25)
- Transparency (p. 3)

APPLICATION NOTE

EXERCISE 31 Banks generally pay compound interest instead of simple interest, but compound interest models are not linear.

GUIDED PRACTICE

Vocabulary Check ✓

1. Explain what it means to *evaluate a variable expression.* **to replace each variable in the expression by a number**

Concept Check ✓

2. What operation is indicated by the expression?

a. $4y$ **multiplication** **b.** $\frac{7}{d}$ **division** **c.** $t + 8$ **addition** **d.** $3 - t$ **subtraction**

3. Write a variable expression for "5 divided by r." $\frac{5}{r}$ **or** $5 \div r$

4. How is unit analysis helpful in solving real-life problems? **See margin.**

4. *Sample answer:* Unit analysis can help you determine if your original model is correct. For example, in Example 5 on page 5, the answer should be in units of time. An answer in miles or in miles per hour would indicate a problem with the model or the solution.

Skill Check ✓

Evaluate the expression when $y = 6$.

5. $5y$ **30** 6. $\frac{24}{y}$ **4** 7. $y + 19$ **25** 8. $y - 2$ **4**

9. $y \div 3$ **2** 10. $27 - y$ **21** 11. $2.5y$ **15** 12. $3.2 + y$ **9.2**

Simplify the real-life expression and show unit analysis.

13. time $= \frac{15 \text{ mi}}{5 \text{ mi/h}}$ **3 h;** $\frac{\text{mi}}{\text{mi/h}} = \text{mi} \cdot \frac{\text{h}}{\text{mi}} = \text{h}$

14. perimeter = 3 cm + 4 cm + 5 cm **12 cm; cm + cm + cm = cm**

15. distance = (60 mi/h)(2.3 h) **138 mi;** $\frac{\text{mi}}{\text{h}} \cdot \text{h} = \text{mi}$

16. simple interest = $(\$100)\left(\frac{0.05}{\text{year}}\right)(2.5 \text{ years})$ **\$12.50; dollars** $\cdot \frac{\text{percent}}{\text{year}} \cdot$ **years = dollars**

HIKING In Exercises 17 and 18, you want to hike a round-trip distance of 10 miles from the Hiouchi Information Center along the Little Bald Hills Trail and back. Calculate how long it will take if you hike at a rate of 1.25 miles per hour.

17. Write a verbal model, provide labels, and write an algebraic model. **time =** $\frac{\text{distance}}{\text{rate}}$**; time = t h, distance = 10 mi, rate = 1.25 mi/h;** $t = \frac{10}{1.25}$**; 8 h**

18. Show unit analysis. $\frac{\text{mi}}{\text{mi/h}} = \text{mi} \cdot \frac{\text{h}}{\text{mi}} = \text{h}$

PRACTICE AND APPLICATIONS

STUDENT HELP
Extra Practice to help you master skills is on p. 797.

EVALUATING EXPRESSIONS Evaluate the expression for the given value of the variable.

19. $12x$ when $x = 5$ **60** 20. $b - 7$ when $b = 24$ **17** 21. $0.5d$ when $d = 0.5$ **0.25**

22. $9 + p$ when $p = 11$ **20** 23. $r(10)$ when $r = 8.2$ **82** 24. $3.67a$ when $a = 2$ **7.34**

25. $\frac{6.3}{x}$ when $x = 3$ **2.1** 26. $\frac{2}{3} \cdot x$ when $x = \frac{1}{3}$ $\frac{2}{9}$ 27. $\frac{d}{12}$ when $d = 60$ **5**

28. $\frac{18}{t}$ when $t = 3$ **6** 29. $\frac{5}{8} - p$ when $p = \frac{3}{16}$ $\frac{7}{16}$ 30. $\frac{1}{2} + t$ when $t = \frac{1}{2}$ **1**

STUDENT HELP
HOMEWORK HELP
Example 1: Exs. 19–30
Example 2: Exs. 35–38
Example 3: Ex. 32
Example 4: Exs. 31, 33, 34
Example 5: Ex. 39

31. **INTEREST EARNED** You invest \$80 at a simple annual interest rate of 2%. How much simple interest would you earn in 1.5 years? Use unit analysis to check the units in your response. **\$2.40**

32. **GEOMETRY CONNECTION** The perimeter of a square is equal to $4s$, where s is the length of one side. Find the perimeter of the square at the right. Show unit analysis. **28 ft; ft + ft + ft + ft = ft**

ERROR ANALYSIS **In Exercises 33 and 34, Samantha deposits $300 in an account that earns an annual interest rate of 2.5%. After 9 months, she computes the simple interest.**

~~$(\$300)\left(\frac{2.5}{\text{year}}\right)(0.75 \text{ year}) = \$562.50$~~

33. What mistake did Samantha make? **See margin.**

34. What is the correct amount of simple interest earned after 9 months? **$5.625, or $5.63**

CALCULATING AVERAGE SPEEDS **In Exercises 35–38, find the average speed for the given distance and time. Show unit analysis to check units.**

35. A train travels 75 miles in 55 minutes. **$1\frac{4}{11}$ mi/min ≈ 1.36 mi/min**

36. In 5 seconds, an athlete runs 40 feet. **8 ft/sec**

37. A horse gallops 4 kilometers in 30 minutes. **$\frac{2}{15}$ km/min ≈ 0.13 km/min**

38. Dick Rutan and Jeana Yeager flew nonstop around the world, a distance of 24,986.73 miles, in 216.06 hours. ▶ Source: National Air and Space Museum **approximately 115.65 mi/h**

39. DRIVING TIME If you are driving at a constant speed of 96 kilometers per hour, how long will it take you to travel 288 kilometers? **3 h**

CALORIES BURNED **In Exercises 40–42, use the following information. The number of calories burned while doing an activity can be expressed by *rm*, where *r* is the rate of calories burned and *m* is the number of minutes spent doing the activity.**

▶ Source: *Home & Garden* Bulletin No. 72, U.S. Government Printing Office

40. A 120-pound student playing volleyball burns 2.7 calories per minute. If the student plays for 30 minutes, how many calories does the student burn? **81 Cal**

41. An in-line skater, who is the same weight as the student in Exercise 40, burns 387 calories in 90 minutes. How many calories does the in-line skater burn per minute? **4.3 Cal/min**

42. Which activity burns more calories per minute, volleyball or in-line skating? **in-line skating**

43. MOUNTAIN CLIMBING Heidi Zimmer plans to climb the highest peak in each continent. She has already climbed summits in North America, Europe, Africa, and South America. Copy and complete the table. Convert *h*, the height in meters, to the height in feet by dividing each given value of *h* by 0.3048. The last column shows how much shorter each mountain is than Mt. Everest, which is 29,029 feet high. **See margin.**

Mountain, Continent	h	$\frac{h}{0.3048}$	$29{,}029 - \frac{h}{0.3048}$
Mt. McKinley, North America	6194	?	?
Mt. Elbrus, Europe	5633	?	?
Mt. Kilimanjaro, Africa	5963	?	?
Mt. Aconcagua, South America	6959	?	?

▶ Source: Peakware™ World Mountain Encyclopedia

STUDENT HELP

HOMEWORK HELP
Visit our Web site www.mcdougallittell.com for help with Exs. 35–39.

43. Distances are given to the nearest foot. Mt. McKinley: 20,322 ft and 8707 ft; Mt. Elbrus: 18,481 ft and 10,548 ft; Mt. Kilimanjaro: 19,564 and 9465 ft; Mt. Aconcagua: 22,831 ft and 6198 ft

FOCUS ON PEOPLE

HEIDI ZIMMER is the first deaf woman to climb Mt. McKinley. It took her 18 days to reach the top of Mt. McKinley.

APPLICATION LINK
www.mcdougallittell.com

! COMMON ERROR

EXAMPLE 33 This is a very common error. Remind students to convert interest rates to a decimal by multiplying by 0.01.

STUDENT HELP NOTES

Application Link Additional information about mountain climbing is available at **www.mcdougallittell.com.**

APPLICATION NOTE

EXERCISE 44 Point out that batting average is normally expressed in thousandths. Some students can confirm this if they know the batting averages of their favorite players.

33. She did not use the correct decimal equivalent of 2.5%, 0.025.

ADDITIONAL PRACTICE AND RETEACHING

For Lesson 1.1:

- Practice Levels A, B, and C (*Chapter 1 Resource Book*, p. 15)
- Reteaching with Practice (*Chapter 1 Resource Book*, p. 18)
- See Lesson 1.1 of the *Personal Student Tutor*

For more Mixed Review:

- Search the *Test and Practice Generator* for key words or specific lessons.

4 ASSESS

Daily Homework Quiz

Transparency Available

1. Evaluate the expression.
   a. $6 + y$ when $y = 15$ **21**
   b. $\frac{24}{p}$ when $p = 4$ **6**
2. You deposit $150 in a bank account that pays an annual interest rate of 4%. How much simple interest will you earn in 1.5 years? **$9**
3. Find the area of the rectangle.

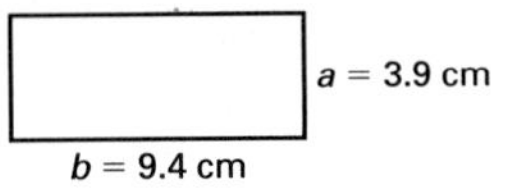

36.66 cm²

4. If you are driving at a constant rate of 108 kilometers per hour, how long will it take you to travel 270 kilometers? **2.5 h**
5. Haresh drove 120 miles in 2 hours 30 minutes. Find the average speed. **48 mi/h**

EXTRA CHALLENGE NOTE

Challenge problems for Lesson 1.1 are available in **blackline** format in the *Chapter 1 Resource Book,* p. 22 and at **www.mcdougallittell.com.**

Additional Test Preparation

1. **OPEN ENDED** Write and solve your own real-life problem that uses one of the formulas in the lesson. ***Sample answer:*** **Ian must ride his bike 3 miles to get to piano lessons. If he can pedal at a rate of 6 miles per hour, how long will it take him to get to his lesson? Answer: $\frac{1}{2}$ hour**

44. BATTING AVERAGE A baseball player's batting average is found by dividing the number of hits h by the official times at bat b. During the 1998 baseball season, Alex Rodriguez of the Seattle Mariners had 686 official times at bat and made 213 hits. Use a verbal model to find his batting average. Round your answer to the nearest thousandth. **0.310**

Test Preparation

45. MULTIPLE CHOICE You drove 200 miles in 3 hours 20 minutes. Which expression represents your average speed if d represents distance and t represents time? **B**

(A) dt (B) $\frac{d}{t}$ (C) $\frac{t}{d}$ (D) $d + t$

46. MULTIPLE CHOICE You invest $300 at a simple annual interest rate of 4.5%. How much simple interest will you earn in 10 years? **C**

(A) $115 (B) $120 (C) $135 (D) $150

47. MULTIPLE CHOICE A rectangular computer screen measures 28 centimeters by 21 centimeters. What is the perimeter of the screen? **B**

(A) 49 cm (B) 98 cm (C) 294 cm (D) 588 cm

★ Challenge

PITCHING SPEED **In baseball the pitcher's mound is 60.5 feet from home plate. The strike zone, or distance across the plate, is 17 inches. The time it takes for a baseball to reach home plate can be determined by dividing the distance the ball travels by the speed at which the pitcher throws the baseball.**

48. If a pitcher throws a baseball at 90 miles per hour, how many seconds does it take for the baseball to reach home plate? **about 0.46 sec**

EXTRA CHALLENGE
www.mcdougallittell.com

49. For Exercise 48, how long is the ball in the strike zone? **about 0.01 sec**

Mixed Review

FRACTIONS **Perform the indicated operation. Simplify your answer, if possible.** (Skills Review, pp. 781–783)

50. $1\frac{7}{8} + \frac{3}{4}$ **$2\frac{5}{8}$** **51.** $\frac{2}{5} - \frac{1}{10}$ **$\frac{3}{10}$** **52.** $5\frac{3}{4} \cdot \frac{1}{3}$ **$1\frac{11}{12}$** **53.** $2\frac{1}{5} \div \frac{4}{5}$ **$2\frac{3}{4}$**

54. $3\frac{2}{3} \cdot 3$ **11** **55.** $\frac{6}{5} \div \frac{3}{10}$ **4** **56.** $4\frac{5}{9} - \frac{1}{5}$ **$4\frac{16}{45}$** **57.** $8\frac{9}{10} + 1\frac{1}{5}$ **$10\frac{1}{10}$**

GEOMETRY CONNECTION **Find the area of the rectangle.** (Skills Review, p. 791)

58.

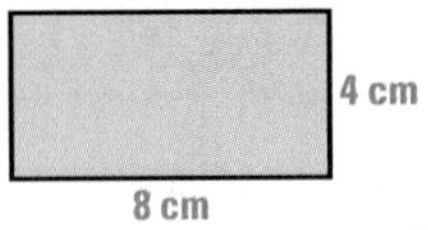

32 cm²

59.

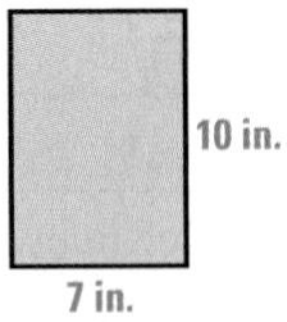

70 in.²

60.

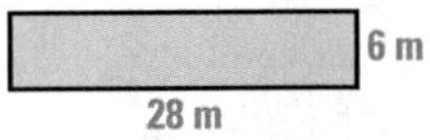

168 m²

EXPRESSIONS **Evaluate the expression when $x = 2$.** (Review 1.1 for Lesson 1.2)

61. $2x$ **4** **62.** $(x)(x)$ **4** **63.** $3x$ **6** **64.** $(x)(x)(x)$ **8**

65. $4x$ **8** **66.** $(x)(x)(x)(x)$ **16** **67.** $5x$ **10** **68.** $(x)(x)(x)(x)(x)$ **32**

1.2 Exponents and Powers

What you should learn

GOAL 1 Evaluate expressions containing exponents.

GOAL 2 Use exponents in **real-life** problems such as finding the volume of an aquarium in **Example 6**.

Why you should learn it

▼To solve **real-life** problems, such as finding the volume of a glass cube in **Ex. 64**.

REAL LIFE

GOAL 1 EXPRESSIONS CONTAINING EXPONENTS

An expression like 4^6 is called a **power**. The **exponent** 6 represents the number of times the **base** 4 is used as a factor.

$$4^6 = 4 \cdot 4 \cdot 4 \cdot 4 \cdot 4 \cdot 4$$

base → 4 (exponent 6); 4^6: power; $4 \cdot 4 \cdot 4 \cdot 4 \cdot 4 \cdot 4$: 6 factors of 4

EXAMPLE 1 *Reading and Writing Powers*

Express the meaning of the power in words and then with numbers or variables.

SOLUTION

EXPONENTIAL FORM	WORDS	MEANING
a. 10^1	ten to the first power	10
b. 4^2	four to the second power, or four squared	$4 \cdot 4$
c. 5^3	five to the third power, or five cubed	$5 \cdot 5 \cdot 5$
d. 7^6	seven to the sixth power	$7 \cdot 7 \cdot 7 \cdot 7 \cdot 7 \cdot 7$
e. x^n	x to the nth power	$x \cdot x \cdot x \cdot x \cdot \ldots \cdot x$ (and so on)

For a number raised to the first power, you usually do not write the exponent 1. For instance, you write 5^1 simply as 5.

EXAMPLE 2 *Evaluating Powers*

Evaluate the expression x^3 when $x = 5$.

SOLUTION

$x^3 = 5^3$ Substitute 5 for x.

$= 5 \cdot 5 \cdot 5$ Write factors.

$= 125$ Multiply.

▶ The value of the expression is 125.

1 PLAN

PACING

Basic: 2 days
Average: 2 days
Advanced: 2 days
Block Schedule: 0.5 block with 1.1
0.5 block with 1.3

LESSON OPENER CALCULATOR ACTIVITY

An alternative way to approach Lesson 1.2 is to use the Calculator Activity Lesson Opener:

- Blackline Master (*Chapter 1 Resource Book*, p. 26)
- Transparency (p. 2)

MEETING INDIVIDUAL NEEDS

- ***Chapter 1 Resource Book***
  Prerequisite Skills Review (p. 5)
  Practice Level A (p. 29)
  Practice Level B (p. 30)
  Practice Level C (p. 31)
  Reteaching with Practice (p. 32)
  Absent Student Catch-Up (p. 34)
  Challenge (p. 36)
- ***Resources in Spanish***
- ***Personal Student Tutor***

NEW-TEACHER SUPPORT

See the Tips for New Teachers on pp. 1–2 of the *Chapter 1 Resource Book* for additional notes about Lesson 1.2.

WARM-UP EXERCISES

Transparency Available

Evaluate the expression.

1. $2 \cdot 2 \cdot 2 \cdot 2$ **16**
2. $3x$ when $x = 6$ **18**
3. $(x)(x)(x)$ when $x = 4$ **64**
4. $x - 8$ when $x = 17$ **9**
5. $y \cdot y \cdot y \cdot y \cdot y$ when $y = 3$ **243**

2 TEACH

MOTIVATING THE LESSON
Ask students how they would evaluate $5 \cdot 5 \cdot 5 \cdot 5$ on their calculator. Some students may know that scientific calculators have power keys like y^x to help evaluate expressions containing exponents, the focus of today's lesson.

EXTRA EXAMPLE 1
Express the meaning of the power in words and then with numbers or variables.
a. 8^1 8 to the first power; 8
b. 6^2 6 to the second power, or 6 squared; $6 \cdot 6$
c. 4^3 4 to the third power, or four cubed; $4 \cdot 4 \cdot 4$
d. 9^7 nine to the seventh power; $9 \cdot 9 \cdot 9 \cdot 9 \cdot 9 \cdot 9 \cdot 9$
e. y^n y to the nth power; $y \cdot y \cdot y \cdot y \cdot \ldots \cdot y$

EXTRA EXAMPLE 2
Evaluate the expression x^4 when $x = 3$. 81

EXTRA EXAMPLE 3
Evaluate the expression when $a = 3$ and $b = 4$.
a. $(a + b)^3$ 343 **b.** $a + (b^3)$ 67

EXTRA EXAMPLE 4
Evaluate the expression when $x = 5$.
a. $3x^2$ 75 **b.** $(3x)^2$ 225

CHECKPOINT EXERCISES
For use after Example 1:
1. Express the meaning of 6^4 with numbers. $6 \cdot 6 \cdot 6 \cdot 6$
For use after Example 2:
2. Find x^5 when $x = 4$. 1024
For use after Example 3:
3. Evaluate the expression when $a = 2$ and $b = 3$.
a. $(a + b)^2$ 25
b. $a^2 + b^2$ 13
For use after Example 4:
4. Evaluate the expression when $x = 2$.
a. $4x^3$ 32 **b.** $(4x)^3$ 512

GROUPING SYMBOLS For problems that have more than one operation, it is important to know which operation to do first. **Grouping symbols**, such as parentheses ( ) or brackets [ ], indicate the order in which the operations should be performed. Operations within the innermost set of grouping symbols are done first. For instance, the value of the expression $(3 \cdot 4) + 7$ is not the same as the value of the expression $3 \cdot (4 + 7)$.

Multiply. Then add.	Add. Then multiply.
$(3 \cdot 4) + 7 = 12 + 7 = 19$	$3 \cdot (4 + 7) = 3 \cdot 11 = 33$

EXAMPLE 3 Evaluating an Exponential Expression

Evaluate the expression when $a = 1$ and $b = 2$.

a. $(a + b)^2$ **b.** $(a^2) + (b^2)$

SOLUTION

a. $(a + b)^2 = (1 + 2)^2$ Substitute 1 for *a* and 2 for *b*.
$= 3^2$ Add within parentheses.
$= 3 \cdot 3$ Write factors.
$= 9$ Multiply.

b. $(a^2) + (b^2) = (1^2) + (2^2)$ Substitute 1 for *a* and 2 for *b*.
$= 1 + 4$ Evaluate power.
$= 5$ Add.

STUDENT HELP
HOMEWORK HELP
Visit our Web site www.mcdougallittell.com for extra examples.

An exponent applies only to the number, variable, or expression immediately to its left. In the expression $2x^3$, the base is x, not $2x$. In the expression $(2x)^3$, the base is $2x$, as indicated by the parentheses.

EXAMPLE 4 Exponents and Grouping Symbols

Evaluate the expression when $x = 4$.

a. $2x^3$ **b.** $(2x)^3$

SOLUTION

a. $2x^3 = 2(4^3)$ Substitute 4 for *x*.
$= 2(64)$ Evaluate power.
$= 128$ Multiply.

b. $(2x)^3 = (2 \cdot 4)^3$ Substitute 4 for *x*.
$= 8^3$ Multiply within parentheses.
$= 512$ Evaluate power.

GOAL 2 REAL-LIFE APPLICATIONS OF EXPONENTS

Exponents often are used in the formulas for area and volume. In fact, the words *squared* and *cubed* come from the formula for the area of a square, $A = s^2$, and the formula for the volume of a cube, $V = s^3$.

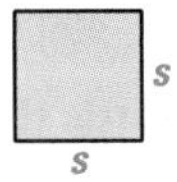

Area of Square: $A = s^2$

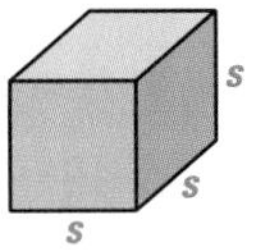

Volume of Cube: $V = s^3$

Units of area, such as square feet, ft^2, can be written using a second power. Units of volume, such as cubic centimeters, cm^3, can be written using a third power.

EXAMPLE 5 *Making a Table*

You can find the volume of cubes that have edge lengths of 1 inch, 2 inches, 3 inches, 4 inches, and 5 inches by using the formula $V = s^3$.

Edge, s	1	2	3	4	5
s^3	1^3	2^3	3^3	4^3	5^3
Volume, V	1 $in.^3$	8 $in.^3$	27 $in.^3$	64 $in.^3$	125 $in.^3$

REAL LIFE *Aquarium*

EXAMPLE 6 *Finding Volume*

The aquarium has the shape of a cube. Each edge is 2.5 feet long.

a. Find the volume in cubic feet.

b. How many gallons of water will the cubic aquarium hold? Convert to liquid volume, where one cubic foot holds 7.48 gallons.

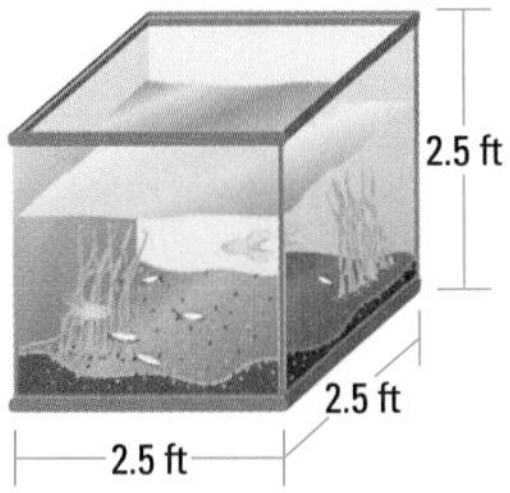

SOLUTION

a. $V = s^3$ — Write formula for volume.

$= 2.5^3$ — Substitute 2.5 for s.

$= 15.625$ — Evaluate power.

▶ The volume of the aquarium is 15.625 ft^3.

b. $V = 15.625 \text{ ft}^3 \left(7.48 \text{ gal}/1 \text{ ft}^3\right)$ — Write conversion factor.

$= 116.875$ gal — Multiply.

▶ A 15.625 cubic foot aquarium will hold 116.875 gallons of water.

STUDENT HELP

KEYSTROKE HELP
Your calculator may have a [y^x] key or a [^] key that you can use to evaluate powers.

Closure Question *Sample answer:*

To evaluate 5 • (8 – 2), you do 8 – 2 first and your answer is 30. To evaluate (5 • 8) – 2, you do 5 • 8 first and your answer is 38.

EXTRA EXAMPLE 5

Make a table showing the area of a square with side lengths 30 cm, 40 cm, 50 cm by using the formula $A = s^2$.

Side, s	s^2	Area
30	900	900 cm^2
40	1600	1600 cm^2
50	2500	2500 cm^2

EXTRA EXAMPLE 6

A tank has the shape of a cube. Each edge is 4.5 feet long.

a. Find the volume in cubic feet. **91.125 ft^3**

b. How many gallons of water will the cubic tank hold? Convert to liquid volume where one cubic foot holds 7.48 gallons. **681.615 gal**

CHECKPOINT EXERCISES

For use after Examples 5 and 6:

1. Find the total volume of 5 cubes that all have the edge length 3.1 cm. **148.955 cm^3**

ENGLISH LEARNERS

Draw students' attention to the number 15.625, write it on the board, and write 15,625 below it. Explain that in the United States, a point separates whole numbers from decimal parts, while a comma separates hundreds from thousands, thousands from millions, and so on.

STUDENT HELP NOTES

Keystroke Help Keystrokes for several models of calculators are available in **blackline** format in the *Chapter 1 Resource Book,* p. 27.

CLOSURE QUESTION

Describe the difference in evaluating 5 • (8 – 2) and (5 • 8) – 2.
See margin.

DAILY PUZZLER

The variables x and y represent different positive integers. Find values for x and y such that $x^y = y^x$.
$x = 2, y = 4$ or $x = 4, y = 2$

3 APPLY

ASSIGNMENT GUIDE

BASIC
Day 1: pp. 12–13 Exs.17–48
Day 2: pp. 13–14 Exs. 55–65, 68, 72–84

AVERAGE
Day 1: pp. 12–13 Exs. 17–48
Day 2: pp. 13–14 Exs.55–66, 68, 72–84

ADVANCED
Day 1: pp. 12–13 Exs. 17–48
Day 2: pp. 13–14 Exs. 58–84

BLOCK SCHEDULE
pp. 12–13 Exs. 17–48 (with 1.1)
pp. 13–14 Exs. 55–66, 68, 72–84 (with 1.3)

EXERCISE LEVELS

Level A: *Easier*
17–62, 72, 73, 75–84
Level B: *More Difficult*
63–66, 68, 74
Level C: *Most Difficult*
67, 69–71

✓ HOMEWORK CHECK

To quickly check student understanding of key concepts, go over the following exercises: Exs. 24, 25, 32, 38, 42, 58, 64. See also the Daily Homework Quiz:

- Blackline Master (*Chapter 1 Resource Book*, p. 39)
- Transparency (p. 4)

GUIDED PRACTICE

Vocabulary Check ✓

1. In the expression 15^3, what is 15 called? What is 3 called? What is the expression called? **the base; the exponent; a power**

Concept Check ✓

2. The expressions $3x^2$ and $(3x)^2$ do not have the same meaning. Explain the difference. **In the first expression, only the x is squared. In the second the product, $3x$, is squared: $(3x)^2 = 9x^2$.**

3. Evaluate the expressions $3x^2$ and $(3x)^2$ when $x = 4$. **48, 144**

Skill Check ✓

Match the power with the words that describe it.

A. five to the sixth power
B. two to the fifth power
C. five squared
D. five cubed

4. 5^2 **C** **5.** 5^3 **D** **6.** 2^5 **B** **7.** 5^6 **A**

Evaluate the expression when $x = 3$.

8. x^2 **9** **9.** $(x + 1)^3$ **64** **10.** $2x^2$ **18** **11.** $(2x)^3$ **216**

12. $(x - 1)^4$ **16** **13.** 5^x **125** **14.** $(3x)^4$ **6561** **15.** 10^x **1000**

16. STEREO SPEAKERS A cubical stereo speaker measures 35 centimeters along each edge. The expression for finding the surface area of a cube is $6s^2$, where s is the length of each edge. Find the surface area of the stereo speaker. **7350 cm²**

PRACTICE AND APPLICATIONS

STUDENT HELP
Extra Practice to help you master skills is on p. 797.

EXPONENTIAL FORM **Write the expression in exponential form.**

17. two cubed **2^3** **18.** p squared **p^2** **19.** nine to the yth power **9^y**

20. b to the eighth power **b^8** **21.** $3 \cdot 3 \cdot 3 \cdot 3 \cdot y$ **3^4y** **22.** $t \cdot t$ **t^2**

23. $c \cdot c \cdot c \cdot c \cdot c \cdot c$ **c^6** **24.** $5 \cdot x \cdot x \cdot x \cdot x \cdot x$ **$5x^5$** **25.** $4x \cdot 4x \cdot 4x$ **$(4x)^3$**

EVALUATING POWERS **Evaluate the power.**

26. 10^2 **100** **27.** 5^2 **25** **28.** 8^2 **64**

29. 6^4 **1296** **30.** 10^5 **100,000** **31.** 7^4 **2401**

32. 4^6 **4096** **33.** 9^3 **729** **34.** 2^5 **32**

EVALUATING EXPRESSIONS **Evaluate the expression for the given value of the variable.**

35. 4^n when $n = 5$ **1024** **36.** b^4 when $b = 9$ **6561** **37.** x^6 when $x = 10$ **1,000,000**

38. c^6 when $c = 2$ **64** **39.** w^3 when $w = 13$ **2197** **40.** p^2 when $p = 2.5$ **6.25**

STUDENT HELP
HOMEWORK HELP
Example 1: Exs. 17–25
Example 2: Exs. 26–40
Example 3: Exs. 41–46
Example 4: Exs. 55–60
Example 5: Ex. 61
Example 6: Exs. 63–66

EXPONENTIAL EXPRESSIONS **Evaluate the expression for the given values of the variables.**

41. $(x + y)^2$ when $x = 5$ and $y = 3$ **64** **42.** $m - n^2$ when $m = 25$ and $n = 4$ **9**

43. $(a - b)^4$ when $a = 4$ and $b = 2$ **16** **44.** $c^3 + d$ when $c = 4$ and $d = 16$ **80**

45. $(d - 3)^2$ when $d = 13$ **100** **46.** $16 + x^3$ when $x = 2$ **24**

EVALUATING POWERS **Use a calculator to evaluate the power. For keystroke help see Student Help box on page 11.**

47. 9^5 **59,049** **48.** 2^{10} **1024** **49.** 5^9 **1,953,125** **50.** 3^{11} **177,147**

51. 8^6 **262,144** **52.** 12^7 **35,831,808** **53.** 6^8 **1,679,616** **54.** 13^5 **371,293**

EXPONENTIAL EXPRESSIONS **Evaluate the expression for the given value of the variable.**

55. $(5w)^3$ when $w = 5$ **15,625** **56.** $6t^4$ when $t = 3$ **486** **57.** $7b^2$ when $b = 7$ **343**

58. $2x^2$ when $x = 15$ **450** **59.** $(8x)^3$ when $x = 2$ **4096** **60.** $5y^5$ when $y = 2$ **160**

61. **USING A TABLE** The area of a square is s^2. Show the relationship between the side length of a square and its area by copying and completing the following table. **1, 4, 9, 16, 25**

Side length, s	1	2	3	4	5
Area, s^2	?	?	?	?	?

62. **CRITICAL THINKING** Copy and complete the table. What pattern do you see?

Power	10^2	100^2	1000^2	$10{,}000^2$
Evaluate	100	?	?	?

10,000; 1,000,000; 100,000,000; the simplified power has twice as many zeroes as the base.

63. **INTERIOR DESIGN** One room in Jean's apartment is a square measuring 12.2 feet along the base of each wall. How many square feet of wall-to-wall carpet does Jean need to carpet the room? **148.84 ft²**

64. **ART CONNECTION** In 1997 the artist Jon Kuhn of North Carolina created a cubic sculpture called Crystal Victory, shown at the left. Each edge of the solid glass cube is 9.5 inches in length. How much liquid glass did Kuhn need to make the cube? **857.375 in.³**

65. **VOLUME OF A SAFE** Each dimension of the cubical storage space inside a fireproof safe is 12 inches. What is the volume of the storage space? **1728 in.³**

66. **SWIMMING POOL** A swimming pool is 50 meters long, 19.5 meters wide, and 3 meters deep. Use the formula for the volume of a rectangular prism to find the volume of water in the pool. The formula is the length times the width times the height. **2925 m³**

67. **RAIN FOREST PYRAMID** The formula for the volume of a pyramid is $\frac{1}{3}$ times the height times the area of the base. The Rain Forest Pyramid in Moody Gardens near Galveston, Texas, is 100 feet high and 200 feet along each side of its square base. What is the volume of space inside the Rain Forest Pyramid?

▶Source: Morris Architects **about 1,333,333 ft³**

FOCUS ON APPLICATIONS

GLASS SCULPTURE Artist John Kuhn used lead crystal and colored glass powders to make the Crystal Victory.

APPLICATION LINK www.mcdougallittell.com

! COMMON ERROR

EXERCISES 55–60 Some students may ignore the parentheses in Exercises 55 and 59. Remind students that operations in parentheses must be done first.

CAREER NOTE

EXERCISE 64 Additional information about artists is available at **www.mcdougallittell.com.**

TEACHING TIPS

EXERCISE 67 If students have difficulty with this exercise suggest that they write out the formula $V = \frac{1}{3}hs^2$, substitute the values for the variables, and then evaluate the expression.

ADDITIONAL PRACTICE AND RETEACHING

For Lesson 1.2:

- Practice Levels A, B, and C (*Chapter 1 Resource Book,* p. 29)
- Reteaching with Practice (*Chapter 1 Resource Book,* p. 32)
- See Lesson 1.2 of the *Personal Student Tutor*

For more Mixed Review:

- Search the *Test and Practice Generator* for key words or specific lessons.

4 ASSESS

DAILY HOMEWORK QUIZ

Transparency Available

1. Write eight cubed in exponential form. 8^3
2. Evaluate 7^3. 343
3. Evaluate the expression for the given values.
   a. 6^n when $n = 3$ 216
   b. $(x + y)^2$ when $x = 4$ and $y = 5$ 81
   c. $d - e^2$ when $d = 16$ and $e = 3$ 7

EXTRA CHALLENGE NOTE

Challenge problems for Lesson 1.2 are available in **blackline** format in the *Chapter 1 Resource Book,* p. 36 and at **www.mcdougallittell.com.**

ADDITIONAL TEST PREPARATION

1. **WRITING** Explain the difference between 3^4 and 4^3. Which is greater? $3^4 = 3 \cdot 3 \cdot 3 \cdot 3 = 81$ and $4^3 = 4 \cdot 4 \cdot 4 = 64$; 3^4 is greater

Test Preparation

69. 9; 81; 729; 6561; 59,049; 531,441; 4,782,969; 43,046,721; the last digit of each number is 1 or 9. The digit is 9 if the exponent is odd and 1 if the exponent is even.

70. $8^1 = 8$; $8^2 = 64$; $8^3 = 512$; $8^4 = 4096$; $8^5 = 32,768$; $8^6 = 262,144$; $8^7 = 2,097,152$; $8^8 = 16,777,216$; the last digit repeats in the following sequence: 8, 4, 2, 6, 8, 4, 2, 6, ...

★ Challenge

68. MULTI-STEP PROBLEM You are making candles to sell at your school's art festival. You melt paraffin wax in a cubic container. Each edge of the container is 6 inches in length. The container is one half full.

a. What is the volume of the wax in the container? **108 in.^{3}**

b. Which of the candle molds could hold all of the melted wax? **A**

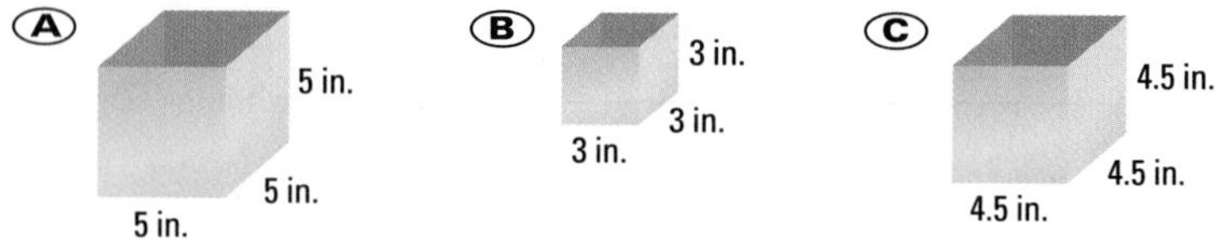

c. *Writing* Design a cubic candle mold different from those given that will hold all of the melted wax. Draw a diagram of the mold. Explain why your mold will hold all of the melted wax. ***Sample answer:*** **any cube with sides longer than 4.76 in.**

FINDING A PATTERN **Copy the table.**

Power	9^1	9^2	9^3	9^4	9^5	9^6	9^7	9^8
Evaluate	?	?	?	?	?	?	?	?

69. Evaluate the powers of 9 in the table. What pattern do you see for the last digit of each product? **See margin.**

70. Make a table like the one shown for powers of 8. Describe any patterns. **See margin.**

EXTRA CHALLENGE

www.mcdougallittell.com

71. Make a table for powers of 7. Describe any patterns. **See margin.**

MIXED REVIEW

71. $7^1 = 7$; $7^2 = 49$; $7^3 = 343$; $7^4 = 2401$; $7^5 = 16,807$; $7^6 = 117,649$; $7^7 = 823,543$; $7^8 = 5,764,801$; the last digit repeats in the following sequence: 7, 9, 3, 1, 7, 9, 3, 1, ...

GEOMETRY CONNECTION **Find the perimeter of the figure when $x = 1.7$.** **(Skills Review, pp. 790–791)**

72. 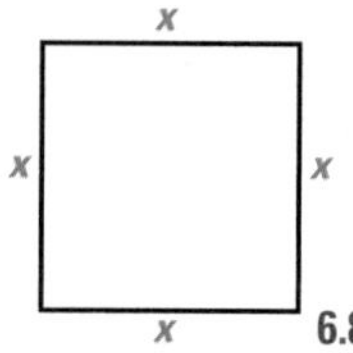

6.8

73. 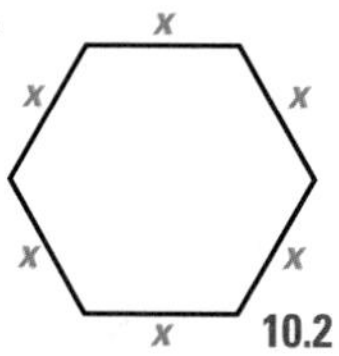

10.2

74. 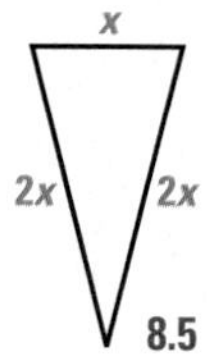

8.5

FRACTIONS, DECIMALS, AND PERCENTS **Write the fraction as a decimal and as a percent.** **(Skills Review, pp. 784–785)**

75. $\frac{5}{8}$ **0.625, 62.5%** 76. $\frac{3}{4}$ **0.75, 75%** 77. $\frac{11}{20}$ **0.55, 55%** 78. $\frac{4}{25}$ **0.16, 16%**

EVALUATING VARIABLE EXPRESSIONS **Evaluate the expression for the given value of the variable.** **(Review 1.1 for 1.3)**

79. $7x$ when $x = 3$ **21** 80. $y + 2$ when $y = 10$ **12** 81. $\frac{a}{2}$ when $a = 8$ **4**

82. $m - 5$ when $m = 17$ **12** 83. $\frac{9}{b}$ when $b = 4$ **$2\frac{1}{4}$** 84. $9b$ when $b = 4$ **36**

Making a Table

Using a graphing calculator to create a table can make it easier to evaluate an expression for many different values of the variable.

▶ EXAMPLE

You are offered a two penny salary that doubles every week. How much is your salary in the 20th week? Use a graphing calculator to make the table.

Week, x	1	2	4	6	8	10	12	14	16	18	20
Salary, 2^x	2	4	16	?	?	?	?	?	?	?	?

▶ SOLUTION

64; 256; 1024; 4096; 16,384; 65,536; 262,144; 1,048,576; $10,485.76

1 Press Y= and enter the expression 2^x as Y_1.

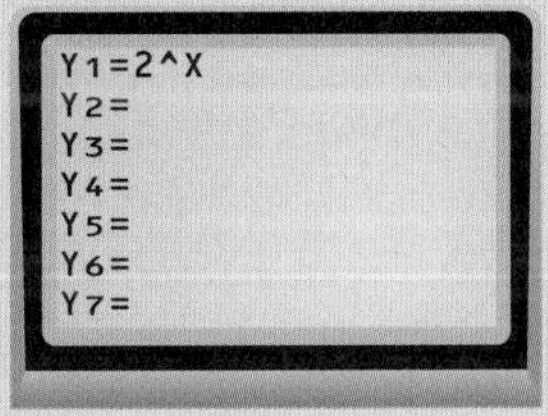

2 Use the Table Setup function to choose values beginning at 6 and increasing by 2.

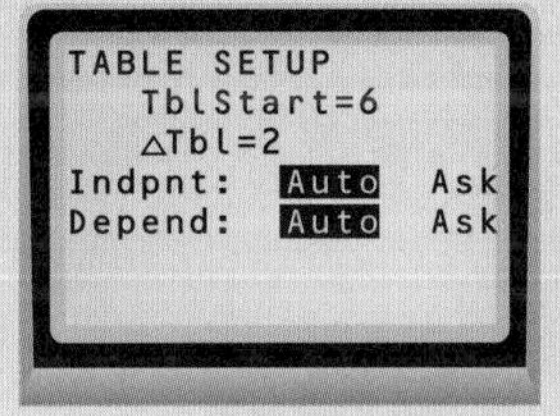

3 View your table. Scroll down to read the twentieth week in the table.

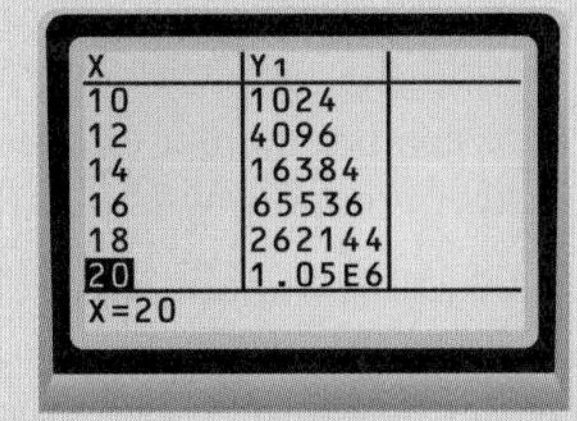

4 The value for the twentieth week is 1.05 E6. This means 1.05×10^6 or 1,050,000.

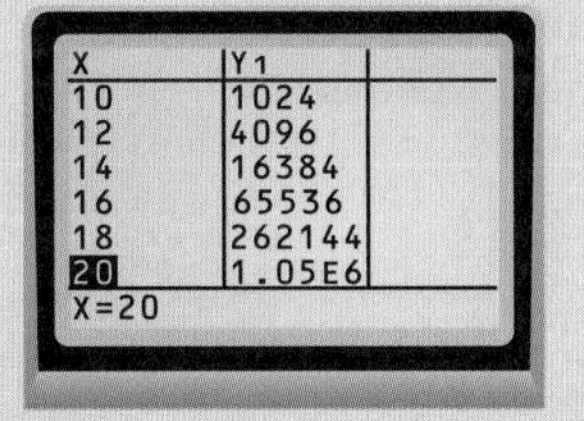

STUDENT HELP

INTERNET KEYSTROKE HELP
See keystrokes for several models of calculators at www.mcdougallittell.com

▶ EXERCISES

EVALUATING EXPRESSIONS **Use the table feature on a graphing calculator to evaluate the exponential expression for the given values of *x*.**

1–6. See margin.

1. 2^x for $x = 25, 50, 75, 100$

2. 3^x for $x = 5, 7, 9, 11, 13$

3. 4^x for $x = 2, 3, 4, 5, 6$

4. 5^x for $x = 4, 8, 12, 16$

5. 6^x for $x = 2, 4, 6, 8, 10$

6. 10^x for $x = 4, 7, 10, 13, 16$

1. 33,554,432; 1.13×10^{15}; 3.78×10^{22}; 1.27×10^{30}
2. 243; 2187; 19,683; 177,147; 1,594,323
3. 16; 64; 256; 1024; 4096
4. 625; 390,625; 244,140,625; 1.53×10^{11}
5. 36; 1296; 46,656; 1,679,616; 60,466,176
6. 10,000; 10,000,000; 1×10^{10}; 1×10^{13}; 1×10^{16}

1 Planning the Activity

PURPOSE
To have students use the table feature of a graphing calculator to evaluate expressions.

MATERIALS
- Graphing calculators
- Keystroke blackline (*Chapter 1 Resource Book,* p. 28)

PACING
- Solution — 15 min
- Exercises — 10 min

LINK TO LESSON
Students can go back to Example 5 and Exercises 61, 62, and 69–71 of Lesson 1.2 and use the table feature to check their work.

2 Managing the Activity

COOPERATIVE LEARNING
If not all students are familiar with graphing calculator techniques, have students work in groups where at least one student does understand how the calculator operates.

ALTERNATIVE APPROACH
To save time, demonstrate the activity with a graphing calculator overhead-view-screen-panel.

3 Closing the Activity

KEY DISCOVERY
Exponential expressions in which the exponent is a variable and the base is greater than 1 increase in value very rapidly.

ACTIVITY ASSESSMENT
JOURNAL Describe how to use a graphing calculator to compare values of 7^x and x^7. Enter 7^x as Y_1 and x^7 as Y_2. Set up the table to display x, Y_1 and Y_2. Compare values of Y_1 and Y_2 for each x-value.

1 PLAN

PACING
Basic: 2 days
Average: 2 days
Advanced: 2 days
Block Schedule: 0.5 block with 1.2
0.5 block with 1.4

LESSON OPENER APPLICATION
An alternative way to approach Lesson 1.3 is to use the Application Lesson Opener:
- Blackline Master (*Chapter 1 Resource Book*, p. 40)
- Transparency (p. 3)

MEETING INDIVIDUAL NEEDS
- ***Chapter 1 Resource Book***
  Prerequisite Skills Review (p. 5)
  Practice Level A (p. 44)
  Practice Level B (p. 45)
  Practice Level C (p. 46)
  Reteaching with Practice (p. 47)
  Absent Student Catch-Up (p. 49)
  Challenge (p. 51)
- ***Resources in Spanish***
- ***Personal Student Tutor***

NEW-TEACHER SUPPORT
See the Tips for New Teachers on pp. 1–2 of the *Chapter 1 Resource Book* for additional notes about Lesson 1.3.

WARM-UP EXERCISES
Transparency Available
Evaluate the expression.
1. $(3 \cdot 5) + 4$ 19
2. $3 \cdot (5 + 4)$ 27
3. $(9 - 3)^2$ 36
4. $9 - (3^2)$ 0
5. $4w^2$ when $w = 5$ 100
6. $(4w)^2$ when $w = 5$ 400

1.3 Order of Operations

What *you should learn*

GOAL 1 Use the order of operations to evaluate algebraic expressions.

GOAL 2 Use a calculator to evaluate **real-life** expressions, such as calculating sales tax in **Example 5**.

Why *you should learn it*

▼ To solve **real-life** problems, such as calculating the cost of admission for a family to a state fair in **Exs. 50** and **51**.

REAL LIFE

GOAL 1 USING THE ORDER OF OPERATIONS

Mathematicians have established an **order of operations** to evaluate an expression involving more than one operation. Start with operations within grouping symbols. Then evaluate powers. Then do multiplications and divisions from left to right. Finally, do additions and subtractions from left to right.

EXAMPLE 1 *Evaluating Without Grouping Symbols*

Evaluate the expression when $x = 4$.

a. $3x^2 + 1$ **b.** $32 \div x^2 - 1$

SOLUTION

a. $3x^2 + 1 = 3 \cdot 4^2 + 1$ Substitute 4 for *x*.
$= 3 \cdot 16 + 1$ Evaluate power.
$= 48 + 1$ Evaluate product.
$= 49$ Evaluate sum.

b. $32 \div x^2 - 1 = 32 \div 4^2 - 1$ Substitute 4 for *x*.
$= 32 \div 16 - 1$ Evaluate power.
$= 2 - 1$ Evaluate quotient.
$= 1$ Evaluate difference.

When you want to change the established order of operations for an expression, you *must* use parentheses or other grouping symbols.

ACTIVITY Developing Concepts

Investigating Grouping Symbols

Without grouping symbols, the expression $3 \cdot 4^2 + 8 \div 4$ has a value of 50. You can insert grouping symbols to produce a different value. For example:

$$3 \cdot (4^2 + 8) \div 4 = 3 \cdot (16 + 8) \div 4 = 3 \cdot 24 \div 4 = 72 \div 4 = 18$$

Insert grouping symbols in the expression $3 \cdot 4^2 + 8 \div 4$ to produce the indicated values.

1. 146 $(3 \cdot 4)^2 + (8 \div 4)$
2. 54 $3 \cdot (4^2 + 8 \div 4)$
3. 38 $((3 \cdot 4)^2 + 8) \div 4$

THE LEFT-TO-RIGHT RULE Operations that have the same priority, such as multiplication and division *or* addition and subtraction, are performed using the *left-to-right rule*, as shown in Example 2.

EXAMPLE 2 Using the Left-to-Right Rule

a. $24 - 8 - 6 = (24 - 8) - 6$ Work from left to right.

$= 16 - 6$

$= 10$

b. $15 \cdot 2 \div 6 = (15 \cdot 2) \div 6$ Work from left to right.

$= 30 \div 6$

$= 5$

c. $16 + 4 \div 2 - 3 = 16 + (4 \div 2) - 3$ Divide first.

$= 16 + 2 - 3$

$= (16 + 2) - 3$ Work from left to right.

$= 18 - 3$

$= 15$

STUDENT HELP

Study Tip
In part (c) of Example 2, you do not perform the operations from left to right because division has a higher priority than addition and subtraction.

ORDER OF OPERATIONS

1. First do operations that occur within grouping symbols.
2. Then evaluate powers.
3. Then do multiplications and divisions from left to right.
4. Finally, do additions and subtractions from left to right.

A fraction bar can act as a grouping symbol: $(1 + 2) \div (4 - 1) = \frac{1 + 2}{4 - 1}$

EXAMPLE 3 Using a Fraction Bar

$\frac{7 \cdot 4}{8 + 7^2 - 1} = \frac{7 \cdot 4}{8 + 49 - 1}$ Evaluate power.

$= \frac{28}{8 + 49 - 1}$ Simplify the numerator.

$= \frac{28}{57 - 1}$ Work from left to right.

$= \frac{28}{56}$ Subtract.

$= \frac{1}{2}$ Simplify.

STUDENT HELP

HOMEWORK HELP
Visit our Web site www.mcdougallittell.com for extra examples.

Skills Review
For help with writing fractions in lowest terms, see pp. 781–783.

2 TEACH

MOTIVATING THE LESSON
Real world activities follow a certain order. For example, you put on your socks before you put on your shoes. The other way, shoes first socks second, just doesn't work. The same is true with mathematics. Today's lesson focuses on order of operations in mathematics.

EXTRA EXAMPLE 1
Evaluate the expression when $x = 3$.
a. $4x^2 - 8$ 28 **b.** $27 \div x^2 + 3$ 6

EXTRA EXAMPLE 2
Simplify the expression.
a. $32 - 16 + 2$ 18
b. $48 \div 3 \cdot 2$ 32
c. $12 + 6 \div 3 - 2$ 12

EXTRA EXAMPLE 3
Simplify the expression $\frac{24 \cdot 3}{5 + 3^2 - 2}$. 6

CHECKPOINT EXERCISES

For use after Example 1:
1. Evaluate the expression when $x = 5$.
a. $2x^2 + 8$ 58
b. $100 \div x^2 + 6$ 10

For use after Example 2:
2. Evaluate the expression.
a. $25 + 10 - 8$ 27
b. $24 \div 2 \cdot 3$ 36
c. $36 - 6 \div 3 + 3$ 37

For use after Example 3:
3. Simplify the expression $\frac{16 \cdot 2 + 8}{10 + 2^2 - 4}$. 4

EXTRA EXAMPLE 4

When you enter the following in your calculator, does the calculator display 5.8 or 1.3?

12 [−] 4.8 [÷] 1.6 [−] 3.2 [Enter]

If the calculator uses order of operations, it will display 5.8. If it displays 1.3, it performs the operations as they are entered.

EXTRA EXAMPLE 5

Suppose $80 is spent on clothing that is taxed at 2% and $45 on a computer game that is taxed at 8%. Which of the following keystroke sequences show the correct amounts owed? Explain why.

a. [8] [0] [+] [4] [5] [+] [8] [0] [(] [0] [.] [0] [2] [)] [Enter]

b. [8] [0] [+] [4] [5] [+] [8] [0] [(] [0] [.] [0] [2] [)] [+] [4] [5] [(] [0] [.] [0] [8] [)] [Enter]

c. [8] [0] [+] [4] [5] [+] [4] [5] [(] [0] [.] [0] [8] [)] [Enter]

b, $130.20, Tax is added on both items.

CHECKPOINT EXERCISES

For use after Examples 4 and 5:

1. For which of the following expressions will your calculator display 8? b

a. $10.5 - 4.2 \div 1.4 + 3.5$

b. $(10.5 - 4.2) \div 1.4 + 3.5$

c. $(10.5 - 4.2) \div (1.4 + 3.5)$

FOCUS ON VOCABULARY

State the order of operations.

First do operations within grouping symbols. Then evaluate powers. Then do multiplications and divisions from left to right. Finally, do additions and subtractions from left to right.

CLOSURE QUESTION

Would you evaluate $\frac{(4+6)^2}{4}$ as 13, 25, or 40? Why? **25; you add 4 + 6 to get 10, square it to get 100, and divide by 4 to get 25.**

GOAL 2 EVALUATING EXPRESSIONS WITH A CALCULATOR

Many calculators use the established order of operations, but some do not. See if your calculator follows the established order of operations in Example 4. If your calculator does not follow the order of operations, you must enter the operations in the correct order.

EXAMPLE 4 *Using a Calculator*

When you enter the following in your calculator, does the calculator display 6.1 or 0.6?

10.5 [−] 6.3 [÷] 2.1 [−] 1.4 [ENTER]

SOLUTION

a. If your calculator uses order of operations, it will display 6.1.

$$10.5 - 6.3 \div 2.1 - 1.4 = 10.5 - (6.3 \div 2.1) - 1.4$$
$$= 10.5 - 3 - 1.4$$
$$= 6.1$$

b. If it displays 0.6, it performs the operations as they are entered.

$$[(10.5 - 6.3) \div 2.1] - 1.4 = (4.2 \div 2.1) - 1.4$$
$$= 2 - 1.4$$
$$= 0.6$$

REAL LIFE *Retail Sales*

EXAMPLE 5 *Calculating Sales Tax*

Suppose you live in a state that charges no sales tax on essentials, such as clothes or food, but charges 6% sales tax on nonessentials, such as CDs or games. You decide to buy a sweatshirt for $24 and a video game for $39. Your calculator follows the established order of operations. Which of the following keystroke sequences will show the correct amount you owe? Explain why.

a. 24 [+] 39 [+] 39 [×] 0.06 [ENTER]

b. 39 [+] 24 [+] 24 [×] 0.06 [ENTER]

c. 39 [+] 24 [+] [( ] 24 [+] 39 [ )] [×] 0.06 [ENTER]

SOLUTION

a. $65.34 is correct. Tax is added on the video game, but not the sweatshirt.

b. $64.44 is wrong. Tax is added on the sweatshirt, but not the video game.

c. $66.78 is wrong. Tax is added on both items.

GUIDED PRACTICE

Vocabulary Check ✓

1. Describe the order of operations agreed upon by mathematicians. **See margin.**

Concept Check ✓

2. If an expression without grouping symbols includes addition and an exponent, which operation should you do first? **evaluating the expression that involves the exponent**
3. If an expression without grouping symbols includes multiplication and division, which operation should you do first? **the first of the two in order from left to right**

Skill Check ✓

Evaluate the expression for the given value of the variable.

1. First do operations that occur within grouping symbols. Then evaluate powers. Then do multiplications and divisions from left to right. Finally, do additions and subtractions from left to right.

4. $x^4 - 3$ when $x = 2$ **13**
5. $5 \cdot 6y$ when $y = 5$ **150**
6. $a^3 + 10a$ when $a = 3$ **57**
7. $\frac{16}{x} - 2$ when $x = 4$ **2**
8. $\frac{22}{x} \div 2 + 16$ when $x = 11$ **17**
9. $\frac{16}{n} + 2^3 - 10$ when $n = 8$ **0**
10. $(x + 5) \div 4$ when $x = 9$ $3\frac{1}{2}$
11. $b + 6 \div 4$ when $b = 1.5$ **3**
12. ERROR ANALYSIS Julio's calculator displayed 7 when he evaluated the following expression. Did his calculator use the established order of operations? If not, how can he use grouping symbols to find the correct value? **no; 72 + (12 ÷ 4) − 14**

72 [+] 12 [÷] 4 [−] 14 [ENTER]   [7]

PRACTICE AND APPLICATIONS

STUDENT HELP

Extra Practice to help you master skills is on p. 797.

EVALUATING VARIABLE EXPRESSIONS **Evaluate the expression for the given value of the variable.**

13. $3 + 2x^3$ when $x = 2$ **19**
14. $y^4 \div 8$ when $y = 4$ **32**
15. $6 \cdot 2p^2$ when $p = 5$ **300**
16. $t^5 - 10t$ when $t = 3$ **213**
17. $13 + 3b$ when $b = 7$ **34**
18. $3r^2 - 17$ when $r = 6$ **91**
19. $\frac{x}{7} + 16$ when $x = 14$ **18**
20. $27 - \frac{24}{b}$ when $b = 8$ **24**
21. $\frac{4}{5} \div n + 13$ when $n = \frac{1}{5}$ **17**
22. $\frac{9}{10} \cdot y - \frac{3}{10}$ when $y = \frac{1}{2}$ $\frac{3}{20}$

EVALUATING NUMERICAL EXPRESSIONS **Evaluate the expression.**

STUDENT HELP

HOMEWORK HELP
Example 1: Exs. 13–22
Example 2: Exs. 23–37
Example 3: Exs. 38–40, 42
Example 4: Exs. 43–46
Example 5: Ex. 41

23. $4 + 9 - 1$ **12**
24. $3 \cdot 2 + \frac{5}{9}$ $6\frac{5}{9}$
25. $6 \div 3 + 2 \cdot 7$ **16**
26. $5 + 8 \cdot 2 - 4$ **17**
27. $16 \div 8 \cdot 2^2$ **8**
28. $2 \cdot 3^2 \div 7$ $2\frac{4}{7}$
29. $10 \div (3 + 2) + 9$ **11**
30. $7[(18 - 6) - 6]$ **42**
31. $[(7 - 4)^2 + 3] + 15$ **27**
32. $3(2.7 \div 0.9) - 5$ **4**
33. $6(5 - 3)^2 + 3$ **27**
34. $[10 + (5^2 \cdot 2)] \div 6$ **10**
35. $\frac{1}{3}(9 \cdot 3) + 18$ **27**
36. $\frac{1}{2} \cdot 26 - 3^2$ **4**
37. $2.5 \cdot 0.5^2 \div 5$ **0.125**

3 APPLY

ASSIGNMENT GUIDE

BASIC
Day 1: pp. 19–20 Exs.13–40
Day 2: pp. 20–22 Exs.47–49, 55, 58–72 even, Quiz 1 Exs. 1–26

AVERAGE
Day 1: pp. 19–20 Exs.14–40 even, 41–49
Day 2: pp. 21–22 Exs.50–55, 58–72 even, Quiz 1 Exs. 1–26

ADVANCED
Day 1: pp. 19–20 Exs.14–40 even, 41–49
Day 2: pp. 21–22 Exs. 50–57, 58–72 even, Quiz 1 Exs. 1–26

BLOCK SCHEDULE
pp. 19–20 Exs. 14–40 even, 41–49 (with 1.2) pp. 21–22 Exs. 50–55, 58–72 even, Quiz 1 Exs. 1–26 (with 1.4)

EXERCISE LEVELS

Level A: ***Easier***
13–40, 58–69, Quiz 1–16

Level B: ***More Difficult***
43–55, 70–72, Quiz 17–26

Level C: ***Most Difficult***
41, 42, 56

✓ HOMEWORK CHECK

To quickly check student understanding of key concepts, go over the following exercises: Exs. 14, 22, 30, 36, 38, 48. See also the Daily Homework Quiz:

- Blackline Master (*Chapter 1 Resource Book*, p. 55)
- Transparency (p. 5)

! COMMON ERROR

EXERCISES 38–40 Some students may try to divide parts of the numerator by parts of the denominator or vice versa. Remind students that the fraction bar is a grouping symbol and that they must simplify the numerator and denominator separately before dividing.

TEACHING TIPS

EXERCISE 47 Point out that grouping symbols are not actually necessary for this expression if one follows the order of operations correctly.

APPLICATION NOTE

EXERCISE 48 Football is played in both the United States and Canada. However, Canadian rules are different. For example, there are 12 players on a side in Canada, instead of 11 as in the United States. Also the Canadian football field is larger than the U.S. field.

EXERCISE 55 Stores have sales on items all the time. Finding how much you save based on the percent of discount on an item is a common real-life experience.

ADDITIONAL PRACTICE AND RETEACHING

For Lesson 1.3:

- Practice Levels A, B, and C (*Chapter 1 Resource Book,* p. 44)
- Reteaching with Practice (*Chapter 1 Resource Book,* p. 47)
- See Lesson 1.3 of the *Personal Student Tutor*

For more Mixed Review:

- Search the *Test and Practice Generator* for key words or specific lessons.

EXPRESSIONS WITH FRACTION BARS **Evaluate the expression.**

38. $\dfrac{9 \cdot 2}{4 + 3^2 - 1}$ $1\frac{1}{2}$ **39.** $\dfrac{13 - 4}{18 - 4^2 + 1}$ 3 **40.** $\dfrac{5^3 \cdot 2}{1 + 6^2 - 8}$ $8\frac{18}{29}$

41. *Writing* You decide to buy two rings from an outdoor vendor. One ring costs \$10.89. The other ring costs \$12.48. The sales tax is 8%. The vendor uses a calculator to obtain the price including sales tax for both rings and gets \$24.37. What mistake did the vendor make? **The vendor taxed only the \$12.48 ring instead of taxing both rings.**

42. LOGICAL REASONING Which is correct? Explain.

A. $\dfrac{(9-7)^2+3}{5} = (9-7)^2 + 3 \div 5$

B. $\dfrac{(9-7)^2+3}{5} = [(9-7)^2 + 3] \div 5$

42. B; the brackets represent the grouping symbol indicated in the original equation by the fraction bar.

ORDER OF OPERATIONS **In Exercises 43–46, two calculators were used to evaluate the expression. They gave different results. Which calculator used the established order of operations? Rewrite the calculator steps with grouping symbols so that both calculators give the correct result.**

43. 15 [–] 6 [÷] 3 [×] 4 [ENTER] **Calculator A: 12; Calculator B: 7**
Calculator B; 15 – [(6 ÷ 3) × 4]

44. 15 [–] 9 [÷] 3 [+] 7 [ENTER] **Calculator A: 19; Calculator B: 9**
Calculator A; 15 – (9 ÷ 3) + 7

45. 15 [+] 10 [÷] 5 [+] 4 [ENTER] **Calculator A: 21; Calculator B: 9**
Calculator A; 15 + (10 ÷ 5) + 4

46. 4 [×] 3 [+] 6 [÷] 2 [ENTER] **Calculator A: 9; Calculator B: 15**
Calculator B; (4 × 3) + (6 ÷ 2)

47. HOTEL RATES A hotel charges \$49.99 per room per night for adults and \$44.10 per room per night for senior citizens. The expression 2 × \$49.99 + 3 × \$44.10 represents the total cost of five rooms for two adults and three senior citizens for an overnight stay. Where in the expression can you put grouping symbols to make sure it is evaluated correctly?
(2 × \$49.99) + (3 × \$44.10)

FOOTBALL UNIFORMS **In Exercises 48 and 49, use the following information.** The table shows the cost of parts of a professional football player's uniform. A sporting goods company offers a \$3200 discount for orders of 30 or more complete professional football player uniforms.

48. Write an expression that represents the cost for an order of 35 complete professional football player uniforms.

48. 35(\$230 + \$300 + \$50 + \$25 + \$100 + \$200) – \$3200

49. Evaluate the expression you wrote in Exercise 48. **\$28,475**

Part of uniform	Jersey and pants	Shoulder pads	Lower body pads	Knee pads	Cleats	Helmet
Cost	\$230	\$300	\$50	\$25	\$100	\$200

55c. \$3.35; one method is to determine the original price, \$25.50, and subtract the discounts to get the sale price of \$21.65, then subtract that amount from \$25. Another method is to calculate the difference between \$25 and the sale price directly: \$25 – ((0.9)(5)(\$2.50) + (0.8)(10)(\$1.30)) = \$3.35.

ADMISSION PRICES In Exercises 50 and 51, use the table below. It shows the admission prices for the California State Fair in 1998. Suppose a family of 2 adults, 1 senior, and 3 children go to the State Fair. The children's ages are 13 years, 10 years, and 18 months.

50. Write an expression that represents the admission price for the family. **3(\$7) + \$5 + \$4**

51. Evaluate the expression you wrote in Exercise 50. **\$30**

California State Fair Admission Prices	
Age	**Admission price**
General Admission (13–61 years of age)	\$7.00
Seniors (62 years and above)	\$5.00
Children (5–12 years)	\$4.00
Children (4 years and under)	Free

▶ Source: *Sacramento Bee*

STUDENT HELP

Study Tip
The expression $b_1 + b_2$ is read "*b* sub 1 plus *b* sub 2".

52. GEOMETRY CONNECTION The area A of a trapezoid with parallel bases of lengths b_1 and b_2 and height h is $A = \frac{1}{2}h(b_1 + b_2)$. Find the area of a trapezoid whose height is 2 meters and whose bases are 6 meters and 10 meters. **16 m²**

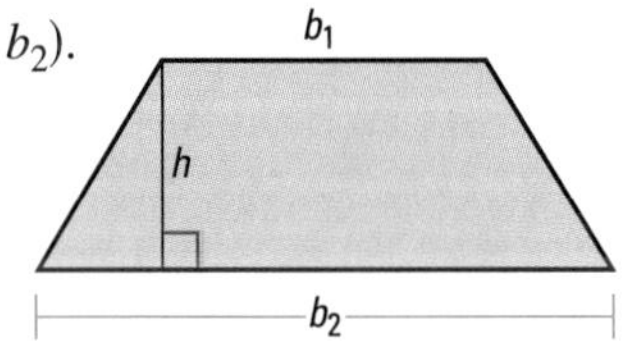

GEOMETRY CONNECTION **In Exercises 53 and 54, use the following information. The surface area of a cylinder equals the lateral surface area ($2\pi r \cdot h$) plus the area of the two bases ($2 \cdot \pi r^2$).**

53. Write the expression for the surface area of a cylinder. **$2\pi rh + 2\pi r^2$ or $2\pi r(h + r)$**

54. Evaluate the expression when $h = 10.5$ centimeters and $r = 2.5$ centimeters. Use 3.14 as an approximation for π. **204.1 cm²**

Test Preparation

55. MULTI-STEP PROBLEM You are shopping for school supplies. A store is offering a 10% discount on binders and a 20% discount on packages of paper. You want to buy 5 binders originally marked \$2.50 each and 10 packages of paper originally marked \$1.30 each.

a. Write an expression that shows how much you will save after the discounts. **0.1(5)(\$2.50) + 0.2(10)(\$1.30)**

b. Evaluate the expression. **\$3.85**

c. *Writing* If you have \$25 to spend on supplies, how much money will you have left over? Explain how you arrived at your answer. **See margin.**

★ Challenge

56. CRITICAL THINKING Without grouping symbols, the expression $2 \cdot 3^3 + 4$ has a value of 58. Insert grouping symbols in the expression $2 \cdot 3^3 + 4$ to produce the indicated values.

a. 62 **$2(3^3 + 4)$** **b.** 220 **$(2 \cdot 3)^3 + 4$** **c.** 4374 **$2 \cdot 3^{3+4}$** **d.** 279,936 **$(2 \cdot 3)^{3+4}$**

57. Create a math puzzle like the one in Exercise 56 with an expression that produces different values when grouping symbols are inserted in different places. **Check puzzles.**

EXTRA CHALLENGE
www.mcdougallittell.com

4 ASSESS

DAILY HOMEWORK QUIZ

Transparency Available

1. Evaluate the expression.

a. $4x^2 - 16$ when $x = 5$ **84**

b. $\frac{7}{8}y - \frac{1}{4}$ when $y = \frac{1}{2}$ **$\frac{3}{16}$**

c. $12 \div 3 + 2 \cdot 8$ **20**

d. $[(9 - 7)^2 + 5] + 26$ **35**

e. $\frac{8 \cdot 2 + 5}{12 + 2^2 - 9}$ **3**

2. A T-shirt company charges \$15 per shirt and offers a \$200 rebate for orders higher than 50 shirts. Write an expression that represents the cost for an order of 75 shirts. Evaluate the expression. **75 • 15 – 200; \$925**

EXTRA CHALLENGE NOTE
Challenge problems for Lesson 1.3 are available in **blackline** format in the *Chapter 1 Resource Book*, p. 51 and at **www.mcdougallittell.com.**

ADDITIONAL TEST PREPARATION

1. OPEN ENDED Using each of the numbers 1, 2, 3, 4, and 5 only once, any or all of the four symbols +, –, ×, and ÷, and grouping symbols if necessary, write expressions that equal 1, 2, 3, 4, and 5. ***Sample answers:***
$5 + 2 + 1 - 4 - 3 = 1$
$5 + (3 \times 1) - (4 + 2) = 2$
$4 + 3 - 5 + 2 - 1 = 3$
$(5 + 3) \div (4 \div 2) \times 1 = 4$
$5 \times (4 - 3) \div (2 - 1) = 5$

ADDITIONAL RESOURCES
An alternative Quiz for Lessons 1.1–1.3 is available in the *Chapter 1 Resource Book,* p.52.

MIXED REVIEW

STUDENT HELP
Skills Review
For help with fractions, see pp. 781–783.

EVALUATING EXPRESSIONS Evaluate the expression for the given value of the variable. (Review 1.1)

58. $8a$ when $a = 4$ **32** **59.** $\frac{24}{x}$ when $x = 3$ **8** **60.** $c + 15$ when $c = 12.5$ **27.5**

61. $\frac{4}{3} \cdot x$ when $x = \frac{1}{6}$ $\frac{2}{9}$ **62.** $9d$ when $d = 0.5$ **4.5** **63.** $\frac{5}{16} - p$ when $p = \frac{3}{8}$ $-\frac{1}{16}$

EVALUATING EXPONENTIAL EXPRESSIONS Evaluate the expression for the given value of the variable. (Review 1.2)

64. $(6w)^2$ when $w = 5$ **900** **65.** $4(t^3)$ when $t = 3$ **108** **66.** $9b^2$ when $b = 8$ **576**

67. $5x^2$ when $x = 16$ **1280** **68.** $(7x)^3$ when $x = 2$ **2744** **69.** $6y^5$ when $y = 4$ **6144**

70. INTEREST EARNED You deposit \$120 in a savings account that pays an annual interest rate of 2.5%. How much simple interest would you earn in 2 years? (Review 1.1) **\$6**

71. DRIVING TIME If you are driving at a constant speed of 68 miles per hour, how long will it take you to travel 170 miles? (Review 1.1) $2\frac{1}{2}$**h**

72. GEOMETRY CONNECTION Find the volume of the cube shown at the right and find the area of one of its sides. (Review 1.2 for 1.4)
64 cm^3, 16 cm^2

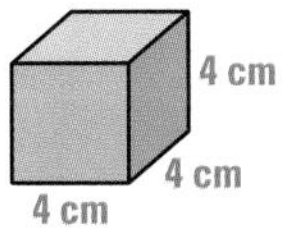

QUIZ 1

Self-Test for Lessons 1.1–1.3

Evaluate the expression when $x = 3$. (Lesson 1.1)

1. $6x$ **18** **2.** $\frac{36}{x}$ **12** **3.** $x + 29$ **32** **4.** $x - 2$ **1**

5. $x \div 3$ **1** **6.** $21 - x$ **18** **7.** $4.5x$ **13.5** **8.** $13.7 + x$ **16.7**

TRAVEL **Find the average speed.** (Lesson 1.1)

9. 120 miles in 3 hours **40 mi/h** **10.** 90 miles in 1.5 hours **60 mi/h** **11.** 360 miles in 6 hours **60 mi/h**

Write the expression in exponential form. (Lesson 1.2)

12. six cubed 6^3 **13.** $4 \cdot 4 \cdot 4 \cdot 4 \cdot 4$ 4^5 **14.** $5y \cdot 5y \cdot 5y$ $(5y)^3$

15. five squared 5^2 **16.** $3 \cdot 3 \cdot 3 \cdot 3 \cdot 3 \cdot 3 \cdot t$ $3^6 \cdot t$ **17.** $7x \cdot 7x$ $(7x)^2$

Evaluate the expression when $x = 6$. (Lesson 1.2)

18. x^5 **7776** **19.** $2x^3$ **432** **20.** $(2x)^3$ **1728** **21.** $x^2 - 3$ **33**

Evaluate the expression. (Lesson 1.3)

22. $\frac{6 \cdot 3}{7 + (2^3 - 1)}$ $1\frac{2}{7}$ **23.** $\frac{2^5 - 12}{2(5^2 - 5)}$ $\frac{1}{2}$ **24.** $\frac{(3^2 - 3)}{2 \cdot 9}$ $\frac{1}{3}$ **25.** $\frac{2(17 + 2 \cdot 4)}{6^2 - 11}$ **2**

26. PACKING BOXES A cubic packing box has dimensions of 1.2 feet on each side. What is the volume of the box? (Lesson 1.2) **1.728 ft^3**

ACTIVITY 1.4

Developing Concepts

Finding Patterns

SET UP
Work in a small group.

MATERIALS
- graph paper
- pencil

▶ QUESTION **How can you use algebra to describe a pattern?**

▶ EXPLORING THE CONCEPT

1 Copy the first four figures on graph paper. Then draw the fifth and sixth figures of the sequence. **See margin.**

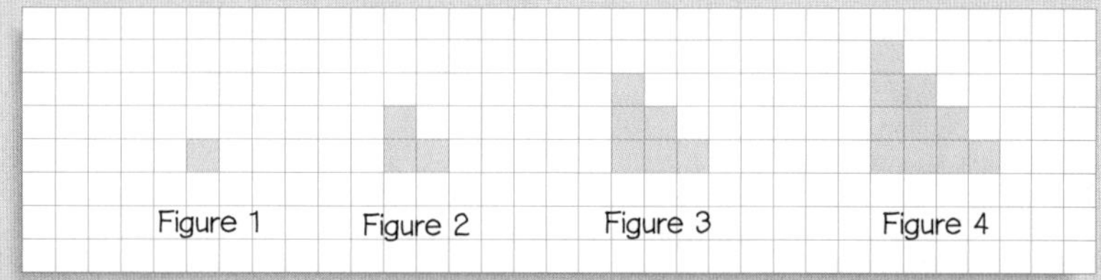

Step 1. Check drawings. Figure 5 is like Figure 4 with another row of 5 squares on the bottom. Figure 6 is like Figure 5 with another row of 6 squares on the bottom.

2 The table shows the mathematical pattern for the perimeters of the first four figures. Copy and complete the table. **figure 5: 20, 4 · 5; figure 6: 24, 4 · 6**

Figure	1	2	3	4	5	6
Perimeter	4	8	12	16	?	?
Pattern	4 • 1	4 • 2	4 • 3	4 • 4	4 • ?	4 • ?

3 Use the pattern in the table to predict the perimeter of the 25th figure. **100**

4 What is the perimeter of the nth figure? Explain your reasoning.
$4n$; *Sample answer:* for each figure the perimeter is 4 times the figure number.

▶ DRAWING CONCLUSIONS

1. Copy the four figures shown below on graph paper. Then draw the fifth and sixth figures for the sequence.

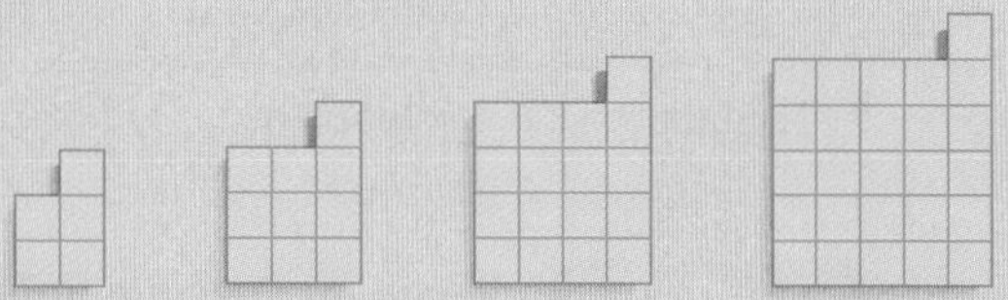

Figure 1 **Figure 2** **Figure 3** **Figure 4**

Check drawings. Figure 5 is a 6 × 6 square with one extra. Figure 6 is a 7 × 7 square with one extra

2. Calculate the perimeters for all six figures. Organize your results in a table. **See margin.**

3. What is the perimeter of the nth figure? Explain your reasoning. **See margin.**

4. Repeat Exercises 1–3 for the sequence below.
See margin; $8n - 4$; *Sample answer:* for each figure, the perimeter is 4 less than 8 times the figure number.

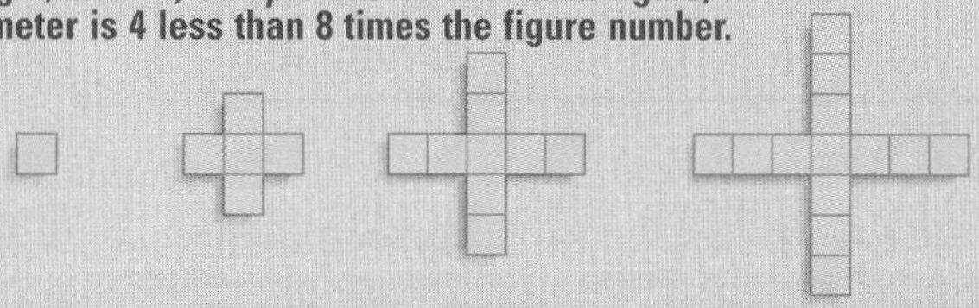

Figure 1 **Figure 2** **Figure 3** **Figure 4**

2.

Fig.	Perim.	Pattern
1	10	4 · 1 + 6
2	14	4 · 2 + 6
3	18	4 · 3 + 6
4	22	4 · 4 + 6
5	26	4 · 5 + 6
6	30	4 · 6 + 6

3. $4n + 6$; *Sample answer:* for each figure, the perimeter is 6 more than 4 times the figure number.

4. Check drawings; Figure 5 is a 9 × 9 cross and Figure 6 is an 11 × 11 cross;

Fig.	Perim.	Pattern
1	4	8 · 1 − 4
2	12	8 · 2 − 4
3	20	8 · 3 − 4
4	28	8 · 4 − 4
5	36	8 · 5 − 4
6	44	8 · 6 − 4

1 Planning the Activity

PURPOSE
To have students find algebraic patterns in block figures.

MATERIALS
- Graph paper
- Activity Support Master (*Chapter 1 Resource Book,* p. 56)

PACING
- Exploring the Concept — 20 min
- Drawing Conclusions — 15 min

▶ *LINK TO LESSON*
The patterns students are finding for the nth figure are actually equations. For example, the pattern in Step 4 is $P = 4n$. When students check possible solutions in Examples 1 and 2 of Lesson 1.4, they are really looking for the "pattern that works."

2 Managing the Activity

CLASSROOM MANAGEMENT
Some groups may have trouble discovering the patterns. Offer some helpful hints to these groups to avoid student frustration.

3 Closing the Activity

★ *KEY DISCOVERY*
Once the pattern for the perimeters is found, an algebraic expression can be written to express the pattern. The algebraic expression can then be used to predict the perimeter of any figure in the sequence.

ACTIVITY ASSESSMENT
Find the perimeter of the nth figure of this sequence of shapes. **$8n - 4$**

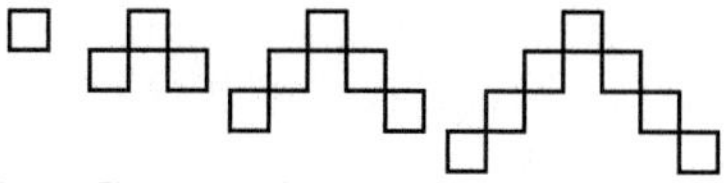

Figure 1 Figure 2 Figure 3 Figure 4

1 PLAN

PACING
Basic: 2 days
Average: 2 days
Advanced: 2 days
Block Schedule: 0.5 block with 1.3
0.5 block with 1.5

LESSON OPENER APPLICATION
An alternative way to approach Lesson 1.4 is to use the Application Lesson Opener:
- Blackline Master (*Chapter 1 Resource Book*, p. 57)
- Transparency (p. 4)

MEETING INDIVIDUAL NEEDS
- ***Chapter 1 Resource Book***
  Prerequisite Skills Review (p. 5)
  Practice Level A (p. 58)
  Practice Level B (p. 59)
  Practice Level C (p. 60)
  Reteaching with Practice (p. 61)
  Absent Student Catch-Up (p. 63)
  Challenge (p. 65)
- ***Resources in Spanish***
- ***Personal Student Tutor***

NEW-TEACHER SUPPORT
See the Tips for New Teachers on pp. 1–2 of the *Chapter 1 Resource Book* for additional notes about Lesson 1.4.

WARM-UP EXERCISES

Transparency Available
Evaluate the expression.

1. $\frac{x}{3} - 6$ when $x = 24$ 2
2. $3x^3$ when $x = 4$ 192
3. $3k - k$ when $k = 12$ 24
4. Tell if the statement is true or false.
  a. $3x$ equals 12 when $x = 2$. False
  b. $6 + b$ equals 8 when $b = 2$. True

1.4 Equations and Inequalities

What you should learn

GOAL 1 Check solutions and solve equations using mental math.

GOAL 2 Check solutions of inequalities in a **real-life** problem, such as regulating your cat's caloric intake in **Example 5**.

Why you should learn it

▼ To solve a **real-life** problem such as how long you must save money to buy a violin in **Ex. 68**.

GOAL 1 CHECKING AND SOLVING EQUATIONS

An **equation** is formed when an equal sign (=) is placed between two expressions creating a left and right side of the equation. An equation that contains one or more variables is an **open sentence**. Here are some examples.

$4 - b = 3$ $\qquad$ $3x + 1 = 7$ $\qquad$ $a + 3 = 3 + a$

When the variable in a single-variable equation is replaced by a number, the resulting statement can be true or false. If the statement is true, the number is a **solution of an equation**.

Substituting a number for a variable in an equation to see whether the resulting statement is true or false is called *checking* a possible solution.

EXAMPLE 1 *Substituting to Check Possible Solutions*

Check whether the numbers 2 and 3 are solutions of the equation $3x + 1 = 7$.

SOLUTION

To check the possible solutions, substitute them into the equation. If both sides of the equation have the same value, then the number is a solution.

x	$3x + 1 = 7$	Result	Conclusion
2	$3(2) + 1 \stackrel{?}{=} 7$	$7 = 7$	2 is a solution
3	$3(3) + 1 \stackrel{?}{=} 7$	$10 \neq 7$ (is not equal to)	3 is not a solution

▶ The number 2 is a solution of $3x + 1 = 7$. The number 3 is *not* a solution.

EXAMPLE 2 *Checking Possible Solutions*

Check whether the numbers 2, 3, and 4 are solutions of the equation $4x - 2 = 10$.

SOLUTION

x	$4x - 2 = 10$	Result	Conclusion
2	$4(2) - 2 \stackrel{?}{=} 10$	$6 \neq 10$	2 is not a solution
3	$4(3) - 2 \stackrel{?}{=} 10$	$10 = 10$	3 is a solution
4	$4(4) - 2 \stackrel{?}{=} 10$	$14 \neq 10$	4 is not a solution

▶ The number 3 is a solution, and 2 and 4 are *not* solutions of the equation.

Finding all the solutions of an equation is called **solving the equation**. Later in the book, you will study several ways to systematically solve equations.

Some equations are simple enough to solve in your head with mental math. For instance, to solve $x + 2 = 5$, ask yourself the question

"What number can be added to 2 to obtain 5?"

If you can see that the answer is 3, then you have solved the equation!

ACTIVITY Developing Concepts

Using Mental Math to Solve Equations

1. Match the equation with the question that can be used to find a solution of the equation.
2. Then use mental math to solve the equation.

EQUATION	MENTAL MATH QUESTION
1. $x + 2 = 6$ E; 4	**A.** 2 times what number gives 10?
2. $x - 3 = 4$ C; 7	**B.** What number divided by 3 gives 1?
3. $2x = 10$ A; 5	**C.** What number minus 3 gives 4?
4. $\frac{x}{3} = 1$ B; 3	**D.** What number cubed gives 8?
5. $x^3 = 8$ D; 2	**E.** What number plus 2 gives 6?

Using mental math to solve equations is something that you already do in everyday life. You probably don't think of it as solving equations, but the mental math process is the same.

REAL LIFE
Grocery Shopping

EXAMPLE 3 *Using Mental Math to Solve a Real-Life Equation*

You need to buy ingredients for nachos. In the supermarket you find that a bag of tortilla chips costs \$2.99, beans cost \$.99, cheese costs \$3.99, two tomatoes cost \$1.00, and olives cost \$1.49. You have a ten-dollar bill. About how much more money do you need?

SOLUTION

You can ask the question: The total cost equals 10 plus what number of dollars? Let x represent any additional money you may need. Use rounding to estimate the total cost.

$$3 + 1 + 4 + 1 + 1.5 = 10 + x$$

$$10.5 = 10 + x$$

▶ Because the total cost of the ingredients is approximately 10.5 or \$10.50, you can see that you need about \$.50 more to purchase all the ingredients.

2 TEACH

MOTIVATING THE LESSON

Many people jog to stay fit. Jogging burns about 4.2 calories per hour for every pound you weigh. If you weigh 150 lb and jog for 2 hours, you can use mental math to figure that you burn about 1200 calories. Today's lesson focuses on checking solutions and solving equations using mental math.

EXTRA EXAMPLE 1

Check whether the numbers 3 and 4 are solutions of the equation $5x - 7 = 8$. **3 is, 4 is not.**

EXTRA EXAMPLE 2

Check whether the numbers 5, 10, and 15 are solutions of the equation $12s + 5 = 125$. **10 is, 5 and 15 are not.**

EXTRA EXAMPLE 3

You need to buy ingredients to make a pizza. Cheese costs \$5.99, pizza dough costs \$2.39, tomato sauce costs \$2.49, and pepperoni costs \$2.98. You have a twenty-dollar bill. About how much change will you receive? **\$6**

CHECKPOINT EXERCISES

For use after Examples 1 and 2:

1. Check whether the numbers 3, 6, and 9 are solutions of the equation $8s - 20 = 52$. **9 is, 3 and 6 are not.**

For use after Example 3:

2. You need to buy supplies for school. A box of pencils costs \$2.98, a package of pens costs \$3.95, a notebook costs \$4.49 and a ruler costs \$.89. You have \$10. How much more money do you need? **about \$2.50**

EXTRA EXAMPLE 4

Decide whether 3 is a solution of the inequality.

a. $3x - 2 > 5$ **yes** **b.** $x + 5 < 8$ **no**
c. $x - 1 \geq 2$ **yes** **d.** $4x^2 \leq 32$ **no**

EXTRA EXAMPLE 5

The average number of calories a 120 lb person burns while walking briskly is greater than 240 times the number of hours the person walks. Using the following values of x, will a 120 lb person walking for x hours burn at least 500 calories?

a. 2 hours **no** **b.** 3 hours **yes**

CHECKPOINT EXERCISES

For use after Example 4:

1. Decide whether 4 is a solution of the inequality.
a. $2x - 1 > 5$ **yes**
b. $x - 2 > 6$ **no**
c. $x + 3 \geq 5$ **yes**
d. $3x^2 \leq 52$ **yes**

For use after Example 5:

2. Your scores on the last three tests were 82, 87, and 74. Your score on the next test is x points. Do the following values for x give you at least 320 total points?
a. 75 **no** **b.** 81 **yes**

FOCUS ON VOCABULARY

Emphasize that an *open sentence* is neither true nor false until values for the variables have been substituted into the sentence.

CLOSURE QUESTION

Tell how to check if a number is a solution to an equation or an inequality. **Substitute the number for the variable. If the result is a true statement then the number is a solution.**

DAILY PUZZLER

A square and an equilateral triangle have the same perimeter. Which has the longer sides? **the triangle**

GOAL 2 CHECKING SOLUTIONS OF INEQUALITIES

STUDENT HELP

Skills Review For help with comparing numbers, see p. 779.

Another type of open sentence is an **inequality**. An inequality is formed when an inequality symbol, such as <, is placed between two expressions.

SYMBOL	MEANING
$<$	is less than
$\leq$	is less than or equal to
$>$	is greater than
$\geq$	is greater than or equal to

A **solution of an inequality** is a number that produces a true statement when it is substituted for the variable in the inequality.

EXAMPLE 4 *Checking Solutions of Inequalities*

Decide whether 4 is a solution of the inequality.

a. $2x - 1 < 8$ **b.** $x + 4 > 9$ **c.** $x - 3 \geq 1$

SOLUTION

INEQUALITY	SUBSTITUTION	RESULT	CONCLUSION
a. $2x - 1 < 8$	$2(4) - 1 \stackrel{?}{<} 8$	$7 < 8$	4 is a solution.
b. $x + 4 > 9$	$4 + 4 \stackrel{?}{>} 9$	$8 \not> 9$	4 is not a solution.
c. $x - 3 \geq 1$	$4 - 3 \stackrel{?}{\geq} 1$	$1 \geq 1$	4 is a solution.

EXAMPLE 5 *Checking Solutions in Real Life*

FOCUS ON CAREERS

VETERINARIAN Vets specialize in the health care of either small animals, such as cats and dogs or large animals, such as horses or elephants.

CAREER LINK www.mcdougallittell.com

VETERINARIAN Your vet told you to restrict your cat's caloric intake to no more than 500 calories each day. Three times a day, you give your cat a serving of food containing x calories. Do the following values of x meet the vet's restriction?

a. 170 calories **b.** 165 calories

SOLUTION

a. $3x \leq 500$ **Write original inequality.**
$3(170) \leq 500$ **Substitute 170 for x.**
$510 \not\leq 500$ **Simplify.**

▶ No. This is too many calories per serving.

b. $3x \leq 500$ **Write original inequality.**
$3(165) \leq 500$ **Substitute 165 for x.**
$495 \leq 500$ **Simplify.**

▶ Yes. If each serving has 165 calories, you will meet your goal.

GUIDED PRACTICE

Vocabulary Check ✓

Decide whether the following is an *expression*, an *equation*, or an *inequality*. Explain your decision. 1–6. Check explanations.

1. $3x + 1 = 14$ Equation
2. $7y - 6$ Expression
3. $5(y^2 + 4) - 7$ Expression
4. $5x - 1 = 3 + x$ Equation
5. $3x + 2 \leq 8$ Inequality
6. $5x > 20$ Inequality
7. Identify the left side and the right side of the equation $8 + 3x = 5x - 9$. left: $8 + 3x$, right: $5x - 9$

Concept Check ✓

8. **ERROR ANALYSIS** Jan says her work shows that 6 is not a solution of $3x - 4 = 14$. What is a likely explanation for her error? The subtraction is done before the multiplication; $3(6) - 4 = 18 - 4 = 14$

~~$3 \cdot (6) - 4 = 6$~~

Skill Check ✓

9. Which question could be used to find the solution of the equation $5 - x = 1$? B
   - **A.** What number can 5 be subtracted from to get 1?
   - **B.** What number can be subtracted from 5 to get 1?
   - **C.** What number can 1 be subtracted from to get 5?

ELECTIONS **The number of votes received by the new student council president is represented by *x*. Match the sentence with the equation or inequality that represents it.**

A. $x = 125$ **B.** $x < 125$ **C.** $x \geq 125$ **D.** $x \leq 125$

10. She received no more than 125 votes. D
11. She received at least 125 votes. C
12. She received exactly 125 votes. A
13. She received less than 125 votes. B

PRACTICE AND APPLICATIONS

STUDENT HELP
Extra Practice to help you master skills is on p. 797.

CHECKING SOLUTIONS OF EQUATIONS **Check whether the given number is a solution of the equation.**

14. $3b + 1 = 13$; 4 solution
15. $5 + x^2 = 17$; 3 not a solution
16. $4c + 2 = 2c + 8$; 2 not a solution
17. $2y^3 + 3 = 5$; 1 solution
18. $5r - 10 = 11$; 5 not a solution
19. $4s - 4 = 30 - s$; 7 not a solution
20. $6d - 5 = 20$; 5 not a solution
21. $\frac{x}{4} - 9 = 9$; 36 not a solution
22. $m + 4m = 60 - 2m$; 10 not a solution
23. $10 + \frac{a}{7} = 12$; 14 solution
24. $p^2 - 5 = 20$; 6 not a solution
25. $4h - h = \frac{12}{h}$; 3 not a solution

STUDENT HELP
HOMEWORK HELP
Example 1: Exs. 14–25
Example 2: Exs. 14–25
Example 3: Exs. 26–37, 65, 66
Example 4: Exs. 38–55
Example 5: Exs. 67, 68

MENTAL MATH **Write a question that could be used to solve the equation. Then use mental math to solve the equation.** 26–37. See margin.

26. $x + 3 = 8$ 5
27. $n + 6 = 11$ 5
28. $p - 11 = 20$ 31
29. $3y = 12$ 4
30. $\frac{x}{4} = 5$ 20
31. $4p = 36$ 9
32. $4r - 1 = 11$ 3
33. $2t - 1 = 9$ 5
34. $m^2 = 144$ 12
35. $\frac{x}{7} = 3$ 21
36. $5q - 2 = 3$ 1
37. $y^3 = 125$ 5

31. What number can be multiplied by 4 to get 36?
32. One less than four times what number is 11?
33. One less than two times what number is 9?
34. What number squared gives 144?
35. What number when divided by 7 gives 3?
36. 3 is 2 less than the product of 5 and what number?
37. What number cubed gives 125?

3 APPLY

ASSIGNMENT GUIDE

BASIC
Day 1: pp. 27–28 Exs. 14–26, 28, 35, 38–54 even
Day 2: pp. 28–30 Exs. 56–65, 69–76, 79–81, 84–100 even

AVERAGE
Day 1: pp. 27–28 Exs. 14–26, 28, 35, 38–54 even
Day 2: pp. 28–30 Exs. 56–65, 69–77, 79–81, 84–100 even

ADVANCED
Day 1: pp. 27–28 Exs. 14–26, 28, 35, 38–54 even
Day 2: pp. 28–30 Exs. 56–64, 67, 69–83, 84–100 even

BLOCK SCHEDULE
pp. 27–28 Exs. 14–26, 28, 35, 38–54 even (with 1.3)
pp. 28–30 Exs. 56–65, 69–77, 79–81, 84–100 even (with 1.5)

EXERCISE LEVELS

Level A: ***Easier***
14–31, 38–64, 69–81
Level B: ***More Difficult***
32–37, 65–68, 84–100
Level C: ***Most Difficult***
82, 83

✓ HOMEWORK CHECK

To quickly check student understanding of key concepts, go over the following exercises: Exs. 18, 28, 42, 52, 58, 65, 70, 74. See also the Daily Homework Quiz:

- Blackline Master (*Chapter 1 Resource Book*, p. 68)
- Transparency (p. 6)

26. What number can be added to 3 to get 8?
27. What number can be added to 6 to get 11?
28. What number can 11 be subtracted from to get 20?
29. What number can be multiplied by 3 to get 12?
30. What number can be divided by 4 to get 5?

! COMMON ERROR

EXERCISE 53 Remind students to substitute 5 for both values of d before simplifying the fraction. Some students may make the mistake of "canceling" the d's first.

APPLICATION NOTE

EXERCISE 65 The price of personal computers has dropped drastically just as their speed and memory have increased drastically. Many computer systems for home and school use are available for less than \$1000.

TEACHING TIPS

EXERCISES 56–64 If students have difficulty with these exercises, tell them to focus on the words that represent the symbols =, <, >, ≤, and ≥.

EXERCISES 69–76 If students have difficulty with these exercises, tell them to focus on the activity taking place to help choose the correct formula.

MATHEMATICAL REASONING

EXERCISES 56–64 Many verbal sentences have a similar structure. To translate them into mathematical models, students need to determine the verb or verb phrase that separates the other phrases. Ask your students to find the verb phrase and the phrases it separates in the sentence "A number is greater than 6". **Verb phrase: is greater than; other phrases: a number; 6**

ADDITIONAL PRACTICE AND RETEACHING

For Lesson 1.4:

- Practice Levels A, B, and C (*Chapter 1 Resource Book,* p. 58)
- Reteaching with Practice (*Chapter 1 Resource Book,* p. 61)
- See Lesson 1.4 of the *Personal Student Tutor*

For more Mixed Review:

- Search the *Test and Practice Generator* for key words or specific lessons.

CHECKING SOLUTIONS OF INEQUALITIES **Check whether the given number is a solution of the inequality.**

38. $n - 2 < 6; 3$ **solution**

39. $5 + s > 8; 4$ **solution**

40. $5 + 5x \geq 10; 1$ **solution**

41. $4p - 1 \geq 8; 2$ **not a solution**

42. $3r - 15 < 0; 5$ **not a solution**

43. $11x \leq x - 7; 9$ **not a solution**

44. $6 + y \leq 8; 3$ **not a solution**

45. $29 - 4b > 5; 7$ **not a solution**

46. $t^2 + 6 > 40; 6$ **solution**

47. $a - 7 \geq 15; 22$ **solution**

48. $6x - 16 < 20; 7$ **not a solution**

49. $y^3 - 2 \leq 8; 2$ **solution**

50. $r + 2r < 30; 9$ **solution**

51. $a(3a + 2) > 50; 4$ **solution**

52. $\frac{c + 5}{3} \leq 4; 3$ **solution**

53. $\frac{25 - d}{d} \geq 4; 5$ **solution**

54. $x^2 - 10 > 16; 6$ **solution**

55. $n(21 - n) < 100; 8$ **not a solution**

EQUATIONS AND INEQUALITIES **Match the verbal sentence with its mathematical representation.**

56. The sum of x and 16 is less than 32. **C**

57. The product of 16 and x is equal to 32. **G**

58. The difference of x and 16 is 32. **F**

59. The quotient of x and 16 is greater than or equal to 32. **H**

60. The product of 16 and x is less than 32. **I**

61. The fourth power of x is 16. **B**

62. The sum of 16 and x is less than or equal to 32. **D**

63. The quotient of x and 16 is greater than 32. **A**

64. The product of 16 and x is greater than 32. **E**

A. $\frac{x}{16} > 32$

B. $x^4 = 16$

C. $x + 16 < 32$

D. $16 + x \leq 32$

E. $16x > 32$

F. $x - 16 = 32$

G. $16x = 32$

H. $\frac{x}{16} \geq 32$

I. $16x < 32$

65. **COMPUTER CENTER** Your school is building a new computer center. Four hundred square feet of the center will be available for computer stations. Each station requires 20 square feet. You want to find how many computer stations can be placed in the new center. You write the equation $20x = 400$ to model the situation. What do 20, x, and 400 represent? Solve the equation. Check your solution. **the number of ft^2 per station, the number of stations, the total number of ft^2 available; 20 stations**

66. **PLAYING A COMPUTER GAME** You are playing a new computer game. For every eight screens you complete, you receive a bonus. You want to know how many bonuses you will receive after completing 96 screens. You write the equation $8x = 96$ to model the situation. What do 8, x, and 96 represent? Solve the equation. Check your solution. **the number of screens per bonus, the number of bonuses, the total number of screens completed; 12 bonuses**

67. **BUYING GAS** You are taking a trip by automobile with the family of a friend. You have \$65 to help pay for gas. It costs \$15 to fill the tank. Can you completely fill the gas tank four times? You use the inequality $15x \leq 65$ to model the situation. What do 15, x, and 65 represent?

67. **cost each time you fill the tank, number of times you fill the tank, total amount to be paid for gas; yes**

68. **BUYING A VIOLIN** You are budgeting money to buy a violin and bow that cost \$250 including tax. If you save \$5 per week, will you have enough money in a year? You write the inequality $5n \geq 250$ to model the situation. What do 5, n, and 250 represent? Solve the inequality. **savings per week, number of weeks, total savings; at least 50 weeks**

STUDENT HELP

HOMEWORK HELP
Visit our Web site www.mcdougallittell.com for help with problem solving in Exs. 69–76.

APPLYING FORMULAS **In Exercises 69–76, match the problem with the formula needed to solve the problem. Then use the Guess, Check, and Revise strategy or another problem-solving strategy to solve the problem.**

Area of a rectangle	$A = lw$	**Distance**	$d = rt$
Simple interest	$I = Prt$	**Volume of a cube**	$V = s^3$
Temperature	$C = \frac{5}{9}(F - 32)$	**Surface area of a cube**	$S = 6s^2$

69. What is the average speed of a runner who completes a 10,000-meter race in 25 minutes? **400 m/min**

70. How long must \$1000 be invested at an annual interest rate of 3% to earn \$300 in simple interest? **10 years**

71. A car travels 60 miles per hour for a distance of 300 miles. How long did the trip take? **5 hours**

72. Carpeting costs \$20 per square yard. You carpet a room that has a width of 15 feet for \$800. What is the length of the room in feet? **24 ft**

73. You measure the temperature of a substance in a chemistry lab at 32°F. What is the temperature in degrees Celsius? **0°C**

74. A piece of cheese cut in the shape of a cube has a volume of 27 cubic inches. What is the length of each edge of the piece of cheese? **3 in.**

75. You want to construct a patio of 80 square feet with a length of 10 feet. What is the width of the patio? **8 ft**

76. A cubic storage box is made with 96 square feet of wood. What is the length of each edge? **4 ft**

77. **AIRCRAFT DESIGN** In the diagram, let x represent the length of a passenger jet and let y represent the jet's wingspan.

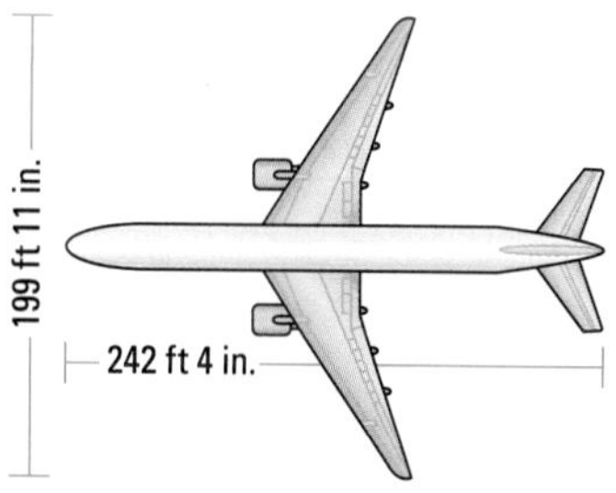

a. What does the equation $1.212y = x$ say about the relationship between the length and the wingspan of the passenger jet? **The length of the jet is 1.212 times the wingspan.**

b. Is the passenger jet longer than it is wide or wider than it is long? Explain your reasoning. **It is longer than it is wide; since $y > 0$, $1.212y > y$.**

78. **MACH NUMBERS** The *Mach number* of an aircraft is the ratio of its maximum speed to the speed of sound. Copy and complete the table. Use the equation $v = 660m$, where v is the speed (in miles per hour) of the aircraft and m is the Mach number, to find the speed of each aircraft. **4422 mi/h; 1452 mi/h; 594 mi/h**

Airplane type	X-15A-2	Supersonic transport	Commercial jet
Mach number, m	6.7	2.2	0.9
Speed, v	?	?	?

FOCUS ON PEOPLE

CHUCK YEAGER was the first person to fly faster than the speed of sound (Mach 1) or about 660 miles per hour.

APPLICATION LINK
www.mcdougallittell.com

4 ASSESS

DAILY HOMEWORK QUIZ

Transparency Available

1. Check whether the number 3 is a solution to $3x - x = 12 - 2x$. **yes**
2. Use mental math to solve $8r + 4 = 36$. **4**
3. Check whether the number 5 is a solution to $x(2x - 3) > 32$. **yes**
4. Write a mathematical representation for "the quotient of y and 6 is less than or equal to 18." $\frac{y}{6} \le 18$
5. Which formula would you use to find the average speed of a car that travels 80 km in 2 hours? **b**
   a. $I = Prt$ **b.** $d = r \cdot t$
   c. $S = 6s^2$

STUDENT HELP NOTES

Homework Help Students can find help for Exs. 69–76 at **www.mcdougallittell.com.** The information can be printed out for students who do not have access to the Internet.

Application Link, p. 29 Additional information about the Mach numbers of an aircraft can be found at **www.mcdougallittell.com.**

EXTRA CHALLENGE NOTE

Challenge problems for Lesson 1.4 are available in **blackline** format in the *Chapter 1 Resource Book*, p. 65 and at **www.mcdougallittell.com.**

ADDITIONAL TEST PREPARATION

1. OPEN ENDED Write a real world situation that can be described by $1.35x \le 10$.
Sample answer: **How many cans of juice that cost $1.35 each can you buy if you have at most $10 to spend?**

APPLICATION NOTE

EXERCISES 82 AND 83

Mention that, in business, the number of items you must sell in order to equal or exceed expenses is normally called the *break-even point.*

Test Preparation

79. MULTIPLE CHOICE You tune and restore pianos. As a piano tuner you charge $75 per tuning. The expenses for your piano restoration business are $2600 per month. Which of the following inequalities could you use to find the number of pianos p you must tune per month in order to at least meet your business expenses? **B**

Ⓐ $75p \le 2600$ Ⓑ $75p \ge 2600$ Ⓒ $\frac{75}{p} \ge 2600$ Ⓓ $75p = 2600$

80. MULTIPLE CHOICE The width of a soccer field cannot be greater than 100 yards. The area cannot be greater than 13,000 square yards. Which of the following would you use to find the possible lengths of a soccer field? **B**

Ⓐ $100x \ge 13{,}000$ Ⓑ $100x \le 13{,}000$

Ⓒ $100 + x \le 13{,}000$ Ⓓ $100x = 130{,}000$

81. MULTIPLE CHOICE For which inequality is $x = 238$ a solution? **A**

Ⓐ $250 \ge x + 12$ Ⓑ $250 < x + 12$ Ⓒ $250 > x + 12$

★ Challenge

BUSINESS You plan to start your own greeting card business. Your startup cost of buying a computer and color printer is $1400. You also want to run an ad for $50 a week for 4 weeks. You plan to sell each card for $1.79. How many cards must you sell to equal or exceed your initial costs?

82. Write an inequality that models the situation. Solve the inequality to find the minimum number of cards you must sell. **at least 894 cards**

83. *Writing* Try raising the price per card. How does the price increase change the number of cards you must sell to equal or exceed your initial costs? What are some factors you need to consider in choosing a price?
As the selling price increases, the number of cards you need to sell decreases. However, you need to be sure the price is not so high as to decrease sales. Some issues to consider are how many cards you can reasonably expect to sell and the price of cards already available.

EXTRA CHALLENGE

www.mcdougallittell.com

MIXED REVIEW

EVALUATING EXPRESSIONS Evaluate the expression for the given value of the variable. **(Review 1.1)**

84. $1.2n$ when $n = 4.8$ **5.76**

85. $\frac{1}{12} + x$ when $x = \frac{1}{6}$ **$\frac{1}{4}$**

86. $b - 12$ when $b = 43$ **31**

87. $\frac{4}{5} \cdot y$ when $y = \frac{1}{5}$ **$\frac{4}{25}$**

EXPONENTIAL FORM Write the expression in exponential form. **(Review 1.2)**

88. three cubed **3^3**

89. y squared **y^2**

90. $6 \cdot 6 \cdot 6 \cdot 6 \cdot 6$ **6^5**

91. $c \cdot c \cdot c \cdot c$ **c^4**

92. five to the fourth power **5^4**

93. $5y \cdot 5y \cdot 5y \cdot 5y$ **$(5y)^4$**

94. three squared **3^2**

95. $9a \cdot 9a \cdot 9a \cdot 9a \cdot 9a \cdot 9a$ **$(9a)^6$**

96. seven squared **7^2**

97. one to the third power **1^3**

98. $6 \cdot x \cdot x \cdot x \cdot x \cdot x \cdot x$ **$6x^6$**

99. $t \cdot t \cdot t \cdot t \cdot t \cdot t$ **t^6**

100. WATER TEMPERATURE The temperature of the water in a swimming pool is 78° Fahrenheit. What is the temperature of the water in degrees Celsius? Use the formula $C = \frac{5}{9}(F - 32)$, where F is the Fahrenheit temperature and C is the Celsius temperature. **(Review 1.3 for 1.5)** **$25\frac{5}{9}$°C, or about 25.56°C**

ACTIVITY 1.5

Developing Concepts

SETUP
Work with a partner.

MATERIALS
- paper
- pencil

Group Activity for use with Lesson 1.5

Exploring Problem Solving

QUESTION **How can problem solving strategies help you find solutions to problems?**

When solving a problem, you need to understand two things: what you are asked to find and what information you need to solve the problem. Then you can use algebra or a problem solving strategy.

EXPLORING THE CONCEPT

1. State what you are asked to find in the following problem.

   In science class you have five 100-point tests and one final 200-point test that counts as two tests. You need an average of at least 90 points to get an A. Your scores for the five 100-point tests are 88, 92, 87, 98, and 81. What is the lowest score you can get on the final test to get an A?
   the lowest score you can get on the final test to get an A

2. List the information you need to solve the problem. **your scores, the number of 100-point test blocks, the lowest average you need to get an A**

3. Use the strategy Guess, Check, and Revise to solve the problem.

Test Average: $\frac{88 + 92 + 87 + 98 + 81 + ?}{7}$

Sample First Guess: $\frac{88 + 92 + 87 + 98 + 81 + 149}{7} = 85$ Too low

Sample Second Guess: $\frac{88 + 92 + 87 + 98 + 81 + 191}{7} = 91$ Too high

Your Guess: $\frac{88 + 92 + 87 + 98 + 81 + ?}{7} = ?$

The lowest score you can get on the final test to get an A is ?. **184**

DRAWING CONCLUSIONS

Solve the real-life problem. Explain your strategy.

1. A plumber charges a basic service fee plus a labor charge for each hour of service. A 2-hour job costs \$120 and a 4-hour job costs \$180. Find the plumber's basic service fee. **\$60**

2. Jerry has received the following math test scores: 80, 75, 79, 86. His final exam is worth 200 points and counts as two tests. What is the lowest score he can get to keep an average of at least 80 points? **160**

3. A Native American art museum charges \$8 per visit to nonmembers. Members pay a \$40 membership fee and get in for a reduced rate of \$5. How many times must a person visit the museum for membership to be less expensive than paying for each visit as a nonmember? **14 visits**

STUDENT HELP

Skills Review
For help with problem solving strategies, see pp. 795–796.

1 Planning the Activity

PURPOSE
To have students find solutions to problems using problem solving strategies.

MATERIALS
- paper
- pencil

PACING
- Exploring the Concept — 10 min
- Drawing Conclusions — 15 min

LINK TO LESSON
When students learn to write algebraic models in Examples 2 and 3 of Lesson 1.5, have them recall that they need to understand what they are being asked to find and what information they need to find it.

2 Managing the Activity

COOPERATIVE LEARNING
As they work through the activity, partners can take turns telling what they are asked to find in each case and what they need to be able to find it.

ALTERNATIVE APPROACH
You may wish to do the Exploring the Concept section as a class demonstration and then have students work individually or in groups to do the problems.

3 Closing the Activity

KEY DISCOVERY
Before attempting to solve a problem, it is important to understand what you are trying to find and what information you need to solve the problem.

ACTIVITY ASSESSMENT
Carol received the following math test scores: 67, 82, 81, 74. What is the lowest test score she can get to have an average of at least 75? **71**

1 PLAN

PACING
Basic: 2 days
Average: 2 days
Advanced: 2 days
Block Schedule: 0.5 block with 1.4
0.5 block with 1.6

LESSON OPENER APPLICATION
An alternative way to approach Lesson 1.5 is to use the Application Lesson Opener:
- Blackline Master (*Chapter 1 Resource Book*, p. 69)
- Transparency (p. 5)

MEETING INDIVIDUAL NEEDS
- ***Chapter 1 Resource Book***
  Prerequisite Skills Review (p. 5)
  Practice Level A (p. 70)
  Practice Level B (p. 71)
  Practice Level C (p. 72)
  Reteaching with Practice (p. 73)
  Absent Student Catch-Up (p. 75)
  Challenge (p. 78)
- ***Resources in Spanish***
- ***Personal Student Tutor***

NEW-TEACHER SUPPORT
See the Tips for New Teachers on pp. 1–2 of the *Chapter 1 Resource Book* for additional notes about Lesson 1.5.

WARM-UP EXERCISES
Transparency Available
Match the equation with the question that can be used to solve it.
1. $x - 3 = 9$ C
2. $3x = 9$ A
3. $9 - x = 3$ B
A. 3 times what number gives 9?
B. 9 minus what number gives 3?
C. What number minus 3 gives 9?

1.5 A Problem Solving Plan Using Models

What you should learn

GOAL 1 Translate verbal phrases into algebraic expressions.

GOAL 2 Use a verbal model to write an algebraic equation or inequality to solve a **real-life** problem, such as making a decision about an airplane's speed in **Example 3**.

Why you should learn it

▼ To solve **real-life** problems such as finding out how many plates of dim sum were ordered for lunch in **Example 2**.

REAL LIFE

GOAL 1 TRANSLATING VERBAL PHRASES

To translate verbal phrases into algebra, look for words that indicate operations.

Operation	Verbal Phrase	Expression
Addition	The *sum* of six and a number	$6 + x$
	Eight *more than* a number	$y + 8$
	A number *plus* five	$n + 5$
	A number *increased* by seven	$x + 7$
Subtraction	The *difference* of five and a number	$5 - y$
	Four *less than* a number	$x - 4$
	Seven *minus* a number	$7 - n$
	A number *decreased* by nine	$n - 9$
Multiplication	The *product* of nine and a number	$9x$
	Ten *times* a number	$10n$
	A number *multiplied* by three	$3y$
Division	The *quotient* of a number and four	$\frac{n}{4}$
	Seven *divided* by a number	$\frac{7}{x}$

Order is important for subtraction and division, but *not* for addition and multiplication. "Four less than a number" is written as $x - 4$, *not* $4 - x$. On the other hand, "the sum of six and a number" can be written as $6 + x$ or $x + 6$.

EXAMPLE 1 *Translating Verbal Phrases into Algebra*

Translate the phrase into an algebraic expression.

SOLUTION

a. Three more than the quantity five times a number n

$5n + 3$ **Think: 3 more than what?**

b. Two less than the sum of six and a number m

$(6 + m) - 2$ **Think: 2 less than what?**

c. A number x decreased by the sum of 10 and the square of a number y

$x - (10 + y^2)$ **Think: x decreased by what?**

GOAL 2 USING A VERBAL MODEL

STUDENT HELP

Study Tip
When you translate verbal expressions, the words "the quantity" tell you what to group. In Example 1, you write $5n + 3$, not $(5 + 3)n$.

In English there is a difference between a phrase and a sentence. Verbal phrases translate into mathematical expressions and verbal sentences translate into equations or inequalities.

Phrase	The sum of six and a number	Expression	$6 + x$
Sentence	The sum of six and a number is twelve.	Equation	$6 + x = 12$
Sentence	Seven times a number is less than fifty.	Inequality	$7x < 50$

Sentences that translate into equations have words that tell how one quantity relates to another. In the first sentence, the word "is" says that one quantity is equal to another. In the second sentence, the words "is less than" indicate an inequality.

Writing algebraic expressions, equations or inequalities that represent real-life situations is called **modeling**. The expression, equation or inequality is a **mathematical model** of the real-life situation. When you write a mathematical model, we suggest that you use three steps.

WRITE A VERBAL MODEL. → **ASSIGN LABELS.** → **WRITE AN ALGEBRAIC MODEL.**

EXAMPLE 2 *Writing an Algebraic Model*

REAL LIFE
Dining Out

You and three friends are having a dim sum lunch at a Chinese restaurant that charges $2 per plate. You order lots of plates of wontons, egg rolls, and dumplings. The waiter gives you a bill for $25.20, which includes tax of $1.20. Use mental math to solve the equation for how many plates your group ordered.

SOLUTION

Be sure that you understand the problem situation before you begin. For example, notice that the tax is added after the cost of the plates of dim sum is figured.

PROBLEM SOLVING STRATEGY

VERBAL MODEL [Cost per plate] · [Number of plates] = [Bill] − [Tax]

LABELS

Cost per plate = **2** (dollars)

Number of plates = ***p*** (plates)

Amount of bill = **25.20** (dollars)

Tax = **1.20** (dollars)

ALGEBRAIC MODEL

$$2p = 25.20 - 1.20$$

$$2p = 24.00$$

$$p = 12$$

▶ Your group ordered 12 plates of food costing $24.00.

2 TEACH

MOTIVATING THE LESSON
Ask students if they think it is possible to know the height of a football at any given moment after a kick-off. Tell students that it is possible to know this because the path of the ball can be described algebraically. Writing and using algebraic models is the focus of today's lesson.

EXTRA EXAMPLE 1
Translate the phrase into an algebraic expression.
a. Six less than 4 times a number y $4y - 6$
b. Three more than the difference of five and a number n $(5 - n) + 3$
c. A number y decreased by the sum of 8 and the square of another number x $y - (8 + x^2)$

EXTRA EXAMPLE 2
You and your friends go to a music store to buy CD's on sale for $6 each. Together you spent $77.76, which included a tax of $5.76. Use mental math to solve the equation for how many CD's you bought together. 12

CHECKPOINT EXERCISES
For use after Example 1:
1. Translate the phrase into an algebraic expression.
a. Eight more than the sum of two times a number y and three. $(2y + 3) + 8$
b. Four less than the sum of five and twice a number n $(5 + 2n) - 4$

For use after Example 2:
2. Eight friends went to a restaurant for dinner. The waiter gave them a bill for $130. At the register a tax and tip of $30 was added to the bill. Use mental math to find about how much each person should contribute to pay an equal share of the bill. $20

STUDENT HELP NOTES

Career Note Additional information about being a pilot is available at **www.mcdougallittell.com.**

EXTRA EXAMPLE 3

An investor finds a mutual fund that has an annual return of 10%. The investor wants to earn $250 in simple interest by the end of one year. Use the formula $I = Prt$.

a. What should be the minimum amount invested? **$2500**

b. If the investor wants to earn twice the amount of simple interest, should the investment or the amount of time be doubled? Explain. **Either method will earn the investor $500 in interest.**

CHECKPOINT EXERCISES

For use after Example 3:

1. A salesperson drives at a speed of 50 miles per hour. When he is 175 miles from his destination, he remembers he has a meeting in 3 hours.

a. At his current speed, will he be on time for his meeting? **No**

b. At what minimum speed should he travel from this point on if he wants to be on time for the meeting? **about 60 mi/h**

FOCUS ON VOCABULARY

Explain *modeling.* **Modeling is the writing of algebraic expressions, equations, or inequalities that represent real-life situations.**

CLOSURE QUESTION

How do you find a mathematical model for a real-life situation? **You write a verbal model to match the situation, assign labels, and then write an algebraic model to match the verbal model.**

DAILY PUZZLER

Find a number whose double is ten less than its triple. **10**

FOCUS ON CAREERS

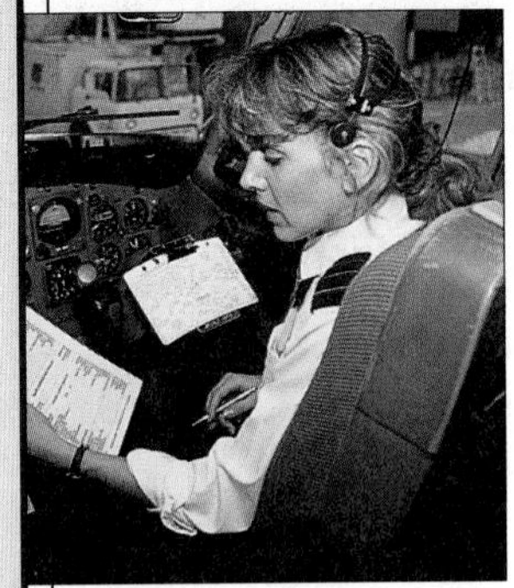

REAL LIFE **JET PILOT** Pilots select a route, an altitude, and a speed that will provide the fastest and safest flight.

INTERNET CAREER LINK www.mcdougallittell.com

A PROBLEM SOLVING PLAN USING MODELS

VERBAL MODEL	Ask yourself what you need to know to solve the problem. Then write a verbal model that will give you what you need to know.
LABELS	Assign labels to each part of your verbal model.
ALGEBRAIC MODEL	Use the labels to write an algebraic model based on your verbal model.
SOLVE	Solve the algebraic model and answer the original question.
CHECK	Check that your answer is reasonable.

Verbal and algebraic modeling can be used as part of a general problem solving plan that includes solving and checking to see that an answer is reasonable.

EXAMPLE 3 *Using a Verbal Model*

JET PILOT A jet pilot is flying from Los Angeles, CA to Chicago, IL at a speed of 500 miles per hour. When the plane is 600 miles from Chicago, an air traffic controller tells the pilot that it will be 2 hours before the plane can get clearance to land. The pilot knows that at its present altitude the speed of the jet must be greater than 322 miles per hour or the jet will stall.

a. At what speed would the jet have to fly to arrive in Chicago in 2 hours?

b. Is it reasonable for the pilot to fly directly to Chicago at the reduced speed from part (a) or must the pilot take some other action?

SOLUTION

a. You can use the formula (rate)(time) = (distance) to write a verbal model.

PROBLEM SOLVING STRATEGY

VERBAL MODEL Speed of jet · Time = Distance to travel

LABELS

Speed of jet = x (miles per hour)

Time = **2** (hours)

Distance to travel = **600** (miles)

ALGEBRAIC MODEL

$2x = 600$ **Write algebraic model.**

$x = 300$ **Solve with mental math.**

▶ To arrive in 2 hours, the pilot would have to slow the jet down to a speed of 300 miles per hour.

b. It is not reasonable for the pilot to fly at 300 miles per hour, because the jet will stall. The pilot should take some other action, such as circling in a holding pattern, to use some of the time.

GUIDED PRACTICE

Vocabulary Check ✓

In Exercises 1 and 2, consider the verbal phrase: *the difference of 7 and a number n.*

1. What operation does the word *difference* indicate? subtraction

2. Translate the verbal phrase into an algebraic expression. $7 - n$

Concept Check ✓

3. Is order important in the expression in Exercise 2? yes

4. Describe how to use a verbal model to solve a problem. See margin.

Skill Check ✓

Match the verbal phrase with its corresponding algebraic expression.

5. Eleven decreased by the quantity four times a number x C
6. Four increased by the quantity eleven times a number x D
7. Four times the quantity of a number x minus eleven B
8. Four times a number x decreased by eleven A

A. $4x - 11$
B. $4(x - 11)$
C. $11 - 4x$
D. $11x + 4$

Write the verbal sentence as an equation or an inequality.

9. A number x increased by ten is 24. $x + 10 = 24$
10. The product of seven and a number y is 42. $7y = 42$
11. Twenty divided by a number n is less than or equal to two. $\frac{20}{n} \le 2$
12. Ten more than a number x is greater than fourteen. $x + 10 > 14$

4. Describe a real-life situation in words, then use the words to write an algebraic model. Note words that indicate operations (*plus, decrease,* or *product,* for example) or tell how one quantity relates to another (is or *is greater than,* for example). Apply labels to the parts of the verbal model to form an algebraic model.

PRACTICE AND APPLICATIONS

STUDENT HELP

Extra Practice to help you master skills is on p. 797.

15. $\frac{1}{2}x + 3$

TRANSLATING PHRASES **Write the verbal phrase as an algebraic expression. Use x for the variable in your expression.**

13. Nine more than a number $x + 9$
14. One half multiplied by a number $\frac{1}{2}x$
15. Three more than half of a number See margin.
16. A number increased by seven $x + 7$
17. Quotient of a number and two tenths $\frac{x}{0.2}$
18. Product of four and a number $4x$
19. Two cubed divided by a number $\frac{2^3}{x}$
20. Difference of ten and a number $10 - x$
21. Five squared minus a number $5^2 - x$
22. Twenty-nine decreased by a number $29 - x$

TRANSLATING VERBAL SENTENCES **Write the verbal sentence as an equation or an inequality.**

23. Nine is greater than three times a number s. $9 > 3s$
24. Twenty-five is the quotient of a number y and 3.5. $25 = \frac{y}{3.5}$
25. The product of 14 and a number x is one. $14x = 1$
26. Nine less than the product of ten and a number d is eleven. $10d - 9 = 11$
27. Three times the quantity two less than a number x is ten. $3(x - 2) = 10$
28. Five decreased by eight is four times y. $5 - 8 = 4y$

STUDENT HELP

HOMEWORK HELP
Example 1: Exs. 13–38
Example 2: Exs. 39–48
Example 3: Exs. 49–57

3 APPLY

ASSIGNMENT GUIDE

BASIC
Day 1: pp. 35–36 Exs. 13–38
Day 2: pp. 36–38 Exs. 39–53, 62–64, 68–78 even, Quiz 2 Exs. 1–13

AVERAGE
Day 1: pp. 35–36 Exs. 13–38
Day 2: pp. 36–38 Exs. 40–48 even, 49–57, 62–64, 68–78 even, Quiz 2 Exs. 1–13

ADVANCED
Day 1: pp. 35–36 Exs. 13–38
Day 2: pp. 36–38 Exs. 40–48 even, 54–66, 68–78 even, Quiz 2 Exs. 1–13

BLOCK SCHEDULE
pp. 35–36 Exs. 13–38 (with 1.4)
pp. 36–38 Exs. 40–48 even, 49–57, 62–64, 68–78 even, Quiz 2 Exs. 1–13 (with 1.6)

EXERCISE LEVELS

Level A: ***Easier***
13–48, Quiz 7–12

Level B: ***More Difficult***
49–57, 62–64, 67–78, Quiz 1–6, 13

Level C: ***Most Difficult***
58–61, 65–66

✓ HOMEWORK CHECK

To quickly check student understanding of key concepts, go over the following exercises: Exs. 20, 26, 30, 44, 50. See also the Daily Homework Quiz:

- Blackline Master (*Chapter 1 Resource Book,* p. 82)
- Transparency (p. 7)

ENGLISH LEARNERS

Note that English language learners often have difficulty translating words and word problems into the correct equations. Explain that the order in which numbers should be placed in an equation does not necessarily match the order in which they appear in the verbal sentence, as in Exercises 24, 26, and 27.

! COMMON ERROR

EXERCISES 29–38 Students often confuse *less than* with *is less than* when writing verbal phrases. Reminds students that *less than* refers to subtraction and that *is less than* refers to the inequality symbol <.

TEACHING TIPS

EXERCISES 39–48 Stress to students the importance of being able to do these exercises correctly. Writing an algebraic model is an extremely important part of the problem solving plan.

TRANSLATING VERBAL SENTENCES In Exercises 29–38, write the verbal sentence as an equation, or an inequality.

29. Twenty-three less than the difference of thirty-eight and a number n is less than eight. $(38 - n) - 23 < 8$

30. A number t increased by the sum of seven and the square of another number s is ten. $t + (7 + s^2) = 10$

31. Five less than the difference of twenty and a number x is greater than or equal to ten. $(20 - x) - 5 \geq 10$

32. Fourteen plus the product of twelve and a number y is less than or equal to fifty. $14 + 12y \leq 50$

33. Nine plus the quotient of a number b and ten is greater than or equal to eleven. $9 + \frac{b}{10} \geq 11$

34. Seventy divided by the product of seven and a number p is equal to one. $\frac{70}{7p} = 1$

35. A number q is equal to or greater than one hundred. $q \geq 100$

36. A number x squared plus forty-four is equal to the number x to the fourth power times three. $x^2 + 44 = 3x^4$

37. The quotient of thirty-five and a number t is less than or equal to seven. $\frac{35}{t} \leq 7$

38. Fifty multiplied by the quantity twenty divided by a number n is greater than or equal to two hundred fifty. $50\left(\frac{20}{n}\right) \geq 250$

ALGEBRAIC MODELING In Exercises 39–48, write an equation or an inequality to model the real-life situation.

39. Ben's hourly wage b at his after school job is \$1.50 less than Eileen's hourly wage e. $b = e - 1.50$

40. The distance s to school is $\frac{1}{5}$ mile more than the distance c to the Community Center swimming pool. $s = c + \frac{1}{5}$

41. The length c of the Colorado River is three times the length r of the Connecticut River, plus 229 miles. $c = 3r + 229$

42. Pi (π) is the quotient of the circumference C and the diameter d of a circle. $\pi = \frac{C}{d}$

43. The volume V of a cube with a side length s is less than or equal to thirty minus three. $V \leq 30 - 3$ or $s^3 \leq 30 - 3$

44. The product of \$25 and the number m of club memberships is greater than or equal to \$500. $25m \geq 500$

45. The perimeter P of a square is equal to four times the difference of a number s and two. $P = 4(s - 2)$

46. The simple interest earned on a principal of three hundred dollars at an annual interest rate of x percent is less than or equal to seventy-two dollars. $300 \cdot \frac{x}{100} \cdot t \leq 72$

47. The area A of a trapezoid is equal to one half times the sum of seven and nine, times a number h plus seven. $A = \frac{1}{2}(7 + 9)(h + 7)$

48. The square of the length c of the hypotenuse of a right triangle is equal to four squared plus three squared. $c^2 = 4^2 + 3^2$

49. The number of subscriptions times the cost of each subscription equals the amount the club needs to raise.

57. $\frac{\text{dollars}}{\text{mi/h}} \cdot \text{mi/h} = \text{dollars}$

STUDENT HELP

HOMEWORK HELP

Visit our Web site www.mcdougallittell.com for help with problem solving in Exs. 54–57.

59. The number of weeks times the total of $20 plus the additional amount saved per week equals the total saved; let x = the additional amount saved per week; $12(x + 20) = 480$.

FOCUS ON APPLICATIONS

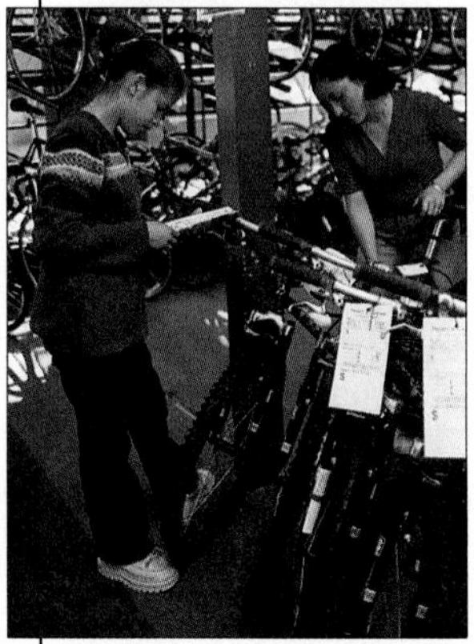

MOUNTAIN BIKES More than 50% of all bicycles sold in the United States are mountain bikes.

DATA UPDATE of Bicycle Product Suppliers Association data at www.mcdougallittell.com

FUNDRAISING In Exercises 49–53, use the following information.
The science club is selling magazine subscriptions at $15 each. The club wants to raise $315 for science equipment.

49. Write a verbal model that relates the number of subscriptions, the cost of each subscription, and the amount of money the club needs to raise.

50. Assign labels and write an algebraic model based on your verbal model.
Let n = the number of subscriptions; $15n = 315$.

51. Use mental math to solve the equation. 21

52. How many subscriptions does the science club need to sell to raise $315? 21 subscriptions

53. Check to see if your answer is reasonable.
$21 \cdot 15 = 315$; the answer is reasonable.

LAW ENFORCEMENT In Exercises 54–57, use the following information.
Jeff lives in a state in which speeders are fined $20 for each mile per hour (mi/h) over the speed limit. Jeff was given a ticket for $260 for speeding on a road where the speed limit is 45 miles per hour. Jeff wants to know how fast he was driving.

Fine per mi/h over speed limit	·	Miles per hour over speed limit	=	Amount of ticket

54. Assign labels to the three parts of the verbal model.
fine per mi/h over speed limit = 20; m = mi/h over speed limit; amount of ticket = 260

55. Use the labels to translate the verbal model into an algebraic model.
$20m = 260$

56. Use mental math to solve the equation. What does the solution represent?
13; the number of mi/h over the speed limit Jeff was driving

57. Perform unit analysis to check that the equation is set up correctly. See margin.

58. **CLASS ELECTION** You are running for class president. At 2:30 on election day you have 95 votes and your opponent has 120 votes. Forty-five more students will be voting. Let x represent the number of students (of the 45) who vote for you.

a. Write an inequality that shows the values of x that will allow you to win the election. $95 + x > 120 + (45 - x)$

b. What is the smallest value of x that is a solution of the inequality? 36

MOUNTAIN BIKES In Exercises 59 and 60, use the following information.
You are shopping for a mountain bike. A store sells two different models. The model that has steel wheel rims costs $220. The model with aluminum wheel rims costs $480. You have a summer job for 12 weeks. You save $20 per week, which would allow you to buy the model with the steel wheel rims. You want to know how much more money you would have to save each week to be able to buy the model with the aluminum wheel rims.

59. Write a verbal model and an algebraic model for how much more money you would have to save each week. See margin.

60. Use mental math to solve the equation. What does the solution represent? See margin.

61. **BASKETBALL** The girls' basketball team scored 544 points in 17 games last year. This year the coach has set a goal for the team to score at least 5 more points per game. If 18 games are scheduled for this year, write an inequality that represents the total number of points the team must equal or exceed to meet their season goal. See margin.

TEACHING TIPS

EXERCISE 53 Stress the importance of checking to see that an answer is reasonable. This is especially important when using a calculator, since the margin for error is increased simply by pushing an incorrect key.

STUDENT HELP NOTES

Homework Help Students can find help for Exs. 54–57 at **www.mcdougallittell.com.** The information can be printed out for students who do not have access to the Internet.

DATA UPDATE

Updated data for Exs. 59–60 is available at **www.mcdougallittell.com.**

APPLICATION NOTE

EXERCISE 61 Students may want to discuss how points are scored in basketball, although this knowledge is not needed to answer the question.

60. $20; the amount over $20 you have to save each week to buy the model with the aluminum wheel rims

61. Let x = the total number of points the team must equal or exceed; $x \geq 18\left(\frac{544}{17} + 5\right)$.

ADDITIONAL PRACTICE AND RETEACHING

For Lesson 1.5:

- Practice Levels A, B, and C (*Chapter 1 Resource Book*, p. 70)
- Reteaching with Practice (*Chapter 1 Resource Book*, p. 73)
- See Lesson 1.5 of the *Personal Student Tutor*

For more Mixed Review:

- Search the *Test and Practice Generator* for key words or specific lessons.

4 ASSESS

DAILY HOMEWORK QUIZ

Transparency Available

1. Write the verbal sentence as an equation or inequality.
   a. Four times the quantity three more than a number n is twelve. $4(n + 3) = 12$
   b. Six less than the sum of the square of a number x and 18 equals twenty. $(x^2 + 18) - 6 = 20$
   c. The product of seven and a number r is less than or equal to forty-nine. $7r \leq 49$
   d. Eighty multiplied by twelve divided by a number n is less than or equal to three hundred forty. $\frac{80 \cdot 12}{n} \leq 340$
   e. The area of the circle is equal to three squared times pi. $A = 3^2\pi$

EXTRA CHALLENGE NOTE

Challenge problems for Lesson 1.5 are available in **blackline** format in the *Chapter 1 Resource Book*, p. 78 and at **www.mcdougallittell.com.**

ADDITIONAL TEST PREPARATION

1. **OPEN ENDED** Give an example of a real world situation and write an algebraic model for it. *Sample answer:* **Juanita received a raise. Her new hourly wage w is \$.75 more per hour than her old hourly wage r. $w = r + 0.75$**

73. Fruits and Vegetables Eaten by California Adults

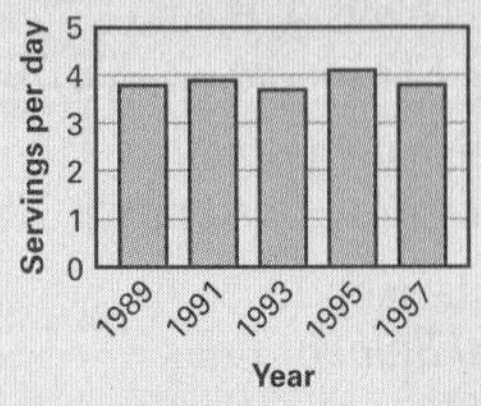

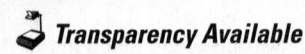

Test Preparation

62. **MULTIPLE CHOICE** Translate the phrase "a number decreased by the quotient of three and four." **A**

    (A) $n - \frac{3}{4}$ (B) $\frac{3}{4} - n$ (C) $\frac{n-3}{4}$ (D) $\frac{3}{4-n}$

63. **MULTIPLE CHOICE** Give the correct algebraic translation of "Howard's hourly wage h is \$2 greater than Marla's hourly wage m." **B**

    (A) $h < m + 2$ (B) $h = m + 2$ (C) $m = h + 2$ (D) $h > m + 2$

64. **MULTIPLE CHOICE** A jet is flying nonstop from Baltimore, Maryland, to Jacksonville, Florida, at a speed r of 500 miles per hour. The distance d between the two cities is about 680 miles. Which equation models the number of hours t the flight will take? **A**

    (A) $t = \frac{680}{500}$ (B) $t = 680(500)$ (C) $680 = \frac{t}{500}$ (D) $t = \frac{500}{680}$

★ Challenge

SCHOOL DANCE In Exercises 65 and 66, use the following information. You are in charge of the music for a school dance. The school's budget allows only \$300 for music, which is enough to hire a disc jockey for 4 hours. You would rather hire a live band, but the band charges \$135 per hour. Your school does not allow students to be charged an admission fee. To raise the money for a live band, you obtain permission for a voluntary contribution of \$1.25 per person.

EXTRA CHALLENGE
www.mcdougallittell.com

65. How much extra money do you need to raise? **\$240**

66. How many students must contribute \$1.25 to cover the cost of a live band? **192 students**

MIXED REVIEW

COMPARING DECIMALS Compare using <, =, or >. (Skills Review, p. 779)

67. 0.3 _?_ 0.30 =
68. 21.1 _?_ 20.99 >
69. 6.7 _?_ 6.079 >
70. 5.68 _?_ 5.678 >
71. 0.333 _?_ 0.3333 <
72. 18.45 _?_ 18.5 <

73. **EATING HABITS** The table gives the number of servings per day of fruits and vegetables consumed by adults in California. Create a bar graph of the data. (Skills Review, pp. 792–794) **See margin.**

Servings of Fruits and Vegetables Eaten by California Adults					
Year	1989	1991	1993	1995	1997
Servings per day	3.8	3.9	3.7	4.1	3.8

▶ Source: California Department of Health Services

FOCUS ON APPLICATIONS

PLANT GROWTH Kudzu was introduced to the United States in 1876. Today, kudzu covers over 7 million acres of the southern United States.

74. **PLANT GROWTH** Kudzu is a vine native to Japan. Kudzu can grow a foot per day during the summer months. Write an expression that shows how much a 20-foot kudzu vine can grow during August. (Review 1.1) **31 ft**

EVALUATING VARIABLE EXPRESSIONS Evaluate the expression. (Review 1.3)

75. $4 + 3x$ when $x = 2$ **10**
76. $y \div 8$ when $y = 32$ **4**
77. $5 \cdot 2p^2$ when $p = 6$ **360**
78. $t^4 - t$ when $t = 7$ **2394**

QUIZ 2

Evaluate the expression. (Lesson 1.3)

1. $12 \div (7 - 3)^2 + 2$ **$2\frac{3}{4}$**

2. $32 - 5 \cdot (2 + 1) + 4$ **21**

3. $x^2 + 4 - x$ when $x = 6$ **34**

4. $y \div 3 + 2$ when $y = 30$ **12**

5. $\frac{r}{s} \cdot 7$ when $r = 30$ and $s = 5$ **42**

6. $5x^2 - y$ when $x = 4$ and $y = 26$ **54**

Check whether the given number is a solution of the equation or inequality. (Lesson 1.4)

7. $2x + 6 = 18; 9$ **not a solution**

8. $13 - 3x = 7; 2$ **solution**

9. $4y + 7 = 5 + 5y; 2$ **solution**

10. $2s + s = 4s; 6$ **not a solution**

11. $3x - 4 > 0; 2$ **solution**

12. $8 - 2y > 4; 3$ **not a solution**

13. PIZZA PARTY You and three friends bought a pizza. You paid \$2.65 for your share $\left(\frac{1}{4} \text{ of the pizza}\right)$. Write an equation that models the situation. What was the total cost of the pizza? (Lesson 1.5) **Let x = the total cost of the pizza; $\frac{1}{4}x = 2.65$; \$10.60.**

MATH & History

Problem Solving

APPLICATION LINK www.mcdougallittell.com

THEN

IN THE NINTH CENTURY, mathematician al-Khwarizmi studied at the House of Wisdom in Baghdad. He developed and published important concepts in algebra. In the 1100s, al-Khwarizmi's text was translated from Arabic into Latin, making his ideas available to Western scholars. The step-by-step problem solving techniques invented by al-Khwarizmi are called *algorithms*—a Latinized form of al-Khwarizmi.

NOW

TODAY, computer programmers use algorithms to write new programs. Algorithms are also an important part of every algebra course. In this chapter you learned the following algorithm.

Write a verbal model. → **Assign labels.** → **Write an algebraic model.**

1. Write a word problem and use the algorithm shown to solve the problem. **Check problems.**

2. Give two more examples of step-by-step methods you have learned in your study of mathematics. Show a worked-out problem using each method.
Sample answer: **order of operations, $(3 + 5)^2 - 2 \cdot 3 = 8^2 - 6 = 64 - 6 = 58$**

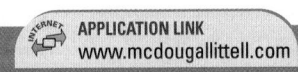

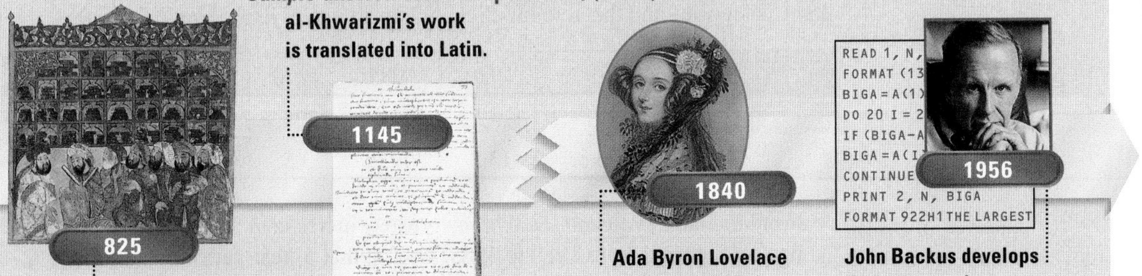
825
1145
1840
1956
READ 1, N,
FORMAT (13
BIGA = A(1)
DO 20 I = 2
IF (BIGA-A
BIGA = A(I
CONTINUE
PRINT 2, N, BIGA
FORMAT 922H1 THE LARGEST

al-Khwarizmi's work is translated into Latin.

Scholars study at the House of Wisdom.

Ada Byron Lovelace writes the first computer program.

John Backus develops the programming language Fortran.

ADDITIONAL RESOURCES

An alternative Quiz for Lessons 1.4 and 1.5 is available in the *Chapter 1 Resource Book,* p. 79.

A **blackline** master with additional Math & History exercises is available in the *Chapter 1 Resource Book,* p. 77.

1 PLAN

PACING
Basic: 1 day
Average: 1 day
Advanced: 1 day
Block Schedule: 0.5 block with 1.5

LESSON OPENER APPLICATION
An alternative way to approach Lesson 1.6 is to use the Application Lesson Opener:

- Blackline Master (*Chapter 1 Resource Book*, p. 83)
- Transparency (p. 6)

MEETING INDIVIDUAL NEEDS

- ***Chapter 1 Resource Book***
  Prerequisite Skills Review (p. 5)
  Practice Level A (p. 84)
  Practice Level B (p. 85)
  Practice Level C (p. 86)
  Reteaching with Practice (p. 87)
  Absent Student Catch-Up (p. 89)
  Challenge (p. 92)
- ***Resources in Spanish***
- ***Personal Student Tutor***

NEW-TEACHER SUPPORT
See the Tips for New Teachers on pp. 1–2 of the *Chapter 1 Resource Book* for additional notes about Lesson 1.6.

WARM-UP EXERCISES

Transparency Available

The table represents U.S. population projections for 2020–2040, in millions.

1. What is the projected population in 2020? 322,700,000
2. What is the projected population in 2040? 370,000,000

Year	Population
2020	322.7
2030	346.9
2040	370.0

EXPLORING DATA AND STATISTICS

1.6

What you should learn

GOAL 1 Use tables to organize data.

GOAL 2 Use graphs to organize **real-life** data, such as the amounts of various foods consumed in **Example 2**.

Why you should learn it

▼ To help you see relationships among **real-life** data, such as the average cost of making a movie in **Example 3**.

REAL LIFE

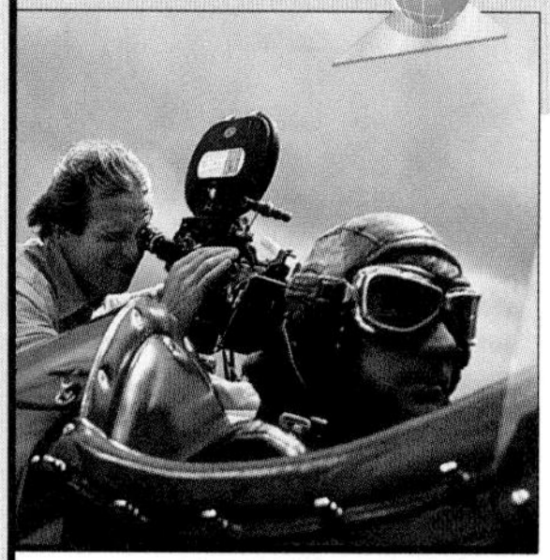

Tables and Graphs

GOAL 1 USING TABLES TO ORGANIZE DATA

Almost every day you have the chance to interpret **data** that describe real-life situations. The word *data* is plural and it means information, facts, or numbers that describe something.

A collection of data is easier to understand when the data are organized in a table or in a graph. There is no "best" way to organize data, but there are many good techniques. Often it helps to put numbers in either increasing or decreasing order. It also helps to group numbers so that patterns or trends are more apparent.

EXAMPLE 1 *Using a Table*

The data in the table were taken from a study on what people in the United States eat. The study grouped what people eat into over twenty categories. The three top categories (listed in pounds per person per year) are shown in the table.

Top Categories of Food Consumed by Americans (lb per person per year)

Year	1970	1975	1980	1985	1990	1995	2000
Dairy	563.8	539.1	543.2	593.7	568.4	584.4	590.0
Vegetables	335.4	337.0	336.4	358.1	382.8	405.0	410.0
Fruit	237.7	252.1	262.4	269.4	273.5	285.4	290.0

DATA UPDATE of U.S. Department of Agriculture at www.mcdougallittell.com
Year 2000 data are projected by the authors.

In which 5-year period did the total consumption per person of dairy, fruit, and vegetables increase the most? When was there a decrease in total consumption?

SOLUTION

Add two more rows to the table. Enter the total consumption per person of all three categories and the amount of change from one 5-year period to the next.

Year	1970	1975	1980	1985	1990	1995	2000
Total	1136.9	1128.2	1142.0	1221.2	1224.7	1274.8	1290.0
Change	——	−8.7	13.8	79.2	3.5	50.1	15.2

▶ From the table, you can see that the greatest increase in total consumption per person occurred from 1980 to 1985. During that 5-year period, total yearly consumption of dairy, vegetables, and fruit increased by 79.2 pounds per person. The minus sign indicates a decrease of 8.7 pounds from 1970 to 1975.

GOAL 2 USING GRAPHS TO ORGANIZE DATA

One way to organize data is with a **bar graph**. The bars can be either vertical or horizontal. Example 2 shows a vertical bar graph of the data from Example 1.

REAL LIFE
Eating Habits

EXAMPLE 2 *Interpreting a Bar Graph*

The bar graph shows the total amount of dairy products, vegetables, and fruit consumed by Americans in a given year. If you glance at the graph, it appears Americans ate seven times the amount of dairy products, vegetables, and fruit in 1995 as compared with 1970. If you study the data in Example 1, you can see that the bar graph could be misleading.

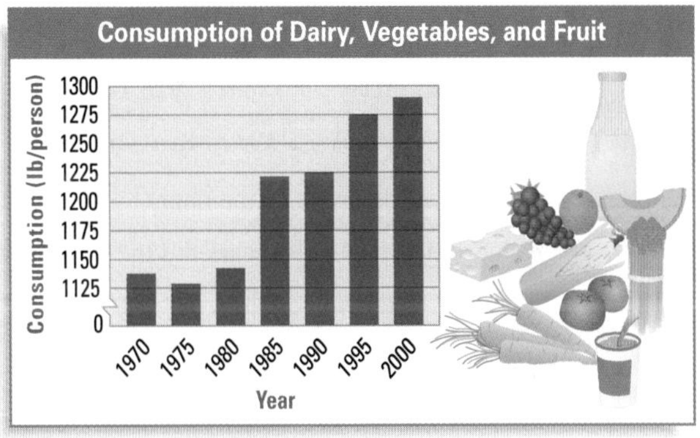

a. Explain why the graph could be misleading.

b. Draw a new bar graph that would not be misleading.

SOLUTION

a. The bar graph could be misleading because the vertical scale is not consistent. A gap exists between 0 and the first tick mark on the vertical axis. This could give the visual impression that consumption for 1970, 1975, and 1980 is much lower than for the other years.

b. To make a bar graph that would not be misleading, you must eliminate the gap and make sure that each tick mark represents the same amount.

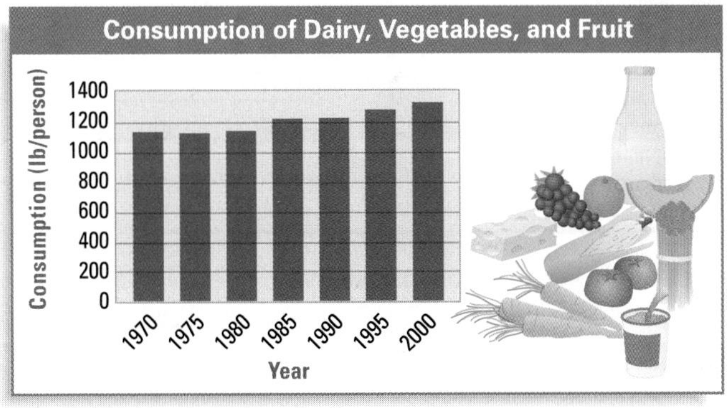

STUDENT HELP
Skills Review
For help with drawing bar graphs, see pp. 792–794.

STUDENT HELP
INTERNET
HOMEWORK HELP
Visit our Web site www.mcdougallittell.com for extra examples.

2 TEACH

MOTIVATING THE LESSON
Tables and graphs are used daily in real life.

EXTRA EXAMPLE 1
The data in the table represents the number of worldwide shipments of personal computers, in millions. During which 2-year period did the number of shipments increase the most?

Year	'90	'92	'94	'96	'98
PCs	23.7	32.4	47.9	70.8	98.4

1996 to 1998

EXTRA EXAMPLE 2
A graph for the data in Extra Example 1 is shown below. Is the graph misleading? Explain.

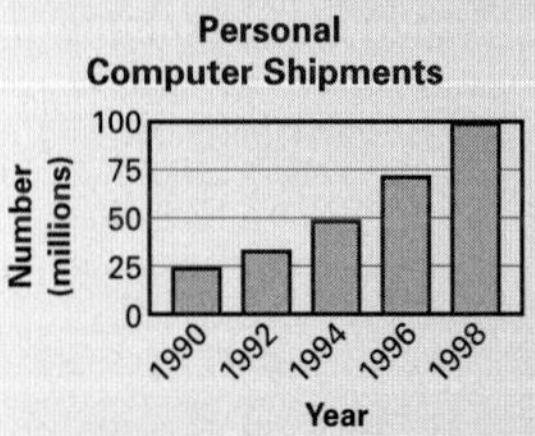

The bar graph is not misleading because the vertical scale is consistent.

CHECKPOINT EXERCISES
For use after Examples 1 and 2:

1. The data in the table represents the number of households in the U.S. in millions.

Year	'60	'70	'80	'90	'00
No.	52.8	63.4	80.8	93.3	103.2

a. During which 10-year period did the number of households increase the most? **1970 to 1980**

b. Draw a bar graph.

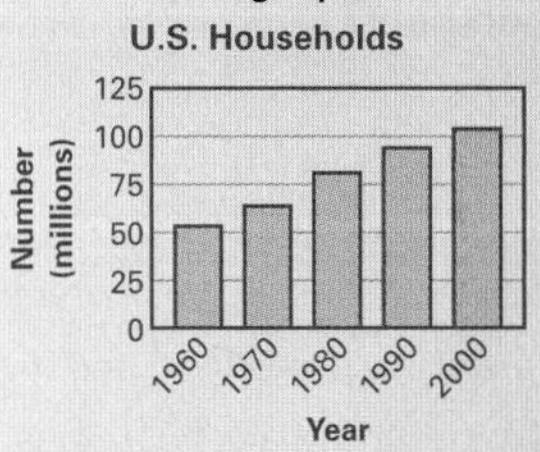

EXTRA EXAMPLE 3

The data in the table represents the number of people over age 65 in the U.S. in millions.

Year	'60	'70	'80	'90	'00
No.	17.4	21.4	27.8	34.3	39.0

a. Draw a line graph.

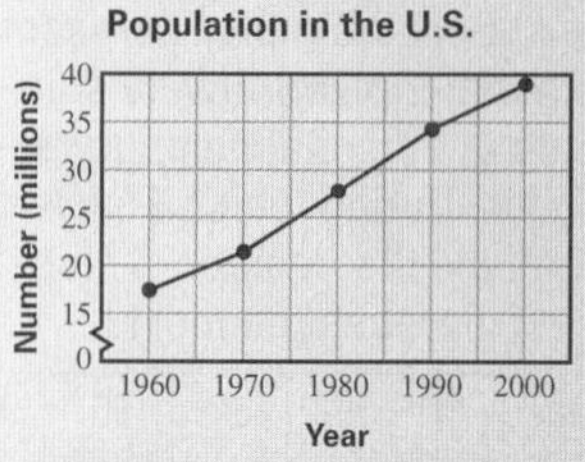

b. During which 10-year period was there the least increase? **between 1960 and 1970**

CHECKPOINT EXERCISES

For use after Example 3:

1. a. Draw a line graph for the data in Extra Example 1.

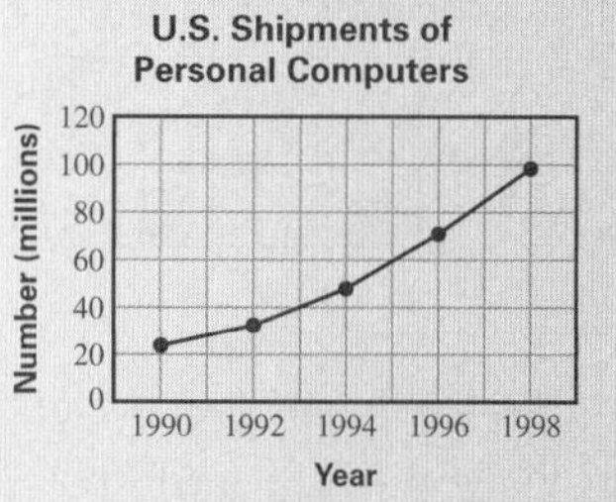

b. During which 2-year period did the number of shipments increase the least? **1990–1992**

CLOSURE QUESTION

What is the difference between a bar graph and a line graph?
A bar graph uses bars to display data. A line graph uses lines to show changes in data over time.

DAILY PUZZLER

What could the graph below represent?

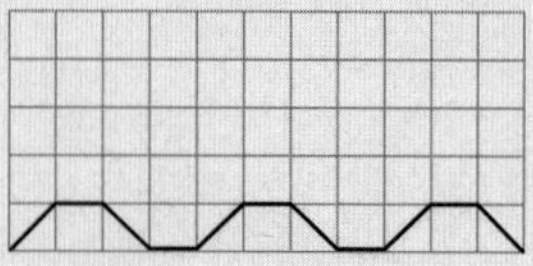

Sample answer: **the speed of a bus or train that stops to pick up and drop off passengers.**

Another way to organize data is with a **line graph**. Line graphs are especially useful for showing changes in data over time.

REAL LIFE

Movie Making

EXAMPLE 3 *Making a Line Graph*

From 1983 to 1996, the average cost (in millions of dollars) of making a movie is given in the table.

Average Cost of Making a Movie (millions of dollars)							
Year	1983	1984	1985	1986	1987	1988	1989
Average cost	$11.8	$14.0	$16.7	$17.5	$20.0	$18.1	$23.3
Year	1990	1991	1992	1993	1994	1995	1996
Average cost	$26.8	$26.1	$28.9	$29.9	$34.3	$36.4	$33.6

▶ Source: *International Motion Picture Almanac*

a. Draw a line graph of the data.

b. During which three years did the average cost decrease from the prior year?

c. Which year showed the greatest decrease?

SOLUTION

a. Draw the vertical scale from 0 to 40 million dollars. Use units of 5 for the vertical axis to represent intervals of five million dollars. Draw the horizontal axis and mark the number of years starting with 1983.

For each average cost in the table, draw a point on the graph.

Draw a line from each point to the next point.

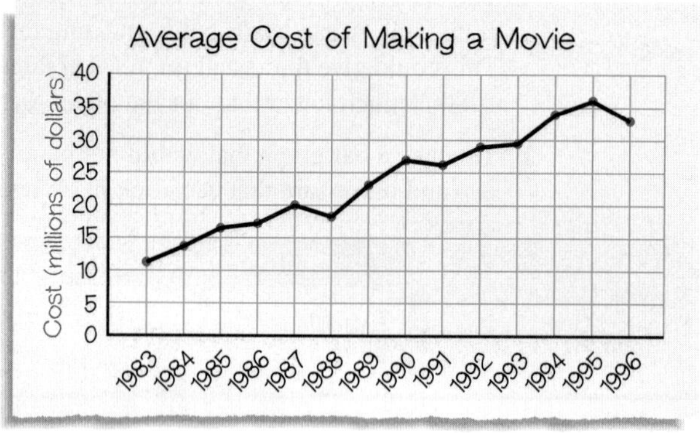

b. In 1988, 1991, and 1996 the average cost of making a movie decreased from the prior year.

c. The greatest decrease in the average cost of making a movie occurred in 1996.

FOCUS ON CAREERS

REAL LIFE **DIRECTOR OF PHOTOGRAPHY**

The composition of the movie shots, type of film, and equipment used are chosen by the director of photography who works closely with the director and producer.

INTERNET CAREER LINK
www.mcdougallittell.com

GUIDED PRACTICE

Vocabulary Check ✓

Concept Check ✓

1. information, facts, or numbers that describe something; sample answer: the heights of all the students in your class

Skill Check ✓

1. Explain what *data* are and give an example. **See margin.**
2. What kind of graph is useful for showing changes over time? **line graph**
3. OLYMPIC EVENTS For a report about the Olympic Games, you want to include the winning times for women running the 100-meter, 200-meter, 400-meter, 800-meter, 1500-meter, and 3000-meter races in 1992, 1996, and 2000. Make a table that you could use to organize the data. **See margin.**

WEATHER **Based on the graph, decide whether each statement is *true* or *false*.**

4. Rainfall increased each month. **false**
5. The amount of rainfall was about the same in May and July. **true**
6. The greatest amount of rainfall occurred in June. **false**
7. The amount of rainfall was the same in January and July. **false**

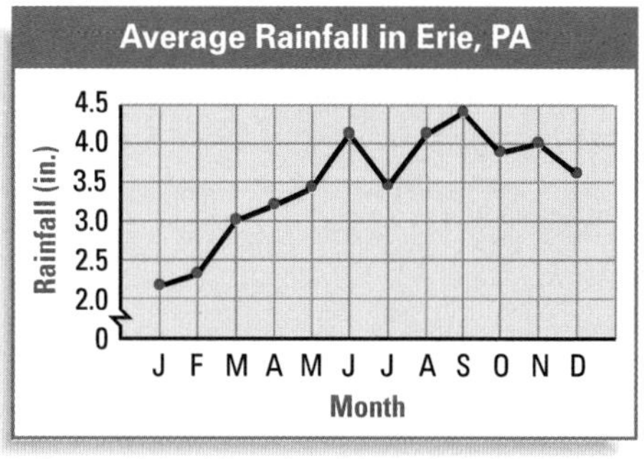

▶ Source: National Oceanic and Atmospheric Administration

PRACTICE AND APPLICATIONS

STUDENT HELP

Extra Practice to help you master skills is on p. 797.

SALARIES **In Exercises 8–10, use the table. It shows average salaries for females of different ages with different numbers of years of education. Based on the table, decide whether each statement is *true* or *false*.**

Age (years)	9–11 years of school	High school graduate	Associate degree	Bachelor's degree
18–24	$2948	$7758	$11,804	$15,245
25–34	$9838	$15,017	$20,835	$25,800
35–44	$11,044	$15,720	$21,807	$26,831
45–54	$11,415	$16,603	$21,944	$27,716

▶ Source: U.S. Bureau of the Census

8. As the years of education increase, the average salary increases. **true**
9. As the years of education increase, the average salary decreases. **false**
10. As the age increases, the average salary increases. **true**
11. WATER NEEDS The table shows the number of gallons of water needed to produce one pound of some foods. Make a bar graph of the data. **See margin.**

Food (1 lb)	lettuce	tomatoes	melons	broccoli	corn
Water (gallons)	21	29	40	42	119

▶ Source: Water Education Foundation

STUDENT HELP

HOMEWORK HELP
Example 1: Exs. 8–10
Example 2: Exs. 11–14
Example 3: Exs. 15–19

See Additional Answers beginning on page AA1.

3 APPLY

ASSIGNMENT GUIDE

BASIC
Day 1: pp. 43–45 Exs. 8–18, 20, 22–28

AVERAGE
Day 1: pp. 43–45 Exs. 8–20, 22–28

ADVANCED
Day 1: pp. 43–45 Exs. 8–28

BLOCK SCHEDULE WITH 1.5
pp. 43–45 Exs. 8–20, 22–28

EXERCISE LEVELS

Level A: ***Easier***
8–17

Level B: ***More Difficult***
18, 19, 22–28

Level C: ***Most Difficult***
20, 21

✓ HOMEWORK CHECK

To quickly check student understanding of key concepts, go over the following exercises: Exs. 8, 10, 14, 18. See also the Daily Homework Quiz:

- Blackline Master (*Chapter 1 Resource Book,* p. 95)
- Transparency (p. 8)

MATHEMATICAL REASONING

EXERCISE 17 Interpreting graphs correctly requires an understanding of what coordinates represent as well as what they are. The difference in the vertical coordinates of consecutive points of a line graph represents the increase or decrease between these points.

3. *Sample answer:*

Times for Women's Olympic Track Events

Year	'92	'96	'00
100 m			
200 m			
400 m			
800 m			
1500 m			
3000 m			

TEACHING TIPS

EXERCISES 12–14 Point out that this is an example of two horizontal bar graphs overlaid on one chart so that the data can be compared.

APPLICATION NOTE

EXERCISE 19 The fuel efficiency of passenger cars has been increasing. This has been due to the use of lighter materials, more aerodynamic designs, and better technology such as multi-port fuel injectors and on-board computers.

18.

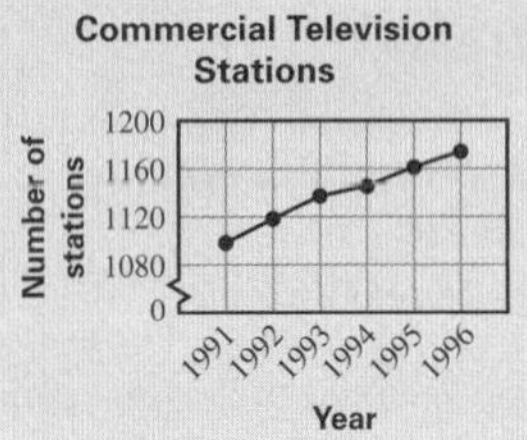

19.

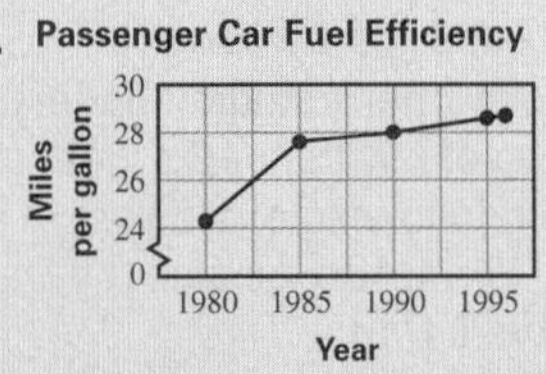

21. ***Sample answer:*** **The graph on the left could be used to support an argument that the number of employees did not change significantly over the six-year period. The graph on the right could be used to exaggerate the change in the number of employees. The graph on the right is misleading because the vertical scale is not consistent.**

ADDITIONAL PRACTICE AND RETEACHING

For Lesson 1.6:

- Practice Levels A, B, and C (*Chapter 1 Resource Book,* p. 84)
- Reteaching with Practice (*Chapter 1 Resource Book,* p. 87)
- See Lesson 1.6 of the *Personal Student Tutor*

For more Mixed Review:

- Search the *Test and Practice Generator* for key words or specific lessons.

MUNICIPAL WASTE In Exercises 12–14, use the double bar graph, which shows the amount of different kinds of waste generated and recycled by city and town dwellers in the United States in 1995.

12–14. Estimates may vary.

12. Which was the most common waste generated by city dwellers? How many millions of tons of this waste were generated? **paper; about 80 million tons**

13. Which kind of waste had the greatest amount recycled? How many millions of tons of this waste were recycled? **paper; about 33 million tons**

14. Which kind of waste had the least amount recycled? How many millions of tons of this waste were recycled? **plastics; about 1 million tons**

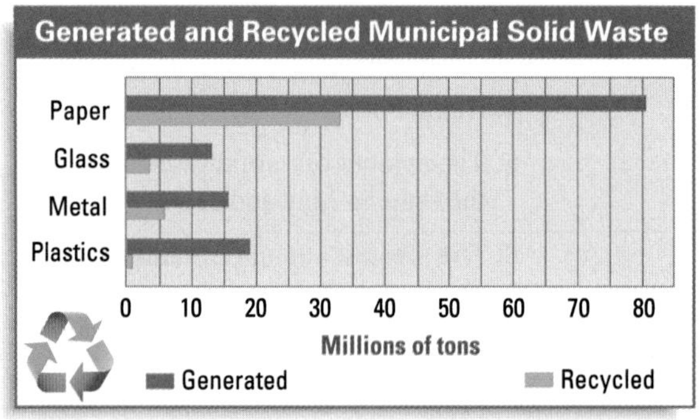

▶ Source: Franklin Associates

MINIMUM WAGE In Exercises 15–17, use the line graph, which shows the minimum wage for different years.

15. For how many years did the minimum wage remain the same? **6 years**

16. What was the minimum wage during the time when it remained the same? **about $4.25**

17. In which year did the minimum wage increase to over $4? **1991**

▶ Source: U.S. Bureau of Labor Statistics

18. *Sample answer:* **The number of television stations increased consistently for the period shown. See margin for graph.**

18. **TELEVISION STATIONS** The table shows the number of commercial television stations for different years. Make a line graph of the data. Discuss what the line graph shows.

Year	1991	1992	1993	1994	1995	1996
Number of stations	1098	1118	1137	1145	1161	1174

▶ Source: *Television Digest*

19. **FUEL EFFICIENCY** The table shows the average fuel efficiency for passenger cars for different years. Make a line graph of the data. **See margin.**

Year	1980	1985	1990	1995	1996
Fuel efficiency (miles per gallon)	24.3	27.6	28.0	28.6	28.7

▶ Source: National Highway Traffic Safety Administration

Test Preparation

20 c. A good answer will include the fact that the standard bar graph is useful for comparing changes for one or more types of data from year to year, or among types of data in a given year, while the stacked bar graph is useful for comparing totals.

20. MULTI-STEP PROBLEM Use the graphs below.

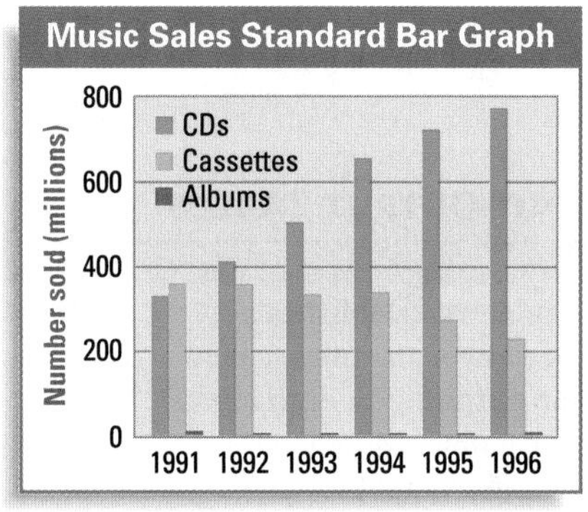

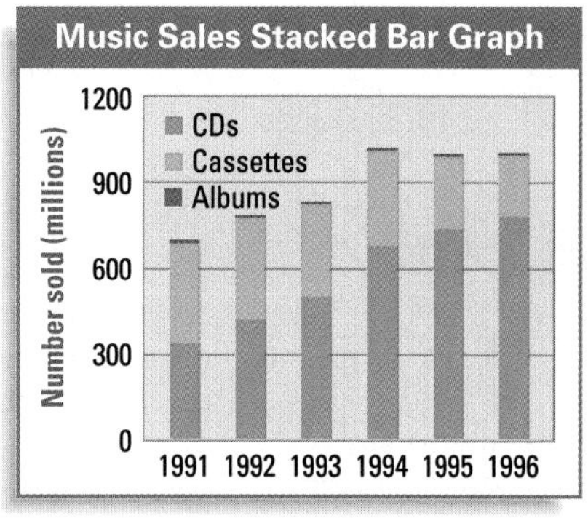

▶ Source: Recording Industry Association of America

a. From 1991 to 1996 which type of recording showed an increase in the number sold? Which showed a decrease? Which remained about the same? **CDs; cassettes; albums**

b. In which two years did the total number of all types of recordings sold remain about the same? **1995 and 1996**

c. CHOOSING A DATA DISPLAY Compare the usefulness of the two graphs.

★ **Challenge**

21. *Writing* The graphs show the same data. Write a statement supported by one graph but *not* the other. Could either graph be misleading? Why? **See margin.**

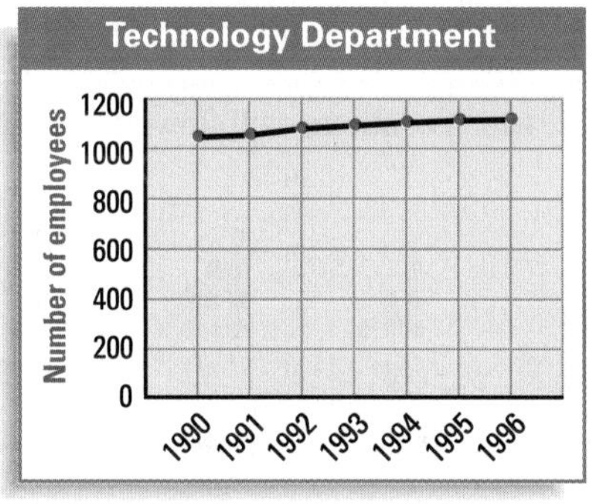

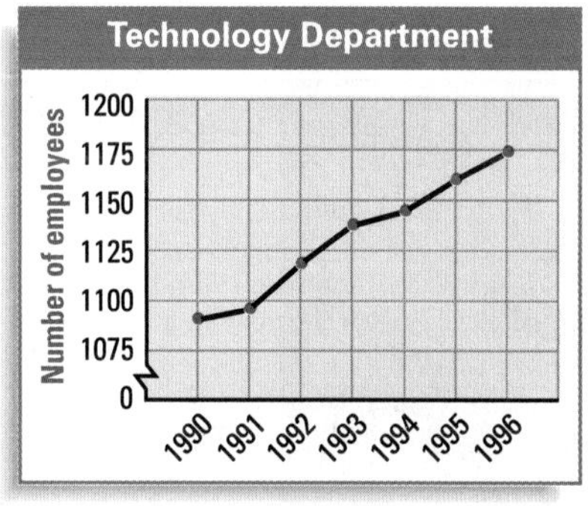

EXTRA CHALLENGE
www.mcdougallittell.com

MIXED REVIEW

SOLUTIONS **Is the number given a solution of the equation?** (Review 1.4 for 1.7)

22. $5x + 2 = 17; 3$ **solution**

23. $12 - 2y = 6; 4$ **not a solution**

24. $3x - 4 = 12 - 5x; 2$ **solution**

25. $2y + 8 = 4y - 2; 5$ **solution**

TRANSLATING **Translate the verbal sentence into an equation.** (Review 1.5)

26. 3 more than a number is 5. $n + 3 = 5$

27. Twelve is the quotient of a number and 3. $12 = \frac{n}{3}$

28. STAMP COLLECTION In your collection of 53 stamps, 37 cost less than \$.25. Let y be the number of stamps that cost \$.25 or more. Which equation models the situation? (Review 1.5) **A**

A. $53 - y = 37$ **B.** $53 + y = 37$ **C.** $53 + 37 = y$

4 ASSESS

DAILY HOMEWORK QUIZ

Transparency Available

1. The number of families in millions with and without children is shown below.

Year	With Children	Without Children
'95	32.6	35.8
'00	33.1	38.6
'05	32.7	42.0

a. Make a bar graph.

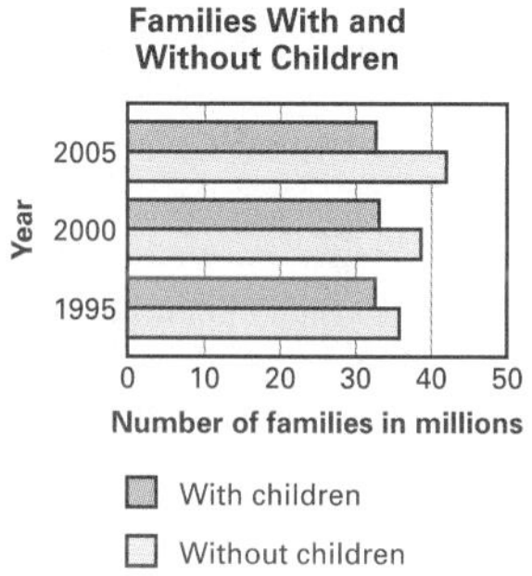

b. Make a line graph showing the number of families with children.

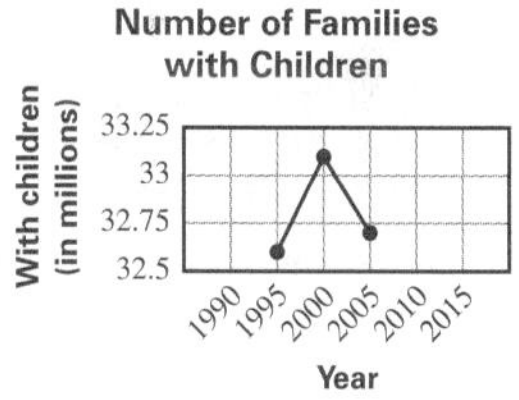

EXTRA CHALLENGE NOTE

Challenge problems for Lesson 1.6 are available in **blackline** format in the *Chapter 1 Resource Book*, p. 92 and at **www.mcdougallittell.com.**

ADDITIONAL TEST PREPARATION

1. WRITING Compare the usefulness of displaying data in a table versus in a graph.
Sample answer: **Both tables and graphs organize the data, but a graph may also give visual cues about how the data are related.**

1 PLAN

PACING

Basic: 1 day
Average: 1 day
Advanced: 1 day
Block Schedule: 0.5 block with Ch. Rev.

LESSON OPENER
VISUAL APPROACH

An alternative way to approach Lesson 1.7 is to use the Visual Approach Lesson Opener:

- Blackline Master (*Chapter 1 Resource Book*, p. 96)
- Transparency (p. 7)

MEETING INDIVIDUAL NEEDS

- ***Chapter 1 Resource Book***
  Prerequisite Skills Review (p. 5)
  Practice Level A (p. 98)
  Practice Level B (p. 99)
  Practice Level C (p. 100)
  Reteaching with Practice (p. 101)
  Absent Student Catch-Up (p. 103)
  Challenge (p. 106)
- ***Resources in Spanish***
- ***Personal Student Tutor***

NEW-TEACHER SUPPORT

See the Tips for New Teachers on pp. 1–2 of the *Chapter 1 Resource Book* for additional notes about Lesson 1.7.

WARM-UP EXERCISES

Transparency Available

Evaluate the expression for the given value of the variable.

1. $12 + 3x$ when $x = 0$ 12
2. $12 + 3x$ when $x = 1$ 15
3. $12 + 3x$ when $x = 2$ 18
4. $12 + 3x$ when $x = 3$ 21
5. Describe a pattern in the answers for Exs. 1–4. *Sample answer:* each increases by 3

1.7 An Introduction to Functions

What you should learn

GOAL 1 Identify a function and make an input-output table for a function.

GOAL 2 Write an equation for a **real-life** function, such as the relationship between water pressure and depth in **Example 3**.

Why you should learn it

▼ To represent **real-life** relationships between two quantities such as time and altitude for a rising hot-air balloon in **Example 2**.

GOAL 1 INPUT-OUTPUT TABLES

A **function** is a rule that establishes a relationship between two quantities, called the **input** and the **output**. For each input, there is exactly one output. More than one input can have the same output.

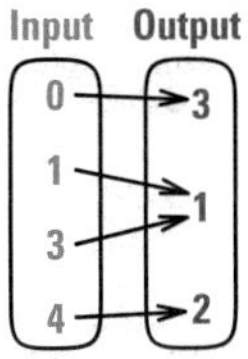

One way to describe a function is to make an **input-output table.** This table lists the inputs and outputs shown in the diagram.

Input	0	1	3	4
Output	3	1	1	2

The collection of all input values is the **domain** of the function and the collection of all output values is the **range** of the function. The table shows that the domain of the function above is 0, 1, 3, 4 and the range is 1, 2, 3.

EXAMPLE 1 *Making an Input-Output Table*

GEOMETRY CONNECTION The diagram shows the first six triangular numbers, 1, 3, 6, 10, 15, 21, which continue on following the same pattern.

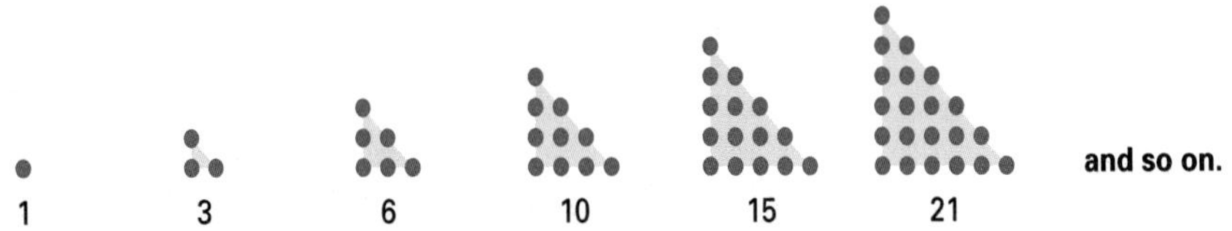

Figure 1 Figure 2 Figure 3 Figure 4 Figure 5 Figure 6 and so on.

a. Make an input-output table in which the input is the Figure number n and the output is the corresponding triangular number T.

b. Does the table represent a function? Justify your answer.

c. Describe the domain and the range.

SOLUTION

a. Use the diagram to make an input-output table.

b. Yes, because for each input there is exactly one output.

c. The collection of all input values is the domain: 1, 2, 3, 4, 5, 6, ... The collection of all output values is the range: 1, 3, 6, 10, 15, 21, ...

Input n	Output T
1	1
2	3
3	6
4	10
5	15
6	21
...	...

EXAMPLE 2 *Using a Table to Graph a Function*

You are at an altitude of 250 feet in a hot-air balloon. You turn on the burner and rise at a rate of 20 feet per minute for 5 minutes. Your altitude h after you have risen for t minutes is given by the function

$h = 250 + 20t$, where $0 \le t \le 5$.

a. For several inputs t, use the function to calculate an output h.

b. Organize the data into an input-output table.

c. Graph the data in the table. Then use this graph to draw a graph of the function.

d. What is the domain of the function? What part of the domain is shown in the table?

SOLUTION

a. List an output for each of several inputs.

INPUT	FUNCTION	OUTPUT
$t = 0$	$h = 250 + 20(0)$	$h = 250$
$t = 1$	$h = 250 + 20(1)$	$h = 270$
$t = 2$	$h = 250 + 20(2)$	$h = 290$
$t = 3$	$h = 250 + 20(3)$	$h = 310$
$t = 4$	$h = 250 + 20(4)$	$h = 330$
$t = 5$	$h = 250 + 20(5)$	$h = 350$

b. Make an input-output table.

t	0	1	2	3	4	5
h	250	270	290	310	330	350

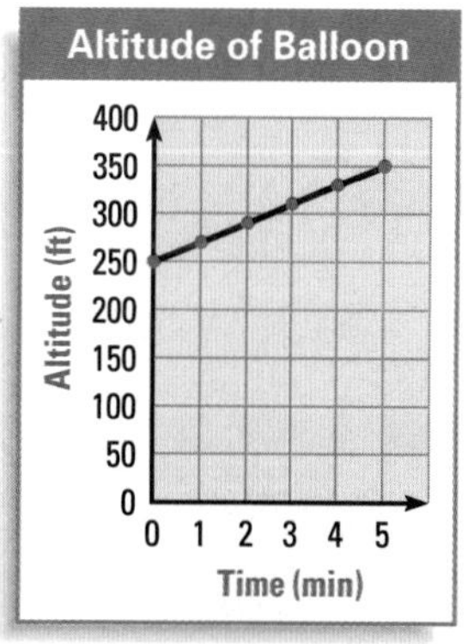

c. The points in blue are the graph of the table data. The black line is the graph of the function.

d. The domain of the function is all values of t such that $0 \le t \le 5$. The values of t shown in the table are 0, 1, 2, 3, 4, 5.

CONCEPT SUMMARY — DESCRIBING A FUNCTION

Example 1 and Example 2 illustrate that functions can be described in these ways:

- Input-Output Table
- Description in Words
- Equation
- Graph

2 TEACH

EXTRA EXAMPLE 1

The profit on the school play is \$4 per ticket minus \$280, the expense to build the set. There are 300 seats in the theater. The profit for n tickets sold is $p = 4n - 280$ for $70 \le n \le 300$.

a. Make an input-output table.

n	70	71	72	73	...	300
p	0	4	8	12	...	920

b. Is this a function? **yes**

c. Describe the domain and range. **Domain: 70, 71, 72, 73, ..., 300; Range: 0, 4, 8, 12, ..., 920**

EXTRA EXAMPLE 2

You bicycle 4 mi and decide to ride for 2.5 more hours at 6 mi/hr. The distance you have traveled d after t hours is given by $d = 4 + 6t$, where $0 \le t \le 2.5$.

a. For several inputs t, calculate an output h.
Sample answers:
$t = 0$, $h = 4$; $t = 0.5$, $h = 7$; $t = 1$, $h = 10$; $t = 1.5$, $h = 13$; $t = 2$, $h = 16$; $t = 2.5$, $h = 19$

b. Make an input-output table.

t	0	0.5	1	1.5	2	2.5
d	4	7	10	13	16	19

c. Graph the data, and then draw a graph of the function.

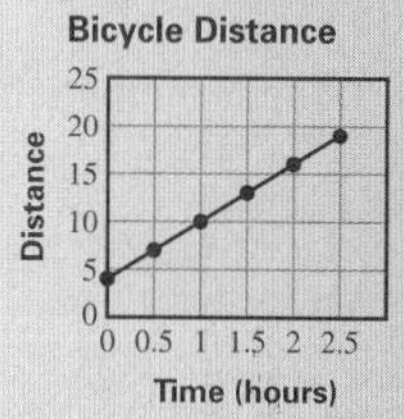

MULTIPLE REPRESENTATIONS

Functions can be represented in words, in symbols, with a table, and with a graph. The focus of Lesson 1.7 is to help students move comfortably among all four representations, and give students a variety of ways to model real-life situations.

Checkpoint Exercises on next page.

✓ CHECKPOINT EXERCISES

For use after Examples 1 and 2:

1. A plane is at 2000 ft. It climbs at a rate of 1000 ft/min for 4 minutes. The altitude h for t minutes is given by $h = 2000 + 1000t$ for $0 \le t \le 4$.

a. Make a table.

t	0	1	2	3	4
d	2000	3000	4000	5000	6000

b. Draw a line graph.

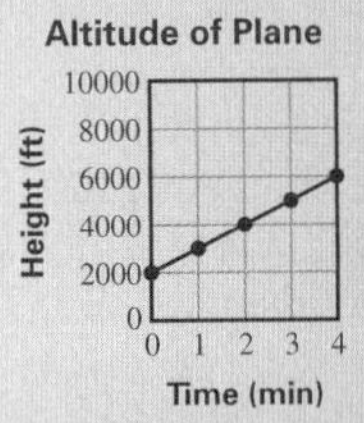

c. Describe the domain and range. **domain: all numbers between and including 0 and 4; range: all numbers between and including 2000 and 6000**

EXTRA EXAMPLE 3

An internet service provider charges $9.00 for the first 10 hours and $.95 per hour for any hours above 10 hours. Represent the cost c as a function of the number of hours h after the first 10 hours.

a. Write an equation.

$c = 9 + 0.95h$ where $h \ge 0$

b. Create an input-output table.

h	c
0	9
1	9.95
2	10.90
3	11.85
...	...

c. Make a line graph.

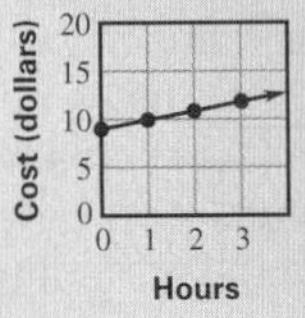

Checkpoint Exercises on next page.

GOAL 2 WRITING EQUATIONS FOR FUNCTIONS

EXAMPLE 3 *Writing an Equation*

SCUBA DIVING As you dive deeper and deeper into the ocean, the pressure of the water on your body steadily increases. The pressure at the surface of the water is 14.7 pounds per square inch (psi). The pressure increases at a rate of 0.445 psi for each foot you descend. Represent the pressure P as a function of the depth d for every 20 feet you descend until you reach a depth of 120 feet.

a. Write an equation for the function.

b. Create an input-output table for the function.

c. Make a line graph of the function.

SOLUTION

a. To write an equation for the function, apply the problem solving strategy you learned in Lesson 1.5. Write a verbal model, assign labels, and write an algebraic model.

PROBLEM SOLVING STRATEGY

VERBAL MODEL **Pressure at given depth** = **Pressure at surface** + **Rate of change in pressure** · **Diving depth**

LABELS

Pressure at given depth = P	(psi)
Pressure at surface = **14.7**	(psi)
Rate of change in pressure = **0.445**	(psi per foot of depth)
Diving depth = d	(feet)

ALGEBRAIC MODEL $P = \mathbf{14.7} + \mathbf{0.445}\, d$ where $0 \le d \le 120$

b. Use the equation to make an input-output table for the function. For inputs, use the depths for every 20 feet you descend down to 120 feet.

Depth d	Pressure P
0	14.7
20	23.6
40	32.5
60	41.4
80	50.3
100	59.2
120	68.1

c. You can represent the data in the table with a line graph.

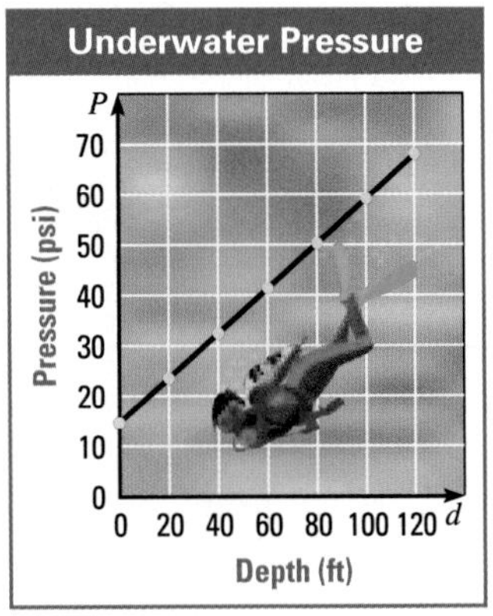

FOCUS ON CAREERS

REAL LIFE SCUBA DIVING INSTRUCTOR
Physics and physiology are very important to scuba diving instructors teaching diver safety.

INTERNET CAREER LINK
www.mcdougallittell.com

GUIDED PRACTICE

Vocabulary Check ✓

Complete the sentence.

1. A function is a relationship between two quantities, called the _?_ and the _?_. **input; output**
2. The collection of all input values is the _?_ of the function. The collection of all output values is the _?_ of the function. **domain; range**

Concept Check ✓

3. Four ways to represent a function are (1) _?_, (2) _?_, (3) _?_, and (4) _?_. **words; input-output table; equation; graph**

For a relationship to be a function, it must be true that for each input, there is exactly one output. Does the table represent a function? Explain. See margin.

4.

Input	Output
1	3
2	4
3	5
4	6

5.

Input	Output
1	3
2	3
3	4
4	4

6.

Input	Output
1	3
1	4
2	5
3	6

4. Yes; for each input, there is exactly one output.
5. Yes; for each input, there is exactly one output.
6. No; there are two output values that correspond to the input value 1.

Skill Check ✓

CAR RACING **The fastest winning speed in the Daytona 500 is about 178 miles per hour. Calculate the distance traveled d (in miles) after time t (in hours) using the equation $d = 178t$.** ▶Source: NASCAR

7. Copy and complete the input-output table. **44.5; 89; 133.5; 178; 222.5; 267**

Time (hours)	0.25	0.50	0.75	1.00	1.25	1.50
Distance traveled (miles)	?	?	?	?	?	?

8. Describe the domain and the range of the function whose values are shown in the table. **domain: 0.25, 0.50, 0.75, 1.00, 1.25, 1.50; range: 44.5, 89, 133.5, 178, 222.5, 267**
9. Graph the data in the table. Use this graph and the fact that when $t = 0$, $d = 0$ and when $t = 2.81$, $d = 500$ to graph the function. See margin.

PRACTICE AND APPLICATIONS

STUDENT HELP

Extra Practice to help you master skills is on p. 797.

CRITICAL THINKING **Does the table represent a function? Explain.** 10–12. See margin.

10.

Input	Output
5	3
6	4
7	5
8	6

11.

Input	Output
1	3
2	6
3	11
4	18

12.

Input	Output
9	5
9	4
8	3
7	2

10. Yes; for each input, there is exactly one output.
11. Yes; for each input, there is exactly one output.
12. No; there are two output values that correspond to the input value 9.

STUDENT HELP

HOMEWORK HELP
Example 1: Exs. 10–25
Example 2: Exs. 26–32
Example 3: Exs. 33–36

INPUT-OUTPUT TABLES **Make an input-output table for the function. Use 0, 1, 2, and 3 as the domain.**

13. $y = 3x + 2$ Output: 2, 5, 8, 11
14. $y = 21 - 2x$ Output: 21, 19, 17, 15
15. $y = 5x$ Output: 0, 5, 10, 15
16. $y = 6x + 1$ Output: 1, 7, 13, 19
17. $y = 2x + 1$ Output: 1, 3, 5, 7
18. $y = x + 4$ Output: 4, 5, 6, 7

CHECKPOINT EXERCISES

For use after Example 3:

1. The temperature at 6:00 A.M. was 62°F and rose 3°F every hour until 9:00 A.M. Represent the temperature T as a function of the number of hours h.
   a. Write an equation. $T = 62 + 3h$ where $0 \le h \le 3$
   b. Make an input-output table.

h	0	1	2	3
T	62	65	68	71

   c. Make a line graph.

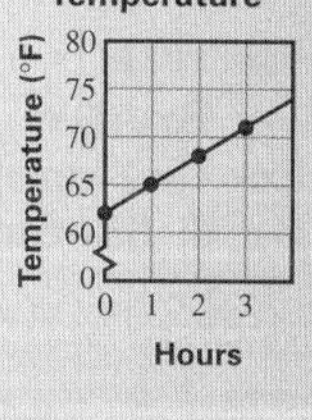

FOCUS ON VOCABULARY

Name four different ways to represent a function. **words, equation, input-output, graph**

CLOSURE QUESTION

Describe the relationship between the domain and the range of a function. **The domain is the set of input values that go into a function which result in the range or set of all output values. For each domain value there is exactly one range value.**

DAILY PUZZLER

Find the next three numbers in the pattern: 10, 12, 15, 19, 24, 30, ... **37, 45, 54**

9.

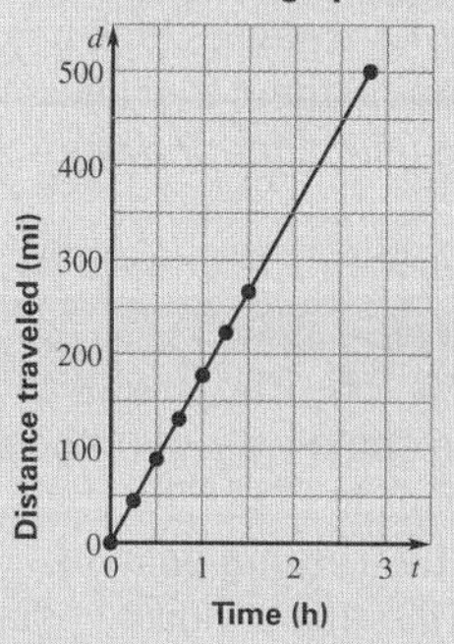

3 APPLY

ASSIGNMENT GUIDE

BASIC
Day 1: pp. 49–52 Exs. 10–24 even, 25, 37–42, 46–58 even, Quiz 3 Exs. 1–5

AVERAGE
Day 1: pp. 49–52 Exs. 10–24 even, 25, 33–36, 39–42, 46–58 even, Quiz 3 Exs. 1–5

ADVANCED
Day 1: pp. 49–52 Exs. 10–26 even, 33–44, 46–58 even, Quiz 3 Exs. 1–5

BLOCK SCHEDULE (WITH CH. 1 REVIEW)
pp. 49–52 Exs. 10–24 even, 25, 33–36, 39–42, 46–58 even, Quiz 3 Exs. 1–5

EXERCISE LEVELS

Level A: *Easier*
10–18, 45–48

Level B: *More Difficult*
19–27, 30–42, 49–59, Quiz 1–5

Level C: *Most Difficult*
28, 29, 43, 44

✔ HOMEWORK CHECK

To quickly check student understanding of key concepts, go over the following exercises: Exs. 10, 12, 14, 20, 25, 37, 38. See also the Daily Homework Quiz:

- Blackline Master (*Chapter 2 Resource Book*, p. 11)
- Transparency (p. 10)

STUDENT HELP NOTES

Look Back As students look back to p. 42, remind them that line graphs are useful for showing changes in data over time.

APPLICATION NOTE

EXERCISES 26–29 The Celsius temperature scale was invented by Anders Celsius in 1742. The scale uses 0° as the freezing point of water and 100° as the boiling point.

MAKING INPUT-OUTPUT TABLES Make an input-output table for the function. Use 1, 1.5, 3, 4.5, and 6 as the domain. 19–24. See margin.

19. Output: 6.5, 8.5, 14.5, 20.5, 26.5
20. Output: 29, 27.5, 23, 18.5, 14
21. Output: 19, 16, 13, 12, 11.5
22. Output: 4, 5, 8, 11, 14
23. Output: 0.5, 1.75, 8.5, 19.75, 35.5
24. Output: 2.5, 3.75, 10.5, 21.75, 37.5

19. $y = 4x + 2.5$ **20.** $y = 32 - 3x$ **21.** $y = \frac{9}{x} + 10$

22. $y = 2 + \frac{x}{0.5}$ **23.** $y = x^2 - 0.5$ **24.** $y = 1.5 + x^2$

25. APPLE TREES A large apple tree may absorb 360 liters of water from the soil per day. The amount of water W absorbed over a short period of time is modeled by the function $W = 360d$, where d represents the number of days. Copy and complete the table.
360 · 2, 720; 360 · 3, 1080; 360 · 4, 1440; 360 · 5; 1800

Input	Function	Output
$d = 1$	$W = 360 \cdot 1$	$W = 360$
$d = 2$	$W = ?$	$W = ?$
$d = 3$	$W = ?$	$W = ?$
$d = 4$	$W = ?$	$W = ?$
$d = 5$	$W = ?$	$W = ?$

TEMPERATURE **In Exercises 26–29, use the following information.**
You use a Celsius thermometer when measuring temperature in science class. Your teacher asks you to show the relationship between Celsius temperature and Fahrenheit temperature with a graph. Use the equation $F = \frac{9}{5}C + 32$ to convert degrees Celsius C to degrees Fahrenheit F.

Input	Function	Output
$C = 0$	$F = \frac{9}{5}(0) + 32$	$F = 32$
$C = 5$	$F = ?$	$F = ?$
$C = 10$	$F = ?$	$F = ?$
$C = 15$	$F = ?$	$F = ?$
$C = 20$	$F = ?$	$F = ?$

26. Copy and complete the input-output table for the function.
41; 50; 59; 68; 77; 86; 95; 104

Input C	0	5	10	15	20	25	30	35	40
Output F	32	?	?	?	?	?	?	?	?

27. Graph the data in the table. Label the vertical axis Temperature (°C) and the horizontal axis Temperature (°F). Use this graph to graph the function. See margin.

STUDENT HELP

Look Back
For help with drawing line graphs, see p. 42.

28. *Writing* Explain why $F = \frac{9}{5}C + 32$ represents a function.
For each Celsius temperature, there is exactly one Fahrenheit temperature.

29. Describe the domain and the range of the function whose values are shown in the table.
domain: 0, 5, 10, 15, 20, 25, 30, 35, 40; range: 32, 41, 50, 59, 68, 77, 86, 95, 104

DRIVE-IN MOVIE **In Exercises 30–32, use the following information.**
A drive-in movie theater charges $5 admission per car and $1.25 admission for each person in the car. The total cost C at the drive-in movie theater is given by $C = 5 + 1.25n$, where n represents the number of people in the car. (Assume a maximum of 7 people per car.)

30. Make an input-output table for the function.
Input: 1, 2, 3, 4, 5, 6, 7; Output: 6.25, 7.50, 8.75, 10.00, 11.25, 12.50, 13.75

31. Describe the domain and range of the function whose values are shown in the table. domain: 1, 2, 3, 4, 5, 6, 7; range: 6.25, 7.50, 8.75, 10.00, 11.25, 12.50, 13.75

32. Graph the data shown in the input-output table. See margin.

STUDENT HELP

HOMEWORK HELP Visit our Web site www.mcdougallittell.com for help with problem solving in Exs. 33–36.

ADVERTISING In Exercises 33–36, use the following information. You are in charge of advertising for the drama club's next performance. You can make two signs from a poster board that is 36 inches long and 24 inches wide. Each poster board costs \$.75.

33. Write an equation that shows the relationship between the number of signs and the total cost of the poster board. Let n represent the number of signs and let C represent the total cost of the poster board. **$C = 0.375n$**

34. Evaluate the equation for $n = 6, 8, 10,$ and 12. Organize your results in an input-output table. **2.25; 3.00; 3.75; 4.50**

35. What do the values in the table represent? **See margin.**

35. Input values represent the number of posters; corresponding output numbers represent the cost of the posters.

36. Suppose you have \$5 to spend on poster board. Do you have enough money to make 14 signs? If not, how much more money do you need? **No; you need \$.25 more.**

HOUSESITTING In Exercises 37 and 38, use the following information. As a summer job, you start a housesitting service in your neighborhood. You agree to get the mail, pick up newspapers, water plants, and feed pets for an initial fee of \$5, plus \$2 per day.

37. Write an equation that shows the relationship between the number of days you housesit and the amount of money you earn for each house. **Let n = the number of days, and a = the amount in dollars that you earn; $a = 2n + 5$.**

38. How much money will you earn housesitting one house for one week? **\$19**

Test Preparation

MULTI-STEP PROBLEM In Exercises 39–43, use the following information. You start a portable catering business. One of your specialties is a barbecue sandwich plate that costs \$.85 to prepare. Suppose you cater an auction where you sell each sandwich plate for \$2.00.

39. Write a function that gives the profit you expect from catering the auction.

39. Let n = the number of sandwiches they prepare, and p = the profit in dollars; $p = 1.15n$.

40. You must also spend \$50 on equipment and supplies to cater the auction. Write a function that includes this cost. **$p = 1.15n - 50$**

41. Use your equation from Exercise 40 to find the profit you will earn if you sell 75 barbecue sandwich plates. **\$36.25**

42. Suppose the cost of sandwich rolls increased by \$.05 each. What effect do you think this will have on the profit? Write a function that includes this cost. **The profit per sandwich will decrease by \$.05; $p = 1.10n - 50$.**

43. Use your equation from Exercise 42 to find how many barbecue sandwich plates you must sell to make a \$100 profit. **137 barbecue sandwich plates**

★ Challenge

44. **PERSONAL RECREATION** The amount of money spent on personal recreation in billions of dollars M can be modeled by $M = 116.3 + 28.5t$, where t is the number of years since 1985. The bar graph shows the amount of money spent on personal recreation in 1985, 1990, and 1995.

Use a calculator to predict the amount of money spent in billions of dollars in the years 2000 and 2005. **\$543.8 billion, \$686.3 billion**

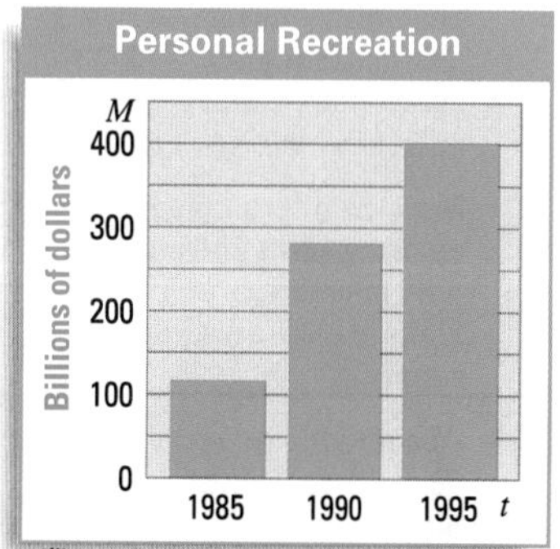

DATA UPDATE U.S. Bureau of Economic Analysis at www.mcdougallittell.com

EXTRA CHALLENGE www.mcdougallittell.com

STUDENT HELP NOTES

Homework Help Students can find help for Exs. 33–36 at **www.mcdougallittell.com.** The information can be printed out for students who do not have access to the Internet.

! COMMON ERROR

EXERCISE 42 Some students may add 0.05 to 1.15 and write the new profit function as $p = 1.20n - 50$. Remind students that as costs go up, profit goes down and that 0.05 should be subtracted from 1.15.

27.

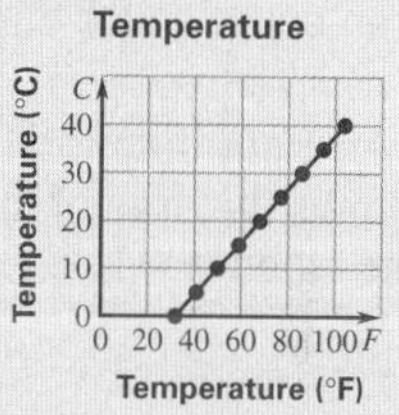

32.

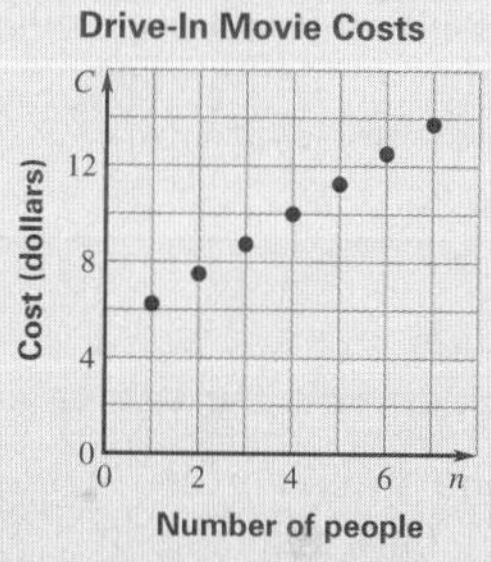

ADDITIONAL PRACTICE AND RETEACHING

For Lesson 1.7:

- Practice Levels A, B, and C (*Chapter 1 Resource Book*, p. 98)
- Reteaching with Practice (*Chapter 1 Resource Book*, p. 101)
- See Lesson 1.7 of the *Personal Student Tutor*

For more Mixed Review:

- Search the *Test and Practice Generator* for key words or specific lessons.

4 ASSESS

DAILY HOMEWORK QUIZ

Transparency Available

1. Does the table represent a function? **Yes**

Input	0	1	3	5	9
Output	3	7	15	23	39

2. Make an input-output table for $y = x^2 + 2.5$. Use 0, 1, 2, and 3 as the domain. **x: 0, 1, 2, 3; y: 2.5, 3.5, 6.5, 11.5**
3. You have carpet remnants that have dimensions of w and $3w$.
   - **a.** Write an equation for the area. **$A = 3w^2$**
   - **b.** Make an input-output table that shows the area of the carpets when w = 1, 2, 3, 4, and 5 ft. **w: 1, 2, 3, 4, 5; A: 3, 12, 27, 48, 75**
   - **c.** Describe the domain and range of the function. **domain: 1, 2, 3, 4, 5; range: 3, 12, 27, 48, 75**

EXTRA CHALLENGE NOTE

Challenge problems for Lesson 1.7 are available in **blackline** format in the *Chapter 1 Resource Book,* p. 106 and at **www.mcdougallittell.com.**

ADDITIONAL TEST PREPARATION

1. **OPEN ENDED** Make up your own example of a function in words. Then give its equation and describe its domain and range. ***Sample answer:* Bill can type 48 words per minute; $w = 48m$; domain is $m \geq 0$; range is $w \geq 0$.**

1. Arts Activity Attendance

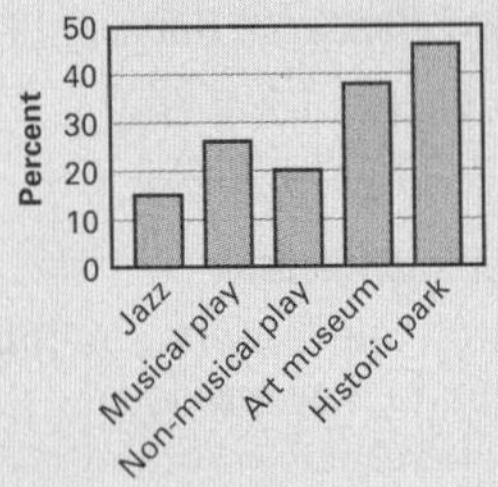

2. Conclusions may include the following:
   - Nearly half of all 18-to-24-year-olds attended a historic park.
   - More of the 18-to-24-year-olds attended a musical play than attended a non-musical play.
   - The events in increasing order of popularity are jazz, non-musical play, musical play, art museum, historic park.
   - More than three times as many 18-to-24-year-olds attended a historic park as a jazz event.
   - About twice as many 18-to-24-year-olds attended an art museum as attended a non-musical play.

MIXED REVIEW

Write the expression in exponential form. (Review 1.2)

45. x raised to the sixth power **x^6**

46. nine cubed **9^3**

47. $y \cdot y \cdot y \cdot y \cdot y \cdot y \cdot y$ **y^7**

48. $15 \cdot 15 \cdot 15 \cdot 15$ **15^4**

Evaluate the expression. (Review 1.3)

49. $\frac{3b + 3c}{5}$ when $b = 1$ and $c = 9$ **6**

50. $6 + \frac{x + 2}{y}$ when $x = 7$ and $y = 3$ **9**

51. $3y^2 + w$ when $y = 8$ and $w = 27$ **219**

52. $2s^3 - 3t$ when $s = 4$ and $t = 6$ **110**

Check whether the number is a solution of the equation or the inequality. (Review 1.4 for 2.1)

53. $7y + 2 = 4y + 8$; 2 **solution**

54. $5x - 1 = 3x + 2$; 4 **not a solution**

55. $32 - 5y > 11$; 3 **solution**

56. $7m + 2 < 12$; 1 **solution**

57. $s - 7 \geq 12 - s$; 9 **not a solution**

58. $n + 2 \leq 2n - 2$; 4 **solution**

59. SHOE SALES A major athletic footwear company had about \$3750 million in sales of athletic footwear during the year ending May 31, 1997. The company's sales fell to about \$3500 million in 1998. By how much did the company's sales decrease? (Review 1.5) **\$250 million**

QUIZ 3

Self-Test for Lessons 1.6 and 1.7

ARTS ACTIVITIES **In Exercises 1 and 2, use the table which shows the percent of 18-to-24-year-olds that attended various arts activities at least once in 1997.** (Lesson 1.6)

Arts Activities Attended by 18-to-24-year-olds				
Jazz	Musical play	Non-musical play	Art museum	Historic park
15%	26%	20%	38%	46%

▶ Source: Statistical Abstract of the United States, 1998

1. Make a bar graph of the data. **See margin.**
2. What conclusions can you draw from the bar graph? **See margin.**
3. Which equation describes the function that contains all of the data points shown in the table? (Lesson 1.7) **B**

Input x	0	1	2	3	4	5
Output y	1	2	5	10	17	26

A. $y = x^2$ **B.** $y = x^2 + 1$ **C.** $y = 2x + 3$ **D.** $y = x^2 - 1$

HOT-AIR BALLOONING **You are at an altitude of 200 feet in a hot-air balloon. You turn on the burner and rise at a rate of 25 feet per minute for 6 minutes.** (Lesson 1.7)

4. Write an equation for your altitude as a function of time. **Let t = the number of minutes after you turn on the heat and a = your altitude; $a = 25t + 200$.**
5. Make an input-output table for the function. **Input: 0, 1, 2, 3, 4, 5, 6; Output: 200, 225, 250, 275, 300, 325, 350**

CHAPTER 1

Chapter Summary

WHAT did you learn?	*WHY did you learn it?*
Evaluate and write variable expressions. (1.1)	Estimate the time it will take you to hike the length of a trail and back. (p. 5)
Evaluate and write expressions containing exponents. (1.2)	Calculate the volume of water that a cubic aquarium holds. (p. 11)
Use the order of operations. (1.3)	Calculate sales tax on a purchase. (p. 18)
Use mental math to solve equations. (1.4)	Estimate the amount you owe for groceries. (p. 25)
Check solutions of inequalities. (1.4)	Check a pet's caloric intake. (p. 26)
Translate verbal phrases and sentences into expressions, equations, and inequalities. (1.5)	Calculate the total number of dim sum plates ordered at a restaurant. (p. 33)
Translate verbal models into algebraic models to solve problems. (1.5)	Model the decision-making of a commercial jet pilot. (p. 34)
Organize data using tables and graphs. (1.6)	Interpret information about eating habits in the United States. (pp. 40 and 41)
Use functions to show the relationship between inputs and outputs. (1.7)	Represent the rise of a hot-air balloon as a function. (p. 47)

How does Chapter 1 fit into the BIGGER PICTURE of algebra?

In this chapter you were introduced to many of the terms and goals of algebra. Communication is a very important part of algebra, so it is important that algebraic terms become part of your vocabulary. Algebra is a language that you can use to solve real-life problems.

STUDY STRATEGY

How did you use the notes in your notebook?

The notes you made, using the **Study Strategy** on page 2, may include this one about order of operations.

Remembering Order of Operations

1. First do operations that occur within grouping symbols.
2. Then evaluate powers.
3. Then do multiplications and divisions from left to right.
4. Finally, do additions and subtractions from left to right.

53

ADDITIONAL RESOURCES

The following resources are available to help review the material in this chapter.

- Chapter Review Games and Activities (*Chapter 1 Resource Book*, p. 107)
- *Instant Replay: Video Review Games*
- *Personal Student Tutor*
- Cumulative Review, Ch. 1 (*Chapter 1 Resource Book*, p. 119)

CHAPTER 1

Chapter Review

VOCABULARY

- variable, p. 3
- values, p. 3
- variable expression, p. 3
- evaluating the expression, p. 3
- unit analysis, p. 5
- verbal model, p. 5
- power, p. 9
- exponent, p. 9
- base, p. 9
- grouping symbols, p. 10
- order of operations, p. 16
- equation, p. 24
- open sentence, p. 24
- solution of an equation, p. 24
- solving the equation, p. 25
- inequality, p. 26
- solution of an inequality, p. 26
- modeling, p. 33
- mathematical model, p. 33
- data, p. 40
- bar graph, p. 41
- line graph, p. 42
- function, p. 46
- input, p. 46
- output, p. 46
- input-output table, p. 46
- domain, p. 47
- range, p. 47

1.1 VARIABLES IN ALGEBRA

Examples on pp. 3–5

EXAMPLES Evaluate the expression when $y = 4$.

$10 - y = 10 - 4$
$= 6$

$11y = 11(4)$
$= 44$

$\frac{16}{y} = \frac{16}{4}$
$= 4$

$y + 9 = 4 + 9$
$= 13$

Evaluate the expression for the given value of the variable.

1. $a + 14$ when $a = 23$ **37**

2. $1.8x$ when $x = 10$ **18**

3. $\frac{m}{1.5}$ when $m = 15$ **10**

4. $\frac{15}{y}$ when $y = 7.5$ **2**

5. $p - 12$ when $p = 22$ **10**

6. $b(0.5)$ when $b = 9$ **4.5**

7. How long will it take to walk 6 miles if you walk at a rate of 3 miles per hour? **2 h**

1.2 EXPONENTS AND POWERS

Examples on pp. 9–11

EXAMPLES Evaluate the expression when $b = 3$.

$b^2 = 3^2$
$= 3 \cdot 3$
$= 9$

$(10 - b)^3 = (10 - 3)^3$
$= 7^3$
$= 7 \cdot 7 \cdot 7$
$= 343$

$12(5^b) = 12(5^3)$
$= 12(5 \cdot 5 \cdot 5)$
$= 12(125)$
$= 1500$

$b^4 + 18 = 3^4 + 18$
$= (3 \cdot 3 \cdot 3 \cdot 3) + 18$
$= 81 + 18$
$= 99$

Evaluate the expression.

8. eight to the fourth power **4096**

9. $(2 + 3)^5$ **3125**

10. s^2 when $s = 1.5$ **2.25**

11. $6 + (b^3)$ when $b = 3$ **33**

12. $2x^4$ when $x = 2$ **32**

13. $(5x)^3$ when $x = 5$ **15,625**

1.3 ORDER OF OPERATIONS

Examples on pp. 16–18

EXAMPLE Evaluate $550 - 4(3 + 5)^2$.

$550 - 4(3 + 5)^2 = 550 - 4(8)^2$ **Evaluate within grouping symbols.**

$= 550 - 4 \cdot 64$ **Evaluate powers.**

$= 550 - 256$ **Multiply or divide.**

$= 294$ **Add or subtract.**

Evaluate the expression.

14. $4 + 21 \div 3 - 3^2$ **2**

15. $(14 \div 7)^2 + 5$ **9**

16. $\frac{6 + 2^2}{17 - 6 \cdot 2}$ **2**

17. $\frac{x - 3y}{6}$ when $x = 15$ and $y = 2$ **1.5**

1.4 EQUATIONS AND INEQUALITIES

Examples on pp. 24–26

EXAMPLE You can check whether the number 4 is a solution of $5x + 3 = 18$.

4 is not a solution, because $5(4) + 3 \neq 18$.

Check whether the given number is a solution of the equation or inequality.

18. $2a - 3 = 2; 4$ **not a solution**

19. $x^2 - x = 2; 2$ **solution**

20. $9y - 3 > 24; 3$ **not a solution**

21. $5x + 2 \leq 27; 5$ **solution**

1.5 A PROBLEM SOLVING PLAN USING MODELS

Examples on pp. 32–34

EXAMPLE You can model problems like the following: If you can save \$5.25 a week, how many weeks must you save to buy a CD that costs \$15.75?

VERBAL MODEL Number of weeks = $\frac{\text{Cost of CD}}{\text{Amount saved per week}}$

LABELS Number of weeks = w (weeks)

Cost of CD = **15.75** (dollars)

Amount saved per week = **5.25** (dollars per week)

ALGEBRAIC MODEL $w = \frac{15.75}{5.25}$ **Write algebraic model**

$w = 3$ **Solve with mental math**

▶ You must save for 3 weeks.

22. You are given \$75 to buy juice for the school dance. Each bottle of juice costs \$.75. Write a verbal and an algebraic model to find how many bottles of juice you can buy. Write an equation and use mental math to solve the equation. **Number of bottles times price per bottle = \$75; let n = the number of bottles of juice; $0.75n = 75$; 100 bottles.**

25. 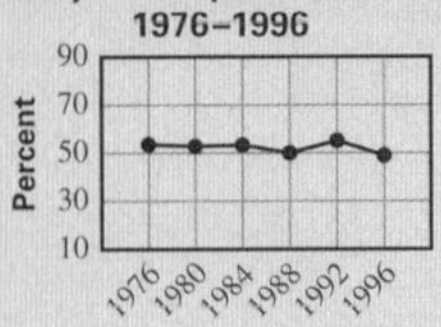

1.6 TABLES AND GRAPHS

Examples on pp. 40–42

EXAMPLES Using the bar graph you can tell the following. From the vertical axis of the graph, you can see that male and female tennis players from the United States both have won the Australian Open 14 times.

You can also see that United States men have won the French Open 10 times, while United States women have won 25 times. So United States women have won 15 more French Open titles than the United States men.

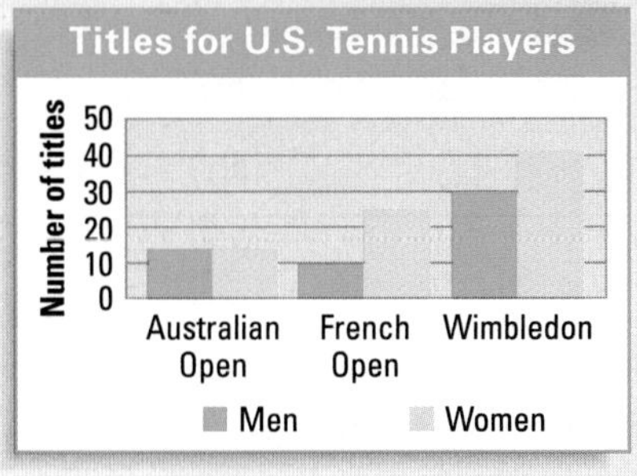

▶ Source: USA Today

In Exercises 23 and 24, use the graph above.

23. Compare the number of titles won by United States women at Wimbledon with the number won by United States men.

24. Find the total number of titles won by United States men and compare with the total for women. **total for men: 54; total for women: 80**

23. ***Sample answer:*** **The women have won 41 Wimbledon titles, while the men have won 30, so the women have won 11 more Wimbledon titles than the men have won.**

Use the data in the table.

25. Make a line graph of the data. **See margin.**

26. What can you conclude from the line graph?

Percent of Voting-Age Population That Voted in Yearly Municipal Referendum, 1976–1996						
Year	1976	1980	1984	1988	1992	1996
Percent	53.5	52.8	53.3	50.3	55.1	48.9

Sample answer: **The percent of the voting age population that voted from 1976 to 1996 varied from just below 50% to just over 55%. The percent did not increase or decrease steadily but rose and fell during that period.**

1.7 AN INTRODUCTION TO FUNCTIONS

Examples on pp. 46–48

EXAMPLES Line for your fishing reel costs $.02 per yard. One lure costs $3.50. Make an input-output table that shows the total cost of buying one lure and 100, 200, 300, or 400 yards of fishing line. The equation is $C = .02n + 3.50$, where n is the number of yards of fishing line. To find C Substitute the given values of n.

Fishing line (yards), n	100	200	300	400
Total (dollars), C	5.50	7.50	9.50	11.50

In Exercises 27–29, you are buying rectangular picture frames that have side lengths of 2*w* and 3*w*.

27. Write an equation for the perimeter, starting with a verbal model. **Perimeter = Length + Width + Length + Width; $P = 10w$**

28. Make an input-output table that shows the perimeter of the frames when $w = 1, 2, 3, 4,$ and 5.

29. Describe the domain and range of the function whose values are shown in the table. **domain = 1, 2, 3, 4, 5; range = 10, 20, 30, 40, 50**

28.

Input	Output
1	10
2	20
3	30
4	40
5	50

CHAPTER 1

Chapter Test

Evaluate the expression when $y = 3$ and $x = 5$.

1. $5y + x^2$ **40**
2. $\frac{24}{y} - x$ **3**
3. $2y + 9x - 7$ **44**
4. $(5y + x) \div 4$ **5**
5. $2x^3 + 4y$ **262**
6. $8(x^2) \div 25$ **8**
7. $(x - y)^3$ **8**
8. $x^4 + 4(y - 2)$ **629**

In Exercises 9–11, write the expression in exponential form.

9. $5y \cdot 5y \cdot 5y \cdot 5y$ **$(5y)^4$**
10. nine cubed **9^3**
11. six to the nth power **6^n**

12. Insert grouping symbols in $5 \cdot 4 + 6 \div 2$ so that the value of the expression is 25. **$5(4 + 6) \div 2$**

13. TRAVEL TIME If you can travel only 35 miles per hour, is $2\frac{1}{2}$ hours enough time to get to a concert that is 85 miles away? Give the expression you used to find the answer. **yes; $2\frac{1}{2}(35) = 87\frac{1}{2}$**

Write an algebraic expression.

14. seven times a number **$7n$**
15. x is at least ninety **$x \ge 90$**
16. the quotient of m and two **$\frac{m}{2}$ or $m \div 2$**

In Exercises 17–22, decide whether the statement is *true* or *false*.

17. $(2 \cdot 3)^2 = 2 \cdot 3^2$ **false**
18. quotient of 3 and 12 is 4 **false**
19. $8 - 6 = 6 - 8$ **false**
20. 10% of \$38 is \$.38 **false**
21. $8 \le y^2 + 3$ when $y = 3$ **true**
22. $9x > x^3$ when $x = 3$ **false**

23. The senior class is planning a trip that will cost \$35 per student. If \$3920 has been collected from the seniors for the trip, how many have paid for the trip? **112 students**

MARCHING BAND In Exercises 24 and 25 members of the marching band are making their own color-guard flags. Each rectangular flag is 1.2 yards by 0.5 yard. The material costs \$1.75 per square yard.

24. Write a function showing the relationship between the number of flags and the cost of the material. **Let x = the number of flags and y = the cost in dollars of the material; $y = 1.05x$.**

25. How much will it cost to make 20 flags? **\$21**

SCHOOL In Exercises 26–29 the bar graph shows the number of students enrolled in schools in the United States in 1995 and the number of students expected to be enrolled in 2000.

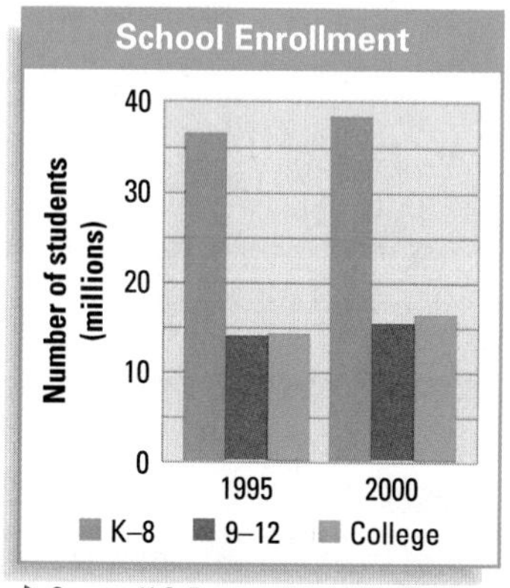

▶ Source: U.S. Bureau of the Census

26. How many students are expected to be in kindergarten through eighth grade in 2000? **about 38 million students**

27. Describe why the K–8 category might be so much larger than 9–12 or College. **The K–8 category includes nine grades while the 9–12 and College categories only include four grades.**

28. What group of students is expected to show the smallest change in enrollment from 1995 to 2000? **9–12**

29. Is the number of students enrolled in school higher in 1995 or in the year 2000? How do you know? **See margin.**

ADDITIONAL RESOURCES

- ***Chapter 1 Resource Book***
  Chapter Test (3 levels) (p. 108)
  SAT/ACT Chapter Test (p. 114)
  Alternative Assessment (p. 115)
- ***Test and Practice Generator***

29. 2000; each of the bars is longer for 2000 than for 1995, so the number of students in each group increased. So the total enrolled for 2000 is higher than in 1995.

ADDITIONAL RESOURCES

- ***Chapter 1 Resource Book***
  Chapter Test (3 levels) (p. 108)
  SAT/ACT Chapter Test (p. 114)
  Alternative Assessment (p. 115)
- ***Test and Practice Generator***

CHAPTER 1

Chapter Standardized Test

TEST-TAKING STRATEGY Avoid spending too much time on one question. Skip questions that are too difficult for you, and spend no more than a few minutes on each question.

1. MULTIPLE CHOICE What is the average speed of a car that traveled 209.2 km in 2 hours? **B**

(A) 20.92 km/h (B) 104.6 km/h
(C) 20.92 h/km (D) 104.6 h/km
(E) 70 mi/h

2. MULTIPLE CHOICE What is the perimeter of the figure? **A**

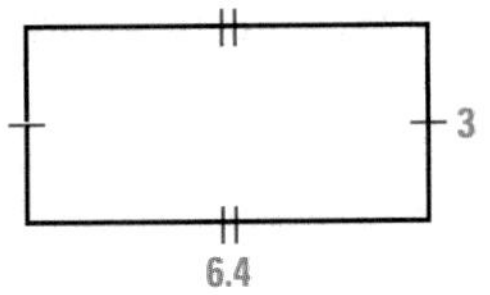

(A) 18.8 (B) 19.2 (C) 27.6
(D) 38.4 (E) 192

3. MULTIPLE CHOICE What is the value of the expression $[(5 \cdot 9) \div x] + 6$ when $x = 3$? **D**

(A) 5 (B) 15 (C) 18
(D) 21 (E) 45

4. MULTIPLE CHOICE If $4t + 5 = 21$, then $t^2 - 3 = \underline{?}$. **B**

(A) 6 (B) 13 (C) 18
(D) 22 (E) 39

5. MULTIPLE CHOICE In the table, what is the value of m? **B**

b	$7b - 2$
2	12
4	26
m	47

(A) 6 (B) 7 (C) 8
(D) 40 (E) 49

6. MULTIPLE CHOICE What is the value of $6 \cdot (15 + 8) - [(2 \cdot 7) - 4]^2$? **B**

(A) −462 (B) 38 (C) 62
(D) 102 (E) 134

QUANTITATIVE COMPARISON In Exercises 7–9, choose the statement below that is true about the given numbers.

(A) The number in column A is greater.
(B) The number in column B is greater.
(C) The two numbers are equal.
(D) The relationship cannot be determined from the information given.

	COLUMN A	COLUMN B	
7.	$3 \cdot 5 - 4$	$3 \cdot (5 - 4)$	**A**
8.	$2x \div 5 + 7$	$2x \div (5 + 7)$	**D**
9.	$42 - (5^2 + 2)$	$42 - (5^2) + 2$	**B**

10. MULTIPLE CHOICE What is the area of a square with sides that are 6.3 cm in length? **E**

(A) 3.969 m^2 (B) 25.2 cm
(C) 25.2 cm^2 (D) 39.69 cm
(E) 39.69 cm^2

11. MULTIPLE CHOICE Which of the following numbers is a solution of the equation $50 - x^2 = 1$?

(A) 5 (B) 6 (C) 7 (D) 8 (E) 9 **C**

12. MULTIPLE CHOICE Which of the following numbers is a solution of the inequality $20 - x \geq x + 2$? **A**

(A) 9 (B) 9.5 (C) 10 (D) 10.5 (E) 11

13. MULTIPLE CHOICE What is the value of $3.4x - 2.3y$, when $x = 11$ and $y = 12$? **D**

(A) 1.1 (B) 2.1 (C) 9 (D) 9.8 (E) 28.7

14. MULTIPLE CHOICE You have decided to save $6 a week to buy an electric guitar costing $150. Which expression shows how much money you still need to save after n weeks? **B**

(A) $150 + 6n$ (B) $150 - 6n$
(C) $(150 + 6)n$ (D) $(150 - 6)n$
(E) $150n + 6n$

15. **MULTIPLE CHOICE** Which algebraic expression is a translation of "five times the difference of eight and a number x"? **A**

Ⓐ $5(8 - x)$ Ⓑ $x - 5 \cdot 8$ Ⓒ $5 \cdot 8 - x$ Ⓓ $5 - 8x$ Ⓔ $5x - 8$

16. **MULTIPLE CHOICE** The number of students on the football team is two more than three times the number of students on the basketball team. If the basketball team has y students, how many students are on the football team? **E**

Ⓐ $3y$ Ⓑ $3y - 2$ Ⓒ $6y$ Ⓓ $2y + 3$ Ⓔ $2 + 3y$

17. **MULTIPLE CHOICE** Which equation represents the function in the table? **D**

Input x	Output y
0	3
1	4
2	7

Ⓐ $y = x + 5$ Ⓑ $y = 2x + 3$
Ⓒ $y = x^2 + 5$ Ⓓ $y = x^2 + 3$
Ⓔ $y = 3x + 1$

18. **MULTIPLE CHOICE** Which of the following represents a function? **B**

I.

Input	Output
1	4
2	4
3	6
4	6

II.

Input	Output
1	3
2	3
3	4
4	4

III.

Input	Output
1	3
1	−3
2	4
3	5

Ⓐ All Ⓑ I and II Ⓒ I and III Ⓓ II and III Ⓔ None

MULTI-STEP PROBLEM Use the graph to compare the amount of chocolate eaten in different countries.

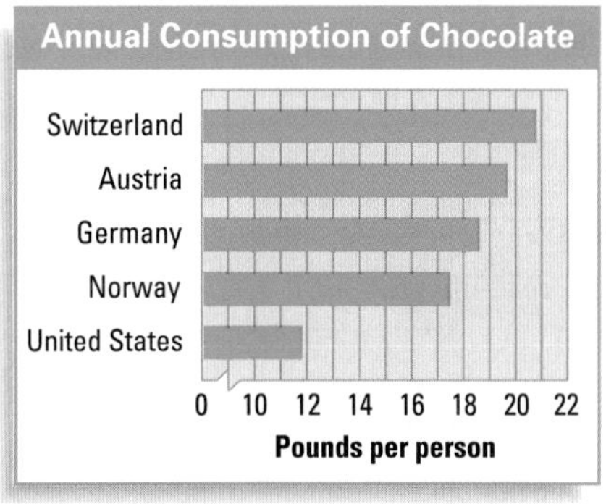

▶ Source: Chocolate Manufacturers Association

19. About how much more chocolate per person is consumed in Switzerland than in the United States? **about 9 lb**
20. About how much more chocolate per person is consumed in Norway than in the United States? **about 5.5 lb**
21. How could the bar graph be misleading? **The horizontal scale is not consistent.**
22. Draw a bar graph representing the same information that would not be misleading. **See margin.**

MULTI-STEP PROBLEM If you place one marble in a measuring cup that contains 200 milliliters of water, the measure on the cup indicates that there is a one millimeter increase in volume. How much does the volume increase when you place from 1 to 10 marbles in the measuring cup?

23. Write an equation to represent the function. **Let x = the number of marbles and y = the rise in milliliter measure of the water; $y = 200 + x$.**
24. Complete an input-output table for the function with domain 0, 1, 2, 3, 4, 5, 6, 7, 8, 9, 10. **See margin.**
25. Describe the domain and range of the function whose values are shown in the table. **domain: 0, 1, 2, 3, 4, 5, 6, 7, 8, 9, 10; range: 200, 201, 202, 203, 204, 205, 206, 207, 208, 209, 210**
26. Graph the data in the table. Use this graph to graph the function. **See margin.**

22.

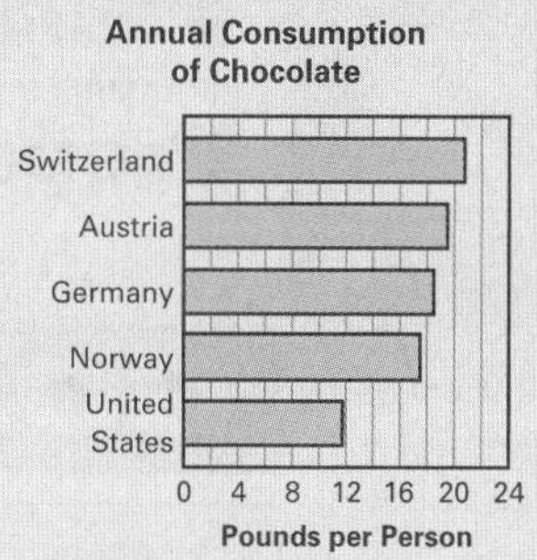

24.

Input	Output
0	200
1	201
2	202
3	203
4	204
5	205
6	206
7	207
8	208
9	209
10	210

26.

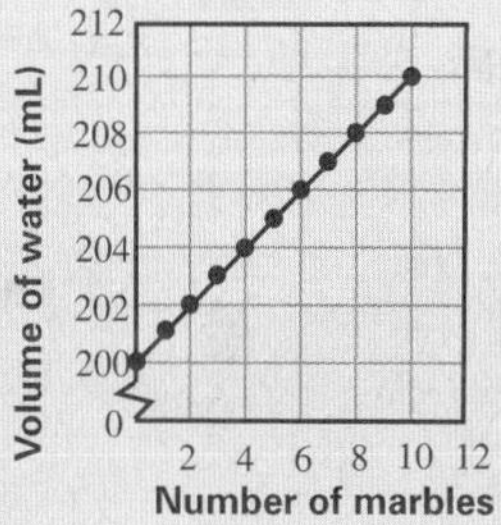

CHAPTER 2

PLANNING THE CHAPTER

Properties of Real Numbers

LESSON	GOALS	NCTM	ITED	SAT9	Terra-Nova	Local
2.1 *pp. 63–70*	GOAL 1 Graph and compare real numbers using a number line. GOAL 2 Find the opposite and the absolute value of a number in real-life applications.	1, 2, 6, 9, 10	IIG, RQWN	6	10, 47, 48, 49	
2.2 *pp. 71–77*	CONCEPT ACTIVITY: 2.2 *Investigate how to model the addition of integers with algebra tiles.* GOAL 1 Add real numbers using a number line or addition rules. GOAL 2 Use addition of real numbers to solve real-life problems.	1, 2, 6, 9, 10	RQWN, RQF	2	10, 11, 17, 18, 47, 48, 49	
2.3 *pp. 78–85*	CONCEPT ACTIVITY: 2.3 *Investigate how to model the subtraction of integers.* GOAL 1 Subtract real numbers using the subtraction rule. GOAL 2 Use subtraction of real numbers to solve real-life problems.	1, 6, 9	RQWN, RQF	2	11, 17, 18, 47, 48, 49	
2.4 *pp. 86–92*	GOAL 1 Organize data in a matrix. GOAL 2 Add and subtract two matrices. TECHNOLOGY ACTIVITY: 2.4 *Find the sum and difference of a matrix on a graphing calculator.*	2, 10	RQWN	34	11, 18	
2.5 *pp. 93–98*	GOAL 1 Multiply real numbers using properties of multiplication. GOAL 2 Multiply real numbers to solve real-life problems.	1, 6, 9	RQWN, RQF	2	11, 17, 18, 47, 48, 49	
2.6 *pp. 99–107*	CONCEPT ACTIVITY: 2.6 *Investigate how to model equivalent expressions using algebra tiles.* GOAL 1 Use the distributive property. GOAL 2 Simplify expressions by combining like terms.	1, 2		2	18, 52	
2.7 *pp. 108–113*	GOAL 1 Divide real numbers. GOAL 2 Use division to simplify algebraic expressions.	1, 2		2	11, 47, 48, 49, 52	
2.8 *pp. 114–120*	GOAL 1 Find the probability of an event. GOAL 2 Find the odds of an event.	5	SAPP, MIP	13, 14, 15, 16	15	

RESOURCES

Chapter Resource Booklets

Chapter Support

Tips for New Teachers	p. 1	Prerequisite Skills Review	p. 5
Parent Guide for Student Success	p. 3	Strategies for Reading Mathematics	p. 7

Lesson Support

Lesson Support	2.1	2.2	2.3	2.4	2.5	2.6	2.7	2.8
Lesson Plans (regular and block)	p. 9	p. 24	p. 38	p. 54	p. 68	p. 80	p. 95	p. 109
Warm-Up Exercises and Daily Quiz	p. 11	p. 26	p. 40	p. 56	p. 70	p. 82	p. 97	p. 111
Activity Support Masters		p. 27	p. 41			p. 83		
Lesson Openers	p. 12	p. 28	p. 42	p. 57	p. 71	p. 84	p. 98	p. 112
Graphing Calculator Activities & Keystrokes	p. 13	p. 29	p. 43	p. 58			p. 99	
Practice (3 levels)	p. 15	p. 30	p. 45	p. 60	p. 72	p. 85	p. 101	p. 113
Reteaching with Practice	p. 18	p. 33	p. 48	p. 63	p. 75	p. 88	p. 104	p. 116
Quick Catch-Up for Absent Students	p. 20	p. 35	p. 50	p. 65	p. 77	p. 90	p. 106	p. 118
Cooperative Learning Activities	p. 21							p. 119
Interdisciplinary Applications	p. 22		p. 51				p. 107	p. 120
Real-Life Applications		p. 36		p. 66	p. 78	p. 91		
Math & History Applications						p. 92		
Challenge: Skills and Applications	p. 23	p. 37	p. 52	p. 67	p. 79	p. 93	p. 108	p. 121

Review and Assessment

Quizzes	pp. 53, 94	Alternative Assessment with Math Journal	p. 130
Chapter Review Games and Activities	p. 122	Project with Rubric	p. 132
Chapter Test (3 levels)	pp. 123–128	Cumulative Review	p. 134
SAT/ACT Chapter Test	p. 129	Resource Book Answers	p. A1

Transparencies

	2.1	2.2	2.3	2.4	2.5	2.6	2.7	2.8
Warm-Up Exercises and Daily Quiz	p. 10	p. 11	p. 12	p. 13	p. 14	p. 15	p. 16	p. 17
Alternative Lesson Opener Transparencies	p. 8	p. 9	p. 10	p. 11	p. 12	p. 13	p. 14	p. 15
Examples/Standardized Test Practice	✓	✓	✓	✓	✓	✓	✓	✓
Answer Transparencies	✓	✓	✓	✓	✓	✓	✓	✓

Technology

- EasyPlanner CD
- EasyPlanner Plus Online
- Classzone.com
- Personal Student Tutor CD
- Test and Practice Generator CD
- Instant Replay: Video Review Games
- Electronic Lesson Presentations
- Chapter Audio Summaries CD

Additional Resources

- Basic Skills Workbook
- Explorations and Projects
- Worked-Out Solution Key
- Data Analysis Sourcebook
- Spanish Resources
- Standardized Test Practice Workbook
- Practice Workbook with Examples
- Notetaking Guide

CHAPTER 2

PACING THE CHAPTER

Resource Key
- STUDENT EDITION
- CHAPTER 2 RESOURCE BOOK
- TEACHER'S EDITION

REGULAR SCHEDULE

Day 1

2.1

STARTING OPTIONS
- Prereq. Skills Review
- Strategies for Reading
- Warm-Up

TEACHING OPTIONS
- Motivating the Lesson
- Les. Opener (Visual)
- Graphing Calc. Activity
- Examples 1–9
- Closure Question
- Guided Practice Exs.

APPLY/HOMEWORK
- See Assignment Guide.
- See the CRB: Practice, Reteach, Apply, Extend

ASSESSMENT OPTIONS
- Checkpoint Exercises
- Daily Quiz (2.1)
- Stand. Test Practice

Day 2

2.2

STARTING OPTIONS
- Homework Check
- Warm-Up or Daily Quiz

TEACHING OPTIONS
- Motivating the Lesson
- Concept Act. & Wksht.
- Les. Opener (Activity)
- Graphing Calc. Activity
- Examples 1–5
- Closure Question
- Guided Practice Exs.

APPLY/HOMEWORK
- See Assignment Guide.
- See the CRB: Practice, Reteach, Apply, Extend

ASSESSMENT OPTIONS
- Checkpoint Exercises
- Daily Quiz (2.2)
- Stand. Test Practice

Day 3

2.3

STARTING OPTIONS
- Homework Check
- Warm-Up or Daily Quiz

TEACHING OPTIONS
- Motivating the Lesson
- Concept Act. & Wksht.
- Les. Opener (Visual)
- Graphing Calc. Activity
- Examples 1–4
- Guided Practice Exs. 1–15

APPLY/HOMEWORK
- See Assignment Guide.
- See the CRB: Practice, Reteach, Apply, Extend

ASSESSMENT OPTIONS
- Checkpoint Exercises, p. 80

Day 4

2.3 *(cont.)*

STARTING OPTIONS
- Homework Check

TEACHING OPTIONS
- Examples 5–6
- Closure Question

APPLY/HOMEWORK
- See Assignment Guide.
- See the CRB: Practice, Reteach, Apply, Extend

ASSESSMENT OPTIONS
- Checkpoint Exercises, p. 81
- Daily Quiz (2.3)
- Stand. Test Practice
- Quiz (2.1–2.3)

Day 5

2.4

STARTING OPTIONS
- Homework Check
- Warm-Up or Daily Quiz

TEACHING OPTIONS
- Les. Opener (Application)
- Graphing Calc. Activity
- Examples 1–4
- Closure Question
- Guided Practice Exs.

APPLY/HOMEWORK
- See Assignment Guide.
- See the CRB: Practice, Reteach, Apply, Extend

ASSESSMENT OPTIONS
- Checkpoint Exercises
- Daily Quiz (2.4)
- Stand. Test Practice

Day 6

2.5

STARTING OPTIONS
- Homework Check
- Warm-Up or Daily Quiz

TEACHING OPTIONS
- Motivating the Lesson
- Les. Opener (Application)
- Examples 1–5
- Closure Question
- Guided Practice Exs.

APPLY/HOMEWORK
- See Assignment Guide.
- See the CRB: Practice, Reteach, Apply, Extend

ASSESSMENT OPTIONS
- Checkpoint Exercises
- Daily Quiz (2.5)
- Stand. Test Practice

Day 9

2.7

STARTING OPTIONS
- Homework Check
- Warm-Up or Daily Quiz

TEACHING OPTIONS
- Motivating the Lesson
- Les. Opener (Activity)
- Graphing Calc. Activity
- Examples 1–5
- Closure Question
- Guided Practice Exs.

APPLY/HOMEWORK
- See Assignment Guide.
- See the CRB: Practice, Reteach, Apply, Extend

ASSESSMENT OPTIONS
- Checkpoint Exercises
- Daily Quiz (2.7)
- Stand. Test Practice

Day 10

2.8

STARTING OPTIONS
- Homework Check
- Warm-Up or Daily Quiz

TEACHING OPTIONS
- Les. Opener (Calculator)
- Examples 1–2
- Guided Practice Exs. 1, 2, 4, 7

APPLY/HOMEWORK
- See Assignment Guide.
- See the CRB: Practice, Reteach, Apply, Extend

ASSESSMENT OPTIONS
- Checkpoint Exercises, p. 115

Day 11

2.8 *(cont.)*

STARTING OPTIONS
- Homework Check

TEACHING OPTIONS
- Examples 3–4
- Closure Question
- Guided Practice Exs. 3, 5, 6, 8

APPLY/HOMEWORK
- See Assignment Guide.
- See the CRB: Practice, Reteach, Apply, Extend

ASSESSMENT OPTIONS
- Checkpoint Exercises, p. 116
- Daily Quiz (2.8)
- Stand. Test Practice

Day 12

Review

DAY 12 START OPTIONS
- Homework Check

REVIEWING OPTIONS
- Chapter 2 Summary
- Chapter 2 Review
- Chapter Review Games and Activities

APPLY/HOMEWORK
- Chapter 2 Test (practice)
- Ch. Standardized Test (practice)

Day 13

Assess

DAY 13 START OPTIONS
- Homework Check

ASSESSMENT OPTIONS
- Chapter 2 Test
- SAT/ACT Ch. 2 Test
- Alternative Assessment

APPLY/HOMEWORK
- Skill Review, p. 130

Day 7

2.6

Starting Options
- Homework Check
- Warm-Up or Daily Quiz

Teaching Options
- Motivating the Lesson
- Concept Act. & Wksht.
- Les. Opener (Calculator)
- Examples 1–4
- Guided Practice Exs. 1–12

Apply/Homework
- See Assignment Guide.
- See the CRB: Practice, Reteach, Apply, Extend

Assessment Options
- Checkpoint Exercises, p. 101

Day 8

2.6 *(cont.)*

Starting Options
- Homework Check

Teaching Options
- Examples 5–6
- Closure Question
- Guided Practice Exs. 13–18

Apply/Homework
- See Assignment Guide.
- See the CRB: Practice, Reteach, Apply, Extend

Assessment Options
- Checkpoint Exercises, p. 102
- Daily Quiz (2.6)
- Stand. Test Practice
- Quiz (2.4–2.6)

BLOCK SCHEDULE

Day 1

Assess & 2.1

(Day 1 = Ch. 1 Day 7)

Assessment Options
- Chapter 1 Test
- SAT/ACT Ch. 1 Test
- Alternative Assessment

Ch. 2 Start Options
- Skill Review, p. 62
- Prereq. Skills Review
- Strategies for Reading

Teaching 2.1 Options
- Warm-Up (Les. 2.1)
- Motivating the Lesson
- Les. Opener (Visual)
- Graphing Calc. Activity
- Examples 1–9
- Closure Question
- Guided Practice Exs.

Apply/Homework
- See Assignment Guide.
- See the CRB: Practice, Reteach, Apply, Extend

Assessment Options
- Checkpoint Exercises
- Daily Quiz (Les. 2.1)
- Stand. Test Practice

Day 2

2.2 & 2.3

Day 2 Start Options
- Homework Check
- W-Up 2.2 or D. Quiz 2.1

Teaching 2.2 Options
- Motivating the Lesson
- Concept Act. & Wksht.
- Les. Opener (Activity)
- Graphing Calc. Activity
- Examples 1–5
- Closure Question
- Guided Practice Exs.

Beginning 2.3 Options
- Warm-Up (Les. 2.3)
- Motivating the Lesson
- Concept Act. & Wksht.
- Les. Opener (Visual)
- Graphing Calc. Activity
- Examples 1–4
- Guided Prac. Exs. 1–15

Apply/Homework
- See Assignment Guide.
- See the CRB: Practice, Reteach, Apply, Extend

Assessment Options
- Checkpoint Exercises
- Daily Quiz (Les. 2.2)

Day 3

2.3 & 2.4

Day 3 Start Options
- Homework Check
- Daily Quiz (Les. 2.2)

Finishing 2.3 Options
- Examples 5–6
- Closure Question

Teaching 2.4 Options
- Warm-Up (Les. 2.4)
- Les. Opener (Appl.)
- Graphing Calc. Activity
- Examples 1–4
- Closure Question
- Guided Practice Exs.

Apply/Homework
- See Assignment Guide.
- See the CRB: Practice, Reteach, Apply, Extend

Assessment Options
- Checkpoint Exercises
- Daily Quiz (Les. 2.3, 2.4)
- Stand. Test Practice
- Quiz (2.1–2.3)

Day 4

2.5 & 2.6

Day 4 Start Options
- Homework Check
- W-Up 2.5 or D. Quiz 2.4

Teaching 2.5 Options
- Motivating the Lesson
- Les. Opener (Appl.)
- Examples 1–5
- Closure Question
- Guided Practice Exs.

Beginning 2.6 Options
- Warm-Up (Les. 2.6)
- Motivating the Lesson
- Concept Act. & Wksht.
- Les. Opener (Calc.)
- Examples 1–4
- Guided Practice Exs. 1–12

Apply/Homework
- See Assignment Guide.
- See the CRB: Practice, Reteach, Apply, Extend

Assessment Options
- Checkpoint Exercises
- Daily Quiz (Les. 2.5)
- Stand. Test Prac. (2.5)

Day 5

2.6 & 2.7

Day 5 Start Options
- Homework Check
- Daily Quiz (Les. 2.5)

Finishing 2.6 Options
- Examples 5–6
- Closure Question
- Guided Practice Exs. 13–18

Teaching 2.7 Options
- Warm-Up (Les. 2.7)
- Motivating the Lesson
- Les. Opener (Activity)
- Graphing Calc. Activity
- Examples 1–5
- Closure Question
- Guided Practice Exs.

Apply/Homework
- See Assignment Guide.
- See the CRB: Practice, Reteach, Apply, Extend

Assessment Options
- Checkpoint Exercises
- Daily Quiz (Les. 2.6, 2.7)
- Stand. Test Practice
- Quiz (2.4–2.6)

Day 6

2.8

Starting Options
- Homework Check
- Warm-Up or Daily Quiz

Teaching 2.8 Options
- Les. Opener (Calc.)
- Examples 1–4
- Closure Question
- Guided Practice Exs.

Apply/Homework
- See Assignment Guide.
- See the CRB: Practice, Reteach, Apply, Extend

Assessment Options
- Checkpoint Exercises
- Daily Quiz (Les. 2.8)
- Stand. Test Practice

Day 7

Review/Assess

Day 7 Start Options
- Homework Check

Reviewing Options
- Chapter 2 Summary
- Chapter 2 Review
- Chapter Review Games and Activities
- Chapter 2 Test (practice)
- Ch. Standardized Test (practice)

Assessment Options
- Chapter 2 Test
- SAT/ACT Ch. 2 Test
- Alternative Assessment

Apply/Homework
- Skill Review, p. 130

MEETING INDIVIDUAL NEEDS

BEFORE THE CHAPTER

The *Chapter 2 Resource Book* has the following materials to distribute and use before the chapter:

- **Parent Guide for Student Success**
- **Prerequisite Skills Review**
- **Strategies for Reading Mathematics (pictured below)**

STRATEGIES FOR READING *Pages 7–8*

CHAPTER 2

NAME ______________________ DATE __________

Strategies for Reading Mathematics

For use with Chapter 2

Chapter Support

Strategy: Reading Vocabulary and Taking Notes in Algebra

One thing you can count on in algebra is the need to understand mathematical terms. Make your studying easier by looking for highlighted vocabulary terms in **heavy type** like this. Then write definitions and examples in your notebook to help you remember the terms. The notebook below shows a sample of the notes you might take about the vocabulary in the next two paragraphs.

Numbers *greater* than zero are **positive numbers.** They are to the right of the origin on a number line. Numbers *less* than zero are **negative numbers.** They are to the left of the origin on a number line.

Integers are the numbers . . . –3, –2, –1, 0, 1, 2, 3, ... On a number line, numbers between the tick marks are not integers.

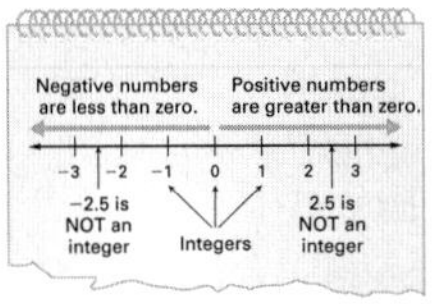

Negative numbers are less than zero.
Positive numbers are greater than zero.
–3
–2
–1
0
1
2
3
–2.5 is NOT an integer
Integers
2.5 is NOT an integer

STUDY TIP
Reading Vocabulary
When you see highlighted words in heavy type, that means the terms are important. Read the sentence that the term is in for its definition. Check sentences just before and after the term to make sure that you have a complete definition.

STUDY TIP
Taking Notes
Make your notes brief. Write important words that will help you remember the meaning of each term. You might also include pictures, diagrams, or examples.

Questions

1. Which of the labeled numbers on the number line shown on the notebook above are positive numbers? How can you tell? Which are negative?
2. Name all the integers labeled on the number line. If you extended the number line, what would be the next negative integer? Explain how you decided.
3. How can you tell that zero is neither positive nor negative?
4. Which numbers in the following list are integers: –2, –3.2, 0, 1.0, 1.08? How could you use the number line on the notebook to help you decide?

STRATEGIES FOR READING MATHEMATICS The first page gives students tips about reading vocabulary and taking notes as they prepare for Chapter 2. The second page (not shown) provides a visual glossary of key vocabulary words in the chapter, such as opposites and the distributive property.

DURING EACH LESSON

The *Chapter 2 Resource Book* has the following alternatives for introducing the lesson:

- **Lesson Openers (pictured below)**
- **Graphing Calculator Activities with Keystrokes**

LESSON OPENER *Page 71*

LESSON 2.5

NAME ______________________ DATE __________

Available as a transparency

Application Lesson Opener

For use with pages 93–98

The temperature fell 2° each hour for 6 hours. Let –2 represent this change in temperature.

1. Use addition to find the total change in temperature.
2. Explain how you can use multiplication instead of addition to find the change.

The scores for your last 3 rounds in a golf tournament were recorded as –1.

3. Use addition to find your total score for these 3 rounds.
4. Explain how you can use multiplication instead of addition to find your score.

A submarine descended 6 feet per second for 8 seconds. Let –6 represent the change in the position of the submarine each second.

5. Use addition to find the total change in position.
6. Explain how you can use multiplication instead of addition to find the change.

Your watch has lost 5 seconds a day for the last 7 days. Let –5 represent the change in the time each day.

7. Use addition to find the total change in time.
8. Explain how you can use multiplication instead of addition to find the change.
9. Use your answers to Questions 2, 4, 6, and 8 to make a conjecture about multiplying a positive number by a negative number.

Lesson 2.5

APPLICATION LESSON OPENER This Lesson Opener provides an alternative way to start Lesson 2.5 in the form of a real-life application. Students see how adding negative numbers and multiplying positive and negative numbers relates to decreases in temperature, golf scores, position, and time.

TECHNOLOGY RESOURCE

Teachers can use the Electronic Lesson Presentations CD-ROM to present Lesson 2.6 using computer animation to step through the Examples.

FOLLOWING EACH LESSON

The *Chapter 2 Resource Book* has a variety of materials to follow-up each lesson. They include the following:

- **Practice (3 levels)**
- **Reteaching with Practice**
- **Quick Catch-Up for Absent Students (pictured below)**
- **Interdisciplinary Applications**
- **Real-Life Applications**

QUICK CATCH-UP *Page 90*

LESSON 2.6

NAME ______________________ DATE ________

Quick Catch-Up for Absent Students

For use with pages 99–107

Lesson 2.6

The items checked below were covered in class on (date missed) ________

Activity 2.6: Modeling the Distributive Property (p. 99)

____ **Goal:** Model equivalent expressions using algebra tiles.

Lesson 2.6: The Distributive Property

____ **Goal 1:** Use the distributive property. (p. 100–101)

Material Covered:

____ Example 1: Using an Area Model

____ Student Help: Study Tip

____ Example 2: Using the Distributive Property

____ Example 3: Using the Distributive Property

____ Example 4: Mental Math Calculations

Vocabulary:

distributive property, p. 100

____ **Goal 2:** Simplify expressions by combining like terms. (p. 102)

Material Covered:

____ Example 5: Simplifying by Combining Like Terms

____ Example 6: Using the Distributive Property to Simplify a Function

Vocabulary:

coefficient, p. 102	like terms, p. 102
constant terms, p. 102	simplified expression, p. 102

____ Other (specify) ______________________

Homework and Additional Learning Support

____ Textbook (specify) pp. 103–107

____ *Reteaching with Practice* worksheet (specify exercises) ________

____ *Personal Student Tutor* for Lesson 2.6

QUICK CATCH-UP FOR ABSENT STUDENTS You can use this form to let students know what they have missed when they've been absent from class. It allows you to quickly check off which Examples and other elements of Lesson 2.6 were covered on a given day and provides space for filling in the homework assignment.

TECHNOLOGY RESOURCE

Students who have missed class can find extra examples for Lessons 2.2, 2.3, 2.4, 2.5, and 2.7 on the Internet at www.mcdougallittell.com.

ASSESSING THE CHAPTER

The *Chapter 2 Resource Book* has the following review and assessment materials:

- **Quizzes**
- **Chapter Review Games and Activities**
- **Chapter Test (3 levels)**
- **SAT/ACT Chapter Test**
- **Alternative Assessment with Math Journal (below)**
- **Project with Rubric**
- **Cumulative Review**

ALTERNATIVE ASSESSMENT *Page 130–131*

CHAPTER 2

NAME ______________________ DATE ________

Alternative Assessment and Math Journal

For use after Chapter 2

JOURNAL 1. Expand and simplify the following expressions: $-2^2, -2^3, -2^4, -2^5, -2^6, (-2)^2, (-2)^3, (-2)^4, (-2)^5$, and $(-2)^6$.

a. Explain how you know the sign of the answer. Describe the patterns that you found with respect to the sign of your answers. Why are parentheses important when simplifying exponents with a negative number base?

b. If n is a positive integer and a is an integer, write a rule for the sign of a^n. Be sure to include all cases.

MULTI-STEP PROBLEM 2. You teacher gives you the following directions: "Choose any integer. Multiply it by negative four. Add ten. Divide your answer by two. Add triple your original number. Subtract twelve from that answer. Add negative ten. Then add 32." Your teacher then asks you for your final number. You reply "18", and your teacher says, "Your original integer was 3." How did your teacher know the original number?

The following expression represents your teacher's directions where the original integer is represented by x.

$$\frac{x \cdot (-4) + 10}{2} + 3x - 12 + (-10) + 32$$

a. State the definition of an integer in your own words.

b. Choose a positive integer for x and evaluate the expression, showing all work.

c. Choose a negative integer for x and evaluate the expression, showing all work.

d. Predict the value of the expression when $x = 40$. Check your answer.

e. Simplify the expression shown. Make a conclusion based on your answer. What did your teacher do to determine your original number?

f. Create an algebraic expression. Write a set of directions that correspond to the expression. Show all work in simplifying the expression, then explain how you can find the original number based on the expression.

3. ***Critical Thinking*** Suppose that your teacher gives the following directions: "Choose a number. Subtract eight. Multiply your answer by two. Take the opposite of your answer. Subtract seventeen. Add four times your number. Then add negative four." Write an expression for this and simplify. When a student gives your teacher an answer, what should your teacher do to find the original number?

4. ***Writing*** Create an algebraic expression for which a student will always get the same answer, no matter what number is chosen.

Review and Assess

ALTERNATIVE ASSESSMENT WITH RUBRIC AND MATH JOURNAL The journal exercise asks students to demonstrate their understanding of operations with positive and negative integers in writing. The Multi-Step Problem has students pull together a variety of concepts from Chapter 2 to solve a problem about a number game. Answers and a scoring rubric are provided on a separate sheet.

CHAPTER OVERVIEW

CHAPTER GOALS

The main content of Chapter 2 is addition, subtraction, multiplication, and division of real numbers. Graphing and comparing real numbers using a number line, finding the opposite and the absolute value of a real number, and solving absolute value equations are also introduced. Students will use the distributive property to evaluate and simplify variable expressions, organize data in a matrix, and add and subtract two matrices. They will also find the domain of a function and study some introductory concepts of probability.

The properties of real numbers will be used to solve real-life problems such as finding velocity and speed or distance, the profit or loss of a business, the difference in stock market prices, or the balance due on a loan.

APPLICATION NOTE

Leonardo Da Vinci first sketched and described a helicopter in 1483, but it was not until the late 1930s that helicopters were actually in use. Unlike airplanes or jets, helicopters can fly vertically, hover, or go slowly in any direction. They can achieve a maximum speed of about 250 miles per hour and are typically flown at between 80 and 150 miles per hour.

Although the military is the largest user of helicopters, they have many commercial uses as well. Commuters depend on traffic reporters who use helicopters to see the traffic flow in a particular area. Cartographers use helicopters to take aerial photographs of regions. Helicopters are used by emergency technicians and hospitals to rescue and transport injured people. Farmers use helicopters for crop dusting and fire-fighters use them to fight fires in remote locations.

Additional information about helicopters is available at **www.mcdougallittell.com.**

PROPERTIES OF REAL NUMBERS

▶ *Why are helicopters able to take off and land without runways?*

CHAPTER

2

APPLICATION: Helicopters

Helicopters are capable of vertical flight—flying straight up and straight down. Rotor blades generate an upward force (lift) as they whirl through the air.

Mathematics provides a useful way of distingushing between upward and downward motion. In this chapter we will use positive numbers to measure the velocity of upward motion and negative numbers to measure the velocity of downward motion.

Think & Discuss

1. Describe some real-life situations that you might represent with negative numbers. *Sample answers:* below-zero temperature readings, golf scores, debt, elevation below sea level
2. Describe the average speed and direction of each helicopter's movement if it travels the given distance in 15 seconds. 8 ft/sec downward; $10\frac{2}{3}$ ft/sec upward

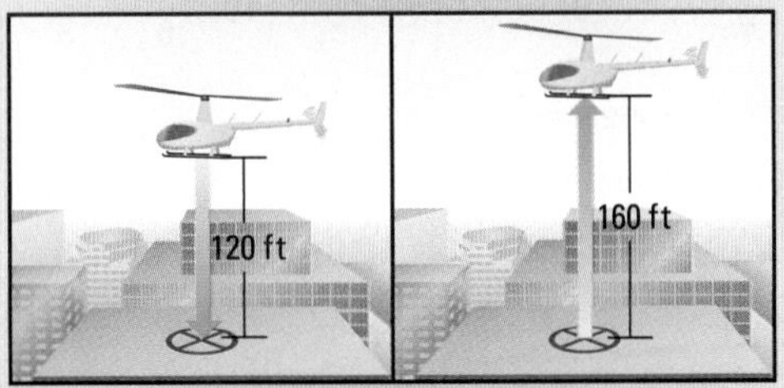

Not drawn to scale

Learn More About It

You will find the speed and velocity of different objects in Exercises 76–78 on p. 69.

APPLICATION LINK Visit www.mcdougallittell.com for more information about helicopters.

ADDITIONAL RESOURCES

Another way to begin the chapter is to show the video clip of a real-life motivator for Chapter 2 on the *Instant Replay: Video Review Games.*

PROJECTS

A project covering Chapters 1–3 appears on pages 198–199 of the Student Edition. An additional project for Chapter 2 is available in the *Chapter 2 Resource Book,* p. 132.

TECHNOLOGY

Software

- *Electronic Teaching Tools*
- *Online Lesson Planner*
- *Personal Student Tutor*
- *Test and Practice Generator*
- *Electronic Lesson Presentations* (Lesson 2.6)

Video

- *Instant Replay: Video Review Games*

Internet Connections
www.mcdougallittell.com

- **Application Links**
  61, 66, 107, 110
- **Data Updates**
  64, 87
- **Student Help**
  69, 73, 80, 81, 87, 92, 94, 104, 109, 118
- **Career Links**
  90, 115
- **Extra Challenge**
  70, 77, 84, 91, 106, 113, 119

ENGLISH LEARNERS

Keep in mind that while practical applications are intended to help students understand the relationship between algebra and the world around them, the vocabulary contained in these applications can be unfamiliar to English language learners. In this application, make sure students understand the meanings of the words *vertical, straight up, straight down, rotor blades, upward force, lift, upward,* and *downward.*

PREPARE

DIAGNOSTIC TOOLS

The **Skill Review** exercises can help you diagnose whether students have the following skills needed in Chapter 2:

- Ordering positive decimals and fractions.
- Operations with fractions.
- Operations with decimals.

The following resources are available for students who need additional help with these skills:

- Prerequisite Skills Review (*Chapter 2 Resource Book,* p. 5; *Warm-Up Transparencies,* p. 9)
- Reteaching with Practice (Chapter Resource Book for Lesson 1.3.)
- *Personal Student Tutor*

ADDITIONAL RESOURCES

The following resources are provided to help you prepare for the upcoming chapter and customize review materials:

- ***Chapter 2 Resource Book***
  Tips for New Teachers (p. 1)
  Parent Guide (p. 3)
  Lesson Plans (every lesson)
  Lesson Plans for Block Scheduling (every lesson)
- ***Electronic Teaching Tools***
- ***Online Lesson Planner***
- ***Test and Practice Generator***

Study Guide

PREVIEW

What's the chapter about?

Chapter 2 is about **real numbers**. In Chapter 2 you'll learn

- how to add, subtract, multiply, and divide real numbers.
- how to determine the likelihood of an event using probability and odds.

KEY VOCABULARY

▶ Review	▶ New	
• variable, p. 3	• real numbers, p. 63	• matrix, p. 86
• order of operations, p. 16	• integers, p. 63	• distributive property, p. 100
• data, p. 40	• absolute value, p. 65	• reciprocal, p. 108
• function, p. 46	• counterexample, p. 66	• probability of an event, p. 114

PREPARE

Are you ready for the chapter?

SKILL REVIEW Do these exercises to review key skills that you'll apply in this chapter. See the given reference page if there is something you don't understand.

STUDENT HELP

Study Tip
"Student Help" boxes throughout the chapter give you study tips and tell you where to look for extra help in this book and on the Internet.

Put the numbers in order from least to greatest. (Skills Review, pp. 779–780)

1. 4.6, 6.1, 5.5, 5.06, 4.07, 5.46 **4.07, 4.6, 5.06, 5.46, 5.5, 6.1**

2. $2\frac{5}{6}, 2\frac{2}{3}, \frac{11}{4}, 2\frac{5}{8}, \frac{25}{11}, 2\frac{3}{5}$ **See margin.**

2. $\frac{25}{11}, 2\frac{3}{5}, 2\frac{5}{8}, 2\frac{2}{3}, \frac{11}{4}, 2\frac{5}{6}$

Find the value of the expression. Write the answer as a fraction or mixed number in lowest terms. (Skills Review, pp. 781–782)

3. $\frac{9}{14} - \frac{3}{7}$ $\frac{3}{14}$ **4.** $\frac{5}{6} + \frac{2}{3}$ $1\frac{1}{2}$ **5.** $1\frac{3}{4} + 2\frac{1}{8}$ $3\frac{7}{8}$ **6.** $8\frac{7}{15} - 2\frac{4}{5}$ $5\frac{2}{3}$

7. $\frac{5}{9} \div \frac{2}{9}$ $2\frac{1}{2}$ **8.** $\frac{7}{9} \cdot \frac{3}{14}$ $\frac{1}{6}$ **9.** $6\frac{6}{7} \cdot 4\frac{1}{12}$ 28 **10.** $9\frac{1}{6} \div 1\frac{3}{8}$ $6\frac{2}{3}$

Evaluate the expression. (Review Example 2, p. 17)

11. $4 + 7 - 3 \cdot 2$ 5 **12.** $64 \div 16 + 5 - 7$ 2 **13.** $9 - 8 \div 4 + 12$ 19

14. $16 - 9 \cdot 5 \div 3$ 1 **15.** $8 \cdot 3 - 6 \cdot 4$ 0 **16.** $12.8 - 7.5 \div 1.5 + 3.2$ 11

STUDY STRATEGY

Here's a study strategy!

Studying a Lesson

- Read the goals and study tips.
- Review the examples and highlighted vocabulary.
- Be sure you understand the rules and properties given in boxes.
- Take notes. Add to your list of vocabulary, rules, and properties in your notebook.

2.1 The Real Number Line

What *you should learn*

GOAL 1 Graph and compare real numbers using a number line.

GOAL 2 Find the opposite and the absolute value of a number in **real-life** applications such as the speed and velocity of an elevator in **Example 8**.

Why *you should learn it*

▼ To help you understand negative amounts in **real life**, such as "below zero" temperature in the wind-chill table in **Exs. 79–81**.

REAL LIFE

GOAL 1 GRAPHING REAL NUMBERS

The numbers used in this algebra book are **real numbers**. They can be pictured as points on a horizontal line called a **real number line**.

The point labeled 0 is the **origin**. Points to the left of zero represent **negative numbers** and points to the right of zero represent **positive numbers**. Zero is neither positive nor negative.

REAL NUMBER LINE

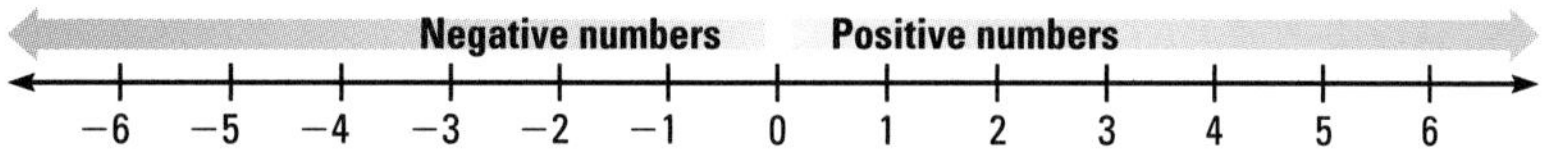

The scale marks are equally spaced and represent **integers**. An integer is either negative, zero, or positive.

..., −3, −2, −1,	0,	1, 2, 3, ...
Negative integers	Zero	Positive integers

The three dots on each side in the list above indicate that the list continues in both directions without end. For instance, the integer immediately to the left of −3 is −4, read as *negative four.*

The real number line has points that represent fractions and decimals, as well as integers. The point that corresponds to a number is the **graph** of the number, and drawing the point is called graphing the number or **plotting** the point.

EXAMPLE 1 *Graphing Real Numbers*

Graph the numbers $\frac{1}{2}$ and -2.3 on a number line.

SOLUTION

The point that corresponds to $\frac{1}{2}$ is one half unit to the *right* of zero. The point that corresponds to -2.3 is 2.3 units to the *left* of zero.

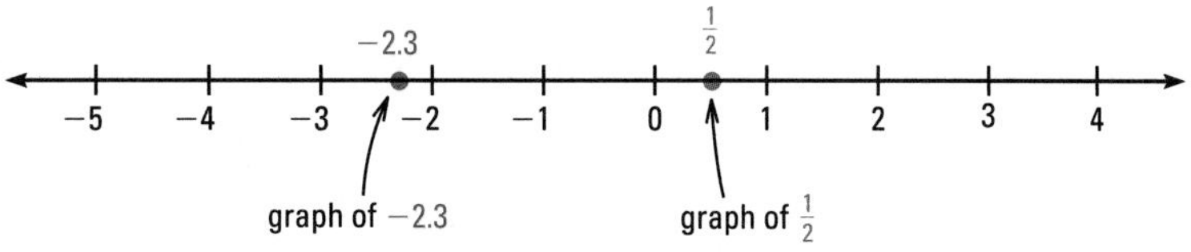

1 PLAN

PACING
Basic: 1 day
Average: 1 day
Advanced: 1 day
Block Schedule: 0.5 block with Ch. 1 Assess.

LESSON OPENER
VISUAL APPROACH
An alternative way to approach Lesson 2.1 is to use the Visual Approach Lesson Opener:
- Blackline Master (*Chapter 2 Resource Book,* p. 12)
- Transparency (p. 8)

MEETING INDIVIDUAL NEEDS
- ***Chapter 2 Resource Book***
  Prerequisite Skills Review (p. 5)
  Practice Level A (p. 15)
  Practice Level B (p. 16)
  Practice Level C (p. 17)
  Reteaching with Practice (p. 18)
  Absent Student Catch-Up (p. 20)
  Challenge (p. 23)
- ***Resources in Spanish***
- ***Personal Student Tutor***

NEW-TEACHER SUPPORT
See the Tips for New Teachers on pp. 1–2 of the *Chapter 2 Resource Book* for additional notes about Lesson 2.1.

WARM-UP EXERCISES

Transparency Available

Complete the statement using > or <.

1. 4 ___ 1 > **2.** $\frac{3}{2}$ ___ $\frac{1}{2}$ >

Write the numbers in increasing order.

3. 3.2, $\frac{1}{2}$, 0, 1.7, 5
$0, \frac{1}{2}, 1.7, 3.2, 5$

4. 1.6, 0.6, 1.06, 1.66, 0.7
0.6, 0.7, 1.06, 1.6, 1.66

2 TEACH

MOTIVATING THE LESSON
Ask students what the opposite of winning \$100 or of losing 5 pounds would be. Explain that the numbers they study in this lesson will allow them to represent such situations.

EXTRA EXAMPLE 1
Graph the numbers $-1\frac{3}{4}$ and 2.7.

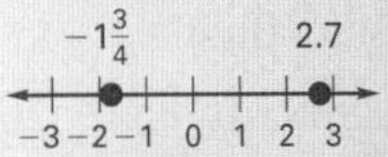

EXTRA EXAMPLE 2
Write two inequalities that compare $-\frac{1}{2}$ and $-\frac{3}{2}$.
$-\frac{3}{2} < -\frac{1}{2}, -\frac{1}{2} > -\frac{3}{2}$

EXTRA EXAMPLE 3
Write the following numbers in increasing order: $-1.5, \frac{1}{3}, -3, 2.5, 0, -1$ $-3, -1.5, -1, 0, \frac{1}{3}, 2.5$

EXTRA EXAMPLE 4
The table shows the balance of a checking account that has overdraft protection each day for four days in August.

Date	Balance (\$)
Aug. 20	−5
Aug. 21	195
Aug. 22	150
Aug. 23	−11

a. Which balance was the lowest? **−\$11**
b. Which dates had a balance above −\$10?
Aug. 20, Aug. 21, Aug. 22

CHECKPOINT EXERCISES
For use after Examples 1–4:
1. a. Write the numbers in increasing order: $\frac{5}{2}, -4, 0, -0.5$, and 3. $-4, -0.5, 0, \frac{5}{2}, 3$
b. Write two inequalities to compare the two least numbers. $-4 < -0.5$ and $-0.5 > -4$

STUDENT HELP
Skills Review
For help with comparing and ordering numbers, see pp. 779–780.

EXAMPLE 2 *Comparing Real Numbers*

Graph −4 and −5 on a number line. Then write two inequalities that compare the two numbers.

SOLUTION

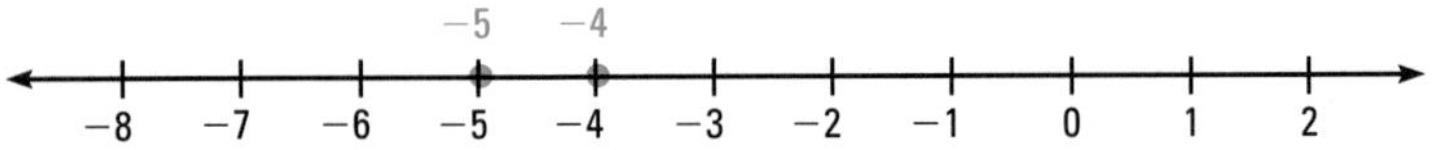

On the graph, −5 is to the left of −4, so −5 is **less than** −4.

$-5 < -4$

On the graph, −4 is to the right of −5, so −4 is **greater than** −5.

$-4 > -5$

EXAMPLE 3 *Ordering Real Numbers*

Write the following numbers in increasing order: $-2, 4, 0, 1.5, \frac{1}{2}, -\frac{3}{2}$.

SOLUTION
Graph the numbers on a number line.

▶ From the graph, you can see that the order is $-2, -\frac{3}{2}, 0, \frac{1}{2}, 1.5, 4$.

FOCUS ON APPLICATIONS

REAL LIFE **NOME, ALASKA** Around December 21, the northern hemisphere has its shortest day and longest night. On that day, Nome has only 3 hours and 54 minutes of daylight.

DATA UPDATE
Visit our Web site www.mcdougallittell.com

EXAMPLE 4 *Comparing Real Numbers*

NOME, ALASKA The table at the right shows the low temperatures recorded in Nome, Alaska, each day for five days in December.

a. Which low temperature reading was the coldest?

b. Which dates had low temperatures above 10°F?

c. Which dates had low temperatures below −5°F?

Date	Temp. (°F)
Dec. 18	−10°F
Dec. 19	−11°F
Dec. 20	16°F
Dec. 21	3°F
Dec. 22	2°F

SOLUTION

a. The coldest temperature was −11°F on December 19.

b. On December 20, the low temperature, 16°F, was above 10°F.

c. Low temperatures were below −5°F on December 18 (−10°F) and December 19 (−11°F).

GOAL 2 FINDING OPPOSITES AND ABSOLUTE VALUES

Two points that are the same distance from the origin but on opposite sides of the origin are **opposites**.

EXAMPLE 5 *Finding the Opposite of a Number*

The numbers -3 and 3 are opposites because each is 3 units from the origin.

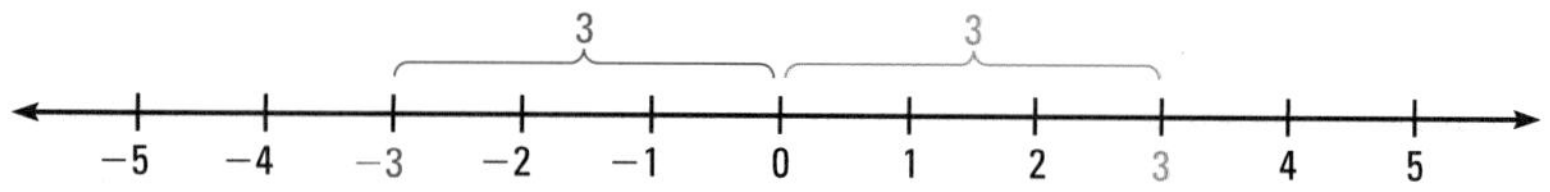

The expression -3 can be stated as "negative 3" or "the opposite of 3." You can read the expression $-a$ as "the opposite of a." You should not assume that $-a$ is a negative number. For instance, if $a = -2$, then $-a = -(-2) = 2$.

The **absolute value** of a real number is the distance between the origin and the point representing the real number. The symbol $|a|$ represents the absolute value of a number a.

CONCEPT SUMMARY THE ABSOLUTE VALUE OF A NUMBER

- If a is a positive number, then $|a| = a$. **Example:** $|3| = 3$
- If a is zero, then $|a| = 0$. **Example:** $|0| = 0$
- If a is a negative number, then $|a| = -a$. **Example:** $|-3| = 3$

The absolute value of a number is *never* negative. If the number a is negative, then its absolute value, $-a$, is positive. For instance, $|-6| = -(-6) = 6$.

EXAMPLE 6 *Finding Absolute Values*

Evaluate the expression.

a. $|2.3|$ **b.** $\left|-\frac{1}{2}\right|$ **c.** $-|-8|$

SOLUTION

a. $|2.3| = 2.3$ **If a is positive, then $|a| = a$.**

b. $\left|-\frac{1}{2}\right| = -\left(-\frac{1}{2}\right) = \frac{1}{2}$ **If a is negative, then $|a| = -a$.**

c. $-|-8| = -(8)$ **The absolute value of -8 is 8.**

$= -8$ **Use definition of opposites.**

EXTRA EXAMPLE 5

The numbers $\frac{1}{2}$ and $-\frac{1}{2}$ are opposites because each is $\frac{1}{2}$ unit from the origin.

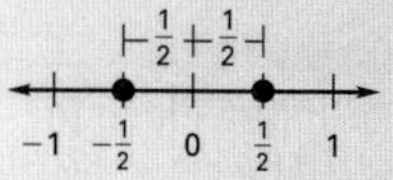

EXTRA EXAMPLE 6

Evaluate the expression.

a. $\left|\frac{1}{2}\right|$ $\frac{1}{2}$

b. $|-1.6|$ 1.6

c. $-|3|$ -3

CHECKPOINT EXERCISES

For use after Example 5:

1. Find the opposite of the number.

a. 0 0 **b.** 1.2 -1.2

For use after Example 6:

2. Evaluate the expression.

a. $|0.9|$ 0.9 **b.** $\left|-\frac{3}{2}\right|$ $\frac{3}{2}$

COMMON ERROR

EXAMPLE 6 In part (c), students may assume that any expression involving an absolute value symbol is positive. Explain that the absolute value must be evaluated before the first negative sign is applied.

STUDENT HELP NOTES

Skills Review As students refer to pp. 779–780, remind them that when comparing fractions, it is often useful to convert them to a common denominator or to convert them to decimal form.

MATHEMATICAL REASONING

In connection with Example 9, discuss the fact that a false statement containing the word *sometimes* cannot be disproved with a counterexample. Consider the statement: "$|a|$ is sometimes negative." In this case you need to use a proof and the definition of absolute value to prove the statement is false.

EXTRA EXAMPLE 7
Use mental math to solve.

a. $|x| = 4.3$ **$-4.3, 4.3$**

b. $|x| = \frac{3}{4}$ **$-\frac{3}{4}, \frac{3}{4}$**

EXTRA EXAMPLE 8
An object is dropped close to Earth's surface. The rate at which it is falling 1 second later is 32 ft/sec (if air resistance is ignored). What are the object's velocity and speed 1 second after it is dropped? **−32 ft/sec, 32 ft/sec**

EXTRA EXAMPLE 9
Decide whether the statement is *true* or *false*. If it is false, give a counterexample.

a. The expression $|a|$ is *always* positive. **False; counterexample: $|0| = 0$ is not positive.**

b. The absolute value of a number is *always* greater than the number. **False; counterexample: $|5| = 5$, and $5 = 5$.**

c. The symbol $-a$ is *sometimes* equal to $|a|$. **True.**

CHECKPOINT EXERCISES
For use after Example 7:

1. Use mental math to solve the equation $|x| = -1$. **no solution**

For use after Example 8:

2. After two bumper cars collide, one moves backwards at a rate of 15 miles per hour. What are its velocity and speed? **−15 mi/h, 15 mi/h**

For use after Example 9:

3. Determine whether the statement is *true* or *false*. If it is *false*, give a counterexample. The expression $-|a|$ is *never* positive. **True.**

FOCUS ON VOCABULARY
Ask students to define the opposite and the absolute value of a number in their own words.

CLOSURE QUESTION
When does a number have a positive absolute value and a negative opposite? **when it is a positive number**

STUDENT HELP

Look Back
For help with the solution of an equation, see p. 24.

EXAMPLE 7 *Solving an Absolute Value Equation*

Use mental math to solve the equation.

a. $|x| = 7$ **b.** $|x| = -5$

SOLUTION

a. Ask "what numbers are 7 units from the origin?" Both 7 and -7 are 7 units from the origin, so there are two solutions: 7 and -7.

b. The absolute value of a number is never negative, so there is no solution.

....

Velocity indicates both speed and direction (up is positive and down is negative). The speed of an object is the absolute value of its velocity.

FOCUS ON APPLICATIONS

SPACE SHUTTLE The space shuttle's elevators have been designed to endure the stress of 320,000 gallons of water released at lift off to lessen heat and vibration.

APPLICATION LINK
www.mcdougallittell.com

EXAMPLE 8 *Finding Velocity and Speed*

SPACE SHUTTLE ELEVATORS A space shuttle launch pad elevator drops at a rate of 10 feet per second. What are its velocity and speed?

SOLUTION

Velocity $= -10$ feet per second — Motion is downward.

Speed $= |-10| = 10$ feet per second — Speed is positive.

....

In mathematics, to prove that a statement is true, you need to show that it is true for all examples. To prove that a statement is false, you need to show that it is *not* true for only one example, called a **counterexample.**

EXAMPLE 9 *Using a Counterexample*

Decide whether the statement is *true* or *false*. If it is false, give a counterexample.

a. The opposite of a number is *always* negative.

b. The absolute value of a number is *never* negative.

c. The expression $-a$ is *never* positive.

d. The expression $-a$ is *sometimes* greater than a.

SOLUTION

a. False. Counterexample: The opposite of -5 is 5, which is positive.

b. True, by definition.

c. False. Counterexample: If $a = -2$, then $-a = -(-2) = 2$, which is positive.

d. True. For instance, if $a = -8$, then $-a = -(-8) = 8$, and $8 > -8$.

GUIDED PRACTICE

Vocabulary Check ✓

1. Write an inequality for the sentence: *Three is greater than negative five.* $3 > -5$

Concept Check ✓

2. Copy the number line. Use it to explain why $-2.5 > -3.5$.

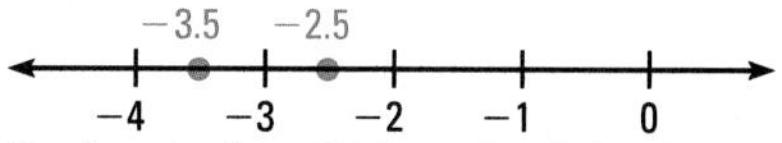

On the graph, -2.5 is to the right of -3.5, so $-2.5 > -3.5$.

3. Use a counterexample to show that the following statement is false. *The opposite of a number is never positive.* Any negative number may be given as a counterexample.

Skill Check ✓

Complete the statement using > or <.

4. $-3 \underline{?} -5$ $>$ 5. $-3 \underline{?} 0$ $<$ 6. $-2 \underline{?} -\frac{1}{2}$ $<$ 7. $-8 \underline{?} 7$ $<$

Graph the numbers on a number line. Then write the numbers in increasing order. 8–11. See margin for graphs.

8. $3, -5, 0$ — $-5, 0, 3$
9. $2, 3, -4$ — $-4, 2, 3$
10. $-\frac{1}{2}, -\frac{3}{4}, \frac{1}{4}$ — $-\frac{3}{4}, -\frac{1}{2}, \frac{1}{4}$
11. $-0.1, -1.1, -1$ — $-1.1, -1, -0.1$

Evaluate the expression.

12. $|-12|$ 12 13. $|4.1|$ 4.1 14. $\left|-\frac{1}{5}\right|$ $\frac{1}{5}$ 15. $|-103|$ 103

Tell whether you would use a positive number or a negative number to represent the velocity.

16. The velocity of a rising rocket positive
17. The velocity of a falling raindrop negative

PRACTICE AND APPLICATIONS

STUDENT HELP

Extra Practice to help you master skills is on p. 798.

WRITING INEQUALITIES **Graph the numbers on a number line. Then write two inequalities that compare the two numbers.** 18–26. See margin for graphs.

18. -6 and 4 — $-6 < 4, 4 > -6$
19. -6.4 and -6.3 — $-6.4 < -6.3, -6.3 > -6.4$
20. -7 and 2 — $-7 < 2, 2 > -7$
21. -0.1 and -0.11 — $-0.1 > -0.11, -0.11 < -0.1$
22. 5.7 and -4.2 — $5.7 > -4.2, -4.2 < 5.7$
23. -2.8 and 3.7 — $-2.8 < 3.7, 3.7 > -2.8$
24. -2.7 and $\frac{3}{4}$ — $-2.7 < \frac{3}{4}, \frac{3}{4} > -2.7$
25. -0.5 and $-\frac{1}{3}$ — $-0.5 < -\frac{1}{3}, -\frac{1}{3} > -0.5$
26. $-1\frac{5}{6}$ and $-1\frac{7}{9}$ — $-1\frac{5}{6} < -1\frac{7}{9}, -1\frac{7}{9} > -1\frac{5}{6}$

ORDERING NUMBERS **Write the numbers in increasing order.**

27. 4.66, 0.7, 4.6, −1.8, 3, −0.66 — −1.8, −0.66, 0.7, 3, 4.6, 4.66
28. −0.03, 0.2, 0, 2.0, −0.2, −0.02 — −0.2, −0.03, −0.02, 0, 0.2, 2.0
29. $4.8, -2.6, 0, -3, \frac{1}{2}, -\frac{1}{2}$ — See margin.
30. $3\frac{1}{2}, 3.4, 4.1, -5, -5.1, -4\frac{1}{2}$ — $-5.1, -5, -4\frac{1}{2}, 3.4, 3\frac{1}{2}, 4.1$
31. $7, -\frac{1}{2}, 2.4, -\frac{3}{4}, -5.8, \frac{1}{3}$ — $-5.8, -\frac{3}{4}, -\frac{1}{2}, \frac{1}{3}, 2.4, 7$
32. 6.03, −6.08, −6.1, −6.11, −6.02, 6.07 — −6.11, −6.1, −6.08, −6.02, 6.03, 6.07

29. $-3, -2.6, -\frac{1}{2}, 0, \frac{1}{2}, 4.8$

OPPOSITES **Find the opposite of the number.**

33. 8 −8 34. −3 3 35. 3.8 −3.8 36. −2.5 2.5

37. $-\frac{5}{6}$ $\frac{5}{6}$ 38. $3\frac{4}{5}$ $-3\frac{4}{5}$ 39. −2.01 2.01 40. $-\frac{1}{9}$ $\frac{1}{9}$

STUDENT HELP

HOMEWORK HELP
Example 1: Exs. 18–26
Example 2: Exs. 18–26
Example 3: Exs. 27–32
Example 4: Exs. 66–70
Example 5: Exs. 33–40

continued on p. 68

3 APPLY

ASSIGNMENT GUIDE

BASIC
Day 1: pp. 67–70 Exs. 18–52 even, 54–65, 83–85, 90, 95, 100–102

AVERAGE
Day 1: pp. 67–70 Exs. 18–52 even, 54–65, 83–85, 90, 95, 100–102

ADVANCED
Day 1: pp. 67–70 Exs. 18–70 even, 83–87, 90, 95, 100–102

BLOCK SCHEDULE
pp. 67–70 Exs. 8–52 even, 54–65, 83–85, 88–102 even (with Ch. 1 Assess.)

EXERCISE LEVELS

Level A: *Easier*
18–26, 33–40

Level B: *More Difficult*
27–32, 41–75, 83–85, 88–102

Level C: *Most Difficult*
76–82, 86, 87

✓ HOMEWORK CHECK

To quickly check student understanding of key concepts, go over the following exercises: Exs. 22, 30, 40, 48, 58, 64. See also the Daily Homework Quiz:

- Blackline Master (*Chapter 2 Resource Book,* p. 26)
- Transparency (p. 11)

8.
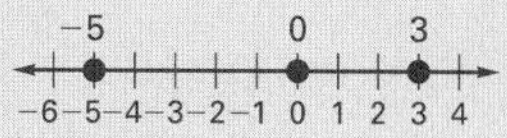

9.
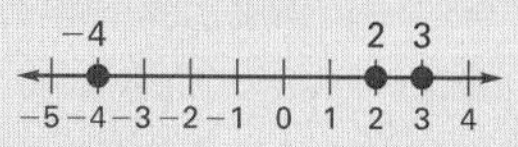

10.
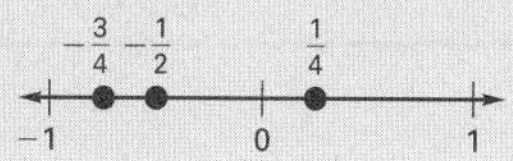

11.
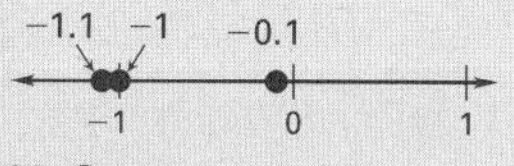

18–26. See next page.

! COMMON ERROR

EXERCISES 46–49 Students may use an incorrect order of operations in exercises like these. Explain that absolute value signs, like parentheses, are grouping symbols and expressions inside them must be evaluated before performing other operations.

18.
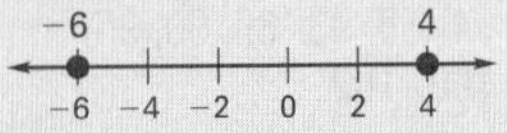

19.
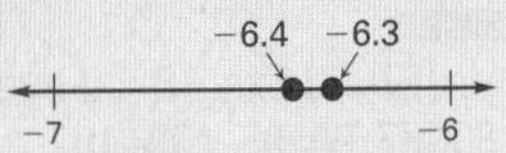

20.
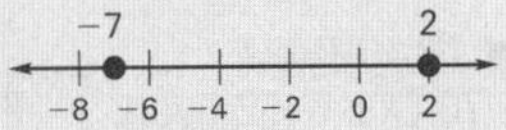

21.
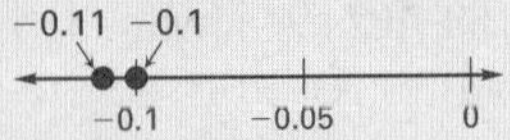

22.
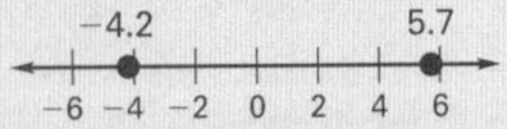

23.
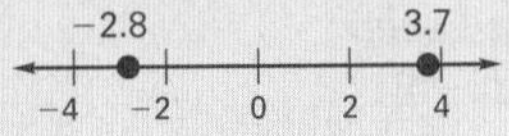

24.
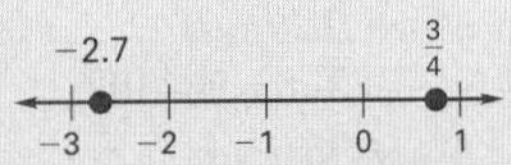

25.
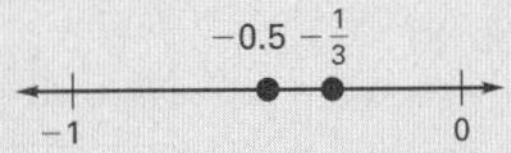

26.
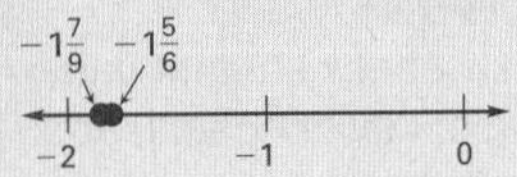

STUDENT HELP

HOMEWORK HELP
continued from p. 67

Example 6: Exs. 41–49
Example 7: Exs. 50–53
Example 8: Exs. 76–78
Example 9: Exs. 54–57

FINDING ABSOLUTE VALUES **Evaluate the expression.**

41. $|7|$ **7**　**42.** $|-4|$ **4**　**43.** $|-4.5|$ **4.5**

44. $\left|\frac{2}{3}\right|$ $\frac{2}{3}$　**45.** $\left|-\frac{4}{5}\right|$ $\frac{4}{5}$　**46.** $|0| + 2$ **2**

47. $|6.3| - 2$ **4.3**　**48.** $-\left|-\frac{8}{9}\right|$ $-\frac{8}{9}$　**49.** $|-6.1| - 6.01$ **0.09**

SOLVE AN EQUATION **Use mental math to solve the equation.**

50. $|x| = 4$ **−4, 4**　**51.** $|x| = 0$ **0**　**52.** $x = |-3.8|$ **3.8**　**53.** $|-x| = 1$ **−1, 1**

COUNTEREXAMPLES **Decide whether the statement is *true* or *false*. If it is false, give a counterexample.**

54. The absolute value of a negative number is always negative. **False; all negative numbers are counterexamples.**

55. The opposite of $-a$ is always positive.

55. False; 0 and all negative numbers are counterexamples.

56. The opposite of $|a|$ is never positive. **True**

57. The value of $|-a|$ is sometimes negative. **False; all real numbers except 0 are counterexamples.**

LOGICAL REASONING **Complete the statement using >, <, ≥, or ≤.**

58. If $x > -4$, then -4 _?_ x. **<**　**59.** If $3 \le y$, then y _?_ 3. **≥**

60. If $m \le 8$, then 8 _?_ m. **≥**　**61.** If $-7 \ge w$, then w _?_ -7. **≤**

ELEVATION **In Exercises 62–65, write a positive number, a negative number, or zero to represent the elevation of the location.**
Elevation is represented by comparing a location to sea level, which is given a value of zero. A location above sea level has a positive elevation, and a location below sea level has a negative elevation.

62. Granite Peak, Montana, 12,799 feet above sea level **12,799**

63. New Orleans, Louisiana, 8 feet below sea level **−8**

64. Death Valley, California, 282 feet below sea level **−282**

65. Long Island Sound, Connecticut, sea level **0**

FOCUS ON APPLICATIONS

ASTRONOMY A star may *appear* to be dim because of its great distance from Earth. Such a star may burn brighter than another star that looks very bright because it is closer to Earth.

ASTRONOMY **In Exercises 66–70, use the table showing the apparent magnitude of several stars.**
A star's brightness as it appears to a person on Earth is measured by its *apparent magnitude*. A bright star has a lesser apparent magnitude than a dim star.

Star	Apparent magnitude
Canopus	−0.72
Procyon	0.38
Pollux	1.14
Altair	0.77
Spica	0.98
Vega	0.03
Regulus	1.35
Sirius	−1.46
Arcturus	−0.04
Deneb	1.25

66. Which star looks the brightest? **Sirius**

67. Which star looks the dimmest? **Regulus**

68. Which stars look dimmer than Altair? **Deneb, Regulus, Pollux, and Spica**

69. Which stars look brighter than Procyon? **Canopus, Vega, Sirius, and Arcturus**

70. Write the stars in order from dimmest to brightest. **Regulus, Deneb, Pollux, Spica, Altair, Procyon, Vega, Arcturus, Canopus, and Sirius**

FOCUS ON APPLICATIONS

REAL LIFE **GOLF SCORES** If you complete a round of golf in 68 strokes at a course with a par of 71 strokes, you have shot "3 under par," or −3.

GOLF In Exercises 71–75, use the table, which lists several players and their final scores at a 1998 Ladies Professional Golf Association tournament.

In golf the total score is given as the number of strokes above or below *par*, the expected score. If you are at "even par," your score is zero. The player with the lowest score wins.

Player	Score
Luciana Bemvenuti	+6
Liz Early	+15
Michelle Estill	−9
Hiromi Kobayashi	+1
Jenny Lidback	−4
Joan Pitcock	−11
Nancy Scranton	+4
Annika Sörenstam	−19

▶ Source: Ladies Professional Golf Association

71. Which player scored closest to par? **Kobayashi**

72. Which player scored farthest from par? **Sörenstam**

73. Which players scored above par? **Bemvenuti, Early, Kobayashi, Scranton**

74. Which players scored below par? **Estill, Lidback, Pitcock, Sörenstam**

75. One of the players listed won the tournament. Who was it? **Sörenstam**

SPEED AND VELOCITY Find the speed and the velocity of the object.

76. A helicopter is descending for a landing at a rate of 6 feet per second. **6 ft/sec, −6 ft/sec**

77. An elevator in the Washington Monument in Washington, DC, climbs about 429 feet per minute. ▶ Source: National Park Service **429 ft/min, 429 ft/min**

78. A diver plunges to the ocean floor at a rate of 3 meters per second. **3 m/sec, −3 m/sec**

WIND CHILL The faster the wind blows, the colder you feel. In Exercises 79–81, use the wind-chill index table to identify the combination of temperature and wind speed that feels colder.

To use the table, find the temperature in the top row. Read down until you are in the row for the appropriate wind speed. For example, when the temperature is 0°F and the wind speed is 10 mi/h, the temperature feels like −22°F.

Wind-Chill Index

Wind speed (mi/h)	Temperature (°F)				
0	−10	−5	0	5	10
5	−15	−10	−5	0	6
10	−34	−27	−22	−15	−9
15	−45	−38	−31	−25	−18
20	−53	−46	−39	−31	−24
25	−59	−51	−44	−36	−29
30	−64	−56	−49	−41	−33

79. A temperature of 10°F with a wind speed of 25 mi/h, or a temperature of −5°F with a wind speed of 10 mi/h **10°F, 25 mi/h**

80. A temperature of 5°F with a wind speed of 15 mi/h, or a temperature of −5°F with a wind speed of 5 mi/h **5°F, 15 mi/h**

81. A temperature of −10°F with a wind speed of 5 mi/h, or a temperature of 10°F with a wind speed of 20 mi/h **10°F, 20 mi/h**

STUDENT HELP

INTERNET **HOMEWORK HELP** Visit our Web site www.mcdougallittell.com for help with Exs. 79–81.

STUDENT HELP NOTES

Homework Help Students can find help for Exs. 79–81 at **www.mcdougallittell.com.** The information can be printed out for students who don't have access to the Internet.

ADDITIONAL PRACTICE AND RETEACHING

For Lesson 2.1:

- Practice Levels A, B, and C (*Chapter 2 Resource Book,* p. 15)
- Reteaching with Practice (*Chapter 2 Resource Book,* p. 18)
- See Lesson 2.1 of the *Personal Student Tutor*

For more Mixed Review:

- Search the *Test and Practice Generator* for key words or specific lessons.

4 ASSESS

DAILY HOMEWORK QUIZ

Transparency Available

1. Write two inequalities that compare −1.8 and 0.4. **−1.8 < 0.4, 0.4 > −1.8**
2. Write the following numbers in increasing order: 2, −0.5, −0.25, 0.75, −5, 1.95. **−5, −0.5, −0.25, 0.75, 1.95, 2**
3. What is the opposite of 6.2? **−6.2**
4. Evaluate −| 1.8 | − 1.8. **−3.6**
5. Write a positive number, a negative number, or zero for the greatest known depth in the Pacific Ocean, which is 36,198 ft. **−36,198 ft**

EXTRA CHALLENGE NOTE

Challenge problems for Lesson 2.1 are available in **blackline** format in the *Chapter 2 Resource Book,* p. 23 and at **www.mcdougallittell.com.**

ADDITIONAL TEST PREPARATION

1. **WRITING** Suppose two numbers have the same absolute value, *k*. How can you determine the distance between the two numbers on the number line? Give at least two examples to support your answer. **Multiply *k* by 2; the distance between −6 and 6 is 2 • 6 = 12, and the distance between −1.8 and 1.8 is 2 • 1.8 = 3.6.**

82. *Writing* You are writing a quiz for this lesson. Write a problem to test whether students can write a set of real numbers in increasing order. Include positive and negative decimals and fractions. Show the answer to your problem on a number line. **Answers may vary.**

Test Preparation

QUANTITATIVE COMPARISON **In Exercises 83–85, choose the statement that is true about the given numbers.**

Ⓐ The number in column A is greater.

Ⓑ The number in column B is greater.

Ⓒ The two numbers are equal.

Ⓓ The relationship cannot be determined from the given information.

	Column A	Column B	
83.	$-\left\|\frac{2}{3}\right\|$	$\left\|-\frac{2}{3}\right\|$	**B**
84.	$x + \|2\|$	$x + \|-2\|$	**C**
85.	$\|x\| + 8$	$x + 8$	**D**

★ Challenge

86. LOGICAL REASONING Is the opposite of the absolute value of a number ever the same as the absolute value of the opposite of the number? In other words, is it ever true that $-|x| = |-x|$? Explain. **Yes; $-|x| = |-x|$ is true only for $x = 0$. Otherwise, $|-x|$ is positive and $-|x|$ is negative.**

EXTRA CHALLENGE

www.mcdougallittell.com

87. LOGICAL REASONING Is it *always, sometimes,* or *never* true that $|x| = |-x|$? **always**

MIXED REVIEW

FRACTION OPERATIONS **Find the sum.** **(Skills Review, p. 781)**

88. $\frac{2}{4} + \frac{3}{8}$ **$\frac{7}{8}$**

89. $\frac{2}{3} + \frac{1}{6}$ **$\frac{5}{6}$**

90. $\frac{2}{5} + \frac{1}{4}$ **$\frac{13}{20}$**

91. $\frac{5}{8} + \frac{1}{3}$ **$\frac{23}{24}$**

92. $3\frac{2}{7} + 4\frac{1}{2}$ **$7\frac{11}{14}$**

93. $1\frac{7}{9} + 4\frac{3}{7}$ **$6\frac{13}{63}$**

EVALUATING EXPRESSIONS **Evaluate the expression for the given value(s) of the variable(s).** **(Review 1.1)**

94. $5x + 3$ when $x = 2$ **13**

95. $2a - 7$ when $a = 6$ **5**

96. $3y + 12$ when $y = 0$ **12**

97. $6b - 39 + c$ when $b = 15$ and $c = 2$ **53**

98. $\frac{5}{6}a - b$ when $a = 6$ and $b = 5$ **0**

99. $\frac{x}{5} - 2y$ when $x = 12$ and $y = \frac{4}{5}$ **$\frac{4}{5}$**

TRANSLATING VERBAL SENTENCES **Write the verbal sentence as an equation or an inequality.** **(Review 1.5)**

100. Five less than z is eight. **$z - 5 = 8$**

101. Eight more than r is seventeen. **$r + 8 = 17$**

102. Nine more than three fourths of y is less than six times w. **$\frac{3}{4}y + 9 < 6w$**

ACTIVITY 2.2

Developing Concepts

Group Activity for use with Lesson 2.2

Modeling Addition of Integers

GROUP ACTIVITY
Work with a partner.

MATERIALS
algebra tiles

▶ QUESTION How can you model addition of integers with algebra tiles?

▶ EXPLORING THE CONCEPT

To model addition problems involving positive and negative integers, you can use tiles labeled + and −. Each + represents positive 1, and each − represents negative 1. Combining a + with a − gives 0. You can use algebra tiles to find the sum of −8 and 3.

1 Model negative 8 and positive 3 using algebra tiles.

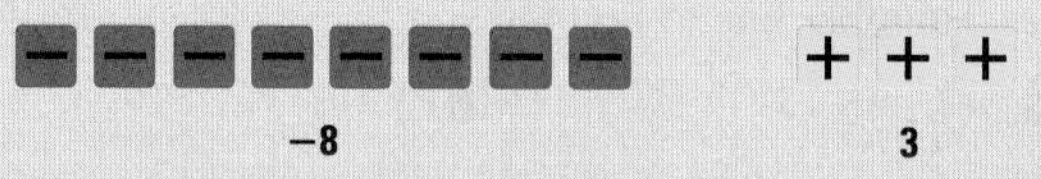

2 Group pairs of positive and negative tiles. Count the remaining tiles.

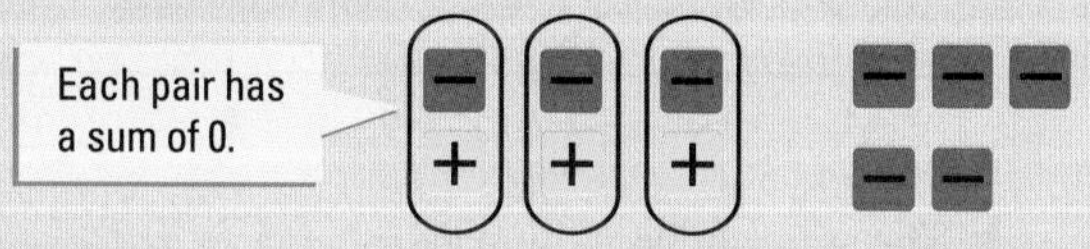

3 The remaining tiles show the sum of −8 and 3.

Complete the statement: −8 + 3 = _?_. −5

▶ DRAWING CONCLUSIONS

Use algebra tiles to find the sum. Sketch your solution.

1. 4 + 5 9 **2.** 3 + 3 6 **3.** −4 + (−2) −6

4. −1 + (−7) −8 **5.** −3 + 2 −1 **6.** 5 + (−2) 3

7. −6 + 6 0 **8.** 2 + (−2) 0 **9.** −6 + (−3) −9

Use your results from Exercises 1–8 to help you decide whether the statement is *true* or *false*. Give an example that supports your answer.

10. The sum of a positive integer and a positive integer is always a positive integer. true; *Sample answer:* 2 + 3 = 5

11. The sum of a negative integer and a negative integer is always a positive integer. false; *Sample answer:* −2 + (−2) = −4

12. The sum of a positive integer and a negative integer is sometimes a negative integer. true; *Sample answer:* 5 + (−7) = −2

13. The sum of a negative integer and a negative integer is never zero. true; *Sample answer:* −3 + (−1) = −4

Activity Assessment *Sample answer:*
when they are both negative, or when the one with the greater absolute value is negative

1 Planning the Activity

PURPOSE
To model integer addition with algebra tiles.

MATERIALS
- A set of 8 positive and 8 negative algebra tiles per group
- Activity Support Master (*Chapter 2 Resource Book,* p. 27)

PACING
- Exploring the Concept — 10 min
- Drawing Conclusions — 10 min

▶ *LINK TO LESSON*
After discussing Example 1 of Lesson 2.2, have students check the answers by drawing algebra tile models of each sum.

2 Managing the Activity

COOPERATIVE LEARNING
For each sum in Exercises 1–9, have one student make the algebra tile model while the other makes the sketch. Have students switch roles for each exercise.

ALTERNATIVE APPROACH
This activity can be done as a demonstration by drawing the algebra tile models on an overhead transparency. After demonstrating −8 + 3, have the class sketch the models for Exercises 1–9 at their desks while students take turns modeling them on the overhead.

3 Closing the Activity

★ *KEY DISCOVERY*
The sum of two positive integers is positive, the sum of two negative integers is negative, and the sum of a positive integer and a negative integer can be positive, negative, or zero.

ACTIVITY ASSESSMENT
When will the sum of two integers be negative?
See sample answer at left.

1 PLAN

PACING
Basic: 1 day
Average: 1 day
Advanced: 1 day
Block Schedule: 0.5 block with 2.3

LESSON OPENER ACTIVITY
An alternative way to approach Lesson 2.2 is to use the Activity Lesson Opener:

- Blackline Master (*Chapter 2 Resource Book,* p. 28)
- Transparency (p. 9)

MEETING INDIVIDUAL NEEDS

- ***Chapter 2 Resource Book***
  Prerequisite Skills Review (p. 5)
  Practice Level A (p. 30)
  Practice Level B (p. 31)
  Practice Level C (p. 32)
  Reteaching with Practice (p. 33)
  Absent Student Catch-Up (p. 35)
  Challenge (p. 37)
- ***Resources in Spanish***
- ***Personal Student Tutor***

NEW-TEACHER SUPPORT
See the Tips for New Teachers on pp. 1–2 of the *Chapter 2 Resource Book* for additional notes about Lesson 2.2.

WARM-UP EXERCISES

Transparency Available

Find the sum.

1. $11 + 23 + 89$ **123**

2. $3 + 1\frac{1}{2}$ $4\frac{1}{2}$

Find the sum. Use a calculator if you wish.

3. $2.87 + 11.01 + 1.23$ **15.11**

4. $7322.6 + 43.21$ **7365.81**

2.2 Addition of Real Numbers

***What** you should learn*

GOAL 1 Add real numbers using a number line or addition rules.

GOAL 2 Use addition of real numbers to solve **real-life** problems such as finding the profit of a business in **Example 5**.

***Why** you should learn it*

▼ To solve **real-life** problems, such as finding the number of yards gained by a football team in **Ex. 51**.

REAL LIFE

GOAL 1 ADDING REAL NUMBERS

Addition can be modeled with movements on a number line.

- You add a positive number by moving to the right.
- You add a negative number by moving to the left.

EXAMPLE 1 *Adding Two Real Numbers*

Use a number line to find the sum.

a. $-2 + 5$ **b.** $2 + (-6)$

SOLUTION

a.

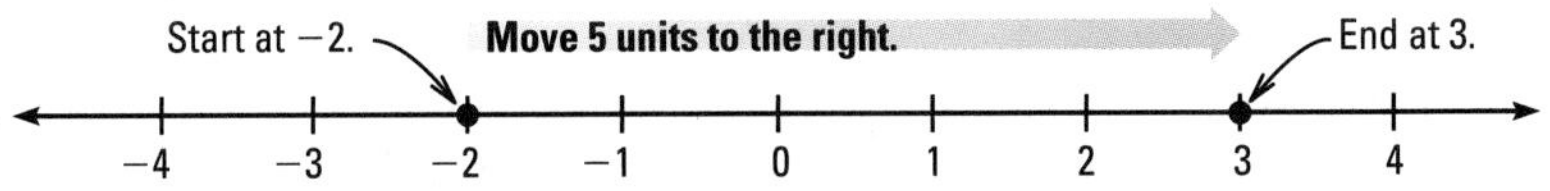

Start at −2.
Move 5 units to the right.
End at 3.
−4 −3 −2 −1 0 1 2 3 4

▶ The sum can be written as $-2 + 5 = 3$.

b.

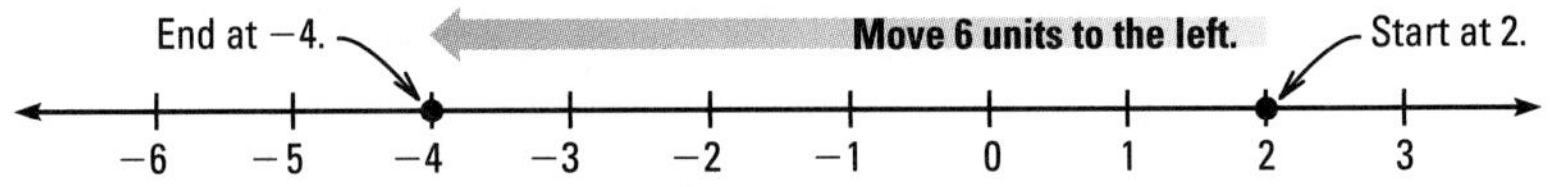

End at −4.
Move 6 units to the left.
Start at 2.
−6 −5 −4 −3 −2 −1 0 1 2 3

▶ The sum can be written as $2 + (-6) = -4$.

EXAMPLE 2 *Adding Three Real Numbers*

Use a number line to find the sum: $-3 + 5 + (-6)$.

SOLUTION

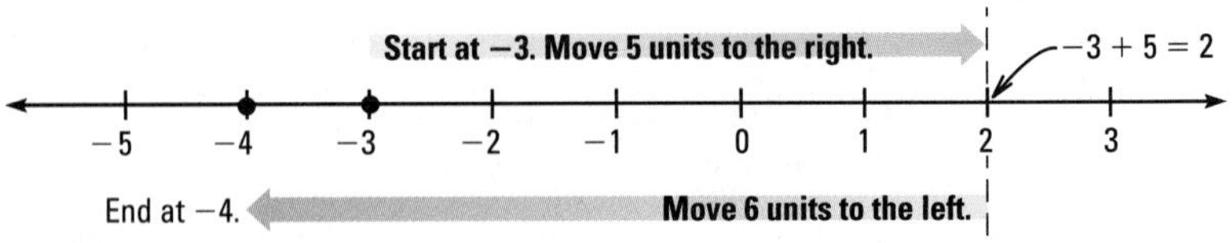

Start at −3. Move 5 units to the right.
−3 + 5 = 2
−5 −4 −3 −2 −1 0 1 2 3
End at −4.
Move 6 units to the left.

▶ The sum can be written as $-3 + 5 + (-6) = -4$.

The rules of addition show how to add two real numbers without a number line.

RULES OF ADDITION

TO ADD TWO NUMBERS WITH THE *SAME SIGN*:

STEP 1 Add their absolute values.

STEP 2 Attach the common sign.

Example: $-4 + (-5)$ **Step 1** $|-4| + |-5| = 9$ **Step 2** -9

TO ADD TWO NUMBERS WITH *OPPOSITE SIGNS*:

STEP 1 Subtract the smaller absolute value from the larger absolute value.

STEP 2 Attach the sign of the number with the larger absolute value.

Example: $3 + (-9)$ **Step 1** $|-9| - |3| = 6$ **Step 2** -6

The above rules of addition will help you find sums of positive and negative numbers. It can be shown that these rules are a consequence of the following Properties of Addition.

PROPERTIES OF ADDITION

COMMUTATIVE PROPERTY

The order in which two numbers are added does not change the sum.

$a + b = b + a$ **Example:** $3 + (-2) = -2 + 3$

ASSOCIATIVE PROPERTY

The way you group three numbers when adding does not change the sum.

$(a + b) + c = a + (b + c)$ **Example:** $(-5 + 6) + 2 = -5 + (6 + 2)$

IDENTITY PROPERTY

The sum of a number and 0 is the number.

$a + 0 = a$ **Example:** $-4 + 0 = -4$

PROPERTY OF ZERO (INVERSE PROPERTY)

The sum of a number and its opposite is 0.

$a + (-a) = 0$ **Example:** $5 + (-5) = 0$

EXAMPLE 3 *Finding a Sum*

a. $1.4 + (-2.6) + 3.1 = 1.4 + (-2.6 + 3.1)$ Use associative property.

$= 1.4 + 0.5$ Simplify.

$= 1.9$

b. $-\frac{1}{2} + 3 + \frac{1}{2} = -\frac{1}{2} + \frac{1}{2} + 3$ Use commutative property.

$= \left(-\frac{1}{2} + \frac{1}{2}\right) + 3$ Use associative property.

$= 0 + 3 = 3$ Use identity property and property of zero.

STUDENT HELP

HOMEWORK HELP Visit our Web site www.mcdougallittell.com for extra examples.

Look Back For help with fraction operations, see pp. 781–783.

2 TEACH

MOTIVATING THE LESSON

In a quiz show game, contestants win money for correct answers and lose money for incorrect answers. Ask students if they know how to determine how much money they have if they answer four questions correctly for a total of $350 and answer two questions incorrectly for a loss of $200. Tell students they can use addition of positive and negative numbers to determine the answer.

EXTRA EXAMPLE 1

Use a number line to find the sum.

a. $7 + (-3)$ **4**

b. $-8 + 2$ **−6**

EXTRA EXAMPLE 2

Use a number line to find the sum: $-2 + (-3) + 6$. **1**

EXTRA EXAMPLE 3

Find the sum.

a. $-0.7 + 1.9 + 0.7$ **1.9**

b. $-3 + \frac{2}{3} + \frac{1}{3}$ **−2**

CHECKPOINT EXERCISES

For use after Example 1:

1. Use a number line to find the sum: $6 + (-9)$. **−3**

For use after Example 2:

2. Use a number line to find the sum: $5 + (-3) + (-2)$. **0**

For use after Example 3:

3. Find the sum. $-3.2 + 4 + (-0.1) + (-4)$ **−3.3**

CONCEPT QUESTION

EXAMPLE 3 Where are the identity property and the property of zero used in this problem?

The property of zero is used to add $-\frac{1}{2}$ and $\frac{1}{2}$. The identity property is used to add 0 and 3.

EXTRA EXAMPLE 4

On a standardized test, students receive a score of +1 for each correct answer and −1 for each incorrect answer. Those questions that are left blank are given a zero. In order to get a grade above average on the test, a student's total score must be above 35. Would the following students score above average on the test?

a. Sylvia: 7 incorrect, 12 blank, 41 correct. **No**

b. Nathan: 8 incorrect, 5 blank, 47 correct. **Yes**

EXTRA EXAMPLE 5

A child care center had the following monthly results after comparing income and expenses. Add the monthly profits and losses to find the overall profit or loss during the four-month period.

March	\$7,982.43
April	−\$1,820.62
May	−\$9,219.56
June	\$12,043.65

The center had a profit of \$8985.90.

CHECKPOINT EXERCISES

For use after Examples 4 and 5:

1. The yearly profit or losses for a bakery are shown for its first 3 years of operation. What is the total profit or loss for the three years? **\$7168.33**

1996	−\$12,931.56
1997	−\$1,290.54
1998	\$21,390.43

CLOSURE QUESTION

Without adding, how can you tell if the sum of two numbers will be positive, negative, or zero?

If both numbers are positive, the sum will be positive. If both are negative, the sum will be negative. If they have different signs, the sign of the sum will be the sign of the number with the greater absolute value. If they are opposites, the sum will be zero.

FOCUS ON APPLICATIONS

REAL LIFE **CRYSTALS** are often made up of ions. The attraction of positive and negative ions to each other results in a regular geometric pattern.

GOAL 2 USING ADDITION IN REAL LIFE

EXAMPLE 4 *Adding Real Numbers*

SCIENCE CONNECTION Atoms are composed of electrons, neutrons, and protons. Each electron has a charge of -1, each neutron has a charge of 0, and each proton has a charge of $+1$. The total charge of an atom is the sum of all the charges of its electrons, neutrons, and protons. An atom is an ion if it has a positive or negative charge. If an atom has a charge of zero, it is *not* an ion. Are the following atoms ions?

a. Aluminum: 13 electrons, 13 neutrons, 13 protons

b. Aluminum: 10 electrons, 13 neutrons, 13 protons

SOLUTION

a. The total charge is $-13 + 0 + 13 = 0$, so the atom is not an ion. In chemistry this aluminum atom is written as Al.

b. The total charge is $-10 + 0 + 13 = 3$, so the atom is an ion. In chemistry this aluminum ion is written as Al^{3+}.

PROFIT AND LOSS A company has a *profit* if its income is greater than its expenses. It has a *loss* if its income is less than its expenses. Business losses can be indicated by negative numbers.

EXAMPLE 5 *Finding the Total Profit*

A consulting company had the following monthly results after comparing income and expenses. Add the monthly profits and losses to find the overall profit or loss during the six-month period.

JANUARY	FEBRUARY	MARCH
-\$13,142.50	-\$6,783.16	-\$4,734.86
APRIL	MAY	JUNE
\$3,825.01	\$7,613.17	\$12,932.54

SOLUTION With this many large numbers, you may want to use a calculator.

13142.50 [+/-] [+] 6783.16 [+/-] [+] 4734.86 [+/-]

[+] 3825.01 [+] 7613.17 [+] 12932.54 [=] -289.8

▶ The display is -289.8. This means the company had a loss of \$289.80.

STUDENT HELP

KEYSTROKE HELP

To enter −5 on a calculator with a [+/-] key, enter 5 [+/-]. To enter −5 on a calculator with a [(-)] key, enter [(-)] 5.

GUIDED PRACTICE

Vocabulary Check ✓

1. Would a business represent *profits* or *losses* with negative numbers? **losses**

Concept Check ✓

2. Is the order in which you add two numbers important? Make a sketch to help explain your answer. What property does this illustrate? **no; commutative property of addition**
3. Show how to model the sum of -3, 2, and -1 in two ways. Make a sketch to illustrate both ways. **See margin.**

Skill Check ✓

4. Write an addition equation to represent the sum modeled on the number line. **$-5 + 9 = 4$**

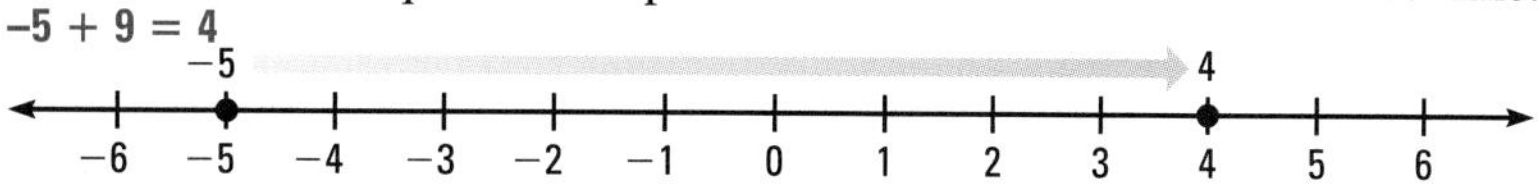

3. Check sketches; *Sample answers:* Start at -3 and move 2 units to the right then 1 unit to the left, ending at -2; start at 2 and move 3 units to the left then 1 more unit to the left, ending at -2.

Find the sum.

5. $-2 + 0$ **-2**
6. $4 + (-3)$ **1**
7. $-2 + (-3)$ **-5**
8. $-7 + 7$ **0**
9. $-1 + 6$ **5**
10. $-4 + \frac{1}{2}$ **$-3\frac{1}{2}$**
11. TEMPERATURE The highest recorded temperature in Hawaii is 100°F. The highest recorded temperature in Colorado is 18°F higher than that of Hawaii. What is the highest recorded temperature in Colorado? **118°F**

▶ Source: National Oceanographic and Atmospheric Administration

PRACTICE AND APPLICATIONS

STUDENT HELP

Extra Practice to help you master skills is on p. 798.

NUMBER LINE SUMS **Use a number line to find the sum.**

12. $-8 + 12$ **4**
13. $2 + (-5)$ **-3**
14. $-3 + (-3)$ **-6**
15. $-3 + (-7)$ **-10**
16. $-4 + 5$ **1**
17. $-10 + 4$ **-6**
18. $-5 + 8 + (-2)$ **1**
19. $2 + (-9) + 3$ **-4**
20. $-5 + 8 + \left(-3\frac{1}{2}\right)$ **$-\frac{1}{2}$**

RULES OF ADDITION **Find the sum.**

21. $-4 + 6$ **2**
22. $17 + 35$ **52**
23. $19 + 0$ **19**
24. $0 + (-5)$ **-5**
25. $-13 + (-6)$ **-19**
26. $14 + (-11)$ **3**
27. $-5 + 10 + (-3)$ **2**
28. $-4 + 10 + (-6)$ **0**
29. $-11.6 + 6.4 + (-3.0)$ **-8.2**
30. $5.7 + (-9.5) + 5.2$ **1.4**
31. $6.8 + 3.3 + (-4.1)$ **6**
32. $9.8 + (-6.3) + (-7.2)$ **-3.7**

ADDITION PROPERTIES **Name the property that makes the statement true.**

33. $-8 + 0 = -8$ **identity property of addition**
34. $2 + (-3) = -3 + 2$ **commutative property of addition**
35. $-2 + 2 = 0$ **property of zero**
36. $(-4 + 3) + 1 = -4 + (3 + 1)$ **associative property of addition**

STUDENT HELP

HOMEWORK HELP

Example 1: Exs. 12–17
Example 2: Exs. 18–20
Example 3: Exs. 21–36
Example 4: Exs. 52–54
Example 5: Exs. 37–42, 55

FINDING SUMS **Find the sum. Use a calculator if you wish.**

37. $-2.95 + 5.76 + (-88.6)$ **-85.79**
38. $10.97 + (-51.14) + (-40.97)$ **-81.14**
39. $20.37 + 190.8 + (-85.13)$ **126.04**
40. $300.3 + (-22.24) + 78.713$ **356.773**
41. $-1.567 + (-2.645) + 5308.34$ **5304.128**
42. $-7344.28 + 2997.65 + (-255.11)$ **-4601.74**

3 APPLY

ASSIGNMENT GUIDE

BASIC
Day 1: pp. 75–77 Exs. 12–50 even, 56–60, 65, 70, 75, 80–82

AVERAGE
Day 1: pp. 75–77 Exs. 12–50 even, 51, 55–60, 65, 70, 75, 80–82

ADVANCED
Day 1: pp. 75–77 Exs. 12–50 even, 51–54, 58–61, 65, 70, 75, 80–82

BLOCK SCHEDULE WITH 2.3
pp. 75–77 Exs. 12–50 even, 51, 55–60, 65, 70, 75, 80–82

EXERCISE LEVELS

Level A: ***Easier***
12–32, 37–42

Level B: ***More Difficult***
33–36, 43–51, 56–60, 62–82

Level C: ***Most Difficult***
52–55, 61

✓ HOMEWORK CHECK

To quickly check student understanding of key concepts, go over the following exercises: Exs. 18, 30, 36, 44. See also the Daily Homework Quiz:

- Blackline Master (*Chapter 2 Resource Book,* p. 40)
- Transparency (p. 12)

STUDENT HELP NOTES

Look Back As students look back to p.3, remind them to replace the variable with the given value and then use the associative and commutative properties to simplify the resulting expression.

ADDITIONAL PRACTICE AND RETEACHING

For Lesson 2.2:

- Practice Levels A, B, and C (*Chapter 2 Resource Book*, p. 30)
- Reteaching with Practice (*Chapter 2 Resource Book*, p. 33)
- See Lesson 2.2 of the *Personal Student Tutor*

For more Mixed Review:

- Search the *Test and Practice Generator* for key words or specific lessons.

STUDENT HELP

Look Back
For help with evaluating expressions, see p. 3.

EVALUATING EXPRESSIONS **Evaluate the expression for the given value of *x*.**

43. $5 + x + (-8); x = 2$ **−1**

44. $4 + x + 10 + (-10); x = 3$ **7**

45. $-24 + 6 + x; x = 8$ **−10**

46. $-6 + x + 4; x = -3$ **−5**

47. $2 + (-5) + x + 14; x = -8$ **3**

48. $-11 + (-2) + 11 + x; x = -10$ **−12**

49. $x + (-6) + (-11); x = -7$ **−24**

50. $9 + x + (-8) + (-3); x = -12$ **−14**

51. **CHAMPIONSHIP GAME** In the game that decides the high school football championship, your team needs to gain 14 yards to score a touchdown and win. Your team's final four plays result in a 9-yard gain, a 5-yard loss, a 4-yard gain, and a 5-yard gain as time runs out. Use a number line to model the gains and losses. Did your team win? **no; $9 + (-5) + 4 + 5 = 13$**

FOCUS ON APPLICATIONS

SODIUM is one of the two elements that make salt, NaCl. Salt, or sodium chloride, is a naturally occurring mineral mined from the ground.

SCIENCE CONNECTION **In Exercises 52–54, use the information in Example 4 on page 74 and in the table below. The table shows the number of electrons and protons in atoms of sodium (Na) and fluorine (F).**

Atom	Electrons	Protons
Na atom 1	10	11
Na atom 2	11	11
F atom 1	9	9
F atom 2	10	9

52. Find the total charge of each sodium atom. Then decide whether the atom is an ion. Which atom has a symbol of Na^+? **Na atom 1: 1, ion; Na atom 2: 0, not an ion; Na atom 1**

53. Find the total charge of each fluorine atom. Then decide whether the atom is an ion. Which atom has a symbol of F^-? **F atom 1: 0, not an ion; F atom 2: −1, ion; F atom 2**

54. **CRITICAL THINKING** You do not need to know the number of neutrons in an atom to find its total charge. Explain why not. Which property of addition supports your answer? **Neutrons have no charge, so they add to and subtract nothing from the total charge; identity property.**

55. **PROFIT AND LOSS** A pest control company had a profit of \$3,514.65 in April, a profit of \$5,674.25 in May, a loss of \$8,992.88 in June, and a loss of \$1,207.03 in July. Did the company make a profit during the 4-month period? Explain. **No; the sum of the losses is greater than the sum of the profits by \$1011.01.**

TIME ZONES **In Exercises 56 and 57, use the time zones map. It shows the time in different zones of the country when it is 1:00 P.M. in California.**

56. A Thanksgiving Day parade in New York City is scheduled to begin at 9:00 A.M. and will be televised live. If you live in Nevada, at what time can you see the parade begin on television? **6:00 A.M.**

57. A New Year's Day parade in California is scheduled to begin at 8:00 A.M. and will be televised live. If you live in Illinois, at what time can you see the parade begin on television? **10:00 A.M.**

Time Zones in the Continental United States

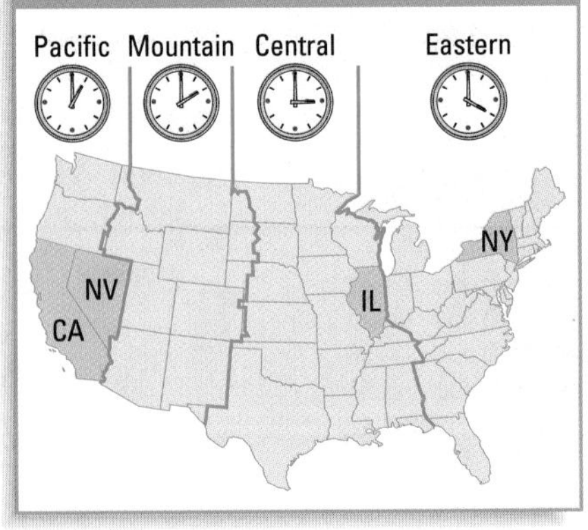

Test Preparation

MULTIPLE CHOICE In Exercises 58–60, use the table, which shows a 6-month record of a household budget used to calculate monthly savings.

Month	\$ Earned	\$ Spent	\$ Saved
Jan.	1676.05	−1427.37	?
Feb.	1554.52	−1771.89	?
Mar.	1851.89	−1556.44	?
Apr.	1567.96	−1874.72	?
May	1921.03	−1602.19	?
June	1667.67	−1989.82	?

58. In which month did the household save the most money? **C**

Ⓐ January Ⓑ March Ⓒ May Ⓓ June

59. In which month did the household's spending most exceed its earnings? **D**

Ⓐ January Ⓑ February Ⓒ April Ⓓ June

60. How much money did the household save during the 6-month period? **B**

Ⓐ \$83.31 Ⓑ \$16.69 Ⓒ \$116.69 Ⓓ \$183.31

★ Challenge

EXTRA CHALLENGE

www.mcdougallittell.com

61. LOGICAL REASONING Formulate the following statement in terms of variables. Then decide whether it is *true* or *false*. *The opposite of the sum of two numbers is equal to the sum of the opposites of the numbers.* If false, give a counterexample. If true, give two examples involving negative numbers.
$-(a+b) = -a + (-b)$; true; *Sample answers:* $-(-2+(-3))$ and $-(-2)+(-(-3))$ are both equal to 5; $-(-4+(-9))$ and $-(-4)+(-(-9))$ are both equal to 13.

MIXED REVIEW

FRACTION OPERATIONS Find the difference. (Skills Review, p. 781)

62. $\frac{4}{5} - \frac{2}{5}$ **$\frac{2}{5}$** **63.** $\frac{8}{9} - \frac{2}{3}$ **$\frac{2}{9}$** **64.** $\frac{3}{4} - \frac{5}{12}$ **$\frac{1}{3}$** **65.** $\frac{7}{8} - \frac{1}{4}$ **$\frac{5}{8}$**

66. $4\frac{2}{3} - 2\frac{1}{5}$ **$2\frac{7}{15}$** **67.** $\frac{5}{6} - \frac{1}{9}$ **$\frac{13}{18}$** **68.** $7\frac{9}{10} - 5\frac{3}{7}$ **$2\frac{33}{70}$** **69.** $\frac{5}{12} - \frac{3}{16}$ **$\frac{11}{48}$**

VARIABLE EXPRESSIONS Evaluate the expression. (Review 1.3)

70. $a^4 + 8$ when $a = 10$ **10,008** **71.** $79 - v^3$ when $v = 4$ **15**

72. $t^2 - 7t + 12$ when $t = 8$ **20** **73.** $2x^2 + 8x - 5$ when $x = 3$ **37**

CHECKING SOLUTIONS Check whether the given number is a solution of the equation. (Review 1.4)

74. $x + 5 = 11$; 7 **no** **75.** $12 - 2a = 18$; 4 **no** **76.** $7y - 15 = 6$; 3 **yes**

77. $3 + 2d = 9 + d$; 6 **yes** **78.** $3w - 7 = w + 1$; 5 **no** **79.** $6z + 5 = 8z - 12$; 8.5 **yes**

PIZZA In Exercises 80–82, use the following information. A pizzeria charges \$6.00 for a large cheese pizza, and \$.85 for each additional topping. The total cost C of a large cheese pizza with n additional toppings is given by $C = 6 + 0.85n$. (Review 1.7)

80. Write an input-output table that shows the total cost of a pizza with 0, 1, 2, 3, 4, and 5 additional toppings.

80.

Input	Output
0	6.00
1	6.85
2	7.70
3	8.55
4	9.40
5	10.25

81. Describe the domain and range of the function.
domain = 0, 1, 2, 3, 4, 5; range = 6.00, 6.85, 7.70, 8.55, 9.40, 10.25

82. Graph the data in the input-output table. **See margin.**

4 ASSESS

DAILY HOMEWORK QUIZ

Transparency Available

Find the sum.

1. $5 + (-10) + 3$ **−2**

2. $-2.4 + 7.6 + (-4.9)$ **0.3**

3. What property of addition makes the following statement true? $-9.5 + (9.5 + (-7.8)) = (-9.5 + 9.5) + (-7.8)$ **associative property**

4. Evaluate the following expression for $x = -5$: $7 + x + (-3)$ **−1**

EXTRA CHALLENGE NOTE

Challenge problems for Lesson 2.2 are available in **blackline** format in the *Chapter 2 Resource Book*, p. 37 and at **www.mcdougallittell.com.**

ADDITIONAL TEST PREPARATION

1. WRITING Explain why you might find the associative and commutative properties helpful when adding three numbers. Give at least one example to demonstrate your reasoning.
When you are adding three or more numbers with different signs, the work is easier if you add all positives together and all negatives together before you find the final sum; $-8.1 + 6.2 + 2.4 + (-7.6) + 15.5 + (-3.4) = (-8.1 + (-7.6) + (-3.4)) + (6.2 + 2.4 + 15.5) = -19.1 + 24.1 = 5$

82. Pizza Costs

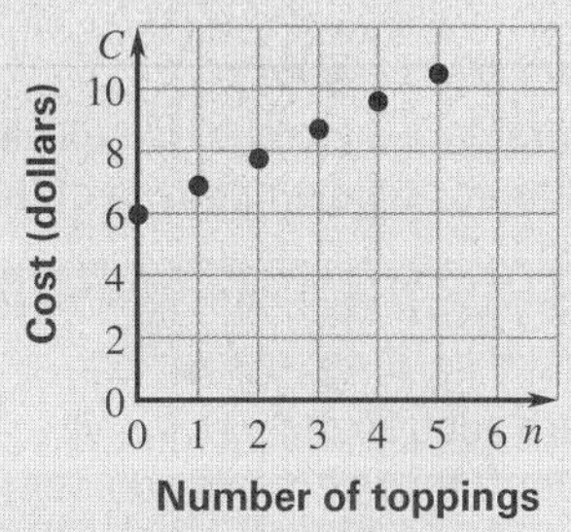

1 Planning the Activity

PURPOSE
To model integer subtraction with algebra tiles.

MATERIALS
- A set of 8 positive and 8 negative algebra tiles per group
- Activity Support Master (*Chapter 2 Resource Book,* p. 41)

PACING
- Exploring the Concept — 10 min
- Drawing Conclusions — 10 min

LINK TO LESSON
After discussing Example 1 of Lesson 2.3, have students check the answers by drawing algebra tile models for parts (*a*) – (*c*).

2 Managing the Activity

COOPERATIVE LEARNING
For each difference in Exercises 1–9, have one student make the algebra tile model while the other makes the sketch. They should switch roles for each exercise.

ALTERNATIVE APPROACH
This activity can be done as a demonstration by drawing the algebra tile models on an overhead sheet. After demonstrating –6 – (–2) and 3 – 6, have the class sketch the models for Exercises 1–9 at their desks while one student does them on the overhead.

3 Closing the Activity

★ ***KEY DISCOVERY***
Subtracting an integer is equivalent to adding the opposite of the integer.

ACTIVITY ASSESSMENT
You are subtracting a negative number from a negative number. Under what circumstances will the difference be positive? Give examples to illustrate your answer.
See sample answer at right.

ACTIVITY 2.3
Developing Concepts

GROUP ACTIVITY
Work with a partner.

MATERIALS
algebra tiles

STUDENT HELP
Look Back
For help with using algebra tiles, see p. 71.

Modeling Subtraction of Integers

QUESTION How can you model the subtraction of integers?

EXPLORING THE CONCEPT: MODELING −6 − (−2)

You can use algebra tiles to find differences.

1. Use algebra tiles to model −6.

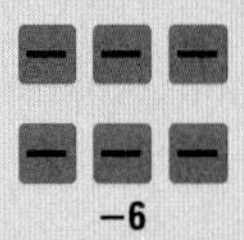
−6

2. Subtract −2 by taking away two −1-tiles.

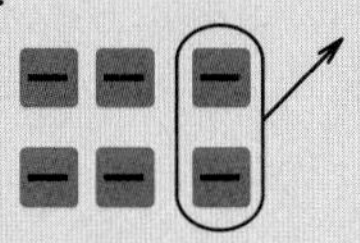
−6 − (−2)

3. The remaining tiles show the difference of −6 and −2.
Complete the statement: −6 − (−2) = _?_. **−4**

EXPLORING THE CONCEPT: MODELING 3 − 6

4. Use algebra tiles to model 3.

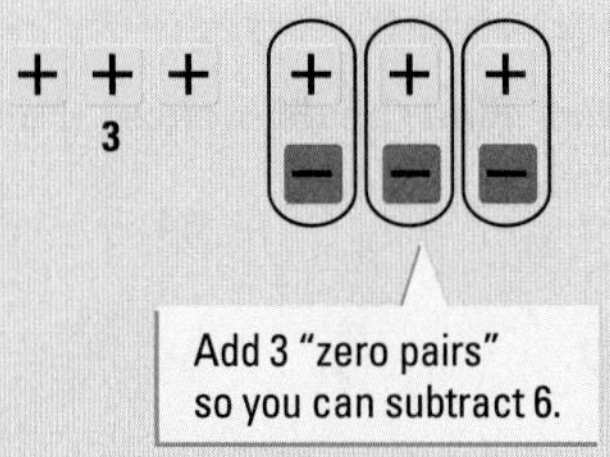
3
Add 3 "zero pairs" so you can subtract 6.

5. Subtract 6 by taking away six 1-tiles.

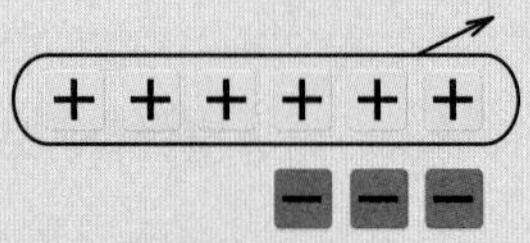
3 − 6

6. The remaining tiles show the difference of 3 and 6.
Complete the statement: 3 − 6 = _?_. **−3**

DRAWING CONCLUSIONS

Use algebra tiles to find the difference. Sketch your solution.

1. 7 − 2 **5**
2. −7 − (−3) **−4**
3. −5 − (−1) **−4**
4. −6 − (−6) **0**
5. 2 − 3 **−1**
6. 4 − 7 **−3**
7. 3 − 5 **−2**
8. 5 − 8 **−3**
9. 0 − (−5) **5**

In Exercises 10 and 11, decide whether the statement is *true* or *false*. Use algebra tiles to show an example that supports your answer. Sketch your example.

10. To subtract a positive integer, add the opposite of the positive integer. **True; check sketches.**
11. To subtract a negative integer, add the opposite of the negative integer. **True; check sketches.**

Activity Assessment *Sample answer:*
The difference will be positive when the absolute value of the original number is less than the absolute value of the number you are subtracting. For example, −3 – (–5) = 2.

2.3 Subtraction of Real Numbers

What you should learn

GOAL 1 Subtract real numbers using the subtraction rule.

GOAL 2 Use subtraction of real numbers to solve **real-life** problems such as finding the differences in stock market prices in **Example 5**.

Why you should learn it

▼ To solve **real-life** problems such as analyzing data on visitors to a historical site in **Ex. 81**.

REAL LIFE

Native American dance at Nez Perce National Historical Park

GOAL 1 SUBTRACTING REAL NUMBERS

Some addition expressions can be evaluated using subtraction.

ADDITION PROBLEM	EQUIVALENT SUBTRACTION PROBLEM
$5 + (-3) = 2$	$5 - 3 = 2$
$9 + (-6) = 3$	$9 - 6 = 3$

Adding the opposite of a number is equivalent to subtracting the number.

SUBTRACTION RULE

To subtract b from a, add the opposite of b to a.

$a - b = a + (-b)$ **Example:** $3 - 5 = 3 + (-5)$

The result is the difference of a and b.

EXAMPLE 1 *Using the Subtraction Rule*

Find the difference.

a. $-4 - 3$ **b.** $10 - 11$ **c.** $11 - 10$ **d.** $-\frac{3}{2} - \left(-\frac{1}{2}\right)$

SOLUTION

a. $-4 - 3 = -4 + (-3)$ Add the opposite of 3.

$= -7$ Use rules of addition.

b. $10 - 11 = 10 + (-11)$ Add the opposite of 11.

$= -1$ Use rules of addition.

c. $11 - 10 = 11 + (-10)$ Add the opposite of 10.

$= 1$ Use rules of addition.

d. $-\frac{3}{2} - \left(-\frac{1}{2}\right) = -\frac{3}{2} + \frac{1}{2}$ Add the opposite of $-\frac{1}{2}$.

$= -1$ Use rules of addition.

....

The commutative property of addition, $a + b = b + a$, shows that the order in which two numbers are added does not affect the result. But as you may have noticed in parts (b) and (c) of Example 1, the order in which two numbers are subtracted does affect the result. Subtraction is *not* commutative.

1 PLAN

PACING

Basic: 2 days
Average: 2 days
Advanced: 2 days
Block Schedule: 0.5 block with 2.2
0.5 block with 2.4

LESSON OPENER
VISUAL APPROACH

An alternative way to approach Lesson 2.3 is to use the Visual Approach Lesson Opener:

- Blackline Master (*Chapter 2 Resource Book*, p. 42)
- Transparency (p. 10)

MEETING INDIVIDUAL NEEDS

- ***Chapter 2 Resource Book***
  Prerequisite Skills Review (p. 5)
  Practice Level A (p. 45)
  Practice Level B (p. 46)
  Practice Level C (p. 47)
  Reteaching with Practice (p. 48)
  Absent Student Catch-Up (p. 50)
  Challenge (p. 52)
- ***Resources in Spanish***
- ***Personal Student Tutor***

NEW-TEACHER SUPPORT

See the Tips for New Teachers on pp. 1–2 of the *Chapter 2 Resource Book* for additional notes about Lesson 2.3.

WARM-UP EXERCISES

Transparency Available

Find the sum.

1. $2 + (-7)$ -5

2. $-3.1 + (-4.6)$ -7.7

Evaluate each function for these values of x: $-1, 0, 1$, and 2. Organize your results in a table.

3. $y = x + 5$

x	y
−1	4
0	5
1	6
2	7

4. $y = -x + 0.1$

x	y
−1	1.1
0	0.1
1	−0.9
2	−1.9

2 TEACH

MOTIVATING THE LESSON
Suppose you owe a friend $25. You can represent this debt as –25. You pay her back $15. You can use subtraction of integers to determine the amount of debt that remains.

EXTRA EXAMPLE 1
Find the difference.
a. $-6-3$ -9
b. $\frac{1}{2}-\frac{2}{3}$ $-\frac{1}{6}$
c. $-15-(-6)$ -9

EXTRA EXAMPLE 2
Evaluate the expression $10-5+14-(-18)$. 37

EXTRA EXAMPLE 3
Find the terms of $3x-5$.
$3x$ and -5

EXTRA EXAMPLE 4
Evaluate the function $y=-x-7$ for these values of x: –2, –1, 0, and 1. Organize your results in a table and describe the pattern.

x	–2	–1	0	1
y	–5	–6	–7	–8

Each time x increases by 1, y decreases by 1.

CHECKPOINT EXERCISES

For use after Examples 1 and 2:

1. Evaluate the expression $-5-(-3)+4-1$. 1

For use after Example 3:

2. Find the terms of $-7-(-2x)$. -7 and $2x$

For use after Example 4:

3. Evaluate the function $y=1-x$ for these values of x: –2, –1, 0, and 1. Organize your results in a table and describe the pattern.

x	–2	–1	0	1
y	3	2	1	0

Each time x is increased by 1, y decreases by 1.

Expressions containing more than one subtraction can also be evaluated by "adding the opposite." To do this, use the left-to-right rule for order of operations.

STUDENT HELP
HOMEWORK HELP
Visit our Web site www.mcdougallittell.com for extra examples.

EXAMPLE 2 *Evaluating Expressions with More than One Subtraction*

Evaluate the expression $3-(-4)-2+8$.

SOLUTION

$3-(-4)-2+8=3+4+(-2)+8$ Add the opposites of -4 and 2.
$=7+(-2)+8$ Add 3 and 4.
$=5+8$ Add 7 and -2.
$=13$ Add 5 and 8.

When an expression is written as a sum, the parts that are added are the **terms** of the expression. For instance, you can write the expression $5-x$ as the sum $5+(-x)$. The terms are 5 and $-x$. You can use the subtraction rule to find the terms of an expression.

EXAMPLE 3 *Finding the Terms of an Expression*

Find the terms of $-9-2x$.

SOLUTION Use the subtraction rule.

$-9-2x=-9+(-2x)$ Rewrite the difference as a sum.

▶ In this form, you can see that the terms of the expression are -9 and $-2x$.

STUDENT HELP
Look Back
For help with functions, see p. 46.

EXAMPLE 4 *Evaluating a Function*

Evaluate the function $y=-5-x$ for these values of x: -2, -1, 0, and 1.

Organize your results in a table and describe the pattern.

SOLUTION

Input	Function	Output
$x=-2$	$y=-5-(-2)$	$y=-3$
$x=-1$	$y=-5-(-1)$	$y=-4$
$x=0$	$y=-5-(0)$	$y=-5$
$x=1$	$y=-5-(1)$	$y=-6$

▶ From the table, you can see that each time x increases by 1, y decreases by 1.

FOCUS ON APPLICATIONS

REAL LIFE STOCK MARKET The Dow Jones utilities average is an average price of the shares of 15 large gas and electric companies. The average is used to show the general trend in the stock prices of these types of companies.

GOAL 2 USING SUBTRACTION IN REAL LIFE

EXAMPLE 5 *Subtracting Real Numbers*

STOCK MARKET You are writing a report about the Dow Jones utilities average. The daily closing averages for one week are given in the table. Find the change in the closing average since the previous day to complete the table.

Date	Feb. 9	Feb. 10	Feb. 11	Feb. 12	Feb. 13
Closing average	266.13	268.08	267.11	269.37	268.02
Change	——	?	?	?	?

▶ Source: *Wall Street Journal*

SOLUTION

Subtract each day's closing average from the closing average for the previous day.

DATE	CLOSING AVERAGE	CHANGE
Feb. 9	266.13	——
Feb. 10	268.08	$268.08 - 266.13 = +1.95$
Feb. 11	267.11	$267.11 - 268.08 = -0.97$
Feb. 12	269.37	$269.37 - 267.11 = +2.26$
Feb. 13	268.02	$268.02 - 269.37 = -1.35$

EXAMPLE 6 *Using a Calculator*

Use the solution from Example 5 to answer the following questions.

a. Write the changes you found in order.

b. Find the difference of the greatest number and the least number.

SOLUTION

a. Written in order from least to greatest, the numbers are as follows.
$-1.35, -0.97, 1.95, 2.26$

b. The difference of the greatest number and the least number is
$2.26 - (-1.35)$.

With a calculator, you can find the difference.

KEYSTROKES	DISPLAY
2.26 [–] 1.35 [+/–] [=]	3.61

▶ The greatest number is 3.61 more than the least number.

STUDENT HELP

INTERNET **KEYSTROKE HELP** Visit our Web site www.mcdougallittell.com to see keystrokes for several models of calculators.

EXTRA EXAMPLE 5
The daily closing prices of a utilities stock for one week are given below. Find the change in the closing price from day to day.

Date	Closing Price	Change
July 12	121.42	—
July 13	122.76	?
July 14	123.51	?
July 15	121.19	?
July 16	117.06	?

+1.34, +0.75, –2.32, –4.13

EXTRA EXAMPLE 6
Use the solution to Extra Example 5.

a. Write the changes you found in order. **–4.13, –2.32, 0.75, 1.34**

b. Find the difference of the greatest number and the least number. **5.47**

CHECKPOINT EXERCISES

For use after Examples 5 and 6:

1. The daily closing prices of a software company's stock for one week are given below.

Date	Closing Price	Change
Nov 2	43.1	—
Nov 3	37.4	?
Nov 4	38.9	?
Nov 5	36.8	?
Nov 6	32.9	?

a. Find the change in the closing price from day to day. **–5.7, +1.5, –2.1, –3.9**

b. Write the changes in order. **–5.7, –3.9, –2.1, 1.5**

c. Find the difference of the greatest number and the least number. **7.2**

CLOSURE QUESTION
How are the addition and subtraction of integers related? **Adding the opposite of a number is equivalent to subtracting the number.**

3 APPLY

ASSIGNMENT GUIDE

BASIC

Day 1: p. 82 Exs. 16–58 even

Day 2: pp. 83–85 Exs. 60–70 even, 76–81, 85, 88, 90, 93–95, Quiz 1 Exs. 1–20

AVERAGE

Day 1: p. 82 Exs. 16–58 even

Day 2: pp. 83–85 Exs. 60–70 even, 71, 72, 76–81, 85, 88, 90, 93–95, Quiz 1 Exs. 1–20

ADVANCED

Day 1: p. 82 Exs. 16–58 even

Day 2: pp. 83–85 Exs. 60–70 even, 73–83, 85, 88, 90, 93–95, Quiz 1 Exs. 1–20

BLOCK SCHEDULE

p. 82 Exs. 16–58 even (with 2.2)
pp. 83–85 Exs. 60–70 even, 71, 72, 76–81, 85, 88, 90, 93–95, Quiz 1 Exs. 1–20 (with 2.4)

EXERCISE LEVELS

Level A: *Easier*
16–29, 65–72

Level B: *More Difficult*
30–64, 73–80, 84–95

Level C: *Most Difficult*
81–83

✔ HOMEWORK CHECK

To quickly check student understanding of key concepts, go over the following exercises: Exs. 22, 26, 32, 42, 52, 62, 77. See also the Daily Homework Quiz:

- Blackline Master (*Chapter 2 Resource Book*, p. 56)
- Transparency (p. 13)

GUIDED PRACTICE

Vocabulary Check ✔

1. Is $7x$ a term of the expression $4y^2 - 7x - 9$? Explain.
No; the terms are the parts that are added, so $-7x$ is a term.

Concept Check ✔

2. Use the number line to complete: $-2 - 5 = \underline{\ ?\ }$. **−7**

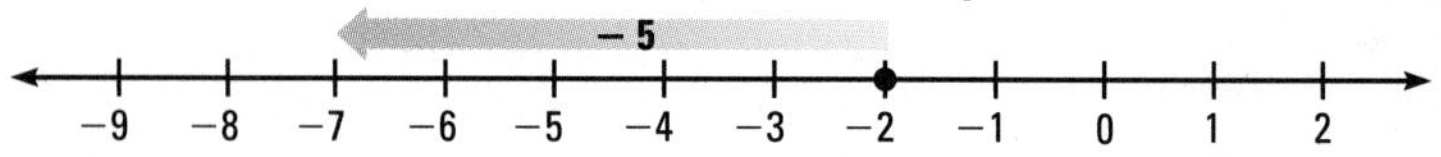

3. Explain the steps you would take to evaluate the expression $5 - 7 - (-4)$. **See margin.**

3. First, use the left-to-right rule: $5 - 7 - (-4) = -2 - (-4)$. Next, add the opposite of -4: $-2 - (-4) = -2 + 4 = 2$.

Skill Check ✔

Use the subtraction rule to rewrite the subtraction expression as an equivalent addition expression. Then evaluate the expression.

4. $4 - 5$ **$4 + (-5)$; -1**
5. $3 - (-8)$ **$3 + 8$; 11**
6. $0 - 7$ **$0 + (-7)$; -7**
7. $2 - (-3) - 6$ **$2 + 3 + (-6)$; -1**
8. $-2.4 - 3$ **$-2.4 + (-3)$; -5.4**
9. $-3.6 - (-6)$ **$-3.6 + 6$; 2.4**
10. $\frac{1}{2} - \frac{1}{4}$ **$\frac{1}{2} + \left(-\frac{1}{4}\right)$; $\frac{1}{4}$**
11. $\frac{2}{3} - \left(-\frac{1}{6}\right) - \frac{1}{3}$ **$\frac{2}{3} + \frac{1}{6} + \left(-\frac{1}{3}\right)$; $\frac{1}{2}$**

Find the terms of the expression.

12. $12 - 5x$ **12 and $-5x$**
13. $-7y^2 + 12y - 6$ **$-7y^2$, $12y$, and -6**
14. $23 - 5w - 8y$ **23, $-5w$, and $-8y$**
15. Evaluate the function $y = -x - 3$ for these values of x: $-2, -1, 0, 1$, and 2. Organize your results in a table. Describe the pattern. **See margin.**

15.

Input	Function	Output
−2	$y = -(-2) - 3$	−1
−1	$y = -(-1) - 3$	−2
0	$y = -0 - 3$	−3
1	$y = -1 - 3$	−4
2	$y = -2 - 3$	−5

Each time x increases by 1, y decreases by 1.

PRACTICE AND APPLICATIONS

STUDENT HELP

Extra Practice to help you master skills is on p. 798.

DIFFERENCES **Find the difference.**

16. $4 - 9$ **−5**
17. $6 - 13$ **−7**
18. $8 - (-5)$ **13**
19. $-2 - (-7)$ **5**
20. $-11 - 8$ **−19**
21. $24 - 39$ **−15**
22. $137 - 355$ **−218**
23. $-65 - (-59)$ **−6**
24. $12.5 - 9.8$ **2.7**
25. $-3.2 - 1.7$ **−4.9**
26. $5.4 - (-3.8)$ **9.2**
27. $-6.6 - (-16.1)$ **9.5**
28. $\frac{4}{3} - \frac{7}{3}$ **−1**
29. $\frac{3}{4} - \left(-\frac{9}{4}\right)$ **3**
30. $-\frac{5}{8} - \frac{3}{4}$ **$-1\frac{3}{8}$**
31. $-\frac{9}{10} - \frac{1}{4}$ **$-1\frac{3}{20}$**
32. $6 - |-2|$ **4**
33. $15 - |-6|$ **9**
34. $|5| - 7.9$ **−2.9**
35. $34.1 - |-57.2|$ **−23.1**

EVALUATING EXPRESSIONS **Evaluate the expression.**

36. $2 - (-4) - 7$ **−1**
37. $8 - 11 - (-6)$ **3**
38. $4 + (-3) - (-5)$ **6**
39. $3 - (-8) + (-9)$ **2**
40. $14 + (-7) - 12$ **−5**
41. $-7 + 42 - 63$ **−28**
42. $-8 - (-12) + 3$ **7**
43. $6 - 1 + 10 - (-8)$ **23**
44. $14 - 8 + 17 - (-23)$ **46**
45. $2.3 + (-9.1) - 1.2$ **−8**
46. $1.3 + (-1.3) - 4.2$ **−4.2**
47. $-8.5 - 3.9 + (-16.2)$ **−28.6**
48. $8.4 - 5.2 - (-4.7)$ **7.9**
49. $-\frac{4}{9} - \frac{2}{3} + \left(-\frac{5}{6}\right)$ **$-1\frac{17}{18}$**
50. $\frac{7}{12} - \left(-\frac{3}{4}\right) + \left(-\frac{1}{8}\right)$ **$1\frac{5}{24}$**

STUDENT HELP

HOMEWORK HELP
Example 1: Exs. 16–35
Example 2: Exs. 36–50
Example 3: Exs. 51–58
Example 4: Exs. 59–64
Example 5: Exs. 71–75
Example 6: Exs. 65–70

FINDING TERMS **Find the terms of the expression.**

51. $-4 - y$ **-4 and $-y$**
52. $-x - 7$ **$-x$ and -7**
53. $-3x + 6$ **$-3x$ and 6**
54. $9 - 28x$ **9 and $-28x$**
55. $-9 + 4b$ **-9 and $4b$**
56. $a + 3b - 5$ **a, $3b$, and -5**
57. $x - y - 7$ **x, $-y$, and -7**
58. $-3x + 5 - 8y$ **$-3x$, 5, and $-8y$**

EVALUATING FUNCTIONS Evaluate the function for these values of x: -2, -1, 0, and 1. Organize your results in a table. 59–64. See margin.

59. $y = x - 8$	**60.** $y = 12 - x$	**61.** $y = -x + 12.1$
62. $y = -8.5 - (-x)$	**63.** $y = 27 + x$	**64.** $y = -x + 13 - x$

EVALUATING EXPRESSIONS Evaluate the expression. Use estimation to check your answer.

65. $5.3 - (-2.5) - 4.7$ **3.1**	**66.** $8.9 - (-2.1) - 7.3$ **3.7**
67. $-4.89 + 2.69 - (-3.74)$ **1.54**	**68.** $-7.85 + 5.96 - (-2.49)$ **0.6**
69. $-13.87 - (-13.87) + 5.8$ **5.8**	**70.** $-15.7 + 0.01 + (-34.44)$ **-50.13**

71. SUBMARINE DEPTH A submarine is at a depth of 725 feet below sea level. Five minutes later, it is at a depth of 450 feet below sea level. What is the change in depth of the submarine? Did it go up or down? **275 ft; up**

72. GASOLINE PRICES Last month the price of gasoline was \$1.19 per gallon. This month the price of gasoline is \$1.13 per gallon. What was the change in the price per gallon of gasoline? **−\$.06/gal**

GOLD PRICES IN LONDON **At 9 A.M., an ounce of gold sells for \$287.56. At noon, gold sells for \$286.90 per ounce. At 4 P.M., the final price for the day is \$287.37 per ounce.**

73. What is the change in the price per ounce of gold from 9 A.M. to noon? **−\$.66/oz**

74. What is the change in the price per ounce of gold from noon to 4 P.M.? **\$.47/oz**

75. What is the change in the price per ounce of gold from 9 A.M. to 4 P.M.? **−\$.19/oz**

HIKING **Use the diagram for Exercises 76–78. It shows the elevation above sea level (in feet) of various points along the Mount Langdon Trail in the White Mountain National Forest in New Hampshire.**

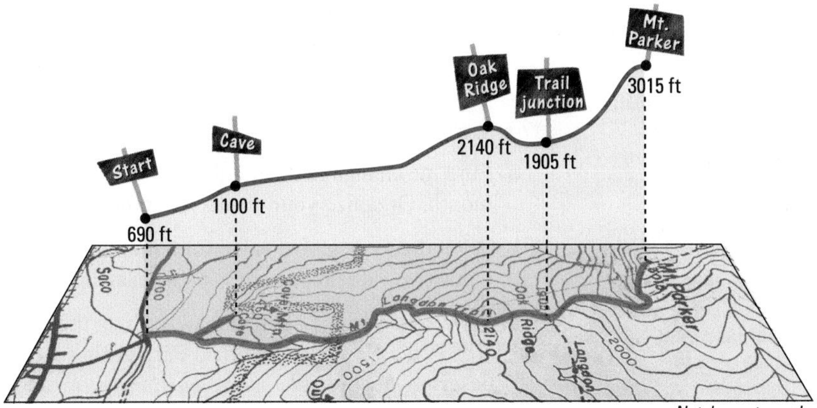

Not drawn to scale

76. How much higher is the cave than the start of the trail? **410 ft**

77. Determine the change in elevation from each point to the next in the diagram. **See margin.**

78. You hike along the trail, visit a cave, and then climb Oak Ridge. From there, you reach a trail junction and continue to the top of Mt. Parker. Find the total of these changes to determine your overall change in elevation. **2325 ft**

White Mountain National Forest, NH

! COMMON ERROR
EXERCISES 32–35 Students may treat absolute value bars as if they were parentheses and distribute the negative sign through them. Remind them to evaluate within the bars first. In Exercise 32, for example, $6 - |-2|$ is equivalent to $6 + (-2)$, not $6 + 2$.

! COMMON ERROR
EXERCISES 59–64 To avoid mistakes when substituting a negative number for a variable, encourage students to replace the variable with the number in parentheses and then simplify the resulting expression in the next step.

TEACHING TIP
EXERCISES 65–70 Stress the importance of estimating to check the calculator answer. When evaluating expressions like the ones in these exercises, it is very easy to enter a negative number as positive. Estimations will help detect this type of error.

ENGLISH LEARNERS
EXERCISES 71–75 While English language learners may be familiar with *feet* as a unit of measurement, they may not be familiar with *gallons* and *ounces*. Point out that *gallons* are used to measure amounts of liquids, while *ounces* measure weight.

59–64, 77. See Additional Answers beginning on page AA1.

ADDITIONAL PRACTICE AND RETEACHING

For Lesson 2.3:
- Practice Levels A, B, and C (*Chapter 2 Resource Book,* p. 45)
- Reteaching with Practice (*Chapter 2 Resource Book,* p. 48)
- See Lesson 2.3 of the *Personal Student Tutor*

For more Mixed Review:
- Search the *Test and Practice Generator* for key words or specific lessons.

4 ASSESS

DAILY HOMEWORK QUIZ

Transparency Available

Find the difference.

1. $-7.2-(-18.5)$ 11.3
2. $25-|-32|$ −7
3. Evaluate $6-12-(-10)+(-5)$. −1
4. What are the terms of $3m-2n-5$? $3m, -2n, -5$
5. Evaluate $y=15-(-x)$ for $x=-2, -1, 0$ and 1. 13, 14, 15, 16
6. A cat jumps from a tree branch 18 ft above the ground to a branch 10 ft up, to the top of a fence 6 ft above the ground, and then to the ground. Find the change in elevation from each point to the next. −8 ft, −4 ft, −6 ft

EXTRA CHALLENGE NOTE

Challenge problems for Lesson 2.3 are available in **blackline** format in the *Chapter 2 Resource Book*, p. 52 and at **www.mcdougallittell.com.**

ADDITIONAL TEST PREPARATION

1. **OPEN ENDED** Describe a real life situation which could be modeled by a negative number subtracted from a positive number.
   See sample answer, below.
2. **WRITING** If a and b are both negative integers, explain how to determine whether the result of the subtraction $a-b$ is positive, negative, or zero.
   See sample answer, below.

79. from cloud to mountain top: −7301 ft; from mountain top to mountain stream: −662 ft; from mountain stream to base of tree: −1883 ft; from base of tree to floating leaf: 77 ft; from floating leaf to lake: −1311 ft; from lake to cloud: 8021 ft

81a.

Month	Change
Jan.	—
Feb.	−5,000
Mar.	5,000
Apr.	1,000
May	5,000
June	13,000
July	3,000
Aug.	−5,000
Sept.	−7,000
Oct.	−7,000
Nov.	10,000
Dec.	−12,000

Test Preparation

81c. A good answer should include several of the following:

- Nez Perce Park has the most visitors in July and the fewest in February.
- The number of visitors increases each month from its low point in February to its high point in July.
- The number of visitors decreases steadily from July to February with the exception of an increase in the number in November.
- The greatest increase occurs from May to June and the greatest decrease from November to December.

★ **Challenge**

EXTRA CHALLENGE

www.mcdougallittell.com

SCIENCE CONNECTION **In Exercises 79 and 80, use the diagram, which shows the journey of a water molecule from left to right.**

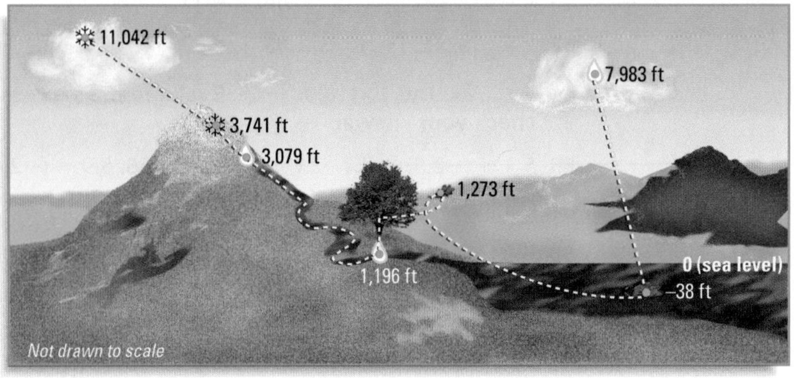

79. Find the change in elevation from each point to the next point. **See margin.**

80. Write an expression using addition and subtraction that models the journey of the water molecule. Then evaluate the expression.
11,042 − 7,301 − 662 − 1,883 + 77 − 1, 311 + 8,021

81. **MULTI-STEP PROBLEM** Use the bar graph, which shows the number of visitors to Nez Perce National Historical Park during 1997.

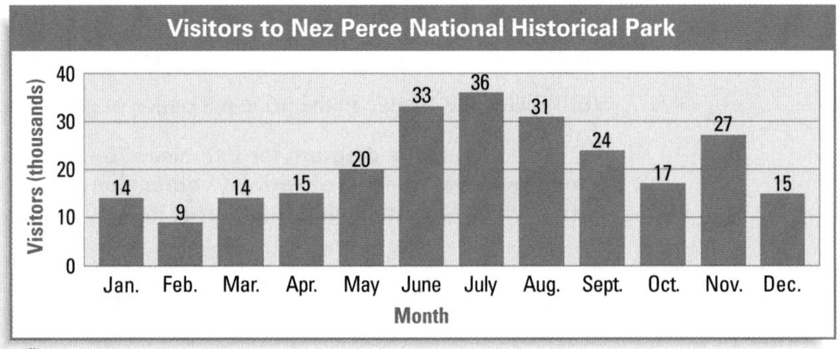

DATA UPDATE of National Park Service data at www.mcdougallittell.com

a. Find the change in the number of visitors from each month to the next month. Organize your results in a table. **See margin.**

b. What does a negative value for a monthly change represent? What does a positive value represent? **a decrease in the number of visitors; an increase in the number of visitors**

c. *Writing* Write a paragraph describing the pattern of visitors to Nez Perce National Historical Park during the year. Use mathematically descriptive words like *most*, *least*, *increase*, and *decrease*. **See margin.**

LOGICAL REASONING **Decide whether the statement is *true* or *false*. Use the subtraction rule or a number line to support your answer.**

82. If you subtract a negative number from a positive number, the result is always a positive number. **True; to subtract the negative number, you add its opposite, which is positive. The sum of two positive numbers is positive.**

83. If you subtract a positive number from a negative number, the result is always a negative number. **True; to subtract the positive number, you add its opposite, which is negative. The sum of two negative numbers is negative.**

Additional Test Preparation *Sample answers:*

1. You want to find the number of days between two days from now and five days ago. This would be modeled by $2-(-5)=7$.
2. If the absolute value of a is greater than the absolute value of b, $a-b$ is negative, if $a=b$, $a-b$ is zero. Otherwise, $a-b$ is positive.

MIXED REVIEW

EVALUATING NUMERIC EXPRESSIONS **Evaluate the expression.** (Review 1.3)

84. $89 - 8 \cdot 5 - 27$ **22** **85.** $\frac{10}{3} - \frac{2}{3} \cdot 4 + 5$ **$5\frac{2}{3}$** **86.** $12 \cdot 9 \div 6 - 13.5$ **4.5**

87. $17 + 100 \div 25 - 5$ **16** **88.** $5 \cdot \frac{8}{9} - \frac{6}{9} + 51 \div 3$ **$20\frac{7}{9}$** **89.** $13 + 11 \cdot 7 - 6 \div 3$ **88**

90. $25 - \left[\frac{3}{10}(6 \cdot 5) - 2\right]$ **18** **91.** $(27 \div 9) \div (7 - 5)$ **$1\frac{1}{2}$** **92.** $[(12 \cdot 9) \div 6] - 13.5$ **4.5**

SPORTS **In Exercises 93 and 94, use the table, which shows the number of male and female participants in high school sports for three years. Based on the table, decide whether the statement is *true* or *false*.** (Review 1.6)

High School Sports Participants (millions)		
Year	Male	Female
1994–95	3.54	2.24
1995–96	3.63	2.37
1996–97	3.71	2.24

▶ Source: National Federation of State High School Associations

93. There were more than six million total participants during 1994–1995. **false**

94. During the three-year period, female participation grew more than male participation. **false**

95. PROFIT AND LOSS A landscaping company had a loss of \$5,126.55 in March. It then had a profit of \$2,943.21 in April, a profit of \$4,988.97 in May, and a loss of \$1,807.81 in June. Did the company make a profit during the four-month period? Explain. (Review 2.2 for 2.4) **Yes; the sum of the profits, \$7932.18, exceeds the sum of the losses, \$6934.36, by \$997.82.**

QUIZ 1

Self-Test for Lessons 2.1–2.3

Write the numbers in increasing order. (Lesson 2.1)

1. 5.31, 5.04, −5.32, −6.2, 6.3, 5.3 **−6.2, −5.32, 5.04, 5.3, 5.31, 6.3**

2. −1.07, 1.06, 1.16, −1.6, 0.18, −0.28 **−1.6, −1.07, −0.28, 0.18, 1.06, 1.16**

3. $7.3, -7.5, 7\frac{2}{3}, -7\frac{1}{3}, 7.5, 7\frac{1}{3}$ **$-7.5, -7\frac{1}{3}, 7.3, 7\frac{1}{3}, 7.5, 7\frac{2}{3}$**

4. $-6\frac{2}{5}, 6.42, \frac{33}{5}, -6.3, -\frac{33}{5}, 6.05$ **$-\frac{33}{5}, -6\frac{2}{5}, -6.3, 6.05, 6.42, \frac{33}{5}$**

Evaluate the expression. (Lesson 2.1)

5. $|3.76|$ **3.76** **6.** $|-75|$ **75** **7.** $-|345|$ **−345**

8. $|-27.5|$ **27.5** **9.** $14 - |-7|$ **7** **10.** $|-75| - 7.6$ **67.4**

Evaluate the expression. (Lessons 2.2 and 2.3)

11. $-11 + 35$ **24** **12.** $29 - 501$ **−472** **13.** $-17 - (-14)$ **−3**

14. $-8 + 12 + (-5)$ **−1** **15.** $-32 - (-27) - 9$ **−14** **16.** $35 - 0 - (-19)$ **54**

17. $12 + (-1.2) + (-7)$ **3.8** **18.** $-143 - (-60) - 8$ **−91** **19.** $14.1 + (-75.2) + 60.7$ **−0.4**

20. POOL DEPTH The water in a pool is 47.3 inches deep on Monday. On Tuesday, 2.1 inches of depth is splashed out. On Wednesday, the depth decreases 11.3 inches due to a leak. On Thursday night, the leak is fixed and 12.9 inches of depth is added overnight. Write and evaluate an expression to find the depth of the water in the pool on Friday morning. (Lesson 2.3) **47.3 + (−2.1) + (−11.3) + 12.9; 46.8 in.**

ADDITIONAL RESOURCES

An alternative Quiz for Lessons 2.1–2.3 is available in the *Chapter 2 Resource Book*, p. 53.

PACING
Basic: 1 day
Average: 1 day
Advanced: 1 day
Block Schedule: 0.5 block with 2.3

➡ **LESSON OPENER**
APPLICATION
An alternative way to approach
Lesson 2.4 is to use the
Application Lesson Opener:

• Blackline Master (*Chapter 2
Resource Book,* p. 57)
• 🖳 Transparency (p. 11)

MEETING INDIVIDUAL NEEDS
• *Chapter 2 Resource Book*
Prerequisite Skills Review (p. 5)
Practice Level A (p. 60)
Practice Level B (p. 61)
Practice Level C (p. 62)
Reteaching with Practice (p. 63)
Absent Student Catch-Up (p. 65)
Challenge (p. 67)
• *Resources in Spanish*
• 🖳 *Personal Student Tutor*

NEW-TEACHER SUPPORT
See the Tips for New Teachers on
pp. 1–2 of the *Chapter 2 Resource
Book* for additional notes about
Lesson 2.4.

WARM-UP EXERCISES

🖳 *Transparency Available*

1. Evaluate the expression
 a. −45,261 + (−1065) −46,326
 b. −2 − 15 −17
2. The monthly profit or loss for
 a store during a three month
 period is given below. What
 was the total profit or loss
 during this time? **$2797**

January	$1,341
February	−$1,105
March	$2,561

**EXPLORING DATA
AND STATISTICS**

2.4

What you should learn

GOAL 1 Organize data in
a matrix.

GOAL 2 Add and subtract
two matrices.

Why you should learn it

To organize data, such as
the number of male and
female members of the
United States Congress in
Example 3.

Adding and Subtracting Matrices

GOAL 1 **ORGANIZING DATA IN A MATRIX**

A **matrix** is a rectangular arrangement of numbers into horizontal rows and
vertical columns. Each number in the matrix is called an **entry** or an **element**.
(The plural of *matrix* is *matrices*.)

$$
\text{row} \rightarrow
\begin{bmatrix}
3 & 1 & 0 & 8 \\
-1 & 2 & 4 & -2 \\
5 & -7 & 3 & 6
\end{bmatrix}
$$

column

The entry in the second row
and third column is 4.

The size of a matrix is described as follows.

$$(\text{the number of rows}) \times (\text{the number of columns})$$

The matrix above is a 3×4 (read "3 by 4") matrix, because it has three rows
and four columns. Think of a matrix as a type of table that can be used to
organize data.

Two matrices are equal if the entries in corresponding positions are equal.

$$
\begin{bmatrix} 3 & -2 \\ \frac{1}{2} & 0 \end{bmatrix} =
\begin{bmatrix} 3 & -2 \\ 0.5 & 0 \end{bmatrix}
\qquad
\begin{bmatrix} -4 & 7 \\ 0 & -1 \end{bmatrix} \neq
\begin{bmatrix} 7 & -4 \\ 0 & -1 \end{bmatrix}
$$

EXAMPLE 1 *Writing a Matrix*

Write a matrix to organize the following information about your CD collection.

Country: 4 groups, 6 solo artists, 0 collections

Rock: 8 groups, 3 solo artists, 3 collections

Blues: 1 group, 5 solo artists, 2 collections

SOLUTION

Country, Rock, and *Blues* can be labels for the rows or for the columns.

AS ROW LABELS:

	Group	Solo artist	Collection
Country	4	6	0
Rock	8	3	3
Blues	1	5	2

AS COLUMN LABELS:

	Country	Rock	Blues
Group	4	8	1
Solo artist	6	3	5
Collection	0	3	2

Extra Example 1

	Entrees	Desserts	Drinks
High School	5	3	5
Middle School	3	2	4

Extra Example 2

a. $\begin{bmatrix} 0 & 0 & 6 \\ 4 & 1 & -13 \end{bmatrix}$ **b.** $\begin{bmatrix} 4 & -6 & 3 \\ 4 & -2 & -1 \end{bmatrix}$

GOAL 2 ADDING AND SUBTRACTING MATRICES

To add or subtract matrices, you add or subtract corresponding entries. Each matrix must have the same number of rows and columns. For instance, you cannot add a matrix that has three rows to a matrix that has only two rows.

STUDENT HELP

INTERNET HOMEWORK HELP Visit our Web site www.mcdougallittell.com for extra examples.

EXAMPLE 2 *Adding and Subtracting Matrices*

a. $$\begin{bmatrix} 4 & 2 \\ 0 & -3 \\ -5 & 1 \end{bmatrix} + \begin{bmatrix} 1 & 0 \\ 2 & -1 \\ 6 & -4 \end{bmatrix} = \begin{bmatrix} 4+1 & 2+0 \\ 0+2 & -3+(-1) \\ -5+6 & 1+(-4) \end{bmatrix}$$

$$= \begin{bmatrix} 5 & 2 \\ 2 & -4 \\ 1 & -3 \end{bmatrix}$$

b. $$\begin{bmatrix} 10 & -6 \\ 5 & 0 \end{bmatrix} - \begin{bmatrix} 4 & 5 \\ -3 & 2 \end{bmatrix} = \begin{bmatrix} 10-4 & -6-5 \\ 5-(-3) & 0-2 \end{bmatrix}$$

$$= \begin{bmatrix} 6 & -11 \\ 8 & -2 \end{bmatrix}$$

EXAMPLE 3 *Political Composition of U.S. Congress*

CONGRESS The United States Congress is composed of the House of Representatives and the Senate. The matrices below show the number of men and women in the Senate and the House at the 1999 start of the 106th Congress. Write and label a single matrix that shows the number of men and women in Congress in 1999. **INTERNET DATA UPDATE** of United States Congress data at www.mcdougallittell.com

HOUSE

	Men	Women
Democrat	172	39
Republican	206	17
Other	1	0

SENATE

	Men	Women
Democrat	39	6
Republican	52	3
Other	0	0

SOLUTION Add the two matrices. Then label the result.

$$\begin{bmatrix} 172 & 39 \\ 206 & 17 \\ 1 & 0 \end{bmatrix} + \begin{bmatrix} 39 & 6 \\ 52 & 3 \\ 0 & 0 \end{bmatrix} = \begin{bmatrix} 211 & 45 \\ 258 & 20 \\ 1 & 0 \end{bmatrix}$$

▶ The result can be written as follows.

CONGRESS

	Men	Women
Democrat	211	45
Republican	258	20
Other	1	0

FOCUS ON PEOPLE

REAL LIFE MARGARET CHASE SMITH was the first woman to serve in Congress in both the Senate and the House.

2 TEACH

EXTRA EXAMPLE 1

Write a matrix to organize the following information.
High school: 5 entrees, 3 desserts, 5 drinks
Middle school: 3 entrees, 2 desserts, 4 drinks
See margin page 86.

EXTRA EXAMPLE 2

a. $\begin{bmatrix} 3 & -1 & 0 \\ 2 & 1 & -5 \end{bmatrix} + \begin{bmatrix} -3 & 1 & 6 \\ 2 & 0 & -8 \end{bmatrix}$

b. $\begin{bmatrix} 4 & -3 & 4 \\ 3 & 0 & -2 \end{bmatrix} - \begin{bmatrix} 0 & 3 & 1 \\ -1 & 2 & -1 \end{bmatrix}$

See margin page 86.

EXTRA EXAMPLE 3

The matrices show the House and the Senate in 1917. Write and label a single matrix that shows the number of men and women in Congress in 1917.

House	Men	Women
Democrat	210	0
Republican	215	1
Other	9	0

Senate	Men	Women
Democrat	53	0
Republican	42	0
Other	1	0

See answer below.

CHECKPOINT EXERCISES

For use after Examples 1–3:

1. The matrices below show the inventory of an auto dealership before a delivery and the cars that were delivered. Write and label a single matrix that gives the auto dealership's inventory after the delivery.

Before	SUVs	Cars
Compact	15	18
Full Size	21	29
Luxury	12	2

Delivered	SUVs	Cars
Compact	5	7
Full Size	9	16
Luxury	3	1

See answer at left.

Congress	Men	Women
Democrat	263	0
Republican	257	1
Other	10	0

1.

After Delivery	SUVs	Cars
Compact	20	25
Full Size	30	45
Luxury	15	3

EXTRA EXAMPLE 4

You own a restaurant that has two different locations. The matrices below show revenue and expenses for 3 months at each location.

Revenue ($)

$$\begin{array}{l} \\ \text{June} \\ \text{July} \\ \text{August} \end{array} \begin{array}{c} \text{Loc. 1} \quad \text{Loc. 2} \\ \begin{bmatrix} 31{,}042 & 46{,}712 \\ 42{,}793 & 40{,}670 \\ 33{,}061 & 42{,}543 \end{bmatrix} \end{array}$$

Expenses ($)

$$\begin{array}{l} \\ \text{June} \\ \text{July} \\ \text{August} \end{array} \begin{array}{c} \text{Loc. 1} \quad \text{Loc. 2} \\ \begin{bmatrix} 30{,}114 & 43{,}998 \\ 42{,}997 & 37{,}940 \\ 32{,}003 & 41{,}056 \end{bmatrix} \end{array}$$

a. Write a matrix that shows the monthly profit for each location. **See below.**

b. Which location had higher overall profits during the 3-month period? **Location 2**

c. Which location lost money? In which month? **Location 1; July**

CHECKPOINT EXERCISES

For Use after Example 4:

1. The Revenue and Expenses for two medical offices for a 2-month period are shown below.

Revenue ($)

$$\begin{array}{l} \\ \text{April} \\ \text{May} \end{array} \begin{array}{c} \text{Office 1} \quad \text{Office 2} \\ \begin{bmatrix} 11{,}041 & 9614 \\ 14{,}397 & 8462 \end{bmatrix} \end{array}$$

Expenses ($)

$$\begin{array}{l} \\ \text{April} \\ \text{May} \end{array} \begin{array}{c} \text{Office 1} \quad \text{Office 2} \\ \begin{bmatrix} 10{,}914 & 9026 \\ 13{,}142 & 6489 \end{bmatrix} \end{array}$$

a. Write a matrix that shows the monthly profit for each office. **See below.**

b. Which office has the higher overall profit during the 2-month period? **Office 2**

CLOSURE QUESTION

Two matrices are added. Their sum is a matrix with all zero entries. What would need to be true about the matrices? What would need to be true about the corresponding entries of the matrices? **They are the same size; the corresponding entries are opposites.**

ORGANIZING DATA As you learned in Example 1, matrices are a useful way to organize and keep track of data. For example, if you have a business, it is important to keep track of revenue (or income) and expenses (or costs). You can find out how much profit has been made by subtracting expenses from revenue. If the profit is a negative number, you lost money.

REAL LIFE *Business*

EXAMPLE 4 *Finding a Profit Matrix*

You own two stores that sell household appliances. The matrices below show revenue and expenses for three months at each store.

REVENUE ($)

$$\begin{array}{r} \\ \text{January} \\ \text{February} \\ \text{March} \end{array} \begin{array}{c} \text{Store 1} \quad \text{Store 2} \\ \begin{bmatrix} 78{,}432 & 109{,}345 \\ 82{,}529 & 120{,}429 \\ 94{,}311 & 118{,}782 \end{bmatrix} \end{array}$$

EXPENSES ($)

$$\begin{array}{r} \\ \text{January} \\ \text{February} \\ \text{March} \end{array} \begin{array}{c} \text{Store 1} \quad \text{Store 2} \\ \begin{bmatrix} 59{,}426 & 98{,}459 \\ 64{,}372 & 104{,}972 \\ 85{,}456 & 120{,}833 \end{bmatrix} \end{array}$$

a. Write a matrix that shows the monthly profit for each store.

b. Which store had higher overall profits during the three-month period?

c. Which store lost money? In which month?

SOLUTION

Profit is the difference of revenue and expenses.

a. To find the *profit matrix,* you can subtract the *expenses matrix* from the *revenue matrix.*

$$\begin{bmatrix} 78{,}432 & 109{,}345 \\ 82{,}529 & 120{,}429 \\ 94{,}311 & 118{,}782 \end{bmatrix} - \begin{bmatrix} 59{,}426 & 98{,}459 \\ 64{,}372 & 104{,}972 \\ 85{,}456 & 120{,}833 \end{bmatrix} = \begin{bmatrix} 19{,}006 & 10{,}886 \\ 18{,}157 & 15{,}457 \\ 8{,}855 & -2{,}051 \end{bmatrix}$$

▶ Label the matrix to identify the monthly profit at each store.

PROFIT ($)

$$\begin{array}{r} \\ \text{January} \\ \text{February} \\ \text{March} \end{array} \begin{array}{c} \text{Store 1} \quad \text{Store 2} \\ \begin{bmatrix} 19{,}006 & 10{,}886 \\ 18{,}157 & 15{,}457 \\ 8{,}855 & -2{,}051 \end{bmatrix} \end{array}$$

b. Add the entries in each column of the profit matrix to find the total profit for each store during the three-month period.

Store 1: $19{,}006 + 18{,}157 + 8{,}855 = \$46{,}018$

Store 2: $10{,}886 + 15{,}457 + (-2{,}051) = \$24{,}292$

▶ Store 1 had higher overall profits.

c. Store 2 had a negative profit of $-2{,}051$ in March. This means the store lost \$2,051 during March.

Extra Example 4

a. Profit ($)

$$\begin{array}{l} \\ \text{June} \\ \text{July} \\ \text{August} \end{array} \begin{array}{c} \text{Location 1} \quad \text{Location 2} \\ \begin{bmatrix} 928 & 2714 \\ -204 & 2730 \\ 1058 & 1487 \end{bmatrix} \end{array}$$

Checkpoint Exercises

1a. Profit ($)

$$\begin{array}{l} \\ \text{April} \\ \text{May} \end{array} \begin{array}{c} \text{Office 1} \quad \text{Office 2} \\ \begin{bmatrix} 127 & 588 \\ 1255 & 1973 \end{bmatrix} \end{array}$$

GUIDED PRACTICE

Vocabulary Check ✓

1. How many rows are there in the matrix at the right? How many columns? **2 rows; 3 columns**

$$\begin{bmatrix} 5 & -7 & 3 \\ 2 & -2 & -4 \end{bmatrix}$$

Concept Check ✓

2. Is the matrix at the right a 3×2 matrix or a 2×3 matrix? **2×3 matrix**

3. CONGRESS Use the matrix showing the number of Democratic and Republican members of the House of Representatives from Arkansas, Delaware, and North Dakota. What is the entry in the first row and second column? What does the number represent? **2; the number of Republican members of the House of Representatives from Arkansas**

	Democrat	Republican
AR	2	2
DE	1	0
ND	0	1

Skill Check ✓

4. VIDEO RENTALS Write a matrix to organize the information about a video store's movies. Label each row and column.

Comedy: 25 new releases, 215 regular selections

Drama: 30 new releases, 350 regular selections

Horror: 26 new releases, 180 regular selections

	New	Reg.
Comedy	25	215
Drama	30	350
Horror	26	180

Find the sum and the difference of the matrices.

5. $\begin{bmatrix} -3 & 0 \\ -6 & 4 \\ 1 & -4 \end{bmatrix}, \begin{bmatrix} 2 & -4 \\ 1 & -3 \\ -1 & 9 \end{bmatrix}$ **See margin.**

6. $\begin{bmatrix} 1 & 8 & -2 \\ -4 & -5 & 6 \end{bmatrix}, \begin{bmatrix} -1 & 9 & 2 \\ 3 & 3 & -5 \end{bmatrix}$ $\begin{bmatrix} 0 & 17 & 0 \\ -1 & -2 & 1 \end{bmatrix}; \begin{bmatrix} 2 & -1 & -4 \\ -7 & -8 & 11 \end{bmatrix}$

Margin answers:

5. $\begin{bmatrix} -1 & -4 \\ -5 & 1 \\ 0 & 5 \end{bmatrix}; \begin{bmatrix} -5 & 4 \\ -7 & 7 \\ 2 & -13 \end{bmatrix}$

11. $\begin{bmatrix} 7 & -5 \\ -3 & -1 \end{bmatrix}$

12. $\begin{bmatrix} -2 & -4 \\ -3 & -12 \end{bmatrix}$

13. $\begin{bmatrix} 4 & -6 & 7 \\ -8 & -2 & 10 \end{bmatrix}$

14. $\begin{bmatrix} -5.2 & 7 & -5.5 \\ 12.6 & 5.3 & 1.7 \end{bmatrix}$

15. $\begin{bmatrix} 7.7 & 8 \\ 4.1 & -6.6 \\ 9.1 & -12.9 \end{bmatrix}$

16. $\begin{bmatrix} 0 & 3 & 1 \\ 0 & 6 & 3 \\ -1 & -1 & 1 \end{bmatrix}$

PRACTICE AND APPLICATIONS

STUDENT HELP

Extra Practice to help you master skills is on p. 798.

STUDENT HELP

HOMEWORK HELP
Example 1: Exs. 23, 24
Example 2: Exs. 7–20
Example 3: Exs. 25, 26
Example 4: Exs. 25, 26

MATRIX OPERATIONS **Tell whether the matrices can be added.**

7. $\begin{bmatrix} 4 & -1 \\ 7 & 5 \end{bmatrix}, \begin{bmatrix} 2 & -2 \\ 5 & -6 \end{bmatrix}$ **yes**

8. $\begin{bmatrix} 3 & -2 & 0 \\ -4 & 1 & -8 \end{bmatrix}, \begin{bmatrix} -4 & 5 \\ 10 & 5 \end{bmatrix}$ **no**

9. $\begin{bmatrix} 4 & 2 \\ -6 & 3 \\ -1 & -2 \end{bmatrix}, \begin{bmatrix} 6 & 4 & -3 \\ 7 & -8 & 1 \end{bmatrix}$ **no**

10. $\begin{bmatrix} 8 & 5 & -8 \\ 4 & -1 & 2 \end{bmatrix}, \begin{bmatrix} -2 & -9 & 1 \\ -6 & 0 & 4 \end{bmatrix}$ **yes**

ADDING MATRICES **Find the sum of the matrices.** **11–16. See margin.**

11. $\begin{bmatrix} 3 & -2 \\ 5 & 1 \end{bmatrix} + \begin{bmatrix} 4 & -3 \\ -8 & -2 \end{bmatrix}$

12. $\begin{bmatrix} 4 & -1 \\ -5 & -9 \end{bmatrix} + \begin{bmatrix} -6 & -3 \\ 2 & -3 \end{bmatrix}$

13. $\begin{bmatrix} 1 & -2 & 2 \\ 0 & -3 & 4 \end{bmatrix} + \begin{bmatrix} 3 & -4 & 5 \\ -8 & 1 & 6 \end{bmatrix}$

14. $\begin{bmatrix} -2.4 & 1.6 & -7.8 \\ 14.3 & 1.1 & -3.9 \end{bmatrix} + \begin{bmatrix} -2.8 & 5.4 & 2.3 \\ -1.7 & 4.2 & 5.6 \end{bmatrix}$

15. $\begin{bmatrix} 6.2 & -1.2 \\ -2.5 & -4.4 \\ 3.4 & -5.8 \end{bmatrix} + \begin{bmatrix} 1.5 & 9.2 \\ 6.6 & -2.2 \\ 5.7 & -7.1 \end{bmatrix}$

16. $\begin{bmatrix} 2 & 9 & -3 \\ 1 & 8 & -2 \\ -3 & -1 & -7 \end{bmatrix} + \begin{bmatrix} -2 & -6 & 4 \\ -1 & -2 & 5 \\ 2 & 0 & 8 \end{bmatrix}$

3 APPLY

ASSIGNMENT GUIDE

BASIC
Day 1: pp. 89–91 Exs. 8–20 even, 21, 22, 29, 35, 40, 45, 50

AVERAGE
Day 1: pp. 89–91 Exs. 8–20 even, 21–24, 29, 35, 40, 45, 50

ADVANCED
Day 1: pp. 89–91 Exs. 8–20 even, 21–31, 35, 40, 45, 50

BLOCK SCHEDULE WITH 2.3
pp. 89–91 Exs. 8–20 even, 21–24, 29, 35, 40, 45, 50

EXERCISE LEVELS

Level A: ***Easier***
7–12, 17, 18, 25

Level B: ***More Difficult***
13–16, 19, 20, 27–29, 32–52

Level C: ***Most Difficult***
21–24, 26, 30, 31

✓ HOMEWORK CHECK

To quickly check student understanding of key concepts, go over the following exercises: Exs. 8, 10, 12, 18, 21. See also the Daily Homework Quiz:

- Blackline Master (*Chapter 2 Resource Book*, p. 70)
- Transparency (p. 14)

! COMMON ERROR
EXERCISES 17–20 When performing matrix subtraction, students frequently subtract corresponding entries in the first columns but then add corresponding entries in the second and third columns. Encourage students to write a subtraction expression for each entry in the difference matrix prior to performing the subtractions.

CAREER NOTE
EXERCISES 27 AND 28
Additional information about a career as a kennel owner is available at **www.mcdougallittell.com.**

ADDITIONAL PRACTICE AND RETEACHING

For Lesson 2.4:

- Practice Levels A, B, and C (*Chapter 2 Resource Book,* p. 60)
- Reteaching with Practice (*Chapter 2 Resource Book,* p. 63)
- See Lesson 2.4 of the *Personal Student Tutor*

For more Mixed Review:

- Search the *Test and Practice Generator* for key words or specific lessons.

19. $\begin{bmatrix} 2 & -2 \\ 5 & -21 \\ 0 & -1 \end{bmatrix}$

20. $\begin{bmatrix} -2 & 11 & -4 \\ -18 & 9 & 7 \end{bmatrix}$

25. $\begin{bmatrix} 76 & 59 \\ 50 & 46 \\ 97 & 66 \end{bmatrix}$

27. $\begin{bmatrix} 0.20 & 0.20 & 0.20 \\ 0.35 & 0.35 & 0.35 \\ 0.45 & 0.45 & 0.45 \end{bmatrix}$

28. $\begin{bmatrix} 5.70 & 6.70 & 7.20 \\ 6.85 & 8.35 & 9.85 \\ 7.95 & 9.20 & 11.45 \end{bmatrix}$

FOCUS ON CAREERS

KENNEL OWNERS provide care for pets in the absence of the pets' owners. Kennels provide daily care, food, grooming, and veterinary services.

CAREER LINK
www.mcdougallittell.com

SUBTRACTING MATRICES **Find the difference of the matrices.**

17. $\begin{bmatrix} 8 & -3 \\ 4 & -1 \end{bmatrix} - \begin{bmatrix} 7 & 7 \\ -2 & -5 \end{bmatrix}$ $\begin{bmatrix} 1 & -10 \\ 6 & 4 \end{bmatrix}$

18. $\begin{bmatrix} 4 & 3 \\ -12 & -10 \end{bmatrix} - \begin{bmatrix} -6 & 1 \\ -4 & 2 \end{bmatrix}$ $\begin{bmatrix} 10 & 2 \\ -8 & -12 \end{bmatrix}$

19. $\begin{bmatrix} -4 & 1 \\ 0 & -13 \\ 2 & -8 \end{bmatrix} - \begin{bmatrix} -6 & 3 \\ -5 & 8 \\ 2 & -7 \end{bmatrix}$ **See margin.**

20. $\begin{bmatrix} -5 & 11 & -2 \\ -10 & 4 & 6 \end{bmatrix} - \begin{bmatrix} -3 & 0 & 2 \\ 8 & -5 & -1 \end{bmatrix}$ **See margin.**

MENTAL MATH **Use mental math to find *a, b, c,* and *d.***

21. $\begin{bmatrix} 3a & 5b \\ c-6 & d \end{bmatrix} = \begin{bmatrix} -12 & -5 \\ 1 & -3 \end{bmatrix}$ **−4, −1, 7, −3**

22. $\begin{bmatrix} 4a & b+3 \\ c & d-3 \end{bmatrix} = \begin{bmatrix} 8 & -1 \\ 0 & -6 \end{bmatrix}$ **2, −4, 0, −3**

WRITING A MATRIX **Write and label a matrix to organize the information.**

23. Music Store Inventory:

CDs: 52 sale price titles, 3300 regular price titles

Tapes: 28 sale price titles, 1600 regular price titles

	Sale	Reg.
CD	52	3300
Tape	28	1600

24. Team Uniform Order:

Shirts: 3 small, 7 medium, 10 large, 5 extra large

Shorts: 7 small, 4 medium, 2 large, 2 extra large

	S	M	L	XL
Shirts	3	7	10	5
Shorts	7	4	2	2

DOG KENNEL **The owner of a kennel keeps records of all the dogs she cares for each year. In Exercises 25 and 26, use the matrices below, which show her records for 3 breeds of dogs cared for in 1999 and 2000.**

1999

	Male	Female
Beagle	36	28
Dalmatian	24	26
Bulldog	51	32

2000

	Male	Female
Beagle	40	31
Dalmatian	26	20
Bulldog	46	34

25. Find the sum of the two matrices. **See margin.**

26. *Writing* Explain what the data represent in the matrix you wrote in Exercise 25. **For the two years, the kennel cared for 76 male beagles, 59 female beagles, 50 male dalmatians, 46 female dalmatians, 97 male bulldogs, and 66 female bulldogs.**

WAGES AND RAISES **In Exercises 27 and 28, use the matrix, which shows employees' hourly wage rates at a grocery store. The wage rates depend on the job and the number of years of experience.**

WAGE RATES

	0–1 year	2–3 years	4+ years
Service Clerk	5.50	6.50	7.00
Cashier	6.50	8.00	9.50
Deli Clerk	7.50	8.75	11.00

27. The store is giving \$.20 raises to all services clerks, \$.35 raises to all cashiers, and \$.45 raises to all deli clerks. Write a matrix that you can add to the matrix above to find the new wage rates after raises are given. **See margin.**

28. Write a matrix that shows the new wage rates after raises are given. **See margin.**

Test Preparation

29. MULTI-STEP PROBLEM Use the table, which shows the monthly average high and low temperatures in degrees Fahrenheit in three cities.

	May		June		July	
City	High	Low	High	Low	High	Low
Atlanta, GA	79.6	58.7	85.8	66.2	88.0	69.5
San Francisco, CA	66.5	49.7	70.3	52.6	71.6	53.9
Anchorage, AK	54.4	38.8	61.6	47.2	65.2	51.7

▶Source: National Climatic Data Center

a. Write a matrix for the average high temperatures in each city during May, June, and July. **See margin.**

b. Write a matrix for the average low temperatures in each city during May, June, and July. **See margin.**

c. **CRITICAL THINKING** Explain how to find the difference of average temperatures in each city during May, June, and July. Use the matrices you wrote in parts (a) and (b). Then write a matrix for the difference of average temperatures in each city during May, June, and July. **See margin.**

★ Challenge

PROPERTIES OF MATRIX ADDITION **In Exercises 30 and 31, recall the properties of addition you learned on page 73.**

30. a. Does $\begin{bmatrix} 5 & -9 \\ -4 & 1 \\ -1 & 4 \end{bmatrix} + \begin{bmatrix} 2 & 8 \\ 1 & 9 \\ -3 & -1 \end{bmatrix} = \begin{bmatrix} 2 & 8 \\ 1 & 9 \\ -3 & -1 \end{bmatrix} + \begin{bmatrix} 5 & -9 \\ -4 & 1 \\ -1 & 4 \end{bmatrix}$? Explain.

Yes; when you add corresponding entries, each of those additions is commutative.

b. Does the commutative property apply when adding matrices? **yes**

EXTRA CHALLENGE
www.mcdougallittell.com

31. Does the associative property apply when adding matrices? Give an example to support your answer.

yes; *Sample answer:*

$\left(\begin{bmatrix} 1 & 2 \\ 3 & 4 \end{bmatrix} + \begin{bmatrix} 1 & 1 \\ 1 & 1 \end{bmatrix}\right) + \begin{bmatrix} 0 & 2 \\ 0 & 2 \end{bmatrix} = \begin{bmatrix} 2 & 5 \\ 4 & 7 \end{bmatrix}$ and $\begin{bmatrix} 1 & 2 \\ 3 & 4 \end{bmatrix} + \left(\begin{bmatrix} 1 & 1 \\ 1 & 1 \end{bmatrix} + \begin{bmatrix} 0 & 2 \\ 0 & 2 \end{bmatrix}\right) = \begin{bmatrix} 2 & 5 \\ 4 & 7 \end{bmatrix}$

MIXED REVIEW

EXPONENTIAL FORM **Write the expression in exponential form.** (Review 1.2)

32. $2y \cdot 2y \cdot 2y$ $(2y)^3$ **33.** five squared 5^2 **34.** four to the sixth power 4^6

35. $x \cdot x \cdot y \cdot y \cdot y$ x^2y^3 **36.** two to the xth power 2^x **37.** $3 \cdot (t \cdot t \cdot t \cdot t)$ $3t^4$

FINDING ABSOLUTE VALUES **Evaluate the expression.** (Review 2.1)

38. $|-82|$ 82 **39.** $-|43.7|$ −43.7 **40.** $-|-4.5|$ −4.5

41. $|-29| + 7$ 36 **42.** $-|13 - 12.1|$ −0.9 **43.** $14 + |-11| - 10$ 15

RULES OF ADDITION **Find the sum.** (Review 2.2 for 2.5)

44. $-19 + (-6)$ −25 **45.** $-12 + (-9)$ −21 **46.** $-3 + 0 + (-29)$ −32

47. $0 + (-5) + 2$ −3 **48.** $-3 + (-6) + (-2)$ −11 **49.** $-5 + (-6) + (-3)$ −14

50. $-7 + (-8) + (-9)$ −24 **51.** $-1 + (-1) + (-1)$ −3 **52.** $-4 + (-4) + (-4)$ −12

2. **Bird Sightings**

Bird Sightings	Charles	Ari	Annette	Miguel
American Goldfinch	2	4	1	5
American Robin	8	10	9	12
Baltimore Oriole	1	3	2	1

4 ASSESS

DAILY HOMEWORK QUIZ

Transparency Available

1. Find the sum and the difference of the given matrices.
$\begin{bmatrix} 75.82 & -10.18 \\ 16.90 & 9.63 \end{bmatrix}$ and $\begin{bmatrix} 19.67 & 4.11 \\ -10.50 & -3.22 \end{bmatrix}$.
$\begin{bmatrix} 95.49 & -6.07 \\ 6.40 & 6.41 \end{bmatrix}$; $\begin{bmatrix} 56.15 & -14.29 \\ 27.40 & 12.85 \end{bmatrix}$

EXTRA CHALLENGE NOTE
Challenge problems for Lesson 2.4 are available in **blackline** format in the *Chapter 2 Resource Book,* p. 67 and at **www.mcdougallittell.com.**

ADDITIONAL TEST PREPARATION

1. WRITING Explain why matrices that are different sizes cannot be added or subtracted. **To add or subtract matrices, you add or subtract corresponding entries. Matrices of different sizes don't have corresponding entries.**

2. OPEN ENDED Give an example of a real-life situation that can be modeled by a 3 by 4 matrix. Write and label the matrix. **See sample answer, below.**

29. See Additional Answers beginning on page AA1.

1 Planning the Activity

PURPOSE
To use a graphing calculator to perform matrix addition and subtraction.

MATERIALS
- graphing calculator per student
- Keystroke blackline (*Chapter 2 Resource Book*, p. 58)

PACING
- Exploring the Concept — 10 min
- Drawing Conclusions — 10 min

LINK TO LESSON
Have students refer to Example 2 on page 87. Ask how many number additions they needed to perform to add the two matrices. **[6]** Explain that by using a graphing calculator, these additions can be made with a single matrix addition.

2 Managing the Activity

COOPERATIVE LEARNING
Students can work through this activity in pairs with a single graphing calculator. Have one student read the entries in the matrix while the other enters them in the calculator. Switch roles for each problem.

CLASSROOM MANAGEMENT
Explain that matrices are always read by rows. That is, each row is read from left to right before moving to the next row. Entries of a matrix are entered into the calculator in the same order that they are read.

3 Closing the Activity

KEY DISCOVERY
Graphing calculators can perform matrix addition and subtraction accurately and quickly.

ACTIVITY ASSESSMENT
Enter any 2 by 2 matrix A in your calculator. Find [A] + [A]. How are the entries of the sum related to the entries of A? **Each entry of A is doubled.**

ACTIVITY 2.4
Using Technology

Adding and Subtracting Matrices

EXAMPLE

Use a graphing calculator to find the sum and difference of $\begin{bmatrix} -5 & 0 \\ -8 & 4 \\ 1 & 5 \end{bmatrix}$ and $\begin{bmatrix} -3 & -1 \\ 4 & 0 \\ 1 & -7 \end{bmatrix}$.

SOLUTION

1. Follow your calculator's procedures to name the first matrix [A].

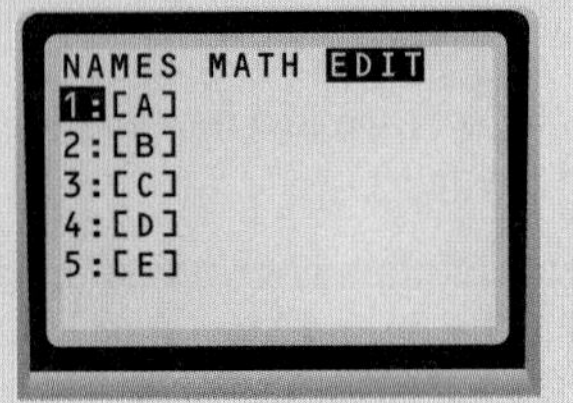

2. Enter the size of the matrix, and then enter each entry in the appropriate position.

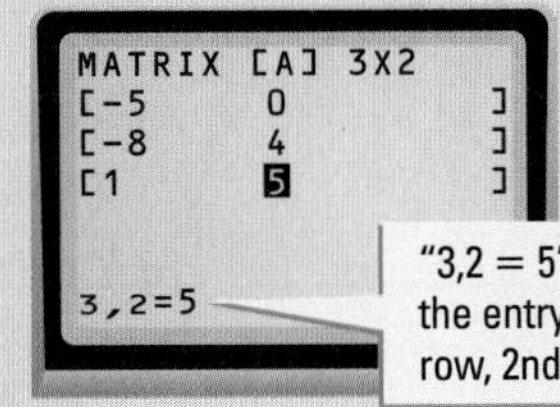

"3,2 = 5" means 5 is the entry in the 3rd row, 2nd column.

STUDENT HELP
KEYSTROKE HELP
See keystrokes for several models of calculators at www.mcdougallittell.com

3. Name the second matrix [B]. Enter the size of the matrix and the entries.

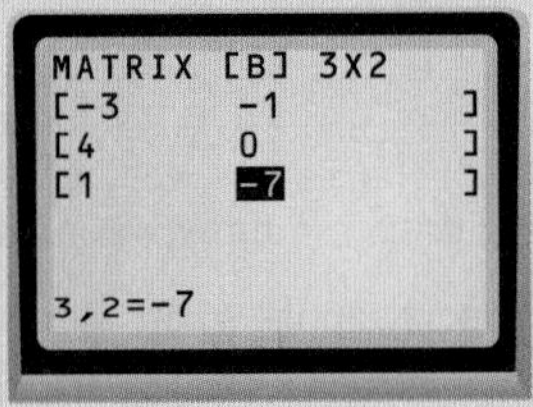

4. Enter [A] + [B] to find the sum, and [A] − [B] to find the difference.

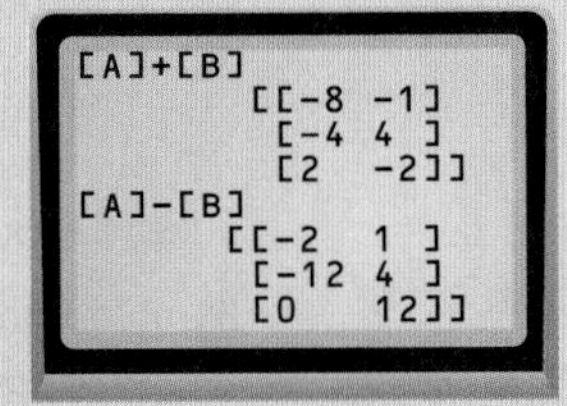

The sum is $\begin{bmatrix} -8 & -1 \\ -4 & 4 \\ 2 & -2 \end{bmatrix}$ and the difference is $\begin{bmatrix} -2 & 1 \\ -12 & 4 \\ 0 & 12 \end{bmatrix}$.

EXERCISES

Use a graphing calculator to find the sum and difference of the matrices.

1–4. See margin.

1. $\begin{bmatrix} 7 & -5 \\ 5 & -3 \\ -9 & -8 \end{bmatrix}, \begin{bmatrix} 0 & -1 \\ 1 & 0 \\ -1 & -6 \end{bmatrix}$

2. $\begin{bmatrix} 52 & 79 & -61 \\ -84 & 21 & -76 \end{bmatrix}, \begin{bmatrix} 10 & 45 & -62 \\ -16 & 98 & -28 \end{bmatrix}$

3. $\begin{bmatrix} 7.4 & -5.1 \\ -0.6 & -4.6 \\ 9.9 & -9.4 \\ 2.1 & -3.6 \\ -0.5 & 7.1 \end{bmatrix}, \begin{bmatrix} 3.3 & 3.2 \\ -3.2 & 2.7 \\ -8.2 & -1.7 \\ 5.1 & -4.8 \\ -3.9 & 6.1 \end{bmatrix}$

4. $\begin{bmatrix} 44 & -51 & 17 \\ 76 & 63 & 25 \\ -20 & -60 & 42 \end{bmatrix}, \begin{bmatrix} -53 & -18 & 11 \\ 29 & -27 & -62 \\ 29 & -35 & -28 \end{bmatrix}$

1. $\begin{bmatrix} 7 & -6 \\ 6 & -3 \\ -10 & -14 \end{bmatrix}; \begin{bmatrix} 7 & -4 \\ 4 & -3 \\ -8 & -2 \end{bmatrix}$

2. $\begin{bmatrix} 62 & 124 & -123 \\ -100 & 119 & -104 \end{bmatrix};$ $\begin{bmatrix} 42 & 34 & 1 \\ -68 & -77 & -48 \end{bmatrix}$

3. $\begin{bmatrix} 10.7 & -1.9 \\ -3.8 & -1.9 \\ 1.7 & -11.1 \\ 7.2 & -8.4 \\ -4.4 & 13.2 \end{bmatrix};$ $\begin{bmatrix} 4.1 & -8.3 \\ 2.6 & -7.3 \\ 18.1 & -7.7 \\ -3 & 1.2 \\ 3.4 & 1 \end{bmatrix}$

4. $\begin{bmatrix} -9 & -69 & 28 \\ 105 & 36 & -37 \\ 9 & -95 & 14 \end{bmatrix};$ $\begin{bmatrix} 97 & -33 & 6 \\ 47 & 90 & 87 \\ -49 & -25 & 70 \end{bmatrix}$

2.5

Multiplication of Real Numbers

What you should learn

GOAL 1 Multiply real numbers using properties of multiplication.

GOAL 2 Multiply real numbers to solve **real-life** problems like finding how much money is owed on a loan in **Exs. 64 and 65**.

Why you should learn it

To solve **real-life** problems like finding how much money a grocery store loses in **Example 5**.

GOAL 1 MULTIPLYING REAL NUMBERS

Remember that multiplication can be modeled as repeated addition. For example, $3(-2) = (-2) + (-2) + (-2) = -6$. The product of a positive number and a negative number is a negative number.

ACTIVITY Developing Concepts

Investigating Multiplication Patterns

Identify and extend the patterns to complete the lists. What generalizations can you make about the sign of a product of real numbers?

FACTOR OF −3	FACTOR OF −2	FACTOR OF −1
$(3)(-3) = -9$	$(3)(-2) = -6$	$(3)(-1) = -3$
$(2)(-3) = -6$	$(2)(-2) = -4$	$(2)(-1) = -2$
$(1)(-3) = -3$	$(1)(-2) = -2$	$(1)(-1) = -1$
$(0)(-3) = 0$	$(0)(-2) = 0$	$(0)(-1) = 0$
$(-1)(-3) = ?$ **3**	$(-1)(-2) = ?$ **2**	$(-1)(-1) = ?$ **1**
$(-2)(-3) = ?$ **6**	$(-2)(-2) = ?$ **4**	$(-2)(-1) = ?$ **2**

The results of the activity can be extended to determine the sign of a product of more than two factors.

STUDENT HELP

Study Tip
- A product is negative if it has an *odd* number of negative factors.
- A product is positive if it has an *even* number of negative factors.

EXAMPLE 1 ***Multiplying Real Numbers***

a. $(-3)(4)(-2) = (-12)(-2) = 24$ — Two negative factors; positive product

b. $\left(-\frac{1}{2}\right)(-2)(-3) = (1)(-3) = -3$ — Three negative factors; negative product

c. $(-1)^4 = (-1)(-1)(-1)(-1) = (1)(-1)(-1)$
$= (-1)(-1) = 1$ — Four negative factors; positive product

In Example 1(c), be sure you understand that $(-1)^4$ is not the same as -1^4.

MULTIPLYING REAL NUMBERS

The product of two real numbers with the same sign is the product of their absolute values. The product of two real numbers with different signs is the *opposite* of the product of their absolute values.

1 PLAN

PACING
Basic: 1 day
Average: 1 day
Advanced: 1 day
Block Schedule: 0.5 block with 2.6

LESSON OPENER
APPLICATION
An alternative way to approach Lesson 2.5 is to use the Application Lesson Opener:
- Blackline Master (*Chapter 2 Resource Book*, p. 71)
- Transparency (p. 12)

MEETING INDIVIDUAL NEEDS
- ***Chapter 2 Resource Book***
  Prerequisite Skills Review (p. 5)
  Practice Level A (p. 72)
  Practice Level B (p. 73)
  Practice Level C (p. 74)
  Reteaching with Practice (p. 75)
  Absent Student Catch-Up (p. 77)
  Challenge (p. 79)
- ***Resources in Spanish***
- ***Personal Student Tutor***

NEW-TEACHER SUPPORT
See the Tips for New Teachers on pp. 1–2 of the *Chapter 2 Resource Book* for additional notes about Lesson 2.5.

WARM-UP EXERCISES

Transparency Available

Write each addition sentence as a multiplication sentence.

1. $2 + 2 + 2$ (3)(2)
2. $5 + 5 + 5 + 5$ (4)(5)

Find the product.

3. $\left(\frac{3}{5}\right)(15)$ 9
4. $(4)\left(\frac{5}{12}\right)$ $1\frac{2}{3}$

2 TEACH

MOTIVATING THE LESSON
Suppose that you borrow \$2500 per year to attend a 4-year college. Assume that you do not pay back any of the loan until after graduation and that no interest is due on the loan during the time you are in school. How can you use integers to model the debt at the time of graduation? **4(−2500)**

ACTIVITY NOTE
Have students rewrite some of the products in the first 3 rows of this activity as repeated addition to verify the answers given. They should then follow the pattern down each column to find the answers for the product of two negative numbers.

EXTRA EXAMPLE 1
Find the product.
a. $(9)(-3)$ **−27**
b. $(8)\left(-\frac{1}{2}\right)(-6)$ **24**
c. $(-3)^3$ **−27**
d. $(-2)\left(-\frac{1}{2}\right)(-3)(-5)$ **15**

EXTRA EXAMPLE 2
Find the product.
a. $(-n)(-n)$ n^2
b. $(-4)(-x)(-x)(x)$ $-4x^3$
c. $-(b)^3$ $-b^3$
d. $(-y)^4$ y^4

EXTRA EXAMPLE 3
Evaluate the expression when $x = -7$.
a. $2(-x)(-x)$ **98** b. $(-5 \cdot x)\left(-\frac{2}{7}\right)$ **−10**

CHECKPOINT EXERCISES
For use after Examples 1–2:
Find the product.
1. $(-2)(4.5)(-10)$ **90**
2. $(-4)(-x)^2$ $-4x^2$

For use after Example 3:
Evaluate the expression when $x = -3$.
3. $(-1 \cdot x)(x)$ **−9**

EXAMPLE 2 ***Products with Variable Factors***

a. $(-2)(-x) = 2x$ — Two negative signs
b. $3(-n)(-n)(-n) = -3n^3$ — Three negative signs
c. $(-1)(-a)^2 = (-1)(-a)(-a) = -a^2$ — Three negative signs
d. $-(y)^4 = -(y \cdot y \cdot y \cdot y) = -y^4$ — One negative sign

CONCEPT SUMMARY — **PROPERTIES OF MULTIPLICATION**

COMMUTATIVE PROPERTY The order in which two numbers are multiplied does not change the product.
$a \cdot b = b \cdot a$ — **Example:** $3 \cdot (-2) = (-2) \cdot 3$

ASSOCIATIVE PROPERTY The way you group three numbers when multiplying does not change the product.
$(a \cdot b) \cdot c = a \cdot (b \cdot c)$ — **Example:** $(-6 \cdot 2) \cdot 3 = -6 \cdot (2 \cdot 3)$

IDENTITY PROPERTY The product of a number and 1 is the number.
$1 \cdot a = a$ — **Example:** $(-4) \cdot 1 = -4$

PROPERTY OF ZERO The product of a number and 0 is 0.
$a \cdot 0 = 0$ — **Example:** $(-2) \cdot 0 = 0$

PROPERTY OF OPPOSITES The product of a number and −1 is the opposite of the number.
$(-1) \cdot a = -a$ — **Example:** $(-1) \cdot (-3) = 3$

STUDENT HELP
HOMEWORK HELP
Visit our Web site www.mcdougallittell.com for extra examples.

EXAMPLE 3 ***Evaluating a Variable Expression***

Evaluate the expression when $x = -5$.

a. $-4(-1)(-x)$ b. $(-9.7 \cdot x)(-2)$

SOLUTION You can simplify the expression first, or substitute for x first.

a. $-4(-1)(-x) = -4x$ — Simplify expression first.
$= -4(-5)$ — Substitute −5 for x.
$= 20$ — Product of two negatives

b. $(-9.7 \cdot x)(-2) = [-9.7 \cdot (-5)] \cdot (-2)$ — Substitute −5 for x first.
$= -9.7 \cdot [-5 \cdot (-2)]$ — Use associative property.
$= -9.7 \cdot 10$ — $-5 \cdot (-2) = 10$
$= -97.0$ — Simplify.

FOCUS ON APPLICATIONS

REAL LIFE **FLYING SQUIRRELS** can't really fly as birds do. But they can glide through the air by using "gliding membranes," flaps of loose skin that extend from wrist to ankle.

GOAL 2 USING MULTIPLICATION IN REAL LIFE

Displacement is the change in the position of an object. Unlike distance, displacement can be positive, negative, or zero.

EXAMPLE 4 *Application of Products of Negatives*

FLYING SQUIRRELS A flying squirrel descends from a tree with a velocity of -6 feet per second. Find its vertical displacement in 3.5 seconds.

SOLUTION

VERBAL MODEL Vertical displacement = Velocity · Time

LABELS Vertical displacement = d (feet)

Velocity = -6 (feet per second)

Time = 3.5 (seconds)

ALGEBRAIC MODEL $d = -6 \cdot 3.5$

$= -21$

▶ The negative displacement indicates downward motion. The squirrel traveled a vertical displacement of -21 feet or 21 feet downward.

UNIT ANALYSIS: Check that *feet* are the units of the solution.

$\frac{\text{feet}}{\cancel{\text{second}}} \cdot \cancel{\text{seconds}} = \text{feet}$

REAL LIFE *Grocery Stores*

EXAMPLE 5 *Application of Products of Negatives*

A grocery store sells pint baskets of strawberries as loss leaders, which means the store is willing to lose money selling them. The store hopes to make up the loss with additional sales to customers attracted to the store. The store loses \$.17 per pint. How much will the store lose if it sells 3450 pints?

SOLUTION

Use a calculator. Multiply the number of pints sold by the loss per pint to find the total loss.

3450 [×] [.] 17 [+/−] [=] → −586.50

▶ The store loses \$586.50 on strawberry sales.

UNIT ANALYSIS: Check that *dollars* are the units of the solution.

$\cancel{\text{pints}} \cdot \frac{\text{dollars}}{\cancel{\text{pint}}} = \text{dollars}$

EXTRA EXAMPLE 4
A leaf floats down from a tree at a velocity of −12 cm/sec. Find the vertical displacement in 4.2 sec. **−50.4 cm**

EXTRA EXAMPLE 5
A grocery store runs a sale where customers can get two bags of spinach for the price of one. The store normally charges \$1.69 per bag. How much will they be losing in sales if they give away 798 free bags?
The store loses \$1348.62.

CHECKPOINT EXERCISES

For use after Example 4:

1. A paper airplane descends at a velocity of −14 in./sec. Find the vertical displacement in 7.5 sec. **−105 in.**

For use after Example 5:

2. After a recent flood, a store owner sells 294 cans of green beans at a reduced price because their labels were ruined. Because the price the store originally paid for the beans is more than the reduced price, the store loses \$.12 on each can of beans sold. How much will the store lose if it sells all 294 cans of beans? **\$35.28**

CLOSURE QUESTION
Suppose you have a product of several nonzero factors. How can you determine if the product will be positive or negative? **Count the negative factors: if there is an even number, the product will be positive; if there is an odd number, the product will be negative.**

DAILY PUZZLER
If x is a positive real number and n is a positive integer, for what values of n will the expression $(-x)^n$ be positive? Negative? Explain. **Since n is the number of negative factors in the expression, $(-x)^n$ will be positive if n is even and negative if n is odd.**

3 APPLY

ASSIGNMENT GUIDE

BASIC
Day 1: pp. 96–98 Exs. 16–56 even, 58–63, 68–70, 72, 73, 80, 85, 95, 97

AVERAGE
Day 1: pp. 96–98 Exs.16–56 even, 58–63, 68–70, 72, 73, 80, 85, 95, 97

ADVANCED
Day 1: pp. 96–98 Exs. 16–56 even, 58–63, 68–75, 80, 85, 95, 97

BLOCK SCHEDULE WITH 2.6
pp. 96–98 Exs. 16–56 even, 58–63, 68–70, 72, 73, 80, 85, 95, 97

EXERCISE LEVELS

Level A: ***Easier***
16–29, 50–57

Level B: ***More Difficult***
30–49, 61–70, 72, 73, 76–97

Level C: ***Most Difficult***
58–60, 71, 74, 75

✔ HOMEWORK CHECK

To quickly check student understanding of key concepts, go over the following exercises: Exs. 24, 34, 44, 60, 68. See also the Daily Homework Quiz:

- Blackline Master (*Chapter 2 Resource Book*, p. 82)
- Transparency (p. 15)

GUIDED PRACTICE

Vocabulary Check ✔

1. Multiplying -12 by -97 produces the same result as multiplying -97 by -12. What property is this? **commutative property of multiplication**

Concept Check ✔

2. Is the product of an odd number of factors always a negative number? **no; only if an odd number of the factors are negative**

3. Is the product of an even number of factors always a positive number? **no; only if an even number of the factors are negative**

4. ERROR ANALYSIS Describe the error shown at the right. **There are 3 negative factors, −1, −5, and −5, so the product is negative.**

~~$-(-5)^2 = 5^2$ 2 negative factors~~
~~$= 25$ Simplify.~~

Skill Check ✔

Find the product.

5. $8 \cdot (-1)$ **−8** **6.** $-3 \cdot 0$ **0** **7.** $-5 \cdot (-7)$ **35** **8.** $-7 \cdot (-5)$ **35**

9. $-12 \cdot (-6)$ **72** **10.** $-30 \cdot 8$ **−240** **11.** $-5 \cdot 2 \cdot (-7)$ **70** **12.** $-(-1)^5$ **1**

Evaluate the expression.

13. $2(-6)(-x)$ when $x = 4$ **48** **14.** $5(x - 4)$ when $x = -3$ **−35**

15. HAWKS A hawk dives from a tree with a velocity of -10 feet per second. Find the vertical displacement in 4.7 seconds. **−47 ft**

PRACTICE AND APPLICATIONS

STUDENT HELP
Extra Practice to help you master skills is on p. 798.

PRODUCTS **Find the product.**

16. $(-8)(3)$ **−24** **17.** $(4)(-4)$ **−16** **18.** $(20)(-65)$ **−1300** **19.** $(-1)(-5)$ **5**

20. $(-7)(-1.2)$ **8.4** **21.** $(-11)\left(\frac{1}{8}\right)$ $-\frac{11}{8}$ **22.** $(-15)\left(\frac{3}{5}\right)$ **−9** **23.** $|(-12)(2)|$ **24**

24. $(-3)(-1)(-6)$ **−18** **25.** $(13)(-2)(-3)$ **78** **26.** $(5)(-2)(7)$ **−70**

27. $(-4)(-7)\left(\frac{3}{7}\right)$ **12** **28.** $(-3)(-1)(4)(-6)$ **−72** **29.** $(-13)(-2)(-2)\left(-\frac{2}{13}\right)$ **8**

SIMPLIFYING EXPRESSIONS **Simplify the variable expression.**

30. $(-3)(-y)$ $3y$ **31.** $(7)(-x)$ $-7x$ **32.** $5(-a)(-a)(-a)$ $-5a^3$

33. $(-4)(-x)(x)(-x)$ $-4x^3$ **34.** $-(-b)^3$ b^3 **35.** $-(-4)^2(y)$ $-16y$

36. $|(8)(-z)(-z)(-z)|$ $8|z^3|$ **37.** $-(y^4)(y)$ $-y^5$ **38.** $(-b^2)(-b^3)(-b^4)$ $-b^9$

39. $-\frac{1}{2}(-2x)$ x **40.** $\frac{2}{3}\left(-\frac{3}{2}x\right)$ $-x$ **41.** $-\frac{3}{7}(-w^2)(7w)$ $3w^3$

STUDENT HELP
HOMEWORK HELP
Example 1: Exs. 16–29
Example 2: Exs. 30–41
Example 3: Exs. 42–49
Example 4: Exs. 61–65
Example 5: Exs. 50–57, 61–63

EVALUATING EXPRESSIONS **Evaluate the expression.**

42. $-8x$ when $x = 6$ **−48** **43.** $y^3 - 4$ when $y = -2$ **−12**

44. $3x^2 - 5x$ when $x = -2$ **22** **45.** $4a + a^2$ when $a = -7$ **21**

46. $-4(|y - 12|)$ when $y = 5$ **−28** **47.** $-2(|x - 5|)$ when $x = -5$ **−20**

48. $-2x^2 + 3x - 7$ when $x = 4$ **−27** **49.** $9r^3 - (-2r)$ when $r = 2$ **76**

EVALUATING EXPRESSIONS Use a calculator to evaluate the expression. Round your answer to two decimal places.

50. $(-7.39)(4.41)(-2.9)$ **94.51**

51. $(4.67)(-8.01)(1.89)$ **−70.70**

52. $(3.6)(-2.67)^3(-9.41)$ **644.80**

53. $(-6.3)^2(9.5)(4.8)$ **1809.86**

54. $x^3 - 8.29$ when $x = -2.47$ **−23.36**

55. $8.3 + y^3$ when $y = -4.6$ **−89.04**

56. $4.7b - (-b^2)$ when $b = 1.99$ **13.31**

57. $x^2 + x - 27.2$ when $x = -7$ **14.8**

COUNTEREXAMPLES Decide whether the statement is *true* or *false*. If it is false, give a counterexample.

58. $(-a) \cdot (-b) = (-b) \cdot (-a)$ **true**

59. The product $(-a) \cdot (-1)$ is always positive.
False; any nonpositive number *a* is a counterexample.

60. If $a > b$, then for any real number c, $a \cdot c > b \cdot c$.
False; let $c = 0$. $a \cdot 0 = 0$ and $b \cdot 0 = 0$ and $0 \cdot 0 = 0$.

LOSS LEADERS To promote sales, a grocery store advertises bananas for \$.25 per pound. The store loses \$.11 on each pound of bananas it sells.

61. Write a verbal model that you can use to find the amount of money that the store loses depending on the number of pounds of bananas it sells. **See margin.**

62. The store sells 2956 pounds of bananas. How much money does the store lose on banana sales? **\$325.16**

63. The store also advertises apple juice for \$1.19 per 64-ounce bottle, and loses \$.08 per bottle sold. Use a verbal model to find how much the store loses on sales of 3107 bottles of apple juice. **\$248.56**

PAYING BACK A LOAN Your aunt lends you \$175 to buy a guitar. She will decrease the amount you owe by \$25 for each day you help her by doing odd jobs.

64. Write a verbal model that you can use to find the decrease in the amount you owe your aunt depending on the number of days you help her out. **See margin.**

65. What is the change in the amount you owe your aunt after helping her out for 5 days? How much do you still owe her? **−\$125; \$50**

VACATION TRAVEL You and your family take a summer vacation to Ireland. You discover that the number of Americans visiting Ireland is increasing by 80,000 visitors per year. Let *x* represent the number of visitors in 1997.

66. Write an expression for the number of visitors in 2000. **$x + 240{,}000$**

67. If the number of visitors in 1997 was 700,000, how many visitors were expected in 2000? Use unit analysis to check your answer. **940,000 visitors**

EXTENSION: SCALAR MULTIPLICATION Multiply the matrix by the real number. **68–70. See margin.**

Sample: $-3\begin{bmatrix} 1 & -2 \\ -4 & 0 \end{bmatrix} = \begin{bmatrix} -3(1) & -3(-2) \\ -3(-4) & -3(0) \end{bmatrix} = \begin{bmatrix} -3 & 6 \\ 12 & 0 \end{bmatrix}$

68. $-8\begin{bmatrix} -4 & -7 \\ 3 & 3 \end{bmatrix}$

69. $-7\begin{bmatrix} 6 & -4 & 3 \\ -1 & 2^2 & -9 \end{bmatrix}$

70. $-5x\begin{bmatrix} 2x & -6y \\ 4b & -8a \end{bmatrix}$

STUDENT HELP

Look Back
For help with counterexamples, see p. 66.

61. The total amount lost is the loss per pound times the number of pounds of bananas sold.

64. The total decrease is equal to the daily decrease times the number of days you help.

68. $\begin{bmatrix} 32 & 56 \\ -24 & -24 \end{bmatrix}$

69. $\begin{bmatrix} -42 & 28 & -21 \\ 7 & -28 & 63 \end{bmatrix}$

70. $\begin{bmatrix} -10x^2 & -30xy \\ -20bx & 40ax \end{bmatrix}$

FOCUS ON APPLICATIONS

REAL LIFE **IRELAND** More people of Irish descent live in New England (4 million) than in Ireland (3.5 million)!

▶ Source: Irish Tourist Board

! COMMON ERROR

EXERCISES 50–57 To avoid rounding error in these problems, students should enter the complete expression in their calculators in one step rather than evaluating part of the expression, rounding, and then using the result at the next step. In exercises that contain variables, encourage students to write the expression after substituting the number and then enter the expression in the calculator.

STUDENT HELP NOTES

Look Back As students review counterexamples on page 66, remind them that a single counterexample is sufficient to show that a statement is false.

APPLICATION NOTE

EXERCISES 66 AND 67
Caution students that the model developed in Exercise 66 assumes the number of visitors from the U.S. to Ireland continues to grow at the same rate. Generally models such as this are valid only for short intervals of time. For example, the model may not give a realistic estimate of the number of visitors from the U.S. to Ireland in 2020.

ADDITIONAL PRACTICE AND RETEACHING

For Lesson 2.5

- Practice Levels A, B, and C (*Chapter 2 Resource Book*, p. 72)
- Reteaching with Practice (*Chapter 2 Resource Book*, p. 75)
- See Lesson 2.5 of the *Personal Student Tutor*

For more Mixed Review:

- Search the *Test and Practice Generator* for key words or specific lessons.

4 ASSESS

DAILY HOMEWORK QUIZ

Transparency Available

1. Find the product.
   a. $(-2)(8)(-1)$ 16
   b. $(7)(-3)(-5)\left(-\frac{1}{3}\right)$ −35
2. Evaluate $a^3 + 2a^2 - a$ when $a = -2$. 2
3. If $a \cdot b > 0$ and $a < 0$, what must be true of b? $b < 0$
4. Find the scalar product.
   $-1\begin{bmatrix} 6 & -3 \\ -6 & 3 \end{bmatrix}$ $\begin{bmatrix} -6 & 3 \\ 6 & -3 \end{bmatrix}$

EXTRA CHALLENGE NOTE

Challenge problems for Lesson 2.5 are available in **blackline** format in the *Chapter 2 Resource Book*, p. 79 and at **www.mcdougallittell.com**.

ADDITIONAL TEST PREPARATION

1. **WRITING** Explain the difference between the expressions -4^4 and $(-4)^4$.
   $-4^4 = -(4^4) = -(4 \cdot 4 \cdot 4 \cdot 4) = -64$; $(-4)^4 = -4 \cdot (-4) \cdot (-4) \cdot (-4) = 64$

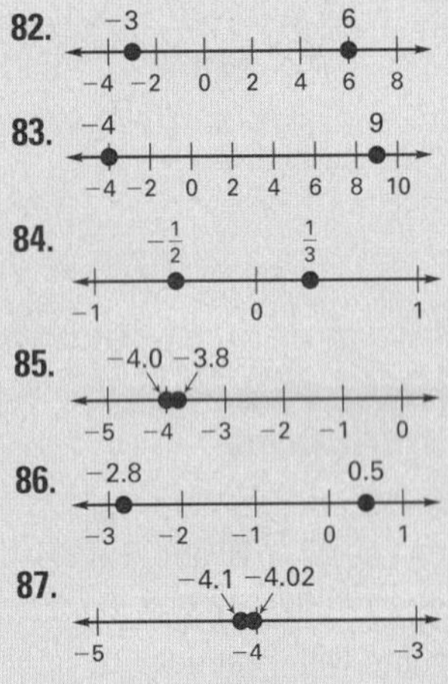

71b. $-\frac{1}{6}(x) + 3.5x$

71. COMPARING METHODS As parts (a) and (b) of Example 3 show, it is sometimes easier to evaluate an expression by simplifying it before substituting, and sometimes easier if you substitute for the variable first. **Sample answers are given.**

   a. Write an expression that is easier to evaluate if you simplify *before* substituting 12 for x. $\frac{1}{3}(x)(-24)$

   b. Write an expression that is easier to evaluate if you substitute 12 for x first. **See margin.**

Test Preparation

72. MULTIPLE CHOICE Which of the following statements is *not* true? **C**

   Ⓐ The product of any number and zero is zero.

   Ⓑ The order in which two numbers are multiplied does not change the product.

   Ⓒ The product of any number and −1 is a negative number.

   Ⓓ The product of any number and −1 is the opposite of the number.

73. MULTIPLE CHOICE Which of the following has the least value? **A**

   Ⓐ $\left[\frac{3}{8}(8 - 6) + \frac{1}{4}\right] \cdot (-12)$ Ⓑ $\frac{3}{8} \cdot 8 - 6 + \frac{1}{4} \cdot (-12)$

   Ⓒ $-\frac{3}{8} \cdot 8 - 6 + \frac{1}{4} \cdot 12$ Ⓓ $-\frac{3}{8} \cdot \left(8 - 6 + \frac{1}{4}\right) \cdot (-12)$

★ Challenge

GROUPING SYMBOLS **Evaluate the expression.**

74. $\frac{3}{4} \cdot [-7 \cdot (-4 - 6) + 30] - 11$ **64**

75. $-3 \cdot \left[\left(2\frac{9}{14} - 3\frac{3}{7}\right) \cdot \frac{28}{11}\right] + 5\left(-9\frac{1}{5} - 9\right)$ **−85**

MIXED REVIEW

MENTAL MATH **Write a question that can be represented by the equation. Then use mental math to solve the equation.** (Review 1.4)

76. $x + 4 = 9$ **5** 77. $y - 7 = 3$ **10** 78. $6x = 18$ **3**

79. $\frac{y}{8} = 4$ **32** 80. $2x + 1 = 7$ **3** 81. $x^2 = 121$ **11, −11**

GRAPHING NUMBERS **Graph the numbers on a number line. Then write two inequalities that compare the two numbers.** (Review 2.1)

82–87. See margin for graphs.

84. $-\frac{1}{2} < \frac{1}{3}, \frac{1}{3} > -\frac{1}{2}$

87. $-4.1 < -4.02$, $-4.02 > -4.1$

82. 6 and −3 **$6 > -3, -3 < 6$** 83. −4 and 9 **$-4 < 9, 9 > -4$** 84. $-\frac{1}{2}$ and $\frac{1}{3}$ **See margin.**

85. −3.8 and −4.0 **$-3.8 > -4.0, -4.0 < -3.8$** 86. −2.8 and 0.5 **$-2.8 < 0.5, 0.5 > -2.8$** 87. −4.1 and −4.02 **See margin.**

FINDING TERMS **Find the terms of the expression.** (Review 2.3 for 2.6)

88. $12 - z$ **12 and $-z$** 89. $-t + 5$ **$-t$ and 5** 90. $4w - 11$ **$4w$ and −11**

91. $31 - 15n$ **31 and $-15n$** 92. $-7 + 4x$ **−7 and $4x$** 93. $m - 2n - t^2$ **m, $-2n$, and $-t^2$**

94. $c^2 - 3c - 4$ **c^2, $-3c$, and −4** 95. $y + 6 - 8x$ **y, 6, and $-8x$** 96. $-9a^2 + 4 - 2a^3$ **$-9a^2$, 4, and $-2a^3$**

97. FEDERAL BUDGET In 1997 the federal government reported a budget deficit of \$21.9 billion. In 1998 the deficit was \$10 billion. What was the change in the deficit? ▶ Source: U.S. Office of Management and Budget (Review 2.3) **−\$11.9 billion**

ACTIVITY 2.6

Developing Concepts

GROUP ACTIVITY
Work with a partner.

MATERIALS
algebra tiles

Group Activity for use with Lesson 2.6

Modeling the Distributive Property

QUESTION How can you model equivalent expressions using algebra tiles?

EXPLORING THE CONCEPT

You can use algebra tiles to model algebraic expressions.

1 **1-tile**
1

This 1-by-1 square tile has an area of 1 square unit.

1 ***x*-tile**
x

This 1-by-*x* rectangular tile has an area of *x* square units.

1 Model $3(x + 4)$.

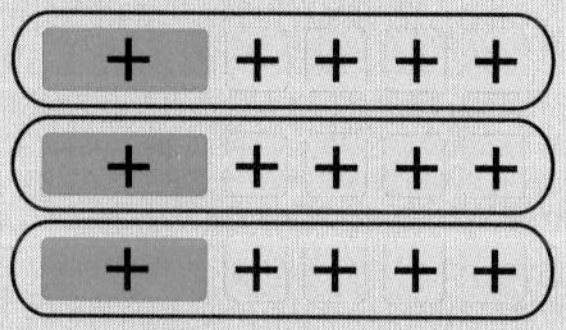

2 Model $3x + 12$.

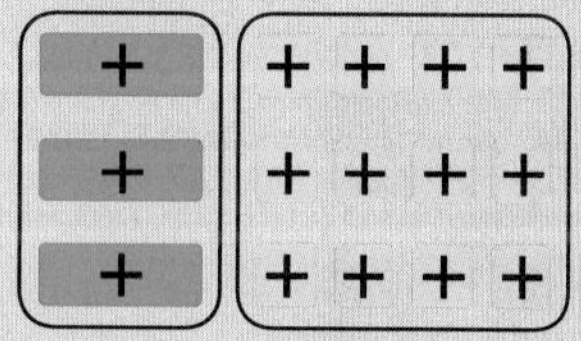

The models show that $3(x + 4) = 3(x) + 3(4) = 3x + 12$. So $3(x + 4)$ and $3x + 12$ are equivalent. This is an example of the distributive property.

DRAWING CONCLUSIONS

1. Use the algebra tiles shown below. Write the expression shown by the tiles in two ways. **$5(x + 2)$; $5x + 10$**

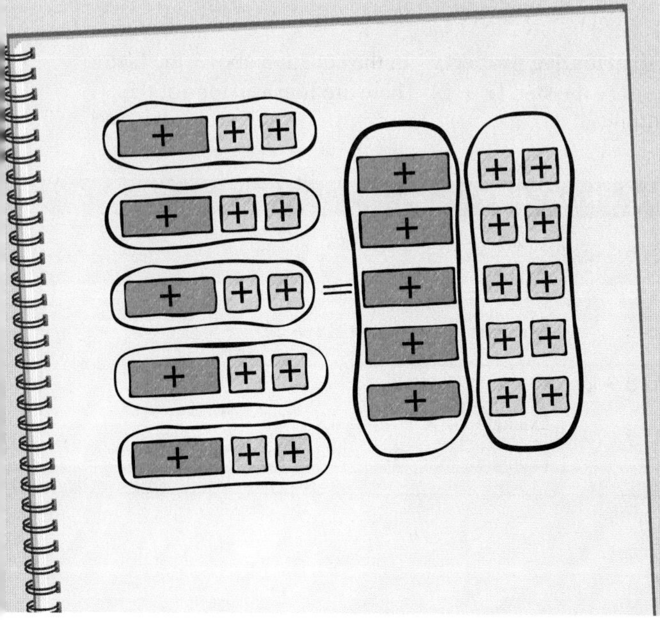

Each equation illustrates the distributive property. Use algebra tiles to model the equation. Draw a sketch of your models. **2–5. See margin.**

2. $2(x + 6) = 2x + 12$ **3.** $4(x + 2) = 4x + 8$

4. $4(x + 4) = 4x + 16$ **5.** $3(x + 5) = 3x + 15$

ERROR ANALYSIS In Exercises 6 and 7, tell whether the equation is *true* or *false*. If false, explain the error and correct the right-hand side of the equation.

6. $4(x + 6) \stackrel{?}{=} 4x + 6$ **false; $4x + 24$** **7.** $3(x + 5) \stackrel{?}{=} 3x + 15$ **true**

8. *Writing* Use your own words to explain the distributive property. Then use *a*, *b*, and *c* to represent the distributive property algebraically. **See margin.**

8. ***Sample answer:*** **When you multiply a number by a sum (or a difference), the result is the same whether you add (or subtract) first and then multiply, or multiply each of the terms by the number and then add (or subtract); for real numbers *a*, *b*, and *c*, $a(b + c) = ab + ac$ and $a(b - c) = ab - ac$.**

1 Planning the Activity

PURPOSE
To model the distributive property using algebra tiles.

MATERIALS
- A set of at least five *x*-tiles and sixteen 1-tiles for each group.
- Activity Support Master (*Chapter 2 Resource Book*, p. 83)

PACING
- Exploring the Concept — 10 min
- Drawing Conclusions — 10 min

LINK TO LESSON
After examining the area model in Example 1, ask students to draw an algebra tile model for this problem as well.

2 Managing the Activity

COOPERATIVE LEARNING
For each equation in Exercises 2–5, have one student make the algebra tile model while the other makes the sketch. They should switch roles for each exercise.

ALTERNATIVE APPROACH
This activity can be done as a demonstration by drawing the algebra tile models on an overhead sheet. After demonstrating $3(x + 4) = 3x + 12$, have the class sketch the models for Exercises 2–5.

3 Closing the Activity

KEY DISCOVERY
The distributive property lets you write equivalent expressions.

ACTIVITY ASSESSMENT
JOURNAL You have a set of two *x*-tiles and eight 1-tiles. Explain how these tiles can be used to model the distributive property. Draw a sketch to illustrate your work.
See sample answer at left.

2–5. See Additional Answers beginning on page AA1.

Activity Assessment *Sample answer:* Check student sketches. They should illustrate $2(x + 4) = 2x + 8$.

1 PLAN

PACING
Basic: 2 days
Average: 2 days
Advanced: 2 days
Block Schedule: 0.5 block with 2.5
0.5 block with 2.7

LESSON OPENER
CALCULATOR ACTIVITY
An alternative way to approach Lesson 2.6 is to use the Calculator Activity Lesson Opener:

- Blackline Master (*Chapter 2 Resource Book,* p. 84)
- Transparency (p. 13)

MEETING INDIVIDUAL NEEDS
- ***Chapter 2 Resource Book***
  Prerequisite Skills Review (p. 5)
  Practice Level A (p. 85)
  Practice Level B (p. 86)
  Practice Level C (p. 87)
  Reteaching with Practice (p. 88)
  Absent Student Catch-Up (p. 90)
  Challenge (p. 93)
- ***Resources in Spanish***
- ***Personal Student Tutor***

NEW-TEACHER SUPPORT
See the Tips for New Teachers on pp.1–2 of the *Chapter 2 Resource Book* for additional notes about Lesson 2.6.

WARM-UP EXERCISES
Transparency Available

1. Find the area of each rectangle.

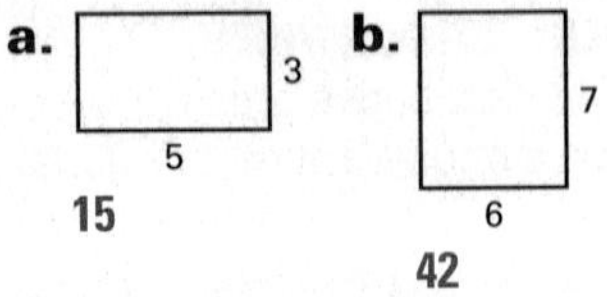

2. Write an expression for the area of each rectangle.

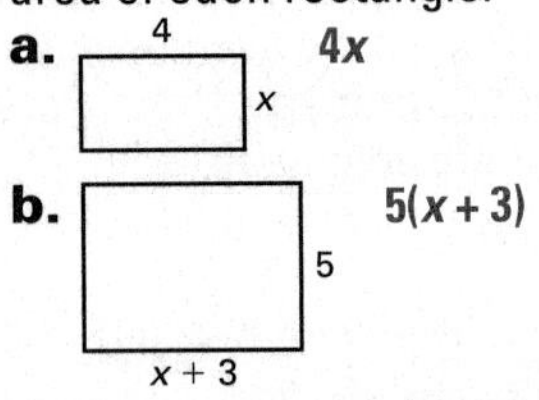

2.6 The Distributive Property

What you should learn

GOAL 1 Use the distributive property.

GOAL 2 Simplify expressions by combining like terms.

Why you should learn it

▼ To solve **real-life** problems such as finding how much you can spend on jeans in **Exs. 70 and 71.**

GOAL 1 USING THE DISTRIBUTIVE PROPERTY

To multiply 3(68) mentally, you could think of 3(68) as

$$3(60 + 8) = 3(60) + 3(8) = 180 + 24 = 204.$$

This is an example of the *distributive property.*

The distributive property is a very important algebraic property. Before discussing the property, study an example that suggests why the property is true.

EXAMPLE 1 *Using an Area Model*

Find the area of a rectangle whose width is 3 and whose length is $x + 2$.

SOLUTION

You can find the area in two ways.

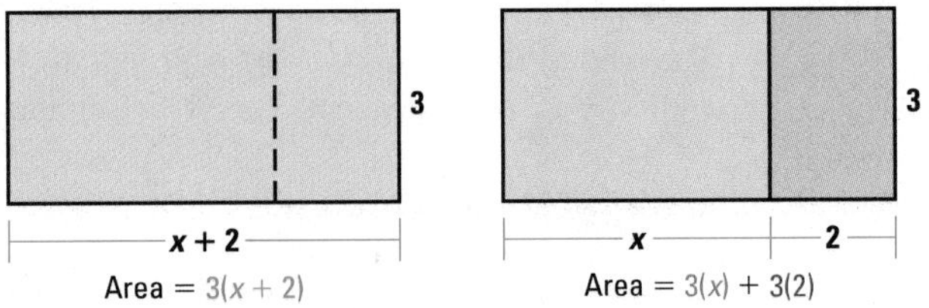

▶ Because both ways produce the same area, the following statement is true.

$$3(x + 2) = 3(x) + 3(2)$$

Example 1 suggests the **distributive property**. In the equation above, the factor 3 is *distributed* to each term of the sum $(x + 2)$. There are four versions of the distributive property, as follows.

THE DISTRIBUTIVE PROPERTY

The product of a and $(b + c)$:

$a(b + c) = ab + ac$ **Example:** $5(x + 2) = 5x + 10$

$(b + c)a = ba + ca$ **Example:** $(x + 4)8 = 8x + 32$

The product of a and $(b - c)$:

$a(b - c) = ab - ac$ **Example:** $4(x - 7) = 4x - 28$

$(b - c)a = ba - ca$ **Example:** $(x - 5)9 = 9x - 45$

STUDENT HELP

Study Tip
In Example 2, the middle steps are usually done as mental math, without writing out the steps. For instance, part (a) can be written as $2(x + 5) = 2x + 10$.

EXAMPLE 2 *Using the Distributive Property*

a. $2(x + 5) = 2(x) + 2(5) = 2x + 10$

b. $(x - 4)x = (x)x - (4)x = x^2 - 4x$

c. $(1 + 2x)8 = (1)8 + (2x)8 = 8 + 16x$

d. $y(1 - y) = y(1) - y(y) = y - y^2$

Study the next problems carefully. Remember that a factor with a negative sign must multiply *each* term of an expression. Forgetting to distribute the negative sign to each term is a common error.

EXAMPLE 3 *Using the Distributive Property*

a. $-3(x + 4) = -3(x) + (-3)(4)$ — Distribute the −3.

$= -3x - 12$ — Simplify.

b. $(y + 5)(-4) = (y)(-4) + (5)(-4)$ — Distribute the −4.

$= -4y - 20$ — Simplify.

c. $-(6 - 3x) = (-1)(6 - 3x)$ — $-a = -1 \cdot a$

$= (-1)(6) - (-1)(3x)$ — Distribute the −1.

$= -6 + 3x$ — Simplify.

d. $(x - 1)(-9x) = (x)(-9x) - (1)(-9x)$ — Distribute the $-9x$.

$= -9x^2 + 9x$ — Simplify.

REAL LIFE
Shopping

EXAMPLE 4 *Mental Math Calculations*

You are shopping for compact discs. You want to buy six compact discs for \$11.95 each. Use the distributive property to calculate the total cost mentally.

SOLUTION

If you think of \$11.95 as \$12.00 − \$.05, the mental math is easier.

$6(11.95) = 6(12 - 0.05)$ — Write 11.95 as a difference.

$= 6(12) - 6(0.05)$ — Use the distributive property.

$= 72 - 0.30$ — Find the products mentally.

$= 71.70$ — Find the difference mentally.

▶ The total cost of 6 compact discs at \$11.95 each is \$71.70.

2 TEACH

MOTIVATING THE LESSON
Suppose you buy 3 pens that cost 49 cents each. What is a quick way to find the total cost of the pens? Some students may realize that you can multiply 50 by 3 and then subtract 3 times 1 to get \$1.47. Explain that they are using the distributive property to make this mental calculation.

EXTRA EXAMPLE 1
Find the area of a rectangle whose width is 4 and whose length is $x + 2$. $4(x + 2) = 4x + 8$

EXTRA EXAMPLE 2
a. $3(x + 1)$ $3x + 3$
b. $(x - 2)7$ $7x - 14$
c. $(2 + 5x)x$ $2x + 5x^2$
d. $d(2 - d)$ $2d - d^2$

EXTRA EXAMPLE 3
a. $-6(x + 3)$ $-6x - 18$
b. $(t + 2)(-3)$ $-3t - 6$
c. $-1(1 - x)$ $-1 + x$
d. $(2x - 3)(-4x)$ $-8x^2 + 12x$

EXTRA EXAMPLE 4
You want to buy 4 new tires for \$44.98 each. Use the distributive property to calculate the total cost mentally. **\$179.92**

CHECKPOINT EXERCISES

For use after Example 1:
1. Find the area of a rectangle whose width is 5 and whose length is $x + 7$.
$5(x + 7) = 5x + 35$

For use after Examples 2 and 3:
2. Use the distributive property to simplify the expression.
a. $-4(x + 3)$ $-4x - 12$
b. $(y - 5)(y)$ $y^2 - 5y$
c. $3x(1 - 2x)$ $3x - 6x^2$
d. $(5 + x)(-5x)$ $-25x - 5x^2$

For use after Example 4:
3. The Drama Club is selling candy bars for \$1.20 each. You buy seven. Use the distributive property to calculate the total cost mentally. **\$8.40**

EXTRA EXAMPLE 5

Simplify the expression.

a. $4x - x$ $3x$

b. $5x^2 - 7 + 3x^2$ $8x^2 - 7$

c. $1 - 4(1 - 2x)$ $-3 + 8x$

EXTRA EXAMPLE 6

It takes you 30 min to get to school. You spend t min walking to the subway stop, and the rest of the time riding the subway. You walk 0.04 mi/min and the subway travels 0.8 mi/min. The total distance you travel is given by the function $D = 0.04t + 0.8(30 - t)$. Simplify this function.
$D = -0.76t + 24$

CHECKPOINT EXERCISES

For use after Example 5:

1. Simplify the expression.
   a. $-4x - 5x$ $-9x$
   b. $3x - 4x^2 + x$ $-4x^2 + 4x$
   c. $-2 - (x + 5)$ $-x - 7$

For use after Example 6:

2. Your phone company charges you \$.25 for the first 2 minutes of any phone call and \$.10 for each additional minute. The total cost of a call that lasts t minutes, where $t \geq 2$, is given by the function $C = 0.25 + 0.10(t - 2)$. Simplify this function. $C = 0.10t + 0.05$

FOCUS ON VOCABULARY

Have students think of the meaning of the word *distribute* to help them remember the various forms of the distributive property. In each of the 4 forms, the factor outside the parentheses is being distributed, or passed out, to each term inside the parentheses.

CLOSURE QUESTION

What are two uses for the distributive property? Give an example of each. **See sample answer, below.**

DAILY PUZZLER

Does the distributive property hold for division? **Yes and no; the property holds for $(b + c) \div a$ and $(b - c) \div a$; it does not hold for $a \div (b + c)$ or $a \div (b - c)$.**

GOAL 2 SIMPLIFYING BY COMBINING LIKE TERMS

In a term that is the product of a number and a variable, the number is the **coefficient** of the variable.

$$-x + 3y^2$$

−1 is the coefficient of *x*. **3 is the coefficient of y^2.**

Like terms are terms in an expression that have the same variable raised to the same power. In the expression below, $5x$ and $-3x$ are like terms, but $5x$ and $-x^2$ are *not* like terms. The **constant terms** -4 and 2 are also like terms.

$$-x^2 + 5x + (-4) + (-3x) + 2$$

The distributive property allows you to *combine like terms* that have variables by adding coefficients. An expression is **simplified** if it has no grouping symbols and if all the like terms have been combined.

EXAMPLE 5 *Simplifying by Combining Like Terms*

a. $8x + 3x = (8 + 3)x$ — Use the distributive property.
$= 11x$ — Add coefficients.

b. $4x^2 + 2 - x^2 = 4x^2 - x^2 + 2$ — Group like terms.
$= 3x^2 + 2$ — Combine like terms.

c. $3 - 2(4 + x) = 3 + (-2)(4 + x)$ — Rewrite as an addition expression.
$= 3 + [(-2)(4) + (-2)(x)]$ — Distribute the −2.
$= 3 + (-8) + (-2x)$ — Multiply.
$= -5 + (-2x) = -5 - 2x$ — Combine like terms and simplify.

FOCUS ON APPLICATIONS

BUSES Some cities offer discount fares to students who use public transportation to get to school.

EXAMPLE 6 *Using the Distributive Property to Simplify a Function*

GETTING TO SCHOOL It takes you 45 minutes to get to school. You spend t minutes walking to the bus stop, and the rest of the time riding the bus. You walk 0.06 mi/min and the bus travels 0.5 mi/min. The total distance you travel is given by the function $D = 0.06t + 0.5(45 - t)$. Simplify this function.

SOLUTION

$D = 0.06t + 0.5(45 - t)$ — Write original function.
$= 0.06t + 22.5 - 0.5t$ — Use the distributive property.
$= 0.06t - 0.5t + 22.5$ — Group like terms.
$= -0.44t + 22.5$ — Combine like terms.

▶ The total distance you travel can be given by $D = -0.44t + 22.5$.

Closure Question ***Sample answer:***
to add like terms in an expression such as $2x + 5x = 7x$ and to simplify mental calculations such as $4(1.24) = 4(1.25) - 4(0.01) = 4.96$.

GUIDED PRACTICE

Vocabulary Check ✓

1. In the expression $4x^2 - 7y - 8$, what is the coefficient of the x^2-term? What is the coefficient of the y-term? **4; −7**

2. What are the like terms in the expression $-6 - 3x^2 + 3y + 7 - 4y + 9x^2$? **−6 and 7, $-3x^2$ and $9x^2$, 3y and −4y**

Concept Check ✓

ERROR ANALYSIS **Describe the error. Then simplify the expression correctly.**

3. ~~$-2(6.5 - 2.1) = 13 - 4.2$~~
~~$= 8.8$~~
See margin.

4. ~~$5 - 4(3 + x) = 5 - 12 + x$~~
~~$= -7 + x$~~
See margin.

Skill Check ✓

Use the distributive property to rewrite the expression without parentheses.

5. $5(w - 8)$ **$5w - 40$**
6. $(y + 19)7$ **$7y + 133$**
7. $(12 - x)y$ **$12y - xy$**
8. $-4(u + 2)$ **$-4u - 8$**
9. $(b - 6)\left(-\frac{5}{6}\right)$ **See margin.**
10. $-\frac{2}{3}(t - 24)$ **$-\frac{2}{3}t + 16$**

Use the distributive property and mental math to simplify the expression.

11. $6(1.15) = 6(1 + 0.15)$ **6.9**
$= \underline{?}$
12. $9(1.95) = 9(\underline{?} - \underline{?})$ **2; 0.05; 17.55**
$= \underline{?}$

Simplify the expression by combining like terms if possible.

13. $9x + 2$ **simplified**
14. $6x^2 - 4x^2$ **$2x^2$**
15. $-3y - 2x$ **simplified**
16. $3x^2 + 4x + 8 - 7x^2$ **$-4x^2 + 4x + 8$**
17. $8t^2 - 2t + 5t - 4$ **$8t^2 + 3t - 4$**
18. $-6w - 12 - 3w + 2x^2$ **$-9w - 12 + 2x^2$**

3. **Mistakes were made when distributing the negative; −8.8.**

4. **The distributive property was applied incorrectly; $-7 - 4x$.**

9. **$-\frac{5}{6}b + 5$**

PRACTICE AND APPLICATIONS

STUDENT HELP

Extra Practice to help you master skills is on p. 798.

LOGICAL REASONING **Decide whether the statement is *true* or *false*. If false, rewrite the right-hand side of the equation so the statement is true.**

19. $3(2 + 7) \stackrel{?}{=} 3(2) + 7$ **false; 3(2) + 3(7)**
20. $(3 + 8)4 \stackrel{?}{=} 3(4) + 8(4)$ **true**
21. $7(15 - 6) \stackrel{?}{=} 7(15) - 7(6)$ **true**
22. $(9 - 2)13 \stackrel{?}{=} 9 - 2(13)$ **false; 9(13) − 2(13)**
23. $\frac{2}{9}\left(\frac{1}{3} - \frac{4}{9}\right) \stackrel{?}{=} \frac{2}{9}\left(\frac{1}{3}\right) - \frac{2}{9}\left(\frac{4}{9}\right)$ **true**
24. $-3.5(6.1 + 8.2) \stackrel{?}{=} -3.5(6.1) - 3.5(8.2)$ **true**

DISTRIBUTIVE PROPERTY **Use the distributive property to rewrite the expression without parentheses.**

STUDENT HELP

HOMEWORK HELP
Example 1: Exs. 81, 82
Example 2: Exs. 19–48
Example 3: Exs. 19–48
Example 4: Exs. 83–85
Example 5: Exs. 49–69
Example 6: Exs. 72–78

25. $3(x + 4)$ **$3x + 12$**
26. $(w + 6)4$ **$4w + 24$**
27. $5(y - 2)$ **$5y - 10$**
28. $(7 - m)4$ **$28 - 4m$**
29. $-(y - 9)$ **$-y + 9$**
30. $-3(r + 8)$ **$-3r - 24$**
31. $-4(t - 8)$ **$-4t + 32$**
32. $(x + 6)(-2)$ **$-2x - 12$**
33. $x(x + 1)$ **$x^2 + x$**
34. $(3 - y)y$ **$3y - y^2$**
35. $-r(r - 9)$ **$-r^2 + 9r$**
36. $-s(7 + s)$ **$-7s - s^2$**
37. $2(3x - 1)$ **$6x - 2$**
38. $(4 + 3y)5$ **$20 + 15y$**
39. $(2x - 4)(-3)$ **$-6x + 12$**
40. $-9(a + 6)$ **$-9a - 54$**
41. $4x(x + 8)$ **$4x^2 + 32x$**
42. $-2t(12 - t)$ **$-24t + 2t^2$**
43. $(3y - 2)5y$ **$15y^2 - 10y$**
44. $-2x(x - 8)$ **$-2x^2 + 16x$**
45. $-9(-t - 3)$ **$9t + 27$**
46. $(6 - 3w)(-w^2)$ **$-6w^2 + 3w^3$**
47. $5\left(\frac{1}{2}x - \frac{2}{3}\right)$ **$\frac{5}{2}x - \frac{10}{3}$**
48. $-y(-y^2 + y)$ **$y^3 - y^2$**

3 APPLY

ASSIGNMENT GUIDE

BASIC
Day 1: pp. 103–104 Exs. 20–68 even
Day 2: pp. 104–107 Exs. 70–76, 79–82, 86–89, 91, 95, 100, 105, 110, Quiz 2 Exs. 1–19

AVERAGE
Day 1: pp. 103–104 Exs. 20–68 even
Day 2: pp. 104–107 Exs. 70–76, 79–82, 86–89, 91, 95, 100, 105, 110, Quiz 2 Exs. 1–19

ADVANCED
Day 1: pp. 103–104 Exs. 20–68 even
Day 2: pp. 104–107 Exs. 70–76, 79–82, 86–89, 91–93, 95, 100, 105, 110, Quiz 2 Exs. 1–19

BLOCK SCHEDULE
pp. 103–105 Exs. 20–68 even, 70–76, 79–82 (with 2.5)
pp. 104–107 Exs. 77, 78, 86–89, 91, 95, 100, 105, 110, Quiz 2 Exs. 1–19 (with 2.7)

EXERCISE LEVELS

Level A: ***Easier***
19–36, 49–60, 79, 80, 94–101
Level B: ***More Difficult***
37–48, 61–78, 81–89, 91, 102–113
Level C: ***Most Difficult***
90, 92, 93

✓ HOMEWORK CHECK

To quickly check student understanding of key concepts, go over the following exercises: Exs. 30, 34, 46, 50, 60, 66, 74, 80. See also the Daily Homework Quiz:

- Blackline Master (*Chapter 2 Resource Book*, p. 97)
- Transparency (p. 16)

! COMMON ERROR

EXERCISES 61–69 Caution students to continue to use the order of operations in problems such as Exercise 65. Students may erroneously subtract $3x$ from $7x$ before using the distributive property to eliminate the parentheses. Also, some students may have trouble deciding which terms to add or subtract after grouping like terms. For these students, suggest that they change all the subtractions to equivalent additions before rearranging.

STUDENT HELP NOTES

Homework Help Students can find help for Exs. 75 and 76 at **www.mcdougallittell.com.** The information can be printed out for students who don't have access to the Internet.

78.

Input	Output
0	240
1	285
2	330
3	375
4	420
5	465
6	510
7	555
8	600
9	645
10	690
11	735
12	780
13	825
14	870
15	915
16	960

SIMPLIFYING EXPRESSIONS **Simplify the expression by combining like terms.**

49. $15x + (-4x)$ **$11x$**

50. $-12y + 5y$ **$-7y$**

51. $-8b - 9b$ **$-17b$**

52. $5 - x + 2$ **$7 - x$**

53. $-3 + y + 7$ **$y + 4$**

54. $4 + a + a$ **$4 + 2a$**

55. $1.3t - 2.1t$ **$-0.8t$**

56. $\frac{7}{9}w + \left(-\frac{2}{3}\right)w$ **$\frac{1}{9}w$**

57. $107a - 208a$ **$-101a$**

58. $3x^2 + 2x^2 - 7$ **$5x^2 - 7$**

59. $9x^3 - 4x^3 - 2$ **$5x^3 - 2$**

60. $8b + 5 - 3b$ **$5b + 5$**

COMBINING LIKE TERMS **Apply the distributive property. Then simplify by combining like terms.**

61. $(3y + 1)(-2) + y$ **$-5y - 2$**

62. $4(2 - a) - a$ **$8 - 5a$**

63. $12s + (7 - s)2$ **$10s + 14$**

64. $(5 - 2x)(-x) + x^2$ **$-5x + 3x^2$**

65. $7x - 3x(x + 1)$ **$4x - 3x^2$**

66. $-4(y + 2) - 6y$ **$-10y - 8$**

67. $3t(t - 5) + 6t^2$ **$9t^2 - 15t$**

68. $-x^3 + 2x(x - x^2)$ **$-3x^3 + 2x^2$**

69. $4w^2 - w(2w - 3)$ **$2w^2 + 3w$**

BUYING JEANS **You have \$58, and you want to buy a pair of jeans and a \$20 T-shirt. There is a 6% sales tax. If x represents the cost of the jeans, then the following inequality is a model that shows how much you can spend on the jeans.**

$$x + 20 + 0.06(x + 20) \le 58$$

70. Simplify the left side of the inequality. **$1.06x + 21.2$**

71. If the jeans cost \$35, can you buy both the T-shirt and the jeans? **no**

FOCUS ON APPLICATIONS

FREIGHT TRAINS A 150-car freight train is so heavy that it takes 1.5 miles to come to a complete stop if traveling 50 miles per hour.

FREIGHT TRAINS **A train with 150 freight cars is used to haul two types of grain. Each freight car can haul 97.3 tons of barley or 114 tons of corn. Let n represent the number of freight cars containing corn.**

72. Which function correctly represents the total weight the train can haul? **A**

A. $W = 97.3(150 - n) + 114n$ **B.** $W = 97.3n + 114(150 - n)$

73. If 90 freight cars contain corn, what is the total weight the train is hauling? **16,098 tons**

74. If 72 freight cars contain barley, what is the total weight the train is hauling? **15,897.6 tons**

INVESTING MONEY **You receive \$5000. You decide to invest the money in a one-year bond paying 2% interest and in a one-year certificate of deposit paying 6% interest.**

75. Let m represent the amount of money invested in the one-year bond. Write a function that represents the total amount of money T that you have after one year. Simplify the function. **$T = 5300 - 0.04s$**

76. If you invest \$2000 in the one-year bond, how much money do you have after one year? What if you invest \$3000 in the one-year bond? **\$5220; \$5180**

MOVING **The van that you are using to move can hold 16 moving boxes. Each box can hold 60 pounds of books or 15 pounds of clothes.**

77. Let b represent the number of boxes filled with books. Write a function that represents the total weight w of the boxes in the van. **$w = 45b + 240$**

78. Use the function you wrote in Exercise 77 to make an input-output table that shows the total weight of the boxes for each combination of boxes of books and clothes. **See margin.**

STUDENT HELP

HOMEWORK HELP Visit our Web site www.mcdougallittell.com for help with Exs. 75–76.

STUDENT HELP

Skills Review
For help with perimeter and area, see p. 790.

GEOMETRY CONNECTION **Write and simplify an expression for the perimeter of the figure.**

79.

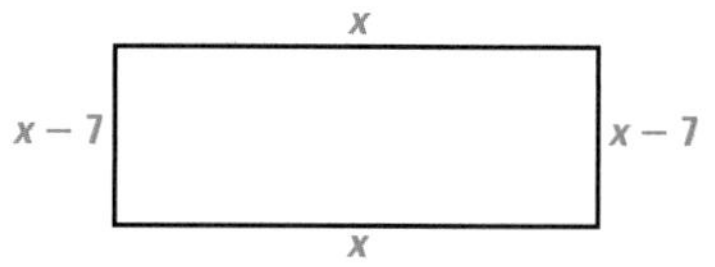

$4x - 14$

80.

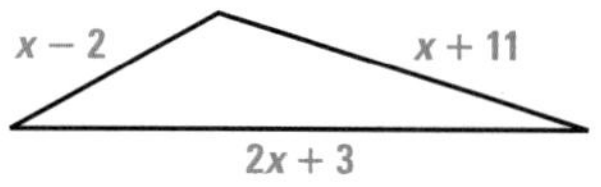

$4x + 12$

GEOMETRY CONNECTION **Write an expression modeling the area of the large rectangle as the product of its length and width. Then write another expression modeling this area as the sum of the areas of the two smaller rectangles. Simplify each expression.**

81.

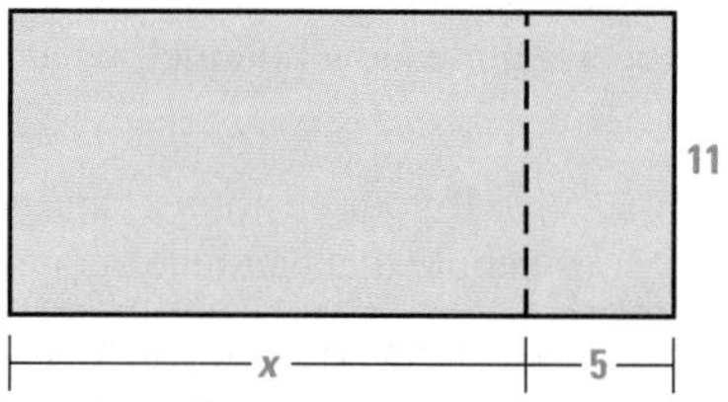

$(x + 5)11 = 11x + 55;\ 11x + 55$

82.

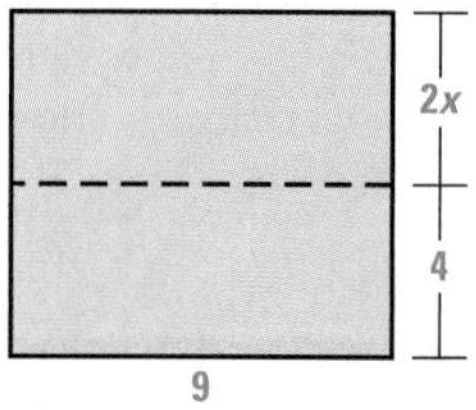

$9(2x + 4) = 18x + 36;\ 18x + 36$

LEAVING A TIP **In Exercises 83–85, use the following information.**
You and a friend decide to leave a 15% tip for restaurant service. You compute the tip, T, as $T = 0.15C$, where C represents the cost of the meal. Your friend claims that an easier way to mentally compute the tip is to calculate 10% of the cost of the meal plus one half of 10% of the cost of the meal.

83. Write an equation that represents your friend's method of computing the tip. $T = 0.10C + 0.5(0.10C)$

84. Simplify the equation. What property did you use to simplify the equation? $T = 0.15C$; distributive property

85. Will both methods give the same results? Explain. See margin.

85. Yes; the simplified version of your friend's equation is identical to your equation.

PLANTING PLAN **In Exercises 86–89, use the sketch below. You are a farmer planting corn and soybeans in a rectangular field measuring 200 yards by 300 yards. It costs $.06 per square yard to plant corn and $.05 per square yard to plant soybeans.**

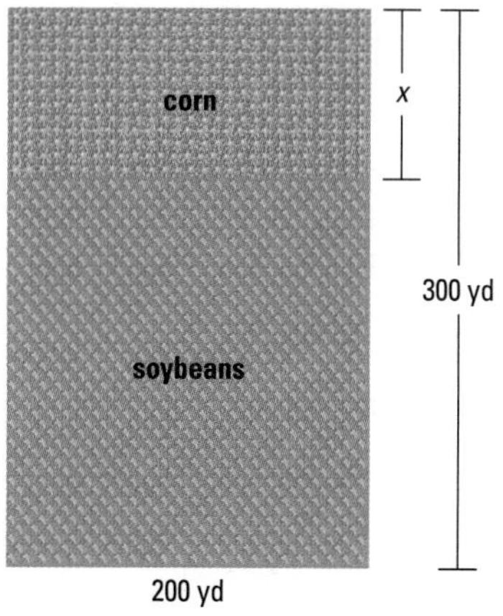

86. Let x represent the width of the field you plant with corn. Write a function that represents the cost of planting the entire field. Simplify.
Let C = the cost; $C = 2x + 3000$.

87. If the width of the field you plant with corn is 75 yards, what is the cost of planting the entire field? $3150

88. If the width of the field you plant with soybeans is 125 yards, what is the cost of planting the entire field? $3350

89. You have $3400 to plant the entire field. Do you have enough money to plant a field of soybeans that is 90 yards wide? no

STUDENT HELP NOTES

Skills Review As students review perimeter and area on page 790, remind them that perimeter is measured in units while area is measured in square units.

ADDITIONAL PRACTICE AND RETEACHING

For Lesson 2.6:

- Practice Levels A, B, and C (*Chapter 2 Resource Book,* p. 85)
- Reteaching with Practice (*Chapter 2 Resource Book,* p. 88)
- See Lesson 2.6 of the *Personal Student Tutor*

For more Mixed Review:

- Search the *Test and Practice Generator* for key words or specific lessons.

1. Use the distributive property to write the expression without parentheses.

 a. $-4(x + 2)$ **b.** $(5 - 2n)(-n)$
 $-4x - 8$ $2n^2 - 5n$

2. Simplify the expression $4y(2 - y) + 6y^2$. $8y + 2y^2$

3. At a used book sale, paperbacks are priced at \$.75 and hardcover books are \$2. You have \$10 to spend.

 a. Write a function to model the situation.
 $0.75p + 2h = 10$

 b. If you buy 6 paperbacks, how many hardcover books can you buy? How much will you have left? 2; \$1.50

ADDITIONAL TEST PREPARATION

1. **OPEN ENDED** Write a real-word problem that could be modeled by the function $y = 8x + 6(15 - x)$. Use the distributive property to simplify the function. *Sample answer:* You bought 15 geraniums to plant in your garden. Large geraniums cost \$8 apiece, and smaller ones cost \$6 apiece. What is the total cost of the geraniums? $y = 2x + 90$

Test Preparation 🔍

90. *Sample answer:* Let a rectangle have one dimension $x + 2$ and the other 7. The area of the rectangle is $(x + 2)7$. The sum of the areas of the smaller rectangles is $7x + 14$.

★ **Challenge**

92. *Sample answer:* Suppose that neither x nor y is zero. If $2(xy) = 2x \cdot 2y$, then $2xy = 4xy$, and $\dfrac{2xy}{xy} = \dfrac{4xy}{xy}$ which implies that $2 = 4$. Clearly this is not true. (If $x = 0$ or $y = 0$, then $2(xy)$ is equal to $2x \cdot 2y$, since both are equal to 0.)

90. ✐ *Writing* Write a problem like Exercise 81 that deals with the area of a rectangle. Explain how you can use the distributive property to model the total area of a rectangle as the sum of two rectangles. **See margin.**

91. **MULTI-STEP PROBLEM** A customer of your flower shop wants to send flowers to 23 people. Each person will receive an \$11.99 "sunshine basket" or a \$16.99 "meadow bouquet."

 a. Let s represent the number of people who will receive a sunshine basket. Which function can you use to find C, the total cost of sending flowers to all 23 people, depending on how many of each arrangement is sent? **A**

 Ⓐ $C = 16.99(23 - s) + 11.99s$ Ⓑ $C = 11.99s + 16.99(23)$

 b. If 8 people receive a sunshine basket, what is the total cost of the flowers? **\$350.77**

 c. If 13 people receive a meadow bouquet, what is the total cost of the flowers? **\$340.77**

 d. **CRITICAL THINKING** If your customer can spend only \$300, what is the greatest number of people that can receive a meadow bouquet? **4 people**

92. **LOGICAL REASONING** You are tutoring a friend in algebra. After learning the distributive property, your friend attempts to apply this property to multiplication and gets $2(xy) = 2x \cdot 2y$.

 Write a convincing argument to show your friend that this is incorrect. **See margin.**

93. **LOGICAL REASONING** Your friend does not understand how the product of a and $(b + c)$ is given by both $ab + ac$ and $ba + ca$.

 Use the properties of multiplication to write a convincing argument to show your friend that both statements are correct. **See below.**

 $a(b + c) = ab + ac$ distributive property
 $ = ba + ca$ commutative property of multiplication
 $ = (b + c)a$ distributive property

MIXED REVIEW

RECIPROCALS **Find the reciprocal.** (Skills Review, p. 782)

94. $\dfrac{2}{7}$ $\dfrac{7}{2}$ 95. $\dfrac{6}{11}$ $\dfrac{11}{6}$ 96. 7 $\dfrac{1}{7}$ 97. 1 1

98. $\dfrac{1}{121}$ 121 99. 435 $\dfrac{1}{435}$ 100. $4\dfrac{1}{2}$ $\dfrac{2}{9}$ 101. $3\dfrac{3}{8}$ $\dfrac{8}{27}$

EXPRESSIONS WITH FRACTION BARS **Evaluate the expression.** (Review 1.3)

102. $\dfrac{10 \cdot 8}{4^2 + 4}$ 4 103. $\dfrac{6^2 - 12}{3^2 + 15}$ 1 104. $\dfrac{75 - 5^2}{11 + (3 \cdot 4)}$ $2\dfrac{4}{23}$

105. $\dfrac{(3 \cdot 7) + 9}{2^3 + 5 - 3}$ 3 106. $\dfrac{(2 + 5)^2}{3^2 - 2}$ 7 107. $\dfrac{6 + 7^2}{3^3 - 9 - 7}$ 5

EVALUATING EXPRESSIONS **Evaluate the expression.** (Review 2.2, 2.3)

108. $6 - (-8) - 11$ 3 109. $4 - 8 - 3$ -7 110. $6 + (-13) + (-5)$ -12

111. $-7 + 9 - 8$ -6 112. $20 + (-16) + (-3)$ 1 113. $12.4 - 9.7 - (-6.1)$ 8.8

QUIZ 2

Self-Test for Lessons 2.4–2.6

1. $\begin{bmatrix} 5 & -13 \\ -4 & 9 \\ -9 & -8 \end{bmatrix}$; $\begin{bmatrix} -1 & 1 \\ -6 & 1 \\ -5 & 8 \end{bmatrix}$

1. Find the sum and difference of the matrices. $\begin{bmatrix} 2 & -6 \\ -5 & 5 \\ -7 & 0 \end{bmatrix}$, $\begin{bmatrix} 3 & -7 \\ 1 & 4 \\ -2 & -8 \end{bmatrix}$ (Lesson 2.4)

Find the product. (Lesson 2.5)

2. $(-7)(9)$ **−63**
3. $(3)(-7)$ **−21**
4. $(-2)(-7)$ **14**
5. $(35)(-80)$ **−2800**
6. $(-15)\left(\frac{1}{5}\right)$ **−3**
7. $(-1.8)(-6)$ **10.8**
8. $(11)(-5)(-2)$ **110**
9. $(-10)(-3)(9)$ **270**

Simplify the variable expression. (Lessons 2.5, 2.6)

10. $(-t)(-7.1)$ **$7.1t$**
11. $(13)(-x)$ **$-13x$**
12. $(-5)(-b)(2b)(-b)$ **$-10b^3$**
13. $(-4)(-x)^3(x)\left(-\frac{1}{8}\right)$ **$-\frac{1}{2}x^4$**
14. $-28x + (-15x)$ **$-43x$**
15. $11y + (-9y)$ **$2y$**
16. $-17t + 9t$ **$-8t$**
17. $-3(8 - b)$ **$-24 + 3b$**
18. $(-4y - 2)(-5)$ **$20y + 10$**

19. SKYSCRAPERS The heights (in feet and in meters) of the three tallest buildings in the United States are shown in the table. Organize the data in a matrix. (Lesson 2.4) **See margin.**

Building	Height (ft)	Height (m)
Sears Tower	1454	443
Empire State Building	1250	381
Aon Center	1136	346

▶ Source: Council on Tall Buildings and Urban Habitat

MATH & History — History of Negative Numbers

APPLICATION LINK www.mcdougallittell.com

THEN

THE 7TH CENTURY Hindu mathematician Brahmagupta explained operations with negative numbers. As late as the 16th century, many mathematicians still considered the idea of a number with a value less than zero absurd.

Hindu math manuscript

1. As early as 200 B.C., negative numbers were used in China. About how many years was it between the time the Chinese displayed a concept of negative numbers and when they were used in 628 A.D. in India? **828 years**

B.C.
A.D.
200 B.C.
0
628 A.D.

NOW

TODAY, negative numbers are used in many ways, including financial statements, altitude measurements, temperature readings, golf scores, and atomic charges.

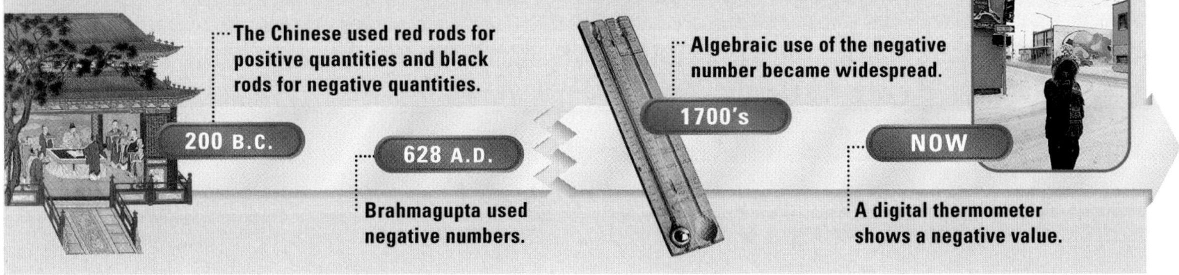

The Chinese used red rods for positive quantities and black rods for negative quantities.
200 B.C.
628 A.D.
Brahmagupta used negative numbers.
1700's
Algebraic use of the negative number became widespread.
NOW
A digital thermometer shows a negative value.

ADDITIONAL RESOURCES

An alternative Quiz for Lessons 2.4–2.6 is available in the *Chapter 2 Resources Book,* p. 94.

A **blackline** master with additional Math & History exercises is available in the *Chapter 2 Resources Book,* p. 92.

19.

	ft	m
Sears Tower	1454	443
Empire State Building	1250	381
Aon Center	1136	346

1 PLAN

PACING
Basic: TK
Average: TK
Advanced: TK
Block Schedule: 0.5 with 2.6

LESSON OPENER ACTIVITY

An alternative way to approach Lesson 2.7 is to use the Activity Lesson Opener:

- Blackline Master (*Chapter 2 Resource Book*, p. 98)
- Transparency (p.14)

MEETING INDIVIDUAL NEEDS

- ***Chapter 2 Resource Book***
  Prerequisite Skills Review (p. 5)
  Practice Level A (p. 101)
  Practice Level B (p. 102)
  Practice Level C (p. 103)
  Reteaching with Practice (p. 104)
  Absent Student Catch-Up (p. 106)
  Challenge (p. 108)
- ***Resources in Spanish***
- ***Personal Student Tutor***

NEW-TEACHER SUPPORT

See the Tips for New Teachers on pp. 1–2 of the *Chapter 2 Resource Book* for additional notes about Lesson 2.7.

WARM-UP EXERCISES

Transparency Available

Find the product.

1. $-1 \cdot \frac{1}{2}$ $-\frac{1}{2}$
2. $3 \cdot \frac{1}{9}$ $\frac{1}{3}$
3. $-12 \cdot \frac{2}{3}$ -8
4. $(-14x) \cdot \left(-\frac{1}{7}\right)$ $2x$
5. $-\frac{1}{3} \cdot 1\frac{1}{5}$ $-\frac{2}{5}$

2.7 Division of Real Numbers

What *you should learn*

GOAL 1 Divide real numbers.

GOAL 2 Use division to simplify algebraic expressions.

Why *you should learn it*

▼ To solve **real-life** problems like finding the average velocity of a model rocket in **Exs. 68–70**.

REAL LIFE

GOAL 1 DIVIDING REAL NUMBERS

For every real number other than zero there exists a number called its *reciprocal*. The product of a number and its **reciprocal** is 1. For instance, the product of the number -3 and its reciprocal $-\frac{1}{3}$ is 1. This property is sometimes referred to as the *inverse property of multiplication*.

- The reciprocal of a is $\frac{1}{a}$. $(a \neq 0)$ **Example:** -8 and $-\frac{1}{8}$
- The reciprocal of $\frac{a}{b}$ is $\frac{b}{a}$. $(a \neq 0, b \neq 0)$ **Example:** $-\frac{2}{5}$ and $-\frac{5}{2}$

Zero is the only real number that has no reciprocal. There is no number that when multiplied by zero gives a product of 1. Because zero does not have a reciprocal, you cannot divide by zero.

You can use a reciprocal to write a division expression as a product.

DIVISION RULE

To divide a number a by a nonzero number b, multiply a by the reciprocal of b.

$a \div b = a \cdot \frac{1}{b}$ **Example:** $-1 \div 3 = -1 \cdot \frac{1}{3} = -\frac{1}{3}$

The result is the quotient of a and b.

EXAMPLE 1 ***Dividing Real Numbers***

Find the quotient.

a. $10 \div (-2)$ **b.** $-39 \div \left(-4\frac{1}{3}\right)$ **c.** $\dfrac{-\frac{1}{3}}{4}$ **d.** $\dfrac{1}{-\frac{3}{4}}$

SOLUTION

a. $10 \div (-2) = 10 \cdot \left(-\frac{1}{2}\right) = -5$

b. $-39 \div \left(-4\frac{1}{3}\right) = -39 \div \left(-\frac{13}{3}\right) = -39 \cdot \left(-\frac{3}{13}\right) = 9$

c. $\dfrac{-\frac{1}{3}}{4} = -\frac{1}{3} \div 4 = -\frac{1}{3} \cdot \frac{1}{4} = -\frac{1}{12}$

d. $\dfrac{1}{-\frac{3}{4}} = 1 \div \left(-\frac{3}{4}\right) = 1 \cdot \left(-\frac{4}{3}\right) = -\frac{4}{3}$

GOAL 2 WORKING WITH ALGEBRAIC EXPRESSIONS

In Example 1, notice that applying the division rule results in the following patterns for finding the sign of a quotient.

THE SIGN OF A QUOTIENT

- The quotient of two numbers with the same sign is positive.

  $-a \div (-b) = \frac{a}{b}$ **Example:** $-20 \div (-4) = 5$

- The quotient of two numbers with opposite signs is negative.

  $-a \div b = -\frac{a}{b}$ **Example:** $-20 \div 4 = -5$

EXAMPLE 2 *Using the Distributive Property to Simplify*

Simplify the expression $\frac{32x - 8}{4}$.

SOLUTION

$$\frac{32x - 8}{4} = (32x - 8) \div 4$$ Rewrite fraction as division expression.

$$= (32x - 8)\left(\frac{1}{4}\right)$$ Multiply by reciprocal.

$$= (32x)\left(\frac{1}{4}\right) - (8)\left(\frac{1}{4}\right)$$ Use distributive property.

$$= 8x - 2$$ Simplify.

The order of operations you learned in Chapter 1 also applies to real numbers and to algebraic expressions.

EXAMPLE 3 *Evaluating an Expression*

Evaluate the expression when $a = -2$ and $b = -3$.

a. $\frac{-2a}{a + b}$ **b.** $\frac{a^2 - 4}{b}$ **c.** $\frac{a}{3 + b}$

SOLUTION

a. $\frac{-2a}{a + b} = \frac{-2(-2)}{-2 + (-3)} = \frac{4}{-5} = -\frac{4}{5}$

b. $\frac{a^2 - 4}{b} = \frac{(-2)^2 - 4}{-3} = \frac{4 - 4}{-3} = \frac{0}{-3} = 0$

c. $\frac{a}{3 + b} = \frac{-2}{3 + (-3)} = \frac{-2}{0}$ (crossed out) (Undefined)

STUDENT HELP

HOMEWORK HELP Visit our Web site www.mcdougallittell.com for extra examples.

STUDENT HELP

Study Tip
Remember that 0 can be divided by any nonzero number. The result will always be zero.

Division *by* zero is undefined.

2 TEACH

MOTIVATING THE LESSON
Suppose you owe your brother \$50. This can be represented by −50. You need to find out how much to pay him each month if he wants you to pay it off in 4 equal monthly payments. This problem can be done using division of real numbers.

EXTRA EXAMPLE 1
Find the quotient.

a. $15 \div (-5)$ **−3**

b. $-45 \div \frac{5}{6}$ **−54**

c. $\frac{2\frac{1}{3}}{-3}$ $-\frac{7}{9}$

d. $\frac{-1}{\frac{2}{3}}$ $-\frac{3}{2}$

EXTRA EXAMPLE 2
Simplify the expression $\frac{9 - 27x}{-3}$.
$9x - 3$

EXTRA EXAMPLE 3
Evaluate the expression when $x = -5$ and $y = -1$.

a. $\frac{x + y}{y + 1}$ **undefined**

b. $\frac{2y}{x - 1}$ $\frac{1}{3}$

c. $\frac{5 + x}{5}$ **0**

CHECKPOINT EXERCISES
For use after Examples 1–3:

1. Simplify the expression $\frac{25d - 125}{-5}$. **$-5d + 25$**

2. Evaluate the expression $\frac{s - 3t}{2s}$ when $s = -3$ and $t = -1$. **0**

STUDENT HELP NOTES

Homework Help Students can find extra examples at **www.mcdougallittell.com** that parallel the examples in the student edition.

EXTRA EXAMPLE 4
A mountain climber descends 300 feet in 50 minutes. What is her velocity? **−6 ft/min**

EXTRA EXAMPLE 5
Find the domain of the function $y = \frac{2-x}{4x}$. **all real numbers except $x = 0$**

CHECKPOINT EXERCISES

For use after Example 4:

1. An elevator descends 120 feet in 1.5 minutes. What is its velocity? **−80 ft/min**

For use after Example 5:

2. Find the domain of the function $y = \frac{6}{4 + x}$ **all real numbers except $x = -4$**

APPLICATION NOTE
EXAMPLE 4 Point out to students that the verbal model given in this example is valid only when the object is moving at a constant velocity. When the velocity of the object varies, the expression $\frac{\text{displacement}}{\text{time}}$ gives the average velocity of the object.

FOCUS ON VOCABULARY
What is the difference between the reciprocal of a number and the opposite of a number? **The product of a number and its reciprocal is 1; whereas the sum of a number and its opposite is zero.**

CLOSURE QUESTION
Why are the rules for finding the sign of a quotient the same as the rules for finding the sign of a product? **You write the quotient as a product in order to evaluate it.**

DAILY PUZZLER
Suppose $a \div b = c$. What do you know about b if c is negative and $c > a$. **b is positive and greater than 1.**

FOCUS ON APPLICATIONS

REAL LIFE HOT-AIR BALLOONING
A hot-air balloon pilot fires the burners of the balloon as it nears the ground, to slow the descent rate and land the balloon gently.

INTERNET APPLICATION LINK
www.mcdougallittell.com

EXAMPLE 4 *Finding a Velocity*

HOT-AIR BALLOONING You are descending in a hot-air balloon. You descend 500 feet in 40 seconds. What is your velocity?

SOLUTION

VERBAL MODEL Velocity = $\frac{\text{Displacement}}{\text{Time}}$

LABELS
Velocity = v (ft/sec)
Displacement = −500 (feet)
Time = 40 (seconds)

ALGEBRAIC MODEL
$v = \frac{-500}{40}$
$= -12.5$

▶ Your velocity, −12.5 ft/sec, is negative because you are descending.

....

When a function is defined by an equation, its domain is restricted to real numbers for which the function can be evaluated. Division by zero is undefined, so input values that make you divide by zero must be excluded from the domain. For instance, the function $y = \frac{5}{x-2}$ does not have $x = 2$ in its domain.

EXAMPLE 5 *Finding the Domain of a Function*

Find the domain of the function $y = \frac{-x}{1-x}$.

STUDENT HELP
Look Back
For help with domain and range, see p. 47.

SOLUTION

To find the domain of $y = \frac{-x}{1-x}$, you must exclude any values for which the denominator, $1 - x$, is equal to zero. To do this, you need to solve the equation $1 - x = 0$. Use mental math.

You can ask: What number subtracted from 1 equals 0?
The answer is 1, because $1 - 1 = 0$.

▶ Since $1 - x = 0$ when $x = 1$, you can see that $x = 1$ is not in the domain of the function, because you cannot divide by zero. All other real numbers are in the domain because there are no other values of x that will make the denominator zero. The domain is all real numbers *except* $x = 1$.

✓**CHECK** When $x = 1$, $y = \frac{-x}{1-x}$

$= \frac{-(1)}{1-(1)}$

$= \frac{-1}{0}$ ← Undefined

Guided Practice

Vocabulary Check ✓

1. The product of a number and _?_ is 1. **its reciprocal**

Concept Check ✓

2. Is dividing by a number the same as multiplying by the opposite of the number? Explain your reasoning. **See margin.**
3. Is the reciprocal of a negative number *sometimes, always*, or *never* positive? **never**
4. A friend tells you that $-\frac{a}{b} = \frac{-a}{b} = \frac{a}{-b}$. Is your friend correct? Use examples or counterexamples to support your answer. **See margin.**
5. Describe the error shown at the right. **$\frac{3}{2}$ is not the reciprocal of $-\frac{2}{3}$.**

~~$6 \div -\frac{2}{3} = 6 \cdot \frac{3}{2} = 9$~~

2. No; dividing by a number is the same as multiplying by its reciprocal. If the given statement were true, then $4 \div 2$ would be equal to $4 \cdot (-2)$. It is not true that $2 = -8$.

Skill Check ✓

Find the reciprocal of the number.

6. 34 **$\frac{1}{34}$**
7. -8 **$-\frac{1}{8}$**
8. $-\frac{3}{4}$ **$-\frac{4}{3}$**
9. $-2\frac{1}{5}$ **$-\frac{5}{11}$**

4. Yes; *Sample answer:* Let $a = 4$ and $b = 2$. Then $-\frac{a}{b}, \frac{-a}{b}$, and $\frac{a}{-b}$ are all equal to -2.

Find the quotient.

10. $7 \div \frac{1}{2}$ **14**
11. $10 \div \left(-\frac{1}{5}\right)$ **−50**
12. $-12 \div 3$ **−4**
13. $\dfrac{16}{\frac{-2}{9}}$ **−72**

14. Find the domain of the function $y = \frac{x+1}{x+2}$. **all real numbers except −2**

15. STOCK You own 18 shares of stock in a computer company. The total value of the shares changes by −\$3.06. By how much does the value of each share of stock change? **−\$.17**

Practice and Applications

STUDENT HELP

▸ **Extra Practice** to help you master skills is on p. 798.

FINDING QUOTIENTS **Find the quotient.**

16. $-51 \div (-17)$ **3**
17. $45 \div (-9)$ **−5**
18. $-24 \div 4$ **−6**
19. $49 \div (-7)$ **−7**
20. $64 \div (-8)$ **−8**
21. $-99 \div 9$ **−11**
22. $-35 \div (-70)$ **$\frac{1}{2}$**
23. $18 \div (-54)$ **$-\frac{1}{3}$**
24. $-90 \div \left(-\frac{2}{3}\right)$ **135**
25. $56 \div \left(-2\frac{4}{7}\right)$ **$-21\frac{7}{9}$**
26. $\dfrac{-26}{-\frac{1}{2}}$ **52**
27. $\dfrac{36}{-\frac{5}{6}}$ **$-43\frac{1}{5}$**
28. $60 \div (-10)$ **−6**
29. $-12.6 \div 1.8$ **−7**
30. $-18 \div \frac{3}{8}$ **−48**
31. $-87 \div \left(-\frac{3}{5}\right)$ **145**

SIMPLIFYING EXPRESSIONS **Simplify the expression.**

32. $42y \div \frac{1}{7}$ **$294y$**
33. $6t \div \left(-\frac{1}{2}\right)$ **$-12t$**
34. $58z \div \left(-\frac{2}{5}\right)$ **$-145z$**
35. $-\frac{x}{12} \div 3$ **$-\frac{x}{36}$**
36. $\frac{d}{4} \div 6$ **$\frac{d}{24}$**
37. $\frac{3y}{4} \div \frac{1}{2}$ **$\frac{3y}{2}$**
38. $-\frac{2b}{7} \div \frac{7}{9}$ **$-\frac{18b}{49}$**
39. $33x \div \frac{3}{11}$ **$121x$**
40. $49x \div 3\frac{1}{2}$ **$14x$**
41. $8x^2 \div \left(-\frac{4}{5}\right)$ **$-10x^2$**
42. $68x \div \left(-\frac{17}{9}\right)$ **$-36x$**
43. $-54x^2 \div \frac{-9}{5}$ **$30x^2$**
44. $\frac{42t}{-14z} \div \frac{-6}{7t}$ **$\frac{7t^2}{2z}$**
45. $8 \cdot \frac{x}{8}$ **x**
46. $3 \cdot \left(-\frac{y}{3}\right)$ **$-y$**
47. $-7 \cdot \left(-\frac{2w}{-7}\right)$ **$-2w$**

STUDENT HELP

▸ HOMEWORK HELP
Example 1: Exs. 16–47
Example 2: Exs. 48–51
Example 3: Exs. 52–57
Example 4: Exs. 62–64
Example 5: Exs. 58–61

3 APPLY

Assignment Guide

BASIC
Day 1: pp. 111–113 Exs. 16–56 even, 58–63, 72, 73, 75, 80, 85, 90, 95, 100

AVERAGE
Day 1: pp. 111–113 Exs. 16–56 even, 58–67, 72, 73, 75, 80, 85, 90, 95, 100

ADVANCED
Day 1: pp. 111–113 Exs. 16–56 even, 65–74, 75, 80, 85, 90, 95, 100

BLOCK SCHEDULE WITH 2.6
pp. 111–113 Exs. 16–56 even, 58–67, 72, 73, 75, 80, 85, 90, 95, 100

Exercise Levels

Level A: ***Easier***
16–47, 62, 63, 75–82

Level B: ***More Difficult***
48–57, 64–70, 72, 73, 83–100

Level C: ***Most Difficult***
58–61, 71, 74

✓ Homework Check

To quickly check student understanding of key concepts, go over the following exercises: Exs. 18, 26, 38, 50, 56, 60. See also the Daily Homework Quiz:

- Blackline Master (*Chapter 2 Resource Book*, p. 111)
- Transparency (p. 17)

! Common Error

EXERCISES 16–47 Students often forget to include the sign of the solution. Encourage them to determine the sign of the solution first and then do the computation.

TEACHING TIP

EXERCISES 65–67 Students may need some help to see how to convert the fraction $\frac{x}{0.1}$ to the product $10x$. Tell them to convert 0.1 to a fraction first and then use division of real numbers.

MATHEMATICAL REASONING

EXERCISES 65–67 Provide a good example of inductive reasoning. With inductive reasoning, you draw a conclusion from examining several examples. Ask students how inductive reasoning is being used in these exercises. ***Sample answer:*** **You find the pattern in the products and then use inductive reasoning to extend the pattern.**

APPLICATION NOTE

EXERCISES 68–70 Point out to students that due to the force of gravity, the rocket's velocity slows as it goes up. In this case, the expression $\frac{\text{displacement}}{\text{time}}$ gives the average velocity of the rocket. On the way down, the parachute controls the speed of the rocket so that it is constant and thus $\frac{\text{displacement}}{\text{time}}$ is the actual velocity of the rocket.

ADDITIONAL PRACTICE AND RETEACHING

For Lesson 2.7:

- Practice Levels A, B, and C (*Chapter 2 Resource Book,* p. 101)
- Reteaching with Practice (*Chapter 2 Resource Book,* p. 104)
- See Lesson 2.7 of the *Personal Student Tutor*

For more Mixed Review:

- Search the *Test and Practice Generator* for key words or specific lessons.

58. all real numbers except 0

59. all real numbers except 2

60. all real numbers except –2

61. all real numbers except 0

66. *Sample answer:* The power of 10 in the product is the same as the number of digits after the decimal point in the quotient.

FOCUS ON PEOPLE

KAREEM ABDUL-JABBAR holds the NBA career record for most points scored. He was a member of 6 NBA championship teams.

DISTRIBUTIVE PROPERTY **Simplify the expression.**

48. $\frac{18x - 9}{3}$ **6x – 3** **49.** $\frac{22x + 10}{-2}$ **–11x – 5** **50.** $\frac{-56 + x}{-8}$ **$7 - \frac{x}{8}$** **51.** $\frac{45 - 5x}{5}$ **9 – x**

EVALUATING EXPRESSIONS **Evaluate the expression for the given value(s) of the variable(s).**

52. $\frac{x - 5}{6}$ when $x = 30$ **$4\frac{1}{6}$**

53. $\frac{3r - 7}{11}$ when $r = 17$ **4**

54. $\frac{3a - b}{a}$ when $a = \frac{1}{3}$ and $b = -3$ **12**

55. $\frac{15x^2 + 10}{y}$ when $x = -3$ and $y = \frac{2}{3}$ **$217\frac{1}{2}$**

56. $\frac{28 - 4x}{y}$ when $x = 2$ and $y = \frac{1}{2}$ **40**

57. $\frac{3a - 4b}{ab}$ when $a = -\frac{1}{3}$ and $b = \frac{1}{4}$ **24**

FINDING THE DOMAIN **Find the domain of the function.** **See margin.**

58. $y = \frac{1}{3x}$ **59.** $y = \frac{3}{2 - x}$ **60.** $y = \frac{1}{x + 2}$ **61.** $y = \frac{4}{x^2}$

62. **COOKING** *Tom Kar Gai* is a soup from Thailand. You need $13\frac{3}{4}$ ounces of chicken broth to make one serving. If you have $82\frac{1}{2}$ ounces of chicken broth, how many servings can you make? **6 servings**

63. **SCUBA DIVING** You are scuba diving in the ocean. You dive down 22.5 feet in 9 seconds. What is your average velocity? **–2.5 ft/sec**

64. **BASKETBALL** Kareem Abdul-Jabbar scored 38,387 points and grabbed 17,440 rebounds in 1560 National Basketball Association games. How many points did he average per game? How many rebounds did he average per game? (Round your answers to the nearest tenth.) ▶ Source: NBA **24.6 points per game; 11.2 rebounds per game**

FINDING A PATTERN **In Exercises 65–67, use the table below.**

Fraction	$\frac{x}{1}$	$\frac{x}{0.1}$	$\frac{x}{0.01}$	$\frac{x}{0.001}$	$\frac{x}{0.0001}$	$\frac{x}{0.00001}$
Product	$1x$	$10x$	?	?	?	?

65. Copy and complete the table by expressing the fraction as a product that has no decimal factor. **100x; 1000x; 10,000x; 100,000x**

66. Describe the pattern in the products. **See margin.**

67. Use the pattern to simplify the expression $\frac{x}{0.0000001}$. **10,000,000x**

MODEL ROCKET **You launch a model rocket that rises 550 feet in 2.75 seconds. It then opens a parachute and falls at a rate of 11 feet per second.**

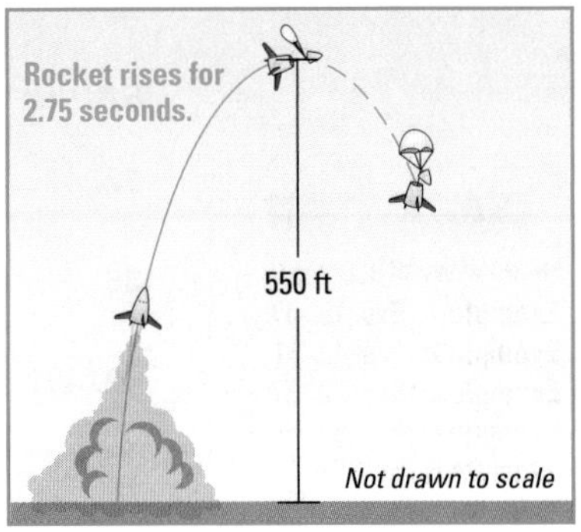

68. What is the rocket's average velocity going up? **200 ft/sec**

69. What is the rocket's velocity coming down? **–11 ft/sec**

70. In how many seconds after the launch will the rocket reach the ground? **50 sec**

Test Preparation

71. Yes; if $a < b$, then $a - b < 0$. Since a and b are positive, ab is positive, so $\frac{a-b}{ab} < \frac{0}{ab}$. Then $\frac{1}{b} - \frac{1}{a} < 0$, which means that $\frac{1}{b} < \frac{1}{a}$. *Sample answer:* $2 < 3$ and $\frac{1}{3} < \frac{1}{2}$.

★ Challenge

EXTRA CHALLENGE

www.mcdougallittell.com

71. *Writing* If a and b are positive and $a < b$, is it true that $\frac{1}{b} < \frac{1}{a}$? Explain and show an example or a counterexample. **See margin.**

72. MULTIPLE CHOICE Which of the following expressions is equivalent to the expression $\frac{6x-4}{-\frac{2}{3}}$? **C**

Ⓐ $2(3x-2) \cdot \frac{3}{2}$ Ⓑ $(6x-4) \cdot \left(-\frac{2}{3}\right)$ Ⓒ $-9x+6$ Ⓓ $9x-6$

73. MULTIPLE CHOICE Which of the following statements is false? **D**

Ⓐ The reciprocal of any negative number is a negative number.

Ⓑ Dividing by a number is the same as multiplying by the reciprocal of the number.

Ⓒ The reciprocal of any positive number is a positive number.

Ⓓ The reciprocal of any number is greater than zero and less than 1.

74. EXTENSION: CLOSURE A set of numbers is *closed* under an operation if applying the operation to any two numbers in the set results in another number in the set. For instance, positive integers are closed under addition because the sum of any two positive integers is a positive integer. Decide whether the set is closed under the given operation.

a. positive integers; subtraction **no**
b. integers; addition and subtraction **yes**
c. integers; multiplication **yes**
d. integers; division **no**

MIXED REVIEW

CONVERTING FRACTIONS **Write the fraction as a decimal. Round to the nearest hundredth if necessary.** (Skills Review, pp. 784–785)

75. $\frac{3}{4}$ **0.75**
76. $\frac{3}{5}$ **0.6**
77. $\frac{1}{10}$ **0.1**
78. $\frac{7}{25}$ **0.28**
79. $\frac{1}{6}$ **0.17**
80. $\frac{2}{7}$ **0.29**
81. $\frac{30}{40}$ **0.75**
82. $\frac{5}{11}$ **0.45**

EXPONENTIAL EXPRESSIONS **Evaluate the expression for the given value of the variable.** (Review 1.2, 2.5)

83. $3x^2$ when $x = 7$ **147**
84. $4(b^3)$ when $b = \frac{1}{2}$ **$\frac{1}{2}$**
85. $2(y^3)$ when $y = -5$ **–250**
86. $5y^3$ when $y = 4$ **320**
87. $(6w)^4$ when $w = 2$ **20,736**
88. $12d^2$ when $d = 9$ **972**
89. $5x^2$ when $x = 0.3$ **0.45**
90. $32x^7$ when $x = -1$ **–32**
91. $(7t)^3$ when $t = -\frac{3}{7}$ **–27**

NUMERICAL EXPRESSIONS **Evaluate the expression.** (Review 1.3, 2.5)

92. $12 - 9 + 7$ **10**
93. $42 \div 6 + 8$ **15**
94. $2(11 - 7) \div 3$ **$2\frac{2}{3}$**
95. $8 + (91 \div 13) \cdot \frac{4}{7}$ **12**
96. $\frac{3}{4} \cdot 8 - 6$ **0**
97. $23 - [(12 \div 3)^2 + 8]$ **–1**
98. $11 \cdot (-5) + 20$ **–35**
99. $-8 \cdot (-9) - 80$ **–8**
100. $-16 + (-6) \cdot (-8)$ **32**

4 ASSESS

DAILY HOMEWORK QUIZ

Transparency Available

1. Find the quotient.
   a. $-35 \div 3\frac{1}{2}$ **–10**
   b. $\frac{11}{\frac{1}{3}}$ **33**
2. Evaluate $\frac{3x-8}{y}$ when $x = 2$ and $y = -4$. **$\frac{1}{2}$**
3. What is the domain of the function $y = \frac{3x}{x-3}$? **all real numbers except $x = 3$**

EXTRA CHALLENGE NOTE

Challenge problems for Lesson 2.7 are available in **blackline** format in the *Chapter 2 Resource Book*, p. 108 and at **www.mcdougallittell.com.**

ADDITIONAL TEST PREPARATION

1. **WRITING** Give a real life example that could be modeled by the expression $-450 \div 5$. ***Sample answer:*** **You are in a cable car descending down a mountain. You descend 450 feet in 5 minutes.**

1 PLAN

PACING
Basic: 2 days
Average: 2 days
Advanced: 2 days
Block Schedule: 1 block

LESSON OPENER
GRAPHING CALCULATOR
An alternative way to approach Lesson 2.8 is to use the Graphing Calculator Lesson Opener:

- Blackline Master (*Chapter 2 Resource Book,* p. 112)
- Transparency (p. 15)

MEETING INDIVIDUAL NEEDS

- ***Chapter 2 Resource Book***
  Prerequisite Skills Review (p. 5)
  Practice Level A (p. 113)
  Practice Level B (p. 114)
  Practice Level C (p. 115)
  Reteaching with Practice (p. 116)
  Absent Student Catch-Up (p. 118)
  Challenge (p. 121)
- ***Resources in Spanish***
- ***Personal Student Tutor***

NEW-TEACHER SUPPORT
See the Tips for New Teachers on pp. 1–2 of the *Chapter 2 Resource Book* for additional notes about Lesson 2.8.

WARM-UP EXERCISES
Transparency Available
Write each fraction as a decimal. Round your answer to the nearest hundredth if necessary.

1. $\frac{3}{4}$ 0.75 **2.** $\frac{1}{3}$ 0.33
3. $\frac{2}{5}$ 0.4 **4.** $\frac{19}{31}$ 0.61
5. $\frac{243}{549}$ 0.44

EXPLORING DATA AND STATISTICS

2.8

What you should learn

GOAL 1 Find the probability of an event.

GOAL 2 Find the odds of an event.

Why you should learn it

▼ To represent **real-life** situations, like the probability that a baby sea turtle reaches the ocean in **Ex. 22.**

REAL LIFE

Probability and Odds

GOAL 1 FINDING THE PROBABILITY OF AN EVENT

The **probability of an event** is a measure of the likelihood that the event will occur. It is a number between 0 and 1, inclusive.

$P = 0$	$P = 0.25$	$P = 0.5$	$P = 0.75$	$P = 1$
Impossible	Unlikely	Occurs half the time	Quite likely	Certain

When you do a probability experiment, the different possible results are called **outcomes**. When an experiment has N *equally likely* outcomes, each of them occurs with probability $\frac{1}{N}$. For example, in the roll of a six-sided number cube, the possible outcomes are 1, 2, 3, 4, 5, and 6, and the probability associated with each outcome is $\frac{1}{6}$.

An **event** consists of a collection of outcomes. In the roll of a six-sided number cube, an "even roll" consists of the outcomes 2, 4, and 6. The **theoretical probability** of an even roll is $\frac{3}{6} = \frac{1}{2}$. The outcomes for an event you wish to have happen are called **favorable outcomes**, and the theoretical probability of this event is calculated by the following rule.

$$\text{Theoretical probability } P = \frac{\text{Number of favorable outcomes}}{\text{Total number of outcomes}}$$

EXAMPLE 1 *Finding the Probability of an Event*

a. You toss two coins. What is the probability P that both are heads?

b. An algebra class has 17 boys and 16 girls. One student is chosen at random from the class. What is the probability P that the student is a girl?

SOLUTION

a. There are four possible outcomes that are equally likely.

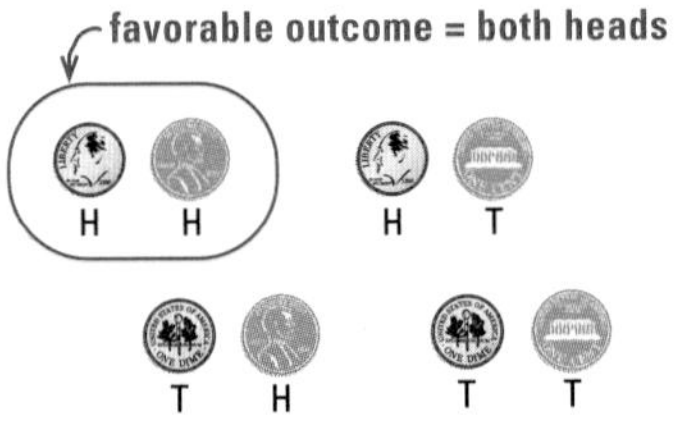

$$P = \frac{\text{Number of favorable outcomes}}{\text{Total number of outcomes}} = \frac{1}{4} = 0.25$$

b. Because the student is chosen at random, it is equally likely that any of the 33 students will be chosen. Let "number of girls" be the favorable outcome. Let "number of students" be the total number of outcomes.

$$P = \frac{\text{Number of favorable outcomes}}{\text{Total number of outcomes}} = \frac{\text{Number of girls}}{\text{Number of students}} = \frac{16}{33} \approx 0.485$$

Another type of probability is **experimental probability**. This type of probability is based on repetitions of an actual experiment and is calculated by the following rule.

$$\text{Experimental probability } P = \frac{\text{Number of favorable outcomes observed}}{\text{Total number of trials}}$$

ACTIVITY
Developing Concepts

Investigating Experimental Probability

Partner Activity Toss three coins 20 times and record the number of heads for each of the 20 tosses.

1. Use your results to find the experimental probability of getting three heads when three coins are tossed.
2. Combine your results with those of all the other pairs in your class. Then use the combined results to find the experimental probability of getting three heads when three coins are tossed.
3. Find the theoretical probability of getting three heads when three coins are tossed. How does it compare with your results from **Step 2**?

A survey is a type of experiment. Probabilities based on the outcomes of a survey can be used to make predictions.

EXAMPLE 2 *Using a Survey to Find a Probability*

SURVEYS Use the circle graph at the right showing the responses of 500 teens to a survey asking "Where would you like to live?"

If you were to ask a randomly chosen teen this question, what is the experimental probability that the teen would say "large city?"

Where Would You Like to Live?

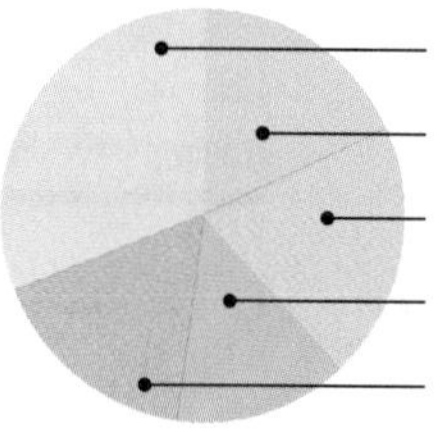

▶ Source: *Youth Views*

SOLUTION

Let "choosing large city" be the favorable outcome. Let "number surveyed" be the total number of trials.

$$\text{Experimental probability } P = \frac{\text{Number choosing large city}}{\text{Number surveyed}}$$

$$= \frac{155}{500}$$

$$= 0.31$$

FOCUS ON CAREERS

REAL LIFE **MARKET RESEARCHERS** use surveys to predict the needs and interests of their potential customers. Then they can make their products better.

INTERNET CAREER LINK
www.mcdougallittell.com

2 TEACH

EXTRA EXAMPLE 1

a. You have 2 red and 2 black socks in a drawer. You reach in and pick two without looking. What is the probability P that they do not match? **0.5**

b. In a group of students, 12 ride the bus to school, 8 are driven to school, and 5 walk. One of the students is chosen at random from the group. What is the probability P that the student walks to school? **0.2**

EXTRA EXAMPLE 2

Use the circle graph below showing the responses of 250 college students to a survey asking "Which factor is most likely to influence your job choice after graduation?" If you were to ask a randomly chosen college student this question, what is the experimental probability that the student would say "type of company"? **about 0.15**

Influences on Job Choices

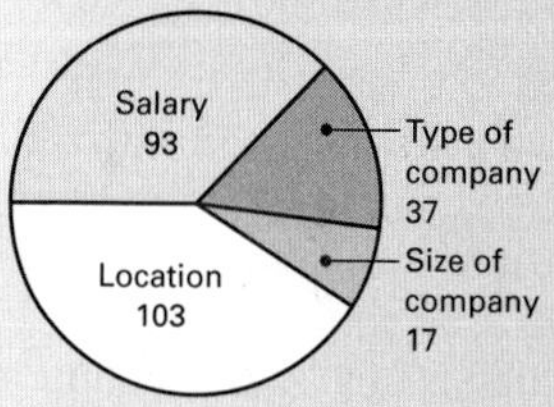

CHECKPOINT EXERCISES

For use after Examples 1 and 2:

1. Last January it snowed 7 days, was sunny 18 days, and was cloudy 6 days.

a. Based on this information, what is the probability that it will snow on a randomly chosen day in January? **about 0.23**

b. Is the probability in part (a) experimental or theoretical? **experimental**

EXTRA EXAMPLE 3
You randomly choose a letter from the word SUMMER. What are the odds that the letter is a vowel? **1 to 2**

EXTRA EXAMPLE 4
The probability that a randomly chosen 4 digit security code contains at least one zero is 0.34. What are the odds that a 4 digit security code contains at least one zero? **17 to 33**

CHECKPOINT EXERCISES

For use after Example 3:

1. A candy dish contains 12 miniature chocolate bars and 21 hard candies. What are the odds that a candy picked at random from the dish will be a hard candy? **7 to 4**

For use after Example 4:

2. In a shoe factory, the probability of a defective shoe being produced is 0.005. What are the odds of producing a defective shoe? **1 to 199**

FOCUS ON VOCABULARY
You have a 4-sided solid with triangular faces numbered from 1 to 4. You want to know the probability of rolling a 3; so you roll the die 100 times and divide the number of times a 3 was rolled by 100. Is this an experimental or theoretical probability? **experimental**

CLOSURE QUESTION
Give an example of an event that would have a probability of 0 and one that would have a probability of 0.75. **See sample answer, below.**

DAILY PUZZLER
The odds of getting your first choice English elective at Jordan's school are 12 to 5. For students who don't get their first choice, the odds of getting their second choice are 2 to 1. If there are 510 students in the school, how many will get their second choice? **100**

GOAL 2 FINDING THE ODDS OF AN EVENT

STUDENT HELP

Study Tip
Odds are always read as the ratio of one quantity *to* another. For example, $\frac{4}{3}$ is read as "four *to* three," not as "four thirds."

THE ODDS OF AN EVENT

When all outcomes are equally likely, the **odds** that an event will occur are given by the formula below.

$$\text{Odds} = \frac{\text{Number of favorable outcomes}}{\text{Number of unfavorable outcomes}}$$

EXAMPLE 3 *Finding the Odds of an Event*

You randomly choose an integer from 0 through 9. What are the odds that the integer is 4 or more?

SOLUTION There are 6 favorable outcomes: 4, 5, 6, 7, 8, and 9. There are 4 unfavorable outcomes: 0, 1, 2, and 3.

$$\text{Odds} = \frac{\text{Number of favorable outcomes}}{\text{Number of unfavorable outcomes}} = \frac{6}{4} = \frac{3}{2}$$

▶ The odds that the integer is 4 or more are 3 to 2.

If you know the probability that an event will occur, then you can find the odds.

$$\text{Odds} = \frac{\text{Probability event will occur}}{\text{Probability event will not occur}} = \frac{\text{Probability event will occur}}{1 - (\text{Probability event will occur})}$$

EXAMPLE 4 *Finding Odds from Probability*

REAL LIFE *Pets*

The probability that a randomly chosen household has a cat is 0.27. What are the odds that a household has a cat? ▶Source: American Veterinary Medical Association

SOLUTION

$$\text{Odds} = \frac{\text{Probability event will occur}}{1 - (\text{Probability event will occur})}$$

$= \frac{0.27}{1 - 0.27}$ **Substitute for probabilities.**

$= \frac{0.27}{0.73}$ **Simplify denominator.**

$= \frac{27}{73}$ **Multiply numerator and denominator by 100.**

▶ The odds are 27 to 73 that a randomly chosen household has a cat.

Closure Question ***Sample answer:***
a person seeing a live dinosaur; choosing a blue marble out of a box that has three blue marbles and one red marble.

GUIDED PRACTICE

Vocabulary Check ✓

1. Using the results of a student lunch survey, you determine that the probability that a randomly chosen student likes green beans is 0.38. Is this probability theoretical or experimental? **experimental**

Concept Check ✓

2. The probability that an event will occur is 0.4. Is it more likely that the event will occur, or is it more likely that the event will *not* occur? **more likely that it will not occur**
3. The odds that an event will occur are 3 to 4. Is it more likely that the event will occur, or is it more likely that the event will *not* occur? **more likely that it will not occur**

Skill Check ✓

Tell whether the event is best described as *impossible, unlikely, likely to occur half the time, quite likely,* or *certain.* Explain your reasoning.

4. The probability of rain is 80%, or 0.8. **Quite likely; the probability is greater than 0.75.**
5. The odds in favor of winning a race are $\frac{1}{3}$. **Unlikely; the probability is 0.25.**
6. The odds of being chosen for a committee are 1 to 1. **Likely to occur half the time; the probability is 0.5.**

TEST PLANNING **Suppose it is equally likely that a teacher will chose any day from Monday, Tuesday, Wednesday, Thursday, and Friday to have the next test.**

7. What is the probability that the next test will be on a Friday? **0.2**
8. What are the odds that the next test will be on a day starting with the letter T? **2 to 3**

PRACTICE AND APPLICATIONS

STUDENT HELP

Extra Practice to help you master skills is on p. 798.

PROBABILITY **Find the probability of randomly choosing a red marble from the given bag of red and white marbles.**

9. Number of red marbles: 16 **0.25**
Total number of marbles: 64
10. Number of red marbles: 8 **0.2**
Total number of marbles: 40
11. Number of white marbles: 7 **0.65**
Total number of marbles: 20
12. Number of white marbles: 24 **0.25**
Total number of marbles: 32

ODDS **Find the odds of randomly choosing the indicated letter from a bag that contains the letters in the name of the given state.**

13. S; MISSISSIPPI **4 to 7**
14. N; PENNSYLVANIA **1 to 3**
15. A; NEBRASKA **1 to 3**
16. G; VIRGINIA **1 to 7**

STUDENT HELP

HOMEWORK HELP
Example 1: Exs. 9–12
Example 2: Exs. 21–24, 26, 28, 29
Example 3: Exs. 13–16, 25, 27, 30
Example 4: Exs. 17, 18, 20

PROBABILITY TO ODDS **Given the probability, find the odds.**

17. The probability of randomly choosing a club from a deck of cards is 0.25. **1 to 3**
18. The probability of tossing a total of 7 using two number cubes is $\frac{1}{6}$. **1 to 5**

COIN TOSS **You toss two coins.**

19. What is the theoretical probability that only one is tails? **0.5**
20. Use the theoretical probability to find the odds that only one is tails. **1 to 1, or even**

3 APPLY

ASSIGNMENT GUIDE

BASIC
Day 1: pp. 117–118 Exs. 9–27
Day 2: pp. 118–120 Exs. 28–30, 33, 34, 40, 45–48, 50, 55, Quiz 3 Exs. 1–21

AVERAGE
Day 1: pp. 117–118 Exs. 9–27
Day 2: pp. 118–120 Exs. 28–31, 33, 34, 40, 45–48, 50, 55, Quiz 3 Exs. 1–21

ADVANCED
Day 1: pp. 117–118 Exs. 9–27
Day 2: pp. 118–120 Exs. 28–38, 40, 45–48, 50, 55, Quiz 3 Exs. 1–21

BLOCK SCHEDULE
pp. 117–118 Exs. 9–27;
pp. 118–120 Exs. 28–31, 33–38, 40, 45–48, 50, 55, Quiz 3 Exs. 1–21

EXERCISE LEVELS

Level A: ***Easier***
9–12, 19

Level B: ***More Difficult***
13–18, 20–25, 28–31, 33, 34, 39–57

Level C: ***Most Difficult***
26, 27, 32, 35–38

✓ HOMEWORK CHECK

To quickly check student understanding of key concepts, go over the following exercises: Exs. 12, 14, 19, 24, 30. See also the Daily Homework Quiz:

- Blackline Master (*Chapter 3 Resource Book*, p. 11)
- Transparency (p. 19)

! COMMON ERROR

EXERCISES 23–30 In problems such as these where some of the numerical information is extraneous, students may have trouble deciding which numbers to use to calculate the probability. Have them give a verbal description of the probability ratio

$$P = \frac{\text{Number of favorable outcomes}}{\text{Total number of outcomes}}$$

before assigning any numerical values. For example, in Exercise 23, they could write $P = \frac{\text{Number of households owning dogs}}{\text{Number of pet-owning households}}$.

ADDITIONAL PRACTICE AND RETEACHING

For Lesson 2.8

- Practice Levels A, B, and C (*Chapter 2 Resource Book,* p. 113)
- Reteaching with Practice (*Chapter 2 Resource Book,* p. 116)
- See Lesson 2.8 of the *Personal Student Tutor*

For more Mixed Review:

- Search the *Test and Practice Generator* for key words or specific lessons.

21. NUMBER CUBES You toss a six-sided number cube 20 times. For twelve of the tosses the number tossed was 3 or more.

a. What is the experimental probability that a number tossed is 3 or more? **0.6**

b. What are the odds that a number tossed is 3 or more? **3 to 2**

22. BABY SEA TURTLES A sea turtle buries 90 eggs in the sand. From the 50 eggs that hatch, 37 turtles do not make it to the ocean. What is the probability that an egg chosen at random hatched and the baby turtle made it to the ocean? $\frac{13}{90} \approx 0.14$

STUDENT HELP

HOMEWORK HELP Visit our Web site www.mcdougallittell.com for help with Exs. 23–27.

PETS **In Exercises 23–25, use the graph.**

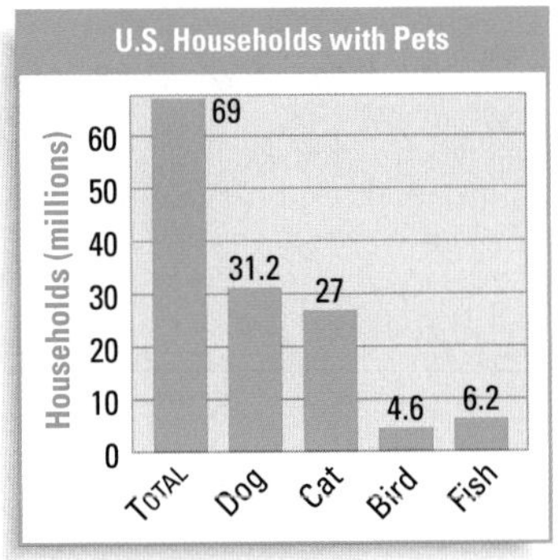

▶ Source: American Veterinary Medical Assoc.

23. What is the probability that a pet-owning household chosen at random owns a dog? **0.45**

24. What is the probability that a pet-owning household chosen at random does *not* own a fish? **0.91**

25. There are approximately 98.8 million households in the United States. If a household is chosen at random, what are the odds that the household owns a pet? **690 to 298, or about 7 to 3**

MOVING **In Exercises 26 and 27, use the table, which shows the percent of citizens from various age groups who changed homes within the United States from 1995 to 1996.**

26. What is the probability that a citizen from the 15–19 age group changed homes? **0.15**

27. What are the odds that a citizen from the 25–29 age group moved to a home in a different state? **1 to 19**

Percent of U.S. Citizens of Given Ages who Moved

Age group	Total	Same county	Different county, same state	Different state
15 to 19	15	10	3	2
20 to 24	33	21	6	5
25 to 29	32	20	7	5
30 to 44	16	10	3	3

▶ Source: U.S. Bureau of the Census

FOCUS ON APPLICATIONS

FAULT LINES A fault line, like the one shown above, may appear in the earth's surface in a region that is prone to earthquakes.

EARTHQUAKES **In Exercises 28–30, use the table, which shows the number of earthquakes of magnitude 4.0 or greater in the western United States since 1900. The magnitude of an earthquake indicates its severity.**

28. What is the probability that the magnitude of an earthquake is from 6.0 to 6.9? $\frac{43}{1310} \approx 0.03$

29. What is the probability that the magnitude of an earthquake is *not* from 4.0 to 4.9? $\frac{253}{1310} \approx 0.19$

30. What are the odds that the magnitude of an earthquake is from 7.0 to 7.9? **3 to 652**

Magnitude	Number of earthquakes
8 and higher	1
7.0–7.9	18
6.0–6.9	129
5.0–5.9	611
4.0–4.9	3171

▶ Source: U.S. Geological Survey

32. The probability you choose a yellow marble is $\frac{\text{no. of favorable outcomes}}{\text{total no. of outcomes}} = \frac{5}{20} = 0.25$. The odds are $\frac{\text{no. of favorable outcomes}}{\text{no. of unfavorable outcomes}} = \frac{5}{15}$ or 1 to 3. Both indicate an event that is unlikely to occur.

31. WARDROBE Your cousin spills spaghetti sauce on her shirt and asks to borrow a clean shirt from you for the rest of the day. You decide to let her choose from a selection of 4 sweatshirts, 1 hockey shirt, 8 T-shirts, and 3 tank tops. If it is equally likely that your cousin will choose any shirt, what are the odds that she will choose a sweatshirt? **1 to 3**

32. *Writing* Suppose you randomly choose a marble from a bag holding 11 green, 4 blue, and 5 yellow marbles. Use probability *and* odds to express how likely it is that you choose a yellow marble. Compare the two ways of expressing how likely it is to choose a yellow marble. If you find one (probability or odds) easier to understand or more useful than the other, explain why. **See margin.**

Test Preparation

MULTIPLE CHOICE **Use the table, which shows the number of students in each grade for three high schools and the number that get to each school by riding the bus.**

3rd District High School Student Bus Riders						
	Marshall		Jordan		King	
Grade	Total	Bus riders	Total	Bus riders	Total	Bus riders
9th	247	198	326	195	265	151
10th	232	176	311	205	273	142
11th	194	141	304	196	264	139
12th	211	142	285	173	231	106

33. If you are in the tenth grade, what is the probability that you ride a bus to get to school? **C**

Ⓐ $\frac{141}{194} \approx 0.73$ Ⓑ $\frac{196}{304} \approx 0.64$ Ⓒ $\frac{523}{816} \approx 0.64$ Ⓓ $\frac{139}{264} \approx 0.53$

34. If you go to Marshall High School, what are the odds that you do *not* ride the bus to get to school? **C**

Ⓐ 657 to 227 Ⓑ 657 to 884 Ⓒ 227 to 657 Ⓓ 227 to 884

★ Challenge

BULL'S-EYE **In Exercises 35–38, use the drawing of the dart board below, and the following information.**

Assume that when a dart is thrown and hits the board, the dart is equally likely to hit any point on the board. The probability that a dart lands within the red bull's-eye circle is given by the following equation.

$$P = \frac{\text{Area of bull's-eye}}{\text{Area of board}}$$

(*Hint*: See the Table of Formulas on p. 813.)

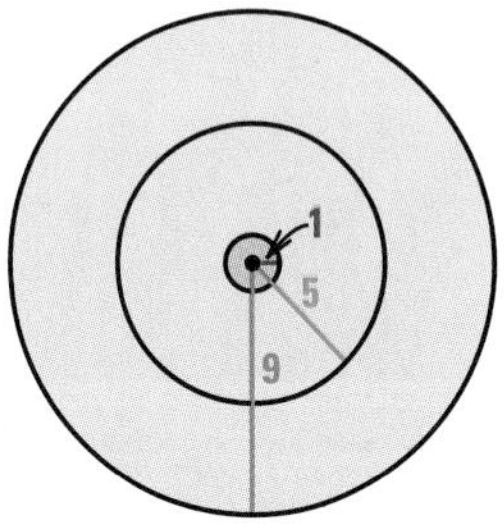

35. What is the area of the bull's-eye? **π square units, or about 3.14 square units**

36. What is the probability that a dart lands within the middle circle? **See margin.**

37. What is the probability that a dart lands between the middle and outer circles? **See margin.**

38. What are the odds that a dart lands on the bull's-eye? **π to 80π, or 1 to 80**

36. $\frac{25\pi}{81\pi}$, or about 0.31

37. $\frac{56\pi}{81\pi}$, or about 0.69

EXTRA CHALLENGE
www.mcdougallittell.com

4 ASSESS

DAILY HOMEWORK QUIZ

Transparency Available

1. What is the probability of choosing a blue marble from a bag containing 20 blue and white marbles if 6 of the marbles are white? **0.7**
2. If a letter is chosen at random from a bag containing the letters in the word *ALGEBRA*, what are the odds that the letter is *A*? **2 to 5**
3. From 1886 to 1997, during the month of September, 319 tropical storms were recorded in the United States. Of these storms, 202 reached hurricane intensity. Is it more likely than not that a September tropical storm will become a hurricane? Explain. **Yes; the probability is about 0.63 that a tropical storm will become a hurricane.**

EXTRA CHALLENGE NOTE

Challenge problems for Lesson 2.8 are available in **blackline** format in the *Chapter 2 Resource Book*, p. 121 and at **www.mcdougallittell.com.**

ADDITIONAL TEST PREPARATION

1. **OPEN ENDED** Give an example of an experimental probability. **You want to know the probability that students prefer their math course to their science course. You ask 25 students who take both math and science. The probability that a student prefers math is $P = \frac{\text{number of students who prefer math}}{25}$.**
2. **WRITING** Explain how to use odds to determine if an event is impossible, unlikely, will occur half the time, quite likely, or certain. **impossible: 0 to 1; unlikely: 1 to 3; will occur half the time: 1 to 1; quite likely: 3 to 1; certain: 1 to 0**

MIXED REVIEW

MENTAL MATH Write a question that can be used to solve the equation. Then use mental math to solve the equation. (Review 1.4)

39–44. Solutions are given. Check students' questions.

39. $x + 17 = 25$ **8** **40.** $a - 5 = 19$ **24** **41.** $2b - 1 = 10$ $5\frac{1}{2}$

42. $11t = 110$ **10** **43.** $\frac{y}{15} = 8$ **120** **44.** $3x + 15 = 24.6$ **3.2**

TRANSLATING VERBAL SENTENCES Translate the sentence into an equation or an inequality. (Review 1.5, 2.1)

45. 17 less than a number z is 9. $z - 17 = 9$

46. 8 more than a number r is less than 17. $r + 8 < 17$

47. -3 is the sum of a number y and -6. $-3 = y + (-6)$

48. -9 is equal to a number y decreased by 21. $-9 = y - 21$

EVALUATING EXPRESSIONS Evaluate the expression. (Review 2.3 for 3.1)

49. $-8 + 4 - 9$ **−13** **50.** $12 - (-8) - 5$ **15** **51.** $20 + (-17) - 8$ **−5**

52. $-17 + 25 - 34$ **−26** **53.** $-6.3 + 4.1 - 9.5$ **−11.7** **54.** $2 - 11 + 5 - (-16)$ **12**

55. $-29.4 - (-8) + 4$ **−17.4** **56.** $\frac{1}{2} + \left(-\frac{4}{5}\right) - \frac{2}{3}$ $-\frac{29}{30}$ **57.** $-1\frac{3}{8} + 4\frac{3}{4} - 7\frac{1}{2}$ $-4\frac{1}{8}$

QUIZ 3

Self-Test for Lessons 2.7 and 2.8

Find the quotient. (Lesson 2.7)

1. $-28 \div \frac{4}{7}$ **−49** **2.** $\frac{36}{\frac{2}{3}}$ **54** **3.** $\frac{32}{\frac{1}{4}}$ **128** **4.** $48 \div (-12)$ **−4**

5. $75 \div (-15)$ **−5** **6.** $-144 \div 9$ **−16** **7.** $-120 \div \frac{3}{8}$ **−320** **8.** $-\frac{13}{27} \div \left(-1\frac{4}{9}\right)$ $\frac{1}{3}$

In Exercises 9–16, simplify the division expression. (Lesson 2.7)

9. $42x \div (-6)$ $-7x$ **10.** $-56 \div (-8x)$ $\frac{7}{x}$ **11.** $9x \div \frac{1}{2}$ $18x$ **12.** $20 \div \frac{4}{x}$ $5x$

13. $25x \div \frac{5}{7}$ $35x$ **14.** $15t \div \left(-\frac{3}{4}\right)$ $-20t$ **15.** $66y \div \left(-\frac{6}{5}\right)$ $-55y$ **16.** $-\frac{2x}{18} \div (-4)$ $\frac{x}{36}$

17. What is the probability that you randomly choose a purple marble from a bag containing 11 red, 6 green, and 8 purple marbles? (Lesson 2.8) **0.32**

Tell whether the event is best described as *impossible, unlikely, likely to occur half the time, quite likely,* or *certain.* Explain your reasoning. (Lesson 2.8)

18. The probability of snow is 20%, or 0.2.
Unlikely; the probability is less than 0.25.

19. The probability of taking the test is 1. **Certain; the definition of a certain event is an event with probability 1.**

20. The odds in favor of winning a car are $\frac{1}{30,000}$.

20. Unlikely; the probability is $\frac{1}{30,001}$, which is very small but not 0.

21. The odds of being chosen for the softball team are 15 to 6.
Quite likely; the probability is about 0.7.

CHAPTER 2

Chapter Summary

WHAT did you learn?	*WHY did you learn it?*
Graph and compare real numbers. (2.1)	Find the coldest temperature recorded. (p. 64)
Find the opposite and the absolute value of a real number. (2.1)	Find the speed and velocity of an elevator. (p. 66)
Add real numbers. (2.2)	Decide whether an atom is an ion. (p. 74)
Subtract real numbers. (2.3)	Find the price change of an ounce of gold. (p. 83)
Organize data in a matrix. (2.4)	Organize music store price data. (p. 90)
Add and subtract two matrices. (2.4)	Find employee wage rates after a raise. (p. 90)
Multiply real numbers. (2.5)	Find the vertical distance a hawk travels. (p. 96)
Use the distributive property. (2.6)	Find the total weight a train is hauling. (p. 104)
Combine like terms. (2.6)	Simplify algebraic expressions. (p. 102)
Divide real numbers. (2.7)	Find the price change of a share of stock. (p. 111)
Express the likelihood of an event as a probability or as odds. (2.8)	Find how likely it is that a test is given on a Friday. (p. 117)

How does Chapter 2 fit into the BIGGER PICTURE of algebra?

The numbers you worked with in Chapter 1 were zero or positive. In this chapter you studied rules for adding, subtracting, multiplying, and dividing with both positive and negative numbers. You are building a sense for the type of problems associated with negative numbers, such as losses, deficits, and low temperatures.

This chapter introduces some of the rules of algebra. By learning to create, simplify, and evaluate algebraic expressions and matrices, you will be able to solve many real-life problems. These skills will also prepare you for the algebra that you will study later in this course.

STUDY STRATEGY

How did you use the lessons in this chapter?

The notes you made as you studied Chapter 2, using the **Study Strategy** on page 62, may include these properties.

Lesson 2.2

Commutative Property

$a + b = b + a$ $\quad -4 + 6 = 6 + (-4)$

Associative Property

$(a + b) + c = a + (b + c)$

$(-8 + 5) + 3 = -8 + (5 + 3)$

Identity Property

$a + 0 = a$ $\quad -9 + 0 = -9$

Property of Zero

$a + -a = 0$ $\quad -7.5 + 7.5 = 0$

ADDITIONAL RESOURCES

The following resources are available to help review the material in this chapter.

- Chapter Review Games and Activities (*Chapter 2 Resource Book*, p. 122)
- *Instant Replay: Video Review Games*
- *Personal Student Tutor*
- Cumulative Review, Chs. 1–2 (*Chapter 2 Resource Book*, p. 134)

1. 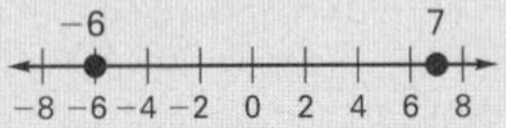

2. 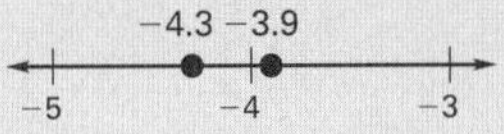

3. 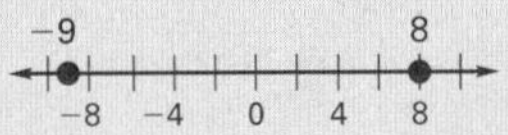

4. 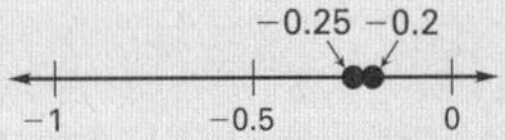

5. 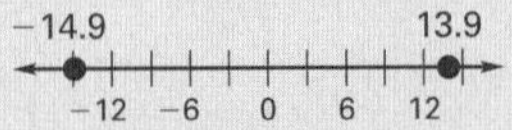

6. 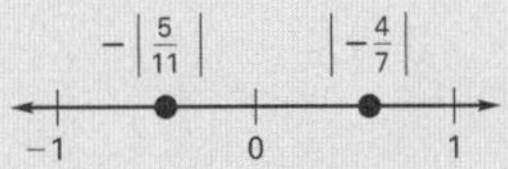

CHAPTER 2

Chapter Review

VOCABULARY

- real number line, p. 63
- real numbers, p. 63
- origin, p. 63
- negative numbers, p. 63
- positive numbers, p. 63
- integers, p. 63
- graph of a number, p. 63
- plotting a point, p. 63
- opposites, p. 65
- absolute value, p. 65
- velocity, p. 66
- counterexample, p. 66
- terms of an expression, p. 80
- matrix, p. 86
- entry, or element, of a matrix, p. 86
- distributive property, p. 100
- coefficient, p. 102
- like terms, p. 102
- constant terms, p. 102
- simplified expression, p. 102
- reciprocal, p. 108
- probability of an event, p. 114
- outcomes, p. 114
- theoretical probability, p. 114
- favorable outcomes, p. 114
- experimental probability, p. 115
- odds, p. 116

2.1 THE REAL NUMBER LINE

Examples on pp. 63–66

EXAMPLES You can use a number line to graph the real numbers 2, $-\frac{1}{2}$, 0.8, $-1\frac{1}{4}$, −2, and −0.8, and then put them in order from least to greatest.

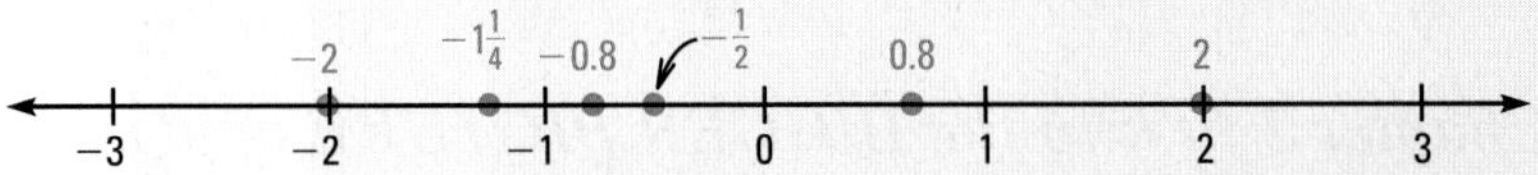

In order from least to greatest, the numbers are −2, $-1\frac{1}{4}$, −0.8, $-\frac{1}{2}$, 0.8, and 2.

The *opposite* of 0.8 is −0.8. Both 2 and −2 have an *absolute value* of 2.

Graph the numbers on a number line. Then write two inequalities that compare the two numbers. **1–6. See margin for graphs.**

1. 7 and −6 **7 > −6, −6 < 7**
2. −3.9 and −4.3 **−3.9 > −4.3, −4.3 < −3.9**
3. −9 and 8 **−9 < 8, 8 > −9**
4. −0.2 and −0.25 **−0.2 > −0.25, −0.25 < −0.2**
5. 13.9 and −14.9 **13.9 > −14.9, −14.9 < 13.9**
6. $\left|-\frac{4}{7}\right|$ and $-\left|\frac{5}{11}\right|$ **$\left|-\frac{4}{7}\right| > -\left|\frac{5}{11}\right|$, $-\left|\frac{5}{11}\right| < \left|-\frac{4}{7}\right|$**

2.2 ADDITION OF REAL NUMBERS

Examples on pp. 72–74

EXAMPLES Find −8 + (−6).

$|-8| + |-6| = 14$ **Same sign rule**

$-8 + (-6) = -14$ **Attach common sign.**

Find 4 + (−7).

$|-7| - |4| = 3$ **Opposite sign rule**

$4 + (-7) = -3$ **Attach sign of −7.**

Find the sum.

7. 12 + (−7) **5**
8. −24 + (−16) **−40**
9. 2.4 + (−3.1) **−0.7**
10. 9 + (−10) + (−3) **−4**
11. −35 + 41 + (−18) **−12**
12. $-2\frac{3}{4} + 5\frac{3}{8} + \left(-4\frac{1}{2}\right)$ **$-1\frac{7}{8}$**

2.3 SUBTRACTION OF REAL NUMBERS

Examples on pp. 79–81

EXAMPLES To subtract a real number, add its opposite.

$3 - (-5) = 3 + 5$ Add the opposite. $-8 - 7 = -8 + (-7)$ Add the opposite.

$= 8$ Find the sum. $= -15$ Find the sum.

Evaluate the expression.

13. $-2 - 7 - (-8)$ **−1** **14.** $5 - 11 - (-6)$ **0** **15.** $-18 - 14 - (-15)$ **−17**

16. $-5.7 + 3.1 - 8.6$ **−11.2** **17.** $-\frac{7}{16} + \left(-\frac{3}{8}\right) - \frac{13}{4}$ **$-4\frac{1}{16}$** **18.** $-\frac{23}{36} - \left|-\frac{4}{9}\right| + \left(-\frac{7}{12}\right)$ **$-1\frac{2}{3}$**

2.4 ADDING AND SUBTRACTING MATRICES

Examples on pp. 86–88

EXAMPLES To add or subtract matrices, add or subtract corresponding entries.

$$\begin{bmatrix} -1 & 3 \\ 2 & -2 \end{bmatrix} + \begin{bmatrix} 2 & 0 \\ -5 & 1 \end{bmatrix} = \begin{bmatrix} -1+2 & 3+0 \\ 2+(-5) & -2+1 \end{bmatrix} = \begin{bmatrix} 1 & 3 \\ -3 & -1 \end{bmatrix}$$

$$\begin{bmatrix} 6 & -6 \\ -2 & 4 \\ -2 & 1 \end{bmatrix} - \begin{bmatrix} -3 & 3 \\ -2 & 5 \\ 7 & -4 \end{bmatrix} = \begin{bmatrix} 6-(-3) & -6-3 \\ -2-(-2) & 4-5 \\ -2-7 & 1-(-4) \end{bmatrix} = \begin{bmatrix} 9 & -9 \\ 0 & -1 \\ -9 & 5 \end{bmatrix}$$

Find the sum and the difference of the matrices.

19. $\begin{bmatrix} -3 & -2 \\ 8 & 4 \end{bmatrix}, \begin{bmatrix} 4 & -2 \\ -7 & 5 \end{bmatrix}$ **20.** $\begin{bmatrix} -2 & 5 & 9 \\ -3 & 10 & 0 \end{bmatrix}, \begin{bmatrix} -1 & -6 & 11 \\ -2 & -7 & 1 \end{bmatrix}$

19. $\begin{bmatrix} 1 & -4 \\ 1 & 9 \end{bmatrix}; \begin{bmatrix} -7 & 0 \\ 15 & -1 \end{bmatrix}$

20. $\begin{bmatrix} -3 & -1 & 20 \\ -5 & 3 & 1 \end{bmatrix}; \begin{bmatrix} -1 & 11 & -2 \\ -1 & 17 & -1 \end{bmatrix}$

2.5 MULTIPLICATION OF REAL NUMBERS

Examples on pp. 93–95

EXAMPLES A product is negative if it has an odd number of negative factors. A product is positive if it has an even number of negative factors.

a. $(-4)(5) = -20$ One negative factor

b. $\left(-\frac{1}{3}\right)(-3)(2) = 2$ Two negative factors

c. $(-2)(-1)(-3) = -6$ Three negative factors

Find the product.

21. $(-3)(12)$ **−36** **22.** $(5)(-8)$ **−40** **23.** $(-40)(-15)$ **600** **24.** $(-1)(9)$ **−9**

25. $(-17)\left(\frac{2}{9}\right)$ **$-3\frac{7}{9}$** **26.** $(-14)(-0.3)$ **4.2** **27.** $|(9)(-5.5)|$ **49.5** **28.** $(-3.2)(-10)(2)$ **64**

29. $(-7)(-6)(-2)$ **−84** **30.** $(-12)(2)(-0.5)$ **12** **31.** $(-24)\left(-\frac{7}{12}\right)$ **14** **32.** $(11)(-1)(-7)(-3)$ **−231**

2.6 THE DISTRIBUTIVE PROPERTY

Examples on pp. 100–102

EXAMPLES You can use the distributive property to rewrite expressions without parentheses.

$8(x + 3) = 8x + 24$ $\qquad a(b + c) = ab + ac$ or $(b + c)a = ba + ca$

$3(x - 6) = 3x - 18$ $\qquad a(b - c) = ab - ac$ or $(b - c)a = ba - ca$

Use the distributive property to rewrite the expression without parentheses.

33. $5(x + 12)$ $5x + 60$ **34.** $(y + 6)9$ $9y + 54$ **35.** $5.5(b - 10)$ $5.5b - 55$ **36.** $(3.2 - w)2$ $6.4 - 2w$

37. $(t + 11)(-3)$ $-3t - 33$ **38.** $-2(s + 13)$ $-2s - 26$ **39.** $-2.5(z - 5)$ $-2.5z + 12.5$ **40.** $-x\left(\frac{3}{7} + y\right)$ $-\frac{3x}{7} - xy$

2.7 DIVISION OF REAL NUMBERS

Examples on pp. 108–110

EXAMPLE To divide 7 by −6, multiply 7 by the reciprocal of −6.

$$7 \div (-6) = 7 \cdot \left(-\frac{1}{6}\right) = -\frac{7}{6} \qquad b \div a = b \cdot \frac{1}{a}$$

Find the quotient.

41. $48 \div (-12)$ -4 **42.** $-34 \div 2$ -17 **43.** $39 \div (-13)$ -3 **44.** $-57 \div (-19)$ 3

45. $55 \div (-1.1)$ -50 **46.** $-63 \div 4\frac{1}{5}$ -15 **47.** $\frac{48}{-\frac{3}{4}}$ -64 **48.** $-\frac{-84}{\frac{7}{8}}$ 96

2.8 PROBABILITY AND ODDS

Examples on pp. 114–116

EXAMPLES Find how likely it is that you will randomly choose a green marble from a bag with 11 green, 2 blue, and 7 red marbles.

Probability

$$P = \frac{\text{Number of favorable outcomes}}{\text{Total number of outcomes}}$$

$$P = \frac{11}{20} = 0.55$$

The probability is 0.55.

Odds

$$\text{Odds} = \frac{\text{Number of favorable outcomes}}{\text{Number of unfavorable outcomes}}$$

$$= \frac{11}{9}$$

The odds are 11 to 9.

Find the probability and the odds of randomly choosing a red marble from a bag of red and white marbles.

49. Number of red marbles: 12
Total number of marbles: 48 **0.25, 1 to 3**

50. Number of red marbles: 9
Total number of marbles: 81 $\frac{1}{9} \approx 0.11$; **1 to 8**

51. Number of white marbles: 36
Total number of marbles: 40 **0.1, 1 to 9**

52. Number of white marbles: 17
Total number of marbles: 68 **0.75, 3 to 1**

CHAPTER 2

Chapter Test

Find the value of the expression.

1. $|8|$ **8**
2. $|-2.7|$ **2.7**
3. $|-6.4| - 3.1$ **3.3**
4. $-(-4.5)$ **4.5**
5. $4 + (-9)$ **−5**
6. $-4.5 + (-9.1)$ **−13.6**
7. $-2\frac{5}{6} + 3\frac{1}{4}$ **$\frac{5}{12}$**
8. $9 + (-10) + 2$ **1**

Add or subtract the matrices. **9–11. See margin.**

9. $\begin{bmatrix} 3 & -7 \\ \frac{1}{2} & 6 \\ 0 & 2 \end{bmatrix} + \begin{bmatrix} 4 & 2 \\ \frac{1}{2} & 4 \\ 5 & \frac{1}{2} \end{bmatrix}$

10. $\begin{bmatrix} -5 & -1 \\ 5 & 8 \\ \frac{7}{8} & 2\frac{1}{2} \end{bmatrix} - \begin{bmatrix} 6 & 3 \\ \frac{1}{2} & -2 \\ 0 & 2 \end{bmatrix}$

11. $\begin{bmatrix} 2 & \frac{1}{2} & 4 \\ 5 & 16 & -7 \end{bmatrix} + \begin{bmatrix} 5 & \frac{1}{2} & -1 \\ -2 & 8 & 4 \end{bmatrix}$

Find the product or quotient.

12. $(-6)(4)$ **−24**
13. $72 \div (-12)$ **−6**
14. $-36 \div (-4)$ **9**
15. $(-8.4)(-100)$ **840**
16. $-56 \div \left(-\frac{7}{8}\right)$ **64**
17. $(-9)(8)\left(-\frac{5}{6}\right)$ **60**
18. $-3740 \div (-10)$ **374**
19. $-(-18)(-5)\left(\frac{7}{15}\right)$ **−42**
20. $-\frac{3}{8} \div \frac{1}{2}$ **$-\frac{3}{4}$**
21. $\left(1\frac{2}{7}\right)\left(1\frac{5}{9}\right)$ **2**
22. $\left(-\frac{1}{2}\right)\left(\frac{3}{5}\right)\left(-\frac{2}{3}\right)\left(\frac{5}{8}\right)$ **$\frac{1}{8}$**
23. $-7\frac{4}{5} \div \left(-1\frac{3}{10}\right)$ **6**

Simplify the expression.

24. $(-5)(-w)(w)(w)$ **$5w^3$**
25. $(-8)^2(-x)^3$ **$-64x^3$**
26. $8(4 - q)$ **$32 - 8q$**
27. $(6 + y)(-12)$ **$-72 - 12y$**
28. $(35 - 10q)\left(-\frac{2}{5}\right)$ **$-14 + 4q$**
29. $-9(y + 11)$ **$-9y - 99$**
30. $5(3 - z) - z$ **$15 - 6z$**
31. $14p + 2(5 - p)$ **$12p + 10$**

In Exercises 32–37, evaluate the expression for the given value of *x*.

32. $5 - (-8) + x$ when $x = -5$ **8**
33. $|-9| - 2x + 5$ when $x = 6$ **2**
34. $-9x + 12$ when $x = -2$ **30**
35. $1 - 2x^2$ when $x = -2$ **−7**
36. $\frac{4 - x}{-3}$ when $x = -1$ **$-\frac{5}{3}$**
37. $-4x^2 - 8x + 9$ when $x = -5$ **−51**

38. **PROFIT** A company had a first-quarter profit of \$2189.70, a second-quarter profit of \$1527.11, a third-quarter loss of \$2502.18, and a fourth-quarter loss of \$266.54. What was the company's profit or loss for the year? **\$948.09 profit**

39. **EAGLES** An eagle dives from its nest with a velocity of $-8\frac{1}{3}$ feet per second. Find the vertical displacement in $4\frac{1}{2}$ seconds. **$-37\frac{1}{2}$ ft**

BEVERAGE SURVEY **You do a survey asking students to identify their favorite beverage from the following categories: *soda, juice, water,* and *other*. You get the following results: 132 students choose soda, 59 choose juice, 43 choose water, and 26 choose other.**

40. What is the probability that a randomly chosen student's favorite drink is juice? **$\frac{59}{260} \approx 0.23$**
41. What are the odds that a randomly chosen student's favorite drink is water? **43 to 217**

ADDITIONAL RESOURCES

- ***Chapter 2 Resource Book***
  Chapter Test (3 levels) (p. 123)
  SAT/ACT Chapter Test (p. 129)
  Alternative Assessment (p. 130)
- ***Test and Practice Generator***

9. $\begin{bmatrix} 7 & -5 \\ 1 & 10 \\ 5 & 2\frac{1}{2} \end{bmatrix}$

10. $\begin{bmatrix} -11 & -4 \\ 4\frac{1}{2} & 10 \\ \frac{7}{8} & \frac{1}{2} \end{bmatrix}$

11. $\begin{bmatrix} 7 & 1 & 3 \\ 3 & 24 & -3 \end{bmatrix}$

ADDITIONAL RESOURCES

- ***Chapter 2 Resource Book***
  Chapter Test (3 levels) (p. 123)
  SAT/ACT Chapter Test (p. 129)
  Alternative Assessment (p. 130)
- ***Test and Practice Generator***

CHAPTER 2

Chapter Standardized Test

TEST-TAKING STRATEGY If you can, check an answer using a method that is different from the one you used originally, to avoid making the same mistake twice.

1. **MULTIPLE CHOICE** Which integer is between $-\frac{55}{4}$ and $-\frac{37}{3}$? **A**

   (A) -13 (B) -12 (C) 2
   (D) $\frac{7}{3}$ (E) $-\frac{69}{5}$

2. **MULTIPLE CHOICE** What is the value of $|x| + |y| - 10$ when $x = -9$ and $y = 2$? **D**

   (A) 17 (B) -43 (C) -1
   (D) 1 (E) 21

3. **MULTIPLE CHOICE** Which of the following is a counterexample that proves the statement $|x| = -x$ is *not* true? **E**

   (A) $x = -9.5$ (B) $x = -2$ (C) $x = -\frac{1}{2}$
   (D) $x = 0$ (E) $x = 4.7$

4. **MULTIPLE CHOICE** What is the value of the expression $-9 + 3 + (-14)$? **B**

   (A) -26 (B) -20 (C) -8
   (D) -2 (E) 26

QUANTITATIVE COMPARISON In Questions 5–7, choose the statement that is true about the given quantities.

(A) The quantity in column A is greater.
(B) The quantity in column B is greater.
(C) The two quantities are equal.
(D) The relationship cannot be determined from the given information.

	Column A	Column B	
5.	$2 - 5 + (-3)$	$2 + (-5) - 3$	**C**
6.	$-\frac{7}{9} - \frac{1}{3}$	$-\frac{11}{12} - \frac{1}{6}$	**B**
7.	x	$-2x$	**D**

8. **MULTIPLE CHOICE** What is the value of the expression $-4 - 6 - (-10)$? **D**

   (A) -20 (B) -12 (C) -8
   (D) 0 (E) 12

9. **MULTIPLE CHOICE** What is the value of the expression $9 - (-13) + (-17) + (-10)$? **C**

   (A) -41 (B) -31 (C) -5
   (D) 5 (E) 15

10. **MULTIPLE CHOICE** Find the sum. **C**

$$\begin{bmatrix} -5 & -4 \\ 2 & -4 \\ 1 & -3 \end{bmatrix} + \begin{bmatrix} 8 & 9 \\ -6 & 1 \\ -1 & 4 \end{bmatrix} = \underline{?}$$

   (A) $\begin{bmatrix} 3 & 5 \\ -6 & -3 \\ 0 & 1 \end{bmatrix}$ (B) $\begin{bmatrix} -3 & 5 \\ 4 & -3 \\ 0 & -1 \end{bmatrix}$

   (C) $\begin{bmatrix} 3 & 5 \\ -4 & -3 \\ 0 & 1 \end{bmatrix}$ (D) $\begin{bmatrix} 3 & 5 \\ -4 & -3 \\ 0 & -1 \end{bmatrix}$

   (E) None of these

11. **MULTIPLE CHOICE** Find the difference. **E**

$$\begin{bmatrix} -1 & 3 \\ -6 & 4 \\ 2 & -5 \end{bmatrix} - \begin{bmatrix} -7 & 3 \\ 2 & 6 \\ 0 & -8 \end{bmatrix} = \underline{?}$$

   (A) $\begin{bmatrix} 6 & 0 \\ -8 & -2 \\ 2 & -3 \end{bmatrix}$ (B) $\begin{bmatrix} -8 & 0 \\ -8 & -2 \\ 2 & 3 \end{bmatrix}$

   (C) $\begin{bmatrix} 6 & 0 \\ -8 & 2 \\ -2 & -3 \end{bmatrix}$ (D) $\begin{bmatrix} 6 & 0 \\ -8 & -2 \\ 0 & 3 \end{bmatrix}$

   (E) None of these

12. **MULTIPLE CHOICE** $-\frac{1}{2} \cdot \frac{-2}{3} \cdot \frac{3}{-4} \cdot \frac{4}{5} = \underline{?}$ **B**

   (A) $-\frac{24}{60}$ (B) $-\frac{1}{5}$ (C) $-\frac{1}{6}$
   (D) $\frac{1}{5}$ (E) $\frac{5}{6}$

13. **MULTIPLE CHOICE** What is the value of $-2m^6 \div 4m^3$ when $m = -2$? **D**

Ⓐ -32 Ⓑ -16 Ⓒ -4 Ⓓ 4 Ⓔ 16

QUANTITATIVE COMPARISON In Questions 14 and 15, choose the statement that is true about the given quantities.

Ⓐ The quantity in column A is greater.

Ⓑ The quantity in column B is greater.

Ⓒ The two quantities are equal.

Ⓓ The relationship cannot be determined from the given information.

	Column A	Column B	
14.	$(-10)^4$	$(-10)^5$	**A**
15.	$-3 \cdot 24 \div (-9)$	$-12 \cdot 8 \div 6$	**A**

16. **MULTIPLE CHOICE** The expression $6(x + 3) - 2(4 - x)$ can be simplified to _?_. **B**

Ⓐ $5x + 5$ Ⓑ $8x + 10$ Ⓒ $5x - 5$ Ⓓ $8x + 1$ Ⓔ $4x + 10$

17. **MULTIPLE CHOICE** What is the value of $\frac{4p + 6pq}{p^2q}$ when $p = -2$ and $q = -3$? **D**

Ⓐ $-\frac{11}{3}$ Ⓑ $\frac{28}{12}$ Ⓒ 2 Ⓓ $-\frac{7}{3}$ Ⓔ $\frac{11}{3}$

18. **MULTIPLE CHOICE** On March 1, you own 15.5 shares of mutual fund stock that are worth \$142.91. On May 1, the shares are worth \$135.16. What was the change in the price per share of stock? **B**

Ⓐ $-\$7.75$ Ⓑ $-\$.50$ Ⓒ $-\$.05$ Ⓓ $\$.05$ Ⓔ $\$.50$

MULTIPLE CHOICE In Questions 19 and 20, you randomly choose a marble from a bag holding 5 red, 7 blue, and 13 yellow marbles.

19. What is the probability that you choose a blue marble? **C**

Ⓐ 0.2 Ⓑ 0.07 Ⓒ 0.28 Ⓓ 0.52 Ⓔ 0.7

20. What are the odds that you choose a red marble? **A**

Ⓐ 1 to 4 Ⓑ 5 to 13 Ⓒ 7 to 13 Ⓓ 5 to 7 Ⓔ 13 to 5

21. **MULTI-STEP PROBLEM** The spinner at the right is evenly divided into 6 sections. Each of the players spins the spinner. Their results are added together. Player A wins if the sum is negative. Otherwise, Player B wins.

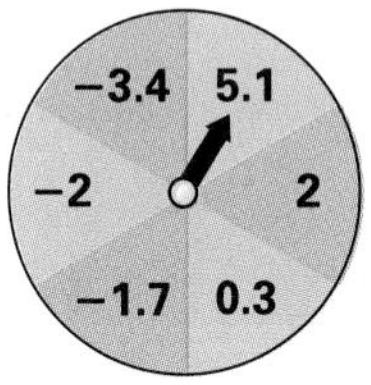

a. List all the possible outcomes for one pair of spins. **See margin.**

b. What are the odds of Player A winning? **17 to 19**

c. **CRITICAL THINKING** A game is fair when all the players have the same probability of winning. Is this a fair game? Explain your reasoning. **See margin.**

d. **CREATING A FAIR GAME** Create your own game like the one described above. Your game should be fair. You may change the numbers on the spinner or the rules of the game. Describe your game and explain why it is fair. **See margin.**

21a.

	−3.4	5.1	2	0.3	−1.7	−2
−3.4	A	B	A	A	A	A
5.1	B	B	B	B	B	B
2	A	B	B	B	B	B
0.3	A	B	B	B	A	A
−1.7	A	B	B	A	A	A
−2	A	B	B	A	A	A

21c. No; the probability that Player A wins is $\frac{17}{36}$, and the probability that Player B wins is $\frac{19}{36}$.

21d. *Sample answer:* Let the spinner have six sections numbered 1 through 6. The players take turns. On a player's turn, he or she wins if the spinner lands on an odd number. If the spinner lands on an even number, the other player wins. On each turn, each player has an equal probability (0.5) of winning.

CHAPTER 3

Planning the Chapter

Solving Linear Equations

Lesson	Goals	NCTM	ITED	SAT9	Terra-Nova	Local
3.1 pp. 131–137	CONCEPT ACTIVITY: 3.1 *Investigate how to use algebra tiles to solve one-step equations.* GOAL 1 Solve linear equations using addition and subtraction. GOAL 2 Use linear equations to solve real-life problems.	1, 2, 6, 8, 9, 10	RQWN, RQF	3, 7	11, 16, 17, 52	
3.2 pp. 138–144	GOAL 1 Solve linear equations using addition and subtraction. GOAL 2 Use multiplication and division equations to solve real-life and geometric problems.	1, 2, 3, 4, 6, 8, 9, 10	RQWN, RQF	3, 7	11, 15, 16, 17, 52	
3.3 pp. 145–152	GOAL 1 Use two or more transformations to solve an equation. GOAL 2 Use multi-step equations to solve real-life problems.	2, 6, 8, 9, 10	RQWN, RQF	1, 7	16, 17, 52	
3.4 pp. 153–159	CONCEPT ACTIVITY: 3.4 *Investigate how to use algebra tiles to solve an equation with a variable on both the left side and right side of an equation.* GOAL 1 Collect variables on one side of an equation. GOAL 2 Use equations to solve real-life problems.	1, 2, 6, 8, 9, 10	RQWN, RQF	2, 3, 7	16, 17	
3.5 pp. 160–165	GOAL 1 Draw a diagram to help you understand real-life problems. GOAL 2 Use tables and graphs to check your answers.	2, 5, 6, 10	RQM, RQG, IIT, IIG	1, 21	15, 17	
3.6 pp. 166–173	GOAL 1 Find exact and approximate solutions of equations that contain decimals. GOAL 2 Solve real-life problems that use decimals. TECHNOLOGY ACTIVITY: 3.6 *Solve multi-step equations on a graphing calculator to generate a table of values.*	1, 2, 6, 9		3	11, 16, 17, 47, 52	
3.7 pp. 174–179	GOAL 1 Solve a formula or literal equation for one of its variables. GOAL 2 Rewrite an equation in function form.	2, 3, 4, 10		4, 21	16, 52	
3.8 pp. 180–186	GOAL 1 Use rates and ratios to model and solve real-life problems. GOAL 2 Use percents to model and solve real-life problems.	1, 2, 6, 8, 9, 10	RQPE, APRA, RQRA	1	17, 18, 50	
Extension pp. 187–188	GOAL 1 Identify and use inductive and deductive reasoning.	7, 8			17	

RESOURCES

Chapter Resource Booklets

Chapter Support

Tips for New Teachers	p. 1
Parent Guide for Student Success	p. 3
Prerequisite Skills Review	p. 5
Strategies for Reading Mathematics	p. 7

Lesson Support

Lesson Support	3.1	3.2	3.3	3.4	3.5	3.6	3.7	3.8
Lesson Plans (regular and block)	p. 9	p. 22	p. 34	p. 50	p. 64	p. 78	p. 94	p. 107
Warm-Up Exercises and Daily Quiz	p. 11	p. 24	p. 36	p. 52	p. 66	p. 80	p. 96	p. 109
Activity Support Masters	p. 12			p. 53				
Lesson Openers	p. 13	p. 25	p. 37	p. 54	p. 67	p. 81	p. 97	p. 110
Graphing Calculator Activities & Keystrokes			p. 38		p. 68	p. 82	p. 98	
Practice (3 levels)	p. 14	p. 26	p. 40	p. 55	p. 70	p. 85	p. 99	p. 111
Reteaching with Practice	p. 17	p. 29	p. 43	p. 58	p. 73	p. 88	p. 102	p. 114
Quick Catch-Up for Absent Students	p. 19	p. 31	p. 45	p. 60	p. 75	p. 90	p. 104	p. 116
Cooperative Learning Activities				p. 61				p. 117
Interdisciplinary Applications		p. 32	p. 46			p. 91	p. 105	
Real-Life Applications	p. 20			p. 62	p. 76			p. 118
Math & History Applications			p. 47					
Challenge: Skills and Applications	p. 21	p. 33	p. 48	p. 63	p. 77	p. 92	p. 106	p. 119

Review and Assessment

Quizzes	pp. 49, 93
Chapter Review Games and Activities	p. 120
Chapter Test (3 levels)	pp. 121–126
SAT/ACT Chapter Test	p. 127
Alternative Assessment with Math Journal	p. 128
Project with Rubric	p. 130
Cumulative Review	p. 132
Resource Book Answers	p. A1

Transparencies

Transparencies	3.1	3.2	3.3	3.4	3.5	3.6	3.7	3.8
Warm-Up Exercises and Daily Quiz	p. 19	p. 20	p. 21	p. 22	p. 23	p. 24	p. 25	p. 26
Alternative Lesson Opener Transparencies	p. 16	p. 17	p. 18	p. 19	p. 20	p. 21	p. 22	p. 23
Examples/Standardized Test Practice	✓	✓	✓	✓	✓	✓	✓	✓
Answer Transparencies	✓	✓	✓	✓	✓	✓	✓	✓

Technology

- EasyPlanner CD
- EasyPlanner Plus Online
- Classzone.com
- Personal Student Tutor CD
- Test and Practice Generator CD
- Instant Replay: Video Review Games
- Electronic Lesson Presentations
- Chapter Audio Summaries CD

Additional Resources

- Basic Skills Workbook
- Explorations and Projects
- Worked-Out Solution Key
- Data Analysis Sourcebook
- Spanish Resources
- Standardized Test Practice Workbook
- Practice Workbook with Examples
- Notetaking Guide

CHAPTER 3

Pacing the Chapter

Resource Key
- ● Student Edition
- ● Chapter 3 Resource Book
- ● Teacher's Edition

REGULAR SCHEDULE

Day 1

3.1

Starting Options
- Prereq. Skills Review
- Strategies for Reading
- Warm-Up

Teaching Options
- Motivating the Lesson
- Concept Act. & Wksht.
- Les. Opener (Activity)
- Examples 1–4
- Closure Question
- Guided Practice Exs.

Apply/Homework
- See Assignment Guide.
- See the CRB: Practice, Reteach, Apply, Extend

Assessment Options
- Checkpoint Exercises
- Daily Quiz (3.1)
- Stand. Test Practice

Day 2

3.2

Starting Options
- Homework Check
- Warm-Up or Daily Quiz

Teaching Options
- Les. Opener (Activity)
- Examples 1–5
- Closure Question
- Guided Practice Exs.

Apply/Homework
- See Assignment Guide.
- See the CRB: Practice, Reteach, Apply, Extend

Assessment Options
- Checkpoint Exercises
- Daily Quiz (3.2)
- Stand. Test Practice

Day 3

3.3

Starting Options
- Homework Check
- Warm-Up or Daily Quiz

Teaching Options
- Motivating the Lesson
- Les. Opener (Application)
- Graphing Calc. Activity
- Examples 1–5
- Guided Practice Exs. 1–8

Apply/Homework
- See Assignment Guide.
- See the CRB: Practice, Reteach, Apply, Extend

Assessment Options
- Checkpoint Exercises, p. 146

Day 4

3.3 *(cont.)*

Starting Options
- Homework Check

Teaching Options
- Examples 6–7
- Closure Question
- Guided Practice Ex. 9

Apply/Homework
- See Assignment Guide.
- See the CRB: Practice, Reteach, Apply, Extend

Assessment Options
- Checkpoint Exercises, p. 147
- Daily Quiz (3.3)
- Stand. Test Practice
- Quiz (3.1–3.3)

Day 5

3.4

Starting Options
- Homework Check
- Warm-Up or Daily Quiz

Teaching Options
- Motivating the Lesson
- Concept Act. & Wksht.
- Les. Opener (Activity)
- Examples 1–4
- Guided Practice Exs. 1–9

Apply/Homework
- See Assignment Guide.
- See the CRB: Practice, Reteach, Apply, Extend

Assessment Options
- Checkpoint Exercises, p. 155

Day 6

3.4 *(cont.)*

Starting Options
- Homework Check

Teaching Options
- Example 5
- Closure Question
- Guided Practice Exs. 10–11

Apply/Homework
- See Assignment Guide.
- See the CRB: Practice, Reteach, Apply, Extend

Assessment Options
- Checkpoint Exercises, p. 156
- Daily Quiz (3.4)
- Stand. Test Practice

Day 9

3.6 *(cont.)*

Starting Options
- Homework Check

Teaching Options
- Example 5
- Closure Question
- Guided Practice Ex. 13

Apply/Homework
- See Assignment Guide.
- See the CRB: Practice, Reteach, Apply, Extend

Assessment Options
- Checkpoint Exercises, p. 168
- Daily Quiz (3.6)
- Stand. Test Practice
- Quiz (3.4–3.6)

Day 10

3.7

Starting Options
- Homework Check
- Warm-Up or Daily Quiz

Teaching Options
- Les. Opener (Application)
- Graphing Calc. Activity
- Examples 1–6
- Closure Question
- Guided Practice Exs.

Apply/Homework
- See Assignment Guide.
- See the CRB: Practice, Reteach, Apply, Extend

Assessment Options
- Checkpoint Exercises
- Daily Quiz (3.7)
- Stand. Test Practice

Day 11

3.8

Starting Options
- Homework Check
- Warm-Up or Daily Quiz

Teaching Options
- Les. Opener (Application)
- Examples 1–6
- Closure Question
- Guided Practice Exs.

Apply/Homework
- See Assignment Guide.
- See the CRB: Practice, Reteach, Apply, Extend

Assessment Options
- Checkpoint Exercises
- Daily Quiz (3.8)
- Stand. Test Practice

Day 12

Review

Day 12 Start Options
- Homework Check

Reviewing Options
- Chapter 3 Summary
- Chapter 3 Review
- Chapter Review Games and Activities

Apply/Homework
- Chapter 3 Test (practice)
- Ch. Standardized Test (practice)

Day 13

Assess

Day 13 Start Options
- Homework Check

Assessment Options
- Chapter 3 Test
- SAT/ACT Ch. 3 Test
- Alternative Assessment

Apply/Homework
- Skill Review, p. 202

Day 7

3.5

Starting Options
- Homework Check
- Warm-Up or Daily Quiz

Teaching Options
- Les. Opener (Visual)
- Graphing Calc. Activity
- Examples 1–3
- Closure Question
- Guided Practice Exs.

Apply/Homework
- See Assignment Guide.
- See the CRB: Practice, Reteach, Apply, Extend

Assessment Options
- Checkpoint Exercises
- Daily Quiz (3.5)
- Stand. Test Practice

Day 8

3.6

Starting Options
- Homework Check
- Warm-Up or Daily Quiz

Teaching Options
- Motivating the Lesson
- Les. Opener (Calculator)
- Graphing Calc. Activity
- Examples 1–4
- Guided Practice Exs. 1–12

Apply/Homework
- See Assignment Guide.
- See the CRB: Practice, Reteach, Apply, Extend

Assessment Options
- Checkpoint Exercises, p. 167

BLOCK SCHEDULE

Day 1

3.1 & 3.2

Day 1 Start Options
- Prereq. Skills Review
- Strategies for Reading
- Warm-Up (Les. 3.1)

Teaching 3.1 Options
- Motivating the Lesson
- Concept Act. & Wksht.
- Les. Opener (Activity)
- Examples 1–4
- Closure Question
- Guided Practice Exs.

Teaching 3.2 Options
- Warm-Up (Les. 3.2)
- Les. Opener (Activity)
- Examples 1–5
- Closure Question
- Guided Practice Exs.

Apply/Homework
- See Assignment Guide.
- See the CRB: Practice, Reteach, Apply, Extend

Assessment Options
- Checkpoint Exercises
- Daily Quiz (Les. 3.1, 3.2)

Day 2

3.3

Day 2 Start Options
- Homework Check
- Warm-Up or Daily Quiz

Teaching 3.3 Options
- Motivating the Lesson
- Les. Opener (Appl.)
- Graphing Calc. Activity
- Examples 1–7
- Closure Question
- Guided Practice Exs.

Apply/Homework
- See Assignment Guide.
- See the CRB: Practice, Reteach, Apply, Extend

Assessment Options
- Checkpoint Exercises
- Daily Quiz (Les. 3.3)
- Stand. Test Practice
- Quiz (3.1–3.3)

Day 3

3.4

Day 3 Start Options
- Homework Check
- Warm-Up or Daily Quiz

Teaching 3.4 Options
- Motivating the Lesson
- Concept Act. & Wksht.
- Les. Opener (Activity)
- Examples 1–5
- Closure Question
- Guided Practice Exs.

Apply/Homework
- See Assignment Guide.
- See the CRB: Practice, Reteach, Apply, Extend

Assessment Options
- Checkpoint Exercises
- Daily Quiz (Les. 3.4)
- Stand. Test Practice

Day 4

3.5 & 3.6

Day 4 Start Options
- Homework Check
- W-Up 3.5 or D. Quiz 3.4

Teaching 3.5 Options
- Les. Opener (Visual)
- Graphing Calc. Activity
- Examples 1–3
- Closure Question
- Guided Practice Exs.

Beginning 3.6 Options
- Warm-Up (Les. 3.6)
- Motivating the Lesson
- Les. Opener (Calc.)
- Graphing Calc. Activity
- Examples 1–4
- Guided Practice Exs. 1–12

Apply/Homework
- See Assignment Guide.
- See the CRB: Practice, Reteach, Apply, Extend

Assessment Options
- Checkpoint Exercises
- Daily Quiz (Les. 3.5)
- Stand. Test Prac. (3.5)

Day 5

3.6 & 3.7

Day 5 Start Options
- Homework Check
- Daily Quiz (Les. 3.5)

Finishing 3.6 Options
- Example 5
- Closure Question
- Guided Practice Ex. 13

Teaching 3.7 Options
- Warm-Up (Les. 3.7)
- Les. Opener (Appl.)
- Graphing Calc. Activity
- Examples 1–6
- Closure Question
- Guided Practice Exs.

Apply/Homework
- See Assignment Guide.
- See the CRB: Practice, Reteach, Apply, Extend

Assessment Options
- Checkpoint Exercises
- Daily Quiz (Les. 3.6, 3.7)
- Stand. Test Practice
- Quiz (3.4–3.6)

Day 6

3.8 & Review

Day 6 Start Options
- Homework Check
- Warm-Up or Daily Quiz

Teaching 3.8 Options
- Les. Opener (Appl.)
- Examples 1–6
- Closure Question
- Guided Practice Exs.

Reviewing Options
- Chapter 3 Summary
- Chapter 3 Review
- Chapter Review Games and Activities

Apply/Homework
- See Assignment Guide.
- See the CRB: Practice, Reteach, Apply, Extend
- Chapter 3 Test (practice)
- Ch. Standardized Test (practice)

Assessment Options
- Checkpoint Exercises
- Daily Quiz (Les. 3.8)
- Stand. Test Prac. (3.8)

Day 7

Assess & 4.1

(Day 7 = Ch. 4 Day 1)

Assessment Options
- Chapter 3 Test
- SAT/ACT Ch. 3 Test
- Alternative Assessment

Ch. 4 Start Options
- Skill Review, p. 202
- Strategies for Reading

Teaching 4.1 Options
- Warm-Up (Les. 4.1)
- Motivating the Lesson
- Les. Starter (Appl.)
- Technology Activities
- Examples 1–3
- Closure Question
- Guided Practice Exs.

Apply/Homework
- See Assignment Guide.
- See the CRB: Practice, Reteach, Apply, Extend

Assessment Options
- Checkpoint Exercises
- Daily Quiz (Les. 4.1)
- Stand. Test Practice

CHAPTER 3

MEETING INDIVIDUAL NEEDS

BEFORE THE CHAPTER

The *Chapter 3 Resource Book* has the following materials to distribute and use before the chapter:

- **Parent Guide for Student Success (pictured below)**
- **Prerequisite Skills Review**
- **Strategies for Reading Mathematics**

PARENT GUIDE *Pages 3–4*

CHAPTER 3 NAME ____________ DATE ________

Parent Guide for Student Success

For use with Chapter 3: Solving Linear Equations

Chapter Support

Chapter Overview One way that you can help your student succeed in Chapter 3 is by discussing the lesson goals in the chart below. When a lesson is completed, ask your student to interpret the lesson goals for you and to explain how the mathematics of the lesson relates to one of the key applications listed in the chart.

Lesson Title	*Lesson Goals*	*Key Applications*
3.1: Solving Equations Using Addition and Subtraction	Solve linear equations using addition and subtraction. Use linear equations to solve real-life problems.	• Temperature Changes • Computer Time • City Parks
3.2: Solving Equations Using Multiplication and Division	Solve linear equations using multiplication and division. Use multiplication and division equations to solve real-life and geometric problems.	• Restoring Movies • Thunderstorms • Bald Eagles
3.3: Solving Multi-Step Equations	Use two or more transformations to solve an equation. Use multi-step equations to solve real-life problems.	• Medical Rescue • Firefighting • Height of a Fountain
3.4: Solving Equations with Variables on Both Sides	Collect variables on one side of an equation. Use equations to solve real-life problems.	• Membership Fees • Rock Climbing • Tall Buildings
3.5: Linear Equations and Problem Solving	Draw a diagram to help you understand real-life problems. Use tables and graphs to check your answers.	• Yearbook Design • Gazelles and Cheetahs • Package Size
3.6: Solving Decimal Equations	Find exact and approximate solutions of equations that contain decimals. Solve real-life problems that use decimals.	• Retail Purchases • Cocoa Consumption • Bridge Expansion
3.7: Formulas and Functions	Solve a formula or literal equation for one of its variables. Rewrite an equation in function form.	• Mars Pathfinder • Scuba Diving • Tree Measure
3.8: Rates, Ratios, and Percents	Use rates, ratios, and percents to model and solve real-life problems.	• Exchange Rates • Glass Recycling • Outdoor Recreation

Test-Taking Strategy

Always **check your solution** against the original problem. When possible **use a different method** to check so you don't repeat an error. Ask your student why going back to the original problem is especially important in Lesson 3.5. Have your student show you an example from the chapter of how to check using different steps than those used to solve the problem.

PARENT GUIDE FOR STUDENT SUCCESS The first page summarizes the content of Chapter 3. Parents are encouraged to have their students explain how the material relates to key applications in the chapter, such as exchange rates. The second page (not shown) provides exercises and an activity that parents can do with their students. In the activity, parents and students compare rates of two long-distance phone services.

DURING EACH LESSON

The *Chapter 3 Resource Book* has the following alternatives for introducing the lesson:

- **Lesson Openers (pictured below)**
- **Graphing Calculator Activities with Keystrokes**

LESSON OPENER *Page 67*

LESSON 3.5 NAME ____________ DATE ________

Available as a transparency

Visual Approach Lesson Opener

For use with pages 160–165

Lesson 3.5

1. The graph shows the total amount of money you have earned so far this year. For example, the point (4, 200) shows that in the first 4 weeks you earned a total of \$200.

   a. What does the graph tell you about how much you earn each week?

   b. You want to find how many weeks it will take you to earn \$750. Write an equation to model this situation.

2. The diagram shows the layout of a rectangular dog run that needs to be fenced in.

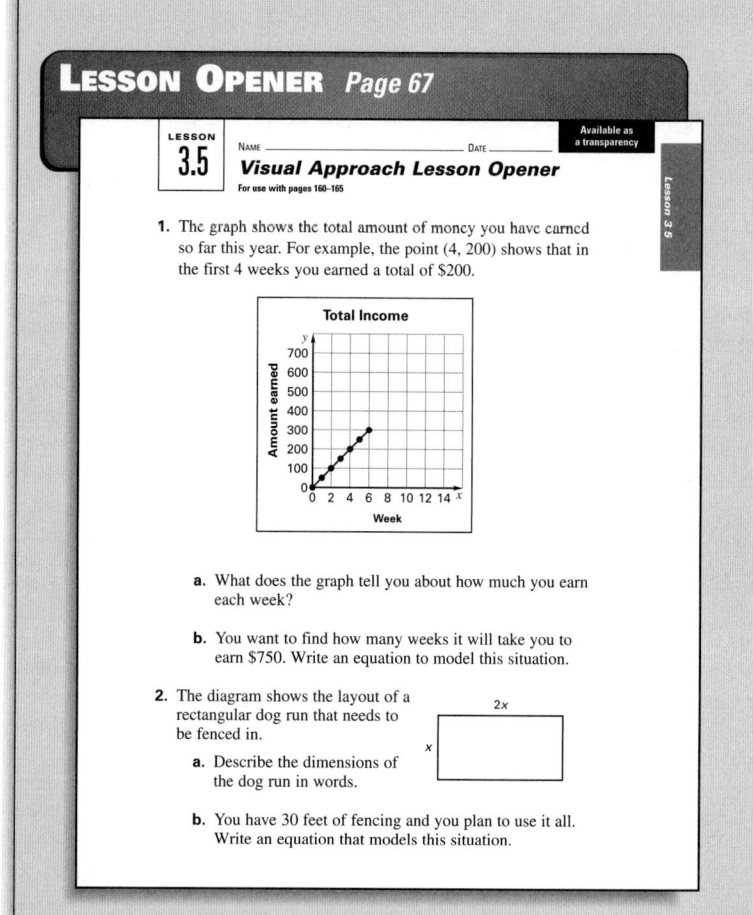

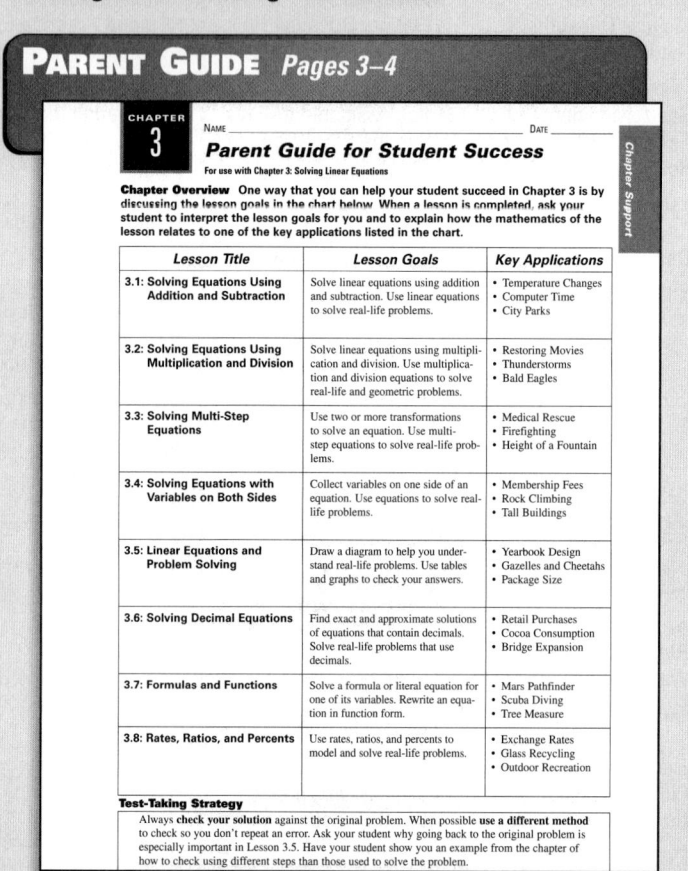

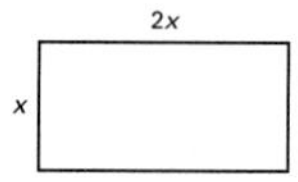

   a. Describe the dimensions of the dog run in words.

   b. You have 30 feet of fencing and you plan to use it all. Write an equation that models this situation.

VISUAL APPROACH LESSON OPENER This Lesson Opener uses visuals as an alternative way to start Lesson 3.5. Students analyze a graph of money earned as an introduction to writing an equation to model a situation.

Following Each Lesson

The *Chapter 3 Resource Book* has a variety of materials to follow-up each lesson. They include the following:

- **Practice (3 levels)**
- **Reteaching with Practice (pictured below)**
- **Quick Catch-Up for Absent Students**
- **Interdisciplinary Applications**
- **Real-Life Applications**

Reteaching with Practice *Pages 43–44*

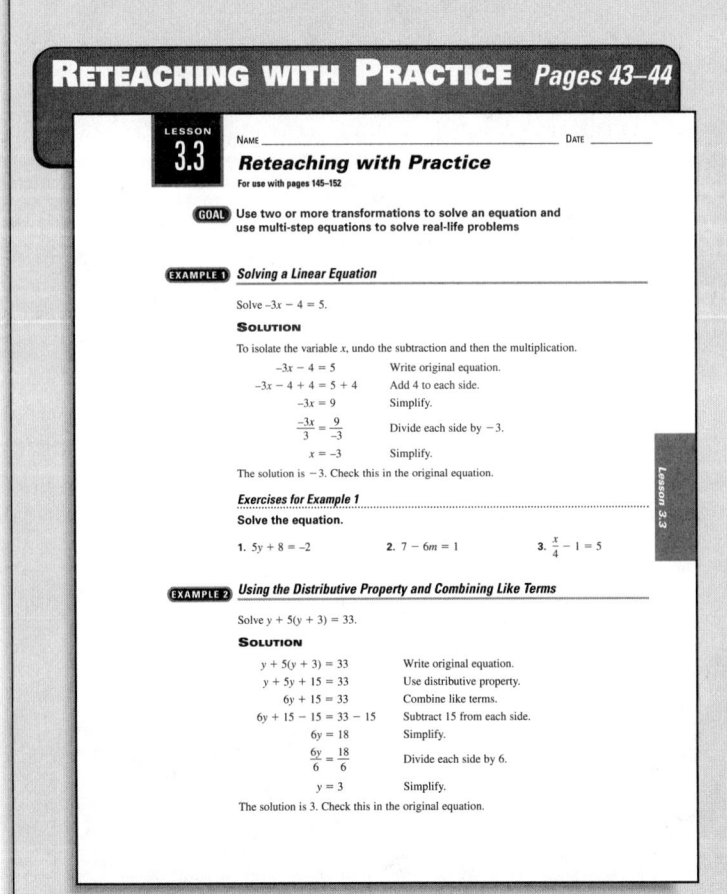

LESSON 3.3

NAME ______________________ DATE __________

Reteaching with Practice

For use with pages 145–152

GOAL Use two or more transformations to solve an equation and use multi-step equations to solve real-life problems

EXAMPLE 1 *Solving a Linear Equation*

Solve $-3x - 4 = 5$.

Solution

To isolate the variable x, undo the subtraction and then the multiplication.

$-3x - 4 = 5$	Write original equation.
$-3x - 4 + 4 = 5 + 4$	Add 4 to each side.
$-3x = 9$	Simplify.
$\frac{-3x}{3} = \frac{9}{-3}$	Divide each side by -3.
$x = -3$	Simplify.

The solution is -3. Check this in the original equation.

Exercises for Example 1

Solve the equation.

1. $5y + 8 = -2$
2. $7 - 6m = 1$
3. $\frac{x}{4} - 1 = 5$

Lesson 3.3

EXAMPLE 2 *Using the Distributive Property and Combining Like Terms*

Solve $y + 5(y + 3) = 33$.

Solution

$y + 5(y + 3) = 33$	Write original equation.
$y + 5y + 15 = 33$	Use distributive property.
$6y + 15 = 33$	Combine like terms.
$6y + 15 - 15 = 33 - 15$	Subtract 15 from each side.
$6y = 18$	Simplify.
$\frac{6y}{6} = \frac{18}{6}$	Divide each side by 6.
$y = 3$	Simplify.

The solution is 3. Check this in the original equation.

Reteaching with Practice On this two-page worksheet, additional examples and practice reinforce the key concepts of Lesson 3.3: solving a linear equation, using the distributive property, and applying techniques to real-life applications.

Technology Resource

Students can use the Personal Student Tutor to find additional reteaching and practice for the skills in Lesson 3.3 and in the rest of Chapter 3.

Assessing the Chapter

The *Chapter 3 Resource Book* has the following review and assessment materials:

- **Quizzes**
- **Chapter Review Games and Activities (pictured below)**
- **Chapter Test (3 levels)**
- **SAT/ACT Chapter Test**
- **Alternative Assessment with Rubric and Math Journal**
- **Project with Rubric**
- **Cumulative Review**

Chapter Review Games *Page 120*

CHAPTER 3

NAME ______________________ DATE __________

Chapter Review Games and Activities

For use after Chapter 3

Solve the following problems in the space provided, and find the answer in the boxes at the bottom of the page. Cross out the box containing each correct answer. Place the remaining letters on the lines below the boxes to find the answer to the riddle.

What do you call emergency relief to Antarctica?

1. $-4 = x - (-5)$
2. $20 = -\frac{2}{3}x$
3. $7(x - 3) = 56$
4. $5x - 16 = 3x$
5. $15 - 9 = -x$
6. $12 + \frac{x}{5} = -7$
7. $17 = 33 - 8x$
8. $-6(8 - x) + 15 = -51$
9. $-4 + x = 15$
10. $\frac{7}{8}x - 22 = 27$
11. $\frac{x}{3} = 8$
12. $\frac{x}{6} = \frac{16}{3}$
13. $6x + 5(x - 2) = 34$
14. $\frac{7}{8}(32 + 16x) = -7x - 14$
15. $4x - 17 = 11$
16. $\frac{3}{2}x + 5 = 20$
17. $\frac{3}{8}x + 21 = 0$

$x = -30$ G	$x = 4$ I	$x = -2$ N	$x = 30$ C	$x = 19$ A	$x = -56$ F
$x = 32$ E	$x = -11$ O	$x = 2$ B	$x = 56$ R	$x = 6$ O	$x = -95$ C
$x = -3$ U	$x = -9$ D	$x = 3$ L	$x = -24$ A	$x = 11$ N	$x = 7$ H
$x = -32$ I	$x = 24$ G	$x = 8$ P	$x = 10$ T	$x = -6$ E	$x = 9$ D

___ ___ ___ ___   ___ ___ ___!!

Review and Assess

Chapter Review Games and Activities On this worksheet, students solve linear equations to find the answer to a riddle as a motivating means of reviewing the material in Chapter 3.

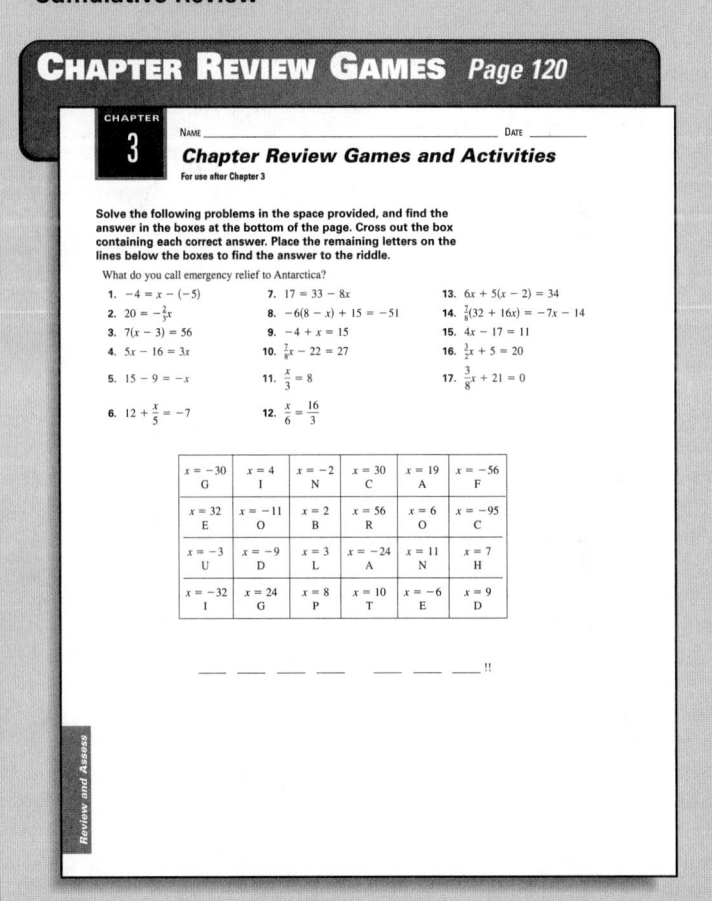

Technology Resource

Teachers can use the video Instant Replay: Video Review Games as another motivating way to review the material in Chapter 3.

CHAPTER OVERVIEW

CHAPTER GOALS

Chapter 3 begins by examining techniques for solving linear equations in one variable. Later in the chapter, students study formulas and learn to solve formulas for a specified variable. They then apply this skill to write equations in function form.The chapter stresses the use of equations (including equations that involve ratios, rates, and percents) to model and solve real-world problems.

APPLICATION NOTE

Bald eagles, like all eagles, are members of the hawk family. Bald eagles have been revered by many cultures including Native Americans who used their feathers to signify strength and courage. The bald eagle received its name because of its white head. Though the birds may not seem very large, they can have wing spans approaching 6 feet.

Even though the bald eagle is the national emblem of the United States, it has been on the list of endangered species for several years. Strong pesticides used in the eagles' habitat made their way into rivers, lakes, and streams and consequently into the fish that are a food source for the eagles. As a result, the egg shells of the birds were weak and few new eagles were being born. The eagle has made a comeback in recent years and has now been taken off the endangered species list.

Additional information about bald eagles is available at **www.mcdougallittell.com.**

SOLVING LINEAR EQUATIONS

Why do you think the bald eagle was chosen as a national symbol?

CHAPTER 3

APPLICATION: Bald Eagles

A bald eagle is featured on the Great Seal of the United States. The bald eagle is considered a symbol of power, strength, grace, and freedom. These birds, found only in North America, can fly up to 30 miles per hour and dive at speeds up to 100 miles per hour.

To solve problems involving flying rates of bald eagles, you can use equations containing variables for distance, rate, and time. In Chapter 3 you will use linear equations to describe many real-world situations.

Think & Discuss

1. Find the distance a bald eagle can travel for the given flying rate and time. **30 mi; 15 mi; 10 mi; $2\frac{1}{2}$ mi**

Flying Rate (miles per hour)	Time (hours)	Distance (miles)
30	1	?
30	$\frac{1}{2}$	?
30	$\frac{1}{12}$	?

2. Rewrite the flying rate in miles per minute. **$\frac{1}{2}$ mi/min**

Learn More About It

You will use equations to find flying and diving times of bald eagles in Exercise 57 on p. 143.

APPLICATION LINK Visit www.mcdougallittell.com for more information about bald eagles.

ADDITIONAL RESOURCES

Another way to begin the chapter is to show the video clip of a real-life motivator for Chapter 3 on the *Instant Replay: Video Review Games.*

PROJECTS

A project covering Chapters 1–3 appears on pages 198–199 of the Student Edition. An additional project for Chapter 3 is available in the *Chapter 3 Resource Book,* p. 130.

TECHNOLOGY

Software

- *Electronic Teaching Tools*
- *Online Lesson Planner*
- *Personal Student Tutor*
- *Test and Practice Generator*
- *Electronic Lesson Presentations* (Lesson 3.3)

Video

- *Instant Replay: Video Review Games*

Internet Connections
www.mcdougallittell.com

- **Application Links**
  162, 184
- **Data Updates**
  182
- **Student Help**
  143, 150, 155, 162, 170, 176, 184
- **Career Links**
  136, 147, 156, 158
- **Extra Challenge**
  159, 165, 179, 185

PREPARE

DIAGNOSTIC TOOLS
The **Skill Review** exercises can help you diagnose whether students have the following skills needed in Chapter 3:

- Find a percent of a number.
- Check the solution of an equation.
- Use the distributive property.
- Identify terms and combine like terms.

The following resources are available for students who need additional help with these skills:

- Prerequisite Skills Review (*Chapter 3 Resource Book,* p. 5; *Warm-Up Transparencies,* p. 18)
- Reteaching with Practice (Chapter Resource Book for Lessons 3.1–3.8)
- *Personal Student Tutor*

ADDITIONAL RESOURCES
The following resources are provided to help you prepare for the upcoming chapter and customize review materials:

- ***Chapter 3 Resource Book***
  Tips for New Teachers (p. 1)
  Parent Guide (p. 3)
  Lesson Plans (every lesson)
  Lesson Plans for Block Scheduling (every lesson)
- ***Electronic Teaching Tools***
- ***Online Lesson Planner***
- ***Test and Practice Generator***

CHAPTER 3

Study Guide

PREVIEW

What's the chapter about?

Chapter 3 is about **solving linear equations**. In Chapter 3, you'll learn

- techniques for solving linear equations systematically.
- ways to apply ratios, rates, percents, and problem solving strategies.

KEY VOCABULARY

▶ **Review**
- solution of an equation, p. 24
- opposites, p. 65
- distributive property, p. 100
- reciprocal, p. 108

▶ **New**
- equivalent equations, p. 132
- inverse operations, p. 132
- linear equation in one variable, p. 133
- ratio of *a* to *b*, p. 140
- identity, p. 155
- formula, p. 174
- rate of *a* per *b*, p. 180

PREPARE

Are you ready for the chapter?

SKILL REVIEW Do these exercises to review key skills that you'll apply in this chapter. See the given **reference page** if there is something you don't understand.

Find the percent of the number. (Skills Review, p. 786)

1. 11% of 650 — 71.5
2. 41% of 71.5 — 29.315
3. 6% of 250 — 15
4. 4.2% of 60 — 2.52

Check whether the given number is a solution of the equation. (Review Examples 1 and 2, p. 24)

5. $9 - x = 7; -2$ — no
6. $-11 = -4y + 1; 3$ — yes
7. $4 - 3w + 5w = 2; -1$ — yes

Use the distributive property to rewrite the expression without parentheses. (Review Examples 2 and 3, p. 101)

8. $3(4r + 6)$ — $12r + 18$
9. $-6(3 - 5z)$ — $-18 + 30z$
10. $(-x + 7)(-3)$ — $3x - 21$

Find the terms of the expression. Then simplify the expression by combining like terms. (Review Example 3, p. 80 and Example 5, p. 102)

11. $-2 + 3x - 7$ — $-9 + 3x$
12. $4a + 2 - a$ — $3a + 2$
13. $1 - 3x + 4y + 3x$ — $1 + 4y$

STUDENT HELP

Study Tip
"Student Help" boxes throughout the chapter give you study tips and tell you where to look for extra help in this book and on the Internet.

STUDY STRATEGY

Here's a study strategy!

Learning to Use Formulas

Be sure you know what each variable means in a formula. To study for tests, it is helpful to write a formula on one side of a 3 × 5 card and a sample problem using that formula on the other side. There is also a helpful Table of Formulas on page 813.

ACTIVITY 3.1

Developing Concepts

Modeling One-Step Equations

GROUP ACTIVITY
Work with a partner.

MATERIALS
algebra tiles

▶ QUESTION **How can you use algebra tiles to solve one-step equations?**

▶ EXPLORING THE CONCEPT: SOLVING $x + 5 = -2$

1. Model the equation $x + 5 = -2$. A [+] represents the unknown value x.

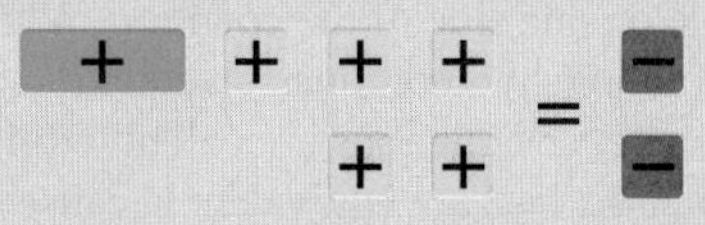

2. To find the value of x, get the x-tile by itself on the left side of the equation. You can undo the addition by subtracting 5. Be sure to subtract the same amount on each side of the equation, so that the two sides stay equal.

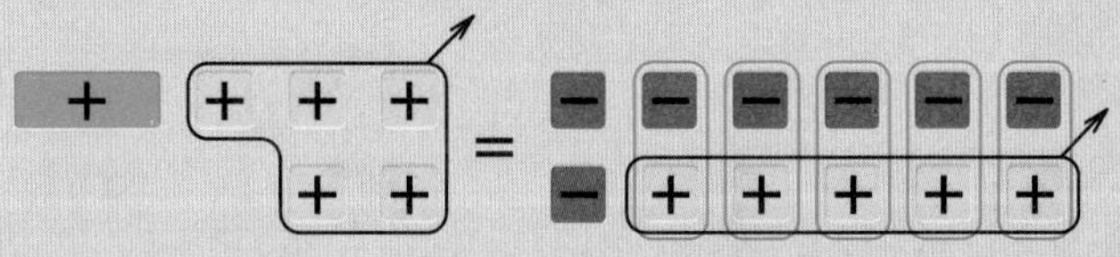

First add 5 zero pairs so you can subtract 5.

STUDENT HELP

Look Back
For help with zero pairs, see p. 78.

3. The resulting equation shows the value of x. So, $x =$ _?_. **−7**

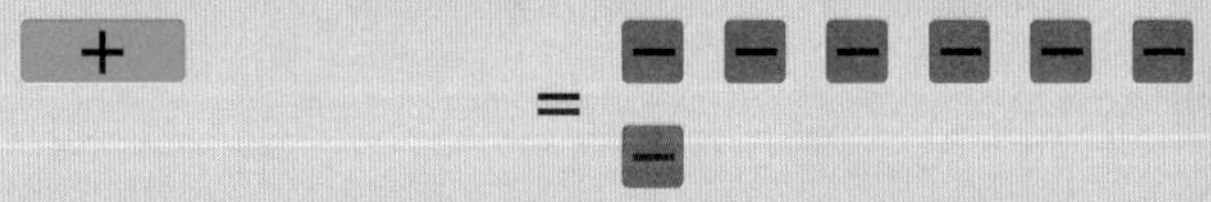

▶ EXPLORING THE CONCEPT: SOLVING $x - 3 = 2$

4. Model the equation $x - 3 = 2$.

You can model the subtraction with −1-tiles because $x - 3 = x + (-3)$.

5. To get the x-tile by itself, add 3 to each side to undo the subtraction.

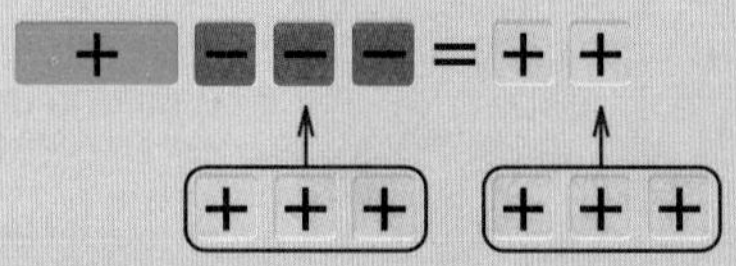

6. Group positive and negative tiles to make zero pairs. What are the remaining tiles on each side of the equation? So, $x =$ _?_. **one x-tile on the left and five +1-tiles on the right; 5**

▶ DRAWING CONCLUSIONS

Use algebra tiles to model and solve the equation. Sketch each step you use.

1. $x + 4 = 6$ **2**　**2.** $x + 5 = 3$ **−2**　**3.** $x + 6 = -1$ **−7**　**4.** $x + 3 = -3$ **−6**

5. $x - 1 = 5$ **6**　**6.** $x - 6 = 2$ **8**　**7.** $x - 7 = -4$ **3**　**8.** $x - 5 = -5$ **0**

9. What operation do you use to solve an addition equation? to solve a subtraction equation? **subtraction; addition**

10. A student solved the equation $x + 3 = -4$ by subtracting 3 on the left side of the equation and got $x = -4$. Is this the correct solution? Explain. **No; you must subtract 3 from each side to get $x = -7$.**

Activity Assessment ***Sample answer:***

$x + 1 = 4$;

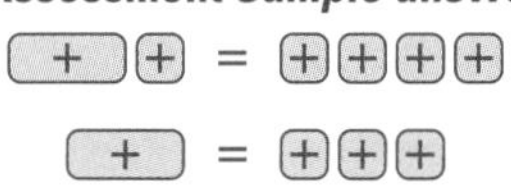

Solution 3

1 Planning the Activity

PURPOSE
To model and solve one-step equations using algebra tiles.

MATERIALS
- algebra tiles
- Activity Support Master (*Chapter 3 Resource Book,* p. 12)

PACING
- Exploring the Concept — 10 min
- Drawing Conclusions — 10 min

▶ LINK TO LESSON
The equation $x - 3 = 2$ used in the Exploring the Concept section is similar to the equation used in Example 1 of Lesson 3.1.

2 Managing the Activity

CLASSROOM MANAGEMENT
When students look back to page 78, you may need to review the meaning of "zero pairs" and how they are used in modeling equations.

ALTERNATIVE APPROACH
Before students begin the activity, model equations such as $x + 3 = 7$ and $x + 2 = 5$ that do not require a zero pair to solve. Then have students use algebra tiles to solve the Exploring the Concept equations.

3 Closing the Activity

★ KEY DISCOVERY
To solve one-step linear equations involving addition, you subtract the same number from both sides of the equation. To solve a one-step linear equation involving subtraction, you add the same number to both sides of the equation.

ACTIVITY ASSESSMENT
JOURNAL Write a one-step equation involving addition or subtraction. Then draw a picture of algebra tiles that models the equation and shows its solution. Then write the solution. **See sample answer at left.**

1 PLAN

PACING
Basic: 1 day
Average: 1 day
Advanced: 1 day
Block Schedule: 0.5 block with 3.2

LESSON OPENER ACTIVITY
An alternative way to approach Lesson 3.1 is to use the Activity Lesson Opener:
- Blackline Master (*Chapter 3 Resource Book*, p. 13)
- Transparency (p. 16)

MEETING INDIVIDUAL NEEDS
- ***Chapter 3 Resource Book***
  Prerequisite Skills Review (p. 5)
  Practice Level A (p. 14)
  Practice Level B (p. 15)
  Practice Level C (p. 16)
  Reteaching with Practice (p. 17)
  Absent Student Catch-Up (p. 19)
  Challenge (p. 21)
- ***Resources in Spanish***
- ***Personal Student Tutor***

NEW-TEACHER SUPPORT
See the Tips for New Teachers on pp. 1–2 of the *Chapter 3 Resource Book* for additional notes about Lesson 3.1.

WARM-UP EXERCISES
Transparency Available
Add or subtract.
1. 7 – 12 **–5**
2. –3 – 4 **–7**
3. –6 + 2 **–4**
4. 9 – (–1) **10**
5. –12 – (–18) **6**

3.1

What you should learn

GOAL 1 Solve linear equations using addition and subtraction.

GOAL 2 Use linear equations to solve **real-life** problems such as finding a record temperature change in **Example 3**.

Why you should learn it

▼ To model **real-life** situations, such as comparing the sizes of city parks in **Exs. 52 and 53.**

REAL LIFE

Solving Equations Using Addition and Subtraction

GOAL 1 ADDITION AND SUBTRACTION EQUATIONS

You can solve an equation by using the *transformations* below and on page 138 to isolate the variable on one side of the equation. When you rewrite an equation using these transformations, you produce an equation with the same solutions as the original equation. These equations are called **equivalent** equations. For example, $x + 3 = 5$ and $x = 2$ are equivalent because $x + 3 = 5$ can be transformed into $x = 2$ and each equation has 2 as its only solution.

To change, or *transform*, an equation into an equivalent equation, think of an equation as having two sides that are "in balance."

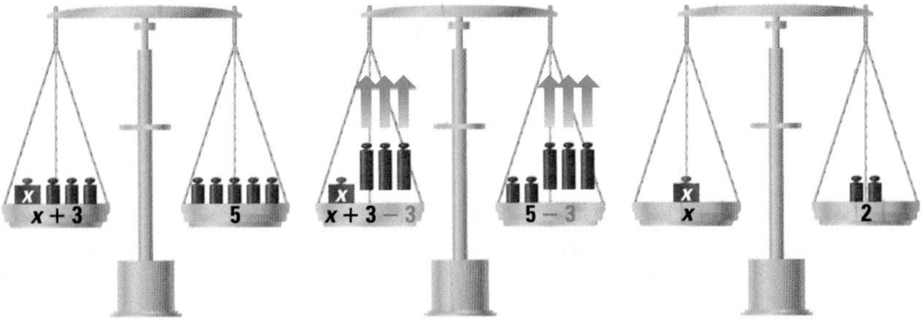

Original equation: $x + 3 = 5$

Subtract 3 from both sides to isolate x on the left. Scale stays in balance.

Simplify both sides. Equivalent equation: $x = 2$

Any transformation you apply to an equation must keep the equation in balance. For instance, if you subtract 3 from one side of the equation, you must also subtract 3 from the other side of the equation. Simplifying one side of the equation does not affect the balance, so you don't have to change the other side.

TRANSFORMATIONS THAT PRODUCE EQUIVALENT EQUATIONS

	ORIGINAL EQUATION		EQUIVALENT EQUATION
• Add the same number to *each* side.	$x - 3 = 5$	Add 3.	$x = 8$
• Subtract the same number from *each* side.	$x + 6 = 10$	Subtract 6.	$x = 4$
• Simplify one or both sides.	$x = 8 - 3$	Simplify.	$x = 5$
• Interchange the sides.	$7 = x$	Interchange.	$x = 7$

Inverse operations are operations that undo each other, such as addition and subtraction. Inverse operations help you to *isolate the variable* in an equation.

STUDENT HELP

Look Back
For help with checking whether a number is a solution, see p. 24.

HOMEWORK HELP
Visit our Web site www.mcdougallittell.com for extra examples.

EXAMPLE 1 *Adding to Each Side*

Solve $x - 5 = -13$.

SOLUTION

On the left side of the equation, 5 is subtracted from x. To isolate x, you need to undo the subtraction by applying the inverse operation of adding 5. Remember, to keep the balance you must add 5 to *each* side.

$x - 5 = -13$	Write original equation.
$x - 5 + 5 = -13 + 5$	Add 5 to each side.
$x = -8$	Simplify.

▶ The solution is -8. Check by substituting -8 for x in the original equation.

✓CHECK

$x - 5 = -13$	Write original equation.
$-8 - 5 \stackrel{?}{=} -13$	Substitute -8 for x.
$-13 = -13$	Solution is correct.

Each time you apply a transformation to an equation, you are writing a **solution step**. Solution steps are written one below the other with the equals signs aligned.

EXAMPLE 2 *Simplifying First*

Solve $-8 = n - (-4)$.

SOLUTION

$-8 = n - (-4)$	Write original equation.
$-8 = n + 4$	Simplify.
$-8 - 4 = n + 4 - 4$	Subtract 4 from each side.
$-12 = n$	Simplify.

STUDENT HELP

Study Tip
After you solve an equation, check by substituting into the *original* equation. To check Example 2, substitute -12 for n in the equation $-8 = n - (-4)$, not in the simplified equation $-8 = n + 4$.

The equations in this chapter are called *linear equations*. In a **linear equation** the variable is raised to the *first* power and does not occur in a denominator, inside a square root symbol, or inside absolute value symbols.

LINEAR EQUATION	NOT A LINEAR EQUATION
$x + 5 = 9$	$x^2 + 5 = 9$
$-4 + n = 2n - 6$	$\lvert x + 3 \rvert = 7$

In Chapter 4 you will see that *linear equations* get their name from the fact that their graphs are straight lines.

2 TEACH

MOTIVATING THE LESSON
You find $10 on the sidewalk and put it in your pocket. Later that day you empty your pocket and find that you have $28. How much did you have before you found the $10? You could find the answer by using the model $x + 10 = 28$ or $28 - 10 = x$.

CONCEPT QUESTION
EXAMPLE 1 Why was 5 instead of 13 added to both sides of the equation to solve for x?
Sample answer: **You are solving for x, so the subtraction on the side of the equation that contains the variable must be undone.**

EXTRA EXAMPLE 1
Solve $x - 9 = -17$. **−8**

EXTRA EXAMPLE 2
Solve $-11 = n - (-2)$ **−13**

CHECKPOINT EXERCISES
For use after Examples 1 and 2:
1. Solve $5 + x = -3$. Check your solution. **−8**

STUDENT HELP NOTES

Homework Help Students can find extra examples at **www.mcdougallittell.com** that parallel the examples in the student edition.

EXTRA EXAMPLE 3

The normal high temperature in January in Bismarck, North Dakota, is 20°F and the normal low temperature is –2°F. How many degrees apart are the normal high and low temperatures?
22°F

EXTRA EXAMPLE 4

Match the real life problem with the equation.

$x + 7 = 9$ $x - 2 = 7$ $9 - x = 7$

a. You have x dollars and your friend repays you the $7 he owes you. You now have $9. How much did you have originally? **$x + 7 = 9$**

b. The temperature was x°F. It fell 2°F and is now 7°F. What was the original temperature? **$x - 2 = 7$**

c. A 9 foot post extends x feet below ground and 7 feet above ground. What is the length x buried below ground? **$9 - x = 7$**

CHECKPOINT EXERCISES

For use after Examples 3 and 4:

1. Your bank balance is $42. If you write a check to buy a pair of shoes, your balance would be –$5. How much do the shoes cost? Does the equation $x + 42 = -5$ model the situation? If not, write an equation that does. Then solve the problem to find the cost of the shoes.
No; $42 - x = -5$; $47

CLOSURE QUESTION

Describe in words how you would solve the equation $x + 6 = -4$ for x using inverse operations.
***Sample answer:* Subtract 6 from both sides of the equation to get $x = -10$.**

DAILY PUZZLER

Suppose you make equations by using 5, 10, 15, and x to fill the boxes in □ – □ = □ + □. What are the greatest and least solutions possible for the equations obtained in this way? **30, –20**

GOAL 2 USING LINEAR EQUATIONS IN REAL-LIFE PROBLEMS

REAL LIFE
Temperature

EXAMPLE 3 *Modeling a Real-Life Problem*

Several record temperature changes have taken place in Spearfish, South Dakota. On January 22, 1943, the temperature in Spearfish fell from 54°F at 9:00 A.M. to −4°F at 9:27 A.M. By how many degrees did the temperature fall?

SOLUTION

PROBLEM SOLVING STRATEGY

VERBAL MODEL

[Temperature at 9:27 A.M.] = [Temperature at 9:00 A.M.] − **Degrees fallen**

LABELS

Temperature at 9:27 A.M.	= **−4**	(degrees Fahrenheit)
Temperature at 9:00 A.M.	= **54**	(degrees Fahrenheit)
Degrees fallen	= ***T***	(degrees Fahrenheit)

ALGEBRAIC MODEL

$\mathbf{-4 = 54 - T}$	**Write algebraic model.**
$-4 - 54 = 54 - T - 54$	**Subtract 54 from each side.**
$-58 = -T$	**Simplify.**
$58 = T$	***T* is the opposite of −58.**

▶ The temperature fell by 58°. Check this in the original statement of the problem.

STUDENT HELP

Study Tip
When you solve an equation, if you get a solution such as $-x = 9$, remember that this is the same as saying "the opposite of x is 9" or "x is the opposite of 9," which you can write as $x = -9$.

EXAMPLE 4 *Translating Verbal Statements*

Match the real-life problem with an equation.

$x - 4 = 16$ $x + 16 = 4$ $16 - x = 4$

a. You owe $16 to your cousin. You paid x dollars back and you now owe $4. How much did you pay back?

b. The temperature was x°F. It rose 16°F and is now 4°F. What was the original temperature?

c. A telephone pole extends 4 feet below ground and 16 feet above ground. What is the total length x of the pole?

SOLUTION

a. Original amount owed (16) − Amount paid back (x) = Amount now owed (4)

b. Original temperature (x) + Degrees risen (16) = Temperature now (4)

c. Total length (x) − Length below ground (4) = Length above ground (16)

Guided Practice

Vocabulary Check ✓

1. Two equations that have the same solutions are called _?_ equations. **equivalent**

Concept Check ✓

2. Show how to check the solution found in Example 3. **Substitute 58 for *T* in the equation: $-4 = 54 - 58$. Since $-4 = -4$, the solution is correct.**
3. Explain each solution step for the equation $-3 - x = 1$. Use the Study Tip and Example 3 on page 134 to help. **Add 3 to each side to get $-x = 4$. Then the opposite of x is 4, so $x = -4$.**

Skill Check ✓

Solve the equation.

4. $r + 3 = 2$ **−1**
5. $9 = x - 4$ **13**
6. $7 + c = -10$ **−17**
7. $-3 = b - 6$ **3**
8. $8 - x = 4$ **4**
9. $r - (-2) = 5$ **3**

SPENDING MONEY **You started with some money in your pocket. All you spent was $4.65 on lunch. You ended up with $7.39 in your pocket.**

10. Write an equation to find how much money you started with. **Let x = the amount of money in dollars; $x - 4.65 = 7.39$.**
11. Describe the transformation you would use to solve the equation. **Add 4.65 to both sides.**
12. Solve the equation. What does the solution mean? **12.04; you started with $12.04 in your pocket.**

Practice and Applications

STUDENT HELP

Extra Practice to help you master skills is on p. 799.

STATING THE INVERSE **State the inverse operation.**

13. Add 28. **Subtract 28.**
14. Add 17. **Subtract 17.**
15. Subtract 15. **Add 15.**
16. Subtract 3. **Add 3.**
17. Add −3. **See margin.**
18. Add −12. **See margin.**
19. Subtract −45. **See margin.**
20. Subtract $-2\frac{1}{2}$. **See margin.**

17. Subtract −3 (or add 3).
18. Subtract −12 (or add 12).
19. Add −45 (or subtract 45).
20. Add $-2\frac{1}{2}$ (or subtract $2\frac{1}{2}$).

SOLVING EQUATIONS **Solve the equation.**

21. $x = 4 - 7$ **−3**
22. $x + 5 = 10$ **5**
23. $t - 2 = 6$ **8**
24. $11 = r - 4$ **15**
25. $-9 = 2 + y$ **−11**
26. $n - 5 = -9$ **−4**
27. $-3 + x = 7$ **10**
28. $\frac{2}{5} = a - \frac{1}{5}$ **$\frac{3}{5}$**
29. $r + 3\frac{1}{4} = 2\frac{1}{2}$ **$-\frac{3}{4}$**
30. $t - (-4) = 4$ **0**
31. $|-6| + y = 11$ **5**
32. $|-8| + x = -3$ **−11**
33. $19 - (-y) = 25$ **6**
34. $|2| - (-b) = 6$ **4**
35. $x + 4 - 3 = 6 \cdot 5$ **29**
36. $-b = 8$ **−8**
37. $12 - 6 = -n$ **−6**
38. $3 - a = 0$ **3**
39. $4 = -b - 12$ **−16**
40. $-3 = -a + (-4)$ **−1**
41. $-r - (-7) = 16$ **−9**

MATCHING AN EQUATION **In Exercises 42–44, match the real-life problem with an equation. Then solve the problem.**

A. $x + 15 = 7$ B. $15 - x = 7$ C. $15 + 7 = x$ D. $x + 15 = -7$

42. You own 15 CDs. You buy 7 more. How many CDs do you own now? **C; 22 CDs**
43. There are 15 members of a high school band brass section. After graduation there are only 7 members. How many members graduated? **B; 8 members**
44. The temperature rose 15 degrees to 7°F. What was the original temperature? **A; −8°F**

STUDENT HELP

HOMEWORK HELP
Example 1: Exs. 13–29
Example 2: Exs. 30–35
Example 3: Exs. 36–49
Example 4: Exs. 42–49

3 APPLY

Assignment Guide

Basic
Day 1: p. 135 Exs. 14–40 even, 42–46, 50, 51, 54, 56–59, 60, 65, 70, 75

Average
Day 1: p. 135 Exs. 14–40 even, 42–48, 50, 51, 54, 56–59, 60, 65, 70, 75

Advanced
Day 1: p. 135 Exs. 14–40 even, 49–55, 56–59, 60, 65, 70, 75

Block Schedule with 3.2
p. 135 Exs. 14–40 even, 42–48, 50, 51, 54, 56–59, 60, 65, 70, 75

Exercise Levels

Level A: *Easier*
13–21

Level B: *More Difficult*
22–48, 50, 51, 54, 56–75

Level C: *Most Difficult*
49, 52, 53, 55

✓ Homework Check

To quickly check student understanding of key concepts, go over the following exercises: Exs. 16, 26, 40, 44, 46, 50. See also the Daily Homework Quiz:

- Blackline Master (*Chapter 3 Resource Book*, p. 24)
- Transparency (p. 20)

English Learners

EXERCISES 45–49 The ability to translate word problems into equations is an essential skill in algebra, but can be extremely difficult for students learning English. When solving word problems, make sure English language learners can identify and understand words and phrases such as *more than, fewer than, added, bonus,* and *per* in Exercises 45–49 successfully.

Career Note

EXERCISES 52 AND 53
Additional information about a career as a park ranger is available at **www.mcdougallittell.com.**

Additional Practice and Reteaching

For Lesson 3.1:

- Practice Levels A, B, and C (*Chapter 3 Resource Book,* p. 14)
- Reteaching with Practice (*Chapter 3 Resource Book,* p. 17)
- See Lesson 3.1 of the *Personal Student Tutor*

For more Mixed Review:

- Search the *Test and Practice Generator* for key words or specific lessons.

SOLVING EQUATIONS **Write and solve an equation to answer the question.**
45–49. Equations may vary.

45. VIDEO PRICES The selling price of a certain video is \$7 more than the price the store paid. If the selling price is \$24, find the price the store paid. **$x + 7 = 24$; \$17**

46. TRACK Jackie Joyner-Kersee won a gold medal in the Olympic Heptathlon in 1988 and in 1992. Her 1992 score was 7044 points. This was 247 fewer points than her 1988 score. What was her 1988 score? **$x - 247 = 7044$; 7291 points**

47. BASEBALL STADIUMS Turner Field in Atlanta, GA, has 49,831 seats. Jacobs Field in Cleveland, OH, has 43,368 seats. How many seats would need to be added to Jacobs Field for it to have as many seats as Turner Field? **$43{,}368 + x = 49{,}831$; 6463 seats**

48. TEST SCORES The last math test that you took had 100 regular points and 10 bonus points. You received a total score of 87, which included 9 bonus points. What would your score have been without any bonus points? **$x + 9 = 87$; 78**

49. COMPUTER TIME The average 12-to-17-year-old spends 645 minutes per month on a personal computer. This is 732 fewer minutes per month than the average 18-to-24-year-old spends. How many minutes per month does the average 18-to-24-year-old spend on a personal computer?
▶ Source: Media Matrix **$x - 732 = 645$; 1377 min**

GEOMETRY CONNECTION **Find the length of the side marked *x*.**

50. The perimeter is 12 feet. **5 ft**

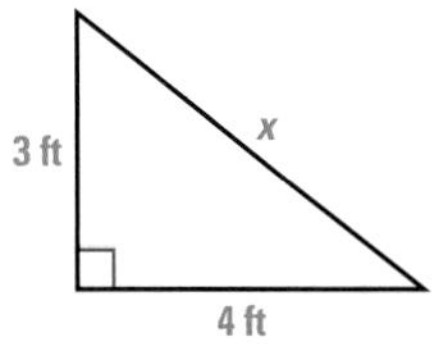

51. The perimeter is 43 centimeters. **20 cm**

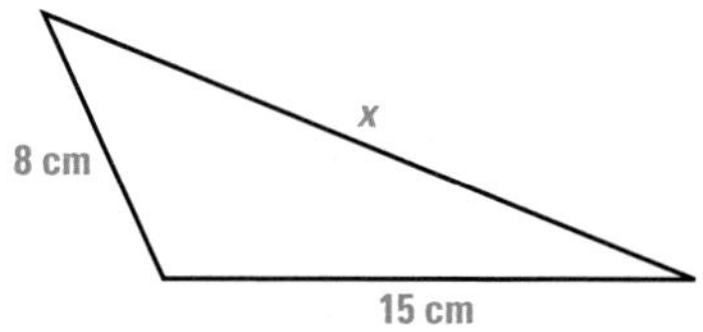

FOCUS ON CAREERS

PARK RANGER Possible duties of a park ranger include gathering and presenting historical or scientific information, managing resources, guiding tours, and maintaining order and safety.

CAREER LINK
www.mcdougallittell.com

CITY PARKS **In Exercises 52 and 53, use the table that shows the sizes (in acres) of the largest city parks in the United States.**

Park (location)	Size (in acres)
Cullen Park (Houston, TX)	?
Fairmount Park (Philadelphia, PA)	8700
Griffith Park (Los Angeles, CA)	4218
Eagle Creek Park (Indianapolis, IN)	?
Pelham Bay Park (Bronx, NY)	2764

▶ Source: The Trust for Public Land

52. You are researching United States city parks. You note that Griffith Park is only 418 acres larger than Eagle Creek Park. Write an equation that models the size of Eagle Creek Park. Solve the equation. ***Sample equation:* $4218 = e + 418$; 3800 acres**

53. You note that the largest city park, Cullen Park, is only 248 acres smaller than the sum of the sizes of Griffith Park, Eagle Creek Park, and Pelham Bay Park. Write an equation that models the size of Cullen Park. Solve the equation. (*Hint:* Use the result of Exercise 52.)
***Sample equation:* $c + 248 = 4218 + 3800 + 2764$; 10,534 acres**

Test Preparation

Sales equation
——
5828 + x = 13,198
13,198 + x = 22,254
22,254 + x = 31,580
31,580 + x = 40,972
40,972 + x = 51,401
51,401 + x = 68,702
68,702 + x = 83,494
83,494 + x = 108,897
108,897 + x = 158,068

54. MULTI-STEP PROBLEM The table shows the number of Digital Versatile Disc (DVD) players sold in the first ten months after their release in 1997.

Month	Cumulative sales	Sales equation	Monthly sales	Monthly sales
March	5,828	——	5,828	——
April	13,198	5,828 + x = 13,198	7,370	——
May	22,254	?	?	9056
June	31,580	?	?	9326
July	40,972	?	?	9392
Aug.	51,401	?	?	10,429
Sept.	68,702	?	?	17,301
Oct.	83,494	?	?	14,792
Nov.	108,897	?	?	25,403
Dec.	158,068	?	?	49,171

▶ Source: INTELECT ASW

a. Copy the table. For each month, write a sales equation relating cumulative and monthly sales. Let x represent the number of players sold that month.

b. Solve your sales equations to fill in the monthly sales column.

c. *Writing* Suppose a DVD player manufacturer started an advertising campaign in September. Use your table to judge the campaign's effect on sales. Write a report explaining whether the campaign was successful. **See margin.**

★ Challenge

55. LOGICAL REASONING You decide to see if you can ride the elevator to street level (Floor 0) without pushing any buttons. The elevator takes you up 4 floors, down 6 floors, up 1 floor, down 8 floors, down 3 floors, up 1 floor, and then down 6 floors to street level. Write and solve an equation to find your starting floor. ***Sample equation***: $x + 4 - 6 + 1 - 8 - 3 + 1 - 6 = 0$; **17**

MIXED REVIEW

54c. Sample answer: It could be argued that the sales campaign was successful, because after an initial decrease in sales, sales increased dramatically. However, it should be noted that advertising alone may not have been responsible. For example, production may have been limited before September, or the distribution system may not have been completely established.

TRANSLATING VERBAL SENTENCES **Write the verbal sentence as an equation.** (Review 1.5 for 3.2)

56. Five times a number is 160. $5n = 160$ **57.** A number divided by 6 is 32. $\frac{n}{6} = 32$

58. One fourth of a number is 36. $\frac{1}{4}n = 36$ **59.** A number multiplied by $\frac{2}{3}$ is 8. $\frac{2}{3}n = 8$

SIMPLIFYING EXPRESSIONS **Simplify the expression.** (Review 2.5 and 2.7 for 3.2)

60. $8\left(\frac{x}{8}\right)$ x **61.** $\frac{1}{8}y \cdot 8$ y **62.** $-\frac{3}{5}\left(-\frac{5}{3}x\right)$ x **63.** $-4x \div (-4)$ x

64. $6\left(-\frac{1}{6}x\right)$ $-x$ **65.** $\frac{7}{12}y \cdot \frac{12}{7}$ y **66.** $\frac{t}{-4}(-4)$ t **67.** $-19a \div 19$ $-a$

DISTRIBUTIVE PROPERTY **Apply the distributive property.** (Review 2.6)

68. $4(x + 2)$ $4x + 8$ **69.** $7(3 - 2y)$ $21 - 14y$ **70.** $-5(-y - 7)$ $5y + 35$ **71.** $(3x + 8)(-2)$ $-6x - 16$

72. $-x(x - 6)$ $-x^2 + 6x$ **73.** $-5x(y + 3)$ $-5xy - 15x$ **74.** $2y(8 - 7y)$ $16y - 14y^2$ **75.** $(-3x - 9y)(-6y)$ $18xy + 54y^2$

4 ASSESS

DAILY HOMEWORK QUIZ

Transparency Available

1. Solve $k + 9 = 21$. **12**
2. Solve $8 = t - (-34)$. **−26**
3. In September, Leonardo spent 780 minutes using the Internet. This was 196 minutes more than in August. Write and solve an equation to find how many minutes he used the Internet in August. $a + 196 = 780$; **584 min**
4. The perimeter of the figure is 52 in. Find the length of the side marked x. **13 in.**

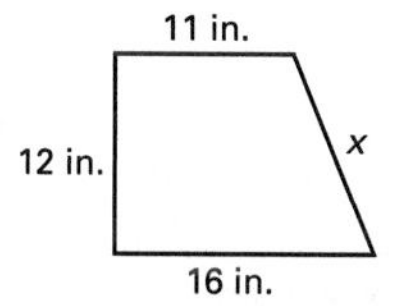

EXTRA CHALLENGE NOTE

Challenge problems for Lesson 3.1 are available in **blackline** format in the *Chapter 3 Resource Book*, p. 21 and at **www.mcdougallittell.com.**

ADDITIONAL TEST PREPARATION

1. **WRITING** Explain why −3 is the only solution to the equation $x + 5 = 2$. ***Sample answer:*** **Because it is the only number that makes the equation true.**

1 PLAN

PACING
Basic: 1 day
Average: 1 day
Advanced: 1 day
Block Schedule: 0.5 block with 3.1

LESSON OPENER ACTIVITY
An alternative way to approach Lesson 3.2 is to use the Activity Lesson Opener:
- Blackline Master (*Chapter 3 Resource Book*, p. 25)
- Transparency (p. 17)

MEETING INDIVIDUAL NEEDS
- ***Chapter 3 Resource Book***
  Prerequisite Skills Review (p. 5)
  Practice Level A (p. 26)
  Practice Level B (p. 27)
  Practice Level C (p. 28)
  Reteaching with Practice (p. 29)
  Absent Student Catch-Up (p. 31)
  Challenge (p. 33)
- ***Resources in Spanish***
- ***Personal Student Tutor***

NEW-TEACHER SUPPORT
See the Tips for New Teachers on pp. 1–2 of the *Chapter 3 Resource Book* for additional notes about Lesson 3.2.

WARM-UP EXERCISES
Transparency Available
Multiply or divide.

1. $\left(\frac{4}{5}\right)\left(\frac{3}{2}\right)$ $\frac{6}{5}$
2. $\frac{1}{2} \cdot 2$ 1
3. $\frac{21}{3}$ 7
4. $\frac{2}{3} \div \frac{-5}{3}$ $-\frac{2}{5}$
5. $\left(-\frac{5}{9}\right)\left(-\frac{5}{9}\right)$ $\frac{25}{81}$

3.2 Solving Equations Using Multiplication and Division

What you should learn

GOAL 1 Solve linear equations using multiplication and division.

GOAL 2 Use multiplication and division equations to solve **real-life** and geometric problems as in **Example 4**.

Why you should learn it

▼ To model and solve **real-life** problems, such as finding how far away you are from a thunderstorm in **Exercise 54**.

REAL LIFE

GOAL 1 MULTIPLICATION AND DIVISION EQUATIONS

In this lesson you will study equations that can be solved by multiplying or dividing each side by the same nonzero number.

Remember, when you solve a linear equation your goal is to isolate the variable on one side of the equation. In the last lesson you used the fact that addition and subtraction are inverse operations to solve linear equations. In this lesson you will use the fact that multiplication and division are inverse operations.

TRANSFORMATIONS THAT PRODUCE EQUIVALENT EQUATIONS

	ORIGINAL EQUATION		EQUIVALENT EQUATION
• Multiply *each* side of the equation by the same nonzero number.	$\frac{x}{2} = 3$	Multiply by 2.	$x = 6$
• Divide *each* side of the equation by the same nonzero number.	$4x = 12$	Divide by 4.	$x = 3$

EXAMPLE 1 *Dividing Each Side of an Equation*

Solve $-4x = 1$.

SOLUTION

On the left side of the equation, x is multiplied by -4. To isolate x, you need to undo the multiplication by applying the inverse operation of dividing by -4.

$-4x = 1$ Write original equation.

$\frac{-4x}{-4} = \frac{1}{-4}$ Divide each side by -4.

$x = -\frac{1}{4}$ Simplify.

▶ The solution is $-\frac{1}{4}$. Check this in the original equation.

✓**CHECK**

$-4x = 1$ Write original equation.

$(-4)\left(-\frac{1}{4}\right) \stackrel{?}{=} 1$ Substitute $-\frac{1}{4}$ for x.

$1 = 1$ Solution is correct.

EXAMPLE 2 Multiplying Each Side of an Equation

Solve $\frac{x}{5} = -30$.

SOLUTION

On the left side of the equation, x is divided by 5. You can isolate x by multiplying each side by 5 to undo the division.

$\frac{x}{5} = -30$ — Write original equation.

$5\left(\frac{x}{5}\right) = 5(-30)$ — Multiply each side by 5.

$x = -150$ — Simplify.

STUDENT HELP

Look Back For help with reciprocals, see page 108.

When you solve an equation with a fractional coefficient, such as $10 = -\frac{2}{3}m$, you can isolate the variable by multiplying by the reciprocal of the fraction.

EXAMPLE 3 Multiplying Each Side by a Reciprocal

Solve $10 = -\frac{2}{3}m$.

SOLUTION

$10 = -\frac{2}{3}m$ — Write original equation.

$\left(-\frac{3}{2}\right)10 = \left(-\frac{3}{2}\right)\left(-\frac{2}{3}m\right)$ — Multiply each side by $-\frac{3}{2}$.

$-15 = m$ — Simplify.

The transformations used to isolate the variable in Lessons 3.1 and 3.2 are based on rules of algebra called **properties of equality**.

CONCEPT SUMMARY — PROPERTIES OF EQUALITY

ADDITION PROPERTY OF EQUALITY	If $a = b$, then $a + c = b + c$.
SUBTRACTION PROPERTY OF EQUALITY	If $a = b$, then $a - c = b - c$.
MULTIPLICATION PROPERTY OF EQUALITY	If $a = b$, then $ca = cb$.
DIVISION PROPERTY OF EQUALITY	If $a = b$ and $c \neq 0$, then $\frac{a}{c} = \frac{b}{c}$.

You have been using these properties to keep equations in balance as you solve them. For instance, in Example 2 you used the multiplication property of equality when you multiplied each side by 5.

2 TEACH

EXTRA EXAMPLE 1

Solve $-3x = 2$. $-\frac{2}{3}$

EXTRA EXAMPLE 2

Solve $\frac{x}{9} = -25$. -225

EXTRA EXAMPLE 3

Solve $14 = -\frac{7}{8}m$. -16

CHECKPOINT EXERCISES

For use after Example 1:

1. Solve $-22 = -2x$. **11**

For use after Examples 2 and 3:

2. Solve the equation $\frac{1}{7}x = -\frac{3}{8}$. $-\frac{21}{8}$

STUDENT HELP NOTES

Homework Help Students can find extra examples at **www.mcdougallittell.com** that parallel the examples in the student edition.

Look Back When working with reciprocals, remind students that to multiply fractions they should multiply numerators and multiply denominators. To divide fractions, they should multiply the first fraction by the reciprocal of the second fraction.

MATHEMATICAL REASONING

Remind students that dividing by a number is equivalent to multiplying by its reciprocal. This means that the methods in Example 2 and 3 are essentially the same.

ENGLISH LEARNERS

Preteach the concept of *similar triangles* by discussing the non-mathematical meaning of the word *similar* ("sharing many qualities, but not exactly the same"). Understanding other uses for mathematical terms can often help English language learners remember their mathematical uses.

EXTRA EXAMPLE 4

a. A strobe light emits 14 flashes per second. Find the total number of flashes it emits in a 5-min period. **4200 flashes**

b. A time-lapse camera photographed a growing plant at the rate of 25 frames per hour over several days. For how many days was the plant photographed if there were a total of 17,400 frames? **29 days**

EXTRA EXAMPLE 5

The two triangles are similar. What is the length of side $\overline{RS}$? **3 cm**

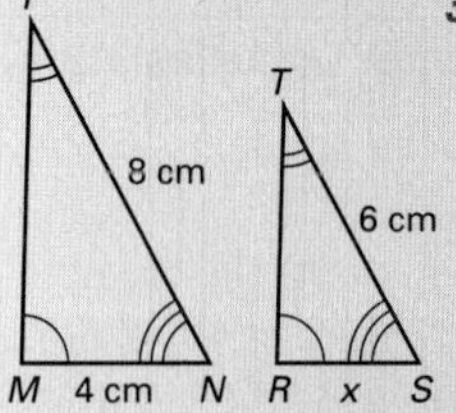

CHECKPOINT EXERCISES

For use after Examples 4 and 5:

1. For the triangles in Extra Example 5, find the length of $\overline{PM}$ given that the length of $\overline{TR}$ is 5 cm. **$6\frac{2}{3}$ cm**

FOCUS ON VOCABULARY

When discussing the properties of equality, emphasize to students that the word *equality* is related to the word *equal* and that all the properties of equality involve an *equal* sign.

CLOSURE QUESTION

How can you tell if you need to use multiplication to solve an equation? if you need to use division?

You solve an equation using multiplication if the variable is being divided by a nonzero integer or multiplied by a fraction. You solve an equation using division if the variable is being multiplied by a nonzero integer.

FOCUS ON APPLICATIONS

REAL LIFE **MOVIE FRAMES** A motion picture camera takes a series of separate pictures as frames. These are projected in rapid sequence when the movie is shown.

GOAL 2 SOLVING REAL-LIFE PROBLEMS

EXAMPLE 4 *Modeling a Real-Life Problem*

RESTORING MOVIES A single picture on a roll of movie film is called a frame. Motion picture studios try to save some older films from decay by cleaning and restoring the film frame by frame. This process is very expensive and time-consuming.

a. The usual rate for taking and projecting professional movies is 24 frames per second. Find the total number of frames in a movie that is 90 minutes long.

b. If a worker can restore 8 frames per hour, how many hours of work are needed to restore all of the frames in a 90-minute movie?

SOLUTION

a. Let x = the total number of frames in the movie. To find the total number of seconds in the movie, multiply $90 \cdot 60$ because each minute is 60 seconds.

$$\frac{\text{Total number of frames in the movie}}{\text{Total number of seconds in the movie}} = \text{Number of frames per second}$$

$$\frac{x}{5400} = 24$$

▶ The solution is $x = 129{,}600$, so a 90-minute movie has 129,600 frames.

b. Let y = the number of hours of work and use the result from part (a).

$$\text{Number of frames restored per hour} \cdot \text{Number of hours of work} = \text{Total number of frames in the movie}$$

$$8 \cdot y = 129{,}600$$

▶ The solution is $y = 16{,}200$, so 16,200 hours of work are needed.

STUDENT HELP

Skills Review
For help with ratios and rates, see p. 787.

You can model some real-life situations with an equation that sets two *ratios* equal. If a and b are two quantities measured in the *same* units, then the **ratio of a to b** is $\frac{a}{b}$.

Two triangles are **similar triangles** if they have equal corresponding angles. It can be shown that this is equivalent to the ratios of the lengths of corresponding sides being equal. Corresponding angles are marked with the same symbol.

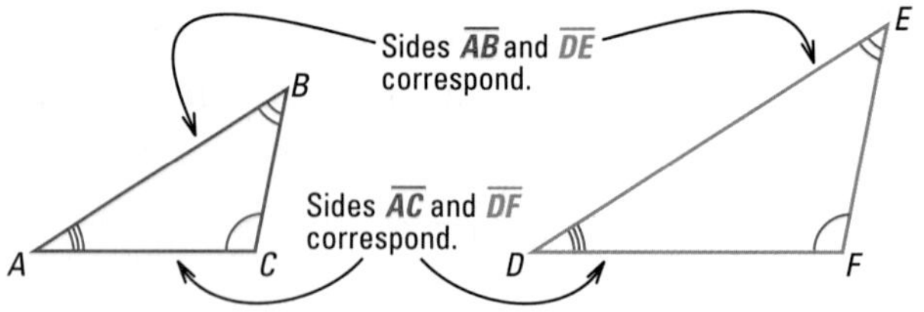

The ratio $\frac{\text{length of } \overline{AB}}{\text{length of } \overline{DE}}$ is equal to the ratio $\frac{\text{length of } \overline{AC}}{\text{length of } \overline{DF}}$.

EXAMPLE 5 *Solving Problems with Similar Triangles*

GEOMETRY CONNECTION The two triangles are similar. What is the length of side $\overline{AB}$?

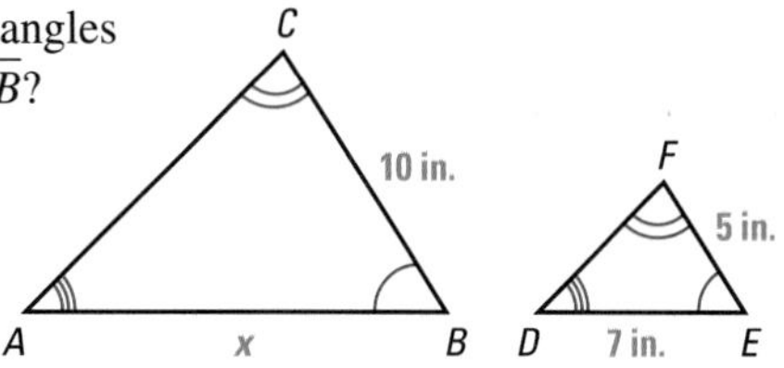

SOLUTION

Set up equal ratios to find the unknown side length.

PROBLEM SOLVING STRATEGY

VERBAL MODEL $\dfrac{\text{Length of } \overline{AB}}{\text{Length of } \overline{DE}} = \dfrac{\text{Length of } \overline{BC}}{\text{Length of } \overline{EF}}$

LABELS
Length of $\overline{AB}$ = x (inches)
Length of $\overline{DE}$ = 7 (inches)
Length of $\overline{BC}$ = 10 (inches)
Length of $\overline{EF}$ = 5 (inches)

ALGEBRAIC MODEL

$\dfrac{x}{7} = \dfrac{10}{5}$ Write algebraic model.

$7\left(\dfrac{x}{7}\right) = 7\left(\dfrac{10}{5}\right)$ Multiply each side by 7.

$x = 14$ Simplify.

▶ The length of $\overline{AB}$ is 14 inches.

GUIDED PRACTICE

Vocabulary Check ✓

1. Name two pairs of inverse operations.
addition and subtraction, multiplication and division

Concept Check ✓

2. Describe six ways to transform an equation into an equivalent equation.
See margin.
3. What is the first step you would use to solve each equation in Exercises 4–7?
See margin.

Skill Check ✓

Solve the equation.

4. $6x = 18$ **3**
5. $\frac{y}{4} = 8$ **32**
6. $\frac{r}{-5} = 20$ **−100**
7. $\frac{5}{6}a = -10$ **−12**
8. $-7b = -4$ $\mathbf{\frac{4}{7}}$
9. $-3x = 5$ $\mathbf{-\frac{5}{3}}$
10. $-\frac{3}{8}t = -6$ **16**
11. $\frac{1}{7}x = \frac{5}{7}$ **5**
12. **CAR TRIPS** Write and solve an equation to find your average speed on a trip from St. Louis to Dallas. You drove 630 miles in $10\frac{1}{2}$ hours.
See margin.
13. The two triangles are similar triangles. Write and solve an equation to find the unknown side length. ***Sample equation:*** $\mathbf{\frac{x}{3} = \frac{4}{6}; 2}$

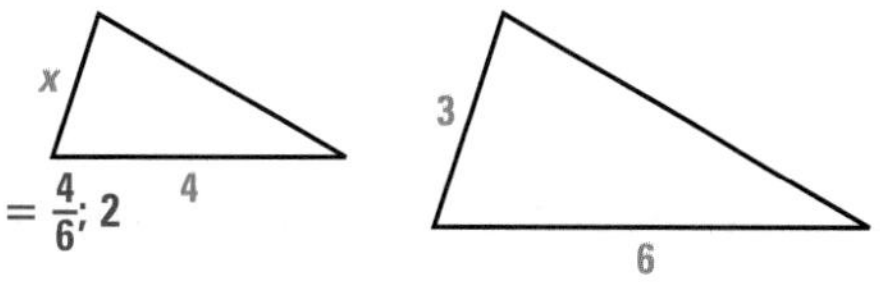

2. Add the same number to each side, subtract the same number from each side, simplify one or both sides, interchange the two sides, multiply each side by the same nonzero number, divide each side by the same nonzero number.

3. *Sample answers:* (4) Divide each side by 6; (5) Multiply each side by 4; (6) Multiply each side by −5; (7) Multiply each side by $\frac{6}{5}$.

12. *Sample equation:* $10\frac{1}{2}x = 630$; 60 mi/h

3 APPLY

ASSIGNMENT GUIDE

BASIC
Day 1: p. 142 Exs. 14–52 even, 58, 59, 62–64, 66–69, 70, 75, 80, 82

AVERAGE
Day 1: p. 142 Exs. 14–52 even, 58, 59, 62–64, 66–69, 70, 75, 80, 82

ADVANCED
Day 1: p. 142 Exs. 14–52 even, 54, 58–61, 62–65, 66–69, 70, 75, 80, 82

BLOCK SCHEDULE WITH 3.1
p. 142 Exs. 14–52 even, 58, 59, 62–64, 66–69, 70, 75, 80, 82

EXERCISE LEVELS

Level A: ***Easier***
14–23

Level B: ***More Difficult***
24–53, 58–64, 66–82

Level C: ***Most Difficult***
54–57, 65

✓ HOMEWORK CHECK

To quickly check student understanding of key concepts, go over the following exercises: Exs. 16, 22, 28, 46, 50, 58. See also the Daily Homework Quiz:

- Blackline Master (*Chapter 3 Resource Book*, p. 36)
- Transparency (p. 21)

! COMMON ERROR
EXERCISES 45–47 Some students may multiply or divide as the first step to solving these equations. Although this method will work, remind students that it is easier first to simplify the right side by taking the absolute value and then to multiply or divide.

PRACTICE AND APPLICATIONS

STUDENT HELP
Extra Practice to help you master skills is on p. 799.

STATING INVERSES State the inverse operation.

14. Divide by 6. **Multiply by 6.**
15. Multiply by -2. **Divide by -2.**
16. Divide by -4. **Multiply by -4.**
17. Multiply by $\frac{2}{3}$. **See margin.**
18. Multiply by $-\frac{9}{4}$. **See margin.**
19. Divide by $-\frac{4}{3}$. **See margin.**

17. Divide by $\frac{2}{3}$ $\left(\text{or multiply by } \frac{3}{2}\right)$.
18. Divide by $-\frac{9}{4}$ $\left(\text{or multiply by } -\frac{4}{9}\right)$.
19. Multiply by $-\frac{4}{3}$.

EQUIVALENT EQUATIONS Tell whether the equations are equivalent.

20. $-4x = 44$ and $x = 11$ **no**
21. $21x = 7$ and $x = 3$ **no**
22. $\frac{x}{10} = -4$ and $x = -40$ **yes**
23. $\frac{2}{3}x = 24$ and $x = 16$ **no**

SOLVING EQUATIONS Solve the equation.

24. $10x = 110$ **11**
25. $-21m = 42$ **-2**
26. $18 = -2a$ **-9**
27. $30b = 5$ **$\frac{1}{6}$**
28. $-4n = -24$ **6**
29. $288 = 16t$ **18**
30. $7r = -56$ **-8**
31. $8x = 3$ **$\frac{3}{8}$**
32. $-10x = -9$ **$\frac{9}{10}$**
33. $\frac{y}{7} = 12$ **84**
34. $\frac{z}{2} = -5$ **-10**
35. $\frac{1}{2}x = -20$ **-40**
36. $\frac{1}{3}y = 82$ **246**
37. $\frac{m}{-4} = -\frac{3}{4}$ **3**
38. $0 = \frac{4}{5}d$ **0**
39. $-\frac{4}{5}x = 72$ **-90**
40. $-\frac{1}{5}y = -6$ **30**
41. $\frac{t}{-2} = \frac{1}{2}$ **-1**
42. $-\frac{2}{3}t = -16$ **24**
43. $\frac{3}{4}z = -5\frac{1}{2}$ **$-7\frac{1}{3}$**
44. $\frac{1}{3}y = 5\frac{2}{3}$ **17**
45. $\frac{3}{4}t = |-15|$ **20**
46. $-\frac{1}{2}b = -|-8|$ **16**
47. $-6y = -|27|$ **$4\frac{1}{2}$**

FOCUS ON APPLICATIONS

NEWSPAPERS Each week the average household reads five pounds of newspapers.
▶ Source: Newspaper Association of America

48. **BUNDLING NEWSPAPERS** You are loading a large pile of newspapers onto a truck. You divide the pile into four equal-size bundles. You find that one bundle weighs 37 pounds. You want to find the weight x of the original pile. Which equation models the situation? Solve the correct equation. **A; 148 lb**

A. $\frac{x}{4} = 37$ **B.** $4x = 37$

MODELING REAL-LIFE PROBLEMS In Exercises 49–53, write and solve an equation to answer the question. **49–53. Equations may vary.**

49. Each household in the United States receives about 676 pieces of junk mail per year. About how many pieces does a household receive per week? **$52x = 676$; 13 pieces**

50. About one eighth of the population is left-handed. In what size school would you expect to find about 50 left-handed students? **$\frac{x}{8} = 50$; 400 students**

51. You ate three of the eight pizza slices and you paid \$3.30 as your share of the cost. How much did the whole pizza cost? **$\frac{3}{8}x = 3.3$; \$8.80**

52. It takes 45 peanuts to make one ounce of peanut butter. How many peanuts will be needed to make a 12-ounce jar of peanut butter? **$\frac{x}{12} = 45$; 540 peanuts**

53. A 10,000-square-foot pizza was created on October 11, 1987. This pie was eaten by about 30,000 people. On average, how much did each person eat? **$30{,}000x = 10{,}000$; $\frac{1}{3}$ ft^2**

STUDENT HELP
HOMEWORK HELP
Example 1: Exs. 14–47
Example 2: Exs. 14–47
Example 3: Exs. 14–47
Example 4: Exs. 48–54
Example 5: Exs. 58–59

FOCUS ON APPLICATIONS

REAL LIFE **THUNDERSTORMS** You see a flash of lightning instantly since light travels so rapidly (186,000 miles/second). You hear the thunder later since sound takes about 5 seconds to travel a mile near the ground.

54. THUNDERSTORMS You can tell how many miles you are from a thunderstorm by counting the seconds between seeing the lightning and hearing the thunder, and then dividing by five. How many seconds would you count for a thunderstorm nine miles away? **45 sec**

55. CRITICAL THINKING Look back at Example 4. For each situation, write a different model than the one shown. The solutions should stay the same. **See margin.**

56. GARDENS A homeowner is installing a fence around the garden at the right. The garden has a perimeter of 216 feet. Write and solve an equation to find the garden's dimensions.

Sample equation: **$10x = 216$; 43.2 ft by 64.8 ft**

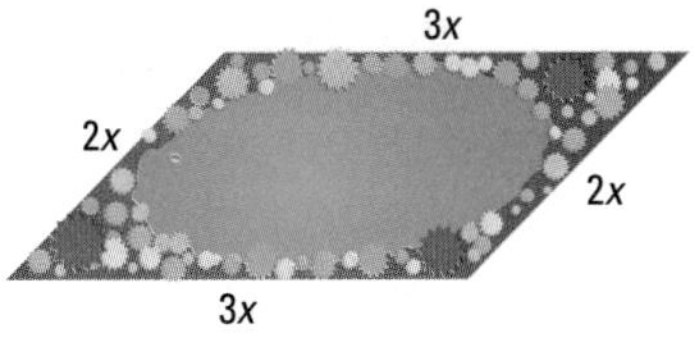

57. BALD EAGLES On page 129 you learned that bald eagles fly up to 30 miles per hour and dive at speeds up to 100 miles per hour. Using this information, write and solve an equation to answer each question.

a. What is the least amount of time that an eagle could take to fly 6 miles? **12 min**

b. An eagle a mile above the water spots a fish. What is the shortest time it would take the eagle to dive for the fish? Express your answer in seconds. **36 sec**

GEOMETRY CONNECTION **In Exercises 58 and 59, the two triangles are similar. Write and solve an equation to find the length of the side marked *x*.**

58.

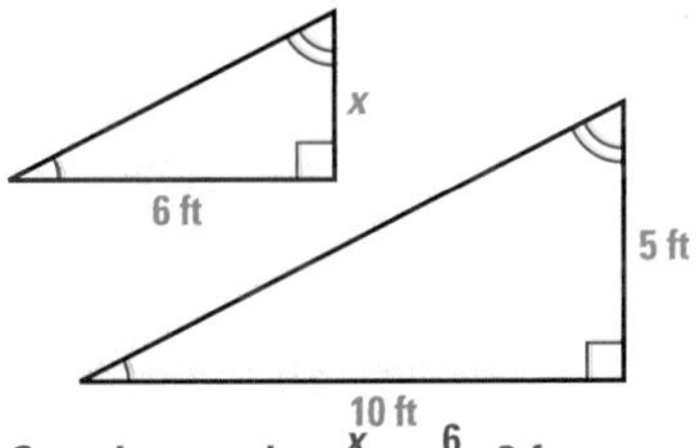

Sample equation: $\frac{x}{5} = \frac{6}{10}$; **3 ft**

59.

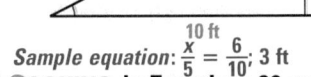

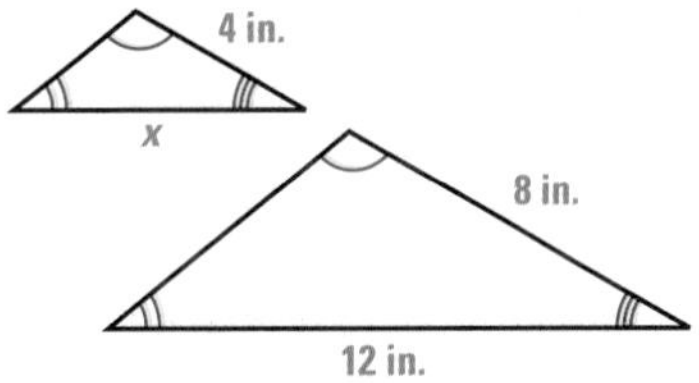

Sample equation: $\frac{x}{12} = \frac{4}{8}$; **6 in.**

COOKING **In Exercises 60 and 61, use the recipe shown below.**

60. You have only 3 cups of rice, so you decrease the recipe. To find the amount of each ingredient, you can write an equation that sets two ratios equal. For the rice you can use the ratio $\frac{3 \text{ cups}}{4 \text{ cups}} = \frac{3}{4}$ that compares the reduced amount to the original amount.

Choose and solve the equation you can use to find the number of teaspoons of soy sauce.

A. $\frac{3}{4} = \frac{2}{x}$ **B.** $\frac{3}{4} = \frac{x}{2}$ **B; $1\frac{1}{2}$ tsp**

FRIED RICE

- 1 tablespoon vegetable oil
- 1/2 cup chopped bamboo shoots
- 1/2 cup chopped mushrooms
- 1/4 cup chopped scallions
- 4 cups cooked rice
- 1 cup cooked diced chicken
- 1/2 cup chicken broth
- 2 teaspoons soy sauce

61. You have 5 cups of rice, so you increase the recipe. Write and solve an equation to find the amount of chicken.

Sample equation: $\frac{5}{4} = \frac{x}{1}$; **$1\frac{1}{4}$ c**

STUDENT HELP

INTERNET **HOMEWORK HELP** Visit our Web site www.mcdougallittell.com for help with Exs. 60 and 61.

! *COMMON ERROR*

EXERCISES 58–59 Some students may not set up the ratios correctly when they write equations for these exercises. You may find it useful to help students in identifying corresponding sides of similar triangles. Stress that it is important to compare the length of a side in the first triangle to the length of its corresponding side in the second triangle.

STUDENT HELP NOTES

Homework Help Students can find help for Exs. 60 and 61 at **www.mcdougallittell.com.** The information can be printed out for students who don't have access to the Internet.

55.a. Let x = the number of frames in the movie; total number of frames in the movie = number of frames per sec × total number of seconds in the movie; $x = 24(5400)$; 129,600 frames.

55b. Let x = the number of hours; number of hours = $\frac{\text{total number of frames restored}}{\text{number of frames restored in an hour}}$, $x = \frac{129{,}600}{8}$; 16,200 h

ADDITIONAL PRACTICE AND RETEACHING

For Lesson 3.2:

- Practice Levels A, B, and C (*Chapter 3 Resource Book,* p. 26)
- Reteaching with Practice (*Chapter 3 Resource Book,* p. 29)
- See Lesson 3.2 of the *Personal Student Tutor*

For more Mixed Review:

- Search the *Test and Practice Generator* for key words or specific lessons.

4 ASSESS

DAILY HOMEWORK QUIZ

Transparency Available

1. What is the inverse operation for "multiply by -6"? **divide by -6**
2. Are the equations $-\frac{3}{5}x = 24$ and $x = 40$ equivalent? **no**
3. Solve $14m = -42$. **-3**
4. Solve $\frac{8}{5}k = |-12|$. **$7\frac{1}{2}$**
5. About $\frac{3}{40}$ of the business calls that Ms. Rodriguez makes are long distance calls. How many business calls would you expect her to make in a month when she makes 24 long distance calls? Write and solve an equation to answer the question. **$\frac{3}{40}c = 24$; 320 business calls**

EXTRA CHALLENGE NOTE

Challenge problems for Lesson 3.2 are available in **blackline** format in the *Chapter 3 Resource Book,* p. 33 and at **www.mcdougallittell.com.**

ADDITIONAL TEST PREPARATION

1. **OPEN ENDED** Describe a real-life situation where you could use a division equation to model and solve a problem involving similar rectangles. ***Sample answer:* A photo that is 3 in. wide and 5 in. long is enlarged to be 8 in. wide. How long is the photo?**

Test Preparation

62. MULTIPLE CHOICE What is the first step you would use to solve $\frac{1}{4} = -7x$?
D

Ⓐ Divide by 4. Ⓑ Multiply by 4.

Ⓒ Multiply by -7. Ⓓ Divide by -7.

63. MULTIPLE CHOICE Solve $-\frac{5}{7}x = -2$.
A

Ⓐ $\frac{14}{5}$ Ⓑ $-\frac{14}{5}$ Ⓒ $\frac{10}{7}$ Ⓓ $\frac{7}{5}$

64. MULTIPLE CHOICE Which of the triangles below are similar triangles?
B

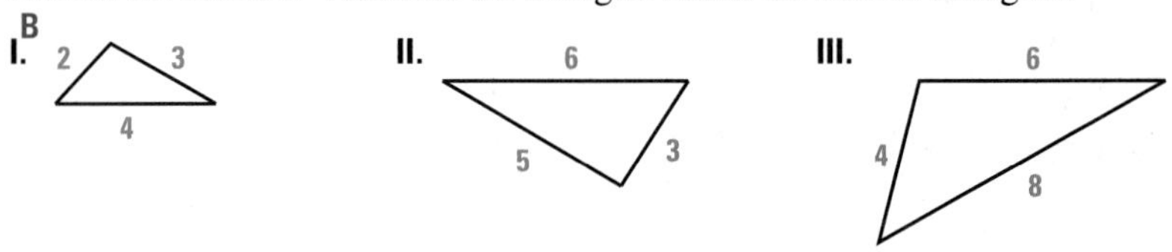

Ⓐ I and II Ⓑ I and III Ⓒ II and III Ⓓ None

★ Challenge

65. ATTENDANCE In 1997, the average home attendance at New York Yankees' baseball games was a little more than 150% of the average home attendance at Anaheim Angels' baseball games. The average home attendance at New York Yankees' baseball games was about 33,000. Write and solve an equation to estimate the average home attendance at Anaheim Angels' baseball games. ▶ Source: *ESPN 1998 Information Please® Sports Almanac*

***Sample equation:* $\frac{150}{100}x = 33{,}000$; 22,000**

EXTRA CHALLENGE
www.mcdougallittell.com

MIXED REVIEW

TRANSLATING VERBAL SENTENCES Write the verbal sentence as an equation. (Review 1.5 for 3.3)

66. The sum of 18 and five times a number is 108. **$18 + 5n = 108$**

67. Twelve less than nine times a number is 60. **$9n - 12 = 60$**

68. Five more than two thirds of a number is 11. **$\frac{2}{3}n + 5 = 11$**

69. Eleven is two fifths of the quantity n decreased by thirteen. **$11 = \frac{2}{5}(n - 13)$**

SIMPLIFYING EXPRESSIONS Simplify the expression. (Review 2.6 for 3.3)

70. $15 - 8x + 12$ **$27 - 8x$** **71.** $4y - 9 + 3y$ **$7y - 9$** **72.** $(x + 8)(-2) - 36$ **$-2x - 52$**

73. $5(y + 3) + 7y$ **$12y + 15$** **74.** $12x - (x - 2)(2)$ **$10x + 4$** **75.** $-25y - 6(-y - 9)$ **$-19y + 54$**

SOLVING EQUATIONS Solve the equation. (Review 3.1)

76. $4 + y = 12$ **8** **77.** $t - 2 = 1$ **3** **78.** $5 - (-t) = 14$ **9**

79. $x - 2 = 28$ **30** **80.** $19 - x = 37$ **-18** **81.** $-9 - (-a) = -2$ **7**

82. PHOTOGRAPHY You take 24 pictures. Seven of the pictures cannot be developed because of bad lighting. Let x represent the number of pictures that can be developed successfully. Which of the following is a correct model for the situation? Solve the correct equation. (Review 3.1) **A; 17 pictures**

A. $x + 7 = 24$ **B.** $x - 7 = 24$

3.3 Solving Multi-Step Equations

GOAL 1 Using Two or More Transformations

What you should learn

GOAL 1 Use two or more transformations to solve an equation.

GOAL 2 Use multi-step equations to solve **real-life** problems, such as finding the temperature inside the Earth in **Example 7**.

Why you should learn it

▼ To model and solve **real-life** problems, such as the fountain problem in **Exercise 58**.

REAL LIFE

Solving a linear equation may require two or more transformations. Here are some guidelines.

- Simplify one or both sides of the equation (if needed).
- Use inverse operations to isolate the variable.

EXAMPLE 1 *Solving a Linear Equation*

Solve $\frac{1}{3}x + 6 = -8$.

SOLUTION To isolate the variable, undo the addition and then the multiplication.

$\frac{1}{3}x + 6 = -8$	Write original equation.
$\frac{1}{3}x + 6 - 6 = -8 - 6$	Subtract 6 from each side.
$\frac{1}{3}x = -14$	Simplify.
$3\left(\frac{1}{3}x\right) = 3(-14)$	Multiply each side by 3.
$x = -42$	Simplify.

EXAMPLE 2 *Combining Like Terms First*

Solve $7x - 3x - 8 = 24$.

SOLUTION

$7x - 3x - 8 = 24$	Write original equation.
$4x - 8 = 24$	Combine like terms.
$4x - 8 + 8 = 24 + 8$	Add 8 to each side.
$4x = 32$	Simplify.
$\frac{4x}{4} = \frac{32}{4}$	Divide each side by 4.
$x = 8$	Simplify.

▶ The solution is 8. When you check the solution, substitute 8 for *each x* in the original equation.

1 PLAN

PACING
Basic: 2 days
Average: 2 days
Advanced: 2 days
Block Schedule: 1 block

LESSON OPENER APPLICATION
An alternative way to approach Lesson 3.3 is to use the Application Lesson Opener:
- Blackline Master (*Chapter 3 Resource Book*, p. 37)
- Transparency (p. 18)

MEETING INDIVIDUAL NEEDS
- ***Chapter 3 Resource Book***
  Prerequisite Skills Review (p. 5)
  Practice Level A (p. 40)
  Practice Level B (p. 41)
  Practice Level C (p. 42)
  Reteaching with Practice (p. 43)
  Absent Student Catch-Up (p. 45)
  Challenge (p. 48)
- ***Resources in Spanish***
- ***Personal Student Tutor***

NEW-TEACHER SUPPORT
See the Tips for New Teachers on pp. 1–2 of the *Chapter 3 Resource Book* for additional notes about Lesson 3.3.

WARM-UP EXERCISES
Transparency Available
Simplify.
1. $2(3 - 9x)$ $6 - 18x$
2. $3x - 7 - 8x$ $-5x - 7$
3. $-7(2x - 11)$ $-14x + 77$
4. $7x - 5x + 3$ $2x + 3$
5. $6x - 2(4x - 1)$ $-2x + 2$

2 TEACH

MOTIVATING THE LESSON
A student has a summer job that pays $6.50 an hour. He wants to save half his earnings minus the $10 he spends each week on transportation. His goal is to save $55 a week. Today's lesson focuses on equations that can help us find how many hours he must work in a week to meet his goal.

EXTRA EXAMPLE 1
Solve $\frac{1}{2}x - 5 = 10$. 30

EXTRA EXAMPLE 2
Solve $2x - 9x + 17 = -4$ 3

CHECKPOINT EXERCISES
For use after Examples 1 and 2:
1. Solve $\frac{2}{5}x - \frac{1}{5}x + 9 = -1$. −50

STUDENT HELP NOTES
Look Back As students look back to page 100, remind them that the value outside the parentheses must be multiplied by each term inside the parentheses. Also, as in Example 4, remind students to watch for a subtraction sign to the left of the value outside the parentheses. This affects what value is used to multiply each term inside the parentheses.

EXTRA EXAMPLE 3
Solve $4x + 12(x - 3) = 28$. 4

EXTRA EXAMPLE 4
Solve $2x - 5(x - 9) = 27$. 6

EXTRA EXAMPLE 5
Solve $12 = \frac{3}{10}(x + 2)$. 38

CHECKPOINT EXERCISES
For use after Examples 3 and 4:
1. Solve $10x - 6(2x + 5) = 20$. −25
For use after Example 5:
2. Solve $-24 = \frac{4}{3}(x - 7)$. −11

STUDENT HELP
Look Back
For help with the distributive property and combining like terms, see pp. 100–102.

EXAMPLE 3 *Using the Distributive Property*

Solve $5x + 3(x + 4) = 28$.

SOLUTION

Method 1 Show All Steps

$$5x + 3(x + 4) = 28$$
$$5x + 3x + 12 = 28$$
$$8x + 12 = 28$$
$$8x + 12 - 12 = 28 - 12$$
$$8x = 16$$
$$\frac{8x}{8} = \frac{16}{8}$$
$$x = 2$$

Method 2 Do Some Steps Mentally

$$5x + 3(x + 4) = 28$$
$$5x + 3x + 12 = 28$$
$$8x + 12 = 28$$
$$8x = 16$$
$$x = 2$$

EXAMPLE 4 *Distributing a Negative*

Solve $4x - 3(x - 2) = 21$.

SOLUTION

$4x - 3(x - 2) = 21$	Write original equation.
$4x - 3x + 6 = 21$	Use distributive property.
$x + 6 = 21$	Combine like terms.
$x = 15$	Subtract 6 from each side.

EXAMPLE 5 *Multiplying by a Reciprocal First*

Solve $66 = -\frac{6}{5}(x + 3)$.

SOLUTION

It is easier to solve this equation if you don't distribute $-\frac{6}{5}$.

$66 = -\frac{6}{5}(x + 3)$	Write original equation.
$\left(-\frac{5}{6}\right)(66) = \left(-\frac{5}{6}\right)\left(-\frac{6}{5}\right)(x + 3)$	Multiply by reciprocal of $-\frac{6}{5}$.
$-55 = x + 3$	Simplify.
$-58 = x$	Subtract 3 from each side.

STUDENT HELP
Study Tip
Before you apply the distributive property, examine the equation. Although you can distribute in Example 5, multiplying by the reciprocal is easier, because it clears the equation of fractions.

Solving equations systematically is an example of deductive reasoning. Notice how each solution step is based on number properties or properties of equality.

FOCUS ON CAREERS

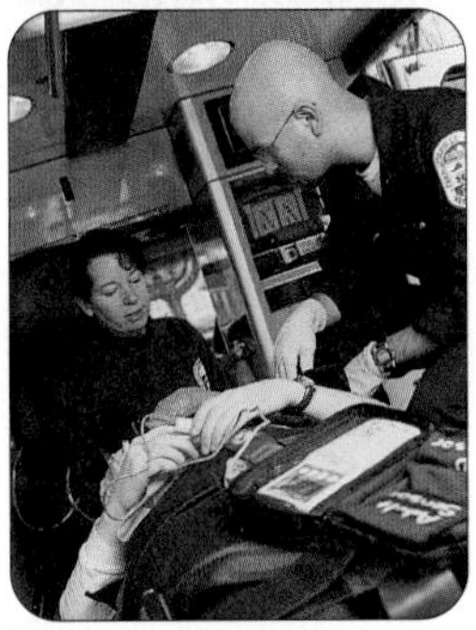

EMERGENCY MEDICAL TECHNICIANS
EMTs quickly evaluate and respond to a patient's condition. For instance, quick treatment of hypothermia may prevent heart and breathing rates from becoming dangerously low.

CAREER LINK
www.mcdougallittell.com

GOAL 2 SOLVING REAL-LIFE PROBLEMS

EXAMPLE 6 *Using a Known Formula*

CHECKING VITAL SIGNS A body temperature of 95°F or lower may indicate the medical condition called hypothermia. What temperature in the Celsius scale may indicate hypothermia?

The Fahrenheit and Celsius scales are related by the equation $F = \frac{9}{5}C + 32$.

SOLUTION Use the formula to convert 95°F to Celsius.

$F = \frac{9}{5}C + 32$	Write known formula.
$95 = \frac{9}{5}C + 32$	Substitute 95° for F.
$63 = \frac{9}{5}C$	Subtract 32 from each side.
$35 = C$	Multiply by $\frac{5}{9}$, the reciprocal of $\frac{9}{5}$.

▶ A temperature of 35°C or lower may indicate hypothermia.

EXAMPLE 7 *Using a Verbal Model*

SCIENCE CONNECTION The temperature within Earth's crust increases about 30° Celsius for each kilometer of depth beneath the surface. If the temperature at Earth's surface is 24°C, at what depth would you expect the temperature to be 114°C?

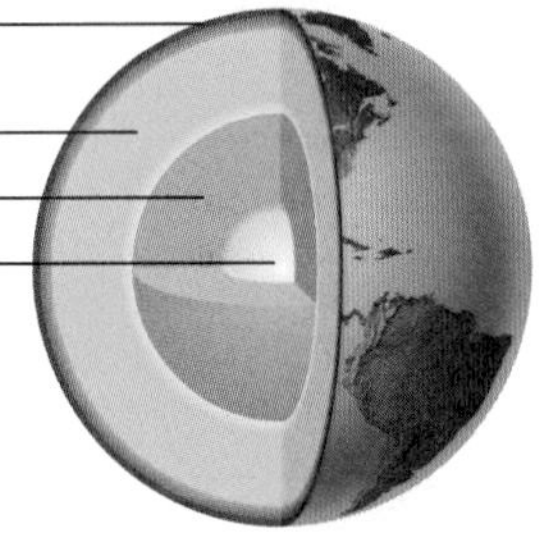

SOLUTION

PROBLEM SOLVING STRATEGY

VERBAL MODEL

[Temperature inside Earth] = [Temperature at Earth's surface] + [Rate of temperature increase] · [Depth below surface]

LABELS

Temperature inside Earth = **114**	(degrees Celsius)
Temperature at Earth's surface = **24**	(degrees Celsius)
Rate of temperature increase = **30**	(degrees Celsius per kilometer)
Depth below surface = d	(kilometers)

ALGEBRAIC MODEL

$114 = 24 + 30 \cdot d$

▶ The solution is $d = 3$. The temperature will be 114°C at 3 kilometers deep.

EXTRA EXAMPLE 6
What Celsius temperature would indicate a fever of 104°F? **40°C**

EXTRA EXAMPLE 7
Air temperature drops about 3°F for each 1000 ft increase in altitude. If the air temperature at sea level is 77°F, at what altitude would you expect the temperature to be 53°F? **8000 ft**

CHECKPOINT EXERCISES

For use after Example 6:

1. Use $F = \frac{9}{5}C + 32$ to convert 212°F to Celsius. **100°C**

For use after Example 7:

2. The speed of a falling object increases 32 ft/sec for each second it falls. How many seconds will it take the speed of a rock dropped from a very high cliff to increase from 50 ft/sec to 210 ft/sec? **5 sec**

CAREER NOTE
EXAMPLE 6 Additional information about Emergency Medical Technicians is available at **www.mcdougallittell.com.**

FOCUS ON VOCABULARY
What is a multi-step equation?
an equation that requires two or more transformations to solve

CLOSURE QUESTION
Describe in words how to solve a multi-step linear equation.
First simplify one or both sides of the equation if needed. Then use inverse operations to isolate the variable.

DAILY PUZZLER
One third of a stack of paper and a 1-pound book will balance one fourth of the stack of paper and 3 of the books. How much does half the stack of paper weigh? **12 lb**

3 APPLY

ASSIGNMENT GUIDE

BASIC

Day 1: p. 148 Exs. 10–42 even

Day 2: p. 148 Exs. 44, 46–54, 56, 61–63, 65, 70, 75, 82–85, Quiz 1 Exs. 1–11

AVERAGE

Day 1: p. 148 Exs. 10–42 even

Day 2: p. 148 Exs. 44, 46–54, 56, 58, 61–65, 70, 75, 82–85, Quiz 1 Exs. 1–11

ADVANCED

Day 1: p. 148 Exs. 10–42 even

Day 2: p. 148 Exs. 44, 46–54, 56, 58, 59–66, 70, 75, 82–85, Quiz 1 Exs. 1–11

BLOCK SCHEDULE

p. 148 Exs. 10–42 even, 44, 46–54, 56, 58, 61–65, 70, 75, 82–85, Quiz 1 Exs. 1–11

EXERCISE LEVELS

Level A: *Easier*
10–15, 67–72

Level B: *More Difficult*
16–40, 44, 50–58, 61, 65, 73–85

Level C: *Most Difficult*
41–43, 45–49, 59, 60, 62–64, 66

✔ HOMEWORK CHECK

To quickly check student understanding of key concepts, go over the following exercises: Exs. 12, 20, 28, 38, 50, 52, 54. See also the Daily Homework Quiz:

- Blackline Master (*Chapter 3 Resource Book*, p. 52)
- Transparency (p. 22)

GUIDED PRACTICE

Concept Check ✔

1. Subtract 3 from each side; multiply each side by 2; divide each side by 5.
2. The distributive property could be used first in any of Exs. 6–8. However, it is simpler in Ex. 6 to divide each side by 5 first and in Ex. 7 to multiply each side by $\frac{4}{3}$.

1. LOGICAL REASONING Copy the solution steps shown. Then, in the right-hand column, describe the transformation that was used in each step. (Lists of types of transformations are shown on pages 132 and 138.) See margin.

Solution Steps	Explanation
$\frac{5x}{2} + 3 = 6$	Original Equation
$\frac{5x}{2} = 3$	?
$5x = 6$	?
$x = \frac{6}{5}$	?

2. For which equations in Exercises 6–8 would you use the distributive property first when solving? Explain. See margin.

Skill Check ✔

In Exercises 3–8, solve the equation. Show how to check your solution.

3. $4x + 3 = 11$ 2
4. $\frac{1}{2}x - 9 = 11$ 40
5. $3x - x + 15 = 41$ 13

6. $5(x - 7) = 90$ 25
7. $\frac{3}{4}(x + 6) = 12$ 10
8. $6x - 4(-3x + 2) = 10$ 1

9. SUMMER JOB You have a summer job running errands for a local business. You earn \$5 per day, plus \$2 for each errand. Write and solve an equation to find how many errands you need to run to earn \$17 in one day. 6 errands

PRACTICE AND APPLICATIONS

STUDENT HELP
Extra Practice to help you master skills is on p. 799.

CHECKING SOLUTIONS **Check whether the given number is a solution of the equation.**

10. $9x - 5x - 19 = 21; -10$ no
11. $\frac{3}{4}x + 1 = -8; -12$ yes

12. $6x - 4(9 - x) = 106; 7$ no
13. $7x - 15 = -1; -2$ no

14. $\frac{1}{2}x - 7 = -4; 6$ yes
15. $\frac{x}{4} - 7 = 13; 24$ no

SOLVING EQUATIONS **Solve the equation.**

16. $2x + 7 = 15$ 4
17. $3x - 1 = 8$ 3
18. $\frac{x}{3} - 5 = -1$ 12

19. $\frac{x}{2} + 13 = 20$ 14
20. $30 = 16 + \frac{1}{5}x$ 70
21. $6 = 14 - 2x$ 4

22. $7 + \frac{2}{3}x = -1$ −12
23. $3 - \frac{3}{4}x = -6$ 12
24. $22 = 18 - \frac{1}{4}x$ −16

25. $8x - 3x = 10$ 2
26. $-7x + 4x = 9$ −3
27. $x + 5x - 5 = 1$ 1

28. $3x - 7 + x = 5$ 3
29. $3(x - 2) = 18$ 8
30. $12(2 - x) = 6$ $1\frac{1}{2}$

31. $\frac{9}{2}(x + 3) = 27$ 3
32. $-\frac{4}{9}(2x - 4) = 48$ −52
33. $17 = 2(3x + 1) - x$ 3

34. $\frac{4x}{3} + 3 = 23$ 15
35. $-10 = 4 - \frac{7x}{4}$ 8
36. $-10 = \frac{1}{2}x + x$ $-6\frac{2}{3}$

STUDENT HELP
HOMEWORK HELP
Example 1: Exs. 16–36
Example 2: Exs. 10–40
Example 3: Exs. 16–40
Example 4: Exs. 37–40
Example 5: Exs. 16–36
Example 6: Exs. 57–58
Example 7: Exs. 51–56

SOLVING EQUATIONS Solve the equation.

37. $5m - (4m - 1) = -12$ **−13**

38. $55x - 3(9x + 12) = -64$ **−1**

39. $22x + 2(3x + 5) = 66$ **2**

40. $9x - 5(3x - 12) = 30$ **5**

ERROR ANALYSIS In Exercises 41–43, find and correct the error. **See margin.**

41.
$2(x - 3) = 5$
$2x - 3 = 5$
$2x = 8$
$x = 4$

42.
$5 - 3x = 10$
$2x = 10$
$x = 5$

43.
$\frac{1}{4}x - 2 = 7$
$x - 2 = 28$
$x = 30$

44. LOGICAL REASONING Solve the equation $-6x + 3(4x - 1) = 9$. Organize your work into two columns. In the left-hand column show the solution steps. In the right-hand column explain the transformation you used in each step. **See margin.**

45. *Writing* Solve the equation $\frac{1}{9}x + 1 = 4$ by following each sequence of steps described in parts (a) and (b). Which method do you prefer? Explain. **Preferences may vary.**

a. Multiply and then subtract. **$x + 9 = 36$; 27**

b. Subtract and then multiply. **$\frac{1}{9}x = 3$; 27**

CHOOSING A METHOD In Exercises 46–49, solve the equation two different ways as in Exercise 45. Explain which method you prefer and why. **Preferences may vary.**

46. $\frac{1}{8}x - 5 = 3$ **64**

47. $\frac{x}{3} + 6 = -2$ **−24**

48. $-4x - 2 = 4$ **$-1\frac{1}{2}$**

49. $\frac{2}{3}x + 1 = \frac{1}{3}$ **−1**

50. CONSECUTIVE INTEGERS *Consecutive integers* are integers that follow each other in order (for example, 5, 6, and 7). You want to find three consecutive integers whose sum is 84.

a. Why does the equation $n + (n + 1) + (n + 2) = 84$ model the situation? **If n is an integer, the next two consecutive integers are $n + 1$ and $n + 2$.**

b. Solve the equation in part (a). Then find the three consecutive integers. **27, 28, and 29**

51. FIND THE NUMBERS The sum of three numbers is 123. The second number is 9 less than two times the first number. The third number is 6 more than three times the first number. Find the three numbers. **21, 33, and 69**

52. FARMING PROJECT You have a 90-pound calf you are raising for a 4-H project. You expect the calf to gain 65 pounds per month. In how many months will the animal weigh 1000 pounds? **14 months**

53. HOURS OF LABOR The bill (parts and labor) for the repair of a car was \$458. The cost of parts was \$339. The cost of labor was \$34 per hour. Write and solve an equation to find the number of hours of labor. ***Sample equation:* $34x + 339 = 458$; 3.5 h**

54. GEOMETRY CONNECTION In any triangle, the sum of the measures of the angles is 180°. In triangle ABC, $\angle A$ is four times as large as $\angle B$. Angle C measures 20° less than $\angle B$. Find the measure of each angle. **See margin.**

55. STUDENT THEATER Your school's drama club charges \$4 per student for admission to *Our Town*. The club borrowed \$400 from parents to pay for costumes, props, and the set. After paying back the parents, the drama club has \$100. How many students attended the play? **125 students**

41. The distributive property was applied incorrectly; $2(x - 3) = 2x - 6$; $5\frac{1}{2}$.

42. The left side was simplified incorrectly; $5 - 3x \neq 2x$; $-1\frac{2}{3}$.

43. The left side was simplified incorrectly; $4\left(\frac{1}{4}x - 2\right) = x - 8$; 36.

44. *Sample answer:*
$-6x + 3(4x - 1) = 9$
Write the original equation.
$-6x + 12x - 3 = 9$
Use the distributive property.
$6x - 3 = 9$; Simplify.
$6x = 12$; Add 3 to each side.
$x = 2$; Divide each side by 6.

54. $m\angle A = 133\frac{1}{3}°$, $m\angle B = 33\frac{1}{3}°$, $m\angle C = 13\frac{1}{3}°$

FOCUS ON APPLICATIONS

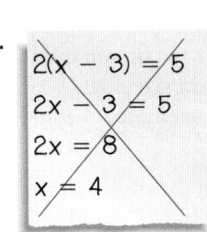

STUDENT THEATER Shown above is one of the original posters for the play *Our Town.* This play has been the second most produced high school play in the last sixty years.

▶ Source: Educational Theatre Association

! COMMON ERROR

EXERCISES 41–43 These exercises demonstrate three common errors that students make in solving equations. All students should understand what these errors are so that they can avoid making them as they solve equations.

TEACHING TIP

EXERCISE 50 Some students may have trouble understanding the concept of consecutive integers. To help students see that n, $n + 1$, and $n + 2$ represent three consecutive integers, you may wish to use the example $5, 5 + 1 = 6, 5 + 2 = 7$.

STUDENT HELP NOTES

Homework Help Students can find help for Exs. 59–64 at **www.mcdougallittell.com.** The information can be printed out for students who don't have access to the Internet.

TEACHING TIP

EXERCISES 62–64 Some students may have trouble setting up equations to solve coin problems. The most common mistake is not multiplying the value of the coin times the number of coins of that type. Remind students that if they have *x* nickels, *y* dimes, and *z* quarters, the respective values are $5x$, $10y$, and $25z$.

62. There are twice as many nickels as quarters, and each nickel has a value of 5 cents; $5(2q) = 10q$

ADDITIONAL PRACTICE AND RETEACHING

For Lesson 3.3:

- Practice Levels A, B, and C (*Chapter 3 Resource Book,* p. 40)
- Reteaching with Practice (*Chapter 3 Resource Book,* p. 43)
- See Lesson 3.3 of the *Personal Student Tutor*

For more Mixed Review:

- Search the *Test and Practice Generator* for key words or specific lessons.

56. HOLIDAY PAY You earn \$9 per hour. On major holidays, such as Thanksgiving, you earn twice as much per hour. You earned a total of \$405 for the week including Thanksgiving. Write and solve an equation to find how many hours you worked on Thanksgiving if you worked 35 hours during the rest of the week. ***Sample equation:*** **$18x + 315 = 405$; 5 h**

57. 48 pounds per square inch

57. FIREFIGHTING The formula $d = \frac{n}{2} + 26$ relates nozzle pressure n (in pounds per square inch) and the maximum distance the water reaches d (in feet) for a fire hose with a certain size nozzle. How much pressure is needed if such a hose is held 50 feet from a fire? ▶ Source: *Fire Department Hydraulics*

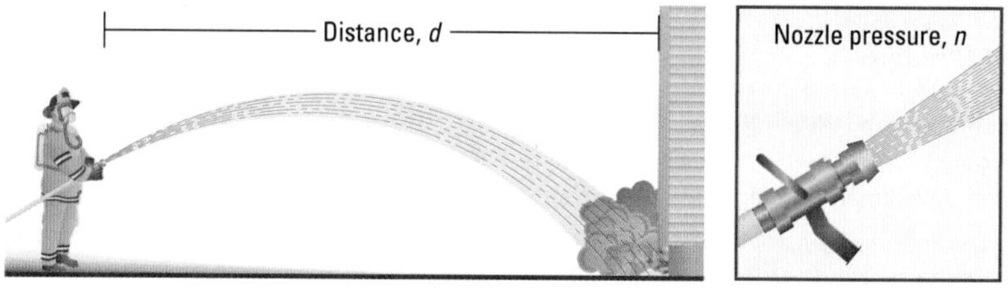

Ex. 58

58. HEIGHT OF A FOUNTAIN Neglecting air resistance, the upward velocity of the water in the stream of a particular fountain is given by the formula $v = -32t + 28$, where t is the number of seconds after the water leaves the fountain. While going upward, the water slows down until, at the top of the stream, the water has a velocity of 0 feet per second. How long does it take a droplet of water to reach the maximum height? **0.875 sec**

59. PHOTOCOPYING An office needs 20 copies of a 420-page report. One photocopier can copy 1800 pages per hour. Another copier is faster and can copy 2400 pages per hour. To find the time it would take the two copiers together to complete the project, you can use this verbal model. Assign labels to the model to form an equation. Then solve the equation. **See margin.**

Rate of first machine	•	Time to complete	+	Rate of second machine	•	Time to complete	=	Total pages

60. CRITICAL THINKING Explain how you can use unit analysis to check that the verbal model in Exercise 59 is correct. **See margin.**

59. Let t = time to complete; $1800t + 2400t = 8400$; 2 h

60. Examine the units in the equation:

$$\frac{\text{pages}}{\text{hour}} \cdot \text{hours} + \frac{\text{pages}}{\text{hour}} \cdot \text{hours} = \text{pages}.$$

61. DATA ENTRY A publishing company needs to enter 910 pages it received from an author into its word processing system, so the writing can be edited and formatted. One person can type 15 pages per hour. Another person can type 20 pages per hour. Write and solve an equation to find how long it will take the two people working together to enter all of the pages. ***Sample equation:*** **$15x + 20x = 910$; 26 h**

REPRESENTING COIN PROBLEMS **In Exercises 62 and 63, use the following information. A person has quarters, dimes, and nickels with a total value of 500 cents (\$5.00). The number of nickels is twice the number of quarters. The number of dimes is four less than the number of quarters.**

62. Explain why the expression $5(2q)$ represents the value of the nickels if q represents the number of quarters. How can you simplify the expression? **See margin.**

63. Write and solve an equation to find the number of each type of coin. ***Sample equation:*** **$25q + 10(q - 4) + 5(2q) = 500$; 12 quarters, 24 nickels, and 8 dimes**

64. COIN PROBLEM There are 4 times as many nickels as dimes in a coin bank. The coins have a total value of 600 cents (\$6.00). Find the number of nickels. **80 nickels**

STUDENT HELP

HOMEWORK HELP Visit our Web site www.mcdougallittell.com for help with work problems and coin problems in Exs. 62–64.

Test Preparation

65. d. 1 represents the entire job, so, letting t = the time it takes to do the whole job, the fraction of the work Luis can do in that time $\left(\frac{2}{7}t\right)$ plus the fraction of the work Mei can do $\left(\frac{3}{14}t\right)$ equals the whole job.

65. MULTI-STEP PROBLEM Two student volunteers are stuffing envelopes for a local food pantry. The mailing will be sent to 560 possible contributors. Luis can stuff 160 envelopes per hour and Mei can stuff 120 envelopes per hour.

a. Working alone, what fraction of the job can Luis complete in one hour? in t hours? Write the fraction in lowest terms. $\frac{2}{7}$; $\frac{2}{7}t$

b. Working alone, what fraction of the job can Mei complete in t hours? $\frac{3}{14}t$

c. Write an expression for the fraction of the job that Luis and Mei can complete in t hours if they work together. $\frac{2}{7}t + \frac{3}{14}t$

d. CRITICAL THINKING To find how long it will take Luis and Mei to complete the job if they work together, you can set the expression you wrote in part (c) equal to 1 and solve for t. Explain why this will work. **See margin.**

e. How long will it take Luis and Mei to complete the job if they work together? Check your solution. **2 h**

★ Challenge

66. ODD INTEGERS Consecutive odd integers are odd integers listed in order such as 5, 7, and 9. Find three consecutive odd integers whose sum is 111. Are there three consecutive odd integers whose sum is 1111? Explain. **35, 37, and 39; no, if x = the smallest of the three odd integers, an equation to solve is $3x + 6 = 1111$, which has no integer solution.**

MIXED REVIEW

WRITING POWERS **Write the expression in exponential form.** (Review 1.2)

67. four squared 4^2 **68.** b cubed b^3 **69.** $(a)(a)(a)(a)(a)(a)$ a^6

70. 10 10^1 **71.** $2 \cdot 2 \cdot 2 \cdot 2$ 2^4 **72.** $3x \cdot 3x \cdot 3x \cdot 3x \cdot 3x$ $(3x)^5$

EVALUATING EXPRESSIONS **Evaluate the expression.** (Review 1.3, 2.7)

73. $5 + 8 - 3$ **10** **74.** $3^2 \cdot 4 + 8$ **44** **75.** $16.9 - 1.5(1.8 + 0.2)$ **13.9**

76. $5 \cdot (12 - 4) + 7$ **47** **77.** $-6 \div 3 - 4 \cdot 5$ **−22** **78.** $10 - [4.3 + 2(6.4 \div 8)]$ **4.1**

79. $\frac{-5 \cdot 4}{3 - 7^2 + 6}$ $\frac{1}{2}$ **80.** $2 - 8 \div \frac{-2}{3}$ **14** **81.** $\frac{(3-6)^2 + 6}{-5}$ **−3**

82. BICYCLING If you ride a bicycle 5 miles per hour, how many miles will you ride in 45 minutes? (Review 1.1) $3\frac{3}{4}$ **mi**

PROFIT **In Exercises 83–85, use the following information. You open a snack stand at a fair. The income and expenses (in dollars) for selling each type of food are shown in the matrices.** (Review 2.4)

Day 1

	Income	Expenses
Hamburgers	72	14
Hot dogs	85	18
Tacos	46	19

Day 2

	Income	Expenses
Hamburgers	62	10
Hot dogs	52	11
Tacos	72	26

83. What were your total income and expenses for selling each type of food for the two days of the fair? **hamburgers: $134 income, $24 expense; hot dogs: $137 income, $29 expense; tacos: $118 income, $45 expense**

84. Which type of food had the largest profit? **hamburgers**

85. Which type of food had the smallest profit? **tacos**

4 ASSESS

DAILY HOMEWORK QUIZ

Transparency Available

1. Solve $\frac{5}{3}y - 7 = 53$. **36**
2. Solve $-12x + 8 + 5x = 14$. $-\frac{6}{7}$
3. Solve $46n - 2(3n - 15) = 60$. $\frac{3}{4}$
4. Victor has $60 in savings. He plans to save $45 a month from money he earns baby sitting. In how many months will his savings be $600? **12 months**
5. The longest sides of a rectangle are 3 in. less than six times the length of the shorter sides. The perimeter of the rectangle is 50 in. Find the measures of the sides of the rectangle. **4 in. and 21 in.**

EXTRA CHALLENGE NOTE

Challenge problems for Lesson 3.3 are available in **blackline** format in the *Chapter 3 Resource Book,* p. 48 and at **www.mcdougallittell.com.**

ADDITIONAL TEST PREPARATION

1. **WRITING** Explain how to solve the equation $2x - 5(x + 4) = 22$. **Distribute −5 to get $2x - 5x - 20 = 22$. Simplify $2x - 5x$ to get $-3x$, and add 20 to both sides of the equation to get $-3x = 42$. Finally, divide both sides by −3 to get $x = -14$. The solution is −14.**

ADDITIONAL RESOURCES
An alternative Quiz for Lessons 3.1–3.3 is available in the *Chapter 3 Resource Book,* p. 49.
A **blackline** master with additional Math & History exercises is available in the *Chapter 3 Resource Book,* p. 47.

QUIZ 1

Self-Test for Lessons 3.1–3.3

Solve the equation. **(Lessons 3.1, 3.2, and 3.3)**

1. $8 - y = -9$ **17**

2. $x + \frac{1}{2} = 5$ $4\frac{1}{2}$

3. $|-14| + z = 12$ **−2**

4. $8b = 5$ $\frac{5}{8}$

5. $\frac{3}{4}q = 24$ **32**

6. $\frac{n}{-8} = -\frac{3}{8}$ **3**

7. $\frac{x}{5} + 10 = \frac{4}{5}$ **−46**

8. $\frac{1}{4}(y + 8) = 5$ **12**

9. $25x - 4(4x + 6) = -69$ **−5**

10. HISTORY TEST You take a history test that has 100 regular points and 8 bonus points. You receive a total score of 91, which includes 4 bonus points. Write and solve an equation to find the score you would have had without the bonus points. **(Lesson 3.1)** ***Sample equation:*** $x + 4 = 91$; **87**

11. TEMPERATURE The temperature is 72°F. What is the temperature in degrees Celsius? $\left(\text{Use the equation } F = \frac{9}{5}C + 32.\right)$ **(Lesson 3.3)** $22\frac{2}{9}$°C

MATH & History — History of Stairs

APPLICATION LINK www.mcdougallittell.com

THEN

IN **1675,** François Blondel, director of France's Royal Academy of Architecture, decided upon a formula for building stairs to accommodate the human gait. He came up with $2R + T = 24$ inches (25.5 in today's inches) where R = riser height and T = tread width.

NOW

IN THE 1990S, a group of building code officials adopted the "7-11" rule which states that a riser should be no taller than 7 inches and no shorter than 4 inches, with a minimum tread width of 11 inches.

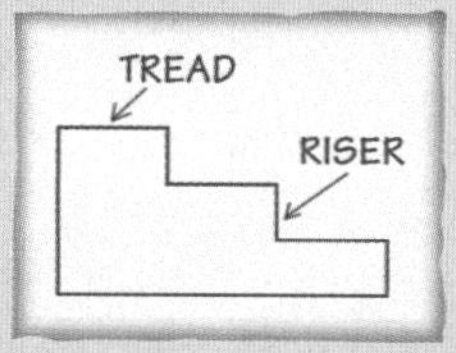

In Exercises 1 and 2 you will compare a traditional staircase with an 8.25-inch riser and a 9-inch tread and a staircase built using the 7-11 rule that has a 7-inch riser and an 11-inch tread. Each of the staircases has 12 steps.

1. Sketch and label an individual stair for each staircase. **Check sketches.**
2. Sketch and label each staircase. Which staircase is steeper? Which design will require more stairs if both need to reach the same height? **Check sketches; the traditional staircase; the staircase built using the 7-11 rule.**

900–1200
The pyramid *El Castillo* in Chichen Itza, Mexico

First escalators installed
1900

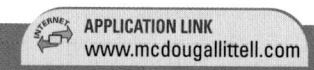

Now
Modern escalator in New York City

ACTIVITY 3.4

Developing Concepts

Modeling Equations with Variables on Both Sides

GROUP ACTIVITY
Work with a partner.

MATERIALS
algebra tiles

QUESTION How can you use algebra tiles to solve an equation with a variable on both the left side and the right side of the equation?

EXPLORING THE CONCEPT

1. Use algebra tiles to model the equation $4x + 5 = 2x + 9$.

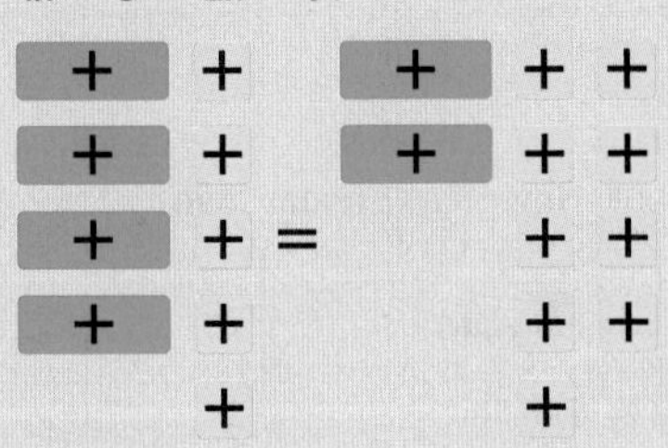

2. You want to have x-tiles on only one side of the equation, so subtract 2 x-tiles from each side. Write the new equation. **$2x + 5 = 9$**

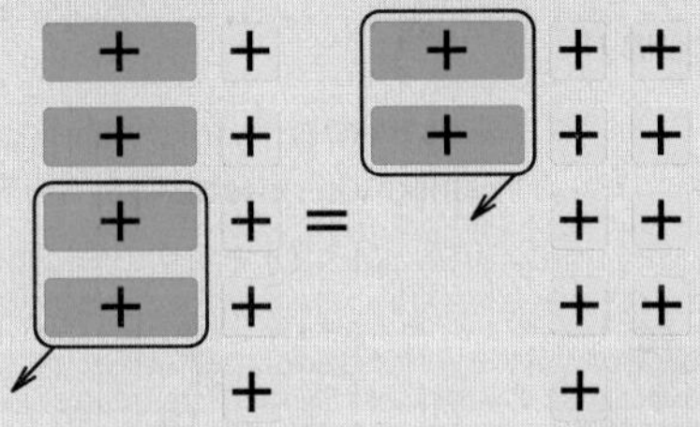

3. To isolate the x-tiles, subtract five 1-tiles from each side. Write the new equation. **$2x = 4$**

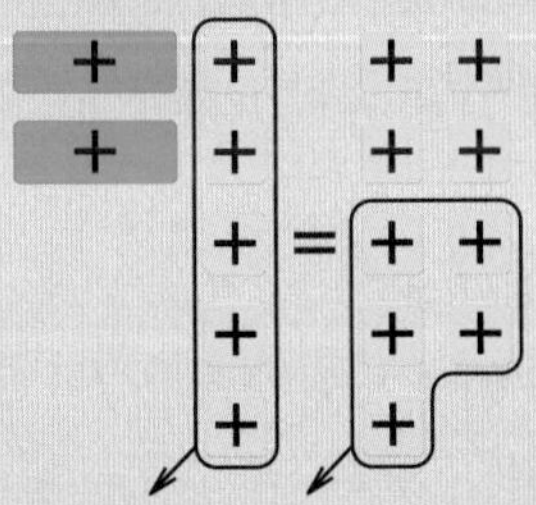

4. You know the value of $2x$. To find the value of x, split the tiles on each side of the equation in half to get x = __?__. **2**

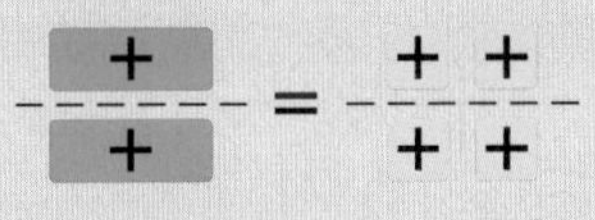

DRAWING CONCLUSIONS

Use algebra tiles to model and solve the equation.

1. $4x + 4 = 3x + 7$ **3**
2. $2x + 3 = 6 + x$ **3**
3. $6x + 5 = 3x + 14$ **3**
4. $5x + 2 = 10 + x$ **2**
5. $8x + 3 = 7x + 3$ **0**
6. $x + 9 = 1 + 3x$ **4**
7. The model at the left shows the solution of an equation. Copy the model. Write the solution step and an explanation of the step beside each part of the model. **See margin.**
8. Look back at **Step 2**. To get x-tiles on only one side of the equation you can subtract 4 x-tiles from each side instead of subtracting 2 x-tiles from each side. How is the result different? Which method do you prefer? Why? **See margin.**

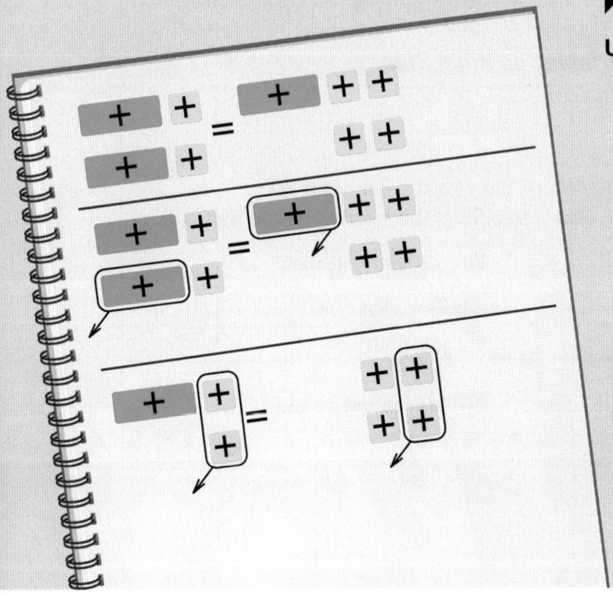

7. $2x + 2 = x + 4$ (model original equation); $x + 2 = 4$ (subtract an x-tile from each side); $x = 2$ (subtract 2 1-tiles from each side).

8. The resulting equation is $5 = -2x + 9$; the original method is simpler, because you don't have to add zero pairs, and because you find a value for x, not the opposite of x.

1 Planning the Activity

PURPOSE
Use algebra tiles to model and solve equations with a variable on both sides.

MATERIALS
- algebra tiles
- Activity Support Master (*Chapter 3 Resource Book*, p. 53)

PACING
- Exploring the Concept — 10 min
- Drawing Conclusions — 10 min

LINK TO LESSON
The four-step approach to solving equations with variables on both sides can be used to solve the equation in Example 1.

2 Managing the Activity

CLASSROOM MANAGEMENT
Remind students that they can only take away tiles that are equal in size from each side. Also, if you take 2 x-tiles away from one side you *must* take away 2 x-tiles from the other.

ALTERNATIVE APPROACH
Have 20 students model $4x + 5 = 2x + 9$. Give 6 students a card with an "X" and 14 students a card with a "1." Put 4 students with an "X" and 5 students with a "1" on one side and 2 students with an "X" and 9 students with a "1" on another side. Then solve for one student with an "X".

3 Closing the Activity

KEY DISCOVERY
To solve equations with variables on both sides, add or subtract to get all x-values on one side and all numbers on the other side. Then divide if necessary to solve.

ACTIVITY ASSESSMENT
Use algebra tiles to model and solve $8x + 1 = 4x + 5$. $x = 1$

1 PLAN

PACING
Basic: 2 days
Average: 2 days
Advanced: 2 days
Block Schedule: 1 block

LESSON OPENER ACTIVITY
An alternative way to approach Lesson 3.4 is to use the Activity Lesson Opener:

- Blackline Master (*Chapter 3 Resource Book,* p. 54)
- Transparency (p. 19)

MEETING INDIVIDUAL NEEDS

- ***Chapter 3 Resource Book***
  Prerequisite Skills Review (p. 5)
  Practice Level A (p. 55)
  Practice Level B (p. 56)
  Practice Level C (p. 57)
  Reteaching with Practice (p. 58)
  Absent Student Catch-Up (p. 60)
  Challenge (p. 63)
- ***Resources in Spanish***
- ***Personal Student Tutor***

NEW-TEACHER SUPPORT
See the Tips for New Teachers on pp. 1–2 of the *Chapter 3 Resource Book* for additional notes about Lesson 3.4.

WARM-UP EXERCISES

Transparency Available

Simplify.

1. $2(4x-5)$ $8x-10$
2. $6x-3x+5$ $3x+5$
3. $-7(2-5x)$ $-14+35x$
4. $3(x-2)+5x$ $8x-6$
5. $9x-(4-3x)$ $12x-4$

3.4 Solving Equations with Variables on Both Sides

What you should learn

GOAL 1 Collect variables on one side of an equation.

GOAL 2 Use equations to solve **real-life** problems such as renting video games in **Example 5**.

Why you should learn it

▼ To solve **real-life** problems, such as deciding whether to join a rock-climbing gym in **Exercise 46**.

GOAL 1 COLLECTING VARIABLES ON ONE SIDE

Some equations have variables on both sides. To solve such equations, you may first collect the variable terms on the side with the *greater* variable coefficient.

EXAMPLE 1 *Collect Variables on Left Side*

Solve $7x + 19 = -2x + 55$.

SOLUTION Look at the coefficients of the x-terms. Since 7 is greater than -2, collect the x-terms on the left side.

$7x + 19 = -2x + 55$	Write original equation.
$7x + 19 + 2x = -2x + 55 + 2x$	Add $2x$ to each side.
$9x + 19 = 55$	Simplify.
$9x + 19 - 19 = 55 - 19$	Subtract 19 from each side.
$9x = 36$	Simplify.
$\frac{9x}{9} = \frac{36}{9}$	Divide each side by 9.
$x = 4$	Simplify.

✓CHECK

$7x + 19 = -2x + 55$	Write original equation.
$7(4) + 19 \stackrel{?}{=} -2(4) + 55$	Substitute 4 for each x.
$47 = 47$	Solution is correct.

EXAMPLE 2 *Collect Variables on Right Side*

Solve $80 - 9y = 6y$.

SOLUTION Look at the coefficients of the y-terms. Think of $80 - 9y$ as $80 + (-9y)$. Since 6 is greater than -9, collect the y-terms on the right side.

$80 - 9y = 6y$	Write original equation.
$80 - 9y + 9y = 6y + 9y$	Add $9y$ to each side.
$80 = 15y$	Simplify.
$\frac{80}{15} = \frac{15y}{15}$	Divide each side by 15.
$\frac{16}{3} = y$	Simplify.

STUDENT HELP

Look Back
For help with terms of an expression, see page 80.

Linear equations do not always have one solution. An **identity** is an equation that is true for all values of the variable. Some linear equations have no solution.

STUDENT HELP

HOMEWORK HELP Visit our Web site www.mcdougallittell.com for extra examples.

EXAMPLE 3 *Many Solutions or No Solution*

Solve the equation.

a. $3(x + 2) = 3x + 6$ **b.** $x + 2 = x + 4$

SOLUTION

a. $3(x + 2) = 3x + 6$ Write original equation.

$3x + 6 = 3x + 6$ Use distributive property.

$6 = 6$ Subtract $3x$ from each side.

▶ All values of x are solutions, because $6 = 6$ is always true. The original equation is an *identity*.

b. $x + 2 = x + 4$ Write original equation.

$2 \neq 4$ Subtract x from each side.

▶ The original equation has no solution, because $2 \neq 4$ for any value of x.

EXAMPLE 4 *Solving More Complicated Equations*

Solve the equation.

a. $4(1 - x) + 3x = -2(x + 1)$ **b.** $\frac{1}{4}(12x + 16) = 10 - 3(x - 2)$

SOLUTION Simplify the equation before you decide whether to collect the variable terms on the right side or the left side of the equation.

a. $4(1 - x) + 3x = -2(x + 1)$ Write original equation.

$4 - 4x + 3x = -2x - 2$ Use distributive property.

$4 - x = -2x - 2$ Add like terms.

$4 + x = -2$ Add $2x$ to each side.

$x = -6$ Subtract 4 from each side.

b. $\frac{1}{4}(12x + 16) = 10 - 3(x - 2)$ Write original equation.

$3x + 4 = 10 - 3x + 6$ Use distributive property.

$3x + 4 = 16 - 3x$ Simplify.

$6x + 4 = 16$ Add $3x$ to each side.

$6x = 12$ Subtract 4 from each side.

$x = 2$ Divide each side by 6.

STUDENT HELP

Study Tip
You can multiply by the reciprocal first in part (b) of Example 4, but distributing the $\frac{1}{4}$ is easier here. Before you solve, look to see which method seems easier.

2 TEACH

MOTIVATING THE LESSON

Ask students how they choose sides for a basketball game. Do they try to make teams with an equal number of players on each team? Ask them what happens if two more people want to play. Point out that to keep the teams equal, they have to add the same number of players to each team. Relate this idea of keeping teams even to keeping both sides of an equation balanced.

EXTRA EXAMPLE 1

Solve $6x + 22 = -3x + 31$. **1**

EXTRA EXAMPLE 2

Solve $64 - 12w = 6w$. $\frac{32}{9}$

EXTRA EXAMPLE 3

Solve the equation.

a. $4(x - 5) = 4x - 20$ **identity**

b. $3x - 9 = 3x + 10$ **no solution**

EXTRA EXAMPLE 4

Solve the equation.

a. $10(2 - x) + 4x = -\frac{3}{10}(x + 3)$ $\frac{11}{3}$

b. $\frac{2}{5}(10x + 15) = 18 - 4(x - 3)$ **3**

CHECKPOINT EXERCISES

For use after Examples 1–4:

Solve the equation.

1. $17 - 2x = 14 + 4x$ $\frac{1}{2}$

2. $2(3x + 4) = 6x + 9$ **no solution**

3. $\frac{1}{2}(12 - 2x) - 4 = 5x + 2(x - 7)$ **2**

CONCEPT QUESTION

EXAMPLE 3 When is a linear equation an identity and when does a linear equation have no solutions? *Sample answer:* **A linear equation is an identity when the solved equation has no variable and the equation is true. A linear equation has no solutions when the solved equation has no variable and the equation is false.**

STUDENT HELP NOTES

Homework Help Students can find extra examples at **www.mcdougallittell.com** that parallel the examples in the student edition.

EXTRA EXAMPLE 5

A gym offers two packages for yearly membership. The first plan costs $50 to be a member. Then each visit to the gym is $5. The second plan costs $200 for a membership fee plus $2 per visit. Which membership is more economical? **fewer than 50 visits: first plan; more than 50 visits: second plan**

CHECKPOINT EXERCISES

For use after Example 5:

1. One CD music club charges a $30 membership fee plus $10 per CD. Another club charges $15 per CD but has no membership fee. Which club is more economical? **fewer than 6 CDs: second club; more than 6 CDs: first club**

FOCUS ON VOCABULARY

When discussing the idea of *identity,* point out that both sides of the equation must have the same, or *identical,* values for all values of *x.*

CLOSURE QUESTION

When solving equations with variables on both sides of the equation, how do you decide which variable term to add or subtract? **Add or subtract the variable term with the lesser coefficient.**

DAILY PUZZLER

Javier said: "I have a number in mind. If I add 5 to it, then subtract 5 from it, the average of the results is the number I started with." Can you tell for sure what number he has in mind? Why? **No; if x is Javier's number, then $\frac{1}{2}[(x+5)+(x-5)] = x$ is an identity, which is true for all values of x.**

GOAL 2 SOLVING REAL-LIFE PROBLEMS

EXAMPLE 5 *Using a Verbal Model*

REAL LIFE *Membership Fees*

A video store charges $8 to rent a video game for five days. You must be a member to rent from the store, but the membership is free. A video game club in town charges only $3 to rent a game for five days, but membership in the club is $50 per year. Which rental plan is more economical?

SOLUTION

Find the number of rentals for which the two plans would cost the same.

PROBLEM SOLVING STRATEGY

VERBAL MODEL

Store rental fee · Number rented = Club rental fee · Number rented + Club membership fee

(Video store: Store rental fee · Number rented; Video club: Club rental fee · Number rented + Club membership fee)

LABELS

Store rental fee = **8** (dollars per game)

Number rented = x (games)

Club rental fee = **3** (dollars per game)

Club membership fee = **50** (dollars)

ALGEBRAIC MODEL

$8 \cdot x = 3 \cdot x + 50$ Write algebraic model.

$5x = 50$ Subtract $3x$ from each side.

$x = 10$ Divide each side by 5.

If you rent 10 video games in a year, the cost would be the same at either the store or the club. If you rent more than 10, the club is more economical. If you rent fewer than 10, the store is more economical.

UNIT ANALYSIS Check that *dollars* are the units of the solution.

$$\frac{\text{dollars}}{\cancel{\text{game}}} \cdot \cancel{\text{games}} = \frac{\text{dollars}}{\cancel{\text{game}}} \cdot \cancel{\text{games}} + \text{dollars}$$

STUDENT HELP

INTERNET CAREER LINK Visit our Web site www.mcdougallittell.com for information about video game designers.

✓**CHECK** A table can help you check the result of Example 5.

Number rented	2	4	6	8	10	12	14	16
Cost at store	$16	$32	$48	$64	$80	$96	$112	$128
Cost at club	$56	$62	$68	$74	$80	$86	$92	$98

Store is cheaper. (2–8) Club is cheaper. (12–16)

GUIDED PRACTICE

Vocabulary Check ✓

Concept Check ✓

1. Is the equation $-2(4 - x) = 2x - 8$ an identity? Explain why or why not. **See margin.**
2. Decide if the statement is *true* or *false*. *The solution of* $x = 2x$ *is zero.* **true**
3. Solve $9(9 - x) = 4x - 10$. Explain what you are doing at each step. **See margin.**

Skill Check ✓

Solve the equation if possible. Does the equation have *one solution*, is it an *identity*, or does it have *no solution*?

4. $2x + 3 = 7x$ **$\frac{3}{5}$; one solution**
5. $12 - 2a = -5a - 9$ **−7; one solution**
6. $x - 2x + 3 = 3 - x$ **all real numbers; an identity**
7. $5x + 24 = 5(x - 5)$ **no solution**
8. $\frac{2}{3}(6c + 3) = 6(c - 3)$ **10; one solution**
9. $6y - (3y - 6) = 5y - 4$ **5; one solution**
10. FUNDRAISING You are making pies to sell at a fundraiser. It costs \$3 to make each pie, plus a one-time cost of \$20 for a pastry blender and a rolling pin. You plan to sell the pies for \$5 each. Which equation should you use to find the number of pies you need to sell to break even? **B**

    A. $3x = 20 + 5x$ **B.** $3x + 20 = 5x$

    C. $3x - 20 = 5x$ **D.** $20 - 5x = 3x$

11. Find the number of pies you need to sell to break even in Exercise 10. **10 pies**

1. Yes; the given equation is true for all values of x since it is equivalent to $2x - 8 = 2x - 8$.

3. $81 - 9x = 4x - 10$ (Use the distributive property.); $81 = 13x - 10$ (Add $9x$ to each side.); $91 = 13x$ (Add 10 to each side.); $7 = x$ (Divide each side by 13.)

PRACTICE AND APPLICATIONS

STUDENT HELP

Extra Practice to help you master skills is on p. 799.

WRITING **Solve the equation and describe each step you use.**

12. $7 - 4c = 10c$ **$\frac{1}{2}$**
13. $-8x + 7 = 4x - 5$ **1**
14. $x + 2 = 3x - 1$ **$1\frac{1}{2}$**
15. $7(1 - y) = -3(y - 2)$ **$\frac{1}{4}$**
16. $\frac{1}{5}(10a - 15) = 3 - 2a$ **$1\frac{1}{2}$**
17. $5(y - 2) = -2(12 - 9y) + y$ **1**

SOLVING EQUATIONS **Solve the equation if possible.**

18. $4x + 27 = 3x$ **−27**
19. $12y + 21 = 9y$ **−7**
20. $-2m = 16m - 9$ **$\frac{1}{2}$**
21. $4n = -28n - 3$ **$-\frac{3}{32}$**
22. $12c - 4 = 12c$ **no solution**
23. $-30d + 12 = 18d$ **$\frac{1}{4}$**
24. $6 - (-5r) = 5r - 3$ **no solution**
25. $6s - 11 = -2s + 5$ **2**
26. $12p - 7 = -3p + 8$ **1**
27. $-12q + 4 = 8q - 6$ **$\frac{1}{2}$**
28. $-7 + 4m = 6m - 5$ **−1**
29. $-7 + 11g = 9 - 5g$ **1**
30. $8 - 9t = 21t - 17$ **$\frac{5}{6}$**
31. $24 - 6r = 6(4 - r)$ **all real numbers**
32. $3(4 + 4x) = 12x + 12$ **all real numbers**
33. $-4(x - 3) = -x$ **4**
34. $10(-4 + y) = 2y$ **5**
35. $8a - 4(-5a - 2) = 12a$ **$-\frac{1}{2}$**
36. $9(b - 4) - 7b = 5(3b - 2)$ **−2**
37. $-2(6 - 10n) = 10(2n - 6)$ **no solution**
38. $-(8n - 2) = 3 + 10(1 - 3n)$ **$\frac{1}{2}$**
39. $\frac{1}{2}(12n - 4) = 14 - 10n$ **1**
40. $\frac{1}{4}(60 + 16s) = 15 + 4s$ **all real numbers**
41. $\frac{3}{4}(24 - 8b) = 2(5b + 1)$ **1**

STUDENT HELP

HOMEWORK HELP

Example 1: Exs. 12–41
Example 2: Exs. 12–41
Example 3: Exs. 18–41
Example 4: Exs. 12–41
Example 5: Exs. 45, 46

3 APPLY

ASSIGNMENT GUIDE

BASIC

Day 1: p. 157 Exs. 12–40 even

Day 2: p. 157 Exs. 31–41 odd, 42, 45, 46, 50–52, 55, 60, 65, 69, 70

AVERAGE

Day 1: p. 157 Exs. 12–40 even

Day 2: p. 157 Exs. 31–41 odd, 42, 43, 45, 46, 50–52, 55, 60, 65, 69, 70

ADVANCED

Day 1: p. 157 Exs. 12–40 even

Day 2: p. 157 Exs. 31–41 odd, 42–54 even, 55, 60, 65, 69, 70

BLOCK SCHEDULE

p. 157 Exs. 12–40 even, 31–41 odd, 42, 43, 45, 46, 50–52, 55, 60, 65, 69, 70

EXERCISE LEVELS

Level A: *Easier*
55–64

Level B: *More Difficult*
12–41, 45, 46, 50–52, 65–70

Level C: *Most Difficult*
42–44, 47–49, 53, 54

✓ HOMEWORK CHECK

To quickly check student understanding of key concepts, go over the following exercises: Exs. 12, 16, 24, 36, 40. See also the Daily Homework Quiz:

- Blackline Master (*Chapter 3 Resource Book*, p. 66)
- Transparency (p. 23)

! COMMON ERROR

EXAMPLES 12–41 Some students may want to write equivalent equations side by side when they solve. Point out that this can easily result in miscommunication, since the right side of an equation and the left side of the next equation may appear to form a single expression. Vertical form avoids this danger and helps reduce the likelihood of errors.

TEACHING TIPS

EXERCISES 24–41 Some students will try to solve these equations by doing too many steps mentally. Emphasize the importance of writing each step in order to communicate the thinking that was used to solve the equation.

! COMMON ERROR

EXERCISES 42–43 These exercises demonstrate some common errors students make in solving equations. All students should understand what these error are to avoid them in their work.

MATHEMATICAL REASONING

In Exercise 44, students may find it helpful to note that they can start with $y = y$ and work up to the top equation. Since $y = y$ is clearly true for all real numbers, so is the top equation.

ADDITIONAL PRACTICE AND RETEACHING

For Lesson 3.4:

- Practice Levels A, B, and C (*Chapter 3 Resource Book*, p. 55)
- Reteaching with Practice (*Chapter 3 Resource Book*, p. 58)
- See Lesson 3.4 of the *Personal Student Tutor*

For more Mixed Review:

- Search the *Test and Practice Generator* for key words or specific lessons.

42. The distributive property is applied improperly on both sides; $7(c - 6) = 7c - 42$ and $(4 - c)(3) = 12 - 3c$; 5.4.

43. Since $-6 = -6$, you can conclude that the given equation is true for all real numbers, not that $b = -6$.

44. The given equation is true for all real numbers.

45. *Sample equation*: $3x + 25 = 5x$; 12.5 sessions; It is worthwhile to become a member if you expect to attend 13 or more sessions.

47. The possible number is $3\frac{1}{2}$ more than the actual number, 50.

FOCUS ON CAREERS

REAL LIFE ARCHITECT Designing a building that is functional, attractive, safe, and economical requires teamwork and many stages of planning. Most architects use computers as they write reports and create plans.

INTERNET CAREER LINK www.mcdougallittell.com

ERROR ANALYSIS **In Exercises 42 and 43, describe the errors.**

42.
$7(c - 6) = (4 - c)(3)$
$7c - 6 = 4 - 3c$
$10c = 10$
$c = 1$

See margin.

43.
$2(4b - 3) = 8b - 6$
$8b - 6 = 8b - 6$
$-6 = -6$
$b = -6$

See margin.

44. **CRITICAL THINKING** A student was confused by the results when solving this equation. Explain what the result means. **See margin.**

$7(y - 2) = -y + 8y - 14$
$7y - 14 = 7y - 14$
$7y = 7y$
$y = y$

45. **COMPUTER TIME** A local computer center charges nonmembers \$5 per session to use the media center. If you pay a membership fee of \$25, you pay only \$3 per session. Write an equation that can help you decide whether to become a member. Then solve the equation and interpret the solution. **See margin.**

46. **ROCK CLIMBING** A rock-climbing gym charges nonmembers \$16 per day to use the gym and \$8 per day for equipment rental. Members pay a yearly fee of \$450 for unlimited climbing and \$6 per day for equipment rental. Write and solve an equation to find how many times you must use the gym to justify becoming a member. ***Sample equation***: **$16x + 8x = 6x + 450$; 25 days**

TALL BUILDINGS **In Exercises 47–49, use the following information.**
In designing a tall building, many factors affect the height of each story. How the space will be used is important. At the Grand Gateway at Xu Hui in Shanghai, the lowest 7 stories have a combined height of about 126 feet. These stories next to a shopping mall are unusually tall. The building's other 43 stories have a more typical height. If the average height of the other stories had been used for the lowest 7 stories, then the building could have fit about $3\frac{1}{2}$ more stories.

▶ Source: Callison Architecture. (Actual dimensions have been simplified.)

Combined height of lowest 7 stories + Number of other stories · Average height of other stories = Possible number of stories · Average height of other stories

47. Explain why you would use $53\frac{1}{2}$ for the "possible number of stories" in the verbal model. **See margin.**

48. Let h represent the average height of the other stories. Finish assigning labels and write the algebraic model. **$126 + 43h = 53\frac{1}{2}(h)$**

49. Solve the equation. Be sure to simplify the answer. About how many feet tall are the stories above the 7th floor? **about 12 ft**

penthouse

other stories

lowest 7 stories

Test Preparation

50. **MULTIPLE CHOICE** Which equations are equivalent? **A**

I. $7x - 9 = 7$ II. $-9 = 7 - 5x$ III. $3(2x - 3) = 7 - x$

Ⓐ I and III Ⓑ II and III Ⓒ All Ⓓ None

51. **MULTIPLE CHOICE** Which equation has more than one solution? **C**

Ⓐ $18y + 13 = 12y - 25$ Ⓑ $6y - (3y - 6) = -14 + 3y$

Ⓒ $-\frac{1}{2}(30x - 18) = 9 - 15x$ Ⓓ $\frac{1}{5}(2x - 5) = 3x + 7$

52. **MULTIPLE CHOICE** Solve $\frac{1}{3}(7x + 5) = 3x - 5$. **C**

Ⓐ 15 Ⓑ -5 Ⓒ 10 Ⓓ $\frac{-5}{2}$

★ Challenge

SUBSCRIPTIONS In Exercises 53 and 54, use the table. It shows monthly expenses and income of a magazine for different numbers of subscribers.

Number of subscribers	50	100	150	200	250	300	350	400
Income	\$75	\$150	\$225	\$300	\$375	\$450	\$525	\$600
Expenses	\$150	\$200	\$250	\$300	\$350	\$400	\$450	\$500

53. **EXPLAINING A PATTERN** Look for patterns in the table. Write an equation that you can use to find how many subscribers the magazine needs for its income to equal its expenses. **See margin.**

54. Explain how to use the table to check your answer in Exercise 53. ***Sample answer:*** **Find the column in the table where income equals expenses. The number of subscribers for that column (200) is the answer to Exercise 53.**

53. Let x = the number of subscribers; income = $1.5x$ and expenses = $x + 100$; $1.5x = x + 100$; 200 subscribers.

EXTRA CHALLENGE

www.mcdougallittell.com

MIXED REVIEW

EQUIVALENTS Write the percent as a decimal. (Skills Review, pages 784–785)

55. 28% **0.28** 56. 40% **0.4** 57. 3% **0.03** 58. 19.5% **0.195**

PERCENT OF A NUMBER Find the number. (Skills Review, page 786)

59. 45% of 84 **37.8** 60. 7% of 28.5 **1.995** 61. 76% of 540 **410.4**

62. 16.3% of 132 **21.516** 63. 8% of \$928.50 **\$74.28** 64. 5.5% of \$74 **\$4.07**

UNIT ANALYSIS Find the resulting unit of measure. (Review 1.1 for 3.5)

65. (dollars per hour) • (hours) **dollars** 66. (years) • (people per year) **people**

67. (miles) ÷ (miles per hour) **hours** 68. (meters) • (kilometers per meter) **kilometers**

GEOMETRY The two triangles are similar. Find the length of the side marked x. (Review 3.2)

69. 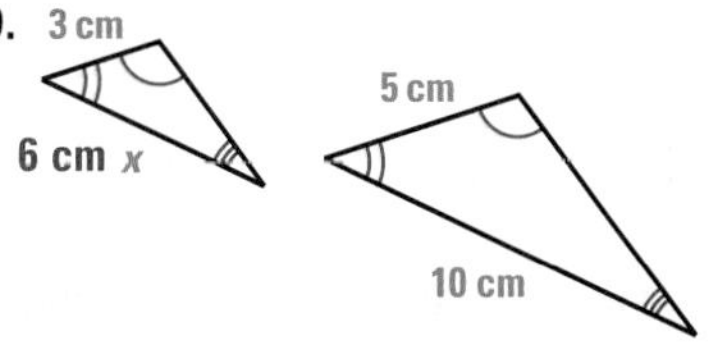

6 cm

70. 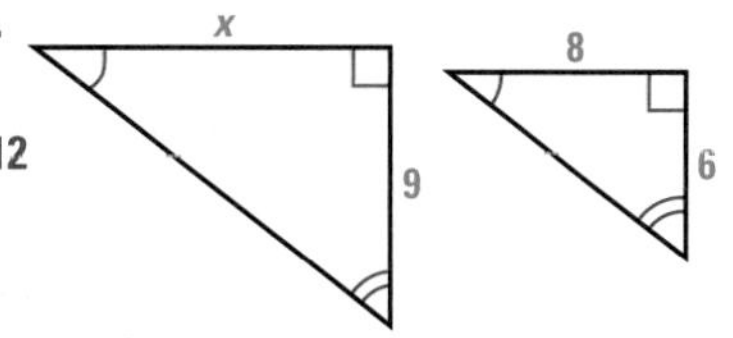

12

4 ASSESS

DAILY HOMEWORK QUIZ

Transparency Available

1. Solve $48 - 9y = 3y$. Describe each step you use. **Add $9y$ to each side. Divide each side by 12. The solution is 4.**

Solve the equation if possible.

2. $\frac{1}{3}(7n - 33) = 8 - 4n$ **3**
3. $4(x + 12) = 40 + 2(4 + 2x)$ **all real numbers**
4. $5(x - 3) - 9x = -2(x + 17)$ $\mathbf{9\frac{1}{2}}$
5. $8k - 2(k + 7) = 15 + 6k$ **no solution**

EXTRA CHALLENGE NOTE

Challenge problems for Lesson 3.4 are available in **blackline** format in the *Chapter 3 Resource Book*, p. 63 and at **www.mcdougallittell.com.**

ADDITIONAL TEST PREPARATION

1. **WRITING** What would be your first and second steps in solving the equation $5x - 3 = 2x + 6$? ***Sample answer:*** **Add 3 to both sides, then subtract $2x$ from both sides (or subtract $2x$ from both sides, then add 3 to both sides).**

1 PLAN

PACING
Basic: 1 day
Average: 1 day
Advanced: 1 day
Block Schedule: 0.5 block with 3.6

LESSON OPENER
VISUAL APPROACH
An alternative way to approach Lesson 3.5 is to use the Visual Approach Lesson Opener:

- Blackline Master (*Chapter 3 Resource Book,* p. 67)
- Transparency (p. 20)

MEETING INDIVIDUAL NEEDS
- ***Chapter 3 Resource Book***
  Prerequisite Skills Review (p. 5)
  Practice Level A (p. 70)
  Practice Level B (p. 71)
  Practice Level C (p. 72)
  Reteaching with Practice (p. 73)
  Absent Student Catch-Up (p. 75)
  Challenge (p. 77)
- ***Resources in Spanish***
- ***Personal Student Tutor***

NEW-TEACHER SUPPORT
See the Tips for New Teachers on pp. 1–2 of the *Chapter 3 Resource Book* for additional notes about Lesson 3.5.

WARM-UP EXERCISES

Transparency Available

Solve.

1. $8x + 3 - 5x = 18$ 5
2. $4(x + 2) = -16$ −6
3. $7(3 - 2x) + 9x = 6$ 3
4. $25x + 10 = 30x + 7$ $\frac{3}{5}$

3.5

What you should learn

GOAL 1 Draw a diagram to help you understand **real-life** problems, such as planning where bicyclists should meet in **Exercise 22**.

GOAL 2 Use tables and graphs to check your answers.

Why you should learn it

▼ To solve **real-life** problems, such as designing a high school yearbook in **Example 1**.

REAL LIFE

Linear Equations and Problem Solving

GOAL 1 DRAWING A DIAGRAM

You have learned to use a verbal model to solve real-life problems. Verbal models can be used with other problem solving strategies, such as drawing a diagram.

EXAMPLE 1 *Visualizing a Problem*

YEARBOOK DESIGN A page of your school yearbook is $8\frac{1}{2}$ inches by 11 inches. The left margin is $\frac{3}{4}$ inch and the space to the right of the pictures is $2\frac{7}{8}$ inches. The space between pictures is $\frac{3}{16}$ inch. How wide can each picture be to fit three across the width of the page?

SOLUTION

DRAW A DIAGRAM The diagram shows that the page width is made up of the width of the left margin, the space to the right of the pictures, two spaces between pictures, and three picture widths.

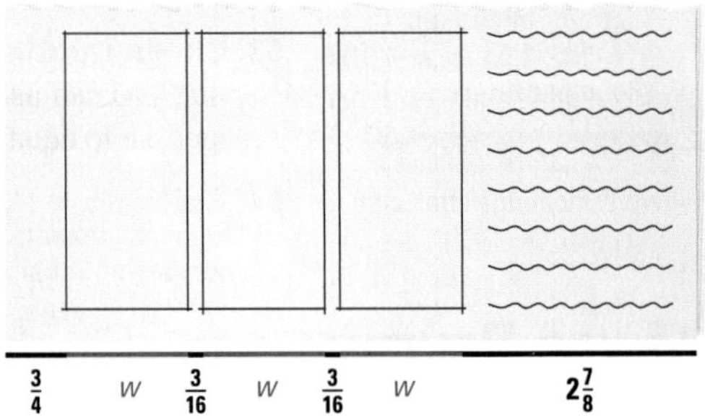

VERBAL MODEL $\boxed{\text{Left margin}} + \boxed{\text{Space to right of pictures}} + 2 \cdot \boxed{\text{Space between pictures}} + 3 \cdot \boxed{\text{Picture width}} = \boxed{\text{Page width}}$

LABELS

Left margin $= \frac{3}{4}$ (inches)

Space to right of pictures $= 2\frac{7}{8}$ (inches)

Space between pictures $= \frac{3}{16}$ (inches)

Picture width $= w$ (inches)

Page width $= 8\frac{1}{2}$ (inches)

ALGEBRAIC MODEL $\frac{3}{4} + 2\frac{7}{8} + 2 \cdot \frac{3}{16} + 3 \cdot w = 8\frac{1}{2}$

▶ When you solve for w, you will find that each picture can be $1\frac{1}{2}$ inches wide.

GOAL 2 USING TABLES AND GRAPHS AS CHECKS

REAL LIFE
Language Study

EXAMPLE 2 *Using a Table as a Check*

At East High School, 579 students take Spanish. This number has been increasing at a rate of about 30 students per year. The number of students taking French is 217 and has been decreasing at a rate of about 2 students per year. At these rates, when will there be three times as many students taking Spanish as taking French?

SOLUTION

First write an expression for the number of students taking each language after some number of years. Then set the expression for students taking Spanish equal to three times the expression for students taking French. Subtraction is used in the expression for students taking French, because there is a decrease.

PROBLEM SOLVING STRATEGY

VERBAL MODEL

[Number taking Spanish now] + [Rate of increase for Spanish] • [Number of years] =

3 • ([Number taking French now] − [Rate of decrease for French] • [Number of years])

LABELS

Number of students taking Spanish now = **579** (students)

Rate of *increase* for Spanish = **30** (students per year)

Number of years = x (years)

Number of students taking French now = **217** (students)

Rate of *decrease* for French = **2** (students per year)

ALGEBRAIC MODEL

$579 + 30 \cdot x = 3(217 - 2 \cdot x)$ Write algebraic model.

$579 + 30x = 651 - 6x$ Use distributive property.

$579 + 36x = 651$ Add $6x$ to each side.

$36x = 72$ Subtract 579 from each side.

$x = 2$ Divide each side by 36.

▶ At these rates, the number of students taking Spanish will be three times the number of students taking French in two years.

✓**CHECK** Use the information in the original problem statement to make a table.

For example, the number of students taking Spanish in Year 1 is 579 + 30 = 609.

Year	0 (now)	1	2
Number taking Spanish	579	609	639
Number taking French	217	215	213
3 · Number taking French	651	645	639

equal

2 TEACH

EXTRA EXAMPLE 1
A page of pictures for a yearbook is $8\frac{1}{2}$ inches by 11 inches. The top margin is $\frac{3}{4}$ inch and the bottom margin is $1\frac{1}{4}$ inches. The space between the pictures is $\frac{3}{16}$ inch. How high can each picture be to fit seven down the length of the page? **$1\frac{1}{8}$ in.**

EXTRA EXAMPLE 2
At West High School, 362 students take Spanish. This number has been increasing at a rate of 20 per year. The number of students taking French is 259 and has been decreasing at a rate of about 3 per year. At these rates, when will there be two times as many students taking Spanish as taking French? **in 6 years**

CHECKPOINT EXERCISES

For use after Example 1:

1. Three pictures that are each $2\frac{5}{8}$ inches wide are to be placed on a page that is $11\frac{1}{2}$ inches wide. If you leave equal space between pictures, how much space can you put between them if the left margin is $\frac{3}{4}$ inch and the right margin is $1\frac{5}{8}$ inches? **$\frac{5}{8}$ in.**

For use after Example 2:

2. At North High School, 176 students buy non-spiral notebooks from the school store. This number has been increasing at a rate of about 24 per year. The number of students buying spiral notebooks is 331 and has been decreasing at a rate of about 15 per year. At these rates, when will the number of spiral notebooks sold be half the number of non-spiral notebooks sold? **9 years**

STUDENT HELP NOTES

Homework Help Students can find extra examples at **www.mcdougallittell.com** that parallel the examples in the student edition.

EXTRA EXAMPLE 3

Jim can run 11 ft/sec and Sarah can run 13 ft/sec. How far ahead of Sarah must Jim be not to fall behind Sarah in the first 15 seconds that they run? **30 ft**

CHECKPOINT EXERCISES

For use after Example 3:

1. The winning speed at the Indianapolis 500 auto race in 1990 was about 186 mi/h. The winning speed in 1997 was 146 mi/h. About how many hours longer was the race in 1997 than in 1990? **about 0.74 h**

FOCUS ON VOCABULARY

Example 3 provides a good opportunity to review basic vocabulary for graphs. Point out the *horizontal axis,* the *vertical axis,* and the *origin* in the graph. Discuss what the *coordinates* of the point of intersection are for the two line graphs.

CLOSURE QUESTION

Explain how drawing a diagram, using a table, and using a graph can help with problem solving. ***Sample answer:*** **Drawing a diagram can help you visualize problems involving spatial relations. Using a table or a graph can provide a numerical or visual check of an answer.**

DAILY PUZZLER

Arrange the digits 1, 2, 3, and 4 in the boxes in the following expression so that the fractions are in simplest form and the smallest possible product is created.

$\frac{\square}{\square} \cdot \frac{\square}{\square}$ $\frac{1}{4} \cdot \frac{2}{3}$ or $\frac{2}{3} \cdot \frac{1}{4}$

STUDENT HELP

HOMEWORK HELP Visit our Web site www.mcdougallittell.com for extra examples.

EXAMPLE 3 *Using a Graph as a Check*

GAZELLE AND CHEETAH A gazelle can run 73 feet per second for several minutes. A cheetah can run faster (88 feet per second) but can only sustain its top speed for about 20 seconds before it is worn out. How far away from the cheetah does the gazelle need to stay for it to be safe?

SOLUTION

You can use a diagram to picture the situation. If the gazelle is too far away for the cheetah to catch it within 20 seconds, the gazelle is probably safe.

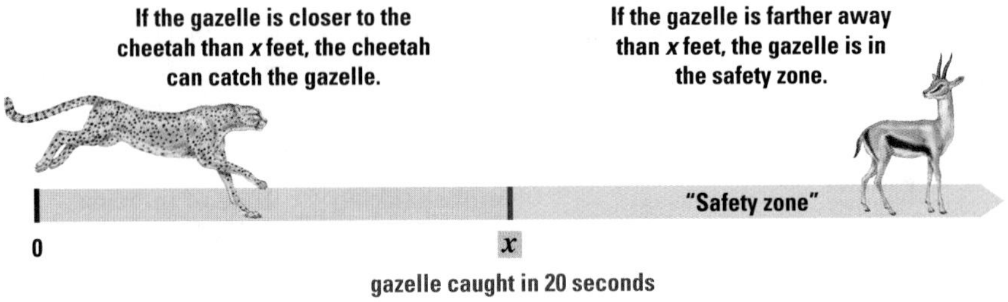

If the gazelle is closer to the cheetah than x feet, the cheetah can catch the gazelle.
If the gazelle is farther away than x feet, the gazelle is in the safety zone.
"Safety zone"
0
x
gazelle caught in 20 seconds

To find the "safety zone," you can first find the starting distance for which the cheetah reaches the gazelle in 20 seconds.

To find the distance each animal runs in 20 seconds, use the formula $d = rt$.

PROBLEM SOLVING STRATEGY

VERBAL MODEL: Distance gazelle runs in 20 sec + Gazelle's starting distance from cheetah = Distance cheetah runs in 20 sec

LABELS:

Gazelle's distance = **73 · 20** (feet)

Gazelle's starting distance from cheetah = x (feet)

Cheetah's distance = **88 · 20** (feet)

ALGEBRAIC MODEL: $73 \cdot 20 + x = 88 \cdot 20$

▶ You find that $x = 300$, so the gazelle needs to stay more than 300 feet away from the cheetah to be safe.

✓CHECK You can use a table or a graph to check your answer. A graph helps you see the relationships. To make a graph, first make a table and then plot the points. You can just find and mark the points for every 5 seconds.

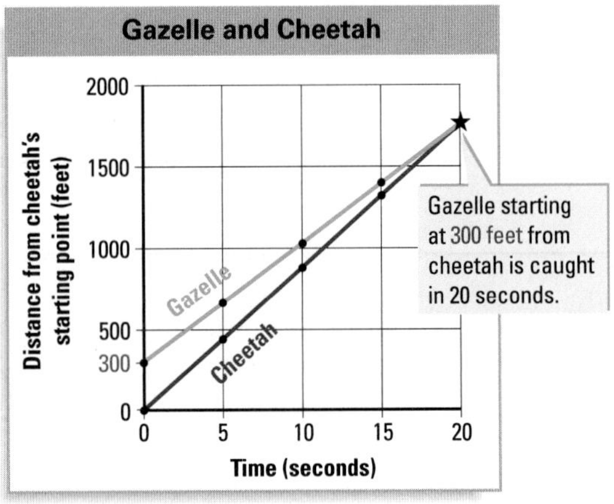

Gazelle and Cheetah
Distance from cheetah's starting point (feet)
2000
1500
1000
500
300
0
0
5
10
15
20
Time (seconds)
Gazelle
Cheetah
Gazelle starting at 300 feet from cheetah is caught in 20 seconds.

FOCUS ON APPLICATIONS

REAL LIFE **GAZELLES** seem to have an instinct for the distance they must keep from predators. They usually will not run from a prowling cheetah until it enters their "safety zone."

APPLICATION LINK www.mcdougallittell.com

Guided Practice

Concept Check ✓

1. **Yearbook Design** In Example 1, the top and bottom margins are each $\frac{7}{8}$ inch. There are 5 rows of pictures, with $\frac{3}{16}$ inch of vertical space between pictures. Draw a diagram and label all the heights on your diagram. Then write and solve an equation to find the height of each picture.
1.7 in.; See margin for sample diagram and equation.
2. Show how unit analysis can help you check the verbal model for Example 3.
Examine the units in the equation: feet + feet = feet.

Skill Check ✓

Distance Problem In Exercises 3–5, use the following problem.
You and your friend are each driving 379 miles from Los Angeles to San Francisco. Your friend leaves first, driving 52 miles per hour. She is 32 miles from Los Angeles when you leave, driving 60 miles per hour. How far do each of you drive before you are side by side? Use the following verbal model.

Rate your friend is driving	·	Time after you start driving	+	Friend's distance when you leave	=	Rate you are driving	·	Time after you start driving

3. Assign labels to each part of the verbal model. Use the labels to translate the verbal model into an algebraic model. **See margin.**
4. Solve the equation. Interpret your solution. How far do each of you drive before you are side by side? **See margin.**
5. Make a table or a graph to check your solution. **See margin.**

3. Rate your friend is driving = 52 (mi/h); Time after you start driving = t (h); Friend's distance when you leave = 32 (mi); rate you are driving = 60 (mi/h); $52t + 32 = 60t$

4. 4; 4 hours after you leave, you and your friend are side by side. Her distance in miles from Los Angeles is $4(52) + 32 = 240$ and yours is $60(4) = 240$.

Practice and Applications

Student Help
Extra Practice to help you master skills is on p. 799.

Package Size In Exercises 6–8, use this U.S. postal regulation: a rectangular package can have a combined length and *girth* of 108 inches. Suppose a package that is 36 inches long and as wide as it is high just meets the regulation.

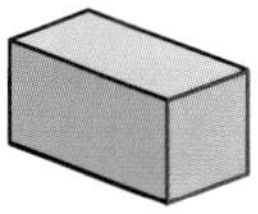

6. **Using a Diagram** Which diagram do you find most helpful for understanding the relationships in the problem? Explain. **See margin.**

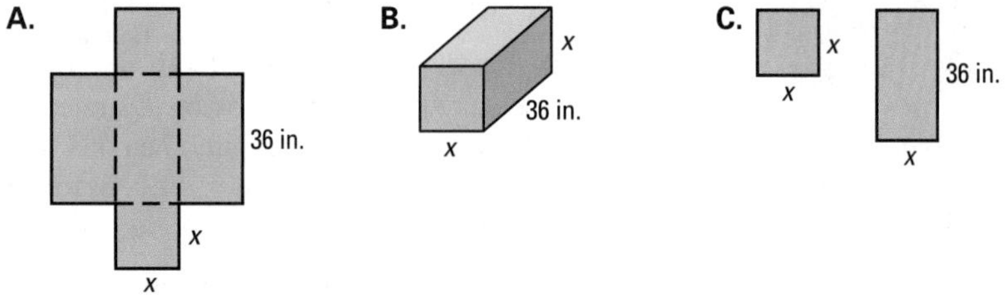

7. **Choosing a Model** Choose the equation you would use to find the width of the package. Then find its width. **C; 18 in.**

A. $x + 36 = 108$ **B.** $2x + 36 = 108$ **C.** $4x + 36 = 108$

8. **Making a Table** Make a table showing possible dimensions, girth, and combined length and girth for a "package that is 36 inches long and as wide as it is high." Which package in your table just meets the postal regulation?
See margin.

6. *Sample answer*: B since it shows the length, width, and height of the 3-dimensional box.

Student Help
Homework Help
Example 1: Exs. 6–7, 14–16
Example 2: Exs. 8–12, 17–21
Example 3: Exs. 9–13, 17–21

3 Apply

Assignment Guide

Basic
Day 1: p. 163 Exs. 6–16, 22, 30–40

Average
Day 1: p. 163 Exs. 6–17, 22, 30–40

Advanced
Day 1: p. 163 Exs. 6–26, 30–40

Block Schedule with 3.6
p. 163 Exs. 6–17, 22, 30–40

Exercise Levels

Level A: *Easier*
27–30

Level B: *More Difficult*
6–16, 22, 31–40

Level C: *Most Difficult*
17–21, 23–26

✓ Homework Check

To quickly check student understanding of key concepts, go over the following exercises: Exs. 6, 8, 12, 13, 16. See also the Daily Homework Quiz:

- Blackline Master (*Chapter 3 Resource Book*, p. 80)
- Transparency (p. 24)

Teaching Tip

Exercise 5 You may wish to have students who created a table to check their solution also create a graph and vice versa. Students could then compare the two methods of checking solutions.

1, 5, 8. See Additional Answers beginning on page AA1.

TEACHING TIPS

EXERCISE 13 Most students will see from the graph that the solution of 25 weeks is incorrect. Ask students to go a step further and find the error in the computation. This may help them avoid making similar errors in their work.

EXERCISES 17 AND 21 As with Exercise 5, you may wish to have students who checked their solutions with a table go back and draw a graph and vice versa

14. See Additional Answers beginning on page AA1.

ADDITIONAL PRACTICE AND RETEACHING

For Lesson 3.5:

- Practice Levels A, B, and C (*Chapter 3 Resource Book*, p. 70)
- Reteaching with Practice (*Chapter 3 Resource Book*, p. 73)
- See Lesson 3.5 of the *Personal Student Tutor*

For more Mixed Review:

- Search the *Test and Practice Generator* for key words or specific lessons.

13. No; according to the graph the solution is 5 weeks, not 25 weeks.

16. $2\frac{1}{4}$; each photo should be $2\frac{1}{4}$ in. wide.

19. $\frac{1}{2}$ mi; $x + \frac{1}{2}$

20. *Sample equation:* $x + \frac{1}{2} = \frac{5}{6}$; $\frac{1}{3}$ mi

FOCUS ON APPLICATIONS

ELEPHANTS need a supply of fresh water, some shade, and plentiful food such as grasses and leaves. To find these, elephants may make yearly migrations of a few hundred miles.

UNIT ANALYSIS **In Exercises 9–12, find the resulting unit of measure.**

9. (hours per day) • (days) **hours**

10. (feet) ÷ (minute) **ft/min**

11. (inches) ÷ (inches per foot) **ft**

12. (miles per hour) • (hours per minute) **mi/min**

13. A VISUAL CHECK Use the graph to check the answer to the problem below. Is the solution correct? Explain.

You have $320 and save $10 each week. Your brother has $445 and spends his income, plus $15 of his savings each week. When will you and your brother have the same amount in savings?

$$320 + 10t = 445 - 15t$$
$$320 = 445 - 5t$$
$$-125 = -5t$$
$$25 = t \quad \text{(in 25 weeks)}$$

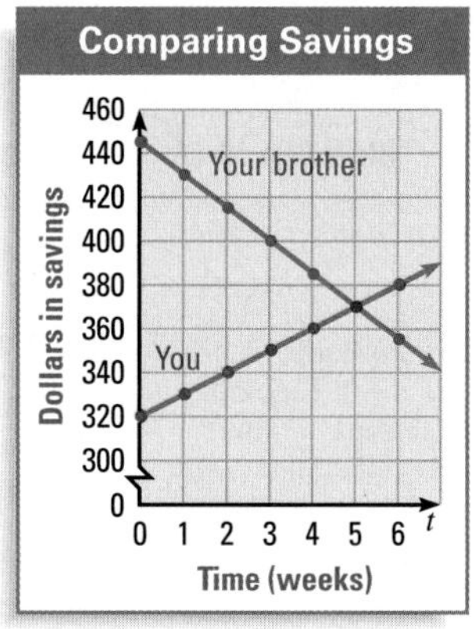

COVER DESIGN **In Exercises 14–16, you want the cover of a sports media guide to show two photos across its width. The cover is $6\frac{1}{2}$ inches wide, and the left and right margins are each $\frac{3}{4}$ inch. The space between the photos is $\frac{1}{2}$ inch. How wide should you make the photos?**

14. Draw a diagram of the cover. **See margin.**

15. Write an equation to model the problem. Use your diagram to help. **Let x = the width in inches of the photos; $2x + 2 = 6\frac{1}{2}$.**

16. Solve the equation and answer the question. **See margin.**

17. LANGUAGE CLASSES At Barton High School, 45 students are taking Japanese. This number has been increasing at a rate of 3 students per year. The number of students taking German is 108 and has been decreasing at a rate of 4 students per year. At these rates, when will the number of students taking Japanese equal the number taking German? Write and solve an equation to answer the question. Check your answer with a table or a graph. ***Sample equation:* $45 + 3x = 108 - 4x$; in 9 years**

BIOLOGY CONNECTION **In Exercises 18–21, use the following information.**

A migrating elephant herd started moving at a rate of 6 miles per hour. One elephant stood still and was left behind. Then this stray elephant sensed danger and began running at a rate of 10 miles per hour to reach the herd. The stray caught up in 5 minutes.

18. How long (in hours) did the stray run to catch up? How far did it run? **$\frac{1}{12}$ h; $\frac{5}{6}$ mi**

19. Find the distance that the herd traveled while the stray ran to catch up. Then write an expression for the total distance the herd traveled. Let x represent the distance (in miles) that the herd traveled while the stray elephant stood still. **See margin.**

20. Use the distances you found in Exercises 18 and 19. Write and solve an equation to find how far the herd traveled while the stray stood still. **See margin.**

21. Make a table or a graph to check your answer. **Check tables and graphs.**

22. **MULTI-STEP PROBLEM** Two friends are 60 miles apart. They decide to ride their bicycles to meet each other. Sally starts from the college and heads east, riding at a rate of 21 miles per hour. At the same time Teresa starts from the river and heads west, riding at a rate of 15 miles per hour.

a. Draw a diagram showing the college and the river 60 miles apart and the two cyclists riding toward each other. **Check diagrams.**

b. How far does each cyclist ride in t hours? **Sally: $21t$ mi; Teresa: $15t$ mi**

c. When the two cyclists meet, what must be true about the distances they have ridden? Write and solve an equation to find when they meet. **See margin.**

d. **CHOOSING A DATA DISPLAY** Would a park that is 26 miles west of the river be a good meeting place? Explain. Use a diagram, a table, or a graph to support your answer.

RABBIT AND COYOTE

In Exercises 23–26, use this information. A rabbit is 30 feet from its burrow. It can run 25 feet per second. A coyote that can run 50 feet per second spots the rabbit and starts running toward it.

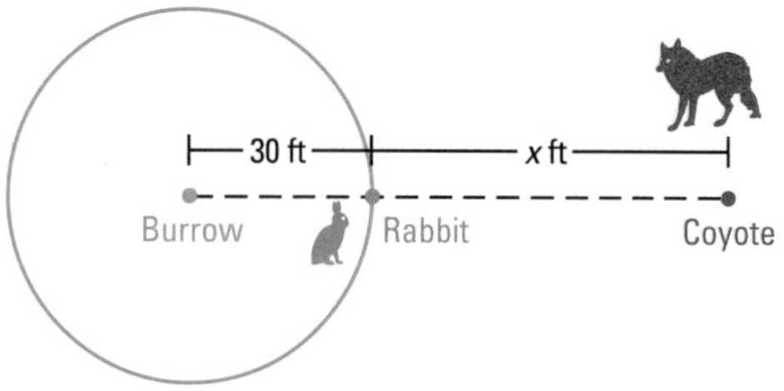

23. Write expressions for the time it will take the rabbit to get to its burrow and for the time it will take the coyote to get to the burrow. **See margin.**

24. Write and solve an equation that relates the two expressions in Exercise 23. **See margin.**

25. **CRITICAL THINKING** Interpret your results from Exercise 24. Include a description of what happens for different values of x. **See margin.**

26. Use a table or a graph to check your conclusions. **Check tables and graphs.**

MIXED REVIEW

FRACTIONS, DECIMALS, AND PERCENTS Write as a decimal rounded to the nearest hundredth. Then write as a percent. (Skills Review, pages 784–785)

27. $\frac{3}{11}$ **0.27; 27%** 28. $\frac{11}{12}$ **0.92; 92%** 29. $\frac{25}{31}$ **0.81; 81%** 30. $\frac{100}{201}$ **0.50; 50%**

EVALUATING EXPRESSIONS Evaluate the expression. (Review 2.5 and 2.7)

31. $\frac{-6}{q} - 2r$ when $q = 2$ and $r = 11$ **−25**

32. $\frac{3}{5}x - y$ when $x = -25$ and $y = -10$ **−5**

33. $\frac{x + y}{12}$ when $x = -24$ and $y = 6$ **−1.5**

34. $\frac{ab}{3a - 10}$ when $a = 8$ and $b = 7$ **4**

SOLVING EQUATIONS Solve the equation. (Review 3.3 and 3.4 for 3.6)

35. $18 + \frac{x}{3} = 9$ **−27**

36. $8 = -\frac{2}{3}(2x - 6)$ **−3**

37. $4(2 - n) = 1$ **$1\frac{3}{4}$**

38. $-3y + 14 = -5y$ **−7**

39. $-4a - 3 = 6a + 2$ **$-\frac{1}{2}$**

40. $5x - (6 - x) = 2(x - 7)$ **−2**

Test Preparation

22. c. The sum of the distances they've ridden must be 60 mi; $21t + 15t = 60$; $1\frac{2}{3}$ h.

d. *Sample answer:* It would be relatively convenient in that it would take Sally about 1.6 h to get there and Teresa about 1.7 h.; Check data displays.

★ Challenge

23. rabbit: 1.2 sec; coyote: $\frac{x + 30}{50}$ sec

24. $\frac{x + 30}{50} = 1.2$; 30 ft

25. If the coyote is 30 ft from the rabbit, the rabbit and the coyote will reach the burrow at the same time. If x is less than 30 ft, the coyote should be able to overtake the rabbit. If x is greater than 30 ft, the rabbit should be able to escape the coyote.

EXTRA CHALLENGE

www.mcdougallittell.com

4 ASSESS

DAILY HOMEWORK QUIZ

Transparency Available

1. Train A leaves the downtown station headed for the other end of the line at 55 mi/h. Train B leaves the other end of the line on a parallel route and heads downtown at 65 mi/h. Use the graph to tell how many minutes it will be before the trains pass one another. **6 min**

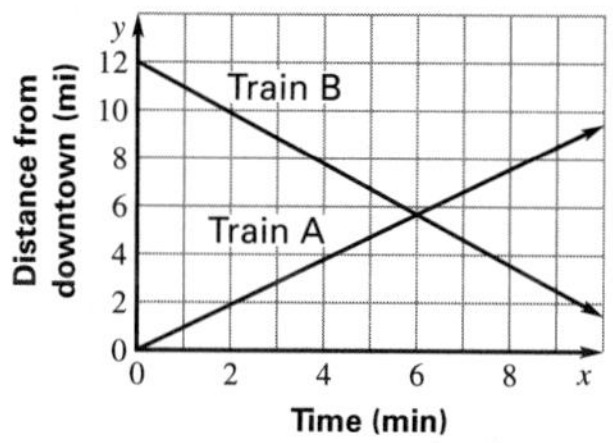

2. Write and solve an equation to check your answer for Ex. 1. **$\frac{55}{60}t = 12 - \frac{65}{60}t$; 6 min**

EXTRA CHALLENGE NOTE

Challenge problems for Lesson 3.5 are available in **blackline** format in the *Chapter 3 Resource Book,* p. 77 and at **www.mcdougallittell.com**.

ADDITIONAL TEST PREPARATION

1. **WRITING** You are personalizing sweatshirts to sell at a fund raiser. Each sweatshirt requires $\frac{3}{4}$ h to paint and $1\frac{1}{2}$ h to dry. It will also require a total of $2\frac{3}{4}$ h to prepare the sweatshirts. If you have a total of 32 h to spend on the sweatshirts and you can work on only one at a time, how many can you make to sell? Write an equation to model the problem. Then solve the problem. **$\frac{3}{4}x + 1\frac{1}{2}x + 2\frac{3}{4} = 32$; 13 sweatshirts**

1 PLAN

PACING
Basic: 2 days
Average: 2 days
Advanced: 2 days
Block Schedule: 0.5 block with 3.5
0.5 block with 3.7

LESSON OPENER
CALCULATOR
An alternative way to approach Lesson 3.6 is to use the Calculator Lesson Opener:

- Blackline Master (*Chapter 3 Resource Book,* p. 81)
- Transparency (p. 21)

MEETING INDIVIDUAL NEEDS
- ***Chapter 3 Resource Book***
  Prerequisite Skills Review (p. 5)
  Practice Level A (p. 85)
  Practice Level B (p. 86)
  Practice Level C (p. 87)
  Reteaching with Practice (p. 88)
  Absent Student Catch-Up (p. 90)
  Challenge (p. 92)
- ***Resources in Spanish***
- ***Personal Student Tutor***

NEW-TEACHER SUPPORT
See the Tips for New Teachers on pp. 1–2 of the *Chapter 3 Resource Book* for additional notes about Lesson 3.6.

WARM-UP EXERCISES
Transparency Available
Round each number to the indicated place value.
1. 1041; tens **1040**
2. 14.256; hundredths **14.26**
3. 22.0401; tenths **22.0**
4. −14.691; ones **−15**
5. −3.755; hundredths **−3.76**

3.6 Solving Decimal Equations

***What** you should learn*

GOAL 1 Find exact and approximate solutions of equations that contain decimals.

GOAL 2 Solve **real-life** problems that use decimals such as calculating sales tax in **Example 5**.

***Why** you should learn it*

▼ To solve **real-life** problems, such as finding the expansion gap for a bridge in **Exs. 56–58**.

REAL LIFE

GOAL 1 SOLVING DECIMAL EQUATIONS

So far in this chapter you have worked with equations that have exact solutions. In real life, exact solutions are not always practical and you must use rounded solutions. Rounded solutions can lead to **round-off error**, as in Example 1.

EXAMPLE 1 *Rounding for a Practical Answer*

Three people want to share equally in the cost of a pizza. The pizza costs \$12.89. You can find each person's share by solving $3x = 12.89$.

$3x = 12.89$	**Write original equation.**
$x = 4.29666...$	**Exact answer is a repeating decimal.**
$x \approx 4.30$	**Round to nearest cent.**

▶ The exact answer is not practical, but three times the rounded answer does not correspond to the price of the pizza. It is one cent too much due to round-off error.

EXAMPLE 2 *Rounding for the Final Answer*

Solve $-38x - 39 = 118$. Round to the nearest hundredth.

SOLUTION

$-38x - 39 = 118$	**Write original equation.**
$-38x = 157$	**Add 39 to each side.**
$x = \frac{157}{-38}$	**Divide each side by −38.**
$x \approx -4.131578947$	**Use a calculator.**
$x \approx -4.13$	**Round to nearest hundredth.**

▶ The solution is approximately −4.13.

✓**CHECK** When you substitute a rounded answer into the original equation, the two sides of the equation may not be exactly equal, but they should be almost equal.

$-38x - 39 = 118$	**Write original equation.**
$-38(\mathbf{-4.13}) - 39 \stackrel{?}{=} 118$	**Substitute −4.13 for *x*.**
$117.94 \approx 118$	**Rounded answer is reasonable.**

STUDENT HELP

Study Tip
When rounding, think about the situation. For example, dollar amounts are usually rounded to hundredths. If no situation is given, use the same number of decimal places in the answer as are used in the original equation.

EXAMPLE 3 ***Original Equation Involving Decimals***

Solve $3.58x - 37.40 = 0.23x + 8.32$. Round to the nearest hundredth.

SOLUTION

Use the same methods you learned for solving equations without decimals.

$3.58x - 37.40 = 0.23x + 8.32$	Write original equation.
$3.35x - 37.40 = 8.32$	Subtract $0.23x$ from each side.
$3.35x = 45.72$	Add 37.40 to each side.
$x = \frac{45.72}{3.35}$	Divide each side by 3.35.
$x \approx 13.64776119$	Use a calculator.
$x \approx 13.65$	Round to nearest hundredth.

▶ The solution is approximately 13.65.

✓**CHECK**

$3.58x - 37.40 = 0.23x + 8.32$	Write original equation.
$3.58(13.65) - 37.40 \stackrel{?}{=} 0.23(13.65) + 8.32$	Substitute 13.65 for each x.
$11.467 \approx 11.4595$	Rounded answer is reasonable.

....

An equation involving decimals can be rewritten as an equivalent equation with only integer terms and coefficients.

EXAMPLE 4 ***Changing Decimal Coefficients to Integers***

Solve $4.5 - 7.2x = 3.4x - 49.5$. Round to the nearest tenth.

SOLUTION

Because the coefficients and constant terms each have only one decimal place, you can rewrite the equation without decimals by multiplying each side by 10.

$4.5 - 7.2x = 3.4x - 49.5$	Write original equation.
$45 - 72x = 34x - 495$	Multiply each side by 10.
$45 = 106x - 495$	Add $72x$ to each side.
$540 = 106x$	Add 495 to each side.
$\frac{540}{106} = x$	Divide each side by 106.
$5.094339623 \approx x$	Use a calculator.
$5.1 \approx x$	Round to nearest tenth.

▶ The solution is approximately 5.1. Check this in the original equation.

2 TEACH

MOTIVATING THE LESSON
To play laser tag, a group of people pay a fixed amount to play for a specific amount of time. Suppose it costs \$24.99 to play laser tag for 30 minutes. How much would it cost each person if 5 people were going to play? **\$5 per person** If 6 people were going to play? **\$4.17 per person** If 2 to 10 people can play laser tag at one time, what is the range of the costs per person? **\$2.50 to \$12.50**

ACTIVITY NOTE
GRAPHING CALCULATOR
Some students may use a calculator for the computations. Advise those students against using a two-place accuracy Mode setting. With such a setting the results would be misleading, because the calculator will be performing the calculations without intermediate rounding.

EXTRA EXAMPLE 1
If four people are sharing the cost of a monthly phone bill of \$58.25, what is each person's share of the bill? **about \$14.56**

EXTRA EXAMPLE 2
Solve $26x - 32 = 99$. Round to the nearest hundredth. **5.04**

EXTRA EXAMPLE 3
Solve $9.92x - 6.13 = 5.96 - 7.28x$. Round to the nearest hundredth. **0.70**

EXTRA EXAMPLE 4
Solve $3.7x - 2.5 = 6.1x - 12.2$. Round to the nearest tenth. **4.0**

CHECKPOINT EXERCISES
For use after Examples 1 and 2:
1. Solve $-12x - 51 = 14$. Round to the nearest hundredth. **−5.42**

For use after Examples 3 and 4:
2. Solve $9.45 - 5.2x = -8.5x + 15.78$. Round to the nearest hundredth. **1.92**

EXTRA EXAMPLE 5

You are shopping for a necklace to match your earrings. The sales tax is 6%. You have a total of \$21.25 to spend. What is the price limit for the necklace? **\$20.04**

CHECKPOINT EXERCISES

For use after Example 5:

1. You are shopping for a new bicycle. The sales tax is 5.5%. You have a total of \$100 to spend. What is your price limit for the bicycle? **\$94.78**

STUDENT HELP NOTES

Skills Review As students look back to the skill review exercises on percents, remind them that percent means "out of one hundred," so students can divide the number before the percent symbol by one hundred to write the percent as a decimal.

FOCUS ON VOCABULARY

When discussing *round-off error,* point out that the earlier the rounding is done in the solution process, the greater the round-off error. You can use the results of the activity to make this clear to students.

CLOSURE QUESTION

How is solving equations involving decimal coefficients the same as solving equations with integer coefficients? How is it different? **See margin.**

DAILY PUZZLER

When the digits of 819 are written upsidedown and written in reverse order, the result is 618. What 4-digit whole numbers can you write upsidedown, reverse, and add to the original number to get a sum of 9699? **1608 and 8091**

GOAL 2 SOLVING PROBLEMS WITH DECIMALS

EXAMPLE 5 *Using a Verbal Model*

Retail Purchase

You are shopping for earrings. The sales tax is 5%. You have a total of \$18.37 to spend. What is your price limit for the earrings?

SOLUTION

The total cost is your price limit plus the tax figured on the price limit.

PROBLEM SOLVING STRATEGY

VERBAL MODEL **Price limit** + **Sales tax rate** · **Price limit** = **Total cost**

LABELS

Price limit = x (dollars)

Sales tax rate = **0.05** (no units)

Total cost = **18.37** (dollars)

ALGEBRAIC MODEL

$x + 0.05 \cdot x = 18.37$	Write algebraic model.
$1.05x = 18.37$	Combine like terms.
$x = \frac{18.37}{1.05}$	Divide each side by 1.05.
$x \approx 17.4952381$	Use a calculator.
$x \approx 17.49$	Round down.

STUDENT HELP

Skills Review
For help with finding a percent of a number, see p. 786.

▶ Because you have a limited amount to spend, the answer is rounded *down* to \$17.49. If you round up to \$17.50, you will be a penny short because the sales tax is rounded up to \$.88.

ACTIVITY Developing Concepts

Investigating Round-Off Error

1. Work with a partner. Each of you will solve the equation below in a different way.

$$18.42x - 12.75 = (5.32x - 6.81)3.46$$

Student 1: Round to the nearest hundredth in intermediate steps.

Student 2: Round to the nearest hundredth only in the last step.

2. Compare the answers you get using both methods. Which method do you think is more accurate? Why? **See margin.**

Student 1: −1081; Student 2: −844.73; the second is more accurate because the first involves three round-off errors.

When solving an equation that has decimals, you need to realize that rounding in the intermediate steps can increase the round-off error.

Closure Question ***Sample answer:***
Solving equations with decimals is the same as solving equations with integers in that you have to isolate the variable on one side of the equation and the numbers on the other side of the equation. It is different in that you may have to round an answer or part of the equation to find a solution. You can also multiply an equation involving decimals by a power of 10 to eliminate the decimals in the equation.

GUIDED PRACTICE

Vocabulary Check ✓

1. What special notation do you need to use when you are giving an approximate answer? **the symbol "≈" (is approximately equal to)**

Concept Check ✓

2. Look back at Example 3. What number would you multiply the equation by to rewrite the equation without decimals? **100**

FIELD TRIP In Exercises 3 and 4, school buses that hold 68 people will be used to transport 377 students and 65 teachers.

3. Write and solve an equation to find the number of buses needed. ***Sample equation:* $68x = 442$; 6.5 buses; they will need a seventh bus.**

4. Is the exact answer in Exercise 3 practical? Explain. **No; the exact answer is 6.5 and it is not possible to rent half a bus.**

Skill Check ✓

Round to the nearest tenth.

5. 23.4459 **23.4** **6.** 108.2135 **108.2** **7.** −13.8953 **−13.9** **8.** 62.9788 **63.0**

Solve the equation. Round the result to the nearest hundredth. Check the rounded solution.

9. $2.2x = 15$ **6.82**

10. $14 - 9x = 37$ **−2.56**

11. $2(3b - 14) = -9$ **3.17**

12. $2.69 - 3.64x = 8.37 + 23.78x$ **−0.21**

13. BUYING A SWEATSHIRT You have \$35.72 to spend for a sweatshirt. The sales tax is 7%. What is your price limit for the sweatshirt? **\$33.38**

PRACTICE AND APPLICATIONS

STUDENT HELP

Extra Practice to help you master skills is on p. 799.

ROUNDING **Perform any indicated operation. Round the result to the nearest tenth and then to the nearest hundredth.**

14. −35.1923 **−35.2; −35.19**

15. 5.34(6.79) **36.3; 36.26**

16. −7.895 + 4.929 **−3.0; −2.97**

17. 47.0362 − 39.7204 **7.3; 7.32**

18. 5.349 ÷ 46.597 **0.1; 0.11**

19. −25.349(−1.369) **34.7; 34.70**

SOLVING AND CHECKING **Solve the equation. Round the result to the nearest hundredth. Check the rounded solution.**

20. $13x - 7 = 27$ **2.62**

21. $18 - 3y = 5$ **4.33**

22. $-7n + 17 = -6$ **3.29**

23. $38 = -14 + 9a$ **5.78**

24. $47 = 28 - 12x$ **−1.58**

25. $14r + 8 = 32$ **1.71**

26. $358 = 39c - 17$ **9.62**

27. $37 - 58b = 204$ **−2.88**

28. $3(31 - 12t) = 82$ **0.31**

29. $4(-7y + 13) = 49$ **0.11**

30. $2(-5a + 7) = -a$ **1.56**

31. $-(d - 3) = 2(3d + 1)$ **0.14**

SOLVING EQUATIONS **Solve the equation. Round the result to the nearest hundredth.**

32. $12.67 + 42.35x = 5.34x + 26.58$ **0.38**

33. $4.65x - 4.79 = 13.57 - 6.84x$ **1.60**

34. $7.45x - 8.81 = 5.29 + 9.47x$ **−6.98**

35. $39.21x + 2.65 = -31.68 + 42.03x$ **12.17**

36. $5.86x - 31.94 = 27.51x - 3.21$ **−1.33**

37. $-2(4.36 - 6.92x) = 9.27x + 3.87$ **2.75**

38. $6.1(3.1 + 2.5x) = 15.3x - 3.9$ **456.2**

39. $4.21x + 5.39 = 12.07(2.01 - 4.72x)$ **0.31**

STUDENT HELP

HOMEWORK HELP
Example 1: Exs. 14–39, 44
Example 2: Exs. 14–39
Example 3: Exs. 32–39
Example 4: Exs. 40–43
Example 5: Exs. 49–53

3 APPLY

ASSIGNMENT GUIDE

BASIC
Day 1: p. 169 Exs. 14–42 even
Day 2: p. 169 Exs. 44, 49–53, 59, 61–71, Quiz 2 Exs. 1–12

AVERAGE
Day 1: p. 169 Exs. 14–42 even
Day 2: p. 169 Exs. 44, 49–55, 59, 61–71, Quiz 2 Exs. 1–12

ADVANCED
Day 1: p. 169 Exs. 14–42 even
Day 2: p. 169 Exs. 44, 46, 48, 50–60, 66–71, Quiz 2 Exs. 1–12

BLOCK SCHEDULE
p. 169 Exs. 14–42 even (with 3.5)
p.169 Exs. 44, 49–55, 59, 61–71, Quiz 2 Exs. 1–12 (with 3.7)

EXERCISE LEVELS

Level A: *Easier*
14–19, 50, 54, 62–65

Level B: *More Difficult*
20–44, 51–53, 55, 59, 61, 66–71

Level C: *Most Difficult*
45–49, 56–58, 60

✓ HOMEWORK CHECK

To quickly check student understanding of key concepts, go over the following exercises: Exs. 16, 24, 28, 34, 42, 44, 52. See also the Daily Homework Quiz:

- Blackline Master (*Chapter 3 Resource Book*, p. 96)
- Transparency (p. 25)

STUDENT HELP NOTES

Homework Help Students can find help for Exs. 51 and 52 at **www.mcdougallittell.com.** The information can be printed out for students who don't have access to the internet.

APPLICATION NOTE

EXERCISES 55 AND 56 These exercises discuss an interesting scientific application. Students can think about other materials that expand and contract in cold and warm weather, such as asphalt or concrete on roads, and items made of wood such as decks or cabinets. Discuss with students the fact that molecules have less energy of movement and fit closer together when it is cold than when it is warm.

44, 48. See Additional Answers beginning on page AA1.

59c. The hot water should run for 5.5 min and the cold water for 3.5 min. The numbers are rounded to the nearest tenth because the numbers in the problem have one decimal place.

60. See Additional Answers beginning on page AA1.

ADDITIONAL PRACTICE AND RETEACHING

For Lesson 3.6:

- Practice Levels A, B, and C (*Chapter 3 Resource Book,* p. 85)
- Reteaching with Practice (*Chapter 3 Resource Book,* p. 88)
- See Lesson 3.6 of the *Personal Student Tutor*

For more Mixed Review:

- Search the *Test and Practice Generator* for key words or specific lessons.

FOCUS ON APPLICATIONS

REAL LIFE **COCOA** is the principle ingredient in chocolate. Typically, it takes an entire year's harvest from a mature cocoa tree to produce 1 kilogram of cocoa.

CHANGING TO INTEGER COEFFICIENTS Multiply the equation by a power of 10 to write an equivalent equation with integer coefficients.

40. $2.5x + 0.7 = 4.6 - 1.3x$
$25x + 7 = 46 - 13x$

41. $-0.625y - 0.184 = 2.506y$
$-625y - 184 = 2506y$

42. $1.67 + 2.43x = 3.29(x - 5)$
$167 + 243x = 329(x - 5)$

43. $4.5n - 0.375 = 0.75n + 2.0$
$4500n - 375 = 750n + 2000$

44. COCOA CONSUMPTION The 267.9 million people in the United States consumed 639.4 million kilograms of the cocoa produced in the 1996–1997 growing year. Which choice better represents the amount of cocoa consumed per person that year? Explain. ▶ Source: International Cocoa Organization **See margin.**

A. 2.38671146 kilograms **B.** about 2.4 kilograms

CRITICAL THINKING Describe a situation where you would round as indicated. **45–47. See margin.**

45. to nearest whole **46.** to nearest tenth **47.** to nearest hundredth

45. a problem involving physical objects that cannot be divided

46. a problem involving metric measures

47. a problem involving money

48. COMPARING ROUNDING METHODS At the 1998 winter Olympics, Marianne Timmer of the Netherlands won the women's 1000-meter speed skating race with a time of 76.51 seconds. Use a calculator to find Timmer's rates in parts (a)–(d) following each method below. Do the two methods give the same final result after rounding to the nearest tenth in part (d)? **See margin.**

Method 1 Round any decimal answers to the nearest tenth before going on.
Method 2 Use the full calculator display until you round for the final answer.

a. speed in meters per second **b.** speed in meters per minute

c. speed in meters per hour **d.** speed in kilometers per hour

49. BUYING DINNER You have \$8.39 to spend for dinner. You want to leave a 15% tip. What is your price limit for the dinner with tax included? **\$7.29**

INTERPRETING DATA **In Exercises 50–52, use the data from a survey about T-shirts.** The survey found that 93 out of 100, or 0.93, of the adults responding own at least one T-shirt. The bar graph shows where those adults who own T-shirts got their oldest T-shirt.

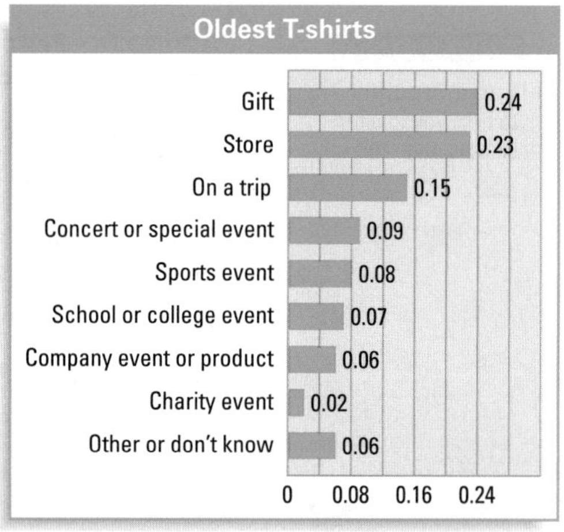

▶ Source: Opinion Research

50. Suppose the survey takers counted 186 people who own at least one T-shirt. Choose the equation you could use to find the number of people in the survey. About how many people were surveyed?

A. $x + 0.93 = 186$ **B.** $0.93 \div x = 186$ **C.** $0.93x = 186$

50. C; about 200 people

51. PROBABILITY CONNECTION What is the probability that an adult chosen at random from a group of adults who own at least one T-shirt received his or her oldest T-shirt as a gift? **0.24 (or 24%)**

52. Suppose a group of adults who own T-shirts includes 42 people who got their oldest T-shirt from a charity event. Estimate the total size of the group. **about 2100 people**

STUDENT HELP

INTERNET **HOMEWORK HELP** Visit our Web site www.mcdougallittell.com for help with Exs. 51 and 52.

FOCUS ON APPLICATIONS

BOTTLE-NOSED WHALES usually stay under the water from 15–20 minutes at a time, but they can stay under for just over an hour. Observers have even reported a two-hour dive!

53. **WHALES** A bottle-nosed whale can dive 440 feet per minute. Suppose a bottle-nosed whale is 500 feet deep and dives at this rate. Write and solve an equation to find how long it will take to reach a depth of 2975 feet. Round to the nearest whole minute. ***Sample equation*: $500 + 440x = 2975$; 6 min**

INTERNET SERVICE In Exercises 54 and 55, your Internet service provider charges \$4.95 per month for the first 3 hours of service, plus \$2.50 for each additional hour. Your total charges last month were \$21.83. Let *x* represent the number of hours you used the service.

54. Which equation models the situation? **B**

A. $4.95 + 2.5x - 3 = 21.83$ **B.** $4.95 + 2.5(x - 3) = 21.83$

55. Solve the equation you chose in Exercise 54 and round the result to the nearest hundredth. What does the result represent?
9.75 h; the total number of hours that you used your Internet service last month

SCIENCE CONNECTION In Exercises 56–58, use the following information.

Bridge sections expand as the temperature goes up, so a small *expansion gap* is left between sections when a bridge is built. As the sections expand, the width of the gap gets smaller.

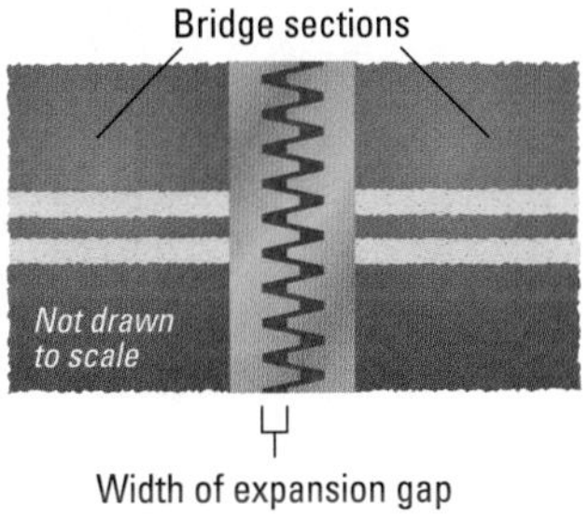

Suppose that for some bridge the expansion gap is 16.8 millimeters wide at 10°C and decreases by 0.37 millimeter for every 1°C rise in temperature.

56. If the temperature is 18°C, by how many degrees did the temperature rise? By how much would the width of the gap decrease? What would the new width of the gap be? Round to the nearest tenth of a millimeter.
8°C; 3.0 mm; 13.8 mm

57. The temperature rises to t°C. Write expressions for the temperature rise, the decrease in the width of the gap, and the new width of the gap.
$(t - 10)$°C, $0.37(t - 10)$ mm, $16.8 - 0.37(t - 10)$

58. Use the expressions from Exercise 57. Write and solve an equation to find the temperature at which the gap decreases to 9.4 millimeters.
$16.8 - 0.37(t - 10) = 9.4$; 30°C

Test Preparation

59. **MULTI-STEP PROBLEM** You are running water into a laundry sink to get a mixture that is one-half hot water and one-half cold water. The hot water flows more slowly, at a rate of 7.8 liters per minute, so you turn it on first. Two minutes later, you also turn on the cold water which flows at a rate of 12.3 liters per minute. You want to know how long to wait before turning the two faucets off.

a. Let t represent the number of minutes the hot water is on. Write a variable expression for the amount of time the cold water is on. **$t - 2$ min**

b. Write an equation that models the situation. Then solve the equation.
$7.8t = 12.3(t - 2)$; $5.4\overline{6}$ min.

c. *Writing* Interpret your solution. If the solution is a decimal, decide what form of the number is most appropriate. Explain your choice. **See margin.**

★ Challenge

60. **ROUND-OFF ERROR** For some decimal equations, you get almost the same solution whether you round early in the solution process or only at the end. For other equations, such as the one in the Activity on page 168, rounding early has a big effect on the solution. Write a decimal equation for which rounding early produces a significant difference in the answer. **See margin.**

EXTRA CHALLENGE
www.mcdougallittell.com

4 ASSESS

DAILY HOMEWORK QUIZ

Transparency Available

Solve the equation. Round the result to the nearest hundredth.

1. $53 = 89 + 14m$ **−2.57**

2. $7(12 - 9x) = 40$ **0.70**

3. $5.31t + 8.25 = 8.54t - 1.19$ **2.92**

4. In a certain rural area, the probability that an adult watches less than 10 h of TV per year is 0.013. A survey in this area found 4 adults who reported watching less than 10 h of TV per year. Estimate the size of the group surveyed. **300 people**

EXTRA CHALLENGE NOTE

Challenge problems for Lesson 3.6 are available in **blackline** format in the *Chapter 3 Resource Book*, p. 92 and at **www.mcdougallittell.com.**

ADDITIONAL TEST PREPARATION

1. OPEN ENDED Describe a situation that would require rounding to each place value:
Sample answers are given.

a. tenths **when estimating timing on a stopwatch**

b. hundredths **when calculating sales tax on an item**

c. ones **when estimating the number of people on a bus**

ADDITIONAL RESOURCES
An alternative Quiz for Lessons 3.4–3.6 is available in the *Chapter 3 Resource Book*, p. 93.

MIXED REVIEW

61.

Input	Output
2	13
3	15.5
4	18
5	20.5
6	23

domain: 2, 3, 4, 5, 6; range: 13, 15.5, 18, 20.5, 23

66. $\frac{1}{26} \approx 0.038$

61. INPUT-OUTPUT TABLES Make an input-output table for the function $A = 8 + 2.5t$ when $t = 2, 3, 4, 5$, and 6. Describe the domain and thc range of the function whose values are shown in the table. (Review 1.7 for 3.7) **See margin.**

OPPOSITES **Find the opposite of the number.** (Review 2.1)

62. 8 **−8** **63.** −3 **3** **64.** −4.9 **4.9** **65.** 7.9 **−7.9**

PROBABILITY **Find the probability of the event.** (Review 2.8)

66. Choosing the letter N from a bag that contains all 26 letters of the alphabet **See margin.**

67. Choosing a blue marble from a bag that contains 16 blue marbles and 14 white marbles $\frac{8}{15} \approx 0.533$

68. Rolling an even number using a six-sided number cube $\frac{1}{2} = 50\%$

ACCOUNT ACTIVITY **In Exercises 69–71, use the table. It shows all of the activity in a checking account during June. Deposits are positive and withdrawals are negative.** (Review 2.2)

Day	Activity
June 6	−\$225.00
June 10	+\$310.25
June 17	+\$152.33
June 25	−\$72.45
June 30	−\$400.00

69. How did the amount of money in the account change from the beginning of June through June 10? **It increased by \$85.25.**

70. Find the total amount withdrawn in June. **\$697.45**

71. What was the total change in the account balance over the course of the month? **−\$234.87**

QUIZ 2

Self-Test for Lessons 3.4–3.6

Solve the equation if possible. (Lesson 3.4)

1. $3x + 1 = 5x$ $\frac{1}{2}$

2. $8 - 2y = 21 - 6y$ $3\frac{1}{4}$

3. $3n = (6 - n)(-3)$ **no solution**

4. $\frac{1}{2}(14 + 8a) = 9a$ $1\frac{2}{5}$

5. $7 - 6d = 3(5 - 2d)$ **no solution**

6. $-7(b + 1) = 5(b - 2)$ $\frac{1}{4}$

Solve the equation. Round the result to the nearest hundredth. (Lesson 3.6)

7. $15y - 8 = 4y - 3$ **0.45**

8. $2(2n + 11) = 31 - 3n$ **1.29**

9. $7.6x + 3.7 = 2.8 - 1.6x$ **−0.10**

10. $4.9(3.7x + 1.4) = 34.7x$ **0.41**

11. PET SHOW The city is sponsoring a pet show. The number of cats, birds, and hamsters equals the number of dogs and turtles. There are 3 birds, 3 turtles, and 5 hamsters. There are twice as many dogs as cats. Let n be the number of cats. Draw a diagram to decide which equation models the situation. (Lesson 3.5) **C**

A. $2n + n = 3 + 3 + 5$

B. $n + 3 = 2n + 3 + 5$

C. $n + 3 + 5 = 2n + 3$

D. $2n + 5 = 3 + 3 + n$

12. Use the results of Exercise 11 to find the number of dogs. (Lesson 3.5) **10 dogs**

ACTIVITY 3.6

Using Technology

Graphing Calculator Activity for use with Lesson 3.6

Using a Table or Spreadsheet to Solve Equations

One way to solve multi-step equations is to use a graphing calculator or spreadsheet software on a computer to generate a table of values. The table can show a value of *x* for which the two sides of the equation are approximately equal.

EXAMPLE

Use a table on a graphing calculator to solve $4.29x + 3.89(8 - x) = 2.65x$. Round your answer to the nearest tenth.

SOLUTION

1 Use the *Table Setup* function on your graphing calculator to set up a table. Choose values beginning at 0 and increasing by 1.

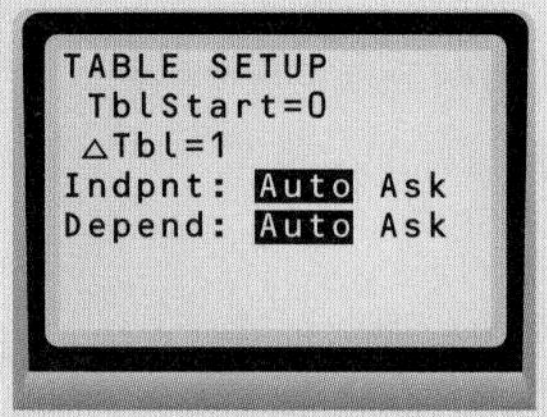

2 You want the table to show the value of the left-hand side of the equation in the second column and the value of the right-hand side of the equation in the third column.

Press Y=. Enter the left-hand side of the equation as Y_1 and the right-hand side of the equation as Y_2.

Enter × for each multiplication. It prints as *.

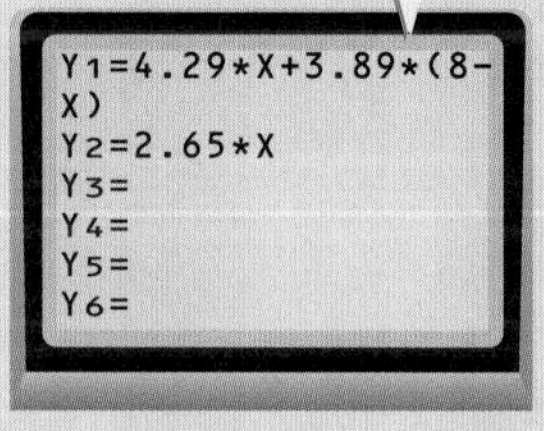

3 View your table. The first column of the table should show values of *x*. Scroll down until you find values in the second and third columns that are approximately equal. The values are closest to being equal when $x = 14$, so the solution must be greater than 13 and less than 15.

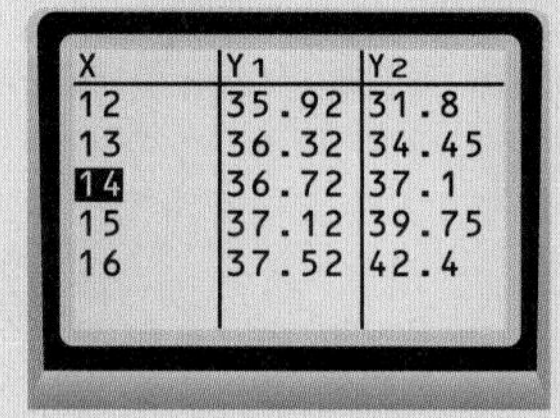

X	Y1	Y2
12	35.92	31.8
13	36.32	34.45
14	36.72	37.1
15	37.12	39.75
16	37.52	42.4

4 To find the solution to the nearest tenth, go back to the table setup and change it so that *x* starts at 13.1 and increases by 0.1. Then view your adjusted table. The values in the second and third columns are closest to being equal when $x = 13.8$.

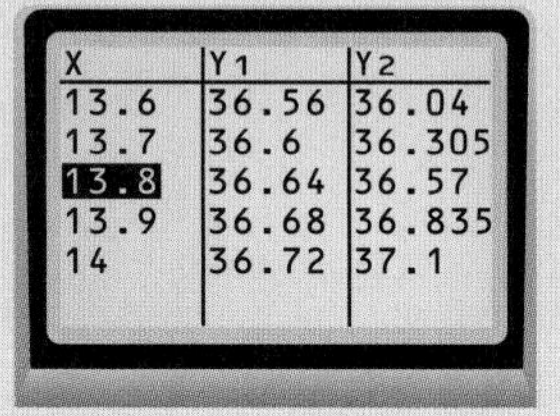

X	Y1	Y2
13.6	36.56	36.04
13.7	36.6	36.305
13.8	36.64	36.57
13.9	36.68	36.835
14	36.72	37.1

▶ The solution to the nearest tenth is 13.8.

EXERCISES

Use a graphing calculator to solve the equation. Round to the nearest tenth.

1. $19.65x + 2.2(x - 6.05) = 255.65$ **12.3**

2. $16.2(3.1 - x) - 31.55x = -19.5$ **1.5**

3. $3.56x + 2.43 = 6.17x - 11.40$ **5.3**

4. $3.5(x - 5.6) + 0.03x = 4.2x - 25.5$ **8.8**

STUDENT HELP

SOFTWARE HELP

See steps for using a computer spreadsheet as an alternative approach at www.mcdougallittell.com

1 Planning the Activity

PURPOSE

To use a graphing calculator or spreadsheet software to generate a table of values to solve equations.

MATERIALS

- graphing calculator or spreadsheet software
- Technology support blackline (*Chapter 3 Resource Book*, p. 82)

PACING

- Exploring the Concept — 15 min
- Drawing Conclusions — 10 min

LINK TO LESSON

You can find solutions correct to a specific place value with a smaller round-off error using a graphing calculator.

2 Managing the Activity

ALTERNATIVE APPROACH

Students may rewrite the equation so that 0 is on one side. Then they need to enter only the equation $Y_1 = 4.29X + 3.89(8 - X) - 2.65X$. They can then use the table feature to find where Y_1 is closest to 0.

CLASSROOM MANAGEMENT

Students may want to work in pairs to check one another's work as they enter the equations and set the table parameters. Explain that the "ΔTbl" symbol indicates how much the *x*-values in the table are changing from one row to the next.

3 Closing the Activity

KEY DISCOVERY

The table feature of a graphing calculator can be used to find approximate solutions to multi-step equations involving decimals.

ACTIVITY ASSESSMENT

Use a graphing calculator to solve the equation $6.2x + 10.5(3.9 - 0.6x) = 22.59$. **183.6**

1 PLAN

PACING
Basic: 1 day
Average: 1 day
Advanced: 1 day
Block Schedule: 0.5 block with 3.6

LESSON OPENER
APPLICATION
An alternative way to approach Lesson 3.7 is to use the Application Lesson Opener:

- Blackline Master (*Chapter 3 Resource Book*, p. 97)
- Transparency (p. 22)

MEETING INDIVIDUAL NEEDS
- ***Chapter 3 Resource Book***
  Prerequisite Skills Review (p. 5)
  Practice Level A (p. 99)
  Practice Level B (p. 100)
  Practice Level C (p. 101)
  Reteaching with Practice (p. 102)
  Absent Student Catch-Up (p. 104)
  Challenge (p. 106)
- ***Resources in Spanish***
- ***Personal Student Tutor***

NEW-TEACHER SUPPORT
See the Tips for New Teachers on pp. 1–2 of the *Chapter 3 Resource Book* for additional notes about Lesson 3.7.

WARM-UP EXERCISES
Transparency Available
Find the value of each expression when $a = 3$ and $b = -4$.

1. $a - b$ **7**
2. $2ab$ **−24**
3. $6a + 12a - b$ **58**
4. $b(5 - a)$ **−8**
5. $(-3b)(-b)$ **48**

3.7 Formulas and Functions

What you should learn

GOAL 1 Solve a formula for one of its variables.

GOAL 2 Rewrite an equation in function form.

Why you should learn it

▼ To solve **real-life** problems about rate, time, and distance as in *Pathfinder*'s flight to Mars in **Example 4**.

GOAL 1 SOLVING FORMULAS

In this lesson you will work with equations that have more than one variable. A **formula** is an algebraic equation that relates two or more real-life quantities.

EXAMPLE 1 ***Solving and Using an Area Formula***

Use the formula for the area of a rectangle, $A = lw$.

a. Find a formula for l in terms of A and w.

b. Use the new formula to find the length of a rectangle that has an area of 35 square feet and a width of 7 feet.

SOLUTION

a. Solve for length l.

$A = lw$ — **Write original formula.**

$\frac{A}{w} = l$ — **To isolate *l*, divide each side by *w*.**

b. Substitute the given values into the new formula.

$$l = \frac{A}{w} = \frac{35}{7} = 5$$

▶ The length of the rectangle is 5 feet.

EXAMPLE 2 ***Solving a Temperature Conversion Formula***

Solve the temperature formula $C = \frac{5}{9}(F - 32)$ for F.

SOLUTION

$C = \frac{5}{9}(F - 32)$ — **Write original formula.**

$\frac{9}{5} \cdot C = \frac{9}{5} \cdot \frac{5}{9}(F - 32)$ — **Multiply each side by $\frac{9}{5}$.**

$\frac{9}{5}C = F - 32$ — **Simplify.**

$\frac{9}{5}C + 32 = F - 32 + 32$ — **Add 32 to each side.**

$\frac{9}{5}C + 32 = F$ — **Simplify.**

EXAMPLE 3 Solving and Using an Interest Formula

a. The simple interest I on an investment of P dollars at an interest rate r for t years is given by $I = Prt$. Solve this formula for r.

b. Find the interest rate r for an investment of \$1500 that earned \$54 in simple interest in one year.

SOLUTION

a. $I = Prt$ — Write original formula.

$\frac{I}{Pt} = \frac{Prt}{Pt}$ — Divide each side by Pt.

$\frac{I}{Pt} = r$ — Simplify.

b. $r = \frac{I}{Pt} = \frac{54}{1500 \cdot 1} = \frac{54}{1500} = 0.036$, or 3.6%

EXAMPLE 4 Solving and Using a Distance Formula

PATHFINDER The *Pathfinder* was launched on December 4, 1996. During its 212-day flight to Mars, it traveled about 310 million miles. Estimate the time the *Pathfinder* would have taken to reach Mars traveling at the following speeds.

a. 30,000 miles per hour

b. 40,000 miles per hour

c. 60,000 miles per hour

d. 80,000 miles per hour

Which of the four speeds is the best estimate of *Pathfinder's* average speed?

SOLUTION

Use the formula $d = rt$ and solve for the time t.

$d = rt$ — Write original formula.

$\frac{d}{r} = \frac{rt}{r}$ — Divide each side by r.

$\frac{d}{r} = t$ — Simplify.

a. $t = \frac{d}{r} = \frac{310{,}000{,}000}{30{,}000} = 10{,}333$ hours ≈ 431 days

b. $t = \frac{d}{r} = \frac{310{,}000{,}000}{40{,}000} = 7{,}750$ hours ≈ 323 days

c. $t = \frac{d}{r} = \frac{310{,}000{,}000}{60{,}000} \approx 5{,}167$ hours ≈ 215 days

d. $t = \frac{d}{r} = \frac{310{,}000{,}000}{80{,}000} = 3{,}875$ hours ≈ 161 days

▶ The trip took 212 days, so *Pathfinder*'s average speed was about 60,000 mi/h.

FOCUS ON APPLICATIONS

REAL LIFE **PATHFINDER** Since its landing on July 4, 1997, the Mars Pathfinder has returned 2.6 billion bits of information, including more than 16,550 images. ▶ Source: NASA

2 TEACH

EXTRA EXAMPLE 1

a. Solve $A = \ell w$ for w. $w = \frac{A}{\ell}$

b. Find the width of a rectangle that has an area of 42 ft^2 and a length of 6 ft. **7 ft**

EXTRA EXAMPLE 2

Solve $K = \frac{5}{9}(F - 32) + 273$ for F.

$\frac{9}{5}(K - 273) + 32 = F$

EXTRA EXAMPLE 3

a. Solve $I = prt$ for t. $t = \frac{I}{Pr}$

b. Find the number of years t that \$2800 was invested to earn \$504 at 4.5%. **4 years**

EXTRA EXAMPLE 4

Voyager 1 was launched on September 5, 1977, heading for Jupiter and reached the planet on March 5, 1979. If the average distance from Earth to Jupiter is about 390 million miles, how long would it take *Voyager 1* to reach Jupiter at the following speeds:

a. 30,000 mi/h **13,000 h ≈ 542 days**

b. 40,000 mi/h **9,750 h ≈ 406 days**

c. 60,000 mi/h **6,500 h ≈ 271 days**

Which is the best estimate of its speed? **30,000 mi/h**

CHECKPOINT EXERCISES

For use after Examples 1–3:

1. a. Solve $I = Prt$ for P. $P = \frac{I}{rt}$

b. Find the amount you need to invest to get \$1950 in interest at 6.5% for 6 years. **\$5000**

For use after Example 4:

2. Refer to Extra Example 4. It took *Voyager 1* 547 days to reach Jupiter. Find the average speed in mi/h. **29,707 mi/h**

EXTRA EXAMPLE 5
Rewrite the equation $2x - y = 9$ so that y is a function of x.
$y = 2x - 9$

EXTRA EXAMPLE 6
a. Write the equation $2x - y = 9$ so that x is a function of y.
$x = \frac{9+y}{2}$
b. Use the result to find x when $y = -2, -1, 0$, and 1.
$\frac{7}{2}, 4, \frac{9}{2}, 5$

CHECKPOINT EXERCISES
For use after Examples 5 and 6:
1. a. Write the equation $5x + y = -8$ so that y is a function of x. $y = -5x - 8$
b. Use the result to find y when $x = -2, -1, 0$, and 1.
2, −3, −8, −13

STUDENT HELP NOTES
Homework Help Students can find extra examples at **www.mcdougallittell.com** that parallel the examples in the student edition.

CLOSURE QUESTION
Explain the steps in solving $p = 2\ell + 2w$ for w. *Sample answer:* First subtract 2ℓ from both sides to get $p - 2\ell = 2w$. Then divide both sides by 2 to get $\frac{p-2\ell}{2} = w$.

DAILY PUZZLER
Jason said: "When I solve my equation for x, I get the same equation as when I solve for y, but the sides are switched." If his equation contains only two variables, what equation did he get when he solved for x? $x = y$

GOAL 2 REWRITING EQUATIONS IN FUNCTION FORM

A two-variable equation is written in *function form* if one of its variables is isolated on one side of the equation. The isolated variable is the output and is a function of the input. For instance, the equation $P = 4s$ describes the perimeter P of a square as a function of its side length s.

STUDENT HELP
Look Back For help with functions, see page 46.

HOMEWORK HELP
Visit our Web site www.mcdougallittell.com for extra examples.

EXAMPLE 5 *Rewriting an Equation in Function Form*

Rewrite the equation $3x + y = 4$ so that y is a function of x.

SOLUTION

$3x + y = 4$	Write original equation.
$3x + y - 3x = 4 - 3x$	Subtract $3x$ from each side.
$y = 4 - 3x$	Simplify.

▶ The equation $y = 4 - 3x$ represents y as a function of x.

EXAMPLE 6 *Rewriting an Equation in Function Form*

a. Rewrite the equation $3x + y = 4$ so that x is a function of y.
b. Use the result to find x when $y = -2, -1, 0$, and 1.

SOLUTION

a.

$3x + y = 4$	Write original equation.
$3x + y - y = 4 - y$	Subtract y from each side.
$3x = 4 - y$	Simplify.
$\frac{3x}{3} = \frac{4-y}{3}$	Divide each side by 3.
$x = \frac{4-y}{3}$	Simplify.

▶ The equation $x = \frac{4-y}{3}$ represents x as a function of y.

b. It is helpful to organize your work in columns.

INPUT	SUBSTITUTE	OUTPUT
$y = -2$	$x = \frac{4-(-2)}{3}$	$x = 2$
$y = -1$	$x = \frac{4-(-1)}{3}$	$x = \frac{5}{3}$
$y = 0$	$x = \frac{4-(0)}{3}$	$x = \frac{4}{3}$
$y = 1$	$x = \frac{4-(1)}{3}$	$x = 1$

Guided Practice

Vocabulary Check ✓

1. Give an example of a formula. State what real-life quantity each variable represents. ***Sample answer:*** $A = \frac{1}{2}bh$**; *A* represents the area of a triangle, *b* represents the base, and *h* represents the height.**

Concept Check ✓

Tell whether the formula shows correctly the relationships among perimeter, length, and width of a rectangle.

2. $P = 2l + 2w$ **yes** **3.** $P = l + w \cdot 2$ **no** **4.** $P = 2(l + w)$ **yes**

5. $l = \frac{P - 2w}{2}$ **yes** **6.** $l = \frac{P}{2w} - 2$ **no** **7.** $w = \frac{P}{2} - 2l$ **no**

8. Write a formula that describes the side length s of a square as a function of its perimeter P. $s = \frac{1}{4}P$

Skill Check ✓

9. Rewrite the equation $2x + 2y = 10$ so that y is a function of x. $y = 5 - x$

10. Use the result of Exercise 9 to find y when $x = -2, -1, 0, 1,$ and 2. **7, 6, 5, 4, 3**

Practice and Applications

Student Help

Extra Practice to help you master skills is on p. 799.

Solving a Formula **Solve for the indicated variable.**

11. Area of a Triangle

Solve for b: $A = \frac{1}{2}bh$ $b = \frac{2A}{h}$

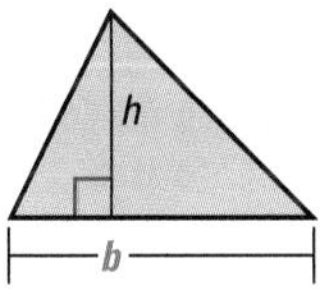

12. Volume of a Rectangular Prism

Solve for h: $V = \ell wh$ $h = \frac{V}{lw}$

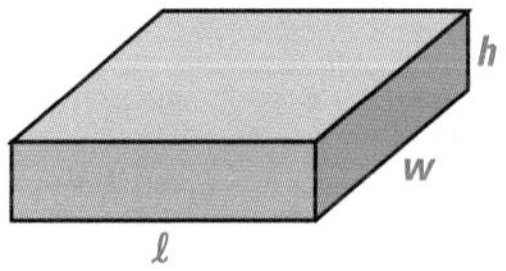

13. Volume of a Circular Cylinder

Solve for h: $V = \pi r^2 h$ $h = \frac{V}{\pi r^2}$

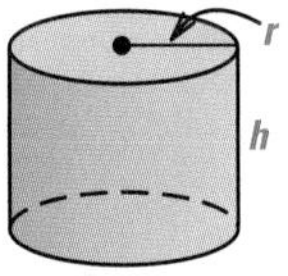

14. Area of a Trapezoid $b_2 = \frac{2A}{h} - b_1$

Solve for b_2: $A = \frac{1}{2}h(b_1 + b_2)$

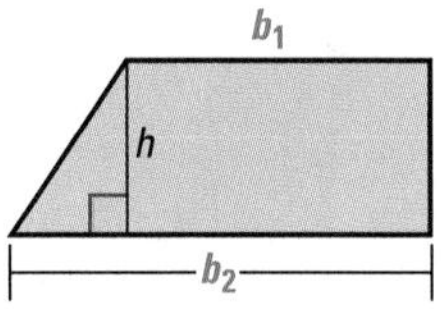

Rewriting Equations **Rewrite the equation so that *y* is a function of *x*.**

15. $2x + y = 5$ $y = 5 - 2x$

16. $3x + 5y = 7$ $y = \frac{7}{5} - \frac{3}{5}x$

17. $13 = 12x - 2y$ $y = 6x - \frac{13}{2}$

18. $2x = -3y + 10$ $y = -\frac{2}{3}x + \frac{10}{3}$

19. $9 - y = 1.5x$ $y = -1.5x + 9$

20. $1 + 7y = 5x - 2$ $y = \frac{5}{7}x - \frac{3}{7}$

21. $\frac{y}{5} - 7 = -2x$ $y = -10x + 35$

22. $\frac{1}{4}y + 3 = -5x$ $y = -20x - 12$

23. $-3x + 4y - 5 = -14$ $y = \frac{3}{4}x - \frac{9}{4}$

24. $7x - 4x + 12 = 36 - 5y$ $y = -\frac{3}{5}x + \frac{24}{5}$

25. $\frac{1}{3}(y + 2) + 3x = 7x$ $y = 12x - 2$

26. $5(y - 3x) = 8 - 2x$ $y = \frac{8}{5} + \frac{13}{5}x$

27. $4x - 3(y - 2) = 15 + y$ $y = x - \frac{9}{4}$

28. $3(x - 2y) = -12(x + 2y)$ $y = -\frac{5}{6}x$

29. $\frac{1}{5}(25 - 5y) = 4x - 9y + 13$ $y = \frac{1}{2}x + 1$

Student Help

Homework Help

Examples 1–4: Exs. 11–14, 34–38
Example 5: Exs. 15–29
Example 6: Exs. 30–33

3 Apply

Assignment Guide

Basic
Day 1: p. 177 Exs. 11–14, 16–34 even, 35, 37–39, 45, 46 50–58 even, 59–61

Average
Day 1: p. 177 Exs. 11–14, 16–34 even, 35, 37–39, 45, 46 50–58 even, 59–61

Advanced
Day 1: p. 177 Exs. 11–14, 16–34 even, 35–39, 45–49

Block Schedule
p. 177 Exs. 11–14, 16–34 even, 35, 37–39, 45, 46, 50–58 even, 59–61

Exercise Levels

Level A: ***Easier***
50–61

Level B: ***More Difficult***
11–13, 15–35, 37–39, 45, 46

Level C: ***Most Difficult***
14, 36, 40–44, 47–49

✓ Homework Check

To quickly check student understanding of key concepts, go over the following exercises: Exs. 12, 14, 18, 28, 34, 38. See also the Daily Homework Quiz:

- Blackline Master (*Chapter 3 Resource Book,* p. 109)
- Transparency (p. 26)

STUDENT HELP NOTES

- **Skills Review** As students work on the exercises on p. 784, remind them that they can multiply the decimal by 100, or move the decimal point two places to the right, to find the number that belongs in front of the percent symbol.
- **Keystroke Help** Keystrokes for several models of calculators are available in **blackline** format in the *Chapter 3 Resource Book,* p. 98 and at **www.mcdougallittell.com.**

ADDITIONAL PRACTICE AND RETEACHING

For Lesson 3.7:

- Practice Levels A, B, and C (*Chapter 3 Resource Book,* p. 99)
- Reteaching with Practice (*Chapter 3 Resource Book,* p. 102)
- See Lesson 3.7 of the *Personal Student Tutor*

For more Mixed Review:

- Search the *Test and Practice Generator* for key words or specific lessons.

32. $x = \frac{17}{14} - \frac{2}{7}y$; $1\frac{11}{14}$, $1\frac{1}{2}$, $1\frac{3}{14}$, $\frac{13}{14}$

33. $x = \frac{5}{2}y - 3$; -8, $-5\frac{1}{2}$, -3, $-\frac{1}{2}$

STUDENT HELP

Skills Review
For help with writing decimals as percents, see pages 784–785.

STUDENT HELP

KEYSTROKE HELP
Visit our Web site www.mcdougallittell.com to see keystrokes for several models of calculators.

44. $C = 2\pi r = \pi d$, so $d = \frac{C}{\pi} = \frac{102.6}{\pi} \approx 32.7$ ft

SUBSTITUTION Rewrite the equation so that *x* is a function of *y*. Then use the result to find *x* when *y* = −2, −1, 0, and 1.

30. $2x + y = 5$ $x = \frac{5}{2} - \frac{1}{2}y$; $3\frac{1}{2}$, 3, $2\frac{1}{2}$, 2

31. $3y - x = 12$ $x = 3y - 12$; -18, -15, -12, -9

32. $4(5 - y) = 14x + 3$ See margin.

33. $5y - 2(x - 7) = 20$ See margin.

DISCOUNTS In Exercises 34 and 35, use the relationship between the sale price *S*, the list price *L*, and the discount rate *r*.

34. Solve for *r* in the formula $S = L - rL$. $r = 1 - \frac{S}{L}$

35. Use the new formula to find the discount rate as a decimal and as a percent.

a. Sale price: \$80 List price: \$100 0.2; 20%

b. Sale price: \$72 List price: \$120 0.4; 40%

c. Sale price: \$12.50 List price: \$25 0.5; 50%

36. *Writing* You can solve the formula $d = rt$ for any of its variables. This means that you need to learn only one formula to solve a problem asking for distance, rate, or time. Explain.
Given *r* and *t*, use $d = rt$; given *d* and *t*, use $r = \frac{d}{t}$; given *d* and *r*, use $t = \frac{d}{r}$.

SCUBA DIVING In Exercises 37–39, use the following information.
If a scuba diver starts at sea level, the pressure on the diver at a depth of *d* feet is given by the formula $P = 64d + 2112$, where *P* represents the total pressure in pounds per square foot. Suppose the current pressure on a diver is 4032 pounds per square foot.

37. Solve the formula for *d*. $d = \frac{P - 2112}{64}$

38. What is the diver's current depth? 30 ft

39. In the original equation, is *P* a function of *d* or is *d* a function of *P*? Explain.
P is a function of *d*; *P* is the isolated variable.

CIRCLE RELATIONSHIPS In Exercises 40–44, use the information at the right. Use the *Table* feature on a graphing calculator (or spreadsheet software) as indicated.

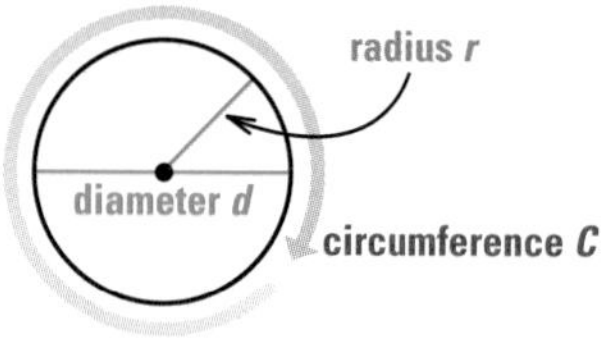

$C = 2\pi r$

40. In the formula, is the circumference a function of the radius or is the radius a function of the circumference?
Circumference is a function of the radius.

41. If *X* is the radius, what expression gives the circumference? On your graphing calculator enter this expression as Y_1. $2\pi X$

42. Solve the circumference formula for *r*. If Y_1 is the circumference, what expression involving Y_1 equals the radius? On your graphing calculator enter this expression as Y_2. To enter Y_1 into your expression, use the *Variables* menu. $Y_1/(2\pi)$

43. LOGICAL REASONING Enter 10 different values for the radius of a circle in the *X*-column. (You may need to choose "Ask" on the Table Setup menu to do this.) View your table and then explain the relationship between the three columns. The first and third columns are identical.

44. GIANT SEQUOIA The *General Sherman* tree, a sequoia tree in Sequoia National Park, has the greatest volume of any tree in the world. The distance around the tree (the circumference) measures 102.6 feet. Show how to find the distance straight through the center of the tree (the diameter) without measuring it directly. Then find this value to the nearest tenth of a foot.
See margin.

Test Preparation

45. **MULTIPLE CHOICE** Which of the choices shows how to rewrite the equation $2y - 3(y - 2x) = 8(x - 1) + 3$ so that y is a function of x? **D**

Ⓐ $x = \frac{5 - y}{2}$ Ⓑ $x = \frac{-y + 5}{10}$

Ⓒ $y = -14x + 5$ Ⓓ $y = -2x + 5$

46. **MULTIPLE CHOICE** The formula $A = \frac{1}{2}h(b_1 + b_2)$ relates the area and the dimensions of a trapezoid. Which way can you rewrite this relationship? **B**

Ⓐ $A = \frac{1}{2}hb_1 + b_2$ Ⓑ $h = \frac{2A}{b_1 + b_2}$

Ⓒ $h = \frac{A}{b_1 + b_2}$ Ⓓ $b_1 = 2A - b_2$

★ Challenge

ROPE AROUND THE WORLD In Exercises 47–49, imagine a rope wrapped around Earth at the equator (Earth's circumference *C*). Then think of adding *d* feet to the rope's length so it can now circle Earth at a distance *h* feet above the equator at all points.

Length of rope = $C + d$

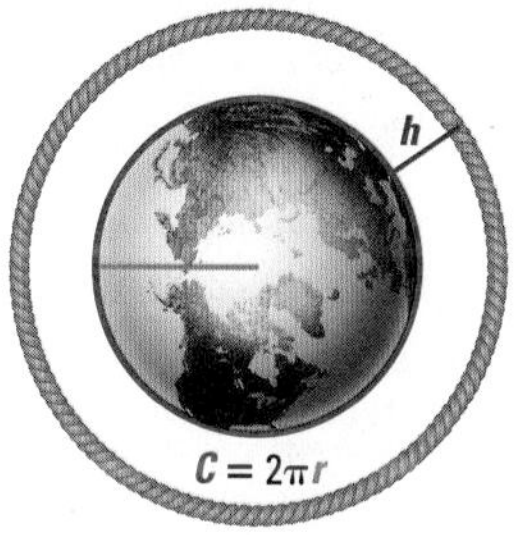

47. Write an equation to model this situation.
$C + d = 2\pi(r + h)$ or $d = 2\pi h$

48. Solve your equation in Exercise 47 for d. **$d = 2\pi h$**

49. How much rope do you need to insert if you want the rope to circle Earth 500 feet above the equator? Use 3.14 as an approximation for π. **1000π ft ≈ 3141.6 ft**

EXTRA CHALLENGE
www.mcdougallittell.com

MIXED REVIEW

CHECKING SOLUTIONS Check whether the given number is a solution of the inequality. (Review 1.4)

50. $y - 5 > 4; 9$ **no** 51. $4 + 7y > 12; 1$ **no** 52. $3x - 3 \geq x + 3; 5$ **yes**

SIMPLIFYING EXPRESSIONS Simplify the variable expression. (Review 2.5)

53. $-(2)^3(b)$ **$-8b$** 54. $(-2)^3(b)$ **$-8b$** 55. $3(-x)(-x)(-x)$ **$-3x^3$**

56. $(-4)(-c)(-c)(-c)$ **$4c^3$** 57. $(-4)^2(t)(t)$ **$16t^2$** 58. $(-5)^2(-y)(-y)$ **$25y^2$**

TEEN ATTENDANCE Each bar shows the percent of teens that attend the activity in a 12-month period. (Review 1.6 for 3.8)

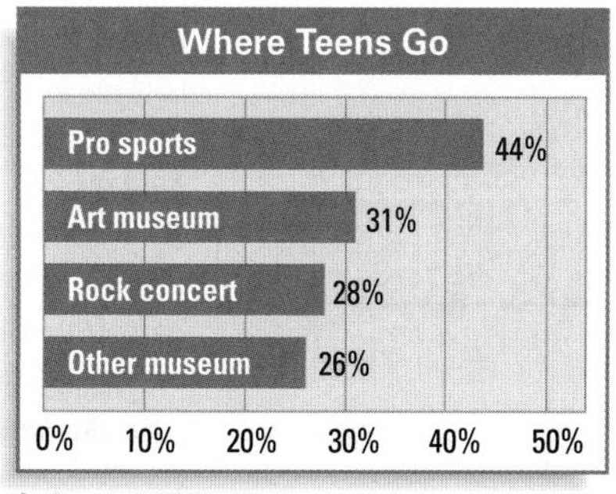

▶ Source: *YOUTHviews*

59. Which activity shown is attended by the most teens? **pro sports**

60. What percent attend rock concerts? **28%**

61. How many out of 100 attend art museums? **31**

4 ASSESS

DAILY HOMEWORK QUIZ

Transparency Available

1. Solve the formula for the volume of a square pyramid for h.
$V = \frac{1}{3}s^2h$ $h = \frac{3V}{s^2}$

Write the equation so that y is a function of x.

2. $4x + 6 = 5y - 10$ $y = \frac{4x + 16}{5}$

3. $5(1 + y) = -12(x - y)$ $y = \frac{12x + 5}{7}$

4. The length L of a piece of steel that was ℓ ft long at a certain temperature is given by $L = \ell + 0.000012\ell t$ if the temperature changes by t degrees Celsius. If a piece of steel that was 5 ft long is later measured to be 5.0009 ft, by how much did the temperature increase? **15°C**

EXTRA CHALLENGE NOTE

Challenge problems for Lesson 3.7 are available in **blackline** format in the *Chapter 3 Resource Book*, p. 106 and at **www.mcdougallittell.com.**

ADDITIONAL TEST PREPARATION

1. **WRITING** One definition of π is the ratio of the circumference of a circle, C, to the diameter of the the circle. How can this definition be used to find the circumference of a circle? How can the definition be used to find the diameter of a circle?
$C = \pi d$, or multiply the diameter by π to find the circumference; $d = \frac{C}{\pi}$, or divide the circumference of the circle by π to find the diameter

1 PLAN

PACING
Basic: 1 day
Average: 1 day
Advanced: 1 day
Block Schedule: 0.5 block with Ch. Rev.

LESSON OPENER
APPLICATION
An alternative way to approach Lesson 3.8 is to use the Application Lesson Opener:

- Blackline Master (*Chapter 3 Resource Book*, p. 110)
- Transparency (p. 23)

MEETING INDIVIDUAL NEEDS

- ***Chapter 3 Resource Book***
  Prerequisite Skills Review (p. 5)
  Practice Level A (p. 111)
  Practice Level B (p. 112)
  Practice Level C (p. 113)
  Reteaching with Practice (p. 114)
  Absent Student Catch-Up (p. 116)
  Challenge (p. 119)
- ***Resources in Spanish***
- ***Personal Student Tutor***

NEW-TEACHER SUPPORT
See the Tips for New Teachers on pp. 1–2 of the *Chapter 3 Resource Book* for additional notes about Lesson 3.8.

DATA UPDATE
Updated data for Example 1 is available at **www.mcdougallittell.com.**

WARM-UP EXERCISES

Transparency Available

Write each fraction as a decimal and as a percent.

1. $\frac{4}{5}$ 0.8; 80% **2.** $\frac{8}{4}$ 2; 200%

3. $\frac{9}{8}$ 1.125; 112.5%

Write each ratio as a fraction in simplest form.

4. 14 to 16 $\frac{7}{8}$ **5.** 25:35 $\frac{5}{7}$

EXPLORING DATA AND STATISTICS

3.8

Rates, Ratios, and Percents

***What** you should learn*

GOAL 1 Use rates and ratios to model and solve **real-life** problems, as in **Example 3**.

GOAL 2 Use percents to model and solve **real-life** problems, as in **Example 6**.

***Why** you should learn it*

▼ To solve **real-life** problems such as interpreting the results of a survey on outdoor recreation in **Exs. 40–42.**

GOAL 1 USING RATES AND RATIOS

In this lesson you will apply rates, ratios, and percents and use your equation solving skills to solve problems. If a and b are two quantities measured in different units, then the **rate of a per b** is $\frac{a}{b}$. Rates are often expressed as *unit rates.* A **unit rate** is a rate per one given unit, such as 60 miles per 1 gallon.

EXAMPLE 1 *Interpreting Large Numbers*

SPENDING How can you relate this information to individual spending?

Estimated Spending for Clothes and Food in the United States in 1996		
Clothing & Shoes	Restaurants	Food at home
\$183 billion	\$190 billion	\$300 billion

DATA UPDATE of U.S. Bureau of Labor Statistics at www.mcdougallittell.com

SOLUTION One solution is to find the average rate of spending *per person* so that people can compare themselves to the average. You can do this by dividing by the total population, which was about 266 million in 1996.

Clothing and shoes: about \$688 per person
Restaurants: about \$714 per person
Food at home: about \$1128 per person

EXAMPLE 2 *Using Collected Data*

MILEAGE You have recorded your car mileage and gasoline use for 2 months. Estimate the number of miles you can drive on a full 18-gallon tank of gasoline.

Number of miles	290	242	196	237	184
Number of gallons	12.1	9.8	8.2	9.5	7.8

SOLUTION You drove a total of 1149 miles and used 47.4 gallons of gasoline, so your average rate was 1149 ÷ 47.4, or about 24.2 miles per gallon. Let x represent the number of miles you can drive on 18 gallons. To estimate x, you can solve the following equation.

Number of miles → x
Number of gallons → 18

$$\frac{x}{18} = 24.2$$ ← **Average rate in miles per gallon**

▶ The solution is $x = 435.6$, so your estimate is 436 miles on an 18-gallon tank.

Exchange Rates

EXAMPLE 3 *Applying Unit Analysis*

a. You are visiting Mexico and you want to exchange \$180 for pesos. The rate of currency exchange is 9.990 pesos per United States dollar. You want to find out how many pesos you will receive.

UNIT ANALYSIS You can use unit analysis to write an equation.

$$\cancel{\text{dollars}} \cdot \frac{\text{pesos}}{\cancel{\text{dollar}}} = \text{pesos}$$

$$D \cdot \frac{9.990}{1} = P \qquad \textbf{Write equation.}$$

$$180 \cdot \frac{9.990}{1} = P \qquad \textbf{Substitute 180 for } \boldsymbol{D} \textbf{ dollars.}$$

$$1798.2 = P \qquad \textbf{Simplify.}$$

▶ You will receive 1798 pesos.

b. When you leave Mexico, you are surprised to find that you have exactly 180 pesos left. You want to find out how many United States dollars you will receive. The rate of currency exchange is now 9.987 pesos per dollar.

UNIT ANALYSIS $\cancel{\text{pesos}} \cdot \frac{\text{dollar}}{\cancel{\text{pesos}}} = \text{dollars}$

$$P \cdot \frac{1}{9.987} = D \qquad \textbf{Write equation.}$$

$$180 \cdot \frac{1}{9.987} = D \qquad \textbf{Substitute 180 for } \boldsymbol{P} \textbf{ pesos.}$$

▶ After simplifying, you find that $D \approx 18.02343046$. You will receive \$18.

EXAMPLE 4 *Using Ratios to Write an Equation*

TULIPS You tested a sample of 50 tulip bulbs from a shipment of 4500 bulbs for resistance to a type of mold. You found that 32 of the bulbs had a natural resistance, and 18 did not. Use these results to estimate the number of bulbs in the shipment that will have a natural resistance to the mold.

SOLUTION

One way to answer the question is to write a ratio. Let x represent the number of resistant bulbs in the shipment.

$$\frac{\textbf{Resistant bulbs in sample}}{\textbf{Total bulbs in sample}} = \frac{\textbf{Resistant bulbs in shipment}}{\textbf{Total bulbs in shipment}}$$

$$\frac{\mathbf{32}}{\mathbf{50}} = \frac{\boldsymbol{x}}{\mathbf{4500}} \qquad \textbf{Write equation.}$$

$$4500 \cdot \frac{32}{50} = x \qquad \textbf{Multiply each side by 4500.}$$

$$2880 = x \qquad \textbf{Simplify.}$$

▶ About 2880 of the 4500 bulbs in the shipment will be resistant to the mold.

FOCUS ON APPLICATIONS

TULIPS About 25% of tulip bulbs sold in the United States are grown in the United States. The Skagit Valley area, shown above, is the largest U.S. producer of tulips.

2 TEACH

EXTRA EXAMPLE 1

Estimated Spending in U.S in 1996

Medical Care	\$913 billion
Housing	\$787 billion
Transportation	\$602 billion

Find the cost per person, assuming the U.S. population in 1996 was about 266 million.
Medical and dental care: about \$3432 per person; Housing: about \$2959 per person; Transportation: about \$2263 per person

EXTRA EXAMPLE 2

You have recorded your car mileage and gasoline use for 5 weeks. Estimate the number of miles you can drive on a full 12-gallon tank of gasoline.

Number of Miles	Number of Gallons
159	5.0
237	6.9
195	6.0
215	6.5
183	5.7

about 395 mi on a 12-gal tank

EXTRA EXAMPLE 3

a. You are visiting Canada and you want to exchange \$150 for Canadian dollars. The rate of currency exchange is 1.4 Canadian dollars per United States dollar. You want to find how many Canadian dollars you will receive.
210 Canadian dollars

b. When you leave Canada you find that you have 15 Canadian dollars left. You want to find how much you can get in United States dollars. The rate of currency exchange is now 1.3 Canadian dollars per U.S. dollar. **\$11.54**

EXTRA EXAMPLE 4
You tested a sample of 100 packages from a shipment of 6000 packages. You found that 70 packages were under weight. Estimate the number of packages that will be under weight. **about 4200**

EXTRA EXAMPLE 5
What percent of the votes for Bob Dole came from California? **9.8%**

	Votes	CA Votes
Dole	39,197,469	3,828,380

EXTRA EXAMPLE 6
About 74% of water used in houses is used in bathrooms, about 5% in kitchens, and about 21% in other places. About 9 gal is used in the kitchen each day. Estimate the total gallons used each day. **about 180 gal**

CHECKPOINT EXERCISES
For use after Examples 1 and 2:

1. Car sales in the U.S. in the millions are shown below.

Year	Domestic	Total Sales
1994	7.26	8.99
1995	7.13	8.63
1996	7.25	8.53

Estimate the number of domestic sales when the total sales are 8.75 million. **about 7.24 million cars**

For use after Examples 3 and 4:

2. If $225 are equivalent to 1350 French francs, how many dollars are equivalent to 1000 French francs? **$166.67**

For use after Examples 5 and 6:

3. About 78% of the air is nitrogen, 21% is oxygen, and 1% is carbon dioxide and other gases. In 450 mL of air, how much is oxygen? **94.5 mL**

CLOSURE QUESTION
How are ratios and rates alike? How are they different? **See margin.**

GOAL 2 USING PERCENTS

REAL LIFE Survey

EXAMPLE 5 *Finding Percents*

What percent of 15–17-year-olds said that they are dating?

A Survey of Teen Dating		
Age Group	Number Surveyed	Number Who Are Dating
12–14-year-olds	1074	216
15–17-year-olds	992	482
18–19-year-olds	307	205

▶ Source: Simmons Market Research Bureau

SOLUTION To find the percent, divide the number who are dating by the number surveyed and then write the result as a percent.

$\frac{482}{992} \approx 0.485887097$, so about 49% of 15–17-year-olds said that they are dating.

EXAMPLE 6 *Using a Graph*

GLASS RECYCLING The circle graph shows the percent of glass containers recycled in the United States in a recent year. About 3.2 million tons of glass containers were recycled. Estimate the total weight of glass containers used that year.

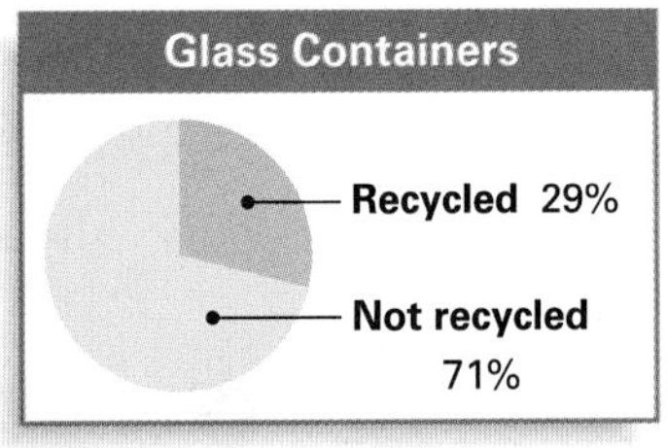

INTERNET DATA UPDATE of U.S. Environmental Protection Agency data at www.mcdougallittell.com

SOLUTION

VERBAL MODEL: Percent of glass containers recycled (in decimal form) · Total weight of glass containers used = Weight of glass containers recycled

LABELS:
Percent of glass containers recycled = **0.29** (no units)
Total weight of glass containers used = x (tons)
Weight of glass containers recycled = **3,200,000** (tons)

ALGEBRAIC MODEL:

$0.29 \cdot x = 3{,}200{,}000$ — Write algebraic model.

$x = \frac{3{,}200{,}000}{0.29}$ — Divide each side by 0.29.

$x \approx 11{,}034{,}482.76$ — Simplify.

▶ About 11.0 million tons of glass containers were used that year.

FOCUS ON APPLICATIONS

REAL LIFE **GLASS RECYCLING** When one glass bottle is recycled, the energy saved could light a 100-watt light bulb for four hours.

Closure Question *Sample answer:*
Both ratios and rates are a comparison of two numbers. Ratios use two numbers that are measured using the same units. Rates use two numbers that are measured using different units.

GUIDED PRACTICE

Vocabulary Check ✓

Concept Check ✓

1. A ratio compares quantities measured in the same units; a rate compares quantities measured in different units.

Skill Check ✓

3. Assuming that the ratio will be about the same for the entire school as for the class, let x = the number of students in the school who have a pet at home and use the equation $\frac{18}{31} = \frac{x}{1746}$.

1. Explain the difference between a rate and a ratio. **See margin.**

2. Which model would you use to change 14 yards to feet? Explain.

A. $14 \text{ yards} \cdot \frac{1 \text{ yard}}{3 \text{ feet}}$ **B.** $14 \text{ yards} \cdot \frac{3 \text{ feet}}{1 \text{ yard}}$ **B; yards $\cdot \frac{\text{feet}}{\text{yard}}$ = feet**

3. SURVEY You took a survey of your class and found that 18 out of the 31 students have a pet at home. Explain how you can use your results to make a prediction for the 1746 students in your school. **See margin.**

4. Look back at Example 1. Find the per person rate of spending for a month for each of the given categories. **clothing and shoes: about $57.33 per person; restaurants: about $59.50 per person; food at home: about $94 per person**

EXCHANGE RATES **Convert the currency using the given exchange rate.**

5. Convert 200 euros to United States dollars. (1 euro is 1.066 dollar.) **$213**

6. Convert 340 U.S. dollars to South African rand. (1 rand is 0.607 dollar.) **560 rand**

In Exercises 7–9, write and solve an equation to find the unknown number.

7. 45% of 280 = ? **0.45(280) = x; 126**

8. 7.5% of 340 = ? **0.075(340) = x; 25.5**

9. 20% of ? = 15 **0.2x = 15; 75**

10. TIPPING What percent was the server's tip if the customer left $1.75 for a $12.50 meal? **14%**

PRACTICE AND APPLICATIONS

STUDENT HELP

Extra Practice to help you master skills is on p. 799.

19. the $1.59 box; the price per bar for the smaller box is $.24, the price per bar for the larger box is about $.23.

STUDENT HELP

HOMEWORK HELP

Example 1: Exs. 11–21
Example 2: Exs. 11–21, 27–29
Example 3: Exs. 23–26, 30–31
Example 4: Ex. 32
Example 5: Exs. 33–37
Example 6: Exs. 38–42

UNIT RATES **Find the unit rate.**

11. $2 for 5 cans of dog food **$.40/can**

12. $121.50 for working 18 hours **$6.75/h**

13. $1.39 for $1\frac{1}{2}$ quarts of juice **about $.93/quart**

14. 6 ounces for 2.5 servings **2.4 oz/serving**

FINDING SPEEDS **In Exercises 15–18, find the average speed.**

15. Fly 1200 miles in 4 hours **300 mi/h**

16. Hike 52 miles in 3 days **$17\frac{1}{3}$ mi/day**

17. Swim 2 miles in 40 minutes **0.05 mi/min**

18. Drive 69 kilometers in $\frac{3}{4}$ hour **92 km/h**

19. *Writing* A store sells a box of 5 frozen yogurt bars for $1.20. The store also sells a box containing 7 of the same frozen yogurt bars for $1.59. Which is the better buy? Explain how you decided. **See margin.**

IN ONE YEAR **In Exercises 20 and 21, find the average rate as indicated.**

20. The United States postal service receives 38 million changes of address per year. Find the rate per day. Round to the nearest thousand. **104,000 changes of address/day**

21. 1,100,000,000 miles are driven to move the mail in the United States per year. Find the rate per week. Round to the nearest million miles. **21,000,000 mi/wk**

22. EQUATION SOLVING Give a calculator keystroke sequence that you could use to find the value of x in Example 4.

Sample answer: **4500 × 32 ÷ 50 =**

3 APPLY

ASSIGNMENT GUIDE

BASIC
Day 1: p. 183 Exs. 12–28 even, 29, 33–35, 39, 44–46, 59, Quiz 3 Exs. 1–12

AVERAGE
Day 1: p. 183 Exs. 12–28 even, 29, 33–35, 39, 40–42, 44–46, 59, Quiz 3 Exs. 1–12

ADVANCED
Day 1: p. 183 Exs. 12–28 even, 29, 33–35, 39, 40–48, 59, Quiz 3 Exs. 1–12

BLOCK SCHEDULE
p. 183 Exs. 12–28 even, 29, 33–35, 39, 40–48, 59, Quiz 3 Exs. 1–12 (with Ch. 3 Review)

EXERCISE LEVELS

Level A: *Easier*
11–18

Level B: *More Difficult*
20–42, 44–46, 49–59

Level C: *Most Difficult*
19, 43, 47, 48

✓ HOMEWORK CHECK

To quickly check student understanding of key concepts, go over the following exercises: Exs. 12, 18, 20, 24, 34, 39. See also the Daily Homework Quiz:

- Blackline Master (*Chapter 4 Resource Book*, p. 11)
- Transparency (p. 28)

! COMMON ERROR

EXERCISES 30–32 Students may set up their proportions incorrectly. Emphasize the importance of writing words before using numbers or variables in the proportions, and explain again the importance of keeping related information in the correct position in the ratios or rates.

ADDITIONAL PRACTICE AND RETEACHING

For Lesson 3.8:

- Practice Levels A, B, and C (*Chapter 3 Resource Book,* p. 111)
- Reteaching with Practice (*Chapter 3 Resource Book,* p. 114)
- See Lesson 3.8 of the *Personal Student Tutor*

For more Mixed Review:

- Search the *Test and Practice Generator* for key words or specific lessons.

MEASUREMENT In Exercises 23–26, convert the measure. Round your answer to the nearest tenth.

23. 12 fluid ounces to cups (1 cup = 8 fluid ounces) **$1\frac{1}{2}$ c, or 1.5 c**

24. 21 inches to centimeters (1 inch = 2.54 centimeters) **53.3 cm**

25. 56 miles to kilometers (1 mile = 1.609 kilometers) **90.1 km**

26. length of a football field (100 yards, not including the end zones) to meters (1 yard = 0.9144 meter) **91.4 m**

27. CAR MILEAGE A car uses fuel at a rate of 15 miles per gallon. Estimate how many miles the car can travel on 20 gallons of fuel. **about 300 mi**

BOOKS A library has 14,588 books which fill its 313 equal-size shelves.

28. What is the average number of books per shelf? **about 47 books**

29. The library plans to install 50 new shelves of this size. Write and solve an equation to estimate how many more books the library will be able to hold. ***Sample equation*: $50 \cdot 47 = x$; about 2350 more books**

EXCHANGE RATES In Exercises 30 and 31, use 114.520 yen per United States dollar as the rate of currency exchange. You are visiting Japan and have taken $325 to spend on your trip.

30. If you exchange the entire amount, how many yen will you receive? **37,219 yen**

31. You have 605 yen left after your trip. How many dollars will you get back? **$5**

32. CALLING FROM THE AIR You are conducting a survey on the use of airplane phones. You survey 320 adults and find that 288 of them never made a phone call from an airplane. If you surveyed 3500 adults, how many of them would you predict *have* made a phone call from an airplane? Explain. **See margin.**

32. about 350 people; the number of people who did make calls was $320 - 288 = 32$; let x = the number of people in the larger survey who made calls; $\frac{32}{320} = \frac{x}{3500}$; $x = 350$

FINDING PERCENTS Find the percent. Round to the nearest whole percent.

33. $2.25 tip on a cab fare of $14 **16%**

34. Tax of $.68 on an item priced at $11.29 **6%**

35. 292 people in favor out of 450 people surveyed **65%**

MOON ROCKS In Exercises 36–38, round to the nearest tenth. A total of 382 kilograms of lunar samples (rocks, dust, and so on) were collected during the six Apollo moon landings between 1969 and 1972.

STUDENT HELP

HOMEWORK HELP Visit our Web site www.mcdougallittell.com for help with Exs. 36–38.

APPLICATION LINK Visit our Web site www.mcdougallittell.com for more information about moon rocks.

36. The largest rock collected weighs 11.7 kilograms. This single rock is what percent of the total weight of the samples? **about 3.1%**

37. The 110.5 kilograms of lunar samples collected by the Apollo 17 astronauts represent what percent of the total weight of the samples? **about 28.9%**

38. About 7.5% of the lunar samples (by weight) have been analyzed and then returned for storage in the Return Sample Vault at NASA's Johnson Space Center. What is the combined weight of the samples in this vault? **about 28.7 kilograms**

39. CONSUMER ACTION Americans ages 18 years and older were asked, *"Have you ever stopped buying a product specifically because the manufacturer of that product pollutes the environment?"* Of those surveyed, 719 said no. About how many people took part in the survey? **about 1013 people**

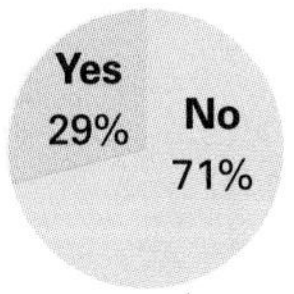

▶Source: Wirthlin Worldwide

In the survey, Basketball was the most popular team sport.

INTERPRETING DATA In Exercises 40–42, use the graph. It shows the percent of 16-to-24-year-olds participating in the top seven categories of outdoor recreation.

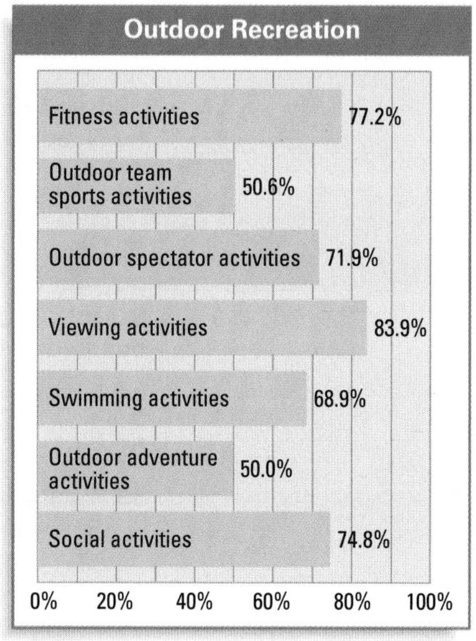

▶Source: National Survey on Recreation and the Environment

40, 41. Estimates may vary.

40. Viewing activities include sightseeing, visiting a nature center, and so on. Estimate how many students participate in viewing activities in a college with 1400 students. **about 1175 students**

41. Estimate the survey size if a survey of 16-to-24-year-olds found that 270 of them participate in fitness activities. **about 350 students**

42. Write your own question based on the survey data. Then find the answer.

42. *Sample answer*: If 299 students say they participate in social activities, estimate the number of students surveyed; 400 students.

43. **CRITICAL THINKING** Compare Example 2 and Example 4 in Goal 1. How are the methods shown alike? How are they different? **See margin.**

43. *Sample answer*: Both involve setting a ratio equal to another number. In Example 4, the other number is also a ratio.

Test Preparation

QUANTITATIVE COMPARISON In Exercises 44–46, choose the statement below that is true about the given values of *x*.

Ⓐ The value of x in column A is greater.

Ⓑ The value of x in column B is greater.

Ⓒ The two values of x are equal.

Ⓓ The relationship cannot be determined from the given information.

	Column A	Column B	
44.	The solution of the equation 160% of 400 = x	The solution of the equation 16% of 4000 = x	C
45.	The solution of the equation $\frac{x}{3} = \frac{7}{8}$	The solution of the equation $\frac{x}{8} = \frac{7}{3}$	B
46.	The solution of the equation 15% of x = 30	The solution of the equation 30% of x = 15	A

★ Challenge

LITERATURE CONNECTION Use this information in Exercises 47 and 48.

In *Citizen of the Galaxy* by Robert Heinlein, Thorby learns that he owns "an interest in a company . . . through a chain of six companies—18% of 31% of 43% of 19% of 44% of 27%, a share so microscopic that he lost track." Let x represent the total worth of the company in which Thorby owns a share.

47, 48. Answers may vary depending on rounding.

47. Write and simplify an expression to show Thorby's share of the company. **See margin.**

48. Suppose Thorby's share of the company is worth 1000 credits. Write and solve an equation to find the total worth of the company. **$0.0005x = 1000$; about 2,000,000 credits**

47. $(0.18)(0.31)(0.43)(0.19) \cdot (0.44)(0.27)x$, or about $0.0005x$

EXTRA CHALLENGE

www.mcdougallittell.com

4 ASSESS

DAILY HOMEWORK QUIZ

Transparency Available

1. What is the average speed of a plane that flies 2629 mi in 5.5 h? **478 mi/h**
2. A moving company one year had an income of $930,500. The number of moving jobs was 882. What was the average rate per job? **$1055/job**
3. Find the percent. Round to the nearest whole percent. Tip of $5.50 on a restaurant bill of $32.50. **17%**
4. A survey of telephone customers in one city asked, "If you could save $15 per month on long distance calls, would you change your long distance company?" Of those surveyed, 76 answered yes. About how many customers took part in the survey? **about 633 people**

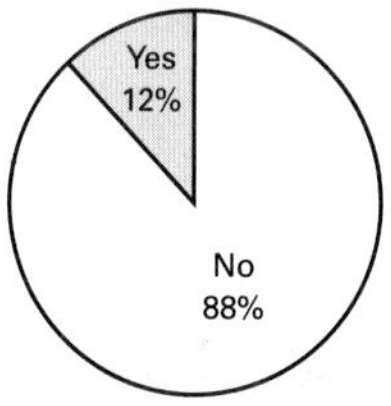

EXTRA CHALLENGE NOTE

Challenge problems for Lesson 3.8 are available in **blackline** format in the *Chapter 3 Resource Book*, p. 119 and at **www.mcdougallittell.com.**

ADDITIONAL TEST PREPARATION

1. **WRITING** "Percent" means "out of 100," and all percents can be written as fractions. Relate how ratios, rates, and percents all use fractions. **See margin.**

Additional Test Preparation ***Sample answer:***
Ratios are fractions that compare two numbers using the same unit of measure; rates are fractions using two numbers with different units of measure; percents are fractions that write the percentage out of 100.

49–54, 59. See Additional Answers beginning on page AA1.

MIXED REVIEW

WRITING INEQUALITIES Graph the numbers on a number line. Then write two inequalities that compare the two numbers. (Review 2.1)
49–54. See margin for graphs.

49. 4 and -3
$4 > -3$; $-3 < 4$

50. -2.4 and 3.2
$-2.4 < 3.2$; $3.2 > -2.4$

51. -0.2 and -0.21
$-0.2 > -0.21$; $-0.21 < -0.2$

52. -1.8 and -2
$-1.8 > -2$; $-2 < -1.8$

53. $\frac{3}{4}$ and $-\frac{5}{6}$
$\frac{3}{4} > -\frac{5}{6}$; $-\frac{5}{6} < \frac{3}{4}$

54. $-1\frac{1}{3}$ and -1.75
$-1\frac{1}{3} > -1.75$; $-1.75 < -1\frac{1}{3}$

SOLVING EQUATIONS Solve the equation. Round the result to the nearest hundredth. Check the rounded solution. (Review 3.6)

55. $-11a - 16 = 5$ **−1.91**

56. $2(4 - 3x) = -x$ **1.6**

57. $2.62x - 4.03 = 7.65 - 5.34x$ **1.47**

58. $2.6(4.2y - 6.5) = -7.1y - 2.8$ **0.78**

59. **POPULATION PROJECTIONS** The table shows the projected number (in millions) of people 85 years and older in the United States for different years. Make a line graph of the data. (Review 1.6 for 4.1) See margin.

Year	2000	2010	2020	2030	2040	2050
Number of people 85 years and older (in millions)	4.1	5.0	5.0	5.8	8.3	9.6

▶Source: *Statistical Abstract of the United States, 1997*

QUIZ 3

Self-Test for Lessons 3.7 and 3.8

Write *y* as a function of *x* and evaluate when *x* is −2, 0, and 2. (Lesson 3.7)

1. $3x + 2y = 12$
$y = -\frac{3}{2}x + 6$; 9, 6, 3

2. $\frac{y}{3} - 4 = x$
$y = 3x + 12$; 6, 12, 18

3. $\frac{1}{2}(2x + 10y) = 8$
$y = -\frac{1}{5}x + 1\frac{3}{5}$; 2, $1\frac{3}{5}$, $1\frac{1}{5}$

Solve the given formula for the specified variable. (Lesson 3.7)

4. Solve $V = lwh$ for w.
$w = \frac{V}{lh}$

5. Solve $A = \frac{1}{2}bh$ for h.
$h = \frac{2A}{b}$

6. Solve $P = C + rC$ for r.
$r = \frac{P}{C} - 1$

Convert the given currency or measure. (Lesson 3.8)

7. 125 Swiss francs to United States dollars (1 Swiss franc is 0.746 dollar.) **$93**

8. 50.8 centimeters to inches (1 inch = 2.54 centimeters) **20 in.**

In Exercises 9–11, write and solve an equation to find the answer. (Lesson 3.8)
9–12. Equations may vary.

9. **CAR MILEAGE** A car uses fuel at a rate of 33 miles per gallon. Estimate how many miles the car will travel on 12 gallons of fuel. $33 \cdot 12 = x$; 396 mi

10. **POPULATION** A school's population increased by 42 students. This was a 3% increase. About how many students attended before the increase?
$0.03x = 42$; 1400 students

11. **SURVEY** Out of 525 people surveyed, 210 answered yes. Estimate how many people would answer yes if 900 people were surveyed.

11. $900 \cdot \frac{210}{525} = x$; about 360 people

12. Use the information in Exercise 11. What percent of the 525 people surveyed answered yes? (Lesson 3.8) **40%**

CHAPTER 3

Extension

Inductive and Deductive Reasoning

What you should learn

GOAL Identify and use inductive and deductive reasoning.

Why you should learn it

To use logical reasoning in mathematics and in everyday life as in **Examples 3 and 4**.

GOAL USING INDUCTIVE AND DEDUCTIVE REASONING

Reasoning is used in mathematics, science, and everyday life. When you reach a conclusion, called a **generalization**, based on several observations, you are using **inductive reasoning**. Such a generalization is not always true. If a counterexample is found, then the generalization is proved to be false.

EXAMPLE 1 *Inductive Reasoning in Everyday Life*

You go to the school library every day for a week to do research for a paper. You notice that the computer you plan to use is not in use at 3:00 each day.

a. Using inductive reasoning, what would you conclude about the computer?

b. What counterexample would show that your generalization is false?

SOLUTION

a. From your observations, you conclude that the computer is available every weekday at 3:00.

b. A counterexample would be to find the computer in use one day at 3:00.

STUDENT HELP

Look Back
For help with counterexamples, see p. 66.

You use inductive reasoning when you make generalizations about number patterns. You assume the next number in the sequence will follow the pattern.

EXAMPLE 2 *Inductive Reasoning and Sequences*

Using inductive reasoning, name the next three numbers of the sequence 1, 5, 9, 13, 17, 21, ...

SOLUTION

The first step is to look for a pattern in the list of numbers. In this case, each number in the list is 4 more than the previous number ($1 + 4 = \mathbf{5}$, $5 + 4 = \mathbf{9}$, $9 + 4 = \mathbf{13}$, $13 + 4 = \mathbf{17}$). The next three numbers would be $21 + 4 = \mathbf{25}$, $25 + 4 = \mathbf{29}$, and $29 + 4 = \mathbf{33}$.

When you use facts, definitions, rules, or properties to reach a conclusion, you are using **deductive reasoning**. A conclusion reached in this way is proved to be true. Deductive reasoning often uses **if-then statements**. The *if* part is called the **hypothesis** and the *then* part is called the **conclusion**. When deductive reasoning has been used to *prove* an if-then statement then the fact that the hypothesis is true implies that the conclusion is true.

EXTRA EXAMPLE 1

Scott took his younger brother Michael to seven professional baseball games during July. They sat in the same section each time and caught at least one fly ball during the pre-game warm-up.

a. Using inductive reasoning, what would Scott and Michael conclude about catching fly balls? **They will always catch a fly ball during warm-ups.**

b. What counterexample would show that their generalization is false? **attending a professional baseball game and not catching a fly ball**

EXTRA EXAMPLE 2

Using inductive reasoning, name the next three numbers in the sequence 1, 3, 7, 15, 31, 63, ... *Sample answer:* **127, 255, and 511**

CHECKPOINT EXERCISES

For use after Examples 1 and 2:

1. Scott tells Michael that he is thinking of a sequence of numbers and that the first three numbers are 1, 2, and 3. He asks Michael what number comes next.

a. Using inductive reasoning, what number will Michael probably say is the next number? **4**

b. Scott says the sequence is the sequence of digits after the decimal point in the decimal for $\frac{123}{999}$ and that the fourth number for the sequence is not 4 but 1. Is this a counterexample for Michael's generalization? **yes**

EXTRA EXAMPLE 3

Scott tells Michael, "If you catch two fly balls during one game, then I will pay you $10." Michael catches two fly balls during one game.

a. Identify the hypothesis and conclusion of the if-then statement. **hypothesis: you catch two fly balls during one game; conclusion: I will pay you $10**

b. Use deductive reasoning to determine whether Michael gets $10. **Michael catches two fly balls during one game, so he will get paid $10.**

EXTRA EXAMPLE 4

Use deductive reasoning and the properties of addition to show that $(-1 + s) + 1 = s$.

Sample answer: **You can use the following steps. The properties used are given below the steps.**

$$\begin{aligned}(-1 + s) + 1 &= (s + (-1)) + 1\\ &= s + ((-1) + 1)\\ &= s + (0)\\ &= s\end{aligned}$$

commutative property; associative property; addition property of opposites; identity property of 0

CHECKPOINT EXERCISES

For use after Examples 3 and 4:

1. The following steps show that $(a - b) + (b - a) = 0$. Give a reason for each step.

$$\begin{aligned}(a - b) + (b - a) &=\\ (a + (-b)) + (b + (-a))\\ &= a + ((-b) + b + (-a))\\ &= a + (0 + (-a))\\ &= a + (-a)\\ &= 0\end{aligned}$$

See margin.

MATHEMATICAL REASONING

You may wish to discuss more examples like the one used in the Checkpoint Exercise for Examples 1 and 2. By examining several such examples, students may come to realize that listing a few terms of a number sequence can never fully specify the sequence. There needs instead to be a specific general rule describing how the terms of the sequence are generated.

EXAMPLE 3 Use of If-Then Statements

During the last week of a class, your teacher tells you that if you receive an A on the final exam, you will have earned an A average in the course.

a. Identify the hypothesis and the conclusion of the if-then statement.

b. Supose you receive an A on the final exam. What will your final grade be?

SOLUTION

a. The hypothesis is "you receive an A on the final exam." The conclusion is "you will have earned an A average in the course."

b. You can conclude that you have earned an A in the course.

EXAMPLE 4 Deductive Reasoning in Mathematics

STUDENT HELP

Look Back
For help with the properties of addition, see p. 73.

To show that the statement "If a, b, and c are real numbers, then $(a + b) + c = (c + b) + a$" is true by deductive reasoning, you can write each step and justify it using the properties of addition.

$(a + b) + c = c + (a + b)$	**Commutative property**
$= c + (b + a)$	**Commutative property**
$= (c + b) + a$	**Associative property**

EXERCISES

In Exercises 1–4, identify the reasoning as inductive or deductive. Explain.

1. The tenth number in the list 1, 4, 9, 16, 25, . . . is 100.
inductive reasoning; a conclusion is reached based on observation of a pattern.

2. You know that in your neighborhood, if it is Sunday, then no mail is delivered. It is Sunday, so you conclude that the mail will not be delivered.
deductive reasoning; a conclusion is reached because the hypothesis is true.

3. If the last digit of a number is 2, then the number is divisible by 2. You conclude that 98,765,432 is divisible by 2.
deductive reasoning; a conclusion is reached because the hypothesis is true.

4. You notice that for several values of x the value of x^2 is greater than x. You conclude that the square of a number is greater than the number itself.
inductive reasoning; a conclusion is reached based on observation of a pattern.

5. Find a counterexample to show that the conclusion in Exercise 4 is false.
Sample counterexample: **The square of 1 is equal to itself.**

6. Give an example of inductive reasoning and an example of deductive reasoning. Do not use any of the examples already given. **See margin.**

7. Name the next three numbers of the sequence: 0, 3, 6, 9, 12, 15, 18, . . . **21, 24, 27**

In Exercises 8 and 9, identify the hypothesis and the conclusion of the if-then statement.

8. If you add two odd numbers, then the answer will be an even number.
hypothesis: "you add two odd numbers"; conclusion: "the answer will be an even number"

9. If you are in Minnesota in January, then you will be cold.
hypothesis: "you are in Minnesota in January"; conclusion: "you will be cold"

10. Use deductive reasoning and the properties of addition to show that if z is a real number, then $(z + 2) + (-2) = z$. **See margin.**

6. **Answers may vary. *Sample answer:* Inductive reasoning: You see your neighbor leave the house every morning at 6 A.M. for 5 consecutive days. You conclude that your neighbor will leave the house every morning at 6 A.M. Deductive reasoning: You know that if the temperature is −20°F or lower, then you should wear a heavy hat and coat. The temperature outside is −30°F, so you conclude you should wear a heavy hat and coat.**

Checkpoint Exercises *Sample answer:*
Subtracting a number is equivalent to adding its opposite; associative property; addition property of opposites; identity property of 0; addition property of opposites

CHAPTER 3

Chapter Summary

10. $(z + 2) + (-2) = z + (2 + (-2))$
[Associative property]
$= z + 0$
[Inverse property]
$= z$
[Identity property]

WHAT did you learn?	*WHY did you learn it?*
Solve a linear equation	
• using addition and subtraction. (3.1)	Model and find increases and decreases. (p. 134)
• using multiplication and division. (3.2)	Find your distance from a thunderstorm. (p. 143)
• using two or more transformations. (3.3)	Find the temperature below Earth's surface. (p. 147)
• with variables on both sides. (3.4)	Decide whether a membership is economical. (p. 156)
• involving decimals. (3.6)	Find the expansion gap for a bridge. (p. 171)
Write an equation using equal ratios. (3.2)	Find unknown side lengths in similar triangles. (p. 141)
Solve a formula for one of its variables. (3.7)	Estimate the speed on Pathfinder's flight to Mars. (p. 175)
Rewrite an equation in function form, and find the output given the input. (3.7)	Prepare for graphing linear equations in two variables. (p. 176)
Use linear equations to model and solve real-life problems	
• using a diagram. (3.5)	Design a high school yearbook. (p. 160)
• using a table or a graph to check. (3.5)	Understand when a gazelle is a safe distance from a cheetah. (p. 162)
• involving rates, ratios, and percents. (3.8)	Estimate how far a car can travel on a tank of gasoline. (p. 180)

How does Chapter 3 fit into the BIGGER PICTURE of algebra?

In Chapters 1 and 2 you used mental math to solve simple equations. In this chapter you learned systematic equation-solving techniques that allow you to solve more complicated equations such as $0.5(x - 4) - (x - 8) = 16$. These techniques are based on the rules of algebra and the methods for simplifying you learned in Chapter 2.

You will need to solve linear equations when you solve linear inequalities in Chapter 6 and systems of equations in Chapter 7.

STUDY STRATEGY

How did you use your formula cards?

One of the formula cards you made, using the **Study Strategy** on p. 130, may look like this one.

The Distance Formula

$D = rt$

D = distance (in feet, miles, kilometers, etc.)
t = time (in minutes, hours, etc.)
r = rate (in feet per minute, miles per hour, etc.)

For example: If D is in kilometers and t is in hours, then r is in kilometers per hour.

ADDITIONAL RESOURCES

The following resources are available to help review the material in this chapter.

- Chapter Review Games and Activities (*Chapter 3 Resource Book*, p. 120)
- *Instant Replay: Video Review Games*
- *Personal Student Tutor*
- Cumulative Review, Chs. 1–3 (*Chapter 3 Resource Book*, p. 132)

CHAPTER 3

Chapter Review

VOCABULARY

- equivalent equations, p. 132
- inverse operations, p. 132
- solution step, p. 133
- linear equation in one variable, p. 133
- properties of equality, p. 139
- ratio of *a* to *b*, p. 140
- similar triangles, p. 140
- identity, p. 155
- round-off error, p. 166
- formula, p. 174
- rate of *a* per *b*, p. 180
- unit rate, p. 180

3.1–3.2 SOLVING EQUATIONS USING ONE OPERATION

Examples on pp. 132–134, 138–141

EXAMPLES Use inverse operations to isolate the variable.

$y - 4 = -6$ Write original equation.

$y = -2$ Add 4 to each side.

$2 - x = 12$ Write original equation.

$-x = 10$ Subtract 2 from each side.

$x = -10$ x is the opposite of 10.

$\frac{1}{8}m = -5$ Write original equation.

$m = -40$ Multiply each side by 8.

$-7n = 28$ Write original equation.

$n = -4$ Divide each side by -7.

Solve the equation.

1. $y - 15 = -4$ 11 **2.** $-7 + x = -3$ 4 **3.** $25 = -35 - c$ −60 **4.** $-11 = z - (-15)$ −26

5. $36 = \frac{h}{-12}$ −432 **6.** $-\frac{2}{3}w = -70$ 105 **7.** $6m = -72$ −12 **8.** $\frac{y}{4} = \frac{15}{6}$ 10

3.3 SOLVING MULTI-STEP EQUATIONS

Examples on pp. 145–147

EXAMPLE Solving some equations requires two or more transformations.

$-2p - (-5) - 2p = 13$ Write original equation.

$-4p + 5 = 13$ Simplify.

$-4p = 8$ Subtract 5 from each side.

$p = -2$ Divide each side by -4.

Solve the equation.

9. $26 - 9p = -1$ 3 **10.** $\frac{4}{5}c - 12 = -32$ −25 **11.** $\frac{y}{4} + 2 = 0$ −8 **12.** $-2(4 - x) - 7 = 5$ 10

13. $6r - 2 - 9r = 1$ −1 **14.** $16 = 5(1 - x)$ $-\frac{11}{5}$ **15.** $-\frac{2}{3}(6 - 2a) = 6$ $\frac{15}{2}$ **16.** $n - 4(1 + 5n) = -2$ $-\frac{2}{19}$

3.4 SOLVING EQUATIONS WITH VARIABLES ON BOTH SIDES

Examples on pp. 154–156

EXAMPLES To solve, try to collect the variable terms on one side of the equation.

An equation with one solution:	An equation with no solution:	An equation with many solutions:
$-21d + 15 = -5d + 7$	$-3(2x - 5) = -(15 + 6x)$	$2s - 5s + 11 = 2 - 3s + 9$
$15 = 16d + 7$	$-6x + 15 = -15 - 6x$	$-3s + 11 = 11 - 3s$
$8 = 16d$	$15 = -15$	$11 = 11$
$\frac{1}{2} = d$	$15 \neq -15$ for any value of x, so the original equation has no solution.	The equation $11 = 11$ is always true, so all values of s are solutions.

Solve the equation if possible.

17. $9z + 24 = -3z$ **−2**

18. $12 - 4h = -18 + 11h$ **2**

19. $24a - 8 - 10a = -2(4 - 7a)$ **all real numbers**

20. $9(-5 - r) = -10 - 2r$ **−5**

21. $\frac{2}{3}(3x - 9) = 4(x + 6)$ **−15**

22. $6m - 3 = 10 - 6(2 - m)$ **no solution**

3.5 LINEAR EQUATIONS AND PROBLEM SOLVING

Examples on pp. 160–162

EXAMPLE A diagram can help you to understand a problem. Two runners leave the starting line at the same time. The first runner crosses the finish line in 25 minutes and averages 12 kilometers per hour. The second runner crosses the finish line 5 minutes after the first runner. Find the second runner's speed.

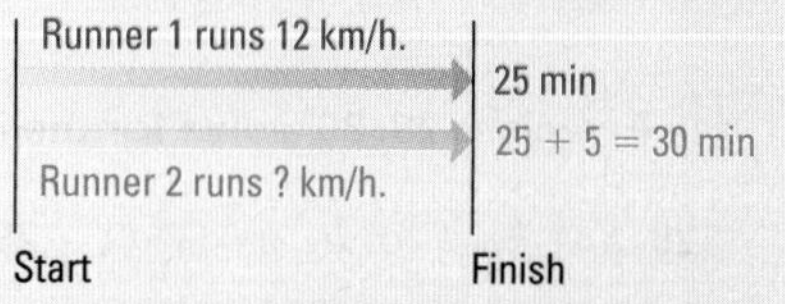

VERBAL MODEL

Speed of first runner · Time of first runner = Speed of second runner · Time of second runner

LABELS

Speed of first runner = **12** (kilometers per hour)

Time of first runner = $\mathbf{\frac{5}{12}}$ (hour)

Speed of second runner = x (kilometers per hour)

Time of second runner = $\mathbf{\frac{1}{2}}$ (hour)

23. Write and solve the equation for the example. Then answer the question. $12\left(\frac{5}{12}\right) = x\left(\frac{1}{2}\right)$; **10 km/h**

24. **TOMATOES** One tomato plant is 12 inches tall and grows $1\frac{1}{2}$ inches per week. Another tomato plant is 6 inches tall and grows 2 inches per week. When will the plants be the same height? Use a table or a graph to check. **12 weeks**

3.6 SOLVING DECIMAL EQUATIONS

Examples on pp. 166–168

EXAMPLE For some equations, you need to give an approximate solution.

$3.45m = -2.93m - 2.95$ **Write original equation.**

$6.38m = -2.95$ **Add 2.93*m* to each side.**

$m \approx -0.462382445$ **Using a calculator, divide each side by 6.38.**

$m \approx -0.46$ **Round answer. (Use hundredths, as in original equation.)**

Solve the equation. Round the result to the nearest tenth.

25. $13.7t - 4.7 = 9.9 + 8.1t$ **2.6** **26.** $4.6(2a + 3) = 3.7a - 0.4$ **−2.6** **27.** $-6(5.61x - 3.21) = 4.75$ **0.4**

3.7 FORMULAS AND FUNCTIONS

Examples on pp. 174–176

EXAMPLE You can solve the formula for the volume of a cone for the height *h*.

$V = \frac{1}{3}\pi r^2 h$ **Write original formula.**

$3V = \pi r^2 h$ **Multiply each side by 3.**

$\frac{3V}{\pi r^2} = h$ **Divide each side by πr^2.**

In Exercises 28–30, solve for the indicated variable.

28. $S = 2\pi rh$ for h $h = \frac{S}{2\pi r}$ **29.** $S = \pi s(R + r)$ for r $r = \frac{S}{\pi s} - R$ **30.** $R = \frac{pV}{nT}$ for p $p = \frac{RnT}{V}$

31. Rewrite the equation $3x + 2y - 4 = 2(5 - y)$ so that y is a function of x. Then use the result to find y when $x = -2, 0, 1,$ and 5. $y = -\frac{3}{4}x + \frac{7}{2}$; $5, \frac{7}{2}, 2\frac{3}{4}, -\frac{1}{4}$

3.8 RATES, RATIOS, AND PERCENTS

Examples on pp. 180–182

EXAMPLE For a United States flag, the ratio $\frac{\text{length}}{\text{width}}$ is $\frac{1.9}{1}$. How long is a 5-inch-wide flag?

length of small flag → $\frac{x}{5} = \frac{1.9}{1}$ ← **standard length to width ratio**
width of small flag →

The solution of the equation is $x = 9.5$, so the flag is 9.5 inches long.

32. You earn \$210 in 40 hours. At this rate, how much do you earn in 55 hours? **\$288.75**

33. A cab driver typically receives about 15% of the fare charged as a tip. To earn a total of \$30 in tips, about how much would a driver need to collect in fares? **\$200**

34. Convert 470 Ethiopian birrs to U.S. dollars. (1 dollar is 7.821 birrs.) **\$60**

CHAPTER 3

Chapter Test

Solve the equation if possible.

1. $2 + x = 8$ **6** **2.** $19 = a - 4$ **23** **3.** $-3y = -18$ **6**

4. $\frac{x}{4} = 5$ **20** **5.** $17 = 5 - 3p$ **−4** **6.** $-\frac{3}{4}x - 2 = -8$ **8**

7. $\frac{5}{3}(9 - w) = -10$ **15** **8.** $-3(x - 2) = x$ $1\frac{1}{2}$

9. $-5r - 6 + 4r = -r + 2$ **no solution** **10.** $-4y - (5y + 6) = -7y + 3$ $-4\frac{1}{2}$

Solve the equation. Round the result to the nearest hundredth.

11. $13.2x + 4.3 = 2(2.7x - 3.6)$ **−1.47** **12.** $-4(2.5x + 8.7) = (1.4 - 9.2x)(6)$ **0.96**

In Exercises 13 and 14, solve for the indicated variable.

13. $C = 2\pi r, r$ $r = \frac{C}{2\pi}$ **14.** $S = B + \frac{1}{2}Pl, l$ $l = \frac{2(S - B)}{P}$

15. Rewrite $3x + 4y = 15 + 6y$ so that y is a function of x. $y = \frac{3}{2}x - \frac{15}{2}$

16. Use the result in Exercise 15 to find y when $x = -1$, 0, and 2. $-9, -7\frac{1}{2}, -4\frac{1}{2}$

17. How many feet are in 3.5 kilometers? (Hint: 1 km ≈ 3281 ft) **approximately 11,483.5 ft**

18. SHOVELING SNOW You shovel snow. You charge \$7 per driveway and earn \$42. Let x represent the number of driveways you shoveled. Which of the following equations is an algebraic model for the situation? **C**

A. $42x = 7$ **B.** $\frac{1}{7}x = 42$ **C.** $7x = 42$ **D.** $\frac{1}{42}x = 7$

EARNINGS **In Exercises 19 and 20, your cousin earns about \$25 per week baby-sitting and receives one \$5 bonus. You earn about \$15 per week mowing lawns and \$12 per week running errands. After working the same number of weeks, you have \$11 more than your cousin.**

19. Write and solve an equation to find how many weeks you worked.
***Sample equation*: $25x + 5 = 15x + 12x - 11$; 8 weeks**

20. Check your solution in Exercise 19 with a table or a graph.
Check graphs and tables.

21. SAVINGS INTEREST You invest \$400. After one year, the total of the investment is \$414.40. Use the formula $A = P + Prt$ to find the annual simple interest rate for the investment, where A is the total of the investment, P is the principal (amount invested), r is the annual simple interest rate, and t is the time in years. **3.6%**

In Exercises 22 and 23, write and solve an equation to answer the question.

22. VOLUNTEER WORK You stuffed 108 envelopes in 45 minutes. At this rate, how many envelopes can you stuff in 2 hours? **288 envelopes**

23. WAGES After an 8% increase in your wages, you receive \$.94 more per hour. About how much did you receive per hour before the increase in your wages? **\$11.75**

ADDITIONAL RESOURCES

- ***Chapter 3 Resource Book***
  Chapter Test (3 levels) (p. 121)
  SAT/ACT Chapter Test (p. 127)
  Alternative Assessment (p. 128)
- ***Test and Practice Generator***

ADDITIONAL RESOURCES

- ***Chapter 3 Resource Book***
  Chapter Test (3 levels) (p. 121)
  SAT/ACT Chapter Test (p. 127)
  Alternative Assessment (p. 128)
- ***Test and Practice Generator***

CHAPTER 3

Chapter Standardized Test

TEST TAKING STRATEGY As soon as the testing time begins, start working. Keep moving and stay focused on the test.

1. **MULTIPLE CHOICE** Solve $4 - x = -5$. **D**

(A) -1 (B) 1 (C) -9
(D) 9 (E) $\frac{5}{4}$

2. **MULTIPLE CHOICE** Which of these steps can you use to solve the equation $\frac{3}{5}x = 12$? **D**

I. Multiply by $\frac{3}{5}$. **II.** Divide by $\frac{5}{3}$.

III. Divide by $\frac{3}{5}$. **IV.** Multiply by $\frac{5}{3}$.

(A) **I** only (B) **III** only
(C) **I** and **II** (D) **III** and **IV**
(E) None of the above

3. **MULTIPLE CHOICE** The perimeter of the rectangle is 30. Find the value of x. **D**

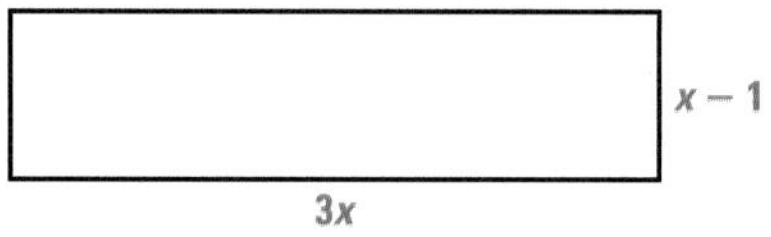

(A) 3 (B) $\frac{31}{4}$ (C) 10
(D) 4 (E) 31

4. **MULTIPLE CHOICE** If $9x - 4(3x - 2) = 4$, then $x = \underline{\ ?\ }$. **A**

(A) $\frac{4}{3}$ (B) $-\frac{4}{3}$ (C) -4
(D) 4 (E) 2

5. **MULTIPLE CHOICE** How many solutions does the equation $-2y + 3(4 - y) = 12 - 5y$ have? **E**

(A) none (B) one (C) two
(D) three (E) an infinite number

6. **MULTIPLE CHOICE** Solve the equation $\frac{1}{3}(27x + 18) = 12 + 6(x - 4)$. **A**

(A) -6 (B) -3 (C) -2
(D) 2 (E) 14

7. **MULTIPLE CHOICE** If $0.75t = 12$, then $t = \underline{\ ?\ }$. **E**

(A) 2 (B) 8 (C) 9
(D) 12 (E) 16

8. **MULTIPLE CHOICE** You sell hot dogs for \$1.25 each. Which equation can you use to find how many hot dogs you sold if you have \$60 from selling hot dogs? **C**

(A) $\frac{x}{60} = 1.25$ (B) $60x = 1.25$
(C) $1.25x = 60$ (D) $\frac{x}{1.25} = 60$
(E) $(1.25)(60) = x$

9. **MULTIPLE CHOICE** Find the value of y if $13.56y - 14.76 = 3(4.12y - 6.72)$. **A**

(A) -4.5 (B) -2.56 (C) -1.2
(D) 1.2 (E) 2.70

10. **MULTIPLE CHOICE** Use the equation $2(2 - y) = 3x$. What are the values of y when $x = -4, 0, \frac{2}{3}$, and 3? **C**

(A) $4, 2, 1, -\frac{5}{2}$ (B) $4, \frac{4}{3}, \frac{8}{9}, -\frac{2}{3}$
(C) $8, 2, 1, -\frac{5}{2}$ (D) $16, 4, 2, -5$
(E) $-8, -2, -1, \frac{5}{2}$

11. **MULTIPLE CHOICE** The sales tax is 6%. Which is the total charge for your meal before tax if the sales tax came to \$.87? **E**

(A) \$14.30 (B) \$5.22 (C) \$6.90
(D) \$1.45 (E) \$14.50

12. **MULTIPLE CHOICE** If $x = -5$ is a solution of $2x + tx - 5 = 30$, what is the value of t? **A**

(A) -9 (B) -7 (C) -5
(D) 3 (E) 9

QUANTITATIVE COMPARISON In Exercises 13 and 14, choose the statement that is true about the value of the variable.

Ⓐ The value in column A is greater.

Ⓑ The value in column B is greater.

Ⓒ The two values are equal.

Ⓓ The relationship cannot be determined from the given information.

	Column A	Column B	
13.	The value of b in the formula $A = \frac{1}{2}bh$ when $A = 39$ and $h = 6$	The value of b in the formula $A = \frac{1}{2}bh$ when $A = 78$ and $h = 12$	C
14.	The value of A in the formula $A = \frac{1}{2}bh$ when $h = 32$	The value of A in the formula $A = \frac{1}{2}bh$ when $h = 5$	D

15. MULTIPLE CHOICE Out of 225 people surveyed, 183 said yes. Which equation can you use to estimate how many people would say yes if 600 people were surveyed? **B**

Ⓐ $\frac{225}{183} = \frac{x}{600}$ Ⓑ $\frac{x}{600} = \frac{183}{225}$ Ⓒ $600 = x \cdot \frac{183}{225}$ Ⓓ $\frac{225}{600} = \frac{x}{183}$ Ⓔ none of them

16. MULTI-STEP PROBLEM Karim and his three children enjoy ice skating. The town ice skating rink charges \$10.00 for adults and \$6.50 for children per visit. The rink also offers a one-year family membership for \$650.00. The membership offers a family unlimited ice skating throughout the year.

Karim's schedule is so busy that he has at most two days each month to go ice skating with his children. Karim wants to decide whether it is worthwhile for him to purchase the one-year membership.

a. Let r represent the number of times Karim's family (Karim and his children) visits the ice skating rink during the year. Write a variable expression for each of the following costs: the cost of one child's visits to the rink during the year, the cost of Karim's visits to the rink during the year, the total cost of the family's visits to the rink during the year. **$6.5r$; $10r$; $29.5r$**

b. Karim decides to compare the total cost of the family's visits during the year to the membership cost. Write an equation that models the situation. Then solve the equation. **$29.5r = 650$; 22**

c. How much money will Karim lose or save by purchasing the membership if he and his children find the time to visit the skating rink 27 times in the year? if they only visit the skating rink once a month? **save \$146.50; lose \$296**

d. *Writing* Based on your results in parts (b) and (c), make a recommendation to Karim. Should he buy the one-year membership with unlimited visits or should he pay for each visit? Explain your reasoning. **See margin.**

16d. According to the given information, Karim is able to visit at most twice per month. If he is sure that he will be able to visit that often, he will save money by purchasing the annual membership. His decision should be based on a realistic estimate of how often he can get to the rink.

ADDITIONAL RESOURCES
A Cumulative Review covering Chapters 1–3 is available in the *Chapter 3 Resource Book*, p. 132.

29. $\begin{bmatrix} 26 & 14 & -6 \\ -54 & -3 & 35 \end{bmatrix}$

CHAPTER 3

Cumulative Practice

Evaluate the numerical expression. (1.3, 2.1, 2.7)

1. $5 + 3(18 \div 6)$ **14**
2. $3.6 \div (2 + 4) + 1.2$ **1.8**
3. $[2(9 - 5) + 1] \cdot 3^2$ **81**
4. $|6 - 14| - 2.5(7)$ **−9.5**
5. $2(3.5) - |13.2 - 7.21|$ **1.01**
6. $4(12.7 - 31.2) + 3.6$ **−70.4**
7. $(20 - 3 + 7) \div (-8)$ **−3**
8. $\frac{4(9-2)}{(7)(8)}$ $\frac{1}{2}$
9. $\frac{2^3 - 2(9)}{-5}$ **2**

Evaluate the variable expression. (1.3, 2.5, 2.7)

10. $-20 - 4y$ when $y = -3$ **−8**
11. $r + 6(5 - r)$ when $r = 10$ **−20**
12. $\frac{x-y}{3}$ when $x = 22$ and $y = 7$ **5**
13. $\frac{15x - 21}{y}$ when $x = 3$ and $y = 6$ **4**
14. $\frac{10x^2 - 8}{y}$ when $x = -2$ and $y = 11$ $\frac{32}{11}$
15. $\frac{x^2y^2}{4x - 10}$ when $x = 10$ and $y = -\frac{1}{2}$ $\frac{5}{6}$

Check whether the number is a solution of the equation or inequality. (1.4)

16. $4 + 2x = 12$; 2 **no**
17. $6x - 5 = 13$; 3 **yes**
18. $3y + 7 = 4y - 2$; 8 **no**
19. $x - 4 < 6$; 9 **yes**
20. $5x + 3 > 8$; 1 **no**
21. $9 - x \le x + 3$; 3 **yes**

Write the verbal phrase or sentence as an expression, an equation, or an inequality. (1.5)

22. A number cubed minus eight $n^3 - 8$
23. The sum of four times a number and seventeen $4n + 17$
24. Four less than twice a number is equal to ten. $2n - 4 = 10$
25. The product of negative three and a number is greater than twelve. $-3n > 12$
26. The quotient of twenty and a number is less than one. $\frac{20}{n} < 1$

Perform the indicated matrix operation. (2.4)

27. $\begin{bmatrix} 8 & -2 \\ 1 & -5 \end{bmatrix} + \begin{bmatrix} -3 & 6 \\ -7 & 0 \end{bmatrix}$ $\begin{bmatrix} 5 & 4 \\ -6 & -5 \end{bmatrix}$
28. $\begin{bmatrix} -5 & 5 \\ -2 & -1 \\ 8 & 4 \end{bmatrix} - \begin{bmatrix} -2 & 0 \\ 4 & -5 \\ 1 & -10 \end{bmatrix}$ $\begin{bmatrix} -3 & 5 \\ -6 & 4 \\ 7 & 14 \end{bmatrix}$
29. $\begin{bmatrix} 23 & -6 & 1 \\ -47 & 15 & 4 \end{bmatrix} + \begin{bmatrix} 3 & 20 & -7 \\ -7 & -18 & 31 \end{bmatrix}$ **See margin.**
30. $[4 \;\; -2 \;\; -7 \;\; 1] - [6 \;\; -4 \;\; -8 \;\; -1]$ $[-2 \;\; 2 \;\; 1 \;\; 2]$

Find the probability of choosing the indicated letter from a bag that contains the letters of the word. (2.8)

31. **Letter:** A
Word: MATHEMATICS $\frac{2}{11} \approx 0.18$
32. **Letter:** I
Word: DIAMETER $\frac{1}{8} = 0.125$
33. **Letter:** P
Word: PHILADELPHIA $\frac{1}{6} \approx 0.17$
34. **Letter:** A
Word: ALEXANDRIA $\frac{3}{10} = 0.3$

Solve the equation. (3.1–3.4)

35. $x + 11 = 19$ **8**
36. $-7 - x = -2$ **−5**
37. $9b = 135$ **15**
38. $35 = 3c - 19$ **18**
39. $\frac{p}{2} - 9 = -1$ **16**
40. $4(2x - 9) = 6(10x - 6)$ **0**
41. $3(q - 12) = 5q + 2$ **−19**
42. $-\frac{3}{4}(2x + 5) = 6$ **$-\frac{13}{2}$**
43. $9(2p + 1) - 3p = 4p - 6$ **$-1\frac{4}{11}$**

Solve the equation. Round the result to the nearest hundredth. (3.6)

44. $-3.46y = -5.78$ **1.67**
45. $4.17n + 3.29 = 2.74n$ **−2.30**
46. $4.2(0.3 + x) = 8.7$ **1.77**
47. $23.5a + 12.5 = 5.2(9.3a - 4.8)$ **1.51**

In Exercises 48–50, rewrite the equation so that *x* is a function of *y*. Then use the result to find *x* when *y* = −2, 0, 1.5, and 3. (3.7) **48–50. See margin.**

48. $x + \frac{1}{2}y = -3$
49. $2(3y - 1) = 4x$
50. $-3(x + y) + 4 = 7y$

51. **TEMPERATURE** Suppose the temperature outside is 82°F. What is the temperature in degrees Celsius? Use the formula $C = \frac{5}{9}(F - 32)$. (1.1) **$27\frac{7}{9}$°C**

52. **SILVER PRODUCTION** The table shows the amount (in metric tons) of silver produced in the United States for different years. Make a line graph of the data. (1.6) **See margin.**

Year	1991	1992	1993	1994	1995	1996
Amount (metric tons)	1860	1800	1640	1480	1560	1570

▶ Source: U.S. Geological Survey

53. **FUND RAISER** Your school band is planning to attend a competition. The total cost for the fifty band members to attend is $750. Each band member will pay $3 toward this cost and the rest of the money will be raised by selling wrapping paper. For each roll of wrapping paper sold, the band makes $2. Write and solve an equation to find how many rolls the band members need to sell. (3.3) **300 rolls**

54. GEOMETRY CONNECTION The volume of a circular cylinder with a radius of 1.5 inches is about 42.4 cubic inches. Find the cylinder's height. Round to the nearest tenth. (*Hint:* Volume of a circular cylinder = $\pi r^2 h$) (3.7) **about 6.0 in.**

VACATION IN SWEDEN In Exercises 55–57, use the following information. You are vacationing in Sweden and have taken $620 to spend. The rate of currency exchange in Sweden is 7.827 kronor (plural of krona) per United States dollar. (3.8)

55. If you exchange $\frac{3}{4}$ of the entire amount, how many kronor will you receive? **3639 kronor**

56. After your vacation, you have 1255 kronor left. If the exchange rate is the same, how many dollars will you get back? **$160**

57. If you exchanged the entire amount at the start of your vacation, how many kronor would you have received? **4852 kronor**

48. $x = -\frac{1}{2}y - 3$; $-2, -3, -3\frac{3}{4}, -4\frac{1}{2}$

49. $x = \frac{3}{2}y - \frac{1}{2}$; $-3\frac{1}{2}, -\frac{1}{2}, 1\frac{3}{4}, 4$

50. $x = -\frac{10}{3}y + \frac{4}{3}$; $8, \frac{4}{3}, -3\frac{2}{3}, -8\frac{2}{3}$

52. See Additional Answers beginning on page AA1.

PROJECT OVERVIEW

PROJECT GOALS

- Use linear equations to model cost and income for a business.
- Use linear equations to compare and find profit.
- Use results from solving linear equations to determine when a business is feasible.

MANAGING THE PROJECT

CLASSROOM MANAGEMENT Students can work individually or in small groups to complete the project. If working in groups, have them work on Exercises 1–8 together. Then have each student explain a part of their results in the presentation. Each student in the group should understand and be able to explain the results.

GUIDING STUDENTS' WORK If students are having difficulty with Exercises 1 and 2, have them find the cost to clean one side of 1, 2, 3, 4, and 5 windows. Then have them look for a pattern and use it to answer Exercise 2. Also, have students show their work when they present their results.

If students are finding the cost per paper towel, they will get an answer of \$.015. This particular answer should not be rounded to \$.02 because the round-off error will be too great. Since they need two paper towels for each side of a window, the cost is 2(\$.015) = \$.03, not 2(\$.02) = \$.04.

CONCLUDING THE PROJECT Students should present their results including any charts or diagrams that they used to solve the exercises. You can use questions like the following for class discussion:

- Do most window washers you have seen use paper towels or a squeegee? How do you think using a squeegee to clean windows will change your cost, income, or profit?

PROJECT
Applying Chapters 1–3

Running a Business

OBJECTIVE **Comparing the expenses and income of a business to determine profitability.**

Materials: paper, pencil, graphing calculator or computer (optional)

INVESTIGATION

In Exercises 1–8, use the financial plan below.

My summer window-cleaning business: Financial Plans

1. Expenses:
   - Paper towels cost \$.90 for a 60-sheet roll.
   - Spray window cleaner costs \$1.98 for a 33-ounce bottle.
   - A step ladder costs \$25.00.
2. Income (per window):

Number of windows	1-14	15 or more
Outside only	\$.60	\$.50
Outside and inside	\$1.10	\$1.00

1. You use two paper towels and 0.5 ounce of window cleaner to clean one side of a window. What is the total cost of materials to clean one side of a window? **\$.06**
2. Write an equation to find the cost C of cleaning one side of x windows. $C = 0.06x$
3. Copy and complete the table. You may want to use a spreadsheet.

Number of windows	5	9	15	25
Cost for one side	?	?	?	?
Cost for both sides	?	?	?	?

\$.30, \$.54, \$.90, \$1.50

\$.60 \$1.08 \$1.80 \$3.00

4. Write an equation for the income I for cleaning both sides of x windows when $x < 15$. $I = 1.1x$
5. How much profit will you make for cleaning both sides of 15 windows? **\$13.20**
6. What is your total profit for cleaning the windows of 6 houses (inside and outside) with 12, 17, 13, 9, 14, and 10 windows respectively? **\$11.76, \$14.96, \$12.74, \$8.82, \$13.72, \$9.80**
7. How many windows do you have to clean to pay for your step ladder? Explain your reasoning. **Answers may vary. *Sample answers*: See margin.**
8. What would it mean if you found a negative profit? **The rates you decided to charge are not high enough to cover your expenses for paper towels and window cleaner.**

7. 26 windows (both sides, at 0–14 rate); 29 windows (both sides, at 15-or-more rate); or combinations such as 30 windows (outside only, at 15-or-more rate) and 13 windows (both sides, at 0–14 rate)

PRESENT YOUR RESULTS

Write a report about your window cleaning plans.

- Include a discussion of income and expenses.
- Include your answers to Exercises 1–8.
- Explain how the cost of the ladder is a different kind of expense from the cost of paper towels and the window cleaner.

You may want to find out whether this business would work in your neighborhood.

- Survey some of your neighbors to find the price they would be willing to pay to have their windows cleaned.
- Do some research to find out what local businesses pay workers for similar work.
- Find out what the minimum hourly wage is. How many windows would you have to clean each hour to earn that amount?

EXTENSION

Think of a business that you would like to start.

1. If you can, interview someone who has experience with the type of business you are interested in. What advice can that person give you?
2. Do some research to find out what kind of supplies or equipment you would need to get started and the cost of these items.
3. Decide what you could charge customers for this service.
4. Survey some of your friends and neighbors to find the price they would be willing to pay for such a product or service.
5. Write a financial plan for your business.

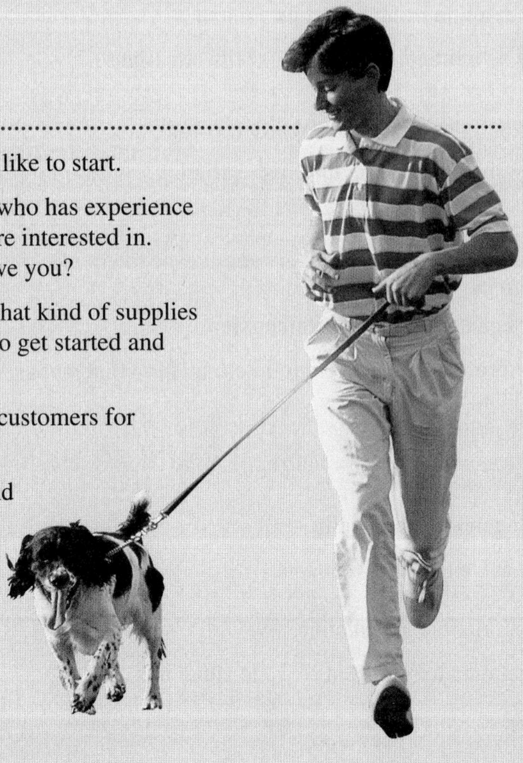

- What group of people do you think will be most interested in your services? Do you think you would have more success from families and houses or from businesses?
- Being a window washer may require travel. What transportation costs might you encounter?
- What other concerns would you need to consider in your business?

GRADING THE PROJECT

4 Students complete Exercises 1–8 correctly and include their solutions in their report. The report is thorough and answers all questions about the project. Students also include an extension to the project and are able to explain their ideas clearly in their presentation.

3 Students complete the Exercises and include them in their report, which answers all of the questions and shows an effort to complete the survey questions. They also make a good presentation to the class, but do not have a complete understanding of the concepts that were being developed in the project. Their report is complete, but students do not make a good effort to extend the project.

2 Students make a report, but it is incomplete, with some answers for the Investigation incorrect or not provided. Students do not answer all the questions asked. Also, students show a lack of understanding of the concepts and ideas that are the focus of the project. Their presentation shows that they do not have an understanding of the ideas being explored.

1 Students attempt to complete the exercises, but many of their answers are incorrect or incomplete. Students show an inadequate grasp of the ideas and concepts being explored in the project. Students should speak with the teacher to review their work and to make a new start on the project.

CHAPTER 4

PLANNING THE CHAPTER

Graphing Linear Equations and Functions

LESSON	GOALS	NCTM	ITED	SAT9	Terra-Nova	Local
4.1 *pp. 203–209*	GOAL 1 Plot points in a coordinate plane. GOAL 2 Draw a scatter plot and make predictions about real-life situations. TECHNOLOGY ACTIVITY: 4.1 *Create a scatter plot on a graphing calculator.*	1, 3, 5, 6, 9, 10	MIG, IIG	1, 8, 9, 10	14, 15	
4.2 *pp. 210–217*	GOAL 1 Graph a linear equation using a table or a list of values. GOAL 2 Graph horizontal and vertical lines.	1, 3, 9, 10	MIG, IIG	9, 10, 34	14, 16	
4.3 *pp. 218–224*	GOAL 1 Find the intercepts of the graph of a linear equation. GOAL 2 Use intercepts to make a quick graph of a linear equation.	1, 3, 6, 9, 10	MIG, IIG		14, 15	
4.4 *pp. 225–233*	CONCEPT ACTIVITY: 4.4 *Investigate how changes in rise and run affect the slope of a ramp.* GOAL 1 Find the slope of a line using two of its points. GOAL 2 Interpret slope as a rate of change in real-life situations.	1, 3, 6, 9, 10	MIG, IIG, APRA	1	14, 15, 16, 17	
4.5 *pp. 234–239*	GOAL 1 Write linear equations that represent direct variation. GOAL 2 Use a ratio to write an equation for direct variation.	1, 3, 6, 9, 10	APRA	1	16	
4.6 *pp. 240–249*	CONCEPT ACTIVITY: 4.6 *Investigate the relationships that exist between members of a family of equations.* GOAL 1 Graph a linear equation in slope-intercept form. GOAL 2 Graph and interpret equations in slope-intercept form that model real-life situations. TECHNOLOGY ACTIVITY: 4.6 *Graph a linear equation on a graphing calculator.*	1, 3, 6, 9, 10	MIG, IIG, APRA	1	14, 16	
4.7 *pp. 250–255*	GOAL 1 Solve a linear equation graphically. GOAL 2 Use a graph to solve real-life problems.	1, 3, 6, 9, 10	MIG, IIG	1, 7, 10	14, 16, 17	
4.8 *pp. 256–262*	GOAL 1 Identify when a relation is a function. GOAL 2 Use function notation to represent real-life situations.	1, 2, 3, 6, 9, 10		20	16	

RESOURCES

Chapter Resource Booklets

Chapter Support

Tips for New Teachers	p. 1	Prerequisite Skills Review	p. 5
Parent Guide for Student Success	p. 3	Strategies for Reading Mathematics	p. 7

Lesson Support

Lesson Support	4.1	4.2	4.3	4.4	4.5	4.6	4.7	4.8
Lesson Plans (regular and block)	p. 9	p. 24	p. 36	p. 50	p. 64	p. 76	p. 95	p. 108
Warm-Up Exercises and Daily Quiz	p. 11	p. 26	p. 38	p. 52	p. 66	p. 78	p. 97	p. 110
Activity Support Masters				p. 53		p. 79		
Lesson Openers	p. 12	p. 27	p. 39	p. 54	p. 67	p. 80	p. 98	p. 111
Graphing Calculator Activities & Keystrokes	p. 13					p. 81	p. 99	
Practice (3 levels)	p. 16	p. 28	p. 40	p. 55	p. 68	p. 86	p. 100	p. 112
Reteaching with Practice	p. 19	p. 31	p. 43	p. 58	p. 71	p. 89	p. 103	p. 115
Quick Catch-Up for Absent Students	p. 21	p. 33	p. 45	p. 60	p. 73	p. 91	p. 105	p. 117
Cooperative Learning Activities				p. 61				
Interdisciplinary Applications	p. 22			p. 62		p. 92		p. 118
Real-Life Applications		p. 34	p. 46		p. 74		p. 106	
Math & History Applications			p. 47					
Challenge: Skills and Applications	p. 23	p. 35	p. 48	p. 63	p. 75	p. 93	p. 107	p. 119

Review and Assessment

Quizzes	pp. 49, 94	Alternative Assessment with Math Journal	p. 128
Chapter Review Games and Activities	p. 120	Project with Rubric	p. 130
Chapter Test (3 levels)	pp. 121–126	Cumulative Review	p. 132
SAT/ACT Chapter Test	p. 127	Resource Book Answers	p. A1

Transparencies

	4.1	4.2	4.3	4.4	4.5	4.6	4.7	4.8
Warm-Up Exercises and Daily Quiz	p. 28	p. 29	p. 30	p. 31	p. 32	p. 33	p. 34	p. 35
Alternative Lesson Opener Transparencies	p. 24	p. 25	p. 26	p. 27	p. 28	p. 29	p. 30	p. 31
Examples/Standardized Test Practice	✓	✓	✓	✓	✓	✓	✓	✓
Answer Transparencies	✓	✓	✓	✓	✓	✓	✓	✓

Technology

- EasyPlanner CD
- EasyPlanner Plus Online
- Classzone.com
- Personal Student Tutor CD
- Test and Practice Generator CD
- Instant Replay: Video Review Games
- Electronic Lesson Presentations
- Chapter Audio Summaries CD

Additional Resources

- Basic Skills Workbook
- Explorations and Projects
- Worked-Out Solution Key
- Data Analysis Sourcebook
- Spanish Resources
- Standardized Test Practice Workbook
- Practice Workbook with Examples
- Notetaking Guide

CHAPTER 4

PACING THE CHAPTER

Resource Key
- STUDENT EDITION
- CHAPTER 4 RESOURCE BOOK
- TEACHER'S EDITION

REGULAR SCHEDULE

Day 1

4.1

STARTING OPTIONS
- Prereq. Skills Review
- Strategies for Reading
- Warm-Up or Daily Quiz

TEACHING OPTIONS
- Motivating the Lesson
- Les. Opener (Appl.)
- Technology Activities
- Examples 1–3
- Closure Question
- Guided Practice Exs.

APPLY/HOMEWORK
- See Assignment Guide.
- See the CRB: Practice, Reteach, Apply, Extend

ASSESSMENT OPTIONS
- Checkpoint Exercises
- Daily Quiz (4.1)
- Stand. Test Practice

Day 2

4.2

STARTING OPTIONS
- Homework Check
- Warm-Up or Daily Quiz

TEACHING OPTIONS
- Motivating the Lesson
- Les. Opener (Activity)
- Examples 1–3
- Guided Practice Exs. 1–10

APPLY/HOMEWORK
- See Assignment Guide.
- See the CRB: Practice, Reteach, Apply, Extend

ASSESSMENT OPTIONS
- Checkpoint Exercises, p. 211

Day 3

4.2 *(cont.)*

STARTING OPTIONS
- Homework Check

TEACHING OPTIONS
- Examples 4–6
- Closure Question
- Guided Practice Ex. 11

APPLY/HOMEWORK
- See Assignment Guide.
- See the CRB: Practice, Reteach, Apply, Extend

ASSESSMENT OPTIONS
- Checkpoint Exercises, pp. 212–213
- Daily Quiz (4.2)
- Stand. Test Practice

Day 4

4.3

STARTING OPTIONS
- Homework Check
- Warm-Up or Daily Quiz

TEACHING OPTIONS
- Motivating the Lesson
- Les. Opener (Activity)
- Examples 1–3
- Guided Practice Exs. 1–12

APPLY/HOMEWORK
- See Assignment Guide.
- See the CRB: Practice, Reteach, Apply, Extend

ASSESSMENT OPTIONS
- Checkpoint Exercises, p. 219

Day 5

4.3 *(cont.)*

STARTING OPTIONS
- Homework Check

TEACHING OPTIONS
- Example 4
- Closure Question
- Guided Practice Ex. 13

APPLY/HOMEWORK
- See Assignment Guide.
- See the CRB: Practice, Reteach, Apply, Extend

ASSESSMENT OPTIONS
- Checkpoint Exercises, p. 220
- Daily Quiz (4.3)
- Stand. Test Practice
- Quiz (4.1–4.3)

Day 6

4.4

STARTING OPTIONS
- Homework Check
- Warm-Up or Daily Quiz

TEACHING OPTIONS
- Motivating the Lesson
- Concept Act. & Wksht.
- Les. Opener (Visual)
- Examples 1–5
- Guided Practice Exs.

APPLY/HOMEWORK
- See Assignment Guide.
- See the CRB: Practice, Reteach, Apply, Extend

ASSESSMENT OPTIONS
- Checkpoint Exercises, pp. 227–228

Day 9

4.6

STARTING OPTIONS
- Homework Check
- Warm-Up or Daily Quiz

TEACHING OPTIONS
- Motivating the Lesson
- Concept Act. & Wksht.
- Les. Opener (Application)
- Graphing Calc. Activity
- Examples 1–3
- Guided Practice Exs. 1–10

APPLY/HOMEWORK
- See Assignment Guide.
- See the CRB: Practice, Reteach, Apply, Extend

ASSESSMENT OPTIONS
- Checkpoint Exercises, p. 242

Day 10

4.6 *(cont.)*

STARTING OPTIONS
- Homework Check

TEACHING OPTIONS
- Example 4
- Technology Activity
- Closure Question
- Guided Practice Exs. 11–12

APPLY/HOMEWORK
- See Assignment Guide.
- See the CRB: Practice, Reteach, Apply, Extend

ASSESSMENT OPTIONS
- Checkpoint Exercises, p. 243
- Daily Quiz (4.6)
- Stand. Test Practice
- Quiz (4.4–4.6)

Day 11

4.7

STARTING OPTIONS
- Homework Check
- Warm-Up or Daily Quiz

TEACHING OPTIONS
- Motivating the Lesson
- Les. Opener (Calculator)
- Graphing Calc. Activity
- Examples 1–5
- Closure Question
- Guided Practice Exs.

APPLY/HOMEWORK
- See Assignment Guide.
- See the CRB: Practice, Reteach, Apply, Extend

ASSESSMENT OPTIONS
- Checkpoint Exercises
- Daily Quiz (4.7)
- Stand. Test Practice

Day 12

4.8

STARTING OPTIONS
- Homework Check
- Warm-Up or Daily Quiz

TEACHING OPTIONS
- Motivating the Lesson
- Les. Opener (Application)
- Examples 1–3
- Guided Practice Exs. 1–9

APPLY/HOMEWORK
- See Assignment Guide.
- See the CRB: Practice, Reteach, Apply, Extend

ASSESSMENT OPTIONS
- Checkpoint Exercises, p. 257

Day 13

4.8 *(cont.)*

STARTING OPTIONS
- Homework Check

TEACHING OPTIONS
- Example 4
- Closure Question
- Guided Practice Ex. 10

APPLY/HOMEWORK
- See Assignment Guide.
- See the CRB: Practice, Reteach, Apply, Extend

ASSESSMENT OPTIONS
- Checkpoint Exercises, p. 258
- Daily Quiz (4.8)
- Stand. Test Practice
- Quiz (4.7–4.8)

Day 14

Review

DAY 14 START OPTIONS
- Homework Check

REVIEWING OPTIONS
- Chapter 4 Summary
- Chapter 4 Review
- Chapter Review Games and Activities

APPLY/HOMEWORK
- Chapter 4 Test (practice)
- Ch. Standardized Test (practice)

Day 7

4.4 (cont.)

Starting Options
- Homework Check

Teaching Options
- Example 6
- Closure Question

Apply/Homework
- See Assignment Guide.
- See the CRB: Practice, Reteach, Apply, Extend

Assessment Options
- Checkpoint Exercises, p. 229
- Daily Quiz (4.4)
- Stand. Test Practice

Day 8

4.5

Starting Options
- Homework Check
- Warm-Up or Daily Quiz

Teaching Options
- Motivating the Lesson
- Les. Opener (Activity)
- Examples 1–4
- Closure Question
- Guided Practice Exs.

Apply/Homework
- See Assignment Guide.
- See the CRB: Practice, Reteach, Apply, Extend

Assessment Options
- Checkpoint Exercises
- Daily Quiz (4.5)
- Stand. Test Practice

Day 15

Assess

Day 15 Start Options
- Homework Check

Assessment Options
- Chapter 4 Test
- SAT/ACT Ch. 4 Test
- Alternative Assessment

Apply/Homework
- Skill Review, p. 272

BLOCK SCHEDULE

Day 1

Assess & 4.1

(Day 1 = Ch. 3 Day 7)

Assessment Options
- Chapter 3 Test
- SAT/ACT Ch. 3 Test
- Alternative Assessment

Ch. 4 Start Options
- Skill Review, p. 202
- Strategies for Reading

Teaching 4.1 Options
- Warm-Up (Les. 4.1)
- Motivating the Lesson
- Les. Opener (Appl.)
- Technology Activities
- Examples 1–3
- Closure Question
- Guided Practice Exs.

Apply/Homework
- See Assignment Guide.
- See the CRB: Practice, Reteach, Apply, Extend

Assessment Options
- Checkpoint Exercises
- Daily Quiz (Les. 4.1)
- Stand. Test Practice

Day 2

4.2

Day 2 Start Options
- Homework Check
- Warm-Up or Daily Quiz

Teaching 4.2 Options
- Motivating the Lesson
- Les. Opener (Activity)
- Examples 1–6
- Closure Question
- Guided Practice Exs.

Apply/Homework
- See Assignment Guide.
- See the CRB: Practice, Reteach, Apply, Extend

Assessment Options
- Checkpoint Exercises
- Daily Quiz (Les. 4.2)
- Stand. Test Practice

Day 3

4.3

Day 3 Start Options
- Homework Check
- Warm-Up or Daily Quiz

Teaching 4.3 Options
- Motivating the Lesson
- Les. Opener (Activity)
- Examples 1–4
- Closure Question
- Guided Practice Exs.

Apply/Homework
- See Assignment Guide.
- See the CRB: Practice, Reteach, Apply, Extend

Assessment Options
- Checkpoint Exercises
- Daily Quiz (Les. 4.3)
- Stand. Test Practice
- Quiz (4.1–4.3)

Day 4

4.4

Day 4 Start Options
- Homework Check
- Warm-Up or Daily Quiz

Teaching 4.4 Options
- Motivating the Lesson
- Concept Act. & Wksht.
- Les. Opener (Visual)
- Examples 1–6
- Closure Question
- Guided Practice Exs.

Apply/Homework
- See Assignment Guide.
- See the CRB: Practice, Reteach, Apply, Extend

Assessment Options
- Checkpoint Exercises
- Daily Quiz (Les. 4.4)
- Stand. Test Practice

Day 5

4.5 & 4.6

Day 5 Start Options
- Homework Check
- W-Up 4.5 or D. Quiz 4.4

Teaching 4.5 Options
- Motivating the Lesson
- Les. Opener (Activity)
- Examples 1–4
- Closure Question
- Guided Practice Exs.

Beginning 4.6 Options
- Warm-Up (Les. 4.6)
- Motivating the Lesson
- Concept Act. & Wksht.
- Les. Opener (Appl.)
- Graphing Calc. Activity
- Examples 1–3
- Guided Pract. Exs. 1–10

Apply/Homework
- See Assignment Guide.
- See the CRB: Practice, Reteach, Apply, Extend

Assessment Options
- Checkpoint Exercises
- Daily Quiz (Les. 4.5)

Day 6

4.6 & 4.7

Day 6 Start Options
- Homework Check
- Daily Quiz (Les. 4.5)

Finishing 4.6 Options
- Example 4
- Technology Activity
- Closure Question
- Guided Practice Exs. 11–12

Teaching 4.7 Options
- Motivating the Lesson
- Les. Opener (Calc.)
- Graphing Calc. Activity
- Examples 1–5
- Closure Question
- Guided Practice Exs.

Apply/Homework
- See Assignment Guide.
- See the CRB: Practice, Reteach, Apply, Extend

Assessment Options
- Checkpoint Exercises
- Daily Quiz (4.6, 4.7)
- Stand. Test Practice
- Quiz (4.4–4.6)

Day 7

4.8

Day 7 Start Options
- Homework Check
- Warm-Up or Daily Quiz

Teaching 4.8 Options
- Motivating the Lesson
- Les. Opener (Appl.)
- Examples 1–4
- Closure Question
- Guided Practice Exs.

Apply/Homework
- See Assignment Guide.
- See the CRB: Practice, Reteach, Apply, Extend

Assessment Options
- Checkpoint Exercises
- Daily Quiz (Les. 4.8)
- Stand. Test Practice
- Quiz (4.7–4.8)

Day 8

Review/Assess

Day 8 Start Options
- Homework Check

Reviewing Options
- Chapter 4 Summary
- Chapter 4 Review
- Chapter Review Games and Activities
- Chapter 4 Test (practice)
- Ch. Standardized Test (practice)

Assessment Options
- Chapter 4 Test
- SAT/ACT Ch. 4 Test
- Alternative Assessment

Apply/Homework
- Skill Review, p. 272

CHAPTER 4

Meeting Individual Needs

Before the Chapter

The *Chapter 4 Resource Book* has the following materials to distribute and use before the chapter:

- **Parent Guide for Student Success**
- **Prerequisite Skills Review (pictured below)**
- **Strategies for Reading Mathematics**

Prerequisite Skills *Pages 5–6*

CHAPTER 4

NAME ______ DATE ______

Prerequisite Skills Review

For use before Chapter 4

Chapter Support

EXAMPLE 1 *Writing Percents as Fractions and Decimals*

Write each percent as a fraction in lowest terms and as a decimal.

a. 34% **b.** 5%

Solution

% means "divided by 100": Move the decimal point **left** two places:

a. $34\% = \frac{34}{100} = \frac{17}{50}$ and $34\% = 34\% = 0.34$

b. $5\% = \frac{5}{100} = \frac{1}{20}$ and $5\% = 05\% = 0.05$

Exercises for Example 1

Write each percent as a fraction in lowest terms and as a decimal.

1. 25% **2.** 60% **3.** 8% **4.** 114%

EXAMPLE 2 *Working with Functions*

Use the function $y = \frac{1}{2}x + 10$, where $-10 \le x \le 10$.

a. Make an input-output table for the function.

b. Draw a line graph to represent the data in the input-output table.

c. Describe the domain and range of the function.

Solution

a. You can choose any input values from –10 to 10 to make the input-output table. You should at least choose the input values –10 and 10.

Input	*Function*	*Output*
$x = -10$	$y = \frac{1}{2}(-10) + 10$	$y = 5$
$x = -5$	$y = \frac{1}{2}(-5) + 10$	$y = 7\frac{1}{2}$
$x = 0$	$y = \frac{1}{2}(0) + 10$	$y = 10$
$x = 5$	$y = \frac{1}{2}(5) + 10$	$y = 12\frac{1}{2}$
$x = 10$	$y = \frac{1}{2}(10) + 10$	$y = 15$

Once you have evaluated each input value, you can use a table to summarize your results.

Input, x	–10	–5	0	5	10
Output, y	5	$7\frac{1}{2}$	10	$12\frac{1}{2}$	15

Prerequisite Skills Review These two pages support the Study Guide on page 202. They help students prepare for Chapter 4 by providing worked-out examples and practice for the following skills needed in the chapter:

- **Write percents in various forms.**
- **Graph a linear function and describe the domain and range.**
- **Evaluate rational expressions with two variables.**

Technology Resource

Students can use the Personal Student Tutor to find additional reteaching and practice for skills from earlier chapters that are used in Chapter 4.

During Each Lesson

The *Chapter 4 Resource Book* has the following alternatives for introducing the lesson:

- **Lesson Openers (pictured below)**
- **Graphing Calculator Activities with Keystrokes**

Lesson Opener *Page 39*

LESSON 4.3

NAME ______ DATE ______

Available as a transparency

Activity Lesson Opener

For use with pages 218–224

SET UP: Work with your class.

1. Create a human coordinate plane with members of your class. Each square on the diagram below shows where one person should sit.

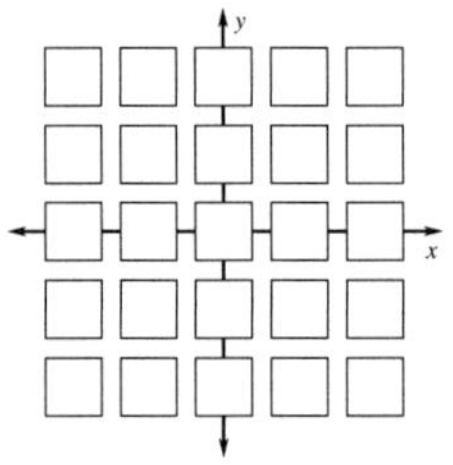

2. Each person on your grid represents a point. What are your coordinates?

3. Consider the equation $x + y = 1$. Find the coordinates of the point where the graph of this equation will cross the x-axis. Stand up if these match your coordinates. Find the coordinates of the point where the graph of this equation will cross the y-axis. Stand up if these match your coordinates.

4. Look at the two class members who are standing. If you are not standing, decide if you are in a position that lines up with the two students who are standing. If you are, then stand up. If not, remain seated. What is true about the people standing up?

5. Repeat Steps 3 and 4 for the equation $x - y = 2$.

6. Use your results to make a conjecture about a quick way to graph a linear equation.

Lesson 4.3

Activity Lesson Opener This Lesson Opener provides an alternative way to start Lesson 4.3 in the form of an activity. In this activity, students form a human coordinate plane and physically graph a line using the x- and y-intercepts.

FOLLOWING EACH LESSON

The *Chapter 4 Resource Book* has a variety of materials to follow-up each lesson. They include the following:

- **Practice (3 levels)**
- **Reteaching with Practice**
- **Quick Catch-Up for Absent Students**
- **Interdisciplinary Applications (pictured below)**
- **Real-Life Applications**

APPLICATION *Page 62*

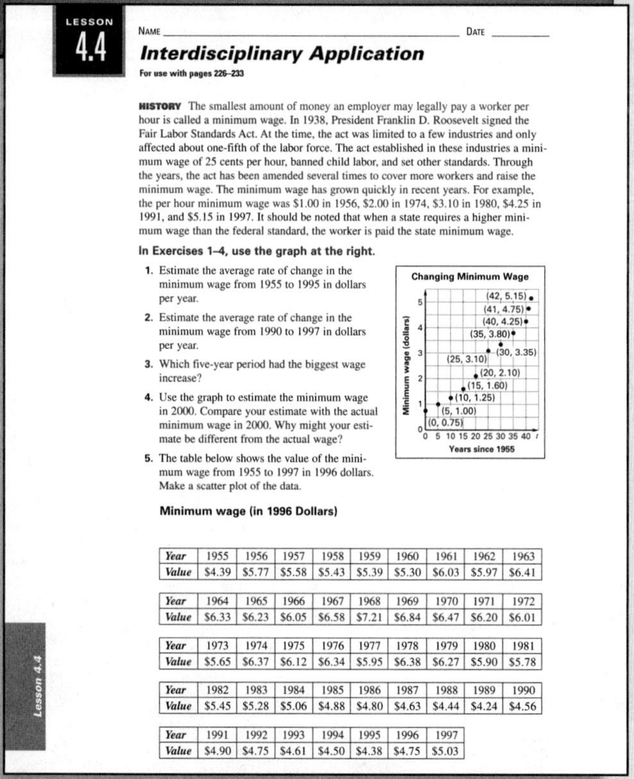

LESSON 4.4

NAME ______________________ DATE ________

Interdisciplinary Application

For use with pages 226–233

HISTORY The smallest amount of money an employer may legally pay a worker per hour is called a minimum wage. In 1938, President Franklin D. Roosevelt signed the Fair Labor Standards Act. At the time, the act was limited to a few industries and only affected about one-fifth of the labor force. The act established in these industries a minimum wage of 25 cents per hour, banned child labor, and set other standards. Through the years, the act has been amended several times to cover more workers and raise the minimum wage. The minimum wage has grown quickly in recent years. For example, the per hour minimum wage was $1.00 in 1956, $2.00 in 1974, $3.10 in 1980, $4.25 in 1991, and $5.15 in 1997. It should be noted that when a state requires a higher minimum wage than the federal standard, the worker is paid the state minimum wage.

In Exercises 1–4, use the graph at the right.

1. Estimate the average rate of change in the minimum wage from 1955 to 1995 in dollars per year.
2. Estimate the average rate of change in the minimum wage from 1990 to 1997 in dollars per year.
3. Which five-year period had the biggest wage increase?
4. Use the graph to estimate the minimum wage in 2000. Compare your estimate with the actual minimum wage in 2000. Why might your estimate be different from the actual wage?
5. The table below shows the value of the minimum wage from 1955 to 1997 in 1996 dollars. Make a scatter plot of the data.

Minimum wage (in 1996 Dollars)

Year	1955	1956	1957	1958	1959	1960	1961	1962	1963
Value	$4.39	$5.77	$5.58	$5.43	$5.39	$5.30	$6.03	$5.97	$6.41

Year	1964	1965	1966	1967	1968	1969	1970	1971	1972
Value	$6.33	$6.23	$6.05	$6.58	$7.21	$6.84	$6.47	$6.20	$6.01

Year	1973	1974	1975	1976	1977	1978	1979	1980	1981
Value	$5.65	$6.37	$6.12	$6.34	$5.95	$6.38	$6.27	$5.90	$5.78

Year	1982	1983	1984	1985	1986	1987	1988	1989	1990
Value	$5.45	$5.28	$5.06	$4.88	$4.80	$4.63	$4.44	$4.24	$4.56

Year	1991	1992	1993	1994	1995	1996	1997
Value	$4.90	$4.75	$4.61	$4.50	$4.38	$4.75	$5.03

Lesson 4.4

INTERDISCIPLINARY APPLICATION This application examines the minimum wage and its rate of change over time, a topic of Lesson 4.4.

ASSESSING THE CHAPTER

The *Chapter 4 Resource Book* has the following review and assessment materials:

- **Quizzes**
- **Chapter Review Games and Activities**
- **Chapter Test (3 levels)**
- **SAT/ACT Chapter Test (pictured below)**
- **Alternative Assessment with Rubric and Math Journal**
- **Project with Rubric**
- **Cumulative Review**

SAT/ACT CHAPTER TEST *Page 127*

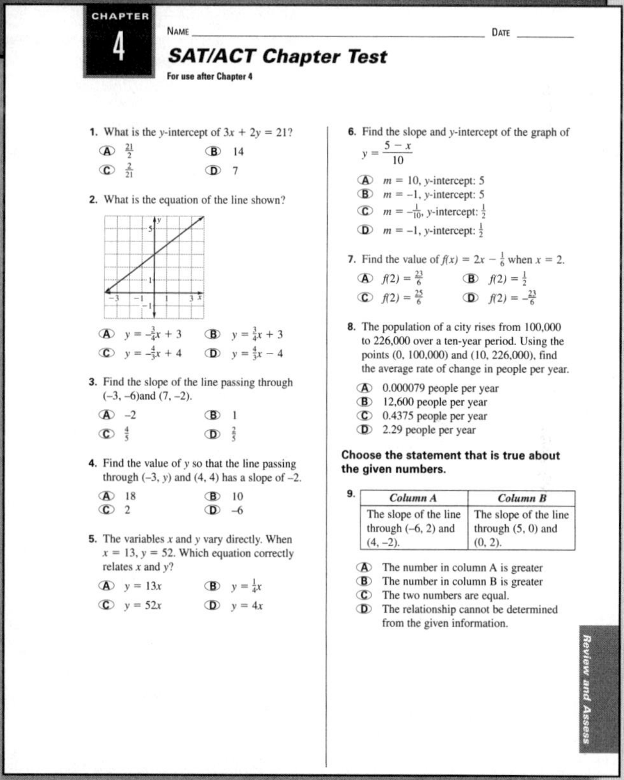

CHAPTER 4

NAME ______________________ DATE ________

SAT/ACT Chapter Test

For use after Chapter 4

1. What is the y-intercept of $3x + 2y = 21$?

   Ⓐ $\frac{21}{2}$ Ⓑ 14
   Ⓒ $\frac{2}{21}$ Ⓓ 7

2. What is the equation of the line shown?

   Ⓐ $y = -\frac{3}{4}x + 3$ Ⓑ $y = \frac{3}{4}x + 3$
   Ⓒ $y = -\frac{4}{3}x + 4$ Ⓓ $y = \frac{4}{3}x - 4$

3. Find the slope of the line passing through $(-3, -6)$and $(7, -2)$.

   Ⓐ −2 Ⓑ 1
   Ⓒ $\frac{4}{5}$ Ⓓ $\frac{2}{5}$

4. Find the value of y so that the line passing through $(-3, y)$ and $(4, 4)$ has a slope of −2.

   Ⓐ 18 Ⓑ 10
   Ⓒ 2 Ⓓ −6

5. The variables x and y vary directly. When $x = 13$, $y = 52$. Which equation correctly relates x and y?

   Ⓐ $y = 13x$ Ⓑ $y = \frac{1}{4}x$
   Ⓒ $y = 52x$ Ⓓ $y = 4x$

6. Find the slope and y-intercept of the graph of $y = \frac{5 - x}{10}$

   Ⓐ $m = 10$, y-intercept: 5
   Ⓑ $m = -1$, y-intercept: 5
   Ⓒ $m = -\frac{1}{10}$, y-intercept: $\frac{1}{2}$
   Ⓓ $m = -1$, y-intercept: $\frac{1}{2}$

7. Find the value of $f(x) = 2x - \frac{1}{6}$ when $x = 2$.

   Ⓐ $f(2) = \frac{23}{6}$ Ⓑ $f(2) = \frac{1}{2}$
   Ⓒ $f(2) = \frac{25}{6}$ Ⓓ $f(2) = -\frac{23}{6}$

8. The population of a city rises from 100,000 to 226,000 over a ten-year period. Using the points (0, 100,000) and (10, 226,000), find the average rate of change in people per year.

   Ⓐ 0.000079 people per year
   Ⓑ 12,600 people per year
   Ⓒ 0.4375 people per year
   Ⓓ 2.29 people per year

Choose the statement that is true about the given numbers.

9.

Column A	*Column B*
The slope of the line through (−6, 2) and (4, −2).	The slope of the line through (5, 0) and (0, 2).

   Ⓐ The number in column A is greater
   Ⓑ The number in column B is greater
   Ⓒ The two numbers are equal.
   Ⓓ The relationship cannot be determined from the given information.

Review and Assess

SAT/ACT CHAPTER TEST This test covers the material in Chapter 4 in a standardized test format. Students taking this form of the test should be reminded to read all of the answer choices before deciding which is the correct one.

TECHNOLOGY RESOURCE

Teachers can use the Time-Saving Test and Practice Generator to create customized review and assessment materials for Chapter 4.

CHAPTER OVERVIEW

CHAPTER GOALS

Students will use linear equations and their graphs to model real-life situations. To do so, they will draw and make predictions from scatter plots. The students will graph linear equations using a variety of techniques, including a table of values, two points, and a point and the slope of the line. They will be able to identify equations of vertical, horizontal, and parallel lines, and equations that represent direct variation. Students will identify and evaluate functions. They will also use their graphs of linear functions to model and solve real-life problems.

APPLICATION NOTE

The invention by Andrew Smith Hallidie of a gripper system on the underside of a streetcar that could grab a moving cable helped replace horse-drawn streetcars with cable cars. San Francisco's first cable car was carrying passengers in 1873.

Streetcars, including trolleys and cable cars, created the first commuter rail lines for suburban communities. The expense of maintaining streetcar systems and the popularity of the automobile caused many cities to replace their streetcar lines with gasoline-powered buses and electric buses. Subsurface cars, which are hybrids of electric trolley cars and subway trains, have also made an appearance in American cities.

San Francisco however, recognizing cable cars' energy efficiency, low emissions, and their role in alleviating traffic congestion, as well as their popularity with tourists, has continued to maintain its cable car system.

Additional information about Transportation is available at **www.mcdougallittell.com**

GRAPHING LINEAR EQUATIONS AND FUNCTIONS

▶ *How do you solve the problem of transporting people up steep hills?*

CHAPTER 4

APPLICATION: Transportation

Before 1873, public transportation in San Francisco was provided by horse-drawn wagons.

To design a transportation system, people needed a mathematical way to describe and measure steepness.

Think & Discuss 1–2. See margin.

1. What are some other real-life situations when steepness is important? When is steepness helpful? When is steepness a problem?
2. How would you describe the steepness of the sections of street shown below?

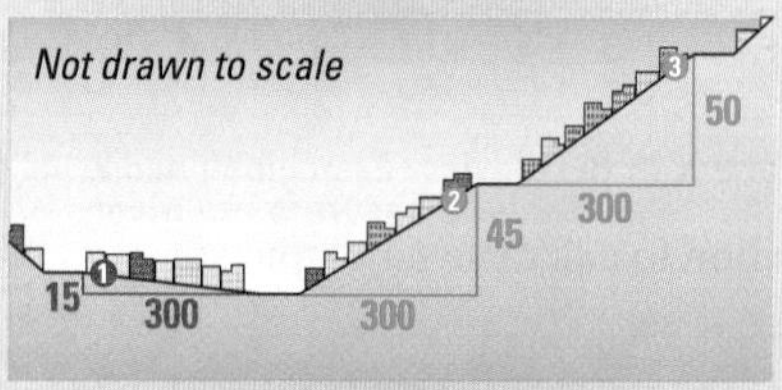

(cross section of side view)

Learn More About It

Slope is a measure of steepness. You will calculate the slope of this street in Exercise 55 on p. 231.

APPLICATION LINK Visit www.mcdougallittell.com for more information about cable cars.

ADDITIONAL RESOURCES

Another way to begin the chapter is to show the video clip of a real-life motivator for Chapter 4 on the *Instant Replay: Video Review Games.*

PROJECTS

A project covering Chapters 4–6 appears on pages 392–393 of the Student Edition. An additional project for Chapter 4 is available in the *Chapter 4 Resource Book,* p. 130.

TECHNOLOGY

Software
- *Electronic Teaching Tools*
- *Online Lesson Planner*
- *Personal Student Tutor*
- *Test and Practice Generator*
- *Electronic Lesson Presentations* (Lesson 4.4)

Video
- *Instant Replay: Video Review Games*

Internet Connections
www.mcdougallittell.com
- **Application Links**
  201, 216, 220, 224, 235, 246, 254
- **Data Updates**
  205, 223, 232, 247, 260
- **Student Help**
  209, 211, 228, 232, 238, 245, 248, 252, 260
- **Career Links**
  231, 236, 243, 252
- **Extra Challenge**
  208, 217, 233, 239, 246, 255, 261

1. ***Sample answer:*** **Steepness is an important factor in ski slopes and mountain roads. It is helpful for downhill skiers. It is a problem for motorists on icy roads.**
2. ***Sample answer:*** **Section 1 is not very steep and slants down from left to right; section 2 is very steep and slants up; section 3 is even steeper and slants up.**

PREPARE

Diagnostic Tools

The **Skill Review** exercises can help you diagnose whether students have the following skills needed in Chapter 4:

- Write percents in various forms.
- Represent linear data with a line graph.
- Evaluate variable expressions.

The following resources are available for students who need additional help with these skills.

- Prerequisite Skills Review (*Chapter 4 Resource Book,* p. 5; *Warm-Up Transparencies,* p. 27)
- Reteaching with Practice (Chapter Resource Books for Lessons 1.7 and 2.2.)
- *Personal Student Tutor*

Additional Resources

The following resources are provided to help you prepare for the upcoming chapter and customize review materials:

- ***Chapter 4 Resource Book***
  Tips for New Teachers (p. 1)
  Parent Guide (p. 3)
  Lesson Plans (every lesson)
  Lesson Plans for Block Scheduling (every lesson)
- ***Electronic Teaching Tools***
- ***Online Lesson Planner***
- ***Test and Practice Generator***

6.

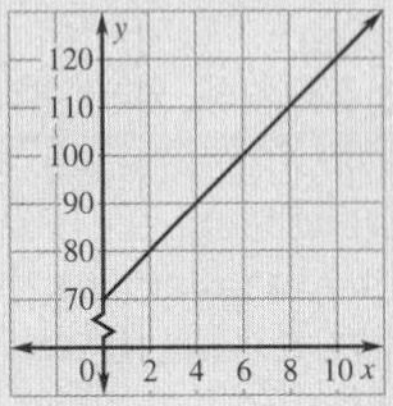

CHAPTER 4

Study Guide

PREVIEW

What's the chapter about?

Chapter 4 is about **graphing linear equations**. In Chapter 4, you'll learn

- how to graph linear equations.
- two ways to graph linear equations quickly.
- how to tell whether an equation or a graph represents a function.

KEY VOCABULARY

▶ **Review**
- **variable expression, p. 3**
- **solution of an equation, p. 24**
- **function, p. 46**
- **linear equation, p. 133**

▶ **New**
- **coordinate plane, p. 203**
- **scatter plot, p. 204**
- **graph of a linear equation, p. 210**
- ***x*-intercept, p. 218**
- ***y*-intercept, p. 218**
- **slope, p. 226**
- **rate of change, p. 229**
- **slope-intercept form, p. 241**
- **function notation, p. 257**

PREPARE

Are you ready for the chapter?

SKILL REVIEW Do these exercises to review key skills that you'll apply in this chapter. See the given **reference page** if there is something you don't understand.

Rewrite as a decimal and as a fraction in lowest terms. (Skills Review, p. 784)

1. 50% **0.5, $\frac{1}{2}$** **2.** 75% **0.75, $\frac{3}{4}$** **3.** 1% **0.01, $\frac{1}{100}$** **4.** 20% **0.2, $\frac{1}{5}$**

Use the function $y = 5x + 70$, where $x \geq 0$. (Review Examples 1 and 2, pp. 46–47)

5. Sample answers are given.

5. For several inputs x, use the function to calculate an output y.

6. Represent the data with a line graph. **See margin.**

7. Describe the domain and range of the function.
domain: $x \geq 0$; range: $y \geq 70$

x	y
0	70
1	75
2	80
5	95
10	120

Evaluate the expression for the given values of the variables. (Review Example 3, p. 109)

8. $\frac{x - y}{2}$ when $x = -3$ and $y = -1$ **−1** **9.** $\frac{x + 2y}{x}$ when $x = 6$ and $y = 3$ **2**

STUDENT HELP

Study Tip
"Student Help" boxes throughout the chapter give you study tips and tell you where to look for extra help in this book and on the Internet.

STUDY STRATEGY

Here's a study strategy!

Getting Your Questions Answered

Each day after you finish your math homework, write a list of questions about things you don't understand. Ask your teacher or another student to answer your questions and write the explanations in your notebook.

EXPLORING DATA AND STATISTICS

4.1 Coordinates and Scatter Plots

What you should learn

GOAL 1 Plot points in a coordinate plane.

GOAL 2 Draw a scatter plot and make predictions about **real-life** situations as in **Example 3**.

Why you should learn it

▼ To solve **real-life** problems such as predicting how much film will be developed for the Winter Olympics in the year 2002 in **Ex. 38**.

REAL LIFE

GOAL 1 PLOTTING POINTS IN A COORDINATE PLANE

A **coordinate plane** is formed by two real number lines that intersect at a right angle. Each point in the plane corresponds to an **ordered pair** of real numbers.

The first number in an ordered pair is the ***x*-coordinate** and the second number is the ***y*-coordinate**. The ordered pair $(3, -2)$ has an x-coordinate of 3 and a y-coordinate of -2 as shown in the graph at the left below.

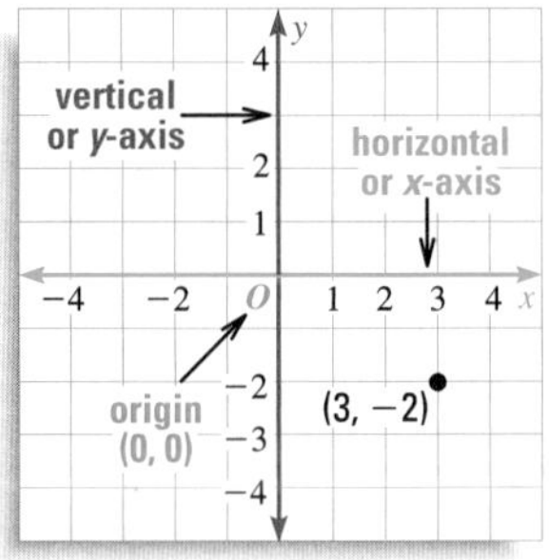

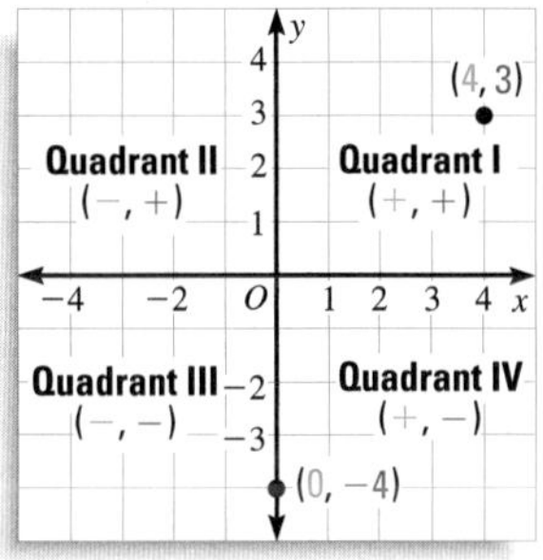

The x-axis and the y-axis (the *axes*) divide the coordinate plane into four quadrants. You can tell which quadrant a point is in by looking at the signs of its coordinates.

In the diagram at the right above, the point $(4, 3)$ is in Quadrant I. The point $(0, -4)$ is on the y-axis and is not inside any of the four quadrants.

The point in the plane that corresponds to an ordered pair (x, y) is called the **graph** of (x, y). To plot a point, you draw the point in the coordinate plane that corresponds to an ordered pair of numbers.

EXAMPLE 1 *Plotting Points in a Coordinate Plane*

a. To plot the point $(3, 4)$, start at the origin. Move 3 units to the right and 4 units up.

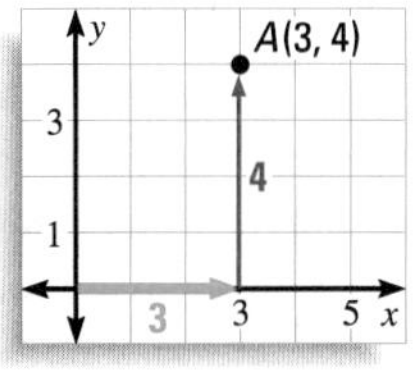

b. To plot the point $(-2, -3)$, start at the origin. Move 2 units to the left and 3 units down.

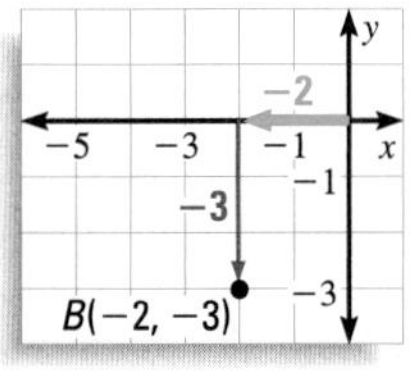

c. To plot the point $(2, 0)$, start at the origin. Move 2 units to the right and 0 units up.

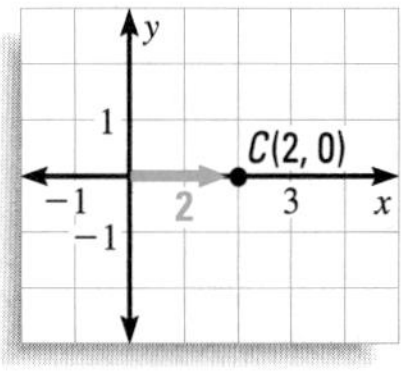

1 PLAN

PACING
Basic: 1 day
Average: 1 day
Advanced: 1 day
Block Schedule: 0.5 block with Ch. 3 Assess.

LESSON OPENER APPLICATION
An alternative way to approach Lesson 4.1 is to use the Application Lesson Opener:
- Blackline Master (*Chapter 4 Resource Book,* p. 12)
- Transparency (p. 24)

MEETING INDIVIDUAL NEEDS
- ***Chapter 4 Resource Book***
  Prerequisite Skills Review (p. 5)
  Practice Level A (p. 16)
  Practice Level B (p. 17)
  Practice Level C (p. 18)
  Reteaching with Practice (p. 19)
  Absent Student Catch-Up (p. 21)
  Challenge (p. 23)
- ***Resources in Spanish***
- ***Personal Student Tutor***

NEW-TEACHER SUPPORT
See the Tips for New Teachers on pp. 1–2 of the *Chapter 4 Resource Book* for additional notes about Lesson 4.1.

WARM-UP EXERCISES
Transparency Available
Use the formula $F = \frac{9}{5}C + 32$ to find the following temperatures.
1. What Celsius temperature is equivalent to 32°F? **0°**
2. What Fahrenheit temperature is equivalent to 225°C? **437°F**
3. What Celsius temperature is equivalent to −40°F? **−40°C**

2 TEACH

MOTIVATING THE LESSON
Ask students if they believe that there is a relationship between the size of a house in a particular town and the selling price of the house. Explain to students that scatter plots will help them to answer this kind of question.

EXTRA EXAMPLE 1
Plot and label the points $A(2, 1)$, $B(5, -3)$, $C(-3, 0)$, $D(-2, -2)$, $E(0, 4)$, and $F(-2, 3)$.

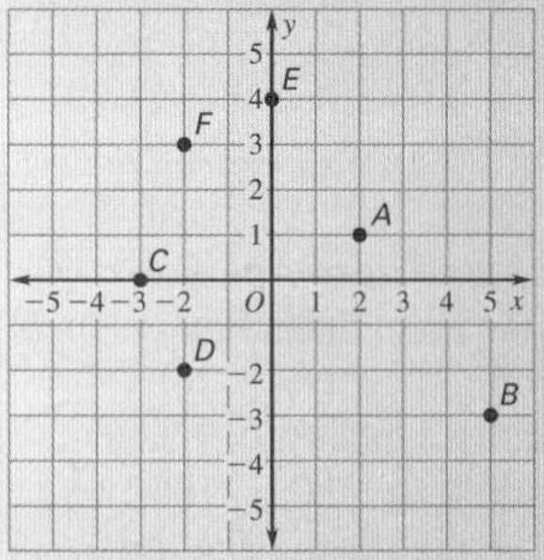

EXTRA EXAMPLE 2
You want to know the distance that five tires with different diameters d cover in one revolution. Your estimates of the circumferences C of the tires are given in the chart. Use a scatter plot to find the incorrect estimate of circumference. **130 in.**

d (in.)	48	23	20	24	30
C (in.)	151	72	63	130	94

CHECKPOINT EXERCISES
For use after Examples 1 and 2:
1. Plot the points in Sets I and II to see which points lie on a straight line. **Set I**

I	II
(−4, 9)	(−2, −6)
(1, −1)	(0, −3)
(2, −3)	(4, 6)
(6, −11)	(7, 14)

GOAL 2 USING A SCATTER PLOT

Many real-life situations can be described in terms of pairs of numbers. Medical charts record both the height and weight of the patient, while weather reports may include both temperature and windspeed. One way to analyze the relationships between two quantities is to graph the pairs of data on a coordinate axis. Such a graph is called a **scatter plot**.

REAL LIFE *Sports*

EXAMPLE 2 *Making a Scatter Plot*

You are the student manager of your high school soccer team. You are working on the team's program guide and have recorded the height and weight of the eleven starting players in the given table.

Height (in.)	72	70	71	70	69	70	69	73	66	70	76
Weight (lb)	190	170	180	175	160	160	150	180	150	150	200

a. Make a scatter plot of the data. Put height h on the horizontal axis and weight w on the vertical axis.

b. Use the scatter plot to estimate the weight of a player who is 69 inches tall and of one who is 71 inches tall.

c. In general, how does weight change as height changes?

d. What would you expect a player who is 74 inches tall to weigh?

(66, 150), (69, 150), (70, 150)
(69, 160), (70, 160)
(70, 170), (70, 175)
(71, 180), (73, 180)
(72, 190)
(76, 200)

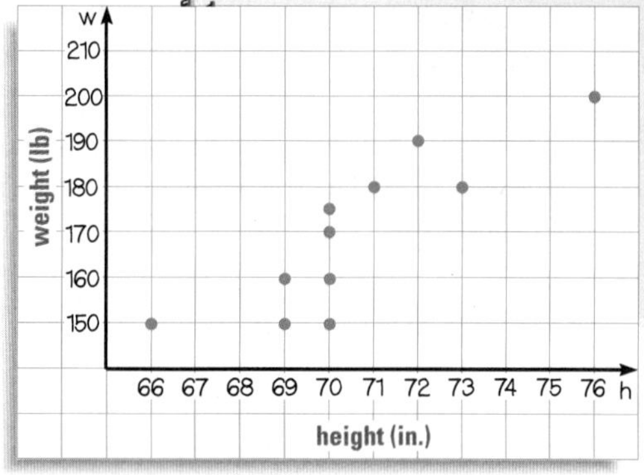

SOLUTION

a. Plot each ordered pair, as shown at the right.

b. The two players who are 69 inches tall weigh 150 pounds and 160 pounds, so a good estimate would be about 155 pounds. The player who is 71 inches tall weighs 180 pounds, so a possible estimate would be about 180 pounds.

c. As height increases, it appears that weight increases.

d. There is no given data value for a player who is 74 inches tall. However, the 72-inch-tall player weighs 190 pounds, the 73-inch-tall player weighs 180 pounds, and the 76-inch-tall player weighs 200 pounds, so a good estimate of the weight of a 74-inch-tall player would be between 190 and 195 pounds.

STUDENT HELP
INTERNET
HOMEWORK HELP
Visit our Web site www.mcdougallittell.com for extra examples.

You can use scatter plots to see trends in data and to make predictions about the future.

EXAMPLE 3 *Describing Patterns from a Scatter Plot*

WINTER SPORTS EQUIPMENT The amount (in millions of dollars) spent in the United States on snowmobiles and ski equipment is shown in the table.

a. Draw a scatter plot of each set of data in the same coordinate plane.

b. Describe the pattern of the amount spent on snowmobiles.

c. Describe the pattern of the amount spent on ski equipment.

Year	1990	1991	1992	1993	1994	1995	1996
Snowmobiles	322	362	391	515	715	924	970
Ski equipment	606	577	627	611	652	607	644

DATA UPDATE of National Sporting Goods Association data at www.mcdougallittell.com

SOLUTION

a. Because you want to see how spending changes over time, put time t on the horizontal axis and spending s on the vertical axis. Let t be the number of years since 1990. The scatter plot is shown below.

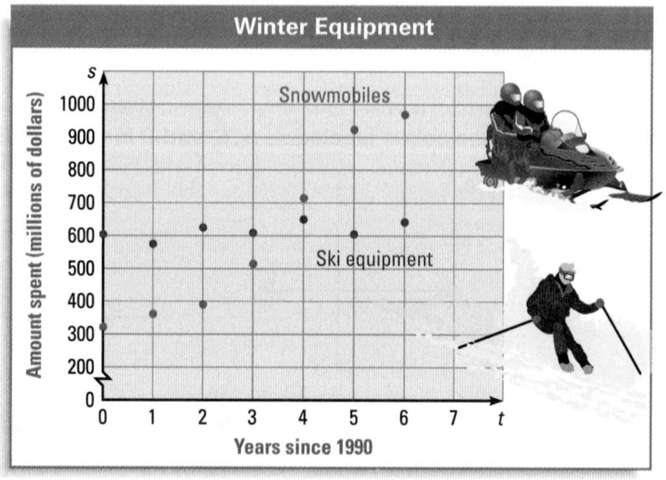

b. From the scatter plot, you can see that the amount spent on snowmobiles has been increasing rapidly since 1990.

c. The amount spent on ski equipment has been fairly constant since 1990. The amount spent has stayed around $600 million.

FOCUS ON APPLICATIONS

REAL LIFE **SNOWMOBILES** have proven essential in the search and rescue field. They can often reach mountainous areas more quickly than an aircraft.

EXTRA EXAMPLE 3

The number of U.S. citizens 7 years or older (in millions) who played soccer is shown below. Draw a scatter plot. Describe the pattern, and predict the number of U.S. citizens who will play soccer in 2007.

Year	1987	1989	1991
Number	9.8	11.2	10.0
Year	1993	1995	1997
Number	10.3	12.0	13.7

With the exception of 1991, the participation has increased each year; about 16 million.

CHECKPOINT EXERCISES

For use after Example 3:

1. The data below represent the weight and height of a male. At age 13, he was 63 in. tall. Predict his weight at age 13. **about 123 lb**

Age	Height (in.)	Weight (lb)
5	50	78
10	60	112
15	67	135
25	70	150

FOCUS ON VOCABULARY

The order of the coordinates in an ordered pair is extremely important. Just as x precedes y in the alphabet, the x-coordinate is always placed first in an ordered pair.

CLOSURE QUESTION

If the data points on a scatter plot rise from left to right, how can you describe the relationship between the x-coordinates and the y-coordinates?
As x increases, y increases.

DAILY PUZZLER

List the coordinates of three points that, when connected by line segments, are the vertices of an isosceles right triangle. ***Sample answer:*** **(−2, −3), (−2, 2), and (3, −3)**

3 APPLY

ASSIGNMENT GUIDE

BASIC
Day 1: pp. 206–208 Exs. 10–26 even, 27–32, 35–38, 42–52 even

AVERAGE
Day 1: pp. 206–208 Exs. 10–26 even, 27–32, 35–38, 42–52 even

ADVANCED
Day 1: pp. 206–208 Exs. 10–26 even, 27–32, 35–38, 42–52 even

BLOCK SCHEDULE
pp. 206–208 Exs. 10–26 even, 27–32, 35–38, 42–52 even (with Ch. 3 Assess.)

EXERCISE LEVELS

Level A: ***Easier***
10–18

Level B: ***More Difficult***
19–31, 38, 42–52

Level C: ***Most Difficult***
32–37, 39–41

✓ HOMEWORK CHECK

To quickly check student understanding of key concepts, go over the following exercises: Exs. 10, 14, 20, 28, 35, 36. See also the Daily Homework Quiz:

- Blackline Master (*Chapter 4 Resource Book*, p. 26)
- Transparency (p. 29)

4.

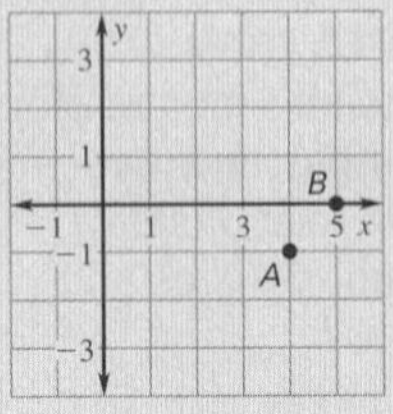

5.

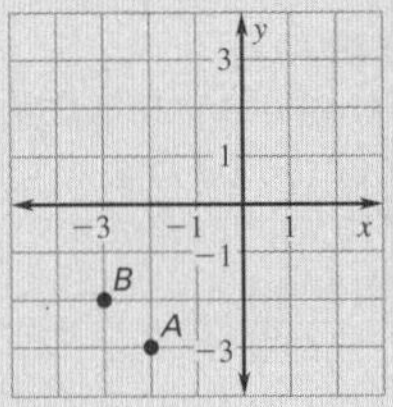

GUIDED PRACTICE

Vocabulary Check ✓

1. In the ordered pair (4, 9), the x-coordinate is _?_. **4**

Concept Check ✓

2. Decide whether the following statement is *true* or *false*. *Each point in a coordinate plane corresponds to an ordered pair of real numbers.* **true**

Skill Check ✓

3. Write the ordered pairs that correspond to the points labeled A, B, and C in the coordinate plane at the right.
***A*(1, 1), *B*(−2, 1), *C*(0, −2)**

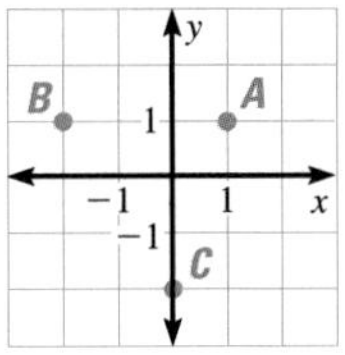

Ex. 3

Plot the ordered pairs in a coordinate plane. **See margin.**

4. $A(4, -1), B(5, 0)$ **5.** $A(-2, -3), B(-3, -2)$

6. The point (−2, 5) lies in Quadrant _?_. **II**

SCATTER PLOT In Exercises 7 and 8, draw a scatter plot of the given data.
See margin.

7.

Time	1:00	3:00	5:00	7:00
Temp.	71°	74°	68°	63°

8.

TV size	19 in.	27 in.	32 in.	36 in.
Price	\$179	\$349	\$499	\$659

9. **SNOWMOBILE SALES** Use the scatter plot in Example 3. If you were given the amounts spent on snowmobiles from 1990 through 1992 only, would your description of the pattern be different than the one given in the Example?
***Sample answer:* yes, the amount spent on snowmobiles increased but not rapidly.**

PRACTICE AND APPLICATIONS

STUDENT HELP
Extra Practice to help you master skills is on p. 800.

IDENTIFYING ORDERED PAIRS Write the ordered pairs that correspond to the points labeled *A*, *B*, *C*, and *D* in the coordinate plane.

10.

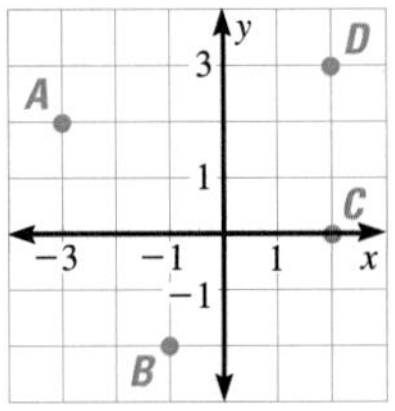

***A*(−3, 2), *B*(−1, −2), *C*(2, 0), *D*(2, 3)**

11.

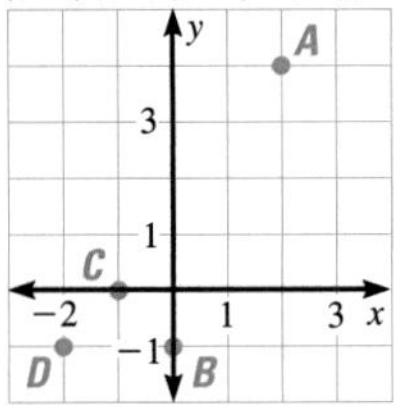

***A*(2, 4), *B*(0, −1), *C*(−1, 0), *D*(−2, −1)**

12.

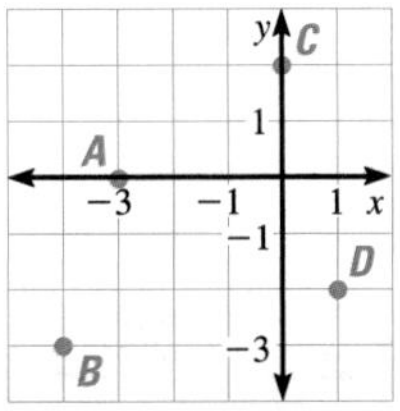

***A*(−3, 0), *B*(−4, −3), *C*(0, 2), *D*(1, −2)**

PLOTTING POINTS Plot and label the ordered pairs in a coordinate plane.
13–18. See margin.

13. $A(0, 3), B(-2, -1), C(2, 0)$ **14.** $A(5, 2), B(4, 3), C(-2, -4)$

15. $A(4, 1), B(0, -3), C(3, 3)$ **16.** $A(0, 0), B(2, -2), C(-2, 0)$

17. $A(-4, 1), B(-1, 5), C(0, -4)$ **18.** $A(3, -5), B(1.5, 3), C(-3, -1)$

STUDENT HELP
HOMEWORK HELP
Example 1: Exs. 10–26
Example 2: Exs. 33–37
Example 3: Exs. 27–32

IDENTIFYING QUADRANTS Without plotting the point, tell whether the point is in Quadrant I, Quadrant II, Quadrant III, or Quadrant IV.

19. (5, −3) **Quadrant IV** **20.** (−2, 7) **Quadrant II** **21.** (6, 17) **Quadrant I** **22.** (14, −5) **Quadrant IV**

23. (−4, −2) **Quadrant III** **24.** (3, 9) **Quadrant I** **25.** (−5, −2) **Quadrant III** **26.** (−5, 6) **Quadrant II**

13–18. See Additional Answers beginning on page AA1.

STUDENT HELP

Look Back
For help with breaks in the scale of a graph, see p. 41.

32. **The gas mileage of a car should decrease as length increases, since weight increases as length increases, and mileage decreases as weight increases.**

34. **Check scatter plots. *Sample answer:* if the points appear to lie on a line that rises from left to right, you and your friend have similar tastes. If the points appear to lie near a line that falls from left to right, you and your friend have very opposite tastes. If the points do not appear to lie near any line, your tastes are probably not comparable.**

CAR COMPARISONS **In Exercises 27–32, use the scatter plots.**

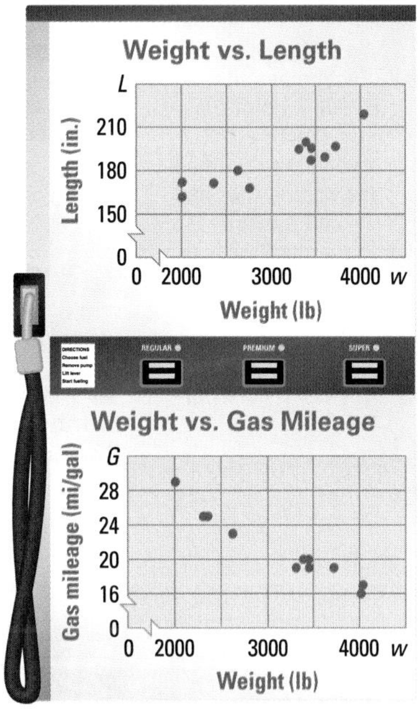

27. In the Weight vs. Length graph, what are the units on the horizontal axis? What are the units on the vertical axis?
pounds; inches

28. In the Weight vs. Length graph, estimate the coordinates of the point for a car that weighs about 4000 pounds.
***Sample answer:* (4000, 210)**

29. Which of the following is true? **C**

A. Length tends to decrease as weight increases.

B. Length is constant as weight increases.

C. Length tends to increase as weight increases.

D. Length is not related to weight.

30. In the Weight vs. Gas Mileage graph, what is the value of w in the ordered pair (2010, 29)? What is the value of G?
2010 lb; 29 mi/gal

31. INTERPRETING DATA In the Weight vs. Gas Mileage graph, how does gas mileage tend to change as weight increases?
It decreases.

32. CRITICAL THINKING How would you expect the length of a car to affect its gas mileage? Explain your reasoning. **See margin.**

33. GATHERING DATA Write a list of ten songs in alphabetical order. Put songs you like on the list and some you don't like. Make two copies. On one copy, rank the songs from 1 to 10 (1 for the most liked and 10 for the least liked). Then ask a friend to rank the songs without looking at your rankings.
Check lists and rankings.

34. INTERPRETING A GRAPH Make a scatter plot with your song ratings from Exercise 33 on the horizontal axis and your friend's ratings on the vertical axis. What conclusions (if any) can you draw from the scatter plot? **See margin.**

FOCUS ON APPLICATIONS

AMERICAN FLAMINGOES have distinctive and unusual features. Their extremely long neck and legs give the impression that the bird could fly backwards. When the flamingo's wings are fully extended, you can see the black in its wings.

BIOLOGY CONNECTION **In Exercises 35–37, the table shows the wing length in millimeters and the wing-beat rate in beats per second.**

Bird	Flamingo	Shellduck	Velvet Scoter	Fulmar	Great Egret
Wing length	400	375	281	321	437
Wing-beat rate	2.4	3.0	4.3	3.6	2.1

35. Make a scatter plot that shows the wing lengths and wing-beat rates for the six birds. Use the horizontal axis to represent the wing length. **See margin.**

36. What is the slowest wing-beat rate shown on the scatter plot? What is the fastest? Where are these located on your scatter plot?
2.1 beats/second; 4.3 beats/second; farthest right and farthest left

37. INTERPRETING DATA Describe the relationship between the wing length and the wing-beat rate. **As wing length increases, wing-beat rate decreases.**

APPLICATION NOTE

EXERCISES 35–37 Representing wing-beat rate on the horizontal axis will also produce a scatter plot with points that fall from left to right. In these exercises, however, the students are analyzing how wing-beat rate is dependent on wing length. Although either variable can be associated with the horizontal axis, it is customary to associate the variable whose values are used to determine the values of the other variable with the horizontal axis.

7.

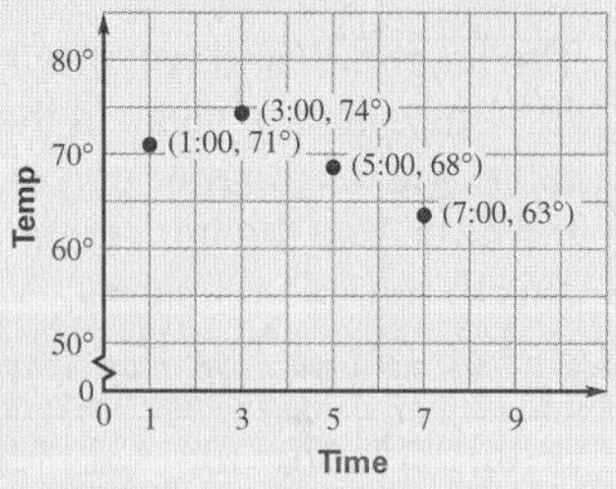

8.

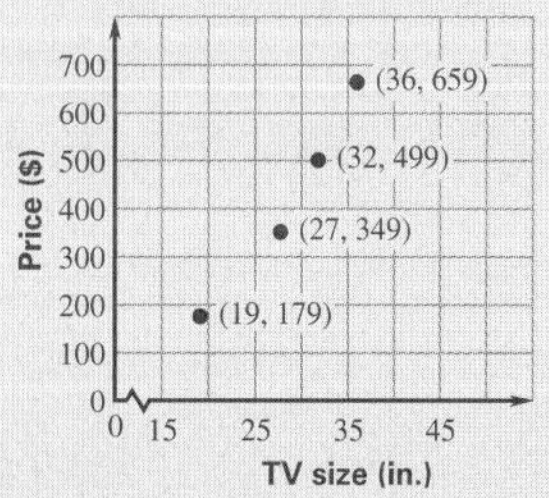

35. Wing Length vs. Wing-beat Rate

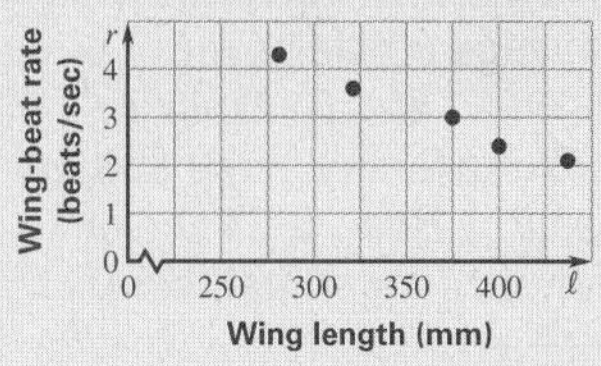

ADDITIONAL PRACTICE AND RETEACHING

For Lesson 4.1:

- Practice Levels A, B, and C (*Chapter 4 Resource Book,* p. 16)
- Reteaching with Practice (*Chapter 4 Resource Book,* p. 19)
- See Lesson 4.1 of the *Personal Student Tutor*

For more Mixed Review:

- Search the *Test and Practice Generator* for key words or specific lessons.

4 ASSESS

DAILY HOMEWORK QUIZ

Transparency Available

1. Use the graph. Write the ordered pairs that correspond to points *A*, *C*, and *E*.
(2, 1), (−2, 2), (3, −2)

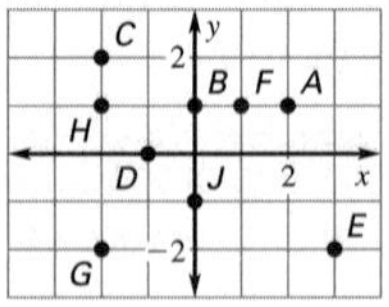

2. The heights *H* and weights *W* of 5 students are given in the table. Make a scatter plot of the data. Use the horizontal axis to represent heights.

H (in.)	63	75	74	66	70
W (lb)	112	176	186	131	138

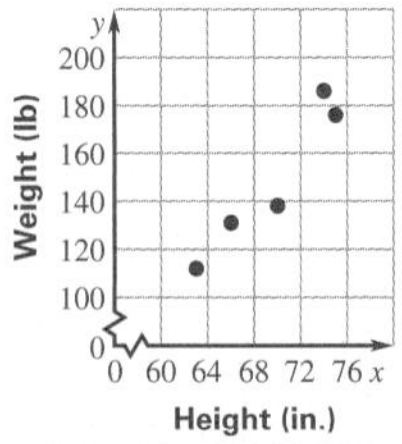

EXTRA CHALLENGE NOTE

Challenge problems for Lesson 4.1 are available in **blackline** format in the *Chapter 4 Resource Book*, p. 23 and at **www.mcdougallittell.com.**

ADDITIONAL TEST PREPARATION

1. WRITING How you can predict values using a scatter plot? **See sample answer at right.**

2. OPEN ENDED Describe a real-life situation that can be modeled by a scatter plot. **See sample answer at right.**

38a, 39, 40. See Additional Answers beginning on page AA1.

Test Preparation

38. b. ***Sample answer:*** **About 83,000 rolls of film; if you sketch a line through the points in the scatter plot and extend it to $x = 22$, the y-coordinate is about 83,000.**

38. MULTI-STEP PROBLEM The table below shows the number of rolls of film developed for the United States media at the Winter Olympics.

Number of years since 1980, *t*	4	8	12	14	18
Rolls of film, *f*	48,200	53,750	60,500	67,500	75,000

a. Construct a scatter plot of the data. Describe the pattern of the number of rolls of film developed for the Winter Olympics from 1984 to 1998.
See margin for plot. The number of rolls developed increased steadily.

b. *Writing* Predict the number of rolls of film that will be developed for the Winter Olympics in the year 2002. Explain how you made your prediction. **See margin.**

★ Challenge

MUSIC **Sara thought that there might be a relationship between the number of songs and the number of minutes of music recorded on a CD. She collected the following data from five CDs she had on her desk.**

Number of songs, *s*	15	12	13	16	18
Number of minutes, *m*	44.9	31.5	40.6	50.5	71.6

39. Make a scatter plot of Sara's data. What relationship do the data suggest?
See margin; as the number of songs increases, the number of minutes increases.

40. To check her findings, Sara collected data from five more CDs. Make a new scatter plot of these data. What relationship do the data suggest?
See margin; as the number of songs increases, the number of minutes decreases.

Number of songs, *s*	12	13	9	11	10
Number of minutes, *m*	46.8	40.6	67.0	53.0	56.1

EXTRA CHALLENGE
www.mcdougallittell.com

41. CRITICAL THINKING Do you think there is a relationship between the number of songs and the number of minutes on a CD? Explain your reasoning. How could you test whether there is a relationship?
No; if there were a relationship, the two scatter plots would show the same trend; sample answer: test a larger sample.

MIXED REVIEW

EVALUATING EXPRESSIONS **Evaluate the expression for the given value of the variable.** **(Review 1.3 and 2.5 for 4.2)**

42. $3x + 9$ when $x = 2$ **15**

43. $13 - (y + 2)$ when $y = 4$ **7**

44. $4.2t + 17.9$ when $t = 3$ **30.5**

45. $-3x - 9y$ when $x = -2$ and $y = -1$ **15**

USING EXPONENTS **Evaluate the expression.** **(Review 1.2)**

46. $x^2 - 3$ when $x = 4$ **13**

47. $12 + y^3$ when $y = 3$ **39**

ABSOLUTE VALUE **In Exercises 48–51, evaluate the expression.** **(Review 2.1)**

48. $|-2.6|$ **2.6**

49. $|1.07|$ **1.07**

50. $\left|\frac{9}{10}\right|$ $\frac{9}{10}$

51. $\left|\frac{-2}{3}\right|$ $\frac{2}{3}$

52. **TEEN SAVINGS** In a survey of 500 teens, 345 said they have a savings account. What is the experimental probability that a randomly chosen teen in the survey has a savings account? **(Review 2.8)** **0.69, or 69%**

Additional Test Preparation *Sample answers:*

1. Look for a pattern in the data points. If the points lie close to a line, locate the x-coordinate of the value on this line. The y-coordinate of this point is the predicted value.

2. You can plot the weights of turkeys in pounds and the number of minutes needed to cook them.

ACTIVITY 4.1

Using Technology

Graphing a Scatter Plot

> **STUDENT HELP**
>
> INTERNET **KEYSTROKE HELP**
>
> See keystrokes for several models of calculators at www.mcdougallittell.com

▶ EXAMPLE

The table below shows the maximum time allowed for boys in the 1-mile run to qualify for the President's Physical Fitness Award. Use a graphing calculator or a computer to make a scatter plot of the data.

Age (years)	10	11	12	13	14	15	16	17
Time (minutes)	7.95	7.53	7.18	6.83	6.43	6.33	6.13	6.10

▶ SOLUTION

1. Write the data as a set of ordered pairs. Use age as the x-coordinate and time as the y-coordinate, for example (10, 7.95). Use the STAT EDIT feature to enter the ordered pairs as List 1 and List 2.

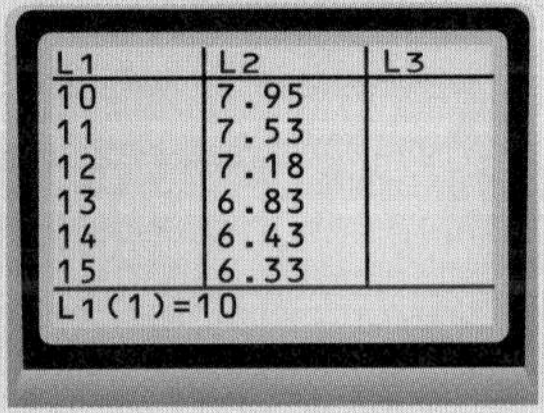

2. Use WINDOW to describe the size of the graph. The x-values are between 0 and 20. The y-values are between 0 and 10.

The x-scale is the number of units per mark on the x-axis of the graph.

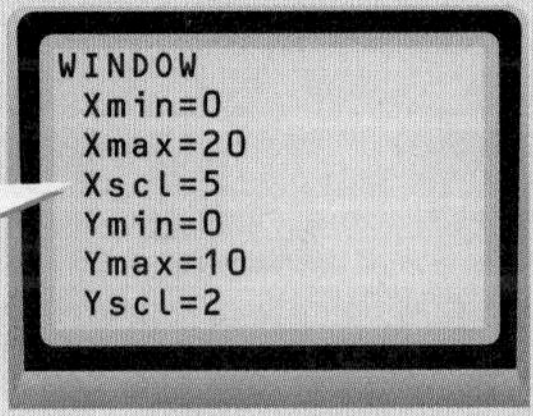

3. Use STAT PLOT. In this window, select scatter plot, List 1 for the x-values, and List 2 for the y-values.

4. Use GRAPH to draw the scatter plot.

Tick marks will be 5 units apart on the x-axis.

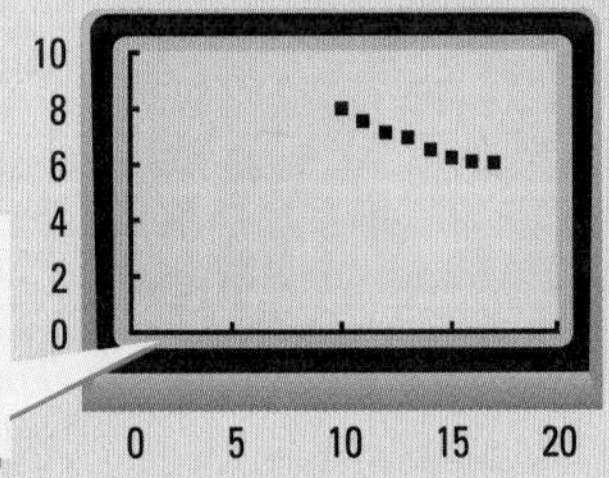

▶ EXERCISES

1. **LOOK FOR A PATTERN** Describe any patterns you see in the scatter plot you made in the example above. **As age increases, time decreases.**

The table below shows the maximum time allowed for girls in the 1-mile run to qualify for the President's Physical Fitness Award.

Age (years)	10	11	12	13	14	15	16	17
Time (minutes)	9.32	9.03	8.38	8.22	7.98	8.13	8.38	8.25

2. Use a graphing calculator to make a scatter plot of the data. **See margin.**
3. **LOOK FOR A PATTERN** Describe any patterns you see in the scatter plot. **See margin.**
4. How does this scatter plot differ from the scatter plot for the boys' times? **See margin.**

4. ***Sample answer:*** **The times do not steadily decrease as in the scatter plot for the boys' times. Also, the times are greater.**

> **STUDENT HELP**
>
> **Study Tip**
>
> You will use the skills from this activity to help with the activity on page 299.

1 Planning the Activity

PURPOSE
To use a graphing calculator or a computer to show how a scatter plot can be used to model the relationship between two variables.

MATERIALS
- graphing calculator per student or a computer with spreadsheet software for each pair of students
- Keystroke blackline (*Chapter 4 Resource Book,* p. 15)

PACING
- Exploring the Concept — 10 min
- Drawing Conclusions — 10 min

▶ *LINK TO LESSON*
Students can use a graphing calculator to demonstrate Example 2.

2 Managing the Activity

CLASSROOM MANAGEMENT
Students whose scatter plots show a different pattern from the one in step 4 should check to make sure they entered the data correctly.

ALTERNATIVE APPROACH
To save time, demonstrate the activity to the class using an overhead projector.

3 Closing the Activity

★ ***KEY DISCOVERY***
A scatter plot models the relationship between two variables and allows patterns to be examined.

ACTIVITY ASSESSMENT
JOURNAL Select 10 people from your class and measure their height and the length of their forearm. Plot the ordered pairs on a scatter plot with height on the horizontal axis. Do the points on the scatter plot rise or fall from left to right? **Scatter plots will vary; rise**

2, 3. See Additional Answers beginning on page AA1.

1 PLAN

PACING
Basic: 2 days
Average: 2 days
Advanced: 2 days
Block Schedule: 1 block

LESSON OPENER ACTIVITY
An alternative way to approach Lesson 4.2 is to use the Activity Lesson Opener:
- Blackline Master (*Chapter 4 Resource Book*, p. 27)
- Transparency (p. 25)

MEETING INDIVIDUAL NEEDS
- ***Chapter 4 Resource Book***
  Prerequisite Skills Review (p. 5)
  Practice Level A (p. 28)
  Practice Level B (p. 29)
  Practice Level C (p. 30)
  Reteaching with Practice (p. 31)
  Absent Student Catch-Up (p. 33)
  Challenge (p. 35)
- ***Resources in Spanish***
- ***Personal Student Tutor***

NEW-TEACHER SUPPORT
See the Tips for New Teachers on pp. 1–2 of the *Chapter 4 Resource Book* for additional notes about Lesson 4.2.

WARM-UP EXERCISES

Transparency Available

Rewrite each equation in function form by solving for y.

1. $2x + y = 10$
$y = -2x + 10$

2. $6x - 3y = -3$
$y = 2x + 1$

Find the value of y when $x = -3$.

3. $y = x - 7$
-10

4. $y = -5x + 1$
16

4.2 Graphing Linear Equations

What you should learn

GOAL 1 Graph a linear equation using a table or a list of values.

GOAL 2 Graph horizontal and vertical lines.

Why you should learn it

▼ To solve **real-life** problems such as finding the amount of calories burned while training for Triathlon in **Exs. 71–73.**

REAL LIFE

GOAL 1 GRAPHING A LINEAR EQUATION

A **solution of an equation** in two variables x and y is an ordered pair (x, y) that makes the equation true. The **graph of an equation** in x and y is the set of *all* points (x, y) that are solutions of the equation. In this lesson you will see that the graph of a *linear* equation is a line.

EXAMPLE 1 *Verifying Solutions of an Equation*

Use the graph to decide whether the point lies on the graph of $x + 3y = 6$. Justify your answer algebraically.

a. $(1, 2)$

b. $(-3, 3)$

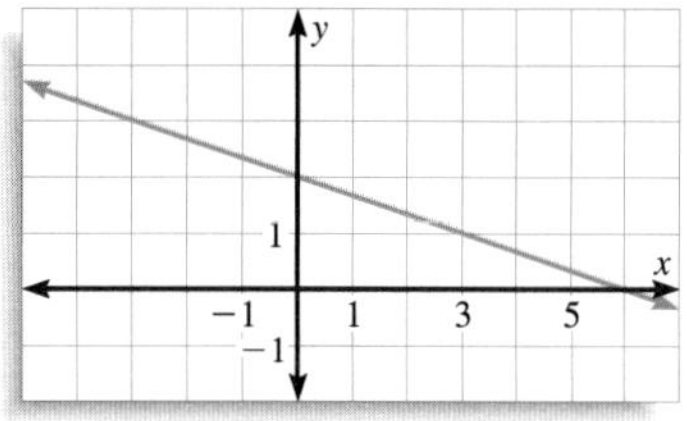

SOLUTION

a. The point $(1, 2)$ is *not* on the graph of $x + 3y = 6$. This means that $(1, 2)$ is not a solution. You can check this algebraically.

$x + 3y = 6$ **Write original equation.**

$1 + 3(2) \stackrel{?}{=} 6$ **Substitute 1 for *x* and 2 for *y*.**

$7 \neq 6$ **Simplify. Not a true statement**

▶ $(1, 2)$ is not a solution of the equation $x + 3y = 6$, so it is not on the graph.

b. The point $(-3, 3)$ is on the graph of $x + 3y = 6$. This means that $(-3, 3)$ is a solution. You can check this algebraically.

$x + 3y = 6$ **Write original equation.**

$-3 + 3(3) \stackrel{?}{=} 6$ **Substitute −3 for *x* and 3 for *y*.**

$6 = 6$ **Simplify. True statement**

▶ $(-3, 3)$ is a solution of the equation $x + 3y = 6$, so it is on the graph.

In Example 1 the point $(-3, 3)$ is on the graph of $x + 3y = 6$, but how many points does the graph have in all? The answer is that most graphs have too many points to list. Then how can you ever graph an equation? One way is to make a table or a list of a few values, plot enough solutions to recognize a pattern, and then connect the points. Even then, the graph extends without limit to the left of the smallest input and to the right of the largest input.

When you make a table of values to graph an equation, you may want to choose values for x that include negative values, zero, and positive values. This way you will see how the graph behaves to the left and right of the y-axis.

EXAMPLE 2 *Graphing an Equation*

Use a table of values to graph the equation $y + 2 = 3x$.

SOLUTION

STUDENT HELP

Look Back
For help with rewriting equations in function form, see p. 176.

Rewrite the equation in function form by solving for y.

$y + 2 = 3x$ — **Write original equation.**

$y = 3x - 2$ — **Subtract 2 from each side.**

Choose a few values for x and make a table of values.

Choose x.	Substitute to find the corresponding y-value.
-2	$y = 3(-2) - 2 = -8$
-1	$y = 3(-1) - 2 = -5$
0	$y = 3(0) - 2 = -2$
1	$y = 3(1) - 2 = 1$
2	$y = 3(2) - 2 = 4$

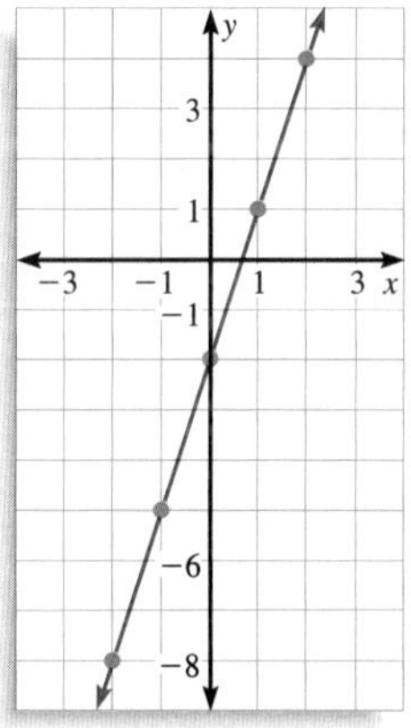

With this table of values you have found the five solutions $(-2, -8)$, $(-1, -5)$, $(0, -2)$, $(1, 1)$, and $(2, 4)$.

Plot the points. Note that they appear to lie on a straight line.

▶ The line through the points is the graph of the equation.

....

STUDENT HELP

HOMEWORK HELP
Visit our Web site www.mcdougallittell.com for extra examples.

In Lesson 3.1 you saw examples of linear equations in one variable. The solution of an equation such as $2x - 1 = 3$ is a real number. Its graph is a point on the real number line. The equation $y + 2 = 3x$ in Example 2 is a linear equation in *two variables*. Its graph is a straight line.

GRAPHING A LINEAR EQUATION

STEP 1 Rewrite the equation in function form, if necessary.

STEP 2 Choose a few values of x and make a table of values.

STEP 3 Plot the points from the table of values. A line through these points is the graph of the equation.

2 TEACH

MOTIVATING THE LESSON

Ask students how many of them have bought CDs recently. Then ask two students for the base price of the CDs to the nearest dollar. Have students find the total cost of the CD with a 5% sales tax. Represent the base cost and the total cost as an ordered pair. Plot the two points and draw a line through them. Tell students that this line is the graph of a linear equation.

EXTRA EXAMPLE 1

Use the graph to decide whether the point lies on the graph of $4x + y = 8$. Justify your answer algebraically.

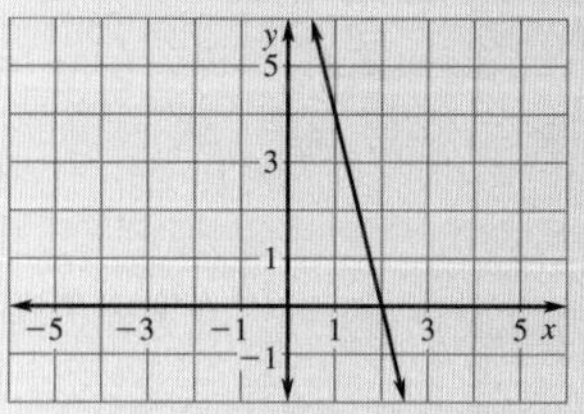

a. (1, 4) yes; $4(1) + 4 = 8$
b. (3, 4) no; $4(3) + 4 \neq 8$

EXTRA EXAMPLE 2

Use a table of values to graph the equation $5x = y - 2$.

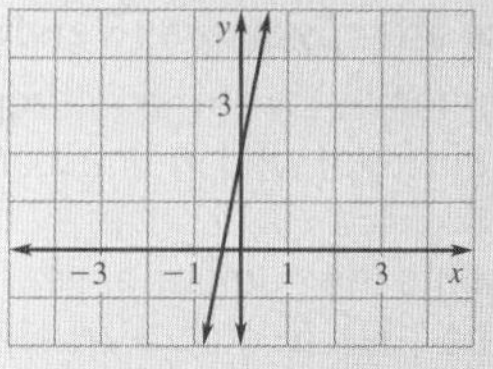

CHECKPOINT EXERCISES

For use after Examples 1 and 2:

1. Use the graph to decide whether the point lies on the graph of $x - 2y = 5$. Justify your answer algebraically.

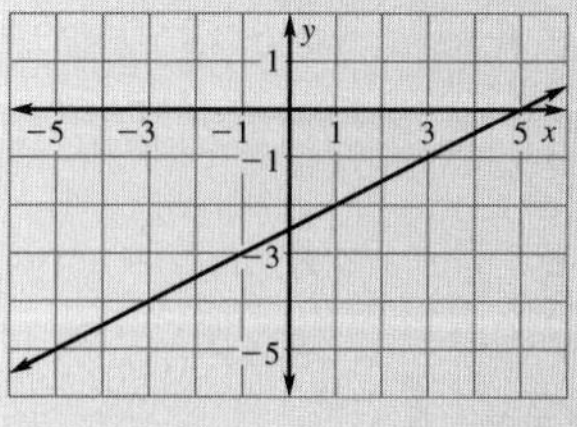

a. (3, −4) no; $3 - 2(-4) \neq 5$
b. (5, 0) yes; $5 - 2(0) = 5$

Extra Example 3

Use a table of values to graph the equation $3x - 2y = 6$.

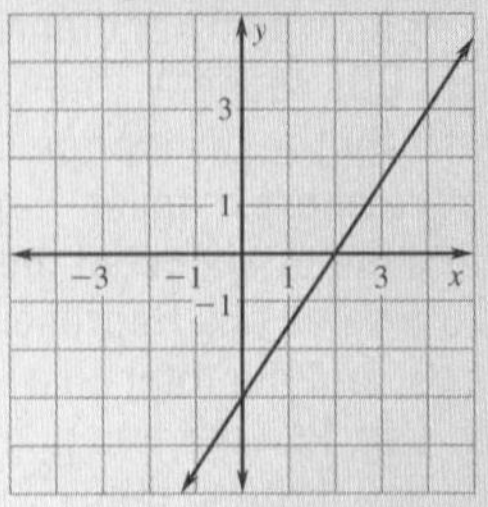

Extra Example 4

The number n of U.S. citizens (in millions) who exercise walk can be modeled by $n = 1.38t + 60.6$, where t represents the number of years since 1987. Sketch a graph of this model. Describe the graph in the context of the real-life situation.

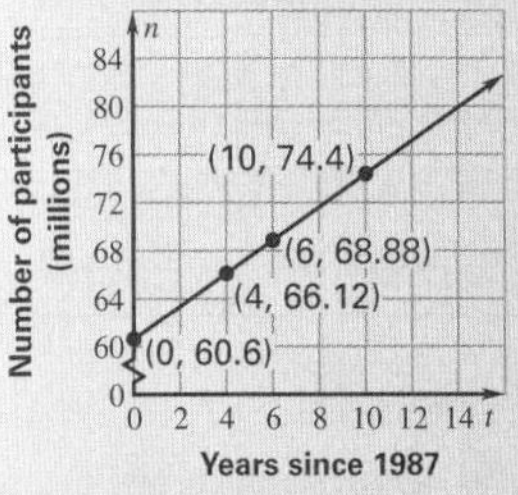

The graph shows the number of participants increasing by about 1.4 million per year.

Checkpoint Exercises

For use after Examples 3 and 4:

1. The number n of U.S. citizens (in millions) who in-line skate is $n = 3.47t + 7.14$, where t represents the number of years since 1991. Sketch a graph of this model. Describe the graph of this line in the context of the real-life situation.

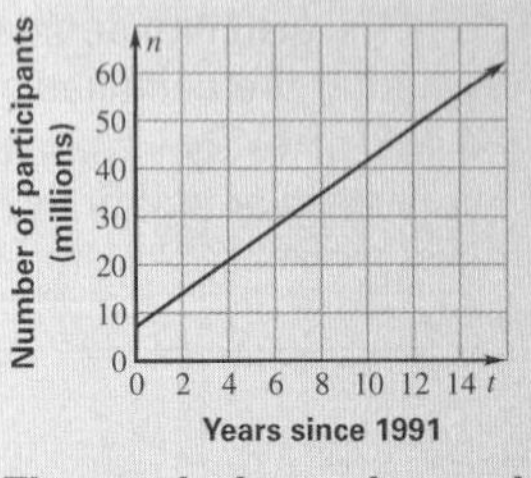

The graph shows the number of participants increasing by about 3.5 million per year.

EXAMPLE 3 *Graphing a Linear Equation*

Use a table of values to graph the equation $3x + 2y = 1$.

SOLUTION

1 Rewrite the equation in function form by solving for y. This will make it easier to make a table of values.

$3x + 2y = 1$	Write original equation.
$2y = -3x + 1$	Subtract $3x$ from each side.
$y = -\frac{3}{2}x + \frac{1}{2}$	Divide each side by 2.

STUDENT HELP

Skills Review
For help with fraction operations, see pp. 780–782.

2 Choose a few values of x and make a table of values.

Choose x.	-2	0	2
Evaluate y.	$\frac{7}{2}$	$\frac{1}{2}$	$-\frac{5}{2}$

With this table of values you have found three solutions.

$$\left(-2, \frac{7}{2}\right), \left(0, \frac{1}{2}\right), \left(2, -\frac{5}{2}\right)$$

3 Plot the points and draw a line through them.

▶ The graph of $3x + 2y = 1$ is shown at the right.

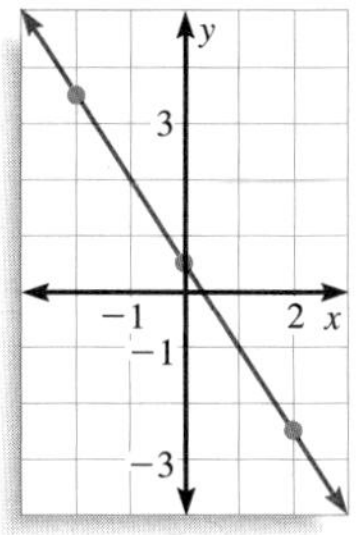

EXAMPLE 4 *Using the Graph of a Linear Model*

REAL LIFE
Internet Access

An Internet Service Provider estimates that the number of households h (in millions) with Internet access can be modeled by $h = 6.76t + 14.9$, where t represents the number of years since 1996. Graph this model. Describe the graph in the context of the real-life situation.

SOLUTION

Make a table of values. Use $0 \le t \le 6$ for 1996–2002.

t	h
0	14.90
1	21.66
2	28.42
3	35.18
4	41.94
5	48.70
6	55.46

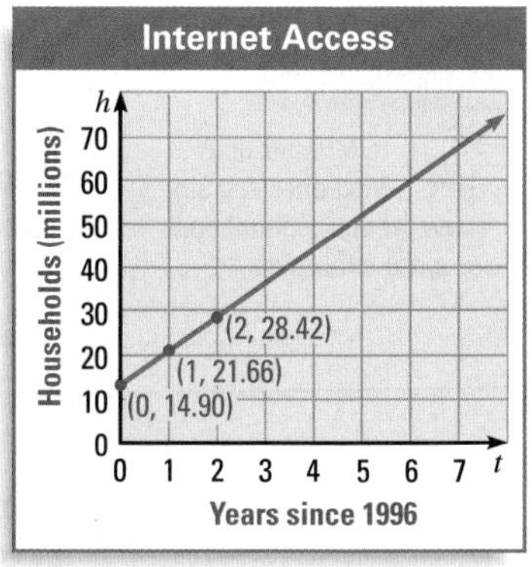

▶ From the table and the graph, you can see that the number of households with Internet access is projected to increase by about 7 million households per year.

GOAL 2 HORIZONTAL AND VERTICAL LINES

All linear equations in x and y can be written in the form $Ax + By = C$. When $A = 0$ the equation reduces to $By = C$ and the graph is a horizontal line. When $B = 0$ the equation reduces to $Ax = C$ and the graph is a vertical line.

EQUATIONS OF HORIZONTAL AND VERTICAL LINES

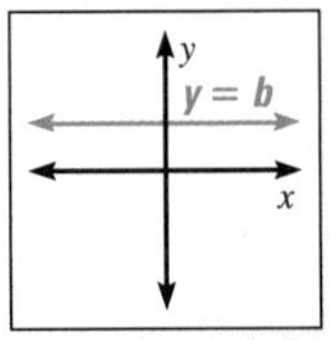

In the coordinate plane, the graph of $y = b$ is a horizontal line.

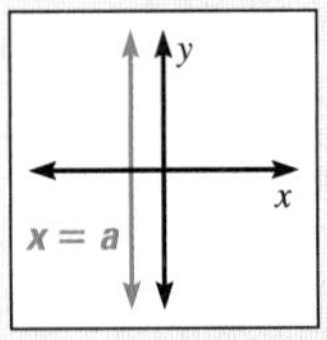

In the coordinate plane, the graph of $x = a$ is a vertical line.

EXAMPLE 5 *Graphing $y = b$*

Graph the equation $y = 2$.

SOLUTION

The equation does not have x as a variable. The y-value is always 2, regardless of the value of x. For instance, here are some points that are solutions of the equation:

$(-3, 2), (0, 2), (3, 2).$

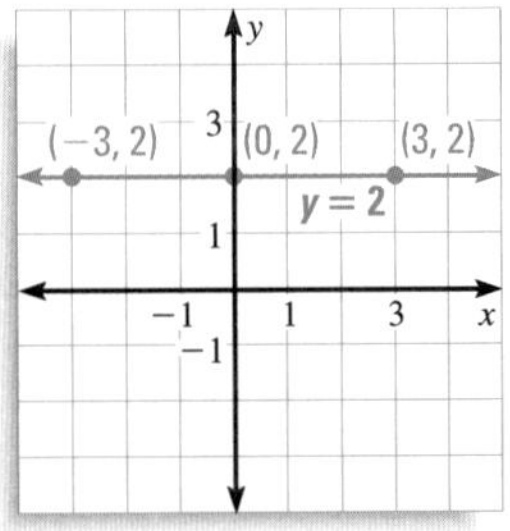

▶ The graph of the equation is a horizontal line 2 units above the x-axis.

EXAMPLE 6 *Graphing $x = a$*

Graph the equation $x = -3$.

SOLUTION

The x-value is always -3, regardless of the value of y. For instance, here are some points that are solutions of the equation:

$(-3, -2), (-3, 0), (-3, 3).$

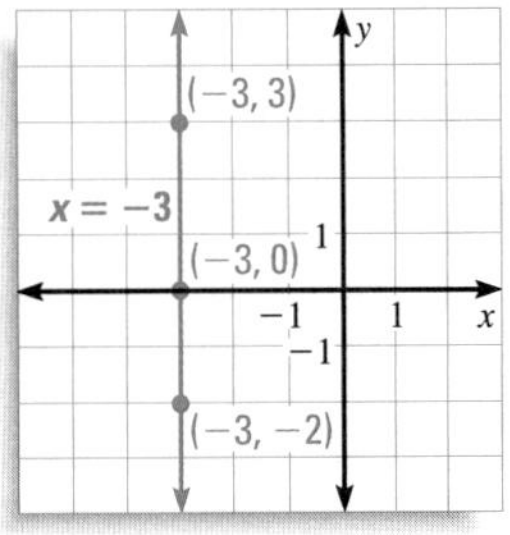

▶ The graph of the equation is a vertical line 3 units to the left of the y-axis.

STUDENT HELP

Study Tip
The equations $y = 2$ and $0x + 1y = 2$ are equivalent. For any value of x, the ordered pair $(x, 2)$ is a solution of $y = 2$.

EXTRA EXAMPLE 5
Sketch the graph of $y = -1$.

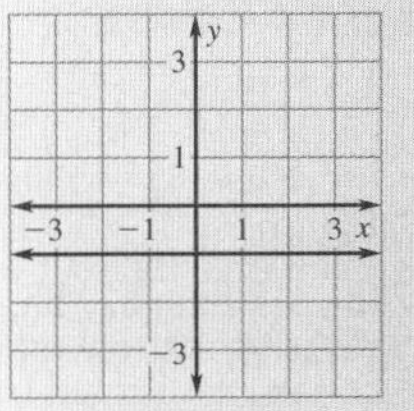

EXTRA EXAMPLE 6
Sketch the graph of $x = 2$.

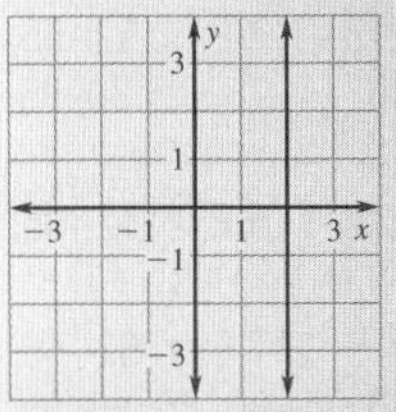

CHECKPOINT EXERCISES

For use after Examples 5 and 6:

1. Describe the graph of $y = -3$ **a horizontal line through (0, −3)**
2. Describe the graph of $x = 0$. **the vertical axis**

CLOSURE QUESTION
What kind of a figure is the graph of the equation $y = -2x + 5$? **a line**

DAILY PUZZLER
Graph four equations that intersect to form a square with center at the origin. **Equations graphed should be of the form $x = a$, $x = -a$, $y = a$, $y = -a$. For example, $x = 5$, $x = -5$, $y = 5$, $y = -5$**

! COMMON ERROR
EXAMPLE 5 Students will commonly graph $y = a$ parallel to the y-axis. Suggest to students that they should plot the point $(0, a)$ on the y-axis and then draw a line through this point perpendicular to the y-axis.

! COMMON ERROR
EXAMPLE 6 Students will commonly graph $x = a$ parallel to the x-axis. Suggest to students that they should plot the point $(a, 0)$ on the x-axis and then draw a line through this point perpendicular to the x-axis.

3 APPLY

ASSIGNMENT GUIDE

BASIC
Day 1: pp. 214–215 Exs. 12–35, 36–50 even
Day 2: pp. 215–217 Exs. 52–55, 60–70, 74–78, 85, 90, 95

AVERAGE
Day 1: pp. 214–215 Exs. 12–35, 36–50 even
Day 2: pp. 215–217 Exs. 52–55, 60–70, 74–78, 85, 90, 95

ADVANCED
Day 1: pp. 214–215 Exs. 12–35, 36–50 even
Day 2: pp. 215–217 Exs. 52–56, 60–66, 71–80, 85, 90, 95

BLOCK SCHEDULE
pp. 214–217 Exs. 12–35, 36–50 even, 52–55, 60–70, 74–78, 85, 90, 95

EXERCISE LEVELS

Level A: *Easier*
12–20, 81–94

Level B: *More Difficult*
21–55, 57, 58, 60–66, 74–78, 95–98

Level C: *Most Difficult*
56, 59, 67–73, 79, 80

✓ HOMEWORK CHECK

To quickly check student understanding of key concepts, go over the following exercises: Exs. 12, 16, 18, 22, 32, 38, 60, 68. See also the Daily Homework Quiz:

- Blackline Master (*Chapter 4 Resource Book,* p. 38)
- Transparency (p. 30)

4.

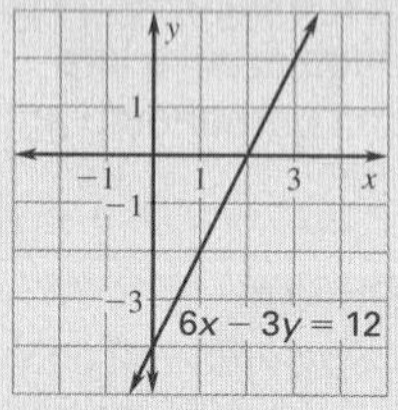

GUIDED PRACTICE

Vocabulary Check ✓

1. Complete the following sentence: An ordered pair that makes an equation in two variables true is called a(n) ? . **solution of the equation**

Concept Check ✓

2. In Example 2, if you choose a value of x different from those in the table of values, will you find a solution that lies on the same line? Explain. **See margin.**

3. Decide whether the following statement is *true* or *false. The graph of the equation $x = 3$ is a horizontal line.* Explain.
False; the graph of $x = 3$ is a vertical line.

Skill Check ✓

2. Yes; for every value of x, you will solve the equation to find the appropriate value for y so that (x, y) is on the line.

Use a table of values to graph the equation. **4–6. See margin.**

4. $6x - 3y = 12$ **5.** $x = 1.5$ **6.** $y = -2$

Tell whether the point is a solution of the equation $4x - y = 1$.

7. $(-1, 3)$ **no** **8.** $(1, 3)$ **yes** **9.** $(1, 0)$ **no** **10.** $(1, 4)$ **no**

11. INTERNET ACCESS Using the graph or the model in Example 4, estimate the number of households that had Internet access in 1999.
about 35,180,000 households

PRACTICE AND APPLICATIONS

STUDENT HELP
Extra Practice to help you master skills is on p. 800.

VERIFYING SOLUTIONS **Use the graph to decide whether the point lies on the graph of the line. Justify your answer algebraically.**

12. $3x - 4y = 10$ **See margin.**
a. $(2, -1)$
b. $(-1, 2)$

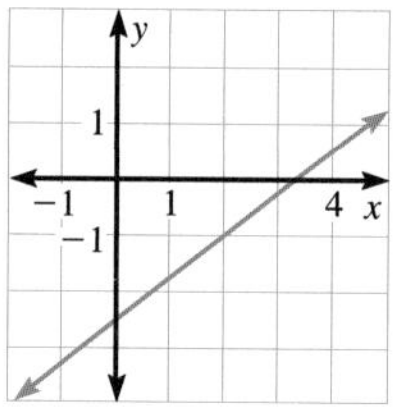

12 a. yes; $3(2) - 4(-1) = 10$
b. no; $3(-1) - 4(2) = -11 \neq 10$

13. $y = 5$
a. $(5, 0)$ **no; $y \neq 5$**
b. $(0, 5)$ **yes; $y = 5$**

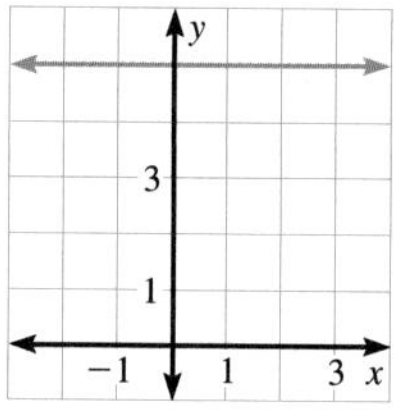

14. $x = 0$
a. $(0, 14)$ **yes; $x = 0$**
b. $(14, 0)$ **no; $x \neq 0$**

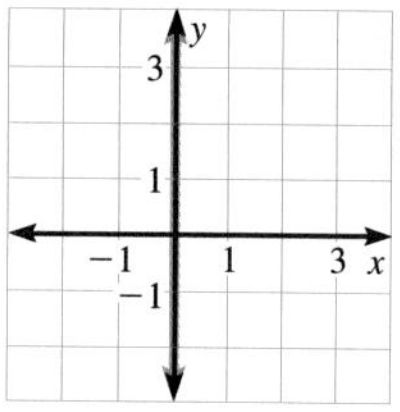

CHECKING SOLUTIONS **Decide whether the given ordered pair is a solution of the equation.**

15. $2y - 4x = 8, (-2, 8)$ **no**
16. $-5x - 8y = 15, (-3, 0)$ **yes**
17. $y = -2, (-2, -2)$ **yes**
18. $x = -4, (1, -4)$ **no**
19. $6y - 3x = -9, (2, -1)$ **no**
20. $-2x - 9y = 7, (-1, -1)$ **no**

FINDING SOLUTIONS **Find three different ordered pairs that are solutions of the equation.** **21–29. Sample answers are given.**

21. $y = 3x - 5$ **(−1, −8), (0, −5), (1, −2)**
22. $y = 7 - 4x$ **See margin.**
23. $y = -2x - 6$ **(−1, −4), (0, −6), (1, −8)**
24. $x = 2$ **See margin.**
25. $x = \frac{1}{2}$ **See margin.**
26. $y = -6$ **See margin.**
27. $y = \frac{1}{2}(4 - 2x)$ **(−1, 3), (0, 2), (1, 1)**
28. $y = 3(6x - 1)$ **(−1, −21), (0, −3), (1, 15)**
29. $y = 4\left(\frac{1}{2}x - 1\right)$ **(−1, −6), (0, −4), (1, −2)**

22. (−1, 11), (0, 7), (1, 3)
24. (2, −1), (2, 0), (2, 1)
25. $\left(\frac{1}{2}, -1\right), \left(\frac{1}{2}, 0\right), \left(\frac{1}{2}, 1\right)$
26. (−1, −6), (0, −6), (1, −6)

STUDENT HELP
HOMEWORK HELP

Example 1:	Exs. 12–20
Example 2:	Exs. 21–55
Example 3:	Exs. 21–55
Example 4:	Exs. 67–70
Example 5:	Exs. 44–55, 60–65
Example 6:	Exs. 44–55, 60–65

31. $y = -\frac{2}{3}x + 2$

32. $y = -\frac{1}{4}x + 12$

33. $y = -x + \frac{19}{5}$

34. $y = -\frac{1}{5}x + \frac{2}{5}$

FUNCTION FORM **Rewrite the equation in function form.**

30. $-3x + y = 12$ $y = 3x + 12$ **31.** $2x + 3y = 6$ See margin. **32.** $x + 4y = 48$ See margin.

33. $5x + 5y = 19$ See margin. **34.** $\frac{1}{2}x + \frac{5}{2}y = 1$ See margin. **35.** $-x - y = 5$ $y = -x - 5$

GRAPHING EQUATIONS **Use a table of values to graph the equation.** 36–51. See margin.

36. $y = -x + 4$ **37.** $y = -2x + 5$ **38.** $y = -(3 - x)$ **39.** $y = -2(x - 6)$

40. $y = 3x + 2$ **41.** $y = 4x - 1$ **42.** $y = \frac{4}{3}x + 2$ **43.** $y = -\frac{3}{4}x + 1$

44. $x = 9$ **45.** $y = -1$ **46.** $y = 0$ **47.** $y = -3x + 1$

48. $x = 0$ **49.** $x - 2y = 6$ **50.** $4x + 4y = 2$ **51.** $x + 2y = -8$

MATCHING EQUATIONS WITH GRAPHS **Match the equation with its graph.**

A. $x = -2$ **B.** $2x - y = 3$

C. $6x + 3y = 0$ **D.** $-x + 2y = 6$

52.
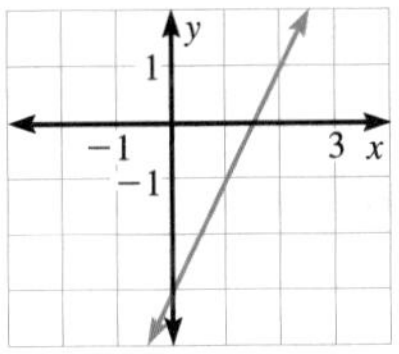
B

53.
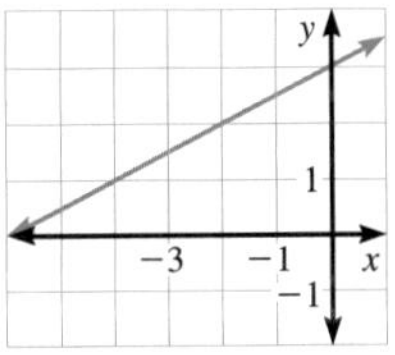
D

54.
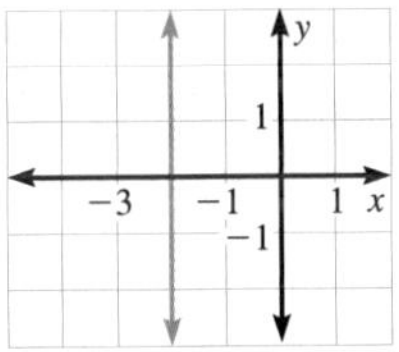
A

55.
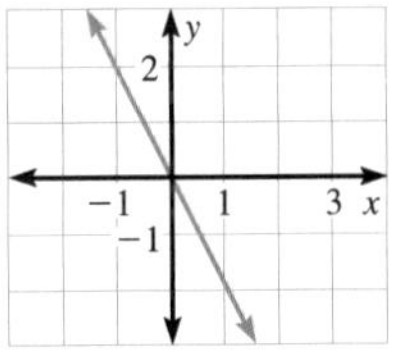
C

56. **CRITICAL THINKING** Robin always finds at least three different ordered pairs when making a table to graph an equation. Why do you think she does this? as a check

STUDENT HELP

Look Back
For help with scatter plots, see p. 204.

PLOTTING POINTS **In Exercises 57–59, use the ordered pairs below.**
(−3, −9), (−2, −7), (−1.6, −5.5), (0.4, −2.2), (3.1, 3.2), (5.2, 7.4)

57. Use a graphing calculator or a computer to graph the ordered pairs. See margin.

58. The line that fits most of the points is the graph of $y = 2x - 3$. One of the points does not appear to fall on the line with the others. Which point is it? (−1.6, −5.5)

59. Show algebraically that one point does not lie on this line. $x = -1.6$; $2(-1.6) - 3 = -6.2$, but $y = -5.5$.

POINTS OF INTERSECTION **In Exercises 60–65, graph the two lines in the same coordinate plane. Then find the coordinates of the point at which the lines cross.** 60–65. Check graphs.

60. $x = -5, y = 2$ (−5, 2) **61.** $y = 11, x = -8$ (−8, 11) **62.** $y = -6, x = 1$ (1, −6)

63. $x = 4, y = -4$ (4, −4) **64.** $y = -1, x = 0$ (0, −1) **65.** $x = 3, y = 0$ (3, 0)

66. **VISUAL THINKING** Name a point that would be on the graphs of both $x = -3$ and $y = 2$. Sketch both graphs to verify your answer. (−3, 2); Check graphs.

STUDENT HELP NOTES

Look Back When students look back to p. 204, remind them that the *x*-coordinates should be entered in List 1 and the *y*-coordinates should be entered in List 2 in the STAT EDIT menu.

GRAPHING CALCULATOR NOTE

EXERCISES 57–58 The graphing calculator will be used to graph decimal values and introduce students to the notion of a line of fit for a set of data.

5.
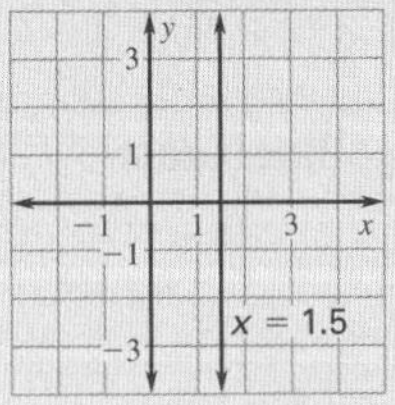

6.
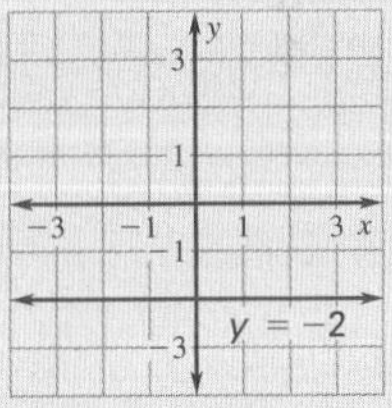

36.
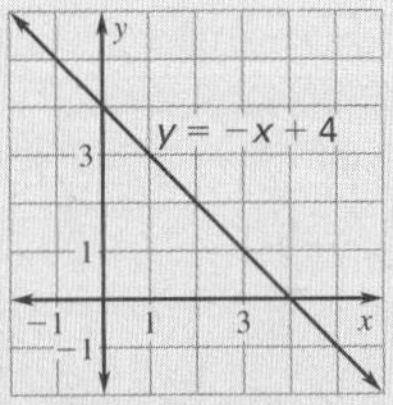

37–51, 57. See Additional Answers beginning on page AA1.

APPLICATION NOTE

EXERCISES 72–73 Students should understand that numbers given for calories burned per minute are averages. The body weight and metabolism of each triathlete are factors affecting the number of calories he or she will burn during each activity. For example, a person who weighs 140 pounds will burn 508 calories while bicycling for one hour at about 12–13 mi/h and exerting a moderate effort. A person weighing 195 pounds will burn 708 calories in the same activity.

Additional information about the Ironman Triathlon World Championship is available at **www.mcdougallittell.com.**

72.

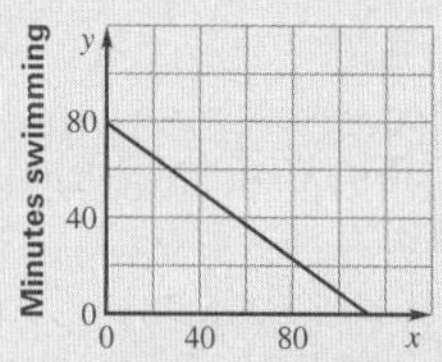

79. See Additional Answers beginning on page AA1.

ADDITIONAL PRACTICE AND RETEACHING

For Lesson 4.2:

- Practice Levels A, B, and C (*Chapter 4 Resource Book*, p. 28)
- Reteaching with Practice (*Chapter 4 Resource Book*, p. 31)
- See Lesson 4.2 of the *Personal Student Tutor*

For more Mixed Review:

- Search the *Test and Practice Generator* for key words or specific lessons.

LANDSCAPING BUSINESS In Exercises 67–70, use the following information.

One summer you charge \$20 to mow a lawn and \$10 to trim bushes. You want to make \$300 in one week. An algebraic model for your earnings is $20x + 10y = 300$, where x is the number of lawns you mow and y is the number of bushes you trim.

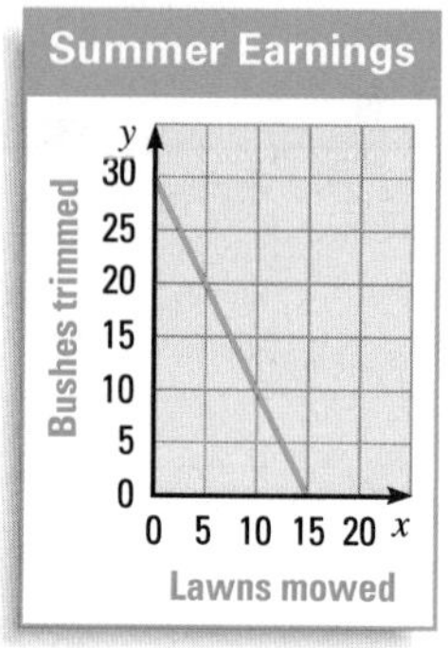

67. Solve the equation for y. **$y = -2x + 30$**

68. Use the equation in function form from Exercise 67 to make a table of values for $x = 5$, $x = 10$, and $x = 15$.

68.

x	$y = -2x + 30$
5	20
10	10
15	0

69. Look at the graph at the right. Does the line shown appear to pass through the points from your table of values? **yes**

70. If you do not trim any bushes during the week, how many lawns will you have to mow to earn \$300? **15 lawns**

TRAINING FOR A TRIATHLON In Exercises 71–73, Mary Gordon is training for a triathlon. Like most triathletes she regularly trains in two of the three events every day. On Saturdays she expects to burn about 800 calories during her workout by running and swimming.

Running: 7.1 calories per minute

Swimming: 10.1 calories per minute

Bicycling: 6.2 calories per minute

71. Copy and complete the model below. Let x represent the number of minutes she spends running, and let y represent the number of minutes she spends swimming. **See margin.**

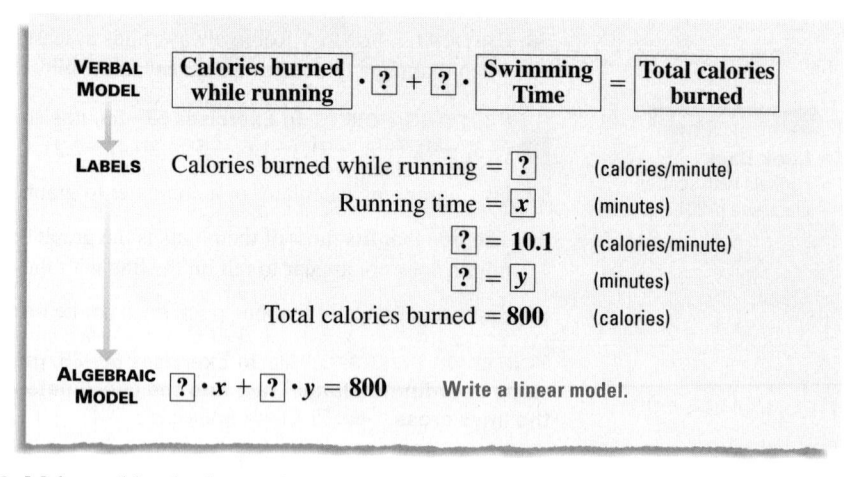

71. Running time; Calories burned while swimming; 7.1; Calories burned while swimming; Swimming time; $7.1x + 10.1y = 800$

VERBAL MODEL [Calories burned while running] · [?] + [?] · [Swimming Time] = [Total calories burned]

LABELS

Calories burned while running = [?] (calories/minute)

Running time = [x] (minutes)

[?] = 10.1 (calories/minute)

[?] = [y] (minutes)

Total calories burned = 800 (calories)

ALGEBRAIC MODEL $[?] \cdot x + [?] \cdot y = 800$ **Write a linear model.**

72. Make a table of values and graph the equation from Exercise 71. **See margin.**

73. If Mary Gordon spends 45 minutes running, about how many minutes will she have to spend swimming to burn 800 calories? **about 48 min**

FOCUS ON APPLICATIONS

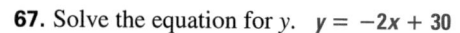

TRIATHLON In 1998, over 1500 people competed in Hawaii's Ironman Triathlon World Championship.

APPLICATION LINK www.mcdougallittell.com

CHOOSING A MODEL In Exercises 74 and 75, decide whether a graph of the data would be points that appear to lie on a *horizontal line*, a *vertical line*, or *neither*. Explain. Let the *x*-axis represent time.

74. The number of books read by Avery each year from 1999 to 2005. **See margin.**

75. The number of senators in the United States Congress each year from 1991 to 2000. **Horizontal; the number of senators does not vary.**

Test Preparation

76. MULTIPLE CHOICE Which point lies on the graph of $2x + 5y = 6$? **D**

(A) $(-2, 2)$ (B) $(3, 0)$ (C) $(0, 3)$ (D) both A and B

77. MULTIPLE CHOICE Which point does *not* lie on the graph of $y = 3$? **C**

(A) $(0, 3)$ (B) $(-3, 3)$ (C) $(3, -3)$ (D) $\left(\frac{1}{3}, 3\right)$

78. MULTIPLE CHOICE The ordered pair $(-3, 5)$ is a solution of __?__. **A**

(A) $y = 5$ (B) $x = 5$ (C) $y = \frac{1}{2}x - 2$ (D) $y = -\frac{1}{2}x - 2$

★ Challenge

MISINTERPRETING GRAPHS Even though they provide an accurate representation of data, some graphs are easy to misinterpret. In Exercises 79 and 80, use the following.

Based on actual and projected data from 1995 to 2000, a linear model for a company's profit P is $P = 3{,}005{,}000 - 900t$, where t represents the number of years since 1995.

79. Sketch two graphs of the model. In the first graph, divide the vertical axis into \$1,000,000 intervals. In the second graph start at \$3,000,000 and use \$1000 intervals. **See margin.**

80. CRITICAL THINKING What are the advantages and disadvantages of each graph? ***Sample answer:* The first graph might be used to imply that profits remained steady. (The actual decrease was about 0.030% per year.) The second graph might be used to imply that profits decreased steeply. The second graph is more likely to be misinterpreted.**

EXTRA CHALLENGE
www.mcdougallittell.com

MIXED REVIEW

EVALUATING EXPRESSIONS Evaluate the expression. (Review 2.2)

81. $5 + 2 + (-3)$ **4** **82.** $-6 + (-14) + 8$ **−12** **83.** $-18 + (-10) + (-1)$ **−29**

84. $-\frac{1}{3} + 6 + \frac{1}{3}$ **6** **85.** $\frac{4}{7} + \frac{3}{7} - 1$ **0** **86.** $-\frac{7}{9} + \frac{1}{3} + 2$ $\frac{14}{9}$

MATRICES Find the sum of the matrices. (Review 2.4) **87–90. See margin.**

87. $\begin{bmatrix} 1 & 6 \\ -4 & 2 \end{bmatrix} + \begin{bmatrix} 15 & -3 \\ 0 & 16 \end{bmatrix}$ **88.** $\begin{bmatrix} 2 & 5 \\ 3 & 10 \end{bmatrix} + \begin{bmatrix} -14 & -5 \\ 12 & 7 \end{bmatrix}$

89. $\begin{bmatrix} 4 & 10 \\ -1 & 9 \end{bmatrix} + \begin{bmatrix} -20 & 40 \\ -8 & 10 \end{bmatrix}$ **90.** $\begin{bmatrix} 5 & -2 \\ 5 & -1 \end{bmatrix} + \begin{bmatrix} -1 & -16 \\ 11 & -3 \end{bmatrix}$

SOLVING EQUATIONS Solve the equation. (Review 3.2 for 4.3)

91. $-2z = -26$ **13** **92.** $9x = 3$ $\frac{1}{3}$ **93.** $6p = -96$ **−16** **94.** $24 = 8c$ **3**

95. $\frac{2}{3}t = -10$ **−15** **96.** $-\frac{p}{7} = -9$ **63** **97.** $\frac{n}{15} = \frac{3}{5}$ **9** **98.** $\frac{c}{6} = \frac{2}{3}$ **4**

74. ***Sample answer:*** **neither; a vertical line makes no sense in this situation and a horizontal line is unlikely since Avery probably does not read the same number of books each year.**

87. $\begin{bmatrix} 16 & 3 \\ -4 & 18 \end{bmatrix}$

88. $\begin{bmatrix} -12 & 0 \\ 15 & 17 \end{bmatrix}$

89. $\begin{bmatrix} -16 & 50 \\ -9 & 19 \end{bmatrix}$

90. $\begin{bmatrix} 4 & -18 \\ 16 & -4 \end{bmatrix}$

4 ASSESS

DAILY HOMEWORK QUIZ

Transparency Available

1. Decide whether the given ordered pair is a solution of $2x - 3y = 8$.

a. $(-2, -4)$ **yes**

b. $(7, -2)$ **no**

2. Rewrite $4x - 2y = 18$ in function form. $y = 2x - 9$

3. Use a table of values to graph $y = 2x + 2$.

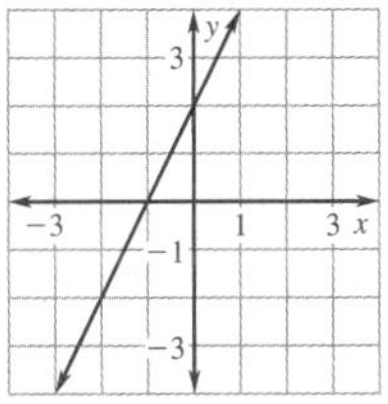

4. Find the coordinates of the point at which the lines cross.

a. $x = -3$, $y = 8$ **(−3, 8)**

b. $y = 0$, $x = 5$ **(5, 0)**

EXTRA CHALLENGE NOTE

Challenge problems for Lesson 4.2 are available in **blackline** format in the *Chapter 4 Resource Book*, p. 35 and at **www.mcdougallittell.com.**

ADDITIONAL TEST PREPARATION

1. WRITING Explain why the point $(3, -1)$ is a solution of both $x = 3$ and $y = -1$. **The solutions of $x = 3$ are all the points whose x-coordinate is 3. The solutions of $y = -1$ are all the points whose y-coordinate is −1. The point $(3, -1)$ satisfies both conditions.**

2. OPEN ENDED Describe a real-life situation that can be modeled by the equation $3x + 5y = 500$. **See sample answer at left.**

Additional Test Preparation *Sample answer:*

2. A store makes a \$3 profit on each short-sleeve T-shirt and a \$5 profit on each long-sleeve T-shirt. If x represents the number of short-sleeve shirts sold and y represents the number of long-sleeve shirts sold, then the solutions of $3x + 5y = 500$ represent the number of each kind of T-shirt sold to make a profit of \$500.

1 PLAN

PACING
Basic: 2 days
Average: 2 days
Advanced: 2 days
Block Schedule: 1 block

***LESSON OPENER* ACTIVITY**
An alternative way to approach Lesson 4.3 is to use the Activity Lesson Opener:

- Blackline Master (*Chapter 4 Resource Book* p. 39)
- Transparency (p. 26)

MEETING INDIVIDUAL NEEDS

- ***Chapter 4 Resource Book***
  Prerequisite Skills Review (p. 5)
  Practice Level A (p. 40)
  Practice Level B (p. 41)
  Practice Level C (p. 42)
  Reteaching with Practice (p. 43)
  Absent Student Catch-Up (p. 45)
  Challenge (p. 48)
- ***Resources in Spanish***
- ***Personal Student Tutor***

NEW-TEACHER SUPPORT
See the Tips for New Teachers on pp. 1–2 of the *Chapter 4 Resource Book* for additional notes about Lesson 4.3.

WARM-UP EXERCISES

Transparency Available

1. Which point lies on the x-axis, (3, 0) or (0, 3)? (3, 0)
2. Which point lies on the y-axis, (–5, 0) or (0, –5)? (0, –5)

The point (x, y) is a solution of $2x + y = 8$.

3. What is the value of y when $x = 0$? 8
4. What is the value of x when $y = 0$? 4

4.3 Quick Graphs Using Intercepts

***What** you should learn*

GOAL 1 Find the intercepts of the graph of a linear equation.

GOAL 2 Use intercepts to make a quick graph of a linear equation as in **Example 4**.

***Why** you should learn it*

▼ To solve **real-life** problems, such as finding numbers of school play tickets to be sold to reach a fundraising goal in **Ex. 63**.

GOAL 1 FINDING THE INTERCEPTS OF A LINE

In Lesson 4.2 you graphed a linear equation by writing a table of values, plotting the points, and drawing a line through the points.

In this lesson, you will learn a quicker way to graph a linear equation. To do this, you need to realize that only two points are needed to determine a line. Two points that are usually convenient to use are points where a graph crosses the axes.

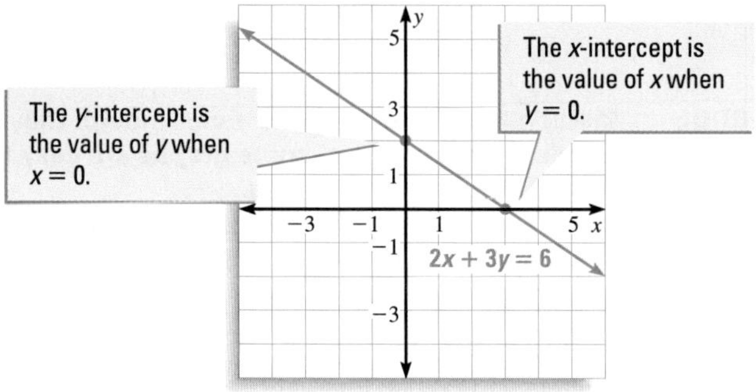
The y-intercept is the value of y when $x = 0$.
The x-intercept is the value of x when $y = 0$.
y
5
3
1
−3
−1
1
5
x
−1
−3
$2x + 3y = 6$

An ***x*-intercept** is the x-coordinate of a point where a graph crosses the x-axis. A ***y*-intercept** is the y-coordinate of a point where a graph crosses the y-axis.

EXAMPLE 1 *Finding Intercepts*

Find the x-intercept and the y-intercept of the graph of the equation $2x + 3y = 6$.

SOLUTION

To find the x-intercept of $2x + 3y = 6$, let $y = 0$.

$2x + 3y = 6$	Write original equation.
$2x + 3(0) = 6$	Substitute 0 for y.
$x = 3$	Solve for x.

▶ The x-intercept is 3. The line crosses the x-axis at the point (3, 0).

To find the y-intercept of $2x + 3y = 6$, let $x = 0$.

$2x + 3y = 6$	Write original equation.
$2(0) + 3y = 6$	Substitute 0 for x.
$y = 2$	Solve for y.

▶ The y-intercept is 2. The line crosses the y-axis at the point (0, 2).

GOAL 2 USING INTERCEPTS TO GRAPH EQUATIONS

EXAMPLE 2 Making a Quick Graph

Graph the equation $3.5x + 7y = 14$.

SOLUTION

Find the intercepts.

$3.5x + 7y = 14$ Write original equation.

$3.5x + 7(0) = 14$ Substitute 0 for *y*.

$x = \frac{14}{3.5} = 4$ The *x*-intercept is 4.

$3.5x + 7y = 14$ Write original equation.

$3.5(0) + 7y = 14$ Substitute 0 for *x*.

$y = \frac{14}{7} = 2$ The *y*-intercept is 2.

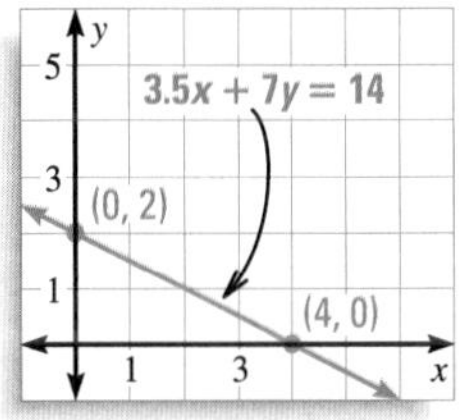

Draw a coordinate plane that includes the points (4, 0) and (0, 2).

Plot the points (4, 0) and (0, 2) and draw a line through them.

STUDENT HELP

INTERNET HOMEWORK HELP Visit our Website www.mcdougallittell.com for extra examples.

EXAMPLE 3 Drawing Appropriate Scales

Graph the equation $y = 4x + 40$.

SOLUTION

Find the intercepts, by substituting 0 for *y* and then 0 for *x*.

$y = 4x + 40$	$y = 4x + 40$
$0 = 4x + 40$	$y = 4(0) + 40$
$-40 = 4x$	$y = 40$
$-10 = x$	The *y*-intercept is 40.
The *x*-intercept is -10.	

Draw a coordinate plane that includes the points $(-10, 0)$ and $(0, 40)$. With these values, it is reasonable to use tick marks at 10-unit intervals.

You may want to draw axes with at least two tick marks to the left of -10 and to the right of 0 on the *x*-axis and two tick marks below 0 and above 40 on the *y*-axis.

Plot the points $(-10, 0)$ and $(0, 40)$ and draw a line through them.

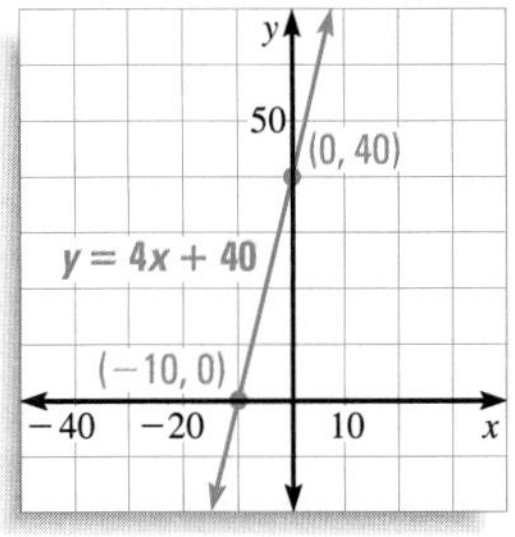

STUDENT HELP

Study Tip
When you make a quick graph, find the intercepts *before* you draw the coordinate plane. This will help you find an appropriate scale on each axis.

2 TEACH

MOTIVATING THE LESSON

You run the 100 meter dash in 20 seconds. The graph of $d = -5t + 100$ models your distance from the finish line. What do the points (0, 100) and (20, 0) represent?
(0, 100) is your initial position at the starting line; (20, 0) is your position at the finish line.

EXTRA EXAMPLE 1

Find the *x*- and *y*-intercepts of the graph of $4x + 3y = 12$.
3 and 4

EXTRA EXAMPLE 2

Graph the equation $x - 2 = 4y$.

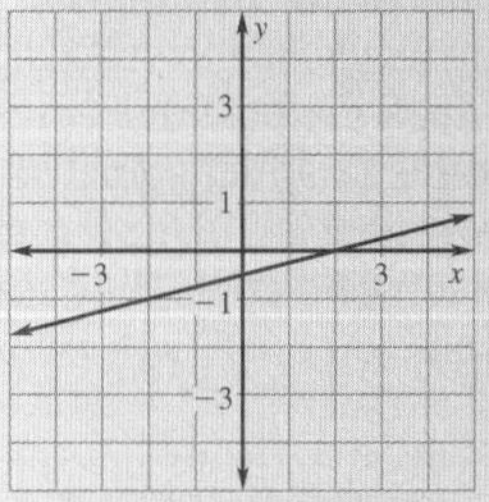

EXTRA EXAMPLE 3

Graph the equation $y = -2x + 25$.

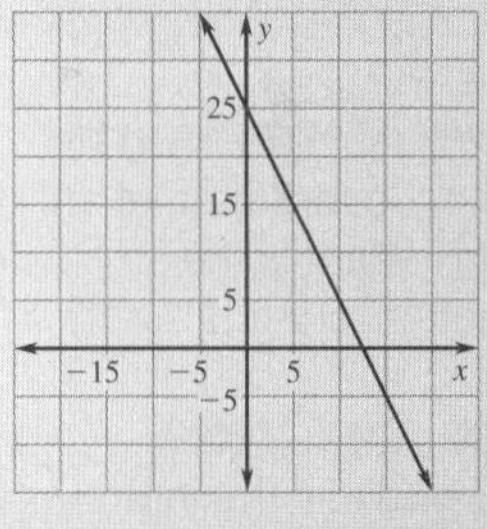

CHECKPOINT EXERCISES

For use after Examples 1–3:

1. Graph the equation $2x + 5y = -10$.

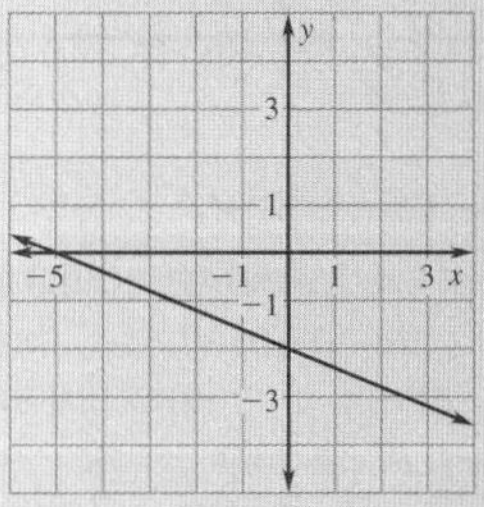

EXTRA EXAMPLE 4

A nutritionist has suggested that you take an additional 2400 mg of iron over the next six months. Two brands of vitamins supply the necessary iron, but Brand A is more expensive. Brand A supplies 40 mg of iron in each pill and Brand B supplies 10 mg. How many of each kind of vitamin might you take? **All points on the line $40x + 10y = 2400$ with integer coordinates are solutions to this problem. For example, you might take 30 Brand A pills and 120 Brand B pills.**

CHECKPOINT EXERCISES

For use after Example 4:

1. The Student Senate is selling tickets to the fall carnival. They would like to sell \$2000 worth. If the cost of a student ticket is \$5 and the cost of an adult ticket is \$10, how many of each kind could they sell to reach their goal? **All points on the line $5x + 10y = 2000$ with integer coordinates are solutions. For example, they could sell 300 student tickets and 50 adult tickets.**

FOCUS ON VOCABULARY

What is the y-coordinate of an x-intercept? What is the x-coordinate of a y-intercept? **0; 0**

CLOSURE QUESTION

Explain how to use the intercepts to graph a linear equation. **Find the two intercepts by substituting 0 for y and then 0 for x. Draw a coordinate plane that includes the two intercepts. Plot the two points on the coordinate plane and draw a line through them.**

DAILY PUZZLER

Three of the following equations have a common property. Identify the equations and describe the property. A. $3x = 2y$ B. $3x + 2y = 5$ C. $2x - 3y = 0$ D. $-2x = -3y$ E. $3y = 2x - 1$ **A, C, and D have a common property; both intercepts are 0.**

EXAMPLE 4 *Writing and Using a Linear Model*

ZOO FUNDRAISING You are organizing the annual spaghetti dinner to raise funds for a zoo. Your goal is to sell \$1500 worth of tickets. Assuming 200 adults and 100 students will attend the dinner, how much should you charge for an adult ticket and for a student ticket?

SOLUTION

PROBLEM SOLVING STRATEGY

VERBAL MODEL: Number of adults · Adult ticket price + Number of students · Student ticket price = Total sales

LABELS

Number of adults	= **200**	(people)
Adult ticket price	= x	(dollars per person)
Number of students	= **100**	(people)
Student ticket price	= y	(dollars per person)
Total salcs	= **1500**	(dollars)

ALGEBRAIC MODEL

$$200x + 100y = 1500$$ **Write linear model.**

$$2x + y = 15$$ **Divide by 100 to simplify model.**

This equation has many solutions. To get a better idea of the possible prices, make a quick graph.

Find the intercepts.

$2x + y = 15$ $\qquad$ $2x + y = 15$

$2x + 0 = 15$ $\qquad$ $2(0) + y = 15$

$x = 7.5$ ← *x*-intercept $\qquad$ $y = 15$ ← *y*-intercept

Draw a coordinate plane that includes the points (7.5, 0) and (0, 15).

Plot the points (7.5, 0) and (0, 15) and draw a line through them. From the graph, you can determine several possible prices to charge.

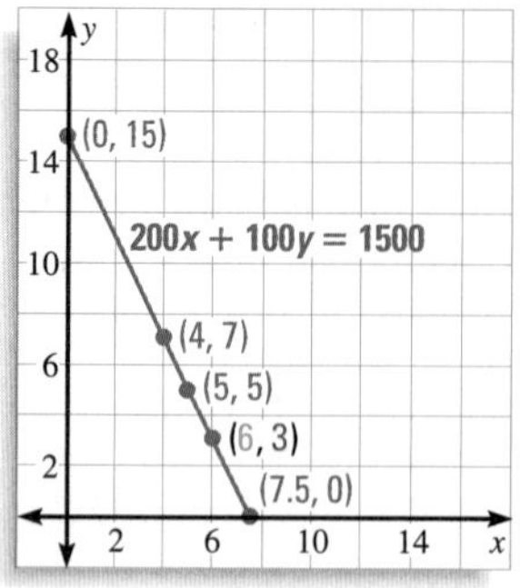

Possible Prices to Raise \$1500

Adult	Student
\$0.00	\$15.00
\$4.00	\$7.00
\$5.00	\$5.00
\$6.00	\$3.00
\$7.50	\$0.00

▶ One reasonable price to charge is \$6 for adults and \$3 for students.

FOCUS ON APPLICATIONS

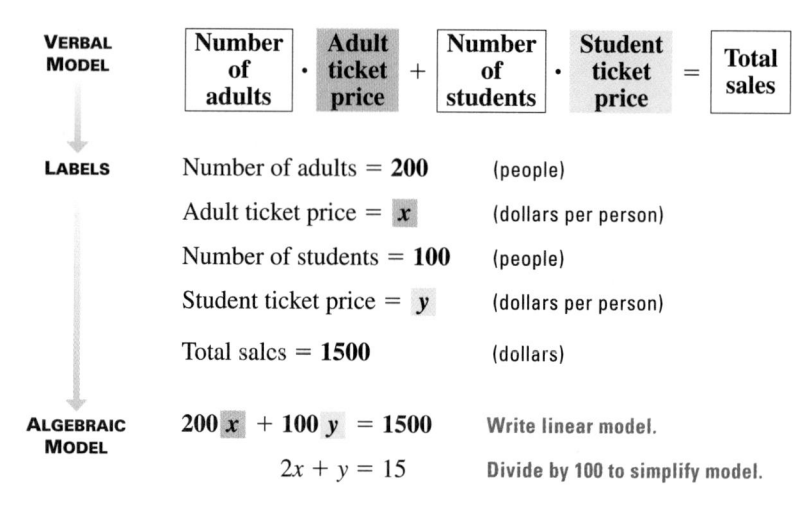

ZOO EXPENSES The American Zoo and Aquarium Institute estimates that it costs a zoo about \$22,000 per year to house, feed, and care for a lion and about \$18,500 for a polar bear.

APPLICATION LINK www.mcdougallittell.com

GUIDED PRACTICE

Vocabulary Check ✓

1. Decide whether 2 is the *x-intercept* or the *y-intercept* of the line $y = 2x + 2$. Explain your choice. **y-intercept; when $x = 0$, $y = 2(0) + 2 = 2$.**

Concept Check ✓

2. How many points are needed to determine a line? **two points**

3. Describe a line that has no x-intercept.
a line with equation $y = a$ for any number a except 0

Skill Check ✓

Find the x-intercept of the graph of the equation.

4. $y = 2x + 20$ **−10** **5.** $y = 0.1x + 0.3$ **−3** **6.** $y = x - \frac{1}{4}$ **$\frac{1}{4}$**

In Exercises 7–12, find the x-intercept and the y-intercept of the graph of the equation. Graph the equation. **7–12. See margin for graphs.**

7. $y = x + 2$ **−2; 2** **8.** $y - 2x = 3$ **$-\frac{3}{2}$; 3** **9.** $2x - y = 4$ **2; −4**

10. $3y = -6x + 3$ **$\frac{1}{2}$; 1** **11.** $5y = 5x + 15$ **−3; 3** **12.** $x - y = 1$ **1; −1**

13. FUNDRAISING If your goal for the fundraising dinner in Example 4 is \$2000, find reasonable prices for adult tickets and student tickets.
***Sample answers:* \$8 for adult tickets and \$4 for student tickets or \$9 for adult tickets and \$2 for student tickets**

PRACTICE AND APPLICATIONS

STUDENT HELP

Extra Practice to help you master skills is on p. 800.

USING GRAPHS TO FIND INTERCEPTS **Use the graph to find the x-intercept and the y-intercept of the line.**

14. **2; 3**

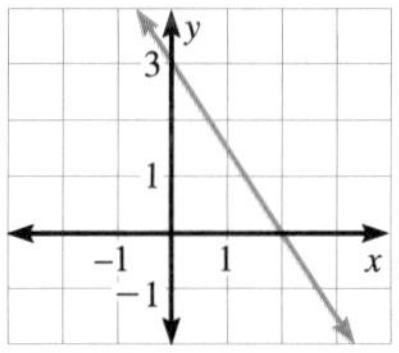

15. **−2; 4**

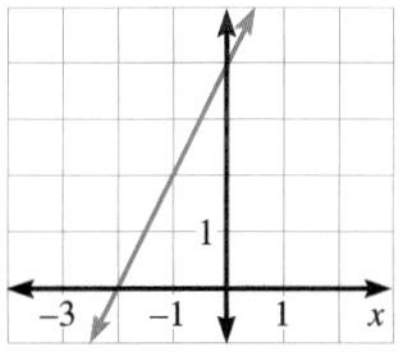

16. **−4; −1**

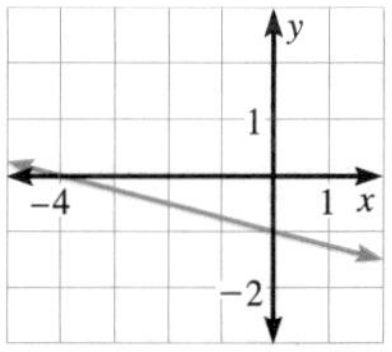

FINDING X-INTERCEPTS **Find the x-intercept of the graph of the equation.**

17. $x + 3y = 5$ **5** **18.** $x - 2y = 6$ **6** **19.** $2x + 2y = -10$ **−5**

20. $3x + 4y = 12$ **4** **21.** $5x - y = 45$ **9** **22.** $-x + 3y = 27$ **−27**

23. $-7x - 3y = 42$ **−6** **24.** $2x + 6y = -24$ **−12** **25.** $-12x - 20y = 60$ **−5**

FINDING Y-INTERCEPTS **Find the y-intercept of the graph of the equation.**

26. $y = -2x + 5$ **5** **27.** $y = 3x - 4$ **−4** **28.** $y = 8x + 27$ **27**

29. $y = 7x - 15$ **−15** **30.** $4x - 5y = -35$ **7** **31.** $6x - 9y = 72$ **−8**

32. $3x + 12y = -84$ **−7** **33.** $-x + 1.7y = 5.1$ **3** **34.** $2x - 6y = -18$ **3**

USING INTERCEPTS **Graph the line that has the given intercepts.**
35–40. See margin.

35. x-intercept: −2; y-intercept: 5
36. x-intercept: 4; y-intercept: 6
37. x-intercept: −7; y-intercept: −3
38. x-intercept: −3; y-intercept: −7
39. x-intercept: −12; y-intercept: −8
40. x-intercept: −7; y-intercept: 15

STUDENT HELP

HOMEWORK HELP
Example 1: Exs. 17–34
Example 2: Exs. 35–55
Example 3: Exs. 44–55
Example 4: Exs. 60–63

3 APPLY

ASSIGNMENT GUIDE

BASIC
Day 1: pp. 221–222 Exs. 14–40 even, 41–43
Day 2: pp. 222–224 Exs. 44–58 even, 60–63, 69, 75, 80, 85, 87, Quiz 1 Exs. 1–16

AVERAGE
Day 1: pp. 221–222 Exs. 14–40 even, 41–43
Day 2: pp. 222–224 Exs. 44–58 even, 60–63, 69, 75, 80, 85, 87, Quiz 1 Exs. 1–16

ADVANCED
Day 1: pp. 221–222 Exs. 14–40 even, 41–43
Day 2: pp. 222–224 Exs. 44–58 even, 60–63, 67–70, 75, 80, 85, 87, Quiz 1 Exs. 1–16

BLOCK SCHEDULE
pp. 221–224 Exs. 14–40 even, 41–43, 44–58 even, 60–63, 69, 75, 80, 85, 87, Quiz 1 Exs. 1–16

EXERCISE LEVELS

Level A: ***Easier***
14–16, 71–84

Level B: ***More Difficult***
17–59, 69, 85–87

Level C: ***Most Difficult***
60–68, 70

✓ HOMEWORK CHECK

To quickly check student understanding of key concepts, go over the following exercises: Exs. 20, 30, 36, 46, 56, 62. See also the Daily Homework Quiz:

- Blackline Master (*Chapter 4 Resource Book*, p. 52)
- Transparency (p. 31)

7.

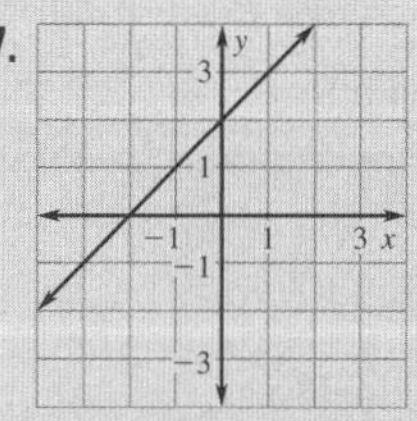

8–12, 35–40. See Additional Answers beginning on page AA1.

! COMMON ERROR

EXERCISE 59 Students often incorrectly identify the graph of an equation of the form $x = a$ as a horizontal line. Remind them that the line $x = 4$ has x-intercept 4, but no y-intercept, thus it cannot be a horizontal line.

APPLICATION NOTE

EXERCISE 69 Though the graph of the equation extends into three quadrants, only the intercepts and the points that lie in the first quadrant model the problem.

44.
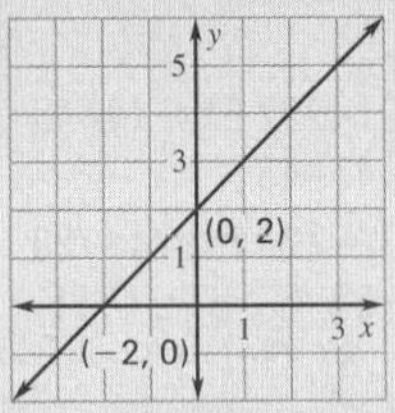

45.
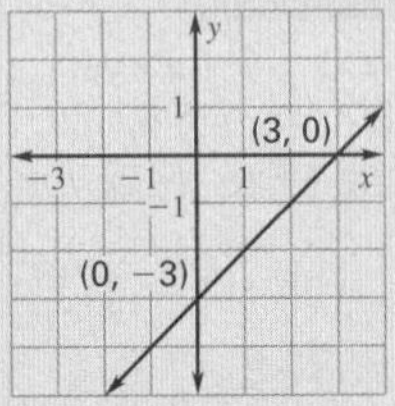

46–55, 60, 65. See Additional Answers beginning on page AA1.

ADDITIONAL PRACTICE AND RETEACHING

For Lesson 4.3:

- Practice Levels A, B, and C (*Chapter 4 Resource Book*, p. 40)
- Reteaching with Practice (*Chapter 4 Resource Book*, p. 43)
- See Lesson 4.3 of the *Personal Student Tutor*

For more Mixed Review:

- Search the *Test and Practice Generator* for key words or specific lessons.

63. The possible pairs (x, y) are (0, 120), (5, 112), (10, 104), (15, 96), (20, 88), (25, 80), (30, 72), (35, 64), (40, 56), (45, 48), (50, 40), (55, 32), (60, 24), (65, 16), (70, 8), and (75, 0).

FOCUS ON APPLICATIONS

MARATHON In 1896 the Olympic Games introduced the "marathon" race to honor a Greek soldier. The legend says a soldier ran 26.2 miles from Marathon to Athens in 3 hours to announce the victory over the Persians.

MATCHING GRAPHS AND EQUATIONS **Match the equation with its graph.**

A. $8x - 2y = 16$ **B.** $8x - 2y = 8$ **C.** $8x - 2y = -8$

41. A

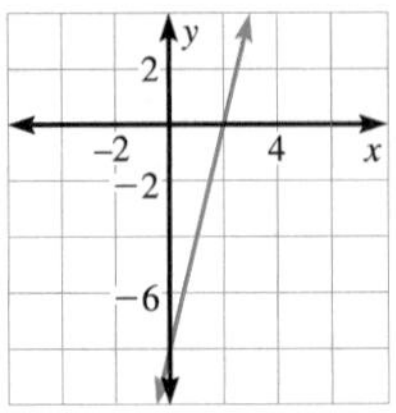

42. C

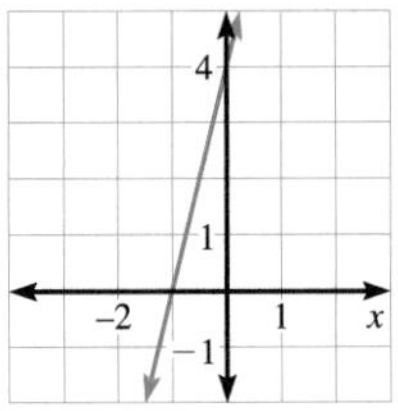

43. B

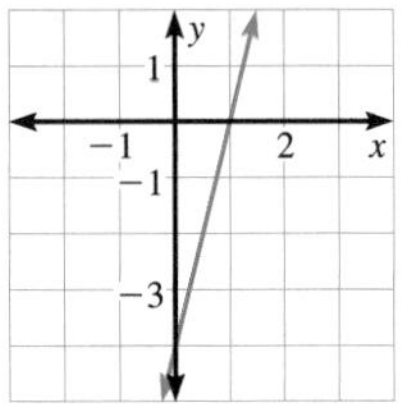

GRAPHING LINES **Find the x-intercept and the y-intercept of the line. Graph the equation. Label the points where the line crosses the axes.**

44–55. See margin for graphs.

44. $y = x + 2$ −2; 2 **45.** $y = x - 3$ 3; −3 **46.** $y = 4x + 8$ −2; 8

47. $y = -6 + 3x$ 2; −6 **48.** $y = 5x + 15$ −3; 15 **49.** $2x + 4y = 16$ 8; 4

50. $-4x + 3y = 24$ −6; 8 **51.** $x - 7y = 21$ 21; −3 **52.** $6x - y = 36$ 6; −36

53. $2x + 9y = -36$ −18; −4 **54.** $4x + 5y = 20$ 5; 4 **55.** $0.5y = -2x + 8$ 4; 16

LOGICAL REASONING **In Exercises 56–59, tell whether the statement is *true* or *false*. Justify your answer.**

56. The y-intercept of the graph of $3x + 5y = 30$ is 10.
False; the y-intercept is 6; when $x = 0$, $5y = 30$ and $y = 6$.

57. The x-intercept of the graph of $3x + 5y = 30$ is 10.
True; when $y = 0$, $3x = 30$ and $x = 10$.

58. The point (3, 5) is on the graph of $3x + 5y = 30$. False; $3(3) + 5(5) = 34$.

59. The graph of the equation $x = 4$ is a horizontal line.
False; the graph is a vertical line.

SCHOOL PLAY **In Exercises 60–63, use the following information.**

Your school drama club is putting on a play next month. By selling tickets for the play, the club hopes to raise \$600 for the drama fund for new costumes, scripts, and scenery for future plays. Let x represent the number of adult tickets they sell at \$8 each, and let y represent the number of student tickets they sell at \$5 each.

60. Graph the linear function $8x + 5y = 600$. See margin.

61. What is the x-intercept? What does it represent in this situation?
75; the number of adult tickets that must be sold if no student tickets are sold

62. What is the y-intercept? What does it represent in this situation?
120; the number of student tickets that must be sold if no adult tickets are sold

63. What are three possible numbers of adult and student tickets to sell that will make the drama club reach its goal? See margin.

MARATHON **In Exercises 64–66, you are running in a marathon. You either run 8 miles per hour or walk 4 miles per hour.**

64. Write an equation to show the relationship between time run and time walked during the 26.2-mile course. Let x = the number of hours running and y = the number of hours walking; $8x + 4y = 26.2$.

65. Graph the equation from Exercise 64. What are some possible running and walking times if you complete the 26.2-mile course? See margin.
Sample answers: (1.5, 3.55), (2.25, 2.05), (2.5, 1.55), (2.75, 1.05), (3, 0.55)

66. If you walk for a total of 1 hour during the course, how long will you have spent running when you cross the finish line of the marathon? 2.775 h

MOVIE PRICES In Exercises 67 and 68, a theater charges \$4 per person before 6:00 P.M. and \$7 per person after 6:00 P.M. The total ticket sales for Saturday were \$11,228.

67. Make a graph showing the possible number of people who attended the theater before and after 6:00 P.M. **See margin.**

68. Suppose no one attended the theater before 6:00 P.M. How many people attended the theater after 6:00 P.M.? Explain how you know.
1604 people; this is the y-intercept. When $x = 0$, $4(0) + 7y = 11{,}228$, so $y = 1604$.

Test Preparation

69. MULTI-STEP PROBLEM The number of people who worked for the railroads in the United States each year from 1989 to 1995 can be modeled by the equation $y = -6.61x + 229$, where x represents the number of years since 1989 and y represents the number of railroad employees (in thousands).

DATA UPDATE of U.S. Bureau of the Census data at www.mcdougallittell.com

a. Find the y-intercept of the line. What does it represent?

b. Find the x-intercept of the line. What does it represent?

c. About how many people worked for the railroads in 1995?

d. *Writing* Do you think the line in the graph will continue to be a good model for the next 50 years? Explain.

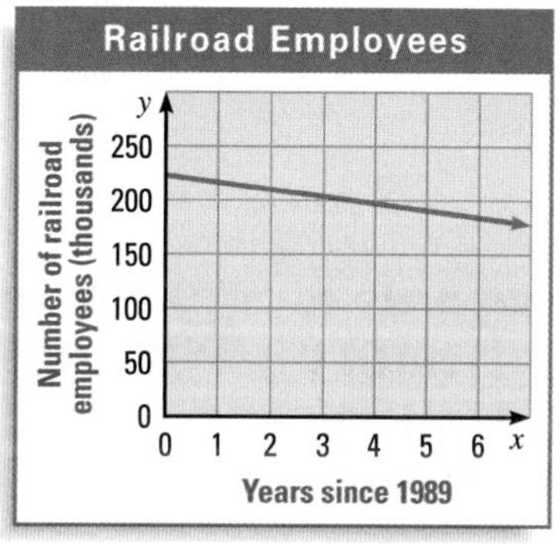

69. a. 229; the number of railroad employees (in thousands) in the United States in 1989

b. approximately 34.6; the number of years since 1989 when the number of railroad employees in the United States will be zero

c. 189,340 people

d. No; after 2024, the number of employees will be negative.

★ Challenge

70. CRITICAL THINKING Consider the equation $6x + 8y = k$. What numbers could replace k so that the x-intercept and the y-intercept are both integers? Explain.
Common multiples of 6 and 8, for example, 24 and 48; when such numbers are divided by 6 or 8, the result is an integer.

MIXED REVIEW

EVALUATING DIFFERENCES Find the difference. (Review 2.3 for 4.4)

71. $5 - 9$ **−4** **72.** $17 - 6$ **11** **73.** $-8 - 9$ **−17** **74.** $|8| - 12.6$ **−4.6**

75. $-\frac{2}{3} - \left(-\frac{7}{3}\right)$ $\frac{5}{3}$ **76.** $13.8 - 6.9$ **6.9** **77.** $7 - |-1|$ **6** **78.** $-4.1 - (-5.1)$ **1**

EVALUATING QUOTIENTS In Exercises 79–86, find the quotient. (Review 2.7 for 4.4)

79. $54 \div 9$ **6** **80.** $-72 \div 8$ **−9** **81.** $12 \div \left(-\frac{1}{5}\right)$ **−60** **82.** $3 \div \frac{1}{4}$ **12**

83. $26 \div (-13)$ **−2** **84.** $-1 \div 8$ $-\frac{1}{8}$ **85.** $-20 \div \left(-2\frac{1}{2}\right)$ **8** **86.** $\frac{1}{8} \div \frac{1}{2}$ $\frac{1}{4}$

87. SCHOOL BAKE SALE You have one hour to make cookies for your school bake sale. You spend 20 minutes mixing the dough. It then takes 12 minutes to bake each tray of cookies. If you bake one tray at a time, which model can you use to find how many trays you can bake? (Review 3.3 and 3.6) **B**

A. $x(20 + 12) = 60$ **B.** $12x + 20 = 60$

Additional Test Preparation ***Sample answer:***
179; the fact that the population of the United States was approximately 179,000,000 in 1960

4 ASSESS

DAILY HOMEWORK QUIZ

Transparency Available

1. Give the x- and y-intercepts of the graph of $2x - y = -4$.
−2 and 4

2. Graph $2x - 3y = 6$. Label the points where the line crosses the axes.

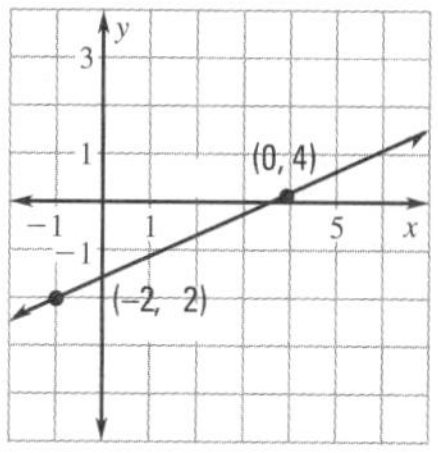

3. An animal shelter has a total of 60 cats and dogs available for adoption.

a. Write an equation to model the situation. **$c + d = 60$**

b. On a graph of the equation the horizontal axis represents the number of dogs available. What are the intercepts? What do they represent?
(60, 0) and (0, 60); 60 dogs, 0 cats and 0 dogs, 60 cats

EXTRA CHALLENGE NOTE

Challenge problems for Lesson 4.3 are available in **blackline** format in the *Chapter 4 Resource Book*, p. 48 and at **www.mcdougallittell.com.**

ADDITIONAL TEST PREPARATION

1. WRITING The population of the United States from 1960 to 1990 can be approximated using the equation $y = 2.3x + 179$, where y is the population in millions and x is the number of years since 1960. What is the y-intercept of the graph of this equation? What does it represent? **See sample answer at left.**

67. See Additional Answers beginning on page AA1.

ADDITIONAL RESOURCES

An alternative Quiz for Lessons 4.1–4.3 is available in the *Chapter 4 Resource Book,* p. 49.

A **blackline** master with additional Math & History exercises is available in the *Chapter 4 Resource Book,* p. 47.

1. 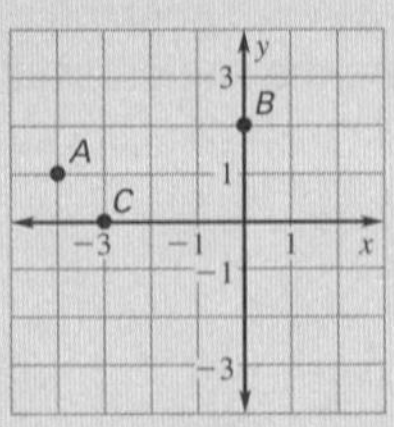

2. 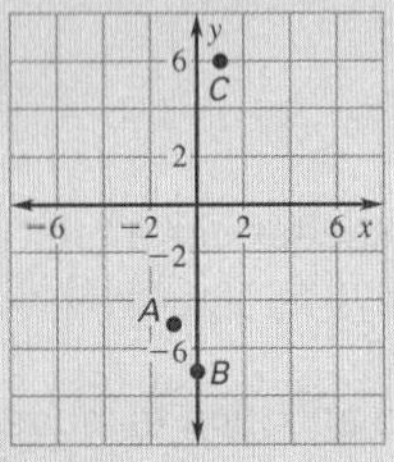

3. 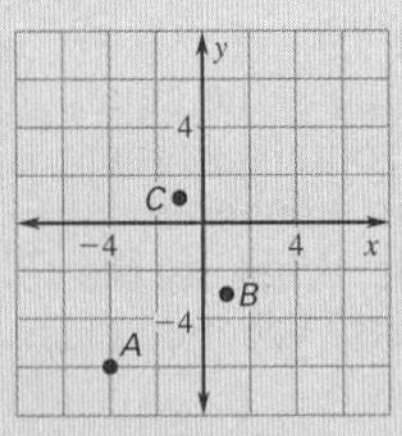

4. 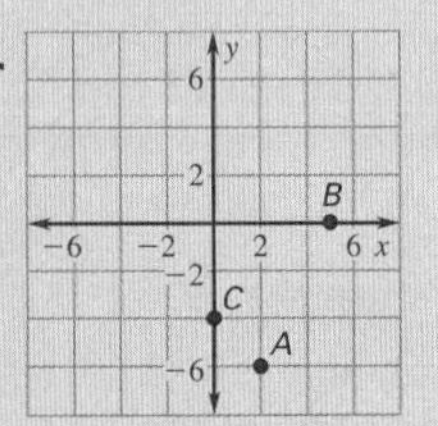

5–16. See Additional Answers beginning on page AA1.

6. (0, −1), $\left(\frac{1}{2}, 1\right)$, $\left(-\frac{1}{2}, -3\right)$

QUIZ 1

Self-Test for Lessons 4.1–4.3

Plot and label the ordered pairs in a coordinate plane. (Lesson 4.1) See margin.

1. $A(-4, 1), B(0, 2), C(-3, 0)$ **2.** $A(-1, -5), B(0, -7), C(1, 6)$

3. $A(-4, -6), B(1, -3), C(-1, 1)$ **4.** $A(2, -6), B(5, 0), C(0, -4)$

Find three different ordered pairs that are solutions of the equation. Graph the equation. (Lesson 4.2) 5–10. See margin for graphs.

5. $y = 2x - 6$ (0, −6), (3, 0), (2, −2) **6.** $y = 4x - 1$ See margin. **7.** $y = 2(-3x + 1)$ (0, 2), (1, −4), (−1, 8)

8. $x = 3$ (3, −1), (3, 0), (3, 1) **9.** $y = -3(x - 4)$ (0, 12), (1, 9), (4, 0) **10.** $y = -5$ (−3, −5), (9, −5), (2, −5)

Find the *x*-intercept and the *y*-intercept of the line. Graph the line. Label the points where the line crosses the axes. (Lesson 4.3) 11–16. See margin for graphs.

11. $y = 4 - x$ 4; 4 **12.** $y = -5 + 2x$ $2\frac{1}{2}$; −5 **13.** $y = 3x + 12$ −4; 12

14. $3x + 3y = 27$ 9; 9 **15.** $-6x + y = -3$ $\frac{1}{2}$; −3 **16.** $y = 10x + 50$ −5; 50

MATH & History — Graphing Mathematical Relationships

APPLICATION LINK www.mcdougallittell.com

THEN

400 YEARS AGO, this chapter would not have been in your algebra book. In 1637 René Descartes published a method for drawing algebraic equations on a coordinate grid—introducing a new connection between algebra and geometry that linked the two branches more closely than ever before.

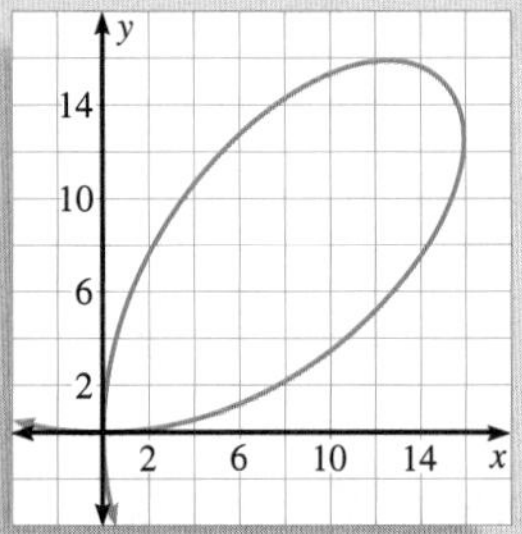

Descartes was interested in studying curves that are more complicated than the lines you have seen in this chapter.

Descartes was interested in equations like $x^3 + y^3 = 3axy$ when $a = 10$, which is graphed at the right. Use algebra to show whether the point is on the graph of the equation.

1. (15, 15) yes **2.** (0, 0) yes **3.** (−1, −1) no **4.** (3, 0.3) no

NOW

TODAY, using a coordinate plane to graph an equation is one of the first things you learn in Algebra 1. Computers that can process large amounts of data have enabled mathematicians to study mathematical systems whose outputs are too complicated to graph by hand.

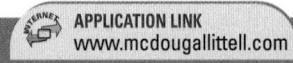

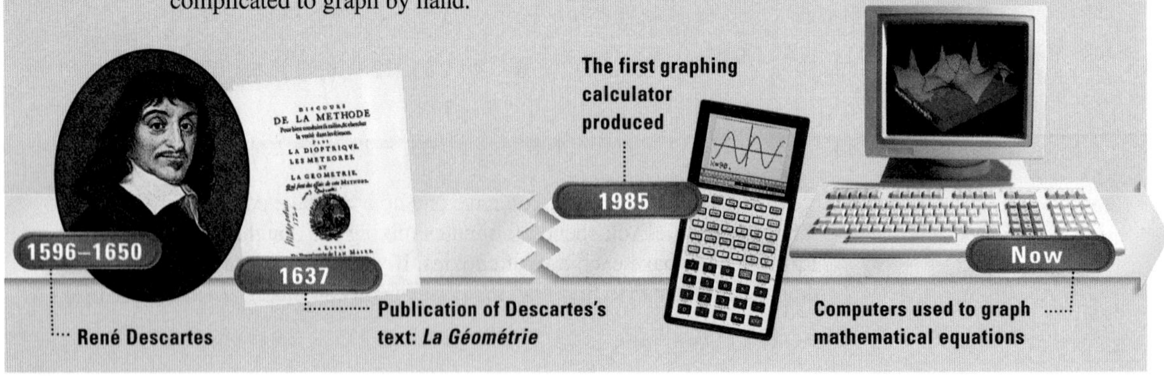
1596–1650
René Descartes
DE LA METHODE
1637
Publication of Descartes's text: *La Géométrie*
The first graphing calculator produced
1985
Now
Computers used to graph mathematical equations

ACTIVITY 4.4

Developing Concepts

Investigating Slope

GROUP ACTIVITY
Work with a partner.

MATERIALS
- several books
- two metric rulers

QUESTION How can you use numbers to describe the steepness of a ramp?

The *slope* of a ramp describes its steepness. For the ramp at the right:

$$\text{slope} = \frac{\text{rise}}{\text{run}} = \frac{2\text{ m}}{5\text{ m}} = \frac{2}{5}$$

Height of books	Distance between ruler and books	Rise	Run	Slope
7 cm	12 cm	7	12	$\frac{7}{12}$
11 cm	5.5 cm	?	?	?

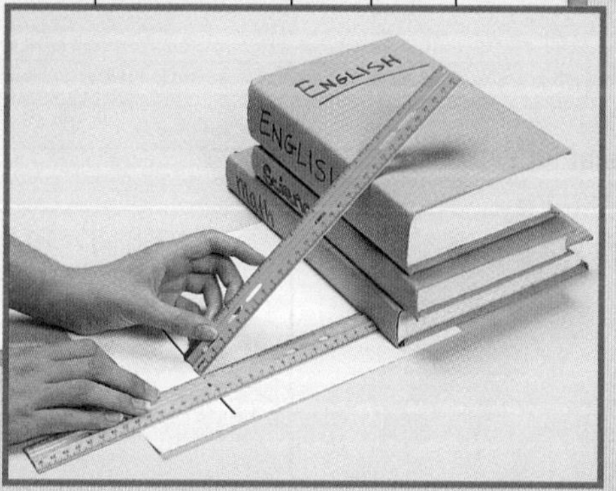

EXPLORING THE CONCEPT

1. Copy and complete the table at the left. **Rise = 11 cm; Run = 5.5 cm; Slope = 2**
2. Make a pile of three books. Measure the height of the pile of books. Use one ruler to create a ramp. Measure the distance from the bottom of the ruler to the books. Use this information to add a row to your table.
3. Without changing the rise, create at least four ramps that have different runs. Record information about these ramps in your table.
4. Place a piece of paper under the edge of your pile of books and make a mark on it that is 15 cm away from the books. Place the end of your ramp on the mark. Add this ramp to your table.
5. Change the rise by adding or removing books. Without changing the run, create at least four ramps that have different rises. Add them to your table.

DRAWING CONCLUSIONS

In Exercises 1 and 2, give examples to support your answer.

1. If the run of a ramp increases and the rise stays the same, how does the slope of the ramp change?
2. If the rise of a ramp increases and the run stays the same, how does the slope of the ramp change? **See margin.**

In Exercises 3–5, you will summarize what you have learned about slope.

3. If a ramp has a slope of 1, what is the relationship between its rise and run? **They are equal.**
4. If a ramp has a slope greater than 1, is it steeper than a ramp with a slope of 1? Describe the relationship between its rise and run. **Yes; the rise is greater than the run.**
5. If a ramp has a slope less than 1, is it steeper than a ramp with a slope of 1? Describe the relationship between its rise and run. **No; the rise is less than the run.**

1. The slope decreases; if the rise is 1 and the run is 4, the slope is $\frac{1}{4}$. If the run increases to 8, the slope decreases to $\frac{1}{8}$.

2. The slope increases; if the rise is 1 and the run is 4, the slope is $\frac{1}{4}$. If the rise increases to 3, the slope increases to $\frac{3}{4}$.

1 Planning the Activity

PURPOSE
To use a physical model to explore slope.

MATERIALS
- 2 metric rulers per group
- 3–5 books per group
- Activity Support Master (*Chapter 4 Resource Book,* p. 53)

PACING
- Exploring the Concept — 15 min
- Drawing Conclusions — 10 min

LINK TO LESSON
When students learn to find the slope of a line between two points in Example 1 of Lesson 4.4, have them imagine a ramp between the two points.

2 Managing the Activity

CLASSROOM MANAGEMENT
Make sure that the students understand how to measure the rise and the run of each ramp using one of the two rulers. Emphasize that they are to use the same stack of books in Steps 2 and 3.

ALTERNATIVE APPROACH
To save time, demonstrate the activity to the class with help from individual students.

3 Closing the Activity

KEY DISCOVERY
The ratio "rise to run" describes the steepness of a ramp, or its slope. As a ramp gets steeper, its slope increases.

ACTIVITY ASSESSMENT
JOURNAL A proposed ramp has a rise of 3 ft and a run of 6 ft. Describe how to alter one measurement to make a steeper ramp. Do the same for a shallower ramp. Give examples. **See sample answer at left.**

Activity Assessment *Sample answer:*

The slope of the proposed ramp is $\frac{1}{2}$. Keeping the same run but increasing the rise produces a steeper ramp, e.g., a rise of 6 ft and a run of 6 ft gives a slope of 1. Keeping the same rise but increasing the run produces a shallower ramp, e.g., a rise of 3 ft and a run of 9 ft gives a slope of $\frac{1}{3}$.

1 PLAN

PACING
Basic: 2 days
Average: 2 days
Advanced: 2 days
Block Schedule: 1 block

LESSON OPENER
VISUAL APPROACH
An alternative way to approach Lesson 4.4 is to use the Visual Approach Lesson Opener:
- Blackline Master (*Chapter 4 Resource Book*, p. 54)
- Transparency (p. 27)

MEETING INDIVIDUAL NEEDS
- ***Chapter 4 Resource Book***
  Prerequisite Skills Review (p. 5)
  Practice Level A (p. 55)
  Practice Level B (p. 56)
  Practice Level C (p. 57)
  Reteaching with Practice (p. 58)
  Absent Student Catch-Up (p. 60)
  Challenge (p. 63)
- ***Resources in Spanish***
- ***Personal Student Tutor***

NEW-TEACHER SUPPORT
See the Tips for New Teachers on pp. 1–2 of the *Chapter 4 Resource Book* for additional notes about Lesson 4.4.

WARM-UP EXERCISES

Transparency Available

Identify the x-coordinate and the y-coordinate of each ordered pair.

1. (–3, 4)
x-coordinate –3
y-coordinate 4

2. (0, –7)
x-coordinate 0
y-coordinate –7

3. (5, 0)
x-coordinate 5
y-coordinate 0

4. (–9, –2)
x-coordinate –9
y-coordinate –2

4.4 The Slope of a Line

***What** you should learn*

GOAL 1 Find the slope of a line using two of its points.

GOAL 2 Interpret slope as a rate of change in **real-life** situations like the parachute problem in **Example 6**.

***Why** you should learn it*

▼ To solve **real-life** problems about the steepness of a hill or the grade of a road in **Ex. 51**.

REAL LIFE

GOAL 1 FINDING THE SLOPE OF A LINE

The **slope** m of a nonvertical line is the number of units the line rises or falls for each unit of horizontal change from left to right.

The slanted line at the right rises 3 units for each 2 units of horizontal change from left to right. So, the slope m of the line is $\frac{3}{2}$.

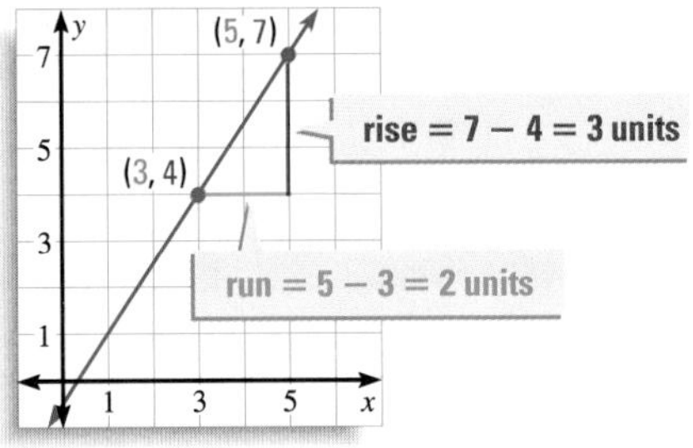

$$\text{slope} = \frac{\text{rise}}{\text{run}} = \frac{7-4}{5-3} = \frac{3}{2}$$

Two points on a line are all that is needed to find its slope.

FINDING THE SLOPE OF A LINE

The slope m of the nonvertical line passing through the points (x_1, y_1) and (x_2, y_2) is

$$m = \frac{\text{rise}}{\text{run}} = \frac{\text{change in } y}{\text{change in } x} = \frac{y_2 - y_1}{x_2 - x_1}.$$

Read x_1 as "x sub one." Think "x-coordinate of the first point."

Read y_2 as "y sub two." Think "y-coordinate of the second point."

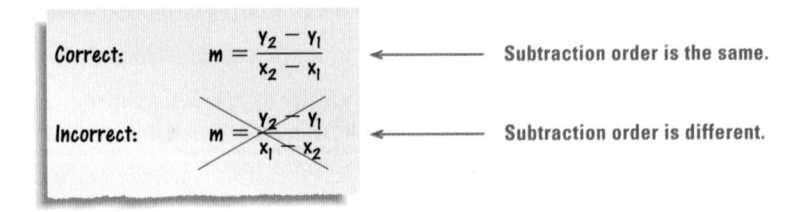

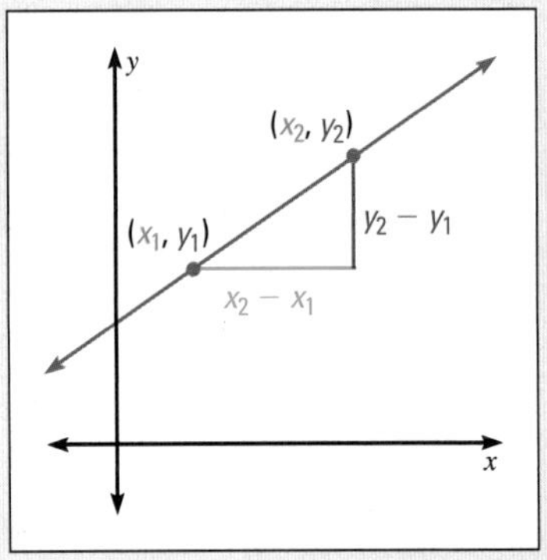

When you use the formula for slope, the order of the subtraction is important. Given two points on a line, you can label either point as (x_1, y_1) and the other point as (x_2, y_2). After doing this, however, you must form the numerator and the denominator using the *same* order of subtraction.

Correct: $m = \frac{y_2 - y_1}{x_2 - x_1}$ ← **Subtraction order is the same.**

Incorrect: ~~$m = \frac{y_2 - y_1}{x_1 - x_2}$~~ ← **Subtraction order is different.**

EXAMPLE 1 *A Line with a Positive Slope Rises*

Find the slope of the line passing through $(-2, 2)$ and $(3, 4)$.

SOLUTION

Let $(x_1, y_1) = (-2, 2)$ and $(x_2, y_2) = (3, 4)$.

$$m = \frac{y_2 - y_1}{x_2 - x_1}$$ ← Rise: Difference of y-values; ← Run: Difference of x-values

$$= \frac{4 - 2}{3 - (-2)}$$ Substitute values.

$$= \frac{2}{3 + 2}$$ Simplify.

$$= \frac{2}{5}$$ Slope is positive.

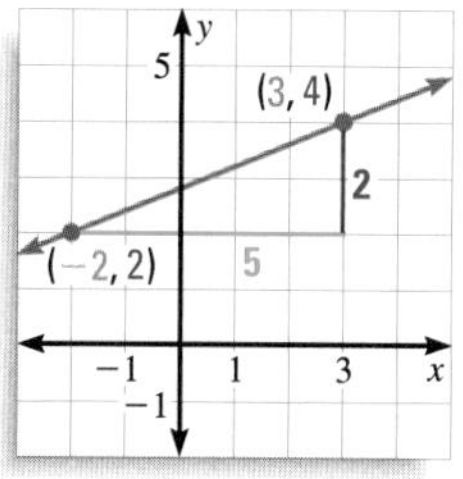

Positive slope: line rises from left to right.

> **STUDENT HELP**
> **Look Back**
> For help with evaluating expressions, see p. 79.

EXAMPLE 2 *A Line with a Zero Slope is Horizontal*

Find the slope of the line passing through $(-1, 2)$ and $(3, 2)$.

SOLUTION

Let $(x_1, y_1) = (-1, 2)$ and $(x_2, y_2) = (3, 2)$.

$$m = \frac{y_2 - y_1}{x_2 - x_1}$$ ← Rise: Difference of y-values; ← Run: Difference of x-values

$$= \frac{2 - 2}{3 - (-1)}$$ Substitute values.

$$= \frac{0}{4}$$ Simplify.

$$= 0$$ Slope is zero.

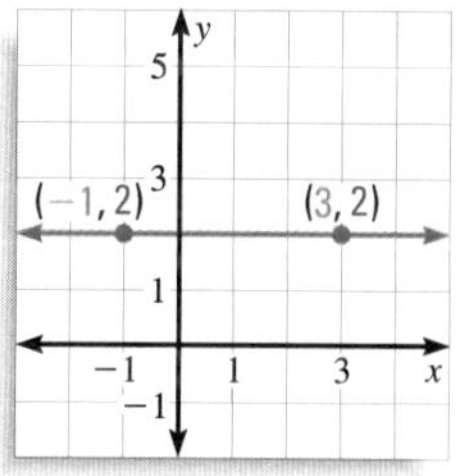

Zero slope: line is horizontal.

EXAMPLE 3 *A Line with a Negative Slope Falls*

Find the slope of the line passing through $(0, 0)$ and $(3, -3)$.

SOLUTION

Let $(x_1, y_1) = (0, 0)$ and $(x_2, y_2) = (3, -3)$

$$m = \frac{y_2 - y_1}{x_2 - x_1}$$ ← Rise: Difference of y-values; ← Run: Difference of x-values

$$= \frac{-3 - 0}{3 - 0}$$ Substitute values.

$$= \frac{-3}{3}$$ Simplify.

$$= -1$$ Slope is negative.

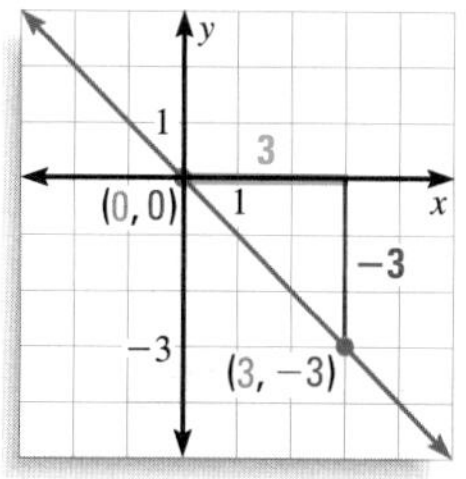

Negative slope: line falls from left to right.

> **STUDENT HELP**
> **Skills Review**
> For help with simplifying fractions, see p. 780.

2 TEACH

MOTIVATING THE LESSON
Signs are often posted to indicate how steep a road is. Ask students which road they think will be harder to climb on a bicycle, one with a 9% grade or one with a 7% grade. [9% grade] Slope, or the steepness of a line, is the topic of this lesson.

EXTRA EXAMPLE 1
Find the slope of the line passing through $(-3, 0)$ and $(-1, 6)$. **3**

EXTRA EXAMPLE 2
Find the slope of the line passing through $(-1, -3)$ and $(5, -3)$. **0**

EXTRA EXAMPLE 3
Find the slope of the line passing through $(-2, 1)$ and $(1, -3)$. $-\frac{4}{3}$

CHECKPOINT EXERCISES

For use after Examples 1–3:

1. Find the slope of the line passing through $(5, 0)$ and $(0, 2)$. Tell whether the line rises or falls from left to right.
$-\frac{2}{5}$; falls

STUDENT HELP NOTES

Look Back As students look back to p. 79, remind them that in expressions with more than one variable each variable must be replaced with the correct value.

Skills Review As students review simplifying fractions on p. 780, remind them that the greatest common factor should be removed from both numerator and denominator.

ENGLISH LEARNERS
Tell students that "rise" refers to movement up or down, and that "run" refers to movement left or right. Say that the subscript number tells them whether to refer to the first or the second ordered pair.

EXTRA EXAMPLE 4
Find the slope of the line passing through (0, −1) and (0, 4). **undefined**

EXTRA EXAMPLE 5
Find the value of y so that the line passing through the points (3, y) and (1, 5) has a slope of −2. **1**

CHECKPOINT EXERCISES

For use after Example 4:

1. Find the slope of the line passing through (−3, 5) and (−3, −2). **undefined**

For use after Example 5:

2. Find the value of y so that the line passing through the points (3, y) and (5, 8) has a slope of 2. **4**

! COMMON ERROR
EXAMPLE 4 Some students confuse a slope of 0 with an undefined slope. Remind students that it is easy to walk on a horizontal surface because the slope is 0. Tell them that since it is impossible to walk straight up a wall, the slope of a vertical line must be undefined.

ENGLISH LEARNERS
Tell students to follow along a line from left to right—the way you read a sentence in English—to tell whether the slope is positive, negative, zero, or undefined.

MULTIPLE REPRESENTATIONS
The slope of a line is represented by a number. However, students need to use algebraic or graphical representations to calculate slope. Eventually, students should feel comfortable using the algebraic formula, but a graph can help build understanding of the concept.

STUDENT HELP NOTES
Homework Help Students can find extra examples at **www.mcdougallittell.com** that parallel the examples in the student edition.

EXAMPLE 4 *Slope of a Vertical Line is Undefined*

Find the slope of the line passing through (2, 4) and (2, 1).

SOLUTION Let $(x_1, y_1) = (2, 1)$ and $(x_2, y_2) = (2, 4)$.

$m = \dfrac{y_2 - y_1}{x_2 - x_1}$ ← **Rise: Difference of y-values** ← **Run: Difference of x-values**

$= \dfrac{4 - 1}{2 - 2}$ **Substitute values.**

$= \dfrac{3}{0}$ (crossed out) **Division by 0 is undefined.**

▶ Because division by 0 is undefined, the expression $\frac{3}{0}$ has no meaning. The slope of a vertical line is undefined.

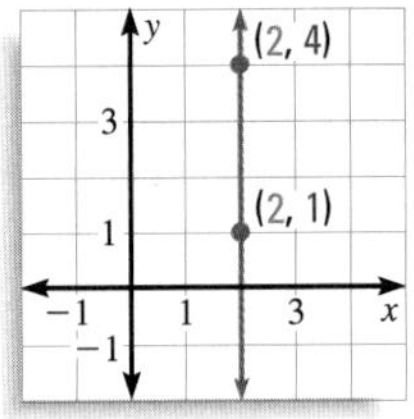

Undefined slope: line is vertical.

CONCEPT SUMMARY CLASSIFICATION OF LINES BY SLOPE

A line with positive slope *rises* from left to right.

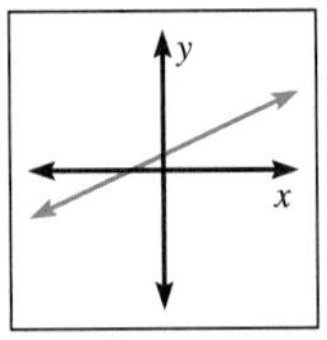

$m > 0$

A line with negative slope *falls* from left to right.

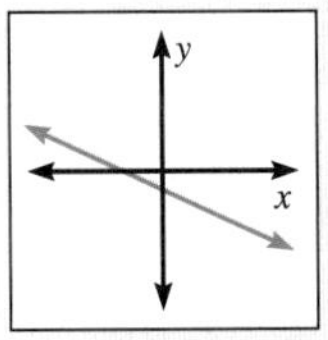

$m < 0$

A line with zero slope is *horizontal*.

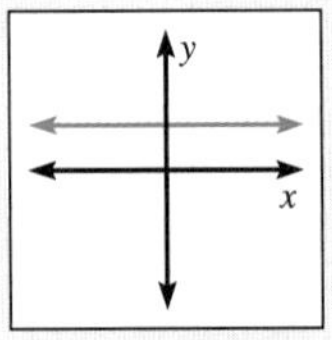

$m = 0$

A line with undefined slope is *vertical*.

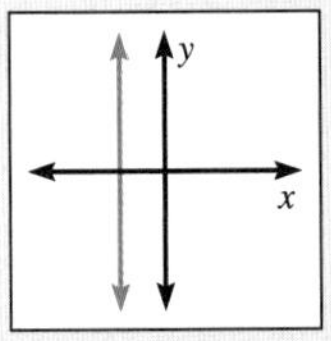

m **is undefined.**

EXAMPLE 5 *Given the Slope, Find a y-Coordinate*

Find the value of y so that the line passing through the points (−2, 1) and (4, y) has a slope of $\frac{2}{3}$.

SOLUTION Let $(x_1, y_1) = (-2, 1)$ and $(x_2, y_2) = (4, y)$.

$m = \dfrac{y_2 - y_1}{x_2 - x_1}$	**Write formula for slope.**
$\dfrac{2}{3} = \dfrac{y - 1}{4 - (-2)}$	**Substitute values for m, x_1, y_1, x_2, and y_2.**
$\dfrac{2}{3} = \dfrac{y - 1}{6}$	**Simplify.**
$6 \cdot \dfrac{2}{3} = 6 \cdot \dfrac{y - 1}{6}$	**Multiply each side by 6.**
$4 = y - 1$	**Simplify.**
$5 = y$	**Add 1 to each side.**

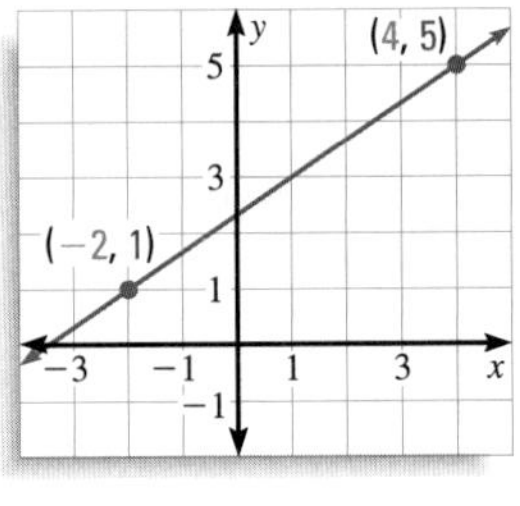

STUDENT HELP
HOMEWORK HELP Visit our Web site www.mcdougallittell.com for extra examples.

GOAL 2 INTERPRETING SLOPE AS A RATE OF CHANGE

A **rate of change** compares two different quantities that are changing. For example, the rate at which distance is changing with time is called *velocity*. Slope provides an important way of visualizing a rate of change.

REAL LIFE
Parachuting

EXAMPLE 6 *Slope as a Rate of Change*

You are parachuting. At time $t = 0$ seconds, you open your parachute at height $h = 2500$ feet above the ground. At time $t = 35$ seconds, you are at height $h = 2115$ feet.

a. What is your rate of change in height?

b. About when will you reach the ground?

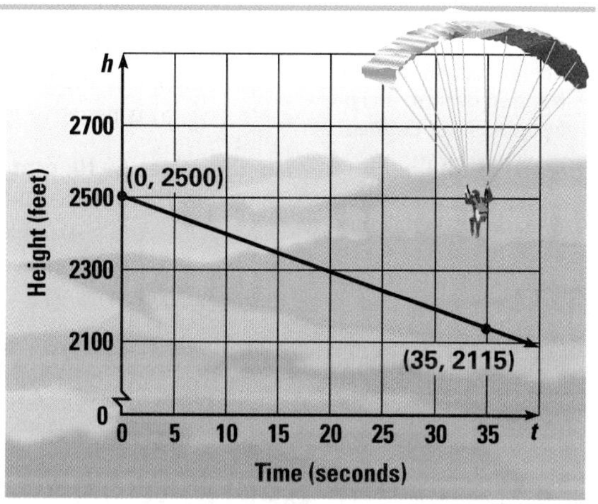

SOLUTION

a. Use the formula for slope to find the rate of change. The change in time is $35 - 0 = 35$ seconds. Subtract in the same order. The change in height is $2115 - 2500 = -385$ feet.

PROBLEM SOLVING STRATEGY

VERBAL MODEL Rate of change $=$ $\dfrac{\text{Change in height}}{\text{Change in time}}$

LABELS Rate of change $= m$ (ft/sec)

Change in height $= \mathbf{-385}$ (ft)

Change in time $= \mathbf{35}$ (sec)

ALGEBRAIC MODEL $m = \dfrac{-385}{35}$

▶ Your rate of change is -11 ft/sec. The negative value indicates that you are falling.

b. Falling at a rate of 11 ft/sec, find the time it will take you to fall 2500 feet.

$$\text{Time} = \frac{\text{Distance}}{\text{Rate}} = \frac{2500 \text{ ft}}{11 \text{ ft/sec}} \approx 227 \text{ sec}$$

▶ You will reach the ground about 227 seconds after opening your parachute.

UNIT ANALYSIS Check that *seconds* are the units of the solution.

$$\frac{\text{ft}}{\text{ft/sec}} = \cancel{\text{ft}} \cdot \frac{\text{sec}}{\cancel{\text{ft}}} = \text{sec}$$

EXTRA EXAMPLE 6

You are traveling by car and leave home at 8:00 A.M. By 8:45 A.M., you are 36 mi from home. Find the average speed in miles per hour. **48 mi/h**

CHECKPOINT EXERCISES

For use after Example 6:

1. You are typing a paper. At 5:15 P.M. you have typed 275 words. By 5:27 P.M. you have typed 660 words. Find the average rate of change in words per minute. **about 32 words per minute**

MATHEMATICAL REASONING

Suppose you turned on a hot water faucet. Explain how the temperature *t* after *m* minutes can be modeled by two linear equations. **From the time the water is turned on until it reaches its maximum temperature, the temperature can be modeled by a linear equation with positive slope. After the water has reached its maximum temperature, the temperature can be modeled by a linear equation with slope 0.**

FOCUS ON VOCABULARY

The graph in Example 6 models uniform motion. In uniform motion, the rate of change of distance with respect to time is constant. Thus the graph of distance versus time is a straight line whose slope is equal to the rate of change of distance with respect to time.

CLOSURE QUESTION

Without drawing a graph, how can you tell from two coordinates that the slope of the line between them is zero? undefined? **The *y*-values are equal; the *x*-values are equal.**

DAILY PUZZLER

You and a friend are playing a game on a grid. You have markers at three *x*-intercepts. Your friend has markers at (1, 3), (3, 3), and (3, 2). Who wins the game? What game are you playing? **you win; tic-tac-toe**

3 APPLY

ASSIGNMENT GUIDE

BASIC

Day 1: pp. 230–231 Exs. 16–20, 21–35 odd, 36, 39–47 odd

Day 2: pp. 230–233 Exs. 37, 42, 48, 52–59, 67–69, 75, 80, 85, 90, 95

AVERAGE

Day 1: pp. 230–231 Exs. 16–20, 21–45 mult. of 3, 47–49

Day 2: pp. 230–233 Exs. 37, 42, 51–61, 67–69, 75, 80, 85, 90, 95

ADVANCED

Day 1: pp. 230–231 Exs. 16–20, 21–45 mult. of 3, 47–50

Day 2: pp. 230–233 Exs. 51–56, 58–61, 67–71, 75, 80, 85, 90, 95

BLOCK SCHEDULE

pp. 230–233 Exs. 12–48 even, 35, 37, 49, 51–55, 58–60, 67–69, 75, 80, 85, 90, 95

EXERCISE LEVELS

Level A: *Easier*
12–19, 72–83

Level B: *More Difficult*
20–49, 52–57, 67–69, 84–95

Level C: *Most Difficult*
50, 51, 58–66, 70–71

✔ HOMEWORK CHECK

To quickly check student understanding of key concepts, go over the following exercises: Exs. 19, 20, 36, 45, 47, 56. See also the Daily Homework Quiz:

- Blackline Master (*Chapter 4 Resource Book*, p. 66)
- Transparency (p. 32)

GUIDED PRACTICE

Vocabulary Check ✓

1. Draw a ramp and label its rise and run. Explain what is meant by the slope of the ramp. **See margin.**

Concept Check ✓

2. ERROR ANALYSIS Describe the error at the right. Then calculate the correct slope. **See margin.**

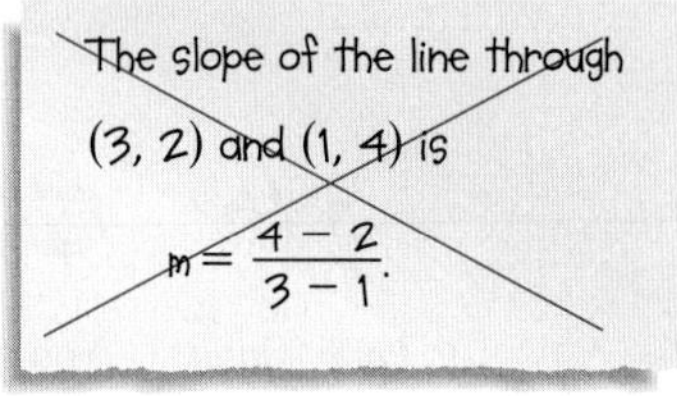

Ex. 2

3. Explain what happens when the formula for slope is applied to a vertical line. **See margin.**

4. CRITICAL THINKING How can you tell that the slope of the line through (2, 2) and (−3, 5) is negative without calculating? **See margin.**

2. The subtractions in the numerator and denominator are not in the same order; −1.

3. The denominator of the fraction is zero; division by zero is not defined.

4. *Sample answer:* Let (x_1, y_1) be (2, 2) and (x_2, y_2) be (−3, 5). Notice that $x_2 < x_1$ and $y_2 > y_1$. Then the numerator and denominator have different signs and the slope is negative.

Skill Check ✓

In Exercises 5–10, plot the points and draw a line through them. Find the slope of the line passing through the points. 5–10. Check graphs.

5. (0, 0), (1, 2) **2** **6.** (0, 0), (−1, −1) **1** **7.** (1, 2), (2, 1) **−1**

8. (3, 2), (1, 4) **−1** **9.** (2, 2), (3, 5) **3** **10.** (−3, −2), (1, 6) **2**

11. Find the value of *y* so that the line passing through the points (0, 3) and (4, *y*) has a slope of −3. **−9**

PRACTICE AND APPLICATIONS

STUDENT HELP
Extra Practice to help you master skills is on p. 800.

DESCRIBING SLOPE Plot the points and draw a line through them. Without calculating, state whether the slope of the line is *positive, negative, zero,* or *undefined.* Explain your reasoning. 12–19. See margin.

12. (6, 9), (4, 3) **13.** (7, 4), (−1, 8) **14.** (5, 10), (5, −4) **15.** (1, 1), (4, −3)

16. (−2, 5), (3, 5) **17.** (0, 0), (−5, 3) **18.** (1, 3), (−2, 1) **19.** (2, −2), (2, −6)

GRAPH AND CALCULATE Plot the points and find the slope of the line passing through the points. 20–34. Check graphs.

20. (4, 5), (2, 3) **1** **21.** (1, 5), (5, 2) $-\frac{3}{4}$ **22.** (2, 3), (−3, 0) $\frac{3}{5}$

23. (0, −6), (8, 0) $\frac{3}{4}$ **24.** (0, 6), (8, 0) $-\frac{3}{4}$ **25.** (2, 4), (4, −4) **−4**

26. (−6, −1), (−6, 4) **undefined** **27.** (0, −10), (−4, 0) $-\frac{5}{2}$ **28.** (1, −2), (−2, 2) $-\frac{4}{3}$

29. (3, 6), (3, 0) **undefined** **30.** (−6, 2), (4, −2) $-\frac{2}{5}$ **31.** (−1, −1), (−3, −6) $\frac{5}{2}$

32. $\left(0, \frac{1}{2}\right)$, (0, 0) **undefined** **33.** (2, 2), (−3, 5) $-\frac{3}{5}$ **34.** (4, 1), (6, 1) **0**

STUDENT HELP
HOMEWORK HELP
Example 1: Exs. 12–37
Example 2: Exs. 12–37
Example 3: Exs. 12–37
Example 4: Exs. 12–37
Example 5: Exs. 38–46
Example 6: Exs. 56, 57

GRAPHICAL REASONING Find the slope of each line.

35.

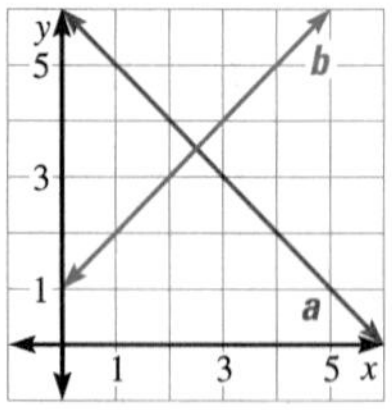

a: −1; *b*: 1

36.

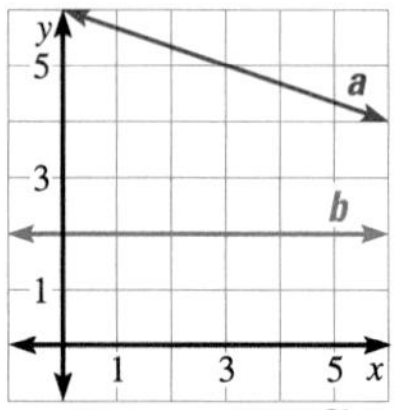

a: $-\frac{1}{3}$; *b*: 0

37.

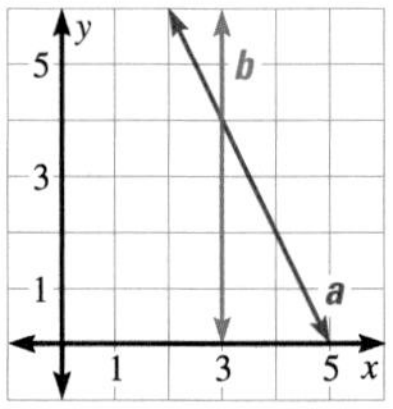

a: −2; *b*: undefined

47. No matter which pairs of points are chosen, the slope is the same, $\frac{1}{2}$.

48. No matter which pairs of points are chosen, the slope is the same, $-\frac{1}{3}$.

49. No; the slope of the line that passes through (−2, 4) and (2, −2) is $-\frac{3}{2}$. The slope of the line that passes through (2, −2) and (6, 0) is $\frac{1}{2}$. The slope of the line that passes through (−2, 4) and (6, 0) is $-\frac{1}{2}$. Three distinct lines, each with a different slope, are determined by the three points; check diagrams.

50. Yes; sample answer: if it made a difference which points you chose, then the line would bend; choose points with integer coordinates.

FOCUS ON CAREERS

SURVEYORS Surveying land is the responsibility of surveyors, survey technicians, and mapping scientists. They measure distances, angles between points, and elevations as in Ex. 51.

CAREER LINK
www.mcdougallittell.com

FINDING A COORDINATE **Find the value of y so that the line passing through the two points has the given slope.**

38. $(2, y), (4, 5), m = 2$ **1**

39. $(0, -2), (2, y), m = 3$ **4**

40. $(-5, 3), (3, y), m = -\frac{1}{2}$ **−1**

41. $(0, y), (2, 5), m = 2$ **1**

42. $(-1, 5), (3, y), m = 5$ **25**

43. $(-1, 3), (5, y), m = -1$ **−3**

44. $(5, 7), (8, y), m = \frac{4}{3}$ **11**

45. $(3, y), (1, 4), m = -\frac{1}{2}$ **3**

46. $(2, -15), (5, y), m = \frac{4}{5}$ **$-\frac{63}{5}$**

INDUCTIVE REASONING **Choose three different pairs of points on the line. Find the slope of the line using each pair. What do you notice?** See margin.

47.

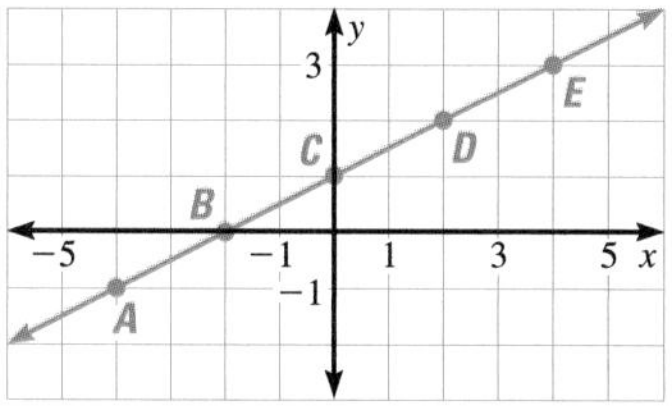

48.

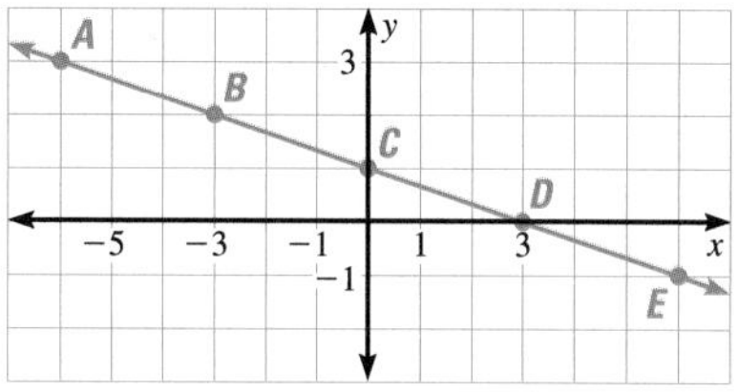

49. GEOMETRY CONNECTION Use the concept of slope to decide whether the points (−2, 4), (2, −2), and (6, 0) are on the same line. Explain your reasoning and include a diagram. **See margin.**

50. **FINDING SLOPE** To find the slope of a line, can you choose any two points on the line? Explain your reasoning. What are some guidelines that you could use to choose convenient points for calculating slope? **See margin.**

51. **ROAD GRADES** The U.S. Department of Transportation requires surveyors to place signs on steep sections of roads. The grade of a road is measured as a percent. For instance, the grade of the road shown below is 6%. What is the slope of the road? Explain the relationship between road grade and slope. **$\frac{3}{50}$; the road grade is the slope of the road expressed as a percent.**

WHAT ARE THE UNITS? **In Exercises 52–54, find the rate of change between the two points. Give the units of measure for the rate.**

52. (2, 2) and (9, 23); x in minutes, y in inches **3 in./min**

53. (3, 5) and (11, 69); x in years, y in dollars. **$8/year**

54. (53, 44) and (32, 14); x in seconds, y in liters. **$\frac{10}{7}$ L/sec**

55. **CABLE CARS** In the 1870s a cable car system was built in San Francisco to climb the steep streets. San Francisco was the site of the first cable car used for public transportation in America. Calculate each labeled slope from left to right in the diagram. **See margin.**

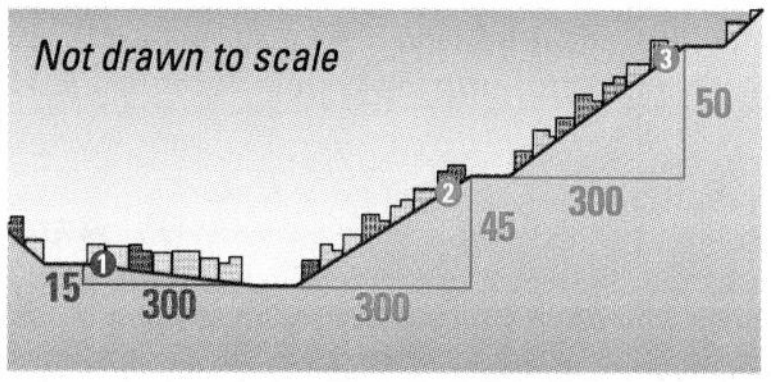

(cross section of side view)

! COMMON ERROR
EXERCISES 5–10 Some students will compute $\frac{\text{run}}{\text{rise}}$ for slope. Remind them that when finding slope, you must "rise to the top."

CAREER NOTE
EXERCISE 51 Additional information about road surveyors is available at **www.mcdougallittell.com**.

1.

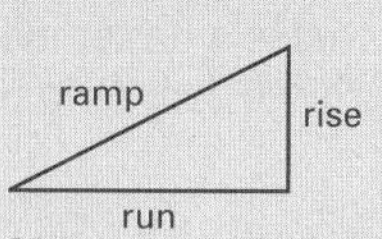

Slope describes the ramp's steepness.

12–19. Check graphs.

12. Positive; the line rises from left to right.

13. Negative; the line falls from left to right.

14. Undefined; the line is vertical.

15. Negative; the line falls from left to right.

16. Zero; the line is horizontal.

17. Negative; the line falls from left to right.

18. Positive; the line rises from left to right.

19. Undefined; the line is vertical.

55. Section 1: $-\frac{1}{20}$; Section 2: $\frac{3}{20}$; Section 3: $\frac{1}{6}$

STUDENT HELP NOTES

Homework Help Students can find help for Exs. 62–64 at **www.mcdougallittell.com.** The information can be printed out for students who don't have access to the Internet.

APPLICATION NOTE

EXERCISES 62–65 Population growth sometimes appears linear over small intervals. However, an exponential model is a better model of population growth over long periods of time.

RESEARCH NOTE

EXERCISE 65 Population data can be found at **www.census.gov** and in the *Information Please Almanac.*

ADDITIONAL PRACTICE AND RETEACHING

For Lesson 4.4:

- Practice Levels A, B, and C (*Chapter 4 Resource Book,* p. 55)
- Reteaching with Practice (*Chapter 4 Resource Book,* p. 58)
- See Lesson 4.4 of the *Personal Student Tutor*

For more Mixed Review:

- Search the *Test and Practice Generator* for key words or specific lessons.

FOCUS ON APPLICATIONS

REAL LIFE **MOVIE PRICES** In 1954 when the Japanese movie *Gojira* was filmed, the average cost of a ticket in the United States was 49¢. In 1998 when the remake *Godzilla* was filmed, the average cost was \$4.69, although some theaters charged more than \$8.00.

56. SPACE SHUTTLE A space shuttle achieves orbit at 9:23 A.M. At 9:31 A.M. it has traveled 2,309.6 miles in orbit. Find the rate of change in miles per minute. **288.7 mi/min**

57. PROFIT In 1990, a company had a profit of \$173,000,000. In 1996, the profit was \$206,000,000. If the profit increased the same amount each year, find the rate of change of the company's profit in dollars per year. **\$5,500,000/year**

MOVIE PRICES **In Exercises 58–61, the graph shows the price of a movie ticket at the Midtown Theater for the years 1960–1995.**

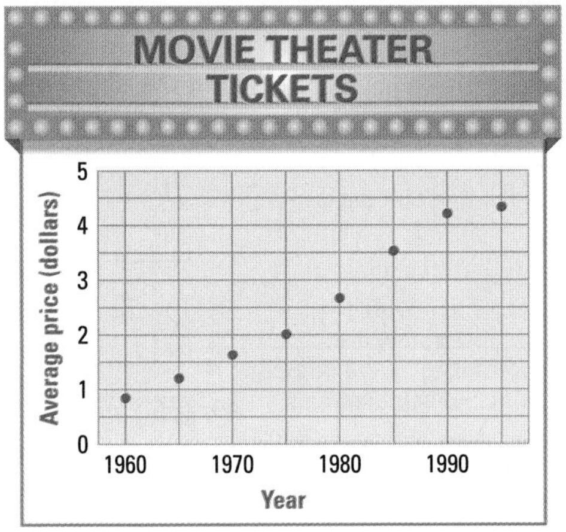

58, 59. Estimates may vary.

58. Estimate the rate of change from 1960 to 1995 in the price in dollars per year. **about \$.10/year**

59. Estimate the rate of change from 1985 to 1990 in the price in dollars per year. **about \$.15/year**

60. Which five-year period had the smallest price increase? **1990 to 1995**

61. *Writing* Use the graph to estimate the cost of going to a movie at the Midtown Theater this year. Why might your estimate be different from the actual cost? **Answers may vary.**

POPULATION RATES **In Exercises 62–65, use the line graph below.**

STUDENT HELP

INTERNET **HOMEWORK HELP** Visit our Web site www.mcdougallittell.com for help with problem solving in Exs. 62–64.

62. During which decade did the population of the South increase the most? Estimate the rate of change for this decade in people per year.

63. During which decade did the population of the Midwest increase the most? Estimate the rate of change for this decade in people per year.

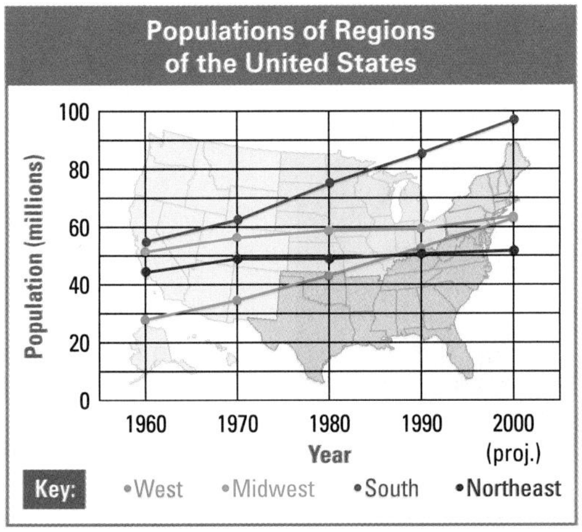

INTERNET **DATA UPDATE** of U.S. Bureau of the Census data at www.mcdougallittell.com

64. INTERPRETING A GRAPH The line graph for the Northeast appears to be horizontal between 1970 and 1980. Write a sentence about what this means in terms of the real-life situation.

65. RESEARCH Find the population of your state in 1980 and in 1990. Find the annual rate of change of this population. **Answers may vary.**

62, 63. Estimates may vary.

62. 1970–1980; about 1.3 million people/year

63. 1960–1970; about 0.5 million people/year

64. The horizontal part of the graph indicates a period in which population growth was zero, or nearly so.

66. **Mathematical Reasoning** When you use the formula $\frac{y_2 - y_1}{x_2 - x_1}$ to find the slope of a line passing through two points, does it matter which point you choose to use as (x_1, y_1)? Give three different examples to support your answer. **See margin.**

Test Preparation

66. No; *Sample answer:* Given (2, 4) and (3, 5), $\frac{5-4}{3-2} = \frac{4-5}{2-3} = 1$; given (1, 4) and (6, 3), $\frac{3-4}{6-1} = \frac{4-3}{1-6} = -\frac{1}{5}$; given (−2, −4) and (−3, 6), $\frac{6-(-4)}{-3-(-2)} = \frac{-4-6}{-2-(-3)} = -10$.

67. **Multiple Choice** The slope of the line passing through the points (0, 5) and (0, −3) is __?__. **D**

Ⓐ positive Ⓑ negative Ⓒ zero Ⓓ undefined

68. **Multiple Choice** The slope of the line passing through the points (8, 0) and (0, 8) is __?__. **B**

Ⓐ positive Ⓑ negative Ⓒ zero Ⓓ undefined

69. **Multiple Choice** What is the slope of the line passing through the points (−3, 4) and (5, −11)? **C**

Ⓐ $\frac{-7}{8}$ Ⓑ $\frac{-7}{2}$ Ⓒ $\frac{15}{-8}$ Ⓓ $\frac{15}{8}$

★ Challenge

70. **Finding an x-Coordinate** Write an expression for the slope of a line passing through the points (0, 3) and (x, −9). What value of x will make the fraction equivalent to −2? $-\frac{12}{x}$; 6

71. Use your answers to Exercise 70 to find the value of x so that the line passing through the points (0, 3) and (x, −9) will have a slope of −3. **4**

Extra Challenge
www.mcdougallittell.com

Mixed Review

Evaluating Expressions **Evaluate the expression for the given value of the variable.** (Review 1.3)

72. $12x + 3$ when $x = 5$ **63**
73. $\frac{3}{4}p$ when $p = 16$ **12**
74. $8n$ when $n = \frac{3}{16}$ **$1\frac{1}{2}$**
75. $\frac{7}{2}y - 3$ when $y = 4$ **11**
76. $5x + x$ when $x = 7$ **42**
77. $\frac{6}{5}z + 2$ when $z = 5$ **8**

Simplifying Expressions **Simplify the variable expression.** (Review 2.5)

78. $(-6)(y)$ **$-6y$**
79. $(-1)(x)(-x)$ **x^2**
80. $-(-3)^2(-y)$ **$9y$**
81. $\left(\frac{2}{5}\right)(5)(-x)(x)$ **$-2x^2$**
82. $(y)(-23)(-y^2)$ **$23y^3$**
83. $\left(\frac{1}{8}\right)(-4)(-x)(-x)$ **$-\frac{1}{2}x^2$**

Writing in Function Form **Solve for y.** (Review 3.7 for 4.5) **84–89. See margin.**

84. $9x - 2y = 14$
85. $5x + 9y = 18$
86. $-2x - 2y = 7$
87. $-x + 4y = 36$
88. $6x - 3y = 21$
89. $3x + 5y = 17$

84. $y = \frac{9}{2}x - 7$
85. $y = -\frac{5}{9}x + 2$
86. $y = -x - \frac{7}{2}$
87. $y = \frac{1}{4}x + 9$
88. $y = 2x - 7$
89. $y = -\frac{3}{5}x + \frac{17}{5}$

Finding Intercepts **Find the x-intercept and the y-intercept of the graph of the equation.** (Review 4.3)

90. $y = x + 8$ **−8; 8**
91. $y = 2x + 12$ **−6; 12**
92. $y = 3x + 9$ **−3; 9**
93. $y = -6 + 2x$ **3; −6**
94. $y = -2x + 16$ **8; 16**
95. $3y = 6x + 3$ **$-\frac{1}{2}$; 1**

4 Assess

Daily Homework Quiz

Transparency Available

1. Plot the points and draw the line through them. Calculate the slope of the line.
   a. (−2, 2), (0, 4) **1**
   b. (1, 1), (4, 2) **$\frac{1}{3}$**

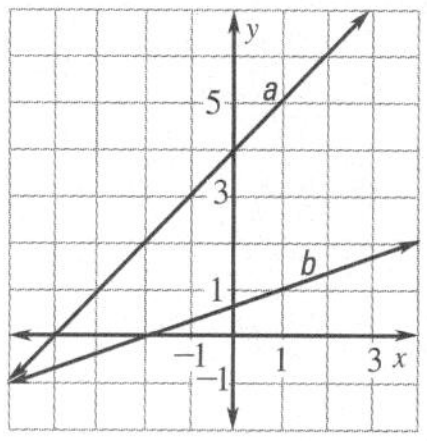

2. The points (−4, −2), (0, 0), and (4, y) are on the same line. What is the value of y? **2**
3. Find the rate of change between the given points. Give the units of measure for the rate. (4, 24) and (20, 120), x in hours, y in dollars **$6/hour**

Extra Challenge Note
Challenge problems for Lesson 4.4 are available in **blackline** format in the *Chapter 4 Resource Book*, p. 63 and at **www.mcdougallittell.com.**

Additional Test Preparation

1. **Writing** Explain how the slope of a line can be 0.
   When a line is horizontal, the rise is always 0 no matter what the run is, and 0 divided by any number except 0 is 0.
2. **Open Ended** Describe a real-life situation where the rate of change is 0.
   the attendance per performance for an opera company during an 80-performance season if the theater is filled to capacity for each performance

1 PLAN

PACING
Basic: 1 day
Average: 1 day
Advanced: 1 day
Block Schedule: 0.5 block with 4.6

LESSON OPENER ACTIVITY
An alternative way to approach Lesson 4.5 is to use the Activity Lesson Opener:

- Blackline Master (*Chapter 4 Resource Book* p. 66)
- Transparency (p. 28)

MEETING INDIVIDUAL NEEDS

- ***Chapter 4 Resource Book***
  Prerequisite Skills Review (p. 5)
  Practice Level A (p. 68)
  Practice Level B (p. 69)
  Practice Level C (p. 70)
  Reteaching with Practice (p. 71)
  Absent Student Catch-Up (p. 73)
  Challenge (p. 75)
- ***Resources in Spanish***
- ***Personal Student Tutor***

NEW-TEACHER SUPPORT
See the Tips for New Teachers on pp. 1–2 of the *Chapter 4 Resource Book* for additional notes about Lesson 4.5.

WARM-UP EXERCISES

Transparency Available

Identify the slope and y-intercept of the graph of each equation.

1. $y = 2x$
2; 0

2. $-9x = 3y$
−3; 0

3. You are driving to a national park for a family vacation. If you travel at 60 mi/h, write an equation that describes the relationship between the number of miles d you drive and the number of hours t you spend driving. **$d = 60t$**

4.5 Direct Variation

What you should learn

GOAL 1 Write linear equations that represent direct variation.

GOAL 2 Use a ratio to write an equation for direct variation, such as the ratio of tail length to body length in alligators in **Example 4**.

Why you should learn it

▼ To solve **real-life** problems such as lengths of several stringed instruments in **Exs. 36 and 37.**

REAL LIFE

GOAL 1 RECOGNIZING AND USING DIRECT VARIATION

Two variables x and y *vary directly* if there is a nonzero number k such that the following is true.

$$y = kx \quad \longleftarrow \text{ Model for direct variation}$$

The number k is the **constant of variation.** Two quantities that vary directly are said to have **direct variation.**

EXAMPLE 1 *Graphs of Direct Variation Models*

Find the constant of variation and the slope of each direct variation model.

a. $y = 2x$

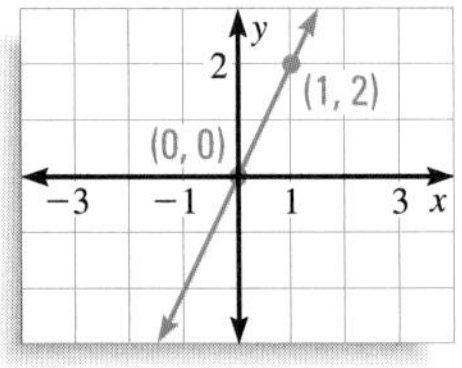

b. $y = -\frac{1}{2}x$

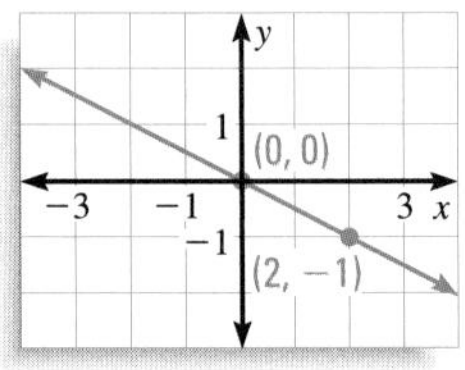

SOLUTION

a. For the equation $y = 2x$, the constant of variation is $k = 2$.

To find the slope of the line, use the slope formula.

$$m = \frac{y_2 - y_1}{x_2 - x_1}$$

$$m = \frac{2 - 0}{1 - 0} = 2$$

b. For the equation $y = -\frac{1}{2}x$, the constant of variation is $k = -\frac{1}{2}$.

To find the slope of the line, use the slope formula.

$$m = \frac{y_2 - y_1}{x_2 - x_1}$$

$$m = \frac{-1 - 0}{2 - 0} = -\frac{1}{2}$$

PROPERTIES OF GRAPHS OF DIRECT VARIATION MODELS

- The graph of $y = kx$ is a line through the origin.
- The slope of the graph of $y = kx$ is k.

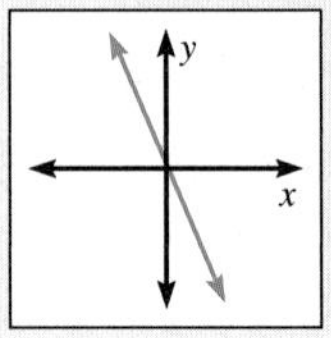

$k < 0$

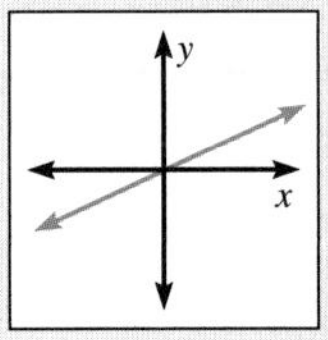

$k > 0$

EXAMPLE 2 *Writing a Direct Variation Equation*

The variables x and y vary directly. When $x = 5$, $y = 20$.

a. Write an equation that relates x and y.

b. Find the value of y when $x = 10$.

SOLUTION

a. Because x and y vary directly, the equation is of the form $y = kx$. You can solve for k as follows.

$y = kx$	Write model for direct variation.
$20 = k(5)$	Substitute 5 for x and 20 for y.
$4 = k$	Divide each side by 5.

▶ An equation that relates x and y is $y = 4x$.

b. $y = 4(10)$ Substitute 10 for x in $y = 4x$.

$y = 40$

▶ When $x = 10$, $y = 40$.

EXAMPLE 3 *Writing a Direct Variation Model*

ZEPPELIN In 1852 Henri Giffard built the first airship successfully used for transportation. It had a volume of 88,000 cubic feet and could support 5650 pounds. The *Graf Zeppelin II*, built in 1937, had a volume of 7,063,000 cubic feet, making it one of the two largest airships ever built. The weight an airship can support varies directly with its volume. How much weight could the *Graf Zeppelin II* support?

SOLUTION

Use the data about the first airship. Find a model that relates the volume V (in cubic feet) to the weight w (in pounds) that an airship can support.

$w = kV$	Write model for direct variation.
$5650 = k(88{,}000)$	Substitute 5650 for w and 88,000 for V.
$\frac{5650}{88{,}000} = k$	Divide each side by 88,000.
$0.064 \approx k$	Solve for k.

A direct variation model for the weight an airship can support is $w = 0.064V$.

Use the model to find the weight the *Graf Zeppelin II* could support.

$w = 0.064V$	Write the model for direct variation.
$w = 0.064(7{,}063{,}000)$	Substitute 7,063,000 for V.
$w = 452{,}032$	Simplify.

▶ The *Graf Zeppelin II* could support about 452,000 pounds.

FOCUS ON APPLICATIONS

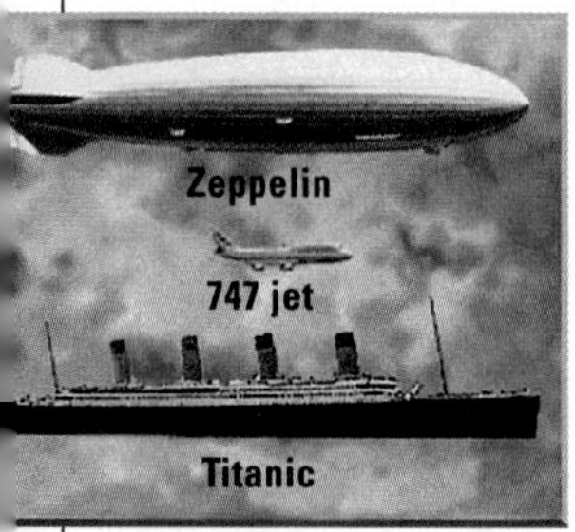

REAL LIFE **ZEPPELIN SIZE** The enormous *Graf Zeppelin II* was 804 ft long. The *Titanic* was similar in length (882.5 ft) but traveled at $\frac{1}{3}$ of the speed.

INTERNET APPLICATION LINK www.mcdougallittell.com

2 TEACH

MOTIVATING THE LESSON
Many situations in everyday life involve a relationship between two measurable quantities. For example, your pay from a part-time job increases as you work more hours. In this lesson we will study a particular kind of relationship between two quantities.

EXTRA EXAMPLE 1
Find the constant of variation and the slope for the direct variation model $y = -3x$. **−3; −3**

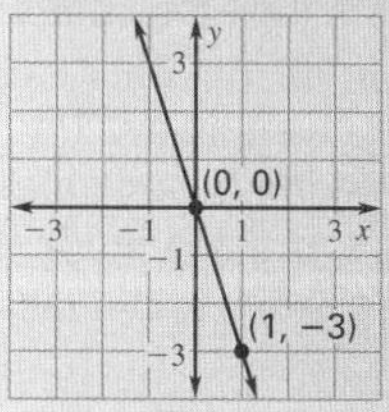

EXTRA EXAMPLE 2
The variables x and y vary directly. When $x = -3$, $y = -30$.

a. Write an equation that relates x and y. **$y = 10x$**

b. Find the value of y when $x = 8$. **80**

EXTRA EXAMPLE 3
The income at a store varies directly as the advertising budget. When the owners spent \$3000 per month to advertise, their monthly gross income was \$120,000. If the owners increase their advertising budget to \$5000 per month, how much gross monthly income can they expect? **\$200,000**

CHECKPOINT EXERCISES

For use after Examples 1–3:

1. The length that a spring will stretch S varies directly with the weight w attached to the spring. If a spring stretches 1.4 inches when a 20 pound weight is attached, how far will it stretch when a 10 pound weight is attached? **0.7 inches**

EXTRA EXAMPLE 4

The forearm lengths and body heights in inches of 5 people are shown below.

a. Write a direct variation model that relates body length *B* to forearm length *F*. **$B = 7.5F$**

b. Estimate the body length of a person whose forearm length is 10 inches. **75 in.**

F	7.0	7.5	8.0	8.5	9.0
B	52.5	55.5	58.4	63.5	68.4

CHECKPOINT EXERCISES

For use after Example 4:

1. The total scores of 4 students on the tests they took before the final exam and their exam score are given below. Use direct variation to estimate the exam score of a student whose pre-exam total was 353. **about 81**

pre-exam	final exam
304	70
325	70
332	73
380	95

FOCUS ON VOCABULARY

Since *y* is dependent on *x* in the direct variation equation $y = kx$, we say *y varies directly as x* or *y is directly proportional to x.*

CLOSURE QUESTION

How can you tell if a graph represents a direct variation equation? **See sample answer, below.**

DAILY PUZZLER

Let $a = 10$ and $b = 5$. Write the number of the month in which you were born. Multiply by *a*. Add *b*. Multiply by *a*. Add the day of the month in which you were born. Multiply by *a*. Add *b*. Multiply by *a*. Add the last two digits of the year in which you were born. Subtract 5050. What does the result represent?
the month, day, and year of your birth

GOAL 2 USING A RATIO TO MODEL DIRECT VARIATION

The model $y = kx$ for direct variation can be rewritten as follows.

$$k = \frac{y}{x}$$ ← **Ratio form of direct variation model**

The ratio form tells you that if *x* and *y* have *direct variation*, then the ratio of *y* to *x* is the same for all values of *x* and *y*. Sometimes real-life data can be approximated by a direct variation model, even though the data may not fit the model exactly.

EXAMPLE 4 *Using a Ratio Based on Data to Write a Model*

ANIMAL STUDIES The tail and body lengths (in feet) of 8 alligators are shown below. The ages range from 2 years to over 50 years. ▶ Source: St. Augustine Alligator Farm

Body length, *B* — Tail length, *T*

Tail, *T*	1.41	2.04	2.77	2.77	3.99	4.67	4.69	5.68
Body, *B*	1.50	2.41	3.08	3.23	4.28	5.04	5.02	6.38

a. Write a model that relates the tail length *T* to the body length *B*.

b. Estimate the body length of an alligator whose tail length is 4.5 feet.

SOLUTION

a. Begin by finding the ratio of tail length to body length for each alligator.

Tail, *T*	1.41	2.04	2.77	2.77	3.99	4.67	4.69	5.68
Body, *B*	1.50	2.41	3.08	3.23	4.28	5.04	5.02	6.38
Ratio	0.94	0.85	0.90	0.86	0.93	0.93	0.93	0.89

The ratio is about 0.9 for each alligator, so use a direct variation model.

$k = \frac{T}{B}$ **Write ratio model for direct variation.**

$0.9 = \frac{T}{B}$ **Substitute 0.9 for *k*.**

$0.9B = T$ **Multiply each side by *B*.**

b. Use the model $0.9B = T$ to estimate the body length of the alligator.

$0.9B = 4.5$ **Substitute 4.5 for *T*.**

$B = 5$ **Divide each side by 0.9.**

▶ You estimate that the alligator's body is about 5 feet long.

FOCUS ON CAREERS

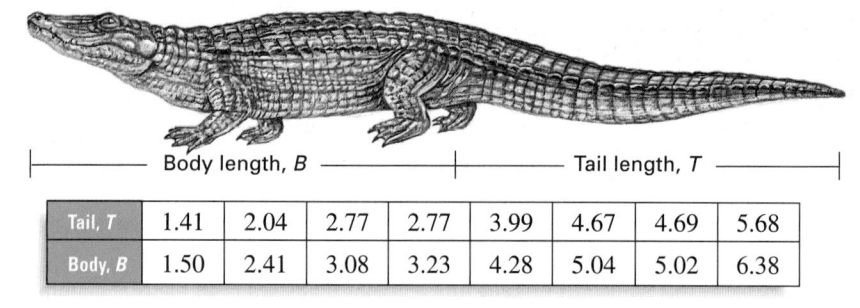

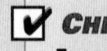

REAL LIFE **ZOOLOGISTS** are biologists who study animals in natural or controlled surroundings.

INTERNET CAREER LINK www.mcdougallittell.com

Closure Question *Sample answer:*
If the graph is a nonvertical straight line that goes through the origin, then it represents a direct variation equation.

GUIDED PRACTICE

Vocabulary Check ✓

1. Explain what it means for x and y to vary directly.
It means that there is a nonzero number k such that $y = kx$.

Concept Check ✓

2. In a direct variation equation, how are the constant of variation and the slope related? **The slope is the constant of variation.**

Skill Check ✓

Graph the equation. State whether the two quantities have direct variation. If they have direct variation, find the constant of variation and the slope of the direct variation model. **3–8. Check graphs.**

3. $y = x$ **direct variation. 1; 1**

4. $y = 4x$ **direct variation. 4; 4**

5. $y = \frac{1}{2}x$ **direct variation. $\frac{1}{2}$; $\frac{1}{2}$**

6. $y = 2x$ **direct variation. 2; 2**

7. $y = x - 4$ **not direct variation.**

8. $y - 0.1x = 0$ **direct variation. 0.1; 0.1**

SALARY In Exercises 9–11, you work a different number of hours each day. The table shows your total pay p and the number of hours h you worked.

Total pay, p	\$18	\$42	\$48	\$30
Hours worked, h	3	7	8	5
Ratio	?	?	?	?

9. Copy and complete the table by finding the ratio of your total pay each day to the number of hours you worked that day. **6; 6; 6; 6**

10. Write a model that relates the variables p and h. **$p = 6h$**

11. If you work 6 hours on the fifth day, what will your total pay be? **\$36**

PRACTICE AND APPLICATIONS

STUDENT HELP

Extra Practice to help you master skills is on p. 800.

DIRECT VARIATION MODEL Find the constant of variation and the slope.

12. $y = 3x$ **3; 3**

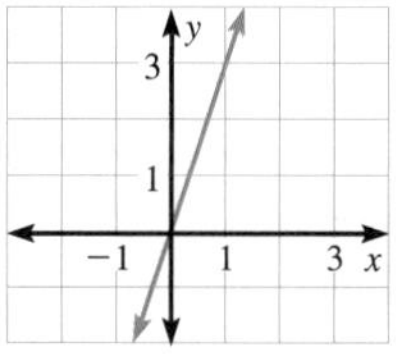

13. $y = -\frac{2}{5}x$ **$-\frac{2}{5}$; $-\frac{2}{5}$**

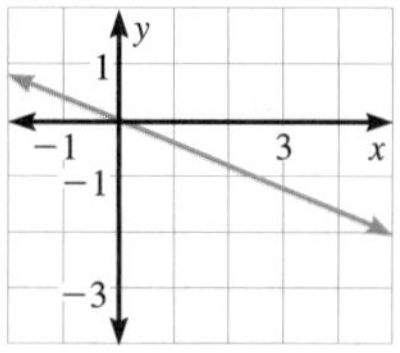

14. $y = 0.75x$ **0.75; 0.75**

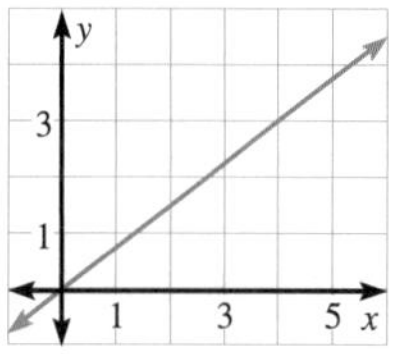

CONSTANT OF VARIATION Graph the equation. Find the constant of variation and the slope of the direct variation model. **15–20. Check graphs.**

15. $y = -3x$ **−3; −3**

16. $y = -5x$ **−5; −5**

17. $y = 0.4x$ **0.4; 0.4**

18. $y = -x$ **−1; −1**

19. $y = \frac{5}{4}x$ **$\frac{5}{4}$; $\frac{5}{4}$**

20. $y = -\frac{1}{5}x$ **$-\frac{1}{5}$; $-\frac{1}{5}$**

STUDENT HELP

HOMEWORK HELP
Example 1: Exs. 12–20
Example 2: Exs. 23–31
Example 3: Exs. 21–22, 32
Example 4: Exs. 36–37

RECOGNIZING DIRECT VARIATION In Exercises 21 and 22, state whether the two quantities have direct variation.

21. **BICYCLING** You ride your bike at an average speed of 14 miles per hour. The number of miles m you ride during h hours is modeled by $m = 14h$. **yes**

22. **GEOMETRY CONNECTION** The circumference C of a circle and its diameter d are related by the equation $C = \pi d$. **yes**

3 APPLY

ASSIGNMENT GUIDE

BASIC
Day 1: pp. 237–239 Exs. 12–20 even, 21–31, 33, 38–41, 44–62 even

AVERAGE
Day 1: pp. 237–239 Exs. 12–20 even, 21–31, 33, 38–41, 44–62 even

ADVANCED
Day 1: pp. 237–239 Exs. 12–20 even, 21–31, 33, 36–43, 44–62 even

BLOCK SCHEDULE WITH 4.6
pp. 237–239 Exs. 12–20 even, 21–31, 33, 38–41, 44–62 even

EXERCISE LEVELS

Level A: ***Easier***
12–14

Level B: ***More Difficult***
15–31, 33, 36–41, 44–63

Level C: ***Most Difficult***
32, 34, 35, 42, 43

✓ HOMEWORK CHECK

To quickly check student understanding of key concepts, go over the following exercises: Exs. 12, 16, 21, 24, 32. See also the Daily Homework Quiz:

- Blackline Master (*Chapter 4 Resource Book*, p. 78)
- Transparency (p. 33)

MATHEMATICAL REASONING

Marlene's salary varies directly with the number of hours she works. If Marlene's salary is graphed against the number of hours she works, what does the slope of the line represent?
her hourly wage

APPLICATION NOTE

EXERCISE 33 Since $k = \frac{\text{weight on Earth}}{\text{weight on moon}}$ is the ratio that describes the weight relationship for both astronauts, students can also solve the equation $\frac{360}{60} = \frac{x}{54}$ to find Tereshkova's total weight on Earth.

STUDENT HELP NOTES

Homework Help Students can find help for Exs. 34–35 at **www.mcdougallittell.com.** The information can be printed out for students who don't have access to the Internet.

APPLICATION NOTE

EXERCISES 36–37 A direct variation function has many uses as a mathematical model. In these exercises, the direct variation function only approximates the total body length of the violins. This is evident because a constant of proportionality can only be approximated. By comparison, the direct variation function in Exercises 34–35 represents the exact relationship between the two variables.

34. See Additional Answers beginning on page AA1.

ADDITIONAL PRACTICE AND RETEACHING

For Lesson 4.5:

- Practice Levels A, B, and C (*Chapter 4 Resource Book*, p. 68)
- Reteaching with Practice (*Chapter 4 Resource Book*, p. 71)
- See Lesson 4.5 of the *Personal Student Tutor*

For more Mixed Review:

- Search the *Test and Practice Generator* for key words or specific lessons.

FOCUS ON PEOPLE

VALENTINA V. TERESHKOVA
In 1963 the Soviet astronaut Valentina Tereshkova was the first woman to travel in space.

FINDING EQUATIONS **In Exercises 23–31, the variables *x* and *y* vary directly. Use the given values to write an equation that relates *x* and *y*.**

23. $x = 4, y = 12$ **$y = 3x$** **24.** $x = 7, y = 35$ **$y = 5x$** **25.** $x = 18, y = 4$ **$y = \frac{2}{9}x$**

26. $x = 22, y = 11$ **$y = \frac{1}{2}x$** **27.** $x = 5.5, y = 1.1$ **$y = 0.2x$** **28.** $x = 16.5, y = 3.3$ **$y = 0.2x$**

29. $x = -1, y = -1$ **$y = x$** **30.** $x = 7\frac{1}{5}, y = -9$ **$y = -\frac{5}{4}x$** **31.** $x = -9, y = 3$ **$y = -\frac{1}{3}x$**

32. SCIENCE CONNECTION The volume V of blood pumped from your heart each minute varies directly with your pulse rate p. Each time your heart beats, it pumps approximately 0.06 liter of blood.

a. Find an equation that relates V and p. **$V = 0.06p$**

b. COLLECTING DATA Take your pulse and find out how much blood your heart pumps per minute. **Answers may vary.**

33. ASTRONAUTS Weight varies directly with gravity. With his equipment Buzz Aldrin weighed 360 pounds on Earth but only 60 pounds on the moon. If Valentina V. Tereshkova had landed on the moon with her equipment and weighed 54 pounds, how much would she have weighed on Earth with equipment? **324 lb**

SWIMMING POOLS **In Exercises 34 and 35, use the following information about several sizes of pools. The amount of chlorine needed in a pool varies with the amount of water in the pool.**

Amount of water (gallons)	6250	5000	12,500	15,000
Chlorine (pounds)	0.75	0.60	1.50	1.80

STUDENT HELP

HOMEWORK HELP Visit our Web site www.mcdougallittell.com for help with scatter plots in Exs. 34–35.

34. Use a graphing calculator or a computer to make a scatter plot of the data. Sketch the graph. **See margin.**

35. Does the scatter plot you made in Exercise 34 show direct variation? **yes**

VIOLIN FAMILY **In Exercises 36 and 37, use the following information.**
The violin family includes the bass, the cello, the viola, and the violin. The size of each instrument determines its tone. The shortest produces the highest tone, while the longest produces the deepest (lowest) tone.

Violin family	Bass	Cello	Viola	Violin
Total length t (inches)	72	47	26	23
Body length b (inches)	44	30	16	14

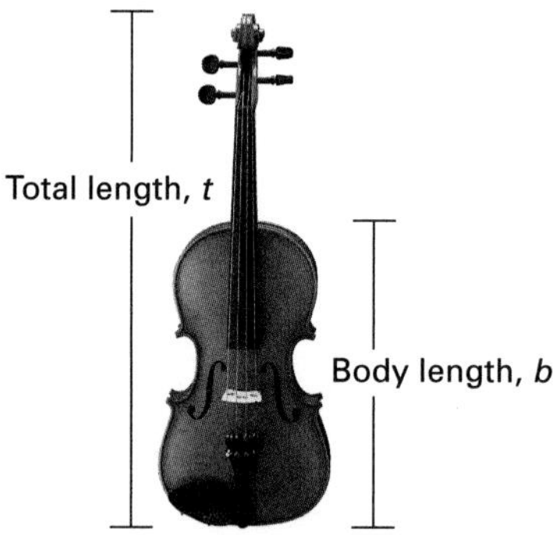

36. Write a direct variation model that you can use to relate the body length of a member of the violin family to its total length. **$b = 0.62t$**

37. Another instrument in the violin family has a body length of 28 inches. Use your model from Exercise 36 to estimate the total length of the instrument. Is its tone higher or lower than a viola's tone? **45 in.; lower**

Test Preparation

38. **MULTIPLE CHOICE** The variables x and y vary directly. When $x = 4$, $y = 24$. Which equation correctly relates x and y? **D**

Ⓐ $x = 4y$ Ⓑ $y = 4x$ Ⓒ $x = 6y$ Ⓓ $y = 6x$

39. **MULTIPLE CHOICE** Which equation models the ratio form of direct variation? **A**

Ⓐ $\frac{-4}{3} = \frac{y}{x}$ Ⓑ $-3y = 4x - 1$ Ⓒ $\frac{4}{-3} = x + y$ Ⓓ $y = -\frac{4}{3}x + 6$

40. **MULTIPLE CHOICE** Find the constant of variation of the direct variation model $3x = y$. **A**

Ⓐ 3 Ⓑ $\frac{1}{3}$ Ⓒ y Ⓓ x

41. **MULTIPLE CHOICE** You buy some oranges for \$.80 per pound. Which direct variation model relates the total cost x to the number of pounds of oranges y? **B**

Ⓐ $y = 8x$ Ⓑ $x = 0.8y$ Ⓒ $y = 0.8x$ Ⓓ $x = 8y$

★ **Challenge**

42. **GEOMETRY CONNECTION** Write an equation for the perimeter of the regular hexagon at the right. Does your equation model direct variation? **$P = 6x$; yes**

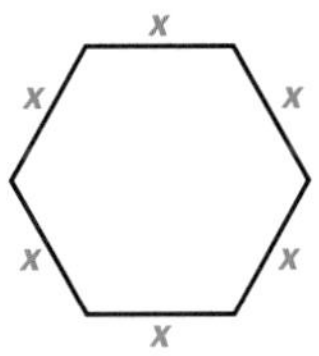

EXTRA CHALLENGE
www.mcdougallittell.com

43. **LOGICAL REASONING** If a varies directly with b, then does b vary directly with a? Explain your reasoning. **Yes; if a varies directly with b, then $a = kb$ for some nonzero k, and $b = \frac{1}{k}a$, so b varies directly with a.**

MIXED REVIEW

SOLVING EQUATIONS **Solve the equation.** (Review 3.4)

44. $7z + 30 = -5$ **−5** 45. $4b = 26 - 9b$ **2** 46. $2(w - 2) = 2$ **3**

47. $9x + 65 = -4x$ **−5** 48. $55 - 5y = 9y + 27$ **2** 49. $7c - 3 = 4(c - 3)$ **−3**

FUNCTIONS **In Exercises 50 and 51, rewrite the equation so that x is a function of y.** (Review 3.7)

50. $15 = 7(x - y) + 3x$ **$x = 0.7y + 1.5$** 51. $3x + 12 = 5(x + y)$ **$x = -\frac{5}{2}y + 6$**

52. **HOURLY WAGE** You get paid \$152.25 for working 21 hours. Find your hourly rate of pay. (Review 3.8) **\$7.25/h**

53. **ORDERED PAIRS** Plot and label the points $R(2, 4)$, $S(0, -1)$, $T(3, 6)$, and $U(-1, -2)$ in a coordinate plane. (Review 4.1) **See margin.**

CHECKING SOLUTIONS **Decide whether the given point lies on the line. Justify your answer both algebraically and graphically.** (Review 4.2) **54–57. See margin for graphs.**

54. $x - y = 10$; $(5, -5)$ **Yes; $5 - (-5) = 5 + 5 = 10$**

55. $3x - 6y = -2$; $(-4, -2)$ **No; $3(-4) - 6(-2) = 0$ and $0 \neq -2$**

56. $5x + 6y = -1$; $(1, -1)$ **Yes; $5(1) + 6(-1) = -1$**

57. $-4x - 3y = -8$; $(-4, 2)$ **No; $-4(-4) - 3(2) = 10$ and $10 \neq -8$**

FINDING INTERCEPTS **Find the x-intercept and the y-intercept of the graph of the equation.** (Review 4.3 for 4.6)

58. $2x + y = 6$ **3; 6** 59. $x - 1.1y = 10$ **10; $-9\frac{1}{11}$** 60. $x - 6y = 4$ **4; $-\frac{2}{3}$**

61. $x + 5y = 10$ **10; 2** 62. $y = x - 5$ **5; −5** 63. $y = \frac{1}{3}x - \frac{7}{3}$ **7; $-\frac{7}{3}$**

4 ASSESS

DAILY HOMEWORK QUIZ

Transparency Available

1. Find the constant of variation and the slope of the direct variation model.
   a. $y = \frac{1}{4}x$ **$\frac{1}{4}$; $\frac{1}{4}$**
   b. $y = -2.5x$ **−2.5; −2.5**
2. The perimeter of a square with side length s is modeled by $p = 4s$. Do the side length and perimeter have direct variation? **yes**
3. The variables x and y vary directly. Use the given values to write an equation that relates x and y.
   a. $x = 5, y = 18$ **$y = \frac{18}{5}x$**
   b. $x = -4, y = 6.8$ **$y = -1.7x$**
4. Sound travels about 12.4 miles in one minute. How long does it take sound to travel 5 miles? **0.4 min, or 24 sec**

EXTRA CHALLENGE NOTE

Challenge problems for Lesson 4.5 are available in **blackline** format in the *Chapter 4 Resource Book*, p. 75 and at **www.mcdougallittell.com.**

ADDITIONAL TEST PREPARATION

1. **WRITING** Explain how the direct variation function $y = kx$ is a special case of the linear function $y = mx + b$. **$y = kx$ is a linear function with slope equal to k and y-intercept 0.**
2. **OPEN ENDED** Describe a real-life situation that can be modeled by a direct variation function. **For a fixed interest rate and period of time, the amount of simple interest varies directly with the principal amount.**

53–57. See Additional Answers beginning on page AA1.

1 Planning the Activity

PURPOSE
To explore the graphs of lines that have the same y-intercept and the graphs of parallel lines.

MATERIALS
- desk per group
- textbooks per group
- graph paper per group
- metric ruler per group
- Activity Support Master (*Chapter 4 Resource Book,* p. 79)

PACING
- Exploring the Concept — 15 min
- Drawing Conclusions — 15 min

LINK TO LESSON
In Example 1 of Lesson 4.6, have students associate mx with the total height of the books. They can also associate b with desk height.

2 Managing the Activity

CLASSROOM MANAGEMENT
Make sure that students graph the equations in the same plane.

ALTERNATIVE APPROACH
Have students work in groups of 4. Ask each group to use a different book to model the height. Then graph each group's results on the same coordinate plane.

3 Closing the Activity

KEY DISCOVERY
The graphs of equations with the same y-intercept intersect at the point associated with the intercept. Equations with the same value of m are parallel when graphed.

ACTIVITY ASSESSMENT
JOURNAL What is true of the graphs of the equations $y = 2x + 1$ and $y = 2x - 1$? **They are parallel lines.**

ACTIVITY 4.6
Developing Concepts

Graphing Families of Linear Equations

GROUP ACTIVITY
Work in a small group.

MATERIALS
- desk
- textbooks
- graph paper
- meter stick or metric ruler

QUESTION What are some relationships that exist between members of a family of equations?

EXPLORING THE CONCEPT

You can use a linear equation $y = mx + b$ to model the height from the floor to the top of a stack of books that are m centimeters thick sitting on a desk b centimeters high.

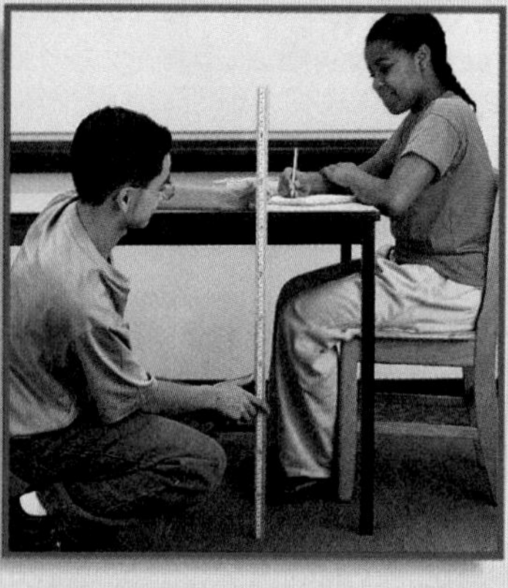

1. Measure the thickness of your algebra textbook. Measure the height of the top of your desk to the floor.
2. Write a model for the height y from the floor to the top of a stack of x algebra books the same size as yours sitting on your desk.
3. Graph and label your model from **Step 2**.
4. Measure the thickness of your English textbook. Write a model for the height y from the floor to the top of a stack of x English books the same size as yours sitting on your desk. Graph this model in the coordinate plane you drew in **Step 3**.
5. Repeat **Step 4** using another book.

DRAWING CONCLUSIONS

1. Equations that have characteristics in common can be thought of as a *family of equations*. List all of the characteristics that the equations have in common. List all of the characteristics that their graphs have in common.
2. Suppose in **Step 2** you used the same book but on a desk or a table of a different height. Write models for the height of a stack of algebra books on a desk 74 cm tall and on a computer table 68 cm tall. Graph these models in the same coordinate plane. What characteristics do the equations share? What characteristics do their graphs share?
3. What characteristics are shared by the family of equations in which $m = 4$? **All the graphs have slope 4.**
4. What is true about the family of linear equations with graphs passing through the point (0, 5)? **When written in the form $y = mx + b$, the equations all have 5 as their constant term b.**

1. Both are in the form $y = mx + b$, where b is the height in centimeters of the desk; their graphs are lines that pass through the point $(0, b)$.

2. Let h be the height of the book; $y = hx + 74$ and $y = hx + 68$; in both equations, the coefficient of x is the same; the graphs have the same slope.

4.6 Quick Graphs Using Slope-Intercept Form

What you should learn

GOAL 1 Graph a linear equation in slope-intercept form.

GOAL 2 Graph and interpret equations in slope-intercept form that model **real-life** situations, such as the snowstorm in **Exs. 56–59**.

Why you should learn it

▼ To model **real-life** situations, such as a flooding river in **Example 4**.

REAL LIFE

Mississippi and Ohio Rivers flooding

GOAL 1 GRAPHING USING SLOPE-INTERCEPT FORM

In Lesson 4.4 you learned to find the slope of a line given two points on the line. There is also a method for finding the slope given an equation of a line.

ACTIVITY
Developing Concepts

Investigating Slope-Intercept Form

Copy and complete the table. Then write a generalization about the meaning of m and b in $y = mx + b$. 1. (0, 1), $\left(-\frac{1}{2}, 0\right)$; 2; 1   2. (0, −3), $\left(-\frac{3}{2}, 0\right)$; −2, −3

	Equation	Two solutions	Slope	y-Intercept
1.	$y = 2x + 1$	(0, ?), (?, 0)	?	?
2.	$y = -2x - 3$	(0, ?), (?, 0)	?	?
3.	$y = x + 4$	(0, ?), (?, 0)	?	?
4.	$y = 0.5x - 2.5$	(0, ?), (?, 0)	?	?

3. (0, 4), (−4, 0); 1; 4   4. (0, −2.5), (5, 0); 0.5; −2.5

In this activity, you may have discovered the following idea, which applies only when an equation is in slope-intercept form.

SLOPE-INTERCEPT FORM OF THE EQUATION OF A LINE

The linear equation $y = mx + b$ is written in **slope-intercept form**. The slope of the line is m. The y-intercept is b.

$y = 2x + 3$
Slope is 2.
y-intercept is 3.

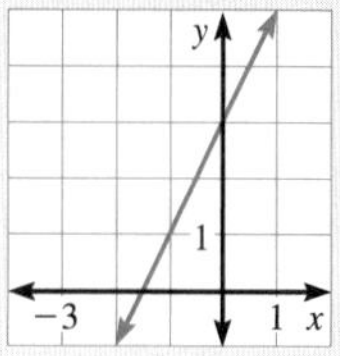

EXAMPLE 1 *Writing Equations in Slope-Intercept Form*

	EQUATION	SLOPE-INTERCEPT FORM	SLOPE	y-INTERCEPT
a.	$y = -x + 2$	$y = (-1)x + 2$	$m = -1$	$b = 2$
b.	$y = \frac{x + 3}{2}$	$y = \frac{1}{2}x + \frac{3}{2}$	$m = \frac{1}{2}$	$b = \frac{3}{2}$
c.	$y = -4$	$y = 0x - 4$	$m = 0$	$b = -4$
d.	$2x - 4y = 16$	$y = 0.5x - 4$	$m = 0.5$	$b = -4$

1 PLAN

PACING
Basic: 2 days
Average: 2 days
Advanced: 2 days
Block Schedule: 0.5 block with 4.5
0.5 block with 4.7

LESSON OPENER ACTIVITY
An alternative way to approach Lesson 4.6 is to use the Activity Lesson Opener:
- Blackline Master (*Chapter 4 Resource Book* p. 80)
- Transparency (p. 29)

MEETING INDIVIDUAL NEEDS
- ***Chapter 4 Resource Book***
  Prerequisite Skills Review (p. 5)
  Practice Level A (p. 86)
  Practice Level B (p. 87)
  Practice Level C (p. 88)
  Reteaching with Practice (p. 89)
  Absent Student Catch-Up (p. 91)
  Challenge (p. 93)
- ***Resources in Spanish***
- ***Personal Student Tutor***

NEW-TEACHER SUPPORT
See the Tips for New Teachers on pp. 1–2 of the *Chapter 4 Resource Book* for additional notes about Lesson 4.6.

WARM-UP EXERCISES
Transparency Available
Use a graph to identify the slope and y-intercept of each equation.

1. $y = -6x$
−6; 0
2. $y = 3x - 7$
3; −7
3. $4x - y = 5$
4; −5
4. $7x - 3y = 4$
$\frac{7}{3}$; $-\frac{4}{3}$

2 TEACH

MOTIVATING THE LESSON

Your family will be traveling to Mexico. When you leave, each U.S. dollar is worth 9.6 pesos. You already have 100 pesos, and will exchange x more U.S. dollars for pesos when you get to the airport. What equation models the number y of pesos you will have after you make the exchange? **$y = 9.6x + 100$**

ACTIVITY NOTE

Using Manipulatives Suggest that students graph the equations to verify the slopes and intercepts they determine algebraically.

EXTRA EXAMPLE 1

Write the equation $3x + 4y = 8$ in slope-intercept form. Then identify the slope and the y-intercept.
$y = -\frac{3}{4}x + 2; -\frac{3}{4}; 2$

EXTRA EXAMPLE 2

Graph $-2x + y = 5$.

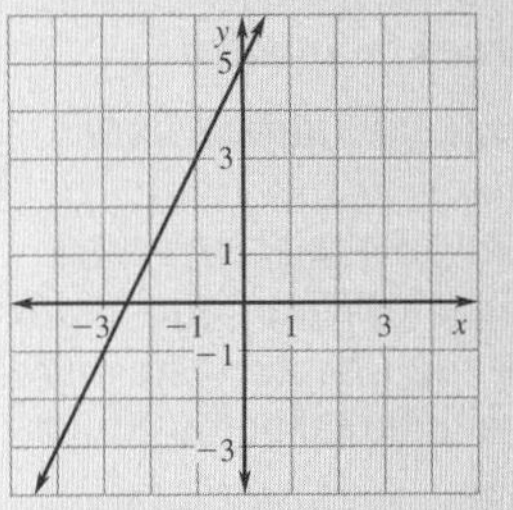

EXTRA EXAMPLE 3

Which of the following lines are parallel?

a. $3y = -9x - 5$
b. $2y - 6x = -5$
c. $12x + 4y = 1$ **a and c**

CHECKPOINT EXERCISES

For use after Examples 1–3:

1. Are the lines $-2x + y = 5$ and $4y = 8x - 1$ parallel? **yes**

! COMMON ERROR

EXAMPLE 3 Students initially may not understand that the coefficient of x represents the slope only when the equation is in slope-intercept form.

EXAMPLE 2 *Graphing Using Slope and y-Intercept*

Graph the equation $3x + y = 2$.

SOLUTION

Write the equation in slope-intercept form.

$y = -3x + 2$

Find the slope and the y-intercept.

$m = -3 \qquad b = 2$

Plot the point $(0, b)$.

Draw a slope triangle to locate a second point on the line.

$m = \frac{-3}{1} = \frac{\text{rise}}{\text{run}}$

Draw a line through the two points.

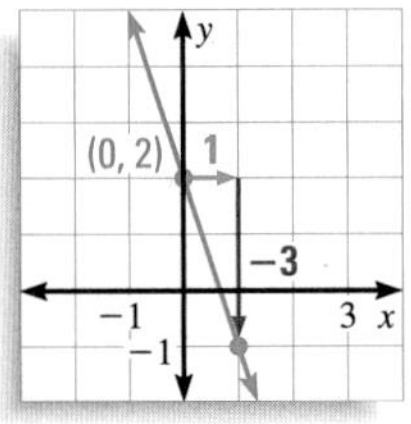

....

Two different lines in the same plane are **parallel** if they do not intersect.

In a coordinate plane, any two vertical lines are parallel. Any two nonvertical lines are parallel if and only if they have the same slope.

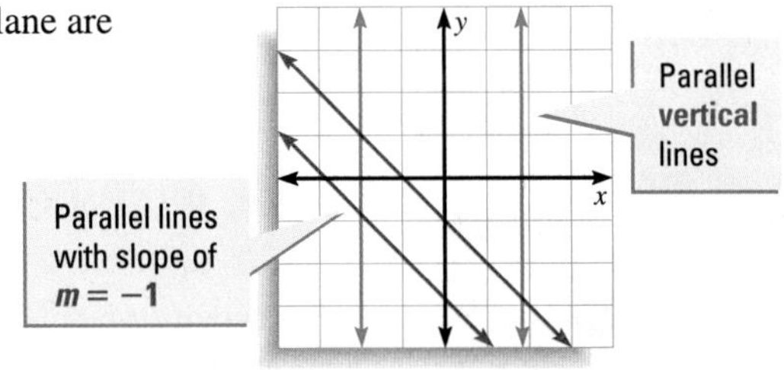

EXAMPLE 3 *Identifying a Family of Parallel Lines*

Which of the following lines are parallel?

line *a*: $-x + 2y = 6$ **line *b*:** $-x + 2y = -2$ **line *c*:** $x + 2y = 4$

SOLUTION Begin by writing each equation in slope-intercept form.

line *a*: $y = \frac{1}{2}x + 3$ **line *b*:** $y = \frac{1}{2}x - 1$ **line *c*:** $y = -\frac{1}{2}x + 2$

Lines a and b are parallel because each has a slope of $\frac{1}{2}$.

Line c is not parallel to either of the other two lines because it has a slope of $-\frac{1}{2}$.

✓**CHECK** The graph gives you a visual check. It shows that line c intersects each of the two parallel lines.

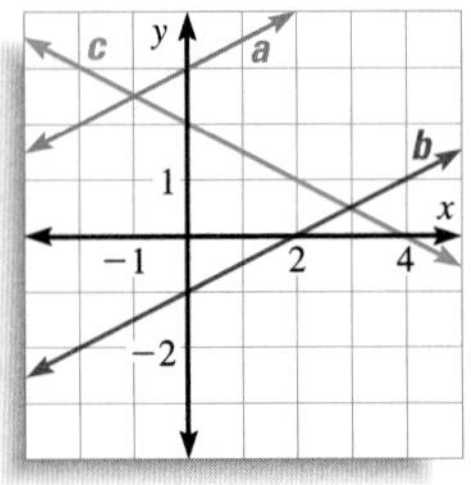

▶ Line a and line b are parallel.

GOAL 2 SOLVING REAL-LIFE PROBLEMS

EXAMPLE 4 *Graphing Using Slope-Intercept Form*

CITY PLANNING You are an intern at a city planner's office and are asked to create a graph for a planning board meeting. The graph will represent different heights of a local river during a flood.

You are given the following equations to model the changing river heights. In each equation, h represents the height of the river (in feet) and t represents the time (in hours) since the flooding began.

Stage 1 (first 60 hours):	$h = \frac{1}{2}t + 24$	$0 \le t \le 60$
Stage 2 (next 12 hours):	$h = 54$	$60 < t \le 72$
Stage 3 (last 48 hours):	$h = -\frac{5}{8}t + 99$	$72 < t \le 120$

SOLUTION

You decide to graph all three equations in the same coordinate plane and label the graph to describe the three stages of flooding. You notice that each equation is in slope-intercept form. The slope in each equation represents the rate at which the river's height is changing during a particular period of time.

Stage 1: During the first 60 hours, the slope is $\frac{1}{2}$. The river height increased $\frac{1}{2}$ foot per hour, from 24 feet to 54 feet.

Stage 2: During the next 12 hours, the river stayed at a constant peak height of 54 feet.

Stage 3: During the last 48 hours, the slope is $-\frac{5}{8}$. The river height decreased $-\frac{5}{8}$ foot per hour, from 54 feet to a normal 24 feet.

To graph each model, you plot each stage within each domain.

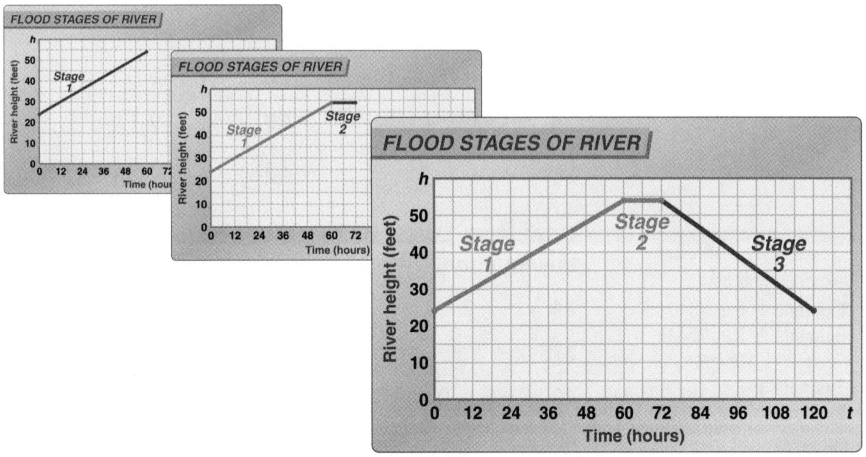

FOCUS ON CAREERS

CITY PLANNERS develop strategies for use of a community's land and resources. They also predict future needs of a community and suggest protection methods for ecologically sensitive areas.

CAREER LINK
www.mcdougallittell.com

EXTRA EXAMPLE 4

The equations below model the changing speeds of a car as it enters an expressway, travels on the expressway, and then exits the expressway. Create a graph to represent the car's different speeds.

Stage 1 (first 2 minutes):
$s = 15t + 25$ Domain: $0 \le t < 2$
Stage 2 (next 10 minutes):
$s = 55$ Domain: $2 \le t < 12$
Stage 3 (next 2 minutes):
$s = -15t + 235$ Domain: $12 \le t < 14$

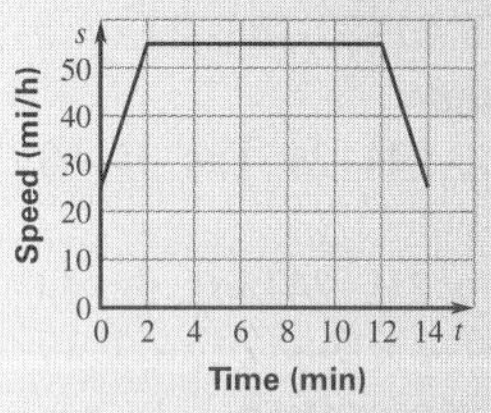

CHECKPOINT EXERCISES

For use after Example 4:

1. Graph each line for the given domain on the same coordinate axes.

a. $y = 2x + 10$
Domain: $-5 \le x < -3$

b. $y = 4$ Domain: $-3 \le x < -1$

c. $y = -4x$ Domain: $-1 \le x < 0$

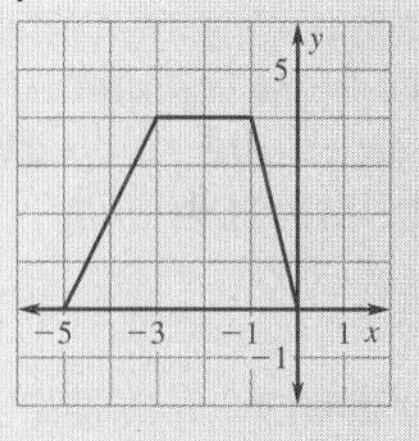

ENGLISH LEARNERS

Paraphrase paragraph 2 orally to help students understand the task described: *You are given the three equations shown below. They can be used to model changes in the river's height that happen during the three stages of the flood. In each equation, h represents the height of the river in feet and t represents the time in hours since the flooding began.*

CLOSURE QUESTION

How do you write an equation in the form $Ax + By + C = 0$ in slope-intercept form?
Solve the equation for y.

3 APPLY

ASSIGNMENT GUIDE

BASIC

Day 1: pp. 244–245 Exs. 14–50 even

Day 2: pp. 245–247 Exs. 56–59, 62–65, 69, 70, 81–89, Quiz 2 Exs. 1–19

AVERAGE

Day 1: pp. 244–245 Exs. 14–50 even

Day 2: pp. 245–247 Exs. 56–65, 70, 81–89, Quiz 2 Exs. 1–19

ADVANCED

Day 1: pp. 244–245 Exs. 14–50 even

Day 2: pp. 245–247 Exs. 62–77, 81–89, Quiz 2 Exs. 1–19

BLOCK SCHEDULE

pp. 244–245 Exs. 14–50 even (with 4.5),
pp. 245–247 Exs. 56–65, 69, 70, 78–88 even,
Quiz 2 Exs. 1–19 (with 4.7)

EXERCISE LEVELS

Level A: ***Easier***
13–21

Level B: ***More Difficult***
22–56, 62–65, 69, 70, 78–89

Level C: ***Most Difficult***
57–61, 66–68, 71–77

✓ HOMEWORK CHECK

To quickly check student understanding of key concepts, go over the following exercises: Exs. 16, 26, 38, 46, 52, 62. See also the Daily Homework Quiz:

- Blackline Master (*Chapter 4 Resource Book*, p. 97)
- Transparency (p. 34)

12, 22–45. See Additional Answers beginning on page AA1.

GUIDED PRACTICE

Vocabulary Check ✓

1. Why is the equation $y = mx + b$ called the *slope-intercept form*? **The slope and y-intercept can be read directly from the equation.**

Concept Check ✓

2. Explain how to graph $2x - 3y = 6$ using slope-intercept form. **See margin.**

3. Decide whether the graphs of $y = x + 2$ and $y = x - 4$ are parallel lines. **yes**

Skill Check ✓

4. Choose the equation whose graph is shown at the right. **B**

   A. $y = \frac{1}{2}x - 2$ **B.** $y = 2x - 2$ **C.** $y = 2x$

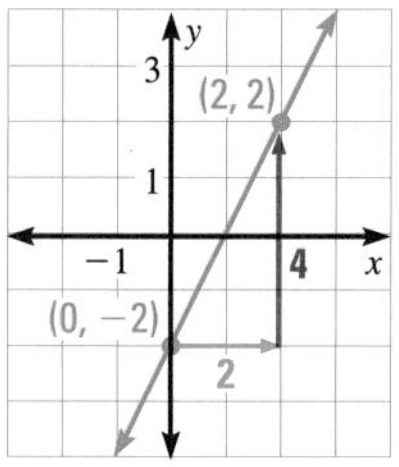

2. Write the equation in slope-intercept form, $y = \frac{2}{3}x - 2$. Plot the y-intercept. Draw a slope triangle to plot a second point. Draw a line through the points you plotted.

Find the slope and the y-intercept of the graph of the equation.

5. $y = 2x + 1$ **2; 1**
6. $y = 11x$ **11; 0**
7. $y = x + 3$ **1; 3**
8. $y = -1.5x$ **−1.5; 0**
9. $y = 5x - 3$ **5; −3**
10. $y = -x - 2$ **−1; −2**

SAVINGS ACCOUNT In Exercises 11 and 12, you have $50 in your savings account at the beginning of the year. Each month you save $30. Assuming no interest is paid, the equation $s = 30m + 50$ models the amount of money s in your savings account after m months.

11. What is the slope of the graph of the line? What is the y-intercept? **30; 50**

12. Graph your total savings. **See margin.**

PRACTICE AND APPLICATIONS

STUDENT HELP

Extra Practice to help you master skills is on p. 800.

SLOPE AND y-INTERCEPT Find the slope and the y-intercept of the graph of the equation.

13. $y = 6x + 4$ **6; 4**
14. $y = 3x + 1$ **3; 1**
15. $y = 2x - 3$ **2; −3**
16. $y - 9x = 0$ **9; 0**
17. $y = -2$ **0; −2**
18. $3x + 4y = 16$ **$-\frac{3}{4}$; 4**
19. $y = \frac{x + 2}{4}$ **$\frac{1}{4}$; $\frac{1}{2}$**
20. $y = \frac{6 - x}{3}$ **$-\frac{1}{3}$; 2**
21. $12x + 4y - 2 = 0$ **−3; $\frac{1}{2}$**

GRAPHING LINES Graph the equation. 22–33. See margin.

22. $y = x + 3$
23. $y = 2x - 1$
24. $y = x + 5$
25. $y = -x + 4$
26. $y = 6 - x$
27. $y = 3x + 7$
28. $y = 4x + 4$
29. $y = x + 9$
30. $y = 4x + 9$
31. $y = \frac{2}{3}x$
32. $y = -\frac{1}{2}x - 3$
33. $y = 2$

41. $y = -\frac{1}{3}x + 1$

42. $y = \frac{2}{3}x - 2$

STUDENT HELP

HOMEWORK HELP
Example 1: Exs. 13–21
Example 2: Exs. 22–45, 52–55
Example 3: Exs. 46–51
Example 4: Exs. 56–59

GRAPHING LINES Write the equation in slope-intercept form. Then graph the equation. 34–45. See margin for graphs.

34. $y = -2$ **$y = -2$**
35. $3x - 6y = 9$ **$y = \frac{1}{2}x - \frac{3}{2}$**
36. $4x + 5y = 15$ **$y = -\frac{4}{5}x + 3$**
37. $4x - y - 3 = 0$ **$y = 4x - 3$**
38. $x - y + 4 = 0$ **$y = x + 4$**
39. $x + y = 0$ **$y = -x$**
40. $x - y = 0$ **$y = x$**
41. $x + 3y - 3 = 0$ **See margin.**
42. $2x - 3y - 6 = 0$ **See margin.**
43. $y - 0.5 = 0$ **$y = 0.5$**
44. $5(x + 3 + y) = 10x$ **$y = x - 3$**
45. $2x + 3y - 4 = x + 5$ **$y = -\frac{1}{3}x + 3$**

PARALLEL LINES **Decide whether the graphs of the two equations are parallel lines. Explain your answer.**

46. $y = -3x + 2, y + 3x = -4$ Yes; both have slope -3.

47. $2x - 12 = y, y = 10 + 2x$ Yes; both have slope 2.

48. $y + 6x - 8 = 0, 2y = 12x - 4$ No; one has slope -6, the other has slope 6

49. $3y - 4x = 3, 3y = -4x + 9$ See margin.

50. $y + 3x = 6x + 1, 9x - 3y = 7$ Yes; both have slope 3.

51. $-x + 3 + y = 2x + 3, y + 4 = 3x$ Yes; both have slope 3.

MATCHING EQUATIONS AND GRAPHS **Match the equation with its graph.**

A. $y = \frac{1}{2}x + 1$ **B.** $y = \frac{1}{2}x - 1$ **C.** $y = x + 2$ **D.** $y = -x - 1$

52. 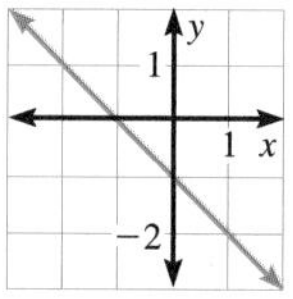 D

53. 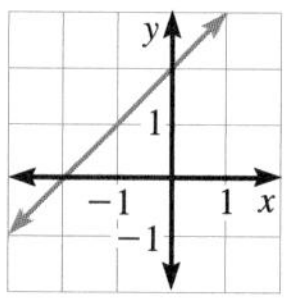 C

54. 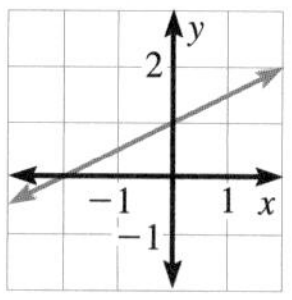 A

55. 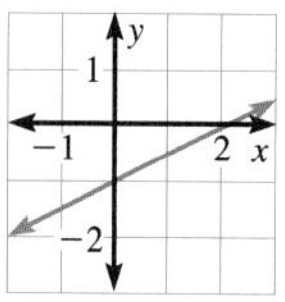 B

SNOWSTORM **In Exercises 56–59, snow fell for 9 hours at a rate of $\frac{1}{2}$ inch per hour. Before the snowstorm began, there were already 6 inches of snow on the ground. The equation $y = \frac{1}{2}x + 6$ models the depth y of snow on the ground after x hours.**

56. What is the slope of $y = \frac{1}{2}x + 6$? What is the y-intercept? $\frac{1}{2}$; 6

57. CRITICAL THINKING Explain what the slope and the y-intercept mean in the snowstorm model. See margin.

58. Graph the amount of snow on the ground during the storm. See margin.

59. *Writing* In the Apple County School District, school is canceled if there is a foot or more of snow on the ground. If this 9-hour storm occurred in Apple County, would school be canceled? Explain your reasoning. No; for $x = 9$, $y = \frac{1}{2}(9) + 6 = 10\frac{1}{2}$.

PHONE CALL **In Exercises 60 and 61, the cost of a long-distance telephone call is \$.87 for the first minute and \$.15 for each additional minute.**

60. PLOTTING POINTS Let c represent the total cost of a call that lasts t minutes. Plot points for the costs of calls that last 1, 2, 3, 4, 5, and 6 minutes. See margin.

61. CRITICAL THINKING Draw a line through the points you plotted in Exercise 60. Find the slope. What does the slope represent? 0.15; the cost per additional minute

GEOMETRY CONNECTION **In Exercises 62–63, use the following linear equations.**

line *a*: $y = -x + 2$ **line *b*:** $y = -1$ **line *c*:** $y = x + 2$

62. Which of the lines are parallel? Explain. No two of the lines are parallel; no two have the same slope.

63. Graph each equation in the same coordinate plane. What type of polygon do they form? Find the area of the polygon. Justify your method. Check graphs; A triangle; 9 square units; the base is 6 units long and the height is 3 units.

MODELING REAL LIFE **In Exercises 64 and 65, graph the situation.**

64. You start from home and drive 55 miles per hour for 3 hours, where d is your distance from home. See margin.

65. You start 165 miles from home and drive toward home at 55 miles per hour for 3 hours, where d is your distance from home. See margin.

64, 65. See Additional Answers beginning on page AA1.

49. No; one has slope $\frac{4}{3}$, the other has slope $-\frac{4}{3}$.

57. The slope is the rate (in inches per hour) at which snow fell; the y-intercept is the amount of snow (in inches) on the ground when the storm began.

FOCUS ON APPLICATIONS

SNOWSTORMS Each year an average of 105 snowstorms affect the continental United States.

60. Check graphs. The points are (1, 0.87), (2, 1.02), (3, 1.17), (4, 1.32), (5, 1.47), and (6, 1.62).

STUDENT HELP

HOMEWORK HELP Visit our Web site www.mcdougallittell.com for help with problem solving in Ex. 63.

APPLICATION NOTE

EXERCISES 11 AND 12 The money students deposit in a savings account would probably earn interest; in which case the equation would not model exactly the savings over time. These exercises present a simplification of a real-world situation.

COMMON ERROR

EXERCISES 34–45 If the numerical coefficient of y is different from 1, students often forget that *both* the coefficient of x *and* the constant term must be divided by this number. Suggest that they first write an equation like $3x - 6y = 9$ in the form $y = \frac{-3x + 9}{-6}$.

COMMON ERROR

EXERCISES 46–51 Students may need reminding that they should rewrite each equation in slope-intercept form to determine whether a pair of lines are parallel.

STUDENT HELP NOTES

Homework Help Students can find help for Ex. 63 at **www.mcdougallittell.com.** The information can be printed out for students who don't have access to the Internet.

COMMON ERROR

EXERCISE 63 Students often forget to multiply the height of a triangle by *one-half* its base length. You may wish to ask students to sketch a rectangle having the same base as the triangle. Ask students what the area of the rectangle is. They should see that the triangle and the rectangle cannot have the same area.

58.

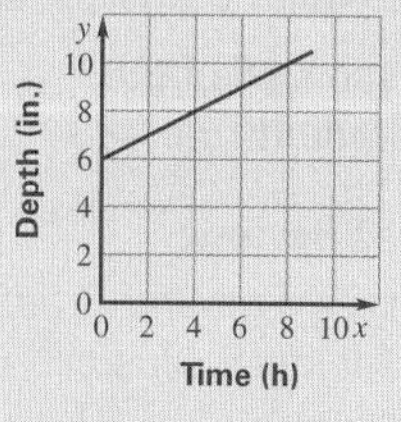

88. Cost of Raising a Child

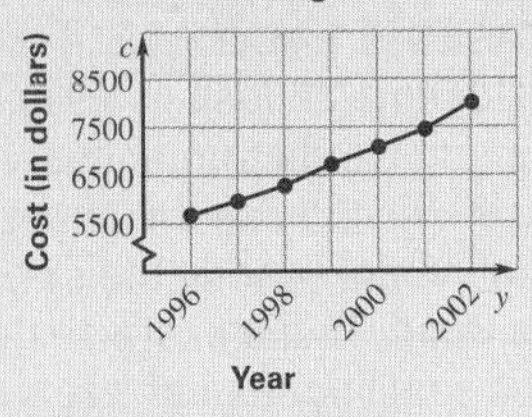

13.

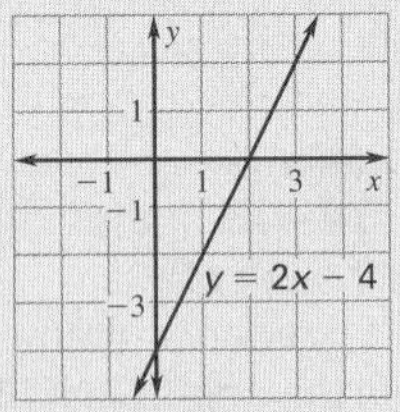

14–18. See Additional Answers beginning on page AA1.

Additional Practice and Reteaching

For Lesson 4.6:

- Practice Levels A, B, and C (*Chapter 4 Resource Book,* p. 86)
- Reteaching with Practice (*Chapter 4 Resource Book,* p. 89)
- See Lesson 4.6 of the *Personal Student Tutor*

For more Mixed Review:

- Search the *Test and Practice Generator* for key words or specific lessons.

STUDENT HELP

INTERNET APPLICATION LINK Visit our Web site www.mcdougallittell.com for more information about roller coasters.

ROLLER COASTER In Exercises 66–68, use the following information. You are supervising the construction of a roller coaster for young children. For the first 20 feet of horizontal distance, the track must rise off the ground at a constant rate. After your crew has constructed 5 feet of horizontal distance, the track is 1 foot off the ground.

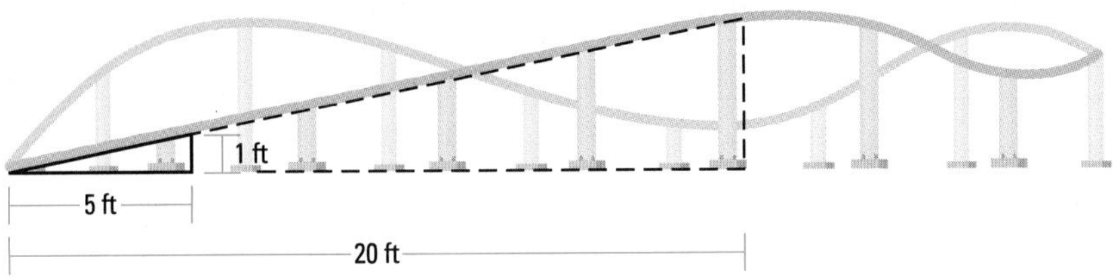

66. $\frac{1}{5}$; the rate at which the track rises off the ground

66. Plot points for the heights of the track at 5-foot intervals. Draw a line through the points. Find the slope of the line. What does the slope represent? **See margin.**

67. After 20 feet of horizontal distance is constructed, you are at the highest point of your roller coaster. How high off the ground is the track? **4 ft**

68. During construction, the park passes a regulation that the track for any roller coaster for young children must not be higher than 5 feet off the ground. Will your roller coaster pass inspection when it is completed? **yes**

Test Preparation

69. MULTIPLE CHOICE What is the slope of a line parallel to the graph of the equation $1.6x - 3.2y = 16$? **B**

Ⓐ −5 Ⓑ 0.5 Ⓒ −0.5 Ⓓ 5

70. MULTIPLE CHOICE If $y + 8 = 0$, what is the slope of the graph? **C**

Ⓐ Undefined Ⓑ 1 Ⓒ 0 Ⓓ −1

★ Challenge

GEOMETRY CONNECTION In Exercises 71–73, graph the two equations in the same coordinate plane. Use the following information to decide whether the lines are perpendicular.

It can be shown that two different nonvertical lines in the same plane with slopes m_1 and m_2 are perpendicular if and only if m_2 is the negative reciprocal of m_1.

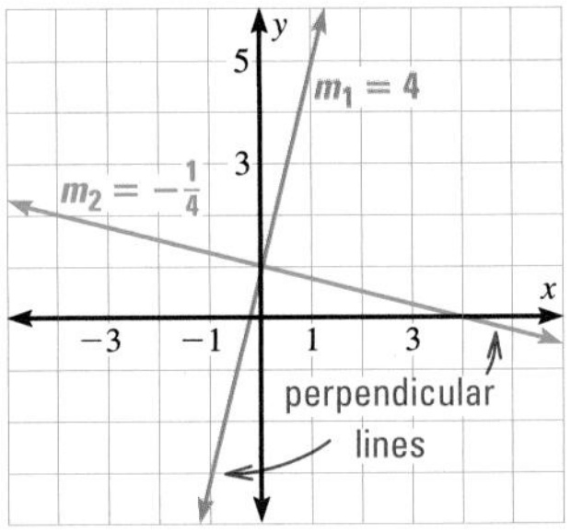

71–73. Check graphs.

71. $y = 2x + 4$
$y = -\frac{1}{2}x + 2$ **yes**

72. $y = -\frac{1}{3}x - 4$
$y = 3x + 2$ **yes**

73. $y = 1 + \frac{2}{3}x$
$y = \frac{3}{2}x - 2$ **no**

PERPENDICULAR LINES In Exercises 74–76, use Exercises 71–73 to help you write an equation of a line that is perpendicular to the given line.

74. $y = 3x - 4$ **75.** $y = -2x + 1$ **76.** $y = \frac{4}{3}x - 3$ **See margin.**

74. $y = -\frac{1}{3}x + k$, for any number k

75. $y = \frac{1}{2}x + k$, for any number k

76. $y = -\frac{3}{4}x + k$, for any number k

77. PARALLEL LINES Is it possible to find another line that is parallel to $3x + 9y = 1$ and passes through the point $\left(0, \frac{1}{9}\right)$? Explain your reasoning.

77. No; the given line passes through $\left(0, \frac{1}{9}\right)$.

EXTRA CHALLENGE www.mcdougallittell.com

MIXED REVIEW

SOLVING EQUATIONS Solve the equation. (Review 3.3, 3.4 for 4.7)

78. $x + 6 = 14$ **8** **79.** $9 - y = 4$ **5** **80.** $7b = 21$ **3**

81. $\frac{a}{4} = 3$ **12** **82.** $\frac{1}{3}g - 2 = 1$ **9** **83.** $3p - 12 = 6$ **6**

84. $2(v + 1) = 4$ **1** **85.** $3(r - 1) = 2(r - 2)$ **−1** **86.** $5(w - 5) = 25$ **10**

87. **COIN COLLECTION** You have 32 coins in a jar. Each coin is either copper or silver. You have 8 more copper coins than silver coins. Let c be the number of copper coins. Which equation correctly models the situation? (Review 1.5) **A**

A. $(c - 8) + c = 32$ **B.** $c + (c + 8) = 32$

COST OF RAISING A CHILD In Exercises 88 and 89, the table shows estimated costs of raising a child born in 1996 to a low income family for the child's first seven years. (Review 1.6)

Year	1996	1997	1998	1999	2000	2001	2002
Cost (in dollars)	5670	5960	6280	6730	7080	7450	8000

DATA UPDATE of USDA data at www.mcdougallittell.com

88. Make a line graph of the data. **See margin.**

89. During which year did the cost increase the most? **from 2001 to 2002**

QUIZ 2

Self-Test for Lessons 4.4–4.6

Find the slope of the line passing through the points. (Lesson 4.4)

1. (0, 0), (5, 2) $\frac{2}{5}$ **2.** (1, −3), (−4, −5) $\frac{2}{5}$ **3.** (3, 3), (−6, −4) $\frac{7}{9}$

4. (−3, 2), (−5, −2) **2** **5.** (0, −4), (5, −4) **0** **6.** (1, −2), (−7, 6) **−1**

The variables x and y vary directly. Given one pair of values for x and y, find an equation that relates the variables. (Lesson 4.5)

7. $x = 2, y = 10$ $y = 5x$ **8.** $x = 8, y = 64$ $y = 8x$ **9.** $x = 12, y = 72$ $y = 6x$

10. $x = 16, y = 12$ $y = \frac{3}{4}x$ **11.** $x = 10, y = 7$ $y = 0.7x$ **12.** $x = 18, y = 8$ $y = \frac{4}{9}x$

Write the equation in slope-intercept form. Then graph the equation. (Lesson 4.6) **13–18. See margin for graphs.**

13. $y = 2x - 4$ **14.** $6x + 12y + 4 = 0$ **15.** $x - 2y = 12$

16. $x - y + 2 = 0$ **17.** $x + 2y - 2 = 0$ **18.** $\frac{3}{2}x + \frac{3}{2}y = \frac{3}{4}$

19. **BUSINESS** On January 1, a company had \$365,800 in its account. On June 1, the company had \$215,400 in its account. If the amount in the company's account changed by an equal amount each month, find the rate of change. Give your answer in dollars per month. (Lesson 4.4) **−\$30,080/month**

13. $y = 2x - 4$
14. $y = -\frac{1}{2}x - \frac{1}{3}$
15. $y = \frac{1}{2}x - 6$
16. $y = x + 2$
17. $y = -\frac{1}{2}x + 1$
18. $y = -x + \frac{1}{2}$

4 ASSESS

DAILY HOMEWORK QUIZ

Transparency Available

1. Find the slope and the y-intercept of $4x - 2y = 8$. **2; −4**

2. Write $x + y + 3 = 0$ in slope-intercept form. Then graph the equation. $y = -x - 3$

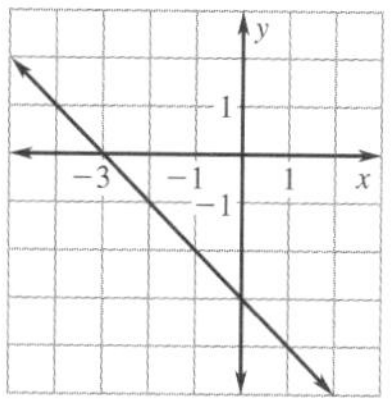

3. Are the graphs of the equations given below parallel: Explain your answer.
$y - 2x = 5$ and $2y = 4x$
Yes; both have slope 2.

4. Describe the figure formed by the intersection of the graphs of the equations. What is its area? $x = -3, x = 3, y = -3, y = 3$
a square; 36 square units

EXTRA CHALLENGE NOTE

Challenge problems for Lesson 4.6 are available in **blackline** format in the *Chapter 4 Resource Book,* p. 93 and at **www.mcdougallittell.com.**

ADDITIONAL TEST PREPARATION

1. WRITING Explain how you can determine algebraically if two lines are parallel. **Write the equations in slope-intercept form. If both lines have the same slope but different y-intercepts, then they are parallel.**

2. OPEN ENDED Graph a family of three lines with the same y-intercept. **Graphs will vary.**

ADDITIONAL RESOURCES

An alternative Quiz for Lessons 4.4–4.6 is available in the *Chapter 4 Resource Book,* p. 94

1 Planning the Activity

PURPOSE
To use a graphing calculator to sketch the graph of a linear equation and find solutions.

MATERIALS
- graphing calculator per student (or a computer with graphing software for each pair of students)
- Keystroke blackline (*Chapter 4 Resource Book*, p. 84)

PACING
- Exploring the Concept — 15 min
- Drawing Conclusions — 15 min

LINK TO LESSON
The standard viewing window would be appropriate for graphing the equations in Examples 2 and 3, but Example 4 would require adjusting the viewing window.

2 Managing the Activity

COOPERATIVE LEARNING
As students complete each step of the activity, have them compare their results with a partner. When the results are different, students should reexamine the problem together to determine the correct result.

CLASSROOM MANAGEMENT
Students may be initially confused by what they see in Step 4. Ask them what they will need to do in order to see the intercepts of the graph.

3 Closing the Activity

KEY DISCOVERY
Writing an equation in slope-intercept form allows a linear equation to be easily graphed.

ACTIVITY ASSESSMENT
JOURNAL How would you adjust the viewing window in Step 5 of Example 1 to show both intercepts of the equation? **Adjust the viewing window to Xmin = –5, Xmax = 20, Ymin = –20, Ymax = 5.**

ACTIVITY 4.6
Using Technology

Graphing a Linear Equation

In Lesson 4.6 you learned how to graph a linear equation using the slope and the *y*-intercept. With a graphing calculator or a computer, you can graph a linear equation and find solutions.

STUDENT HELP
INTERNET KEYSTROKE HELP
See keystrokes for several models of calculators at www.mcdougallittell.com

▶ EXAMPLE 1

Use a graphing calculator to graph $2x - 3y = 33$.

▶ SOLUTION

1 Rewrite the equation in terms of x and y if necessary. Then solve the equation for y.

$$2x - 3y = 33$$
$$-3y = -2x + 33$$
$$y = \frac{2}{3}x - 11$$

2 Press Y= (2 ÷ 3) x – 11 ENTER .
Without parentheses, the calculator may interpret the fraction as $\frac{2}{3x}$.

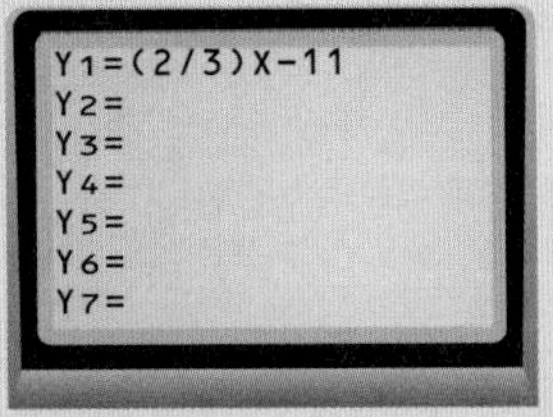

3 Think of the screen as a "window" that lets you look at part of a coordinate plane. Press WINDOW to set the size of the graph.

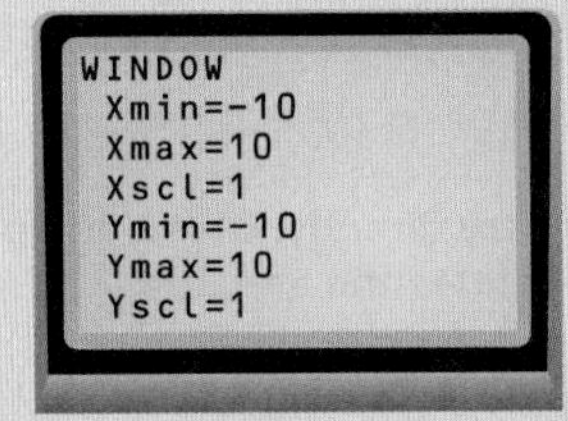

4 Press GRAPH to graph the equation. A standard viewing window is shown.

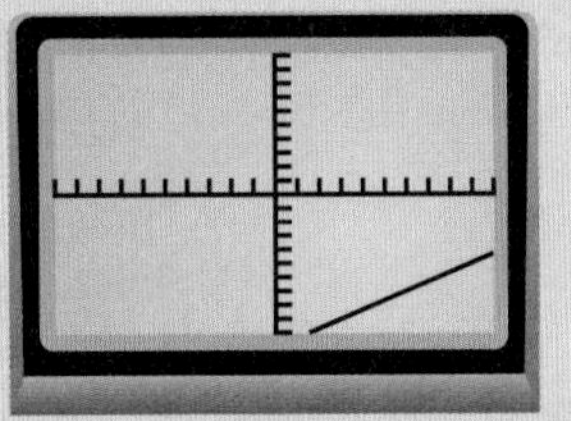

5 To see the point where the graph crosses the x-axis, you can adjust the viewing window. Press WINDOW and enter new values. Then press GRAPH to graph the equation.

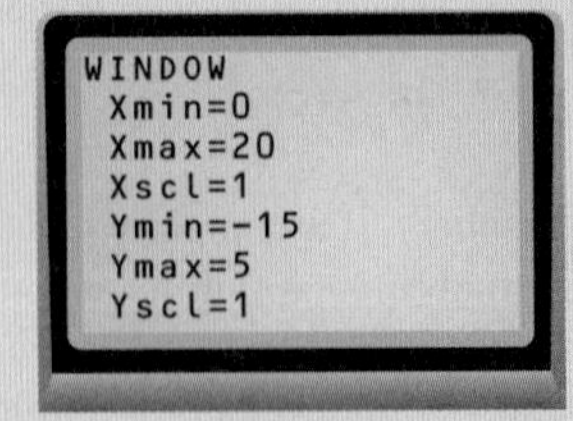

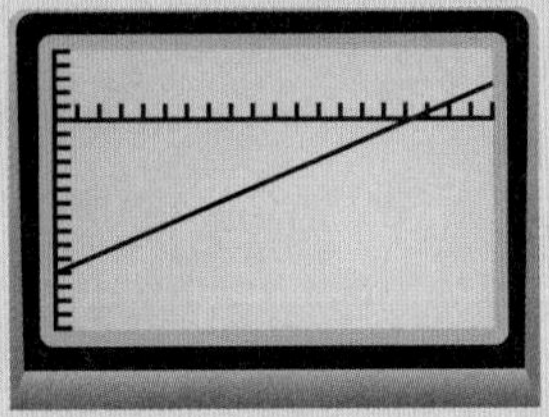

▶ EXAMPLE 2

Estimate the value of y when $x = -7$ in the equation $y = \frac{2}{3}x - \frac{45}{8}$.

▶ SOLUTION

1 Graph the equation $y = \frac{2}{3}x - \frac{45}{8}$ using a viewing window that will show the graph when $x \approx -7$.

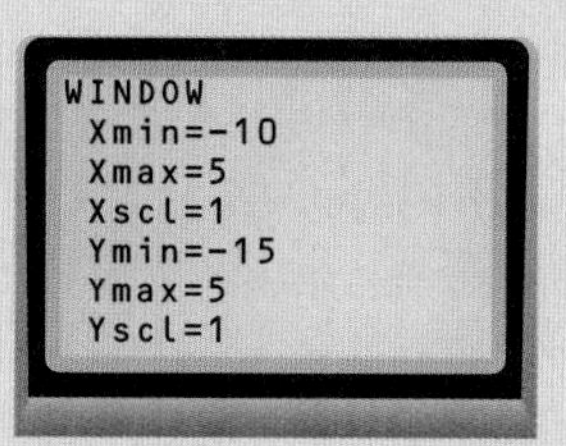

2 Press TRACE and a flashing cursor appears. The x-coordinate and y-coordinate of the cursor's location are at the bottom of the screen. Press right and left arrows to move it. Move the trace cursor until the x-coordinate of the point is at about -7.

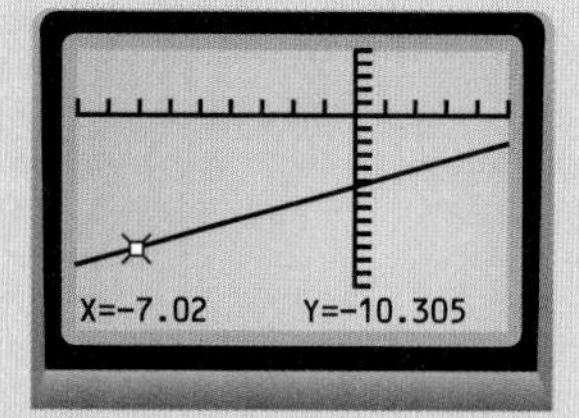

3 Use the ZOOM feature to get a more accurate estimate. A common way to zoom is to press ZOOM and select *Zoom In*. You now have a closer look at the graph at that point. Repeat **Step 2**.

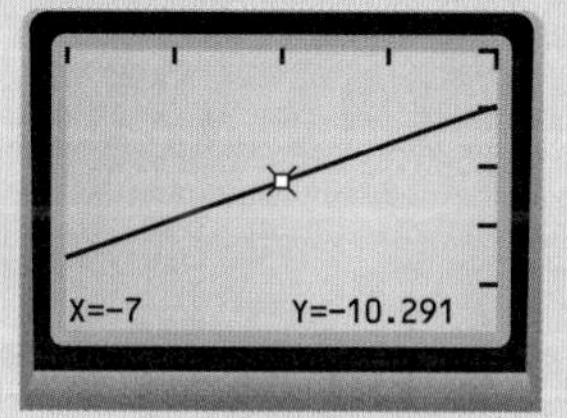

▶ When $x = -7$, $y \approx -10.3$.

STUDENT HELP

Study Tip
You can continue to use zoom until the y-coordinate is to the nearest tenth, hundredth, or any other decimal place you need.

▶ EXERCISES

Use the standard viewing window to graph the equation. **See margin.**

1. $y = -2x - 3$ **2.** $y = 2x + 2$ **3.** $x + 2y = -1$ **4.** $x - 3y = 3$

Use the indicated viewing window to graph the equation. **See margin.**

5. $y = x + 25$	**6.** $y = 0.1x$	**7.** $y = 100x + 2500$
Xmin = –10	Xmin = –10	Xmin = 0
Xmax = 10	Xmax = 10	Xmax = 100
Xscl = 1	Xscl = 1	Xscl = 10
Ymin = –5	Ymin = –5	Ymin = 0
Ymax = 35	Ymax = 1	Ymax = 15000
Yscl = 5	Yscl = 0.1	Yscl = 1000

Determine an appropriate viewing window for the graph of the equation.

8. $y = x - 330$ **9.** $y = x - 0.3$ **10.** $y = 120x$ **11.** $y = 40{,}000 - 1500x$

Use a graph of the equation to estimate the value of y for the given value of x. **12–15. Estimates may vary.**

12. $y = -9x$ when $x = -1.05$ **about 9.45**

13. $y = 5x + 651$ when $x = 2.3$ **about 662.5**

14. $y + \frac{1}{3}x = \frac{1}{5}$ when $x = 19$ **about –6.1$\overline{3}$**

15. $y = -2x - 3$ when $x = 954$ **about –1911**

1.

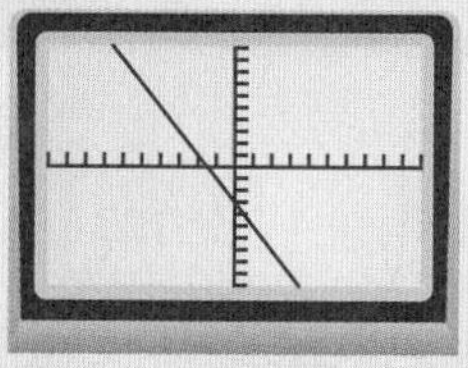

2.

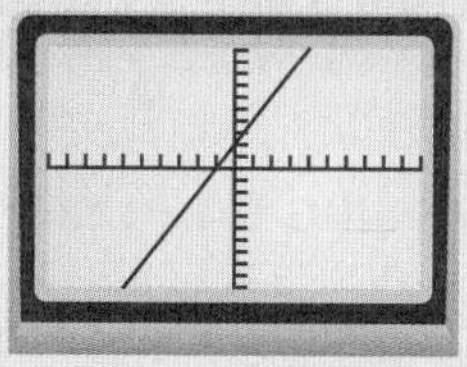

3.

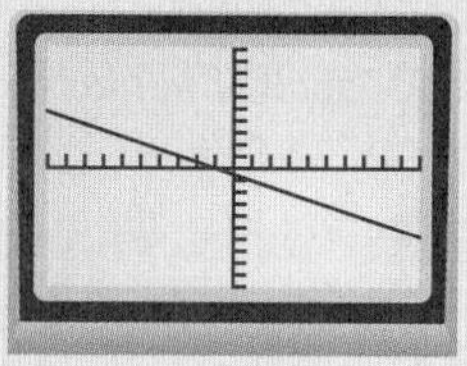

4.

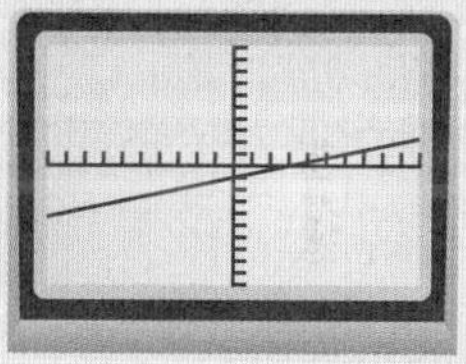

5–7. See Additional Answers beginning on page AA1.

8–11. Sample answers are given.

8. Xmin = –500, Xmax = 500, Xscl = 50, Ymin = –500, Ymax = 500, Yscl = 50

9. standard viewing window

10. Xmin = –10, Xmax = 10, Xscl = 1, Ymin = –1200, Ymax = 1200, Yscl = 100

11. Xmin = –10, Xmax = 30, Xscl = 5, Ymin = –10000, Ymax = 50000, Yscl = 5000

249

1 PLAN

PACING
Basic: 1 day
Average: 1 day
Advanced: 1 day
Block Schedule: 0.5 block with 4.6

LESSON OPENER
GRAPHING CALCULATOR
An alternative way to approach Lesson 4.7 is to use the Graphing Calculator Lesson Opener:
- Blackline Master (*Chapter 4 Resource Book,* p. 98)
- Transparency (p. 30)

MEETING INDIVIDUAL NEEDS
- ***Chapter 4 Resource Book***
  Prerequisite Skills Review (p. 5)
  Practice Level A (p. 100)
  Practice Level B (p. 101)
  Practice Level C (p. 102)
  Reteaching with Practice (p. 103)
  Absent Student Catch-Up (p. 105)
  Challenge (p. 107)
- ***Resources in Spanish***
- ***Personal Student Tutor***

NEW-TEACHER SUPPORT
See the Tips for New Teachers on pp. 1–2 of the *Chapter 4 Resource Book* for additional notes about Lesson 4.7.

WARM-UP EXERCISES
Transparency Available
Solve each equation.

1. $-4x + 1 = 17$
-4

2. $5 - 3y = 8$
-1

What is the x-intercept of each function?

3. $4x - 8 = y$
2

4. $y = -x - 3$
-3

5. $x + y = -5$
-5

4.7 Solving Linear Equations Using Graphs

What you should learn

GOAL 1 Solve a linear equation graphically.

GOAL 2 Use a graph to solve **real-life** problems, such as estimating when there will be 2.5 million registered nurses in **Example 4.**

Why you should learn it

▼ To solve **real-life** problems such as finding production costs of a small business in **Ex. 44.** REAL LIFE

GOAL 1 SOLVING LINEAR EQUATIONS GRAPHICALLY

Up to this point, you have been solving linear equations and checking the solutions *algebraically* (by using algebra). In this lesson you will learn how to solve linear equations and check solutions *graphically* (by using a graph).

STEPS FOR SOLVING LINEAR EQUATIONS GRAPHICALLY

The solution of a linear equation can be found graphically using the following steps:

STEP 1 Write the equation in the form $ax + b = 0$.

STEP 2 Write the related function $y = ax + b$.

STEP 3 Graph the equation $y = ax + b$.

The solution of $ax + b = 0$ is the x-intercept of $y = ax + b$.

EXAMPLE 1 *Using a Graphical Check for a Solution*

Solve $2x - 5 = 1$ algebraically. Check your solution graphically.

SOLUTION

$2x - 5 = 1$	**Write original equation.**
$2x = 6$	**Add 5 to each side.**
$x = 3$	**Divide each side by 2.**

✓**CHECK** Solve $2x - 5 = 1$ graphically as a check.

1 Write the equation in the form $ax + b = 0$.

$2x - 5 = 1$

$2x - 6 = 0$

2 Write the related function in two variables.

$y = 2x - 6$

3 Graph the related function

$y = 2x - 6.$

Notice that the x-intercept is 3, which is the solution of $2x - 6 = 0$. This confirms that 3 is the solution of $2x - 5 = 1$.

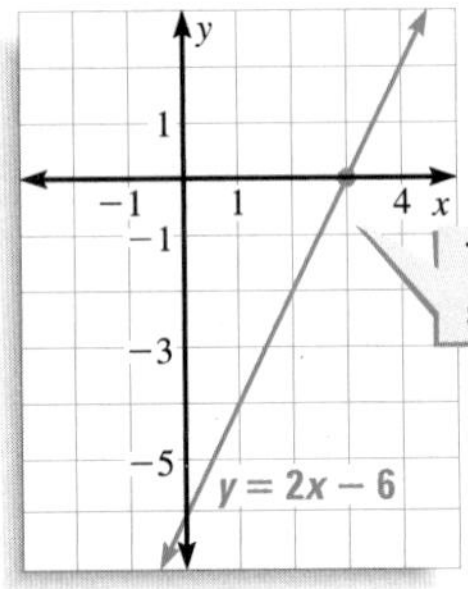

The x-intercept is 3, so the solution is 3.

STUDENT HELP

Study Tip
When you know two ways to solve a problem, it's a good idea to use one method to get a solution and the other method to check the solution. This is done in Examples 1 and 2.

EXAMPLE 2 *Solving an Equation Graphically*

Solve $\frac{4}{3}x + 3 = 2x$ graphically. Check your solution algebraically.

SOLUTION

1. Write the equation in the form $ax + b = 0$.

$\frac{4}{3}x + 3 = 2x$ **Write original equation.**

$-\frac{2}{3}x + 3 = 0$ **Subtract 2*x* from each side.**

2. Write the related function $y = -\frac{2}{3}x + 3$.

3. Graph the equation $y = -\frac{2}{3}x + 3$. The x-intercept appears to be $\frac{9}{2}$.

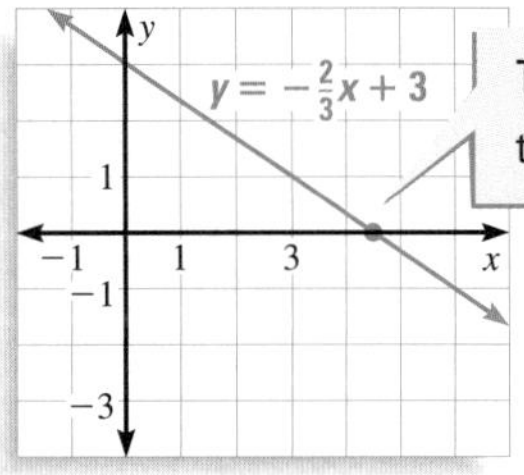

The x-intercept is $\frac{9}{2}$, so the solution is $\frac{9}{2}$.

✓**CHECK** Use substitution.

$\frac{4}{3}x + 3 = 2x$

$\frac{4}{3}\left(\frac{9}{2}\right) + 3 \stackrel{?}{=} 2\left(\frac{9}{2}\right)$

$6 + 3 = 9$ ← **True statement**

▶ The solution of $\frac{4}{3}x + 3 = 2x$ is $\frac{9}{2}$.

EXAMPLE 3 *Using a Graphing Calculator*

Use a graphing calculator to approximate the solution of the linear equation $2.65(4x - 9) = -8.85 + 7.6x$.

SOLUTION

When you graph an equation using a graphing calculator, you do not need to simplify the equation before entering it.

$2.65(4x - 9) + 8.85 - 7.6x = 0$ **Rewrite equation so one side is 0.**

Then use a graphing calculator (or a computer) to graph the related function

$y = 2.65(4x - 9) + 8.85 - 7.6x.$

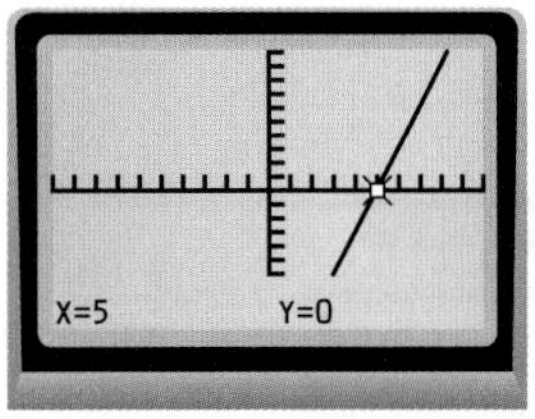

▶ The x-intercept appears to be 5. Check that 5 is a solution of the original equation.

2 TEACH

MOTIVATING THE LESSON
Some of the roads in the Mojave Desert are rough 4-wheel drive routes. You are on one of these routes. The graph shows how the amount of gas you have left depends on how many hours you have been driving. It shows that you can drive for about 2 hours before you run out of gas. In this lesson you'll learn how to determine solutions to linear equations graphically.

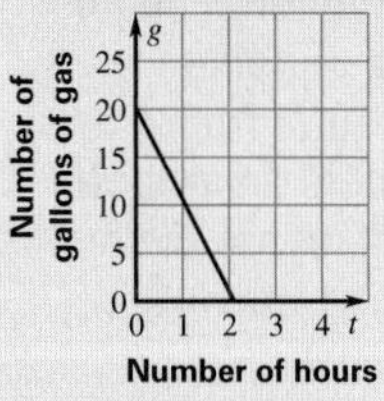

EXTRA EXAMPLE 1
Solve $-x - 3 = 0.5x$ algebraically. $x = -2$

EXTRA EXAMPLE 2
Solve $\frac{5}{2}x - 1 = \frac{9}{2}x$ graphically.
$x = -\frac{1}{2}$

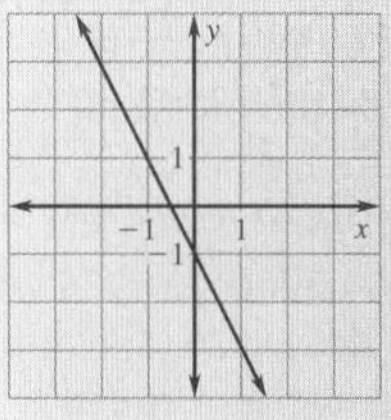

EXTRA EXAMPLE 3
Use a graphing calculator to approximate the solution of the linear equation $1.42(5 - 2x) = 4.75x - 10.2$. $x \approx 2.3$

CHECKPOINT EXERCISES
For use after Examples 1–3:
1. Solve $2(x - 3) - 5x = 9$ algebraically. $x = -5$.

EXTRA EXAMPLE 4

Based on census data from 1987 to 1995, a model for the hourly wage w of people employed in the production of computers and related goods in the United States is $w = 0.409t + 10.74$, where t is the number of years since 1987. According to this model, in what year will the hourly wage for these workers be approximately $18.10? **2005**

EXTRA EXAMPLE 5

Write and graph a function for each side of the equation $18.10 = 0.409t + 10.74$ from Extra Example 4. What is the t-coordinate of the point where the two graphs intersect? **about 18**

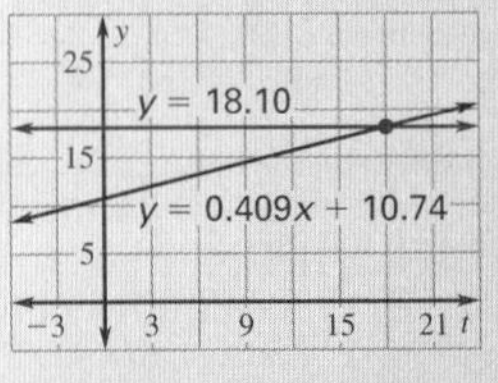

CHECKPOINT EXERCISES

For use after Examples 4 and 5:

1. The average speeds in miles per hour of the first six winners (1911–1916) of the Indianapolis 500 can be modeled by the equation $y = 1.84x + 80$, where x is the number of years since 1911 and y is the average speed in miles per hour. Use this model to find the average speed of Ralph DePalma, the 1915 winner. **about 87.4 mph**

CLOSURE QUESTION

What linear function is related to the equation $4x + 3 = 5$? $y = 4x - 2$

DAILY PUZZLER

A family of lines all have the same x-intercept. What, if anything, does this tell you? **The equations related to this family of lines all have the same solution.**

FOCUS ON CAREERS

REAL LIFE **NURSES** Registered nurses are often responsible for measuring vital signs and giving correct doses of medication.

INTERNET CAREER LINK www.mcdougallittell.com

GOAL 2 APPROXIMATING SOLUTIONS IN REAL LIFE

Example 4 shows how to apply the graphical approach of Examples 1–3 to a real-life problem. Example 5 shows another graphical approach to this problem.

EXAMPLE 4 *Approximating a Real-Life Solution*

NURSES Based on data from 1980 to 1997, a model for the number n (in millions) of registered nurses in the United States is $n = 0.055t + 1.26$, where t is the number of years since 1980. According to this model, in what year will the United States have 2.5 million registered nurses? ▶ Source: U.S. Public Health Service

SOLUTION

Substitute 2.5 for n in the linear model. You can answer the question by solving the resulting linear equation, $2.5 = 0.055t + 1.26$.

Write the equation in the form $ax + b = 0$.

$$2.5 = 0.055t + 1.26$$

$$0 = 0.055t - 1.24$$

Graph the related function.

$$y = 0.055t - 1.24$$

The t-intercept is about 22.5.

The t-intercept is about 22.5. This is where $0.055t - 1.24 = 0$.

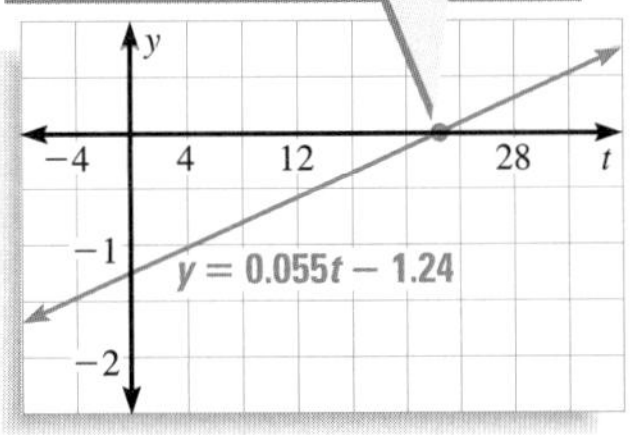

▶ Because t is the number of years since 1980, you can estimate that there will be 2.5 million registered nurses sometime in 2002.

EXAMPLE 5 *Solving by Graphing Both Sides*

Another graphical approach that you can use to solve the equation $2.5 = 0.055t + 1.26$ in Example 4 is to write and graph a function for each side of the equation.

$2.5 = 0.055t + 1.26$	Original equation.
$y = 2.5$	Function for left side of equation.
$y = 0.055t + 1.26$	Function for right side of equation.

Then graph both functions in the same coordinate plane. The graphs cross where the value of y for both functions is about 2.5.

▶ The t-coordinate where the graphs intersect is about 22.5, the same value found by the method in Example 4.

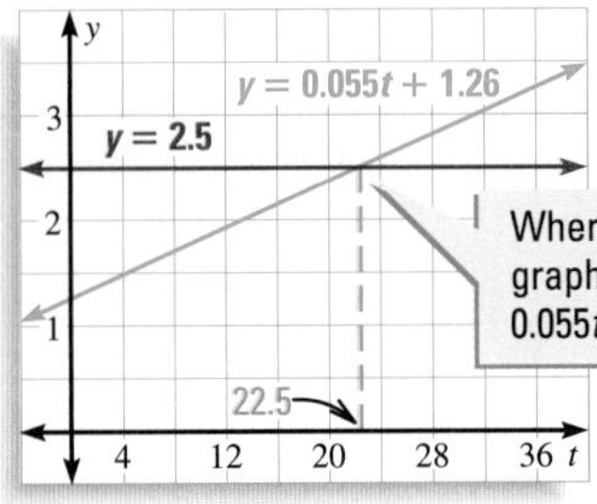

Where the two graphs intersect, $0.055t + 1.26 = 2.5$.

STUDENT HELP

INTERNET HOMEWORK HELP Visit our Web site www.mcdougallittell.com for extra examples.

Guided Practice

Vocabulary Check ✓

1. Explain how to solve a linear equation in one variable *graphically*. **See margin.**

Concept Check ✓

2. In Example 2, explain why the x-intercept of the line $y = -\frac{2}{3}x + 3$ is the solution of the equation $0 = -\frac{2}{3}x + 3$. **The x-intercept is the value of x when $y = 0$.**

Solve the equation graphically. Check your answer algebraically.

3. $x - 3 = 7$ **10**
4. $2 - x = -5$ **7**
5. $5x + 6 = -9$ **−3**

Skill Check ✓

In Exercises 6–10, match the one-variable equation with its related function.

6. $2x = 10$ **C**
7. $2x - 10 = 6$ **A**
8. $5x + 4 = 3x + 5$ **D**
9. $-4x - 6 = -2x + 5$ **B**
10. $19x + 3 = x$ **E**

A. $y = 2x - 16$
B. $y = -2x - 11$
C. $y = 2x - 10$
D. $y = 2x - 1$
E. $y = 18x + 3$

1. ***Sample answer:*** **Write the equation in the form $ax + b = 0$. Then write the related function in two variables, $y = ax + b$, and graph the related function. The x-intercept of the graph is the solution of the equation.**

Practice and Applications

STUDENT HELP

Extra Practice to help you master skills is on p. 800.

MATCHING EQUATIONS AND GRAPHS **Match the equation with the line whose x-intercept is the solution of the equation.**

A. $2x - 4 = 0$ B. $x + 6 = 0$ C. $x - 6 = 0$

11.
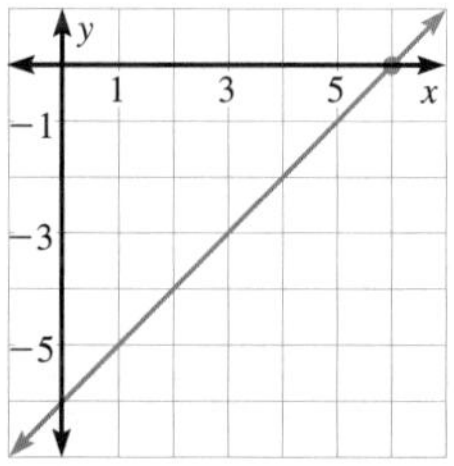

C

12.
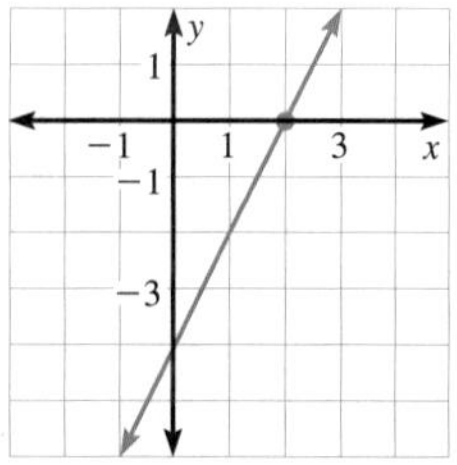

A

13.
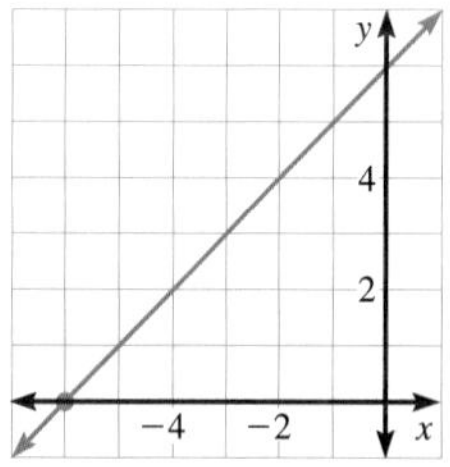

B

RELATED FUNCTIONS **Write the equation in the form $ax + b = 0$. Then write the related function $y = ax + b$.**

14. $7x - 3 = 3$
$7x - 6 = 0$; $y = 7x - 6$
15. $6 - 4x = 13$
$-4x - 7 = 0$; $y = -4x - 7$
16. $9 + 5x = 19$
$5x - 10 = 0$; $y = 5x - 10$
17. $12x + 5 = 8x$
$4x + 5 = 0$; $y = 4x + 5$
18. $-4x - 5 = 3x + 8$
$-7x - 13 = 0$; $y = -7x - 13$
19. $6 - \frac{4}{7}x = 13 + \frac{3}{7}x$
$x + 7 = 0$; $y = x + 7$

SOLVING EQUATIONS **Solve the equation algebraically. Check your solution graphically.**

20. $5x + 3 = -2$ **−1**
21. $-3x + 11 = 2$ **3**
22. $-x = -2$ **2**
23. $2x + 7 = 10$ **$1\frac{1}{2}$**
24. $\frac{1}{4}x + 1 = -\frac{1}{2}$ **−6**
25. $\frac{1}{2}x + 5 = 3$ **−4**
26. $-\frac{2}{3}x - 6 = -4$ **−3**
27. $-\frac{3}{4}x + 3 = \frac{9}{4}$ **1**
28. $\frac{2}{3}x - \frac{2}{3} = 2$ **4**

STUDENT HELP

HOMEWORK HELP
Example 1: Exs. 11–28
Example 2: Exs. 11–37
Example 3: Exs. 38–43
Example 4: Exs. 44, 45
Example 5: Exs. 44, 45

3 APPLY

Assignment Guide

BASIC
Day 1: pp. 253–255 Exs. 11–13, 14–36 even, 44, 45, 50, 55, 60, 65, 66

AVERAGE
Day 1: pp. 253–255 Exs. 11–13, 14–36 even, 50, 55, 60, 65, 66

ADVANCED
Day 1: pp. 253–255 Exs. 11–13, 14–42 even, 46–51, 55, 60, 65, 66

BLOCK SCHEDULE WITH 4.6
pp. 253–255 Exs. 11–13, 14–36 even, 44–48, 50, 55, 60, 65, 66

Exercise Levels

Level A: *Easier*
52–56

Level B: *More Difficult*
11–50, 57–65

Level C: *Most Difficult*
51, 66

✓ Homework Check

To quickly check student understanding of key concepts, go over the following exercises: Exs. 12, 16, 21, 26, 30, 35, 44. See also the Daily Homework Quiz:

- Blackline Master (*Chapter 4 Resource Book,* p. 110)
- Transparency (p. 35)

! Common Error

EXERCISES 3–5 Some students may forget to solve the equation for 0. Remind them that they are solving for x, so they need to determine where the equation crosses the x-axis. That is, they must determine the value of x when $y = 0$. The first step, then, is to write the equation in the form $ax + b = 0$.

GRAPHING CALCULATOR NOTE

EXERCISES 38–43 These exercises would require much more algebraic manipulation if done without the aid of a graphing calculator.

APPLICATION NOTE

EXERCISE 44 If Chai's income from the sale of her hats can also be represented by a linear function, her cost and revenue functions can be graphed together. The point at which the two graphs meet is called the break-even point. At this point, Chai will neither sustain a loss nor make a profit.

EXERCISES 47 AND 48
By graphing the linear functions for these exercises on the same coordinate axes, students can see that as game shows continue their decline, talk shows will increase in popularity. The events described in these two exercises will occur at about the same time, according to the models.

ENGLISH LEARNERS

In Exercises 46–49, explain that *Viewing Habits* means "habits of people who view TV," not "observing people's habits." Note that it can be difficult for a non-native speaker to sense whether an *-ing* form preceding a noun has the function of a transitive verb (for example, running laps) or an adjective (for example, running shoes).

ADDITIONAL PRACTICE AND RETEACHING

For Lesson 4.7:

- Practice Levels A, B, and C (*Chapter 4 Resource Book,* p. 100)
- Reteaching with Practice (*Chapter 4 Resource Book,* p. 103)
- See Lesson 4.7 of the *Personal Student Tutor*

For more Mixed Review:

- Search the *Test and Practice Generator* for key words or specific lessons.

SOLVING GRAPHICALLY Solve the equation graphically. Check your solution algebraically.

29. $2x - 7 = -5$ **1** **30.** $5x - 2 = 8$ **2** **31.** $-4x = -12$ **3**

32. $6x + 9 = 3$ **−1** **33.** $7 - 9x = -11$ **2** **34.** $3x + 10 = -2x$ **−2**

35. $-5x + 4 = 12 - 3x$ **−4** **36.** $\frac{1}{3}x + 1 = 4$ **9** **37.** $\frac{1}{2}x + 5 = 3$ **−4**

SOLVING GRAPHICALLY Use a graphing calculator to find the solution of the equation. Check your solution algebraically.

38. $4(x + 2) = 3(x + 5)$ **7** **39.** $-1.6(1.5x + 7.5) = 0.6(6x + 30)$ **−5**

40. $\frac{1}{2}(x + 7) = \frac{1}{3}(10x + 2)$ **1** **41.** $0.7(3x - 20) = 22 - 3.9x$ **6**

42. $\frac{3}{4}(4x - 15) = -\frac{3}{2}(4x - 18)$ **4.25** **43.** $\frac{1}{3}(10x + 27) = -\frac{1}{5}\left(\frac{55}{3}x + 25\right)$ **−2**

44. **PRODUCTION COSTS** Chai has a small business making decorated hats. She calculates her monthly cost y of producing x hats using the function $y = 1.9x + 350$. In January, her cost was $458.30. How many hats did she make that month? Solve algebraically and graphically. **57 hats**

STUDENT HELP
INTERNET **APPLICATION LINK**
Visit our Web site www.mcdougallittell.com for more information about tourism in Ex. 45.

45. **TOURISM** Based on data from 1991–1995, a model for the annual number of visitors v from the United States to Europe is

$$v = 559{,}100t + 6{,}423{,}000$$

where t is the number of years since 1991. According to this model, in what year will the number of visitors to Europe reach 11,000,000? **1999**

▶ Source: U.S. Bureau of the Census

Venice, Italy

VIEWING HABITS In Exercises 46–49, use the models below. They can be used to estimate the average number of hours h per week that a household in the United States watches television programs. In all three models, t is the number of years since 1985. ▶ Source: Media Dynamics, Inc.

Hours spent watching television:	$h = 0.57t + 54.85$
Hours spent watching talk shows:	$h = 0.35t + 4.06$
Hours spent watching game shows:	$h = -0.2t + 3.4$

46. According to the model, in what year will a household watch television an average of 70 hours per week? **2011**

47. According to the model, in what year will a household watch an average of 10 hours of talk shows per week? **2001**

48. According to the model, in what year will the average number of hours per week that a household watches game shows be zero? **2002**

49. *Writing* Did you solve Exercises 46–48 algebraically or graphically? Give a reason for your choice. **See margin.**

49. ***Sample answer:*** **The arithmetic involved makes Exs. 46 and 47 easier to solve graphically. Ex. 48 may be solved either way.**

Test Preparation

50\. a. $800{,}000 = 13{,}272t + 736{,}000$; $800{,}000 = 9455t + 736{,}000$; $800{,}000 = 11{,}455t - 736{,}000$

b. Model 1: 1998; Model 2: 2000; Model 3: 1999

c. Model 1; Model 2; the x-intercept of the graph tells you how soon the number of mechanics will reach 800,000. The graph with the lowest x-intercept (Model 1) reaches this number first. The graph with the highest x-intercept (Model 2) reaches this number last.

★ Challenge

51\. Yes; *Sample answer:* $x - 2 = 0$ and $-x + 2 = 0$ are equivalent equations.

EXTRA CHALLENGE
www.mcdougallittell.com

50. MULTI-STEP PROBLEM The U.S. Bureau of Labor Statistics projects job growth by using three models to make low, moderate, and high estimates. The equations below model the projected number of auto mechanics m from 1994 to 2005. In all three models, t is the number of years since 1994. **50. a–c. See margin.**

Model 1: $m = 13{,}272t + 736{,}000$

Model 2: $m = 9455t + 736{,}000$

Model 3: $m = 11{,}455t + 736{,}000$

a. For each model, write an equation that enables you to predict the year in which the number of auto mechanics will reach 800,000.

b. In the same coordinate plane, graph the related function for each equation that you found in part (a). According to each model, in what year will the number of auto mechanics reach 800,000?

c. VISUAL THINKING Which model gives a high estimate of the number of mechanics? a low estimate? How can you tell this from the graphs of the models?

51. COMPARING METHODS Below, Tanya and Dwight solve $3x - 5 = 2x - 3$ graphically. Does each method give a correct solution? Explain. **See margin.**

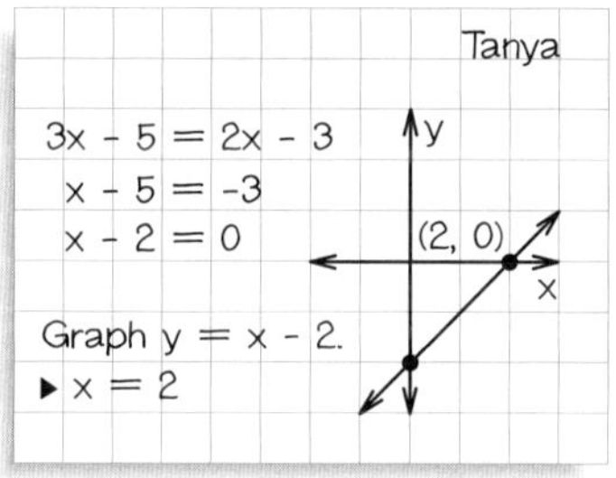

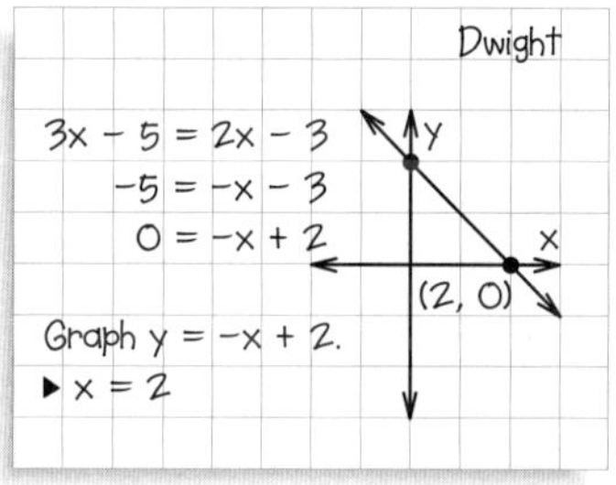

MIXED REVIEW

CHECKING SOLUTIONS **Is the given number a solution of the inequality?** (Review 1.4)

52. $x + 3 > 5; 7$ yes **53.** $-2 - x \leq -9; -7$ no **54.** $3x - 11 \geq 2; 4$ no

55. $8 + 5x < -2; -2$ no **56.** $4x - 1 > 10; 3$ yes **57.** $2x + 4 \leq 3x - 3; 5$ no

DISTRIBUTIVE PROPERTY **Apply the distributive property.** (Review 2.6) **58–65. See margin.**

58. $4(x + 7)$ **59.** $-8(8 - y)$ **60.** $(3 - b)9$ **61.** $(q + 4)(-3q)$

62. $10(6x + 2)$ **63.** $-2(d - 5)$ **64.** $(7 - 2a)(4a)$ **65.** $-5w(-3 + 2w)$

66. FLOWER SERVICE You start a daily flower club and charge \$10 to join and \$.50 per day. Every day each member of the club gets a fresh flower. Let n represent the number of club members and let I represent your income for four weeks. A model for the situation is $I = [10 + 4 \cdot 7(0.5)]n$. Write an input-output table that shows your income for 2, 4, 6, 8, and 10 club members. (Review 1.7) **See margin.**

58\. $4x + 28$
59\. $-64 + 8y$
60\. $27 - 9b$
61\. $-3q^2 - 12q$
62\. $60x + 20$
63\. $-2d + 10$
64\. $28a - 8a^2$
65\. $15w - 10w^2$

Additional Test Preparation *Sample answer:*

1\. At the point where the graph of $y = 3x - 5$ crosses the x-axis, the value of y is 0. Therefore, the x-coordinate of this point is the value of x that satisfies $3x - 5 = 0$.

4 ASSESS

DAILY HOMEWORK QUIZ

Transparency Available

1\. Solve the equations algebraically.
a. $-7x + 5 = -9$ **2**
b. $\frac{3}{2}x + 5 = 20$ **10**

2\. Solve the equations graphically.
a. $2x + 3 = 5$ **1**
b. $4x - 2 = 5x$ **−2**

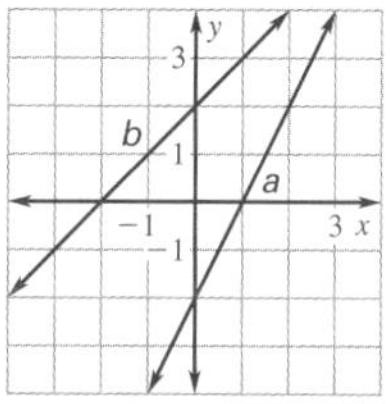

3\. Mike has a lawn care service. He calculates his monthly cost y of caring for x lawns using the function $y = 1.8x + 25$. In July, his cost was \$133. How many lawns did he care for that month? **60 lawns**

EXTRA CHALLENGE NOTE

Challenge problems for Lesson 4.7 are available in **blackline** format in the *Chapter 4 Resource Book*, p. 107 and at **www.mcdougallittell.com.**

ADDITIONAL TEST PREPARATION

1\. **WRITING** Explain why the function $y = 3x - 5$ can be graphed to determine the solution of the equation $3x - 5 = 0$. **See sample answer, below.**

2\. **WRITING** Explain why the graphs of the linear functions $y = 3x + 10$ and $y = -2x$ will determine the solution of the equation $3x + 10 = -2x$. **Since the expressions $3x + 10$ and $-2x$ are equal, the x-coordinate of the intersection point of the graphs of the linear functions is the solution of the equation.**

66\. See Additional Answers beginning on page AA1.

1 PLAN

PACING
Basic: 2 days
Average: 2 days
Advanced: 2 days
Block Schedule: 1 block

LESSON OPENER
APPLICATION
An alternative way to approach Lesson 4.8 is to use the Application Lesson Opener:
- Blackline Master (*Chapter 4 Resource Book,* p. 111)
- Transparency (p. 31)

MEETING INDIVIDUAL NEEDS
- ***Chapter 4 Resource Book***
  Prerequisite Skills Review (p. 5)
  Practice Level A (p. 112)
  Practice Level B (p. 113)
  Practice Level C (p. 114)
  Reteaching with Practice (p. 115)
  Absent Student Catch-Up (p. 117)
  Challenge (p. 119)
- ***Resources in Spanish***
- ***Personal Student Tutor***

NEW-TEACHER SUPPORT
See the Tips for New Teachers on pp. 1–2 of the *Chapter 4 Resource Book* for additional notes about Lesson 4.8.

WARM-UP EXERCISES
Transparency Available
State the domain and range of each function.
1. A store sells video cassettes for \$5 each. The cost c of buying n video cassettes can be modeled by $c = 5n$.
   Domain: whole numbers
   Range: nonnegative multiples of 5.
2. The function $F = 1.8C + 32$ converts Celsius temperatures to Fahrenheit temperatures.
   Domain: real numbers
   Range: real numbers

4.8 Functions and Relations

***What** you should learn*

GOAL 1 Identify when a relation is a function.

GOAL 2 Use function notation to represent **real-life** situations, such as modeling butterfly migration in **Example 4**.

***Why** you should learn it*

▼ To solve **real-life** problems such as projecting the number of high school students and college students in **Exs. 50–52**.

GOAL 1 IDENTIFYING FUNCTIONS

In defining functions, you learned that for every input there corresponds exactly one output, as in the case of linear functions defined by the rule $y = mx + b$. However, there are rules that associate more than one output with an input, as in $x = y^2$ where the input $x = 4$ corresponds to the outputs $y = 2$ and $y = -2$. A rule of this type, where an input can have more than one output, is a *relation*. A **relation** is defined to be any set of ordered pairs.

EXAMPLE 1 *Identifying Functions*

Decide whether the relation is a function. If so, give the domain and the range.

a. 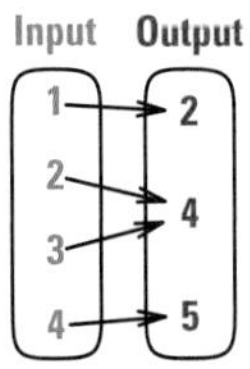

b. 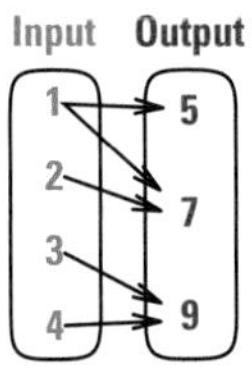

SOLUTION

a. The relation is a function. For each input there is exactly one output. The domain of the function is the set of input values 1, 2, 3, and 4. The range is the set of output values 2, 4, and 5.

b. The relation is not a function because the input 1 has two outputs: 5 and 7.

....

When you graph a relation, the input is given by the horizontal axis and the output is given by the vertical axis.

STUDENT HELP
Look Back
For help with domain and range, see p. 47.

VERTICAL LINE TEST FOR FUNCTIONS

A relation is a function of the horizontal-axis variable if and only if no vertical line passes through two or more points on the graph.

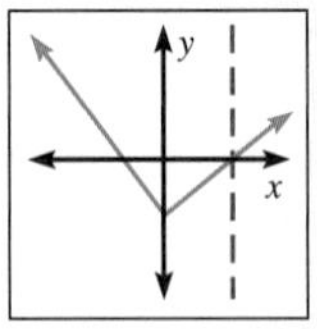

function

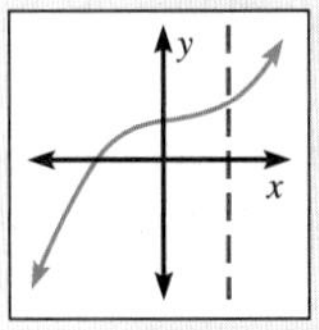

function

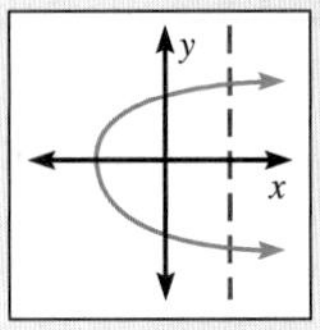

not a function

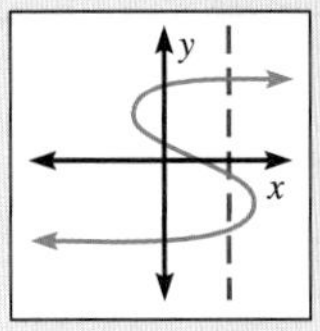

not a function

GOAL 2 USING FUNCTION NOTATION

When a function is defined by an equation, it is convenient to give the function a name. In making a table of values for the equation $y = 3x + 2$, you apply the rule *multiply by three and add two* to transform an input x into the output y. By giving the rule *multiply by three and add two* the name f, you have a shorthand notation for applying this rule to different numbers and variables. The fact that this rule transforms the input 1 into the output 5 can now be written $f(1) = 5$.

In general, the symbol $f(x)$ replaces y and is read as "the value of f at x" or simply as "f of x." It does not mean f times x. $f(x)$ is called **function notation**.

EXAMPLE 2 *Evaluating a Function*

> **STUDENT HELP**
> **Study Tip**
> You don't have to use f to name a function. You can use other letters, such as g and h.

Evaluate the function for the given value of the variable.

a. $f(x) = 2x - 3$ when $x = -2$ **b.** $g(x) = -5x$ when $x = 0$

SOLUTION

a. $f(x) = 2x - 3$ **Write original function.**

$f(-2) = 2(-2) - 3$ **Substitute −2 for *x*.**

$= -7$ **Simplify.**

b. $g(x) = -5x$ **Write original function.**

$g(0) = -5(0)$ **Substitute 0 for *x*.**

$= 0$ **Simplify.**

....

Since function notation allows you to write $f(x)$ in place of y, the **graph of a function** f is the set of all points $(x, f(x))$, where x is in the domain of the function. To graph a linear function, you may want to rewrite the function using x-y notation.

EXAMPLE 3 *Graphing a Linear Function*

Graph $f(x) = -\frac{1}{2}x + 3$.

SOLUTION

Rewrite the function as $y = -\frac{1}{2}x + 3$.

The y-intercept is 3, so plot (0, 3).

The slope is $-\frac{1}{2}$.

Draw a slope triangle to locate a second point on the line.

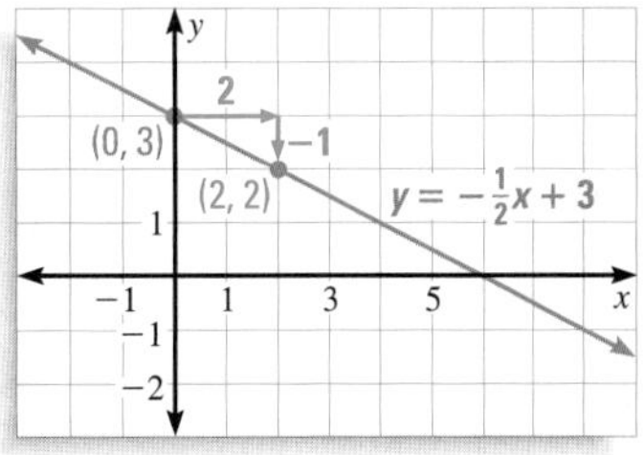

2 TEACH

MOTIVATING THE LESSON
Do you know how to say *hello* in Spanish? in Italian? in Japanese? Knowing a language enables you to communicate in that language. Mathematics is the language of science. It is important for scientists to understand mathematical notation. In this lesson, you will learn function notation.

EXTRA EXAMPLE 1
Decide whether the relation shown is a function. If it is, give the domain and the range.

a. Input → Output **b.** Input → Output

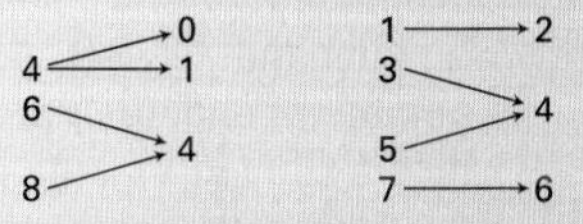

a. no; b. yes; domain: 1, 3, 5, 7; range: 2, 4, 6

EXTRA EXAMPLE 2
Evaluate the function for the given value of the variable.

a. $f(x) = 7 - 3x$ when $x = 4$ **−5**

b. $f(x) = 10x + 3$ when $x = -2$ **−17**

EXTRA EXAMPLE 3
Graph $f(x) = 5x - 4$.

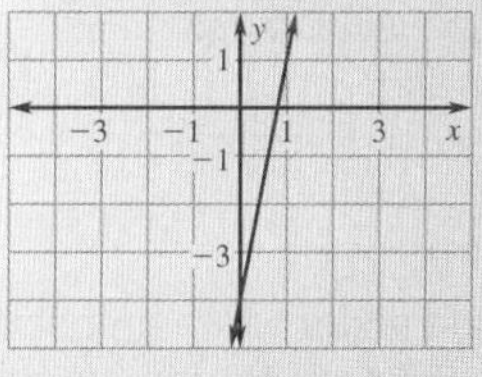

CHECKPOINT EXERCISES

For use after Example 1:

1. Is the set of ordered pairs (−4, 1), (−3, 2), (−2, 5), (−1, 1) a function? **yes**

For use after Example 2:

2. Evaluate $f(x) = -100x$ for $x = -3$. **300**

For use after Example 3:

3. Graph $f(x) = -2x + 3$.

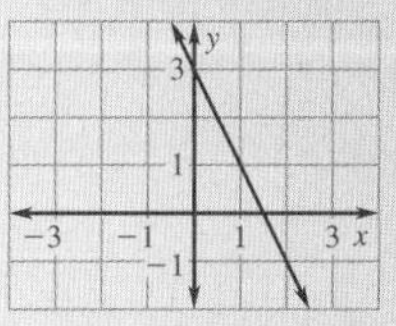

EXTRA EXAMPLE 4

A paving company is resurfacing some roads. It takes 150 h to resurface 30 mi of road.

a. Write a linear function that models the length of road in miles d that can be resurfaced in t hours. **$d = 0.2t$**

b. Use the model to find the length of road that can be resurfaced in 500 h. **100 mi**

c. Give the point that represents the length of road that can be resurfaced in 500 h. **(500, 100)**

CHECKPOINT EXERCISES

For use after Example 4

1. A manufacturing company can produce 1320 jet engines in 6 months. Write a linear function that models the number of engines y produced by the plant in x months. Give the point that represents the number of engines produced in 2 months. **$y = 220x$; (2, 440)**

CLOSURE QUESTION

Why is a relation represented by a graph not a function of the horizontal-axis variable if a vertical line passes through two or more points of the graph? ***Sample answer:*** **Some value of the horizontal-axis variable corresponds to more than one vertical-axis value.**

DAILY PUZZLER

Plot the last digits of the first seven multiples of 6 on a coordinate grid. Connect the points in order to form six segments. If you continue the pattern, what will be the slopes of the next three segments? **6, −4, −4**

EXAMPLE 4 *Writing and Using a Linear Function*

BUTTERFLIES Use the diagram about monarch butterfly migration.

a. Write a linear function that models the distance traveled by a migrating butterfly.

b. Use the model to estimate the distance traveled after 30 days of migration.

c. Graph your model and label the point that represents the distance traveled after 30 days.

Monarch butterflies migrate from the northern United States to Mexico.

SOLUTION

a. You can see in the photo caption below that the butterflies travel about 2000 miles in 40 days. Their average speed is 50 miles per day. To create a linear model, we approximate the actual speed with this constant.

PROBLEM SOLVING STRATEGY

VERBAL MODEL Distance traveled = Average speed • Time

LABELS

Time	= t	(days)
Average speed	= **50**	(miles per day)
Distance traveled	= $f(t)$	(miles)

ALGEBRAIC MODEL $f(t) = 50\,t$

b. To estimate the distance traveled after 30 days, substitute 30 for t.

$f(t) = 50t$ **Write linear function.**

$f(30) = 50 \cdot 30$ **Substitute 30 for t.**

$= 1500$ **Simplify.**

The distance traveled after 30 days is 1500 miles.

UNIT ANALYSIS You can use unit analysis to check that *miles* are the units of the solution.

$$\frac{\text{miles}}{\cancel{\text{day}}} \cdot \cancel{\text{days}} = \text{miles}$$

c. From the graph, you can see that distance traveled is a function of time.

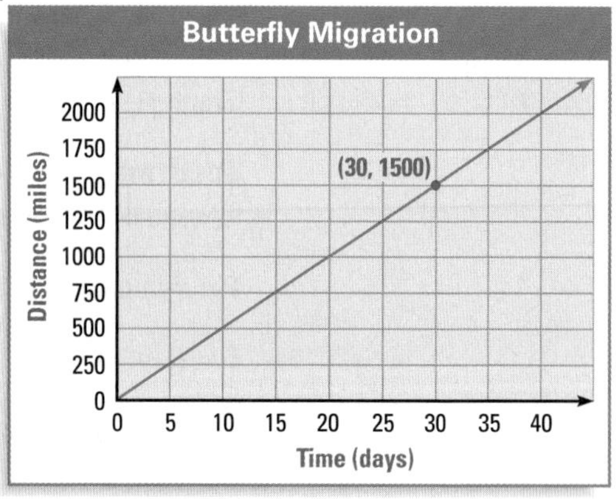

FOCUS ON APPLICATIONS

MONARCH BUTTERFLY
Thousands of monarch butterflies migrate in one large flock. The 2000-mile trip takes about 40 days.

GUIDED PRACTICE

Vocabulary Check ✓
Concept Check ✓
Skill Check ✓

1. Is every function a relation? Is every relation a function? Explain. **See margin.**

2. Describe a line that cannot be the graph of a linear function. **a vertical line**

Evaluate the function $f(x) = 3x - 10$ for the given value of x.

3. $x = 0$ **−10** **4.** $x = 20$ **50** **5.** $x = -2$ **−16** **6.** $x = \frac{2}{3}$ **−8**

In Exercises 7–9, decide whether the relation is a function. If it is a function, give the domain and the range.

7.
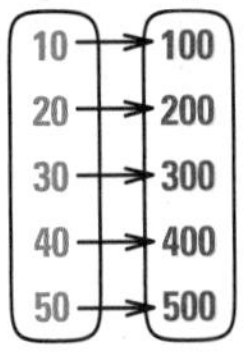

yes; 10, 20, 30, 40, 50; 100, 200, 300, 400, 500

8.
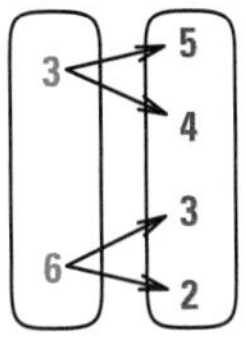

no

9.
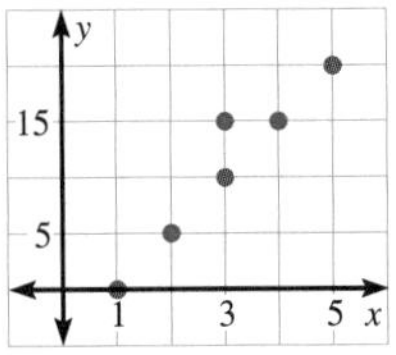

no

10. CAR TRAVEL Your average speed during a trip is 40 miles per hour. Write a linear function that models the distance you travel $d(t)$ as a function of t, the time spent traveling. **$d(t) = 40t$**

1. Yes; no; every function is a set of ordered pairs and, so, is a relation. A relation is a function only if for each input there is exactly one output.

PRACTICE AND APPLICATIONS

STUDENT HELP

Extra Practice to help you master skills is on p. 800.

GRAPHICAL REASONING **Decide whether the graph represents y as a function of x. Explain your reasoning.** **11–13. See margin.**

11.
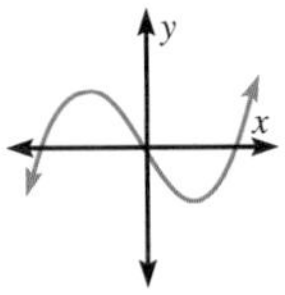

12.
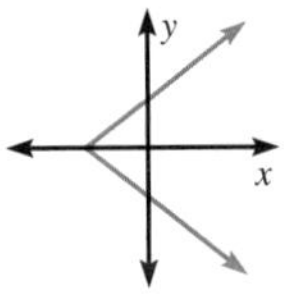

13.
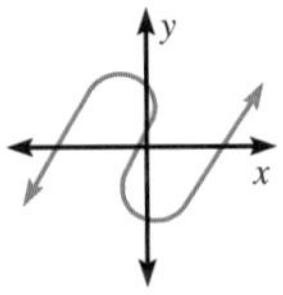

11. Yes; no vertical line passes through two points on the graph.

12. No; there are vertical lines that pass through two points on the graph.

13. No; there are vertical lines that pass through two or more points on the graph.

RELATIONS AND FUNCTIONS **In Exercises 14–19, decide whether the relation is a function. If it is a function, give the domain and the range.**

14.
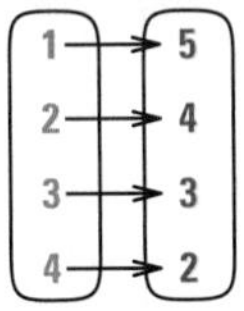

yes; 1, 2, 3, 4; 5, 4, 3, 2

15.
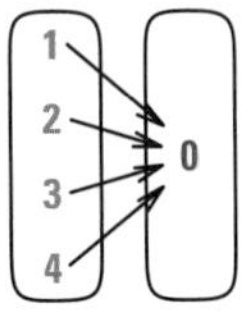

yes; 1, 2, 3, 4; 0

16.
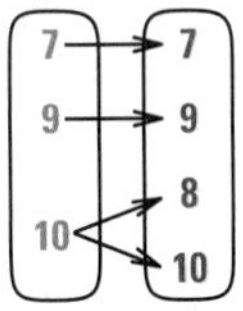

no

17. **yes; 0, 1, 2, 3; 2, 4, 6, 8**

Input	Output
0	2
1	4
2	6
3	8

18. **no**

Input	Output
0	1
2	2
4	3
2	4

19. **yes; 1, 3, 5, 7; 1, 2, 3**

Input	Output
1	1
3	2
5	3
7	1

STUDENT HELP

HOMEWORK HELP
Example 1: Exs. 11–19
Example 2: Exs. 20–28
Example 3: Exs. 29–40
Example 4: Exs. 50–52

3 APPLY

ASSIGNMENT GUIDE

BASIC
Day 1: pp. 259–260 Exs. 11–19, 20–28 even, 29–31, 32–48 even
Day 2: pp. 260–262 Exs. 33–49 odd, 50–52, 56–58, 65, 70, 72–79, Quiz 3 Exs. 1–19

AVERAGE
Day 1: pp. 259–260 Exs. 11–19, 20–28 even, 29–31, 32–48 even
Day 2: pp. 260–262 Exs. 33–49 odd, 50–53, 56–58, 65, 70, 72–79, Quiz 3 Exs. 1–19

ADVANCED
Day 1: pp. 259–260 Exs. 11–19, 20–28 even, 29–31, 32–48 even
Day 2: pp. 260–262 Exs. 33–49 odd, 53–55, 59–62, 65, 70, 72–79, Quiz 3 Exs. 1–19

BLOCK SCHEDULE
pp. 259–262 Exs. 11–19, 20–28 even, 29–31, 32–49, 50–53, 56–58, 65, 70, 72–79, Quiz 3 Exs. 1–19

EXERCISE LEVELS

Level A: ***Easier***
11–13, 29–31, 63–66

Level B: ***More Difficult***
14–28, 32–40, 42–52, 56–58, 67–79

Level C: ***Most Difficult***
41, 53–55, 59–62

✓ HOMEWORK CHECK

To quickly check student understanding of key concepts, go over the following exercises: Exs. 12, 16, 18, 22, 32, 29, 42, 50. See also the Daily Homework Quiz:

- Blackline Master (*Chapter 5 Resource Book,* p. 11)
- Transparency (p. 37)

STUDENT HELP NOTES

Homework Help Students can find help for Exs. 42–49 at **www.mcdougallittell.com.** The information can be printed out for students who don't have access to the Internet.

APPLICATION NOTE

EXERCISES 50–52 Although the relationships between years and enrollments represent functions, they cannot be modeled by simple linear functions. Each function could be modeled by the union of a set of linear functions over restricted domains.

32.

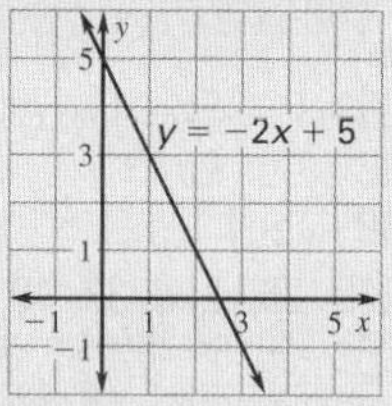

33.

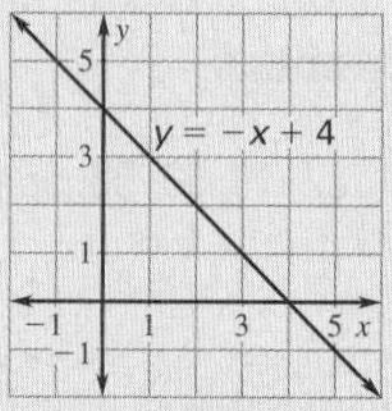

34.

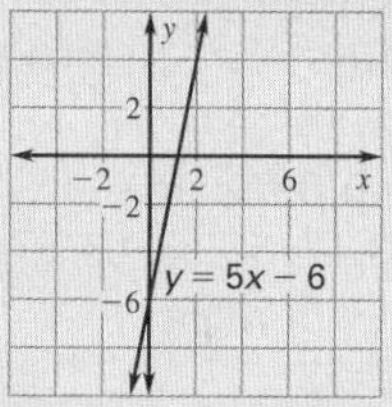

35.

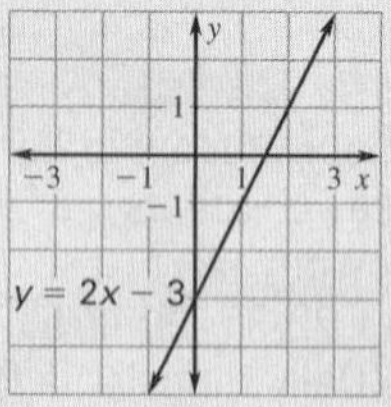

36.

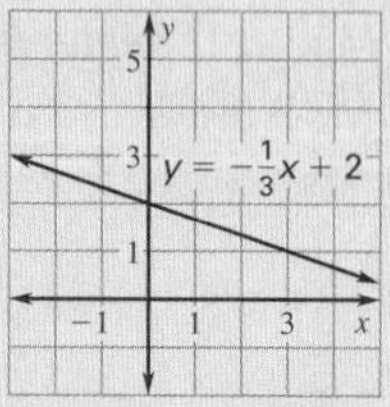

37–40, 59. See Additional Answers beginning on page AA1.

41. The points (1, 2) and (3, −1) are on the graph, so the slope is $\frac{-1-2}{3-1} = -\frac{3}{2}$.

STUDENT HELP

HOMEWORK HELP Visit our Web site www.mcdougallittell.com for help with Exs. 42–49.

50. Yes; yes; in both cases, each input is paired with exactly one output.

EVALUATING FUNCTIONS Evaluate the function when $x = 2$, $x = 0$, and $x = -3$.

20. $f(x) = 10x + 1$ **21; 1; −29**

21. $g(x) = 8x - 2$ **14; −2; −26**

22. $h(x) = 3x + 6$ **12; 6; −3**

23. $g(x) = 1.25x$ **2.5; 0; −3.75**

24. $h(x) = 0.75x + 8$ **9.5; 8; 5.75**

25. $f(x) = 0.33x - 2$ **−1.34; −2; −2.99**

26. $h(x) = \frac{3}{4}x - 4$ **−2.5; −4; −6.25**

27. $g(x) = \frac{2}{5}x + 7$ **7.8; 7; 5.8**

28. $f(x) = \frac{2}{7}x + 4$ **$4\frac{4}{7}$; 4; $3\frac{1}{7}$**

GRAPHICAL REASONING Match the function with its graph.

A. $f(x) = 3x - 2$ **B.** $f(x) = 2x + 2$ **C.** $f(x) = \frac{1}{2}x - 2$

29.

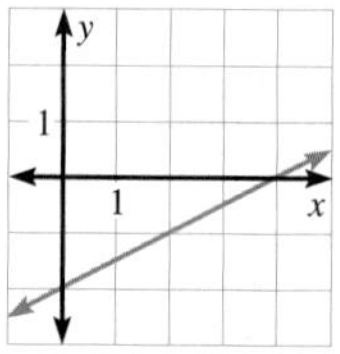

C

30.

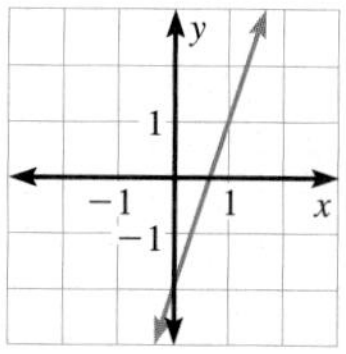

A

31.

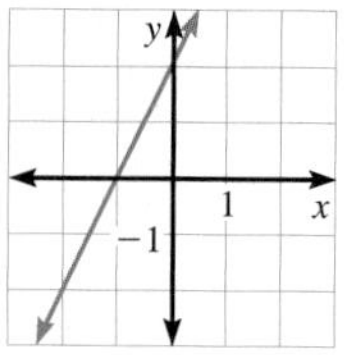

B

GRAPHING FUNCTIONS In Exercises 32–40, graph the function. **32–40. See margin.**

32. $f(x) = -2x + 5$ **33.** $g(x) = -x + 4$ **34.** $h(x) = 5x - 6$

35. $g(x) = 2x - 3$ **36.** $h(x) = -\frac{1}{3}x + 2$ **37.** $f(x) = -\frac{1}{2}x + 1$

38. $f(x) = 4x + 1$ **39.** $h(x) = 5$ **40.** $g(x) = -3x - 2$

41. *Writing* Describe how to find the slope of the graph of linear function f given that $f(1) = 2$ and $f(3) = -1$. **See margin.**

FINDING SLOPE Find the slope of the graph of the linear function f.

42. $f(2) = -3, f(-2) = 5$ **−2**

43. $f(0) = 4, f(4) = 0$ **−1**

44. $f(-3) = -9, f(3) = 9$ **3**

45. $f(6) = -1, f(3) = 8$ **−3**

46. $f(9) = -1, f(-1) = 2$ **−0.3**

47. $f(-2) = -1, f(2) = 6$ **1.75**

48. $f(2) = 2, f(3) = 3$ **1**

49. $f(-1) = 2, f(3) = 2$ **0**

HIGH SCHOOL AND COLLEGE STUDENTS In Exercises 50–52, use the graph below. It shows the projected number of high school and college students in the United States for different years. Let x be the number of years since 1980. **51–52. Estimates may vary.**

50. Is the projected high school enrollment a function of the year? Is the projected college enrollment a function of the year? Explain.

51. Let $f(x)$ represent the projected number of high school students in year x. Estimate $f(2005)$. **about 16 million**

52. Let $g(x)$ represent the projected number of college students in year x. Estimate $g(2005)$. **about 17 million**

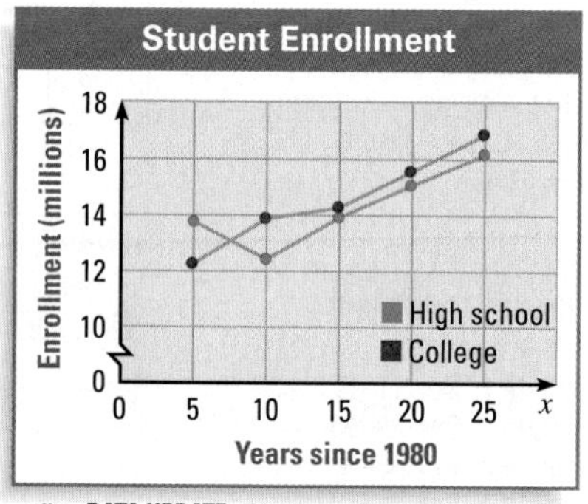

DATA UPDATE of National Center for Education Statistics data at www.mcdougallittell.com

FOCUS ON APPLICATIONS

REAL LIFE **ZYDECO MUSIC** blends elements from a variety of cultures. The accordion has German origins. The rub board was invented by Louisiana natives whose roots go back to France and Canada.

53. ZYDECO MUSIC The graph shows the number of people who attended the Southwest Louisiana Zydeco Music Festival for different years, where k is the number of years since 1980. Is the number of people who attended the festival a function of the year? Explain.

▶ Source: Louisiana Zydeco Music Festival

Yes; each input is paired with exactly one output.

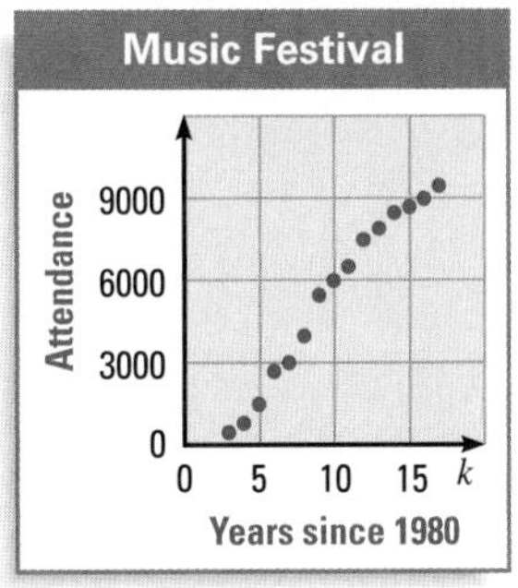

54. MASTERS TOURNAMENT The table shows information about the top 7 winners of the 1997 Masters Tournament in Augusta, Georgia. Graph the relation. Is the money earned a function of the score? Explain. ▶ Source: Golfweb

Check graphs; yes; each input is paired with exactly one output.

Score	270	282	283	284	285	285	286
Prize ($)	486,000	291,600	183,600	129,600	102,600	102,600	78,570

55. SCIENCE CONNECTION It takes 4.25 years for starlight to travel 25 trillion miles. Let t be the number of years and let $f(t)$ be trillions of miles traveled. Write a linear function $f(t)$ that expresses the distance traveled as a function of time. **$f(t) = \frac{100}{17}t$, or $f(t) \approx 5.88t$**

Test Preparation

QUANTITATIVE COMPARISON **In Exercises 56–58, choose the statement that is true about the given quantities.**

Ⓐ The number in column A is greater.

Ⓑ The number in column B is greater.

Ⓒ The two numbers are equal.

Ⓓ The relationship cannot be determined from the given information.

	Column A	Column B	
56.	$f(x) = -7x + 4$ when $x = -4$	$f(x) = -7x + 4$ when $x = -5$	**B**
57.	$f(x) = \frac{1}{2}x - 3$ when $x = 7$	$f(x) = \frac{1}{2}x - 5$ when $x = 11$	**C**
58.	$g(x) = 6x - \frac{1}{7}$ when $x = \frac{2}{7}$	$h(t) = 6t - \frac{1}{7}$ when $t = \frac{2}{7}$	**C**

60. If $f(x) = \frac{1}{x}$, there is no output value corresponding to the input value 0; if $f(x) = \frac{1}{x+3}$, there is no output value corresponding to the input value -3.

61. Let the domain of $f(x) = \frac{1}{x}$ be all real numbers except 0; let the domain of $f(x) = \frac{1}{x+3}$ be all real numbers except -3.

★ Challenge

RESTRICTING DOMAINS **In Exercises 59–62, use the equations $y = \frac{1}{x}$ and $y = \frac{1}{x+3}$.** **60–62. See margin.**

59. Use a graphing calculator or computer to graph both equations. **See margin for graph.**

60. Explain why each equation is not a function for all real values of x.

61. How can you restrict the domain of each equation so that it is a function?

62. Describe what happens to the graph of $y = \frac{1}{x}$ near the value(s) of x for which the graph is not a function.

62. For $x < 0$ but close to 0, the graph falls sharply. For $x > 0$ but close to 0, the graph rises sharply.

EXTRA CHALLENGE

www.mcdougallittell.com

APPLICATION NOTE

EXERCISE 53 Students should recognize that the data do not represent a linear function, although a linear function can be used for modeling purposes. Additional information about Zydeco Music is available at **www.mcdougallittell.com.**

ADDITIONAL PRACTICE AND RETEACHING

For Lesson 4.8:

- Practice Levels A, B, and C (*Chapter 4 Resource Book,* p. 112)
- Reteaching with Practice (*Chapter 4 Resource Book,* p. 115)
- See Lesson 4.8 of the *Personal Student Tutor*

For more Mixed Review:

- Search the *Test and Practice Generator* for key words or specific lessons.

4 ASSESS

DAILY HOMEWORK QUIZ

Transparency Available

1. Does the graph represent a function? Explain your answer.

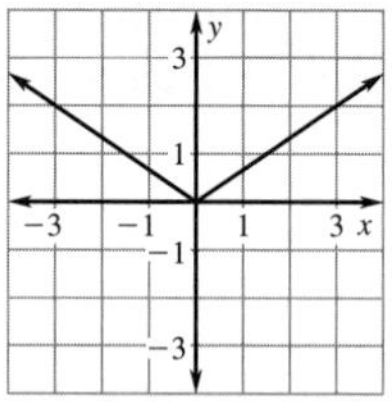

Yes; no vertical line passes through two points of the graph.

2. Evaluate $f(x) = -3x + 2$ when $x = -1, 0,$ and 3. **5, 2, −7**

3. Find the slope of the graph of the linear function f if $f(3) = -2$ and $f(0) = 1$. **−1**

EXTRA CHALLENGE NOTE

Challenge problems for Lesson 4.8 are available in **blackline** format in the *Chapter 4 Resource Book,* p. 119 and at **www.mcdougallittell.com.**

ADDITIONAL TEST PREPARATION

1. WRITING Explain the relationship between a relation and a function. **A function is a relation in which each value in the domain is associated with exactly one value in the range.**

2. OPEN ENDED Give an example of a linear function that models a real-world problem. Explain the significance of one point on the graph of the function. **See sample answer at right.**

73–79. See Additional Answers beginning on page AA1.

63. $\begin{bmatrix} 9 & -11 \\ 12 & 7 \end{bmatrix}$

64. $\begin{bmatrix} -4.1 & -9.3 \\ 4.3 & 0.7 \end{bmatrix}$

65. $\begin{bmatrix} 9.8 & -17.9 \\ -12.3 & -0.1 \\ 12.4 & -13.2 \end{bmatrix}$

66. $\begin{bmatrix} 3 & 4 & 1 \\ -6 & -3 & -3 \\ -5 & 5 & 0 \end{bmatrix}$

MIXED REVIEW

MATRICES **Find the sum or the difference of the matrices.** (Review 2.4) **63–66. See margin.**

63. $\begin{bmatrix} 4 & -8 \\ 7 & 0 \end{bmatrix} - \begin{bmatrix} -5 & 3 \\ -5 & -7 \end{bmatrix}$

64. $\begin{bmatrix} -6.5 & -4.2 \\ 0 & 3.7 \end{bmatrix} + \begin{bmatrix} 2.4 & -5.1 \\ 4.3 & -3 \end{bmatrix}$

65. $\begin{bmatrix} 6.2 & -12 \\ -2.5 & -4.4 \\ 3.4 & -5.8 \end{bmatrix} - \begin{bmatrix} -3.6 & 5.9 \\ 9.8 & -4.3 \\ -9 & 7.4 \end{bmatrix}$

66. $\begin{bmatrix} 9 & 1 & 6 \\ -4 & -7 & 1 \\ -5 & 0 & -1 \end{bmatrix} + \begin{bmatrix} -6 & 3 & -5 \\ -2 & 4 & -4 \\ 0 & 5 & 1 \end{bmatrix}$

SOLVING EQUATIONS **Solve the equation if possible.** (Review 3.4)

67. $4x + 8 = 24$ **4** **68.** $3n = 5n - 12$ **6** **69.** $9 - 5z = -8z$ **−3**

70. $-5y + 6 = 4y + 3$ $\frac{1}{3}$ **71.** $3b + 8 = 9b - 7$ $2\frac{1}{2}$ **72.** $-7q - 13 = 4 - 7q$ **no solution**

GRAPHING LINES **In Exercises 73–78, write the equation in slope-intercept form. Then graph the equation.** (Review 4.6 for 5.1) **73–78. See margin.**

73. $2x - y + 3 = 0$ **74.** $x + 2y - 6 = 0$ **75.** $y - 2x = -7$

76. $5x - y = 4$ **77.** $x - 2y + 4 = 2$ **78.** $4y + 12 = 0$

79. **CHARITY WALK** You start 5 kilometers from the finish line and walk $1\frac{1}{2}$ kilometers per hour for 2 hours, where d is your distance from the finish line. Graph the situation. (Review 4.6 for 5.1) **See margin.**

QUIZ 3

Self-Test for Lessons 4.7 and 4.8

Solve the equation graphically. Check your solution algebraically. (Lesson 4.7)

1. $4x + 3 = -5$ **−2** **2.** $6x - 12 = -9$ $\frac{1}{2}$ **3.** $8x - 7 = x$ **1**

4. $-5x - 4 = 3x$ $-\frac{1}{2}$ **5.** $\frac{1}{3}x + 5 = -\frac{2}{3}x - 8$ **−13** **6.** $\frac{3}{4}x + 2 = -\frac{3}{4}x - 6$ $-5\frac{1}{3}$

Evaluate the function when $x = 3$, $x = 0$, and $x = -4$. (Lesson 4.8)

7. $h(x) = 5x - 9$ **6; −9; −29** **8.** $g(x) = -4x + 3$ **−9; 3; 19** **9.** $f(x) = 1.75x - 2$ **3.25; −2; −9**

10. $h(x) = -1.4x$ **−4.2; 0; 5.6** **11.** $f(x) = \frac{1}{4}x + 9$ $9\frac{3}{4}$; 9; 8 **12.** $g(x) = \frac{4}{7}x + \frac{2}{7}$ 2; $\frac{2}{7}$; −2

In Exercises 13–18, graph the function. (Lesson 4.8) **13–18. See margin.**

13. $f(x) = -5x$ **14.** $h(x) = 4x - 7$ **15.** $f(x) = 3x - 2$

16. $h(x) = \frac{2}{5}x - 1$ **17.** $f(x) = \frac{1}{4}x + \frac{1}{2}$ **18.** $g(x) = -6x + 5$

19. **LUNCH BOX BUSINESS** You have a small business making and delivering box lunches. You calculate your average weekly cost y of producing x lunches using the function $y = 2.1x + 75$. Last week your cost was \$600. How many lunches did you make last week? Solve algebraically and graphically. (Lesson 4.7) **250 lunches**

Additional Test Preparation *Sample answer:*

2. I send my film away for processing. The company charges me \$4.10 for shipping and \$3.75 for processing each roll of film. The equation $y = 3.75x + 4.10$, where x = the number of rolls of film that are processed and y = the total cost, models the situation; the point (4, 19.1) represents the cost (\$19.10) of processing 4 rolls of film.

CHAPTER 4

Chapter Summary

WHAT did you learn?	*WHY did you learn it?*
Plot points and draw scatter plots. **(4.1)**	See relationships between two real-life quantities and make predictions. **(p. 205)**
Graph a linear equation in two variables	
• using a table of values. **(4.2)**	Model earnings from a business. **(p. 216)**
• using intercepts. **(4.3)**	Make a quick graph to help plan a fundraiser. **(p. 220)**
• using slope-intercept form. **(4.6)**	Make a quick graph of a river's heights. **(p. 243)**
Find the slope of a line passing through two points. **(4.4)**	Represent the steepness of a road. **(p. 231)**
Interpret slope as a rate of change. **(4.4)**	Describe the rate of change of a parachutist's height above the ground. **(p. 229)**
Write and graph direct variation equations. **(4.5)**	Model the relationship between lengths of stringed instruments. **(p. 238)**
Use a graph to check or approximate the solution of a linear equation. **(4.7)**	Model production costs for a business. **(p. 254)**
Identify, evaluate, and graph functions. **(4.8)**	Model projected school enrollments. **(p. 260)**

How does Chapter 4 fit into the BIGGER PICTURE of algebra?

In this chapter you saw that relationships between variables may be expressed in algebraic form as an equation or in geometric form as a graph. Recognizing and using the connection between equations and graphs is one of the most important skills you can acquire to help you solve real-life problems.

STUDY STRATEGY

How did you use your list of questions?

The list of questions and answers you made, using the **Study Strategy** on page 202, may resemble this one.

Getting Questions Answered

1. In Lesson 4.1, Exercises 19–26, how can I find the quadrant without plotting the points?

Answer: Mary told me that I just need to look at each coordinate to see the direction I move.

For example, with (4, −5), I go right and then down, so the point is in Quadrant IV.

13.

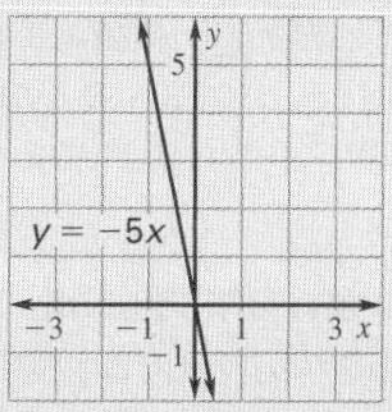

14.

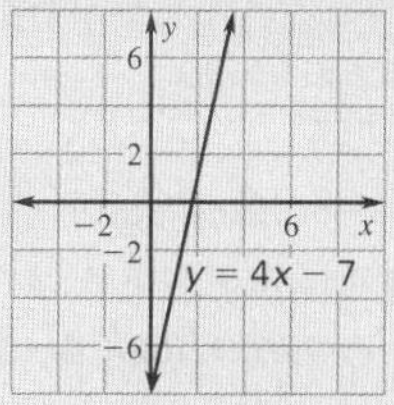

15.

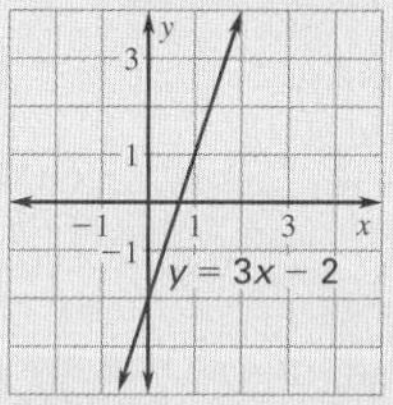

16.

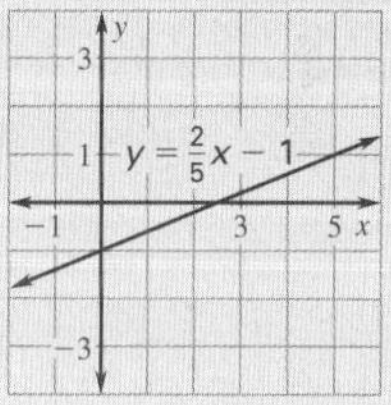

17–18. See Additional Answers beginning on page AA1.

ADDITIONAL RESOURCES

The following resources are available to help review the material in this chapter.

- Chapter Review Games and Activities (*Chapter 4 Resource Book,* p. 120)
- *Instant Replay: Video Review Games*
- *Personal Student Tutor*
- Cumulative Review, Chs. 1–4 (*Chapter 4 Resource Book,* p. 132)

1.

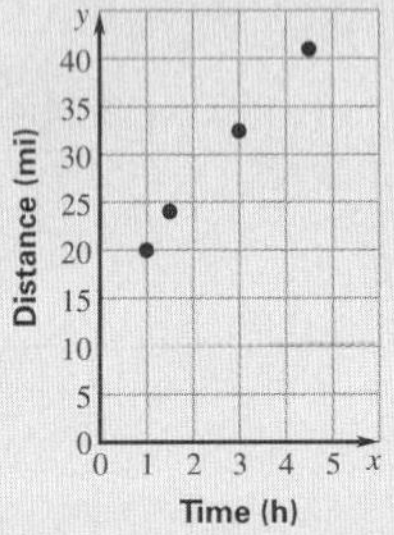

2–5.

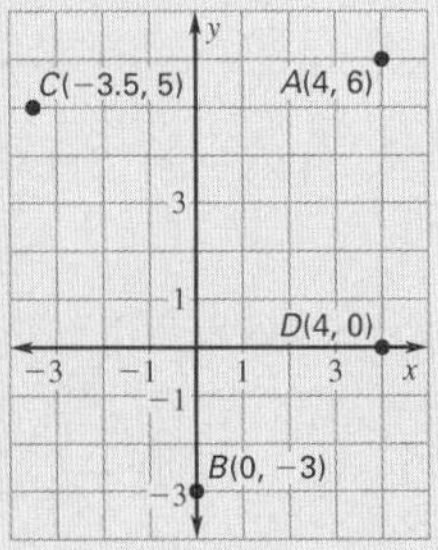

CHAPTER 4

Chapter Review

VOCABULARY

- coordinate plane, p. 203
- ordered pair, p. 203
- *x*-coordinate, p. 203
- *y*-coordinate, p. 203
- *x*-axis, p. 203
- *y*-axis, p. 203
- origin, p. 203
- quadrants, p. 203
- graph of an ordered pair, p. 203
- scatter plot, p. 204
- solution of an equation, p. 210
- graph of an equation, p. 210
- *x*-intercept, p. 218
- *y*-intercept, 218
- slope, p. 226
- rate of change, p. 229
- constant of variation, p. 234
- direct variation, p. 234
- slope-intercept form, p. 241
- parallel lines, p. 242
- relation, p. 256
- function notation, p. 257
- graph of a function, p. 257

4.1 COORDINATES AND SCATTER PLOTS

Examples on pp. 203–205

EXAMPLES

To plot the point (4, −2), start at the origin. Move 4 units to the right and 2 units down.

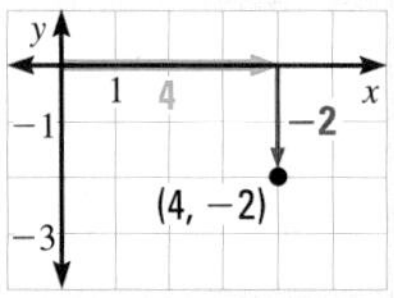

To plot the point (−3, 1), start at the origin. Move 3 units to the left and 1 unit up.

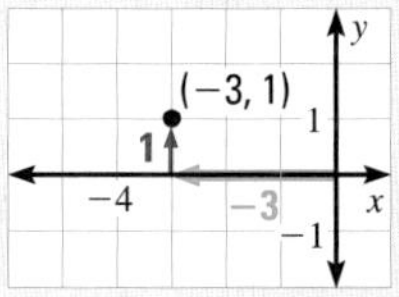

1. Make a scatter plot of the data in the table at the right. **See margin.**

Time (h)	1	1.5	3	4.5
Distance (mi)	20	24	32.5	41

Plot and label the ordered pair in a coordinate plane. **2–5. See margin.**

2. $A(4, 6)$ **3.** $B(0, -3)$ **4.** $C(-3.5, 5)$ **5.** $D(4, 0)$

4.2 GRAPHING LINEAR EQUATIONS

Examples on pp. 210–213

EXAMPLE To graph $3y = x - 6$, solve the equation for y, make a table of values, and plot the points.

x	-1	0	1
y	$-2\frac{1}{3}$	-2	$-1\frac{2}{3}$

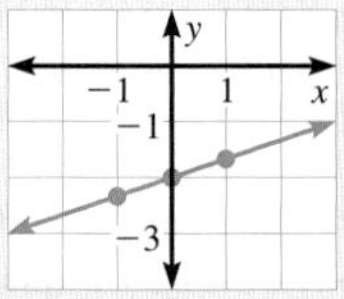

$3y = x - 6$ **Write original equation.**

$y = \frac{x - 6}{3}$ **Divide each side by 3.**

Graph the equation. **6–9. See margin.**

6. $y = 2x + 2$ **7.** $y = 7 - \frac{1}{2}x$ **8.** $y = -4(x + 1)$ **9.** $x - 10 = 2y$

4.3 QUICK GRAPHS USING INTERCEPTS

Examples on pp. 218–220

EXAMPLE To graph $y + 2x = 10$, first find the intercepts.

$y + 2x = 10$ $\qquad$ $y + 2x = 10$

$0 + 2x = 10$ $\qquad$ $y + 2(0) = 10$

$x = 5$ $\qquad$ $y = 10$

Plot (5, 0) and (0, 10). Then draw a line through the points.

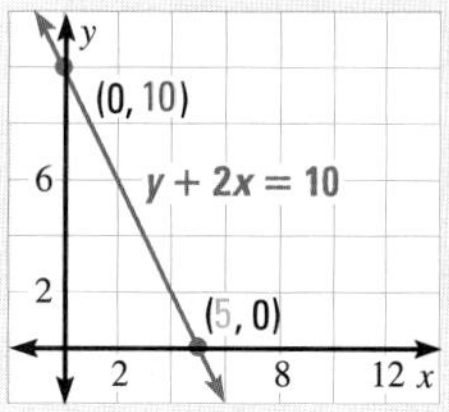

Graph the equation. Label the intercepts. **10–13. See margin.**

10. $-x + 4y = 8$ $\qquad$ **11.** $3x + 5y = 15$ $\qquad$ **12.** $4x - 5y = -20$ $\qquad$ **13.** $2x + 3y = 12$

4.4 THE SLOPE OF A LINE

Examples on pp. 227–229

EXAMPLE To find the slope of the line passing through the points $(-2, 5)$ and $(4, -7)$, let $(x_1, y_1) = (-2, 5)$ and $(x_2, y_2) = (4, -7)$.

$m = \frac{y_2 - y_1}{x_2 - x_1}$ **Write formula for slope.**

$m = \frac{-7 - 5}{4 - (-2)}$ **Substitute values.**

$m = \frac{-12}{6}$ **Simplify.**

$m = -2$ **Slope is negative.**

Plot the points and find the slope of the line passing through the points.

14. (2, 1), (3, 4) **3** $\qquad$ **15.** (0, 8), (−1, 2) **6** $\qquad$ **16.** (2, 4), (5, 0) $-\frac{4}{3}$ $\qquad$ **17.** (0, 5), (−4, 5) **0**

4.5 DIRECT VARIATION

Examples on pp. 234–236

EXAMPLE If x and y vary directly, the equation that relates x and y is of the form $y = kx$. If $x = 3$ when $y = 18$, then you can write an equation that relates x and y.

$y = kx$ **Write model for direct variation.**

$18 = k(3)$ **Substitute 3 for x and 18 for y.**

$6 = k$ **Divide each side by 3.**

An equation that relates x and y is $y = 6x$.

The variables *x* and *y* vary directly. Use the given values of the variables to write an equation that relates *x* and *y*.

18. $x = 7, y = 35$ $\quad$ $y = 5x$ $\qquad$ **19.** $x = 12, y = -4$ $\quad$ $y = -\frac{1}{3}x$ $\qquad$ **20.** $x = 4, y = -16$ $\quad$ $y = -4x$ $\qquad$ **21.** $x = 3, y = 10.5$ $\quad$ $y = 3.5x$

6.

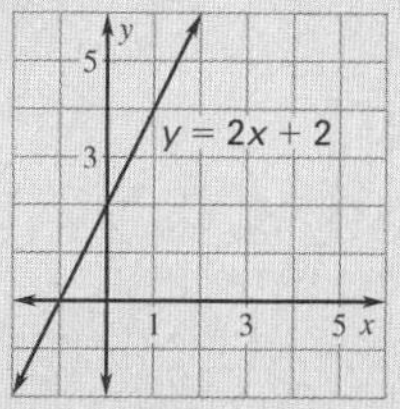

7.

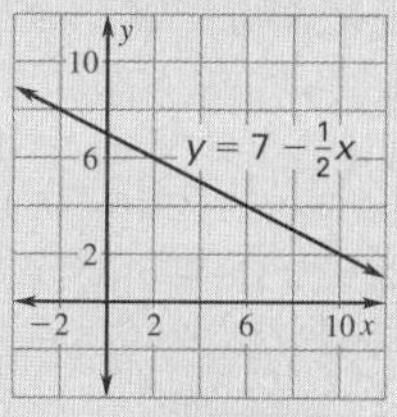

8.

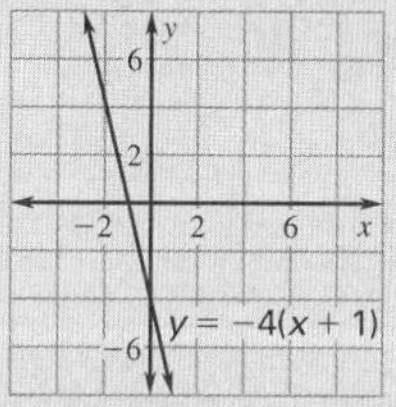

9.

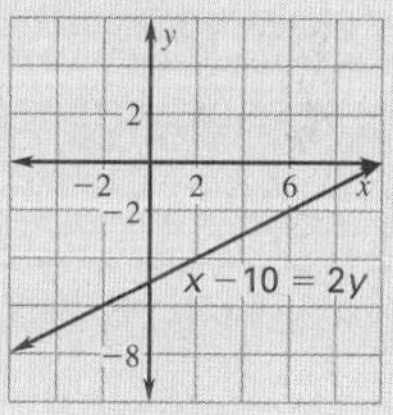

10–13. See Additional Answers beginning on page AA1.

22.

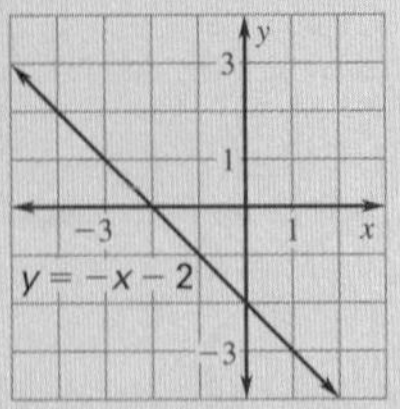

23.

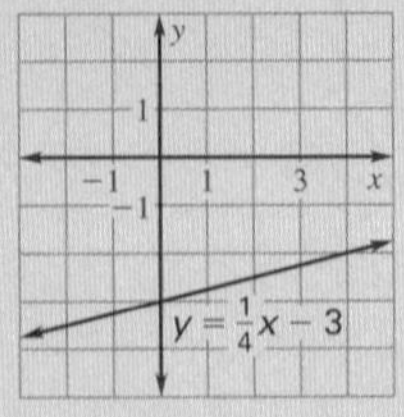

24.

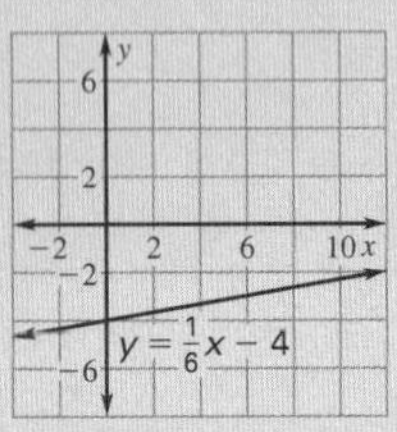

29.

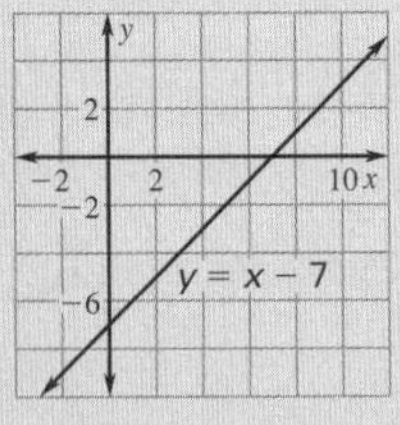

30–32. See Additional Answers beginning on page AA1.

4.6 QUICK GRAPHS USING SLOPE-INTERCEPT FORM

Examples on pp. 241–243

EXAMPLE Use the following steps to graph $4x + y = 0$.

STEP 1 Write the equation in $y = mx + b$ form: $y = -4x$.

STEP 2 Find the slope and the y-intercept: $m = -4$, $b = 0$.

STEP 3 Plot the point $(0, 0)$. Draw a slope triangle to locate a second point on the line. Draw a line through the two points.

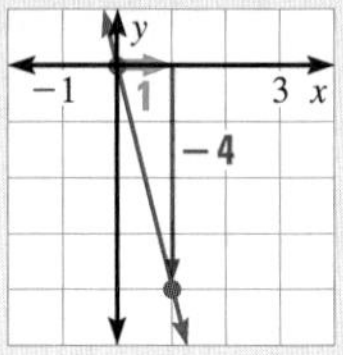

Graph the equation. **22–24. See margin.**

22. $y = -x - 2$ **23.** $x - 4y = 12$ **24.** $-x + 6y = -24$

4.7 SOLVING LINEAR EQUATIONS USING GRAPHS

Examples on pp. 250–252

EXAMPLE You can solve the equation $2x - 4 = 2$ graphically.

$2x - 4 = 2$	**Write original equation.**
$2x - 6 = 0$	**Write in form $ax + b = 0$.**
$y = 2x - 6$	**Write related function $y = ax + b$.**

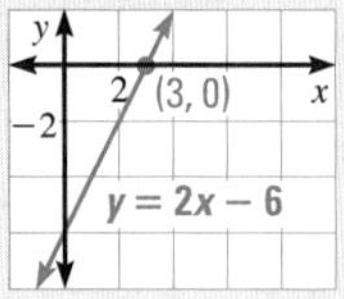

Graph $y = 2x - 6$. The x-intercept is 3, so the solution is $x = 3$.

✓CHECK

$2x - 4 = 2$	**Write original equation.**
$2(3) - 4 \stackrel{?}{=} 2$	**Substitute 3 for x.**
$2 = 2$	**True statement.**

Solve the equation graphically. Check your solution algebraically.

25. $3x - 6 = 0$ **2** **26.** $-5x - 3 = 0$ $-\frac{3}{5}$ **27.** $3x + 8 = 4x$ **8** **28.** $-4x - 1 = 7$ **−2**

4.8 FUNCTIONS AND RELATIONS

Examples on pp. 256–258

EXAMPLE To evaluate the function $f(x) = -\frac{1}{5}x + 1$ when $x = 15$, substitute the given value for x.

$f(x) = -\frac{1}{5}x + 1$	**Write original function.**
$f(15) = -\frac{1}{5}(15) + 1$	**Substitute 15 for x.**
$f(15) = -2$	**Simplify.**

Evaluate the function for the given value of x. Then graph the function. **29–32. See margin for graphs.**

29. $f(x) = x - 7$ when $x = -2$ **−9** **30.** $f(x) = -x + 4$ when $x = 4$ **0**

31. $f(x) = 2x + 6$ when $x = -3$ **0** **32.** $f(x) = 1.5x - 4.2$ when $x = -9$ **−17.7**

CHAPTER 4 Chapter Test

Plot and label the points in a coordinate plane. **1–4. See margin.**

1. $A(2, 6), B(-4, -1), C(-1, 4), D(3, -5)$

2. $A(-5, 1), B(0, 3), C(-1, -5), D(4, 6)$

3. $A(7, 3), B(-2, -2), C(0, 4), D(6, -2)$

4. $A(0, -1), B(-5, -6), C(7, -2), D(2, 4)$

Graph the line that has the given intercepts. **5–8. See margin.**

5. x-intercept: 3
y-intercept: -1

6. x-intercept: -5
y-intercept: 4

7. x-intercept: 6
y-intercept: 6

8. x-intercept: $-\frac{1}{2}$
y-intercept: -3

Use a table of values to graph the equation. **9–12. See margin.**

9. $y = -x + 3$

10. $y = 4$

11. $y = -(5 - x)$

12. $x = 6$

Graph the equation. Tell which method you used. **13–16. Methods may vary. Samples are given; See margin.**

13. $2x + y - 11 = 0$ **using slope-intercept form**

14. $3x - 2y - 2 = 0$ **using intercepts**

15. $-7x - y + 49 = 0$ **using intercepts**

16. $\frac{2}{3}x + y - 32 = 0$ **using slope-intercept form**

Plot the points and find the slope of the line passing through the points.

17. $(0, 1), (-2, -6)$ **$\frac{7}{2}$**

18. $(-4, -1), (5, -7)$ **$-\frac{2}{3}$**

19. $(-3, 5), (2, -2)$ **$-\frac{7}{5}$**

20. $(-3, -1), (2, -1)$ **0**

The variables *x* and *y* vary directly. Use the given values of the variables to write an equation that relates *x* and *y*.

21. $x = -2, y = -2$ **$y = x$**

22. $x = 2, y = 10$ **$y = 5x$**

23. $x = -3, y = 7$ **$y = -\frac{7}{3}x$**

24. $x = \frac{1}{2}, y = 6$ **$y = 12x$**

25. $x = 1.3, y = 3.9$ **$y = 3x$**

26. $x = 16, y = 3.2$ **$y = 0.2x$**

In Exercises 27 and 28, decide whether the graphs of the two equations are parallel lines. Explain your answer.

27. $y = 4x + 3, y = -4x - 5$
No; the lines have different slopes.

28. $10y + 20 = 6x, 5y = 3x + 35$
Yes; both lines have slope $\frac{3}{5}$.

29. Solve $x - 2 = -3x$ graphically. Check your solution algebraically. **$\frac{1}{2}$**

In Exercises 30–32, evaluate the function when $x = 3$, $x = 0$, and $x = -4$.

30. $f(x) = 6x$ **18; 0; −24**

31. $f(x) = -(x - 2)$ **−1; 2; 6**

32. $g(x) = 3.2x + 2.8$ **12.4; 2.8; −10**

33. **FLOOD WATERS** A river has risen 6 feet above flood stage. Beginning at time $t = 0$, the water level drops at a rate of two inches per hour. The number of feet above flood stage y after t hours is given by $y = 6 - \frac{1}{6}t$. Graph the equation over the 12-hour period from $t = 0$ to $t = 12$. **See margin.**

34. **SHOE SIZES** The table below shows how foot length relates to women's shoe sizes. Is shoe size a function of foot length? Why or why not? **Yes; each input is paired with exactly one output.**

Foot length (in inches), x	$9\frac{1}{4}$	$9\frac{1}{2}$	$9\frac{5}{8}$	$9\frac{3}{4}$	$9\frac{15}{16}$	$10\frac{1}{4}$	$10\frac{1}{2}$
Shoe size, y	$6\frac{1}{2}$	7	7	8	8	$9\frac{1}{2}$	$9\frac{1}{2}$

ADDITIONAL RESOURCES
- ***Chapter 4 Resource Book***
  Chapter Test (3 levels) (p. 121)
  SAT/ACT Chapter Test (p. 127)
  Alternative Assessment (p. 128)
- ***Test and Practice Generator***

1.
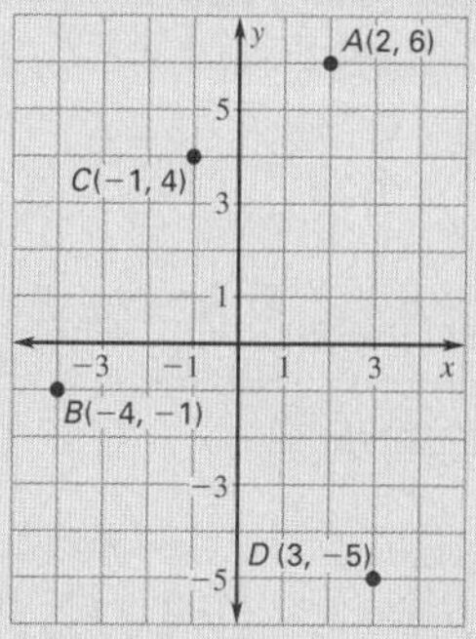

2.
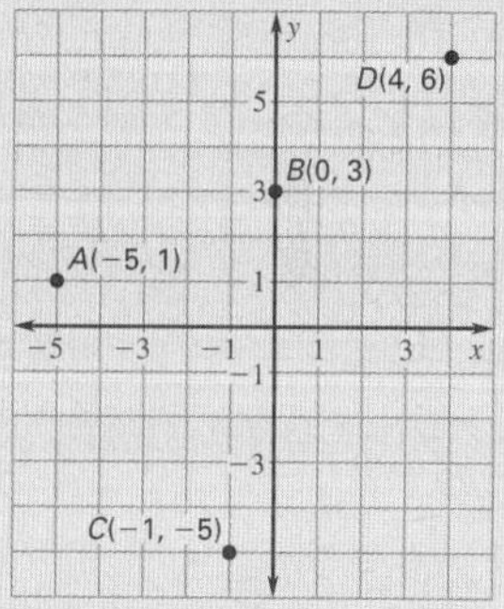

3.
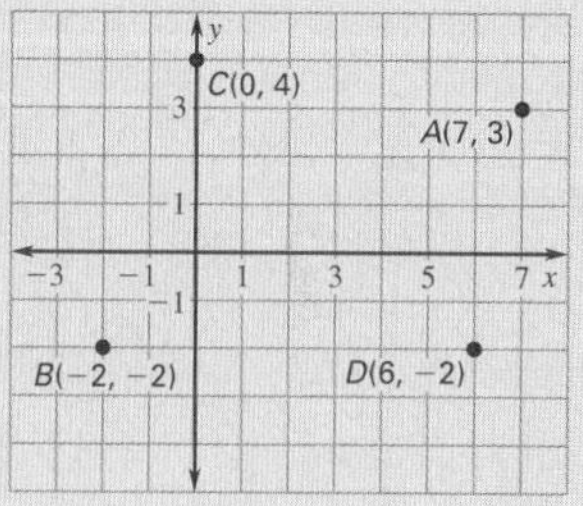

4.
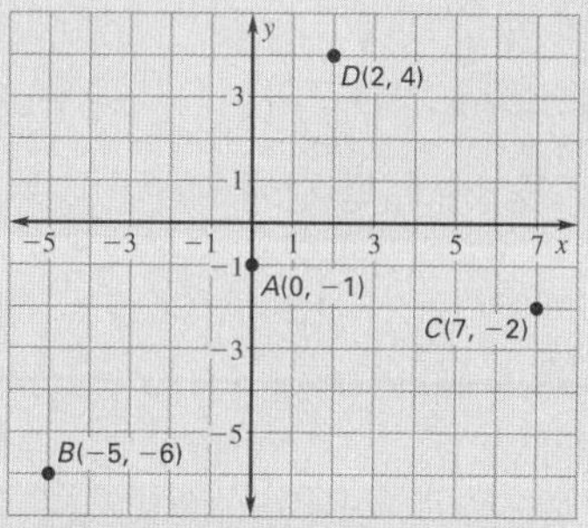

5–16, 33. See Additional Answers beginning on page AA1.

ADDITIONAL RESOURCES

- ***Chapter 4 Resource Book***
  Chapter Test (3 levels) (p. 121)
  SAT/ACT Chapter Test (p. 127)
  Alternative Assessment (p. 128)
- ***Test and Practice Generator***

CHAPTER 4

Chapter Standardized Test

TEST-TAKING STRATEGY Read all of the answer choices before deciding which is the correct one.

1. MULTIPLE CHOICE What is the equation of the line shown? **E**

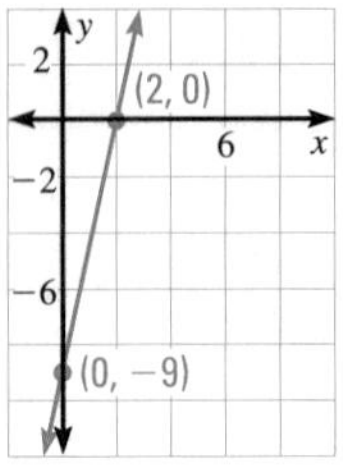

Ⓐ $9x - 2y = -18$

Ⓑ $-9x - 2y = 18$

Ⓒ $9x + 2y = 18$

Ⓓ $9x + 2y = -18$

Ⓔ $-9x + 2y = -18$

2. MULTIPLE CHOICE What is the y-intercept of the line $-4x - \frac{1}{2}y = 10$? **A**

Ⓐ -20 Ⓑ -4

Ⓒ $-\frac{5}{2}$ Ⓓ 5

Ⓔ 20

3. MULTIPLE CHOICE Write the equation $3x - 4y = 20$ in slope-intercept form. **C**

Ⓐ $y = -\frac{3}{4}x - 5$

Ⓑ $y = -\frac{3}{4}x + 5$

Ⓒ $y = \frac{3}{4}x - 5$

Ⓓ $y = \frac{3}{4}x + 5$

Ⓔ $y = 20 - 3x$

4. MULTIPLE CHOICE Find the slope of the line passing through the points $(1, 2)$ and $(2, 1)$. **D**

Ⓐ 1 Ⓑ 3

Ⓒ 2 Ⓓ -1

Ⓔ -2

5. MULTIPLE CHOICE What is the slope of the line shown? **C**

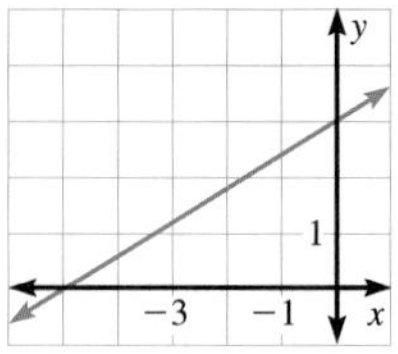

Ⓐ -5 Ⓑ $-\frac{3}{5}$

Ⓒ $\frac{3}{5}$ Ⓓ $\frac{5}{3}$

Ⓔ 3

6. MULTIPLE CHOICE What is the slope of the graph of the equation $5x - y = -2$? **B**

Ⓐ -5 Ⓑ 5

Ⓒ 1 Ⓓ -2

Ⓔ 2

7. MULTIPLE CHOICE Which point does *not* lie on the graph of $x = -12$? **D**

Ⓐ $(-12, 0)$

Ⓑ $(-12, -12)$

Ⓒ $(-12, 1)$

Ⓓ $(-1, -12)$

Ⓔ $(-12, 12)$

8. MULTIPLE CHOICE What is the x-intercept of $-13x - y = -65$? **D**

Ⓐ -65 Ⓑ -5

Ⓒ 0 Ⓓ 5

Ⓔ 65

9. MULTIPLE CHOICE Find the value of $f(x) = -x^2 - 6x - 7$ when $x = -2$. **C**

Ⓐ -23 Ⓑ -15

Ⓒ 1 Ⓓ 7

Ⓔ 9

QUANTITATIVE COMPARISON In Exercises 10–12, choose the statement below that is true about the given numbers.

Ⓐ The number in column A is greater.

Ⓑ The number in column B is greater.

Ⓒ The two numbers are equal.

Ⓓ The relationship cannot be determined from the given information.

	Column A	Column B	
10.	The slope of the line through $(4, -3)$ and $(-12, -3)$	0	**C**
11.	The slope of the line through $(4.5, 6)$ and $(-7, 4)$	The slope of the line through $(-6, 4.5)$ and $(4, -7)$	**A**
12.	The slope of the line through $(3.5, y)$ and $(6.8, 4)$	The slope of the line through $(3.5, q)$ and $(6.8, 4)$	**D**

13. MULTI-STEP PROBLEM An Internet provider offers three different levels of monthly service.

Standard: \$10 for the first 10 hours and \$1 for each additional hour.

Upgrade: \$15 for the first 20 hours and \$1 for each additional hour.

Unlimited: \$20 per month with no hourly charge.

a. Tell whether each graph represents Standard, Upgrade, or Unlimited service. Explain your reasoning.

13. a. Graph I: Upgrade; Graph II: Unlimited; Graph III: Standard; explanations will vary.

I.

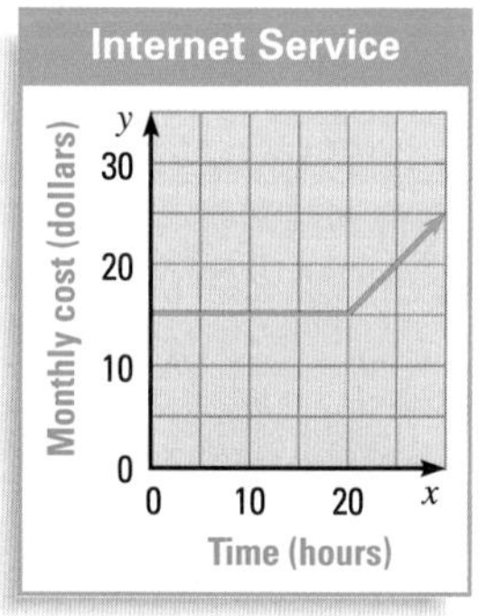

II.

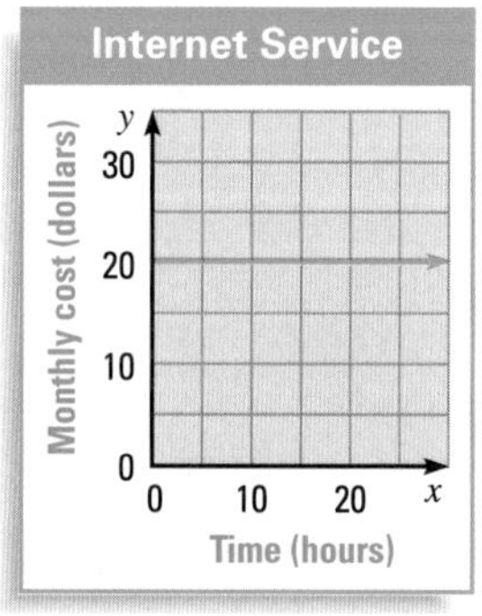

III.

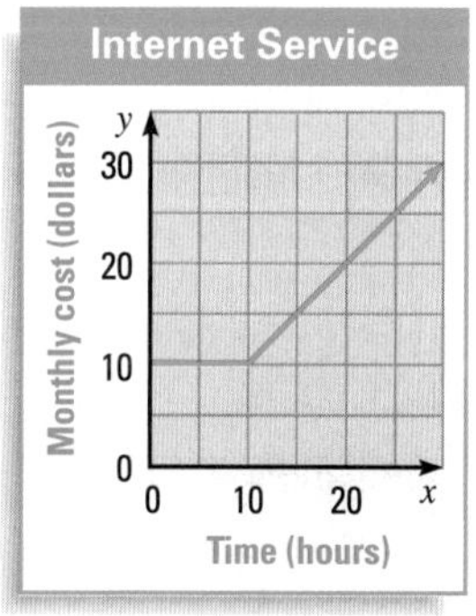

b. Write an equation for the total cost T per month for Upgrade service as a function of the number of additional hours used b. **$T = 15 + b$**

c. If you use the Internet 13 hours per month, which service will cost the least? Explain. **Standard; the cost for Standard service is less for any number of hours up to 15 hours.**

d. If you use the Internet 24 hours per month, which service will cost the least? Explain. **Upgrade at \$19; Standard would cost \$24 and Unlimited would cost \$20.**

e. The equation $C = 10 + a$ gives the total cost C per month for Standard service as a function of the number of additional hours used a. Explain how this model is different from the one you labeled Standard in part (a). **See margin.**

e. *Sample answer:* the model in part (a) gives cost as a function of total time. The model $C = 10 + a$ gives cost as a function of time beyond 10 hours.

CHAPTER 5

PLANNING THE CHAPTER

Writing Linear Equations

LESSON	GOALS	NCTM	ITED	SAT9	Terra-Nova	Local
5.1 *pp. 273–278*	GOAL 1 Use the slope-intercept form to write the equation of a line. GOAL 2 Model a real-life situation with a linear function.	1, 2, 4, 6, 8, 9, 10	APRA, RQRA		16	
5.2 *pp. 279–284*	GOAL 1 Use slope and any point on a line to write an equation of the line. GOAL 2 Use a linear model to make predictions about a real-life situation.	1, 2, 3, 4, 6, 9, 10	APRA, RQRA		16, 18	
5.3 *pp. 285–291*	GOAL 1 Write an equation of a line given two points on the line. GOAL 2 Use a linear equation to model a real-life problem.	1, 2, 3, 4, 6, 8, 9, 10	APRA, RQRA		16, 18	
5.4 *pp. 292–299*	GOAL 1 Find a linear equation that approximates a set of data points. GOAL 2 Determine whether there is a positive or negative correlation in a set of real-life data. TECHNOLOGY ACTIVITY: 5.4 *Find the best-fitting line for data on a graphing calculator.*	2, 3, 4, 6, 9, 10	MIG	9, 21	14, 15, 16	
5.5 *pp. 300–306*	GOAL 1 Use the point-slope form to write an equation of a line. GOAL 2 Use the point-slope form to model a real-life situation.	1, 2, 3, 4, 6, 8, 9, 10	APRA, RQRA		16, 18	
5.6 *pp. 307–314*	CONCEPT ACTIVITY: 5.6 *Investigate the standard form of a linear equation.* GOAL 1 Write a linear equation in standard form. GOAL 2 Use the standard form of an equation to model real-life situations.	2, 6, 8, 9, 10	APRA, RQRA	34	16, 18	
5.7 *pp. 315–322*	CONCEPT ACTIVITY: 5.7 *Investigate linear modeling.* GOAL 1 Determine whether a linear model is appropriate. GOAL 2 Use a linear model to make a real-life prediction.	2, 3, 6, 9, 10	IIG, MIG		18	

RESOURCES

CHAPTER RESOURCE BOOKLETS

CHAPTER SUPPORT

Tips for New Teachers	p. 1	Prerequisite Skills Review	p. 5
Parent Guide for Student Success	p. 3	Strategies for Reading Mathematics	p. 7

LESSON SUPPORT

LESSON SUPPORT	5.1	5.2	5.3	5.4	5.5	5.6	5.7
Lesson Plans (regular and block)	p. 9	p. 21	p. 34	p. 49	p. 63	p. 77	p. 92
Warm-Up Exercises and Daily Quiz	p. 11	p. 23	p. 36	p. 51	p. 65	p. 79	p. 94
Activity Support Masters						p. 80	p. 95
Lesson Openers	p. 12	p. 24	p. 37	p. 52	p. 66	p. 81	p. 96
Graphing Calculator Activities & Keystrokes		p. 25	p. 38	p. 53		p. 82	p. 97
Practice (3 levels)	p. 13	p. 26	p. 40	p. 54	p. 67	p. 84	p. 98
Reteaching with Practice	p. 16	p. 29	p. 43	p. 57	p. 70	p. 87	p. 101
Quick Catch-Up for Absent Students	p. 18	p. 31	p. 45	p. 59	p. 72	p. 89	p. 103
Cooperative Learning Activities				p. 60			
Interdisciplinary Applications	p. 19		p. 46		p. 73		
Real-Life Applications		p. 32		p. 61		p. 90	p. 104
Math & History Applications					p. 74		
Challenge: Skills and Applications	p. 20	p. 33	p. 47	p. 62	p. 75	p. 91	p. 105

REVIEW AND ASSESSMENT

Quizzes	pp. 48, 76	Alternative Assessment with Math Journal	p. 114
Chapter Review Games and Activities	p. 106	Project with Rubric	p. 116
Chapter Test (3 levels)	pp. 107–112	Cumulative Review	p. 118
SAT/ACT Chapter Test	p. 113	Resource Book Answers	p. A1

TRANSPARENCIES

	5.1	5.2	5.3	5.4	5.5	5.6	5.7
Warm-Up Exercises and Daily Quiz	p. 37	p. 38	p. 39	p. 40	p. 41	p. 42	p. 43
Alternative Lesson Opener Transparencies	p. 32	p. 33	p. 34	p. 35	p. 36	p. 37	p. 38
Examples/Standardized Test Practice	✓	✓	✓	✓	✓	✓	✓
Answer Transparencies	✓	✓	✓	✓	✓	✓	✓

TECHNOLOGY

- EasyPlanner CD
- EasyPlanner Plus Online
- Classzone.com
- Personal Student Tutor CD
- Test and Practice Generator CD
- Instant Replay: Video Review Games
- Electronic Lesson Presentations
- Chapter Audio Summaries CD

ADDITIONAL RESOURCES

- Basic Skills Workbook
- Explorations and Projects
- Worked-Out Solution Key
- Data Analysis Sourcebook
- Spanish Resources
- Standardized Test Practice Workbook
- Practice Workbook with Examples
- Notetaking Guide

CHAPTER 5

PACING THE CHAPTER

Resource Key
- Student Edition
- Chapter 5 Resource Book
- Teacher's Edition

REGULAR SCHEDULE

Day 1

5.1

STARTING OPTIONS
- Prereq. Skills Review
- Strategies for Reading
- Warm-Up or Daily Quiz

TEACHING OPTIONS
- Les. Opener (Calculator)
- Examples 1–4
- Closure Question
- Guided Practice Exs.

APPLY/HOMEWORK
- See Assignment Guide.
- See the CRB: Practice, Reteach, Apply, Extend

ASSESSMENT OPTIONS
- Checkpoint Exercises
- Daily Quiz (5.1)
- Stand. Test Practice

Day 2

5.2

STARTING OPTIONS
- Homework Check
- Warm-Up or Daily Quiz

TEACHING OPTIONS
- Motivating the Lesson
- Les. Opener (Activity)
- Graphing Calc. Activity
- Examples 1–3
- Closure Question
- Guided Practice Exs.

APPLY/HOMEWORK
- See Assignment Guide.
- See the CRB: Practice, Reteach, Apply, Extend

ASSESSMENT OPTIONS
- Checkpoint Exercises
- Daily Quiz (5.2)
- Stand. Test Practice

Day 3

5.3

STARTING OPTIONS
- Homework Check
- Warm-Up or Daily Quiz

TEACHING OPTIONS
- Motivating the Lesson
- Les. Opener (Application)
- Graphing Calc. Activity
- Examples 1–2
- Guided Practice Exs. 1–17

APPLY/HOMEWORK
- See Assignment Guide.
- See the CRB: Practice, Reteach, Apply, Extend

ASSESSMENT OPTIONS
- Checkpoint Exercises, p. 286

Day 4

5.3 *(cont.)*

STARTING OPTIONS
- Homework Check

TEACHING OPTIONS
- Example 3
- Closure Question

APPLY/HOMEWORK
- See Assignment Guide.
- See the CRB: Practice, Reteach, Apply, Extend

ASSESSMENT OPTIONS
- Checkpoint Exercises, p. 287
- Daily Quiz (5.3)
- Stand. Test Practice
- Quiz (5.1–5.3)

Day 5

5.4

STARTING OPTIONS
- Homework Check
- Warm-Up or Daily Quiz

TEACHING OPTIONS
- Les. Opener (Application)
- Graphing Calc. Activity
- Examples 1–3
- Closure Question
- Guided Practice Exs.

APPLY/HOMEWORK
- See Assignment Guide.
- See the CRB: Practice, Reteach, Apply, Extend

ASSESSMENT OPTIONS
- Checkpoint Exercises
- Daily Quiz (5.4)
- Stand. Test Practice

Day 6

5.5

STARTING OPTIONS
- Homework Check
- Warm-Up or Daily Quiz

TEACHING OPTIONS
- Motivating the Lesson
- Les. Opener (Activity)
- Examples 1–2
- Guided Practice Exs. 1–11

APPLY/HOMEWORK
- See Assignment Guide.
- See the CRB: Practice, Reteach, Apply, Extend

ASSESSMENT OPTIONS
- Checkpoint Exercises, p. 301

Day 9

5.6 *(cont.)*

STARTING OPTIONS
- Homework Check

TEACHING OPTIONS
- Example 4
- Closure Question
- Guided Practice Exs. 16–17

APPLY/HOMEWORK
- See Assignment Guide.
- See the CRB: Practice, Reteach, Apply, Extend

ASSESSMENT OPTIONS
- Checkpoint Exercises, p. 310
- Daily Quiz (5.6)
- Stand. Test Practice

Day 10

5.7

STARTING OPTIONS
- Homework Check
- Warm-Up or Daily Quiz

TEACHING OPTIONS
- Motivating the Lesson
- Concept Act. & Wksht.
- Les. Opener (Visual)
- Graphing Calc. Activity
- Example 1
- Guided Practice Exs. 1–2

APPLY/HOMEWORK
- See Assignment Guide.
- See the CRB: Practice, Reteach, Apply, Extend

ASSESSMENT OPTIONS
- Checkpoint Exercises, p. 317

Day 11

5.7 *(cont.)*

STARTING OPTIONS
- Homework Check

TEACHING OPTIONS
- Examples 2–3
- Closure Question
- Guided Practice Exs. 3–10

APPLY/HOMEWORK
- See Assignment Guide.
- See the CRB: Practice, Reteach, Apply, Extend

ASSESSMENT OPTIONS
- Checkpoint Exercises, pp. 317–318
- Daily Quiz (5.7)
- Stand. Test Practice
- Quiz (5.6–5.7)

Day 12

Review

DAY 12 START OPTIONS
- Homework Check

REVIEWING OPTIONS
- Chapter 5 Summary
- Chapter 5 Review
- Chapter Review Games and Activities

APPLY/HOMEWORK
- Chapter 5 Test (practice)
- Ch. Standardized Test (practice)

Day 13

Assess

DAY 13 START OPTIONS
- Homework Check

ASSESSMENT OPTIONS
- Chapter 5 Test
- SAT/ACT Ch. 5 Test
- Alternative Assessment

APPLY/HOMEWORK
- Skill Review, p. 332

Day 7

5.5 *(cont.)*

Starting Options
- Homework Check

Teaching Options
- Example 3
- Closure Question
- Guided Practice Exs. 12–17

Apply/Homework
- See Assignment Guide.
- See the CRB: Practice, Reteach, Apply, Extend

Assessment Options
- Checkpoint Exercises, p. 302
- Daily Quiz (5.5)
- Stand. Test Practice
- Quiz (5.4–5.5)

Day 8

5.6

Starting Options
- Homework Check
- Warm-Up or Daily Quiz

Teaching Options
- Motivating the Lesson
- Concept Act. & Wksht.
- Les. Opener (Activity)
- Graphing Calc. Activity
- Examples 1–3
- Guided Practice Exs. 1–15

Apply/Homework
- See Assignment Guide.
- See the CRB: Practice, Reteach, Apply, Extend

Assessment Options
- Checkpoint Exercises, p. 309

BLOCK SCHEDULE

Day 1

5.1 & 5.2

Day 1 Start Options
- Prereq. Skills Review
- Strategies for Reading
- W-Up 5.1 or D. Quiz 4.8

Teaching 5.1 Options
- Les. Opener (Calc.)
- Examples 1–4
- Closure Question
- Guided Practice Exs.

Teaching 5.2 Options
- Warm-Up (Les. 5.2)
- Motivating the Lesson
- Les. Opener (Activity)
- Graphing Calc. Activity
- Examples 1–3
- Closure Question
- Guided Practice Exs.

Apply/Homework
- See Assignment Guide.
- See the CRB: Practice, Reteach, Apply, Extend

Assessment Options
- Checkpoint Exercises
- Daily Quiz (Les. 5.1, 5.2)
- Stand. Test Practice

Day 2

5.3

Day 2 Start Options
- Homework Check
- Warm-Up or Daily Quiz

Teaching 5.3 Options
- Motivating the Lesson
- Les. Opener (Appl.)
- Graphing Calc. Activity
- Examples 1–3
- Closure Question
- Guided Practice Exs.

Apply/Homework
- See Assignment Guide.
- See the CRB: Practice, Reteach, Apply, Extend

Assessment Options
- Checkpoint Exercises
- Daily Quiz (5.3)
- Stand. Test Practice
- Quiz (5.1–5.3)

Day 3

5.4 & 5.5

Day 3 Start Options
- Homework Check
- W-Up 5.4 or D. Quiz 5.3

Teaching 5.4 Options
- Les. Opener (Appl.)
- Graphing Calc. Activity
- Examples 1–3
- Closure Question
- Guided Practice Exs.

Beginning 5.5 Options
- Warm-Up (Les. 5.5)
- Motivating the Lesson
- Les. Opener (Activity)
- Examples 1–2
- Guided Practice Exs. 1–11

Apply/Homework
- See Assignment Guide.
- See the CRB: Practice, Reteach, Apply, Extend

Assessment Options
- Checkpoint Exercises
- Daily Quiz (Les. 5.4)
- Stand. Test Prac. (5.4)

Day 4

5.5 & 5.6

Day 4 Start Options
- Homework Check
- Daily Quiz (Les. 5.4)

Finishing 5.5 Options
- Example 3
- Closure Question
- Guided Practice Exs. 12–17

Beginning 5.6 Options
- Warm-Up (Les. 5.6)
- Motivating the Lesson
- Concept Act. & Wksht.
- Les. Opener (Activity)
- Graphing Calc. Activity
- Examples 1–3
- Guided Practice Exs. 1–15

Apply/Homework
- See Assignment Guide.
- See the CRB: Practice, Reteach, Apply, Extend

Assessment Options
- Checkpoint Exercises
- Daily Quiz (Les. 5.5)
- Stand. Test Prac. (5.5)
- Quiz (5.4–5.5)

Day 5

5.6 & 5.7

Day 5 Start Options
- Homework Check
- Daily Quiz (Les. 5.5)

Finishing 5.6 Options
- Example 4
- Closure Question
- Guided Practice Exs. 16–17

Beginning 5.7 Options
- Warm-Up (Les. 5.7)
- Motivating the Lesson
- Concept Act. & Wksht.
- Les. Opener (Visual)
- Graphing Calc. Activity
- Example 1
- Guided Practice Exs. 1–2

Apply/Homework
- See Assignment Guide.
- See the CRB: Practice, Reteach, Apply, Extend

Assessment Options
- Checkpoint Exercises
- Daily Quiz (Les.5.6)
- Stand. Test Prac. (5.6)

Day 6

5.7 & Review

Starting Options
- Homework Check
- Daily Quiz (Les. 5.6)

Finishing 5.7 Options
- Examples 2–3
- Closure Question
- Guided Practice Exs. 3–10

Reviewing Options
- Chapter 5 Summary
- Chapter 5 Review
- Chapter Review Games and Activities

Apply/Homework
- See Assignment Guide.
- See the CRB: Practice, Reteach, Apply, Extend
- Chapter 5 Test (practice)
- Ch. Standardized Test (practice)

Assessment Options
- Checkpoint Exercises
- Daily Quiz (5.7)
- Stand. Test Practice
- Quiz (5.6–5.7)

Day 7

Assess & 6.1

(Day 7 = Ch. 6 Day 1)

Assessment Options
- Chapter 5 Test
- SAT/ACT Ch. 5 Test
- Alternative Assessment

Ch. 6 Start Options
- Skill Review, p. 332
- Prereq. Skills Review
- Strategies for Reading

Teaching Options
- Warm-Up (Les. 6.1)
- Motivating the Lesson
- Les. Opener (Appl.)
- Examples 1–4
- Closure Question
- Guided Practice Exs.

Apply/Homework
- See Assignment Guide.
- See the CRB: Practice, Reteach, Apply, Extend

Assessment Options
- Checkpoint Exercises
- Daily Quiz (6.1)
- Stand. Test Practice

CHAPTER 5

MEETING INDIVIDUAL NEEDS

BEFORE THE CHAPTER

The *Chapter 5 Resource Book* has the following materials to distribute and use before the chapter:

- **Parent Guide for Student Success**
- **Prerequisite Skills Review**
- **Strategies for Reading Mathematics (pictured below)**

STRATEGIES FOR READING *Pages 7–8*

CHAPTER 5 NAME ______________________ DATE ________

Strategies for Reading Mathematics

For use with Chapter 5

Chapter Support

Strategy: Translating Words into Symbols

When you use mathematics to solve real-life problems, you will often translate the words into symbols. Read the following problem.

A family pays a one-time registration fee of \$35 to enroll their child in daycare. The weekly charge thereafter is \$200 per week. How much will the family pay for daycare?

The steps outlined below can help you interpret this problem and other problems like it.

Verbal Model: [Total cost] = [Weekly charge] · [Number of weeks] + [Registration fee]

Labels:

Total cost	$= y$	(dollars)
Weekly charge	$= 200$	(dollars per week)
Number of weeks	$= x$	(weeks)
Registration fee	$= 35$	(dollars)

Algebraic Model: $y = 200x + 35$

STUDY TIP
Use a Problem-Solving Model
To model a real-life situation using mathematics, begin by writing a verbal model based on the relationships in the situation. Next assign labels to the pieces of your model. Then use your labels to write an algebraic model for the situation.

Questions

1. How much does it cost for three weeks of daycare? for five weeks of daycare? What is the rate of change in dollars per week?
2. If you graphed the equation, what would be the y-intercept?
3. Suppose the registration fee was \$40 and the weekly charge was \$225. Write an equation to represent this situation.
4. Write a verbal model, labels, and an algebraic model for the following situation.

Rachel receives a weekly allowance of \$3.00 for doing certain household chores. She can earn more money by doing additional chores for \$1.50 per chore. How much money can Rachel earn in a week?

STRATEGIES FOR READING MATHEMATICS The first page gives students tips about representing word problems with equations as they prepare for Chapter 5. The second page (not shown) provides a visual glossary of key vocabulary words in the chapter, such as correlation, slope-intercept form, point-slope form, and standard form.

DURING EACH LESSON

The *Chapter 5 Resource Book* has the following alternatives for introducing the lesson:

- **Lesson Openers (pictured below)**
- **Graphing Calculator Activities with Keystrokes**

LESSON OPENER *Page 52*

LESSON 5.4 NAME ______________________ DATE ________

Available as a transparency

Application Lesson Opener

For use with pages 292–298

The data in the graph show the resting heart rates and ages of 20 students in an aerobics class.

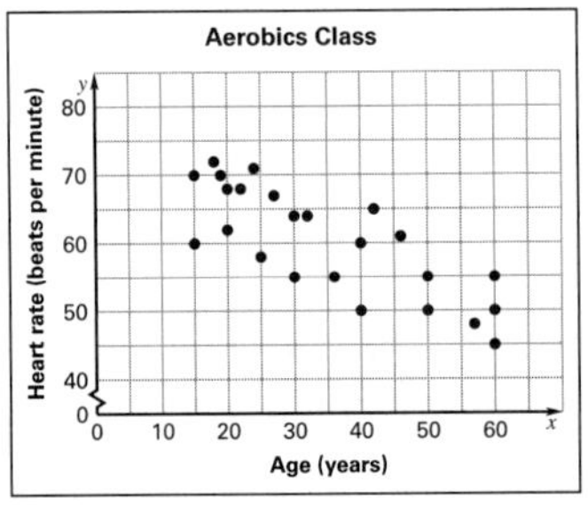

1. Graph the equations on the graph above.
   a. $y = -\frac{2}{5}x + 80$
   b. $y = -\frac{2}{5}x + 75$
   c. $y = 60$
   d. $y = -\frac{3}{5}x + 75$
2. Which equation seems to best represent the data? Why?

Lesson 5.4

APPLICATION LESSON OPENER This Lesson Opener provides an alternative way to start Lesson 5.4 in the form of a real-life application. Students see how graphing equations relates to plotting data for heart rate and age.

TECHNOLOGY RESOURCE
Teachers can use the Electronic Lesson Presentations CD-ROM to present Lesson 5.4 using computer animation to step through the Examples.

FOLLOWING EACH LESSON

The *Chapter 5 Resource Book* has a variety of materials to follow-up each lesson. They include the following:

- **Practice (3 levels)**
- **Reteaching with Practice**
- **Quick Catch-Up for Absent Students**
- **Interdisciplinary Applications (pictured below)**
- **Real-Life Applications**

APPLICATION *Page 46*

LESSON 5.3

NAME ______________________ DATE ________

Interdisciplinary Application

For use with pages 285–291

Bald Eagles (Science)

The bald eagle, our national symbol, is making a comeback from the brink of extinction. Although it has been illegal to hunt bald eagles since 1940, when the Bald Eagle Protection Act made it illegal to kill, harm, harass, or possess bald eagles, the eagle was listed as endangered in 1973 in most of the lower 48 states.

The eagle population was decimated after World War II when the pesticide DDT went into widespread use. The pesticide caused the birds to lay thin-shelled eggs that broke during incubation. DDT was banned in the U.S. on June 14, 1972. Since then, the eagle has steadily increased in numbers, and was down listed from endangered to threatened in August 1995. In July 1999, Congress and the president proposed removing it from the threatened list under the Endangered Species Act.

The raptor research and technical assistance center coordinates the bald eagle survey throughout the lower 48 states. The survey reported 15,896 eagles in 1994. In 1995, 16289 eagles were counted.

1. Write a linear equation to model the eagle population. Let x represent the number of years since 1990 and y the eagle population.
2. Graph the equation from Exercise 1.
3. Use the equation from Exercise 1 to estimate the eagle population in the year 2000.
4. In 1996, the number of eagles sighted declined slightly. What factors could have influenced the bald eagle count?
5. If the number of eagles counted in 1999 is 18,362, write a linear equation using year 1999 and year 1995 as your points.
6. Graph the equation from Exercise 5.
7. Use the equation from Exercise 5 to estimate the eagle population in the year 2000.

Lesson 5.3

INTERDISCIPLINARY APPLICATION This application makes a connection between the bald eagle population of the U.S. and writing, graphing, and analyzing a linear equation, the topic of Lesson 5.3.

ASSESSING THE CHAPTER

The *Chapter 5 Resource Book* has the following review and assessment materials:

- **Quizzes**
- **Chapter Review Games and Activities**
- **Chapter Test (3 levels)**
- **SAT/ACT Chapter Test**
- **Alternative Assessment with Rubric and Math Journal**
- **Project with Rubric**
- **Cumulative Review (pictured below)**

CUMULATIVE REVIEW *Pages 118–119*

CHAPTER 5

NAME ______________________ DATE ________

Cumulative Review

For use after Chapters 1–5

Evaluate the expression for the given values of the variable. (1.2)

1. $(2s + t)^3$ when $s = -1$ and $t = -5$
2. $a - \frac{1}{2}b^3$ when $a = \frac{1}{3}$ and $b = -\frac{2}{3}$
3. $-10.9(k - 0.36)$ when $k = 0.45$
4. $8 - (x^2) - (y^3)$ when $x = -3$ and $y = 6$

Write an equation or an inequality to model the situation. (1.5)

5. The length l of a table is three times its width w.
6. The time t it takes to go to work from home is less than one half the time s it takes to go to the mall.
7. The height h of a triangle is equal to the quotient of two times the area a and its base b.

Find the sum or difference. (2.2–2.3)

8. $|-3| - |2^2| - (-2)$
9. $-23.8 - 0.327 + 45.96$
10. $|-3.677| + 4.279 - 5.698$
11. $-\frac{7}{3} - \left|-\frac{5}{6}\right| + \left(\frac{2}{3}\right)^2$

Solve the equation. (3.1–3.4)

12. $-14x - 5 = 93$
13. $-15a + 30 = -89$
14. $-(x - 2.3) - 4(5.9 - x)$
15. $\frac{2}{9}\left(x - \frac{6}{7}\right) = \left(\frac{1}{63}x + \frac{40}{63}\right)$

Find the *x*- and *y*-intercepts of the equation. (4.3)

16. $y = 4x - 14$
17. $\frac{8}{9}x + \frac{40}{27} = 10y$

Find the slope of the line passing through the given points. (4.4)

18. $(0, -18), (-3.5, -18.99)$
19. $\left(\frac{8}{3}, \frac{4}{7}\right), \left(\frac{50}{3}, \frac{6}{21}\right)$

Write an equation of the line with the given slope and *y*-intercept. (5.1)

20. $m = 3, b = -2$
21. $m = -5, b = 4$
22. $m = 7, b = 11$
23. $m = \frac{1}{5}, b = -6$
24. $m = 0, b = 8$
25. $m = -6.5, b = 4.5$

Write an equation of the line that is parallel to the given line and passes through the given point. (5.2)

26. $y = 4x + 6, (4, -2)$
27. $y = -\frac{1}{4}x - 1, (4, 1)$
28. $y = -3x + 9, (3, -2)$
29. $y = \frac{4}{3}x - 6, (3, 1)$

Review and Assess

CUMULATIVE REVIEW The content of Chapters 1–5 is reviewed in this two-page cumulative review. Lesson references with each cluster of exercises guide students to the places in the textbook where they can go to review each concept.

TECHNOLOGY RESOURCE

Students can use the Personal Student Tutor to find additional examples that review the material covered in the Cumulative Review.

CHAPTER OVERVIEW

Chapter Goals

The primary goal of Chapter 5 is for students to learn to write and use linear equations. They will write the equation of a line when given a graph, the slope and a point on the line, or two points on the line. The standard form, point-slope form and slope-intercept form of a linear equation are all presented. Equations for horizontal and vertical lines, as well as parallel and perpendicular lines are investigated. Students develop and use linear models for real-life situations and find a linear equation that approximates a set of data points. The concept of correlation is also introduced.

Application Note

Carbon-14 is a radioactive form of carbon that is found in all plants and animals. When the organism is alive, the amount of carbon-14 is fairly constant, but after death, the amount of carbon-14 remaining decreases over time in a regular pattern. This pattern can be used to find the age of the plant or animal fossil. Carbon-14 dating is most commonly used to date bones, shells, wood, and seeds.

Additional information about radiocarbon dating is available at **www.mcdougallittell.com.**

WRITING LINEAR EQUATIONS

How can you figure out how old an artifact is?

270

CHAPTER 5

APPLICATION: Archaeology

Archaeologists use radiocarbon dating to estimate the age of an artifact. To do this, they use an instrument to measure the amount of carbon-14 remaining in a sample.

Think & Discuss

In Exercises 1–3, use the graph below.

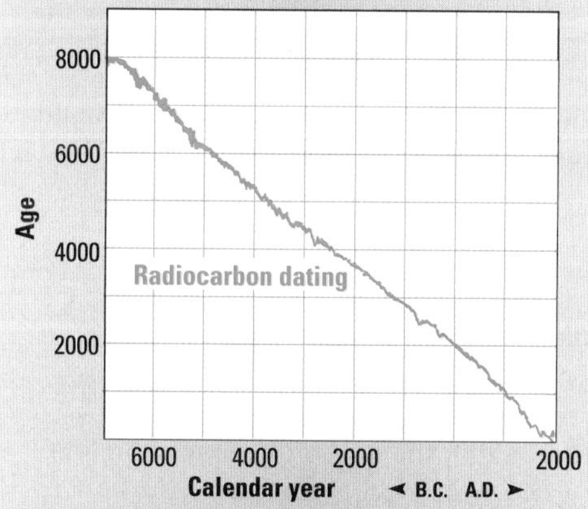

1. What calendar year corresponds to an artifact that has a radiocarbon age of 7000? **about 5800 B.C.**
2. What is the radiocarbon age of an object from 4000 B.C.? **about 5200 years**
3. For which calendar years does radiocarbon dating produce the most accurate results? **from about 1000 B.C. to A.D. 2000**

Learn More About It

You will learn more about archaeological dating in the Math & History on p. 306.

APPLICATION LINK Visit www.mcdougallittell.com for more information about archaeology.

ADDITIONAL RESOURCES

Another way to begin the chapter is to show the video clip of a real-life motivator for Chapter 5 on the *Instant Replay: Video Review Games.*

PROJECTS

A project covering Chapters 4–6 appears on pages 392–393 of the Student Edition. An additional project for Chapter 5 is available in the *Chapter 5 Resource Book,* p. 116.

TECHNOLOGY

Software
- *Electronic Teaching Tools*
- *Online Lesson Planner*
- *Personal Student Tutor*
- *Test and Practice Generator*
- *Electronic Lesson Presentations* (Lesson 5.4)

Video
- *Instant Replay: Video Review Games*

Internet Connections
www.mcdougallittell.com
- **Application Links**
  271, 295, 306
- **Data Updates**
  277, 281, 294, 320
- **Student Help**
  277, 280, 286, 297, 299, 301, 313, 318
- **Career Links**
  287, 318
- **Extra Challenge**
  278, 284, 290, 298, 314, 321

PREPARE

DIAGNOSTIC TOOLS

The **Skill Review** exercises can help you diagnose whether students have the following skills needed in Chapter 5:

- Solve one-variable equations.
- Plot points in the plane.
- Graph lines using *x*- and *y*-intercepts.

The following resources are available for students who need additional help with these skills:

- Prerequisite Skills Review (*Chapter 5 Resource Book*, p. 5; Warm-Up Transparencies, p. 36)
- Reteaching with Practice (Chapter Resource Books for Lessons 3.4, 4.1, and 4.3)
- *Personal Student Tutor*

ADDITIONAL RESOURCES

The following resources are provided to help you prepare for the upcoming chapter and customize review materials:

- ***Chapter 5 Resource Book***
  Tips for New Teachers (p. 1)
  Parent Guide (p. 3)
  Lesson Plans (every lesson)
  Lesson Plans for Block Scheduling (every lesson)
- ***Electronic Teaching Tools***
- ***Online Lesson Planner***
- ***Test and Practice Generator***

4–8. See Additional Answers beginning on page AA1.

CHAPTER 5

Study Guide

PREVIEW

What's the chapter about?

Chapter 5 is about writing **different forms of linear equations**. You'll learn

- three forms of linear equations.
- to write a linear equation given a slope and a point, or given two points.
- to write an equation of a line perpendicular to another line.
- to fit a line to data and use linear interpolation or linear extrapolation.

KEY VOCABULARY

▶ **Review**
- **scatter plot, p. 204**
- ***x*-intercept, p. 218**
- ***y*-intercept, p. 218**
- **slope, p. 226**

▶ **New**
- **slope-intercept form, p. 273**
- **best-fitting line, p. 292**
- **correlation, p. 295**
- **point-slope form, p. 300**
- **standard form, p. 308**
- **linear interpolation, p. 318**
- **linear extrapolation, p. 318**

PREPARE

Are you ready for the chapter?

SKILL REVIEW Do these exercises to review key skills that you'll apply in this chapter. See the given **reference page** if there is something you don't understand.

STUDENT HELP

Study Tip
"Student Help" boxes throughout the chapter give you study tips and tell you where to look for extra help in this book and on the Internet.

Solve the equation. **(Review Example 3, p. 155)**

1. $4(x + 8) = 20x$ **2** **2.** $2x + 2 = 3x - 8$ **10** **3.** $9a = 4a - 24 + a$ **−6**

Plot the ordered pairs in a coordinate plane. **(Review Example 1, p. 203)**
4, 5. See margin.

4. $A(5, -7), B(-6, -8), C(-9, 0)$ **5.** $A(8, -5), B(7, 3), C(-2, 0)$

Find the *x*-intercept and the *y*-intercept in the equation. Graph the line.
(Review Example 2, p. 219)

6. $4x + 9y = 18$
4.5, 2; See margin.

7. $-x - 5y = 15 + 2x$
−5, −3; See margin.

8. $-6x - 2y = 8$
$-\frac{4}{3}$, **−4; See margin.**

STUDY STRATEGY

Here's a study strategy!

Practice Test

Writing and taking a practice test are good ways to determine how ready you are for an actual test. Review the chapter. Create a test and exchange it with a classmate. After taking the tests, correct and discuss them. You should study any topics that gave you difficulty.

5.1 Writing Linear Equations in Slope-Intercept Form

What you should learn

GOAL 1 Use the slope-intercept form to write an equation of a line.

GOAL 2 Model a **real-life** situation with a linear function, such as the population of California in **Example 3**.

Why you should learn it

▼ To model **real-life** situations, such as finding the cost of renting a bicycle in **Exs. 9–11**.

GOAL 1 USING SLOPE-INTERCEPT FORM

In this lesson you will learn to write an equation of a line using its slope and y-intercept. To do this, you need to use the **slope-intercept form** of the equation of a line.

$y = mx + b$ ⟵ Slope-intercept form

In the equation, m is the slope and b is the y-intercept. Recall that the y-intercept is the y-coordinate of the point where the line crosses the y-axis.

EXAMPLE 1 *Writing an Equation of a Line*

Write an equation of the line whose slope m is -2 and whose y-intercept b is 5.

SOLUTION

You are given the slope $m = -2$ and the y-intercept $b = 5$.

$y = mx + b$ Write slope-intercept form.

$y = -2x + 5$ Substitute -2 for m and 5 for b.

EXAMPLE 2 *Writing an Equation of a Line from a Graph*

Write an equation of the line shown in the graph.

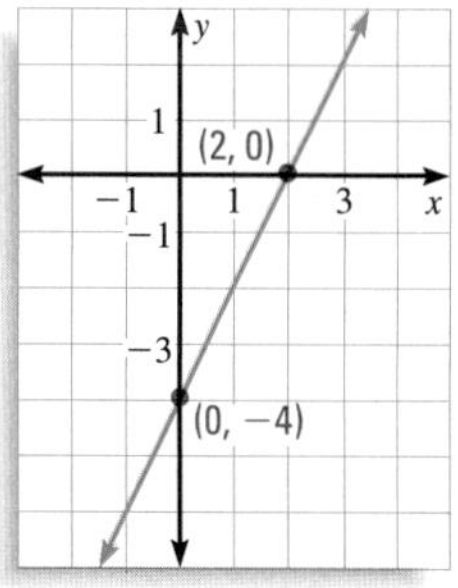

SOLUTION

The line intersects the y-axis at $(0, -4)$, so the y-intercept b is -4. The line also passes through the point $(2, 0)$. Find the slope of the line:

$$m = \frac{\text{rise}}{\text{run}} = \frac{-4 - 0}{0 - 2} = \frac{-4}{-2} = 2$$

Knowing the slope and y-intercept, you can write an equation of the line.

$y = mx + b$ Write slope-intercept form.

$y = 2x + (-4)$ Substitute 2 for m and -4 for b.

$y = 2x - 4$ Simplify.

STUDENT HELP

Study Tip
In Example 2, you can also find the slope of the line by reading the graph.
$\frac{\text{rise}}{\text{run}} = \frac{4}{2} = 2$

1 PLAN

PACING
Basic: 1 day
Average: 1 day
Advanced: 1 day
Block Schedule: 0.5 block with 5.2

LESSON OPENER
GRAPHING CALCULATOR
An alternative way to approach Lesson 5.1 is to use the Graphing Calculator Lesson Opener:
- Blackline Master (*Chapter 5 Resource Book*, p. 12)
- Transparency (p. 32)

MEETING INDIVIDUAL NEEDS
- ***Chapter 5 Resource Book***
  Prerequisite Skills Review (p. 5)
  Practice Level A (p. 13)
  Practice Level B (p. 14)
  Practice Level C (p. 15)
  Reteaching with Practice (p. 16)
  Absent Student Catch-Up (p. 18)
  Challenge (p. 20)
- ***Resources in Spanish***
- ***Personal Student Tutor***

NEW-TEACHER SUPPORT
See the Tips for New Teachers on pp. 1–2 of the *Chapter 5 Resource Book* for additional notes about Lesson 5.1.

WARM-UP EXERCISES
Transparency Available
Simplify each expression.
1. $\frac{2-3}{-4+1}$ $\frac{1}{3}$ **2.** $\frac{-2-1}{2+(-3)}$ 3
Evaluate for $x = -1$, 0, and 2.
3. $f(x) = 3x - 7$ −10, −7, −1
4. $f(x) = 205 + 32x$ 173, 205, 269

ENGLISH LEARNERS
Draw English learners' attention to the hyphen in *slope-intercept form.* Explain that the hyphen connects two words, *slope* and *intercept,* and makes them into one adjective describing *form.*

2 TEACH

EXTRA EXAMPLE 1

Write an equation of the line whose slope m is $\frac{3}{4}$ and whose y-intercept b is -4. $y = \frac{3}{4}x - 4$

EXTRA EXAMPLE 2

Write an equation of the line shown in the graph.

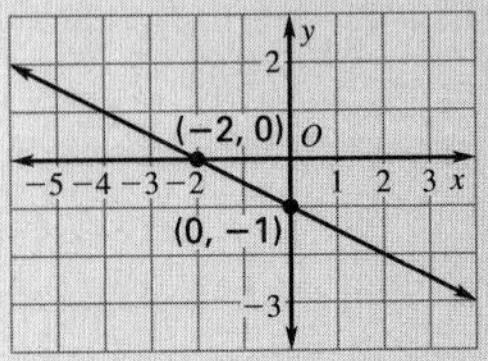

$y = -\frac{1}{2}x - 1$

EXTRA EXAMPLE 3

The number of trout in a lake was estimated to be 44.5 thousand in 1995. During the next 10 years, the trout population is expected to increase by about 3.2 thousand per year.

a. Write an equation to model the trout population in any year between 1995 and 2005. Let t be the number of years since 1995. $P = 44.5 + 3.2t$

b. Predict the trout population in 2002. **about 66.9 thousand trout**

CHECKPOINT EXERCISES

For use after Examples 1–3:

1. A mosquito population is currently estimated at 27.8 million mosquitoes per mi^2. A spraying program will decrease the number of mosquitoes by about 1.8 million mosquitoes per mi^2 every hour for the first 12 hours.

a. Write an equation to model the mosquito population for any of the 12 hours. Let P represent the number of mosquitoes per mi^2 and t represent the number of hours since spraying. $P = 27.8 - 1.8t$

b. Predict the mosquito population 5 hours after the spraying. **18.8 million mosquitoes per square mile**

FOCUS ON APPLICATIONS

REAL LIFE POPULATION DATA In 1990, California had a population of about 29.76 million. During the next 15 years, the state's population was expected to increase by an average of about 0.31 million people per year.

▶ Source: U.S. Bureau of the Census

GOAL 2 MODELING A REAL-LIFE SITUATION

In Lesson 4.8 you saw examples of linear functions.

$f(x) = mx + b$ ← **Linear function**

A *linear model* is a linear function that is used to model a real-life situation. In a linear model, m is the rate of change and b is the initial amount. In such models, sometimes line graphs can be used to approximate change that occurs in discrete steps.

EXAMPLE 3 *A Linear Model for Population*

POPULATION CHANGE Linear functions can approximate population change.

a. Write a linear equation to approximate the expected population of California in any year between 1990 and 2005. Use the information below the photo.

b. Use the equation to predict the population of California in 2005.

SOLUTION

a. **VERBAL MODEL** California Population = 1990 Population + Increase per year · Years since 1990

LABELS

California Population = P		(million people)
1990 Population = **29.76**		(million people)
Increase per year = **0.31**		(million people per year)
Years since 1990 = t		(years)

ALGEBRAIC MODEL $P = 29.76 + 0.31 \cdot t$ **Linear model**

UNIT ANALYSIS Check that *people* are the units in the algebraic model:

$$people = \text{people} + \frac{\text{people}}{\cancel{\text{year}}} \cdot \cancel{\text{year}}$$

b. When $t = 15$, the year will be 2005. Substitute in your equation.

$P = 29.76 + 0.31(15)$

$= 29.76 + 4.65$

$= 34.41$

▶ In 2005 the population will be about 34.4 million people. The graph represents this prediction.

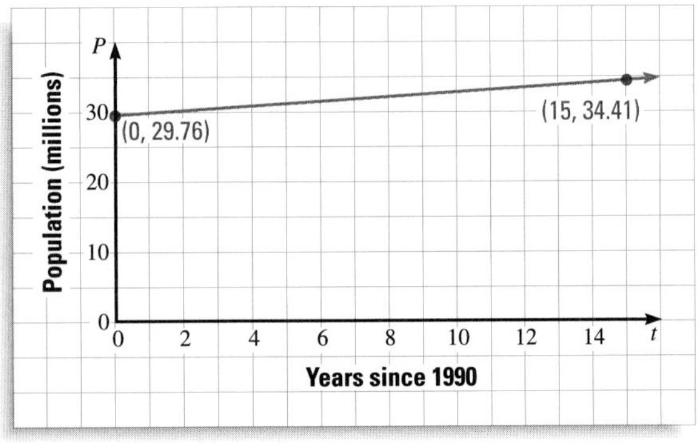

STUDENT HELP

Look Back
For help with graphing linear equations, see p. 219.

Telephone Charges

EXAMPLE 4 A Linear Model for Telephone Charges

You can approximate collect telephone call charges using a linear function.

a. Write a linear equation to approximate the cost of collect telephone calls. Use the information shown on the telephone bill to create your model.

b. How much would a 9-minute collect telephone call cost?

Date	Time	Area Number	Ist Min.	Rate	Add'l Min	Amount
Jan 7	857PM	360 555-1829	$3.44	$0.24	4	$4.40
Jan 16	558PM	360 555-1829	$3.44	$0.24	1	$3.68
Jan 17	837PM	360 555-1829	$3.44	$0.24	17	$7.52
Jan 22	601PM	360 555-1829	$3.44	$0.24	9	$5.60
Feb 3	859PM	360 555-1829	$3.44	$0.24	10	$5.84

SOLUTION

PROBLEM SOLVING STRATEGY

a. **VERBAL MODEL** Total cost = Cost for first minute + Rate per additional minute • Number of additional minutes

LABELS

Total cost = C (dollars)

Cost for first minute = **3.44** (dollars)

Rate per additional minute = **0.24** (dollars per minute)

Number of additional minutes = t (minutes)

ALGEBRAIC MODEL $C = 3.44 + 0.24 \cdot t$ **Linear model**

UNIT ANALYSIS Check that *dollars* are the units in the algebraic model:

$$dollars = \text{dollars} + \frac{\text{dollars}}{\cancel{\text{min}}} \cdot \cancel{\text{min}}$$

b. A 9-minute call would involve 8 additional minutes. Substitute in your equation.

$C = 3.44 + 0.24 \cdot t$

$= 3.44 + 0.24(8)$

$= 5.36$

▶ A 9-minute call would cost $5.36. The graph represents this prediction.

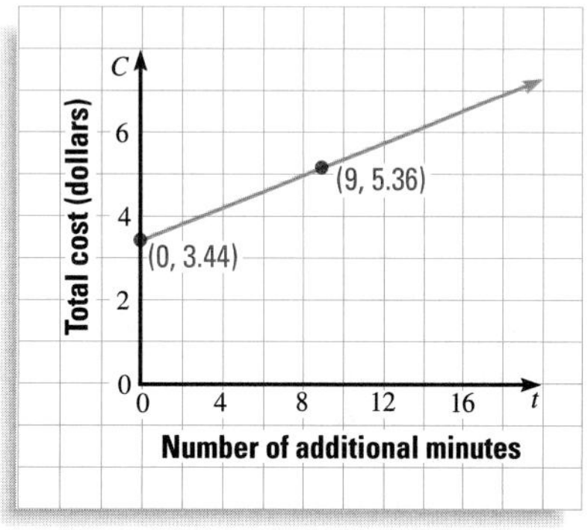

EXTRA EXAMPLE 4

a. Write a linear model that reflects the cost of an Internet connection. Use the information shown on the Internet bill below to create your model. **$C = 1.63 + 0.87t$, where t is the number of additional hours**

b. How much would a 5-hour Internet connection cost?

Internet Connection			
Date	3/1	3/12	3/17
1st Hour	$1.63	$1.63	$1.63
Rate	$0.87	$0.87	$0.87
Add'l Hours	2	9	7
Amount	$3.37	$9.46	$7.72

$5.11

CHECKPOINT EXERCISES

For use after Example 4:

1. A plumber charges $66 for the first hour of a service call and $45 for any additional hours. Write a linear model to reflect the plumber's charges and use the equation to determine the cost of a 3-hour service call. **$C = 66 + 45t$; where t is the number of additional hours; $156**

FOCUS ON VOCABULARY

Explain why the equation $y = mx + b$ is called the *slope-intercept form* of the equation of a line.
Sample answer: **$y = mx + b$ is called the slope-intercept form because the slope m and the y-intercept b of the graph can be read directly from the equation.**

CLOSURE QUESTION

In real-life situations that can be modeled by a linear equation in slope-intercept form, such as Example 3 and Example 4, what clues can you use to determine the equation? ***Sample answer:*** **The y-intercept is the starting value, or fixed number; the slope is the rate of change from the starting value.**

3 APPLY

ASSIGNMENT GUIDE

BASIC
Day 1: pp. 276–278 Exs. 12–24 even, 26–32, 38, 39, 42, 44–51

AVERAGE
Day 1: pp. 276–278 Exs. 12–24 even, 26–33, 38, 39, 42, 44–51

ADVANCED
Day 1: pp. 276–278 Exs. 12–24 even, 30–40, 42, 44–51

BLOCK SCHEDULE WITH 5.2
pp. 276–278 Exs. 12–24 even, 26–33, 38, 39, 42, 44–51

EXERCISE LEVELS

Level A: *Easier*
12–19

Level B: *More Difficult*
20–33, 38, 39, 41–57

Level C: *Most Difficult*
34–37, 40

✔ HOMEWORK CHECK

To quickly check student understanding of key concepts, go over the following exercises: Exs. 12, 14, 22, 24, 26, 27, 32. See also the Daily Homework Quiz:

- Blackline Master (*Chapter 5 Resource Book,* p. 23)
- Transparency (p. 38)

! COMMON ERROR

EXERCISES 20–25 Encourage students to use a visual check of each graph to see that the sign of the slope in their equations matches that of the graph.

10. Bike Rental Costs

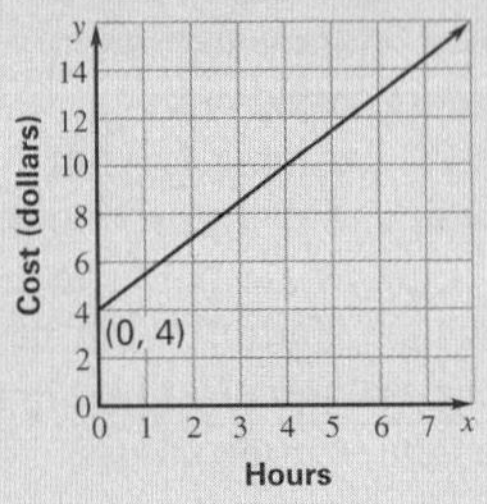

GUIDED PRACTICE

Vocabulary Check ✔
Concept Check ✔

1. Is 5 the *x-intercept* or the *y-intercept* of the line $y = 3x + 5$? Explain. **See margin.**

2. Describe the rate of change in the context of Example 3 on page 274. **the increase in the number of millions of people per year; 0.31**

Skill Check ✔

Write an equation of the line in slope-intercept form.

3. The slope is 1; the y-intercept is 0. $y = x$

4. The slope is -7; the y-intercept is $-\frac{2}{3}$. $y = -7x - \frac{2}{3}$

5. The slope is -1; the y-intercept is 3. $y = -x + 3$

6. The slope is -2; the y-intercept is 0. $y = -2x$

7. The slope is -3; the y-intercept is $\frac{1}{2}$. $y = -3x + \frac{1}{2}$

8. The slope is 4; the y-intercept is -6. $y = 4x - 6$

1. The y-intercept; *Sample answer:* the equation is in slope-intercept form. The slope is 3 and the y-intercept is 5.

RENTING A BIKE **Suppose that bike rentals cost \$4 plus \$1.50 per hour.**

9. Write an equation to model the total cost y of renting a bike for x hours. $y = 1.5x + 4$

10. Graph the equation. Label the y-intercept. **See margin.**

11. Use the equation to find the cost of renting a bike for 12 hours. **\$22**

PRACTICE AND APPLICATIONS

STUDENT HELP
Extra Practice to help you master skills is on p. 801.

WRITING EQUATIONS **Write an equation of the line in slope-intercept form.**

12. The slope is 3; the y-intercept is -2. $y = 3x - 2$

13. The slope is 1; the y-intercept is 2. $y = x + 2$

14. The slope is 0; the y-intercept is 4. $y = 4$

15. The slope is 2; the y-intercept is -1. $y = 2x - 1$

16. The slope is $\frac{3}{2}$; the y-intercept is 3. $y = \frac{3}{2}x + 3$

17. The slope is $-\frac{1}{4}$; the y-intercept is 1. $y = -\frac{1}{4}x + 1$

18. The slope is -6; the y-intercept is $\frac{3}{4}$. $y = -6x + \frac{3}{4}$

19. The slope is -3; the y-intercept is $-\frac{1}{2}$. $y = -3x - \frac{1}{2}$

GRAPHICAL REASONING **Write an equation of the line shown in the graph.**

20.

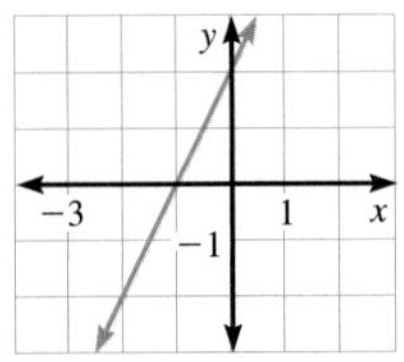

$y = 2x + 2$

21.

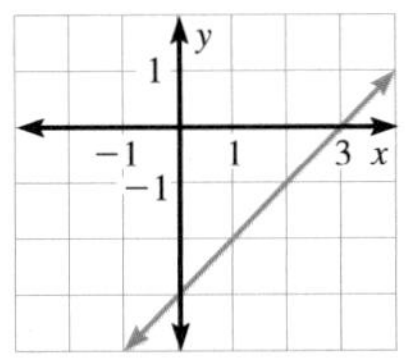

$y = x - 3$

22.

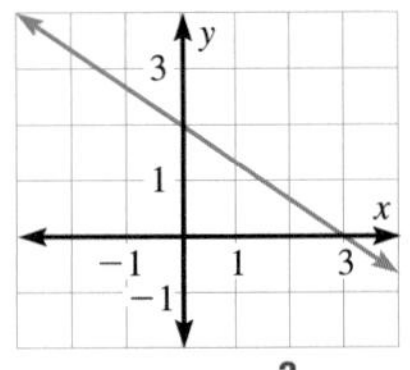

$y = -\frac{2}{3}x + 2$

PARALLEL LINES **Write an equation of each line in slope-intercept form.** **23–25. See margin.**

23. a: $y = 2x + 3$; b: $y = 2x$; c: $y = 2x - 4$

24. a: $y = -x + 4$; b: $y = -x + 2$; c: $y = -x$

25. a: $y = 4$; b: $y = 1$; c: $y = -2$

23.

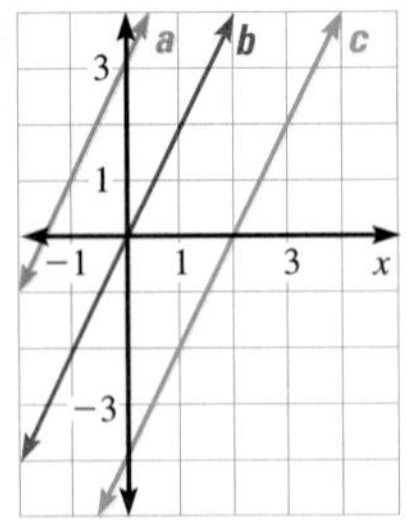

24.

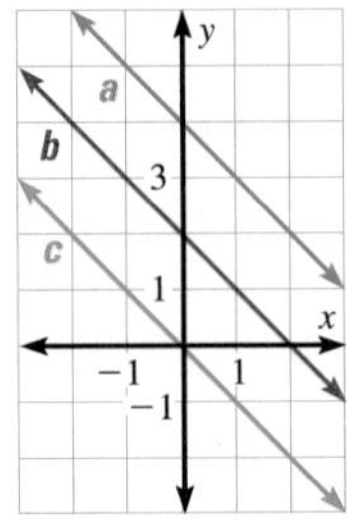

25.

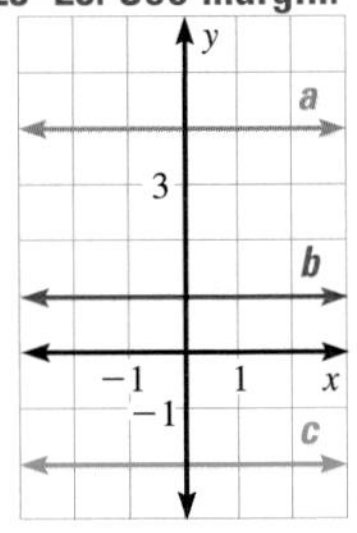

STUDENT HELP
HOMEWORK HELP
Example 1: Exs. 12–19
Example 2: Exs. 20–25
Example 3: Exs. 26, 27
Example 4: Exs. 28, 29

SOUTH CAROLINA POPULATION In Exercises 26 and 27, use the following information. In 1990 the population of South Carolina was approximately 3,486,000. During the next five years, the population increased by approximately 37,400 people per year.

DATA UPDATE of U.S. Bureau of the Census data at www.mcdougallittell.com

26. Write an equation to model the population P of South Carolina in terms of t, the number of years since 1990. $P = 37{,}400t + 3{,}486{,}000$

27. Estimate the population of South Carolina in 1996. **about 3,710,400**

RENTING A MOVING VAN In Exercises 28 and 29, a rental company charges a flat fee of \$30 and an additional \$.25 per mile to rent a moving van.

28. Write an equation to model the total charge y (in dollars) in terms of x, the number of miles driven. $y = 0.25x + 30$

29. Copy and complete the table using the equation you found in Exercise 28. **\$36.25; \$42.50; \$48.75; \$55**

Miles (x)	25	50	75	100
Cost (y)	?	?	?	?

UNIT ANALYSIS In Exercises 30–32, write a linear equation to model the situation. Use unit analysis to check your model. **30–32. See margin.**

30. You borrow \$40 from your sister. To repay the loan, you pay her \$5 a week.

31. Your uncle weighed 180 pounds. He has lost 2 pounds a month for 8 months.

32. You have walked 5 miles on a hiking trail. You continue to walk at the rate of 2 miles per hour for 6 hours.

33. TRAVELING HOME You are traveling home in a bus whose speed is 50 miles per hour. At noon you are 200 miles from home. Write an equation that models your distance y from home in terms of t, the number of hours since noon. Why does the line given by this equation have a negative slope?
$y = 200 - 50t$; The distance from home decreases as time increases.

FUNDRAISING In Exercises 34–37, you are designing a calendar as a fund-raising project for your Biology Club. The cost of printing is \$500, plus \$2.50 per calendar.

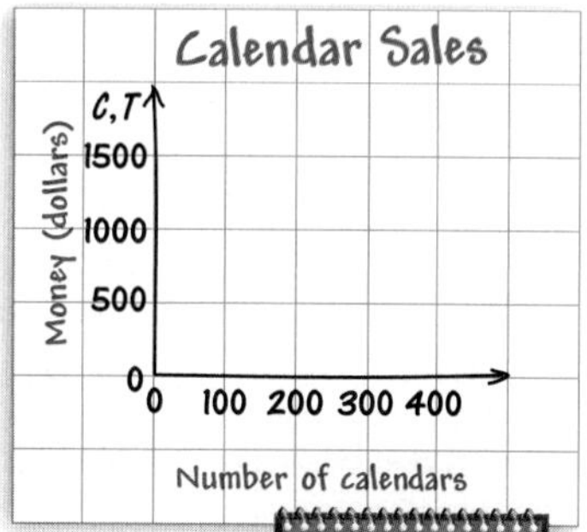

34. Write an equation to model the total cost C of printing x calendars. $C = 2.5x + 500$

35. You sell each calendar for \$5.00. Write an equation to model the total income T for selling x calendars. $T = 5x$

36. Graph both equations in the same coordinate plane. Estimate the point at which the two lines intersect. Explain the significance of this point of intersection in the context of the real-life problem. **See margin.**

37. ANALYZING DATA You estimate the club can sell 200 calendars. Write an analysis of how effective this fundraising project will be for the club. **See margin.**

30. Let y = the amount you owe and x = the number of weeks; $y = 40 - 5x$; dollars = dollars $-$ $\frac{\text{dollars}}{\text{week}}$ · weeks.

31. Let y = your uncle's weight in lb and x = the number of months from the start of his weight loss; $y = 180 - 2x$; lb = lb $-$ $\frac{\text{lb}}{\text{month}}$ · months.

32. Let y = the total distance in mi and x = the number of hours since you started; $y = 2x + 5$; mi = mi + $\frac{\text{mi}}{\text{hour}}$ · hours.

STUDENT HELP

HOMEWORK HELP Visit our Web site www.mcdougallittell.com for help with Exs. 34–37.

36. Check graphs; (200, 1000); the point of intersection (200, 1000) represents the number of calendars, 200, for which the printing cost, \$1000, is equal to the income.

37. *Sample answer:* If the club members can sell only 200 calendars, the fund-raising project will not raise any money at all; therefore, it is not a very effective project.

! COMMON ERROR

EXERCISES 20–25 Some students may incorrectly use the x-intercept when writing their equations. Remind students that they should use the point where the line crosses the y-axis to find b.

DATA UPDATE

Updated data for Exs. 26 and 27 is available at **www.mcdougallittell.com.**

STUDENT HELP NOTES

Homework Help Students can find help for Exs. 34–37 at **www.mcdougallittell.com.** The information can be printed out for students who don't have access to the Internet.

ADDITIONAL PRACTICE AND RETEACHING

For Lesson 5.1:

- Practice Levels A, B, and C (*Chapter 5 Resource Book*, p. 13)
- Reteaching with Practice (*Chapter 5 Resource Book*, p. 16)
- See Lesson 5.1 of the *Personal Student Tutor*

For more Mixed Review:

- Search the *Test and Practice Generator* for key words or specific lessons.

4 ASSESS

DAILY HOMEWORK QUIZ

Transparency Available

Write an equation of the line in slope-intercept form.

1. slope: $-\frac{3}{2}$; y-intercept: 2
   $y = -\frac{3}{2}x + 2$
2. 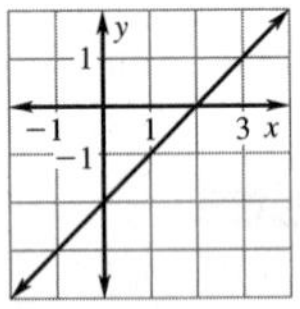

   $y = x - 2$
3. You rent a bike for a flat rate of \$15 plus an additional charge of \$2.50 per hour. Write an equation to model the total cost C in terms of the number of hours h you rent the bike.
   $C = 2.5h + 15$

EXTRA CHALLENGE NOTE

Challenge problems for Lesson 5.1 are available in **blackline** format in the *Chapter 5 Resource Book*, p. 20 and at **www.mcdougallittell.com.**

ADDITIONAL TEST PREPARATION

1. **OPEN ENDED** Describe a real-life situation that could be modeled using a linear equation in slope-intercept form. Write an equation to model the situation. Answers will vary.
2. **WRITING** You are given the equation $y = 56.3 - 4x$. Explain how you could use the equation to find the slope of its graph. What is the slope?
   Sample answer: **You can write it in slope-intercept form $y = -4x + 56.3$, to see that the slope is −4.**

40, 47–57. See Additional Answers beginning on page AA1.

Test Preparation

FLORIDA POPULATION In Exercises 38 and 39, use the following information. The U.S. Bureau of the Census predicted that the population of Florida would be about 17.4 million in 2010 and then would increase by about 0.22 million per year until 2025.

38. MULTIPLE CHOICE Choose the linear model that predicts the population P of Florida (in millions) in terms of t, the number of years since 2010. **C**

(A) $P = 17.4t + 0.22$ (B) $P = -0.22t + 17.4$

(C) $P = 0.22t + 17.4$ (D) $P = -17.4t + 0.22$

39. MULTIPLE CHOICE According to the correct model, the population of Florida in 2011 will be about __?__ million. **C**

(A) 17.18 (B) 3.8 (C) 17.62 (D) 39.4

★ Challenge

40. CLASS SIZE In January, your ceramics class begins with 12 students. In every month after January, three new students join and one student drops out.

a. Write a linear equation to model the situation. **Let y = the number of students and x = the number of months since January; $y = 2x + 12$.**

b. Graph the model. **See margin.**

c. Predict the size of your ceramics class in May and June. **May: 20 students, June: 22 students**

EXTRA CHALLENGE

www.mcdougallittell.com

MIXED REVIEW

EVALUATING EXPRESSIONS **Evaluate the expression when $x = -3$ and $y = 6$.** (Review 2.7)

41. $\frac{3x}{x + y}$ **−3** **42.** $\frac{x}{x + 2}$ **3** **43.** $\frac{y^2 + x}{x}$ **−11**

SCHOOL NEWSPAPER **You are designing a newspaper page with three photos. The page is $13\frac{1}{4}$ inches wide with 1 inch margins on both sides. You need to allow $\frac{3}{4}$ inch between photographs. How wide should you make the photos if they are of equal size?** (Review 3.5)

44. Sketch a diagram of the newspaper page. **Check drawings.**

45. Write an equation to model the problem. Use the diagram to help.

46. Solve the equation and answer the question. **$3\frac{1}{4}$ in.**

45. Let x = the width of each photograph; $3x + 3\frac{1}{2} = 13\frac{1}{4}$.

WINDSURFING **Match the description with the linear model $y = 10$ or the linear model $y = 10x$. Graph the model.** (Review 4.2)

47. You rent a sailboard for \$10 per hour. **$y = 10x$; See margin.**

48. You rent a life jacket for a flat fee of \$10. **$y = 10$; See margin.**

GRAPHING LINEAR EQUATIONS **Find the slope and the y-intercept of the graph of the equation. Then graph the equation.** (Review 4.6 for 5.2)

49–57. See margin.

49. $y + 2x = 2$ **−2, 2** **50.** $3x - y = -5$ **3, 5** **51.** $2y - 3x = 6$ **$\frac{3}{2}$, 3**

52. $4x + 2y = 6$ **−2, 3** **53.** $4y + 12x = 16$ **−3, 4** **54.** $25x - 5y = 30$ **5, −6**

55. $x + 3y = 15$ **$-\frac{1}{3}$, 5** **56.** $x + 6y = 12$ **$-\frac{1}{6}$, 2** **57.** $x - y = 10$ **1, −10**

5.2 Writing Linear Equations Given the Slope and a Point

What you should learn

GOAL 1 Use slope and any point on a line to write an equation of the line.

GOAL 2 Use a linear model to make predictions about a **real-life** situation, such as the number of vacation trips in **Example 3**.

Why you should learn it

▼ To model a **real-life** situation, such as finding the total cost of a taxi ride in **Ex. 48**.

REAL LIFE

GOAL 1 USING SLOPE AND A POINT ON A LINE

In Lesson 5.1 you learned how to write an equation of a line when given the slope and the y-intercept of the line. You used the slope-intercept form of the equation of a line, $y = mx + b$.

In this lesson you will learn to write an equation of a line when given its slope and *any* point on the line.

EXAMPLE 1 Writing an Equation of a Line

Write an equation of the line that passes through the point $(6, -3)$ and has a slope of -2.

SOLUTION

Find the y-intercept. Because the line has a slope of $m = -2$ and passes through the point $(x, y) = (6, -3)$, you can substitute the values $m = -2$, $x = 6$, and $y = -3$ into the slope-intercept form and solve for b.

$y = mx + b$	Write slope-intercept form.
$-3 = (-2)(6) + b$	Substitute -2 for m, 6 for x, and -3 for y.
$-3 = -12 + b$	Simplify.
$9 = b$	Solve for b.

The y-intercept is $b = 9$.

Write an equation of the line. Because you now know both the slope and the y-intercept, you can use the slope-intercept form.

$y = mx + b$	Write slope-intercept form.
$y = -2x + 9$	Substitute -2 for m and 9 for b.

✓CHECK You can check your result by graphing $y = -2x + 9$.

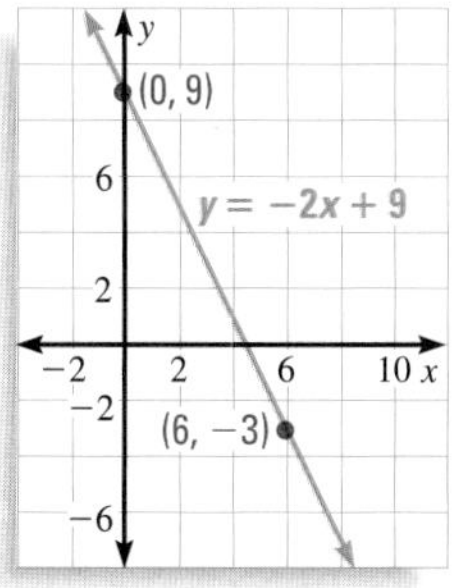

Note that the line crosses the y-axis at the point $(0, 9)$ and passes through the given point $(6, -3)$.

You can count the number of units in the rise and the run to check that the slope is -2.

1 PLAN

PACING
Basic: 1 day
Average: 1 day
Advanced: 1 day
Block Schedule: 0.5 block with 5.1

LESSON OPENER ACTIVITY
An alternative way to approach Lesson 5.2 is to use the Activity Lesson Opener:
- Blackline Master (*Chapter 5 Resource Book,* p. 24)
- Transparency (p. 33)

MEETING INDIVIDUAL NEEDS
- ***Chapter 5 Resource Book***
  Prerequisite Skills Review (p. 5)
  Practice Level A (p. 26)
  Practice Level B (p. 27)
  Practice Level C (p. 28)
  Reteaching with Practice (p. 29)
  Absent Student Catch-Up (p. 31)
  Challenge (p. 33)
- ***Resources in Spanish***
- ***Personal Student Tutor***

NEW-TEACHER SUPPORT
See the Tips for New Teachers on pp. 1–2 of the *Chapter 5 Resource Book* for additional notes about Lesson 5.2.

WARM-UP EXERCISES

Transparency Available

Write an equation of the line in slope-intercept form.

1. The slope is 5 and the y-intercept is -2. $y = 5x - 2$
2. The slope is $-\frac{2}{3}$ and the y-intercept is 1. $y = -\frac{2}{3}x + 1$
3. The slope is 0 and the y-intercept is $-\frac{1}{2}$. $y = -\frac{1}{2}$

Solve each equation for b.

4. $8 = (-2)(3) + b$ 14
5. $-2 = \frac{1}{4}(-3) + b$ $-\frac{5}{4}$

2 TEACH

MOTIVATING THE LESSON

All the employees of a garden center are given a \$0.40 per hour raise each year. At the start of your fourth year as an employee, you are earning \$7.15 per hour. What was your starting pay? Your starting pay will be the y-intercept of the line with slope 0.4 that contains (3, 7.15).

EXTRA EXAMPLE 1

Write an equation of the line that passes through the point (–3, 0) and has a slope of $\frac{1}{3}$. $y = \frac{1}{3}x + 1$

EXTRA EXAMPLE 2

Write an equation of the line that is parallel to the line $y = -3x - 2$ through the point (3, –4).
$y = -3x + 5$

CHECKPOINT EXERCISES

For use after Example 1:

1. Write an equation of the line that passes through the point (–2, –1) and has a slope of –3.
$y = -3x - 7$

For use after Example 2:

2. Write an equation of the line that is parallel to the line $y = -\frac{1}{4}x + 2$ and passes through the point (3, –2).
$y = -\frac{1}{4}x - \frac{5}{4}$

STUDENT HELP NOTES

Look Back As students look back to p. 242, remind them that vertical lines have undefined slope. The case of vertical lines is not included in this lesson.

MATHEMATICAL REASONING

In Example 2, the relationship between parallel lines and slope is given as a *biconditional.* Explain to students that the expression "if and only if" means the implication can be read in either direction. Ask them to write the biconditional as two conditional statements.

WRITING AN EQUATION OF A LINE GIVEN ITS SLOPE AND A POINT

STEP 1 **First find the y-intercept.** Substitute the slope m and the coordinates of the given point (x, y) into the slope-intercept form, $y = mx + b$. Then solve for the y-intercept b.

STEP 2 **Then write an equation of the line.** Substitute the slope m and the y-intercept b into the slope-intercept form, $y = mx + b$.

EXAMPLE 2 *Writing Equations of Parallel Lines*

STUDENT HELP

Look Back
For help with parallel lines, see p. 242.

Two nonvertical lines are parallel if and only if they have the same slope. Write an equation of the line that is parallel to the line $y = \frac{2}{3}x - 2$ and passes through the point $(-2, 1)$.

SOLUTION

1 The given line has a slope of $m = \frac{2}{3}$. Because parallel lines have the same slope, a parallel line through $(-2, 1)$ must also have a slope of $m = \frac{2}{3}$. Use this information to find the y-intercept.

$y = mx + b$	Write slope-intercept form.
$1 = \frac{2}{3}(-2) + b$	Substitute $\frac{2}{3}$ for m, -2 for x, and 1 for y.
$1 = -\frac{4}{3} + b$	Simplify.
$\frac{7}{3} = b$	Solve for b.

The y-intercept is $b = \frac{7}{3}$.

2 Write an equation using the slope-intercept form.

$y = mx + b$	Write slope-intercept form.
$y = \frac{2}{3}x + \frac{7}{3}$	Substitute $\frac{2}{3}$ for m and $\frac{7}{3}$ for b.

✓**CHECK** You can check your result graphically. The original line $y = \frac{2}{3}x - 2$ is parallel to $y = \frac{2}{3}x + \frac{7}{3}$.

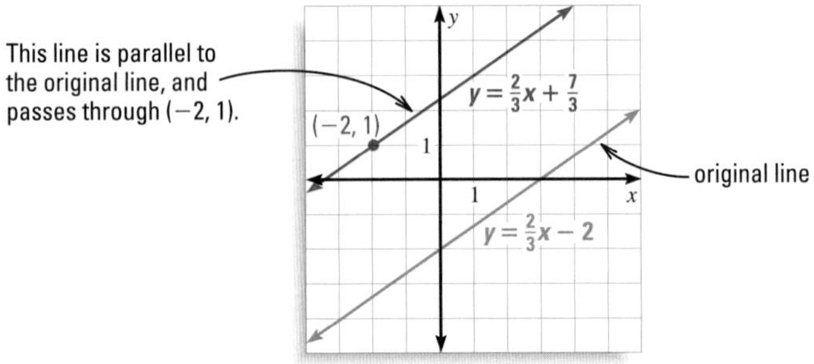

STUDENT HELP

HOMEWORK HELP
Visit our Web site www.mcdougallittell.com for extra examples.

Mathematical Reasoning ***Sample answer:***
If two nonvertical lines are parallel, then their slopes are equal. If the slopes of two nonvertical lines are equal, then the lines are parallel.

FOCUS ON APPLICATIONS

VACATIONS One favorite vacation destination is Washington, DC, with 22 million visitors in 1997. ▶ Source: Washington DC Convention and Visitors Association

GOAL 2 WRITING AND USING REAL-LIFE MODELS

EXAMPLE 3 *Writing and Using a Linear Model*

VACATION TRIPS Between 1985 and 1995, the number of vacation trips in the United States taken by United States residents increased by about 26 million per year. In 1993, United States residents went on 740 million vacation trips within the United States. DATA UPDATE of Travel Industry of America at www.mcdougallittell.com

a. Write a linear equation that models the number of vacation trips y (in millions) in terms of the year t. Let t be the number of years since 1985.

b. Estimate the number of vacation trips in the year 2005.

SOLUTION

a. The number of trips increased by about 26 million per year, so you know the slope is $m = 26$. You also know that $(t, y) = (8, 740)$ is a point on the line, because 740 million trips were taken in 1993, 8 years after 1985.

$y = mt + b$ **Write slope-intercept form.**

$740 = (26)(8) + b$ **Substitute 26 for m, 8 for t, and 740 for y.**

$740 = 208 + b$ **Simplify.**

$532 = b$ **The y-intercept is $b = 532$.**

Write an equation of the line using $m = 26$ and $b = 532$.

$y = mt + b$ **Write slope-intercept form.**

$y = 26t + 532$ **Substitute 26 for m and 532 for b.**

b. You can estimate the number of vacation trips in the year 2005 by substituting $t = 20$ into the linear model.

$y = 26t + 532$ **Write linear model.**

$= 26(20) + 532$ **Substitute 20 for t.**

$= 1052$ **Simplify.**

▶ You can estimate that United States residents will take about 1052 million vacation trips in the year 2005. A graph can help check this result. The line goes through (20, 1052).

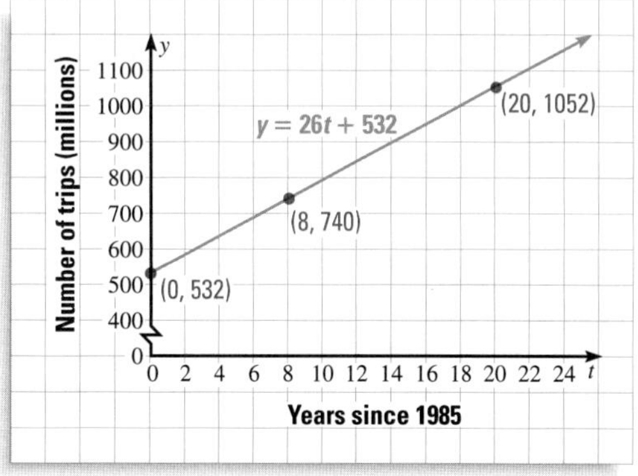

EXTRA EXAMPLE 3

To learn more about water evaporation, you measure the depth of a puddle after a rain storm. Your first measurement taken 3 hours after the storm, was 2.25 in. You take another measurement an hour later to determine that the rate of evaporation of the puddle is 0.25 in. per hour.

a. Write a linear equation that models the depth of the puddle d in inches in terms of t, the number of hours since the storm. $d = -0.25t + 3$

b. Estimate the depth of the puddle 7 hours after the storm. **1.25 in.**

CHECKPOINT EXERCISES

For use after Example 3:

1. All the employees of a garden center are given a \$0.40 per hour raise each year. You make \$7.15 per hour after three years as an employee. Write a linear equation that models your salary per hour S in terms of n the number of years you have worked at the garden center. Then find your hourly salary after 6 years. $S = 0.4n + 5.95$; **\$8.35**

DATA UPDATE

Updated data for Example 3 is available at **www.mcdougallittell.com.**

FOCUS ON VOCABULARY

You are given the slope-intercept equations for two nonvertical lines. How do you know if they are parallel? **If the slopes are equal, the lines are parallel.**

CLOSURE QUESTION

How is the procedure used to find the equation of a line given its slope and a point related to the procedure used to find the equation of a line given its slope and y-intercept? *Sample answer:* **The second step in the procedure used to find the equation of a line given its slope and a point is the same as the procedure used to find the equation of a line given its slope and y-intercept.**

3 APPLY

ASSIGNMENT GUIDE

BASIC

Day 1: pp. 282–284 Exs. 12–40 even, 44, 45, 48–50, 55, 65, 75

AVERAGE

Day 1: pp. 282–284 Exs. 12–40 even, 42–50, 55, 65, 75

ADVANCED

Day 1: pp. 282–284 Exs. 12–40 even, 42–51, 55, 65, 75

BLOCK SCHEDULE WITH 5.1

pp. 282–284 Exs. 12–40 even, 42–50, 55, 65, 75

EXERCISE LEVELS

Level A: ***Easier***
52–75

Level B: ***More Difficult***
12–50

Level C: ***Most Difficult***
51

✔ HOMEWORK CHECK

To quickly check student understanding of key concepts, go over the following exercises: Exs. 14, 22, 24, 26, 28, 30, 36, 44, 49. See also the Daily Homework Quiz:

- Blackline Master (*Chapter 5 Resource Book,* p. 36)
- Transparency (p. 39)

! COMMON ERROR

EXERCISES 12–23 Some students may reverse the values when substituting for x and y in the equation $y = mx + b$. It may help students to write x and y above the coordinates and write out the equation $y = mx + b$ before doing any substitution. After students find the equation, have them check that it is satisfied by the given coordinates.

GUIDED PRACTICE

Vocabulary Check ✓

1. Two nonvertical lines that have the same slope must be _?_. **parallel**

Concept Check ✓

2. Explain how to find the equation of a line given its slope and a point on the line. **See margin.**

Skill Check ✓

Write an equation of the line that passes through the point and has the given slope. Write the equation in slope-intercept form.

3. $(3, 4), m = \frac{1}{2}$ **$y = \frac{1}{2}x + \frac{5}{2}$**
4. $(2, -4), m = -5$ **$y = -5x + 6$**
5. $(10, -10), m = \frac{2}{3}$ **See margin.**
6. $(-12, 2), m = -2$ **$y = -2x - 22$**
7. $(4, 8), m = 5$ **$y = 5x - 12$**
8. $(0, -5), m = 0$ **$y = -5$**

Write an equation of the line that is parallel to the given line and passes through the point.

9. $y = \frac{1}{2}x + 8, (-6, 4)$ **$y = \frac{1}{2}x + 7$**
10. $y = -3x - 3, (-4, -3)$ **$y = -3x - 15$**

11. **BANKING** Christina has her savings in a bank account. She withdraws \$8.25 per week from her account. After 10 weeks, the balance is \$534. Write an equation that models the balance y of Christina's account in terms of the number of weeks x. Do not consider any interest earned by the account. **$y = -8.25x + 616.5$**

2. Substitute the slope and the coordinates of the given point in the equation $y = mx + b$ and solve for b, the y-intercept. Substitute b and m into the equation $y = mx + b$ to get an equation of the line.

5. $y = \frac{2}{3}x - \frac{50}{3}$

PRACTICE AND APPLICATIONS

STUDENT HELP

Extra Practice to help you master skills is on p. 801.

WRITING EQUATIONS **Write an equation of the line that passes through the point and has the given slope. Write the equation in slope-intercept form.**

12. $(1, 4), m = 3$ **$y = 3x + 1$**
13. $(2, -3), m = 2$ **$y = 2x - 7$**
14. $(-4, 2), m = -1$ **$y = -x - 2$**
15. $(1, -3), m = -4$ **$y = -4x + 1$**
16. $(-3, -5), m = -2$ **$y = -2x - 11$**
17. $(1, 3), m = 4$ **$y = 4x - 1$**
18. $(2, 5), m = \frac{1}{2}$ **$y = \frac{1}{2}x + 4$**
19. $(-3, 2), m = \frac{1}{3}$ **$y = \frac{1}{3}x + 3$**
20. $(0, 2), m = 3$ **$y = 3x + 2$**
21. $(0, -2), m = 4$ **$y = 4x - 2$**
22. $(3, 4), m = 0$ **$y = 4$**
23. $(-2, 4), m = 0$ **$y = 4$**

GRAPHICAL REASONING **Write an equation of the line shown in the graph.**

24.
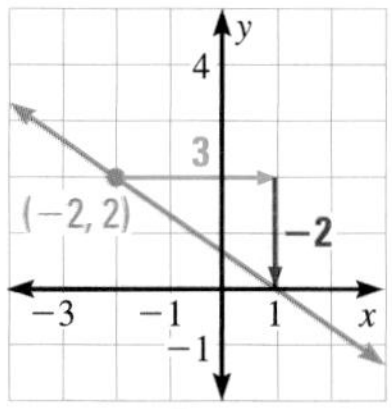

See margin.

25.
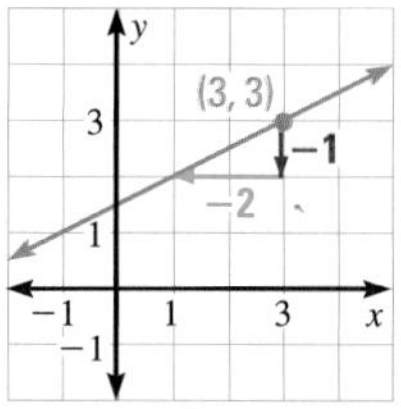

See margin.

26.
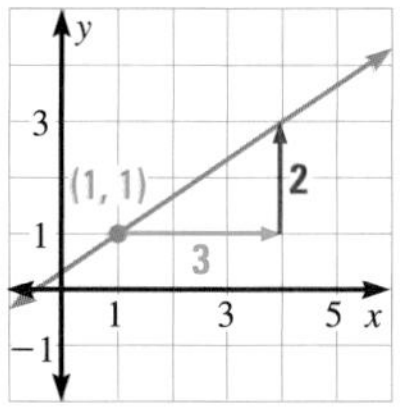

See margin.

27.
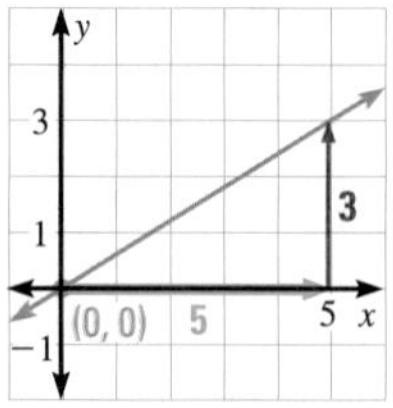

See margin.

28.
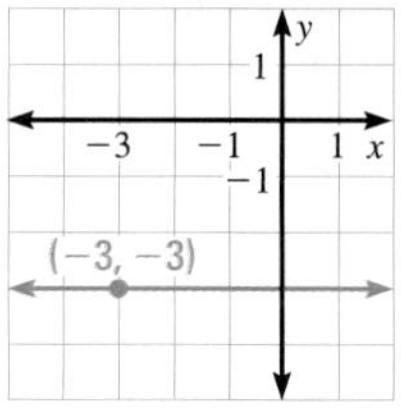

$y = -3$

29.
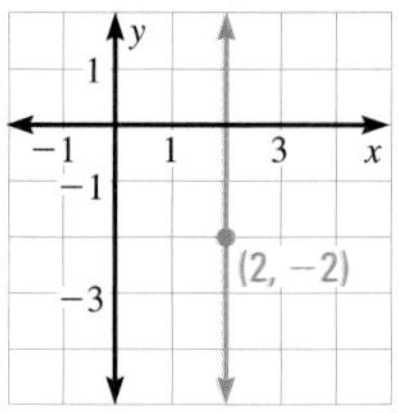

$x = 2$

24. $y = -\frac{2}{3}x + \frac{2}{3}$

25. $y = \frac{1}{2}x + \frac{3}{2}$

26. $y = \frac{2}{3}x + \frac{1}{3}$

27. $y = \frac{3}{5}x$

STUDENT HELP

HOMEWORK HELP
Example 1: Exs. 12–29
Example 2: Exs. 32–41
Example 3: Exs. 42–45

USING x-INTERCEPTS Write an equation of the line that has the given x-intercept and slope.

30. x-intercept $= 2$, $m = -\frac{2}{3}$ $\quad$ **$y = -\frac{2}{3}x + \frac{4}{3}$**

31. x-intercept $= 4$, $m = 3$ $\quad$ **$y = 3x - 12$**

PARALLEL LINES Write an equation of the line that is parallel to the given line and passes through the given point.

32. $y = 2x + 2$, $(3, 2)$ $\quad$ **$y = 2x - 4$**

33. $y = x + 4$, $(-2, 0)$ $\quad$ **$y = x + 2$**

34. $y = -2x + 3$, $(4, 4)$ $\quad$ **$y = -2x + 12$**

35. $y = -3x + 1$, $(4, 2)$ $\quad$ **$y = -3x + 14$**

36. $y = \frac{2}{3}x - 2$, $(2, 1)$ $\quad$ **See margin.**

37. $y = -\frac{1}{3}x - 1$, $(4, 1)$ $\quad$ **See margin.**

38. $y = -4x - 2$, $(5, 3)$ $\quad$ **$y = -4x + 23$**

39. $y = 6x + 9$, $(5, -3)$ $\quad$ **$y = 6x - 33$**

40. $y = -9x - 8$, $(7, -2)$ $\quad$ **$y = -9x + 61$**

36. $y = \frac{2}{3}x - \frac{1}{3}$

37. $y = -\frac{1}{3}x + \frac{7}{3}$

41. GEOMETRY CONNECTION $\overline{AB}$ is part of a line with a slope of $\frac{1}{2}$. To begin to draw a parallelogram, you want to draw a line parallel to $\overline{AB}$ through point C. Write an equation of the line. $\quad$ **$y = \frac{1}{2}x - 2$**

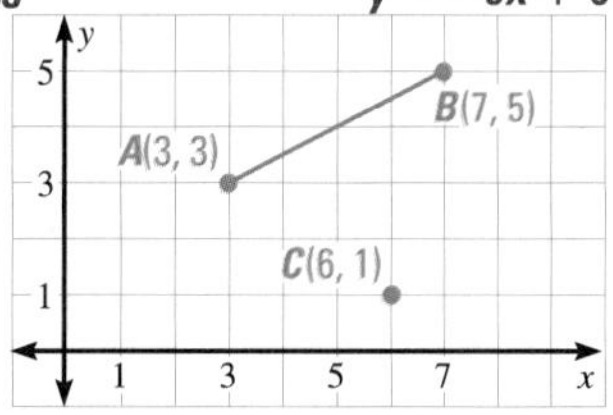

FOCUS ON APPLICATIONS

REAL LIFE **KENOSHA** borders Lake Michigan and originally had an economy dominated by the shipping industry.

POPULATION In Exercises 42 and 43, use the following information.

In 1991, the population of Kenosha, Wisconsin, was 132,000. Between 1991 and 1996, the population of Kenosha increased by approximately 2000 people per year. ▶ Source: U.S. Bureau of the Census

42. Write an equation that models the population y of Kenosha in terms of x, where x represents the number of years since 1991. $\quad$ **$y = 2000x + 132{,}000$**

43. Use the model to estimate the population of Kenosha in 2006. $\quad$ **about 162,000**

SALARY In Exercises 44 and 45, use the following information.

At the start of your second year as a veterinary technician, you receive a raise of \$750. You expect to receive the same raise every year. Your total yearly salary after your first raise is \$18,000 per year.

44. Write an equation that models your total salary s in terms of the number of years n since you started as a technician. $\quad$ **$s = 750n + 17{,}250$**

45. Calculate your yearly salary after six years as a veterinary technician. $\quad$ **\$21,750**

CELLULAR RATES In Exercises 46 and 47, use the following information.

You are moving to Houston, Texas, and are switching your cellular phone company. Your new peak air time rate in Houston is \$.23 per minute. Your bill also includes a monthly access charge. For 110 minutes of peak air time your bill is \$51.30.

46. Write an equation that models the cost C of your monthly bill in terms of the number of minutes m used. (All of your minutes are during peak air time.) $\quad$ **$C = 0.23m + 26$**

47. How much is your monthly bill for 60 minutes of peak air time? $\quad$ **\$39.80**

TAXI RIDE In Exercises 48 and 49, the cost of a taxi ride is an initial fee plus \$1.50 for each mile. Your fare for 9 miles is \$15.50.

48. Write an equation that models the total cost y of a taxi ride in terms of the number of miles x. $\quad$ **$y = 1.5x + 2$**

49. How much is the initial fee? $\quad$ **\$2**

TEACHING TIPS

EXERCISE 29 Students should understand that the formula $y = mx + b$ cannot be used in the case of a vertical line, since the slope of the line is undefined. Stress that the procedure outlined for writing the equation of a line given its slope and a point is valid only for *nonvertical lines.*

APPLICATION NOTE

EXERCISES 42 AND 43 Point out to students that the estimate found in Exercise 43 assumes that the population growth rate remained fairly consistent from 1991 to 2006. This assumption is generally valid for short intervals of time only. The model would not be likely to give a good estimate of the population in 2050, for example.

ADDITIONAL PRACTICE AND RETEACHING

For Lesson 5.2:

- Practice Levels A, B, and C (*Chapter 5 Resource Book,* p. 26)
- Reteaching with Practice (*Chapter 5 Resource Book,* p. 29)
- See Lesson 5.2 of the *Personal Student Tutor*

For more Mixed Review:

- Search the *Test and Practice Generator* for key words or specific lessons.

4 ASSESS

DAILY HOMEWORK QUIZ

Transparency Available

Write an equation of the line that passes through the point and has the given slope.

1. a. (6, −2), $m = -1$ **$y = -x + 4$**

b. (2, 0), $m = 4$ **$y = 4x - 8$**

2. Write an equation of the line through (4, −1) that is parallel to the line $y = 2x - 3$. **$y = 2x - 9$**

3. Between 1990 and 1998, the population of Riverton decreased by about 150 people per year. In 1997, the population was 75,450.

a. Write a linear equation that models the population P of Riverton in terms of the year t, where t is the number of years since 1990. **$P = -150t + 76{,}500$**

b. Estimate Riverton's population in 2003. **74,550**

EXTRA CHALLENGE NOTE

Challenge problems for Lesson 5.2 are available in **blackline** format in the *Chapter 5 Resource Book*, p. 33 and at **www.mcdougallittell.com.**

ADDITIONAL TEST PREPARATION

1. WRITING Keesha claimed the equation of the line through the point (4, −3) with slope $\frac{1}{12}$ is $y = \frac{1}{12}x + 8$. Her work is shown below. Explain her mistake and find the correct equation.

$y = mx + b$

$4 = \frac{1}{12}(-3) + b$

$4 = -4 + b$

$8 = b$

$y = \frac{1}{12}x + 8$

See sample answer at right.

51b. See Additional Answers beginning on page AA1.

Test Preparation

50c. 127 mi; *Sample answer:* No; it is not reasonable to assume the 2 mi per week increase can continue indefinitely.

★ Challenge

50. MULTI-STEP PROBLEM After 6 weeks on a fitness program, Greg jogs 35 miles per week. His average mileage gain has been 2 miles per week.

a. Write an equation that models Greg's weekly mileage m in terms of the number of weeks n that he stays on the program. **$m = 2n + 23$**

b. When will Greg jog over 45 miles per week? **after 11 weeks on the program**

c. *Writing* According to the equation, what will be Greg's weekly mileage after 52 weeks? Do you think this is realistic? Explain.

51. RENTING A CAR You are comparing the costs of car rental agencies for a one-day car rental. Car Rental Agency A charges \$30 a day plus \$.08 per mile. Car Rental Agency B charges a flat fee of \$40 a day.

a. Write an equation in slope-intercept form that models the cost of renting a car from each car rental agency. **$y = 0.08x + 30$ and $y = 40$**

b. Use a graphing calculator to graph the two equations on the same coordinate plane. **See margin.**

c. Under what conditions would the cost of renting a car from either rental agency be the same? **if you drive exactly 125 miles**

d. Under what conditions would the cost of renting a car from Car Rental Agency A be the best deal? Under what conditions would it be cheaper for you to rent a car from Car Rental Agency B?
If you drive less than 125 miles, Car Rental Agency A is less expensive. If you drive more than 125 miles, Car Rental Agency B is less expensive.

EXTRA CHALLENGE
www.mcdougallittell.com

MIXED REVIEW

EVALUATING EXPONENTIAL EXPRESSIONS Use a calculator to evaluate the power. (Review 1.2)

52. 5^7 **78,125** **53.** 8^5 **32,768** **54.** 9^6 **531,441** **55.** 3^{12} **531,441**

56. 2^8 **256** **57.** 4^3 **64** **58.** 8^4 **4096** **59.** 10^4 **10,000**

60. 7^8 **5,764,801** **61.** 6^9 **10,077,696** **62.** 5^{11} **48,828,125** **63.** 9^{13} **about 2,541,865,828,000**

RULES OF ADDITION Find the sum. (Review 2.2)

64. $-7 + (-1)$ **−8** **65.** $4 + \left(-4\frac{1}{2}\right)$ **$-\frac{1}{2}$** **66.** $-9 + 11$ **2**

67. $-10 + (-1)$ **−11** **68.** $2 + (-6)$ **−4** **69.** $-18 + (-2)$ **−20**

70. $6 + (-8) - 4$ **−6** **71.** $4 - (-7) + 3$ **14** **72.** $5 + (-3) + (-5)$ **−3**

FINDING SLOPE Find the slope of the line. (Review 4.4 for 5.3)

73. 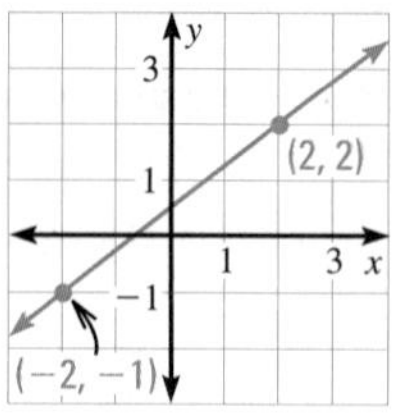

$\frac{3}{4}$

74. 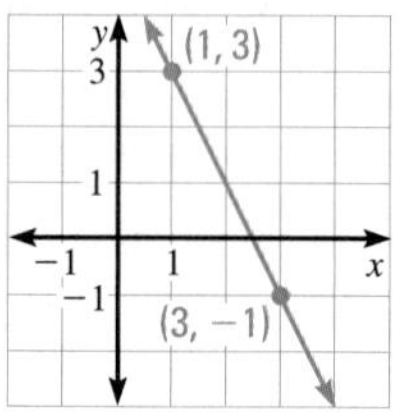

−2

75. 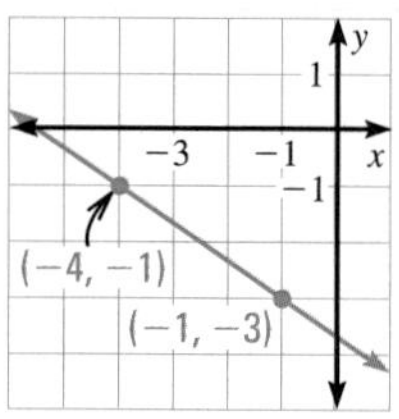

$-\frac{2}{3}$

Additional Test Preparation *Sample answer:* She switched the coordinates for x and y when she substituted in $y = mx + b$. The correct equation is $y = \frac{1}{12}x - \frac{10}{3}$.

5.3 Writing Linear Equations Given Two Points

What you should learn

GOAL 1 Write an equation of a line given two points on the line.

GOAL 2 Use a linear equation to model a **real-life** problem, such as estimating height in **Example 3**.

Why you should learn it

▼ To model a **real-life** situation, such as the distance an echo travels across a canyon in **Ex. 57**.

REAL LIFE

GOAL 1 USING TWO POINTS TO WRITE AN EQUATION

So far in this chapter, you have been writing equations of lines for which you were given the slope. In this lesson you will work with problems in which you must first find the slope. To do this, you can use the formula given in Chapter 4.

$$m = \frac{\text{rise}}{\text{run}} = \frac{y_2 - y_1}{x_2 - x_1} \longleftarrow \text{Slope of a line through } (x_1, y_1) \text{ and } (x_2, y_2)$$

EXAMPLE 1 *Writing an Equation Given Two Points*

Write an equation of the line that passes through the points (1, 6) and (3, −4).

SOLUTION

1. Find the slope of the line. Let $(x_1, y_1) = (1, 6)$ and $(x_2, y_2) = (3, -4)$.

$m = \frac{y_2 - y_1}{x_2 - x_1}$ Write formula for slope.

$= \frac{-4 - 6}{3 - 1}$ Substitute.

$= \frac{-10}{2} = -5$ Simplify.

2. Find the y-intercept. Let $m = -5$, $x = 1$, and $y = 6$ and solve for b.

$y = mx + b$ Write slope-intercept form.

$6 = (-5)(1) + b$ Substitute −5 for m, 1 for x, and 6 for y.

$6 = -5 + b$ Simplify.

$11 = b$ Solve for b.

3. Write an equation of the line.

$y = mx + b$ Write slope-intercept form.

$y = -5x + 11$ Substitute −5 for m and 11 for b.

✓**CHECK** You can use a graph to check your work. Plot the two given points, (1, 6) and (3, −4), and draw a line through them. Then check that the line has a y-intercept of 11 and a slope of −5.

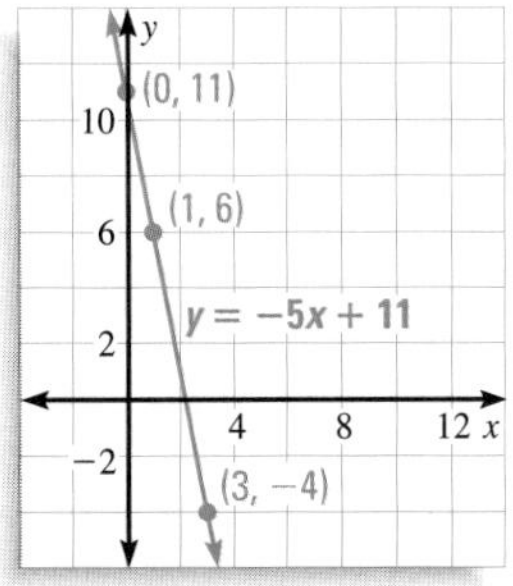

1 PLAN

PACING
Basic: 2 days
Average: 2 days
Advanced: 2 days
Block Schedule: 1 block

LESSON OPENER
APPLICATION
An alternative way to approach Lesson 5.3 is to use the Application Lesson Opener:
- Blackline Master (*Chapter 5 Resource Book*, p. 37)
- Transparency (p. 34)

MEETING INDIVIDUAL NEEDS
- ***Chapter 5 Resource Book***
  Prerequisite Skills Review (p. 5)
  Practice Level A (p. 40)
  Practice Level B (p. 41)
  Practice Level C (p. 42)
  Reteaching with Practice (p. 43)
  Absent Student Catch-Up (p. 45)
  Challenge (p. 47)
- ***Resources in Spanish***
- ***Personal Student Tutor***

NEW-TEACHER SUPPORT
See the Tips for New Teachers on pp. 1–2 of the *Chapter 5 Resource Book* for additional notes about Lesson 5.3.

WARM-UP EXERCISES

Transparency Available

Write the equation in slope-intercept form of the line that passes through the point and has the given slope.

1. (−2, 5), $m = \frac{1}{2}$ $y = \frac{1}{2}x + 6$

2. (6, −3), $m = -2$ $y = -2x + 9$

Write an equation of the line that is parallel to the given line and passes through the given point.

3. $y = -3x + 2$, (2, 1) $y = -3x + 7$

4. $y = \frac{1}{4}x + 1$, (−2, 0) $y = \frac{1}{4}x + \frac{1}{2}$

2 TEACH

MOTIVATING THE LESSON
During an unusually hot summer, the temperature in Paris, France, reached a high of 40°C. You know that the freezing point of water is 0°C or 32°F and the boiling point of water is 100°C or 212°F. How can you use this information to determine the Fahrenheit equivalent of 40°C?

EXTRA EXAMPLE 1
Write an equation of the line that passes through the points (–3, 1) and (5, 5). $y = \frac{1}{2}x + \frac{5}{2}$

EXTRA EXAMPLE 2

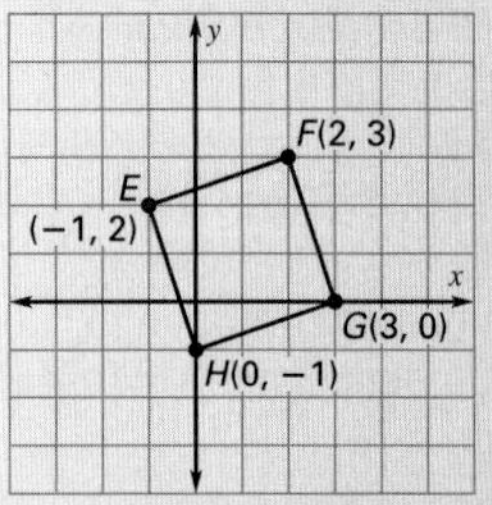

a. Show that $\overline{FG}$ and $\overline{GH}$ are perpendicular sides of figure *EFGH*. **The slope of $\overline{FG}$ is –3 and the slope of $\overline{GH}$ is $\frac{1}{3}$. $\overline{FG}$ and $\overline{GH}$ are perpendicular, since –3 is the negative reciprocal of $\frac{1}{3}$.**

b. Write equations for the lines containing $\overline{FG}$ and $\overline{GH}$.
$y = -3x + 9$, $y = \frac{1}{3}x - 1$

CHECKPOINT EXERCISES
For use after Examples 1 and 2:

1. Triangle *ABC* has vertices *A*(0, 1), *B*(1, 3), and *C*(5, 1).
  a. Is triangle *ABC* a right triangle? Explain your answer. **Yes; the slopes of sides $\overline{AB}$ and $\overline{BC}$ are negative reciprocals. Therefore $\overline{AB}$ and $\overline{BC}$ are perpendicular, and angle *B* is a right angle.**
  b. Write an equation for the line containing $\overline{BC}$.
  $y = -\frac{1}{2}x + \frac{7}{2}$

STUDENT HELP
INTERNET
HOMEWORK HELP
Visit our Web site www.mcdougallittell.com for extra examples.

WRITING AN EQUATION OF A LINE GIVEN TWO POINTS

STEP 1 **Find the slope.** Substitute the coordinates of the two given points into the formula for slope, $m = \frac{y_2 - y_1}{x_2 - x_1}$.

STEP 2 **Find the *y*-intercept.** Substitute the slope *m* and the coordinates of one of the points into the slope-intercept form, $y = mx + b$, and solve for the *y*-intercept *b*.

STEP 3 **Write an equation of the line.** Substitute the slope *m* and the *y*-intercept *b* into the slope-intercept form, $y = mx + b$.

EXAMPLE 2 *Writing Equations of Perpendicular Lines*

GEOMETRY CONNECTION Two different nonvertical lines are perpendicular if and only if their slopes are negative reciprocals of each other.

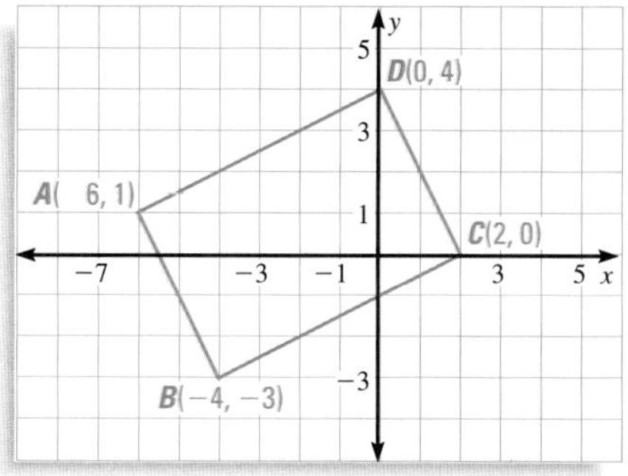

a. Show that $\overline{AB}$ and $\overline{BC}$ are perpendicular sides of figure *ABCD*.

b. Write equations for the lines containing $\overline{AB}$ and $\overline{BC}$.

SOLUTION

a. Find the slopes.

Slope of $\overline{AB}$: $m = \frac{1 - (-3)}{-6 - (-4)} = -2$

Slope of $\overline{BC}$: $m = \frac{0 - (-3)}{2 - (-4)} = \frac{1}{2}$

▸ $\overline{AB}$ and $\overline{BC}$ are perpendicular because $\frac{1}{2}$ is the negative reciprocal of -2.

b. Find the *y*-intercepts of the lines containing $\overline{AB}$ and $\overline{BC}$. Substitute the slopes from part (a) and the coordinates of one point into $y = mx + b$.

For $\overline{AB}$

$y = mx + b$
$-3 = (-2)(-4) + b$
$-11 = b$

For $\overline{BC}$

$y = mx + b$
$-3 = \left(\frac{1}{2}\right)(-4) + b$
$-1 = b$

Write an equation in slope-intercept form by substituting for *m* and *b*.

Equation of line containing $\overline{AB}$

$y = mx + b$
$y = -2x - 11$

Equation of line containing $\overline{BC}$

$y = mx + b$
$y = \frac{1}{2}x - 1$

GOAL 2 MODELING A REAL-LIFE SITUATION

EXAMPLE 3 *Writing and Using a Linear Model*

ARCHAEOLOGY While working at an archaeological dig, you find an upper leg bone (femur) that belonged to an adult human male. The bone is 43 centimeters long. In humans, femur length is linearly related to height. To estimate the height of the person, you measure the femur and height of two complete adult male skeletons found at the same excavation.

Person 1: 40-centimeter femur, 162-centimeter height

Person 2: 45-centimeter femur, 173-centimeter height

Estimate the height of the person whose femur was found.

SOLUTION

Write a linear equation to model the height of a person in terms of the femur length. Let x represent femur length (in centimeters) and let y represent the height of the person (in centimeters).

Find the slope of the line through the points (40, 162) and (45, 173).

$$m = \frac{173 - 162}{45 - 40}$$

$$= \frac{11}{5} = 2.2$$

Find the y-intercept.

$y = mx + b$	**Write slope-intercept form.**
$173 = 2.2(45) + b$	**Substitute for y, m, and x.**
$74 = b$	**Solve for b.**

Write a linear equation.

$y = mx + b$	**Write slope-intercept form.**
$y = 2.2x + 74$	**Substitute 2.2 for m and 74 for b.**

Estimate the height of the person whose femur is 43 centimeters by substituting 43 for x in the equation you wrote.

$$y = 2.2x + 74$$

$$y = 2.2(43) + 74$$

$$y = 168.6$$

▶ The person was about 169 centimeters tall.

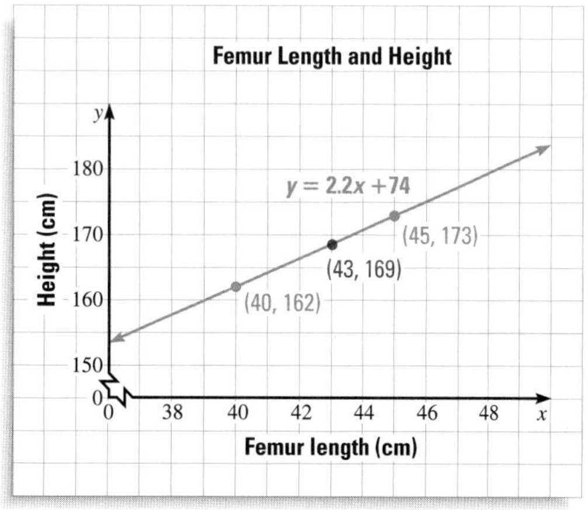

FOCUS ON CAREERS

ANTHROPOLOGIST
Physical anthropologists examine bones from archaeological digs to investigate the height, weight, and diet of members of a culture.

CAREER LINK
www.mcdougallittell.com

Closure Question *Sample answer:*
Use the slope formula to find the slope of the line. Use the slope-intercept formula with the computed slope and one of the given points to find the y-intercept. Substitute the slope and y-intercept into the slope-intercept form to get the equation of the line.

EXTRA EXAMPLE 3
For science class, you need to know the Celsius equivalent of a normal room temperature of 70°F. You know that Celsius temperature is linearly related to Fahrenheit temperature. To estimate the Celsius equivalent, you use what you know about the Fahrenheit and Celsius temperatures of the freezing and boiling point of water.
Freezing point: 32°F, 0°C
Boiling point: 212°F, 100°C
Estimate the Celsius equivalent of normal room temperature.
about 21°C

CHECKPOINT EXERCISES
For use after Example 3:

1. A camp program charges a registration fee and a daily amount. If the total bill for one camper was $338 for 12 days and the total bill for another camper was $506 for 19 days, how much will the bill be for a camper who enrolls for 30 days? **$770**

CONCEPT QUESTION
EXAMPLE 3 How do you decide which variable is the independent variable and which is the dependent variable when writing a linear model? ***Sample answer:* The unknown quantity is the dependent variable and the known quantity is the independent variable.**

FOCUS ON VOCABULARY
What is the relationship between the slopes of perpendicular lines?
The slopes are negative reciprocals.

CLOSURE QUESTION
Describe the procedure used to find the equation of a line through two points.
See sample answer at left.

DAILY PUZZLER
ABCD has vertices $A(-1, 2)$, $B(1, k)$, $C(3, 5)$ and $D(1,1)$. Find the value of k if polygon *ABCD* is a parallelogram.
6

3 APPLY

ASSIGNMENT GUIDE

BASIC
Day 1: pp. 288–289
Exs. 18–44 even
Day 2: pp. 289–291 Exs. 45–52, 58–60, 66–76 even, Quiz 1 Exs. 1–13

AVERAGE
Day 1: pp. 288–289
Exs. 18–44 even
Day 2: pp. 289–291 Exs. 45–55, 58–60, 66–76 even, Quiz 1 Exs. 1–13

ADVANCED
Day 1: pp. 288–289
Exs. 18–44 even
Day 2: pp. 289–291 Exs. 45–47, 51–65, 66–76 even, Quiz 1 Exs. 1–13

BLOCK SCHEDULE
pp. 288–291 Exs. 18–44 even, 45–55, 58–60, 66–76 even, Quiz 1 Exs. 1–13

EXERCISE LEVELS
Level A: *Easier*
18–20
Level B: *More Difficult*
21–55, 58–60, 66–76
Level C: *Most Difficult*
56, 57, 61–55

✔ HOMEWORK CHECK
To quickly check student understanding of key concepts, go over the following exercises: Exs. 18, 24, 31, 37, 40, 46, 52. See also the Daily Homework Quiz:

- Blackline Master (*Chapter 5 Resource Book,* p. 51)
- Transparency (p. 40)

GUIDED PRACTICE

1–2. Sample answers are given. See margin.

Vocabulary Check ✓

1. Compare writing linear equations given the slope and a point with writing linear equations given two points.

Concept Check ✓

2. Explain why you should find the slope m first when finding the equation of a line passing through two points.

Skill Check ✓

1. If you are given two points, you must first use the coordinates to determine the slope of the line. You then choose one of the points and proceed as you would if you were given that point and the slope.
2. You can use the slope and the coordinates of one of the given points in the equation $y = mx + b$ to determine the y-intercept.

Give the slope of a line perpendicular to the given line.

3. $y = -4x + 2$ $\frac{1}{4}$ **4.** $y = \frac{1}{2}x - 3$ -2 **5.** $y = x - 3$ -1

Write an equation in slope-intercept form of the line shown in the graph.

6.
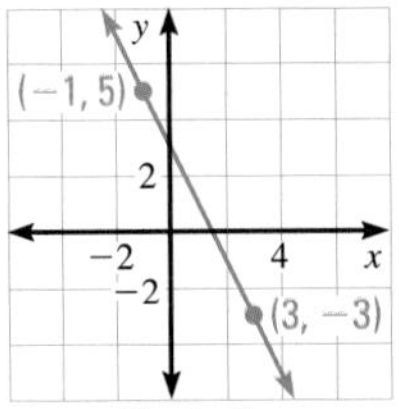

$y = -2x + 3$

7.
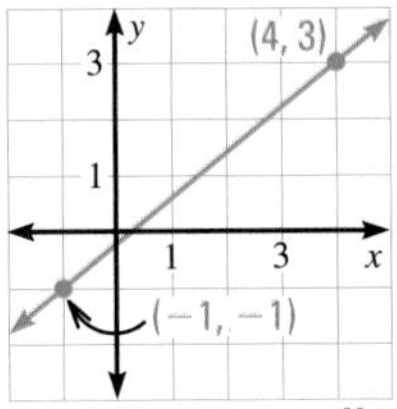

$y = \frac{4}{5}x - \frac{1}{5}$

8.
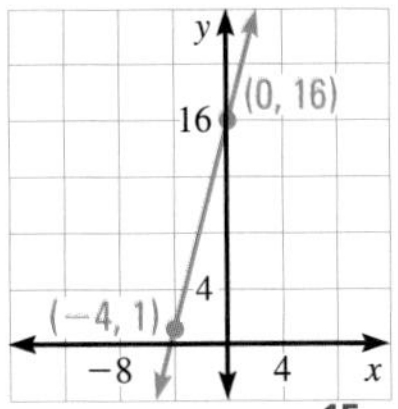

$y = \frac{15}{4}x + 16$

Write an equation in slope-intercept form of the line that passes through the points.

9. (−1, 1), (4, 5) $y = \frac{4}{5}x + \frac{9}{5}$ **10.** (3, −2), (−6, 4) $y = -\frac{2}{3}x$ **11.** (−4, 3), (−1, −7) See margin.

12. (−2, 5), (−6, −8) See margin. **13.** (−8, −4), (4, 2) $y = \frac{1}{2}x$ **14.** (−1, −3), (−8, −9) $y = \frac{6}{7}x - \frac{15}{7}$

15. (5, 3), (4, −3) $y = 6x - 27$ **16.** (6, −10), (−3, −8) $y = -\frac{2}{9}x - \frac{26}{3}$ **17.** (12, 2), (7, 2) $y = 2$

11. $y = -\frac{10}{3}x - \frac{31}{3}$
12. $y = \frac{13}{4}x + \frac{23}{2}$
21. $y = \frac{8}{3}x + \frac{2}{3}$
22. $y = -\frac{5}{4}x + \frac{21}{4}$
24. $y = -\frac{3}{2}x + 3$

PRACTICE AND APPLICATIONS

STUDENT HELP
Extra Practice to help you master skills is on p. 801.

27. $y = -\frac{6}{7}x + \frac{8}{7}$
28. $y = \frac{9}{10}x + \frac{4}{5}$
32. $y = \frac{7}{2}x + \frac{11}{2}$
33. $y = -\frac{1}{11}x + \frac{17}{11}$
34. $y = \frac{9}{2}x + 29$
35. $y = -\frac{14}{13}x + \frac{116}{13}$

STUDENT HELP
HOMEWORK HELP
Example 1: Exs. 18–44
Example 2: Exs. 45–49
Example 3: Exs. 53–55

WRITING EQUATIONS **Write an equation in slope-intercept form of the line shown in the graph.**

18.
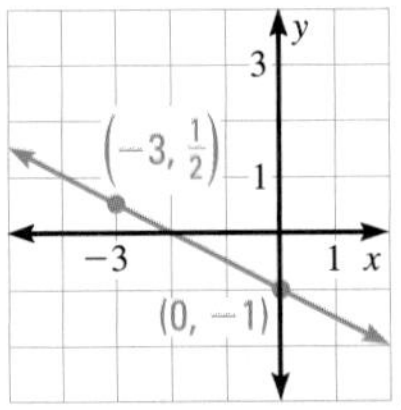

$y = -\frac{1}{2}x - 1$

19.
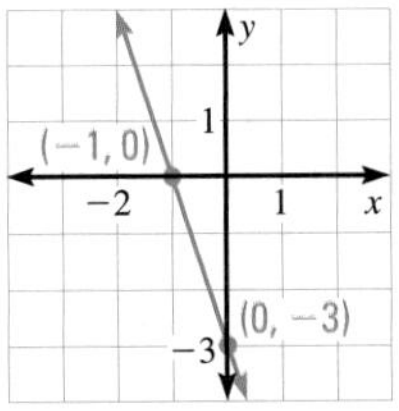

$y = -3x - 3$

20.
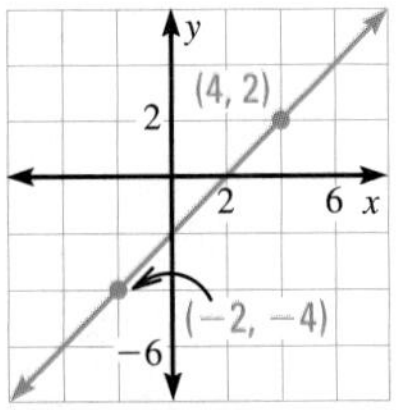

$y = x - 2$

GRAPHING **Graph the points and draw a line through them. Write an equation in slope-intercept form of the line that passes through the points.**

21–35. Check graphs.

21. (−1, −2), (2, 6) See margin. **22.** (1, 4), (5, −1) See margin. **23.** (−1, −2), (3, −2) $y = -2$

24. (2, 0), (−2, 6) See margin. **25.** (2, −3), (−3, 7) $y = -2x + 1$ **26.** (0, −5), (3, 4) $y = 3x - 5$

27. (6, −4), (−1, 2) See margin. **28.** (−2, −1), (8, 8) See margin. **29.** (1, 1), (7, 4) $y = \frac{1}{2}x + \frac{1}{2}$

30. (2, 4), (1, −2) $y = 6x - 8$ **31.** (5, −6), (5, −3) $x = 5$ **32.** (−3, −5), (1, 9) See margin.

33. (−5, 2), (6, 1) See margin. **34.** (−6, 2), (−4, 11) See margin. **35.** (−1, 10), (12, −4) See margin.

36. $y = \frac{9}{7}x + \frac{19}{7}$

37. $y = 3$

38. $y = \frac{3}{7}x - \frac{85}{7}$

39. $y = -\frac{3}{5}x + \frac{27}{5}$

40. $y = -\frac{1}{4}x + \frac{29}{4}$

41. $y = -\frac{2}{3}x + \frac{11}{3}$

42. $y = \frac{16}{63}x + \frac{122}{63}$

43. $y = -\frac{15}{7}x + \frac{4}{7}$

44. $y = -\frac{1650}{1183}x - \frac{24,159}{4732}$

48. $\overline{WX}$ and $\overline{XY}$, $\overline{ZY}$ and $\overline{XY}$; since the slope of $\overline{WX}$ and $\overline{ZY}$ is $-\frac{4}{3}$ and the slope of $\overline{XY}$ is $\frac{3}{4}$, the slopes are negative reciprocals.

49. $\overline{WX}$: $y = -\frac{4}{3}x + 4$; $\overline{XY}$: $y = \frac{3}{4}x - \frac{9}{4}$; $\overline{ZY}$: $y = -\frac{4}{3}x - \frac{13}{3}$

50. $\overline{WX}$: $y = -\frac{4}{3}x + 4$; $\overline{ZY}$: $y = -\frac{4}{3}x - \frac{13}{3}$; they have the same slope.

52. $y = \frac{35}{3}x - \frac{1600}{3}$; $\frac{35}{3}$; the French side

SLOPE-INTERCEPT FORM Write an equation in slope-intercept form of the line that passes through the points. 36–44. See margin.

36. $(-6, -5), (1, 4)$ 37. $(2, 3), (4, 3)$ 38. $(5, -10), (12, -7)$

39. $(14, -3), (-6, 9)$ 40. $(-7, 9), (-3, 8)$ 41. $(-8, 9), (10, -3)$

42. $\left(\frac{1}{4}, 2\right), \left(-5, \frac{2}{3}\right)$ 43. $\left(\frac{1}{2}, -\frac{1}{2}\right), \left(\frac{1}{9}, \frac{3}{9}\right)$ 44. $(-8.5, 6.75), (3.33, -9.75)$

45. **PERPENDICULAR LINES** Which of the lines are perpendicular? Explain. **Lines p and r; their slopes are negative reciprocals.**

line p: $y = \frac{1}{5}x + 2$ line q: $y = 5x - \frac{1}{2}$ line r: $y = -5x + 3$

46. Write an equation of a line through $(0, 2)$ that is perpendicular to $y = -4x + 6$. $y = \frac{1}{4}x + 2$

47. Write an equation of a line through $(4, 5)$ that is perpendicular to $y = \frac{1}{2}x + 3$. $y = -2x + 13$

GEOMETRY CONNECTION In Exercises 48–50, use the graph.

48. Find the perpendicular sides of trapezoid $WXYZ$. How do you know mathematically that these sides are perpendicular?

49. Write equations of the lines containing the perpendicular sides.

50. Write equations of the lines containing the two parallel sides. How do you know these sides are parallel?

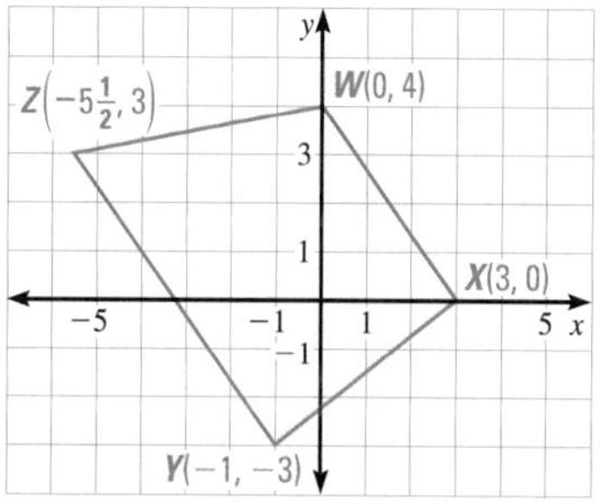

CHUNNEL In Exercises 51 and 52, use the diagram of the *Chunnel,* a railroad tunnel built under the English channel. It is one of the most ambitious engineering feats of the twentieth century.

51. Write an equation of the line from point A to point B. What is the slope of the line? $y = -\frac{26}{3}x + 60$; $-\frac{26}{3}$

52. Write an equation of the line from point C to point D. What is the slope of the line? Is the Chunnel steeper on the French side or on the English side?

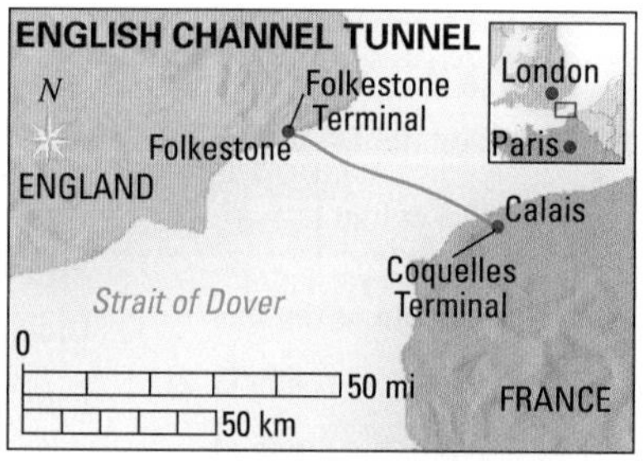

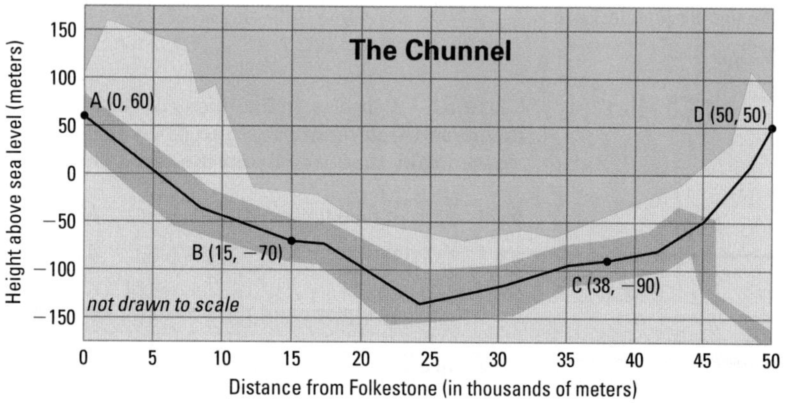

! COMMON ERROR

EXERCISES 6–35 Computational and algebraic errors are very common in these problems. Encourage students to always check their equation by testing the point not used to find the *y*-intercept.

! COMMON ERROR

EXERCISES 53–55 Students may confuse the independent and dependent variable in this type of problem. Help students look for key words such as *in terms of* to help them identify the variables. Often, the dependent variable is what they are looking for; whereas the independent variable is known. Note that if the variables are switched, the equation is still a valid expression of the relationship between the two variables, but it will not match the answer given.

ADDITIONAL PRACTICE AND RETEACHING

For Lesson 5.3:

- Practice Levels A, B, and C (*Chapter 5 Resource Book,* p. 40)
- Reteaching with Practice (*Chapter 5 Resource Book,* p. 43)
- See Lesson 5.3 of the *Personal Student Tutor*

For more Mixed Review:

- Search the *Test and Practice Generator* for key words or specific lessons.

4 ASSESS

DAILY HOMEWORK QUIZ

Transparency Available

Write an equation in slope-intercept form of the line that passes through the points.

1. **a.** (−3, 1), (2, −2) $y = -\frac{3}{5}x - \frac{4}{5}$
   b. (2, 3), (−4, 0) $y = \frac{1}{2}x + 2$
2. Write an equation of the line through (2, −1) that is perpendicular to $y = \frac{1}{3}x + 1$.
   $y = -3x + 5$
3. Show that $\triangle ABC$ is a right triangle.

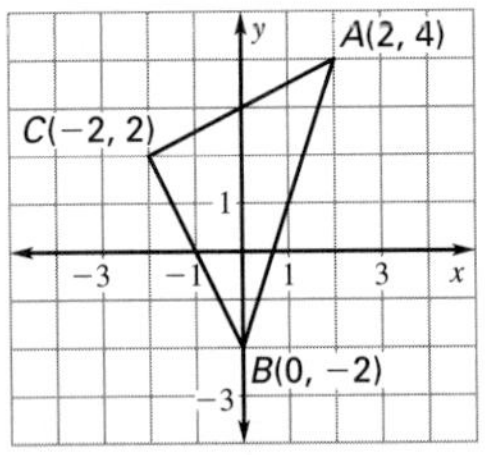

The slope of $\overline{AC}$ is $\frac{1}{2}$; the slope of $\overline{BC}$ is −2. Therefore, $\overline{AC}$ is perpendicular to $\overline{BC}$.

EXTRA CHALLENGE NOTE

Challenge problems for Lesson 5.3 are available in **blackline** format in the *Chapter 5 Resource Book,* p. 47 and at **www.mcdougallittell.com.**

ADDITIONAL TEST PREPARATION

1. **WRITING** Gregg claimed that if two lines are perpendicular, the product of their slopes is −1. Do you agree? Justify your answer with examples. **See sample answer below.**
2. **OPEN ENDED** Does it matter which point you use to find the *y*-intercept in Step 2 of the procedure outlined on page 286? Illustrate your answer with an example. **No; check students' examples.**

SCIENCE CONNECTION **In Exercises 53–55, use the following information.**
At sea level, the speed of sound in air is linearly related to the air temperature. If it is 35°C, sound will travel at a rate of 352 meters per second. If it is 15°C, sound will travel at a rate of 340 meters per second.

53. Write a linear equation that models speed of sound *s* in terms of air temperature *T*. $s = \frac{3}{5}T + 331$

54. How fast will sound travel at sea level if it is 25°C outside? **346 m/sec**

55. If sound travels at a rate of 346 meters per second at sea level, what is the temperature? **25°C**

ECHOES **In Exercises 56 and 57, use the following information.**
You yell across a canyon. Your sound travels at a rate of 343 meters per second.

56. Use your equation from Exercise 53 to determine the air temperature. **20°C**

57. After you shout, it takes 4 seconds to hear your echo. How far are you from the canyon wall? **686 m**

Test Preparation

58. MULTIPLE CHOICE What is the equation of the line that passes through the points (7, 4) and (−5, −2)? **D**

Ⓐ $y = \frac{1}{2}x - \frac{1}{2}$ Ⓑ $y = -\frac{1}{2}x + \frac{1}{2}$

Ⓒ $y = -\frac{1}{2}x - \frac{1}{2}$ Ⓓ $y = \frac{1}{2}x + \frac{1}{2}$

59. MULTIPLE CHOICE What is the equation of the line shown in the graph? **D**

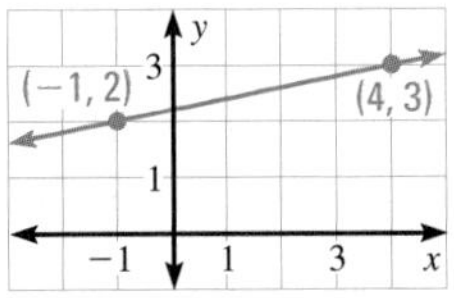

Ⓐ $y = 5x + 11$ Ⓑ $y = 5x - 17$

Ⓒ $y = \frac{11}{5}x + \frac{1}{5}$ Ⓓ $y = \frac{1}{5}x + \frac{11}{5}$

60. MULTIPLE CHOICE Choose which lines are perpendicular. **C**

Line *p* passes through (4, 0) and (6, 4)

Line *q* passes through (0, 4) and (6, 4)

Line *r* passes through (0, 4) and (0, 0)

Ⓐ line *p* and line *q* Ⓑ line *p* and line *r*

Ⓒ line *q* and line *r* Ⓓ none of these

★ Challenge

CRITICAL THINKING **In Exercises 61–64, use a graph to determine whether the given three points seem to lie on the same line. If they do, prove algebraically that they lie on the same line and write an equation of the line.**

61. (−3, −1), (0, 1), (12, 9) **See margin.** **62.** (4, −2), (−1, 2), (−8, 9) **See margin.**

63. (−2, −1), (3, 2), (7, 5) **See margin.** **64.** (3, −3), (−1, 13), (1, 5) **yes; $y = -4x + 9$**

65. What is an equation of the line that passes through the points $\left(3\frac{1}{2}, 4\right)$ and $\left(-5, -\frac{9}{2}\right)$? $y = x + \frac{1}{2}$

61. yes; $y = \frac{2}{3}x + 1$

62. No; the slope of the line through (4, −2) and (−1, 2) is $-\frac{4}{5}$, while the slope of the line through (−1, 2) and (−8, 9) is −1.

63. No; the slope of the line through (−2, −1) and (3, 2) is $\frac{3}{5}$, while the slope of the line through (3, 2) and (7, 5) is $\frac{3}{4}$.

EXTRA CHALLENGE
www.mcdougallittell.com

Additional Test Preparation ***Sample answer:***
1. No; If one line is horizontal, then the line perpendicular to it is vertical, which has no slope.

MIXED REVIEW

SIMPLIFYING EXPRESSIONS Simplify the expression. (Review 2.7)

66. $\frac{6x + 12y}{24}$ $\frac{x + 2y}{4}$ **67.** $\frac{48a - 56b}{16}$ $\frac{6a - 7b}{2}$ **68.** $\frac{14x}{-28y - 1}$ in simplest form

SOLVING EQUATIONS Solve the equation. (Review 3.4)

69. $4x - 11 = -31$ -5 **70.** $5x - 7 + x = 19$ $4\frac{1}{3}$ **71.** $18 = 4 - \frac{2x}{5}$ -35

72. $2x - 6 = 20$ 13 **73.** $\frac{1}{2}a + 8\frac{1}{2}a = 3$ $\frac{1}{3}$ **74.** $7y = 9y - 8$ 4

PLOTTING POINTS Plot the points in the table. Determine whether the slope of the line given by the points is *positive* or *negative*. (Review 4.1, 4.4 for 5.4)

75.

x	−1	0	1	2	3	4
y	−7	−4	−1	2	5	8

positive

76.

x	−2	−1	0	1	2	3
y	7	5	3	1	−1	−3

negative

QUIZ 1

Self-Test for Lessons 5.1–5.3

Write an equation of the line. (Lesson 5.1)

1. 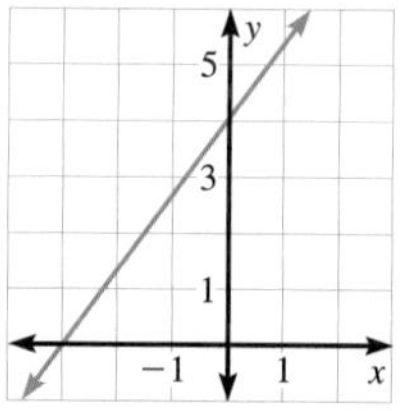
$y = \frac{4}{3}x + 4$

2. 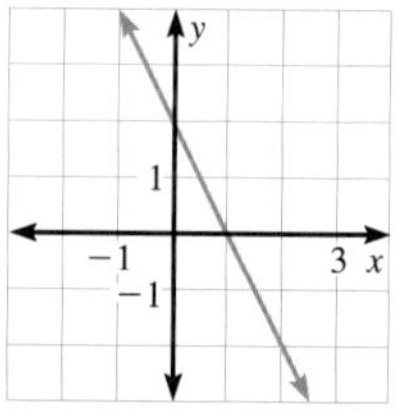
$y = -2x + 2$

3. 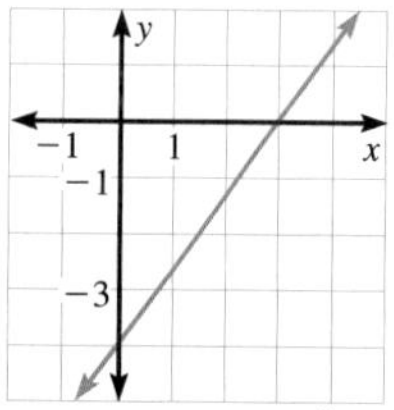
$y = \frac{4}{3}x - 4$

Write an equation of the line that passes through the point and has the given slope. Write the equation in slope-intercept form. (Lesson 5.2)

4. $(-5, 4)$, $m = 2$ $y = 2x + 14$ **5.** $(2, 1)$, $m = -3$ $y = -3x + 7$ **6.** $(-3, -6)$, $m = 1$ $y = x - 3$

7. Find an equation of the line that is perpendicular to the line $y = -x + 6$ and passes through $(-2, -4)$. (Lesson 5.3) $y = x - 2$

Write an equation in slope-intercept form of the line that passes through the points. (Lesson 5.3)

9. $y = \frac{13}{3}x + \frac{44}{3}$

8. $(8, -3)$, $(5, -2)$ $y = -\frac{1}{3}x - \frac{1}{3}$ **9.** $(-2, 6)$, $(-5, -7)$ See margin. **10.** $(4, 4)$, $(-7, 4)$ $y = 4$

CANOE RENTAL In Exercises 11–13, suppose your family rents a canoe for a deposit of \$10 plus \$28 per day. (Lesson 5.1)

11. Write an equation to model the total cost y of renting a canoe for x days. $y = 28x + 10$

12. Graph the equation. Label the y-intercept. See margin

13. Use the equation to find the cost of renting a canoe for 3 days. \$94

ADDITIONAL RESOURCES
An alternative Quiz for Lessons 5.1–5.3 is available in the *Chapter 5 Resource Book*, p. 48.

12. 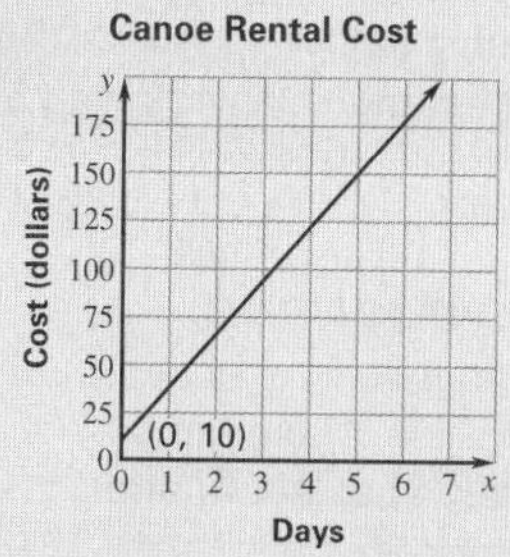

1 PLAN

PACING
Basic: 1 day
Average: 1 day
Advanced: 1 day
Block Schedule: 0.5 block with 5.5

LESSON OPENER APPLICATION
An alternative way to approach Lesson 5.4 is to use the Application Lesson Opener:
- Blackline Master (*Chapter 5 Resource Book,* p. 52)
- Transparency (p. 35)

MEETING INDIVIDUAL NEEDS
- ***Chapter 5 Resource Book***
  Prerequisite Skills Review (p. 5)
  Practice Level A (p. 54)
  Practice Level B (p. 55)
  Practice Level C (p. 56)
  Reteaching with Practice (p. 57)
  Absent Student Catch-Up (p. 59)
  Challenge (p. 62)
- ***Resources in Spanish***
- ***Personal Student Tutor***

NEW-TEACHER SUPPORT
See the Tips for New Teachers on pp. 1–2 of the *Chapter 5 Resource Book* for additional notes about Lesson 5.4.

WARM-UP EXERCISES

Transparency Available

Write an equation in slope-intercept form of the line that passes through the points.

1. (5, 32), (7, 16) $y = -8x + 72$
2. (0, 160), (25, 610) $y = 18x + 160$
3. (–12, –15), (–18, –12)
$y = -\frac{1}{2}x - 21$

1–4. See Additional Answers beginning on page AA1.

EXPLORING DATA AND STATISTICS

5.4 Fitting a Line to Data

What you should learn

GOAL 1 Find a linear equation that approximates a set of data points.

GOAL 2 Determine whether there is a positive or negative correlation or no correlation in a set of **real-life** data, like the Olympic data in **Example 3.**

Why you should learn it

To investigate trends in **real-life** data and to make predictions such as future football salaries in **Exs. 23** and **24.**

GOAL 1 FITTING A LINE TO DATA

In this lesson you will learn how to write a linear model to represent a collection of data points.

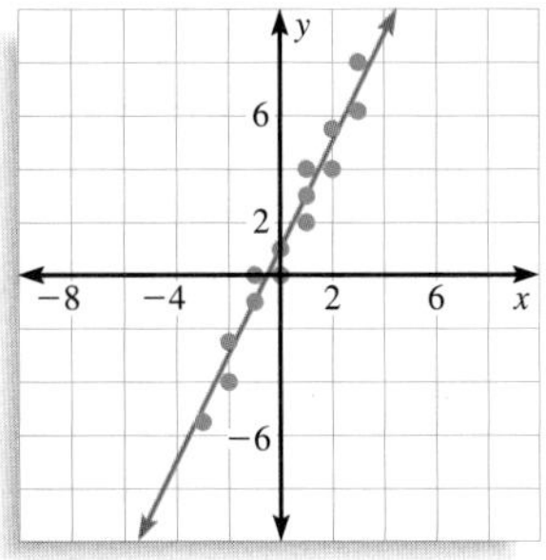

Usually there is no single line that passes through all of the data points, so you try to find the line that best fits the data, as shown at the right. This is called the **best-fitting line**.

There is a mathematical definition of the best-fitting line that is called *least squares approximation*. Many calculators have a built-in program for finding the equation of the best-fitting line. Since least squares approximation is not part of Algebra 1, you will be asked to use a graphical approach for drawing a line that is probably close to the best-fitting line.

ACTIVITY
Developing Concepts

Approximating a Best-Fitting Line

With your group, use the following steps to approximate a best-fitting line.

1. Carefully plot the following points on graph paper.
   (0, 3.3), (0, 3.9), (1, 4.2), (1, 4.5), (1, 4.8), (2, 4.7), (2, 5.1), (3, 4.9), (3, 5.6), (4, 6.1), (5, 6.4), (5, 7.1), (6, 6.8), (7, 7.5), (8, 7.8)
2. Use a ruler to sketch the line that you think best approximates the data points. Describe your strategy.
3. Locate two points on the line. Approximate the x-coordinate and the y-coordinate for each point. (These do not have to be two of the original data points.)
4. Use the method from Lesson 5.3 to find an equation of the line that passes through the two points.

Biology

EXAMPLE 1 *Approximating a Best-Fitting Line*

The data in the table show the forearm lengths and foot lengths (without shoes) of 18 students in an algebra class. After graphing these data points, draw a line that corresponds closely to the data. Write an equation of your line.

Forearm length	Foot length
22 cm	24 cm
20 cm	19 cm
24 cm	24 cm
21 cm	23 cm
25 cm	23 cm
18 cm	18 cm
20 cm	21 cm
23 cm	23 cm
24 cm	25 cm
20 cm	22 cm
19 cm	19 cm
25 cm	25 cm
23 cm	22 cm
22 cm	23 cm
18 cm	19 cm
24 cm	23 cm
21 cm	24 cm
22 cm	22 cm

SOLUTION

Let x represent the forearm length and let y represent the foot length. To begin, plot the points given by the ordered pairs. Then sketch the line that appears to best fit the points.

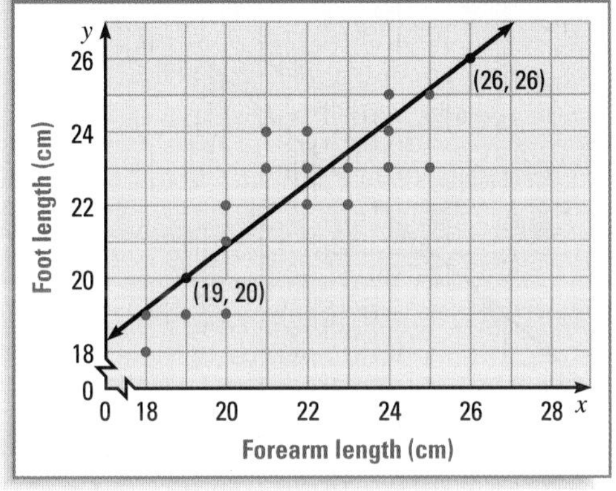

Next, find two points that lie on the line. You might choose the points (19, 20) and (26, 26). Find the slope of the line through these two points.

$m = \frac{y_2 - y_1}{x_2 - x_1}$ Write slope formula.

$m = \frac{26 - 20}{26 - 19}$ Substitute.

$m = \frac{6}{7}$ Simplify.

≈ 0.86 Decimal approximation.

To find the y-intercept of the line, substitute the values $m = 0.86$, $x = 19$, and $y = 20$ in the slope-intercept form.

$y = mx + b$ Write slope-intercept form.

$20 = (0.86)(19) + b$ Substitute 0.86 for m, 19 for x, and 20 for y.

$20 = 16.34 + b$ Simplify.

$3.66 = b$ Solve for b.

▶ An approximate equation of the best-fitting line is $y = 0.86x + 3.66$. In general, if a student has a long forearm, then that student also has a long foot.

STUDENT HELP

Study Tip
If you choose different points, you might get a different linear model. Be sure to choose points that will give you a line that is close to most of the points.

2 TEACH

ACTIVITY NOTE

USING MANIPULATIVES Clear rulers are very useful for this activity so that students can see all the points when drawing the line. A useful strategy for Step 2 is to have about half the points lie above the line and half the points lie below the line. Students may come up with other equally valid strategies that they can share with the class. Stress that there is not a unique answer to this problem. Each group might get a different line that corresponds closely to the data.

EXTRA EXAMPLE 1

You are studying the way a tadpole turns into a frog. You collect data to make a table that shows the ages and the lengths of the tails of 8 tadpoles.Draw a line that corresponds closely to the data. Write an equation of your line.

Age (days)	Length of Tail (mm)
5	14
2	15
9	3
7	8
12	1
10	3
3	12
6	9

Sample answer: $y = -1.7x + 19.9$

CHECKPOINT EXERCISES

For use after Example 1:

1. Draw a line that corresponds closely to the data in the table. Write an equation of the line.

x	2	9	1	8	5
y	11	27	9	23	17

Sample answer: $y = 2.2x + 6.6$

EXTRA EXAMPLE 2

The winning Olympic times for the women's 100 meter run from 1948 to 1996 are shown in the table. Draw a line that corresponds closely to these times. Write an equation of your line.

Olympic Year	Winning Time
1948	11.9 s
1952	11.5 s
1956	11.5 s
1960	11.0 s
1964	11.4 s
1968	11.0 s
1972	11.1 s
1976	11.1 s
1980	11.1 s
1984	11.0 s
1988	10.5 s
1992	10.8 s
1996	10.9 s

Check students' graphs. *Sample answer:* $y = -0.02x + 11.75$, where x is the number of years after 1940.

CHECKPOINT EXERCISES

For use after Example 2:

1. Draw a line that corresponds closely to the data in the table below. Write an equation of your line.

x	24	46	27	33	52
y	63	55	61	57	51

Sample answer: **$y = -0.38x + 71.29$**

CONCEPT QUESTION

EXAMPLE 2 Why was the point (96, 230) used to find the equation of the line when the winning throw for 1996 was 227.7 ft? **To find the equation of the approximation, you use points on the line. These points are not necessarily data points.**

FOCUS ON APPLICATIONS

DISCUS EVENT The discus throw is one of the original Olympic events. New records for the distance an athlete can throw the discus continue to be set.

EXAMPLE 2 Approximating a Best-Fitting Line

DISCUS THROWS The winning Olympic discus throws from 1908 to 1996 are shown in the table. After graphing these data points, draw a line that corresponds closely to the data. Write an equation of your line.

Olympic year	Winning throw
1908	134.1 ft
1912	148.3 ft
1920	146.6 ft
1924	151.4 ft
1928	155.3 ft
1932	162.4 ft
1936	165.6 ft
1948	173.2 ft
1952	180.5 ft
1956	184.9 ft
1960	194.2 ft
1964	200.1 ft
1968	212.5 ft
1972	211.3 ft
1976	221.5 ft
1980	218.7 ft
1984	218.5 ft
1988	225.8 ft
1992	213.7 ft
1996	227.7 ft

DATA UPDATE of *Information Please Almanac* at www.mcdougallittell.com

SOLUTION

Let x represent the years since 1900. Let y represent the winning throw. To begin, plot the points given by the ordered pairs. Then sketch the line that appears to best fit the points.

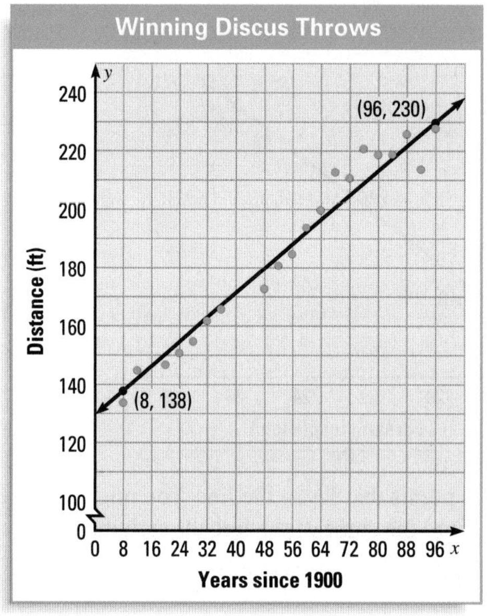

Next, find two points that lie on the line, such as (8, 138) and (96, 230). Find the slope of the line through these points.

$$m = \frac{y_2 - y_1}{x_2 - x_1} = \frac{230 - 138}{96 - 8} = \frac{92}{88} \approx 1.05$$

To find the y-intercept of the line, substitute the values $m = 1.05$, $x = 8$, and $y = 138$ in the slope-intercept form.

$y = mx + b$ — Write slope-intercept form.

$138 = (1.05)(8) + b$ — Substitute 1.05 for m, 8 for x, and 138 for y.

$138 = 8.4 + b$ — Simplify.

$129.6 = b$ — Solve for b.

▶ An approximate equation of the best-fitting line is $y = 1.05x + 129.6$. In most years, the winner of the discus throw was able to throw the discus farther than the previous winner.

GOAL 2 DETERMINING THE CORRELATION OF *X* AND *Y*

Correlation is a number r satisfying $-1 \le r \le 1$ that indicates how well a particular set of data can be approximated by a straight line. The Activity on page 299 shows how to find an r-value using a calculator. In this course you will describe a correlation without actually finding r-values, as described below.

When the points on a scatter plot can be approximated by a line with positive slope, x and y have a **positive correlation**. When the points can be approximated by a line with negative slope, x and y have a **negative correlation**. When the points cannot be well approximated by a straight line, we say that there is **relatively no correlation.**

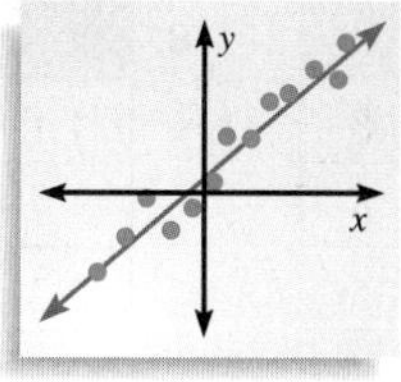

Positive correlation

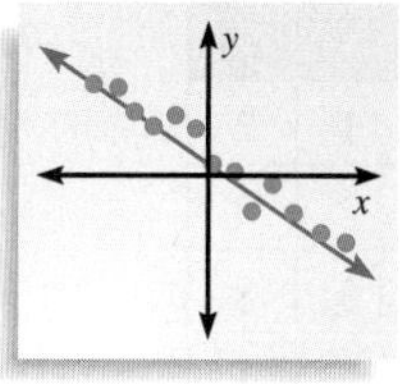

Negative correlation

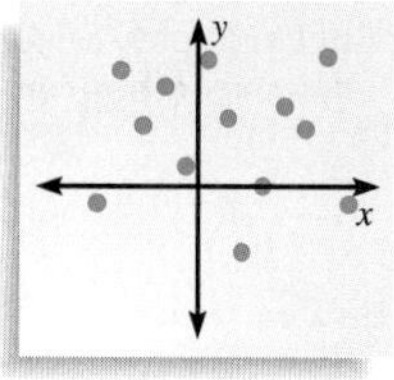

Relatively no correlation

EXAMPLE 3 *Using Correlation*

In swimming events, performance is positively correlated with time.

a. The two graphs show the winning 100-meter women's freestyle swimming times and the winning women's long jump distances for the Olympics from 1948 through 1996. Which is which? Explain your reasoning.

b. Describe the correlation of each set of data.

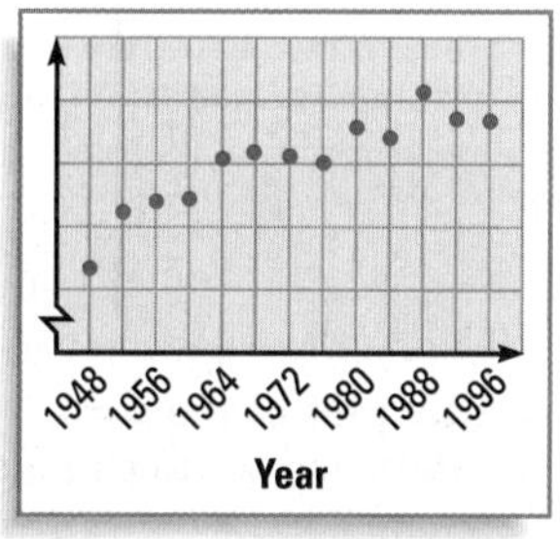

▶ Source: *Information Please Almanac*

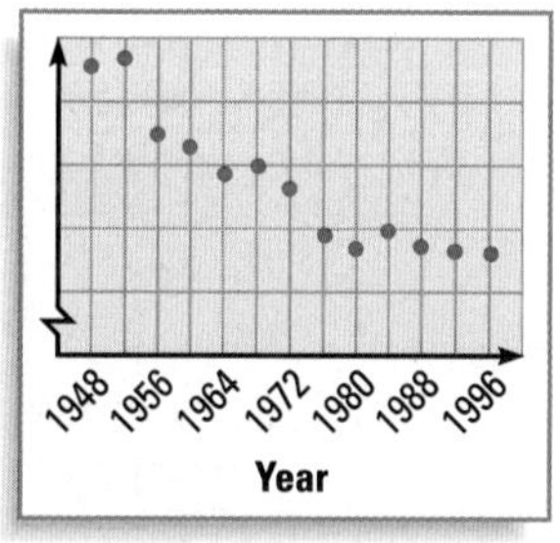

▶ Source: *Information Please Almanac*

SOLUTION

a. The first graph must represent the long jump distances because the winners have tended to jump farther with each Olympic year. The second graph must represent the swimming times because the winners have tended to swim faster with each Olympic year, so their times have been decreasing.

b. The first graph shows a positive correlation between the year and the winning distance. The second graph shows a negative correlation between the year and the winning time.

FOCUS ON APPLICATIONS

REAL LIFE FREESTYLE SWIMMING In a freestyle swimming event, the swimmer can choose the stroke. Usually, the swimmer chooses the Australian Crawl.

INTERNET **APPLICATION LINK**
www.mcdougallittell.com

EXTRA EXAMPLE 3

The Hernandez family spent 6 hours traveling by car.

a. The two graphs show the gallons of gas that remain in the gas tank and the miles driven for each of the 6 hours. Which is which? Explain. **See answer, below.**

b. Describe the correlation of each set of data. **See answer, below.**

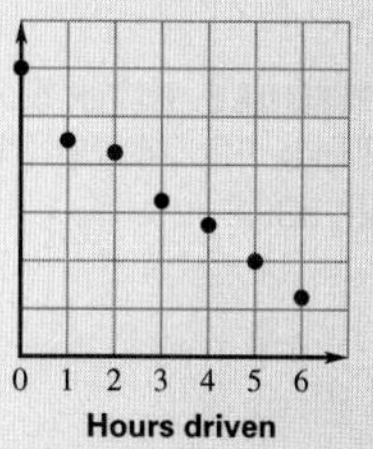

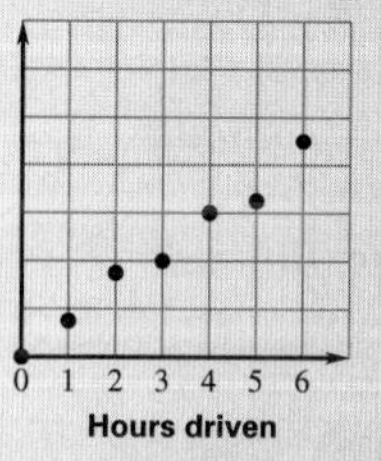

CHECKPOINT EXERCISES

For use after Example 3:

1. High school students are asked how many hours they watch television per week and how many hours they spend reading. Do you think the data will show a negative or positive correlation? Explain. **It will show a negative correlation because the number of hours a person reads will probably decrease as the number of hours they watch television increases.**

CLOSURE QUESTION

Describe the procedure used to approximate a best-fitting line. **Make a scatter plot. Draw a line to correspond closely to the data. Pick two points on the line and use the procedure from Lesson 5.3 to find the equation.**

Extra Example 3

a. **The first graph is the number of gallons left in the gas tank because the gas should decrease as the hours increase. The second graph is the miles driven, as this should increase as the hours increase.**

b. **The first graph shows a negative correlation between hours driven and gallons of gas remaining. The second graph shows a positive correlation between hours driven and number of miles driven.**

3 APPLY

ASSIGNMENT GUIDE

BASIC
Day 1: pp. 296–298 Exs. 10–24, 28–31, 35–38, 40, 45, 50

AVERAGE
Day 1: pp. 296–298 Exs. 10–24, 28–31, 35–38, 40, 45, 50

ADVANCED
Day 1: pp. 296–298 Exs. 10–40, 45, 50

BLOCK SCHEDULE WITH 5.5
pp. 296–298 Exs. 10–24, 28–31, 35–38, 40, 45, 50

EXERCISE LEVELS

Level A: *Easier*
10–12, 17–22

Level B: *More Difficult*
13–16, 23–30, 35–50

Level C: *Most Difficult*
31–34

✓ HOMEWORK CHECK

To quickly check student understanding of key concepts, go over the following exercises: Exs. 10, 14, 18, 22, 24. See also the Daily Homework Quiz:

- Blackline Master (*Chapter 5 Resource Book,* p. 65)
- Transparency (p. 41)

! COMMON ERROR

EXERCISES 6–16 Students often use incorrect points when finding the equation of a line that corresponds closely to the data. Stress that the points used must be on the line itself. They do not need to be data points.

GUIDED PRACTICE

Vocabulary Check ✓

1. The line that most closely fits a set of data is called the _?_. **best-fitting line**

Concept Check ✓

Draw a scatter plot of data that have the given correlation.
2–4. Check graphs. Graphs might resemble those on page 295 of the text.

2. Positive **3.** Negative **4.** Relatively no

5. **DISCUS THROWS** Predict the distance of the winning throw in the year 2005 from the data in Example 2. Explain your reasoning.
about 239.85 ft; for the year 2005, $x = 2005 - 1900 = 105$, so $y = 1.05(105) + 129.6 = 239.85$.

Skill Check ✓

Draw a scatter plot of the data. State whether x and y have a *positive correlation*, a *negative correlation*, or *relatively no correlation*. If possible, draw a line that closely fits the data and write an equation of the line.

6–16. Check graphs. Sample equations are given.

6.

x	y
1	2
2	9
3	8
4	1
5	4
6	8

relatively no correlation

7.

x	y
−3	8
−2	6
−1	5
0	3
1	2
2	0

negative correlation; $y = -1.54x + 3.23$

8.

x	y
1.1	5.1
1.7	5.5
2.2	5.9
2.6	6.3
3.3	7.5
3.5	7.6

positive correlation; $y = 1.10x + 3.68$

9.

x	y
5.5	0.4
6.2	1.0
7.7	2.5
8.1	2.9
9.2	4.3
9.7	5.5

positive correlation; $y = 1.16x - 6.22$

PRACTICE AND APPLICATIONS

STUDENT HELP
Extra Practice to help you master skills is on p. 801.

FITTING LINES **Copy the graph and draw a line that corresponds closely to the data. Write an equation of your line.**

10.
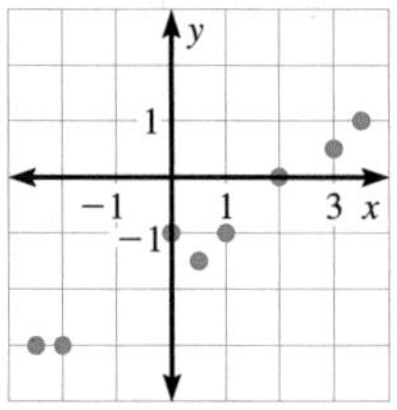

$y = 0.67x - 1.46$

11.
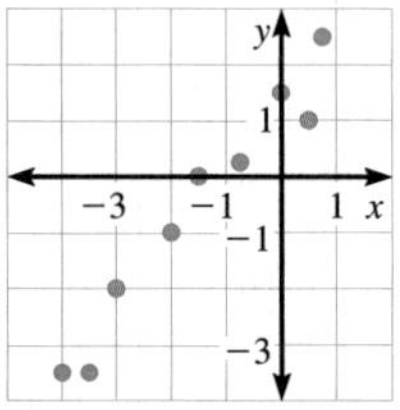

$y = 1.19x + 1.19$

12.
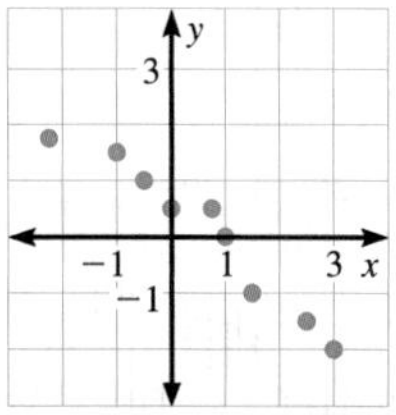

$y = -0.87x + 0.66$

FITTING LINES **Draw a scatter plot of the data. Draw a line that corresponds closely to the data and write an equation of the line.**
13–16. See margin.

13. $y = 1.64x + 2.57$
14. $y = 1.18x + 3.83$
15. $y = -1.79x + 15.45$
16. $y = -1.40x + 13.25$

13.

x	y
1.0	3.8
1.5	4.2
1.7	5.3
2.0	5.8
2.0	5.5
1.5	6.7

14.

x	y
3.0	7.1
3.4	8.1
4.0	8.5
4.1	8.9
4.8	9.6
5.2	9.8

15.

x	y
3.0	9.9
3.5	9.7
3.7	8.6
4.0	8.1
4.0	8.4
4.5	7.4

16.

x	y
5.0	6.8
5.4	5.8
6.0	5.6
6.1	5.2
6.8	4.3
7.2	3.5

STUDENT HELP
HOMEWORK HELP
Example 1: Exs. 10–16, 23
Example 2: Exs. 10–16, 23
Example 3: Exs. 17–22

GRAPHICAL REASONING **State whether *x* and *y* have a *positive correlation*, a *negative correlation*, or *relatively no correlation*.**

17. 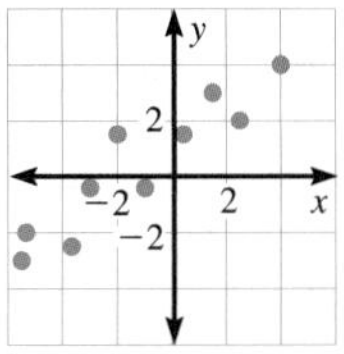
positive correlation

18. 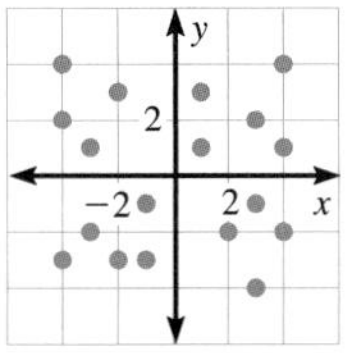
relatively no correlation

19. 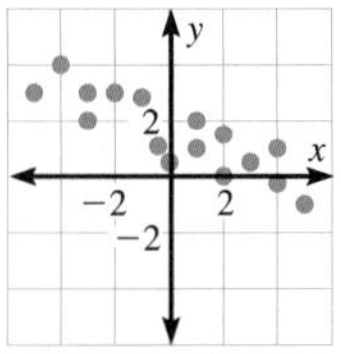
negative correlation

20. 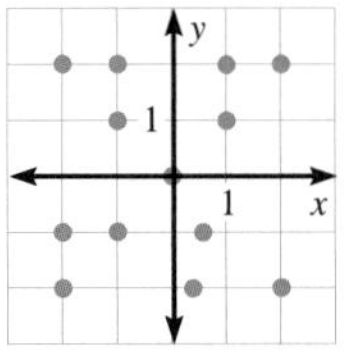
relatively no correlation

21. 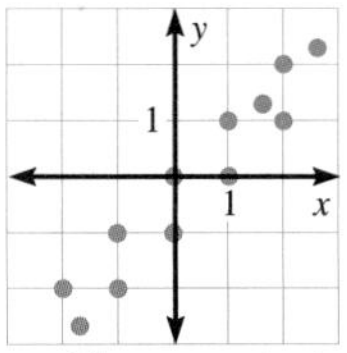
positive correlation

22. 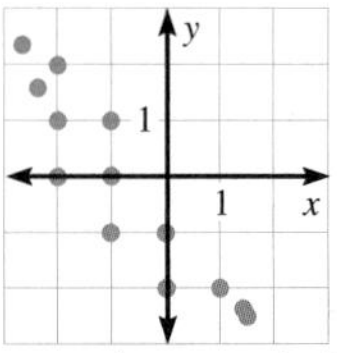
negative correlation

FOOTBALL SALARIES

In Exercises 23 and 24, use the following information. The median base salary for players in the National Football League from 1983 to 1997 is shown in the scatter plot at the right. In the scatter plot, *y* represents the salary and *x* represents the number of years since 1980.

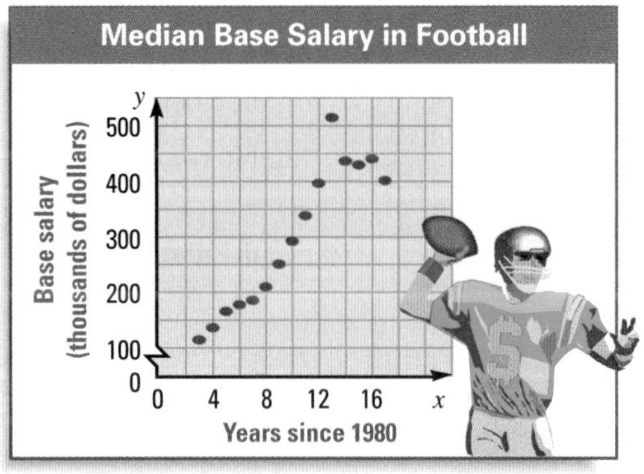

▶ Source: National Football League Players Association

23. Find an equation of the line that you think closely fits the data. ***Sample answer:* $y = 25x + 50$**

24. Use the equation from Exercise 23 to approximate the median base salary in the year 2010. **about $800,000**

BIOLOGY CONNECTION **In Exercises 25–27, use the following information.** As people grow older, the size of their pupils tends to get smaller. The average diameter (in millimeters) of one person's pupils is given in the table.

Sample Pupil Diameters		
Age (years)	Day	Night
20	4.7	8.0
30	4.3	7.0
40	3.9	6.0
50	3.5	5.0
60	3.1	4.1
70	2.7	3.2
80	2.3	2.5

25. Draw a scatter plot of the day diameters and another of the night diameters. Let *x* represent the person's age and let *y* represent pupil diameters. **See margin.**

26. Find an equation of the line that closely fits the day and the night sets of data for pupil diameters. **See margin.**

27. **CRITICAL THINKING** Do the two lines have the same slope? Explain your answer in the context of the real-life data. **See margin.**

STUDENT HELP

HOMEWORK HELP Visit our Web site www.mcdougallittell.com for help with Exs. 23 and 24.

26. daylight: $y = -0.04x + 5.5$; night: $y = -0.09x + 9.76$ (sample equations determined using a graphing calculator and rounding to the nearest hundredth)

27. No; the graph of the night diameters is much steeper (that is, the absolute value of the slope is much greater) than the graph of the daylight diameters. This indicates that the night diameter of the pupil decreases with age much more dramatically than the daylight diameter, which means that the eye's ability to adjust to lower light decreases with age.

! *COMMON ERROR*

EXERCISES 17–19 Students may assume that for *x* and *y* to have relatively no correlation they must have no relationship. In fact, *x* and *y* can have any relationship other than a linear one. For example, in the graph below, the data exhibit a quadratic relationship, but as this is not a linear relationship, *x* and *y* have no linear correlation.

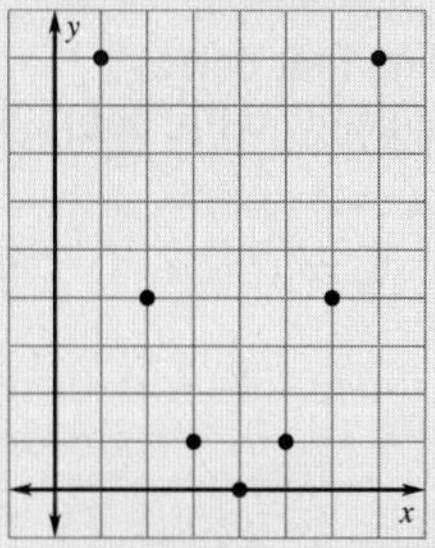

STUDENT HELP NOTES

Homework Help Students can find help for Exs. 23 and 24 at **www.mcdougallittell.com.** The information can be printed out for students who don't have access to the Internet.

ENGLISH LEARNERS

EXERCISES 25–27 Keep in mind that the language contained in word problems is often specialized. Students who are unfamiliar with the biological definition of the word *pupil* will have difficulty understanding the problem. Whenever possible, clarify specialized language.

25. See Additional Answers beginning on page AA1.

ADDITIONAL PRACTICE AND RETEACHING

For Lesson 5.4:

- Practice Levels A, B, and C (*Chapter 5 Resource Book*, p. 54)
- Reteaching with Practice (*Chapter 5 Resource Book*, p. 57)
- See Lesson 5.4 of the *Personal Student Tutor*

For more Mixed Review:

- Search the *Test and Practice Generator* for key words or specific lessons.

4 ASSESS

DAILY HOMEWORK QUIZ

Transparency Available

Use the table for Exercises 1–3.

x	1	2	3	4	5
y	10	12	15	17	20

1. a. Draw a scatter plot of the data.

b. Draw a line that corresponds closely to the data .

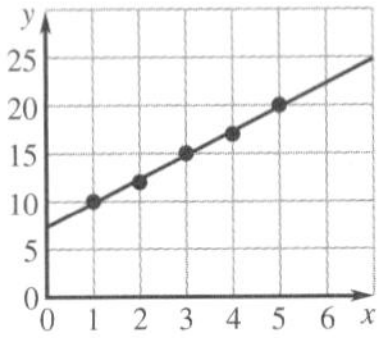

2. Write an equation of the line. *Sample answer:* $y = 2.5x + 7.3$

3. Do x and y have a *positive correlation,* a *negative correlation,* or *relatively no correlation*? **positive correlation**

EXTRA CHALLENGE NOTE

Challenge problems for Lesson 5.4 are available in **blackline** format in the *Chapter 5 Resource Book,* p. 62 and at **www.mcdougallittell.com.**

ADDITIONAL TEST PREPARATION

1. OPEN ENDED Give a real-life situation in which two variables would have a negative correlation. **Check students' work.**

2. WRITING A researcher is studying the relationship between the number of players on a soccer team and the number of minutes that a player spends in the game rather than on the sideline. What type of correlation would you expect to find between these two quantities? Explain. **See sample answer, below.**

32, 33. See Additional Answers beginning on page AA1.

Test Preparation

MULTI-STEP PROBLEM In Exercises 28–31, use the scatter plots. They show a library's annual book budget *B* and the average price *P* of a book.
28, 29. Sample answers are given.

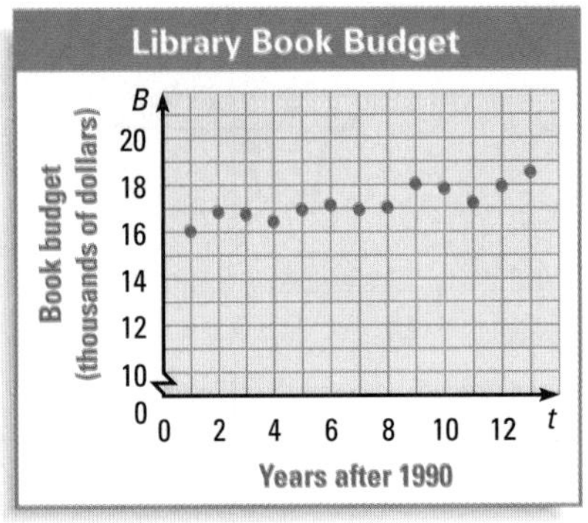

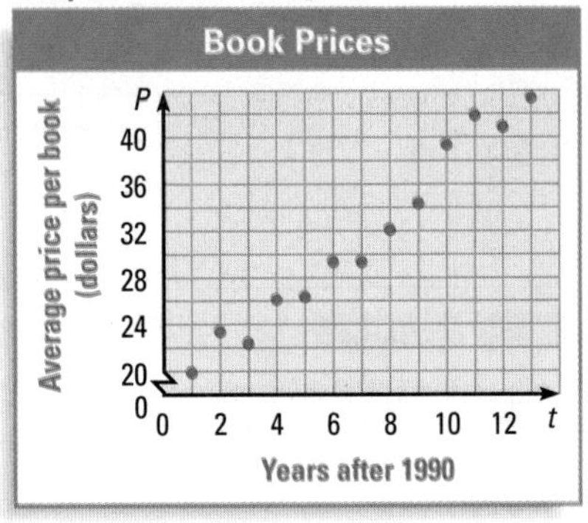

28. Find an equation of the line that you think best represents the library's budget for purchasing books. $B = \frac{1}{7}t + \frac{117}{7}$

29. Find an equation of the line that you think best represents the average price of a new book purchased by the library. $P = 2t + 18$

30. Interpret the slopes of the two lines in the context of the problem. **The average book price is increasing much more rapidly than the library's purchasing budget.**

31. CRITICAL THINKING Suppose that you want to ask town officials to increase the library's annual book budget. Use your results and the information above to write a letter that explains why the library budget should be increased. **See margin.**

ANALYZING DATA In Exercises 32–34, use the Book Prices scatter plot above.

32. Explain why you would not want to choose the points for the years 1992 and 1997 to write an equation that represents the data. **See margin.**

33. Will choosing the two points that are the farthest apart always give you the closest line? Explain why or why not. If not, sketch a counterexample. **See margin.**

34. Explain how to choose a good pair of points to find a line that is probably close to the best-fitting line.
Sample answer: **Once you have chosen the line, try to choose two points on it whose coordinates can be closely approximated and, if possible, will simplify the arithmetic of determining the equation.**

★ Challenge

EXTRA CHALLENGE

www.mcdougallittell.com

31. *Sample letter:* I depend on the town library as a source for references as well as for reading pleasure. I believe it is important to maintain a well-stocked and up to date library. I have enclosed copies of graphs that I made showing the library's annual book budget and the average price per book for the 13-year period beginning in 1991. As you will notice, the average price per book increased much more rapidly than the budget. In 1991, the annual budget allowed for the purchase of about 800 books. In 2003, the budget allowed for the purchase of only about 430 books. Please support an increase in the library budget to help remedy this situation.

MIXED REVIEW

HORIZONTAL AND VERTICAL LINES Decide whether the line is *horizontal* or *vertical.* Then graph the line. (Review 4.2) **35–38. See margin.**

35. $y = -2$ **36.** $y = 3$ **37.** $x = 4$ **38.** $x = -5$

VISUAL REASONING Without calculating, state whether the slope of the line through the points is *positive, negative, zero,* or *undefined.* (Review 4.4)

39. (1, 4), (3, 5) **positive**
40. (2, 5), (4, 1) **negative**
41. (6, 1), (3, 1) **zero**
42. (5, 2), (4, 3) **negative**

WRITING EQUATIONS OF LINES Write an equation of the line that passes through the points. (Review 5.3 for 5.5)

43. (2, 5), (6, 4) **See margin.**
44. (1, 4), (3, 7) **See margin.**
45. (3, 7), (7, 3) $y = -x + 10$
46. (5, 2), (4, 3) $y = -x + 7$
47. (−3, 1), (4, −2) **See margin.**
48. (−2, −3), (0, 4) **See margin.**
49. (5, 1), (3, −6) **See margin.**
50. (0, −6), (−1, 7) $y = -13x - 6$

35. horizontal
36. horizontal
37. vertical
38. vertical
43. $y = -\frac{1}{4}x + \frac{11}{2}$
44. $y = \frac{3}{2}x + \frac{5}{2}$
47. $y = -\frac{3}{7}x - \frac{2}{7}$
48. $y = \frac{7}{2}x + 4$
49. $y = \frac{7}{2}x - \frac{33}{2}$

Additional Test Preparation *Sample answer:*
2. Negative correlation; you would expect that as the number of players increases, the number of minutes would decrease as more players would be used.

Best-Fitting Lines

A graphing calculator can be used to find a best-fitting line. One way to tell how well a line fits a set of data is to look at the *r*-value. The closer the absolute value of *r* is to 1, the better the line fits the data.

STUDENT HELP

Look Back
For help with scatter plots, see p. 209.

KEYSTROKE HELP
See keystrokes for several models of calculators at www.mcdougallittell.com

EXAMPLE

Use a graphing calculator to find the best-fitting line for the data.
(38, 62), (28, 46), (56, 102), (56, 88), (24, 36), (77, 113), (40, 69), (46, 60)

SOLUTION

1. Enter the ordered pairs into the graphing calculator. Make a scatter plot of the data.

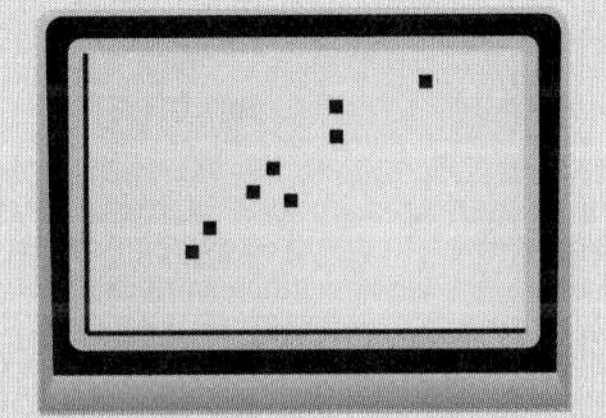

2. Use linear regression to find the best-fitting line. Select L_1 as the x list and L_2 as the y list.

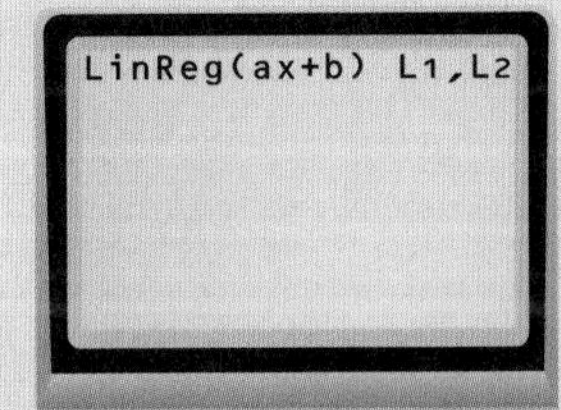

3. The equation $y = 1.49x + 3.85$ is the line of best fit with an r-value of approximately 0.95.

4. Graph the equation $y = 1.49x + 3.85$ with the data points.

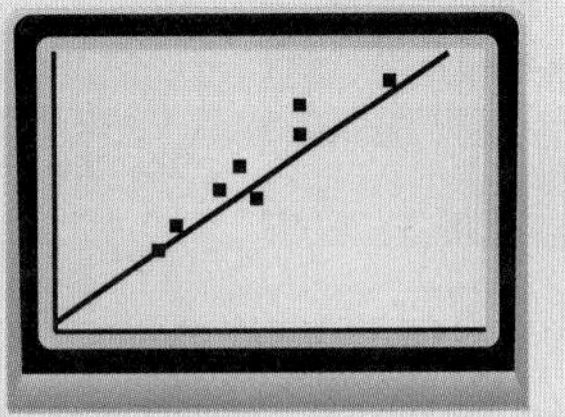

The r-value of 0.95 is close to 1. The equation $y = 1.49x + 3.85$ fits the data points well.

EXERCISES

Find the best-fitting line for the points. 1–4. Values of m and b are rounded to the nearest hundredth.

1. (0.1, 2.1), (1.0, 2.5), (2.2, 2.9), (2.9, 3.4), (4.0, 4.0), (4.9, 4.3) $y = 0.47x + 2.01$
2. (31, 114), (40, 136), (49, 165), (62, 177), (70, 185), (78, 209) $y = 1.86x + 61.88$
3. (0, 1), (1, 2), (1, 3), (2, 3), (2, 3.5), (3, 4), (3, 4.5), (4, 5.5), (4, 6), (5, 5), (5, 6), (5, 6.5), (6, 7), (6, 8), (7, 7.5) $y = 0.95x + 1.41$
4. (0, 8), (1, 7.5), (1, 6), (2, 6.5), (2, 6), (3, 5.5), (3, 5), (4, 4), (4, 3.5), (5, 3), (5, 2.5), (6, 2), (6, 1.5), (7, 1), (7, 0) $y = -1.07x + 8.13$

1 Planning the Activity

PURPOSE
To have students learn to use calculators to find the best-fitting line and to introduce them to *r*-value.

MATERIALS
- Graphing calculator per group
- Keystroke blackline (*Chapter 5 Resource Book,* p. 53)

PACING
- Exploring the Concept — 10 min
- Drawing Conclusions — 15 min

LINK TO LESSON
Have students refer back to Example 1 on page 293. They should find the best-fitting line for the data using the calculator and see how this compares to the approximation given in the example.

2 Managing the Activity

COOPERATIVE LEARNING
This activity can be done in pairs. Have one student read the data while the other enters the values into the calculator and finds the best-fitting line. Students should switch roles for each set of data.

ALTERNATIVE APPROACH
This activity can be done as a class using an overhead projector with a graphing calculator display.

3 Closing the Activity

KEY DISCOVERY
The *r*-value indicates how well the best-fitting line fits the data.

ACTIVITY ASSESSMENT
Predict the *r*-value of the best-fitting line for these points: (0, 0), (1, 1), (2, 2). Find the best-fitting line and *r*-value to test your prediction.
$y = x$ with r-value of 1.

1 PLAN

PACING
Basic: 2 days
Average: 2 days
Advanced: 2 days
Block Schedule: 0.5 block with 5.4
0.5 block with 5.6

LESSON OPENER ACTIVITY
An alternative way to approach Lesson 5.5 is to use the Activity Lesson Opener:

- Blackline Master (*Chapter 5 Resource Book,* p. 66)
- Transparency (p. 36)

MEETING INDIVIDUAL NEEDS

- ***Chapter 5 Resource Book***
  Prerequisite Skills Review (p. 5)
  Practice Level A (p. 67)
  Practice Level B (p. 68)
  Practice Level C (p. 69)
  Reteaching with Practice (p. 70)
  Absent Student Catch-Up (p. 72)
  Challenge (p. 75)
- ***Resources in Spanish***
- ***Personal Student Tutor***

NEW-TEACHER SUPPORT
See the Tips for New Teachers on pp. 1–2 of the *Chapter 5 Resource Book* for additional notes about Lesson 5.5.

WARM-UP EXERCISES
Transparency Available
Find the slope of the line through the points.

1. (–2, 3), (5, –1) $-\frac{4}{7}$
2. (6, –2), (–1, –5) $\frac{3}{7}$

Solve each equation for *y*.

3. $y - 21 = \frac{3}{5}(x + 10)$ $y = \frac{3}{5}x + 27$
4. $y + 1 = -2(x - 3)$ $y = -2x + 5$

5.5 Point-Slope Form of a Linear Equation

What you should learn

GOAL 1 Use the point-slope form to write an equation of a line.

GOAL 2 Use the point-slope form to model a **real-life** situation, such as running pace in **Example 3.**

Why you should learn it

▼ To estimate the **real-life** time it will take a mountain climber to reach the top of a cliff in **Ex. 63.**

GOAL 1 USING THE POINT-SLOPE FORM

In Lesson 5.2 you learned one strategy for writing a linear equation when given the slope and a point on the line. In this lesson you will learn a different strategy—one that uses the **point-slope form** of an equation of a line.

EXAMPLE 1 *Developing the Point-Slope Form*

Write an equation of the line. Use the points (2, 5) and (*x*, *y*).

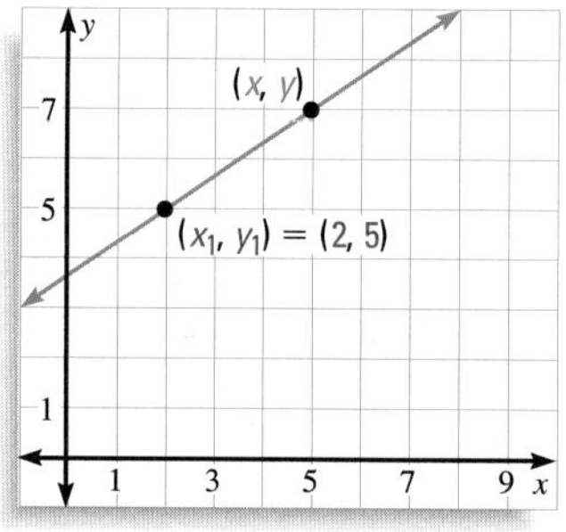

SOLUTION

You are given one point on the line. Let (*x*, *y*) be any point on the line. Because (2, 5) and (*x*, *y*) are two points on the line, you can write the following expression for the slope of the line.

$m = \frac{y - 5}{x - 2}$ **Use formula for slope.**

The graph shows that the slope is $\frac{2}{3}$. Substitute $\frac{2}{3}$ for *m* in the formula for slope.

$\frac{y - 5}{x - 2} = \frac{2}{3}$ **Substitute $\frac{2}{3}$ for *m*.**

$y - 5 = \frac{2}{3}(x - 2)$ **Multiply each side by $(x - 2)$.**

The equation $y - 5 = \frac{2}{3}(x - 2)$ is written in point-slope form.

POINT-SLOPE FORM OF THE EQUATION OF A LINE

The **point-slope form** of the equation of the nonvertical line that passes through a given point (x_1, y_1) with a slope of *m* is

$y - y_1 = m(x - x_1)$.

You can use the point-slope form when you are given the slope and a point on the line. In the point-slope form, (x_1, y_1) is the given point and (x, y) is any other point on the line. You can also use the point-slope form when you are given two points on the line. First find the slope. Then use either given point as (x_1, y_1).

EXAMPLE 2 *Using the Point-Slope Form*

STUDENT HELP

INTERNET HOMEWORK HELP
Visit our Web site www.mcdougallittell.com for extra examples.

Write an equation of the line shown at the right.

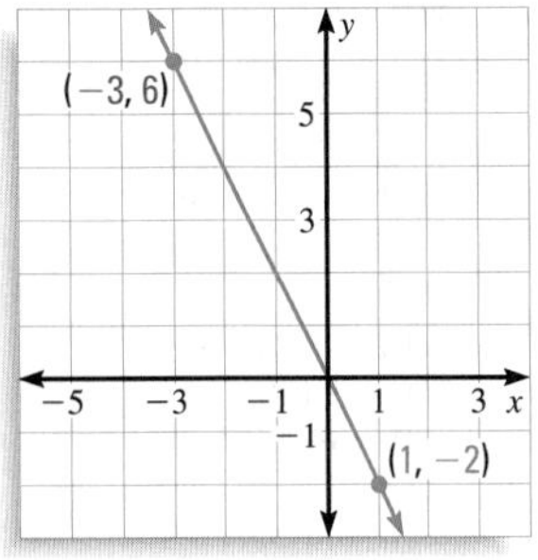

SOLUTION

First find the slope. Use the points $(x_1, y_1) = (-3, 6)$ and $(x_2, y_2) = (1, -2)$.

$$m = \frac{y_2 - y_1}{x_2 - x_1} = \frac{-2 - 6}{1 - (-3)} = \frac{-8}{4} = -2$$

Then use the slope to write the point-slope form. Choose either point as (x_1, y_1).

$y - y_1 = m(x - x_1)$	Write point-slope form.
$y - 6 = -2[x - (-3)]$	Substitute for m, x_1, and y_1.
$y - 6 = -2(x + 3)$	Simplify.
$y - 6 = -2x - 6$	Use distributive property.
$y = -2x$	Add 6 to each side.

STUDENT HELP

Study Tip
The point-slope form $y - y_1 = m(x - x_1)$ has two minus signs. Be sure to account for these signs when the point (x_1, y_1) has negative coordinates.

ACTIVITY Developing Concepts

Investigating the Point-Slope Form

The line shown at the right is labeled with four points.

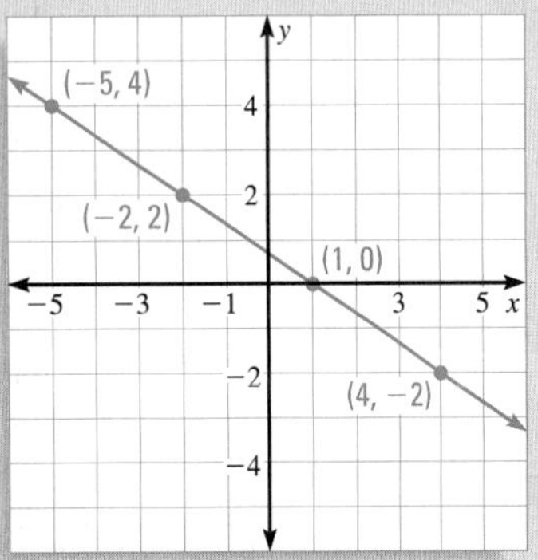

1. Each person in your group should use a different pair of points to write an equation of the line in point-slope form. $y = -\frac{2}{3}x + \frac{2}{3}$

2. Compare the equations of the people in your group. Discuss whether the following statement seems to be always, sometimes, or never true. *Any two points on a line can be used to find an equation of the line.*

2. Always true; *Sample answer:* No matter which two points you use, you can rewrite the equations in slope-intercept form and you'll get the same equation.

2 TEACH

MOTIVATING THE LESSON
Water can take on three different forms: solid ice, liquid water, or gas steam. Each form has its own advantages and disadvantages depending on what you want to use the water for. Ice is good for cooling your drink but not for watering your garden. Similarly, linear equations can have different forms, and each form has its own advantages and disadvantages depending on the application.

ACTIVITY NOTE
The point-slope forms of the equations will be different for each pair of points used in Step 1. Students will need to convert the equations to slope-intercept form in Step 2 to see that the different point-slope forms are equivalent.

EXTRA EXAMPLE 1
Write an equation of the line. Use the points (−2, 3) and (−1, 1).

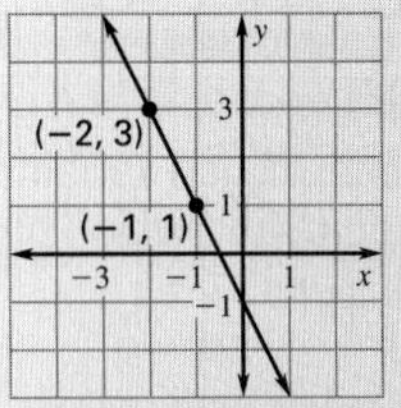

$y - 3 = -2(x + 2)$

EXTRA EXAMPLE 2
Write an equation of the line.

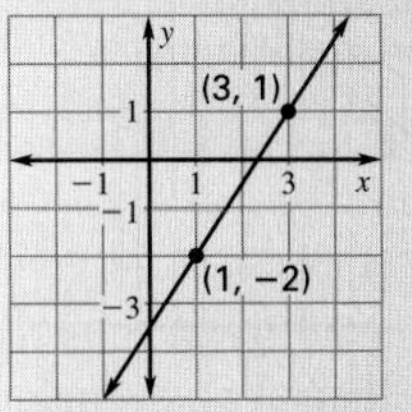

$y = \frac{3}{2}x - \frac{7}{2}$

CHECKPOINT EXERCISES
For use after Examples 1 and 2:
1. Write an equation of the line in point-slope form that passes through the point (−1, −6) with slope $-\frac{1}{3}$. $y + 6 = -\frac{1}{3}(x + 1)$

EXTRA EXAMPLE 3

The base rent on a 5000 ft^2 retail space is $40,000 per year. For larger spaces, any increase of 50 ft^2 above the 5000 ft^2 results in an additional $300 per year in rent.

a. Use the information to write a linear model for rent.
$R = 6a + 10{,}000$, where R is the rent per year, and a is the area of the retail space in square feet

b. Use the model to find the rent on a 6400 ft^2 retail space.
$48,400

CHECKPOINT EXERCISES

For use after Example 3:

1. A crate can hold up to 30 oranges. The crate with 16 oranges weighs 8.8 lb. Assuming that each orange weighs about 0.4 lb, write a linear model to represent the weight of the crate with x oranges. How much will a full crate weigh?
$W = 2.4 + 0.4x$; 14.4 lb

FOCUS ON VOCABULARY

Write the *point-slope form* of the equation of a line. Which variables represent the point and which one represents the slope?
$y - y_1 = m(x - x_1)$; x_1, and y_1 represent the point and m represents the slope.

CLOSURE QUESTION

Give the slope and a point on the line whose equation is $y - 3 = \frac{2}{3}(x + 1)$. **$\frac{2}{3}$ and $(-1, 3)$**

DAILY PUZZLER

On a balance scale, 9 cylinders balance one rectangular prism and 4 cubes. Without the prism, 18 cylinders balance 12 cubes. Write an equation for the number of cylinders y needed to balance the prism and x cubes. **$y = \frac{3}{2}x + 3$**

GOAL 2 MODELING A REAL-LIFE SITUATION

FOCUS ON PEOPLE

LAMECK AGUTA of Kenya won the Boston Marathon in 1997. The high temperature on that day was 52° and the low temperature was 42°.

EXAMPLE 3 *Writing and Using a Linear Model*

MARATHON The information below was taken from an article that appeared in a newspaper.

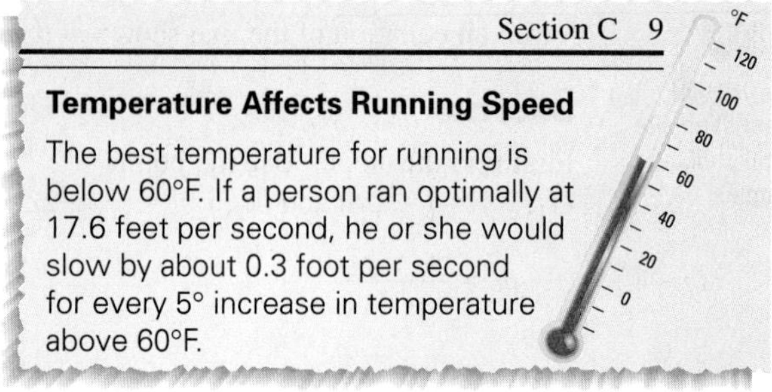

Section C 9

Temperature Affects Running Speed

The best temperature for running is below 60°F. If a person ran optimally at 17.6 feet per second, he or she would slow by about 0.3 foot per second for every 5° increase in temperature above 60°F.

a. Use the information to write a linear model for optimal running pace.

b. Use the model to find the optimal running pace for a temperature of 80°F.

SOLUTION

a. Let T represent the temperature in degrees Fahrenheit. Let P represent the optimal pace in feet per second.

From the article, you know that the optimal running pace at 60° F is 17.6 feet per second so one point on the line is $(T_1, P_1) = (60, 17.6)$. Find the slope of the line.

$$m = \frac{\text{change in } P}{\text{change in } T} = \frac{-0.3}{5} = -0.06.$$

Use the point-slope form to write the model.

$P - P_1 = m(T - T_1)$	Write point-slope form.
$P - 17.6 = (-0.06)(T - 60)$	Substitute for m, T_1, and P_1.
$P - 17.6 = -0.06T + 3.6$	Use distributive property.
$P = -0.06T + 21.2$	Add 17.6 to each side.

b. Use the model to find the optimal pace at 80°F.

$$P = -0.06(80) + 21.2$$
$$= 16.4$$

▶ At 80°F the optimal running pace is 16.4 feet per second.

✓**CHECK** A graph can help you check this result. You can see that as the temperature increases the optimal running pace decreases.

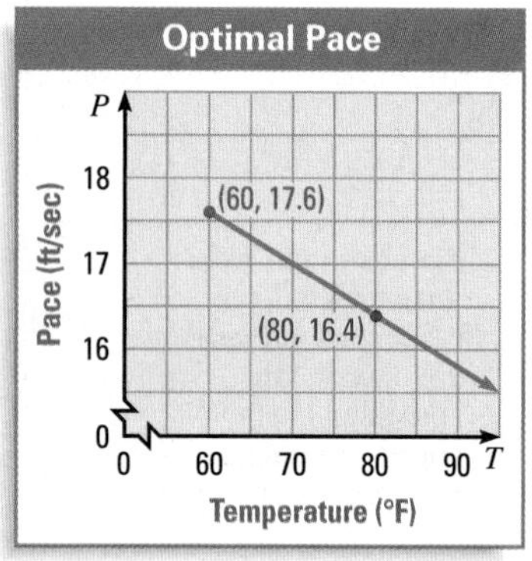

GUIDED PRACTICE

Vocabulary Check ✓

Concept Check ✓

Skill Check ✓

1. Name the following form of the equation of a line: $y - y_1 = m(x - x_1)$.
 point-slope form

2. What are the values of m, x_1, and y_1 in the equation $y - (-1) = 3(x - 2)$?
 3, 2, −1

Write an equation of the line in point-slope form. 3–5. See margin.

3. $y - 4 = \frac{1}{2}(x - 3)$

4. $y + 5 = \frac{2}{3}(x - 1)$

5. $y + 2 = 0$

7. $y - 4 = \frac{1}{2}(x - 3)$

12. $y - 12 = -14(x - 5)$
 or $y + 2 = -14(x - 6)$

13. $y + 1 = -\frac{3}{5}(x + 4)$ or
 $y - 2 = -\frac{3}{5}(x + 9)$

14. $y + 8 = \frac{2}{5}(x + 17)$ or
 $y + 4 = \frac{2}{5}(x + 7)$

15. $y - 12 = -2(x - 2)$ or
 $y - 2 = -2(x - 7)$

16. $y + 12 = \frac{16}{5}(x - 3)$ or
 $y - 4 = \frac{16}{5}(x - 8)$

3.

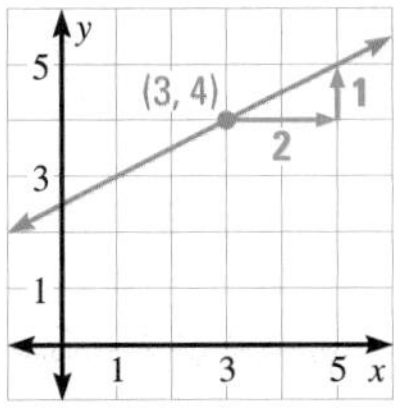

4.

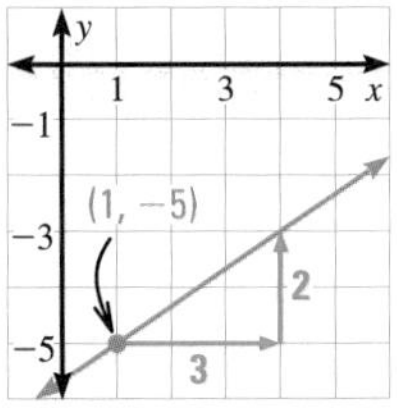

5.

Write an equation of the line in point-slope form that passes through the given point and has the given slope.

6. $(2, -1)$, $m = 3$
 $y + 1 = 3(x - 2)$

7. $(3, 4)$, $m = \frac{1}{2}$
 See margin.

8. $(-5, -7)$, $m = -2$
 $y + 7 = -2(x + 5)$

9. $(-3, 10)$, $m = 4$
 $y - 10 = 4(x + 3)$

10. $(-2, -9)$, $m = 4$
 $y + 9 = 4(x + 2)$

11. $(7, -7)$, $m = -8$
 $y + 7 = -8(x - 7)$

Write an equation of the line that passes through the given points.
12–17. See margin.

12. $(5, 12)$, $(6, -2)$

13. $(-4, -1)$, $(-9, 2)$

14. $(-17, -8)$, $(-7, -4)$

15. $(2, 12)$, $(7, 2)$

16. $(3, -12)$, $(8, 4)$

17. $(-4, -7)$, $(2, 2)$

PRACTICE AND APPLICATIONS

STUDENT HELP

► **Extra Practice**
to help you master
skills is on p. 801.

17. $y + 7 = \frac{3}{2}(x + 4)$ or
 $y - 2 = \frac{3}{2}(x - 2)$

19. $y + 3 = -\frac{1}{2}(x + 1)$

21. $y - 5 = 5(x - 1)$ or
 $y + 5 = 5(x + 1)$

22. $y + 5 = -\frac{1}{9}(x + 2)$ or
 $y + 6 = -\frac{1}{9}(x - 7)$

23. $y - 10 = -\frac{13}{5}(x + 9)$
 or $y + 3 = -\frac{13}{5}(x + 4)$

STUDENT HELP

► **HOMEWORK HELP**
Example 1: Exs. 18–20
Example 2: Exs. 21–29
Example 3: Exs. 57–61

USING A GRAPH **Write an equation of the line in point-slope form.**

18.

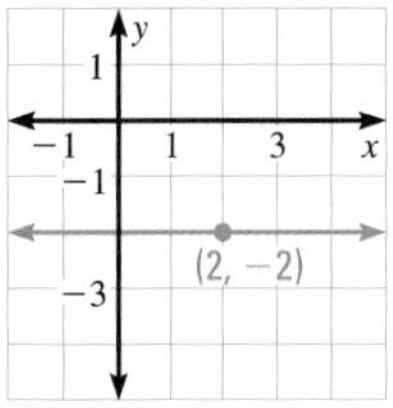

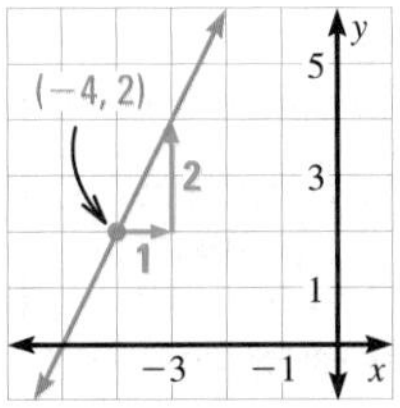

19.

20.

$y - 2 = 2(x + 4)$

See margin.

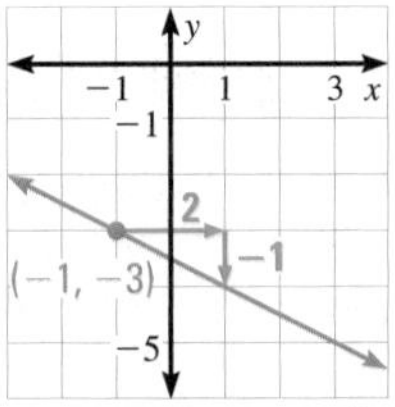

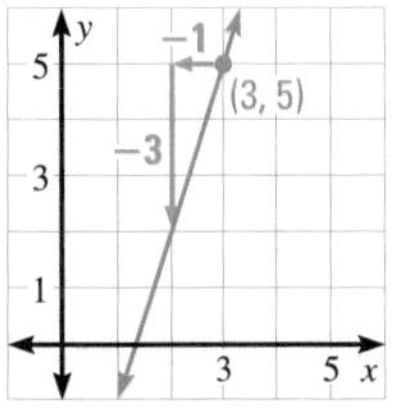

$y - 5 = 3(x - 3)$

POINT-SLOPE FORM **Write an equation in point-slope form of the line that passes through the given points.** 21–29. See margin.

21. $(1, 5)$, $(-1, -5)$

22. $(-2, -5)$, $(7, -6)$

23. $(-9, 10)$, $(-4, -3)$

24. $(4, -5)$, $(-2, -7)$

25. $(-5, 10)$, $(-4, -2)$

26. $(-3, -2)$, $(-3, -8)$

27. $(-3, -9)$, $(-6, -8)$

28. $(-3, -7)$, $(-4, -8)$

29. $(1, -7)$, $(-1, -5)$

WRITING EQUATIONS **Write an equation in point-slope form of the line that passes through the given point and has the given slope.**

30. $(-1, -3)$, $m = 4$
 $y + 3 = 4(x + 1)$

31. $(-6, 2)$, $m = -5$
 $y - 2 = -5(x + 6)$

32. $(-10, 0)$, $m = 2$
 $y = 2(x + 10)$

33. $(-8, -2)$, $m = 2$
 $y + 2 = 2(x + 8)$

34. $(-4, 3)$, $m = -6$
 $y - 3 = -6(x + 4)$

35. $(-3, 4)$, $m = 6$
 $y - 4 = 6(x + 3)$

36. $(12, 2)$, $m = -7$
 $y - 2 = -7(x - 12)$

37. $(8, -1)$, $m = 0$
 $y + 1 = 0$

38. $(5, -12)$, $m = -11$
 $y + 12 = -11(x - 5)$

5.5 *Point-Slope Form of a Linear Equation* **303**

3 APPLY

○ **ASSIGNMENT GUIDE**

BASIC
Day 1: pp. 303–304 Exs. 18–20,
 22–46 even
Day 2: pp. 304–306 Exs. 48–56
 even, 57–59, 64–66, 75, 80,
 85–87, Quiz 2 Exs. 1–12

AVERAGE
Day 1: pp. 303–304 Exs. 18–20,
 22–46 even
Day 2: pp. 304–306 Exs. 48–56
 even, 57–61, 64–66, 75, 80,
 85–87, Quiz 2 Exs. 1–12

ADVANCED
Day 1: pp. 303–304 Exs. 18–20,
 22–46 even
Day 2: pp. 304–306 Exs. 48–56
 even, 57–70, 75, 80, 85–87,
 Quiz 2 Exs. 1–12

BLOCK SCHEDULE
pp. 303–304 Exs. 18–20, 22–46
even (with 5.4),
pp. 304–306 Exs. 48–56 even,
57–61, 64–66, 75, 80, 85–87, Quiz 2
Exs. 1–12 (with 5.6)

EXERCISE LEVELS
Level A: *Easier*
18–29
Level B: *More Difficult*
30–61, 64–66, 71–87
Level C: *Most Difficult*
62, 63, 67–70

✓ **HOMEWORK CHECK**
To quickly check student understanding of key concepts, go
over the following exercises:
Exs. 18, 22, 26, 34, 42, 52. See also
the Daily Homework Quiz:

• Blackline Master (*Chapter 5
 Resource Book*, p. 79)
• 🖨 Transparency (p. 42)

24–29. See next page.

! COMMON ERROR

EXERCISES 57–63 Caution students to pay close attention to what the dependent variable represents in each of these problems. In Exs. 57–59, 62, and 63, the dependent variable represents the distance that still needs to be traveled to reach the destination rather than the distance traveled. In Exs. 60 and 61, the dependent variable represents the time remaining rather than the time that has elapsed.

24. $y + 5 = \frac{1}{3}(x - 4)$ or $y + 7 = \frac{1}{3}(x + 2)$
25. $y - 10 = -12(x + 5)$ or $y + 2 = -12(x + 4)$
26. $x = -3$
27. $y + 9 = -\frac{1}{3}(x + 3)$ or $y + 8 = -\frac{1}{3}(x + 6)$
28. $y + 7 = x + 3$ or $y + 8 = x + 4$
29. $y + 7 = -(x - 1)$ or $y + 5 = -(x + 1)$

ADDITIONAL PRACTICE AND RETEACHING

For Lesson 5.5:

- Practice Levels A, B, and C (*Chapter 5 Resource Book*, p. 67)
- Reteaching with Practice (*Chapter 5 Resource Book*, p. 70)
- See Lesson 5.5 of the *Personal Student Tutor*

For more Mixed Review:

- Search the *Test and Practice Generator* for key words or specific lessons.

39. $y - 4 = 2(x - 1)$; $y = 2x + 2$
40. $y - 4 = 3(x + 2)$; $y = 3x + 10$
41. $y - 2 = \frac{1}{2}(x - 6)$; $y = \frac{1}{2}x - 1$
42. $y + 1 = 4(x + 3)$; $y = 4x + 11$
43. $y + 1 = -1(x - 5)$; $y = -x + 4$
44. $y + 5 = -2(x + 5)$; $y = -2x - 15$
45. $y - 1 = -\frac{1}{8}(x + 1)$; $y = -\frac{1}{8}x + \frac{7}{8}$
46. $y + 2 = \frac{1}{4}(x - 4)$; $y = \frac{1}{4}x - 3$
47. $y + 6 = -\frac{2}{3}(x + 9)$; $y = -\frac{2}{3}x - 12$
48. $y = -\frac{4}{3}x + 2$
49. $y = -\frac{1}{3}x - \frac{4}{3}$
50. $y = \frac{3}{5}x - 3$
51. $y - 3 = 2(x - 1)$ or $y - 5 = 2(x - 2)$
52. $y - 3 = -2(x + 2)$ or $y + 5 = -2(x - 2)$
53. $y + 10 = -\frac{3}{2}(x - 7)$ or $y + 22 = -\frac{3}{2}(x - 15)$
54. $y - 6 = -\frac{5}{3}(x - 2)$ or $y - 1 = -\frac{5}{3}(x - 5)$
55. $y + 4 = \frac{3}{4}(x + 3)$ or $y + 1 = \frac{3}{4}(x - 1)$
56. $y + 2 = -\frac{7}{13}(x - 4)$ or $y - 5 = -\frac{7}{13}(x + 9)$

COMPARING FORMS Write the point-slope form of the equation of the line that passes through the point and has the given slope. Then rewrite the equation in slope-intercept form. 39–47. See margin.

39. $(1, 4), m = 2$ **40.** $(-2, 4), m = 3$ **41.** $(6, 2), m = \frac{1}{2}$

42. $(-3, -1), m = 4$ **43.** $(5, -1), m = -1$ **44.** $(-5, -5), m = -2$

45. $(-1, 1), m = -\frac{1}{8}$ **46.** $(4, -2), m = \frac{1}{4}$ **47.** $(-9, -6), m = -\frac{2}{3}$

WRITING EQUATIONS Write an equation of the line. 48–50. See margin.

48.

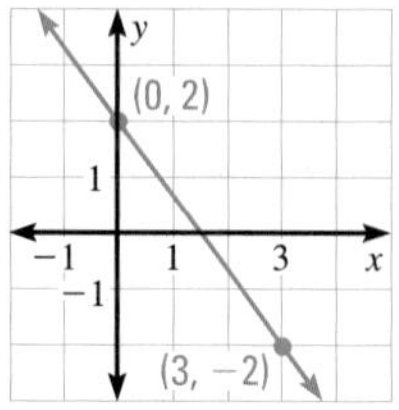

49.

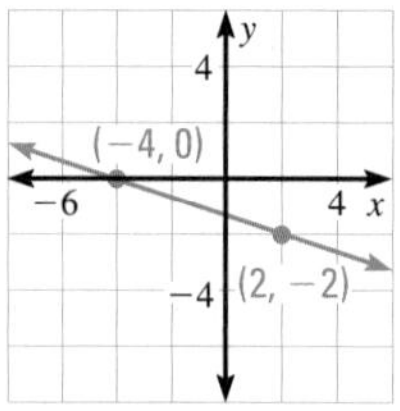

50.

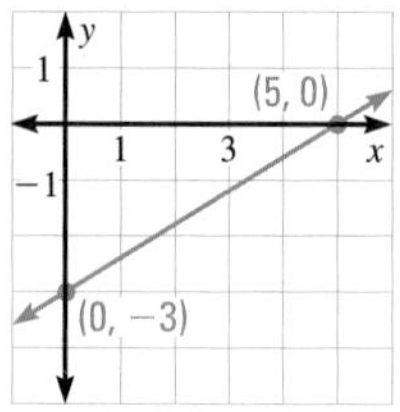

WRITING EQUATIONS GIVEN TWO POINTS Write the point-slope form of the equation of the line that passes through the two points. 51–56. See margin.

51. $(1, 3), (2, 5)$ **52.** $(-2, 3), (2, -5)$ **53.** $(7, -10), (15, -22)$

54. $(2, 6), (5, 1)$ **55.** $(-3, -4), (1, -1)$ **56.** $(4, -2), (-9, 5)$

FIELD TRIP In Exercises 57–59, you are going on a trip to the Natural History Museum. At 9:00 A.M., you leave for the museum, which is 120 miles away. At 10:15 A.M., you are 63 miles away from the museum.

57. Write a linear equation that gives the distance d (in miles) from the museum in terms of the time t. Let t represent the number of minutes since 9:00 A.M. **$d = -0.76t + 120$**

58. Find the distance you are from the museum after you have traveled 2 hours. **28.8 mi**

59. According to your equation, when will you reach the museum? **at about 11:38 A.M.**

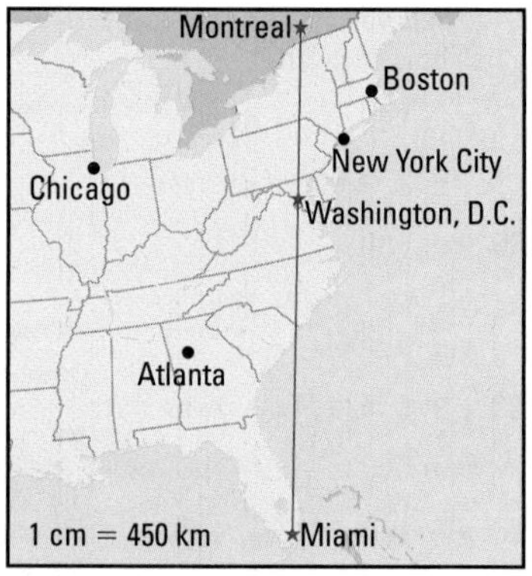

TRAVEL In Exercises 60 and 61, you are flying from Montreal to Miami in a plane. The plane leaves Montreal at 8:00 A.M. and at 9:15 A.M. is flying over Washington, DC.

60. Use the map to write a linear equation that gives the distance y (in kilometers) remaining in the flight in terms of the number of minutes x since 8:00 A.M. **$y = -10.2x + 2340$**

61. When will you reach Miami? (Assume that the plane is flying at a constant speed.) **11:49 A.M.**

MOUNTAIN CLIMBING In Exercises 62 and 63, a mountain climber is scaling a 300-foot cliff at a constant rate. The climber starts at the bottom at 12:00 P.M. By 12:30 P.M., the climber has moved 62 feet up the cliff.

62. Write an equation that gives the distance d (in feet) remaining in the climb in terms of the time t (in hours). What is the slope of the line? **$d = -124t + 300$**

63. At what time will the mountain climber reach the top of the cliff? **at about 2:25 P.M.**

Test Preparation

64. **MULTIPLE CHOICE** The time t (in hours) needed to produce x units of a product is modeled by $t = px + s$. If it takes 265 hours to produce 200 units and 390 hours to produce 300 units, what is the value of s? **B**

Ⓐ 1.25　　Ⓑ 15　　Ⓒ 100　　Ⓓ 125

65. **MULTIPLE CHOICE** What is the equation in point-slope form of the line in the graph? **A**

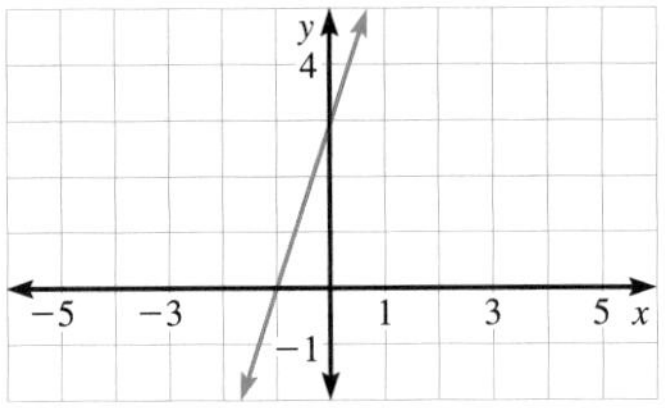

Ⓐ $y - 3 = 3(x - 0)$

Ⓑ $y = 3x + 3$

Ⓒ $y - (-1) = 3(x - 3)$

Ⓓ $y - 3 = 3[x - (-1)]$

66. **MULTIPLE CHOICE** What is the equation in point-slope form of the line perpendicular to the line in the graph and through $(2, -1)$? **B**

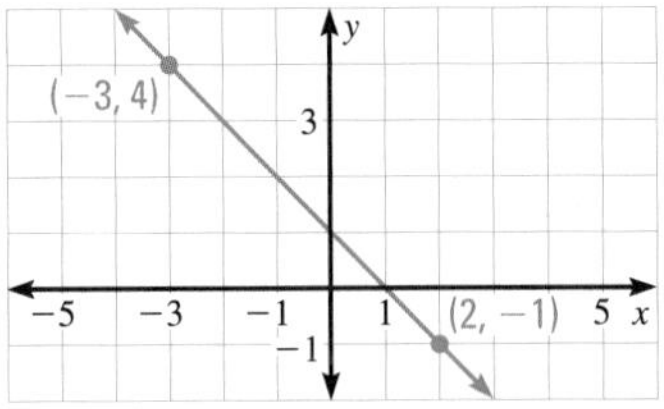

Ⓐ $y = -x + 1$

Ⓑ $y - (-1) = -1(x - 2)$

Ⓒ $y - 4 = -1(x + 3)$

Ⓓ $y - (-1) = 1(x - 2)$

★ **Challenge**

Find the value of k so that the line through the given points has slope m.

67. $(2k, 3), (1, k); m = 2$ **1**

68. $(k + 1, k - 1), (k, -k); m = k + 1$ **2**

69. $(k, k + 1), (3, 2); m = 3$ **4**

70. $(k + 1, 3 + 2k), (k - 1, 1 - k); m = k$ **−2**

MIXED REVIEW

CHECKING SOLUTIONS OF INEQUALITIES **Check whether the given number is a solution of the inequality.** (Review 1.4)

71. $2x < 24; 8$ **solution**

72. $7y + 6 \geq 10; 3$ **solution**

73. $16p - 9 \leq 71; 5$ **solution**

74. $12a \leq a - 9; -2$ **solution**

75. $4x \leq 28; 7$ **solution**

76. $6c - 4 > 14; 3$ **not a solution**

GRAPHING FUNCTIONS **Graph the function.** (Review 4.8 for 5.6) **77–85. See margin.**

77. $f(x) = -3x + 4$

78. $g(x) = -x - 7$

79. $h(x) = 2x - 8$

80. $f(x) = -5x - 7$

81. $f(x) = 6x + 7$

82. $f(x) = 3 - 7x$

83. $g(x) = -\frac{1}{2}x - 3$

84. $f(x) = 8x + \frac{2}{3}$

85. $h(x) = \frac{6}{5}x + 5$

86. **TEEN SURVEY** In a school of 500 teens, 40 have taken drum lessons. What are the odds that a randomly chosen teen has not taken drum lessons? (Review 2.8) **23 to 2**

87. **PROBABILITY CONNECTION** A jar contains 10 blue marbles, 7 red marbles, and 4 yellow marbles. Without looking, you reach into the jar and choose one marble. What is the probability that the marble is red? (Review 2.8) **$33\frac{1}{3}$%**

4 ASSESS

DAILY HOMEWORK QUIZ

Transparency Available

Write an equation in point-slope form of the line that passes through the points.

1. (5, 3), (−1, 0)
$y - 3 = \frac{1}{2}(x - 5)$ or $y = \frac{1}{2}(x + 1)$
2. (−4, −2), (1, −17)
$y + 2 = -3(x + 4)$ or $y + 17 = -3(x - 1)$
3. What is an equation in point-slope form of the line through (6, 4) that is perpendicular to the line $y - 2 = \frac{2}{5}(x - 5)$?
$y - 4 = -\frac{5}{2}(x - 6)$

EXTRA CHALLENGE NOTE

Challenge problems for Lesson 5.5 are available in **blackline** format in the *Chapter 5 Resource Book*, p. 75 and at **www.mcdougallittell.com.**

ADDITIONAL TEST PREPARATION

1. **WRITING** Show how to find an equation of the line through the points (1, −4) and (3, 2) using the slope-intercept form. Then find an equation of the same line using the point-slope form. Rewrite this equation in slope-intercept form to show that the methods produce equivalent equations. **Slope-intercept form: $y = 3x - 7$; point-slope form: $y + 4 = 3(x - 1)$, which is $y = 3x - 7$ in slope-intercept form.**

77–79. See next page.
80–85. See Additional Answers beginning on page AA1.

ADDITIONAL RESOURCES

An alternative Quiz for Lessons 5.4–5.5 is available in the *Chapter 5 Resource Book*, p. 76.

A **blackline** master with additional Math & History exercises is available in the *Chapter 5 Resource Book*, p. 74.

77.
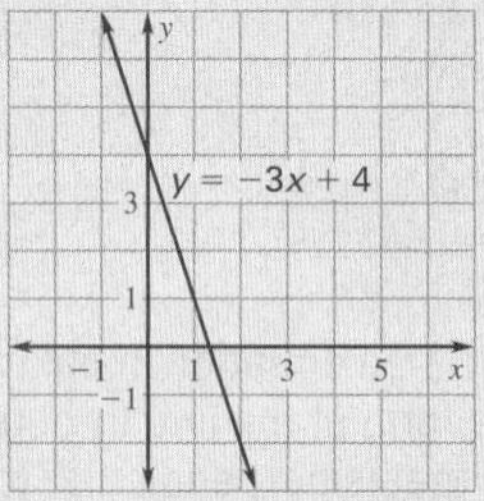

78.
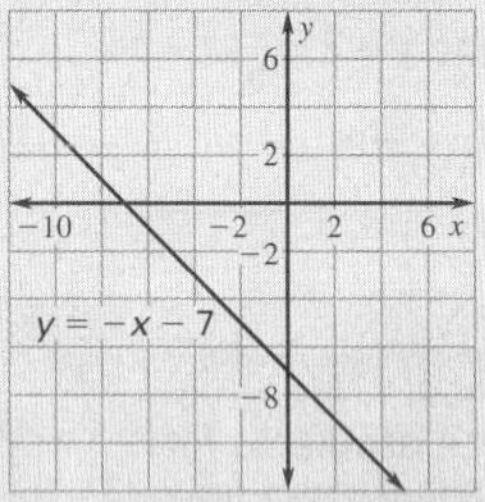

79.
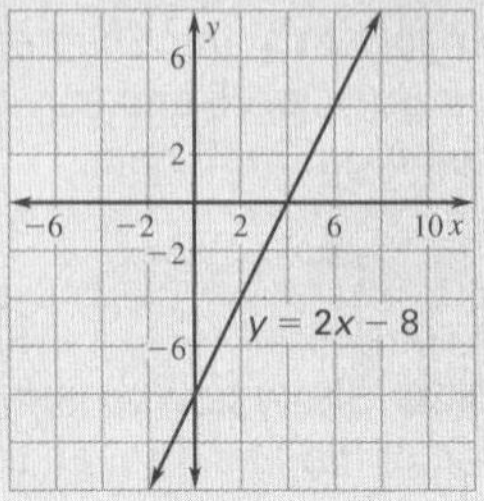

QUIZ 2

Self-Test for Lessons 5.4 and 5.5

1–3. Check graphs. Sample equations given were determined using a graphing calculator and rounding to the nearest hundredth.

Copy the graph and draw a best-fitting line for the scatter plot. Find an equation of the line. State whether x and y have a *positive correlation* or a *negative correlation*. (Lesson 5.4)

1.
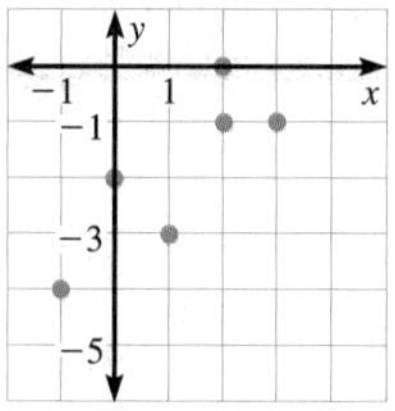

$y = 0.82x - 2.78$; positive

2.
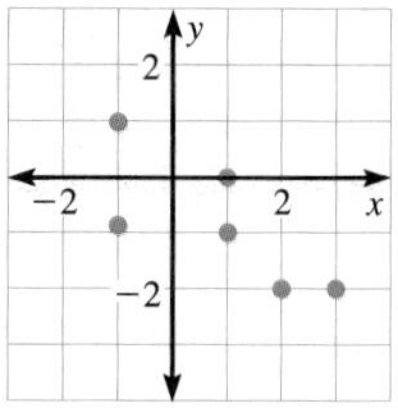

$y = -0.53x - 0.73$; negative

3.
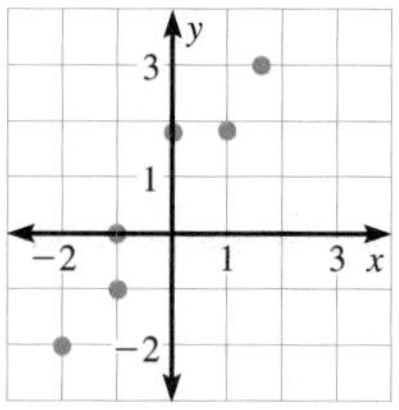

$y = 1.38x + 0.99$; positive

Write an equation in point-slope form of the line that passes through the given point and has the given slope. (Lesson 5.5)

4. $(1, -4), m = -1$ $y + 4 = -1(x - 1)$

5. $(2, 5), m = 2$ $y - 5 = 2(x - 2)$

6. $(-3, -2), m = -2$ $y + 2 = -2(x + 3)$

7. $(-3, -3), m = 0$ $y + 3 = 0$

8. $(-6, 6), m = \frac{1}{2}$ See margin.

9. $(3, 8), m = 7$ $y - 8 = 7(x - 3)$

8. $y - 6 = \frac{1}{2}(x + 6)$

Write an equation in point-slope form of the line that passes through the given points. (Lesson 5.5)

10. $(0, 0), (-3, -1)$ $y = \frac{1}{3}x$ or $y + 1 = \frac{1}{3}(x + 3)$

11. $(-5, -2), (-6, -2)$ $y + 2 = 0$

12. $(-5, -3), (1, 6)$ $y + 3 = \frac{3}{2}(x + 5)$ or $y - 6 = \frac{3}{2}(x - 1)$

MATH & History — Archaeological Dating

APPLICATION LINK www.mcdougallittell.com

THEN BY THE MIDDLE OF THE TWENTIETH CENTURY, radiocarbon dating was being used to find the ages of objects. During the 1950s, increasing differences emerged between the radiocarbon dates and the historical dates of civilizations.

NOW TODAY, archaeologists calibrate tree ring dates with radiocarbon dates to accurately determine the age of an object.

1. Radiocarbon dates calibrate well with tree ring dates after approximately what year? about 1000 BC
2. Does radiocarbon dating produce calendar years that are earlier or later than those based on tree ring dating? later

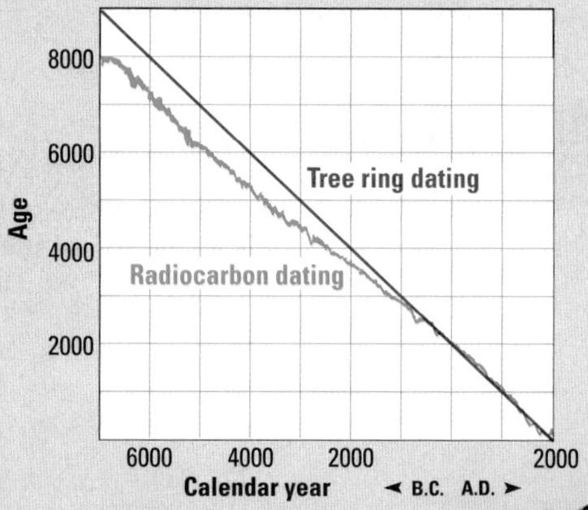

An excavation unearths pottery at Mesa Verde.

A. E. Douglass uses tree rings to date Mesa Verde sites.

Modern archaeologists use a variety of dating methods.

ACTIVITY 5.6

Developing Concepts

Group Activity for use with Lesson 5.6

Investigating the Standard Form of a Linear Equation

GROUP ACTIVITY
Work in a small group.

MATERIALS
graph paper

QUESTION **How can you model possible combinations of two stocks you can buy with a limited amount of money?**

EXPLORING THE CONCEPT

Your social studies class is learning about the stock market. Each person in class "invests" up to \$80 in either GFE stock, which costs \$8 per share, or JIH stock, which costs \$20 per share.

Step 1: 20, 28, 36, 44, 52, 60

1. Copy and complete the table.

Number of GFE shares	0	1	2	3	4	5
Number of JIH shares	1	1	1	1	1	1
Total cost (dollars)	?	?	?	?	?	?

Step 2: They are on the line $y = 1$.

2. Total costs that represent different combinations of the two stocks are written at points on the graph. Locate the values from your table on the graph.

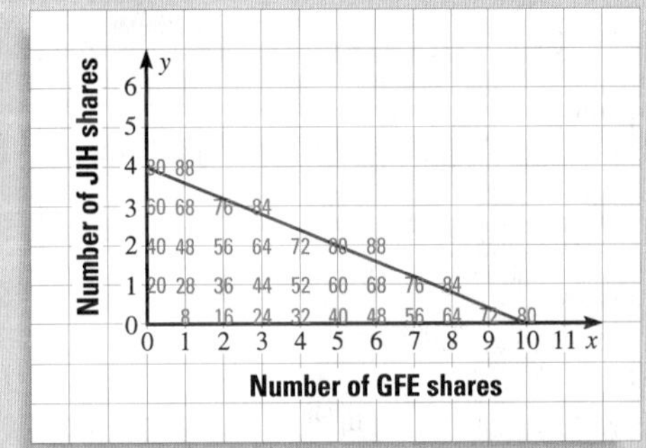

Step 3: the points that satisfy $8x + 20y = 80$

3. Find the cost of 5 shares of GFE stock and 2 shares of JIH stock. What does the red line on the graph represent?

Step 4: $y = -\frac{2}{5}x + 4$

4. Write an equation of the line in slope-intercept form.

DRAWING CONCLUSIONS

1. Each member of the group should copy the graph and choose a different amount to invest from the following list: \$48, \$64, \$96, and \$108. **Check students' work.**
2. Draw a line that represents the combinations of stocks you can invest in using all of your money. Write an equation of the line in slope-intercept form. **See margin. For each amount n, the equation is $y = -\frac{2}{5}x + \frac{n}{20}$.**
3. Use the verbal model below to write an equation that models the combinations of stocks you can buy using your investment amount as the total cost. **For each amount n, the equation is $8x + 20y = n$.**

GFE stock price · Number of GFE shares + JIH stock price · Number of JIH shares = Total cost

4. *Writing* You have modeled your investments using a graph and two forms of equations. Compare the results of each person in the group. What do you notice about the lines? about the equations from Exercises 2 and 3?

4. The graphs are parallel; in Question 2, m is the same for each equation; in Question 3, the coefficients of x and y are the same for each equation.

1 Planning the Activity

PURPOSE
To investigate a real-life situation that can be modeled by a linear equation in standard form.

MATERIALS
- Activity Support Master (*Chapter 5 Resource Book*, p. 80)

PACING
- Exploring the Concept — 10 min
- Drawing Conclusions — 15 min

LINK TO LESSON
Refer back to this activity when discussing Example 4 on page 310.

2 Managing the Activity

CLASSROOM MANAGEMENT
In Step 4, students should read the y-intercept from the graph. They can find the slope from the graph or by choosing two points on the line and using the slope formula.

ALTERNATIVE APPROACH
This activity can be done as a demonstration. Use the chalkboard or an overhead to display the graph while completing Steps 1–4. Have students complete Questions 1–3 individually. Discuss Question 4 as a class and then have each student write his/her own response.

3 Closing the Activity

KEY DISCOVERY
Standard form linear equations can be useful for modeling situations involving combinations of items.

ACTIVITY ASSESSMENT
Derek has \$100 to spend on school supplies. He plans to buy notebooks at \$8 each and paper at \$4 per package. Write an equation that models the combinations of notebooks and paper he can buy.
$8x + 4y = 100$, where x = notebooks and y = packages of paper.

2. See Additional Answers beginning on page AA1.

1 PLAN

PACING

Basic: 2 days
Average: 2 days
Advanced: 2 days
Block Schedule: 0.5 block with 5.5
0.5 block with 5.7

LESSON OPENER ACTIVITY

An alternative way to approach Lesson 5.6 is to use the Activity Lesson Opener:

- Blackline Master (*Chapter 5 Resource Book,* p. 81)
- Transparency (p. 37)

MEETING INDIVIDUAL NEEDS

- ***Chapter 5 Resource Book***
  Prerequisite Skills Review (p. 5)
  Practice Level A (p. 84)
  Practice Level B (p. 85)
  Practice Level C (p. 86)
  Reteaching with Practice (p. 87)
  Absent Student Catch-Up (p. 89)
  Challenge (p. 91)
- ***Resources in Spanish***
- ***Personal Student Tutor***

NEW-TEACHER SUPPORT

See the Tips for New Teachers on pp. 1–2 of the *Chapter 5 Resource Book* for additional notes about Lesson 5.6.

WARM-UP EXERCISES

Transparency Available

Write an equation for each line in point-slope form.

1. the line through (–1, 3) with slope –2 $y - 3 = -2(x + 1)$
2. the line through (3, –5) and (2, –4) $y + 5 = -(x - 3)$

Convert each equation in point-slope form to slope-intercept form.

3. $y + 1 = -2(x - 5)$ $y = -2x + 9$
4. $y - 3 = \frac{1}{2}(x + 6)$ $y = \frac{1}{2}x + 6$

1, 2. See Additional Answers beginning on page AA1.

5.6 The Standard Form of a Linear Equation

What you should learn

GOAL 1 Write a linear equation in standard form.

GOAL 2 Use the standard form of an equation to model **real-life** situations, such as the amount of food to purchase for a barbecue in **Example 4**.

Why you should learn it

▼ To model **real-life** situations, such as buying combinations of bird seeds in **Exs. 64–66**.

REAL LIFE

GOAL 1 WRITING AN EQUATION IN STANDARD FORM

Most of the linear equations in this chapter are written in slope-intercept form.

Slope-intercept form: $y = mx + b$

Another commonly used form is the **standard form** of the equation of a line.

Standard (or general) form: $Ax + By = C$

In this form, notice that the variable terms are on the left side of the equation and the constant term is on the right. A, B, and C represent real numbers and A and B are not both zero.

ACTIVITY
Developing Concepts

Investigating Forms of Equations

1, 2. See margin.

1. Graph each of these equations.

   a. $y = \frac{4}{3}x - 2$ **b.** $-4x + 3y = -6$ **c.** $4x - 3y = 6$

2. Which, if any, of the equations are equivalent? How do you know?

EXAMPLE 1 *Writing an Equation in Standard Form*

Write $y = \frac{2}{5}x - 3$ in standard form with integer coefficients.

SOLUTION

To write the equation in standard form, isolate the variable terms on the left and the constant term on the right.

$y = \frac{2}{5}x - 3$	Write original equation.
$5y = 5\left(\frac{2}{5}x - 3\right)$	Multiply each side by 5.
$5y = 2x - 15$	Use distributive property.
$-2x + 5y = -15$	Subtract $2x$ from each side.

A linear equation can have more than one standard form. For instance, if you multiply each side of the standard form above by -1, you get $2x - 5y = 15$, which is also in standard form.

STUDENT HELP

Look Back
For help with solving linear equations, see p. 155.

EXAMPLE 2 *Writing a Linear Equation*

Write the standard form of an equation of the line passing through $(-4, 3)$ with a slope of -2.

SOLUTION

You are given a point on the line and its slope, so you can write the point-slope form of the equation of the line.

$y - y_1 = m(x - x_1)$	**Write point-slope form.**
$y - 3 = -2[x - (-4)]$	**Substitute for y_1, m, and x_1.**
$y - 3 = -2(x + 4)$	**Simplify.**
$y - 3 = -2x - 8$	**Use distributive property.**
$2x + y = -5$	**Add $2x$ and 3 to each side.**

▶ The equation $2x + y = -5$ is in standard form.

EXAMPLE 3 *Horizontal and Vertical Lines*

Write the standard form of an equation

a. of the horizontal line.

b. of the vertical line.

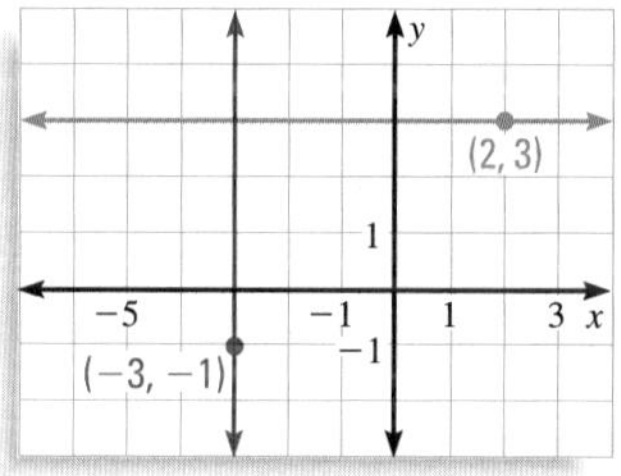

SOLUTION

a. The horizontal line has a point with a y-coordinate of 3. The standard form of the equation is $y = 3$.

b. The vertical line has a point with an x-coordinate of -3. The standard form of the equation is $x = -3$.

. . . .

You have now studied all of the commonly used forms of linear equations. They are summarized in the following list.

CONCEPT SUMMARY EQUATIONS OF LINES

SLOPE-INTERCEPT FORM:	$y = mx + b$
POINT-SLOPE FORM:	$y - y_1 = m(x - x_1)$
VERTICAL LINE (UNDEFINED SLOPE):	$x = a$
HORIZONTAL LINE (ZERO SLOPE):	$y = b$
STANDARD FORM:	$Ax + By = C$, where A, B, and C are real numbers; A and B are not both zero.

2 TEACH

MOTIVATING THE LESSON

Ask students to imagine that they have $50 to spend at a used book sale. They can buy paperbacks for $1 and hardcover books for $4. Ask for combinations of paperbacks and hardcovers they might buy for $50. Explain that another form of a linear equation that can be used to model this situation is the focus of today's lesson.

ACTIVITY NOTE

This activity helps students recognize the graph of a linear equation written in slope-intercept form and an equivalent standard form.

EXTRA EXAMPLE 1

Write $-5x + 11 = \frac{1}{2}y$ in standard form with integer coefficients.
$10x + y = 22$

EXTRA EXAMPLE 2

Write the standard form of the equation of the line passing through $(-5, 1)$ with a slope of $\frac{3}{4}$.
$-3x + 4y = 19$

EXTRA EXAMPLE 3

Write the standard form of the equation:

a. of the horizontal line through $(6, -5)$ **$y = -5$**

b. of the vertical line through $(-2, -7)$ **$x = -2$**

CHECKPOINT EXERCISES

For use after Examples 1 and 2:

1. Write the standard form of the equation of the line passing though the points $(3, 0)$ and $(-5, 3)$. **$3x + 8y = 9$**

For use after Example 3:

2. Write the standard form of the equation of the vertical line through the point $(2, 11)$.
$x = 2$

EXTRA EXAMPLE 4

You are in charge of buying apples and oranges for the school picnic. A bag of oranges costs $4.00 and a bag of apples costs $3.00. You have $48 to spend.

a. Model the possible combinations of bags of oranges and apples. $4x + 3y = 48$

b. Make a table.

Oranges	Apples
0	16
3	12
6	8
9	4
12	0

CHECKPOINT EXERCISES

For use after Example 4:

1. You are starting a garden and want to plant marigolds and zinnias. Marigold plants cost $1.50 each while zinnia plants cost $2.50 each. You have $15 to spend. Write a linear model for the possible combinations of marigold and zinnia plants. How many zinnia plants can you buy if you buy 5 marigold plants? $15x + 25y = 150$; 3

CLOSURE QUESTION

How do you write the standard form equation of a line if you are given a point on the line and its slope?
See sample answer, below.

DAILY PUZZLER

A square in the coordinate plane has a perimeter of 36 and its lower left vertex at (–1, 4). Write an equation for the line though each one of the sides of the square if each side is parallel to one of the two coordinate axes. $x = -1$, $x = 8$, $y = 4$, $y = 13$

GOAL 2 MODELING A REAL-LIFE SITUATION

EXAMPLE 4 *Writing and Using a Linear Model*

BARBECUE You are in charge of buying the hamburger and boned chicken for a barbecue. The hamburger costs $2 per pound and the boned chicken costs $3 per pound. You have $30 to spend.

a. Write an equation that models the different amounts of hamburger and chicken that you can buy.

b. Make a table and a graph that illustrate several different amounts of hamburger and chicken that you can buy.

SOLUTION

a. Model the possible combinations of hamburger and chicken.

PROBLEM SOLVING STRATEGY

VERBAL MODEL: Price of hamburger · Weight of hamburger + Price of chicken · Weight of chicken = Total cost

LABELS:

Price of hamburger = **2** (dollars per pound)

Weight of hamburger = x (pounds)

Price of chicken = **3** (dollars per pound)

Weight of chicken = y (pounds)

Total cost = **30** (dollars)

ALGEBRAIC MODEL: $2x + 3y = 30$ Linear model

b. Use the algebraic model to make a table.

Hamburger (lb), x	0	3	6	9	12	15
Chicken (lb), y	10	8	6	4	2	0

The points corresponding to these different amounts are shown in the graph. Note that as the number of pounds of hamburger you purchase increases, the number of pounds of chicken you purchase decreases, and vice versa.

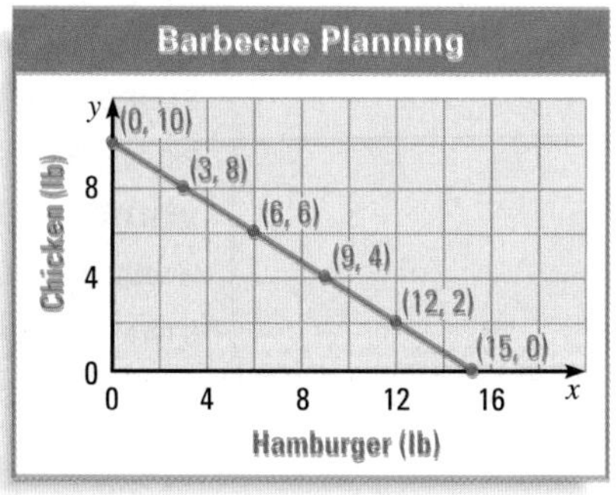

FOCUS ON APPLICATIONS

REAL LIFE **BARBECUE** The International Bar-B-Que Festival in Owensboro, Kentucky, uses over 5000 chickens in its annual cookout.

Closure Question ***Sample answer:***
Use either the point-slope form or slope-intercept form to write an equation of the line, then use algebra to convert it to standard form.

GUIDED PRACTICE

Vocabulary Check ✓

1. Name the following form of the equation of a line: $y = mx + b$. **slope-intercept form**
2. Name the following form of the equation of a line: $Ax + By = C$. **standard form**

Concept Check ✓

3. The equations $y = -\frac{2}{3}x + 3$ and $2x + 3y = 9$ are equivalent. Describe how you would use each equation to graph the equation. **See margin.**

Skill Check ✓

3. ***Sample answer:*** **To graph $y = -\frac{2}{3}x + 3$, I would first use the y-intercept to plot (0, 3), then use a slope triangle to plot a second point. To graph $2x + 3y = 9$, I would make a table of values as in Example 4 on page 310.**

Write the equation in standard form with integer coefficients.
4–16. Sample answers are given.

4. $y = 2x - 9$ **$2x - y = 9$**
5. $y = 6 - 5x$ **$5x + y = 6$**
6. $y = 9 + 1x$ **$1x - y = -9$**
7. $y = \frac{1}{2}x + 8$ **$x - 2y = -16$**
8. $y = 7x + 3$ **$7x - y = -3$**
9. $y = -2 + \frac{3}{2}x$ **$3x - 2y = 4$**

Write an equation in standard form of the line that passes through the given point and has the given slope.

10. $(-3, 4), m = -4$ **$4x + y = -8$**
11. $(1, -2), m = 5$ **$5x - y = 7$**
12. $(2, 5), m = 3$ **$3x - y = 1$**
13. $(5, -8), m = -1$ **$x + y = -3$**
14. $(-5, -6), m = 3$ **$3x - y = -9$**
15. $(4, -3), m = -2$ **$2x + y = 5$**

16. **VEGETABLES** You have \$10 to buy tomatoes and carrots for a salad. Tomatoes cost \$2 per pound and carrots cost \$1.25 per pound. Write a linear equation that models the different amounts of tomatoes x and carrots y that you can buy. ***Sample answer:*** **$8x + 5y = 40$**

17. Copy and complete the table using the linear model in Exercise 16. **8; 6; 4; 2; 0**

Tomatoes (lb), x	0	1.25	2.5	3.75	5
Carrots (lb), y	?	?	?	?	?

PRACTICE AND APPLICATIONS

Sample answers with integer coefficients are given for equations in standard form.

STUDENT HELP

Extra Practice to help you master skills is on p. 801.

EQUATIONS **Write the equation in standard form with integer coefficients.**

18. $4x - y - 7 = 0$ **$4x - y = 7$**
19. $x + 3y - 4 = 0$ **$x + 3y = 4$**
20. $-4x + 5y + 16 = 0$ **$4x - 5y = 16$**
21. $2x - 3y - 14 = 0$ **$2x - 3y = 14$**
22. $x - 5 = 0$ **$x = 5$**
23. $y + 3 = 0$ **$y = -3$**
24. $y = -5x + 2$ **$5x + y = 2$**
25. $y = 3x - 8$ **$3x - y = 8$**
26. $y = -0.4x + 1.2$ **$2x + 5y = 6$**
27. $3x + 9 = \frac{7}{2}y$ **$6x - 7y = -18$**
28. $y = 9x + \frac{1}{2}$ **$18x - 2y = -1$**
29. $y = \frac{5}{2}x + 9$ **$5x - 2y = -18$**
30. $y = -\frac{3}{4}x + \frac{5}{4}$ **$3x + 4y = 5$**
31. $y = -\frac{1}{7}x + \frac{6}{7}$ **$x + 7y = 6$**
32. $y = -\frac{1}{3}x - 4$ **$x + 3y = -12$**

STUDENT HELP

HOMEWORK HELP
Example 1: Exs. 18–32
Example 2: Exs. 33–50
Example 3: Exs. 51–61
Example 4: Exs. 64–69

WRITING EQUATIONS **Write an equation in standard form of the line that passes through the given point and has the given slope.**

33. $(-8, 3), m = 2$ **$2x - y = -19$**
34. $(-2, 7), m = -4$ **$4x + y = -1$**
35. $(-1, 4), m = -3$ **$3x + y = 1$**
36. $(-6, -7), m = -1$ **$x + y = -13$**
37. $(3, -2), m = 5$ **$5x - y = 17$**
38. $(10, 6), m = 7$ **$7x - y = 64$**
39. $(2, 9), m = -7$ **$7x + y = 23$**
40. $(5, -8), m = 10$ **$10x - y = 58$**
41. $(7, 3), m = -2$ **$2x + y = 17$**
42. $(-4, 2), m = -2$ **$2x + y = -6$**
43. $(0, 3), m = 1$ **$x - y = -3$**
44. $(-3, 3), m = 4$ **$4x - y = -15$**

3 APPLY

ASSIGNMENT GUIDE

BASIC
Day 1: pp. 311–312 Exs. 18–52 even
Day 2: pp. 312–314 Exs. 54–69, 75–80, 86, 91, 96, 101

AVERAGE
Day 1: pp. 311–312 Exs. 18–52 even
Day 2: pp. 312–314 Exs. 54–69, 75–80, 86, 91, 96, 104

ADVANCED
Day 1: pp. 311–312 Exs. 18–52 even
Day 2: pp. 312–314 Exs. 54–69, 75–83, 86, 91, 96, 101

BLOCK SCHEDULE
pp. 311–312 Exs. 18–52 even (with 5.5)
pp. 312–314 Exs. 54–69, 75–80, 86, 91, 96, 101 (with 5.7)

EXERCISE LEVELS

Level A: ***Easier***
18–32

Level B: ***More Difficult***
33–69, 75–80, 84–108

Level C: ***Most Difficult***
70–74, 81–83

✓ HOMEWORK CHECK

To quickly check student understanding of key concepts, go over the following exercises: Exs. 20, 30, 36, 44, 46, 52, 56, 64. See also the Daily Homework Quiz:

- Blackline Master (*Chapter 5 Resource Book*, p. 94)
- Transparency (p. 43)

! COMMON ERROR
EXERCISES 45–50
Computational and algebraic errors are very common in problems like these. Encourage students to always check their work by testing one point on the line in the equation.

65.

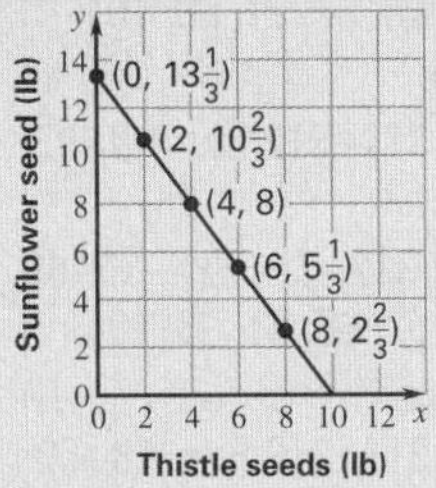

66. See graph for Exercise 65.

WRITING EQUATIONS **Write an equation in standard form of the line that passes through the two points.**

45. (1, 4), (5, 7) **$3x - 4y = -13$**
46. (−3, −3), (7, 2) **$x - 2y = 3$**
47. (−4, 1), (2, −5) **$x + y = -3$**
48. (9, −2), (−3, 2) **$x + 3y = 3$**
49. (0, 0), (2, 0) **$y = 0$**
50. (4, −7), (5, −1) **$6x - y = 31$**

HORIZONTAL AND VERTICAL LINES **Write an equation in standard form of the horizontal line and the vertical line.**

51.

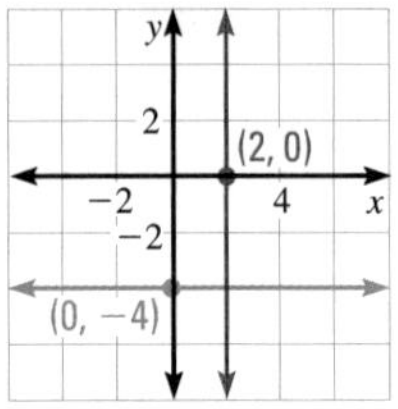

$y = -4, x = 2$

52.

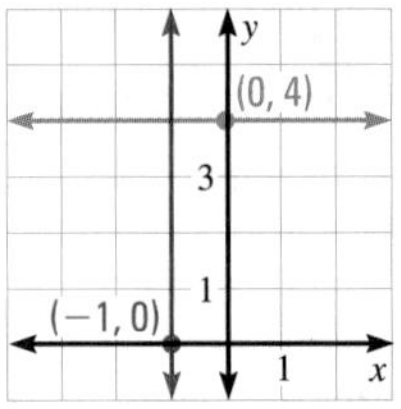

$y = 4, x = -1$

53.

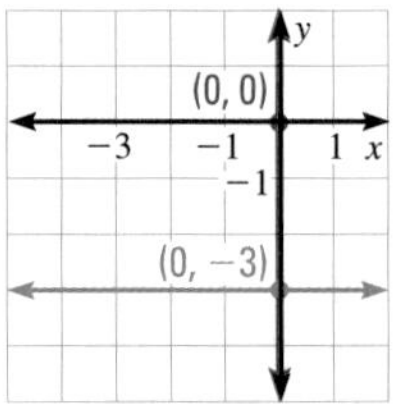

$y = -3, x = 0$

HORIZONTAL AND VERTICAL LINES **Write an equation in standard form of the horizontal line and the vertical line that pass through the point.**

54. (1, 3) **$y = 3, x = 1$**
55. (−4, 4) **$y = 4, x = -4$**
56. (−2, −5) **$y = -5, x = -2$**
57. (6, −1) **$y = -1, x = 6$**
58. (0, 7) **$y = 7, x = 0$**
59. (−9, 0) **$y = 0, x = -9$**
60. (−3, 7) **$y = 7, x = -3$**
61. (10, −3) **$y = -3, x = 10$**

ERROR ANALYSIS **In Exercises 62 and 63, describe the error.**

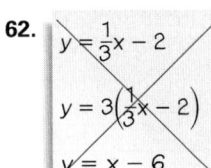

62.
$y = \frac{1}{3}x - 2$
$y = 3\left(\frac{1}{3}x - 2\right)$
$y = x - 6$

62. The left side of the equation was not multiplied by 3.

63.
$y - 4 = -3[x - (-6)]$
$y - 4 = -3(x - 6)$
$y - 4 = -3x + 18$
$y = -3x + 22$

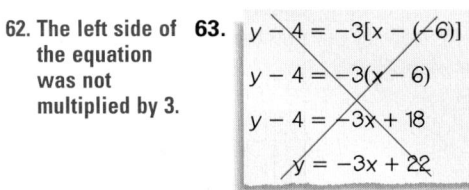

63. The expression inside the brackets was simplified incorrectly; $x - (-6) = x + 6$.

BIRD SEED MIXTURE **In Exercises 64–66, use the following information.** You are buying $20 worth of bird seed that consists of two types of seed. Thistle seed attracts finches and costs $2 per pound. Dark oil sunflower seed attracts many kinds of song birds and costs $1.50 per pound.

64. Write an equation that represents the different amounts of $2 thistle seed x and $1.50 dark oil sunflower seed y that you could buy. ***Sample answer:* $4x + 3y = 40$**

65. Graph the line representing the possible seed mixtures in Exercise 64. **65, 66. See margin.**

66. Copy and complete the table. Label the points from the table on the graph created in Exercise 65.

Thistle seed (lb), x	0	2	4	6	8
Dark oil sunflower seed (lb), y	?	?	?	?	?

FOCUS ON APPLICATIONS

FEEDING HABITS The American Goldfinch uses its feet to extract seeds. Before a storm, American Goldfinches increase the amount they eat tremendously and gain weight. ▶ Source: Canadian Wildlife Service

PRIZES **In Exercises 67–69, you are in charge of buying prizes for a school contest. A one-strand rosette ribbon costs $2.00 and a three-strand rosette ribbon costs $3.00. You have $10 to spend.**

67. Write an equation that represents the different numbers of one-strand and three-strand rosette ribbons that you can purchase. ***Sample answer:* $2x + 3y = 10$**

68. What is the greatest number of students that can receive a ribbon? **5 students**

69. What is the least number of students that can receive a ribbon? **3 students**

STUDENT HELP

HOMEWORK HELP Visit our Web site www.mcdougallittell.com for help with Exs. 70–74.

COLOR PRINTING In Exercises 70–74, use the following information.
In printing, different shades of color are often formed by *overprinting* different percents of two or more primary colors such as magenta (red) and cyan (blue). For instance, combining a 40% magenta screen with a 20% cyan screen produces a more reddish purple than combining a 20% magenta screen with a 40% cyan screen. This process reduces the number of colors of ink needed.

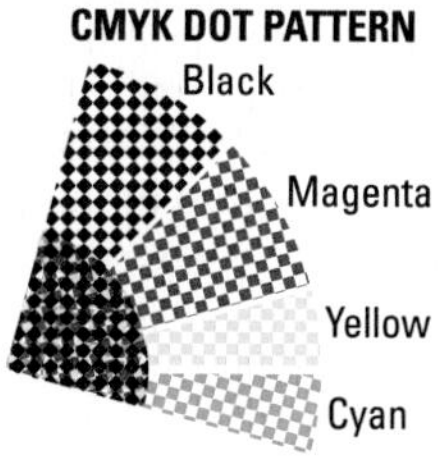

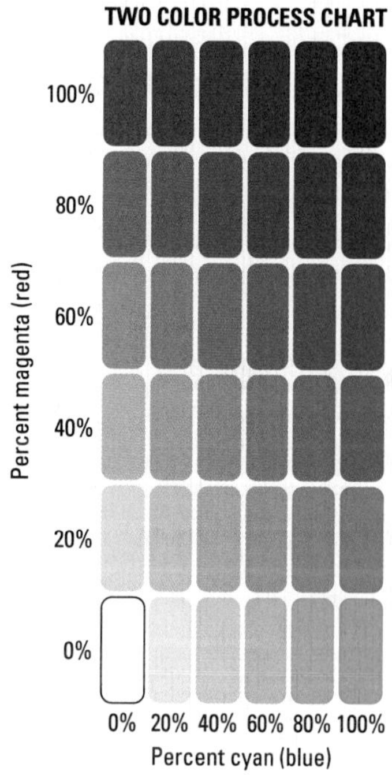

70. You are designing an advertising piece. You want the *combined* percent of magenta and cyan screens for a background color to be 60%. Write a linear equation that models the different percents of magenta and cyan you can use. Let x represent the percent of magenta, and let y represent the percent of cyan. ***Sample answer:*** $x + y = 60$

71. Trace the magenta-cyan color palette matrix at the right and identify the possible shades you can use for a combined percent of 60%.

71. 0% blue, 60% red; 10% blue, 50% red; 20% blue, 40% red; 30% blue, 30% red; 40% blue, 20% red; 50% blue, 10% red; 60% blue, 0% red

72. Write a linear equation that models the different percents of magenta and cyan you can use for a background color of 80%. ***Sample answer:*** $x + y = 80$

73. Trace the magenta-cyan color palette matrix and identify the possible shades you can use for a combined percent of 80%. **See margin.**

73. 0% blue, 80% red; 10% blue, 70% red; 20% blue, 60% red; 30% blue, 50% red; 40% blue, 40% red; 50% blue, 30% red; 60% blue, 20% red; 70% blue, 10% red; 80% blue, 0% red

74. How do the patterns formed by the possible shades relate to the lines whose equations you found in Exercises 70 and 72? **See margin.**

74. ***Sample answer:*** For each equation, if the palette were a coordinate grid with each square representing a vertex, each color square would be a point on the graph of the equation.

Test Preparation

75. MULTIPLE CHOICE Choose the equation in standard form of the line that passes through the point $(-1, -4)$ and has a slope of 2. **B**

Ⓐ $2x + y = -2$ Ⓑ $-2x + y = -2$

Ⓒ $2x + y = 2$ Ⓓ $2x - y = -2$

76. MULTIPLE CHOICE Choose the standard form of the equation shown in the graph. **D**

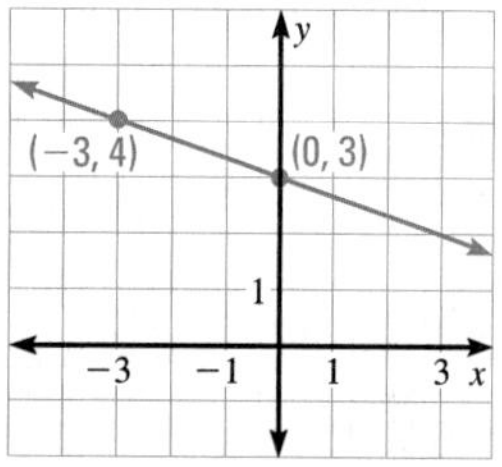

Ⓐ $3x + y = 3$

Ⓑ $-3y = x + (-9)$

Ⓒ $-x - y = 9$

Ⓓ $-x - 3y = -9$

STUDENT HELP NOTES

Homework Help Students find help for Exs. 70–74 at **www.mcdougallittell.com.** This information can be printed out for students who don't have access to the Internet.

ADDITIONAL PRACTICE AND RETEACHING

For Lesson 5.6:

- Practice Levels A, B, and C (*Chapter 5 Resource Book,* p. 84)
- Reteaching with Practice (*Chapter 5 Resource Book,* p. 87)
- See Lesson 5.6 of the *Personal Student Tutor*

For more Mixed Review:

- Search the *Test and Practice Generator* for key words or specific lessons.

4 ASSESS

DAILY HOMEWORK QUIZ

Transparency Available

1. Write the equation $y = \frac{7}{10}x - \frac{3}{2}$ in standard form.
Sample answer: $-7x + 10y = -15$

2. Write an equation in standard form of the line that passes through (5, −1) and (−1, 3).
Sample answer: $2x + 3y = 7$

3. You are buying apples and oranges for picnic lunches. Apples cost 20¢ apiece and oranges cost 25¢. You have \$5 to spend for the fruit. Write an equation in standard form to represent the number of apples x and oranges y you could buy. *Sample answer:* $20x + 25y = 500$

EXTRA CHALLENGE NOTE

Challenge problems for Lesson 5.6 are available in **blackline** format in the *Chapter 5 Resource Book,* p. 91 and at **www.mcdougallittell.com.**

ADDITIONAL TEST PREPARATION

1. WRITING In the equation $Ax + By = C$ is C the y-intercept? Explain your answer using at least one example.
No, the y-intercept is $\frac{C}{B}$; *sample answer:* the y-intercept of $2x + 3y = 5$ is $\frac{5}{3}$.

90–95. See Additional Answers beginning on page AA1.

EXPLORING EQUATIONS **In Exercises 77–80, use the equation $2x + 7y = 14$.**

77. What is the x-intercept? **7**

78. What is the y-intercept? **2**

79. Multiply each term in the equation by $\frac{1}{14}$. $\frac{1}{7}x + \frac{1}{2}y = 1$

80. Find a relationship between the denominators on the left side of the equation and the x-intercept and the y-intercept. **The denominator of the x-term is the x-intercept; the denominator of the y-term is the y-intercept.**

★ Challenge

FINDING EQUATIONS **In Exercises 81–83, use the following information.**
The equation represents the intercept form of the equation of a line. In the equation, the x-intercept is a and the y-intercept is b.

$$\frac{x}{a} + \frac{y}{b} = 1$$

81. Write the intercept form of the equation of the line whose x-intercept is 2 and y-intercept is 3. $\frac{x}{2} + \frac{y}{3} = 1$

82. Write the equation in Exercise 81 in standard form. $3x + 2y = 6$

83. Write the equation in Exercise 82 in slope-intercept form. $y = -\frac{3}{2}x + 3$

EXTRA CHALLENGE
www.mcdougallittell.com

MIXED REVIEW

SOLVING EQUATIONS **Solve the equation. Check the result.** (Review 3.4)

84. $8 + y = 3$ **−5**
85. $y - 9 = 2$ **11**
86. $-6(q + 22) = 5q$ **−12**
87. $2(x + 5) = 18$ **4**
88. $7 - 2a = -14$ $10\frac{1}{2}$
89. $-2 + 4c = 19$ $5\frac{1}{4}$

GRAPHING EQUATIONS **Use a table of values to graph the equation. Label the x-intercept and the y-intercept.** (Review 4.2, 4.3) **90–95. See margin.**

90. $y = -x + 8$
91. $y = 4x - 4$
92. $y = x + 5$
93. $y = -9 + 3x$
94. $y = -x - 1$
95. $y = 10 - x$

SLOPE-INTERCEPT FORM **Graph the line that passes through the points. Write its equation in slope-intercept form.** (Review 5.3 for 5.7) **96–104. Check graphs. See margin.**

96. (2, 3), (−4, 1)
97. (7, 6), (5, −9)
98. (−5, −2), (−1, 12)
99. (6, −2), (−5, −8)
100. (2, 9), (4, 7)
101. (19, −2), (4, −1)
102. (5, −8), (7, −3)
103. (14, −9), (−5, −3)
104. (−5, 9), (5, 4)

96. $y = \frac{1}{3}x + \frac{7}{3}$
97. $y = \frac{15}{2}x - \frac{93}{2}$
98. $y = \frac{7}{2}x + \frac{31}{2}$
99. $y = \frac{6}{11}x - \frac{58}{11}$
100. $y = -x + 11$
101. $y = -\frac{1}{15}x - \frac{11}{15}$
102. $y = \frac{5}{2}x - \frac{41}{2}$
103. $y = -\frac{6}{19}x - \frac{87}{19}$
104. $y = -\frac{1}{2}x + \frac{13}{2}$

105. **BUYING A CD PLAYER** You have \$125.79 to spend for a CD player. The sales tax is 6%. What is your price limit for the CD player? (Review 3.6) **\$118.67**

106. **BUYING A CALENDAR** You have \$15 to spend for a calendar. The sales tax is 8.5%. What is your price limit for the calendar? (Review 3.6) **\$13.82**

107. **LEAVING A TIP** You leave a \$1.50 tip for a lunch of \$6.00. What percent tip did you leave? (Review 3.8) **25%**

108. **SHOPPING** A coat regularly sells for \$84. You buy the coat on sale and save \$10. What percent is the discount? (Review 3.8) **11.9% (to the nearest tenth of a percent)**

Investigating Linear Modeling

GROUP ACTIVITY
Work in a small group.

MATERIALS
- toothpicks
- graph paper

▶ QUESTION **How can you decide whether data can be represented by a linear model?**

▶ EXPLORING THE CONCEPT

The following figures represent the first five stages in a sequence.

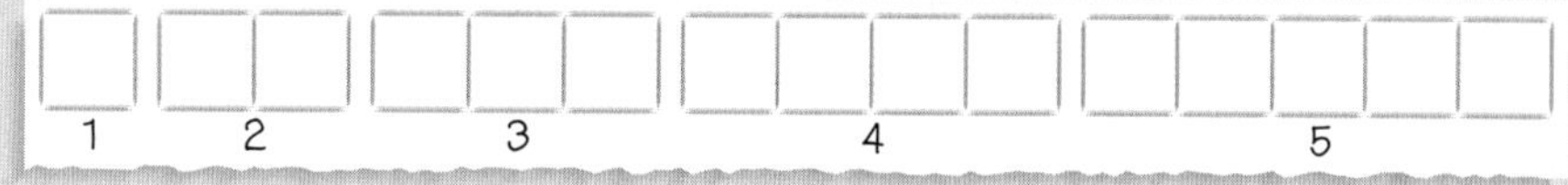

1. Copy and complete the table showing the number of toothpicks used to make each figure. **7, 10, 13, 16**

Number of squares in figure	1	2	3	4	5
Number of toothpicks	4	?	?	?	?

2. Make a scatter plot of the data. Put the number of squares on the horizontal axis. What do you observe about the data points? **The points all lie on a line; See margin.**
3. Choose two points on the line and calculate the slope. **3**
4. Write the equation of the line in slope-intercept form. $y = 3x + 1$
5. Use the equation to predict the number of toothpicks necessary to construct a figure with 10 squares. Check your result. **31**

▶ DRAWING CONCLUSIONS

1–4. See margin.

The figures show the first three stages in a sequence. Each group member should choose one sequence and answer Exercises 1–4 individually.

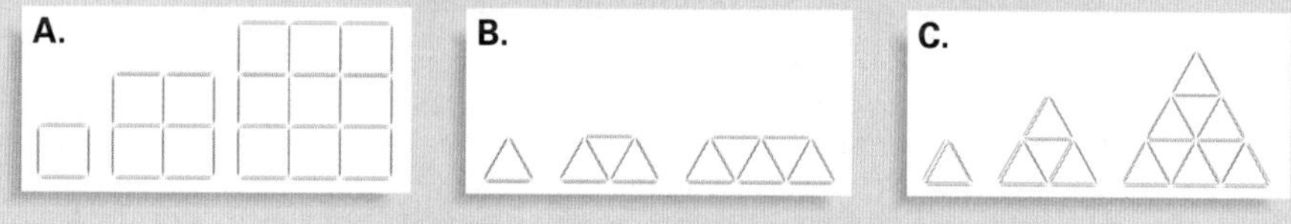

1. Use toothpicks to make the next three figures in the sequence.
2. Make a table of the results. For each stage, give the number of small squares or small triangles in the figure and the number of toothpicks used.
3. Make a scatter plot of the number of small squares or small triangles as a function of the stage number.
4. Make a scatter plot of the number of toothpicks as a function of the number of small squares or small triangles.
5. Compare your scatter plots with those of others in your group. Which of the sequences can be modeled with a linear model? Which cannot? Explain.
6. If the data appear to fit a linear model, find an equation of the line.

5. Answers may vary. The scatter plots for A and C of Question 3 cannot be modeled with linear models. All the others can.

6. For sequences 4A and 4C, the sample answer given was obtained using a graphing calculator.

3B: $y = 2x - 1$

4A: $y = 2.24x + 3.22$

4B: $y = 2x + 1$

4C: $y = 1.68x + 2.42$

1 Planning the Activity

PURPOSE
To investigate the appropriateness of a linear model to represent data.

MATERIALS
- box of toothpicks per group
- Activity Support Master (*Chapter 5 Resource Book,* p. 95)

PACING
- Exploring the Concept — 10 min
- Drawing Conclusions — 15 min

▶ *LINK TO LESSON*
When discussing Example 1 on page 316, refer back to Exercise 5 in this activity. Point out that when using data from real-life, the points may not lie exactly on a line, but the general shape of the scatter plot might be linear.

2 Managing the Activity

COOPERATIVE LEARNING
When working through Steps 1–5, have students take turns completing the steps.

CLASSROOM MANAGEMENT
You may want to have students make the figures on a tray to keep the toothpicks from rolling onto the floor and to make clean up easier.

3 Closing the Activity

★ ***KEY DISCOVERY***
Not all data fit a linear model.

ACTIVITY ASSESSMENT
JOURNAL Draw a scatter plot that could be represented with a linear model and a scatter plot that should not be represented by a linear model. Write about how you can tell from a graph if data can be represented by a linear model.
See sample answer at left; check scatter plots.
Step 2, 1–4. See Additional Answers beginning on page AA1.

Activity Assessment *Sample answer:*
If the points in the scatter plot appear to lie along or close to a best-fitting line, then the data can be represented by a linear model.

1 PLAN

PACING
Basic: 2 days
Average: 2 days
Advanced: 2 days
Block Schedule: 0.5 block with Ch. Rev.

LESSON OPENER
VISUAL APPROACH
An alternative way to approach Lesson 5.7 is to use the Visual Approach Lesson Opener:

- Blackline Master (*Chapter 5 Resource Book,* p. 96)
- Transparency (p. 38)

MEETING INDIVIDUAL NEEDS

- ***Chapter 5 Resource Book***
  Prerequisite Skills Review (p. 5)
  Practice Level A (p. 98)
  Practice Level B (p. 99)
  Practice Level C (p. 100)
  Reteaching with Practice (p. 101)
  Absent Student Catch-Up (p. 103)
  Challenge (p. 105)
- ***Resources in Spanish***
- ***Personal Student Tutor***

NEW-TEACHER SUPPORT
See the Tips for New Teachers on pp. 1–2 of the *Chapter 5 Resource Book* for additional notes about Lesson 5.7.

WARM-UP EXERCISES

Transparency Available

Write a linear equation for the line through the two points.

1. (21, 14,256) and (45, 39,561)
$y = 1054x - 7886$

2. (12.3, 4162) and (14.7, 3219)
$y = -393x + 8995$

Tell whether each scatter plot has positive, negative, or relatively no correlation.

3. the points cannot be approximated by a line **relatively no correlation**

4. the points can be approximated by a line with positive slope **positive correlation**

EXPLORING DATA AND STATISTICS

5.7 Predicting with Linear Models

What you should learn

GOAL 1 Determine whether a linear model is appropriate.

GOAL 2 Use a linear model to make a **real-life** prediction, such as spending on advertising in **Example 3**.

Why you should learn it

▼ To predict the cost of college tuition in the future in **Exs. 29** and **30**.

REAL LIFE

GOAL 1 DECIDING WHEN TO USE A LINEAR MODEL

Two of your major goals in this course are to learn about different types of models and to learn which type of model to use in a specific real-life situation.

In this lesson you will learn how to decide when a linear model can be used to represent real-life data.

EXAMPLE 1 *Which Data Set is More Linear?*

The amount (in millions of dollars) spent on advertising in broadcast television and on the Internet from 1995 through 2001 is given in the table. Which data are better modeled with a linear model?

Year	1995	1997	1999	2001
Broadcast television advertising expenditures (in millions)	32,720	36,893	41,230*	46,140*
Internet advertising expenditures (in millions)	24	1300	15,000*	46,000*

▶ Sources: Veronis, Suhler & Associates Communications Industry Forecast and *Activ* Media

*Projected data

SOLUTION

A good way to decide whether data can be represented by a linear model is to draw a scatter plot of the data.

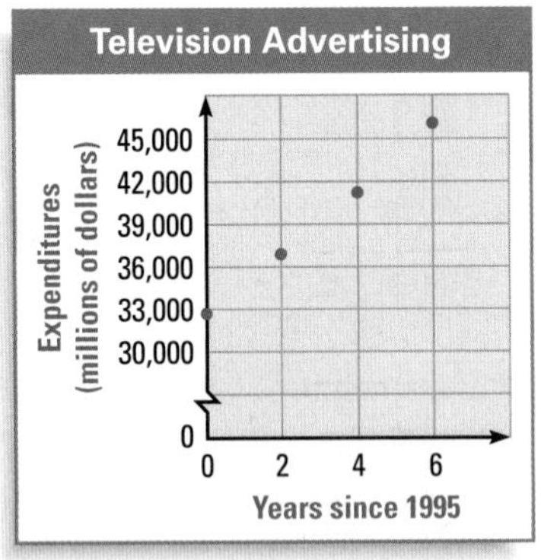

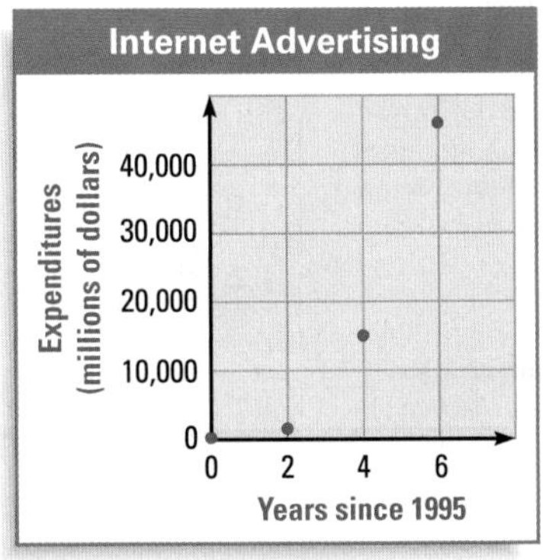

From the two scatter plots, you can see that the broadcast television data fall almost exactly on a line. The Internet data are much less linear. That is, the data points don't lie as close to a straight line.

GOAL 2 PREDICTING WITH LINEAR MODELS

REAL LIFE
Advertising

EXAMPLE 2 *Writing a Linear Model*

Write a linear model for the amounts spent on television advertising given in the table in Example 1.

SOLUTION

METHOD 1 Make a scatter plot of the data and draw the line that best fits the points. The line does not need to pass through any of the data points.

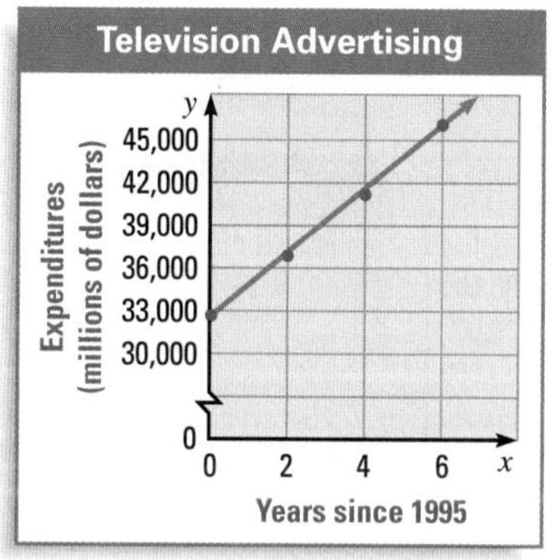

Find two points on the best-fitting line such as (0, 32,720) and (6, 46,140). Use these points to find the slope of the best-fitting line.

$$m = \frac{y_2 - y_1}{x_2 - x_1} = \frac{46{,}140 - 32{,}720}{6 - 0} \approx 2236.67 \approx 2237$$

Using a y-intercept of $b = 32{,}720$ and a slope of $m = 2237$ you can write an equation of the line.

$y = mx + b$ — Write slope-intercept form.

$y = 2237x + 32{,}720$ — Substitute 2237 for m and 32,720 for b.

▶ A linear model for the amounts spent on television advertising is $y = 2237x + 32{,}720$.

METHOD 2 You can use a graphing calculator and follow the procedure given on page 299.

The linear regression program gives you the results shown at the right. A linear model for spending on television advertising is $y = 2229.85x + 32{,}556.2$. The r-value of about 0.999 indicates that the linear model is a very good fit.

STUDENT HELP

Look Back
For help with best-fitting lines, see p. 293.

STUDENT HELP

Study Tip
In Example 2, notice that the two methods give slightly different results. The equation found using the graphing calculator is the more accurate model.

2 TEACH

MOTIVATING THE LESSON
You are a restaurant owner and are making a menu with new pricing. You want the menu to last 3 years. By how much should you increase the prices so that they will keep up with increases in costs over the next 3 years? Investigating whether prices can be predicted by a linear model will help you decide.

EXTRA EXAMPLE 1
The manager of a restaurant made the following table.

Average Price Per Pound ($)		
Year	Fish	Meat
1991	3.75	2.50
1993	4.25	2.70
1995	5.25	3.00
1997	6.75	3.30
1999	8.75	3.50

Which data are better modeled by a linear model? **meat prices**

EXTRA EXAMPLE 2
Write a linear model for the average meat prices given in Extra Example 1.
Sample answer: $y = 0.13x + 2.35$

CHECKPOINT EXERCISES

For use after Examples 1 and 2:

1. set 1

x	1	2	3	4
y	12	25	36	48

set 2

x	1	2	3	4
y	10	5	2	1

a. Which data are better modeled using a linear model? **set 1**

b. Write a linear model for the set you choose in part (a). ***Sample answer:*** $y = 11.9x + 0.5$

EXTRA EXAMPLE 3

Use the linear model you found in Extra Example 2 to estimate the average price per pound of meat in the given year. Tell whether you use linear interpolation or linear extrapolation.

a. 2002 **$3.91; extrapolation**

b. 1998 **$3.39; interpolation**

CHECKPOINT EXERCISES

For use after Example 3:

1. A linear model for the data below is $y = 32x + 27$, where x is the number of years after 1985. Use the model to estimate the value of y in 2002. Tell whether you use linear interpolation or linear extrapolation. **571; extrapolation**

x	3	7	9	11
y	120	256	315	376

STUDENT HELP NOTES

Homework Help Students can find extra examples at **www.mcdougallittell.com** that parallel the examples in the student edition.

CAREER NOTE

EXAMPLE 3 Additional information about graphic design is available at **www.mcdougallittell.com.**

MATHEMATICAL REASONING

In Example 1 on page 316, ask students to compare how the values in each row of the table change from year to year. They should notice that the Broadcast Television row has increases that are fairly constant, while the increases in the Internet row are increasing. Point out how this difference relates to the graphs. This exercise should help students see that a linear function is one with a constant rate of change that is the slope of the graph.

CLOSURE QUESTION

Explain how to use a linear model to make predictions from given data. **See sample answer at right.**

STUDENT HELP

HOMEWORK HELP Visit our Web site www.mcdougallittell.com for extra examples.

INTERPOLATION AND EXTRAPOLATION When you use a linear model to estimate data points that are not given, you are using *linear interpolation* or *linear extrapolation*. **Linear interpolation** is a method of estimating the coordinates of a point that lies between two given data points. **Linear extrapolation** is a method of estimating the coordinates of a point that lies to the right or left of all of the given data points.

EXAMPLE 3 *Linear Interpolation and Linear Extrapolation*

ADVERTISING Use the linear model you found in Example 2 to estimate the amount spent on advertising in broadcast television in the given year. Tell whether you use *linear interpolation* or *linear extrapolation.*

a. 1996 **b.** 2015

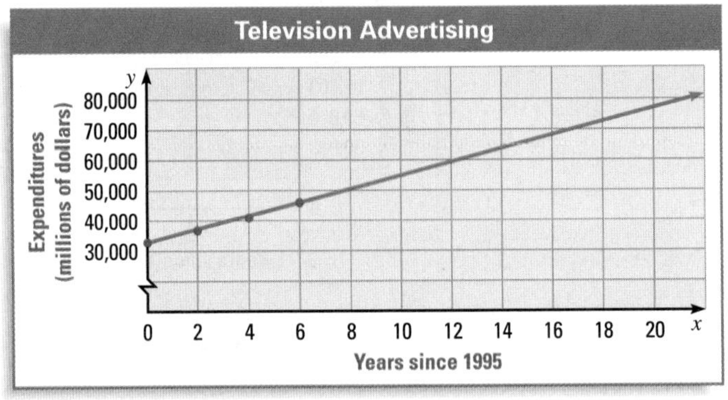

a. You used data for 1995 and 2001 to write the model. Because 1996 is between these two dates, you will use *linear interpolation.* Because x represents the number of years since 1995, you can estimate the amount spent in 1996 by substituting $x = 1$ into the linear model from Example 2.

$y = 2237x + 32{,}720$ **Write linear model.**

$= 2237(1) + 32{,}720$ **Substitute 1 for x.**

$= 34{,}957$ **Simplify.**

▶ The model predicts that almost $35,000 million, or $35 billion, was spent in 1996.

b. Because 2015 is to the right of all of the given data, you will use *linear extrapolation.* You can estimate the amount spent in 2015 by substituting $x = 20$ into the linear model from Example 2.

$y = 2237x + 32{,}720$ **Write linear model.**

$= 2237(20) + 32{,}720$ **Substitute 20 for x.**

$= 77{,}460$ **Simplify.**

▶ The model predicts that about $77,000 million, or about $77 billion, will be spent in 2015.

FOCUS ON CAREERS

GRAPHIC DESIGNERS
Graphic designers choose the art and design layouts. They may work on advertising for television, newspapers, and other media.

CAREER LINK www.mcdougallittell.com

Closure Question ***Sample answer:***
First graph the data to see if a linear model is a good fit. Then find the equation of a best-fitting line. This equation can be used to make predictions.

GUIDED PRACTICE

Throughout, linear models are determined using a graphing calculator and rounding to the nearest hundredth.

Vocabulary Check ✓

1. Explain the difference between linear interpolation and linear extrapolation. **See margin.**

Concept Check ✓

2. Explain how to decide whether a data set can be represented by a linear model. ***Sample answer:* Make a scatter plot and decide whether the data points lie close to a straight line.**

1. Linear interpolation is a method of estimating the coordinates of a point that lies between two given data points. Linear extrapolation is a method of estimating the coordinates of a point that lies to the right or left of all the given data points.

Tell whether it is reasonable for the graph to be represented by a linear model.

3. 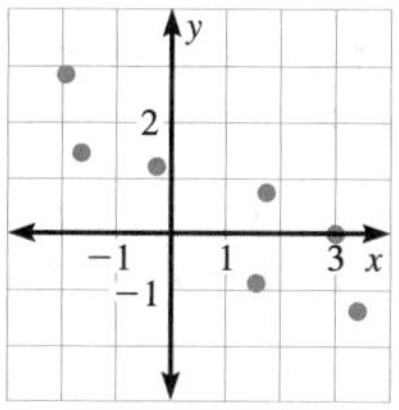 **yes**

4. 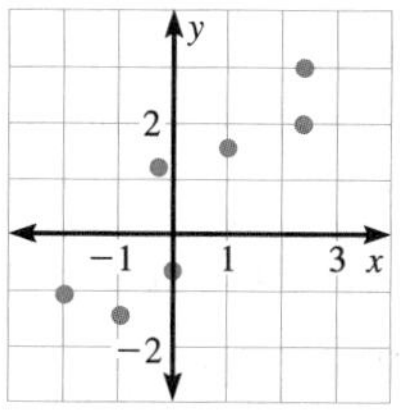 **yes**

5. 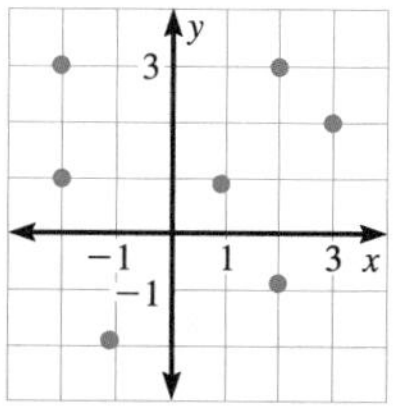 **no**

Skill Check ✓

BOTTLED WATER In Exercises 6–10, use the data on bottled water consumption per person.

6. Make a scatter plot of the data. **See margin.**

7. Write a linear model for the data. $y = 0.62x + 1.96$

Year	Water (in gallons)
1980	2.4
1985	4.5
1990	8.0
1995	11.6

▶ Source: *Statistical Abstract of the United States*

Use your model to estimate the consumption of bottled water in the given year. Tell whether you use *linear interpolation* or *linear extrapolation*.

8. 1994 **about 10.64 gal; interpolation**

9. 2010 **about 20.56 gal; extrapolation**

10. 1979 **1.34 gal; extrapolation**

PRACTICE AND APPLICATIONS

STUDENT HELP

Extra Practice to help you master skills is on p. 801.

REPRESENTING DATA Tell whether it is reasonable for the graph to be represented by a linear model.

11. 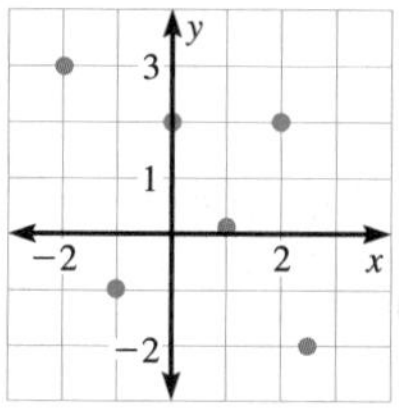 **no**

12. 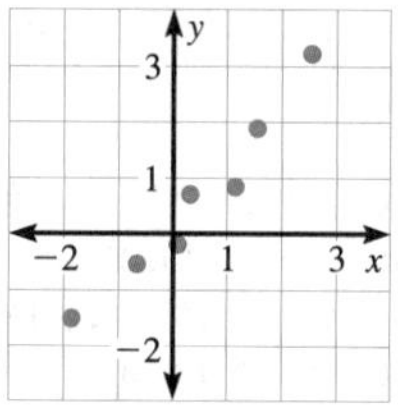 **yes**

13. 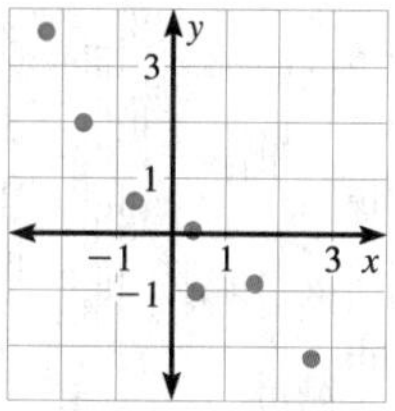 **yes**

14. 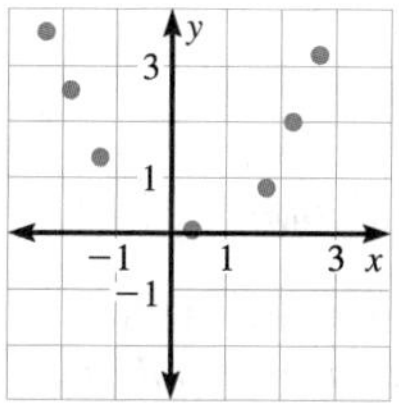 **no**

15. 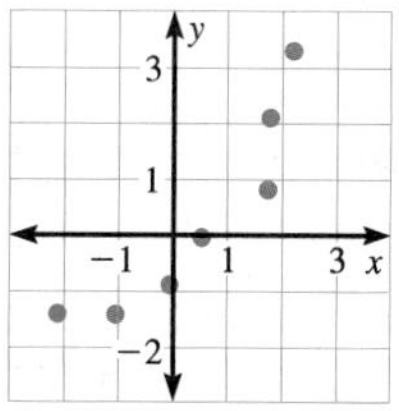 **yes**

16. 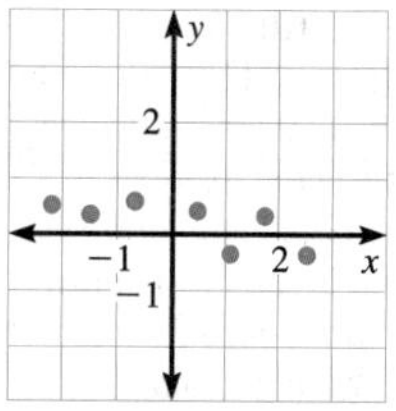 **yes**

STUDENT HELP

HOMEWORK HELP
Example 1: Exs. 11–16, 17–19
Example 2: Exs. 20, 23, 25
Example 3: Exs. 21, 22, 24, 26, 27

3 APPLY

ASSIGNMENT GUIDE

BASIC
Day 1: pp. 319–320 Exs. 11–24
Day 2: pp. 320–322 Exs. 25–34, 38–43, 44–54 even, Quiz 3 Exs. 1–15

AVERAGE
Day 1: pp. 319–320 Exs. 11–24
Day 2: pp. 320–322 Exs. 25–34, 38–43, 44–54 even, Quiz 3 Exs. 1–15

ADVANCED
Day 1: pp. 319–320 Exs. 11–24
Day 2: pp. 320–322 Exs. 25–43, 44–54 even
Quiz 3 Exs. 1–15

BLOCK SCHEDULE
pp. 319–320 Exs. 11–24 (with 5.6)
pp. 320–322 Exs. 25–34, 38–43, 44–54 even, Quiz 3 Exs. 1–15 (with Ch. 5 Review)

EXERCISE LEVELS

Level A: ***Easier***
11–16

Level B: ***More Difficult***
17–34, 38–54

Level C: ***Most Difficult***
35–37

✓ HOMEWORK CHECK

To quickly check student understanding of key concepts, go over the following exercises: Exs. 12, 14, 17, 20, 21, 25, 26. See also the Daily Homework Quiz:

- Blackline Master (*Chapter 6 Resource Book*, p. 11)
- Transparency (p. 45)

6. **Bottled Water Consumption**

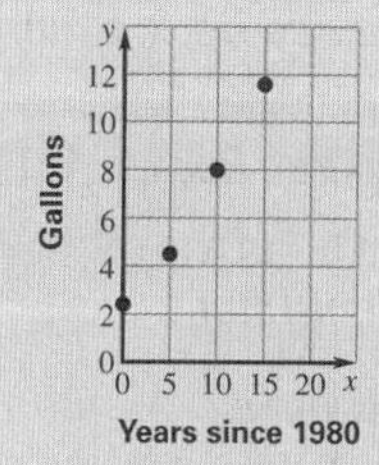

Data Update

Updated data for Exs. 17–22 is available at **www.mcdougallittell.com.**

Application Note

Exercises 25–27 Caution students that extrapolation is less reliable than interpolation. Extrapolation is generally used only for points close to the other data points.

17.

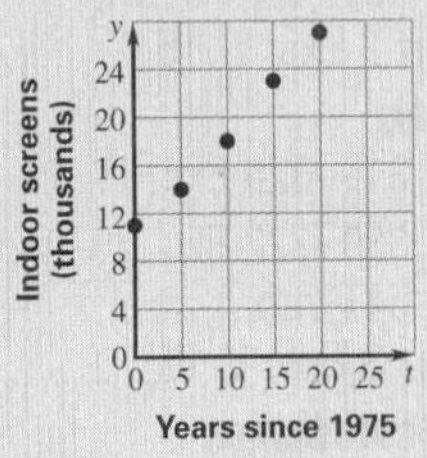

18.

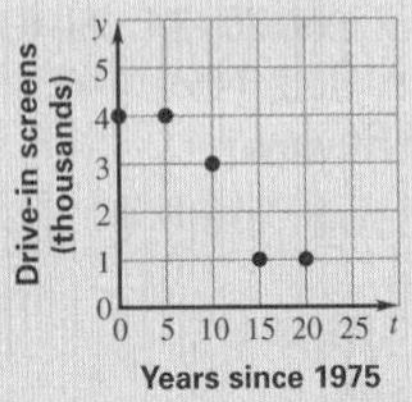

Additional Practice and Reteaching

For Lesson 5.7:

- Practice Levels A, B, and C (*Chapter 5 Resource Book,* p. 98)
- Reteaching with Practice (*Chapter 5 Resource Book,* p. 101)
- See Lesson 5.7 of the *Personal Student Tutor*

For more Mixed Review:

- Search the *Test and Practice Generator* for key words or specific lessons.

Movie Theaters **In Exercises 17–22, use the table which shows the number of movie theater screens (in thousands) from 1975 to 1995.**

Year	1975	1980	1985	1990	1995
Indoor screens (in thousands)	11	14	18	23	27
Drive-in screens (in thousands)	4	4	3	1	1

DATA UPDATE of Motion Picture Association of America, Inc. data at www.mcdougallittell.com

17. Make a scatter plot of the number of indoor movie screens in terms of the year t. Let t represent the number of years since 1975. **See margin.**

18. Make a scatter plot of the number of drive-in movie screens in terms of the year t. Let t represent the number of years since 1975. **See margin.**

19. Which data are better modeled with a linear model? **indoor screens**

20. Write a linear model for the number of indoor movie screens. $y = 0.82t + 10.4$

21. Use the linear model to estimate the number of indoor movie screens in 2005. **about 35,000 theaters**

22. Use the linear model to estimate the number of indoor movie screens in 1989. Did you use *linear interpolation* or *linear extrapolation*? **about 21,880 theaters; linear interpolation**

Books and Maps **In Exercises 23 and 24, use the table which shows the number of dollars (in billions) spent on books and maps in the United States from 1990 through 1995.**

Years since 1990	0	1	2	3	4	5
Billions of dollars	16.5	16.9	17.7	19.0	20.1	20.9

▶ Source: U.S. Bureau of Economic Analysis

23. Write a linear model for the amount spent on books and maps. $y = 0.94x + 16.17$

24. Use the linear model to estimate the amount spent in 2005. Did you use *linear interpolation* or *linear extrapolation*? **about $30.27 billion; linear extrapolation**

FOCUS ON APPLICATIONS

Real Life **SPORTS EQUIPMENT** Life jackets should be chosen by the type of water activities, as well as the expected water conditions. Type I can be used in rough, offshore water while Type III are appropriate for calm, inland waters.

Toys and Sport Supplies **In Exercises 25–27, use the table which shows the number of dollars (in billions) spent on toys and sport supplies in the United States from 1990 through 1995.**

Years since 1990	Billions of dollars
0	31.6
1	32.8
2	?
3	36.5
4	40.1
5	42.7

▶ Source: U.S. Bureau of Economic Analysis

25. Write a linear model for the amount spent on toys and sport supplies. $y = 2.24x + 30.91$

26. Use the linear model to estimate the amount spent in 1992. Did you use *linear interpolation* or *linear extrapolation*? **about $35.39 billion; linear interpolation**

27. Use the linear model to estimate the amount spent in 2005. Did you use *linear interpolation* or *linear extrapolation*? **about $64.51 billion; linear extrapolation**

COLLEGE TUITION In Exercises 28–30, use the table which shows the average tuition for attending a private and a public four-year college.

Year	Public college	Private college
1990	$2,035	$10,348
1991	$2,159	$11,379
1992	$2,410	$12,192
1993	$2,604	$13,055
1994	$2,820	$13,874
1995	$2,977	$14,537
1996	$3,151	$15,581

▶ Source: U.S. National Center for Education Statistics

28. Write a linear model of the tuition for attending a public and of the tuition for attending a private college.
public: $y = 192.64x + 2015.79$; private: $y = 846.32x + 10456.18$

29. Use linear extrapolation to estimate the tuition for attending a public college when you graduate from high school. **Answers may vary.**

30. Use linear extrapolation to estimate the tuition for attending a private college when you graduate from high school. **Answers may vary.**

Test Preparation

MULTI-STEP PROBLEM In Exercises 31–34, use the following information. You are the produce manager at a new grocery store. It is your job to decide how much fruit to order for the week of the grand opening. You find a table that shows yearly per person consumption of pounds of bananas.

Year	Pounds of bananas
1990	24.4
1991	25.1
1992	27.3
1993	26.8
1994	28.1
1995	27.4

▶ Source: U.S. Department of Agriculture

31. What was the average *weekly* consumption per person of bananas in 1995? What was the average *monthly* consumption?
about 0.53 lb; about 2.28 lb

32. Write a linear model for the average weekly consumption per person of bananas in pounds. **$y = 0.01x + 0.48$, where x is the number of years since 1990.**

33. Use the linear model to estimate the average weekly consumption per person of bananas in pounds for this year. **Answers may vary.**

34. If you expect 1750 customers for the week of the grand opening, how many pounds of bananas should you order? **Answers may vary.**

★ Challenge

In Exercises 35–37, use the data from Exercises 28–30 and a graphing calculator. Let x represent years since 1990 and let y represent the average tuition at a public college.

35. Enter the data for the tuition for public colleges and graph the equation. **Check graphs.**

36. Locate the x-intercept of the graph. What does this point represent?
The x-intercept represents the year when the tuition would be $0.

37. *Writing* What are the reasonable limits of linear extrapolation in terms of college tuition? **See margin.**

37. The absolute lower limit is -10, since tuition values for $x < -10$ are negative. Reasonable limits might be -5 and 10, since most economic trends do not continue indefinitely.

EXTRA CHALLENGE
www.mcdougallittell.com

4 ASSESS

DAILY HOMEWORK QUIZ

Transparency Available

The table shows the number of U.S. high schools with computer networks from 1994 to 1997.

Years since 1994	Number of schools
0	6576
1	8159
2	9166
3	9565

1. Write a linear model for the number of high schools having computer networks.
***Sample answer:* $y = 940x + 7000$**

2. Use the linear model to estimate the number of high schools with computer networks in 2002. Did you use linear interpolation or linear extrapolation? **about 14,500 schools; linear extrapolation**

EXTRA CHALLENGE NOTE

Challenge problems for Lesson 5.7 are available in **blackline** format in the *Chapter 5 Resource Book*, p. 105 and at **www.mcdougallittell.com.**

ADDITIONAL TEST PREPARATION

1. OPEN ENDED Write two sets of 5 data points so that a linear model is a good fit for one but not for the other. Find a linear model for the appropriate set. Use extrapolation to predict a sixth data point in the set.

Sample answer:

Set A:

1	2	3	4	5
17	19	21	26	25

Set B:

1	2	3	4	5
1	8	27	64	125

A linear model for set A is $y = 2.1x + 14.9$.
A prediction for the sixth data point is (6, 27.5).

MIXED REVIEW

38. 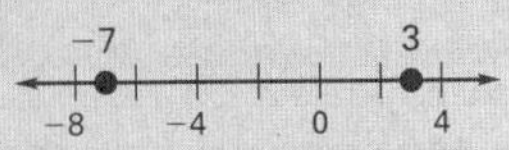

39. 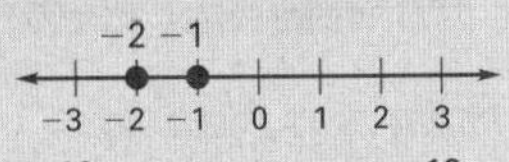

40. 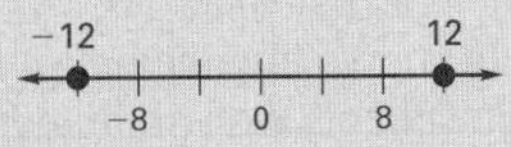

41. 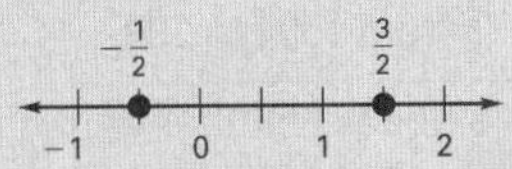

42. 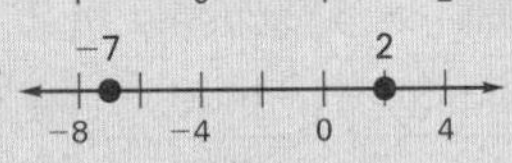

43. 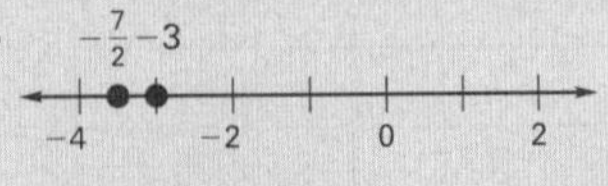

WRITING INEQUALITIES Graph the numbers on a number line. Then write two inequalities that compare the two numbers. (Review 2.1 for 6.1)
38–43. See margin.

38. -7 and 3 **39.** -2 and -1 **40.** 12 and -12

41. $-\frac{1}{2}$ and $\frac{3}{2}$ **42.** -7 and 2 **43.** -3 and $-\frac{7}{2}$

38. $-7 < 3, 3 > -7$; See margin.

39. $-2 < -1, -1 > -2$; See margin.

40. $12 > -12, -12 < 12$; See margin.

41. $-\frac{1}{2} < \frac{3}{2}, \frac{3}{2} > -\frac{1}{2}$; See margin.

42. $-7 < 2, 2 > -7$; See margin.

43. $-3 > -\frac{7}{2}, -\frac{7}{2} < -3$; See margin.

SLOPE-INTERCEPT FORM Write an equation of the line in slope-intercept form. (Review 5.1)

44. The slope is 2; the y-intercept is -5. $y = 2x - 5$

45. The slope is $\frac{1}{2}$; the y-intercept is -8. $y = \frac{1}{2}x - 8$

46. The slope is -4; the y-intercept is 7. $y = -4x + 7$

47. The slope is -2; the y-intercept is -6. $y = -2x - 6$

WRITING EQUATIONS Write an equation in slope-intercept form of the line that passes through the point and has the given slope. (Review 5.2)

48. $(6, 5), m = 2$ $y = 2x - 7$

49. $(-3, 7), m = 7$ $y = 7x + 28$

50. $(8, 2), m = 7$ $y = 7x - 54$

51. $(8, 4), m = -5$ $y = -5x + 44$

52. $(3, 1), m = 7$ $y = 7x - 20$

53. $(-8, 7), m = 2$ $y = 2x + 23$

54. FEATHER GROWTH At nine days old, a flight feather of a night heron is 17 millimeters. At 30 days, the flight feather is 150 millimeters. Write a linear model that gives feather length f in terms of age a. (Review 5.5) $f = \frac{19}{3}a - 40$

QUIZ 3

Self-Test for Lessons 5.6 and 5.7

Write the equation in standard form with integer coefficients. (Lesson 5.6)
1–6. Sample answers are given.

1. $y = \frac{1}{2}x + 3$ $x - 2y = -6$

2. $3y + 9x - 18 = 0$ $9x + 3y = 18$

3. $6y - 2x = 8$ $-2x + 6y = 8$

4. $y = 9x + 12$ $9x - y = -12$

5. $y = \frac{2}{3}x + 6$ $2x - 3y = -18$

6. $-8y = \frac{1}{2} - 7x$ $14x - 16y = 1$

Write an equation in standard form of the line that passes through the point and has the given slope. (Lesson 5.6)

7. $(6, -8), m = -2$ $2x + y = 4$

8. $(4, 1), m = -\frac{1}{2}$ $x + 2y = 6$

9. $(1, 5), m = -3$ $3x + y = 8$

10. $(-3, -1), m = 2$ $2x - y = -5$

11. $(-12, 4), m = 3$ $3x - y = -40$

12. $(-16, -2), m = 6$ See margin.

12. $6x - y = -94$

LIFE EXPECTANCY In Exercises 13–15, use the table on projected life expectancy at birth. (Lesson 5.7)

Birth year	Life expectancy
1985	74.7
1990	75.4
1995	76.3
2000*	76.7
2005*	77.3
2010*	77.9

▶ Source: U.S. National Center for Health Statistics *Projected

13. Write a linear equation of the data.

13. Let x = the number of years since 1985; $y = 0.13x + 74.80$.

14. Interpolate the life expectancy for a baby born in 1989. **about 75.32 years**

15. Extrapolate the life expectancy for a baby born in 2015. **about 78.7 years**

CHAPTER 5

Chapter Summary

WHAT did you learn?	*WHY did you learn it?*
Write an equation of a line in slope-intercept form.	
• given its slope and its y-intercept (5.1)	Calculate the cost of a phone call. (p. 275)
• given its slope and one of its points (5.2)	Predict the number of vacations trips in 2005. (p. 281)
• given two of its points (5.3)	Estimate height at an archaeological dig. (p. 287)
Write equations of perpendicular lines. (5.3)	Write equations for sides of geometric figures. (p. 286)
Fit a line to data. (5.4)	Represent winning Olympic discus throws. (p. 294)
Write an equation of a line in point-slope form. (5.5)	Find the optimal running pace for a given temperature. (p. 302)
Write an equation of a line in standard form. (5.6)	Plan possible purchases. (p. 310)
Create a linear model for a real-life situation. (5.1–5.6)	Make predictions and estimates. (pp. 274 and 281)
Predict using a linear model. (5.7)	Estimate the amount spent in 2015 on advertising. (p. 318)

How does Chapter 5 fit into the BIGGER PICTURE of algebra?

In this chapter you studied techniques for writing equations of lines. Using the slope-intercept form that you saw in Chapter 4, you learned the point-slope form and standard form of an equation. You used many techniques you learned in earlier chapters: the problem-solving strategies in Chapter 1; the rules of simplification in Chapter 2; the equation solving in Chapter 3; and the graphing methods in Chapter 4.

Knowing which form of a linear equation is best to use is a basic problem solving strategy you will continue to apply in the following chapters.

STUDY STRATEGY

How did you use your practice test?

A practice test you may have written using the **Study Strategy** on page 272 may have begun like this one.

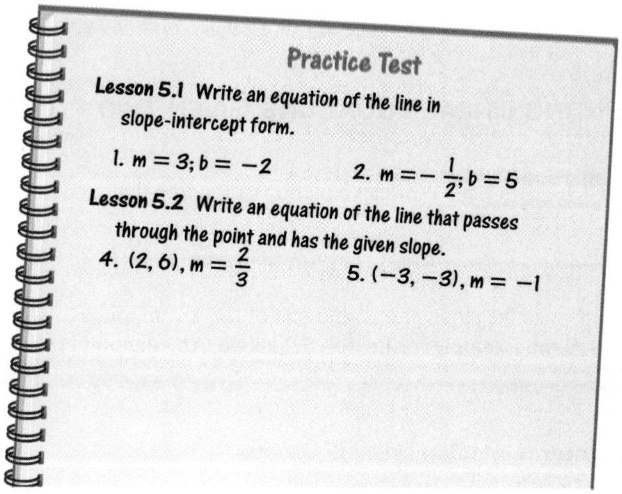
Practice Test

Lesson 5.1 Write an equation of the line in slope-intercept form.

1. $m = 3; b = -2$ 2. $m = -\frac{1}{2}; b = 5$

Lesson 5.2 Write an equation of the line that passes through the point and has the given slope.

4. $(2, 6), m = \frac{2}{3}$ 5. $(-3, -3), m = -1$

323

ADDITIONAL RESOURCES

The following resources are available to help review the material in this chapter.

- Chapter Review Games and Activities (*Chapter 5 Resource Book*, p. 106)
- *Instant Replay: Video Review Games*
- *Personal Student Tutor*
- Cumulative Review, Chs. 1–5 (*Chapter 5 Resource Book*, p. 118)

CHAPTER 5

Chapter Review

VOCABULARY

- slope-intercept form, p. 273
- best-fitting line, p. 292
- positive correlation, p. 295
- negative correlation, p. 295
- point-slope form, p. 300
- standard form, p. 308
- linear interpolation, p. 318
- linear extrapolation, p. 318

5.1 WRITING LINEAR EQUATIONS: SLOPE-INTERCEPT FORM

Examples on pp. 273–275

EXAMPLES Write an equation of the line with a slope of $\frac{1}{2}$ and a y-intercept of -2.

$y = mx + b$ — Write slope-intercept form.

$y = \frac{1}{2}x - 2$ — Substitute $\frac{1}{2}$ for m and -2 for b.

Write an equation of the line with the given slope and y-intercept.

1. $m = 2; b = -2$ $y = 2x - 2$ **2.** $m = -\frac{1}{2}; b = 5$ $y = -\frac{1}{2}x + 5$ **3.** $m = -8; b = -3$ $y = -8x - 3$

5.2 WRITING LINEAR EQUATIONS GIVEN THE SLOPE AND A POINT

Examples on pp. 279–281

EXAMPLES Write an equation of the line through $(2, -1)$ with a slope of 3.

$y = mx + b$ — Write slope-intercept form.

$-1 = 3(2) + b$ — Substitute -1 for y, 2 for x, and 3 for m.

$-7 = b$ — Simplify and solve for b.

$y = 3x - 7$ — Substitute 3 for m and -7 for b in slope-intercept form.

Write an equation of the line that passes through the point and has the given slope.

4. $(4, -3), m = 6$ $y = 6x - 27$ **5.** $(-9, 4), m = 2$ $y = 2x + 22$ **6.** $(-3, 2), m = -1$ $y = -x - 1$

5.3 WRITING LINEAR EQUATIONS GIVEN TWO POINTS

Examples on pp. 285–287

EXAMPLES Write an equation of the line that passes through $(-2, -6)$ and $(3, 4)$.

$$m = \frac{y_2 - y_1}{x_2 - x_1} = \frac{4 - (-6)}{3 - (-2)} = \frac{10}{5} = 2$$ Find the slope of the line.

Now use the slope $m = 2$ and one of the given points to write an equation of the line as in the Example for Lesson 5.2 above. An equation is $y = 2x - 2$.

EXAMPLES If two lines are perpendicular their slopes are negative reciprocals.

A line perpendicular to the line $y = -5x + 3$ has a slope of $\frac{1}{5}$.

Write the slope-intercept form of an equation of the line that passes through the points.

7. $(4, -9), (-3, 2)$ See margin. **8.** $(1, 8), (-2, -1)$ $y = 3x + 5$ **9.** $(2, 5), (-8, 2)$ $y = \frac{3}{10}x + \frac{22}{5}$

10. Write the slope-intercept form of an equation of the line perpendicular to the line in Exercise 7 with a y-intercept of -3. $y = \frac{7}{11}x - 3$

5.4 FITTING A LINE TO DATA

Examples on pp. 292–295

EXAMPLES On the scatter plot, sketch a line to approximate the data. Choose two points on your line, say $(11, 5)$ and $(6, -5)$. Find the slope.

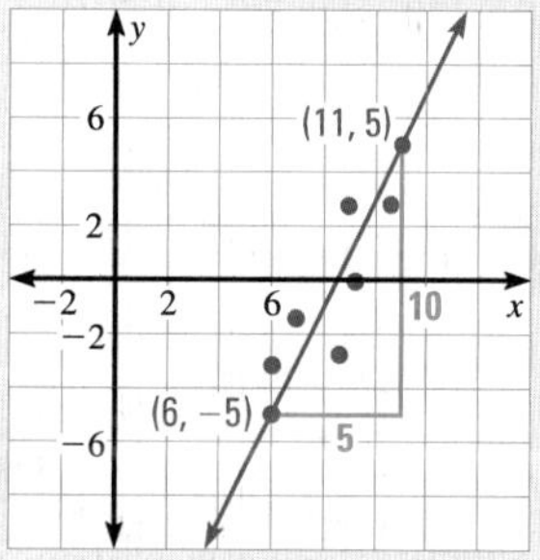

$$m = \frac{y_2 - y_1}{x_2 - x_1} = \frac{5 - (-5)}{11 - 6} = \frac{10}{5} = 2$$

$y = mx + b$ **Write slope-intercept form.**

$-5 = 2(6) + b$ **Substitute for y, x, and m.**

$-17 = b$ **Solve for b.**

An equation that approximates the best-fitting line is $y = 2x - 17$.

Draw a scatter plot of the data. Find an equation of the line that corresponds closely to the data and that best fits the data.

11, 12. Check graphs. Sample equations are given.

11.

x	−1	−3	5	9	12
y	−1	−4	9	20	25

$y = 1.98x + 1.11$

12.

x	2	1	−2	−4	−6
y	10	5	0	−5	−10

$y = 2.34x + 4.22$

5.5 POINT-SLOPE FORM OF A LINEAR EQUATION

Examples on pp. 300–302

EXAMPLES Write an equation of the line through $(5, -4)$ and $(-3, 4)$.

$$m = \frac{y_2 - y_1}{x_2 - x_1} = \frac{4 - (-4)}{-3 - 5} = \frac{8}{-8} = -1$$

$y - y_1 = m(x - x_1)$ **Write point-slope form.**

$y - (-4) = (-1)(x - 5)$ **Substitute for m, x_1, and y_1.**

$y = -x + 1$ **Simplify.**

Write an equation in point-slope form of the line that passes through the two points. Then rewrite the equation in slope-intercept form. 13–15. See margin.

13. $(-4, 4), (2, 5)$ **14.** $(-2, 3), (5, 0)$ **15.** $(1, -2), (-1, 8)$

7. $y = -\frac{11}{7}x - \frac{19}{7}$

13. $y - 4 = \frac{1}{6}(x + 4)$ or $y - 5 = \frac{1}{6}(x - 2)$; $y = \frac{1}{6}x + \frac{14}{3}$

14. $y - 3 = -\frac{3}{7}(x + 2)$ or $y = -\frac{3}{7}(x - 5)$; $y = -\frac{3}{7}x + \frac{15}{7}$

15. $y + 2 = -5(x - 1)$ or $y - 8 = -5(x + 1)$; $y = -5x + 3$

22. about 2 million; linear extrapolation
23. about 242 million; linear extrapolation
24. about 114 million; linear extrapolation
25. about 178 million; linear extrapolation

5.6 THE STANDARD FORM OF A LINEAR EQUATION

Examples on pp. 308–310

EXAMPLES The standard form of a linear equation is $Ax + By = C$.

Write the equation $y = -\frac{2}{3}x + 6$ in standard form with integer coefficients.

$y = -\frac{2}{3}x + 6$	**Write given form.**
$3y = -2x + 18$	**Multiply each side by 3.**
$2x + 3y = 18$	**Add 2*x* to each side.**

Rewrite the equation in standard form with integer coefficients. **16–21. Sample answers are given.**

16. $y = 2x + 9$ **$2x - y = -9$** **17.** $3y = -8x + 2$ **$8x + 3y = 2$** **18.** $2y = -2x + 6$ **$2x + 2y = 6$**

19. $y = -\frac{1}{3}x + \frac{2}{3}$ **$x + 3y = 2$** **20.** $y = \frac{3}{4}x + \frac{1}{2}$ **$3x - 4y = -2$** **21.** $\frac{1}{2}y = \frac{2}{3}x - 2$ **$4x - 3y = 12$**

5.7 PREDICTING WITH LINEAR MODELS

Examples on pp. 316–318

EXAMPLES You can use linear models to make predictions. Use the graph showing the number of World Wide Web users (in millions) for different years, with x representing the number of years since 1996. Predict the number of World Wide Web users in the year 2005.

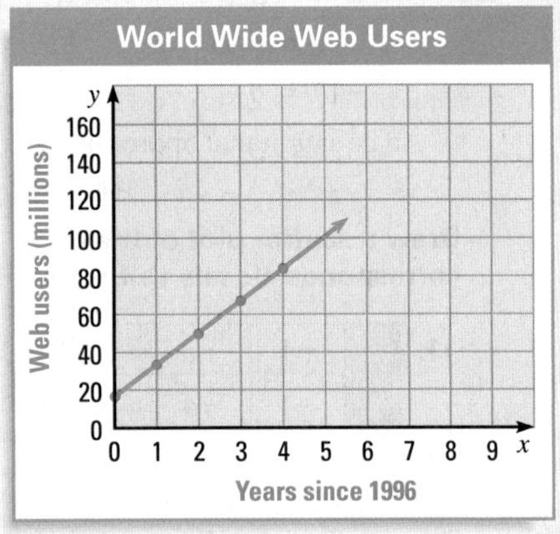

▶ Source: IDC Link

$m = \frac{66 - 34}{3 - 1} = 16$	**Find slope using (1, 34) and (3, 66).**
$34 = (16)1 + b$	**Substitute 34 for *y*, 1 for *x*, and 16 for *m* in *y* = *mx* + *b*.**
$b = 18$	**Simplify and solve for *b*.**
$y = 16x + 18$	**Write equation with *m* and *b*.**
$y = 16(9) + 18$	**Substitute 9 to represent 2005.**
$y = 162$	**Simplify and solve.**

Using the linear model $y = 16x + 18$, you can estimate the number of World Wide Web users in 2005 to be 162 million.

Use the linear model in the Example above to estimate the number of World Wide Web users in the given year. Tell whether you use a *linear interpolation* or *linear extrapolation*. **22–25. See margin.**

22. 1995 **23.** 2010 **24.** 2002 **25.** 2006

CHAPTER 5

Chapter Test

Write an equation of the line with the given slope and y-intercept. Write the equation in slope-intercept form.

1. $m = 2, b = -1$ $y = 2x - 1$ **2.** $m = -4, b = 3$ $y = -4x + 3$ **3.** $m = 6, b = 9$ $y = 6x + 9$

4. $m = \frac{1}{4}, b = -3$ $y = \frac{1}{4}x - 3$ **5.** $m = -3, b = 3$ $y = -3x + 3$ **6.** $m = 0, b = 4$ $y = 4$

Write an equation of the line that passes through the given point and has the given slope. Write the equation in slope-intercept form.

7. $(2, 6), m = 2$ $y = 2x + 2$ **8.** $(3, -9), m = -5$ $y = -5x + 6$ **9.** $(-5, -6), m = -3$ $y = -3x - 21$

10. $(1, 8), m = -4$ $y = -4x + 12$ **11.** $(4, -2), m = \frac{1}{2}$ $y = \frac{1}{2}x - 4$ **12.** $\left(\frac{1}{3}, -5\right), m = 8$ $y = 8x - \frac{23}{3}$

Graph the line that passes through the points. Then write an equation of the line in slope-intercept form. 13–18. Check sketches.

13. $(-3, 2), (4, -1)$ $y = -\frac{3}{7}x + \frac{5}{7}$ **14.** $(6, 2), (8, -4)$ $y = -3x + 20$ **15.** $(-2, 5), (2, 4)$ $y = -\frac{1}{4}x + \frac{9}{2}$

16. $(-2, -8), (-1, 0)$ $y = 8x + 8$ **17.** $(-5, 2), (2, 4)$ $y = \frac{2}{7}x + \frac{24}{7}$ **18.** $(9, -1), (1, -9)$ $y = x - 10$

19. Write an equation of a line that is perpendicular to $y = -2x + 6$ and passes through $(-4, 7)$. $y = \frac{1}{2}x + 9$

In Exercises 20–25, rewrite the equation in standard form with integer coefficients. 20–27. Sample answers are given.

20. $4y = 24 + 2x$ $2x - 4y = -24$ **21.** $y = 7x + 8$ $7x - y = -8$ **22.** $6y = -18x + 3$ $18x + 6y = 3$

23. $\frac{1}{2} - x = 9y$ $2x + 18y = 1$ **24.** $5y = 25x$ $25x - 5y = 0$ **25.** $-2y + \frac{1}{2}x = 4$ $x - 4y = 8$

26. Rewrite the equation $y = \frac{5}{13}x + 4$ in standard form with integer coefficients. $5x - 13y = -52$

27. NICKELS AND DIMES Maria needs \$2.20 to buy a magazine. The only money she has is a jar of nickels and dimes. Write an equation in standard form for the different amounts of nickels x and dimes y she could use. $5x + 10y = 220$

28. MONTHLY PAY A salesperson for an appliance store earns a monthly pay of \$1250 plus a 4% commission on the sales. Write an equation in slope-intercept form that gives the total monthly pay y in terms of sales x. $y = 0.04x + 1250$

CELLULAR PHONE INDUSTRY **In Exercises 29–33, use the following information. The table shows the number of employees in the cellular telephone industry in the United States from 1990 through 1995.**

▶ Source: Cellular Telecommunications Industry Association

Cellular Telephone Industry	
Years since 1990	Employees
0	21,400
1	26,300
2	34,300
3	39,800
4	?
5	68,200

29. Make a scatter plot and fit a line to the data. Check graphs.

30. Write an equation of the line in slope-intercept form. See margin.

31. Use the linear model to estimate the number of employees in 1994. Did you use *linear interpolation* or *linear extrapolation*? See margin.

32. Use the linear model to estimate the number of employees in 2004. Did you use *linear interpolation* or *linear extrapolation*? See margin.

33. Use the linear model to estimate the year in which the number of employees was 0. Did you use *linear interpolation* or *linear extrapolation*? Is this a realistic prediction? Why or why not? See margin.

ADDITIONAL RESOURCES

- ***Chapter 5 Resource Book***
  Chapter Test (3 levels) (p. 107)
  SAT/ACT Chapter Test (p. 113)
  Alternative Assessment (p. 114)
- ***Test and Practice Generator***

30. $y = 9277.03x + 17{,}590.54$ (sample answer)
31. about 54,699 employees; linear interpolation
32. about 147,469 employees; linear extrapolation
33. 1988; linear extrapolation; no; Cellular phones were available before 1988, so clearly there were employees at that time.

ADDITIONAL RESOURCES

- ***Chapter 5 Resource Book***
  Chapter Test (3 levels) (p. 107)
  SAT/ACT Chapter Test (p. 113)
  Alternative Assessment (p. 114)
- ***Test and Practice Generator***

CHAPTER 5

Chapter Standardized Test

TEST-TAKING STRATEGY Avoid spending too much time on one question. Skip questions that are too difficult for you, and spend no more than a few minutes on each question.

1. **MULTIPLE CHOICE** What is an equation of the line that passes through the points $(-4, 2)$ and $(6, 6)$? **A**

   (A) $y = \frac{2}{5}x + \frac{18}{5}$ (B) $y = \frac{2}{5}x - \frac{12}{5}$

   (C) $y = 2x - 6$ (D) $y = \frac{2}{5}x - \frac{18}{5}$

   (E) $y = 2x + 18$

2. **MULTIPLE CHOICE** An equation of the line perpendicular to the line $y = -2x - 3$ with a y-intercept of $-\frac{3}{4}$ is $\underline{?}$. **D**

   (A) $y = -2x + \frac{3}{4}$ (B) $y = -2x - \frac{3}{4}$

   (C) $y = 2x - \frac{3}{4}$ (D) $y = \frac{1}{2}x - \frac{3}{4}$

   (E) $y = -\frac{1}{2}x + \frac{3}{4}$

3. **MULTIPLE CHOICE** A line with a slope of -1 passes through the point $(2, -1)$. If $(-4, p)$ is another point on the line, what is the value of p? **E**

   (A) -5 (B) -1 (C) 1

   (D) 2 (E) 5

4. **MULTIPLE CHOICE** What is an equation of a line that best fits the scatter plot? **A**

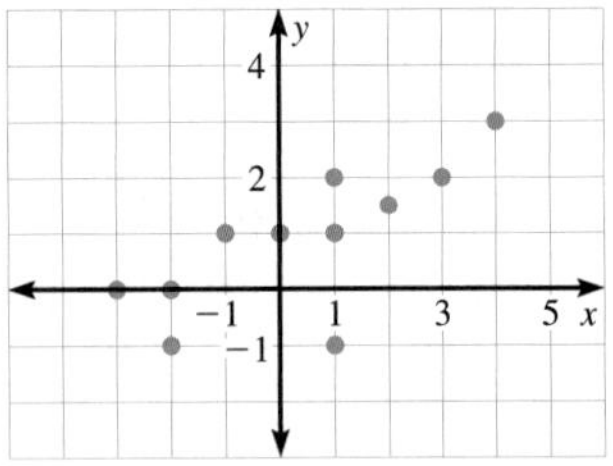

   (A) $y = \frac{1}{2}x + 1$ (B) $y = 1$

   (C) $y = -\frac{1}{2}x + 1$ (D) $y = x + 1$

   (E) $y = x - 1$

5. **MULTIPLE CHOICE** A bike rental shop charges \$8 to rent a bike, plus \$1.50 for every half hour you ride. If the shop charges you and your friend a total of \$25, how many hours did you each ride? (Assume that you each rode a separate bike for an equal amount of time.) **B**

   (A) 1 (B) 1.5

   (C) 2 (D) 2.5

   (E) 3

6. **MULTIPLE CHOICE** What is an equation of the line that passes through the point $(4, -5)$ and has a slope of $\frac{1}{2}$? **E**

   (A) $y = x - 5$ (B) $y = -\frac{1}{2}x + 7$

   (C) $y = \frac{1}{2}x + 7$ (D) $y = -\frac{1}{2}x - 7$

   (E) $y = \frac{1}{2}x - 7$

7. **MULTIPLE CHOICE** An equation of the line whose x-intercept is 3 and whose y-intercept is 5 is $\underline{?}$. **C**

   (A) $y = \frac{5}{3}x + 5$ (B) $y = -\frac{3}{5}x + 5$

   (C) $y = -\frac{5}{3}x + 5$ (D) $y = \frac{3}{5}x + 5$

   (E) $y = -\frac{5}{3}x - 5$

8. **MULTIPLE CHOICE** Which two points lie on the line $y = -2x + 7$? **D**

   (A) $(0, 7), (1, -5)$ (B) $(4, 0), (-2, -8)$

   (C) $(-3, -4), (2, 6)$ (D) $(-1, 9), (3, 1)$

   (E) $(2, -1), (-3, -11)$

9. **MULTIPLE CHOICE** Which equation is in standard form with integer coefficients? **E**

   (A) $x - \frac{1}{2}y = \frac{5}{2}$ (B) $y = 2x + -5$

   (C) $y = -5 + 2x$ (D) $x = \frac{1}{2}y + \frac{5}{2}$

   (E) $-2x + y = -5$

10. **MULTIPLE CHOICE** An equation in standard form of the line that passes through the point $(-6, 1)$ and has a slope of -2 is ___?___. **C**

Ⓐ $2x + y = 13$ Ⓑ $2x - y = 11$

Ⓒ $2x + y = -11$ Ⓓ $2x + y = -13$

Ⓔ $2x - y = -13$

11. **MULTIPLE CHOICE** An equation in standard form of a line that is perpendicular to the line that passes through the points $(3, -4)$ and $(6, 1)$ is ___?___. **D**

Ⓐ $5x + 3y - 6 = 0$ Ⓑ $x + 3y - 6 = 0$

Ⓒ $y = \frac{-3}{5}x + 6$ Ⓓ $3x + 5y = 6$

Ⓔ $5y = -3x + 6$

QUANTITATIVE COMPARISON In Exercises 12–14, choose the statement that is true.

Ⓐ The number in Column A is greater.

Ⓑ The number in Column B is greater.

Ⓒ The two numbers are equal.

Ⓓ The relationship cannot be determined from the information given.

	Column A	Column B	
12.	slope of $2x + 3y = 12$	slope of $-5y = 6 + 10x$	A
13.	y-intercept of $2x + 3y = 12$	y-intercept of $-5y = 6 + 10x$	A
14.	x-intercept of $\frac{2}{3}x + 6y = 8$	x-intercept of $3y = -7 + 2x$	A

MULTI-STEP PROBLEM In Exercises 15–20, all students in a class were surveyed after they took a chapter test. The teacher wanted to know if studying at home produced good test grades. After the survey was taken, the following data were recorded in a table.

Hours spent studying the chapter	0	.25	.5	.75	1	1.5	2	3	5	7
Average grade on the chapter test	29	32	35	38	40	47	54	66	79	89

15. Make a scatter plot of the data. **See margin.**

16. Make a linear model of the average grade on the chapter test based on the number of hours spent studying the chapter at home. **See margin.**

17. If you study for four hours, approximately what grade can you expect to earn on the chapter test according to the model? **about 68**

18. Use linear extrapolation to estimate the number of hours needed to earn a grade of 93 on the chapter test. **about 7 h**

19. Use linear interpolation to find the average grade on the chapter test by students who study for 4.5 hours. **about 72**

20. *Writing* If you were the parent of a child studying this chapter, what advice would you give your child about studying at home for the chapter test? Explain your reasoning.

20. ***Sample answer:*** **The survey shows that this material must be pretty difficult. We could discuss it and that might give us both some idea of how well you have learned it. Then, if you are having problems, you might have an opportunity to get extra help before you try studying alone.**

15.

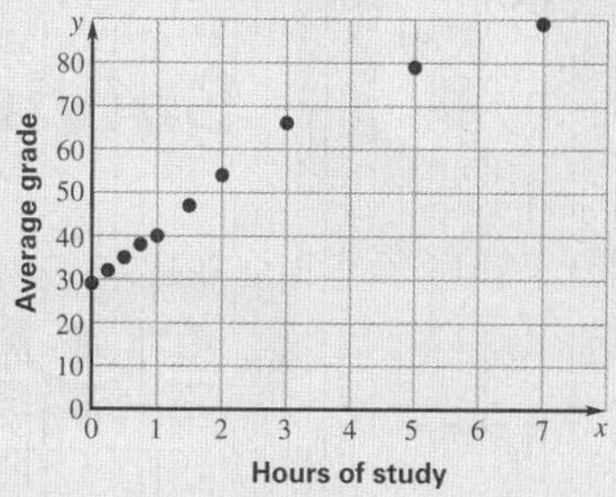

16. $y = 8.92x + 32.16$ **(sample answer obtained using a graphing calculator and rounding to nearest hundredth)**

CHAPTER 6

PLANNING THE CHAPTER

Solving and Graphing Linear Inequalities

LESSON	GOALS	NCTM	ITED	SAT9	Terra-Nova	Local
6.1 *pp. 333–339*	CONCEPT ACTIVITY: 6.1 *Investigate inequalities.* GOAL 1 Graph linear inequalities in one variable. GOAL 2 Solve one-step linear inequalities.	1, 2, 3, 10	IIG, MIG	6	14, 16, 52	
6.2 *pp. 340–345*	GOAL 1 Solve multi-step linear inequalities. GOAL 2 Use linear inequalities to model and solve real-life problems.	1, 2, 6, 8, 9, 10	RQWN, RQF	3, 6	16, 17, 18, 52	
6.3 *pp. 346–352*	GOAL 1 Write, solve, and graph compound inequalities. GOAL 2 Model a real-life situation with a compound inequality.	1, 2, 3, 6, 8, 9, 10	RQWN, RQF	3, 6	14, 16, 18, 52	
6.4 *pp. 353–359*	GOAL 1 Solve absolute-value equations. GOAL 2 Solve absolute-value inequalities. TECHNOLOGY ACTIVITY: 6.4 *Solve absolute-value equations on a graphing calculator.*	1, 2	IIG, MIG	6	14, 16, 52	
6.5 *pp. 360–367*	GOAL 1 Graph a linear inequality in two variables. GOAL 2 Model a real-life situation using a linear inequality in two variables. TECHNOLOGY ACTIVITY: 6.5 *Graph inequalities on a graphing calculator.*	1, 2, 3, 6, 8, 9, 10	IIG, MIG	3	14, 16	
6.6 *pp. 368–374*	GOAL 1 Make and use a stem-and-leaf plot to put data in order. GOAL 2 Find the mean, median, and mode of data.	1, 5, 10	IIG, MIG	12	15	
6.7 *pp. 375–382*	GOAL 1 Draw a box-and-whisker plot to organize real-life data. GOAL 2 Read and interpret a box-and-whisker plot of real-life data. TECHNOLOGY ACTIVITY: 6.7 *Draw a box-and-whisker plot on a graphing calculator.*	1, 5, 9, 10	IIG, MIG		15	

RESOURCES

Chapter Resource Booklets

Chapter Support

Tips for New Teachers	p. 1	Prerequisite Skills Review	p. 5
Parent Guide for Student Success	p. 3	Strategies for Reading Mathematics	p. 7

Lesson Support

Lesson Support	6.1	6.2	6.3	6.4	6.5	6.6	6.7
Lesson Plans (regular and block)	p. 9	p. 21	p. 33	p. 49	p. 62	p. 76	p. 91
Warm-Up Exercises and Daily Quiz	p. 11	p. 23	p. 35	p. 51	p. 64	p. 78	p. 93
Activity Support Masters							
Lesson Openers	p. 12	p. 24	p. 36	p. 52	p. 65	p. 79	p. 94
Graphing Calculator Activities & Keystrokes			p. 37	p. 53	p. 66	p. 80	p. 95
Practice (3 levels)	p. 13	p. 25	p. 39	p. 54	p. 67	p. 82	p. 96
Reteaching with Practice	p. 16	p. 28	p. 42	p. 57	p. 70	p. 85	p. 99
Quick Catch-Up for Absent Students	p. 18	p. 30	p. 44	p. 59	p. 72	p. 87	p. 101
Cooperative Learning Activities						p. 88	
Interdisciplinary Applications		p. 31			p. 73	p. 89	
Real-Life Applications	p. 19		p. 45	p. 60			p. 102
Math & History Applications			p. 46				
Challenge: Skills and Applications	p. 20	p. 32	p. 47	p. 61	p. 74	p. 90	p. 103

Review and Assessment

Quizzes	pp. 48, 75	Alternative Assessment with Math Journal	p. 112
Chapter Review Games and Activities	p. 104	Project with Rubric	p. 114
Chapter Test (3 levels)	pp. 105–110	Cumulative Review	p. 116
SAT/ACT Chapter Test	p. 111	Resource Book Answers	p. A1

Transparencies

	6.1	6.2	6.3	6.4	6.5	6.6	6.7
Warm-Up Exercises and Daily Quiz	p. 45	p. 46	p. 47	p. 48	p. 49	p. 50	p. 51
Alternative Lesson Opener Transparencies	p. 39	p. 40	p. 41	p. 42	p. 43	p. 44	p. 45
Examples/Standardized Test Practice	✓	✓	✓	✓	✓	✓	✓
Answer Transparencies	✓	✓	✓	✓	✓	✓	✓

Technology

- EasyPlanner CD
- EasyPlanner Plus Online
- Classzone.com
- Personal Student Tutor CD
- Test and Practice Generator CD
- Instant Replay: Video Review Games
- Electronic Lesson Presentations
- Chapter Audio Summaries CD

Additional Resources

- Basic Skills Workbook
- Explorations and Projects
- Worked-Out Solution Key
- Data Analysis Sourcebook
- Spanish Resources
- Standardized Test Practice Workbook
- Practice Workbook with Examples
- Notetaking Guide

CHAPTER 6

PACING THE CHAPTER

Resource Key
- STUDENT EDITION
- CHAPTER 6 RESOURCE BOOK
- TEACHER'S EDITION

REGULAR SCHEDULE

Day 1

6.1

STARTING OPTIONS
- Prereq. Skills Review
- Strategies for Reading
- Warm-Up or Daily Quiz

TEACHING OPTIONS
- Motivating the Lesson
- Les. Opener (Application)
- Examples 1–4
- Closure Question
- Guided Practice Exs.

APPLY/HOMEWORK
- See Assignment Guide.
- See the CRB: Practice, Reteach, Apply, Extend

ASSESSMENT OPTIONS
- Checkpoint Exercises
- Daily Quiz (6.1)
- Stand. Test Practice

Day 2

6.2

STARTING OPTIONS
- Homework Check
- Warm-Up or Daily Quiz

TEACHING OPTIONS
- Motivating the Lesson
- Les. Opener (Visual)
- Examples 1–4
- Closure Question
- Guided Practice Exs.

APPLY/HOMEWORK
- See Assignment Guide.
- See the CRB: Practice, Reteach, Apply, Extend

ASSESSMENT OPTIONS
- Checkpoint Exercises
- Daily Quiz (6.2)
- Stand. Test Practice

Day 3

6.3

STARTING OPTIONS
- Homework Check
- Warm-Up or Daily Quiz

TEACHING OPTIONS
- Motivating the Lesson
- Les. Opener (Activity)
- Graphing Calc. Activity
- Examples 1–6
- Closure Question
- Guided Practice Exs.

APPLY/HOMEWORK
- See Assignment Guide.
- See the CRB: Practice, Reteach, Apply, Extend

ASSESSMENT OPTIONS
- Checkpoint Exercises
- Daily Quiz (6.3)
- Stand. Test Practice
- Quiz (6.1–6.3)

Day 4

6.4

STARTING OPTIONS
- Homework Check
- Warm-Up or Daily Quiz

TEACHING OPTIONS
- Motivating the Lesson
- Les. Opener (Calculator)
- Graphing Calc. Activity
- Examples 1–2
- Guided Practice Exs. 1–3

APPLY/HOMEWORK
- See Assignment Guide.
- See the CRB: Practice, Reteach, Apply, Extend

ASSESSMENT OPTIONS
- Checkpoint Exercises, p. 354

Day 5

6.4 *(cont.)*

STARTING OPTIONS
- Homework Check

TEACHING OPTIONS
- Examples 3–5
- Closure Question
- Guided Practice Exs. 4–18

APPLY/HOMEWORK
- See Assignment Guide.
- See the CRB: Practice, Reteach, Apply, Extend

ASSESSMENT OPTIONS
- Checkpoint Exercises, pp. 354–355
- Daily Quiz (6.4)
- Stand. Test Practice

Day 6

6.5

STARTING OPTIONS
- Homework Check
- Warm-Up or Daily Quiz

TEACHING OPTIONS
- Les. Opener (Activity)
- Graphing Calc. Activity
- Examples 1–4
- Guided Practice Exs. 1–12

APPLY/HOMEWORK
- See Assignment Guide.
- See the CRB: Practice, Reteach, Apply, Extend

ASSESSMENT OPTIONS
- Checkpoint Exercises, p. 361

Day 9

6.6 *(cont.)*

STARTING OPTIONS
- Homework Check

TEACHING OPTIONS
- Examples 3–4
- Closure Question
- Guided Practice Ex. 10

APPLY/HOMEWORK
- See Assignment Guide.
- See the CRB: Practice, Reteach, Apply, Extend

ASSESSMENT OPTIONS
- Checkpoint Exercises, p. 370
- Daily Quiz (6.6)
- Stand. Test Practice

Day 10

6.7

STARTING OPTIONS
- Homework Check
- Warm-Up or Daily Quiz

TEACHING OPTIONS
- Motivating the Lesson
- Les. Opener (Activity)
- Graphing Calc. Activity
- Examples 1–2
- Guided Practice Exs. 1–8

APPLY/HOMEWORK
- See Assignment Guide.
- See the CRB: Practice, Reteach, Apply, Extend

ASSESSMENT OPTIONS
- Checkpoint Exercises, p. 376

Day 11

6.7 *(cont.)*

STARTING OPTIONS
- Homework Check

TEACHING OPTIONS
- Example 3
- Closure Question
- Guided Practice Exs. 9–10

APPLY/HOMEWORK
- See Assignment Guide.
- See the CRB: Practice, Reteach, Apply, Extend

ASSESSMENT OPTIONS
- Checkpoint Exercises, p. 377
- Daily Quiz (6.7)
- Stand. Test Practice
- Quiz (6.6–6.7)

Day 12

Review

DAY 12 START OPTIONS
- Homework Check

REVIEWING OPTIONS
- Chapter 6 Summary
- Chapter 6 Review
- Chapter Review Games and Activities

APPLY/HOMEWORK
- Chapter 6 Test (practice)
- Ch. Standardized Test (practice)

Day 13

Assess

DAY 13 START OPTIONS
- Homework Check

ASSESSMENT OPTIONS
- Chapter 6 Test
- SAT/ACT Ch. 6 Test
- Alternative Assessment

APPLY/HOMEWORK
- Skill Review, p. 396

Day 7

6.5 *(cont.)*

Starting Options
- Homework Check

Teaching Options
- Example 5
- Closure Question
- Guided Practice Exs. 13–14

Apply/Homework
- See Assignment Guide.
- See the CRB: Practice, Reteach, Apply, Extend

Assessment Options
- Checkpoint Exercises, p. 362
- Daily Quiz (6.5)
- Stand. Test Practice
- Quiz (6.4–6.5)

Day 8

6.6

Starting Options
- Homework Check
- Warm-Up or Daily Quiz

Teaching Options
- Motivating the Lesson
- Les. Opener (Application)
- Graphing Calc. Activity
- Examples 1–2
- Guided Practice Exs. 1–9

Apply/Homework
- See Assignment Guide.
- See the CRB: Practice, Reteach, Apply, Extend

Assessment Options
- Checkpoint Exercises, p. 369

BLOCK SCHEDULE

Day 1

Assess & 6.1

(Day 1 = Ch. 5 Day 7)

Assessment Options
- Chapter 5 Test
- SAT/ACT Ch. 5 Test
- Alternative Assessment

Ch. 6 Start Options
- Skill Review, p. 332
- Prereq. Skills Review
- Strategies for Reading

Teaching Options
- Warm-Up (Les. 6.1)
- Motivating the Lesson
- Les. Opener (Appl.)
- Examples 1–4
- Closure Question
- Guided Practice Exs.

Apply/Homework
- See Assignment Guide.
- See the CRB: Practice, Reteach, Apply, Extend

Assessment Options
- Checkpoint Exercises
- Daily Quiz (6.1)
- Stand. Test Practice

Day 2

6.2 & 6.3

Day 2 Start Options
- Homework Check
- W-Up 6.2 or D. Quiz 6.1

Teaching 6.2 Options
- Motivating the Lesson
- Les. Opener (Visual)
- Examples 1–4
- Closure Question
- Guided Practice Exs.

Teaching 6.3 Options
- Warm-Up (Les. 6.3)
- Motivating the Lesson
- Les. Opener (Activity)
- Graphing Calc. Activity
- Examples 1–6
- Closure Question
- Guided Practice Exs.

Apply/Homework
- See Assignment Guide.
- See the CRB: Practice, Reteach, Apply, Extend

Assessment Options
- Checkpoint Exercises
- Daily Quiz (Les. 6.2, 6.3)
- Stand. Test Practice
- Quiz (6.1–6.3)

Day 3

6.4

Day 3 Start Options
- Homework Check
- Warm-Up or Daily Quiz

Teaching 6.4 Options
- Motivating the Lesson
- Les. Opener (Calc.)
- Graphing Calc. Activity
- Examples 1–5
- Closure Question
- Guided Practice Exs.

Apply/Homework
- See Assignment Guide.
- See the CRB: Practice, Reteach, Apply, Extend

Assessment Options
- Checkpoint Exercises
- Daily Quiz (Les. 6.4)
- Stand. Test Practice

Day 4

6.5

Day 4 Start Options
- Homework Check
- Warm-Up or Daily Quiz

Teaching 6.5 Options
- Les. Opener (Activity)
- Graphing Calc. Activity
- Examples 1–5
- Closure Question
- Guided Practice Exs.

Apply/Homework
- See Assignment Guide.
- See the CRB: Practice, Reteach, Apply, Extend

Assessment Options
- Checkpoint Exercises
- Daily Quiz (Les. 6.5)
- Stand. Test Practice
- Quiz (6.4–6.5)

Day 5

6.6

Day 5 Start Options
- Homework Check
- Warm-Up or Daily Quiz

Teaching 6.6 Options
- Motivating the Lesson
- Les. Opener (Appl.)
- Graphing Calc. Activity
- Examples 1–4
- Closure Question
- Guided Practice Exs.

Apply/Homework
- See Assignment Guide.
- See the CRB: Practice, Reteach, Apply, Extend

Assessment Options
- Checkpoint Exercises
- Daily Quiz (Les. 6.6)
- Stand. Test Practice

Day 6

6.7

Day 6 Start Options
- Homework Check
- Warm-Up or Daily Quiz

Teaching 6.7 Options
- Motivating the Lesson
- Les. Opener (Activity)
- Graphing Calc. Activity
- Examples 1–3
- Closure Question
- Guided Practice Exs.

Apply/Homework
- See Assignment Guide.
- See the CRB: Practice, Reteach, Apply, Extend

Assessment Options
- Checkpoint Exercises
- Daily Quiz (Les. 6.7)
- Stand. Test Practice
- Quiz (6.6–6.7)

Day 7

Review/Assess

Day 7 Start Options
- Homework Check

Reviewing Options
- Chapter 6 Summary
- Chapter 6 Review
- Chapter Review Games and Activities
- Chapter 6 Test (practice)
- Ch. Standardized Test (practice)

Assessment Options
- Chapter 6 Test
- SAT/ACT Ch. 6 Test
- Alternative Assessment

Apply/Homework
- Skill Review, p. 396

CHAPTER 6

MEETING INDIVIDUAL NEEDS

BEFORE THE CHAPTER

The *Chapter 6 Resource Book* has the following materials to distribute and use before the chapter:

- **Parent Guide for Student Success (pictured below)**
- **Prerequisite Skills Review**
- **Strategies for Reading Mathematics**

PARENT GUIDE *Pages 3–4*

CHAPTER 6 NAME ______________________ DATE __________

Parent Guide for Student Success

For use with Chapter 6: Solving and Graphing Linear Inequalities

Chapter Support

Chapter Overview One way that you can help your student succeed in Chapter 6 is by discussing the lesson goals in the chart below. When a lesson is completed, ask your student to interpret the lesson goals for you and to explain how the mathematics of the lesson relates to one of the key applications listed in the chart.

Lesson Title	*Lesson Goals*	*Key Applications*
6.1: Solving One-Step Linear Inequalities	Graph linear inequalities in one variable. Solve one-step linear inequalities.	• Running a Race • Steel Arch Bridge • Musical Instruments
6.2: Solving Multi-Step Linear Inequalities	Solve multi-step linear inequalities. Use linear inequalities to model and solve real-life problems.	• Population • Fly Fishing Business • Animated Films
6.3: Solving Compound Inequalities	Write, solve, and graph compound inequalities. Model a real-life situation with a compound inequality.	• Elevations • Prices of Fine Art • Solar System
6.4: Solving Absolute-Value Equations and Inequalities	Solve absolute-value equations and inequalities.	• Quality Control • Boxing • Fireworks
6.5: Graphing Linear Inequalities in Two Variables	Graph a linear inequality in two variables. Model a real-life situation using a linear inequality in two variables.	• Treasure Diving • Football • Nutrition
6.6: Stem-and-Leaf Plots and Mean, Median, and Mode	Make and use a stem-and-leaf plot to put data in order. Find the mean, median, and mode of data.	• Age Distribution • Manufacturing • Number of Households
6.7: Box-and-Whisker Plots	Draw a box-and-whisker plot to organize real-life data. Read and interpret a box-and-whisker plot of real-life data.	• Meteorology • Cellular Telephones • Skyscrapers

Study Strategy

Showing Your Work is the study strategy featured in Chapter 6 (see page 332). Encourage your student to record all steps on assignments since this helps to prevent errors and makes it easier to find and correct mistakes that do occur. When an answer is incorrect, you may be able to help your student in identifying the mistake and deciding how to correct the steps.

PARENT GUIDE FOR STUDENT SUCCESS The first page summarizes the content of Chapter 6. Parents are encouraged to have their students explain how the material relates to key applications in the chapter, such as nutrition. The second page (not shown) provides exercises and an activity that parents can do with their students. In the activity, parents and students use the results of coin tosses to make a box-and-whisker plot.

DURING EACH LESSON

The *Chapter 6 Resource Book* has the following alternatives for introducing the lesson:

- **Lesson Openers (pictured below)**
- **Graphing Calculator Activities with Keystrokes**

LESSON OPENER *Page 52*

LESSON 6.4 NAME ______________________ DATE __________ Available as a transparency

Graphing Calculator Lesson Opener

For use with pages 353–358

In Questions 1–4, graph each equation using a standard viewing window.

1. Enter the equation $y = |x - 1|$ into your calculator.

a. Estimate the x-intercept of the graph.

b. Use the *Table* feature to check your estimate.

c. Substitute the x-intercept into the equation $|x - 1| = 0$ What do you notice?

2. Enter the equation $y = |x| - 1$ into your calculator.

a. Estimate the x-intercepts of the graph.

b. Use the *Table* feature to check your estimate.

c. Substitute the x-intercepts into the equation $|x| - 1 = 0$. What do you notice?

3. Enter the equation $y = |2x|$ into your calculator.

a. Estimate the x-intercept of the graph.

b. Use the *Table* feature to check your estimate.

c. Substitute the x-intercept into the equation $y = |2x| = 0$. What do you notice?

4. Enter the equation $y = |2x| - 2$ into your calculator.

a. Estimate the x-intercepts of the graph.

b. Use the *Table* feature to check your estimate.

c. Substitute the x-intercepts into the equation $|2x| - 2 = 0$. What do you notice?

Lesson 6.4

GRAPHING CALCULATOR LESSON OPENER This Lesson Opener provides an alternative way to start Lesson 6.4 through the use of a graphing calculator. Students graph equations containing the absolute value function to develop an understanding of finding the x-intercept.

TECHNOLOGY RESOURCE

Teachers can use the Electronic Lesson Presentations CD-ROM to present Lesson 6.4 using computer animation to step through the Examples.

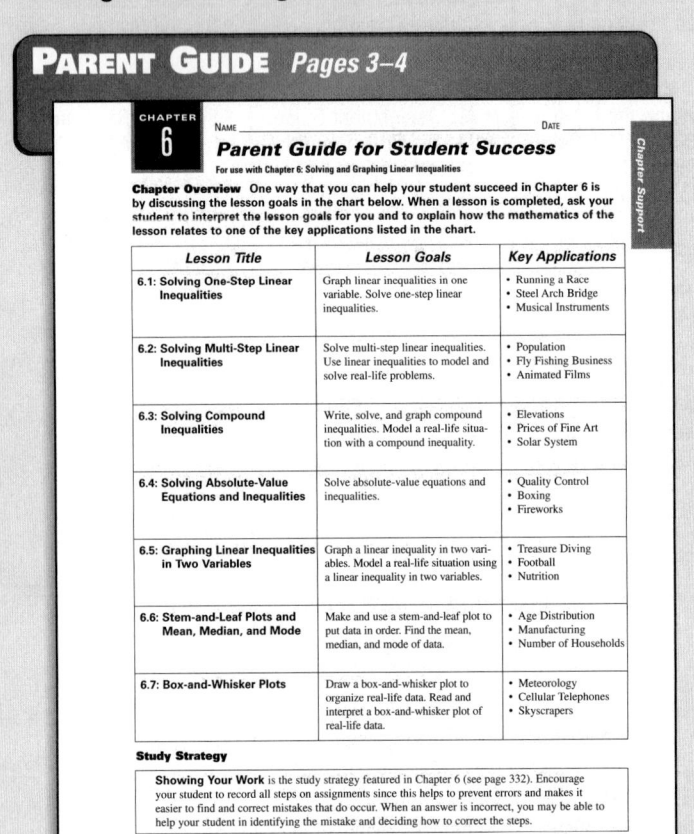

FOLLOWING EACH LESSON

The *Chapter 6 Resource Book* has a variety of materials to follow-up each lesson. They include the following:

- **Practice (3 levels)**
- **Reteaching with Practice**
- **Quick Catch-Up for Absent Students**
- **Interdisciplinary Applications**
- **Real-Life Applications (pictured below)**

APPLICATION *Page 45*

LESSON 6.3

NAME ______________________ DATE __________

Real-Life Application

For use with pages 346–352

The Value and Cost of Education

Your future income will be greatly influenced by the level of education you attain. According to 1994 income statistics published by the U.S. Bureau of the Census, the average annual earnings by level of education were as follows.

Level of Education	*Average Annual Earnings*
Did not finish high school	\$12,809
High school diploma	\$18,737
Associate's degree	\$24,398
Bachelor's degree	\$32,629
Master's degree	\$40,368
Professional	\$74,560

In Exercises 1–3, write a compound inequality that describes the situation.

1. An individual who has a Master's degree is earning more than average but less than a professional.
2. A high school graduate is earning at least the national average but less than someone with a Bachelor's degree.
3. A person with an Associate's degree is earning less than average but more than a high school graduate.

College Costs

In Exercises 4–5, use the following information.

If continuing your education is one of your goals, planning ahead and saving money is important. It is never too late to start saving. Education costs continue to increase each year and expenses range widely from one institution to another. When planning for education expenses, items such as books, room and board, transportation, and incidentals must also be taken into consideration. Four-year institutions range from \$6,000 per year to \$30,000 per year, while two-year business and trade schools range from \$5,000 per year to \$15,000 per year.

4. Write compound inequalities that describe each type of school.
5. Graph the inequalities from Exercise 4.

Lesson 6.3

REAL-LIFE APPLICATION This application examines the relationship between level of education and annual income and relates this to writing and graphing inequalities, the topic of Lesson 6.3.

ASSESSING THE CHAPTER

The *Chapter 6 Resource Book* has the following review and assessment materials:

- **Quizzes (pictured below)**
- **Chapter Review Games and Activities**
- **Chapter Test (3 levels)**
- **SAT/ACT Chapter Test**
- **Alternative Assessment with Rubric and Math Journal**
- **Project with Rubric**
- **Cumulative Review**

QUIZ *Page 48*

CHAPTER 6

NAME ______________________ DATE __________

Quiz 1

For use after Lessons 6.1–6.3

1. Solve $-11 + x \geq -3$ and graph the solution. (Lesson 6.1)
2. Solve $-\frac{x}{4} < 7$and graph the solution. (Lesson 6.1)
3. The highest waterfall in the world is Angel Falls in Venezuela at 1000 meters. Write an inequality that describes the height h (in meters) of every other waterfall. Graph the inequality. (Lesson 6.1)
4. Solve $-4x - 3 > -5x + 8$.(Lesson 6.2)
5. Write an inequality for the value of x. (Lesson 6.2)
   Area $\leq$ 48 square centimeters

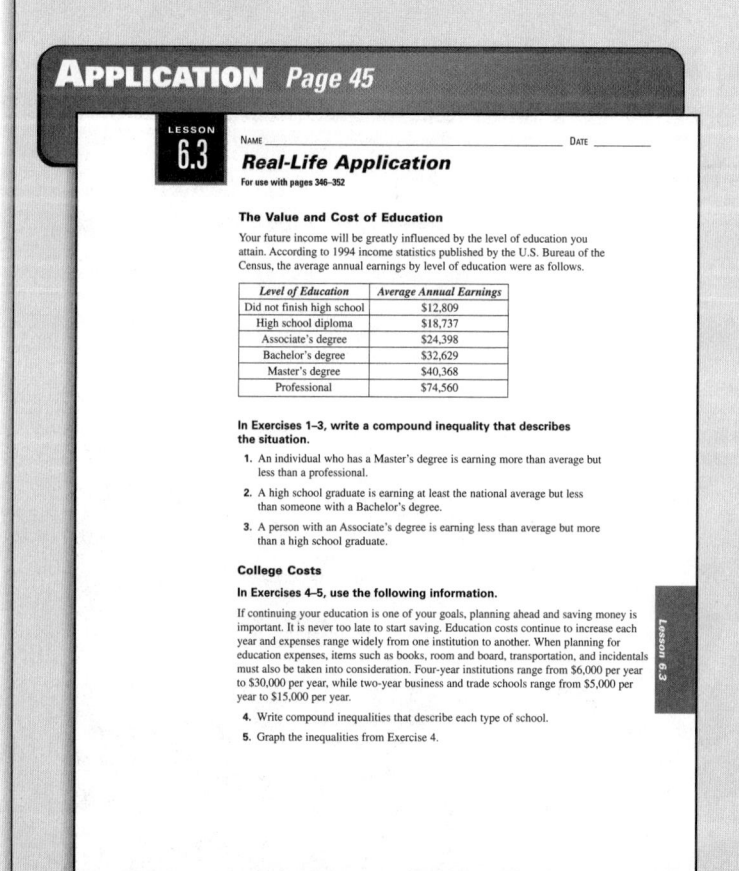

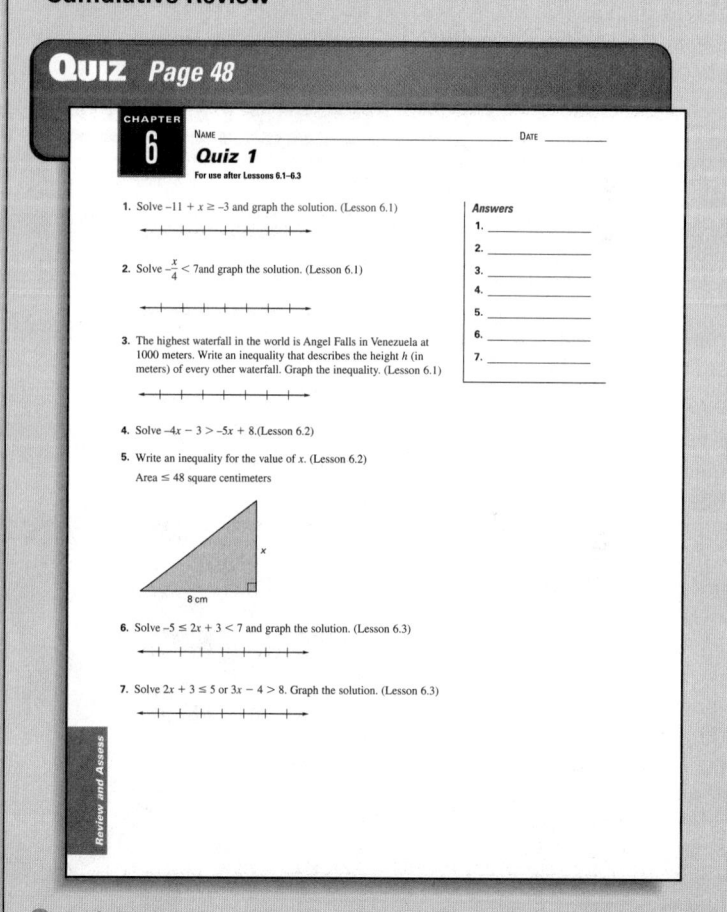

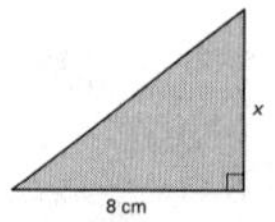

6. Solve $-5 \leq 2x + 3 < 7$ and graph the solution. (Lesson 6.3)
7. Solve $2x + 3 \leq 5$ or $3x - 4 > 8$. Graph the solution. (Lesson 6.3)

Answers

1. ________
2. ________
3. ________
4. ________
5. ________
6. ________
7. ________

Review and Assess

QUIZ Lessons 6.1–6.3 are covered on this quiz, which is an alternate form of the Quiz on page 352 of the textbook. You might want to assign the quiz in the textbook for homework and then use this version for assessment.

TECHNOLOGY RESOURCE

Teachers can use the Time-Saving Test and Practice Generator to create a customized quiz for Lessons 6.1 through 6.3, or for any other combination of lessons in Chapter 6.

CHAPTER OVERVIEW

CHAPTER GOALS

Students will write, solve, and graph linear inequalities in one variable, including compound inequalities. They will solve absolute value equations and inequalities, and will graph linear inequalities in two variables. Along the way, students will use their new skills to model real-life situations. Students will complete the chapter by organizing data in stem-and-leaf and box-and-whisker plots. They will learn to interpret these plots and to determine the mean, median, and mode of a data set.

APPLICATION NOTE

Sound travels in repeating waves. The distance between successive crests or troughs is the *wavelength*, λ. The number of cycles of the wave that are completed in one second is the wave's *frequency, f.* The product of the wavelength and the frequency gives the speed of the wave, V. So, $f\lambda = V$, or $f = \frac{V}{\lambda}$.

When you play an instrument, the sound speed is constant. So, the higher the frequency of sound that you make, the shorter its wavelength.

Flutes are the oldest members of the family of instruments called *aerophones,* which produce sound by causing air to vibrate. The flute in the photograph is called a *transverse* flute because it is held horizontally. The flutist (*flautist*) opens or closes the holes in the flute wall to create different pitches.

Additional information about music is available at **www.mcdougallittell.com.**

SOLVING AND GRAPHING LINEAR INEQUALITIES

▶ *How are sounds produced?*

CHAPTER
6

APPLICATION: Music

Musical instruments produce vibrations in the air that we hear as music. Not all instruments produce these vibrations in the same way, and different instruments produce sounds in different frequencies.

A flute player blows across an opening in the flute, which causes the air inside to vibrate. A clarinet or saxophone player blows on a wooden reed whose vibrations cause the air inside to vibrate.

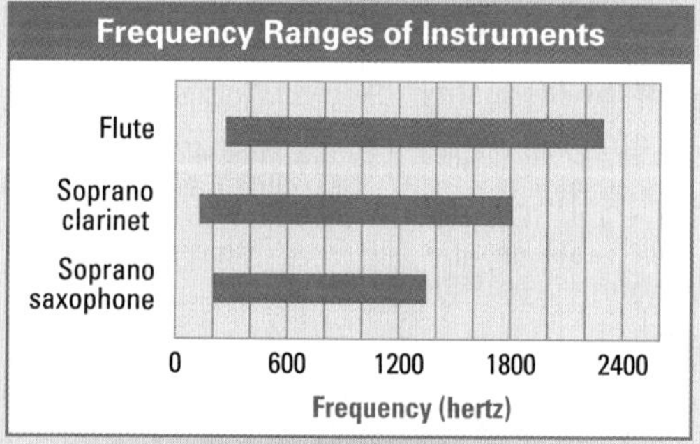

Think & Discuss

1. Estimate the frequency range of each instrument. **See margin.**
2. Which of these instruments has the greatest frequency range? **flute**
3. Estimate the frequency range that a flute can play that a soprano clarinet cannot. ***Sample answer:*** **$1800 \leq x \leq 2300$**

Learn More About It

You will write inequalities to describe frequency ranges in Exercises 66 and 67 on p. 338.

APPLICATION LINK Visit www.mcdougallittell.com for more information about music.

331

ADDITIONAL RESOURCES

Another way to begin the chapter is to show the video clip of a real-life motivator for Chapter 6 on the *Instant Replay: Video Review Games.*

PROJECTS

A project covering Chapters 4–6 appears on pages 392–393 of the Student Edition. An additional project for Chapter 6 is available in the *Chapter 6 Resource Book,* p.114.

TECHNOLOGY

Software

- *Electronic Teaching Tools*
- *Online Lesson Planner*
- *Personal Student Tutor*
- *Test and Practice Generator*
- *Electronic Lesson Presentations* (Lesson 6.4)

Video

- *Instant Replay: Video Review Games*

Internet Connections
www.mcdougallittell.com

- **Application Links** 338, 352, 357, 362, 379
- **Data Updates** 341, 372, 379
- **Student Help** 336, 340, 344, 347, 355, 359, 364, 367, 372, 377, 379, 382
- **Career Links** 355, 365
- **Extra Challenge** 339, 345, 351, 358, 365, 374, 380

1. ***Sample answer:***
flute: $260 \leq x \leq 2300$;
soprano clarinet: $135 \leq x \leq 1800$;
soprano saxophone: $200 \leq x \leq 1350$

PREPARE

DIAGNOSTIC TOOLS

The **Skill Review** exercises can help you diagnose whether students have the following skills needed in Chapter 6:

- Identify the solutions of an inequality.
- Solve linear equations.
- Graph linear equations in two variables.

The following resources are available for students who need additional help with these skills:

- Prerequisite Skills Review (*Chapter 6 Resource Book,* p. 5; *Warm-Up Transparencies,* p. 44)
- Reteaching with Practice (Chapter Resource Books for Lessons 1.4, 3.3, 4.2, 4.3, and 4.6)
- *Personal Student Tutor*

ADDITIONAL RESOURCES

The following resources are provided to help you prepare for the upcoming chapter and customize review materials:

- ***Chapter 6 Resource Book***
  Tips for New Teachers (p. 1)
  Parent Guide (p. 3)
  Lesson Plans (every lesson)
  Lesson Plans for Block Scheduling (every lesson)
- ***Electronic Teaching Tools***
- ***Online Lesson Planner***
- ***Test and Practice Generator***

13.

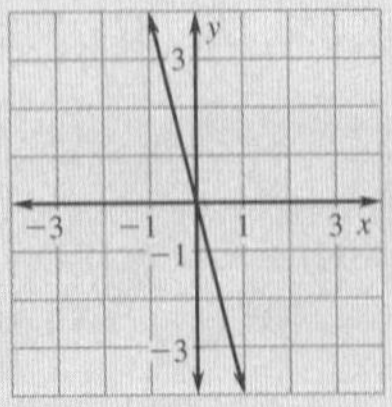

14.

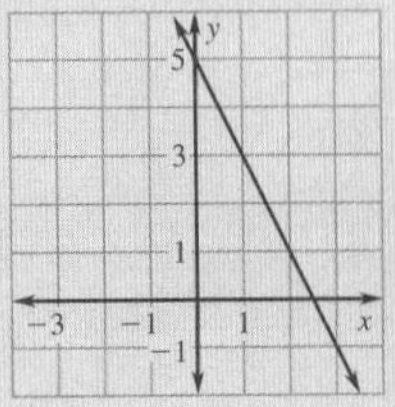

15–16. See Additional Answers beginning on page AA1.

CHAPTER 6

Study Guide

PREVIEW

What's the chapter about?

Chapter 6 is about **solving and graphing linear inequalities**, many of which model real-life applications. In Chapter 6 you'll learn

- how to solve and graph inequalities.
- how to solve and graph absolute-value equations and inequalities.
- how to use measures of central tendency and statistical plots.

KEY VOCABULARY

▶ **Review**
- **equation, p. 24**
- **inequality, p. 26**
- **absolute value, p. 65**
- **graph of an equation, p. 210**

▶ **New**
- **graph of a linear inequality, p. 334**
- **compound inequality, p. 346**
- **solution of a linear inequality, p. 360**
- **stem-and-leaf plot, p. 368**
- **mean, median, mode, p. 369**
- **box-and-whisker plot, p. 375**

PREPARE

Are you ready for the chapter?

SKILL REVIEW Do these exercises to review skills that you'll apply in this chapter. See the given **reference page** if there is something you don't understand.

STUDENT HELP

Study Tip
"Student Help" boxes throughout the chapter give you study tips and tell you where to look for extra help in this book and on the Internet.

Decide whether 5 is a solution of the inequality. **(Review Example 4, p. 26)**

1. $3x + 2 < 18$ **solution**
2. $x - 3 > 2$ **not a solution**
3. $4 + 4x \geq 24$ **solution**
4. $1 + 2x \leq 5$ **not a solution**
5. $8 + x < 12$ **not a solution**
6. $2x^2 > 45$ **solution**

Solve the equation. **(Review Examples 1–3, pp. 145–146)**

7. $\frac{1}{2}x + 10 = 15$ **10**
8. $\frac{2}{3}x - 4 = 12$ **24**
9. $20 = 2(x + 1)$ **9**
10. $4x + 5x - 1 = 53$ **6**
11. $6x - 2x + 8 = 36$ **7**
12. $4(5 - x) = 12$ **2**

Graph the equation. **(Review Examples, pp. 211, 219, and 242)** **13–16. See margin.**

13. $y = -4x$
14. $y = -2x + 5$
15. $2y - x = 4$
16. $-2x + y = 3$

STUDY STRATEGY

Here's a study strategy!

Showing Your Work

Show all the steps when you do your homework. Showing your work makes it easier to find the place where you made a mistake. Write corrections next to the errors.

ACTIVITY 6.1

Developing Concepts

Group Activity for use with Lesson 6.1

Investigating Inequalities

GROUP ACTIVITY
Work in a small group.

MATERIALS
- paper
- pencil

QUESTION How do operations change an inequality?

EXPLORING THE CONCEPT

1. Each member of your group should write a different inequality by choosing two numbers and placing $>$ or $<$ between them to show which is greater. *Sample answer:* $4 > -8$
2. Apply each rule below to both sides of your inequality. Write the correct inequality symbol between the two resulting numbers. Sample answers are given.

   a. Add 4. $4 + 4 > -8 + 4;\ 8 > -4$
   b. Subtract 4. $4 - 4 > -8 - 4;\ 0 > -12$
   c. Multiply by 4. $4 \cdot 4 > -8 \cdot 4;\ 16 > -32$
   d. Divide by 4. $\frac{4}{4} > \frac{-8}{4};\ 1 > -2$
   e. Multiply by -4. $4 \cdot (-4) < -8 \cdot (-4);\ -16 < 32$
   f. Divide by -4. $\frac{4}{-4} < \frac{-8}{-4};\ -1 < 2$

3. In **Step 2**, when did you have to change the direction of the inequality symbol? in parts (e) and (f)
4. Use your inequality from **Step 1**. Repeat **Step 2**, but change 4 and -4 to some other positive and negative numbers. When did you have to change the direction of the inequality symbol? *Sample answer:* when multiplying or dividing by a negative number

DRAWING CONCLUSIONS

In Exercises 1–6, predict whether the direction of the inequality symbol will change when you apply the given rule. Check your prediction.

1. $4 < 9$; add 7 no
2. $15 > 12$; subtract -4 no
3. $4 > -3$; multiply by 5 no
4. $2 > -11$; add -7 no
5. $-6 < 2$; divide by -3 yes
6. $1 < 8$; multiply by -10 yes

7. Copy and complete the table. no, no; no, no; no, yes; no, yes

Does the inequality symbol change directions?		
	a positive number	a negative number
Add	?	?
Subtract	?	?
Multiply by	?	?
Divide by	?	?

Apply the given rule to solve the inequality.

8. $x + 3 > 9$; subtract 3 $x > 6$
9. $x + 7 \le 12$; add -7 $x \le 5$
10. $4x \ge 15$; divide by 4 $x \ge \frac{15}{4}$
11. $-3x > 11$; divide by -3 $x < -\frac{11}{3}$
12. $2x < 11$; multiply by $\frac{1}{2}$ $x < \frac{11}{2}$
13. $-\frac{1}{3}x \le 12$; multiply by -3 $x \ge -36$
14. $x + 6 < 15$; subtract 6 $x < 9$
15. $x - 2 \ge 90$; add 2 $x \ge 92$
16. $5x \le 25$; divide by 5 $x \le 5$
17. $-6x > 30$; multiply by $-\frac{1}{6}$ $x < -5$

1 Planning the Activity

PURPOSE
To investigate the effects on the inequality sign when arithmetic operations are performed on a linear inequality.

PACING
- Exploring the Concept — 10 min
- Drawing Conclusions — 10 min

LINK TO LESSON
Students should notice that the "yes" responses that they recorded in the table in Exercise 7 coincide with the last two statements in the chart preceding Example 4 on page 336.

2 Managing the Activity

COOPERATIVE LEARNING
In Steps 1 and 2, have students exchange inequalities and perform the same operations to confirm each student's results.

CLASSROOM MANAGEMENT
To help students check their results in Exercises 8–17, ask them to choose a value for the variable that satisfies their inequality. Then ask them to substitute the same value in the original inequality. Both statements should be true.

3 Closing the Activity

KEY DISCOVERY
The inequality symbol is reversed when both sides of an inequality are multiplied or divided by a negative number.

ACTIVITY ASSESSMENT
JOURNAL Write a summary of the similarities and differences between solving linear inequalities and linear equations.
See sample answer at left.

Activity Assessment *Sample answer:* A good answer should include the following:
- In solving both, you must do to one side what you do to the other; that is, any number that you add, subtract, multiply by, or divide by on one side of the equals or inequality sign, you must do likewise with on the other side.
- Only in solving inequalities can the sign itself change—it reverses direction when you multiply or divide both sides by a negative number.

1 PLAN

PACING
Basic: 1 day
Average: 1 day
Advanced: 1 day
Block Schedule: 0.5 block with Ch. 5 Assess

LESSON OPENER APPLICATION
An alternative way to approach Lesson 6.1 is to use the Application Lesson Opener:
- Blackline Master (*Chapter 6 Resource Book*, p. 12)
- Transparency (p. 39)

MEETING INDIVIDUAL NEEDS
- ***Chapter 6 Resource Book***
  Prerequisite Skills Review (p. 5)
  Practice Level A (p. 13)
  Practice Level B (p. 14)
  Practice Level C (p. 15)
  Reteaching with Practice (p. 16)
  Absent Student Catch-Up (p. 18)
  Challenge (p. 20)
- ***Resources in Spanish***
- ***Personal Student Tutor***

NEW-TEACHER SUPPORT
See the Tips for New Teachers on pp. 1–2 of the *Chapter 6 Resource Book* for additional notes about Lesson 6.1.

WARM-UP EXERCISES
Transparency Available
Solve each equation.
1. $5x - 7 = -12$ -1
2. $6 = 3 - x$ -3
3. $\frac{x}{4} = 12$ 48
4. $8x = -32$ -4

ENGLISH LEARNERS
Write the symbols <, >, ≤, and ≥ on the board. Review with English learners equivalent verbal expressions. Include *more than* and *greater than* for >, *less than* and *fewer than* for <, and so on.

6.1 Solving One-Step Linear Inequalities

What you should learn

GOAL 1 Graph linear inequalities in one variable.

GOAL 2 Solve one-step linear inequalities.

Why you should learn it

▼ To describe **real-life** situations, such as the speeds of runners in **Example 1**.

REAL LIFE

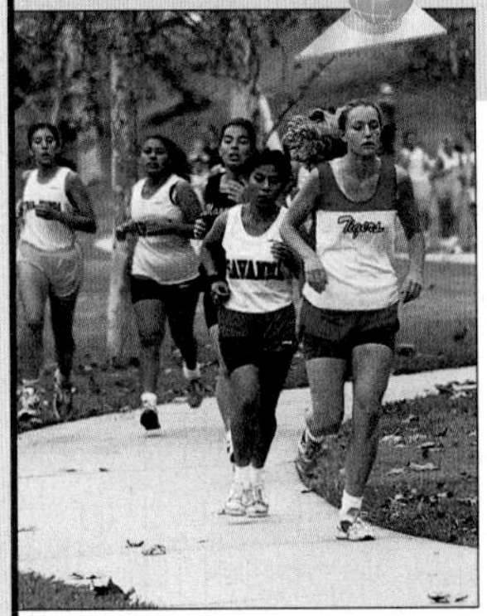

GOAL 1 GRAPHING LINEAR INEQUALITIES

The **graph** of a linear inequality in one variable is the set of points on a number line that represent all solutions of the inequality.

VERBAL PHRASE	INEQUALITY	GRAPH
All real numbers less than 2	$x < 2$	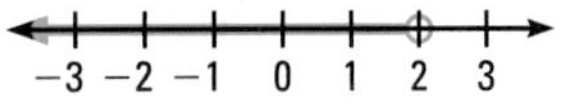
All real numbers greater than -2	$x > -2$	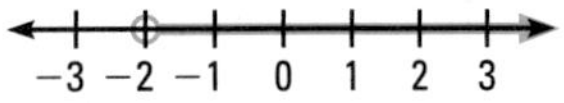
All real numbers less than or equal to 1	$x \le 1$	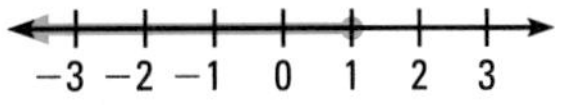
All real numbers greater than or equal to 0	$x \ge 0$	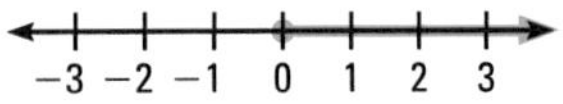

An open dot is used for < or > and a solid dot for ≤ or ≥.

EXAMPLE 1 *Write and Graph a Linear Inequality*

Sue ran a 2-kilometer race in 8 minutes. Write an inequality to describe the average speeds of runners who were faster than Sue. Graph the inequality.

SOLUTION Average speed is $\frac{\text{distance}}{\text{time}}$. A faster runner's average speed must be greater than Sue's average speed.

VERBAL MODEL $\boxed{\text{Faster average speed}} > \frac{\text{Distance}}{\text{Sue's time}}$

LABELS
Faster average speed = s (kilometers per minute)
Distance = **2** (kilometers)
Sue's time = **8** (minutes)

ALGEBRAIC MODEL
$s > \frac{2}{8}$ Write linear inequality.
$s > \frac{1}{4}$ Simplify.

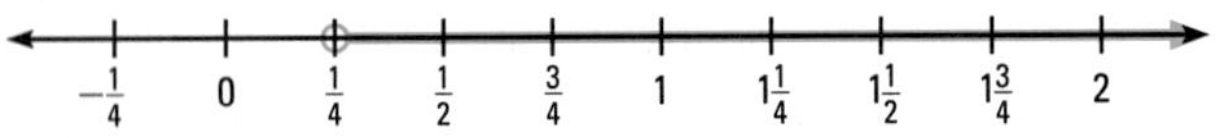

STUDENT HELP
Look Back
For help with inequalities, see p. 26.

STUDENT HELP

Look Back
For help with solving equations with addition and subtraction, see pp. 132–134.

GOAL 2 SOLVING ONE-STEP LINEAR INEQUALITIES

Solving a linear inequality in one variable is much like solving a linear equation in one variable. To solve the inequality, you isolate the variable on one side using transformations that produce **equivalent inequalities**, which have the same solution(s).

TRANSFORMATIONS THAT PRODUCE EQUIVALENT INEQUALITIES

	Original Inequality		Equivalent Inequality
• Add the same number to *each* side.	$x - 3 < 5$	Add 3.	$x < 8$
• Subtract the same number from *each* side.	$x + 6 \geq 10$	Subtract 6.	$x \geq 4$

EXAMPLE 2 *Using Subtraction to Solve an Inequality*

Solve $x + 5 \geq 3$. Graph the solution.

SOLUTION

$x + 5 \geq 3$ — Write original inequality.

$x + 5 - 5 \geq 3 - 5$ — Subtract 5 from each side.

$x \geq -2$ — Simplify.

▶ The solution is all real numbers greater than or equal to -2. Check several numbers that are greater than or equal to -2 in the original inequality.

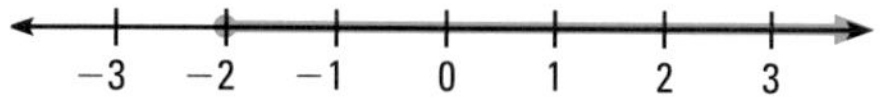

EXAMPLE 3 *Using Addition to Solve an Inequality*

Solve $-2 > n - 4$. Graph the solution.

SOLUTION

$-2 > n - 4$ — Write original inequality.

$-2 + 4 > n - 4 + 4$ — Add 4 to each side.

$2 > n$ — Simplify.

▶ The solution is all real numbers less than 2. Check several numbers that are less than 2 in the original inequality.

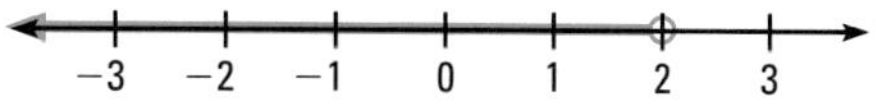

STUDENT HELP

Study Tip
Check several possible solutions by substituting them in the original inequality. Also try some numbers that are *not* solutions to make sure that they do *not* satisfy the original inequality.

2 TEACH

MOTIVATING THE LESSON
Ask students to imagine waiting with several friends for an elevator. The doors on the old, 1200 lb capacity elevator open to reveal two football players, who must together weigh 500 lb, already aboard. The friends look at each other, wondering who all should get on. Tell students that a linear inequality provides a good model for this situation, where "less than or equal to" is very important.

EXTRA EXAMPLE 1
Abu was sure that he didn't score less than 73 on his algebra test. Write an inequality to describe Abu's possible score. Graph the inequality. $q \geq 73$

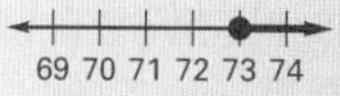

EXTRA EXAMPLE 2
Solve $x + 8 \geq 1$. Graph the solution. $x \geq -7$

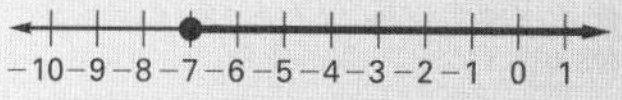

EXTRA EXAMPLE 3
Solve $3 < m - 5$. Graph the solution. $m > 8$

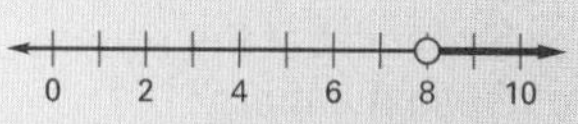

CHECKPOINT EXERCISES

For use after Example 1:

1. Karim hopes that his next biology test grade will be higher than his average to date. His first three test grades were 77, 83, and 86. Write and simplify an inequality to express the scores that will meet Karim's hopes. $g > \frac{77 + 83 + 86}{3}$, or $g > 82$

For use after Examples 2 and 3:

2. Solve $5 + x \leq 3$. Graph the solution. $x \leq -2$

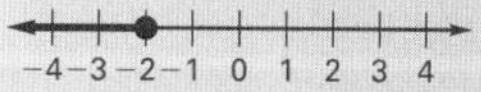

3. Solve $3 > m - 2$. Graph the solution. $m < 5$

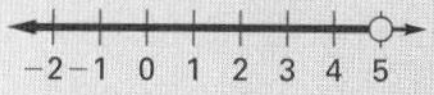

EXTRA EXAMPLE 4

Solve the inequality and graph its solution.

a. $\frac{p}{2} \leq -2$ $p \leq -4$

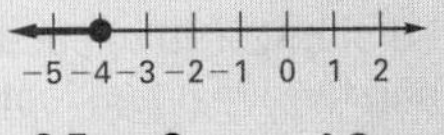

b. $-2.5y > 3$ $y < -1.2$

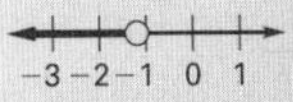

CHECKPOINT EXERCISES

For use after Example 4:

1. Which two inequalities have the same solution?

a. $\frac{n}{5} > -2$ **b.** $-2n < 20$

c. $-\frac{n}{2} < -5$ **a and b**

FOCUS ON VOCABULARY

What is the graph of a linear inequality in one variable? **the set of points on the number line that represents all solutions of the inequality**

CLOSURE QUESTION

How do you produce equivalent inequalities? **See the Transformations boxes on pp. 335 and 336.**

DAILY PUZZLER

The sum of three consecutive perfect squares is at most 302. What is the maximum sum of the square roots of these numbers? **30**

STUDENT HELP

Look Back
For help with solving equations with multiplication and division, see p. 138.

USING MULTIPLICATION AND DIVISION The operations used to solve linear inequalities are similar to those used to solve linear equations, but there are important differences. When you multiply or divide each side of an inequality by a *negative* number, you must *reverse* the inequality symbol to maintain a true statement. For instance, to reverse $>$, replace it with $<$.

TRANSFORMATIONS THAT PRODUCE EQUIVALENT INEQUALITIES

	Original inequality		Equivalent inequality
• Multiply each side by the same *positive* number.	$\frac{1}{2}x > 3$	Multiply by 2.	$x > 6$
• Divide each side by the same *positive* number.	$3x \leq 9$	Divide by 3.	$x \leq 3$
• Multiply each side by the same *negative* number and *reverse* the inequality symbol.	$-x < 4$	Multiply by (−1).	$x > -4$
• Divide each side by the same *negative* number and *reverse* the inequality symbol.	$-2x \leq 6$	Divide by (−2).	$x \geq -3$

EXAMPLE 4 *Using Multiplication or Division to Solve an Inequality*

a. $\frac{a}{3} \leq 12$ Original inequality

$3 \cdot \frac{a}{3} \leq 3 \cdot 12$ Multiply each side by positive 3.

$a \leq 36$ Simplify.

▶ The solution is all real numbers less than or equal to 36. Check several numbers that are less than or equal to 36 in the original inequality.

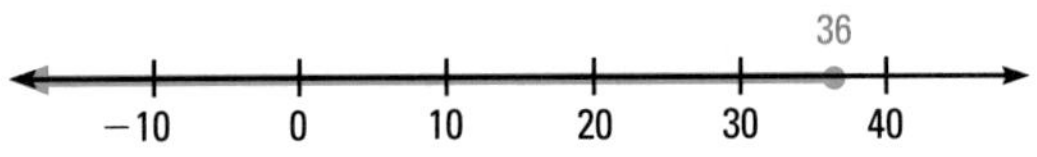

b. $-4.2m > 6.3$ Original inequality

$\frac{-4.2m}{-4.2} < \frac{6.3}{-4.2}$ Divide each side by −4.2 and reverse inequality symbol.

$m < -1.5$ Simplify.

▶ The solution is all real numbers less than -1.5. Check several numbers that are less than -1.5 in the original inequality.

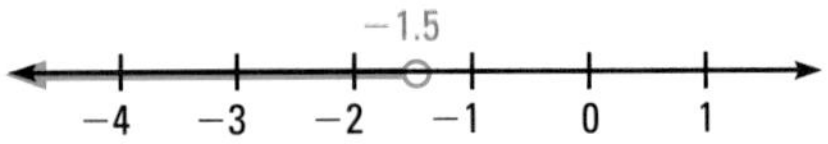

STUDENT HELP

INTERNET **HOMEWORK HELP**
Visit our Web site www.mcdougallittell.com for extra examples.

GUIDED PRACTICE

Vocabulary Check ✓

1. Since $x - 3 < 5$ and $x < 8$ have the same solution, they are _?_ inequalities. equivalent

Concept Check ✓

Write a verbal phrase that describes the inequality. 2–5. See margin.

2. $x > -7$ 3. $x < 1$ 4. $9 \le x$ 5. $10 \ge x$

2. all real numbers greater than -7
3. all real numbers less than 1
4. all real numbers greater than or equal to 9
5. all real numbers less than or equal to 10

ERROR ANALYSIS **Describe and correct the error.** 6, 7. See margin.

6. (crossed out)
$-3x \ge 15$
$\frac{-3x}{-3} \ge \frac{15}{-3}$
$x \ge -5$

7. (crossed out)
$x + 4 < 1$
$x + 4 - 4 > 1 - 4$
$x > -3$

6. The inequality symbol was not reversed when each side of the inequality was divided by a negative number; $x \le -5$.
7. The inequality symbol should not have been reversed when a number was subtracted from each side of the inequality; $x < -3$.

Tell whether you should use an *open dot* or a *closed dot* on the graph of the inequality.

8. $x < 3$ open dot
9. $x > 10$ open dot
10. $x \ge 5$ closed dot
11. $3x + 5 < 4$ open dot
12. $5x - 3 \ge 12$ closed dot
13. $-2x - 1 \le 3$ closed dot

Skill Check ✓

Solve the inequality and graph its solution. 14–19. See margin for graphs.

14. $-2 + x < 5$ $x < 7$
15. $-3 \le y + 2$ $y \ge -5$
16. $-2b \le -8$ $b \ge 4$
17. $x - 4 > 10$ $x > 14$
18. $5x > -45$ $x > -9$
19. $-6y \le 36$ $y \ge -6$

COLLECTIBLES **In 1997 a software executive bid \$19,550 for Superman's cape from the 1978 movie *Superman*. This was the highest bid.**

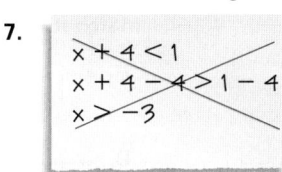

20. Let b represent the amount of a bid for the cape. Write an inequality for b. $b \le 19{,}550$
21. Graph the inequality. See margin.

PRACTICE AND APPLICATIONS

STUDENT HELP

Extra Practice to help you master skills is on p. 802.

GRAPHING **Graph the inequality.** 22–30. See margin for graphs.

22. $x > -2$
23. $x < 16$
24. $x \le 7$
25. $x \ge -6$
26. $2 \ge x$
27. $2.5 \le x$
28. $1 \ge -x$
29. $-10 \le -x$
30. $x < -0.5$

STUDENT HELP

HOMEWORK HELP
Example 1: Exs. 22–30, 61–63
Examples 2, 3: Exs. 31–45, 55–60
Example 4: Exs. 46–60

SOLVING AND GRAPHING **Solve the inequality and graph its solution.** 31–45. See margin for graphs.

31. $x + 6 < 8$ $x < 2$
32. $-5 < 4 + x$ $x > -9$
33. $-4 + x < 20$ $x < 24$
34. $8 + x \le -9$ $x \le -17$
35. $p - 12 \ge -1$ $p \ge 11$
36. $-2 > b - 5$ $b < 3$
37. $x - 3 > 2$ $x > 5$
38. $x - 5 \ge 1$ $x \ge 6$
39. $6 \le c + 2$ $c \ge 4$
40. $-8 \le x - 14$ $x \ge 6$
41. $m + 7 \ge -10$ $m \ge -17$
42. $-6 > x - 4$ $x < -2$
43. $-2 + x < 0$ $x < 2$
44. $-10 > a - 6$ $a < -4$
45. $5 + x \ge -5$ $x \ge -10$

3 APPLY

ASSIGNMENT GUIDE

BASIC
Day 1: pp. 337–339 Exs. 22–54 even, 55–63, 66–68, 72, 76, 80, 86, 87

AVERAGE
Day 1: pp. 337–339 Exs. 22–54 even, 55–63, 66–68, 72, 76, 80, 86, 87

ADVANCED
Day 1: pp. 337–339 Exs. 22–54 even, 55–63, 66–70, 72, 76, 80, 86, 87

BLOCK SCHEDULE
pp. 337–339 Exs. 22–54 even, 55–63, 66–68, 72, 76, 80, 86, 87 (with Ch. 5 Assess.)

EXERCISE LEVELS

Level A: *Easier*
22–30

Level B: *More Difficult*
31–68, 71–87

Level C: *Most Difficult*
69–70

✓ HOMEWORK CHECK

To quickly check student understanding of key concepts, go over the following exercises: Exs. 26, 36, 48, 52, 66. See also the Daily Homework Quiz:

- Blackline Master (*Chapter 6 Resource Book,* p. 23)
- Transparency (p. 46)

APPLICATION NOTE

EXERCISES 20 AND 21
Minimum bids are generally established for auction items that will bring a high price, such as Superman's cape. The actual inequality and its graph are likely to include a range of values between the minimum bid and the highest bid.

14–19, 21. See the next page.
22–45. See Additional Answers beginning on page AA1.

! COMMON ERROR

EXERCISES 22–54 Included in these exercises are some with the variable on the right. Some students tend to draw an incorrect graph for these types of exercises by reversing the directions of the arrow. Refer students back to Example 3 on p. 335 to help remediate this error.

APPLICATION NOTE

EXERCISE 62 Additional information about mercury is available at **www.mcdougallittell.com.**

EXERCISE 65 For a real-life bridge, only the positive values that satisfy the inequality are appropriate. A typical span for a modern bridge is between 200 ft and 800 ft.

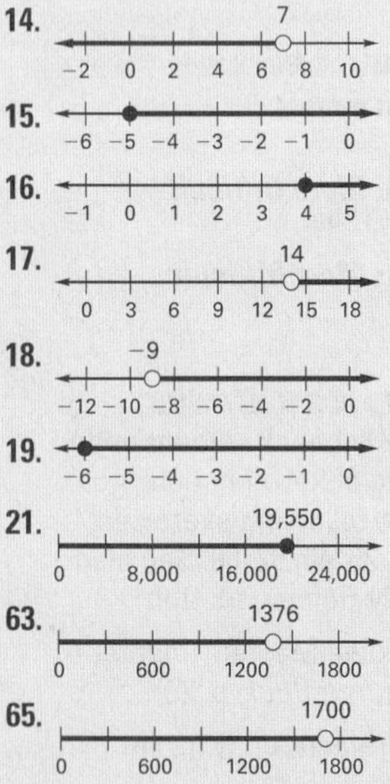

ADDITIONAL PRACTICE AND RETEACHING

For Lesson 6.1:

- Practice Levels A, B, and C (*Chapter 6 Resource Book,* p. 13)
- Reteaching with Practice (*Chapter 6 Resource Book,* p. 16)
- See Lesson 6.1 of the *Personal Student Tutor*

For more Mixed Review:

- Search the *Test and Practice Generator* for key words or specific lessons.

SOLVING INEQUALITIES Solve the inequality.

46. $15p < 60$ **$p < 4$** **47.** $-10a > 100$ **$a < -10$** **48.** $-\frac{n}{5} < 17$ **$n > -85$**

49. $11 \geq -2.2m$ **$m \geq -5$** **50.** $-18.2x \geq -91$ **$x \leq 5$** **51.** $2.1x \leq -10.5$ **$x \leq -5$**

52. $13 \leq -\frac{x}{3}$ **$x \leq -39$** **53.** $-\frac{a}{10} \leq -2$ **$a \geq 20$** **54.** $\frac{x}{4} \leq -9$ **$x \leq -36$**

SOLVING AND MATCHING In Exercises 55–60, solve the inequality. Then match its solution with its graph.

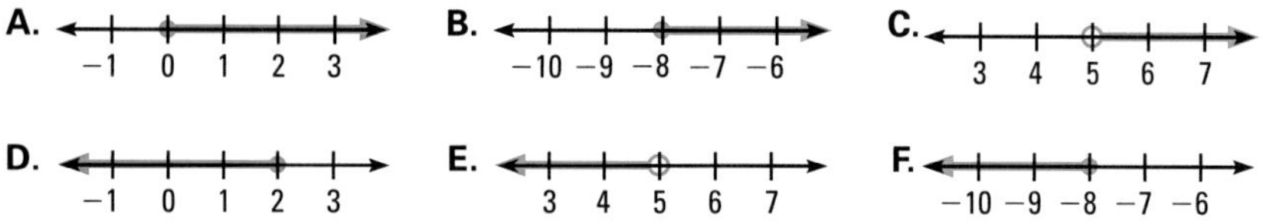

55. $x - 3.2 < 1.8$ **$x < 5$; E** **56.** $x + 4 \leq 6$ **$x \leq 2$; D** **57.** $10x > 50$ **$x > 5$; C**

58. $x - 2 \geq -10$ **$x \geq -8$; B** **59.** $-\frac{x}{2} \leq 0$ **$x \geq 0$; A** **60.** $-3.5x \geq 28$ **$x \leq -8$; F**

61. WALKING RACE You finish a three-mile walking race in 27.5 minutes. Write an inequality that describes the average speed of a walker who finished after you did. **$x < \frac{6}{55}$**

62. SCIENCE CONNECTION Mercury is the metallic element with the lowest melting point, −38.87°C. Write an inequality that describes the melting point p (in degrees Celsius) of any other metallic element. **$p > -38.87$**

FOCUS ON APPLICATIONS

MERCURY A melting point is the temperature at which a solid becomes a liquid. Mercury is used in thermometers. It is the only metal that is a liquid at room temperature.

APPLICATION LINK www.mcdougallittell.com

63. LARGEST MARLIN The world record for the largest Pacific blue marlin is 1376 pounds. It was caught in Kaaiwi Point, Kona, Hawaii. Let M represent the weight of a Pacific blue marlin that has been caught. Write an inequality for M. Graph the inequality. ▶ Source: International Game Fish Association **$M < 1376$; See margin for graph.**

64. BOWLING TOURNAMENT After two games of bowling, Brenda has a total score of 475. To win the tournament, she needs a total score of 684 or higher. Let x represent the score she needs for her third game to win the tournament. Write an inequality for x. What is the lowest score she can get for her third game and win the tournament? **$x + 475 \geq 684$; 209**

65. STEEL ARCH BRIDGE The longest steel arch bridge in the world is the New River Gorge Bridge near Fayetteville, West Virginia, at 1700 feet. Write an inequality that describes the length l (in feet) of any other steel arch bridge. Graph the inequality. **$l < 1700$; See margin for graph.**

MUSICAL INSTRUMENTS **Write an inequality to describe the frequency range f of the instrument.**

66. The frequency range of a guitar is from 73 hertz to 698 hertz. **$73 \leq f \leq 698$**

67. The frequency range of cymbals is from 131 hertz to 587 hertz. **$131 \leq f \leq 587$**

Test Preparation

68. **MULTI-STEP PROBLEM** You start your own business making wind chimes. Your equipment costs \$48. It costs \$3.50 to make each wind chime. You sell each wind chime for \$7.50. The graph shows your total costs.

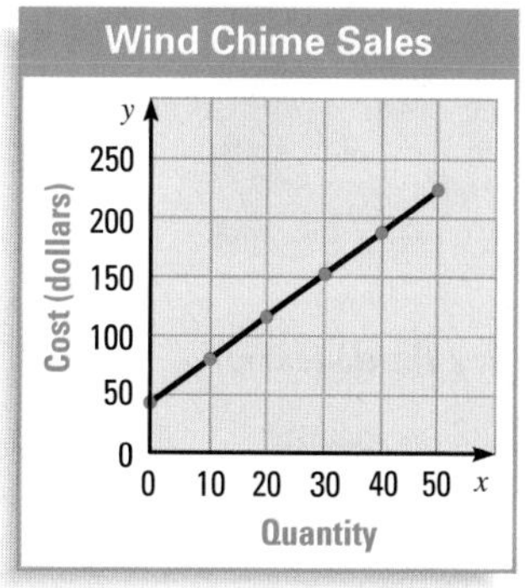

a. Copy the graph. On the same graph draw the line that represents the total revenue for different numbers of wind chimes sold. **See margin.**

b. How many wind chimes must you sell before you start making a profit? **12 wind chimes**

c. Write an inequality that describes the number of wind chimes you must sell before you start making a profit. ***Sample answer:*** **$48 + 3.5x \leq 7.5x$**

69. *Sample answer:* An inequality such as $x < y$ means that x is to the left of y on a number line. Since a number and its opposite are on opposite sides of zero on a number line, if $x < y$, then $-x > -y$. For example, $2 < 3$ and $-2 > -3$; $-4 < -1$ and $4 > 1$.

★ Challenge

69. **REVERSING INEQUALITIES** Explain why you must reverse the direction of the inequality symbol when multiplying by a negative number. Give two examples to support your reasoning. **See margin.**

70. **LOGICAL REASONING** Explain why multiplying by x to solve the inequality $\frac{4}{x} > 2$ might lead to an error.
Sample answer: **You don't know whether x is positive or negative, so you don't know whether or not to reverse the inequality symbol. (Note that x must be positive or negative and not zero.)**

EXTRA CHALLENGE
www.mcdougallittell.com

MIXED REVIEW

SOLVING EQUATIONS **Solve the equation.** (Review 3.3 for 6.2)

71. $45b - 3 = 2$ **$\frac{1}{9}$**
72. $-5x + 50 = 300$ **-50**
73. $-3s - 2 = -44$ **14**
74. $\frac{1}{3}x + 5 = -4$ **-27**
75. $\frac{x}{4} + 4 = 18$ **56**
76. $-8 = \frac{3}{5}a - 5$ **-5**
77. $x + 2x + 5 = 14$ **3**
78. $3(x - 6) = 12$ **10**
79. $9 = -\frac{3}{2}(x - 2)$ **-4**

WRITING EQUATIONS **Write the slope-intercept form of the equation of the line that passes through the two points. Graph the line. Label the points where the line crosses the axes.** (Review 4.3, 5.3) **80–85. See margin for graphs.**

80. (1, 2), (4, −1)
81. (2, 0), (−4, −3)
82. (1, 1), (−3, 5)
83. (−1, 4), (2, 4)
84. (−1, −3), (2, 3)
85. (5, 0), (5, −2)

80. $y = -x + 3$; (0, 3) and (3, 0)
81. $y = \frac{1}{2}x - 1$; (0, −1) and (2, 0)
82. $y = -x + 2$; (0, 2) and (2, 0)
83. $y = 4$; the line does not cross the x-axis and crosses the y-axis at (0, 4).
84. $y = 2x - 1$; (0, −1) and $\left(\frac{1}{2}, 0\right)$
85. $x = 5$; the line crosses the x-axis at (5, 0) and does not cross the y-axis.

86. **EXAM GRADES** There are 20 questions on an exam, each worth 5 points. Your percent grade p varies directly with the number n of correct answers. Find an equation that relates p and n. What is your percent grade if you get 17 correct answers? (Review 4.5) **$p = 5n$; 85**

87. **BIKE RIDING** The table shows the time t (in minutes) and the distance d (in miles) that Maria rode her bike each week. Write a model that relates the variables d and t. (Review 4.5) **$d = 0.12t$**

Time (minutes), t	3	5	10	12
Distance (miles), d	0.36	0.60	1.20	1.44

4 ASSESS

DAILY HOMEWORK QUIZ

Transparency Available

1. Graph $-1.5 < x$.

Solve the inequality and graph its solution.

2. $x - 4 \leq -1$ **$x \leq 3$**

3. $2 > -1 - x$ **$x > -3$**

Solve the inequality.

4. $1.8x \geq -8.1$ **$x \geq -4.5$**
5. $-6 < -\frac{x}{7}$ **$x < 42$**

EXTRA CHALLENGE NOTE

Challenge problems for Lesson 6.1 are available in **blackline** format in the *Chapter 6 Resource Book*, p. 20 and at **www.mcdougallittell.com.**

ADDITIONAL TEST PREPARATION

1. **WRITING** What does the solid dot in the graph of $x \geq 6$ represent? What does the empty circle in the graph of $x > 6$ represent? **See sample answer below.**
2. **OPEN ENDED** Describe a real-life situation that can be modeled by the inequality $x > 70$. **See sample answer below.**

68a.

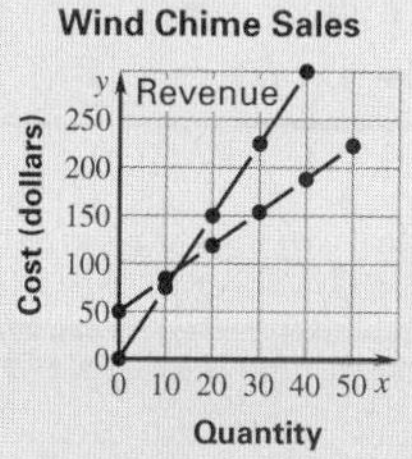

80–85. See Additional Answers beginning on page AA1.

Additional Test Preparation *Sample answers:*

1. The solid dot indicates that 6 is included in the solution set of the inequality. The empty circle indicates that 6 is not included in the solution set of the inequality.
2. If the speed limit is 70 mi/h, then $x > 70$ describes the speeds that exceed the speed limit, and for which you might be ticketed.

1 PLAN

PACING
Basic: 1 day
Average: 1 day
Advanced: 1 day
Block Schedule: 0.5 block with 6.3

LESSON OPENER
VISUAL APPROACH
An alternative way to approach Lesson 6.2 is to use the Visual Approach Lesson Opener:

- Blackline Master (*Chapter 6 Resource Book,* p. 24)
- Transparency (p. 40)

MEETING INDIVIDUAL NEEDS

- ***Chapter 6 Resource Book***
  Prerequisite Skills Review (p. 5)
  Practice Level A (p. 25)
  Practice Level B (p. 26)
  Practice Level C (p. 27)
  Reteaching with Practice (p. 28)
  Absent Student Catch-Up (p. 30)
  Challenge (p. 32)
- ***Resources in Spanish***
- ***Personal Student Tutor***

NEW-TEACHER SUPPORT
See the Tips for New Teachers on pp. 1–2 of the *Chapter 6 Resource Book* for additional notes about Lesson 6.2.

WARM-UP EXERCISES

Transparency Available

Solve the linear equation.

1. $2x - 1 = 7$ 4
2. $5 = 3x - 7$ 4
3. $2.5 = 3.3 - 4x$ 0.2
4. $5 + \frac{2}{3}x = 1$ −6

6.2 Solving Multi-Step Linear Inequalities

What you should learn

GOAL 1 Solve multi-step linear inequalities.

GOAL 2 Use linear inequalities to model and solve **real-life** problems, as in the business problem in **Example 4**.

Why you should learn it

▼ To describe **real-life** problems, such as city population growth in Nashville, Tennessee, in **Example 3**.

GOAL 1 SOLVING MULTI-STEP INEQUALITIES

In Lesson 6.1 you solved inequalities using one step. In this lesson you will learn how to solve linear inequalities using two or more steps.

EXAMPLE 1 *Using More than One Step*

Solve $2y - 5 < 7$.

SOLUTION

$2y - 5 < 7$	Write original inequality.
$2y < 12$	Add 5 to each side.
$y < 6$	Divide each side by 2.

▶ The solution is all real numbers less than 6.

EXAMPLE 2 *Multiplying or Dividing by a Negative Number*

Solve the inequality.

a. $5 - x > 4$ **b.** $2x - 4 \geq 4x - 1$

SOLUTION

a. $5 - x > 4$	Write original inequality.
$-x > -1$	Subtract 5 from each side.
$(-x)(-1) < (-1)(-1)$	Multiply each side by −1. *Reverse* inequality symbol.
$x < 1$	Simplify.

▶ The solution is all real numbers less than 1.

b. $2x - 4 \geq 4x - 1$	Write original inequality.
$2x \geq 4x + 3$	Add 4 to each side.
$-2x \geq 3$	Subtract $4x$ from each side.
$x \leq -\frac{3}{2}$	Divide each side by −2. *Reverse* inequality symbol.

▶ The solution is all real numbers less than or equal to $-\frac{3}{2}$.

STUDENT HELP
HOMEWORK HELP
Visit our Web site www.mcdougallittell.com for extra examples.

In part (b) of Example 2, you could begin by subtracting $2x$ from each side. Your solution would still be all real numbers less than or equal to $-\frac{3}{2}$.

GOAL 2 MODELING AND SOLVING REAL-LIFE PROBLEMS

REAL LIFE
Population

EXAMPLE 3 *Writing a Linear Model*

In 1990 Nashville, Tennessee, had a population of 985,000. From 1990 through 1996, the population increased at an average rate of about 22,000 per year. Write a linear model for the population of Nashville.

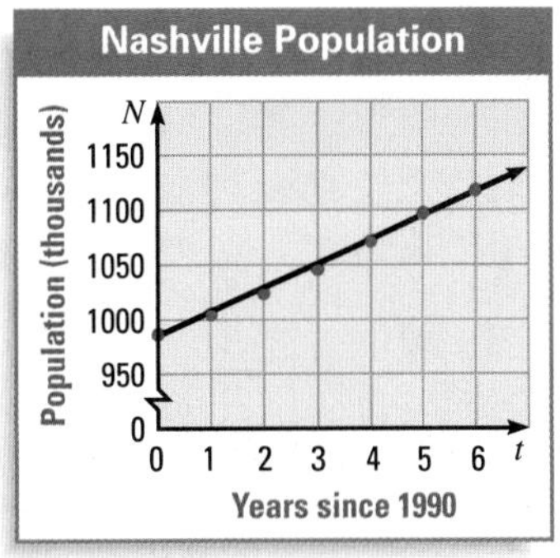

DATA UPDATE of U.S. Bureau of the Census data at www.mcdougallittell.com

SOLUTION

To write a linear model of the population of Nashville, express the population as the 1990 population plus 22,000 per year.

Let N represent the population of Nashville.

Let t represent the number of years since 1990.

Population = 1990 population + 22,000 • Years since 1990

$$N = 985{,}000 + 22{,}000t$$

In the Activity below, you will compare populations of Nashville and Las Vegas.

ACTIVITY Developing Concepts

Investigating Problem Solving

1. In 1990 Las Vegas, Nevada, had a population of 853,000. From 1990 through 1996, the population increased at an average rate of about 58,000 per year. Write a linear model for the population of Las Vegas. Let L represent the population of Las Vegas. **$L = 58{,}000t + 853{,}000$**
2. Graph your linear model from **Step 1**. Graph the linear model from Example 3 in the same coordinate plane. According to the models, in what year will the populations be the same? **1993**
3. Write and solve a linear inequality to find the year that the population of Las Vegas exceeded the population of Nashville. **$58{,}000t + 853{,}000 > 22{,}000t + 985{,}000$; 1993**
4. In which years was the population of Las Vegas less than the population of Nashville? **1990, 1991, 1992**
5. Write an inequality to represent the numbers of years y (since 1990) in which the population of Las Vegas was less than the population of Nashville. **$58{,}000y + 853{,}000 < 22{,}000y + 985{,}000$**

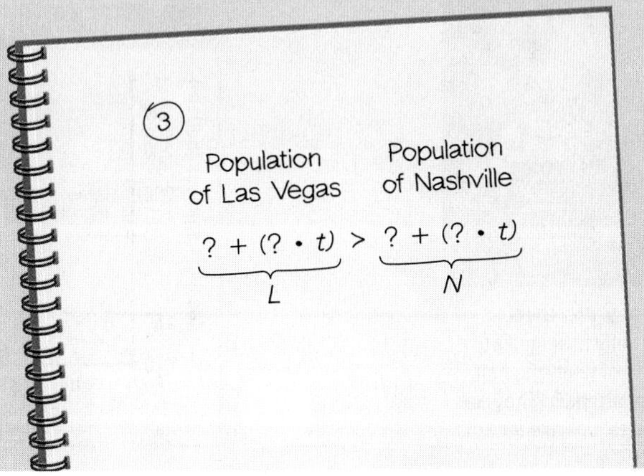

2 TEACH

MOTIVATING THE LESSON

You are going to an amusement park with $25. Admittance costs $5, with each ride costing $1.25. Ask students how they would find out how many rides they could go on. Point out that one way to do this is to write and solve an inequality that takes more than one step to solve.

ACTIVITY NOTE

GRAPHING CALCULATOR By graphing $y = 22{,}000x + 985{,}000$ and $y = 58{,}000x + 853{,}000$ on a graphing calculator with a window such as [0, 10] × [800000, 1200000], students can investigate the problem more precisely. By using an "intersect" command or by tracing and zooming, students can see Las Vegas' population passing Nashville's after $3\frac{2}{3}$ years, or after August, 1993.

EXTRA EXAMPLE 1

Solve $5x - 3 > 12$. $x > 3$

EXTRA EXAMPLE 2

Solve the inequality.

a. $-3x + 1 \le 22$ $x \ge -7$

b. $-x + 1 \le 3x + 21$ $x \ge -5$

EXTRA EXAMPLE 3

In 1988, the number of U.S. citizens who had backpacked or wilderness camped was about 9.1 million. From 1988 to 1998, this number increased at a yearly rate of about 850,000. Write a linear model for the number in millions of U.S. citizens who had backpacked or wilderness camped during this period. $y = 9.1 + 0.85x$

CHECKPOINT EXERCISES

For use after Examples 1 and 2:

1. Solve $-3x - 5 < 4$. $x > -3$
2. Solve $6x + 7 \ge 7x - 9$. $x \le 16$

For use after Example 3:

3. Write a linear model for a population that begins at 110,000 and decreases by 2500 per year. $N = 110{,}000 - 2500t$

EXTRA EXAMPLE 4
You start growing geraniums from seed and plan to sell potted plants for Mother's day. It costs you \$1.75 to pot and grow each plant. If you sell each plant for \$3.25, how many plants will you need to sell if you want to make a profit of at least \$150?
at least 100 plants

CHECKPOINT EXERCISES
For use after Example 4:

1. You plan to publish an online newsletter that reports the results of mountain bike competitions. You do not want your monthly costs to exceed \$2370. Your fixed monthly costs are \$1200. You must also pay \$130 per month to each article writer. How many writers can you afford to hire in a month? **nine or fewer**

APPLICATION NOTE
EXAMPLE 4 The point at which the costs equal the income is called the *break-even* point. For about 35 flies sold, the expenses are equal to the income from sales.

FOCUS ON VOCABULARY
Why is the graph in Example 3 called a linear model? **The graph, which is a line, can be used to predict, or model, the population in each year.**

CLOSURE QUESTION
How is solving a multi-step inequality different from solving a multi-step equation? **When both sides of the inequality are multiplied or divided by a negative number, the inequality symbol reverses.**

DAILY PUZZLER
You're operating an arts stand at a fair. You're out of coins and dollar bills. You have one \$5 bill left, but plenty of tens and twenties. A customer hands you a \$100 bill for a \$34 purchase. For the customer not to overpay, how much extra change must you return? **at least \$4**

REAL LIFE *Fly Fishing*

EXAMPLE 4 *Writing and Using a Linear Model*

You see an advertisement for instructions on how to tie flies for fishing. The cost of materials for each fly is \$.15. You plan to sell each fly for \$.58, and you want to make a profit of at least \$200. How many flies will you need to tie and sell?

SOLUTION

Your total expenses will be \$.15 per fly, plus \$15 for the instruction book.

PROBLEM SOLVING STRATEGY

VERBAL MODEL: Price per fly · Number of flies sold − Total expenses ≥ Desired profit

LABELS

Price per fly = **0.58**	(dollars per fly)
Number of flies sold = x	(flies)
Total expenses = $\mathbf{0.15x + 15}$	(dollars)
Desired profit = **200**	(dollars)

ALGEBRAIC MODEL

$0.58x - (0.15x + 15) \geq 200$	**Write algebraic model.**
$0.58x - 0.15x - 15 \geq 200$	**Use distributive property.**
$0.43x - 15 \geq 200$	**Combine like terms.**
$0.43x \geq 215$	**Add 15 to each side.**
$x \geq 500$	**Divide each side by 0.43.**

▶ You need to tie and sell at least 500 flies.

✓**CHECK** You can check your result graphically by graphing equations for sales and for expenses separately.

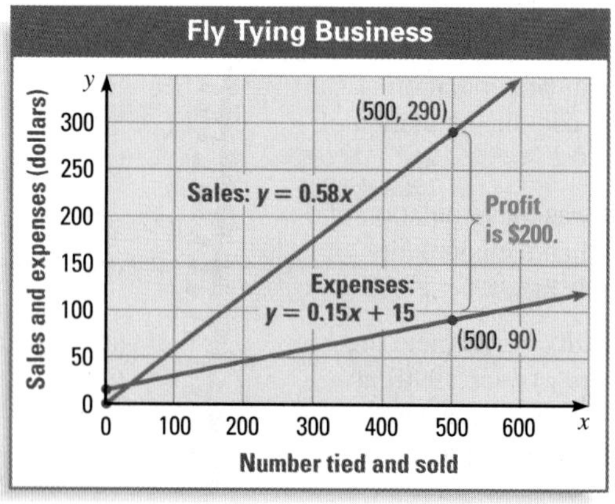

FOCUS ON APPLICATIONS

FLY FISHING In fly fishing, flies are made of feathers, hair, or other materials. They are intended to look like insects.

GUIDED PRACTICE

Vocabulary Check ✓

1. Explain why an inequality such as $3a + 6 \geq 0$ is called a *multi-step* inequality. Give another example of a multi-step inequality. **There are multiple steps involved in the solution; sample: $3x - 4 < 8$.**

Concept Check ✓

2. Describe the steps you could use to solve the inequality $-3y + 2 > 11$. **Subtract 2 from each side, then divide each side by −3 and reverse the inequality symbol.**

ERROR ANALYSIS Describe and correct the error. **See margin.**

3.
$-4y + 10 < 15$
$-4y < 5$
$\frac{-4y}{-4} < \frac{5}{-4}$
$y < -\frac{5}{4}$

4.
$6x - 4 \geq 2x + 1$
$6x \geq 2x - 3$
$4x \geq -3$
$x \geq -\frac{3}{4}$

3. The inequality symbol was not reversed when each side of the inequality was divided by −4; $y > -\frac{5}{4}$.
4. 4 was added to the left side of the inequality but subtracted from the right side; $x \geq \frac{5}{4}$.

Skill Check ✓

Solve the inequality.

5. $y + 2 > -1$ **$y > -3$**
6. $-2x < -14$ **$x > 7$**
7. $-4x \geq -12$ **$x \leq 3$**
8. $4y - 3 < 13$ **$y < 4$**
9. $5x + 12 \leq 62$ **$x \leq 10$**
10. $10 - c \geq 6$ **$c \leq 4$**
11. $2 - x > 6$ **$x < -4$**
12. $3x + 2 \leq 7x$ **$x \geq \frac{1}{2}$**
13. $2x - 1 > 6x + 2$ **$x < -\frac{3}{4}$**
14. BUSINESS In Example 4, how many flies will you need to tie and sell to make a profit of at least $150? **384 flies**

PRACTICE AND APPLICATIONS

STUDENT HELP
Extra Practice to help you master skills is on p. 802.

SOLVING INEQUALITIES Solve the inequality.

15. $x + 5 > -13$ **$x > -18$**
16. $15 - x < 7$ **$x > 8$**
17. $-5 \leq 6x - 12$ **$x \geq \frac{7}{6}$**
18. $-6 + 5x < 19$ **$x < 5$**
19. $7 - 3x \leq 16$ **$x \geq -3$**
20. $-3x - 0.4 > 0.8$ **$x < -0.4$**
21. $6x + 5 < 23$ **$x < 3$**
22. $-17 > 5x - 2$ **$x < -3$**
23. $-x + 9 \geq 14$ **$x \leq -5$**
24. $4x - 1 \leq -17$ **$x \leq -4$**
25. $12 > -2x - 6$ **$x > -9$**
26. $-x - 4 > 3x - 2$ **$x < -\frac{1}{2}$**
27. $\frac{2}{3}x + 3 \geq 11$ **$x \geq 12$**
28. $6 \geq \frac{7}{3}x - 1$ **$x \leq 3$**
29. $-\frac{1}{2}x + 3 < 7$ **$x > -8$**
30. $3x + 1.2 < -7x - 1.3$ **$x < -0.25$**
31. $x + 3 \leq 2(x - 4)$ **$x \geq 11$**
32. $2x + 10 \geq 7(x + 1)$ **$x \leq \frac{3}{5}$**
33. $-x + 4 < 2(x - 8)$ **$x > 6\frac{2}{3}$**
34. $-x + 6 > -(2x + 4)$ **$x > -10$**
35. $-2(x + 3) < 4x - 7$ **$x > \frac{1}{6}$**
36. POPULATION In 1990 Tampa, Florida, had a population of 280,015. From 1990 through 1996, the population increased at an average rate of about 850 per year. Write a linear model for the population of Tampa.
▶ Source: U.S. Bureau of the Census **Let p = population and n = number of years after 1990; $p = 850n + 280{,}015$.**
37. AMUSEMENT PARK An amusement park charges $5 for admission and $1.25 for each ride. You go to the park with $25. Write an inequality that represents the possible number of rides you can go on. What is the maximum number of rides you can go on? **$1.25r + 5 \leq 25$; 16 rides**

STUDENT HELP
HOMEWORK HELP
Example 1: Exs. 15–35
Example 2: Exs. 15–35
Example 3: Ex. 36
Example 4: Exs. 37, 39

3 APPLY

ASSIGNMENT GUIDE

BASIC
Day 1: pp. 343–345 Exs. 16–34 even, 37, 39, 43–51, 55, 60, 61, 64, 65

AVERAGE
Day 1: pp. 343–345 Exs. 16–34 even, 37, 39, 43–51, 55, 60, 61, 64, 65

ADVANCED
Day 1: pp. 343–345 Exs. 16–34 even, 37, 39, 43–53, 55, 60, 61, 64, 65

BLOCK SCHEDULE WITH 6.3
pp. 343–345 Exs. 16–34 even, 37, 39, 43–51, 55, 60, 61, 64, 65

EXERCISE LEVELS

Level A: *Easier*
15–29
Level B: *More Difficult*
30–39, 43–51, 54–65
Level C: *Most Difficult*
40–42, 52, 53

✓ HOMEWORK CHECK

To quickly check student understanding of key concepts, go over the following exercises: Exs. 16, 30, 34, 39, 46. See also the Daily Homework Quiz:
- Blackline Master (*Chapter 6 Resource Book,* p. 35)
- Transparency (p. 47)

! COMMON ERROR

EXERCISES 34 AND 35 Some students may forget to distribute the negative in these exercises. Remind students that when multiplying by a negative number in the distributive property, the negative is distributed to both terms inside the parentheses.

MATHEMATICAL REASONING

Students can use a graphing calculator to solve a linear inequality in the form $ax + b < c$ (or $\leq$, $>$, or $\geq$), but they must be able to interpret the results. For example, to find the solutions of $3x + 2 < 5$, students begin by graphing $y_1 = 3x + 2$ and $y_2 = 5$. Ask them how the graphs indicate that the solution is $x < 1$. ***Sample answer:* For $x < 1$, the graph of $y_1 = 3x + 2$ will be below the graph of $y_2 = 5$.**

APPLICATION NOTE

EXERCISE 39 Students should understand that the domain for the inequality is the whole numbers. Point out that the maximum number of toppings would still be six even if the initial amount were \$19. A graph of the solution set would show a set of discreet points with a lower bound of zero.

STUDENT HELP NOTES

Homework Help Students can find help for Ex. 39 at **www.mcdougallittell.com.** The information can be printed out for students who don't have access to the Internet.

ADDITIONAL PRACTICE AND RETEACHING

For Lesson 6.2:

- Practice Levels A, B, and C (*Chapter 6 Resource Book,* p. 25)
- Reteaching with Practice (*Chapter 6 Resource Book,* p. 28)
- See Lesson 6.2 of the *Personal Student Tutor*

For more Mixed Review:

- Search the *Test and Practice Generator* for key words or specific lessons.

STUDENT HELP

HOMEWORK HELP Visit our Web site www.mcdougallittell.com for help with problem solving in Ex. 39.

38. **SWEETENERS** From 1980 to 1995, the per capita consumption of sugar sweeteners dropped from 83.6 pounds to 65.5 pounds. The per capita consumption of corn syrup rose from 38.2 pounds to 83.2 pounds. For which years did the per capita consumption of corn syrup exceed sugar sweeteners? **1991 to 1995**

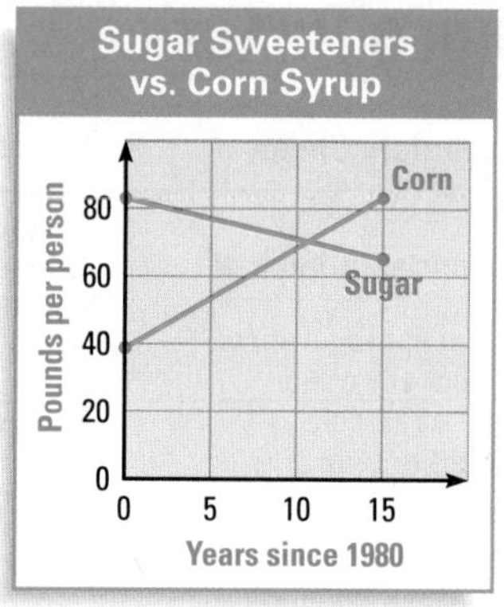

▶ Source: U.S. Department of Agriculture

39. **PIZZA** You have \$18.50 to spend on pizza. It costs \$14 plus \$.75 for each additional topping, tax included. Solve an inequality to find the maximum number of toppings the pizza can have. **$0.75t + 14 \leq 18.50$; 6 toppings**

40. $n \geq \left(\frac{24}{1 \text{ sec}}\right)\left(\frac{3 \text{ h}}{2}\right)\left(\frac{60 \text{ min}}{1 \text{ h}}\right)\left(\frac{60 \text{ sec}}{1 \text{ min}}\right)$, or $n \geq 129{,}600$

FOCUS ON APPLICATIONS

ANIMATED FILMS begin with animators' drawings. Artists trace the drawings onto sheets of acetate called *cels* and then paint the cels.

42. **If everyone worked the same number of hours per year as those on the original team, the time would be cut in half.**

ANIMATED FILMS In Exercises 40–42, you are working as an animator. Each frame in an animated feature film takes at least one hour to draw. When projected, 35-millimeter film runs at 24 frames per second. A $2\frac{1}{2}$-hour movie has about 216,000 frames.

40. Write an inequality that describes the number of hours it would take to draw the frames needed for a $1\frac{1}{2}$-hour animated feature film. **See margin.**

41. You are part of a team of 36 artists, each working 40 hours per week for 45 weeks per year. How many years would it take your team to draw the $1\frac{1}{2}$-hour animated feature film? **2 years**

42. **LOGICAL REASONING** If you doubled the size of your team, how would that affect the time it would take to draw the film? **See margin.**

GEOMETRY CONNECTION **Write an inequality for the values of *x*.**

43. Area > 36 square meters

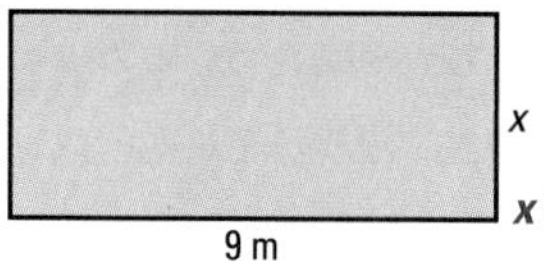

$x > 4$ m

44. Area ≥ 25 square meters

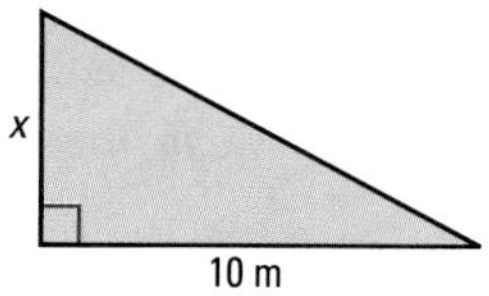

$x \geq 5$ m

45. Area < 30 square feet

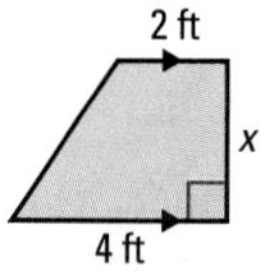

$x < 10$ ft

46. Area < 144 square inches

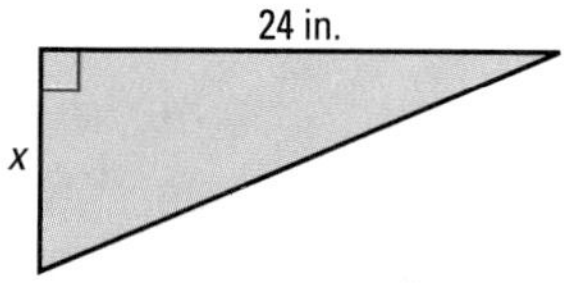

$x < 12$ in.

47. Area ≤ 25 square inches

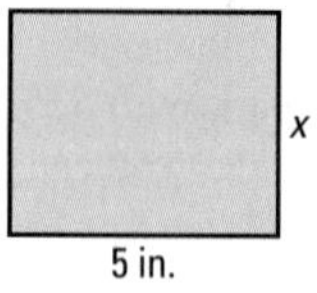

$x \leq 5$ in.

48. Area > 60 square centimeters

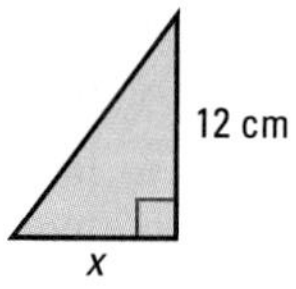

$x > 10$ cm

49. **MULTIPLE CHOICE** At a grocery store, four oranges cost \$2.39 and kiwi fruit costs \$.69 each. If you have \$5 to spend and you buy four oranges, which inequality represents the number of kiwi fruit you can buy? **A**

Ⓐ $x \le 3$ Ⓑ $x \ge 3$ Ⓒ $x > 3$ Ⓓ $x < 7$

50. **MULTIPLE CHOICE** The Glee Club budgeted \$250 for food for the annual Spaghetti Supper. Each meal costs \$1.75 to prepare. Which inequality represents the number of meals that can be prepared without going over the budget? **C**

Ⓐ $x \le 143$ Ⓑ $x \ge 143$ Ⓒ $x \le 142$ Ⓓ $x \ge 142$

51. **MULTIPLE CHOICE** For Park College's basketball games, it costs \$15 to attend. A season's pass costs \$170. At most, how many games could you attend at the \$15 price before spending more than the cost of a season's pass? **B**

Ⓐ 10 Ⓑ 11 Ⓒ 15 Ⓓ 17

★ **Challenge**

52. **GRAPHICAL REASONING** Use the graph of $y = \frac{2}{3}x - 2$ to solve the inequality $\frac{2}{3}x - 2 < 0$. Explain your reasoning. **See margin.**

53. **GRAPHICAL REASONING** Use the graph of $y = -2x + 4$ to solve the inequality $-2x + 4 > 0$. Explain your reasoning. **See margin.**

Test Preparation

52. *Sample answer*: To find the solution of $\frac{2}{3}x - 2 = 0$, graph the equations relating to both sides of the equation, $y = \frac{2}{3}x - 2$ and $y = 0$, and locate their intersection, (3, 0). The x-coordinate, 3, is the solution of the original equation. Notice that on the graph of $y = \frac{2}{3}x - 2$, for points to the left of (3, 0), $y < 0$ and for points to the right of (3, 0), $y > 0$. Then the solution of $\frac{2}{3}x - 2 < 0$ is $x < 3$.

EXTRA CHALLENGE

www.mcdougallittell.com

MIXED REVIEW

EVALUATING EXPRESSIONS **Evaluate the expression.** (Review 1.3, 2.7)

54. $(a + 4) - 8$ when $a = 7$ **3**

55. $3x + 2$ when $x = -4$ **−10**

56. $b^3 - 5$ when $b = 2$ **3**

57. $2(r + s)$ when $r = 2$ and $s = 4$ **12**

58. $x^2 - 3x$ when $x = 5$ **10**

59. $\frac{a^2}{8 - b}$ when $a = 4$ and $b = -9$ $\frac{16}{17}$

TRANSLATING VERBAL SENTENCES **Translate the verbal sentence into an equation or an inequality.** (Review 1.5 for 6.3)

60. Sarah's height S is 4 inches more than Joanne's height J. $S = J + 4$

61. The number of tickets t sold this week is more than twice the number of tickets l sold last week. $t > 2l$

WRITING INTEGERS **In Exercises 62 and 63, write an integer to represent the situation.** (Review 2.1)

62. a loss of \$12 **−12**

63. three feet above ground **3**

64. **SHOPPING TRIP** You buy a pair of shoes for \$42.99, a shirt for \$14.50, and a pair of jeans for \$29.99. You have a coupon for \$10 off your purchase. How much did you spend? (Review 1.3) **\$77.48**

65. **PROBABILITY** You put 20 slips of paper in a hat. Eight slips are yellow, five slips are blue, and the rest are red. Without looking, you reach into the hat and choose a slip of paper. What is the probability that the slip of paper is not blue? (Review 2.8) **0.75**

53. *Sample answer*: To find the solution of $-2x + 4 = 0$, graph the equations relating to both sides of the equation, $y = -2x + 4$ and $y = 0$, and locate their intersection, (2, 0). The x-coordinate, 2, is the solution of the original equation. Notice that on the graph of $y = -2x + 4$, for points to the left of (2, 0), $y > 0$ and for points to the right of (2, 0), $y < 0$. Then the solution of $-2x + 4 > 0$ is $x < 2$.

4 ASSESS

DAILY HOMEWORK QUIZ

Transparency Available

Solve the inequality.

1. $4 - 3x \ge 7$ $x \le -1$
2. $2x - 3 < x + 6$ $x < 9$
3. $-2x + 1 > 3(x - 8)$ $x < 5$
4. A triangle has a height of 15 cm and a base of x cm. Write and solve an inequality for the values of x that give an area of at most 105 cm^2. $\frac{1}{2}x(15) \le 105$; $x \le 14$ cm

EXTRA CHALLENGE NOTE

Challenge problems for Lesson 6.2 are available in **blackline** format in the *Chapter 6 Resource Book*, p. 32 and at **www.mcdougallittell.com.**

ADDITIONAL TEST PREPARATION

1. **WRITING** The graph shows how the cost of producing two types of sandals is dependent on the demand for each type. Explain the significance of the dashed segment.

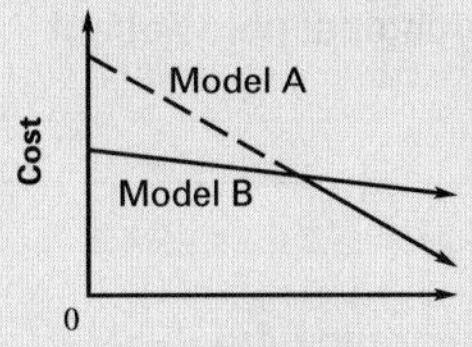

Sample answer: This section of the graph represents the levels of demand for which the cost of Model A is higher than the cost of Model B.

1 PLAN

PACING
Basic: 1 day
Average: 1 day
Advanced: 1 day
Block Schedule: 0.5 block with 6.2

LESSON OPENER ACTIVITY
An alternative way to approach Lesson 6.3 is to use the Activity Lesson Opener:
- Blackline Master (*Chapter 6 Resource Book,* p. 36)
- Transparency (p. 41)

MEETING INDIVIDUAL NEEDS
- ***Chapter 6 Resource Book***
  Prerequisite Skills Review (p. 5)
  Practice Level A (p. 39)
  Practice Level B (p. 40)
  Practice Level C (p. 41)
  Reteaching with Practice (p. 42)
  Absent Student Catch-Up (p. 44)
  Challenge (p. 47)
- ***Resources in Spanish***
- ***Personal Student Tutor***

NEW-TEACHER SUPPORT
See the Tips for New Teachers on pp. 1–2 of the *Chapter 6 Resource Book* for additional notes about Lesson 6.3.

WARM-UP EXERCISES
Transparency Available
Use the numbers –5, –4, –3, –2, –1, 0, 1, 2, 3, 4, 5.
1. Which numbers are less than or equal to –1 and greater than or equal to –2? **–2, –1**
2. Which numbers are greater than 1 or less than –3? **–5, –4, 2, 3, 4, 5**
3. Which numbers are less than or equal to –2 and less than or equal to 2? **–5, –4, –3, –2**
4. Which numbers are greater than –1 or greater than 3? **0, 1, 2, 3, 4, 5**

6.3 Solving Compound Inequalities

What you should learn
GOAL 1 Write, solve, and graph compound inequalities.
GOAL 2 Model a **real-life** situation with a compound inequality, such as the distances in **Example 6**.

Why you should learn it
▼ To describe **real-life** situations, such as elevations on Mount Rainier in **Example 2**.

REAL LIFE

GOAL 1 SOLVING COMPOUND INEQUALITIES

In Lesson 6.1 you studied four types of simple inequalities. In this lesson you will study *compound inequalities.* A **compound inequality** consists of two inequalities connected by *and* or *or*.

EXAMPLE 1 *Writing Compound Inequalities*

Write an inequality that represents the set of numbers and graph the inequality.

a. All real numbers that are greater than zero *and* less than or equal to 4.

b. All real numbers that are less than -1 *or* greater than 2.

SOLUTION

a. $0 < x \le 4$

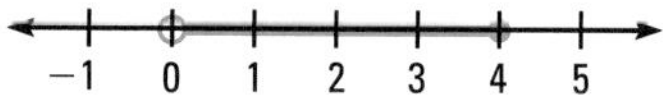

This inequality is also written as $0 < x$ *and* $x \le 4$.

b. $x < -1$ *or* $x > 2$

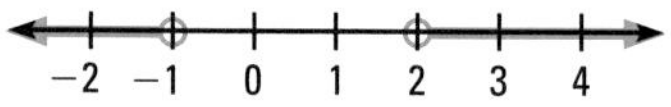

EXAMPLE 2 *Compound Inequalities in Real Life*

Write an inequality that describes the elevations of the regions of Mount Rainier.

a. Timber region below 6000 ft

b. Alpine meadow region below 7500 ft

c. Glacier and permanent snow field region

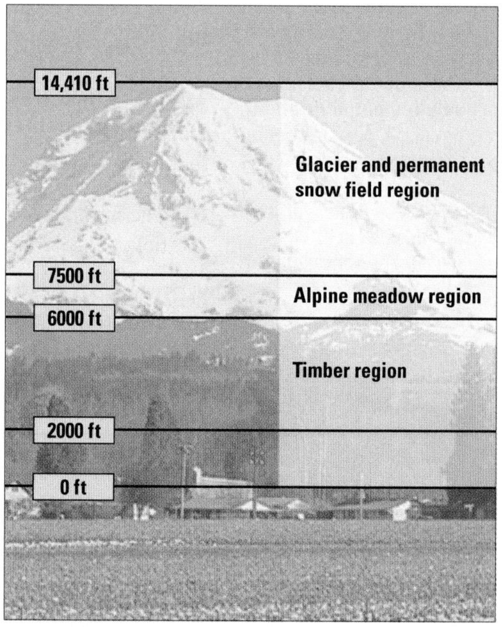

SOLUTION Let y represent the approximate elevation (in feet).

a. Timber region:
$2000 \le y < 6000$

b. Alpine meadow region:
$6000 \le y < 7500$

c. Glacier and permanent snow field region:
$7500 \le y \le 14{,}410$

STUDENT HELP

HOMEWORK HELP
Visit our Web site www.mcdougallittell.com for extra examples.

EXAMPLE 3 *Solving a Compound Inequality with And*

Solve $-2 \le 3x - 8 \le 10$. Graph the solution.

SOLUTION
Isolate the variable x between the two inequality symbols.

$-2 \le 3x - 8 \le 10$ **Write original inequality.**

$6 \le 3x \le 18$ **Add 8 to each expression.**

$2 \le x \le 6$ **Divide each expression by 3.**

▶ The solution is all real numbers that are greater than or equal to 2 *and* less than or equal to 6.

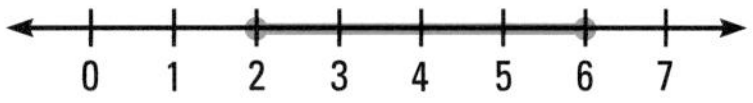

EXAMPLE 4 *Solving a Compound Inequality with Or*

Solve $3x + 1 < 4$ *or* $2x - 5 > 7$. Graph the solution.

SOLUTION A solution of this inequality is a solution of either of its simple parts. You can solve each part separately.

$3x + 1 < 4$ *or* $2x - 5 > 7$

$3x < 3$ *or* $2x > 12$

$x < 1$ *or* $x > 6$

▶ The solution is all real numbers that are less than 1 *or* greater than 6.

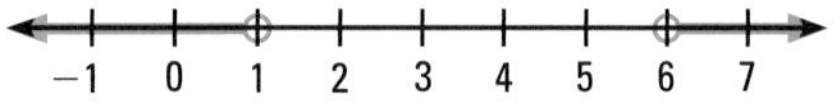

STUDENT HELP

Study Tip
When you multiply or divide by a negative number to solve a compound inequality, remember that you have to reverse *both* inequality symbols.

EXAMPLE 5 *Reversing Both Inequality Symbols*

Solve $-2 < -2 - x < 1$. Graph the solution.

SOLUTION
Isolate the variable x between the two inequality signs.

$-2 < -2 - x < 1$ **Write original inequality.**

$0 < -x < 3$ **Add 2 to each expression.**

$0 > x > -3$ **Multiply each expression by −1 and *reverse* both inequality symbols.**

▶ To match the order of numbers on a number line, this compound inequality is usually written as $-3 < x < 0$. The solution is all real numbers that are greater than -3 *and* less than 0.

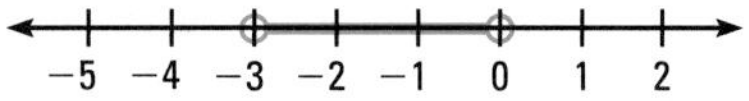

2 TEACH

MOTIVATING THE LESSON
Have students imagine they are helping plan the spring dance. It will take at least 100 people to make the dance profitable. The building safety code prohibits more than 300 occupants. Tell students that the number of students must satisfy two inequalities at once. It takes a *compound* inequality to model this situation.

EXTRA EXAMPLE 1
Write an inequality that represents the set of numbers and sketch its graph.

a. All real numbers that are greater than or equal to −2 *and* less than 3. $-2 \le x < 3$

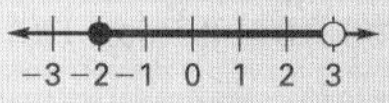

b. All real numbers that are less than or equal to −3 *or* greater than 0. $x \le -3$ or $x > 0$

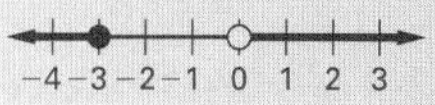

EXTRA EXAMPLE 2
Write an inequality that describes each condition.

a. Water is a nonliquid when the temperature is 32°F or below, or is at least 212°F. $T \le 32$ or $T \ge 212$

b. A refrigerator is designed to work on an electric line carrying from 115 volts to 120 volts. $115 \le V \le 120$

EXTRA EXAMPLE 3
Solve $-5 \le 2x + 3 < 7$. $-4 \le x < 2$

EXTRA EXAMPLE 4
Solve $6x - 5 < 7$ *or* $8x + 1 > 25$. $x < 2$ or $x > 3$

EXTRA EXAMPLE 5
Solve $-3 < -1 - 2x \le 5$. $-3 \le x < 1$

CHECKPOINT EXERCISES
For use after Examples 1–5:

1. Solve $4 \ge 3x + 1 > -2$. $-1 < x \le 1$

2. Solve $-3x + 5 > 8$ *or* $-x < -4$. $x < -1$ or $x > 4$

EXTRA EXAMPLE 6

The park is $1\frac{1}{2}$ miles from your house. The mall is $2\frac{1}{4}$ miles from your house.

a. Find the minimum distance between the park and the mall. **$\frac{3}{4}$ mi**

b. Find the maximum distance between the park and the mall. **$3\frac{3}{4}$ mi**

c. Write an inequality that describes the possible distances *d* between the park and the mall. **$\frac{3}{4} \le d \le 3\frac{3}{4}$**

CHECKPOINT EXERCISES

For use after Example 6:

1. Your family's new house is 45 miles from a ski area and 25 miles from a lake. What is the farthest you would have to drive to go from the ski area to the lake? **70 mi**

FOCUS ON VOCABULARY

Another way to express the compound inequality $a < x < b$ is to say that *x* is *between a* and *b.* When this kind of inequality is graphed, the number line is shaded between *a* and *b,* with open circles at *a* and *b.*

CLOSURE QUESTION

What is the difference between the graphs of "and" inequalities and "or" inequalities? ***Sample answer:*** **The graph of an "and" inequality is a segment. The graph of an "or" inequality is two rays going in opposite directions.**

DAILY PUZZLER

It's 0.7 mi from your home to the bakery. It's 0.5 mi from the bakery to the florist, 1.1 mi from the florist to the bookstore, and 1.0 mi from the bookstore to the shoe store. If all the stores lie in a line, what is the shortest possible distance you would have to walk from your home to the shoe store and back? **0.2 mi**

GOAL 2 MODELING REAL-LIFE PROBLEMS

EXAMPLE 6 *Modeling with a Compound Inequality*

REAL LIFE *Distances*

You have a friend Bill who lives three miles from school and another friend Mary who lives two miles from the same school. You wish to estimate the distance *d* that separates their homes.

a. What is the smallest value *d* might have?

b. What is the largest value *d* might have?

c. Write an inequality that describes all the possible values that *d* might have.

PROBLEM SOLVING STRATEGY

SOLUTION

DRAW A DIAGRAM A good way to begin this problem is to draw a diagram with the school at the center of a circle.

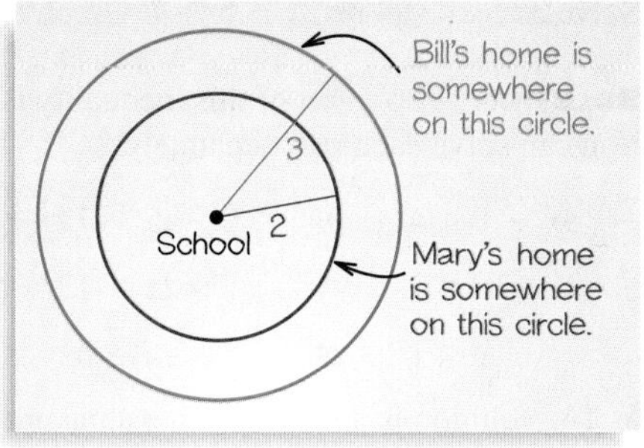

Bill's home is somewhere on the circle with radius 3 miles and center at the school.

Mary's home is somewhere on the circle with radius 2 miles and center at the school.

a. If both homes are on the same line going toward school, the distance is 1 mile.

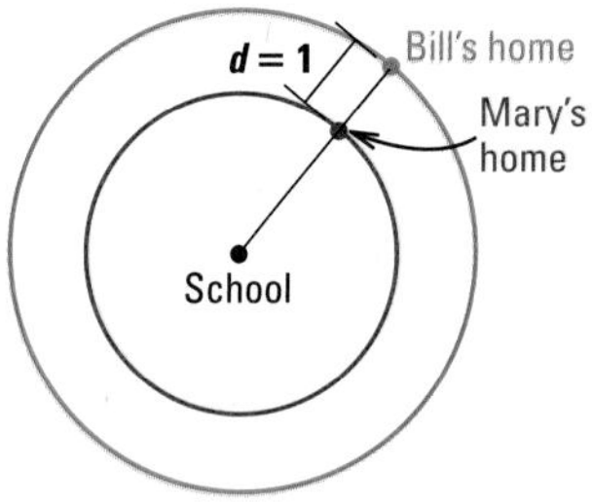

b. If both homes are on the same line but in opposite directions from school, the distance is 5 miles.

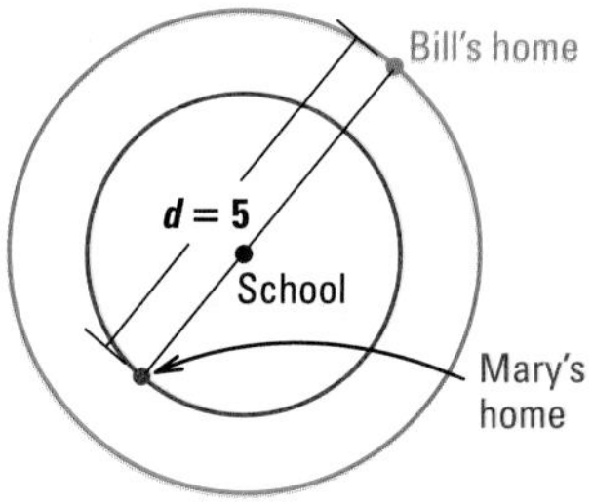

c. The values of *d* can be described by the inequality $1 \le d \le 5$.

Guided Practice

Vocabulary Check ✓

1. Write a compound inequality and explain why it is a compound inequality. **See margin.**

Concept Check ✓

2. Write a compound inequality that describes the graph at the right. **$-4 \le x < 0$**

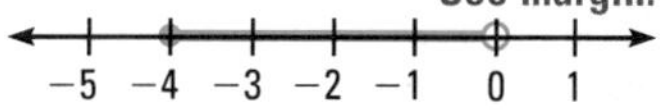

Skill Check ✓

WRITING INEQUALITIES Write an inequality that represents the statement and graph the inequality. 3, 4. See margin for graphs.

3. x is less than 5 and greater than 2. **$2 < x < 5$**
4. x is greater than 3 or less than -1. **$x > 3$ or $x < -1$**

Solve the inequality.

5. $7 < 4 + x < 8$ **$3 < x < 4$**
6. $-1 < 2x + 3 \le 13$ **$-2 < x \le 5$**
7. $6 < 4x - 2 \le 14$ **$2 < x \le 4$**
8. $4x - 1 > 7$ or $5x - 1 < -6$ **$x > 2$ or $x < -1$**
9. $2x + 3 < -1$ or $3x - 5 > -2$ **$x < -2$ or $x > 1$**
10. $4 \le -8 - x < 7$ **$-15 < x \le -12$**
11. In Example 6, write an inequality that describes the situation if you live two miles from school and your friend lives one mile from the same school. **$1 \le d \le 3$**

1. *Sample answers*: $x < -2$ or $x > 4$ (a compound inequality because it consists of two inequalities connected by *or*); $-12 \le x \le 9$ (a compound inequality because it consists of two inequalities connected by *and*).

Practice and Applications

STUDENT HELP

Extra Practice to help you master skills is on p. 802.

WRITING INEQUALITIES Write an inequality that represents the statement and graph the inequality. 12–17. See margin for graphs.

12. x is less than 8 and greater than 3. **$3 < x < 8$**
13. x is greater than 7 or less than 5. **$x > 7$ or $x < 5$**
14. x is less than 5 and is at least 0. **$0 \le x < 5$**
15. x is less than 4 and is at least -9. **$-9 \le x < 4$**
16. x is less than -2 and is at least -4. **$-4 \le x < -2$**
17. x is greater than -6 and less than -1. **$-6 < x < -1$**

SOLVING INEQUALITIES Solve the inequality. Write a sentence that describes the solution.

18. $6 < x - 6 \le 8$ **x is greater than 12 and less than or equal to 14.**
19. $-5 < x - 3 < 6$ **x is greater than -2 and less than 9.**
20. $-4 < 2 + x < 1$ **x is greater than -6 and less than -1.**
21. $8 \le 2x + 6 \le 18$ **x is greater than or equal to 1 and less than or equal to 6.**
22. $6 + 2x > 20$ or $8 + x \le 0$ **x is greater than 7 or less than or equal to -8.**
23. $-3x - 7 \ge 8$ or $-2x - 11 \le -31$ **See margin.**
24. $-4 \le -3x - 13 \le 26$ **See margin.**
25. $-13 \le 5 - 2x < 9$ **x is greater than -2 and less than or equal to 9.**

23. **x is less than or equal to -5 or greater than or equal to 10.**

24. **x is greater than or equal to -13 and less than or equal to -3.**

GRAPHING INEQUALITIES Solve the inequality and graph the solution. 26–31. See margin for graphs.

26. $3 \le 2x < 7$ **$\frac{3}{2} \le x < \frac{7}{2}$**
27. $-25 < 5x < -20$ **$-5 < x < -4$**
28. $-5 < 3x + 4 < 19$ **$-3 < x < 5$**
29. $-4 \le 9x - 1 < 5$ **$-\frac{1}{3} \le x < \frac{2}{3}$**
30. $2x + 7 < 3$ or $5x + 5 \ge 10$ **$x < -2$ or $x \ge 1$**
31. $3x + 8 > 17$ or $2x + 5 \le 7$ **$x > 3$ or $x \le 1$**

CHECKING SOLUTIONS Solve the inequality and graph the solution. Then check graphically whether the given x-value is a solution by graphing the x-value on the same number line. 32–35. See margin for graphs.

32. $-4 < 4x - 8 < 12$; $x = 1$ **$1 < x < 5$; not a solution**
33. $-1 < 7x - 15 \le 20$; $x = 5$ **$2 < x \le 5$; solution**
34. $-2x \ge 6$ or $2x + 1 > 5$; $x = 0$ **$x \le -3$ or $x > 2$; not a solution**
35. $7 \le -2x + 21 < 31$; $x = 0$ **$-5 < x \le 7$; solution**

STUDENT HELP

HOMEWORK HELP

Example 1: Exs. 12–17
Example 2: Exs. 36–38
Example 3: Exs. 18–35
Example 4: Exs. 18–35
Example 5: Exs. 18–35
Example 6: Exs. 41–44

12–17. See the next page.
26–35. See Additional Answers beginning on page AA1.

3 APPLY

Assignment Guide

BASIC
Day 1: pp. 349–351 Exs. 12–38 even, 39, 40, 45, 50, 62, 65, Quiz 1 Exs. 1–15

AVERAGE
Day 1: pp. 349–351 Exs. 12–38 even, 39, 40, 45, 50, 62, 65, Quiz 1 Exs. 1–15

ADVANCED
Day 1: pp. 349–351 Exs. 12–38 even, 39, 40, 45, 46–48, 50, 62, 65, Quiz 1 Exs. 1–15

BLOCK SCHEDULE WITH 6.2
pp. 349–351 Exs. 12–38 even, 39, 40, 45, 50, 62, 65, Quiz 1 Exs. 1–15

Exercise Levels

Level A: *Easier*
12–17

Level B: *More Difficult*
18–38, 49–69

Level C: *Most Difficult*
39–48

✓ Homework Check

To quickly check student understanding of key concepts, go over the following exercises: Exs. 14, 20, 24, 30, 34. See also the Daily Homework Quiz:

- Blackline Master (*Chapter 6 Resource Book*, p. 51)
- Transparency (p. 48)

! Common Error

EXERCISES 15 AND 16 Some students may have difficulty representing the expression *at least.* Have them choose several points from their graphs and verify that they satisfy both conditions of the inequality. Help students compare and contrast *at least, at most, not more than,* and *not less than.*

3. 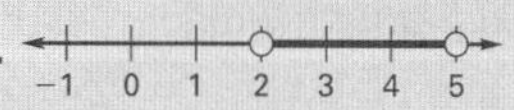

4. 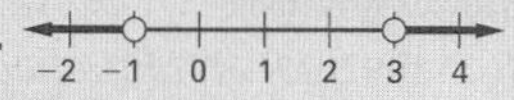

MATHEMATICAL REASONING

To check students' understanding of *and* compound inequalities, have them rewrite the inequalities $a < x < b$ and $a > x > b$ each as two simple inequalities joined by *and,* with x on the left side of each inequality. Ask them if for any two real numbers a and b both compound inequalities can be true. **$x > a$ and $x < b$; $x > b$ and $x < a$; no**

APPLICATION NOTE

EXERCISES 41–44 You may want to discuss the logarithmic scale used in the diagram. Explain that the planets are so far apart that it is otherwise difficult to show them together in one picture.

12.
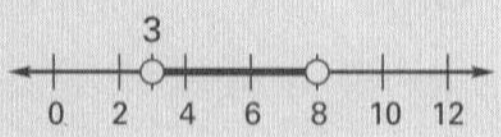

13.
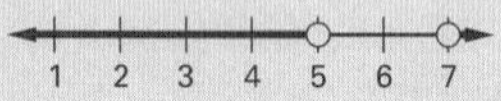

14.
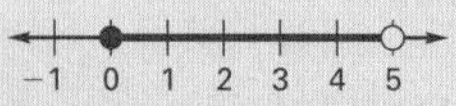

15.
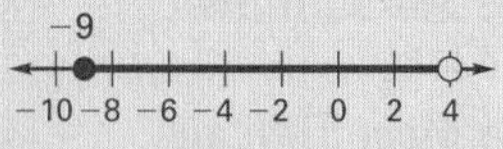

16.
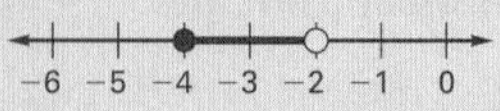

17.
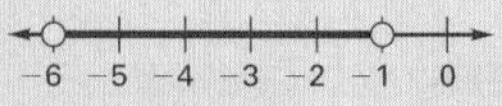

ADDITIONAL PRACTICE AND RETEACHING

For Lesson 6.3:

- Practice Levels A, B, and C (*Chapter 6 Resource Book,* p. 39)
- Reteaching with Practice (*Chapter 6 Resource Book,* p. 42)
- See Lesson 6.3 of the *Personal Student Tutor*

For more Mixed Review:

- Search the *Test and Practice Generator* for key words or specific lessons.

FOCUS ON APPLICATIONS

FINE ART The painting above is *Still Life with Apples* by Paul Cézanne. Another of Cézanne's paintings, *Still Life with Curtain, Pitcher and Bowl of Fruit,* sold at auction for $60.5 million in 1999.

36. **PRICES OF FINE ART** In 1958, the painting *Still Life with Apples* by Paul Cézanne sold at auction for \$252,000. The painting was auctioned again in 1993 and sold for \$28.6 million. Write a compound inequality that represents the different values that the painting probably was worth between 1958 and 1993. **$252{,}000 \le x \le 28{,}600{,}000$**

37. **TELEVISION ADVERTISING** In 1967 a 60-second TV commercial during the first Super Bowl cost \$85,000. In 1998 advertisers paid \$2.6 million for 60 seconds of commercial time (two 30-second spots). Write a compound inequality that represents the different prices that 60 seconds of commercial time during the Super Bowl probably cost between 1967 and 1998. **$85{,}000 \le x \le 2{,}600{,}000$**

38. **ANTELOPES** The table gives the weights of some adult antelopes. The eland is the largest antelope in Africa. The royal antelope is the smallest of all. Write a compound inequality that represents the different weights of these adult antelopes. **$7 \le x \le 2000$**

Antelope	Weight (lb)
Eland	2000
Kudu	700
Nyala	280
Springbok	95
Royal	7

39–40. Sample answers are given.

39. **LOGICAL REASONING** Explain why the inequality $3 < x < 1$ has no solution. **A solution would be a number on the graph of both inequalities. Since the graphs do not overlap, there are no such numbers.**

40. **LOGICAL REASONING** Explain why the inequality $x < 2$ *or* $x > 1$ has every real number as a solution.

40. A solution is a number that is on the graph of either inequality. Since every point that is not on the graph of one inequality is on the graph of the other, every real number is a solution.

SCIENCE CONNECTION **Use the diagram of distances in our solar system.**

41. Which compound inequality best describes the distance d (in miles) between the Sun and any of the nine planets? **A**

A. $10^7 < d < 10^{10}$

B. $10^6 < d < 10^{11}$

C. $10^8 < d < 10^{11}$

42. Write an inequality to describe an estimate of the distance d (in miles) between the Sun and Mercury. **$10^7 < d < 10^8$**

43. Write an inequality to describe an estimate of the distance d (in miles) between the Sun and Saturn. **$10^8 < d < 10^9$**

44. The Moon's distance d (in miles) from Earth varies from about 220,000 miles to about 250,000 miles. Write an inequality to represent this fact. **$220{,}000 < d < 250{,}000$**

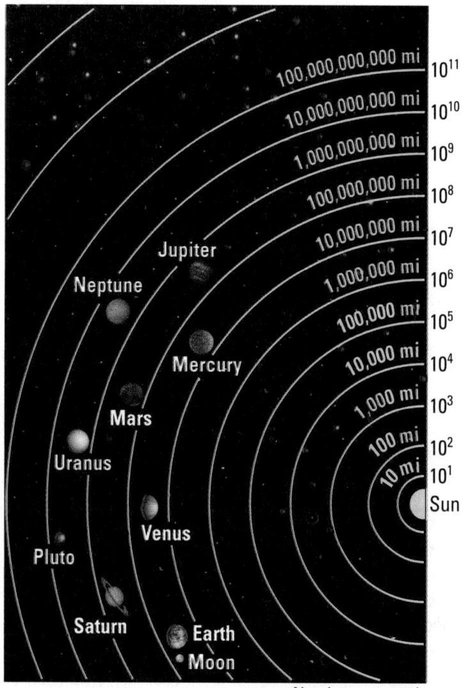

Not drawn to scale

Test Preparation

45. Multi-Step Problem You have a friend Kiko who lives one mile from a pool. Another friend D'evon lives 0.5 mile from Kiko. Kiko walks from her home to the pool, where she meets D'evon. They both walk from the pool to D'evon's home and then to Kiko's home.

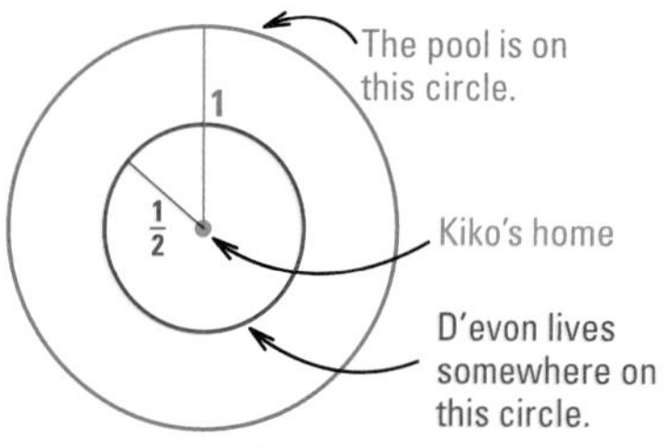

a. Write an inequality that describes the values, d, of the possible distances that D'evon could live from the pool. **$\frac{1}{2} \le d \le 1\frac{1}{2}$**

b. Write an inequality that describes the values, D, of the possible distances that Kiko could have walked. (Assume that each part of the walk followed a straight path.) **$2 \le d \le 3$**

★ Challenge

Geometry Connection **Write a compound inequality that must be satisfied by the length of the side labeled x. Use the fact that the sum of the lengths of any two sides of a triangle is greater than the length of the third side.**

46.

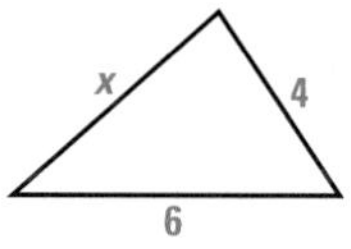

$2 < x < 10$

47.

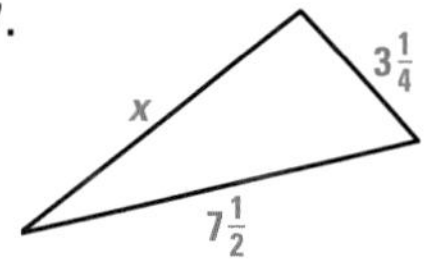

$4\frac{1}{4} < x < 10\frac{3}{4}$

48.

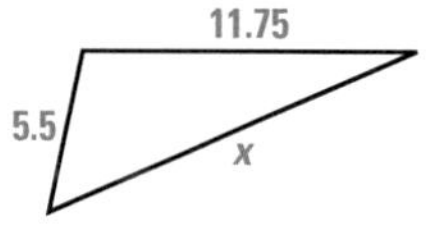

$6.25 < x < 17.25$

Extra Challenge

www.mcdougallittell.com

Mixed Review

Evaluating Expressions **Evaluate the expression.** (Review 1.1)

49. $x + 5$ when $x = 2$ **7**
50. $6.5a$ when $a = 4$ **26**
51. $m - 20$ when $m = 30$ **10**
52. $\frac{x}{15}$ when $x = 30$ **2**
53. $5x$ when $x = 3.3$ **16.5**
54. $4.2p$ when $p = 4.1$ **17.22**

Solving Equations **Solve the equation.** (Review 3.1, 3.2 for 6.4)

55. $x + 17 = 9$ **−8**
56. $8 = x + 2\frac{1}{2}$ **$5\frac{1}{2}$**
57. $x - 4 = 12$ **16**
58. $x - (-9) = 15$ **6**
59. $\frac{1}{2}x = -6$ **−12**
60. $-3x = -27$ **9**
61. $4x = -28$ **−7**
62. $-\frac{3}{4}x = 21$ **−28**
63. $x + 3.2 = 11$ **7.8**
64. $x - 4 = 16.7$ **20.7**
65. $\frac{x}{-6} = -\frac{1}{2}$ **3**
66. $\frac{5}{6}x = -25$ **−30**

67. Ice Skating An ice skating rink charges \$4.75 for admission and skate rental. If you bring your own skates, the admission is \$3.25. You can buy a pair of ice skates for \$45. How many times must you go ice skating to justify buying your own skates? (Review 3.4) **at least 30 times**

Hiking **You are hiking a six-mile trail at a constant rate in Topanga Canyon. You begin at 10 A.M. At noon you are two miles from the end of the trail.** (Review 5.5)

68. Write a linear equation that gives the distance d (in miles) from the end of the trail in terms of time t. Let t represent the number of hours since 10 A.M. **$d = 6 - 2t$**

69. Find the distance you are from the end of the trail at 11 A.M. **4 mi**

4 ASSESS

Daily Homework Quiz

Transparency Available

1. Write an inequality that represents the statement "x is at most −4 or at least 1." **$x \le -4$ or $x \ge 1$**

Solve the inequality. Write a sentence that describes the solution.

2. $-11 < -3x - 2 < 1$
$-1 < x < 3$; x is greater than −1 and less than 3.

3. $-2x - 3 > 5$ or $3x + 2 \ge -1$
$x < -4$ or $x \ge -1$; x is less than −4 or x is at least −1.

Solve the inequality. Graph the solution. Is the indicated value of x a solution?

4. $3 \le 2x - 7 < 11$; $x = 9$
$5 \le x < 9$; no

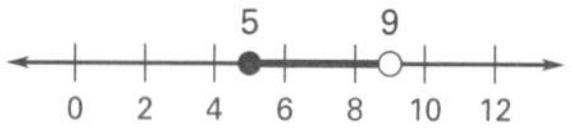

5. $5x - 3 \le 6$ or $7 < 4x - 9$; $x = 1.9$
$x \le 1.8$ or $x > 4$; no

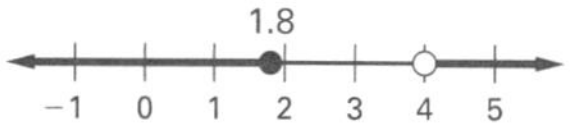

Extra Challenge Note

Challenge problems for Lesson 6.3 are available in **blackline** format in the *Chapter 6 Resource Book*, p. 47 and at **www.mcdougallittell.com.**

Additional Test Preparation

1. Writing Write two different sentences that describe the compound inequality $-4 \le x \le 6$. ***Sample sentences:*** **x is greater than or equal to −4 and less than or equal to 6; x is at least −4 and at most 6; x is between −4 and 6, inclusive; x is not less than −4 and is not greater than 6.**

In Exercises 1–13, solve the inequality and graph the solution.
(Lessons 6.2 and 6.3) 1–13. See margin for graphs.

1. $x + 2 < 7$ $x < 5$

2. $-3 + x \le -11$
 $x \le -8$

3. $3.4x \le 13.6$
 $x \le 4$

4. $5 \le -\frac{x}{2}$ $x \le -10$

5. $-4x - 2 \ge 14$
 $x \le -4$

6. $-5 < x - 8 < 4$
 $3 < x < 12$

7. $-x - 4 > 3x - 12$
 $x < 2$

8. $x + 3 \le 2(x - 7)$
 $x \ge 17$

9. $-10 \le -4x - 18 \le 30$
 $-12 \le x \le -2$

10. $-3 < x + 6$ or $-\frac{x}{3} > 4$
 $x > -9$ or $x < -12$

11. $2 - x < -3$ or $2x + 14 < 12$
 $x > 5$ or $x < -1$

12. $2x - 6 < -8$ or $10 - 5x < -19$

13. $6x - 2 > -7$ or $-3x - 1 > 11$

14. 🌐 **AMUSEMENT PARK** A person must be at least 52 inches tall to ride the *Power Tower* ride at Cedar Point in Ohio. Write an inequality that describes the required heights. (Lesson 6.1) $x \ge 52$

15. 🌐 **TEMPERATURES** The lowest temperature ever recorded was $-128.6°F$ at the Soviet station Vostok in Antarctica. The highest temperature ever recorded was $136°F$ at Azizia, Libya. Write a compound inequality whose solution includes all of the other temperatures T ever recorded. (Lesson 6.3)
▶ Source: National Climatic Data Center
$-128.6 < T < 136$

Margin answers (graphs):

1. (number line) 0 1 2 3 4 5 6
2. (number line) -12 -8 -4 0
3. (number line) -1 0 1 2 3 4 5
4. (number line) -12 -8 -4 0
5. (number line) -8 -6 -4 -2 0 2 4
6. (number line) -3 0 3 6 9 12 15
7. (number line) -2 -1 0 1 2 3 4
8. (number line) 17; 0 8 16 24
9. (number line) -12 -8 -4 0
10. (number line) -15 -12 -9 -6 -3 0 3
11. (number line) -1 0 1 2 3 4 5
12. $x < -1$ or $x > 5\frac{4}{5}$; (number line) -1, $5\frac{4}{5}$; -2 0 2 4 6 8 10
13. $x > -\frac{5}{6}$ or $x < -4$; (number line) $-\frac{5}{6}$; -5 -4 -3 -2 -1 0 1

MATH & History **History of Communication**

🌐 APPLICATION LINK
www.mcdougallittell.com

THEN ▶ **IN 1860** you could send a message from St. Joseph, Missouri, to Sacramento, California, in 10 to 11 days via Pony Express. By 1884 you could send a message in at least $4\frac{1}{2}$ days by transcontinental railroad.

NOW ▶ **BY 1972** it took less than a second to send a message across the country via e-mail.

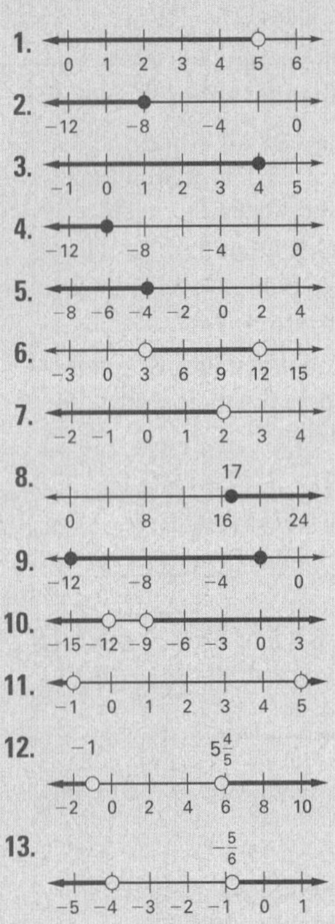

Write an inequality to represent the time it took to send the message.

1. Via Pony Express. Let t represent the time (in days). $10 \le t \le 11$

2. Via transcontinental railroad. Let t represent the time (in hours). $t \ge 108$

3. Via e-mail. Let t represent the time (in seconds). $t < 1$

Pony Express

Transcontinental Railroad

Air mail

e-mail

1860 1884 1920 1972

6.4 Solving Absolute-Value Equations and Inequalities

What you should learn

GOAL 1 Solve absolute-value equations.

GOAL 2 Solve absolute-value inequalities.

Why you should learn it

▼ To solve **real-life** problems such as finding the wavelengths of different colors of fireworks in **Exs. 65–68**.

REAL LIFE

GOAL 1 SOLVING ABSOLUTE-VALUE EQUATIONS

You can solve some absolute-value equations using mental math. For instance, you learned in Lesson 2.1 that the equation $|x| = 8$ has *two* solutions: 8 and -8.

To solve absolute-value equations, you can use the fact that the expression inside the absolute value symbols can be either positive or negative.

EXAMPLE 1 *Solving an Absolute-Value Equation*

Solve $|x - 2| = 5$.

SOLUTION

Because $|x - 2| = 5$, the expression $x - 2$ can be equal to 5 or to -5.

***x* − 2 IS POSITIVE**	***x* − 2 IS NEGATIVE**
$\lvert x - 2\rvert = 5$	$\lvert x - 2\rvert = 5$
$x - 2 = +5$	$x - 2 = -5$
$x = 7$	$x = -3$

▶ The equation has two solutions: 7 and -3.

✓CHECK Substitute both values into the original equation.

$|7 - 2| = |5| = 5$ $\qquad$ $|-3 - 2| = |-5| = 5$

EXAMPLE 2 *Solving an Absolute-Value Equation*

Solve $|2x - 7| - 5 = 4$.

SOLUTION

Isolate the absolute-value expression on one side of the equation.

2*x* − 7 IS POSITIVE	**2*x* − 7 IS NEGATIVE**
$\lvert 2x - 7\rvert - 5 = 4$	$\lvert 2x - 7\rvert - 5 = 4$
$\lvert 2x - 7\rvert = 9$	$\lvert 2x - 7\rvert = 9$
$2x - 7 = +9$	$2x - 7 = -9$
$2x = 16$	$2x = -2$
$x = 8$	$x = -1$

▶ The equation has two solutions: 8 and -1. Check these solutions in the original equation.

1 PLAN

PACING

Basic: 2 days
Average: 2 days
Advanced: 2 days
Block Schedule: 1 block

LESSON OPENER CALCULATOR ACTIVITY

An alternative way to approach Lesson 6.4 is to use the Calculator Activity Lesson Opener:

- Blackline Master (*Chapter 6 Resource Book,* p. 52)
- Transparency (p. 42)

MEETING INDIVIDUAL NEEDS

- ***Chapter 6 Resource Book***
  Prerequisite Skills Review (p. 5)
  Practice Level A (p. 54)
  Practice Level B (p. 55)
  Practice Level C (p. 56)
  Reteaching with Practice (p. 57)
  Absent Student Catch-Up (p. 59)
  Challenge (p. 61)
- ***Resources in Spanish***
- ***Personal Student Tutor***

NEW-TEACHER SUPPORT

See the Tips for New Teachers on pp. 1–2 of the *Chapter 6 Resource Book* for additional notes about Lesson 6.4.

WARM-UP EXERCISES

Transparency Available

Which values of *x* make the statement true?

1. $|x| = 5$ **5 and −5**
2. $-|x| = -9$ **9 and −9**

Solve.

3. $x - 7 < -2$ *or* $x - 7 > 7$ **$x < 5$ or $x > 14$**
4. $10 \geq 2x + 4 \geq -10$ **$-7 \leq x \leq 3$**

2 TEACH

MOTIVATING THE LESSON
Ask students if they have seen news survey results on television. Point out that there is usually a note shown saying something like "results accurate to ±4 percentage points." Tell students that an absolute-value inequality represents the expected actual values.

ACTIVITY NOTE
Partner Activity Have students work in pairs. One student can graph the solution set. The other student can write the compound inequality that describes the solution set. Have partners switch roles.

EXTRA EXAMPLE 1
Solve $|x - 4| = 8$. **−4 and 12**

EXTRA EXAMPLE 2
Solve $|5x + 1| + 3 = 14$. **−2.4 and 2**

EXTRA EXAMPLE 3
Solve $|x + 6| < 8$. **$-14 < x < 2$**

CHECKPOINT EXERCISES
For use after Examples 1 and 2:
1. Solve $|9 + x| = 15$. **−24 and 6**
2. Solve $|7 - 4x| - 4 = 14$. **−2.75 and 6.25**
For use after Example 3:
3. Solve $|x - 9| \le 1$. **$8 \le x \le 10$**

CONCEPT QUESTION
EXAMPLE 1 How many units from 2 will the two values of x be? **5**

COMMON ERROR
EXAMPLE 2 Make sure students understand that the definition of absolute value should be applied to this equation only after the absolute-value expression is isolated on one side of the equation.

1–3. See Additional Answers beginning on page AA1.

GOAL 2 SOLVING ABSOLUTE-VALUE INEQUALITIES

Recall that $|x|$ is the distance between x and 0. If $|x| < 8$, then any number between -8 and 8 is a solution of the inequality.

ACTIVITY Developing Concepts

Investigating Absolute-Value Inequalities

Use Guess, Check, and Revise to find values of x that satisfy each absolute-value inequality. Graph the solution set on a number line. Then use a compound inequality to describe the solution set.

1. $|x| < 2$ **$-2 < x < 2$** **2.** $|x + 2| \ge 1$ **$x \le -3$ or $x \ge -1$** **3.** $|x - 3| \le 2$ **$1 \le x \le 5$**

1–3. See margin for graphs.

You can use these properties to solve absolute-value inequalities and equations.

SOLVING ABSOLUTE-VALUE EQUATIONS AND INEQUALITIES

Each absolute-value equation or inequality is rewritten as two equations or two inequalities joined by *and* or *or*.

- $|ax + b| < c$ means $ax + b < c$ *and* $ax + b > -c$.
- $|ax + b| \le c$ means $ax + b \le c$ *and* $ax + b \ge -c$.
- $|ax + b| = c$ means $ax + b = c$ *or* $ax + b = -c$.
- $|ax + b| > c$ means $ax + b > c$ *or* $ax + b < -c$.
- $|ax + b| \ge c$ means $ax + b \ge c$ *or* $ax + b \le -c$.

Notice that when an absolute value is *less than* a number, the inequalities are connected by *and*. When an absolute value is *greater than* a number, the inequalities are connected by *or*.

EXAMPLE 3 *Solving an Absolute-Value Inequality*

Solve $|x - 4| < 3$.

SOLUTION

$x - 4$ IS POSITIVE	$x - 4$ IS NEGATIVE
$\lvert x - 4\rvert < 3$	$\lvert x - 4\rvert < 3$
$x - 4 < +3$	$x - 4 > -3$ ← **Reverse inequality symbol.**
$x < 7$	$x > 1$

▶ The solution is all real numbers greater than 1 *and* less than 7, which can be written as $1 < x < 7$.

STUDENT HELP
Study Tip
When you rewrite an absolute-value inequality, reverse the inequality symbol in the inequality involving $-c$.

STUDENT HELP

HOMEWORK HELP
Visit our Web site www.mcdougallittell.com for extra examples.

EXAMPLE 4 *Solving an Absolute-Value Inequality*

Solve $|2x + 1| - 3 \geq 6$. Then graph the solution.

SOLUTION

2x + 1 IS POSITIVE	**2x + 1 IS NEGATIVE**	
$\|2x + 1\| - 3 \geq 6$	$\|2x + 1\| - 3 \geq 6$	
$\|2x + 1\| \geq 9$	$\|2x + 1\| \geq 9$	
$2x + 1 \geq +9$	$2x + 1 \leq -9$	← **Reverse inequality symbol.**
$2x \geq 8$	$2x \leq -10$	
$x \geq 4$	$x \leq -5$	

▶ The solution of $|2x + 1| - 3 \geq 6$ is all real numbers greater than or equal to 4 *or* less than or equal to -5, which can be written as the compound inequality $x \leq -5$ *or* $x \geq 4$.

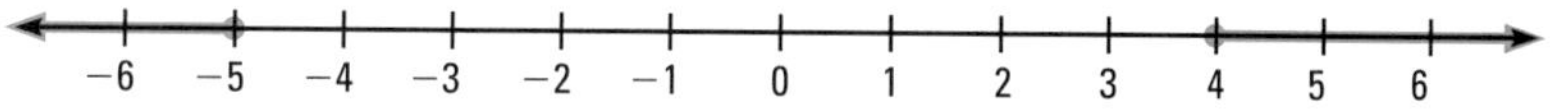

FOCUS ON CAREERS

QUALITY CONTROL INSPECTOR
A quality control inspector usually works in a manufacturing company. Experienced workers usually advance to more complex inspecting positions.

CAREER LINK
www.mcdougallittell.com

EXAMPLE 5 *Writing an Absolute-Value Inequality*

You work in the quality control department of a manufacturing company. The diameter of a drill bit must be between 0.62 inch and 0.63 inch.

a. Write an absolute-value inequality to represent this requirement.

b. A bit has a diameter of 0.623 inch. Does it meet the requirement?

SOLUTION

a. Let d represent the diameter (in inches) of the drill bit.

Write a compound inequality.

$$0.62 \leq d \leq 0.63$$

Find the halfway point: 0.625.

Subtract 0.625 from each part of the compound inequality.

$$0.62 - 0.625 \leq d - 0.625 \leq 0.63 - 0.625$$

$$-0.005 \leq d - 0.625 \leq 0.005$$

Rewrite as an absolute-value inequality:

$$|d - 0.625| \leq 0.005$$

This inequality can be read as "the actual diameter must differ from 0.625 inch by no more than 0.005 inch."

b. Because $|0.623 - 0.625| \leq 0.005$, the bit does meet the requirement.

EXTRA EXAMPLE 4
Solve $|3x - 3| + 4 \geq 10$. Then graph the solution. $x \leq -1$ **or** $x \geq 3$

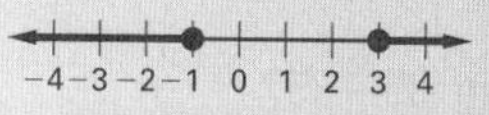

EXTRA EXAMPLE 5
To pass inspection, a computer part must measure from 0.0049 mm to 0.0051 mm.

a. Write an absolute-value inequality to represent this requirement.
$|d - 0.0050| \leq 0.0001$

b. A part measures 0.00495 mm. Will it pass inspection? **yes**

CHECKPOINT EXERCISES

For use after Example 4:

1. Solve $|2x + 5| - 2 \geq 7$.
$x \leq -7$ **or** $x \geq 2$

For use after Example 5:

2. If a machine part must measure from 0.0366 in. to 0.0378 in., what is the margin for error? **0.0006 in.**

FOCUS ON VOCABULARY
Though students may think of the *absolute value* of a number as "the positive of the number and its opposite," the algebraic operations in this lesson should reinforce the definition of absolute value as magnitude or distance from zero.

CLOSURE QUESTION
Describe the procedure to solve absolute-value equations and inequalities.
See sample answer below.

DAILY PUZZLER
Asieh's first shot on 6 golf holes varied less than 15 yd from 161 yd. Benna's first shot varied less than 18 yd from 128 yd. Is it possible that Benna hit as far as Asieh on any first shot? **no**

Closure Question *Sample answer:*
Isolate the absolute-value expression on one side of the equals sign or inequality symbol. Drop the absolute-value signs and set the expression equal to the positive and negative values. For inequalities, reverse the symbol for negative values. Solve the two resulting equations or inequalities.

3 APPLY

ASSIGNMENT GUIDE

BASIC

Day 1: pp. 356–357 Exs. 20–44 even

Day 2: pp. 357–358 Exs. 46–66 even, 69–71, 76–85

AVERAGE

Day 1: pp. 356–357 Exs. 20–44 even

Day 2: pp. 357–358 Exs. 46–68 even, 69–71, 76–85

ADVANCED

Day 1: pp. 356–357 Exs. 20–44 even

Day 2: pp. 357–358 Exs. 46–68 even, 69–73, 76–85

BLOCK SCHEDULE

pp. 356–358 Exs. 20–68 even, 69–71, 76–85

EXERCISE LEVELS

Level A: *Easier*
19–21, 74, 75

Level B: *More Difficult*
22–71, 76–85

Level C: *Most Difficult*
72, 73

✔ HOMEWORK CHECK

To quickly check student understanding of key concepts, go over the following exercises: Exs. 20, 26, 36, 40, 44, 56, 62. See also the Daily Homework Quiz:

- Blackline Master (*Chapter 6 Resource Book*, p. 64)
- Transparency (p. 49)

TEACHING TIPS

EXERCISE 16 This is the same as saying that the distance between x and 2 is greater than 9. Have students locate 2 on a number line. Then ask them to graph two points that are 9 units from 2. The points on the number line outside these two points are the solutions to the inequality.

GUIDED PRACTICE

Vocabulary Check ✔

1. $|x + 3| = 8$ is an example of a(n) _?_ and $|x + 3| \leq 8$ is an example of a(n) _?_. **absolute-value equation; absolute-value inequality**

Concept Check ✔

2. Explain each step you should use to solve $|x + 3| = 8$.
3. Explain why $|x - 5| < 2$ means that $x - 5$ is between -2 and 2. **See margin.**

2. Write two equations, $x + 3 = 8$ and $x + 3 = -8$. Solve each equation by subtracting 3 from each side: $x = 5$ and $x = -11$.

Complete the sentence using the word *and* or the word *or*.

4. $|x - 5| < 2$ means $x - 5 < 2$ _?_ $x - 5 > -2$. **and**
5. $|x - 5| > 2$ means $x - 5 > 2$ _?_ $x - 5 < -2$. **or**

Skill Check ✔

Solve the equation.

6. $|n| = 5$ **−5, 5**
7. $|a| = 0$ **0**
8. $|x + 3| = 6$ **−9, 3**
9. $|x - 4| = 10$ **−6, 14**
10. $|2n - 3| + 4 = 8$ **See margin.**
11. $|3x + 2| + 2 = 5$ **$\frac{1}{3}, -1\frac{2}{3}$**

3. $|x - 5| < 2$ means that $x - 5 > -2$ and $x - 5 < 2$, or $-2 < x - 5 < 2$, which is the same as saying that $x - 5$ is between -2 and 2.

10. $-\frac{1}{2}, 3\frac{1}{2}$

Match the inequality with the graph of its solution.

12. $|x + 2| \geq 1$ **B**
13. $|x + 2| \leq 1$ **C**
14. $|x + 2| > 1$ **A**

A. 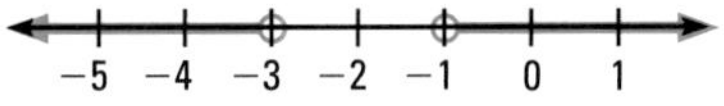

B. 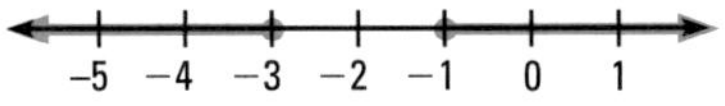

C. 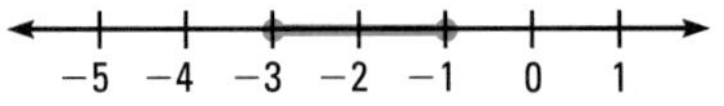

Solve the inequality.

15. $|x + 6| < 4$ **$-10 < x < -2$**
16. $|x - 2| > 9$ **$x < -7$ or $x > 11$**
17. $|3x + 1| \leq 5$ **$-2 \leq x \leq 1\frac{1}{3}$**
18. **MANUFACTURING** In Example 5, suppose the diameter of the drill bit could be 0.5 inch, plus or minus as much as 0.005 inch. Write an absolute value inequality to represent this requirement. **$|d - 0.5| \leq 0.005$**

PRACTICE AND APPLICATIONS

STUDENT HELP

Extra Practice to help you master skills is on p. 802.

SOLVING EQUATIONS **Solve the equation.**

19. $|x| = 7$ **−7, 7**
20. $|x| = 10$ **−10, 10**
21. $|x| = 25$ **−25, 25**
22. $|x - 4| = 6$ **−2, 10**
23. $|x + 5| = 11$ **−16, 6**
24. $|x + 8| = 9$ **−17, 1**
25. $|x - 5| = 2$ **3, 7**
26. $|x - 1| = 4$ **−3, 5**
27. $|x + 3| = 9$ **−12, 6**
28. $|x + 1| = 6$ **−7, 5**
29. $|x - 3.2| = 7$ **−3.8, 10.2**
30. $|x + 5| = 6.5$ **−11.5, 1.5**
31. $|4x - 2| = 22$ **−5, 6**
32. $|6x - 4| = 2$ **$\frac{1}{3}, 1$**
33. $|3x + 5| = 23$ **$-9\frac{1}{3}, 6$**
34. $|5 - 4x| - 3 = 4$ **$-\frac{1}{2}, 3$**
35. $|2x - 4| - 8 = 10$ **−7, 11**
36. $|7x + 3| + 2 = 33$ **$-4\frac{6}{7}, 4$**
37. $|x + 3.6| = 4.6$ **−8.2, 1**
38. $|x - 1.2| - 2 = 5$ **−5.8, 8.2**
39. $\left|x - \frac{1}{2}\right| = \frac{5}{2}$ **−2, 3**

STUDENT HELP

HOMEWORK HELP
Example 1: Exs. 19–39
Example 2: Exs. 19–39

continued on p. 357

STUDENT HELP

HOMEWORK HELP
continued from p. 356
Example 3: Exs. 40–60
Example 4: Exs. 40–60
Example 5: Exs. 61, 62

SOLVING INEQUALITIES **Solve the inequality.**

40. $|x + 3| < 8$ **$-11 < x < 5$**
41. $|2x - 9| \le 11$ **$-1 \le x \le 10$**
42. $|x + 10| \ge 20$ **$x \le -30$ or $x \ge 10$**
43. $|x - 2.2| > 3$ **$x < -0.8$ or $x > 5.2$**
44. $|4x + 2| - 1 < 5$ **$-2 < x < 1$**
45. $|5x - 15| - 4 \ge 21$ **$x \le -2$ or $x \ge 8$**

SOLVING AND GRAPHING **Solve the inequality. Then graph the solution.** **46–60. See margin for graphs.**

46. $|9 + x| \le 7$
47. $|4 - x| < 5$
48. $|x + 12| > 36$
49. $|x - 3| \le 17$
50. $|x + 5| \ge 1$
51. $|x + 3| \ge 8$
52. $|10 - 4x| \le 2$
53. $|2x + 3| > 4$
54. $|x + 2| - 5 \ge 8$
55. $|3 + x| + 7 < 10$
56. $|3x + 2| - 1 \ge 10$
57. $|5x + 1| - 8 \le 16$
58. $|5x + 3| - 4 \ge 9$
59. $|2x + 5| - 1 < 6$
60. $|3x - 9| + 2 > 7$

46. $-16 \le x \le -2$
47. $-1 < x < 9$
48. $x < -48$ or $x > 24$
49. $-14 \le x \le 20$
50. $x \le -6$ or $x \ge -4$
51. $x \le -11$ or $x \ge 5$
52. $2 \le x \le 3$
53. $x < -3\frac{1}{2}$ or $x > \frac{1}{2}$
54. $x \le -15$ or $x \ge 11$
55. $-6 < x < 0$
56. $x \le -4\frac{1}{3}$ or $x \ge 3$
57. $-5 \le x \le 4\frac{3}{5}$
58. $x \le -3\frac{1}{5}$ or $x \ge 2$
59. $-6 < x < 1$
60. $x < 1\frac{1}{3}$ or $x > 4\frac{2}{3}$

61. **BASKETBALL** On your basketball team, the starting players' scoring averages are between 8 and 22 points per game. Write an absolute-value inequality describing the scoring averages for the players. **$|x - 15| \le 7$**

62. **TEST SCORES** The test scores in your class range from 60 to 100. Write an absolute-value inequality describing the range of the test scores. **$|x - 80| \le 20$**

63. **CAR MILEAGE** Your car averages 28 miles per gallon in the city. The actual mileage varies from the average by at most 4 miles per gallon. Write an absolute-value inequality that shows the range for the mileage your car gets. **$|x - 28| \le 4$**

64. **BOXING** The cruiser weight division in boxing is centered at 183 pounds. A boxer's weight can be as much as 7 pounds more than or less than 183 pounds. Write an absolute-value inequality for this weight requirement. **$|x - 183| \le 7$**

FIREWORKS **When a firework star bursts, the color of the "stars" is determined by the chemical compounds in the firework. The wavelengths for different colors in the spectrum are shown below.**

Color	Wavelength, w
Ultraviolet	$w < 400$
Violet	$400 \le w < 424$
Blue	$424 \le w < 491$
Green	$491 \le w < 575$
Yellow	$575 \le w < 585$
Orange	$585 \le w < 647$
Red	$647 \le w < 700$
Infrared	$700 \le w$

65. A firework star contains strontium. When it is burned, strontium emits light at wavelengths given by $|w - 643| < 38$. What colors could the star be? **orange or red**

66. A firework star contains a copper compound. When it is burned, the compound emits light at wavelengths given by $|w - 455| < 23$. Determine the color of the star. **blue**

67. A firework star contains barium chlorate. When it is burned, barium chlorate emits light at wavelengths given by $|w - 519.5| < 12.5$. What color is the star? **green**

68. A firework star contains a sodium compound. When it is burned, the compound emits light at wavelengths given by $|w - 600| < 5$. Determine the color of the star. **orange**

FOCUS ON APPLICATIONS

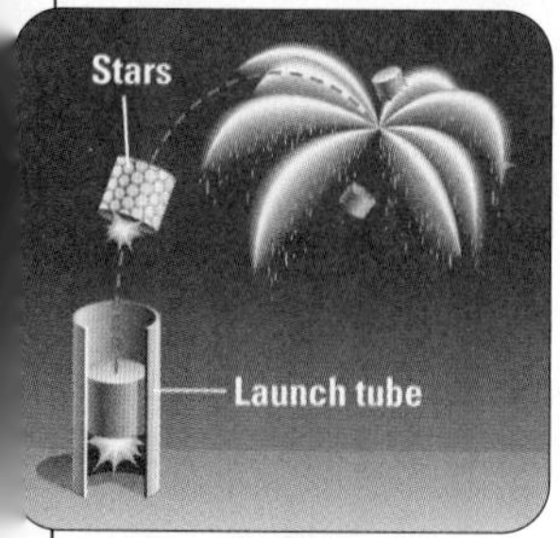

REAL LIFE **FIREWORKS** The diagram above shows what happens when a firework is launched.

INTERNET APPLICATION LINK
www.mcdougallittell.com

APPLICATION NOTE

EXERCISE 62 Unlike the inequalities in Exercises 40–60, the domain of the inequality in this exercise is the set of positive integers. The graph of the solution, therefore, would include dots rather than shading of the number line.

EXERCISE 63 Mileage is affected by several conditions. For example, the speed at which a car travels, the number of stops and starts it makes, the outside temperature and wind speed, the pavement type and condition, and how well the engine is tuned can all cause variations in gas mileage.

EXERCISES 65–68 The range of wavelengths from about 400 nanometers to about 700 nanometers occurs in the visible part of the electromagnetic spectrum. Thus, we can see energy emitted whose wavelength is in this range. Special infrared film allows us to "see" in the infrared portion of the spectrum, as do night-vision goggles.

Additional information about fireworks is available at **www.mcdougallittell.com.**

46–60. See Additional Answers beginning on page AA1.

ADDITIONAL PRACTICE AND RETEACHING

For Lesson 6.4:
- Practice Levels A, B, and C (*Chapter 6 Resource Book,* p. 54)
- Reteaching with Practice (*Chapter 6 Resource Book,* p. 57)
- See Lesson 6.4 of the *Personal Student Tutor*

For more Mixed Review:
- Search the *Test and Practice Generator* for key words or specific lessons.

4 ASSESS

DAILY HOMEWORK QUIZ

Transparency Available

Solve the equation or inequality.

1. $|x-3|=1$ **2, 4**
2. $|3x-7|=8$ **$-\frac{1}{3}$, 5**
3. $|2x+7|>5$ **$x<-6$ or $x>-1$**
4. Solve the inequality $|3-5x|\le 7$. Then graph the solution. **$-0.8\le x\le 2$**

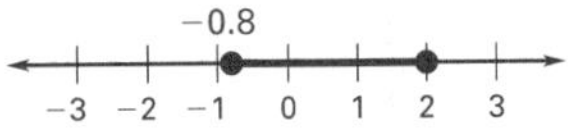

5. Write an absolute-value inequality representing the possible values for a stock that had a low for the year of 23 and a high of 41. **$|x-32|\le 9$**

EXTRA CHALLENGE NOTE

Challenge problems for Lesson 6.4 are available in **blackline** format in the *Chapter 6 Resource Book*, p. 61 and at **www.mcdougallittell.com.**

ADDITIONAL TEST PREPARATION

1. **WRITING** Explain why $x-5=3$ and $-(x-5)=3$ both describe solutions to $|x-5|=3$. **See sample answer below.**
2. **OPEN ENDED** Describe a real-life situation that can be modeled by the inequality $\left|x-3\frac{1}{2}\right|\le\frac{1}{2}$. **See sample answer below.**

79.

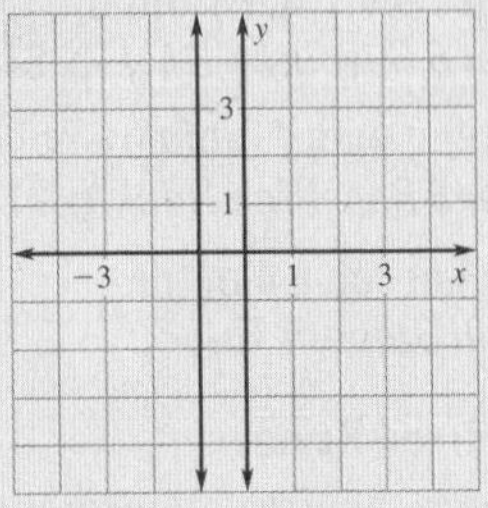

80 and 81. See Additional Answers beginning on page AA1.

Test Preparation

69. MULTIPLE CHOICE Solve $|x-7|<6$. **C**

Ⓐ $-6<x<6$ Ⓑ $-7<x<7$ Ⓒ $1<x<13$ Ⓓ $-1<x<13$

70. MULTIPLE CHOICE Solve $|3x+3|>12$. **C**

Ⓐ $-5<x<3$ Ⓑ $3<x<-5$ Ⓒ $x>3$ or $x<-5$ Ⓓ $x>3$

71. MULTIPLE CHOICE Solve $|2x-4|\le 3$. **B**

Ⓐ $2\le x\le 7$ Ⓑ $\frac{1}{2}\le x\le\frac{7}{2}$ Ⓒ $-\frac{7}{2}\le x\le\frac{1}{2}$ Ⓓ $-\frac{7}{2}\le x\le\frac{7}{2}$

★ Challenge

SCIENCE CONNECTION **In Exercises 72 and 73, use the diagram below. It shows how light is absorbed in seawater.**

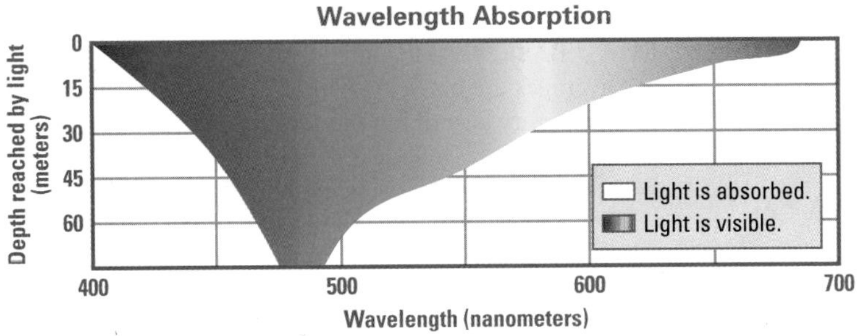

72. Write an absolute-value inequality approximating the wavelengths (w) of light that reach a depth of 30 meters in seawater. ***Sample answer:* $|w-505|\le 65$**

EXTRA CHALLENGE
www.mcdougallittell.com

73. Write an absolute-value inequality approximating the wavelengths (w) of light that reach a depth of 60 meters in seawater. ***Sample answer:* $|w-490|\le 20$**

MIXED REVIEW

MATRICES **Find the sum or difference of the matrices.** (Review 2.4)

74. $\begin{bmatrix}-2 & 7\\ 0 & 4\end{bmatrix}+\begin{bmatrix}3 & -6\\ -1 & -5\end{bmatrix}$ **$\begin{bmatrix}1 & 1\\ -1 & -1\end{bmatrix}$**

75. $\begin{bmatrix}-4 & -9\\ -1 & 0\end{bmatrix}-\begin{bmatrix}-12 & 8\\ -10 & -5\end{bmatrix}$ **$\begin{bmatrix}8 & -17\\ 9 & 5\end{bmatrix}$**

SLOPE-INTERCEPT FORM **Rewrite the equation in slope-intercept form.** (Review 3.7, 4.6 for 6.5)

76. $x+5y=20$ **77.** $6x+9y=36$ **78.** $3x-7y=42$ **$y=\frac{3}{7}x-6$**

76. $y=-\frac{1}{5}x+4$
77. $y=-\frac{2}{3}x+4$

EQUATIONS **Graph the equation.** (Review 4.6) **79–81. See margin.**

79. $x=-1$ **80.** $3y=15$ **81.** $x+y=7$

WRITING EQUATIONS **Write the point-slope form of the equation of the line that passes through the point and has the given slope. Then rewrite the equation in slope-intercept form.** (Review 5.5) **82–84. See margin.**

82. $(0, 4)$, $m=3$ **83.** $(2, -5)$, $m=-2$ **84.** $(-3, 1)$, $m=\frac{2}{3}$

82. $y-4=3(x-0)$; $y=3x+4$
83. $y+5=-2(x-2)$; $y=-2x-1$
84. $y-1=\frac{2}{3}(x+3)$; $y=\frac{2}{3}x+3$

85. **TRAVELING** It is 368 miles from New York City to Pittsburgh. You make the trip in $6\frac{1}{2}$ hours. What was your average speed? Round your answer to the nearest mile per hour. (Review 1.1) **57 mi/h**

Additional Test Preparation *Sample answers:*

1. For any real number n, $|n|=3$ has two solutions, n and the opposite of n.
2. You spend $3\frac{1}{2}$ hours each school night on homework, give or take a half hour.

ACTIVITY 6.4

Using Technology

Graphing Calculator Activity for use with Lesson 6.4

Graphing Absolute-Value Equations

▶ EXAMPLE

Solve the equation $|2x + 1| - 2 = 5$ using a graphing calculator.

▶ SOLUTION

To solve the equation, graph both sides of the equation and find the intersections.

STUDENT HELP

INTERNET KEYSTROKE HELP

See keystrokes for several models of calculators at www.mcdougallittell.com

❶ Enter the equations
$y = |2x + 1| - 2$ and
$y = 5$.

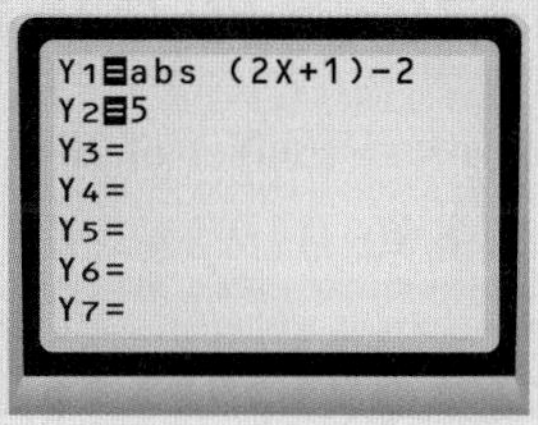

❷ Using a standard viewing window, graph both equations.

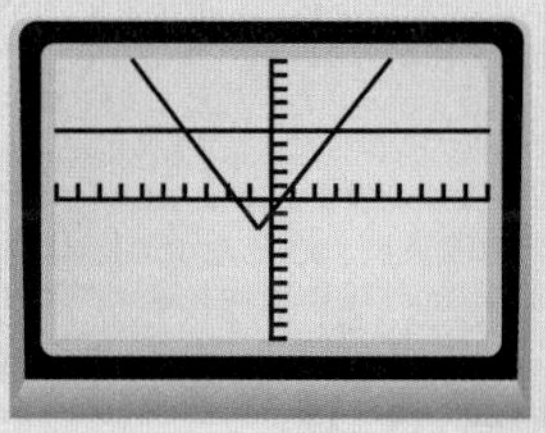

❸ Use the *Intersect* feature to estimate a point where the graphs intersect. If you need to select a graph, use the direction keys to move the cursor to one of the graphs.

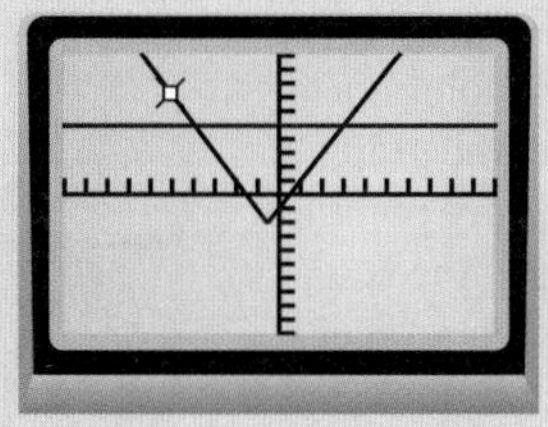

❹ Move the cursor to the approximate intersection point. Follow your calculator's procedure to display the coordinate values. Repeat the steps to find the other intersection point.

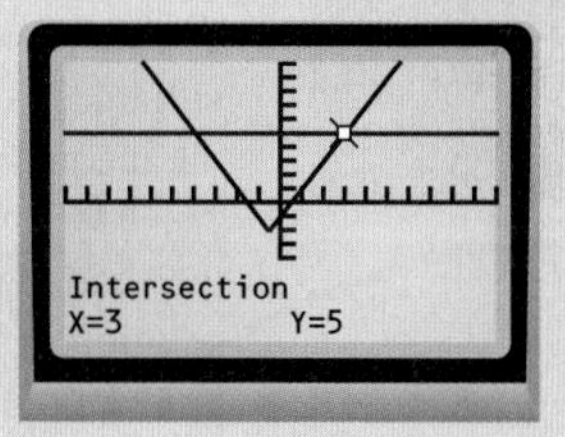

▶ The solutions are 3 and −4.

▶ EXERCISES

Use a graphing calculator or computer to solve the absolute-value equation. Check your solutions algebraically. Use a standard viewing window.

1. $|x + 3| - 1 = 3$ **−7, 1**

2. $|4x - 4| + 1 = 6$ **−0.25, 2.25**

3. $\left|\frac{1}{2}x - 2\right| + 1 = 8$ **−10, 18**

4. $|x - 1.67| - 3.24 = -1.1$ **−0.47, 3.81**

5. $|2x - 60| - 55 = 13$ **−4, 64**

6. $\left|\frac{1}{3}x + 1\right| + 2 = 4$ **−9, 3**

7. $|x - 4.3| + 2.8 = 5.3$ **1.8, 6.8**

8. $|x - 7.2| - 7.6 = 10.9$ **−11.3, 25.7**

9. $|x + 36| + 35 = 49$ **−50, −22**

1 Planning the Activity

PURPOSE

To solve absolute-value equations using a graphing calculator.

MATERIALS

- graphing calculator for each student
- Keystroke blackline (*Chapter 6 Resource Book,* p. 53)

PACING

- Example — 5 min
- Exercises — 10 min

▶ *LINK TO LESSON*

Example 2 of Lesson 6.4 showed how to solve an absolute-value equation algebraically. In this activity, absolute-value equations are solved by graphing.

2 Managing the Activity

CLASSROOM MANAGEMENT

Use an overhead display for your calculator if one is available. Perform one keystroke at a time, making sure that students can compare their displays to yours. If no overhead is available, circulate among the students to make sure they understand the keystrokes.

3 Closing the Activity

★ *KEY DISCOVERY*

When each side of an absolute-value equation is graphed separately, the intersections are the solutions of the original absolute-value equation.

ACTIVITY ASSESSMENT

Describe the appearance of the solution on a graphing calculator screen when solving $|x - 1| = 8$ by graphing. **The solutions are the points of intersection of a "V"-shaped graph (the left side of the equation) and the horizontal line $y = 8$.**

1 PLAN

PACING
Basic: 2 days
Average: 2 days
Advanced: 2 days
Block Schedule: 1 block

LESSON OPENER ACTIVITY
An alternative way to approach Lesson 6.5 is to use the Activity Lesson Opener:

- Blackline Master (*Chapter 6 Resource Book*, p. 65)
- Transparency (p. 43)

MEETING INDIVIDUAL NEEDS

- ***Chapter 6 Resource Book***
  Prerequisite Skills Review (p. 5)
  Practice Level A (p. 67)
  Practice Level B (p. 68)
  Practice Level C (p. 69)
  Reteaching with Practice (p. 70)
  Absent Student Catch-Up (p. 72)
  Challenge (p. 74)
- ***Resources in Spanish***
- ***Personal Student Tutor***

NEW-TEACHER SUPPORT
See the Tips for New Teachers on pp. 1–2 of the *Chapter 6 Resource Book* for additional notes about Lesson 6.5.

WARM-UP EXERCISES

Transparency Available

Write the equation in slope-intercept form.

1. $x - 2y = -8$ $\quad y = \frac{1}{2}x + 4$

2. $-y - x = 4$ $\quad y = -x - 4$

Find the x- and y-intercepts of the graph of the equation.

3. $6x + 8y = 48$ $\quad$ 8, 6

4. $2.5x + 0.75y = 82.5$ $\quad$ 33; 110

6.5 Graphing Linear Inequalities in Two Variables

***What** you should learn*

GOAL 1 Graph a linear inequality in two variables.

GOAL 2 Model a **real-life** situation using a linear inequality in two variables, such as purchasing produce in **Ex. 64**.

***Why** you should learn it*

▼ To model **real-life** situations, such as salvaging coins from a shipwreck in **Example 5**.

GOAL 1 GRAPHING LINEAR INEQUALITIES

A **linear inequality** in x and y is an inequality that can be written as follows.

$ax + by < c \qquad ax + by \le c \qquad ax + by > c \qquad ax + by \ge c$

An ordered pair (x, y) is a **solution** of a linear inequality if the inequality is true when the values of x and y are substituted into the inequality.

EXAMPLE 1 *Checking Solutions of a Linear Inequality*

Check whether the ordered pair is a solution of $2x - 3y \ge -2$.

a. (0, 0) **b.** (0, 1) **c.** (2, −1)

SOLUTION

(x, y)	$2x - 3y \ge -2$	Conclusion
a. (0, 0)	$2(0) - 3(0) = 0 \ge -2$	(0, 0) is a solution.
b. (0, 1)	$2(0) - 3(1) = -3 \not\ge -2$	(0, 1) is not a solution.
c. (2, −1)	$2(2) - 3(-1) = 7 \ge -2$	(2, −1) is a solution.

The **graph** of a linear inequality in two variables is the graph of the solutions of the inequality. For instance, the graph of $2x - 3y \ge -2$ is shown. Every point in the shaded region and on the line is a solution of the inequality. Every other point in the plane is not a solution.

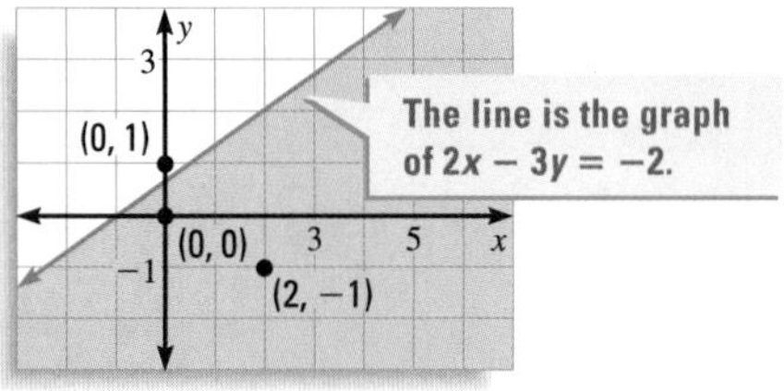

GRAPHING A LINEAR INEQUALITY

STEP 1 Graph the corresponding equation. Use a *dashed* line for inequalities with > or < to show that the points on the line are not solutions. Use a *solid* line for inequalities with ≥ or ≤ to show that the points on the line are solutions.

STEP 2 The line you drew separates the coordinate plane into two **half-planes.** Test a point in one of the half-planes to find whether it is a solution of the inequality.

STEP 3 If the test point is a solution, shade the half-plane it is in. If not, shade the other half-plane.

EXAMPLE 2 Graphing a Linear Inequality

Sketch the graph of $x < -2$.

SOLUTION

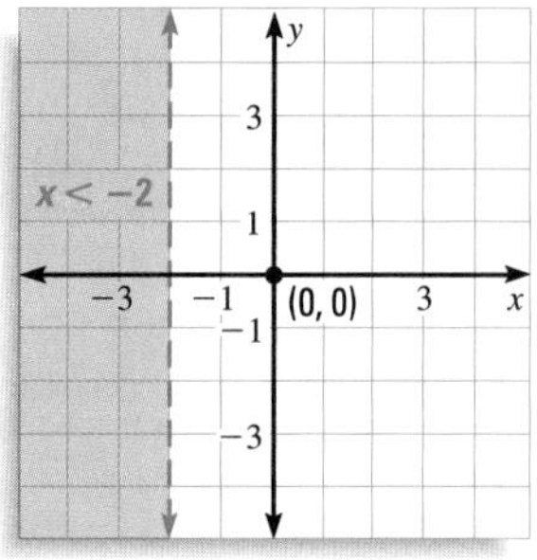

1. Graph the corresponding equation $x = -2$, a vertical line. Use a dashed line.
2. Test a point. The origin (0, 0) is *not* a solution and it lies to the right of the line. So, the graph of $x < -2$ is all points to the left of the line $x = -2$.
3. Shade the region to the left of the line.

EXAMPLE 3 Graphing a Linear Inequality

Sketch the graph of $y \le 1$.

SOLUTION

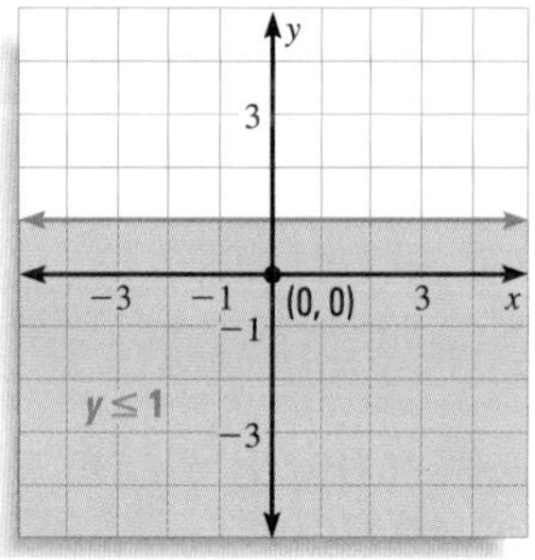

1. Graph the corresponding equation $y = 1$, a horizontal line. Use a solid line.
2. Test a point. The origin (0, 0) *is* a solution and it lies below the line. So, the graph of $y \le 1$ is all points on or below the line $y = 1$.
3. Shade the region below the line.

EXAMPLE 4 Writing in Slope-Intercept Form

Sketch the graph of $x + y > 3$.

SOLUTION

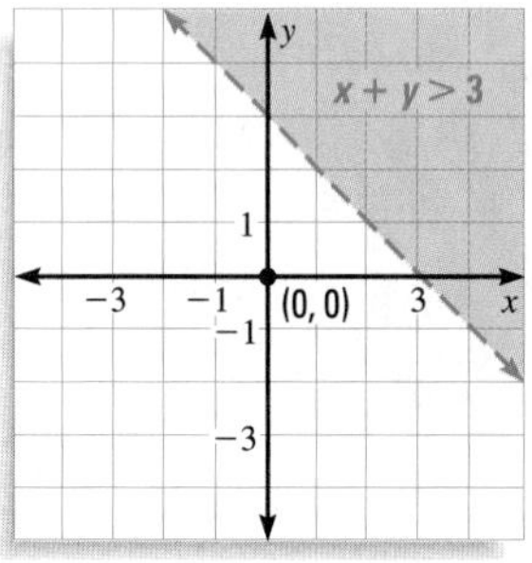

The corresponding equation is $x + y = 3$. To graph this line, you can first write the equation in slope-intercept form.

$$y = -x + 3$$

Then graph the line that has a slope of -1 and a y-intercept of 3. Use a dashed line.

The origin (0, 0) is *not* a solution and it lies below the line. So, the graph of $x + y > 3$ is all points above the line $y = -x + 3$.

✓**CHECK** Test any point above the line. Any point you choose will satisfy the inequality.

STUDENT HELP

Study Tip
You can use any point that is not on the line as a test point. The origin is often used because it is convenient to evaluate expressions in which 0 is substituted for each variable.

2 TEACH

EXTRA EXAMPLE 1
Check whether the ordered pair is a solution of $4x + 5y \le 12$.
a. (−3, 5) no **b.** (0, 2) yes
c. (6, −8) yes

EXTRA EXAMPLE 2
Sketch the graph of $x \ge 1$.

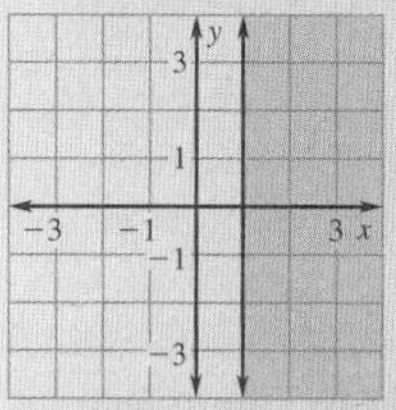

EXTRA EXAMPLE 3
Sketch the graph of $y > -3$.

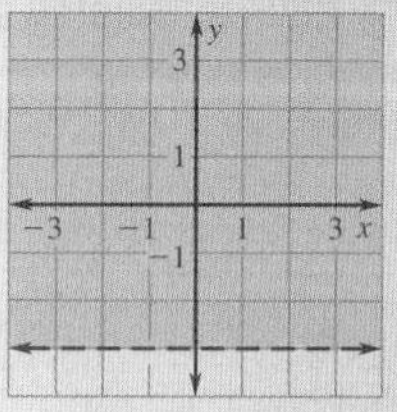

EXTRA EXAMPLE 4
Sketch the graph of $x - y \le 2$.

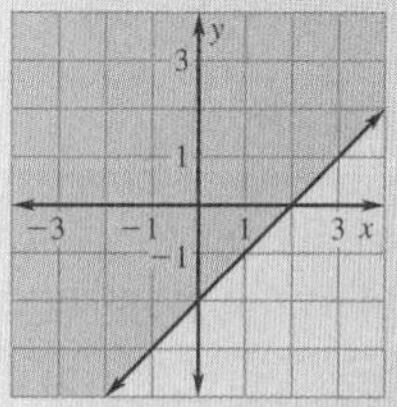

CHECKPOINT EXERCISES

For use after Example 1:

1. Is (−2, 3) a solution of $-2x + y > 0$? yes

For use after Example 2:

2. Describe the graph of $x > -5$. *Sample answer:* a dashed line through $x = -5$ with shading to the right of the line

For use after Examples 3 and 4:

3. Sketch the graph of $y + x > -1$.

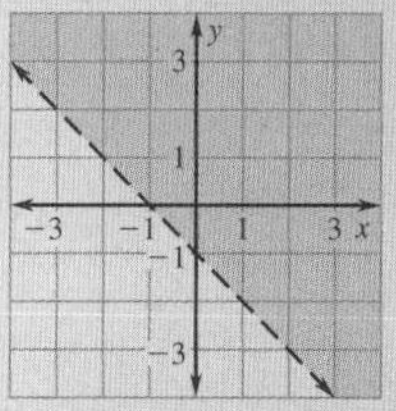

EXTRA EXAMPLE 5

On school nights you are allowed a maximum of 40 minutes total to surf the Internet or to hang out in chat rooms. What are the different amounts of time you can spend surfing or chatting? Display your answer with a graph.

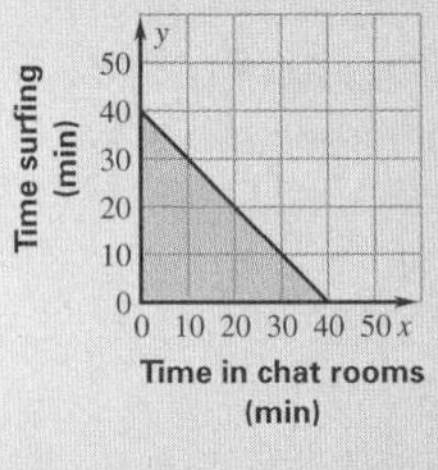

CHECKPOINT EXERCISES

For use after Example 5:

1. Sam has 30 ft of fencing that she can use to enclose a rectangular sandbox. Display the possible dimensions of the sandbox with a graph.

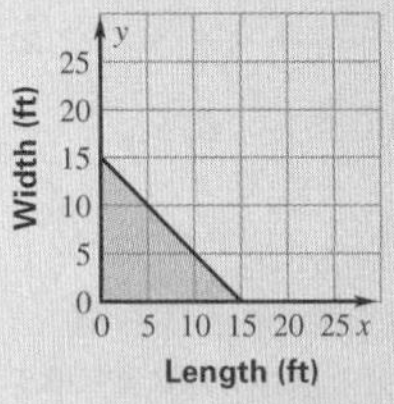

CLOSURE QUESTION

How do you know whether to shade above or below the boundary line (or to its left or right) when graphing an inequality? ***Sample answer:*** **Choose one point not on the boundary line. If it satisfies the inequality, shade the half-plane containing it. Otherwise, shade the other half-plane.**

DAILY PUZZLER

Your cat weighs more than 11 lb. It weighs less than your dog weighs, but more than three fourths what the dog weighs. The sum of the weights is less than 29 lb. What are whole-number possible weights of the animals? **Possible answers are: cat 12 lb, dog 13 lb; cat 12 lb, dog 14 lb; cat 12 lb, dog 15 lb; cat 13 lb, dog 14 lb; cat 13 lb, dog 15 lb**

GOAL 2 MODELING A REAL-LIFE SITUATION

EXAMPLE 5 *Modeling with a Linear Inequality*

REAL LIFE *Treasure Diving*

You are on a treasure-diving ship that is hunting for gold and silver coins. Objects collected by the divers are placed in a wire basket. One of the divers signals you to reel in the basket. It feels as if it contains no more than 50 pounds of material.

If each gold coin weighs about 0.5 ounce and each silver coin weighs about 0.25 ounce, what are the different amounts of coins that could be in the basket?

SOLUTION

Find the number of ounces in the basket. There are 16 ounces in a pound.

$16 \cdot 50 = 800$

Write an algebraic model.

PROBLEM SOLVING STRATEGY

VERBAL MODEL Weight per gold coin $\cdot$ Number of gold coins $+$ Weight per silver coin $\cdot$ Number of silver coins $\leq$ Weight in basket

LABELS

Weight per gold coin = **0.5**	(ounces per coin)
Number of gold coins = x	(coins)
Weight per silver coin = **0.25**	(ounces per coin)
Number of silver coins = y	(coins)
Maximum weight in basket = **800**	(ounces)

ALGEBRAIC MODEL $0.5x + 0.25y \leq 800$ Write algebraic model.

Graph the inequality to see the possible solutions. To make a quick graph of the corresponding equation, find the x-intercept and the y-intercept. The x-intercept is (1600, 0). The y-intercept is (0, 3200). Then graph the line, test the origin, and shade the graph of the inequality.

FOCUS ON APPLICATIONS

REAL LIFE **GOLD** found in the ocean can look brand new. Most other metals deteriorate quickly in salt water.

APPLICATION LINK www.mcdougallittell.com

The graph shows all the solutions of the inequality. The possible numbers of gold and silver coins, however, are only the ordered pairs of integers in the graph.

One solution is all gold coins.

(1600, 0)

Another solution is all silver coins.

(0, 3200)

There are many other solutions, including no gold or silver coins.

(0, 0)

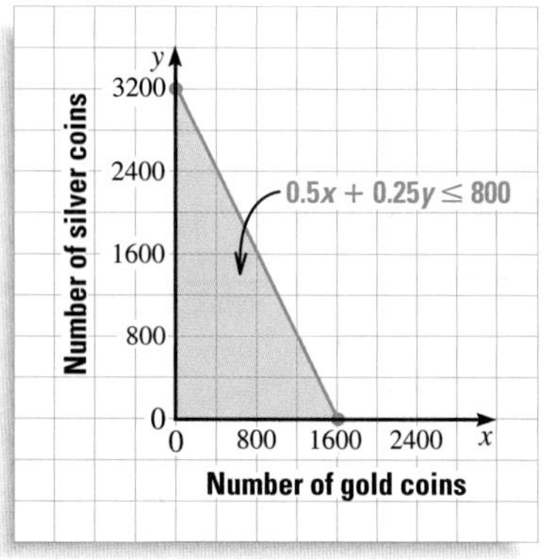

GUIDED PRACTICE

Vocabulary Check ✓

1. Explain what a *solution* of a linear inequality in x and y is. **an ordered pair (x, y) for which substituting the values of x and y in the inequality produces a true statement**

Concept Check ✓

2. In the graph in Example 5, why is shading shown only in Quadrant 1? Use the real-life context to explain your reasoning. **The real-life values of x and y must both be positive, since they are numbers of coins.**

3. Choose the inequality whose solution is shown by the graph. Explain your reasoning.

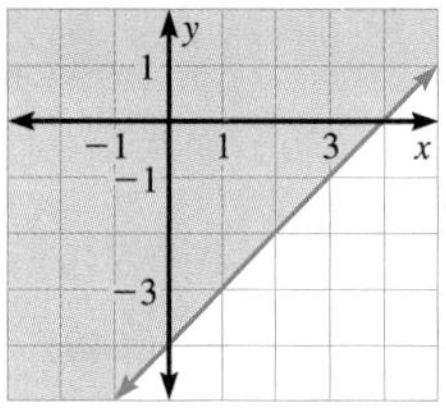

A. $x - y > 4$

B. $x - y < 4$

C. $x - y \geq 4$

D. $x - y \leq 4$

D; the solid line is the graph of $x - y = 4$. Since (0, 0) is in the shaded region and (0, 0) satisfies the inequality $x - y < 4$, the shaded region with the solid boundary line is the graph of $x - y \leq 4$.

Skill Check ✓

Check whether (0, 0) is a solution. Then sketch the graph of the inequality.
4–12. See margin for graphs.

4. $y < -2$ **not a solution** **5.** $x > -2$ **solution** **6.** $x + y \geq -1$ **solution**

7. $x - y \leq -2$ **not a solution** **8.** $x + y < 4$ **solution** **9.** $x - y \leq 5$ **solution**

10. $x + y > 3$ **not a solution** **11.** $3x - y < 3$ **solution** **12.** $x - 3y \geq 12$ **not a solution**

BASKETBALL SCORES **With two minutes left in a basketball game, your team is 12 points behind. What are two different numbers of 2-point and 3-point shots your team could score to earn at least 12 points?**

13. Write a verbal model for the situation. Assign labels to each part of the verbal model and write an inequality. **Let x = the number of 2-point shots and y = the number of 3-point shots; $2x + 3y \geq 12$.**

14. Sketch the graph of the inequality. Then name two ways your team could score at least 12 points. **The table shows all possible ways to score exactly 12 points. Some combinations that score more than 12 points (in this case, 13) are three 3-point shots and two 2-point shots or one 3-point shot and five 2-point shots. See margin for graph.**

14.

2-point shots	3-point shots
0	4
3	2
6	0

PRACTICE AND APPLICATIONS

STUDENT HELP

Extra Practice to help you master skills is on p. 802.

STUDENT HELP

HOMEWORK HELP
Example 1: Exs. 15–22
Example 2: Exs. 23–42
Example 3: Exs. 23–42
Example 4: Exs. 43–60
Example 5: Exs. 64–66

CHECKING SOLUTIONS **Is each ordered pair a solution of the inequality?**

15. $x + y > -3$; (0, 0), (−6, 3) **solution, not a solution**

16. $2x + 2y \leq 0$; (−1, −1), (1, 1) **solution, not a solution**

17. $2x + 5y \geq 10$; (1, 2), (6, 1) **solution, solution**

18. $4x + 7y \leq 26$; (3, 2), (2, 3) **solution, not a solution**

19. $0.6x + 0.6y > 2.4$; (2, 2), (3, −3) **not a solution, not a solution**

20. $1.8x - 3.8y \geq 5$; (0, 0), (1, −1) **not a solution, solution**

21. $\frac{3}{4}x - \frac{3}{4}y < 2$; (8, 8), (8, −8) **solution, not a solution**

22. $\frac{5}{6}x + \frac{5}{3}y > 4$; (6, −12), (8, −8) **not a solution, not a solution**

SKETCHING GRAPHS **Sketch the graph of the inequality.** **23–42. See margin for graphs.**

23. $x \geq 4$ **24.** $x \leq 5$ **25.** $y > -3$ **26.** $y < 9$

27. $x + 3 > -2$ **28.** $7 - x \leq 16$ **29.** $y + 6 > 5$ **30.** $8 - y \leq 0$

31. $4x < -12$ **32.** $-2x \geq 10$ **33.** $5y \leq -25$ **34.** $8y > 24$

35. $-3x \geq 15$ **36.** $6y \leq 24$ **37.** $-4y < 8$ **38.** $-5x > -10$

39. $x < \frac{1}{2}$ **40.** $y \geq 1.5$ **41.** $2y > 1$ **42.** $x \leq 3.5$

3 APPLY

ASSIGNMENT GUIDE

BASIC
Day 1: p. 363 Exs. 16–42 even
Day 2: pp. 364–365 Exs. 43–52, 57–64, 70, 71, 76, 78, 82, 88, 90, Quiz 2 Exs. 1–23

AVERAGE
Day 1: p. 363 Exs. 16–42 even
Day 2: pp. 364–366 Exs. 43–52, 57–65, 70, 71, 76, 78, 82, 88, 90, Quiz 2 Exs. 1–23

ADVANCED
Day 1: p. 363 Exs. 16–42 even
Day 2: pp. 364–366 Exs. 43–52, 57–65, 68, 70–73, 76, 78, 82, 88, 90, Quiz 2 Exs. 1–23

BLOCK SCHEDULE
pp. 363–366 Exs. 16–42 even, 43–52, 57–65, 70, 71, 76, 78, 82, 88, 90, Quiz 2 Exs. 1–23

EXERCISE LEVELS

Level A: *Easier*
15–20, 74–76

Level B: *More Difficult*
21–66, 70, 71, 77–91

Level C: *Most Difficult*
67–69, 72, 73

✓ HOMEWORK CHECK

To quickly check student understanding of key concepts, go over the following exercises: Exs. 18, 24, 26, 38, 46, 52, 62. See also the Daily Homework Quiz:

- Blackline Master (*Chapter 6 Resource Book*, p. 78)
- Transparency (p. 50)

APPLICATION NOTE

EXERCISE 14 Students should recognize that discrete points, rather than a shaded region, are solutions to this problem.

4–12, 14, 23–42.
See Additional Answers beginning on page AA1.

! COMMON ERROR

EXERCISES 43–48 Some students may make the erroneous assumption that the solutions to the inequalities in these exercises lie below the graphed boundary lines if the given inequality contains a less than symbol. Similarly, they may assume that the solutions lie above the graphed lines for those given inequalities with a greater than symbol. Students can use the inequality symbol to determine the solution set only if the inequality is first solved for *y*.

APPLICATION NOTE

EXERCISE 64 Encourage students to interpret the graph before they use a mechanical test for determining the solution set. Since the points above the line represent a total cost greater than $12, those points should be excluded from the solution set.

STUDENT HELP NOTES

Homework Help Students can find help for Ex. 65 at **www.mcdougallittell.com.** The information can be printed out for students who don't have access to the Internet.

49–60. See Additional Answers beginning on page AA1.

64.

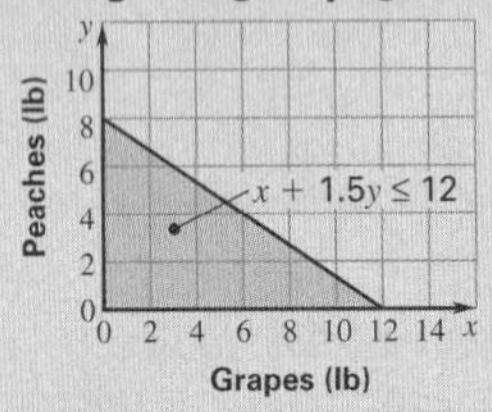

65a.

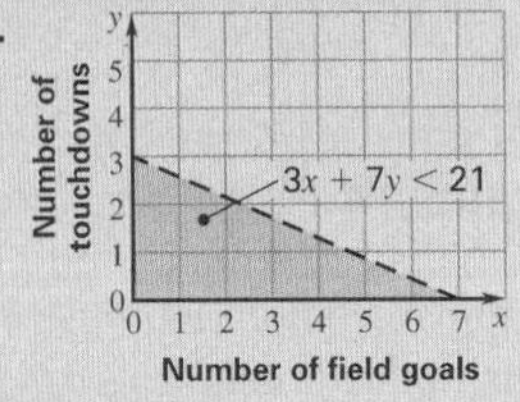

MATCHING GRAPHS **Match the inequality with its graph.**

A. $2x - y \le 1$ **B.** $-2x - y < 1$ **C.** $2x - y \ge 1$

D. $2x - y > 1$ **E.** $2x - y < 1$ **F.** $-2x - y > 1$

43. 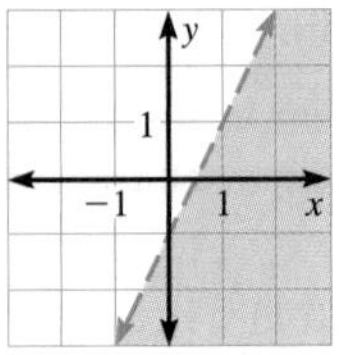 D

44. 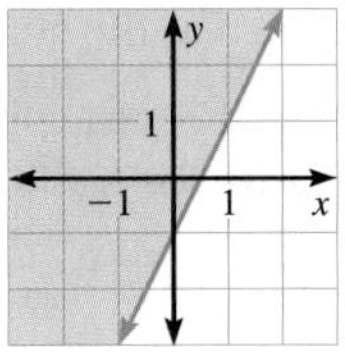 A

45. 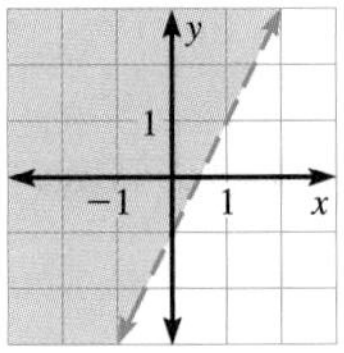 E

46. 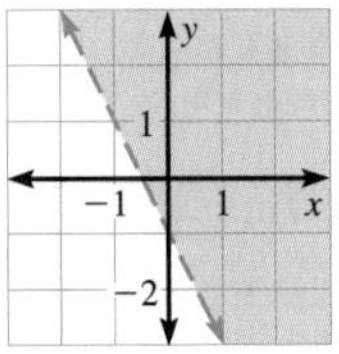 B

47. 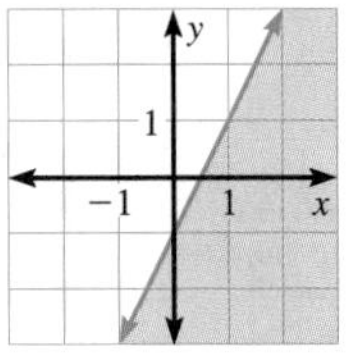 C

48. 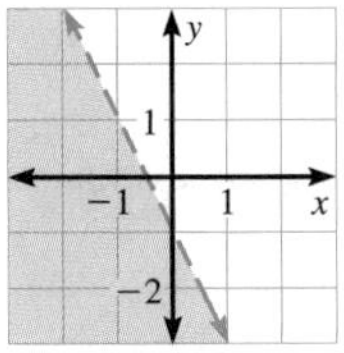 F

SKETCHING GRAPHS **Sketch the graph of the inequality.** **49–60. See margin for graphs.**

49. $x + y > -8$ **50.** $x - y \ge 4$ **51.** $y - x \le 11$ **52.** $-x - y < 3$

53. $x + 6y \le 12$ **54.** $3x - y \ge 5$ **55.** $2x + y > 6$ **56.** $y - 5x < 0$

57. $4x + 3y \le 24$ **58.** $9x - 3y \ge 18$ **59.** $\frac{1}{4}x + \frac{1}{2}y < 1$ **60.** $\frac{1}{3}x - \frac{2}{3}y \ge 2$

WRITING INEQUALITIES FROM GRAPHS **Write an inequality whose solution is shown in the graph.**

61. 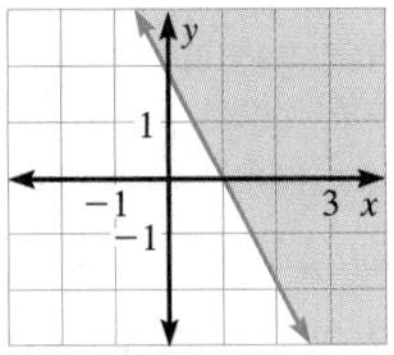

$y \ge -2x + 2$ or $2x + y \ge 2$

62. 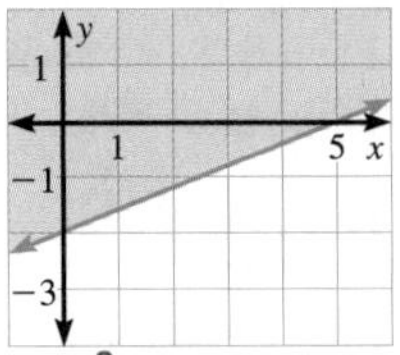

$y \ge \frac{2}{5}x - 2$ or $2x - 5y \le 10$

63. 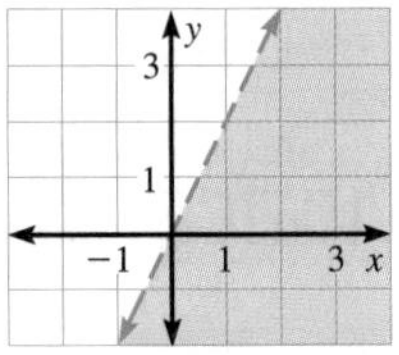

$y < 2x$ or $2x - y > 0$

64. **PURCHASING PRODUCE** You have $12 to spend on fruit for a meeting. Grapes cost $1 per pound and peaches cost $1.50 per pound. Let *x* represent the number of pounds of grapes you can buy. Let *y* represent the number of pounds of peaches you can buy. Write and graph an inequality to model the amounts of grapes and peaches you can buy. **$x + 1.5y \le 12$; See margin for graph.**

65b. **No; only points for which both coordinates are nonnegative integers represent solutions, since only a whole number of touchdowns or field goals can be scored. The real-life solutions are (0, 0), (0, 1), (0, 2), (1, 0), (1, 1), (1, 2), (2, 0), (2, 1), (2, 2), (3, 0), (3, 1), (4, 0), (4, 1), (5, 0), and (6, 0).**

STUDENT HELP

HOMEWORK HELP Visit our Web site www.mcdougallittell.com for help with Ex. 65.

65. **FOOTBALL** In the last quarter of a high school football game, your team is behind by 21 points. A field goal is 3 points and a touchdown (with the point-after-touchdown) is 7 points. Let *x* represent the number of field goals scored. Let *y* represent the number of touchdowns scored.

a. Write and graph an inequality that models the different numbers of field goals and touchdowns your team could score and still not win or tie. (Assume the other team scores no more points.) **$3x + 7y < 21$; See margin for graph.**

b. Does every point on the graph represent a solution of the real-life problem? Give examples to support your answer. **See margin.**

FOCUS ON CAREERS

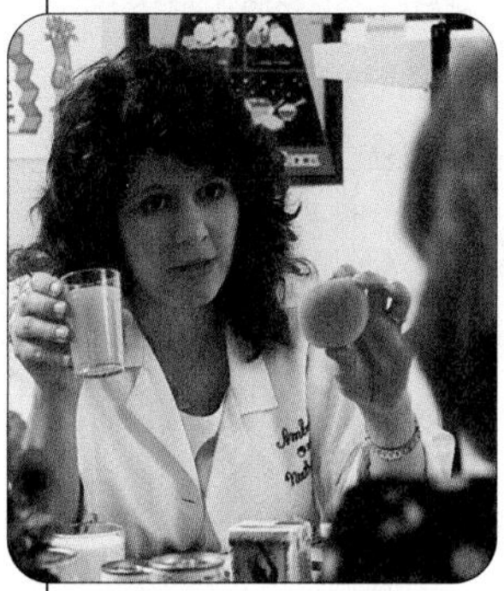

NUTRITIONIST Nutritionists plan food programs and supervise the preparing of meals.

CAREER LINK www.mcdougallittell.com

66. CAR SALES You are a car dealer. You have \$1,408,000 available to purchase compact cars and sport utility vehicles for your lot. The compact car costs \$11,000 and the sport utility vehicle costs \$22,000. Let x represent the number of compact cars and let y represent the number of sport utility vehicles you purchase. Write an inequality that models the different numbers of compact cars and sport utility vehicles that you could purchase.
$11{,}000x + 22{,}000y \leq 1{,}408{,}000$ or $11x + 22y \leq 1408$

NUTRITION **In Exercises 67–69, use the following information. You are planning a breakfast that supplies at most 500 calories. The table shows the numbers of calories in certain foods.**

Breakfast food	Calories
Bagel, 1 plain	200
Cereal, 1 cup	88
Juice, Apple, 1 glass	120
Juice, Tomato, 1 glass	43
Egg	75
Milk, skim, 1 cup	85
Margarine, 1 tsp	33

67. a. You select tomato juice, some cereal, and milk. What different amounts of cups of cereal and milk could you choose along with one glass of tomato juice to supply at most 500 calories? **See margin.**

b. Describe a reasonable meal of tomato juice, cereal, and milk that supplies at most 500 calories. **See margin.**

c. Find reasonable amounts of cups of cereal and milk, with a glass of tomato juice, to supply at most 600 calories. **See margin.**

68. a. Another morning you choose a bagel with some margarine, and apple juice. Find different numbers of teaspoons of margarine and glasses of apple juice, with a bagel, to supply at most 500 calories. **See margin.**

b. Describe a reasonable meal of a bagel with margarine, and apple juice, that supplies at most 500 calories. ***Sample answer:* 1 bagel, 2 tsp of margarine, and 1 glass of apple juice**

69. *Writing* Describe two methods you could use to solve part (a) of Exercise 68. ***Sample answer:* Graph the inequality $33x + 120y + 200 \leq 500$ and find specific solutions from the graph or make a table and try different combinations.**

67a. Let (x, y) represent x cups of cereal and y cups of skim milk; the possible solutions are: (1, 1), (1, 2), (1, 3), (1, 4), (2, 1), (2, 2), (2, 3), (3, 1), (3, 2), (4, 1).

Test Preparation

70. MULTIPLE CHOICE Choose the inequality whose graph is shown. **C**

Ⓐ $2y - 6x < -4$

Ⓑ $2y - 6x \leq -4$

Ⓒ $2y - 6x > -4$

Ⓓ $6x + 2y > 4$

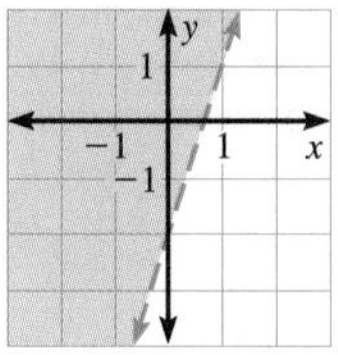

71. MULTIPLE CHOICE Choose the inequality whose graph is shown. **B**

Ⓐ $2y - 6x < -4$

Ⓑ $2y - 6x \leq -4$

Ⓒ $2y - 6x \geq -4$

Ⓓ $6x + 2y > 4$

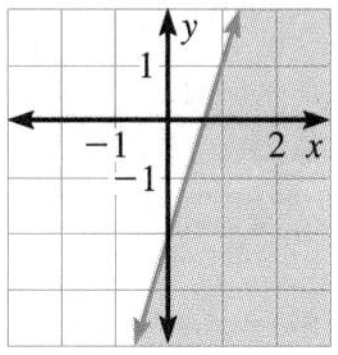

67b. *Sample answer:* 1 glass of tomato juice, 2 cups of cereal, and 2 or 3 cups of milk

67c. The possible solutions are all those in part (a) as well as (1, 5), (2, 4), (3, 3), (4, 2), and (5, 1). *Sample answer:* one glass of tomato juice, 2 cups of cereal, and 3 cups of milk

★ Challenge

72. Sketch the graph of $|y| < 2$ in a coordinate plane. Find the domain and the range.
See margin; domain: all real numbers; range: all real numbers between −2 and 2

73. Sketch the graph of $|x + 3| \leq y$ in a coordinate plane. **See margin.**

APPLICATION NOTE

EXERCISE 66 The dealer can buy many more compact cars with the budget, but must consider the popularity of sport utility vehicles. And although the dealer can buy fewer sport utility vehicles, there is likely a greater profit on the sale of each of these vehicles than on the sale of each compact car.

EXERCISE 67 The U.S. Department of Agriculture suggests that a variety of foods will supply necessary nutrients, but that most of our food servings should come from grain products such as bread, cereal, rice, and pasta, and from fruits and vegetables.

CAREER NOTE

EXERCISES 67–69 Additional information about a career as a nutritionist is available at **www.mcdougallittell.com.**

68a. Let (x, y) represent x tsp of margarine and y glasses of apple juice; the possible solutions are (1, 1), (1, 2), (2, 1), (3, 1), (4, 1), and (5, 1).

72.

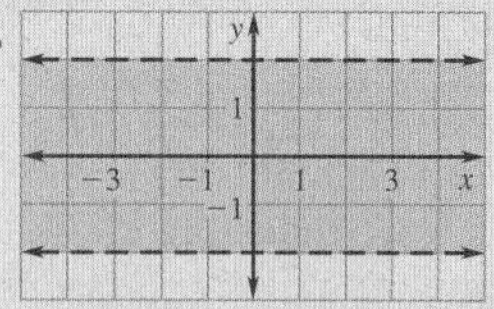

73.

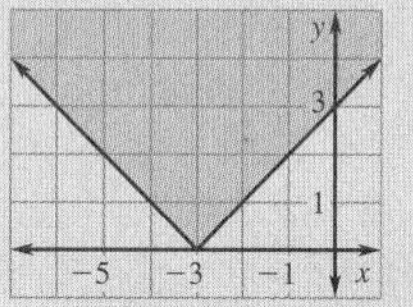

ADDITIONAL PRACTICE AND RETEACHING

For Lesson 6.5:

- Practice Levels A, B, and C (*Chapter 6 Resource Book,* p. 67)
- Reteaching with Practice (*Chapter 6 Resource Book,* p. 70)
- See Lesson 6.5 of the *Personal Student Tutor*

For more Mixed Review:

- Search the *Test and Practice Generator* for key words or specific lessons.

4 ASSESS

DAILY HOMEWORK QUIZ

Transparency Available

1. Tell whether the ordered pairs $(1, -1)$ and $(4, 4)$ are solutions of $3x - 2y < 5$. **no, yes**

Sketch the graph of the inequality.

2. $x + 3 \leq 5$

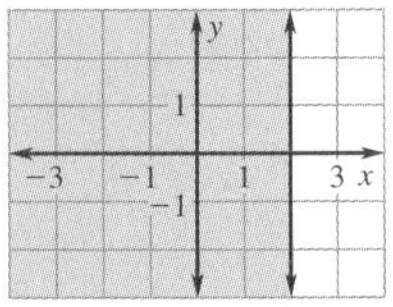

3. $x - y \geq 2$

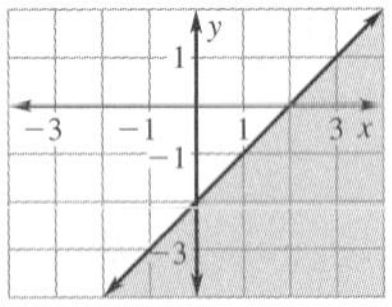

4. $3y + 2x > 1$

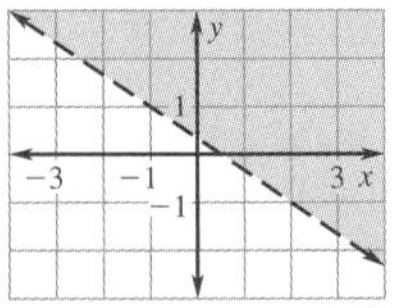

EXTRA CHALLENGE NOTE

Challenge problems for Lesson 6.5 are available in **blackline** format in the *Chapter 6 Resource Book*, p. 74 and at **www.mcdougallittell.com.**

ADDITIONAL TEST PREPARATION

1. WRITING How does the graph of a linear inequality represent the solutions to the inequality? **See sample answer below.**

2. OPEN ENDED Use an inequality in two variables to model a real-life situation. **See sample answer below.**

ADDITIONAL RESOURCES

An alternative Quiz for Lessons 6.4 and 6.5 is available in the *Chapter 6 Resource Book*, p. 75.

79–82, 7–12, 17–22. See Additional Answers beginning on page AA1.

MIXED REVIEW

EVALUATING EXPRESSIONS Evaluate the expression. (Review 1.3 for 6.6)

74. $\frac{16 + 11 + 18}{3}$ **15** **75.** $\frac{20 + 15 + 22 + 19}{4}$ **19** **76.** $\frac{37 + 65 + 89 + 72 + 82}{5}$ **69**

SOLVING A FORMULA Solve for the indicated variable. (Review 3.7)

77. Distance traveled
Solve for r: $d = rt$ $r = \frac{d}{t}$

78. Volume of a pyramid
Solve for h: $V = \frac{1}{3}Bh$ $h = \frac{3V}{B}$

GRAPHING EQUATIONS Use a table of values to graph the equation. (Review 4.2) **79–82. See margin.**

79. $y = -6x + 7$

80. $y = 3x - 7$

81. $-2x + 2y = 5$

82. $3x + 4y = 20$

FINDING SLOPES AND *y*-INTERCEPTS Find the slope and the *y*-intercept of the line. (Review 4.6)

83. $y = -5x + 2$ **−5, 2** **84.** $y = \frac{x}{2} - 2$ $\frac{1}{2}, -2$ **85.** $5x - 5y = 1$ $1, -\frac{1}{5}$

86. $6x + 2y = 14$ **−3, 7** **87.** $y = -2$ **0, −2** **88.** $y = 5$ **0, 5**

89. $y = 3x - 6.5$ **3, −6.5** **90.** $y = -4x + \frac{1}{2}$ $-4, \frac{1}{2}$ **91.** $3x + 2y = 10$ $-\frac{3}{2}, 5$

QUIZ 2

Self-Test for Lessons 6.4 and 6.5

Solve the equation. (Lesson 6.4)

1. $|x| = 15$ **−15, 15** **2.** $|x| = 22$ **−22, 22** **3.** $|x + 3| = 6$ **−9, 3**

4. $|x - 2| = 10$ **−8, 12** **5.** $|2x + 7| = 7$ **−7, 0** **6.** $|3x - 2| - 2 = 5$ $-\frac{5}{3}, 3$

Solve the inequality. Then graph its solution on a number line. (Lesson 6.4)
7–12. See margin for graphs.

7. $|x - 4| > 1$ $x < 3$ or $x > 5$ **8.** $|x + 7| < 2$ $-9 < x < -5$ **9.** $|x - 12| \leq 9$ $3 \leq x \leq 21$

10. $\left|x - \frac{1}{4}\right| > \frac{9}{4}$ $x < -2$ or $x > \frac{5}{2}$ **11.** $|2x + 7| \leq 25$ $-16 \leq x \leq 9$ **12.** $|4x + 2| - 5 > 17$ $x < -6$ or $x > 5$

Is each ordered pair a solution of the inequality? (Lesson 6.5)

13. $x + y \leq 4$; $(0, -1)$, $(2, 2)$ **solution, solution**

14. $y - 3x > 0$; $(0, 0)$, $(-4, 1)$ **not a solution, solution**

15. $-2x + 5y \geq 5$; $(2, 1)$, $(-1, 2)$ **not a solution, solution**

16. $-x - 2y < 4$; $(1, -1)$, $(2, -3)$ **solution, not a solution**

Sketch the graph of the inequality in a coordinate plane. (Lesson 6.5)
17–22. See margin for graphs.

17. $y - 5x > 0$ **18.** $y \geq 3$ **19.** $x \leq -4$

20. $y < -2x$ **21.** $2x - y \geq 10$ **22.** $3x + y > 15$

23. **STOCK MARKET** You purchase shares of a technology company's stock and hold onto them for one year. During that time period, the price per share ranged from \$65 to \$143. Write an absolute-value inequality that shows the range of stock prices for the year. (Lesson 6.4) $|x - 104| \leq 39$

1. If the boundary separating the half-planes is a solid line, then all the points on the line and the points in the shaded half-plane are solutions. If the boundary line is dashed, then all the points in the shaded half-plane, but not those on the boundary line, are solutions.

2. If canned peas are \$.75 per can and canned beans are \$.60 per can, then the numbers of cans of peas and beans you can buy without exceeding \$12 is given by $0.75p + 0.60b \leq 12$, where p is the number of cans of peas and b is the number of cans of beans.

Graphing Inequalities

EXAMPLE

Graph the solution of the inequality $x - 2y \le -6$.

SOLUTION

1 Rewrite the inequality so that y is isolated on the left.

$x - 2y \le -6$	**Write original inequality.**
$-2y \le -x - 6$	**Subtract x from each side.**
$y \ge \frac{x}{2} + 3$	**Divide each side by −2 and reverse inequality symbol.**

2 Use your calculator's procedure for graphing and shading an inequality. You may need to decide whether the graph of the corresponding equation is part of the solution, because that may not be clear on the screen.

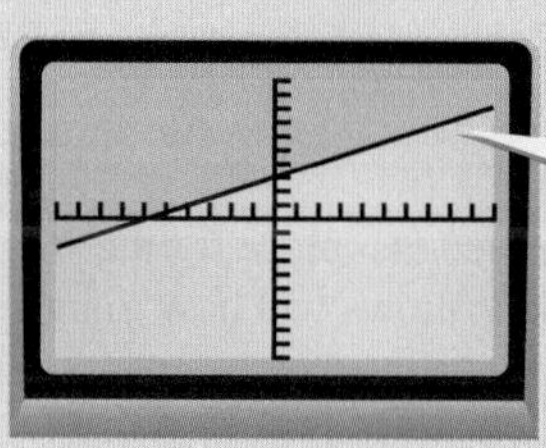

The graph of $y = \frac{x}{2} + 3$ is part of the solution.

3 The corresponding equation should be shown with a solid line, because in this case the inequality is "less than or equal to."

STUDENT HELP

INTERNET KEYSTROKE HELP

See keystrokes for several models of calculators at www.mcdougallittell.com

EXERCISES

In Exercises 1–12, use a graphing calculator or a computer to graph the inequality. (For Exercises 1–8, use a standard viewing rectangle. For Exercises 9–12, use the indicated viewing rectangle.) 1–12. See margin for graphs.

1. $y < -2x - 3$ **2.** $y > 2x + 2$ **3.** $x + 2y \le -1$ **4.** $x - 3y \ge 3$

5. $y > 0.5x + 2$ **6.** $y < 3.2x - 3$ **7.** $\frac{3}{4}x + y \ge 1$ **8.** $\frac{x}{2} - 2y \le 2$

9. $y < x + 25$	**10.** $y > -x + 25$	**11.** $y \le 0.1x$	**12.** $y \ge 100x + 2500$
Xmin = −10	Xmin = −10	Xmin = −10	Xmin = 0
Xmax = 10	Xmax = 10	Xmax = 10	Xmax = 100
Xscl = 1	Xscl = 1	Xscl = 1	Xscl = 10
Ymin = −5	Ymin = −5	Ymin = −1	Ymin = 0
Ymax = 35	Ymax = 35	Ymax = 1	Ymax = 15000
Yscl = 5	Yscl = 5	Yscl = 0.1	Yscl = 1000

13. Write an inequality that represents all points that lie above the line $y = x$. Use a graphing calculator or a computer to check your answer. $y > x$

14. Write an inequality that represents all points that lie below the line $y = x + 2$. Use a graphing calculator or a computer to check your answer. $y < x + 2$

1–12. See Additional Answers beginning on page AA1.

1 Planning the Activity

PURPOSE
To graph a linear inequality in two variables using a graphing calculator.

MATERIALS
- graphing calculator for each student
- Keystroke blackline (*Chapter 6 Resource Book,* p. 66)

PACING
- Example — 10 min
- Exercises — 15 min

LINK TO LESSON
Students should recognize that the first step for graphing the inequality on a calculator involves isolating y, as is the case in Example 4.

2 Managing the Activity

CLASSROOM MANAGEMENT
Shading often involves choosing a *lower function* and an *upper function.* Help students see that when they shade below a boundary line, then the boundary line is the upper function. Any horizontal line at or below the bottom of the window (such as Ymin) will make an appropriate lower function. When students shade above a boundary line, then it is the lower function. Any horizontal line at or above the top of the window (such as Ymax) will make an appropriate upper function.

3 Closing the Activity

KEY DISCOVERY
When graphing a linear inequality on a graphing calculator, you first isolate y. Then you make sure that the region you shade corresponds to the solution set.

ACTIVITY ASSESSMENT
What will the graph of $y - 2x \le 8$ look like on your graphing calculator?
the graph of $y = 2x + 8$ with the half-plane below this line shaded

1 PLAN

PACING
Basic: 2 days
Average: 2 days
Advanced: 2 days
Block Schedule: 1 block

LESSON OPENER APPLICATION
An alternative way to approach Lesson 6.6 is to use the Application Lesson Opener:
- Blackline Master (*Chapter 6 Resource Book,* p. 79)
- Transparency (p. 44)

MEETING INDIVIDUAL NEEDS
- ***Chapter 6 Resource Book***
  Prerequisite Skills Review (p. 5)
  Practice Level A (p. 82)
  Practice Level B (p. 83)
  Practice Level C (p. 84)
  Reteaching with Practice (p. 85)
  Absent Student Catch-Up (p. 87)
  Challenge (p. 90)
- ***Resources in Spanish***
- ***Personal Student Tutor***

NEW-TEACHER SUPPORT
See the Tips for New Teachers on pp. 1–2 of the *Chapter 6 Resource Book* for additional notes about Lesson 6.6.

WARM-UP EXERCISES
Transparency Available
Solve.

1. $x = \frac{98 + 92 + 88 + 95}{4}$ 93.25

2. $x = \frac{3(35) + 3(27) + 2(28)}{3 + 3 + 2}$ 30.25

3. $158 = \frac{x + 123 + 178}{3}$ 173

ENGLISH LEARNERS
The definitions for *mean, median,* and *mode* may confuse English learners. After they have learned the concepts, encourage them to take notes describing how each measure relates to the numbers in a set.

EXPLORING DATA AND STATISTICS

6.6

What you should learn

GOAL 1 Make and use a stem-and-leaf plot to put data in order.

GOAL 2 Find the mean, median, and mode of data, such as tire usage in **Example 3**.

Why you should learn it

▼ To describe **real-life** data, such as the ages of the residents of Alcan, Alaska, in **Example 1**.

REAL LIFE

Stem-and-Leaf Plots and Mean, Median, and Mode

GOAL 1 MAKING STEM-AND-LEAF PLOTS

A **stem-and-leaf plot** is an arrangement of digits that is used to display and order numerical data.

EXAMPLE 1 *Making a Stem-and-Leaf Plot*

The following data show the ages of the 27 residents of Alcan, Alaska. Make a stem-and-leaf plot to display the data. ▶ Source: U.S. Bureau of the Census

45	1	52	42	10	40	50	40	7
46	19	35	3	11	31	6	41	12
43	37	8	41	48	42	55	30	58

SOLUTION

Use the digits in the tens' place for the *stems* and the digits in the ones' place for the *leaves*. The key shows you how to interpret the digits.

Unordered stem-and-leaf plot

Stems	Leaves
0	1 7 3 6 8
1	0 9 1 2
2	
3	5 1 7 0
4	5 2 0 0 6 1 3 1 8 2
5	2 0 5 8

Key: 4|5 = 45

You can order the leaves to make an *ordered* stem-and-leaf plot.

Ordered stem-and-leaf plot

Stems	Leaves
0	1 3 6 7 8
1	0 1 2 9
2	
3	0 1 5 7
4	0 0 1 1 2 2 3 5 6 8
5	0 2 5 8

Key: 4|5 = 45

Your choice of place values for the stem and the leaves will depend on the data. For data between 0 and 100, the leaves are the digits in the ones' place. Be sure to include a key that explains how to interpret the digits.

GOAL 2 USING THE MEAN, MEDIAN, AND MODE

A **measure of central tendency** is a number that is used to represent a typical number in a data set. What value is "most typical" may depend on the type of data, and it may be a matter of opinion. There are three commonly used measures of central tendency of a collection of numbers: the *mean*, the *median*, and the *mode*.

CONCEPT SUMMARY MEASURES OF CENTRAL TENDENCY

- The **mean,** or **average,** of n numbers is the sum of the numbers divided by n.
- The **median** of n numbers is the middle number when the numbers are written in order. If n is even, the median is taken to be the average of the two middle numbers.
- The **mode** of n numbers is the number that occurs most frequently. A set of data can have more than one mode or no mode.

REAL LIFE *Population*

EXAMPLE 2 *Finding the Mean, Median, and Mode*

Find the measure of central tendency of the ages of the residents of Alcan, Alaska, given in Example 1.

a. mean **b.** median **c.** mode

SOLUTION

a. To find the mean, add the 27 ages and divide by 27.

$$\text{mean} = \frac{1 + 3 + \cdots + 55 + 58}{27} \quad \textbf{Definition of mean}$$

$$= \frac{853}{27} \quad \textbf{Add ages.}$$

The mean is $\frac{853}{27}$, or about 32.

b. To find the median, write the ages in order and find the middle number. To order the ages, use the ordered stem-and-leaf plot in Example 1.

1	3	6	7	8	10	11	12	19
30	31	35	37	**40**	40	41	41	42
42	43	45	46	48	50	52	55	58

From this list, you can see that the median age is 40. Half of the ages fall below 40 and half of the ages are 40 or older.

c. To find the mode, use the ordered list in part (b).

There are three modes, 40, 41, and 42.

2 TEACH

MOTIVATING THE LESSON
Ask students what it means to be "better than average" or "below average." Does it always mean the same thing? What would students think, for example, if the bowling coach told one of them that her bowling score was "below average" when she finished third on a team of eight? Tell students that there is more than one way to describe what it means to be "average" or "typical."

EXTRA EXAMPLE 1
The data show the ages of 24 patients who were treated in one day at a health clinic. Make a stem-and-leaf plot.
25 12 30 20 15 21 16 13
18 27 43 23 19 22 20 67
53 19 25 66 35 11 20 37

1	1 2 3 5 6 8 9 9
2	0 0 0 1 2 3 5 5 7
3	0 5 7
4	3
5	3
6	6 7

Key: 4 | 3 = 43

EXTRA EXAMPLE 2
Find the measures of central tendency of the ages of the patients given in Extra Example 1.
a. mean **$\frac{219}{8}$, or about 27**
b. median **21.5** **c.** mode **20**

CHECKPOINT EXERCISES
For use after Examples 1 and 2:
1. Find the measures of central tendency of the data.
46 31 20 28 62 39 42 46
55 19 19 23 48 29 39 40
37 50 56 40 18 31 35 34
a. mean **$\frac{887}{24}$, or about 37**
b. median **38**
c. mode **19, 31, 39, 40, and 46**

APPLICATION NOTE
EXAMPLE 2 Because only 11 of 27 people are younger than the mean, students should see that the median is more representative of a typical age in the town.

EXTRA EXAMPLE 3

A ski area surveyed 131 season pass holders to see how many days they had skied the previous season. Responses are shown in the histogram. Find the median and the mode of the data. **22, 22**

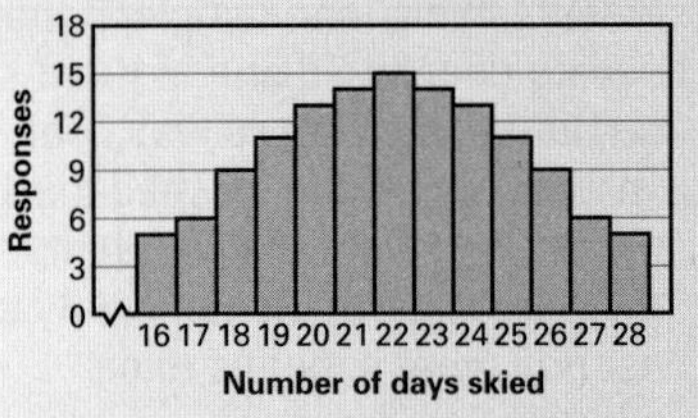

EXTRA EXAMPLE 4

A teacher presents a lesson two ways, each with a different group of students. The first group scores 8, 3, 5, 3, and 9 on a quiz; the second group scores 5, 6, 1, 5, and 4. Using measures of central tendency, do you think one group did better than the other? Explain. ***Sample answer:* The first group scored better. Both groups had the same median, but the mean was 5.6 for the first group, and only 4.2 for the second.**

CHECKPOINT EXERCISES

For use after Examples 3 and 4:

1. Two students scored 100 (the mode) on a quiz. The other eight scores varied from 65 to 80 (the mean). What measure of central tendency likely best represents the "typical" score? **the median**

CLOSURE QUESTION

Explain how to find the mean, the median, and the mode of a set of data. **See the Concept Summary box on p. 369.**

DAILY PUZZLER

The range of ages of three sisters is 5 years. The mean of their ages is one more than the middle sister's age. What are their possible ages if one sister is 22 years old?
17, 18, 22; 21, 22, 26; 22, 23, 27

Many collections of numbers in real life have graphs that are *bell shaped.* For such collections, the mean, the median, and the mode are about equal.

REAL LIFE

Manufacturing

EXAMPLE 3 *A Bell-Shaped Distribution*

A tire manufacturer conducted a survey of 118 people who had purchased a particular type of tire. Each person was asked to report the number of miles driven before replacing the tires. Responses are shown in the histogram. Find the median and the mode of the data.

STUDENT HELP

Look Back
For help with graphs, see p. 41.

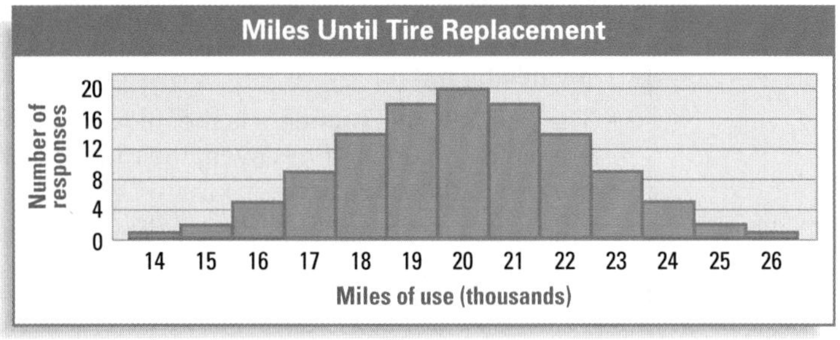

SOLUTION The tallest bar is at 20,000 miles, so that is the mode. The number of responses to the left of that bar is the same as the number of responses to the right, so 20,000 miles is also the median. The mean is also 20,000 miles.

EXAMPLE 4 *Interpreting Measures of Central Tendency*

80-Year-Old Graduates at Elliot School

Elliot School had 28 graduates this year. The town turned out to witness the graduation of John Wilson, age 80. Of the other graduates, seven were 17 years old, nineteen were 18 years old, and one was 19 years old. One person remarked, "The average age of the graduates must be about 35 years."

Was the person's estimate of the average age correct? Would the mean be the most representative measure of central tendency?

SOLUTION

$$\text{Mean} = \frac{7(17) + 19(18) + 1(19) + 1(80)}{28} = \frac{560}{28} = 20$$

The person's estimate was wrong, because the mean is 20.

Because both the median and the mode of the ages are 18 years, the mean is not the most representative measure of central tendency.

....

People usually use the mean to represent a typical number in a data set. If some data are very different, the median may be more representative of the data.

GUIDED PRACTICE

Vocabulary Check ✓

1. Define the *mean*, the *median*, and the *mode* of a collection of numbers. **See margin.**

Concept Check ✓

2. Explain why you might make a stem-and-leaf plot for a data collection. **See margin.**
3. Give an example in which the mean of a collection of numbers is *not* representative of a typical number in the collection. ***Sample answer:*** **3, 4, 5, 5, 4, 3, 2, 2, 152; the mean is 20.**

Skill Check ✓

4. Make a stem-and-leaf plot for the data at the right. **See margin.** 22, 34, 11, 55, 13, 22, 30, 21, 39, 48, 38, 46

11–16. Check plots.

11. 50, 52, 54, 60, 63, 65, 70, 74, 74, 74, 75, 78, 78

12. 17, 18, 19, 20, 22, 23, 24, 29, 30, 39, 40, 45, 50, 51

13. 1, 2, 3, 4, 5, 20, 20, 22, 24, 26, 28, 28, 28, 30, 31, 37, 39

14. 1, 9, 10, 12, 13, 15, 17, 23, 30, 31, 32, 32, 32, 38, 39, 41, 45, 46

15. 30, 33, 37, 39, 44, 48, 52, 54, 61, 61, 62, 68, 68, 76, 76, 76, 77, 79, 80, 81, 82, 84, 86, 87, 87

Find the mean, the median, and the mode of the collection of numbers.

5. 2, 2, 2, 2, 4, 4, 5 **3, 2, 2**
6. 5, 10, 15, 1, 2, 3, 7, 8 **6.375, 6, no mode**
7. −4, 8, 9, −6, −2, 1 **1, $-\frac{1}{2}$, no mode**
8. 6.5, 3.2, 1.7, 10.1, 3.2 **4.94, 3.2, 3.2**
9. Use the stem-and-leaf plot below to order the data set. **11, 13, 13, 15, 16, 17, 18, 20, 20, 22, 25, 27, 27, 28, 31, 33, 34, 37, 38**

1	6 3 8 1 7 5 3
2	8 2 5 0 7 7 0
3	8 3 1 4 7

Key: 3 | 8 = 38

10. AGES If someone said that the mean age of everyone in your algebra class is about $16\frac{1}{2}$ years, do you think the age of the teacher was included in the calculation? Explain. ***Sample answer:*** **If the class is fairly large and the teacher is very young, the teacher's age may have been included. If the class is small and the teacher is much older than the students, the teacher's age was probably not included.**

PRACTICE AND APPLICATIONS

STUDENT HELP

Extra Practice to help you master skills is on p. 802.

MAKING STEM-AND-LEAF PLOTS **Make a stem-and-leaf plot for the data. Use the result to list the data in increasing order.** **11–16. See margin.**

11. 60, 74, 75, 63, 78, 70, 50, 74, 52, 74, 65, 78, 54
12. 24, 29, 17, 50, 39, 51, 19, 22, 40, 45, 20, 18, 23, 30
13. 4, 31, 22, 37, 39, 24, 2, 28, 1, 26, 28, 30, 28, 3, 20, 20, 5
14. 15, 39, 13, 31, 46, 9, 38, 17, 32, 10, 12, 45, 30, 1, 32, 23, 32, 41
15. 87, 61, 54, 77, 79, 86, 30, 76, 52, 44, 48, 76, 87, 68, 82, 61, 84, 33, 39, 68, 37, 80, 62, 81, 76
16. 48, 10, 48, 25, 40, 42, 44, 23, 21, 13, 50, 17, 18, 19, 21, 57, 35, 33, 25, 50, 13, 12, 46

16. 10, 12, 13, 13, 17, 18, 19, 21, 21, 23, 25, 25, 33, 35, 40, 42, 44, 46, 48, 48, 50, 50, 57

STUDENT HELP

HOMEWORK HELP
Example 1: Exs. 11–16, 32–34
Example 2: Exs. 17–27
Example 3: Exs. 36, 37
Example 4: Exs. 29, 30

FINDING MEAN, MEDIAN, AND MODE **Find the mean, the median, and the mode of the collection of numbers.**

17. 4, 2, 10, 6, 10, 7, 10 **7, 7, 10**
18. 1, 2, 1, 2, 1, 3, 3, 4, 3 **$2\frac{2}{9}$, 2, 1 and 3**
19. 8, 5, 6, 5, 6, 6 **6, 6, 6**
20. 4, 4, 4, 4, 4, 4 **4, 4, 4**
21. 5, 3, 10, 13, 8, 18, 5, 17, 2, 7, 9, 10, 4, 1 **8, 7.5, 5 and 10**
22. 12, 5, 6, 15, 12, 9, 13, 1, 4, 6, 8, 14, 12 **9, 9, 12**
23. 3.56, 4.40, 6.25, 1.20, 8.52, 1.20 **4.19, 3.98, 1.2**
24. 161, 146, 158, 150, 156, 150 **153.5, 153, 150**

3 APPLY

ASSIGNMENT GUIDE

BASIC
Day 1: p. 371 Exs. 11–20
Day 2: pp. 371–374 Exs. 21–27, 29, 33, 36, 37, 41, 42, 44–46, 50–58 even

AVERAGE
Day 1: p. 371 Exs. 11–20
Day 2: pp. 371–374 Exs. 21–27, 29, 33, 34, 36, 37, 41, 42, 44–46, 50–58 even

ADVANCED
Day 1: p. 371 Exs. 11–20
Day 2: pp. 371–374 Exs. 21–27, 29, 30–34, 36, 37, 41–46, 50–58 even

BLOCK SCHEDULE
pp. 371–374 Exs. 11–27, 29, 33, 34, 36, 37, 41, 42, 44–46, 50–58 even

EXERCISE LEVELS

Level A: ***Easier***
44–49

Level B: ***More Difficult***
11–27, 29, 33–37, 41, 42, 50–59

Level C: ***Most Difficult***
28, 30–32, 38–40, 43

✓ HOMEWORK CHECK

To quickly check student understanding of key concepts, go over the following exercises: Exs. 12, 18, 22, 25, 29, 36. See also the Daily Homework Quiz:

- Blackline Master (*Chapter 6 Resource Book,* p. 93)
- Transparency (p. 51)

1. ***Sample answer:*** **Given a set of *n* numbers, the mean is the sum of the numbers divided by *n*, the median is the middle number when the numbers are written in order, and the mode is the number(s) that occurs most frequently. If *n* is even, the median is the average of the two middle numbers.**

2 and 4. See Additional Answers beginning on page AA1.

APPLICATION NOTE

EXERCISE 29 A histogram of the salaries would probably show most salaries clustered around $25,000 (the mean excluding the millionaire). The multimillionaire's salary would stand alone at the distant right of the histogram. This indicates that the millionaire's salary pulled the mean toward itself. Income data are often pulled, or skewed, to the right because there are likely to be many moderate incomes in a group, some large incomes, and a small number of very large incomes.

STUDENT HELP NOTES

Homework Help Students can find help for Ex. 29 at **www.mcdougallittell.com.** The information can be printed out for students who don't have access to the Internet.

DATA UPDATE

Updated data for Exs. 30 and 31 are available at **www.mcdougallittell.com.**

APPLICATION NOTE

EXERCISE 31 If the census data were graphed on a scatter plot, the number of households in California would appear as an *outlier,* that is, it falls outside the overall pattern of the graph. When outliers exist, it is helpful also to describe the spread of the data, using, for example, the range.

EXERCISE 32 A histogram that is skewed slightly to the left could represent the first set of data, with the lower mean age. A histogram that is skewed slightly to the right could represent the second set of data, with the higher mean age.

33. See Additional Answers beginning on page AA1.

28. You could add any of the numbers (other than 61) in the list. There would be two modes, the number you added and 61.

29. No; the mean salary of the other 99 adults is $25,000. The median and mode (both $25,000) are much more representative of the "average" salary.

30. *Sample answer:* The median (619) and the mode (619) are both more representative of the data than the mean (1902), due to the high population in California.

STUDENT HELP

HOMEWORK HELP Visit our Web site www.mcdougallittell.com for help with problem solving in Ex. 29.

31. *Sample answer:* The median (619) and the mode (619) are both slightly more representative than the mean (753) because they are closer to more scores in the lower three fourths of the data.

32. *Sample answers:* (1) 10, 12, 12, 12, 18, 18, 18, 20, 20, 20

(2) 12, 12, 15, 15, 16, 16, 18, 18, 29, 29

BASEBALL **In Exercises 25–28, use the following information.**

The table shows the number of shutouts that ten baseball pitchers had in their careers. A shutout is a complete game pitched without allowing a run.

Pitcher	Shutouts
Warren Spahn	63
Christy Mathewson	80
Eddie Plank	69
Nolan Ryan	61
Bert Blyleven	60
Don Sutton	58
Grover Alexander	90
Walter Johnson	110
Cy Young	76
Tom Seaver	61

▶ Source: *The Baseball Encyclopedia*

25. Find the mean and the median for the set of data. **72.8, 66**

26. Write the numbers in decreasing order. **110, 90, 80, 76, 69, 63, 61, 61, 60, 58**

27. Does the set of data have a mode? If so, what is it? **yes; 61**

28. CRITICAL THINKING What number could you add to the set of data to make it have more than one mode? Explain why you chose the number. **See margin.**

29. AVERAGE SALARY A small town has an adult population of 100. One person is a multimillionaire who makes $5,000,000 per year. The mean annual salary in the town is $74,750. Is the mean a fair measure of a typical salary in the town? What is the mean annual salary of the other 99 adults? **See margin.**

30. NUMBER OF HOUSEHOLDS The U.S. Bureau of the Census reports on the number of households (in thousands) in each state. Some of the data are shown in the table. Is one measure of central tendency the most representative of the typical number of households in these states? Explain your reasoning. **See margin.**

State	Households
Arizona	1,687
California	11,101
Colorado	1,502
Idaho	430
Montana	341
Nevada	619
New Mexico	619
Utah	639
Wyoming	184

31. Use the data in Exercise 30. If you exclude California from the data, is one measure of central tendency the most representative of the data? Explain your reasoning.

DATA UPDATE Visit our Web site www.mcdougallittell.com

32. CRITICAL THINKING Create two different sets of data, each having 10 ages. Create one set so that the mean age is 16 and the median age is 18. Create the other set so that the median age is 16 and the mean age is 18. **See margin.**

33. BIRTHDAYS Use a stem-and-leaf plot (months as stems, days as leaves) to write the birthdays in order from earliest in the year to latest (1 = January, 2 = February, and so on). Include a key with your stem-and-leaf plot. **See margin.**

10-11	4-14	7-31	12-28	4-17
2-22	8-21	1-24	9-12	1-3
4-30	10-17	6-5	1-25	5-10
12-9	4-1	8-26	12-15	3-17
4-30	2-3	11-11	6-13	11-4
6-24	6-3	4-8	2-20	11-28

34. **WEIGHTS** The numbers shown below represent the weights (in pounds) of twelve high school students. Use a stem-and-leaf plot to order the weights of the students from least to greatest. (*Hint:* Use 17 | 7 = 177 for your key.)

177, 164, 108, 158, 195, 206, 106, 126, 143, 118, 148, 146 **See margin for plot.**

35. **ASTEROIDS** The numbers below are the mean distances from the sun (in millions of miles) of the first ten asteroids. Use a stem-and-leaf plot to order the distances from least to greatest. (*Hint:* Use 25 | 70 = 257.0 for your key.)

257.0, 257.4, 247.8, 219.3, 239.3, 225.2, 221.4, 204.4, 221.7, 222.6 **See margin for plot.**

VACATIONS In Exercises 36–38, use the following information.

A business conducted a survey of its employees who go on vacation every year. Each person was asked to report the number of miles traveled (to the nearest hundred) on the last vacation. The responses are shown in the histogram.

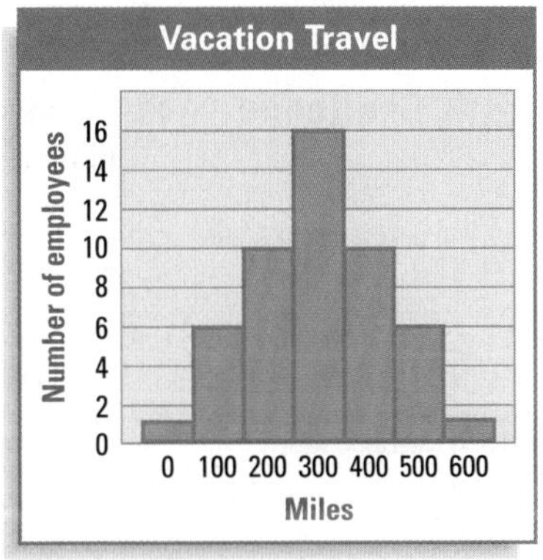

36. Find the median of the data. **300**

37. Find the mode of the data. **300**

38. *Writing* Suppose a new employee traveled 400 miles on the last vacation. Decide whether this would affect the median or the mode. Explain your thinking.

EXTENSION: DOUBLE STEM-AND-LEAF PLOT In Exercises 39 and 40, use the following information.

Another type of visual model that you can use is a *double stem-and-leaf plot.* This type of visual model is used to compare two sets of data. The Stem column is in the middle of the plot, so that the "stem" can apply to both sets of data.

The double stem-and-leaf plot below shows the number of all-time tournament wins by the top 17 male and female professional golfers.

▶ Source: Professional Golfers Association

Male	Stem	Female
1	8	8 2
0	7	
3 0	6	
2 1	5	7 5 0
0 0	4	4 2 2
0 1 8 6 3 2 1	3	8 5 2 1 1 1
9 9	2	9 9 6

Key: 1 | 3 | 8 = 31, 38

39. What conclusions can be drawn from the double stem-and-leaf plot? **See margin.**

40. Use the data in the double stem-and-leaf plot to make two conventional stem-and-leaf plots. Explain what changes you need to make. **See margin for plots.**

34. 106, 108, 118, 126, 143, 146, 148, 158, 164, 177, 195, 206

35. 204.4, 219.3, 221.4, 221.7, 222.6, 225.2, 239.3, 247.8, 257.0, 257.4

38. *Sample answer:* It would not affect either. There would be 51 numbers in the data set and the median would be the twenty-sixth, which would be 300. The number 400 would still appear fewer times than the number 300, which would still be the mode.

39. *Sample answers:*
- About half of the top 17 male and female golfers had fewer than 40 tournament wins.
- For both males and females, by far the most common stem was 3, meaning between 30 and 39 tournament wins.
- There were 4 male and 2 female golfers with 60 or more tournament wins.

40. *Sample answer:* The positions of the stems and leaves on the left-hand plot must be reversed. To make standard ordered stem-and-leaf plots, the order of the stems must be reversed and the leaves must be arranged in increasing order.

APPLICATION NOTE

EXERCISE 38 The median can be determined in terms of the area of a histogram: the area to the left of the median equals the area to the right. Adding the new data value adds a little area to the right side of the histogram, so the median will shift to the right to a position that equalizes the areas of the histogram on both sides of the median.

! COMMON ERROR

EXERCISE 39 Students can easily be confused by a double stem-and-leaf plot because they must read the left side of the plot from right to left. Remind students that this plot has single stems, which are shared by leaves on both sides.

34.

10	6 8
11	8
12	6
13	
14	3 6 8
15	8
16	4
17	7
18	
19	5
20	6

Key: 17 | 7 = 177

35.

20	44
21	93
22	14 17 26 52
23	93
24	78
25	70 74

Key: 25 | 70 = 257.0

40. See Additional Answers beginning on page AA1.

ADDITIONAL PRACTICE AND RETEACHING

For Lesson 6.6:
- Practice Levels A, B, and C (*Chapter 6 Resource Book,* p. 82)
- Reteaching with Practice (*Chapter 6 Resource Book,* p. 85)
- See Lesson 6.6 of the *Personal Student Tutor*

For more Mixed Review:
- Search the *Test and Practice Generator* for key words or specific lessons.

4 ASSESS

DAILY HOMEWORK QUIZ

Transparency Available

Make a stem-and-leaf plot of the data. Find the mean, median, and mode.

1. High temperatures (°F)

55 47 52 55 59 61 70
56 48 45 54 63 57 62

4	5 7 8
5	2 4 5 5 6 7 9
6	1 2 3
7	0

Key: 6 | 1 = 61

56, 55.5, 55

2. 100 meter race results (sec)

11.35 11.42 11.19 11.56 11.69
11.61 11.56 11.58 11.48 11.28
11.62 11.67 11.57 11.39 11.37

11.1	9
11.2	8
11.3	5 7 9
11.4	2 8
11.5	6 6 7 8
11.6	1 2 7 9

Key: 11.5 | 6 = 11.56

11.49, 11.56, 11.56

EXTRA CHALLENGE NOTE

Challenge problems for Lesson 6.6 are available in **blackline** format in the *Chapter 6 Resource Book,* p. 90 and at **www.mcdougallittell.com.**

ADDITIONAL TEST PREPARATION

1. WRITING Explain how you can use a stem-and-leaf plot to determine the median of a set of data. **See sample answer below.**

2. OPEN ENDED Describe a real-life situation in which the mean is lower than the median. ***Sample answer:*** **You just bought 5 CDs, four that cost \$14.99 and one that was on sale for \$5. The mean price of the CDs is lower than the median price.**

54–59. See Additional Answers beginning on page AA1.

Test Preparation

43. a. $\frac{84 + 92 + 76 + x}{4}$, or $\frac{x + 252}{4}$

b. $\frac{x + 252}{4} \geq 85$; 88

c. ***Sample answer:*** **The average of the three tests is 84. Subtract 3 × 84 from 4 × 85; the result is 88.**

QUANTITATIVE COMPARISON **In Exercises 41 and 42, choose the statement below that is true about the given numbers.**

(A) The number in column A is greater.

(B) The number in column B is greater.

(C) The two numbers are equal.

(D) The relationship cannot be determined from the given information.

	Column A	Column B	
41.	The mean of 1, 2, 3, 4, 5, 5, 10	The mean of 2, 3, 4, 5, 5, 10, 11	**B**
42.	The mode of −2, −2, −2, 4	The mode of −2, −2, −3, −3, −3	**A**

★ Challenge

43. **TEST SCORES** You have test scores of 84, 92, and 76. There will be one more test in the marking period. You want your mean test score for the marking period to be 85 or higher. **See margin.**

a. Let x represent your last test score. Write an expression for the mean of your test scores for the marking period.

b. Write and solve an inequality to find what you must score on the last test.

c. Solve the problem without using algebra. Describe your method. Do your answers agree?

EXTRA CHALLENGE
www.mcdougallittell.com

MIXED REVIEW

EVALUATING EXPRESSIONS **Evaluate the expression.** **(Review 1.3 for 6.7)**

44. $35 + 40 \div 2$ **55** **45.** $120 \times 3 - 2$ **358** **46.** $237 - 188 \div 4$ **190**

CHECKING SOLUTIONS **Check whether the given number is a solution of the inequality.** **(Review 1.4)**

47. $x + 5 > 11$; 6 **not a solution** **48.** $2z - 3 < 4$; 4 **not a solution**

49. **ROAD TRIP** You and two friends each agree to drive $\frac{1}{3}$ of a 75-mile trip. Which equation would you use to find the number of miles each of you must drive? **(Review 1.5)** **A**

A. $3x = 75$ **B.** $\frac{1}{3}x = 75$

FINDING SLOPE **Find the slope of the graph of the linear function f.** **(Review 4.8)**

50. $f(2) = -1, f(5) = 5$ **2** **51.** $f(0) = 5, f(5) = 0$ **−1**

52. $f(-3) = -12, f(3) = 12$ **4** **53.** $f(0) = 1, f(4) = -1$ $-\frac{1}{2}$

SOLVING INEQUALITIES **Solve the inequality and graph the solution.** **(Review 6.2–6.4)** **54–59. See margin for graphs.**

54. $-x + 7 > 13$ $x < -6$ **55.** $16 - x \leq 7$ $x \geq 9$ **56.** $-4 < x + 3 < 7$ $-7 < x < 4$

57. $7 < 3 - 2x \leq 19$ $-8 \leq x < -2$ **58.** $|x - 3| < 12$ $-9 < x < 15$ **59.** $|x + 16| \geq 9$ $x \leq -25$ or $x \geq -7$

Additional Test Preparation *Sample answer:*

1. Create an ordered stem-and-leaf plot for the data. Count either from the top or the bottom to locate the middle number in the leaves if there is an odd number of data values. Read this leaf with its stem. This is the median. If there is an even number of data values, locate the middle two numbers in the leaves and find their mean. This number with its stem is the median.

EXPLORING DATA AND STATISTICS

6.7 Box-and-Whisker Plots

What you should learn

GOAL 1 Draw a box-and-whisker plot to organize **real-life** data.

GOAL 2 Read and interpret a box-and-whisker plot of **real-life** data.

Why you should learn it

▼ To interpret **real-life** data, such as data about the lengths of commuters' trips to work, as in **Exs. 31–33**.

REAL LIFE

GOAL 1 DRAWING A BOX-AND-WHISKER PLOT

A **box-and-whisker plot** is a data display that divides a set of data into four parts. The median or **second quartile** separates the set into two halves: the numbers that are below the median and the numbers that are above the median. The **first quartile** is the median of the lower half. The **third quartile** is the median of the upper half.

EXAMPLE 1 *Finding Quartiles*

Use this set of data: 11, 19, 5, 34, 9, 25, 28, 16, 17, 11, 12, 7.

a. Find the first, second, and third quartiles of the data.

b. Draw a box-and-whisker plot of the data.

SOLUTION

a. Begin by writing the numbers in increasing order. You must find the second quartile before you find the first and third quartiles.

Second quartile: $\frac{12 + 16}{2} = 14$

5, 7, 9, | 11, 11, 12, | 16, 17, 19, | 25, 28, 34

First quartile: $\frac{9 + 11}{2} = 10$ **Third quartile:** $\frac{19 + 25}{2} = 22$

b. ***Draw*** a number line that includes the least number and the greatest number in the data set.

Plot the least number, the first quartile, the second quartile, the third quartile, and the greatest number. Draw a line from the least number to the greatest number below your number line. Plot the same points on that line.

The "box" extends from the first to the third quartile. Draw a vertical line in the box at the second quartile. The "whiskers" connect the box to the least and greatest numbers.

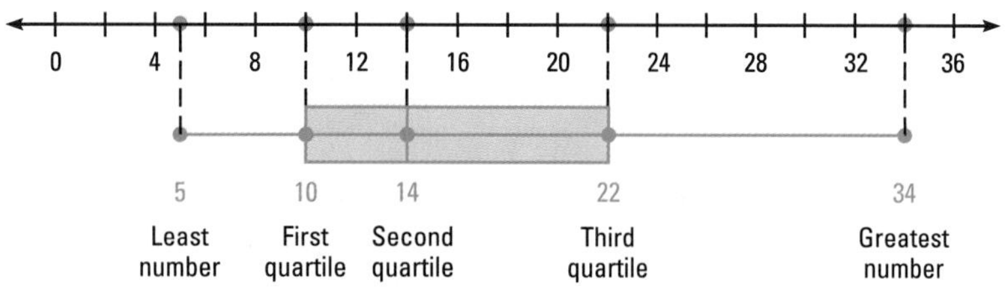

1 PLAN

PACING
Basic: 2 days
Average: 2 days
Advanced: 2 days
Block Schedule: 1 block

LESSON OPENER ACTIVITY
An alternative way to approach Lesson 6.7 is to use the Activity Lesson Opener:
- Blackline Master (*Chapter 6 Resource Book,* p. 94)
- Transparency (p. 45)

MEETING INDIVIDUAL NEEDS
- ***Chapter 6 Resource Book***
  Prerequisite Skills Review (p. 5)
  Practice Level A (p. 96)
  Practice Level B (p. 97)
  Practice Level C (p. 98)
  Reteaching with Practice (p. 99)
  Absent Student Catch-Up (p. 101)
  Challenge (p. 103)
- ***Resources in Spanish***
- ***Personal Student Tutor***

NEW-TEACHER SUPPORT
See the Tips for New Teachers on pp. 1–2 of the *Chapter 6 Resource Book* for additional notes about Lesson 6.7.

WARM-UP EXERCISES

Transparency Available

In its first seven games, your football team scored the following numbers of points in the first two quarters:
Quarter 1: 14, 10, 10, 17, 21, 16, 13
Quarter 2: 6, 7, 14, 13, 3, 15, 10

1. What was the median number of points scored in the first quarter? **14**

2. What was the median number of points scored in the second quarter? **10**

3. What was the median number of points scored in either quarter? **13**

2 TEACH

MOTIVATING THE LESSON
Ask students how many of them would like to try to find the median height of all male students in their grade or school from a stem-and-leaf plot. Most probably won't care much for the counting involved. Tell students that a new kind of plot, a *box-and-whisker plot,* makes it easy to find the median and other important information about a data set.

EXTRA EXAMPLE 1
Use the data: 39 32 27 42 21 35 38
a. Find the first, second, and third quartiles of the data.
27, 35, 39
b. Draw a box-and-whisker plot.

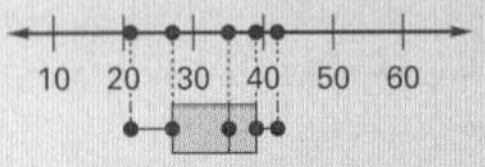

EXTRA EXAMPLE 2
You found the following prices (in dollars) of 20 different styles of bicycle helmets.
52 67 102 85 68 89 99
92 80 89 59 79 79 90
75 78 63 94 79 64
a. Draw a stem-and-leaf plot.

Stem	Leaves
5	2 9
6	3 4 7 8
7	5 8 9 9 9
8	0 5 9 9
9	0 2 4 9
10	2

Key: 7 | 5 = 75

b. Draw a box-and-whisker plot.

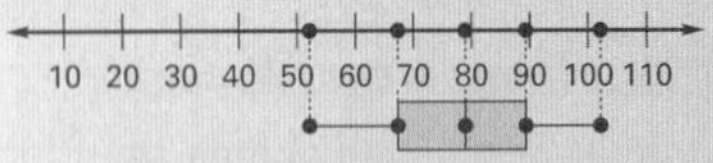

c. Describe the results.
Sample answer: **One half of the prices are between \$67.50 and \$89.50. The median price is \$79. Three fourths of the helmets cost \$89.50 or less. The least price is \$52 and the greatest price is \$102.**

CHECKPOINT EXERCISES
For use after Examples 1 and 2:
1. Find the first, second, and third quartiles of these data:
65 56 66 43 72 81 78 74 62
59, 66, 76

REAL LIFE *Retail Prices*

EXAMPLE 2 *Organizing Data*

You found the following prices of 24 different brands of backpacks.

a. Draw a stem-and-leaf plot to order the data.

b. Draw a box-and-whisker plot of the data.

c. Describe the results.

\$15	\$25	\$50	\$42	\$29
\$50	\$33	\$27	\$35	\$32
\$40	\$35	\$28	\$40	\$40
\$35	\$35	\$75	\$40	\$40
\$25	\$50	\$35	\$50	

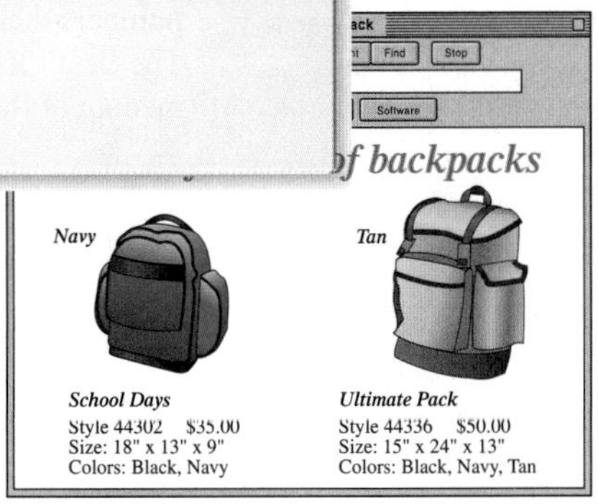

STUDENT HELP
Study Tip
A stem with no leaves shows that there are no data in the set that begin with that stem number.

SOLUTION

a. Draw an ordered stem-and-leaf plot.

Stem	Leaves
1	5
2	5 5 7 8 9
3	2 3 5 5 5 5 5
4	0 0 0 0 0 2
5	0 0 0 0
6	
7	5

Key: 5 | 0 = 50

From this plot, you can order the data.

\$15	\$25	\$25	\$27	\$28	\$29	\$32	\$33
\$35	\$35	\$35	\$35	\$35	\$40	\$40	\$40
\$40	\$40	\$42	\$50	\$50	\$50	\$50	\$75

b. Use the ordered data to find the quartiles.

Second quartile: $\frac{35 + 35}{2} = 35$

First quartile: $\frac{29 + 32}{2} = 30.5$

Third quartile: $\frac{40 + 42}{2} = 41$

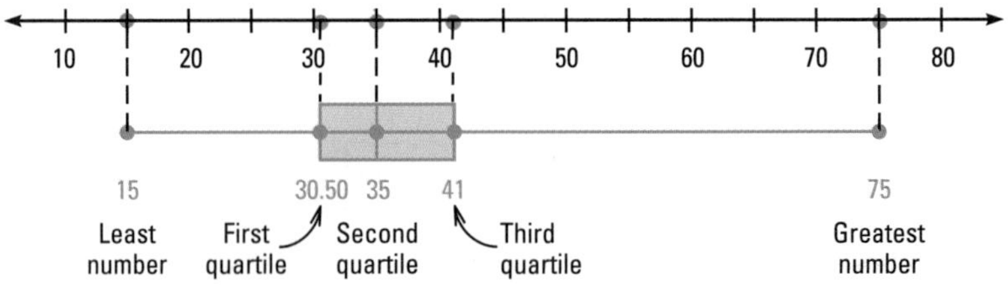

c. One half of the prices are between \$30.50 and \$41. The median price is \$35. Three fourths of the backpacks cost \$41 or less. The least price is \$15 and the greatest price is \$75.

GOAL 2 READING BOX-AND-WHISKER PLOTS

You can use a box-and-whisker plot to see how data are grouped within a set of data. You can also use a box-and-whisker plot to compare two different sets of data.

REAL LIFE
Meteorology

EXAMPLE 3 *Interpreting a Box-and-Whisker Plot*

The box-and-whisker plots below show the annual precipitation (in inches) in Chicago, Illinois, and in San Diego, California, recorded over the last 30 years.

Precipitation in Chicago, Illinois

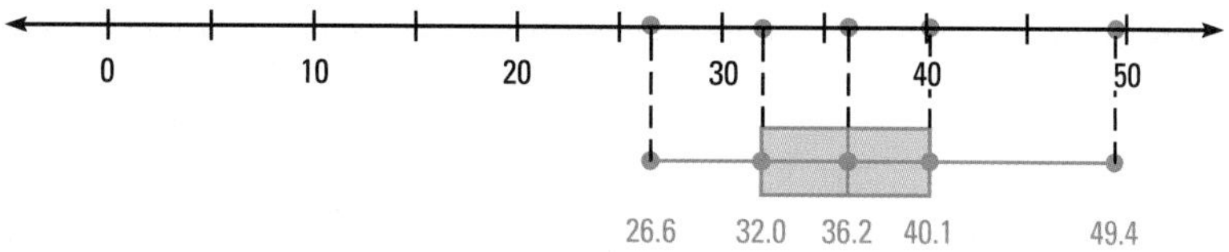

Precipitation in San Diego, California

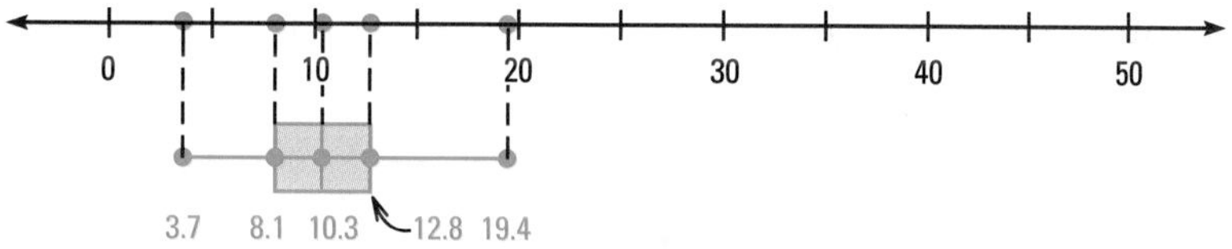

a. What is the median amount of precipitation in Chicago? in San Diego?

b. Which city has had more precipitation in the last 30 years?

c. *Writing* How else do the data sets differ?

SOLUTION

a. The median amount of precipitation in Chicago is about 36.2 inches. The median amount of precipitation in San Diego is about 10.3 inches.

b. Chicago has had more precipitation than San Diego. The lowest "whisker" in the Chicago plot represents a greater amount than the highest "whisker" in the San Diego plot.

c.

Chicago has had between 26.6 inches and 49.4 inches of precipitation. San Diego has had between 3.7 inches and 19.4 inches of precipitation. The difference between the extremes of precipitation in Chicago is about 22.8 inches. The difference between the extremes of precipitation in San Diego is about 15.7 inches.

In about half of the years, Chicago had a precipitation between 32 inches and 40 inches. In about half of the years, San Diego had a precipitation between 8 inches and 13 inches.

STUDENT HELP

INTERNET HOMEWORK HELP
Visit our Web site www.mcdougallittell.com for extra examples.

EXTRA EXAMPLE 3

The box-and-whisker plot shows how a group of teenagers rated two fruit drinks on a scale of 1 to 10.

Brand A

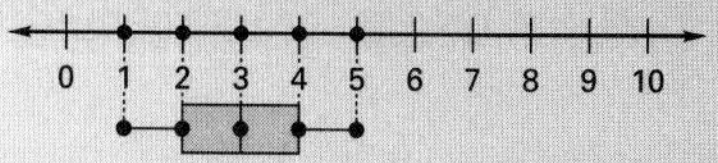

Brand B

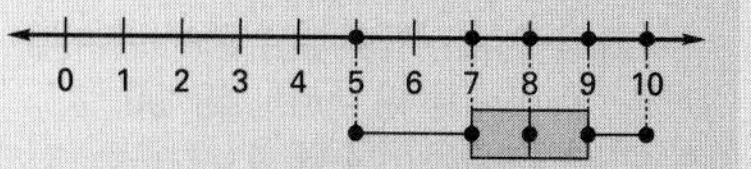

a. What is the median rating for Brand A? for Brand B? **3; 8**

b. Are all of Brand B's ratings higher than Brand A's? **no**

c. How else do the data sets differ? ***Sample answer:*** **Brand A had ratings between 1 and 5. Brand B had ratings between 5 and 10. The difference between the extremes for Brand A is 4, and for Brand B is 5. Half the ratings for Brand A were between 2 and 4. Half the ratings for Brand B were between 7 and 9.**

CHECKPOINT EXERCISES

For use after Example 3:

1. Here are measures for two data sets:
Set 1: min: 20; first quartile: 27; median: 34; third quartile: 50; max: 86
Set 2: min: 5; first quartile: 10; median: 18.5; third quartile: 29; max: 33
Which data set has a greater distance between the ends of the whiskers? For which data set are the left and right ends of the box closer to equally distant from the median?
Set 1; Set 2

CLOSURE QUESTION

How do you locate the first and third quartiles from a box-and-whisker plot? ***Sample answer:*** **The first quartile is the value at the left end of the box. The third quartile is the value at the right end of the box.**

3 APPLY

ASSIGNMENT GUIDE

BASIC
Day 1: p. 378 Exs. 11–22
Day 2: pp. 379–381 Exs. 25–36, 40–48 even, 49, 50, Quiz 3 Exs. 1–9

AVERAGE
Day 1: p. 378 Exs. 11–22
Day 2: pp. 379–381 Exs. 25–36, 40–48 even, 49, 50, Quiz 3 Exs. 1–9

ADVANCED
Day 1: p. 378 Exs. 11–22
Day 2: pp. 379–381 Exs. 25–37, 40–48 even, 49, 50, Quiz 3 Exs. 1–9

BLOCK SCHEDULE
pp. 378–381 Exs. 11–22, 25–36, 40–48 even, 49, 50, Quiz 3 Exs. 1–9

EXERCISE LEVELS

Level A: ***Easier***
11–18

Level B: ***More Difficult***
19–36, 38–50

Level C: ***Most Difficult***
37

✓ HOMEWORK CHECK

To quickly check student understanding of key concepts, go over the following exercises: Exs. 12, 16, 20, 25, 26, 32, 34. See also the Daily Homework Quiz:

- Blackline Master (*Chapter 7 Resource Book,* p. 11)
- Transparency (p. 53)

! COMMON ERROR

EXERCISES 4–7 Make sure that students do not include the median in the top and/or bottom halves of the data values when they are determining the first and third quartiles.

8, 15–22. See Additional Answers beginning on page AA1.

GUIDED PRACTICE

Vocabulary Check ✓

1. Another expression for the median of an ordered set of numbers is ___?___. **second quartile**

Concept Check ✓

2. Describe the "box" of a box-and-whisker plot. What does it represent? **See margin.**

3. Describe the "whiskers" of a box-and-whisker plot. What do they represent? **See margin.**

2. The box extends from the first quartile to the third quartile and represents the middle half of the data.

3. The lower whisker connects the least number to the first quartile and the upper whisker connects the greatest number to the third quartile. Each whisker represents a quarter of the data.

Skill Check ✓

In Exercises 4–7, find the first, second, and third quartiles of the data.

4. 1, 2, 3, 1, 2, 1, 2, 1, 2 **1, 2, 2**

5. 5, 6, 7, 2, 1, 3, 4 **2, 4, 6**

6. 12, 30, 19, 15, 18, 22 **15, 18.5, 22**

7. 4, 20, 14, 6, 8, 18, 2, 10, 12, 16 **6, 11, 16**

8. Draw a box-and-whisker plot of the following ages of horses in a stable. **See margin.**

2, 5, 12, 3, 18, 23, 7, 8, 11, 24, 4, 10, 3, 5, 12

9. Tell which set of data is shown by the box-and-whisker plot. **C**

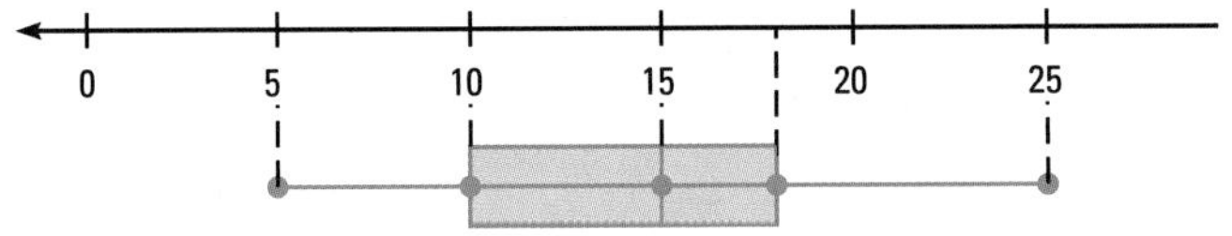

A. 0, 10, 15, 18, 25 **B.** 5, 10, 15, 18, 30 **C.** 5, 10, 15, 18, 25

10. Use the box-and-whisker plots in Example 3. Is it true that 50% of the yearly precipitation in San Diego is less than 8.1 inches? Explain.
No; the precipitation was less than 8.1 inches in 25% of the years described.

PRACTICE AND APPLICATIONS

STUDENT HELP
Extra Practice to help you master skills is on p. 802.

FINDING QUARTILES **Find the first, second, and third quartiles of the data.**

11. 12, 5, 3, 8, 10, 7, 6, 5 **5, 6.5, 9**

12. 20, 73, 31, 53, 22, 64, 47 **22, 47, 64**

13. 1, 12, 6, 5, 4, 7, 5, 10, 3, 4 **4, 5, 7**

14. 2.3, 5.6, 3.4, 4.5, 3.8, 1.2, 9.7 **2.3, 3.8, 5.6**

DRAWING BOX-AND-WHISKER PLOTS **In Exercises 15–22, draw a box-and-whisker plot of the data.** **15–22. See margin for plots.**

15. 6, 7, 10, 6, 2, 8, 7, 7, 8

16. 10, 5, 9, 50, 10, 3, 4, 15, 20, 6

17. 12, 13, 7, 6, 25, 25, 5, 10, 15, 10, 16, 14, 29

18. 8, 8, 10, 10, 1, 12, 8, 6, 5, 1, 9, 10

19. Number of days of rainfall in a year: 39, 46, 26, 12, 34, 57, 38, 37, 69, 15, 44, 47, 38, 58, 75, 29, 40, 35, 22, 69, 22, 37, 51, 55, 46, 27, 19, 36, 72, 49

20. Dogs' weights (in pounds): 88, 23, 75, 46, 77, 22, 83, 75, 97, 89, 46, 79, 69, 72, 91, 20, 75, 78, 83, 35, 39, 98, 59, 77, 84, 69, 82, 79, 57, 91

21. Attendance at early showing of a movie: 48, 60, 40, 68, 51, 47, 57, 41, 65, 61, 20, 65, 49, 34, 63, 53, 52, 35, 45, 35, 65, 65, 48, 36, 24, 53, 64, 48, 40

22. Weekly tips earned (in dollars): 167, 191, 190, 154, 188, 174, 192, 166, 180, 155, 157, 161, 163, 172, 169, 167, 182, 184, 158, 160, 176

STUDENT HELP
HOMEWORK HELP
Example 1: Exs. 11–22
Example 2: Exs. 25–29
Example 3: Exs. 31–33

FOCUS ON APPLICATIONS

REAL LIFE **CELLULAR TELEPHONES** generated over $26 billion in operating revenue in 1996. An average monthly bill was about $48.

INTERNET APPLICATION LINK www.mcdougallittell.com

23. *Sample answer*: 10, 11, 11, 12, 14, 15, 16, 18, 20, 21, 25, 27, 27, 29, 30, 38

24. *Sample answer*: 106, 110, 118, 124, 124, 130, 140, 149, 151, 155, 159, 160, 164, 180, 182, 193

30. *Sample answer*: The feed change increased milk production significantly. Notice that all three quartiles and the greatest number increased. The quartiles increased by amounts ranging from 6 to 7.5. The part of the graph from the first quartile to the right represents 75% of the data as does the part of the graph from the third quartile to the left. The second graph indicates that 75% of the data from the second group are higher than 75% of the data from the first.

33. No. *Sample answer*: More people travel 0.5–1 hour; the box represents about 50% of the data, each whisker about 25%.

STUDENT HELP

INTERNET HOMEWORK HELP Visit our Web site www.mcdougallittell.com for help with Exs. 31–33.

CREATING DATA SETS Create a collection of 16 numbers that could be represented by the box-and-whisker plot. 23, 24. See margin.

23.

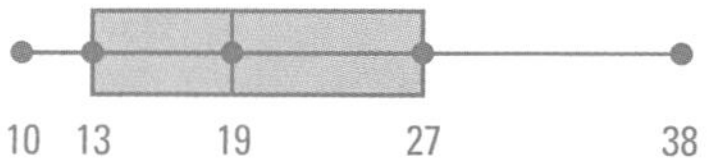

24.

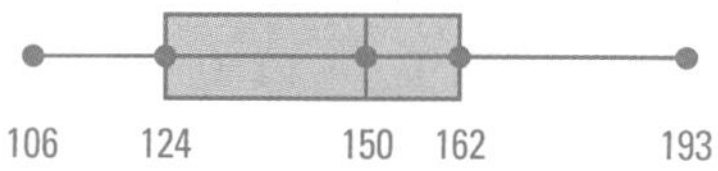

CELLULAR TELEPHONES In Exercises 25–29, use the table that lists the weights (in ounces) and operating times (in minutes) for 10 cellular telephones. 25–28. See margin for plots.

Weight	Operating Time
8.8	140
7.7	80
4.8	90
7.0	90
7.1	90
8.4	70
3.8	60
3.6	260
6.7	75
6.2	135

INTERNET DATA UPDATE Visit our Web site www.mcdougallittell.com

25. Make a stem-and-leaf plot to order the weights of the phones. 3.6, 3.8, 4.8, 6.2, 6.7, 7.0, 7.1, 7.7, 8.4, 8.8

26. Draw a box-and-whisker plot of the weights of the phones.

27. Make a stem-and-leaf plot to order the operating times of the phones. 60, 70, 75, 80, 90, 90, 90, 135, 140, 260

28. Draw a box-and-whisker plot of the operating times of the phones.

29. *Writing* Using the plot in Exercise 28, describe the distribution of the operating time data. What does this mean in terms of operating time? See margin.

30. MILK PRODUCTION To increase the amount of milk produced by the herd, a dairy farmer changes the cows' feed. The data show the average daily milk yield (in pints) for 10 cows before the feed change and one month after the feed change. Did the feed change increase the average daily milk yield of a cow? Use box-and-whisker plots to support your answer. See margin for plots.

Before	39	42	43	44	39	40	42	51	40	47
After	52	53	50	46	39	49	50	49	51	44

COMMUTING In Exercises 31–33, use the box-and-whisker plot that shows the lengths (in hours) of commuters' trips to work.

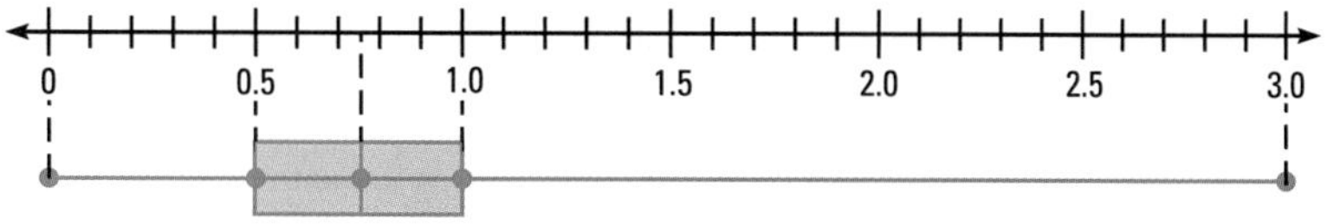

31. How long is the median trip to work? 0.75 h

32. Compare the number of people whose trip is 0–0.5 hour long to the number of people whose trip is 1–3 hours long. Explain your reasoning. *Sample answer*: The numbers are about the same; each whisker represents about 25% of the data.

33. Do more people travel 1–3 hours than travel 0.5–1 hour? Explain. See margin.

MATHEMATICAL REASONING
Box-and-whisker plots can at times exhibit unusual behavior or mislead. The box-and-whisker plot for the data 15, 17, 19, 19, 19, 19, 19, 22, 24, 24, 24, 26, 28 has a box with no center segment because the first quartile and median are the same. Ask students to explain how two identical box-and-whisker plots could represent data sets with very different means. See sample answer below.

! COMMON ERROR
EXERCISES 15–22 When first learning to draw box-and-whisker plots, students may think that the longer a whisker or a half of the box, the more data values it represents. Remind them that each whisker and each division of the box always represent one quarter of the data, no matter how the lengths compare.

APPLICATION NOTE
EXERCISES 25–29 Additional information about cellular telephones is available at **www.mcdougallittell.com.**

DATA UPDATE
Updated data for Exs. 25–29 are available at **www.mcdougallittell.com.**

APPLICATION NOTE
EXERCISES 31–33 Though the maximum suggests that the data are skewed to the right, the extremes are only single observations. In fact, they may say little about the distribution of the data. A histogram or a stem-and-leaf plot would reveal whether or not the maximum is outside the general distribution of the other data.

STUDENT HELP NOTES
Homework Help Students can find help for Exs. 31–33 at **www.mcdougallittell.com.** The information can be printed out for students who don't have access to the Internet.

Mathematical Reasoning *Sample answer:*
For example, in one data set the upper quarter of values could cluster near the outside end of the right whisker, while in the second set the upper quarter of values could cluster near the right end of the box. In the first data set, the distribution would tend to pull the mean higher.

25–30. See Additional Answers beginning on page AA1.

! COMMON ERROR

EXERCISE 35 Because the middle bar is at 40, students may interpret this value as the median. Suggest that they count the total frequencies represented both by the bars to the left of 40 and those to the right of 40.

APPLICATION NOTE

EXERCISE 36 Box-and-whisker plots can be drawn either horizontally or vertically. Some students may be able to interpret this data more easily from a vertical plot. In either case, numerical scales are necessary.

34.

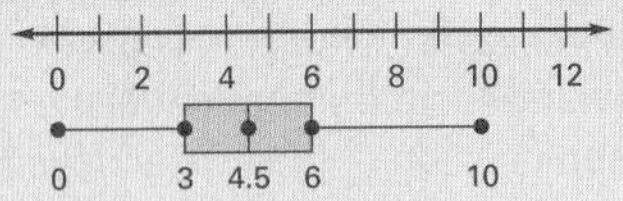

35.

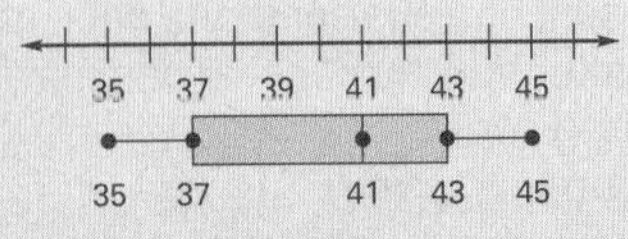

36a.

Stem	Leaves
27	4 5 9
28	0 7 8 9
29	
30	0 5 5
31	0
32	
33	
34	3 6
35	
36	
37	
38	1

Key: 34 | 3 = 343

b.

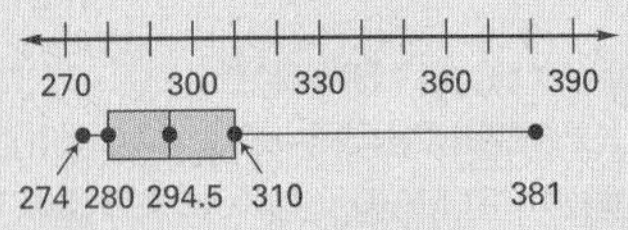

ADDITIONAL PRACTICE AND RETEACHING

For Lesson 6.7:

- Practice Levels A, B, and C (*Chapter 6 Resource Book*, p. 96)
- Reteaching with Practice (*Chapter 6 Resource Book*, p. 99)
- See Lesson 6.7 of the *Personal Student Tutor*

For more Mixed Review:

- Search the *Test and Practice Generator* for key words or specific lessons.

36. c. ***Sample answer***: It is easier to determine the median from the box-and-whisker plot, since it is clearly indicated. You can determine the median from the stem-and-leaf plot, but you must count to find the middle entry.

37. ***Sample answer***: The comparison should include the fact that conclusions about specific data can be drawn from a stem-and-leaf plot, but not a box-and-whisker plot (except for the least and greatest numbers.) Conclusions about how the data are distributed are more easily obtained from a box-and-whisker plot. The median can be read directly from the box-and-whisker plot, while it can be found by counting on the stem-and-leaf plot to locate the middle item. The mean can be calculated from the stem-and-leaf plot and the mode can be observed. Neither the mean nor the mode can be derived from a box-and-whisker plot.

DRAWING BOX-AND-WHISKER PLOTS **Draw a box-and-whisker plot of the data.** 34, 35. See margin.

34.

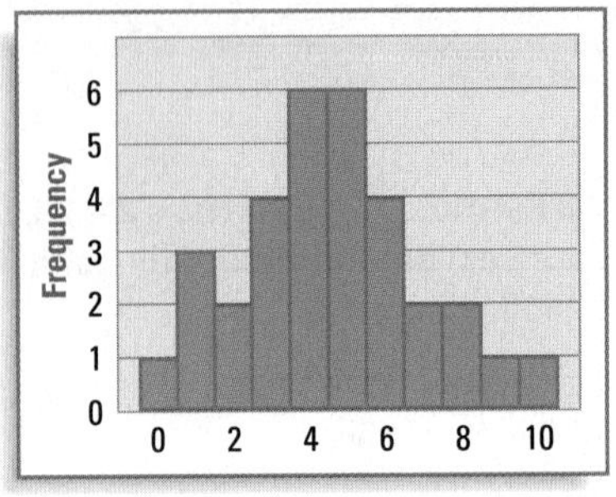

35.

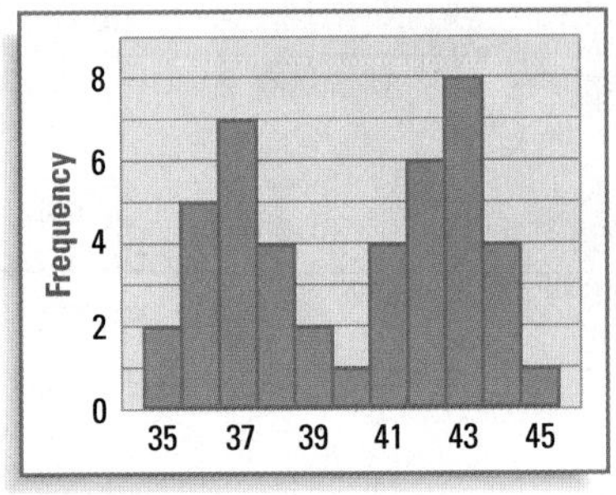

Test Preparation

36. MULTI-STEP PROBLEM **The heights (in meters) of some of the world's tallest skyscrapers are given in the table below.** a, b. See margin for plots.

Skyscraper, location	Height (m)
One Liberty Place, Philadelphia	288
Columbia Seafirst Center, Seattle	287
First Interstate World Center, Los Angeles	310
Society Tower, Cleveland	289
One Peachtree Center, Atlanta	275
Texas Commerce Tower, Houston	305
Two Prudential Center, Chicago	274
First Interstate Plaza, Houston	300
Bank of China, Hong Kong	305
Citicorp Center, New York City	279
Empire State, New York City	381
John Hancock Center, Chicago	343
Overseas Union Bank, Singapore	280
Aon Center, Chicago	346

a. Make a stem-and-leaf plot of the heights. List the heights from largest to smallest. **381, 346, 343, 310, 305, 305, 300, 289, 288, 287, 280, 279, 275, 274**

b. Draw a box-and-whisker plot of the heights.

c. CRITICAL THINKING Explain which type of plot you would prefer to use to answer this question: What is the median height of the skyscrapers that are listed in the table? **See margin.**

★ Challenge

37. ANALYZING DATA In Exercises 36(a) and 36(b) you made a stem-and-leaf plot and a box-and-whisker plot. Compare the types of plots. Include a discussion of how to find the measures of central tendency with each plot. Also include an explanation of the types of conclusions you can draw from the plots. **See margin.**

EXTRA CHALLENGE
www.mcdougallittell.com

MIXED REVIEW

FUNCTIONS Rewrite the equation so that *x* is a function of *y*. Then use the result to find *x* when *y* = −2, −1, 0, and 1. (Review 3.7)

38. $x = -\frac{1}{3}y + \frac{7}{3}; 3, \frac{8}{3}, \frac{7}{3}, 2$

38. $3x + y = 7$

39. $4x - y = 4$ $x = \frac{1}{4}y + 1; \frac{1}{2}, \frac{3}{4}, 1, 1\frac{1}{4}$

40. $x = \frac{1}{3}y + \frac{2}{9}; -\frac{4}{9}, -\frac{1}{9}, \frac{2}{9}, \frac{5}{9}$

40. $3(2 - y) = -9x + 8$

41. $6y - 4(x + 3) = -2$

$x = \frac{3}{2}y - \frac{5}{2}; -\frac{11}{2}, -4, -2\frac{1}{2}, -1$

UNIT RATES Find the unit rate. (Review 3.8)

42. 12 ounces for 4 servings **3 oz/serving** **43.** $420 for 40 hours **$10.50/h**

FINDING POINTS Decide whether the given ordered pair is a solution of the equation. (Review 4.2 for 7.1)

44. $3x - 2y = 2; (1, 3)$ **not a solution** **45.** $5x + 4y = 6; (-2, 4)$ **solution**

46. $-2x - 2y = -6; (4, -1)$ **solution** **47.** $x = 3; (-3, 3)$ **not a solution**

🌐 **UNITED STATES HOUSEHOLDS** Use the table that shows the number of households in the United States from 1970 to 1995. (Review 5.5)

Years since 1970	Households (in thousands)
0	63,401
5	71,120
10	80,776
15	86,789
20	93,347
25	98,990

▶ Source: *Statistical Abstract of the United States*

48. Find the year in the table in which the number of households first exceeded 80,000,000. **1980**

49. Write a linear model for the number of households in the United States.
Sample answer: $y = 1430x + 64,500$ (in thousands)

50. Use the linear model to estimate the number of households in 2005.
Sample answer: about 115,000,000 households

QUIZ 3 *Self-Test for Lessons 6.6 and 6.7*

1.
```
0 | 5 7
1 | 0 6
2 | 0 3 4 5 9
3 | 1 2 7 8
```
Key: 3|1 = 31

2.
```
1 | 2 8
2 | 6 7
3 | 3 3
4 | 2 4 6 7
5 | 9
6 | 1
```
Key: 6|1 = 61

Make a stem-and-leaf plot for the data. Use the result to list the data in increasing order. (Lesson 6.6) 1–2. See margin for plots.

1. 23, 16, 24, 31, 38, 10, 32, 29, 5, 25, 37, 7, 20 **5, 7, 10, 16, 20, 23, 24, 25, 29, 31, 32, 37, 38**

2. 33, 59, 27, 44, 47, 12, 61, 46, 18, 33, 42, 26 **12, 18, 26, 27, 33, 33, 42, 44, 46, 47, 59, 61**

Find the mean, the median, and the mode of the data. (Lesson 6.6)

3. 9, 9, 5, 10, 7, 10, 6 **8, 9, 9 and 10** **4.** 42, 43, 50, 20, 30, 40 **37.5, 41, no mode**

Find the first, second, and third quartiles of the data. Then draw a box-and-whisker plot of the data. (Lesson 6.7) 5–8. See margin for plots.

5. 13, 10, 16, 25, 5, 12, 20, 7, 15 **8.5, 13, 18** **6.** 50, 70, 75, 60, 30, 10, 40, 80, 15, 20 **20, 45, 70**

7. 57, 48, 9, 27, 8, 80, 56, 55, 6, 33 **9, 40.5, 56** **8.** 2, 6, 11, 16, 10, 7, 3, 12, 20, 5, 1, 14 **4, 8.5, 13**

9. 🌐 **BASKETBALL WINS** The data represent the number of wins by some NBA basketball teams for the 1996–1997 season. Find the mean, the median, and the mode of the data to the nearest whole number. (Lesson 6.6)

64, 57, 24, 20, 14, 40, 21, 15, 61, 57, 22, 26, 45, 44, 57, 30, 34, 56, 49

39, 40, 57

6.7 Box-and-Whisker Plots **381**

Additional Test Preparation *Sample answers:*
1. A box-and-whisker plot displays the minimum value, the first quartile, the second quartile (the median), the third quartile, and the maximum value.

2. If the data set were large, such as the median income in each of the 50 United States, a stem-and-leaf plot would be impractical or cumbersome.

4 ASSESS

DAILY HOMEWORK QUIZ

📖 *Transparency Available*

Make a box-and-whisker plot of the data.

1. Waiting times at a doctor's office (min)
22 45 52 37 43 11 62 29
39 28 44 33 28 41 39 28

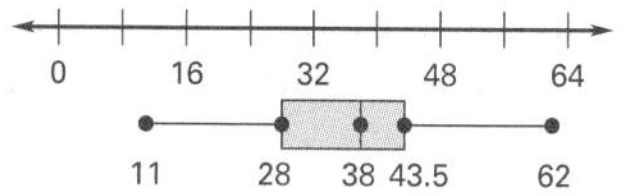

2. Highway fuel efficiencies (mi/gal)
21 18 26 31 25 33 41 20 18
19 24 28 34 55 37 28 30 25

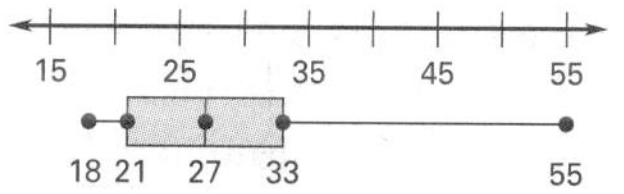

3. In the plot from Question 2 above, compare the spread of the data in the lowest fourth and highest fourth of the data values. *Sample answer:* The lowest fourth of the values vary by only 3 mi/gal, while the highest fourth vary by 22 mi/gal.

EXTRA CHALLENGE NOTE

↳ Challenge problems for Lesson 6.7 are available in **blackline** format in the *Chapter 6 Resource Book*, p. 103 and at **www.mcdougallittell.com.**

ADDITIONAL TEST PREPARATION

1. WRITING Explain why the information included in a box-and-whisker plot is called a "five-number summary" of the data.
See sample answer at left.

2. OPEN ENDED Describe a set of data that would be better suited to displaying on a box-and-whisker plot than on a stem-and-leaf plot. Explain.
See sample answer at left.

5–8. See Additional Answers beginning on page AA1.

1 Planning the Activity

PURPOSE
To use a graphing calculator or a computer to graph a box-and-whisker plot.

MATERIALS
- a graphing calculator for each student or a computer with graphing software for each pair of students
- Keystroke blackline (*Chapter 6 Resource Book,* p. 95)

PACING
- Example — 10 min
- Exercises — 15 min

▶ LINK TO LESSON
Have students look back to Example 2 of Lesson 6.7. Point out to them that by graphing the data using a graphing calculator, they can enter the data as they read it, without first ordering the data.

2 Managing the Activity

COOPERATIVE LEARNING
As students complete each step of the activity, have them compare results with a partner. If results are different, students can reexamine the keystrokes together to determine the correct sequence.

3 Closing the Activity

★ KEY DISCOVERY
Using technology to graph box-and-whisker plots, you can read the extremes and quartiles directly from the graph by tracing.

ACTIVITY ASSESSMENT
Display the plots for Exs.1 and 2 in the same window. How do the medians compare? Which plot shows a wider variation in the middle half of the data?
Sample answer: They are about the same.; Exercise 1

1–4. See Additional Answers beginning on page AA1.

▷ ACTIVITY 6.7

Using Technology

Box-and-Whisker Plots

You can draw a box-and-whisker plot using a graphing calculator or computer. Technology is especially useful for large groups of data or multiple sets of data.

▶ EXAMPLE
Draw a box-and-whisker plot of the following data.

32, 24, 39, 55, 67, 35, 23, 61, 11, 44, 43, 30, 29

▶ SOLUTION

STUDENT HELP

INTERNET **KEYSTROKE HELP**

See keystrokes for several models of calculators at www.mcdougallittell.com

Calculators vary but follow the same general instructions for graphing box-and-whisker plots. Be sure to set reasonable WINDOW settings.

1 Clear all lists and enter the given data.

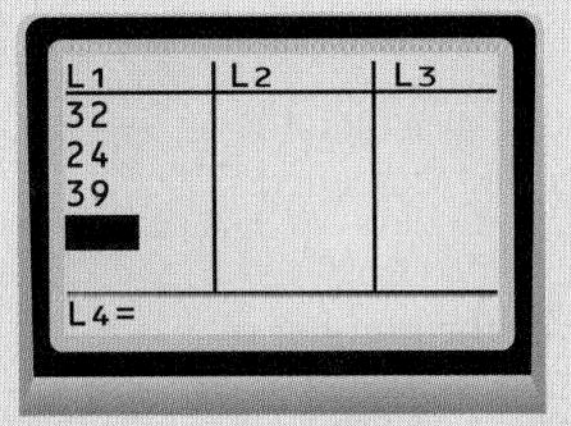

2 Choose to sketch a box-and-whisker plot.

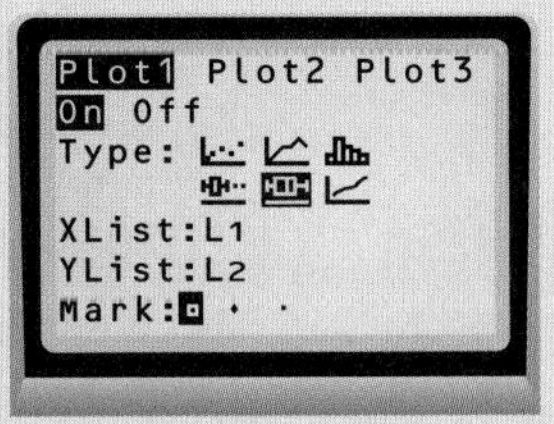

3 Use ZOOM to set a statistics viewing window. Press GRAPH to sketch the box-and-whisker plot.

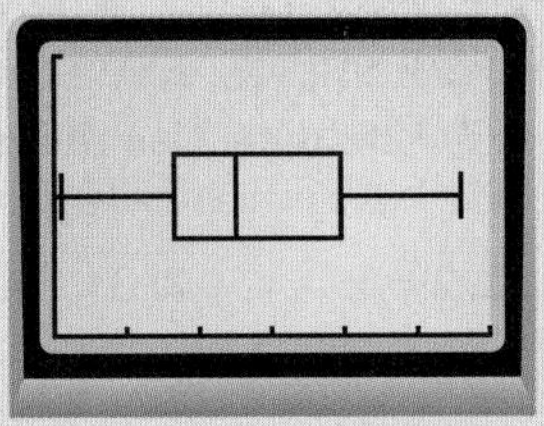

4 Use TRACE and the direction arrows to find quartile values.

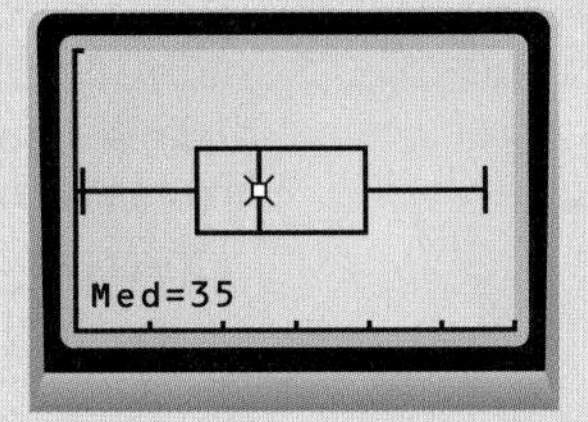

▶ EXERCISES

Use a graphing calculator or computer to draw a box-and-whisker plot of the data. List the first, second, and third quartiles.

1–4. See margin for plots.

1. Ages of students' brothers: 19, 12, 14, 9, 14, 4, 17, 13, 19, 11, 6, 19 **10, 13.5, 18** (least: 4; greatest: 19)
2. Cost of CDs (in dollars): 11, 10, 12, 14, 17, 16, 14, 6, 8, 14, 20 **10, 14, 16** (least: 6; greatest: 20)
3. Temperatures in February (in degrees Fahrenheit): 20.1, 43.4, 34.9, 23.9, 33.5, 24.1, 22.5, 42.4, 25.7 **23.2, 25.7, 38.65** (least: 20.1 greatest: 43.4)
4. Miles traveled on weekends: 78.4, 76.3, 107.5, 78.5, 93.2, 90.3, 77.8, 37.1, 97.1, 75.5, 58.8, 65.6, 3.2 **62.2, 77.8, 91.75** (least: 3.2 greatest: 107.5)

CHAPTER 6 Chapter Summary

WHAT did you learn?	*WHY did you learn it?*
Write, solve, and graph an inequality in one variable.	
• Simple linear inequalities (6.1)	Summarize the melting points of metallic elements. (p. 338)
• Multi-step linear inequalities (6.2)	Model and predict city population growth. (p. 341)
• Compound inequalities (6.3)	Describe distances between planets. (p. 350)
• Absolute-value inequalities (6.4)	Determine wavelengths of colors of fireworks. (p. 357)
Write, solve, and graph a linear inequality in two variables. (6.5)	Find different amounts of coins collected on a treasure-diving ship. (p. 362)
Use an inequality to model a real-life situation. (6.1–6.5)	Analyze real-life situations. (p. 342)
Make a stem-and-leaf plot for data. (6.6)	Display and order population data. (p. 368)
Use measures of central tendency to represent data. (6.6)	Describe the average age of a high school graduate. (p. 370)
Draw a box-and-whisker plot of data. (6.7)	Organize and compare backpack prices. (p. 376)

How does Chapter 6 fit into the BIGGER PICTURE of algebra?

You studied solving and graphing linear inequalities in one and two variables and absolute-value inequalities. Many real-life situations can be described with phrases like “at most” or “less than,” which can be modeled using inequalities.

You also organized data using stem-and-leaf plots, measures of central tendency, and box-and-whisker plots. When reviewing this chapter, look for connections between solving and graphing equations and solving and graphing inequalities.

STUDY STRATEGY

How did showing all your work help you make corrections?

When you show all your work, using the **Study Strategy** on p. 332, the solution steps may be like these.

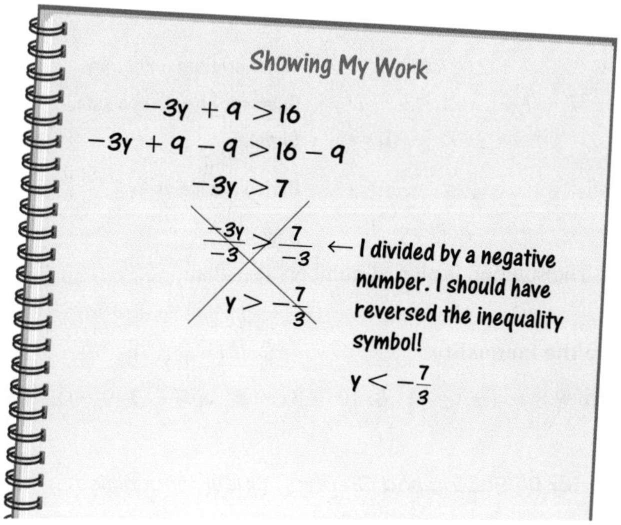

Showing My Work
$-3y + 9 > 16$
$-3y + 9 - 9 > 16 - 9$
$-3y > 7$
$\frac{-3y}{-3} > \frac{7}{-3}$
← I divided by a negative number. I should have reversed the inequality symbol!
$y > -\frac{7}{3}$
$y < -\frac{7}{3}$

383

ADDITIONAL RESOURCES

The following resources are available to help review the material in this chapter.

- Chapter Review Games and Activities (*Chapter 6 Resource Book*, p. 104)
- *Instant Replay: Video Review Games*
- *Personal Student Tutor*
- Cumulative Review, Chs. 1–6 (*Chapter 6 Resource Book*, p. 116)

1. 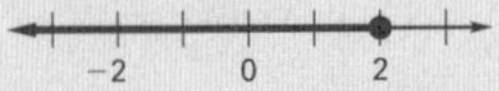
2. 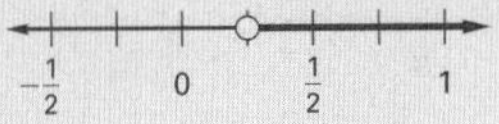
3. 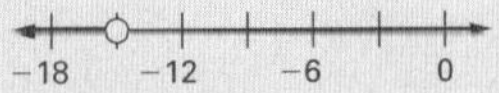
4. 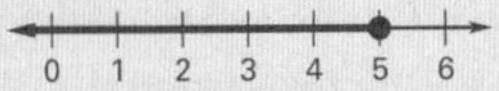

CHAPTER 6

Chapter Review

VOCABULARY

- graph of a linear inequality in one variable, p. 334
- equivalent inequalities, p. 335
- compound inequality, p. 346
- linear inequality in *x* and *y*, p. 360
- solution of a linear inequality, p. 360
- graph of a linear inequality in two variables, p. 360
- half-plane, p. 360
- stem-and-leaf plot, p. 368
- measure of central tendency, p. 369
- mean, or average, p. 369
- median, p. 369
- mode, p. 369
- box-and-whisker plot, p. 375
- quartiles, p. 375

6.1 SOLVING ONE-STEP LINEAR INEQUALITIES

Examples on pp. 334–336

EXAMPLE Solve $-0.9x \geq 3.6$. Graph the solution.

$-0.9x \geq 3.6$	**Write original inequality.**
$\frac{-0.9x}{-0.9} \leq \frac{3.6}{-0.9}$	**Divide each side by −0.9 and *reverse* the inequality symbol.**
$x \leq -4$	**Simplify.**

▶ The solution is all real numbers less than or equal to -4.

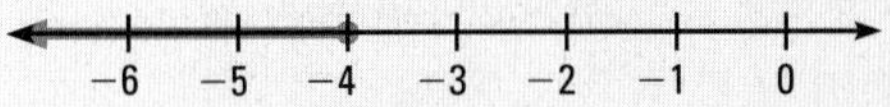

Solve the inequality and graph its solution. **1–4. See margin for graphs.**

1. $x - 5 \leq -3$ $x \leq 2$ **2.** $\frac{3}{4} + x > 1$ $x > \frac{1}{4}$ **3.** $\frac{x}{3} < -5$ $x < -15$ **4.** $-6x \geq -30$ $x \leq 5$

6.2 SOLVING MULTI-STEP LINEAR INEQUALITIES

Examples on pp. 340–342

EXAMPLE Solve $7 + 2x < -3$.

$7 + 2x < -3$	**Write original inequality.**
$7 - 7 + 2x < -3 - 7$	**Subtract 7 from each side.**
$2x < -10$	**Simplify.**
$\frac{2x}{2} < \frac{-10}{2}$	**Divide each side by 2.**
$x < -5$	**Simplify.**

▶ The solution is all real numbers less than -5.

Solve the inequality.

5. $6x + 8 > 4$ $x > -\frac{2}{3}$ **6.** $10 - 3x < 5$ $x > \frac{5}{3}$ **7.** $9 - 4x \geq -11$ $x \leq 5$ **8.** $5 - \frac{1}{2}x \leq -3$ $x \geq 16$

6.3 SOLVING COMPOUND INEQUALITIES

Examples on pp. 346–348

EXAMPLE Solve $1 < -2x + 3 < 9$.

$1 < -2x + 3 < 9$	**Write original inequality.**
$-2 < -2x < 6$	**Subtract 3 from each expression.**
$1 > x > -3$	**Divide each expression by -2. Reverse both inequality symbols.**

▶ The solution is all real numbers greater than -3 and less than 1.

Solve the inequality. Write a sentence that describes the solution.

9. $3 < x + 1 < 6$
x is greater than 2 and less than 5.

10. $6 \le 2x + 3 \le 10$
See margin.

11. $x + \frac{1}{2} \le -\frac{3}{4}$ or $\frac{1}{4}x - 1 > \frac{3}{4}$
See margin.

6.4 SOLVING ABSOLUTE-VALUE EQUATIONS AND INEQUALITIES

Examples on pp. 353–355

EXAMPLE To solve both absolute-value equations and inequalities, remember that expressions inside absolute-value symbols can be either positive or negative.

Solve $|x - 4| > 3$.

$x - 4$ is positive.	**$x - 4$ is negative.**
$x - 4 > 3$	$x - 4 < -3$
$x > 7$	$x < 1$

▶ The solution is all real numbers greater than 7 or less than 1, or $x < 1$ or $x > 7$.

Solve the equation or the inequality.

12. $|x| = 12$ **$-12, 12$**

13. $|x - 5| = 11$ **$-6, 16$**

14. $|x + 0.5| = 0.25$ **$-0.25, -0.75$**

15. $|5 + x| \le 6$ **$-11 \le x \le 1$**

16. $|7 - x| > 2$ **$x < 5$ or $x > 9$**

17. $|3x + 2| - 4 \le 8$ **$-4\frac{2}{3} \le x \le 3\frac{1}{3}$**

6.5 GRAPHING LINEAR INEQUALITIES IN TWO VARIABLES

Examples on pp. 360–362

EXAMPLE To graph $2x + y < 1$, first sketch the graph of the corresponding equation $2x + y = 1$. Use a dashed line because the inequality symbol is $<$. Next, test a point. The origin $(0, 0)$ *is* a solution, so the graph of $2x + y < 1$ includes $(0, 0)$ and all points below the line.

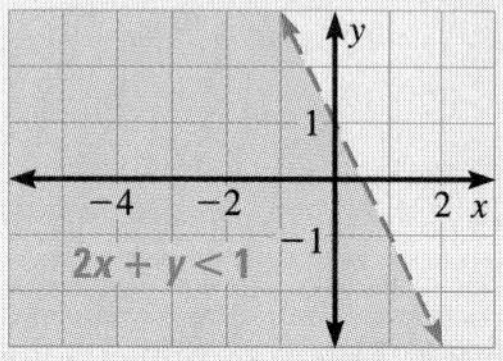

Graph the inequality. **18–21. See margin.**

18. $x < 2$

19. $-2x + y > 4$

20. $\frac{1}{3}x - 3y \le 3$

21. $2y - 6x \ge -2$

10. ***Sample answer:*** **x is greater than or equal to $\frac{3}{2}$ and less than or equal to $\frac{7}{2}$.**

11. **x is less than or equal to $-1\frac{1}{4}$ or x is greater than 7.**

18.

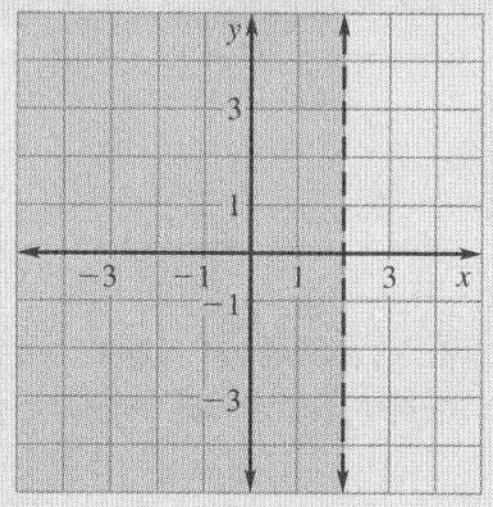

19.

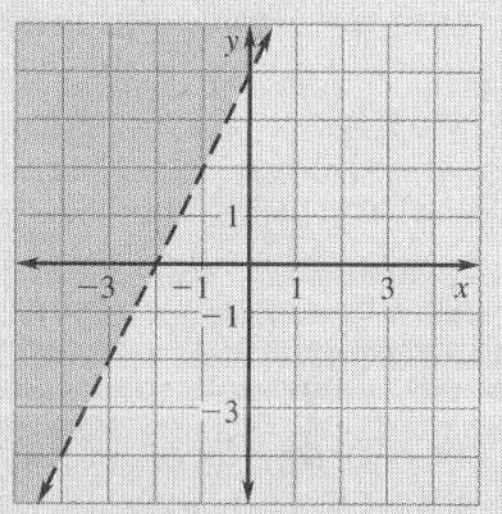

20.

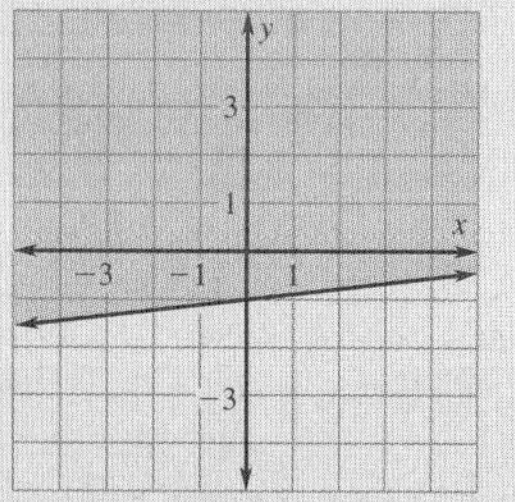

21.

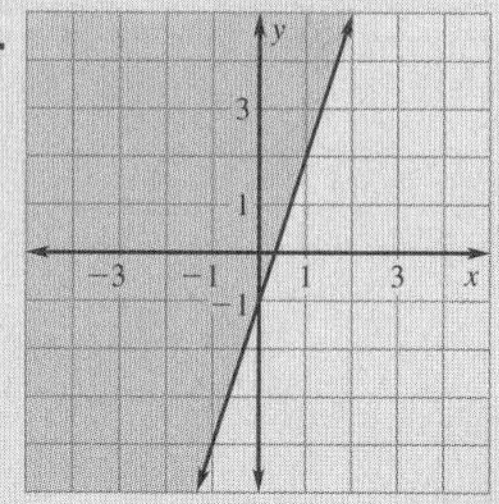

22.

Stem	Leaves
5	6 8 8
6	3 5
7	1 8 8
8	2 4 5 6

Key: 8 | 2 = 82

24. Paris Temperatures

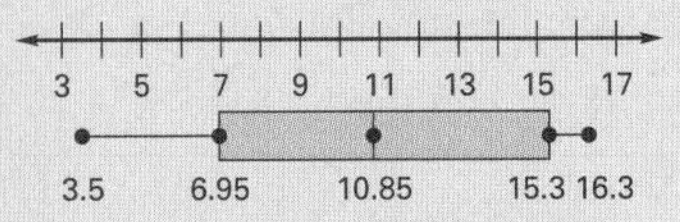

25. ***Sample answer:*** **The temperatures in Tokyo are much higher. The highest Paris temperature is only slightly higher than the median Tokyo temperature. Only 25% of the Paris temperatures are at least 15.3°, while half of the Tokyo temperatures are at least 15.85°.**

6.6 STEM-AND-LEAF PLOTS AND MEAN, MEDIAN, AND MODE

Examples on pp. 368–370

EXAMPLE To make a stem-and-leaf plot for the following data, use the digits in the tens' place for the stem and the digits in the ones' place for the leaves.

40, 60, 34
43, 68, 45
61, 64, 54
51, 64, 52

Stems	Leaves
3	4
4	0 3 5
5	1 2 4
6	0 1 4 4 8

Key: 6 | 0 = 60

The mean of the data is the sum of the numbers divided by 12, which is 53.

The median of the data is the average of the two middle numbers, which is 53.

The mode of the data is the number that occurs most often, which is 64.

22. The data below show the monthly water temperature (in degrees Fahrenheit) of the Gulf of Mexico near Pensacola, Florida. Make a stem-and-leaf plot of the data: 56, 58, 63, 71, 78, 84, 85, 86, 82, 78, 65, 58. **See margin.**

23. Find the mean, the median, and the mode of the data in Exercise 22. **72, 74.5, 58 and 78**

6.7 BOX-AND-WHISKER PLOTS

Examples on pp. 375–377

EXAMPLE The data show the average temperature (in degrees Celsius) for each month in Tokyo: 5.2, 5.6, 8.5, 14.1, 18.6, 21.7, 25.2, 27.1, 23.2, 17.6, 12.6, 7.9.

Write the numbers in order and find the quartiles of the data.

5.2 5.6 7.9 8.5 12.6 14.1 17.6 18.6 21.7 23.2 25.2 27.1

$$\text{Second quartile} = \frac{14.1 + 17.6}{2} = 15.85$$

$$\text{First quartile} = \frac{7.9 + 8.5}{2} = 8.2 \qquad \text{Third quartile} = \frac{21.7 + 23.2}{2} = 22.45$$

To make a box-and-whisker plot, draw a box that extends from the first to the third quartiles. Connect the least and greatest numbers to the box as the "whiskers."

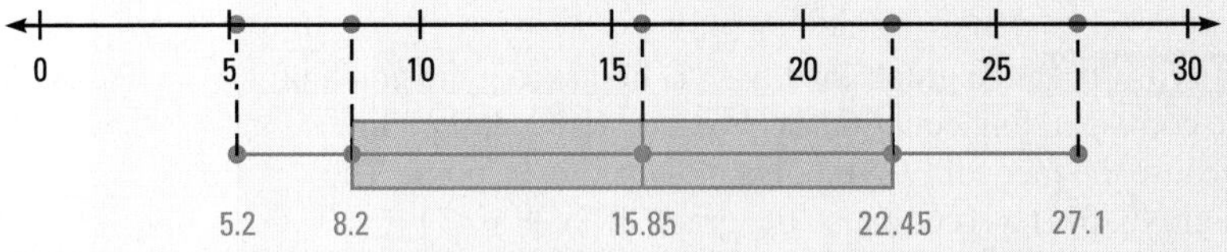

24. The data below show the average temperature (in degrees Celsius) for each month in Paris. Make a box-and-whisker plot of the data. **See margin.**

3.5, 4.2, 6.6, 9.5, 13.2, 16.3, 15.9, 16.0, 14.7, 12.2, 8.9, 7.3

25. Compare the box-and-whisker plot in Exercise 24 to the one in the Example. How do the data sets differ? **See margin.**

CHAPTER 6

Chapter Test

Solve the inequality. Graph the solution on a number line. **1–9. See margin for graphs.**

1. $x - 3 < 10$ **$x < 13$** **2.** $-6 > x + 5$ **$x < -11$** **3.** $-32x > 64$ **$x < -2$**

4. $\frac{x}{4} \le 8$ **$x \le 32$** **5.** $\frac{2}{3}x + 2 \le 4$ **$x \le 3$** **6.** $6 - x > 15$ **$x < -9$**

7. $3x + 5 \le 2x - 1$ **$x \le -6$** **8.** $(x + 6) \ge 2(1 - x)$ **$x \ge -\frac{4}{3}$** **9.** $-2x + 8 > 3x + 10$ **$x < -\frac{2}{5}$**

Solve the inequality. Write a sentence that describes the solution.

10. $-15 \le 5x < 20$ **x is at least -3 and is less than 4.**
11. $-3 \le 4x + 5 \le 7$ **x is at least -2 and at most $\frac{1}{2}$.**
12. $8x - 11 < 5$ or $4x - 7 > 13$ **x is less than 2 or x is greater than 5.**
13. $-2x > 8$ or $3x + 1 \ge 7$ **x is less than -4 or x is at least 2.**
14. $12 > 4 - x > -5$ **x is between -8 and 9.**
15. $6x + 9 \ge 21$ or $9x - 5 \le 4$ **x is at least 2 or x is at most 1.**

Write a compound inequality that describes the graph.

16. 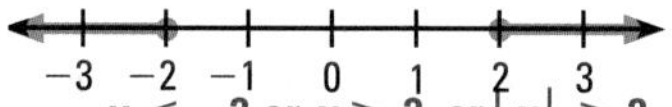

$x \le -2$ or $x \ge 2$, or $|x| \ge 2$

17. 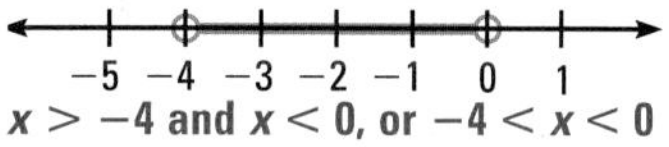

$x > -4$ and $x < 0$, or $-4 < x < 0$

Solve the equation or the inequality.

18. $|x + 7| = 11$ **$-18, 4$**
19. $|x - 8| - 3 \le 10$ **$-5 \le x \le 21$**
20. $|x + 4.2| + 3.6 = 16.2$ **$-16.8, 8.4$**
21. $|2x - 6| > 14$ **$x < -4$ or $x > 10$**
22. $|4x + 5| - 6 \le 1$ **$-3 \le x \le \frac{1}{2}$**
23. $|3x - 9| + 6 = 18$ **$-1, 7$**

Sketch the graph of the inequality. **24–29. See margin.**

24. $x > -1$ **25.** $x - 1 \le -3$ **26.** $-4x \le 8$

27. $x + 2y > 6$ **28.** $3x + 4y \ge 12$ **29.** $7y - 2x + 3 < 17$

EXERCISE BICYCLE PRICES In Exercises 30–33, use these prices:
\$1130, \$695, \$900, \$220, \$350, \$500, \$630, \$180, \$170, \$145, \$185, \$140.

30. Make a stem-and-leaf plot for the data. List the data in increasing order. **See margin for plot.**
140, 145, 170, 180, 185, 220, 350, 500, 630, 695, 900, 1130

31. Find the mean, the median, and the mode of the data. **437.08, 285, no mode**

32. Find the first, second, and third quartiles. Which quartile is the median? **175, 285, 662.5; the second quartile**

33. Draw a box-and-whisker plot of the data. **See margin.**

34. PAPER MAKING One kind of machine makes paper in a roll that can be as wide as 33 feet or as narrow as 12 feet. Write a compound inequality for the possible widths of a roll of paper that this machine can produce. **$12 \le x \le 33$**

35. WALKING DISTANCE Walking at a rate of 210 feet per minute, you take 12 minutes to walk from your home to school. Your uncle's home is closer to school than your home is. Write an inequality for the distance d (in feet) that your uncle lives from your school. **$d < 2520$**

36. RAFTING Members of an outdoor club rent several rafts and launch all of the rafts at the same time. The first raft finishes in 36 minutes. The last raft finishes in 48 minutes. Write an absolute-value inequality that describes the finishing times. **$|t - 42| \le 6$**

ADDITIONAL RESOURCES

- ***Chapter 6 Resource Book***
  Chapter Test (3 levels) (p. 105)
  SAT/ACT Chapter Test (p. 111)
  Alternative Assessment (p. 112)
- ***Test and Practice Generator***

1. 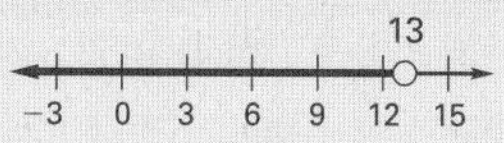

2. 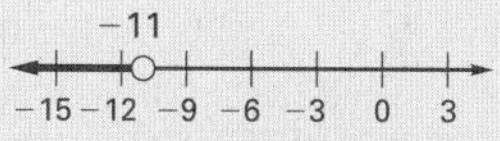

3. 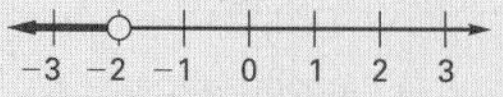

4. 

5. 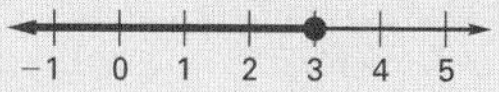

6. 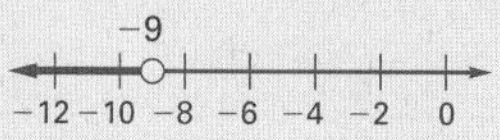

7. 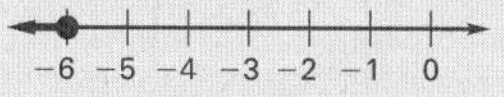

8. 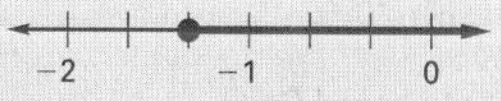

9. 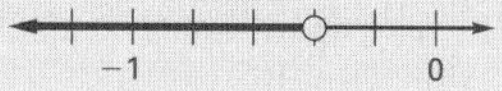

24. 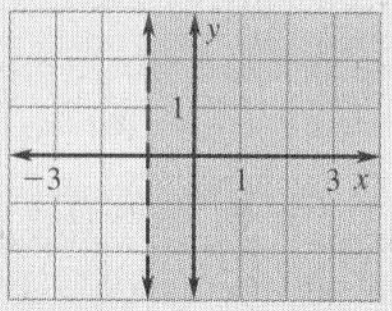

25. 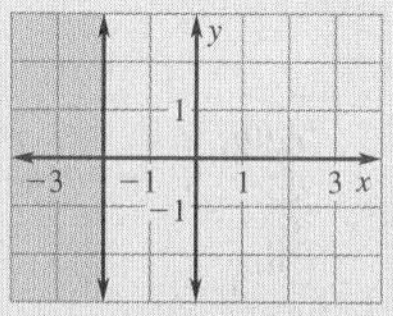

26. 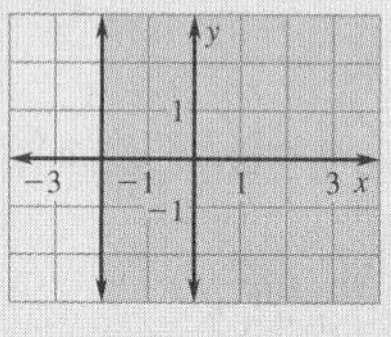

27–30, 33. See the next page.

ADDITIONAL RESOURCES

- ***Chapter 6 Resource Book***
  Chapter Test (3 levels) (p. 105)
  SAT/ACT Chapter Test (p. 111)
  Alternative Assessment (p. 112)
- ***Test and Practice Generator***

27. 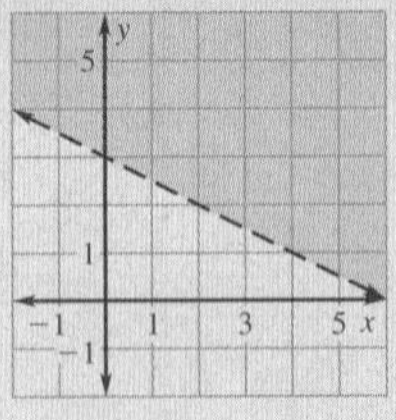

28. 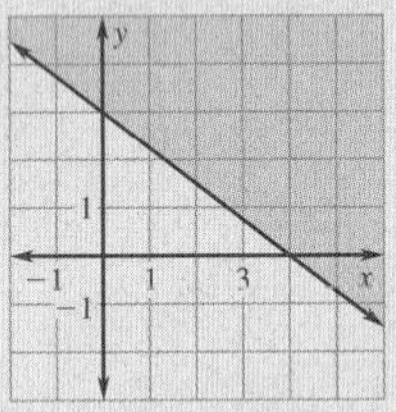

29. 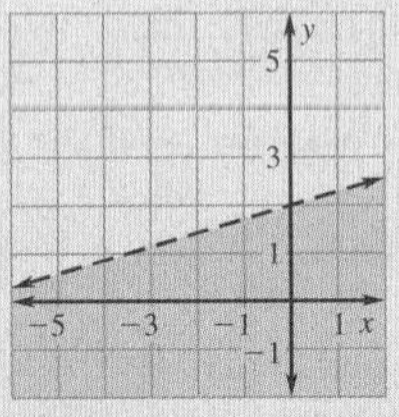

30.

Stem	Leaves
1	40 45 70 80 85
2	20
3	50
4	
5	00
6	30 95
7	
8	
9	00
10	
11	30

Key: 11 | 30 = 1130

33. Costs of Exercise Bikes

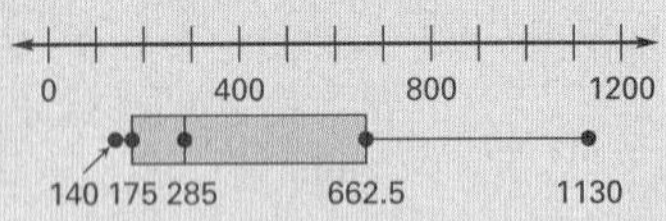

CHAPTER 6

Chapter Standardized Test

TEST-TAKING STRATEGY Work as fast as you can through the easier problems, but not so fast that you are careless.

1. MULTIPLE CHOICE Which graph represents the solution of $x + 10 < 17$? **D**

Ⓐ 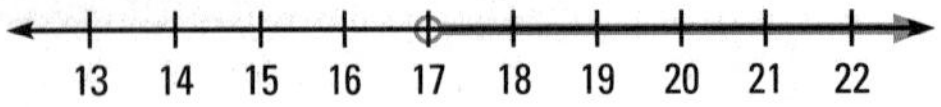

Ⓑ 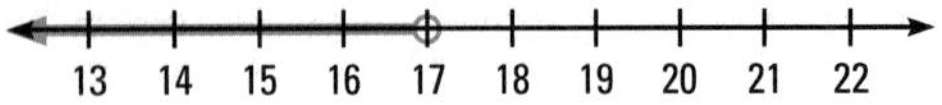

Ⓒ 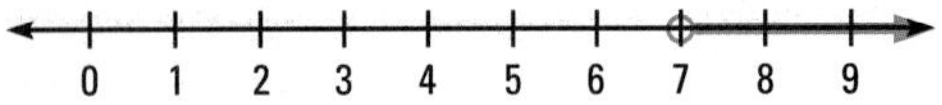

Ⓓ 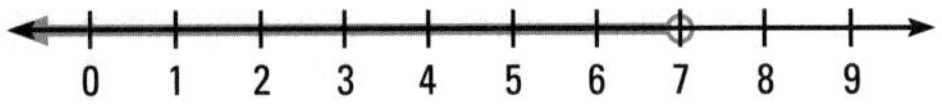

Ⓔ 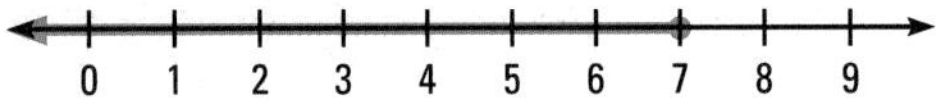

2. MULTIPLE CHOICE Describe the solution of the inequality $x + 3 \le 7$. **C**

Ⓐ All real numbers less than 4

Ⓑ All real numbers less than or equal to 10

Ⓒ All real numbers less than or equal to 4

Ⓓ All real numbers less than or equal to -4

Ⓔ None of these

3. MULTIPLE CHOICE Which inequality is equivalent to $2 - 3x \ge -4$? **B**

Ⓐ $x \ge 2$ Ⓑ $x \le 2$

Ⓒ $x \le -\frac{2}{3}$ Ⓓ $x \ge -2$

Ⓔ None of these

4. MULTIPLE CHOICE Describe the solution of the compound inequality $-3x + 2 > 11$ or $5x + 1 > 6$. **A**

Ⓐ All real numbers less than -3 or greater than 1

Ⓑ All real numbers greater than 3 or less than 1

Ⓒ All real numbers less than 3 or greater than -1

Ⓓ All real numbers less than -3 and greater than 1

Ⓔ None of these

5. MULTIPLE CHOICE For which values of x is the inequality $5(3x + 4) \le 5x - 10$ true? **B**

Ⓐ $x \ge -3$ Ⓑ $x \le -3$

Ⓒ $x \le -1$ Ⓓ $x \le 1$

Ⓔ $x \le 3$

6. MULTIPLE CHOICE Which graph represents the solution of $|2x - 10| \ge 6$? **D**

Ⓐ 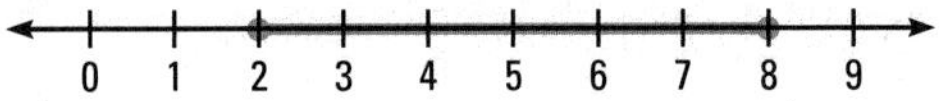

Ⓑ 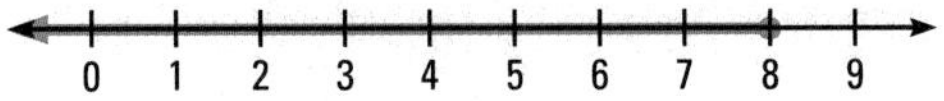

Ⓒ 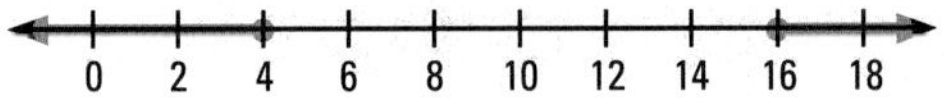

Ⓓ 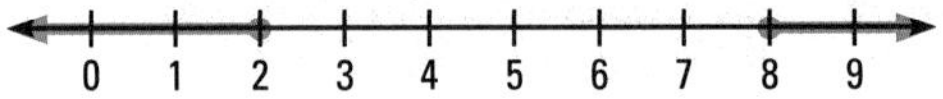

Ⓔ 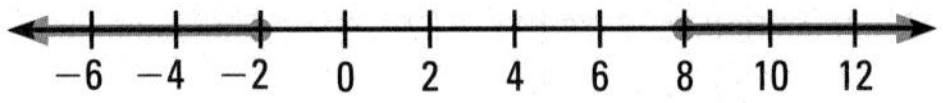

7. MULTIPLE CHOICE Which numbers are solutions to the absolute-value equation $|x - 7| + 5 = 17$? **E**

Ⓐ 5 and -19 Ⓑ 5 and -5

Ⓒ 12 and -5 Ⓓ 19 and -19

Ⓔ 19 and -5

8. MULTIPLE CHOICE Which point *is not* a solution of $y < x + 5$? **C**

Ⓐ $(4, -4)$ Ⓑ $(1, 4)$

Ⓒ $(-1, 4)$ Ⓓ $(-3, 1)$

Ⓔ $(4, 6)$

9. MULTIPLE CHOICE Choose the inequality whose solution is shown in the graph. **C**

Ⓐ $2x + y < 4$

Ⓑ $2x + y \ge 4$

Ⓒ $2x + y > 4$

Ⓓ $y - 2x > 4$

Ⓔ $y - 2x \le 4$

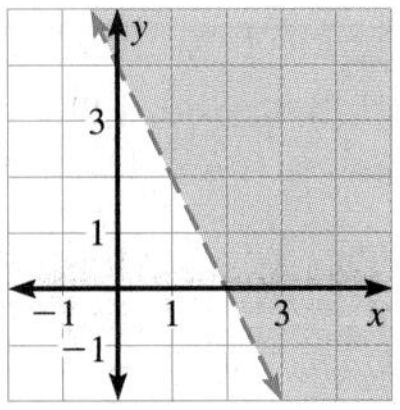

10. **MULTIPLE CHOICE** The stem-and-leaf plot shows the ages of 20 people. What percent of people are under 20 years old? **E**

Ⓐ 7% Ⓑ 8% Ⓒ 10% Ⓓ 20% Ⓔ 40%

0	5 7 1 2
1	1 3 3 8
2	2 4 5 9 4 2
3	0 3 3 5 3 8

Key: 3 | 0 = 30

11. **MULTIPLE CHOICE** What is the mean of the collection of numbers: 25, 29, 33, 38, 40, 45, 51, 53, 66, 73? **D**

Ⓐ 40 Ⓑ 42.5 Ⓒ 45 Ⓓ 45.3 Ⓔ 73

12. **MULTIPLE CHOICE** Which set of numbers is represented by the box-and-whisker plot? **D**

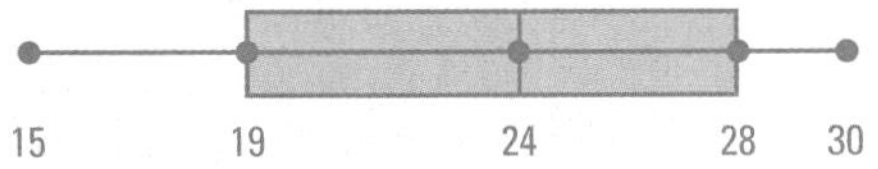

Ⓐ 20, 24, 15, 20, 25, 18, 19

Ⓑ 19, 18, 15, 23, 25, 28, 30

Ⓒ 24, 22, 18, 16, 29, 28, 18, 20, 21, 20, 19

Ⓓ 25, 23, 17, 15, 19, 21, 28, 28, 30, 26

Ⓔ 20, 18, 14, 13, 22, 26, 22, 25, 29

QUANTITATIVE COMPARISON In Exercises 13 and 14, choose the statement below that is true about the given numbers.

Ⓐ The number in column A is greater.

Ⓑ The number in column B is greater.

Ⓒ The two numbers are equal.

Ⓓ The relationship cannot be determined from the given information.

	Column A	Column B	
13.	The mean of 2, 3, 4, 5, 6	The median of 2, 3, 4, 5, 6	**C**
14.	$\lvert a + 2 \rvert$	$\lvert a - 2 \rvert$	**D**

15. **MULTI-STEP PROBLEM** You own a printing shop and order paper each week to meet your customers' printing needs. Company A charges $4.50 for a ream of 500 sheets of paper plus a $25 delivery fee for each order. Company B charges only $4.00 for a ream but charges a $75 delivery fee for each order.

a. Let x represent the number of reams of paper you purchase. Write an expression that represents the total cost of purchasing paper from Company A. Write an expression that represents the total cost of purchasing paper from Company B. **$4.5x + 25$; $4x + 75$**

b. Choose two different amounts of paper between 50 reams and 150 reams that you could purchase. Calculate the charges from each supplier.

c. *Writing* Draw a graph to show the costs of ordering up to 200 reams of paper from each supplier. Explain how to use the graph to determine which company charges the lower price for the same order.

d. Write an inequality that represents orders for which Company A offers the lower price. Write an inequality that represents orders for which Company B offers the lower price. **$x < 100$; $x > 100$**

15. b. ***Sample answer*: For 75 reams, Company A charges $362.50. Company B charges $375.**

15. c. **For any number of reams, the graph that is lower represents the company that charges the lower price. See margin for graph.**

15c.

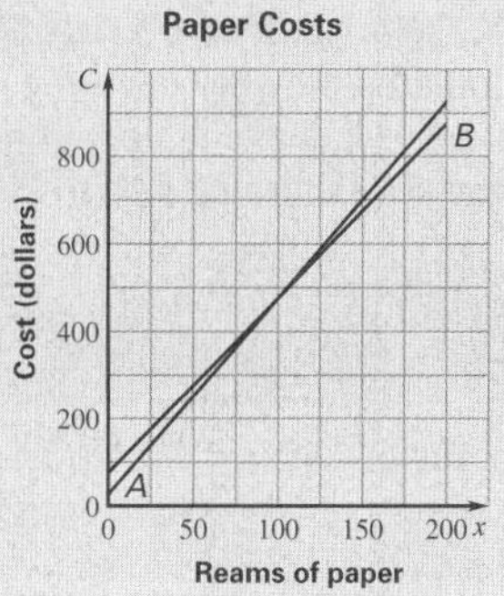

ADDITIONAL RESOURCES
A Cumulative Review covering Chapters 1–6 is available in the *Chapter 6 Resource Book,* p. 116.

37.
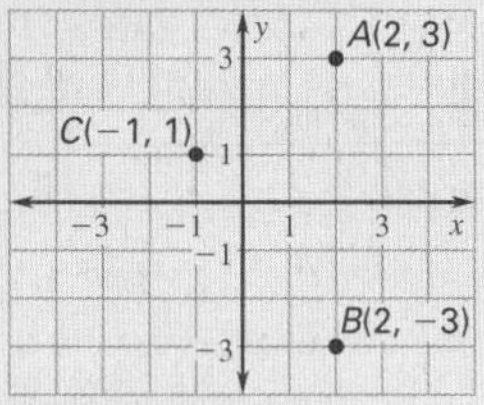

38.
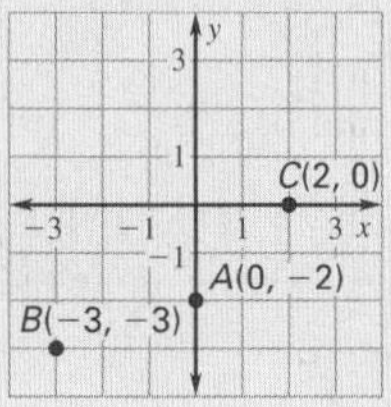

39.
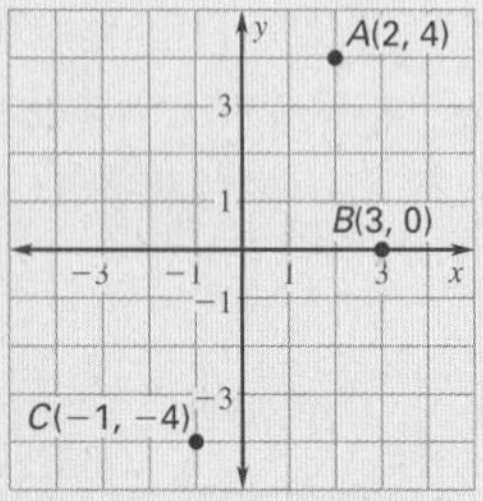

40.
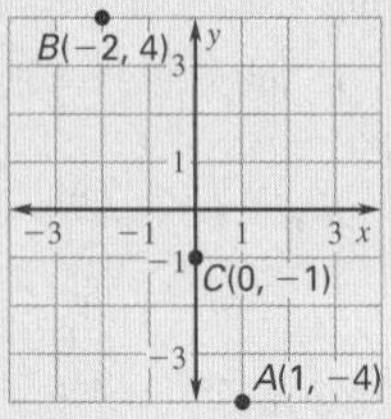

41.
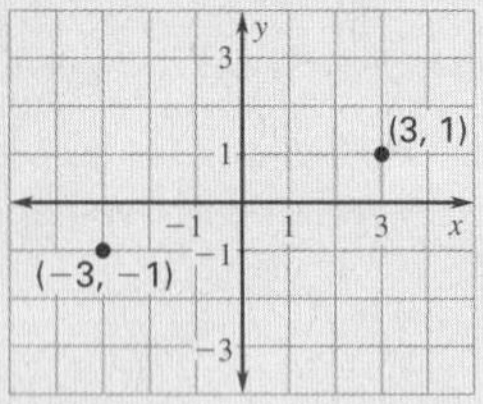

42.
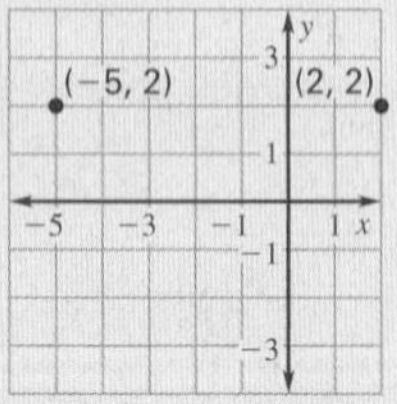

43.
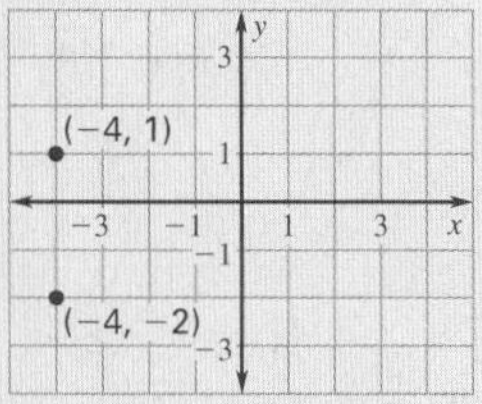

44–50. See the next page.

CHAPTER 6

Cumulative Practice

for Chapters 1–6

Evaluate the expression. (1.1–1.3)

1. $x + 8$ when $x = -1$ **7**
2. $3x - 2$ when $x = 7$ **19**
3. $x(4 + x)$ when $x = 5$ **45**
4. $(x - 5)^2$ when $x = 1$ **16**
5. $2\left(\frac{x + 8}{x}\right)$ when $x = 4$ **6**
6. $x^3 - 3x + 1$ when $x = 2$ **3**

Evaluate the expression. (2.2 and 2.3)

7. $2 - (-6) + (-14)$ **−6**
8. $3.1 + (-3.3) - 1.8$ **−2**
9. $20 - |-5.5|$ **14.5**

Find the sum or the difference of the matrices. (2.4)

10. $\begin{bmatrix} 3 & 2 \\ 8 & -2 \end{bmatrix} + \begin{bmatrix} -4 & -2 \\ -1 & 0 \end{bmatrix}$ $\begin{bmatrix} -1 & 0 \\ 7 & -2 \end{bmatrix}$

11. $\begin{bmatrix} -2 & 7 & -3 \\ 5 & 4 & 1 \end{bmatrix} - \begin{bmatrix} -8 & 0 & 3 \\ 10 & 5 & -4 \end{bmatrix}$ $\begin{bmatrix} 6 & 7 & -6 \\ -5 & -1 & 5 \end{bmatrix}$

Simplify the expression. (2.5 and 2.6)

12. $4(y - 4)$ $4y - 16$
13. $3(6 + x)$ $18 + 3x$
14. $2y(5 + y)$ $10y + 2y^2$
15. $-5t(3 - t)$ $-15t + 5t^2$
16. $20x - 17x$ $3x$
17. $4b + 7 + 7b$ $11b + 7$
18. $5x^2 - 3x^2$ $2x^2$
19. $4.2y + 1.1y$ $5.3y$

Solve the equation. (3.1–3.4, 3.6)

20. $x + 4 = -1$ **−5**
21. $-3 = n - 15$ **12**
22. $5b = -25$ **−5**
23. $\frac{x}{4} = 6$ **24**
24. $3x + 4 = 13$ **3**
25. $6 + \frac{2}{3}x = 14$ **12**
26. $5(x - 2) = 15$ **5**
27. $14x + 50 = 75$ $1\frac{11}{14}$
28. $x + 8 = 3(x - 4)$ **10**
29. $-(x - 7) = \frac{1}{2}x + 1$ **4**
30. $3x - 15.6 = 75.3$ **30.3**
31. $5.5x + 2.1 = 7.6$ **1**

Rewrite the equation so that *y* is a function of *x*. (3.7)

32. $x = y + 3$ $y = x - 3$
33. $4x - 5y = 13$ $y = \frac{4}{5}x - \frac{13}{5}$
34. $3(y - x) = 10 - 4x$ $y = -\frac{1}{3}x + \frac{10}{3}$

Find the unit rate. (3.8)

35. \$1 for two cans of dog food **\$.50/can**
36. \$440 for working 40 hours **\$11/h**

Plot and label the ordered pairs in a coordinate plane. (4.1) **See margin.**

37. $A(2, 3), B(2, -3), C(-1, 1)$
38. $A(0, -2), B(-3, -3), C(2, 0)$
39. $A(2, 4), B(3, 0), C(-1, -4)$
40. $A(1, -4), B(-2, 4), C(0, -1)$

Plot the points and find the slope of the line passing through the points. (4.4) **41–44. See margin for graphs.**

41. $(3, 1), (-3, -1)$ $\frac{1}{3}$
42. $(2, 2), (-5, 2)$ **0**
43. $(-4, 1), (-4, -2)$ **undefined**
44. $(-2, 0), (0, -4)$ **−2**

Graph the equation. (4.2, 4.3, and 4.6) **45–50. See margin.**

45. $x - y = 4$
46. $2x - y + 1 = 0$
47. $x + 2y - 4 = 0$
48. $x + 3y = 7x$
49. $x + 4y - 1 = 0$
50. $y - 2.5 = 0$

Write an equation of the line in slope-intercept form. (5.1)

51. The slope is 1; the *y*-intercept is −3. $y = x - 3$
52. The slope is −2; the *y*-intercept is 5. $y = -2x + 5$

Write an equation of the line that passes through the point and has the given slope. Write the equation in slope-intercept form. (5.2)

53. $(-1, 1), m = 2$ **$y = 2x + 3$** **54.** $(3, -1), m = \frac{1}{4}$ **$y = \frac{1}{4}x - \frac{7}{4}$** **55.** $(-3, 6), m = -5$ **$y = -5x - 9$**

Write an equation in slope-intercept form of the line that passes through the points. (5.3)

56. $(-1, -7), (-2, 1)$ **$y = -8x - 15$**

57. $(0, 3), (2, 4)$ **$y = \frac{1}{2}x + 3$**

58. $(4.2, -3.6), (7.0, 3.4)$ **$y = 2.5x - 14.1$**

Solve the inequality. (6.1–6.4)

59. $6 > 3x$ **$x < 2$**

60. $-6 \le x + 12$ **$x \ge -18$**

61. $-\frac{x}{6} \ge 8$ **$x \le -48$**

62. $-4 - 5x \le 31$ **$x \ge -7$**

63. $-4x + 3 > -21$ **$x < 6$**

64. $-x + 2 < 2(x - 5)$ **$x > 4$**

65. $-3.2 + x \ge 6.9$ **$x \ge 10.1$**

66. $-4 \le -2x \le 10$ **$-5 \le x \le 2$**

67. $5 < 4x - 11 < 13$ **$4 < x < 6$**

68. $-2 < x - 7 \le 15$ **$5 < x \le 22$**

69. $|x - 8| > 10$ **$x < -2$ or $x > 18$**

70. $|2x + 5| \le 7$ **$-6 \le x \le 1$**

Find the mean, the median, and the mode of the collection of numbers. (6.6)

71. 10, 5, 25, 5, 10, 15, 20, 50, 5, 15 **16, 12.5, 5**

72. 8, 7, 5, 2, 3, 5, 2, 3, 2, 7, 1, 2 **3.92, 3, 2**

73. PURCHASES You have \$25. You buy two CDs that cost \$9.99 each, tax included. Do you have enough money left over to buy a cassette that costs \$5.95? Explain. (1.4) **No. *Sample answer:* By estimation, you have about \$25 − 2(\$10) = \$5 left. The actual amount is \$25 − 2(\$9.99) = \$5.02, which is less than \$5.95.**

74. PHOTO COSTS A photography studio charges \$65 for a basic graduation package of photos, plus \$3 for each additional wallet photo. Use the equation $65 + 3n = C$, where n represents the number of additional wallet photos and C represents the total cost, to make an input-output table of the costs for ordering 0 through 6 additional wallet photos. (1.7) **See margin.**

75. VELOCITY Recall that the speed of an object is the absolute value of its velocity. A hot-air balloon drops at a rate of 100 feet per minute. What are its velocity and speed? (2.1) **−100 ft/min, 100 ft/min**

76. TEMPERATURES On February 21, 1918, the temperature in Granville, North Dakota, rose from −33°F to 50°F. By how many degrees did the temperature rise? (3.1) **83°F**

77. SALES TAX You are shopping for a pen. The sales tax is 6%. You have a total of \$10.50 to spend. What is your price limit for the pen? (3.6) **\$9.90**

AMUSEMENT PARKS **In Exercises 78–80, use the table that shows the number of dollars (in millions) spent at amusement parks in the United States from 1991 through 1995.** (4.1 and 5.4)

Years since 1991	0	1	2	3	4
Dollars (in millions)	4820	5366	5663	5905	6376

▶ Source: U.S. Bureau of the Census

78. Draw a scatter plot of the data. **See margin.**

79. Write a linear model for the amount spent at amusement parks. **Sample answer: $y = 365x + 4896$**

80. Use the linear model to estimate the amount spent in 2005. **Sample answer: about 10 billion dollars**

44.

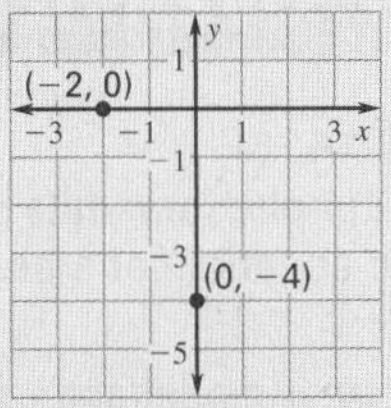

45.

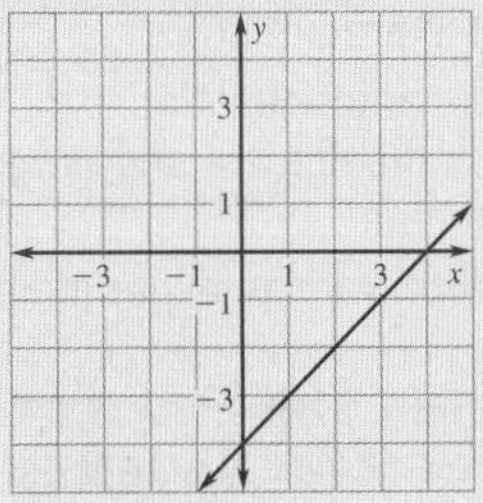

46.

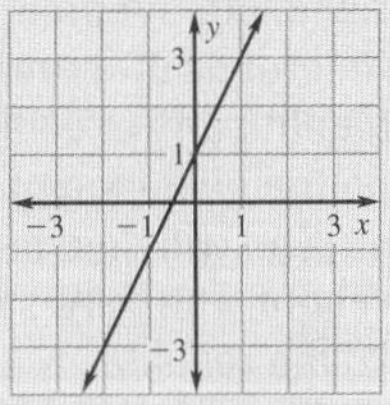

47.

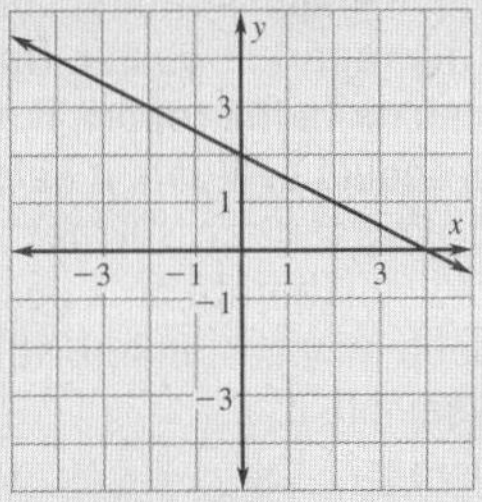

48.

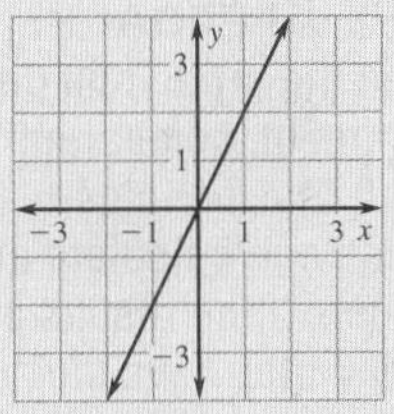

49.

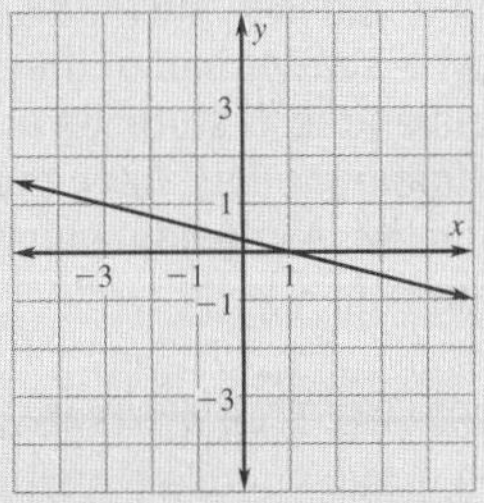

50.

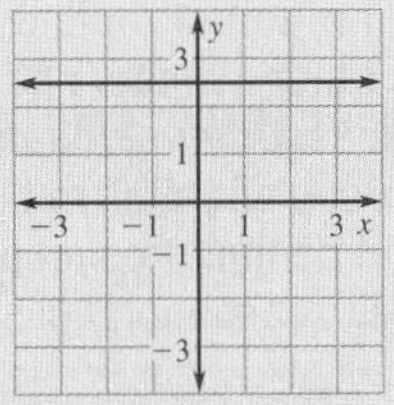

74.

Input	Output
0	65
1	68
2	71
3	74
4	77
5	80
6	83

78. See Additional Answers beginning on page AA1.

PROJECT OVERVIEW

PROJECT GOALS
- Draw a scatter plot for a data set and write an equation for a best-fitting line.
- Interpret a linear model and use it to make predictions.
- Write linear inequalities to describe the domain and range of real-life data.
- Generalize from an experiment.

MANAGING THE PROJECT

CLASSROOM MANAGEMENT
Students can complete the project individually or in groups of three. In groups, one student can set up the experiment with the guidance of a second student, who will then perform the experiment. The third student can record the results.

All group members should contribute to the report or poster. One member can write the report, but the others should evaluate it before the final draft is written. All members should collaborate on a poster, but the group may assign one student to assemble it.

ALTERNATIVE APPROACH
Each student in the group can perform the experiment, comparing results for each trial. For close results, students can calculate an average distance. If the results are not close for any trial, have students repeat that trial. Have the group draw a scatter plot of the average distances. If students find their own linear models, have them compare models to decide which fits the data best.

CONCLUDING THE PROJECT
Have each group choose a member to present their report or project to the class. The other members can answer questions. Other students should pay special attention to the predictions and conjectures made to evaluate their reasonableness.

PROJECT
Applying Chapters 4–6

Investigating Elasticity

OBJECTIVE **Explore the relationship between the length of a rubber band and the weight suspended from it.**

Materials: paper cup, paper clip, masking tape, metric ruler, string, rubber band, hole punch, scissors, 100 pennies, graph paper or graphing calculator (optional)

COLLECTING THE DATA

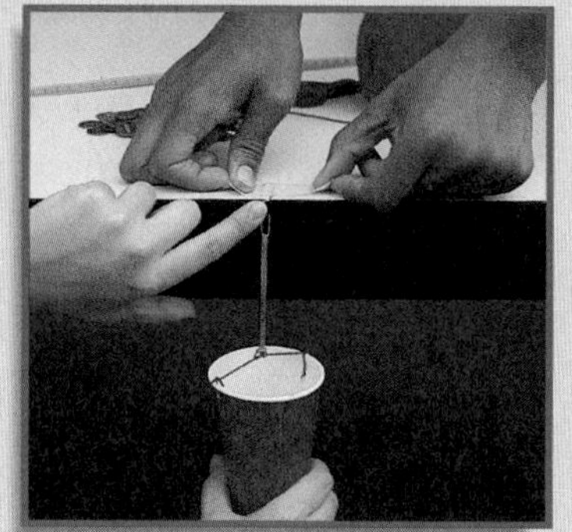

1. Punch two holes in the paper cup about $\frac{1}{2}$ inch from the top of the rim directly across from each other. Thread the string through the two holes. Tie off the string on both sides of the cup. Cut off any extra string on either end. Then attach the rubber band to the string.

2. Tape the paper clip to the edge of a table or desk so that one end hangs over the edge. Attach the rubber band to the end of the paper clip so the cup is hanging on the rubber band over the side of the table.

3. Tape the ruler to the edge of the table next to the rubber band with 0 on the ruler next to the top of the rubber band. Record the distance from the top of the rubber band to the bottom of the cup. This is the initial distance.

4. Add 10 pennies to the cup. Record the number of pennies you added to the cup and the distance from the top of the rubber band to the bottom of the cup.

5. Repeat **Step 4** several more times, each time increasing the number of pennies placed in the cup until you have 50 pennies in the cup.

3. ***Sample answer*: The y-intercept is the distance from the top of the rubber band to the bottom of the cup before any pennies are added. The slope is the change in distance per added penny.**

4. ***Sample answer:* The domain is all integers greater than or equal to zero and less than or equal to 50 (the number of pennies): $0 \le x \le 50$. The range is all real numbers greater than or equal to the initial distance from the top of the rubber band to the bottom of the cup and less than or equal to the maximum distance.**

INVESTIGATING THE DATA

1. Make a scatter plot of the data you have collected. Describe any patterns you see. **Check plots. *Sample answer:* As the number of pennies increases, the distance from the top of the rubber band to the bottom of the cup increases.**
2. Find an equation for a best-fitting line. You may want to use a graphing calculator or a computer. **Answers may vary.**
3. Explain what the y-intercept and the slope mean in terms of your data. **See margin.**
4. Describe a reasonable domain and the range of your equation. Write inequalities to represent both. **See margin.**

5. Check answers. Predictions can be made using the scatter plot in Ex. 1, the equation in Ex. 2, or the relationship described in the answer to Ex. 6.

6. *Sample answer:* The initial distance stays the same, but the additional distance due to stretching is multiplied by the same factor that the number of pennies is multiplied by. For example, this stretching is twice as much for 100 pennies as for 50 pennies.

7. *Sample answer:* The initial distance will be greater for a longer rubber band, so the equation will be different. Examples may vary.

8. *Sample answer:* The graph would have a negative slope rather than a positive slope because the distance from the floor to the bottom of the rubber band decreases rather than increases as pennies are added.

MAKING CONJECTURES

5. Make a conjecture about what the distance from the top of the rubber band to the bottom of the cup would be if you placed 100 pennies in the cup.
6. Test your conjecture. What relationship do you see between the number of pennies used and the length of the stretch? **See margin.**

In Exercises 7 and 8, discuss your results with others in your class.

7. Make a conjecture about how the length of the rubber band might affect the total distances. Would the equations be the same or different? Give examples to support your answer. **See margin.**
8. How would your graph be different if you measured from the floor up to the bottom of the rubber band? Explain your answer. **See margin.**

PRESENTING YOUR RESULTS

Write a report or make a poster to report your results.

- Include a table with your data.
- Include your answers to Exercises 1–4.
- Describe the conjectures that you made in Exercises 5–8 and your reasons for believing them to be true.
- Describe any patterns you found when you discussed results with others in your class.
- What advice would you give to someone else who is going to do this project?

EXTENDING THE PROJECT

- How does the thickness of the rubber band affect the distance it stretches? Do a second experiment with a thicker rubber band to find out.
- A grocery store scale operates in a similar way. When you put fruits or vegetables on a scale, the spring inside the scale stretches. The heavier the item, the larger the stretch. Can you think of other items that work in a similar way?

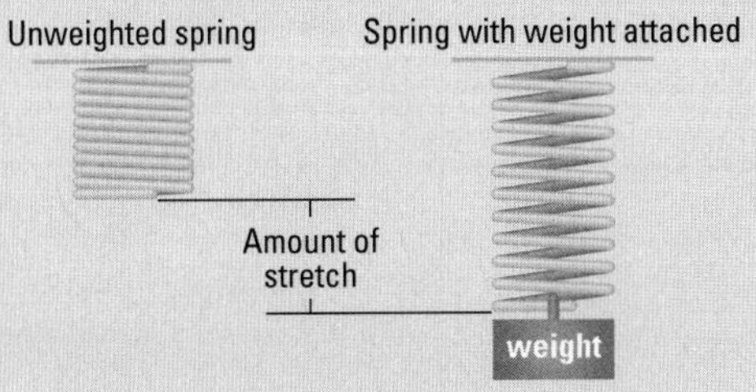

GRADING THE PROJECT

4 Students complete the experiment and record the results correctly. They make an accurate scatter plot of the data with proper axes and scale. They include an appropriate best-fitting line. Their interpretations of their models and consideration of domain and range correctly relate the data with the mathematical ideas. Conjectures are reasonable, and tested as possible. Students extend the project by repeating the experiment with a thicker rubber band. Their report or poster is well thought out, clearly presented, and contains a clear and concise table of values.

3 Students complete the experiment, make a scatter plot, and find a best-fitting line, but their plot may not be well-labeled, or their line may not be the very best model. They do not completely relate the mathematical ideas to the data or support their conjectures fully. The report or poster includes all required information and shows considerable effort. Students extend the project.

2 Students complete the experiment but may not accurately record, describe, or model the data with a best-fitting line. They neither answer all of the questions correctly nor complete any extensions. Their conjectures are incomplete or not well-reasoned. A report or a poster is created, but may be missing a clear data summary, and does not include all of the ideas needed to convey the meaning of the activity.

1 Students' work is incomplete, showing little understanding of the mathematical ideas. The scatter plot and/or best-fitting line are incomplete. Conjectures show little understanding of the relationships between weight and stretch or are missing. The report or poster does not present the data accurately, does not present the ideas clearly, and shows little effort.

CHAPTER 7

Planning the Chapter

Systems of Linear Equations and Inequalities

Lesson	Goals	NCTM	ITED	SAT9	Terra-Nova	Local
7.1 *pp. 397–404*	CONCEPT ACTIVITY: 7.1 *Investigate a system of linear equations by graphing.* GOAL 1 Solve a system of linear equations by graphing. GOAL 2 Model a real-life problem using a linear system. TECHNOLOGY ACTIVITY: 7.1 *Solve linear systems on a graphing calculator.*	1, 2, 3, 6, 8, 9, 10	IIG, MIG, RQWN, RQF, RQGE	1, 7	14, 16, 17, 18, 52	
7.2 *pp. 405–410*	GOAL 1 Use substitution to solve a linear system. GOAL 2 Model a real-life situation using a linear system.	1, 2, 6, 8, 9, 10	RQWN, RQF	7	16, 18, 52	
7.3 *pp. 411–417*	GOAL 1 Use linear combinations to solve a system of linear equations. GOAL 2 Model a real-life problem using a system of linear equations.	1, 2, 6, 8, 9, 10	RQWN, RQF	7	16, 17, 18, 52	
7.4 *pp. 418–424*	GOAL 1 Choose the best method to solve a system of linear equations. GOAL 2 Use a system to model real-life problems.	1, 2, 3, 6, 8, 9, 10	RQWN, RQF, RQGE	7	16, 17, 18, 52	
7.5 *pp. 425–431*	CONCEPT ACTIVITY: 7.5 *Investigate special types of linear systems.* GOAL 1 Identify linear systems as having one solution, no solution, or infinitely many solutions. GOAL 2 Model real-life problems using a linear system.	1, 2, 6, 8, 9, 10	RQWN, RQF	1, 7	16, 17, 18, 52	
7.6 *pp. 432–438*	GOAL 1 Solve a system of linear inequalities by graphing. GOAL 2 Use a system of linear inequalities to model a real-life situation.	1, 2, 3, 6, 8, 9, 10	IIG, MIG, RQWN, RQF, RQGE	1, 6, 7	14, 16, 18, 52	

RESOURCES

Chapter Resource Booklets

Chapter Support

Tips for New Teachers	p. 1	Prerequisite Skills Review	p. 5
Parent Guide for Student Success	p. 3	Strategies for Reading Mathematics	p. 7

Lesson Support	7.1	7.2	7.3	7.4	7.5	7.6
Lesson Plans (regular and block)	p. 9	p. 23	p. 36	p. 50	p. 63	p. 78
Warm-Up Exercises and Daily Quiz	p. 11	p. 25	p. 38	p. 52	p. 65	p. 80
Activity Support Masters	p. 12					
Lesson Openers	p. 13	p. 26	p. 39	p. 53	p. 66	p. 81
Graphing Calculator Activities & Keystrokes	p. 14	p. 27			p. 67	p. 82
Practice (3 levels)	p. 15	p. 28	p. 40	p. 54	p. 70	p. 85
Reteaching with Practice	p. 18	p. 31	p. 43	p. 57	p. 73	p. 88
Quick Catch-Up for Absent Students	p. 20	p. 33	p. 45	p. 59	p. 75	p. 90
Cooperative Learning Activities				p. 60		
Interdisciplinary Applications		p. 34		p. 61		p. 91
Real-Life Applications	p. 21		p. 46		p. 76	
Math & History Applications			p. 47			
Challenge: Skills and Applications	p. 22	p. 35	p. 48	p. 62	p. 77	p. 92

Review and Assessment

Quizzes	pp. 49, 93	Alternative Assessment with Math Journal	p. 102
Chapter Review Games and Activities	p. 94	Project with Rubric	p. 104
Chapter Test (3 levels)	pp. 95–100	Cumulative Review	p. 106
SAT/ACT Chapter Test	p. 101	Resource Book Answers	p. A1

Transparencies

	7.1	7.2	7.3	7.4	7.5	7.6
Warm-Up Exercises and Daily Quiz	p. 53	p. 54	p. 55	p. 56	p. 57	p. 58
Alternative Lesson Opener Transparencies	p. 46	p. 47	p. 48	p. 49	p. 50	p. 51
Examples/Standardized Test Practice	✓	✓	✓	✓	✓	✓
Answer Transparencies	✓	✓	✓	✓	✓	✓

Technology

- EasyPlanner CD
- EasyPlanner Plus Online
- Classzone.com
- Personal Student Tutor CD
- Test and Practice Generator CD
- Instant Replay: Video Review Games
- Electronic Lesson Presentations
- Chapter Audio Summaries CD

Additional Resources

- Basic Skills Workbook
- Explorations and Projects
- Worked-Out Solution Key
- Data Analysis Sourcebook
- Spanish Resources
- Standardized Test Practice Workbook
- Practice Workbook with Examples
- Notetaking Guide

Pacing the Chapter

Resource Key
- Student Edition
- Chapter 7 Resource Book
- Teacher's Edition

Regular Schedule

Day 1

7.1

Starting Options
- Prereq. Skills Review
- Strategies for Reading
- Warm-Up or Daily Quiz

Teaching Options
- Motivating the Lesson
- Concept Act. & Wksht.
- Les. Opener (Visual)
- Graphing Calc. Activity
- Examples 1–3
- Closure Question
- Guided Practice Exs.

Apply/Homework
- See Assignment Guide.
- See the CRB: Practice, Reteach, Apply, Extend

Assessment Options
- Checkpoint Exercises
- Daily Quiz (7.1)
- Stand. Test Practice

Day 2

7.2

Starting Options
- Homework Check
- Warm-Up or Daily Quiz

Teaching Options
- Motivating the Lesson
- Les. Opener (Activity)
- Graphing Calc. Activity
- Examples 1–3
- Closure Question
- Guided Practice Exs.

Apply/Homework
- See Assignment Guide.
- See the CRB: Practice, Reteach, Apply, Extend

Assessment Options
- Checkpoint Exercises
- Daily Quiz (7.2)
- Stand. Test Practice

Day 3

7.3

Starting Options
- Homework Check
- Warm-Up or Daily Quiz

Teaching Options
- Motivating the Lesson
- Les. Opener (Application)
- Examples 1–3
- Guided Practice Exs. 1–8

Apply/Homework
- See Assignment Guide.
- See the CRB: Practice, Reteach, Apply, Extend

Assessment Options
- Checkpoint Exercises, p. 412

Day 4

7.3 *(cont.)*

Starting Options
- Homework Check

Teaching Options
- Example 4
- Closure Question

Apply/Homework
- See Assignment Guide.
- See the CRB: Practice, Reteach, Apply, Extend

Assessment Options
- Checkpoint Exercises, p. 413
- Daily Quiz (7.3)
- Stand. Test Practice
- Quiz (7.1–7.3)

Day 5

7.4

Starting Options
- Homework Check
- Warm-Up or Daily Quiz

Teaching Options
- Motivating the Lesson
- Les. Opener (Application)
- Example 1
- Guided Practice Exs. 1–5

Apply/Homework
- See Assignment Guide.
- See the CRB: Practice, Reteach, Apply, Extend

Assessment Options
- Checkpoint Exercises, p. 419

Day 6

7.4 *(cont.)*

Starting Options
- Homework Check

Teaching Options
- Examples 2–3
- Closure Question
- Guided Practice Exs. 6–9

Apply/Homework
- See Assignment Guide.
- See the CRB: Practice, Reteach, Apply, Extend

Assessment Options
- Checkpoint Exercises, pp. 419–420
- Daily Quiz (7.4)
- Stand. Test Practice

Day 9

7.6 *(cont.)*

Starting Options
- Homework Check

Teaching Options
- Example 4
- Closure Question

Apply/Homework
- See Assignment Guide.
- See the CRB: Practice, Reteach, Apply, Extend

Assessment Options
- Checkpoint Exercises, p. 434
- Daily Quiz (7.6)
- Stand. Test Practice
- Quiz (7.4–7.6)

Day 10

Review

Day 10 Start Options
- Homework Check

Reviewing Options
- Chapter 7 Summary
- Chapter 7 Review
- Chapter Review Games and Activities

Apply/Homework
- Chapter 7 Test (practice)
- Ch. Standardized Test (practice)

Day 11

Assess

Day 11 Start Options
- Homework Check

Assessment Options
- Chapter 7 Test
- SAT/ACT Ch. 7 Test
- Alternative Assessment

Apply/Homework
- Skill Review, p. 448

BLOCK SCHEDULE

Day 1

7.1 & 7.2

Day 1 Start Options
- Prereq. Skills Review
- Strategies for Reading
- W-Up 7.1 or D. Quiz 6.7

Teaching 7.1 Options
- Motivating the Lesson
- Concept Act. & Wksht.
- Les. Opener (Visual)
- Graphing Calc. Activity
- Examples 1–3
- Closure Question
- Guided Practice Exs.

Teaching 7.2 Options
- Warm-Up (Les. 7.2)
- Motivating the Lesson
- Les. Opener (Activity)
- Graphing Calc. Activity
- Examples 1–3
- Closure Question
- Guided Practice Exs.

Apply/Homework
- See Assignment Guide.
- See the CRB: Practice, Reteach, Apply, Extend

Assessment Options
- Checkpoint Exercises
- D. Quiz (Les. 7.1, 7.2)

Day 2

7.3

Starting Options
- Homework Check
- Warm-Up or Daily Quiz

Teaching Options
- Motivating the Lesson
- Les. Opener (Appl.)
- Examples 1–4
- Closure Question
- Guided Practice Exs.

Apply/Homework
- See Assignment Guide.
- See the CRB: Practice, Reteach, Apply, Extend

Assessment Options
- Checkpoint Exercises
- Daily Quiz (7.3)
- Stand. Test Practice
- Quiz (7.1–7.3)

Day 3

7.4

Day 3 Start Options
- Homework Check
- Warm-Up or Daily Quiz

Teaching 7.4 Options
- Motivating the Lesson
- Les. Opener (Appl.)
- Examples 1–3
- Closure Question
- Guided Practice Exs.

Apply/Homework
- See Assignment Guide.
- See the CRB: Practice, Reteach, Apply, Extend

Assessment Options
- Checkpoint Exercises
- Daily Quiz (7.4)
- Stand. Test Practice

Day 4

7.5 & 7.6

Day 4 Start Options
- Homework Check
- W-Up 7.5 or D. Quiz 7.4

Teaching 7.5 Options
- Motivating the Lesson
- Les. Opener (Calc.)
- Graphing Calc. Activity
- Examples 1–4
- Closure Question
- Guided Practice Exs.

Beginning 7.6 Options
- Warm-Up (Les. 7.6)
- Les. Opener (Activity)
- Graphing Calc. Activity
- Examples 1–3
- Guided Practice Exs. 1–8

Apply/Homework
- See Assignment Guide.
- See the CRB: Practice, Reteach, Apply, Extend

Assessment Options
- Checkpoint Exercises
- Daily Quiz (Les. 7.5)
- Stand. Test Practice

Day 5

7.6 & Review

Day 5 Start Options
- Homework Check
- Daily Quiz (Les. 7.5)

Finishing 7.6 Options
- Example 4
- Closure Question

Reviewing Options
- Chapter 7 Summary
- Chapter 7 Review
- Chapter Review Games and Activities

Apply/Homework
- See Assignment Guide.
- See the CRB: Practice, Reteach, Apply, Extend
- Chapter 7 Test (practice)
- Ch. Standardized Test (practice)

Assessment Options
- Checkpoint Exercises
- Daily Quiz (7.6)
- Stand. Test Practice
- Quiz (7.4–7.6)

Day 6

Assess & 8.1

(Day 6 = Ch. 8 Day 1)

Assessment Options
- Chapter 7 Test
- SAT/ACT Ch. 7 Test
- Alternative Assessment

Ch. 8 Start Options
- Skill Review, p. 448
- Prereq. Skills Review
- Strategies for Reading

Beginning 8.1 Options
- Warm-Up (Les. 8.1)
- Motivating the Lesson
- Concept Act. & Wksht.
- Les. Opener (Activity)
- Graphing Calc. Activity
- Examples 1–4
- Guided Practice Exs. 1–21

Apply/Homework
- See Assignment Guide.
- See the CRB: Practice, Reteach, Apply, Extend

Assessment Options
- Checkpoint Exercises

Day 7

7.5

Starting Options
- Homework Check
- Warm-Up or Daily Quiz

Teaching Options
- Motivating the Lesson
- Les. Opener (Calculator)
- Graphing Calc. Activity
- Examples 1–4
- Closure Question
- Guided Practice Exs.

Apply/Homework
- See Assignment Guide.
- See the CRB: Practice, Reteach, Apply, Extend

Assessment Options
- Checkpoint Exercises
- Daily Quiz (7.5)
- Stand. Test Practice

Day 8

7.6

Starting Options
- Homework Check
- Warm-Up or Daily Quiz

Teaching Options
- Les. Opener (Activity)
- Graphing Calc. Activity
- Examples 1–3
- Guided Practice Exs. 1–8

Apply/Homework
- See Assignment Guide.
- See the CRB: Practice, Reteach, Apply, Extend

Assessment Options
- Checkpoint Exercises, p. 433

CHAPTER 7

Meeting Individual Needs

Before the Chapter

The *Chapter 7 Resource Book* has the following materials to distribute and use before the chapter:

- **Parent Guide for Student Success**
- **Prerequisite Skills Review (pictured below)**
- **Strategies for Reading Mathematics**

Prerequisite Skills *Pages 5–6*

CHAPTER 7

NAME ______________________ DATE ________

Prerequisite Skills Review

For use before Chapter 7

Chapter Support

EXAMPLE 1 ***Simplifying by Combining Like Terms***

Simplify the expression.

$9 - 3(5 - x)$

SOLUTION

$9 - 3(5 - x) = 9 + (-3)(5 - x)$	Rewrite as an addition expression.
$= 9 + [(-3)(5) + (-3)(-x)]$	Distribute the -3.
$= 9 + (-15) + (-3x)$	Multiply.
$= -6 + 3x$	Combine like terms and simplify.

Exercises for Example 1

Simplify the expression.

1. $8r + (-64r)$ **2.** $\frac{1}{48}g - \frac{3}{168}g$

3. $\frac{9}{10}s + 2\frac{1}{5}s$ **4.** $0.1t - 0.02t$

EXAMPLE 2 ***Solving Equations with Variables on Both Sides***

Solve the equation if possible.

$9 - 3(5 - x)$

SOLUTION

$9x + 35 = -x - 6$	Write original equation.
$9x + 34 + x = -x - 6 + x$	Add x to each side.
$10x + 34 = -6$	Simplify.
$10x = -40$	Subtract 34 from both sides.
$10x = -40$	Simplify.

The solution is –4.

When you check the solution, substitute –4 for each x in the equation.

Exercises for Example 2

Solve the equation if possible.

5. $3x + 7(x - 1) = 23$ **6.** $8x + 4x = 12x - 1$

7. $1 + 7x - 3.1 = 0$ **8.** $12(8 - y) - 11y + y + 16$

PREREQUISITE SKILLS REVIEW These two pages support the Study Guide on page 396. They help students prepare for Chapter 7 by providing worked-out examples and practice for the following skills needed in the chapter:

- **Combine like terms.**
- **Solve equations with variables on both sides.**
- **Verify solutions of an equation.**

TECHNOLOGY RESOURCE

Students can use the Personal Student Tutor to find additional reteaching and practice for skills from earlier chapters that are used in Chapter 7.

During Each Lesson

The *Chapter 7 Resource Book* has the following alternatives for introducing the lesson:

- **Lesson Openers**
- **Graphing Calculator Activities with Keystrokes (below)**

Calculator Activity *Pages 82–83*

LESSON 7.6

NAME ______________________ DATE ________

Graphing Calculator Activity

For use with pages 432–438

Lesson 7.6

GOAL **To predict which area of a graph represents the solution of a system of linear inequalities.**

By observing the linear inequalities in a system of linear inequalities, it is possible to predict which area of the graph represents the solution of the system. The graph of the solution of two linear inequalities is shown at the left.

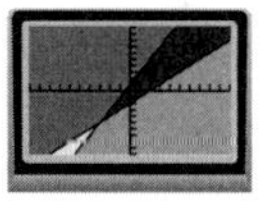

Activity

1. Listed below is a system of linear inequalities.

   $y \geq x + 3$ Inequality 1

   $y \geq 3$ Inequality 2

2. Predict the area that will be shaded by the graph representing the system of linear inequalities given in Step 1.
3. Use your graphing calculator to enter the two inequalities.
4. Graph the two inequalities using the shading feature of your calculator.
5. Does your graph verify your prediction?

Exercises

1. Predict the area of the graph that will be shaded by the system of linear inequalities. Use your graphing calculator to check your prediction.

a. $y \leq -3$, $y \leq x - 5$ **b.** $y \geq x + 5$, $y \leq x + 7$ **c.** $y \geq -2x - 3$, $y \geq 3x - 3$

2. Using the information in the graph, complete the system of inequalities with $\leq$ or $\geq$.

a. y_1 __?__ -2, y_2 __?__ $-3x + 3$ **b.** y_1 __?__ $-x - 4$, y_2 __?__ $-5x - 9$ **c.** y_1 __?__ $-4x - 7$, y_2 __?__ $-6x + 5$

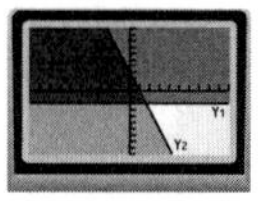

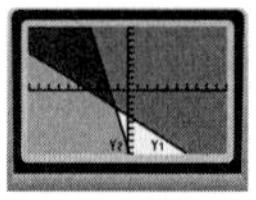

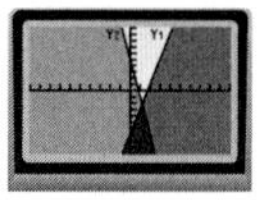

GRAPHING CALCULATOR ACTIVITY This activity uses a graphing calculator to develop an understanding of solving systems of linear inequalities, one of the concepts covered in Lesson 7.6. Keystrokes for this activity are provided on a separate sheet.

Following Each Lesson

The *Chapter 7 Resource Book* has a variety of materials to follow-up each lesson. They include the following:

- **Practice (3 levels)**
- **Reteaching with Practice**
- **Quick Catch-Up for Absent Students (pictured below)**
- **Interdisciplinary Applications**
- **Real-Life Applications**

Quick Catch-Up *Page 20*

LESSON 7.1 NAME ____________ DATE ________

Quick Catch-Up for Absent Students

For use with pages 203–209

The items checked below were covered in class on (date missed) __________

Activity 7.1: Investigating Graphs of Linear Systems (p. 397)

____ **Goal:** Determine if two different linear equations have a solution in common.

____ Student Help: Look Back

Lesson 7.1: Solving Linear Systems by Graphing

____ **Goal 1:** Solve a system of linear equations by graphing (pp. 398–399)

Material Covered:

____ Student Help: Look Back

____ Example 1: Checking the Intersection Point

____ Example 2: Using the Graph-and-Check Method

Vocabulary:

system of linear equations, p. 398 linear system, p. 398

solution of a system of linear equations, p. 398

____ **Goal 2:** Model a real-life problem using a linear system. (p. 400)

Material Covered:

____ Example 3: Writing and Using a Linear System

Activity 7.1: Solving Linear Systems by Graphing (p. 404)

____ **Goal:** Solve a linear system using a graphing calculator.

____ Student Help: Keystroke Help

____ Other (specify) ____________

Homework and Additional Learning Support

____ Textbook (specify) pp. 401–403

____ Internet: Extra Examples at www.mcdougallittell.com

____ *Reteaching with Practice* worksheet (specify exercises) ________

____ *Personal Student Tutor* for Lesson 7.1

Lesson 7.1

Quick Catch-Up for Absent Students You can use this form to let students know what they have missed when they've been absent from class. It allows you to quickly check off which Examples and other elements of Lesson 7.1 were covered on a given day and provides space for filling in the homework assignment.

Technology Resource

Students who have missed class can find keystrokes for the technology activity in Lesson 7.1 on the Internet at www.mcdougallittell.com.

Assessing the Chapter

The *Chapter 7 Resource Book* has the following review and assessment materials:

- **Quizzes**
- **Chapter Review Games and Activities**
- **Chapter Test (3 levels)**
- **SAT/ACT Chapter Test**
- **Alternative Assessment with Rubric and Math Journal**
- **Project with Rubric (pictured below)**
- **Cumulative Review**

Project with Rubric *Pages 104–105*

CHAPTER 7 NAME ____________ DATE ________

Project: Going Up!

For use with Chapter 7

OBJECTIVE **Explore how relationships among stair dimensions affect safety and comfort.**

MATERIALS inch ruler, paper, pencil, graph paper

INVESTIGATION Have you found that some stairways are more difficult to climb or easier to trip on than others? Have you thought about why this happens? Stairway safety and comfort can be influenced by the length of the riser (the vertical part between treads), the length of the tread (horizontal surface we step on), and the ratio of these lengths.

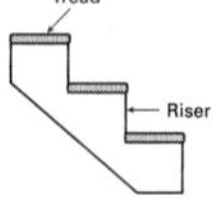

1. Make a table with heads shown below. Collect and fill in the data for at least eight different stairways. Try for a good variety of step sizes.

Stairway Location	*Tread Length* (in.)	*Riser Length* (in.)	*Riser/Tread ratio*	*How hard or easy to climb?*	*How safe does it feel?*

2. Based on the data you collected, what seems to be a good riser/tread ratio. Explain your decision. Did the stairways with tread lengths or riser lengths within a certain range seem safer or easier to climb? Explain.
3. Use your answers to Question 2. Design your own stairway to go up a total distance of 120 inches from one floor to the next. Determine the riser and tread lengths and the number of steps. Then make a scale drawing.
4. Below are two generally accepted rules for stairway construction. To find the riser and tread lengths that satisfy these rules, first write the relationships as four inequalities involving tread length t and riser length r. Then graph the solution of that system. Since the boundary lines are close together, use small intervals such as $\frac{1}{2}$ or $\frac{1}{4}$ along axes. (You can make the graph very large or you can use breaks in the axes and focus on the part of the graph near the solution region.)

   Rule 1: The sum of one riser length and one tread length should be from 17 inches to 18 inches.

   Rule 2: The sum of two riser lengths and one tread length should be from 24 inches to 25 inches.
5. On the same graph, plot the ordered pairs from the data you collected and from the stairway you designed. Do the stairways all satisfy the two rules?

PRESENT YOUR RESULTS Make a poster presenting your results. Include your graph and the scale drawing of the stairway you designed. Also include an analysis of the safety of the stairs for which you collected data and of the one you designed.

Review and Assess

Project with Rubric The Project for Chapter 7 provides students with the opportunity to apply the concepts they have learned in the chapter in a new way. In this project, students use what they have learned about systems of linear inequalities to explore how relationships among stair dimensions affect safety and comfort. Teacher's notes and a scoring rubric are provided on a separate sheet.

CHAPTER OVERVIEW

CHAPTER GOALS

Chapter 7 introduces systems of linear equations and systems of inequalities. Students will solve a system of two linear equations by using graphing, substitution, and linear combinations. The choice of method for solving a particular linear system is considered and linear systems that have one solution, no solution, or infinitely many solutions are identified. Systems of linear inequalities are solved by graphing. Throughout the chapter, real-life problems are modeled using linear systems.

APPLICATION NOTE

The average household in the United States spends about $\frac{1}{4}$ of its family budget on housing. Families in the lower income brackets, however, often spend over half their family budget on housing. You might ask students to imagine a family of 4 that makes $12,000 per year but spends $6,000 on housing. What expenses would they have other than housing? You might break these expenses down into monthly amounts.

You could also ask students what happens to those families that need low-income housing and don't get it? Students could do some research to find data about the number of homeless individuals or homeless families during this same period from 1970. Students can make a linear model and compare it with the models for needed and available low-income housing.

Additional information about low-income housing is available at **www.mcdougallittell.com.**

SYSTEMS OF LINEAR EQUATIONS AND INEQUALITIES

▶ *How can you analyze the need for low-income housing?*

CHAPTER 7

APPLICATION: Housing

To see how the need for low-income rental housing has changed over time, you can look at a model. The graph below shows the number of households with annual earnings of $12,000 or less that need to rent housing and the number of rental units available that they can afford.

In this chapter, you will learn how to use pairs of linear models to analyze problems.

Low-Income Rental Housing

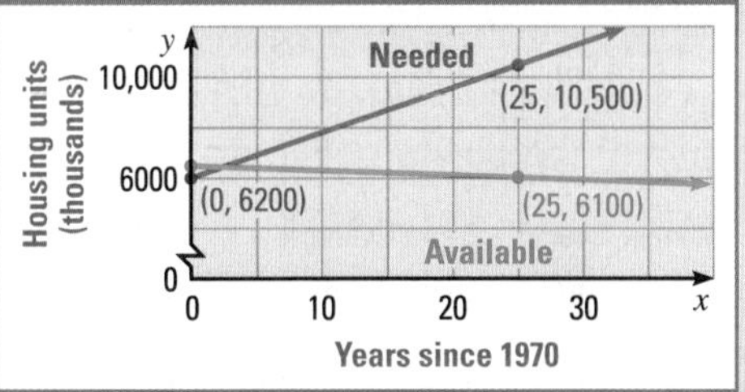

Think & Discuss

Use the graph to answer the following questions.

1. How many low-cost housing units were available in 1995? **6,100,000 units**
2. In 1995, how many more low-income renters were there than low-cost units? **4,400,000 renters**

Learn More About It

You will use a linear system to analyze the need for low-income housing in Ex. 51 on p. 422.

APPLICATION LINK www.mcdougallittell.com

ADDITIONAL RESOURCES

Another way to begin the chapter is to show the video clip of a real-life motivator for Chapter 7 on the *Instant Replay: Video Review Games.*

PROJECTS

A project covering Chapters 7–9 appears on pages 570–571 of the Student Edition. An additional project for Chapter 7 is available in the *Chapter 7 Resource Book,* p. 104.

TECHNOLOGY

Software
- *Electronic Teaching Tools*
- *Online Lesson Planner*
- *Personal Student Tutor*
- *Test and Practice Generator*
- *Electronic Lesson Presentations* (Lesson 7.5)

Video
- *Instant Replay: Video Review Games*

Internet Connections
www.mcdougallittell.com
- **Application Links**
  395, 407, 413, 415, 417, 419, 422, 423
- **Data Updates**
  402, 424
- **Student Help**
  399, 404, 409, 412, 430, 433
- **Career Links**
  400, 430, 434, 436
- **Extra Challenge**
  403, 410, 416, 424, 431, 437

PREPARE

DIAGNOSTIC TOOLS
The **Skill Review** exercises can help you diagnose whether students have the following skills needed in Chapter 7:

- Simplify expressions.
- Solve linear equations.
- Identify solutions to linear inequalities.

The following resources are available for students who need additional help with these skills:

- Prerequisite Skills Review (*Chapter 7 Resource Book,* p. 5; *Warm-Up Transparencies,* p. 52)
- Reteaching with Practice (Chapter Resource Books for Lessons 2.6, 3.4, 4.2, 6.5)
- *Personal Student Tutor*

ADDITIONAL RESOURCES
The following resources are provided to help you prepare for the upcoming chapter and customize review materials:

- ***Chapter 7 Resource Book***
  Tips for New Teachers (p. 1)
  Parent Guide (p. 3)
  Lesson Plans (every lesson)
  Lesson Plans for Block Scheduling (every lesson)
- ***Electronic Teaching Tools***
- ***Online Lesson Planner***
- ***Test and Practice Generator***

CHAPTER 7

Study Guide

PREVIEW

What's the chapter about?

Chapter 7 is about **systems of equations**. In Chapter 7, you'll learn

- three methods for solving a system of linear equations.
- to determine the number of solutions of a linear system.
- to graph and solve a system of linear inequalities.

KEY VOCABULARY

▶ Review
- coefficient, p. 102
- linear equation, p. 210
- ordered pair, p. 203
- parallel lines, p. 242

▶ New
- linear system, p. 398
- solution of a linear system, p. 398
- linear combination, p. 411
- system of linear inequalities, p. 432
- solution of a system of inequalities, p. 432
- graph of a system of inequalities, p. 432

PREPARE

Are you ready for the chapter?

SKILL REVIEW Do these exercises to review key skills that you'll apply in this chapter. See the given **reference page** if there is something you don't understand.

Simplify the expression. (Review Example 5, p. 102)

1. $13x + 8x$ **$21x$**
2. $9r + (-45r)$ **$-36r$**
3. $0.1d - 1.1d$ **$-d$**
4. $\frac{1}{2}w + \frac{3}{4}w$ **$\frac{5}{4}w$**
5. $-\frac{3}{7}g + \left(-\frac{1}{3}\right)g$ **$-\frac{16}{21}g$**
6. $-\frac{3}{10}y + \frac{7}{8}y$ **$\frac{23}{40}y$**

Solve the equation if possible. (Review Example 3 and Example 4, p. 155)

7. $2x + 6(x + 1) = -2$ **-1**
8. $5y + 8 = 5y$ **no solution**
9. $1 + 4x = 4x + 1$ **all real numbers**
10. $2(4y - 1) + 2y = 3$ **$\frac{1}{2}$**

Decide whether the given ordered pair is a solution of the equation or inequality. (Review Example 1, p. 210 and Example 1, p. 360)

11. $2x + y = 1, (1, -1)$ **solution**
12. $5x + 5y = 10, (2, 1)$ **not a solution**
13. $3y - x \geq 9, (3, -2)$ **not a solution**
14. $7y - 8x > 56, (112, 7)$ **not a solution**

STUDENT HELP

Study Tip
"Student Help" boxes throughout the chapter give you study tips and tell you where to look for extra help in this book and on the Internet.

STUDY STRATEGY

Here's a study strategy!

Solving Problems

It's important to sort out the useful information in a problem. Then you can organize the information and plan how to answer the question.

In your notebook, keep a list of different types of problems you find and how to solve them. Once you have a plan for finding a solution, you can calculate the answer.

ACTIVITY 7.1

Developing Concepts

Group Activity for use with Lesson 7.1

Investigating Graphs of Linear Systems

GROUP ACTIVITY
Work as a class.

MATERIALS
graph paper

▶ QUESTION **Can two different linear equations have a solution in common?**

In this activity, use the following linear equations.

$x - y = -2$ **Equation 1**
$x + y = 4$ **Equation 2**

▶ EXPLORING THE CONCEPT

1 Turn your classroom into a coordinate grid by arranging your desks into rows. Every position in your grid should be occupied.

2 Choose one student's position to represent the origin. Write down the ordered pair you represent. A sample classroom grid is shown.

3 Substitute the x-value and y-value that you represent into Equation 1. If you get a true statement, stand up.

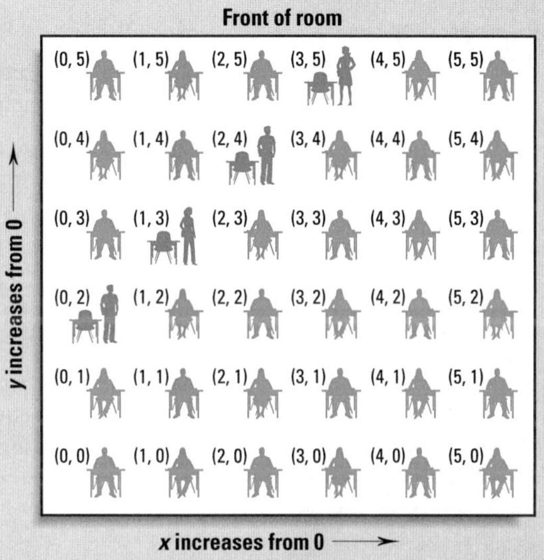

STUDENT HELP

Look Back
For help with checking solutions, see p. 210.

4 What do you notice about the positions of the students standing up? On graph paper, draw axes and plot a point for each student who is standing.

5 Have everyone sit down. Then repeat **Step 3** and **Step 4** using Equation 2.

▶ DRAWING CONCLUSIONS

1. Look at your graph and describe what you observe. Which student's ordered pair do you think will make both equations true? Is this the only ordered pair that is a solution of both equations? Explain.

1. All the students whose ordered pairs are solutions of Equation 1 are in a line, as are all students whose ordered pairs are solutions of Equation 2. The ordered pair belonging to the student in both lines is the only ordered pair that is a solution of both equations. If two lines intersect, they intersect in only one point, so there is no other solution of both equations.

In Exercises 2–5, use the class as a coordinate grid to find an ordered pair that is a solution of both equations.

2. $x - y = -1$ **(1, 2)**
$x + y = 3$

3. $x + y = 4$ **(3, 1)**
$-x + y = -2$

4. $-x + y = 0$ **(3, 3)**
$x + y = 6$

5. $x = 3$ **(3, 2)**
$y = 2$

6. *Writing* In this activity your class has been modeling the solutions of pairs of linear equations. What have you learned about the solutions of two linear equations whose graphs intersect at one point? **The solution of such a system is the ordered pair whose graph is the intersection of the graphs of the linear equations.**

1 Planning the Activity

PURPOSE
To model a system of linear equations using the classroom as a coordinate grid.

MATERIALS
- graph paper
- Activity Support Master (*Chapter 7 Resource Book,* p. 12)

PACING
- Exploring the Concept — 10 min
- Drawing Conclusions — 10 min

▶ *LINK TO LESSON*
When discussing Example 1 of Lesson 7.1, have students think about how these equations would be modeled by the class.

2 Managing the Activity

CLASSROOM MANAGEMENT
In Step 4, you can have the students standing up call out their coordinates for other students to record on their graph paper.

ALTERNATIVE APPROACH
This activity can be done by groups of four using large grid graph paper. Have each student test the points in one quadrant of the graph and mark those that satisfy the first equation using one color. They can then repeat this process for the second equation using a different color.

3 Closing the Activity

★ *KEY DISCOVERY*
The solution to a system of linear equations is the ordered pair for the point of intersection of their graphs.

ACTIVITY ASSESSMENT
Two lines intersect in a point. What do the coordinates of this point tell you? **They tell what ordered pair is a solution of both equations of the lines.**

1 PLAN

PACING
Basic: 1 day
Average: 1 day
Advanced: 1 day
Block Schedule: 0.5 block with 7.2

LESSON OPENER
VISUAL APPROACH
An alternative way to approach Lesson 7.1 is to use the Visual Approach Lesson Opener:

- Blackline Master (*Chapter 7 Resource Book,* p. 13)
- Transparency (p. 46)

MEETING INDIVIDUAL NEEDS

- ***Chapter 7 Resource Book***
  Prerequisite Skills Review (p. 5)
  Practice Level A (p. 15)
  Practice Level B (p. 16)
  Practice Level C (p. 17)
  Reteaching with Practice (p. 18)
  Absent Student Catch-Up (p. 20)
  Challenge (p. 22)
- ***Resources in Spanish***
- ***Personal Student Tutor***

NEW-TEACHER SUPPORT
See the Tips for New Teachers on pp. 1–2 of the *Chapter 7 Resource Book* for additional notes about Lesson 7.1.

WARM-UP EXERCISES

Transparency Available

Is the ordered pair a solution of the equation $2x - 3y = 5$?

1. (1, 0) no **2.** (−1, 1) no
3. (1, −1) yes **4.** (4, 1) yes

7.1 Solving Linear Systems by Graphing

What you should learn

GOAL 1 Solve a system of linear equations by graphing.

GOAL 2 Model a **real-life** problem using a linear system, such as predicting the number of visits at Internet sites in **Example 3**.

Why you should learn it

▼ To solve **real-life** problems, such as comparing the number of people who live inland to the number of people living on the coastline in **Exs. 37–39**.

REAL LIFE

GOAL 1 GRAPHING A LINEAR SYSTEM

In this chapter you will study *systems of linear equations* in two variables. Here are two equations that form a **system of linear equations** or simply a **linear system**.

$x + 2y = 5$ **Equation 1**
$2x - 3y = 3$ **Equation 2**

A **solution of a system of linear equations** in two variables is an ordered pair (x, y) that satisfies each equation in the system.

Because the solution of a linear system satisfies each equation in the system, the solution must lie on the graph of both equations. When the solution has integer values, it is possible to find the solution by graphical methods.

EXAMPLE 1 *Checking the Intersection Point*

Use the graph at the right to solve the system of linear equations. Then check your solution algebraically.

$3x + 2y = 4$ **Equation 1**
$-x + 3y = -5$ **Equation 2**

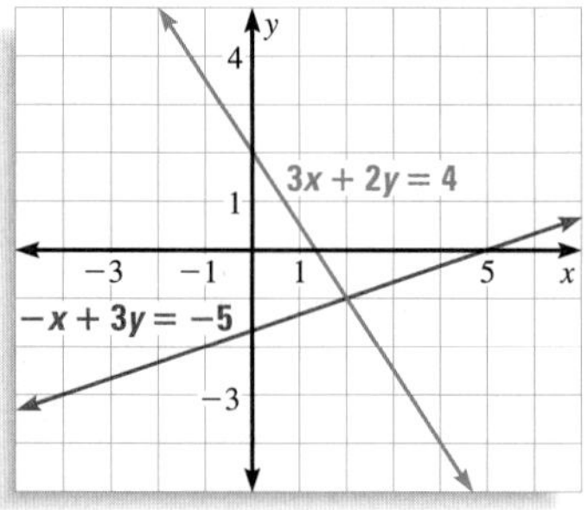

SOLUTION

The graph gives you a visual model of the solution.

The lines appear to intersect once at $(2, -1)$.

✓**CHECK** To check $(2, -1)$ as a solution algebraically, substitute 2 for x and -1 for y in each equation.

EQUATION 1	EQUATION 2
$3x + 2y = 4$	$-x + 3y = -5$
$3(2) + 2(-1) \stackrel{?}{=} 4$	$-(2) + 3(-1) \stackrel{?}{=} -5$
$6 - 2 \stackrel{?}{=} 4$	$-2 - 3 \stackrel{?}{=} -5$
$4 = 4$	$-5 = -5$

▶ Because $(2, -1)$ is a solution of each equation, $(2, -1)$ is the solution of the system of linear equations. Because the lines in the graph of this system intersect at only one point, $(2, -1)$ is the only solution of the linear system.

STUDENT HELP
Look Back
For help with checking solutions, see p. 210.

SOLVING A LINEAR SYSTEM USING GRAPH-AND-CHECK

To use the graph-and-check method to solve a system of linear equations in two variables, use the following steps.

STEP 1 Write each equation in a form that is easy to graph.

STEP 2 Graph both equations in the same coordinate plane.

STEP 3 Estimate the coordinates of the point of intersection.

STEP 4 Check the coordinates algebraically by substituting into each equation of the original linear system.

STUDENT HELP

HOMEWORK HELP Visit our Web site www.mcdougallittell.com for extra examples.

EXAMPLE 2 *Using the Graph-and-Check Method*

Solve the linear system graphically. Check the solution algebraically.

$x + y = -2$ **Equation 1**

$2x - 3y = -9$ **Equation 2**

SOLUTION

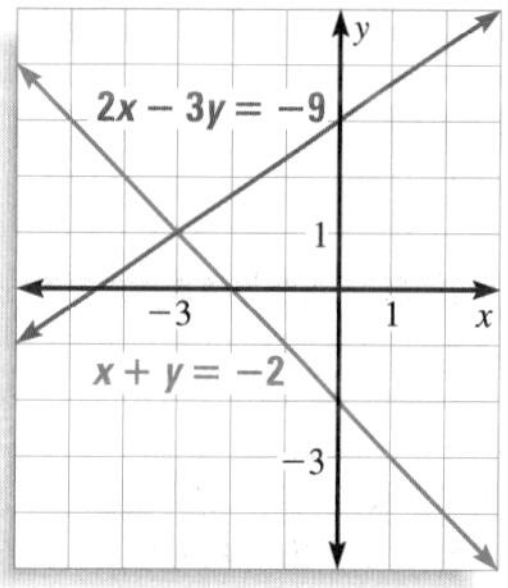

1. Write each equation in a form that is easy to graph, such as slope-intercept form.

   $y = -x - 2$ Slope: -1, y-intercept: -2

   $y = \frac{2}{3}x + 3$ Slope: $\frac{2}{3}$, y-intercept: 3

2. Graph these equations.

3. The two lines appear to intersect at $(-3, 1)$.

4. To check $(-3, 1)$ as a solution algebraically, substitute -3 for x and 1 for y in each original equation.

EQUATION 1

$x + y = -2$

$-3 + 1 \stackrel{?}{=} -2$

$-2 = -2$

EQUATION 2

$2x - 3y = -9$

$2(-3) - 3(1) \stackrel{?}{=} -9$

$-6 - 3 \stackrel{?}{=} -9$

$-9 = -9$

▶ Because $(-3, 1)$ is a solution of each equation in the linear system, it is a solution of the linear system.

....

In Example 2 you used the slope-intercept form of each equation in the system to graph the lines. However, there are other methods. For instance, in Lesson 4.3 you learned how to make a quick graph of a linear equation using intercepts.

In the Graphing Calculator Activity at the end of this lesson, you will learn how to graph a system of linear equations using a graphing calculator.

2 TEACH

MOTIVATING THE LESSON

Suppose you have won a prize where you must spend exactly \$1000 in 10 minutes on music cassettes. The store sells new releases for \$10.00 each and other cassettes for \$9.00 each. You want to know how many of each type you need to buy if you want a total of 104 cassettes. Explain that you can use a system of linear equations to solve this problem.

EXTRA EXAMPLE 1

Use the graph to solve the system of linear equations. Then check your solution algebraically.

$y = 3x - 12$

$y = -2x + 3$ **(3, −3)**

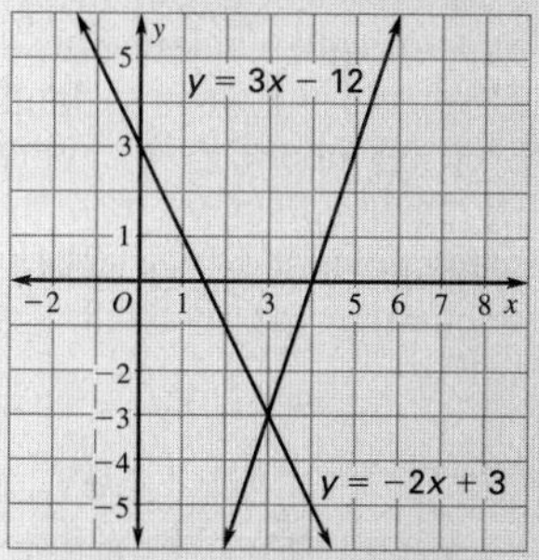

EXTRA EXAMPLE 2

Solve the linear system graphically. Check the solution algebraically.

$3x + y = 11$

$x - 2y = 6$ **(4, −1)**

CHECKPOINT EXERCISES

For use after Examples 1 and 2:

1. Solve the linear system graphically. Check the solution algebraically.

$-2x + y = 2$

$x + y = -1$ **(−1, 0)**

STUDENT HELP NOTES

Homework Help Students can find extra examples at **www.mcdougallittell.com** that parallel the examples in the student edition.

EXTRA EXAMPLE 3

You are an ecologist studying the population of two types of fish in a lake.

Type A: There are currently 5750 fish and they are decreasing at a rate of 250 per year.

Type B: There are currently 3500 fish and they are increasing at a rate of 500 per year.

Predict how many years it will take for there to be an equal number of the two types of fish.

3 years

CHECKPOINT EXERCISES

For use after Example 3:

1. Bryce and his sister Julia each have a savings account for college. In April, Bryce's account had a balance of $2000 and he plans to deposit an additional $50 per month. At the same time, Julia's account had a balance of $1600 and she plans to deposit $100 per month. Predict when Bryce and Julia will have the same amount in their college accounts. **in December**

FOCUS ON VOCABULARY

Suppose your friend gives you a system of linear equations and an ordered pair he claims is the *solution* to the system. What must be true about this ordered pair? **The ordered pair must satisfy each equation in the system.**

CLOSURE QUESTION

Explain how to use the graph-and-check method to solve a system of linear equations. **Graph each equation. Write the ordered pair for the point where the lines intersect. Check by substituting the coordinates in each equation.**

DAILY PUZZLER

Ask students to solve the situation presented in Motivating the Lesson. **64 new releases and 40 other cassettes**

GOAL 2 MODELING A REAL-LIFE PROBLEM

EXAMPLE 3 *Writing and Using a Linear System*

INTERNET In the fall, the math club and the science club each created an Internet site. You are the webmaster for both sites. It is now January and you are comparing the number of times each site is visited each day.

Science Club: There are currently 400 daily visits and the visits are increasing at a rate of 25 daily visits per month.

Math Club: There are currently 200 daily visits and the visits are increasing at a rate of 50 daily visits per month.

Predict when the number of visits at the two sites will be the same.

SOLUTION

PROBLEM SOLVING STRATEGY

VERBAL MODEL

Daily visits = Current visits to science site + Monthly increase (sci) · Number of months

Daily visits = Current visits to math site + Monthly increase (math) · Number of months

LABELS

Daily visits = V	(daily visits)
Current visits (science) = 400	(daily visits)
Increase (science) = 25	(daily visits per month)
Number of months = t	(months)
Current visits (math) = 200	(daily visits)
Increase (math) = 50	(daily visits per month)

ALGEBRAIC MODEL

$V = 400 + 25t$ **Equation 1 (science)**

$V = 200 + 50t$ **Equation 2 (math)**

FOCUS ON CAREERS

WEBMASTER Webmasters build Web sites for clients. They often update their technical skills to meet the demands for first-class Web sites. Some also help design Web pages and are responsible for updating the content.

INTERNET CAREER LINK www.mcdougallittell.com

Use the graph-and-check method to solve the system. The point of intersection of the two lines appears to be (8, 600). Check this solution in Equation 1 and in Equation 2.

$600 = 400 + 25(8)$

$600 = 200 + 50(8)$

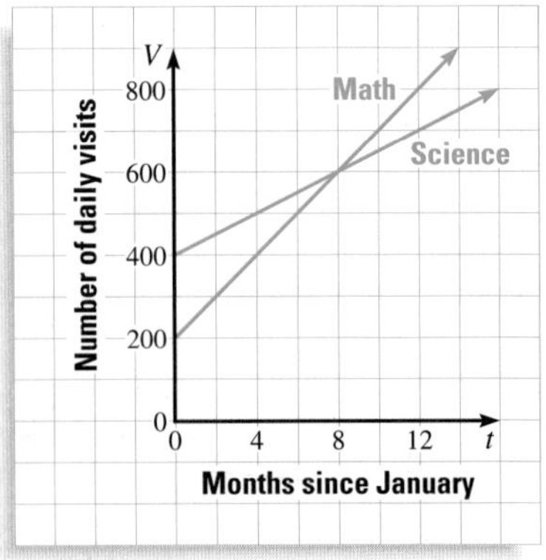

▶ If the monthly increases continue at the same rates, the sites will have the same number of visits by the eighth month after January, which is September.

Guided Practice

Vocabulary Check ✓

1. Explain what it means to solve a system of linear equations. **See margin.**

Concept Check ✓

2. Explain how to use the graph at the right to solve the system of linear equations. **Determine the coordinates of the point of intersection and substitute them into each equation. The intersection is (0, 2). Since 2 = −0 + 2 and 2 = 0 + 2, (0, 2) is the solution of the system.**

$y = -x + 2$
$y = x + 2$

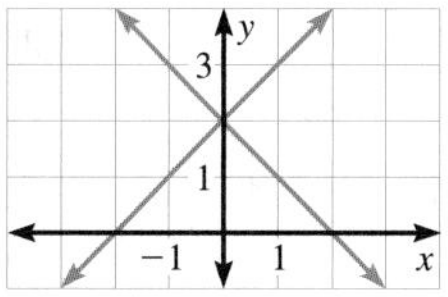

Ex. 2

1. to find an ordered pair that satisfies each of the equations in the system

Skill Check ✓

Graph the linear system below. Then decide if the ordered pair is a solution of the system. See margin for graph.

$-x + y = -2$
$2x + y = 10$

3. (−4, −2) **not a solution**
4. (4, −2) **not a solution**
5. (−4, 2) **not a solution**
6. (4, 2) **solution**
7. Confirm your answer for Exercise 3 algebraically. **See margin.**

7. −(−4) + (−2) = 2; 2(−4) + (−2) = −10; (−4, −2) is not a solution of either equation; −4 + (−2) = −6; 2(4) + (−2) = 6; (4, −2) is not a solution of either equation; −(−4) + 2 = 6; 2(−4) + 2 = −6; (−4, 2) is not a solution of either equation; −4 + 2 = −2; 2(4) + 2 = 10; (4, 2) is a solution of both equations, so it is a solution of the system.

Use the graph-and-check method to solve the system of linear equations.

8. $y = 2x - 1$ **(2, 3)**
$y = x + 1$

9. $y = -2x + 3$ **(2, −1)**
$y = x - 3$

10. $y = \frac{1}{2}x + 2$ **(2, 3)**
$y = -x + 5$

Practice and Applications

Student Help

Extra Practice to help you master skills is on p. 803.

Checking for Solutions **Decide whether the ordered pair is a solution of the system of linear equations.**

11. $3x - 2y = 11$, $-x + 6y = 7$; (5, 2) **solution**
12. $6x - 3y = -15$, $2x + y = -3$; (−2, 1) **solution**
13. $x + 3y = 15$, $4x + y = 6$; (3, −6) **not a solution**
14. $-5x + y = 19$, $x - 7y = 3$; (−4, −1) **solution**
15. $-15x + 7y = 1$, $3x - y = 1$; (3, 5) **not a solution**
16. $-2x + y = -11$, $-x - 9y = -15$; (6, 1) **solution**

Solving Systems Graphically **Use the graph to solve the linear system. Check your solution algebraically.**

17. $-x + 2y = 6$ **(4, 5)**
$x + 4y = 24$

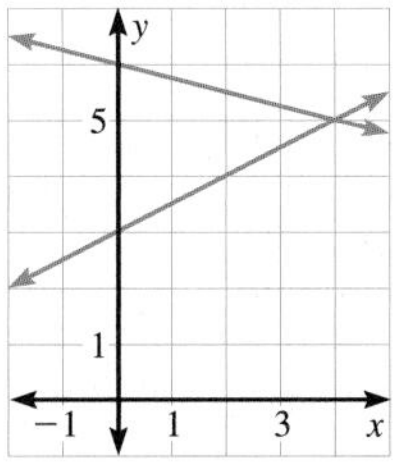

18. $2x - y = -2$ **(−2, −2)**
$4x - y = -6$

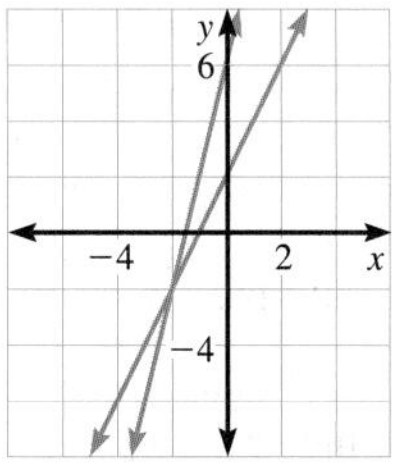

19. $x + y = 3$ **(3, 0)**
$-2x + y = -6$

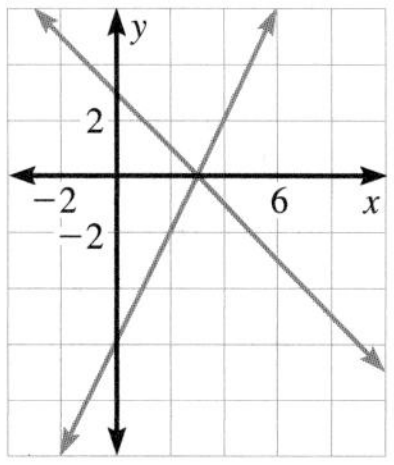

Student Help

Homework Help
Example 1: Exs. 11–19
Example 2: Exs. 20–34
Example 3: Exs. 35, 36

3 Apply

Assignment Guide

Basic
Day 1: pp. 401–403 Exs. 12–34 even, 35, 41, 42, 50, 52, 56, 59

Average
Day 1: pp. 401–403 Exs. 12–34 even, 35, 36, 41, 42, 50, 52, 56, 59

Advanced
Day 1: pp. 401–403 Exs. 12–34 even, 36–46, 50, 52, 56, 59

Block Schedule with 7.2
pp. 401–403 Exs. 12–34 even, 35, 36, 41, 42, 50, 52, 56, 59

Exercise Levels

Level A: *Easier*
11–16

Level B: *More Difficult*
17–36, 41, 42, 47–59

Level C: *Most Difficult*
37–40, 43–46

✓ Homework Check

To quickly check student understanding of key concepts, go over the following exercises: Exs. 16, 18, 26, 32, 35. See also the Daily Homework Quiz:

- Blackline Master (*Chapter 7 Resource Book,* p. 25)
- Transparency (p. 54)

3–6.

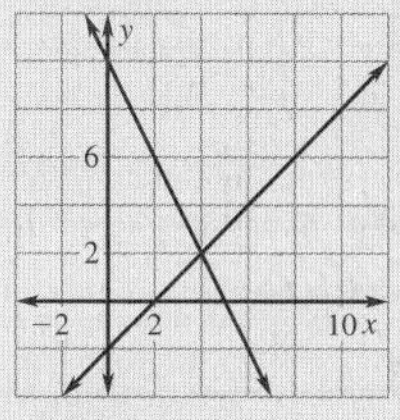

ENGLISH LEARNERS

EXERCISES 35–40 Although English language learners may understand the mathematical concepts presented in Chapter 7, many will have difficulty translating the challenging language of the word problems into the appropriate equations. Consider pairing students with English proficient partners for the duration of the chapter to help them understand unfamiliar vocabulary and interpret word problems.

APPLICATION NOTE

EXERCISES 37–39 This application provides an opportunity to discuss the limits of the graphing method for finding the solution to a system of linear equations. In this case, the intersection point has coordinates that are not lattice points on the graph and thus an exact solution is not possible using the graphing method. This application will lead nicely into the next lesson since students will realize that a method other than graphing is needed for some systems.

38. U.S. Population

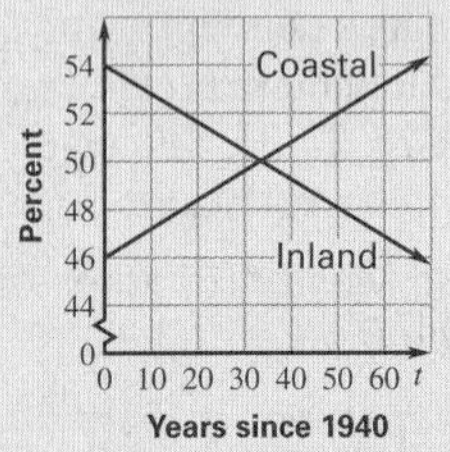

59.

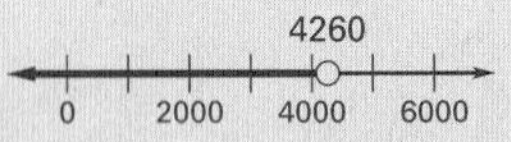

ADDITIONAL PRACTICE AND RETEACHING

For Lesson 7.1:

- Practice Levels A, B, and C (*Chapter 7 Resource Book,* p. 15)
- Reteaching with Practice (*Chapter 7 Resource Book,* p. 18)
- See Lesson 7.1 of the *Personal Student Tutor*

For more Mixed Review:

- Search the *Test and Practice Generator* for key words or specific lessons.

GRAPH AND CHECK **Graph and check to solve the linear system.**

20. $y = -x + 3$; $y = x + 1$ **(1, 2)**

21. $y = -6$; $x = 6$ **(6, −6)**

22. $y = 2x - 4$; $y = -\frac{1}{2}x + 1$ **(2, 0)**

23. $2x - 3y = 9$; $x = -3$ **(−3, −5)**

24. $5x + 4y = 16$; $y = -16$ **(16, −16)**

25. $x - y = 1$; $5x - 4y = 0$ **(−4, −5)**

26. $3x + 6y = 15$; $-2x + 3y = -3$ **(3, 1)**

27. $7y = -14x + 42$; $7y = 14x + 14$ **(1, 4)**

28. $0.5x + 0.6y = 5.4$; $-x + y = 9$ **(0, 9)**

29. $15x - 10y = -80$; $6x + 8y = -80$ **(−8, −4)**

30. $-3x + y = 10$; $7x + y = 20$ **(1, 13)**

31. $x - 8y = -40$; $-5x + 8y = 8$ **(8, 6)**

32. $\frac{1}{5}x + \frac{3}{5}y = \frac{12}{5}$; $-\frac{1}{5}x + \frac{3}{5}y = \frac{6}{5}$ **(3, 3)**

33. $\frac{3}{4}x - \frac{1}{4}y = -\frac{1}{2}$; $\frac{1}{4}x - \frac{3}{4}y = \frac{3}{2}$ $\left(-1\frac{1}{2}, -2\frac{1}{2}\right)$

34. $2.8x + 1.4y = 1.4$; $0.7x - 0.7y = 1.4$ **(1, −1)**

35. **COMPARING CARS** Car model X costs \$22,000 to purchase and an average of \$.12 per mile to maintain. Car model Y costs \$24,500 to purchase and an average of \$.10 per mile to maintain. Use the graph to determine how many miles must be driven for the total cost of the two models to be the same. **125,000 mi**

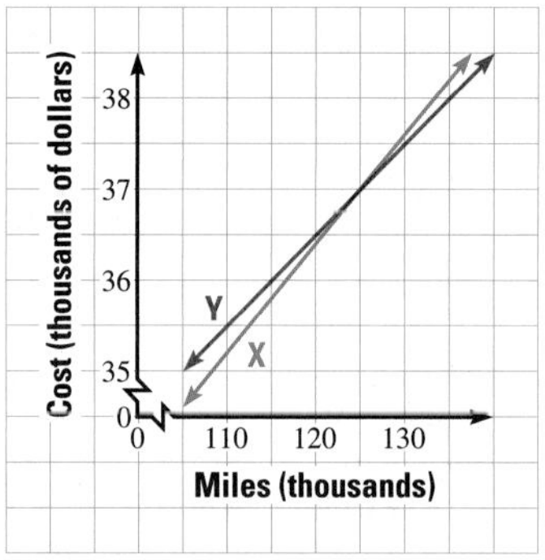

Ex. 35

36. **AEROBICS CLASSES** A fitness club offers two water aerobics classes. There are currently 40 people regularly going to the morning class, and attendance is increasing at a rate of 2 people per month. There are currently 22 people regularly going to the evening class, and attendance is increasing at a rate of 8 people per month. Predict when the number of people in each class will be the same. **in 3 months**

COASTAL POPULATION **In Exercises 37–39, use the table below, which gives the percents of people in the contiguous United States living within 50 miles of a coastal shoreline and those living further inland.**

Population in United States	1940	1997
Living within 50 miles of a coastal shoreline	46%	53%
Living farther inland	54%	47%

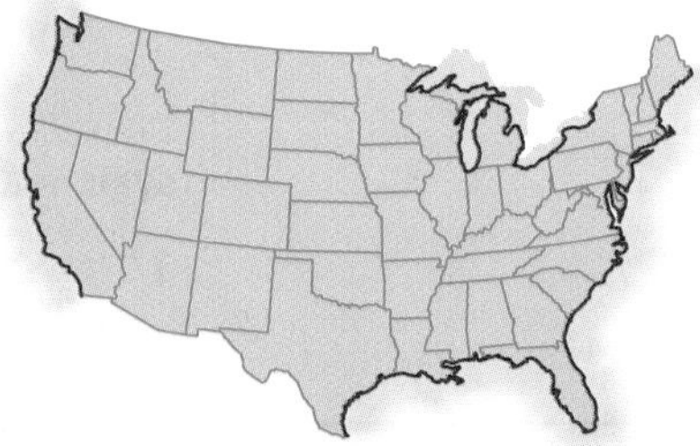

DATA UPDATE U.S. Bureau of the Census at www.mcdougallittell.com

37–39. Sample answers are given.

37. For each location, write a linear model to represent the percent at time t, where t represents the number of years since 1940. **coastal: $y = 0.12t + 46$; inland: $y = -0.12t + 54$**

38. Graph the linear equations you wrote. **See margin.**

39. From your graphs, estimate when the percent of people living near the coast equaled the percent living inland. **1973**

40. **LAUNDRY** You do 4 loads of laundry each week at a launderette where each load costs \$1.25. You could buy a washing machine that costs \$400. Washing 4 loads at home will cost about \$1 per week for electricity. How many loads of laundry must you do in order for the costs to be equal? **400 loads**

Test Preparation

41. **MULTIPLE CHOICE** Which ordered pair is a solution of the linear system? **C**

$x + y = 0.5$
$x + 2y = 1$

(A) (0, −2) (B) (−0.5, 0) (C) (0, 0.5) (D) (0, −0.5)

42. **MULTIPLE CHOICE** If $y - x = -3$ and $4x + y = 2$, then $x = \underline{?}$. **A**

(A) 1 (B) −1 (C) 5 (D) −5

★ Challenge

43. **SOLVE USING A SYSTEM** You know how to solve the equation $\frac{1}{2}x + 2 = \frac{3}{2}x - 12$ algebraically. This equation can also be solved graphically by solving the linear system.

$y = \frac{1}{2}x + 2$

$y = \frac{3}{2}x - 12$

a. Explain how the linear system is related to the original equation.

b. Solve the system graphically. **(14, 9)**

c. Check that the x-coordinate from part (b) satisfies the original equation $\frac{1}{2}x + 2 = \frac{3}{2}x - 12$ by substituting the x-coordinate for x. **See margin.**

43. a. The linear system represents the equations relating to both sides of the original equation. The x-coordinate of the intersection of the graphs is the solution of the original equation.

Use the method shown in Exercise 43 to solve the equation graphically.

44. $\frac{4}{5}x = 3x - 11$ **5**

45. $x + 1 = \frac{3}{2}x + 2$ **−2**

46. $1.2x - 2 = 3.4x - 13$ **5**

EXTRA CHALLENGE
www.mcdougallittell.com

MIXED REVIEW

43. c. $\frac{1}{2}(14) + 2 \stackrel{?}{=} \frac{3}{2}(14) - 12$; $9 = 9$; **the x-coordinate, 14, satisfies the original equation.**

58. $y = \frac{2}{3}x + \frac{17}{3}$

SOLVING EQUATIONS **Solve the equation.** (Review 3.3 for 7.2)

47. $3x + 7 = -2$ **−3**

48. $15 - 2a = 7$ **4**

49. $21 = 7(w - 2)$ **5**

50. $2y + 5 = 3y$ **5**

51. $2(z - 3) = 12$ **9**

52. $-(3 - x) = -7$ **−4**

WRITING EQUATIONS **Write an equation of the line that passes through the point and has the given slope. Use slope-intercept form.** (Review 5.2)

53. (3, 0), $m = -4$ **$y = -4x + 12$**

54. (−4, 3), $m = 1$ **$y = x + 7$**

55. (1, −5), $m = 4$ **$y = 4x - 9$**

56. (−4, −1), $m = -2$ **$y = -2x - 9$**

57. (2, 3), $m = 2$ **$y = 2x - 1$**

58. (−1, 5), $m = \frac{2}{3}$ **See margin.**

Verrazano-Narrows Bridge

59. **SUSPENSION BRIDGES** The Verrazano-Narrows Bridge in New York City is the longest suspension bridge in North America, with a main span of 4260 feet. Write an inequality that describes the length in feet x of every other suspension bridge in North America. Graph the inequality on a number line. (Review 6.1) **$x < 4260$; See margin for graph.**

4 ASSESS

DAILY HOMEWORK QUIZ

Transparency Available

1. Decide whether (−3, 5) is a solution of the linear system. $2x - 5y = -31$; $-3x + y = 14$ **yes**

2. Use the graph to solve the linear system. $y = 2x + 3$; $3x + 2y = -1$ **(−1, 1)**

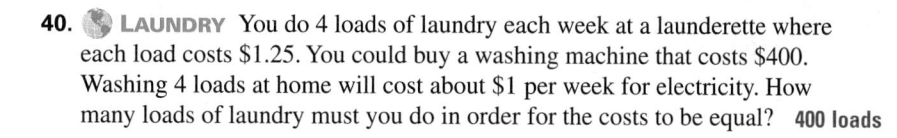

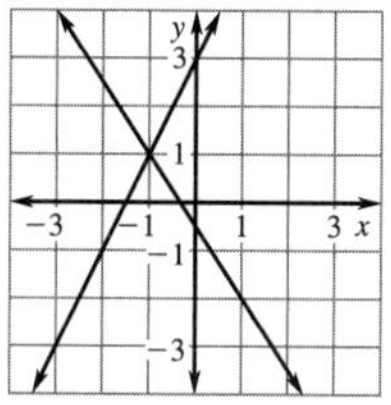

Graph and check to solve the linear system. **Check students' graphs.**

3. $2x + 4y = -2$
$-x + 2y = -7$ **(3, −2)**

4. $\frac{3}{7}x + \frac{1}{7}y = -\frac{5}{7}$
$-\frac{2}{7}x + \frac{3}{7}y = -\frac{4}{7}$ **(−1, −2)**

5. The Smith family made an \$800 downpayment and pays \$75 a month for new furniture. At the same time, the Cooper family made a \$500 downpayment and pays \$95 a month for its new furniture. Use a graph to determine how many months it will be before the amounts they have paid are equal. **15 months**

EXTRA CHALLENGE NOTE

Challenge problems for Lesson 7.1 are available in **blackline** format in the *Chapter 7 Resource Book*, p. 22 and at **www.mcdougallittell.com.**

ADDITIONAL TEST PREPARATION

1. **OPEN ENDED** Write a system of linear equations that has (2, −1) as its solution.
Sample answer:
$x + 2y = 0$
$-3x + y = -7$

1 Planning the Activity

PURPOSE
To familiarize students with the procedure of using a graphing calculator to solve a system of equations.

MATERIALS
- graphing calculator
- Keystroke blackline (*Chapter 7 Resource Book*, p. 14)

PACING
- Exploring the Concept — 10 min
- Drawing Conclusions — 10 min

LINK TO LESSON
Have students use graphing calculators to check their solutions to Exercise 39 on page 402.

2 Managing the Activity

COOPERATIVE LEARNING
Students can complete this activity in pairs. Have one student enter the equations and find the intersection while the other reads the equations and records the intersection.

ALTERNATIVE APPROACH
This activity can be done as a demonstration by using an overhead projection device.

3 Closing the Activity

★ ***KEY DISCOVERY***
Graphing calculators can be used to find an approximation to the solution of a system of equations.

ACTIVITY ASSESSMENT
Write a series of steps to follow when using a graphing calculator to solve this system of equations.

$2x - y = 5$
$x - y = -6$

Sample answer: **Step 1: Solve the equations for *y*. Step 2: Enter the equations in the calculator. Step 3: Adjust the viewing window. Step 4: Use the intersect feature to find the coordinates of the solution.**

ACTIVITY 7.1
Using Technology

Solving Linear Systems by Graphing

STUDENT HELP
INTERNET KEYSTROKE HELP
See keystrokes for several models of calculators at www.mcdougallittell.com

▶ EXAMPLE

Solve the linear system by graphing.

$y = -0.3x + 1.8$ **Equation 1**
$y = 0.6x - 1.5$ **Equation 2**

▶ SOLUTION

1. Enter the equations.

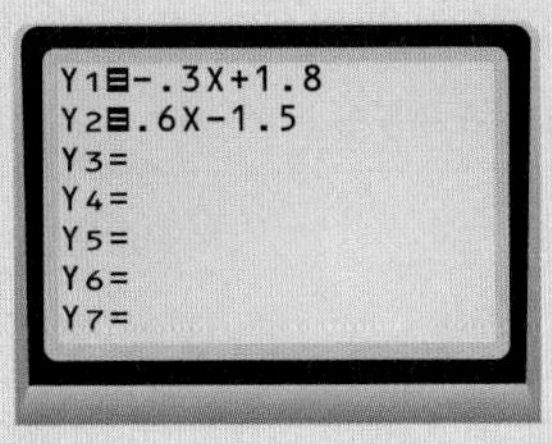

2. Set an appropriate viewing window to graph both equations.

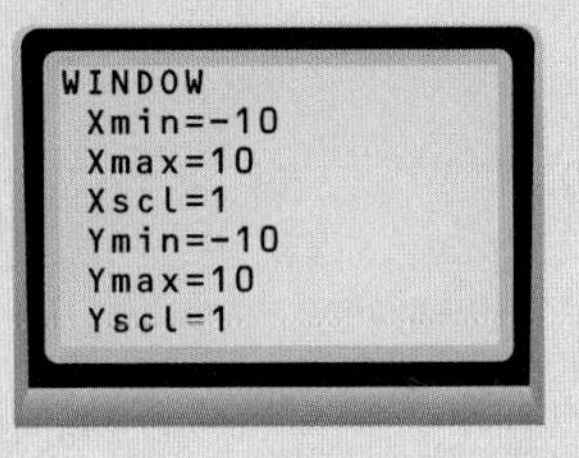

3. Graph both equations. You can use the direction keys to move the cursor to the approximate intersection point.

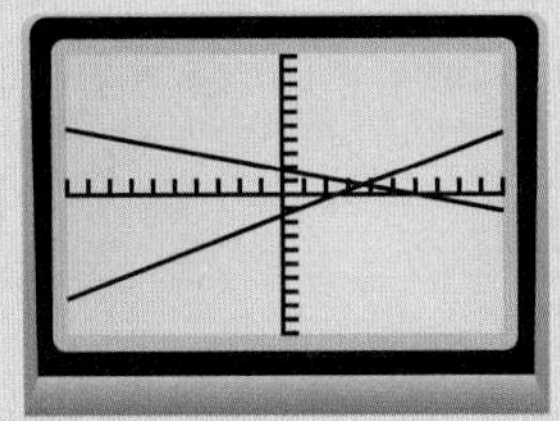

4. Use the *Intersection* feature to estimate a point where the graphs intersect. Follow your calculator's procedure to display the coordinate values.

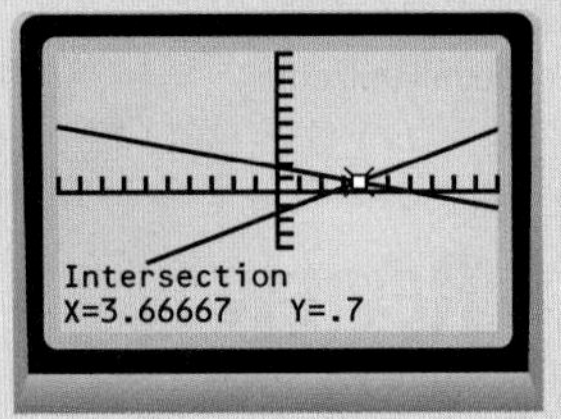

▶ The solution of the system of linear equations is approximately (3.7, 0.7).

▶ EXERCISES

In Exercises 1–4, solve the linear system. Check the result in each of the original equations.

1. $y = x + 6$
$y = -x - 1$ **(−3.5, 2.5)**

2. $3x + y = -2$
$x - y = -8$ $\left(-\frac{5}{2}, \frac{11}{2}\right)$

3. $-0.25x - y = 2.25$
$-1.25x + 1.25y = -1.25$ **(−1, −2)**

4. $-0.8x + 0.6y = -12.0$
$1.25x - 1.50y = 12.75$ **approximately (23, 10.7)**

5. Graph the linear system at the right. $3x + 9y = 8$, $2x + 6y = 7$ Describe the lines, and explain why the linear system has no solution. **The lines are parallel. Since the lines do not intersect, the system has no solution.**

7.2 Solving Linear Systems by Substitution

What you should learn

GOAL 1 Use substitution to solve a linear system.

GOAL 2 Model a **real-life** situation using a linear system, such as the average number of visitors to a Museum in **Example 3**.

Why you should learn it

▼ To solve **real-life** problems such as finding a distance run in **Exs. 46** and **47**.

REAL LIFE

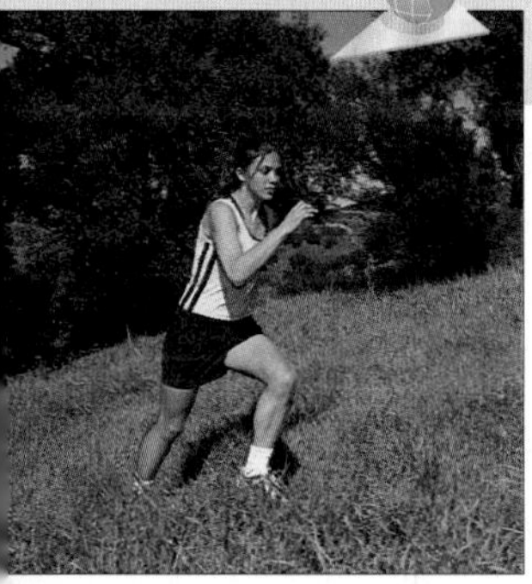

GOAL 1 USING SUBSTITUTION

In this lesson you will study an algebraic method for solving a linear system.

SOLVING A LINEAR SYSTEM BY SUBSTITUTION

STEP 1 Solve one of the equations for one of its variables.

STEP 2 Substitute the expression from Step 1 into the other equation and solve for the other variable.

STEP 3 Substitute the value from Step 2 into the revised equation from Step 1 and solve.

STEP 4 Check the solution in each of the original equations.

EXAMPLE 1 *The Substitution Method*

Solve the linear system.

$-x + y = 1$ **Equation 1**

$2x + y = -2$ **Equation 2**

SOLUTION

1. Solve for y in Equation 1.

$y = x + 1$ **Revised Equation 1**

2. Substitute $x + 1$ for y in Equation 2 and solve for x.

$2x + y = -2$	Write Equation 2.
$2x + (x + 1) = -2$	Substitute $x + 1$ for y.
$3x + 1 = -2$	Simplify.
$3x = -3$	Subtract 1 from each side.
$x = -1$	Solve for x.

3. To find the value of y, substitute -1 for x in the revised Equation 1.

$y = x + 1$	Write revised Equation 1.
$y = -1 + 1$	Substitute -1 for x.
$y = 0$	Solve for y.

4. Check that $(-1, 0)$ is a solution by substituting -1 for x and 0 for y in each of the original equations.

▶ The solution is $(-1, 0)$.

1 PLAN

PACING

Basic: 1 day
Average: 1 day
Advanced: 1 day
Block Schedule: 0.5 block with 7.1

LESSON OPENER ACTIVITY

An alternative way to approach Lesson 7.2 is to use the Activity Lesson Opener:

- Blackline Master (*Chapter 7 Resource Book*, p. 26)
- Transparency (p. 47)

MEETING INDIVIDUAL NEEDS

- ***Chapter 7 Resource Book***
  Prerequisite Skills Review (p. 5)
  Practice Level A (p. 28)
  Practice Level B (p. 29)
  Practice Level C (p. 30)
  Reteaching with Practice (p. 31)
  Absent Student Catch-Up (p. 33)
  Challenge (p. 35)
- ***Resources in Spanish***
- ***Personal Student Tutor***

NEW-TEACHER SUPPORT

See the Tips for New Teachers on pp. 1–2 of the *Chapter 7 Resource Book* for additional notes about Lesson 7.2.

WARM-UP EXERCISES

Transparency Available

Is $(-3, 2)$ a solution of the equation?

1. $3x - y = 4$ no
2. $x + 2y = 1$ yes
3. Is $(4, 0)$ a solution of the system of equations?
$2x + y = 8$
$-x + 3y = 4$ no

Solve each equation for y.

4. $2x + y = 12$ $y = -2x + 12$
5. $-3x - 2y = -2$ $y = -\frac{3}{2}x + 1$

2 TEACH

MOTIVATING THE LESSON

Tell students that on a standardized test, 5 points are awarded for each correct answer and 1 point is subtracted for each incorrect answer. Problems left unanswered are given a score of zero. If there are 120 questions on the test and you get a score of 345, how many questions did you leave blank? Explain that this is the type of question that can be solved using a system of linear equations. The focus of today's lesson is solving systems of linear equations using substitution.

EXTRA EXAMPLE 1

Solve the linear system.
$3x + y = 5$
$2x - y = 10$ (3, –4)

EXTRA EXAMPLE 2

Solve the linear system.
$2x + 6y = 15$
$x = 2y$ $\left(3, \frac{3}{2}\right)$

CHECKPOINT EXERCISES

For use after Examples 1 and 2:

1. Solve the linear system.
$x + 2y = 4$
$-x + y = -7$ (6, –1)

ENGLISH LEARNERS

To help English language learners understand and remember the mathematical concept of *substitution,* discuss the non-mathematical meanings of the word. You may want to give the example of a substitute teacher to help them understand the general concept. Then relate the idea of substitution to the algebraic context.

MULTIPLE REPRESENTATIONS

EXAMPLE 2 Help students make the connection between the graphical and algebraic methods of solving a system of equations by first writing both equations in terms of *y*.

STUDENT HELP

Study Tip
When using substitution, you will get the same solution whether you solve for *y* first or *x* first. You should begin by solving for the variable that is easier to isolate.

EXAMPLE 2 ***The Substitution Method***

Solve the linear system.

$2x + 2y = 3$ **Equation 1**
$x - 4y = -1$ **Equation 2**

SOLUTION

Solve for x in Equation 2 because it is easy to isolate x.

$x = 4y - 1$ **Revised Equation 2**

Substitute $4y - 1$ for x in Equation 1 and solve for y.

$2x + 2y = 3$	Write Equation 1.
$2(4y - 1) + 2y = 3$	Substitute $4y - 1$ for x.
$8y - 2 + 2y = 3$	Distribute the 2.
$10y - 2 = 3$	Simplify.
$10y = 5$	Add 2 to each side.
$y = \frac{1}{2}$	Solve for y.

Substitute $\frac{1}{2}$ for y in the revised Equation 2 to find the value of x.

$x = 4y - 1$	Write revised Equation 2.
$x = 4\left(\frac{1}{2}\right) - 1$	Substitute $\frac{1}{2}$ for y.
$x = 1$	Solve for x.

Check by substituting 1 for x and $\frac{1}{2}$ for y in each of the original equations.

EQUATION 1	EQUATION 2
$2x + 2y = 3$	$x - 4y = -1$
$2(1) + 2\left(\frac{1}{2}\right) \stackrel{?}{=} 3$	$1 - 4\left(\frac{1}{2}\right) \stackrel{?}{=} -1$
$2 + 1 \stackrel{?}{=} 3$	$1 - 2 \stackrel{?}{=} -1$
$3 = 3$	$-1 = -1$

▶ The solution is $\left(1, \frac{1}{2}\right)$.

. . . .

When you use the substitution method, you can still use a graph to check the reasonableness of your solution. For instance, the graph at the right shows a graphic check for Example 2.

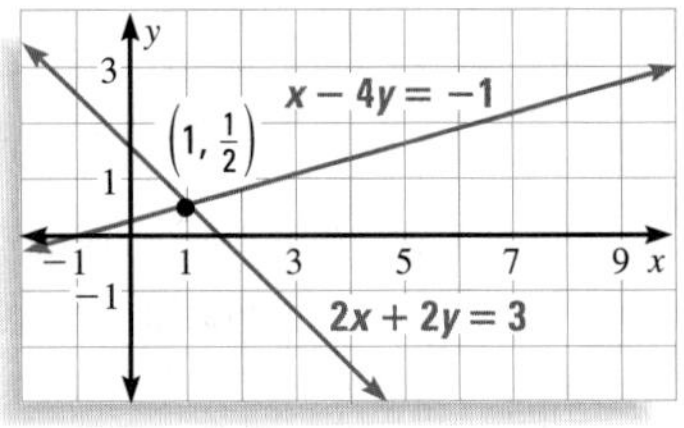

GOAL 2 MODELING A REAL-LIFE PROBLEM

EXAMPLE 3 *Writing and Using a Linear System*

MUSEUM ADMISSIONS In one day the National Civil Rights Museum in Memphis, Tennessee, collected \$1590 from 321 people admitted to the museum. The price of each adult admission is \$6. People with the ages of 4–17 pay the child admission, \$4. Estimate how many adults and how many children were admitted that day.

SOLUTION

Use a verbal model to find the number of adults and children admitted to the museum that day.

PROBLEM SOLVING STRATEGY

VERBAL MODEL

[Number of adults] + [Number of children] = [Total number admitted]

[Number of adults] · [Price of adult admission] + [Number of children] · [Price of child admission] = [Total amount collected]

LABELS

Number of adults	= x	(people)
Number of children	= y	(people)
Total number admitted	= **321**	(people)
Price of adult admission	= **6**	(dollars per person)
Price of child admission	= **4**	(dollars per person)
Total amount collected	= **1590**	(dollars)

ALGEBRAIC MODEL

$x + y = 321$ **Equation 1 (Number admitted)**

$6x + 4y = 1590$ **Equation 2 (Amount collected)**

Using the substitution method, you can determine that $x = 153$ when $y = 168$. The solution is the ordered pair (153, 168).

The graphs of the two equations appear to intersect at (153, 168). The solution checks graphically.

▶ You can conclude that 153 adults were admitted and 168 children were admitted to the National Civil Rights Museum that day.

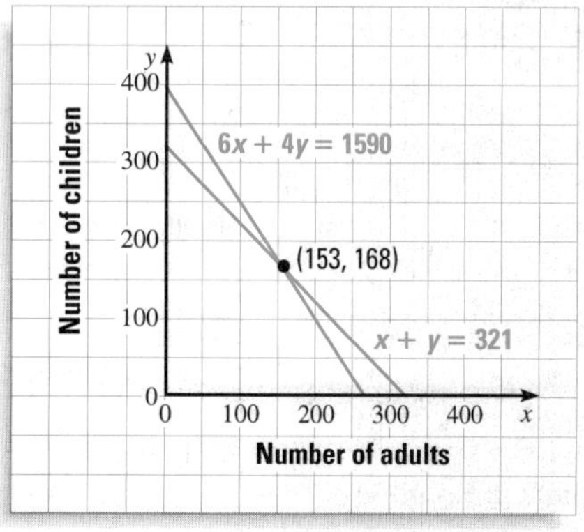

FOCUS ON APPLICATIONS

NATIONAL CIVIL RIGHTS MUSEUM
The National Civil Rights Museum educates people about the history of the civil rights movement through its unique collections and powerful exhibits.

APPLICATION LINK
www.mcdougallittell.com

Closure Question ***Sample answer:***
Solve one of the equations for one of the variables. Replace the variable by the resulting expression in the other equation. Solve the equation you obtain to find the value of one of the variables. Substitute this value back into the revised first equation to find the value of the other variable. Check your solution in both of the original equations.

EXTRA EXAMPLE 3
An office supply company sells two types of fax machines. They charge \$150 for one of the machines and \$225 for the other. If the company sold 22 fax machines for a total of \$3900 last month, how many of each type were sold?
14 of the \$150 machines and 8 of the \$225 machines

CHECKPOINT EXERCISES
For use after Example 3:

1. A quilt maker sews both large and small quilts. A large quilt requires 8 yards of fabric while the small quilt requires 3 yards. How many of each size quilt did she make if she used a total of 90 yards of fabric to make 15 quilts?
6 small quilts and 9 large quilts

APPLICATION NOTE
EXAMPLE 3 Additional information about the National Civil Rights Museum is available at **www.mcdougallittell.com.**

FOCUS ON VOCABULARY
Why is the method for solving linear systems used in this lesson called the *substitution method*?
Sample answer: **You solve one of the equations for one of the variables and then *substitute* the resulting expression for the variable in the other equation.**

CLOSURE QUESTION
Describe in your own words how to solve a system of linear equations using substitution. **See margin.**

DAILY PUZZLER
For a certain 3-digit number, interchanging the first and last digits gives a 3-digit number 297 less than the original number. Doubling the first and last digits of the original number gives a 3-digit number and increases the sum of the digits to 15. What is the number? **451**

3 APPLY

ASSIGNMENT GUIDE

BASIC
Day 1: pp. 408–410 Exs. 14–44 even, 35, 48–51, 55, 60, 65, 68

AVERAGE
Day 1: pp. 408–410 Exs. 14–44 even, 35, 43, 48–51, 55, 60, 65, 68

ADVANCED
Day 1: pp. 408–410 Exs. 14–44 even, 35, 43, 48–53, 55, 60, 65, 68

BLOCK SCHEDULE WITH 7.1
pp. 408–410 Exs. 14–44 even, 35, 43, 48–51, 55, 60, 65, 68

EXERCISE LEVELS

Level A: *Easier*
14–16, 54–57

Level B: *More Difficult*
17–45, 48–50, 58–68

Level C: *Most Difficult*
46, 47, 51–53

✔ HOMEWORK CHECK

To quickly check student understanding of key concepts, go over the following exercises: Exs. 16, 20, 26, 30, 36, 42. See also the Daily Homework Quiz:

- Blackline Master (*Chapter 7 Resource Book,* p. 38)
- Transparency (p. 55)

14. Answers may vary. *Sample answer:* I would isolate *y* in the second equation because it is easy to do so and the value is easy to substitute in the other equation.
15. Answers may vary. *Sample answer:* I would isolate *m* in the second equation because it is easy to do so and the value is easy to substitute in the other equation.
16. See Additional Answers beginning on page AA1.

GUIDED PRACTICE

Concept Check ✔

In Exercises 1–5, use the linear system below.

$-x + y = 5$ **Equation 1**

$\frac{1}{2}x + y = 8$ **Equation 2**

1. Which equation would you choose to solve for y? Why?
2. Solve for y in the equation that you chose. **See margin.**
3. Substitute the expression into the other equation and solve for x. **See margin.**
4. Substitute the value of x into your equation from Exercise 2. What is the solution of the linear system? **(2, 7)**
5. Explain how you can check the solution algebraically and graphically. **See margin.**

1. You may choose either equation; it is equally simple to isolate *y* in either equation.

2. Equation 1: $y = x + 5$; Equation 2: $y = -\frac{1}{2}x + 8$

3. Equation 2: $\frac{1}{2}x + (x + 5) = 8$, $x = 2$; Equation 1: $-x + \left(-\frac{1}{2}x + 8\right) = 5$

Skill Check ✔

Use substitution to solve the linear system.

6. $3x + y = 3$, $7x + 2y = 1$ **(−5, 18)**
7. $2x - y = -1$, $2x + y = -7$ **(−2, −3)**
8. $3x - y = 0$, $5y = 15$ **(1, 3)**
9. $2x + y = 4$, $-x + y = 1$ **(1, 2)**
10. $x - y = 0$, $x + y = 2$ **(1, 1)**
11. $x + y = 1$, $2x - y = 2$ **(1, 0)**
12. $-x + 4y = 10$, $x - 3y = 11$ **(74, 21)**
13. $x + y = 1$, $x - y = 2$ **$\left(1\frac{1}{2}, -\frac{1}{2}\right)$**

PRACTICE AND APPLICATIONS

STUDENT HELP
▸ Extra Practice to help you master skills is on p. 803.

REASONING **Tell which equation you would use to isolate a variable. Explain your reasoning.** **14–16. Sample answers are given. See margin.**

14. $2x + y = -10$, $3x - y = 0$
15. $m + 4n = 30$, $m - 2n = 0$
16. $5c + 3d = 11$, $5c - d = 5$

5. To check the solution algebraically, substitute 2 for *x* and 7 for *y* in each of the original equations. To check graphically, graph both equations and check that the graphs intersect at (2, 7).

28. $\left(-3\frac{1}{2}, -6\frac{1}{2}\right)$

SOLVING LINEAR SYSTEMS **Use the substitution method to solve the linear system.**

17. $y = x - 4$, $4x + y = 26$ **(6, 2)**
18. $s = t + 4$, $2t + s = 19$ **(9, 5)**
19. $2c - d = -2$, $4c + d = 20$ **(3, 8)**
20. $2a = 8$, $a + b = 2$ **(4, −2)**
21. $2x + 3y = 31$, $y = x + 7$ **(2, 9)**
22. $p + q = 4$, $4p + q = 1$ **(−1, 5)**
23. $x - 2y = -25$, $3x - y = 0$ **(5, 15)**
24. $u - v = 0$, $7u + v = 0$ **(0, 0)**
25. $x - y = 0$, $12x - 5y = -21$ **(−3, −3)**
26. $m + 2n = 1$, $5m + 3n = -23$ **(−7, 4)**
27. $x - y = -5$, $x + 4 = 16$ **(12, 17)**
28. $-3a + b = 4$, $-9a + 5b = -1$ **See margin.**
29. $3w - 2u = 12$, $w - u = 60$ **(−168, −108)**
30. $y = 3x$, $x = 3y$ **(0, 0)**
31. $x + y = 5$, $0.5x + 6.0y = 8.0$ **(4, 1)**
32. $x + y = 12$, $x + \frac{3}{2}y = \frac{3}{2}$ **(33, −21)**
33. $7g + h = -2$, $g - 2h = 9$ **$\left(\frac{1}{3}, -4\frac{1}{3}\right)$**
34. $\frac{1}{8}p + \frac{3}{4}q = 7$, $\frac{3}{2}p - q = 4$ **(8, 8)**

STUDENT HELP
▸ HOMEWORK HELP
Example 1: Exs. 14–34
Example 2: Exs. 14–34
Example 3: Exs. 42–44

35. **The variables are eliminated, leaving a statement that is always true. You solved for one variable in one equation and substituted for that variable in the *same* equation; substitute $y = -3x + 9$ into Equation 2 to get (1, 6) for the solution.**

35. ERROR ANALYSIS You are helping a friend with tonight's math homework. Answer your friend's questions.

$3x + y = 9$ ← Equation 1
$-2x + y = 4$ ← Equation 2

$3x + y = 9$
$y = -3x + 9$

$3x + (-3x + 9) = 9$
$3x - 3x + 9 = 9$
$9 = 9$

What does this mean?

How do I find the answer?

FOCUS ON APPLICATIONS

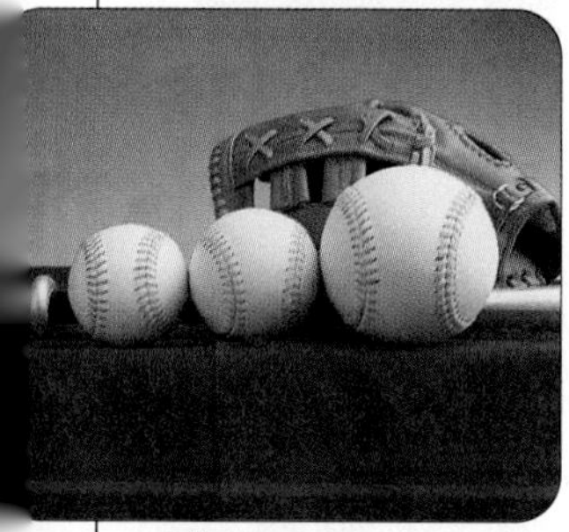

SOFTBALL SIZES Softballs are measured by their circumference. There are three official softball sizes: 11 inches, 12 inches, and 16 inches.

SOLVING LINEAR SYSTEMS **In Exercises 36–41, use substitution to solve the linear system. Then use a graphing calculator or a computer to check your solution.**

36. $x - y = 2$
$2x + y = 1$ **(1, −1)**

37. $2y = x$
$4y = 300 - x$ **(100, 50)**

38. $x - 2y = 9$
$1.5x + 0.5y = 6.5$ **(5, −2)**

39. $0.50x + 0.25y = 2.00$
$x + y = 1$ **(7, −6)**

40. $x + y = 20$
$\frac{1}{5}x + \frac{1}{2}y = 8$ $\left(6\frac{2}{3}, 13\frac{1}{3}\right)$

41. $1.5x - y = 40.0$
$0.5x + 0.5y = 10.0$ **(24, −4)**

42. TICKET SALES You are selling tickets for a high school play. Student tickets cost \$4 and general admission tickets cost \$6. You sell 525 tickets and collect \$2876. How many of each type of ticket did you sell? **137 student tickets, 388 general admission tickets**

43. ORDERING SOFTBALLS You are ordering softballs for two softball leagues. The Pony League uses an 11-inch softball priced at \$2.75. The Junior League uses a 12-inch softball priced at \$3.25. The bill smeared in the rain, but you know the total was 80 softballs for \$245. How many of each size did you order? **30 11-in. softballs, 50 12-in. softballs**

44. MATH TEST Your math teacher tells you that next week's test is worth 100 points and contains 38 problems. Each problem is worth either 5 points or 2 points. Because you are studying systems of linear equations, your teacher says that for extra credit you can figure out how many problems of each value are on the test. How many of each value are there? **8 5-point questions, 30 2-point questions**

45. INVESTING IN STOCKS The value of your EFG stock is three times the value of your PQR stock. If the total value of the stocks is \$4500, how much is invested in each company? **\$3375 in EFG, \$1125 in PQR**

RUNNING **In Exercises 46 and 47, you can run 250 meters per minute downhill and 180 meters per minute uphill. One day you run a total of 1557 meters in 7.6 minutes on a route that goes both uphill and downhill.**

46. **Let u = meters uphill and d = meters downhill; $u + d = 1557$; $\frac{u}{180} + \frac{d}{250} = 7.6$.**

46. Assign labels to the verbal model below. Then write an algebraic model. **See margin.**

Meters uphill + Meters downhill = Total meters

$\frac{\text{Meters uphill}}{\text{Rate uphill}} + \frac{\text{Meters downhill}}{\text{Rate downhill}} = \text{Total time}$

47. Find the number of meters you ran uphill and the number of meters you ran downhill. **882 m uphill, 675 m downhill**

STUDENT HELP

HOMEWORK HELP Visit our Web site www.mcdougallittell.com for help with problem solving in Exs. 46 and 47.

! COMMON ERROR

EXERCISES 6–34 After finding the value of one variable, students often forget to substitute again to find the value of the second variable. Students should make a habit of checking the solution to avoid this problem.

MATHEMATICAL REASONING

In Ex. 35, 9 = 9 suggests that the system has infinitely many solutions. (Students study such systems in Lesson 7.5.) This system, however, has only one solution, since the graphs of the equations are intersecting lines. So there must have been a mistake in the procedure used. Brief inspection shows that $-3x + 9$ was substituted for y in the wrong equation.

GRAPHING CALCULATOR NOTE

EXERCISES 36–41 Remind students to solve both equations for y prior to entering them into a graphing calculator or computer. Students may need to review the procedure for using a graphing calculator to solve a system of linear equations in Activity 7.1 on p. 404.

STUDENT HELP NOTES

Homework Help Students can find help for Exs. 46 and 47 at **www.mcdougallittell.com.** The information can be printed out for students who don't have access to the Internet.

ADDITIONAL PRACTICE AND RETEACHING

For Lesson 7.2:
- Practice Levels A, B, and C (*Chapter 7 Resource Book,* p. 28)
- Reteaching with Practice (*Chapter 7 Resource Book,* p. 31)
- See Lesson 7.2 of the *Personal Student Tutor*

For more Mixed Review:
- Search the *Test and Practice Generator* for key words or specific lessons.

4 ASSESS

DAILY HOMEWORK QUIZ

Transparency Available

Solve the linear system.

1. $-3x + 2y = -11$
   $5x - y = 23$ **(5, 2)**
2. $a + 4b = 8$
   $2a + 3b = 1$ **(−4, 3)**
3. $5p - 4q = -3$
   $2p - q = -3$ **(−3, −3)**
4. $m = 7n$
   $3m + 8n = -29$ **(−7, −1)**
5. A sightseeing boat charges $5 for children and $8 for adults. On its first trip of the day, it collected $439 for 71 paying passengers. How many children and how many adults were there? **43 children, 28 adults**

EXTRA CHALLENGE NOTE

Challenge problems for Lesson 7.2 are available in **blackline** format in the *Chapter 7 Resource Book*, p. 35 and at **www.mcdougallittell.com.**

ADDITIONAL TEST PREPARATION

1. **OPEN ENDED** Write a system of equations that you would solve using substitution. Solve the system, explaining your work at each step.
   Check students' work.
2. **WRITING** Solve the system of equations below using both the graphing and substitution methods. Which method works better for this problem? Explain.
   $x + 5y = 8$
   $y = 3x$
   See margin.

48, 58–68. See Additional Answers beginning on page AA1.

Test Preparation

48. The graphs of the equations intersect at (0, 3), which means that (0, 3) is the solution of the system.

49. (0, 3); the point (0, 3) is a solution of each equation in the linear system.

★ Challenge

50. *Sample answer:* substitution

51. The graphing method provides a visual interpretation of the solution, but if the coordinates of the solution are not integers, they may be difficult to estimate without a graphing calculator. The substitution method yields an exact solution and may be quicker and simpler in some cases (as in the system in Exs. 48–51), but may involve complicated arithmetic in others.

EXTRA CHALLENGE

www.mcdougallittell.com

MULTI-STEP PROBLEM In Exercises 48–51, use the linear system below.

$y = x + 3$
$y = 2x + 3$

48. Graph the system. Explain what the graph shows. **See margin for graph.**

49. Solve the linear system using substitution. What does the solution mean? **See margin.**

50. Which method do you think is easier for solving this linear system? **See margin.**

51. *Writing* Describe the advantages and disadvantages of each method. **See margin.**

HISTORY CONNECTION In Exercises 52 and 53, use the following information. On May 1, 1976, a team of adventurers set sail to discover how the ancient Polynesians regularly navigated the 3000-nautical-mile route shown at the right. They sailed from Hawaii to Tahiti on a traditional twin-hulled canoe called the *Hokule'a.*

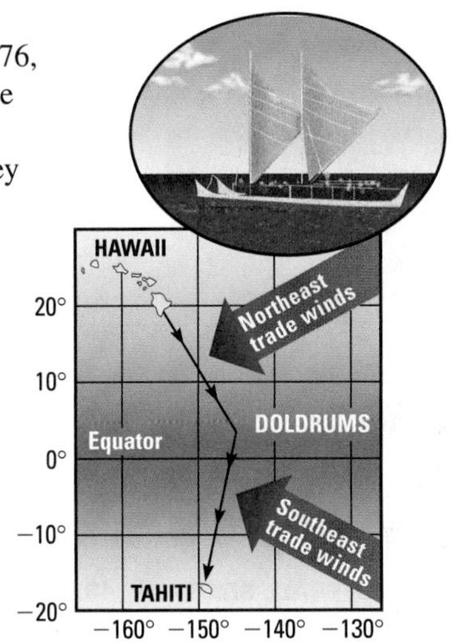

Sailing into northeast trade winds, the crew maintained a course represented by

$$y = -\frac{3}{2}x - 215.$$

Sailing into southeast trade winds, the crew maintained a course represented by

$$y = 7x + 1026.$$

At the point of intersection, the team became caught in the "doldrums" and made little headway for 5 to 6 days.

52. Find the coordinates of the point where the team was caught in the doldrums. **(−146°, 4°)**

53. The equation of the straight line passing through Hawaii and Tahiti is $36x + 7y = -5483$. Find the coordinates of Hawaii and Tahiti. **Hawaii: (−156°, 19°), Tahiti: (−149°, −17°)**

MIXED REVIEW

SIMPLIFYING EXPRESSIONS Simplify the variable expression. (Review 2.6 for 7.3)

54. $4g + 3h + 2g - 3h$ **6g**

55. $3x + 2y - (5x + 2y)$ **−2x**

56. $6(2p - m) - 3m - 12p$ **−9m**

57. $4(3x + 5y) + 3(-4x + 2y)$ **26y**

GRAPHING LINES Write the equation in slope-intercept form. Then graph the equation. (Review 4.6) **58–63. See margin for graphs.**

58. $6x + y = 0$ **$y = -6x$**

59. $8x - 4y + 16 = 0$ **$y = 2x + 4$**

60. $3x + y + 5 = 0$ **$y = -3x - 5$**

61. $5x + 3y = 3$ **61. $y = -\frac{5}{3}x + 1$**

62. $x + y = 0$ **$y = -x$**

63. $y = -2$ **$y = -2$**

SOLVING AND GRAPHING Solve the inequality. Then graph its solution. (Review 6.3, 6.4) **64–68. See margin for graphs.**

64. $-5 < -x \le -1$ **$1 \le x < 5$**

65. $|x + 5| \le 14$ **$-19 \le x \le 9$**

66. $3 > -x > -1$ **$-3 < x < 1$**

67. $2x - 6 < -7$ or $2x - 6 > 5$ **67. $x < -\frac{1}{2}$ or $x > 5\frac{1}{2}$**

68. $3x - 2 > 4$ or $3x - 2 < -5$ **$x > 2$ or $x < -1$**

Additional Test Preparation *Sample answers:*

2. Substitution works better, since one of the equations is already solved for *y*. Also, the coordinates of the solution are not integers, so they would be hard to read from the graph.

7.3 Solving Linear Systems by Linear Combinations

What you should learn

GOAL 1 Use linear combinations to solve a system of linear equations.

GOAL 2 Model a **real-life** problem using a system of linear equations, such as the mixture problem in **Example 4**.

Why you should learn it

▼ To solve **real-life** problems such as finding the speed of the current in a river in **Ex. 48**.

GOAL 1 USING LINEAR COMBINATIONS

Sometimes when you want to solve a linear system, it is not easy to isolate one of the variables. In that case, you can solve the system by *linear combinations*. A **linear combination** of two equations is an equation obtained by adding one of the equations (or a multiple of one of the equations) to the other equation.

SOLVING A LINEAR SYSTEM BY LINEAR COMBINATIONS

STEP 1 Arrange the equations with like terms in columns.

STEP 2 Multiply one or both of the equations by a number to obtain coefficients that are opposites for one of the variables.

STEP 3 Add the equations from Step 2. Combining like terms will eliminate one variable. Solve for the remaining variable.

STEP 4 Substitute the value obtained in Step 3 into either of the original equations and solve for the other variable.

STEP 5 Check the solution in each of the original equations.

EXAMPLE 1 *Using Addition*

Solve the linear system. $4x + 3y = 16$ **Equation 1**
$2x - 3y = 8$ **Equation 2**

SOLUTION

1 The equations are already arranged.

2 The coefficients for y are already opposites.

3 Add the equations to get an equation in one variable.

$4x + 3y = 16$ **Write Equation 1.**
$2x - 3y = 8$ **Write Equation 2.**
$6x = 24$ **Add equations.**
$x = 4$ **Solve for *x*.**

4 Substitute 4 for x in the first equation and solve for y.

$4(4) + 3y = 16$ **Substitute 4 for *x*.**
$y = 0$ **Solve for *y*.**

5 Check by substituting 4 for x and 0 for y in each of the original equations.

▶ The solution is (4, 0).

1 PLAN

PACING
Basic: 2 days
Average: 2 days
Advanced: 2 days
Block Schedule: 1 block

LESSON OPENER APPLICATION
An alternative way to approach Lesson 7.3 is to use the Application Lesson Opener:
- Blackline Master (*Chapter 7 Resource Book*, p. 39)
- Transparency (p. 48)

MEETING INDIVIDUAL NEEDS
- ***Chapter 7 Resource Book***
  Prerequisite Skills Review (p. 5)
  Practice Level A (p. 40)
  Practice Level B (p. 41)
  Practice Level C (p. 42)
  Reteaching with Practice (p. 43)
  Absent Student Catch-Up (p. 45)
  Challenge (p. 48)
- ***Resources in Spanish***
- ***Personal Student Tutor***

NEW-TEACHER SUPPORT
See the Tips for New Teachers on pp. 1–2 of the *Chapter 7 Resource Book* for additional notes about Lesson 7.3.

WARM-UP EXERCISES
Transparency Available
Give the opposite of each number.
1. 3 −3 **2.** 1 −1
3. $-\frac{1}{2}$ $\frac{1}{2}$
Simplify.
4. $3(2x - 7y)$ $6x - 21y$
5. $-2(5x + y)$ $-10x - 2y$

2 TEACH

MOTIVATING THE LESSON
Tell students that copiers with a limited number of settings can be used to get reduced copies that one might think are not possible. For example, a copier that will make a copy 120% of the original size or 75% of the original size can be used to get a copy that is 90% of the original size. Simply combine a reduction with an enlargement. Similarly, one can often find the value of a variable in a system of equations by working not with just one equation of the system but with a combination of both of the equations.

EXTRA EXAMPLE 1
Solve the linear system.
$-x+2y=-8$
$x+6y=-16$ (2, –3)

EXTRA EXAMPLE 2
Solve the linear system.
$5x-4y=3$
$2x+8y=-2$ $\left(\frac{1}{3}, -\frac{1}{3}\right)$

EXTRA EXAMPLE 3
Solve the linear system.
$3x=-6y+12$
$-x+3y=6$ (0, 2)

CHECKPOINT EXERCISES
For use after Example 1:
1. Solve the linear system.
$11x+3y=1$
$-5x-3y=-7$ (–1, 4)

For use after Examples 2 and 3:
2. Solve the linear system.
$2u=4v+8$
$3v=5u-13$ (2, –1)

STUDENT HELP NOTES
Homework Help Students can find extra examples at **www.mcdougallittell.com** that parallel the examples in the student edition.

STUDENT HELP
Study Tip
Sometimes you can add the original equations because the coefficients are already opposites as in Example 1. In Example 2 you need to multiply both equations by appropriate numbers first.

EXAMPLE 2 *Using Multiplication First*

Solve the linear system. $3x + 5y = 6$ **Equation 1**
$-4x + 2y = 5$ **Equation 2**

SOLUTION

The equations are already arranged. You can get the coefficients of x to be opposites by multiplying the first equation by 4 and the second equation by 3.

$$\begin{aligned} 3x + 5y &= 6 \quad \text{Multiply by 4.} & 12x + 20y &= 24 \\ -4x + 2y &= 5 \quad \text{Multiply by 3.} & -12x + 6y &= 15 \\ & & 26y &= 39 \quad \text{Add equations.} \\ & & y &= 1.5 \quad \text{Solve for } y. \end{aligned}$$

Substitute 1.5 for y in the second equation and solve for x.

$$\begin{aligned} -4x + 2y &= 5 && \text{Write Equation 2.} \\ -4x + 2(1.5) &= 5 && \text{Substitute 1.5 for } y. \\ -4x + 3 &= 5 && \text{Simplify.} \\ x &= -0.5 && \text{Solve for } x. \end{aligned}$$

▶ The solution is (–0.5, 1.5). Check this in the original equations.

EXAMPLE 3 *Arranging Like Terms in Columns*

Solve the linear system. $3x + 2y = 8$ **Equation 1**
$2y = 12 - 5x$ **Equation 2**

SOLUTION

First arrange the equations.

$$\begin{aligned} 3x + 2y &= 8 && \text{Write Equation 1.} \\ 5x + 2y &= 12 && \text{Rearrange Equation 2.} \end{aligned}$$

You can get the coefficients of y to be opposites by multiplying the second equation by -1.

$$\begin{aligned} 3x + 2y &= 8 & & 3x + 2y = 8 \\ 5x + 2y &= 12 \quad \text{Multiply by } -1. & & -5x - 2y = -12 \\ & & & -2x = -4 \quad \text{Add equations.} \\ & & & x = 2 \quad \text{Solve for } x. \end{aligned}$$

Substitute 2 for x into the first equation and solve for y.

$$\begin{aligned} 3x + 2y &= 8 && \text{Write Equation 1.} \\ 3(2) + 2y &= 8 && \text{Substitute 2 for } x. \\ 6 + 2y &= 8 && \text{Simplify.} \\ y &= 1 && \text{Solve for } y. \end{aligned}$$

▶ The solution is (2, 1). Check this in the original equations.

STUDENT HELP
INTERNET
HOMEWORK HELP
Visit our Web site www.mcdougallittell.com for extra examples.

GOAL 2 MODELING A REAL-LIFE SITUATION

Legend has it that Archimedes used the relationship between the weight of an object and its volume to prove that his king's gold crown was not pure gold. If the crown was all gold, it would have had the same volume as an equal amount of gold.

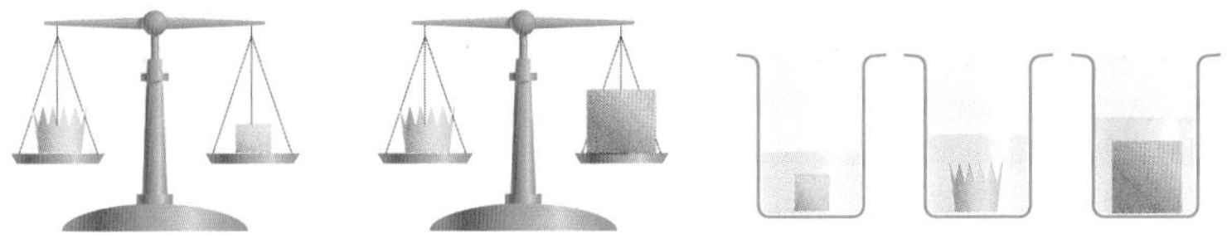

Gold block has same weight as crown. | Silver block has same weight as crown. | Gold block, crown, and silver block displace different amounts of water.

EXAMPLE 4 *Writing and Using a Linear System*

GOLD AND SILVER A gold crown, suspected of containing some silver, was found to have a weight of 714 grams and a volume of 46 cubic centimeters. The density of gold is about 19 grams per cubic centimeter. The density of silver is about 10.5 grams per cubic centimeter. What percent of the crown is silver?

SOLUTION

VERBAL MODEL

Gold volume + Silver volume = Total volume

Gold density · Gold volume + Silver density · Silver volume = Total weight

LABELS

Volume of gold = G	(cubic centimeters)
Volume of silver = S	(cubic centimeters)
Total volume = 46	(cubic centimeters)
Density of gold = 19	(grams per cubic centimeter)
Density of silver = 10.5	(grams per cubic centimeter)
Total weight = 714	(grams)

ALGEBRAIC MODEL

$G + S = 46$ **Equation 1**

$19G + 10.5S = 714$ **Equation 2**

Use linear combinations to solve for S.

$-19G - 19S = -874$ Multiply Equation 1 by -19.

$19G + 10.5S = 714$ Write Equation 2.

$-8.5S = -160$ Add equations.

▶ The volume of silver is about 19 cm^3. The crown has a volume of 46 cm^3, so the crown is $\frac{19}{46} \approx 41\%$ silver by volume.

FOCUS ON PEOPLE

ARCHIMEDES In the third century B.C., Archimedes supposedly used the relationship between the weight of an object and its volume to uncover a fraud in the manufacture of a gold crown.

APPLICATION LINK www.mcdougallittell.com

EXTRA EXAMPLE 4
A landscape designer uses two types of gravel for walkways. The larger gravel weighs an average of 40 pounds per cubic foot and the smaller weighs an average of 48 pounds per cubic foot. In his truck, the designer has a 12-cubic foot load of gravel that weighs 520 pounds. Is there a mixture of small and large gravel in the truck? If so, how many cubic feet of each type of gravel are in the truck?
yes; 7 ft^3 of large and 5 ft^3 of small

CHECKPOINT EXERCISES
For use after Example 4:

1. A toy maker produces wooden trains and wooden airplanes. Each train requires 3 ounces of paint and each airplane requires 5 ounces of paint. The toy maker has a gallon can of paint (64 ounces). If he wants to use it to paint 14 toys, how many of each can he paint?
3 trains and 11 airplanes

APPLICATION NOTE
EXAMPLE 4 Additional information about Archimedes is available at **www.mcdougallittell.com.**

FOCUS ON VOCABULARY
Tell students that the word *combine* often means to bring together some quantity of one item with some quantity of another. In the case of linear combinations, you are combining, or adding, a multiple of one equation with a multiple of another.

CLOSURE QUESTION
When solving a linear system using combinations, how do you determine the numbers used to multiply each equation? **You choose the numbers so that after multiplying, the coefficients of one of the variables are opposites.**

3 APPLY

ASSIGNMENT GUIDE

BASIC

Day 1: p. 414 Exs. 8–30 even

Day 2: pp. 415–417 Exs. 32–42 even, 43, 44, 49–51, 53–55, 60, 65, 70, Quiz 1 Exs. 1–10

AVERAGE

Day 1: p. 414 Exs. 8–30 even

Day 2: pp. 415–417 Exs. 32–42 even, 43–51, 53–55, 60, 65, 70, Quiz 1 Exs. 1–10

ADVANCED

Day 1: p. 414 Exs. 8–30 even

Day 2: pp. 415–417 Exs. 32–42 even, 43–56, 60, 65, 70, Quiz 1 Exs. 1–10

BLOCK SCHEDULE

pp. 414–417 Exs. 8–42 even, 43–51, 53–55, Quiz 1 Exs. 1–10

EXERCISE LEVELS

Level A: ***Easier***
8–15

Level B: ***More Difficult***
16–44, 49–51, 53–55, 57–72

Level C: ***Most Difficult***
45–48, 52, 56

✔ HOMEWORK CHECK

To quickly check student understanding of key concepts, go over the following exercises: Exs. 10, 18, 24, 28, 34, 42, 44, 50. See also the Daily Homework Quiz:

- Blackline Master (*Chapter 7 Resource Book,* p. 52)
- Transparency (p. 56)

GUIDED PRACTICE

Vocabulary Check ✔

1. When you use the linear combinations method to solve a linear system, what is the purpose of using multiplication as the first step? **to obtain coefficients that are opposites for one of the variables, so that variable can be eliminated by addition**

Concept Check ✔

ERROR ANALYSIS **Find and describe the error. Then correctly solve the linear system by using linear combinations.**

2.

$x + y = 1 \rightarrow 10x + 10y = 10$

$5x + 4y = 14 \rightarrow -10x + 8y = -28$

$18y = -18$

$y = -1$

$-2(5x + 4y) = -10x - 8y$; (10, −9)

3.

$3x + y = 24 \rightarrow 9x + 3y = 24$

$7x - 3y = 8 \rightarrow 7x - 3y = 8$

$2x = 32$

$x = 16$

See margin.

3. The right side of the first equation was not multiplied by 3; (5, 9)

4. Add the equations, solve for *x*, then substitute in either original equation to solve for *y*; (9, −1).

Skill Check ✔

Explain the steps you would use to solve the system of equations using linear combinations. Then solve the system. **4–7. Sample answers are given. See margin.**

4. $x + 3y = 6$
$x - 3y = 12$

5. $x - 3y = 0$
$x + 10y = 13$

6. $3x - 4y = 7$
$2x - y = 3$

7. $2y = -2 + 2x$
$2x + 3y = 12$

5. Multiply either equation by −1, solve for *y*, then substitute in either original equation to solve for *x*; (3, 1).

PRACTICE AND APPLICATIONS

STUDENT HELP

Extra Practice to help you master skills is on p. 803.

6. Multiply the second equation by −4, solve for *x*, then substitute in the second original equation to solve for *y*; (1, −1).

7. Subtract 2*x* from each side of the first equation, solve for y, then substitute in either original equation to solve for *x*; (3, 2).

STUDENT HELP

HOMEWORK HELP
Example 1: Exs. 8–15
Example 2: Exs. 16–24, 31–42
Example 3: Exs. 25–42
Example 4: Ex. 43

USING ADDITION **Use linear combinations to solve the system of linear equations.**

8. $2x + y = 4$
$x - y = 2$ **(2, 0)**

9. $a - b = 8$
$a + b = 20$ **(14, 6)**

10. $y - 2x = 0$
$6y + 2x = 0$ **(0, 0)**

11. $m + 3n = 2$
$-m + 2n = 3$ **(−1, 1)**

12. $p + 4q = 23$
$-p + q = 2$ **(3, 5)**

13. $3v - 2w = 1$
$2v + 2w = 4$ **(1, 1)**

14. $\frac{1}{2}g + h = 2$
$-g - h = 2$ **(−8, 6)**

15. $6.5x - 2.5y = 4.0$
$1.5x + 2.5y = 4.0$ **(1, 1)**

USING MULTIPLICATION FIRST **Use linear combinations to solve the system of linear equations.**

16. $x - y = 0$
$-3x - y = 2$ **$\left(-\frac{1}{2}, -\frac{1}{2}\right)$**

17. $v - w = -5$
$v + 2w = 4$ **(−2, 3)**

18. $x + 3y = 3$
$x + 6y = 3$ **(3, 0)**

19. $2g - 3h = 0$
$3g - 2h = 5$ **(3, 2)**

20. $2p - q = 2$
$2p + 3q = 22$ **$\left(3\frac{1}{2}, 5\right)$**

21. $2a + 6z = 4$
$3a - 7z = 6$ **(2, 0)**

22. $5e + 4f = 9$
$4e + 5f = 9$ **(1, 1)**

23. $10m + 16n = 140$
$5m - 8n = 60$ **$\left(13, \frac{5}{8}\right)$**

24. $9x - 3z = 20$
$3x + 6z = 2$ **$\left(2, -\frac{2}{3}\right)$**

ARRANGING LIKE TERMS **Use linear combinations to solve the system of linear equations.**

25. $x + 3y = 12$
$-3y + x = 30$ **(21, −3)**

26. $3b + 2c = 46$
$5c + b = 11$ **(16, −1)**

27. $y = x - 9$
$x + 8y = 0$ **(8, −1)**

28. $2q = 7 - 5p$
$4p - 16 = q$ **(3, −4)**

29. $2v = 150 - u$
$2u = 150 - v$ **(50, 50)**

30. $0.1g - h + 4.3 = 0$
$3.6 = 0.2g + h$ **(7, 5)**

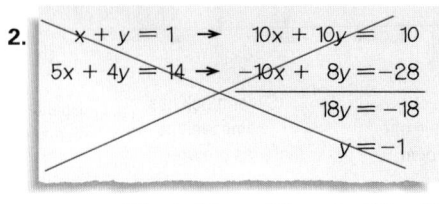

STUDENT HELP

Look Back
For help with simplifying expressions, see p. 102.

SOLVING LINEAR SYSTEMS In Exercises 31–42, use linear combinations to solve the system of linear equations.

31. $x + 2y = 5$, $5x - y = 3$ **(1, 2)**

32. $3p - 2 = -q$, $-q + 2p = 3$ **(1, −1)**

33. $3g - 24 = -4h$, $-2 + 2h = g$ **(4, 3)**

34. $t + r = 1$, $2r - t = 2$ **(1, 0)**

35. $x + 1 - 3y = 0$, $2x = 7 - 3y$ **(2, 1)**

36. $3a + 9b = 8b - a$, $5a - 10b = 4a - 9b + 5$ **(1, −4)**

37. $2m - 4 = 4n$, $m - 2 = n$ **(2, 0)**

38. $3y = -5x + 15$, $-y = -3x + 9$ **(3, 0)**

39. $3j + 5k = 19$, $4j - 8k = -4$ **(3, 2)**

40. $1.5v - 6.5w = 3.5$, $0.5v + 2w = -3$ **(−2, −1)**

41. $5y - 20 = -4x$, $y = -\frac{5}{4}x + 4$ **(0, 4)**

42. $9g - 7h = \frac{2}{3}$, $3g + h = \frac{1}{3}$ $\left(\frac{1}{10}, \frac{1}{30}\right)$

43. WEIGHT OF GOLD A gold and copper bracelet weighs 238 grams. The volume of the bracelet is 15 cubic centimeters. Gold weighs 19.3 grams per cubic centimeter, and copper weighs 9 grams per cubic centimeter. How many grams of copper are mixed with the gold? **5 g**

44. HONEY BEE PATHS A farmer is tracking two wild honey bees in his field. He maps the first bee's path back to the hive on the line $y = \frac{9}{7}x$. The second bee's path follows the line $y = -3x + 12$. Their paths cross at the hive. At what coordinates will the farmer find the hive? $\left(\frac{14}{5}, \frac{18}{5}\right)$

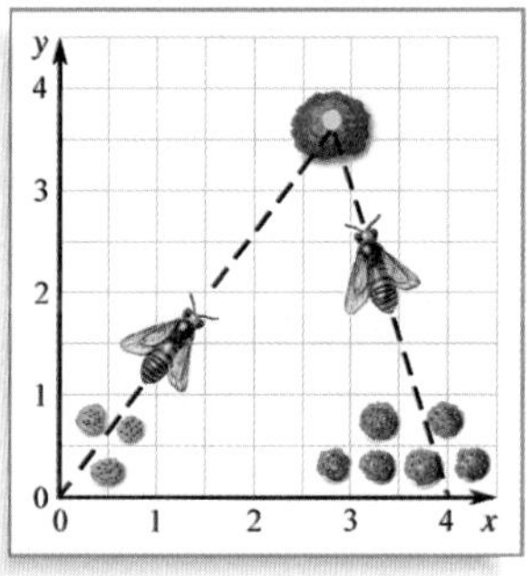

FOCUS ON APPLICATIONS

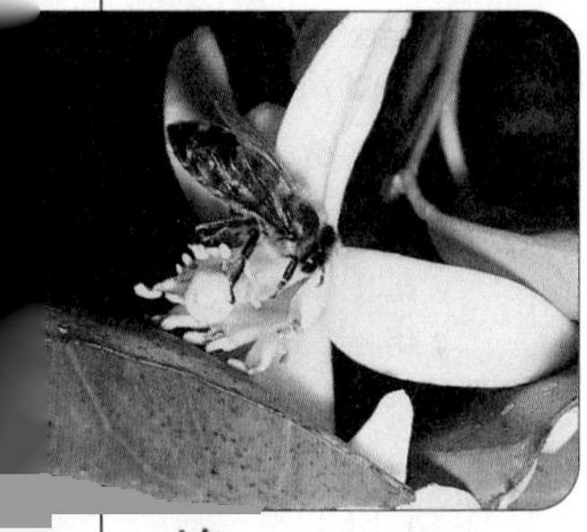

HONEY BEES In 1990, unusually cold weather and predators destroyed 90% of the wild honey bee colonies in the United States. If a colony is destroyed, farmers may need to rent bee hives to pollinate their crops.

APPLICATION LINK
www.mcdougallittell.com

AIRPLANE SPEED **In Exercises 45–47, use the following information.**
It took 3 hours for a plane, flying against the wind, to travel 900 miles from Alabama to Minnesota. The "ground speed" of the plane is 300 miles per hour. On the return trip, the flight took only 2 hours with a ground speed of 450 miles per hour. During both flights the speed and the direction of the wind were the same. The plane's speed decreases or increases because of the wind as the verbal model below shows.

Speed in still air − Wind speed = Ground speed against wind

Speed in still air + Wind speed = Ground speed with wind

45. Assign labels to the verbal model shown above. Use the labels to translate the verbal model into a system of linear equations. **Let s = speed in still air, and w = wind speed; $s - w = 300$, $s + w = 450$**

46. Solve the linear system. **(375, 75)**

47. What was the speed of the plane in still air? What was the speed of the wind? **375 mi/h; 75 mi/h**

48. STEAMBOAT SPEED A steamboat went 8 miles upstream in 1 hour. The return trip took only 30 minutes. Assume that the speed of the current and the direction were constant during both parts of the trip. Find the speed of the boat in still water and the speed of the current. **12 mi/h, 4 mi/h**

! COMMON ERROR
EXERCISES 16–24 Students may multiply only the side of the equation that contains the variable by the number. Remind them to multiply each term on both sides of the equation by the same number.

STUDENT HELP NOTES
Look Back As students review simplifying on p. 102, remind them that they can simplify fractions, decimals, and percents by multiplying the entire equation by a constant to eliminate the fractions or decimals. For example in Ex. 40, multiply both equations by 10 to eliminate the decimals, then solve the system using linear combinations. Multiply equations containing fractions by the least common denominator of the fractions.

APPLICATION NOTE
EXERCISE 44 Additional information about honey bees is available at **www.mcdougallittell.com.**

ADDITIONAL PRACTICE AND RETEACHING

For Lesson 7.3:

- Practice Levels A, B, and C (*Chapter 7 Resource Book,* p. 40)
- Reteaching with Practice (*Chapter 7 Resource Book,* p. 43)
- See Lesson 7.3 of the *Personal Student Tutor*

For more Mixed Review:

- Search the *Test and Practice Generator* for key words or specific lessons.

4 ASSESS

DAILY HOMEWORK QUIZ

Transparency Available

Solve the system.

1. $-3x + 5y = -7$
   $3x - 8y = 1$ $\left(\frac{17}{3}, 2\right)$
2. $7m + n = -2$
   $3m + n = 2$ $(-1, 5)$
3. $8e - 3f = 12$
   $2e + 7f = -28$ $(0, -4)$
4. $10g + 3h = 10$
   $-12g - 6h = -24$ $\left(-\frac{1}{2}, 5\right)$
5. $5p = 1 + 3q$
   $-4p + 6q = 10$ $(2, 3)$
6. $5x + 9y = -\frac{17}{5}$
   $2x - 7y = 5$ $\left(\frac{2}{5}, -\frac{3}{5}\right)$
7. Two toy robots are placed on a coordinate grid. Both are moving down from the x-axis. One is traveling along the line $x + 2y = \frac{2}{7}$, and the other is traveling along the line $x = -\frac{12}{5}y$. At what point will the robots meet? $\left(\frac{12}{7}, -\frac{5}{7}\right)$

EXTRA CHALLENGE NOTE

Challenge problems for Lesson 7.3 are available in **blackline** format in the *Chapter 7 Resource Book,* p. 48 and at **www.mcdougallittell.com.**

ADDITIONAL TEST PREPARATION

1. **OPEN ENDED** Make up a word problem that you could solve using a linear system. Solve the problem using linear combinations.
   Check students' work.

49. *x*; only one equation must be multiplied.

50. *x*; only one equation must be multiplied.

51. *x* or *y*: only one equation must be multiplied.

SOLVING EFFICIENTLY **In Exercises 49–51, decide which variable to eliminate when using linear combinations to solve the system. Explain your thinking.** 49–51. See margin.

49. $2x + 3y = 1$
$4x - 2y = 10$

50. $5y - 3x = 7$
$x + 3y = 7$

51. $\frac{1}{3}x + 6y = 6$
$-x + 3y = 3$

52. *Writing* Describe a general method for deciding which variable to eliminate when using linear combinations. See margin.

Test Preparation

52. *Sample answer:* First determine whether one variable may be eliminated without any multiplication. If not, determine whether either variable may be eliminated by multiplying only one equation. If not, you may eliminate either variable.

QUANTITATIVE COMPARISON **In Exercises 53–55, solve the system. Then choose the statement below that is true about the solution of the system.**

Ⓐ The value of x is greater than the value of y.

Ⓑ The value of y is greater than the value of x.

Ⓒ The values of x and y are equal.

Ⓓ The relationship cannot be determined from the given information.

53. $x + y = 4$ **A**
$x - 2y = 1$

54. $x - 5y + 1 = 6$ **A**
$x = 12y$

55. $3x + 5y = -8$ **C**
$x - 2y = 1$

★ Challenge

56. **SYSTEM OF THREE EQUATIONS** Solve for x, y, and z in the system of equations. Explain each of your solution steps. See margin.

$3x + 2y + z = 42$
$2y + z + 12 = 3x$
$x - 3y = 0$

EXTRA CHALLENGE
www.mcdougallittell.com

MIXED REVIEW

56. (9, 3, 9); rearrange the second equation so like terms in the first and second equations are aligned in columns. Multiply the second equation by −1. Add the first two equations to eliminate the variable *z*. In doing so, the variable *y* is also eliminated and you can solve to find $x = 9$. Then substitute 9 for *x* in the third equation to find $y = 3$. Next, substitute the values of *x* and *y* in one of the first two equations to find $z = 9$. Finally, check that (9, 3, 9) is a solution of all three equations.

WRITING EQUATIONS **Write an equation of the line in slope-intercept form that passes through the two points, or passes through the point and has the given slope.** (Review 5.2, 5.3)

57. $(-2, -1), (4, 2)$ $y = \frac{1}{2}x$

58. $(-2, 4), m = 3$ $y = 3x + 10$

59. $(6, 5), (2, 1)$ $y = x - 1$

60. $(5, 1), m = 5$ $y = 5x - 24$

61. $(9, 3), m = -\frac{1}{3}$ $y = -\frac{1}{3}x + 6$

62. $(4, -5), (-1, -3)$ $y = -\frac{2}{5}x - \frac{17}{5}$

CHECKING SOLUTIONS **Check whether each ordered pair is a solution of the inequality.** (Review 6.5)

63. $3x - 2y < 2$; $(1, 3), (2, 0)$ solution; not a solution

64. $5x + 4y \geq 6$; $(-2, 4), (5, 5)$ solution; solution

65. $5x + y > 5$; $(5, 5), (-5, -5)$ solution; not a solution

66. $12y - 3x \leq 3$; $(-2, 4), (1, -1)$ not a solution; solution

SOLVING SYSTEMS **Use substitution to solve the linear system.** (Review 7.2 for 7.4)

67. $-6x - 5y = 28$ $(-3, -2)$
$x - 2y = 1$

68. $m + 2n = 1$ $(-3, 2)$
$5m - 4n = -23$

69. $g - 5h = 20$ $(10, -2)$
$4g + 3h = 34$

70. $p + 4q = -9$ $(-1, -2)$
$2p - 3q = 4$

71. $\frac{3}{5}b - a = 0$ $(3, 5)$
$1 + b = 2a$

72. $d - e = 8$ $(5, -3)$
$\frac{1}{5}d = e + 4$

QUIZ 1

Self-Test for Lessons 7.1–7.3

Graph and check to solve the linear system. (Lesson 7.1)

1. $3x + y = 5$, $-x + y = -7$ (3, −4)
2. $\frac{1}{2}x + \frac{3}{4}y = 9$, $-2x + y = -4$ (6, 8)
3. $x - 2y = 0$, $3x - y = 0$ (0, 0)

Use substitution to solve the linear system. (Lesson 7.2)

4. $4x + 3y = 31$, $y = 2x + 7$ (1, 9)
5. $-12x + y = 15$, $3x + 2y = 3$ (−1, 3)
6. $x + \frac{1}{2}y = 7$, $3x + 2y = 18$ (10, −6)

Use linear combinations to solve the linear system. (Lesson 7.3)

7. $x + 7y = 12$, $3x - 5y = 10$ (5, 1)
8. $3x - 5y = -4$, $-9x + 7y = 8$ $\left(-\frac{1}{2}, \frac{1}{2}\right)$
9. $\frac{2}{3}x + \frac{1}{6}y = \frac{2}{3}$, $-y = 12 - 2x$ $\left(2\frac{2}{3}, -6\frac{2}{3}\right)$

10. **COMPACT DISC SALE** A store is selling compact discs for \$10.50 and \$8.50. You buy 10 discs and spend a total of \$93. How many compact discs did you buy that cost \$10.50? that cost \$8.50? (Lessons 7.1–7.3) **4 CDs; 6 CDs**

MATH & History

Systems of Linear Equations

APPLICATION LINK www.mcdougallittell.com

THEN

THE FIRST KNOWN SYSTEM of linear equations appeared in a Chinese book about 2000 years ago. The problem below appeared in the book *Shu-shu chiu-chang* in 1247.

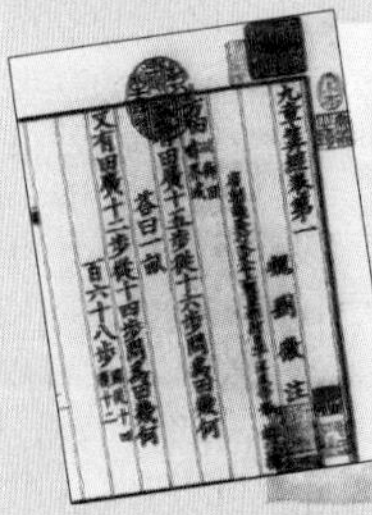

A storehouse has three kinds of stuff: cotton, floss silk, and raw silk. They take inventory of the materials and wish to cut out and make garments for the army. As for the cotton, if we use 8 rolls for 6 men, we have a shortage of 160 rolls; if we use 9 rolls for 7 men, there is a surplus of 560 rolls. . . . We wish to know the number of men [we can clothe] and the amounts of cotton [we will use]. . .

$$-x = -\frac{8y}{6} + 160$$
$$x = \frac{9y}{7} + 560$$
$$0 = \frac{9y}{7} - \frac{8y}{6} + 720$$

1. In the equations above, what does x represent? What does y represent? **x = rolls of cotton and y = number of men**
2. Solve the linear system and interpret the solution. **(20,000, 15,120); 20,000 rolls of cotton will be used and there are 15,120 men.**

NOW

CASH REGISTERS can keep track of inventory and notify a store's manager when inventory needs to be ordered.

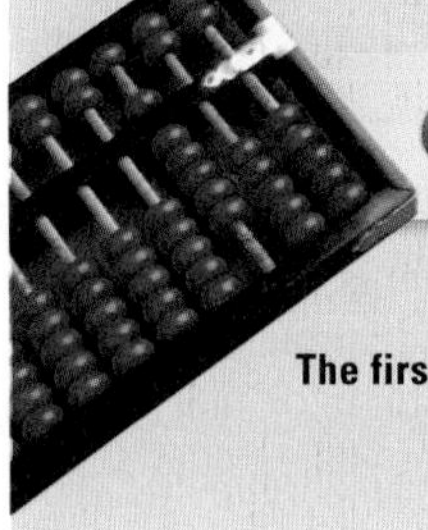

c. 3000 B.C.

The first abacus

1879

Cash register is invented.

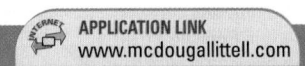

Now

Modern cash registers have scanners.

ADDITIONAL RESOURCES
An alternative Quiz for Lessons 7.1–7.3 is available in the *Chapter 7 Resource Book*, p. 49.

MATH & HISTORY NOTE
See the enrichment master Math & History Applications on p. 47 of the *Chapter 7 Resource Book* for additional exercises.

1 PLAN

PACING
Basic: 2 days
Average: 2 days
Advanced: 2 days
Block Schedule: 1 block

LESSON OPENER APPLICATION
An alternative way to approach Lesson 7.4 is to use the Application Lesson Opener:
- Blackline Master (*Chapter 7 Resource Book*, p. 53)
- Transparency (p. 49)

MEETING INDIVIDUAL NEEDS
- ***Chapter 7 Resource Book***
  Prerequisite Skills Review (p. 5)
  Practice Level A (p. 54)
  Practice Level B (p. 55)
  Practice Level C (p. 56)
  Reteaching with Practice (p. 57)
  Absent Student Catch-Up (p. 59)
  Challenge (p. 62)
- ***Resources in Spanish***
- ***Personal Student Tutor***

NEW-TEACHER SUPPORT
See the Tips for New Teachers on pp. 1–2 of the *Chapter 7 Resource Book* for additional notes about Lesson 7.4.

WARM-UP EXERCISES
Transparency Available
1. Solve the linear system using substitution.
   $2x - y = 7$
   $3x + 3y = -3$ (2, –3)
2. Solve the linear system using the graph-and-check method.
   $y = x$
   $y = -2x + 3$ (1, 1)
3. Solve the linear system using linear combinations.
   $-3x + 4y = -4$
   $3x - 6y = 6$ (0, –1)

EXPLORING DATA AND STATISTICS

7.4 Applications of Linear Systems

What you should learn

GOAL 1 Choose the best method to solve a system of linear equations.

GOAL 2 Use a system to model **real-life** problems, such as the mixture problem in **Example 2**.

Why you should learn it

▼ To solve **real-life** problems such as choosing between two jobs in **Example 3**.

GOAL 1 CHOOSING A SOLUTION METHOD

CONCEPT SUMMARY — WAYS TO SOLVE A SYSTEM OF LINEAR EQUATIONS

GRAPHING: A useful method for approximating a solution, checking the reasonableness of a solution, and providing a visual model. **(Examples 1–3, pp. 398–400)**

SUBSTITUTION: A useful method when one of the variables has a coefficient of 1 or –1. **(Examples 1–3, pp. 405–407)**

LINEAR COMBINATIONS: A useful method when none of the variables has a coefficient of 1 or –1. **(Examples 1–4, pp. 411–413)**

EXAMPLE 1 *Choosing a Solution Method*

SELLING SHOES A store sold 28 pairs of cross-trainer shoes for a total of \$2220. Style A sold for \$70 per pair and Style B sold for \$90 per pair. How many of each style were sold?

SOLUTION

PROBLEM SOLVING STRATEGY

VERBAL MODEL

Number of Style A + Number of Style B = Total number sold

Price of Style A · Number of Style A + Price of Style B · Number of Style B = Total receipts

LABELS

Number of Style A = x		(pairs of shoes)
Number of Style B = y		(pairs of shoes)
Total number sold = **28**		(pairs of shoes)
Price of Style A = **70**		(dollars per pair)
Price of Style B = **90**		(dollars per pair)
Total receipts = **2220**		(dollars)

ALGEBRAIC MODEL

$x + y = 28$ **Equation 1**

$70x + 90y = 2220$ **Equation 2**

Because the coefficients of x and y are 1 in Equation 1, substitution is most convenient. Solve Equation 1 for x and substitute the result into Equation 2. Simplify to obtain $y = 13$. Substitute 13 for y in Equation 1 and solve for x.

▶ The solution is 15 pairs of Style A and 13 pairs of Style B.

STUDENT HELP

Study Tip
Examples 1 and 2 are called *mixture problems.* Mixture problems often have one equation of the form
$x + y =$ amount
and another equation in which the coefficients of x and y are not 1.

GOAL 2 SOLVING REAL-LIFE PROBLEMS

EXAMPLE 2 *Solving a Mixture Problem*

CAR MAINTENANCE Your car's manual recommends that you use at least 89-octane gasoline. Your car's 16-gallon gas tank is almost empty. How much regular gasoline (87-octane) do you need to mix with premium gasoline (92-octane) to produce 16 gallons of 89-octane gasoline?

SOLUTION

An octane rating is the percent of isooctane in the gasoline, so 16 gallons of 89-octane gasoline contains 89% of 16, or 14.24, gallons of isooctane.

PROBLEM SOLVING STRATEGY

VERBAL MODEL

Volume of regular + **Volume of premium** = **Volume of 89-octane**

Isooctane in regular + **Isooctane in premium** = **Isooctane in 89-octane**

LABELS

Volume of regular	= x	(gallons)
Volume of premium	= y	(gallons)
Volume of 89-octane	= **16**	(gallons)
Isooctane in regular	= $0.87x$	(gallons)
Isooctane in premium	= $0.92y$	(gallons)
Isooctane in 89-octane	= **14.24**	(gallons)

ALGEBRAIC MODEL

$x + y = 16$ **Equation 1**

$0.87x + 0.92y = 14.24$ **Equation 2**

Use substitution. Solve Equation 1 for y and substitute into Equation 2.

$y = 16 - x$ Solve Equation 1 for y.

$0.87x + 0.92(16 - x) = 14.24$ Substitute $16 - x$ for y in Equation 2.

$-0.05x = -0.48$ Simplify.

$x = 9.6$ Solve for x.

$9.6 + y = 16$ Substitute 9.6 for x in Equation 1.

$y = 6.4$ Solve for y.

✓**CHECK** Substitute in each equation to check your result.

EQUATION 1	EQUATION 2
$9.6 + 6.4 \stackrel{?}{=} 16$	$0.87(9.6) + 0.92(6.4) \stackrel{?}{=} 14.24$
$16.0 = 16$	$8.352 + 5.888 = 14.24$

▶ You need to mix 9.6 gallons of regular gasoline with 6.4 gallons of premium gasoline to get 16 gallons of 89-octane gasoline.

FOCUS ON APPLICATIONS

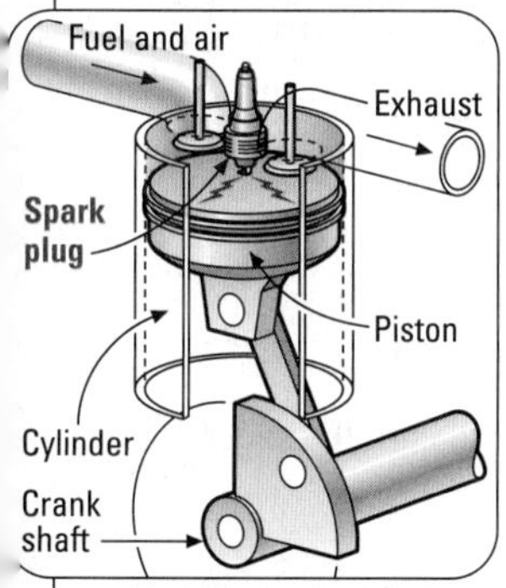

REAL LIFE CAR ENGINE In a car engine, fuel and air in the cylinder are ignited by a spark to provide power to drive the engine. Using a fuel with the correct octane rating makes the engine run more smoothly.

APPLICATION LINK
www.mcdougallittell.com

2 TEACH

MOTIVATING THE LESSON

Ask students to consider the route they take when they walk, drive, or take public transportation to school, the library, or some other location. For most students, there are several different routes they could have taken. Why did they choose the one they did? Was it the shortest? Did they choose the route to pick up a friend or stop at a store? We often choose to do something a certain way to make things faster, easier, or to achieve a certain goal. Today's lesson will focus on the choices possible for solving linear systems.

EXTRA EXAMPLE 1

A health food store mixes granola that costs them $4 per pound and raisins that cost them $2 per pound together to make 25 pounds of raisin granola. How many pounds of raisins should they include if they want the mixture to cost them a total of $80? **10 lb**

EXTRA EXAMPLE 2

A chemist needs to make 30 ounces of a 25% alcohol solution by mixing together a 15% alcohol solution with a 40% alcohol solution. How much of each solution should they use? **18 oz of 15% solution and 12 oz of 40% solution**

CHECKPOINT EXERCISES

For use after Examples 1 and 2:

1. A group of 40 children attended a baseball game on a field trip. Each child received either a hot dog or bag of popcorn. Hot dogs were $2.25 and popcorn was $1.75. If the total bill was $83.50, how many hot dogs were bought? **27 hot dogs**

EXTRA EXAMPLE 3

You are offered two different credit cards. Card A charges \$75 per year and 15% interest on your average balance. Card B charges \$15 per year and 18% interest on your average balance.

a. What would your average balance need to be for the cards to cost you the same amount? **\$2000**

b. You believe that your average balance is between \$1000 and \$2000. Which credit card should you choose? **card B**

CHECKPOINT EXERCISES

For use after Example 3:

1. You have a choice of two different Internet service companies. Company A charges \$12 per month plus \$2.00 per hour. Company B charges \$27 per month plus \$.50 per hour.

a. How many hours would you need to use the Internet per month for the services to be the same price? **10 h**

b. You estimate that you will spend between 15 and 20 hours per month on the Internet. Which Internet service company should you choose? **company B**

FOCUS ON VOCABULARY

Tell students to think of the words *substitute* and *combine* to help them remember the different methods for solving systems of linear equations.

CLOSURE QUESTION

What things do you consider when deciding which method to use to solve a linear system? **See margin.**

DAILY PUZZLER

Suppose 2 cubes and 1 cone balance on a scale with 12 marbles. If 1 cube balances with 1 cone and 3 marbles, then how many marbles balance with 1 cube? **5**

REAL LIFE

Career Decisions

EXAMPLE 3 *Making a Decision*

You are offered two different sales jobs. Job A offers an annual salary of \$30,000 plus a year-end bonus of 1% of your total sales. Job B offers an annual salary of \$24,000 plus a year-end bonus of 2% of your total sales.

a. How much would you have to sell to earn the same amount in each job?

b. You believe you can sell between \$500,000 and \$800,000 of merchandise per year. Which job should you choose?

SOLUTION

a. First find a linear system that models the situation.

PROBLEM SOLVING STRATEGY

VERBAL MODEL

Total earnings = Job A salary + 1% · **Total sales**

Total earnings = Job B salary + 2% · **Total sales**

LABELS

Total earnings = y (dollars)

Total sales = x (dollars)

Salary for Job A = **30,000** (dollars)

Salary for Job B = **24,000** (dollars)

ALGEBRAIC MODEL

$y = 30{,}000 + 0.01x$ **Equation 1 (Job A)**

$y = 24{,}000 + 0.02x$ **Equation 2 (Job B)**

Use linear combinations to solve this system of linear equations because you can easily get the coefficients of y to be opposites.

$y = 30{,}000 + 0.01x$	Write Equation 1.
$-y = -24{,}000 - 0.02x$	Multiply Equation 2 by −1.
$0 = 6000 - 0.01x$	Add equations.
$0.01x = 6000$	Add $0.01x$ to each side.
$x = 600{,}000$	Solve for x.

The solution of the linear system is $x = 600{,}000$ and $y = 36{,}000$.

▶ You would have to sell \$600,000 worth of merchandise to earn the same amount of money in each job.

b. One way to decide which job to choose is to draw a graph of the linear system.

▶ From the graph you can see that if your sales are greater than \$600,000, Job B would pay better than Job A.

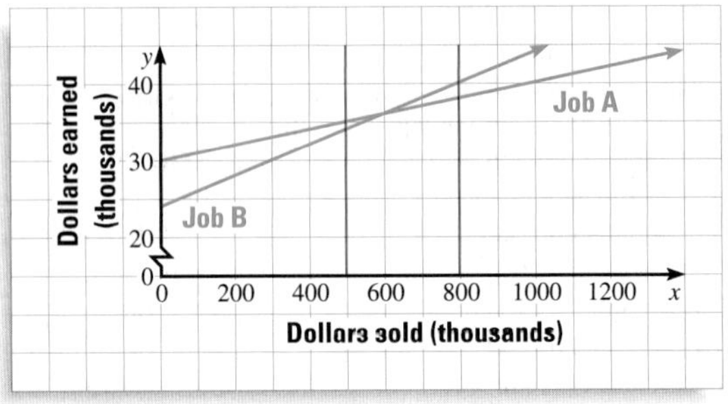

Closure Question *Sample answer:*

I look to see if the coefficient of one of the variables is 1 or −1, in which case I use substitution. If I only want an approximate answer or a visual model, I use graphing. Otherwise, I use linear combinations.

GUIDED PRACTICE

Concept Check ✓

1. Solve the linear system in Example 3 using the substitution method. **See margin.**
2. You have seen three ways to solve the linear system in Example 3. Which method or methods do you prefer for solving the linear system in Example 3? Why? **Personal preferences may vary.**

Skill Check ✓

Choose a method to solve the linear system. Explain your choice. **3–5. Sample answers are given. See margin.**

3. $x + y = 300$
   $x + 3y = 18$
4. $3x + 5y = 25$
   $2x - 6y = 12$
5. $2x + y = 0$
   $x + y = 5$

3. Substitution or linear combinations; it would be simple to write either variable in terms of the other or to eliminate x by multiplying either equation by -1.

4. Linear combinations; neither variable has a coefficient of 1 or -1 in either equation.

PRICE PER GALLON **In Exercises 6–9, use the following problem.** The total cost of 10 gallons of regular gasoline and 15 gallons of premium gasoline is \$32.75. Premium costs \$.20 more per gallon than regular. What is the cost per gallon of each type of gasoline?

6. Write a verbal model for the problem. **See margin.**
7. Assign labels to the verbal model. **Let x = price per gallon of regular and y = price per gallon of premium.**
8. Write a linear system as an algebraic model. **$10x + 15y = 32.75$, $y = x + 0.20$**
9. Solve the system and answer the question. **regular: \$1.19; premium: \$1.39**

PRACTICE AND APPLICATIONS

STUDENT HELP

Extra Practice to help you master skills is on p. 803.

COMPARING METHODS **Solve the linear system using all three methods on page 418.**

10. $x + y = 2$ **(0, 2)**
    $6x + y = 2$
11. $x - y = 1$ **(3, 2)**
    $x + y = 5$
12. $3x - y = 3$ **(3, 6)**
    $-x + y = 3$

CHOOSING A SOLUTION METHOD **Choose a method to solve the linear system. Explain your choice.** **13–18. Sample answers are given. See margin.**

13. $6x + y = 2$
    $9x - y = 5$
14. $2x + 3y = 3$
    $5x + 5y = 10$
15. $-3x = 36$
    $-6x + y = 1$
16. $2x - 5y = 0$
    $x - y = 3$
17. $3x + 2y = 10$
    $2x + 5y = 3$
18. $6.2x - 0.5y = -27.8$
    $0.3x + 0.4y = 68.7$

5. Any of the three methods would be reasonable; it would be simple to write either variable in terms of the other, both would be simple to graph, and y could be eliminated by multiplying either equation by -1.

SOLVING LINEAR SYSTEMS **Choose a method to solve the linear system. Explain your choice, and then solve the system.** **19–30. Methods chosen may vary.**

19. $2x + y = 5$ **(2, 1)**
    $x - y = 1$
20. $2x - y = 3$ **(3, 3)**
    $4x + 3y = 21$
21. $x - 2y = 4$ **(2, −1)**
    $6x + 2y = 10$
22. $3x + 6y = 8$ $\left(\frac{4}{15}, 1\frac{1}{5}\right)$
    $-6x + 3y = 2$
23. $x + y = 0$ **(1, −1)**
    $3x + 2y = 1$
24. $2x - 3y = -7$ **(−2, 1)**
    $3x + y = -5$
25. $3y + 4x = 5$ **(2, −1)**
    $x + y = 1$
26. $8x + y = 15$ $\left(1\frac{1}{2}, 3\right)$
    $9 = 2y + 2x$
27. $100 - 9x = 5y$ $\left(5\frac{5}{9}, 10\right)$
    $0 = 5y - 9x$
28. $x + 2y = 2$ **(6, −2)**
    $x + 4y = -2$
29. $-y = -4$ **(−4, 4)**
    $x + 2y = 4$
30. $0.2x - 0.5y = -3.8$
    $0.3x + 0.4y = 10.4$ **(16, 14)**

STUDENT HELP

HOMEWORK HELP
Example 1: Exs. 10–48
Example 2: Exs. 10–48
Example 3: Exs. 10–48

3 APPLY

ASSIGNMENT GUIDE

BASIC
Day 1: pp. 421–422 Exs. 10–38 even
Day 2: pp. 422–424 Exs. 40–48 even, 49, 50, 61–66, 68–75

AVERAGE
Day 1: pp. 421–422 Exs. 10–38 even
Day 2: pp. 422–424 Exs. 40–48 even, 49–51, 61–66, 68–75

ADVANCED
Day 1: pp. 421–422 Exs. 10–38 even
Day 2: pp. 421–422 Exs. 40–48 even, 49–51, 61–75

BLOCK SCHEDULE
pp. 421–424 Exs. 10–48 even, 49–51, 61–66, 68–75

EXERCISE LEVELS

Level A: ***Easier***
10–12
Level B: ***More Difficult***
13–54, 61–75
Level C: ***Most Difficult***
55–60

✓ HOMEWORK CHECK

To quickly check student understanding of key concepts, go over the following exercises: Exs. 10, 14, 20, 26, 36, 44, 50, 62. See also the Daily Homework Quiz:

- Blackline Master (*Chapter 7 Resource Book*, p. 65)
- Transparency (p. 57)

1, 6, 13–18. See next page for answers.

! COMMON ERROR
EXERCISES 41–45 Remind students that to avoid computational errors with decimals and fractions, they can multiply each equation by an appropriate constant as a first step.

APPLICATION NOTE
EXERCISE 46 Additional information about urban gardens is available at **www.mcdougallittell.com.**

1. $30{,}000 + 0.01x = 24{,}000 + 0.02x$; $6000 = 0.01x$; $x = 600{,}000$; $y = 30{,}000 + 0.01(600{,}000) = 30{,}600$; $(600{,}000, 36{,}000)$
6. The number of gallons of regular multiplied by the price per gallon of regular plus the number of gallons of premium multiplied by the price per gallon of premium is equal to \$32.75; the price per gallon of premium is \$.20 more than the price per gallon of regular.
13. Substitution or linear combinations; it would be simple to write y in terms of x, and y could be eliminated by adding the equations.
14. Linear combinations; no variable has a coefficient of 1 or –1.
15. Substitution; either variable may be easily eliminated using one of the equations.
16. Substitution; it would be simple to solve the second equation for x or for y.
17. Linear combinations; no variable has a coefficient of 1 or –1.
18. Linear combinations; no variable has a coefficient of 1 or –1.

FOCUS ON APPLICATIONS

URBAN GARDENS Cities are blossoming with thousands of gardens. They provide green areas of plant life amid acres of steel and concrete.

APPLICATION LINK
www.mcdougallittell.com

42. $\left(5\frac{1}{4}, -\frac{1}{4}\right)$

51. $y = 172x + 6200$ and $y = -16x + 6500$; $\left(1\frac{28}{47}, 6474\frac{22}{47}\right)$; the point of intersection represents the number of years after 1970 (about 1.6) when the demand for low-income housing and the availability were equal (about 6,474,500).

CHOOSING A METHOD In Exercises 31–45, solve the linear system.

31. $6x - y = 18$
$8x + y = 24$ **(3, 0)**

32. $x - y = -4$
$2y + x = 5$ **(–1, 3)**

33. $x + 2y = 8$
$3x - 2y = 8$ **(4, 2)**

34. $x + 2y = 1$
$5x - 4y = -23$ **(–3, 2)**

35. $8x + 4y = 8$
$-2x + 3y = 12$ $\left(-\frac{3}{4}, 3\frac{1}{2}\right)$

36. $3x - 5y = 8$
$-2x + 3y = 3$ **(–39, –25)**

37. $2x - y = 6$
$y - x = 0$ **(6, 6)**

38. $8x + 9y = 42$
$6x - y = 16$ **(3, 2)**

39. $7x + 4y = 22$
$-5x - 9y = 15$ **(6, –5)**

40. $3x - 5y = 3$
$9x - 20y = 6$ $\left(2, \frac{3}{5}\right)$

41. $0.5x + 2.2y = 9$
$6x + 0.4y = -22$ **(–4, 5)**

42. $1.5x - 2.5y = 8.5$
$6x + 30y = 24$ **See margin.**

43. $3x + 9y = 1$ $\left(\frac{1}{3}, 0\right)$
$2x + 3y = \frac{2}{3}$

44. $3x - 2y = 8$ **(8, 8)**
$x + \frac{3}{2}y = 20$

45. $\frac{1}{2}x - y = -5$ **(2, 6)**
$x - \frac{1}{3}y = 0$

46. **GARDEN PLANTS** In early spring, you buy 6 potted tomato plants for your container garden. The plants contained in 8-inch pots sell for \$5 and the plants contained in 10-inch pots sell for \$8. If you spend \$36, how many of each size are you buying? **4 8-in. pots, 2 10-in. pots**

Number of 8-inch pots + Number of 10-inch pots = Total number buying

Cost of all 8-inch pots + Cost of all 10-inch pots = Total cost

47. **SCIENCE CONNECTION** In your chemistry class you have a bottle of 5% boric acid solution and a bottle of 2% boric acid solution. You need 60 milliliters of a 3% boric acid solution for an experiment. How much of each solution do you need to mix together? **20 mL of the 5% solution, 40 mL of the 2% solution**

48. **TREE GROWTH** You plant a 14-inch hemlock tree in your backyard that grows at a rate of 4 inches per year and an 8-inch blue spruce tree that grows at a rate of 6 inches per year. In how many years after you plant the trees will the two trees be the same height? How tall will each tree be? **3 years; 26 in.**

PARTY PLANNING In Exercises 49 and 50, you are planning a birthday party for your 8-year-old cousin. You can have the party at a pizza place for \$8 per person plus \$30 for favors and a small cake or at a taco place for \$12 per person plus \$14 for a large cake.

49. How many children would you have to invite to the party for the cost to be the same for both places? **4 children**

50. How many children would you have to invite to the party for the cost to be less at the pizza place? **more than 4 children**

51. **LOW-INCOME HOUSING** The graph at the right represents the demands for low-income rental housing in the United States and the number of affordable rental units. Use the coordinates given to write a linear system. Solve the system. What does the point of intersection represent in the context of the situation?

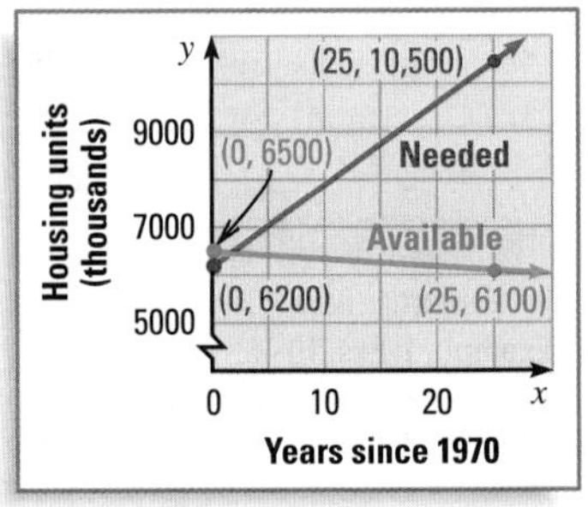

DRAWING PORTRAITS In Exercises 52–54, you are earning extra money drawing sketches for your friends and family. You receive \$12 per sketch. Your expenses for art supplies are \$50.

52. Write an equation that models only your income. **$y = 12x$**

53. Suppose you plan to spend an additional \$7 per sketch for a frame. Write an equation that models only your expenses. **$y = 7x + 50$**

54. How many framed sketches do you have to sell in order for your income to equal your expenses? **10 sketches**

55. 1 h at 4 mi/h and $\frac{1}{2}$ h at 6 mi/h

55. TREADMILLS You exercised on a treadmill for 1.5 hours. You ran at 4 miles per hour, then you sprinted at 6 miles per hour. If the treadmill monitor says that you ran and sprinted 7 miles, how long did you run at each speed?

STUDENT HELP

HOMEWORK HELP Visit our Web site www.mcdougallittell.com for help with Ex. 55.

56. TRAVEL TIME You are driving to a concert. To arrive on time, you plan to travel at an average speed of 45 miles per hour. For the first two hours of the trip you travel at 40 miles per hour. Find the expression for the total time you traveled. Then use the following verbal models to find how many hours you traveled at 55 miles per hour to arrive on time.

Distance at 40 mi/h + Distance at 55 mi/h = Total distance

Average speed • Total time = Total distance **$\frac{\text{Total distance}}{45}$; 1 h**

SCIENCE CONNECTION **In Exercises 57–60, a chemist is calculating absolute zero using the procedure below. Absolute zero is the lowest possible temperature for all substances. It is believed to be the temperature at which the volume of an ideal gas (at constant pressure) contracts to zero.**

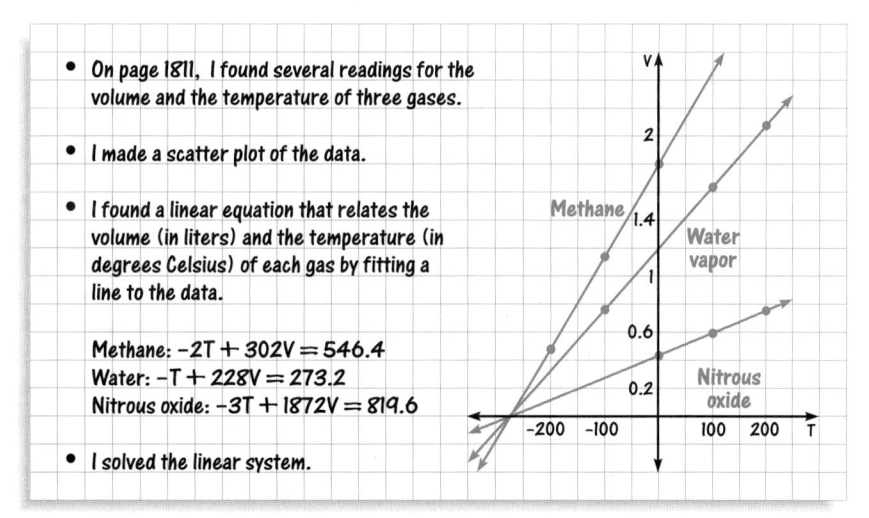

- On page 1811, I found several readings for the volume and the temperature of three gases.
- I made a scatter plot of the data.
- I found a linear equation that relates the volume (in liters) and the temperature (in degrees Celsius) of each gas by fitting a line to the data.

Methane: $-2T + 302V = 546.4$
Water: $-T + 228V = 273.2$
Nitrous oxide: $-3T + 1872V = 819.6$

- I solved the linear system.

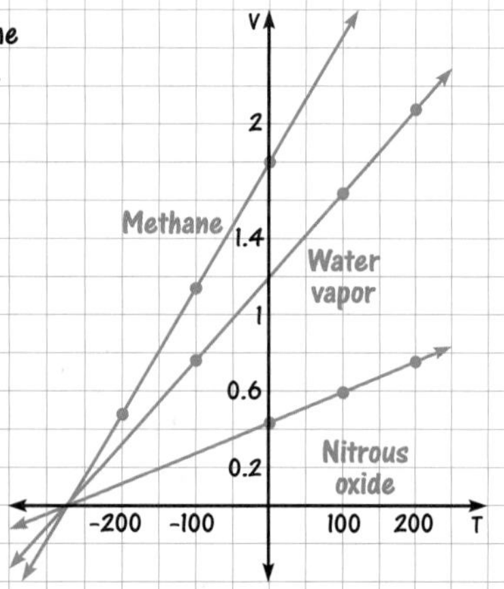

59. No; solving the system involving any two of the equations will produce the desired solution.

STUDENT HELP

APPLICATION LINK Visit our Web site www.mcdougallittell.com for more information about absolute zero.

57. What methods could the chemist use to solve the linear system? **substitution, graphing, linear combinations**

58. Use the chemist's graph to predict what absolute zero is on the Celsius scale. ***Sample answer:* about −275°C**

59. Does the chemist need to solve all three equations to find T? Explain. **See margin.**

60. Solve the system algebraically. **−273.2°C**

STUDENT HELP NOTES

Homework Help Students can find help for Ex. 55 at **www.mcdougallittell.com.**

71.

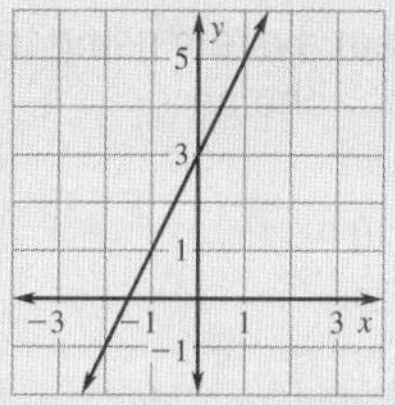

72.

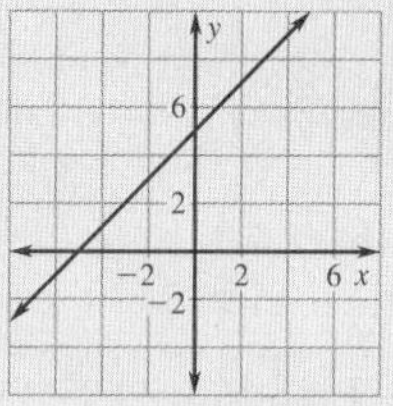

73.

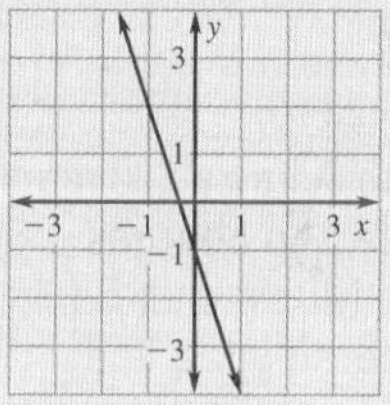

ADDITIONAL PRACTICE AND RETEACHING

For Lesson 7.4:

- Practice Levels A, B, and C (*Chapter 7 Resource Book,* p. 54)
- Reteaching with Practice (*Chapter 7 Resource Book,* p. 57)
- See Lesson 7.4 of the *Personal Student Tutor*

For more Mixed Review:

- Search the *Test and Practice Generator* for key words or specific lessons.

4 ASSESS

DAILY HOMEWORK QUIZ

Transparency Available

Solve the linear system using the method of your choice.

1. $x - 2y = 5$
   $x + 3y = -10$ **(−1, −3)**
2. $3x + 7y = -29$
   $4x + 3y = -7$ **(2, −5)**
3. $6x - 5y = 0$
   $3x + 2y = 9$ **$\left(\frac{5}{3}, 2\right)$**
4. $4x + y = 7$
   $9x + 2y = 16$ **(2, −1)**
5. $y = 1 - x$
   $4x - 2y = -1$ **$\left(\frac{1}{6}, \frac{5}{6}\right)$**
6. $5x + 3y = 9$
   $-2x - 5y = 23$ **(6, −7)**
7. Music club A has a \$30 membership fee and charges \$8 per CD. Music club B has a \$45 membership fee and charges \$7 per CD. How many CDs would a person need to buy for club A to be more expensive than Club B? **16 or more CDs**

EXTRA CHALLENGE NOTE

Challenge problems for Lesson 7.4 are available in **blackline** format in the *Chapter 7 Resource Book*, p. 62 and at **www.mcdougallittell.com.**

ADDITIONAL TEST PREPARATION

1. **WRITING** For each of the three methods listed below, write a linear system that would be best solved using that method. Explain why the method is best for each system.
   a. graph and check
   b. substitution
   c. linear combinations
   Check students' work.

71–73. See previous page for answers.

MARBLES In Exercises 61–63, consider a bag containing 12 marbles that are either red or blue. A marble is drawn at random. There are three times as many red marbles as there are blue marbles in the bag.

61. Write a linear system to describe this situation. **$r + b = 12$; $r = 3b$**

62. How many red marbles are in the bag? **9 red marbles**

63. PROBABILITY CONNECTION What is the probability of drawing a red marble? **0.75**

Test Preparation

64. **MULTIPLE CHOICE** At what point do the lines $3x - 2y = 0$ and $5x + 2y = 0$ intersect? **D**

(A) (1, 2) (B) (5, 2) (C) (3, 2) (D) (0, 0)

65. **MULTIPLE CHOICE** Find the x-value of the solution for the linear system below. **B**

$y = 2x - 2$
$y = 3x + 1$

(A) −8 (B) −3 (C) 3 (D) 8

★ Challenge

66. RELAY RACE The total time for a two-member team to complete a 5765-meter relay race was 19 minutes. The first runner averaged 285 meters per minute and the second runner averaged 335 meters per minute. Use the verbal model to find how many minutes the first runner ran. **12 min**

Time for 1st runner	+	Time for 2nd runner	=	Total time
Distance for 1st runner	+	Distance for 2nd runner	=	Total distance

EXTRA CHALLENGE
www.mcdougallittell.com

MIXED REVIEW

PARALLEL LINES Decide whether the graphs of the two equations are parallel lines. (Review 4.6 for 7.5)

67. $y = 4x + 3$; $2y - 8x = -3$ **parallel**

68. $4y + 5x = 1$; $10x + 2y = 2$ **not parallel**

69. $3x + 9y + 2 = 0$; $2y = -6x + 3$ **not parallel**

70. $4y - 1 = 5$; $6y + 2 = 8$ **parallel**

GRAPHING FUNCTIONS Graph the function. (Review 4.8) **71–73. See margin for graphs.**

71. $f(x) = 2x + 3$

72. $h(x) = x + 5$

73. $g(x) = -3x - 1$

CD-ROM DRIVES In Exercises 74 and 75, the number of United States households with CD-ROM drives in their computers is shown in the scatter plot at the right. (Review 5.4 and 5.7)

74. $y = \frac{20}{3}x + \frac{10}{3}$

74. Find an equation of the line that you think best fits the data. Notice that x represents the number of years since 1993 and y represents the number of households out of 100.

75. Use a linear model to estimate the number of households that will have computer-installed CD-ROM drives in 2003. **70 out of 100 households**

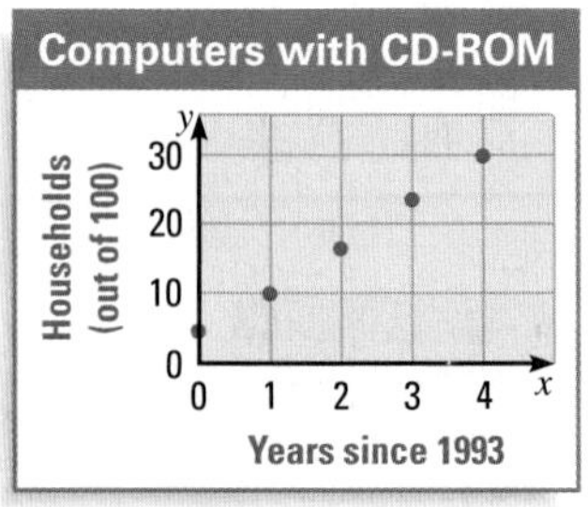

DATA UPDATE
www.mcdougallittell.com

ACTIVITY 7.5

Developing Concepts

Group Activity for use with Lesson 7.5

Investigating Special Types of Linear Systems

GROUP ACTIVITY
Work in a small group.

MATERIALS
- graph paper

QUESTION How can you identify the number of solutions of a linear system by graphing?

EXPLORING THE CONCEPT

1. Each member of your group should choose a different one of the linear systems below and graph it. **a–c. See margin for graphs.**

   a. $x + y = 0$ / $3x - 2y = 10$ **b.** $2x - 4y = 6$ / $x - 2y = 3$ **c.** $x - y = 1$ / $-3x + 3y = 3$

2. Share your graphs. How are the three graphs different? **See margin.**
3. For the system you graphed, write both equations in the form $y = mx + b$. **See margin.**
4. Share your results from **Step 3** with the others in your group. How are the equations within each system alike or different? **In the first pair of equations, both the slopes and the y-intercepts are different; in the second, both the slopes and the y-intercepts are the same; in the third, the slopes are the same and the y-intercepts are different.**

Step 2. The graph for the first system has two lines that intersect in one point; for the second, there is only one line; for the third, there are two parallel lines.

Step 3.
a. $y = -x$, $y = \frac{3}{2}x - 5$
b. $y = \frac{1}{2}x - \frac{3}{2}$, $y = \frac{1}{2}x - \frac{3}{2}$
c. $y = x - 1$, $y = x + 1$

STUDENT HELP
Look Back
For help with graphing linear systems, see p. 399.

DRAWING CONCLUSIONS

1. Repeat **Steps 1** through **4** above using the following systems. **a–c. See margin.**

a. $x - 3y = 9$ / $-2x + 6y = -18$ **b.** $x - \frac{1}{4}y = 5$ / $5x + \frac{1}{4}y = 7$ **c.** $x + 2y = 3$ / $x + 2y = 6$

Write a linear system for the graphical model. If only one line is shown, write two different equations for the line. **2–4. Where possible, equations are given in slope-intercept form. See margin.**

2.
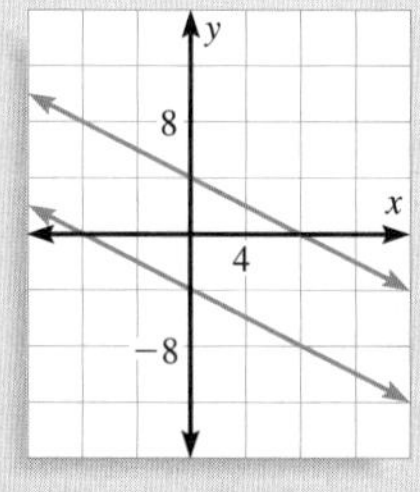

3.
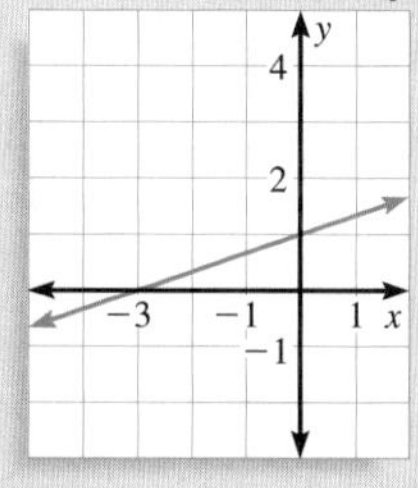

4.
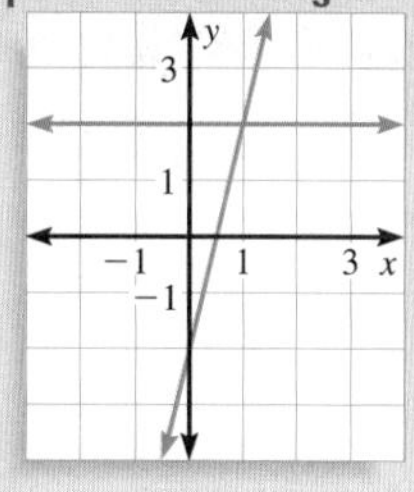

CRITICAL THINKING **In Exercises 5–7, the graph of a linear system is described. Decide whether the system has *no solution, exactly one solution,* or *many solutions.* Explain your reasoning.**

5. The slope and the y-intercept of the lines are the same. **Many solutions; the graphs are the same line, so every point on the line is a solution of both equations.**
6. The lines have different slopes and y-intercepts. **Exactly 1 solution; the graphs intersect in exactly 1 point.**
7. The lines have the same slope but different y-intercepts. **No solution; the graphs are parallel and do not intersect.**

Have each member of your group give an example of a linear system that has the given number of solutions. Compare your results. **8–10. See margin.**

8. No solution **9.** Exactly one solution **10.** Many solutions

2. $y = -\frac{1}{2}x + 4$, $y = -\frac{1}{2}x - 4$
3. any two equations that can be obtained by multiplying both sides of $y = \frac{1}{3}x + 1$ by nonzero constants
4. $y = 2$, $y = 4x - 2$
8. *Sample answer:* any two equations of the form $y = mx + b$ and $y = mx + c$, $b \neq c$
9. *Sample answer:* any two equations of the form $y = mx + b$ and $y = nx + c$, $m \neq n$
10. any two equations that result in the same equation when written in slope-intercept form

1 Planning the Activity

PURPOSE
To explore graphs and equations of linear systems with no solution, one solution, and many solutions.

MATERIALS
- graph paper
- Activity Support Master (*Chapter 7 Resource Book,* p. 67)

PACING
- Exploring the Concept — 10 min
- Drawing Conclusions — 15 min

LINK TO LESSON
Refer back to this activity when discussing the Concept Summary box on page 426.

2 Managing the Activity

COOPERATIVE LEARNING
This activity works best with three students in each group. In Step 1 and Question 1, each of the three students will work with one of the systems of equations. Group members should then exchange papers and check each other's work before moving on to Step 2.

3 Closing the Activity

KEY DISCOVERY
Systems of equations can have one solution, no solution, or many solutions. The number of solutions can be determined from the graph or from the equations.

ACTIVITY ASSESSMENT
JOURNAL Explain how to use equations or a graph to determine if a system of equations has one solution, no solution, or many solutions.
See answer at left.

Activity Assessment ***Sample answer:***
A system of two equations has one solution when the slopes are different. The system has no solution when the lines are parallel (same slope but different y-intercepts). The system has many solutions when the lines coincide (same slope, same y-intercept).

Exploring the Concept, Step 1 a–c and Drawing Conclusions, Step 1 a–c. See Additional Answers beginning on page AA1.

1 PLAN

PACING
Basic: 1 day
Average: 1 day
Advanced: 1 day
Block Schedule: 0.5 block with 7.6

LESSON OPENER
GRAPHING CALCULATOR
An alternative way to approach Lesson 7.5 is to use the Graphing Calculator Lesson Opener:
- Blackline Master (*Chapter 7 Resource Book,* p. 66)
- Transparency (p. 50)

MEETING INDIVIDUAL NEEDS
- ***Chapter 7 Resource Book***
  Prerequisite Skills Review (p. 5)
  Practice Level A (p. 70)
  Practice Level B (p. 71)
  Practice Level C (p. 72)
  Reteaching with Practice (p. 73)
  Absent Student Catch-Up (p. 75)
  Challenge (p. 77)
- ***Resources in Spanish***
- ***Personal Student Tutor***

NEW-TEACHER SUPPORT
See the Tips for New Teachers on pp. 1–2 of the *Chapter 7 Resource Book* for additional notes about Lesson 7.5.

WARM-UP EXERCISES
Transparency Available
Write each equation in slope-intercept form. Then identify the slope and y-intercept.

1. $x + y = 12$ $\quad y = -x + 12; -1, 12$
2. $-x + 3y = 3$ $\quad y = \frac{1}{3}x + 1; \frac{1}{3}, 1$
3. $4x - 6y = 15$ $\quad y = \frac{2}{3}x - \frac{5}{2}; \frac{2}{3}, -\frac{5}{2}$
4. $x - 5y = 10$ $\quad y = \frac{1}{5}x - 2; \frac{1}{5}, -2$

7.5 Special Types of Linear Systems

What you should learn

GOAL 1 Identify linear systems as having one solution, no solution, or infinitely many solutions.

GOAL 2 Model **real-life** problems using a linear system, as in **Ex. 30**.

Why you should learn it

▼ To solve **real-life** problems such as finding the heights of statues on Easter Island in **Example 4**.

REAL LIFE

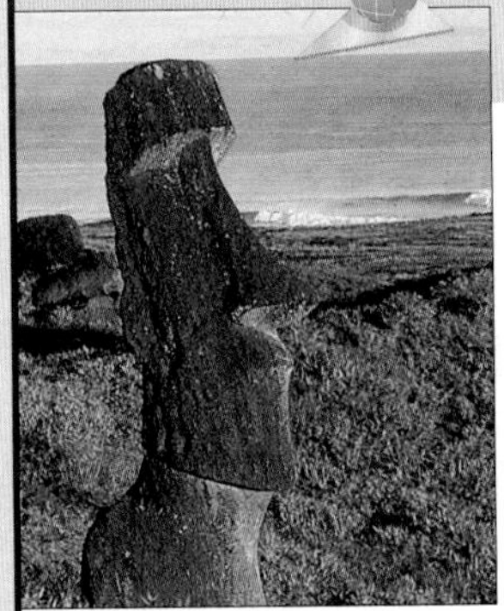

GOAL 1 IDENTIFYING THE NUMBER OF SOLUTIONS

In Lessons 7.1 through 7.4, each linear system in two variables has exactly one solution. The summary below shows that there are two other possibilities.

CONCEPT SUMMARY NUMBER OF SOLUTIONS OF A LINEAR SYSTEM

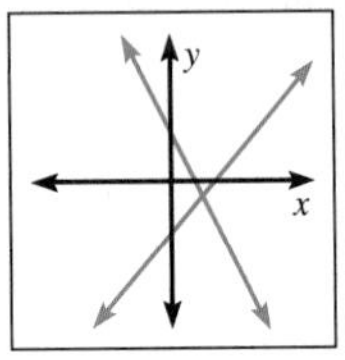

Lines intersect
one solution

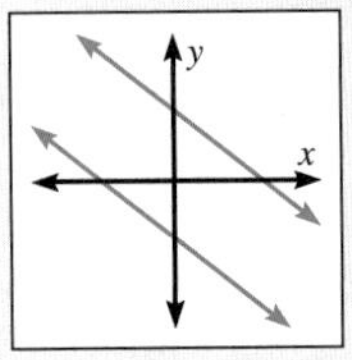

Lines are parallel
no solution

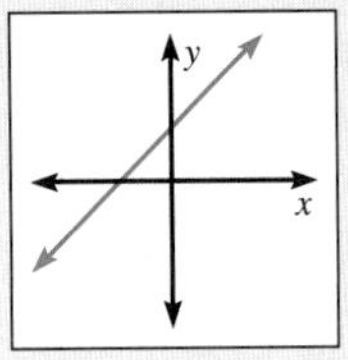

Lines coincide
infinitely many solutions (the coordinates of every point on the line)

EXAMPLE 1 *A Linear System with No Solution*

Show that the linear system has no solution.

$2x + y = 5$ **Equation 1**
$2x + y = 1$ **Equation 2**

SOLUTION

Method 1: **GRAPHING** Rewrite each equation in slope-intercept form. Then graph the linear system.

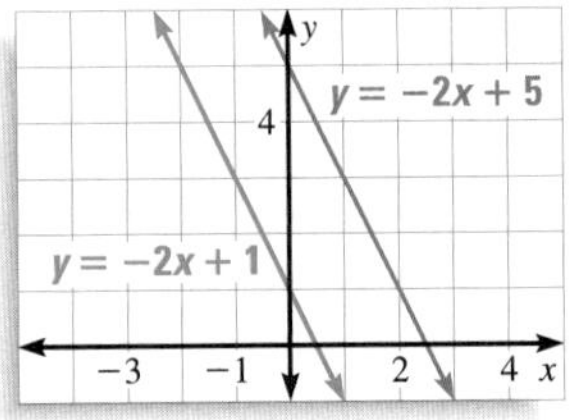

$y = -2x + 5$ **Revised Equation 1**
$y = -2x + 1$ **Revised Equation 2**

▶ Because the lines have the same slope but different y-intercepts, they are parallel. Parallel lines do not intersect, so the system has no solution.

Method 2: **SUBSTITUTION** Because Equation 2 can be revised to $y = -2x + 1$, you can substitute $-2x + 1$ for y in Equation 1.

$2x + y = 5$ **Write Equation 1.**
$2x + (-2x + 1) = 5$ **Substitute $-2x + 1$ for y.**
$1 = 5$ **Simplify. False statement**

▶ The variables are eliminated and you are left with a statement that is not true regardless of the values of x and y. This tells you the system has no solution.

STUDENT HELP
Look Back
For help with equations in one variable with no solution, see p. 155.

EXAMPLE 2 *A Linear System with Many Solutions*

Show that the linear system has infinitely many solutions.

$-2x + y = 3$ **Equation 1**

$-4x + 2y = 6$ **Equation 2**

SOLUTION

Method 1: **GRAPHING** Rewrite each equation in slope-intercept form.

$y = 2x + 3$ **Revised Equation 1**

$y = 2x + 3$ **Revised Equation 2**

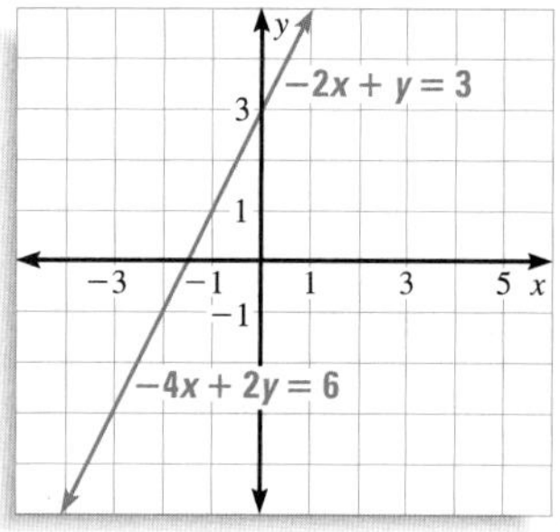

▶ From these equations you can see that the equations represent the same line. Any point on the line is a solution.

> **STUDENT HELP**
>
> **Study Tip**
> When both equations of a system are in the form $y = mx + b$, you can determine the number of solutions.
> - different slopes: one solution
> - same slopes, different y-intercepts: no solution
> - same slopes, same y-intercepts: infinitely many solutions

Method 2: **LINEAR COMBINATIONS** You can multiply Equation 1 by −2.

$4x - 2y = -6$	**Multiply Equation 1 by −2.**
$-4x + 2y = 6$	**Write Equation 2.**
$0 = 0$	**Add equations. True statement**

▶ The variables are eliminated and you are left with a statement that is true regardless of the values of x and y. This result tells you that the linear system has infinitely many solutions.

....

Notice in Example 2 that $-2x + y = 3$ can be transformed to look exactly like $-4x + 2y = 6$ by multiplying by 2. The two equations are *equivalent*. That is why your result is $0 = 0$.

EXAMPLE 3 *Identifying the Number of Solutions*

Solve the system and interpret the results.

a. $3x + y = -1$, $-9x - 3y = 3$

b. $x - 2y = 5$, $-2x + 4y = 2$

c. $2x + y = 4$, $4x - 2y = 0$

SOLUTION

a. Using the substitution method, you get $3 = 3$. This is true regardless of the values of x and y. This linear system has infinitely many solutions.

b. Using linear combinations, you get $0 = 12$. This is not true for any values of x and y. This linear system has no solution.

c. Using the substitution method to eliminate y, you get the equation $x = 1$. This linear system has exactly one solution—the ordered pair (1, 2).

2 TEACH

MOTIVATING THE LESSON
Many real-life problems can have more than one solution. Today's lesson will examine how systems of linear equations can have one solution, many solutions, or no solution at all.

MATHEMATICAL REASONING
Method 2 of Example 1 uses proof by contradiction. You assume something is true, show that the assumption leads to a contradiction or false statement, and conclude that the opposite of what you assumed is true.

EXTRA EXAMPLE 1
Show that this linear system has no solution.
$-6x + 2y = -8$
$-3x + y = 7$
See margin.

EXTRA EXAMPLE 2
Show that the linear system has infinitely many solutions.
$-x + 2y = -2$
$3x - 6y = 6$
The slope-intercept form of both equations is $y = \frac{1}{2}x - 1$. The lines coincide, so there are infinitely many solutions. Alternately, using linear combinations leads to the equation $0 = 0$. So the system has infinitely many solutions.

EXTRA EXAMPLE 3
Solve the system and interpret the results.
$6x - 2y = 10$
$-3x + y = 12$
$0 = 34$; The system has no solution.

CHECKPOINT EXERCISES
For use after Examples 1–3:
1. Solve the system and interpret your results.
$3x + y = 4$
$-6x - 2y = -8$
$0 = 0$; The system has infinitely many solutions.

Example 1 ***Sample answer:***
The slope-intercept equations are: $y = 3x - 4$ and $y = 3x + 7$. Since these are parallel lines, there is no solution. Alternately, solving the second equation for y and substituting leads to the false equation $14 = -8$. So there is no solution.

EXTRA EXAMPLE 4

Sandra and Elio make the following statements about their ages. Write a system of linear equations to represent the situation. Solve the equations and interpret the results. Tell how you know that one of the statements is incorrect.

Sandra: Elio is 2 more than 3 times as old as I am.

Elio: If you take one-third of my age and subtract 1, you will have Sandra's age.

$E = 3S + 2$

$S = \frac{1}{3}E - 1$

$0 = -1$; There is no solution. One of the statements must be incorrect.

CHECKPOINT EXERCISES

For use after Example 4:

1. Trudy bought 2 notebooks and 3 packs of paper for \$6.25. Tran bought 6 packs of paper and 4 notebooks for \$12.50. Write a system of linear equations to represent this situation. Solve the system to explain why it is not possible to determine the price of a pack of paper or a notebook.
   $2x + 3y = 6.25$
   $4x + 6y = 12.50$
   The equations are equivalent, so there are many solutions.

FOCUS ON VOCABULARY

Saying that an equation or system of equations has infinitely many solutions means that the number of solutions is unlimited.

CLOSURE QUESTION

How is the number of solutions of a linear system related to its graph? **See margin.**

DAILY PUZZLER

Rene said, "I am thinking of an equation of the form $Ax + By = C$. Give me any linear equation, and I guarantee my equation and yours will form a system with infinitely many solutions." What is her equation?

$0x + 0y = 0$

FOCUS ON APPLICATIONS

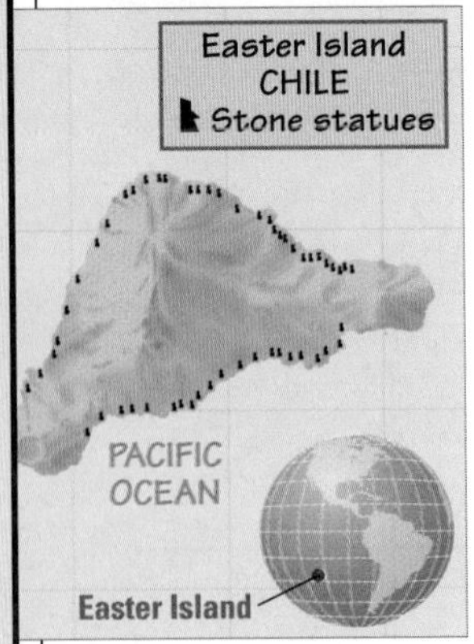

EASTER ISLAND is located in the South Pacific. On the island there are more than 600 massive stone statues called *moai.* The statues were carved hundreds of years ago, and some are over 40 feet high.

GOAL 2 MODELING A REAL-LIFE PROBLEM

EXAMPLE 4 *Error Analysis*

GEOMETRY CONNECTION Two students are visiting the mysterious statues on Easter Island in the South Pacific. To find the heights of two statues that are too tall to measure, they tried a technique involving proportions. They measured the shadow lengths of the statues at 2:00 P.M. and again at 3:00 P.M. Why were they unable to use their measurements to determine the heights of the statues?

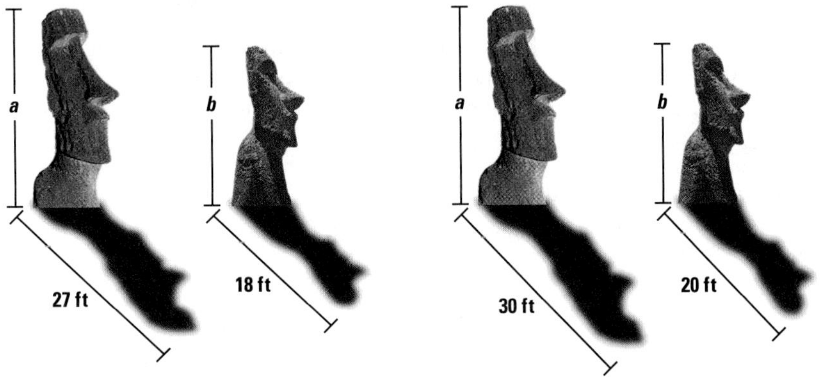

SOLUTION They let a and b represent the heights of the two statues. Because the ratios of corresponding sides of similar triangles are equal, the students wrote the following two equations.

2:00 P.M. $\frac{a}{27} = \frac{b}{18}$

$a = \frac{27}{18}b$

$a = \frac{3}{2}b$

3:00 P.M. $\frac{a}{30} = \frac{b}{20}$

$a = \frac{30}{20}b$

$a = \frac{3}{2}b$

▶ They got stuck because the equations that they wrote are equivalent. All that the students found from these measurements was that $a = \frac{3}{2}b$. In other words, the height of the first statue is one and one half times the height of the second.

To find the heights of the two statues, the students need to measure the shadow length of something else whose height they know or can measure. For instance, if at 2:00 P.M. a student 5 feet tall casts a shadow 3 feet long, the proportion $3:5 = 27:a$ can be used to find the height of the taller statue.

Closure Question *Sample answer:*
If the graph is two intersecting lines, there is one solution. If the graph is parallel lines there is no solution and if the graph is a single line, there are infinitely many solutions.

GUIDED PRACTICE

Vocabulary Check ✓

In Exercises 1–3, describe the graph of a linear system that has the given number of solutions. Sketch an example. **1–3. See margin.**

1. no solution **2.** infinitely many solutions **3.** exactly one solution

Concept Check ✓

1. The graphs of the equations are parallel lines.
2. The graphs of the equations are the same line.
3. The graphs of the equations are intersecting lines.

4. ERROR ANALYSIS Patrick says that the graph of the linear system shown at the right has no solution. Why is he wrong? **If the graphs were extended, they clearly would intersect.**

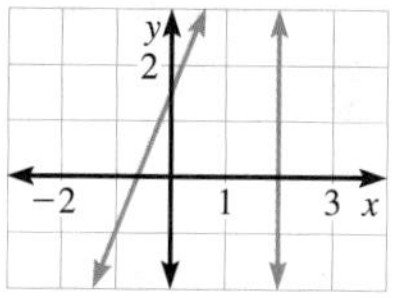

Ex. 4

5. Explain how you can tell if the system of linear equations has a solution. Then solve the system.

$x - y = 2$
$4x - 4y = 8$

Write both equations in slope-intercept form. If the slopes and y-intercepts are both different, the system has exactly one solution. If the slopes and the y-intercepts are both the same, the system has many solutions.

Skill Check ✓

Graph the system of linear equations. Does the system have *exactly one solution, no solution,* or *infinitely many solutions*? **See margin for graphs.**

6. $2x + y = 5$
$-6x - 3y = -15$
infinitely many solutions

7. $-6x + 2y = 4$
$-9x + 3y = 12$
no solution

8. $2x + y = 7$
$3x - y = -2$
exactly one solution

Use the substitution method or linear combinations to solve the linear system and tell how many solutions the system has.

9. $-x + y = 7$
$2x - 2y = -18$
no solution

10. $-4x + y = -8$
$-12x + 3y = -24$
infinitely many solutions

11. $-4x + y = -8$
$2x - 2y = -14$
exactly one solution, (5, 12)

PRACTICE AND APPLICATIONS

STUDENT HELP
Extra Practice to help you master skills is on p. 803.

MATCHING GRAPHS **Match the graph with its linear system. Does the system have *exactly one solution, no solution,* or *infinitely many solutions*?**

A. $-2x + 4y = 1$
$3x - 6y = 9$

B. $2x - 2y = 4$
$-x + y = -2$

C. $2x + y = 4$
$-4x - 2y = -8$

D. $-x + y = 1$
$x - y = 1$

E. $5x + 3y = 17$
$x - 3y = -2$

F. $x - y = 0$
$5x - 2y = 6$

12.
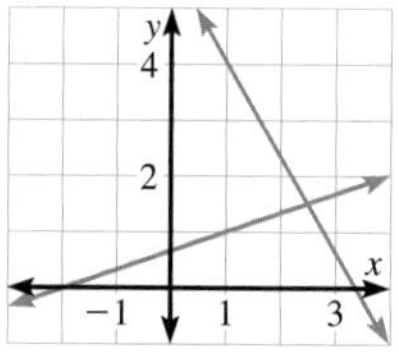

E; exactly one solution

13.
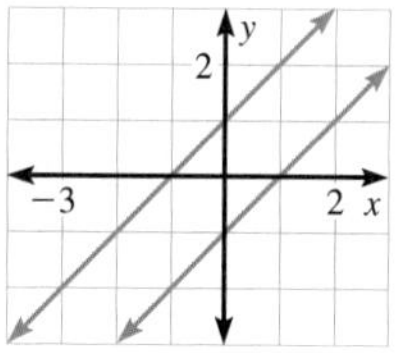

D; no solution

14.
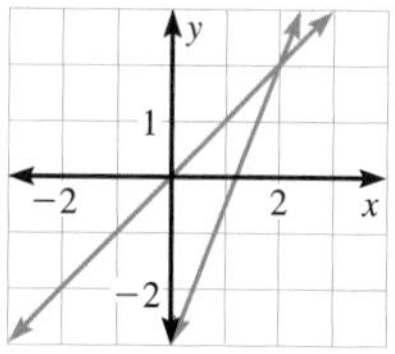

F; exactly one solution

15.
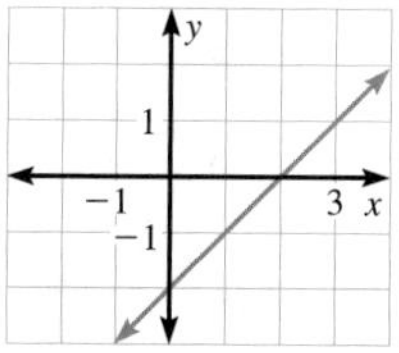

B; infinitely many solutions

16.
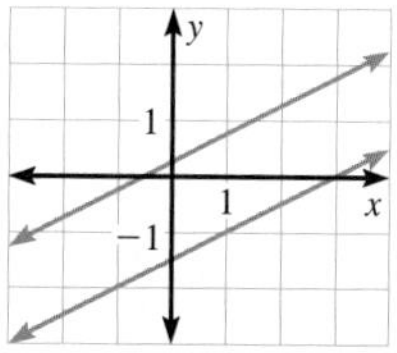

A; no solution

17.
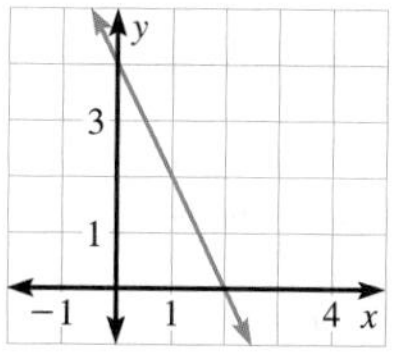

C; infinitely many solutions

STUDENT HELP
HOMEWORK HELP
Example 1: Exs. 12–29
Example 2: Exs. 12–29
Example 3: Exs. 18–29
Example 4: Ex. 30

3 APPLY

ASSIGNMENT GUIDE

BASIC
Day 1: pp. 429–431 Exs. 12–17, 18–28 even, 31, 32, 36–38, 43–48

AVERAGE
Day 1: pp. 429–431 Exs. 12–17, 18–28 even, 30–33, 36–38, 43–48

ADVANCED
Day 1: pp. 429–431 Exs. 12–17, 18–28 even, 30–33, 36–48

BLOCK SCHEDULE WITH 7.6
pp. 429–431 Exs. 12–17, 18–28 even, 30–33, 36–38, 43–48

EXERCISE LEVELS

Level A: *Easier*
12–17

Level B: *More Difficult*
18–32, 34–38, 43–48

Level C: *Most Difficult*
33, 39–43

✓ HOMEWORK CHECK

To quickly check student understanding of key concepts, go over the following exercises: Exs. 14, 16, 18, 22, 28, 32. See also the Daily Homework Quiz:

- Blackline Master (*Chapter 7 Resource Book,* p. 80)
- Transparency (p. 58)

6–8. See next page for answers.

CAREER NOTE

EXERCISES 31 AND 32

Additional information about carpenters is available at **www.mcdougallittell.com.**

STUDENT HELP NOTES

Software Help Instructions for several software packages are available at **www.mcdougallittell.com.**

6.

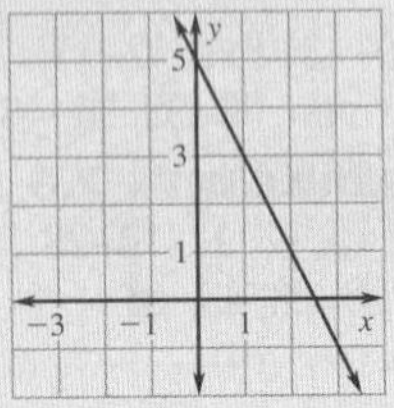

7.

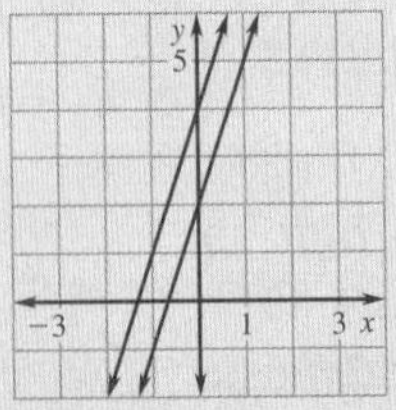

8.

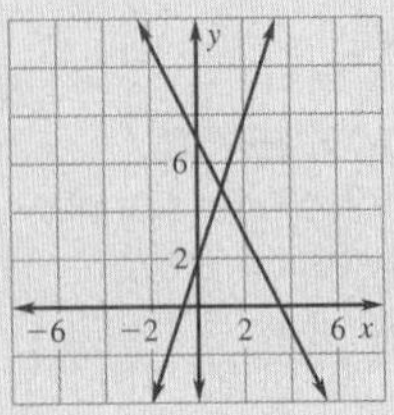

ADDITIONAL PRACTICE AND RETEACHING

For Lesson 7.5:

- Practice Levels A, B, and C (*Chapter 7 Resource Book,* p. 70)
- Reteaching with Practice (*Chapter 7 Resource Book,* p. 73)
- See Lesson 7.5 of the *Personal Student Tutor*

For more Mixed Review:

- Search the *Test and Practice Generator* for key words or specific lessons.

FOCUS ON CAREERS

CARPENTERS Carpenters know which materials will work best on a variety of projects. Their skills include framing walls, installing doors and windows, and tiling.

CAREER LINK www.mcdougallittell.com

30. No; the system of equations that describes the situation is $30x + 6y = 3.6$, $10x + 2y = 1.2$, which has infinitely many solutions.

32. Yes; you can combine the new equation, $8x + y = 139.69$, with either of the original equations to produce a system that has a solution, (14.98, 19.85). A sheet of oak paneling costs $14.98.

34. The system has no solution; the graphs of the equations are parallel; at every value of x, the difference between the y-values is the same.

STUDENT HELP

SOFTWARE HELP Visit our Web site www.mcdougallittell.com to see instructions for several software packages.

INTERPRETING ALGEBRAIC RESULTS **Use the substitution method or linear combinations to solve the linear system and tell how many solutions the system has.**

18. $-7x + 7y = 7$
$2x - 2y = -18$
no solution

19. $4x + 4y = -8$
$2x + 2y = -4$
infinitely many solutions

20. $2x + y = -4$
$4x - 2y = 8$
exactly one solution, (0, −4)

21. $15x - 5y = -20$
$-3x + y = 4$
infinitely many solutions

22. $-6x + 2y = -2$
$-4x - y = 8$
exactly one solution, (−1, −4)

23. $2x + y = -1$
$-6x - 3y = -15$
no solution

INTERPRETING GRAPHING RESULTS **Use the graphing method to solve the linear system and tell how many solutions the system has.**

24. $x + y = 8$
$x + y = -1$
no solution

25. $3x - 2y = 3$
$-6x + 4y = -6$
infinitely many solutions

26. $x - y = 2$
$-2x + 2y = 2$
no solution

27. $-x + 4y = -20$
$3x - 12y = 48$
no solution

28. $6x - 2y = 4$
$-4x + 2y = -\frac{8}{3}$
exactly one solution, $\left(\frac{2}{3}, 0\right)$

29. $\frac{3}{4}x + \frac{1}{2}y = 10$
$-\frac{3}{2}x - y = 4$
no solution

30. **JEWELRY** You have a necklace and matching bracelet with 2 types of beads. There are 30 small beads and 6 large beads on the necklace. The bracelet has 10 small beads and 2 large beads. The necklace weighs 3.6 grams and the bracelet weighs 1.2 grams. If the chain has no significant weight, can you find the weight of one large bead? Explain. **See margin.**

CARPENTRY SUPPLIES **In Exercises 31 and 32, use the following information.**

A carpenter is buying supplies for a job. The carpenter needs 4 sheets of oak paneling and 2 sheets of shower tileboard. The carpenter pays $99.62 for these supplies. For the next job the carpenter buys 12 sheets of oak paneling and 6 sheets of shower tileboard and pays $298.86.

31. Could you find how much the carpenter is spending on 1 sheet of oak paneling? Explain. **No; the system of equations that describes the situation is** $4x + 2y = 99.62$, $12x + 6y = 298.86$, **which has infinitely many solutions.**

32. If the carpenter later spends a total of $139.69 for 1 sheet of shower tileboard and 8 sheets of oak paneling, could you find how much 1 sheet of oak paneling costs? Explain. **See margin.**

33. **CRITICAL THINKING** Use linear systems to determine where the x-coordinate is equal to the y-coordinate on the graph of $2x + 3y = 25$. **(5, 5)**

USING A SPREADSHEET **In Exercises 34 and 35, use the spreadsheet at the right to investigate the linear system.**

$y = 2x + 3$
$y = 2x - 9$

34. What do the results mean? Explain how you know. **See margin.**

35. Use a spreadsheet or a table of values to verify your answers to Exercises 6–8. **Check spreadsheets.**

linear system

	A	B	C	D
1	x	y = 2x + 3	y = 2x – 9	col. B – col. C
2	–3	–3	–15	12
3	–2	–1	–13	12
4	–1	1	–11	12
5	0	3	–9	12
6	1	5	–7	12
7	2	7	–5	12
8	3	9	–3	12
9	4	11	–1	12

Exs. 34 and 35

Test Preparation

36. $y = \frac{1}{2}[2(x + 10) - 18] - x$

37. The system has infinitely many solutions, all ordered pairs whose graphs are on the line $y = 1$. This means that every real number x is a solution.

MULTI-STEP PROBLEM **In Exercises 36–38, use the description shown below.**

> Choose any number.
> Add 10 to the number.
> Multiply the result by 2.
> Subtract 18 from the result.
> Multiply the result by one half.
> Subtract the original number.

36. Write a linear equation for the description. Let x represent the number that is chosen and let y represent the final result. **See margin.**

37. To find the number a person would have chosen to obtain a final result of 1, solve the linear system consisting of the equation you wrote in Exercise 36 and the equation $y = 1$. Explain your result.

38. **LOGICAL REASONING** Write a number trick of your own that will give similar results. Explain why your trick works. **Check results.**

★ Challenge

ALTERING LINEAR SYSTEMS **In Exercises 39–42, perform parts (a)–(c).**

a. Find a value of n so that the linear system has infinitely many solutions.

b. Find a value of n so that the linear system has no solution.

c. Graph both results.

EXTRA CHALLENGE
www.mcdougallittell.com

39. $x - y = 3$
$4x - 4y = n$
a. 12
b. any real number n, $n \neq 12$
c. Check graphs.

40. $x - y = 4$
$-2x + 2y = n$
a. −8
b. any real number n, $n \neq -8$
c. Check graphs.

41. $6x - 9y = n$
$-2x + 3y = 3$
a. −9
b. any real number n, $n \neq -9$
c. Check graphs.

42. $9x + 6y = n$
$1.8x + 1.2y = 3$
a. 15
b. any real number n, $n \neq 15$
c. Check graphs.

MIXED REVIEW

MATCHING GRAPHS **Match the inequality with its graph.** (Review 6.5 for 7.6)

A. $2x + y \leq 3$ **B.** $2x + y < 3$ **C.** $2x + y \geq 3$ **D.** $2x + y > 3$

43. 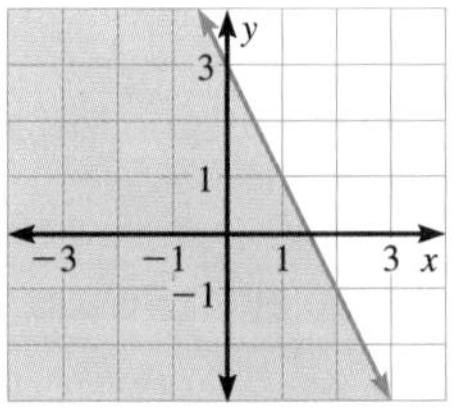
A

44. 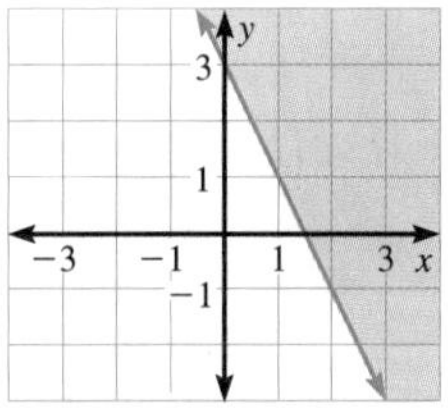
C

45. 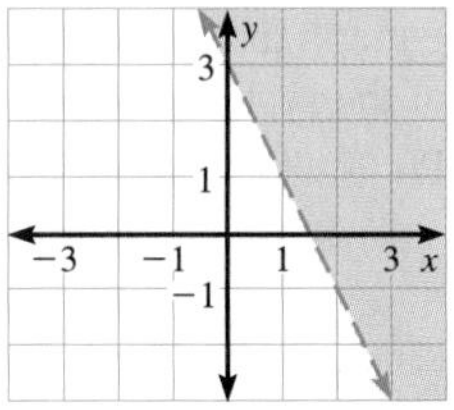
D

46. 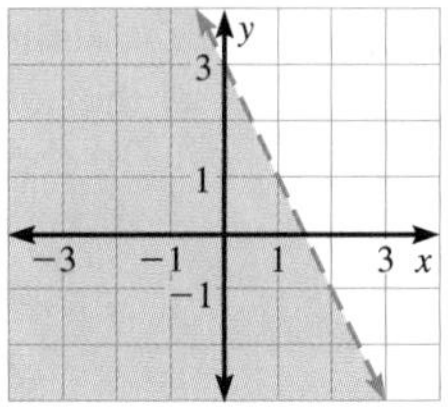
B

ROCK CLIMBING **In Exercises 47 and 48, you are climbing a 500-foot cliff. By 1:00 P.M. you have climbed 125 feet up the cliff. By 4:00 P.M. you have reached a height of 290 feet.** (Review 4.4)

47. What is your rate of change in height? **55 ft/h**

48. If you continue climbing the cliff at the same rate, at what time will you reach the top? **at about 7:49 P.M.**

4 ASSESS

DAILY HOMEWORK QUIZ

Transparency Available

The diagram shows the graphs of the following three equations.
A: $x - 2y = 2$
B: $-2x + 4y = 8$
C: $3x + 2y = 2$

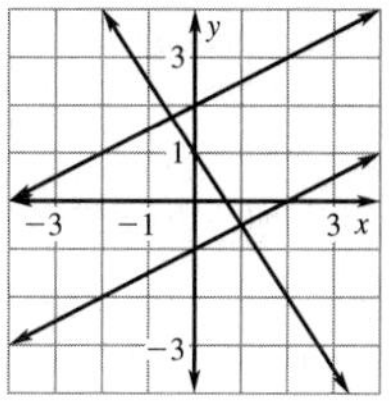

How many solutions are there for the system formed by the pair of equations?

1. A and B **no solution**
2. A and C **1 solution**

Solve the linear system. Tell how many solutions the system has.

3. $3x + 4y = 5$
$-2x + y = 4$ **1 solution, (−1, 2)**

4. $7x - 3y = 4$
$-14x + 6y = -8$
infinitely many solutions

5. For 4 hamburgers and 4 sodas you pay \$8. For 6 hamburgers and 6 sodas you pay \$12. Can you use this information to find the price of 1 hamburger and the price of 1 soda? Explain. **no; the system $4h + 4s = 8$, $6h + 6s = 12$ has infinitely many solutions.**

EXTRA CHALLENGE NOTE
Challenge problems for Lesson 7.5 are available in **blackline** format in the *Chapter 7 Resource Book,* p. 77 and at **www.mcdougallittell.com.**

ADDITIONAL TEST PREPARATION

1. OPEN ENDED Write a system of linear equations that has no solution. Solve the system using both the graphing method and either substitution or linear combinations. **Check students' work.**

1 PLAN

PACING
Basic: 2 days
Average: 2 days
Advanced: 2 days
Block Schedule: 0.5 block with 7.5
0.5 block with Ch. Rev.

LESSON OPENER ACTIVITY

An alternative way to approach Lesson 7.6 is to use the Activity Lesson Opener:

- Blackline Master (*Chapter 7 Resource Book,* p. 81)
- Transparency (p. 51)

MEETING INDIVIDUAL NEEDS

- ***Chapter 7 Resource Book***
  Prerequisite Skills Review (p. 5)
  Practice Level A (p. 85)
  Practice Level B (p. 86)
  Practice Level C (p. 87)
  Reteaching with Practice (p. 88)
  Absent Student Catch-Up (p. 90)
  Challenge (p. 92)
- ***Resources in Spanish***
- ***Personal Student Tutor***

NEW-TEACHER SUPPORT

See the Tips for New Teachers on pp. 1–2 of the *Chapter 7 Resource Book* for additional notes about Lesson 7.6.

WARM-UP EXERCISES

Transparency Available

In Exs. 1 and 2, tell whether the graph of the inequality will have a solid or a dashed boundary line.

1. $y \leq 2x + 3$ **solid**

2. $-2x + 1 > y$ **dashed**

3. Which inequality's graph is included in the graph of one of the other inequalities?
$2x + 3y \leq 8$ $3x + 2y \leq 7$
$4x + 6y \leq 9$ $x + y > 6$
$\mathbf{4x + 6y \leq 9}$

7.6

What you should learn

GOAL 1 Solve a system of linear inequalities by graphing.

GOAL 2 Use a system of linear inequalities to model a **real-life** situation such as earnings from two jobs in **Exs. 38** and **39**.

Why you should learn it

▼ To model **real-life** situations, such as buying spotlights for a concert hall in **Example 4**.

REAL LIFE

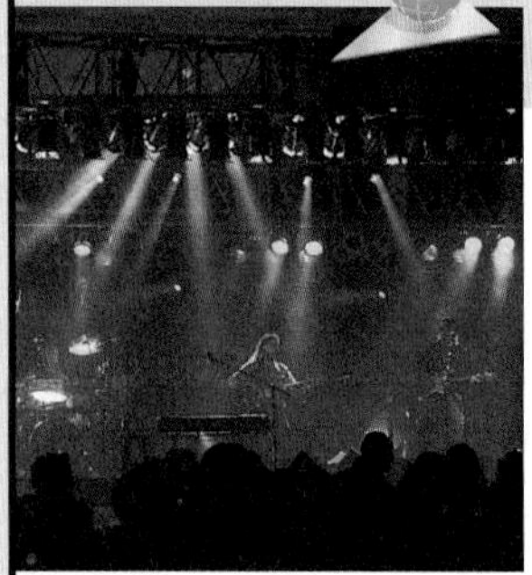

Solving Systems of Linear Inequalities

GOAL 1 SOLVING INEQUALITIES BY GRAPHING

From Lesson 6.5, remember that the graph of a linear inequality in two variables is a half-plane. The boundary line of the half-plane is dashed if the inequality is $<$ or $>$ and solid if the inequality is $\leq$ or $\geq$.

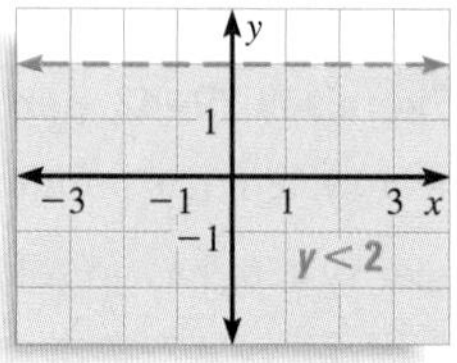

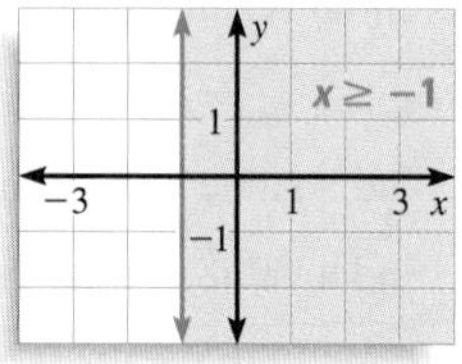

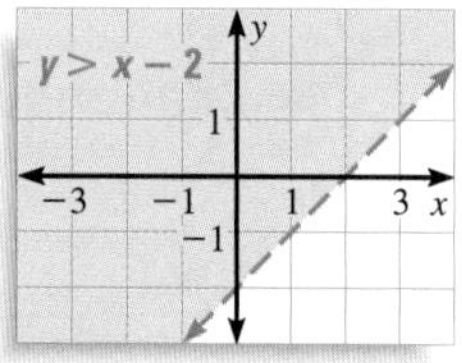

Two or more linear inequalities form a **system of linear inequalities** or simply a *system of inequalities*. A **solution** of a system of linear inequalities is an ordered pair that is a solution of each inequality in the system. The **graph** of a system of linear inequalities is the graph of *all solutions* of the system.

EXAMPLE 1 *A Triangular Solution Region*

Graph the system of linear inequalities.

$y < 2$ **Inequality 1**
$x \geq -1$ **Inequality 2**
$y > x - 2$ **Inequality 3**

SOLUTION

Graph all three inequalities in the same coordinate plane. The graph of the system is the overlap, or *intersection,* of the three half-planes shown.

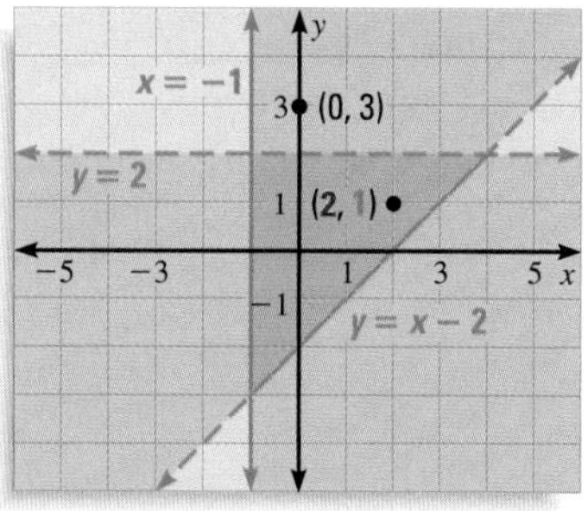

✓**CHECK** You can see from the graph that the point (2, 1) is a solution of the system. To check this, substitute the point into each inequality.

$1 < 2$ **True**
$2 \geq -1$ **True**
$1 > 2 - 2$ **True**

The point (0, 3) is not in the graph of the system. Notice (0, 3) is not a solution of Inequality 1. This point is not a solution of the system.

....

When graphing a system of inequalities, it is helpful to find each corner point (or *vertex*). For instance, the graph in Example 1 has three corner points: (−1, 2), (−1, −3), and (4, 2).

STUDENT HELP

HOMEWORK HELP
Visit our Web site www.mcdougallittell.com for extra examples.

EXAMPLE 2 *Solution Region Between Parallel Lines*

Write a system of inequalities that defines the shaded region at the right.

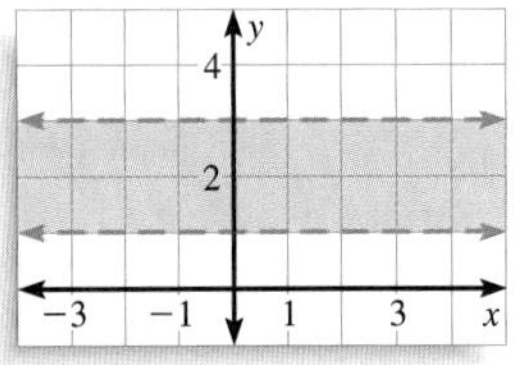

SOLUTION

The graph of one inequality is the half-plane *below* the line $y = 3$.

The graph of the other inequality is the half-plane *above* the line $y = 1$.

The shaded region of the graph is the horizontal band that lies *between* the two horizontal lines, $y = 3$ and $y = 1$, but not *on* the lines.

▶ The system of linear inequalities at the right defines the shaded region.

$y < 3$ **Inequality 1**
$y > 1$ **Inequality 2**

EXAMPLE 3 *A Quadrilateral Solution Region*

STUDENT HELP

Study Tip
When you are graphing a system of inequalities, try shading with several colored pencils. The graph of the system is the region that has been shaded by every color.

Graph the system of linear inequalities. Label each vertex of the solution region. Describe the shape of the region.

$x \geq 0$ **Inequality 1**
$y \geq 0$ **Inequality 2**
$y \leq 2$ **Inequality 3**
$y \leq -\frac{1}{2}x + 3$ **Inequality 4**

SOLUTION

The graph of the first inequality is the half-plane *on and to the right* of the *y*-axis.

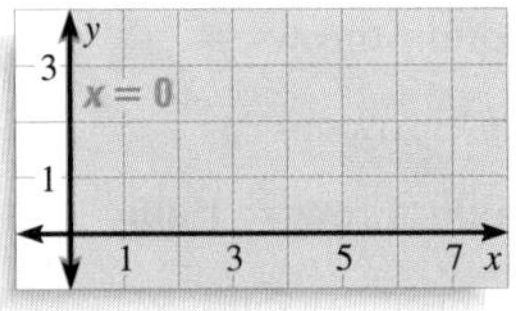

The graph of the second inequality is the half-plane *on and above* the *x*-axis.

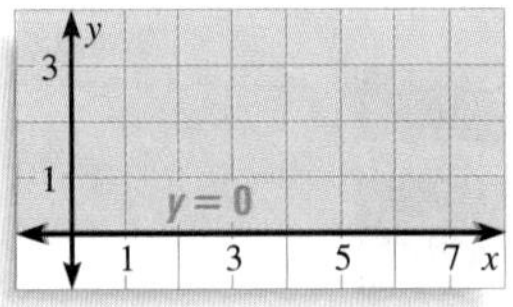

The graph of the third inequality is the half-plane *on and below* the horizontal line $y = 2$.

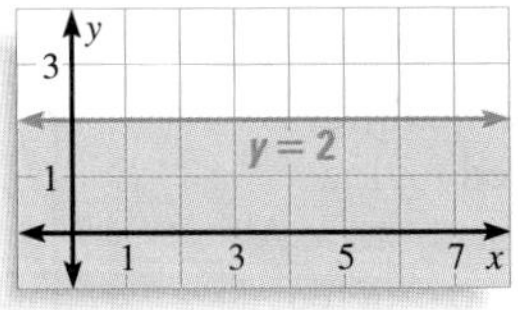

The graph of the fourth inequality is the half-plane *on and below* the line $y = -\frac{1}{2}x + 3$.

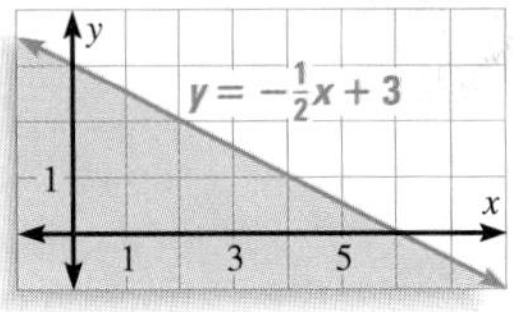

▶ The region that lies in all four half-planes is a *quadrilateral* with vertices at (0, 2), (0, 0), (6, 0), and (2, 2). Note that the point (0, 3) is not a vertex of the solution region even though two boundary lines meet at that point.

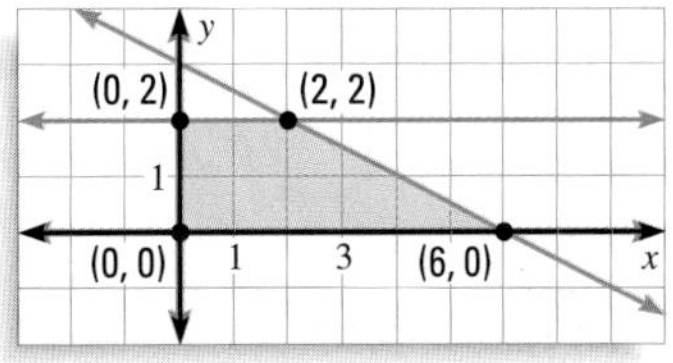

2 TEACH

EXTRA EXAMPLE 1
Graph the system.
$x < 4$; $y \geq -3$; $x - y \geq 4$

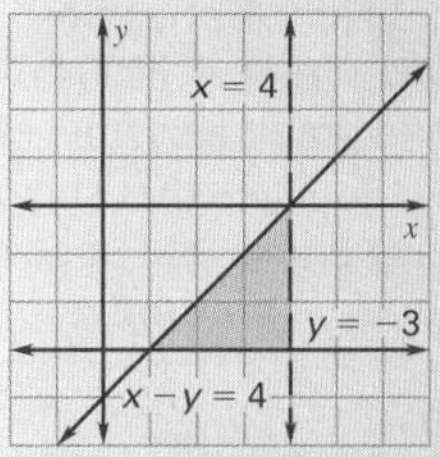

EXTRA EXAMPLE 2
Write a system of inequalities that defines the shaded region below.

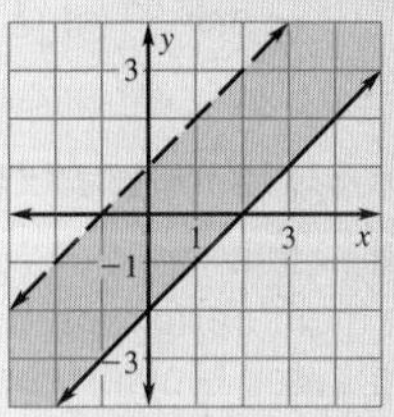

$y \geq x - 2$
$y < x + 1$

EXTRA EXAMPLE 3
Sketch the graph of the system of linear inequalities.
$x \geq 1$; $x \leq 4$; $y \geq -1$; $y \leq \frac{1}{2}x + 1$

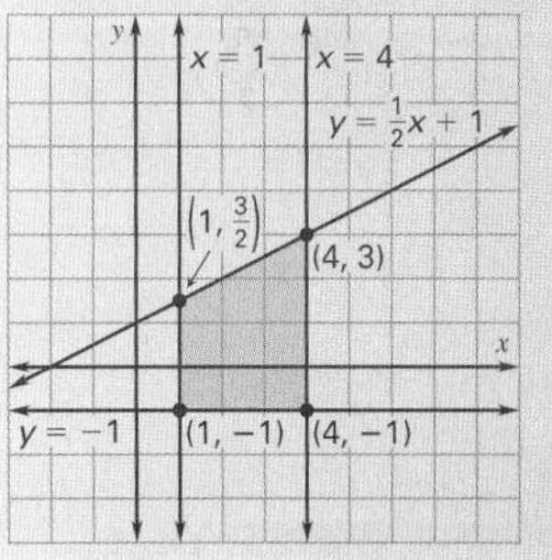

CHECKPOINT EXERCISES
For use after Examples 1–3:
1. Sketch the graph of the system of inequalities.
$x \geq 0$; $-x + 3y \leq 6$; $y > x$

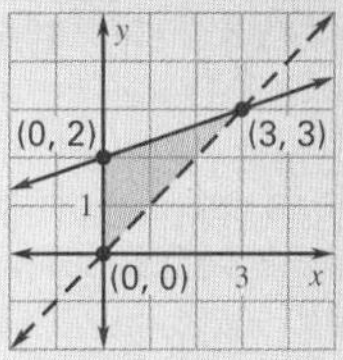

Extra Example 4

You are ordering bricks and sand. You need a minimum of 500 bricks and 10 bags of sand. Bricks weigh 2 lb each and sand weighs 50 lb per bag. The maximum weight that can be delivered in a truck is 3000 lb.

a. Write and graph a system of linear inequalities that shows how many bricks and bags of sand could be ordered.

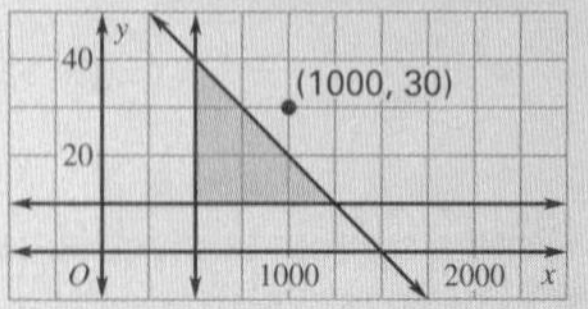

$x \geq 500$, $y \geq 10$, $2x + 50y \leq 3000$

b. Can 1000 bricks and 30 bags of sand be delivered? **no**

Checkpoint Exercises

For use after Example 4:

1. You have a gift certificate to a movie theater. Movies cost \$7 for adults and \$5 for children. You need to spend at least as much as the \$50 certificate. You want at least 2 adults to go. Write and graph a system of linear inequalities.

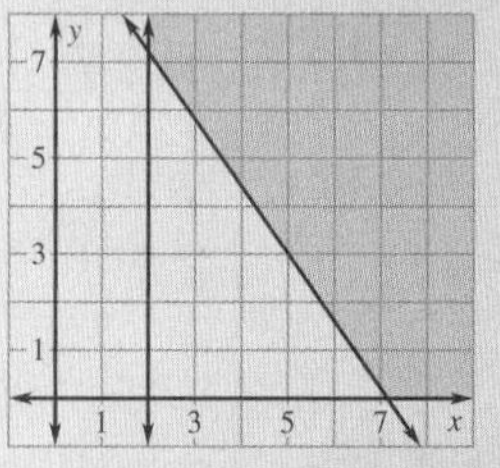

$7x + 5y \geq 50$, $x \geq 2$, $y \geq 0$

Closure Question

How is the solution of a system of linear inequalities similar to the solution of a system of linear equations. How is it different? **The solution to a system of linear inequalities must satisfy each inequality just as the solution to a system of linear equations must satisfy each equation. The solution to a system of linear inequalities is usually a region, whereas a solution to a system of linear equations is a point or a line.**

GOAL 2 Modeling a Real-Life Problem

EXAMPLE 4 *Writing a System of Linear Inequalities*

LIGHTING You are ordering lighting for a theater so the spotlights can follow the performers. The lighting technician needs at least 3 medium-throw spotlights and at least 1 long-throw spotlight. A medium-throw spotlight costs \$1000 and a long-throw spotlight costs \$3500. The minimum order for free delivery is \$10,000.

a. Write and graph a system of linear inequalities that shows how many medium-throw spotlights and long-throw spotlights should be ordered to get the free delivery.

b. Will an order of 4 medium-throw spotlights and 1 long-throw spotlight be delivered free?

Solution

PROBLEM SOLVING STRATEGY

a. VERBAL MODEL

Number of medium-throws $\geq$ **3**

Number of long-throws $\geq$ **1**

Number of medium-throws • **Price of a medium-throw** + **Number of long-throws** • **Price of a long-throw** $\geq$ **10,000**

LABELS

Number of medium-throws = x (no units)

Number of long-throws = y (no units)

Price of a medium-throw = **1000** (dollars)

Price of a long-throw = **3500** (dollars)

ALGEBRAIC MODEL

$x \geq 3$ **Inequality 1**

$y \geq 1$ **Inequality 2**

$1000x + 3500y \geq 10{,}000$ **Inequality 3**

The graph of the system of inequalities is shown. Any point in the shaded region of the graph is a solution of the system. A fraction of a spotlight cannot be ordered, so only ordered pairs of integers in the shaded region will correctly answer the problem.

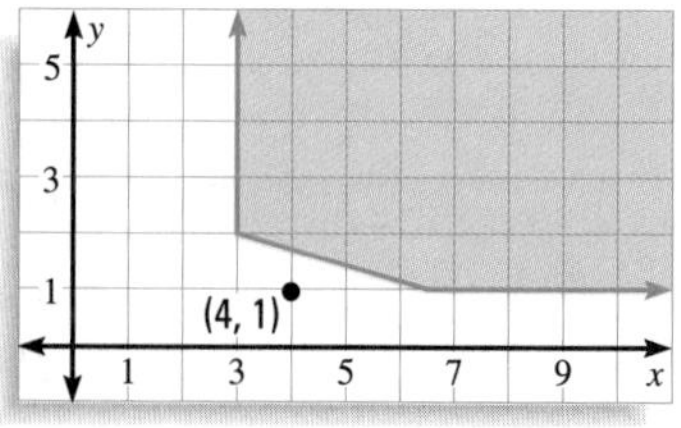

b. The point (4, 1) is outside the solution region, so an order of 4 medium-throw spotlights and 1 long-throw spotlight would not be delivered free.

FOCUS ON CAREERS

REAL LIFE **LIGHTING TECHNICIANS** use extra spotlights on each area of the stage to do color changes for the whole stage. Lighting control boards can be programmed in advance to produce complicated lighting sequences.

INTERNET CAREER LINK
www.mcdougallittell.com

Guided Practice

Vocabulary Check ✓

1. Determine whether the following statement is *true* or *false. A solution of a system of linear inequalities is an ordered pair that is a solution of any one of the inequalities in the system.* Explain. **See margin.**

Concept Check ✓

2. How is graphing a system of linear inequalities the same as graphing a system of linear equations? How is it different? **See margin.**

3. After you graph a system of linear inequalities, how can you use algebra to check whether the correct region is shaded? **See margin.**

Skill Check ✓

In Exercises 4 and 5, use both the student graph at the right and the system of linear inequalities shown below. **4, 5. See margin.**

$y > -1$
$x \geq 2$
$y > x - 4$

4. Describe the errors the student made while graphing the system.

5. Graph the system correctly.

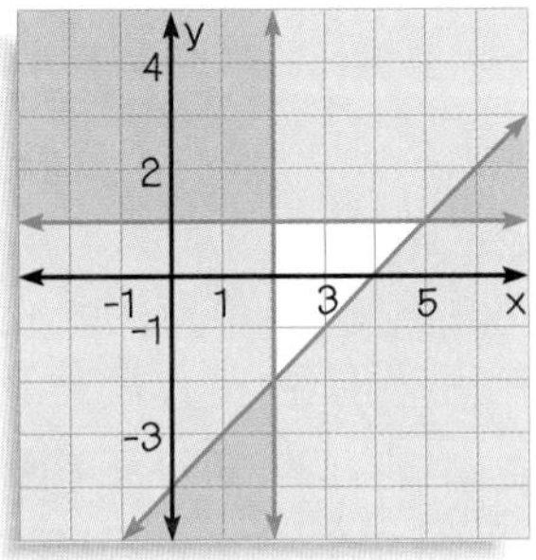

Ex. 4

Graph the system of linear inequalities. **6–8. See margin for graphs.**

6. $y > -2x + 2$
$y \leq -1$

7. $y > x$
$x \leq 1$

8. $x + 1 > y$
$y \geq 0$

1. False; a solution of a system of linear inequalities is an ordered pair that is a solution of each of the inequalities in the system.

2. *Sample answer:* You must graph the boundary lines, which are graphs of linear equations; you must determine whether the boundary lines are dashed or solid and which half-plane determined by the boundary line must be shaded.

Practice and Applications

Student Help

Extra Practice to help you master skills is on p. 803.

Matching Graphs **Match the system of linear inequalities with its graph.**

A. $2x + y < 4$
$-2x + y \leq 4$

B. $2x + y \geq -4$
$x - 2y < 4$

C. $2x + y \leq 4$
$2x + y \geq -4$

9. 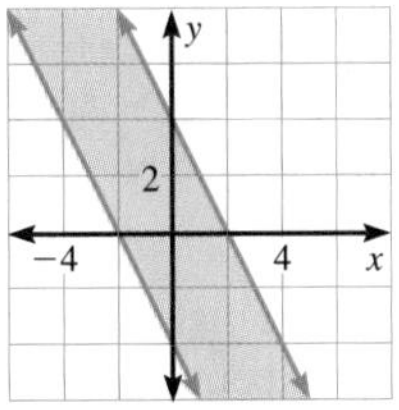

C

10. 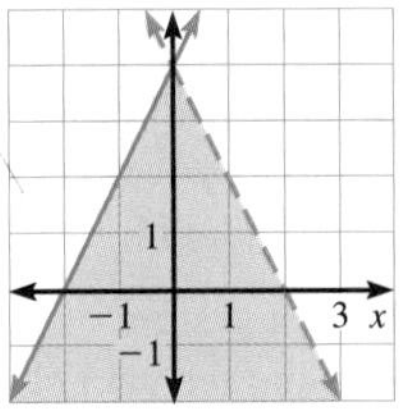

A

11. 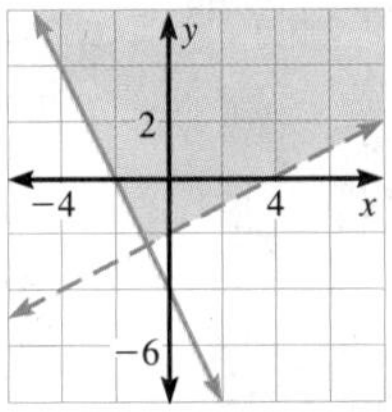

B

12. $y \leq -\frac{5}{2}x + 4$, $y > -\frac{1}{2}x - 2$

13. $y \leq \frac{1}{2}x + 2$, $y \geq \frac{1}{2}x - 2$

14. $y \leq -\frac{1}{2}x + 2$, $y < \frac{1}{2}x + 2$

Writing a System **Write a system of linear inequalities that defines the shaded region.** **12–14. See margin.**

12. 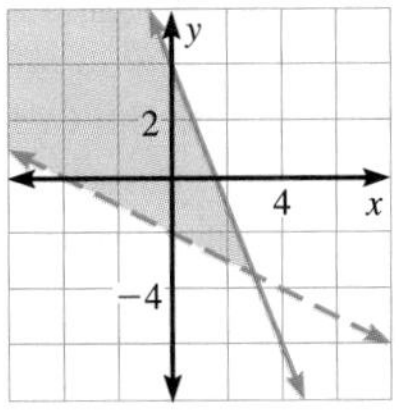

13. 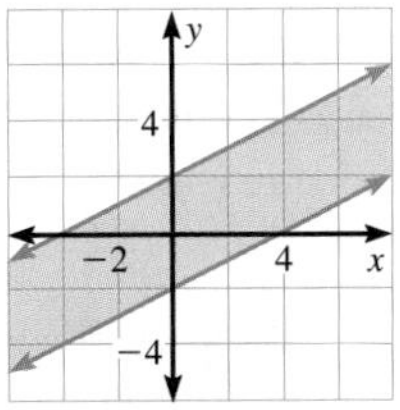

14. 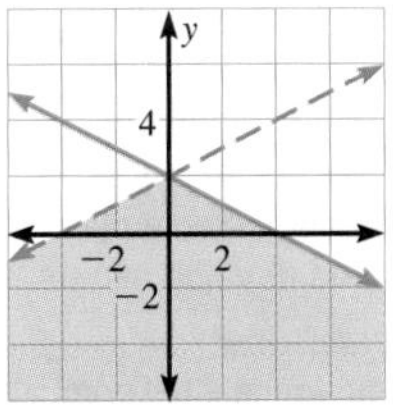

Student Help

Homework Help
Example 1: Exs. 9–11, 15–26
Example 2: Exs. 12–14
Example 3: Exs. 15–26
Example 4: Exs. 34–38

3 Apply

Assignment Guide

Basic
Day 1: pp. 435–436 Exs. 9–14, 16–26 even
Day 2: pp. 436–438 Exs. 30–36, 42, 45–52, 57, 58, Quiz 2 Exs. 1–14

Average
Day 1: pp. 435–436 Exs. 9–14, 16–26 even
Day 2: pp. 436–438 Exs. 30–39, 42, 45–52, 57, 58, Quiz 2 Exs. 1–14

Advanced
Day 1: pp. 435–436 Exs. 9–14, 16–26 even
Day 2: pp. 436–438 Exs. 30–39, 42–52, 57, 58, Quiz 2 Exs. 1–14

Block Schedule
pp. 435–436 Exs. 9–14, 16–26 even (with 7.5)
pp. 436–438 Exs. 30–39, 42, 45–52, 57, 58, Quiz 2 Exs. 1–14 (with Ch. 7 Review)

Exercise Levels

Level A: *Easier*
9–14, 45–56

Level B: *More Difficult*
15–39, 42, 57, 58

Level C: *Most Difficult*
40, 41, 43, 44

✓ Homework Check

To quickly check student understanding of key concepts, go over the following exercises: Exs. 10, 12, 16, 20, 26, 30, 34. See also the Daily Homework Quiz:

- Blackline Master (*Chapter 8 Resource Book,* p. 11)
- Transparency (p. 60)

3–8. See Additional Answers beginning on page AA1.

! COMMON ERROR

EXERCISES 15–26 Students may shade the wrong region of the graph. To help them avoid this error, they should pick a point in the region to test at the completion of each problem. The coordinates of the point should satisfy each of the inequalities in the system. In addition, students should check that they have used solid and dashed lines appropriately.

GRAPHING CALCULATOR NOTE

EXERCISES 27–29 Use of a graphing calculator in these exercises will speed the graphing process and help students focus on the solution to the system of inequalities. They can use the *intersect* feature of the calculator to find the coordinates of the vertex points.

CAREER NOTE

EXERCISES 34 AND 35 Additional information on a career as a chef is available at **www.mcdougallittell.com.**

15–26. See Additional Answers beginning on page AA1.
35–36. See Additional Answers beginning on page AA1.
38. See Additional Answers beginning on page AA1.
40. See Additional Answers beginning on page AA1.

ADDITIONAL PRACTICE AND RETEACHING

For Lesson 7.6:

- Practice Levels A, B, and C (*Chapter 7 Resource Book,* p. 85)
- Reteaching with Practice (*Chapter 7 Resource Book,* p. 88)
- See Lesson 7.6 of the *Personal Student Tutor*

For more Mixed Review:

- Search the *Test and Practice Generator* for key words or specific lessons.

30. $y \geq 0$, $y \leq x + 2$, $y \leq -x + 2$

31. $1 \leq x \leq 7$, $1 \leq y \leq 6$

32. $y \geq 0$, $y \leq \frac{5}{4}x + \frac{35}{4}$, $y \geq -\frac{5}{3}x$

33. $y \leq x + 2$, $y \leq -x + 7$, $1 \leq y \leq 3$

GRAPHING A SYSTEM **Graph the system of linear inequalities.** 15–26. See margin for graphs.

15. $2x + y > 2$
$6x + 3y < 12$

16. $2x - 2y < 6$
$x - y < 9$

17. $x - 3y \geq 12$
$x - 6y \leq 12$

18. $x + y < 4$
$x + y > -2$

19. $x + y \leq 6$
$x \geq 1$
$y \geq 0$

20. $\frac{3}{2}x + y < 3$
$x > 0$
$y > 0$

21. $x \geq 0$
$y \geq 0$
$x \leq 3$
$y \leq 5$

22. $x > -2$
$y \geq -2$
$x \leq 1$
$y \leq 4$

23. $-\frac{3}{2}x + y \leq 3$
$\frac{1}{4}x + y > -\frac{1}{2}$
$4x + y < 2$

24. $2x + y \geq 2$
$x \leq 3$
$y \leq \frac{1}{2}$

25. $x \geq 2$
$x - 2y \geq 3$
$3x + y \geq 9$
$x + y \leq 7$

26. $x \geq 1$
$x - 2y \leq 3$
$3x + 2y \geq 9$
$x + y \leq 6$

GRAPHING A SYSTEM **Use a graphing calculator or a computer to graph the system of inequalities. Give the coordinates of each vertex of the solution region.**

27. $x + y \leq 11$
$5x - 3y \leq 15$
$x \geq 0$
(0, −5), (0, 11), (6, 5)

28. $5x - 3y \geq -7$
$3x + y \leq 7$
$x + 5y \geq -7$
(1, 4), (−2, −1), (3, −2)

29. $x + 2y \geq 0$
$5x - 2y \leq 0$
$-x + y \leq 3$
(0, 0), (−2, 1), (2, 5)

GEOMETRY CONNECTION **In Exercises 30–33, plot the points and draw line segments connecting the points to create the polygon. Then write a system of linear inequalities that defines the polygonal region.** 30–33. See margin.

30. Triangle: (−2, 0), (2, 0), (0, 2)

31. Rectangle: (1, 1), (7, 1), (7, 6), (1, 6)

32. Triangle: (0, 0), (−7, 0), (−3, 5)

33. Trapezoid: (−1, 1), (1, 3), (4, 3), (6, 1)

FOOD BUDGET **In Exercises 34 and 35, use the following information.** You are planning the menu for your restaurant. Each evening you offer two different meals and have at least 240 customers. For Saturday night you plan to offer roast beef and teriyaki chicken. You expect that fewer customers will order the beef than will order the chicken. The beef costs you \$5 per serving and the chicken costs you \$3 per serving. You have a budget of at most \$1200 for meat for Saturday night.

34. Write a system of linear inequalities that shows the various numbers of roast beef and teriyaki chicken meals you could make for Saturday night. $x + y \geq 240$, $y > x$, $5x + 3y \leq 1200$, $x \geq 0$, $y \geq 0$

35 Graph the system of linear inequalities. See margin.

36. **LOCAL MAGAZINE** A monthly magazine is hiring reporters to cover school events and local events. In each magazine, the managing editor wants at least 4 reporters covering local news and at least 1 reporter covering school news. The budget allows for not more than 9 different reporters' articles to be in one magazine. Graph the region that shows the possible combinations of local and school events covered in a magazine. See margin.

37. **GEOMETRY CONNECTION** What is the area of the region described by the system of linear inequalities $x \leq 3$, $y \leq 1$, and $x + y \geq 0$? 8 square units

FOCUS ON CAREERS

REAL LIFE **CHEF** Chefs prepare meals that appeal to both the taste buds and the eye. They often direct kitchen workers, estimate food requirements, and order new supplies. Some plan meals and develop menus.

INTERNET CAREER LINK www.mcdougallittell.com

EARNING MONEY In Exercises 38–39, use the following information. Suppose you can work a total of no more than 20 hours per week at your two jobs. Baby-sitting pays \$5 per hour, and your cashier job pays \$6 per hour. You need to earn at least \$90 per week to cover your expenses.

38. Write a system of inequalities that shows the various numbers of hours you can work at each job. Graph the result. **$x + y \le 20$, $5x + 6y \ge 90$, $x \ge 0$, $y \ge 0$; See margin for graphs.**

39. Give two possible ways you could divide your hours between the two jobs.

39. *Sample answers:* **4 h baby-sitting and 14 h as cashier, 12 h baby-sitting and 6 h as cashier**

EXTENSION: LINEAR PROGRAMMING In Exercises 40 and 41, use the following information.
You own a bottle recycling center that receives bottles that are either sorted by color or unsorted. To sort and recycle all of the bottles, you can use up to 4200 hours of human labor and up to 2400 hours of machine time. The system below represents the number of hours your center spends sorting and recycling bottles where x is the number of tons of unsorted bottles and y is the number of tons of sorted bottles.

$4x + y \le 4200$ — **Human labor: up to 4200 hours**
$2x + y \le 2400$ — **Machine time: up to 2400 hours**
$x \ge 0, y \ge 0$ — **Cannot have negative number of tons**

40. Graph the system of linear inequalities. **See margin.**

41. a. Find the vertices of your graph. **(0, 0), (0, 2400), (900, 600), (1050, 0)**

b. You will earn \$30 per ton for bottles that are sorted by color and earn \$10 per ton for unsorted bottles. Let P be the maximum profit. Substitute the ordered pairs from part (a) into the following equation and solve.

$P = 10x + 30y$ **0; 72,000; 27,000; 10,500**

c. Assuming that the maximum profit occurs at one of the four vertices, what is the maximum profit? **\$72,000**

Test Preparation

42. MULTI-STEP PROBLEM Use the tree farm shown at the right.

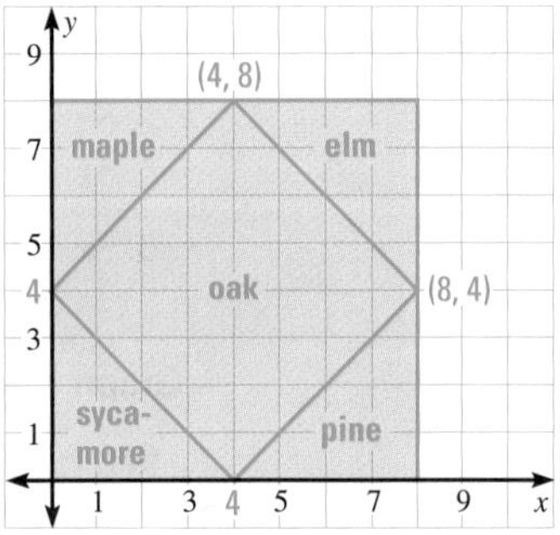

a. Write a system of inequalities that defines the region containing maple trees. **$x \ge 0$, $y \le 8$, $y \ge x + 4$**

b. Write a system of inequalities that defines the region containing sycamore trees. **$x \ge 0$, $y \ge 0$, $y \le -x + 4$**

c. Write a system of inequalities that defines the region containing oak trees. **See margin.**

d. Calculate the area of the oak tree region. Explain your method. **See margin.**

42c. **$y \le x + 4$, $y \ge -x + 4$, $y \le -x + 12$, $y \ge x - 4$**

42d. **32 square units;** ***Sample answer:*** **The area of the large square is 64 square units and the area of each triangular region is 8 square units; $64 - 4(8) = 32$.**

★ Challenge

43. CRITICAL THINKING Explain why the system of inequalities has no solution.

$2x - y > 4$
$y > 2x - 2$

The half-planes determined by the inequalities do not intersect.

44. VISUAL THINKING Describe the graph of the system of inequalities.

$2x + 3y > -6$
$2x + 3y \ge 6$

the line with equation $2x + 3y = 6$ and the half-plane above it

EXTRA CHALLENGE
www.mcdougallittell.com

Additional Test Preparation ***Sample answer:***
2. Check students' graphs. All points that satisfy $x \ge 1$ and $x + y \le 6$ already meet the condition imposed by $y \le 5$.

4 ASSESS

DAILY HOMEWORK QUIZ

Transparency Available

Graph the system of linear inequalities.

1. $y \le x + 1$; $x > 1$; $x < 3$; $y \ge 0$

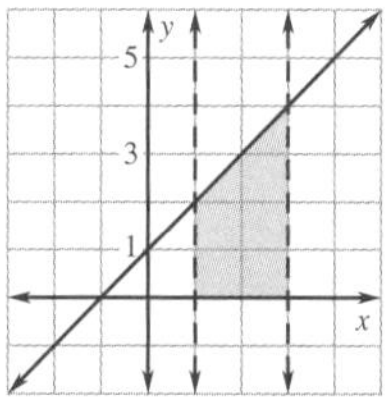

2. $y \ge 0.5x - 1$; $y \le 0.5x + 2$; $x \ge -1$; $x \le 2$

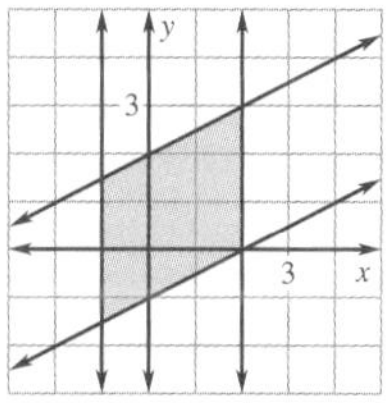

3. Plot the points (–1, 1), (3, 4), and (3, 1). Connect the points with segments. Write a system of linear inequalities to describe the triangular region. **Check students' graphs.; $y \le \frac{3}{4}x + \frac{7}{4}$; $y \ge 1$; $x \le 3$**

EXTRA CHALLENGE NOTE

Challenge problems for Lesson 7.6 are available in **blackline** format in the *Chapter 7 Resource Book,* p. 92 and at **www.mcdougallittell.com.**

ADDITIONAL TEST PREPARATION

1. OPEN ENDED Write a real-life problem that can be solved using a system of linear inequalities. Write and graph the system. **Check students' work.**

2. WRITING Use the system of linear inequalities below. Graph the system and explain why one of the inequalities does not affect the solution.
$x \ge 1$; $y > x$; $y \le 5$; $x + y \le 6$;
See margin.

ADDITIONAL RESOURCES
An alternative Quiz for Lessons 7.4–7.6 is available in the *Chapter 7 Resource Book*, p. 93.

8. 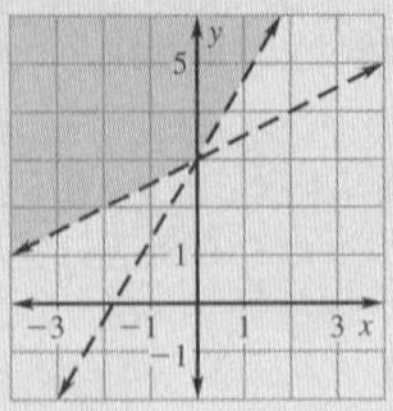

9. 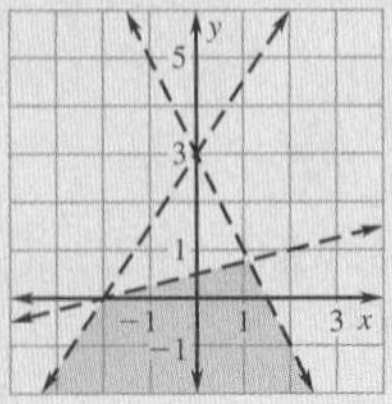

10. 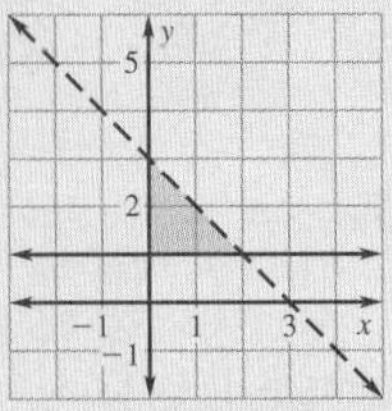

11. 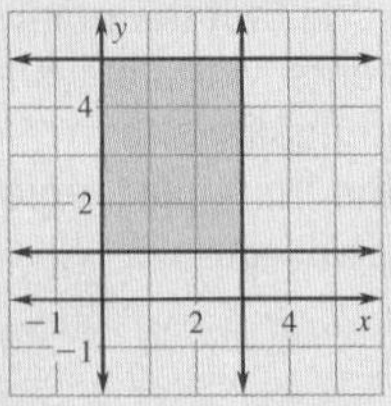

12. 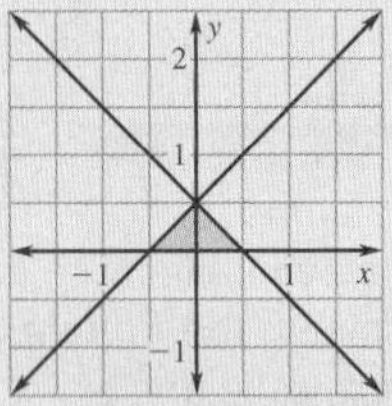

13. 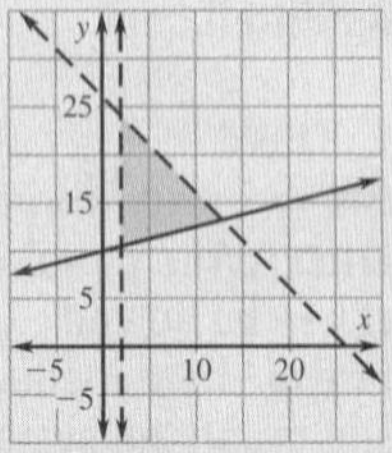

MIXED REVIEW

EVALUATING NUMERICAL EXPRESSIONS **Evaluate the expression.** **(Review 1.2, 1.3 for 8.1)**

45. $3 \cdot 3 \cdot 3 \cdot 3 \cdot 3$ **243** **46.** $8^2 - 17$ **47** **47.** $5^3 + 12$ **137**

48. $(3^3 - 20)^2$ **49** **49.** $2^6 - 31$ **33** **50.** $5 \cdot 5 + 3 \cdot 3 \cdot 3$ **52**

EVALUATING ALGEBRAIC EXPRESSIONS **In Exercises 51–56, evaluate the exponential expression.** **(Review 1.2)**

51. $(x + y)^2$ when $x = 5$ and $y = 2$ **49** **52.** $(b - c)^2$ when $b = 2$ and $c = 1$ **1**

53. $g - h^2$ when $g = 4$ and $h = 8$ **−60** **54.** $x^2 + z$ when $x = 8$ and $z = 12$ **76**

55. $m^3 + 1$ when $m = 2$ **9** **56.** $v^3 + w$ when $v = 3$ and $w = 5$ **32**

57. **MEAN, MEDIAN, AND MODE** Find the mean, the median, and the mode of the collection of numbers shown below. **(Review 6.6)** **47, 56, 57**

57, 40, 57, 57, 30, 56, 40, 30, 56

58. **TEST QUESTIONS** Your teacher is giving a test worth 250 points. There are 68 questions. Some questions are worth 5 points and the rest are worth 2 points. How many of each question are on the test? **(Review 7.3)**
38 5-point questions and 30 2-point questions

QUIZ 2

Self-Test for Lessons 7.4–7.6

1. **GEOMETRY CONNECTION** The perimeter of the rectangle is 22 feet. The perimeter of the triangle is 12 feet. Find the dimensions of the rectangle. **(Lesson 7.4)**
width = 3 ft, length = 8 ft

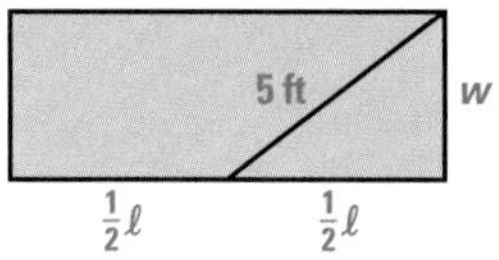

Use the graphing method to solve the linear system and tell how many solutions the system has. **(Lesson 7.5)**

2. $5x - 2y = 0$
$5x - 2y = -4$
no solution

3. $3x + 2y = 12$
$9x + 6y = 18$
no solution

4. $-4x + 11y = 44$
$4x - 11y = -44$
infinitely many solutions

5. $3x + y = 9$
$2x + y = 4$
exactly one solution, (5, −6)

6. $4x + 8y = 8$
$x + y = 1$
exactly one solution, (0, 1)

7. $3x + 6y = 30$
$4x + 8y = 40$
infinitely many solutions

Graph the system of linear inequalities. **(Lesson 7.6)** **8–13. See margin.**

8. $x - 2y < -6$
$5x - 3y < -9$

9. $-3x + 2y < 6$
$x - 4y > -2$
$2x + y < 3$

10. $x + y < 3$
$x \geq 0$
$y \geq 1$

11. $x \geq 0$
$y \geq 1$
$x \leq 3$
$y \leq 5$

12. $x + y \leq \frac{1}{2}$
$-x + y \leq \frac{1}{2}$
$y \geq 0$

13. $x + y < 26$
$-x + 4y \geq 40$
$2x > 4$

14. **INVESTMENTS** A total of \$16,000 is invested in two individual retirement accounts paying 5% and 6% annual interest. The combined annual interest is \$860. How much of the \$16,000 is invested in each account? **(Lesson 7.4)** **\$10,000 at 5%, \$6000 at 6%**

CHAPTER 7

Chapter Summary

WHAT did you learn? / *WHY did you learn it?*

Solve a system of linear equations
- by graphing. **(7.1)** → Predict traffic at two Internet sites. **(p. 400)**
- by substitution. **(7.2)** → Analyze a museum's ticket sales. **(p. 407)**
- by linear combinations. **(7.3)** → Model compositions of chemical mixtures. **(p. 413)**

Choose the best method to solve a linear system. **(7.4)** → Make a decision about two job offers. **(p. 420)**

Identify the number of solutions of a linear system. **(7.5)** → Interpret results when modeling with linear systems. **(p. 428)**

Use a system of linear equations to model a real-life situation. **(7.1–7.5)** → Find the speed of an airplane flying against the wind. **(p. 415)**

Solve a system of linear inequalities. **(7.6)** → Plan a minimum order for free delivery. **(p. 434)**

Use a system of linear inequalities to model a real-life situation. **(7.6)** → Find ways to work within a budget. **(p. 436)**

How does Chapter 7 fit into the BIGGER PICTURE of algebra?

You saw in this chapter how to choose and apply an appropriate method to solve systems of linear equations and systems of linear inequalities. Knowing the advantages and disadvantages of each method is part of becoming an efficient problem solver.

Linear systems often occur as models of real-life situations. The information you gain from studying systems that model real-life problems will help you analyze situations and make decisions.

STUDY STRATEGY

How did you use your list of problem types?

One type of problem that you listed in your notebook, using the **Study Strategy** on page 396, may be this one.

Types of Problems

- **Motion Problems**

S = rate of boat in still water

r = rate of current

rate of boat upstream $= S - r$

rate of boat downstream $= S + r$

How to solve: Use linear combinations because you can quickly eliminate the variable r.

439

ADDITIONAL RESOURCES

The following resources are available to help review the material in this chapter.

- Chapter Review Games and Activities (*Chapter 7 Resource Book*, p. 94)
- *Instant Replay: Video Review Games*
- *Personal Student Tutor*
- Cumulative Review, Chs. 1–7 (*Chapter 7 Resource Book*, p. 106)

CHAPTER 7

Chapter Review

VOCABULARY

- **system of linear equations, or linear system (p. 398)**
- **solution of a system of linear equations (p. 398)**
- **linear combination (p. 411)**
- **system of linear inequalities (p. 432)**
- **solution of a system of linear inequalities (p. 432)**
- **graph of a system of linear inequalities (p. 432)**

7.1 SOLVING LINEAR SYSTEMS BY GRAPHING

Examples on pp. 398–400

EXAMPLE To solve a linear system by graphing, you can graph each equation using a table of values, the intercepts, or slope-intercept form.

Graph and solve the system. $y = x + 3$ **Equation 1**
$y = -x + 7$ **Equation 2**

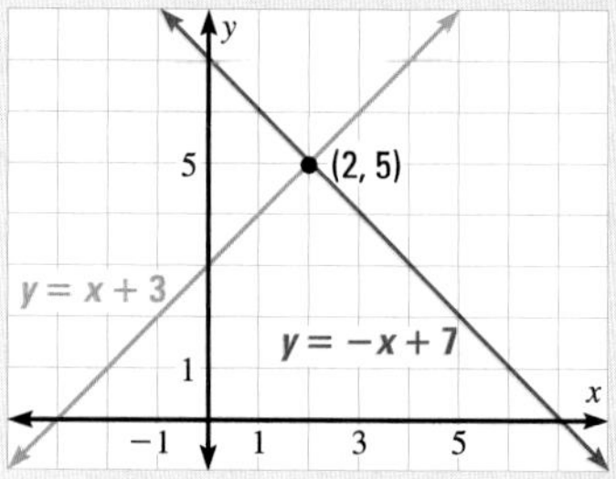

The two lines seem to intersect at the point (2, 5). Check this solution algebraically in both equations.

EQUATION 1:

$y = x + 3$

$5 \stackrel{?}{=} 2 + 3$

$5 = 5$

EQUATION 2:

$y = -x + 7$

$5 \stackrel{?}{=} -(2) + 7$

$5 = 5$

▶ The solution is (2, 5).

Graph and check to solve the linear system. **1–6. See margin.**

1. $x + y = 6$
$x - y = 12$

2. $4x - y = 3$
$3x + y = 4$

3. $x + 9y = 9$
$3x + 6y = 6$

4. $5x - y = -5$
$3x + 6y = -3$

5. $7x + 8y = 24$
$x - 8y = 8$

6. $2x - 3y = -3$
$x + 6y = -9$

7.2 SOLVING LINEAR SYSTEMS BY SUBSTITUTION

Examples on pp. 405–407

EXAMPLE To solve the linear system at the right by substitution, first solve one equation for one of its variables.

$4x - 2y = 0$ **Equation 1**
$8x - 2y = 16$ **Equation 2**

$y = 2x$ **Revised Equation 1**

Then substitute $2x$ for y in Equation 2 and solve for x.

$8x - 2(2x) = 16$ Substitute $2x$ for y.

$x = 4$ Solve for x.

$y = 2(4) = 8$ Substitute 4 for x in $y = 2x$.

▶ The solution is (4, 8). Check the solution in the original equations.

Use the substitution method to solve the linear system.

7. $x + 3y = 9$ **(0, 3)**
$4x - 2y = -6$

8. $-2x - 5y = 7$ **(−1, −1)**
$7x + y = -8$

9. $4x - 3y = -2$ $\left(\frac{5}{8}, 1\frac{1}{2}\right)$
$4x + y = 4$

10. $-x + 3y = 24$ **(−9, 5)**
$5x + 8y = -5$

11. $4x + 9y = 2$ $\left(\frac{1}{2}, 0\right)$
$2x + 6y = 1$

12. $9x + 6y = 3$ $\left(-1\frac{2}{3}, 3\right)$
$3x - 7y = -26$

7.3 SOLVING LINEAR SYSTEMS BY LINEAR COMBINATIONS

Examples on pp. 411–413

EXAMPLE To solve the linear system, get coefficients for x or y that are opposites.

$2x - 15y = -10$ **Equation 1**
$-4x + 5y = -30$ **Equation 2**

$4x - 30y = -20$ **Multiply Equation 1 by 2.**
$-4x + 5y = -30$ **Write Equation 2.**
$-25y = -50$ **Add the equations.**
$y = 2$ **Solve for *y*.**

Then substitute 2 for y in Equation 2 and solve for x.

$-4x + 5y = -30$ **Write Equation 2.**
$-4x + 5(2) = -30$ **Substitute 2 for *y*.**
$x = 10$ **Solve for *x*.**

▶ The solution is (10, 2). Check the solution in the original equations.

Use linear combinations to solve the linear system.

13. $-4x - 5y = 7$ $\left(-5, 2\frac{3}{5}\right)$
$x + 5y = 8$

14. $2x + y = 0$ **(2, −4)**
$5x - 4y = 26$

15. $3x + 5y = -16$ **(3, −5)**
$-2x + 6y = -36$

16. $9x + 6y = 3$ **(11, −16)**
$3y + 6x = 18$

17. $2 - 7x = 9y$ **(−1, 1)**
$2y - 4x = 6$

18. $4x - 9y = 1$ **(−2, −1)**
$-5x + 6y = 4$

7.4 APPLICATIONS OF LINEAR SYSTEMS

Examples on pp. 418–420

EXAMPLES After you model a real-life problem with a linear system, you have a choice of three methods for solving the system.

GRAPHING: Use to approximate a solution.

SUBSTITUTION: Use when one variable has a coefficient of 1 or −1.

LINEAR COMBINATIONS: Use when no variable has a coefficient of 1 or −1.

19. **CARNIVAL** You have 50 tickets to ride the Ferris wheel and the roller coaster. If you ride 12 times, using 3 tickets for each Ferris wheel ride and 5 tickets for each roller coaster ride, how many times did you go on each ride? **Ferris wheel: 5 times, roller coaster: 7 times**

20. **RENTING MOVIES** You spend \$13 to rent five movies for the weekend. Since new releases rent for \$3 and regular movies rent for \$2, how many regular movies did you rent? How many new releases did you rent? **2 regular movies; 3 new releases**

1.
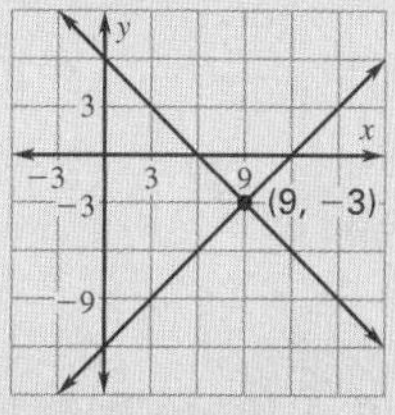

2.
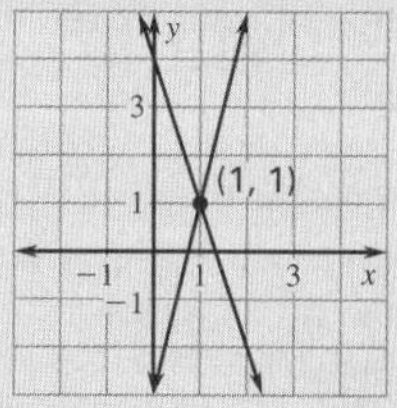

3.
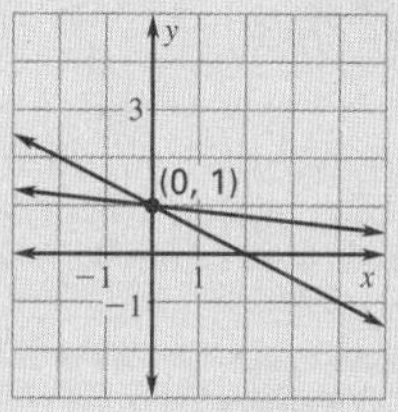

4.
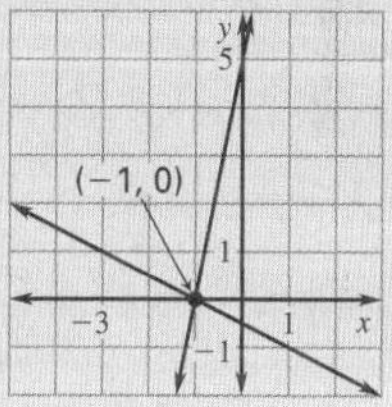

5.
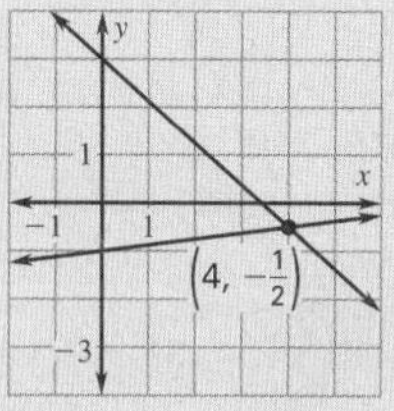

6.
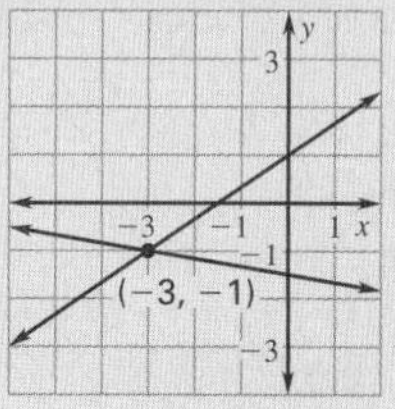

27.

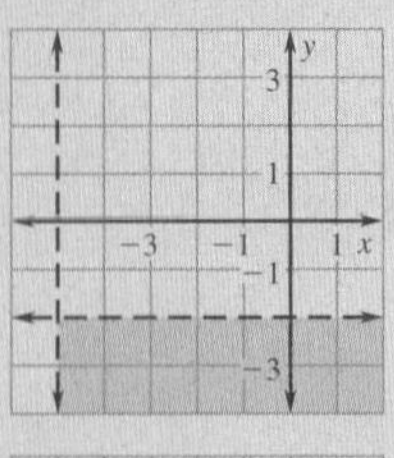

28.

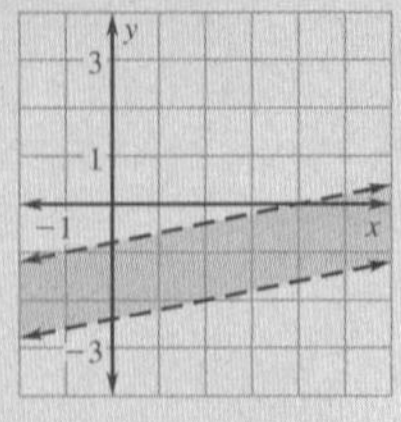

29.

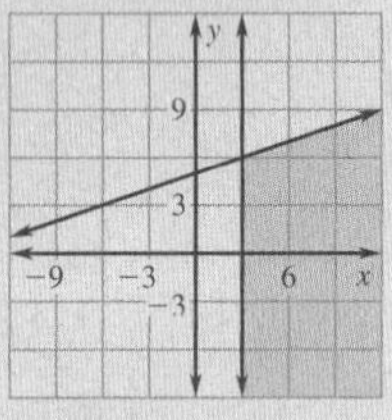

30.

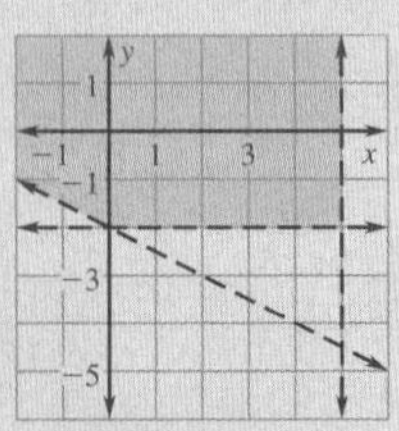

31.

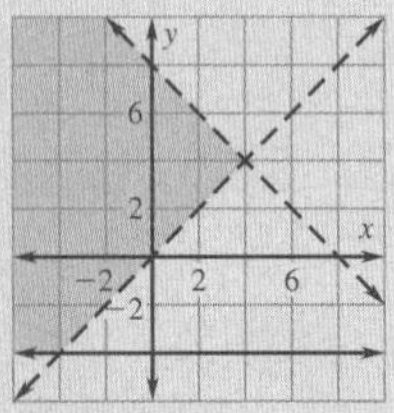

32.

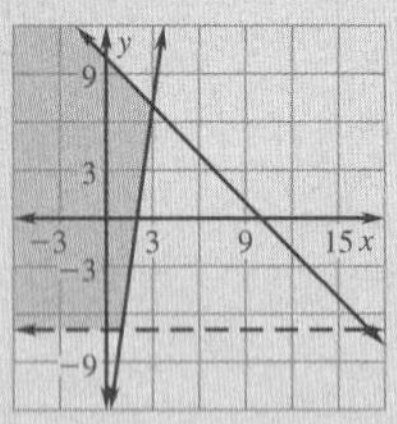

7.5 SPECIAL TYPES OF LINEAR SYSTEMS

Examples on pp. 426–428

EXAMPLES A system of linear equations in two variables can have exactly one solution, no solution, or infinitely many solutions.

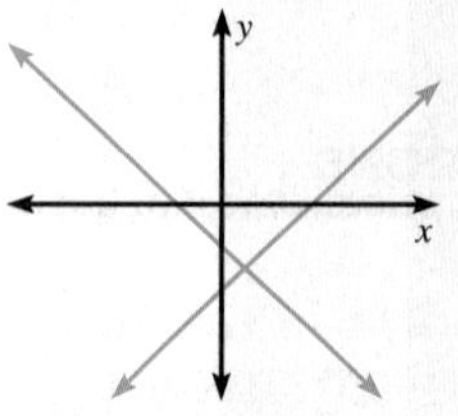

Since the lines intersect, there is one solution.

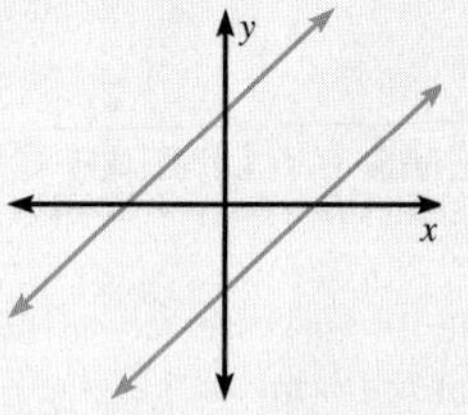

Since the lines are parallel, there is no solution.

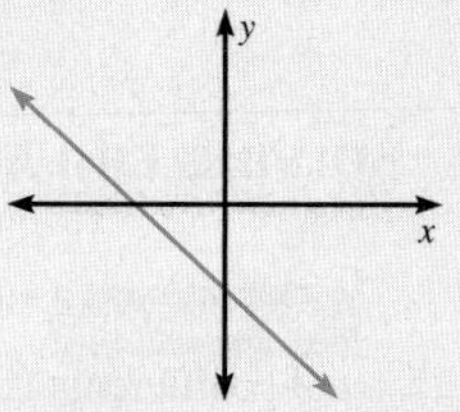

Since the lines coincide, there are infinitely many solutions.

Solve the linear system and tell how many solutions the linear system has.

21. $\frac{1}{3}x + y = 2$
$2x + 6y = 12$
infinitely many solutions

22. $2x - 3y = 1$
$-2x + 3y = 1$
no solution

23. $-6x + 5y = 18$
$7x + 2y = 26$
exactly one solution, (2, 6)

24. $10x + 4y = 25$
$5x + 8y = 11$
exactly one solution, $\left(2\frac{3}{5}, -\frac{1}{4}\right)$

25. $14x + 7y = 0$
$-2x + y = 13$
exactly one solution, $\left(-3\frac{1}{4}, 6\frac{1}{2}\right)$

26. $21x + 28y = 14$
$9x + 12y = 6$
infinitely many solutions

7.6 SOLVING SYSTEMS OF LINEAR INEQUALITIES

Examples on pp. 432–434

EXAMPLE The boundary line of the graph of a linear inequality in two variables is dashed if the inequality is $<$ or $>$ and solid if the inequality is $\le$ or $\ge$.

The graph of the system of linear inequalities below is the intersection of the three half-planes shown at the right.

$x \ge 0$ **Inequality 1**
$3x - y \le 7$ **Inequality 2**
$2x + y < 2$ **Inequality 3**

The graph of the first inequality is the half-plane *on and to the right* of the line $x = 0$.

The graph of the second inequality is the half-plane *on and above* the line $y = 3x - 7$.

The graph of the third inequality is the half-plane *below* the line $y = -2x + 2$.

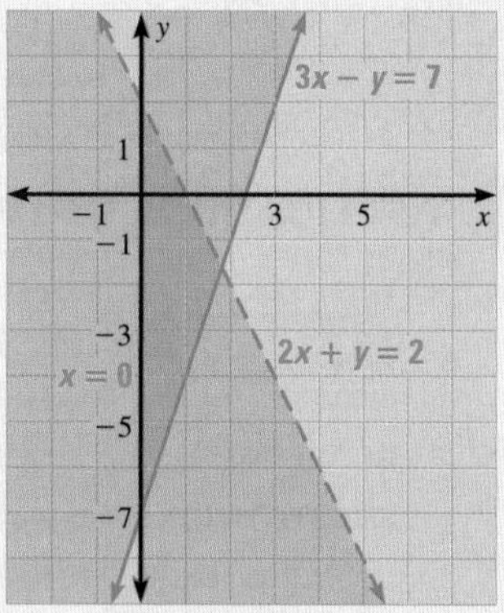

Graph the system of linear inequalities. **27–32. See margin.**

27. $x > -5$
$y < -2$

28. $2x - 10y > 8$
$x - 5y < 12$

29. $-x + 3y \le 15$
$9x \ge 27$

30. $x < 5$
$y > -2$
$x + 2y > -4$

31. $x + y < 8$
$x - y < 0$
$y \ge -4$

32. $7y > -49$
$-7x + y \ge -14$
$x + y \le 10$

CHAPTER 7

Chapter Test

Graph and check to solve the linear system.

1. $y = 2x - 3$ $(1, -1)$
$-y = 2x - 1$

2. $6x + 2y = 16$ $(2, 2)$
$-2x + y = -2$

3. $4x - y = 10$ $(4, 6)$
$-2x + 4y = 16$

4. $-4x + y = -10$ $(3, 2)$
$6x + 2y = 22$

5. $3x + 5y = -10$ $(-10, 4)$
$-x + 2y = 18$

6. $2x - 3y = 12$ $(6, 0)$
$-x - 3y = -6$

Use the substitution method to solve the linear system.

7. $-4x + 7y = -2$ $(-3, -2)$
$-x - y = 5$

8. $7x + 4y = 5$ $(-1, 3)$
$x - 6y = -19$

9. $-3x + 6y = 24$ $(-2, 3)$
$-2x - y = 1$

10. $5x - y = 7$ $(1, -2)$
$4x + 8y = -12$

11. $x + 6y = 9$ $(-3, 2)$
$-x + 4y = 11$

12. $8x + 3y = 0$ $\left(4, -10\frac{2}{3}\right)$
$-x - 9y = 92$

Use linear combinations to solve the linear system.

13. $6x + 7y = 5$ $\left(-1\frac{1}{2}, 2\right)$
$4x - 2y = -10$

14. $-7x + 2y = -5$ $\left(\frac{1}{3}, -1\frac{1}{3}\right)$
$10x - 2y = 6$

15. $-3x + 3y = 12$ $(2, 6)$
$4x + 2y = 20$

16. $3x + 4y = 9$ $\left(1\frac{2}{3}, 1\right)$
$4y - 3x = -1$

17. $8x - 2 + y = 0$ $(13, -102)$
$9x - y = 219$

18. $5y - 3x = 1$ $(18, 11)$
$4y + 2x = 80$

Solve the system using the method of your choice and tell how many solutions the system has.

19. $8x + 4y = -4$
$2x - y = -3$
exactly one solution, $(-1, 1)$

20. $-6x + 3y = -6$
$2x + 6y = 30$
exactly one solution, $(3, 4)$

21. $-x + \frac{1}{3}y = -6$
$3x - y = -16$
no solution

22. $3x + y = 8$
$4x + 6y = 6$
exactly one solution, $(3, -1)$

23. $3x - 4y = 8$
$\frac{9}{2}x - 6y = 12$
infinitely many solutions

24. $6x + y = 12$
$-4x - 2y = 0$
exactly one solution, $(3, -6)$

Graph the system of linear inequalities. 25–27. See margin.

25. $x \le 4$
$y \ge 1$
$y \le x + 2$

26. $x < 5$
$y \le 6$
$y > -2x + 3$

27. $y > \frac{3}{2}x + \frac{3}{2}$
$y < -\frac{1}{4}x - \frac{1}{2}$

Write a system of linear inequalities that defines the shaded region.

28.

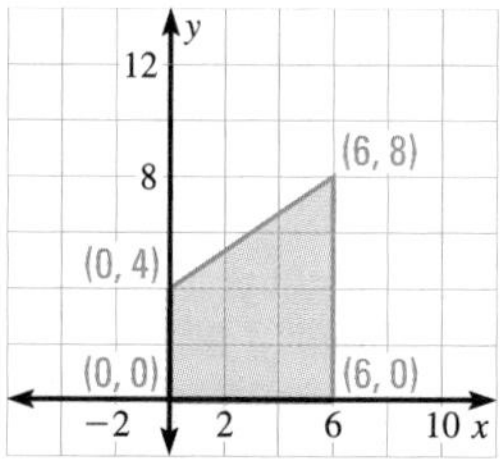

$x \ge 0, y \ge 0, x \le 6, y \le \frac{2}{3}x + 4$

29.

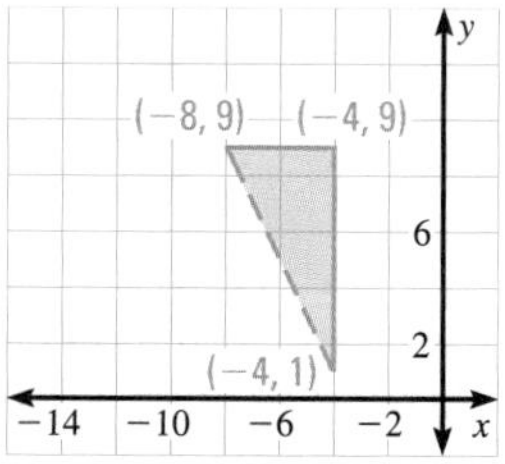

$x \le -4, y \le 9, y > -2x - 7$

30.

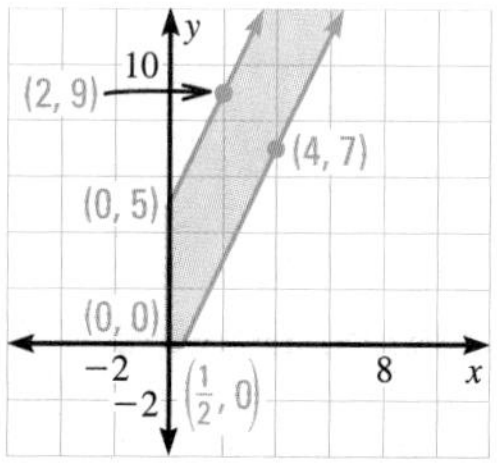

$x \ge 0, y \ge 0, y \le 2x + 5, y \ge 2x - 1$

31. WILD BIRD FOOD You buy six bags of wild bird food to fill the feeders in your yard. Oyster shell grit, a natural calcium source, sells for \$4.00 a bag, and sunflower seeds sell for \$4.45 a bag. If you spend \$25.80, how many bags of each type of feed are you buying? 2 bags of oyster shell grit, 4 bags of sunflower seeds

ADDITIONAL RESOURCES

- ***Chapter 7 Resource Book***
  Chapter Test (3 levels) (p. 95)
  SAT/ACT Chapter Test (p. 101)
  Alternative Assessment (p. 102)
- ***Test and Practice Generator***

25.

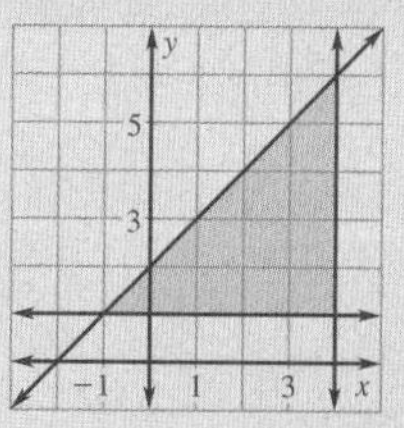

26.

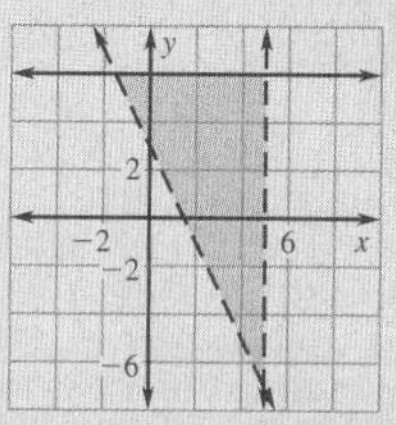

27.

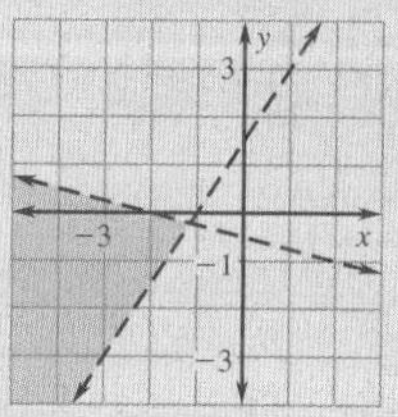

ADDITIONAL RESOURCES

- ***Chapter 7 Resource Book***
  Chapter Test (3 levels) (p. 95)
  SAT/ACT Chapter Test (p. 101)
  Alternative Assessment (p. 102)
- ***Test and Practice Generator***

CHAPTER 7

Chapter Standardized Test

TEST-TAKING STRATEGY Go back and check as much of your work as you can.

1. **MULTIPLE CHOICE** Which point represents the solution of the system of linear equations? **D**

$y = x - 2$
$y = -x - 4$

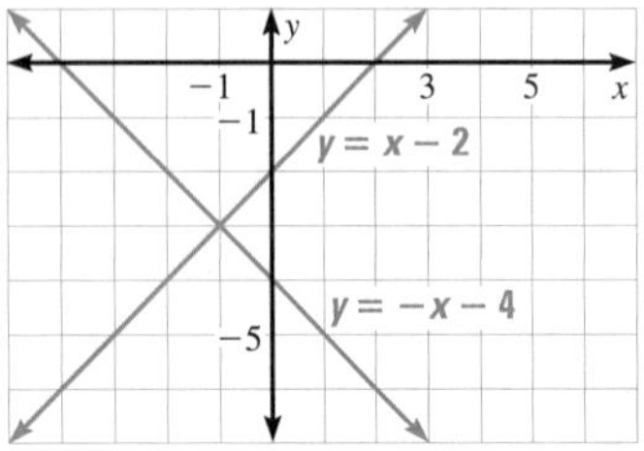

Ⓐ (2, 0) Ⓑ (−3, −1) Ⓒ (1, −1)
Ⓓ (−1, −3) Ⓔ (0, −4)

2. **MULTIPLE CHOICE** What is the solution of the system of linear equations? **C**

$\frac{4}{5}x + \frac{1}{2}y = 16$
$x + y = 24$

Ⓐ (8, 16) Ⓑ (10, 14)
Ⓒ $\left(\frac{40}{3}, \frac{32}{3}\right)$ Ⓓ $\left(\frac{41}{4}, \frac{55}{4}\right)$
Ⓔ (2, 22)

3. **MULTIPLE CHOICE** If $x + y = 15$ and $x - y = -19$, then $xy =$ _?_. **C**

Ⓐ −76 Ⓑ −68 Ⓒ −34
Ⓓ 15 Ⓔ 34

4. **MULTIPLE CHOICE** If $-2x + 7y = -8$ and $x - 6y = 10$, then $x - y =$ _?_. **A**

Ⓐ −2 Ⓑ 0 Ⓒ $\frac{3}{2}$
Ⓓ $\frac{5}{2}$ Ⓔ 4

5. **MULTIPLE CHOICE** If $-\frac{3}{4}x + \frac{7}{10}y = 5$ and $\frac{1}{4}x - \frac{3}{10}y = -5$, then $x + y =$ _?_. **E**

Ⓐ −90 Ⓑ −10 Ⓒ $-\frac{25}{8}$
Ⓓ 10 Ⓔ 90

6. **MULTIPLE CHOICE** If $y = 5x - 2$, then $-3y =$ _?_. **E**

Ⓐ $5x - 6$ Ⓑ $15x - 6$ Ⓒ $15x + 6$
Ⓓ $-15x - 6$ Ⓔ $-15x + 6$

7. **MULTIPLE CHOICE** Your teacher is giving a test worth 150 points. There is a total of 42 five-point and two-point questions. How many two-point questions are on the test? **B**

Ⓐ 18 Ⓑ 20 Ⓒ 22
Ⓓ 24 Ⓔ 26

8. **MULTIPLE CHOICE** How many solutions does the linear system have? **B**

$4x - 2y = 6$
$7x + y = 15$

Ⓐ None Ⓑ Exactly one Ⓒ Two
Ⓓ Infinitely many Ⓔ Cannot be determined

9. **MULTIPLE CHOICE** How many solutions does the linear system have? **B**

$y = 2x + 4$
$y = 2$

Ⓐ None Ⓑ Exactly one Ⓒ Two
Ⓓ Infinitely many Ⓔ Cannot be determined

10. **MULTIPLE CHOICE** The ordered pair (3, 4) is a solution of _?_. **A**

Ⓐ $x + y = 7$, $x + 2y = 11$ Ⓑ $x - y = 1$, $2x - y = 9$
Ⓒ $x - y = 1$, $2x + y = 10$ Ⓓ $x + y = 7$, $2x - 2y = 14$
Ⓔ Cannot be determined

11. **MULTIPLE CHOICE** Which point is a solution of the system of linear inequalities? **D**

$y < -x$
$y < x$

Ⓐ (2, 6) Ⓑ (6, −2) Ⓒ (−2, 6)
Ⓓ (−1, −6) Ⓔ (−6, −1)

12. MULTIPLE CHOICE Which system of inequalities is represented by the graph at the right? **C**

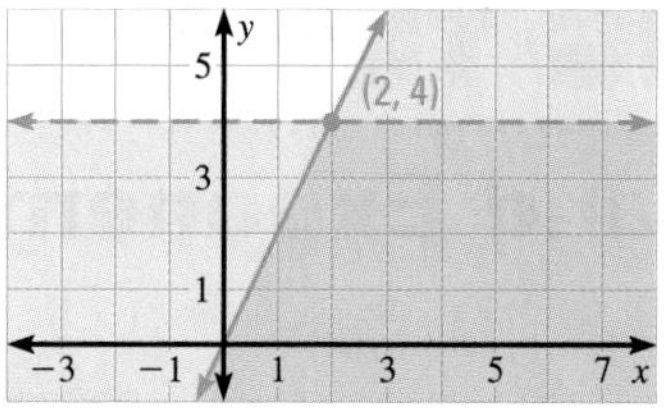

Ex. 12

Ⓐ $y \le 4$, $y < 2x + 1$ Ⓑ $y < 4$, $y \ge 2x$ Ⓒ $y < 4$, $y \le 2x$

Ⓓ $y < 4$, $y < 2x$ Ⓔ $y < 4$, $y \le 2x + 1$

13. MULTIPLE CHOICE A ship is traveling toward a lighthouse following the line $4x - y = 10$. Another ship is traveling toward the lighthouse following the line $x + y = 5$. What are the coordinates of the lighthouse? **E**

Ⓐ $(3, -2)$ Ⓑ $(2, 3)$ Ⓒ $(-2, -3)$ Ⓓ $(-2, 3)$ Ⓔ $(3, 2)$

14. MULTIPLE CHOICE Which point lies on the graph of the system? **B**

$$3y = 3$$
$$x + 2y = 4$$

Ⓐ $(3, 1)$ Ⓑ $(2, 1)$ Ⓒ $(4, 1)$

Ⓓ $\left(\frac{3}{2}, 1\right)$ Ⓔ Cannot be determined

15. QUANTITATIVE COMPARISON Solve the linear system. Then choose the statement below that is true about the solution of the system. **A**

$$2x + 6y = 13$$
$$x - 4y = 3$$

Ⓐ The value of x is greater. Ⓑ The value of y is greater.

Ⓒ The values of x and y are equal.

Ⓓ The relationship cannot be determined from the given information.

16. MULTI-STEP PROBLEM The members of the city cultural center have decided to put on a play once a night for a week. Their auditorium holds 500 people. By selling tickets, the members would like to raise \$3150 every night to cover all expenses. Let x represent the number of adult tickets sold at \$7.50 each. Let y represent the number of student tickets sold at \$4.50 each.

a. Write a linear equation that models the \$3150 income the members hope to raise from the sale of adult and student tickets for one night. **$7.5x + 4.5y = 3150$**

b. Write a second linear equation that models the number of tickets they sell if all 500 seats are filled for a single night's performance. **$x + y = 500$**

c. If all 500 seats are filled for a performance, how many of each type of ticket must have been sold for the members to raise exactly \$3150? **300 adult tickets, 200 student tickets**

d. At one performance there were three times as many adult tickets sold as student tickets. If there were 400 tickets sold, how much below the goal of \$3150 did the ticket sales fall? **\$450**

e. *Writing* At another performance, only adults attended. The members know they raised at least \$3150 that night. Find the possible numbers of tickets that could have been sold. Explain your method.
At least 420 tickets; the possible numbers are solutions of the inequality $7.5x \ge 3150$.

CHAPTER 8

Planning the Chapter

Exponents and Exponential Functions

Lesson	GOALS	NCTM	ITED	SAT9	Terra-Nova	Local
8.1 *pp. 449–455*	CONCEPT ACTIVITY: 8.1 *Investigate using addition and multiplication to multiply exponential expressions.* GOAL 1 Use properties of exponents to multiply exponential expressions. GOAL 2 Use powers to model real-life problems.	1, 2, 6, 8, 9, 10	MCE, APE	2, 3	10, 11, 16, 17, 18	
8.2 *pp. 456–462*	GOAL 1 Evaluate powers that have zero and negative exponents. GOAL 2 Graph exponential functions. TECHNOLOGY ACTIVITY: 8.2 *Graph an exponential function on a graphing calculator.*	1, 2, 3	MCE, MIG, IIG	2	10, 14, 16	
8.3 *pp. 463–469*	GOAL 1 Use the division properties of exponents to evaluate powers and simplify expressions. GOAL 2 Use the division properties of exponents to find a probability.	1, 2, 5	MCE, SAPP, MIP	2	10, 11, 15, 52	
8.4 *pp. 470–475*	GOAL 1 Use scientific notation to represent numbers and perform operations with them. GOAL 2 Use scientific notation to describe real-life situations.	1, 6, 8, 9, 10	RQWN, RQF		10	
8.5 *pp. 476–482*	CONCEPT ACTIVITY: 8.5 *Investigate linear and exponential growth models.* GOAL 1 Write and use models for exponential growth. GOAL 2 Graph models for exponential growth.	1, 2, 3, 8, 10	MCE, MIG, IIG, APE		10, 14, 18, 52	
8.6 *pp. 483–492*	CONCEPT ACTIVITY: 8.6 *Investigate exponential decay.* GOAL 1 Write and use models for exponential decay. GOAL 2 Graph models for exponential decay. TECHNOLOGY ACTIVITY: 8.6 *Find a best-fitting exponential growth or decay model on a graphing calculator.*	1, 2, 3, 8, 10	MCE, MIG, IIG, APE		10, 14, 15, 18, 52	

RESOURCES

Chapter Resource Booklets

Chapter Support

Tips for New Teachers	p. 1	Prerequisite Skills Review	p. 5
Parent Guide for Student Success	p. 3	Strategies for Reading Mathematics	p. 7

Lesson Support

Lesson Support	8.1	8.2	8.3	8.4	8.5	8.6
Lesson Plans (regular and block)	p. 9	p. 23	p. 37	p. 52	p. 66	p. 81
Warm-Up Exercises and Daily Quiz	p. 11	p. 25	p. 39	p. 54	p. 68	p. 83
Activity Support Masters	p. 12				p. 69	p. 84
Lesson Openers	p. 13	p. 26	p. 40	p. 55	p. 70	p. 85
Graphing Calculator Activities & Keystrokes	p. 14	p. 27	p. 41	p. 56	p. 71	p. 86
Practice (3 levels)	p. 15	p. 29	p. 43	p. 57	p. 73	p. 88
Reteaching with Practice	p. 18	p. 32	p. 46	p. 60	p. 76	p. 91
Quick Catch-Up for Absent Students	p. 20	p. 34	p. 48	p. 62	p. 78	p. 93
Cooperative Learning Activities				p. 63		p. 94
Interdisciplinary Applications		p. 35		p. 64		
Real-Life Applications	p. 21		p. 49		p. 79	p. 95
Math & History Applications						p. 96
Challenge: Skills and Applications	p. 22	p. 36	p. 50	p. 65	p. 80	p. 97

Review and Assessment

Quizzes	pp. 51, 98	Alternative Assessment with Math Journal	p. 107
Chapter Review Games and Activities	p. 99	Project with Rubric	p. 109
Chapter Test (3 levels)	pp. 100–105	Cumulative Review	p. 111
SAT/ACT Chapter Test	p. 106	Resource Book Answers	p. A1

Transparencies

	8.1	8.2	8.3	8.4	8.5	8.6
Warm-Up Exercises and Daily Quiz	p. 60	p. 61	p. 62	p. 63	p. 64	p. 65
Alternative Lesson Opener Transparencies	p. 52	p. 53	p. 54	p. 55	p. 56	p. 57
Examples/Standardized Test Practice	✓	✓	✓	✓	✓	✓
Answer Transparencies	✓	✓	✓	✓	✓	✓

Technology

- EasyPlanner CD
- EasyPlanner Plus Online
- Classzone.com
- Personal Student Tutor CD
- Test and Practice Generator CD
- Instant Replay: Video Review Games
- Electronic Lesson Presentations
- Chapter Audio Summaries CD

Additional Resources

- Basic Skills Workbook
- Explorations and Projects
- Worked-Out Solution Key
- Data Analysis Sourcebook
- Spanish Resources
- Standardized Test Practice Workbook
- Practice Workbook with Examples
- Notetaking Guide

CHAPTER 8

PACING THE CHAPTER

Resource Key
- Student Edition
- Chapter 8 Resource Book
- Teacher's Edition

REGULAR SCHEDULE

Day 1	Day 2	Day 3	Day 4	Day 5	Day 6
8.1	**8.1** *(cont.)*	**8.2**	**8.2** *(cont.)*	**8.3**	**8.3** *(cont.)*
Starting Options • Prereq. Skills Review • Strategies for Reading • Warm-Up or Daily Quiz **Teaching Options** • Motivating the Lesson • Concept Act. & Wksht. • Les. Opener (Activity) • Graphing Calc. Activity • Examples 1–4 • Guided Practice Exs. 1–21 **Apply/Homework** • See Assignment Guide. • See the CRB: Practice, Reteach, Apply, Extend **Assessment Options** • Checkpoint Exercises, p. 451	**Starting Options** • Homework Check **Teaching Options** • Examples 5–6 • Closure Question **Apply/Homework** • See Assignment Guide. • See the CRB: Practice, Reteach, Apply, Extend **Assessment Options** • Checkpoint Exercises, p. 452 • Daily Quiz (8.1) • Stand. Test Practice	**Starting Options** • Homework Check • Warm-Up or Daily Quiz **Teaching Options** • Motivating the Lesson • Les. Opener (Visual) • Graphing Calc. Activity • Examples 1–4 • Guided Practice Exs. 1–10 **Apply/Homework** • See Assignment Guide. • See the CRB: Practice, Reteach, Apply, Extend **Assessment Options** • Checkpoint Exercises, p. 457	**Starting Options** • Homework Check **Teaching Options** • Examples 5–6 • Closure Question • Guided Practice Exs. 11–13 **Apply/Homework** • See Assignment Guide. • See the CRB: Practice, Reteach, Apply, Extend **Assessment Options** • Checkpoint Exercises, p. 458 • Daily Quiz (8.2) • Stand. Test Practice	**Starting Options** • Homework Check • Warm-Up or Daily Quiz **Teaching Options** • Motivating the Lesson • Les. Opener (Calculator) • Graphing Calc. Activity • Examples 1–3 • Guided Practice Exs. 1–18 **Apply/Homework** • See Assignment Guide. • See the CRB: Practice, Reteach, Apply, Extend **Assessment Options** • Checkpoint Exercises, p. 464	**Starting Options** • Homework Check **Teaching Options** • Examples 4–5 • Closure Question **Apply/Homework** • See Assignment Guide. • See the CRB: Practice, Reteach, Apply, Extend **Assessment Options** • Checkpoint Exercises, p. 465 • Daily Quiz (8.3) • Stand. Test Practice • Quiz (8.1–8.3)

Day 9	Day 10	Day 11	Day 12
8.6	**8.6** *(cont.)*	**Review**	**Assess**
Starting Options • Homework Check • Warm-Up or Daily Quiz **Teaching Options** • Concept Act. & Wksht. • Les. Opener (Application) • Graphing Calc. Activity • Examples 1–3 • Guided Practice Exs. 1–8 **Apply/Homework** • See Assignment Guide. • See the CRB: Practice, Reteach, Apply, Extend **Assessment Options** • Checkpoint Exercises, p. 485	**Starting Options** • Homework Check **Teaching Options** • Examples 4–6 • Closure Question • Guided Practice Ex. 9 **Apply/Homework** • See Assignment Guide. • See the CRB: Practice, Reteach, Apply, Extend **Assessment Options** • Checkpoint Exercises, pp. 486–487 • Daily Quiz (8.6) • Stand. Test Practice • Quiz (8.4–8.6)	**Day 11 Start Options** • Homework Check **Reviewing Options** • Chapter 8 Summary • Chapter 8 Review • Chapter Review Games and Activities **Apply/Homework** • Chapter 8 Test (practice) • Ch. Standardized Test (practice)	**Day 12 Start Options** • Homework Check **Assessment Options** • Chapter 8 Test • SAT/ACT Ch. 8 Test • Alternative Assessment **Apply/Homework** • Skill Review, p. 502

Day 7

8.4

Starting Options
- Homework Check
- Warm-Up or Daily Quiz

Teaching Options
- Les. Opener (Application)
- Graphing Calc. Activity
- Examples 1–6
- Closure Question
- Guided Practice Exs.

Apply/Homework
- See Assignment Guide.
- See the CRB: Practice, Reteach, Apply, Extend

Assessment Options
- Checkpoint Exercises
- Daily Quiz (8.4)
- Stand. Test Practice

Day 8

8.5

Starting Options
- Homework Check
- Warm-Up or Daily Quiz

Teaching Options
- Motivating the Lesson
- Concept Act. & Wksht.
- Les. Opener (Appl.)
- Graphing Calc. Activity
- Examples 1–5
- Closure Question
- Guided Practice Exs.

Apply/Homework
- See Assignment Guide.
- See the CRB: Practice, Reteach, Apply, Extend

Assessment Options
- Checkpoint Exercises
- Daily Quiz (8.5)
- Stand. Test Practice

BLOCK SCHEDULE

Day 1

Assess & 8.1

(Day 1 = Ch. 7 Day 6)

Assessment Options
- Chapter 7 Test
- SAT/ACT Ch. 7 Test
- Alternative Assessment

Ch. 8 Start Options
- Skill Review, p. 448
- Prereq. Skills Review
- Strategies for Reading

Beginning 8.1 Options
- Warm-Up (Les. 8.1)
- Motivating the Lesson
- Concept Act. & Wksht.
- Les. Opener (Activity)
- Graphing Calc. Activity
- Examples 1–4
- Guided Practice Exs. 1–21

Apply/Homework
- See Assignment Guide.
- See the CRB: Practice, Reteach, Apply, Extend

Assessment Options
- Checkpoint Exercises

Day 2

8.1 & 8.2

Day 2 Start Options
- Homework Check

Finishing 8.1 Options
- Examples 5–6
- Closure Question

Beginning 8.2 Options
- Warm-Up (Les. 8.2)
- Motivating the Lesson
- Les. Opener (Visual)
- Graphing Calc. Activity
- Examples 1–4
- Guided Practice Exs. 1–10

Apply/Homework
- See Assignment Guide.
- See the CRB: Practice, Reteach, Apply, Extend

Assessment Options
- Checkpoint Exercises
- Daily Quiz (Les. 8.1)
- Stand. Test Prac. (8.1)

Day 3

8.2 & 8.3

Day 3 Start Options
- Homework Check
- Daily Quiz (Les. 8.1)

Finishing 8.2 Options
- Examples 5–6
- Closure Question
- Guided Practice Exs. 11–13

Beginning 8.3 Options
- Warm-Up (Les. 8.3)
- Motivating the Lesson
- Les. Opener (Calc.)
- Graphing Calc. Activity
- Examples 1–3
- Guided Practice Exs. 1–18

Apply/Homework
- See Assignment Guide.
- See the CRB: Practice, Reteach, Apply, Extend

Assessment Options
- Checkpoint Exercises
- Daily Quiz (Les. 8.2)
- Stand. Test Prac. (8.2)

Day 4

8.3 & 8.4

Day 4 Start Options
- Homework Check
- Daily Quiz (Les. 8.2)

Finishing 8.3 Options
- Examples 4–5
- Closure Question

Teaching 8.4 Options
- Warm-Up (Les. 8.4)
- Les. Opener (Appl.)
- Graphing Calc. Activity
- Examples 1–6
- Closure Question
- Guided Practice Exs.

Apply/Homework
- See Assignment Guide.
- See the CRB: Practice, Reteach, Apply, Extend

Assessment Options
- Checkpoint Exercises
- Daily Quiz (Les. 8.3, 8.4)
- Stand. Test Practice
- Quiz (8.1–8.3)

Day 5

8.5 & 8.6

Day 5 Start Options
- Homework Check
- W-Up 8.5 or D. Quiz 8.4

Teaching 8.5 Options
- Motivating the Lesson
- Concept Act. & Wksht.
- Les. Opener (Appl.)
- Graphing Calc. Activity
- Examples 1–5
- Closure Question
- Guided Practice Exs.

Beginning 8.6 Options
- Warm-Up (Les. 8.6)
- Concept Act. & Wksht.
- Les. Opener (Appl.)
- Graphing Calc. Activity
- Examples 1–3
- Guided Practice Exs. 1–8

Apply/Homework
- See Assignment Guide.
- See the CRB: Practice, Reteach, Apply, Extend

Assessment Options
- Checkpoint Exercises
- Daily Quiz (Les. 8.5)
- Stand. Test Prac. (8.5)

Day 6

8.6 & Review

Day 6 Start Options
- Homework Check
- Daily Quiz (Les. 8.5)

Finishing 8.6 Options
- Examples 4–6
- Closure Question
- Guided Practice Ex. 9

Reviewing Options
- Chapter 8 Summary
- Chapter 8 Review
- Chapter Review Games and Activities

Apply/Homework
- See Assignment Guide.
- See the CRB: Practice, Reteach, Apply, Extend
- Chapter 8 Test (practice)
- Ch. Standardized Test (practice)

Assessment Options
- Checkpoint Exercises
- Daily Quiz (Les. 8.6)
- Stand. Test Practice
- Quiz (8.4–8.6)

Day 7

Assess & 9.1

(Day 7 = Ch. 9 Day 1)

Assessment Options
- Chapter 8 Test
- SAT/ACT Ch. 8 Test
- Alternative Assessment

Ch. 9 Start Options
- Skill Review, p. 502
- Prereq. Skills Review
- Strategies for Reading

Teaching 9.1 Options
- Warm-Up (Les. 9.1)
- Motivating the Lesson
- Les. Opener (Appl.)
- Graphing Calc. Activity
- Examples 1–7
- Closure Question
- Guided Practice Exs.

Apply/Homework
- See Assignment Guide.
- See the CRB: Practice, Reteach, Apply, Extend

Assessment Options
- Checkpoint Exercises
- Daily Quiz (Les. 9.1)
- Stand. Test Practice

CHAPTER 8

MEETING INDIVIDUAL NEEDS

BEFORE THE CHAPTER

The *Chapter 8 Resource Book* has the following materials to distribute and use before the chapter:

- **Parent Guide for Student Success**
- **Prerequisite Skills Review**
- **Strategies for Reading Mathematics (pictured below)**

STRATEGIES FOR READING *Pages 7–8*

CHAPTER 8 NAME ______ DATE ______

Strategies for Reading Mathematics

For use with Chapter 8

Chapter Support

Strategy: Reading Properties and Rules

Symbols are used when writing algebraic properties and rules in order to make the ideas clearer and easier to see and to remember. Reading these properties is not hard if you take it step by step and think about what each symbol stands for.

The example below shows steps you can follow to read and understand a property involving exponents. Remember that exponential notation is a short way of writing repeated multiplication. For example, to say that a variable x is taken as a factor six times, you can just write the following symbols.

base → x^6 ← exponent (Read as "x to the sixth power.")

Product of Powers Property

Property in symbols:	If a is a number and m and n are positive integers, then				
	a^m	$\cdot$	a^n	$=$	a^{m+n}
Reading the symbols:	a to the mth power	times	a to the nth power	equals	a to the power of the sum of m and n
Visualizing the relationship:	$(a \cdot a \cdot a \cdot \ldots \cdot a)$ base a as a factor m times	$\cdot$	$(a \cdot a \cdot a \cdot \ldots \cdot a)$ base a as a factor n times	$=$	$(a \cdot a \cdot a \cdot \ldots \cdot a)$ base a as a factor $m = n$ times
Property in words:	To multiply powers having the same base, add the exponents and take the base to the power of that sum.				

STUDY TIP
Reading Properties
Read the property part by part. Use a drawing, numerical example, or other representation to visualize what the property is saying. Try to express the ideas of the property in your own words.

STUDY TIP
Using Resources
If you forget what a symbol means, find the table of symbols in your textbook. If you forget what a property is all about, use the glossary or use the index to find pages on which the property is taught or applied.

Questions

1. Identify the exponent and the base of the power that appears in the function $y = ab^x$. Then use arrows and labels to show how to read this function.
2. The distributive property for addition can be written $a(b + c) = ab + ac$. Read and record in words what the symbols say. Next use a numerical example or a diagram to visualize the relationship. Then explain the property in your own words.
3. The formula for the volume of a cube is $V = s^3$. Follow the three steps shown above to interpret this formula. Tell what each variable stands for.

STRATEGIES FOR READING MATHEMATICS The first page gives students tips about reading and interpreting mathematical properties and rules as they prepare for Chapter 8. The second page (not shown) provides a visual glossary of key vocabulary words in the chapter, such as exponential growth and growth factor.

DURING EACH LESSON

The *Chapter 8 Resource Book* has the following alternatives for introducing the lesson:

- **Lesson Openers (pictured below)**
- **Graphing Calculator Activities with Keystrokes**

LESSON OPENER *Page 26*

LESSON 8.2 NAME ______ DATE ______ Available as a transparency

Visual Approach Lesson Opener

For use with pages 456–461

Lesson 8.2

1. The function $y = 2^x$ is graphed below. Find the coordinates of the points shown on the graph.

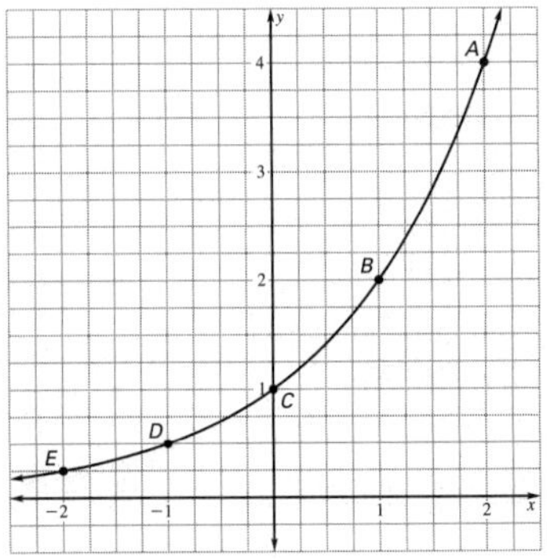

2. What do you notice about the value of y when x is positive?
3. What do you notice about the value of y when x is 0?
4. What do you notice about the value of y when x is negative?

VISUAL APPROACH LESSON OPENER This Lesson Opener uses visuals as an alternative way to start Lesson 8.2. Students look at the graph of $y = 2^x$ as an introduction to values of an exponential equation.

TECHNOLOGY RESOURCE

Teachers can use the Electronic Lesson Presentations CD-ROM to present Lesson 8.5 or Lesson 8.6 using computer animation to step through the Examples.

Following Each Lesson

The *Chapter 8 Resource Book* has a variety of materials to follow-up each lesson. They include the following:

- **Practice (3 levels)**
- **Reteaching with Practice**
- **Quick Catch-Up for Absent Students**
- **Interdisciplinary Applications**
- **Real-Life Applications (pictured below)**

Application *Page 79*

LESSON 8.5

NAME ______________________ DATE __________

Real-Life Application

For use with pages 477–482

Lesson 8.5

Investing for College

Saving enough money for college can seem impossible, especially when many sources claim it will cost $120,000 or more for four years. Though many Ivy League institutions are this expensive, other public and private schools can be as low as $13,000 for four years. Even this lower amount requires financial planning. Fortunately, there are many ways to invest money that can be used for college in the future. The best savings plans are those that are started when a child is born. If this is not possible, the sooner you can begin to save, the more money you will have available to you when you actually begin your journey into higher education.

Roth IRAs, traditionally individual retirement accounts, can be used for qualified higher-education expenses after an account has been open for five successive tax years. Owners may then take tax-free and penalty-free distributions of earnings. Educational IRAs can be established as well. Like a Roth IRA, parents or grandparents can make annual contributions, which are not tax deductible. The money, however, is allowed to grow tax-free until it is withdrawn for higher education. The drawback is that there is presently a limit of only $500 a year per student.

Traditional savings accounts and certificates can also be opened with banks or credit unions.

1. Imagine that your parents established a savings account for you when you were born. They deposited $1500 with an interest rate of 3.54% compounded yearly. How much balance will you have to further your education at age 18?
2. Certificates of deposit (CDs), which are guaranteed for a specific amount of time, generally offer a higher percentage rate than savings accounts. When you entered sixth grade, your parents opened a $2000 seven-year CD with 5.87% interest rate that is compounded yearly. How much would your balance be after high school graduation?
3. Imagine that you could have the same percentage rate in your savings account that you had in your certificate of deposit. How much would your balance be at age 18? (Use your original $1500 deposit from Exercise 1.)
4. Graph the exponential growth model in Exercise 1. Let the horizontal axis represent the number of years, with $t = 0$ being your birth year. Let the vertical axis represent the amount of money in the balance. (Make a table of values using your exponential growth model from Exercise 1 for eighteen years.)
5. Graph the exponential growth model in Exercise 2. Let the horizontal axis represent the number of years, with $t = 0$ being your birth year. Let the vertical axis represent the amount of money in the balance. (Make a table of values using your exponential growth model from Exercise 2 for seven years.)

Real-Life Application When students ask "When will I ever use exponential growth models?," you can have them work through this Real-Life Application in which they apply the content of Lesson 8.5 to saving money for college.

Assessing the Chapter

The *Chapter 8 Resource Book* has the following review and assessment materials:

- **Quizzes**
- **Chapter Review Games and Activities**
- **Chapter Test (3 levels)**
- **SAT/ACT Chapter Test (pictured below)**
- **Alternative Assessment with Rubric and Math Journal**
- **Project with Rubric**
- **Cumulative Review**

SAT/ACT Chapter Test *Page 106*

CHAPTER 8

NAME ______________________ DATE __________

SAT/ACT Chapter Test

For use after Chapter 8

Review and Assess

1. Simplify $(3^4 \cdot 3^3)^2$.
   - Ⓐ 3^9 Ⓑ 3^{14}
   - Ⓒ 3^{16} Ⓓ 3^{24}
2. Simplify $[(1 + x^2)]^3$ when $x = 2$.
   - Ⓐ 33 Ⓑ 65
   - Ⓒ 125 Ⓓ 721
3. What is the equation of the graph?

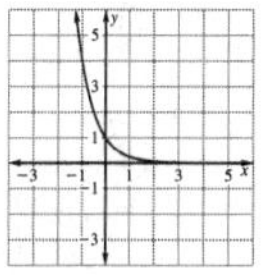

   - Ⓐ $y = 4^x$
   - Ⓑ $y = 2^x$
   - Ⓒ $y = (\frac{1}{2})^x$
   - Ⓓ $y = (\frac{1}{4})^x$
4. Simplify $\frac{3x^2y}{4x^3y^2} \cdot \frac{8x^{-3}y^5}{3x^{-5}y}$.
   - Ⓐ $\frac{2}{x^3y}$ Ⓑ $\frac{x^3y}{2}$
   - Ⓒ $2x^3y$ Ⓓ $2xy^3$
5. Which of the following numbers is *not* written in scientific notation?
   - Ⓐ 61.2×10^4 Ⓑ 4.56×10^{18}
   - Ⓒ 8.642×10^{-3} Ⓓ 1.1987×10^3
6. Rewrite 4.5×10^{-7} in decimal form.
   - Ⓐ 0.00000045 Ⓑ 0.0000045
   - Ⓒ 45,000,000 Ⓓ 450,000,000
7. Which of the models below are exponential decay models?
   - I. $y = 1.19^x$ II. $y = 0.12^x$
   - III. $y = (\frac{5}{3})^x$ IV. $y = (\frac{2}{3})^x$
   - Ⓐ I and II Ⓑ II and III
   - Ⓒ I and III Ⓓ II and IV

In Questions 8 and 9, choose the statement below that is true about the given numbers.

- **A.** The number in column A is greater.
- **B.** The number in column B is greater.
- **C.** The two numbers are equal.
- **D.** The relationship cannot be determined from the given information.

8.

	Column A	*Column B*
When $x = 0$,	$y = (\frac{2}{3})^x$	$y = (\frac{2}{3})^{-x}$

Ⓐ Ⓑ Ⓒ Ⓓ

9.

	Column A	*Column B*
When $x = -1$,	$y = 5x$	$y = 5^x$

Ⓐ Ⓑ Ⓒ Ⓓ

10. You deposit $500 in an account that pays 5% interest compounded yearly. How much money is in the account after 4 years?
    - Ⓐ $541.90 Ⓑ $607.75
    - Ⓒ $1118.03 Ⓓ $2531.25
11. You buy a used truck for $22,000. It depreciates at the rate of 10% per year. What is the value of the truck after 4 years?
    - Ⓐ $13,000 Ⓑ $14,434.20
    - Ⓒ $21,133.11 Ⓓ $21,912.13

SAT/ACT Chapter Test This test covers the material in Chapter 8 in a standardized test format. Students taking this form of the test should be reminded to read all of the answer choices before deciding which is the correct one.

Technology Resource

Teachers can use the Time-Saving Test and Practice Generator to create customized review and assessment materials for Chapter 8.

CHAPTER OVERVIEW

CHAPTER GOALS

In this chapter, students will learn how to multiply and divide expressions with exponents, including zero and negative exponents. They will use scientific notation to represent numbers and solve problems. They will graph exponential functions and use exponential functions to solve real-life problems that involve exponential growth and exponential decay. Students will also learn how to use technology to find a best-fitting exponential growth or decay model.

APPLICATION NOTE

Basic training for hill riding should include gear changes to increase pedaling resistance and improve technique. The units of measure for gears are complicated. For example, a 39 × 23-tooth (39 × 23T) gear is one that can be turned easily while riding. A moderate gear that can be used for warming up is a 39 × 17-tooth (30 × 17T) gear.

Breathing technique is essential to the bicycle racer. Racers must be aware of how breathing patterns are related to pedal strokes and to maximum heart rate. There are four ways to improve breathing control: (1) stretch your diaphragm by using your stomach to help breathe, (2) adopt a new rhythm of breathing, (3) breathe through your nose, and (4) open your nostrils with breathing strips.

Additional information about bicycle racing is available at **www.mcdougallittell.com.**

EXPONENTS AND EXPONENTIAL FUNCTIONS

How does switching gears cause a chain reaction?

CHAPTER

8

APPLICATION: Bicycle Racing

Shifting into high gear helps a racer increase speed on level or downhill surfaces, but pedaling becomes more difficult. When the racer exerts energy the brain sends a message to increase the rate and depth of breathing.

The relationship between breathing rate and bicycle speed can be represented with a type of mathematical model that you'll study in Chapter 8.

Think & Discuss

1. Construct a scatter plot of the data below. Draw a smooth curve through the points. **See margin.**

Bicycle speed, x (miles per hour)	Breathing rate, y (liters per minute)
0	6.4
5	10.7
10	18.1
15	30.5
20	51.4

2. Describe the change in breathing rate after each increase of 5 mi/h in bike speed. Does it increase by the same amount? the same percent? **See margin.**

Learn More About It

You will use an exponential model that relates the breathing rate to bike speed in Ex. 17 on p. 480.

APPLICATION LINK Visit www.mcdougallittell.com for more information about bicycle racing.

ADDITIONAL RESOURCES

Another way to begin the chapter is to show the video clip of a real-life motivator for Chapter 8 on the *Instant Replay: Video Review Games.*

PROJECTS

A project covering Chapter 8 appears on pages 570–571 of the Student Edition. An additional project for Chapter 8 is in the *Chapter 8 Resource Book,* p. 109.

TECHNOLOGY

Software

- *Electronic Teaching Tools*
- *Online Lesson Planner*
- *Personal Student Tutor*
- *Test and Practice Generator*
- *Electronic Lesson Presentations* (Lessons 8.5 and 8.6)

Video

- *Instant Replay: Video Review Games*

Internet Connections
www.mcdougallittell.com

- **Application Links** 447, 481
- **Data Updates** 468
- **Student Help** 462, 464, 471, 474, 492
- **Career Links** 454, 465, 489
- **Extra Challenge** 455, 468, 475, 482, 490

1.

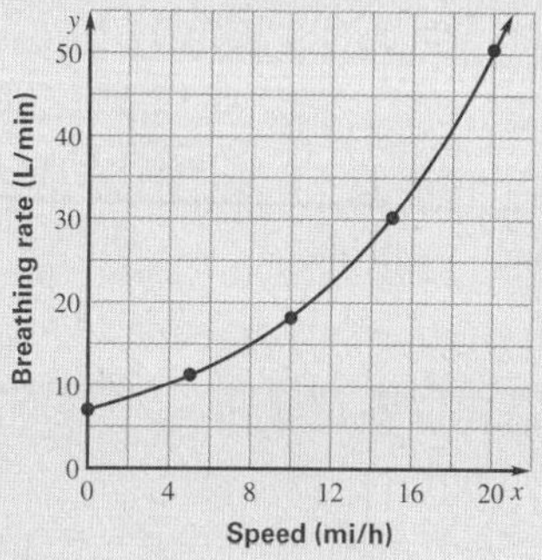

2. ***Sample answer:*** **The rate increases by increasing amounts. The amounts of increase are not the same, but the percents are approximately equal.**

PREPARE

DIAGNOSTIC TOOLS

The **Skill Review** exercises can help you diagnose whether students have the following skills needed in Chapter 8:

- Write expressions in exponential forms.
- Evaluate expressions.
- Solve equations.
- Graph equations.

The following resources are available for students who need additional help with these skills:

- Prerequisite Skills Review (*Chapter 8 Resource Book*, p. 5; *Warm-Up Transparencies*, p. 59)
- Reteaching with Practice (Chapter Resource Books for Lessons 8.1–8.6)
- *Personal Student Tutor*

ADDITIONAL RESOURCES

The following resources are provided to help you prepare for the upcoming chapter and customize review materials:

- ***Chapter 8 Resource Book***
  Tips for New Teachers (p. 1)
  Parent Guide (p. 3)
  Lesson Plans (every lesson)
  Lesson Plans for Block Scheduling (every lesson)
- ***Electronic Teaching Tools***
- ***Online Lesson Planner***
- ***Test and Practice Generator***

CHAPTER 8

Study Guide

PREVIEW

What's the chapter about?

Chapter 8 is about **exponents**. In Chapter 8 you'll learn

- how to multiply and divide expressions with exponents
- how to use scientific notation in problem solving
- how to use exponential growth and decay models to solve real-life problems

KEY VOCABULARY

▶ Review	▶ New	
• power, p. 9	• exponential function, p. 458	• growth factor, p. 477
• base, p. 9	• scientific notation, p. 470	• initial amount, p. 477, 484
• function, p. 46	• exponential growth, p. 477	• exponential decay, p. 484
		• decay factor, p. 484

PREPARE

Are you ready for the chapter?

SKILL REVIEW Do these exercises to review key skills that you'll apply in this chapter. See the given reference page if there is something you don't understand.

STUDENT HELP

Study Tip
"Student Help" boxes throughout the chapter give you study tips and tell you where to look for extra help in this book and on the Internet.

Write the expression in exponential form. (Review Example 1, p. 9)

1. six squared 6^2 **2.** four cubed 4^3 **3.** $2y \cdot 2y \cdot 2y \cdot 2y \cdot 2y$ $(2y)^5$

Find the probability of choosing a blue marble from the given bag of red and blue marbles. (Review Example 1, p. 114)

4. Number of blue marbles: 20
Total number of marbles: 80 $\frac{1}{4}$

5. Number of blue marbles: 6
Total number of marbles: 54 $\frac{1}{9}$

6. Number of red marbles: 12
Total number of marbles: 48 $\frac{3}{4}$

7. Number of red marbles: 25
Total number of marbles: 50 $\frac{1}{2}$

Find the unit rate. (Review pp. 180–182)

8. \$123.75 for working 15 hours **\$8.25/h**

9. \$3.16 for 4 cantaloupes **\$.79/cantaloupe**

10. \$6 for 12 cans of cat food **\$.50/can**

11. 270 miles for 6 gallons of gasoline **45 mi/gal**

STUDY STRATEGY

Here's a study strategy!

Planning Your Time:

A schedule or weekly planner can be a useful tool that allows you to coordinate your study time with time for other activities and responsibilities.

- Make a plan for each day of the week.
- Study every day, not just the day before a test.

ACTIVITY 8.1

Developing Concepts

Investigating Powers

SETUP
Work in a small group.

MATERIALS
- paper
- pencil

▶ **QUESTION** **How can you use addition to multiply exponential expressions? How can you use multiplication to raise an exponential expression to a power?**

▶ EXPLORING THE CONCEPT: PRODUCT OF POWERS

1 Copy and complete the table. To simplify an expression, expand the product. Then count the factors.

Product of powers	Expanded product	Number of factors	Product as a power	
$7^3 \cdot 7^2$	$(7 \cdot 7 \cdot 7) \cdot (7 \cdot 7)$	5	7^5	
$2^4 \cdot 2^4$	$(2 \cdot 2 \cdot 2 \cdot 2) \cdot (2 \cdot 2 \cdot 2 \cdot 2)$	8	?	2^8
$x^4 \cdot x^5$	$(x \cdot x \cdot x \cdot x) \cdot (x \cdot x \cdot x \cdot x \cdot x)$	?	?	9; x^9

Step 2

Sum of Exponents
5
8
9

2 Add a column to your table that shows the sum of the exponents that are in the first column. What pattern do you notice? **The sum of the exponents is the same as the number of factors. See margin.**

▶ EXPLORING THE CONCEPT: POWER OF A POWER

3 Copy and complete the table. To simplify an expression, expand the product. Then count the factors. **$[(-3) \cdot (-3)] \cdot [(-3) \cdot (-3)]$; 4; $(-3)^4$; $(b^2) \cdot (b^2) \cdot (b^2) \cdot (b^2)$; $(b \cdot b) \cdot (b \cdot b) \cdot (b \cdot b) \cdot (b \cdot b)$; 8; b^8**

Power of a power	Expanded product	Expanded product	Number of factors	Product as a power
$(5^2)^3$	$(5^2) \cdot (5^2) \cdot (5^2)$	$(5 \cdot 5) \cdot (5 \cdot 5) \cdot (5 \cdot 5)$	6	5^6
$[(-3)^2]^2$	$[(-3)^2] \cdot [(-3)^2]$	?	?	?
$(b^2)^4$	?	?	?	?

Step 4

Product of Exponents
6
4
8

4 Add a column to your table that shows the product of the exponents that are in the first column. What pattern do you notice? **The product of the exponents is the same as the number of factors. See margin.**

▶ DRAWING CONCLUSIONS

Expand the product. Then write your answer as a power.

1. $6^3 \cdot 6^2$ **6^5** **2.** $(-2) \cdot (-2)^4$ **$(-2)^5$** **3.** $p^4 \cdot p^6$ **p^{10}** **4.** $x^{12} \cdot x^7$ **x^{19}**

5. $(4^2)^6$ **4^{12}** **6.** $[(-5)^2]^4$ **$(-5)^8$** **7.** $(d^5)^5$ **d^{25}** **8.** $[(-n)^3]^8$ **$(-n)^{24}$**

9. What operation do you use to simplify a product of powers? Give examples.

10. What operation do you use to simplify a power of a power? Give examples.

11. CRITICAL THINKING Does $x^3 \cdot y^5 = xy^8$? Explain your answer.
No; $x^3 \cdot y^5 = x \cdot x \cdot x \cdot y \cdot y \cdot y \cdot y \cdot y$, while $xy^8 = x \cdot y \cdot y \cdot y \cdot y \cdot y \cdot y \cdot y \cdot y$.

9. Addition; examples should be like Exercises 1–4.

10. Multiplication; examples should be like Exercises 5–8.

1 Planning the Activity

PURPOSE
To multiply exponential expressions.

MATERIALS
- Activity Support Master (*Chapter 8 Resource Book,* p. 12)

PACING
- Exploring the Concept — 10 min
- Drawing Conclusions — 10 min

▶ ***LINK TO LESSON***
Students may wish to use charts as in Steps 1 and 3 to verify the results of Examples 1 and 2 in Lesson 8.1.

2 Managing the Activity

COOPERATIVE LEARNING
Students can do Steps 1 and 3 individually. After each student has completed his or her tables, the group can get together to do Steps 2 and 4 and discuss their results.

CLASSROOM MANAGEMENT
Some students may confuse when they should add exponents and when they should multiply them. Suggest that if they are not sure, they should expand the product as is done in Steps 1 or 3 of this activity.

3 Closing the Activity

★ ***KEY DISCOVERY***
To multiply powers having the same base, add the exponents. To find the power of a power, multiply the exponents.

ACTIVITY ASSESSMENT
Expand the power of the expression, and write the product as a power.

1. $5^4 \cdot 5^6$ 5^{10}
2. $x^3 \cdot x^3 \cdot x$ x^7
3. $(-3^4)^2$ 3^8
4. $(x^4)^3$ x^{12}

1 PLAN

PACING
Basic: 2 days
Average: 2 days
Advanced: 2 days
Block Schedule: 0.5 block with Ch. 7 Assess.
0.5 block with 8.2

LESSON OPENER ACTIVITY
An alternative way to approach Lesson 8.1 is to use the Activity Lesson Opener:
- Blackline Master (*Chapter 8 Resource Book,* p. 13)
- Transparency (p. 52)

MEETING INDIVIDUAL NEEDS
- ***Chapter 8 Resource Book***
  Prerequisite Skills Review (p. 5)
  Practice Level A (p. 15)
  Practice Level B (p. 16)
  Practice Level C (p. 17)
  Reteaching with Practice (p. 18)
  Absent Student Catch-Up (p. 20)
  Challenge (p. 22)
- ***Resources in Spanish***
- ***Personal Student Tutor***

NEW-TEACHER SUPPORT
See the Tips for New Teachers on pp. 1–2 of the *Chapter 8 Resource Book* for additional notes about Lesson 8.1.

WARM-UP EXERCISES
Transparency Available
Evaluate when $x = 4$.
1. $x \cdot x \cdot x$ 64
2. $(14 - x)^2$ 100
3. $x \cdot x \cdot x^2$ 256
4. $(-x) \cdot (-x)$ 16

ENGLISH LEARNERS
Remembering mathematical terms can be challenging for English learners. Before beginning Section 8.1, briefly review the terms *power, base,* and *exponent* to prepare students.

8.1 Multiplication Properties of Exponents

What you should learn

GOAL 1 Use properties of exponents to multiply exponential expressions.

GOAL 2 Use powers to model **real-life** problems, such as finding the area of crop irrigation circles in **Example 5**.

Why you should learn it

▼ To solve **real-life** problems such as calculating the power generated by a windmill in **Ex. 77**.

REAL LIFE

GOAL 1 MULTIPLYING EXPONENTIAL EXPRESSIONS

To multiply two powers that have the same base, you add exponents. Here is an example.

$$a^2 \cdot a^3 = \overbrace{\underbrace{a \cdot a}_{\text{2 factors}} \cdot \underbrace{a \cdot a \cdot a}_{\text{3 factors}}}^{\text{5 factors}} = a^5 = a^{2+3}$$

To find a power of a power, you multiply exponents. Here is an example.

$$(a^2)^3 = \overbrace{a^2 \cdot a^2 \cdot a^2}^{\text{3 factors}} = \overbrace{a \cdot a \cdot a \cdot a \cdot a \cdot a}^{\text{6 factors}} = a^6 = a^{2 \cdot 3}$$

These two rules for exponents and the rule for raising a product to a power are summarized below.

CONCEPT SUMMARY MULTIPLICATION PROPERTIES OF EXPONENTS

Let a and b be numbers and let m and n be positive integers.

PRODUCT OF POWERS PROPERTY
To multiply powers having the same base, add the exponents.
$a^m \cdot a^n = a^{m+n}$ **Example:** $3^2 \cdot 3^7 = 3^{2+7} = 3^9$

POWER OF A POWER PROPERTY
To find a power of a power, multiply the exponents.
$(a^m)^n = a^{m \cdot n}$ **Example:** $(5^2)^4 = 5^{2 \cdot 4} = 5^8$

POWER OF A PRODUCT PROPERTY
To find a power of a product, find the power of each factor and multiply.
$(a \cdot b)^m = a^m \cdot b^m$ **Example:** $(2 \cdot 3)^6 = 2^6 \cdot 3^6$

EXAMPLE 1 ***Using the Product of Powers Property***

a. $5^3 \cdot 5^6 = 5^{3+6}$
$= 5^9$

b. $x^2 \cdot x^3 \cdot x^4 = x^{2+3+4}$
$= x^9$

c. $3 \cdot 3^5 = 3^1 \cdot 3^5$
$= 3^{1+5}$
$= 3^6$

d. $(-2)(-2)^4 = (-2)^1 \cdot (-2)^4$
$= (-2)^{1+4}$
$= (-2)^5$

STUDENT HELP
Look Back
For help with exponential expressions, see page 9.

STUDENT HELP

HOMEWORK HELP Visit our Web site www.mcdougallittell.com for extra examples.

EXAMPLE 2 *Using the Power of a Power Property*

a. $(3^5)^2 = 3^{5 \cdot 2}$
$= 3^{10}$

b. $(y^2)^4 = y^{2 \cdot 4}$
$= y^8$

c. $[(-3)^3]^2 = (-3)^{3 \cdot 2}$
$= (-3)^6$

d. $[(a+1)^2]^5 = (a+1)^{2 \cdot 5}$
$= (a+1)^{10}$

When you use the power of a power property, it is the quantity within the parentheses that is raised to the power *not* the individual terms.

Correct:	$(a+1)^3 = (a+1)(a+1)(a+1)$	← 3 factors
Incorrect:	~~$(a+1)^3 = a^3 + 1^3$~~	← 2 terms

EXAMPLE 3 *Using the Power of a Product Property*

a. $(6 \cdot 5)^2 = 6^2 \cdot 5^2$ — Raise each factor to a power.
$= 36 \cdot 25$ — Evaluate each power.
$= 900$ — Multiply.

b. $(4yz)^3 = (4 \cdot y \cdot z)^3$ — Identify factors.
$= 4^3 \cdot y^3 \cdot z^3$ — Raise each factor to a power.
$= 64y^3z^3$ — Simplify.

c. $(-2w)^2 = (-2 \cdot w)^2$ — Identify factors.
$= (-2)^2 \cdot w^2$ — Raise each factor to a power.
$= 4w^2$ — Simplify.

d. $-(2w)^2 = -(2 \cdot w)^2$ — Identify factors.
$= -(2^2 \cdot w^2)$ — Raise each factor to a power.
$= -4w^2$ — Simplify.

STUDENT HELP

Study Tip
In parts (c) and (d) of Example 3, notice that the two expressions are different. In $(-2w)^2$, the negative sign is part of the base. In $-(2w)^2$, the negative sign is not part of the base.

EXAMPLE 4 *Using All Three Properties*

Simplify $(4x^2y)^3 \cdot x^5$.

SOLUTION

$(4x^2y)^3 \cdot x^5 = 4^3 \cdot (x^2)^3 \cdot y^3 \cdot x^5$ — Power of a product
$= 64 \cdot x^6 \cdot y^3 \cdot x^5$ — Power of a power
$= 64x^{11}y^3$ — Product of powers

2 TEACH

MOTIVATING THE LESSON
Tell students that a pharmacy uses two-part prescription numbers. There is a 5-digit number, a dash, and then a 3-digit number. The digit 0 is never used. The number of 5-digit numbers is 9 to the fifth power, and the number of 3-digit numbers is 9 to the third power. The total number of possible prescription numbers is the product of these two powers of 9. Today's lesson will focus on multiplying powers of the same number.

EXTRA EXAMPLE 1
a. $4^5 \cdot 4^3$ 4^8
b. $y^3 \cdot y^4 \cdot y^5$ y^{12}
c. $2 \cdot 2^6$ 2^7
d. $(-5)(-5)^3$ $(-5)^4$

EXTRA EXAMPLE 2
a. $(5^2)^3$ 5^6
b. $(x^3)^2$ x^6
c. $[(-2)^3]^4$ $(-2)^{12}$
d. $[(a-2)^3]^2$ $(a-2)^6$

EXTRA EXAMPLE 3
a. $(3 \cdot 4)^2$ 144
b. $(3xy)^4$ $81x^4y^4$
c. $(-3y)^2$ $9y^2$
d. $-(3y)^2$ $-9y^2$

EXTRA EXAMPLE 4
Simplify $(3x^4y)^2 \cdot y^5$. $9x^8y^7$

CHECKPOINT EXERCISES
For use after Example 1:
1. a. $3^5 \cdot 3^2$ 3^7
b. $z \cdot z^2 \cdot z^4$ z^7
c. $5 \cdot 5^5$ 5^6
d. $(-3)(-3)^5$ $(-3)^6$

For use after Example 2:
2. a. $(4^2)^4$ 4^8
b. $(y^3)^5$ y^{15}
c. $[(-4)^3]^5$ $(-4)^{15}$
d. $[(y+2)^4]^2$ $(y+2)^8$

For use after Examples 3 and 4:
3. Simplify $(2x^2y)^5 \cdot x^6$. $32x^{16}y^5$

EXTRA EXAMPLE 5
The diameter of a small pizza is 6 in., and the diameter of a large pizza is 18 in.
a. Find the ratio of the area of the large pizza to the area of the small pizza. $\frac{9}{1}$
b. Write a general statement about tripling the diameter. **Tripling the diameter makes the area *nine* times as large.**

EXTRA EXAMPLE 6
A multiple choice test has two parts. There are 4^3 ways to answer the 3 questions in Part A. There are 4^7 ways to answer the 7 questions in Part B.
a. How many ways are there to answer all 10 questions? **1,048,576 ways**
b. If you guess each answer, what is the probability you will get them all right? $\frac{1}{1{,}048{,}576}$, **or about 0.00000095**

CHECKPOINT EXERCISES
For use after Example 5:
1. A large circle has radius 10 and a small circle has radius 2. Find the ratio of the area of the large circle to the area of the small circle. $\frac{25}{1}$
For use after Example 6:
2. A company uses 6-digit ID numbers. The first 3 digits must be odd, and there are 3^5 ways to make this part of the ID. The last 3 digits are nonzero even numbers, and there are 3^4 ways to make this part of the ID.
a. How many ID numbers are possible? **3^9, or 19,683**
b. If you try to guess an ID number, what is the probability that you guess it on the first try? $\frac{1}{19{,}683}$, **or about 0.000051**

CLOSURE QUESTION
How is finding the product of two powers of the same base different from finding a power of a power of that same base? **See margin.**

GOAL 2 USING POWERS IN REAL LIFE

REAL LIFE *Irrigation*

EXAMPLE 5 *Using the Power of a Product Property*

Some farmers use *center-pivot irrigation,* which is a sprinkler system that revolves around a center pivot. The large irrigation circles help farmers to conserve water, maximize crop yield, and reduce the cost of pesticides.

a. Find the ratio of the area of the larger irrigation circle to the area of the smaller irrigation circle.

b. Write a general statement about *doubling* the radius of an irrigation circle.

SOLUTION

a. Ratio $= \frac{\pi(2r)^2}{\pi r^2} = \frac{\pi \cdot 2^2 \cdot r^2}{\pi \cdot r^2} = \frac{\pi \cdot 4 \cdot r^2}{\pi \cdot r^2} = \frac{4}{1}$

b. Doubling the radius of an irrigation circle makes the area *four* times as large.

EXAMPLE 6 *Using the Product of Powers Property*

PROBABILITY CONNECTION A true-false test has two parts. There are 2^{10} ways to answer the 10 questions in Part A. There are 2^{15} ways to answer the 15 questions in Part B.

a. How many ways are there to answer all 25 questions?

b. If you guess each answer, what is the probability you will get them all right?

STUDENT HELP
Skills Review
For help with the counting principle, see page 788.

SOLUTION

a. For each of the 2^{10} ways to answer the questions in Part A, there are 2^{15} ways to answer the questions in Part B. Use the counting principle to find the total number of ways to answer for both parts. The number of ways to answer the 25 questions is the product of 2^{10} and 2^{15}.

$2^{10} \cdot 2^{15} = 2^{10+15}$ **Use product of powers property.**

$= 2^{25}$ **Add exponents.**

$= 33{,}554{,}432$ **Use a calculator.**

▶ The number of ways to answer the 25 questions is 33,554,432.

b. Probability $= \frac{\text{Ways to get all right}}{\text{Ways to answer}} = \frac{1}{33{,}554{,}432}$

▶ The probability of guessing and getting all answers correct is about 0.00000003.

Closure Question *Sample answer:*
For the product, you add the exponents, but for a power of a power, you multiply them.

GUIDED PRACTICE

Vocabulary Check ✓ 1. In the expression a^5, a is called the _?_ of the expression. base

Concept Check ✓ 2. How are the expressions $x^7 \cdot x^3$ and $(x^7)^3$ different? Explain your answer. See margin.

3. Can $a^3 \cdot b^4$ be simplified? Explain your answer. No; the product of powers property can only be used if the bases of the powers are the same.

Skill Check ✓ **Use the product of powers property to simplify the expression.**

4. $c \cdot c \cdot c$ c^3
5. $m \cdot m^2$ m^3
6. $2^2 \cdot 2^3$ 2^5, or 32
7. $3^2 \cdot 3^5$ 3^7, or 2187
8. $a^4 \cdot a^6$ a^{10}
9. $x^4 \cdot x^5$ x^9

Use the power of a power property to simplify the expression.

10. $(3)^2$ 9
11. $(-2)^2$ 4
12. $(2^4)^3$ 2^{12}, or 4096
13. $(4^3)^3$ 4^9, or 262,144
14. $(y^4)^5$ y^{20}
15. $(m^4)^8$ m^{32}

Use the power of a product property to simplify the expression.

16. $(2m^2)^3$ $8m^6$
17. $(ab^2)^2$ a^2b^4
18. $(5x)^2$ $25x^2$
19. $(x^3y^5)^4$ $x^{12}y^{20}$
20. $(x^3y^8)^5$ $x^{15}y^{40}$
21. $(-2x^3)^3$ $(-2)^3x^9$, or $-8x^9$

2. $x^7 \cdot x^3 = x \cdot x \cdot x \cdot x \cdot x \cdot x \cdot x \cdot x \cdot x \cdot x = x^{10}$; $(x^7)^3 = x^7 \cdot x^7 \cdot x^7 = x \cdot x = x^{21}$

The first expression uses the product of powers property. The second uses the power of a power property.

PRACTICE AND APPLICATIONS

STUDENT HELP

Extra Practice to help you master skills is on p. 804.

SIMPLIFYING EXPRESSIONS **Simplify, if possible. Write your answer as a power or as a product of powers.**

22. $3^4 \cdot 3^6$
23. $5^8 \cdot 5^3$
24. $(2^3)^2$ 2^6, or 64
25. $(7^4)^2$
26. $x \cdot x^6$ x^7
27. $(3 \cdot 7)^2$
28. $(2x)^2$
29. $(-5a)^3$
30. $(-2m^4n^6)^2$
31. $[(-4)^2]^3$
32. $[(-5xy)^2]^5$
33. $[(5+x)^3]^6$ $(5+x)^{18}$
34. $[(2x+3)^3]^2$ $(2x+3)^6$
35. $(3b)^3 \cdot b$ 3^3b^4, or $27b^4$
36. $5^3 \cdot (5a^4)^2$
37. $4x \cdot (x \cdot x^3)^2$ $4x^9$
38. $(-3a)^5 \cdot (4a)^2$
39. $-(3x)^2 \cdot (7x^4)^2$
40. $2x^3 \cdot (3x)^2$
41. $3y^2 \cdot (2y)^3$
42. $(-ab)(a^2b)^2$ $-a^5b^3$
43. $(-rs)(rs^3)^2$ $-r^3s^7$
44. $(-2xy)^3(-x^2)$
45. $(-3cd)^3(-d^2)$ $-(-3)^3c^3d^5$, or $27c^3d^5$
46. $(5b^2)^3\left(\frac{1}{2}b^3\right)^2$
47. $(6a^4)^2\left(\frac{1}{4}a^3\right)^2$
48. $(2t)^3(-t^2)$ -2^3t^5, or $-8t^5$
49. $(-w^3)(3w^2)^2$
50. $(-y)^3(-y)^4(-y)^5$
51. $(-x)^4(-x)^3(-x)^2$ $(-x)^9$, or $-x^9$
52. $(abc^2)^3(a^2b)^2$ $a^7b^5c^6$
53. $-(r^2st^3)^2(s^4t)^3$ $-r^4s^{14}t^9$
54. $(-3xy^2)^3(-2x^2y)^2$

22. 3^{10}, or 59,049
23. 5^{11}, or 48,828,125
25. 7^8, or 5,764,801
27. $3^2 \cdot 7^2$, or 441
28. 2^2x^2, or $4x^2$
29. $(-5)^3a^3$, or $-125a^3$
30. $(-2)^2m^8n^{12}$, or $4m^8n^{12}$
31. $(-4)^6$, or 4096
32. $(-5)^{10}x^{10}y^{10}$, or $9{,}765{,}625x^{10}y^{10}$
36. 5^5a^8, or $3125a^8$
38. $(-3)^5 \cdot 4^2 \cdot a^7$, or $-3888a^7$
39. $-3^2 \cdot 7^2 \cdot x^{10}$, or $-441x^{10}$
40. $2 \cdot 3^2 \cdot x^5$, or $18x^5$
41. $3 \cdot 2^3 \cdot y^5$, or $24y^5$

STUDENT HELP

HOMEWORK HELP
Example 1: Exs. 22–62
Example 2: Exs. 22–62
Example 3: Exs. 22–62
Example 4: Exs. 36–62
Example 5: Exs. 77, 78
Example 6: Ex. 79

EVALUATING EXPRESSIONS **Simplify. Then evaluate the expression when $a = 1$ and $b = 2$.**

55. $a^2 \cdot a^3$ a^5; 1
56. $b \cdot b^4$ b^5; 32
57. $(a^3)^2$ a^6; 1
58. $(-b)^3 \cdot b^2$ $-b^5$; -32
59. $(a \cdot b^2)^2$ a^2b^4; 16
60. $(a^2b)^4$ a^8b^4; 16
61. $-(ab^3)^2$ $-(a^2b^6)$; -64
62. $(b^2 \cdot b^3) \cdot (b^2)^4$ b^{13}; 8192

3 APPLY

ASSIGNMENT GUIDE

BASIC
Day 1: p. 453 Exs. 22–58
Day 2: pp. 453–455 Exs. 59–75, 82, 83, 90, 95, 100, 105

AVERAGE
Day 1: p. 453 Exs. 22–58
Day 2: pp. 453–455 Exs. 59–75, 77, 82, 83, 90, 95, 100, 105

ADVANCED
Day 1: p. 453 Exs. 22–58
Day 2: pp. 453–455 Exs. 59–77, 79, 82–85, 90, 95, 100, 105

BLOCK SCHEDULE
p. 453 Exs. 22–58 (with Ch. 7 Assess)
pp. 453–455 Exs. 59–75, 82, 83, 90, 95, 100, 105 (with 8.2)

EXERCISE LEVELS

Level A: *Easier*
22–26, 55–57, 99–102

Level B: *More Difficult*
27–54, 58–74, 82, 87–98, 103–105

Level C: *Most Difficult*
75–81, 83–85

✓ HOMEWORK CHECK

To quickly check student understanding of key concepts, go over the following exercises: Exs. 22, 24, 28, 30, 34, 46, 54, 58, 68. See also the Daily Homework Quiz:

- Blackline Master (*Chapter 8 Resource Book*, p. 25)
- Transparency (p. 61)

44. $-(-2)^3x^5y^3$, or $8x^5y^3$
46. $5^3 \cdot \left(\frac{1}{2}\right)^2 \cdot b^{12}$, or $\frac{125}{4}b^{12}$
47. $6^2 \cdot \left(\frac{1}{4}\right)^2 \cdot a^{14}$, or $\frac{9}{4}a^{14}$
49. -3^2w^7, or $-9w^7$
50. $(-y)^{12}$, or y^{12}
54. $(-3)^3 \cdot (-2)^2x^7y^8$, or $-108x^7y^8$

! COMMON ERROR

EXERCISES 16–74 Students often take only the power of the rightmost factor when finding a power of a product. Remind them that the power must be taken for each factor.

STUDENT HELP NOTES

Career Link Additional information about engineering technicians is available at **www.mcdougallittell.com.**

82b.

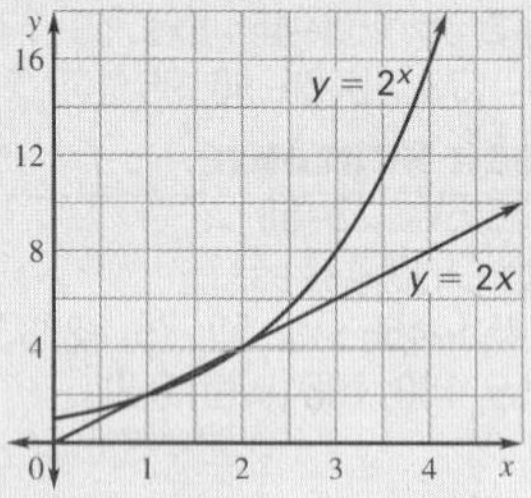

84. $(a^2)^3 = a^2 \cdot a^2 \cdot a^2$ definition of x^3
$= a \cdot a \cdot a \cdot a \cdot a \cdot a$ definition of x^2
$= a^6$ definition of x^6

92–101. See Additional Answers beginning on page AA1.

ADDITIONAL PRACTICE AND RETEACHING

For Lesson 8.1:

- Practice Levels A, B, and C (*Chapter 8 Resource Book*, p. 15)
- Reteaching with Practice (*Chapter 8 Resource Book*, p. 18)
- See Lesson 8.1 of the *Personal Student Tutor*

For more Mixed Review:

- Search the *Test and Practice Generator* for key words or specific lessons.

WRITING INEQUALITIES Complete the statement using > or <.

63. $(5 \cdot 6)^4 \ \underline{?}\ 5 \cdot 6^4$ >
64. $5^2 \cdot 3^6 \ \underline{?}\ (5 \cdot 3)^6$ <
65. $(3^6 \cdot 3^{12}) \ \underline{?}\ 3^{72}$ <
66. $4^2 \cdot 4^8 \ \underline{?}\ 4^{16}$ <
67. $(7^2)^3 \ \underline{?}\ 7^5$ >
68. $(6^2 \cdot 3)^3 \ \underline{?}\ 6^5 \cdot 3^3$ >

EVALUATING POWERS In Exercises 69–74, simplify the expression. Then use a calculator to evaluate the expression. Round the result to the nearest tenth when appropriate.

69. $(2.1 \cdot 4.4)^3$ 788.9
70. $6.5^3 \cdot 6.5^4$ 490,222.8
71. $2.6^4 \cdot 2.6^2$ 308.9
72. $(5.0 \cdot 4.9)^2$ 600.3
73. $(3.7^3)^5$ 333,446,267.9
74. $(8.4^2)^4$ 24,787,589.1

75. GEOMETRY CONNECTION The volume of a sphere is given by $V = \frac{4}{3}\pi r^3$, where r is the radius and π is approximately 3.14. What is the volume of the sphere in terms of a? about $113.04a^3$

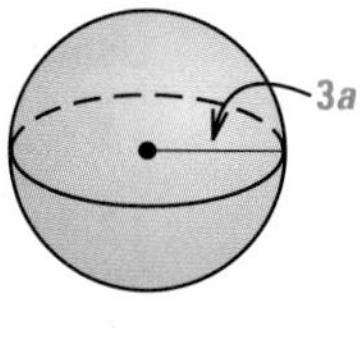

76. GEOMETRY CONNECTION The volume of a cone is given by $V = \frac{1}{3}\pi r^2 h$, where r is the radius of the base, h is the height, and $\pi \approx 3.14$. What is the volume of the cone in terms of b? about $100.48b^4$

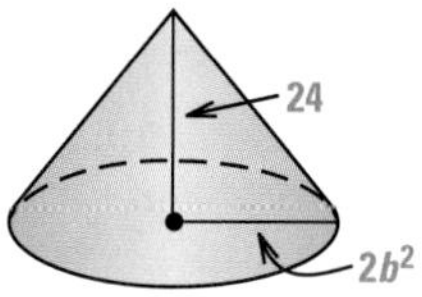

WINDMILLS In Exercises 77 and 78, use the following information. The power generated by a windmill can be modeled by the equation $w = 0.015s^3$, where w is the power measured in watts and s is the wind speed in miles per hour.

77. Find the ratio of the power generated by a windmill when the wind speed is 20 miles per hour to the power generated when the wind speed is 10 miles per hour. 8 to 1

78. *Writing* Write a general statement about how doubling the wind speed affects the amount of power generated by a windmill. See margin.

78. If the wind speed is doubled, the power generated is multiplied by 8.

FOCUS ON CAREERS

ENGINEERING TECHNICIANS use science, engineering, and math to solve technical problems in the development, construction, inspection, and maintenance of industrial machinery.

CAREER LINK
www.mcdougallittell.com

79. PROBABILITY CONNECTION Part A of a test has 10 true-false questions. Part B has 10 multiple-choice questions. Each of the multiple-choice questions has 4 possible answers. There are 2^{10} ways to answer the 10 questions in Part A. There are 4^{10} ways to answer the 10 questions in Part B.

a. How many ways are there to answer all 20 questions? 2^{30} or 1,073,741,824 ways

b. If you guess the answer to each question, what is the probability that you will get them all right? $\left(\frac{1}{2}\right)^{30}$ or about 0.0000000009

PROBABILITY CONNECTION **In Exercises 80 and 81, suppose you put one red marble, one green marble, and one blue marble in each of six bags. There are 3^6 possible orderings of the colors of the marbles you can get when you choose one marble from each bag.** See margin.

80. If you use 8 bags, there are 3^8 possible orderings of colors. What is the probability that the marbles you choose will all be red?

81. If you use 14 bags, how many different orderings of colors are there? What is the probability that the marbles you choose will all be red?

80. $\frac{1}{3^8}$ or about 0.00015

81. 3^{14}; $\frac{1}{3^{14}}$ or about 0.0000002

Test Preparation

MULTI-STEP PROBLEM **Use the results of Exercise 82 for Exercise 83.**

82. a. Copy and complete the table of values.

x	0	1	2	3	4
$2x$	?	?	?	?	?
2^x	?	?	?	?	?

x	0	1	2	3	4
$2x$	0	2	4	6	8
$2x$	1	2	4	8	16

b. Sketch the graphs of $y = 2x$ and $y = 2^x$ in the same coordinate plane. **See margin.**

c. Compare the graphs. How are they the same? How are they different? **See margin.**

82 c. ***Sample answer:*** **Both graphs increase from left to right and both pass through (2, 4); the graph of $y = 2^x$ is a curve, while the graph of $y = 2x$ is a line.**

83. CRITICAL THINKING You are offered a job that pays $2x$ dollars or 2^x dollars for x hours of work. Assuming you must work at least 2 hours, which method of payment would you choose? Explain your reasoning. **See margin.**

★ Challenge

84. LOGICAL REASONING Fill in the blanks and give a reason for each step to complete a convincing argument that the power of a power property is true. **See margin.**

$$(a^2)^3 = a^2 \cdot \underline{?} \cdot \underline{?}$$
$$= \underline{?} \cdot \underline{?} \cdot \underline{?} \cdot \underline{?} \cdot \underline{?} \cdot \underline{?}$$
$$= \underline{?}$$

EXTRA CHALLENGE
www.mcdougallittell.com

85. LOGICAL REASONING Write a convincing argument to show that the power of a product property is true.
Sample answer: **$(a \cdot b)^m = (a \cdot b) \cdot (a \cdot b) \cdot \ldots \cdot (a \cdot b)$, with $a \cdot b$ as a factor m times. The associative and commutative properties of multiplication allow you to rewrite this expression as $a \cdot a \cdot \ldots \cdot a \cdot b \cdot b \cdot \ldots \cdot b$, with each factor appearing m times. This expression is equal to $a^m \cdot b^m$.**

MIXED REVIEW

83. If you work more than 2 h, the pay is much better at the rate of 2^x dollars per hour. For example, at 2^x dollars per hour, you would earn \$256 for 8 hours, while at $2x$ dollars per hour, you would earn \$16.

EXPONENTIAL EXPRESSIONS **Evaluate the expression.** **(Review 1.3 for 8.2)**

86. b^2 when $b = 8$ **64**

87. $(5y)^4$ when $y = 2$ **10,000**

88. $\frac{1}{2}n^3$ when $n = -2$ **−4**

89. $\frac{1}{y^2}$ when $y = 5$ $\mathbf{\frac{1}{25}}$

90. $\frac{24}{x^3}$ when $x = 2$ **3**

91. $\frac{45}{a^2}$ when $a = 2$ $\mathbf{\frac{45}{4}}$

GRAPHING EQUATIONS **Use a table of values to graph the equation.** **(Review 4.2) 92–97. See margin.**

92. $y = x + 2$

93. $y = -(x - 4)$

94. $y = \frac{1}{2}x - 5$

95. $y = \frac{3}{4}x + 2$

96. $y = 2$

97. $x = -3$

GRAPHING INEQUALITIES **Sketch the graph of the inequality.** **(Review 6.1) 98–101. See margin.**

98. $x < 4$

99. $x > 15$

100. $x \geq -9$

101. $x \leq 3$

SOLVING INEQUALITIES **Solve the inequality.** **(Review 6.2)**

102. $-x - 2 < -5$ $\mathbf{x > 3}$

103. $8 + 3x \geq -2$ $\mathbf{x \geq -\frac{10}{3}}$

104. $2 < 2x + 7$ $\mathbf{x > -\frac{5}{2}}$

105. PRICE OF MILK In 1910, a quart of milk cost \$.07. In 1994, a quart of milk cost \$.71. After 1910, the price of milk increased steadily, never falling below \$.07 per quart again. During the period 1910–1994, the maximum price of a quart was \$.71. Write a compound inequality that represents the possible costs of a quart of milk between 1910 and 1994. **(Review 6.3)**
$\$0.07 \leq p \leq \$0.71$, where p is the price of a quart of milk

4 ASSESS

DAILY HOMEWORK QUIZ

Transparency Available

Simplify. Write your answer as a power or as a product of powers.

1. $5^7 \cdot 5^4$ **5^{11}, or 48,828,125**

2. $(3^4)^5$ **3^{20}, or 3,486,784,401**

3. $(7m)^6$ **7^6m^6, or $117{,}649m^6$**

4. $(-4x^7y^2)^3$ **$(-4)^3x^{21}y^6$, or $-64x^{21}y^6$**

5. $[(5p + 2)^2]^5$ **$(5p + 2)^{10}$**

6. $\left(\frac{2}{3}h^4\right)^3(5h^2)^4$ **See below.**

7. $(-6s^3t)^2(9s^2t^4)^3$
$(-6)^2 \cdot 9^3 \cdot s^{12}t^{14}$, or $26{,}244\,s^{12}t^{14}$

8. Simplify $(-b)^4 \cdot a^3 \cdot (b \cdot a)$. Evaluate the resulting expression when $a = -3$ and $b = -2$.
a^4b^5; −2592

9. Complete the statement using > or <. $(4^3 \cdot 5^2)^2$ ____ $4^9 \cdot 5^4$ **<**

6. $\left(\frac{2}{3}\right)^3 \cdot 5^4 \cdot h^{20}$, or $\frac{5000}{27}h^{20}$

EXTRA CHALLENGE NOTE
Challenge problems for Lesson 8.1 are available in **blackline** format in the *Chapter 8 Resource Book*, p. 22 and at **www.mcdougallittell.com.**

ADDITIONAL TEST PREPARATION

1. OPEN ENDED Explain when exponents are added and when they are multiplied in simplifying the expression $(2^4)^3 \cdot 2^5$.
You multiply exponents when you are finding a power of a power. For example, $(2^4)^3 = 2^{12}$. You add exponents when you are multiplying two powers with the same base. To find $2^{12} \cdot 2^5$, you would add 12 and 5 to get 2^{12+5}, or 2^{17}.

1 PLAN

PACING
Basic: 2 days
Average: 2 days
Advanced: 2 days
Block Schedule: 0.5 block with 8.1
0.5 block with 8.3

LESSON OPENER
VISUAL APPROACH
An alternative way to approach Lesson 8.2 is to use the Visual Approach Lesson Opener:
- Blackline Master (*Chapter 8 Resource Book,* p. 26)
- Transparency (p. 53)

MEETING INDIVIDUAL NEEDS
- ***Chapter 8 Resource Book***
  Prerequisite Skills Review (p. 5)
  Practice Level A (p. 29)
  Practice Level B (p. 30)
  Practice Level C (p. 31)
  Reteaching with Practice (p. 32)
  Absent Student Catch-Up (p. 34)
  Challenge (p. 36)
- ***Resources in Spanish***
- ***Personal Student Tutor***

NEW-TEACHER SUPPORT
See the Tips for New Teachers on pp. 1–2 of the *Chapter 8 Resource Book* for additional notes about Lesson 8.2.

WARM-UP EXERCISES
Transparency Available
Evaluate each expression.
1. 4^2 16
2. $(-2)^3$ −8
3. $\left(\frac{1}{2}\right)^3$ $\frac{1}{8}$
4. $2 \cdot 1.05^2$ 2.205
5. $3^4 \cdot 3^3$ 2187

8.2 Zero and Negative Exponents

What you should learn

GOAL 1 Evaluate powers that have zero and negative exponents.

GOAL 2 Graph exponential functions.

Why you should learn it

▼ To solve **real-life** problems such as predicting a player's average score per game in **Example 6**.

REAL LIFE

GOAL 1 USING ZERO AND NEGATIVE EXPONENTS

In this lesson you will study how to use the multiplication properties of exponents when working with negative exponents.

ACTIVITY Developing Concepts

Investigating Zero and Negative Exponents

1. Copy the table and discuss any patterns you see. Use the patterns to complete the table. Write non-integers as fractions in simplest form. **See margin.**

Exponent, n	3	2	1	0	−1	−2	−3
Power, 2^n	8	4	2	?	?	?	?
Power, 3^n	27	9	3	?	?	?	?
Power, 4^n	64	16	4	?	?	?	?

2. What appears to be the value of a^0 for any number a? **1**
3. How can you evaluate an expression of the form a^{-n}? **If $a \neq 0$, $a^{-n} = \frac{1}{a^n}$.**

Step 1.

0	−1	−2	−3
1	$\frac{1}{2}$	$\frac{1}{4}$	$\frac{1}{8}$
1	$\frac{1}{3}$	$\frac{1}{9}$	$\frac{1}{27}$
1	$\frac{1}{4}$	$\frac{1}{16}$	$\frac{1}{64}$

DEFINITION OF ZERO AND NEGATIVE EXPONENTS

Let a be a nonzero number and let n be a positive integer.
- A nonzero number to the zero power is 1: $a^0 = 1$, $a \neq 0$.
- a^{-n} is the reciprocal of a^n: $a^{-n} = \frac{1}{a^n}$, $a \neq 0$.

STUDENT HELP
HOMEWORK HELP
Visit our Web site www.mcdougallittell.com for extra examples.

EXAMPLE 1 *Powers with Zero and Negative Exponents*

a. $2^{-2} = \frac{1}{2^2} = \frac{1}{4}$ — 2^{-2} is the reciprocal of 2^2.

b. $(-2)^0 = 1$ — a^0 is 1.

c. $5^{-x} = \frac{1}{5^x}$ — 5^{-x} is the reciprocal of 5^x.

d. $\left(\frac{1}{3}\right)^{-1} = 3$ — The reciprocal of $\frac{1}{3}$ is 3.

e. $0^{-3} = \frac{1}{0^3}$ (Undefined) — Zero has no reciprocal.

EXAMPLE 2 Simplifying Exponential Expressions

Rewrite with positive exponents.

a. $5(2^{-x})$ **b.** $2x^{-2}y^{-3}$

SOLUTION

a. $5(2^{-x}) = 5\left(\frac{1}{2^x}\right) = \frac{5}{2^x}$

b. $2x^{-2}y^{-3} = 2 \cdot \frac{1}{x^2} \cdot \frac{1}{y^3} = \frac{2}{x^2y^3}$

> **STUDENT HELP**
> **Study Tip**
> Informally, you can think of rewriting expressions with positive exponents as "moving factors" from the denominator to the numerator or vice versa.
> $\frac{1}{3x^{-4}} = \frac{x^4}{3}$

EXAMPLE 3 Evaluating Exponential Expressions

Evaluate the expression.

a. $3^{-2} \cdot 3^2$ **b.** $(2^{-3})^{-2}$ **c.** 3^{-4}

SOLUTION

a. $3^{-2} \cdot 3^2 = 3^{-2+2}$ Use product of powers property.

$= 3^0$ Add exponents.

$= 1$ a^0 is 1.

b. $(2^{-3})^{-2} = 2^{-3 \cdot (-2)}$ Use power of a power property.

$= 2^6$ Multiply exponents.

$= 64$ Evaluate.

c. You might want to evaluate 3^{-4} with a calculator.

KEYSTROKES	DISPLAY
3 [yˣ] 4 [+/-] [=]	.01234568

> **STUDENT HELP**
> **KEYSTROKE HELP**
> Use [yˣ] or [^] to input the exponent −4.

EXAMPLE 4 Simplifying Exponential Expressions

Rewrite with positive exponents.

a. $(5a)^{-2}$ **b.** $\frac{1}{d^{-3n}}$

SOLUTION

a. $(5a)^{-2} = 5^{-2} \cdot a^{-2}$ Use power of a product property.

$= \frac{1}{5^2} \cdot \frac{1}{a^2}$ Write reciprocals of 5^2 and a^2.

$= \frac{1}{25a^2}$ Multiply fractions.

b. $\frac{1}{d^{-3n}} = (d^{-3n})^{-1}$ Use definition of negative exponents.

$= d^{(-3n) \cdot (-1)}$ Use power of a power property.

$= d^{3n}$ Multiply exponents.

2 TEACH

MOTIVATING THE LESSON
A savings account was opened a few years ago. There have been no withdrawals and the only money added to the account has been the interest. If you know the interest rate and the present balance, you can use negative exponents to find how much money was in the past.

Correction — as printed: A savings account was opened a few years ago. There have been no withdrawals and the only money added to the account has been the interest. If you know the interest rate and the present balance, you can use negative exponents to find how much money was in the account at some time in the past.

EXTRA EXAMPLE 1

a. $3^{-4} = ?$ $\frac{1}{81}$ **b.** $(-5.2)^0 = ?$ 1

c. $4^{-y} = ?$ $\frac{1}{4^y}$ **d.** $\left(\frac{3}{5}\right)^{-1} = ?$ $\frac{5}{3}$

e. $0^{-1} = ?$ undefined

EXTRA EXAMPLE 2
Rewrite with positive exponents. Assume k is positive.

a. $4(3^{-k})$ $\frac{4}{3^k}$

b. $5g^{-3} \cdot h^{-4}$ $\frac{5}{g^3h^4}$

EXTRA EXAMPLE 3
Evaluate the expression.

a. $4^{-3} \cdot 4^3$ 1

b. $(5^{-2})^{-3}$ 15,625

c. 2^{-3} $\frac{1}{8}$

EXTRA EXAMPLE 4
Rewrite with positive exponents. Assume n is positive.

a. $(4y)^{-3}$ $\frac{1}{64y^3}$ **b.** $\frac{1}{g^{-2n}}$ g^{2n}

CHECKPOINT EXERCISES

For use after Examples 1 and 2:

1. Rewrite with positive exponents. Assume h is positive.

a. $3(2^{-h})$ $\frac{3}{2^h}$

b. $4k^{-3} \cdot m^{-2}$ $\frac{4}{k^3m^2}$

For use after Examples 3 and 4:

2.a. $6^{-2} \cdot 6^2$ 1

b. $(4n^{-2})^{-3}$ $\frac{n^6}{64}$

EXTRA EXAMPLE 5

Sketch the graph of $y = 4^x$.

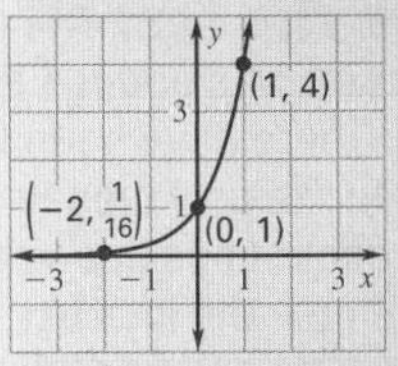

EXTRA EXAMPLE 6

Census figures showed that a town's population can be modeled by $P = 5325(1.02)^t$, where $t = 0$ represents the year 2010.

a. Estimate the population in the year 1999. **4283**

b. Estimate the population in the year 2010. **5325**

CHECKPOINT EXERCISES

For use after Example 5:

1. Sketch the graph of $y = \left(\frac{1}{2}\right)^x$.

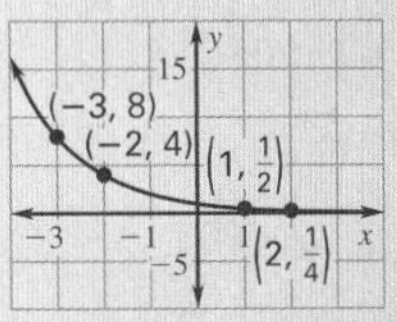

For use after Example 6:

2. A middle range sales price for a house in the U.S. can be modeled by $P = 124(1.03)^t$, where t represents the year 1997 and P is the price in thousands of dollars.

a. Estimate the middle home sales price in the year 1991. **about $103,848**

b. Estimate the price in the year 1997. **about $124,000**

FOCUS ON VOCABULARY

Explain why $y = 3^x$ is an exponential function and $y = x^3$ is not.
In $y = 3^x$, the variable occurs as an exponent and not as a base.

CLOSURE QUESTION

Explain why a^{-n} is not defined when n is positive and $a = 0$.
See margin.

DAILY PUZZLER

a and b are two different nonzero integers. If $(-a)^{-a} = -a$ and $(-b)^{-b} = -b$, what is the value of ab? **±1**

GOAL 2 GRAPHING EXPONENTIAL FUNCTIONS

So far we have used expressions of the form b^n where $b \neq 0$ and n is an integer. To model some situations, we need an **exponential function** of the form $y = a \cdot b^x$ where x is a real number, $b > 0$, and $b \neq 1$. To graph the exponential function $y = a \cdot b^x$, make a table using integer values for x and plot the corresponding points. To complete the graph for all real values of x, connect the points with a smooth curve.

EXAMPLE 5 *Graphing an Exponential Function*

To sketch the graph of $y = 2^x$, make a table that includes negative x-values.

x	−2	−1	0	1	2	3
2^x	$2^{-2} = \frac{1}{4}$	$2^{-1} = \frac{1}{2}$	$2^0 = 1$	2	4	8

Draw a coordinate plane and plot the six points given by the table. Then draw a smooth curve through the points.

Notice that the graph has a y-intercept of 1, and that it gets closer to the negative side of the x-axis as the x-values get smaller.

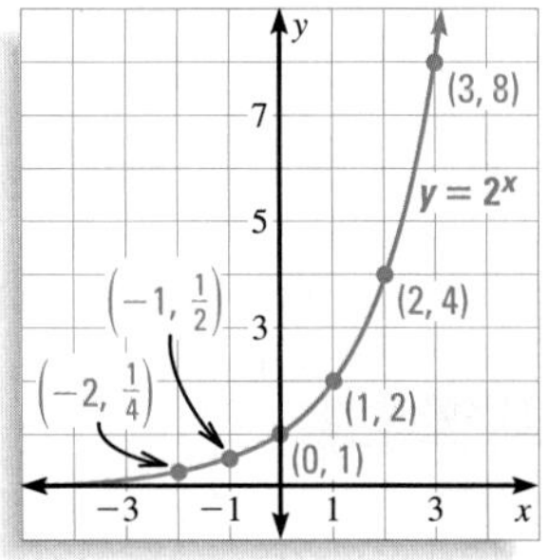

In real-life applications of $y = ab^x$, x is often the time period.

EXAMPLE 6 *Evaluating an Exponential Function*

REAL LIFE *Basketball*

A professional basketball player's first year in the NBA was 1998. Suppose it were estimated that from 1998 to 2010 his average points per game could be modeled by $P = 18.5(1.038)^t$, where $t = 0$ represents the year 2000.

a. Estimate the player's average points per game in 1998.

b. Estimate his average points per game in the year 2000.

SOLUTION

a. $P = 18.5 \cdot 1.038^{-2}$ Substitute −2 for t.

≈ 17.17 Use a calculator.

▶ In the year 1998 his average points per game was about 17.2 points.

b. $P = 18.5 \cdot 1.038^0$ Substitute 0 for t.

$= 18.5$ a^0 is 1.

▶ In the year 2000 his average points per game was about 18.5 points.

Closure Question ***Sample answer:*** **a^{-n} is a short way of writing $\frac{1}{a^n}$. If $a = 0$ and n is positive, $\frac{1}{a^n}$ would equal the quotient of 1 and 0, which is not defined.**

GUIDED PRACTICE

Vocabulary Check ✓

Concept Check ✓

Skill Check ✓

1. The function $y = ab^x$ is a(n) _?_ function. **exponential**

2. **ERROR ANALYSIS** Describe the error at the right.

~~$5x^{-3} = \frac{1}{5x^3}$~~

The exponent applies only to the variable; $5x^{-3} = \frac{5}{x^3}$.

Evaluate the expression.

3. 3^{-1} **$\frac{1}{3}$**
4. 0^{-4} **not defined**
5. 0^0 **not defined**
6. $6 \cdot 3^0$ **6**

Rewrite as an expression with positive exponents.

7. m^{-2} **$\frac{1}{m^2}$**
8. $3c^{-5}$ **$\frac{3}{c^5}$**
9. a^5b^{-8} **$\frac{a^5}{b^8}$**
10. $\frac{1}{(2x)^{-3}}$ **$8x^3$**

11. Does the graph at the right appear to be the graph of the exponential function $y = 3^x$? **yes**

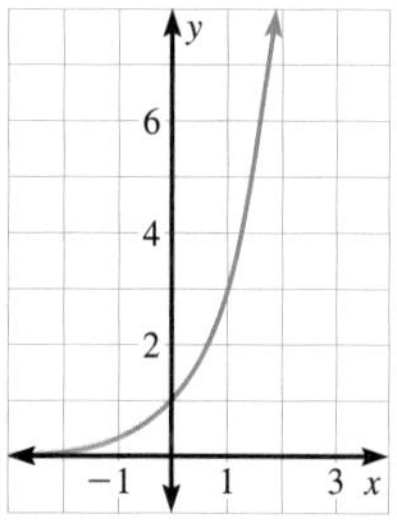

Ex. 11

12. Tell whether the following statement is true.

If a is positive, then a^{-n} is positive.

Explain your reasoning.

12. Yes; *Sample answer:* If *a* is positive, then any power a^n is a product of positive factors and, so, is positive. The reciprocal of a positive number is positive, so a^{-n} is positive for every number *n*.

13. **BASKETBALL** Use the information in Example 6. If the player's final year in the NBA is 2010, estimate his average points per game in his final year. **about 26.9 points per game**

PRACTICE AND APPLICATIONS

STUDENT HELP

Extra Practice to help you master skills is on p. 804.

EVALUATING EXPRESSIONS **Evaluate the exponential expression. Write fractions in simplest form.**

14. 4^{-2} **$\frac{1}{16}$**
15. 3^{-4} **$\frac{1}{81}$**
16. $\left(\frac{1}{5}\right)^{-1}$ **5**
17. $8\left(\frac{1}{4}\right)^{-1}$ **32**
18. $4(4^{-2})$ **$\frac{1}{4}$**
19. $\left(\frac{1}{10}\right)^{-2}$ **100**
20. $-6^0 \cdot \frac{1}{3^{-2}}$ **−9**
21. $2^{-3} \cdot 2^2$ **$\frac{1}{2}$**
22. $8^3 \cdot 0^{-1}$ **not defined**
23. $7^4 \cdot 7^{-4}$ **1**
24. $8^{-7} \cdot 8^7$ **1**
25. $-4 \cdot (-4)^{-1}$ **1**
26. $(5^{-3})^2$ **$\frac{1}{15,625}$**
27. $(-3^{-2})^{-1}$ **−9**
28. $11 \cdot 11^{-1}$ **1**
29. $4^0 \cdot 5^{-3}$ **$\frac{1}{125}$**

STUDENT HELP

HOMEWORK HELP

Example 1: Exs. 14–29
Example 2: Exs. 30–45
Example 3: Exs. 14–29, 49–52
Example 4: Exs. 30–45
Example 5: Exs. 53–63
Example 6: Exs. 64, 65

SIMPLIFYING EXPRESSIONS **Rewrite the expression with positive exponents.**

30. x^{-5} **$\frac{1}{x^5}$**
31. $3x^{-4}$ **$\frac{3}{x^4}$**
32. $\frac{1}{2x^{-5}}$ **$\frac{x^5}{2}$**
33. $x^{-2}y^4$ **$\frac{y^4}{x^2}$**
34. x^4y^{-7} **$\frac{x^4}{y^7}$**
35. $8x^{-2}y^{-6}$ **$\frac{8}{x^2y^6}$**
36. $\frac{1}{9x^{-3}y^{-1}}$ **$\frac{x^3y}{9}$**
37. $\frac{1}{4x^{-10}y^{14}}$ **$\frac{x^{10}}{4y^{14}}$**
38. $(-9)^0x$ **x**
39. $(-4x)^{-3}$ **$-\frac{1}{64x^3}$**
40. $(-10a)^0$ **1**
41. $(3xy)^{-2}$ **$\frac{1}{9x^2y^2}$**
42. $(6a^{-3})^3$ **$\frac{216}{a^9}$**
43. $\frac{8}{m^{-2}}$ **$8m^2$**
44. $\frac{1}{(4x)^{-5}}$ **$1024x^5$**
45. $\left(\frac{-4x^2}{2x^{-1}}\right)^{-1}$ **$-\frac{1}{2x^3}$**

3 APPLY

ASSIGNMENT GUIDE

BASIC

Day 1: pp. 459–460 Exs.14–44 even, 46–48, 50–60 even

Day 2: pp. 459–461 Exs. 15–45 odd, 51–61 odd, 66–69, 71–75

AVERAGE

Day 1: pp. 459–460 Exs.14–44 even, 46–48, 50–62 even

Day 2: pp. 459–461 Exs. 15–45 odd, 51–63 odd, 66–69, 71–75

ADVANCED

Day 1: pp. 459–460 Exs.14–44 even, 46–48, 50–64 even

Day 2: pp. 459–461 Exs. 15–45 odd, 51–63 odd, 66–69, 70–75

BLOCK SCHEDULE

pp. 459–460 Exs.14–44 even, 46–48, 50–62 even (with 8.1)
pp. 459–461 Exs. 15–45 odd, 51–63 odd, 66–69, 71–75 (with 8.3)

EXERCISE LEVELS

Level A: *Easier*
14–29

Level B: *More Difficult*
30–61, 64–69, 71–88

Level C: *Most Difficult*
62, 63, 70

✓ HOMEWORK CHECK

To quickly check student understanding of key concepts, go over the following exercises: Exs. 16,18, 26, 28, 32, 34, 40, 44, 58. See also the Daily Homework Quiz:

- Blackline Master (*Chapter 8 Resource Book*, p. 39)
- Transparency (p. 62)

! COMMON ERROR

EXERCISE 53 A common error is to interpret -3^x to mean negative 3 to the xth power rather than the opposite of the xth power of 3. Point out that "negative 3 to the xth power" would be written in symbols as $(-3)^x$. You might suggest that students evaluate -3^x and $(-3)^x$ for several integer values of x.

MATHEMATICAL REASONING

Students need to understand that while each graph in Exs. 57–60 comes very close to the x-axis, the graph never actually reaches the x-axis. For example, in Ex 57, if x is a large positive integer, $\left(\frac{1}{3}\right)^x$ will have the same value as $\frac{1}{3^x}$. The denominator will be very large and the fraction very small. But *close to 0* and *equal to 0* are not the same.

TEACHING TIP

EXERCISE 64 This application models the formula for an annual growth rate of 6%. Emphasize the use of a negative exponent to find the account balance in previous years.

57–60, 65b, 75–82.
See Additional Answers beginning on page AA1.

ADDITIONAL PRACTICE AND RETEACHING

For Lesson 8.2:

- Practice Levels A, B, and C (*Chapter 8 Resource Book,* p. 29)
- Reteaching with Practice (*Chapter 8 Resource Book,* p. 32)
- See Lesson 8.2 of the *Personal Student Tutor*

For more Mixed Review:

- Search the *Test and Practice Generator* for key words or specific lessons.

MATCHING THE GRAPH Match the equation with its graph.

A.
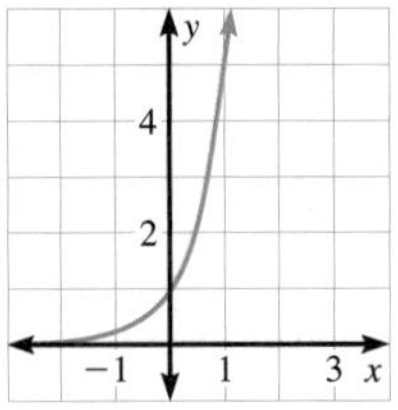

B.
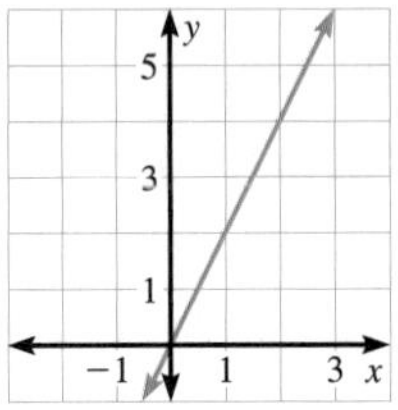

C.
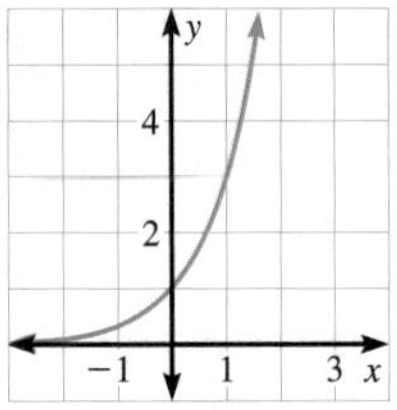

46. $y = 2x$ **B** **47.** $y = 3^x$ **C** **48.** $y = 5^x$ **A**

EVALUATING EXPONENTIAL EXPRESSIONS **Use a calculator to evaluate the expression. Round your answer to the nearest hundred thousandth.**

49. 2^{-5} **0.03125** **50.** $(1.1)^{-2}$ **0.82645** **51.** $55 \cdot 5^{-6}$ **0.00352** **52.** $3^{-5} \div 0.9$ **0.00457**

CHECKING POINTS **Does the graph of the function contain the point (0, 1)?**

53. $y = -3^x$ **no** **54.** $y = 4^x$ **yes** **55.** $y = 4 \cdot 1^x$ **no** **56.** $y = 60^x$ **yes**

GRAPHING FUNCTIONS **In Exercises 57–60, graph the exponential function.** **57–60. See margin.**

57. $y = \left(\frac{1}{3}\right)^x$ **58.** $y = \left(\frac{1}{5}\right)^x$ **59.** $y = 4^{-x}$ **60.** $y = 5^x$

61. VISUAL THINKING Sketch the graphs of $y = 2^x$ and $y = \left(\frac{1}{2}\right)^x$. How are the graphs related? **See margin.**

61. *Sample answer:* Both graphs are curves that pass through (0, 1) and both increase without limit in one direction and get very close to the x-axis in the other. See Additional Answers for graph.

62. CRITICAL THINKING Sketch the graphs of $y = 3^x$ and $y = \left(\frac{1}{3}\right)^x$. Use these graphs and the ones you sketched in Exercise 61 to predict how the graphs of $y = b^x$ and $y = \left(\frac{1}{b}\right)^x$ are related. **See margin.**

62. Both graphs are curves that pass through (0, 1) and are mirror images, or reflections of each other over the y-axis. See Additional Answers for graph.

63. COMMON POINTS What point do all graphs of the form $y = a^x$ have in common? Is there a point that all graphs of the form $y = 2(a)^x$ have in common? If so, name the point. **(0, 1); yes; (0, 2)**

64. SAVINGS ACCOUNT You started a savings account in 1990. The balance A is modeled by $A = 450(1.06)^t$, where $t = 0$ represents the year 2000. What is the balance in the account in 1990? in 2000? in 2010? **$251.28; $450; $805.88**

FOCUS ON APPLICATIONS

SHIPWRECKS In 1996 the excavation of the *Belle* off the coast of Texas uncovered the hull of the ship, three bronze cannons, millions of glass beads, pottery and even the skeleton of a crew member.

APPLICATION LINK
www.mcdougallittell.com

65. SHIPWRECKS Suppose that from 1860 to 1980 the number of shipwrecks in the Gulf of Mexico increased by about the same percent each year and that the number of shipwrecks S for each decade t can be modeled by $S = 292(1.2)^t$, where $t = 0$ represents the decade 1920 to 1929.

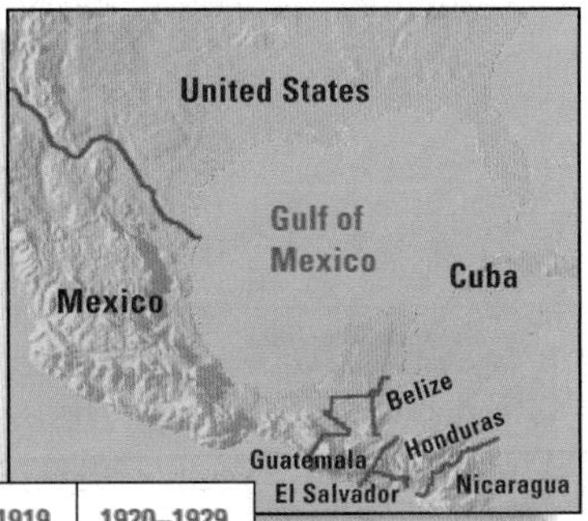

a. Copy and complete the table below. **117; 141; 203; 243; 292**

	1870–1879	1880–1889	1900–1909	1910–1919	1920–1929
t	−5	−4	−2	−1	0
S	?	?	?	?	?

b. Graph the function and check your results. **See margin.**

Test Preparation

QUANTITATIVE COMPARISON In Exercises 66–68, evaluate each function. Then choose the statement below that is true about the given values of y.

Ⓐ The value of y in column A is greater.

Ⓑ The value of y in column B is greater.

Ⓒ The two values of y are equal.

Ⓓ The relationship cannot be determined from the given information.

		Column A	Column B	
66.	When $x = 3$,	$y = 2x$	$y = 2^x$	B
67.	When $x = 1$,	$y = 2^x$	$y = 2^{-x}$	A
68.	When $x = 0$,	$y = \left(\frac{1}{2}\right)^x$	$y = \left(\frac{1}{2}\right)^{-x}$	C

69. **MULTIPLE CHOICE** What is a possible equation of the graph? **C**

Ⓐ $y = 2^x$ Ⓑ $y = 3^x$

Ⓒ $y = \left(\frac{1}{2}\right)^x$ Ⓓ $y = \left(\frac{1}{3}\right)^x$

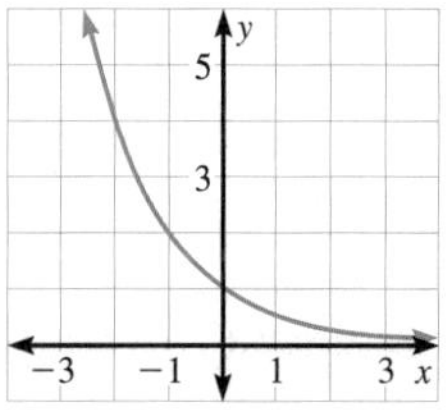

★ Challenge

70. *Writing* Suppose you did not know that for $b \neq 0$, $b^0 = 1$. Based on the equation $b^2 \cdot b^0 = b^{2+0} = b^2$, explain why you might want to make this definition. **We are given $b^2 \cdot b^0 = b^{0+2} = b^2$. We know by the identity property of multiplication that $b^2 \cdot 1 = b^2$. Since $b^2 \cdot b^0 = b^2 = b^2 \cdot 1$, b^0 must be equal to 1.**

MIXED REVIEW

EVALUATING EXPRESSIONS Evaluate the expression. (Review 1.3 for 8.3)

71. $\left(\frac{2}{5}\right)^2$ $\frac{4}{25}$ **72.** $\left(\frac{1}{2}\right)^3$ $\frac{1}{8}$ **73.** $\left(-\frac{9}{10}\right)^3$ $-\frac{729}{1000}$ **74.** $\left(\frac{1}{5}\right)^4$ $\frac{1}{625}$

SOLVE AND GRAPH Solve the inequality. Then sketch a graph of the solution on a number line. (Review 6.4) **75–80. See margin for graphs.**

75. $-12 \le x \le 2$

76. $x > 2$ or $x < -6\frac{2}{3}$

77. $-11 \le x \le 7$

75. $|5 + x| + 4 \le 11$ **76.** $|3x + 7| - 4 > 9$ **77.** $|x + 2| - 1 \le 8$

78. $|3 - x| - 6 > -4$ **$x < 1$ or $x > 5$** **79.** $|9 - 2x| + 3 < 4$ **$4 < x < 5$** **80.** $|3x + 2| + 9 \ge -1$ **all real numbers**

STATISTICS Draw a box-and-whisker plot of the data. (Review 6.7) **81–82. See margin.**

81. 48, 10, 48, 25, 40, 42, 44, 23, 21, 13, 50, 17

82. 85, 61, 55, 78, 79, 86, 30, 76, 76, 87, 68, 82

SOLVING SYSTEMS Use substitution to solve the system. (Review 7.2)

83. $2x - y = -2$, $4x + y = 5$ $\left(\frac{1}{2}, 3\right)$ **84.** $-3x + y = 4$, $-9x + 5y = 10$ $\left(-1\frac{2}{3}, -1\right)$ **85.** $x + 4y = 300$, $x - 2y = 0$ **(100, 50)**

86. $2x - 3y = 10$, $3x + 3y = 15$ **(5, 0)** **87.** $x + 15y = 6$, $-x - 5y = 84$ **(−129, 9)** **88.** $4x - y = 5$, $2x + 4y = 15$ $\left(1\frac{17}{18}, 2\frac{7}{9}\right)$

4 ASSESS

DAILY HOMEWORK QUIZ

Transparency Available

Evaluate the exponential expression. Write fractions in simplest form.

1. $7(7^{-3})$ $\frac{1}{49}$ **2.** $\left(\frac{1}{3}\right)^{-2}$ 9

3. $(4^{-2})^2$ $\frac{1}{256}$ **4.** $12 \cdot 12^{-1}$ 1

Rewrite the expression with positive exponents.

5. $m^{-3}n^5$ $\frac{n^5}{m^3}$ **6.** $\frac{1}{7k^{-3}}$ $\frac{k^3}{7}$

7. $(15r)^0$ 1 **8.** $\frac{2}{(8x)^{-3}}$ $1024x^3$

9. Graph the exponential function $y = \left(\frac{1}{7}\right)^x$.

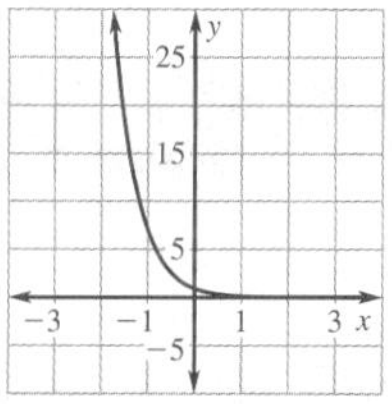

EXTRA CHALLENGE NOTE

Challenge problems for Lesson 8.2 are available in **blackline** format in the *Chapter 8 Resource Book*, p. 36 and at **www.mcdougallittell.com.**

ADDITIONAL TEST PREPARATION

1. WRITING Explain how the expression b^n, where n is a positive integer, can be rewritten as an expression involving only negative exponents. What do you have to assume about the value of the base b?
You can rewrite b^n as $(b^{-n})^{-1}$. You have to assume b is not equal to 0.

1 Planning the Activity

PURPOSE
To use a graphing calculator to graph an exponential function.

MATERIALS
- Graphing calculator
- Keystroke blackline (*Chapter 8 Resource Book,* p. 27)

PACING
- Exploring the Concept — 5 min
- Drawing Conclusions — 10 min

LINK TO LESSON
All exponential functions of the form $y = b^x$ contain the point (0, 1). When students learn to graph an exponential function in Example 5 of Lesson 8.2 and Exs. 47, 48, 57, and 58, they should notice that (0, 1) is a point of each graph.

2 Managing the Activity

CLASSROOM MANAGEMENT
Some students may have trouble graphing exponential functions with a fractional base. Make sure that students enclose the fraction in parentheses when they type it in.

ALTERNATIVE APPROACH
To save time, do this activity as a demonstration and use the table features of the calculator to display coordinates.

3 Closing the Activity

★ ***KEY DISCOVERY***
If the base of an exponential function is less than 1 and greater than zero, the y-values get smaller as the x-values get larger.

ACTIVITY ASSESSMENT
JOURNAL Describe how to use a graphing calculator to graph exponential functions. *Sample answer:* **Step 1: Enter the equation into the calculator. Step 2: Adjust the viewing window. Step 3: Graph.**

ACTIVITY 8.2
Using Technology

Graphing Exponential Functions

You can use a graphing calculator to graph an exponential function.

EXAMPLE

Graph $y = \left(\frac{1}{2}\right)^x$.

SOLUTION

1 To enter the function in your graphing calculator, press Y=. Enter the function as 0 . 5 ^ X,T,θ.

```
Y1=0.5^X
Y2=
Y3=
Y4=
Y5=
Y6=
Y7=
```

2 Adjust the *viewing window* to get the best scale for your graph.

```
WINDOW
 Xmin=-10
 Xmax=10
 Xscl=2
 Ymin=-10
 Ymax=10
 Yscl=2
```

3 Now you are ready to graph the function. Press GRAPH to see the graph.

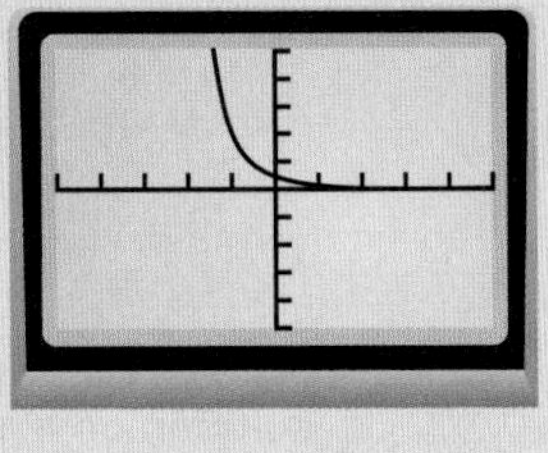

STUDENT HELP
INTERNET KEYSTROKE HELP
See keystrokes for several models of calculators at www.mcdougallittell.com

EXERCISES

Use a graphing calculator to graph the exponential function. **1–6. Check graphs.**

1. $y = 2^x$ **2.** $y = 10^x$ **3.** $y = -3^x$

4. $y = 5^{-x}$ **5.** $y = (0.27)^x$ **6.** $y = -\left(\frac{2}{3}\right)^x$

CRITICAL THINKING **Use your results from Exercises 1–6 to answer the following questions.** **7–10. Sample answers are given. See margin.**

7. If $a > 1$, what does the graph of $y = a^x$ look like?

8. If $0 < a < 1$, what does the graph of $y = a^x$ look like?

9. If $a > 1$, what does the graph of $y = -(a^x)$ look like?

10. If $0 < a < 1$, what does the graph of $y = -(a^x)$ look like?

7. The graph is close to the negative x-axis and curves upward to the right.

8. The graph curves upward to the left and is close to the x-axis on the right.

9. The graph is close to the negative x-axis and curves downward to the right.

10. The graph is close to the positive x-axis and curves downward to the left.

8.3 Division Properties of Exponents

What you should learn

GOAL 1 Use the division properties of exponents to evaluate powers and simplify expressions.

GOAL 2 Use the division properties of exponents to find a probability as in **Example 5**.

Why you should learn it

▼ To solve **real-life** problems such as comparing the top speeds of boats in an Olympic rowing competition in **Ex. 58**.

REAL LIFE

GOAL 1 DIVIDING WITH EXPONENTS

In Lesson 8.1 you learned that you multiply powers with the same base by adding exponents. To divide powers with the same base, you subtract exponents. Here is an example.

$$\frac{4^5}{4^3} = \frac{\overbrace{4 \cdot 4 \cdot 4 \cdot 4 \cdot 4}^{5 \text{ factors}}}{\underbrace{4 \cdot 4 \cdot 4}_{3 \text{ factors}}} = \underbrace{4 \cdot 4}_{2 \text{ factors}} = 4^{5-3} = 4^2$$

CONCEPT SUMMARY — DIVISION PROPERTIES OF EXPONENTS

Let a and b be numbers and let m and n be integers.

QUOTIENT OF POWERS PROPERTY

To divide powers having the same base, subtract exponents.

$\frac{a^m}{a^n} = a^{m-n}, a \neq 0$ **Example:** $\frac{3^7}{3^5} = 3^{7-5} = 3^2$

POWER OF A QUOTIENT PROPERTY

To find a power of a quotient, find the power of the numerator and the power of the denominator and divide.

$\left(\frac{a}{b}\right)^m = \frac{a^m}{b^m}, b \neq 0$ **Example:** $\left(\frac{4}{5}\right)^3 = \frac{4^3}{5^3}$

EXAMPLE 1 *Using the Quotient of Powers Property*

a. $\frac{6^5}{6^4} = 6^{5-4}$
$= 6^1$
$= 6$

b. $\frac{(-5)^2}{(-5)^2} = (-5)^{2-2}$
$= (-5)^0$
$= 1$

c. $\frac{9^4 \cdot 9^2}{9^7} = \frac{9^6}{9^7}$
$= 9^{6-7}$
$= 9^{-1}$
$= \frac{1}{9}$

d. $\frac{1}{y^5} \cdot y^3 = \frac{y^3}{y^5}$
$= y^{3-5}$
$= y^{-2}$
$= \frac{1}{y^2}$

1 PLAN

PACING

Basic: 2 days
Average: 2 days
Advanced: 2 days
Block Schedule: 0.5 block with 8.2
0.5 block with 8.4

LESSON OPENER
CALCULATOR ACTIVITY

An alternative way to approach Lesson 8.3 is to use the Calculator Activity Lesson Opener:

- Blackline Master (*Chapter 8 Resource Book,* p. 40)
- Transparency (p. 54)

MEETING INDIVIDUAL NEEDS

- ***Chapter 8 Resource Book***
  Prerequisite Skills Review (p. 5)
  Practice Level A (p. 43)
  Practice Level B (p. 44)
  Practice Level C (p. 45)
  Reteaching with Practice (p. 46)
  Absent Student Catch-Up (p. 48)
  Challenge (p. 50)
- ***Resources in Spanish***
- ***Personal Student Tutor***

NEW-TEACHER SUPPORT

See the Tips for New Teachers on pp. 1–2 of the *Chapter 8 Resource Book* for additional notes about Lesson 8.3.

WARM-UP EXERCISES

Transparency Available

Simplify.

1. 5^{-2} $\frac{1}{25}$
2. $(3w^2)^2$ $9w^4$
3. $(4vw^3)^2$ $16v^2w^6$
4. $(2xy^3)^4(4x^2y^5)$ $64x^6y^{17}$
5. $(-2r^2s^3)^5$ $-32r^{10}s^{15}$

2 TEACH

MOTIVATING THE LESSON
A book reviewer for a newspaper receives 15 new novels one week and 23 new novels the following week. For the first week there are 2^{15} ways to decide which, if any, of the novels to recommend. For the next week, there are 2^{23} ways. Today's lesson will examine a simple way to use exponents to compare these two sets of possibilities.

EXTRA EXAMPLE 1

a. $\frac{4^5}{4^4}$ 4 b. $\frac{(-3)^4}{(-3)^4}$ 1

c. $\frac{6^3 \cdot 6^5}{6^9}$ $\frac{1}{6}$ d. $\frac{2}{x^4} \cdot x^2$ $\frac{2}{x^2}$

EXTRA EXAMPLE 2
Simplify the expression.

a. $\left(\frac{1}{4}\right)^3$ $\frac{1}{64}$ b. $-\left(\frac{4}{x}\right)^2$ $-\frac{16}{x^2}$

c. $\left(\frac{3}{2}\right)^{-4}$ $\frac{16}{81}$

EXTRA EXAMPLE 3
Simplify the expression.

a. $\frac{3x^3y}{4x} \cdot \frac{12x^2y^2}{y^3}$ $9x^4$

b. $\left(\frac{3x^2}{y}\right)^3$ $\frac{27x^6}{y^3}$

CHECKPOINT EXERCISES

For use after Examples 1 and 2:

1. Simplify the expression.

a. $\frac{7^5}{7^6}$ $\frac{1}{7}$ b. $\frac{(-4)^3}{(-4)^3}$ 1

c. $\left(\frac{-2}{m}\right)^4$ $\frac{16}{m^4}$ d. $\left(\frac{5}{3}\right)^{-3}$ $\frac{27}{125}$

For use after Example 3:

2. Simplify the expression.

a. $\frac{-6m^3n^4}{4n} \cdot \frac{16m^2}{6mn}$ $-4m^4n^2$

b. $\left(\frac{3y}{r^2}\right)^2$ $\frac{9y^2}{r^4}$

STUDENT HELP

HOMEWORK HELP
Visit our Web site www.mcdougallittell.com for extra examples.

EXAMPLE 2 *Using the Power of a Quotient Property*

Simplify the expression.

a. $\left(\frac{2}{3}\right)^2$ b. $\left(-\frac{3}{y}\right)^3$ c. $\left(\frac{7}{4}\right)^{-3}$

SOLUTION

a. $\left(\frac{2}{3}\right)^2 = \frac{2^2}{3^2} = \frac{4}{9}$ Square numerator and denominator and simplify.

b. $\left(-\frac{3}{y}\right)^3 = \left(\frac{-3}{y}\right)^3$ Rewrite fraction.

$= \frac{(-3)^3}{y^3}$ Power of a quotient

$= \frac{-27}{y^3}$ Simplify.

c. $\left(\frac{7}{4}\right)^{-3} = \frac{7^{-3}}{4^{-3}}$ Power of a quotient

$= \frac{4^3}{7^3}$ Definition of negative exponents

$= \frac{64}{343}$ Simplify.

EXAMPLE 3 *Simplifying Expressions*

Simplify the expression.

a. $\frac{2x^2y}{3x} \cdot \frac{9xy^2}{y^4}$ b. $\left(\frac{2x}{y^2}\right)^4$

SOLUTION

a. $\frac{2x^2y}{3x} \cdot \frac{9xy^2}{y^4} = \frac{(2x^2y)(9xy^2)}{(3x)(y^4)}$ Multiply fractions.

$= \frac{18x^3y^3}{3xy^4}$ Product of powers

$= 6x^2y^{-1}$ Quotient of powers

$= \frac{6x^2}{y}$ Definition of negative exponents

b. $\left(\frac{2x}{y^2}\right)^4 = \frac{(2x)^4}{(y^2)^4}$ Power of a quotient

$= \frac{2^4 \cdot x^4}{y^{2 \cdot 4}}$ Product of a product and power of a power

$= \frac{16x^4}{y^8}$ Simplify.

GOAL 2 USING POWERS AS REAL-LIFE MODELS

FOCUS ON CAREERS

FLOOR TRADER Floor traders at the New York Stock Exchange use hand signals that date back to the 1880s to relay information about their trades.

CAREER LINK
www.mcdougallittell.com

EXAMPLE 4 *Using the Quotient of Powers Property*

STOCK EXCHANGE The number of shares N (in billions) listed on the New York Stock Exchange from 1977 through 1997 can be modeled by

$N = 92.56 \cdot (1.112)^t$

where $t = 0$ represents 1990. Find the ratio of shares listed in 1997 to the shares listed in 1977. ▶ Source: Federal Reserve Bank of New York

SOLUTION

Use $t = -13$ for 1977 and $t = 7$ for 1997.

$$\frac{\text{Number listed in 1997}}{\text{Number listed in 1977}} = \frac{92.56 \cdot (1.112)^7}{92.56 \cdot (1.112)^{-13}}$$

$$= 1.112^{7-(-13)}$$

$$= 1.112^{20}$$

$$\approx 8.4$$

▶ The ratio of shares listed in 1997 to the shares listed in 1977 is 8.4 to 1. There were about 8.4 times as many listed in 1997 as in 1977.

EXAMPLE 5 *Using the Power of a Quotient Property*

PROBABILITY CONNECTION You toss a fair coin ten times. Show that the probability that the coin lands heads up each time is about 0.001.

SOLUTION

Probability that the first toss is heads: $\frac{1}{2}$

Probability that the first two tosses are heads: $\left(\frac{1}{2}\right)^2$

Probability that the first three tosses are heads: $\left(\frac{1}{2}\right)^3$

⋮ ⋮

Probability that the first nine tosses are heads: $\left(\frac{1}{2}\right)^9$

Probability that the first ten tosses are heads: $\left(\frac{1}{2}\right)^{10}$

Use the power of a quotient property to evaluate.

$$\left(\frac{1}{2}\right)^{10} = \frac{1}{2^{10}} = \frac{1}{1024} \approx 0.001$$

▶ The probability is $\left(\frac{1}{2}\right)^{10}$ or about 0.001.

STUDENT HELP

Look Back
For help with probability, see page 114.

EXTRA EXAMPLE 4
A middle-range weekly wage for a man in the U.S. from 1980 to 1997 can be modeled by $y = 460(1.035)^t$ where $t = 0$ represents the year 1990. Find the ratio of weekly earnings in 1997 to the earnings in 1980.
about 1.8 to 1

EXTRA EXAMPLE 5
You toss a number cube 5 times in a row. Show that the probability of getting all sixes is about 0.00013. $\left(\frac{1}{6}\right)^5 \approx 0.00013$

CHECKPOINT EXERCISES

For use after Example 4:

1. A middle range weekly wage for a woman in the U.S. from 1980 to 1997 can be modeled by $y = 348(1.035)^t$, where $t = 0$ represents the year 1990. Find the ratio of weekly earnings in 1995 to the earnings in 1985. **about 1.4 to 1**

For use after Example 5:

2. You toss a number cube 5 times in a row. Show that the probability of getting no sixes is about 0.402. $\left(\frac{5}{6}\right)^5 \approx 0.402$

STUDENT HELP NOTES

Career Link Additional information about being a stock broker is available at **www.mcdougallittell.com.**

Look Back As students look back to p. 114, remind them each toss of the coin or number cube is independent of the other tosses.

CLOSURE QUESTION
With respect to the exponents, how is dividing two powers of the same nonzero base different from multiplying these powers? **To divide the powers, you subtract the exponents. To multiply the powers, you add the exponents.**

3 APPLY

ASSIGNMENT GUIDE

BASIC

Day 1: p. 466 Exs. 20–48 even

Day 2: pp. 467–469 Exs. 49–58, 62–65, 70, 75, 80, 85, Quiz 1 Exs. 1–24

AVERAGE

Day 1: p. 466 Exs. 20–48 even

Day 2: pp. 467–469 Exs. 49–58, 62–65, 70, 75, 80, 85, Quiz 1 Exs. 1–24

ADVANCED

Day 1: p. 466 Exs. 20–48 even

Day 2: pp. 467–469 Exs. 49–66, 70, 75, 80, 85, Quiz 1 Exs. 1–24

BLOCK SCHEDULE

pp. 466–467 Exs. 20–48 even (with 8.2)

pp. 467–469 Exs. 49–58, 62–65, 70, 75, 80, 85, Quiz 1 Exs. 1–24 (with 8.4)

EXERCISE LEVELS

Level A: ***Easier***
19, 20, 31–35, 67–75

Level B: ***More Difficult***
21–30, 36–58, 60–64, 76–86

Level C: ***Most Difficult***
59, 65, 66

✓ HOMEWORK CHECK

To quickly check student understanding of key concepts, go over the following exercises: Exs. 20, 28, 30, 38, 40, 46, 54. See also the Daily Homework Quiz:

- Blackline Master (*Chapter 8 Resource Book*, p. 54)
- Transparency (p. 63)

40. $-\frac{27x^3}{y^6}$

42. $-\frac{8x^4}{y}$

GUIDED PRACTICE

Vocabulary Check ✓

1. The expression $\frac{a^4}{a^6}$ can be simplified by using the _?_ property. **quotient of powers**

Concept Check ✓

2. Can $\frac{x^8}{y^3}$ be simplified? Explain your answer. **No; the quotient of powers property only applies when the powers have the same base.**

Skill Check ✓

Use the quotient of powers property to simplify the expression.

3. $\frac{5^4}{5^1}$ **125**
4. $\frac{7^6}{7^9}$ $\mathbf{\frac{1}{343}}$
5. $\frac{a^{12}}{a^9}$ $\mathbf{a^3}$
6. $\frac{m^5}{m^{11}}$ $\mathbf{\frac{1}{m^6}}$
7. $\frac{a^5}{a^2}$ $\mathbf{a^3}$
8. $\frac{(-2)^8}{(-2)^3}$ **−32**
9. $\frac{5^3 \cdot 5^5}{5^9}$ $\mathbf{\frac{1}{5}}$
10. $\frac{x^7 \cdot x}{x^{-2}}$ $\mathbf{x^{10}}$

Use the power of a quotient property to simplify the expression.

11. $\left(\frac{1}{2}\right)^5$ $\mathbf{\frac{1}{32}}$
12. $\left(\frac{3}{5}\right)^3$ $\mathbf{\frac{27}{125}}$
13. $\left(\frac{5}{m}\right)^2$ $\mathbf{\frac{25}{m^2}}$
14. $\left(\frac{2}{b}\right)^4$ $\mathbf{\frac{16}{b^4}}$
15. $\left(\frac{5}{4}\right)^{-3}$ $\mathbf{\frac{64}{125}}$
16. $\left(\frac{x^4}{2^3}\right)^2$ $\mathbf{\frac{x^8}{64}}$
17. $\left(\frac{x^3}{y^5}\right)^6$ $\mathbf{\frac{x^{18}}{y^{30}}}$
18. $\left(\frac{a^6}{b^9}\right)^5$ $\mathbf{\frac{a^{30}}{b^{45}}}$

PRACTICE AND APPLICATIONS

STUDENT HELP

Extra Practice to help you master skills is on p. 804.

EVALUATING EXPRESSIONS **Evaluate the expression. Write fractions in simplest form.**

19. $\frac{5^6}{5^3}$ **125**
20. $\frac{8^3}{8^1}$ **64**
21. $\frac{(-3)^6}{-3^6}$ **−1**
22. $\frac{(-3)^9}{(-3)^9}$ **1**
23. $\frac{3^3}{3^{-4}}$ **2187**
24. $\frac{8^3 \cdot 8^2}{8^5}$ **1**
25. $\frac{5 \cdot 5^4}{5^8}$ $\mathbf{\frac{1}{125}}$
26. $\left(\frac{3}{4}\right)^2$ $\mathbf{\frac{9}{16}}$
27. $\left(\frac{6}{2}\right)^3$ **27**
28. $\left(-\frac{2}{3}\right)^3$ $\mathbf{-\frac{8}{27}}$
29. $\left(-\frac{3}{5}\right)^2$ $\mathbf{\frac{9}{25}}$
30. $\left(\frac{9}{6}\right)^{-1}$ $\mathbf{\frac{2}{3}}$

SIMPLIFYING EXPRESSIONS **Simplify the expression. The simplified expression should have no negative exponents.**

31. $\left(\frac{3}{x}\right)^4$ $\mathbf{\frac{81}{x^4}}$
32. $\frac{x^4}{x^5}$ $\mathbf{\frac{1}{x}}$
33. $\left(\frac{1}{x}\right)^5$ $\mathbf{\frac{1}{x^5}}$
34. $x^3 \cdot \frac{1}{x^2}$ $\mathbf{x}$
35. $x^5 \cdot \frac{1}{x^8}$ $\mathbf{\frac{1}{x^3}}$
36. $\left(\frac{a^9}{a^5}\right)^{-1}$ $\mathbf{\frac{1}{a^4}}$
37. $\left(\frac{y^2}{y^3}\right)^{-2}$ $\mathbf{y^2}$
38. $\frac{m^3 \cdot m^5}{m^2}$ $\mathbf{m^6}$
39. $\frac{(r^3)^4}{(r^3)^8}$ $\mathbf{\frac{1}{r^{12}}}$
40. $\left(\frac{-6x^2y}{2xy^3}\right)^3$ **See margin.**
41. $\left(\frac{2x^3y^4}{3xy}\right)^3$ $\mathbf{\frac{8x^6y^9}{27}}$
42. $\frac{16x^3y}{-4xy^3} \cdot -\frac{2xy}{-x^{-1}}$ **See margin.**
43. $\frac{4x^3y^3}{2xy} \cdot \frac{5xy^2}{2y}$ $\mathbf{5x^3y^3}$
44. $\frac{36a^8b^2}{ab} \cdot \left(\frac{6}{ab^2}\right)^{-1}$ $\mathbf{6a^8b^3}$
45. $\frac{16x^5y^{-8}}{x^7y^4} \cdot \left(\frac{x^3y^2}{8xy}\right)^4$ $\mathbf{\frac{x^6}{256y^8}}$
46. $\frac{6x^{-2}y^2}{xy^{-3}} \cdot \frac{(4x^2y)^{-2}}{xy^2}$ $\mathbf{\frac{3y}{8x^8}}$
47. $\frac{5x^{-3}y^2}{x^5y^{-1}} \cdot \frac{(2xy^3)^{-2}}{xy}$ $\mathbf{\frac{5}{4x^{11}y^4}}$
48. $\left(\frac{2xy^{-2}y^4}{3x^{-1}y}\right)^{-2} \cdot \left(\frac{4xy}{2x^{-1}y^{-3}}\right)^2$ $\mathbf{9y^6}$

STUDENT HELP

HOMEWORK HELP
Example 1: Exs. 19–50
Example 2: Exs. 19–48
Example 3: Exs. 31–50
Example 4: Exs. 51–59
Example 5: Exs. 60, 61

49. $\frac{6^3}{6} = 6^{3-1} = 6^2 = 36$

50. $\frac{x^{-9}}{x^{-3}} = x^{-9-(-3)} =$

$x^{-6} = \frac{1}{x^6}$

ERROR ANALYSIS In Exercises 49 and 50, find and correct the errors.

49. 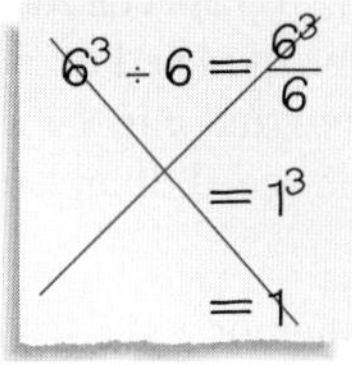

50. 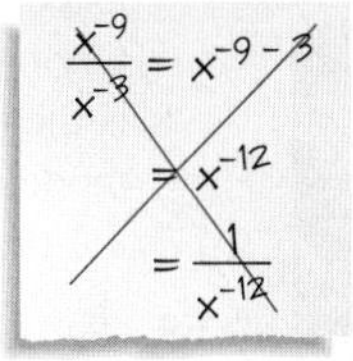

51. **EARTH AND MOON** The volume of a sphere is given by $V = \frac{4}{3}\pi r^3$, where r is the radius of the sphere. Assuming that the radius of the Moon is $\frac{1}{4}$ the radius of Earth, find the ratio of the volume of Earth to the volume of the Moon. Let r represent the radius of Earth.
The volume of the Earth is about 64 times the volume of the Moon.

52. **RETAIL SALES** From 1994 to 1998, the sales for a national clothing store increased by about the same percent each year. The sales S (in millions of dollars) for year t can be modeled by

$$S = 3723\left(\frac{6}{5}\right)^t$$

where $t = 0$ corresponds to 1994. Find the ratio of 1998 sales to 1995 sales. **$\frac{216}{125}$**

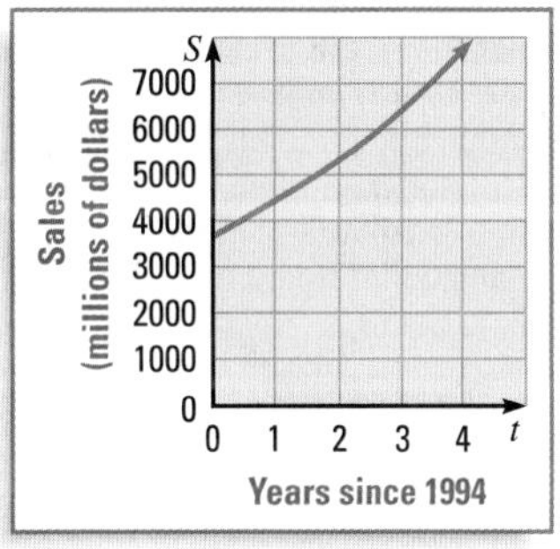

53. **BASEBALL SALARIES** The average salary for a professional baseball player in the United States can be approximated by $y = 283(1.2)^t$, where $t = 0$ represents the year 1984. Using this approximation, find the ratio of an average salary in 1988 to an average salary in 1994. **$\frac{1}{1.2^6} \approx \frac{1}{3}$**

ATLANTIC COD In Exercises 54–56, use the following information. The average weight w (in pounds) of an Atlantic cod t years old can be modeled by the equation $w = 1.16(1.44)^t$. ▶ Source: National Marine Fisheries Service, Northeast Science Center

54. Find the ratio of the weight of a 5-year-old cod to the weight of a 2-year-old cod. Express this ratio as a power of 1.44. **1.44^3**

55. A 5-year-old cod weighs how many times as much as a 2-year-old cod? **about 3 times**

56. According to the model, what is the average weight of an Atlantic cod when it is hatched? How did you get your answer?
1.16 lb; substitute 0 for t and evaluate the expression.

57. **SALES** From 1995 through 1999, the sales for a national furniture store increased by about the same percent each year. The sales s (in millions of dollars) for year t can be modeled by $s = 476(1.13)^t$, where $t = 0$ represents 1995. Find the ratio of 1997 sales to 1999 sales. **$\frac{1}{1.13^2}$ or about 0.8**

FOCUS ON APPLICATIONS

REAL LIFE ATLANTIC COD Adult Atlantic cod average about 3 feet in length and weigh from 10 to 25 pounds, though some may grow much larger.

! COMMON ERROR
EXERCISES 8, 21, 22, 28, 29, AND 40 Some students may simplify expressions such as $\left(-\frac{3}{5}\right)^2$ as $-\left(\frac{3}{5}\right)^2$, or $-\frac{9}{25}$. Students who make this error may find it helpful to first change $\left(-\frac{3}{5}\right)^2$ to $\left[(-1)\left(\frac{3}{5}\right)\right]^2$ and apply the power of a product property.

MATHEMATICAL REASONING
Students have seen situations in which they add, subtract, or multiply exponents. Some students may wonder whether it is ever necessary to divide exponents. They should be able to see that division would be needed to find what should replace the question mark in an equation such as $(2^?)^3 = 2^{12}$.

APPLICATION NOTE
EXERCISE 53 The model given in this application implies that the average salary for a professional baseball player goes up 20% per year.

TEACHING TIP

EXERCISE 59 Ask students if the equation that models this situation is based on any assumptions about the study habits of the student referred to in the exercise. They should be able to see that the model assumes the student does little if any review of the word list.

EXERCISE 62 Point out that this model is based on a 2.08% increase in the population of Arizona every year.

ADDITIONAL PRACTICE AND RETEACHING

For Lesson 8.3:

- Practice Levels A, B, and C (*Chapter 8 Resource Book,* p. 43)
- Reteaching with Practice (*Chapter 8 Resource Book,* p. 46)
- See Lesson 8.3 of the *Personal Student Tutor*

For more Mixed Review:

- Search the *Test and Practice Generator* for key words or specific lessons.

FOCUS ON APPLICATIONS

REAL LIFE **ROWING** Shells with 4 or 8 rowers usually have an additional nonrowing member of the team to direct the rest of the crew. This person is called the coxswain.

59b. 19 weeks; *Sample answer:* Since the number of words remembered each week is 0.8 times the number remembered the previous week, I did repeated multiplications on my calculator, beginning with 0.8(200) and counted the number of times it took to get to 3.

58. OLYMPIC ROWING The racing shells (boats) used in rowing competition usually have 1, 2, 4, or 8 rowers. Top speeds for racing shells in the Olympic 2000-meter races can be modeled by $s = 16.3(1.0285)^n$, where s is the speed in kilometers per hour and n is the number of rowers. Use the model to estimate the ratio of the speed of an 8-rower shell to the speed of a 2-rower shell. **about 1.2**

59. LEARNING SPANISH You memorized a list of 200 Spanish vocabulary words. Unfortunately, each week you forget one fifth of the words you knew the previous week. The number of words S you remember after n weeks can be approximated by the following equation.

Vocabulary words remembered: $S = 200\left(\frac{4}{5}\right)^n$

a. Copy and complete the table showing the number of words you remember after n weeks. **200; 160; 128; 102; 82; 66; 52**

Weeks, n	0	1	2	3	4	5	6
Words, S	?	?	?	?	?	?	?

b. CRITICAL THINKING How many weeks does it take to forget all but three words? Explain your answer. **See margin.**

60. PROBABILITY CONNECTION You roll a die eight times. What is the probability that you will roll eight sixes in a row? $\left(\frac{1}{6}\right)^8 \approx 0.0000006$

61. PROBABILITY CONNECTION You roll a die six times. What is the probability that you will roll six even numbers in a row? $\left(\frac{1}{2}\right)^6 \approx 0.02$

Test Preparation

MULTI-STEP PROBLEM **In Exercises 62–65, use the following information.** You work for a real estate company that wants to build a new apartment complex. A team is formed to decide in which state to build the complex. One team member wants to build in Arizona. Another team member wants to build in Michigan. Your boss asks you to decide where to build the complex.

62. You find that the population P of Arizona (in thousands) in 1995 projected through 2025 can be modeled by $P = 4264(1.0208)^t$, where $t = 0$ represents 1995. Find the ratio of the population in 2025 to the population in 2000.

INTERNET **DATA UPDATE** of U.S. Bureau of the Census data at www.mcdougallittell.com $(1.0208)^{25} \approx 1.67$

63. You find that the population P of Michigan (in thousands) in 1995 projected through 2025 can be modeled by $P = 9540(1.0026)^t$, where $t = 0$ represents 1995. Find the ratio of the population in 2025 to the population in 2000.

▶ Source: U.S. Bureau of the Census $(1.0026)^{25} \approx 1.07$

64. Which population is projected to grow more rapidly? **Arizona**

65. *Writing* Use the results from Exercises 62–64 to decide where to build the complex. Write a memo to your boss explaining your decision. **See margin.**

65. The answer should take into account that, while Arizona's population is growing more rapidly, Michigan's population will be greater than that of Arizona for nearly 50 years. Encourage students who are interested to do more research before choosing. Students may wish to compare state tax rates, construction costs, average family size and income, population density, etc. rather than just basing the decision on population.

★ Challenge

66. STACKING PAPER A piece of notebook paper is about 0.0032 inch thick. If you begin with a stack consisting of a single sheet and double the stack 25 times, how tall will the stack be in inches? How tall will it be in feet? (*Hint:* Write and solve an exponential equation to find the height of the stack in inches. Then use unit analysis to find the height in feet.) **107,374 in.; 8948 ft**

EXTRA CHALLENGE
www.mcdougallittell.com

MIXED REVIEW

POWERS OF TEN **Evaluate the expression.** (Review 1.2, 8.2 for 8.4)

67. 10^5 **100,000** **68.** 10^3 **1000** **69.** 10^{-4} $\frac{1}{10{,}000}$ **70.** 10^{-8} $\frac{1}{100{,}000{,}000}$

SKETCHING GRAPHS **Sketch the graph of the inequality in a coordinate plane.** (Review 6.5) **71–78. See margin.**

71. $x \geq 5$ **72.** $x + 3 < 4$ **73.** $y > -2$ **74.** $y \leq -1.5$

75. $x \geq 2.5$ **76.** $3x - y < 0$ **77.** $y \leq \frac{x}{2}$ **78.** $\frac{3}{4}x + \frac{1}{4}y \geq 1$

CHECKING FOR SOLUTIONS **Decide whether the ordered pair is a solution of the system.** (Review 7.1)

79. $2x + 4y = 2$, $-x + 5y = 13$ $(-3, 2)$ **solution**

80. $x - 5y = 9$, $3x + y = 11$ $(1, -4)$ **not a solution**

81. $8a + 4b = 6$, $4a + b = 3$ $\left(\frac{3}{4}, 0\right)$ **solution**

82. $3c - 8d = 11$, $c + 6d = 8$ $\left(5, -\frac{1}{2}\right)$ **not a solution**

SOLVING LINEAR SYSTEMS **Use linear combinations to solve the system.** (Review 7.3)

83. $x - y = 4$, $x + y = 12$ **(8, 4)**

84. $-x + 2y = 12$, $x + 6y = 20$ **(−4, 4)**

85. $2a + 3b = 17$, $3a + 4b = 24$ **(4, 3)**

QUIZ 1

Self-Test for Lessons 8.1–8.3

Evaluate the expression. (Lessons 8.1, 8.2, and 8.3)

1. $3^3 \cdot 3^4$ **2187** **2.** $(2^2)^4$ **256** **3.** $[(8 + 2)^2]^2$ **10,000** **4.** 7^{-4} $\frac{1}{2401}$

5. $4^{-3} \cdot 4^{-4}$ $\frac{1}{16{,}384}$ **6.** $\left(\frac{6}{7}\right)^{-1}$ $\frac{7}{6}$ **7.** $\frac{5^{-3}}{5^2}$ $\frac{1}{3125}$ **8.** $\frac{3^4 \cdot 3^6}{3^3}$ **2187**

9. $\left(\frac{5}{4}\right)^{-3}$ $\frac{64}{125}$ **10.** $\frac{(-2)^9}{(-2)^2}$ **−128** **11.** $6^0 \cdot \frac{1}{4^{-3}}$ **64** **12.** $\frac{2^3 \cdot 2^{-4}}{2^{-3}}$ **4**

Simplify the expression. Write your answer with no negative exponents. (Lessons 8.1, 8.2, and 8.3)

13. $x^4 \cdot x^5$ x^9 **14.** $(-2x)^5$ $-32x^5$ **15.** $-\frac{3}{a^{-5}}$ $-3a^5$ **16.** $200^0 c^5$ c^5

17. $\frac{x^6}{x^4}$ x^2 **18.** $\frac{x^{-5}}{x^{-6}}$ x **19.** $\left(\frac{-2m^2n}{3mn^2}\right)^4$ $\frac{16m^4}{81n^4}$ **20.** $x^4 \cdot \frac{1}{x^3}$ x

21. $(3a)^3 \cdot (-4a)^3$ $-1728a^6$ **22.** $(8m^3)^2\left(\frac{1}{2}m^2\right)^2$ $16m^{10}$ **23.** $\frac{20x^3y}{4xy^2} \cdot \frac{-6xy}{-x}$ $30x^2$

24. SAVINGS ACCOUNT You started a savings account in 1994. The balance A is given by $A = 250(1.08)^t$, where $t = 0$ represents the year 2001. What is the balance in the account in 1994? in 1999? in 2001? (Lesson 8.2)
\$145.87; \$214.33; \$250

Additional Test Preparation ***Sample answer:***
If \$2000 is invested in an account that earns compound interest at the rate of 5% a year, then the ratio of the amount in the account after 15 years to the amount after 12 years can be found by dividing $2000(1.05)^{15}$ by $2000(1.05)^{12}$.

4 ASSESS

DAILY HOMEWORK QUIZ

Transparency Available

Evaluate the expression. Write fractions in simplest form.

1. $\frac{7^5}{7^3}$ **49** **2.** $\left(-\frac{3}{5}\right)^4$ $\frac{81}{625}$

3. $\left(\frac{9}{4}\right)^{-1}$ $\frac{4}{9}$

Simplify the expression. The simplified expression should have no negative exponents.

4. $b^7 \cdot \frac{b^4}{b^3}$ b^8

5. $\left(-\frac{9m^4n}{27mn^2}\right)^3$ $-\frac{m^9}{27n^3}$

6. $\frac{-5x^{-3}y^4}{x^2y^{-1}} \cdot \frac{(3x^2y^2)^3}{x^3y}$ $\frac{-135y^{10}}{x^2}$

7. The amount of money after t years in an account that begins with \$700 and that earns compound interest at a rate of 4.7% per year can be modeled by the equation $A = 700(1.047)^t$. Find the ratio of the amount after 6 years to the amount after 4 years. Express the ratio as a power of 1.047. 1.047^2

EXTRA CHALLENGE NOTE

Challenge problems for Lesson 8.3 are available in **blackline** format in the *Chapter 8 Resource Book,* p. 50 and at **www.mcdougallittell.com.**

ADDITIONAL TEST PREPARATION

1. OPEN ENDED Describe a real-life situation that might require using the quotient of powers property. **See margin.**

ADDITIONAL RESOURCES

An alternative Quiz for Lessons 8.1–8.3 is available in the *Chapter 8 Resource Book,* p. 51.

71–78. See Additional Answers beginning on page AA1.

1 PLAN

PACING
Basic: 1 day
Average: 1 day
Advanced: 1 day
Block Schedule: 0.5 block with 8.3

LESSON OPENER
APPLICATION
An alternative way to approach Lesson 8.4 is to use the Application Lesson Opener:
- Blackline Master (*Chapter 8 Resource Book,* p. 55)
- Transparency (p. 55)

MEETING INDIVIDUAL NEEDS
- ***Chapter 8 Resource Book***
  Prerequisite Skills Review (p. 5)
  Practice Level A (p. 57)
  Practice Level B (p. 58)
  Practice Level C (p. 59)
  Reteaching with Practice (p. 60)
  Absent Student Catch-Up (p. 62)
  Challenge (p. 65)
- ***Resources in Spanish***
- ***Personal Student Tutor***

NEW-TEACHER SUPPORT
See the Tips for New Teachers on pp. 1–2 of the *Chapter 8 Resource Book* for additional notes about Lesson 8.4.

WARM-UP EXERCISES
Transparency Available
Evaluate each expression.
1. 10^3 **1000**
2. 10^{-4} **0.0001**
3. 100^2 **10,000**
4. 10^{-8} **0.00000001**
5. 10^{10} **10,000,000,000**

EXPLORING DATA AND STATISTICS

8.4

What you should learn

GOAL 1 Use scientific notation to represent numbers.

GOAL 2 Use scientific notation to describe **real-life** situations, such as the price per acre of the Alaska purchase in **Example 6**.

Why you should learn it

▼ To solve **real-life** problems, such as finding the amount of water discharged by the Amazon River each year in **Example 5**.

REAL LIFE

Scientific Notation

GOAL 1 USING SCIENTIFIC NOTATION

A number is written in **scientific notation** if it is of the form $c \times 10^n$, where $1 \le c < 10$ and n is an integer.

ACTIVITY Developing Concepts

Investigating Scientific Notation

1. Rewrite each number in decimal form.

   a. 6.43×10^4 **64,300** **b.** 3.072×10^6 **3,072,000** **c.** 4.2×10^{-2} **0.042** **d.** 1.52×10^{-3} **0.00152**

2. Describe a general rule for writing the decimal form of a number given in scientific notation. How many places do you move the decimal point? Do you move the decimal point left or right?
   Move the decimal point n places to the right if $n > 0$ and $|n|$ places left if $n < 0$.

EXAMPLE 1 *Rewriting in Decimal Form*

Rewrite in decimal form.

a. 2.834×10^2 **b.** 4.9×10^5 **c.** 7.8×10^{-1} **d.** 1.23×10^{-6}

SOLUTION

a. $2.834 \times 10^2 = 283.4$ **Move decimal point right 2 places.**

b. $4.9 \times 10^5 = 490{,}000$ **Move decimal point right 5 places.**

c. $7.8 \times 10^{-1} = 0.78$ **Move decimal point left 1 place.**

d. $1.23 \times 10^{-6} = 0.00000123$ **Move decimal point left 6 places.**

EXAMPLE 2 *Rewriting in Scientific Notation*

a. $34{,}690 = 3.469 \times 10^4$ **Move decimal point left 4 places.**

b. $1.78 = 1.78 \times 10^0$ **Move decimal point 0 places.**

c. $0.039 = 3.9 \times 10^{-2}$ **Move decimal point right 2 places.**

d. $0.000722 = 7.22 \times 10^{-4}$ **Move decimal point right 4 places.**

e. $5{,}600{,}000{,}000 = 5.6 \times 10^9$ **Move decimal point left 9 places.**

STUDENT HELP

HOMEWORK HELP Visit our Web site www.mcdougallittell.com for extra examples.

To multiply, divide, or find powers of numbers in scientific notation, use the properties of exponents.

EXAMPLE 3 *Computing with Scientific Notation*

Evaluate the expression. Write the result in scientific notation.

a. $(1.4 \times 10^4)(7.6 \times 10^3)$

b. $(1.2 \times 10^{-1}) \div (4.8 \times 10^{-4})$

c. $(4.0 \times 10^{-2})^3$

SOLUTION

a. $(1.4 \times 10^4)(7.6 \times 10^3)$

$= (1.4 \cdot 7.6) \times (10^4 \cdot 10^3)$ Associative property of multiplication

$= 10.64 \times 10^7$ Simplify.

$= 1.064 \times 10^8$ Write in scientific notation.

b. $\frac{1.2 \times 10^{-1}}{4.8 \times 10^{-4}} = \frac{1.2}{4.8} \times \frac{10^{-1}}{10^{-4}}$ Rewrite as a product.

$= 0.25 \times 10^3$ Simplify.

$= 2.5 \times 10^2$ Write in scientific notation.

c. $(4.0 \times 10^{-2})^3 = 4^3 \times (10^{-2})^3$ Power of a product

$= 64 \times 10^{-6}$ Power of a power

$= 6.4 \times 10^{-5}$ Write in scientific notation.

....

Many calculators automatically switch to scientific notation to display large or small numbers. Try multiplying 98,900,000 by 500. If your calculator follows standard conventions, it will display the product using scientific notation.

4.945 E10 ← Calculator display for 4.945×10^{10}

EXAMPLE 4 *Using a Calculator*

Use a calculator to multiply 0.000000748 by 2,400,000,000.

SOLUTION

Enter 0.000000748 as 7.48×10^{-7} and 2,400,000,000 as 2.4×10^9.

KEYSTROKES	DISPLAY
7.48 EE 7 +/- × 2.4 EE 9 ENTER	1.7952 E3

▶ The product is 1.7952×10^3, or 1795.2.

STUDENT HELP

KEYSTROKE HELP If your calculator does not have an EE function, you can enter the number in scientific notation as a product. 7.48 × 10 y^x 7 +/-

2 TEACH

EXTRA EXAMPLE 1

Rewrite in decimal form.

a. 3.128×10^3 **3128**

b. 6.4×10^4 **64,000**

c. 3.9×10^{-1} **0.39**

d. 6.12×10^{-5} **0.0000612**

EXTRA EXAMPLE 2

Rewrite in scientific notation.

a. 52,314 $\mathbf{5.2314 \times 10^4}$

b. 3.2 $\mathbf{3.2 \times 10^0}$

c. 0.0471 $\mathbf{4.71 \times 10^{-2}}$

d. 0.0000428 $\mathbf{4.28 \times 10^{-5}}$

e. 602,000,000 $\mathbf{6.02 \times 10^8}$

EXTRA EXAMPLE 3

Evaluate the expression. Write the result in scientific notation.

a. $(2.5 \times 10^4)(5.8 \times 10^2)$ $\mathbf{1.45 \times 10^7}$

b. $(1.82 \times 10^{-1}) \div (1.4 \times 10^{-3})$ $\mathbf{1.3 \times 10^2}$

c. $(1.5 \times 10^{-4})^3$ $\mathbf{3.375 \times 10^{-12}}$

EXTRA EXAMPLE 4

Use a calculator to multiply 0.00000052 by 3,500,000,000. $\mathbf{1.820 \times 10^3}$**, or 1820**

CHECKPOINT EXERCISES

For use after Example 1:

1. Rewrite in decimal form.

a. 6.873×10^4 **68,730**

b. 8.92×10^3 **8920**

c. 9.7×10^{-3} **0.0097**

d. 7.39×10^{-4} **0.000739**

For use after Examples 2 and 3:

2. Evaluate the expression. Write the result in scientific notation.

a. $(3.8 \times 10^3)(4.1 \times 10^4)$ $\mathbf{1.558 \times 10^8}$

b. $(3.69 \times 10^{-1}) \div (8.2 \times 10^{-4})$ $\mathbf{4.5 \times 10^2}$

c. $(2.1 \times 10^{-3})^2$ $\mathbf{4.41 \times 10^{-6}}$

For use after Example 4:

3. Use a calculator to multiply 0.00000382 by 4,700,000,000. $\mathbf{1.7954 \times 10^4}$

EXTRA EXAMPLE 5

Light travels at a speed of 1.863×10^5 mi/sec. How far does it travel in a year?
about 5.88×10^{12} mi

EXTRA EXAMPLE 6

In 1998, the population of the world was about 5.9 billion. In the same year, the population of the U.S. was 2.7×10^8. What percent of the world population was the U.S. population?
about 4.6%

CHECKPOINT EXERCISES

For use after Example 5:

1. The mass of one proton is 1.67×10^{-24} g. One atom of copper has 29 protons. Find the mass of the protons in 2.37×10^{17} atoms of copper. Write your answer in scientific notation. **about 1.15×10^{-5} g**

For use after Example 6:

2. In 1997, cell phone bills totaled about 2.75×10^{10} dollars. There were 55 million subscribers. What was the average monthly cost per subscriber? **$41.67**

FOCUS ON VOCABULARY

Use the words *power* and *exponent* to explain how scientific notation works. **See margin.**

CLOSURE QUESTION

When a number is written in scientific notation, what does the sign of the exponent in the power of 10 tell you about the size of the number?
If the exponent is negative, the number is less than 1; if the exponent is 0, the number is greater than or equal to 1 but less than 10; if the exponent is positive, the number is greater than or equal to 10.

DAILY PUZZLER

I am thinking of a power of 2. When I write it in standard form, it has 16 digits. If I multiply it by 2, the answer has 17 digits. What power of 2 am I thinking of? **53rd power of 2**

FOCUS ON APPLICATIONS

AMAZON RIVER The Amazon River in Brazil contributes more water to Earth's oceans than any other river. Each second, it discharges 4.2×10^6 cubic feet of water into the Atlantic Ocean.

GOAL 2 SOLVING REAL-LIFE PROBLEMS

EXAMPLE 5 *Multiplying with Scientific Notation*

AMAZON RIVER How much water does the Amazon River discharge into the Atlantic Ocean each year?

SOLUTION

First find the number of seconds in a year. Then multiply the amount of water discharged per second by the number of seconds per year to find the total amount of water discharged into the Atlantic Ocean.

Use unit analysis to find the number of seconds in a year.

$$\frac{\text{days}}{\text{year}} \cdot \frac{\text{hours}}{\text{day}} \cdot \frac{\text{minutes}}{\text{hour}} \cdot \frac{\text{seconds}}{\text{minute}} = \frac{\text{seconds}}{\text{year}}$$

$$\frac{365}{1} \cdot \frac{24}{1} \cdot \frac{60}{1} \cdot \frac{60}{1} = 31{,}536{,}000 \approx 3.2 \times 10^7$$

There are about 3.2×10^7 seconds in a year.

Multiply to find the total amount of water discharged by the Amazon River into the Atlantic Ocean in one year.

$$\text{Total amount of water} = \frac{4.2 \times 10^6 \text{ cubic feet}}{1 \text{ second}} \cdot \frac{3.2 \times 10^7 \text{ seconds}}{1 \text{ year}}$$

$$= (4.2 \times 10^6) \cdot (3.2 \times 10^7)$$

$$= 13.44 \times 10^{13}$$

$$= 1.344 \times 10^{14}$$

▶ One trillion is 1×10^{12}, so the total amount of water discharged into the Atlantic Ocean by the Amazon River is about 134 trillion cubic feet per year.

REAL LIFE History

EXAMPLE 6 *Dividing with Scientific Notation*

In 1867, the United States purchased Alaska from Russia for \$7.2 million. The total area of Alaska is about 3.78×10^8 acres. What was the price per acre?

STUDENT HELP

Look Back
For help with unit rates, see p. 180.

SOLUTION The price per acre is a unit rate.

$$\text{Price per acre} = \frac{\text{Total price}}{\text{Number of acres}}$$

$$= \frac{7.2 \times 10^6}{3.78 \times 10^8} \quad \leftarrow 7.2 \text{ million} = 7.2 \times 10^6$$

$$= \frac{7.2}{3.78} \times 10^{-2}$$

$$\approx 0.019$$

▶ The price was about 2¢ per acre.

Focus on Vocabulary *Sample answer:*
Scientific notation represents a number as the product of a power of 10 and a number greater than or equal to 1 and less than 10. The exponent in the power of 10 tells how many places to the left or right you would move the decimal point in the other factor to write the number in decimal form.

GUIDED PRACTICE

Vocabulary Check ✓

Concept Check ✓

1. Is the number 12.38×10^2 in scientific notation? Explain. **No; 12.38 is not between 1 and 10.**
2. Given the number 6.39×10^7, would you move the decimal point to the *left* or to the *right* to rewrite the number in decimal form? **right**

Skill Check ✓

Rewrite in decimal form.

3. 4.3×10^2 **430**
4. 8.11×10^3 **8110**
5. 2.45×10^{-1} **0.245**
6. 9.38×10^5 **938,000**

Rewrite in scientific notation.

7. 39.6 **3.96×10**
8. 0.72 **7.2×10^{-1}**
9. 1200 **1.2×10^3**
10. 0.0003 **3×10^{-4}**
11. 6,900,000 **6.9×10^6**
12. 0.0000205 **2.05×10^{-5}**
13. 72,000,000 **7.2×10^7**
14. 0.000000006 **6×10^{-9}**
15. ASTRONOMY The distance between the ninth planet Pluto and the Sun is 5.9×10^9 kilometers. Light travels at a speed of about 3.0×10^5 kilometers per second. How long does it take light to travel from the Sun to Pluto? **approximately 1.97×10^4 sec, or about 5.5 h**

PRACTICE AND APPLICATIONS

STUDENT HELP

Extra Practice to help you master skills is on p. 804.

DECIMAL FORM **Rewrite in decimal form.**

16. 2.14×10^4 **21,400**
17. 98×10^{-2} **0.98**
18. 7.75×10^0 **7.75**
19. 8.6521×10^3 **8652.1**
20. 4.65×10^{-4} **0.000465**
21. 6.002×10^{-6} **0.000006002**
22. 4.332×10^8 **433,200,000**
23. 1.00012×10^8 **100,012,000**
24. 1.1098×10^{10} **11,098,000,000**

SCIENTIFIC NOTATION **Rewrite in scientific notation.**

25. 0.05 **5×10^{-2}**
26. 95.2 **9.52×10**
27. 0.0422 **4.22×10^{-2}**
28. 370.207 **3.70207×10^2**
29. 700,000,000 **7×10^8**
30. 19.314 **1.9314×10**
31. 0.008551 **8.551×10^{-3}**
32. 2,730,000,000 **2.73×10^9**
33. 0.000459 **4.59×10^{-4}**
34. 0.00032954 **3.2954×10^{-4}**
35. 88,000,000 **8.8×10^7**
36. 0.0000288 **2.88×10^{-5}**

EVALUATING EXPRESSIONS **Evaluate the expression without using a calculator. Write the result in scientific notation and in decimal form.**

37. $(4 \times 10^{-2}) \cdot (3 \times 10^6)$ **1.2×10^5; 120,000**
38. $(7 \times 10^{-3}) \cdot (8 \times 10^{-4})$ **5.6×10^{-6}; 0.0000056**
39. $(6 \times 10^5) \cdot (2.5 \times 10^{-1})$ **1.5×10^5; 150,000**
40. $(1.2 \times 10^{-6}) \cdot (2.3 \times 10^4)$ **2.76×10^{-2}; 0.0276**
41. $\frac{8 \times 10^{-3}}{5 \times 10^{-5}}$ **1.6×10^2; 160**
42. $\frac{1.4 \times 10^{-1}}{3.5 \times 10^{-4}}$ **4.0×10^2; 400**
43. $\frac{6.6 \times 10^{-1}}{1.1 \times 10^{-1}}$ **6.0×10^0; 6**
44. $(3.0 \times 10^{-3})^2$ **9.0×10^{-6}; 0.000009**
45. $(9 \times 10^3)^2$ **8.1×10^7; 81,000,000**
46. $(3 \times 10^{-2})^4$ **8.1×10^{-7}; 0.00000081**

CALCULATOR **Use a calculator to evaluate the expression. Write the result in scientific notation and in decimal form.**

47. $2{,}000{,}000 \cdot 12{,}000$ **2.4×10^{10}; 24,000,000,000**
48. $6{,}000{,}000 \cdot 324{,}000$ **1.944×10^{12}; 1,944,000,000,000**
49. $0.000279 \cdot 3{,}940{,}000{,}000$ **1.09926×10^6; 1,099,260**
50. $654{,}000 \cdot 0.000042$ **2.7468×10; 27.468**
51. $(2.4 \times 10^{-4})^2$ **5.76×10^{-8}; 0.0000000576**
52. 0.000094^3 **8.30584×10^{-13}; 0.000000000000830584**

STUDENT HELP

HOMEWORK HELP
Example 1: Exs. 16–24
Example 2: Exs. 25–36
Example 3: Exs. 37–46
Example 4: Exs. 47–52
Example 5: Exs. 61, 62
Example 6: Exs. 58–60

3 APPLY

ASSIGNMENT GUIDE

BASIC
Day 1: pp. 473–475 Exs. 16–56 even, 64, 65, 70, 75, 78, 80, 82

AVERAGE
Day 1: pp. 473–475 Exs. 16–56 even, 61, 64, 65, 70, 75, 78, 80, 82

ADVANCED
Day 1: pp. 473–475 Exs. 16–56 even, 60–66, 70, 75, 78, 80, 82

BLOCK SCHEDULE WITH 8.3
pp. 473–475 Exs. 16–56 even, 61, 64, 65, 70, 75, 78, 80, 82

EXERCISE LEVELS

Level A: *Easier*
16–36, 67–70

Level B: *More Difficult*
37–59, 61–65, 71–83

Level C: *Most Difficult*
60, 66

✓ HOMEWORK CHECK

To quickly check student understanding of key concepts, go over the following exercises: Exs. 18, 20, 22, 28, 32, 34, 38, 42. See also the Daily Homework Quiz:

- Blackline Master (*Chapter 8 Resource Book*, p. 68)
- Transparency (p. 64)

! COMMON ERROR

EXERCISES 3–6 AND 16–24
When working with scientific notation, students sometimes think that the exponent for 10 tells how many 0s come immediately before or after the decimal point. You can use examples to point out why this is not necessarily the case.

APPLICATION NOTE

EXERCISE 55 Astronomy makes extensive use of scientific notation because of the vast distances in space.

EXERCISE 56 Chemistry makes extensive use of scientific notation because of the extremely small size of an atom.

STUDENT HELP NOTES

Homework Help Students can find help for Ex. 62 at **www.mcdougallittell.com.**

ADDITIONAL PRACTICE AND RETEACHING

For Lesson 8.4:

- Practice Levels A, B, and C (*Chapter 8 Resource Book,* p. 57)
- Reteaching with Practice (*Chapter 8 Resource Book,* p. 60)
- See Lesson 8.4 of the *Personal Student Tutor*

For more Mixed Review:

- Search the *Test and Practice Generator* for key words or specific lessons.

SCIENTIFIC NOTATION IN REAL LIFE In Exercises 53–57, write the number in scientific notation.

53. LIGHTNING The speed of a lightning bolt is 120,000,000 feet per second. **1.2×10^8**

54. WORLD POPULATION In 1997, the population of the world was estimated at 5,852,000,000. **DATA UPDATE** of U.S. Bureau of the Census data at www.mcdougallittell.com **5.852×10^9**

55. ASTRONOMY The star Sirius in the constellation Canis Major is about 50,819,000,000,000 miles from Earth. **5.0819×10^{13}**

56. CHEMISTRY The mass of a carbon atom is 0.00000000000000000000002 gram. **2×10^{-23}**

57. SIZE OF JUPITER Jupiter, the largest planet in our solar system, has a radius of about 4.4×10^4 miles. Use the equation $V = \frac{4}{3}\pi r^3$ to find Jupiter's volume. **about 3.57×10^{14} mi^3**

STUDENT HELP

APPLICATION LINK Visit our Web site www.mcdougallittell.com for more information about the purchase of Alaska, the Louisiana Purchase, and the Gadsden Purchase in Example 6 and Exs. 58–60.

58. Louisiana Purchase: \$18.12/mi^2; Gadsden Purchase: \$340.14/mi^2

59. Louisiana Purchase: \$.03/acre; Gadsden Purchase: \$.53/acre

60. Answers may vary. *Sample answer*: As time passed, land values increased.

HISTORY CONNECTION In Exercises 58–60, use the following information. In 1803, the Louisiana Purchase added 8.28×10^5 square miles to the United States. The cost of this land was \$15 million. In 1853, the Gadsden Purchase added 2.94×10^4 square miles, and the cost was \$10 million.

58. Find the average cost of a square mile for each of the purchases.

59. Find the average cost of an acre for each of the purchases. (*Hint:* There are 640 acres in a square mile.)

60. *Writing* Describe a factor that you think might explain the difference in the price per acre for these two purchases.

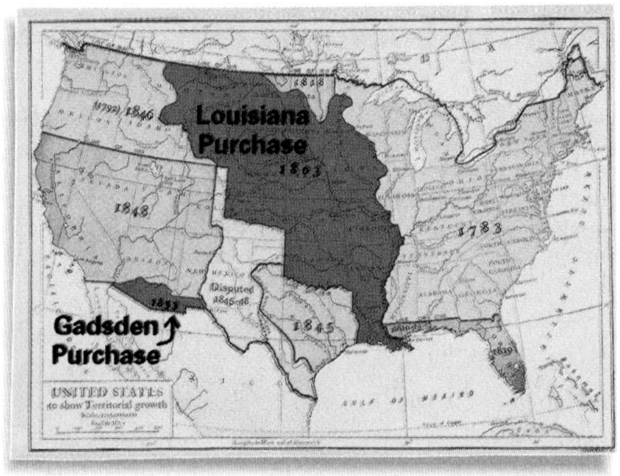

61. WATERFALL Stanley Falls in Congo, Africa, has an average flow of about 1.7×10^4 cubic meters per second. How much water goes over Stanley Falls in a typical 30-day month? **4.4064×10^{10} m^3**

STUDENT HELP

HOMEWORK HELP Visit our Web site www.mcdougallittell.com for help with Ex. 62.

62. HEARTBEATS Consider a person whose heart beats 70 times per minute and who lives to be 85 years old. Estimate the number of times the person's heart beats during his or her life. Do not acknowledge leap years. Write your answer in decimal form and in scientific notation. **3,127,320,000 or 3.12732×10^9 times**

63. TELEPHONE SURVEY Use the table which shows the population and the number of local telephone calls made in five states in 1994 to find the number of local calls made per person in each state. **Answers are rounded to the nearest whole number; TX: 2167; MN: 1522; PA; 1583; VT: 810; CA: 1806**

State	Local Calls	Population
Texas	3.9×10^{10}	1.8×10^7
Minnesota	7.0×10^9	4.6×10^6
Pennsylvania	1.9×10^{10}	1.2×10^7
Vermont	4.7×10^8	5.8×10^5
California	5.6×10^{10}	3.1×10^7

▶ Source: U.S. Bureau of the Census

Test Preparation

64. **MULTIPLE CHOICE** Which number is not in scientific notation? **D**

(A) 1×10^4 (B) 3.4×10^{-3} (C) 9.02×10^2 (D) 12.25×10^{-5}

65. **MULTIPLE CHOICE** Evaluate $\frac{1.1 \times 10^{-1}}{5.5 \times 10^{-5}}$ using scientific notation. **C**

(A) 0.2×10^{-4} (B) 2.0×10^4 (C) 2.0×10^3 (D) 0.2×10^4

★ Challenge

66. **BASEBALL** A baseball pitcher can throw a ball to home plate in about 0.5 second. The distance between the pitcher's mound and home plate is 60.5 feet.

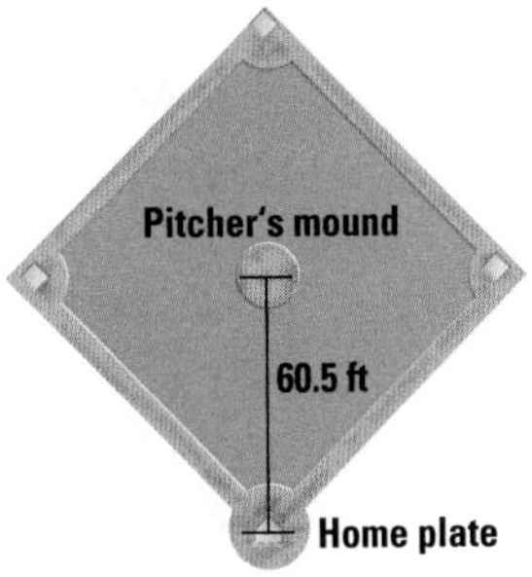

a. Fill in the missing numbers and simplify the following expression to find the rate, in scientific notation, at which the ball is traveling in millimeters per second.

$$\frac{?\text{ feet}}{?\text{ second}} \cdot \frac{1\text{ meter}}{3.3\text{ feet}} \cdot \frac{?\text{ millimeters}}{1\text{ meter}} \approx \underline{?} \times 10^{?}\text{ millimeters per second}$$

b. To hit a home run, a batter has a leeway of about 200 millimeters in the point of contact between the bat and the ball. What is the most the batter's timing could be off and still hit a home run? Explain your calculations. (*Hint:* Find the time it takes the ball to travel 200 millimeters.)

c. **CRITICAL THINKING** You are at bat. The pitcher throws you a pitch. Your timing is off by 0.006 second. Could you hit a home run on this pitch? Explain your answer.

66. a. $\frac{60.5\text{ ft}}{0.5\text{ sec}} \cdot \frac{1\text{ m}}{3.3\text{ ft}} \cdot \frac{1000\text{ mm}}{\text{m}} \approx 3.67 \times 10^4$ **mm/sec**

b. About 0.005 sec; the player has the 0.005 sec it takes the ball to travel those 200 mm.

c. No; in 0.006 sec, the ball would have traveled farther than 200 mm and I would not be able to hit it.

EXTRA CHALLENGE
www.mcdougallittell.com

MIXED REVIEW

PERCENTS AS DECIMALS Write the percent as a decimal. (Skills Review page 784 for 8.5)

67. 22% **0.22** 68. 87.5% **0.875** 69. 0.07% **0.0007** 70. 8.42% **0.0842**

71. $\frac{1}{2}$% **0.005** 72. $\frac{3}{4}$% **0.0075** 73. 255% **2.55** 74. $1\frac{1}{4}$% **0.0125**

GRAPHING LINEAR SYSTEMS Use the graphing method to solve the linear system and describe its solution(s). (Review 7.5)

75. $4x + 2y = 12$ **(1, 4)**
$-6x + 3y = 6$

76. $3x - 2y = 0$ **no solution**
$3x - 2y = -4$

GRAPHING Graph the system of linear inequalities. (Review 7.6) **See margin.**

77. $2x + y \le 1$
$-2x + y \le 1$

78. $x + 2y < 3$
$x - 3y > 1$

79. $2x + y \ge 2$
$x \le 2$

SIMPLIFYING EXPRESSIONS Simplify the expression. (Review 8.2 and 8.3)

80. 2^{-4} **$\frac{1}{16}$** 81. $\left(\frac{1}{10}\right)^{-3}$ **1000** 82. $\frac{1}{(2x)^{-2}}$ **$4x^2$** 83. $\frac{7^4 \cdot 7}{7^7}$ **$\frac{1}{49}$**

4 ASSESS

DAILY HOMEWORK QUIZ

Transparency Available

Rewrite in decimal form.

1. 6.32×10^{-5} **0.0000632**
2. 4.55×10^0 **4.55**
3. 7.168×10^7 **71,680,000**

Rewrite in scientific notation.

4. 3591.76 **3.59176×10^3**
5. 0.0000849 **8.49×10^{-5}**
6. 460,000,000 **4.6×10^8**

Evaluate the expression without using a calculator. Write the result in decimal form.

7. $(9 \times 10^{-12}) \cdot (6 \times 10^3)$ **0.000000054**
8. $\frac{1.8 \times 10^{-2}}{2.4 \times 10^{-7}}$ **75,000**

EXTRA CHALLENGE NOTE

Challenge problems for Lesson 8.4 are available in **blackline** format in the *Chapter 8 Resource Book*, p. 65 and at **www.mcdougallittell.com.**

ADDITIONAL TEST PREPARATION

1. **WRITING** Suppose your calculator will not display answers in scientific notation if the exponent for 10 is greater than 99. Can you get around this difficulty to somehow find the product of 7.495×10^{53} and 8.09×10^{72} in scientific notation? Explain your answer. **Yes; use the calculator to multiply 7.495 and 8.09 to get 60.63455. Write this in scientific notation as 6.063455×10^1. Multiply this result by the product of 10^{53} and 10^{72}, which is 10^{125}, to get 6.063455×10^{126}.**

77–79. See Additional Answers beginning on page AA1.

1 Planning the Activity

PURPOSE

To compare linear growth models and exponential growth models.

MATERIALS

- Graph paper
- Activity Support Master (*Chapter 8 Resource Book*, p. 69)

PACING

- Exploring the Concept — 10 min
- Drawing Conclusions — 10 min

LINK TO LESSON

Remind students of the results of this activity when you discuss and compare the linear and exponential models in Examples 4 and 5 of Lesson 8.5.

2 Managing the Activity

COOPERATIVE LEARNING

Ask groups of students to make up and graph examples of linear and exponential growth models. Have groups compare the methods they use to identify each type of growth model.

ALTERNATIVE APPROACH

Students can use graphing calculators to graph linear and exponential growth models. Use the table feature to show how the *y*-coordinates increase more rapidly for exponential growth models.

3 Closing the Activity

KEY DISCOVERY

Exponential growth models have a variable used as an exponent. Their values eventually change much more rapidly than those of linear models.

ACTIVITY ASSESSMENT

Identify the following models as linear or exponential.

a. $y = 8(2)^x$ **exponential**

b. $y = 8 + 2x$ **linear**

Linear and Exponential Growth Models

SET UP

Work in a small group.

MATERIALS

graph paper

▶ QUESTION How are linear growth models and exponential growth models different?

▶ EXPLORING THE CONCEPT

1 The equation $y = 5x + 20$ is a *linear growth model*. Copy and complete the table. **30; 35; 40; 45**

x	0	1	2	3	4	5
y	20	25	?	?	?	?

2 Graph $y = 5x + 20$. **Check graphs.**

3 The equation $y = 5^x$ is an *exponential growth model*. Copy and complete the table. **25; 125; 625; 3125**

x	0	1	2	3	4	5
y	1	5	?	?	?	?

4 Graph $y = 5^x$. **Check graphs.**

5 Which of the graphs below shows a *linear growth model*? Which shows an *exponential growth model*? Explain how you know. **b; a; *Sample explanation:* The points of graph (b) lie on a straight line while the points of graph (a) lie on an exponential curve.**

a.

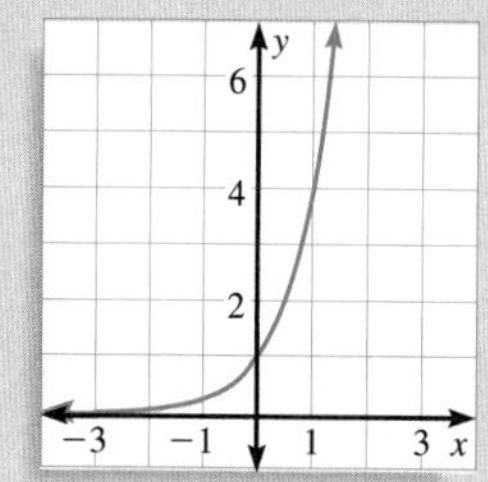

b.

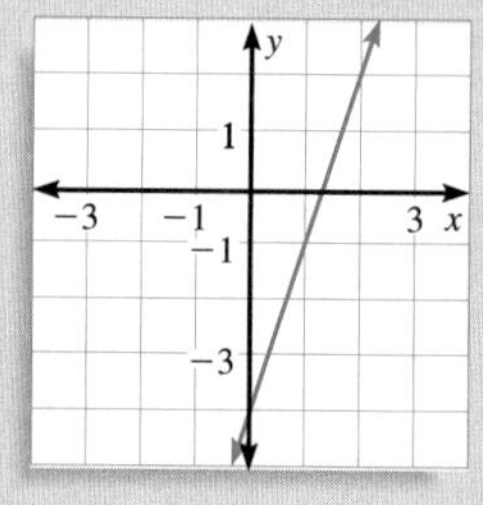

▶ DRAWING CONCLUSIONS

In Exercises 1–6, identify the equation as a *linear* growth model or an *exponential* growth model.

1. $y = x + 5$ **linear growth model**

2. $y = 3^x$ **exponential growth model**

3. $y = 10 + 2x$ **linear growth model**

4. $y = 15 + 2^x$ **exponential growth model**

5. $y = 5(4x - 7)$ **linear growth model**

6. $y = 10(1.2)^x$ **exponential growth model**

7. Look at your data and graph in Steps 1 and 2 to complete the statement. **same**

A linear growth model increases the _?_ amount for each unit on the x-axis.

8. Describe the rate of increase in an exponential growth model. ***Sample answer:* The rate of increase is increasing.**

9. CRITICAL THINKING You accept a job that pays $20,000 your first year. Would you rather receive a raise of $500 each year or a raise of 3% of your current salary each year? Does your answer depend on how long you plan to stay at the job? Explain your reasoning. **See margin.**

9. *Sample answer:* **For every year after the first, the salary is higher with the 3% annual raise. In the first few years after that, the differences are only a few hundred dollars. After 10 years, the difference is nearly $2000.**

8.5 Exponential Growth Functions

What you should learn

GOAL 1 Write and use models for exponential growth.

GOAL 2 Graph models for exponential growth.

Why you should learn it

▼ To solve **real-life** problems such as finding the weight of a channel catfish in **Example 2**.

REAL LIFE

GOAL 1 WRITING EXPONENTIAL GROWTH MODELS

A quantity is *growing exponentially* if it increases by the same percent in each unit of time. This is called **exponential growth.**

EXPONENTIAL GROWTH MODEL

$$y = C(1 + r)^t$$

C is the **initial amount.**

t is the **time period.**

$(1 + r)$ is the **growth factor,** r is the **growth rate.**

The **percent of increase** is $100r$.

EXAMPLE 1 *Finding the Balance in an Account*

COMPOUND INTEREST You deposit \$500 in an account that pays 8% annual interest compounded yearly. What is the account balance after 6 years?

SOLUTION

Method 1 SOLVE A SIMPLER PROBLEM

Find the account balance A_1 after 1 year and multiply by the growth factor to find the balance for each of the following years. The growth rate is 0.08, so the growth factor is $1 + 0.08 = 1.08$.

$A_1 = 500(1.08) = 540$ — Balance after 1 year

$A_2 = 500(1.08)(1.08) = 583.20$ — Balance after 2 years

$A_3 = 500(1.08)(1.08)(1.08) = 629.856$ — Balance after 3 years

⋮

$A_6 = 500(1.08)^6 \approx 793.437$ — Balance after 6 years

▶ The account balance after 6 years will be about \$793.44.

Method 2 USE A FORMULA

Use the exponential growth model to find the account balance A. The growth rate is 0.08. The initial value is 500.

$A = C(1 + r)^t$ — Exponential growth model

$= 500(1 + 0.08)^6$ — Substitute 500 for C, 0.08 for r, and 6 for t.

$= 500(1.08)^6$ — Simplify.

≈ 793.437 — Evaluate.

▶ The balance after 6 years will be about \$793.44.

1 PLAN

PACING

Basic: 1 day
Average: 1 day
Advanced: 1 day
Block Schedule: 0.5 block with 8.6

LESSON OPENER APPLICATION

An alternative way to approach Lesson 8.5 is to use the Application Lesson Opener:

- Blackline Master (*Chapter 8 Resource Book,* p. 70)
- Transparency (p. 56)

MEETING INDIVIDUAL NEEDS

- ***Chapter 8 Resource Book***
  Prerequisite Skills Review (p. 5)
  Practice Level A (p. 73)
  Practice Level B (p. 74)
  Practice Level C (p. 75)
  Reteaching with Practice (p. 76)
  Absent Student Catch-Up (p. 78)
  Challenge (p. 80)
- ***Resources in Spanish***
- ***Personal Student Tutor***

NEW-TEACHER SUPPORT

See the Tips for New Teachers on pp. 1–2 of the *Chapter 8 Resource Book* for additional notes about Lesson 8.5.

WARM-UP EXERCISES

Transparency Available

Evaluate when $x = 5$.

1. $2 + 3.1x$ **17.5**
2. $2 + (3.1)^x$ **288.29151**
3. $-12 + 9x$ **33**
4. $-12 + 9^x$ **59,037**

ENGLISH LEARNERS

Before beginning Lesson 8.5, point out that *growth* is the noun equivalent of the verb *grow.* Then invite volunteers to give a definition for *growth* and to discuss non-mathematical contexts in which they might use the word.

EXTRA EXAMPLE 1
A savings certificate of $1000 pays 6.5% annual interest compounded yearly. What is the balance when the certificate matures after 5 years? **about $1370.09**

EXTRA EXAMPLE 2
An experiment started with 100 bacteria. They double in number every hour.
a. Write a model for the number of bacteria after 8 hours. $B = 100(2)^t$, where B is the number of bacteria, and t is the number of 1-hour periods
b. Find the number of bacteria after 8 hours. **25,600**

EXTRA EXAMPLE 3
A stock with initial cost of $20 per share doubled its price per share every year for 3 years.
a. What is the percent of increase each year? **100%**
b. What is the price per share after 3 years? **$160**

✔ CHECKPOINT EXERCISES

For use after Example 1:
1. You deposit $1000 in a savings account that pays 5.5% annual interest compounded yearly. What is the balance after 3 years? **$1174.24**

For use after Examples 2 and 3:
2. A company starts with 50 employees and after one year has 75. It increases the employees at the same rate every year for 4 years.
a. What is the percent of increase each year? **50%**
b. Find the number of employees after 4 years. **about 253 employees**

478

Catfish Growth

EXAMPLE 2 *Writing an Exponential Growth Model*

A newly hatched channel catfish typically weighs about 0.3 gram. During the first six weeks of life, its growth is approximately exponential, increasing by about 10% each day.

a. Write a model for the weight during the first six weeks.

b. Find the weight at the end of six weeks.

SOLUTION

a. Let W represent the weight in grams, and let t represent the time in days. The initial weight is $C = 0.3$ and the growth rate is 0.10.

$W = C(1 + r)^t$	**Exponential growth model**
$= 0.3(1 + 0.10)^t$	**Substitute 0.3 for C and 0.10 for r.**
$= 0.3(1.1)^t$	**Simplify.**

b. To find the weight at the end of 6 weeks (or 42 days), substitute 42 for t.

$W = 0.3(1.1)^{42}$	**Substitute.**
≈ 16.42910977	**Use a calculator.**
≈ 16.4	**Round to the nearest tenth.**

▶ The weight is about 16.4 grams.

Population Growth

EXAMPLE 3 *Writing an Exponential Growth Model*

A population of 20 rabbits is released into a wild-life region. The population triples each year for 5 years.

a. What is the percent of increase each year?

b. What is the population after 5 years?

SOLUTION

a. The population triples each year, so the growth factor is 3.

$$1 + r = 3$$

▶ So, the growth rate r is 2 and the percent of increase each year is 200%.

STUDENT HELP

↳ Study Tip
Notice that the growth factor and the percent of increase are not the same. In Example 3, the growth factor is 3 but the percent of increase is 200%.

b. After 5 years, the population is

$P = C(1 + r)^t$	**Exponential growth model**
$= 20(1 + 2)^5$	**Substitute for C, r, and t.**
$= 20 \cdot 3^5$	**Simplify.**
$= 4860$	**Evaluate.**

▶ There will be about 4860 rabbits after 5 years.

GOAL 2 GRAPHING EXPONENTIAL GROWTH MODELS

You can graph exponential growth models in the same way you graphed exponential functions in Lesson 8.2.

Compound Interest

EXAMPLE 4 *A Model with a Small Growth Factor*

Graph the exponential growth model in Example 1.

SOLUTION

Use the values found in Method 1 of Example 1 to plot points in a coordinate plane. Then, draw a smooth curve through the points.

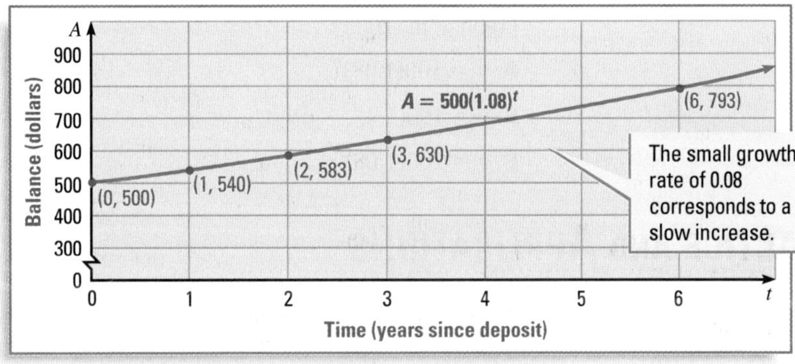

Population Growth

EXAMPLE 5 *A Model with a Large Growth Factor*

Graph the exponential growth model in Example 3.

SOLUTION

Make a table of values, plot the points in a coordinate plane, and draw a smooth curve through the points.

t	0	1	2	3	4	5
P	20	60	180	540	1620	4860

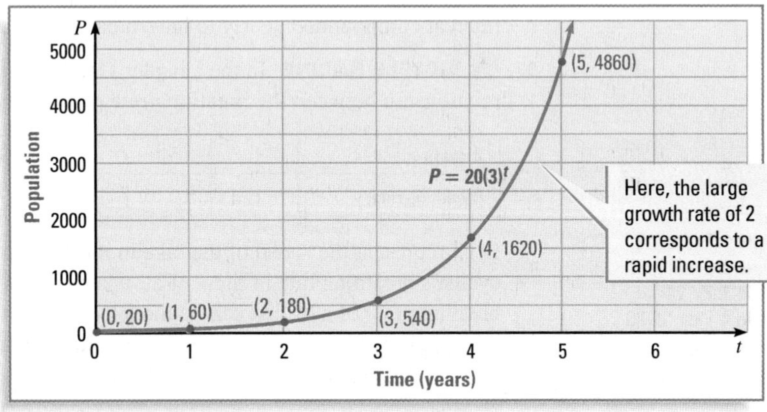

EXTRA EXAMPLE 4

Graph Extra Example 1.

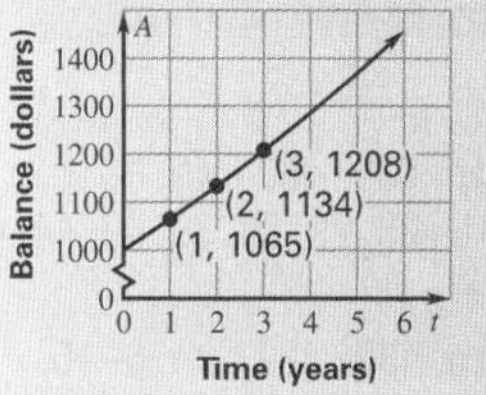

EXTRA EXAMPLE 5

Graph Extra Example 2.

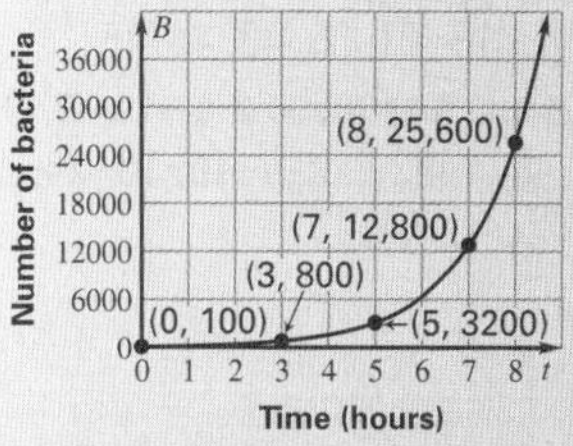

CHECKPOINT EXERCISES

For use after Example 4:

1. Graph $y = 40(1.05)^t$, where t is time in years.

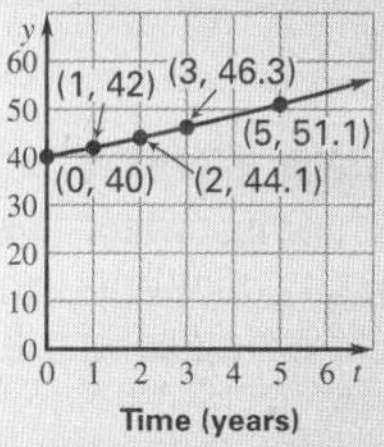

For use with Example 5:

2. Graph Extra Example 3.

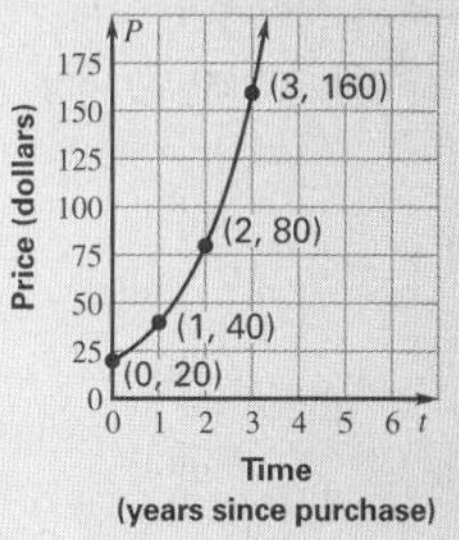

CLOSURE QUESTION

Give a general formula for an exponential growth model.

Sample: **$A = P(1 + r)^t$, where A is the amount, P is the starting amount, r is the growth rate, and t is the number of time periods.**

3 APPLY

ASSIGNMENT GUIDE

BASIC
Day 1: pp. 480–482 Exs. 6–16, 18–23, 29, 30, 35, 40, 44, 45

AVERAGE
Day 1: pp. 480–482 Exs. 6–16, 18–24, 29, 30, 35, 40, 44, 45

ADVANCED
Day 1: pp. 480–482 Exs. 6–16, 18–31, 35, 40, 44, 45

BLOCK SCHEDULE WITH 8.6
pp. 480–482 Exs. 6–16, 18–24, 29, 30, 35, 40, 44, 45

EXERCISE LEVELS

Level A: *Easier*
32–38

Level B: *More Difficult*
6–24, 29, 30, 39–48

Level C: *Most Difficult*
25–28, 31

✓ HOMEWORK CHECK

To quickly check student understanding of key concepts, go over the following exercises: Exs. 8, 12, 14, 16, 18, 23. See also the Daily Homework Quiz:

- Blackline Master (*Chapter 8 Resource Book,* p. 83)
- Transparency (p. 65)

GUIDED PRACTICE

Vocabulary Check ✓

1. In the exponential growth model, $y = C(1 + r)^t$, C is the _?_ and $(1 + r)$ is the _?_. **initial amount; growth factor**

Concept Check ✓

2. Look back at Example 1. Suppose that you got a 12% annual interest rate. Write a new exponential growth model for the balance in the account. $A = 500(1.12)^t$

3. Look back at Example 3. Suppose that the rabbit population doubled every year for 5 years. What was the percent of increase? the growth factor? **100%; 2**

Skill Check ✓

4. ACCOUNT BALANCE You deposit $500 in an account that pays 4% interest compounded yearly. What is the balance after 5 years? after 10 years? **$608.33; $740.12**

5. CHOOSE A MODEL Which model best represents the growth curve shown in the graph at the right? **B**

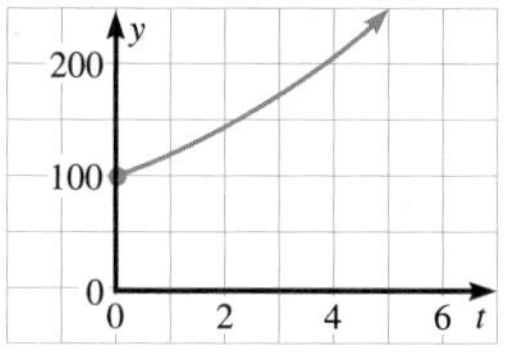

A. $y = 100(1.08)^t$

B. $y = 100(1.2)^t$

C. $y = 200(1.08)^t$

PRACTICE AND APPLICATIONS

STUDENT HELP
Extra Practice to help you master skills is on p. 804.

FIND THE BALANCE **You deposit $1400 in an account that pays 6% interest compounded yearly. Find the balance for the given time period.**

6. 5 years **$1873.52** **7.** 8 years **$2231.39** **8.** 12 years **$2817.08** **9.** 20 years **$4489.99**

FIND THE BALANCE **Find the balance after 5 years of an account that pays 4.8% interest compounded yearly given the following investment amounts.**

10. $250 **$316.04** **11.** $300 **$379.25** **12.** $350 **$442.46** **13.** $400 **$505.67**

14. ACCOUNT BALANCE A principal of $200 is deposited in an account that pays 4.2% interest compounded yearly. Find the balance after 5 years. **$245.68**

15. INVESTING How much must you deposit in an account that pays 3.5% interest compounded yearly to have a balance of $400 after 6 years? **$325.40**

16. INVESTING How much must you deposit in an account that pays 6% interest compounded yearly to have a balance of $1000 after 8 years? **$627.41**

17. BICYCLE RACING In the Chapter Opener you learned that there is a relationship between the breathing rate of a cyclist and the bicycle speed.

Bicycle speed, x	0	5	10	15	20
Breathing rate, y	6.4	10.7	18.1	30.5	51.4

Let x represent the speed of the bike in miles per hour, and let y represent the cyclist's breathing rate in liters of air taken into the lungs per minute. The breathing rate of a cyclist can be modeled by $y = 6.37(1.11)^x$. What is the cyclist's breathing rate if the bike is traveling 19 miles per hour? 25 miles per hour? **about 46.3 L/min; about 86.5 L/min**

STUDENT HELP
HOMEWORK HELP
Example 1: Exs. 6–17
Example 2: Exs. 21, 22
Example 3: Ex. 24
Example 4: Exs. 18–20
Example 5: Exs. 25–28

MATCHING THE GRAPH Match the description with its graph.

A.

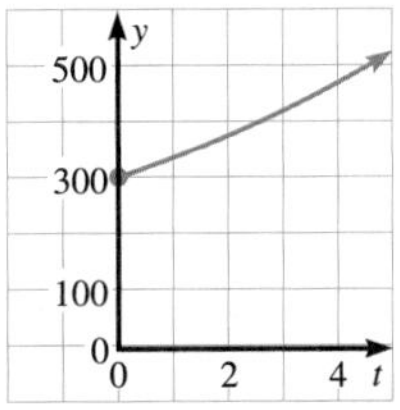

B.

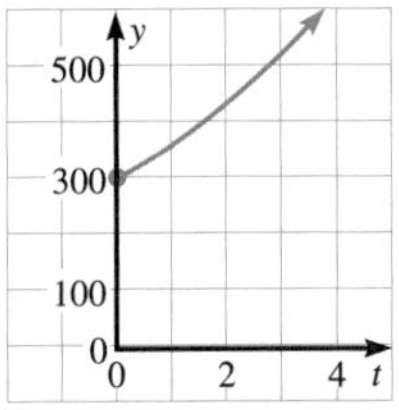

C.

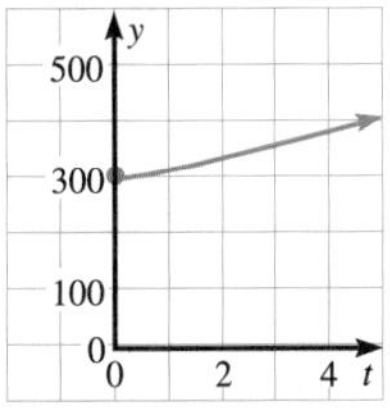

18. Deposit: \$300 **C**
Annual rate: 6%

19. Deposit: \$300 **A**
Annual rate: 12%

20. Deposit: \$300 **B**
Annual rate: 20%

BUSINESS In Exercises 21 and 22, write an exponential growth model.

21. A business had a \$10,000 profit in 1990. Then the profit increased by 25% per year for the next 10 years. **$P = 10{,}000(1.25)^t$**

22. A business had a \$20,000 profit in 1990. Then the profit increased by 20% per year for the next 10 years. **$P = 20{,}000(1.2)^t$**

23. VISUAL THINKING Graph the exponential growth models in Exercises 21 and 22 in the same coordinate plane. Which business would you rather own? Explain. **See margin.**

24. POPULATION GROWTH A population of 30 mice is released in a wildlife region. The population doubles each year for 4 years. What is the population after 4 years? **480**

BIOLOGY CONNECTION In Exercises 25–28, use the following information.

The average of the diameters of the two cross-sections shown can be related to the length of the adult bird. This average for a robin's egg is 25 millimeters (mm); for a bluebird, 19 mm; and for a chickadee, 13 mm. The graph shows the relationship between the average A of the two diameters and the length L (in centimeters) of an adult bird.

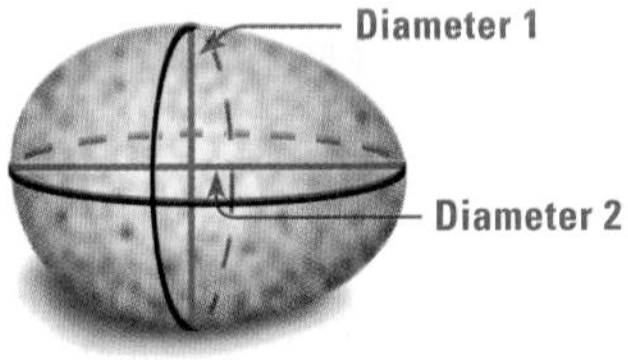

25. CHOOSING A MODEL Which model best represents the growth curve shown in the graph? **B**

A. $L = (0.09)(1.23)^A$

B. $L = 12 + (0.09)(1.23)^A$

C. $L = 12 + [(0.09)(1.23)]^A$

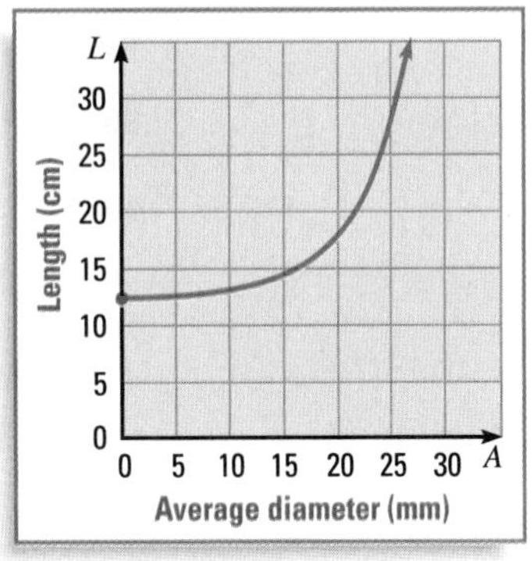

26. Estimate the length of an adult bluebird from the graph. **Estimates may vary; about 17 cm.**

27. Estimate the length of an adult bluebird using the exponential growth model. How does it compare to the length you found in Exercise 26? **About 16.6 cm; comparisons may vary.**

28. Use the exponential growth model to find the ratio of the length of a chickadee to the length of a robin. **about $\frac{1}{2}$**

23. ***Sample answer:*** **I would base my choice on how long I thought I would own the business (and how realistic I thought the continued annual increase was). The profits for the business in Ex. 22 started higher than those for the other business and would remain higher for 16 years. After that, the profits of the other business would increase more rapidly and be much higher. For example, after 20 years the profits would be more than \$100,000 greater.**

FOCUS ON APPLICATIONS

REAL LIFE CHICKADEES Black-capped Chickadees survive freezing weather by lowering their body temperatures at night, entering a state of controlled hypothermia.

INTERNET APPLICATION LINK
www.mcdougallittell.com

! COMMON ERROR

EXERCISES 21 AND 22 Some students may use 10 as the exponent in the equation modeling the situation. Stress that that an exponential growth model must have a variable as the exponent.

TEACHING TIP

EXERCISE 24 Ask students if doubling represents a low growth rate or a large growth rate. **large**

STUDENT HELP NOTES

Application Link Additional information about chickadees is available at **www.mcdougallittell.com.**

23.

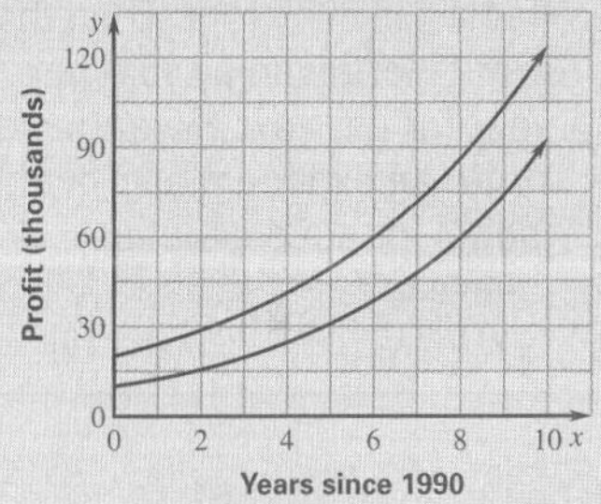

ADDITIONAL PRACTICE AND RETEACHING

For Lesson 8.5:

- Practice Levels A, B, and C (*Chapter 8 Resource Book,* p. 73)
- Reteaching with Practice (*Chapter 8 Resource Book,* p. 76)
- See Lesson 8.5 of the *Personal Student Tutor*

For more Mixed Review:

- Search the *Test and Practice Generator* for key words or specific lessons.

4 ASSESS

DAILY HOMEWORK QUIZ

Transparency Available

1. You deposit \$700 in an account that pays 8% interest compounded yearly. Find the balance after 13 years. **\$1903.74**
2. How much must you deposit in an account that pays 8% interest compounded yearly to have a balance of \$2000 after 6 years? **\$1260.34**
3. Company A had a \$15,000 profit. Then the profit increased by 20% per year for the next 15 years. Company B had a \$25,000 profit. Then the profit increased by 16% per year for the next 15 years. Write an exponential growth model for each situation. **company A: $P = 15{,}000(1.2)^t$; company B: $P = 25{,}000(1.16)^t$**

EXTRA CHALLENGE NOTE

Challenge problems for Lesson 8.5 are available in **blackline** format in the *Chapter 8 Resource Book,* p. 80 and at **www.mcdougallittell.com.**

ADDITIONAL TEST PREPARATION

1. **OPEN ENDED** Describe a real-life application of exponential growth models.
   Sample answer: **the growth of the population of a country**

Test Preparation

29. MULTIPLE CHOICE The hourly rate of your new job is \$5.00 per hour. You expect a raise of 9% each year. At the end of your first year, you receive your first raise. What will your hourly rate be at the end of your fifth year? **C**

Ⓐ \$5.45 Ⓑ \$7.25 Ⓒ \$7.69 Ⓓ \$7.76

30. MULTIPLE CHOICE What is the equation of the graph? **C**

Ⓐ $y = 2(0.88)^x$ Ⓑ $y = 4(0.88)^x$

Ⓒ $y = 2(1.3)^x$ Ⓓ $y = 4(1.3)^x$

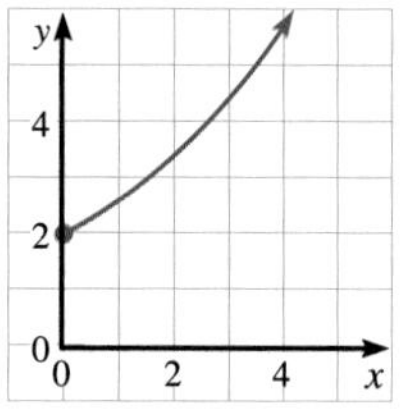

★ Challenge

31. EXTENSION: COMPOUND INTEREST What is the value of an \$8000 investment after 5 years if it earns 8% annual interest compounded quarterly? To solve, use the compound interest formula, $A = P(1 + i)^n$, where P is the original value of the investment, i is the interest rate per compounding period, n is the total number of compounding periods, and A is the value of the investment after n periods.

a. What is the interest rate per quarter? **2%**

b. How many compounding periods (quarters) are there in 5 years? **20 periods**

c. Use the formula $A = P(1 + i)^n$ to find the value of the investment after 5 years. **\$11,887.58**

EXTRA CHALLENGE
www.mcdougallittell.com

MIXED REVIEW

PERCENT OF A NUMBER **Find the percent of a number.** (Skills Review page 786)

32. 12% of 56 **6.72** **33.** 75% of 235 **176.25** **34.** 1.25% of 90 **1.125**

35. 200% of 130 **260** **36.** 2% of 105 **2.1** **37.** 0.8% of 120 **0.96**

EVALUATING VARIABLE EXPRESSIONS **Evaluate the expression.** (Review 1.3)

38. $24 + m^3$ when $m = 5$ **149**

39. $\frac{a^2 - b^2}{ab}$ when $a = 3$ and $b = 5$ **$-\frac{16}{15}$**

40. $x^6 - 1$ when $x = 1.2$ **1.985984**

41. $3y^4 + 15y$ when $y = -0.02$ **−0.29999952**

42. $(1 - x)^t$ when $x = 0.5$ and $t = 3$ **0.125**

43. $\frac{(1 - x)^t}{2}$ when $x = 0.09$ and $t = 2$ **0.41405**

44. BREAKFAST You are in charge of bringing breakfast for your scout troop. You buy 6 bagels and 8 donuts for a total of \$4.10. Then you decide to buy 3 extra of each for a total of \$1.80. How much did each bagel and donut cost? (Review 7.4) **bagel: \$.35, donut: \$.25**

SOLVING EQUATIONS **Solve the equation.** (Review 3.3, 3.4, 3.6)

45. $-2(7 - 5x) = 10$ **2.4**

46. $25 - (6x + 5) = 4(3x - 5) + 4$ **2**

47. $\frac{3}{2}(8m - 30) = -3m$ **3**

48. $1.4(6.4y - 3.5) = -9.54y + 22.85$ **1.5**

Investigating Exponential Decay

SET UP
Work in a small group.

MATERIALS
- 100 pennies
- cup

QUESTION Can an exponential decay model show a decreasing amount?

EXPLORING THE CONCEPT

1. Make a table to record your results.

Number of toss	0	1	2	3	4	5	6	7
Number of pennies remaining	?	?	?	?	?	?	?	?

2. Place the pennies in the cup. Shake the pennies in the cup then spill them onto a flat surface. Remove all of the pennies that land face up. Count and record the number of pennies remaining.

3. Repeat **Step 2** until there are no pennies left in the cup.

4. Make a scatter plot of the data you have collected.

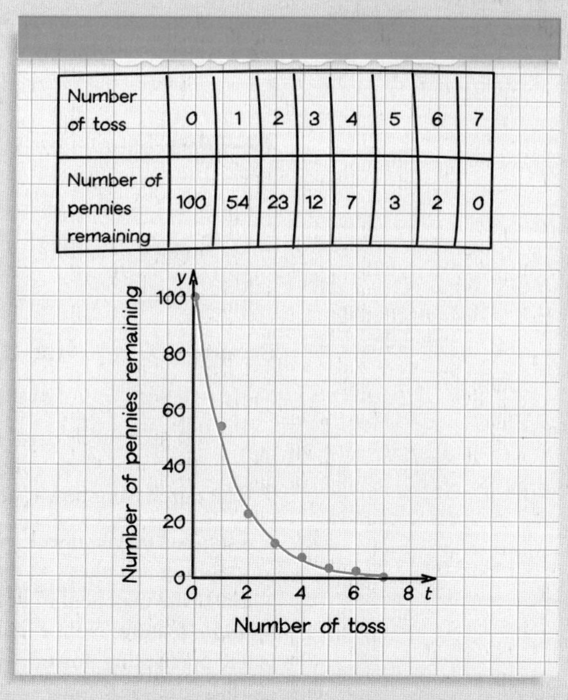

DRAWING CONCLUSIONS

1. Describe any patterns suggested by the scatter plot. ***Sample answer:* The number of pennies remaining appears to decrease exponentially.**
2. Look at your data and scatter plot to complete the following statement. **half**

> Each time the cup is emptied, the number of pennies you remove is about _?_ of the number in the cup.

3. Write an exponential equation to model this situation. $y = 100\left(\frac{1}{2}\right)^x$**, where x is the number of the toss**
4. Compare your equations and graphs with those of other groups. Describe any patterns you see. **See margin.**

Activity Assessment *Sample answer:*
It is a curve that decreases rapidly and approaches the x-axis very closely as x increases.

1 Planning the Activity

PURPOSE
To have students find and graph an exponential decay model.

MATERIALS
- 100 pennies
- Cups
- Graph paper
- Activity Support Master (*Chapter 8 Resource Book*, p. 84)

PACING
- Exploring the Concept — 10 min
- Drawing Conclusions — 10 min

LINK TO LESSON
Students can graph the model for their data with a graphing calculator as is done in Example 5 of Lesson 8.6.

2 Managing the Activity

COOPERATIVE LEARNING
Have students work in groups of 4. One student should make the table. A second student should manage the pennies. The third student should make the scatter plot, and the fourth student should record the conclusions.

ALTERNATIVE APPROACH
To save time, do the exploration as a demonstration. Have the class describe any patterns in your data.

3 Closing the Activity

KEY DISCOVERY
A quantity that decreases by a factor less than 1 can be modeled by an exponential equation that represents exponential decay.

ACTIVITY ASSESSMENT
Describe the graph of an exponential equation $y = ab^x$, where $a > 0$ and b is a positive number less than 1.
See sample answer at left.

4. See Additional Answers beginning on page AA1.

1 PLAN

PACING
Basic: 2 days
Average: 2 days
Advanced: 2 days
Block Schedule: 0.5 block with 8.6
0.5 block with
Ch. Rev.

LESSON OPENER
APPLICATION
An alternative way to approach Lesson 8.6 is to use the Application Lesson Opener:
- Blackline Master (*Chapter 8 Resource Book*, p. 85)
- Transparency (p. 57)

MEETING INDIVIDUAL NEEDS
- ***Chapter 8 Resource Book***
  Prerequisite Skills Review (p. 5)
  Practice Level A (p. 88)
  Practice Level B (p. 89)
  Practice Level C (p. 90)
  Reteaching with Practice (p. 91)
  Absent Student Catch-Up (p. 93)
  Challenge (p. 97)
- ***Resources in Spanish***
- ***Personal Student Tutor***

NEW-TEACHER SUPPORT
See the Tips for New Teachers on pp. 1–2 of the *Chapter 8 Resource Book* for additional notes about Lesson 8.6.

WARM-UP EXERCISES
Transparency Available
Evaluate each expression. Round to the nearest hundredth.
1. $(0.85)^5$ 0.44
2. $(0.97)^4$ 0.89
3. $(0.5)^3$ 0.13
4. $(1 - 0.03)^4$ 0.89
5. $(1 - 0.08)^6$ 0.61

8.6 Exponential Decay Functions

What you should learn

GOAL 1 Write and use models for exponential decay.

GOAL 2 Graph models for exponential decay.

Why you should learn it

▼ To solve **real-life** problems, such as comparing the buying power of a dollar in different years as in **Example 1**.

REAL LIFE

GOAL 1 WRITING EXPONENTIAL DECAY MODELS

In Lesson 8.5, you learned that a quantity is *growing exponentially* if it *increases* by the same percent in each unit of time. A quantity is *decreasing exponentially* if it *decreases* by the same percent in each unit of time. This is called **exponential decay**.

EXPONENTIAL DECAY MODEL

$$y = C(1 - r)^t$$

C is the **initial amount.**

t is the **time period.**

$(1 - r)$ is the **decay factor,** r is the **decay rate.** To assure $1 > (1 - r) > 0$, it is necessary that $0 < r < 1$.

The **percent of decrease** is $100r$.

EXAMPLE 1 *Writing an Exponential Decay Model*

PURCHASING POWER From 1982 through 1997, the purchasing power of a dollar decreased by about 3.5% per year. Using 1982 as the base for comparison, what was the purchasing power of a dollar in 1997?

▶ Source: *Statistical Abstract of the United States:* 1998.

SOLUTION Let y represent the purchasing power and let $t = 0$ represent the year 1982. The initial amount is \$1. Use an exponential decay model.

$y = C(1 - r)^t$ **Exponential decay model**

$= (1)(1 - 0.035)^t$ **Substitute 1 for C and 0.035 for r.**

$= 0.965^t$ **Simplify.**

Because 1997 is 15 years after 1982, substitute 15 for t.

$y = 0.965^{15}$ **Substitute 15 for t.**

≈ 0.59 **Use a calculator.**

▶ The purchasing power of a dollar in 1997 compared to 1982 was \$.59.

✓**CHECK** You can check your result by using a graphing calculator to make a table of values. Enter the exponential decay model and use the *Table* feature. Notice that as the prices *inflate*, the purchasing power of a dollar *deflates*.

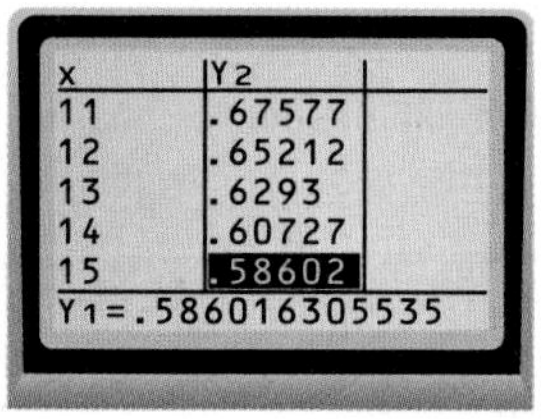

Depreciation

EXAMPLE 2 *Writing an Exponential Decay Model*

You bought a used car for \$18,000. The value of the car will be less each year because of depreciation. The car depreciates (loses value) at the rate of 12% per year.

a. Write an exponential decay model to represent this situation.

b. Estimate the value of your car in 8 years.

SOLUTION

STUDENT HELP

Skills Review
For help with writing a percent as a decimal, see page 784.

a. The initial value C is \$18,000. The decay rate r is 0.12. Let y be the value and let t be the age of the car in years.

$y = C(1 - r)^t$ Exponential decay model

$= 18{,}000(1 - 0.12)^t$ Substitute 18,000 for C and 0.12 for r.

$= 18{,}000(0.88)^t$ Simplify.

▶ The exponential decay model is $y = 18{,}000(0.88)^t$.

b. To find the value in 8 years, substitute 8 for t.

$y = 18{,}000(0.88)^t$ Exponential decay model

$= 18{,}000(0.88)^8$ Substitute 8 for t.

≈ 6473 Use a calculator.

▶ According to this model, the value of your car in 8 years will be about \$6473.

Depreciation

EXAMPLE 3 *Making a List to Verify a Model*

Verify the model you found in Example 2. Find the value of the car for each year by multiplying the value in the previous year by the decay factor.

SOLUTION The decay rate is 0.12. Decay factor = 1 − 0.12 = 0.88.

Year	Value
0	**18,000** ← Initial value of the car
1	0.88(**18,000**) = 15,840
2	0.88(15,840) ≈ 13,939
3	0.88(13,939) ≈ 12,266
4	0.88(12,266) ≈ 10,794
5	0.88(10,794) ≈ 9499
6	0.88(9499) ≈ 8359
7	0.88(8359) ≈ 7356
8	0.88(7356) ≈ 6473

From the list you can see that the value of the car in 8 years will be about \$6473, which is consistent with your model.

2 TEACH

EXTRA EXAMPLE 1
From 1983 to 1997, the ratio of students per computer at a school has dropped by about 16.8% per year. If there were 103 students per computer in 1983 and 1983 is the base for comparison, what was the number of students per computer in 1997? **7.8 students per computer**

EXTRA EXAMPLE 2
You bought a computer for \$1800. The value of the computer will be less each year because of depreciation. The computer depreciates at the rate of 29% per year.

a. Write an exponential decay model to represent this situation. **$y = 1800(0.71)^t$, where y = value and t is the time in years.**

b. Estimate the value of the computer in 2 years. **\$907**

EXTRA EXAMPLE 3
Verify the model you found in Extra Example 2.

Year	0	1	2
Value	1800	1278	907

CHECKPOINT EXERCISES
For use after Examples 1–3:

1. You bought a used boat for \$2300. The value of the boat will be less each year because of depreciation. The boat depreciates at the rate of 8% per year.

a. Write an exponential decay model to represent this situation. **$y = 2300(0.92)^t$**

b. Estimate the value of the boat in 2 years. **\$1947**

c. Verify the model.

Year	0	1	2
Value	2300	2116	1947

EXTRA EXAMPLE 4

Graph the exponential decay model in Extra Example 1.

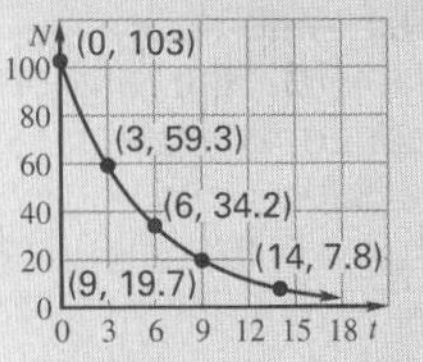

EXTRA EXAMPLE 5

a. Graph the exponential decay model in Extra Example 2.

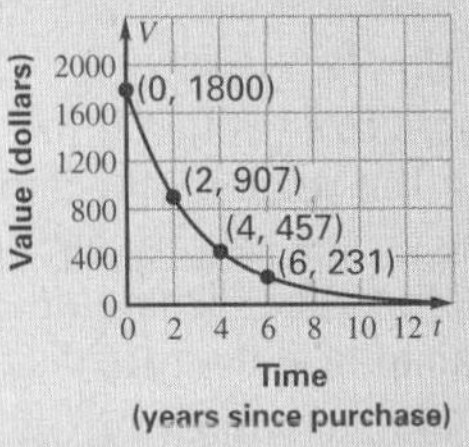

b. Use the graph to estimate the value of your computer in 7 years. ***Sample answer:*** **\$164**

CHECKPOINT EXERCISES

For use after Examples 4 and 5:

1. a. Suppose the value of a \$2300 computer t years after purchase is modeled by $y = 2300(0.92)^t$. Graph the exponential decay model.

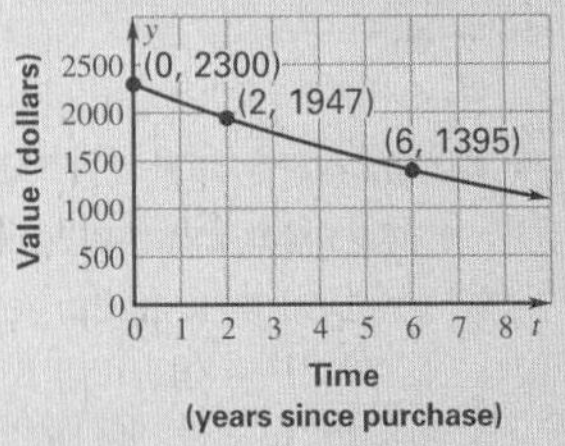

b. Estimate the value in 8 years. **\$1180**

GOAL 2 GRAPHING EXPONENTIAL DECAY MODELS

EXAMPLE 4 *Graphing an Exponential Decay Model*

REAL LIFE
Purchasing Power

Graph the exponential decay model in Example 1.

SOLUTION

Use your graphing calculator to graph the model. Set the viewing rectangle so that $0 \le x \le 20$ and $0 \le y \le 1$.

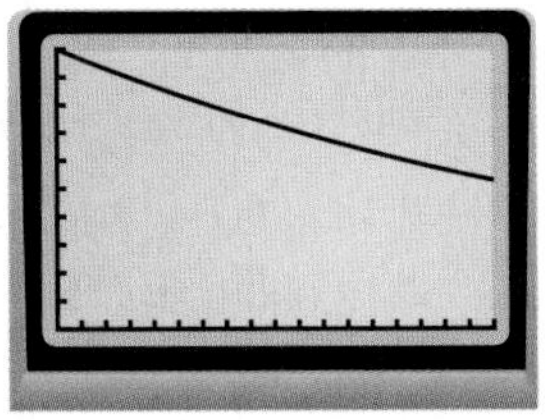

EXAMPLE 5 *Graphing an Exponential Decay Model*

REAL LIFE
Depreciation

a. Graph the exponential decay model in Example 2.

b. Use the graph to estimate the value of your car in 10 years.

SOLUTION

a. Use the table of values in Example 3. Plot the points in a coordinate plane, and draw a smooth curve through the points.

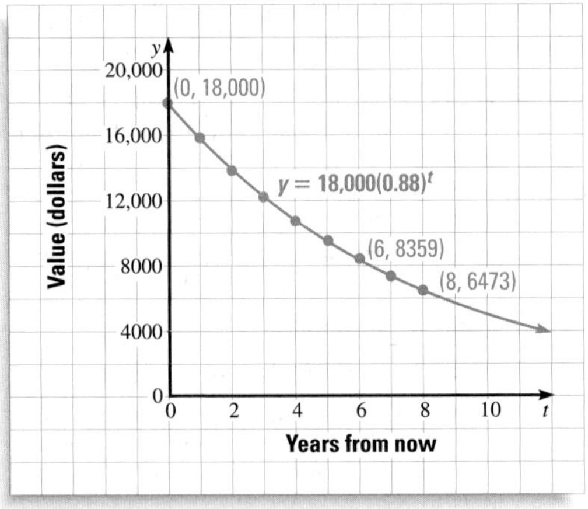

b. From the graph, the value of your car in 10 years will be about \$5000.

....

An exponential model $y = a \cdot b^t$ represents exponential growth if $b > 1$ and exponential decay if $0 < b < 1$.

EXAMPLE 6 Comparing Growth and Decay Models

Classify each model as *exponential growth* or *exponential decay*. In each case identify the growth or decay factor and the percent of increase or percent of decrease per time period. Then graph each model.

a. $y = 30(1.2)^t$ **b.** $y = 30\left(\frac{3}{5}\right)^t$

SOLUTION

a. Because $1.2 > 1$, the model $y = 30(1.2)^t$ is an exponential growth model.

The growth factor $(1 + r)$ is 1.20.

The growth rate is 0.2, so the percent of increase is 20%.

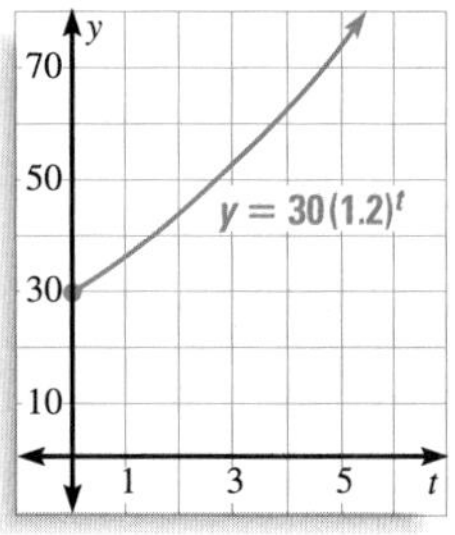

b. Because $0 < \frac{3}{5} < 1$, the model $y = 30\left(\frac{3}{5}\right)^t$ is an exponential decay model.

The decay factor $(1 - r) = \frac{3}{5}$.

The decay rate is $\frac{2}{5}$ or 0.4, so the percent of decrease is 40%.

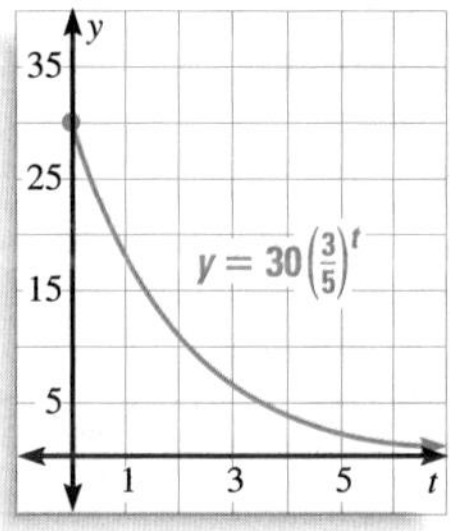

CONCEPT SUMMARY — EXPONENTIAL GROWTH AND DECAY MODELS

EXPONENTIAL GROWTH MODEL

$y = C(1 + r)^t$ ← time period

initial amount: C; growth factor: $(1 + r)$

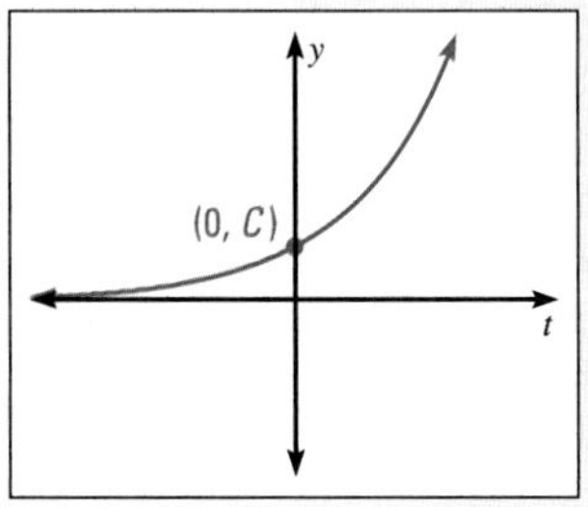

$1 + r > 1$

EXPONENTIAL DECAY MODEL

$y = C(1 - r)^t$ ← time period

initial amount: C; decay factor: $(1 - r)$

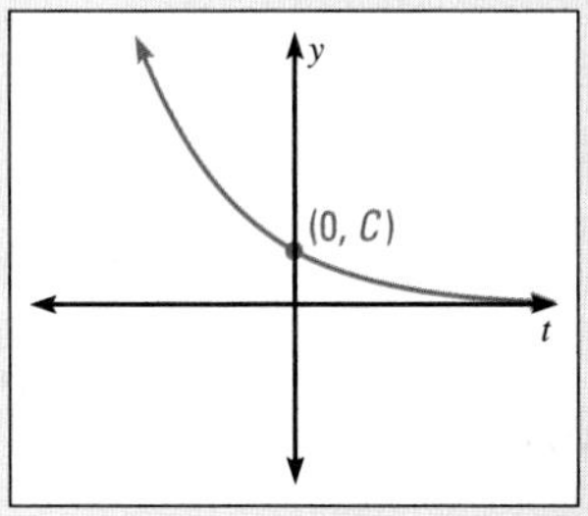

$0 < 1 - r < 1$

EXTRA EXAMPLE 6

Classify each model as *exponential growth* or *exponential decay*. In each case identify the growth or decay factor and the percent of increase or decrease per time period. Then graph the model.

a. $y = 40(0.75)^t$

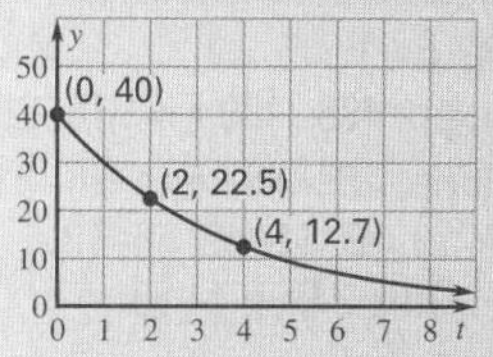

exponential decay; 0.75, 25% decrease

b. $y = 40\left(\frac{6}{5}\right)^t$

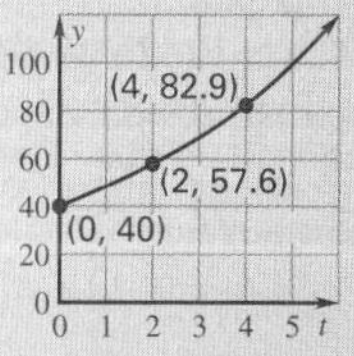

exponential growth; 1.2, 20% increase

CHECKPOINT EXERCISES

For use after Example 6:

1. Classify $y = 100(1.1)^t$ as *exponential growth* or *exponential decay*. Identify the growth or decay factor and the percent of increase or decrease per time period. Then draw the graph.

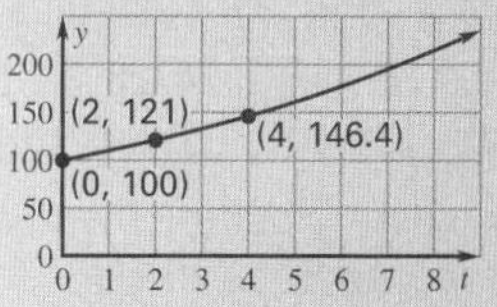

exponential growth; 1.1, 10% increase

CLOSURE QUESTION

Given an exponential model $y = 80(b)^t$, how can you tell if the model represents growth or decay? **If b is between 0 and 1, it is decay. If b is greater than one, it is growth.**

3 APPLY

ASSIGNMENT GUIDE

BASIC
Day 1: pp. 488–489 Exs. 10–25
Day 1: pp. 489–491 Exs. 26–33, 35, 36, 41–43, 45, 48, Quiz 2 Exs. 1–10

AVERAGE
Day 1: pp. 488–489 Exs. 10–25
Day 1: pp. 489–491 Exs. 26–33, 35, 36, 41–43, 45, 48, Quiz 2 Exs. 1–10

ADVANCED
Day 1: pp. 488–489 Exs. 10–25
Day 1: pp. 489–491 Exs. 26–43, 45, 48, Quiz 2 Exs. 1–10

BLOCK SCHEDULE
pp. 488–489 Exs. 10–25 (with 8.5)
pp. 489–491, Exs. 26–33, 35, 36, 41–43, 45, 48, Quiz 2 Exs. 1–10 (with Ch. 8 Review)

EXERCISE LEVELS

Level A: ***Easier***
41–44

Level B: ***More Difficult***
10–33, 35, 36, 45–48

Level C: ***Most Difficult***
34, 37–40

✔ HOMEWORK CHECK

To quickly check student understanding of key concepts, go over the following exercises: Exs. 12, 16, 18, 22, 24, 28, 31. See also the Daily Homework Quiz:

- Blackline Master (*Chapter 9 Resource Book,* p. 11)
- Transparency (p. 67)

18. exponential decay; 0.98; 2%
19. exponential growth; $\frac{5}{4}$; 25%
21. exponential decay; $\frac{2}{5}$; 60%

GUIDED PRACTICE

Vocabulary Check ✔ **1.** In the exponential decay model, $y = C(1 - r)^t$, what is the decay factor? **$1 - r$**

Concept Check ✔ **2.** Is $y = 1.02^t$ an exponential decay model? Explain.
No; it is an exponential growth model; 1.02 is a growth factor, not a decay factor.

3. Look back at Example 2. Suppose that the car was depreciating at the rate of 20% per year. Write a new exponential decay model. **$y = 18{,}000(0.8)^t$**

Skill Check ✔ **CAR VALUE** **You buy a used car for $7000. The car depreciates at the rate of 6% per year. Find the value of the car in the given years.**

4. 2 years **$6185** **5.** 5 years **$5137** **6.** 8 years **$4267** **7.** 10 years **$3770**

8. **BUSINESS DECLINE** A business earned $85,000 in 1990. Then its earnings decreased by 2% each year for 10 years. Write an exponential decay model for the earnings E in year t. Let $t = 0$ represent 1990. **$E = 85{,}000(0.98)^t$**

9. **CHOOSE A MODEL** Which model best represents the decay curve shown in the graph at the right? **C**

A. $y = 60(0.80)^t$

B. $y = 60(1.20)^t$

C. $y = 60(0.40)^t$

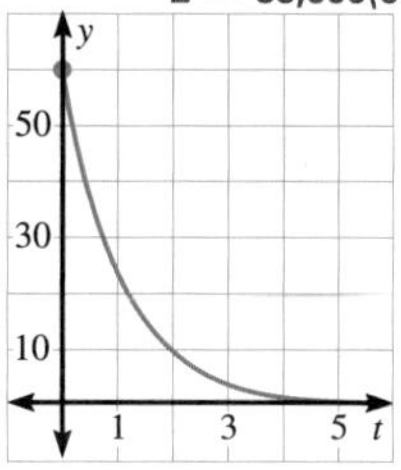

PRACTICE AND APPLICATIONS

STUDENT HELP
Extra Practice
to help you master skills is on p. 804.

TRUCK VALUE **You buy a used truck for $20,000. It depreciates at the rate of 15% per year. Find the value of the truck in the given years.**

10. 3 years **$12,283** **11.** 8 years **$5450** **12.** 10 years **$3937** **13.** 12 years **$2845**

MATCHING THE GRAPH **Match the equation with its graph.**

A.

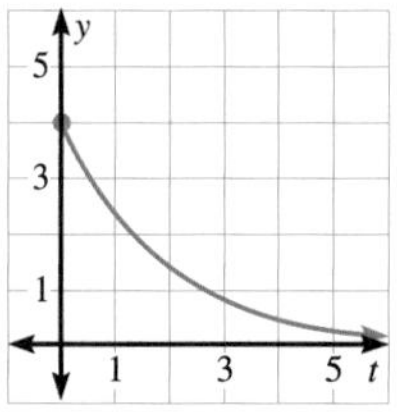

B.

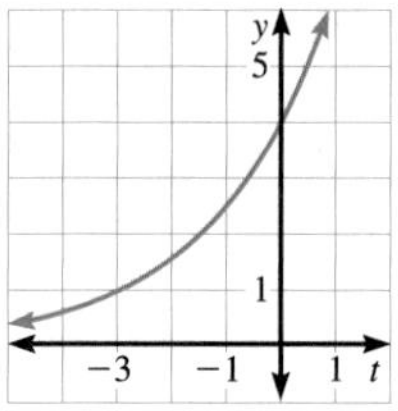

C.

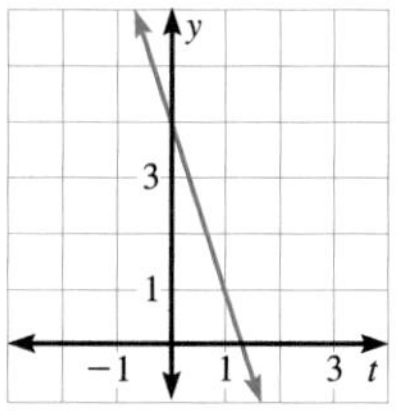

14. $y = 4 - 3t$ **C** **15.** $y = 4(1.6)^t$ **B** **16.** $y = 4(0.6)^t$ **A**

STUDENT HELP
HOMEWORK HELP
Examples 1, 2: Exs. 10–13, 23–29
Example 3: Exs. 30–33
Examples 4, 5: Exs. 14–16, 30, 33
Example 6: Exs. 14–22

RECOGNIZING MODELS **Classify the model as *exponential growth* or *exponential decay*. Identify the growth or decay factor and the percent of increase or decrease per time period.**

17. $y = 24(1.18)^t$ **exponential growth; 1.18; 18%**

18. $y = 14(0.98)^t$ **See margin.**

19. $y = 35\left(\frac{5}{4}\right)^t$ **See margin.**

20. $y = 112(0.4)^t$ **exponential decay; 0.4; 60%**

21. $y = 9\left(\frac{2}{5}\right)^t$ **See margin.**

22. $y = 97(1.01)^t$ **exponential growth; 1.01; 1%**

FOCUS ON CAREERS

PHARMACISTS Pharmacists must understand the use, composition, and effects of drugs.

CAREER LINK
www.mcdougallittell.com

SCIENCE CONNECTION In Exercises 23–25, use the following information. The concentration of aspirin in a person's bloodstream can be modeled by the equation $y = A(0.8)^t$, where y represents the concentration of aspirin in a person's bloodstream in milligrams (mg), A represents the amount of aspirin taken, and t represents the number of hours since the medication was taken. Find the amount of aspirin remaining in a person's bloodstream at the given dosage.

23. Dosage: 250 mg **160 mg**
Time: after 2 hours

24. Dosage: 500 mg **229 mg**
Time: after 3.5 hours

25. Dosage: 750 mg **246 mg**
Time: after 5 hours

26. BUYING A TRUCK You buy a used truck in 1996 for $10,500. Each year the truck depreciates by 10%. Write an exponential decay model to represent this situation. Then estimate the value of the truck in 10 years. **$y = 10{,}500(0.9)^t$; $3661**

BASKETBALL **In Exercises 27–29, use the following information.** Each year in the month of March, the NCAA basketball tournament is held to determine the national champion. At the start of the tournament there are 64 teams, and after each round, one half of the remaining teams are eliminated.

27. Write an exponential decay model showing the number of teams N left in the tournament after round t. $N = 64\left(\frac{1}{2}\right)^t$

28. How many teams remain after 3 rounds? after 4 rounds? **8 teams; 4 teams**

29. CRITICAL THINKING If a team won 6 games in a row in the tournament, does it mean that it won the national championship? Explain your reasoning. **Yes; there are only six rounds so the team won each one, including the final one.**

30. SUMMER CAMP A summer youth camp had a declining enrollment from 1995 to 2000. The enrollment in 1995 was 320 people. Each year for the next five years, the enrollment decreased by 2%. Copy and complete the table showing the enrollment for each year. Sketch a graph of the results. **320; 314; 307; 301; 295; 289; See margin.**

Year	1995	1996	1997	1998	1999	2000
Enrollment	?	?	?	?	?	?

CABLE CARS **In Exercises 31–33, use the following information. From 1894 to 1903 the number of miles of cable car track decreased by about 10% per year. There were 302 miles of track in 1894.**

31. Write an exponential decay model showing the number of miles M of cable car track left in year t. $M = 302(0.9)^t$

32. Copy and complete the table. You may want to use a calculator. **302; 245; 198; 178; 160; 144; 117**

Year	1894	1896	1898	1899	1900	1901	1903
Miles of track	?	?	?	?	?	?	?

33. Sketch a graph of the results. **See margin.**

34. CRITICAL THINKING A store is having a sale on sweaters. On the first day the price of a sweater is reduced by 20%. The price will be reduced another 20% each day until the sweater is sold. Denise thinks that on the fifth day of the sale the sweater will be free. Is she right? Explain. **See margin.**

34. No; *Sample answer:* The sweater will never be free because there is no number *n* other than 0 for which 80% of *n* is 0. If the price reduction actually did continue at 20% each day, and the sweater did not sell, the price would eventually be under $.01. For example a $40 sweater would sell for less than $.01 after 38 days. This is not a realistic possibility however.

APPLICATION NOTE

EXERCISE 23 Since there is still aspirin in the bloodstream after 2 hours, the concentration of aspirin is increased if another dose is taken after the first dose.

COMMON ERROR

EXERCISES 1, 3–8, 10–22, 30, 31, AND 34 Students may confuse the percent of increase or decrease from one time period to the next with the growth factor or decay factor. These students may benefit from a review of what each variable means in the growth models summarized in the Concept Summary box found on page 487.

STUDENT HELP NOTES

Career Link Additional information about pharmacists is available at **www.mcdougallittell.com.**

30.

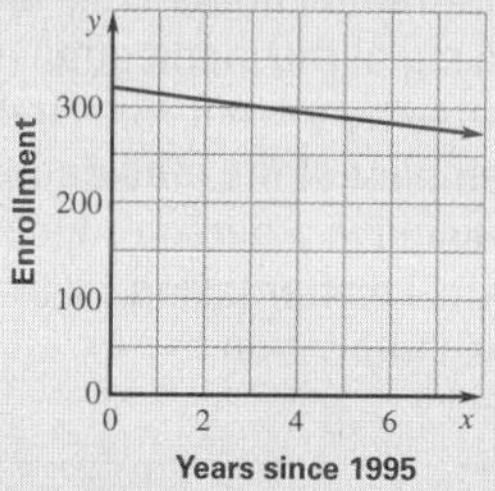

33. See Additional Answers beginning on page AA1.

ADDITIONAL PRACTICE AND RETEACHING

For Lesson 8.6:

- Practice Levels A, B, and C (*Chapter 8 Resource Book,* p. 88)
- Reteaching with Practice (*Chapter 8 Resource Book,* p. 91)
- See Lesson 8.6 of the *Personal Student Tutor*

For more Mixed Review:

- Search the *Test and Practice Generator* for key words or specific lessons.

4 ASSESS

DAILY HOMEWORK QUIZ

Transparency Available

1. In a factory, a piece of machinery that cost \$45,000 depreciates at the rate of 16% per year. Find the value of the machinery 5 years from the date of purchase.
about \$18,820

Classify the model as *exponential growth* or *exponential decay.* Identify the growth or decay factor and the percent of increase or decrease per time period.

2. $y = 120(0.89)^t$
exponential decay; 0.89; 11%

3. $y = 93(1.03)^t$
exponential growth; 1.03; 3%

4. The concentration y of a certain medication in a person's bloodstream t hours after taking A mg of the medication is modeled by $y = A(0.75)^t$. Find the amount of the medication remaining in a person's bloodstream 3 h after taking 1000 mg of the medication.
about 422 mg

EXTRA CHALLENGE NOTE

Challenge problems for Lesson 8.6 are available in **blackline** format in the *Chapter 8 Resource Book,* p. 97 and at **www.mcdougallittell.com.**

ADDITIONAL TEST PREPARATION

1. OPEN ENDED Describe a real-life situation that can be modeled by an exponential decay model. ***Sample answer:*** **The value of a car may decrease by 10% each year.**

Test Preparation

37. $y = 4^t$

t	y
−3	0.015625
−2	0.0625
−1	0.25
0	1
1	4
2	16
3	64

35. MULTIPLE CHOICE In 1995, you purchase a parcel of land for \$8000. The value of the land depreciates by 4% every year. What will the approximate value of the land be in 2002? **C**

Ⓐ \$5760 Ⓑ \$5771 Ⓒ \$6012 Ⓓ \$6262

36. MULTIPLE CHOICE Which model best represents the decay curve shown in the graph at the right? **B**

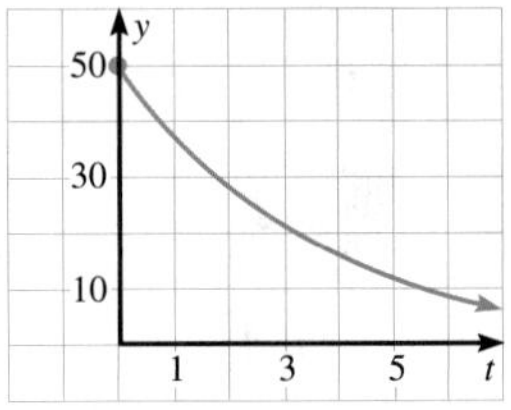

Ⓐ $y = 20(1.16)^t$

Ⓑ $y = 50(0.75)^t$

Ⓒ $y = 20(0.75)^t$

Ⓓ $y = 50 + 20(1.16)^t$

★ Challenge

$y = \left(\frac{1}{4}\right)^t$

t	y
−3	64
−2	16
−1	4
0	1
1	0.25
2	0.0625
3	0.015625

UNDERSTANDING GRAPHS In Exercises 37–39, use a graphing calculator.

37. Make an input-output table for the equations $y = 4^t$ and $y = \left(\frac{1}{4}\right)^t$. Use −3, −2, −1, 0, 1, 2, and 3 as the input. Then sketch the graph of each equation. **See margin for input-output tables. Check graphs.**

38. VISUAL THINKING Interpret the graphs. How are they related?
Sample answer: **Each graph is the mirror image of the other reflected in the *y*-axis.**

39. VISUAL THINKING The graph at the right is $y = 2.26^t$. Based on the relationships between the graphs in Exercise 37, predict the graph of $y = \left(\frac{1}{2.26}\right)^t$.

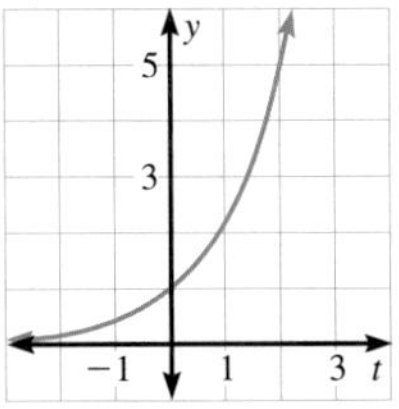

Sample answer: **The graph will be the mirror image of the given graph reflected in the *y*-axis. It will curve up to the left, and come very close to the *t*-axis on the right.**

EXTRA CHALLENGE
www.mcdougallittell.com

MIXED REVIEW

VARIABLE EXPRESSIONS Evaluate the expression. (Review 1.3 for 9.1)

40. $4a^2 + 11$ when $a = 5$ **111**

41. $c^3 + 6cd$ when $c = 2$ and $d = 1$ **20**

42. $b^2 - 4ac$ when $a = 1, b = 3$, and $c = 5$ **−11**

43. $\frac{a^2 - b^2}{2c^2} + 9$ when $a = -3, b = 5$, and $c = -2$ **7**

SOLVING EQUATIONS Solve the equation. Round the result to the nearest tenth if necessary. (Review 3.6)

44. $12m - 9 = 5m - 2$ **1**

45. $5(2x + 2.3) - 11.2 = 6x - 5$ **−1.3**

46. $-1.3y + 3.7 = 4.2 - 5.4y$ **0.1**

47. $2.5(3.5p + 6.4) = 18.2p - 6.5$ **2.4**

48. TELEVISION A TV station's local news program has 50,000 viewers. The managers of the station plan to increase the number of viewers by 2% per month. Write an exponential growth model to represent the number of viewers in t months. (Review 8.5) $y = 50{,}000(1.02)^t$

QUIZ 2

Self-Test for Lessons 8.4–8.6

SCIENTIFIC NOTATION **Rewrite in scientific notation.** (Lesson 8.4)

1. 0.011205 1.1205×10^{-2}
2. 140,000,000 1.4×10^{8}
3. 0.000000067 6.7×10^{-8}
4. 30,720,000,000 3.072×10^{10}

DECIMAL FORM **Rewrite in decimal form.** (Lesson 8.4)

5. 4.82×10^3 4820
6. 5×10^9 5,000,000,000
7. 7.04×10^{-6} 0.00000704
8. 1.112×10^{-2} 0.01112

9. INVESTMENT In 1995, you bought a baseball card for $50 that you expect to increase in value 5% each year for the next 10 years. Write an exponential growth model and estimate the value of the baseball card in 2002. (Lesson 8.5) $y = 50(1.05)^t$; about $70

10. BUYING A CAMPER You buy a used camper in 1995 for $20,000. Each year the camper depreciates by 15%. Write an exponential decay model to represent this situation. Then estimate the value of the camper in 5 years. (Lesson 8.6) $y = 20{,}000(0.85)^t$; about $8874

MATH & History — History of Microscopes

APPLICATION LINK www.mcdougallittell.com

THEN

THE EARLIEST MICROSCOPES consisted of a single, strongly curved lens mounted on a metal plate. These simple microscopes used visible light to illuminate the object being viewed and could magnify objects as much as 400 times.

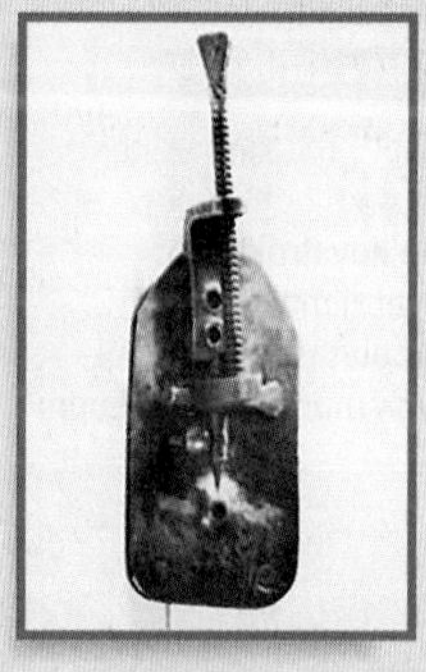

NOW

TODAY, scientists use microscopes that use electrons instead of light. The *transmission electron microscope* (or TEM) can resolve two objects as close together as 0.0000005 millimeter (mm) with magnifications up to 1,000,000 times. At this magnification, a sneaker would be long enough to reach from Boston to New York City.

1. Write 0.0000005 mm in scientific notation. 5×10^{-7}

2. The distance from Boston to New York City is 1,003,200 feet. Write this number in scientific notation. 1.0032×10^6 ft

3. How many times stronger is the magnification of the TEM compared to the earliest microscopes? Write your answer in scientific notation. 2.5×10^3 times

Anton van Leeuwenhoek was the first to see red blood cells.

1670

Max Knoll and Ernst Ruska of Germany developed the first TEM.

1931

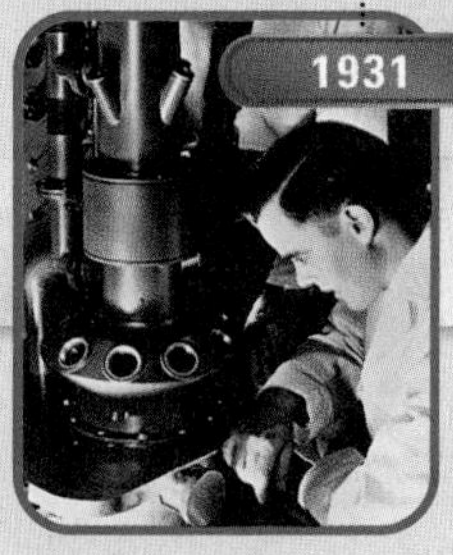

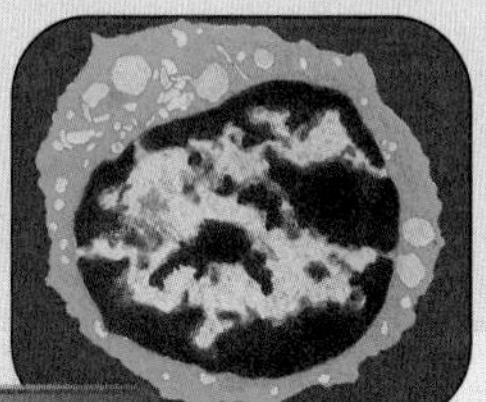

Today

This TEM image shows a white blood cell at a magnification of 9600 times. The image has been colored to show the different parts of the cell.

ADDITIONAL RESOURCES
An alternative Quiz for Lessons 8.4–8.6 is available in the *Chapter 8 Resource Book*, p. 98.

MATH & HISTORY NOTE
See the enrichment master Math & History Applications on p. 96 of the *Chapter 8 Resource Book* for additional exercises.

ADDITIONAL RESOURCES
A **blackline** master with additional Math & History exercises is available in the *Chapter 8 Resource Book*, p. 96.

1 Planning the Activity

PURPOSE
To have students use a graphing calculator to find a best-fitting exponential growth or decay model.

MATERIALS
- Graphing calculator
- Keystroke blackline (*Chapter 8 Resource Book,* p. 86)

PACING
- Exploring the Concept — 15 min
- Drawing Conclusions — 10 min

LINK TO LESSON
Students can use a graphing calculator to confirm that they get the equation in Example 2 part a of Lesson 8.6 if they fit an exponential model to the data in the table for Example 3.

2 Managing the Activity

CLASSROOM MANAGEMENT
Students can work on this activity with a partner.

ALTERNATIVE APPROACH
To save time, do this activity as a demonstration. Discuss the meanings of *a, b,* and *x* in the model $y = a \cdot b^x$.

3 Closing the Activity

KEY DISCOVERY
A graphing calculator can be used to see how well a set of data is modeled by an exponential growth or decay model.

ACTIVITY ASSESSMENT
Use a graphing calculator to find a best-fitting growth or decay model for this data. (0, 102), (1, 98), (2, 95), (3, 91), (4, 87), (5, 82), (6, 80)
$y = 102.5(0.959)^x$

ACTIVITY 8.6

Using Technology

Fitting Exponential Models

In Chapter 5, you learned that you can use a graphing calculator to find a best-fitting line. A graphing calculator can also be used to find a best-fitting exponential growth or decay model.

EXAMPLE

A rubber ball is dropped from a height of 0.82 meter. Using a CBL unit, the height of the ball on each successive bounce was recorded. The *x*-values represent the bounce and the *y*-values represent the height. Use a graphing calculator to find an exponential model for these data.

x	0	1	2	3	4	5	6	7	8	9
y	0.82	0.64	0.50	0.39	0.30	0.24	0.18	0.14	0.11	0.08

SOLUTION

STUDENT HELP
INTERNET KEYSTROKE HELP
See keystrokes for several models of calculators at www.mcdougallittell.com

1 Enter the ordered pairs into the graphing calculator. Select L_1 as the *x* list and L_2 as the *y* list.

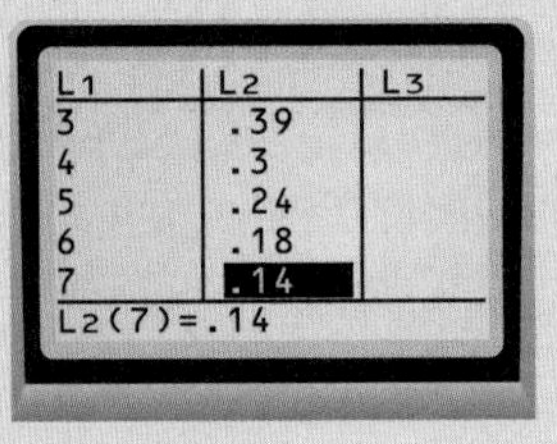

2 Use exponential regression to find an exponential model. The equation $y = 0.8335(0.7744)^x$ is the best-fitting exponential model.

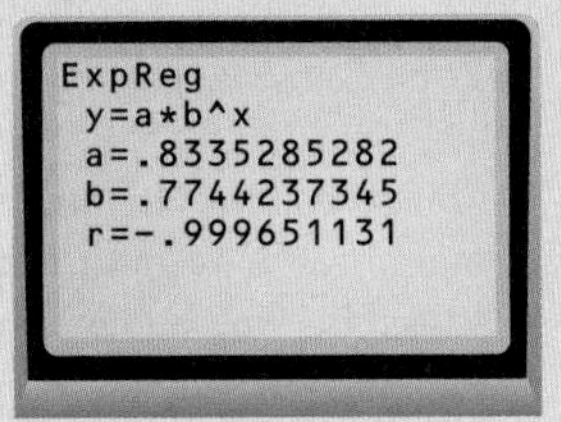

3 Set the viewing rectangle so that $0 \le x \le 20$ and $0 \le y \le 1$.

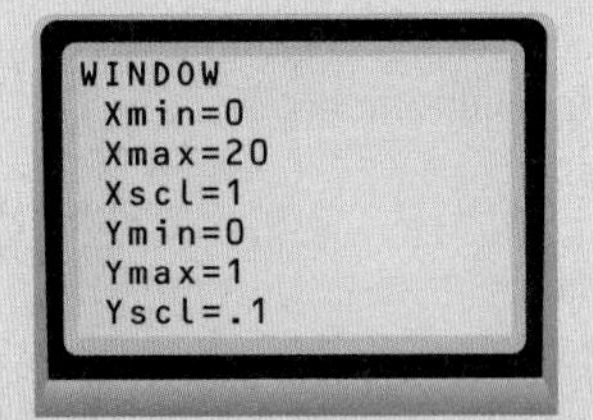

4 Graph the equation $y = 0.8335(0.7744)^x$.

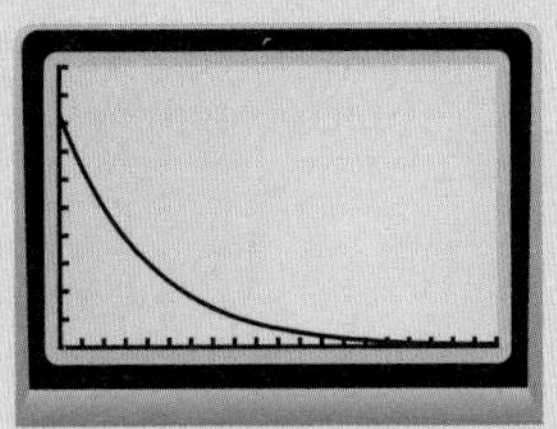

EXERCISES

Use a graphing calculator to find the best-fitting exponential growth model for the points.

1. (0, 1), (1, 1.4), (2, 3), (3, 5), (4, 8), (5, 12), (6, 20), (7, 30), (8, 50), (9, 80)
$y = 1.0339(1.6293)^x$
2. (0, 0.5), (2, 0.8), (3, 1), (4, 1.4), (5, 1.8), (6, 2.7), (7, 3.6), (8, 4.9), (9, 7)
$y = 0.4439(1.3476)^x$

CHAPTER 8 Chapter Summary

WHAT did you learn?	*WHY did you learn it?*
Evaluate exponential expressions	
• using multiplication properties of exponents. **(8.1)**	Find the power generated by a windmill. **(p. 454)**
• that have negative and zero exponents. **(8.2)**	Predict a basketball player's average score per game. **(p. 458)**
• using division properties of exponents. **(8.3)**	Estimate the speed of an Olympic rowing team. **(p. 468)**
Convert numbers from scientific notation to decimal form. **(8.4)**	Find the amount of water discharged by the Amazon River each year. **(p. 472)**
Convert numbers from decimal form to scientific notation. **(8.4)**	Find the price per acre of the Alaska purchase. **(p. 472)**
Perform operations with numbers in scientific notation. **(8.4)**	Find how long it takes light to travel from the Sun to Pluto. **(p. 473)**
Use scientific notation in problem solving. **(8.4)**	Estimate the number of heartbeats in a person's life. **(p. 474)**
Use exponential growth models. **(8.5)**	Find the weight of a channel catfish. **(p. 478)**
Use exponential decay models. **(8.6)**	Find the buying power of a dollar. **(p. 484)**
Sketch the graphs of exponential growth and decay models. **(8.5 and 8.6)**	Find the balance on a savings account. **(p. 480)**

How does Chapter 8 fit into the BIGGER PICTURE of algebra?

In Chapters 3–7, you learned how to solve linear equations such as $4(x - 3y) = 24$. In this chapter you learned how to use the properties of exponents and scientific notation to solve exponential functions such as $y = 42(1.2)^x$. You will need to know how to use the properties of exponents when you solve quadratic equations in Chapter 9 and polynomial equations in Chapter 10.

STUDY STRATEGY

How did you use your schedule?

A schedule that you made using the **Study Strategy** on p. 448 may resemble this one.

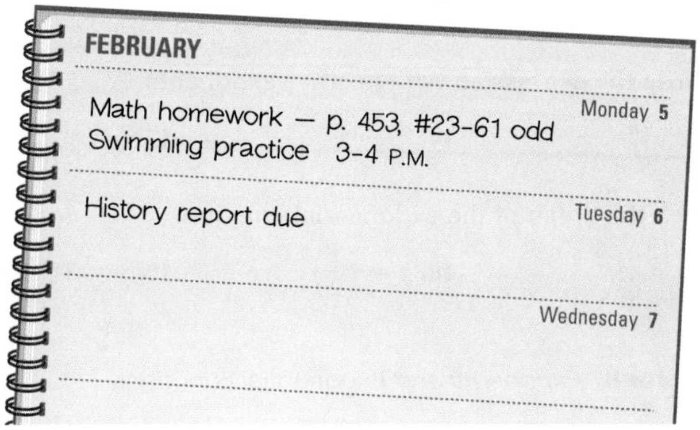

493

ADDITIONAL RESOURCES

The following resources are available to help review the material in this chapter.

- Chapter Review Games and Activities (*Chapter 8 Resource Book*, p. 99)
- *Instant Replay: Video Review Games*
- *Personal Student Tutor*
- Cumulative Review, Chs. 1–8 (*Chapter 8 Resource Book*, p. 111)

17.

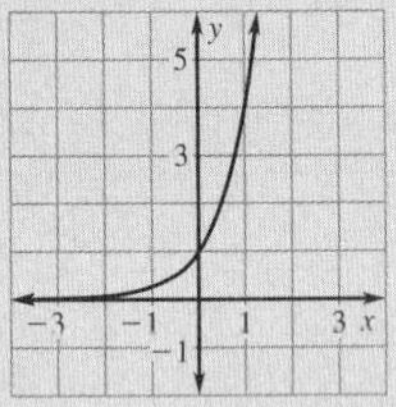

CHAPTER 8

Chapter Review

VOCABULARY

- exponential function, p. 458
- scientific notation, p. 470
- exponential growth, p. 477
- growth factor, p. 477
- initial amount, pp. 477, 484
- time period, pp. 474, 484
- percent of increase, p. 477
- growth rate, p. 477
- exponential decay, p. 484
- decay factor, p. 484
- decay rate, p. 484
- percent of decrease, p. 484

8.1 MULTIPLICATION PROPERTIES OF EXPONENTS

Examples on pp. 450–452

EXAMPLES Use the multiplication properties of exponents.

a. $4^2 \cdot 4^7 = 4^{2+7} = 4^9$ — Product of powers property

b. $(5^2)^3 = 5^{2 \cdot 3} = 5^6$ — Power of a power property

c. $(6 \cdot x)^3 = 6^3 \cdot x^3$ — Power of a product property

Simplify the expression.

1. $2^2 \cdot 2^7$ **2^9, or 512**
2. $(4^3)^2$ **4^6, or 4096**
3. $(3a)^3 \cdot (2a)^2$ **$3^3 \cdot 2^2 \cdot a^5$, or $108a^5$**
4. $(w^3x^4y)^2 \cdot (wx^2y^3)^4$ **$w^{10}x^{16}y^{14}$**

Simplify. Then evaluate the expression when $s = 2$ and $t = 3$.

5. $s^3 \cdot s^4$ **s^7; 128**
6. $s^4 \cdot (-t)^3$ **$-s^4t^3$; −432**
7. $(s^3 \cdot t)^2$ **s^6t^2; 576**
8. $-(st^2)^2$ **$-s^2t^4$; −324**

8.2 ZERO AND NEGATIVE EXPONENTS

Examples on pp. 456–458

EXAMPLES Use zero and negative exponents.

a. $6^0 = 1$ — A nonzero number to the zero power is 1.

b. $7(x^{-3}) = 7\left(\frac{1}{x^3}\right) = \frac{7}{x^3}$ — a^{-n} is the reciprocal of a^n.

Evaluate the expression. Write fractions in simplest form.

9. 5^{-3} **$\frac{1}{125}$**
10. $7^{-4} \cdot 7^6$ **49**
11. $16\left(\frac{1}{2}\right)^{-1}$ **32**
12. $2^0 \cdot \left(\frac{1}{4^{-2}}\right)$ **16**

Rewrite the expression with positive exponents.

13. x^6y^{-6} **$\frac{x^6}{y^6}$**
14. $\frac{1}{5p^8q^{-3}}$ **$\frac{q^3}{5p^8}$**
15. $(a^2b)^0$ **1**
16. $(-2y)^{-4}$ **$\frac{1}{16y^4}$**

Sketch the graph of the exponential function. **17–20. See margin.**

17. $y = 4^x$
18. $y = \left(\frac{1}{2}\right)^x$
19. $y = 3^{-x}$
20. $y = \left(\frac{2}{3}\right)^{-x}$

8.3 DIVISION PROPERTIES OF EXPONENTS

Examples on pp. 463–465

EXAMPLES Using the division properties of exponents.

a. $\frac{6^4}{6^2} = 6^{4-2} = 6^2 = 36$ **Quotient of powers property**

b. $\left(\frac{2}{3}\right)^3 = \frac{2^3}{3^3} = \frac{8}{27}$ **Power of a quotient property**

Evaluate the expression. Write fractions in simplest form.

21. $\frac{3^2}{3^5}$ **$\frac{1}{27}$** **22.** $\frac{5^2}{5^{-2}}$ **625** **23.** $\left(-\frac{4}{9}\right)^2$ **$\frac{16}{81}$** **24.** $\left(\frac{10}{7}\right)^{-1}$ **$\frac{7}{10}$**

Simplify the expression. The simplified expression should have no negative exponents.

25. $\left(\frac{9}{b}\right)^6$ **$\frac{531{,}441}{b^6}$** **26.** $\frac{x^{12}}{x^6}$ **x^6** **27.** $\left(\frac{m^7}{m^4}\right)^2$ **m^6** **28.** $\frac{(p^2)^3}{(p^2)^5}$ **$\frac{1}{p^4}$**

29. $\left(\frac{-9a^2b^2}{3ab}\right)^3$ **$-27a^3b^3$** **30.** $\left(\frac{25a^4b^5}{-5a^2b}\right)^3$ **$-125a^6b^{12}$** **31.** $\frac{32a^4b^{-2}}{2a^3b^3} \cdot \frac{3a^2b^7}{-2a}$ **$-24a^2b^2$** **32.** $\frac{9x^{-3}y^6}{x^4y^{-5}} \cdot \frac{(3x^2y)^{-2}}{xy^3}$ **$\frac{y^6}{x^{12}}$**

33. SALES From 1994 through 1999, the sales for a national book store increased by about the same percent each year. The sales s (in millions of dollars) for year t can be modeled by $s = 1686(1.17)^t$ where $t = 0$ represents 1994. Find the ratio of 1994 sales to 1999 sales. **$\frac{1}{(1.17)^5}$ or about 0.46**

8.4 SCIENTIFIC NOTATION

Examples on pp. 470–472

EXAMPLES Rewriting numbers in decimal form and scientific notation.

a. $1.247 \times 10^2 = 124.7$ **Move decimal point right 2 places.**

b. $1.045 \times 10^{-3} = 0.001045$ **Move decimal point left 3 places.**

c. $79{,}500 = 7.95 \times 10^4$ **Move decimal point left 4 places.**

d. $0.0588 = 5.88 \times 10^{-2}$ **Move decimal point right 2 places.**

Rewrite the number in decimal form.

34. 6.667×10^{-3} **0.006667** **35.** 7.68×10^5 **768,000** **36.** 3.75×10^{-1} **0.375** **37.** 2×10^{-4} **0.0002**

Rewrite the number in scientific notation.

38. 523,000,000 **5.23×10^8** **39.** 0.000679 **6.79×10^{-4}** **40.** 0.0000000233 **2.33×10^{-8}**

41. SPACE TRAVEL Astronaut Shannon W. Lucid holds the United States single-mission, space-flight endurance record. Upon completion of her 1996 mission aboard the Russian Space Station *Mir*, Dr. Lucid had traveled 75,200,000 miles. Write 75,200,000 miles in scientific notation. **7.52×10^7**

42. ASTRONOMY The distance from Earth to the star Alpha Centauri is about 4.07×10^{13} kilometers (km). Light travels at a speed of about 3.0×10^5 km per second. How long does it take light to travel from this star to Earth? **about 1.36×10^8 sec or about 4.3 years**

18.

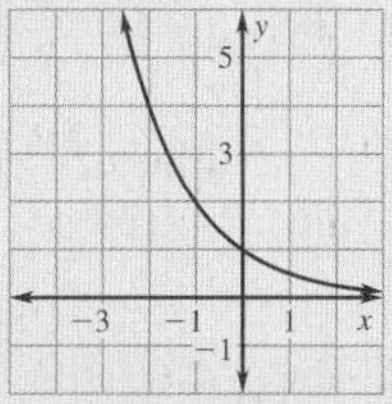

19.

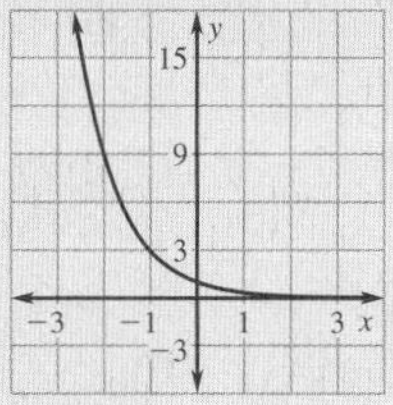

20.

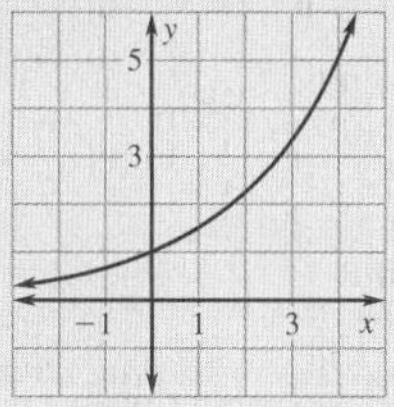

8.5 EXPONENTIAL GROWTH FUNCTIONS

Examples on pp. 477–479

EXAMPLE

You deposited \$1200 in a savings account that pays 9% annual interest compounded yearly. What is the balance after 8 years?

$y = C(1 + r)^t$ **Exponential growth model**

$= 1200(1 + 0.09)^t$ **Substitute 1200 for *C* and 0.09 for *r*.**

$= 1200(1.09)^t$ **Simplify.**

After 8 years, the balance would be

$y = 1200(1.09)^8$ **Substitute 8 for *t*.**

≈ 2391.08 **Use a calculator.**

After 8 years, the balance would be \$2391.08.

FITNESS PROGRAM In Exercises 43 and 44, use the following information. You start a walking program. The first week you walk 2 miles. Over the next 9 weeks, you increase your distance 5% per week.

43. Write an exponential growth model to represent the number of miles w you are walking after x weeks. **$w = 2(1.05)^x$**

44. How far are you walking in the tenth week? **about 3.3 mi**

8.6 EXPONENTIAL DECAY FUNCTIONS

Examples on pp. 484–487

EXAMPLE

In 1995 you bought a 32-inch television for \$600. The television is depreciating at the rate of 8% per year. Write an exponential decay model and estimate the value of the television in 6 years.

$y = C(1 - r)^t$ **Exponential decay model**

$= 600(1 - 0.08)^t$ **Substitute 600 for *C* and 0.08 for *r*.**

$= 600(0.92)^t$ **Simplify.**

After 6 years, the balance would be

$y = 600(0.92)^6$ **Substitute 6 for *t*.**

≈ 363.81 **Use a calculator.**

After 6 years, the television would be worth \$363.81.

TENNIS CLUB In Exercises 45 and 46, use the following information. A tennis club had a declining enrollment from 1993 to 2000. The enrollment in 1993 was 125 people. Each year for 7 years, the enrollment decreased by 3%.

45. Write an exponential decay model to represent enrollment e after x weeks. **$e = 125(0.97)^x$**

46. Estimate the enrollment in 2000. **101 people**

CHAPTER 8

Chapter Test

In Exercises 1–12, simplify the expression. The simplified expression should have no negative exponents.

1. $x^3 \cdot x^4$ **x^7**
2. $a^0 \cdot a^4$ **a^4**
3. $b^2 \cdot b^{-5}$ **$\frac{1}{b^3}$**
4. $5y^{-4}$ **$\frac{5}{y^4}$**
5. $(x^3)^7$ **x^{21}**
6. $(a^{-2})^3$ **$\frac{1}{a^6}$**
7. $\frac{n^3}{n^5}$ **$\frac{1}{n^2}$**
8. $(2b)^3(b^{-4})$ **$\frac{2^3}{b}$, or $\frac{8}{b}$**
9. $(mn)^2 \cdot n^4$ **m^2n^6**
10. $3a^5 \cdot 5a^{-2} \cdot a^3$ **$15a^6$**
11. $\left(\frac{x^3}{xy^4}\right)\left(\frac{y}{x}\right)^5$ **$\frac{y}{x^3}$**
12. $\frac{a^{-1}b^2}{ab} \cdot \frac{a^2b^3}{b^{-2}}$ **b^6**

In Exercises 13–20, evaluate the expression.

13. $5^4 \cdot 5^{-1}$ **125**
14. 4^{-3} **$\frac{1}{64}$**
15. $(425^2)^0$ **1**
16. $\left(\frac{5}{2}\right)^{-2}$ **$\frac{4}{25}$**
17. $\frac{3 \cdot 3^5}{3^4}$ **9**
18. $\left(\frac{3}{4}\right)^3 \cdot 4^2 \cdot 3^0$ **$\frac{27}{4}$**
19. $(5 \cdot 4)^3 \cdot 5^{-2}$ **320**
20. $[(-2)^5]^2$ **1024**

In Exercises 21–24, write the number in decimal form.

21. $4.27 \cdot 10^5$ **427,000**
22. $6.283 \cdot 10^{-9}$ **0.000000006283**
23. 4.56×10^{10} **45,600,000,000**
24. 5×10^{-12} **0.000000000005**

In Exercises 25–28, write the number in scientific notation.

25. 9,875,000 **9.875×10^6**
26. 0.00125 **1.25×10^{-3}**
27. 6,557,000,000 **6.557×10^9**
28. 0.0000000317 **3.17×10^{-8}**

In Exercises 29–31, sketch a graph of the equation. **29–31. See margin.**

29. $y = 2^x$
30. $y = \left(\frac{1}{3}\right)^x$
31. $y = 10(1.4)^x$

32. GEOMETRY CONNECTION The volume of a cube is given by $V = s^3$, where s is the length of a side. The cube has a side of length $3a$. What is the volume of the cube if $a = 2$? **216**

33. SAVINGS ACCOUNT You started a savings account in 1996. The balance A is given by $A = 400(1.1)^t$, where $t = 0$ represents the year 1996. What is the balance in 2000? in 2003? **\$585.64; \$779.49**

34. SALES In 1996, you started your own business. In the first year, your sales totaled \$88,500. Then each year for the next 4 years, your sales increased by 20%. Write an exponential growth model to represent this situation. Then estimate your sales in 2001. **$y = 88{,}500(1.2)^x$; \$220,216**

35. RADIOISOTOPES The amount of time it takes for a radioactive substance to reduce to half of its original amount is called its half-life. The half-life of carbon 11 (^{11}C) is 20 minutes. If you start with 16 grams of ^{11}C, the number of grams remaining after h half-life periods would be $W = 16(0.5)^h$. Copy and complete the table and use the results to sketch the graph. **16; 8; 4; 2; 1; See margin for graph.**

Half-life periods, h	0	1	2	3	4
Grams remaining, W	?	?	?	?	?

ADDITIONAL RESOURCES

- ***Chapter 8 Resource Book***
  Chapter Test (3 levels) (p. 100)
  SAT/ACT Chapter Test (p. 106)
  Alternative Assessment (p. 107)
- ***Test and Practice Generator***

29.

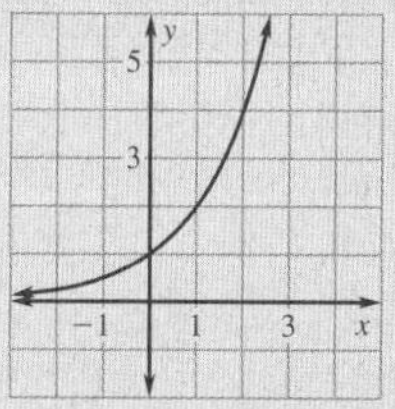

30.

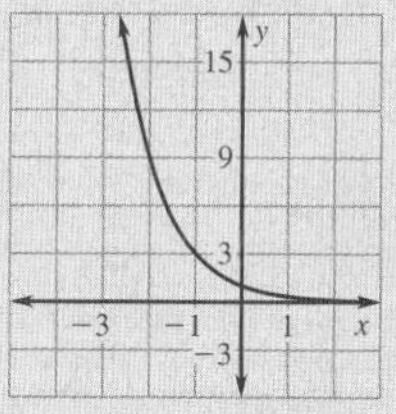

31.

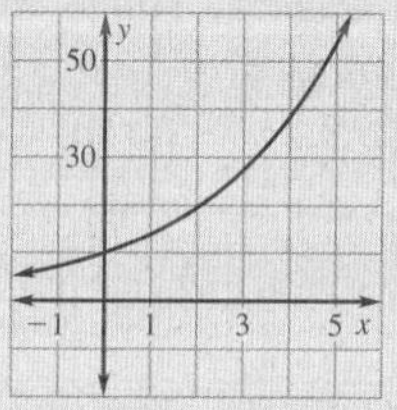

35.

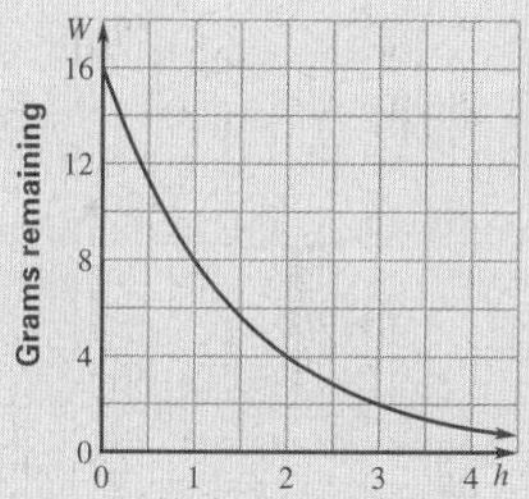

ADDITIONAL RESOURCES

- ***Chapter 8 Resource Book***
  Chapter Test (3 levels) (p. 100)
  SAT/ACT Chapter Test (p. 106)
  Alternative Assessment (p. 107)
- ***Test and Practice Generator***

CHAPTER 8

Chapter Standardized Test

TEST-TAKING STRATEGY Be aware of how much time you have left, but keep focused on your work.

1. MULTIPLE CHOICE Simplify $7^4 \cdot 7^7$. Write the answer as a power. **B**

Ⓐ 7^3 Ⓑ 7^{11} Ⓒ 7^{28}

Ⓓ 49^{11} Ⓔ 49^{28}

2. MULTIPLE CHOICE Evaluate $[(a + 1)^2]^2 \cdot a^3$ when $a = 2$. **E**

Ⓐ 72 Ⓑ 128 Ⓒ 136

Ⓓ 200 Ⓔ 648

3. MULTIPLE CHOICE Evaluate $(5^{-3})^2$. **D**

Ⓐ -3125 Ⓑ $-\frac{1}{15,625}$

Ⓒ $\frac{1}{3125}$ Ⓓ $\frac{1}{15,625}$

Ⓔ 15,625

4. MULTIPLE CHOICE Evaluate $-8^0 \cdot 2^x \cdot 10^y$ when $x = -2$ and $y = -3$. **A**

Ⓐ $-\frac{1}{4000}$ Ⓑ $-\frac{1}{500}$

Ⓒ $\frac{1}{500}$ Ⓓ $\frac{1}{4000}$

Ⓔ 4000

5. MULTIPLE CHOICE Simplify $\frac{4x^2y^2}{4xy} \cdot \frac{8xy^3}{4y}$. **A**

Ⓐ $2x^2y^3$ Ⓑ $4x^2y$

Ⓒ $2xy^2$ Ⓓ $2x^2y^4$

Ⓔ $2xy^3$

6. MULTIPLE CHOICE Which expression simplifies to x^3? **C**

Ⓐ $\frac{x^2}{x^5}$ Ⓑ $\frac{x^5}{x^{-2}}$

Ⓒ $\frac{x^5}{x^2}$ Ⓓ $x^5 \cdot x^2$

Ⓔ $x^5 - x^2$

7. MULTIPLE CHOICE Which of the following numbers *is not* written in scientific notation? **C**

Ⓐ 8.62×10^4 Ⓑ 2.12×10^{-12}

Ⓒ 21.2×10^{-5} Ⓓ 9.9132×10^{-1}

Ⓔ 2.0001×10^{-3}

8. MULTIPLE CHOICE Rewrite 3.6×10^{-6} in decimal form. **E**

Ⓐ 0.000036 Ⓑ 3,600,000

Ⓒ 0.00000036 Ⓓ 36,000,000

Ⓔ 0.0000036

9. MULTIPLE CHOICE Evaluate the product $(6.2 \times 10^4) \cdot (2.4 \times 10^5)$. Write the result in scientific notation. **B**

Ⓐ 1.488×10^8 Ⓑ 1.488×10^{10}

Ⓒ 14.88×10^1 Ⓓ 14.88×10^{10}

Ⓔ 14.88×10^{20}

10. MULTIPLE CHOICE Evaluate $\frac{(2.3622 \times 10^4)}{(3.81 \times 10^{-3})}$. Write the result in scientific notation. **A**

Ⓐ 6.2×10^6 Ⓑ 6.2×10^8

Ⓒ 0.62×10^6 Ⓓ 6.2×10^1

Ⓔ 6.2×10^{-12}

11. MULTIPLE CHOICE Which model best represents the growth curve shown below? **A**

Ⓐ $y = 50(1.4)^t$

Ⓑ $y = 100(1.4)^t$

Ⓒ $y = 50(1.12)^t$

Ⓓ $y = 50(1.4)^{-t}$

Ⓔ $y = 200(1.08)^t$

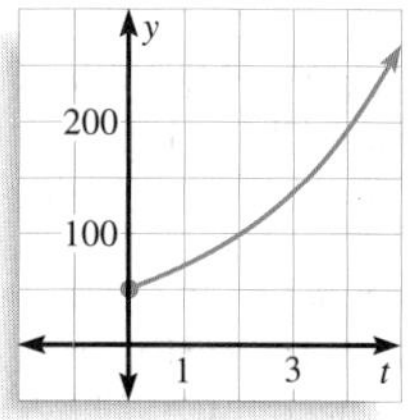

12. MULTIPLE CHOICE You deposit \$450 into a savings account that pays 6% interest compounded yearly. How much money is in the account after 6 years? Assume you make no more deposits or withdrawals. **C**

Ⓐ \$477.00 Ⓑ \$602.20

Ⓒ \$638.33 Ⓓ \$676.63

Ⓔ \$693.33

13. MULTIPLE CHOICE In 1994 you bought a rare stamp for \$500 that you expect to increase in value 12% each year for the next 15 years. Write an exponential growth model and estimate the value of the stamp in 2002. **D**

Ⓐ \$881.17 Ⓑ \$986.91 Ⓒ \$1557.92 Ⓓ \$1237.98 Ⓔ \$2000

QUANTITATIVE COMPARISON In Exercises 14–16, evaluate each function. Then choose the statement below that is true about the given values of *y*.

A. The value of y in column A is greater.

B. The value of y in column B is greater.

C. The two values of y are equal.

D. The relationship cannot be determined from the given information.

		Column A	Column B	
14.	When $x = 3$,	$y = 3x$	$y = 3^x$	**B**
15.	When $x = -2$,	$y = 3^x$	$y = \left(\frac{1}{3}\right)^x$	**B**
16.	When $x = 1$,	$y = 3^{-x}$	$y = \left(\frac{1}{3}\right)^x$	**C**

17. MULTIPLE CHOICE The concentration of an allergy medication in a person's bloodstream in nanograms per milliliter (ng/mL) can be modeled by the equation $y = 263(0.92)^t$, where t represents the number of hours since the medication was taken. What is the concentration of the medication remaining in the person's bloodstream after 4 hours? **D**

Ⓐ 263 ng/mL Ⓑ 222 ng/mL Ⓒ 205 ng/mL Ⓓ 188 ng/mL Ⓔ 170 ng/mL

18. MULTIPLE CHOICE A business had a profit of \$142,000 in 1994. Then its profit decreased by 8% each year for the next 6 years. Which exponential decay model would you use to find how much the business earned in the year 2000? Let E represent the earnings and let t represent the year. **C**

Ⓐ $E = 142{,}000(0.08)^t$ Ⓑ $E = 142{,}000(0.96)^t$ Ⓒ $E = 142{,}000(0.92)^t$

Ⓓ $t = 142{,}000(0.08)^E$ Ⓔ $E = 0.92(142{,}000)^t$

19. MULTIPLE CHOICE Which models below are exponential decay models? **E**

I $y = 1.25^t$ **II** $y = 0.97^t$ **III** $y = \left(\frac{4}{3}\right)^t$ **IV** $y = \left(\frac{2}{3}\right)^t$

Ⓐ I and II Ⓑ III and IV Ⓒ II and III Ⓓ I and III Ⓔ II and IV

20. MULTI-STEP PROBLEM The population in one midwestern town was tracked over several years. Based on the data for the town, population experts determined that the population, in thousands of people, could be represented by the expression $2(1.175)^t$, where t is the number of years from now.

a. What is the population this year? **2000**

b. What is the estimated population 5 years from now? **4479**

c. What was the population 2 years ago? **1449**

d. *Writing* What advice would you give to the city planners who are trying to decide whether or not to build freeways? Include in your advice a population prediction for 10 years and 20 years from now. **Answers may vary.**

CHAPTER 9

PLANNING THE CHAPTER

Quadratic Equations and Functions

LESSON	GOALS	NCTM	ITED	SAT9	Terra-Nova	Local
9.1 *pp. 503–510*	GOAL 1 Evaluate and approximate square roots. GOAL 2 Solve a quadratic equation by finding square roots.	1, 2	APE, MCE	5	52	
9.2 *pp. 511–516*	GOAL 1 Use properties of radicals to simplify radicals. GOAL 2 Use quadratic equations to model real-life problems.	1, 2, 6, 9, 10	APE, MCE	5	17, 18, 52	
9.3 *pp. 517–525*	CONCEPT ACTIVITY: 9.3 *Investigate graphs of quadratic functions.* GOAL 1 Sketch the graph of a quadratic function. GOAL 2 Use quadratic models in real-life settings. TECHNOLOGY ACTIVITY: 9.3 *Graph quadratic curves of best fit on a graphing calculator.*	1, 2, 3, 6, 9, 10	MCE, MIG, IIG, RQGE, APE		14, 16, 18	
9.4 *pp. 526–532*	GOAL 1 Solve a quadratic equation graphically. GOAL 2 Use quadratic models in real-life settings. TECHNOLOGY ACTIVITY: 9.4 *Solve quadratic equations by graphing on a graphing calculator.*	1, 2, 3, 6, 9, 10	MCE, MIG, IIG, RQGE, APE		14, 18, 52	
9.5 *pp. 533–539*	GOAL 1 Use the quadratic formula to solve a quadratic equation. GOAL 2 Use quadratic models in real-life settings. TECHNOLOGY ACTIVITY: 9.5 *Write a program for the quadratic formula on a graphing calculator.*	1, 2, 6, 9, 10	MCE, APE		18, 52	
9.6 *pp. 540–547*	CONCEPT ACTIVITY: 9.6 *Investigate applications of the discriminant.* GOAL 1 Use the discriminant to find the number of solutions of a quadratic equation. GOAL 2 Apply the discriminant to solve real-life problems.	1, 2, 6, 9, 10	MCE, APE		17, 52	
9.7 *pp. 548–553*	GOAL 1 Sketch the graph of a quadratic inequality. GOAL 2 Use quadratic inequalities as real-life models.	1, 2, 3, 6, 9, 10	MCE, MIG, IIG, RQGE, APE	6	14, 16, 18	
9.8 *pp. 554–560*	GOAL 1 Choose a model that best fits a collection of data. GOAL 2 Use models in real-life settings.	1, 2, 5, 6, 9, 10	MCS	21	18	

RESOURCES

Chapter Resource Booklets

Chapter Support

Tips for New Teachers	p. 1
Parent Guide for Student Success	p. 3
Prerequisite Skills Review	p. 5
Strategies for Reading Mathematics	p. 7

Lesson Support

Lesson Support	9.1	9.2	9.3	9.4	9.5	9.6	9.7	9.8
Lesson Plans (regular and block)	p. 9	p. 22	p. 34	p. 52	p. 67	p. 81	p. 96	p. 109
Warm-Up Exercises and Daily Quiz	p. 11	p. 24	p. 36	p. 54	p. 69	p. 83	p. 98	p. 111
Activity Support Masters						p. 84		
Lesson Openers	p. 12	p. 25	p. 37	p. 55	p. 70	p. 85	p. 99	p. 112
Graphing Calculator Activities & Keystrokes	p. 13		p. 38	p. 56	p. 71		p. 100	p. 113
Practice (3 levels)	p. 14	p. 26	p. 43	p. 59	p. 73	p. 86	p. 101	p. 116
Reteaching with Practice	p. 17	p. 29	p. 46	p. 62	p. 76	p. 89	p. 104	p. 119
Quick Catch-Up for Absent Students	p. 19	p. 31	p. 48	p. 64	p. 78	p. 91	p. 106	p. 121
Cooperative Learning Activities								p. 122
Interdisciplinary Applications	p. 20			p. 65	p. 79			p. 123
Real-Life Applications		p. 32	p. 49			p. 92	p. 107	
Math & History Applications						p. 93		
Challenge: Skills and Applications	p. 21	p. 33	p. 50	p. 66	p. 80	p. 94	p. 108	p. 124

Review and Assessment

Quizzes	pp. 51, 95
Chapter Review Games and Activities	p. 125
Chapter Test (3 levels)	pp. 126–131
SAT/ACT Chapter Test	p. 132
Alternative Assessment with Math Journal	p. 133
Project with Rubric	p. 135
Cumulative Review	p. 137
Resource Book Answers	p. A1

Transparencies

Transparencies	9.1	9.2	9.3	9.4	9.5	9.6	9.7	9.8
Warm-Up Exercises and Daily Quiz	p. 67	p. 68	p. 69	p. 70	p. 71	p. 72	p. 73	p. 74
Alternative Lesson Opener Transparencies	p. 58	p. 59	p. 60	p. 61	p. 62	p. 63	p. 64	p. 65
Examples/Standardized Test Practice	✓	✓	✓	✓	✓	✓	✓	✓
Answer Transparencies	✓	✓	✓	✓	✓	✓	✓	✓

Technology

- EasyPlanner CD
- EasyPlanner Plus Online
- Classzone.com
- Personal Student Tutor CD
- Test and Practice Generator CD
- Instant Replay: Video Review Games
- Electronic Lesson Presentations
- Chapter Audio Summaries CD

Additional Resources

- Basic Skills Workbook
- Explorations and Projects
- Worked-Out Solution Key
- Data Analysis Sourcebook
- Spanish Resources
- Standardized Test Practice Workbook
- Practice Workbook with Examples
- Notetaking Guide

CHAPTER 9

PACING THE CHAPTER

Resource Key
- STUDENT EDITION
- CHAPTER 9 RESOURCE BOOK
- TEACHER'S EDITION

REGULAR SCHEDULE

Day 1

9.1

STARTING OPTIONS
- Prereq. Skills Review
- Strategies for Reading
- Warm-Up or Daily Quiz

TEACHING OPTIONS
- Motivating the Lesson
- Les. Opener (Appl.)
- Graphing Calc. Activity
- Examples 1–7
- Closure Question
- Guided Practice Exs.

APPLY/HOMEWORK
- See Assignment Guide.
- See the CRB: Practice, Reteach, Apply, Extend

ASSESSMENT OPTIONS
- Checkpoint Exercises
- Daily Quiz (9.1)
- Stand. Test Practice

Day 2

9.2

STARTING OPTIONS
- Homework Check
- Warm-Up or Daily Quiz

TEACHING OPTIONS
- Motivating the Lesson
- Les. Opener (Activity)
- Examples 1–3
- Closure Question
- Guided Practice Exs.

APPLY/HOMEWORK
- See Assignment Guide.
- See the CRB: Practice, Reteach, Apply, Extend

ASSESSMENT OPTIONS
- Checkpoint Exercises
- Daily Quiz (9.2)
- Stand. Test Practice

Day 3

9.3

STARTING OPTIONS
- Homework Check
- Warm-Up or Daily Quiz

TEACHING OPTIONS
- Motivating the Lesson
- Concept Activity
- Les. Opener (Calculator)
- Graphing Calc. Activity
- Examples 1–2
- Guided Practice Exs. 1–19

APPLY/HOMEWORK
- See Assignment Guide.
- See the CRB: Practice, Reteach, Apply, Extend

ASSESSMENT OPTIONS
- Checkpoint Exercises, p. 519

Day 4

9.3 *(cont.)*

STARTING OPTIONS
- Homework Check

TEACHING OPTIONS
- Example 3
- Technology Activity
- Closure Question
- Guided Practice Ex. 20

APPLY/HOMEWORK
- See Assignment Guide.
- See the CRB: Practice, Reteach, Apply, Extend

ASSESSMENT OPTIONS
- Checkpoint Exercises, p. 520
- Daily Quiz (9.3)
- Stand. Test Practice
- Quiz (9.1–9.3)

Day 5

9.4

STARTING OPTIONS
- Homework Check
- Warm-Up or Daily Quiz

TEACHING OPTIONS
- Motivating the Lesson
- Les. Opener (Visual)
- Graphing Calc. Activity
- Examples 1–4
- Technology Activity
- Closure Question
- Guided Practice Exs.

APPLY/HOMEWORK
- See Assignment Guide.
- See the CRB: Practice, Reteach, Apply, Extend

ASSESSMENT OPTIONS
- Checkpoint Exercises
- Daily Quiz (9.4)
- Stand. Test Practice

Day 6

9.5

STARTING OPTIONS
- Homework Check
- Warm-Up or Daily Quiz

TEACHING OPTIONS
- Motivating the Lesson
- Les. Opener (Activity)
- Graphing Calc. Activity
- Examples 1–2
- Guided Practice Exs. 1–15

APPLY/HOMEWORK
- See Assignment Guide.
- See the CRB: Practice, Reteach, Apply, Extend

ASSESSMENT OPTIONS
- Checkpoint Exercises, p. 534

Day 9

9.6 *(cont.)*

STARTING OPTIONS
- Homework Check

TEACHING OPTIONS
- Example 4
- Closure Question

APPLY/HOMEWORK
- See Assignment Guide.
- See the CRB: Practice, Reteach, Apply, Extend

ASSESSMENT OPTIONS
- Checkpoint Exercises, p. 543
- Daily Quiz (9.6)
- Stand. Test Practice
- Quiz (9.4–9.6)

Day 10

9.7

STARTING OPTIONS
- Homework Check
- Warm-Up or Daily Quiz

TEACHING OPTIONS
- Les. Opener (Application)
- Graphing Calc. Activity
- Examples 1–4
- Closure Question
- Guided Practice Exs.

APPLY/HOMEWORK
- See Assignment Guide.
- See the CRB: Practice, Reteach, Apply, Extend

ASSESSMENT OPTIONS
- Checkpoint Exercises
- Daily Quiz (9.7)
- Stand. Test Practice

Day 11

9.8

STARTING OPTIONS
- Homework Check
- Warm-Up or Daily Quiz

TEACHING OPTIONS
- Les. Opener (Application)
- Graphing Calc. Activity
- Examples 1–2
- Guided Practice Exs. 1–8

APPLY/HOMEWORK
- See Assignment Guide.
- See the CRB: Practice, Reteach, Apply, Extend

ASSESSMENT OPTIONS
- Checkpoint Exercises, p. 555

Day 12

9.8 *(cont.)*

STARTING OPTIONS
- Homework Check

TEACHING OPTIONS
- Example 3
- Closure Question

APPLY/HOMEWORK
- See Assignment Guide.
- See the CRB: Practice, Reteach, Apply, Extend

ASSESSMENT OPTIONS
- Checkpoint Exercises, p. 556
- Daily Quiz (9.8)
- Stand. Test Practice
- Quiz (9.7–9.8)

Day 13

Review

DAY 13 START OPTIONS
- Homework Check

REVIEWING OPTIONS
- Chapter 9 Summary
- Chapter 9 Review
- Chapter Review Games and Activities

APPLY/HOMEWORK
- Chapter 9 Test (practice)
- Ch. Standardized Test (practice)

Day 14

Assess

DAY 14 START OPTIONS
- Homework Check

ASSESSMENT OPTIONS
- Chapter 9 Test
- SAT/ACT Ch. 9 Test
- Alternative Assessment

APPLY/HOMEWORK
- Skill Review, p. 574

Day 7

9.5 *(cont.)*

Starting Options
- Homework Check

Teaching Options
- Examples 3–4
- Technology Activity
- Closure Question
- Guided Practice Exs. 16–22

Apply/Homework
- See Assignment Guide.
- See the CRB: Practice, Reteach, Apply, Extend

Assessment Options
- Checkpoint Exercises, pp. 534–535
- Daily Quiz (9.5)
- Stand. Test Practice

Day 8

9.6

Starting Options
- Homework Check
- Warm-Up or Daily Quiz

Teaching Options
- Motivating the Lesson
- Concept Act. & Wksht.
- Les. Opener (Activity)
- Examples 1–3
- Guided Practice Exs. 1–8

Apply/Homework
- See Assignment Guide.
- See the CRB: Practice, Reteach, Apply, Extend

Assessment Options
- Checkpoint Exercises, p. 542

BLOCK SCHEDULE

Day 1

Assess & 9.1

(Day 1 = Ch. 8 Day 7)

Assessment Options
- Chapter 8 Test
- SAT/ACT Ch. 8 Test
- Alternative Assessment

Ch. 9 Start Options
- Skill Review, p. 502
- Prereq. Skills Review
- Strategies for Reading

Teaching 9.1 Options
- Warm-Up (Les. 9.1)
- Motivating the Lesson
- Les. Opener (Appl.)
- Graphing Calc. Activity
- Examples 1–7
- Closure Question
- Guided Practice Exs.

Apply/Homework
- See Assignment Guide.
- See the CRB: Practice, Reteach, Apply, Extend

Assessment Options
- Checkpoint Exercises
- Daily Quiz (Les. 9.1)
- Stand. Test Practice

Day 2

9.2 & 9.3

Day 2 Start Options
- Homework Check
- W-Up (Les. 9.2) or D. Quiz (Les. 9.1)

Teaching 9.2 Options
- Motivating the Lesson
- Les. Opener (Activity)
- Examples 1–3
- Closure Question
- Guided Practice Exs.

Beginning 9.3 Options
- Warm-Up (Les. 9.3)
- Motivating the Lesson
- Concept Activity
- Les. Opener (Calc.)
- Graphing Calc. Activity
- Examples 1–2
- Guided Prac. Exs. 1–19

Apply/Homework
- See Assignment Guide.
- See the CRB: Practice, Reteach, Apply, Extend

Assessment Options
- Checkpoint Exercises
- Daily Quiz (Les. 9.2)

Day 3

9.3 & 9.4

Day 3 Start Options
- Homework Check
- Daily Quiz (Les. 9.2)

Finishing 9.3 Options
- Example 3
- Technology Activity
- Closure Question
- Guided Practice Ex. 20

Teaching 9.4 Options
- Warm-Up (Les. 9.4)
- Motivating the Lesson
- Les. Opener (Visual)
- Graphing Calc. Activity
- Examples 1–4
- Technology Activity
- Closure Question
- Guided Practice Exs.

Apply/Homework
- See Assignment Guide.
- See the CRB: Practice, Reteach, Apply, Extend

Assessment Options
- Checkpoint Exercises
- Daily Quiz (Les. 9.3, 9.4)
- Quiz (9.1–9.3)

Day 4

9.5

Day 4 Start Options
- Homework Check
- Warm-Up or Daily Quiz

Teaching 9.5 Options
- Motivating the Lesson
- Les. Opener (Activity)
- Graphing Calc. Activity
- Examples 1–4
- Technology Activity
- Closure Question
- Guided Practice Exs.

Apply/Homework
- See Assignment Guide.
- See the CRB: Practice, Reteach, Apply, Extend

Assessment Options
- Checkpoint Exercises,
- Daily Quiz (Les. 9.5)
- Stand. Test Practice

Day 5

9.6

Day 5 Start Options
- Homework Check
- Warm-Up or Daily Quiz

Teaching 9.6 Options
- Motivating the Lesson
- Concept Act. & Wksht.
- Les. Opener (Activity)
- Examples 1–4
- Closure Question
- Guided Practice Exs.

Apply/Homework
- See Assignment Guide.
- See the CRB: Practice, Reteach, Apply, Extend

Assessment Options
- Checkpoint Exercises
- Daily Quiz (Les. 9.6)
- Stand. Test Practice
- Quiz (9.4–9.6)

Day 6

9.7 & 9.8

Day 6 Start Options
- Homework Check
- W-Up (Les. 9.7) or D. Quiz (Les. 9.6)

Teaching 9.7 Options
- Les. Opener (Appl.)
- Graphing Calc. Activity
- Examples 1–4
- Closure Question
- Guided Practice Exs.

Beginning 9.8 Options
- Warm-Up (Les. 9.8)
- Les. Opener (Appl.)
- Graphing Calc. Activity
- Examples 1–2
- Guided Practice Exs. 1–8

Apply/Homework
- See Assignment Guide.
- See the CRB: Practice, Reteach, Apply, Extend

Assessment Options
- Checkpoint Exercises
- Daily Quiz (Les. 9.7)
- Stand. Test Prac. (9.7)

Day 7

9.8 & Review

Day 7 Start Options
- Homework Check
- Daily Quiz (Les. 9.7)

Finishing 9.8 Options
- Example 3
- Closure Question

Reviewing Options
- Chapter 9 Summary
- Chapter 9 Review
- Chapter Review Games and Activities

Apply/Homework
- See Assignment Guide.
- See the CRB: Practice, Reteach, Apply, Extend
- Chapter 9 Test (practice)
- Ch. Standardized Test (practice)

Assessment Options
- Checkpoint Exercises
- Daily Quiz (Les. 9.8)
- Stand. Test Practice
- Quiz (9.7–9.8)

Day 8

Assess & 10.1

(Day 8 = Ch. 10 Day 1)

Assessment Options
- Chapter 9 Test
- SAT/ACT Ch. 9 Test
- Alternative Assessment

Ch. 10 Start Options
- Skill Review, p. 574
- Prereq. Skills Review
- Strategies for Reading

Teaching 10.1 Options
- Warm-Up (Les. 10.1)
- Motivating the Lesson
- Concept Act. & Wksht.
- Les. Opener (Appl.)
- Graphing Calc. Activity
- Examples 1–6
- Technology Activity
- Closure Question
- Guided Practice Exs.

Apply/Homework
- See Assignment Guide.
- See the CRB: Practice, Reteach, Apply, Extend

Assessment Options
- Checkpoint Exercises
- Daily Quiz (Les. 10.1)

MEETING INDIVIDUAL NEEDS

BEFORE THE CHAPTER

The *Chapter 9 Resource Book* has the following materials to distribute and use before the chapter:

- **Parent Guide for Student Success (pictured below)**
- **Prerequisite Skills Review**
- **Strategies for Reading Mathematics**

PARENT GUIDE *Pages 3–4*

CHAPTER 9 NAME ______ DATE ______

Parent Guide for Student Success

For use with Chapter 9: Quadratic Equations and Functions

Chapter Overview One way that you can help your student succeed in Chapter 9 is by discussing the lesson goals in the chart below. When a lesson is completed, ask your student to interpret the lesson goals for you and to explain how the mathematics of the lesson relates to one of the key applications listed in the chart.

Lesson Title	*Lesson Goals*	*Key Applications*
9.1: Solving Quadratic Equations by Finding Square Roots	Evaluate and approximate square roots. Solve a quadratic equation by finding square roots.	• Engineering • Computer Sales • Minerals
9.2: Simplifying Radicals	Use properties of radicals to simplify radicals. Use quadratic equations to model real-life problems.	• Boat Racing • Body Surface Area • Tsunami
9.3: Graphing Quadratic Functions	Sketch the graph of a quadratic function. Use quadratic models in real-life settings.	• Track and Field • Water Arc • Gold Production
9.4: Solving Quadratic Equations by Graphing	Solve a quadratic equation graphically. Use quadratic models in real-life settings.	• Shot Put • Swiss Cheese • RV Sales
9.5: Solving Quadratic Equations by the Quadratic Formula	Use the quadratic formula to solve a quadratic equation. Use quadratic models in real-life settings.	• Balloon Competition • Peregrine Falcon • Baseball
9.6: Applications of the Discriminant	Use the discriminant to find the number of solutions of a quadratic equation. Apply the discriminant to solve real-life problems.	• Camping • Government Payroll • Financial Analysis
9.7: Graphing Quadratic Inequalities	Sketch the graph of a quadratic inequality. Use quadratic inequalities as real-life models.	• Bridge Building • Diamonds • Milk Consumption
9.8: Comparing Linear, Exponential, and Quadratic Models	Choose a model that best fits a collection of data and use models in real-life settings.	• Chambered Nautilus • Pendulums • NHL Attendance

Study Strategy

Explaining Ideas is the study strategy featured in Chapter 9 (see page 502). Having your student try to explain material to you can provide an opportunity for him or her to pull ideas together, to identify and overcome misunderstandings, and to review and prepare for tests.

Chapter Support

PARENT GUIDE FOR STUDENT SUCCESS The first page summarizes the content of Chapter 9. Parents are encouraged to have their students explain how the material relates to key applications in the chapter, such as baseball. The second page (not shown) provides exercises and an activity that parents can do with their students. In the activity, parents and students design a garden with a sidewalk border to work with areas and quadratic equations.

DURING EACH LESSON

The *Chapter 9 Resource Book* has the following alternatives for introducing the lesson:

- **Lesson Openers (pictured below)**
- **Graphing Calculator Activities with Keystrokes**

LESSON OPENER *Page 112*

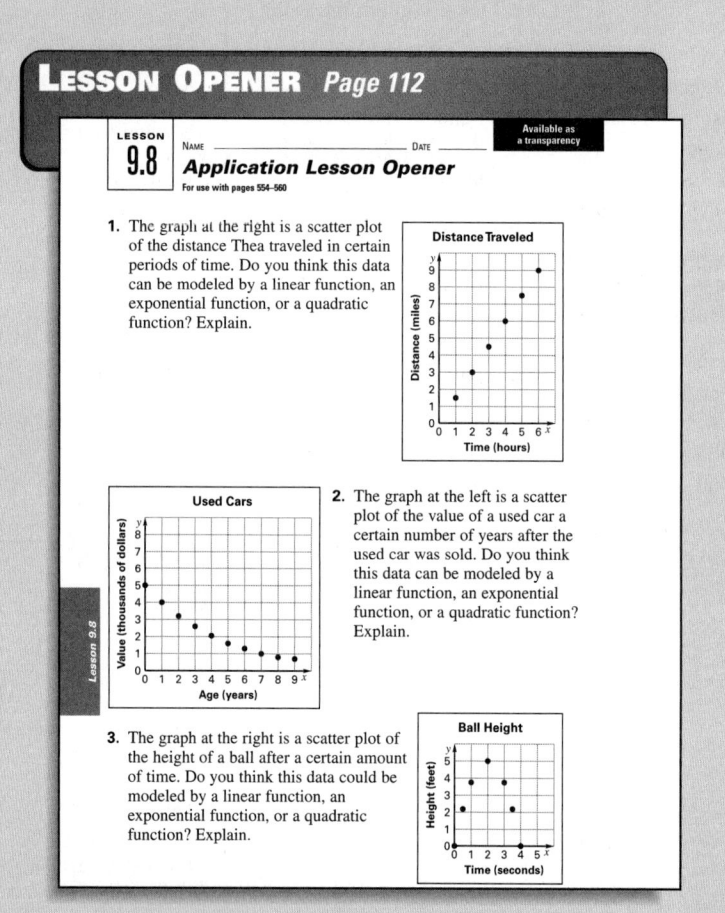

LESSON 9.8 NAME ______ DATE ______ Available as a transparency

Application Lesson Opener

For use with pages 554–560

1. The graph at the right is a scatter plot of the distance Thea traveled in certain periods of time. Do you think this data can be modeled by a linear function, an exponential function, or a quadratic function? Explain.

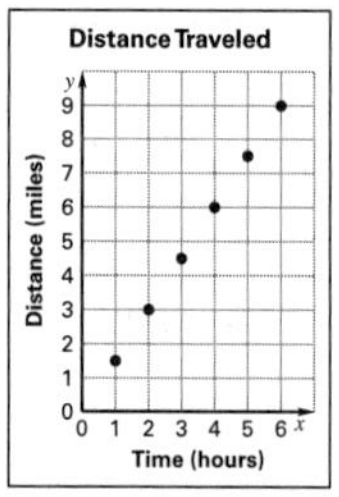

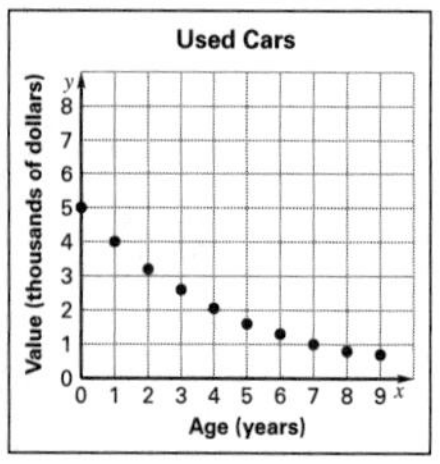

2. The graph at the left is a scatter plot of the value of a used car a certain number of years after the used car was sold. Do you think this data can be modeled by a linear function, an exponential function, or a quadratic function? Explain.

3. The graph at the right is a scatter plot of the height of a ball after a certain amount of time. Do you think this data could be modeled by a linear function, an exponential function, or a quadratic function? Explain.

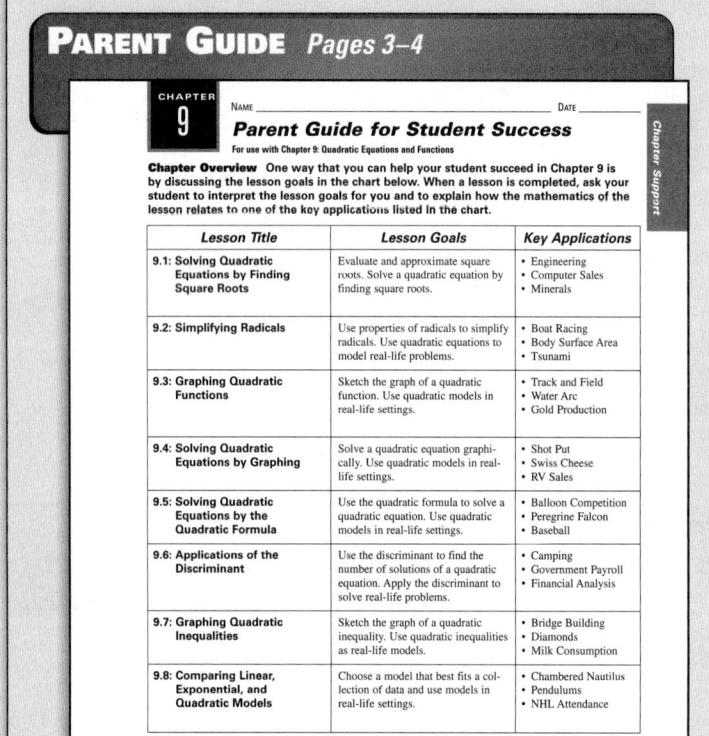

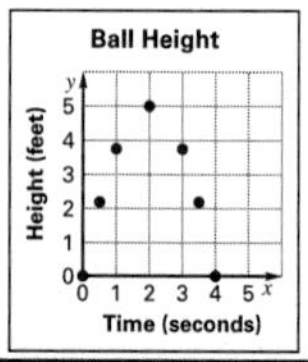

Lesson 9.8

APPLICATION LESSON OPENER This Lesson Opener provides an alternative way to start Lesson 9.8 in the form of a real-life application. Students see how a scatter plot relates to distance traveled, the value of a used car, and the height of a ball.

TECHNOLOGY RESOURCE

Teachers can use the Electronic Lesson Presentations CD-ROM to present Lesson 9.4 using computer animation to step through the Examples.

FOLLOWING EACH LESSON

The *Chapter 9 Resource Book* has the following alternatives for introducing the lesson:

- **Practice (3 levels)**
- **Reteaching with Practice**
- **Quick Catch-Up for Absent Students**
- **Interdisciplinary Applications**
- **Real-Life Applications (pictured below)**

APPLICATION *Page 49*

LESSON 9.3

NAME ______________________ DATE __________

Real-Life Application: When Will I Ever Use This?

For use with pages 518–524

Ballet Recital

Your ballet class has decided to have a recital during December. The most popular choice of ballets voted on by class members is *The Nutcracker* by Pyotr (Peter) Tchaikovsky.

The music, completed in 1892, was commissioned by the Imperial Opera Directorate. Tchaikovsky actually disliked the subject matter of the ballet, which is a story based on a nutcracker and a mouse king. In spite of this, Tchaikovsky accepted the commission and the first performance of the ballet was on December 18, 1892. Critics did not like the finished work, but generations of ballet audiences have since disagreed with them. *The Nutcracker* is still one the most popular holiday ballets.

Your class chooses two primary dancers for the lead roles, one male and one female. One of the male dancer's leaps can be modeled by the equation $h = -8t^2 + 8t$ where h is height in feet and t is time in seconds. One of the female dancer's leaps can be modeled by the equation $h = -10t^2 + 8t$.

1. Estimate the maximum height reached by the male dancer.
2. How many seconds does it take the male dancer to reach his maximum jump height? Round to the nearest tenth.
3. Estimate the maximum height reached by the female dancer. Round to the nearest tenth.
4. How many seconds does it take the female dancer to reach her maximum height? Round to the nearest tenth.
5. Sketch the graph of the equation for the male dancer.

Lesson 9.3

REAL-LIFE APPLICATION When students ask "When will I ever use quadratic models?," you can have them work through this Real-Life Application in which they apply the content of Lesson 9.3 to determine the paths of leaps by ballet dancers.

ASSESSING THE CHAPTER

The *Chapter 9 Resource Book* has the following review and assessment materials:

- **Quizzes**
- **Chapter Review Games and Activities**
- **Chapter Test (3 levels)**
- **SAT/ACT Chapter Test**
- **Alternative Assessment and Math Journal (below)**
- **Project with Rubric**
- **Cumulative Review**

ALTERNATIVE ASSESSMENT *Pages 133–134*

CHAPTER 9

NAME ______________________ DATE __________

Alternative Assessment and Math Journal

For use after Chapter 9

JOURNAL 1. In this chapter, you have learned how to solve vertical motion problems using two models.

a. Create three different word problems: one in which an object is dropped, one in which an object is thrown down, and one where an object is thrown upward. At least two of the three problems should request the solver to find the time traveled in the air. Be sure the objects are thrown in a vertical motion.

b. Solve each problem, showing all work. As part of your solution, explain how you know what number to substitute in for each variable.

MULTI-STEP PROBLEM 2. During testing, it was determined that the horsepower for an engine is represented by $H = -59.3r^2 + 1041r - 3676$,where H represents the horsepower and r represents revolutions per minute (rev/min) in thousands. (Horsepower is a unit for measuring the power of an engine and other devices.)

For problems a–d, round answers to the nearest whole number.

a. Predict what the graph will look like. Include what shape the graph will be and explain how you know. What does the leading coefficient tell you about the graph?

b. Graph the engine's horsepower equation from 7200 rev/min to 9600 rev/min. (Hint: rev/min is in thousands). Sketch the curve and label your axes. Use the curve to explain how revolutions per minute affects horse power.

c. Find the horsepower at 8000 rev/min. Explain how you determined your answer.

d. Estimate the rev/min at which the engine produces 876 horsepower. At how many points does this occur? Describe, in two or three complete sentences, how you could solve this problem both by using your calculator and algebraically.

3. After some adjustments to the engine, another test is run. The engine now develops a maximum of 902 horsepower. What is the percentage improvement in horsepower? Explain how you got your answer.

Review and Assess

ALTERNATIVE ASSESSMENT WITH RUBRIC AND MATH JOURNAL The journal exercise asks students to demonstrate their understanding of vertical motion problems in writing. The Multi-Step Problem has students pull together a variety of concepts from Chapter 9 to solve a problem about horsepower and revolutions per minute of a car engine. Answers and a scoring rubric are provided on a separate sheet.

CHAPTER OVERVIEW

CHAPTER GOALS

Students will approximate and evaluate square roots and simplify radicals. They will use these skills to solve quadratic equations by finding square roots and by using the quadratic formula. Students will graph quadratic functions, and use the x-intercepts of the graphs to solve the related quadratic equations. They will see how the discriminant relates to the number of x-intercepts of the graph of a quadratic function and the number of solutions of a quadratic equation. Students will also graph quadratic inequalities. Finally, students choose a linear, quadratic, or exponential model that best fits a collection of data.

APPLICATION NOTE

A slugger like Mark McGwire of the St. Louis Cardinals, who finished the 1998 season with a record-breaking 70 home runs, or Sammy Sosa of the Chicago Cubs, who finished close behind with 66 home runs, has to hit the ball *hard* to hit a home run. That is, the initial velocity of the ball off the bat must be very high. Using this velocity, the height at which the ball is hit, and the angle at which it is hit, and accounting for gravity, a quadratic model of the path of the ball can be constructed.

Using this model, and knowing the distance to and height of the outfield wall, you can predict whether a hit will be a home run by solving an equation or using a graph. One famous outfield wall that a home-run ball must clear is the "Green Monster," the 37.17 foot high left-field wall in Fenway Park in Boston, which stands from 310 to 370 feet from home plate.

Additional information about baseball is available at **www.mcdougallittell.com.**

QUADRATIC EQUATIONS AND FUNCTIONS

What is the path of a home run ball?

500

CHAPTER 9

APPLICATION: Baseball

A batter usually scores a home run by hitting a ball over the outfield fence. Mathematical models can be used to find how long it takes for a baseball to reach the ground.

You'll use quadratic equations to solve different types of vertical motion problems in Chapter 9.

Think & Discuss

1. Use the graph to approximate the maximum height the ball reaches. **about 100 ft**
2. Use the graph to approximate the maximum horizontal distance it travels. **about 400 ft**

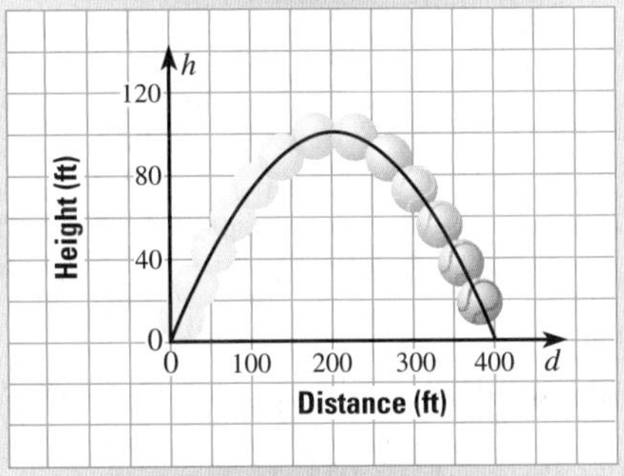

Learn More About It

You will use a quadratic model to learn more about the path of a ball in Exercise 84 on p. 538.

INTERNET **APPLICATION LINK** Visit www.mcdougallittell.com for more information about baseball.

ADDITIONAL RESOURCES

Another way to begin the chapter is to show the video clip of a real-life motivator for Chapter 9 on the *Instant Replay: Video Review Games.*

PROJECTS

A project covering Chapters 7–9 appears on pages 570 and 571 of the Student Edition. An additional project for Chapter 9 is available in the *Chapter 9 Resource Book,* p. 135.

TECHNOLOGY

Software

- *Electronic Teaching Tools*
- *Online Lesson Planner*
- *Personal Student Tutor*
- *Test and Practice Generator*
- *Electronic Lesson Presentations* (Lesson 9.4)

Video

- *Instant Replay: Video Review Games*

INTERNET **Internet Connections**
www.mcdougallittell.com

- **Application Links** 501, 513, 535, 543, 547, 550
- **Data Updates** 530
- **Student Help** 509, 512, 517, 523, 525, 527, 532, 534, 539, 542, 552, 558
- **Career Links** 509, 545
- **Extra Challenge** 510, 516, 523, 531, 538, 546, 553, 559

PREPARE

DIAGNOSTIC TOOLS

The **Skill Review** exercises can help you diagnose whether students have the following skills needed in Chapter 9:

- Evaluate algebraic expressions.
- Use tables to graph linear equations.
- Check whether an ordered pair is a solution of a linear inequality.

The following resources are available for students who need additional help with these skills:

- Prerequisite Skills Review (*Chapter 9 Resource Book,* p. 5; *Warm-Up Transparencies,* p. 66)
- Reteaching with Practice (Chapter Resource Books for Lessons 2.5, 2.7, 4.2, and 6.5)
- *Personal Student Tutor*

ADDITIONAL RESOURCES

The following resources are provided to help you prepare for the upcoming chapter and customize review materials:

- ***Chapter 9 Resource Book***
  Tips for New Teachers (p. 1)
  Parent Guide (p. 3)
  Lesson Plans (every lesson)
  Lesson Plans for Block Scheduling (every lesson)
- ***Electronic Teaching Tools***
- ***Online Lesson Planner***
- ***Test and Practice Generator***

5.

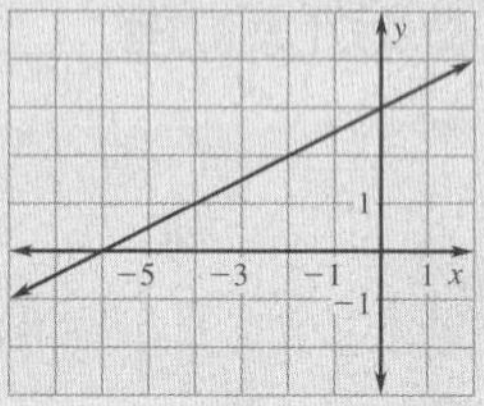

6.

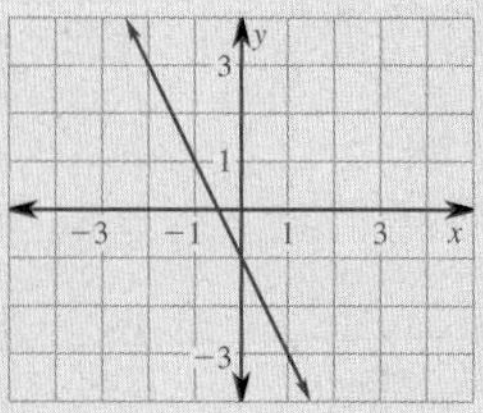

7. See the next page.

CHAPTER 9

Study Guide

PREVIEW

What's the chapter about?

Chapter 9 is about **solving quadratic equations and graphing quadratic functions**, many of which model real-life applications. In Chapter 9 you'll learn

- how to evaluate and approximate square roots.
- how to simplify radicals.
- how to solve a quadratic equation.
- how to sketch the graph of a quadratic function and a quadratic inequality.

KEY VOCABULARY

▶ Review	▶ New	
• linear equation, p. 133	• square root, p. 503	• quadratic function, p. 518
• *x*-intercept, p. 218	• irrational number, p. 504	• parabola, p. 518
• exponential growth, p. 477	• radical expression, p. 504	• roots, p. 526
• exponential decay, p. 484	• quadratic equation, p. 505	• discriminant, p. 541
		• quadratic inequalities, p. 548

PREPARE

Are you ready for the chapter?

SKILL REVIEW Do these exercises to review key skills that you'll apply in this chapter. See the given reference page if there is something you don't understand.

> STUDENT HELP
> **Study Tip**
> "Student Help" boxes throughout the chapter give you study tips and tell you where to look for extra help in this book and on the Internet.

Evaluate the expression. (Review pp. 94 and 109)

1. $3x^2 - 108$ when $x = -4$ **−60**

2. $8x^2 \div \frac{2}{3}$ when $x = -1$ **12**

3. $x^2 - 4xy$ when $x = -2$ and $y = 5$ **44**

4. $-\frac{x}{2y}$ when $x = 12$ and $y = -3$ **2**

Use a table of values to graph the equation. (Review Examples 2 and 3, pp. 211–212) **5–7. See margin.**

5. $y = \frac{1}{2}x + 3$

6. $y + 3 = -2x + 2$

7. $x + 7y = 14$

Check whether the ordered pair is a solution. (Review Example 1, p. 360)

8. $3x + 4y < 5, (-1, 2)$ **not a solution**

9. $\frac{1}{2}x - \frac{2}{3}y \geq -6, (0, 0)$ **solution**

10. $6x - 2y > -8, (2, -3)$ **solution**

STUDY STRATEGY

Here's a study strategy!

Explaining Ideas

Sometimes explaining things to another person can help you understand a topic better. Or, someone else's questions may point out something that you don't fully understand. Talking about math is a good way to check how well you know the material and to work through questions.

9.1 Solving Quadratic Equations by Finding Square Roots

What you should learn

GOAL 1 Evaluate and approximate square roots.

GOAL 2 Solve a quadratic equation by finding square roots.

Why you should learn it

▼ To solve **real-life** problems such as finding the time it takes an egg to hit the ground in an engineering contest in **Example 7**.

REAL LIFE

GOAL 1 EVALUATING SQUARE ROOTS

You know how to find the square of a number. For instance, the square of 3 is $3^2 = 9$. The square of -3 is also 9.

In this lesson you will study the reverse problem: finding a *square root* of a number.

SQUARE ROOT OF A NUMBER

If $b^2 = a$, then b is a **square root** of a.

Example: If $3^2 = 9$, then 3 is a square root of 9.

All positive real numbers have two square roots: a **positive square root** (or principal square root) and a **negative square root**. Square roots are written with a radical symbol $\sqrt{}$. The number or expression inside a radical symbol is the **radicand**. In the following examples, 9 is the radicand.

Meaning	Positive square root	Negative square root	The positive and negative square roots
Symbol	$\sqrt{}$	$-\sqrt{}$	$\pm\sqrt{}$
Example	$\sqrt{9} = 3$	$-\sqrt{9} = -3$	$\pm\sqrt{9} = \pm 3 = +3$ or -3

The symbol $\pm$ is read as "plus or minus" and refers to both the positive square root and the negative square root. Zero has only one square root: zero. Negative numbers have no real square roots, because the square of every real number is either positive or zero.

EXAMPLE 1 *Finding Square Roots of Numbers*

Evaluate the expression.

a. $\sqrt{64}$ **b.** $-\sqrt{64}$ **c.** $\sqrt{0}$ **d.** $\pm\sqrt{0.25}$ **e.** $\sqrt{-4}$

SOLUTION

a. $\sqrt{64} = 8$ — Positive square root

b. $-\sqrt{64} = -8$ — Negative square root

c. $\sqrt{0} = 0$ — Square root of zero

d. $\pm\sqrt{0.25} = \pm 0.5$ — Two square roots

e. $\sqrt{-4}$ (undefined) — No real square root

1 PLAN

PACING

Basic: 1 day
Average: 1 day
Advanced: 1 day
Block Schedule: 0.5 block with Ch. 8 Assess

LESSON OPENER APPLICATION

An alternative way to approach Lesson 9.1 is to use the Application Lesson Opener:

- Blackline Master (*Chapter 9 Resource Book,* p. 12)
- Transparency (p. 58)

MEETING INDIVIDUAL NEEDS

- ***Chapter 9 Resource Book***
  Prerequisite Skills Review (p. 5)
  Practice Level A (p. 14)
  Practice Level B (p. 15)
  Practice Level C (p. 16)
  Reteaching with Practice (p. 17)
  Absent Student Catch-Up (p. 19)
  Challenge (p. 21)
- ***Resources in Spanish***
- ***Personal Student Tutor***

NEW-TEACHER SUPPORT

See the Tips for New Teachers on pp. 1–2 of the *Chapter 9 Resource Book* for additional notes about Lesson 9.1.

WARM-UP EXERCISES

Transparency Available

Simplify.

1. 6^2 36
2. $(-14)^2$ 196
3. -9^2 −81
4. 0^2 0
5. $-4x^2$, for $x = 3$ −36

7.

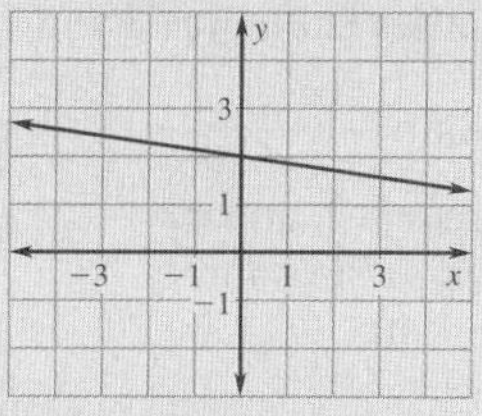

2 TEACH

MOTIVATING THE LESSON

A student might read on a quart can of paint that it will cover 75–100 square feet. Ask students what this means in terms of how big a square the paint will cover. Many will know to find the square root of the area to find the side length. Inform them that what they are really doing is solving a quadratic equation.

EXTRA EXAMPLE 1

Evaluate the expression.

a. $\sqrt{81}$ **9**

b. $-\sqrt{81}$ **−9**

c. $-\sqrt{0}$ **0**

d. $\pm\sqrt{0.0016}$ **±0.04**

e. $\sqrt{-9}$ **undefined**

EXTRA EXAMPLE 2

Evaluate the expression. Use a calculator if necessary.

a. $-\sqrt{36}$ **−6**

b. $\sqrt{1.21}$ **1.1**

c. $-\sqrt{0.04}$ **−0.2**

d. $\sqrt{8}$ **about 2.83**

EXTRA EXAMPLE 3

Evaluate $\sqrt{b^2 - 4ac}$ for $a = 7$, $b = 8$, and $c = 1$. **6**

EXTRA EXAMPLE 4

Evaluate $\frac{1 \pm 2\sqrt{6}}{4}$ to the nearest hundredth. **−0.97, 1.47**

CHECKPOINT EXERCISES

For use after Examples 1 and 2:

1. Evaluate the expression. Use a calculator if necessary.
   - a. $\sqrt{49}$ **7**
   - b. $\pm\sqrt{100}$ **±10**
   - c. $-\sqrt{0.81}$ **−0.9**
   - d. $-\sqrt{-0.04}$ **undefined**
   - e. $-\sqrt{12}$ **about −3.46**

For use after Examples 3 and 4:

2. Evaluate $\frac{-b \pm \sqrt{b^2 - 4ac}}{2a}$ for $a = 6$, $b = -2$, and $c = -1$. Round to the nearest hundredth. **0.61, −0.27**

Numbers whose square roots are integers or quotients of integers are called **perfect squares.** The square roots of numbers that are not perfect squares must be written using the radical symbol or approximated. These numbers are part of the set of *irrational numbers*. An **irrational number** is a number that cannot be written as the quotient of two integers.

STUDENT HELP

Square Root Table
For a table of square roots, see p. 811.

EXAMPLE 2 *Evaluating Square Roots*

Evaluate the expression. Use a calculator if necessary.

a. $-\sqrt{121}$ **b.** $-\sqrt{1.44}$ **c.** $\sqrt{0.09}$ **d.** $\sqrt{7}$

SOLUTION

a. $-\sqrt{121} = -11$ — 121 is a perfect square: $11^2 = 121$.

b. $-\sqrt{1.44} = -1.2$ — 1.44 is a perfect square: $1.2^2 = 1.44$.

c. $\sqrt{0.09} = 0.3$ — 0.09 is a perfect square: $0.3^2 = 0.09$.

d. $\sqrt{7} \approx 2.65$ — Round to the nearest hundredth.

A **radical expression** involves square roots (or *radicals*). If the symbol ± precedes the radical, the expression represents two different numbers, as in Example 4. The square root symbol is a grouping symbol. Operations inside a radical symbol must be performed before the square root is evaluated.

EXAMPLE 3 *Evaluating a Radical Expression*

Evaluate $\sqrt{b^2 - 4ac}$ when $a = 1$, $b = -2$, and $c = -3$.

SOLUTION

$\sqrt{b^2 - 4ac} = \sqrt{(-2)^2 - 4(1)(-3)}$ — Substitute values.

$= \sqrt{4 + 12}$ — Simplify.

$= \sqrt{16}$ — Simplify.

$= 4$ — Positive square root

EXAMPLE 4 *Evaluating an Expression with a Calculator*

Evaluate $\frac{1 \pm 2\sqrt{3}}{4}$ to the nearest hundredth.

SOLUTION This expression represents two numbers.

(1 + 2 × 3 √) ÷ 4 ENTER → 1.116025404

(1 − 2 × 3 √) ÷ 4 ENTER → −0.61602540

STUDENT HELP

KEYSTROKE HELP
To find the square root of 3 on your calculator you may need to press √ 3 or 3 √. Test your calculator to find out which order is correct.

▶ Rounded to the nearest hundredth, the expression represents 1.12 and −0.62.

GOAL 2 SOLVING A QUADRATIC EQUATION

A **quadratic equation** is an equation that can be written in the following **standard form.**

$ax^2 + bx + c = 0$, where $a \neq 0$

In standard form, a is the **leading coefficient.**

When $b = 0$, this equation becomes $ax^2 + c = 0$. To solve quadratic equations of this form, isolate x^2 on one side. Then find the square root(s) of each side.

CONCEPT SUMMARY SOLVING $x^2 = d$ BY FINDING SQUARE ROOTS

- If $d > 0$, then $x^2 = d$ has two solutions: $x = \pm\sqrt{d}$.
- If $d = 0$, then $x^2 = d$ has one solution: $x = 0$.
- If $d < 0$, then $x^2 = d$ has no real solution.

EXAMPLE 5 *Solving Quadratic Equations*

Solve each equation.

a. $x^2 = 4$ **b.** $x^2 = 5$ **c.** $x^2 = 0$ **d.** $x^2 = -1$

SOLUTION

a. $x^2 = 4$ has two solutions: $x = +2$ and $x = -2$.

b. $x^2 = 5$ has two solutions: $x = \sqrt{5}$ and $x = -\sqrt{5}$.

c. $x^2 = 0$ has one solution: $x = 0$.

d. $x^2 = -1$ has no real solution.

EXAMPLE 6 *Rewriting Before Finding Square Roots*

Solve $3x^2 - 48 = 0$.

SOLUTION

$3x^2 - 48 = 0$	Write original equation.
$3x^2 = 48$	Add 48 to each side.
$x^2 = 16$	Divide each side by 3.
$x = \pm\sqrt{16}$	Find square roots.
$x = \pm 4$	16 is a perfect square: $4^2 = 16$, $(-4)^2 = 16$.

▶ The solutions are 4 and −4. Check both solutions in the original equation. Both 4 and −4 make the equation true, so $3x^2 - 48 = 0$ has two solutions.

STUDENT HELP

Study Tip
When checking solutions of quadratic equations, remember to check *all* solutions in the original equation. In Example 6, check both 4 and −4.
$3(4)^2 - 48 \stackrel{?}{=} 0$
$3(-4)^2 - 48 \stackrel{?}{=} 0$

STUDENT HELP NOTES

Keystroke Help Keystrokes for several models of calculators are available in **blackline** format in the *Chapter 9 Resource Book*, p. 13 and at **www.mcdougallittell.com.**

EXTRA EXAMPLE 5
Solve each equation.
a. $x^2 = 16$ 4, −4
b. $x^2 = 11$ $\sqrt{11}, -\sqrt{11}$
c. $x^2 = -4$ no real solution
d. $x^2 = 100$ 10, −10

EXTRA EXAMPLE 6
Solve $120 - 6x^2 = -30$. 5, −5

CHECKPOINT EXERCISES
For use after Example 5:
1. Solve each equation.
a. $x^2 = 25$ 5, −5
b. $x^2 = 81$ 9, −9
c. $x^2 = 7$ $\sqrt{7}, -\sqrt{7}$
d. $x^2 = -12$ no real solution

For use after Example 6:
2. Solve $12x^2 - 60 = 0$. $\sqrt{5}, -\sqrt{5}$

CONCEPT QUESTION
EXAMPLE 5 Explain why the equation $x^2 = -a$ has no real solution if a is positive. ***Sample answer:*** **There is no real number whose square is a negative number.**

EXTRA EXAMPLE 7
Construction waste falls down a 25 foot chute into a bin. The height h of the waste after t seconds is modeled by $h = -16t^2 + s$, where s is the initial height. How long will it take the waste to reach the bin? Assume there is no air resistance. **1.25 sec**

CHECKPOINT EXERCISES

For use after Example 7:

1. Use the falling object model $h = -16t^2 + s$ to find how long it takes a free-fall ride at an amusement park to drop 121 feet. Assume there is no air resistance. **2.75 sec**

FOCUS ON VOCABULARY
Give examples of a perfect square and an irrational number.
Sample answer: **64, $\sqrt{10}$**

CLOSURE QUESTION
Describe how to solve $-16x^2 + 48 = 0$. ***Sample answer:*** **Add −48 to each side. Divide each side by −16. Find the square root of each side. The answer is $\pm\sqrt{3}$.**

DAILY PUZZLER
Name four perfect squares between 25 and 81. ***Sample answer:*** **36, 49, 64, and 72.25**

FALLING OBJECT MODEL When an object is dropped, the speed with which it falls continues to increase. Ignoring air resistance, its height h can be approximated by the falling object model.

$h = -16t^2 + s$ ⟵ **Falling object model**

Here h is measured in feet, t is the number of seconds the object has fallen, and s is the initial height from which the object was dropped.

The object's increase in speed is due to Earth's gravitational pull. On a planet whose gravitational pull is different from that of Earth, the factor -16 would be replaced by another constant.

REAL LIFE
Engineering

EXAMPLE 7 *Using a Falling Object Model*

An engineering student is in an "egg dropping contest." The goal is to create a container for an egg so it can be dropped from a height of 32 feet without breaking the egg. To the nearest tenth of a second, about how long will it take for the egg's container to hit the ground? Assume there is no air resistance.

SOLUTION Write an equation to model the egg container's height h as function of time t, where the initial height $s = 32$.

$h = -16t^2 + s$ **Write falling object model.**

$h = -16t^2 + 32$ **Substitute 32 for *s*.**

The falling object model for the egg container is $h = -16t^2 + 32$.

PROBLEM SOLVING STRATEGY

Method 1 **MAKE A TABLE** One way to solve the problem is to find the height h for different values of time t in the function $h = -16t^2 + 32$. Organize the data in a table. The egg container will hit the ground when $h = 0$.

Time (sec)	0.0	0.5	1.0	1.1	1.2	1.3	1.4	1.5
Height (ft)	32	28	16	12.64	8.96	4.96	0.64	−4

▶ From the table, you can see that $h = 0$ between 1.4 and 1.5 seconds. The egg container will take between 1.4 and 1.5 seconds to hit the ground.

Method 2 **USE AN EQUATION** Another way to approach the problem is to solve the quadratic equation for the time t that gives a height of $h = 0$ feet.

$h = -16t^2 + 32$ **Write falling egg model.**

$0 = -16t^2 + 32$ **Substitute 0 for *h*.**

$-32 = -16t^2$ **Subtract 32 from each side.**

$2 = t^2$ **Divide each side by −16.**

$\sqrt{2} = t$ **Find positive square root.**

$1.4 \approx t$ **Use a calculator.**

▶ The egg container will hit the ground in about 1.4 seconds. You can ignore the negative square root, because -1.4 seconds is not a reasonable solution.

STUDENT HELP
Look Back
For help with using a table of values to graph an equation, see pp. 211–212.

GUIDED PRACTICE

Vocabulary Check ✓

1. State the meanings of the symbols $\sqrt{\ }$, $-\sqrt{\ }$, and $\pm\sqrt{\ }$. **the principal (or positive) square root; the negative square root; both the positive and the negative square root**
2. Give an example of a perfect square and an example of an irrational number. ***Sample answers:* 121; $\sqrt{11}$**

Concept Check ✓

3. Explain how to find solutions of an equation of the form $ax^2 + c = 0$. **See margin.**

3. *Sample answer:* Subtract c from both sides, then divide both sides by a. If $-\frac{c}{a}$ is positive, find both square roots of $-\frac{c}{a}$. If $-\frac{c}{a}$ is zero, the only solution is 0. If $-\frac{c}{a}$ is negative, there are no real solutions.

If the statement is *true,* give an example. If false, give a counterexample.

4. Some numbers have no real square root. **True. *Sample answer:* $\sqrt{-1}$**
5. No number has only one square root. **False. *Sample answer:* $\sqrt{0} = 0$**
6. All positive numbers have two different square roots. **True. *Sample answer:* −5 and 5 are both square roots of 25.**
7. The square root of the sum of two numbers is equal to the sum of the square roots of the numbers. **False. *Sample answer:* $\sqrt{6+3} = 3$; $\sqrt{6} + \sqrt{3} = 4.8$**

Skill Check ✓

Evaluate the expression.

8. $\sqrt{36}$ **6**
9. $\sqrt{0.81}$ **0.9**
10. $-\sqrt{0.04}$ **−0.2**
11. $\pm\sqrt{9}$ **±3**

Evaluate the radical expression when $a = 2$ and $b = 4$.

12. $\sqrt{b^2 + 10a}$ **6**
13. $\dfrac{10 \pm 2\sqrt{b}}{a}$ **3, 7**
14. $\sqrt{b^2 - 8a}$ **0**

Solve the equation. If there is no solution, state the reason.

15. $2x^2 - 8 = 0$ **−2, 2**
16. $x^2 + 25 = 0$ **See margin.**
17. $x^2 - 1.44 = 0$ **−1.2, 1.2**
18. $5x^2 = -15$ **See margin.**

16. There is no solution; negative numbers have no real square roots.

18. There is no solution; negative numbers have no real square roots.

FALLING OBJECTS **If an object is dropped from an initial height s, how long will it take to reach the ground? Assume there is no air resistance.**

19. $s = 48$ **about 1.7 sec**
20. $s = 96$ **about 2.4 sec**
21. $s = 192$ **about 3.5 sec**
22. If you double the height from which the object falls, do you double the falling time? If not, why? **No. *Sample answer:* The height is a function of the square of the falling time rather than of the falling time.**

PRACTICE AND APPLICATIONS

STUDENT HELP

Extra Practice to help you master skills is on p. 805.

FINDING SQUARE ROOTS **Find all square roots of the number or write *no square roots.* Check the results by squaring each root.**

23. 49 **−7, 7**
24. 144 **−12, 12**
25. −4 **no square roots**
26. −25 **no square roots**
27. 0 **0**
28. 100 **−10, 10**
29. 0.09 **−0.3, 0.3**
30. 0.16 **−0.4, 0.4**

STUDENT HELP

HOMEWORK HELP
Example 1: Exs. 23–30
Example 2: Exs. 31–38
Example 3: Exs. 39–44

continued on p. 508

EVALUATING SQUARE ROOTS **Evaluate the expression. Give the exact value if possible. Otherwise, approximate to the nearest hundredth.**

31. $-\sqrt{169}$ **−13**
32. $\sqrt{64}$ **8**
33. $\sqrt{13}$ **3.61**
34. $-\sqrt{125}$ **−11.18**
35. $\sqrt{0.04}$ **0.2**
36. $\pm\sqrt{0.75}$ **±0.87**
37. $-\sqrt{0.1}$ **−0.32**
38. $\pm\sqrt{6.25}$ **±2.5**

EVALUATING EXPRESSIONS **Evaluate $\sqrt{b^2 - 4ac}$ for the given values.**

39. $a = 4, b = 5, c = 1$ **3**
40. $a = -2, b = 8, c = -8$ **0**
41. $a = 3, b = -7, c = 6$ **undefined**
42. $a = 2, b = 4, c = 0.5$ **$2\sqrt{3}$**
43. $a = -3, b = 7, c = 5$ **$\sqrt{109}$**
44. $a = 6, b = -8, c = 4$ **undefined**

3 APPLY

ASSIGNMENT GUIDE

BASIC
Day 1: pp. 507–510 Exs. 23, 26, 29, and every 3rd Ex. through Ex. 77, 79–81, 95, 96, 104, 107, 110, 111

AVERAGE
Day 1: pp. 507–510 Exs. 23, 26, 29, and every 3rd Ex. through Ex. 77, 79–81, 87, 95, 96, 104, 107, 110, 111

ADVANCED
Day 1: pp. 507–510 Exs. 23, 26, 29, and every 3rd Ex. through Ex. 77, 78–81, 87, 95–100, 104, 107, 110, 111

BLOCK SCHEDULE
pp. 507–510 Exs. 23, 26, 29, and every 3rd Ex. through Ex. 77, 79–81, 87, 95, 96, 104, 107, 110, 111 (with Ch. 8 Assess.)

EXERCISE LEVELS

Level A: ***Easier***
23–38, 101–104

Level B: ***More Difficult***
39–77, 79–88, 95, 96, 105–112

Level C: ***Most Difficult***
78, 89–94, 97–100

✓ HOMEWORK CHECK

To quickly check student understanding of key concepts, go over the following exercises: Exs. 26, 38, 44, 50, 62, 74, 79. See also the Daily Homework Quiz:

- Blackline Master (*Chapter 9 Resource Book,* p. 24)
- Transparency (p. 68)

! COMMON ERROR

EXERCISE 29 Some students may quickly write that the square roots of 0.09 are ±0.03. Remind them that when finding square roots of decimals less than one, the positive square root is larger than the original decimal.

! COMMON ERROR

EXERCISE 73 After subtracting 12 from each side, students will likely multiply both sides by $\frac{5}{4}$ before concluding that there is no solution. Point out that by being observant, they can save a step. After subtracting, the right side is negative and the coefficient of x is positive, so they know that the right side will remain negative without actually having to multiply.

STUDENT HELP

HOMEWORK HELP
continued from p. 507

Example 4: Exs. 45–53
Example 5: Exs. 54–59
Example 6: Exs. 60–77
Example 7: Exs. 79–86

EVALUATING EXPRESSIONS **Use a calculator to evaluate the expression. Round the results to the nearest hundredth.**

45. $\frac{3 \pm 4\sqrt{6}}{3}$ **−2.27, 4.27** **46.** $\frac{6 \pm 4\sqrt{2}}{-1}$ **−11.66, −0.34** **47.** $\frac{2 \pm 5\sqrt{3}}{5}$ **−1.33, 2.13**

48. $\frac{1 \pm 6\sqrt{8}}{6}$ **−2.66, 3.00** **49.** $\frac{7 \pm 3\sqrt{2}}{-1}$ **−11.24, −2.76** **50.** $\frac{2 \pm 5\sqrt{6}}{2}$ **−5.12, 7.12**

51. $\frac{5 \pm 6\sqrt{3}}{3}$ **−1.80, 5.13** **52.** $\frac{3 \pm 4\sqrt{5}}{4}$ **−1.49, 2.99** **53.** $\frac{7 \pm 0.3\sqrt{12}}{-6}$ **−1.34, −0.99**

QUADRATIC EQUATIONS **Solve the equation or write *no solution*. Write the solutions as integers if possible. Otherwise write them as radical expressions.**

54. $x^2 = 36$ **−6, 6** **55.** $b^2 = 64$ **−8, 8** **56.** $5x^2 = 500$ **−10, 10**

57. $x^2 = 16$ **−4, 4** **58.** $x^2 = 0$ **0** **59.** $x^2 = -9$ **no solution**

60. $3x^2 = 6$ $-\sqrt{2}, \sqrt{2}$ **61.** $a^2 + 3 = 12$ **−3, 3** **62.** $x^2 - 7 = 57$ **−8, 8**

63. $2s^2 - 5 = 27$ **−4, 4** **64.** $5x^2 + 5 = 20$ $-\sqrt{3}, \sqrt{3}$ **65.** $7x^2 + 30 = 9$ **no solution**

66. $x^2 + 4.0 = 0$ **no solution** **67.** $6x^2 - 54 = 0$ **−3, 3** **68.** $7x^2 - 63 = 0$ **−3, 3**

SOLVING EQUATIONS **Use a calculator to solve the equation or write *no solution*. Round the results to the nearest hundredth.**

69. $4x^2 - 3 = 57$ **−3.87, 3.87** **70.** $6y^2 + 22 = 34$ **−1.41, 1.41** **71.** $2x^2 - 4 = 10$ **−2.65, 2.65**

72. $\frac{2}{3}n^2 - 6 = 2$ **−3.46, 3.46** **73.** $\frac{4}{5}x^2 + 12 = 5$ **no solution** **74.** $\frac{1}{2}x^2 + 3 = 8$ **−3.16, 3.16**

75. $3x^2 + 7 = 31$ **−2.83, 2.83** **76.** $6s^2 - 12 = 0$ **−1.41, 1.41** **77.** $5a^2 + 10 = 20$ **−1.41, 1.41**

78. **CRITICAL THINKING** Write your own example of a quadratic equation in the form $x^2 = d$ for each of the following.

a. An equation that has no real solution. **$x^2 = d$, for any number $d < 0$**

b. An equation that has one solution. **$x^2 = 0$**

c. An equation that has two solutions. **$x^2 = d$, for any number $d > 0$**

ROAD SAFETY **In Exercises 79–82, a boulder falls off the top of a cliff during a storm. The cliff is 60 feet high. Find how long it will take for the boulder to hit the road below.**

79. Write the falling object model for $s = 60$. **$h = -16t^2 + 60$**

80. Try to determine the time when $h = 0$ from an input-output table for this model. **See margin.**

81. Solve the falling object model for $h = 0$. **$\sqrt{3.75} \approx 1.94$ sec**

82. *Writing* Which problem solving method do you prefer? Why? **See margin.**

80. ***Sample answer:***

t	h
0	60
1	44
1.5	24
1.75	11
1.8	8.16
1.9	2.24
1.95	−0.84

about 1.95 sec

82. ***Sample answer:*** **I prefer solving the equation; it is more straightforward and provides an exact answer.**

FALLING OBJECT MODEL **In Exercises 83–86, an object is dropped from a height *s*. How long does it take to reach the ground? Assume there is no air resistance.**

83. $s = 144$ feet **3 sec** **84.** $s = 256$ feet **4 sec** **85.** $s = 400$ feet **5 sec** **86.** $s = 600$ feet **about 6.1 sec**

STUDENT HELP

HOMEWORK HELP
Visit our Web site www.mcdougallittell.com for help with Exs. 87–88.

FOCUS ON CAREERS

MINERALOGISTS
The properties and distribution of minerals are studied by mineralogists. The Vickers scale applies to thin slices of minerals that can be examined with a microscope.

CAREER LINK
www.mcdougallittell.com

87. HOME COMPUTER SALES The sales S (in millions of dollars) of home computers in the United States from 1988 to 1995 can be modeled by $S = 145.63t^2 + 3327.56$, where t is the number of years since 1988. Use this model to estimate the year in which sales of home computers will be \$36,000 million. ▶ Source: Electronic Industries Association **2003**

88. COMPUTER SOFTWARE SALES The sales S (in millions of dollars) of computer software in the United States from 1990 to 1995 can be modeled by $S = 61.98t^2 + 1001.15$, where t is the number of years since 1990. Use this model to estimate the year in which sales of computer software will be \$7200 million. ▶ Source: Electronic Industries Association **2000**

MINERALS **In Exercises 89–94, use the following information.**
Mineralogists use the Vickers scale to measure the hardness of minerals. The hardness H of a mineral can be determined by hitting the mineral with a pyramid-shaped diamond and measuring the depth d of the indentation. The harder the mineral, the smaller the depth of the indentation. A model that relates mineral hardness with the indentation depth (in millimeters) is $Hd^2 = 1.89$.

Use a calculator to find the depth of the indentation for the mineral with the given value of H. Round to the nearest hundredth of a millimeter.

89. Graphite: $H = 12$ **0.40 mm**

90. Gold: $H = 50$ **0.19 mm**

91. Galena: $H = 80$ **0.15 mm**

92. Platinum: $H = 125$ **0.12 mm**

93. Copper: $H = 140$ **0.12 mm**

94. Hematite: $H = 755$ **0.05 mm**

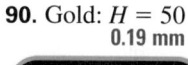

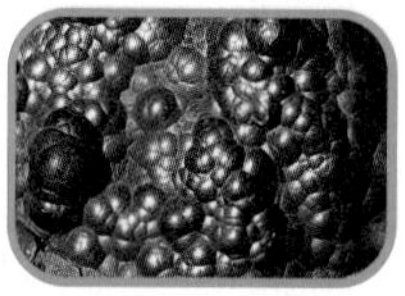

Test Preparation

QUANTITATIVE COMPARISON **In Exercises 95–96, choose the statement below that is true about the given numbers.**

Ⓐ The number in column A is greater.

Ⓑ The number in column B is greater.

Ⓒ The two numbers are equal.

Ⓓ The relationship cannot be determined from the given information.

	Column A	Column B	
95.	$-\sqrt{121}$	$-2\sqrt{121}$	**A**
96.	The positive value of x in $2x^2 - 14 = 18$	The positive value of x in $2x^2 + 14 = 46$	**C**

STUDENT HELP NOTES

Homework Help Students can find help for Exs. 87 and 88 at **www.mcdougallittell.com.** The information can be printed out for students who don't have access to the Internet.

APPLICATION NOTE
EXERCISES 89–94 Although graphite has the lowest Vickers scale hardness number, 12, industrial diamonds may be synthesized by subjecting graphite to very high pressures and temperatures. The mineral chromite is at the high end of the scale, with a Vickers number of 1206.

ENGLISH LEARNERS
EXERCISES 87–94
Understanding the language of word problems can be extremely difficult for English language learners. In Exercises 87 and 88, you may need to clarify the usage of the word *where,* common in mathematical language but different from normal usage. For Exercises 89–94, make sure students understand both the vocabulary and the cause-effect relationship in the sentence *The harder the mineral, the smaller the depth of the indentation.*

CAREER NOTE
EXERCISES 89–94
Additional information about mineralogists is available at **www.mcdougallittell.com.**

ADDITIONAL PRACTICE AND RETEACHING

For Lesson 9.1:

- Practice Levels A, B, and C (*Chapter 9 Resource Book,* p. 14)
- Reteaching with Practice (*Chapter 9 Resource Book,* p. 17)
- See Lesson 9.1 of the *Personal Student Tutor*

For more Mixed Review:

- Search the *Test and Practice Generator* for key words or specific lessons.

4 ASSESS

DAILY HOMEWORK QUIZ

Transparency Available

1. Find all square roots of 0.25. ± 0.5
2. Evaluate $-\sqrt{1.44}$. -1.2
3. Evaluate $\sqrt{b^2 - 4ac}$ for $a = 3$, $b = 3$, and $c = -6$. 9
4. Using a calculator, evaluate $\frac{-3 \pm 2\sqrt{5}}{4}$ to the nearest hundredth. $-1.87, 0.37$
5. Solve $x^2 - 7 = 13$. Write the result as a radical expression. $\pm\sqrt{20}$
6. Solve $3y^2 + 4 = 11$. Write the result to the nearest hundredth. ± 1.53
7. A cliff diver jumps from an 84 ft cliff. Write a falling object model for the time t it will take the diver to hit the water. Assume there is no air resistance. $h = -16t^2 + 84$

EXTRA CHALLENGE NOTE

Challenge problems for Lesson 9.1 are available in **blackline** format in the *Chapter 9 Resource Book*, p. 21 and at **www.mcdougallittell.com.**

ADDITIONAL TEST PREPARATION

1. **WRITING** Describe how to solve the quadratic equation $ax^2 + c = 0$. ***Sample answer:* Add $-c$ to each side. Then divide each side by a to get $x^2 = -\frac{c}{a}$. If $-\frac{c}{a}$ is nonnegative, take the square root of both sides to get the solution $\pm\sqrt{-\frac{c}{a}}$. If $-\frac{c}{a}$ is negative, there is no real solution.**

★ Challenge

99. 4 times; the distance that an object is dropped is a function of the time squared. So $(2t)^2 = 2^2 \cdot t^2 = 4t^2$

100. *Sample answer:* The first equation represents the distance a falling object falls in a given time t, and the second represents the height of the same object at time t. Then $h = s - d$ and so $-16t^2 = -\frac{1}{2}g(t^2)$; that is, the exponential term in both equations is the same.

EXTRA CHALLENGE

www.mcdougallittell.com

SCIENCE CONNECTION **In Exercises 97–100, use the following information.** Scientists simulate a gravity-free environment called *microgravity* in free-fall situations. A similar microgravity environment can be felt on free-fall rides at amusement parks or when stepping off a high diving platform. The distance d (in meters) that an object that is dropped falls in t seconds can be modeled by the equation $d = \frac{1}{2}g(t^2)$, where g is the acceleration due to gravity (9.8 meters per second per second).

97. The NASA Lewis Research Center has two microgravity facilities. One provides a 132-meter drop into a hole and the other provides a 24-meter drop inside a tower. How long will each free-fall period be? **about 5.19 sec, about 2.21 sec**

98. In Japan a 490-meter-deep mine shaft has been converted into a microgravity facility. This creates the longest period of free fall currently available on Earth. How long will a period of free-fall be? **10 sec**

99. If you want to double the free-fall time, how much do you have to increase the height from which the object was dropped? **See margin.**

100. CRITICAL THINKING How are these formulas similar? **See margin.**

$d = \frac{1}{2}g(t^2)$ when d is distance, g is gravity, and t is time

$h = -16t^2 + s$ when h is height, s is initial height, and t is time

MIXED REVIEW

FACTORING INTEGERS **Write the prime factorization.** (Skills Review, p. 777)

101. 11 **11** **102.** 24 $2^3 \times 3$ **103.** 72 $2^3 \times 3^2$ **104.** 108 $2^2 \times 3^3$

GRAPH AND CHECK **Use a graph to solve the linear system. Check your solution algebraically.** (Review 7.1)

105. $-3x + 4y = -5$, $4x + 2y = -8$ **(−1, −2)**

106. $4x + 5y = 20$, $\frac{5}{4}x + y = 4$ **(0, 4)**

107. $\frac{1}{2}x + 3y = 18$, $2x + 6y = -12$ **(−48, 14)**

SOLVING SYSTEMS **In Exercises 109–111, use linear combinations to solve the system.** (Review 7.3)

108. $12x - 4y = -32$, $x + 3y = 4$ **(−2, 2)**

109. $10x - 3y = 17$, $-7x + y = 9$ **(−4, −19)**

110. $8x - 5y = 100$, $2x + \frac{1}{2}y = 4$ **(5, −12)**

111. BASKETBALL TICKETS You are selling tickets at a high school basketball game. Student tickets cost $2 and general admission tickets cost $3. You sell 2342 tickets and collect $5801. How many of each type of ticket did you sell? (Review 7.2) **1225 student tickets, 1117 general admission tickets**

112. FLOWERS You are buying a combination of irises and white tulips for a flower arrangement. The irises are $1 each and the white tulips are $.50. You spend $20 total to purchase an arrangement of 25 flowers. How many of each kind did you purchase? (Review 7.2) **15 irises and 10 white tulips**

9.2 Simplifying Radicals

What you should learn

GOAL 1 Use properties of radicals to simplify radicals.

GOAL 2 Use quadratic equations to model **real-life** problems, such as the speed of a tsunami in **Ex. 55**.

Why you should learn it

▼ To compare the speeds of two different-sized sailboats in a **real-life** boat race in **Example 3**.

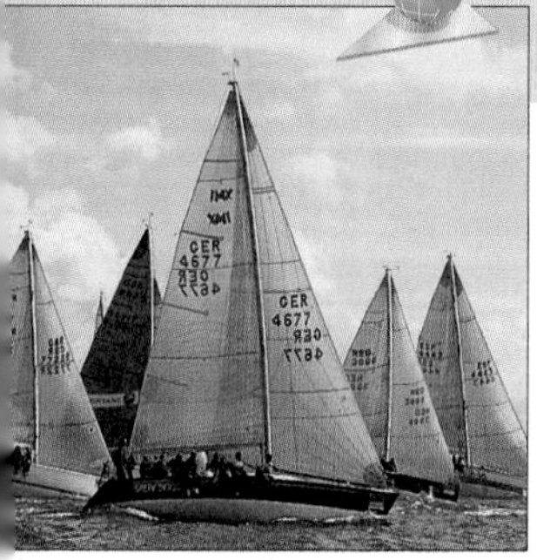

GOAL 1 SIMPLIFYING RADICALS

In this lesson you will learn how to simplify radical expressions and how to recognize when a radical is written in simplest form.

ACTIVITY Developing Concepts

Investigating Properties of Radicals

1. Evaluate the radical expressions $\sqrt{ab}$ and $\sqrt{a} \cdot \sqrt{b}$ for the given values of a and b.

   a. $a = 4, b = 9$ **6, 6** **b.** $a = 64, b = 100$ **80, 80** **c.** $a = 25, b = 4$ **10, 10**

   d. $a = 36, b = 16$ **24, 24** **e.** $a = 100, b = 625$ **250, 250** **f.** $a = 121, b = 49$ **77, 77**

2. What can you conclude? **If a and b are both positive, $\sqrt{ab} = \sqrt{a} \cdot \sqrt{b}$.**

3. Evaluate the radical expressions $\sqrt{\frac{a}{b}}$ and $\frac{\sqrt{a}}{\sqrt{b}}$ for the given values of a and b.

   a. $a = 4, b = 49$ $\frac{2}{7}, \frac{2}{7}$ **b.** $a = 16, b = 64$ $\frac{1}{2}, \frac{1}{2}$ **c.** $a = 25, b = 36$ $\frac{5}{6}, \frac{5}{6}$

   d. $a = 225, b = 4$ $\frac{15}{2}, \frac{15}{2}$ **e.** $a = 144, b = 100$ $\frac{6}{5}, \frac{6}{5}$ **f.** $a = 9, b = 81$ $\frac{1}{3}, \frac{1}{3}$

4. What can you conclude? **If a and b are both positive, $\sqrt{\frac{a}{b}} = \frac{\sqrt{a}}{\sqrt{b}}$.**

In the activity you may have discovered the following two properties that can be used to simplify expressions that contain radicals.

PROPERTIES OF RADICALS

PRODUCT PROPERTY The square root of a product equals the product of the square roots of the factors.

$$\sqrt{ab} = \sqrt{a} \cdot \sqrt{b} \text{ where } a \geq 0 \text{ and } b \geq 0$$

Example: $\sqrt{4 \cdot 100} = \sqrt{4} \cdot \sqrt{100}$

QUOTIENT PROPERTY The square root of a quotient equals the quotient of the square roots of the numerator and denominator.

$$\sqrt{\frac{a}{b}} = \frac{\sqrt{a}}{\sqrt{b}} \text{ where } a \geq 0 \text{ and } b > 0$$

Example: $\sqrt{\frac{9}{25}} = \frac{\sqrt{9}}{\sqrt{25}}$

1 PLAN

PACING

Basic: 1 day
Average: 1 day
Advanced: 1 day
Block Schedule: 0.5 block with 9.3

LESSON OPENER ACTIVITY

An alternative way to approach Lesson 9.2 is to use the Activity Lesson Opener:

- Blackline Master (*Chapter 9 Resource Book*, p. 25)
- Transparency (p. 59)

MEETING INDIVIDUAL NEEDS

- ***Chapter 9 Resource Book***
  Prerequisite Skills Review (p. 5)
  Practice Level A (p. 26)
  Practice Level B (p. 27)
  Practice Level C (p. 28)
  Reteaching with Practice (p. 29)
  Absent Student Catch-Up (p. 31)
  Challenge (p. 33)
- ***Resources in Spanish***
- ***Personal Student Tutor***

NEW-TEACHER SUPPORT

See the Tips for New Teachers on pp. 1–2 of the *Chapter 9 Resource Book* for additional notes about Lesson 9.2.

WARM-UP EXERCISES

Transparency Available

Evaluate each expression.

1. $\sqrt{25}$ 5
2. $-\sqrt{100}$ −10
3. $\sqrt{16} \cdot \sqrt{4}$ 8
4. $\frac{2}{3} \cdot \sqrt{9}$ 2
5. $\frac{\sqrt{100}}{5}$ 2

2 TEACH

MOTIVATING THE LESSON
Ask students how fast they think a tsunami moves (the speed can be more than 500 mi/h). They may be surprised to learn that the speed depends on the water depth—as the water gets shallower near shore, the wave slows down. This model (from Exs. 55–57) involves radical expressions. By simplifying these expressions, students can better picture how the models work.

ACTIVITY NOTE
Students should understand that *a* and *b* must be positive for the conclusions to be valid.

EXTRA EXAMPLE 1
Simplify the expression $\sqrt{48}$.
$4\sqrt{3}$

EXTRA EXAMPLE 2
Simplify the expression.

a. $\sqrt{\frac{7}{16}}$ $\frac{\sqrt{7}}{4}$

b. $\frac{\sqrt{18}}{3}$ $\sqrt{2}$

c. $\sqrt{\frac{80}{45}}$ $\frac{4}{3}$

CHECKPOINT EXERCISES
For use after Examples 1 and 2:
1. Simplify the expression.

a. $\sqrt{125}$ $5\sqrt{5}$ **b.** $\sqrt{\frac{18}{25}}$ $\frac{3\sqrt{2}}{5}$

c. $\frac{\sqrt{72}}{6}$ $\sqrt{2}$ **d.** $\sqrt{\frac{40}{90}}$ $\frac{2}{3}$

STUDENT HELP NOTES
→ **Skills Review** As students review factoring on pp. 777–778, remind them that perfect squares are the squares of 1, 2, 3, and so on.
→ **Homework Help** Students can find extra examples at **www.mcdougallittell.com** that parallel the examples in the student edition.

An expression with radicals is in **simplest form** if the following are true:

- No perfect square factors other than 1 are in the radicand. $\sqrt{8} \rightarrow \sqrt{4 \cdot 2} \rightarrow 2\sqrt{2}$
- No fractions are in the radicand. $\sqrt{\frac{5}{16}} \rightarrow \frac{\sqrt{5}}{\sqrt{16}} \rightarrow \frac{\sqrt{5}}{4}$
- No radicals appear in the denominator of a fraction. $\frac{1}{\sqrt{4}} \rightarrow \frac{1}{2}$

EXAMPLE 1 Simplifying with the Product Property

Simplify the expression $\sqrt{50}$.

SOLUTION

You can use the product property to simplify a radical by removing perfect square factors from the radicand.

$\sqrt{50} = \sqrt{25 \cdot 2}$ **Factor using perfect square factor.**

$= \sqrt{25} \cdot \sqrt{2}$ **Use product property.**

$= 5\sqrt{2}$ **Simplify.**

STUDENT HELP
Skills Review
For help with factoring, see pp. 777–778.

EXAMPLE 2 Simplifying with the Quotient Property

Simplify the expression.

a. $\sqrt{\frac{3}{4}}$ **b.** $\frac{\sqrt{20}}{4}$ **c.** $\sqrt{\frac{32}{50}}$

SOLUTION

a. $\sqrt{\frac{3}{4}} = \frac{\sqrt{3}}{\sqrt{4}}$ **Use quotient property.**

$= \frac{\sqrt{3}}{2}$ **Simplify.**

b. $\frac{\sqrt{20}}{4} = \frac{\sqrt{4 \cdot 5}}{4}$ **Factor using perfect square factor.**

$= \frac{2\sqrt{5}}{4}$ **Remove perfect square factors.**

$= \frac{\sqrt{5}}{2}$ **Divide out common factors.**

c. $\sqrt{\frac{32}{50}} = \sqrt{\frac{2 \cdot 16}{2 \cdot 25}}$ **Factor using perfect square factor.**

$= \sqrt{\frac{16}{25}}$ **Divide out common factors.**

$= \frac{\sqrt{16}}{\sqrt{25}}$ **Use quotient property.**

$= \frac{4}{5}$ **Simplify.**

STUDENT HELP
INTERNET **HOMEWORK HELP**
Visit our Web site www.mcdougallittell.com for extra examples.

FOCUS ON APPLICATIONS

CHASE RACES allow boats with different maximum speeds to race against each other. In Example 3, the smaller boat would start a 10-mile race 40 minutes before the bigger boat to allow for the bigger boat's speed advantage.

APPLICATION LINK
www.mcdougallittell.com

GOAL 2 USING QUADRATIC MODELS IN REAL LIFE

Quadratic equations can be used to model real-life situations, such as comparing the maximum speeds of two sailboats with different water line lengths.

EXAMPLE 3 *Simplifying Radical Expressions*

BOAT RACING The maximum speed s (in knots or nautical miles per hour) that some kinds of boats can travel can be modeled by

$s^2 = \frac{16}{9}x$, where x is the length of the water line in feet.

A sailboat with a 16-foot water line is racing against a sailboat with a 32-foot water line. The smaller sailboat is traveling at a maximum speed of 5.3 knots.

a. Find the maximum speed of the sailboat with the 32-foot water line. Find the speed to the nearest tenth.

b. Is the maximum speed of the sailboat with the 32-foot water line twice that of the sailboat with the 16-foot water line? Explain.

SOLUTION

a. Write the model for maximum speed of a sailboat and let $x = 32$ feet.

$s^2 = \frac{16}{9}x$ — Write quadratic model.

$s^2 = \frac{16}{9} \cdot 32$ — Substitute 32 for x.

$\sqrt{s^2} = \sqrt{\frac{16}{9} \cdot 32}$ — Find square root of both sides.

$s = \frac{\sqrt{16}}{\sqrt{9}} \cdot \sqrt{32}$ — Use quotient and product properties.

$= \frac{\sqrt{16}}{\sqrt{9}} \cdot \sqrt{16} \cdot \sqrt{2}$ — Identify perfect square factors.

$= \frac{4}{3} \cdot 4\sqrt{2}$ — Remove perfect square factors.

$= \frac{16\sqrt{2}}{3}$ — Simplify.

≈ 7.5 — Use a calculator or square root table.

b. No, because 7.5 knots is not twice 5.3 knots. The maximum speed of the smaller sailboat is $s = \sqrt{\frac{16^2}{9}}$, or $\frac{16}{3}$, so the longer sailboat's maximum speed is only $\sqrt{2}$ times the smaller sailboat's maximum speed.

EXTRA EXAMPLE 3
The distance d you can see to the horizon depends on your height h. A model is $d^2 = 1.5h$, with d in miles and h in feet.
a. Find the exact distance you can see from the top of a 400 ft building. **$10\sqrt{6}$ mi**
b. Find the distance in Part (a) to the nearest tenth. **24.5 mi**
c. If you were $3 \times 400 = 1200$ ft up in a skyscraper, how far could you see to the nearest tenth? **42.4 mi**

CHECKPOINT EXERCISES
For use after Example 3:
1. Solve $y^2 = \frac{9}{16}x$ for y if $x = 24$. $\frac{3}{2}\sqrt{6}, -\frac{3}{2}\sqrt{6}$

APPLICATION NOTE
EXAMPLE 3 Additional information about chase races is available at **www.mcdougallittell.com.**

FOCUS ON VOCABULARY
Give an example of the product property of square roots. ***Sample answer:*** $\sqrt{25 \cdot 16} = \sqrt{25} \cdot \sqrt{16}$

CLOSURE QUESTION
When is a radical expression in simplest form? ***Sample answer:*** **when the radicand has no perfect square factors other than one; when there are no fractions in the radicand, and when no radicals appear in the denominator of a fraction.**

DAILY PUZZLER
What's wrong with this "proof" that $4 = -4$?
$4 = \sqrt{4(4)} = \sqrt{-4(-4)} = \sqrt{-4} \cdot \sqrt{-4} = (\sqrt{-4})^2 = -4$
Sample answer: **The product property does not apply to negative numbers. It is not true that $\sqrt{-4(-4)} = \sqrt{-4} \cdot \sqrt{-4}$.**

3 APPLY

ASSIGNMENT GUIDE

BASIC
Day 1: pp. 514–516 Exs. 10–52 even, 58–60, 65, 70, 74, 75

AVERAGE
Day 1: pp. 514–516 Exs. 10–52 even, 53, 54, 58–60, 65, 70, 74, 75

ADVANCED
Day 1: pp. 514–516 Exs. 10–52 even, 55–65, 70, 74, 75

BLOCK SCHEDULE WITH 9.3
pp. 514–516 Exs. 10–52 even, 53, 54, 58–60, 65, 70, 74, 75

EXERCISE LEVELS

Level A: *Easier*
10–21

Level B: *More Difficult*
22–56, 58–60, 65–75

Level C: *Most Difficult*
57, 61–64

✔ HOMEWORK CHECK

To quickly check student understanding of key concepts, go over the following exercises: Exs. 16, 26, 28, 36, 48, 52. See also the Daily Homework Quiz:

- Blackline Master (*Chapter 2 Resource Book*, p. 36)
- Transparency (p. 69)

! COMMON ERROR

EXERCISE 48 Students may need to see a similar problem with integers to remind them that the denominator can be paired with either factor in the numerator when simplifying. For example, since $3 = 3 \cdot 1$, $\frac{6 \cdot 9}{3}$ could be rewritten as $\frac{6}{3} \cdot 9$ or as $6 \cdot \frac{9}{3}$. The factor to choose is the one that makes simplifying the easiest.

GUIDED PRACTICE

Vocabulary Check ✓

1. Is the radical expression in simplest form? Explain. **a–c. See margin.**

a. $\frac{3}{5}\sqrt{2}$ **b.** $\sqrt{\frac{3}{16}}$ **c.** $5\sqrt{40}$

Concept Check ✓

2. Explain how to use the product property of radicals to simplify $\sqrt{3} \cdot \sqrt{15}$.
$\sqrt{3} \cdot \sqrt{15} = \sqrt{3} \cdot \sqrt{3 \cdot 5} = \sqrt{3} \cdot \sqrt{3} \cdot \sqrt{5} = 3\sqrt{5}$

3. Explain how to use the quotient property of radicals to simplify $\sqrt{\frac{4}{25}}$.
$\sqrt{\frac{4}{25}} = \frac{\sqrt{4}}{\sqrt{25}} = \frac{2}{5}$

Skill Check ✓

Match the radical expression with its simplified form.

A. $3\sqrt{6}$ **B.** $9\sqrt{6}$ **C.** $2\sqrt{2}$ **D.** $4\sqrt{2}$

4. $\sqrt{32}$ **D** **5.** $\sqrt{54}$ **A** **6.** $\sqrt{486}$ **B** **7.** $\sqrt{8}$ **C**

8. SAILING Find the maximum speed of a sailboat with a 25-foot water line using the formula in Example 3. **about 6.7 knots**

9. ERROR ANALYSIS Describe the error.

$\sqrt{50} = \sqrt{5 \cdot 10}$
$= 5\sqrt{10}$

Simplify correctly. ***Sample answer:*** **The factor 5 cannot be taken outside the radicand. The first step should be $\sqrt{50} = \sqrt{25 \cdot 2}$. Then, by the product property, this equals $\sqrt{25} \cdot \sqrt{2}$, or $5\sqrt{2}$.**

1. a. Yes; there are no perfect square factors other than 1 in the radicand, there are no fractions in the radicand, and no radicals appear in the denominator of the fraction.

PRACTICE AND APPLICATIONS

STUDENT HELP
Extra Practice to help you master skills is on p. 805.

1. b. No; there is a fraction in the radicand and the denominator of the fraction is a perfect square.

c. No; the radicand contains a perfect square factor.

PRODUCT PROPERTY **Simplify the expression.**

10. $\sqrt{44}$ $2\sqrt{11}$ **11.** $\sqrt{27}$ $3\sqrt{3}$ **12.** $\sqrt{48}$ $4\sqrt{3}$ **13.** $\sqrt{75}$ $5\sqrt{3}$

14. $\sqrt{90}$ $3\sqrt{10}$ **15.** $\sqrt{125}$ $5\sqrt{5}$ **16.** $\sqrt{200}$ $10\sqrt{2}$ **17.** $\sqrt{80}$ $4\sqrt{5}$

18. $\frac{1}{2}\sqrt{112}$ $2\sqrt{7}$ **19.** $\frac{1}{3}\sqrt{54}$ $\sqrt{6}$ **20.** $\sqrt{2} \cdot \sqrt{8}$ 4 **21.** $\sqrt{6} \cdot \sqrt{8}$ $4\sqrt{3}$

QUOTIENT PROPERTY **Simplify the expression.**

22. $\sqrt{\frac{7}{9}}$ $\frac{\sqrt{7}}{3}$ **23.** $\sqrt{\frac{11}{16}}$ $\frac{\sqrt{11}}{4}$ **24.** $2\sqrt{\frac{5}{4}}$ $\sqrt{5}$ **25.** $18\sqrt{\frac{5}{81}}$ $2\sqrt{5}$

26. $2\sqrt{\frac{10}{2}}$ $2\sqrt{5}$ **27.** $3\sqrt{\frac{9}{3}}$ $3\sqrt{3}$ **28.** $8\sqrt{\frac{13}{9}}$ $\frac{8\sqrt{13}}{3}$ **29.** $3\sqrt{\frac{8}{64}}$ $\frac{3\sqrt{2}}{4}$

30. $4\sqrt{\frac{16}{4}}$ 8 **31.** $3\sqrt{\frac{3}{16}}$ $\frac{3\sqrt{3}}{4}$ **32.** $5\sqrt{\frac{6}{2}}$ $5\sqrt{3}$ **33.** $8\sqrt{\frac{20}{4}}$ $8\sqrt{5}$

SIMPLIFYING **Simplify the expression.**

34. $\frac{\sqrt{32}}{\sqrt{25}}$ $\frac{4\sqrt{2}}{5}$ **35.** $\sqrt{\frac{27}{36}}$ $\frac{\sqrt{3}}{2}$ **36.** $\frac{\sqrt{49}}{\sqrt{4}}$ $\frac{7}{2}$ **37.** $\frac{\sqrt{36}}{\sqrt{9}}$ 2

38. $\frac{\sqrt{9}}{\sqrt{49}}$ $\frac{3}{7}$ **39.** $\frac{\sqrt{48}}{\sqrt{81}}$ $\frac{4\sqrt{3}}{9}$ **40.** $\frac{\sqrt{64}}{\sqrt{16}}$ 2 **41.** $\frac{\sqrt{120}}{\sqrt{4}}$ $\sqrt{30}$

42. $\frac{1}{2}\sqrt{32} \cdot \sqrt{2}$ 4 **43.** $3\sqrt{63} \cdot \sqrt{4}$ $18\sqrt{7}$ **44.** $\sqrt{9} \cdot 4\sqrt{25}$ 60 **45.** $-2\sqrt{27} \cdot \sqrt{3}$ −18

46. $\sqrt{7} \cdot \frac{\sqrt{18}}{\sqrt{2}}$ $3\sqrt{7}$ **47.** $-\sqrt{4} \cdot \frac{\sqrt{81}}{\sqrt{36}}$ −3 **48.** $\frac{\sqrt{10} \cdot \sqrt{16}}{\sqrt{5}}$ $4\sqrt{2}$ **49.** $\frac{-2\sqrt{20}}{\sqrt{100}}$ $\frac{-2\sqrt{5}}{5}$

STUDENT HELP
HOMEWORK HELP
Example 1: Exs. 10–21
Example 2: Exs. 22–41
Example 3: Exs. 42–57

STUDENT HELP

Skills Review
For help with area of geometric figures, see pp. 790–791.

GEOMETRY CONNECTION **Find the area of the figure. Give both the exact answer in simplified form and the decimal approximation rounded to the nearest hundredth.**

50.

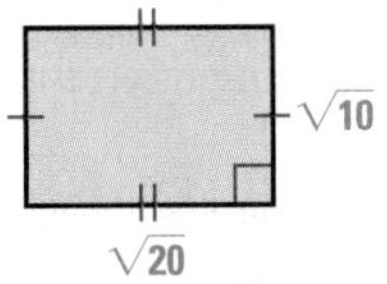

$10\sqrt{2}$; 14.14

51.

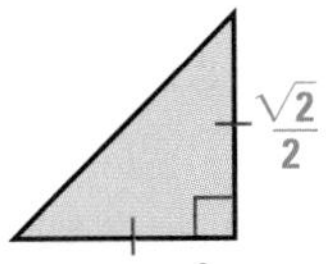

$\frac{1}{4}$; 0.25

52.

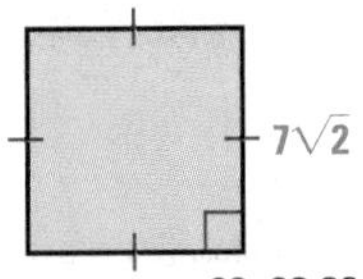

98; 98.00

BODY SURFACE AREA **In Exercises 53 and 54, use the following information.**

Physicians can approximate the Body Surface Area of an adult (in square meters) using an index called *BSA* where *H* is height in centimeters and *W* is weight in kilograms.

Body Surface Area: $\sqrt{\frac{HW}{3600}}$

53. Find the *BSA* of a person who is 180 centimeters tall and weighs 75 kilograms. $\frac{\sqrt{15}}{2} \approx 1.94\ m^2$

54. Find the *BSA* of a person who is 160 centimeters tall and weighs 50 kilograms. $\frac{2\sqrt{5}}{3} \approx 1.49\ m^2$

TSUNAMI **In Exercises 55–57, use the following information.**

A *tsunami* is a destructive, fast-moving ocean wave that is caused by an undersea earthquake, landslide, or volcano. The Pacific Tsunami Warning Center is responsible for monitoring earthquakes that could potentially cause tsunamis in the Pacific Ocean. Through measuring the water level and calculating the speed of a tsunami, scientists can predict arrival times of tsunamis.

The speed s (in meters per second) at which a tsunami moves is determined by the depth d (in meters) of the ocean.

$s = \sqrt{gd}$, where g is 9.8 meters per second per second

55. Find the speed of a tsunami in a region of the ocean that is 1000 meters deep. Write the result in simplified form. $70\sqrt{2}$ m/sec

56. Find the speed of a tsunami in a region of the ocean that is 4000 meters deep. Write the result in simplified form. $140\sqrt{2}$ m/sec

57. **CRITICAL THINKING** Is the speed of a tsunami at a depth of 4000 meters four times the speed of a tsunami at 1000 meters? Explain why or why not.

57. No; the speed is twice the speed of a tsunami at 1000 m deep; the speed is a function of the square root of the depth, not of the depth.

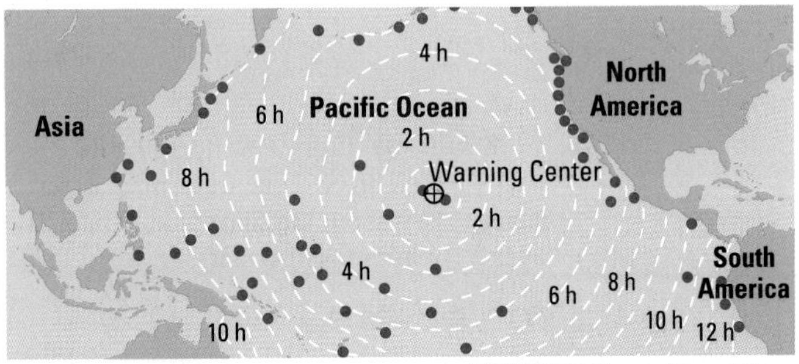

• Water elevation stations

Tsunami travel times (in hours) to Honolulu, Hawaii

STUDENT HELP NOTES

Skills Review As students review the area of geometric figures on pp. 790–791, remind them that area formulas also apply when lengths are irrational.

APPLICATION NOTE

EXERCISES 55–57 Because energy is conserved, when a tsunami slows as it enters shallow water near shore, its height increases. So, a wave that is barely noticeable in deep water may tower as it hits the coastline.

ADDITIONAL PRACTICE AND RETEACHING

For Lesson 9.2:

- Practice Levels A, B, and C (*Chapter 9 Resource Book*, p. 26)
- Reteaching with Practice (*Chapter 9 Resource Book*, p. 29)
- See Lesson 9.2 of the *Personal Student Tutor*

For more Mixed Review:

- Search the *Test and Practice Generator* for key words or specific lessons.

4 ASSESS

DAILY HOMEWORK QUIZ

Transparency Available

Simplify the expression.

1. $\sqrt{112}$ $4\sqrt{7}$
2. $-3\sqrt{\frac{32}{4}}$ $-6\sqrt{2}$
3. $6\sqrt{\frac{28}{9}}$ $4\sqrt{7}$
4. $\frac{\sqrt{76}}{\sqrt{25}}$ $\frac{2\sqrt{19}}{5}$
5. $\sqrt{12} \cdot \frac{\sqrt{18}}{\sqrt{3}}$ $6\sqrt{2}$
6. A rectangle has width $2\sqrt{5}$ and length $\sqrt{20}$. Find its area. 20 units^2

EXTRA CHALLENGE NOTE

Challenge problems for Lesson 9.2 are available in **blackline** format in the *Chapter 9 Resource Book,* p. 33 and at **www.mcdougallittell.com.**

ADDITIONAL TEST PREPARATION

1. **OPEN ENDED** How can you decide if $\sqrt{n}$ is in simplest form? ***Sample answer:*** **Look at the factors of *n*. If there are any repeated factors, then *n* has perfect square factors, and you can apply the product property to simplify. If there are no repeated factors, then it is in simplest form.**

61a.

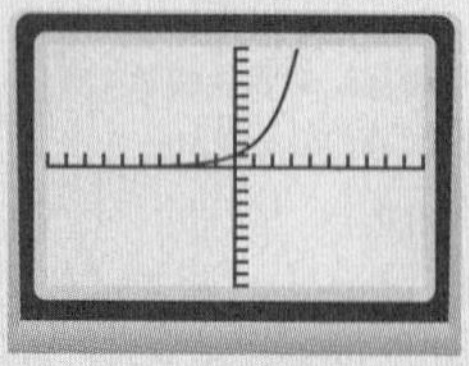

65.

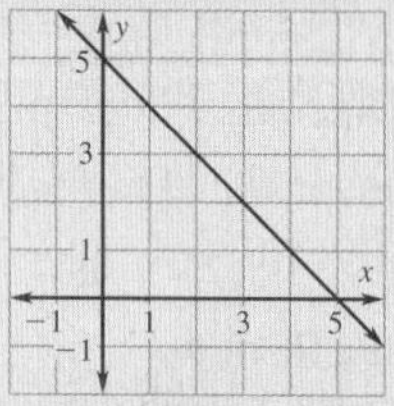

66 and 67. See Additional Answers beginning on page AA1.

Test Preparation

58. **MULTIPLE CHOICE** Which is the simplified form of $4\frac{\sqrt{125}}{\sqrt{25}}$? **B**

   Ⓐ $2\sqrt{5}$ Ⓑ $4\sqrt{5}$ Ⓒ $20\sqrt{5}$ Ⓓ $\frac{4\sqrt{5}}{5}$

59. **MULTIPLE CHOICE** Which is the simplified form of $33\sqrt{\frac{2}{121}}$? **C**

   Ⓐ $\frac{2\sqrt{33}}{\sqrt{11}}$ Ⓑ $33\sqrt{2}$ Ⓒ $3\sqrt{2}$ Ⓓ $\frac{3\sqrt{2}}{\sqrt{11}}$

60. **MULTIPLE CHOICE** Which is the simplified form of $\frac{6\sqrt{52}}{\sqrt{2} \cdot \sqrt{8}}$? **B**

   Ⓐ $\frac{3\sqrt{13}}{\sqrt{2}}$ Ⓑ $3\sqrt{13}$ Ⓒ $3\sqrt{26}$ Ⓓ $\frac{6\sqrt{13}}{\sqrt{2}}$

★ Challenge

EXTENSION: FRACTIONAL EXPONENTS **Sometimes $\sqrt{3}$ is written as $3^{1/2}$.**

61. You can obtain a graphical representation of the relationship $2^{1/2} = \sqrt{2}$ by investigating the graph of $f(x) = 2^x$.

   a. Graph $f(x) = 2^x$. **See margin.**

   b. Use the *Trace* feature to find values of f when $x = \frac{1}{2}$. **1.4142136**

   c. Compare the value from part (b) with the value of $\sqrt{2}$. **The two values are the same.**

EXTRA CHALLENGE
www.mcdougallittell.com

Using the fact that $x^{1/2} = \sqrt{x}$, rewrite in simplest radical form.

62. $6x^{1/2}$ $6\sqrt{x}$
63. $x^{1/2} \cdot 4\sqrt{2}$ $4\sqrt{2x}$
64. $18^{1/2}x \cdot 9x^{1/2}x$ $27x^2\sqrt{2x}$

MIXED REVIEW

GRAPHING EQUATIONS **Use a table of values to graph the equation.** **(Review 4.2 for 9.3)** **65–67. See margin.**

65. $y = -x + 5$
66. $y = x - 7$
67. $y = 3x - 1$

EVALUATING EXPRESSIONS **Simplify the expression. Then evaluate the expression when $a = 1$ and $b = 2$.** **(Review 8.1)**

68. $(a^3)^2$ a^6; 1
69. $b^6 \cdot b^2$ b^8; 256
70. $(a^3b)^4$ $a^{12}b^4$; 16

REWRITING EXPRESSIONS **Rewrite the expression using positive exponents.** **(Review 8.2)**

71. $\frac{1}{2x^{-5}}$ $\frac{x^5}{2}$
72. $\frac{1}{4x^{-7}}$ $\frac{x^7}{4}$
73. $x^{-4}y^3$ $\frac{y^3}{x^4}$
74. $6x^{-2}y^{-6}$ $\frac{6}{x^2y^6}$

75. **BUSINESS** From 1994 to 1999, the sales for a chain of home furnishing stores increased by about the same annual rate. The sales S (in millions of dollars) in year t can be modeled by $S = 455\left(\frac{13}{10}\right)^t$ where t represents years since 1994. Find the ratio of 1999 sales to 1995 sales. **(Review 8.3)** $(1.3)^4 \approx 2.86$

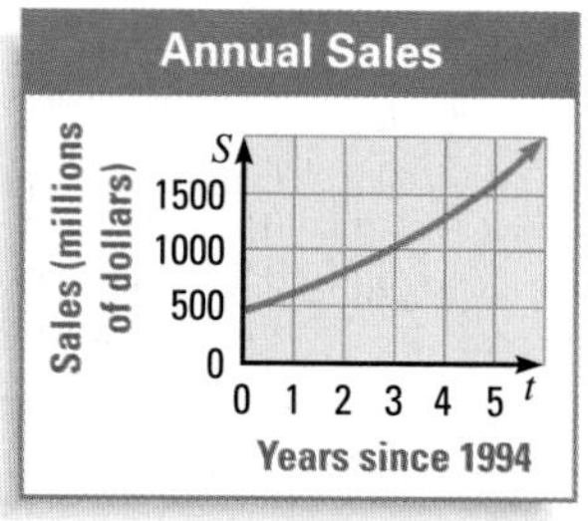

ACTIVITY 9.3

Developing Concepts

Group Activity for use with Lesson 9.3

Investigating Graphs of Quadratic Functions

QUESTION **How do the coefficients *a*, *b*, and *c* affect the shape of the graph of the quadratic function $y = ax^2 + bx + c$?**

EXPLORING THE CONCEPT **Steps 1–3. Sample answers are given. See margin.**

1. Use a graphing calculator to graph $y = ax^2$ using −2, −1, −0.5, 0.5, 1, and 2 as values of *a*. Adjust the viewing window if necessary. Discuss your results with others in your group.

   Write a sentence that describes how the value of *a* affects the graph of $y = ax^2$.

2. Use a graphing calculator to graph $y = x^2 + bx$ using −4, −2, −1, 0, 1, 2, and 4 as values of *b*. Adjust the viewing window if necessary. Discuss your results with others in your group.

   Write a sentence that describes how the value of *b* affects the graph of $y = x^2 + bx$.

3. Use a graphing calculator to graph $y = x^2 + c$ using −5, −3, −1, 1, 3, and 5 as values of *c*. Adjust the viewing window if necessary. Discuss your results with others in your group.

   Write a sentence that describes how the value of *c* affects the graph of $y = x^2 + c$.

DRAWING CONCLUSIONS

ANALYZING GRAPHS **From the graph, what can you tell about the *a*, *b*, and *c* values of the function?**

1.

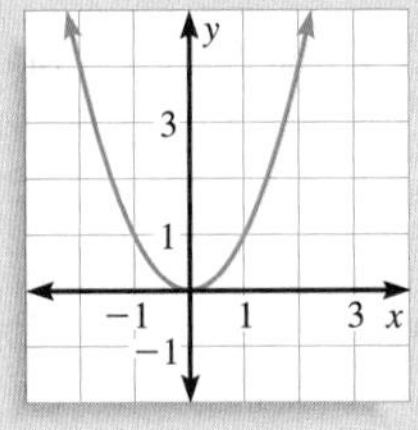

$a = 1$ (or $a > 0$), $b = 0$, $c = 0$

2.

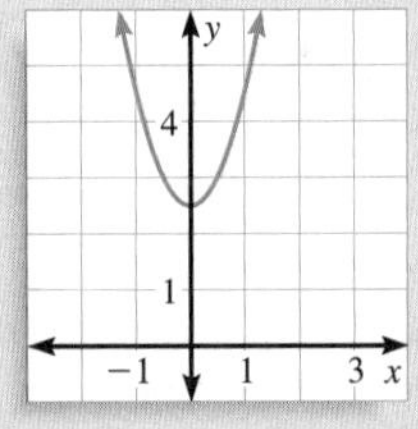

$a = 1$ (or $a > 0$), $b = 0$, $c > 0$

3.

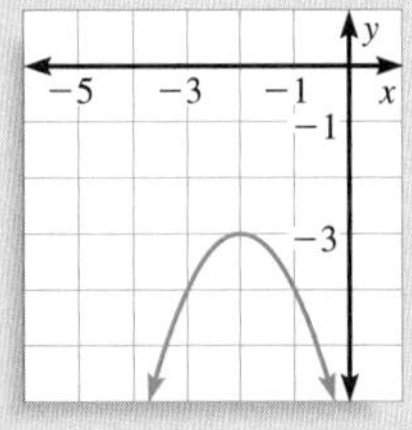

$a < 0$, $b < 0$, $c < 0$

SKETCHING GRAPHS **Sketch the graph of the function. Use the *a*, *b*, and *c* values.** **4–9. See margin.**

4. $y = 5x^2$ **5.** $y = 5x^2 + 10x$ **6.** $y = 5x^2 + 10x - 5$

7. $y = -2x^2$ **8.** $y = -2x^2 - 7x$ **9.** $y = -2x^2 - 7x + 6$

10. **GRAPHING FUNCTIONS** Which of the quadratic functions could be shown by the graph at the right? Explain your reasoning.

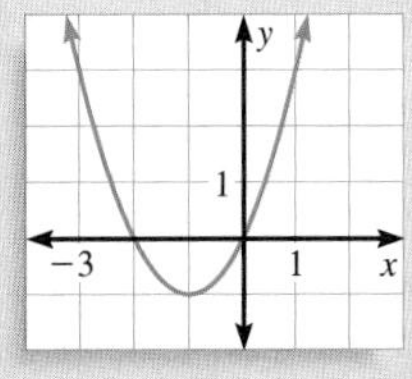

A. $y = x^2 - 2$ **B.** $y = x^2 + 2$

C. $y = 2x^2$ **D.** $y = x^2 + 2x$

D. *Sample answer:* For the graph shown, $b \neq 0$.

GROUP ACTIVITY
Work in a small group.

MATERIALS
graphing calculator

STUDENT HELP

KEYSTROKE HELP
See keystrokes for several models of calculators at www.mcdougallittell.com

Step 1. As $|a|$ increases, the graph of $y = ax^2$ gets narrower, and as $|a|$ decreases, the graph of $y = ax^2$ gets wider. If *a* is positive, the graph of $y = ax^2$ opens up, and if *a* is negative, the graph of $y = ax^2$ opens down.

Step 2. For negative values of *b*, the graph of $y = x^2 + bx$ is the graph of $y = x^2$ moved down and to the right. For positive values of *b*, the graph of $y = x^2 + bx$ is the graph of $y = x^2$ moved down and to the left.

Step 3. For negative values of *c*, the graph of $y = x^2 + c$ is the graph of $y = x^2$ moved down. For positive values of *c*, the graph of $y = x^2 + c$ is the graph of $y = x^2$ moved up.

1 Planning the Activity

PURPOSE
To investigate the shapes of graphs of quadratic functions.

MATERIALS
- graphing calculator for each group
- Keystroke blackline (*Chapter 9 Resource Book*, p. 40)

PACING
- Exploring the Concept — 15 min
- Drawing Conclusions — 15 min

LINK TO LESSON
The effect of changing *b* on the graph of $y = ax^2 + bx + c$ may be unclear to students. When they learn on p. 518 that the *x*-coordinate of the vertex is $-\frac{b}{2a}$, they will see the relationship more clearly.

2 Managing the Activity

CLASSROOM MANAGEMENT
As students explore the effects of changing *a*, *b*, and *c*, encourage them to look at the *shape* of the graphs, the *position* of the graphs (especially that of the vertex) and the *direction* that the graphs open.

ALTERNATE APPROACH
To save time, demonstrate this activity using a graphing calculator display. You can use the table features to display coordinates.

3 Closing the Activity

KEY DISCOVERY
You can predict the shape of the graph of $y = ax^2 + bx + c$, if it opens up or down, and its general position by examining *a*, *b*, and *c*.

ACTIVITY ASSESSMENT
JOURNAL Describe the effect of changing *a* in the quadratic function $y = ax^2 + bx + c$.
See sample answer at left.

Activity Assessment *Sample answer:*
If *a* is positive, the graph opens up. If *a* is negative, it opens down. As the absolute value of *a* increases, the graph narrows. As the absolute value decreases, the graph widens.

4–9. See Additional Answers beginning on page AA1.

1 PLAN

PACING
Basic: 2 days
Average: 2 days
Advanced: 2 days
Block Schedule: 0.5 block with 9.2
0.5 block with 9.4

LESSON OPENER
CALCULATOR ACTIVITY
An alternative way to approach Lesson 9.3 is to use the Calculator Activity Lesson Opener:

- Blackline Master (*Chapter 9 Resource Book,* p. 37)
- Transparency (p. 60)

MEETING INDIVIDUAL NEEDS

- ***Chapter 9 Resource Book***
  Prerequisite Skills Review (p. 5)
  Practice Level A (p. 43)
  Practice Level B (p. 44)
  Practice Level C (p. 45)
  Reteaching with Practice (p. 46)
  Absent Student Catch-Up (p. 48)
  Challenge (p. 50)
- ***Resources in Spanish***
- ***Personal Student Tutor***

NEW-TEACHER SUPPORT
See the Tips for New Teachers on pp. 1–2 of the *Chapter 9 Resource Book* for additional notes about Lesson 9.3.

WARM-UP EXERCISES

Transparency Available

Evaluate $-\frac{b}{2a}$ for the following values.

1. $a = -3$ and $b = 8$ $\frac{4}{3}$

2. $a = 4$ and $b = -6$ $\frac{3}{4}$

Evaluate the expression for $x = -4$ and $x = 3$.

3. $x^2 + 4x - 1$ **−1, 20**

4. $-2x^2 - 7x + 3$ **−1, −36**

9.3 Graphing Quadratic Functions

What you should learn

GOAL 1 Sketch the graph of a quadratic function.

GOAL 2 Use quadratic models in **real-life** settings, such as finding the winning distance of a shot put in **Example 3**.

Why you should learn it

▼ To model **real-life** parabolic situations, such as the height above the water of a jumping dolphin in **Exs. 65** and **66**.

REAL LIFE

GOAL 1 SKETCHING A QUADRATIC FUNCTION

A **quadratic function** is a function that can be written in the **standard form**

$y = ax^2 + bx + c$, where $a \neq 0$.

Every quadratic function has a U-shaped graph called a **parabola**. If the leading coefficient a is positive, the parabola *opens up.* If the leading coefficient is negative, the parabola *opens down* in the shape of an upside down U.

The graph on the left has a positive leading coefficient so it opens up. The graph on the right has a negative leading coefficient so it opens down.

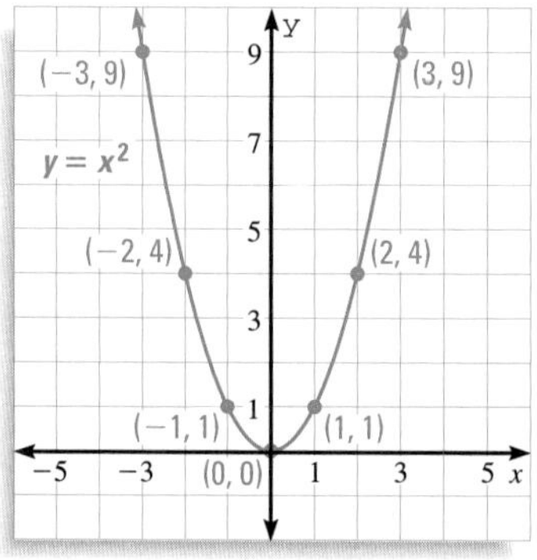

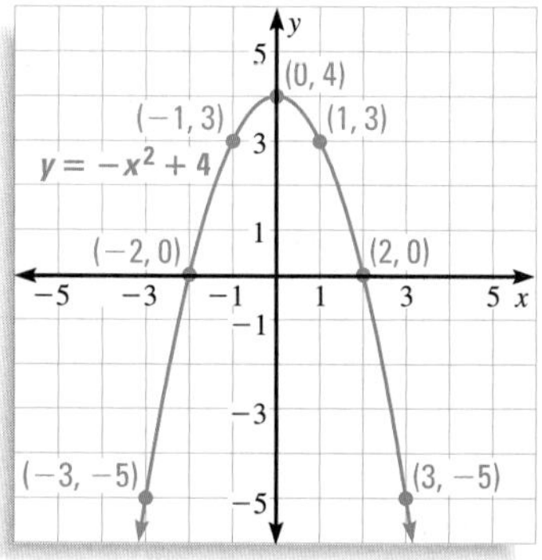

The **vertex** is the lowest point of a parabola that opens up and the highest point of a parabola that opens down. The parabola on the left has a vertex of (0, 0) and the parabola on the right has a vertex of (0, 4).

The line passing through the vertex that divides the parabola into two symmetric parts is called the **axis of symmetry**. The two symmetric parts are mirror images of each other.

GRAPH OF A QUADRATIC FUNCTION

The graph of $y = ax^2 + bx + c$ is a parabola.

- If a is positive, the parabola opens up.
- If a is negative, the parabola opens down.
- The vertex has an x-coordinate of $-\frac{b}{2a}$.
- The axis of symmetry is the vertical line $x = -\frac{b}{2a}$.

GRAPHING A QUADRATIC FUNCTION

STEP 1 Find the x-coordinate of the vertex.

STEP 2 Make a table of values, using x-values to the left and right of the vertex.

STEP 3 Plot the points and connect them with a smooth curve to form a parabola.

EXAMPLE 1 *Graphing a Quadratic Function with a Positive a-value*

Sketch the graph of $y = x^2 - 2x - 3$.

SOLUTION

1. Find the x-coordinate of the vertex when $a = 1$ and $b = -2$.

$$-\frac{b}{2a} = -\frac{-2}{2(1)} = 1$$

2. Make a table of values, using x-values to the left and right of $x = 1$.

x	-2	-1	0	1	2	3	4
y	5	0	-3	-4	-3	0	5

3. Plot the points. The vertex is $(1, -4)$ and the axis of symmetry is $x = 1$. Connect the points to form a parabola that opens up since a is positive.

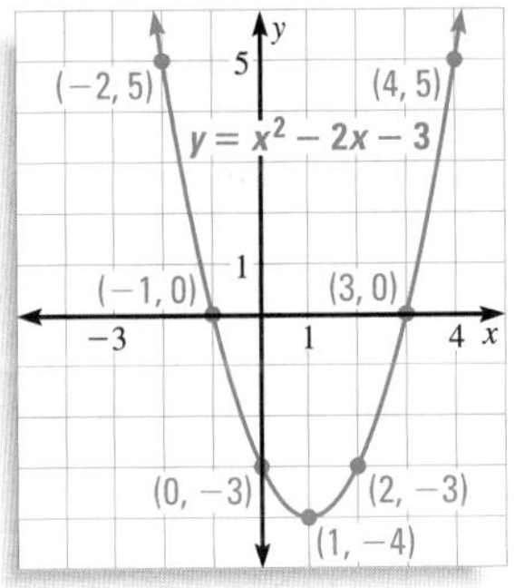

EXAMPLE 2 *Graphing a Quadratic Function with a Negative a-value*

Sketch the graph of $y = -2x^2 - x + 2$.

SOLUTION

1. Find the x-coordinate of the vertex when $a = -2$ and $b = -1$.

$$-\frac{b}{2a} = -\frac{-1}{2(-2)} = -\frac{1}{4}$$

2. Make a table of values, using x-values to the left and right of $x = -\frac{1}{4}$.

x	-2	-1	$-\frac{1}{4}$	0	1	2
y	-4	1	$2\frac{1}{8}$	2	-1	-8

3. Plot the points. The vertex is $\left(-\frac{1}{4}, 2\frac{1}{8}\right)$ and the axis of symmetry is $x = -\frac{1}{4}$. Connect the points to form a parabola that opens down since a is negative.

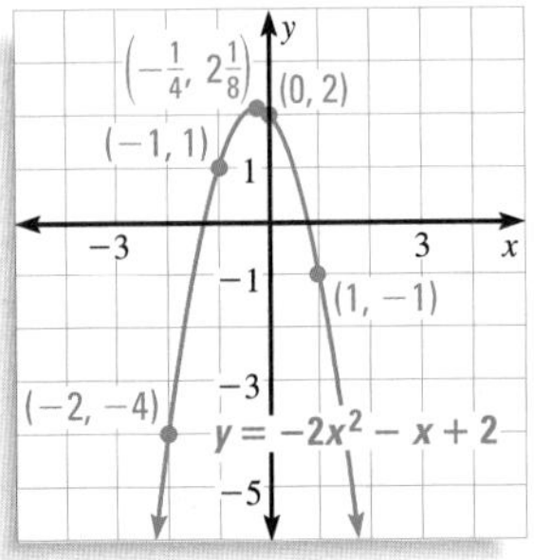

> **STUDENT HELP**
>
> **Study Tip**
> If the x-coordinate of the vertex is a fraction, you can still choose whole numbers when you make a table.

2 TEACH

MOTIVATING THE LESSON
Many sports students participate in involve throwing or kicking balls. Ask students what shape the path of a ball has. They will probably answer that the shape is an arc. Tell students that this arc is called a *parabola,* and that it can be described by a quadratic function.

EXTRA EXAMPLE 1
Sketch the graph of $y = x^2 + 3x - 4$.

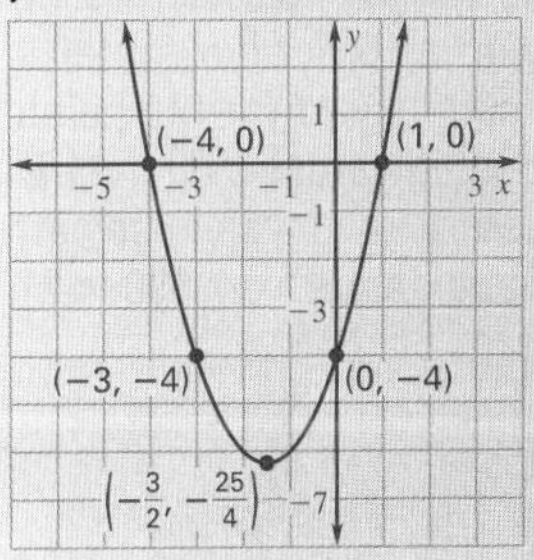

EXTRA EXAMPLE 2
Sketch the graph of $y = -\frac{1}{2}x^2 - x + 1$.

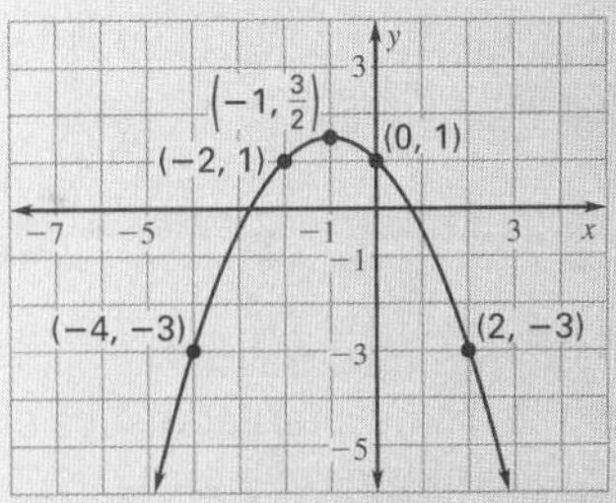

CHECKPOINT EXERCISES
For use after Examples 1 and 2:
Give the coordinates of the vertex of the graph of each quadratic function, and tell whether it opens up or down.

1. $y = x^2 + x - 2$ $\left(-\frac{1}{2}, -\frac{9}{4}\right)$, **up**

2. $y = -2x^2 + 5x - 4$ $\left(\frac{5}{4}, -\frac{7}{8}\right)$, **down**

EXTRA EXAMPLE 3

A struck baseball follows a path modeled by $y = -0.00993x^2 + 1.1106x + 4.1$, where x is the horizontal distance and y is the height (both in feet).

a. To the nearest foot, what is the maximum height of the baseball? **35 ft**

b. How far, to the nearest foot, does the ball travel before hitting the ground? **115 ft**

CHECKPOINT EXERCISES

For use after Example 3:

1. A golf ball follows a path modeled by $y = -0.005x^2 + 1.1x$, where x is the horizontal distance and y is the height (both in yards). Will the ball clear a sand trap in its path that extends from 190 to 210 yards out? **yes**

FOCUS ON VOCABULARY

Describe the axis of symmetry of the graph of a quadratic function. ***Sample answer:*** **The axis of symmetry is the vertical line $\left(\text{with equation } x = -\frac{b}{2a}\right)$ that passes through the vertex of a parabola and divides it into two symmetric halves that are mirror images of each other.**

CLOSURE QUESTION

Describe some characteristics of the graph of $y = -x^2 + 3x - 2$. ***Sample answer:*** **It is a parabola that opens down, has axis of symmetry $x = \frac{3}{2}$, vertex $\left(\frac{3}{2}, \frac{1}{4}\right)$, and contains the points (0, −2), (1, 0), and (2, 0).**

DAILY PUZZLER

A parabola has vertex (0, 1), contains (2, 0), and has the same width as the parabola $y = \frac{1}{4}x^2$. What are a, b, and c in its equation $y = ax^2 + bx + c$? Write the equation. **$a = -\frac{1}{4}$, $b = 0$, $c = 1$; $y = -\frac{1}{4}x^2 + 1$**

GOAL 2 USING QUADRATIC MODELS IN REAL LIFE

When an object has little air resistance, its path through the air can be approximated by a parabola.

FOCUS ON PEOPLE

NATALYA LISOVSKAYA
In 1987, the women's world record in shot put was set by Lisovskaya. She also won an Olympic gold medal in 1988.

EXAMPLE 3 *Using a Quadratic Model*

TRACK AND FIELD Natalya Lisovskaya holds the world record for the women's shot put. The path of her record-breaking throw can be modeled by $y = -0.01347x^2 + 0.9325x + 5.5$, where x is the horizontal distance in feet and y is the height (in feet). The initial height is represented by 5.5, the height at which the shot (a 4-kilogram metal ball) was released.

a. What was the maximum height (in feet) of the shot thrown by Lisovskaya?

b. What was the distance of the throw to the nearest hundredth of a foot?

SOLUTION

a. The maximum height of the throw occurred at the vertex of the parabolic path. Find the x-coordinate of the vertex. Use $a = -0.01347$ and $b = 0.9325$.

$$-\frac{b}{2a} = -\frac{0.9325}{2(-0.01347)} \approx 34.61$$

Substitute 34.61 for x in the model to find the maximum height.

$$y = -0.01347(34.61)^2 + 0.9325(34.61) + 5.5 \approx 21.6$$

▶ The maximum height of the shot was about 21.6 feet.

b. To find the distance of the throw, sketch the parabolic path of the shot. Because a is negative, the graph opens down. Use (35, 22) as the vertex. The y-intercept is 5.5. Sketch a symmetric curve.

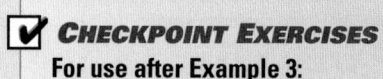

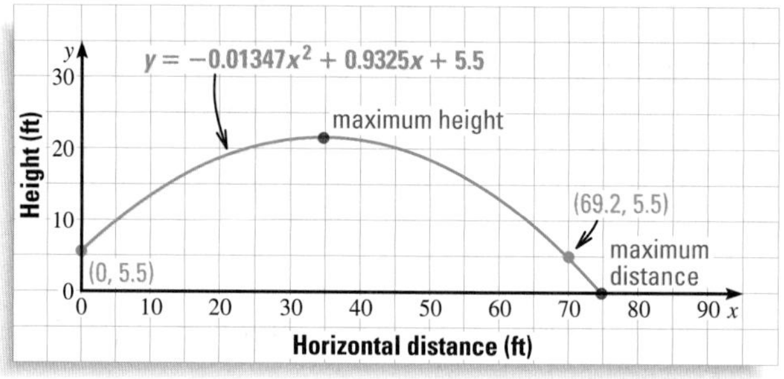

The distance of the throw is the x-value that yields a y-value of 0. The graph shows that the distance was between 74 and 75 feet. You can use a table to refine this estimate.

Distance, x	74.5 ft	74.6 ft	74.7 ft	74.8 ft
Height, y	0.209 ft	0.102 ft	−0.006 ft	−0.114 ft

▶ The shot hit the ground at a distance between 74.6 and 74.7 feet.

Guided Practice

Vocabulary Check ✓

1. Identify the values of a, b, and c for the quadratic function in standard form $y = -5x^2 + 7x - 4$. **$a = -5$, $b = 7$, $c = -4$**

2. Why is the vertical line that passes through the vertex of a parabola called the axis of symmetry? **The axis of symmetry divides the parabola into two parts that are symmetrical, that is, they are mirror images of each other.**

Concept Check ✓

3. Explain how you can decide whether the graph of $y = 3x^2 + 2x - 4$ opens up or down. **The graph opens up because $a = 3$, which is positive.**

4. Find the coordinates of the vertex of the graph of $y = 2x^2 + 4x - 2$. **$(-1, -4)$**

Skill Check ✓

Tell whether the graph opens up or down. Write an equation of the axis of symmetry.

5. $y = x^2 + 4x - 1$ **up; $x = -2$**
6. $y = 3x^2 + 8x - 6$ **up; $x = -\frac{4}{3}$**
7. $y = x^2 + 7x - 1$ **up; $x = -\frac{7}{2}$**
8. $y = -x^2 - 4x + 2$ **down; $x = -2$**
9. $y = 5x^2 - 2x + 4$ **up; $x = \frac{1}{5}$**
10. $y = -x^2 + 4$ **down; $x = 0$**

Sketch the graph of the function. Label the vertex. **11–19. See margin for graphs.**

11. $y = -3x^2$ **(0, 0)**
12. $y = -3x^2 + 6x + 2$ **(1, 5)**
13. $y = -5x^2 + 10$ **(0, 10)**
14. $y = x^2 + 4x + 7$ **(−2, 3)**
15. $y = x^2 - 6x + 8$ **(3, −1)**
16. $y = 5x^2 + 5x - 2$
17. $y = -4x^2 - 4x + 12$
18. $y = 3x^2 - 6x + 1$ **(1, −2)**
19. $y = 2x^2 - 8x + 3$ **(2, −5)**

16. $\left(-\frac{1}{2}, -\frac{13}{4}\right)$

17. $\left(-\frac{1}{2}, 13\right)$

20. **BASKETBALL** You throw a basketball whose path can be modeled by $y = -16x^2 + 15x + 6$, where x represents time (in seconds) and y represents height of the basketball (in feet).
    a. What is the maximum height that the basketball reaches? **about 9.5 ft**
    b. In how many seconds will the basketball hit the ground if no one catches it? **about 1.2 sec**

Practice and Applications

STUDENT HELP

Extra Practice to help you master skills is on p. 805.

PREPARING TO GRAPH **Complete these steps for the function.** **21–43. See margin.**

a. Tell whether the graph of the function opens up or down.

b. Find the coordinates of the vertex.

c. Write an equation of the axis of symmetry.

21. $y = 2x^2$
22. $y = -7x^2$
23. $y = 6x^2$
24. $y = \frac{1}{2}x^2$
25. $y = -5x^2$
26. $y = -4x^2$
27. $y = -16x^2$
28. $y = 5x^2 - x$
29. $y = 2x^2 - 10x$
30. $y = -7x^2 + 2x$
31. $y = -10x^2 + 12x$
32. $y = 6x^2 + 2x + 4$
33. $y = 5x^2 + 10x + 7$
34. $y = -4x^2 - 4x + 8$
35. $y = 2x^2 - 7x - 8$
36. $y = 2x^2 + 7x - 21$
37. $y = -x^2 + 8x + 32$
38. $y = \frac{1}{2}x^2 + 3x - 7$
39. $y = 4x^2 + \frac{1}{4}x - 8$
40. $y = -10x^2 + 5x - 3$
41. $y = 0.78x^2 - 4x - 8$
42. $y = 3.5x^2 + 2x - 8$
43. $y = -10x^2 - 7x + 2.66$

STUDENT HELP

HOMEWORK HELP
Example 1: Exs. 21–64
Example 2: Exs. 21–64
Example 3: Exs. 65–75

3 APPLY

Assignment Guide

BASIC
Day 1: pp. 521–522 Exs. 24–64 multiples of 4
Day 2: pp. 522–524 Exs. 65, 66, 69–71, 76, 84, 86, 90, 94, 100, Quiz 1 Exs. 1–25

AVERAGE
Day 1: pp. 521–522 Exs. 24–64 multiples of 4
Day 2: pp. 522–524 Exs. 65, 66, 69–71, 76, 84, 86, 90, 94, 100, Quiz 1 Exs. 1–25

ADVANCED
Day 1: pp. 521–522 Exs. 24–64 multiples of 4
Day 2: pp. 522–524 Exs. 65, 66, 69–80, 84, 86, 90, 94, 100, Quiz 1 Exs. 1–25

BLOCK SCHEDULE
pp. 521–522 Exs. 24–64 multiples of 4 (with 9.2)
pp. 522–524 Exs. 65, 66, 69–71, 76, 84, 86, 90, 94, 100, Quiz 1 Exs. 1–25 (with 9.4)

Exercise Levels

Level A: *Easier*
96–101
Level B: *More Difficult*
21–74, 76, 81–95
Level C: *Most Difficult*
75, 77–80

✓ Homework Check

To quickly check student understanding of key concepts, go over the following exercises: Exs. 32, 40, 52, 60, 66, 70. See also the Daily Homework Quiz:

- Blackline Master (*Chapter 2 Resource Book,* p. 54)
- Transparency (p. 70)

11–19, 21–43.
See Additional Answers beginning on page AA1.

MATHEMATICAL REASONING

Students can use their knowledge of symmetry to graph a quadratic function. A parabola is symmetric about a vertical line through the vertex, so students need only find function values on one side of the line of symmetry. The graph of $y = 2x^2 - 2x + 1$ has vertex $\frac{1}{2}$ and passes through (1, 1), (2, 5), and (2.5, 8.5). Have students give the corresponding points on the left side of the parabola without substituting. **(0, 1), (−1, 5), (−1.5, 8.5)**

APPLICATION NOTE

EXERCISE 66 Students should see that because the height starts at 0 when the dolphin leaves the water and ends at zero when the dolphin reenters the water, then the total distance of the jump will, by symmetry, be twice the horizontal distance to the vertex.

EXERCISES 69–71 Tell students that the vertex of a parabola is called the "turning point," which is the point where a graph changes from decreasing to increasing, or vice versa. You may want to point out that the "slope of the curve" at this point is 0, a fact that will be very important in future mathematics courses.

44.

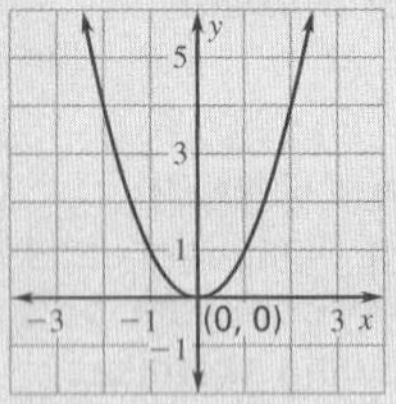

45.

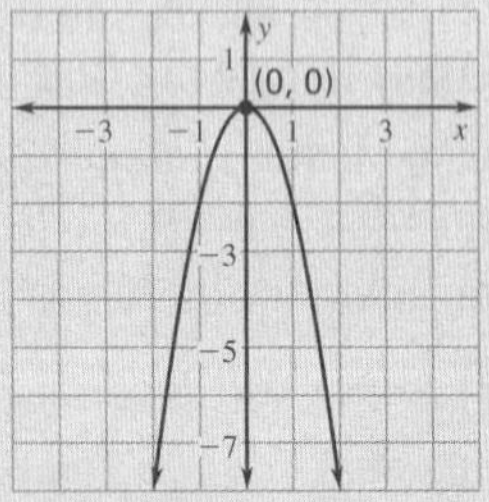

46–64. See Additional Answers beginning on page AA1.

SKETCHING GRAPHS **Sketch the graph of the function. Label the vertex.**
44–64. See margin.

44. $y = x^2$ **45.** $y = -2x^2$ **46.** $y = 4x^2$

47. $y = x^2 + 4x - 1$ **48.** $y = -3x^2 + 6x - 9$ **49.** $y = 4x^2 + 8x - 3$

50. $y = 2x^2 - x$ **51.** $y = 6x^2 - 4x$ **52.** $y = 3x^2 - 2x$

53. $y = x^2 + x + 4$ **54.** $y = x^2 + x + \frac{1}{4}$ **55.** $y = 3x^2 - 2x - 1$

56. $y = 2x^2 + 6x - 5$ **57.** $y = -3x^2 - 2x - 1$ **58.** $y = -4x^2 + 32x - 20$

59. $y = -4x^2 + 4x + 7$ **60.** $y = -3x^2 - 3x + 4$ **61.** $y = -2x^2 + 6x - 5$

62. $y = -\frac{1}{3}x^2 + 2x - 3$ **63.** $y = -\frac{1}{2}x^2 - 4x + 6$ **64.** $y = -\frac{1}{4}x^2 - x - 1$

DOLPHIN **In Exercises 65 and 66, use the following information.**

A bottlenose dolphin jumps out of the water. The path the dolphin travels can be modeled by $h = -0.2d^2 + 2d$, where h represents the height of the dolphin and d represents horizontal distance.

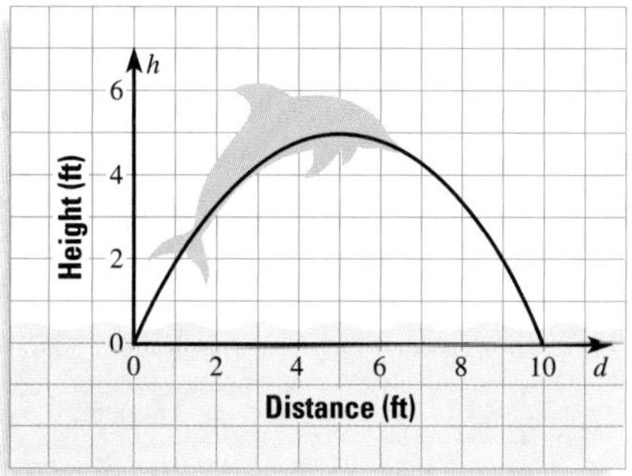

65. What is the maximum height the dolphin reaches? **5 ft**

66. How far did the dolphin jump? **10 ft**

FOCUS ON APPLICATIONS

WATER ARC To celebrate the engineering feat of reversing the flow of the Chicago River, the Water Arc was built on the hundredth anniversary of this event.

WATER ARC **In Exercises 67 and 68, use the following information.**

On one of the banks of the Chicago River, there is a water cannon, called the Water Arc, that sprays recirculated water across the river. The path of the Water Arc is given by the model

$$y = -0.006x^2 + 1.2x + 10$$

where x is the distance (in feet) across the river, y is the height of the arc (in feet), and 10 is the number of feet the cannon is above the river.

67. What is the maximum height of the water sprayed from the Water Arc? **70 ft**

68. How far across the river does the water land? **about 208 ft**

GOLD PRODUCTION **In Exercises 69–71, use the following information.**

In Ghana from 1980 to 1995, the annual production of gold G in thousands of ounces can be modeled by $G = 12t^2 - 103t + 434$, where t is the number of years since 1980.

69. From 1980 to 1995, during which years was the production of gold in Ghana decreasing? **1980 to 1984**

70. From 1980 to 1995, during which years was the production of gold increasing? **1984 to 1995**

71. How are the questions asked in Exercises 69 and 70 related to the vertex of the graph? **See margin.**

71. The vertex is the point at which the graph changes direction, that is, when the values of *G* stop decreasing and begin increasing.

STUDENT HELP

INTERNET HOMEWORK HELP
Visit our Web site www.mcdougallittell.com for help with Exs. 72–75.

75. ***Sample answers:*** **The force with which the ball is hit, the angle at which it is hit, and the spin placed on the ball; a harder hit, a steeper angle, and a favorable spin; a softer hit, a flatter angle, and an unfavorable spin**

76.c. ***Sample answer:*** **yes; the angle would be between 35° and 60° and the reach would be longer than on the graphs shown; 90°.**

Test Preparation

78. A change in the value of a makes the graph of $y = ax^2 + bx + c$ narrower or wider, and affects whether the graph opens up or down.

79. A change in the value of b moves the graph of $y = ax^2 + bx + c$ horizontally and vertically.

80. A change in the value of c moves the graph of $y = ax^2 + bx + c$ up or down.

★ Challenge

EXTRA CHALLENGE
www.mcdougallittell.com

TABLE TENNIS **In Exercises 72–75, use the following information.**
Suppose a table-tennis ball is hit in such a way that its path can be modeled by $h = -4.9t^2 + 2.07t$, where h is the height in meters above the table and t is the time in seconds.

72. Estimate the maximum height reached by the table-tennis ball. Round to the nearest tenth.
about 0.22 m

73. About how many seconds did it take for the table-tennis ball to reach its maximum height after its initial bounce? Round to the nearest tenth. **about 0.21 sec**

74. About how many seconds did it take for the table-tennis ball to travel from the initial bounce to land on the other side of the net? Round to the nearest tenth. **about 0.42 sec**

75. CRITICAL THINKING What factors would change the path of the table-tennis ball? What combination of factors would result in the table-tennis ball bouncing the highest? What combination of factors would result in the table-tennis ball bouncing the lowest? **See margin.**

76. MULTI-STEP PROBLEM A sprinkler can eject water at an angle of 35°, 60°, or 75° with the ground. For these settings, the paths of the water can be modeled by the equations below where x and y are measured in feet.
Answers are rounded to the nearest tenth.

35°: $y = -0.06x^2 + 0.70x + 0.5$

60°: $y = -0.16x^2 + 1.73x + 0.5$

75°: $y = -0.60x^2 + 3.73x + 0.5$

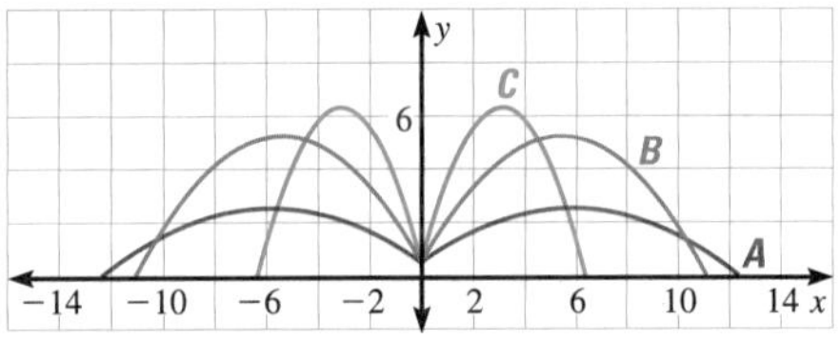

a. Find the maximum height of the water for each setting.
35°: 2.5 ft; 60°: 5.2 ft; 75°: 6.3 ft

b. Find how far from the sprinkler the water reaches for each setting.
35°: 12.3 ft; 60°: 11.1 ft; 75°: 6.3 ft

c. CRITICAL THINKING Do you think there is an angle setting for the sprinkler that will reach farther than any of the settings above? How do the angle and reach represented by the graph of $y = -0.08x^2 + x + 0.5$ compare with the others? What angle setting would reach the least distance?
See margin.

77. UNDERSTANDING GRAPHS Sketch the graphs of the three functions in the same coordinate plane. Describe how the three graphs are related.
See margin.

a. $y = x^2 + x + 1$	**b.** $y = x^2 - 1x + 1$	**c.** $y = x^2 - x + 1$
$y = \frac{1}{2}x^2 + x + 1$	$y = x^2 - 5x + 1$	$y = x^2 - x + 3$
$y = 2x^2 + x + 1$	$y = x^2 - 10x + 1$	$y = x^2 - x - 2$

78. How does a change in the value of a change the graph of $y = ax^2 + bx + c$? **See margin.**

79. How does a change in the value of b change the graph of $y = ax^2 + bx + c$? **See margin.**

80. How does a change in the value of c change the graph of $y = ax^2 + bx + c$? **See margin.**

! COMMON ERROR

EXERCISE 74 Because the height of the ball is given as a function of time rather than a function of the horizontal *distance* traveled by the ball (as in Example 3), students may confuse the time it takes for the ball to land on the other side and the distance to where it lands. Help students understand that the graph shows the height of the ball at various times, but does not show the actual path of the ball in space, though both models are quadratic.

STUDENT HELP NOTES

Homework Help Students can find help for Exs. 72–75 at **www.mcdougallittell.com.** The information can be printed out for students who don't have access to the Internet.

77a.

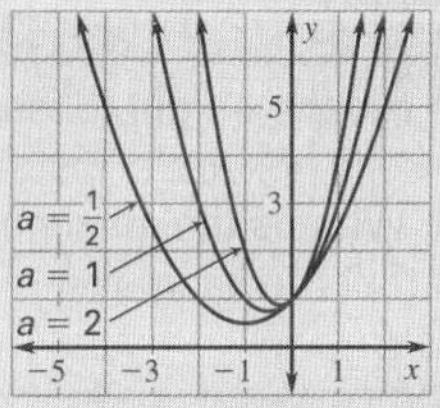

The functions differ in the value of a, and vary in the width of the parabola. When a changes from 1 to $\frac{1}{2}$, the graph becomes wider. When a changes from 1 to 2, the graph becomes narrower.

77b. and 77c. See Additional Answers beginning on page AA1.

ADDITIONAL PRACTICE AND RETEACHING

For Lesson 9.3:

- Practice Levels A, B, and C (*Chapter 9 Resource Book,* p. 43)
- Reteaching with Practice (*Chapter 9 Resource Book,* p. 46)
- See Lesson 9.3 of the *Personal Student Tutor*

For more Mixed Review:

- Search the *Test and Practice Generator* for key words or specific lessons.

4 ASSESS

DAILY HOMEWORK QUIZ

Transparency Available

Tell whether the graph opens up or down. Give the vertex and the axis of symmetry.

1. $y = 3x^2 - 3x + 4$
   up; $\left(\frac{1}{2}, \frac{13}{4}\right)$, $x = \frac{1}{2}$
2. $y = -2x^2 - x + 4$
   down; $\left(-\frac{1}{4}, \frac{33}{8}\right)$, $x = -\frac{1}{4}$
3. Graph $y = -2x^2 + 4x + 1$. Label the vertex.

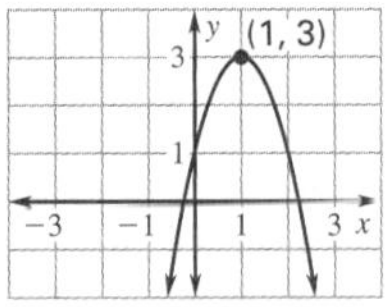

4. A kangaroo's jump follows the path $h = -0.0347d^2 + 0.833d$, where h is the height in feet and d is the horizontal distance in feet. What is the maximum height that the kangaroo reaches? **5 ft**

EXTRA CHALLENGE NOTE

Challenge problems for Lesson 9.3 are available in **blackline** format in the *Chapter 9 Resource Book,* p. 50 and at **www.mcdougallittell.com.**

ADDITIONAL TEST PREPARATION

1. **WRITING** The path of a stone thrown into the air is modeled by $y = -0.0128x^2 + x + 6$, with height y and horizontal distance x in feet. Describe your solution for finding the maximum height. **See sample answer below.**

ADDITIONAL RESOURCES

An alternative Quiz for Lessons 9.1–9.3 is available in the *Chapter 9 Resource Book,* p. 51.

81–87, 23–25.
See Additional Answers beginning on page AA1.

MIXED REVIEW

GRAPHING **Write the equation in slope-intercept form, and then graph the equation. Label the *x*- and *y*-intercepts on the graph.** (Review 4.6 for 9.4) **81–84. See margin for graphs.**

81. $-3x + y + 6 = 0$ **$y = 3x - 6$**
82. $-x + y - 7 = 0$ **$y = x + 7$**
83. $4x + 2y - 12 = 0$ **$y = -2x + 6$**
84. $x + 2y - 7 = 5x + 1$ **$y = 2x + 4$**

GRAPHING LINEAR INEQUALITIES **Graph the system of linear inequalities.** (Review 7.6) **85–87. See margin.**

85. $x - 3y \geq 3$
$x - 3y \leq 12$

86. $x + y \leq 5$
$x \geq 2$
$y \geq 0$

87. $x + y < 10$
$2x + y > 10$
$x - y < 2$

SIMPLIFYING EXPRESSIONS **Simplify. Write your answer as a power or as an expression containing powers.** (Review 8.1)

88. $4^5 \cdot 4^8$
89. $(3^3)^2$
90. $(3^6)^3$
91. $a \cdot a^5$ **a^6**
92. $(3b^4)^2$ **$9b^8$**
93. $6x \cdot (6x)^2$ **$216x^3$**
94. $(3t)^3(-t^4)$ **$-27t^7$**
95. $(-3a^2b^2)^3$ **$-27a^6b^6$**

88. 4^{13}, or 67,108,864
89. 3^6, or 729
90. 3^{18}, or 387,420,489

SCIENTIFIC NOTATION **Rewrite the number in scientific notation.** (Review 8.4)

96. 0.0012 **1.2×10^{-3}**
97. 987,000 **9.87×10^5**
98. 3,984,328 **3.984328×10^6**
99. 1,229,000,000 **1.229×10^9**
100. 0.000432 **4.32×10^{-4}**
101. 0.00999 **9.99×10^{-3}**

QUIZ 1

Self-Test for Lessons 9.1–9.3

Evaluate the expression. Give the exact value if possible. Otherwise, approximate to the nearest hundredth. (Review 9.1)

1. $\sqrt{144}$ **12**
2. $-\sqrt{196}$ **−14**
3. $-\sqrt{676}$ **−26**
4. $-\sqrt{27}$ **−5.20**
5. $\sqrt{6}$ **2.45**
6. $\sqrt{1.5}$ **1.22**
7. $\sqrt{0.16}$ **0.4**
8. $\sqrt{2.25}$ **1.5**

Solve the equation. (Review 9.1)

9. $x^2 = 169$ **−13, 13**
10. $4x^2 = 64$ **−4, 4**
11. $12x^2 = 120$ **$-\sqrt{10}, \sqrt{10}$**
12. $-6x^2 = -48$ **$-2\sqrt{2}, 2\sqrt{2}$**

Simplify the expression. (Review 9.2)

13. $\sqrt{18}$ **$3\sqrt{2}$**
14. $\sqrt{5} \cdot \sqrt{20}$ **10**
15. $\frac{2\sqrt{121}}{\sqrt{4}}$ **11**
16. $\sqrt{\frac{45}{36}}$ **$\frac{\sqrt{5}}{2}$**

Tell whether the graph of the function opens up or down. Find the coordinates of the vertex. Write the equation of the axis of symmetry of the function. (Review 9.3) **17–22. See margin.**

17. $y = x^2 + 2x - 11$
18. $y = 2x^2 - 8x - 6$
19. $y = 3x^2 + 6x - 10$
20. $y = \frac{1}{2}x^2 + 5x - 3$
21. $y = 7x^2 - 7x + 7$
22. $y = x^2 + 9x$

17. up; $(-1, -12)$; $x = -1$
18. up; $(2, -14)$; $x = 2$
19. up; $(-1, -13)$; $x = -1$
20. up; $\left(-5, -15\frac{1}{2}\right)$; $x = -5$
21. up; $\left(\frac{1}{2}, 5\frac{1}{4}\right)$; $x = \frac{1}{2}$
22. up; $\left(-4\frac{1}{2}, -20\frac{1}{4}\right)$; $x = -4\frac{1}{2}$

Sketch a graph of the function. (Review 9.3) **23–25. See margin.**

23. $y = -x^2 + 5x - 5$
24. $y = 3x^2 + 3x + 1$
25. $y = -2x^2 + x - 3$

Additional Test Preparation *Sample answer:*
Substitute $a = -0.0128$ and $b = 1$ into $-\frac{b}{2a}$ to find the x-coordinate of the vertex. Substitute the result into the model to get (39.1, 25.5) as the coordinates of the vertex. The maximum height is the y-coordinate, about 25.5 feet.

Graphing Quadratic Curves of Best Fit

MATERIALS
CBL (optional)

▶ EXAMPLE

A falling object is dropped from a height of 5.23 feet. Using a CBL unit, the height of the object was recorded. Use a graphing calculator to find a quadratic model for the data.

Time	0.0	0.04	0.12	0.16	0.22	0.26	0.32	0.36	0.41
Height	5.23	5.16	4.85	4.63	4.20	3.85	3.23	2.77	1.98

▶ SOLUTION

1 Let L_1 represent time (x) and L_2 represent height (y). Enter the ordered pairs.

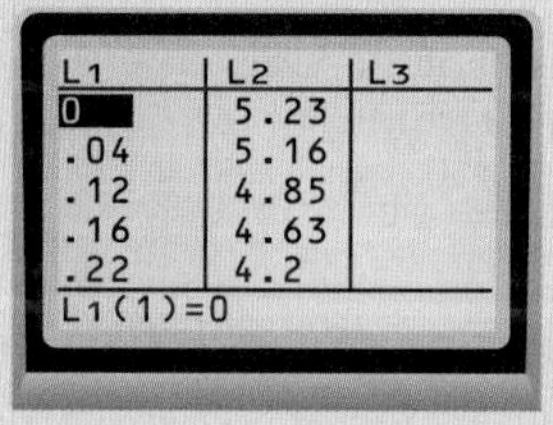

2 Set the viewing window so the range of x is from 0 to 1 and the range of y is from 0 to 6. Make a scatter plot.

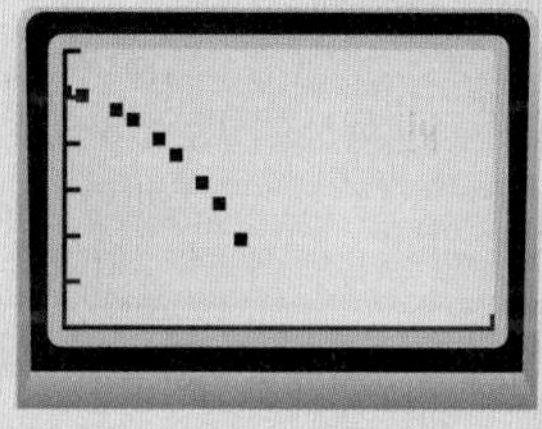

3 Use quadratic regression to find a quadratic model. Approximate to the nearest hundredths place.

4 Graph the equation $y = -16.80x^2 - 0.90x + 5.22$ with the initial points.

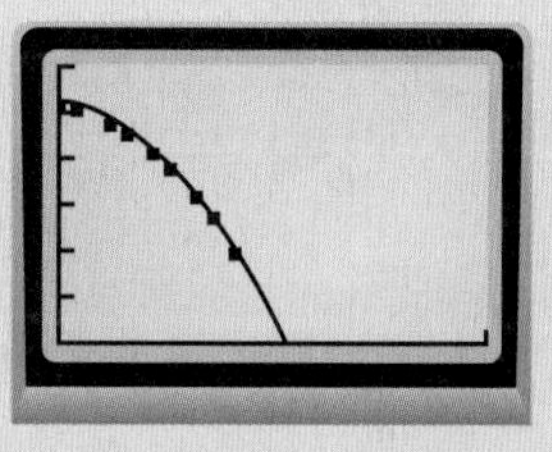

▶ From the scatter plot and the curve, you can see that the data appear to be part of a parabola that opens downward.

▶ EXERCISES

In Exercises 1 and 2, find the best-fitting quadratic model for the points.

1. (1, 16), (3, 138), (5, 366), (7, 652), (9, 970), (11, 1310), (13, 1660) **$y = 5.25x^2 + 67.21x - 81.46$**

2. (3.9, −2.8), (5.1, −8.2), (6.3, −16.3), (−3.1, −16.1), (−2.8, −8.9), (−1.3, −2.5), (0, 2), (1.8, 4.1), (2.7, 2.8) **$y = -0.86x^2 + 2.42x + 2.12$**

3. If you have access to a CBL unit, collect data for three different falling objects such as a basketball, a baseball, and a table-tennis ball. Find a quadratic model for the height of each object. Are the values of a very different for different objects? **Check work.**

STUDENT HELP

INTERNET KEYSTROKE HELP
Visit our Web site www.mcdougallittell.com for information on using a CBL to collect your own data.

1 Planning the Activity

PURPOSE
To find the quadratic model of best fit for a set of data.

MATERIALS
- CBL (optional)
- Keystroke blackline (*Chapter 9 Resource Book*, p. 41)

PACING
- Example — 10 min
- Exercises — 10 min

▶ *LINK TO LESSON*
A quadratic model for data can be found when at least three points are known. Students can verify the model given for Exs. 69–71 of Lesson 9.3 by finding the quadratic regression model for the coordinates of the graph.

2 Managing the Activity

COOPERATIVE LEARNING
If several CBL units are available, divide the class into groups and have them generate their own data to analyze. Ask groups to graph the quadratic model on the same coordinate grid as the scatter plot to see how well the model fits the data.

ALTERNATIVE APPROACH
If time permits, you may want to use a CBL unit to record the height of a falling object. Use a graphing calculator to analyze this new set of data.

3 Closing the Activity

★ ***KEY DISCOVERY***
The height of a falling object as a function of time can be modeled by a quadratic function.

ACTIVITY ASSESSMENT
When should you try to find a quadratic model for a set of data?
Sample answer: **when a graph of the data appears to follow a parabola-shaped curve**

1 PLAN

PACING
Basic: 1 day
Average: 1 day
Advanced: 1 day
Block Schedule: 0.5 block with 9.3

LESSON OPENER
VISUAL APPROACH
An alternative way to approach Lesson 9.4 is to use the Visual Approach Lesson Opener:

- Blackline Master (*Chapter 9 Resource Book,* p. 55)
- Transparency (p. 61)

MEETING INDIVIDUAL NEEDS

- ***Chapter 9 Resource Book***
  Prerequisite Skills Review (p. 5)
  Practice Level A (p. 59)
  Practice Level B (p. 60)
  Practice Level C (p. 61)
  Reteaching with Practice (p. 62)
  Absent Student Catch-Up (p. 64)
  Challenge (p. 66)
- ***Resources in Spanish***
- ***Personal Student Tutor***

NEW-TEACHER SUPPORT
See the Tips for New Teachers on pp. 1–2 of the *Chapter 9 Resource Book* for additional notes about Lesson 9.4.

WARM-UP EXERCISES

Transparency Available

Solve the equation.

1. $3x^2 = 75$ 5, −5
2. $x^2 + 11 = 36$ 5, −5
3. $-2x^2 + 8 = -154$ 9, −9
4. $\frac{1}{4}x^2 = 64$ 16, −16

ENGLISH LEARNERS
To help English learners understand the process of solving quadratic equations using graphs, point out that the word *related* means "having something in common with" or "having the same origin" both in this context and in everyday language.

526

9.4 Solving Quadratic Equations by Graphing

***What** you should learn*

GOAL 1 Solve a quadratic equation graphically.

GOAL 2 Use quadratic models in **real-life** settings, such as the consumption of Swiss cheese in the United States in **Ex. 52**.

***Why** you should learn it*

▼ To make predictions with **real-life** situations, such as the length of a shot put in **Example 4**.

GOAL 1 SOLVING QUADRATIC EQUATIONS GRAPHICALLY

In Lesson 4.7 you learned to use the x-intercept of a linear graph to estimate the solution of a linear equation. A similar procedure applies to quadratic equations.

SOLVING QUADRATIC EQUATIONS USING GRAPHS

The solution of a quadratic equation in one variable x can be solved or checked graphically with the following steps.

STEP 1 Write the equation in the form $ax^2 + bx + c = 0$.

STEP 2 Write the related function $y = ax^2 + bx + c$.

STEP 3 Sketch the graph of the function $y = ax^2 + bx + c$.
The solutions, or **roots,** of $ax^2 + bx + c = 0$ are the x-intercepts.

EXAMPLE 1 *Representing a Solution Using a Graph*

Solve $\frac{1}{2}x^2 = 8$ algebraically. Represent your solutions as the x-intercepts of a graph.

SOLUTION

$\frac{1}{2}x^2 = 8$ **Write original equation.**

$x^2 = 16$ **Multiply each side by 2.**

$x = \pm 4$ **Find the square root of each side.**

Represent these solutions using a graph.

1. Write the equation in the form $ax^2 + bx + c = 0$.

   $\frac{1}{2}x^2 = 8$ **Rewrite original equation.**

   $\frac{1}{2}x^2 - 8 = 0$ **Subtract 8 from each side.**

2. Write the related function $y = ax^2 + bx + c$.

   $y = \frac{1}{2}x^2 - 8$

3. Sketch the graph of $y = \frac{1}{2}x^2 - 8$.
   The x-intercepts are ± 4, which are the algebraic solutions.

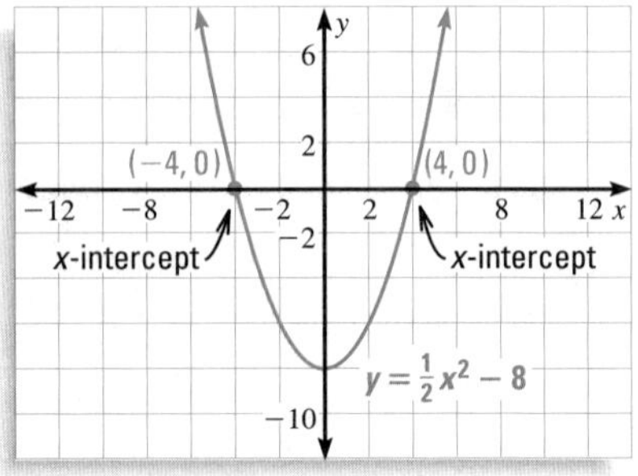

STUDENT HELP

HOMEWORK HELP
Visit our Web site www.mcdougallittell.com for extra examples.

EXAMPLE 2 *Solving an Equation Graphically*

Solve $x^2 - x = 2$ graphically. Check your solution algebraically.

SOLUTION

1. Write the equation in the form $ax^2 + bx + c = 0$.

$x^2 - x = 2$ **Write original equation.**

$x^2 - x - 2 = 0$ **Subtract 2 from each side.**

2. Write the related function $y = ax^2 + bx + c$.

$y = x^2 - x - 2$

3. Sketch the graph of the function $y = x^2 - x - 2$.

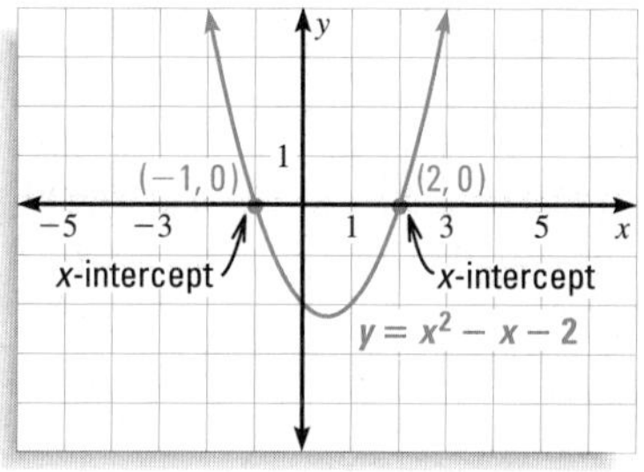

From the graph, the x-intercepts appear to be $x = -1$ and $x = 2$.

✓**CHECK** You can check this by substitution.

Check $x = -1$:	**Check $x = 2$:**
$x^2 - x = 2$	$x^2 - x = 2$
$(-1)^2 - (-1) \stackrel{?}{=} 2$	$2^2 - 2 \stackrel{?}{=} 2$
$1 + 1 = 2$	$4 - 2 = 2$

EXAMPLE 3 *Using a Graphing Calculator*

STUDENT HELP

Look Back
For help with using a graphing calculator to graph equations, see p. 248.

Use a graphing calculator to approximate the solution of $x^2 + 4x - 12 = 0$. Check your solution algebraically.

SOLUTION

$x^2 + 4x - 12 = 0$ **Write original equation.**

$y = x^2 + 4x - 12$ **Write related function.**

Use a graphing calculator to graph the function.

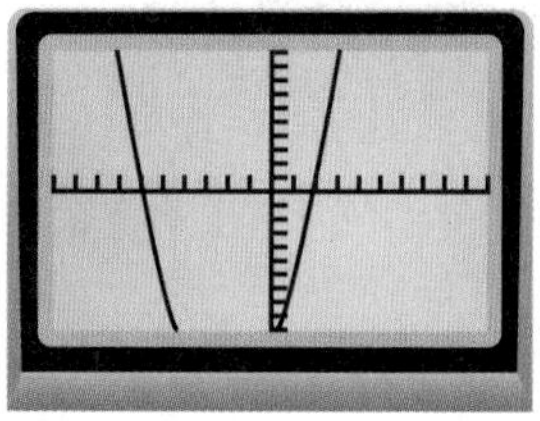

From the graph, you can see that the x-intercepts appear to be $x = 2$ and $x = -6$.

✓**CHECK** You can check this by substitution.

Check $x = 2$:	**Check $x = -6$:**
$x^2 + 4x - 12 = 0$	$x^2 + 4x - 12 = 0$
$2^2 + 4(2) - 12 \stackrel{?}{=} 0$	$(-6)^2 + 4(-6) - 12 \stackrel{?}{=} 0$
$4 + 8 - 12 = 0$	$36 + (-24) - 12 = 0$

2 TEACH

MOTIVATING THE LESSON
Have students imagine throwing a stone as far out onto a lake as they can. Ask them at what angle above the ground they think they should throw the stone for maximum distance. Tell them that they will use quadratic models to answer questions like this.

EXTRA EXAMPLE 1
Solve $3x^2 = 27$ algebraically. Check your solution graphically. **3, −3**

EXTRA EXAMPLE 2
Solve $x^2 + x = 6$ graphically. Check your solution algebraically.

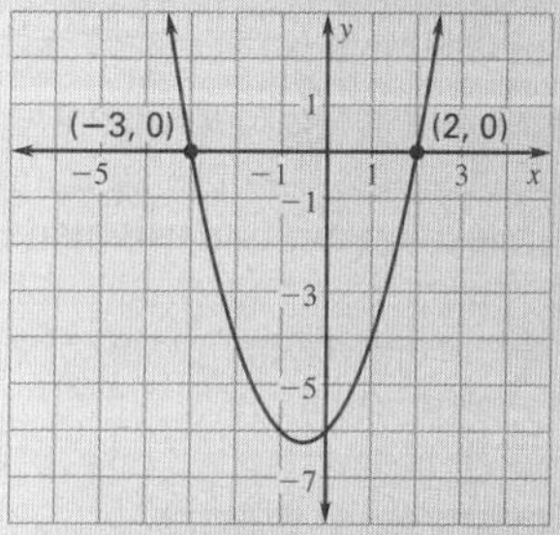

−3, 2

EXTRA EXAMPLE 3
Use a graphing calculator to approximate the solution of $x^2 - 4x - 5 = 0$. Check your solution algebraically. **−1, 5**

CHECKPOINT EXERCISES

For use after Example 1:

1. Solve $\frac{1}{3}x^2 = 12$ algebraically. Check your solution graphically. **6, −6**

For use after Examples 2 and 3:

2. Solve $x^2 + 6x = -8$ graphically. Check your solution algebraically.

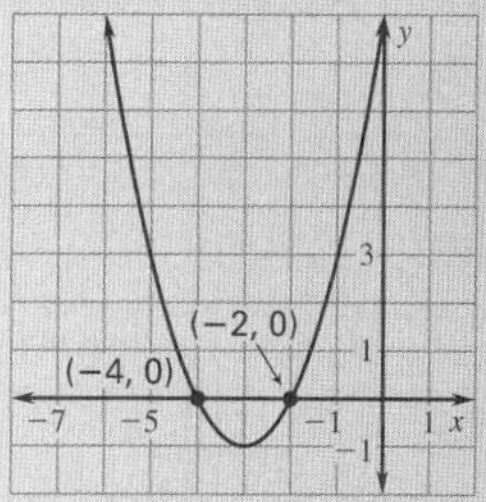

−2, −4

EXTRA EXAMPLE 4

A baseball is thrown at 100 ft/sec from left field toward home plate. The models below give paths of the ball for two initial angles, with height y and horizontal distance x (both in feet).

15°: $y = -0.00171x^2 + 0.268x + 6$

25°: $y = -0.00195x^2 + 0.466x + 6$

If home plate is 236 feet away, which angle(s) have the ball hitting the ground before reaching the plate? **15°**

CHECKPOINT EXERCISES

For use after Example 4:

1. Which model has an x-intercept closer to 26?
A: $y = -0.0356x^2 + 0.625x + 3$
B: $y = -0.0741x^2 + 1.376x + 3$
A

FOCUS ON VOCABULARY

Remind students that the solutions to a quadratic equation are also called "roots," and correspond to the x-intercepts of the graph of a quadratic function.

CLOSURE QUESTION

What do you need to do first in order to solve the equation $2x^2 - x = 15$ by graphing and finding the x-intercepts? Explain. ***Sample answer:*** **Subtract 15 from each side and rewrite the equation in the form $ax^2 + bx + c = 0$. This will give you the related function, $y = ax^2 + bx + c$, which you can then graph.**

DAILY PUZZLER

You graph the quadratic equation $x^2 + 2x + 2 = 0$ and find out that there are no x-intercepts. What does this mean about the algebraic solutions? **There are no real solutions.**

GOAL 2 USING QUADRATIC MODELS IN REAL LIFE

REAL LIFE — *Shot Put*

EXAMPLE 4 *Comparing Two Quadratic Models*

A shot put champion performs an experiment for your math class. Assume that both times he releases the shot with the same initial speed but at different angles. The path of each put can be modeled by one of the equations below, where x represents the horizontal distance (in feet) and y represents the height (in feet).

Initial angle of 35°: $y = -0.010735x^2 + 0.700208x + 6$

Initial angle of 65°: $y = -0.040330x^2 + 2.144507x + 6$

Which angle results in a farther throw?

SOLUTION

Begin by graphing both models in the same coordinate plane. Use only positive x-values because x represents the distance of the throw.

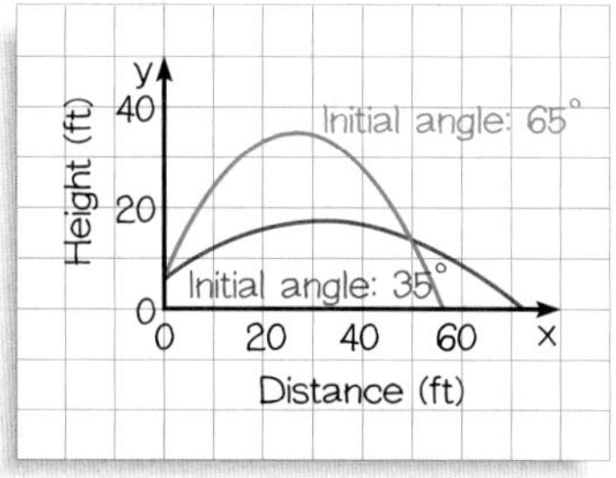

You can see that the 35° angle produced a greater distance.

PARABOLIC MOTION It can be shown that an angle of 45° produces the maximum distance if an object is propelled from ground level. An angle smaller than 45° is better when a shot is released above the ground. If a shot is released from 6 feet above the ground, a 43° angle produces a maximum distance of 74.25 feet.

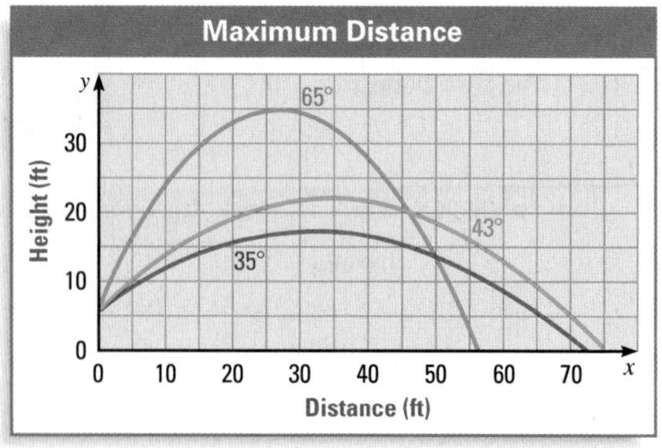

GUIDED PRACTICE

Vocabulary Check ✓ **1.** Define the roots of a quadratic equation. **See margin.**

Concept Check ✓ **2.** Write the steps for solving a quadratic equation using a graph. **See margin.**

1. *Sample answer:* The roots of a quadratic equation are its solutions, which correspond to the *x*-intercepts of its graph.

Match the quadratic equation with the graph of its related function.

A. $x^2 = 3$ **B.** $x^2 + x = 4$ **C.** $-x^2 = -2x - 1$

3. 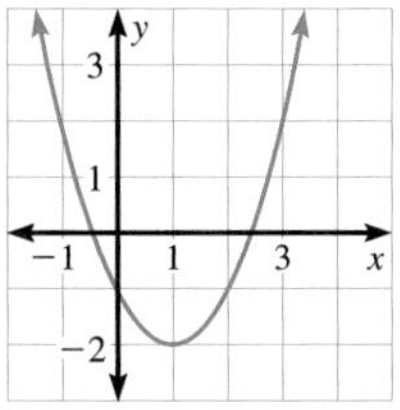

C

4. 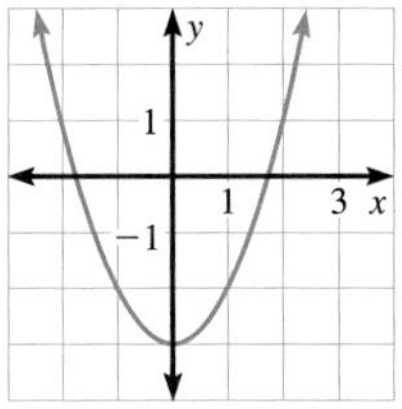

A

5. 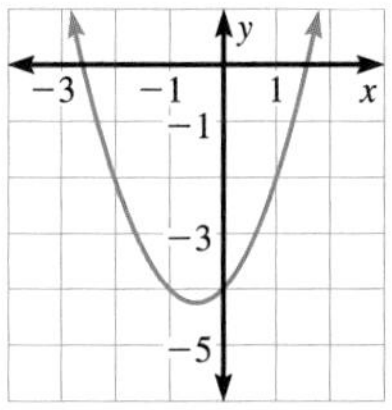

B

Skill Check ✓ **Solve the equation algebraically. Check the solution graphically.**

2. *Sample answer:* Write the equation in the form $ax^2 + bx + c = 0$ and then write the related function $y = ax^2 + bx + c$. Sketch the graph of the function. The solutions of the original equation are the *x*-intercepts of the graph.

6. $3x^2 = 12$ **−2, 2** **7.** $4x^2 = 16$ **−2, 2** **8.** $5x^2 = 125$ **−5, 5**

9. $3x^2 = 27$ **−3, 3** **10.** $8x^2 = 32$ **−2, 2** **11.** $-2x^2 = -18$ **−3, 3**

Solve the equation graphically. Check the solutions algebraically.

12. $3x^2 = 48$ **−4, 4** **13.** $x^2 - 4 = 5$ **−3, 3** **14.** $-x^2 + 7x - 10 = 0$ **2, 5**

15. $2x^2 + 6x = -4$ **−2, −1** **16.** $\frac{1}{3}x^2 + x - 6 = 0$ **−6, 3** **17.** $x - x^2 = -20$ **−4, 5**

PRACTICE AND APPLICATIONS

STUDENT HELP

Extra Practice to help you master skills is on p. 805.

ESTIMATING ROOTS **For each quadratic equation, use the graph to estimate the roots of the equation.**

18. $-x^2 + 3x - 2 = 0$ **1, 2**

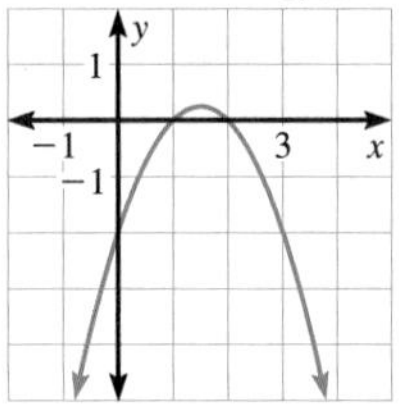

19. $x^2 + 4x - 1 = 0$ **−4.2, 0.2**

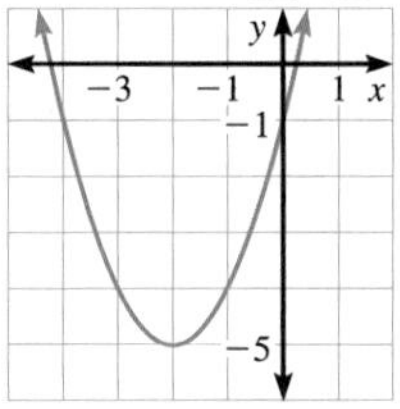

20. $-3x^2 - 2x + 1 = 0$ **$-1, \frac{1}{3}$**

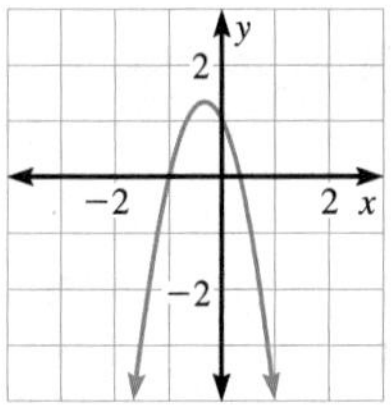

STUDENT HELP

HOMEWORK HELP
Example 1: Exs. 21–32
Example 2: Exs. 18–20, 33–44
Example 3: Exs. 45–50
Example 4: Exs. 51–56

CHECKING GRAPHICALLY **Solve the equation algebraically. Check the solutions graphically.**

21. $2x^2 = 32$ **−4, 4** **22.** $4x^2 = 16$ **−2, 2** **23.** $4x^2 = 100$ **−5, 5**

24. $\frac{1}{3}x^2 = 3$ **−3, 3** **25.** $\frac{1}{4}x^2 = 36$ **−12, 12** **26.** $\frac{1}{2}x^2 = 18$ **−6, 6**

27. $x^2 - 11 = 14$ **−5, 5** **28.** $x^2 - 13 = 36$ **−7, 7** **29.** $x^2 - 4 = 12$ **−4, 4**

30. $x^2 - 53 = 11$ **−8, 8** **31.** $x^2 + 37 = 118$ **−9, 9** **32.** $2x^2 - 89 = 9$ **−7, 7**

3 APPLY

ASSIGNMENT GUIDE

BASIC
Day 1: pp. 529–531 Exs. 18–44 even, 52, 57, 58, 65, 72–82 even, 83

AVERAGE
Day 1: pp. 529–531 Exs. 18–44 even, 52, 57, 58, 65, 72–82 even, 83

ADVANCED
Day 1: pp. 529–531 Exs. 18–44 even, 52, 57–65, 72–82 even, 83

BLOCK SCHEDULE WITH 9.3
pp. 529–531 Exs. 18–44 even, 52, 57, 58, 65, 72–82 even, 83

EXERCISE LEVELS

Level A: *Easier*
18–32

Level B: *More Difficult*
33–58, 65–83

Level C: *Most Difficult*
59–64

✓ HOMEWORK CHECK

To quickly check student understanding of key concepts, go over the following exercises: Exs. 20, 24, 30, 32, 34, 44, 52. See also the Daily Homework Quiz:

- Blackline Master (*Chapter 2 Resource Book*, p. 69)
- Transparency (p. 71)

APPLICATION NOTE

EXERCISES 53–56 Students may wonder why the model doesn't always seem to be true. Point out that the data this model are based on come only from a short period, and that we might be able to make a better model if we looked at more data. Also tell students that factors that affect economic and consumer models change rapidly, so that a good short-term model may be less helpful for long-term predictions.

! *COMMON ERROR*

EXERCISE 58 In standardized tests, time is important. Students may waste time trying one answer at a time. In this exercise, students should immediately notice that the y-intercept of the graph is –5, which means that B and C are the only choices that need be considered.

53.

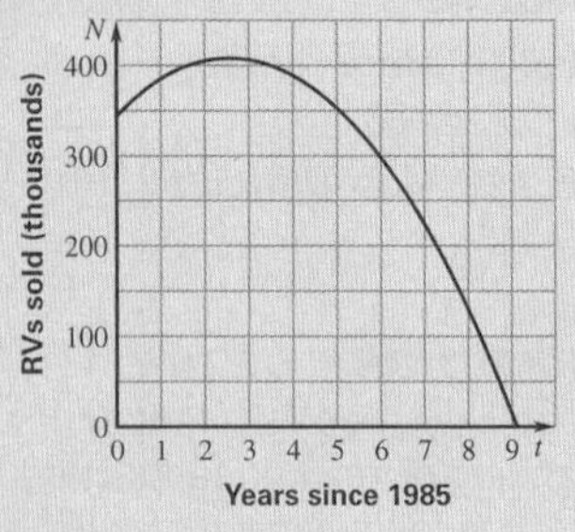

DATA UPDATE

Updated data for Exs. 53–56 is available at **www. mcdougallittell.com**.

ADDITIONAL PRACTICE AND RETEACHING

For Lesson 9.4:

- Practice Levels A, B, and C (*Chapter 9 Resource Book*, p. 59)
- Reteaching with Practice (*Chapter 9 Resource Book*, p. 62)
- See Lesson 9.4 of the *Personal Student Tutor*

For more Mixed Review:

- Search the *Test and Practice Generator* for key words or specific lessons.

51. 69.19 ft; the distance was greater than that for an initial angle of 65°, but less than that for an initial angle of 35°.

56. No. *Sample answer:* An improved economy, lower vehicle prices, and lower gas prices would all tend to increase sales, while opposite conditions would decrease sales.

FOCUS ON APPLICATIONS

RECREATIONAL VEHICLES
Motorized homes, travel and camping trailers, and truck campers are all considered recreational vehicles.

GRAPHICAL REPRESENTATION **Represent the solution graphically. Check the solution algebraically.** **33–44. Check graphs. Solutions are given.**

33. $x^2 - x = 6$ **−2, 3** **34.** $x^2 + 2x = 3$ **−3, 1** **35.** $-x^2 + x = -2$ **−1, 2**

36. $-x^2 + 3x = -4$ **−1, 4** **37.** $2x^2 + 4x = 6$ **−3, 1** **38.** $3x^2 + 3x = 6$ **−2, 1**

39. $x^2 - 4x - 5 = 0$ **−1, 5** **40.** $x^2 - x = 12$ **−3, 4** **41.** $x^2 + 4x = 21$ **−7, 3**

42. $8x^2 - 4 = 4x$ **$-\frac{1}{2}$, 1** **43.** $-7x^2 - 21x = 14$ **−2, −1** **44.** $-2x^2 - 4x = -30$ **−5, 3**

APPROXIMATING SOLUTIONS **Use a graphing calculator to approximate the solution of the equation.**

45. $x^2 - 3x - 4 = 0$ **−1, 4** **46.** $x^2 + 6x - 7 = 0$ **−7, 1**

47. $5x^2 + 5x - 1 = 0$ **−1.17, 0.17** **48.** $-8x^2 - 24x + 32 = 0$ **−4, 1**

49. $\frac{1}{2}x^2 + 2x - 16 = 0$ **−8, 4** **50.** $\frac{5}{4}x^2 + 15x + 40 = 0$ **−8, −4**

51. SHOT PUT A shot put champion releases the shot at a 55° angle. The path of the shot is modeled by the equation below where x represents the horizontal distance (in feet) and y represents the height (in feet).

$$y = -0.021895x^2 + 1.428148x + 6$$

How far does the shot travel? Assume the initial speed is the same as in Example 4. Compare the results with the results in Example 4.

52. SWISS CHEESE The consumption of Swiss cheese in the United States from 1970 to 1996 can be modeled by $P = -0.002t^2 + 0.056t + 0.889$, where P is the number of pounds per person and t is the number of years since 1970. According to the graph of the model, in what year would the consumption of Swiss cheese drop to 0? Is this a realistic prediction? **2009; no**

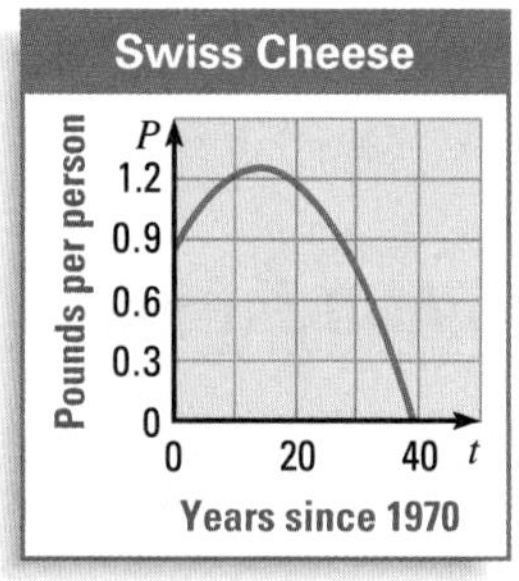

▶ Source: U.S. Department of Agriculture

RV SALES **In Exercises 53–56, use the following information.**
The number of recreational vehicles (RVs) sold in the United States from 1985 to 1991 can be modeled by $N = -9.5t^2 + 48.9t + 343.5$, where N represents the number of vehicles sold (in thousands) and t represents the number of years since 1985.

DATA UPDATE of Recreation Vehicle Industry Association data at www.mcdougallittell.com

53. Sketch a graph of the model for positive values of x and y. **See margin.**

54. Use the graph to estimate a positive root of the equation

$0 = -9.5t^2 + 48.9t + 343.5$. **9.11**

55. According to the model, in what year will the number of RVs sold in the United States drop to 0? **1994**

56. Do you think this prediction is realistic? What factors might explain a decrease or an increase in the number of sales of recreational vehicles? **See margin.**

Test Preparation

57. **MULTIPLE CHOICE** What are the x-intercepts of $y = x^2 - 2x - 3$? **D**

(A) 1 and -3 (B) -2 and -3 (C) 6 and -1 (D) 3 and -1

58. **MULTIPLE CHOICE** Choose the equation whose roots are shown in the graph below. **B**

(A) $5x^2 - 1 = 0$

(B) $\frac{1}{5}x^2 - 5 = 0$

(C) $x^2 - 5 = 0$

(D) $\frac{1}{5}x^2 - 1 = 0$

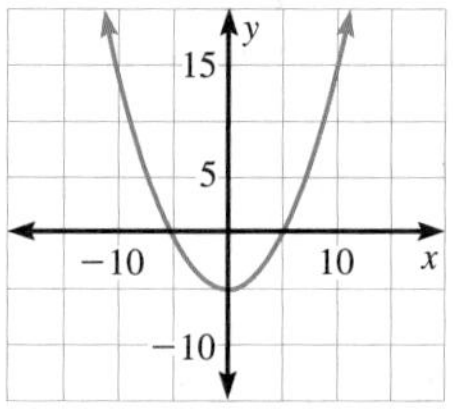

★ Challenge

SOLVING EQUATIONS WITH A GRAPHING CALCULATOR **The solution of a quadratic equation can be found by graphing each side separately and locating the points of intersection. You may wish to consult page 532 for help in approximating solutions.**

59. $3x^2 + 2x + 5 = 6x^2$ **−1, 1.67**

60. $-5x^2 + 4x = 2x^2 - 8$ **−0.82, 1.39**

61. $-2x^2 + 5x = 8x^2 - 2$ **−0.26, 0.76**

62. $-x^2 - 2 = 4x^2 + 6x - 3$ **−1.35, 0.15**

63. $0.75x^2 + 2.67x = 6.22x^2 - 4.1$ **−0.66, 1.14**

64. $-4.87x^2 + 1.44x = 5.22x^2 + 6x$ **−0.45, 0**

EXTRA CHALLENGE
www.mcdougallittell.com

MIXED REVIEW

SOLVING LINEAR SYSTEMS **Solve the system of linear equations if possible. Does the system have exactly *one solution, no solution,* or *infinitely many solutions*?** (Review 7.5)

65. $-2x + 8y = 11$, $x + 6y = 2$ **$\left(-\frac{5}{2}, \frac{3}{4}\right)$; one solution**

66. $4x + 5y = 37$, $-5x + 3y = 0$ **(3, 5); one solution**

67. $-2x + 2y = 4$, $x - y = -2$ **infinitely many solutions**

68. $8x + 4y = -4$, $4x - y = -20$ **$\left(-\frac{7}{2}, 6\right)$; one solution**

69. $10x - 6y = -5$, $3y = 5x + 2$ **no solution**

70. $4y = -5x - 3$, $15x + 12y = 9$ **no solution**

EVALUATING EXPRESSIONS **Evaluate the expression to the nearest hundredth.** (Review 9.1 for 9.5)

71. $\frac{5 \pm 3\sqrt{6}}{2}$ **−1.17, 6.17**

72. $\frac{2 \pm 6\sqrt{3}}{3}$ **−2.80, 4.13**

73. $\frac{-3 \pm 2\sqrt{5}}{-1}$ **−1.47, 7.47**

74. $\frac{-2 \pm 4\sqrt{2}}{-2}$ **−1.83, 3.83**

SIMPLIFYING RADICAL EXPRESSIONS **Simplify the radical expression.** (Review 9.2)

75. $\sqrt{40}$ **$2\sqrt{10}$**

76. $\sqrt{24}$ **$2\sqrt{6}$**

77. $\sqrt{60}$ **$2\sqrt{15}$**

78. $\sqrt{200}$ **$10\sqrt{2}$**

79. $\frac{1}{2}\sqrt{80}$ **$2\sqrt{5}$**

80. $\frac{1}{3}\sqrt{27}$ **$\sqrt{3}$**

81. $\frac{1}{8}\sqrt{32}$ **$\frac{1}{2}\sqrt{2}$**

82. $\frac{2}{3}\sqrt{300}$ **$\frac{20}{3}\sqrt{3}$**

83. **LUNCH TIME** At lunch, you order 2 pasta dishes and 1 type of salad. Your friend orders 1 pasta dish and 2 types of salads. The restaurant charges the same price for each pasta dish and the same price for each salad. Your bill is \$13.85 and your friend's bill is \$9.85. How much did each pasta dish and each salad cost? (Review 7.4) **pasta dishes: \$5.95, salads: \$1.95**

4 ASSESS

DAILY HOMEWORK QUIZ

Transparency Available

1. Use the graph of $y = x^2 + 0.5x - 1.5$ to identify the roots of $x^2 + 0.5x - 1.5 = 0$.

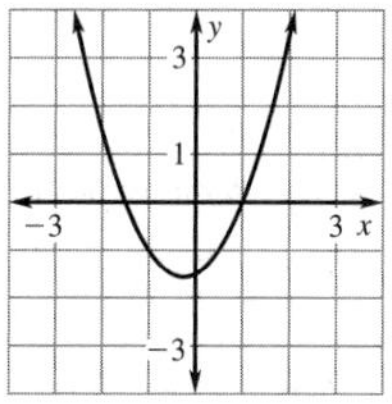

−1.5, 1

Solve the equation algebraically. Check the solutions graphically.

2. $x^2 + 5 = 21$ **−4, 4**

3. $3x^2 - 7 = 68$ **−5, 5**

Solve the equation graphically. Check the solutions algebraically.

4. $x^2 + 3x = 4$ **−4, 1**

5. $-3x^2 - 9x = -30$ **−5, 2**

EXTRA CHALLENGE NOTE

Challenge problems for Lesson 9.4 are available in **blackline** format in the *Chapter 9 Resource Book*, p. 66 and at **www.mcdougallittell.com.**

ADDITIONAL TEST PREPARATION

1. OPEN ENDED Give a quadratic function with x-intercepts 6 and −6.
Sample answer: $y = x^2 - 36$

2. WRITING Explain why the solutions of $ax^2 + bx + c = 0$ are the same as the x-intercepts of the graph of $y = ax^2 + bx + c$.
***Sample answer:* The x-intercepts occur where the y-value of the graph is 0. Substituting 0 for y in the quadratic function gives the original equation.**

1 Planning the Activity

PURPOSE
To find the roots of a quadratic equation by using a graphing calculator's root or zero feature.

MATERIALS
- graphing calculator for each student
- Keystroke blackline (*Chapter 9 Resource Book*, p. 57)

PACING
- Example — 5 min
- Exercises — 15 min

LINK TO LESSON
Students can use the techniques of this activity to find the distances in Example 4 of Lesson 9.4.

2 Managing the Activity

CLASSROOM MANAGEMENT
Students may have difficulty choosing an interval around the root. Point out that if they have chosen appropriate bounds, the sign of the y-value at the left bound will differ from the sign at the right bound.

ALTERNATIVE APPROACH
Using the fact that the sign of a graph changes as it crosses a root, you can also use a calculator's table feature to estimate roots. In the interval where the sign changes, students can choose a smaller delta value to refine their estimates.

3 Closing the Activity

KEY DISCOVERY
By finding the x-intercepts of $y = ax^2 + bx + c$, you find the solutions of $ax^2 + bx + c = 0$.

ACTIVITY ASSESSMENT
Approximate both roots of $3.274x^2 - 0.172x - 1 = 0$ to the nearest hundredth. **−0.53, 0.58**

ACTIVITY 9.4
Using Technology

Approximating Solutions by Graphing

You can use the root or zero feature of a graphing calculator to approximate the solutions, or roots, of a quadratic equation.

STUDENT HELP
KEYSTROKE HELP
See keystrokes for several models of calculators at www.mcdougallittell.com

EXAMPLE

Approximate the roots of $2x^2 + 3x - 4 = 0$.

SOLUTION

The four screens below show the steps in approximating the roots of an equation.

1. Enter the related function $y = 2x^2 + 3x - 4$ into the graphing calculator.

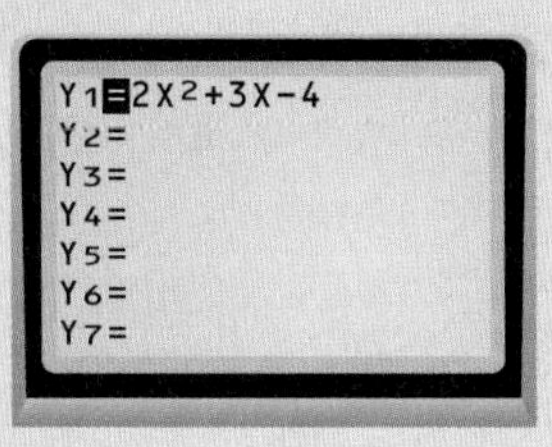

2. Adjust the viewing window so you can see the graph cross the x-axis twice. Graph the function.

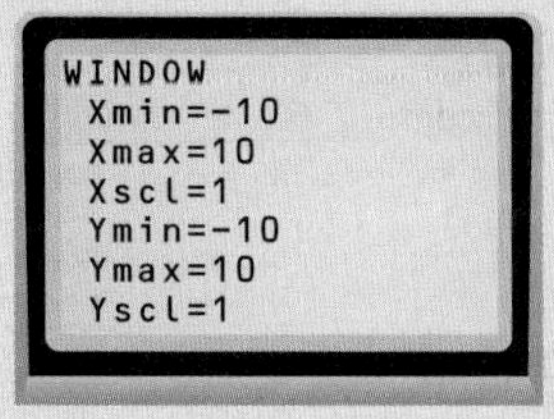

3. Choose the *Root* or *Zero* feature.

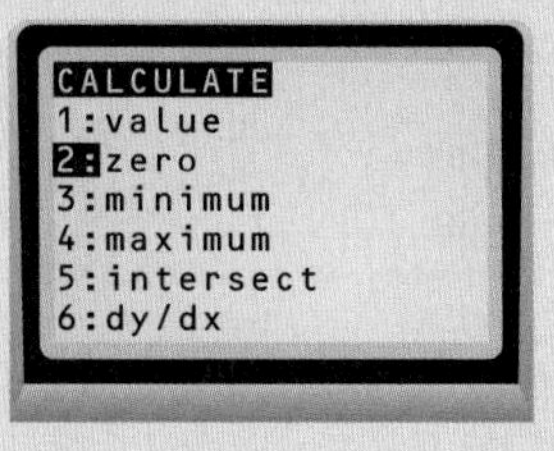

4. Follow your graphing calculator's procedure to find one root.

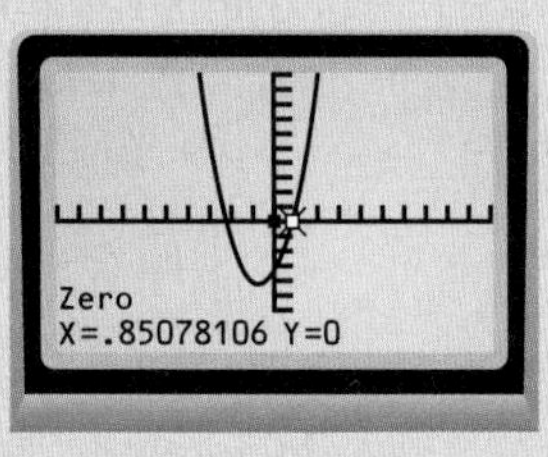

▶ The approximate positive root is 0.85. Follow similar steps to find the negative root, −2.35.

EXERCISES

APPROXIMATING ROOTS Use a graphing calculator to approximate both roots of the quadratic equation to the nearest hundredth.

1. $3x^2 - 20x + 5 = 0$ **0.26, 6.41**
2. $-4x^2 + 6x + 7 = 0$ **−0.77, 2.27**
3. $-x^2 + 5x - 1 = 0$ **0.21, 4.79**
4. $6x^2 + 4x - 5.1 = 0$ **−1.31, 0.65**
5. $-1.4x^2 + 5.2x - 4.8 = 0$ **1.71, 2**
6. $2.87x^2 - 9.43x - 4.53 = 0$ **−0.43, 3.71**
7. $-0.53x^2 + 5x - 10.3 = 0$ **3.04, 6.40**
8. $4.72x^2 + 8x - 7.65 = 0$ **−2.38, 0.68**

9.5

Solving Quadratic Equations by the Quadratic Formula

What you should learn

GOAL 1 Use the quadratic formula to solve a quadratic equation.

GOAL 2 Use quadratic models for **real-life** situations, such as the hot-air balloon competition in **Example 4**.

Why you should learn it

▼ To model **real-life** situations, such as the maximum height of a baseball in **Ex. 84**.

REAL LIFE

GOAL 1 USING THE QUADRATIC FORMULA

In Lesson 9.1 you learned how to solve quadratic equations of the form $ax^2 + c = 0$ by finding square roots. In this lesson you will learn how to use the **quadratic formula** to solve *any* quadratic equation.

THE QUADRATIC FORMULA

The solutions of the quadratic equation $ax^2 + bx + c = 0$ are

$x = \dfrac{-b \pm \sqrt{b^2 - 4ac}}{2a}$ when $a \neq 0$ and $b^2 - 4ac \geq 0$.

You can read this formula as "*x* equals the opposite of *b*, plus or minus the square root of *b* squared minus 4*ac*, all divided by 2*a*."

EXAMPLE 1 *Using the Quadratic Formula*

Solve $x^2 + 9x + 14 = 0$.

SOLUTION

Use the quadratic formula.

$1x^2 + 9x + 14 = 0$	Identify $a = 1$, $b = 9$, and $c = 14$.
$x = \dfrac{-9 \pm \sqrt{9^2 - 4(1)(14)}}{2(1)}$	Substitute values in the quadratic formula.
$x = \dfrac{-9 \pm \sqrt{81 - 56}}{2}$	Simplify.
$x = \dfrac{-9 \pm \sqrt{25}}{2}$	Simplify.
$x = \dfrac{-9 \pm 5}{2}$	Solutions

▶ The equation has two solutions:

$$x = \frac{-9 + 5}{2} = -2 \text{ and } x = \frac{-9 - 5}{2} = -7.$$

✓Check these solutions in the original equation.

Check $x = -2$:	Check $x = -7$:
$x^2 + 9x + 14 = 0$	$x^2 + 9x + 14 = 0$
$(-2)^2 + 9(-2) + 14 \stackrel{?}{=} 0$	$(-7)^2 + 9(-7) + 14 \stackrel{?}{=} 0$
$0 = 0$	$0 = 0$

1 PLAN

PACING

Basic: 2 days
Average: 2 days
Advanced: 2 days
Block Schedule: 1 block

LESSON OPENER ACTIVITY

An alternative way to approach Lesson 9.5 is to use the Activity Lesson Opener:

- Blackline Master (*Chapter 9 Resource Book*, p. 70)
- Transparency (p. 62)

MEETING INDIVIDUAL NEEDS

- ***Chapter 9 Resource Book***
  Prerequisite Skills Review (p. 5)
  Practice Level A (p. 73)
  Practice Level B (p. 74)
  Practice Level C (p. 75)
  Reteaching with Practice (p. 76)
  Absent Student Catch-Up (p. 78)
  Challenge (p. 80)
- ***Resources in Spanish***
- ***Personal Student Tutor***

NEW-TEACHER SUPPORT

See the Tips for New Teachers on pp. 1–2 of the *Chapter 9 Resource Book* for additional notes about Lesson 9.5.

WARM-UP EXERCISES

Transparency Available

Evaluate the expression $\sqrt{b^2 - 4ac}$ for the given values. Give the answer in simplest form.

1. $a = 3, b = 8, c = 0$ 8
2. $a = -2, b = 0, c = 5$ $2\sqrt{10}$
3. $a = 2, b = 6, c = 3$ $2\sqrt{3}$
4. $a = -3, b = 8, c = 1$ $2\sqrt{19}$
5. $a = -3, b = 5, c = 0$ 5

2 TEACH

MOTIVATING THE LESSON
Students have used quadratic models for dropped or thrown objects, such as a shot put. Except for the simplest models, they have used graphing to find the roots. Tell students that using the *quadratic formula,* they can find the roots of any quadratic model without graphing.

EXTRA EXAMPLE 1
Solve $x^2 + 5x - 6 = 0$ using the quadratic formula. **−6, 1**

EXTRA EXAMPLE 2
Solve $-3x^2 + x = -5$.
$\frac{1+\sqrt{61}}{6} \approx 1.47, \frac{1-\sqrt{61}}{6} \approx -1.14$

EXTRA EXAMPLE 3
Find the *x*-intercepts of the graph of $y = x^2 + 3x - 8$.
$\frac{-3+\sqrt{41}}{2} \approx 1.70, \frac{-3-\sqrt{41}}{2} \approx -4.70$

CHECKPOINT EXERCISES

For use after Examples 1 and 2:

1. Solve $4x^2 - x = 7$.
$\frac{1+\sqrt{113}}{8} \approx 1.45,$
$\frac{1-\sqrt{113}}{8} \approx -1.20$

For use after Example 3:

2. Find the *x*-intercepts of the graph of $y = -3x^2 + 4x + 5$.
$\frac{4+\sqrt{76}}{6} \approx 2.12, \frac{4-\sqrt{76}}{6} \approx -0.79$

MULTIPLE REPRESENTATIONS
EXAMPLE 3 points out an important connection between the solutions of a quadratic equation $ax^2 + bx + c = 0$ and the *x*-intercepts of its related quadratic function $y = ax^2 + bx + c$: the solutions to the equation are the *x*-intercepts of its related function. Students can use this relationship to check solutions graphically or to check that graphs have the correct *x*-intercepts.

STUDENT HELP
INTERNET HOMEWORK HELP
Visit our Web site www.mcdougallittell.com for extra examples.

EXAMPLE 2 *Writing in Standard Form*

Solve $2x^2 - 3x = 8$.

SOLUTION

$2x^2 - 3x = 8$	**Write original equation.**
$2x^2 - 3x - 8 = 0$	**Rewrite equation in standard form.**
$x = \frac{-(-3) \pm \sqrt{(-3)^2 - 4(2)(-8)}}{2(2)}$	**Substitute values into the quadratic formula: $a = 2$, $b = -3$, $c = -8$.**
$x = \frac{3 \pm \sqrt{9 + 64}}{4}$	**Simplify.**
$x = \frac{3 \pm \sqrt{73}}{4}$	**Solutions**

▶ The equation has two solutions:

$$x = \frac{3 + \sqrt{73}}{4} \approx 2.89 \text{ and } x = \frac{3 - \sqrt{73}}{4} \approx -1.39.$$

EXAMPLE 3 *Finding the x-Intercepts of a Graph*

Find the *x*-intercepts of the graph of $y = -x^2 - 2x + 5$.

SOLUTION

The *x*-intercepts occur when $y = 0$.

$y = -x^2 - 2x + 5$	**Write original equation.**
$0 = -x^2 - 2x + 5$	**Substitute 0 for *y*.**
$x = \frac{-(-2) \pm \sqrt{(-2)^2 - 4(-1)(5)}}{2(-1)}$	**Substitute values into the quadratic formula: $a = -1$, $b = -2$, and $c = 5$.**
$x = \frac{2 \pm \sqrt{4 + 20}}{-2}$	**Simplify.**
$x = \frac{2 \pm \sqrt{24}}{-2}$	**Solutions**

▶ The two solutions,

$$x = \frac{2 + \sqrt{24}}{-2} \approx -3.45 \text{ and}$$

$$x = \frac{2 - \sqrt{24}}{-2} \approx 1.45,$$

are the *x*-intercepts of the graph of $y = -x^2 - 2x + 5$.

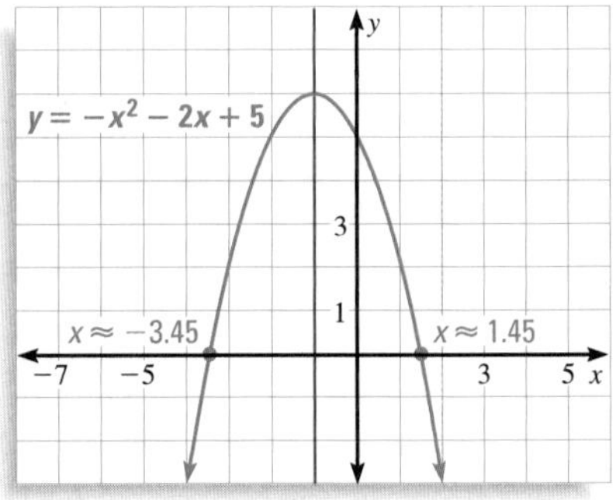

✓ Check your solution graphically. You can see that the graph shows the *x*-intercepts between −3 and −4 and between 1 and 2.

STUDENT HELP
Study Tip
Note that the *x*-intercepts are *x*-coordinates of points that are the same distance from the axis of symmetry.

GOAL 2 USING QUADRATIC MODELS IN REAL LIFE

In Lesson 9.1 you studied the model for the height of a falling object that is *dropped*. For an object that is *thrown* down or up, the model has an extra term to take into account, the initial velocity. Problems involving these two models are called *vertical motion* problems.

VERTICAL MOTION MODELS

OBJECT IS DROPPED: $h = -16t^2 + s$

OBJECT IS THROWN: $h = -16t^2 + vt + s$

h = height (feet) t = time in motion (seconds)

s = initial height (feet) v = initial velocity (feet per second)

In these models the coefficient of t^2 is one half the acceleration due to gravity. On the surface of Earth, this acceleration is approximately 32 feet per second per second.

Remember that velocity v can be positive (for an object moving up), negative (for an object moving down), or zero (for an object that is not moving). Speed is the absolute value of velocity.

EXAMPLE 4 *Modeling Vertical Motion*

BALLOON COMPETITION You are competing in the Field Target Event at a hot-air balloon festival. You throw a marker down from an altitude of 200 feet toward a target. When the marker leaves your hand, its speed is 30 feet per second. How long will it take the marker to hit the target?

SOLUTION

Because the marker is thrown down, the initial velocity is $v = -30$ feet per second. The initial height is $s = 200$ feet. The marker will hit the target when the height is 0.

$h = -16t^2 + vt + s$ — **Choose the vertical motion model for a thrown object.**

$h = -16t^2 + (-30)t + 200$ — **Substitute values for v and s into the vertical motion model.**

$0 = -16t^2 - 30t + 200$ — **Substitute 0 for h. Write in standard form.**

$t = \frac{-(-30) \pm \sqrt{(-30)^2 - 4(-16)(200)}}{2(-16)}$ — **Substitute values for a, b, and c into the quadratic formula.**

$t = \frac{30 \pm \sqrt{13{,}700}}{-32}$ — **Simplify.**

$t \approx 2.72$ or -4.60 — **Solutions**

▶ The weighted marker will hit the target about 2.72 seconds after it was thrown. As a solution, -4.60 doesn't make sense in the problem.

FOCUS ON APPLICATIONS

REAL LIFE **HOT-AIR BALLOONS**
The annual Albuquerque International Balloon Fiesta has several competitive events including the Field Target Event and Prize Grab. There is also a Special Shapes Rodeo and a night time Balloon Glow event.

INTERNET APPLICATION LINK
www.mcdougallittell.com

EXTRA EXAMPLE 4
You jump off a 15 foot diving board with an initial upward speed of 3 ft/sec. How long will it take you to hit the water?
about 1.07 sec

CHECKPOINT EXERCISES
For use after Example 4:
1. From the top of a 40 foot cliff, you throw a stone downward at 20 ft/sec into the water below. How long will it take the stone to hit the water? **about 1.08 sec**

APPLICATION NOTE
EXAMPLE 4 Additional information about hot-air balloons is available at **www.mcdougallittell.com.**

FOCUS ON VOCABULARY
Emphasize to students the need to memorize the quadratic formula because it can be used to solve any quadratic equation.

CLOSURE QUESTION
Describe two ways to solve a quadratic equation of the form $ax^2 + bx + c = 0$. ***Sample answer:*** **Use the quadratic formula to find the solutions $\frac{-b \pm \sqrt{b^2 - 4ac}}{2a}$, or use a graph to find the x-intercepts of the related quadratic function, $y = ax^2 + bx + c$.**

DAILY PUZZLER
You throw a ball from 6 ft high to a friend 40 ft up in the stands at 49 ft/sec. Write and solve a quadratic equation for how long the ball takes to reach your friend. Find and explain the second root.
$-16t^2 + 49t + 6 = 40$, or $-16t^2 + 49t - 34 = 0$; about 1.06 sec; 2 sec; the second root is the time at which the ball passes the friend on the way back down if it is not intercepted.

3 APPLY

ASSIGNMENT GUIDE

BASIC

Day 1: p. 536 Exs. 24–52 even

Day 2: pp. 537–538 Exs. 54–80 even, 84, 87–90, 92–102 even

AVERAGE

Day 1: p. 536 Exs. 24–52 even

Day 2: pp. 537–538 Exs. 54–80 even, 82–84, 87–90, 92–102 even

ADVANCED

Day 1: p. 536 Exs. 24–52 even

Day 2: pp. 537–538 Exs. 54–80 even, 82–90, 92–102 even

BLOCK SCHEDULE

pp. 536–538 Exs. 24–80 even, 82–84, 87–90, 92–102 even

EXERCISE LEVELS

Level A: ***Easier***
87–91

Level B: ***More Difficult***
23–84, 92–103

Level C: ***Most Difficult***
85, 86

✔ HOMEWORK CHECK

To quickly check student understanding of key concepts, go over the following exercises: Exs. 24, 32, 40, 48, 54, 64, 70, 72, 78. See also the Daily Homework Quiz:

- Blackline Master (*Chapter 2 Resource Book*, p. 83)
- Transparency (p. 72)

! COMMON ERROR

EXERCISE 44 Students may look at the equation and quickly assume that c is 30 instead of –30. Remind them to write a quadratic equation in standard form before using the quadratic formula. It may help students also to note that they can multiply through by –1 to remove the negative signs.

36–43. See Additional Answers beginning on page AA1.

GUIDED PRACTICE

Vocabulary Check ✔

1. What formula can you use to solve any quadratic equation? **the quadratic formula**

Concept Check ✔

2. Explain how to use the quadratic formula to solve $-2x^2 + 5x = -7$. **See margin.**
3. What is the difference between the two models for vertical motion? **See margin.**

Skill Check ✔

Use the quadratic formula to solve the equation.

4. $x^2 + 6x - 7 = 0$ **1, –7**
5. $x^2 - 2x - 15 = 0$ **5, –3**
6. $x^2 + 12x + 36 = 0$ **–6**
7. $4x^2 - 8x + 3 = 0$ **0.5, 1.5**
8. $3x^2 + x - 1 = 0$ **See margin.**
9. $x^2 + 6x - 3 = 0$ **$-3 + 2\sqrt{3}, -3 - 2\sqrt{3}$**

Write in standard form. Use the quadratic formula to solve the equation.

10. $2x^2 = -x + 6$ **$2x^2 + x - 6 = 0; -2, \frac{3}{2}$**
11. $6x = -8x^2 + 2$ **$8x^2 + 6x - 2 = 0; -1, \frac{1}{4}$**
12. $3 = 3x^2 + 8x$ **$3x^2 + 8x - 3 = 0; -3, \frac{1}{3}$**
13. $-14x = -2x^2 + 36$ **$2x^2 - 14x - 36 = 0; -2, 9$**
14. $-x^2 + 4x = 3$ **$-x^2 + 4x - 3 = 0; 1, 3$**
15. $4x^2 + 4x = -1$ **$4x^2 + 4x + 1 = 0; -\frac{1}{2}$**

Find the x-intercepts of the graph of the equation.

16. $y = x^2 - 11x + 24$ **3, 8**
17. $y = x^2 + 10x + 16$ **–2, –8**
18. $y = 2x^2 + 4x - 30$ **–5, 3**
19. $y = 2x^2 + 6x - 9$ **See margin.**
20. $y = 5x^2 + 8x - 8$ **See margin.**
21. $y = 4x^2 + 8x - 1$ **See margin.**
22. FIELD TARGET EVENT From a hot-air balloon directly over a target, you throw a marker with an initial downward velocity of –40 feet per second from a height of 180 feet. How long does it take the marker to reach the target? **2.33 sec**

2. ***Sample answer:*** **Rewrite the equation in standard form. Then substitute $a = -2$, $b = 5$, and $c = 7$ into the quadratic formula. The solutions are $\frac{-5 + \sqrt{5^2 - 4(-2)(7)}}{2(-2)}$, or $x = -1$, and $\frac{-5 - \sqrt{5^2 - 4(-2)(7)}}{2(-2)}$, or $x = 3.5$.**

3. In one model, the object is dropped, that is, its initial velocity is 0. In the other, the object is thrown, and, therefore, has a nonzero initial velocity.

8. $\frac{-1 + \sqrt{13}}{6}, \frac{-1 - \sqrt{13}}{6}$

19. $\frac{-3 - 3\sqrt{3}}{2} \approx -4.10$, $\frac{-3 + 3\sqrt{3}}{2} \approx 1.10$

20. $\frac{-4 - 2\sqrt{14}}{5} \approx -2.30$, $\frac{-4 + 2\sqrt{14}}{5} \approx 0.70$

21. $\frac{-2 - \sqrt{5}}{2} \approx -2.12$, $\frac{-2 + \sqrt{5}}{2} \approx 0.12$

PRACTICE AND APPLICATIONS

STUDENT HELP

Extra Practice to help you master skills is on p. 805.

FINDING VALUES **Find the value of $b^2 - 4ac$ for the equation.**

23. $x^2 - 3x - 4 = 0$ **25**
24. $4x^2 + 5x + 1 = 0$ **9**
25. $x^2 - 11x + 30 = 0$ **1**
26. $5w^2 - 3w - 2 = 0$ **49**
27. $s^2 - 13s + 42 = 0$ **1**
28. $3x^2 - 5x - 12 = 0$ **169**
29. $5x^2 + 5x + \frac{1}{5} = 0$ **21**
30. $\frac{1}{2}a^2 + 5a - 8 = 0$ **41**
31. $\frac{1}{4}v^2 - 6v - 3 = 0$ **39**

SOLVING EQUATIONS **Use the quadratic formula to solve the equation.**

32. $4x^2 - 13x + 3 = 0$ **$\frac{1}{4}, 3$**
33. $y^2 + 11y + 10 = 0$ **–10, –1**
34. $7x^2 + 8x + 1 = 0$ **$-1, -\frac{1}{7}$**
35. $-3y^2 + 2y + 8 = 0$ **$-\frac{4}{3}, 2$**
36. $6n^2 - 10n + 3 = 0$ **See margin.**
37. $9x^2 + 14x + 3 = 0$ **See margin.**
38. $8m^2 + 6m - 1 = 0$ **See margin.**
39. $7x^2 + 2x - 1 = 0$ **See margin.**
40. $-6x^2 - 3x + 2 = 0$ **See margin.**
41. $-\frac{1}{2}x^2 + 6x + 13 = 0$ **See margin.**
42. $-10a^2 + 3a + 2 = 0$ **See margin.**
43. $-2d^2 - 5d + 19 = 0$ **See margin.**

STANDARD FORM **Write the quadratic equation in standard form. Solve using the quadratic formula.**

44. $2x^2 = 4x + 30$ **$2x^2 - 4x - 30 = 0; -3, 5$**
45. $2 - 3x + x^2 = 0$ **$x^2 - 3x + 2 = 0; 1, 2$**
46. $16 = -x^2 + 11x$ **See margin.**
47. $5x - 1 = -6x^2$ **$6x^2 + 5x - 1 = 0; -1, \frac{1}{6}$**
48. $2q^2 - 6 = -4q$ **$2q^2 + 4q - 6 = 0; 1, -3$**
49. $5z - 2z^2 + 15 = 8$ **$2z^2 - 5z - 7 = 0; -1, \frac{7}{2}$**
50. $-1 + 3x^2 = 2x$ **$3x^2 - 2x - 1 = 0; -\frac{1}{3}, 1$**
51. $-5c^2 + 9c = 4$ **$5c^2 - 9c + 4 = 0; \frac{4}{5}, 1$**
52. $-16b = -8b^2 - 8$ **$8b^2 - 16b + 8 = 0; 1$**

46. $x^2 - 11x + 16 = 0$; $\frac{11 - \sqrt{57}}{2} \approx 1.73$, $\frac{11 + \sqrt{57}}{2} \approx 9.27$

STUDENT HELP

HOMEWORK HELP
Example 1: Exs. 23–43
Example 2: Exs. 44–52
Example 3: Exs. 53–61
Example 4: Exs. 71–80

56. $\frac{3 - 3\sqrt{3}}{2} \approx -1.10$,

$\frac{3 + 3\sqrt{3}}{2} \approx 4.10$

57. $\frac{-1 - \sqrt{17}}{2} \approx -2.56$,

$\frac{-1 + \sqrt{17}}{2} \approx 1.56$

58. $\frac{-7 - \sqrt{57}}{2} \approx -7.27$,

$\frac{-7 + \sqrt{57}}{2} \approx 0.27$

60. $\frac{2 - \sqrt{2}}{2} \approx 0.29$,

$\frac{2 + \sqrt{2}}{2} \approx 1.71$

61. $\frac{1 - \sqrt{5}}{2} \approx -0.62$,

$\frac{1 + \sqrt{5}}{2} \approx 1.62$

FINDING INTERCEPTS **Find the x-intercepts of the graph of the equation.**

53. $y = 3x^2 - 6x - 24$ 4, −2
54. $y = 2x^2 - 6x - 8$ 4, −1
55. $y = 2x^2 - 2x - 12$ −2, 3
56. $y = -2x^2 + 6x + 9$ See margin.
57. $y = x^2 + x - 4$ See margin.
58. $y = x^2 + 7x - 2$ See margin.
59. $y = -3x^2 - 2x + 1$ $-1, \frac{1}{3}$
60. $y = -4x^2 + 8x - 2$ See margin.
61. $y = -5x^2 + 5x + 5$ See margin.

CHOOSING A METHOD **Solve the quadratic equation by finding square roots or by using the quadratic formula. Explain why you chose the method.** 62–70. See margin.

62. $6x^2 + 20x + 5 = 0$
63. $m^2 = 32$
64. $x^2 - 625 = 0$
65. $4y^2 - 49 = 0$
66. $-2x^2 + 6x + 1 = 0$
67. $h^2 + 18h + 81 = 0$
68. $5x^2 = 25$
69. $9y^2 - 3y = 1$
70. $v^2 = 8v + 2$

FIELD TARGET EVENT **In Exercises 71–76, six balloonists compete in the Field Target Event at a hot-air balloon festival. Calculate the amount of time it takes for the marker to hit the target when thrown down from the given initial height (in feet) with the given initial velocity (in feet per second).**

71. $s = 200$; $v = -50$ 2.30 sec
72. $s = 150$; $v = -25$ 2.38 sec
73. $s = 100$; $v = -10$ 2.21 sec
74. $s = 150$; $v = -33$ 2.20 sec
75. $s = 80$; $v = -40$ 1.31 sec
76. $s = 50$; $v = 0$ 1.77 sec

VERTICAL MOTION **In Exercises 77–80, use a vertical motion model to find how long it will take for the object to reach the ground.**

77. You drop keys from a window 30 feet above ground to your friend below. Your friend does not catch them. 1.37 sec

78. An acorn falls 45 feet from the top of a tree. 1.68 sec

79. A lacrosse player throws a ball upward from her playing stick with an initial height of 7 feet, at an initial speed of 90 feet per second. 5.70 sec

80. You throw a ball downward with an initial speed of 10 feet per second out of a window to a friend 20 feet below. Your friend does not catch the ball. 0.85 sec

81. PEREGRINE FALCON A falcon dives toward a pigeon on the ground. When the falcon is at a height of 100 feet the pigeon sees the falcon, which is diving at 220 feet per second. Estimate the time the pigeon has to escape. 0.44 sec

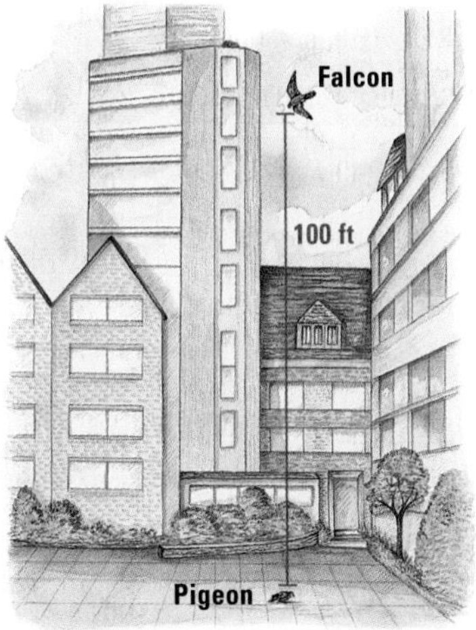

82. RED-TAILED HAWK A hawk dives toward a snake. When the hawk is at a height of 200 feet the snake sees the hawk, which is diving at 105 feet per second. Estimate the time the snake has to escape. 1.54 sec

83. BALD EAGLE An eagle dives toward a mouse. When the eagle is at a height of 125 feet the mouse sees the eagle, which is diving at 145 feet per second. Estimate the time the mouse has to escape. 0.79 sec

FOCUS ON APPLICATIONS

REAL LIFE **URBAN BIRDS** Cities provide a habitat for many species of wildlife including birds of prey such as Peregrine falcons and red-tailed hawks.

APPLICATION NOTE

EXERCISE 81 It is important to specify that the wings are folded so that air resistance is reduced and the dive is more like free fall. Students may have seen films in which sky divers fold in their arms to make a faster descent. Ski jumpers widen their arms and skis so that they fall less quickly.

62–70. Sample explanations are given.

62. $\frac{-10 + \sqrt{70}}{6} \approx -0.27$, $\frac{-10 - \sqrt{70}}{6} \approx -3.06$; **quadratic formula; the expression on the left cannot be written as a perfect square.**

63. $4\sqrt{2} \approx 5.66$, $-4\sqrt{2} \approx -5.66$; **finding square roots; simplest method**

64. −25, 25; **finding square roots; simplest method**

65. $-\frac{7}{2}, \frac{7}{2}$; **finding square roots; simplest method**

66. $\frac{-3 + \sqrt{11}}{-2} \approx -0.16$, $\frac{-3 - \sqrt{11}}{-2} \approx 3.16$; **quadratic formula; the expression on the left cannot be written as a perfect square.**

67. −9; **finding square roots; the expression on the left can be written as $(h + 9)^2$.**

68. $\sqrt{5} \approx 2.24$, $-\sqrt{5} \approx -2.24$; **finding square roots; simplest method**

69 and 70. See Additional Answers beginning on page AA1.

ADDITIONAL PRACTICE AND RETEACHING

For Lesson 9.5:

- Practice Levels A, B, and C (*Chapter 9 Resource Book*, p. 73)
- Reteaching with Practice (*Chapter 9 Resource Book*, p. 76)
- See Lesson 9.5 of the *Personal Student Tutor*

For more Mixed Review:

- Search the *Test and Practice Generator* for key words or specific lessons.

4 ASSESS

DAILY HOMEWORK QUIZ

Transparency Available

1. Find the value of $b^2 - 4ac$ for $-2x^2 + 7x - 6$. **1**
2. Use the quadratic formula to solve $6x^2 - x - 15$. $-\frac{3}{2}, \frac{5}{3}$
3. Write $-17x = -6x^2 - 12$ in standard form. Solve using the quadratic formula. $6x^2 - 17x + 12 = 0; \frac{4}{3}, \frac{3}{2}$
4. Find the x-intercepts of the graph of $y = 2x^2 - 5x + 1$. $\frac{5-\sqrt{17}}{4} \approx 0.22, \frac{5+\sqrt{17}}{4} \approx 2.28$
5. Solve $9x^2 - 36 = 0$ by finding square roots or by using the quadratic formula. **−2, 2**
6. The initial upward velocity of a golf ball is 60 ft/sec. How long will it take the ball to return to the ground? **3.75 sec**

EXTRA CHALLENGE NOTE

Challenge problems for Lesson 9.5 are available in **blackline** format in the *Chapter 9 Resource Book*, p. 80 and at **www.mcdougallittell.com.**

ADDITIONAL TEST PREPARATION

1. **WRITING** Describe how to use the x-intercepts of the graph of a quadratic function to find the vertex of the graph.
Sample answer: **Find the x-coordinate of the point midway between the x-intercepts. This is the x-coordinate of the vertex. Substitute this value into the equation of the function to find the y-coordinate of the vertex.**

91.
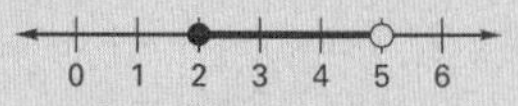

92.
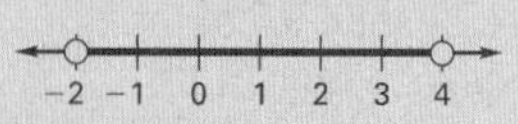

93.

94–102. See Additional Answers beginning on page AA1.

Test Preparation

84. d. *Sample answer:* **The speed of the ball approaching the batter, the speed of the swinging bat, the angle of the bat, and the wind and weather conditions can change the path of the baseball. The dimensions of the ball park also determine whether a hit is a home run.**

84. MULTI-STEP PROBLEM In parts (a)–(d), a batter hits a pitched baseball when it is 3 feet off the ground. After it is hit, the height h (in feet) of the ball at time t (in seconds) is modeled by

$$h = -16t^2 + 80t + 3$$

where t is the time (in seconds).

a. Find the time when the ball hits the ground in the outfield. **5.04 sec**

b. Write a quadratic equation that you can use to find the time when the baseball is at its maximum height of 103 feet. Solve the quadratic equation. $-16t^2 + 80t + 3 = 103$, **or** $-16t^2 + 80t - 100 = 0$; **2.5 sec**

c. Use a graphing calculator to graph the function. Use the zoom feature to approximate the time when the baseball is at its maximum height. Compare your results with those you obtained in part (b). **2.5 sec; this is the same as the value found in part (b).**

d. **CRITICAL THINKING** What factors change the path of a baseball? What factors would contribute to hitting a home run? **See margin.**

★ Challenge

85. $x = 1\frac{1}{2}$; **the x-intercepts are −3 and 6, and the midpoint of (−3, 0) and (6, 0) is** $\left(1\frac{1}{2}, 0\right)$.

EXTRA CHALLENGE
www.mcdougallittell.com

85. VISUAL THINKING Write an equation of the axis of symmetry of the graph and show that it lies halfway between the two x-intercepts. **See margin.**

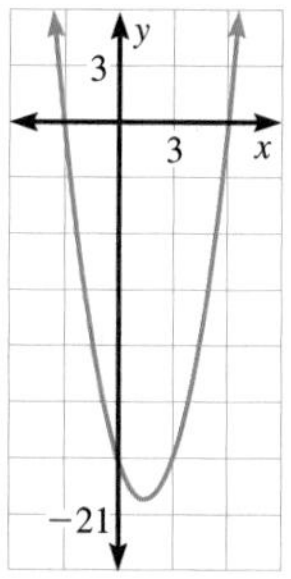

Ex. 85

86. *Writing* Explain how you can use this two-part form of the quadratic formula

$$x = \frac{-b}{2a} \pm \frac{\sqrt{b^2 - 4ac}}{2a}$$

to find the distance between the axis of symmetry of a parabola and either of its x-intercepts. **See margin.**

MIXED REVIEW

86. $x = \frac{-b}{2a}$ **is the equation of the axis of symmetry of the graph of $y = ax^2 + bx + c$. The given values of x are the x-intercepts of the equation, which are equidistant from the axis of symmetry. The distance of each point from the axis of symmetry is** $\frac{\sqrt{b^2 - 4ac}}{2a}$.

97. $\left(\frac{1}{3}, -\frac{5}{3}\right)$

98. $\left(-\frac{5}{6}, \frac{61}{12}\right)$

99. $\left(-\frac{3}{4}, \frac{25}{8}\right)$

101. $\left(\frac{1}{32}, 3\frac{255}{256}\right)$

102. $\left(-\frac{1}{20}, \frac{41}{80}\right)$

EVALUATING EXPRESSIONS Evaluate the expression. (Review 1.3 for 9.6)

87. x^2 when $x = -5$ **25**

88. $-y^2$ when $y = -1$ **−1**

89. $-4xy$ when $x = -2$ and $y = -6$ **−48**

90. $y^2 - y$ when $y = -2$ **6**

INEQUALITIES Solve the inequality and graph the solution. (Review 6.3, 6.4) **91–96. See margin for graphs.**

91. $2 \le x < 5$

92. $8 > 2x > -4$ **$4 > x > -2$**

93. $-12 < 2x - 6 < 4$ **$-3 < x < 5$**

94. $-3 < -x < 1$ **$-1 < x < 3$**

95. $|x + 5| \ge 10$ **$x \le -15$ or $x \ge 5$**

96. $|2x + 9| \le 15$ **$-12 \le x \le 3$**

SKETCHING GRAPHS Sketch the graph of the function. Label the vertex. (Review 9.3) **97–102. See margin for graphs.**

97. $y = 6x^2 - 4x - 1$

98. $y = -3x^2 - 5x + 3$

99. $y = -2x^2 - 3x + 2$

100. $y = \frac{1}{2}x^2 + 2x - 1$ **(−2, −3)**

101. $y = 4x^2 - \frac{1}{4}x + 4$

102. $y = -5x^2 - 0.5x + 0.5$

103. RECREATION There were 1.4×10^7 people who visited Golden Gate Recreation Area in California in 1996. On average, how many people visited per day? per month? ▶ Source: National Park Service (Review 8.4) **about 3.84×10^4 people; about 1.17×10^6 people**

ACTIVITY 9.5

Using Technology

Graphing Calculator Activity for use with Lesson 9.5

Writing a Program for the Quadratic Formula

A graphing calculator or a computer can be programmed to use the quadratic formula to solve a quadratic equation. Both graphing calculators and computers use step-by-step instructions to perform operations.

EXAMPLE

Write a graphing calculator program to solve the equation $5x^2 - 5.5x + 1.5 = 0$.

SOLUTION

An algorithm is a step-by-step model. The algorithm below shows the steps needed when writing a program that uses the quadratic formula to solve a quadratic equation.

1. Enter values for a, b, and c.
2. Calculate the value of $b^2 - 4ac$.
3. If $b^2 - 4ac < 0$, display "No solution."
4. If $b^2 - 4ac \geq 0$, proceed to the next step.
5. Find the first solution: $\dfrac{-b + \sqrt{b^2 - 4ac}}{2a}$.
6. Display the first solution.
7. Find the second solution: $\dfrac{-b - \sqrt{b^2 - 4ac}}{2a}$.
8. Display the second solution.

```
PROGRAM:QUADFORM
:Prompt A,B,C
:B²-4AC→D
:If D<0
:Then
:Disp "NO
SOLUTION"
:Pause
```

Follow your graphing calculator's procedure to enter a program. Run the program. The program produces the solutions 0.6 and 0.5.

✓Check these solutions in the original equation.

```
prgmQUADFORM
A=?5
B=?-5.5
C=?1.5
THE FIRST
SOLUTION IS...
                .6
```

STUDENT HELP

INTERNET KEYSTROKE HELP

Visit our Web site www.mcdougallittell.com to see keystrokes for programming several models of calculators.

EXERCISES

QUADRATIC FORMULA **Use your graphing calculator or a computer program to find the solutions of the quadratic equation.**

1. $x^2 + x - 30 = 0$ **−6, 5**

2. $2x^2 + x - 21 = 0$ **−3.5, 3**

3. $x^2 + 4x + 4 = 0$ **−2**

4. $x^2 - 6x + 10 = 0$ **no solution**

5. $-3x^2 - 6x - 4 = 0$ **no solution**

6. $x^2 - 3x + 3 = 0$ **no solution**

7. $0.25x^2 + 0.35x - 0.60 = 0$ **−2.4, 1**

8. $x^2 - 1.38x - 4.32 = 0$ **−1.5, 2.88**

9. $x^2 + 8.51x + 13.716 = 0$ **−6.35, −2.16**

10. $0.032x^2 + 0.712x - 9 = 0$ **−31.25, 9**

1 Planning the Activity

PURPOSE

To write a program that uses the quadratic formula to solve quadratic equations.

MATERIALS

- graphing calculators for each student
- Keystroke blackline (*Chapter 9 Resource Book,* p. 71)

PACING

- Example — 10 min
- Exercises — 10 min

LINK TO LESSON

Because you need only substitute values for *a*, *b*, and *c* to use the quadratic formula, a program for the formula will produce solutions very efficiently. In Lesson 9.5, Example 4, entering $a = -16$, $b = -30$, and $c = 200$ will save you computational steps.

2 Managing the Activity

COOPERATIVE LEARNING

Ask students already familiar with programming their calculators to help others with the keystrokes.

ALTERNATIVE APPROACH

Have one student in a group enter the program into a calculator and then transfer (link) the program to other students in the group.

3 Closing the Activity

★ ***KEY DISCOVERY***

A calculator or computer can easily be programmed to use the quadratic formula to give you the real roots, if any, of a quadratic equation.

ACTIVITY ASSESSMENT

Find the solutions of $0.5x^2 - 2.83x - 4.5 = 0$ to the nearest hundredth. **6.95, −1.29**

1 Planning the Activity

PURPOSE
To explore how the discriminant relates to the number of solutions of a quadratic equation.

MATERIALS
- graph paper
- graphing calculators (optional)
- Activity Support Master (*Chapter 9 Resource Book,* p. 84)

PACING
- Exploring the Concept — 5 min
- Drawing Conclusions — 10 min

LINK TO LESSON
A nonnegative discriminant indicates real solutions to a quadratic equation. In Example 4 of Lesson 9.6, the discriminant is used to indicate when there are real solutions to an application problem.

2 Managing the Activity

CLASSROOM MANAGEMENT
Organization will help students reach valid conclusions. In Exs. 11 and 12, students can make a table indicating the discriminant value, the number of solutions, and the number of *x*-intercepts.

ALTERNATIVE APPROACH
Some students may be interested in programming their calculators to find the discriminant.

3 Closing the Activity

KEY DISCOVERY
If the discriminant is positive, there are 2 real solutions; if it is 0, there is one real solution; and if it is negative, there are no real solutions.

ACTIVITY ASSESSMENT
Find the number of real solutions of the quadratic equation.

a. $6x^2 - 2x - 5 = 0$ **2**

b. $-2x^2 + 3x - 7 = 0$ **none**

ACTIVITY 9.6

Developing Concepts

GROUP ACTIVITY
Work in a small group.

MATERIALS
- graph paper
- graphing calculator (optional)

1. (1) See margin. **(2)** one *x*-intercept **(3)** one solution

2. (1) See margin. **(2)** two *x*-intercepts **(3)** two solutions

3. (1) See margin. **(2)** no *x*-intercept **(3)** no real solution

4. (1) See margin. **(2)** two *x*-intercepts **(3)** two solutions

5. (1) See margin. **(2)** no *x*-intercept **(3)** no real solution

6. (1) See margin. **(2)** one *x*-intercept **(3)** one solution

STUDENT HELP

Look Back
For help with using a graphing calculator to graph equations, see p. 248.

7. (1) See margin. **(2)** two *x*-intercepts **(3)** two solutions

8. (1) See margin. **(2)** no *x*-intercept **(3)** no real solution

9. (1) See margin. **(2)** two *x*-intercepts **(3)** two solutions

Group Activity for use with Lesson 9.6

Investigating Applications of the Discriminant

QUESTION **How can you determine the number of solutions of a quadratic equation?**

EXPLORING THE CONCEPT

In Lesson 9.4 you learned that the solutions of a quadratic equation $ax^2 + bx + c = 0$ are the *x*-intercepts of the graph of the related function $y = ax^2 + bx + c$.

1. Begin with the equation $x^2 + 3x - 2 = 0$. Graph the related function $y = x^2 + 3x - 2$. **Check graphs.**
2. How many *x*-intercepts does your graph have? **two**
3. How many solutions does the equation $x^2 + 3x - 2 = 0$ have? **two**
4. Repeat **Steps 1–3** for each of the following equations. **Check graphs; one *x*-intercept, no *x*-intercept; one solution, no real solution**

   $x^2 - 6x + 9 = 0$

   $x^2 + 3 = 0$

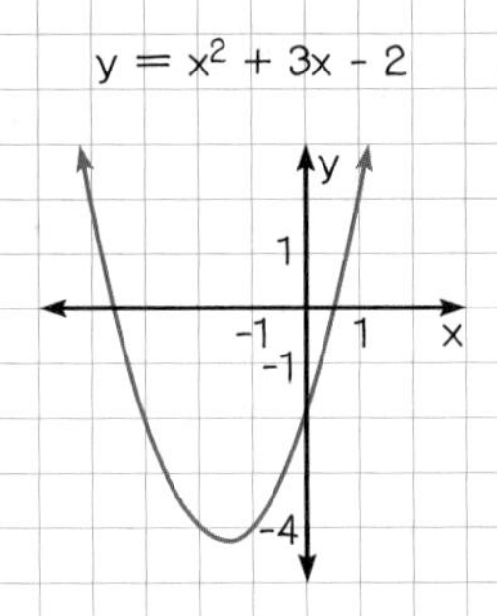

The graph has ______ x-intercepts.

5. Compare the graphs you have drawn. How many intercepts did you find? **Each graph has as many *x*-intercepts as the related equation has solutions.**

DRAWING CONCLUSIONS

In Exercises 1–9, begin with the given quadratic equation. Follow Steps 1–3 for each equation. **1–9. See margin.**

1. $9x^2 - 24x + 16 = 0$ **2.** $2x^2 - 5x - 4 = 0$ **3.** $3x^2 - x + 2 = 0$

4. $-x^2 - 4x + 3 = 0$ **5.** $-5x^2 - 5x - 12 = 0$ **6.** $-2x^2 - 4x - 2 = 0$

7. $4x^2 + 8x - 2 = 0$ **8.** $-6x^2 + 2x - \frac{1}{2} = 0$ **9.** $x^2 + 2x - 1 = 0$

10. Compare your results in Exercises 1–9 with those of others in your group. How are the graphs alike? How are they different? How many intercepts did you find? How are the parabolas positioned with respect to the *x*-axis? **See margin.**

11. Each of the equations in Exercises 1–9 is in the form $ax^2 + bx + c = 0$. For each equation, calculate the value of $b^2 - 4ac$. Write the value next to the sketch of the related graph. **See margin.**

12. CRITICAL THINKING Use your results of Exercises 1–11 to write a rule that relates the number of solutions of a quadratic equation to the value of $b^2 - 4ac$. **The quadratic equation $ax^2 + bx + c = 0$ has two solutions if $b^2 - 4ac > 0$, one solution if $b^2 - 4ac = 0$, and no real solution if $b^2 - 4ac < 0$.**

1–11. See Additional Answers beginning on page AA1.

9.6 Applications of the Discriminant

What you should learn

GOAL 1 Use the discriminant to find the number of solutions of a quadratic equation.

GOAL 2 Apply the discriminant to solve **real-life** problems, such as financial analysis in **Exs. 29 and 30**.

Why you should learn it

▼ To model a **real-life** situation, such as storing your food while camping in **Example 4**.

REAL LIFE

GOAL 1 NUMBER OF SOLUTIONS OF A QUADRATIC

In the quadratic formula, the expression inside the radical is the **discriminant**.

$$x = \frac{-b \pm \sqrt{b^2 - 4ac}}{2a} \longleftarrow \text{Discriminant}$$

The discriminant $b^2 - 4ac$ of a quadratic equation can be used to find the number of solutions of the quadratic equation.

THE NUMBER OF SOLUTIONS OF A QUADRATIC EQUATION

Consider the quadratic equation $ax^2 + bx + c = 0$.

- If $b^2 - 4ac$ is positive, then the equation has two solutions.
- If $b^2 - 4ac$ is zero, then the equation has one solution.
- If $b^2 - 4ac$ is negative, then the equation has no real solution.

EXAMPLE 1 *Finding the Number of Solutions*

Find the value of the discriminant and use the value to tell if the equation has *two solutions*, *one solution*, or *no solution*.

a. $x^2 - 3x - 4 = 0$ **b.** $-x^2 + 2x - 1 = 0$ **c.** $2x^2 - 2x + 3 = 0$

SOLUTION

a. $b^2 - 4ac = (-3)^2 - 4(1)(-4)$ Substitute 1 for a, −3 for b, −4 for c.

$= 9 + 16$ Simplify.

$= 25$ Discriminant is positive.

▶ The discriminant is positive, so the equation has two solutions.

b. $b^2 - 4ac = (2)^2 - 4(-1)(-1)$ Substitute −1 for a, 2 for b, −1 for c.

$= 4 - 4$ Simplify.

$= 0$ Discriminant is zero.

▶ The discriminant is zero, so the equation has one solution.

c. $b^2 - 4ac = (-2)^2 - 4(2)(3)$ Substitute 2 for a, −2 for b, 3 for c.

$= 4 - 24$ Simplify.

$= -20$ Discriminant is negative.

▶ The discriminant is negative, so the equation has no real solution.

1 PLAN

PACING
Basic: 2 days
Average: 2 days
Advanced: 2 days
Block Schedule: 1 block

LESSON OPENER ACTIVITY
An alternative way to approach Lesson 9.6 is to use the Activity Lesson Opener:
- Blackline Master (*Chapter 9 Resource Book,* p. 85)
- Transparency (p. 63)

MEETING INDIVIDUAL NEEDS
- ***Chapter 9 Resource Book***
  Prerequisite Skills Review (p. 5)
  Practice Level A (p. 86)
  Practice Level B (p. 87)
  Practice Level C (p. 88)
  Reteaching with Practice (p. 89)
  Absent Student Catch-Up (p. 91)
  Challenge (p. 94)
- ***Resources in Spanish***
- ***Personal Student Tutor***

NEW-TEACHER SUPPORT
See the Tips for New Teachers on pp. 1–2 of the *Chapter 9 Resource Book* for additional notes about Lesson 9.6.

WARM-UP EXERCISES

Transparency Available

Evaluate the expression $b^2 - 4ac$ for the given values.

1. $a = 2, b = 8, c = 1$ 56
2. $a = 3, b = 6, c = 3$ 0
3. $a = 5, b = 3, c = 6$ −111
4. $a = 2, b = 4, c = 0$ 16
5. $a = 3, b = 0, c = -3$ 36

2 TEACH

MOTIVATING THE LESSON
Students like to avoid unnecessary work. Tell them that the discriminant of a quadratic equation tells them before graphing or trying to solve when there are no real solutions.

EXTRA EXAMPLE 1
Find the value of the discriminant and use the value to tell if the equation has two solutions, one solution, or no real solution.
a. $x^2 - 2x + 4 = 0$ **no real solution**
b. $-3x^2 + 5x - 1 = 0$ **two solutions**
c. $-x^2 - 10x - 25 = 0$ **one solution**

EXTRA EXAMPLE 2
Use the related equation to find the number of x-intercepts of the graph of the function.
a. $y = x^2 + 6x + 3$ **2**
b. $y = x^2 + 6x + 10$ **0**
c. $y = x^2 + 6x + 9$ **1**

EXTRA EXAMPLE 3
Sketch the graph of the equations in Extra Example 2 to check the number of x-intercepts. What is true if $c > 9$?

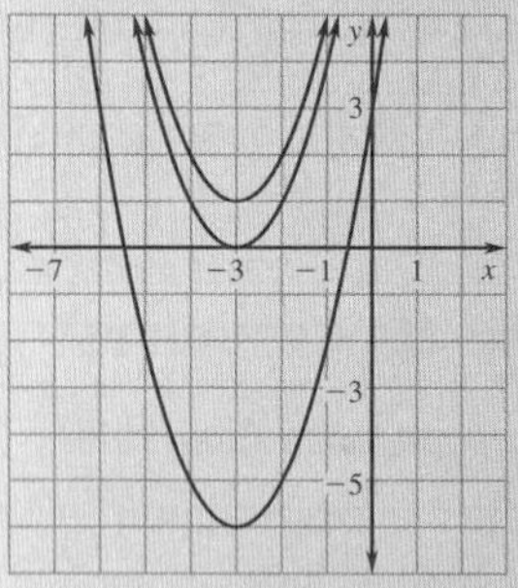

There are no real solutions.

CHECKPOINT EXERCISES
For use after Examples 1–3:
1. Find the number of x-intercepts of the graph of the function.
a. $y = -x^2 + 3x - 2$ **2**
b. $y = x^2 + x + 1$ **0**
c. $y = -x^2 + 2x - 1$ **1**

The number of x-intercepts of the graph of $y = ax^2 + bx + c$ is the same as the number of real solutions of the equation $ax^2 + bx + c = 0$.

EXAMPLE 2 *Finding the Number of x-Intercepts*

Use the related equation to find the number of x-intercepts of the graph of the function.

a. $y = x^2 + 2x - 2$ **b.** $y = x^2 + 2x + 1$ **c.** $y = x^2 + 2x + 3$

STUDENT HELP
HOMEWORK HELP
Visit our Web site www.mcdougallittell.com for extra examples.

SOLUTION

For each function, let $y = 0$. Then find the value of the discriminant.

a. $b^2 - 4ac = 2^2 - 4(1)(-2)$ $a = 1, b = 2, c = -2$
$= 4 + 8$ Simplify.
$= 12$ Discriminant is positive.

▶ The discriminant is positive, so the equation has two solutions *and* the graph has two x-intercepts.

b. $b^2 - 4ac = 2^2 - 4(1)(1)$ $a = 1, b = 2, c = 1$
$= 4 - 4$ Simplify.
$= 0$ Discriminant is zero.

▶ The discriminant is zero, so the equation has one solution *and* the graph has one x-intercept.

c. $b^2 - 4ac = 2^2 - 4(1)(3)$ $a = 1, b = 2, c = 3$
$= 4 - 12$ Simplify.
$= -8$ Discriminant is negative.

▶ The discriminant is negative, so the equation has no real solutions *and* the graph has no x-intercepts.

EXAMPLE 3 *Changing the Value of c*

Sketch the graphs of the equations in Example 2 to check the number of x-intercepts. What effect does changing the value of c have on the graph?

STUDENT HELP
Look Back
For help with sketching a quadratic function, see p. 518.

SOLUTION

By changing the value of c, you can move the graph of $y = x^2 + 2x + c$ up or down in the coordinate plane.

If the graph is moved high enough, it will not have an x-intercept and the equation $x^2 + 2x + c = 0$ will have no real solution.

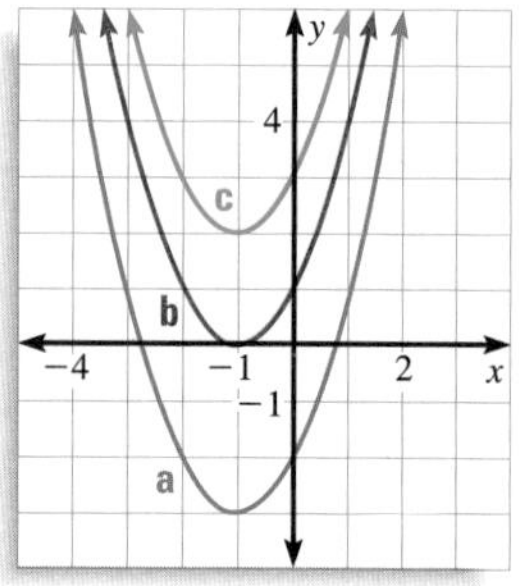

GOAL 2 USING THE DISCRIMINANT IN REAL LIFE

REAL LIFE

Camping

EXAMPLE 4 *Using the Discriminant*

You and a friend are camping in Glacier National Park in Montana. You want to hang a food pack from a high tree branch in order to protect your food from bears. You attach a stick to a rope and your friend is preparing to throw it over a tree branch that is 20 feet from the ground.

a. Your friend can throw the stick upward with an initial velocity of 29 feet per second from an initial height of 6 feet. Will the stick reach the branch when it is thrown?

b. You can throw a stick upward with an initial velocity of 32 feet per second from the same initial height as your friend. Will the stick reach the branch when it is thrown?

Not drawn to scale

SOLUTION

In both situations, you can use a vertical motion model $h = -16t^2 + vt + s$ where h is the height you are trying to reach, t is the time in motion, v is the initial velocity, and s is the initial height.

Because the initial height for you and your friend is 6 feet, the vertical motion model is $h = -16t^2 + vt + 6$. After you substitute for the initial velocity, evaluating the discriminant will tell you whether the stick reaches the tree branch.

a. Solve for an initial velocity v of 29. Use the discriminant to decide if the equation has a solution when $h = 20$.

$$h = -16t^2 + vt + 6 \quad \text{Write vertical motion model.}$$

$$20 = -16t^2 + 29t + 6 \quad \text{Substitute 20 for } h \text{ and 29 for } v.$$

$$0 = -16t^2 + 29t - 14 \quad \text{Rewrite in standard form.}$$

Evaluate $b^2 - 4ac = 29^2 - 4(-16)(-14)$. The discriminant is -55.

▶ Because the discriminant is negative, the equation has no solution. The stick thrown by your friend *will not* reach the branch.

b. Solve for an initial velocity v of 32. Use the discriminant to decide if the equation has a solution when $h = 20$.

$$h = -16t^2 + vt + 6 \quad \text{Write vertical motion model.}$$

$$20 = -16t^2 + 32t + 6 \quad \text{Substitute 20 for } h \text{ and 32 for } v.$$

$$0 = -16t^2 + 32t - 14 \quad \text{Rewrite in standard form.}$$

Evaluate $b^2 - 4ac = 32^2 - 4(-16)(-14)$. The discriminant is 128.

▶ Because the discriminant is positive, the equation has two solutions. The stick that you throw *will* reach the branch.

STUDENT HELP

INTERNET APPLICATION LINK
www.mcdougallittell.com

EXTRA EXAMPLE 4

You throw a ball toward a friend in a window, 15 ft off the ground, on the upper floor of your house.

a. You throw the ball with an initial velocity of 35 ft/sec from a height of 6 ft. Will the ball reach the window? **yes**

b. You throw the ball with an initial velocity of 20 ft/sec from a height of 6 ft. Will the ball reach the window? **no**

CHECKPOINT EXERCISES

For use after Example 4:

1. A basketball is shot from 7 ft high with an upward velocity of 24 ft/sec. Write a vertical motion model to see if the ball reaches a height of 16 ft. Find the discriminant. Does the ball reach a height of 16 ft? **$-16x^2 + 24x + 7 = 16$; 0; yes**

APPLICATION NOTE

EXAMPLE 4 Additional information about camping is available at **www.mcdougallittell.com.**

FOCUS ON VOCABULARY

What is the discriminant, and how can it help when solving a quadratic equation? *Sample answer:* **The discriminant is the value of $b^2 - 4ac$ from the quadratic formula. If the value is negative, there are no real solutions, so you do not need to look for them.**

CLOSURE QUESTION

Describe how to find the number of real solutions of $ax^2 + bx + c$. *Sample answer:* **Calculate $b^2 - 4ac$. A positive result indicates two solutions, a negative result no real solutions, and a zero result one solution.**

DAILY PUZZLER

When does the discriminant tell you the minimum or maximum of a quadratic function? **A zero discriminant means the vertex lies on the x-axis, so the minimum or maximum (depending on the sign of a) is zero.**

3 APPLY

ASSIGNMENT GUIDE

BASIC
Day 1: pp. 544–545 Exs. 9–26, 31
Day 2: pp. 546–547 Exs. 35–48, Quiz 2 Exs. 1–13

AVERAGE
Day 1: pp. 544–545 Exs. 9–26, 31
Day 2: pp. 546–547 Exs. 35–48, Quiz 2 Exs. 1–13

ADVANCED
Day 1: pp. 544–545 Exs. 9–26, 31–34
Day 2: pp. 546–547 Exs. 35–48, Quiz 2 Exs. 1–13

BLOCK SCHEDULE
pp. 544–547 Exs. 9–26, 31, 35–48, Quiz 2 Exs. 1–13

EXERCISE LEVELS

Level A: ***Easier***
9–20
Level B: ***More Difficult***
21–31, 35–48
Level C: ***Most Difficult***
32–34

✓ HOMEWORK CHECK

To quickly check student understanding of key concepts, go over the following exercises: Exs. 10, 14, 18, 20, 22, 26. See also the Daily Homework Quiz:

- Blackline Master (*Chapter 2 Resource Book*, p. 98)
- Transparency (p. 73)

GUIDED PRACTICE

Vocabulary Check ✓

1. Write the quadratic formula and circle the part that is the discriminant. **See margin.**

Concept Check ✓

2. How can you use the discriminant to tell the number of solutions of $ax^2 + bx + c = 0$ and the number of x-intercepts of the graph of the equation? **See margin.**

Skill Check ✓

1. The solutions of a quadratic equation $ax^2 + bx + c = 0$ are $x = \frac{-b \pm \sqrt{b^2 - 4ac}}{2a}$.

Find the discriminant for the equation. Then tell if the equation has *two solutions, one solution,* or *no real solution.*

3. $3x^2 - 2x + 5 = 0$ **−56; no real solution**

4. $-3x^2 + 6x - 3 = 0$ **0; one solution**

5. $x^2 - 5x - 10 = 0$ **65; two solutions**

Match the discriminant with the graph.

A. $b^2 - 4ac = 2$ **B.** $b^2 - 4ac = 0$ **C.** $b^2 - 4ac = -3$

6. 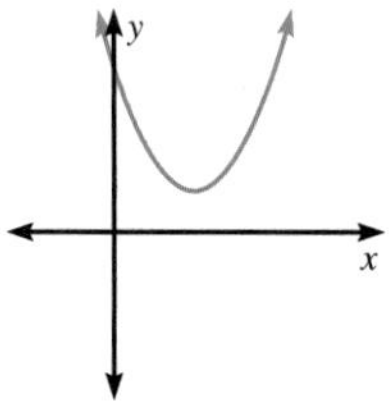

C

7. 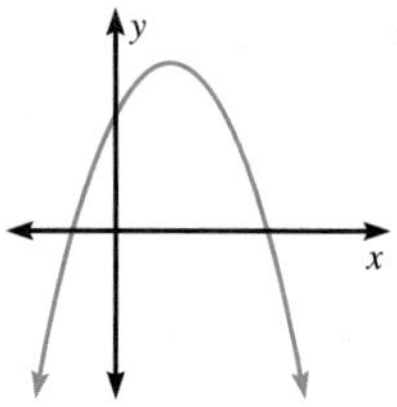

A

8. 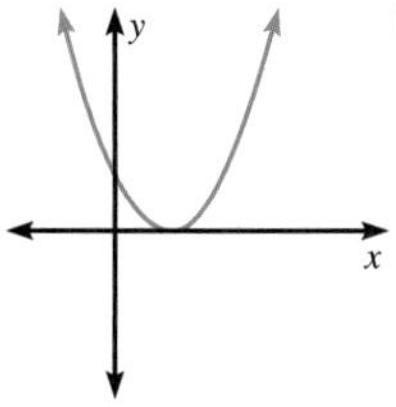

B

PRACTICE AND APPLICATIONS

STUDENT HELP
Extra Practice to help you master skills is on p. 805.

2. ***Sample answer:*** **The quadratic equation $ax^2 + bx + c = 0$ has two solutions if $b^2 - 4ac > 0$, one solution if $b^2 - 4ac = 0$, and no real solution if $b^2 - 4ac < 0$. Similarly, the related function $y = ax^2 + bx + c$ has two x-intercepts if $b^2 - 4ac > 0$, one x-intercept if $b^2 - 4ac = 0$, and no x-intercept if $b^2 - 4ac < 0$.**

USING THE DISCRIMINANT **Tell if the equation has *two solutions, one solution,* or *no real solution.***

9. $x^2 - 3x + 2 = 0$ **two solutions**

10. $2x^2 - 4x + 3 = 0$ **no real solution**

11. $-3x^2 + 5x - 1 = 0$ **two solutions**

12. $-\frac{1}{3}x^2 + x + 4 = 0$ **two solutions**

13. $6x^2 - 2x + 4 = 0$ **no real solution**

14. $3x^2 - 6x + 3 = 0$ **one solution**

NUMBER OF X-INTERCEPTS **Use the related equation to find the number of *x*-intercepts the graph of the function has. Then match the function with its graph.**

A. 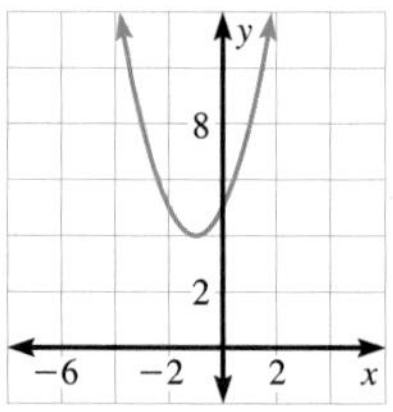

B. 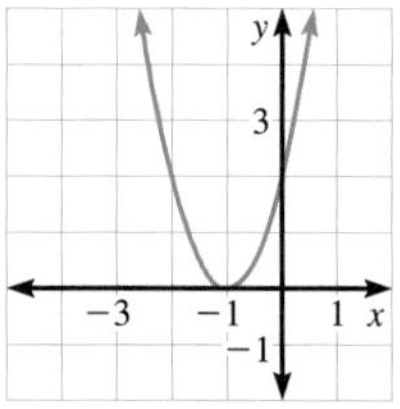

C. 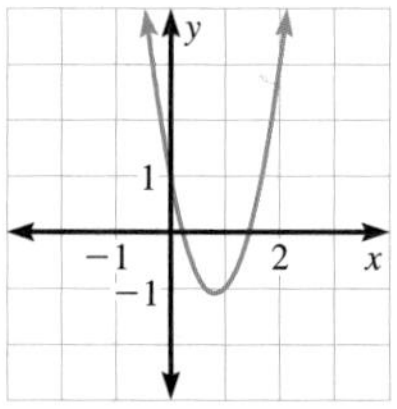

15. $y = 2x^2 + 4x + 2$ **one x-intercept; B**

16. $y = 3x^2 - 5x + 1$ **two x-intercepts; C**

17. $y = x^2 + 2x + 5$ **no x-intercept; A**

INTERPRETING THE DISCRIMINANT **Consider the equation $\frac{1}{2}x^2 + \frac{2}{3}x - 3 = 0$.**

18. Evaluate the discriminant. $\frac{58}{9}$

19. How many solutions does the equation have? **two solutions**

20. What does the discriminant tell you about the graph of $y = \frac{1}{2}x^2 + \frac{2}{3}x - 3$? Does the graph cross the x-axis? **The graph of $y = \frac{1}{2}x^2 + \frac{2}{3}x - 3$ crosses the x-axis at two different points.**

STUDENT HELP
HOMEWORK HELP
Example 1: Exs. 9–14
Example 2: Exs. 15–20
Example 3: Exs. 21–23
Example 4: Exs. 24–26

21. The equation has two solutions for all values of $c < 4$, one solution for $c = 4$, and no real solution for all values of $c > 4$; check graphs.

22. The equation has two solutions for all values of $c < 1$, one solution for $c = 1$, and no real solution for all values of $c > 1$; check graphs.

23. The equation has two solutions for all values of $c < 1\frac{1}{8}$, one solution for $c = 1\frac{1}{8}$, and no real solution for all values of $c > 1\frac{1}{8}$; check graphs.

25. Yes; no. *Sample answer:* Because the discriminant for you is positive, your equation has two solutions; because the discriminant for your friend is negative, your friend's equation has no real solution. That is, there is no time when your friend's jump height will be 3.4 ft.

FOCUS ON CAREERS

REAL LIFE FINANCIAL ANALYSTS
Mathematical models help financial analysts analyze and predict a company's future earnings.

INTERNET CAREER LINK
www.mcdougallittell.com

CHANGING c-VALUES In Exercises 21–23, find values of *c* so that the equation will have two solutions, one solution, and no real solution. Then sketch the graph of the equation for each value of *c* that you chose.
21–23. See margin.

21. $x^2 + 4x + c = 0$ **22.** $x^2 - 2x + c = 0$ **23.** $2x^2 + 3x + c = 0$

24. FIREFIGHTING You see a firefighter aim a fire hose from 4 feet above the ground at a window that is 26 feet above the ground. The equation $h = -0.01d^2 + 1.06d + 4$ models the path of the water when h equals height in feet. Estimate, to the nearest whole number, the possible horizontal distances d (in feet) between the firefighter and the building.
about 28 ft or about 78 ft

BASKETBALL In Exercises 25 and 26, use the vertical motion model $h = -16t^2 + vt + s$ (p. 535) and the following information.
You and a friend are playing basketball. You can jump with an initial velocity of 12 feet per second. You need to jump 2.2 feet to dunk a basketball. Your friend can jump with an initial velocity of 14 feet per second. Your friend needs to jump 3.4 feet to dunk a basketball.

25. Can you dunk the ball? Can your friend? Justify your answers.
See margin.

26. Suppose you can jump with an initial velocity of 11.5 feet per second and your friend can jump with an initial velocity of 15.5 feet per second. How, if at all, would this change your answers to Exercise 25?
In this case, you would not be able to dunk but your friend would be able to dunk.

GOVERNMENT PAYROLL In Exercises 27 and 28, use a graphing calculator and the following information.
For a recent 12-year period, the total government payroll (local, state, and federal) in the United States can be modeled by $P = 26t^2 + 1629t + 19{,}958$, where P is the payroll in millions of dollars and t is the number of years since the beginning of the 12-year period. ▶ Source: U.S. Bureau of the Census

27. Use the discriminant to show that the payroll will reach 80 billion dollars in the future according to the model. See margin.

28. Use a graphing calculator to find out how many years it will take for the total payroll to reach 80 billion dollars according to the model. about 26 yr

FINANCIAL ANALYSIS In Exercises 29 and 30, use a graphing calculator and the following information.
You are a financial analyst for a software company. You have been asked to project the net profit of your company. The net profit of the company from 1993 to 1998 can be modeled by $P = 6.84t^2 - 3.76t + 9.29$ where P is the profit in millions of dollars and t represents the number of years since 1993.

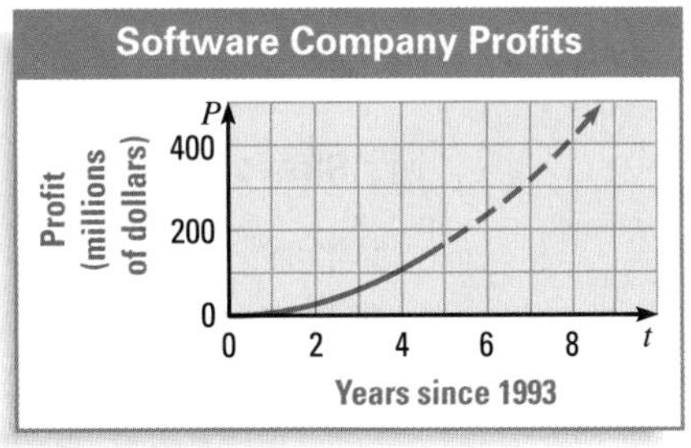

29. Use the model to predict whether the net profit will reach 650 million dollars.
yes

30. Use a graphing calculator to estimate how many years it will take for the company's net profit to reach 475 million dollars according to the model.
about 8.5 yr

MATHEMATICAL REASONING
The discriminant is not a magic number. The reason that it gives the number of roots is very logical. Have students look at $0.5x^2 - x + c = 0$, for which the solutions are $1 \pm \sqrt{1 - 2c}$. Ask them to tell how many solutions there are and explain why for $c = -1$, 0.5, and 1. $c = -1$: 2, because $\sqrt{1 - 2c} = \sqrt{3}$, so $1 \pm \sqrt{1 - 2c}$ gives two distinct numbers; $c = 0.5$: 1, because $\sqrt{1 - 2c} = 0$, so $1 \pm \sqrt{1 - 2c}$ gives only one number; $c = 1$: 0, because $\sqrt{1 - 2c} = \sqrt{-1}$, so $1 \pm \sqrt{1 - 2c}$ cannot be a real number.

APPLICATION NOTE
EXERCISE 24 This is an example of parabolic motion as presented in Lessons 9.3 and 9.4, where h represents the vertical distance and d represents the horizontal distance. This model gives the *path* of the water, while the vertical motion models presented in Lesson 9.5 and used in Example 4 give the height as a function of time, not distance.

! COMMON ERROR
EXERCISES 27 AND 28 Some students may become confused with the units. Remind them that a billion is a thousand million, so that 80 billion is 80,000 million. Also, in Ex. 28, if students graph the related function of the equation used in Ex. 27, remind them they are looking for a zero, not a value of 80,000.

27. See the next page.

ADDITIONAL PRACTICE AND RETEACHING

For Lesson 9.6:
- Practice Levels A, B, and C (*Chapter 9 Resource Book*, p. 86)
- Reteaching with Practice (*Chapter 9 Resource Book*, p. 89)
- See Lesson 9.6 of the *Personal Student Tutor*

For more Mixed Review:
- Search the *Test and Practice Generator* for key words or specific lessons.

4 ASSESS

DAILY HOMEWORK QUIZ

Transparency Available

Tell if the equation has *two solutions, one solution,* or *no real solution.*

1. $4x^2 - 8x + 5 = 0$
   no real solution
2. $-0.5x^2 + 11x - 33 = 0$
   two solutions
3. Evaluate the discriminant of $-\frac{3}{4}x^2 - \frac{5}{2}x - 2$. Does the graph of the related function cross the *x*-axis? **$\frac{1}{4}$, yes**
4. Find values of *c* so that $-x^2 + 4x + c$ will have two solutions, one solution, and no solutions.
   $c > -4$, $c = -4$, $c < -4$
5. An antelope must jump 5.5 ft to clear a fence. It can jump with an initial vertical velocity of 20 ft/sec. Using the discriminant and the vertical motion model $h = -16t^2 + vt + s$, can the antelope make the jump?
   yes

EXTRA CHALLENGE NOTE

Challenge problems for Lesson 9.6 are available in **blackline** format in the *Chapter 9 Resource Book,* p. 94 and at **www.mcdougallittell.com.**

ADDITIONAL TEST PREPARATION

1. **OPEN ENDED** Give an example of a real-world problem that has only one solution even though the discriminant indicates two solutions. **See sample answer below.**

27. ***Sample answer:* The payroll will reach \$80 billion if the discriminant of the equation $26t^2 + 1629t - 60{,}042 = 0$ is nonnegative; since the discriminant is 8,898,009, the payroll will reach that amount.**

35–37, 42–44. See the next page.

45–47. See Additional Answers beginning on page AA1.

Test Preparation

31c. No. *Sample answer:* Solving the equation $0.039x^2 - 0.331x - 218.5 = 0$ gives a solution of approximately 79.2 horizontal feet required. A horizontal distance of 75 feet will allow a height of only 196.4 ft.

32. *Sample answer:* If $b^2 - 4ac > 0$, then $\sqrt{b^2 - 4ac}$ is defined and is a positive number, so the values $\frac{-b + \sqrt{b^2 - 4ac}}{2a}$ and $\frac{-b - \sqrt{b^2 - 4ac}}{2a}$ are distinct solutions of the equation.

★ Challenge

EXTRA CHALLENGE

www.mcdougallittell.com

31. MULTI-STEP PROBLEM You are on a team that is building a roller coaster. The vertical height of the first hill of the roller coaster is supposed to be 220 feet. According to the design, the path of the first hill can be modeled by $y = 0.039x^2 - 0.331x + 1.850$, where *y* is the vertical height in feet and *x* is the horizontal distance in feet. The first hill can use only 75 feet of horizontal distance.

a. Use the model to determine whether the first hill will reach a height of 220 feet. **yes**

b. What minimum horizontal distance is needed for the first hill to reach a vertical height of 220 feet? **79.2 ft**

c. *Writing* Can you build the first hill high enough? Explain your findings. **See margin.**

d. **CRITICAL THINKING** Is the shape of the graph the same as the shape of the hill? Why or why not? **No. *Sample answer:* The horizontal and vertical scales are not the same.**

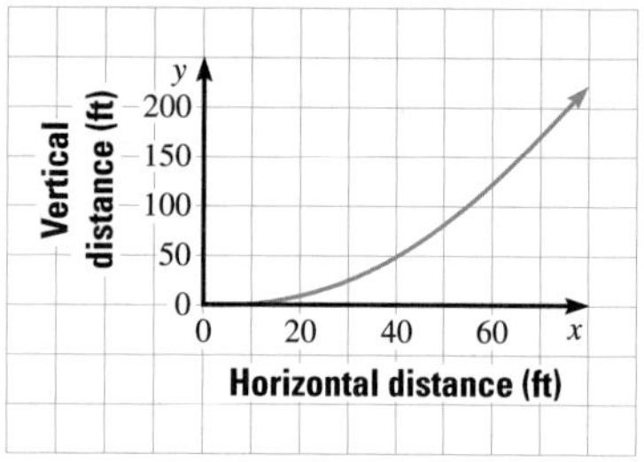

LOGICAL REASONING **Consider the equation $ax^2 + bx + c = 0$ and use the quadratic formula to justify the statement.**

32. If $b^2 - 4ac$ is positive, then the equation has two solutions. **See margin.**

33. If $b^2 - 4ac$ is zero, then the equation has one solution. **See margin.**

34. If $b^2 - 4ac$ is negative, then the equation has no real solution. **See margin.**

MIXED REVIEW

33. *Sample answer:* If $b^2 - 4ac = 0$, then $\sqrt{b^2 - 4ac}$ is 0, so $\frac{-b + \sqrt{b^2 - 4ac}}{2a}$ and $\frac{-b - \sqrt{b^2 - 4ac}}{2a}$ are both equal to $-\frac{b}{2a}$, which is the only solution of the equation.

34. *Sample answer:* If $b^2 - 4ac < 0$, then $\sqrt{b^2 - 4ac}$ is not defined, so $\frac{-b + \sqrt{b^2 - 4ac}}{2a}$ and $\frac{-b - \sqrt{b^2 - 4ac}}{2a}$ are not defined, and there is no real solution of the equation.

GRAPHING FUNCTIONS **Graph the function.** (Review 4.8) **35–37. See margin.**

35. $f(x) = -x + 1$ **36.** $f(x) = -6x + 1$ **37.** $f(x) = 3x - 9$

SOLVING INEQUALITIES **Solve the inequality.** (Review 6.1)

38. $\frac{x}{6} \le -2$ **$x \le -12$** **39.** $-\frac{x}{3} \ge 15$ **$x \le -45$** **40.** $-12.3x > 86.1$ **$x < -7$** **41.** $11.2x \le 134.4$ **$x \le 12$**

SKETCHING GRAPHS **Sketch the graph of the inequality.** (Review 6.5 for 9.7) **42–45. See margin.**

42. $3x + y > 9$ **43.** $y - 4x < 0$ **44.** $-2x - y \ge 4$ **45.** $-y - 3x \le 6$

COMPUTER MODEMS **In Exercises 46–48, use the following data which list the prices of several computer modems.** (Review 6.6, 6.7)

\$230, \$220, \$170, \$215, \$190, \$200, \$200, \$150, \$170

46. Use a stem-and-leaf plot to order the data from least to greatest. **See margin for plot; 150, 170, 170, 190, 200, 200, 215, 220, 230**

47. Make a box-and-whisker plot of the data. **See margin.**

48. Use your results to describe the prices. ***Sample answer:* The prices range from \$150 to \$230, with half the prices between \$170 and \$217.50. The median price is \$200.**

Additional Test Preparation *Sample answer:*
If the area of a square with side *s* is 36, you can solve $s^2 - 36 = 0$ to find the length of a side. The discriminant, 144, indicates two solutions. Solving gives the solutions −6 and 6, but you must reject −6 because side length cannot be negative.

SOLVING GRAPHICALLY **Solve the equation graphically. Check the results algebraically.** (Lesson 9.4)

1. $x^2 - 3x = 10$ **−2, 5**
2. $x^2 - 12x = -36$ **6**
3. $3x^2 + 12x = -9$ **−3, −1**

USING THE FORMULA **Use the quadratic formula to solve the equation.** (Lesson 9.5)

4. $x^2 + 6x + 9 = 0$ **−3**
5. $2x^2 + 13x + 6 = 0$ **−6, −0.5**
6. $-x^2 + 6x + 16 = 0$ **−2, 8**
7. $-2x^2 + 7x - 6 = 0$ **1.5, 2**
8. $-3x^2 - 5x + 12 = 0$ **$-3, 1\frac{1}{3}$**
9. $5x^2 + 8x + 3 = 0$ **−1, −0.6**

USING THE DISCRIMINANT **Tell if the equation has *two solutions, one solution,* or *no real solution.*** (Lesson 9.6)

10. $x^2 - 15x + 56 = 0$ **two solutions**
11. $x^2 + 8x + 16 = 0$ **one solution**
12. $x^2 - 3x + 4 = 0$ **no real solution**

13. **THROWING A BASEBALL** Your friend is standing on a 45-foot balcony. He asks you to throw a baseball up to him. You throw the baseball with an initial upward velocity of 50 feet per second. Assuming that you released the baseball 6 feet above the ground, did it reach your friend? Explain. (Lesson 9.6)
Yes. *Sample answer:* The discriminant of the equation $-16t^2 + 50t + 6 = 45$ is positive, so the equation has two real solutions, meaning that the ball would reach a height of 45 ft.

MATH & History — History of the Quadratic Formula

APPLICATION LINK www.mcdougallittell.com

THEN

OVER 3700 YEARS AGO, in Mesopotamia, math exercises using quadratic equations were written on cuneiform tablets. A cuneiform tablet contains the math exercise:

A region consists of two nonoverlapping squares of total area 1000. The side of one square is two-thirds of the side of the other square, diminished by ten. What are the sides of the two squares?

1. Use the equations $x^2 + y^2 = 1000$ and $y = \frac{2}{3}x - 10$ to write a quadratic equation to find the length of a side x of the figure. **$\frac{13}{9}x^2 - \frac{40}{3}x - 900 = 0$, or $13x^2 - 120x - 8100 = 0$**
2. Use the discriminant to find the number of solutions. How many reasonable solutions are there? **two; one**
3. Solve the quadratic equation. What are the lengths of both sides of the figure? **The larger square is 30 ft by 30 ft, the smaller is 10 ft by 10 ft.**

NOW

TODAY, there are many modern applications of quadratic equations. NASA relies on aircraft that fly in a parabolic arc in which to train astronauts.

c. 1775 B.C.

Cuneiform tablets record knowledge of quadratic equations.

1535 A.D.

Mathematician Niccolo Tartaglia solves cubic equations.

Astronaut Mary Cleave trains in a low gravity environment.

1985

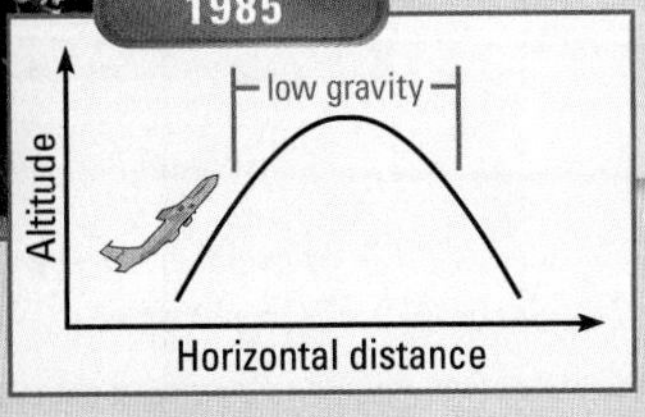

ADDITIONAL RESOURCES

An alternative Quiz for Lessons 9.4–9.6 is available in the *Chapter 9 Resource Book,* p. 95.

A **blackline** master with additional Math & History exercises is available in the *Chapter 9 Resource Book,* p. 93.

35.
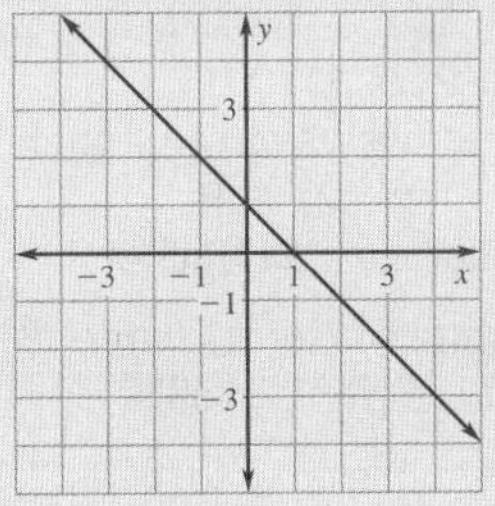

36.
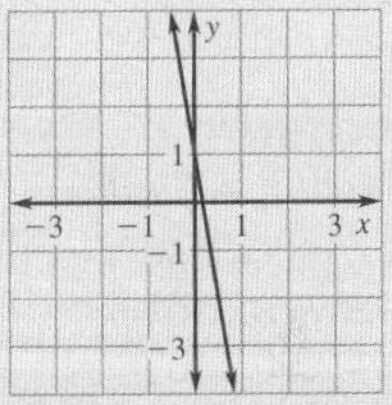

37.
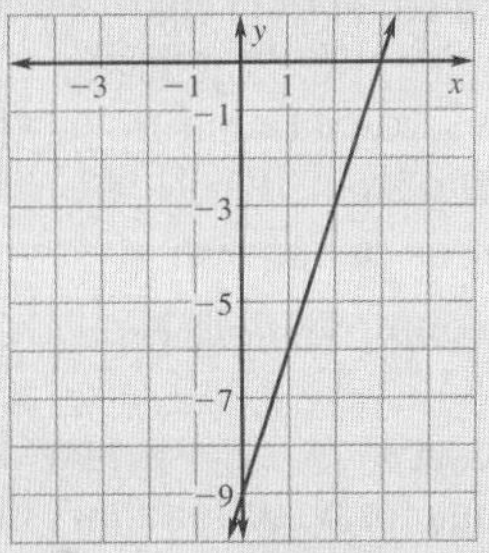

42.
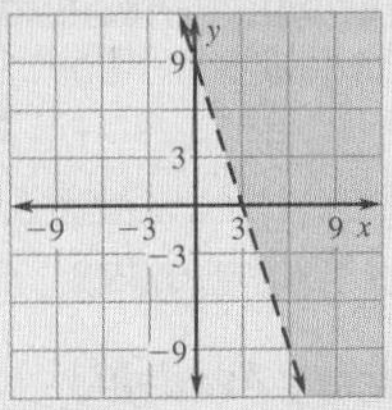

43.
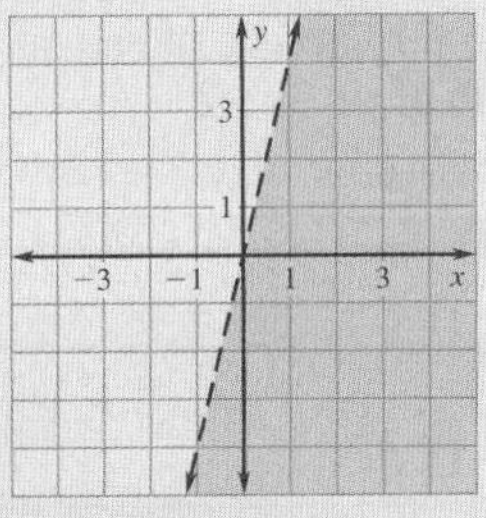

44.
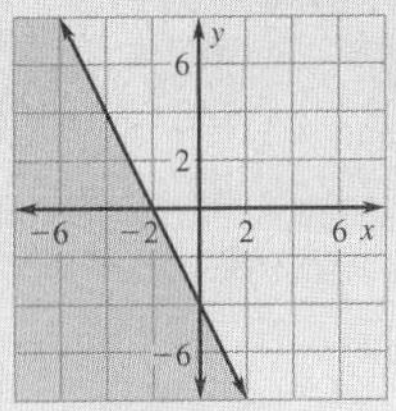

1 PLAN

PACING
Basic: 1 day
Average: 1 day
Advanced: 1 day
Block Schedule: 0.5 block with 9.8

LESSON OPENER APPLICATION
An alternative way to approach Lesson 9.7 is to use the Application Lesson Opener:
- Blackline Master (*Chapter 9 Resource Book*, p. 99)
- Transparency (p. 64)

MEETING INDIVIDUAL NEEDS
- ***Chapter 9 Resource Book***
  Prerequisite Skills Review (p. 5)
  Practice Level A (p. 101)
  Practice Level B (p. 102)
  Practice Level C (p. 103)
  Reteaching with Practice (p. 104)
  Absent Student Catch-Up (p. 106)
  Challenge (p. 108)
- ***Resources in Spanish***
- ***Personal Student Tutor***

NEW-TEACHER SUPPORT
See the Tips for New Teachers on pp. 1–2 of the *Chapter 9 Resource Book* for additional notes about Lesson 9.7.

WARM-UP EXERCISES
Transparency Available
Tell whether each point lies on the graph of $y = 3x + 2$.
1. (0, 2) yes
2. (−1, 4) no
3. (−2, −1) no
4. (3, 11) yes
5. (−1, −1) yes

9.7 Graphing Quadratic Inequalities

What you should learn

GOAL 1 Sketch the graph of a quadratic inequality.

GOAL 2 Use quadratic inequalities as **real-life** models, as in bridge building in **Example 4**.

Why you should learn it

▼ To model a **real-life** situation such as the height of the two towers on the Golden Gate Bridge in **Ex. 32**.

REAL LIFE

GOAL 1 GRAPHING A QUADRATIC INEQUALITY

In this lesson you will study the following types of **quadratic inequalities.**

$y < ax^2 + bx + c$ $\qquad$ $y \le ax^2 + bx + c$

$y > ax^2 + bx + c$ $\qquad$ $y \ge ax^2 + bx + c$

The **graph** of a quadratic inequality consists of the graph of all ordered pairs (x, y) that are solutions of the inequality.

EXAMPLE 1 *Checking Points*

Sketch the graph of $y = x^2 - 3x - 3$. Plot and label the points $A\,(3, 2)$, $B\,(1, 4)$, and $C\,(4, -3)$. Tell whether each point lies inside or outside the parabola.

SOLUTION

1. Sketch the graph of $y = x^2 - 3x - 3$.
2. Plot and label the points $A(3, 2)$, $B(1, 4)$, and $C(4, -3)$.

   A and B lie inside the parabola while C lies outside.

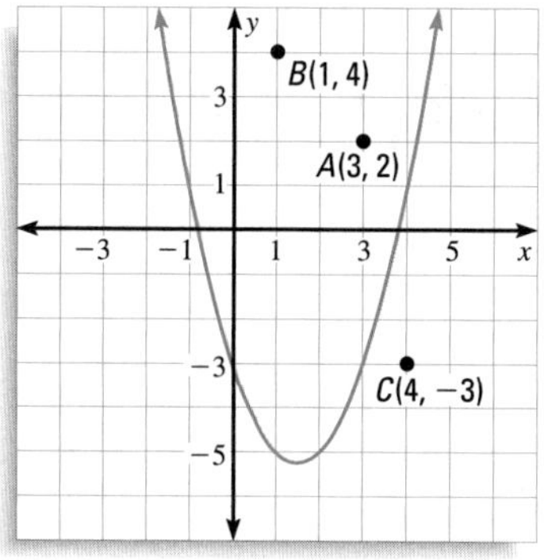

Checking points helps you graph quadratic inequalities by telling you which region to shade. You can use the following guidelines to graph any quadratic inequality.

SKETCHING THE GRAPH OF A QUADRATIC INEQUALITY

STEP 1 Sketch the graph of the equation $y = ax^2 + bx + c$ that corresponds to the inequality. Sketch a dashed parabola for inequalities with < or > to show that the points on the parabola are *not* solutions. Sketch a solid parabola for inequalities with ≤ or ≥ to show that the points on the parabola are solutions.

STEP 2 The parabola you drew separates the coordinate plane into two regions. Test a point that is not on the parabola to find whether it is a solution of the inequality.

STEP 3 If the test point is a solution, shade its region. If not, shade the other region.

EXAMPLE 2 *Graphing a Quadratic Inequality*

Sketch the graph of $y < 2x^2 - 3x$.

SOLUTION

STEP 1 Sketch the equation $y = 2x^2 - 3x$ that corresponds to the inequality $y < 2x^2 - 3x$. Use a dashed line since the inequality is $<$. The parabola opens up.

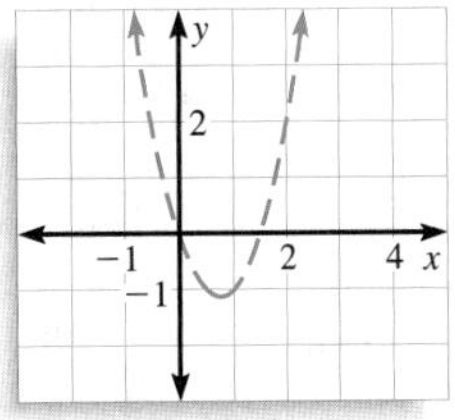

STEP 2 Test a point that is not on the parabola, say (0, 2).

$y < 2x^2 - 3x$	**Write original inequality.**
$2 \overset{?}{<} 2(0)^2 - 3(0)$	**Substitute 0 for *x* and 2 for *y*.**
$2 \not< 0$	**2 is not less than 0.**

Because 2 is not less than 0, the ordered pair (0, 2) is not a solution.

STEP 3 The point (0, 2) is not a solution and it is inside the parabola, so the graph of $y < 2x^2 - 3x$ is all the points outside, but not on, the parabola.

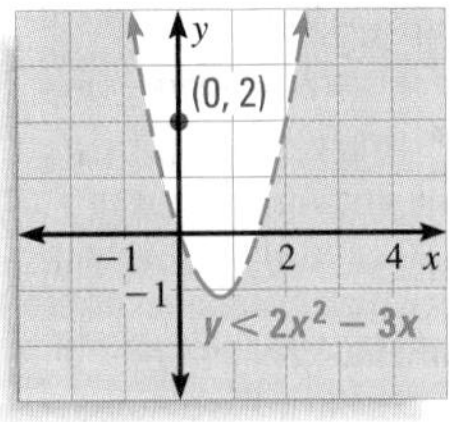

STUDENT HELP

Look Back
For help with checking ordered pairs as solutions, see p. 360.

EXAMPLE 3 *Graphing a Quadratic Inequality*

Sketch the graph of $y \le -x^2 - 5x + 4$.

SOLUTION

STEP 1 Sketch the equation $y = -x^2 - 5x + 4$ that corresponds to the inequality $y \le -x^2 - 5x + 4$. Use a solid line since the inequality is $\le$. The parabola opens down.

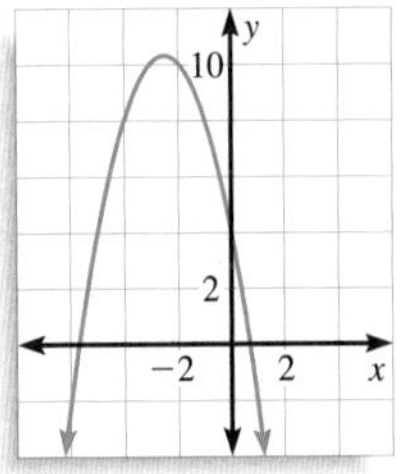

STEP 2 Test a point that is not on the parabola, say (0, 0).

$y \le -x^2 - 5x + 4$	**Write original inequality.**
$0 \overset{?}{\le} -(0)^2 - 5(0) + 4$	**Substitute 0 for *x* and *y*.**
$0 \le 4$	**True.**

Because 0 *is* less than or equal to 4, the ordered pair (0, 0) is a solution.

STEP 3 The point (0, 0) is a solution and it is inside the parabola, so the graph of $y \le -x^2 - 5x + 4$ is all the points inside or on the parabola $y = -x^2 - 5x + 4$.

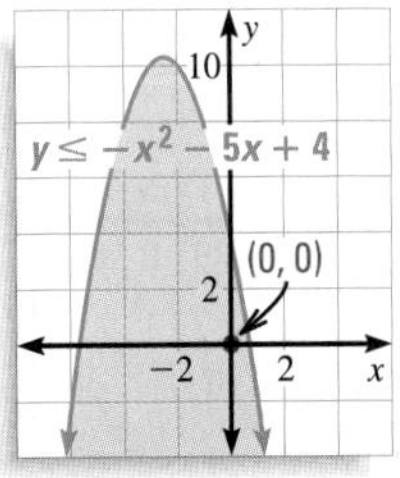

2 TEACH

EXTRA EXAMPLE 1

Sketch the graph of $y = x^2 + 2x - 4$. Plot and label the points $A(-1, -2)$, $B(-2, 3)$, and $C(3, 0)$. Tell whether each point lies inside or outside the parabola.

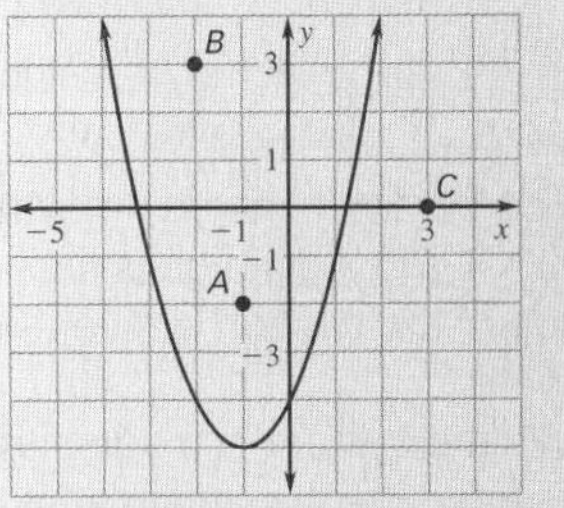

***A*: inside; *B*: inside; *C*: outside**

EXTRA EXAMPLE 2

Sketch the graph of $y \ge 3x^2 + 2x$.

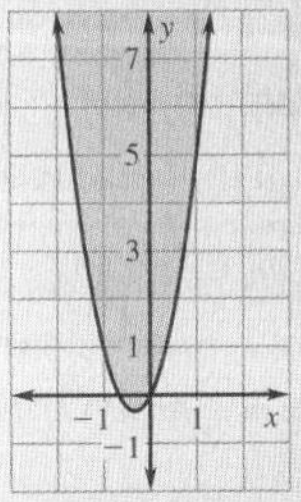

EXTRA EXAMPLE 3

Sketch the graph of $y < -x^2 + 3x - 1$.

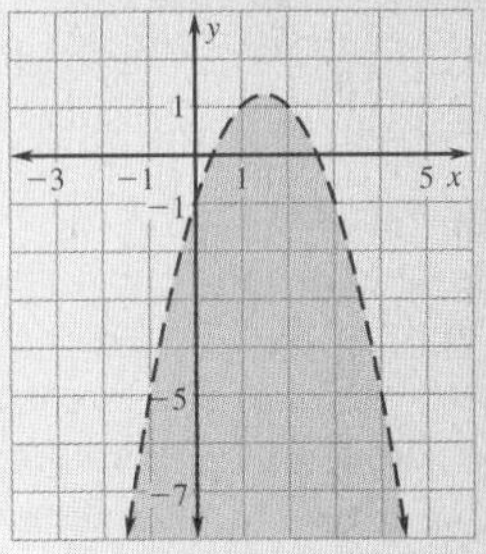

CHECKPOINT EXERCISES

For use after Examples 1–3:

1. Tell whether each point is included in the graph of the inequality $y > -2x^2 - x + 1$.
$A(-2, -4)$ $B(0, -1)$
$C(2, -6)$ $D(3, -20)$
***A*: yes; *B*: no; *C*: yes; *D*: no**

EXTRA EXAMPLE 4

A support over a river is shown below. The parabolic opening can be modeled by $y = -0.006667x^2 - 6$.

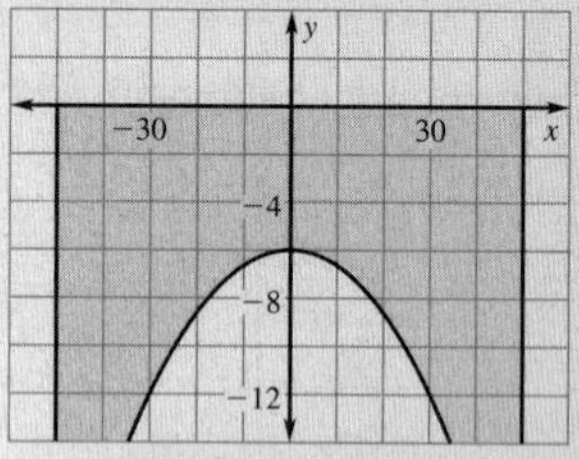

a. The parabola meets the water at $x = \pm 30$. How high is the top of the support above the water? **12 ft**

b. Write a system of inequalities to describe the region above water of the support.
$-50 \le x \le 50$; $-12 \le y \le 0$; $y \ge -0.006667x^2 - 6$

CHECKPOINT EXERCISES

For use after Example 4:

1. Write a system of inequalities to describe the region inside $y = 0.005x^2 - 10$, on the y-axis or in Quadrant I, and below a horizontal line through (3, 10).
$y \ge 0.005x^2 - 10$; $x \ge 0$; $y < 10$

FOCUS ON VOCABULARY

Describe the graph of a quadratic inequality. **See sample answer below.**

CLOSURE QUESTION

Describe how to graph $y \le ax^2 + bx + c$. ***Sample answer:*** **Graph $y = ax^2 + bx + c$. The inequality is $\le$, so use a solid line. Substitute a point not on the parabola into the inequality. If the inequality is true, shade the region containing the point. If not, shade the other region.**

DAILY PUZZLER

What is the only point with integer coordinates that satisfies all three inequalities?

$y \ge x^2 - 2x$ $\quad y \le -x^2 + 2$

$y \ge x^2 + 4x + 1$ $\quad$ **(0, 1)**

GOAL 2 USING QUADRATIC INEQUALITIES IN REAL LIFE

EXAMPLE 4 *Using a Quadratic Inequality Model*

FOCUS ON APPLICATIONS

BRIDGE BUILDING The Akashi Kaikyo Bridge is part of a network of bridges built to connect all of Japan's four main islands. It connects the cities of Kobe and Awaji.

APPLICATION LINK
www.mcdougallittell.com

BRIDGE BUILDING The Akashi Kaikyo Bridge in Japan is the longest suspension bridge in the world. The shape of the main cable between the two main towers can be modeled by

$$y = 0.000188x^2 + 15$$

where x is the distance (in meters) from the center of the bridge and y is the height (in meters) above the road.

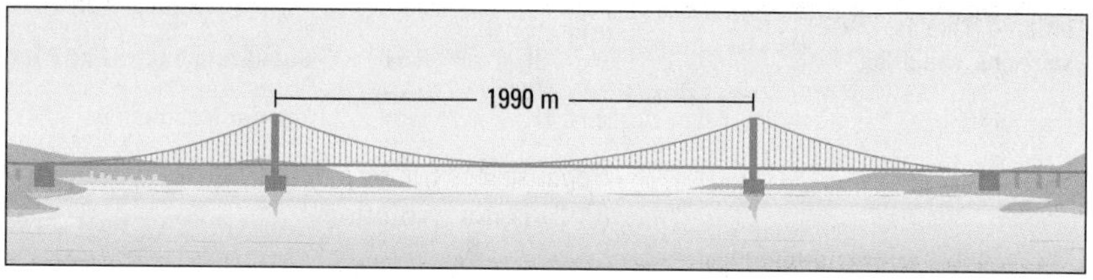

a. How high are the towers above the road?

b. Write a system of inequalities that describes the region under the main cable, above the roadbed, and between the main towers. Assume that the roadbed is modeled by $y = 0$. Graph the system.

SOLUTION

a. The diagram shows that the distance between the two towers is 1990 meters. Because the vertex of the parabola is at the center of the bridge, the x-coordinates of the tops of the towers are ± 995. Substitute either value.

$y = 0.000188x^2 + 15$	**Write quadratic equation.**
$= 0.000188(995)^2 + 15$	**Substitute 995 for x.**
≈ 201.12	**Approximate with a calculator.**

▶ The towers are about 201 meters above the road.

b. The parabola $y = 0.000188x^2 + 15$ opens up. To include the region below the parabola and the main cable, use the inequality $\le$.

Region below the cable:	$y \le 0.00188x^2 + 15$
Region between the towers:	$-995 \le x \le 995$
Region above the roadbed:	$y \ge 0$

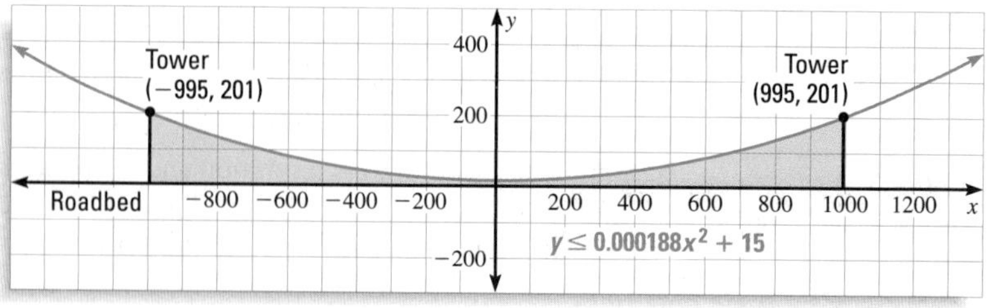

Focus on Vocabulary *Sample answer:*
The graph is a parabola, which may or may not be part of the solution, and one of the regions into which it divides the coordinate plane: one with points outside the parabola and one with points inside.

GUIDED PRACTICE

Vocabulary Check ✓ **1.** Give an example of each of the types of quadratic inequalities. **See margin.**

Concept Check ✓ **2.** Describe the steps used to sketch the graph of a quadratic inequality. **See margin.**

3. Is (0, 3) inside or outside the graph of $y = x^2 + 3x + 2$? **inside**

Skill Check ✓

Decide whether each labeled ordered pair is a solution of the inequality. **4–6. See margin.**

4. $y < -x^2$

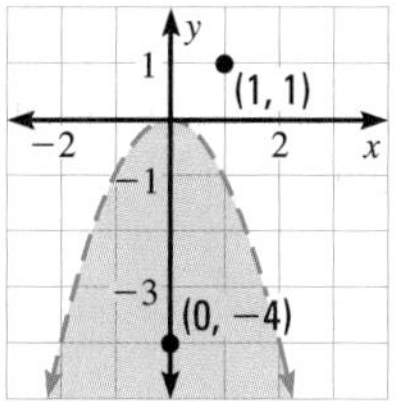

5. $y \geq x^2 - 2$

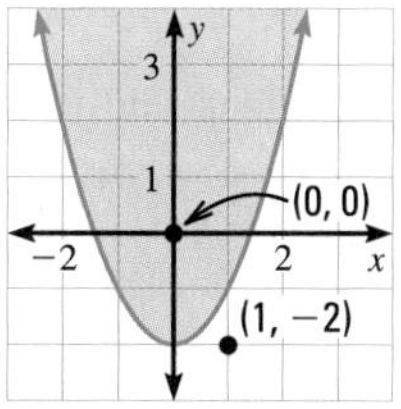

6. $y \leq 2x^2 + 5x$

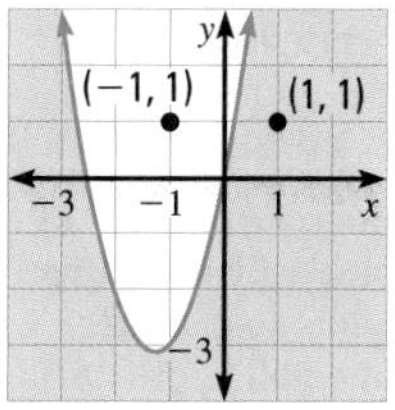

Sketch the graph of the inequality. **7–12. See margin.**

7. $y \leq x^2$ **8.** $y > -x^2 + 3$ **9.** $y < x^2 + 2x$

10. $y \geq x^2 - 2x$ **11.** $y > -2x^2 + 5x$ **12.** $y < 4x^2 - 2x + 1$

1. ***Sample answers:*** $y < 2x^2 + 3x - 1$, $y \leq -2x^2 + 7x - 3$, $y > x^2 + x + 10$, $y \geq -5x^2 - 3x - 2$

2. ***Sample answer:*** Write the corresponding equation by replacing the inequality symbol with an equal sign. Graph the related function. Use a dashed curve for < and > inequalities and a solid curve for ≤ and ≥ inequalities. Test a point in one of the two regions of the plane determined by the curve. If the point is a solution of the inequality, shade the region containing the point. If not, shade the other region.

4. (1, 1) is not a solution; (0, −4) is a solution.

5. (1, −2) is not a solution; (0, 0) is a solution.

6. (−1, 1) is not a solution; (1, 1) is a solution.

PRACTICE AND APPLICATIONS

STUDENT HELP

Extra Practice to help you master skills is on p. 805.

STUDENT HELP

HOMEWORK HELP
Example 1: Exs. 13–22
Example 2: Exs. 17–31
Example 3: Exs. 17–31
Example 4: Exs. 32, 33

SOLUTIONS **Decide whether the ordered pair is a solution of the inequality.**

13. $y \geq 2x^2 - x$, (2, 6) **solution** **14.** $y < x^2 + 9x$, (−3, 10) **not a solution**

15. $y > 4x^2 - 7x$, (2, −10) **not a solution** **16.** $y \geq x^2 - 13x$, (−1, 14) **solution**

MATCHING INEQUALITIES **In Exercises 17–22, match the graph with its inequality.**

A. $y \geq -2x^2 - 2x + 1$ **B.** $y > -2x^2 + 4x + 3$ **C.** $y \leq 2x^2 + x + 1$

D. $y \geq 4x^2 - 6x + 5$ **E.** $y < 2x^2 - x - 3$ **F.** $y > -4x^2 + 6x - 5$

17. **C**

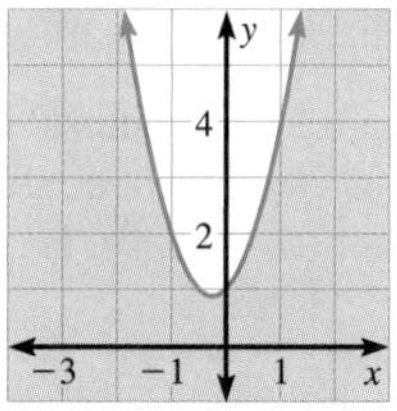

18. **F**

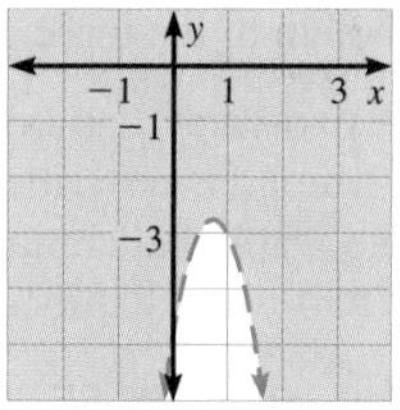

19. **E**

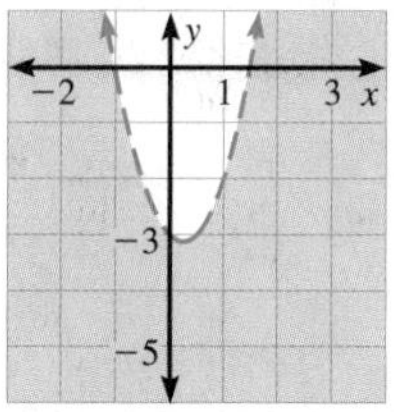

20. **A**

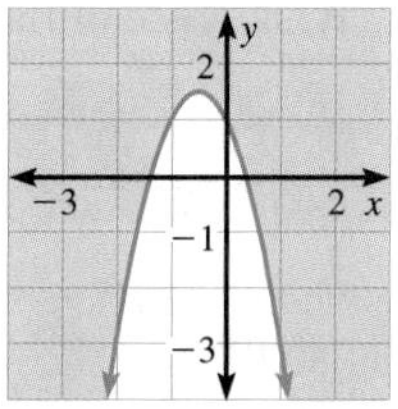

21. **D**

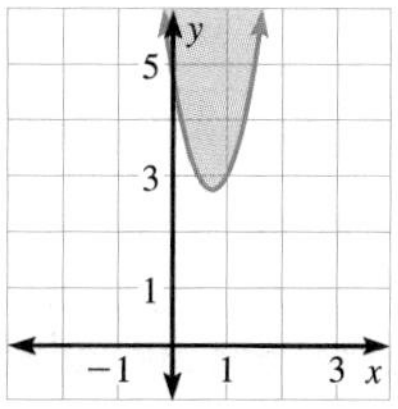

22. **B**

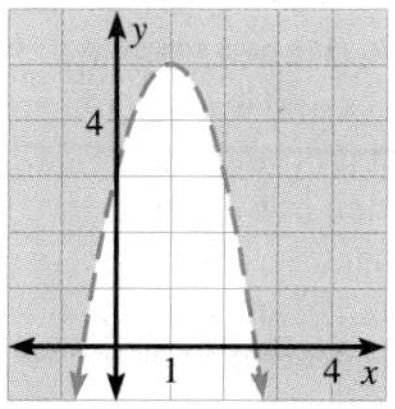

3 APPLY

ASSIGNMENT GUIDE

BASIC
Day 1: pp. 551–553 Exs. 13–33, 39, 40, 44–66 even

AVERAGE
Day 1: pp. 551–553 Exs. 13–33, 39, 40, 44–66 even

ADVANCED
Day 1: pp. 551–553 Exs. 13–33, 39, 40–43, 44–66 even

BLOCK SCHEDULE WITH 9.8
pp. 551–553 Exs. 13–33, 39, 40, 44–66 even

EXERCISE LEVELS
Level A: ***Easier***
13–22
Level B: ***More Difficult***
23–40, 44–67
Level C: ***Most Difficult***
41–43

✓ HOMEWORK CHECK
To quickly check student understanding of key concepts, go over the following exercises: Exs. 14, 16, 17, 22, 24, 30, 32. See also the Daily Homework Quiz:

- Blackline Master (*Chapter 2 Resource Book,* p. 111)
- Transparency (p. 74)

7.

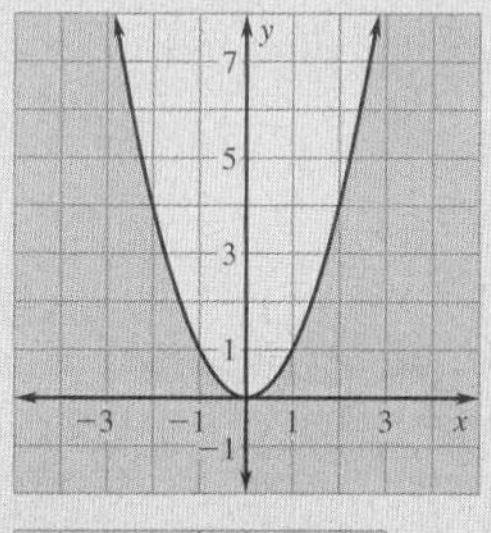

8.

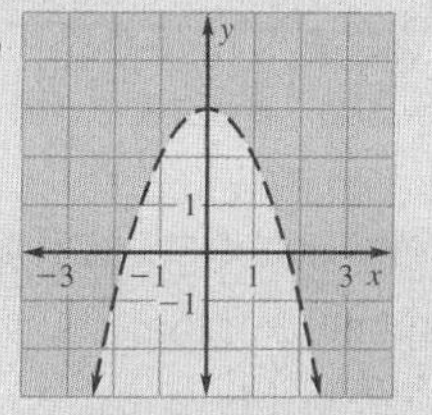

9–12. See Additional Answers beginning on page AA1.

STUDENT HELP NOTES

Homework Help Students can find help for Exs. 32–33 at **www.mcdougallittell.com.** The information can be printed out for students who don't have access to the Internet.

! COMMON ERROR

EXERCISE 38 Remind students that the measure of quality is not the price per carat, since the cost graphs are not linear. If two diamonds of different sizes had the same cost per carat, the larger diamond would be of lower quality. For a given quality, the price per carat increases rapidly as the size increases because high-quality larger diamonds are rarer.

23.

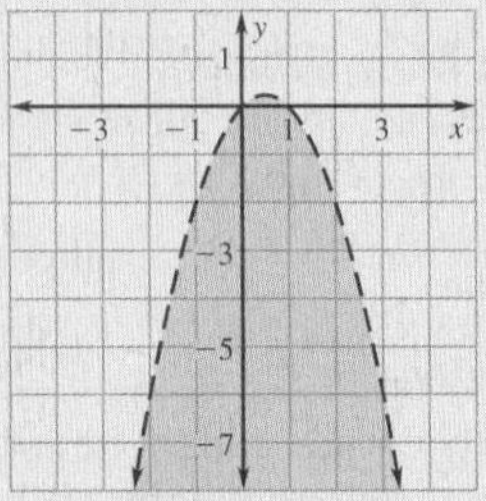

24.

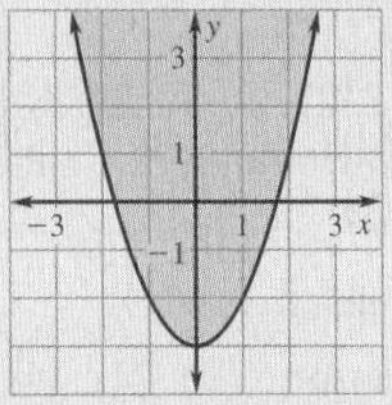

25–31, 33. See Additional Answers beginning on page AA1.

ADDITIONAL PRACTICE AND RETEACHING

For Lesson 9.7:

- Practice Levels A, B, and C (*Chapter 9 Resource Book,* p. 101)
- Reteaching with Practice (*Chapter 9 Resource Book,* p. 104)
- See Lesson 9.7 of the *Personal Student Tutor*

For more Mixed Review:

- Search the *Test and Practice Generator* for key words or specific lessons.

STUDENT HELP

INTERNET **HOMEWORK HELP** Visit our Web site www.mcdougallittell.com for help with Exs. 32–33.

34. IF; for every value of *x*, the value of y (that is, the cost of the diamond) is greater for that grade than for either of the others.

35. I3; for every value of *x*, the value of *y* (that is, the cost of the diamond) is lower for that grade than for either of the others.

36. The region described by $y > 82.12x^2 + 119.13x - 48.15$; yes; the region is bounded below by the equation for an IF diamond.

FOCUS ON APPLICATIONS

REAL LIFE **DIAMONDS** A carat is a unit of weight for precious stones. The price of a diamond is determined by its weight, cut, clarity, and degree of flawlessness.

SKETCHING GRAPHS Sketch the graph of the inequality. **23–31. See margin.**

23. $y < -x^2 + x$	**24.** $y \geq x^2 - 3$	**25.** $y \geq x^2 - 5x$
26. $y < -x^2 - 3x - 1$	**27.** $y < x^2 + x + 8$	**28.** $y \leq -x^2 + 3x + 2$
29. $y > -4x^2 - 8x - 4$	**30.** $y \geq -4x^2 - 3x + 8$	**31.** $y \geq 2x^2 + 5x + 3$

GOLDEN GATE BRIDGE Use the following information for Exercises 32 and 33.

The Golden Gate Bridge connects northern California to San Francisco. The vertical cables that are suspended from the main cable lie in the region given by $0 \leq y \leq 0.000112x^2 + 5$ where x is horizontal distance from the middle of the bridge (in feet) and y is vertical distance (in feet) above the road.

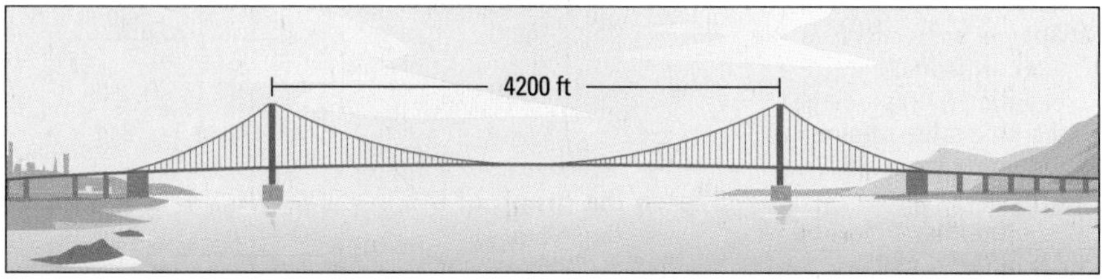

32. How high are the towers above the road? **about 499 ft**

33. Sketch a graph of the region between the towers and under the main cable. **See margin.**

DIAMONDS In Exercises 34–38, use the equations below which represent the sample price *y* (in hundreds of dollars) of diamonds of *x* carats with different grades of flawlessness.

Flawlessness	Equation
IF	$y = 82.12x^2 + 119.13x - 48.15$
VS2	$y = 38.88x^2 + 26.62x - 5.07$
I3	$y = 1.87x^2 + 14.83x - 2.3$

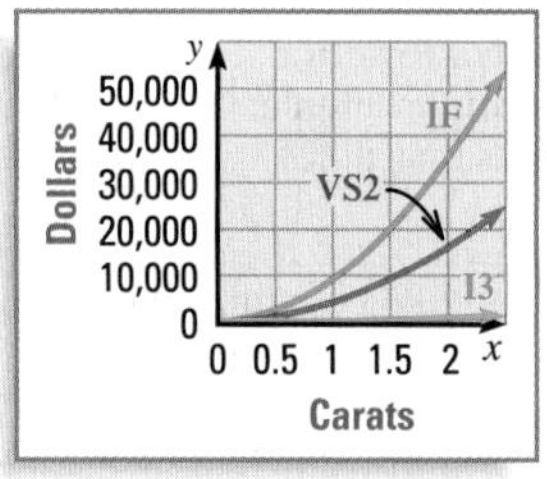

34. Which of the three grades of diamond is the least flawed? How do you know? **See margin.**

35. Which of the three grades of diamond is the most flawed? How do you know? **See margin.**

36. A two-carat diamond is selling for the wholesale price of \$55,000. The diamond is described as having the highest quality. Describe the region in the graph that contains the point (2, 55,000). Is this the highest quality for a two-carat diamond? Explain. **See margin.**

37. The wholesale price of a one-carat diamond is \$12,000. Assume that the price fairly represents the flawlessness of the diamond. Which region contains the point (1, 12,000)? What grade of flawlessness is the diamond? **A; VS2**

A. $y > 1.87x^2 + 14.83x - 2.3$ **B.** $y < 1.87x^2 + 14.83x - 2.3$

38. You see a $\frac{1}{2}$-carat diamond on display for a wholesale price of \$4000. Next to it is a two-carat diamond for a wholesale price of \$12,000. Is the more expensive one necessarily less flawed? **no**

Test Preparation

39. MULTIPLE CHOICE Which ordered pair is *not* a solution of the inequality $y \geq 2x^2 - 7x - 10$? **C**

Ⓐ (0, −4) Ⓑ (−1, −1) Ⓒ (4, −13) Ⓓ (5, 15)

40. MULTIPLE CHOICE Identify the graph of $y \leq 3x^2 + 2x - 5$. **C**

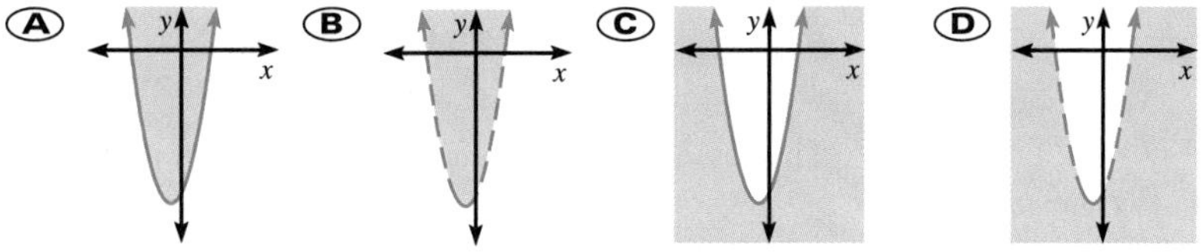

★ Challenge

MILK CONSUMPTION **In Exercises 41–43, use the following information.** The per capita consumption (pounds per person) of whole milk W and reduced fat milk R from 1980 to 1995 can be modeled by the equations below where t represents years since 1980. ▶ Source: U.S. Department of Agriculture

Whole milk: $W = 0.09t^2 - 5.29t + 114.69$

Reduced fat: $R = -0.23t^2 + 2.40t + 71.39$

41. Sketch a graph of each equation on the same coordinate plane. **See margin.**

42. Shade the region where the per capita consumption of reduced fat milk was greater than that of whole milk. What inequalities represent this region? **$y > 0.09t^2 - 5.29t + 114.69$ and $y < -0.23t^2 + 2.40t + 71.39$; See margin for graph.**

43. Find the points where the parabolas intersect. What does each of these points represent? **about (9, 74.4) and (15, 55.5); a year in which per capita consumption of whole milk and reduced fat milk was the same.**

EXTRA CHALLENGE
→ www.mcdougallittell.com

MIXED REVIEW

GRAPHING EQUATIONS **Graph the equation.** (Review 4.3, 9.3 for 9.8) **44–49. See margin.**

44. $y = x + 4$ 45. $3x + 6y = -18$ 46. $2x - 4y = 24$

47. $y = x^2 + x + 2$ 48. $y = -x^2 + 4x + 1$ 49. $y = 3x^2 - 2x + 6$

DIRECT VARIATION **The variables x and y vary directly. Use the given values to write an equation that relates x and y.** (Review 4.5)

50. $x = 6, y = 42$ **$y = 7x$** 51. $x = 54, y = -9$ **$y = -\frac{1}{6}x$** 52. $x = 7, y = 5$ **$y = \frac{5}{7}x$**

53. $x = -13, y = -52$ **$y = 4x$** 54. $x = \frac{3}{4}, y = 3$ **$y = 4x$** 55. $x = 4.6, y = 1.2$ **$y = \frac{6}{23}x$**

GRAPHING EXPONENTIAL EQUATIONS **Sketch the graph of the exponential equation.** (Review 8.5, 8.6) **56–61. See margin.**

56. $y = 2^x$ 57. $y = 0.5^x$ 58. $y = \frac{1}{2}(2)^x$

59. $y = 0.9^x$ 60. $y = 4(1.5)^x$ 61. $y = 5(0.5)^x$

SOLVING QUADRATIC EQUATIONS **Use the quadratic formula to solve the equation.** (Review 9.5)

62. $x^2 - 2x - 3 = 0$ **−1, 3** 63. $2x^2 - 6x + 4 = 0$ **1, 2** 64. $2x^2 - 2x - 12 = 0$ **−2, 3**

65. $-\frac{2}{3}x^2 - 3x + 1 = 0$ **See margin.** 66. $-7x^2 - 2.5x + 3 = 0$ **$-\frac{6}{7}, \frac{1}{2}$** 67. $2x^2 + 4x - 3 = 0$ **See margin.**

65. $\frac{-9 - \sqrt{105}}{4} \approx -4.81$, $\frac{-9 + \sqrt{105}}{4} \approx 0.31$

67. $\frac{-2 - \sqrt{10}}{2} \approx -2.58$; $\frac{-2 + \sqrt{10}}{2} \approx 0.58$

41, 42, 44–49, 56–61. See Additional Answers beginning on page AA1.

4 ASSESS

DAILY HOMEWORK QUIZ

Transparency Available

1. Is the ordered pair (−2, 4) a solution of the inequality $y > 3x^2 + 7x + 5$? **yes**
2. Which inequality describes the graph?
   A. $y > -x^2 + 3x + 1$
   B. $y < -x^2 + 3x + 1$
   C. $y \leq -x^2 + 3x + 1$

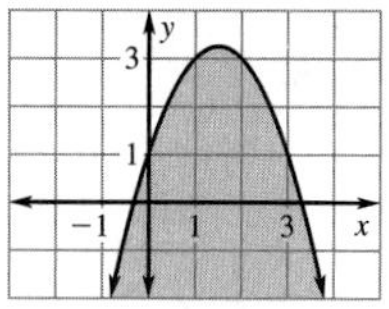

C

3. Graph $y < x^2 - 2x - 2$.

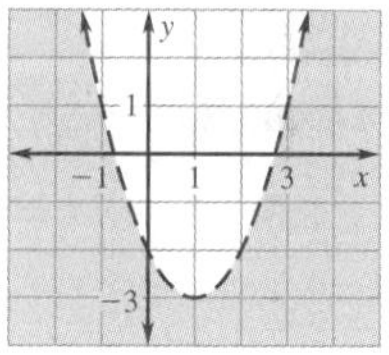

4. The area under a parabolic bridge support is described by $y \leq -0.005x^2 + 120$. The support is 310 ft wide at its base. A road crosses the arch 45 ft in from each side of the base of the support. How high is the road above the base? **59.5 ft**

EXTRA CHALLENGE NOTE
→ Challenge problems for Lesson 9.7 are available in **blackline** format in the *Chapter 9 Resource Book*, p. 108 and at **www.mcdougallittell.com.**

ADDITIONAL TEST PREPARATION

1. **OPEN ENDED** Write a quadratic inequality for which the outside of the graph is below the parabola and another for which the outside of the graph is above the parabola.
   Sample answers:
   below: $y \geq 9x^2 + 2x + 2$;
   above: $y \leq -2x^2 + x - 1$

1 PLAN

PACING
Basic: 2 days
Average: 2 days
Advanced: 2 days
Block Schedule: 0.5 block with 9.7
0.5 block with Ch. Rev.

LESSON OPENER APPLICATION
An alternative way to approach Lesson 9.8 is to use the Application Lesson Opener:
- Blackline Master (*Chapter 9 Resource Book*, p. 112)
- Transparency (p. 65)

MEETING INDIVIDUAL NEEDS
- ***Chapter 9 Resource Book***
  Prerequisite Skills Review (p. 5)
  Practice Level A (p. 116)
  Practice Level B (p. 117)
  Practice Level C (p. 118)
  Reteaching with Practice (p. 119)
  Absent Student Catch-Up (p. 121)
  Challenge (p. 124)
- ***Resources in Spanish***
- ***Personal Student Tutor***

NEW-TEACHER SUPPORT
See the Tips for New Teachers on pp. 1–2 of the *Chapter 9 Resource Book* for additional notes about Lesson 9.8.

WARM-UP EXERCISES
Transparency Available
Find the ratio of the value of the function when $x = 2$ to the value when $x = 1$.
1. $y = 2x^2$ 4
2. $y = 3.5x^2$ 4
3. $y = 3^x$ 3
4. $y = 100(1 - 0.05)^x$ 0.95

EXPLORING DATA AND STATISTICS

9.8 Comparing Linear, Exponential, and Quadratic Models

What you should learn

GOAL 1 Choose a model that best fits a collection of data.

GOAL 2 Use models in **real-life** settings, such as the stretch of a spring in **Ex. 20.**

Why you should learn it

▼ To select the best model for **real-life** data, such as the increase in volume of the chambers of a nautilus in **Example 2**.

Chambered Nautilus

GOAL 1 CHOOSING A MODEL

This lesson will help you choose the type of model that best fits a collection of data.

CONCEPT SUMMARY THREE BASIC MODELS

LINEAR MODEL
$y = mx + b$

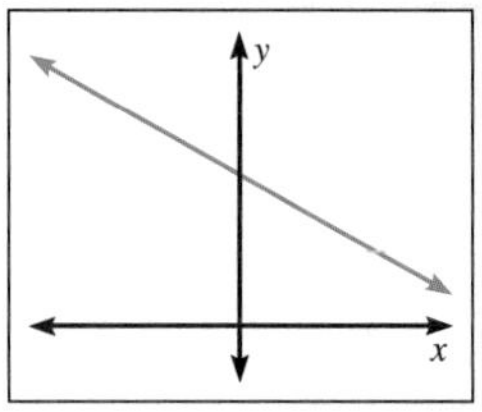

EXPONENTIAL MODEL
$y = C(1 \pm r)^t$

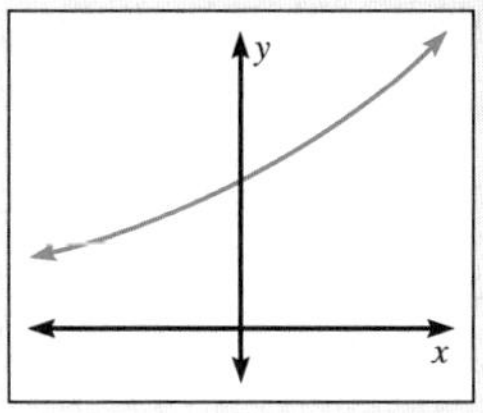

QUADRATIC MODEL
$y = ax^2 + bx + c$

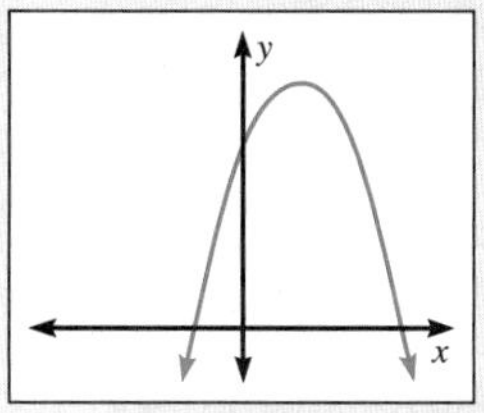

EXAMPLE 1 *Choosing a Model*

Name the type of model that best fits each data collection.

a. $(-3, 4), \left(-2, \frac{7}{2}\right), (-1, 3), \left(0, \frac{5}{2}\right), (1, 2), \left(2, \frac{3}{2}\right), (3, 1)$

b. $(-3, 4), (-2, 2), (-1, 1), \left(0, \frac{1}{2}\right), \left(1, \frac{1}{4}\right), \left(2, \frac{1}{8}\right), \left(3, \frac{1}{16}\right)$

c. $(-3, 4), \left(-2, \frac{7}{3}\right), \left(-1, \frac{4}{3}\right), (0, 1), \left(1, \frac{4}{3}\right), \left(2, \frac{7}{3}\right), (3, 4)$

SOLUTION

Make scatter plots of the data. Then decide whether the points appear to lie on a line (linear model), an exponential curve (exponential model), or a parabola (quadratic model).

a. Linear Model

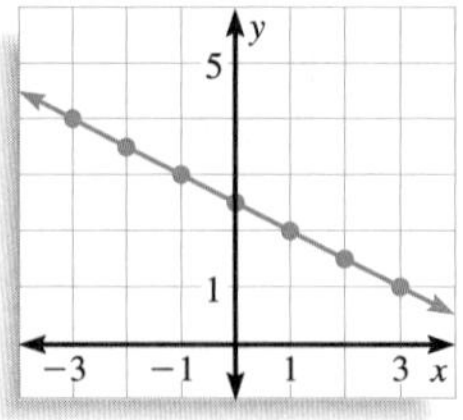

b. Exponential Model

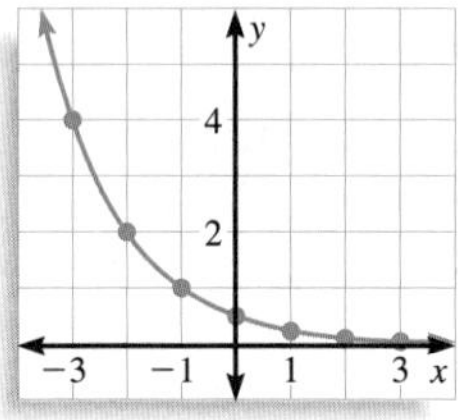

c. Quadratic Model

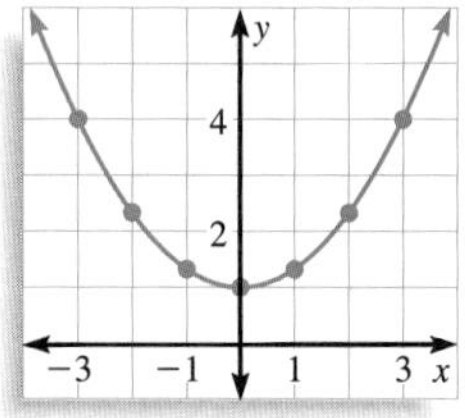

FOCUS ON APPLICATIONS

CHAMBERED NAUTILUS is a marine animal whose soft body is covered with a shell. The nautilus lives in the outermost chamber. Shown here is a cross section of the shell.

GOAL 2 USING MODELS IN REAL-LIFE SETTINGS

EXAMPLE 2 *Writing a Model*

NAUTILUS As a chambered nautilus grows, its chambers get larger and larger. The relationship between the volume and consecutive chambers usually follows the same pattern.

The volumes v (in cubic centimeters) of nine consecutive chambers of a chambered nautilus are given in the table. The chambers are numbered from 0 to 8, with 0 being the first and smallest chamber.

Decide which type of model best fits the data. Write a model.

Chamber, n	Volume (cm^3), v
0	0.787
1	0.837
2	0.889
3	0.945
4	1.005
5	1.068
6	1.135
7	1.207
8	1.283

SOLUTION

Draw a scatter plot of the data. You can see that the graph has a slight curve.

Test whether an exponential model fits the data by finding the ratios of consecutive volumes of the chambers.

$$\frac{\text{Volume of Chamber 1}}{\text{Volume of Chamber 0}} = \frac{0.837}{0.787} \approx 1.064$$

$$\frac{\text{Volume of Chamber 2}}{\text{Volume of Chamber 1}} = \frac{0.889}{0.837} \approx 1.062$$

$$\frac{\text{Volume of Chamber 3}}{\text{Volume of Chamber 2}} = \frac{0.945}{0.889} \approx 1.063$$

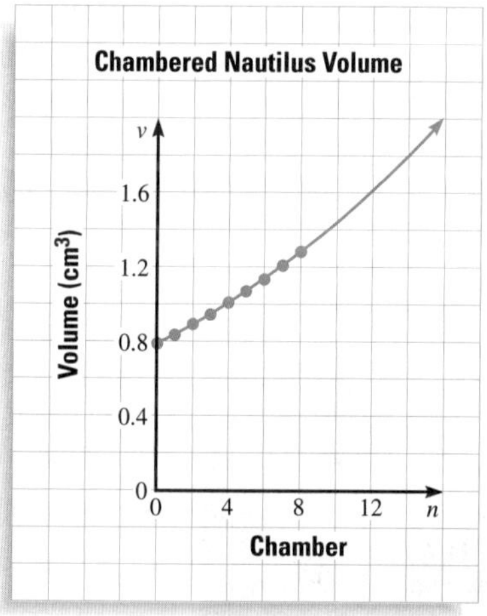

The ratios show that each chamber's volume is about 0.063, or 6.3%, larger than the volume of the chamber just before it. Assuming this pattern applies to all data points, it is appropriate to use an exponential growth model $v = C(1 + r)^n$.

The rate of growth r is 0.063 and the initial volume in chamber 0 is 0.787.

✓**CHECK** Check values of n in the exponential model $v = 0.787(1.063)^n$.

$v = 0.787(1.063)^n$	$v = 0.787(1.063)^n$
$v = 0.787(1.063)^3$	$v = 0.787(1.063)^8$
$v = 0.945$	$v = 1.283$

The exponential model $v = 0.787(1.063)^n$ fits the data.

STUDENT HELP

Look Back
For help writing an exponential growth model, see p. 477.

2 TEACH

EXTRA EXAMPLE 1

Name the type of model that best fits each data set.

a. $\left(-1, -\frac{1}{2}\right), \left(0, -\frac{3}{2}\right), \left(1, -\frac{1}{2}\right), \left(2, \frac{5}{2}\right)$ **quadratic**

b. $\left(-2, \frac{5}{2}\right), (-1, 1), \left(0, -\frac{1}{2}\right), (1, -2)$ **linear**

c. $\left(-1, \frac{5}{4}\right), \left(0, \frac{5}{8}\right), \left(1, \frac{5}{16}\right), \left(2, \frac{5}{32}\right)$ **exponential**

EXTRA EXAMPLE 2

The faster a car's speed, the farther it takes to stop. The table gives the stopping distances of a car at various speeds. Decide which type of model corresponds most closely to the data. Write a model.

Speed (ft/sec)	Distance (ft)
37	45
51	86
66	144
81	217
96	304

quadratic; $y = 0.033x^2$

CHECKPOINT EXERCISES

For use after Examples 1 and 2:

1. Decide which type of model corresponds most closely to the data. Write a model.

Time (hours)	Number of bacteria
0	100
1	250
2	625
3	1560
4	3900

exponential; $y = 100(2.5)^x$

EXTRA EXAMPLE 3

The data show the heights of a bouncing ball dropped from an initial height of 8 feet. Decide which type of model corresponds most closely to the data. Write a model.

Bounce	Height (ft)
0	8
1	7.2
2	6.5
3	5.8
4	5.2

exponential; $H = 8(0.9)^x$

CHECKPOINT EXERCISES

For use after Example 3:

1. Decide which type of model corresponds most closely to the data. Write a model.

x	2	4	6	8	10
y	10	40	90	160	250

quadratic; $y = 2.5x^2$

FOCUS ON VOCABULARY

Give a simple example of a linear model, a quadratic model, and an exponential model.

Sample answers: **linear: $y = 2x$; quadratic: $y = 2x^2$; exponential: $y = 2(0.5)^x$**

CLOSURE QUESTION

Describe how to choose the type of model that corresponds most closely to a collection of data. **See sample answer below.**

DAILY PUZZLER

Given points (2, 2) and (4, 8), find a linear model, an exponential model, and a quadratic model. What does this mean in terms of modeling?

Sample answers: **linear: $y = 3x - 4$; exponential: $y = \frac{1}{2} \cdot 2^x$; quadratic: $y = \frac{1}{2}x^2$; This means you need several data points before you can draw conclusions about a model.**

EXAMPLE 3 *Writing a Model*

PENDULUMS A simple pendulum can be constructed with a string hung from a fixed point and a weight attached at the other end. The time it takes for the pendulum to swing from one side to the other and all the way back again is called its *period*. The period t (in seconds) of the pendulum is related to its length d (in inches). The table gives values of t and d for several different pendulums. Decide which type of model best fits the data. Write a model.

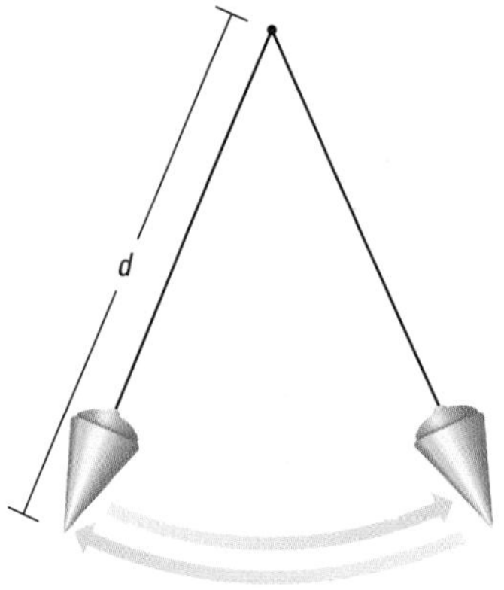

Period (sec), t	Length (in.), d
0.25	0.61
0.50	2.45
0.75	5.50
1.00	9.78
1.25	15.28
1.50	22.00
1.75	29.95
2.00	39.12

SOLUTION

Draw a scatter plot for the data. From the scatter plot, you can see that the pattern is not linear.

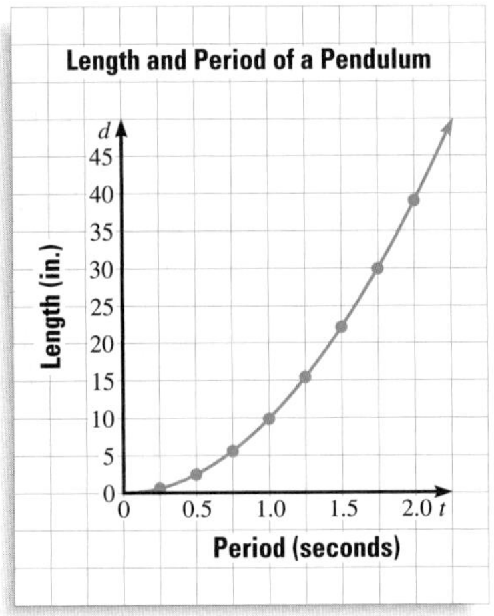

You can test whether an exponential model fits by finding the ratios of consecutive pendulum lengths. Since the lengths do not increase by the same percent, an exponential model does not fit the data.

Test whether a quadratic model fits. Write the simple quadratic model $d = at^2$. To find a, substitute any known values of d and t.

$d = at^2$	**Write quadratic model.**
$9.78 = a(1^2)$	**Substitute 9.78 for *d* and 1 for *t*.**
$9.78 = a$	**Solve for *a*.**
$d = 9.78t^2$	**Substitute *a* in the equation.**

✓**CHECK** Check values of t in the quadratic model $d = 9.78t^2$

$d = 9.78t^2$	$d = 9.78t^2$
$d = 9.78(0.25)^2$	$d = 9.78(2.0)^2$
$d = 0.61$	$d = 39.12$

▶ The quadratic model $d = 9.78t^2$ fits the data.

FOCUS ON APPLICATIONS

REAL LIFE **PENDULUM CLOCKS** One of the first clocks to use a pendulum was invented by the Dutch physicist, astronomer, and mathematician, Christiaan Huygens in the 1650s.

Closure Question *Sample answer:*
Draw a scatter plot for the data. If the points fit on a line, use a linear model. If the points fit on a curve, see if the ratios of consecutive function values are constant. If they are, use an exponential model. Otherwise, try to find a quadratic model.

GUIDED PRACTICE

Vocabulary Check ✓

1. Name, sketch, and write an equation for each of the three types of algebraic models you have studied. **See margin.**

Concept Check ✓

2. Describe the steps you would follow to decide which of the three basic models best fits a collection of data. ***Sample answer:*** **Make a scatter plot. If you cannot determine whether the best fit is an exponential or quadratic model, compare the ratios of consecutive y-coordinates.**

Skill Check ✓

Name the type of model suggested by the graph.

3.

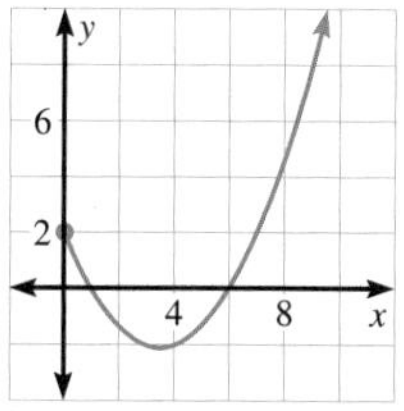

quadratic

4.

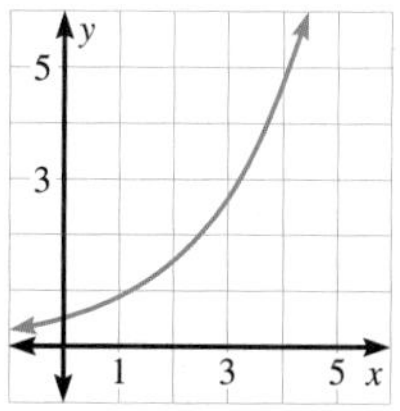

exponential

5.

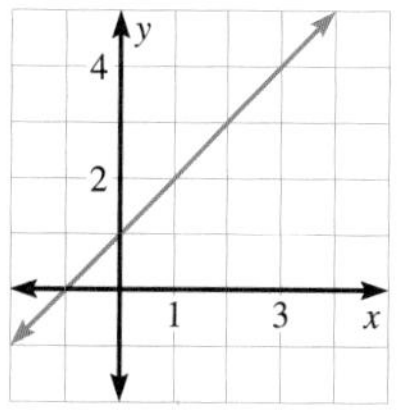

linear

1. Sample answers:
linear: $y = x$
See margin for graph.
exponential: $y = 2^x$
See margin for graph.
quadratic: $y = x^2$
See margin for graph.

Use the data: (0, 1), (1, 1.25), (2, 2), (3, 3.25), (4, 5), (5, 7.25).

6. Draw a scatter plot of the data. **See margin.**

7. Decide which type of model best fits the data. Explain your reasoning. **Quadratic; the graph is a curve but the ratios of consecutive y-coordinates are not the same.**

8. Write a model that fits the data. $y = \frac{1}{4}x^2 + 1$

PRACTICE AND APPLICATIONS

STUDENT HELP
Extra Practice to help you master skills is on p. 805.

CHOOSING MODELS **Make a scatter plot of the data. Then name the type of model that best fits the data.** **9–17. See margin for plots.**

9. $(-1, -6), (-3, 4), (2, 9), (-2, -3), (0, -5), (1, 0)$ **quadratic**

10. $(0, 3), (8, 3), (-4, -1), (4, 4), (-6, -3), (10,1)$ **quadratic**

11. $(1, 3), (2.5, 16.5), (0.5, 1.5), (-2, 0.1), (0, 1), (1.5, 5)$ **exponential**

12. $(-2, -1), (-1, -2.5), (0, -3), (1, -2.5), (2, -1), (3, 1.5)$ **quadratic**

13. $(-3, 2), \left(-2, \frac{5}{2}\right), \left(-1, \frac{7}{2}\right), (0, 5), (1, 7), \left(2, \frac{19}{2}\right)$ **exponential**

14. $(-2, 2), \left(-1, \frac{5}{2}\right), (0, 3), \left(1, \frac{7}{2}\right), (2, 4), \left(3, \frac{9}{2}\right)$ **linear**

STUDENT HELP
HOMEWORK HELP
Example 1: Exs. 9–17
Example 2: Exs. 18–20
Example 3: Exs. 18–20

15.

x	y
5	4
0	−6
7	−6
−1	−14
6	0
3	6

quadratic

16.

x	y
−1	8
1	2
−2	16
3	0.5
0	4
2	1

exponential

17.

x	y
0	5
2	−3
−2	13
1	1
4	−11
3	−7

linear

3 APPLY

ASSIGNMENT GUIDE

BASIC
Day 1: pp. 557–558 Exs. 9–19
Day 2: pp. 558–560 Exs. 20–22, 27–29, 34–48 even, Quiz 3 Exs. 1–13

AVERAGE
Day 1: pp. 557–558 Exs. 9–19
Day 2: pp. 558–560 Exs. 20–22, 27–29, 34–48 even, Quiz 3 Exs. 1–13

ADVANCED
Day 1: pp. 557–558 Exs. 9–19
Day 2: pp. 558–560 Exs. 20–22, 27–32, 34–48 even, Quiz 3 Exs. 1–13

BLOCK SCHEDULE
pp. 557–558 Exs. 9–19 (with 9.7)
pp. 558–560 Exs. 20–22, 27–39, 34–48 even, Quiz 3 Exs. 1–13 (with Ch. 9 Review)

EXERCISE LEVELS

Level A: ***Easier***
9–17
Level B: ***More Difficult***
18–22, 27–29, 33–49
Level C: ***Most Difficult***
23–26, 30–32

✓ HOMEWORK CHECK

To quickly check student understanding of key concepts, go over the following exercises: Exs. 10, 14, 16, 18, 20, 22. See also the Daily Homework Quiz:

- Blackline Master (*Chapter 10 Resource Book,* p. 11)
- Transparency (p. 76)

1. See the next page.
6, 9–17. See Additional Answers beginning on page AA1.

MATHEMATICAL REASONING

To help students see why the ratios of consecutive terms of an exponential function are always the same, ask them to find the ratios of consecutive terms when $x = 2, 3, 4,$ and 5 for the exponential functions $y = 2^x$ and $y = 4(0.6)^x$. Then have them find the same ratios for $y = x^2$.
$y = 2^x$: 2, 2, 2; $y = 4(0.6)^x$: 0.6, 0.6, 0.6; $y = x^2$: $\frac{9}{4}, \frac{16}{9}, \frac{25}{16}$

APPLICATION NOTE

EXERCISE 18 Because the consecutive differences are not the same, students will rule out a linear model. You may want to suggest that students look at the differences of the differences. They will find that these *second differences* are the same when a quadratic model is called for.

STUDENT HELP NOTES

Homework Help Students can find help for Exs. 21 and 22 at **www.mcdougallittell.com.** The information can be printed out for students who don't have access to the Internet.

1a. 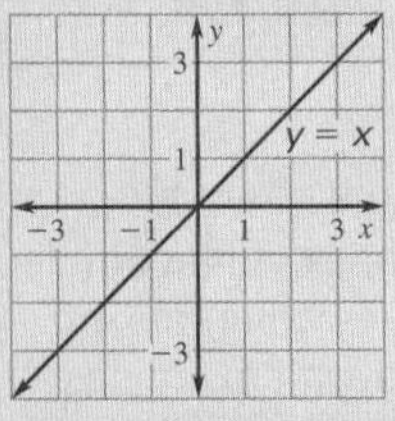

linear

b. 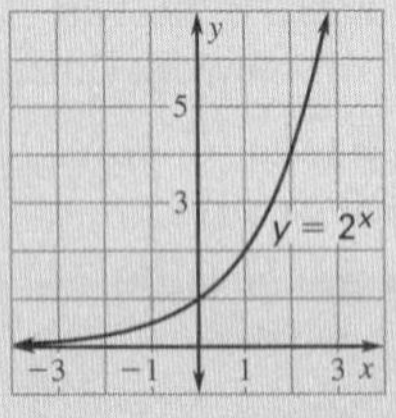

exponential

c. 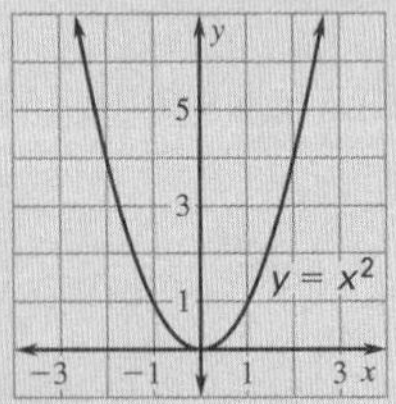

quadratic

21. See Additional Answers beginning on page AA1.

18. GEOMETRY CONNECTION The surface area S of a sphere with radius r is shown in the table. Write a model that best fits the data. **$S = 4\pi(r^2)$**

Radius, r	1	2	3	4	5
Surface area, S	4π	16π	36π	64π	100π

19. BODY MASS INDEX Body mass index is a measure of weight in relation to height. The table shows the body mass index B of a person who is 152 centimeters tall and weighs w kilograms. Which type of model best fits the data? Write a model. **linear; $B = 0.43w$**

Weight, w	45.40	49.94	54.48	59.02	63.56	68.10	72.64
Body mass index, B	19.55	21.50	23.46	25.41	27.37	29.32	31.28

▶ Source: Reuters Health Information

20. SCIENCE CONNECTION Different masses M (in kilograms) are hung from a spring. The distances d (in centimeters) that the spring stretches are shown in the table. Test different models to see which type of model best fits the data. Write a model that accurately represents the data. **linear; $d = 2.6M$**

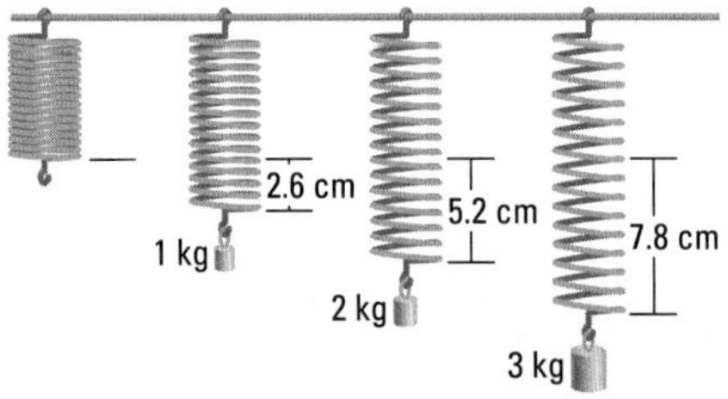

Mass, M	1	2	3	4	5	6	7
Distance, d	2.6	5.2	7.8	10.4	13	15.6	18.2

STUDENT HELP

HOMEWORK HELP
Visit our Web site www.mcdougallittell.com for help with Exs. 21 and 22.

NHL ATTENDANCE **In Exercises 21 and 22, use the following information.** The regular season attendance A (in thousands) of National Hockey League games during a recent five-year period can be modeled by

$$A = 341.93t^2 - 583.61t + 12{,}580.66$$

where t is the number of years since the beginning of the five-year period.

DATA UPDATE of National Hockey League data at www.mcdougallittell.com

21. Copy and complete the table. **See margin.**

Year t	Attendance A	Change in A from previous year
0	12,580,660	———
1	?	?
2	?	?
3	?	?
4	?	?

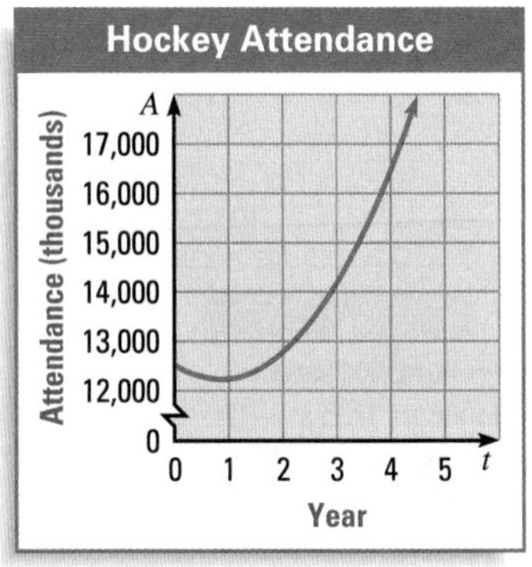

22. Describe the pattern in the change of attendance.
Sample answer: **The change, C, fits a linear model, $C = 683{,}860t - 925{,}540$.**

EXTENSION: ABSOLUTE-VALUE EQUATIONS

In Exercises 23–26, use the V-shaped graph. It is the graph of an absolute-value equation of the form $y = a|x - b| + c$.

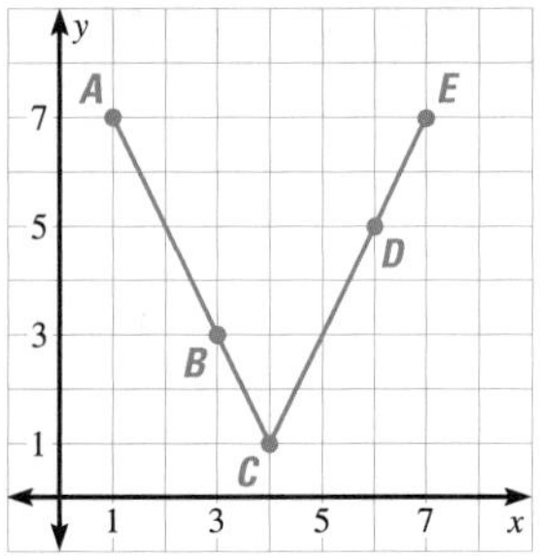

23. Write an equation of the line passing through the points A and B. What is its slope? $y = -2x + 9; -2$

24. Write an equation of the line passing through the points D and E. What is its slope? $y = 2x - 7; 2$

25. The intersection of the two lines in Exercises 23 and 24 is the vertex of the graph. Find the intersection of the two lines. **(4, 1)**

26. Find values of a, b, and c such that the equation of the graph is of the form $y = a|x - b| + c$, where a is the slope of the line to the right of the vertex and (b, c) is the vertex. $a = 2, b = 4, c = 1$

Test Preparation

27. MULTIPLE CHOICE The graph shows the number of NCAA men's college basketball teams during a recent ten-year period, where t is the number of years since the beginning of the period. Which type of model best fits the data? **A**

▶ Source: National College Athletic Association

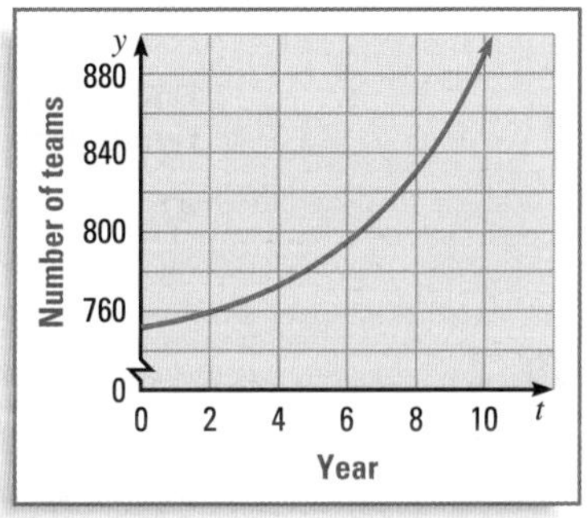

Ⓐ Exponential Ⓑ Quadratic

Ⓒ Linear Ⓓ None

28. MULTIPLE CHOICE In 1970 the population A of Arizona was 1,775,000. Since then, the average annual percent of increase has been about 3.41%. Which model best fits the situation where t is the number of years since 1970?

▶ Source: U.S. Bureau of the Census **D**

Ⓐ $A = 3.41t + 1{,}775{,}000$ Ⓑ $A = 60{,}527.5t + 1{,}775{,}000$

Ⓒ $A = 1{,}835{,}527.5t^2$ Ⓓ $A = 1{,}775{,}000(1.0341)^t$

29. MULTIPLE CHOICE The area of an ellipse whose semi-minor and semi-major axes are r and $r + 1$ respectively is given below. Which model best fits the data? **B**

Radius, r	1	2	3	4	5	6
Area A	2π	6π	12π	20π	30π	42π

Ⓐ $A = \pi r + 2$ Ⓑ $A = \pi r(r + 1)$ Ⓒ $A = \pi(1 + r)^t$ Ⓓ $A = r^2 + \pi r + 2$

★ Challenge

CHOOSING A MODEL Use a graphing calculator to graph the points. Which type of model best fits the data?

30. $(-3, 4), \left(-2, \frac{7}{2}\right), (-1, 3), \left(0, \frac{5}{2}\right), (1, 2), \left(2, \frac{3}{2}\right), (3, 1)$ **linear**

31. $(-3, 4), (-2, 2), (-1, 1), \left(0, \frac{1}{2}\right), \left(1, \frac{1}{4}\right), \left(2, \frac{1}{8}\right), \left(3, \frac{1}{16}\right)$ **exponential**

32. $(-3, 4), \left(-2, \frac{7}{3}\right), \left(-1, \frac{4}{3}\right), (0, 1), \left(1, \frac{4}{3}\right), \left(2, \frac{7}{3}\right), (3, 4)$ **quadratic**

EXTRA CHALLENGE

www.mcdougallittell.com

! COMMON ERROR

EXERCISE 28 Students may spend unnecessary time evaluating the choices. When they see that they are looking for a constant percent of increase, they should immediately consider only an exponential model.

ADDITIONAL PRACTICE AND RETEACHING

For Lesson 9.8:

- Practice Levels A, B, and C (*Chapter 9 Resource Book,* p. 116)
- Reteaching with Practice (*Chapter 9 Resource Book,* p. 119)
- See Lesson 9.8 of the *Personal Student Tutor*

For more Mixed Review:

- Search the *Test and Practice Generator* for key words or specific lessons.

4 ASSESS

DAILY HOMEWORK QUIZ

Transparency Available

Make a scatter plot of the data. Name the type of model that best fits the data.

1. $\left(1, -\frac{15}{4}\right)$, (−2, −2), $\left(2, -\frac{31}{8}\right)$, (−1, −3), (−3, 0), (−4, 4)

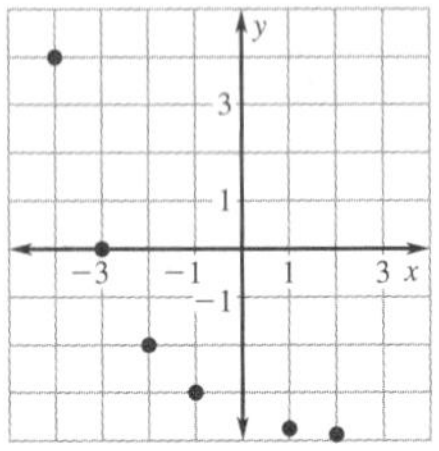

exponential

2. (0, −1.5), (2, 2.5), (−3, 0), (1, 0), (−1, −2), (−2, −1.5)

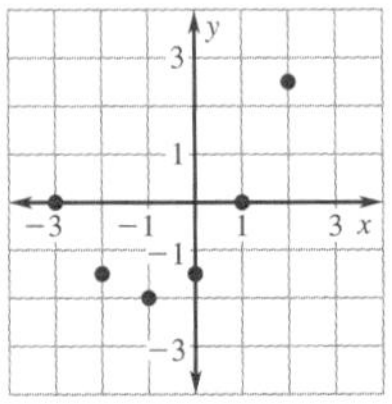

quadratic

3. The ordered pairs give the year and the population in thousands of a town. Write a model that accurately reflects the data. (1995, 24.5), (1996, 27.0), (1997, 29.6), (1998, 32.6), (1999, 35.9)
$P = 24.5 \cdot 1.1^t$, where t is the number of years after 1995

EXTRA CHALLENGE NOTE

Challenge problems for Lesson 9.8 are available in **blackline** format in the *Chapter 9 Resource Book,* p. 124 and at **www.mcdougallittell.com.**

ADDITIONAL TEST PREPARATION

1. WRITING You have decided that an exponential model best fits a data set. Explain how to write a model. **See sample answer at right.**

MIXED REVIEW

FINDING TERMS List the terms of the expression. (Review 2.3)

33. $-3 + x$ **−3, x**
34. $a - 5$ **a, −5**
35. $-5 - 8x$ **−5, −8x**
36. $-4r + s - 1$ **−4r, s, −1**

SIMPLIFYING EXPRESSIONS Simplify the variable expression. (Review 2.5 for 10.1)

37. $-(-5)(y)(-y)$ **$-5y^2$**
38. $(-3)^3(x)(x)$ **$-27x^2$**
39. $(-2)(-r)(-r)(-r)$ **$2r^3$**
40. $(-n)(-n)(-4)$ **$-4n^2$**
41. $-(y)^3$ **$-y^3$**
42. $(-1)(-x)^4$ **$-x^4$**

SOLVING EQUATIONS Solve the equation. Round the result to two decimal places. (Review 3.6)

43. $15.67x + 23.61 = 1.56 + 45.8x$ **0.73**
44. $17.87 - 2.87x = 1.87 - 4.92x$ **−7.80**
45. $6.35x - 9.94 = 3.88 + 40.34x$ **−0.41**
46. $5.6(1.2 + 1.9x) = 20.4x + 6.8$ **−0.01**

WRITING EQUATIONS Write an equation of the line that passes through the two points. (Review 5.3)

47. (3, −2), (5, 4) **$y = 3x - 11$**
48. (−3, −9), (5, 7) **$y = 2x - 3$**
49. (2, 3), (−4, 6) **$y = -\frac{1}{2}x + 4$**

QUIZ 3

Self-Test for Lessons 9.7 and 9.8

Decide whether the ordered pair is a solution of the inequality. (Lesson 9.7)

1. $y \geq 2x^2 - x + 9$, (2, 10) **not a solution**
2. $y > 4x^2 - 64x + 115$, (12, −60) **solution**
3. $y < x^2 + 6x + 12$, (−1, 6) **solution**
4. $y \leq x^2 - 7x + 9$, (−1, 16) **solution**

Match the inequality with its graph. (Lesson 9.7)

A.

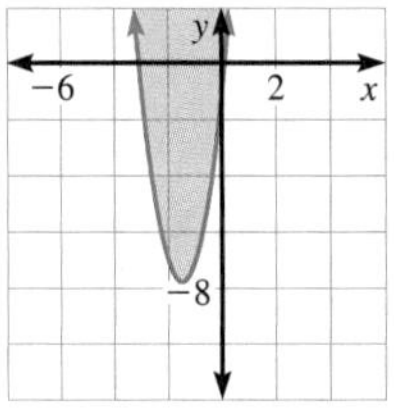

B.

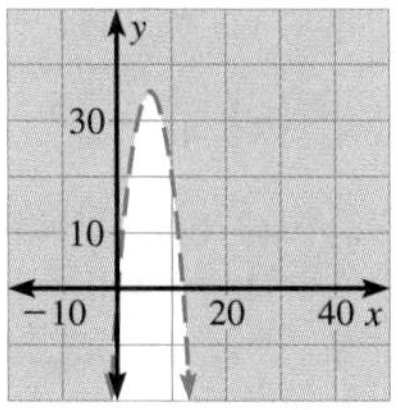

C.

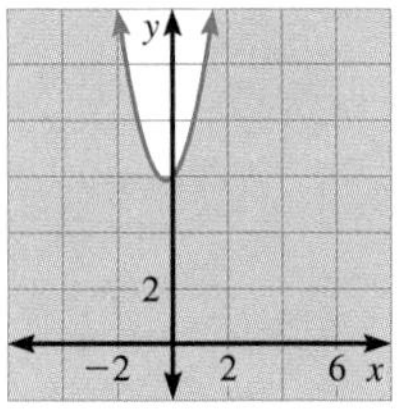

5. $y > -x^2 + 12x - 1$ **B**
6. $y \leq 2x^2 + x + 6$ **C**
7. $y \geq 3x^2 + 9x - 1$ **A**

Graph the equation. Name the model it represents. (Lesson 9.8)
8–10. See margin for graphs.

8. $y = 3x - 6$ **linear**
9. $y = 0.5(1.4)^x$ **exponential**
10. $y = 2x^2 + 8x - 4$ **quadratic**

Make a scatter plot of the data. Then name the type of model that best fits the data and write a model. (Lesson 9.8) **11–13. See margin for plots.**

11. (−8.5, 20), (−8, 19), (−2, 7), (0, 3), (1, 1), (6, −9), (7.5, −12), (10, −17) **linear; $y = -2x + 3$**

12. (−3, 1.5), (−1.5, 2.7), (−0.5, 4), (0, 5), (1, 7.5), (3, 17) **exponential; $y = 5(1.5)^x$**

13. (3.3, 15.0), (2.1, 11.7), (1.0, 6.8), (0, 2.9), (−1.1, −0.9) **linear; $y = 3.72x + 3.16$**

Additional Test Preparation *Sample answer:*
Find the constant ratio of consecutive function values. This gives you $1 \pm r$ in $y = C(1 \pm r)^t$. To find C, substitute data values for y and t and solve for C. Then check other values of y and t to make sure they fit the model.

8–13. See Additional Answers beginning on page AA1.

CHAPTER 9

Chapter Summary

WHAT did you learn?	*Why did you learn it?*
Evaluate and approximate square roots. (9.1)	Find solutions using a quadratic model for mineral hardness. (p. 509)
Solve a quadratic equation.	
• by finding square roots (9.1)	Estimate the time for an object to fall. (p. 506)
• by sketching its graph (9.4)	Compare shot-put throws. (p. 528)
• by using the quadratic formula. (9.5)	Model vertical motion problems. (p. 535)
Simplify radicals. (9.2)	Compare the speeds of two sailboats. (p. 513)
Sketch the graph of a quadratic function. (9.3)	Estimate the maximum height of a water spray. (p. 522)
Find the number of solutions of a quadratic equation by using the discriminant. (9.6)	Decide whether an outcome is possible in a camping situation. (p. 543)
Sketch the graph of a quadratic inequality. (9.7)	Sketch the region between the towers and under the main cable of a suspension bridge. (p. 550)
Choose an algebraic model that best fits a collection of data. (9.8)	Recognize the characteristics of different models in real-life settings. (pp. 555 and 556)

How does Chapter 9 fit into the BIGGER PICTURE of algebra?

A quadratic equation contains the square of a variable. Try to remember the "algebra-geometry" connection between equations in one and two variables. For instance, the equation $x^2 - 9 = 0$ has two solutions, or roots, which correspond to the two x-intercepts of the graph of the function $y = x^2 - 9$. A quadratic function has a U-shaped graph called a *parabola*.

This chapter focuses on quadratic models, which have many applications, such as vertical motion and parabolic path problems. The last lesson helps you decide whether to apply a linear, an exponential, or a quadratic model to fit a collection of real-life data.

STUDY STRATEGY

How did explaining ideas help you to understand a topic?

Talking with a classmate about Lesson 9.4 may have led you to think about ideas that you would understand fully in Lesson 9.6.

Talking about x-intercepts following Lesson 9.4

"Let's see what the graph of $y = x^2$ looks like. It has only one x-intercept. When I solve $x^2 = 0$, I get $x = \sqrt{0} = 0$. It has only one solution, too!"

"If a quadratic function has a graph with no x-intercepts, then I don't think the equation will have any solutions either."

561

ADDITIONAL RESOURCES

The following resources are available to help review the material in this chapter.

- Chapter Review Games and Activities (*Chapter 9 Resource Book*, p. 125)
- *Instant Replay: Video Review Games*
- *Personal Student Tutor*
- Cumulative Review, Chs. 1–9 (*Chapter 9 Resource Book*, p. 137)

CHAPTER 9

Chapter Review

VOCABULARY

- square root, p. 503
- positive square root, p. 503
- negative square root, p. 503
- radicand, p. 503
- perfect square, p. 504
- irrational number, p. 504
- radical expression, p. 504
- quadratic equation in standard form, p. 505
- leading coefficient, p. 505
- simplest form of a radical expression, p. 512
- quadratic function in standard form, p. 518
- parabola, p. 518
- vertex, p. 518
- axis of symmetry, p. 518
- roots, p. 526
- quadratic formula, p. 533
- discriminant, p. 541
- quadratic inequalities, p. 548
- graph of a quadratic inequality, p. 548

9.1 SOLVING QUADRATIC EQUATIONS BY FINDING SQUARE ROOTS

Examples on pp. 503–506

EXAMPLE To find the real solutions of a quadratic equation of the form $ax^2 + c = 0$, isolate x^2 and then find the square root of each side.

$2x^2 - 98 = 0$	Write original equation.
$2x^2 = 98$	Add 98 to each side.
$x^2 = 49$	Divide each side by 2.
$x = \pm\sqrt{49}$	Find square roots.
$x = \pm 7$	49 is a perfect square.

Solve the equation.

1. $x^2 - 144 = 0$ **−12, 12**
2. $8y^2 = 968$ **−11, 11**
3. $4t^2 + 19 = 19$ **0**
4. $16y^2 - 80 = 0$ **$-\sqrt{5}, \sqrt{5}$**
5. $\frac{1}{5}a^2 = 5$ **−5, 5**
6. $\frac{1}{3}x^2 - 7 = -4$ **−3, 3**

9.2 SIMPLIFYING RADICALS

Examples on pp. 511–513

EXAMPLES Use the product property and the quotient property.

a. $\sqrt{28} = \sqrt{4 \cdot 7}$	Factor using perfect square factor.
$= \sqrt{4} \cdot \sqrt{7} = 2\sqrt{7}$	Use product property and simplify.
b. $\sqrt{\frac{169}{625}} = \frac{\sqrt{169}}{\sqrt{625}} = \frac{13}{25}$	Use quotient property and simplify.

Simplify the expression.

7. $\sqrt{45}$ **$3\sqrt{5}$**
8. $\sqrt{441}$ **21**
9. $\sqrt{\frac{36}{64}}$ **$\frac{3}{4}$**
10. $\sqrt{\frac{99}{25}}$ **$\frac{3\sqrt{11}}{5}$**

9.3 GRAPHING QUADRATIC FUNCTIONS

Examples on pp. 518–520

EXAMPLE To sketch the graph of $y = x^2 - 4x - 3$, first find the x-coordinate of the vertex: $-\frac{b}{2a} = -\frac{-4}{2(1)} = 2$. Make a table of values using x-values to the left and right of $x = 2$. Plot the points and connect them to form a parabola.

x	−1	0	1	2	3	4	5
y	2	−3	−6	−7	−6	−3	2

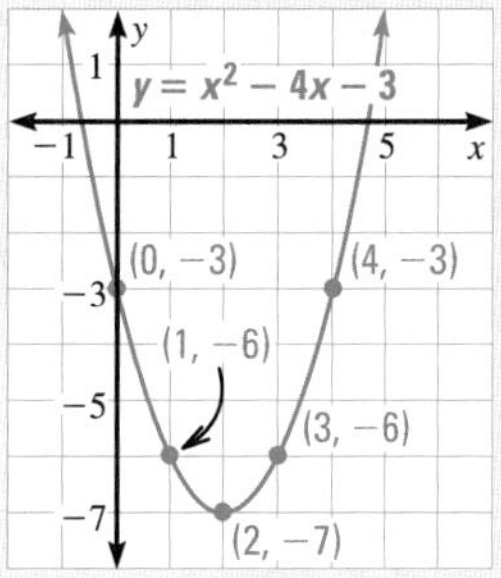

Sketch the graph of the function. Label the vertex. 11–13. See margin.

11. $y = x^2 - x - 5$ **12.** $y = -x^2 - 3x + 2$ **13.** $y = -4x^2 + 6x + 3$

9.4 SOLVING QUADRATIC EQUATIONS BY GRAPHING

Examples on pp. 526–528

EXAMPLE To solve the quadratic equation $-x^2 + 3x = 2$ by graphing, first rewrite the equation in the form $ax^2 + bx + c = 0$.

$-x^2 + 3x - 2 = 0.$

Sketch the graph of the related function $y = -x^2 + 3x - 2$.

▶ From the graph, the x-intercepts appear to be $x = 1$ and $x = 2$. Check these in the original equation.

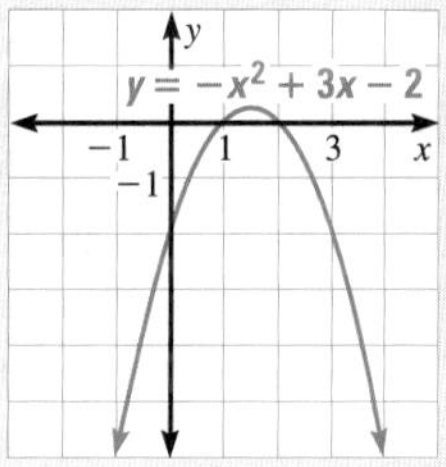

Solve the equation by graphing. Check the solution algebraically.

14. $x^2 + 2 = 3x$ **1, 2** **15.** $x^2 - 2x = 15$ **−3, 5** **16.** $\frac{1}{2}x^2 + 5x = -8$ **−8, −2** **17.** $-x^2 - 2x = -24$ **−6, 4**

9.5 SOLVING QUADRATIC EQUATIONS BY THE QUADRATIC FORMULA

Examples on pp. 533–535

EXAMPLE Solve equations of the form $ax^2 + bx + c = 0$ by substituting the values of a, b, and c into the quadratic formula. Solve $x^2 + 6x - 16 = 0$.

$$x = \frac{-6 \pm \sqrt{6^2 - 4(1)(-16)}}{2(1)} = \frac{-6 \pm \sqrt{36 + 64}}{2} = \frac{-6 \pm \sqrt{100}}{2} = \frac{-6 \pm 10}{2}$$

▶ The equation has two solutions: $x = \frac{-6 + 10}{2} = 2$ and $x = \frac{-6 - 10}{2} = -8$.

Use the quadratic formula to solve the equation.

18. $3x^2 - 4x + 1 = 0$ $\frac{1}{3}, 1$ **19.** $-2x^2 + x + 6 = 0$ $-\frac{3}{2}, 2$ **20.** $10x^2 - 11x + 3 = 0$ $\frac{1}{2}, \frac{3}{5}$

11.

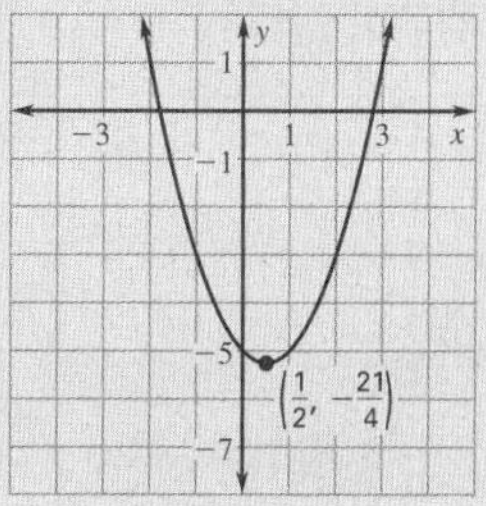

12.

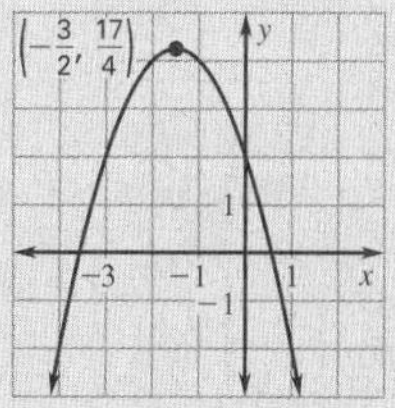

13.

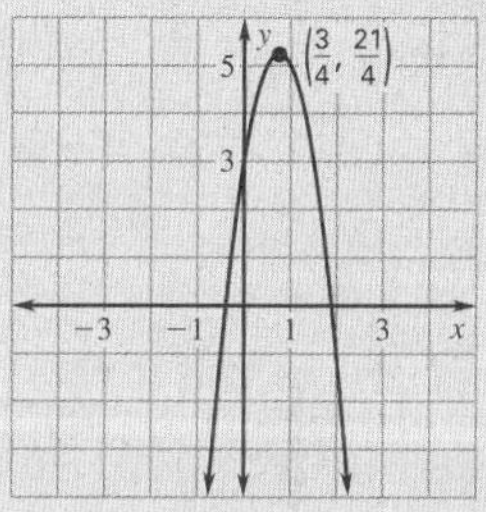

24.
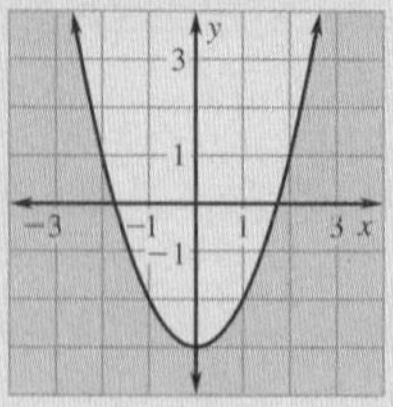

25.
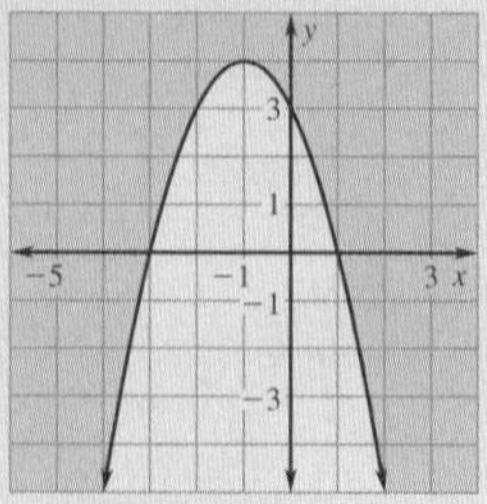

26.
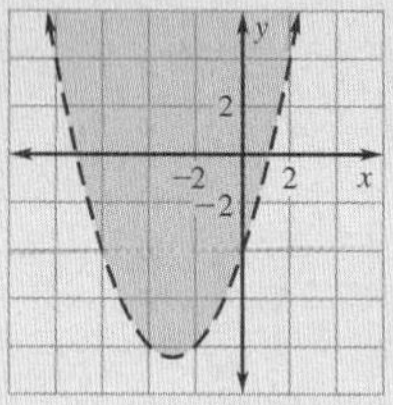

27.
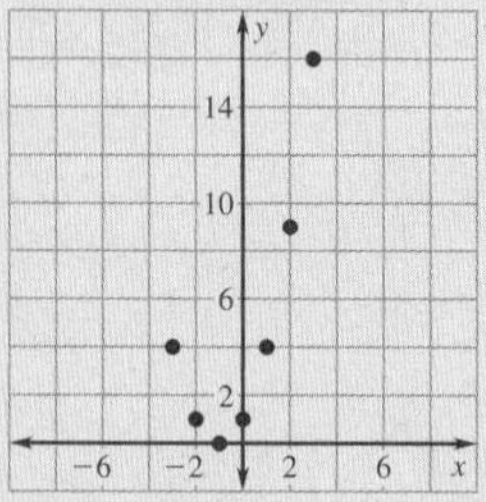

28.
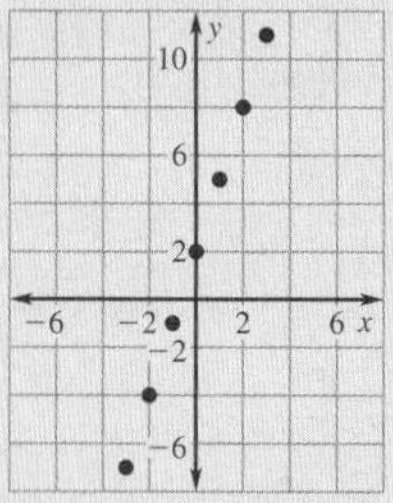

29.
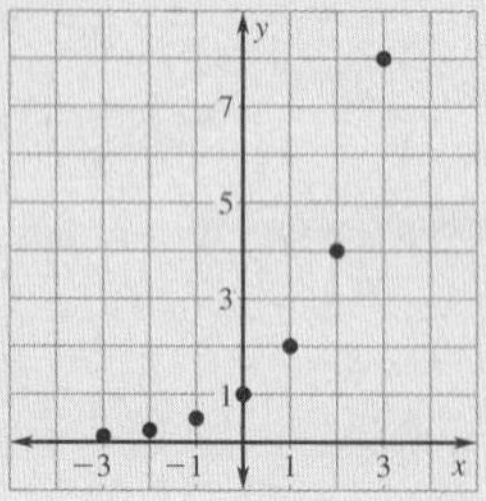

9.6 APPLICATIONS OF THE DISCRIMINANT

Examples on pp. 541–543

EXAMPLES Use the discriminant, $b^2 - 4ac$, to find the number of solutions of a quadratic equation $ax^2 + bx + c = 0$.

EQUATION	DISCRIMINANT	NUMBER OF SOLUTIONS
$3x^2 + 6x + 2 = 0$	$6^2 - 4(3)(2) = 12$	2
$2x^2 + 8x + 8 = 0$	$8^2 - 4(2)(8) = 0$	1
$x^2 + 7x + 15 = 0$	$7^2 - 4(1)(15) = -11$	0

Tell if the equation has *two solutions, one solution,* or *no real solution.*

21. $3x^2 - 12x + 12 = 0$ **one solution**

22. $2x^2 + 10x + 6 = 0$ **two solutions**

23. $-x^2 + 3x - 5 = 0$ **no real solution**

9.7 GRAPHING QUADRATIC INEQUALITIES

Examples on pp. 548–550

EXAMPLE To sketch the graph of the quadratic inequality $y < x^2 - 9$, first sketch the graph of the parabola $y = x^2 - 9$. Use a dashed parabola for inequalities with $>$ or $<$, and a solid parabola for inequalities with $\geq$ or $\leq$.

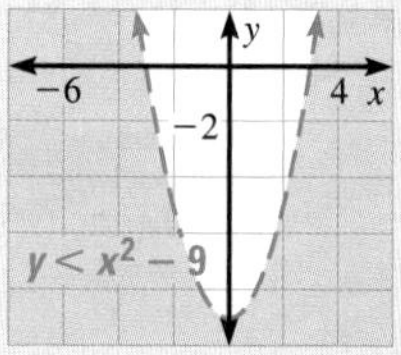

Then test a point that is not on the parabola, say (0, 0). If the test point is a solution, shade its region. If not, shade the other region.

Sketch the graph of the inequality. **24–26. See margin.**

24. $x^2 - 3 \geq y$

25. $-x^2 - 2x + 3 \leq y$

26. $\frac{1}{2}x^2 + 3x - 4 < y$

9.8 COMPARING LINEAR, EXPONENTIAL, AND QUADRATIC MODELS

Examples on pp. 554–556

EXAMPLE To choose the type of model that best fits a set of data, first make a scatter plot of the data. Then decide whether the points lie on a line, an exponential curve, or a parabola.

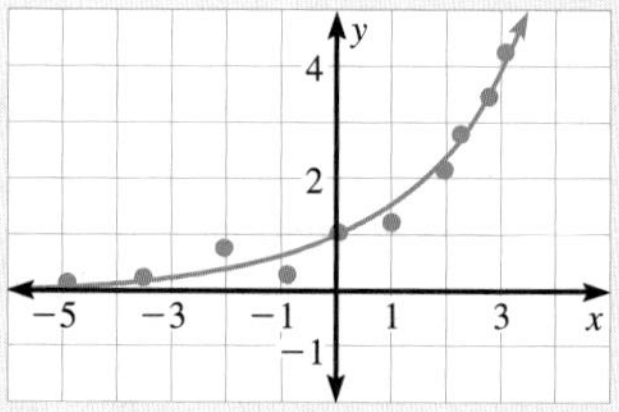

The points in the graph appear to lie on an exponential curve. An exponential model could be used to model the set of data.

Make a scatter plot and name the type of model that best fits the data. **27–29. See margin for plots.**

27. $(-3, 4), (-2, 1), (-1, 0), (0, 1), (1, 4), (2, 9), (3, 16)$ **quadratic**

28. $(-3, -7), (-2, -4), (-1, -1), (0, 2), (1, 5), (2, 8), (3, 11)$ **linear**

29. $\left(-3, \frac{1}{8}\right), \left(-2, \frac{1}{4}\right), \left(-1, \frac{1}{2}\right), (0, 1), (1, 2), (2, 4), (3, 8)$ **exponential**

CHAPTER 9 Chapter Test

Simplify the expression.

1. $-\sqrt{9}$ **−3** **2.** $\sqrt{0.0064}$ **0.08** **3.** $\pm\sqrt{121}$ **±11** **4.** $\sqrt{8^2 - 4(2)(8)}$ **0**

5. $\sqrt{192}$ **$8\sqrt{3}$** **6.** $\sqrt{5} \cdot \sqrt{30}$ **$5\sqrt{6}$** **7.** $\sqrt{\frac{27}{147}}$ **$\frac{3}{7}$** **8.** $\frac{10\sqrt{8}}{\sqrt{16}}$ **$5\sqrt{2}$**

Solve the equation by finding square roots or using the quadratic formula.

9. $8x^2 = 800$ **−10, 10** **10.** $4x^2 + 11 = 12$ **$-\frac{1}{2}, \frac{1}{2}$** **11.** $2x^2 + 5x - 7 = 0$ **$-\frac{7}{2}, 1$**

12. $-2x^2 + 4x + 6 = 0$ **−1, 3** **13.** $64x^2 - 5 = 11$ **$-\frac{1}{2}, \frac{1}{2}$** **14.** $10x^2 + 17x - 11 = 0$ **$-\frac{11}{5}, \frac{1}{2}$**

Match the equation or inequality with its graph.

A.
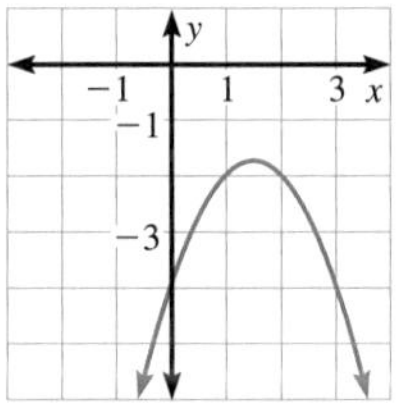

B.
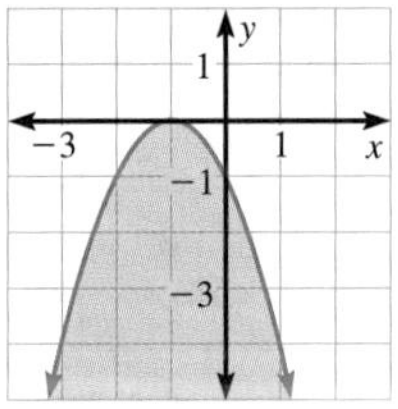

C.
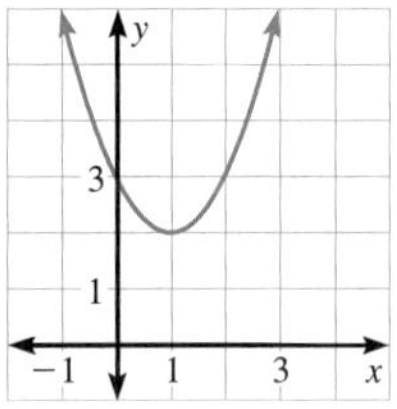

D.
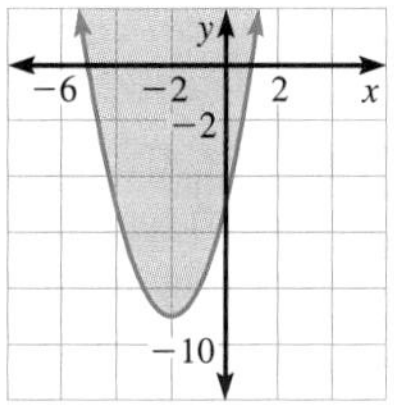

15. $y = x^2 - 2x + 3$ **C** **16.** $y = -x^2 + 3x - 4$ **A** **17.** $y \geq x^2 + 4x - 5$ **D** **18.** $y \leq -x^2 - 2x - 1$ **B**

Decide how many solutions the equation has. Check the results by graphing.

19. $5x^2 - 20x - 60 = 0$ **two solutions** **20.** $-3x^2 + 4x - 2 = 0$ **no real solution** **21.** $x^2 - 4x + 4 = 0$ **one solution**

Sketch the graph of the equation or the inequality. **22–25. See margin.**

22. $y = \frac{1}{3}x^2 + 2x - 3$ **23.** $y = -x^2 + 5x - 6$ **24.** $y < x^2 + 7x + 6$ **25.** $y \leq -\frac{1}{2}x^2 + 2x + \frac{5}{2}$

Name the type of model suggested by the graph.

26.
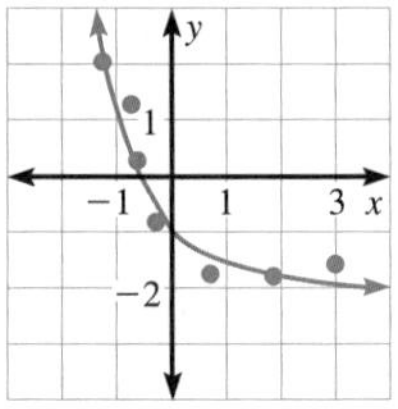

exponential

27.
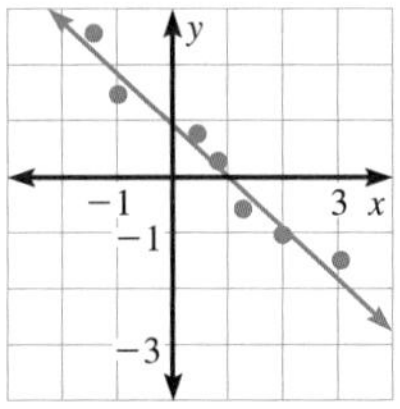

linear

28.
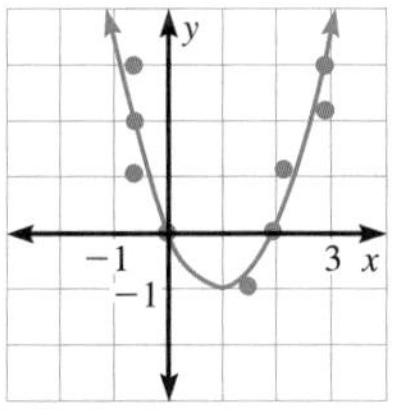

quadratic

VERTICAL MOTION **In Exercises 29–31, you are standing on a bridge over a creek, holding a stone 20 feet above the water.**

29. You release the stone. How long will it take the stone to hit the water? **about 1.12 sec**

30. You take another stone and toss it straight up with an initial velocity of 30 feet per second. How long will it take the stone to hit the water? **about 2.40 sec**

31. If you throw a stone straight up into the air with an initial velocity of 50 feet per second, could the stone reach a height of 60 feet above the water? **no**

ADDITIONAL RESOURCES

- ***Chapter 9 Resource Book***
  Chapter Test (3 levels) (p. 126)
  SAT/ACT Chapter Test (p. 132)
  Alternative Assessment (p. 133)
- ***Test and Practice Generator***

22.
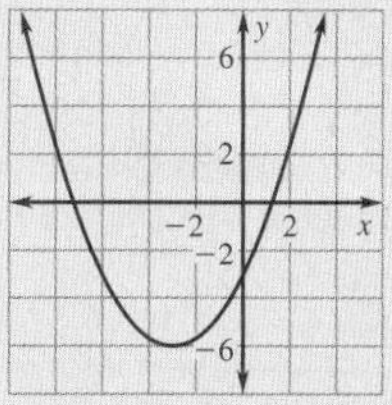

23.
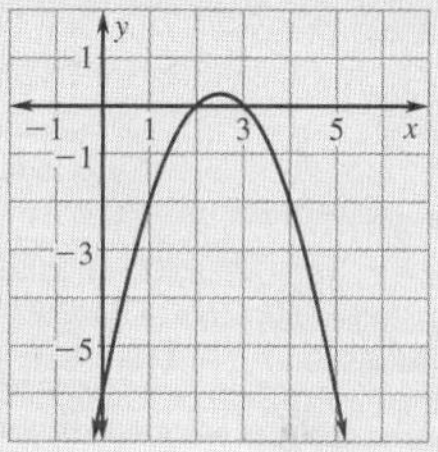

24.
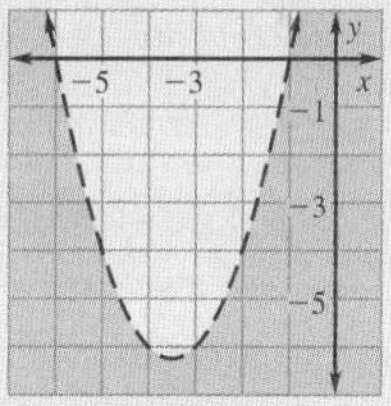

25.
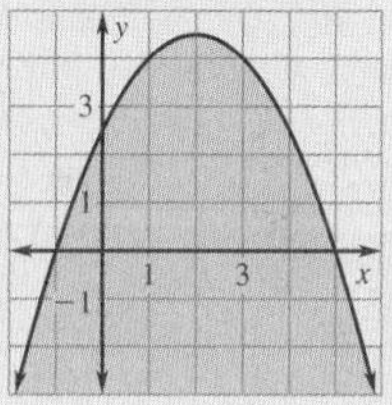

ADDITIONAL RESOURCES

- ***Chapter 9 Resource Book***
  Chapter Test (3 levels) (p. 126)
  SAT/ACT Chapter Test (p. 132)
  Alternative Assessment (p. 133)
- ***Test and Practice Generator***

CHAPTER 9

Chapter Standardized Test

TEST-TAKING STRATEGY Before you give up on a question, try to eliminate some of your choices so you can make an educated guess.

1. **MULTIPLE CHOICE** Which one of the following is *not* a quadratic equation? **D**

   Ⓐ $x^2 - 4 = 0$ Ⓑ $x^2 + 10x + 21 = 0$
   Ⓒ $-9 + x^2 = 0$ Ⓓ $-7x + 12 = 0$
   Ⓔ $-2 + 9x + x^2 = 0$

2. **MULTIPLE CHOICE** Which one of the following is a solution of the equation $\frac{2}{3}t^2 - 7 = 17$? **A**

   Ⓐ -6 Ⓑ -4 Ⓒ 4
   Ⓓ $\sqrt{15}$ Ⓔ $\sqrt{24}$

3. **MULTIPLE CHOICE** You drop a rock from a bridge 320 feet above a river. How long will it take the rock to hit the river? **D**

   Ⓐ 2.5 sec Ⓑ 3.5 sec Ⓒ 3.8 sec
   Ⓓ 4.5 sec Ⓔ 5.5 sec

4. **MULTIPLE CHOICE** The surface area S of a cube is 150 square feet. What is the length (in feet) of each edge of the cube? $(S = 6s^2)$ **C**

   Ⓐ ± 5
   Ⓑ $5\sqrt{6}$
   Ⓒ 5
   Ⓓ 25
   Ⓔ None of these

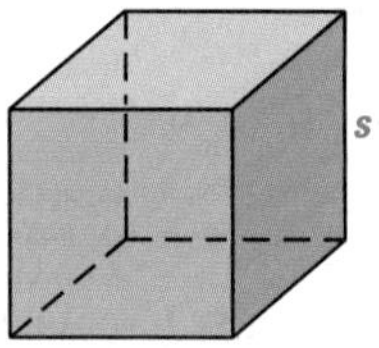

5. **MULTIPLE CHOICE** Find the area of the rectangle. **E**

   Ⓐ $20\sqrt{3}$
   Ⓑ 240
   Ⓒ 60
   Ⓓ $12\sqrt{5}$
   Ⓔ $4\sqrt{15}$

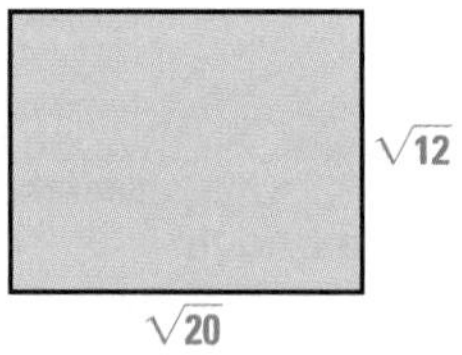

6. **MULTIPLE CHOICE** What is the value of x when $3x^2 - 78 = 114$? **D**

   Ⓐ $\pm 2\sqrt{3}$ Ⓑ $\pm 4\sqrt{3}$ Ⓒ ± 6
   Ⓓ ± 8 Ⓔ $\pm 2\sqrt{29}$

QUANTITATIVE COMPARISON **In Questions 7 and 8, perform the indicated operation and simplify the result. Then choose the statement below that is true about the given numbers.**

Ⓐ The number in column A is greater.
Ⓑ The number in column B is greater.
Ⓒ The two numbers are equal.
Ⓓ The relationship cannot be determined from the given information.

	Column A	Column B	
7.	$\sqrt{24} \cdot \sqrt{6}$	$\sqrt{25} \cdot \sqrt{5}$	**A**
8.	$\frac{\sqrt{27}\sqrt{3}}{\sqrt{9}}$	$\frac{\sqrt{30}\sqrt{6}}{\sqrt{12}}$	**B**

9. **MULTIPLE CHOICE** What is the x-coordinate of the vertex for the graph of the equation $y = -\frac{1}{2}x^2 - x + 8$? **B**

   Ⓐ -2 Ⓑ -1 Ⓒ $-\frac{1}{2}$
   Ⓓ $\frac{1}{2}$ Ⓔ 1

10. **MULTIPLE CHOICE** What are the x-intercepts of the graph of $y = -x^2 - 6x + 40$? **C**

    Ⓐ 4 and 10 Ⓑ −7 and 1
    Ⓒ −10 and 4 Ⓓ −8 and 2
    Ⓔ −11 and 5

11. **MULTIPLE CHOICE** Which one of the following is a solution of $-3x^2 + 22x + 93 = 0$? **D**

    Ⓐ -7 Ⓑ 3 Ⓒ $-\frac{6}{31}$
    Ⓓ -3 Ⓔ $\frac{31}{6}$

12. **MULTIPLE CHOICE** Which one of the following is a solution of $4x^2 - 17x + 13 = 0$? **E**

    Ⓐ $-\frac{17}{13}$ Ⓑ -1 Ⓒ $\frac{4}{13}$
    Ⓓ 13 Ⓔ 1

13. **MULTIPLE CHOICE** What is the discriminant of the equation $-7x^2 - 2x + 5 = 0$? **B**

Ⓐ 12 Ⓑ 144 Ⓒ -12 Ⓓ 136 Ⓔ 1

14. **MULTIPLE CHOICE** Which of the following inequalities is represented by the graph? **E**

Ⓐ $y > -4x^2 + 8x - 5$

Ⓑ $y < 4x^2 + 8x - 5$

Ⓒ $y > 4x^2 + 8x - 5$

Ⓓ $y \leq -4x^2 + 8x - 5$

Ⓔ $y < -4x^2 + 8x - 5$

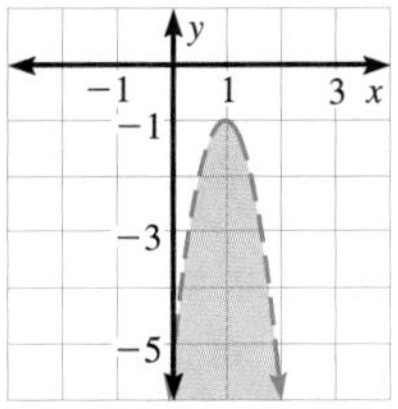

15. **MULTIPLE CHOICE** Name the type of model suggested by the graph. **B**

Ⓐ Quadratic

Ⓑ Exponential decay

Ⓒ Exponential growth

Ⓓ Linear

Ⓔ None of these

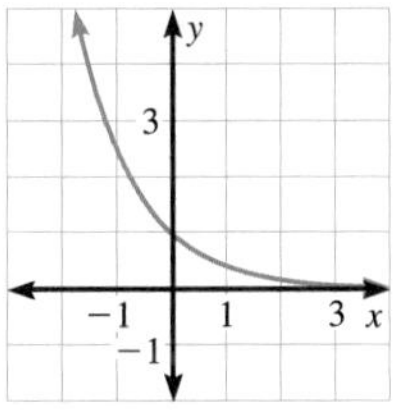

16. **MULTI-STEP PROBLEM** Use the data in the table below.

x	−5	−4	−2	0	1	2
y	0.3	0.5	1.5	5	9	16.2

a. Make a scatter plot of the data. **See margin.**

b. Name the type of model that best fits the data. **exponential**

c. Write a model that best fits the data. $y = 5(1.8)^x$

16a.

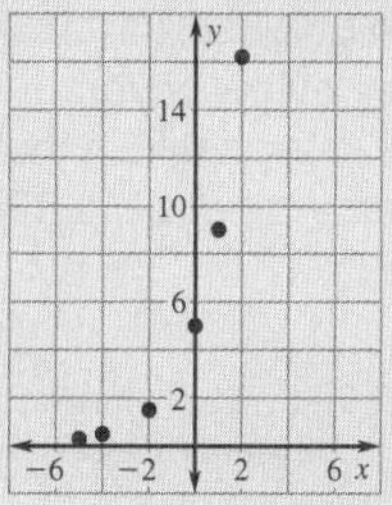

17. **MULTI-STEP PROBLEM** A bird flying above the edge of a cliff sees a fish several meters below the surface of the water in a lake bounded by the cliff. The bird makes a dive to catch the fish. The path of the dive follows a parabolic curve given by the function $y = x^2 - 12x + 32$. At the right is a graph of this function. Imagine that the x-axis runs along the surface of the lake and that the y-axis runs along the edge of the cliff.

a. How far from the side of the cliff is the fish if it is at the vertex of the curve? **6 m**

b. How far below the surface of the water does the bird have to dive to catch the fish? **4 m**

c. How far from the side of the cliff will the bird enter and exit the water (x-intercepts)? **4 m, 8 m**

d. Approximate the height of the cliff (y-intercept). **32 ft**

e. *Writing* If the bird enters the water 4 meters from the side of the cliff and exits it 8 meters from the cliff, write an inequality to represent the region in which the fish can swim safely.

The graph of $y < x^2 - 12x + 32$ represents the parabola below which the fish can swim out of reach of the bird.

Not drawn to scale

ADDITIONAL RESOURCES
A Cumulative Review covering Chapters 1–9 is available in the *Chapter 9 Resource Book,* p. 137.

26.
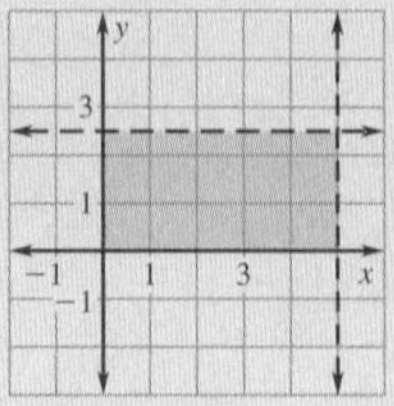

27.
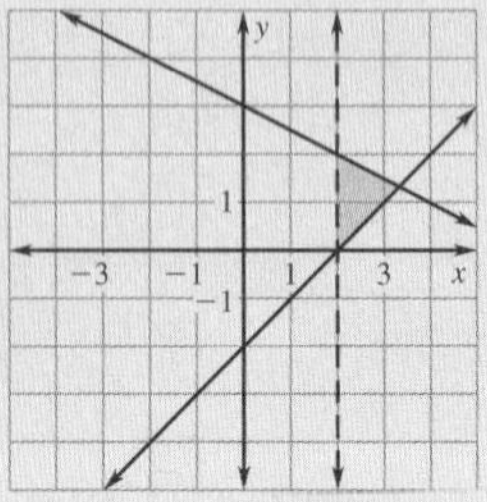

28.
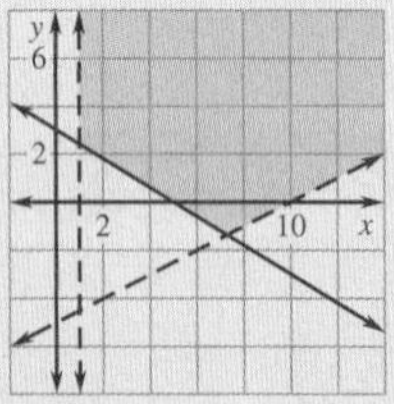

29.
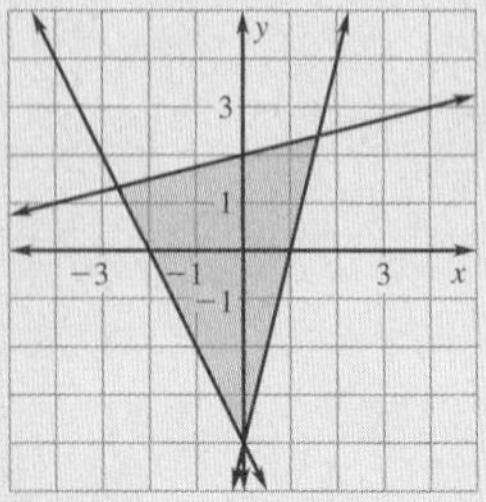

CHAPTER 9 Cumulative Practice

Does the table represent a function? Explain. (1.7)
No; The input value 5 has two different output values.

1.

Input	1	2	3	4
Output	5	8	11	14

2.

Input	5	3	5	2
Output	8	7	4	3

3.

Input	3	6	9	12
Output	5	8	5	8

1, 3. Yes; no input value has two different output values.

Find the probability of choosing a green marble from a bag of green and yellow marbles. Then find the odds of choosing a green marble. (2.8)

4. Number of green marbles: 30 **$\frac{5}{14} \approx 36\%$; 5 to 9**
Total number of marbles: 84

5. Number of green marbles: 16 **$\frac{2}{9} \approx 22\%$; 2 to 7**
Total number of marbles: 72

In Exercises 6 and 7, the two triangles are similar. Write an equation and solve it to find the length of the side marked *x*. (3.2)

6.
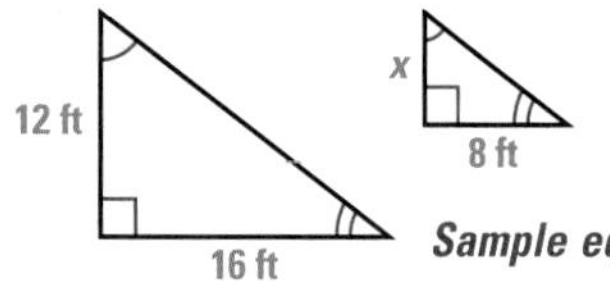

Sample equation: **$\frac{x}{12} = \frac{8}{16}$; 6 ft**

7.
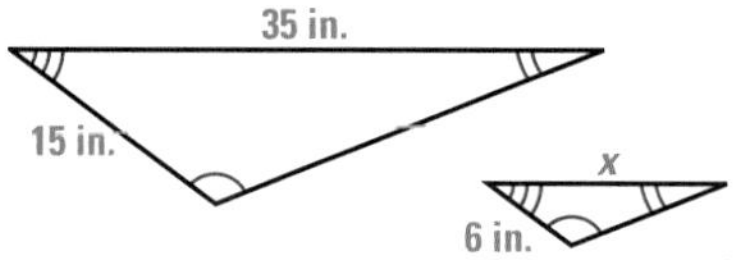

Sample equation: **$\frac{x}{35} = \frac{6}{15}$; 14 in.**

Find the value of *y* so that the line passing through the two points has the given slope. (4.4)

8. $(2, y), (11, 8), m = \frac{1}{3}$ **5**

9. $(7, 3), (6, y), m = -1$ **4**

10. $(5, -10), (8, y), m = 4$ **2**

11. $(5, y), (-5, 7), m = -\frac{1}{2}$ **2**

12. $(0, -12), (3, y), m = 5$ **3**

13. $(-4, y), (3, -8), m = -2$ **6**

Write the equation in standard form with integer coefficients. (5.6)

14. $3x - 5y + 6 = 0$ **$3x - 5y = -6$**

15. $6y = 2x + 4$ **$2x - 6y = -4$**

16. $-2x + 7y - 15 = 0$ **$-2x + 7y = 15$**

Find the first, second, and third quartiles of the data. (6.7)

17. 8, 5, 7, 5, 6, 9, 8, 5, 7 **5, 7, 8**

18. 10, 12, 7, 30, 25, 8, 6, 10, 5, 8 **7, 9, 12**

19. 4, 4, 8, 2, 10, 8, 4, 6, 2, 10, 4 **4, 4, 8**

Solve the linear system. (7.1–7.4)

20. $x + y = 8$
$2x + y = 10$ **(2, 6)**

21. $\frac{1}{4}x - y = 7$
$x + 4y = 0$ **$\left(14, -3\frac{1}{2}\right)$**

22. $-2x + 20y = 10$
$x - 5y = -5$ **(−5, 0)**

23. $3.2x + 1.1y = -19.3$
$-32x + 4y = 148$ **(−5, −3)**

24. $\frac{1}{10}x - \frac{3}{2}y = -1$
$-10x + 3y = 2$ **$\left(0, \frac{2}{3}\right)$**

25. $1.4x + 2.1y = 1.75$
$2.8x - 4.2y = 34.58$ **(6.8, −3.7)**

Sketch the graph of the system of linear inequalities. (7.6) **26–29. See margin.**

26. $x \geq 0$
$y \geq 0$
$x < 5$
$y < \frac{5}{2}$

27. $x > 2$
$x - y \leq 2$
$\frac{1}{2}x + y \leq 3$

28. $3x + 5y \geq 15$
$x - 2y < 10$
$x > 1$

29. $-\frac{1}{4}x + y \leq 2$
$-4x + y \geq -4$
$2x + y \geq -4$

Simplify the expression. (8.1–8.3)

30. $\left(\frac{1}{x^2}\right)^7$ **$\frac{1}{x^{14}}$**

31. $\frac{x^8}{x^{10}}$ **$\frac{1}{x^2}$**

32. $\frac{4}{(2x)^{-3}}$ **$32x^3$**

33. $\left(\frac{-8x^3}{4xy^5}\right)^2$ **$\frac{4x^4}{y^{10}}$**

34. $5x \cdot (x \cdot x^{-4})^2$ **$\frac{5}{x^5}$**

35. $(6a^3)^2\left(\frac{1}{2}a^3\right)^2$ **$9a^{12}$**

36. $(r^2st^5)^0(s^4t^2)^3$ **$s^{12}t^6$**

37. $\frac{6x^4y^4}{3xy} \cdot \frac{5x^2y^3}{2y^2}$ **$5x^5y^4$**

Evaluate the expression. Write the result in scientific notation and in decimal form. (8.4)

38. $(5 \times 10^{-2}) \cdot (3 \times 10^4)$ **$1.5 \times 10^3 = 1500$**

39. $(6 \times 10^{-4}) \cdot (7 \times 10^{-5})$ **$4.2 \times 10^{-8} = 0.000000042$**

40. $(20 \times 10^{-4}) \div (2.5 \times 10^{-8})$ **$8 \times 10^4 = 80{,}000$**

41. $(7 \times 10^3)^{-3}$ **$2.92 \times 10^{-12} = 0.00000000000292$**

42. $(8.8 \times 10^{-1}) \div (11 \times 10^{-1})$ **$8 \times 10^{-1} = 0.8$**

43. $(2.8 \times 10^{-2})^3$ **$2.1952 \times 10^{-5} = 0.000021952$**

Match the graph with its description. (8.5)

A. Deposit: \$300, Annual rate: 2.5%

B. Deposit: \$300, Annual rate: 4%

C. Deposit: \$250, Annual rate: 2%

D. Deposit: \$250, Annual rate: 4.5%

44.

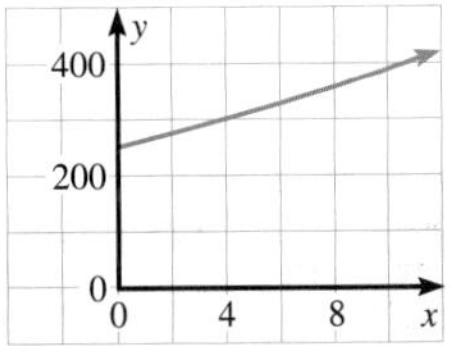

D

45.

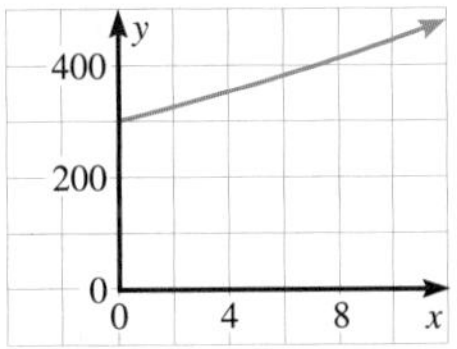

B

46.

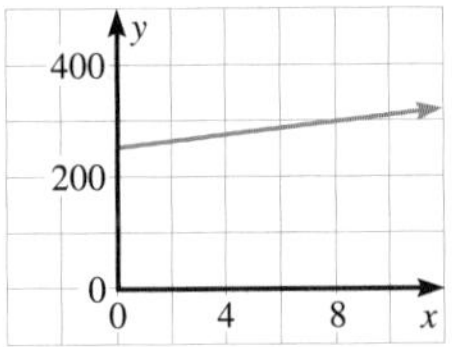

C

47.

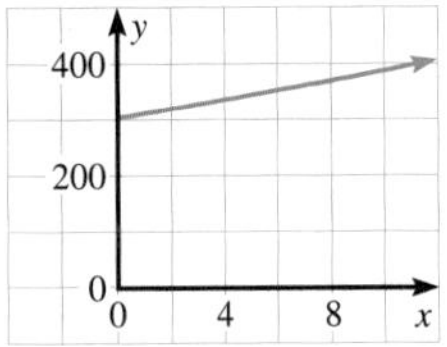

A

Simplify the expression. (9.2)

48. $\sqrt{48}$ **$4\sqrt{3}$**

49. $\sqrt{\frac{28}{36}}$ **$\frac{\sqrt{7}}{3}$**

50. $\frac{1}{4}\sqrt{84}$ **$\frac{1}{2}\sqrt{21}$**

51. $\frac{\sqrt{112}}{\sqrt{49}}$ **$\frac{4\sqrt{7}}{7}$**

52. $\sqrt{12} \cdot \sqrt{63}$ **$6\sqrt{21}$**

53. $\sqrt{9} \cdot \frac{\sqrt{18}}{\sqrt{54}}$ **$\sqrt{3}$**

54. $\frac{-2\sqrt{98}}{\sqrt{7}}$ **$-2\sqrt{14}$**

55. $\frac{\sqrt{33} \cdot \sqrt{75}}{\sqrt{11}}$ **15**

Sketch the graph of the function or the inequality. (9.3 and 9.7) **56–58. See margin.**

56. $y = -3x^2 + 12x - 7$

57. $y \geq 5x^2 + 20x + 13$

58. $y \leq \frac{1}{2}x^2 + 4x + 1$

Use the quadratic formula to solve the equation. (9.5)

59. $x^2 + 10x + 9 = 0$ **$-9, -1$**

60. $-x^2 + 5x - 6 = 0$ **2, 3**

61. $3x^2 + 8x - 5 = 0$ **$\frac{-4 - \sqrt{31}}{3} \approx -3.19, \frac{-4 + \sqrt{31}}{3} \approx 0.52$**

62. $-2x^2 + 5x + 12 = 0$ **$-\frac{3}{2}, 4$**

63. $-\frac{1}{2}x^2 + 3x - \frac{5}{2} = 0$ **1, 5**

64. $7x^2 + 12x - 2 = 0$ **$\frac{-6 - 5\sqrt{2}}{7} \approx -1.87, \frac{-6 + 5\sqrt{2}}{7} \approx 0.15$**

65. COMPOUND INTEREST Your savings account earns 4.8% interest, compounded annually. Another bank in town is offering 5.1% interest, compounded annually. The balance in your account is \$567. How much additional interest could you earn in 5 years by moving your account to the bank with the 5.1% interest? in 10 years? (8.5) **\$10.31; \$26.28**

SENDING UP FLARES **In Exercises 69 and 70, a flare is fired straight up from ground level with an initial velocity of 100 feet per second.** (9.5)

66. How long will it take the flare to reach an altitude of 150 feet? **2.5 sec**

67. Will the flare reach an altitude of 180 feet? Explain. **See margin.**

68. GEOMETRY CONNECTION Is it possible for a rectangle with a perimeter of 52 centimeters to have an area of 148.75 square centimeters? Explain. (9.5–9.6) **See margin.**

56.

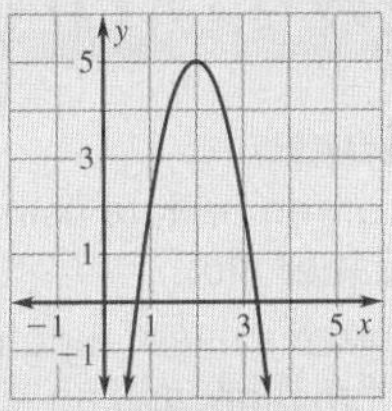

57.

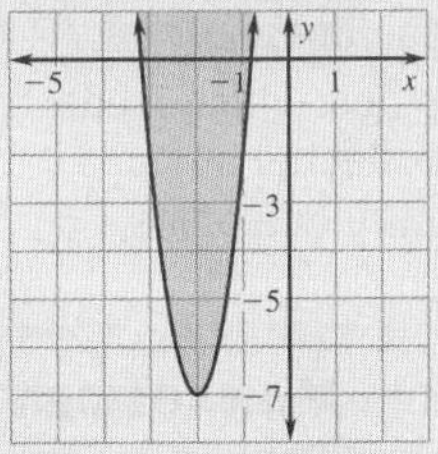

58.

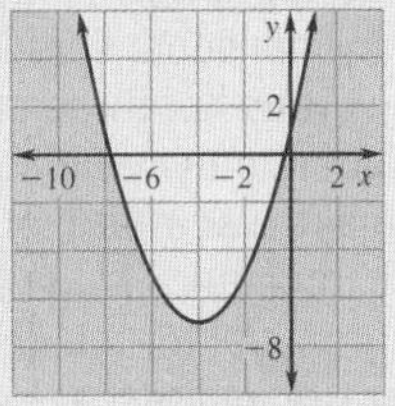

67. No; the discriminant of the equation $-16t^2 + 100t = 180$ is negative, so the equation has no real solution. That is, there is no time t at which the flare will reach a height of 180 ft.

68. Yes; the area of the rectangle is $x(26 - x)$. The discriminant of the equation $x(26 - x) = 148.75$ is positive, so there are two solutions. (The rectangle has dimensions 8.5 cm and 17.5 cm.)

PROJECT OVERVIEW

PROJECT GOALS

- Collect and organize data using a table and scatter plot.
- Model data with a linear function, a quadratic function, and an exponential function.
- Select the best-fitting model for the data collected and use it to predict future outcomes.

MANAGING THE PROJECT

CLASSROOM MANAGEMENT

The Project for Chapters 7–9 may be completed by individual students or in groups. In groups of three students, each student can complete the table and make a scatter plot. Then one group member can write and test a linear model for the data, a second can write and test a quadratic model, and the third can write and test an exponential model. Group members can discuss their models to decide which is most appropriate.

Group members can also work together to find the best-fitting model for the data they collect about another business. All members should collaborate on the report or poster, with each member including his or her findings.

GUIDING STUDENTS' WORK

Students may have difficulty deciding whether a linear, quadratic, or exponential model is the best-fitting model. Ask students who do a paper-and-pencil scatter plot to check if the yearly difference in the total sales is constant, indicating a linear model, or if the yearly ratios of total sales are constant, indicating an exponential model. Encourage students who use graphing calculators or computers to investigate the value of r, the correlation coefficient. The closer this value is to 1 or −1, the better the model fit. Ask them to include a discussion of r in their report.

2. See Additional Answers beginning on page AA1.

PROJECT

Applying Chapters 7–9

Fitting a Model to Data

OBJECTIVE Fitting a model to data and then using the model to predict future outcomes.

Materials: graph paper, graphing calculator or computer (optional)

COLLECTING YOUR DATA

You opened a pizza business eight years ago, and you want to write an equation to represent the growth of the business since you opened.

1. Copy and complete the table. **18,360; 31,230; 53,100; 90,225; 153,360; 260,730; 443,160**

Pizza Business Sales Report

Year	Number of Pizzas Sold	Total Sales ($9 per pizza)
1	1200	$9 \times 1200 = 10{,}800$
2	2040	$9 \times 2040 = ?$
3	3470	?
4	5900	?
5	10,025	?
6	17,040	?
7	28,970	?
8	49,240	?

INVESTIGATING YOUR DATA

2. Make a scatter plot of the data. Describe any patterns you see. **See margin.**
3. To write an equation that represents your data, test a linear model, a quadratic model, and an exponential model to determine which fits the data best. You may want to use a graphing calculator or computer. (See the Graphing Calculator Activity in Chapter 8, page 492, on how to fit a model to a given set of data.) Explain why you think that each model either fits or doesn't fit the data. **See margin.**
4. Make a prediction on what the total sales would be in the tenth year. **about $1,280,000**

2. See margin for plot; *Sample answer:* The more that time increases, the steeper the graph for number of pizzas sold climbs.

3. *Sample answer:* An exponential model, such as $y = 6353(1.7)^x$, fits the data best. From the graph it is clear that a linear model would not fit the data well. An exponential model will fit the data well because if you calculate the ratio of each y-coordinate to the preceding y-coordinate, the ratios are approximately equal.

INVESTIGATING ANOTHER BUSINESS

5–7. Check work.

5. Find the annual sales for at least four years for a business or an industry that interests you. Make a table of the data.

6. Make a scatter plot of the data. Try to fit a linear model, a quadratic model, and an exponential model to the data. Which model fits best?

7. Can you use the best-fitting model from Exercise 6 to predict future sales for this business or industry? If you can, use your model to predict sales five years from now. If you cannot, explain why not.

PRESENT YOUR RESULTS

Write a report or make a poster to report your results.

- Include your answers to Exercises 1–4.
- Include all the curves you tried and your explanations of whether they fit or not.
- Describe the business or industry you looked at in Exercises 5–7. Tell where you found your data. Explain whether you were able to fit a model to your data. Include your answers to Exercises 5–7.
- If you used a graphing calculator or a computer to create your models, give the *r*-value for each model and explain what the values mean.
- Describe what you learned about fitting a model to data. For example, what information or values helped you to determine the best model to use?

EXTENSION

- Find sales data for a business or industry related to the one you used for Exercises 5–7. Compare the two sets of data. Compare the growth in sales in the two businesses or industries.
- Do some research to find a stock market index that you can use to compare the performance of the business or industry you used in the project with the performance of stocks during the same period of time.

CONCLUDING THE PROJECT

Before students present their results, tell them to be prepared to defend their choice of models. You may want students to contact a representative of the business or industry they investigated for comments about their project.

GRADING THE PROJECT

4 Students complete the table, do a scatter plot correctly, and find correct linear, quadratic, and exponential models for the data. They explain clearly why each model either fits or doesn't fit the data and include an appropriate prediction about the total sales in ten years. Students also investigate another business or industry with the same thoroughness. Reports or posters present the data clearly and thoroughly support model choices.

3 Students complete the table, do a scatter plot correctly, and find correct linear, quadratic, and exponential models for the data. They explain clearly why each model either fits or doesn't fit the data and include an appropriate prediction about the total sales in ten years. Students also investigate another business or industry, but their work is less thorough. Reports or posters present the data clearly, but model choices may not always be thoroughly supported.

2 Students complete the table and do the scatter plot correctly, but not all models are appropriate. They include a prediction about the total sales in ten years, and investigate another business or industry. Reports or posters present the data and models, but show haste and do not support model choices.

1 Students complete the table and do the scatter plot correctly. They do not find all the models for the data, but do include a prediction for total sales. They do not investigate another business or industry. Reports or posters are unclear and incomplete, and show little grasp of finding a model for data.

CHAPTER 10

PLANNING THE CHAPTER

Polynomials and Factoring

LESSON	GOALS	NCTM	ITED	SAT9	Terra-Nova	Local
10.1 *pp. 575–583*	CONCEPT ACTIVITY: 10.1 *Investigate how to model the addition of polynomials with algebra tiles.* GOAL 1 Add and subtract polynomials. GOAL 2 Use polynomials to model real-life situations. TECHNOLOGY ACTIVITY: 10.1 *Find the sum and difference of polynomials on a graphing calculator.*	1, 2, 3, 6, 8, 9, 10	RQWN, RQF, IIG, MIG	2, 3	11, 14, 15, 16, 18	
10.2 *pp. 584–589*	GOAL 1 Multiply two polynomials. GOAL 2 Use polynomial multiplication in real-life situations.	1, 2, 6, 8, 9, 10	RQWN, RQF	2, 3	11, 16	
10.3 *pp. 590–596*	GOAL 1 Use special product patterns for the product of a sum and a difference, and for the square of a binomial. GOAL 2 Use special products as real-life models.	1, 2, 6, 8, 9, 10	MCE, APE, RQWN, RQF	2, 3	11, 16	
10.4 *pp. 597–602*	GOAL 1 Solve a polynomial equation in factored form. GOAL 2 Relate factors and x-intercepts.	1, 2	RQWN, RQF	2	14, 18, 52	
10.5 *pp. 603–609*	CONCEPT ACTIVITY: 10.5 *Investigate how to model the factorization of a trinomial of the form $x^2 + bx + c$ using algebra tiles.* GOAL 1 Factor a quadratic expresssion of the form $x^2 + bx + c$. GOAL 2 Solve quadratic equations by factoring.	1, 2, 10	MCE, APE		10, 11, 16, 18, 52	
10.6 *pp. 610–618*	CONCEPT ACTIVITY: 10.6 *Investigate how to model the factorization of a trinomial of the form $ax^2 + bx + c$ using algebra tiles.* GOAL 1 Factor a quadratic expression of $ax^2 + bx + c$. GOAL 2 Solve quadratic equations by factoring.	1, 2, 3, 10	MCE, APE		14, 16, 18, 52	
10.7 *pp. 619–624*	CONCEPT ACTIVITY: 10.7 *Investigate how to model the factorization of a trinomial of the form $(ax^2) + 2abx + b^2$ using algebra tiles.* GOAL 1 Use special product patterns to factor quadratric polynomials. GOAL 2 Solve quadratic equations by factoring.	1, 2, 8	MCE, APE		11, 16, 52	
10.8 *pp. 625–632*	GOAL 1 Use the distributive property to factor a polynomial. GOAL 2 Solve polynomial equations by factoring.	1, 2, 8	RQWN, RQF		10, 11, 16, 52	

RESOURCES

Chapter Resource Booklets

Chapter Support

Tips for New Teachers	p. 1	Prerequisite Skills Review	p. 5
Parent Guide for Student Success	p. 3	Strategies for Reading Mathematics	p. 7

Lesson Support

Lesson Support	10.1	10.2	10.3	10.4	10.5	10.6	10.7	10.8
Lesson Plans (regular and block)	p. 9	p. 23	p. 37	p. 52	p. 64	p. 77	p. 92	p. 107
Warm-Up Exercises and Daily Quiz	p. 11	p. 25	p. 39	p. 54	p. 66	p. 79	p. 94	p. 109
Activity Support Masters	p. 12				p. 67	p. 80		
Lesson Openers	p. 13	p. 26	p. 40	p. 55	p. 68	p. 81	p. 95	p. 110
Graphing Calculator Activities & Keystrokes	p. 14	p. 27				p. 82	p. 96	
Practice (3 levels)	p. 15	p. 29	p. 41	p. 56	p. 69	p. 83	p. 99	p. 111
Reteaching with Practice	p. 18	p. 32	p. 44	p. 59	p. 72	p. 86	p. 102	p. 114
Quick Catch-Up for Absent Students	p. 20	p. 34	p. 46	p. 61	p. 74	p. 88	p. 104	p. 116
Cooperative Learning Activities			p. 47					
Interdisciplinary Applications	p. 21		p. 48		p. 75	p. 89		
Real-Life Applications		p. 35		p. 62			p. 105	p. 117
Math & History Applications			p. 49					
Challenge: Skills and Applications	p. 22	p. 36	p. 50	p. 63	p. 76	p. 90	p. 106	p. 118

Review and Assessment

Quizzes	pp. 51, 91	Alternative Assessment with Math Journal	p. 127
Chapter Review Games and Activities	p. 119	Project with Rubric	p. 129
Chapter Test (3 levels)	pp. 120–125	Cumulative Review	p. 131
SAT/ACT Chapter Test	p. 126	Resource Book Answers	p. A1

Transparencies

	10.1	10.2	10.3	10.4	10.5	10.6	10.7	10.8
Warm-Up Exercises and Daily Quiz	p. 76	p. 77	p. 78	p. 79	p. 80	p. 81	p. 82	p. 83
Alternative Lesson Opener Transparencies	p. 66	p. 67	p. 68	p. 69	p. 70	p. 71	p. 72	p. 73
Examples/Standardized Test Practice	✓	✓	✓	✓	✓	✓	✓	✓
Answer Transparencies	✓	✓	✓	✓	✓	✓	✓	✓

Technology

- EasyPlanner CD
- EasyPlanner Plus Online
- Classzone.com
- Personal Student Tutor CD
- Test and Practice Generator CD
- Instant Replay: Video Review Games
- Electronic Lesson Presentations
- Chapter Audio Summaries CD

Additional Resources

- Basic Skills Workbook
- Explorations and Projects
- Worked-Out Solution Key
- Data Analysis Sourcebook
- Spanish Resources
- Standardized Test Practice Workbook
- Practice Workbook with Examples
- Notetaking Guide

CHAPTER 10

PACING THE CHAPTER

Resource Key
- STUDENT EDITION
- CHAPTER 10 RESOURCE BOOK
- TEACHER'S EDITION

REGULAR SCHEDULE

Day 1

10.1

STARTING OPTIONS
- Prereq. Skills Review
- Strategies for Reading
- Warm-Up or Daily Quiz

TEACHING OPTIONS
- Motivating the Lesson
- Concept Act. & Wksht.
- Les. Opener (Appl.)
- Graphing Calc. Activity
- Examples 1–6
- Technology Activity
- Closure Question
- Guided Practice Exs.

APPLY/HOMEWORK
- See Assignment Guide.
- See the CRB: Practice, Reteach, Apply, Extend

ASSESSMENT OPTIONS
- Checkpoint Exercises
- Daily Quiz (10.1)
- Stand. Test Practice

Day 2

10.2

STARTING OPTIONS
- Homework Check
- Warm-Up or Daily Quiz

TEACHING OPTIONS
- Motivating the Lesson
- Les. Opener (Visual)
- Graphing Calc. Activity
- Examples 1–5
- Closure Question
- Guided Practice Exs.

APPLY/HOMEWORK
- See Assignment Guide.
- See the CRB: Practice, Reteach, Apply, Extend

ASSESSMENT OPTIONS
- Checkpoint Exercises
- Daily Quiz (10.2)
- Stand. Test Practice

Day 3

10.3

STARTING OPTIONS
- Homework Check
- Warm-Up or Daily Quiz

TEACHING OPTIONS
- Motivating the Lesson
- Les. Opener (Application)
- Graphing Calc. Activity
- Examples 1–3
- Guided Practice Exs. 1, 4, 6–14

APPLY/HOMEWORK
- See Assignment Guide.
- See the CRB: Practice, Reteach, Apply, Extend

ASSESSMENT OPTIONS
- Checkpoint Exercises, p. 591

Day 4

10.3 *(cont.)*

STARTING OPTIONS
- Homework Check

TEACHING OPTIONS
- Examples 4–5
- Closure Question
- Guided Practice Exs. 2–3, 5

APPLY/HOMEWORK
- See Assignment Guide.
- See the CRB: Practice, Reteach, Apply, Extend

ASSESSMENT OPTIONS
- Checkpoint Exercises, p. 592
- Daily Quiz (10.3)
- Stand. Test Practice
- Quiz (10.1–10.3)

Day 5

10.4

STARTING OPTIONS
- Homework Check
- Warm-Up or Daily Quiz

TEACHING OPTIONS
- Motivating the Lesson
- Les. Opener (Application)
- Graphing Calc. Activity
- Examples 1–5
- Closure Question
- Guided Practice Exs.

APPLY/HOMEWORK
- See Assignment Guide.
- See the CRB: Practice, Reteach, Apply, Extend

ASSESSMENT OPTIONS
- Checkpoint Exercises
- Daily Quiz (10.4)
- Stand. Test Practice

Day 6

10.5

STARTING OPTIONS
- Homework Check
- Warm-Up or Daily Quiz

TEACHING OPTIONS
- Motivating the Lesson
- Concept Act. & Wksht.
- Les. Opener (Activity)
- Examples 1–6
- Guided Practice Exs. 1–11

APPLY/HOMEWORK
- See Assignment Guide.
- See the CRB: Practice, Reteach, Apply, Extend

ASSESSMENT OPTIONS
- Checkpoint Exercises, pp. 605–606

Day 9

10.6 *(cont.)*

STARTING OPTIONS
- Homework Check

TEACHING OPTIONS
- Example 6
- Technology Activity
- Closure Question
- Guided Practice Ex. 4

APPLY/HOMEWORK
- See Assignment Guide.
- See the CRB: Practice, Reteach, Apply, Extend

ASSESSMENT OPTIONS
- Checkpoint Exercises, p. 613
- Daily Quiz (10.6)
- Stand. Test Practice
- Quiz (10.4–10.6)

Day 10

10.7

STARTING OPTIONS
- Homework Check
- Warm-Up or Daily Quiz

TEACHING OPTIONS
- Motivating the Lesson
- Les. Opener (Activity)
- Graphing Calc. Activity
- Examples 1–5
- Guided Practice Exs. 1–17

APPLY/HOMEWORK
- See Assignment Guide.
- See the CRB: Practice, Reteach, Apply, Extend

ASSESSMENT OPTIONS
- Checkpoint Exercises, pp. 620–621

Day 11

10.7 *(cont.)*

STARTING OPTIONS
- Homework Check

TEACHING OPTIONS
- Example 6
- Closure Question

APPLY/HOMEWORK
- See Assignment Guide.
- See the CRB: Practice, Reteach, Apply, Extend

ASSESSMENT OPTIONS
- Checkpoint Exercises, p. 621
- Daily Quiz (10.7)
- Stand. Test Practice

Day 12

10.8

STARTING OPTIONS
- Homework Check
- Warm-Up or Daily Quiz

TEACHING OPTIONS
- Motivating the Lesson
- Les. Opener (Activity)
- Examples 1–7
- Guided Practice Exs. 1–14

APPLY/HOMEWORK
- See Assignment Guide.
- See the CRB: Practice, Reteach, Apply, Extend

ASSESSMENT OPTIONS
- Checkpoint Exercises, p. 626–627

Day 13

10.8 *(cont.)*

STARTING OPTIONS
- Homework Check

TEACHING OPTIONS
- Examples 8–9
- Closure Question

APPLY/HOMEWORK
- See Assignment Guide.
- See the CRB: Practice, Reteach, Apply, Extend

ASSESSMENT OPTIONS
- Checkpoint Exercises, pp. 627–628
- Daily Quiz (10.8)
- Stand. Test Practice
- Quiz (10.7–10.8)

Day 14

Review

DAY 14 START OPTIONS
- Homework Check

REVIEWING OPTIONS
- Chapter 10 Summary
- Chapter 10 Review
- Chapter Review Games and Activities

APPLY/HOMEWORK
- Chapter 10 Test (practice)
- Ch. Standardized Test (practice)

Day 7

10.5 *(cont.)*

Starting Options
- Homework Check

Teaching Options
- Example 7
- Closure Question

Apply/Homework
- See Assignment Guide.
- See the CRB: Practice, Reteach, Apply, Extend

Assessment Options
- Checkpoint Exercises, p. 606
- Daily Quiz (10.5)
- Stand. Test Practice

Day 8

10.6

Starting Options
- Homework Check
- Warm-Up or Daily Quiz

Teaching Options
- Motivating the Lesson
- Concept Act. & Wksht.
- Les. Opener (Calculator)
- Graphing Calc. Activity
- Examples 1–5
- Guided Practice Exs. 1–3, 5–14

Apply/Homework
- See Assignment Guide.
- See the CRB: Practice, Reteach, Apply, Extend

Assessment Options
- Checkpoint Exercises, pp. 612–613

Day 15

Assess

Day 15 Start Options
- Homework Check

Assessment Options
- Chapter 10 Test
- SAT/ACT Ch. 10 Test
- Alternative Assessment

Apply/Homework
- Skill Review, p. 642

BLOCK SCHEDULE

Day 1

Assess & 10.1

(Day 1 = Ch. 9 Day 8)

Assessment Options
- Chapter 9 Test
- SAT/ACT Ch. 9 Test
- Alternative Assessment

Ch. 10 Start Options
- Skill Review, p. 574
- Prereq. Skills Review
- Strategies for Reading

Teaching 10.1 Options
- Warm-Up (Les. 10.1)
- Motivating the Lesson
- Concept Act. & Wksht.
- Les. Opener (Appl.)
- Graphing Calc. Activity
- Examples 1–6
- Technology Activity
- Closure Question
- Guided Practice Exs.

Apply/Homework
- See Assignment Guide.
- See the CRB: Practice, Reteach, Apply, Extend

Assessment Options
- Checkpoint Exercises
- Daily Quiz (Les. 10.1)
- Stand. Test Practice

Day 2

10.2 & 10.3

Day 2 Start Options
- Homework Check
- W-Up 10.2 or D. Quiz 10.1

Teaching 10.2 Options
- Motivating the Lesson
- Les. Opener (Visual)
- Graphing Calc. Activity
- Examples 1–5
- Closure Question
- Guided Practice Exs.

Beginning 10.3 Options
- Warm-Up (Les. 10.3)
- Motivating the Lesson
- Les. Opener (Appl.)
- Examples 1–3
- Guided Practice Exs. 1, 4, 6–14

Apply/Homework
- See Assignment Guide.
- See the CRB: Practice, Reteach, Apply, Extend

Assessment Options
- Checkpoint Exercises
- Daily Quiz (Les. 10.2)
- Stand. Test Prac. (10.2)

Day 3

10.3 & 10.4

Day 3 Start Options
- Homework Check
- Daily Quiz (Les. 10.2)

Finishing 10.3 Options
- Examples 4–5
- Closure Question
- Guided Practice Exs. 2–3, 5

Teaching 10.4 Options
- Warm-Up (Les. 10.4)
- Motivating the Lesson
- Les. Opener (Appl.)
- Examples 1–5
- Closure Question
- Guided Practice Exs.

Apply/Homework
- See Assignment Guide.
- See the CRB: Practice, Reteach, Apply, Extend

Assessment Options
- Checkpoint Exercises
- Daily Quiz (Les. 10.3, 10.4)
- Stand. Test Practice
- Quiz (10.1–10.3)

Day 4

10.5

Day 4 Start Options
- Homework Check
- Warm-Up or Daily Quiz

Teaching 10.5 Options
- Motivating the Lesson
- Concept Act. & Wksht.
- Les. Opener (Activity)
- Examples 1–7
- Closure Question
- Guided Practice Exs.

Apply/Homework
- See Assignment Guide.
- See the CRB: Practice, Reteach, Apply, Extend

Assessment Options
- Checkpoint Exercises
- Daily Quiz (Les. 10.5)
- Stand. Test Practice

Day 5

10.6

Day 5 Start Options
- Homework Check
- Warm-Up or Daily Quiz

Teaching 10.6 Options
- Motivating the Lesson
- Concept Act. & Wksht.
- Les. Opener (Calc.)
- Graphing Calc. Activity
- Examples 1–6
- Technology Activity
- Closure Question
- Guided Practice Exs.

Apply/Homework
- See Assignment Guide.
- See the CRB: Practice, Reteach, Apply, Extend

Assessment Options
- Checkpoint Exercises
- Daily Quiz (Les. 10.6)
- Stand. Test Practice
- Quiz (10.4–10.6)

Day 6

10.7

Day 6 Start Options
- Homework Check
- Warm-Up or Daily Quiz

Teaching 10.7 Options
- Motivating the Lesson
- Les. Opener (Activity)
- Graphing Calc. Activity
- Examples 1–6
- Closure Question
- Guided Practice Exs.

Apply/Homework
- See Assignment Guide.
- See the CRB: Practice, Reteach, Apply, Extend

Assessment Options
- Checkpoint Exercises
- Daily Quiz (Les. 10.7)
- Stand. Test Practice

Day 7

10.8

Day 7 Start Options
- Homework Check
- Warm-Up or Daily Quiz

Teaching 10.8 Options
- Motivating the Lesson
- Les. Opener (Activity)
- Examples 1–9
- Closure Question
- Guided Practice Exs.

Apply/Homework
- See Assignment Guide.
- See the CRB: Practice, Reteach, Apply, Extend

Assessment Options
- Checkpoint Exercises
- Daily Quiz (Les. 10.8)
- Stand. Test Practice
- Quiz (10.7–10.8)

Day 8

Review/Assess

Day 8 Start Options
- Homework Check

Reviewing Options
- Chapter 10 Summary
- Chapter 10 Review
- Chapter Review Games and Activities
- Chapter 10 Test (practice)
- Ch. Standardized Test (practice)

Assessment Options
- Chapter 10 Test
- SAT/ACT Ch. 10 Test
- Alternative Assessment

Apply/Homework
- Skill Review, p. 642

CHAPTER 10

Meeting Individual Needs

Before the Chapter

The *Chapter 10 Resource Book* has the following materials to distribute and use before the chapter:

- **Parent Guide for Student Success**
- **Prerequisite Skills Review (pictured below)**
- **Strategies for Reading Mathematics**

Prerequisite Skills *Pages 5–6*

CHAPTER 10

NAME ________________ DATE ________

Prerequisite Skills Review

For use before Chapter 10

Chapter Support

EXAMPLE 1 ***Simplifying by Combining Like Terms***

Combine like terms.

$6x - 4(7 + x)$

Solution

$6x - 4(7 + x) = 6x + (-4)(7 + x)$	Rewrite as addition expression.
$= 6x + [(-4)(7) + (-4)(x)]$	Distribute the –4.
$= 6x + (-28) + (-4x)$	Multiply.
$= 2x - 28$	Combine like terms and simplify.

Exercises for Example 1

Combine like terms.

1. $(5 + 10x)2$
2. $-\frac{1}{4}(t - 16)$
3. $-(y - 6) - y$
4. $7s + 2(s - 10)$
5. $3.2(7.2 + x) + x$
6. $3x^2 - x(3 + 2x)$

EXAMPLE 2 ***Multiplying Exponential Expressions***

Simplify the expression.

a. $(-2st^3)(-s^4)^5$ b. $(5x^3y^4)^2 \cdot x^6$

Solution

a. $(-2st^3)(-s^4)^5 = (-2)^3(s)^3(t^3)^3(-s^4)^5$	Raise each factor to a power.
$= -8 \cdot s^3 \cdot t^9 \cdot (-s^{20})$	Evaluate each power.
$= 8s^{23}t^9$	Simplify.
b. $(5x^3y^4)^2 \cdot x^6 = 5^2(x^3)^2(y^4)^2 \cdot x^6$	Raise each factor to a power.
$= 25 \cdot x^6 \cdot y^8 \cdot x^6$	Evaluate each power.
$= 25x^{12}y^8$	Simplify.

Exercises for Example 2

Simplify the expression.

7. $x^2 \cdot y^5 \cdot x^4 \cdot y^{12}$
8. $(-4ab^4)(3a^6b^9)$
9. $(4s^5t^9)^3 \cdot 3t^4$
10. $\left(\frac{1}{2}x^4y^6\right)^5(x^4y)^2$

Prerequisite Skills Review These two pages support the Study Guide on page 574. They help students prepare for Chapter 10 by providing worked-out examples and practice for the following skills needed in the chapter:

- **Simplify by combining like terms.**
- **Multiply exponential expressions.**
- **Find the number of solutions.**

Technology Resource

Students can use the Personal Student Tutor to find additional reteaching and practice for skills from earlier chapters that are used in Chapter 10.

During Each Lesson

The *Chapter 10 Resource Book* has the following alternatives for introducing the lesson:

- **Lesson Openers (pictured below)**
- **Graphing Calculator Activities with Keystrokes**

Lesson Opener *Page 26*

LESSON 10.2

NAME ________________ DATE ________

Available as a transparency

Visual Approach Lesson Opener

For use with pages 511–516

Lesson 10.2

Match each area to the correct model. Then use the model to find the product.

1. $x(x)$
2. $x(x + 1)$
3. $2x(x)$
4. $2x(x + 1)$
5. $(x + 2)(x + 1)$
6. $(x + 3)(x + 1)$

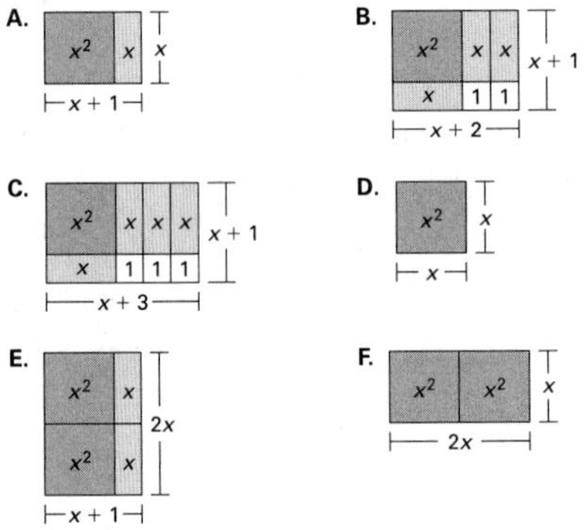

Visual Approach Lesson Opener This Lesson Opener uses visuals as an alternative way to start Lesson 10.2. Students use area models as an introduction to multiplying polynomials.

FOLLOWING EACH LESSON

The *Chapter 10 Resource Book* has a variety of materials to follow-up each lesson. They include the following:

- **Practice (3 levels)**
- **Reteaching with Practice (pictured below)**
- **Quick Catch-Up for Absent Students**
- **Interdisciplinary Applications**
- **Real-Life Applications**

RETEACHING WITH PRACTICE *Pages 32–33*

LESSON 10.2

NAME ______________________ DATE __________

Reteaching with Practice

For use with pages 584–589

GOAL **Multiply two polynomials and use polynomial multiplication in real-life situations.**

VOCABULARY

To multiply two binomials, use a pattern called the **FOIL** pattern. Multiply the First, Outer, Inner, and Last terms.

EXAMPLE 1 ***Multiplying Binomials Using the FOIL Pattern***

Find the product $(4x + 3)(x + 2)$.

SOLUTION

$(4x + 3)(x + 2) = 4x^2 + 8x + 3x + 6$ (F, O, I, L) — Mental math

$= 4x^2 + 11x + 6$ — Simplify.

Exercises for Example 1

Use the FOIL pattern to find the product.

1. $(2x + 3)(x + 1)$ **2.** $(y - 2)(y - 3)$ **3.** $(3a + 2)(2a - 1)$

EXAMPLE 2 ***Multiplying Polynomials Vertically***

Find the product $(x + 3)(4 - 2x^2 + x)$.

SOLUTION

To multiply two polynomials that have three or more terms, you must multiply each term of one polynomial by each term of the other polynomial. Align like terms in columns.

$-2x^2 + x + 4$	Standard form
$x + 3$	Standard form
$-6x^2 + 3x + 12$	$3(-2x^2 + x + 4)$
$-2x^3 + x^2 + 4x$	$x(-2x^2 + x + 4)$
$-2x^3 - 5x^2 + 7x + 12$	Combine like terms.

Exercises for Example 2

Multiply the polynomials vertically.

4. $(a + 4)(a^2 + 3 - 2a)$ **5.** $(2y + 1)(y^2 - 5 + y)$

Lesson 10.2

RETEACHING WITH PRACTICE On this two-page worksheet, additional examples and practice reinforce the key concepts of Lesson 10.2: multiplying polynomials and using polynomials in real-life situations.

ASSESSING THE CHAPTER

The *Chapter 10 Resource Book* has the following review and assessment materials:

- **Quizzes**
- **Chapter Review Games and Activities (pictured below)**
- **Chapter Test (3 levels)**
- **SAT/ACT Chapter Test**
- **Alternative Assessment with Rubric and Math Journal**
- **Project with Rubric**
- **Cumulative Review**

CHAPTER REVIEW GAMES *Page 119*

CHAPTER 10

NAME ______________________ DATE __________

Chapter Review Games and Activities

For use after Chapter 10

Solve the following problems, and find the answer at the right of the page. Place the letter of the answer on the line with the problem number to answer the riddle.

How does a pea farmer keep from getting sunburned at the beach?

1. $(-2x^3 + 2x^2 - x - 1) - (3x^3 + 5x^2 - x + 5) =$

2. $(x + 3)(-3x - 6) =$

3. $(4x - 8)(4x + 8) =$

4. $(5x + 2)^2 =$

5. Solve for x: $(x - 4)(x + 2) = 0$

6. Solve for x: $x^2 + 3x = 54$

7. Factor: $3x^2 - 19x - 40$

8. Factor: $9x^2 - 12x + 4$

(S) $x^3 + 7x^2 - 2x + 4$
(C) $x = -4$ and 2
(S) $(3x - 2)(3x - 2)$
(U) $3x^2 + 15x - 18$
(A) $x = 9$ and -6
(N) $16x^2 - 64$
(O) $25x^2 + 20x + 4$
(S) $16x^2 + 64$
(A) $(3x + 5)(x - 8)$
(R) $(3x - 2)(3x + 2)$
(E) $x = -9$ and 6
(E) $(9x - 2)(x - 2)$
(P) $x = -2$ and 4
(A) $-3x^2 - 15x - 18$
(N) $(-3x - 20)(x + 2)$
(M) $25x^2 + 4$
(C) $-5x^3 - 3x^2 - 6$

He uses ____ (1) ____ (2) ____ (3) ____ (4) ____ (5) ____ (6) ____ (7) ____ (8) !!

Review and Assess

CHAPTER REVIEW GAMES AND ACTIVITIES On this worksheet, students solve problems to answer a riddle as a motivating means of reviewing the material in Chapter 10.

TECHNOLOGY RESOURCE

Teachers can use the video Instant Replay: Video Review Games as another motivating way to review the material in Chapter 10.

CHAPTER OVERVIEW

CHAPTER GOALS

In Chapter 10 students identify, compute with, and factor polynomial expressions; and they solve quadratic equations. Students add, subtract, and multiply polynomial expressions in the first half of the chapter. In the second half of the chapter, they factor polynomial expressions and use factoring to solve quadratic equations.

APPLICATION NOTE

Radio telescopes are used to collect astronomical data beyond the region of visible light. They are different from optical telescopes in that they are not placed on a mountain top or on something high in the air, but are located in remote valleys away from radio and electrical interference. Also, they can be used day or night in any weather, while optical telescopes are used when the night is clear. Radio telescopes can be used to study pulsars, which are stars that emit extremely regular radio signals.

The study and use of radio waves began at Holmdel, New Jersey in 1931 when Karl Jansky discovered radio waves coming from a region in Sagittarius. In the late 1940s and 1950s, Australia, England, Holland, and later the United States all began to use and make radio observatories.

Additional information about the Arecibo Observatory is available at **www.mcdougallittell.com.**

POLYNOMIALS AND FACTORING

Where can scientists dish up astronomical research projects?

CHAPTER 10

APPLICATION: Arecibo Observatory

The Arecibo radio telescope is the largest single-dish radio telescope in the world. Its enormous reflector dish, built in a natural limestone sinkhole in Puerto Rico, is sensitive enough to collect radio waves originating 7000 trillion miles from Earth.

The cross section of many radio telescopes, such as the one shown below, where x and y are in feet, can be modeled by a polynomial equation whose graph is a parabola. You will solve polynomial equations in Chapter 10.

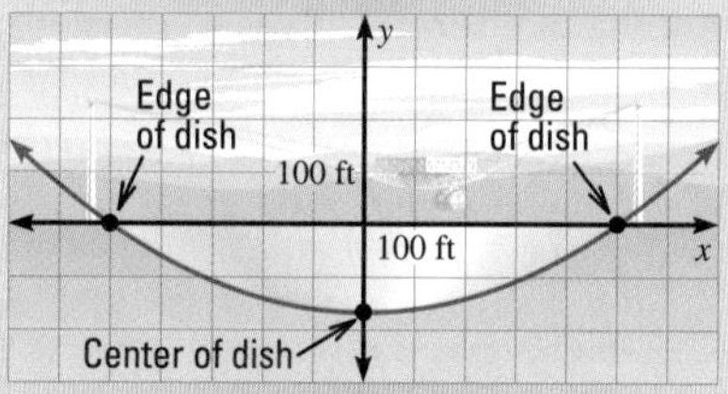

Think & Discuss

Use the graph modeling a cross section of the telescope's reflector dish for Exercises 1 and 2.

1. Find the x-intercepts. How can you use this information to find the diameter of the dish? **−500, 500; The diameter is the distance between the x-intercepts, 1000 ft.**
2. Estimate the depth of the dish. **Estimates may vary; the actual depth is 167 feet.**

Learn More About It

You will use an algebraic model of a radio telescope in Exercises 53 and 54 on page 601.

INTERNET APPLICATION LINK Visit www.mcdougallittell.com for more information about the Arecibo Observatory.

ADDITIONAL RESOURCES
Another way to begin the chapter is to show the video clip of a real-life motivator for Chapter 10 on the *Instant Replay: Video Review Games.*

PROJECTS
A project covering Chapter 10–12 appears on pages 774–775 of the Student Edition. An additional project for Chapter 10 is available in the *Chapter 10 Resource Book,* p. 129.

TECHNOLOGY

Software
- *Electronic Teaching Tools*
- *Online Lesson Planner*
- *Personal Student Tutor*
- *Test and Practice Generator*
- *Electronic Lesson Presentations* (Lesson 10.6)

Video
- *Instant Replay: Video Review Games*

INTERNET **Internet Connections**
www.mcdougallittell.com
- **Application Links**
573, 588, 596, 608, 630
- **Data Updates**
581, 589
- **Student Help**
577, 583, 585, 594, 601, 605, 615, 618, 620, 628
- **Career Links**
581, 594, 606
- **Extra Challenge**
582, 589, 609, 616, 624, 631

PREPARE

DIAGNOSTIC TOOLS

The **Skill Review** exercises can help you diagnose whether students have the following skills needed in Chapter 10:

- Use the distributive property.
- Simplify expressions using the order of operations.
- Find the discriminant of a quadratic equation.

The following resources are available for students who need additional help with these skills:

- Prerequisite Skills Review (*Chapter 10 Resource Book,* p. 5; *Warm-Up Transparencies,* p. 75)
- Reteaching with Practice (Chapter Resource Books for Lessons 1.3, 2.6, and 9.6)
- *Personal Student Tutor*

ADDITIONAL RESOURCES

The following resources are provided to help you prepare for the upcoming chapter and customize review materials:

- ***Chapter 10 Resource Book***
  Tips for New Teachers (p. 1)
  Parent Guide (p. 3)
  Lesson Plans (every lesson)
  Lesson Plans for Block Scheduling (every lesson)
- ***Electronic Teaching Tools***
- ***Online Lesson Planner***
- ***Test and Practice Generator***

ENGLISH LEARNERS

For students to understand algebraic concepts presented in English, they must have a clear understanding of key vocabulary terms. You may want to have English language learners prepare for Chapter 10 by beginning their own vocabulary guides. Encourage them to define each review vocabulary term in their own words, and to add new vocabulary to their guides as they complete the chapter.

CHAPTER 10 Study Guide

PREVIEW — *What's the chapter about?*

Chapter 10 is about **polynomials and factoring**. In Chapter 10 you'll learn

- how to add, subtract, and multiply polynomials.
- how to factor polynomials.
- how to solve polynomial equations by factoring.

KEY VOCABULARY

▶ Review

- distributive property, p. 100
- like terms, p. 102
- *x*-intercept, p. 218
- parabola, p. 518
- quadratic formula, p. 533
- vertical motion model, p. 535
- discriminant, p. 541

▶ New

- polynomial, p. 576
- FOIL pattern, p. 585
- factored form, p. 597
- zero-product property, p. 597
- factor a quadratic expression, p. 604
- factor completely, p. 625

PREPARE — *Are you ready for the chapter?*

SKILL REVIEW Do these exercises to review key skills that you'll apply in this chapter. See the given reference page if there is something you don't understand.

STUDENT HELP

Study Tip
"Student Help" boxes throughout the chapter give you study tips and tell you where to look for extra help in this book and on the Internet.

Apply the distributive property. Combine like terms. (Review Examples 2 and 3, p. 101 and Example 5, p. 102)

1. $3x(x + 6)$ — $3x^2 + 18x$
2. $(2 - 4x)7$ — $14 - 28x$
3. $-4(x + 5)$ — $-4x - 20$
4. $(8 - 2x)(-8)$ — $-64 + 16x$
5. $2(x - 3) + x$ — $3x - 6$
6. $5x + 3(1 - x)$ — $2x + 3$
7. $-(x + 4) - x$ — $-2x - 4$
8. $2x^2 - x(1 + x)$ — $x^2 - x$

Simplify the expression. (Review pp. 450 and 451)

9. $x^5 \cdot x^4 \cdot x$ — x^{10}
10. $(x^6)^2$ — x^{12}
11. $(-2ab)^3$ — $-8a^3b^3$
12. $(3xy^4)^2 \cdot x^2$ — $9x^4y^8$

Find the discriminant. Tell if the equation has *two solutions, one solution,* or *no real solution.* (Review Example 1, p. 541)

13. $3x^2 - 4x + 6 = 0$ — −56; no real solution
14. $5x^2 + 6x - 8 = 0$ — 196; two solutions
15. $2x^2 - 12x + 18 = 0$ — 0; one solution

STUDY STRATEGY — *Here's a study strategy!*

Working on Trouble Spots

Look through the chapter and identify your trouble spots: questions that you had difficulty answering or homework exercises that you solved incorrectly. Review the material covering these topics and then try to answer the questions again.

ACTIVITY 10.1

Developing Concepts

Modeling Addition of Polynomials

SET UP
Work with a partner.

MATERIALS
algebra tiles

▶ QUESTION **How can you model the addition of polynomials with algebra tiles?**

▶ EXPLORING THE CONCEPT

Algebra tiles can be used to model polynomials.

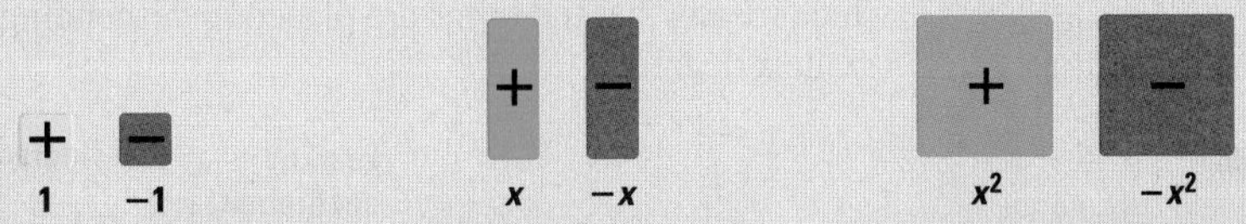

These 1-by-1 square tiles have an area of 1 square unit.

These 1-by-*x* rectangular tiles have an area of *x* square units.

These *x*-by-*x* square tiles have an area of x^2 square units.

1. You can use algebra tiles to add the polynomials $x^2 + 4x + 2$ and $2x^2 - 3x - 1$.

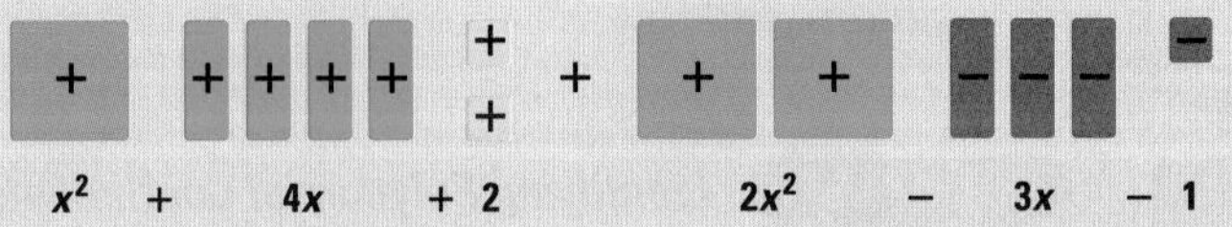

2. To add the polynomials, combine like terms. Group the x^2-tiles, the x-tiles, and the 1-tiles.

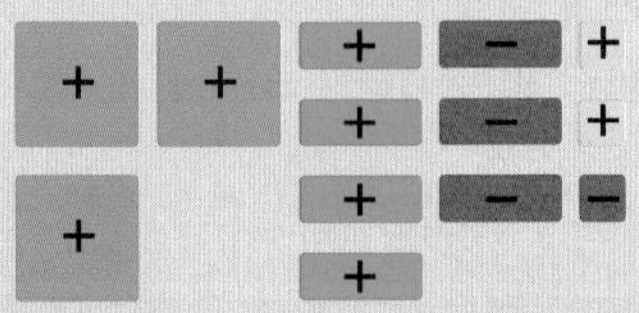

3. Rearrange the tiles to form zero pairs. Remove the zero pairs. The sum is $3x^2 + x + 1$.

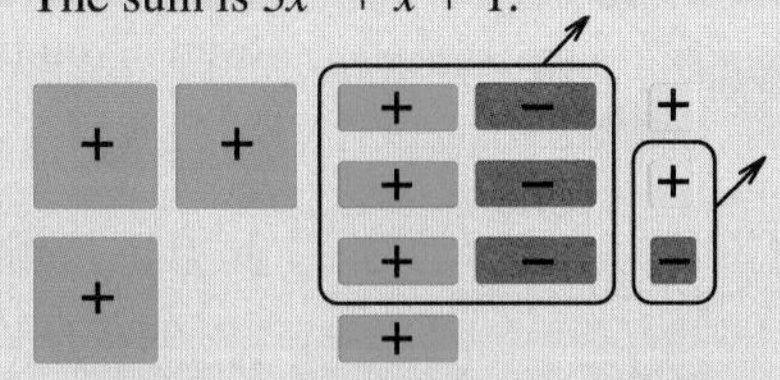

▶ DRAWING CONCLUSIONS

In Exercises 1–6, use algebra tiles to find the sum. Sketch your solution. **Exs. 1–3, 5, 6. See margin for sketches.**

1. $(-x^2 + x - 1) + (4x^2 + 2x - 3)$ **$3x^2 + 3x - 4$**
2. $(3x^2 + 5x - 6) + (-2x^2 - 3x - 6)$ **$x^2 + 2x - 12$**
3. $(5x^2 - 3x + 4) + (-x^2 + 3x - 2)$ **$4x^2 + 2$**
4. $(2x^2 - x - 1) + (-2x^2 + x + 1)$ **0**
5. $(-x^2 + 3x + 7) + (x^2 - 7)$ **$3x$**
6. $(4x^2 + 5) + (4x^2 + 5x)$ **$8x^2 + 5x + 5$**
7. Describe how to use algebra tiles to model *subtraction* of polynomials.

7. *Sample answer:* Use the fact that subtracting a polynomial is the same as adding its opposite. Multiply each term of the second polynomial by −1 and add the resulting polynomial to the first one.

Use algebra tiles to find the difference.

8. $(x^2 + 3x + 4) - (x^2 + 3)$ **$3x + 1$**
9. $(x^2 - 2x + 5) - (3 - 2x)$ **$x^2 + 2$**
10. $(2x^2 + 5) - (-x^2 + 3)$ **$3x^2 + 2$**
11. $(x^2 + 4) - (2x^2 + x)$ **$-x^2 - x + 4$**

1 Planning the Activity

PURPOSE
To model the addition of polynomials using algebra tiles, and to use algebra tiles to find the sum or difference of two polynomials.

MATERIALS
- algebra tiles for each pair of students
- Activity Support Master (*Chapter 10 Resource Book,* p. 12)

PACING
- Exploring the Concept — 5 min
- Drawing Conclusions — 10 min

▶ *LINK TO LESSON*
Before discussing Example 3 in Lesson 10.1, use algebra tiles to demonstrate adding or subtracting two quadratic polynomials.

2 Managing the Activity

COOPERATIVE LEARNING
Have one student model the first polynomial and the other student model the second polynomial. Then the partners can work together to solve the addition or subtraction of the two polynomials.

CLASSROOM MANAGEMENT
Algebra tiles made for the overhead projector can be used to model the polynomials during the Exploring the Concept part of the activity.

3 Closing the Activity

★ *KEY DISCOVERY*
When adding or subtracting polynomials, combine like terms.

ACTIVITY ASSESSMENT
Use algebra tiles to find the sum of $x^2 + 3x + 4$ and $2x^2 - 6x - 1$.
$3x^2 - 3x + 3$

1–3, 5, 6. See Additional Answers beginning on page AA1.

1 PLAN

PACING
Basic: 1 day
Average: 1 day
Advanced: 1 day
Block Schedule: 0.5 block with Ch. 9 Assess

LESSON OPENER APPLICATION
An alternative way to approach Lesson 10.1 is to use the Application Lesson Opener:
- Blackline Master (*Chapter 10 Resource Book,* p. 13)
- Transparency (p. 66)

MEETING INDIVIDUAL NEEDS
- ***Chapter 10 Resource Book***
  Prerequisite Skills Review (p. 5)
  Practice Level A (p. 15)
  Practice Level B (p. 16)
  Practice Level C (p. 17)
  Reteaching with Practice (p. 18)
  Absent Student Catch-Up (p. 20)
  Challenge (p. 22)
- ***Resources in Spanish***
- ***Personal Student Tutor***

NEW-TEACHER SUPPORT
See the Tips for New Teachers on pp. 1–2 of the *Chapter 10 Resource Book* for additional notes about Lesson 10.1.

WARM-UP EXERCISES

Transparency Available

Give the slope and the y-intercept of the graph of the equation.

1. $y = 4x - 2$ $4; -2$
2. $y = \frac{2}{3}x + \frac{4}{3}$ $\frac{2}{3}; \frac{4}{3}$
3. $2x - y = 8$ $2; -8$

Simplify.

4. $2x + 5 - 7x - 7$ $-5x - 2$
5. $7 - (2x + 8) - 3$ $-2x - 4$

10.1 Adding and Subtracting Polynomials

What you should learn

GOAL 1 Add and subtract polynomials.

GOAL 2 Use polynomials to model **real-life** situations, such as energy use in **Exs. 67–69**.

Why you should learn it

▼ To represent **real-life** situations, like mounting a photo on a mat in **Example 5**.

GOAL 1 ADDING AND SUBTRACTING POLYNOMIALS

An expression which is the sum of terms of the form ax^k where k is a nonnegative integer is a **polynomial**. Polynomials are usually written in **standard form**, which means that the terms are placed in descending order, from largest degree to smallest degree.

Polynomial in standard form: $2x^3 + 5x^2 - 4x + 7$ (Leading coefficient: 2; Degree: 3; Constant term: 7)

The **degree** of each term of a polynomial is the exponent of the variable. The **degree of a polynomial** is the largest degree of its terms. When a polynomial is written in standard form, the coefficient of the first term is the **leading coefficient**.

EXAMPLE 1 *Identifying Polynomial Coefficients*

Identify the coefficients of $-4x^2 + x^3 + 3$.

SOLUTION First write the polynomial in standard form. Account for each degree, even if you must use a zero coefficient.

$$-4x^2 + x^3 + 3 = (1)x^3 + (-4)x^2 + (0)x + 3$$

▶ The coefficients are 1, −4, 0, and 3.

....

A polynomial with only one term is called a **monomial**. A polynomial with two terms is called a **binomial**. A polynomial with three terms is called a **trinomial**.

EXAMPLE 2 *Classifying Polynomials*

POLYNOMIAL	DEGREE	CLASSIFIED BY DEGREE	CLASSIFIED BY NUMBER OF TERMS
a. 6	0	constant	monomial
b. $-2x$	1	linear	monomial
c. $3x + 1$	1	linear	binomial
d. $-x^2 + 2x - 5$	2	quadratic	trinomial
e. $4x^3 - 8x$	3	cubic	binomial
f. $2x^4 - 7x^3 - 5x + 1$	4	quartic	polynomial

To add or subtract two polynomials, add or subtract the like terms. You can use a vertical format or a horizontal format.

STUDENT HELP

HOMEWORK HELP Visit our Web site www.mcdougallittell.com for extra examples.

EXAMPLE 3 *Adding Polynomials*

Find the sum. Write the answer in standard form.

a. $(5x^3 - x + 2x^2 + 7) + (3x^2 + 7 - 4x) + (4x^2 - 8 - x^3)$

b. $(2x^2 + x - 5) + (x + x^2 + 6)$

SOLUTION

a. Vertical format: Write each expression in standard form. Align like terms.

$$\begin{array}{rrrr} 5x^3 & +\ 2x^2 & -\ \ x & +\ 7 \\ & 3x^2 & -\ 4x & +\ 7 \\ -x^3 & +\ 4x^2 & & -\ 8 \\ \hline 4x^3 & +\ 9x^2 & -\ 5x & +\ 6 \end{array}$$

b. Horizontal format: Add like terms.

$$(2x^2 + x - 5) + (x + x^2 + 6) = (2x^2 + x^2) + (x + x) + (-5 + 6)$$
$$= 3x^2 + 2x + 1$$

EXAMPLE 4 *Subtracting Polynomials*

Find the difference.

a. $(-2x^3 + 5x^2 - x + 8) - (-2x^3 + 3x - 4)$

b. $(x^2 - 8) - (7x + 4x^2)$

c. $(3x^2 - 5x + 3) - (2x^2 - x - 4)$

SOLUTION

a. Use a vertical format. To subtract, you *add the opposite*. This means that you can multiply each term in the subtracted polynomial by -1 and add.

$$\begin{array}{rr} & (-2x^3 + 5x^2 - x + 8) \\ - & (-2x^3 + 3x - 4) \\ \hline \end{array} \quad \textbf{Add the opposite.} \quad \begin{array}{rrrrr} & -2x^3 & +\ 5x^2 & -\ \ x & +\ 8 \\ + & 2x^3 & & -\ 3x & +\ 4 \\ \hline & & 5x^2 & -\ 4x & +\ 12 \end{array}$$

b.

$$\begin{array}{rr} & (x^2 - 8) \\ - & (7x + 4x^2) \\ \hline \end{array} \quad \textbf{Add the opposite.} \quad \begin{array}{rrrr} & x^2 & & -\ 8 \\ + & -4x^2 & -\ 7x & \\ \hline & -3x^2 & -\ 7x & -\ 8 \end{array}$$

c. Use a horizontal format.

$$(3x^2 - 5x + 3) - (2x^2 - x - 4) = 3x^2 - 5x + 3 - 2x^2 + x + 4$$
$$= (3x^2 - 2x^2) + (-5x + x) + (3 + 4)$$
$$= x^2 - 4x + 7$$

STUDENT HELP

Study Tip
A common mistake in algebra is to forget to change signs correctly when subtracting one expression from another.
~~$(x^2 - 3x) - (2x - 5x + 4)$~~
~~$= x^2 - 3x - 2x - 5x + 4$~~
Wrong signs

2 TEACH

MOTIVATING THE LESSON
Draw pictures like these on the board:

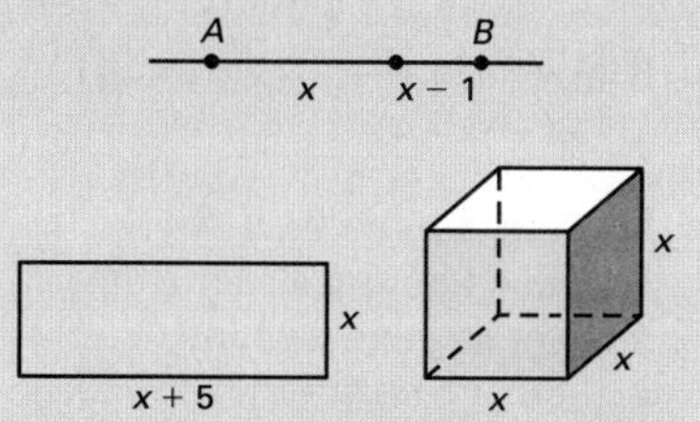

Ask students if they can write algebraic expressions for the length of $\overline{AB}$, the area of the rectangle, and the volume of the cube. Explain that these expressions are *polynomial* expressions, and discuss the differences among them, such as number of terms and types of terms. **($2x - 1$, $x^2 + 5x$, x^3)**

EXTRA EXAMPLE 1
Identify the coefficients of $4 - x + 2x^3$. **2, 0, −1, 4**

EXTRA EXAMPLE 2
Give the degree and classify the polynomial by degree and number of terms.
a. -5 **0; constant; monomial**
b. $\frac{1}{4}x$ **1; linear; monomial**
c. $-9x + 2$ **1; linear; binomial**
d. $x^2 - 6$ **2; quadratic; binomial**
e. $-x^3 + 2x + 1$ **3; cubic; trinomial**
f. $3x^4 + 2x^3 - x^2 + 5x - 8$ **4; quartic; polynomial**

CHECKPOINT EXERCISES
For use after Examples 1 and 2:
1. Identify the coefficients and classify the polynomial $2x + 4 - x^3$. **−1, 0, 2, 4; degree: 3; cubic; trinomial**

STUDENT HELP NOTES
Homework Help Students can find extra examples at **www.mcdougallittell.com** that parallel the examples in the student edition.

EXTRA EXAMPLE 3

Find the sum. Write the answer in standard form.

a. $(-8x^3 + x - 9x^2 + 2) + (8x^2 - 2x + 4) + (4x^2 - 1 - 3x^3)$
$-11x^3 + 3x^2 - x + 5$

b. $(6x^2 - x + 3) + (-2x + x^2 - 7)$
$7x^2 - 3x - 4$

EXTRA EXAMPLE 4

Find the difference. Write the answer in standard form.

a. $(-6x^3 + 5x - 3) - (2x^3 + 4x^2 - 3x + 1)$
$-8x^3 - 4x^2 + 8x - 4$

b. $(4x^2 - 1) - (3x - 2x^2)$
$6x^2 - 3x - 1$

c. $(12x - 8x^2 + 6) - (-8x^2 - 3x + 4)$
$15x + 2$

EXTRA EXAMPLE 5

You are enlarging an 8 inch by 10 inch photo by a scale factor of $2x$ and mounting it on a mat. You want the mat to be 1.5 times as wide as the enlarged photo and 3 inches less than 1.5 times as high as the enlarged photo.

a. Draw a diagram to represent the described situation. Label the dimensions.
See sample answer, below.

b. Write a model for the area of the mat around the photograph as a function of x.
$A = 400x^2 - 72x$

CHECKPOINT EXERCISES

For use after Examples 3 and 4:

1. Simplify the expression.
$(2x^2 + 9x - 4) + (6x - 3x^2 + 1) - (x^2 + x + 1)$ $-2x^2 + 14x - 4$

For use after Example 5:

2. Use the 8 inch by 10 inch photo in Extra Example 5 above. Suppose you want to reduce the picture by a factor of x, where $0 < x < 1$, and are mounting it on a mat. You want the mat to be 1.5 times as wide as the reduced photo and 3 inches less than 1.5 times as high as the reduced photo. Write a model for the area of the mat around the photograph as a function of x.
$100x^2 - 36x$

GOAL 2 USING POLYNOMIALS IN REAL LIFE

REAL LIFE
Geometry

EXAMPLE 5 *Subtracting Polynomials*

You are enlarging a 5-inch by 7-inch photo by a scale factor of x and mounting it on a mat. You want the mat to be twice as wide as the enlarged photo and 2 inches less than twice as high as the enlarged photo.

a. Draw a diagram to represent the described situation. Label the dimensions.

b. Write a model for the area of the mat around the photograph as a function of the scale factor.

SOLUTION

a. Use rectangles to represent the mat and the photo. Use the description of the problem to label the dimensions as shown in the sample diagram below.

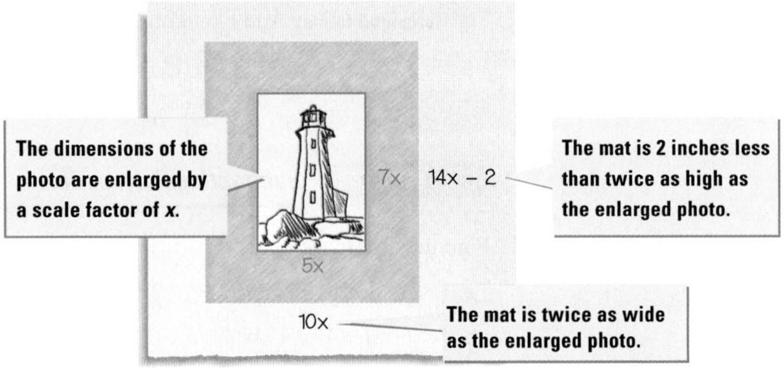

The dimensions of the photo are enlarged by a scale factor of x.
7x
14x − 2
The mat is 2 inches less than twice as high as the enlarged photo.
5x
10x
The mat is twice as wide as the enlarged photo.

b. Use a verbal model. Use the diagram to find expressions for the labels.

PROBLEM SOLVING STRATEGY

VERBAL MODEL — Area of mat = Total area − Area of photo

LABELS
Area of mat $= A$ (square inches)
Total area $= (10x)(14x - 2)$ (square inches)
Area of photo $= (5x)(7x)$ (square inches)

ALGEBRAIC MODEL
$$A = (10x)(14x - 2) - (5x)(7x)$$
$$= 140x^2 - 20x - 35x^2$$
$$= 105x^2 - 20x$$

▶ A model for the area of the mat around the photograph as a function of the scale factor x is $A = 105x^2 - 20x$.

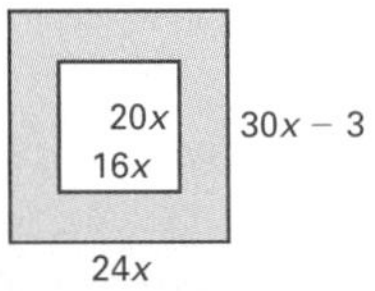

20x
16x
30x − 3
24x

REAL LIFE Internet

EXAMPLE 6 *Adding Polynomials*

From 1991 through 1998, the number of commercial C and education E Internet Web sites can be modeled by the following equations, where t is the number of years since 1991. ▶ Source: Network Wizards

Commercial sites (in millions): $C = 0.321t^2 - 1.036t + 0.698$

Education sites (in millions): $E = 0.099t^2 - 0.120t + 0.295$

Find a model for the total number S of commercial and education sites.

SOLUTION

You can find a model for S by adding the models for C and E.

$$\begin{array}{r} 0.321t^2 - 1.036t + 0.698 \\ +\ 0.099t^2 - 0.120t + 0.295 \\ \hline 0.42t^2 - 1.156t + 0.993 \end{array}$$

▶ The model for the sum is $S = 0.42t^2 - 1.156t + 0.993$.

GUIDED PRACTICE

Vocabulary Check ✓

1. Is $9x^2 + 8x - 4x^3 + 3$ a polynomial with a degree of 2? Explain.
No; in standard form, the polynomial is $-4x^3 + 9x^2 + 8x + 3$; the degree is 3.

Concept Check ✓

In Exercises 2–4, consider the polynomial expression $5x + 6 - 3x^3 - 4x^2$.

2. Write the expression in standard form and name its terms.
$-3x^3 - 4x^2 + 5x + 6$; $-3x^3$, $-4x^2$, $5x$, 6
3. Name the coefficients of the terms. Which is the leading coefficient?
$-3, -4, 5, 6; -3$
4. What is the degree of the polynomial? **3**

ERROR ANALYSIS Describe the error shown.

5.
$$\begin{array}{r} 7x^3 - 3x^2 + 5 \\ +\ 2x^3 - 5x - 7 \\ \hline 9x^3 - 8x^2 - 2 \end{array}$$

5. $-3x^2$ and $-5x$ are not like terms and cannot be added.

6.
$(4x^2 - 9x) - (-8x^2 + 3x - 7)$
$= (4x^2 + 8x^2) + (-9x + 3x) - 7$
$= 12x^2 - 6x - 7$

$-9x - 3x \neq -9x + 3x$ and $0 - (-7) \neq -7$.

Skill Check ✓

Classify the polynomial by degree and by the number of terms.

7. $-9y + 5$ **linear, binomial**
8. $12x^2 + 7x$ **quadratic, binomial**
9. $4w^3 - 8w + 9$ **cubic, trinomial**
10. $\frac{1}{2}x - \frac{3}{4}x^2$ **quadratic, binomial**
11. -4.3 **constant, monomial**
12. $7y + 2y^3 - y^2 + 3y^4$ **quartic, polynomial**

Find the sum or the difference.

13. $(x^2 - 4x + 3) + (3x^2 - 3x - 5)$ **$4x^2 - 7x - 2$**
14. $(-x^2 + 3x - 4) - (2x^2 + x - 1)$ **$-3x^2 + 2x - 3$**
15. $(-3x^2 + x + 8) - (x^2 - 8x + 4)$ **$-4x^2 + 9x + 4$**
16. $(5x^2 - 2x - 1) + (-3x^2 - 6x - 2)$ **$2x^2 - 8x - 3$**
17. $(4x^2 - 2x - 9) + (x - 7 - 5x^2)$ **$-x^2 - x - 16$**
18. $(2x - 3 + 7x^2) - (3 - 9x^2 - 2x)$ **$16x^2 + 4x - 6$**

EXTRA EXAMPLE 6

From 1890 through 1990, the number of men M and women W in the United States labor force can be modeled by the following equations, where t is the number of years since 1890.
Number of men (in millions): $M = 0.0016t^2 + 0.315t + 19.467$
Number of women (in millions): $W = 0.007t^2 - 0.228t + 5.908$
Find a model for the total number S of men and women in the U.S. labor force.
$S = 0.0086t^2 + 0.087t + 25.375$

CHECKPOINT EXERCISES

For use after Example 6:

1. Use the equations in Extra Example 6. Find a model that represents how many more men than women were in the U.S. labor force.
$y = -0.0054t^2 + 0.543t + 13.559$

FOCUS ON VOCABULARY

The vocabulary words in this lesson are all used to describe different kinds of polynomial expressions. To help students understand the vocabulary, write several polynomial expressions and have your students classify them by degree and by the number of terms; then have them state the leading coefficient and the degree.

CLOSURE QUESTION

What are the key ideas when adding and subtracting polynomials?
Sample answer: **Put the terms of the polynomial expressions in order from highest degree to lowest degree; add or subtract only like terms; pay attention to signs and parentheses.**

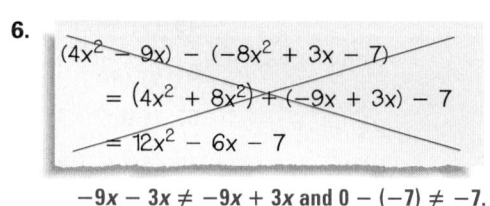

DAILY PUZZLER

A square piece of paper is folded in thirds. If the perimeter of one of the thirds is 24 cm, what is the area of the square? **81 cm²**

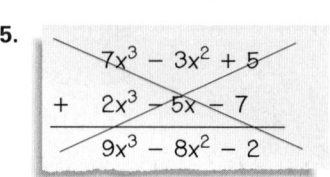

3 APPLY

ASSIGNMENT GUIDE

BASIC
Day 1: pp. 580–582 Exs. 20–62 even, 65, 66, 70, 75, 80, 85

AVERAGE
Day 1: pp. 580–582 Exs. 20–62 even, 63–66, 70, 75, 80, 85

ADVANCED
Day 1: pp. 580–582 Exs. 20–62 even, 63–66, 70–72, 75, 80, 85

BLOCK SCHEDULE
pp. 580–582 Exs. 20–62 even, 63–66, 70, 75, 80, 85 (with Ch. 9 Assess.)

EXERCISE LEVELS

Level A: *Easier*
19–30

Level B: *More Difficult*
31–70, 73–86

Level C: *Most Difficult*
71, 72

✔ HOMEWORK CHECK

To quickly check student understanding of key concepts, go over the following exercises: Exs. 20, 23, 32, 38, 46, 52, 56. See also the Daily Homework Quiz:

- Blackline Master (*Chapter 10 Resource Book,* p. 25)
- Transparency (p. 77)

! COMMON ERROR

EXERCISES 31–62 Students may forget that when they are subtracting a polynomial that is enclosed in parentheses, they must subtract each term inside the parentheses. The end result is that the sign of each term within the parentheses changes. To avoid mistakes, have students include the step of changing the signs of each term and rewriting the subtraction as addition.

PRACTICE AND APPLICATIONS

STUDENT HELP

Extra Practice
to help you master skills is on p. 806.

CLASSIFYING POLYNOMIALS **Identify the leading coefficient, and classify the polynomial by degree and by number of terms.**

19. $-3w + 7$ −3; linear, binomial

20. $-4x^2 + 2x - 1$ −4; quadratic, trinomial

21. $8 + 5t^2 - 3t + t^3$ 1; cubic, polynomial

22. $8 + 5y^2 - 3y$ 5; quadratic trinomial

23. -6 −6; constant, monomial

24. $14w^4 + 9w^2$ 14; quartic, binomial

25. $-\frac{2}{3}x + 5x^4 - \frac{5}{6}$ 5; quartic, trinomial

26. $-4.1b^2 + 7.4b^3$ 7.4; cubic, binomial

27. $-9t^2 + 3t^3 - 4t^4 - 15$ −4; quartic, polynomial

28. $9y^3 - 5y^2 + 4y - 1$ 9; cubic, polynomial

29. $-16x^3$ −16; cubic, monomial

30. $-8z^2 + 74 + 39z - 95z^4$ −95; quartic, polynomial

VERTICAL FORMAT **Use a vertical format to add or subtract.**

31. $(12x^3 + 10) - (18x^3 - 3x^2 + 6)$ $-6x^3 + 3x^2 + 4$

32. $(a + 3a^2 + 2a^3) - (a^4 - a^3)$ $-a^4 + 3a^3 + 3a^2 + a$

33. $(2m - 8m^2 - 3) + (m^2 + 5m)$ $-7m^2 + 7m - 3$

34. $(8y^2 + 2) + (5 - 3y^2)$ $5y^2 + 7$

35. $(3x^2 + 7x - 6) - (3x^2 + 7x)$ -6

36. $(4x^2 - 7x + 2) + (-x^2 + x - 2)$ $3x^2 - 6x$

37. $(8y^3 + 4y^2 + 3y - 7) + (2y^2 - 6y + 4)$ $8y^3 + 6y^2 - 3y - 3$

38. $(7x^4 - x^2 + 3x) - (x^3 + 6x^2 - 2x + 9)$ $7x^4 - x^3 - 7x^2 + 5x - 9$

HORIZONTAL FORMAT **Use a horizontal format to add or subtract.**

39. $(x^2 - 7) + (2x^2 + 2)$ $3x^2 - 5$

40. $(-3a^2 + 5) + (-a^2 + 4a - 6)$ $-4a^2 + 4a - 1$

41. $(x^3 + x^2 + 1) - x^2$ $x^3 + 1$

42. $12 - (y^3 + 4)$ $-y^3 + 8$

43. $(3n^3 + 2n - 7) - (n^3 - n - 2)$ $2n^3 + 3n - 5$

44. $(3a^3 - 4a^2 + 3) - (a^3 + 3a^2 - a - 4)$ $2a^3 - 7a^2 + a + 7$

45. $(6b^4 - 3b^3 - 7b^2 + 9b + 3) + (4b^4 - 6b^2 + 11b - 7)$ $10b^4 - 3b^3 - 13b^2 + 20b - 4$

46. $(x^3 - 6x) - (2x^3 + 9) - (4x^2 + x^3)$ $-2x^3 - 4x^2 - 6x - 9$

POLYNOMIAL ADDITION AND SUBTRACTION **Use a vertical format or a horizontal format to add or subtract.**

47. $(9x^3 + 12) + (16x^3 - 4x + 2)$ $25x^3 - 4x + 14$

48. $(-2t^4 + 6t^2 + 5) - (-2t^4 + 5t^2 + 1)$ $t^2 + 4$

49. $(3x + 2x^2 - 4) - (x^2 + x - 6)$ $x^2 + 2x + 2$

50. $(u^3 - u) - (u^2 + 5)$ $u^3 - u^2 - u - 5$

51. $(-7x^2 + 12) - (6 - 4x^2)$ $-3x^2 + 6$

52. $(10x^3 + 2x^2 - 11) + (9x^2 + 2x - 1)$ $10x^3 + 11x^2 + 2x - 12$

53. $(-9z^3 - 3z) + (13z - 8z^2)$ $-9z^3 - 8z^2 + 10z$

54. $(21t^4 - 3t^2 + 43) - (19t^3 + 33t - 58)$ $21t^4 - 19t^3 - 3t^2 - 33t + 101$

55. $(6t^2 - 19t) - (3 - 2t^2) - (8t^2 - 5)$ $-19t + 2$

56. $(7y^2 + 15y) + (5 - 15y + y^2) + (24 - 17y^2)$ $-9y^2 + 29$

57. $\left(x^4 - \frac{1}{2}x^2\right) + \left(x^3 + \frac{1}{3}x^2\right) + \left(\frac{1}{4}x^2 - 9\right)$ $x^4 + x^3 + \frac{1}{12}x^2 - 9$

58. $(10w^3 + 20w^2 - 55w + 60) + (-25w^2 + 15w - 10) + (-5w^2 + 10w - 20)$ $10w^3 - 10w^2 - 30w + 30$

59. $(9x^4 - x^2 + 7x) + (x^3 - 6x^2 + 2x - 9) - (4x^3 + 3x + 8)$ $9x^4 - 3x^3 - 7x^2 + 6x - 17$

60. $(6.2b^4 - 3.1b + 8.5) + (-4.7 + 5.8b^2 - 2.4b^4)$ $3.8b^4 + 5.8b^2 - 3.1b + 3.8$

61. $(-3.8y^3 + 6.9y^2 - y + 6.3) - (-3.1y^3 + 2.9y - 4.1)$ $-0.7y^3 + 6.9y^2 - 3.9y + 10.4$

62. $\left(\frac{2}{5}a^4 - 2a + 7\right) - \left(-\frac{3}{10}a^4 + 6a^3\right) - (2a^2 - 7)$ $\frac{7}{10}a^4 - 6a^3 - 2a^2 - 2a + 14$

STUDENT HELP

HOMEWORK HELP

Example 1: Exs. 19–30
Example 2: Exs. 19–30
Example 3: Exs. 31–62
Example 4: Exs. 31–62
Example 5: Exs. 63, 64
Example 6: Exs. 65–69

BUILDING A HOUSE In Exercises 63 and 64, you plan to build a house that is $1\frac{1}{2}$ times as long as it is wide. You want the land around the house to be 20 feet wider than the width of the house, and twice as long as the length of the house, as shown at the right.

63. Write an expression for the area of the land surrounding the house. **$1.5x^2 + 60x$**

64. If $x = 30$ feet, what is the area of the house? What is the area of the entire property? **1350 ft^2; 4500 ft^2**

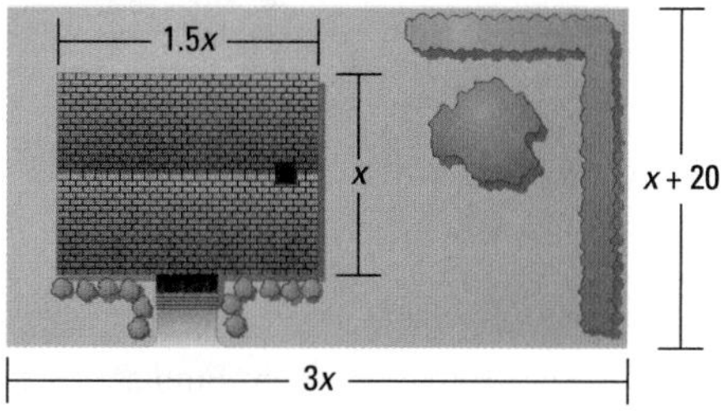

POPULATION In Exercises 65 and 66, use the following information. Projected from 1950 through 2010, the total population P and the male population M of the United States (in thousands) can be modeled by the following equations, where t is the number of years since 1950.

DATA UPDATE of U.S. Bureau of the Census data at www.mcdougallittell.com

Total population model: $P = 2387.74t + 155{,}211.46$

Male population model: $M = 1164.16t + 75{,}622.43$

65. Find a model that represents the female population F of the United States from 1950 through 2010. **$F = 1223.58t + 79{,}589.03$**

66. For the year 2010, the value of P is 298,475.86 and the value of M is 145,472.03. Use these figures to predict the female population in 2010. **153,003,830**

ENERGY USE In Exercises 67–69, use the following information. From 1989 through 1993, the amounts (in billions of dollars) spent on natural gas N and electricity E by United States residents can be modeled by the following equations, where t is the number of years since 1989.

▶ Source: U.S. Energy Information Administration

Gas spending model: $N = 1.488t^2 - 3.403t + 65.590$

Electricity spending model: $E = -0.107t^2 + 6.897t + 169.735$

67. Find a model for the total amount A (in billions of dollars) spent on natural gas *and* electricity by United States residents from 1989 through 1993. **$A = 1.381t^2 + 3.494t + 235.325$**

68. According to the models, will more money be spent on natural gas or on electricity in 2020? **natural gas**

69. The graph at the right shows U.S. energy spending starting in 1989. Models N, E, and A are shown. Copy the graph and label the models N, E, and A. **From top to bottom the models are A, E, and N.**

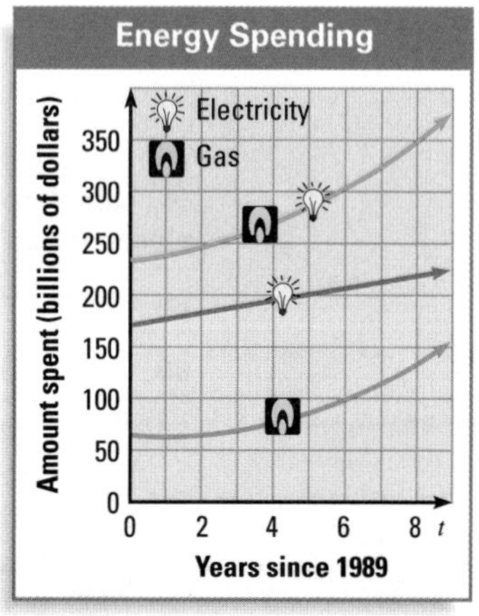

FOCUS ON CAREERS

ELECTRICAL ENGINEERS design, test, and monitor the performance of electrical equipment. This includes equipment used by power utilities to generate and transmit electricity.

CAREER LINK www.mcdougallittell.com

CAREER NOTE

EXERCISES 67–69 Additional information about a career as an electrical engineer is available at **www.mcdougallittell.com.**

ADDITIONAL PRACTICE AND RETEACHING

For Lesson 10.1:

- Practice Levels A, B, and C (*Chapter 10 Resource Book,* p. 15)
- Reteaching with Practice (*Chapter 10 Resource Book,* p. 18)
- See Lesson 10.1 of the *Personal Student Tutor*

For more Mixed Review:

- Search the *Test and Practice Generator* for key words or specific lessons.

4 ASSESS

DAILY HOMEWORK QUIZ

Transparency Available

Identify the leading coefficient; then classify the polynomial by degree and number of terms.

1. $7 - 3n + 5n^2$
5; quadratic, trinomial

2. $-3x$ **−3; linear, monomial**

Add or subtract.

3. $(4x^2 + 6) + (-x^2 + 3x - 4)$
$3x^2 + 3x + 2$

4. $(x^3 - 2x^2 + 3x - 5) - (x^3 - 3x^2 + 6)$ $x^2 + 3x - 11$

5. $(b^4 - 2b^3 - b + 5) + (3b^3 + b - 5)$
$b^4 + b^3$

6. $(8y^2 + 3y) - (6y^2 - 5y - 10)$
$2y^2 + 8y + 10$

EXTRA CHALLENGE NOTE

Challenge problems for Lesson 10.1 are available in **blackline** format in the *Chapter 10 Resource Book,* p. 22 and at **www.mcdougallittell.com.**

ADDITIONAL TEST PREPARATION

1. WRITING Explain the meaning of *like terms.* ***Sample answer:*** **terms that contain the same variable(s) with the same exponent(s)**

2. OPEN ENDED When adding ax^2 and dx^2, what will be the coefficient of the sum? $a + d$

79.

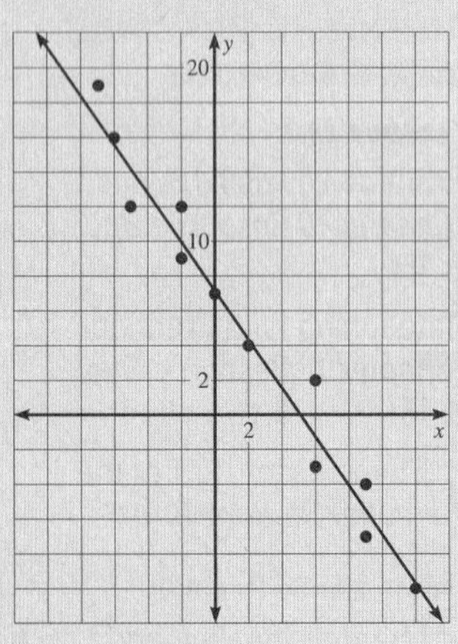

$y = -1.42x + 7.35$

Test Preparation

70. MULTI-STEP PROBLEM The table below shows the amounts that Megan and Sara plan to deposit in their savings accounts to buy a used car. Their savings accounts have the same annual growth rate g.

Date	1/1/00	1/1/01	1/1/02	1/1/03
Megan	\$250	\$400	\$170	\$625
Sara	\$475	\$50	\$300	\$540

a. On January 1, 2003, the value of Megan's account M can be modeled by $M = 250g^3 + 400g^2 + 170g + 625$, where g is the annual growth rate. Find a model for the value of Sara's account S on January 1, 2003.
$S = 475g^3 + 50g^2 + 300g + 540$

b. Find a model for the total value of Megan's and Sara's accounts together on January 1, 2003. $T = 725g^3 + 450g^2 + 470g + 1165$

c. The annual growth rate g is equal to $1 + r$, where r is the annual interest rate. The annual interest rate on both accounts is 2.5% for the three-year period. Find the combined value of the two accounts on January 1, 2003.
\$2900.28

d. If the used car that Megan and Sara want to buy costs \$2500, will they have enough money? **yes**

71. An odd number is a number than can be written in the form $2x + 1$, where x is an integer. Since $x + (x + 1) = 2x + 1$ and x is an integer, the sum of two consecutive integers is odd.

★ Challenge

71. The sum of any two consecutive integers can be written as $(x) + (x + 1)$. Show that the sum of any two consecutive integers is always odd. **See margin.**

EXTRA CHALLENGE
www.mcdougallittell.com

72. Use algebra to show that the sum of any four consecutive integers is always even. **Four consecutive integers can be written x, $x + 1$, $x + 2$ and $x + 3$. Their sum is $x + (x + 1) + (x + 2) + (x + 3) = 4x + 6 = 2(2x + 3)$. Since x is an integer, $2x + 3$ is an integer, and $2(2x + 3)$ is even. Then the sum of four consecutive integers is even.**

MIXED REVIEW

DISTRIBUTIVE PROPERTY Simplify the expression. (Review 2.6 for 10.2)

73. $-3(x + 1) - 2$ $-3x - 5$ **74.** $(2x - 1)(2) + x$ $5x - 2$ **75.** $11x + 3(8 - x)$ $8x + 24$

76. $(5x - 1)(-3) + 6$ $-15x + 9$ **77.** $-4(1 - x) + 7$ $3 + 4x$ **78.** $-12x - 5(11 - x)$ $-7x - 55$

79. *Sample Answer*: $y = -1.42x + 7.35$.

79. BEST-FITTING LINES Draw a scatter plot. Then draw a line that approximates the data and write an equation of the line. (Review 5.4) **See margin for graph.**

$(-7, 19), (-6, 16), (-5, 12), (-2, 12), (-2, 9), (0, 7),$

$(2, 4), (6, -3), (6, 2), (9, -4), (9, -7), (12, -10)$

EXPONENTIAL EXPRESSIONS In Exercises 80–85, simplify. Then use a calculator to evaluate the expression. Round the result to two decimal places when appropriate. (Review 8.1)

80. $(4 \cdot 3^2 \cdot 2^3)^4$ $4^4 \cdot 3^8 \cdot 2^{12}$; **6,879,707,136** **81.** $(2^4 \cdot 2^4)^2$ 2^{16}; **65,536** **82.** $(-6 \cdot 3^4)^3$ $-216 \cdot 3^{12}$; **−114,791,256**

83. $(1.1 \cdot 3.3)^3$ $(1.1^3)(3.3^3)$; **47.83** **84.** $5.5^3 \cdot 5.5^4$ 5.5^7; **152,243.52** **85.** $(2.9^3)^5$ 2.9^{15}; **8,629,188.75**

86. 1.30; *Sample answer*: The ratio of the population in 2025 to the population in 2000 is about 1.23, or 1.0104^{20} times the ratio of the population in 2000 to the population in 1995.

86. ALABAMA The population P of Alabama (in thousands) for 1995 projected through 2025 can be modeled by $P = 4227(1.0104)^t$, where t is the number of years since 1995. Find the ratio of the population in 2025 to the population in 2000. Compare this ratio with the ratio of the population in 2000 to the population in 1995. ▶ Source: U.S. Bureau of the Census (Review 8.3)

ACTIVITY 10.1

Using Technology

Graphing Polynomial Functions

You can use a graphing calculator or a computer to check an answer when finding the sum or difference of polynomials.

STUDENT HELP

INTERNET KEYSTROKE HELP

See keystrokes for several models of calculators at www.mcdougallittell.com

▶ EXAMPLE

One of the expressions below is the difference of $2x^2 + 3x - 1$ and $-x^2 - 2x + 3$. Use a graphing calculator to check which is correct.

A. $x^2 + 5x - 4$ **B.** $3x^2 + 5x - 4$

▶ SOLUTION

1 Enter $(2x^2 + 3x - 1) - (-x^2 - 2x + 3)$ as equation Y1.

Enter $x^2 + 5x - 4$ as equation Y2.

Enter $3x^2 + 5x - 4$ as equation Y3.

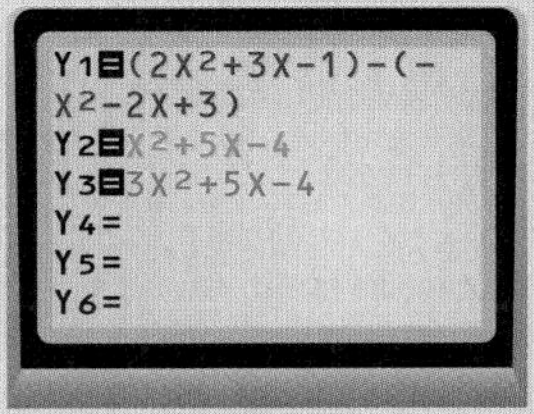

2 Graph equations Y1 and Y2 on the same screen.

The graphs do not coincide, so

$(2x^2 + 3x - 1) - (-x^2 - 2x + 3)$

is not equal to $x^2 + 5x - 4$.

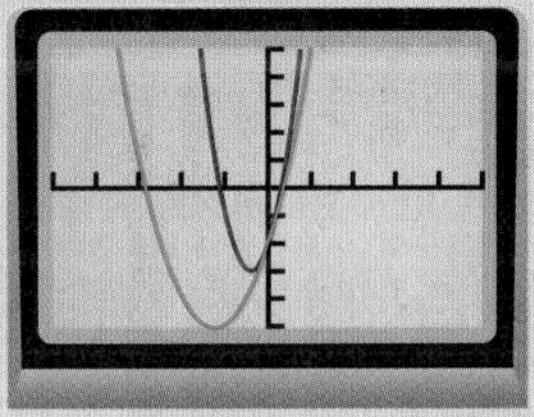

3 Graph equations Y1 and Y3 on the same screen.

The graphs coincide, so

$(2x^2 + 3x - 1) - (-x^2 - 2x + 3)$

equals $3x^2 + 5x - 4$.

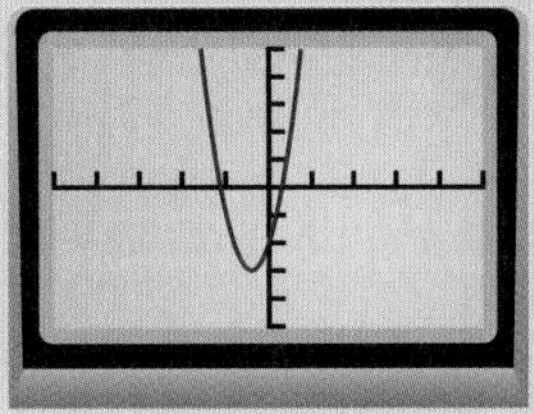

▶ EXERCISES

Tell whether the given answer is a correct sum or difference. If it is incorrect, find the correct answer.

1. $(2x^2 - 3x + 5) + (-x^2 + 2x - 6) \stackrel{?}{=} x^2 - x - 1$ **correct**
2. $(-x^2 - 3x - 1) - (-2x^2 + 4x + 5) \stackrel{?}{=} -3x^2 + x + 4$ **incorrect; $x^2 - 7x - 6$**

Find the sum or difference. Then use a graphing calculator or a computer to check your answer.

3. $(2x^2 - 6x - 3) + (x^2 - 3x + 3)$ **$3x^2 - 9x$**
4. $(x^2 - 14x + 5) + (-2x^2 - 3x + 2)$ **$-x^2 - 17x + 7$**
5. $(x^2 + 12x + 6) - (3x^2 - 2x + 2)$ **$-2x^2 + 14x + 4$**
6. $(-3x^2 + 5x - 8) - (x^2 - 5x - 8)$ **$-4x^2 + 10x$**
7. $(2x^2 + 10x + 3) + (4x^2 + 2x - 4)$ **$6x^2 + 12x - 1$**
8. $(x^2 - 4x + 5) - (-2x^2 - 3x + 7)$ **$3x^2 - x - 2$**

Activity Assessment *Sample answer:*
When the algebraic solution and the expression involving the sum or difference of the two polynomials are graphed together, the two graphs will coincide if the two expressions are equal.

1 Planning the Activity

PURPOSE
To use a graphing calculator to check if two polynomials were added or subtracted correctly.

MATERIALS
- graphing calculator
- Keystroke blackline (*Chapter 10 Resource Book,* p. 14)

PACING
- Exploring the Concept — 5 min
- Drawing Conclusions — 10 min

▶ *LINK TO LESSON*
Have students use a graphing calculator to check the solutions in Examples 3 and 4 on page 577.

2 Managing the Activity

ALTERNATIVE APPROACH
Instead of entering the sum or difference into Y1, students can enter the first polynomial equation into Y1 and the second polynomial equation into Y2. Then, in Y3, use the VARS key and go under Y-VARS to enter Y1 + Y2 or Y1 – Y2. The equations given as solutions can be entered into Y4 or Y5 of the Y = list. Then graph only the polynomial in Y3 and the polynomial solution that you are checking to see if they coincide.

3 Closing the Activity

★ ***KEY DISCOVERY***
The graph of a possible solution of a polynomial sum or difference and the graph of the expression involving the sum or difference will coincide if the two polynomials were added or subtracted correctly.

ACTIVITY ASSESSMENT
JOURNAL Explain how a graphing calculator can show if two polynomials were added or subtracted correctly.
See sample answer at left.

1 PLAN

PACING
Basic: 1 day
Average: 1 day
Advanced: 1 day
Block Schedule: 0.5 block with 10.3

LESSON OPENER
VISUAL APPROACH
An alternative way to approach Lesson 10.2 is to use the Visual Approach Lesson Opener:

- Blackline Master (*Chapter 10 Resource Book,* p. 26)
- Transparency (p. 67)

MEETING INDIVIDUAL NEEDS
- ***Chapter 10 Resource Book***
  Prerequisite Skills Review (p. 5)
  Practice Level A (p. 29)
  Practice Level B (p. 30)
  Practice Level C (p. 31)
  Reteaching with Practice (p. 32)
  Absent Student Catch-Up (p. 34)
  Challenge (p. 36)
- ***Resources in Spanish***
- ***Personal Student Tutor***

NEW-TEACHER SUPPORT
See the Tips for New Teachers on pp. 1–2 of the *Chapter 10 Resource Book* for additional notes about Lesson 10.2.

WARM-UP EXERCISES

Transparency Available

Simplify the expression.

1. $2(x+4)$ $2x+8$
2. $-3(2-5x)$ $15x-6$
3. $7(12+3x)-10$ $21x+74$
4. $(3x)(-4x)$ $-12x^2$
5. $3x+(-4x)$ $-x$

ENGLISH LEARNERS
The FOIL pattern requires knowledge of the English-language word each letter represents. If students have trouble using FOIL, encourage them to come up with a similar acronym in their first language.

10.2 Multiplying Polynomials

What you should learn

GOAL 1 Multiply two polynomials.

GOAL 2 Use polynomial multiplication in **real-life** situations, such as calculating the area of a window in **Example 5**.

Why you should learn it

▼ To solve **real-life** problems such as finding a model for movie ticket revenue in **Exs. 58** and **59**.

REAL LIFE

GOAL 1 MULTIPLYING POLYNOMIALS

In Lesson 2.6 you learned how to multiply a polynomial by a monomial by using the distributive property.

$$(3x)(2x^2 - 5x + 3) = (3x)(2x^2) + (3x)(-5x) + (3x)(3) = 6x^3 - 15x^2 + 9x$$

In the following activity, you will see how an area model illustrates the multiplication of two binomials.

ACTIVITY Developing Concepts

Investigating Binomial Multiplication

The rectangle shown at the right has a width of $(x + 2)$ and a height of $(2x + 1)$.

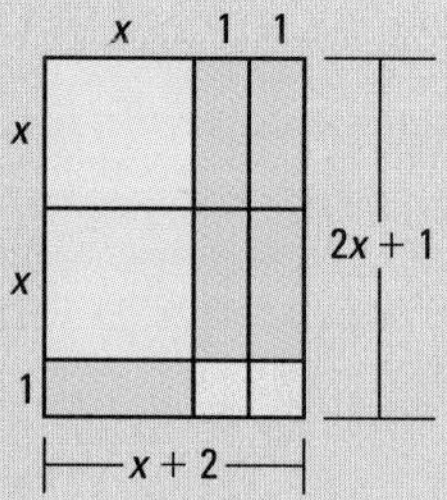

1. Copy the model. What is the area of each part of the rectangle? **The area of the blue part is $2x^2$, the area of the green part is $5x$, and the area of the yellow part is 2.**
2. Find the product of $(x + 2)$ and $(2x + 1)$ by adding the areas of the parts to get an expression for total area. **$2x^2 + 5x + 2$**
3. Copy and complete the equation: $(x + 2)(2x + 1) = \underline{?}$. **$(x + 2)(2x + 1) = 2x^2 + 5x + 2$**
4. Use an area model to multiply $(x + 3)$ and $(2x + 4)$. **$2x^2 + 10x + 12$**

Another way to multiply two binomials is to use the distributive property *twice.*

First use $(a + b)(c + d) = a(c + d) + b(c + d)$.

Then use $a(c + d) = ac + ad$ and $b(c + d) = bc + bd$.

This shows that $(a + b)(c + d) = ac + ad + bc + bd$.

This property can also be applied to binomials of the form $a - b$ or $c - d$.

EXAMPLE 1 ***Using the Distributive Property***

Find the product $(x + 2)(x - 3)$.

SOLUTION

$$(x + 2)(x - 3) = x(x - 3) + 2(x - 3) \quad (b + c)a = ba + ca$$
$$= x^2 - 3x + 2x - 6 \quad a(b - c) = ab - ac$$
$$= x^2 - x - 6 \quad \text{Combine like terms.}$$

STUDENT HELP

Look Back
For help with the distributive property, see p. 100.

When you multiply two binomials, you can remember the results given by the distributive property by means of the **FOIL** pattern. Multiply the **F**irst, **O**uter, **I**nner, and **L**ast terms.

Product of First terms | Product of Outer terms | Product of Inner terms | Product of Last terms

$$(3x + 4)(x + 5) = 3x^2 + 15x + 4x + 20$$
$$= 3x^2 + 19x + 20$$

STUDENT HELP

INTERNET HOMEWORK HELP
Visit our Web site www.mcdougallittell.com for extra examples.

EXAMPLE 2 *Multiplying Binomials Using the FOIL Pattern*

F O I L

$$(3x - 4)(2x + 1) = 6x^2 + 3x - 8x - 4 \quad \textbf{Mental math}$$
$$= 6x^2 - 5x - 4 \quad \textbf{Simplify.}$$

....

To multiply two polynomials that have three or more terms, remember that *each term of one polynomial must be multiplied by each term of the other polynomial.* Use a vertical or a horizontal format. Write each polynomial in standard form.

EXAMPLE 3 *Multiplying Polynomials Vertically*

Find the product $(x - 2)(5 + 3x - x^2)$.

SOLUTION Align like terms in columns.

$-x^2 + 3x + 5$	**Standard form**
$\times \quad x - 2$	**Standard form**
$2x^2 - 6x - 10$	← $-2(-x^2 + 3x + 5)$
$-x^3 + 3x^2 + 5x$	← $x(-x^2 + 3x + 5)$
$-x^3 + 5x^2 - x - 10$	**Combine like terms.**

EXAMPLE 4 *Multiplying Polynomials Horizontally*

Find the product $(4x^2 - 3x - 1)(2x - 5)$.

SOLUTION Multiply $2x - 5$ by each term of $4x^2 - 3x - 1$.

$$(4x^2 - 3x - 1)(2x - 5) = 4x^2(2x - 5) - 3x(2x - 5) - 1(2x - 5)$$
$$= 8x^3 - 20x^2 - 6x^2 + 15x - 2x + 5$$
$$= 8x^3 - 26x^2 + 13x + 5$$

STUDENT HELP

Look Back
For help with multiplying exponential expressions, see p. 451.

2 TEACH

MOTIVATING THE LESSON
Write the polynomial $(x + 4)(x + 5)$ and the polynomial $x^2 + 20$. Take a poll in your class as to how many students think the two polynomials are equal. Ask how you can decide whether they are equal or not equal. Use substitution by putting a value in for x ($x \neq 0$) in both expressions. This shows that the two expressions are not equal, and that you cannot multiply polynomials just by multiplying like terms.

EXTRA EXAMPLE 1
Find the product $(x + 8)(x - 7)$.
$x^2 + x - 56$

EXTRA EXAMPLE 2
Find the product $(2x + 3)(5x + 1)$.
$10x^2 + 17x + 3$

CHECKPOINT EXERCISES
For use after Examples 1 and 2:
1. Find the product $(3x + 4)(x - 3)$. $3x^2 - 5x - 12$

STUDENT HELP NOTES

- **Homework Help** Students can find extra examples at **www.mcdougallittell.com** that parallel the examples in the student edition.
- **Look Back** As students look back to page 100 for help with the distributive property, remind them that the term outside of the parentheses must be multiplied with each term inside the parentheses, and that the terms are separated by "+" and "–" signs.
- **Look Back** As students look back to page 451 for help with multiplying expressions, remind them to add the exponents for variables that are the same, and multiply coefficients. Also, remind them that only like terms can be added or subtracted, but unlike terms can be multiplied or divided.

EXTRA EXAMPLE 3
Find the product
$(x - 4)(5x + 9 - 2x^2)$.
$-2x^3 + 13x^2 - 11x - 36$

EXTRA EXAMPLE 4
Find the product
$(5x^2 - x - 3)(6x - 5)$.
$30x^3 - 31x^2 - 13x + 15$

EXTRA EXAMPLE 5
The glass portion of a sliding glass door has a height-to-width ratio of 2:1. The framework adds 10 inches to the width and 12 inches to the height.
a. Write a polynomial expression that represents the total area of the window, including the framework. $2x^2 + 32x + 120$
b. Find the area when $x = 8, 9, 10, 11$, and 12. 504 in.^{2}; 570 in.^{2}; 640 in.^{2}; 714 in.^{2}; 792 in.^{2}

CHECKPOINT EXERCISES

For use after Examples 3 and 4:

1. Find the product $(2x - 1)(3x^2 - x - 4)$.
$6x^3 - 5x^2 - 7x + 4$

For use after Example 5:

2. You want to make a table top where the length to width ratio is 5:4. The trim of the table top adds 2 inches to the width and 1 inch to the length.
a. Write a polynomial expression that represents the total area of the table top including the trim.
$20x^2 + 14x + 2$
b. Find the area when $x = 3$.
224 in.^{2}

CLOSURE QUESTION
When multiplying two binomials, how is using the distributive property twice equivalent to using the FOIL method for multiplying?
See sample answer, below.

DAILY PUZZLER
Two numbers are mirror numbers if one can be obtained from the other by reversing the order of the digits, for example, 123 and 321. Find two mirror numbers whose product is 53,703. **153 and 351**

GOAL 2 USING POLYNOMIALS IN REAL LIFE

REAL LIFE *Carpentry*

EXAMPLE 5 ***Multiplying Binomials to Find an Area***

The diagram at the right shows the basic dimensions for a window. The glass portion of the window has a height-to-width ratio of 3 : 2. The framework adds 6 inches to the width and 10 inches to the height.

a. Write a polynomial expression that represents the total area of the window, including the framework.

b. Find the area when $x = 10, 11, 12, 13$, and 14.

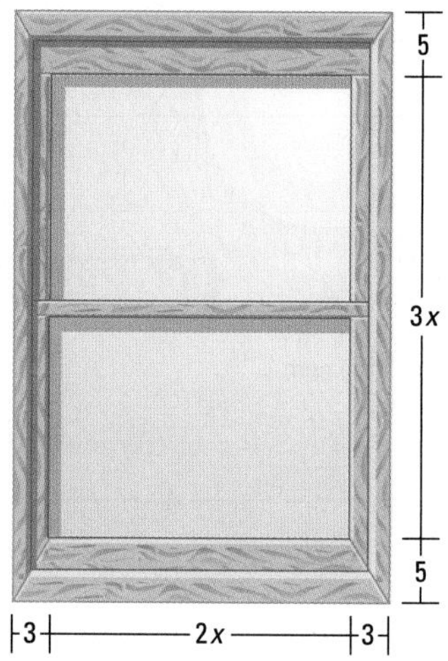

SOLUTION

a. Use a verbal model.

PROBLEM SOLVING STRATEGY

VERBAL MODEL Total area = Height of window · Width of window

LABELS Total area = A (square inches)

Height of window = $3x + 10$ (inches)

Width of window = $2x + 6$ (inches)

ALGEBRAIC MODEL

$A = (3x + 10) \cdot (2x + 6)$ **Area model**

$= 6x^2 + 18x + 20x + 60$ **FOIL pattern**

$= 6x^2 + 38x + 60$ **Combine like terms.**

b. You can evaluate the polynomial expression $6x^2 + 38x + 60$ by substituting x-values. For instance, to find the total area of the window when $x = 10$, substitute 10 for x.

$A = 6x^2 + 38x + 60$

$= 6(\mathbf{10})^2 + 38(\mathbf{10}) + 60$

$= 600 + 380 + 60$

$= 1040$

▶ The areas for all five x-values are listed in the table.

x (in.)	10	11	12	13	14
A (in.^{2})	1040	1204	1380	1568	1768

Closure Question *Sample answer:*
Using the distributive property twice results in multiplying the firsts, the outers, the inners, and the lasts, and then adding these products. This is the same as using the FOIL method.

GUIDED PRACTICE

Vocabulary Check ✓

1. How do the letters in "FOIL" help you remember how to multiply two binomials? **See margin.**

Concept Check ✓

2. Write an equation that represents the product of two binomials as shown in the area model at the right.
$(x + 1)(2x + 1) = 2x^2 + 3x + 1$

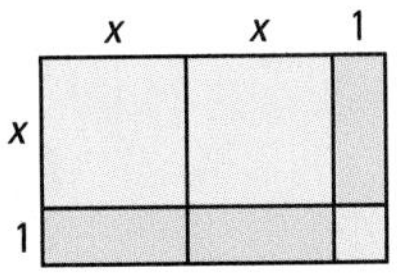

Explain how to use the distributive property to find the product. **3–4. See margin.**

3. $(2x - 3)(x + 4)$
4. $(x + 1)(x^2 - x + 1)$

Skill Check ✓

Use the given method to find the product $(2x + 3)(4x + 1)$.

5. Use an area model.
$8x^2 + 14x + 3$. See margin.
6. Use the distributive property.
$2x(4x + 1) + 3(4x + 1) = 8x^2 + 2x + 12x + 3 = 8x^2 + 14x + 3$
7. Use the FOIL pattern.
$2x(4x) + 2x(1) + 3(4x) + 3(1) = 8x^2 + 2x + 12x + 3 = 8x^2 + 14x + 3$
8. Which of the methods you used in Exercises 5–7 do you prefer? Explain.
Preferences may vary. For example, visual learners might prefer the area model.

Find the product.

9. $(4x + 7)(-2x)$ **$-8x^2 - 14x$**
10. $-4x^2(3x^2 + 2x - 6)$ **$-12x^4 - 8x^3 + 24x^2$**
11. $(-y - 2)(y + 8)$ **$-y^2 - 10y - 16$**
12. $(w - 3)(2w + 5)$ **$2w^2 - w - 15$**
13. $(x + 6)(x + 9)$ **$x^2 + 15x + 54$**
14. $(-2x - 4)(8x + 3)$ **$-16x^2 - 38x - 12$**
15. $(b + 8)(6 - 2b)$ **$-2b^2 - 10b + 48$**
16. $(-4y + 5)(-7 - 3y)$ **$12y^2 + 13y - 35$**
17. $(3w^2 - 9)(5w - 1)$ **$15w^3 - 3w^2 - 45w + 9$**

1. "F" stands for first: multiply the two first terms; "O" stands for outer: next, multiply the two outer terms; "I" stands for inner; next, multiply the two inner terms; "L" stands for last: multiply the two last terms. Finally, add the four products.

3. Multiply $2x(x + 4)$, then $(-3)(x + 4)$, and add the products.

4. Multiply $x(x^2 - x + 1)$, then $1(x^2 - x + 1)$, and add the products.

PRACTICE AND APPLICATIONS

STUDENT HELP

Extra Practice to help you master skills is on p. 806.

MULTIPLYING EXPRESSIONS **Find the product.**

18. $(2x - 5)(-4x)$ **$-8x^2 + 20x$**
19. $3t^2(7t - t^3 - 3)$ **$21t^3 - 3t^5 - 9t^2$**
20. $2x(x^2 - 8x + 1)$ **$2x^3 - 16x^2 + 2x$**
21. $(-y)(6y^2 + 5y)$ **$-6y^3 - 5y^2$**
22. $4w^2(3w^3 - 2w^2 - w)$ **$12w^5 - 8w^4 - 4w^3$**
23. $-b^2(6b^3 - 16b + 11)$ **$-6b^5 + 16b^3 - 11b^2$**

DISTRIBUTIVE PROPERTY **Use the distributive property to find the product.**

24. $(t + 8)(t + 5)$ **$t^2 + 13t + 40$**
25. $(2d + 3)(3d + 1)$ **$6d^2 + 11d + 3$**
26. $(4y^2 + y - 7)(2y - 1)$ **See margin.**
27. $(3s^2 - s - 1)(s + 2)$ **$3s^3 + 5s^2 - 3s - 2$**
28. $(a^2 + 8)(a^2 - a - 3)$ **$a^4 - a^3 + 5a^2 - 8a - 24$**
29. $(x + 6)(x^2 - 6x - 2)$ **$x^3 - 38x - 12$**

USING THE FOIL PATTERN **Use the FOIL pattern to find the product.**

30. $(4q - 1)(3q + 8)$ **$12q^2 + 29q - 8$**
31. $(2z + 7)(3z + 2)$ **$6z^2 + 25z + 14$**
32. $(x + 6)(x - 6)$ **$x^2 - 36$**
33. $(2w - 5)(w + 5)$ **$2w^2 + 5w - 25$**
34. $(x - 9)(2x + 15)$ **$2x^2 - 3x - 135$**
35. $(5t - 3)(2t + 3)$ **$10t^2 + 9t - 9$**

MULTIPLYING EXPRESSIONS **Find the product.**

36. $(d - 5)(d + 3)$ **$d^2 - 2d - 15$**
37. $(4x + 1)(x - 8)$ **$4x^2 - 31x - 8$**
38. $(3b - 1)(b - 9)$ **$3b^2 - 28b + 9$**
39. $(9w + 8)(11w - 10)$ **$99w^2 - 2w - 80$**
40. $(11t - 30)(5t - 21)$ **$55t^2 - 381t + 630$**
41. $(9.4y - 5.1)(7.3y - 12.2)$ **See margin.**
42. $(3x + 4)\left(\frac{2}{3}x + 1\right)$ **See margin.**
43. $\left(n + \frac{6}{5}\right)(4n - 10)$ **See margin.**
44. $\left(x + \frac{1}{8}\right)\left(x - \frac{9}{8}\right)$ **See margin.**
45. $(2.5z - 6.1)(z + 4.3)$ **$2.5z^2 + 4.65z - 26.23$**
46. $(t^2 + 6t - 8)(t - 6)$ **$t^3 - 44t + 48$**
47. $(-4s^2 + s - 1)(s + 4)$ **$-4s^3 - 15s^2 + 3s - 4$**

26. $8y^3 - 2y^2 - 15y + 7$

41. $68.62y^2 - 151.91y + 62.22$

42. $2x^2 + \frac{17}{3}x + 4$

43. $4n^2 - \frac{26}{5}n - 12$

44. $x^2 - x - \frac{9}{64}$

STUDENT HELP

HOMEWORK HELP
Example 1: Exs. 18–29
Example 2: Exs. 30–35
Example 3: Exs. 36–47
Example 4: Exs. 36–47
Example 5: Exs. 52–55

3 APPLY

ASSIGNMENT GUIDE

BASIC
Day 1: pp. 587–589 Exs. 18–46 even, 54, 55, 60, 65, 70, 73, 75, 80, 85

AVERAGE
Day 1: pp. 587–589 Exs. 18–46 even, 54–57, 60, 65, 70, 73, 75, 80, 85

ADVANCED
Day 1: pp. 587–589 Exs. 18–46 even, 54–57, 60, 61, 65, 70, 73, 75, 80, 85

BLOCK SCHEDULE WITH 10.3
pp. 587–589 Exs. 18–46 even, 54–57, 60, 65, 70, 73, 75, 80, 85

EXERCISE LEVELS

Level A: *Easier*
18–23

Level B: *More Difficult*
24–60, 62–85

Level C: *Most Difficult*
61

✓ HOMEWORK CHECK

To quickly check student understanding of key concepts, go over the following exercises: Exs. 20, 24, 28, 30, 32, 42, 54, 55. See also the Daily Homework Quiz:

- Blackline Master (*Chapter 10 Resource Book*, p. 39)
- Transparency (p. 78)

! COMMON ERROR

EXERCISES 30–35 When multiplying binomials, students may be confused about when to multiply and when to add. Stress the words behind the meaning of FOIL, the *product* of the first terms *plus* the *product* of the outer terms *plus* the *product* of the inner terms *plus* the *product* of the last terms. Writing FOIL as F + O + I + L may also help.

5. See Additional Answers beginning on page AA1.

Graphing Calculator Note

Exercises 48–51 Students can review the method of checking equations using a graphing calculator by looking back to Activity 10.1 on page 583.

60a.

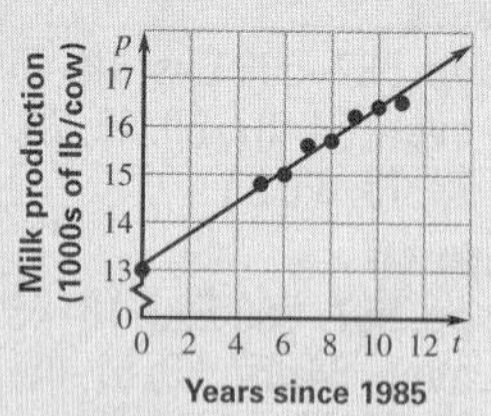

b.

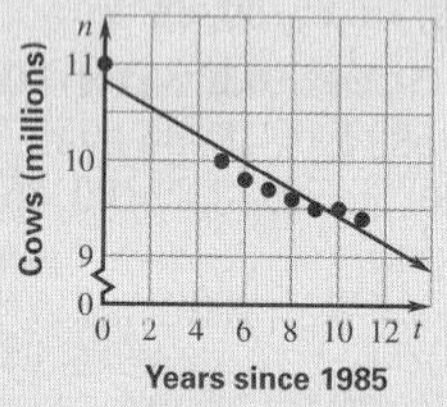

Additional Practice and Reteaching

For Lesson 10.2:

- Practice Levels A, B, and C (*Chapter 10 Resource Book*, p. 29)
- Reteaching with Practice (*Chapter 10 Resource Book*, p. 32)
- See Lesson 10.2 of the *Personal Student Tutor*

For more Mixed Review:

- Search the *Test and Practice Generator* for key words or specific lessons.

Using a Graph to Check **Find the product. Check your result by comparing a graph of the given expression with a graph of the product.**

48. $(x + 5)(x + 4)$ $x^2 + 9x + 20$

49. $(2x + 1)(x - 6)$ $2x^2 - 11x - 6$

50. $(3x^2 - 8x - 1)(9x + 4)$ $27x^3 - 60x^2 - 41x - 4$

51. $(x + 7)(-x^2 - 6x + 2)$ $-x^3 - 13x^2 - 40x + 14$

Geometry Connection **Find an expression for the area of the figure. (*Hint:* the *Table of Formulas* is on p. 813.) Give your answer as a quadratic polynomial.**

52.

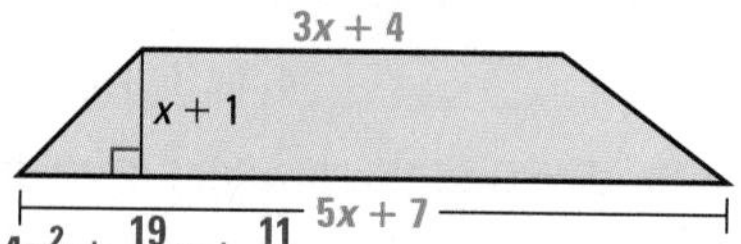

$4x^2 + \frac{19}{2}x + \frac{11}{2}$

53.

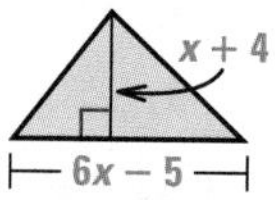

$3x^2 + \frac{19}{2}x - 10$

Football **In Exercises 54 and 55, a football field's dimensions can be represented by a width of $\left(\frac{1}{2}x + 10\right)$ feet and a length of $\left(\frac{5}{4}x - 15\right)$ feet.**

54. Find an expression for the area A of a football field. Give your answer as a quadratic trinomial.

54. $\frac{5}{8}x^2 + 5x - 150$

55. An actual football field is 160 feet wide and 360 feet long. For what value of x does the expression you wrote in Exercise 54 give these dimensions? 300

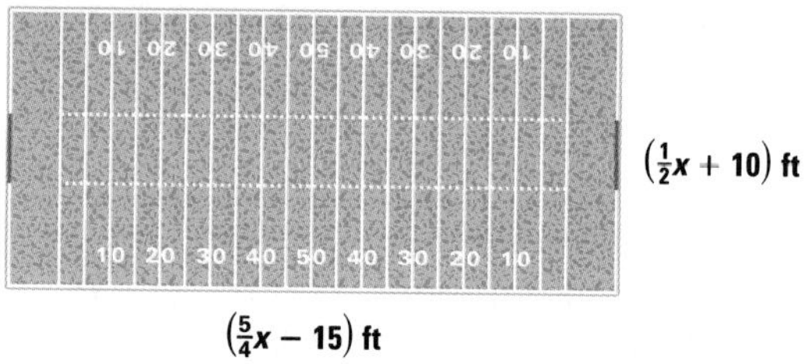

Videocassettes **In Exercises 56 and 57, use the following information about videocassette sales from 1987 to 1996, where *t* is the number of years since 1987.**

The annual number of blank videocassettes B sold in the United States can be modeled by $B = 15t + 281$, where B is measured in millions. The wholesale price P for a videocassette can be modeled by $P = -0.21t + 3.52$, where P is measured in dollars. ▶ Source: EIA Market Research Department

56. Find a model for the revenue from annual sales of blank videocassettes. Give the model as a quadratic trinomial. $R = -3.15t^2 - 6.21t + 989.12$, where R is measured in millions of dollars.

57. Describe what happens to revenue during the period from 1987 to 1996. The revenue decreases from $989.12 million to $678.08 million.

Student Help

Application Link www.mcdougallittell.com

Box Office **In Exercises 58 and 59, use the following information about movie theater admissions from 1980 to 1996, where *t* is the number of years since 1980.**

The annual number of admissions D into movie theaters can be modeled by $D = 13.75t + 1057.36$, where D is measured in millions. The admission price P can be modeled by $P = 0.11t + 2.90$, where P is the price per person.

▶ Source: Motion Picture Association of America

58. Find a model for the annual box office revenue. Give the model as a quadratic trinomial. $R = 1.5125t^2 + 156.1846t + 3066.344$, where R is measured in millions of dollars.

59. What conclusions can you make from your model? *Sample answer:* As time goes on, the revenue will increase, because the value of the model $1.5125t^2 + 156.1846t + 3066.344$ will increase as t increases.

Test Preparation

60. **MULTI-STEP PROBLEM** Use the table, which gives annual data on cows and milk production (in pounds) in the United States from 1985 to 1996, where t is the number of years since 1985. **Sample answers are given.**

Year (t)	0	5	6	7	8	9	10	11
Milk production (1000s of lb/cow)	13	14.8	15	15.6	15.7	16.2	16.4	16.5
Millions of cows	11	10	9.8	9.7	9.6	9.5	9.5	9.4

DATA UPDATE of U.S. Department of Agriculture data at www.mcdougallittell.com

a. Make a scatter plot of the milk production data. **See margin.**

b. Make a scatter plot of the annual data on the number of cows. **See margin.**

c. Write a linear model for each set of data. (Use the line-fitting technique in Lesson 5.4.) $p = 0.33t + 13.09$; $n = -0.14t + 10.82$

d. Find a model for the total amount of milk produced. Check to see whether the model matches the information in the table below. $A = -0.0462t^2 + 1.738t + 141.6338$

Year (t)	0	5	6	7	8	9	10	11
U.S. milk production (billion lb)	143	148	148	151	151	154	155	154

▶ Source: U.S. Department of Agriculture

e. What conclusions can you make based on the information? **See margin.**

60.e. *Sample answer:* Even though the number of cows is decreasing, the actual milk production is increasing because the milk production per cow is increasing.

★ **Challenge**

61. a. **PATTERNS** Find each product: $(x - 1)(x + 1)$, $(x - 1)(x^2 + x + 1)$, and $(x - 1)(x^3 + x^2 + x + 1)$. Find a pattern in the results. **See margin.**

b. Use the pattern to predict the product $(x - 1)(x^4 + x^3 + x^2 + x + 1)$. Verify your guess by multiplying or graphing. $x^5 - 1$

61. $x^2 - 1$, $x^3 - 1$, $x^4 - 1$; $(x - 1)(x^n + x^{n-1} + x^{n-2} + \ldots + x + 1) = x^{n+1} - 1$

EXTRA CHALLENGE

www.mcdougallittell.com

MIXED REVIEW

SIMPLIFYING EXPRESSIONS **Simplify.** (Review 8.1 for 10.3)

62. $(7x)^2$ $49x^2$ **63.** $\left(\frac{1}{3}m\right)^2$ $\frac{1}{9}m^2$ **64.** $\left(\frac{2}{5}y\right)^2$ $\frac{4}{25}y^2$ **65.** $(0.5w)^2$ $0.25w^2$

66. $9^3 \cdot 9^5$ **67.** $(4^2)^4$ **68.** $b^2 \cdot b^5$ b^7 **69.** $(4c^2)^4$ $256c^8$

70. $(2t)^4 \cdot 3^3$ $432t^4$ **71.** $(-w^4)^3$ $-w^{12}$ **72.** $(-3xy)^3(2y)^2$ $-108x^3y^5$ **73.** $(8x^2y^8)^3$ $512x^6y^{24}$

66. 9^8, or 43,046,721

67. 4^8, or 65,536

FINDING SOLUTIONS **Tell how many solutions the equation has.** (Review 9.6)

74. $2x^2 - 3x - 1 = 0$ two solutions **75.** $4x^2 + 4x + 1 = 0$ one solution **76.** $3x^2 - 7x + 5 = 0$ no real solution

77. $7x^2 - 8x - 6 = 0$ two solutions **78.** $10x^2 - 13x - 9 = 0$ two solutions **79.** $6x^2 - 12x - 6 = 0$ two solutions

SKETCHING GRAPHS **Sketch the graph of the inequality.** (Review 9.7)

80–85. See margin.

80. $y \geq 4x^2 - 7x$ **81.** $y < x^2 - 3x - 10$ **82.** $y > -2x^2 + 4x + 16$

83. $y < 6x^2 - 1$ **84.** $y \geq x^2 - 3x + 1$ **85.** $y \leq 8x^2 - 3$

4 ASSESS

DAILY HOMEWORK QUIZ

Transparency Available

Find the product.

1. $(2x)(8 - x - 3x^2)$ $16x - 2x^2 - 6x^3$

2. $(3t + 5)(t - 3)$ $3t^2 - 4t - 15$

3. $(y - 4)(y + 4)$ $y^2 - 16$

4. $(2x + 10)\left(\frac{1}{2}x + 2\right)$ $x^2 + 9x + 20$

5. Write a polynomial expression for the area of the rectangle.

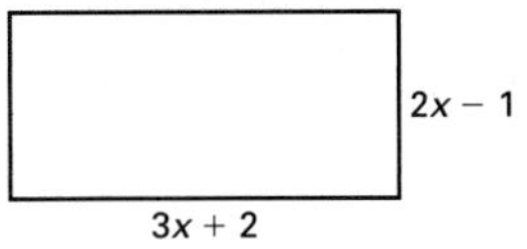

$6x^2 + x - 2$

EXTRA CHALLENGE NOTE

Challenge problems for Lesson 10.2 are available in **blackline** format in the *Chapter 10 Resource Book*, p. 36 and at **www.mcdougallittell.com.**

ADDITIONAL TEST PREPARATION

1. WRITING Explain how to multiply two binomial factors.
Sample answer: **Multiply one term in the first factor by both terms in the second factor. Then multiply the second term in the first factor by both terms in the second factor. Combine any like terms.**

80–85. See Additional Answers beginning on page AA1.

1 PLAN

PACING

Basic: 2 days
Average: 2 days
Advanced: 2 days
Block Schedule: 0.5 block with 10.2
0.5 block with 10.4

LESSON OPENER
APPLICATION

An alternative way to approach Lesson 10.3 is to use the Application Lesson Opener:

- Blackline Master (*Chapter 10 Resource Book*, p. 40)
- Transparency (p. 68)

MEETING INDIVIDUAL NEEDS

- ***Chapter 10 Resource Book***
  Prerequisite Skills Review (p. 5)
  Practice Level A (p. 41)
  Practice Level B (p. 42)
  Practice Level C (p. 43)
  Reteaching with Practice (p. 44)
  Absent Student Catch-Up (p. 46)
  Challenge (p. 50)
- ***Resources in Spanish***
- ***Personal Student Tutor***

NEW-TEACHER SUPPORT

See the Tips for New Teachers on pp. 1–2 of the *Chapter 10 Resource Book* for additional notes about Lesson 10.3.

WARM-UP EXERCISES

Transparency Available

Simplify the expression.

1. $(2x)^2$ $4x^2$
2. $(-8m)^2$ $64m^2$
3. $\left(\frac{1}{4}y\right)^2$ $\frac{1}{16}y^2$
4. $(b^3)^2$ b^6
5. $(3n^2)^4$ $81n^8$

10.3 Special Products of Polynomials

What you should learn

GOAL 1 Use special product patterns for the product of a sum and a difference, and for the square of a binomial.

GOAL 2 Use special products as **real-life** models such as a genetic model in **Example 5**.

Why you should learn it

▼ To solve **real-life** problems like finding the percent of second-generation tigers that are white in **Exs. 49 and 50**.

REAL LIFE

Step 2.

1. The factors in the expression are a sum and difference of two terms. The product is the difference of the squares of the two terms.
2. The expression is the square of a sum of two terms. The product is the square of the first plus twice the product of the terms plus the square of the second.
3. The expression is the square of a difference of two terms. The product is the square of the first minus twice the product of the terms plus the square of the second.

GOAL 1 USING SPECIAL PRODUCT PATTERNS

As you learned in Lesson 10.2, you can always use the FOIL pattern to multiply two binomials. Some pairs of binomials have *special products*. If you learn to recognize these pairs, finding the product of two binomials will sometimes be quicker and easier.

ACTIVITY Developing Concepts

Investigating Special Product Patterns

1. Use the FOIL pattern to find the products in each group.

1. $(x-2)(x+2)$	**2.** $(x+3)^2$	**3.** $(z-2)^2$
$(2n+3)(2n-3)$	$(3m+1)^2$	$(6x-4)^2$
$(4t-1)(4t+1)$	$(5s+2)^2$	$(5p-7)^2$
$(x+y)(x-y)$	$(x+y)^2$	$(x-y)^2$

2. Describe any patterns that you see in each group. **See margin.**

Step 1. 1. x^2-4; $4n^2-9$; $16t^2-1$; x^2-y^2
2. x^2+6x+9; $9m^2+6m+1$; $25s^2+20s+4$; $x^2+2xy+y^2$
3. z^2-4z+4; $36x^2-48x+16$; $25p^2-70p+49$; $x^2-2xy+y^2$

In this activity, you may have noticed that the product of the sum and difference of two terms has no "middle term." Also, you may have noticed that in the square of a binomial the middle term is twice the product of the terms in the binomial.

SPECIAL PRODUCT PATTERNS

SUM AND DIFFERENCE PATTERN

$(a+b)(a-b)=a^2-b^2$ **Example:** $(3x-4)(3x+4)=9x^2-16$

SQUARE OF A BINOMIAL PATTERN

$(a+b)^2=a^2+2ab+b^2$ **Example:** $(x+4)^2=x^2+8x+16$

$(a-b)^2=a^2-2ab+b^2$ **Example:** $(2x-6)^2=4x^2-24x+36$

The area model shown at the right gives a geometric representation of the *square of a binomial* pattern $(a+b)^2=a^2+2ab+b^2$.

The area of the large square is $(a+b)^2$, which is equal to the sum of the areas of the two small squares and two rectangles. Note that the two rectangles with area ab produce the middle term $2ab$.

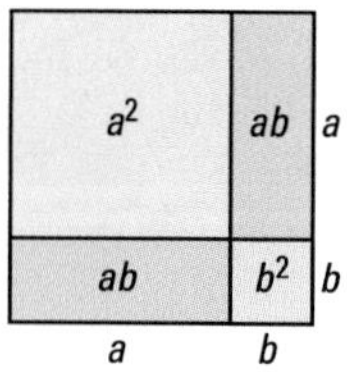

STUDENT HELP

Study Tip
When you use these special product patterns, remember that *a* and *b* can be numbers, variables, or sometimes variable expressions.

EXAMPLE 1 Using the Sum and Difference Pattern

Find the product $(5t - 2)(5t + 2)$.

SOLUTION

$(a - b)(a + b) = a^2 - b^2$ Write pattern.

$(5t - 2)(5t + 2) = (5t)^2 - 2^2$ Apply pattern.

$= 25t^2 - 4$ Simplify.

✓**CHECK** You can use the FOIL pattern to check your answer.

$(5t - 2)(5t + 2) = (5t)(5t) + (5t)(2) + (-2)(5t) + (-2)(2)$ Use FOIL.

$= 25t^2 - 4$ Simplify.

EXAMPLE 2 Squaring a Binomial

Find the product.

a. $(3n + 4)^2$ **b.** $(2x - 7y)^2$

SOLUTION

a. $(a + b)^2 = a^2 + 2ab + b^2$ Write pattern.

$(3n + 4)^2 = (3n)^2 + 2(3n)(4) + 4^2$ Apply pattern.

$= 9n^2 + 24n + 16$ Simplify.

b. $(a - b)^2 = a^2 - 2ab + b^2$ Write pattern.

$(2x - 7y)^2 = (2x)^2 - 2(2x)(7y) + (7y)^2$ Apply pattern.

$= 4x^2 - 28xy + 49y^2$ Simplify.

....

The special product patterns can help you use mental math to find products.

EXAMPLE 3 Special Products and Mental Math

Use mental math to find the product.

a. $17 \cdot 23$ **b.** 29^2

SOLUTION

a. $17 \cdot 23 = (20 - 3)(20 + 3)$ Write as product of difference and sum.

$= 400 - 9$ Apply pattern.

$= 391$ Simplify.

b. $29^2 = (30 - 1)^2$ Write as square of binomial.

$= 900 - 60 + 1$ Apply pattern.

$= 841$ Simplify.

2 TEACH

MOTIVATING THE LESSON

Use area models similar to those used in Example 4 of this lesson to find the area of a square with side lengths of x, then $x + 1$, then $x + 2$, and so on if necessary. Then discuss their areas. The first square has an area of x^2. The second square can be broken up into a square with area x^2, 2 rectangles with area x, and a square with area 1. The total area then is $x^2 + 2x + 1$. Complete this with the other squares, and use it to lead into the "square" of a binomial pattern.

ACTIVITY NOTE

When students are completing Step 2 of the activity, focus their attention and descriptions on the use of addition and subtraction in the binomials, as well as the type of polynomial that you start with or end with (that is, binomial or trinomial). The titles of the special product patterns involve these ideas.

EXTRA EXAMPLE 1

Find the product $(3b - 5)(3b + 5)$. **$9b^2 - 25$**

EXTRA EXAMPLE 2

Find the product.

a. $(7a + 2)^2$ **$49a^2 + 28a + 4$**

b. $(2p - 5q)^2$ **$4p^2 - 20pq + 25q^2$**

EXTRA EXAMPLE 3

Use mental math to find the product.

a. $19 \cdot 21$ **399**

b. 38^2 **1444**

CHECKPOINT EXERCISES

For use after Example 1:

1. Find the product $(2x - 1)(2x + 1)$. **$4x^2 - 1$**

For use after Example 2:

2. Find the product $(3x + 5y)^2$. **$9x^2 + 30xy + 25y^2$**

For use after Example 3:

3. Use mental math to find the product.

a. $18 \cdot 22$ **396**

b. 42^2 **1764**

EXTRA EXAMPLE 4
Find an expression for the area of the unshaded region.

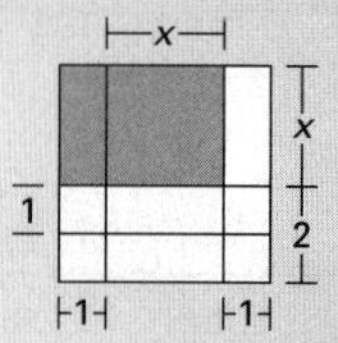

$3x + 4$

EXTRA EXAMPLE 5
Two green snapdragons, each with one yellow gene and one blue gene, are crossed. Each parent passes along only one gene for color to its offspring. Show how the square of a binomial can be used to model the Punnett square. $(0.5Y + 0.5B)^2 = (0.5Y)^2 + 2(0.5Y)(0.5B) + (0.5B)^2 = 0.25Y^2 + 0.5YB + 0.25B^2$; 25% yellow, 50% green, and 25% blue

CHECKPOINT EXERCISES

For use after Example 4:

4. Find an expression for the area of the unshaded region.

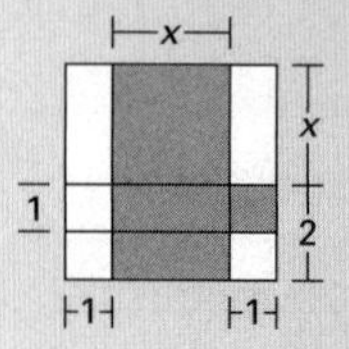

$2x + 3$

For use after Example 5:

5. Two people with wavy hair and having one gene for curly hair and one gene for straight hair become parents. Show how the square of a binomial can be used to model the Punnett square for the type of hair their offspring will have. $(0.5C + 0.5S)^2 = (0.5C)^2 + 2(0.5C)(0.5S) + (0.5S)^2 = 0.25C^2 + 0.5CS + 0.25S^2$; 25% curly, 50% wavy, and 25% straight

CLOSURE QUESTION
Write the pattern for the square of a binomial when the binomial involves subtraction. $(a - b)^2 = a^2 - 2ab + b^2$

GOAL 2 APPLYING SPECIAL PRODUCTS

EXAMPLE 4 *Finding an Area*

GEOMETRY CONNECTION Find an expression for the area of the blue region.

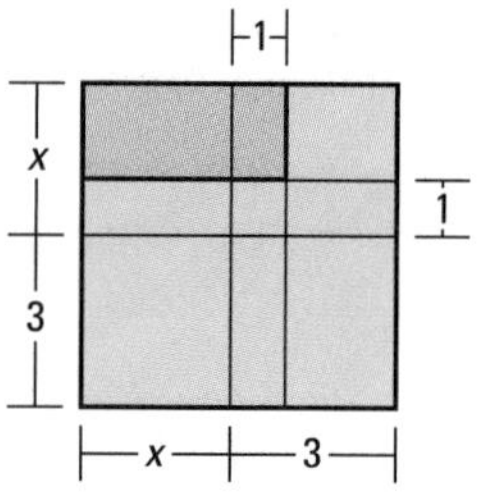

SOLUTION

PROBLEM SOLVING STRATEGY

VERBAL MODEL Area of blue region = Area of entire square − Area of red region

LABELS

Area of blue region = A (square units)

Area of entire square = $(x + 3)^2$ (square units)

Area of red region = $(x + 1)(x - 1)$ (square units)

ALGEBRAIC MODEL

$A = (x + 3)^2 - (x + 1)(x - 1)$ — Write algebraic model.

$= (x^2 + 6x + 9) - (x^2 - 1)$ — Apply pattern.

$= x^2 + 6x + 9 - x^2 + 1$ — Use distributive property.

$= 6x + 10$ — Simplify.

EXAMPLE 5 *Modeling a Punnett Square*

SCIENCE CONNECTION The Punnett square at the right is an area model that shows the possible results of crossing two pink snapdragons, each with one red gene R and one white gene W. Each parent snapdragon passes along only one gene for color to its offspring. Show how the square of a binomial can be used to model the Punnett square.

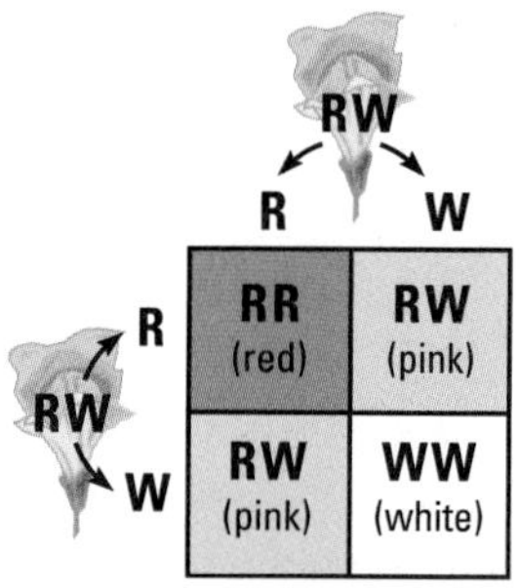

FOCUS ON APPLICATIONS

REAL LIFE **PUNNETT SQUARES** are used in genetics to model the mixing of parents' genes in the resulting offspring.

SOLUTION

Each parent snapdragon has half red genes and half white genes. You can model the genetic makeup of each parent as $0.5R + 0.5W$. The genetic makeup of the offspring can be modeled by the product $(0.5R + 0.5W)^2$.

$$(0.5R + 0.5W)^2 = (0.5R)^2 + 2(0.5R)(0.5W) + (0.5W)^2$$
$$= 0.25R^2 + 0.5RW + 0.25W^2$$

Red — Pink — White

▶ 25% of the offspring will be red, 50% will be pink, and 25% will be white.

GUIDED PRACTICE

Vocabulary Check ✓

Concept Check ✓

1. The square of the sum and difference of two terms is the difference of the squares of the terms.

Skill Check ✓

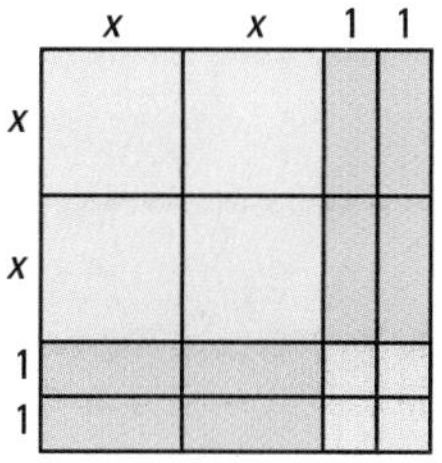

1. What is the sum and difference pattern? **See margin.**
2. Write two expressions for the product shown by the area model at the right. $(2x + 2)(2x + 2)$, $4x^2 + 8x + 4$
3. Tell whether the following statement is *true* or *false*. *The product of* $(a + b)$ *and* $(a + b)$ *is* $a^2 + b^2$. Explain. **False; the product of $(a + b)$ and $(a + b)$ is $a^2 + 2ab + b^2$.**
4. Find the missing term: $(a - b)^2 = a^2 - \boxed{?} + b^2$. **$2ab$**
5. Show the product $(x + 2)^2$ using an area model or algebra tiles. Draw a sketch of the model and use it to help you write $(x + 2)^2$ as a trinomial. **$x^2 + 4x + 4$. See margin.**

Use a special product pattern to find the product.

6. $(x - 6)^2$ **$x^2 - 12x + 36$**
7. $(w + 11)(w - 11)$ **$w^2 - 121$**
8. $(6 + p)^2$ **$36 + 12p + p^2$**
9. $(2y - 3)^2$ **$4y^2 - 12y + 9$**
10. $(3z + 2)^2$ **$9z^2 + 12z + 4$**
11. $(t - 6)(t + 6)$ **$t^2 - 36$**
12. $(x - 3)(x - 3)$ **$x^2 - 6x + 9$**
13. $(2y + 5)(2y - 5)$ **$4y^2 - 25$**
14. $(4n + 3)^2$ **$16n^2 + 24n + 9$**

PRACTICE AND APPLICATIONS

STUDENT HELP

Extra Practice to help you master skills is on p. 806.

SUM AND DIFFERENCE PATTERN **Write the product of the sum and difference.**

15. $(x + 3)(x - 3)$ **$x^2 - 9$**
16. $(y - 1)(y + 1)$ **$y^2 - 1$**
17. $(2m + 2)(2m - 2)$ **$4m^2 - 4$**
18. $(3b - 1)(3b + 1)$ **$9b^2 - 1$**
19. $(3 + 2x)(3 - 2x)$ **$9 - 4x^2$**
20. $(6 - 5n)(6 + 5n)$ **$36 - 25n^2$**

SQUARE OF A BINOMIAL **Write the square of the binomial as a trinomial.**

21. $(x + 5)^2$ **$x^2 + 10x + 25$**
22. $(a + 8)^2$ **$a^2 + 16a + 64$**
23. $(3t + 1)^2$ **$9t^2 + 6t + 1$**
24. $(2s - 4)^2$ **$4s^2 - 16s + 16$**
25. $(4b - 3)^2$ **$16b^2 - 24b + 9$**
26. $(x - 7)^2$ **$x^2 - 14x + 49$**

SPECIAL PRODUCT PATTERNS **Find the product.**

27. $(x + 4)(x - 4)$ **$x^2 - 16$**
28. $(x - 3)(x + 3)$ **$x^2 - 9$**
29. $(3x + 1)(3x - 1)$ **$9x^2 - 1$**
30. $(6x + 5)(6x - 5)$ **$36x^2 - 25$**
31. $(a + 2b)(a - 2b)$ **$a^2 - 4b^2$**
32. $(4n - 8m)(4n + 8m)$ **$16n^2 - 64m^2$**
33. $(3y + 8)^2$ **$9y^2 + 48y + 64$**
34. $(9 - 4t)(9 + 4t)$ **$81 - 16t^2$**
35. $\left(2x + \frac{1}{2}\right)\left(2x - \frac{1}{2}\right)$ **$4x^2 - \frac{1}{4}$**
36. $(-5 - 4x)^2$ **$25 + 40x + 16x^2$**
37. $(3s + 4t)(3s - 4t)$ **$9s^2 - 16t^2$**
38. $(-a - 2b)^2$ **$a^2 + 4ab + 4b^2$**

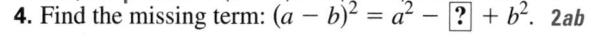

CHECKING SOLUTIONS **Tell whether the statement is *true* or *false*. If the statement is false, rewrite the right-hand side to make the statement true.**

39. $(9x + 8)(9x - 8) \stackrel{?}{=} 81x^2 - 64$ **true**
40. $(6y - 7w)^2 \stackrel{?}{=} 36y^2 - 49w^2$ **false; $36y^2 - 84yw + 49w^2$**
41. $\left(\frac{1}{3}a + 3b\right)^2 \stackrel{?}{=} \frac{1}{9}a^2 + 2ab + 9b^2$ **true**
42. $\left(\frac{2}{7}n - 3m\right)\left(\frac{2}{7}n - 3m\right) \stackrel{?}{=} \frac{4}{49}n^2 - 9m^2$ **false; $\frac{4}{49}n^2 - \frac{12}{7}mn + 9m^2$**

MENTAL MATH **Use mental math to find the product.**

43. $26 \cdot 34$ **884**
44. $45 \cdot 55$ **2475**
45. 16^2 **256**
46. 41^2 **1681**

STUDENT HELP

HOMEWORK HELP
Example 1: Exs. 15–20, 27–42
Example 2: Exs. 21–42
Example 3: Exs. 43–46
Example 4: Exs. 47, 48
Example 5: Exs. 49–52

3 APPLY

ASSIGNMENT GUIDE

BASIC
Day 1: p. 593 Exs. 15–35
Day 2: pp. 593–596 Exs. 36–42, 47–50, 56–58, 65, 70, 75, 80, 81, Quiz 1 Exs. 1–17

AVERAGE
Day 1: p. 593 Exs. 15–35
Day 2: pp. 593–596 Exs. 36–52, 56–58, 65, 70, 75, 80, 81, Quiz 1 Exs. 1–17

ADVANCED
Day 1: p. 593 Exs. 15–35
Day 2: pp. 593–596 Exs. 36–52, 56–62, 65, 70, 75, 80, 81, Quiz 1 Exs. 1–17

BLOCK SCHEDULE
p. 593 Exs. 15–35, (with 10.2)
pp. 593–596 Exs. 36–52, 56, 58, 65, 70, 75, 80, 81, Quiz 1 Exs. 1–17 (with 10.4)

EXERCISE LEVELS

Level A: ***Easier***
15–20

Level B: ***More Difficult***
21–52, 56–58, 63–81

Level C: ***Most Difficult***
53–55, 59–62

✓ HOMEWORK CHECK

To quickly check student understanding of key concepts, go over the following exercises: Exs. 16, 20, 25, 32, 36, 40, 42, 48, 50. See also the Daily Homework Quiz:

- Blackline Master (*Chapter 10 Resource Book*, p. 54)
- Transparency (p. 79)

5.

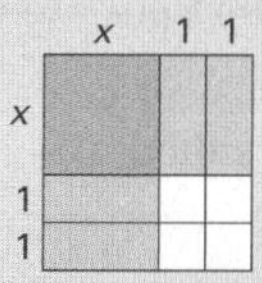

STUDENT HELP NOTES
Homework Help Students can find help for Exs. 49–55 at **www.mcdougallittell.com.** The information can be printed out for students who don't have access to the Internet.

CAREER NOTE

EXERCISES 49–52 Additional information about a career as a geneticist is available at **www.mcdougallittell.com.**

AREA MODELS Write two expressions for the area of the figure. Describe the special product pattern that is represented.

47. 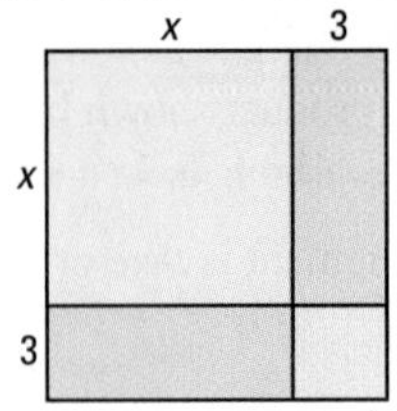

$(x + 3)^2$, $x^2 + 6x + 9$; square of a binomial

48. 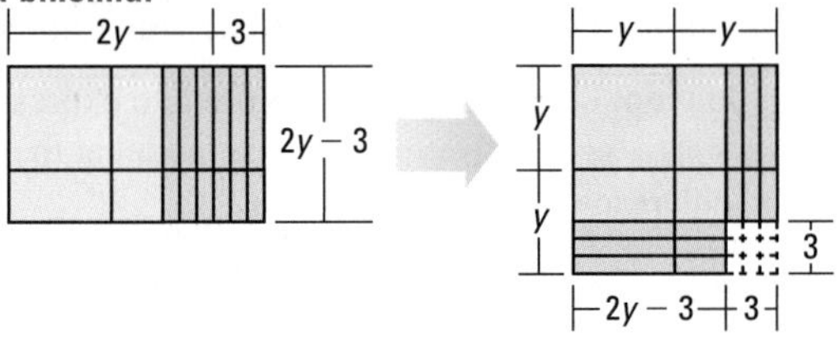

$(2y^2) - 3^2$, $(2y + 3)(2y - 3)$; sum and difference

STUDENT HELP
HOMEWORK HELP Visit our Web site www.mcdougallittell.com for help with Exs. 49–55.

SCIENCE CONNECTION In tigers, the normal color gene *C* is *dominant* and the white color gene *c* is *recessive*. This means that a tiger whose color genes are *CC* or *Cc* will have normal coloring. A tiger whose color genes are *cc* will be white.

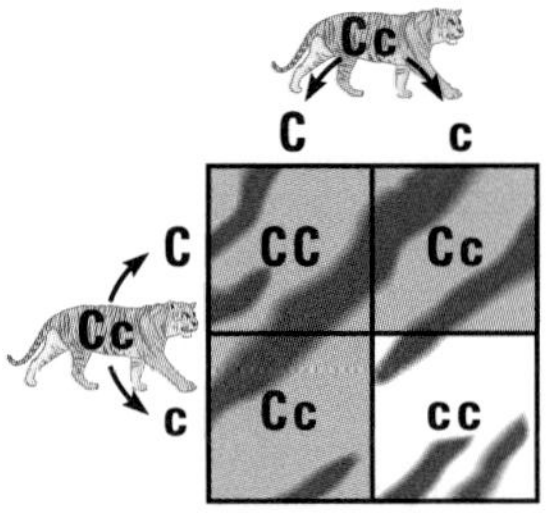

49. The Punnett square at the right shows the possible results of crossing two tigers that have recessive white genes. Find a model that can be used to represent the Punnett square and write the model as a polynomial.
$(0.5C + 0.5c)^2 = 0.25C^2 + 0.5Cc + 0.25c^2$

50. What percent of the offspring will have normal coloring? What percent will be white? 75%; 25%

FOCUS ON CAREERS

GENETICISTS do research on the biological and chemical processes that result in the the physical traits of plants and animals.

CAREER LINK
www.mcdougallittell.com

SCIENCE CONNECTION In Exercises 51 and 52, use the following information.
In chickens, neither the normal-feathered gene *F* nor the frizzle-feathered gene *f* is dominant. So chickens whose feather genes are *FF* will have normal feathers. Chickens with *Ff* will have mildly frizzled feathers. Chickens with *ff* will have extremely frizzled feathers.

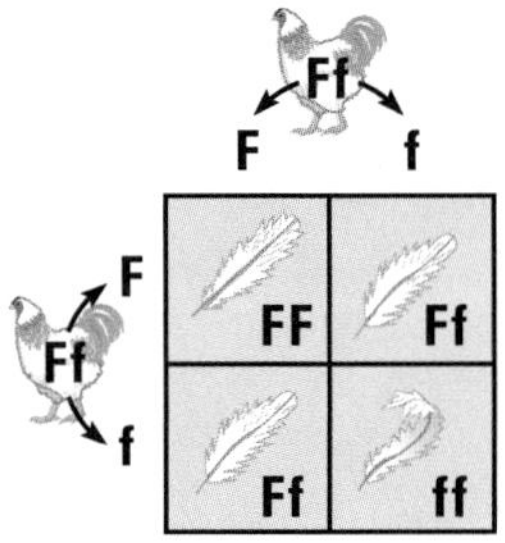

51. The Punnett square at the right shows the possible results of crossing two chickens with mildly frizzled feathers. Find a model that can be used to represent the Punnett square and write the model as a polynomial.
$(0.5F + 0.5f)^2 = 0.25F^2 + 0.5Ff + 0.25f^2$

52. What percent of the offspring will have normal feathers? What percent will have mildly frizzled feathers? What percent will have extremely frizzled feathers? 25%; 50%; 25%

INVESTMENT VALUE In Exercises 53–55, an investment of *P* dollars that gains *r* percent of its value in one year is worth $P(1 + r)$ at the end of that year. An investment that loses *r* percent of its value in one year is worth $P(1 - r)$ at the end of that year.

53. Write a model for the value of an investment *P* that loses *r* percent one year, then gains *r* percent the following year. $P(1 - r)(1 + r) = P(1 - r^2)$

54. According to the model, did the investment increase or decrease in value? By how much? decrease; Pr^2 dollars

55. If the investment gains *r* percent the first year and loses *r* percent the second year, what is the increase or decrease in the value of the investment?
a decrease of Pr^2 dollars

ADDITIONAL PRACTICE AND RETEACHING

For Lesson 10.3:
- Practice Levels A, B, and C (*Chapter 10 Resource Book*, p. 41)
- Reteaching with Practice (*Chapter 10 Resource Book*, p. 44)
- See Lesson 10.3 of the *Personal Student Tutor*

For more Mixed Review:
- Search the *Test and Practice Generator* for key words or specific lessons.

Test Preparation

QUANTITATIVE COMPARISON **In Exercises 56 and 57, choose the statement that is true about the given numbers.**

Ⓐ The number in column A is greater.

Ⓑ The number in column B is greater.

Ⓒ The two numbers are equal.

Ⓓ The relationship cannot be determined from the given information.

	Column A	Column B	
56.	$(3x - 7)(3x + 7)$ when $x = 4$	$(3x - 7)^2$ when $x = 4$	**A**
57.	$a^2 + b^2$ when $a = 1$ and $b = -2$	$(a + b)^2$ when $a = 1$ and $b = -2$	**A**

58. **MULTIPLE CHOICE** Which of the following is equal to $(3x + 2)^2 - (x + 5)(x - 5)$? **D**

Ⓐ $8x^2 + 29$ Ⓑ $8x^2 + 12x - 21$

Ⓒ $10x^2 + 6x + 29$ Ⓓ $8x^2 + 12x + 29$

★ Challenge

WRITING FACTORS **Write a pair of factors that have the given product.**

59. $x^2 + 12x + 36$ **$x + 6$ and $x + 6$**

60. $x^2 - 4$ **$x + 2$ and $x - 2$**

61. $25x^2 - 30x + 9$ **$5x - 3$ and $5x - 3$**

62. $49x^2 - 169$ **$7x + 13$ and $7x - 13$**

MIXED REVIEW

SIMPLIFYING EXPRESSIONS **Simplify the expression. The simplified expression should have no negative exponents.** **(Review 8.3)**

63. $\left(\frac{x}{4}\right)^3$ **$\frac{x^3}{64}$**

64. $\frac{x^3}{x^2}$ **x**

65. $\left(\frac{4x}{y^3}\right)^3$ **$\frac{64x^3}{y^9}$**

66. $x^7 \cdot \frac{1}{x^4}$ **x^3**

67. $\frac{3x^2y}{2x} \cdot \frac{6xy^2}{y^3}$ **$9x^2$**

68. $\frac{5x^4y}{3xy^2} \cdot \frac{9xy}{x^2y}$ **$\frac{15x^2}{y}$**

70. $\left(1\frac{1}{2}, -18\frac{3}{4}\right)$; $x = 1\frac{1}{2}$

FINDING PARTS OF A GRAPH **Find the coordinates of the vertex and write the equation of the axis of symmetry.** **(Review 9.3 for 10.4)**

69. $y = 2x^2 + 3x + 6$ **$\left(-\frac{3}{4}, 4\frac{7}{8}\right)$; $x = -\frac{3}{4}$**

70. $y = 3x^2 - 9x - 12$ **See margin.**

71. $y = -x^2 + 4x + 16$ **(2, 20); $x = 2$**

72. $y = -4x^2 - 2x + 5$ **$\left(-\frac{1}{4}, 5\frac{1}{4}\right)$; $x = -\frac{1}{4}$**

73. $y = -\frac{1}{2}x^2 + 6x - 4$ **(6, 14); $x = 6$**

74. $y = \frac{1}{6}x^2 - \frac{1}{3}x + 2$ **$\left(1, 1\frac{5}{6}\right)$; $x = 1$**

CHECKING GRAPHICALLY **Solve the equation algebraically. Check the solutions graphically.** **(Review 9.4)**

75. $x^2 - 10 = 6$ **−4, 4**

76. $x^2 + 12 = 48$ **−6, 6**

77. $\frac{1}{5}x^2 = 5$ **−5, 5**

78. $3x^2 = 192$ **−8, 8**

79. $\frac{2}{3}x^2 = 6$ **−3, 3**

80. $2x^2 - 66 = 96$ **−9, 9**

81. **GEOMETRY CONNECTION** Is it possible for a rectangle with a perimeter of 52 centimeters to have an area of 148.75 square centimeters? Explain. **(Review 9.6)** **See margin.**

81. Yes; ***Sample answer:*** **Let x be the length of the rectangle and y be the width. Then $2x + 2y = 52$ and $xy = 148.75$. Substituting, $x(26 - x) = 148.75$, which is equivalent to $x^2 - 26x + 148.75 = 0$. Since the discriminant of the equation is 81, the equation has two solutions. The rectangle is 17.5 by 8.5.**

4 ASSESS

DAILY HOMEWORK QUIZ

Transparency Available

Find the product.

1. $(5x + 5)(5x - 5)$ **$25x^2 - 25$**

2. $(x - 11)^2$ **$x^2 - 22x + 121$**

3. $(3n + 4m)^2$ **$9n^2 + 24mn + 16m^2$**

4. The Punnett square shows the possible outcomes when 2 coins are tossed. The outcome on each coin can be modeled by $\frac{1}{2}H + \frac{1}{2}T$. Show how the square of a binomial can be used to model the results when 2 coins are tossed.

	H	T
H	HH	HT
T	TH	TT

$\left(\frac{1}{2}H + \frac{1}{2}T\right)^2 = \frac{1}{4}H^2 + \frac{1}{2}HT + \frac{1}{4}T^2$

EXTRA CHALLENGE NOTE

Challenge problems for Lesson 10.3 are available in **blackline** format in the *Chapter 10 Resource Book*, p. 50 and at **www.mcdougallittell.com.**

ADDITIONAL TEST PREPARATION

1. WRITING Describe the middle term when using the sum and difference pattern and when using the square of a binomial pattern. **There is no middle term when using the sum and difference pattern. The middle term is twice the product of the two terms in the binomial when using the square of a binomial pattern.**

2. WRITING When using the sum and difference pattern for multiplying $(2x - 7)(2x + 7)$, what are the coefficients of the x^2-term, the x-term, and the constant term? **4, 0, and −49**

ADDITIONAL RESOURCES

An alternative quiz for Lessons 10.1–10.3 is available in the *Chapter 10 Resource Book*, p. 51.

A **blackline** master with additional Math & History exercises is available in the *Chapter 10 Resource Book*, p. 49.

17.

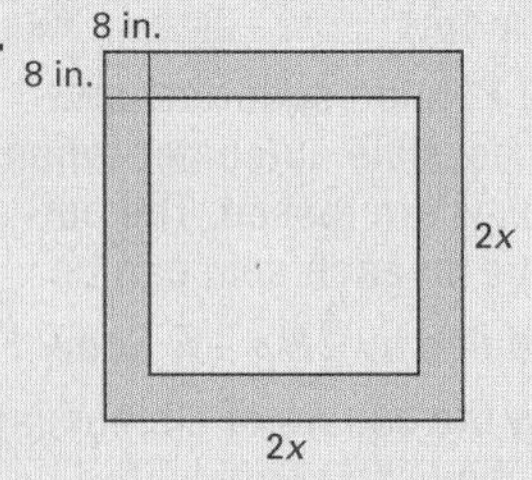

QUIZ 1

Use a vertical format or a horizontal format to add or subtract. (Lesson 10.1)

1. $(2x^2 + 7x + 1) + (x^2 - 2x + 8)$ **$3x^2 + 5x + 9$**
2. $(-4x^3 - 5x^2 + 2x) - (2x^3 + 9x^2 + 2)$ **$-6x^3 - 14x^2 + 2x - 2$**
3. $(7t^2 - 3t + 5) - (4t^2 + 10t - 9)$ **$3t^2 - 13t + 14$**
4. $(5x^3 - x^2 + 3x + 3) + (x^3 - 4x^2 + x)$ **$6x^3 - 5x^2 + 4x + 3$**

Find the product. (Lesson 10.2)

5. $(x + 8)(x - 1)$ **$x^2 + 7x - 8$**
6. $(4n + 7)(4n - 7)$ **$16n^2 - 49$**
7. $(-2x^2 + x - 4)(x - 2)$ **$-2x^3 + 5x^2 - 6x + 8$**
8. $(9m - 4)(3m + 1)$ **$27m^2 - 3m - 4$**
9. $\left(\frac{1}{2}x - 3\right)\left(\frac{1}{2}x + 5\right)$ **$\frac{1}{4}x^2 + x - 15$**
10. $-x^2(12x^3 - 11x^2 + 3)$ **$-12x^5 + 11x^4 - 3x^2$**

Use special products to find the product. (Lesson 10.3)

11. $(x - 6)(x + 6)$ **$x^2 - 36$**
12. $(4x + 3)(4x - 3)$ **$16x^2 - 9$**
13. $(5 + 3b)(5 - 3b)$ **$25 - 9b^2$**
14. $(2x - 7y)(2x + 7y)$ **$4x^2 - 49y^2$**
15. $(3x + 6)^2$ **$9x^2 + 36x + 36$**
16. $(-6 - 8x)^2$ **$36 + 96x + 64x^2$**
17. NEW PATIO You are making a square concrete patio. You want a brick border that is 8 inches wide around the outer edge of the patio, and the total length of a side is $2x$ inches. Draw a diagram and write a polynomial for the area of the floor of the patio not including the brick border. (Lesson 10.3)
$(2x - 16)^2 = (4x^2 - 64x + 256)$ in.^{2} See margin for diagram.

MATH & History — Plant Genetics

APPLICATION LINK www.mcdougallittell.com

THEN

IN 1866, Gregor Mendel identified the pattern in which pea plants passed their characteristics on to the next generation of plants. In Lesson 10.3, you used special products to model genetic patterns, as shown in Punnett squares.

Mendel identified the pattern for pea plants by breeding many plants under identical circumstances. This enabled him to calculate the experimental probability that a given generation of pea plants would have a given characteristic.

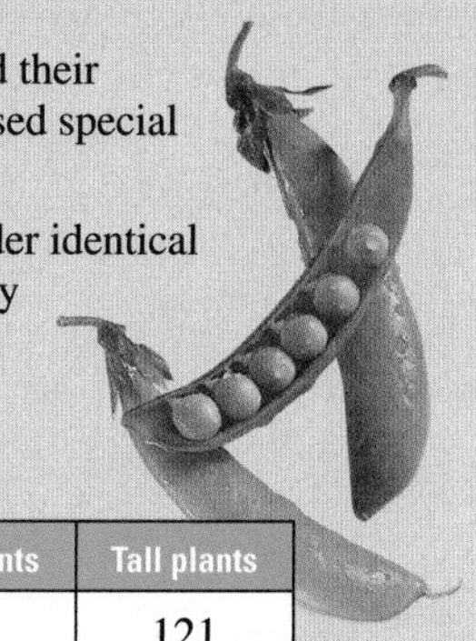

Use the table. Find the experimental probability. (see p. 115)

1. In trial 1, what is the probability that a pea plant is tall? **$\frac{121}{160} \approx 0.76$**
2. In trial 2, what is the probability that a pea plant is short? **$\frac{81}{316} \approx 0.26$**

	Short plants	Tall plants
Trial 1	39	121
Trial 2	81	235

NOW

TODAY, knowledge of genetics helps researchers grow disease-resistant crops, more nutritious vegetables, taller trees, and more varieties of flowers.

Machinery dramatically increases farm production. 1860–1890

1896 **George Washington Carver heads Agriculture Dept. at Tuskegee Institute.**

Barbara McClintock receives Nobel Prize for her work on the genetics of corn. 1983

10.4 Solving Polynomial Equations in Factored Form

What you should learn

GOAL 1 Solve a polynomial equation in factored form.

GOAL 2 Relate factors and x-intercepts.

Why you should learn it

▼ To solve **real-life** problems like estimating the dimensions of the Gateway Arch in **Exs. 55 and 56.**

REAL LIFE

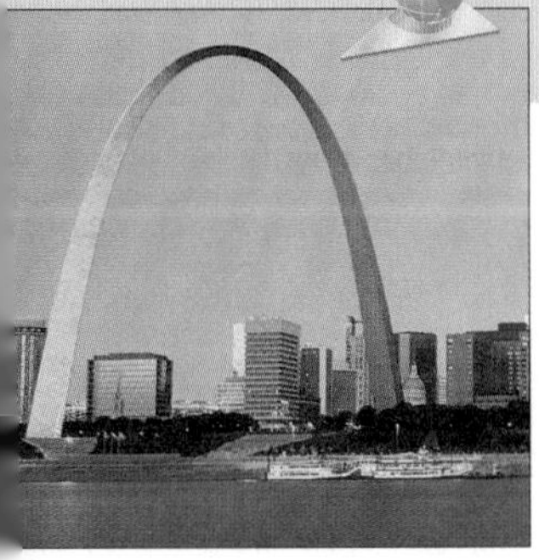

GOAL 1 SOLVING FACTORED EQUATIONS

In Chapter 9 you learned how to solve quadratic equations in standard form.

Standard form: $2x^2 + 7x - 15 = 0$

In this lesson you will learn how to solve quadratic and other polynomial equations in *factored form*. A polynomial is in **factored form** if it is written as the product of two or more linear factors.

Factored form: $(2x - 3)(x + 5) = 0$

Factored form: $(2x - 3)(x + 5)(x - 7) = 0$

ACTIVITY
Developing Concepts

Investigating Factored Equations

1. Copy and complete the table by substituting the x-values into the given expression. **See margin.**

Expression	x-value						
	−3	−2	−1	0	1	2	3
$(x - 3)(x + 2)$	$(-6)(-1) = 6$	?	?	?	?	?	?
$(x + 1)(x + 3)$	?	?	?	?	?	?	?
$(x - 2)(x - 2)$	?	?	?	?	?	?	?
$(x - 1)(x + 2)$	?	?	?	?	?	?	?

2. Use the table to solve each of the following equations.

   a. $(x - 3)(x + 2) = 0$ **3, −2**

   b. $(x + 1)(x + 3) = 0$ **−1, −3**

   c. $(x - 2)(x - 2) = 0$ **2**

   d. $(x - 1)(x + 2) = 0$ **1, −2**

3. Write a general rule for solving an equation that is in factored form. **See margin.**

In this activity, you may have noticed that the product of two factors is zero *only* when at least one of the factors is zero. This is because the real numbers satisfy the **zero-product property**.

ZERO-PRODUCT PROPERTY

Let a and b be real numbers. If $ab = 0$, then $a = 0$ or $b = 0$.

Examples 1, 2, and 3 show how to use this property to solve factored equations.

1 PLAN

PACING
Basic: 1 day
Average: 1 day
Advanced: 1 day
Block Schedule: 0.5 block with 10.3

LESSON OPENER
APPLICATION
An alternative way to approach Lesson 10.4 is to use the Application Lesson Opener:

- Blackline Master (*Chapter 10 Resource Book*, p. 55)
- Transparency (p. 69)

MEETING INDIVIDUAL NEEDS

- ***Chapter 10 Resource Book***
  Prerequisite Skills Review (p. 5)
  Practice Level A (p. 56)
  Practice Level B (p. 57)
  Practice Level C (p. 58)
  Reteaching with Practice (p. 59)
  Absent Student Catch-Up (p. 61)
  Challenge (p. 63)
- ***Resources in Spanish***
- ***Personal Student Tutor***

NEW-TEACHER SUPPORT
See the Tips for New Teachers on pp. 1–2 of the *Chapter 10 Resource Book* for additional notes about Lesson 10.4.

WARM-UP EXERCISES

Transparency Available

Solve the equation.

1. $x^2 = 9$ **±3**

2. $x^2 - 13 = 12$ **±5**

3. $\frac{1}{4}x^2 = 16$ **±8**

Write each product.

4. $(x - 3)(2x - 1)$ $2x^2 - 7x + 3$

5. $(x + 8)^2$ $x^2 + 16x + 64$

1, 3. See Additional Answers beginning on page AA1.

See Activity Note on next page.

2 TEACH

MOTIVATING THE LESSON
Ask your students to find the product of a complex multiplication expression that contains 0 as a factor as you write it on the board or overhead. Tell them they cannot use a calculator. Use numbers in the thousands and millions as you write the expression, and do not include the 0 until you are near the end. If someone realizes that the product will be 0, ask them to explain how they know. Use the converse of this idea to explain the zero product property.

ACTIVITY NOTE
Students should understand that they should not look in the "0" column for solutions, but they should find the entries in the columns that are zeros. When students are writing the general rule for solving an equation, try to get them to use the vocabulary from the chapter, such as *factor* or *polynomial.*

EXTRA EXAMPLE 1
Solve the equation $(x-4)(x+1)=0$. 4, –1

EXTRA EXAMPLE 2
Solve $(x+8)^2=0$. –8

EXTRA EXAMPLE 3
Solve $(3x-2)(4x+3)(x+4)=0$. $\frac{2}{3}, -\frac{3}{4}, -4$

CHECKPOINT EXERCISES

For use after Example 1:
1. Solve the equation $(x-9)(x-7)=0$. 9, 7

For use after Example 2:
2. Solve $(x-2)^2=0$. 2

For use after Example 3:
3. Solve $(x-3)(2x+5)(5x-1)=0$. $3, -\frac{5}{2}, \frac{1}{5}$

EXAMPLE 1 *Using the Zero-Product Property*

Solve the equation $(x-2)(x+3)=0$.

SOLUTION Use the zero-product property: either $x-2=0$ or $x+3=0$.

$(x-2)(x+3)=0$ — Write original equation.

$x-2=0$ — Set first factor equal to 0.

$x=2$ — Solve for *x*.

$x+3=0$ — Set second factor equal to 0.

$x=-3$ — Solve for *x*.

▶ The solutions are 2 and −3. Check these in the original equation.

EXAMPLE 2 *Solving a Repeated-Factor Equation*

Solve $(x+5)^2=0$.

SOLUTION

This equation has a *repeated* factor. To solve the equation you need to set only $x+5$ equal to zero.

$(x+5)^2=0$ — Write original equation.

$x+5=0$ — Set repeated factor equal to 0.

$x=-5$ — Solve for *x*.

▶ The solution is −5. Check this in the original equation.

EXAMPLE 3 *Solving a Factored Cubic Equation*

Solve $(2x+1)(3x-2)(x-1)=0$.

SOLUTION

$(2x+1)(3x-2)(x-1)=0$ — Write original equation.

$2x+1=0$ — Set first factor equal to 0.

$x=-\frac{1}{2}$ — Solve for *x*.

$3x-2=0$ — Set second factor equal to 0.

$x=\frac{2}{3}$ — Solve for *x*.

$x-1=0$ — Set third factor equal to 0.

$x=1$ — Solve for *x*.

▶ The solutions are $-\frac{1}{2}, \frac{2}{3}$, and 1. Check these in the original equation.

STUDENT HELP

Study Tip
A polynomial equation in one variable can have as many solutions as it has linear factors that include the variable. As you saw in Example 2, an equation may have fewer solutions than factors if a factor is repeated.

GOAL 2 RELATING FACTORS AND X-INTERCEPTS

CONCEPT SUMMARY — FACTORS, SOLUTIONS, AND X-INTERCEPTS

For any quadratic polynomial $ax^2 + bx + c$, if one of the following statements is true, then all three statements are true.

- $(x - p)$ is a factor of the quadratic expression $ax^2 + bx + c$.
  Example: $(x - 4)$ and $(x + 3)$ are factors of $x^2 - x - 12$.
- $x = p$ is a solution of the quadratic equation $ax^2 + bx + c = 0$.
  Example: $x = 4$ and $x = -3$ are solutions of $x^2 - x - 12 = 0$.
- p is an x-intercept of the graph of the function $y = ax^2 + bx + c$.
  Example: 4 and -3 are x-intercepts of $y = x^2 - x - 12$.

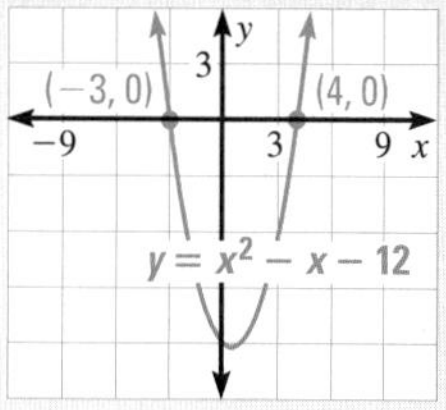

EXAMPLE 4 *Relating x-Intercepts and Factors*

STUDENT HELP

Study Tip
The statements above about factors, solutions, and x-intercepts are *logically equivalent* because each statement implies the truth of the other statements.

To sketch the graph of $y = (x - 3)(x + 2)$:

First solve $(x - 3)(x + 2) = 0$ to find the x-intercepts: 3 and -2.

Then find the coordinates of the vertex.

- The x-coordinate of the vertex is the average of the x-intercepts.

$$x = \frac{3 + (-2)}{2} = \frac{1}{2}$$

- Substitute to find the y-coordinate.

$$y = \left(\frac{1}{2} - 3\right)\left(\frac{1}{2} + 2\right) = -\frac{25}{4}$$

- The coordinates of the vertex are $\left(\frac{1}{2}, -\frac{25}{4}\right)$.

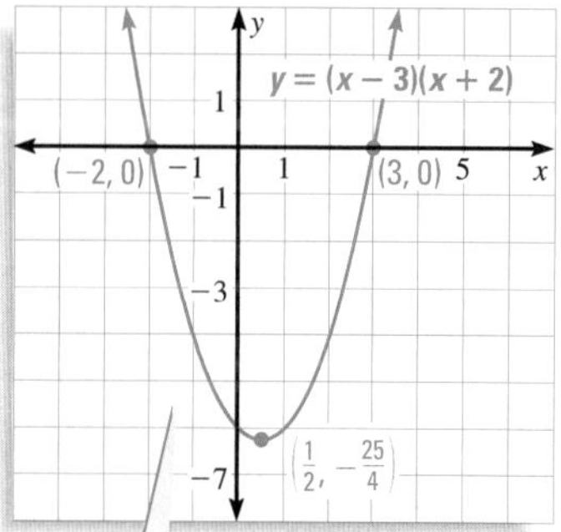

Sketch the graph by using the x-intercepts and vertex coordinates.

EXAMPLE 5 *Using a Quadratic Model*

REAL LIFE *Archways*

An arch is modeled by the equation $y = -0.15(x - 8)(x + 8)$, with x and y measured in feet. How wide is the arch at the base? How high is the arch?

SOLUTION Sketch a graph of the model.

- The x-intercepts are $x = 8$ and $x = -8$.
- The vertex is at (0, 9.6).

▶ The arch is $8 - (-8)$, or 16 feet wide at the base and 9.6 feet high.

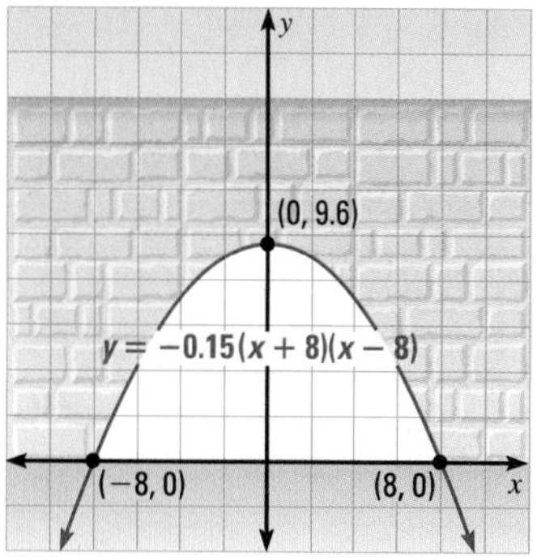

Closure Question ***Sample answer:***
If a polynomial expression is in factored form and equals 0, then each factor of the polynomial expression can be set equal to 0 and solved to find the solutions to the polynomial equation.

EXTRA EXAMPLE 4

Sketch $y = (x - 4)(x + 2)$.

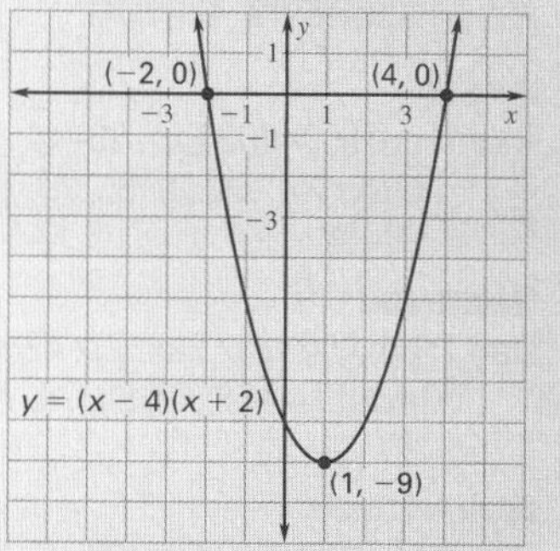

EXTRA EXAMPLE 5

An arched garden trellis is modeled by $y = -1.8(x - 2)(x + 2)$, with x and y measured in feet. How wide is the arch at the base? How high is the arch? **4 ft wide; 7.2 ft high**

CHECKPOINT EXERCISES

For use after Example 4:

1. Find the x-intercepts and the vertex of the graph of $y = (x + 2)(2x - 4)$. **−2, 2; (0, −8)**

For use after Example 5:

2. The entrance of an underpass contains an arch that is modeled by $y = -0.2(x - 10)(x + 10)$, with x and y measured in feet. How wide is the arch at the base? How high is the arch? **20 ft wide; 20 ft high**

CLOSURE QUESTION

Explain how the zero product property can be used to solve polynomial equations.
See sample answer, below.

DAILY PUZZLER

You have three different envelopes: one labeled 1st place, one labeled 2nd place, and one labeled 3rd place. Inside each envelope is a ribbon that corresponds to the correct place, blue, red, and yellow respectively. Someone takes each ribbon out and puts it in a different envelope. You open the 1st place envelope and find a red ribbon. What color ribbon is in the 2nd place envelope? **yellow**

3 APPLY

ASSIGNMENT GUIDE

BASIC
Day 1: pp. 600–602 Exs. 20–42 even, 44–54, 59, 65, 70, 75, 80–82

AVERAGE
Day 1: pp. 600–602 Exs. 20–42 even, 44–56, 59, 65, 70, 75, 80–82

ADVANCED
Day 1: pp. 600–602 Exs. 20–42 even, 44–56, 59, 60, 65, 70, 75, 80–82

BLOCK SCHEDULE WITH 10.3
pp. 600–602 Exs. 20–42 even, 44–56, 59, 65, 70, 75, 80–82

EXERCISE LEVELS

Level A: ***Easier***
61–68

Level B: ***More Difficult***
19–59, 69–84

Level C: ***Most Difficult***
60

HOMEWORK CHECK

To quickly check student understanding of key concepts, go over the following exercises: Exs. 22, 26, 32, 34, 38, 42, 48, 54. See also the Daily Homework Quiz:

- Blackline Master (*Chapter 10 Resource Book*, p. 66)
- Transparency (p. 80)

18.

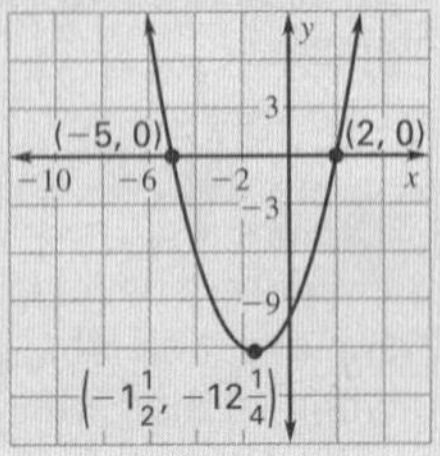

GUIDED PRACTICE

Vocabulary Check ✓

Concept Check ✓

1. If the product of two numbers is zero, at least one of the numbers is zero. (If $ab = 0$, then $a = 0$ or $b = 0$.)

2. No; the solutions are solutions of $x + 1 = 0$ and $x - 4 = 0$, or $x = -1$ and $x = 4$.

Skill Check ✓

3. No; the equation has two linear factors, so it has only two solutions, the solutions of $x - 2 = 0$ and $x + 5 = 0$, or $x = 2$ and $x = -5$.

1. What is the zero-product property? **See margin.**

2. Are 1 and −4 the solutions of $(x + 1)(x - 4) = 0$? Explain. **See margin.**

3. Are −5, 2, and 3 the solutions of $3(x - 2)(x + 5) = 0$? Explain. **See margin.**

Tell whether the statement about $x^2 + 12x + 32$ is *true* or *false*.

4. $x^2 + 12x + 32 = (x + 4)(x + 8)$ **true**

5. $x = -4$ and $x = -8$ are solutions of $x^2 + 12x + 32 = 0$. **true**

6. $(x + 4)$ and $(x + 8)$ are factors of $x^2 + 12x + 32$. **true**

7. The x-intercepts of the graph of $y = x^2 + 12x + 32$ are −4 and −8. **true**

Does the graph of the function have x-intercepts of 4 and −5?

8. $y = 2(x + 4)(x - 5)$ **no**

9. $y = 4(x - 4)(x - 5)$ **no**

10. $y = -(x - 4)(x + 5)$ **yes**

11. $y = 3(x + 5)(x - 4)$ **yes**

In Exercises 12–17, use the zero-product property to solve the equation.

12. $(b + 1)(b + 3) = 0$ **−1, −3**

13. $(t - 3)(t - 5) = 0$ **3, 5**

14. $(x - 7)^2 = 0$ **7**

15. $(y + 9)(y - 2) = 0$ **−9, 2**

16. $(3d + 6)(2d + 5) = 0$ **$-2, -\frac{5}{2}$**

17. $\left(2w + \frac{1}{2}\right)^2 = 0$ **$-\frac{1}{4}$**

18. Sketch the graph of $y = (x - 2)(x + 5)$. Label the vertex and the x-intercepts. **See margin.**

PRACTICE AND APPLICATIONS

STUDENT HELP
Extra Practice to help you master skills is on p. 806.

ZERO-PRODUCT PROPERTY **Use the zero-product property to solve the equation.**

19. $(x + 4)(x + 1) = 0$ **−4, −1**

20. $(y + 3)^2 = 0$ **−3**

21. $(t + 8)(t - 6) = 0$ **−8, 6**

22. $(w - 17)^2 = 0$ **17**

23. $(b - 9)(b + 8) = 0$ **9, −8**

24. $(d + 7)^2 = 0$ **−7**

25. $(y - 2)(y + 1) = 0$ **2, −1**

26. $(z + 2)(z + 3) = 0$ **−2, −3**

27. $(v - 7)(v - 5) = 0$ **7, 5**

28. $\left(t + \frac{1}{2}\right)(t - 4) = 0$ **$-\frac{1}{2}, 4$**

29. $4(c + 9)^2 = 0$ **−9**

30. $(u - 3)\left(u - \frac{2}{3}\right) = 0$ **$3, \frac{2}{3}$**

31. $(y - 5.6)^2 = 0$ **5.6**

32. $(a - 40)(a + 12) = 0$ **40, −12**

33. $7(b - 5)^3 = 0$ **5**

STUDENT HELP
HOMEWORK HELP
Example 1: Exs. 19–43
Example 2: Exs. 19–43
Example 3: Exs. 38–43
Example 4: Exs. 44–52
Example 5: Exs. 53–58

SOLVING FACTORED EQUATIONS **Solve the equation.**

34. $(4x - 8)(7x + 21) = 0$ **2, −3**

35. $(2d + 8)(3d + 12) = 0$ **−4**

36. $5(3m + 9)(5m - 15) = 0$ **−3, 3**

37. $8(9n + 27)(6n - 9) = 0$ **$-3, \frac{3}{2}$**

38. $(6b - 18)(2b + 2)(2b + 2) = 0$ **3, −1**

39. $(4y - 5)(2y - 6)(3y - 4) = 0$ **$\frac{5}{4}, 3, \frac{4}{3}$**

40. $(x + 44)(3x - 2)^2 = 0$ **$-44, \frac{2}{3}$**

41. $(5x - 9.5)^2(3x + 6.3) = 0$ **1.9, −2.1**

42. $\left(\frac{1}{2}x + 2\right)\left(\frac{2}{3}x + 6\right)\left(\frac{1}{6}x - 1\right) = 0$ **−4, −9, 6**

43. $\left(2n - \frac{1}{4}\right)\left(5n + \frac{3}{10}\right)\left(3n - \frac{2}{3}\right) = 0$ **$\frac{1}{8}, -\frac{3}{50}, \frac{2}{9}$**

MATCHING FUNCTIONS AND GRAPHS Match the function with its graph.

A. $y = (x + 2)(x - 4)$ **B.** $y = (x - 2)(x + 4)$ **C.** $y = (x + 4)(x + 2)$

44.
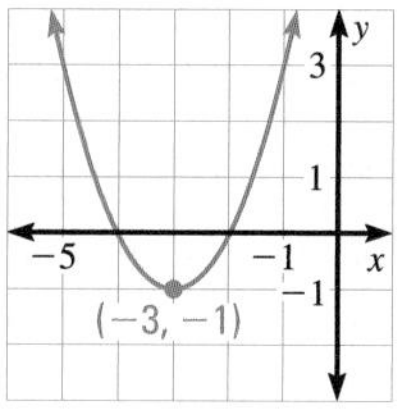

C

45.
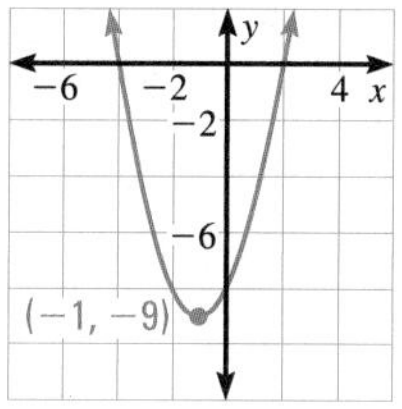

B

46.
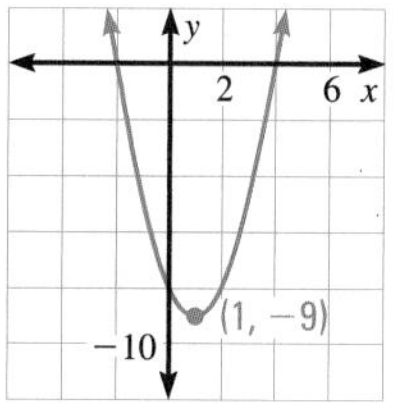

A

SKETCHING GRAPHS Find the *x*-intercepts and the vertex of the graph of the function. Then sketch the graph of the function. 47-52. See margin.

47. $y = (x - 4)(x + 2)$ **48.** $y = (x + 5)(x + 3)$ **49.** $y = (-x - 2)(x + 2)$

50. $y = (-x - 1)(x + 7)$ **51.** $y = (x - 2)(x - 6)$ **52.** $y = (-x + 4)(x + 3)$

STUDENT HELP

INTERNET HOMEWORK HELP Visit our Web site www.mcdougallittell.com for help with Exs. 53–58.

RADIO TELESCOPE In Exercises 53 and 54, use the cross section of the radio telescope dish shown below.

The cross section of the telescope's dish can be modeled by the polynomial function

$$y = \frac{167}{500^2}(x + 500)(x - 500)$$

where x and y are measured in feet, and the center of the dish is where $x = 0$.

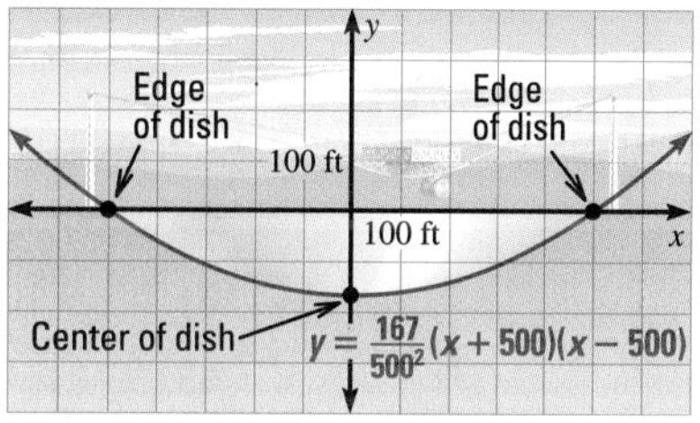

53. Use the distance between the *x*-intercepts at (−500, 0) and (500, 0).

53. Explain how to use the algebraic model to find the width of the dish.

54. Use the model to find the coordinates of the center of the dish. (0, −167)

GATEWAY ARCH In Exercises 55 and 56, use the following information.

The Gateway Arch in St. Louis, Missouri, has the shape of a *catenary* (a U-shaped curve similar to a parabola). It can be approximated by the following model, where x and y are measured in feet. ▶ Source: National Park Service

Gateway Arch model: $y = -\frac{7}{1000}(x + 300)(x - 300)$

55. According to the model, how far apart are the legs of the arch? 600 ft

56. How high is the arch? 630 ft

FOCUS ON APPLICATIONS

REAL LIFE **THE BARRINGER METEOR CRATER** was formed about 49,000 years ago when a nickel and iron meteorite struck the desert at about 25,000 miles per hour.

BARRINGER METEOR CRATER In Exercises 57 and 58, use the following equation, where *x* and *y* are measured in meters, to model a cross section of the Barringer Meteor Crater, near Winslow, Arizona.

▶ Source: Jet Propulsion Laboratory

Barringer Meteor Crater model: $y = \frac{1}{1800}(x - 600)(x + 600)$

57. Assuming the lip of the crater is at $y = 0$, how wide is the crater? 1200 m

58. What is the depth of the crater? 200 m

STUDENT HELP NOTES

Homework Help Students can find help for Exs. 53–58 at **www.mcdougallittell.com.** The information can be printed out for students who don't have access to the Internet.

MATHEMATICAL REASONING

Ask students to evaluate a polynomial expressions such as $(x + 3)(x - 2)$ for $x = -1, 0, 1$. Next ask them to solve the equation $(x + 3)(x - 2) = 0$. Then ask them to describe the difference between evaluating an expression and solving an equation. −6, −6, −4; −3, 2; *Sample answer:* When you evaluate an expression, you find its value by replacing the variable with the given numbers; when you solve an equation, you find the values of the variable that result in true statements.

47–52. See Additional Answers beginning on page AA1.

ADDITIONAL PRACTICE AND RETEACHING

For Lesson 10.4:

- Practice Levels A, B, and C (*Chapter 10 Resource Book,* p. 56)
- Reteaching with Practice (*Chapter 10 Resource Book,* p. 59)
- See Lesson 10.4 of the *Personal Student Tutor*

For more Mixed Review:

- Search the *Test and Practice Generator* for key words or specific lessons.

4 ASSESS

DAILY HOMEWORK QUIZ

Transparency Available

Solve the equation.

1. $(y-7)^2=0$ 7
2. $(x+3)(x-9)=0$ −3, 9
3. $2(2t-2)(3t+5)=0$ $1, -\frac{5}{3}$
4. Find the x-intercepts and the vertex of the graph of $y=(x+1)(x-1)$. Then sketch the graph of the function. −1, 1, (0, −1)

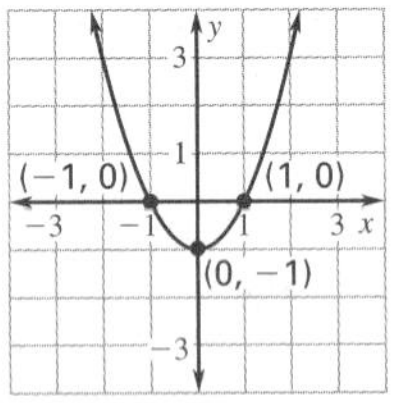

EXTRA CHALLENGE NOTE

Challenge problems for Lesson 10.4 are available in **blackline** format in the *Chapter 10 Resource Book*, p. 63 and at **www.mcdougallittell.com.**

ADDITIONAL TEST PREPARATION

1. **WRITING** Explain what a factored polynomial expression tells you about the graph of the related polynomial equation. *Sample answer:* When each of the factors is set equal to 0, the solutions are the x-intercepts of the graph of the polynomial equation.
2. **OPEN ENDED** Name several important points on the graph of a quadratic equation. *Sample answer:* the vertex, the x-intercept(s), the y-intercept

Test Preparation

59. MULTI-STEP PROBLEM You sell hot dogs for \$1.00 each at your concession stand at a baseball park and have about 200 customers. You want to increase the price of a hot dog. You estimate that you will lose three sales for every \$.10 increase. The following equation models your hot dog sales revenue R, where n is the number of \$.10 increases.

Concession stand revenue model: $R = (1 + 0.1n)(200 - 3n)$

a. To find your revenue from hot dog sales, you multiply the price of each hot dog sold by the number of hot dogs sold. In the formula above, what does $1 + 0.1n$ represent? What does $200 - 3n$ represent? **the price of a hot dog after n \$.10 increases; the number of hot dogs sold after n \$.10 increases**

b. How many times would you have to raise the price by \$.10 to reduce your revenue to zero? Make a graph to help find your answer. **67 times**

c. Decide how high you should raise the price to make the most money. Explain how you got your answer. **See margin.**

59. c. \$3.80; *Sample answer:* I used my graph from part (b) and determined that the maximum value of R occurs at $n = 28\frac{1}{3}$, which is closer to 28 than to 29. I then found the cost by finding $1 + 0.1n$ for $n = 28$.

★ Challenge

60. *Writing* Write your own multi-step problem about selling a product like the one in Exercise 59. Include a model that shows the relationship between price and number of items sold. Explain what each factor in the model represents. **Answer will vary.**

MIXED REVIEW

DECIMAL FORM Rewrite in decimal form. (Review 8.4)

61. 2.1×10^5 210,000 **62.** 4.443×10^{-2} 0.04443 **63.** 8.57×10^8 857,000,000 **64.** 1.25×10^6 1,250,000

65. 3.71×10^{-3} 0.00371 **66.** 9.96×10^6 9,960,000 **67.** 7.22×10^{-4} 0.000722 **68.** 8.17×10^7 81,700,000

MULTIPLYING EXPRESSIONS Find the product. (Review 10.2 for 10.5)

69. $(x-2)(x-7)$ $x^2 - 9x + 14$ **70.** $(x+8)(x-8)$ $x^2 - 64$ **71.** $(x-4)(x+5)$ $x^2 + x - 20$

72. $(x+6)(x-7)$ $x^2 - x - 42$ **73.** $\left(x+\frac{2}{3}\right)\left(x-\frac{1}{3}\right)$ $x^2 + \frac{1}{3}x - \frac{2}{9}$ **74.** $(x-3)\left(x-\frac{1}{6}\right)$ $x^2 - \frac{19}{6}x + \frac{1}{2}$

75. $(2x+7)(3x-1)$ $6x^2 + 19x - 7$ **76.** $(5x-1)(5x+2)$ $25x^2 + 5x - 2$ **77.** $(3x+1)(8x-3)$ $24x^2 - x - 3$

78. $(2x-4)\left(\frac{1}{4}x-2\right)$ $\frac{1}{2}x^2 - 5x + 8$ **79.** $(x+10)(x+10)$ $x^2 + 20x + 100$ **80.** $(3x+5)\left(\frac{2}{3}x-3\right)$ $2x^2 - \frac{17}{3}x - 15$

EXPONENTIAL MODELS Tell whether the situation can be represented by a model of *exponential growth* or *exponential decay*. Then write a model that represents the situation. (Review 8.5, 8.6)

81. COMPUTER PRICES From 1996 to 2000, the average price of a computer company's least expensive home computer system decreased by 16% per year. **Exponential decay; let P be the price in 1996, t the number of years since 1996, and y the price after t years; $y = P(0.84)^t$.**

82. MUSIC SALES From 1995 to 1999, the number of CDs a band sold increased by 23% per year. **Exponential growth; let n be the number sold in 1995, t the number of years since 1995, and y the price after t years; $y = n(1.23)^t$.**

83. COOKING CLUB From 1996 to 2000, the number of members in the cooking club decreased by 3% per year. **See margin.**

83. Exponential decay; let n be the number of members in 1996, t the number of years since 1996, and y the number of members after t years; $y = n(0.97)^t$.

84. INTERNET SERVICE From 1993 to 1998, the total revenues for a company that provides Internet service increased by about 137% per year. **Exponential growth; let R be the revenue in 1993, t the number of years since 1993, and y the revenue after t years; $y = R(2.37)^t$.**

ACTIVITY 10.5

Developing Concepts

Modeling the Factorization of $x^2 + bx + c$

GROUP ACTIVITY
Work with a partner.

MATERIALS
algebra tiles

▶ QUESTION **How can you model the factorization of a trinomial of the form $x^2 + bx + c$ using algebra tiles?**

▶ EXPLORING THE CONCEPT

You can use algebra tiles to create a model that can be used to factor a trinomial that has a leading coefficient of 1. Factor the trinomial $x^2 + 5x + 6$ as follows.

1. Use algebra tiles to model $x^2 + 5x + 6$.

x^2 + $5x$ + 6

2. With the x^2-tile at the upper left, arrange the x-tiles and the 1-tiles around the x^2-tile to form a rectangle.

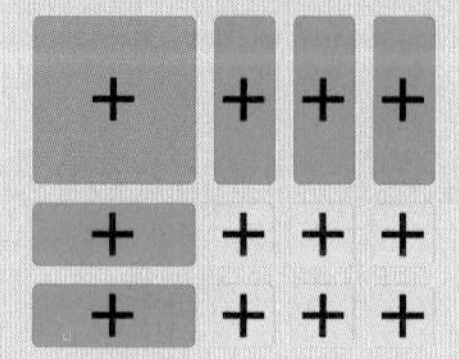

3. The width of the rectangle is _?_, and the length of the rectangle is _?_. Complete the statement: $x^2 + 5x + 6 =$ _?_ • _?_.
$x + 2$; $x + 3$; $(x + 2)(x + 3)$

▶ EXERCISES

Use the model to write the factors of the trinomial.

1. 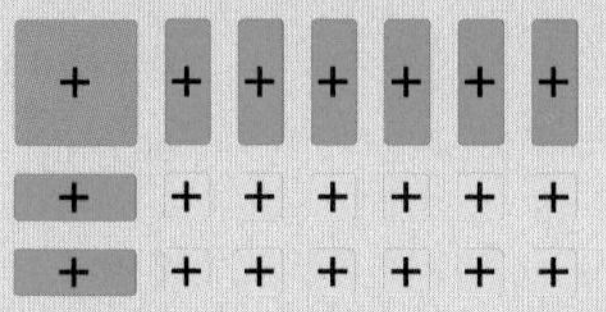
$(x + 2)(x + 6)$

2. 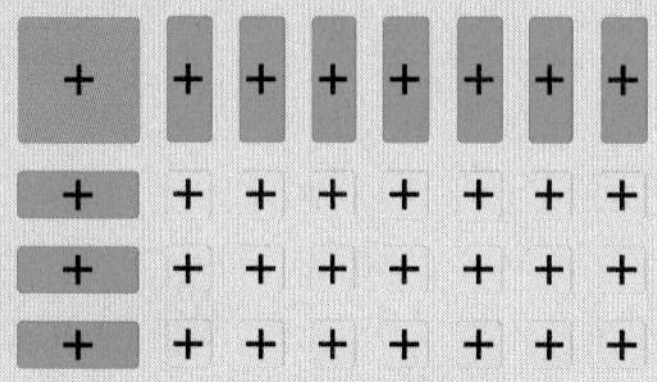
$(x + 3)(x + 7)$

In Exercises 3–8, use algebra tiles to factor the trinomial. Sketch your model.

3. $x^2 + 7x + 6$ $(x + 1)(x + 6)$
4. $x^2 + 6x + 8$ $(x + 4)(x + 2)$
5. $x^2 + 8x + 15$ $(x + 3)(x + 5)$
6. $x^2 + 6x + 9$ $(x + 3)^2$
7. $x^2 + 4x + 4$ $(x + 2)^2$
8. $x^2 + 7x + 10$ $(x + 2)(x + 5)$
9. Use algebra tiles to show why the trinomial $x^2 + 3x + 4$ cannot be factored.
The tiles cannot be formed into a rectangle.

Activity Assessment ***Sample answer:***
Make a rectangle with the tiles (if possible). The product of the binomials will be the product of the length and width of the rectangle.

1 Planning the Activity

PURPOSE
To model trinomials of the form $x^2 + bx + c$ with algebra tiles in order to find the factorization of the trinomial.

MATERIALS
- algebra tiles for each pair
- Activity Support Master (*Chapter 10 Resource Book,* p. 67)

PACING
- Exploring the Concept — 5 min
- Drawing Conclusions — 10 min

LINK TO LESSON
Have students use algebra tiles to model Example 1 of Lesson 10.5.

2 Managing the Activity

CLASSROOM MANAGEMENT
When modeling trinomials with algebra tiles, students with different arrangements may think that they model different rectangles, when actually the length and the width are switched. For example, to model $x^2 + 5x + 6$, the length could be $x + 3$ and width $x + 2$, or the length could be $x + 2$ and the width $x + 3$. You can use this idea to show the commutative property of multiplication, where $x^2 + 5x + 6 = (x + 3)(x + 2) = (x + 2)(x + 3)$.

3 Closing the Activity

★ KEY DISCOVERY
When you factor a quadratic trinomial, you write it as the product of two linear binomial factors that represent the length and width of a rectangle.

ACTIVITY ASSESSMENT
Describe how to model the factorization of a quadratic trinomial into the product of two linear binomials using algebra tiles.
See sample answer at left.

1 PLAN

PACING
Basic: 2 days
Average: 2 days
Advanced: 2 days
Block Schedule: 1 block

LESSON OPENER ACTIVITY
An alternative way to approach Lesson 10.5 is to use the Activity Lesson Opener
- Blackline Master (*Chapter 10 Resource Book,* p. 68)
- Transparency (p. 70)

MEETING INDIVIDUAL NEEDS
- ***Chapter 10 Resource Book***
  Prerequisite Skills Review (p. 5)
  Practice Level A (p. 69)
  Practice Level B (p. 70)
  Practice Level C (p. 71)
  Reteaching with Practice (p. 72)
  Absent Student Catch-Up (p. 74)
  Challenge (p. 76)
- ***Resources in Spanish***
- ***Personal Student Tutor***

NEW-TEACHER SUPPORT
See the Tips for New Teachers on pp. 1–2 of the *Chapter 10 Resource Book* for additional notes about Lesson 10.5.

WARM-UP EXERCISES

Transparency Available

Find the product.

1. $(x+2)(x+6)$ $x^2+8x+12$
2. $(x-9)(x+9)$ x^2-81
3. $(x-3)^2$ x^2-6x+9
4. $(2x-5)(x+5)$ $2x^2+5x-25$
5. $\left(x+\frac{1}{4}\right)\left(x-\frac{1}{4}\right)$ $x^2-\frac{1}{16}$

10.5 Factoring $x^2 + bx + c$

What you should learn

GOAL 1 Factor a quadratic expression of the form $x^2 + bx + c$.

GOAL 2 Solve quadratic equations by factoring.

Why you should learn it

▼ To help solve **real-life** problems, such as finding the width of a stone border in **Example 7**.

REAL LIFE

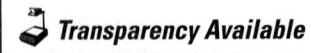

GOAL 1 FACTORING A QUADRATIC TRINOMIAL

To **factor** a quadratic expression means to write it as the product of two linear expressions. In this lesson, you will learn how to factor quadratic trinomials that have a leading coefficient of 1.

FACTORING $x^2 + bx + c$

You know from the FOIL method that $(x + p)(x + q) = x^2 + (p + q)x + pq$. So to factor $x^2 + bx + c$, you need to find numbers p and q such that

$$p + q = b \quad \text{and} \quad pq = c$$

because $x^2 + (p + q)x + pq = x^2 + bx + c$ if and only if $p + q = b$ and $pq = c$.

Example: $x^2 + 6x + 8 = (x + 4)(x + 2)$ $\qquad 4 + 2 = 6$ and $4 \cdot 2 = 8$

EXAMPLE 1 *Factoring when b and c are Positive*

Factor $x^2 + 3x + 2$.

SOLUTION

For this trinomial, $b = 3$ and $c = 2$. You need to find two numbers whose sum is 3 and whose product is 2.

$x^2 + 3x + 2 = (x + p)(x + q)$ $\qquad$ Find p and q when $p + q = 3$ and $pq = 2$.

$= (x + 1)(x + 2)$ $\qquad$ $p = 1$ and $q = 2$

✓**CHECK:** You can check the result by multiplying.

$(x + 1)(x + 2) = x^2 + 2x + x + 2$ $\qquad$ Use FOIL to check.

$= x^2 + 3x + 2$ $\qquad$ Simplify.

EXAMPLE 2 *Factoring when b is Negative and c is Positive*

Factor $x^2 - 5x + 6$.

SOLUTION

Because b is negative and c is positive, both p and q must be negative numbers. Find two numbers whose sum is -5 and whose product is 6.

$x^2 - 5x + 6 = (x + p)(x + q)$ $\qquad$ Find p and q when $p + q = -5$ and $pq = 6$.

$= (x - 2)(x - 3)$ $\qquad$ $p = -2$ and $q = -3$

STUDENT HELP

HOMEWORK HELP
Visit our Web site www.mcdougallittell.com for extra examples.

EXAMPLE 3 *Factoring when b and c are Negative*

Factor $x^2 - 2x - 8$.

SOLUTION For this trinomial, $b = -2$ and $c = -8$. Because c is negative, you know that p and q cannot both have negative values.

$x^2 - 2x - 8 = (x + p)(x + q)$ Find p and q when $p + q = -2$ and $pq = -8$.

$= (x + 2)(x - 4)$ $p = 2$ and $q = -4$

✓**CHECK:** Use a graphing calculator. Graph $y = x^2 - 2x - 8$ and $y = (x + 2)(x - 4)$ on the same screen.

The graphs coincide, so your answer is correct.

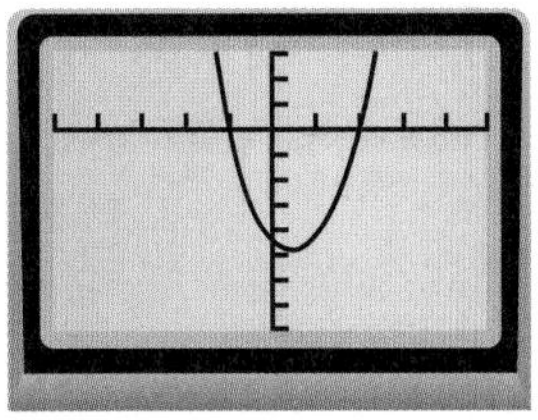

EXAMPLE 4 *Factoring when b is Positive and c is Negative*

Factor $x^2 + 7x - 18$.

SOLUTION For this trinomial, $b = 7$ and $c = -18$. Because c is negative, you know that p and q cannot both have negative values.

$x^2 + 7x - 18 = (x + p)(x + q)$ Find p and q when $p + q = 7$ and $pq = -18$.

$= (x + 9)(x - 2)$ $p = 9$ and $q = -2$

....

STUDENT HELP

Look Back
For help with finding the discriminant, see p. 541.

It is important to realize that *many* quadratic trinomials with integer coefficients cannot be factored into linear factors with integer coefficients. A quadratic trinomial $x^2 + bx + c$ can be factored (using integer coefficients) only if the *discriminant* is a perfect square.

EXAMPLE 5 *Using the Discriminant*

Tell whether the trinomial can be factored.

a. $x^2 + 3x - 4$ **b.** $x^2 + 3x - 6$

SOLUTION Find the discriminant.

a. $b^2 - 4ac = 3^2 - 4(1)(-4)$ $a = 1$, $b = 3$, and $c = -4$

$= 25$ Simplify.

▶ The discriminant is a perfect square, so the trinomial can be factored.

b. $b^2 - 4ac = 3^2 - 4(1)(-6)$ $a = 1$, $b = 3$, and $c = -6$

$= 33$ Simplify.

▶ The discriminant is not a perfect square, so the trinomial cannot be factored.

2 TEACH

MOTIVATING THE LESSON
Write the numbers from 1 to 10 on small cards or pieces of paper. Then give the cards to 10 students. Ask the students if the numbers on their cards are factors of larger numbers, such as 24, 50, or 60. Discuss what it means to be a factor of a number, and then relate this idea to what it means to be a factor of a polynomial.

EXTRA EXAMPLE 1
Factor $x^2 + 8x + 15$. $(x + 5)(x + 3)$

EXTRA EXAMPLE 2
Factor $x^2 - 9x + 20$. $(x - 5)(x - 4)$

EXTRA EXAMPLE 3
Factor $x^2 - 8x - 9$. $(x - 9)(x + 1)$

EXTRA EXAMPLE 4
Factor $x^2 + 3x - 18$. $(x + 6)(x - 3)$

EXTRA EXAMPLE 5
Tell whether the trinomial can be factored.
a. $x^2 + 6x - 7$ yes
b. $x^2 + 6x - 5$ no

CHECKPOINT EXERCISES

For use after Examples 1 and 2:

1. a. Factor $x^2 + 11x + 10$. $(x + 10)(x + 1)$

b. Factor $x^2 - 17x + 60$. $(x - 12)(x - 5)$

For use after Examples 3 and 4:

2. a. Factor $x^2 - 2x - 48$. $(x - 8)(x + 6)$

b. Factor $x^2 + 4x - 12$. $(x + 6)(x - 2)$

For use after Example 5:

3. Tell whether $x^2 - 10x + 25$ can be factored. yes

STUDENT HELP NOTES

Homework Help Students can find extra examples at **www.mcdougallittell.com** that parallel the examples in the student edition.

EXTRA EXAMPLE 6
Solve the equation $x^2 - 5x = 24$.
−3, 8

EXTRA EXAMPLE 7
You are putting a dog run in your backyard. The dog run will be put in along one side and the back of your backyard and will have a constant width. After the dog run is installed, the remaining backyard will be 20 feet by 56 feet. You want the dog run to take up a total of 320 square feet. How wide should the dog run be? **4 feet**

CHECKPOINT EXERCISES
For use after Example 6:
1. Solve the equation $x^2 + 11x = -18$. **−2, −9**

For use after Example 7:
2. Solve the problem in Extra Example 7 above if the total area of the dog run takes up 500 square feet and the remaining yard will be 20 feet by 75 feet. **5 feet**

CAREER NOTE
EXAMPLE 7 Additional information about a career as a Landscape Designer is available at **www.mcdougallittell.com.**

FOCUS ON VOCABULARY
The concept of *factor* is key to understanding Lessons 10.5 through 10.8. Ask students what it means to say that $(x + p)$ and $(x + q)$ are the factors of $x^2 + bx + c$.
$(x + p)(x + q) = x^2 + bx + c$, with $p + q = b$ and $pq = c$.

CLOSURE QUESTION
Describe how to factor $x^2 + bx + c$.
See sample answer, below.

DAILY PUZZLER
How many total squares are in the figure below?

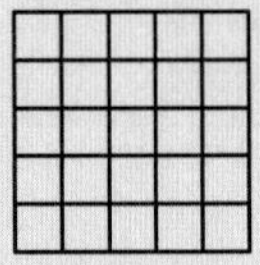

25 + 16 + 9 + 4 + 1 = 55

GOAL 2 SOLVING QUADRATIC EQUATIONS BY FACTORING

EXAMPLE 6 *Solving a Quadratic Equation*

$x^2 - 3x = 10$	Write equation.
$x^2 - 3x - 10 = 0$	Write in standard form.
$(x - 5)(x + 2) = 0$	Factor left side.
$(x - 5) = 0$ or $(x + 2) = 0$	Use zero-product property.
$x - 5 = 0$	Set first factor equal to 0.
$x = 5$	Solve for *x*.
$x + 2 = 0$	Set second factor equal to 0.
$x = -2$	Solve for *x*.

▶ The solutions are 5 and −2. Check these in the original equation.

EXAMPLE 7 *Writing a Quadratic Model*

LANDSCAPE DESIGN You are putting a stone border along two sides of a rectangular Japanese garden that measures 6 yards by 15 yards. Your budget limits you to only enough stone to cover 46 square yards. How wide should the border be?

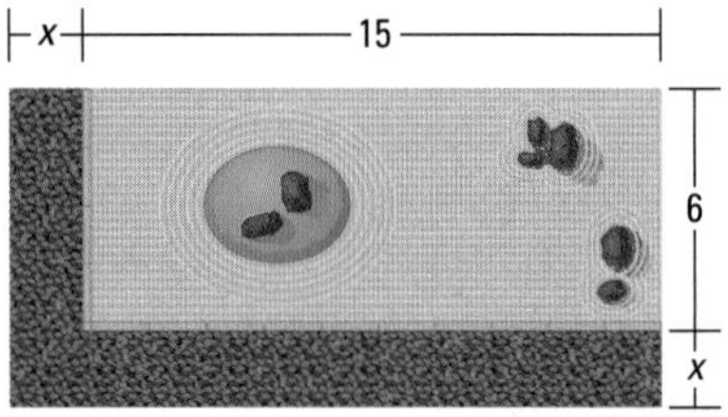

SOLUTION

Begin by drawing and labeling a diagram.

Area of border	=	Total area	−	Garden area

$46 = (x + 15)(x + 6) - (15)(6)$	Write quadratic model.
$46 = x^2 + 21x + 90 - 90$	Multiply.
$0 = x^2 + 21x - 46$	Write in standard form.
$0 = (x + 23)(x - 2)$	Factor.
$(x + 23) = 0$ or $(x - 2) = 0$	Use zero-product property.
$x + 23 = 0$	Set first factor equal to 0.
$x = -23$	Solve for *x*.
$x - 2 = 0$	Set second factor equal to 0.
$x = 2$	Solve for *x*.

▶ The solutions are −23 and 2. Only $x = 2$ is a reasonable solution, because negative values for dimension do not make sense. Make the border 2 yards wide.

FOCUS ON CAREERS

REAL LIFE **LANDSCAPE DESIGNERS** plan and map out the appearance of outdoor spaces like parks, gardens, golf courses, and other recreation areas.

INTERNET **CAREER LINK**
www.mcdougallittell.com

Closure Question ***Sample answer:***
Write the parentheses for the binomials and place *x* as the first term of each. Then find two factors of *c* that add up to *b*. Place those as the second terms in the binomials.

GUIDED PRACTICE

Vocabulary Check ✓

Concept Check ✓

1. What does it mean to factor a quadratic expression?
to write it as the product of two linear expressions
2. Factor $x^2 - 4x + 3$. When testing possible factorizations, why is it unnecessary to test $(x - 1)(x + 3)$ and $(x + 1)(x - 3)$? **See margin.**

2. $(x - 1)(x - 3)$; the product of the constant terms is positive, so both must be positive or both must be negative.

3. Factor $x^2 + 2x - 3$. When testing possible factorizations, why is it unnecessary to test $(x - 1)(x - 3)$ and $(x + 1)(x + 3)$? **$(x + 3)(x - 1)$; the product of the constant terms is negative, so one must be positive and the other negative.**
4. Can a trinomial $x^2 + bx + c$, where b and c are integers, be factored with integer coefficients if its discriminant is 35? Explain.
No; 35 is not a perfect square.

Skill Check ✓

Match the trinomial with a correct factorization.

5. $x^2 - x - 20$ **D**
6. $x^2 + x - 20$ **A**
7. $x^2 + 9x + 20$ **B**
8. $x^2 - 9x + 20$ **C**

A. $(x + 5)(x - 4)$
B. $(x + 4)(x + 5)$
C. $(x - 4)(x - 5)$
D. $(x + 4)(x - 5)$

Use the discriminant to decide whether the equation can be solved by factoring. Explain your reasoning.

9. $x^2 - 4x + 4 = 0$
The equation can be solved by factoring because the discriminant, 0, is a perfect square.
10. $x^2 - 4x - 5 = 0$
The equation can be solved by factoring because the discriminant, 36, is a perfect square.
11. $x^2 - 4x - 6 = 0$
The equation cannot be solved by factoring because the discriminant, 40, is not a perfect square.

PRACTICE AND APPLICATIONS

STUDENT HELP

Extra Practice to help you master skills is on p. 806.

FACTORED FORM **Choose the correct factorization. If neither is correct, find the correct factorization.**

12. $x^2 + 7x + 12$ **B**
A. $(x + 6)(x + 2)$
B. $(x + 4)(x + 3)$

13. $x^2 - 10x + 16$ **B**
A. $(x - 4)(x - 4)$
B. $(x - 8)(x - 2)$

14. $x^2 + 11x - 26$
A. $(x - 13)(x + 2)$
B. $(x - 13)(x - 2)$
$(x + 13)(x - 2)$

FACTORING TRINOMIALS **Factor the trinomial.**

15. $x^2 + 8x - 9$ **$(x + 9)(x - 1)$**
16. $t^2 - 10t + 21$ **$(t - 3)(t - 7)$**
17. $b^2 + 5b - 24$ **$(b + 8)(b - 3)$**
18. $w^2 + 13w + 36$ **$(w + 4)(w + 9)$**
19. $y^2 - 3y - 18$ **$(y - 6)(y + 3)$**
20. $c^2 + 14c + 40$ **$(c + 10)(c + 4)$**
21. $m^2 - 7m - 30$ **$(m - 10)(m + 3)$**
22. $32 + 12n + n^2$ **$(4 + n)(8 + n)$**
23. $44 - 15s + s^2$ **$(11 - s)(4 - s)$**
24. $z^2 + 65z + 1000$ **$(z + 40)(z + 25)$**
25. $x^2 - 45x + 450$ **$(x - 15)(x - 30)$**
26. $d^2 - 33d - 280$ **$(d - 40)(d + 7)$**

STUDENT HELP

HOMEWORK HELP
Examples 1–4: Exs. 12–26, 48–51
Example 5: Exs. 42–47
Example 6: Exs. 27–41
Example 7: Exs. 56–60

SOLVING QUADRATIC EQUATIONS **Solve the equation by factoring.**

27. $x^2 + 7x + 10 = 0$ **−2, −5**
28. $x^2 + 5x - 14 = 0$ **−7, 2**
29. $x^2 - 9x = -14$ **2, 7**
30. $x^2 + 32x = -220$ **−22, −10**
31. $x^2 + 16x = -15$ **−1, −15**
32. $x^2 + 3x = 54$ **−9, 6**
33. $x^2 + 8x = 65$ **−13, 5**
34. $-x + x^2 = 56$ **−7, 8**
35. $x^2 - 20x = -51$ **3, 17**
36. $x^2 - 5x = 84$ **−7, 12**
37. $x^2 + 3x - 31 = -3$ **−7, 4**
38. $x^2 - 2x - 19 = -4$ **−3, 5**
39. $x^2 - x - 8 = 82$ **−9, 10**
40. $x^2 + 42 = 13x$ **6, 7**
41. $x^2 - 9x + 18 = 2x$ **2, 9**

3 APPLY

ASSIGNMENT GUIDE

BASIC
Day 1: pp. 607–608 Exs. 12–46 even, 52, 53, 56–58
Day 2: pp. 607–609 Exs. 13–47 odd, 54, 55, 59, 60, 65–67, 75, 80, 85, 90, 95, 96

AVERAGE
Day 1: pp. 607–608 Exs. 12–46 even, 52, 53, 56–58
Day 2: pp. 607–609 Exs. 13–47 odd, 54, 55, 59–62, 65–67, 75, 80, 85, 90, 95, 96

ADVANCED
Day 1: pp. 607–608 Exs. 12–46 even, 52, 53, 56–58
Day 2: pp. 607–609 Exs. 13–47 odd, 54, 55, 59, 60–62, 65–71, 75, 80, 85, 90, 95, 96

BLOCK SCHEDULE
pp. 607–609 Exs. 12–47, 52–62, 65–67, 75, 80, 85, 90, 95, 96

EXERCISE LEVELS

Level A: *Easier*
12–14
Level B: *More Difficult*
15–67, 72–96
Level C: *Most Difficult*
68–71

✓ HOMEWORK CHECK

To quickly check student understanding of key concepts, go over the following exercises: Exs. 12, 18, 20, 28, 44, 52, 58, 62. See also the Daily Homework Quiz:

- Blackline Master (*Chapter 10 Resource Book*, p. 79)
- Transparency (p. 81)

GRAPHING CALCULATOR NOTE

EXERCISES 48–51 Students can review the method of graphing and solving quadratic equations by looking back to Activity 9.4 on page 532.

APPLICATION NOTE

EXERCISES 63 AND 64 Additional information about the Taj Mahal is available at **www.mcdougallittell.com.**

MATHEMATICAL REASONING

Ask students to tell what the result is when you find the product of all the prime factors of a number. Then ask what the result is when you find the product of the linear factors of a quadratic expression. **the original number; the original expression**

ADDITIONAL PRACTICE AND RETEACHING

For Lesson 10.5:

- Practice Levels A, B, and C (*Chapter 10 Resource Book,* p. 69)
- Reteaching with Practice (*Chapter 10 Resource Book,* p. 72)
- See Lesson 10.5 of the *Personal Student Tutor*

For more Mixed Review:

- Search the *Test and Practice Generator* for key words or specific lessons.

USING THE DISCRIMINANT **Tell whether the quadratic expression can be factored with integer coefficients. If it can, find the factors.**

42. $t^2 + 7t - 144$ yes; $(x + 16)(x - 9)$

43. $y^2 + 19y + 60$ yes; $(y + 15)(y + 4)$

44. $x^2 - 11x + 24$ yes; $(x - 3)(x - 8)$

45. $w^2 - 6w + 16$ no

46. $z^2 - 26z - 87$ yes; $(z + 3)(z - 29)$

47. $b^2 + 14b + 35$ no

CHECKING GRAPHICALLY **Solve the equation by factoring. Then use a graphing calculator to check your answer.**

48. $x^2 - 17x + 30 = 0$ 2, 15

49. $x^2 + 8x = 105$ −15, 7

50. $x^2 - 20x + 21 = 2$ 1, 19

51. $x^2 + 52x + 680 = 40$ −32, −20

WRITING EQUATIONS **Write a quadratic equation with the given solutions.**
52–55. Sample equations given are in standard form with leading coefficients of 1.

52. 12 and −21 $x^2 + 9x - 252 = 0$

53. −5 and −6 $x^2 + 11x + 30 = 0$

54. −41 and 5 $x^2 + 36x - 205 = 0$

55. 427 and 0 $x^2 - 427x = 0$

STUDENT HELP

TABLE OF FORMULAS
For help with area, see the Table of Formulas, p. 813.

GEOMETRY CONNECTION **In Exercises 56–58, consider a rectangle having one side of length $x - 6$ and having an area given by $A = x^2 - 17x + 66$.**

56. Use factoring to find an expression for the other side of the rectangle. $x - 11$

57. If the area of the rectangle is 84 square feet, what are possible values of x? There is only one, 18.

58. For the value of x found in Exercise 57, what are the dimensions of the rectangle? 12 ft by 7 ft

GEOMETRY CONNECTION **Consider a circle whose radius is greater than 9 and whose area is given by $A = \pi(x^2 - 18x + 81)$. (Use $\pi \approx 3.14$.)**

59. Use factoring to find an expression for the radius of the circle. $x - 9$

60. If the area of the circle is 12.56 square meters, what is the value of x? 11

MAKING A SIGN **In Exercises 61 and 62, a triangular sign has a base that is 2 feet less than twice its height. A local zoning ordinance restricts the surface area of street signs to no more than 20 square feet.**

61. Write an inequality involving the height of the triangle that represents the largest triangular sign allowed. $x^2 - x \le 20$

62. Find the base and height of the largest triangular sign that meets the zoning ordinance. The base is 8 ft wide and the triangle is 5 ft high.

THE TAJ MAHAL **In Exercises 63 and 64, refer to the illustration at the right of the Taj Mahal.**

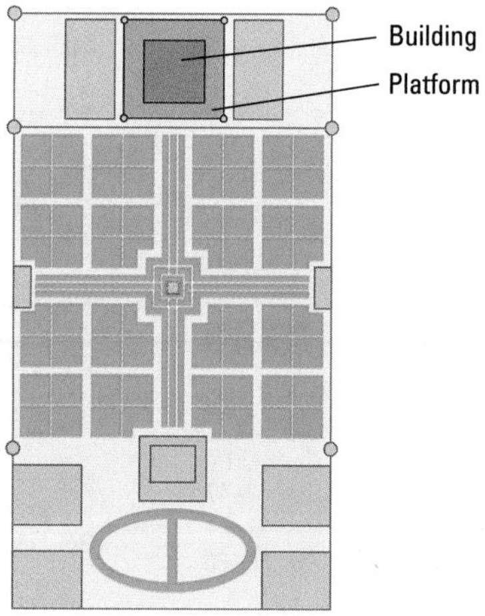

63. The platform is about 38 meters wider than the main building. The total area of the platform is about 9025 square meters. Find the dimensions of the platform and the base of the building. (Assume each is a square.) platform: 95 m by 95 m; building: 57 m by 57 m

64. The entire complex of the Taj Mahal is about 245 meters longer than it is wide. The area of the entire complex is about 167,750 square meters. What are the dimensions of the entire complex? Explain your steps in finding the solution. See margin.

64. 305 m by 550 m; Let x = width of complex, $x + 245$ = length; $x(x + 245) = 167{,}750$; $x^2 + 245x - 167{,}750 = 0$; $(x + 550)(x - 305) = 0$; $x = 305$; $x + 245 = 550$.

FOCUS ON APPLICATIONS

TAJ MAHAL
It took more than 20,000 daily workers 22 years to complete the Taj Mahal around 1643 in India. Constructed primarily of white marble and red sandstone, the Taj Mahal is renowned for its beauty.

APPLICATION LINK
www.mcdougallittell.com

Test Preparation

65. **MULTIPLE CHOICE** The length of a rectangular plot of land with an area of 880 square meters is 24 meters more than its width. A paved area measuring 8 meters by 12 meters is placed on the plot. If w represents the width of the plot of land in meters, which of the following equations can be factored to find possible values of the width of the land? **A**

Ⓐ $w^2 + 24w = 880$ Ⓑ $w^2 - 24w = 880$

Ⓒ $w^2 + 24w = -880$ Ⓓ $w^2 - 24w = -880$

66. **MULTIPLE CHOICE** A triangle's base is 16 feet less than 2 times its height. If h represents the height in feet, and the total area of the triangle is 48 square feet, which of the following equations can be used to determine the height? **B**

Ⓐ $2h + 2(h + 4) = 48$ Ⓑ $h^2 - 8h = 48$

Ⓒ $h^2 + 8h = 48$ Ⓓ $2h^2 - 16h = 48$

67. **MULTIPLE CHOICE** Which of the following equations does *not* have solutions that are integers? **D**

Ⓐ $x^2 + 21x + 100 = -10$ Ⓑ $x^2 - 169 = 0$

Ⓒ $x^2 - 8x - 105 = 0$ Ⓓ $x^2 - 15x - 75 = 0$

★ Challenge

FACTORING CHALLENGE In Exercises 68–71, n is a positive integer. Factor the expression. (*Hint*: $(a^n)^2 = a^{2n}$)

EXTRA CHALLENGE
www.mcdougallittell.com

68. $a^{2n} - b^{2n}$ $(a^n + b^n)(a^n - b^n)$

69. $a^{2n} + 2a^nb^n + b^{2n}$ $(a^n + b^n)^2$

70. $a^{2n} + 18a^nb^n + 81b^{2n}$ $(a^n + 9b^n)^2$

71. $5a^{2n} - 9a^nb^n - 2b^{2n}$ $(5a^n + b^n)(a^n - 2b^n)$

MIXED REVIEW

FINDING THE GCF Find the greatest common factor. (Skills Review, p. 777)

72. 30, 45 **15**
73. 49, 64 **1**
74. 412, 18 **2**
75. 77, 91 **7**
76. 20, 32, 40 **4**
77. 36, 54, 162 **18**

MULTIPLYING EXPRESSIONS Find the product. (Review 10.2, 10.3)

78. $3q(q^3 - 5q^2 + 6)$ $3q^4 - 15q^3 + 18q$
79. $(y + 9)(y - 4)$ $y^2 + 5y - 36$
80. $(7x - 11)^2$ $49x^2 - 154x + 121$
81. $(5 - w)(12 + 3w)$ $60 + 3w - 3w^2$
82. $(3a - 2)(4a + 6)$ $12a^2 + 10a - 12$
83. $(2b - 4)(b^3 + 4b^2 + 5b)$ $2b^4 + 4b^3 - 6b^2 - 20b$
84. $(9x + 8)(9x - 8)$ $81x^2 - 64$
85. $\left(6z + \frac{1}{3}\right)^2$ $36z^2 + 4z + \frac{1}{9}$
86. $(5t - 3)(4t - 10)$ $20t^2 - 62t + 30$

SOLVING FACTORED EQUATIONS Solve the equation. (Review 10.4)

87. $(x + 12)(x + 7) = 0$ $-12, -7$
88. $(z + 2)(z + 3) = 0$ $-2, -3$
89. $(t - 19)^2 = 0$ 19
90. $\left(b - \frac{2}{5}\right)\left(b - \frac{5}{6}\right) = 0$ $\frac{2}{5}, \frac{5}{6}$
91. $(x - 9)(x - 6) = 0$ $9, 6$
92. $(y + 47)(y - 27) = 0$ $-47, 27$
93. $(z - 1)(4z + 2) = 0$ $1, -\frac{1}{2}$
94. $(3a - 8)(a + 5) = 0$ See margin.
95. $(4n - 6)^3 = 0$ $\frac{3}{2}$

94. $\frac{8}{3}, -5$

96. **DISASTER RELIEF** You drop a box of supplies from a helicopter at an altitude of 40 feet above a drop area. Use a vertical motion model to find the time it takes the box to reach the ground. (Review 9.5 for 10.6) **about 1.58 sec**

4 ASSESS

DAILY HOMEWORK QUIZ

Transparency Available

Factor the trinomial.

1. $y^2 + 5y - 14$ $(y + 7)(y - 2)$
2. $x^2 - 3x - 88$ $(x + 8)(x - 11)$
3. $t^2 + 23t + 90$ $(t + 18)(t + 5)$
4. Solve the equation $x^2 - 32 = 4x$. $-4, 8$
5. Can $y^2 + 2y - 120$ be factored with integer coefficients? If so, what are the factors? yes; $(y + 12)(y - 10)$

EXTRA CHALLENGE NOTE

Challenge problems for Lesson 10.5 are available in **blackline** format in the *Chapter 10 Resource Book*, p. 76 and at **www.mcdougallittell.com.**

ADDITIONAL TEST PREPARATION

1. **OPEN ENDED** If a quadratic equation has only positive coefficients and two real solutions, what must be true of the solutions? **Both solutions must be negative.**

1 Planning the Activity

Purpose
To use algebra tiles to model quadratic expressions of the form $ax^2 + bx + c$ and to model their factorization, if possible.

Materials
- algebra tiles for each group
- Activity Support Master (*Chapter 10 Resource Book,* p. 80)

Pacing
- Exploring the Concept — 5 min
- Drawing Conclusions — 10 min

Link to Lesson
Have students use algebra tiles to model the factorization in Example 1 of Lesson 10.6.

2 Managing the Activity

Cooperative Learning
Each student can have a set of algebra tiles when working on the activity, and the group can discuss the combinations they have tried that did not work or show others the correct combination. If each group has only one set, one member could be responsible for the x^2-tiles, one for the x-tiles, and one for the 1-tiles. All students can then work together to model the polynomial.

3 Closing the Activity

★ Key Discovery
Some quadratic expressions $ax^2 + bx + c$ can be factored into the product of binomial linear expressions that represent the length and width of a rectangle.

Activity Assessment
Explain the process of modeling a quadratic trinomial and its factorization with algebra tiles.
See sample answer at right.

7–9. See Additional Answers beginning on page AA1.

ACTIVITY 10.6
Developing Concepts

Modeling the Factorization of $ax^2 + bx + c$

Set Up
Work in a small group.

Materials
algebra tiles

▶ QUESTION **How can you model the factorization of a trinomial of the form $ax^2 + bx + c$ using algebra tiles?**

▶ EXPLORING THE CONCEPT

You can use algebra tiles to create a model that can be used to factor a trinomial that has a leading coefficient other than 1. Factor the trinomial $2x^2 + 5x + 3$ as follows.

1. Use algebra tiles to model $2x^2 + 5x + 3$.

$2x^2$ + $5x$ + 3

2. With the x^2-tiles at the upper left, arrange the x-tiles and the 1-tiles around the x^2-tiles to form a rectangle.

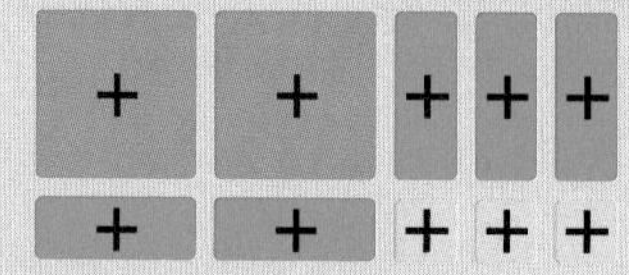

3. The width of the rectangle is _?_, and the length of the rectangle is _?_.

Complete the statement: $2x^2 + 5x + 3 =$ _?_ • _?_ **$x + 1$, $2x + 3$; $(x + 1)(2x + 3)$**

▶ EXERCISES

Use algebra tiles to factor the trinomial. Sketch your model.

1. $2x^2 + 9x + 9$ **$(2x + 3)(x + 3)$**
2. $2x^2 + 7x + 3$ **$(2x + 1)(x + 3)$**
3. $3x^2 + 4x + 1$ **$(3x + 1)(x + 1)$**
4. $3x^2 + 10x + 3$ **$(3x + 1)(x + 3)$**
5. $3x^2 + 10x + 8$ **$(3x + 4)(x + 2)$**
6. $4x^2 + 5x + 1$ **$(4x + 1)(x + 1)$**

Error Analysis **The algebra tile model is incorrect. Sketch the correct model, and use the model to factor the trinomial.**

7. $2x^2 + 3x + 1$

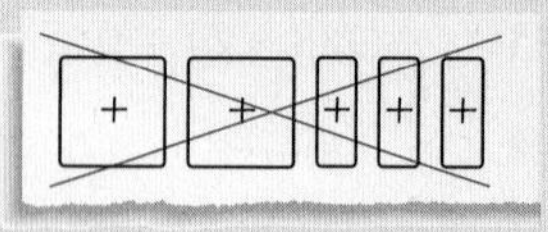

$(2x + 1)(x + 1)$. See margin.

8. $2x^2 + 4x + 2$

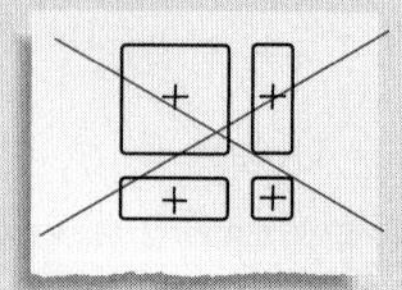

$(2x + 2)(x + 1)$. See margin.

9. $4x^2 + 4x + 1$

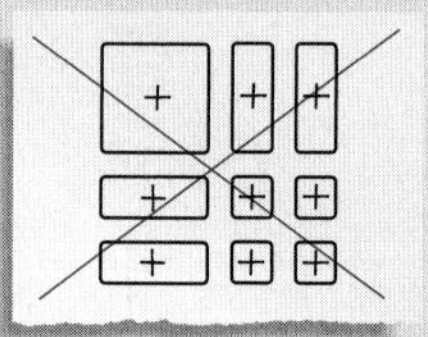

$(2x + 1)(2x + 1)$. See margin.

Activity Assessment ***Sample answer:***
First use algebra tiles to model the trinomial. Then use the tiles to make a rectangle (if possible). The width and length of the rectangle are the binomials into which the trinomial can be factored.

10.6 Factoring $ax^2 + bx + c$

GOAL 1 FACTORING A QUADRATIC TRINOMIAL

What you should learn

GOAL 1 Factor a quadratic expression of the form $ax^2 + bx + c$.

GOAL 2 Solve quadratic equations by factoring.

Why you should learn it

▼ To help solve **real-life** problems, such as finding the time it takes a cliff diver to reach the water in **Example 6**.

REAL LIFE

In this lesson, you will learn to factor quadratic polynomials whose leading coefficient is not 1. Find the factors of a (m and n) and the factors of c (p and q) so that the sum of the outer and inner products (mq and pn) is b.

$$ax^2 + bx + c = (mx + p)(nx + q) \qquad c = pq, \quad a = mn$$

$$b = mq + pn$$

Example: $6x^2 + 22x + 20 = (3x + 5)(2x + 4) \qquad 20 = 5 \cdot 4, \quad 6 = 3 \cdot 2$

$$22 = (3 \cdot 4) + (5 \cdot 2)$$

EXAMPLE 1 *One Pair of Factors for a and c*

Factor $2x^2 + 11x + 5$.

SOLUTION Test the possible factors of a (1 and 2) and c (1 and 5).

Try $a = 1 \cdot 2$ and $c = 1 \cdot 5$.

$(1x + 1)(2x + 5) = 2x^2 + 7x + 5$ **Not correct**

Now try $a = 1 \cdot 2$ and $c = 5 \cdot 1$.

$(1x + 5)(2x + 1) = 2x^2 + 11x + 5$ **Correct**

▶ The correct factorization of $2x^2 + 11x + 5$ is $(x + 5)(2x + 1)$.

EXAMPLE 2 *One Pair of Factors for a and c*

Factor $3x^2 - 4x - 7$.

SOLUTION

FACTORS OF a AND c	PRODUCT	CORRECT?
$a = 1 \cdot 3$ and $c = (-1)(7)$	$(x - 1)(3x + 7) = 3x^2 + 4x - 7$	No
$a = 1 \cdot 3$ and $c = (7)(-1)$	$(x + 7)(3x - 1) = 3x^2 + 20x - 7$	No
$a = 1 \cdot 3$ and $c = (1)(-7)$	$(x + 1)(3x - 7) = 3x^2 - 4x - 7$	Yes
$a = 1 \cdot 3$ and $c = (-7)(1)$	$(x - 7)(3x + 1) = 3x^2 - 20x - 7$	No

▶ The correct factorization of $3x^2 - 4x - 7$ is $(x + 1)(3x - 7)$.

1 PLAN

PACING

Basic: 2 days
Average: 2 days
Advanced: 2 days
Block Schedule: 1 block

LESSON OPENER
GRAPHING CALCULATOR

An alternative way to approach Lesson 10.6 is to use the Graphing Calculator Lesson Opener:

- Blackline Master (*Chapter 10 Resource Book*, p. 81)
- Transparency (p. 71)

MEETING INDIVIDUAL NEEDS

- ***Chapter 10 Resource Book***
  Prerequisite Skills Review (p. 5)
  Practice Level A (p. 83)
  Practice Level B (p. 84)
  Practice Level C (p. 85)
  Reteaching with Practice (p. 86)
  Absent Student Catch-Up (p. 88)
  Challenge (p. 90)
- ***Resources in Spanish***
- ***Personal Student Tutor***

NEW-TEACHER SUPPORT

See the Tips for New Teachers on pp. 1–2 of the *Chapter 10 Resource Book* for additional notes about Lesson 10.6.

WARM-UP EXERCISES

Transparency Available

Factor the trinomial.

1. $x^2 - 10x + 24$ $(x - 6)(x - 4)$
2. $x^2 + 10x + 9$ $(x + 9)(x + 1)$
3. $x^2 + 6x - 40$ $(x + 10)(x - 4)$
4. $x^2 - 6x - 40$ $(x - 10)(x + 4)$
5. $x^2 - 11x + 24$ $(x - 8)(x - 3)$

2 TEACH

MOTIVATING THE LESSON
Use small cards to write all of the possible factors of a trinomial of the form $ax^2 + bx + c$. First write a trinomial such as $4x^2 + 4x - 3$, which factors into $(2x + 3)(2x - 1)$, on the board. Then on the cards, you would write: $4x - 1$, $4x + 1$, $4x - 3$, $4x + 3$, $2x - 1$, $2x + 1$, $2x - 3$, and $2x + 3$. Give the cards to students and ask pairs of students if the product of the expressions on their cards is equal to the trinomial. Use several examples like this one to introduce the factoring process discussed in this lesson.

EXTRA EXAMPLE 1
Factor $3x^2 + 5x + 2$.
$(3x + 2)(x + 1)$

EXTRA EXAMPLE 2
Factor $2x^2 + 21x - 11$.
$(x + 11)(2x - 1)$

EXTRA EXAMPLE 3
Factor $8x^2 - 14x - 15$.
$(2x - 5)(4x + 3)$

EXTRA EXAMPLE 4
Factor $6x^2 + 9x - 27$.
$3(2x - 3)(x + 3)$

CHECKPOINT EXERCISES

For use after Examples 1 and 2:
1. Factor $7x^2 - 4x - 3$.
$(7x + 3)(x - 1)$

For use after Example 3:
2. Factor $6x^2 - 5x - 21$.
$(3x - 7)(2x + 3)$

For use after Example 4:
3. Factor $10x^2 + 25x - 90$.
$5(x - 2)(2x + 9)$

EXAMPLE 3 *Several Pairs of Factors for a and c*

Factor $6x^2 - 19x + 15$.

SOLUTION

Both factors of c must be negative, because b is negative and c is positive. Test the possible factors of a and c.

FACTORS OF *a* AND *c*	PRODUCT	CORRECT?
$a = 1 \cdot 6$ and $c = (-1)(-15)$	$(x - 1)(6x - 15) = 6x^2 - 21x + 15$	No
$a = 1 \cdot 6$ and $c = (-15)(-1)$	$(x - 15)(6x - 1) = 6x^2 - 91x + 15$	No
$a = 1 \cdot 6$ and $c = (-3)(-5)$	$(x - 3)(6x - 5) = 6x^2 - 23x + 15$	No
$a = 1 \cdot 6$ and $c = (-5)(-3)$	$(x - 5)(6x - 3) = 6x^2 - 33x + 15$	No
$a = 2 \cdot 3$ and $c = (-1)(-15)$	$(2x - 1)(3x - 15) = 6x^2 - 33x + 15$	No
$a = 2 \cdot 3$ and $c = (-15)(-1)$	$(2x - 15)(3x - 1) = 6x^2 - 47x + 15$	No
$a = 2 \cdot 3$ and $c = (-3)(-5)$	$(2x - 3)(3x - 5) = 6x^2 - 19x + 15$	Yes
$a = 2 \cdot 3$ and $c = (-5)(-3)$	$(2x - 5)(3x - 3) = 6x^2 - 21x + 15$	No

▶ The correct factorization of $6x^2 - 19x + 15$ is $(2x - 3)(3x - 5)$.

EXAMPLE 4 *A Common Factor for a, b, and c*

Factor $6x^2 - 2x - 8$.

SOLUTION

Begin by factoring out the common factor 2.

$$6x^2 - 2x - 8 = 2(3x^2 - x - 4)$$

Now factor $3x^2 - x - 4$ by testing the possible factors of a and c.

FACTORS OF *a* AND *c*	PRODUCT	CORRECT?
$a = 1 \cdot 3$ and $c = (-1)(4)$	$(x - 1)(3x + 4) = 3x^2 + x - 4$	No
$a = 1 \cdot 3$ and $c = (4)(-1)$	$(x + 4)(3x - 1) = 3x^2 + 11x - 4$	No
$a = 1 \cdot 3$ and $c = (1)(-4)$	$(x + 1)(3x - 4) = 3x^2 - x - 4$	Yes
$a = 1 \cdot 3$ and $c = (-4)(1)$	$(x - 4)(3x + 1) = 3x^2 - 11x - 4$	No
$a = 1 \cdot 3$ and $c = (-2)(2)$	$(x - 2)(3x + 2) = 3x^2 - 4x - 4$	No
$a = 1 \cdot 3$ and $c = (2)(-2)$	$(x + 2)(3x - 2) = 3x^2 + 4x - 4$	No

▶ The correct factorization is $6x^2 - 2x - 8 = 2(x + 1)(3x - 4)$.

In Examples 1 through 4, each test is shown for the sake of completeness. But when you factor, you can stop testing once you find the correct factorization.

STUDENT HELP

Study Tip
You may find it easier to factor quadratic trinomials that have positive leading coefficients. If the leading coefficient is negative, factor out −1 first. For instance, $-2x^2 + 3x - 1 = (-1)(2x^2 - 3x + 1)$.

GOAL 2 SOLVING QUADRATIC EQUATIONS BY FACTORING

EXAMPLE 5 *Solving a Quadratic Equation*

$21n^2 + 14n + 7 = 6n + 11$	Write equation.
$21n^2 + 8n - 4 = 0$	Write in standard form.
$(3n + 2)(7n - 2) = 0$	Factor left side.
$(3n + 2) = 0$ or $(7n - 2) = 0$	Use zero-product property.
$3n + 2 = 0 \qquad 7n - 2 = 0$	
$n = -\frac{2}{3} \qquad n = \frac{2}{7}$	

▶ The solutions are $-\frac{2}{3}$ and $\frac{2}{7}$.

✓**CHECK** Substitute $-\frac{2}{3}$ and $\frac{2}{7}$ for n in $21n^2 + 14n + 7 = 6n + 11$.

$$21\left(-\frac{2}{3}\right)^2 + 14\left(-\frac{2}{3}\right) + 7 = 6\left(-\frac{2}{3}\right) + 11$$

$$7 = 7$$

$$21\left(\frac{2}{7}\right)^2 + 14\left(\frac{2}{7}\right) + 7 = 6\left(\frac{2}{7}\right) + 11$$

$$\frac{89}{7} = \frac{89}{7}$$

REAL LIFE
Cliff Diving

EXAMPLE 6 *Writing a Quadratic Model*

When a diver jumps from a ledge, the vertical component of his motion can be modeled by the vertical motion model. Suppose the ledge is 48 feet above the ocean and the initial upward velocity is 8 feet per second. How long will it take until the diver enters the water?

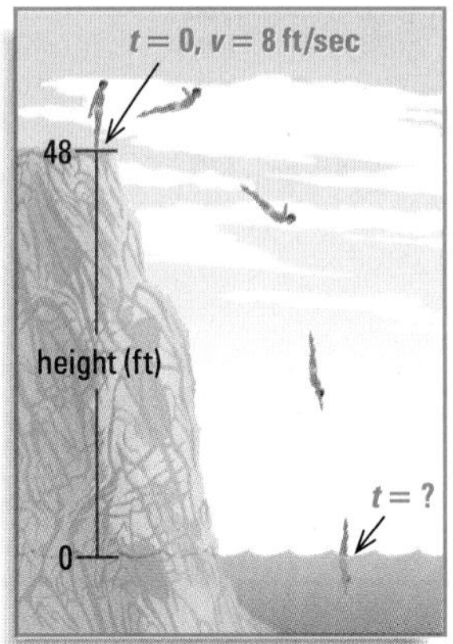

Not drawn to scale

SOLUTION Use a vertical motion model.
Let $v = 8$ and $s = 48$.

$h = -16t^2 + vt + s$	Vertical motion model
$= -16t^2 + 8t + 48$	Substitute values.

To find the time when the diver enters the water, let $h = 0$ and solve the resulting equation for t.

$0 = -16t^2 + 8t + 48$	Write quadratic model.
$0 = (-8)(2t^2 - t - 6)$	Factor out -8.
$0 = (-8)(t - 2)(2t + 3)$	Factor.

▶ The solutions are 2 and $-\frac{3}{2}$. Negative values for time do not make sense, so the only reasonable solution is $t = 2$. It will take the diver 2 seconds.

STUDENT HELP

Look Back
For help with using a vertical motion model, see p. 535.

Closure Question *Sample answer:*
The value of *b* is not the sum of the factors of *c* but it is the sum of the product of factors for *a* and *c*.

EXTRA EXAMPLE 5
Solve $8x^2 + 10x - 11 = 12x + 10$. **$\frac{7}{4}, -\frac{3}{2}$**

EXTRA EXAMPLE 6
You jump off the high dive platform at the city pool which is 12 feet above the water with an initial velocity of 4 feet per second. How long will it take you to enter the water? **1 second**

CHECKPOINT EXERCISES

For use after Example 5:

1. Solve $12x^2 - 8x - 4 = -4x - 3$. **$-\frac{1}{6}, \frac{1}{2}$**

For use after Example 6:

2. If the platform at the city pool is 2 feet above the water and you jump off the platform with an initial velocity of 4 feet per second, how long will it take you to enter the water? **0.5 second**

STUDENT HELP NOTES

Look Back As students look back to p. 535 for help with using a vertical motion model, remind them that the force of gravity (in feet or meters), the initial velocity, and the initial height are needed to write the equation.

FOCUS ON VOCABULARY
Students need to understand the difference between a variable and a letter that represents a coefficient. In the expression $ax^2 + bx + c$, the variable is x, and a, b, and c represent coefficients that *will not* change value within an expression.

CLOSURE QUESTION
How is factoring a trinomial of the form $ax^2 + bx + c$ different from factoring a trinomial of the form $x^2 + bx + c$?
See sample answer, below.

DAILY PUZZLER
Correct this equation by repositioning one number and doing nothing else. 82 – 43 = 1 **$82 - 3^4 = 1$**

3 APPLY

ASSIGNMENT GUIDE

BASIC

Day 1: pp. 614–615 Exs. 16–62 even

Day 2: pp. 614–617 Exs. 15–63 odd, 64, 65, 73–76, 80, 83, 85, 88, 90, Quiz 2 Exs. 1–27

AVERAGE

Day 1: pp. 614–615 Exs. 16–62 even

Day 2: pp. 614–617 Exs. 15–63 odd, 64, 65, 73–76, 80, 83, 85, 88, 90, Quiz 2 Exs. 1–27

ADVANCED

Day 1: pp. 614–615 Exs. 16–62 even

Day 2: pp. 614–617 Exs. 15–63 odd, 64, 65, 73–76, 80, 83, 85, 88, 90, Quiz 2 Exs. 1–27

BLOCK SCHEDULE

pp. 614–617 Exs. 15–65, 73–76, 80, 83, 85, 88, 90, Quiz 2 Exs. 1–27

EXERCISE LEVELS

Level A: *Easier*
15–17

Level B: *More Difficult*
18–68, 73–76, 79–90

Level C: *Most Difficult*
69–72, 77, 78

✔ HOMEWORK CHECK

To quickly check student understanding of key concepts, go over the following exercises: Exs. 16, 18, 24, 30, 38, 46, 50, 52, 58, 62. See also the Daily Homework Quiz:

- Blackline Master (*Chapter 10 Resource Book*, p. 94)
- Transparency (p. 82)

GUIDED PRACTICE

Concept Check ✔

1. False; *Sample answer:* There is no product property of 1 similar to the zero product property.
2. Before factoring, the equation should have been rewritten with one side equal to zero; $-1\frac{1}{2}$, 3.

1. Tell whether the following statement is *true* or *false*. *If* $(5x - 1)(x + 3) = 1$, *then* $5x - 1 = 1$ *or* $x + 3 = 1$. Explain. **See margin.**

2. ERROR ANALYSIS Describe the error at the right. Then solve the equation by factoring correctly. **See margin.**

$2x^2 - 3x + 1 = 10$
$2x^2 - 3x + 1 = (2x - 1)(x - 1)$
$2x - 1 = 0$
$x = \frac{1}{2}$
$x - 1 = 0$
$x = 1$

3. Can a quadratic expression be factored if its discriminant is 1? Explain. **Yes; 1 is a perfect square.**

4. The sum of a number and its square is zero. Write and solve an equation to find numbers that fit this description. $x + x^2 = 0$; $-1, 0$

Skill Check ✔

Match the trinomial with the correct factorization.

5. $3x^2 - 17x - 6$	B	**A.** $(3x + 2)(x + 3)$
6. $3x^2 + 7x - 6$	D	**B.** $(3x + 1)(x - 6)$
7. $3x^2 + 11x + 6$	A	**C.** $(3x - 1)(x + 6)$
8. $3x^2 + 17x - 6$	C	**D.** $(3x - 2)(x + 3)$

Use the discriminant to decide whether the expression can be factored. If it can be factored, factor the expression.

9. $2x^2 - 3x - 2$ $(2x + 1)(x - 2)$

10. $-3y^2 - 4y + 7$ $(-3y - 7)(y - 1)$

11. $6t^2 - 4t - 5$ cannot be factored

Solve the equation.

12. $3b^2 + 26b + 35 = 0$ $-7, -\frac{5}{3}$

13. $2z^2 + 15z = 8$ $-8, \frac{1}{2}$

14. $-7n^2 - 40n = -12$ $-6, \frac{2}{7}$

PRACTICE AND APPLICATIONS

STUDENT HELP

Extra Practice to help you master skills is on p. 806.

FACTORIZATIONS **Choose the correct factorization. If neither is correct, find the correct factorization.**

15. $3x^2 + 2x - 8$ **A**
A. $(3x - 4)(x + 2)$
B. $(3x - 4)(x - 2)$

16. $6y^2 - 29y - 5$
A. $(2y + 1)(3y - 5)$
B. $(6y - 1)(y + 5)$
$(6y + 1)(y - 5)$

17. $4w^2 - 14w - 30$ **A**
A. $(2w + 3)(2w - 10)$
B. $(4w + 15)(w - 2)$

STUDENT HELP

HOMEWORK HELP
Example 1: Exs. 15–32
Example 2: Exs. 15–32
Example 3: Exs. 15–32
Example 4: Exs. 15–32
Example 5: Exs. 33–47
Example 6: Exs. 64–66

FACTORING TRINOMIALS **Factor the trinomial if possible. If it cannot be factored, write *not factorable*.**

18. $3t^2 + 16t + 5$ $(3t + 1)(t + 5)$

19. $6b^2 - 11b - 2$ $(6b + 1)(b - 2)$

20. $4n^2 - 26n - 42$ not factorable

21. $5w^2 - 9w - 2$ $(5w + 1)(w - 2)$

22. $4x^2 + 27x + 35$ $(4x + 7)(x + 5)$

23. $6y^2 - 11y - 10$ $(2y - 5)(3y + 2)$

24. $6x^2 - 21x - 9$ not factorable

25. $3c^2 - 37c + 44$ $(3c - 4)(c - 11)$

26. $10x^2 + 17x + 6$ $(2x + 1)(5x + 6)$

27. $14y^2 - 15y + 4$ $(7y - 4)(2y - 1)$

28. $4z^2 + 32z + 63$ $(2z + 9)(2z + 7)$

29. $6t^2 + t - 70$ $(2t + 7)(3t - 10)$

30. $8b^2 + 2b - 3$ $(2b - 1)(4b + 3)$

31. $2z^2 + 19z - 10$ $(z + 10)(2z - 1)$

32. $12m^2 + 48m + 96$ not factorable

SOLVING EQUATIONS **Solve the equation by factoring.**

33. $2x^2 - 9x - 35 = 0$ $-2\frac{1}{2}, 7$ **34.** $7x^2 - 10x + 3 = 0$ $\frac{3}{7}, 1$ **35.** $3x^2 + 34x + 11 = 0$ $-11, -\frac{1}{3}$

36. $4x^2 - 21x + 5 = 0$ $\frac{1}{4}, 5$ **37.** $2x^2 - 17x - 19 = 0$ $-1, 9\frac{1}{2}$ **38.** $5x^2 - 3x - 26 = 0$ $-2, 2\frac{3}{5}$

39. $2x^2 + 19x = -24$ $-8, -1\frac{1}{2}$ **40.** $4x^2 - 8x = -3$ $\frac{1}{2}, 1\frac{1}{2}$ **41.** $6x^2 - 23x = 18$ $-\frac{2}{3}, 4\frac{1}{2}$

42. $8x^2 - 34x + 24 = -11$ $1\frac{3}{4}, 2\frac{1}{2}$ **43.** $10x^2 + x - 10 = -2x + 8$ $-1\frac{1}{2}, 1\frac{1}{5}$

44. $28x^2 - 9x - 1 = -4x + 2$ $-\frac{1}{4}, \frac{3}{7}$ **45.** $24x^2 + 39x + 15 = 0$ $-1, -\frac{5}{8}$

46. $30x^2 - 80x + 50 = 7x - 4$ $\frac{9}{10}, 2$ **47.** $18x^2 - 30x - 100 = 67x + 30$ $-1\frac{1}{9}, 6\frac{1}{2}$

STUDENT HELP

Look Back
For help with using square roots to solve quadratic equations, see p. 505.

For help with the quadratic formula, see p. 533.

CHOOSING A METHOD **Solve the equation by factoring, by finding square roots, or by using the quadratic formula.**

48. $\frac{1}{16}s^2 = 4$ $-8, 8$ **49.** $x^2 - 10x = -25$ 5 **50.** $y^2 - 7y + 6 = -6$ $3, 4$

51. $4n^2 + 2n = 0$ $-\frac{1}{2}, 0$ **52.** $12t^2 = 0$ 0 **53.** $35b^2 - 61b + 24 = 0$ $\frac{3}{5}, 1\frac{1}{7}$

54. $9x^2 - 19 = -3$ $-\frac{4}{3}, \frac{4}{3}$ **55.** $20d^2 - 10d = 100$ $-2, 2\frac{1}{2}$ **56.** $8c^2 - 27c - 24 = 6$ See margin.

57. $56w^2 - 61w = 22$ $-\frac{2}{7}, 1\frac{3}{8}$ **58.** $3n^2 + 12n + 9 = -1$ See margin. **59.** $24z^2 + 46z - 55 = 10$ See margin.

56. $\frac{27 - 3\sqrt{145}}{16}, \frac{27 + 3\sqrt{145}}{16}$

58. $\frac{-6 - \sqrt{6}}{3}, \frac{-6 + \sqrt{6}}{3}$

59. $\frac{-23 - \sqrt{2089}}{24}, \frac{-23 + \sqrt{2089}}{24}$

MULTIPLY AND SOLVE **Multiply each side of the equation by an appropriate power of ten to obtain integer coefficients. Then solve by factoring.**

60. $0.8x^2 + 3.2x + 2.40 = 0$ $-3, -1$ **61.** $0.23t^2 - 0.54t + 0.16 = 0$ $\frac{8}{23}, 2$

62. $0.3n^2 - 2.2n + 8.4 = 0$ no solution **63.** $0.119y^2 - 0.162y + 0.055 = 0$ $\frac{11}{17}, \frac{5}{7}$

VERTICAL COMPONENT OF MOTION **In Exercises 64–66, use the vertical motion model $h = -16t^2 + vt + s$, where h is the height (in feet), t is the time in motion (in seconds), v is the initial velocity (in feet per second), and s is the initial height (in feet). Solve by factoring.**

STUDENT HELP

INTERNET HOMEWORK HELP Visit our Web site www.mcdougallittell.com for help with Exs. 64–66.

64. GYMNASTICS A gymnast dismounts the uneven parallel bars at a height of 8 feet with an initial upward velocity of 8 feet per second. Find the time t (in seconds) it takes for the gymnast to reach the ground. Is your answer reasonable? **1 sec; yes**

65. CIRCUS ACROBATS An acrobat is shot out of a cannon and lands in a safety net that is 10 feet above the ground. Before being shot out of the cannon, she was 4 feet above the ground. She left the cannon with an initial upward velocity of 50 feet per second. Find the time t (in seconds) it takes for her to reach the net. Explain why only one of the two solutions is reasonable.

65. **3 sec; $t = 0.125$ is not a reasonable solution, because at that time, the acrobat is still rising.**

66. T-SHIRT CANNON At a basketball game, T-shirts are rolled-up into a ball and shot from a "T-shirt cannon" into the crowd. The T-shirts are released from a height of 6 feet with an initial upward velocity of 44 feet per second. If you catch a T-shirt at your seat 30 feet above the court, how long was it in the air before you caught it? Is your answer reasonable?

66. **2 sec; $t = 0.75$ is not a reasonable solution, because at that time the T-shirt is leaving the cannon.**

! COMMON ERROR

EXERCISES 18–32 The first step in factoring is to see if all of the terms have any factors in common and to use the distributive property to factor them out. Many students will forget this step and will not completely factor trinomials. Remind student of the procedure used in Example 4 on page 612 when you assign these exercises.

STUDENT HELP NOTES

- **Look Back** As students look back to p. 505 for help with using square roots, remind them that positive and negative solutions result when the square root is taken of both sides of an equation.
- **Look Back** As students look back to p. 533 for help with the quadratic formula, remind them that they need to know the values of a, b, and c in the equation. The value of a is the coefficient of the x^2-term, the value of b is the coefficient of the x-term, and the value of c is the constant term.
- **Homework Help** Students can find help for Exs. 64–66 at **www.mcdougallittell.com.** The information can be printed out for students who don't have access to the Internet.

ADDITIONAL PRACTICE AND RETEACHING

For Lesson 10.6:

- Practice Levels A, B, and C (*Chapter 10 Resource Book,* p. 83)
- Reteaching with Practice (*Chapter 10 Resource Book,* p. 86)
- See Lesson 10.6 of the *Personal Student Tutor*

For more Mixed Review:

- Search the *Test and Practice Generator* for key words or specific lessons.

4 ASSESS

DAILY HOMEWORK QUIZ

Transparency Available

Factor, if possible. If not, write *not factorable.*

1. $14x^2 + 36x - 18$ **$2(7x - 3)(x + 3)$**
2. $x^2 + 4x - 7$ **not factorable**
3. $4n^2 + 4n - 288$ **$4(n + 9)(n - 8)$**

Solve the equation.

4. $6x^2 + 19x - 36 = 0$ **$\frac{4}{3}, -\frac{9}{2}$**
5. $5y^2 + 22y = 15$ **$-5, 0.6$**
6. $0.4z^2 - 0.6 = z$ **$-\frac{1}{2}, 3$**

EXTRA CHALLENGE NOTE

Challenge problems for Lesson 10.6 are available in **blackline** format in the *Chapter 10 Resource Book,* p. 90 and at **www.mcdougallittell.com.**

ADDITIONAL TEST PREPARATION

1. **WRITING** Explain the difference between factoring a polynomial expression and solving a polynomial equation. ***Sample answer:*** **A polynomial expression is written as the product of linear factors. A polynomial equation is set equal to 0, then factored, and then the factors can be set equal to 0 to find the solutions.**

WARP AND WEFT In Exercises 67 and 68, use the following information.
In every square inch of sailcloth, the warp (lengthwise threads) intersects the weft (crosswise threads) about 9000 times. The density (number of threads per inch) of the weft threads to the warp threads is about 5 to 2.

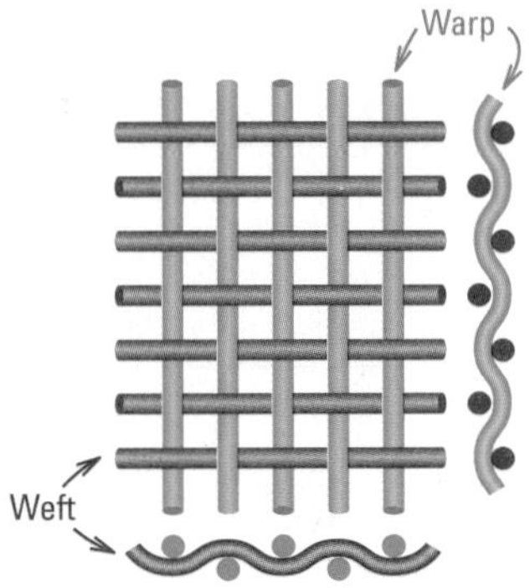

67. Write an equation that you can solve to find the number of threads per square inch. Let x represent the number of warp threads. **$2.5x^2 = 9000$**
68. How many threads are there in each square inch of sailcloth? **210 (60 warp threads and 150 weft threads)**

Test Preparation

EXTENSION: WRITING EQUATIONS In Exercises 69–72, you are tutoring a friend and want to create some quadratic equations that can be solved by factoring. Find a quadratic equation that has the given solutions and explain the procedure you used to obtain the equation. **See margin.**

69. 4 and -3 **$x^2 - x - 12 = 0$**
70. 8 and -8 **$x^2 - 64 = 0$**
71. $-\frac{1}{2}$ and $\frac{1}{3}$ **$6x^2 + x - 1 = 0$**
72. $-\frac{5}{4}$ and $-\frac{8}{3}$ **$12x^2 + 47x + 40 = 0$**

69–72. *Sample procedure*: If integers a and b are solutions of a quadratic equation, then the equation can be written in the form $(x - a)(x - b) = 0$. If fractions $\frac{a}{b}$ and $\frac{c}{d}$ are solutions of a quadratic equation, then the equation can be written in the form $(bx - a)(dx - c) = 0$. The equations given were written first in this form, then the factors were multiplied.

73. **MULTIPLE CHOICE** Which one of the following equations *cannot* be solved by factoring with integer coefficients? **D**

Ⓐ $12x^2 - 15x - 63 = 0$　Ⓑ $12x^2 + 46x - 8 = 0$
Ⓒ $6x^2 - 38x - 28 = 0$　Ⓓ $8x^2 - 49x - 68 = 0$

74. **MULTIPLE CHOICE** Which one of the following is a correct factorization of the expression $-16x^2 + 36x + 52$? **C**

Ⓐ $(-16x + 26)(x - 2)$　Ⓑ $(-4x - 13)(4x + 4)$
Ⓒ $-1(4x - 13)(4x + 4)$　Ⓓ $-1(4x - 4)(4x + 13)$

75. **MULTIPLE CHOICE** Which one of the following is a solution of the equation $7x^2 - 11x - 6 = 0$? **B**

Ⓐ -2　Ⓑ $-\frac{3}{7}$　Ⓒ $-\frac{2}{7}$　Ⓓ $\frac{3}{7}$

76. **MULTIPLE CHOICE** A pebble is thrown upward from the edge of a building 132 feet above the ground with an initial upward velocity of 4 feet per second. How long does it take to reach the ground? **C**

Ⓐ $2\frac{1}{2}$ seconds　Ⓑ $2\frac{3}{4}$ seconds　Ⓒ 3 seconds　Ⓓ 6 seconds

★ Challenge

SAVING FOR A TRIP You are saving money for a trip to Europe that costs \$3600. You make an investment of \$1000 two years before the trip and an investment of \$1800 one year before the trip.

77. Use $(1 + r)$ as a growth factor to write an equation that can be solved to find the growth rate r that you need in order to have \$3600 at the time of the trip to Europe. **$1000(1 + r)^2 + 1800(1 + r) = 3600$**
78. Can the equation be solved by factoring? Solve the equation. What growth rate do you need in order to have \$3600 at the time of the trip to Europe? **yes; 20%**

EXTRA CHALLENGE
www.mcdougallittell.com

MIXED REVIEW

SOLVING SYSTEMS Use linear combinations to solve the system. (Review 7.3)

79. $4x + 5y = 7$ (−2, 3)
$6x - 2y = -18$

80. $6x - 5y = 3$
$-12x + 8y = 5$ $\left(-4\frac{1}{12}, -5\frac{1}{2}\right)$

81. $2x + y = 120$ (40, 40)
$x + 2y = 120$

SOLVING INEQUALITIES Decide whether the ordered pair is a solution of the inequality. (Review 9.7)

82. $y > 2x^2 - x + 7$; (2, 15) solution

83. $y \geq 4x^2 - 64x + 92$; (1, 30) not a solution

SPECIAL PRODUCT PATTERNS Find the product. (Review 10.3 for 10.7)

84. $(4t - 1)^2$ $16t^2 - 8t + 1$

85. $(b + 9)(b - 9)$ $b^2 - 81$

86. $(3x + 5)(3x + 5)$ $9x^2 + 30x + 25$

87. $(2a - 7)(2a + 7)$ $4a^2 - 49$

88. $(11 - 6x)^2$ $121 - 132x + 36x^2$

89. $(100 + 27x)^2$ $10{,}000 + 5400x + 729x^2$

90. BASKETBALL The graph shows the number of women's college basketball teams from 1985 projected through 2000. Which type of model best fits the data?

▶ Source: National Collegiate Athletic Association

(Review 9.8) quadratic

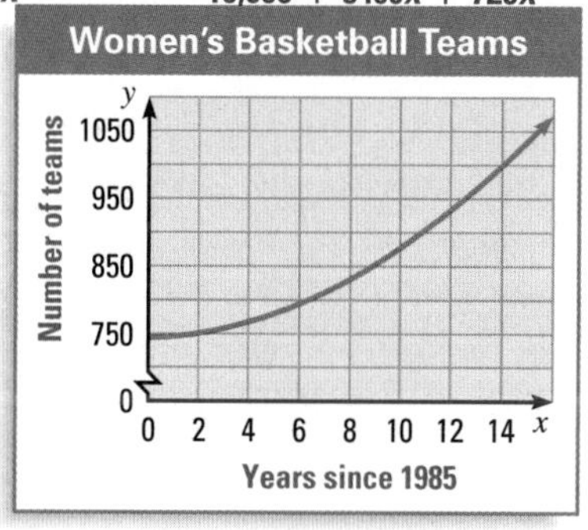

QUIZ 2

Self-Test for Lessons 10.4–10.6

Solve the equation. (Lesson 10.4)

1. $(x + 5)(2x + 10) = 0$ −5

2. $(4x - 6)(5x - 20) = 0$ $\frac{3}{2}, 4$

3. $(2x + 7)(3x - 12) = 0$ $-\frac{7}{2}, 4$

4. $\left(4x + \frac{1}{3}\right)^2 = 0$ $-\frac{1}{12}$

5. $(2x + 8)^2 = 0$ −4

6. $(x - 4)(x + 7)(x + 1) = 0$ 4, −7, −1

Name the *x*-intercepts and the vertex of the graph of the function. Then sketch the graph of the function. (Lesson 10.4) 7–9. See margin for graphs.

7. $y = \left(x - \frac{1}{2}\right)(x + 2)$

8. $y = (x + 4)(x - 4)$

9. $y = (3x - 1)(x + 2)$

Factor the trinomial. (Lessons 10.5 and 10.6)

10. $y^2 + 3y - 4$ $(y + 4)(y - 1)$

11. $w^2 + 13w + 22$ $(w + 2)(w + 11)$

12. $n^2 + 16n - 57$ $(n + 19)(n - 3)$

13. $x^2 + 17x + 66$ $(x + 6)(x + 11)$

14. $t^2 - 41t - 86$ $(t - 43)(t + 2)$

15. $-45 + 14z - z^2$ $(9 - z)(-5 + z)$

16. $12b^2 - 17b - 99$ $(4b + 9)(3b - 11)$

17. $2t^2 + 63t + 145$ $(t + 29)(2t + 5)$

18. $18d^2 - 54d + 28$ $2(3d - 2)(3d - 7)$

Solve the equation by factoring. (Lessons 10.5 and 10.6)

19. $y^2 + 5y - 6 = 0$ −6, 1

20. $n^2 + 26n + 25 = 0$ −25, −1

21. $t^2 - 11t = -18$ 2, 9

22. $w^2 - 29w = 170$ −5, 34

23. $x^2 + 35x + 3 = 77$ −37, 2

24. $2a^2 + 33a + 136 = 0$ See margin.

25. $15n^2 + 41n = -14$ See margin.

26. $18b^2 - 89b = -36$ See margin.

27. $6x^2 + x - 96 = 80$ See margin.

7. $-2, \frac{1}{2}, \left(-\frac{3}{4}, -\frac{25}{16}\right)$

8. −4, 4, (0, −16)

9. $\frac{1}{3}, -2, \left(-\frac{5}{6}, -\frac{49}{12}\right)$

24. $-\frac{17}{2}, -8$

25. $-\frac{7}{3}, -\frac{2}{5}$

26. $\frac{4}{9}, \frac{9}{2}$

27. $-\frac{11}{2}, \frac{16}{3}$

ADDITIONAL RESOURCES

An alternative quiz for Lessons 10.4–10.6 is available in the *Chapter 10 Resource Book*, p. 91.

7.

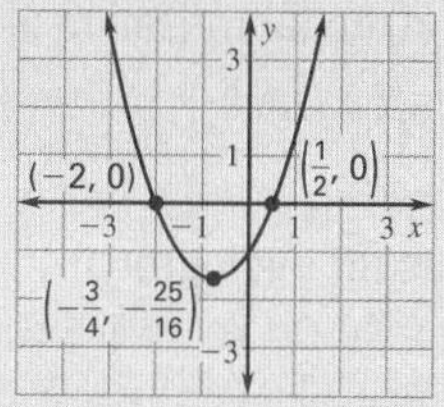

8.

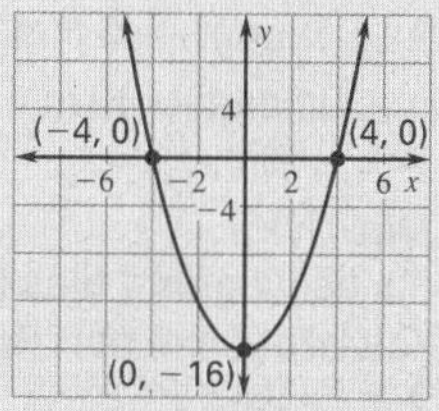

9.

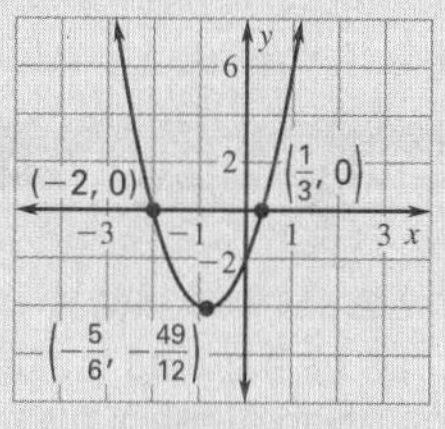

1 Planning the Activity

PURPOSE
To model trinomials of the form $(ax)^2 + 2abx + b^2$ with algebra tiles in order to find the factorization of the trinomial.

MATERIALS
- algebra tiles for each group

PACING
- Exploring the Concept — 5 min
- Drawing Conclusions — 10 min

LINK TO LESSON
Have students use algebra tiles to model the factorization in part (b) of Example 2 in Lesson 10.7.

2 Managing the Activity

ALTERNATIVE APPROACH
This activity can be done as a class demonstration using algebra tiles for the overhead projector. Students can offer suggestions as to how to model the trinomial and how to rearrange the tiles in a square formation.

3 Closing the Activity

KEY DISCOVERY
Trinomials of the form $(ax)^2 + 2abx + b^2$ can be factored as $(ax + b)^2$ using the factoring pattern $a^2 + 2ab + b^2 = (a + b)^2$.

ACTIVITY ASSESSMENT
Use multiplication to show why you can factor $a^2 + 2ab + b^2$ as $(a + b)^2$. Then show why you can factor $a^2 - 2ab + b^2$ as $(a - b)^2$.
$(a + b)^2 = (a + b)(a + b) = a^2 + ab + ab + b^2 = a^2 + 2ab + b^2$; $(a - b)^2 = (a - b)(a - b) = a^2 - ab - ab + b^2 = a^2 - 2ab + b^2$

ACTIVITY 10.7
Developing Concepts

Group Activity for Use with Lesson 10.7

Modeling the Factorization of $(ax)^2 + 2abx + b^2$

SET UP
Work in a small group.

MATERIALS
- algebra tiles

QUESTION

How can you model the factorization of a trinomial of the form $(ax)^2 + 2abx + b^2$ using algebra tiles?

EXPLORING THE CONCEPT

You can use algebra tiles to create a model that can be used to factor a trinomial in which the first and last terms are perfect squares and the middle term is twice the product of the square roots of the first and last terms. Factor the trinomial $x^2 + 6x + 9$ as follows.

1. Use algebra tiles to model $x^2 + 6x + 9$.

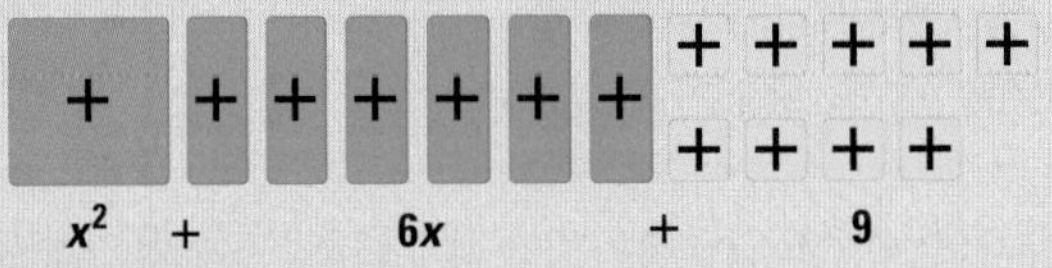

2. Arrange the tiles to form a square.

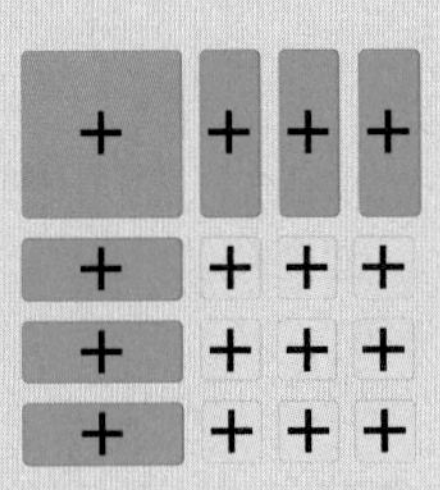

3. The length of each side of the square is __?__. Complete this statement: **$x + 3$**
$x^2 + 6x + 9 = (\underline{?})(\underline{?}) = (\underline{?})^2$ **$x + 3$; $x + 3$; $x + 3$**

DRAWING CONCLUSIONS

Use algebra tiles to factor the trinomial.

1. $x^2 + 4x + 4$ **$(x + 2)^2$**
2. $x^2 + 8x + 16$ **$(x + 4)^2$**
3. $4x^2 + 4x + 1$ **$(2x + 1)^2$**
4. $4x^2 + 12x + 9$ **$(2x + 3)^2$**
5. Look for a pattern in the factorizations in Exercises 1–4. Complete this statement: $a^2 + 2ab + b^2 = (\underline{?})(\underline{?}) = \underline{?}$ **$a + b$; $a + b$; $(a + b)^2$**

Use the pattern from Exercise 5 to factor the trinomial.

6. $9x^2 + 6x + 1$ **$(3x + 1)^2$**
7. $16x^2 + 56x + 49$ **$(4x + 7)^2$**
8. $49x^2 + 28x + 4$ **$(7x + 2)^2$**
9. Make a conjecture about how to complete this statement: $a^2 - 2ab + b^2 = (\underline{?})(\underline{?}) = \underline{?}$ **$(a - b)$; $(a - b)$; $(a - b)^2$**

10.7 Factoring Special Products

What you should learn

GOAL 1 Use special product patterns to factor quadratic polynomials.

GOAL 2 Solve quadratic equations by factoring.

Why you should learn it

▼ To solve **real-life** problems, such as finding the velocity of a pole-vaulter in **Exs. 68 and 69**.

REAL LIFE

GOAL 1 FACTORING QUADRATIC POLYNOMIALS

Each of the special product patterns you studied in Lesson 10.3 can be used to factor polynomials. Note that there are two forms of perfect square trinomials.

FACTORING SPECIAL PRODUCTS

DIFFERENCE OF TWO SQUARES PATTERN	**Example**
$a^2 - b^2 = (a + b)(a - b)$	$9x^2 - 16 = (3x + 4)(3x - 4)$
PERFECT SQUARE TRINOMIAL PATTERN	**Example**
$a^2 + 2ab + b^2 = (a + b)^2$	$x^2 + 8x + 16 = (x + 4)^2$
$a^2 - 2ab + b^2 = (a - b)^2$	$x^2 - 12x + 36 = (x - 6)^2$

EXAMPLE 1 *Factoring the Difference of Two Squares*

a. $m^2 - 4 = m^2 - 2^2$ — Write as $a^2 - b^2$.

$= (m + 2)(m - 2)$ — Factor using pattern.

b. $4p^2 - 25 = (2p)^2 - 5^2$ — Write as $a^2 - b^2$.

$= (2p + 5)(2p - 5)$ — Factor using pattern.

c. $50 - 98x^2 = 2(25 - 49x^2)$ — Factor out common factor.

$= 2[5^2 - (7x)^2]$ — Write as $a^2 - b^2$.

$= 2(5 + 7x)(5 - 7x)$ — Factor using pattern.

EXAMPLE 2 *Factoring Perfect Square Trinomials*

a. $x^2 - 4x + 4 = x^2 - 2(x)(2) + 2^2$ — Write as $a^2 - 2ab + b^2$.

$= (x - 2)^2$ — Factor using pattern.

b. $16y^2 + 24y + 9 = (4y)^2 + 2(4y)(3) + 3^2$ — Write as $a^2 + 2ab + b^2$.

$= (4y + 3)^2$ — Factor using pattern.

c. $3x^2 - 30x + 75 = 3(x^2 - 10x + 25)$ — Factor out common factor.

$= 3[x^2 - 2(x)(5) + 5^2]$ — Write as $a^2 - 2ab + b^2$.

$= 3(x - 5)^2$ — Factor using pattern.

1 PLAN

PACING

Basic: 2 days
Average: 2 days
Advanced: 2 days
Block Schedule: 1 block

LESSON OPENER ACTIVITY

An alternative way to approach Lesson 10.7 is to use the Activity Lesson Opener:

- Blackline Master (*Chapter 10 Resource Book*, p. 95)
- Transparency (p. 72)

MEETING INDIVIDUAL NEEDS

- ***Chapter 10 Resource Book***
  Prerequisite Skills Review (p. 5)
  Practice Level A (p. 99)
  Practice Level B (p. 100)
  Practice Level C (p. 101)
  Reteaching with Practice (p. 102)
  Absent Student Catch-Up (p. 104)
  Challenge (p. 106)
- ***Resources in Spanish***
- ***Personal Student Tutor***

NEW-TEACHER SUPPORT

See the Tips for New Teachers on pp. 1–2 of the *Chapter 10 Resource Book* for additional notes about Lesson 10.7.

WARM-UP EXERCISES

Transparency Available

Find the product.

1. $(2n + 3)^2$ $4n^2 + 12n + 9$
2. $(p - 6)^2$ $p^2 - 12p + 36$
3. $(2y - 1)(2y + 1)$ $4y^2 - 1$

Factor the trinomial.

4. $x^2 - 4x - 5$ $(x - 5)(x + 1)$
5. $x^2 + 2x + 1$ $(x + 1)(x + 1)$

2 TEACH

MOTIVATING THE LESSON
Write down several numerical sequences, such as 1, 3, 5, 7, 9, ... Ask students to continue the sequences and to explain how they find the next terms. Bring in the idea of looking for a pattern, and relate it to the factoring patterns described at the beginning of this lesson.

EXTRA EXAMPLE 1
Factor.
a. $m^2 - 9$ $(m-3)(m+3)$
b. $49q^2 - 81$ $(7q-9)(7q+9)$
c. $12 - 27x^2$ $-3(3x+2)(3x-2)$

EXTRA EXAMPLE 2
Factor.
a. $x^2 - 8x + 16$ $(x-4)^2$
b. $9y^2 + 60y + 100$ $(3y+10)^2$
c. $2x^2 - 12x + 18$ $2(x-3)^2$

EXTRA EXAMPLE 3
Solve the equation
$-3x^2 - 18x - 27 = 0.$ -3

EXTRA EXAMPLE 4
Solve the equation
$x^2 - \frac{2}{3}x + \frac{1}{9} = 0.$ $\frac{1}{3}$

CHECKPOINT EXERCISES
For use after Example 1:
1. Factor $4n^2 - 16m^2$.
$4(n-2m)(n+2m)$
For use after Example 2:
2. Factor $12x^2 + 12x + 3$.
$3(2x+1)^2$
For use after Examples 3 and 4:
3. Solve the equation
$2x^2 - 10x + \frac{25}{2} = 0.$ $\frac{5}{2}$

STUDENT HELP NOTES
Homework Help Students can find extra examples at **www.mcdougallittell.com** that parallel the examples in the student edition.

GOAL 2 SOLVING QUADRATIC EQUATIONS

STUDENT HELP
INTERNET HOMEWORK HELP
Visit our Web site www.mcdougallittell.com for extra examples.

EXAMPLE 3 *Graphical and Analytical Reasoning*

Solve the equation $-2x^2 + 12x - 18 = 0$.

SOLUTION

$-2x^2 + 12x - 18 = 0$	Write original equation.
$-2(x^2 - 6x + 9) = 0$	Factor out common factor.
$-2[x^2 - 2(3x) + 3^2] = 0$	Write as $a^2 - 2ab + b^2$.
$-2(x - 3)^2 = 0$	Factor using pattern.
$x - 3 = 0$	Set repeated factor equal to 0.
$x = 3$	Solve for x.

▶ The solution is 3.

✓**CHECK:** You can check your answer by substitution or by graphing.

Graph $y = -2x^2 + 12x - 18$.

Graph the x-axis, $y = 0$.

Use your graphing calculator's *Intersect* feature to find the x-intercept, where $-2x^2 + 12x - 18 = 0$.

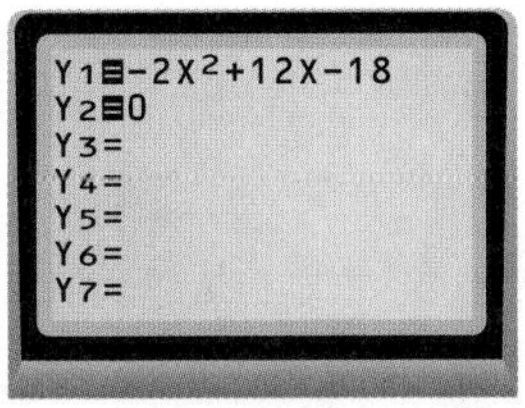

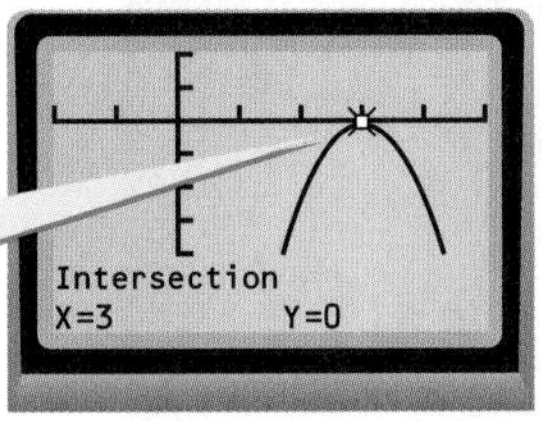

When $x = 3$, $-2x^2 + 12x - 18 = 0$, which appears to confirm your solution.

EXAMPLE 4 *Solving a Quadratic Equation*

Solve $x^2 + x + \frac{1}{4} = 0$.

SOLUTION

$x^2 + x + \frac{1}{4} = 0$	Write original equation.
$x^2 + 2\left(\frac{1}{2}x\right) + \left(\frac{1}{2}\right)^2 = 0$	Write as $a^2 + 2ab + b^2$.
$\left(x + \frac{1}{2}\right)^2 = 0$	Factor using pattern.
$x + \frac{1}{2} = 0$	Set repeated factor equal to 0.
$x = -\frac{1}{2}$	Solve for x.

▶ The solution is $-\frac{1}{2}$. Check this in the original equation.

STUDENT HELP
Study Tip
Another way to solve the equation in Example 4 is to first multiply each side of the equation by 4. This allows you to work less with fractions.

$x^2 + x + \frac{1}{4} = 0$
$4x^2 + 4x + 1 = 0$
$(2x + 1)^2 = 0$
$2x + 1 = 0$
$x = -\frac{1}{2}$

FOCUS ON APPLICATIONS

REAL LIFE **BLOCK AND TACKLE**
A block and tackle makes it easier to lift a heavy object. For instance, using a block and tackle with 4 pulleys, you can lift 1000 pounds with only 250 pounds of applied force.

EXAMPLE 5 *Writing and Using a Quadratic Model*

BLOCK AND TACKLE An object lifted with a rope or wire should not weigh more than the *safe working load* for the rope or wire. The safe working load S (in pounds) for a natural fiber rope is a function of C, the circumference of the rope in inches.

Safe working load model: $150 \cdot C^2 = S$

You are setting up a block and tackle to lift a 1350-pound safe. What size natural fiber rope do you need to have a safe working load?

SOLUTION Use the model to find a safe rope size. Substitute 1350 for S.

$150C^2 = S$	Write model.
$150C^2 = 1350$	Substitute.
$150C^2 - 1350 = 0$	Subtract 1350 from each side.
$150(C^2 - 9) = 0$	Factor out common factor.
$150(C - 3)(C + 3) = 0$	Factor.
$(C - 3) = 0$ or $(C + 3) = 0$	Use zero-product property.
$C - 3 = 0$	Set first factor equal to 0.
$C = 3$	Solve for C.
$C + 3 = 0$	Set second factor equal to 0.
$C = -3$	Solve for C.

▶ The negative solution makes no sense. You need a rope with a circumference of at least 3 inches.

EXAMPLE 6 *Writing and Using a Quadratic Model*

If you project a ball straight up from the ground with an initial velocity of 64 feet per second, will the ball reach a height of 64 feet? If it does, how long will it take to reach that height?

SOLUTION Use a vertical motion model where $v = 64$, $s = 0$, and $h = 64$.

$-16t^2 + vt + s = h$	Vertical motion model
$-16t^2 + 64t + 0 = 64$	Substitute for v, s, and h.
$-16t^2 + 64t - 64 = 0$	Write in standard form.
$-16(t^2 - 4t + 4) = 0$	Factor out common factor.
$-16(t - 2)^2 = 0$	Factor.
$t - 2 = 0$	Set repeated factor equal to 0.
$t = 2$	Solve for t.

▶ Because there is a solution, you know that the ball will reach a height of 64 feet. The solution is $t = 2$, so it will take 2 seconds to reach that height.

EXTRA EXAMPLE 5
The formula for the area of a circle is $A = \pi r^2$. If the wheel of a pulley has an area of 121π cm^2, what are the radius and circumference of the pulley?
11 cm; $22\pi \approx 69.12$ cm

EXTRA EXAMPLE 6
If you kick a ball with an initial velocity of 32 ft/sec, will the ball reach a height of 16 feet? If it does, how long will it take?
yes; 1 sec

CHECKPOINT EXERCISES

For use after Example 5:

1. Suppose the area of a wheel of a pulley is 81π cm^2. Use the formula $A = \pi r^2$, to find the radius of the pulley. **9 cm**

For use after Example 6:

2. If you project a ball straight up from an initial height of 14 ft with an initial velocity of 48 ft/sec, will the ball reach a height of 50 ft? If it does, how long will it take to reach the height? **yes; 1.5 sec**

FOCUS ON VOCABULARY
The meanings of *perfect square, difference,* and *trinomial* are important to understanding the patterns described in this lesson. Reviewing these ideas and discussing how they are used to describe the patterns will help students learn how to apply them.

CLOSURE QUESTION
Explain the two types of factoring patterns that produce special products.
See sample answer at left.

DAILY PUZZLER
If a man is 45 inches tall plus twice one fifth of his own height, how tall is he in feet and inches?
6 feet 3 inches

The difference of two squares pattern starts with a binomial in which both terms are perfect squares and one term is subtracted from the other. The pattern is $a^2 - b^2 = (a - b)(a + b)$. A perfect square trinomial is one where the first and third terms are perfect squares and the middle term is twice the product of the square roots of the first and third terms. It can be factored into the square of a binomial. The patterns are $a^2 - 2ab + b^2 = (a - b)^2$ and $a^2 + 2ab + b^2 = (a + b)^2$.

3 APPLY

ASSIGNMENT GUIDE

BASIC

Day 1: pp. 622–623 Exs. 18–62 even, 63, 64

Day 2: pp. 622–624 Exs. 19–61 odd, 65, 66, 70–72, 75, 80, 85, 90, 95

AVERAGE

Day 1: pp. 622–623: Exs. 18–62 even, 63, 64

Day 2: pp. 622–624 Exs. 19–61 odd, 65–72, 75, 80, 85, 90, 95

ADVANCED

Day 1: pp. 622–623 Exs. 18–62 even, 63, 64

Day 2: pp. 622–624 Exs. 19–61 odd, 65–73, 75, 80, 85, 90, 95

BLOCK SCHEDULE

pp. 622–624 Exs. 18–72, 75, 80, 85, 90, 95

EXERCISE LEVELS

Level A: *Easier*
74–77

Level B: *More Difficult*
15–72, 78–96

Level C: *Most Difficult*
73

✔ HOMEWORK CHECK

To quickly check student understanding of key concepts, go over the following exercises: Exs. 18, 30, 36, 45, 54, 60, 66. See also the Daily Homework Quiz:

- Blackline Master (*Chapter 10 Resource Book,* p. 109)
- Transparency (p. 83)

4, 42–50. See Additional Answers beginning on page AA1.

GUIDED PRACTICE

Vocabulary Check ✔

Concept Check ✔

1. What is the difference of two squares pattern? **See margin.**
2. ERROR ANALYSIS Describe the error at the right. Then solve the equation correctly. **See margin.**
3. Look back to Lesson 10.3. Describe the relationship between the special products shown in Lesson 10.3 and the polynomials you factored in this lesson. **See margin.**
4. Write the three special product factoring patterns. Give an example of each pattern. **See margin.**
5. Factor out the common factor in $3x^2 - 6x + 9$. $3(x^2 - 2x + 3)$

$9x^2 - 6x + 1 = 0$
$(3x + 1)(3x - 1) = 0$
$3x + 1 = 0 \qquad 3x - 1 = 0$
$x = -\frac{1}{3} \qquad x = \frac{1}{3}$
$x = -\frac{1}{3}$ or $\frac{1}{3}$

1. The difference of the squares of two terms is equal to the product of the sum and the difference of the two terms. $[a^2 - b^2 = (a + b)(a - b)]$

2. The wrong special products pattern is used.

Skill Check ✔

Correct solution should use $(a - b)^2$ to get $(3x - 1)^2 = 0$ and $x = \frac{1}{3}$.

3. *Sample answer:* Each of the special products is the basis for one of the rules for factoring.

Factor the expression.

6. $x^2 - 9$ $(x + 3)(x - 3)$
7. $t^2 + 10t + 25$ $(t + 5)^2$
8. $w^2 - 16w + 64$ $(w - 8)^2$
9. $16 - t^2$ $(4 + t)(4 - t)$
10. $6y^2 - 24$ $6(y + 2)(y - 2)$
11. $18 - 2z^2$ $2(3 + z)(3 - z)$

Use factoring to solve the equation.

12. $x^2 + 6x + 9 = 0$ -3
13. $144 - y^2 = 0$ $-12, 12$
14. $s^2 - 14s + 49 = 0$ 7
15. $-25 + x^2 = 0$ $-5, 5$
16. $7x^2 + 28x + 28 = 0$ -2
17. $-4y^2 + 24y - 36 = 0$ 3

PRACTICE AND APPLICATIONS

STUDENT HELP

Extra Practice to help you master skills is on p. 806.

39. $(z + 5)(z - 5)$; difference of two squares

40. $(y + 6)^2$; perfect square trinomial

41. $4(n + 3)(n - 3)$; difference of two squares

STUDENT HELP

HOMEWORK HELP
Example 1: Exs. 18–26, 39–50
Example 2: Exs. 27–50
Example 3: Exs. 51–62
Example 4: Exs. 51–62
Example 5: Exs. 63–66
Example 6: Ex. 67

DIFFERENCE OF TWO SQUARES **Factor the expression.**

18. $n^2 - 16$ $(n + 4)(n - 4)$
19. $100x^2 - 121$ $(10x + 11)(10x - 11)$
20. $6m^2 - 150$ $6(m + 5)(m - 5)$
21. $60y^2 - 540$ $60(y + 3)(y - 3)$
22. $16 - 81r^2$ $(4 + 9r)(4 - 9r)$
23. $98 - 2t^2$ $2(7 + t)(7 - t)$
24. $w^2 - y^2$ $(w + y)(w - y)$
25. $9t^2 - 4q^2$ $(3t - 2q)(3t + 2q)$
26. $-28y^2 + 7t^2$ $-7(2y + t)(2y - t)$

PERFECT SQUARES **Factor the expression.**

27. $x^2 + 8x + 16$ $(x + 4)^2$
28. $x^2 - 20x + 100$ $(x - 10)^2$
29. $y^2 + 30y + 225$ $(y + 15)^2$
30. $b^2 - 14b + 49$ $(b - 7)^2$
31. $9x^2 + 6x + 1$ $(3x + 1)^2$
32. $4r^2 + 12r + 9$ $(2r + 3)^2$
33. $25n^2 - 20n + 4$ $(5n - 2)^2$
34. $36m^2 - 84m + 49$ $(6m - 7)^2$
35. $18x^2 + 12x + 2$ $2(3x + 1)^2$
36. $48y^2 - 72xy + 27x^2$ $3(4y - 3x)^2$
37. $-16w^2 - 80w - 100$ $-4(2w + 5)^2$
38. $-3k^2 + 42k - 147$ $-3(k - 7)^2$

FACTORING EXPRESSIONS **Factor the expression. Tell which special product factoring pattern you used.** 39–50. See margin.

39. $z^2 - 25$
40. $y^2 + 12y + 36$
41. $4n^2 - 36$
42. $32 - 18x^2$
43. $4b^2 - 40b + 100$
44. $-27t^2 - 18t - 3$
45. $-2x^2 + 52x - 338$
46. $169 - x^2$
47. $x^2 - 10{,}000w^2$
48. $-108 + 147x^2$
49. $x^2 + \frac{2}{3}x + \frac{1}{9}$
50. $\frac{3}{4} - 12x^2$

SOLVING EQUATIONS Use factoring to solve the equation. Use a graphing calculator to check your solution if you wish.

51. $2x^2 - 72 = 0$ **−6, 6** **52.** $3x^2 - 24x + 48 = 0$ **4** **53.** $25x^2 - 4 = 0$ $-\frac{2}{5}, \frac{2}{5}$

54. $\frac{1}{5}x^2 - 2x + 5 = 0$ **5** **55.** $27 - 12x^2 = 0$ $-\frac{3}{2}, \frac{3}{2}$ **56.** $50x^2 + 60x + 18 = 0$ $-\frac{3}{5}$

57. $\frac{1}{3}x^2 - 6x + 27 = 0$ **9** **58.** $90x^2 - 120x + 40 = 0$ $\frac{2}{3}$

59. $x^2 - \frac{5}{3}x + \frac{25}{36} = 0$ $\frac{5}{6}$ **60.** $112x^2 - 252 = 0$ $-\frac{3}{2}, \frac{3}{2}$

61. $-16x^2 + 56x - 49 = 0$ $\frac{7}{4}$ **62.** $-\frac{4}{5}x^2 - \frac{4}{5}x - \frac{1}{5} = 0$ $-\frac{1}{2}$

SAFE WORKING LOAD **In Exercises 63 and 64, the safe working load *S* (in tons) for a *wire* rope is a function of *D*, the diameter of the rope in inches.**

Safe working load model for wire rope: $4 \cdot D^2 = S$

63. What diameter of wire rope do you need to lift a 9-ton load and have a safe working load? **1.5 in.**

64. When determining the safe working load *S* of a rope that is old or worn, decrease *S* by 50%. Write a model for *S* when using an old wire rope. What diameter of old wire rope do you need to safely lift a 9-ton load? $2 \cdot D^2 = S$; $\frac{3\sqrt{2}}{2} \approx 2.12$ **in.**

HANG TIME **In Exercises 65 and 66, use the following information about *hang time*, the length of time a basketball player is in the air after jumping.**

The maximum height *h* jumped (in feet) is a function of *t*, where *t* is the hang time (in seconds).

Hang time model: $h = 4t^2$

65. If you jump 1 foot into the air, what is your hang time? $\frac{1}{2}$ **sec**

66. If a professional player jumps 4 feet into the air, what is the hang time? **1 sec**

67. VERTICAL MOTION An object is propelled from the ground with an initial upward velocity of 224 feet per second. Will the object reach a height of 784 feet? If it does, how long will it take the object to reach that height? Solve by factoring. **yes; 7 sec**

POLE-VAULTING **In Exercises 68 and 69, use the following information.**
In the sport of pole-vaulting, the height *h* (in feet) reached by a pole-vaulter is a function of *v*, the velocity of the pole-vaulter, as shown in the model below. The constant *g* is approximately 32 feet per second per second.

Pole-vaulter height model: $h = \frac{v^2}{2g}$

68. To reach a height of 9 feet, what is the pole-vaulter's velocity? **24 ft/sec**

69. To reach a height of 16 feet, what is the pole-vaulter's velocity? **32 ft/sec**

FOCUS ON APPLICATIONS

REAL LIFE **POLE-VAULTERS** must practice to perfect their technique in order to maximize the height vaulted.

! COMMON ERROR
EXERCISES 39–50 When working with factoring special products, some students may have difficulty because they cannot determine *a* and *b* from the patterns in a specific polynomial. Have students use the first and last terms of the given polynomial to help them. Once common factors have been removed, they are the squares of *a* and *b*, so they can be used to find what *a* and *b* are.

! COMMON ERROR
EXERCISES 39–62 The coefficient of the middle term of the perfect square trinomial contains a factor of 2. Students may forget that to fit this pattern, the middle term must be *twice* the product of *a* and *b*.

ADDITIONAL PRACTICE AND RETEACHING

For Lesson 10.7:
- Practice Levels A, B, and C (*Chapter 10 Resource Book*, p. 99)
- Reteaching with Practice (*Chapter 10 Resource Book*, p. 102)
- See Lesson 10.7 of the *Personal Student Tutor*

For more Mixed Review:
- Search the *Test and Practice Generator* for key words or specific lessons.

4 ASSESS

DAILY HOMEWORK QUIZ

Transparency Available

Factor the expression.

1. $36x^2 - 144$ $36(x-2)(x+2)$
2. $r^2 + 18r + 81$ $(r+9)^2$
3. $8y^2 - 40y + 50$ $2(2y-5)^2$

Solve the equation.

4. $4x^2 - 9 = 0$ $-\frac{3}{2}, \frac{3}{2}$
5. $\frac{2}{3}x^2 + 8x + 24 = 0$ -6
6. On his way to the basket, Michael Jordan jumped 3 ft in the air. How long did he stay in the air? Use the hang time model $h = 4t^2$.
$\frac{\sqrt{3}}{2}$ sec, or about 0.87 sec

EXTRA CHALLENGE NOTE

Challenge problems for Lesson 10.7 are available in **blackline** format in the *Chapter 10 Resource Book*, p. 106 and at **www.mcdougallittell.com.**

ADDITIONAL TEST PREPARATION

1. **WRITING** Write the factoring pattern for the difference of two squares using the letters *a* and *b*.
$a^2 - b^2 = (a-b)(a+b)$
2. **OPEN ENDED** What value(s) of *k* will make the expression $4x^2 + kx + 9$ a perfect square trinomial? 12 or −12

Test Preparation

70. MULTIPLE CHOICE Which of the following is a correct factorization of $-12x^2 + 147$? **D**

(A) $-3(2x+7)^2$ (B) $3(2x-7)(2x+7)$
(C) $-2(2x-7)(2x+7)$ (D) $-3(2x-7)(2x+7)$

71. MULTIPLE CHOICE Which of the following is a correct factorization of $72x^2 - 24x + 2$? **C**

(A) $-9(3x-1)^2$ (B) $8\left(9x - \frac{1}{2}\right)^2$
(C) $8\left(3x - \frac{1}{2}\right)\left(3x - \frac{1}{2}\right)$ (D) $-8\left(3x - \frac{1}{2}\right)^2$

72. MULTIPLE CHOICE Which of the following is the solution of the equation $-4x^2 + 24x - 36 = 0$? **D**

(A) −6 (B) −3 (C) 2 (D) 3

★ Challenge

73. SAFE WORKING LOAD Example 5 on page 621 models the safe working load of a natural fiber rope as a function of its *circumference*. Exercises 63 and 64 use a model for the safe working load of a wire rope as a function of its *diameter*. Find a way to use these models to compare the strength of a wire rope to the strength of a natural fiber rope.
If a wire rope and a natural fiber rope have the same radius, the ratio of the safe load of the wire rope to the safe load of the natural fiber rope is $\frac{4}{150\pi^2} \approx \frac{1}{370} \approx 0.003$. That is, the natural fiber rope is about 370 times stronger.

EXTRA CHALLENGE
www.mcdougallittell.com

MIXED REVIEW

FINDING THE GCF **Find the greatest common factor of the numbers.** (Skills Review, p. 777)

74. 9 and 12 **3** 75. 15 and 45 **15** 76. 55 and 132 **11** 77. 14 and 18 **2**

CHECKING FOR SOLUTIONS **Decide whether or not the ordered pair is a solution of the system of linear equations.** (Review 7.1)

78. $x + 9y = -11$, $-4x + y = -30$; $(7, -2)$ **solution**
79. $2x + 6y = 22$, $-x - 4y = -13$; $(-5, -2)$ **not a solution**
80. $-2x + 7y = -41$, $3x + 5y = 15$; $(-10, 3)$ **not a solution**
81. $-5x - 8y = 28$, $9x - 2y = 48$; $(4, -6)$ **solution**

SOLVING LINEAR SYSTEMS **Use the substitution method to solve the linear system.** (Review 7.2)

82. $x - y = 2$, $2x + y = 1$ **(1, −1)**
83. $x - 2y = 10$, $3x - y = 0$ **(−2, −6)**
84. $-x + y = 0$, $2x + y = 0$ **(0, 0)**
85. $2x + 3y = -5$, $x - 2y = -6$ **(−4, 1)**

SIMPLIFYING RADICAL EXPRESSIONS **Simplify the expression.** (Review 9.2)

86. $\sqrt{216}$ $6\sqrt{6}$
87. $\sqrt{5} \cdot \sqrt{15}$ $5\sqrt{3}$
88. $\sqrt{10} \cdot \sqrt{20}$ $10\sqrt{2}$
89. $\sqrt{4} \cdot 3\sqrt{9}$ 18
90. $\sqrt{\frac{28}{49}}$ $\frac{2\sqrt{7}}{7}$
91. $\frac{10\sqrt{8}}{\sqrt{25}}$ $4\sqrt{2}$
92. $\frac{12\sqrt{4}}{\sqrt{9}}$ 8
93. $\frac{-6\sqrt{12}}{\sqrt{4}}$ $-6\sqrt{3}$

SOLVING EQUATIONS **Use the quadratic formula to solve the equation.** (Review 9.5 for 10.8)

94. $9x^2 - 14x - 7 = 0$ $\frac{7 - 4\sqrt{7}}{9}, \frac{7 + 4\sqrt{7}}{9}$
95. $9d^2 - 58d + 24 = 0$ $\frac{4}{9}, 6$
96. $7y^2 - 9y - 17 = 0$ **See margin.**

96. $\frac{9 - \sqrt{557}}{14}, \frac{9 + \sqrt{557}}{14}$

10.8 Factoring Using the Distributive Property

What you should learn

GOAL 1 Use the distributive property to factor a polynomial.

GOAL 2 Solve polynomial equations by factoring.

Why you should learn it

▼ To solve some types of **real-life** problems, such as modeling the effect of gravity in **Exs. 51–54**.

REAL LIFE

Shannon Lucid on Space Station *Mir*

GOAL 1 FACTORING AND THE DISTRIBUTIVE PROPERTY

In dealing with polynomials, you have already been using the distributive property to factor out integers common to the various terms of the expression.

$9x^2 - 15 = 3(3x^2 - 5)$ **Factor out common factor.**

In many situations, it is important to factor out common *variable* factors. To save steps, you should factor out the *greatest common factor* (GCF).

EXAMPLE 1 *Finding the Greatest Common Factor*

Factor the greatest common factor out of $14x^4 - 21x^2$.

SOLUTION First find the greatest common factor. It is the product of all the common factors.

$14x^4 = 2 \cdot 7 \cdot x \cdot x \cdot x \cdot x$

$21x^2 = 3 \cdot 7 \cdot x \cdot x$

$\text{GCF} = 7 \cdot x \cdot x = 7x^2$

Then use the distributive property to factor the greatest common factor out of the polynomial.

$14x^4 - 21x^2 = 7x^2(2x^2 - 3)$

....

In this lesson we restrict the polynomials we consider to polynomials having integer coefficients. A polynomial is **prime** if it is not the product of polynomials having integer coefficients. To **factor a polynomial completely**, write it as the product of these types of factors:

- monomial factors
- prime factors with at least two terms

EXAMPLE 2 *Recognizing Complete Factorization*

Tell whether the polynomial is factored completely.

a. $2x^2 + 8 = 2(x^2 + 4)$ **b.** $2x^2 - 8 = 2(x^2 - 4)$

SOLUTION

a. This polynomial is factored completely because $x^2 + 4$ cannot be factored using integer coefficients.

b. This polynomial is not factored completely because $x^2 - 4$ can be factored as $(x - 2)(x + 2)$.

STUDENT HELP

Skills Review
For help with finding a greatest common factor, see p. 777.

1 PLAN

PACING
Basic: 2 days
Average: 2 days
Advanced: 2 days
Block Schedule: 1 block

LESSON OPENER ACTIVITY
An alternative way to approach Lesson 10.8 is to use the Activity Lesson Opener:
- Blackline Master (*Chapter 10 Resource Book*, p. 110)
- Transparency (p. 73)

MEETING INDIVIDUAL NEEDS
- ***Chapter 10 Resource Book***
  Prerequisite Skills Review (p. 5)
  Practice Level A (p. 111)
  Practice Level B (p. 112)
  Practice Level C (p. 113)
  Reteaching with Practice (p. 114)
  Absent Student Catch-Up (p. 116)
  Challenge (p. 118)
- ***Resources in Spanish***
- ***Personal Student Tutor***

NEW-TEACHER SUPPORT
See the Tips for New Teachers on pp. 1–2 of the *Chapter 10 Resource Book* for additional notes about Lesson 10.8.

WARM-UP EXERCISES
Transparency Available
Find the greatest common factor.
1. 9, 12 3
2. 51, 85 17
3. 30, 45, 105 15
Use the distributive property to simplify.
4. $3x(4x + 9)$ $12x^2 + 27x$
5. $(-3)(2x^3 + 5)$ $-6x^3 - 15$

2 TEACH

MOTIVATING THE LESSON
Ask students to begin listing all of the prime numbers that are less than 10. Then ask them for other prime numbers that they can think of. How they can tell a number is prime? Ask them what they think it means for a polynomial to be prime. **A prime polynomial is one that cannot be written as the product of factors with integer coefficients.**

EXTRA EXAMPLE 1
Factor the greatest common factor out of $33x^5 - 121x^2$. $11x^2(3x^3 - 11)$

EXTRA EXAMPLE 2
Tell whether the polynomial is factored completely.
a. $4x^2 + 36 = 4(x^2 + 9)$ yes
b. $4x^2 - 36 = 4(x^2 - 9)$
No; $x^2 - 9$ can be factored into $(x - 3)(x + 3)$.

EXTRA EXAMPLE 3
Factor $5x^3 - 25x^2 - 30x$ completely. $5x(x - 6)(x + 1)$

EXTRA EXAMPLE 4
Factor $75x^4 - 3x^2$ completely. $3x^2(5x - 1)(5x + 1)$

EXTRA EXAMPLE 5
Factor $x^3 + 4x^2 + 6x + 24$ completely. $(x + 4)(x^2 + 6)$

EXTRA EXAMPLE 6
Factor $x^3 + 2x^2 - 36x - 72$ completely. $(x + 2)(x - 6)(x + 6)$

CHECKPOINT EXERCISES
For use after Examples 1–4:
1. Factor $7x^3 - 7x^2 - 84x$ completely. $7x(x - 4)(x + 3)$
For use after Examples 5 and 6:
2. Factor $2x^3 + 3x^2 - 50x - 75$ completely. $(2x + 3)(x - 5)(x + 5)$

EXAMPLE 3 *Factoring Completely*

Factor $4x^3 + 20x^2 + 24x$ completely.

SOLUTION

$$4x^3 + 20x^2 + 24x = 4x(x^2 + 5x + 6)$$
$$= 4x(x + 2)(x + 3)$$

Monomial factor ($4x$) — Prime factors ($(x + 2)(x + 3)$)

EXAMPLE 4 *Factoring Completely*

Factor $45x^4 - 20x^2$ completely.

SOLUTION

$45x^4 - 20x^2 = 5x^2(9x^2 - 4)$ — Factor out GCF.

$= 5x^2(3x - 2)(3x + 2)$ — Factor difference of squares.

....

GROUPING Another use of the distributive property is in factoring polynomials that have four terms. Sometimes you can factor the polynomial by grouping into two groups of terms and factoring the greatest common factor out of each term.

EXAMPLE 5 *Factoring by Grouping*

Factor $x^3 + 2x^2 + 3x + 6$ completely.

SOLUTION

$x^3 + 2x^2 + 3x + 6 = (x^3 + 2x^2) + (3x + 6)$ — Group terms.

$= x^2(x + 2) + 3(x + 2)$ — Factor each group.

$= (x + 2)(x^2 + 3)$ — Use distributive property.

EXAMPLE 6 *Factoring by Grouping*

Factor $x^3 - 2x^2 - 9x + 18$ completely.

SOLUTION

$x^3 - 2x^2 - 9x + 18 = (x^3 - 2x^2) - (9x - 18)$ — Group terms.

$= x^2(x - 2) - 9(x - 2)$ — Factor each group.

$= (x^2 - 9)(x - 2)$ — Use distributive property.

$= (x - 3)(x + 3)(x - 2)$ — Factor difference of squares.

GOAL 2 SOLVING POLYNOMIAL EQUATIONS

EXAMPLE 7 *Solving a Polynomial Equation*

Solve $8x^3 - 18x = 0$.

SOLUTION

$8x^3 - 18x = 0$	Write original equation.
$2x(4x^2 - 9) = 0$	Factor out GCF.
$2x(2x - 3)(2x + 3) = 0$	Factor difference of squares.

▶ By setting each variable factor equal to zero, you can find that the solutions are 0, $\frac{3}{2}$, and $-\frac{3}{2}$.

EXAMPLE 8 *Writing and Using a Polynomial Model*

The width of a box is 1 inch less than the length. The height is 4 inches greater than the length. The box has a volume of 12 cubic inches. What are the dimensions of the box?

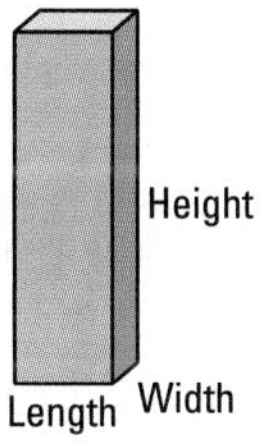

SOLUTION

VERBAL MODEL Length · Width · Height = Volume

LABELS

Length = x	(inches)
Width = $x - 1$	(inches)
Height = $x + 4$	(inches)
Volume = 12	(cubic inches)

ALGEBRAIC MODEL

$x(x - 1)(x + 4) = 12$	Write model.
$x^3 + 3x^2 - 4x = 12$	Multiply.
$(x^3 + 3x^2) - (4x + 12) = 0$	Write in standard form and group terms.
$x^2(x + 3) - 4(x + 3) = 0$	Factor each group of terms.
$(x^2 - 4)(x + 3) = 0$	Use distributive property.
$(x - 2)(x + 2)(x + 3) = 0$	Factor difference of squares.

▶ By setting each factor equal to zero, you can see that the solutions are 2, −2, and −3. The only positive solution is $x = 2$. The dimensions of the box are 2 inches by 1 inch by 6 inches.

EXTRA EXAMPLE 7

Solve $16x^3 - 100x = 0$.
0, $\frac{5}{2}$, $-\frac{5}{2}$

EXTRA EXAMPLE 8

The width of a box is 2 in. less than its length. The height is 8 in. greater than the length. The box has a volume of 96 in.^{3}. What are the dimensions of the box?
2 in. by 4 in. by 12 in.

CHECKPOINT EXERCISES

For use after Example 7:

1. Solve $2x^3 - 10x^2 + 8x = 0$.
0, 1, 4

For use after Example 8:

2. The width of a box is 1 cm less than its length. The height of the box is 9 cm greater than the length. The box has a volume of 72 cm^3. What are the dimensions of the box?
2 cm by 3 cm by 12 cm

CONCEPT QUESTION

Describe the three ways to solve polynomial equations.
(1) solve by graphing: the solutions are the *x*-intercepts of the graph; (2) solve by using the quadratic formula: the solutions are found by substituting the values of *a*, *b*, and *c* into the formula; (3) solve by factoring (if possible): the solutions are the values that make each factor equal to 0.

EXTRA EXAMPLE 9

Solve the equation.

a. $-60x^4 + 114x^3 - 36x^2 = 0$

$0, \frac{3}{2}, \frac{2}{5}$

b. $12x^3 - 44x^2 + 16x = 0$

$0, \approx 3.26, \approx 0.41$

CHECKPOINT EXERCISES

For use after Example 9:

1. Solve the equation $10x^4 - 32x^3 + 18x^2 = 0$.

$0, \approx 2.47, \approx 0.73$

CONCEPT QUESTION

What is the greatest number of solutions a polynomial equation of degree 1 can have? What about a quadratic equation? A cubic equation? Use your answers to make a conjecture about the greatest number of solutions a quartic equation can have. What about a polynomial equation of degree n? **1; 2; 3; 4; n**

FOCUS ON VOCABULARY

The idea of a prime factor of a polynomial equation is similar to the idea of a prime number. If it cannot be written as the product of two other expressions with integer coefficients, then it is prime.

CLOSURE QUESTION

How can the distributive property be used to solve equations by factoring? ***Sample answer:*** **If all of the terms of the polynomial expression have any factors in common, the distributive property can be used to factor out the greatest common factor.**

DAILY PUZZLER

What integer values of a produce no real solutions to either of the equations $x^2 - ax + 4 = 0$ and $x^2 - x - a = 0$? $a = -1, -2, -3$

CONCEPT SUMMARY — **WAYS TO SOLVE POLYNOMIAL EQUATIONS**

GRAPHING: Can be used to solve any equation, but gives only approximate solutions. Examples 2 and 3, p. 527

THE QUADRATIC FORMULA: Can be used to solve any *quadratic* equation. Examples 1–3, pp. 533–534

FACTORING: Can be used with the zero-product property to solve an equation that is factorable.

- **Factoring $x^2 + bx + c$:** Examples 1–6, pp. 604–606

  $x^2 + 9x + 18 = (x + 3)(x + 6)$
- **Factoring $ax^2 + bx + c$:** Examples 1–5, pp. 611–613

  $3x^2 + 10x + 7 = (3x + 7)(x + 1)$
- **Special Products:** Examples 1–4, pp. 619–620

  $a^2 - b^2 = (a + b)(a - b)$ — **Example:** $4x^2 - 36 = (2x + 6)(2x - 6)$

  $a^2 + 2ab + b^2 = (a + b)^2$ — **Example:** $x^2 + 18x + 81 = (x + 9)^2$

  $a^2 - 2ab + b^2 = (a - b)^2$ — **Example:** $x^2 - 16x + 64 = (x - 8)^2$
- **Factoring Completely:** Examples 1–7, pp. 625–627

STUDENT HELP

HOMEWORK HELP Visit our Web site www.mcdougallittell.com for extra examples.

EXAMPLE 9 *Solving Quadratic and Other Polynomial Equations*

Solve the equation.

a. $-30x^4 + 58x^3 - 24x^2 = 0$

b. $18x^3 - 30x^2 = 60x$

SOLUTION

a.

$-30x^4 + 58x^3 - 24x^2 = 0$	**Write original equation.**
$-2x^2(15x^2 - 29x + 12) = 0$	**Factor out $-2x^2$.**
$-2x^2(5x - 3)(3x - 4) = 0$	**Find correct factorization for trinomial.**

▶ Set each variable factor equal to zero. The solutions are 0, $\frac{3}{5}$, and $\frac{4}{3}$.

✓ **CHECK** Graph $y = -30x^4 + 58x^3 - 24x^2$. Use your calculator's TRACE feature to estimate the x-intercepts.

The graph appears to confirm the solutions.

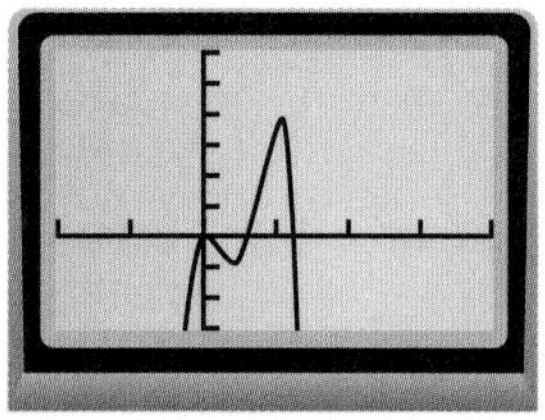

b.

$18x^3 - 30x^2 - 60x = 0$	**Rewrite equation in standard form.**
$6x(3x^2 - 5x - 10) = 0$	**Factor out GCF.**

▶ $6x$ factors out, so one solution is 0. Because $3x^2 - 5x - 10$ is not factorable, use the quadratic formula to find the other solutions, $x \approx 2.84$ and $x \approx -1.17$.

GUIDED PRACTICE

Vocabulary Check ✓

1. In writing a polynomial as the product of polynomials of lesser degrees, what does it mean to say that a factor is *prime*?
that it cannot be factored using integer coefficients

Concept Check ✓

2. To factor a polynomial completely, you must write it as the product of what two types of factors? **monomial factors and prime factors with at least two terms**

3. Name three ways to find the solution(s) of a quadratic equation. Which way do you prefer? Explain. **See margin.**

3. Graphing, the quadratic formula, factoring; *Sample answer:* Graphing may be very convenient, but may give only approximate solutions. The quadratic formula may be used for any equation, but may involve more calculation than is necessary if the polynomial expression is factorable.

ERROR ANALYSIS Describe and correct the factoring error.

4.
$4x^3 + 36x$
$= 4x(x^2 + 9)$
$= 4x(x + 3)(x - 3)$

See margin.

5.
$-2b^3 + 12b^2 - 14b$
$= -2b(b^2 + 6b - 7)$
$= -2b(b + 7)(b - 1)$

See margin.

Skill Check ✓

Factoring can be used only for equations involving factorable expressions.

Find the greatest common factor and factor it out of the expression.

6. $5n^3 - 20n$ **$5n(n^2 - 4)$**
7. $6x^2 + 3x^4$ **$3x^2(2 + x^2)$**
8. $6y^4 + 14y^3 - 10y^2$ **$2y^2(3y^2 + 7y - 5)$**

Tell whether the expression is factored completely. If the expression is not factored completely, write the complete factorization.

9. $7x^3 - 11x$ **no; $x(7x^2 - 11)$**
10. $9t(t^2 + 49)$ **yes**
11. $3w(9w^2 - 16)$ **no; $3w(3w + 4)(3w - 4)$**

Solve the equation. Tell which method you used.

12. $y^2 - 4y - 5 = 0$ **$-1, 5$; *Sample answer:* factoring**
13. $z^2 + 11z + 30 = 0$ **$-5, -6$; *Sample answer:* factoring**
14. $5a^2 + 11a + 2 = 0$ **$-2, -\frac{1}{5}$; *Sample answer:* factoring**

PRACTICE AND APPLICATIONS

STUDENT HELP
Extra Practice to help you master skills is on p. 806.

FACTORING OUT THE GCF Find the greatest common factor and factor it out of the expression.

15. $6v^3 - 18v$ **$6v(v^2 - 3)$**
16. $4q^4 + 12q$ **$4q(q^3 + 3)$**
17. $3x - 9x^2$ **$3x(1 - 3x)$**
18. $24t^5 + 6t^3$ **$6t^3(4t^2 + 1)$**
19. $4a^5 + 8a^3 - 2a^2$ **$2a^2(2a^3 + 4a - 1)$**
20. $18d^6 - 6d^2 + 3d$ **$3d(6d^5 - 2d + 1)$**

4. The sum of two squares is not factorable; $4x(x^2 + 9)$.

FACTORING COMPLETELY Factor the expression completely.

21. $24x^3 + 18x^2$ **$6x^2(4x + 3)$**
22. $-3w^4 + 21w^3$ **$-3w^3(w - 7)$**
23. $2y^3 - 10y^2 - 12y$ **$2y(y - 6)(y + 1)$**
24. $5s^3 + 30s^2 + 40s$ **$5s(s + 4)(s + 2)$**
25. $-7m^3 + 28m^2 - 21m$ **$-7m(m - 3)(m - 1)$**
26. $2d^4 + 2d^3 - 60d^2$ **$2d^2(d + 6)(d - 5)$**
27. $4t^3 - 144t$ **$4t(t + 6)(t - 6)$**
28. $-12z^4 + 3z^2$ **$-3z^2(2z + 1)(2z - 1)$**
29. $c^4 + c^3 - 12c - 12$ **$(c^3 - 12)(c + 1)$**
30. $x^3 - 3x^2 + x - 3$ **$(x^2 + 1)(x - 3)$**
31. $6b^4 + 5b^3 - 24b - 20$ **$(b^3 - 4)(6b + 5)$**
32. $3y^3 - y^2 - 21y + 7$ **$(y^2 - 7)(3y - 1)$**
33. $a^3 + 6a^2 - 4a - 24$ **$(a + 6)(a + 2)(a - 2)$**
34. $t^3 - t^2 - 16t + 16$ **$(t - 1)(t + 4)(t - 4)$**
35. $3m^3 - 15m^2 - 6m + 30$ **$3(m^2 - 2)(m - 5)$**
36. $7n^5 + 7n^4 - 3n^2 - 6n - 3$ **$(7n^4 - 3n - 3)(n + 1)$**

STUDENT HELP
HOMEWORK HELP
Example 1: Exs. 15–20
Examples 2–6: Exs. 21–36
Example 7: Exs. 37–50
Example 8: Exs. 55–57
Example 9: Exs. 37–50

3 APPLY

ASSIGNMENT GUIDE

BASIC
Day 1: pp. 629–630 Exs. 16–50 even
Day 2: pp. 629–632 Exs. 21–51 odd, 52, 61–63, 70, 75, 80–82, Quiz 3 Exs. 1–25

AVERAGE
Day 1: pp. 629–630 Exs. 16–50 even
Day 2: pp. 629–632 Exs. 21–51 odd, 52, 55–57, 61–63, 70, 75, 80–82, Quiz 3 Exs. 1–25

ADVANCED
Day 1: pp. 629–630 Exs. 16–50 even
Day 2: pp. 629–632 Exs. 21–51 odd, 52–57, 61–68, 70, 75, 80–82, Quiz 3 Exs. 1–25

BLOCK SCHEDULE
pp. 629–632 Exs. 16–52, 55–57, 61–63, 70, 75, 80–82, Quiz 3 Exs. 1–25

EXERCISE LEVELS

Level A: *Easier*
15–20
Level B: *More Difficult*
21–52, 55–57, 61–63, 69–82
Level C: *Most Difficult*
53, 54, 58–60, 64–68

✓ HOMEWORK CHECK

To quickly check student understanding of key concepts, go over the following exercises: Exs. 16, 22, 26, 32, 44, 46, 52. See also the Daily Homework Quiz:

- Blackline Master (*Chapter 11 Resource Book,* p. 11)
- Transparency (p. 85)

5. The distributive property was applied incorrectly; $-2b(b^2 - 6b + 7) = -2b(b - 7)(b + 1)$

APPLICATION NOTE

EXERCISES 51–54 Additional information about gravity is available at **www.mcdougallittell.com.**

ADDITIONAL PRACTICE AND RETEACHING

For Lesson 10.8:

- Practice Levels A, B, and C (*Chapter 10 Resource Book,* p. 111)
- Reteaching with Practice (*Chapter 10 Resource Book,* p. 114)
- See Lesson 10.8 of the *Personal Student Tutor*

For more Mixed Review:

- Search the *Test and Practice Generator* for key words or specific lessons.

42. $\frac{21-\sqrt{105}}{2}, \frac{21+\sqrt{105}}{2}$; quadratic formula

44. $0, -\frac{1}{2}, \frac{1}{2}$; *Sample answer:* factoring

45. $0, \frac{29-\sqrt{593}}{2}, \frac{29+\sqrt{593}}{2}$; factoring and quadratic formula

46. $0, 1, \frac{3}{2}$; *Sample answer:* factoring

FOCUS ON APPLICATIONS

GRAVITY varies depending on the mass of a planet. This is why acceleration due to gravity is greater on Jupiter than on Earth.

APPLICATION LINK www.mcdougallittell.com

47. $\frac{-9-\sqrt{305}}{16}, \frac{-9+\sqrt{305}}{16}$; quadratic formula

49. $0, \frac{-3-\sqrt{457}}{8}, \frac{-3+\sqrt{457}}{8}$; factoring and quadratic formula

53. Earth; the coefficient of the t^2-term in the vertical motion model for Earth is 6 times greater than that of the Moon.

56. $h(h-3)(h-9) = 324$ or $h^3 - 12h^2 + 27h - 324 = 0$

SOLVING EQUATIONS Solve the equation. Tell which solution method you used.

37. $y^2 + 7y + 12 = 0$ **−3, −4; *Sample answer:* factoring**

38. $x^2 - 3x - 4 = 0$ **−1, 4; *Sample answer:* factoring**

39. $b^2 + 4b - 117 = 0$ **−13, 9; *Sample answer:* factoring**

40. $t^2 - 16t + 65 = 0$ **no solution**

41. $27 + 6w - w^2 = 0$ **−3, 9; *Sample answer:* factoring**

42. $x^2 - 21x + 84 = 0$ **See margin.**

43. $5x^4 - 80x^2 = 0$ **0, −4, 4; *Sample answer:* factoring**

44. $-16x^3 + 4x = 0$ **See margin.**

45. $10x^3 - 290x^2 + 620x = 0$ **See margin.**

46. $34x^4 - 85x^3 + 51x^2 = 0$ **See margin.**

47. $8x^2 + 9x - 7 = 0$ **See margin.**

48. $18x^2 - 21x + 28 = 0$ **no solution**

49. $24x^3 + 18x^2 - 168x = 0$ **See margin.**

50. $-14x^4 + 118x^3 + 72x^2 = 0$ **$0, -\frac{4}{7}, 9$; *Sample answer:* factoring**

VERTICAL MOTION In Exercises 51–54, use the vertical motion models, where *h* is the initial height (in feet), *v* is the initial velocity (in feet per second) and *t* is the time (in seconds) the object spends in the air. (Note that the acceleration due to gravity on the Moon is $\frac{1}{6}$ that of Earth.)

Model for vertical motion on Earth: $h = 16t^2 - vt$

Model for vertical motion on the Moon: $h = \frac{16}{6}t^2 - vt$

51. EARTH On Earth, you toss a tennis ball from a height of 96 feet with an initial upward velocity of 16 feet per second. How long will it take the tennis ball to reach the ground? **3 sec**

52. MOON On the Moon, you toss a tennis ball from a height of 96 feet with an initial upward velocity of 16 feet per second. How long will it take the tennis ball to reach the surface of the moon? **$(3 + 3\sqrt{5})$ sec (about 9.7 sec)**

53. *Writing* Do objects fall faster on Earth or on the Moon? Use your results from Exercises 51 and 52 and the vertical motion models shown above to support your answer. **See margin.**

54. CRITICAL THINKING The coefficient of the t^2-term in the vertical motion models is one-half the acceleration of a falling object due to gravity. On the surface of Jupiter, the acceleration due to gravity is about 2.4 times that on the surface of Earth. Write a vertical motion model for Jupiter. **$h = 2.4(16t^2) - vt$ or $38.4t^2 - vt$**

PACKAGING In Exercises 55–57, use the following information. Refer to the diagram at the right.

The length ℓ of a box is 3 inches less than the height h. The width w is 9 inches less than the height. The box has a volume of 324 cubic inches.

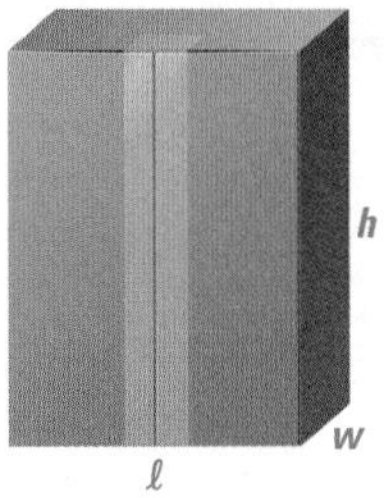

55. Copy and complete the diagram by labeling the dimensions. **$\ell = h - 3, w = h - 9$**

56. Write a model that you can solve to find the length, height, and width of the box.

57. What are the dimensions of the box? **$\ell = 9$ in., $w = 3$ in., $h = 12$ in.**

EXTENSION: BRITISH METHOD Another method for finding the binomial factors of factorable trinomials of the form $ax^2 + bx + c$ is sometimes called the *British method.* To factor $6x^2 - x - 2$ by the British method, use the following steps.

1. Find the "magic" number ac.

   $6 \cdot (-2) = -12$

2. Write the magic number as the product of two factors whose sum is b.

   $-12 = 3 \cdot (-4)$

3. Rewrite the trinomial using the factors.

   $6x^2 - x - 2 = 6x^2 + 3x - 4x - 2$

4. Factor using the distributive property.

   $$6x^2 + 3x - 4x - 2 = 3x(2x + 1) - 2(2x + 1)$$
   $$= (3x - 2)(2x + 1)$$

Use the British method to factor the trinomials.

58. $8x^2 - 2x - 3$ $(2x + 1)(4x - 3)$ **59.** $3x^2 + 13x + 14$ $(3x + 7)(x + 2)$ **60.** $5x^2 + 27x - 18$ $(5x - 3)(x + 6)$

Test Preparation

61. MULTIPLE CHOICE Which of the following is the complete factorization of $x^3 - 5x^2 + 4x - 20$? **C**

(A) $(x + 2)(x + 2)(x - 5)$ (B) $(x + 2)(x - 2)(x - 5)$

(C) $(x^2 + 4)(x - 5)$ (D) $(x - 4)(x - 1)(x - 20)$

62. MULTIPLE CHOICE Which of the following equations have real solutions? **D**

I. $6x^2 + x - 12 = 0$ **II.** $6x^2 + 7x - 12 = 0$ **III.** $6x^2 + 7x - 13 = 0$

(A) I only (B) II only (C) III only (D) All

63. MULTIPLE CHOICE The length l of a box is 3 centimeters more than half the height h. The width w is 2 centimeters more than one fifth of the height. The box has a volume of 3780 cubic centimeters. Which of the following equations can be used to find the height of the box? **A**

(A) $h\left(\frac{1}{5}h + 2\right)\left(\frac{1}{2}h + 3\right) = 3780$ (B) $h\left(\frac{1}{2}h + 2\right)\left(\frac{1}{5}h + 3\right) = 3780$

(C) $\frac{1}{10}h(h + 2)(h + 3) = 3780$ (D) $h\left(\frac{1}{2}w + 2\right)\left(\frac{1}{5}l + 3\right) = 3780$

★ Challenge

CUBIC EQUATIONS Sums and differences of *cubes* can be factored using the following patterns.

Sum of cubes pattern: $a^3 + b^3 = (a + b)(a^2 - ab + b^2)$

Difference of cubes pattern: $a^3 - b^3 = (a - b)(a^2 + ab + b^2)$

In Exercises 64–67, use the patterns above to factor the cubic expression completely. Use the distributive property to verify your results.

64. $y^3 - 125$ 5 $(y - 5)(y^2 + 5y + 25)$ **65.** $b^3 + 27$ −3 See margin. **66.** $\frac{1}{8} + 8x^3$ $-\frac{1}{4}$ See margin. **67.** $216 - 343t^3$ $\frac{6}{7}$ See margin.

68. Set each expression in Exercises 64–67 equal to zero and solve the equation.

65. $(b + 3)(b^2 - 3b + 9)$

66. $\left(\frac{1}{2} + 2x\right)\left(\frac{1}{4} - x + 4x^2\right)$

67. $(6 - 7t)(36 + 42t + 49t^2)$

EXTRA CHALLENGE

www.mcdougallittell.com

4 ASSESS

DAILY HOMEWORK QUIZ

Transparency Available

Factor the expression completely.

1. $2n^3 - 4n^2 - 30n$ $2n(n + 3)(n - 5)$

2. $-6x^3 + 36x^2 - 54x$ $-6x(x - 3)^2$

3. $-y^4 - 3y^3 - 2y^2 - 6y$

$-y(y^2 + 2)(y + 3)$

Solve the equation. Tell which method you used.

4. $5x^2 - 10x = 120$

−4, 6; *Sample answer:* factoring

5. $-3x^2 - 7x + 84 = 0$

$\frac{7 \pm \sqrt{1057}}{-6}$; quadratic formula

EXTRA CHALLENGE NOTE

Challenge problems for Lesson 10.8 are available in **blackline** format in the *Chapter 10 Resource Book,* p. 118 and at **www.mcdougallittell.com.**

ADDITIONAL TEST PREPARATION

1. WRITING What does it mean to factor a quadratic polynomial completely? *Sample answer:* If there are no common factors in all of the terms of the polynomial and the polynomial cannot be written as the product of factors of lesser degree, then the polynomial has been factored completely.

69. 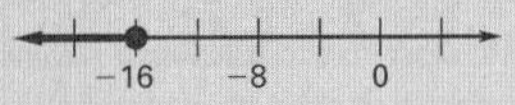

70. 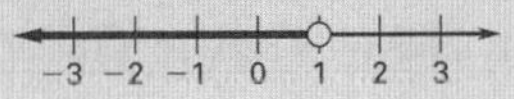

71. 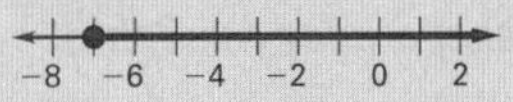

78. 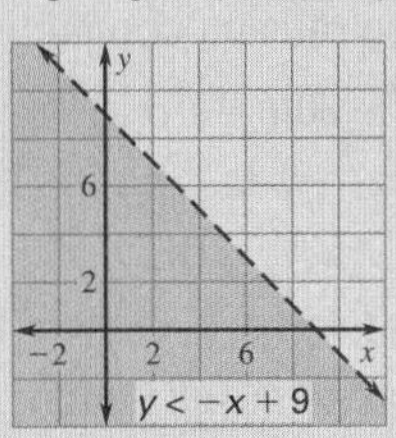

79. 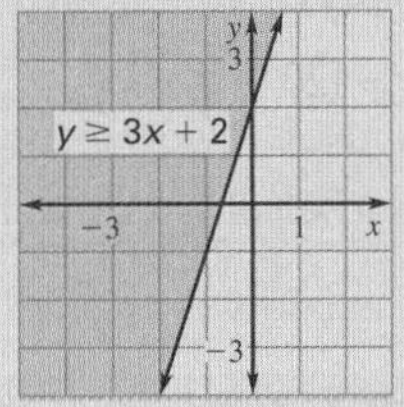

80. 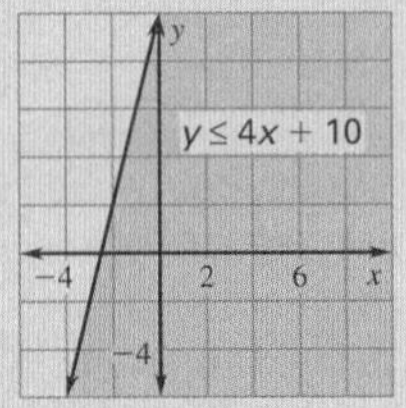

MIXED REVIEW

SOLVING AND GRAPHING INEQUALITIES Solve the inequality and graph its solution. (Review 6.1, 6.2) **69–71. See margin.**

69. $7 + x \leq -9$ **$x \leq -16$** **70.** $-3 > 2x - 5$ **$x < 1$** **71.** $-x + 6 \leq 13$ **$x \geq -7$**

SOLVING EQUATIONS Solve the equation. (Review 6.4)

72. $|x| = 3$ **−3, 3** **73.** $|x - 5| = 7$ **−2, 12** **74.** $|x + 6| = 13$ **−19, 7**

75. $|x + 8| - 2 = -12$ **no solution** **76.** $|2x - 5| + 7 = 16$ **−2, 7** **77.** $\left|x + \frac{3}{4}\right| = \frac{9}{4}$ **$-3, \frac{3}{2}$**

SKETCHING GRAPHS Sketch the graph of the inequality. (Review 6.5) **78–80. See margin.**

78. $x + y < 9$ **79.** $y - 3x \geq 2$ **80.** $y - 4x \leq 10$

81. **HEIGHT** One of the tallest people in the world is Ri Myong-hun of North Korea, who is about 93 inches tall. The average height of a United States male is about 69 inches. Compare Ri Myong-hun's height to the average U.S. male height using a ratio and a percent. (Review 3.8 for 11.1) **$\frac{31}{23} \approx 134.78\%$**

82. **COMPARING ROOMS** One bedroom in a house is 10 feet by 12 feet. The living room is 24 feet by 15 feet. Compare the areas of the two rooms using a ratio. (Review 3.8 for 11.1) **$\frac{1}{3}$**

1. $(7x + 8)(7x - 8)$; difference of two squares

2. $(11 + 3x)(11 - 3x)$; difference of two squares

3. $(2t + 5)^2$; perfect square trinomial

4. $2(6 + 5y)(6 - 5y)$; difference of two squares

5. $(3y + 7)^2$; perfect square trinomial

6. $3(n - 6)^2$; perfect square trinomial

QUIZ 3

Self-Test for Lessons 10.7 and 10.8

20. $0, -\frac{3}{2}, \frac{3}{2}$; factoring special products

21. $-3, \frac{5}{3}$; factoring $ax^2 + bx + c$

22. 2; factoring

23. 0, about 0.77, about 2.26; quadratic formula

24. 0, −3, 4; factoring

25. 0, about −7.15, about −0.35; quadratic formula

Factor the expression. Tell which special product factoring pattern you used. (Lesson 10.7) **1–6. See margin.**

1. $49x^2 - 64$ **2.** $121 - 9x^2$ **3.** $4t^2 + 20t + 25$

4. $72 - 50y^2$ **5.** $9y^2 + 42y + 49$ **6.** $3n^2 - 36n + 108$

Use factoring to solve the equation. (Lesson 10.7)

7. $3x^2 - 192 = 0$ **−8, 8** **8.** $4x^2 + 32x + 64 = 0$ **−4** **9.** $9x^2 + 96x + 256 = 0$ **$-\frac{16}{3}$**

10. $\frac{1}{2}x^2 - 4x + 8 = 0$ **4** **11.** $216x^2 - 96 = 0$ **$-\frac{2}{3}, \frac{2}{3}$** **12.** $-3x^2 - 2x - \frac{1}{3} = 0$ **$-\frac{1}{3}$**

Find the greatest common factor and factor it out of the expression. (Lesson 10.8)

13. $3x^3 + 12x^2$ **$3x^2(x + 4)$** **14.** $6x^2 + 3x$ **$3x(2x + 1)$** **15.** $18x^4 - 9x^3$ **$9x^3(2x - 1)$** **16.** $8x^5 + 4x^2 - 2x$ **$2x(4x^4 + 2x - 1)$**

Factor the expression completely. (Lesson 10.8)

17. $2x^3 - 6x^2 + 4x$ **$2x(x - 1)(x - 2)$** **18.** $48x^3 - 75x$ **$3x(4x - 5)(4x + 5)$** **19.** $x^3 + 3x^2 + 4x + 12$ **$(x^2 + 4)(x + 3)$**

Solve the equation. Tell which solution method you used. (Lesson 10.8) **20–25. See margin.**

20. $12x^4 - 27x^2 = 0$ **21.** $-3x^2 - 4x + 15 = 0$ **22.** $3x^3 - 6x^2 + 5x - 10 = 0$

23. $56x^3 + 98x = 170x^2$ **24.** $17x^6 = 17x^5 + 204x^4$ **25.** $6x^3 + 45x^2 + 15x = 0$

CHAPTER 10 Chapter Summary

WHAT did you learn?	*WHY did you learn it?*
Classify a polynomial by degree and by number of terms. **(10.1)**	Understand how to use polynomials to model real-life situations. **(p. 578)**
Add and subtract polynomials. **(10.1)**	Represent the growth in the number of Internet sites. **(p. 579)**
Multiply polynomials	
• using the FOIL pattern. **(10.2)**	Evaluate a polynomial expression to find the area of a window. **(p. 586)**
• using special product patterns. **(10.3)**	Find products of some polynomials more easily. **(p. 591)**
Solve polynomial equations in factored form. **(10.4)**	Find the dimensions of an archway. **(p. 599)**
Factor quadratic trinomials. **(10.5–10.6)**	Find the time it takes a diver to reach the water. **(p. 613)**
Factor a polynomial	
• using special product patterns. **(10.7)**	Find the size of rope needed to lift a safe. **(p. 621)**
• using the distributive property. **(10.8)**	Find the dimensions of a box. **(p. 627)**
Solve polynomial equations by factoring. **(10.5–10.8)**	Apply factoring skills to write models for real-life situations. **(pp. 604–632)**

How does Chapter 10 fit into the BIGGER PICTURE of algebra?

Polynomials can model a great variety of real-life situations. You were already familiar with three types of polynomial models: $y = a$ (constant model), $y = ax + b$ (linear model), and $y = ax^2 + bx + c$ (quadratic model).

In this chapter you learned how to add, subtract and multiply polynomials. You also learned how to "undo" multiplication by a process called factoring. This chapter has many connections with the mathematics you studied in Chapter 9. For instance, in this chapter you learned how to use factoring to solve quadratic equations.

STUDY STRATEGY

How did identifying trouble spots help you review?

One of your notes about trouble spots, using the **Study Strategy** on p. 574, may look like this.

Trouble Spots:

Lesson 10.1

Remember to change all of the signs when subtracting.

$(x^2 + 4x) - (x^2 - 3x + 1)$

$= x^2 + 4x - x^2 + 3x - 1$

$= 7x - 1$

633

ADDITIONAL RESOURCES

The following resources are available to help review the material in this chapter.

- Chapter Review Games and Activities (*Chapter 10 Resource Book,* p. 119)
- *Instant Replay: Video Review Games*
- *Personal Student Tutor*
- Cumulative Review, Chs. 1–10 (*Chapter 10 Resource Book,* p. 131)

CHAPTER 10

Chapter Review

VOCABULARY

- polynomial, p. 576
- standard form of a polynomial, p. 576
- degree of a term, p. 576
- degree of a polynomial, p. 576
- leading coefficient, p. 576
- monomial, p. 576
- binomial, p. 576
- trinomial, p. 576
- FOIL pattern, p. 585
- factored form of a polynomial, p. 597
- zero-product property, p. 597
- factor a quadratic expression, p. 604
- prime factor, p. 625
- factor a polynomial completely, p. 625

10.1 ADDING AND SUBTRACTING POLYNOMIALS

Examples on pp. 576–579

EXAMPLES To add or subtract polynomials, add or subtract like terms.

Horizontal Format

$$(4x^3 + 6x - 8) - (-x^2 + 7x - 2)$$
$$= 4x^3 + 6x - 8 + x^2 - 7x + 2$$
$$= 4x^3 + x^2 - x - 6$$

Vertical Format

$$\begin{array}{rrrrr} & -2x^3 - 4x^2 - & x & + 5 \\ & 3x^3 + 2x^2 - & 4x & + 9 \\ + & -x^3 + 5x^2 - & x & - 1 \\ \hline & 3x^2 - & 6x & + 13 \end{array}$$

Find the sum or the difference.

1. $(-x^2 + x + 2) + (3x^2 + 4x + 5)$ **$2x^2 + 5x + 7$**
2. $(x^2 + 3x - 1) - (4x^2 - 5x + 6)$ **$-3x^2 + 8x - 7$**
3. $(x^3 + 5x^2 - 4x) - (3x^2 - 6x + 2)$ **$x^3 + 2x^2 + 2x - 2$**
4. $(4x^3 + x^2 - 1) + (2 - x - x^2)$ **$4x^3 - x + 1$**

10.2 MULTIPLYING POLYNOMIALS

Examples on pp. 584–586

EXAMPLES To multiply polynomials, use the distributive property or the FOIL pattern.

a. $(3x + 2)(5x^2 - 4x + 1) = 5x^2(3x + 2) - 4x(3x + 2) + 1(3x + 2)$
$$= 15x^3 + 10x^2 - 12x^2 - 8x + 3x + 2$$
$$= 15x^3 - 2x^2 - 5x + 2$$

b. $(4x + 5)(-3x - 6) = \underset{\text{F}}{-12x^2} - \underset{\text{O}}{24x} - \underset{\text{I}}{15x} - \underset{\text{L}}{30}$
$$= -12x^2 - 39x - 30$$

Find the product.

5. $(-x)(8x^3 - 12x^2)$ **$-8x^4 + 12x^3$**
6. $4x^3(-x^2 + 2x - 7)$ **$-4x^5 + 8x^4 - 28x^3$**
7. $(x + 3)(x + 11)$ **$x^2 + 14x + 33$**
8. $(7x - 1)(5x + 2)$ **$35x^2 + 9x - 2$**
9. $(x + 5)(2x^2 + x - 10)$ **$2x^3 + 11x^2 - 5x - 50$**
10. $(4x^2 + 9x)(-x^2 - 3)$ **$-4x^4 - 9x^3 - 12x^2 - 27x$**

10.3 SPECIAL PRODUCTS OF POLYNOMIALS

Examples on pp. 590–592

EXAMPLES Use special product patterns to multiply some polynomials.

$(a + b)(a - b) = a^2 - b^2$

$(3x + 7)(3x - 7) = (3x)^2 - 7^2$

$= 9x^2 - 49$

$(a + b)^2 = a^2 + 2ab + b^2$

$(5t + 4)^2 = (5t)^2 + 2(5t)(4) + 4^2$

$= 25t^2 + 40t + 16$

Find the product.

11. $(x + 15)(x - 15)$ $x^2 - 225$

12. $(5x - 2)(5x + 2)$ $25x^2 - 4$

13. $(x + 2)^2$ $x^2 + 4x + 4$

14. $(7m - 6)^2$ $49m^2 - 84m + 36$

10.4 SOLVING POLYNOMIAL EQUATIONS IN FACTORED FORM

Examples on pp. 597–599

EXAMPLE Solve the equation $(x + 1)(x - 5) = 0$.

$(x + 1) = 0$ or $(x - 5) = 0$ **Use zero-product property.**

$x + 1 = 0$ $x - 5 = 0$

$x = -1$ $x = 5$

▶ The solutions are -1 and 5. Check these in the original equation.

Solve the equation.

15. $(x + 1)(x + 10) = 0$ $-1, -10$

16. $(x - 3)(x - 2) = 0$ 3, 2

17. $(x + 9)^2 = 0$ -9

18. $(3x + 6)(4x - 1)(x - 4) = 0$ $-2, \frac{1}{4}, 4$

10.5 FACTORING $x^2 + bx + c$

Examples on pp. 604–606

EXAMPLE Solve the equation $x^2 - 9x = -14$ by factoring.

$x^2 - 9x = -14$ **Write original equation.**

$x^2 - 9x + 14 = 0$ **Write in standard form.**

$(x - 2)(x - 7) = 0$ **Factor.**

$(x - 2) = 0$ or $(x - 7) = 0$ **Use zero-product property.**

$x - 2 = 0$ $x - 7 = 0$

$x = 2$ $x = 7$

▶ The solutions are 2 and 7. Check these in the original equation.

Solve the equation by factoring.

19. $x^2 - 21x + 108 = 0$ 9, 12

20. $x^2 + 10x - 200 = 0$ $-20, 10$

21. $x^2 + 26x = -169$ -13

10.6 FACTORING $ax^2 + bx + c$

Examples on pp. 611–613

EXAMPLE To factor $3x^2 - x - 2$, test the possible factors of a (1 and 3) and of c (-2 and 1 or 2 and -1).

$(x - 2)(3x + 1) = 3x^2 - 5x - 2$ Not correct

$(x + 1)(3x - 2) = 3x^2 + x - 2$ Not correct

$(x + 2)(3x - 1) = 3x^2 + 5x - 2$ Not correct

$(x - 1)(3x + 2) = 3x^2 - x - 2$ Correct

Factor the trinomial.

22. $12x^2 + 7x + 1$ **$(3x + 1)(4x + 1)$**

23. $2x^2 + 5x - 12$ **$(2x - 3)(x + 4)$**

24. $6x^2 + 4x - 10$ **$(3x + 5)(2x - 2)$**

25. $4x^2 - 12x + 9$ **$(2x - 3)(2x - 3)$**

10.7 FACTORING SPECIAL PRODUCTS

Examples on pp. 619–621

EXAMPLES Factor using special product patterns to solve the equations.

$a^2 - b^2 = (a + b)(a - b)$

$4x^2 - 64 = 0$

$(2x + 8)(2x - 8) = 0$

$2x + 8 = 0$ or $2x - 8 = 0$

▶ The solutions are -4 and 4.

$a^2 - 2ab + b^2 = (a - b)^2$

$x^2 - 4x + 4 = 0$

$x^2 - 2(x)(2) + 2^2 = 0$

$(x - 2)^2 = 0$

$x - 2 = 0$

▶ The solution is 2.

Use factoring to solve the equation.

26. $100x^2 - 121 = 0$ **$-\frac{11}{10}, \frac{11}{10}$**

27. $16x^2 + 24x + 9 = 0$ **$-\frac{3}{4}$**

28. $9x^2 - 18x + 9 = 0$ **1**

10.8 FACTORING USING THE DISTRIBUTIVE PROPERTY

Examples on pp. 625–628

EXAMPLES Use the distributive property to factor out common variable factors.

$3x^3 - 27x = 3x(x^2 - 9)$

$= 3x(x + 3)(x - 3)$

$x^4 - 4x^3 + x^2 - 4x = (x^4 + x^2) + (-4x^3 - 4x)$

$= x^2(x^2 + 1) - 4x(x^2 + 1)$

$= (x^2 - 4x)(x^2 + 1)$

$= x(x - 4)(x^2 + 1)$

Factor the expression completely.

29. $-2x^5 - 2x^4 + 4x^3$ **$-2x^3(x + 2)(x - 1)$**

30. $2x^4 - 32x^2$ **$2x^2(x + 4)(x - 4)$**

31. $3x^3 + x^2 + 15x + 5$ **$(x^2 + 5)(3x + 1)$**

32. $x^3 + 3x^2 - 4x - 12$ **$(x + 3)(x + 2)(x - 2)$**

CHAPTER 10 Chapter Test

Use a vertical format or a horizontal format to add or subtract.

1. $(x^2 + 4x - 1) + (5x^2 + 2)$ $6x^2 + 4x + 1$
2. $(5t^2 - 9t + 1) - (8t + 13)$ $5t^2 - 17t - 12$
3. $(7n^3 + 2n^2 - n - 4) - (4n^3 - 3n^2 + 8)$ $3n^3 + 5n^2 - n - 12$
4. $(x^4 + 6x^2 + 7) + (2x^4 - 3x^2 + 1)$ $3x^4 + 3x^2 + 8$

Find the product.

5. $(x + 3)(2x + 3)$ $2x^2 + 9x + 9$
6. $(9x - 1)(7x + 4)$ $63x^2 + 29x - 4$
7. $(w - 6)(4w^2 + w - 7)$ $4w^3 - 23w^2 - 13w + 42$
8. $(5t^3 + 2)(4t^2 + 8t - 7)$ $20t^5 + 40t^4 - 35t^3 + 8t^2 + 16t - 14$
9. $(3z^3 - 5z^2 + 8)(z + 2)$ $3z^4 + z^3 - 10z^2 + 8z + 16$
10. $(4x + 1)\left(\frac{1}{2}x - \frac{3}{8}\right)$ $2x^2 - x - \frac{3}{8}$
11. $(x - 12)^2$ $x^2 - 24x + 144$
12. $(7x + 2)^2$ $49x^2 + 28x + 4$
13. $(8x + 3)(8x - 3)$ $64x^2 - 9$

Use the zero-product property to solve the equation.

14. $(6x - 5)(x + 2) = 0$ $\frac{5}{6}, -2$
15. $(x + 8)^2 = 0$ -8
16. $(4x + 3)(x - 1)(3x + 9) = 0$ $-\frac{3}{4}, 1, -3$

Find the *x*-intercepts and the vertex of the graph. Then sketch the graph. 17–19. See margin for graphs.

17. $y = (x + 1)(x - 5)$ $-1, 5; (2, -9)$
18. $y = (x + 2)(x + 6)$ $-6, -2; (-4, -4)$
19. $y = (-x - 4)(x + 7)$ $-7, -4; \left(-\frac{11}{2}, \frac{9}{4}\right)$

Solve the equation by factoring.

20. $x^2 + 13x + 30 = 0$ $-10, -3$
21. $x^2 - 19x + 84 = 0$ $7, 12$
22. $x^2 - 34x - 240 = 0$ $-6, 40$
23. $2x^2 + 15x - 108 = 0$ $-12, \frac{9}{2}$
24. $9x^2 - 9x = 28$ $-\frac{4}{3}, \frac{7}{3}$
25. $18x^2 - 57x = -35$ $\frac{5}{6}, \frac{7}{3}$

Factor the expression.

26. $x^2 - 196$ $(x + 14)(x - 14)$
27. $16x^2 - 36$ $4(2x + 3)(2x - 3)$
28. $128 - 50x^2$ $2(8 + 5x)(8 - 5x)$
29. $x^2 - 6x + 9$ $(x - 3)^2$
30. $4x^2 + 44x + 121$ $(2x + 11)^2$
31. $2x^2 + 28x + 98$ $2(x + 7)^2$
32. $9t^2 - 54$ $9(t^2 - 6)$
33. $4x^2 + 38x + 34$ $2(2x + 17)(x + 1)$
34. $x^3 + 2x^2 - 16x - 32$ $(x + 2)(x + 4)(x - 4)$

Solve the equation by a method of your choice.

35. $x^2 - 60 = -11$ $-7, 7$
36. $2x^2 + 15x - 8 = 0$ $-8, \frac{1}{2}$
37. $x^2 - 13x = -40$ $5, 8$
38. $x(x - 16) = 0$ $0, 16$
39. $12x^2 + 3x = 0$ $-\frac{1}{4}, 0$
40. $8x^2 + 6x = 5$ $-\frac{5}{4}, \frac{1}{2}$
41. $12x^2 + 17x = 7$ $-\frac{7}{4}, \frac{1}{3}$
42. $81x^2 - 6 = 30$ $-\frac{2}{3}, \frac{2}{3}$
43. $16x^2 - 34x - 15 = 0$ $-\frac{3}{8}, \frac{5}{2}$

44. **TIC-TAC-TOE** You are making a tic-tac-toe board. Each square will have sides of x inches. The board will have a border with a width of 1 inch. Draw a diagram and label the dimensions. Write a polynomial expression for the area of the board. $(3x + 2)^2 = 9x^2 + 12x + 4$. See margin for diagram.

45. **ROOM DIMENSIONS** A room's length is 3 feet less than twice its width. The area of the room is 135 square feet. What are the room's dimensions? 9 ft by 15 ft

46. **KICKING A FOOTBALL** You kick a football into the air. The vertical component of this motion can be modeled by the vertical motion model. Suppose the football has an initial upward velocity of 31 feet per second. The ball makes contact with your foot about 2 feet above the ground. Find the time t (in seconds) for the football to reach the ground. Use the vertical motion model on page 621. 2 sec

ADDITIONAL RESOURCES

- ***Chapter 10 Resource Book***
  Chapter Test (3 levels) (p. 120)
  SAT/ACT Chapter Test (p. 126)
  Alternative Assessment (p. 127)
- ***Test and Practice Generator***

17.

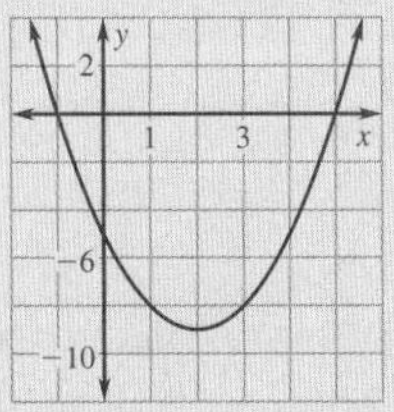

18.

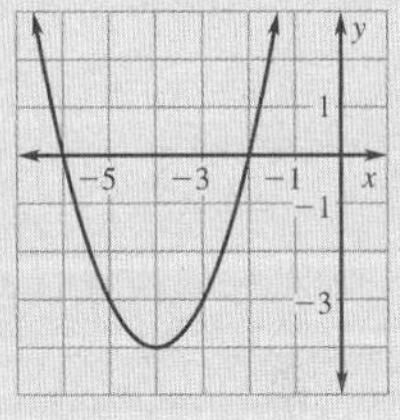

19.

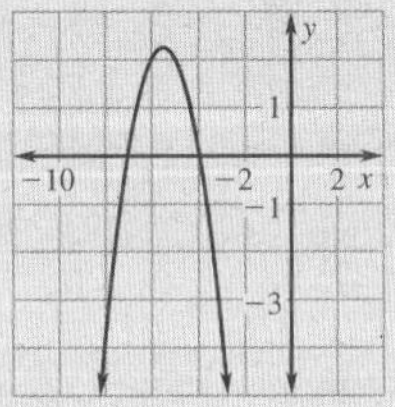

44. See Additional Answers beginning on page AA1.

ADDITIONAL RESOURCES

- ***Chapter 10 Resource Book***
  Chapter Test (3 levels) (p. 120)
  SAT/ACT Chapter Test (p. 126)
  Alternative Assessment (p. 127)
- ***Test and Practice Generator***

CHAPTER 10

Chapter Standardized Test

TEST-TAKING STRATEGY Some questions involve more than one step. Reading too quickly might lead to mistaking the answer to a preliminary step for your final answer.

1. MULTIPLE CHOICE Classify $3x^2 - 7 + 4x^3 - 5x$ by degree and by the number of terms. **B**

Ⓐ quadratic trinomial

Ⓑ cubic polynomial

Ⓒ quartic polynomial

Ⓓ quadratic polynomial

Ⓔ cubic trinomial

2. MULTIPLE CHOICE Which of the following is equal to $(-x^2 - 5x + 7) + (-7x^2 + 5x - 2)$? **E**

Ⓐ $8x^2 - 5$ Ⓑ $-8x^2 + 10x + 5$

Ⓒ $6x^2 + 5$ Ⓓ $-8x^2 - 10x + 5$

Ⓔ $-8x^2 + 5$

3. MULTIPLE CHOICE Which of the following is equal to $(5x^3 + 3x^2 - x + 1) - (2x^3 + x - 5)$? **D**

Ⓐ $3x^3 + 3x^2 - 4$

Ⓑ $3x^3 + 3x^2 - 2x - 4$

Ⓒ $3x^3 + 3x^2 - 2x - 6$

Ⓓ $3x^3 + 3x^2 - 2x + 6$

Ⓔ $7x^3 + 3x^2 - 2x + 6$

4. MULTIPLE CHOICE The base of a triangular sail is x feet and its height is $\frac{1}{2}x + 7$ feet. Which expression represents the sail's area? $\left(\text{The area of a triangle is } A = \frac{1}{2}bh.\right)$ **B**

Ⓐ $\frac{1}{2}x^2 + 7x$ Ⓑ $\frac{1}{4}x^2 + \frac{7}{2}x$

Ⓒ $\frac{1}{2}x^2 + \frac{7}{2}x$ Ⓓ $\frac{7}{2}x^2 + \frac{1}{2}x$

Ⓔ $\frac{1}{4}x^2 + 7x$

5. MULTIPLE CHOICE Which of the following is equal to $(4x - 9)(7x - 2)$? **A**

Ⓐ $28x^2 - 71x + 18$ Ⓑ $28x^2 - 55x + 18$

Ⓒ $28x^2 - 71x - 18$ Ⓓ $28x^2 + 55x + 18$

Ⓔ $28x^2 - 69x + 18$

6. MULTIPLE CHOICE Which trinomial represents the area of the trapezoid? (The area of a trapezoid is $A = \frac{1}{2}h(b_1 + b_2)$.) **C**

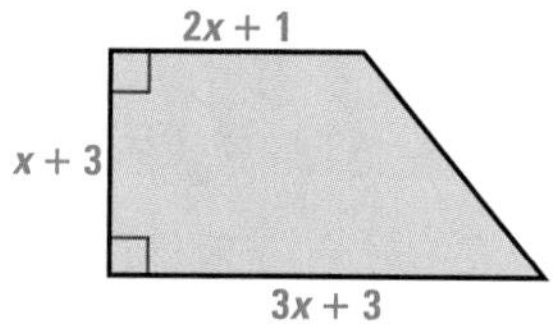

Ⓐ $\frac{5}{2}x^2 + 19x + 12$ Ⓑ $5x^2 + 19x + 6$

Ⓒ $\frac{5}{2}x^2 + \frac{19}{2}x + 6$ Ⓓ $5x^2 + 19x - 6$

Ⓔ $\frac{5}{2}x^2 + \frac{17}{2}x + 6$

7. MULTIPLE CHOICE Which of the following is equal to $(2x - 9)^2$? **E**

Ⓐ $x^2 + 81$ Ⓑ $x^2 - 18x - 81$

Ⓒ $4x^2 + 36x + 81$ Ⓓ $4x^2 - 18x + 81$

Ⓔ $4x^2 - 36x + 81$

8. MULTIPLE CHOICE What are the coordinates of the vertex of the graph of $y = (x - 6)(x + 5)$? **D**

Ⓐ $\left(-\frac{1}{2}, -24\frac{1}{2}\right)$ Ⓑ $\left(\frac{1}{2}, -25\frac{1}{2}\right)$

Ⓒ $(2, -28)$ Ⓓ $\left(\frac{1}{2}, -30\frac{1}{4}\right)$

Ⓔ $\left(-\frac{1}{2}, -24\frac{1}{4}\right)$

9. MULTIPLE CHOICE Which of the following is one of the solutions of the equation $x^2 - 2x = 120$? **B**

Ⓐ -12 Ⓑ -10 Ⓒ 10

Ⓓ 20 Ⓔ 60

10. MULTIPLE CHOICE The area of a circle is given by $A = \pi(9x^2 + 30x + 25)$. Which expression represents the radius of the circle? **A**

Ⓐ $|3x + 5|$ Ⓑ $9x^2 - 25$ Ⓒ $(3x - 5)^2$

Ⓓ $9x^2 + 25$ Ⓔ $(3x + 5)^2$

11. **MULTIPLE CHOICE** If $x = -2$ is a solution of $x^2 - bx - 16 = 0$, what is the value of b? **C**

Ⓐ -8 Ⓑ -6 Ⓒ 6 Ⓓ 8 Ⓔ 10

12. **MULTIPLE CHOICE** A ball is tossed into the air from a height of 10 feet with an initial upward velocity of 12 feet per second. Find the time in seconds for the ball to reach the ground. **C**

Ⓐ $\frac{1}{2}$ Ⓑ $\frac{4}{5}$ Ⓒ $1\frac{1}{4}$ Ⓓ $1\frac{1}{2}$ Ⓔ 2

13. **MULTIPLE CHOICE** Which of the following is a correct factorization of $-45x^2 + 150x - 125$? **D**

Ⓐ $5(-3x + 5)$ Ⓑ $-5(3x + 5)^2$ Ⓒ $-5(3x + 5)(3x - 5)$

Ⓓ $-5(3x - 5)^2$ Ⓔ $-5(-3x + 5)(-3x - 5)$

QUANTITATIVE COMPARISON In Exercises 14 and 15, evaluate the expression for the given values and choose the statement that is true about the results.

Ⓐ The number in Column A is greater. Ⓑ The number in Column B is greater.

Ⓒ The two numbers are equal. Ⓓ The relationship cannot be determined from the information given.

	Column A	Column B	
14.	$(a + b)^2$ when $a = 17$ and $b = -8$	$(a - b)^2$ when $a = 17$ and $b = -8$	**B**
15.	$(a^2 - b^2)$ when $a = 3$ and $b = -4$	$(a - b)^2$ when $a = 3$ and $b = -4$	**B**

16. **MULTIPLE CHOICE** Which of the following is equal to the expression $x^3 - 2x^2 - 11x + 22$? **C**

Ⓐ $(x - 2)(x - 11)$ Ⓑ $(x - 2)(x^2 + 11)$ Ⓒ $(x - 2)(x^2 - 11)$

Ⓓ $(x - 2)(x + 11)$ Ⓔ none of these

17. **MULTI-STEP PROBLEM** You have made clay animals to sell for charity. Each animal is about 6 inches long by 8 inches wide by 8 inches tall. You want to package each animal in a box with the top of its head showing. You will not use a lid for the box. You have received a donation of cardboard sheets that are 24 inches by 20 inches to make the boxes. You must cut out corner regions of x^2 so that the flaps can be folded up to form each box.

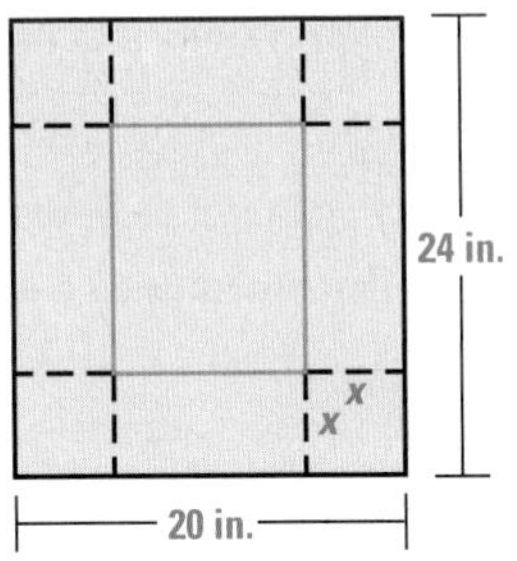

a. Write a polynomial expression for the area of the box bottom. Find the area of the box bottom in terms of x. **$480 - 88x + 4x^2$**

b. Write a polynomial expression for the volume of the box. Find the volume of the box in terms of x. **$x(480 - 88x + 4x^2)$**

c. Is it possible to use squares that are 12 inches for the corners? Explain your reasoning. **No; the combined length of two squares would be longer than the shorter side and the same length as the longer side.**

d. Is it possible to use the donated cardboard sheets to make boxes that will be large enough to hold the clay animals? Explain.

d. Yes; for example, if the squares were $6\frac{1}{2}$ in. on a side, the box would be 7 in. wide, 11 in. long, and $6\frac{1}{2}$ in. high. The animal would fit into the box and the animal's head would show at the top.

CHAPTER 11

PLANNING THE CHAPTER

Rational Equations and Functions

LESSON	GOALS	NCTM	ITED	SAT9	Terra-Nova	Local
11.1 *pp. 643–648*	GOAL 1 Solve proportions. GOAL 2 Use proportions to solve real-life problems.	1, 2, 6, 8, 9, 10	APRA	1, 3	16, 17	
11.2 *pp. 649–655*	GOAL 1 Use equations to solve percent problems. GOAL 2 Use percents in real-life problems.	1, 2, 6, 8, 9, 10	RQPE	3	16, 50, 52	
11.3 *pp. 656–663*	GOAL 1 Use direct and inverse variation. GOAL 2 Use direct and inverse variation to model real-life situations. TECHNOLOGY ACTIVITY: 11.3 *Use a graphing calculator to decide whether quantities vary directly, inversely, or neither.*	1, 2, 3, 6, 8, 9, 10	APRA, MIG	3	14, 15, 17	
11.4 *pp. 664–669*	GOAL 1 Simplify a rational expression. GOAL 2 Use rational expressions to find geometric probability.	1, 2, 3, 5, 6, 8, 10	SAPP, RQF	13	14, 15, 52	
11.5 *pp. 670–675*	GOAL 1 Multiply and divide rational expressions. GOAL 2 Use rational expressions as real-life models.	1, 2, 8, 9, 10	RQF	3	11, 18, 52	
11.6 *pp. 676–682*	GOAL 1 Add and subtract rational expressions that have like denominators. GOAL 2 Add and subtract rational expressions that have unlike denominators.	1, 2			11, 52	
11.7 *pp. 683–689*	CONCEPT ACTIVITY: 11.7 *Investigate how to model division of polynomials using algebra tiles.* GOAL 1 Divide a polynomials by a monomial or by a binomial factor. GOAL 2 Use polynomial long division.	1, 2, 10			11, 18	
11.8 *pp. 690–698*	GOAL 1 Solve rational equations. GOAL 2 Graph rational functions. TECHNOLOGY ACTIVITY: 11.8 *Use a graphing calculator to recognize input values that are not in the domain of a rational function.*	1, 2, 3, 10	MIG, IIG, RQF		14, 15, 52	

RESOURCES

Chapter Resource Booklets

Chapter Support

Tips for New Teachers	p. 1	Prerequisite Skills Review	p. 5
Parent Guide for Student Success	p. 3	Strategies for Reading Mathematics	p. 7

Lesson Support

Lesson Support	11.1	11.2	11.3	11.4	11.5	11.6	11.7	11.8
Lesson Plans (regular and block)	p. 9	p. 21	p. 36	p. 53	p. 65	p. 78	p. 92	p. 105
Warm-Up Exercises and Daily Quiz	p. 11	p. 23	p. 38	p. 55	p. 67	p. 80	p. 94	p. 107
Activity Support Masters							p. 95	
Lesson Openers	p. 12	p. 24	p. 39	p. 56	p. 68	p. 81	p. 96	p. 108
Graphing Calculator Activities & Keystrokes		p. 25	p. 40			p. 82		p. 109
Practice (3 levels)	p. 13	p. 27	p. 43	p. 57	p. 69	p. 83	p. 97	p. 111
Reteaching with Practice	p. 16	p. 30	p. 46	p. 60	p. 72	p. 86	p. 100	p. 114
Quick Catch-Up for Absent Students	p. 18	p. 32	p. 48	p. 62	p. 74	p. 88	p. 102	p. 116
Cooperative Learning Activities		p. 33			p. 75			
Interdisciplinary Applications		p. 34		p. 63	p. 76			p. 117
Real-Life Applications	p. 19		p. 49			p. 89	p. 103	
Math & History Applications			p. 50					
Challenge: Skills and Applications	p. 20	p. 35	p. 51	p. 64	p. 77	p. 90	p. 104	p. 118

Review and Assessment

Quizzes	pp. 52, 91	Alternative Assessment with Math Journal	p. 127
Chapter Review Games and Activities	p. 119	Project with Rubric	p. 129
Chapter Test (3 levels)	pp. 120–125	Cumulative Review	p. 131
SAT/ACT Chapter Test	p. 126	Resource Book Answers	p. A1

Transparencies

	11.1	11.2	11.3	11.4	11.5	11.6	11.7	11.8
Warm-Up Exercises and Daily Quiz	p. 85	p. 86	p. 87	p. 88	p. 89	p. 90	p. 91	p. 92
Alternative Lesson Opener Transparencies	p. 74	p. 75	p. 76	p. 77	p. 78	p. 79	p. 80	p. 81
Examples/Standardized Test Practice	✓	✓	✓	✓	✓	✓	✓	✓
Answer Transparencies	✓	✓	✓	✓	✓	✓	✓	✓

Technology

- EasyPlanner CD
- EasyPlanner Plus Online
- Classzone.com
- Personal Student Tutor CD
- Test and Practice Generator CD
- Instant Replay: Video Review Games
- Electronic Lesson Presentations
- Chapter Audio Summaries CD

Additional Resources

- Basic Skills Workbook
- Explorations and Projects
- Worked-Out Solution Key
- Data Analysis Sourcebook
- Spanish Resources
- Standardized Test Practice Workbook
- Practice Workbook with Examples
- Notetaking Guide

CHAPTER 11

PACING THE CHAPTER

Resource Key
- STUDENT EDITION
- CHAPTER 11 RESOURCE BOOK
- TEACHER'S EDITION

REGULAR SCHEDULE

Day 1

11.1

STARTING OPTIONS
- Prereq. Skills Review
- Strategies for Reading
- Warm-Up or Daily Quiz

TEACHING OPTIONS
- Motivating the Lesson
- Les. Opener (Application)
- Examples 1–5
- Closure Question
- Guided Practice Exs.

APPLY/HOMEWORK
- See Assignment Guide.
- See the CRB: Practice, Reteach, Apply, Extend

ASSESSMENT OPTIONS
- Checkpoint Exercises
- Daily Quiz (11.1)
- Stand. Test Practice

Day 2

11.2

STARTING OPTIONS
- Homework Check
- Warm-Up or Daily Quiz

TEACHING OPTIONS
- Motivating the Lesson
- Les. Opener (Calculator)
- Graphing Calc. Activity
- Examples 1–5
- Closure Question
- Guided Practice Exs.

APPLY/HOMEWORK
- See Assignment Guide.
- See the CRB: Practice, Reteach, Apply, Extend

ASSESSMENT OPTIONS
- Checkpoint Exercises
- Daily Quiz (11.2)
- Stand. Test Practice

Day 3

11.3

STARTING OPTIONS
- Homework Check
- Warm-Up or Daily Quiz

TEACHING OPTIONS
- Motivating the Lesson
- Les. Opener (Application)
- Graphing Calc. Activity
- Examples 1–2
- Guided Practice Exs. 1–11

APPLY/HOMEWORK
- See Assignment Guide.
- See the CRB: Practice, Reteach, Apply, Extend

ASSESSMENT OPTIONS
- Checkpoint Exercises, p. 657

Day 4

11.3 *(cont.)*

STARTING OPTIONS
- Homework Check

TEACHING OPTIONS
- Example 3
- Technology Activity
- Closure Question

APPLY/HOMEWORK
- See Assignment Guide.
- See the CRB: Practice, Reteach, Apply, Extend

ASSESSMENT OPTIONS
- Checkpoint Exercises, p. 658
- Daily Quiz (11.3)
- Stand. Test Practice
- Quiz (11.1–11.3)

Day 5

11.4

STARTING OPTIONS
- Homework Check
- Warm-Up or Daily Quiz

TEACHING OPTIONS
- Les. Opener (Activity)
- Examples 1–5
- Closure Question
- Guided Practice Exs.

APPLY/HOMEWORK
- See Assignment Guide.
- See the CRB: Practice, Reteach, Apply, Extend

ASSESSMENT OPTIONS
- Checkpoint Exercises
- Daily Quiz (11.4)
- Stand. Test Practice

Day 6

11.5

STARTING OPTIONS
- Homework Check
- Warm-Up or Daily Quiz

TEACHING OPTIONS
- Motivating the Lesson
- Les. Opener (Activity)
- Examples 1–6
- Closure Question
- Guided Practice Exs.

APPLY/HOMEWORK
- See Assignment Guide.
- See the CRB: Practice, Reteach, Apply, Extend

ASSESSMENT OPTIONS
- Checkpoint Exercises
- Daily Quiz (11.5)
- Stand. Test Practice

Day 9

11.7

STARTING OPTIONS
- Homework Check
- Warm-Up or Daily Quiz

TEACHING OPTIONS
- Motivating the Lesson
- Concept Act. & Wksht.
- Les. Opener (Visual)
- Examples 1–3
- Guided Practice Exs. 1–4, 6, 9–12, 14

APPLY/HOMEWORK
- See Assignment Guide.
- See the CRB: Practice, Reteach, Apply, Extend

ASSESSMENT OPTIONS
- Checkpoint Exercises, p. 685

Day 10

11.7 *(cont.)*

STARTING OPTIONS
- Homework Check

TEACHING OPTIONS
- Examples 4–5
- Closure Question
- Guided Practice Exs. 5, 7–8, 13

APPLY/HOMEWORK
- See Assignment Guide.
- See the CRB: Practice, Reteach, Apply, Extend

ASSESSMENT OPTIONS
- Checkpoint Exercises, p. 686
- Daily Quiz (11.7)
- Stand. Test Practice

Day 11

11.8

STARTING OPTIONS
- Homework Check
- Warm-Up or Daily Quiz

TEACHING OPTIONS
- Motivating the Lesson
- Les. Opener (Application)
- Graphing Calc. Activity
- Examples 1–5
- Guided Practice Exs. 1–11

APPLY/HOMEWORK
- See Assignment Guide.
- See the CRB: Practice, Reteach, Apply, Extend

ASSESSMENT OPTIONS
- Checkpoint Exercises, pp. 691–692

Day 12

11.8 *(cont.)*

STARTING OPTIONS
- Homework Check

TEACHING OPTIONS
- Examples 6–7
- Technology Activity
- Closure Question
- Guided Practice Exs. 12–13

APPLY/HOMEWORK
- See Assignment Guide.
- See the CRB: Practice, Reteach, Apply, Extend

ASSESSMENT OPTIONS
- Checkpoint Exercises, p. 693
- Daily Quiz (11.8)
- Stand. Test Practice
- Quiz (11.7–11.8)

Day 13

Review

DAY 13 START OPTIONS
- Homework Check

REVIEWING OPTIONS
- Chapter 11 Summary
- Chapter 11 Review
- Chapter Review Games and Activities

APPLY/HOMEWORK
- Chapter 11 Test (practice)
- Ch. Standardized Test (practice)

Day 14

Assess

DAY 14 START OPTIONS
- Homework Check

ASSESSMENT OPTIONS
- Chapter 11 Test
- SAT/ACT Ch. 11 Test
- Alternative Assessment

APPLY/HOMEWORK
- Skill Review, p. 708

Day 7

11.6

Starting Options
- Homework Check
- Warm-Up or Daily Quiz

Teaching Options
- Les. Opener (Activity)
- Graphing Calc. Activity
- Examples 1–4
- Guided Practice Exs. 1–9

Apply/Homework
- See Assignment Guide.
- See the CRB: Practice, Reteach, Apply, Extend

Assessment Options
- Checkpoint Exercises, p. 677

Day 8

11.6 *(cont.)*

Starting Options
- Homework Check

Teaching Options
- Example 5
- Closure Question

Apply/Homework
- See Assignment Guide.
- See the CRB: Practice, Reteach, Apply, Extend

Assessment Options
- Checkpoint Exercises, p. 678
- Daily Quiz (11.6)
- Stand. Test Practice
- Quiz (11.4–11.6)

BLOCK SCHEDULE

Day 1

11.1 & 11.2

Day 1 Start Options
- Prereq. Skills Review
- Strategies for Reading
- Homework Check
- W-Up 11.1 or D. Quiz 10.8

Teaching 11.1 Options
- Motivating the Lesson
- Les. Opener (Appl.)
- Examples 1–5
- Closure Question
- Guided Practice Exs.

Teaching 11.2 Options
- Warm-Up (Les. 11.2)
- Motivating the Lesson
- Les. Opener (Calc.)
- Graphing Calc. Activity
- Examples 1–5
- Closure Question
- Guided Practice Exs.

Apply/Homework
- See Assignment Guide.
- See the CRB: Practice, Reteach, Apply, Extend

Assessment Options
- Checkpoint Exercises
- D. Quiz (Les. 11.1, 11.2)
- Stand. Test Practice

Day 2

11.3

Day 2 Start Options
- Homework Check
- Warm-Up or Daily Quiz

Teaching 11.3 Options
- Motivating the Lesson
- Les. Opener (Appl.)
- Graphing Calc. Activity
- Examples 1–3
- Technology Activity
- Closure Question
- Guided Practice Exs.

Apply/Homework
- See Assignment Guide.
- See the CRB: Practice, Reteach, Apply, Extend

Assessment Options
- Checkpoint Exercises,
- Daily Quiz (Les. 11.3)
- Stand. Test Practice
- Quiz (11.1–11.3)

Day 3

11.4 & 11.5

Day 3 Start Options
- Homework Check
- W-Up 11.4 or D. Quiz 11.3

Teaching 11.4 Options
- Les. Opener (Activity)
- Examples 1–5
- Closure Question
- Guided Practice Exs.

Teaching 11.5 Options
- Warm-Up (Les. 11.5)
- Motivating the Lesson
- Les. Opener (Activity)
- Examples 1–6
- Closure Question
- Guided Practice Exs.

Apply/Homework
- See Assignment Guide.
- See the CRB: Practice, Reteach, Apply, Extend

Assessment Options
- Checkpoint Exercises
- D. Quiz (Les. 11.4. 11.5)
- Stand. Test Practice

Day 4

11.6

Day 4 Start Options
- Homework Check
- Warm-Up or Daily Quiz

Teaching 11.6 Options
- Les. Opener (Activity)
- Graphing Calc. Activity
- Examples 1–5
- Closure Question
- Guided Practice Exs.

Apply/Homework
- See Assignment Guide.
- See the CRB: Practice, Reteach, Apply, Extend

Assessment Options
- Checkpoint Exercises
- Daily Quiz (Les. 11.6)
- Stand. Test Practice
- Quiz (11.4–11.6)

Day 5

11.7

Day 5 Start Options
- Homework Check
- Warm-Up or Daily Quiz

Teaching 11.7 Options
- Motivating the Lesson
- Concept Act. & Wksht.
- Les. Opener (Visual)
- Examples 1–5
- Closure Question
- Guided Practice Exs.

Apply/Homework
- See Assignment Guide.
- See the CRB: Practice, Reteach, Apply, Extend

Assessment Options
- Checkpoint Exercises
- Daily Quiz (Les. 11.7)
- Stand. Test Practice

Day 6

11.8

Day 6 Start Options
- Homework Check
- Warm-Up or Daily Quiz

Teaching 11.8 Options
- Motivating the Lesson
- Les. Opener (Appl.)
- Graphing Calc. Activity
- Examples 1–7
- Technology Activity
- Closure Question
- Guided Practice Exs.

Apply/Homework
- See Assignment Guide.
- See the CRB: Practice, Reteach, Apply, Extend

Assessment Options
- Checkpoint Exercises
- Daily Quiz (Les. 11.8)
- Stand. Test Practice
- Quiz (11.7–11.8)

Day 7

Review/Assess

Day 7 Start Options
- Homework Check

Reviewing Options
- Chapter 11 Summary
- Chapter 11 Review
- Chapter Review Games and Activities
- Chapter 11 Test (practice)
- Ch. Standardized Test (practice)

Assessment Options
- Chapter 11 Test
- SAT/ACT Ch. 11 Test
- Alternative Assessment

Apply/Homework
- Skill Review, p. 708

MEETING INDIVIDUAL NEEDS

BEFORE THE CHAPTER

The *Chapter 11 Resource Book* has the following materials to distribute and use before the chapter:

- **Parent Guide for Student Success**
- **Prerequisite Skills Review**
- **Strategies for Reading Mathematics (pictured below)**

STRATEGIES FOR READING *Pages 7–8*

CHAPTER
11

NAME _____ DATE _____

Strategies for Reading Mathematics
For use with Chapter 11

Chapter Support

Strategy: Reading Proportions

When listing people's names, you can record the last name first or the first name first. However, to avoid misunderstandings, it should be clear which order is being used and it should be used consistently. Similarly, a consistent order must be used when reading and writing the two ratios in a proportion.

Reading Proportions

The example below shows a standard way to read proportions.

Read this ratio first. Read equals sign as "as." Read this ratio last.

Read ratio bar as "is to" $\dfrac{a}{b}$ = $\dfrac{c}{d}$

a is to b as c is to d

Proportions as Models

The arrowed labels can help you think about the order used in the verbal model.

	Previous trip	**Planned trip**

Distance (mi) → Miles you drove / Miles you plan to drive
Gas usage (gal) → Gallons of gas you used = Gallons of gas you expect to use

STUDY TIP
Reading Proportions
When reading proportions, you read top to bottom for each ratio as well as reading left to right for the relationship between the two ratios. You can use the same kind of wording as for verbal analogies such as "Day is to night as light is to dark."

STUDY TIP
Proportions as Models
When reading or writing proportions for a real-life situation, think about what is being compared to what. Writing horizontal and vertical labels can help you to organize your work and to be consistent about the order in which you compare items.

Questions

1. The proportion in the first example can be written in the form $a{:}b = c{:}d$. Use this form to write the proportion "24 is to 9 as the number x is to 6."

2. Record how to read each of the following proportions.

 a. $30{:}60 = 1{:}2$ b. $\dfrac{4}{6} = \dfrac{2}{x}$ c. $\dfrac{7}{9} = \dfrac{x}{18}$ d. $3{:}x = 75{:}100$

3. On a previous trip you used 11 gallons of gas to drive 209 miles. You want to estimate how much gas you will need for a 475-mile trip. Use symbols to model this situation two ways–by writing a proportion that compares miles to gallons and by writing a proportion that compares gallons to miles.

4. A recipe calls for 11 cups of apples to make 2 pies. You are making 3 pies. Write a proportion for this situation using words and one using symbols.

STRATEGIES FOR READING MATHEMATICS The first page gives students tips about reading and writing proportions as they prepare for Chapter 11. The second page (not shown) provides a visual glossary of key vocabulary words in the chapter, such as rational expression and rational equation.

DURING EACH LESSON

The *Chapter 11 Resource Book* has the following alternatives for introducing the lesson:

- **Lesson Openers (pictured below)**
- **Graphing Calculator Activities with Keystrokes**

LESSON OPENER *Page 56*

LESSON
11.4

NAME _____ DATE _____

Available as a transparency

Activity Lesson Opener
For use with pages 664–669

Lesson 11.4

SET UP: Work with a partner.

Tell whether the solution is correct or incorrect. If it is incorrect, identify the error and write the correct solution.

1. $\dfrac{2 \cdot 3}{5 \cdot 3}$ $= \dfrac{2 \cdot \cancel{3}}{5 \cdot \cancel{3}}$

 $= \dfrac{2}{5}$

2. $\dfrac{3 \cdot 2}{3(2 + 5)}$ $= \dfrac{\cancel{3} \cdot 2}{\cancel{3}(2 + 5)}$

 $= \dfrac{1}{5}$

3. $\dfrac{3 + 4}{3}$ $= \dfrac{3 + 4}{\cancel{3}}$

 $= 4$

4. $\dfrac{4 \cdot (5 + 6)}{4 \cdot 6}$ $= \dfrac{\cancel{4} \cdot (5 + 6)}{\cancel{4} \cdot 6}$

 $= \dfrac{5 + 6}{6}$

 $= \dfrac{11}{6}$

5. $\dfrac{2 + 5}{6 + 10}$ $= \dfrac{2 + 5}{2 \cdot 3 + 2 \cdot 5}$

 $= \dfrac{1}{5}$

ACTIVITY LESSON OPENER This Lesson Opener provides an alternative way to start Lesson 11.4 in the form of an activity. In this activity, students work with a partner to check solutions as they learn about the rules of dividing out common factors to simplify fractions.

TECHNOLOGY RESOURCE

Teachers can use the Electronic Lesson Presentations CD-ROM to present Lesson 11.3 using computer animation to step through the Examples.

FOLLOWING EACH LESSON

The *Chapter 11 Resource Book* has a variety of materials to follow-up each lesson. They include the following:

- **Practice (3 levels)**
- **Reteaching with Practice**
- **Quick Catch-Up for Absent Students**
- **Interdisciplinary Applications (pictured below)**
- **Real-Life Applications**

APPLICATION *Page 34*

LESSON 11.2

NAME ______________________ DATE ________

Interdisciplinary Application

For use with pages 649–655

Markup and Cost (Economics)

The price of a good or service must be competitive and attractive to consumers, while also covering expenses and providing a reasonable profit for a business.

Expenses are broken down into two major categories: the cost of goods sold and operating expenses or overhead. Operating expenses and profit are combined and referred to as markup. Markup M is the amount of dollars added to the cost C of the goods. Together they equal the selling price S of the item and can be represented by the equation $S = C + M$.

Companies will often determine the amount of markup dollars using a desired markup percent or rate. There are two ways to calculate the markup rate: one is based on cost and the second is based on selling price.

A company may have a set price for the goods they are buying. If this is the case, they have a fixed cost and may determine the markup rate based on cost. For example, a store purchases cereal for $1.50 a box and desires a markup rate R of 80% based on cost. Using the formula $M = RC$ the markup and then the selling price can be determined.

$M = RC$ *Markup based on cost.* $S = C + M$

$M = (0.80)(1.50) = 1.20$ $S = 1.50 + 1.20 = 2.70$

On the other hand, a company may have a predetermined selling price. For example, a hamburger franchise that must sell hamburgers at $1.29 has a set selling price. If it desires a markup rate of 45% based on selling price, they could then determine the amount of money they could spend to purchase supplies or its cost.

$M = RS$ *Markup based on selling price.* $S = C + M$

$M = (0.45)(1.29) = 0.58$ $1.29 = C + 0.58$

$0.71 = C$

Although the calculations are identical, there are actually two different markup rates and therefore two different markups.

Exercises

1. Your school store is selling pencils. The store paid $.15 per pencil and needs a markup of $.05 each. Determine the selling price.
2. Sweatshirts with your school mascot cost the school $20.00 each. If the store marks up the shirts 35% based on cost, determine the markup and selling price.
3. Backpacks are sold at the store for $35. If the store paid $25 each, determine the markup and the percent of markup based on cost.
4. In Exercise 3, if the markup had been based on the selling price, what would the percent markup be? Round to the nearest tenth.
5. Baseball hats are sold for $15 each. If the store marks them up $7, determine the cost and the percent of the markup based on cost.

Lesson 11.2

INTERDISCIPLINARY APPLICATION This application makes a connection between determining the price of a product and using equations to solve percent problems, the topic of Lesson 11.2.

ASSESSING THE CHAPTER

The *Chapter 11 Resource Book* has the following review and assessment materials:

- **Quizzes**
- **Chapter Review Games and Activities**
- **Chapter Test (3 levels)**
- **SAT/ACT Chapter Test**
- **Alternative Assessment with Rubric and Math Journal**
- **Project with Rubric (pictured below)**
- **Cumulative Review**

PROJECT WITH RUBRIC *Pages 129–130*

CHAPTER 11

NAME ______________________ DATE ________

Project: Miniature Room

For use with Chapter 11

OBJECTIVE **Make a three-dimensional scale model of a room.**

MATERIALS cardboard box, poster board, construction paper, scissors, glue, other decorative materials such as felt or scraps of cloth

INVESTIGATION

1. Measure the dimensions of a room and the positions of fixed objects in the room such as doors and windows. Draw diagrams for later reference.
2. Set up a table to record all necessary measurements and computations. The table below shows some sample items. In the first two columns, fill in the measurements you took in Step 1. Then choose some furnishings you would like to include in your model. Measure their dimensions and add those objects and their actual measurements to your table.

Object	*Actual length*	*Proportion to use*	*Length in model*
Length of the room			
Width of the room			
Height of the room			
Distance from corner to door			
Distance from the bottom of the window to the floor			
Length of table			

3. Decide on an appropriate scale to use for your model. If you want to use a box for the walls of the room, then set up the scale so that the length or the width of the box corresponds to the length or the width of the room. (You can trim the sides of the box to match the other dimensions.) For each object, write and solve a proportion to find its length in the model. Record the proportions and the lengths you find in your table.
4. Use your diagrams and your table to complete a three-dimensional scale model of the room, including some furnishings.

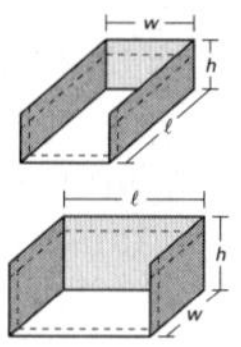

PRESENT YOUR RESULTS Display your model with a poster. On the poster, include your table and an explanation of your procedures, including how you solved your proportions.

Review and Assess

PROJECT WITH RUBRIC The Project for Chapter 11 provides students with the opportunity to apply the concepts they have learned in the chapter in a new way. In this project, students use what they have learned about ratios and proportions to make a scale model of a room. Teacher's notes and a scoring rubric are provided on a separate sheet.

CHAPTER OVERVIEW

CHAPTER GOALS

In this chapter, students work with expressions that involve polynomials and rational expressions. They learn to solve and apply proportions. They set up and solve percent problems. Direct variation is reviewed and inverse variation is introduced. Rational expressions are simplified and used to find geometric probabilities. Students learn to add and subtract rational expressions. They divide polynomials by a monomial or binomial, including the use of polynomial long division. Finally, students solve rational equations and explore the graphs of rational functions.

APPLICATION NOTE

The basic premise behind a scale model is that the dimensions should be in proportion to those of the actual object. That is, the relative sizes of the different parts of the model should be in the same relationship to each other as they are in real life. If a door is twice as high as it is wide in real life, it should be twice as high as it is wide in the model.

One interesting avenue to investigate with students is children's toys. Have students bring in action figures or dolls. Are these scale models of actual people? Have students measure the figures and see if they are proportional to real human beings.

Additional information about scale models is available at **www.mcdougallittell.com.**

RATIONAL EQUATIONS AND FUNCTIONS

▶ *How do scale models fit into the design process?*

CHAPTER

11

APPLICATION: Scale Models

An *architectural scale model* is a smaller, three-dimensional representation of a structure. Models can be built for a single building or an entire city. Architects, builders, and city planners use them.

The dimensions of a scale model can be determined by solving a proportion. In Chapter 11 you'll learn ways to solve proportions and other rational equations.

Think & Discuss

Use the drawing of a scale model for Exercises 1–2.

$x = 336$; 28 ft

1. Solve the equation $\frac{7}{1} = \frac{x}{48}$ to find the height x of the actual house. Write your answer in feet.

2. A shutter on the house is 1 foot wide. How wide is it on the model? Write your answer in inches. **$\frac{1}{4}$ in.**

Learn More About It

You will write and use a proportion for a problem about a scale model in Exercise 43 on page 647.

APPLICATION LINK Visit www.mcdougallittell.com for more information about scale models.

ADDITIONAL RESOURCES

Another way to begin the chapter is to show the video clip of a real-life motivator for Chapter 11 on the *Instant Replay: Video Review Games.*

PROJECTS

A project covering Chapters 10–12 appears on pages 774 and 775 of the Student Edition. An additional project for Chapter 11 is available in the *Chapter 11 Resource Book,* p. 129.

TECHNOLOGY

Software
- *Electronic Teaching Tools*
- *Online Lesson Planner*
- *Personal Student Tutor*
- *Test and Practice Generator*
- *Electronic Lesson Presentations* (Lesson 11.3)

Video
- *Instant Replay: Video Review Games*

Internet Connections
www.mcdougallittell.com
- **Application Links** 641, 645, 651, 662
- **Data Updates** 648, 688
- **Student Help** 647, 650, 660, 663, 665, 671, 680, 686, 695, 698
- **Career Links** 647, 654, 674, 691
- **Extra Challenge** 648, 655, 661, 669, 675, 681, 689, 696

PREVIEW

What's the chapter about?

Chapter 11 is about **rational expressions**. In Chapter 11 you'll learn

• how to solve rational equations.
• how to add, subtract, multiply, and divide rational expressions.
• how to graph rational functions, including inverse variation functions.

KEY VOCABULARY

▶ Review
• equivalent equations, p. 132
• transformations, p. 132
• direct variation, p. 234
• polynomial, p. 576
• zero-product property, p. 597

▶ New
• proportion, p. 643
• extraneous solution, p. 644
• inverse variation, p. 656
• rational number, p. 664
• rational expression, p. 664

• geometric probability, p. 666
• rational equation, p. 690
• rational function, p. 692
• hyperbola, p. 692

PREPARE

Are you ready for the chapter?

SKILL REVIEW These exercises will help you review skills that you'll apply in this chapter. See the given reference page if there is something you don't understand.

Find the value of the expression. Write the answer as a fraction or mixed number in lowest terms. (Skills Review, p. 781–783)

1. $\frac{11}{14} - \frac{4}{14}$ $\frac{1}{2}$ **2.** $\frac{9}{15} + \frac{7}{10}$ $1\frac{3}{10}$ **3.** $\frac{6}{11} - \frac{3}{5}$ $-\frac{3}{55}$ **4.** $\frac{8}{12} + \frac{11}{18}$ $1\frac{5}{18}$

Simplify the expression. (Review Examples 1 and 2, pp. 108–109)

5. $\frac{36x}{15} \div \frac{9}{5}$ $\frac{4x}{3}$ **6.** $49x^2 \div \frac{-7x}{3}$ $-21x$ **7.** $\frac{15x + 25}{5}$ $3x + 5$ **8.** $\frac{16 - 4x}{8}$ $\frac{4 - x}{2}$

Solve the equation by finding square roots or by factoring.
(Review pp. 505 and 613)

9. $3x^2 - 65 = 178$
$-9, 9$

10. $4x^2 - 10x + 6 = 0$
$1, 1\frac{1}{2}$

11. $10x^2 - 30x - 40 = 0$
$-1, 4$

STUDY STRATEGY

Here's a study strategy!

Previewing and Reviewing

Preview the chapter. Write down what you already know about each topic. After studying the chapter, go back to each topic and write down what you then know about it. Compare the two sets of notes. See what you have learned.

11.1 Ratio and Proportion

What you should learn

GOAL 1 Solve proportions.

GOAL 2 Use proportions to solve **real-life** problems, such as making estimates about an archaeological dig in **Example 4**.

Why you should learn it

▼ To solve **real-life** problems such as creating a mural in **Exs. 41 and 42**.

GOAL 1 SOLVING PROPORTIONS

In Chapter 3 you solved problems involving ratios. An equation that states that two ratios are equal is a **proportion**.

$\frac{a}{b} = \frac{c}{d}$ *b* and *d* are nonzero.

This proportion is read as "*a* is to *b* as *c* is to *d*." When the ratios are written in this order, the numbers *a* and *d* are the **extremes** of the proportion and the numbers *b* and *c* are the **means** of the proportion.

PROPERTIES OF PROPORTIONS

RECIPROCAL PROPERTY

If two ratios are equal, their reciprocals, if they exist, are also equal.

If $\frac{a}{b} = \frac{c}{d}$, then $\frac{b}{a} = \frac{d}{c}$. **Example:** $\frac{2}{3} = \frac{4}{6} \Rightarrow \frac{3}{2} = \frac{6}{4}$

CROSS PRODUCT PROPERTY

The product of the extremes equals the product of the means.

If $\frac{a}{b} = \frac{c}{d}$, then $ad = bc$. **Example:** $\frac{2}{3} = \frac{4}{6} \Rightarrow 2 \cdot 6 = 3 \cdot 4$

When a proportion involves a single variable, finding the value of that variable is called **solving the proportion**.

EXAMPLE 1 *Using the Reciprocal Property*

Solve the proportion $\frac{3}{y} = \frac{5}{8}$.

SOLUTION

$\frac{3}{y} = \frac{5}{8}$ Write original proportion.

$\frac{y}{3} = \frac{8}{5}$ Use reciprocal property.

$y = 3 \cdot \frac{8}{5}$ Multiply each side by 3.

$y = \frac{24}{5}$ Simplify.

✓**CHECK** When you substitute to check, $\frac{3}{\frac{24}{5}}$ becomes $3 \cdot \frac{5}{24}$ which simplifies to $\frac{5}{8}$.

1 PLAN

PACING

Basic: 1 day
Average: 1 day
Advanced: 1 day
Block Schedule: 0.5 block with 11.2

LESSON OPENER APPLICATION

An alternative way to approach Lesson 11.1 is to use the Application Lesson Opener:

- Blackline Master (*Chapter 11 Resource Book*, p. 12)
- Transparency (p. 74)

MEETING INDIVIDUAL NEEDS

- ***Chapter 11 Resource Book***
  - Prerequisite Skills Review (p. 5)
  - Practice Level A (p. 13)
  - Practice Level B (p. 14)
  - Practice Level C (p. 15)
  - Reteaching with Practice (p. 16)
  - Absent Student Catch-Up (p. 18)
  - Challenge (p. 20)
- ***Resources in Spanish***
- ***Personal Student Tutor***

NEW-TEACHER SUPPORT

See the Tips for New Teachers on pp. 1–2 of the *Chapter 11 Resource Book* for additional notes about Lesson 11.1.

WARM-UP EXERCISES

Transparency Available

Solve.

1. $y^2 = 16$ −4, 4
2. $3(x + 7) = 4x - 5$ 26
3. $x(x - 2) = 15$ −3, 5
4. $6t = -2t(t + 1)$ −4, 0

2 TEACH

MOTIVATING THE LESSON

Have students recount instances where they've seen or used scale drawings or models. Ask them what we mean when we say that these drawings or models are *proportional* to the actual things they represent. Tell them that by learning the mathematical definition of proportion, they will see how they (and architects and engineers) can use proportions to create scale drawings and scale models.

EXTRA EXAMPLE 1

Solve the proportion $\frac{2}{3} = \frac{3}{w}$. $\frac{9}{2}$

EXTRA EXAMPLE 2

Solve the proportion $\frac{10}{t} = \frac{2t}{5}$.
$-5, 5$

EXTRA EXAMPLE 3

Solve the proportion $\frac{-3}{d} = \frac{d-3}{2d}$.
-3

CHECKPOINT EXERCISES

For use after Example 1:

1. Solve the proportion $\frac{4}{r} = \frac{16}{5}$. $\frac{5}{4}$

For use after Example 2:

2. Solve the proportion $\frac{x+3}{4} = \frac{x}{5}$. -15

For use after Example 3:

3. Solve the proportion $\frac{x+3}{x+5} = \frac{x-3}{-5}$. $-7, 0$

ENGLISH LEARNERS

Real-life applications, such as the archaeological one in Example 4, help students understand practical uses for algebra. However, such examples can be inaccessible to English language learners who are unable to interpret the vocabulary. For this example, briefly describe *archaeology* for students, highlighting the meanings of the words *archaeologist, clay, excavation, sites,* and *tomb.*

EXAMPLE 2 ***Using the Cross Product Property***

Solve the proportion $\frac{x}{3} = \frac{12}{x}$.

SOLUTION

$\frac{x}{3} = \frac{12}{x}$	Write original proportion.	
$x \cdot x = 3 \cdot 12$	Use cross product property.	$\frac{x}{3} \times \frac{12}{x}$
$x^2 = 36$	Simplify.	
$x = \pm 6$	Take square root of each side.	

▶ The solutions are $x = 6$ and $x = -6$. Check these in the original proportion.

STUDENT HELP

Study Tip
Remember to check your solution in the original proportion. Notice that Example 2 has two solutions so you need to check both of them.

Up to now, you have been checking solutions to make sure that you didn't make errors in the solution process. Even with no mistakes, it can happen that a trial solution does not satisfy the original equation. This type of solution is called **extraneous.** An extraneous solution should not be listed as an actual solution.

EXAMPLE 3 ***Checking Solutions***

Solve the proportion $\frac{y^2 - 9}{y + 3} = \frac{y - 3}{2}$.

STUDENT HELP

Look Back
For help with solving an equation by finding square roots or by factoring, see pp. 505 and 613.

SOLUTION

$\frac{y^2 - 9}{y + 3} = \frac{y - 3}{2}$	Write original proportion.
$2(y^2 - 9) = (y + 3)(y - 3)$	Use cross product property.
$2y^2 - 18 = y^2 - 9$	Use distributive property.
$y^2 = 9$	Isolate variable term.
$y = \pm 3$	Take square root of each side.

At this point, the solutions appear to be $y = 3$ and $y = -3$.

✓**CHECK** Check each solution by substituting it into the original proportion.

$y = 3$:

$$\frac{y^2 - 9}{y + 3} = \frac{y - 3}{2}$$

$$\frac{3^2 - 9}{3 + 3} \stackrel{?}{=} \frac{3 - 3}{2}$$

$$\frac{0}{6} \stackrel{?}{=} \frac{0}{2}$$

$$0 = 0$$

$y = -3$:

$$\frac{y^2 - 9}{y + 3} = \frac{y - 3}{2}$$

$$\frac{(-3)^2 - 9}{(-3) + 3} \stackrel{?}{=} \frac{(-3) - 3}{2}$$

$$\frac{0}{0} \neq \frac{-6}{2}$$

▶ You can conclude that $y = -3$ is extraneous because the check results in a false statement. The only solution is $y = 3$.

FOCUS ON APPLICATIONS

ARCHAEOLOGY In 1974, archaeologists discovered the tomb of Emperor Qin Shi Huang (259–210 B.C.) in China. Buried close to the tomb was an entire army of life-sized clay warriors.

APPLICATION LINK
www.mcdougallittell.com

GOAL 2 USING PROPORTIONS IN REAL LIFE

When writing a proportion to model a situation, you can set up your proportion in more than one way.

EXAMPLE 4 *Writing and Using a Proportion*

ARCHAEOLOGY Archaeologists excavated three pits containing the clay army. To estimate the number of warriors in Pit 1 shown below, an archaeologist might excavate three sites. The sites at the ends together contain 450 warriors. The site in the central region contains 282 warriors. This 10-meter-wide site is thought to be representative of the 200-meter central region. Estimate the number of warriors in the central region. Then estimate the total number of warriors in Pit 1.

Not drawn to scale

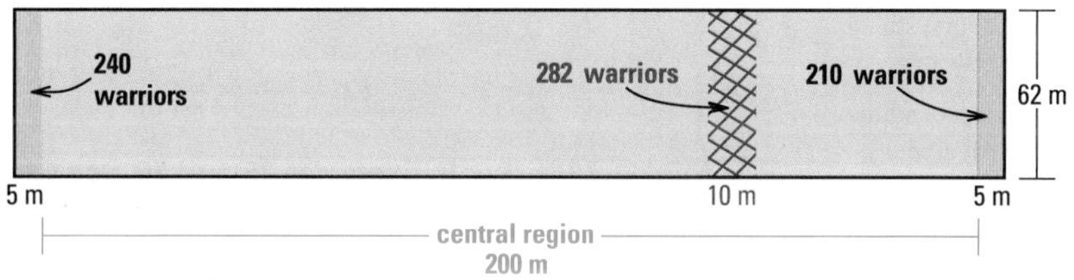

SOLUTION Let n represent the number of warriors in the 200-meter central region. You can find the value of n by solving a proportion.

$$\frac{\text{Number of warriors found}}{\text{Total number of warriors}} = \frac{\text{Number of meters excavated}}{\text{Total number of meters}}$$

$$\frac{282}{n} = \frac{10}{200}$$

▶ The solution is $n = 5640$, indicating that there are about 5640 warriors in the central region. With the 450 warriors at the ends, that makes a total of about 6090 warriors in Pit 1.

REAL LIFE *Model Making*

EXAMPLE 5 *Writing and Using a Proportion*

You want to make a scale model of one of the clay horses found in the tomb. The clay horse is 1.5 meters tall and 2 meters long. Your scale model will be 18 inches long. How tall should it be?

SOLUTION Let h represent the height of the model.

$$\frac{\text{Height of actual statue}}{\text{Length of actual statue}} = \frac{\text{Height of model}}{\text{Length of model}}$$

$$\frac{1.5}{2} = \frac{h}{18}$$

▶ The solution is $h = 13\frac{1}{2}$. Your scale model should be $13\frac{1}{2}$ inches tall.

EXTRA EXAMPLE 4
The route for your town's Fourth of July parade stretches 2 miles (10,560 feet). Spectators are usually evenly distributed along both sides of the route. On one side of a 50 foot section of the route, you count 15 children. Estimate the number of children who attended the parade. **6336**

EXTRA EXAMPLE 5
You want to make a scale model of a parade float. The float is 5.5 feet high and 10 feet long. Your model will be 14 inches long. How tall should it be? **7.7 in.**

CHECKPOINT EXERCISES
For use after Examples 4 and 5:

1. A lemonade recipe calls for $\frac{3}{4}$ cup of lemon juice for 2 quarts of lemonade. How much lemon juice should you use for an 8 quart jug of lemonade? **3 c**

APPLICATION NOTE
EXAMPLE 4 Additional information about archaeology is available at **www.mcdougallittell.com.**

FOCUS ON VOCABULARY
How is an extraneous solution different from an actual solution?
An extraneous solution does not satisfy the original equation.

CLOSURE QUESTION
Can you use the cross product property to solve the equation $\frac{3}{x} + \frac{x}{3} = \frac{3}{15}$ directly? Explain.
No. *Sample answer:* The left side of the equation is not a single ratio, so the equation is not a proportion. The cross product property can be used only with proportions. You must first combine the left side over a common denominator to use the cross product property.

DAILY PUZZLER
On planet Arkoid, 6 quins are worth 48 balzars. How many balzars are 9 quins worth? **72 balzars**

3 APPLY

ASSIGNMENT GUIDE

BASIC
Day 1: pp. 646–648 Exs. 18–40 even, 47, 48, 50, 55, 60

AVERAGE
Day 1: pp. 646–648 Exs. 18–40 even, 43, 47, 48, 50, 55, 60

ADVANCED
Day 1: pp. 646–648 Exs. 18–40 even, 43, 47–50, 55, 60

BLOCK SCHEDULE WITH 11.2
pp. 646–648 Exs. 18–40 even, 43, 47, 48, 50, 55, 60

EXERCISE LEVELS

Level A: *Easier*
17–23, 51–58

Level B: *More Difficult*
24–48, 50, 59–66

Level C: *Most Difficult*
49

HOMEWORK CHECK

To quickly check student understanding of key concepts, go over the following exercises: Exs. 18, 22, 28, 38, 40, 47. See also the Daily Homework Quiz:

- Blackline Master (*Chapter 11 Resource Book*, p. 23)
- Transparency (p. 86)

COMMON ERROR

EXERCISES 17–40 Stress to students that even if their algebra is correct, they still need to check their work to eliminate any extraneous solutions.

GUIDED PRACTICE

Vocabulary Check ✓

1. Write the extremes and the means of the proportion $\frac{3}{4} = \frac{9}{12}$.
extremes: 3 and 12; means: 4 and 9

Concept Check ✓

Write *yes* or *no* to tell whether the equation is a consequence of $\frac{a}{b} = \frac{c}{d}$.

2. $ac = bd$ **no** **3.** $ba = dc$ **no** **4.** $ad = bc$ **yes**

5. $\frac{a}{b} = \frac{d}{c}$ **no** **6.** $\frac{b}{a} = \frac{d}{c}$ **yes** **7.** $\frac{a}{d} = \frac{b}{c}$ **no**

8. Solve the proportion $\frac{4}{x+1} = \frac{7}{2}$ two ways—using the reciprocal property and using the cross product method. Which method do you prefer? Why? **$\frac{1}{7}$; See margin.**

8. *Sample answer:* I prefer the cross product method because you don't have to work with fractions until the last step.

Skill Check ✓

Solve the proportion. Check for extraneous solutions.

9. $\frac{x}{3} = \frac{2}{7}$ **$\frac{6}{7}$** **10.** $\frac{6}{x} = \frac{5}{3}$ **$3\frac{3}{5}$** **11.** $\frac{2}{2x+1} = \frac{1}{5}$ **$4\frac{1}{2}$**

12. $\frac{3}{x} = \frac{x+1}{4}$ **−4, 3** **13.** $\frac{t-2}{t} = \frac{2}{t+3}$ **−2, 3** **14.** $\frac{2u-3}{4u} = \frac{u-1}{u}$ **$\frac{1}{2}$**

MODEL MAKING In Exercises 15 and 16, use Example 5. The proportion used in Example 5 compares height to length, but the situation could be described using other comparisons instead.

15. Set up a different proportion to represent the situation in Example 5.

16. Solve the proportion you wrote in Exercise 15. Do you get the same solution as in Example 5? **Any of the proportions in the answer given for Ex. 15 will give the correct answer.**

15. Other proportions include $\frac{\text{height of actual statue}}{\text{height of model}} = \frac{\text{length of actual statue}}{\text{length of model}}$, $\frac{\text{height of model}}{\text{height of actual statue}} = \frac{\text{length of model}}{\text{length of actual statue}}$, $\frac{\text{length of actual statue}}{\text{height of actual statue}} = \frac{\text{length of model}}{\text{height of model}}$

PRACTICE AND APPLICATIONS

STUDENT HELP
Extra Practice to help you master skills is on p. 807.

SOLVING PROPORTIONS Solve the proportion. Check for extraneous solutions.

17. $\frac{16}{4} = \frac{12}{x}$ **3** **18.** $\frac{4}{2x} = \frac{7}{3}$ **$\frac{6}{7}$** **19.** $\frac{5}{8} = \frac{c}{9}$ **$5\frac{5}{8}$**

20. $\frac{x}{3} = \frac{2}{5}$ **$1\frac{1}{5}$** **21.** $\frac{5}{3c} = \frac{2}{3}$ **$2\frac{1}{2}$** **22.** $\frac{24}{5} = \frac{9}{y+2}$ **$-\frac{1}{8}$**

23. $\frac{6}{3} = \frac{x+8}{-1}$ **−10** **24.** $\frac{r+4}{3} = \frac{r}{5}$ **−10** **25.** $\frac{w+4}{2w} = \frac{-5}{6}$ **$-1\frac{1}{2}$**

26. $\frac{5}{2y} = \frac{7}{y-3}$ **$-1\frac{2}{3}$** **27.** $\frac{x+6}{3} = \frac{x-5}{2}$ **27** **28.** $\frac{x-2}{4} = \frac{x+10}{10}$ **10**

29. $\frac{8}{x+2} = \frac{3}{x-1}$ **$2\frac{4}{5}$** **30.** $\frac{x-3}{18} = \frac{3}{x}$ **−6, 9** **31.** $\frac{-2}{a-7} = \frac{a}{5}$ **2, 5**

32. $\frac{u}{3} = \frac{1}{2u-1}$ **$-1, \frac{3}{2}$** **33.** $\frac{d}{d+4} = \frac{d-2}{d}$ **4** **34.** $\frac{3x}{4x-1} = \frac{1}{x}$ **$\frac{1}{3}, 1$**

35. $\frac{x-3}{x} = \frac{x}{x+6}$ **6** **36.** $\frac{5}{m+1} = \frac{4m}{m}$ **$\frac{1}{4}$** **37.** $\frac{2}{3t} = \frac{t-1}{t}$ **$1\frac{2}{3}$**

38. $\frac{2}{6x+1} = \frac{2x}{1}$ **$-\frac{1}{2}, \frac{1}{3}$** **39.** $\frac{-2}{q} = \frac{q+1}{q^2}$ **$-\frac{1}{3}$** **40.** $\frac{6}{19n} = \frac{-2}{n^2+2}$ **$-6, -\frac{1}{3}$**

STUDENT HELP
HOMEWORK HELP
Example 1: Exs. 17–40
Example 2: Exs. 17–40
Example 3: Exs. 17–40
Example 4: Exs. 41–43
Example 5: Exs. 41–43

STUDENT HELP

HOMEWORK HELP
Visit our Web site www.mcdougallittell.com for help with writing proportions to solve problems.

43. *Sample answer:* By rewriting 1 ft as 12 in., you can set up the proportion $\frac{\frac{1}{16}\text{ in.}}{12\text{ in.}} = \frac{1\text{ in.}}{192\text{ in.}}$. Now that all the units are the same, you can use the cross product property. This gives 12 in. = 12 in., which shows that the proportion is correct.

FOCUS ON CAREERS

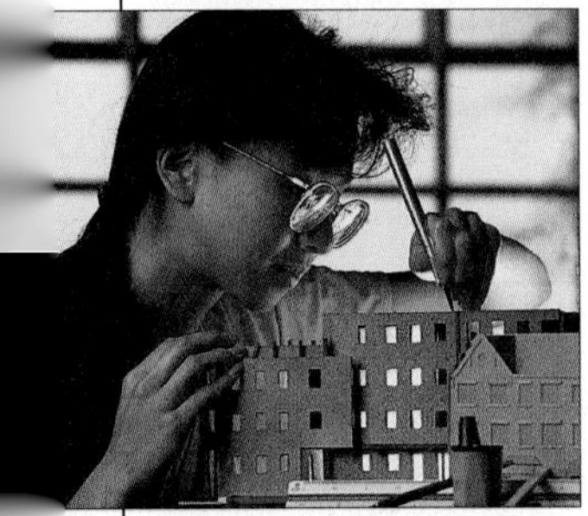

MODEL BUILDER
To make scale models, model builders must pay close attention to detail. They use proportions to find the correct sizes for objects they are modeling as in Exercise 43.

CAREER LINK
www.mcdougallittell.com

MURAL PROJECT In Exercises 41–42, use the following information.
The *Art is the Heart of the City* fence mural project in Charlotte, North Carolina, involved artists Cordelia Williams and Paul Rousso along with 22 students in grades 11 and 12. Students created drawings on paper. Slides of the drawings were made and projected to fit onto 4-foot-wide by 8-foot-long sheets of plywood used for the fence panels. Students traced and later painted the enlarged images.

41. If the paper used for the original drawings was 11 inches wide, how long did it need to be? **22 in.**

42. Suppose the length of the paintbrush on the panel shown is $2\frac{1}{2}$ feet. Use Exercise 41 to find the length of the paintbrush in the student's drawing. **$6\frac{7}{8}$ in.**

43. SCALE MODELS A scale model uses a scale of $\frac{1}{16}$ inch to represent 1 foot. Explain how you can use a proportion and cross products to show that a scale of $\frac{1}{16}$ in. to 1 ft is the same as a scale of 1 in. to 192 in. **See margin.**

WHAT STUDENTS BUY In Exercises 44–46, use the table. It shows the results of a survey in which 100 students were asked how they spent money last week.

How 100 students spent money	
Item	Number of students
Food	78
Clothes, accessories	20
Books, magazines, comics	15
Toys, stickers, games	14
Movie tickets	14
Arcade games	14
Gifts	13
Movie rentals	13
Music	12
Footwear	11
Grooming products	11

44. Estimate the number of students out of 500 that bought clothes or accessories in the last week. **about 100 students**

45. Choose 3 items that were bought by different numbers of students. Based on the survey, how many students out of 20 would you predict to have bought each item? **See margin.**

46. COLLECTING DATA Choose one item from Exercise 45. Ask 20 students whether they bought that item. Compare the results with your prediction. Are the results what you expected? Explain. **Check work.**

47. SAMPLING FISH POPULATIONS Researchers studying fish populations at Dryden Lake in New York caught, marked, and then released 232 Chain pickerel. Later a sample of 329 Chain pickerel were caught and examined. Of these, 16 were found to be marked. Use the proportion below to estimate the total Chain pickerel population in the lake. **about 4771**

$$\frac{\text{Marked pickerel in sample}}{\text{Total pickerel in sample}} = \frac{\text{Marked pickerel in lake}}{\text{Total pickerel in lake}}$$

Mathematical Reasoning *Sample answer:*
Not everyone had phones. Those who did might, for example, have been more affluent than those who didn't. Also, some people have unlisted numbers, some families might have had multiple listings, and so on.

STUDENT HELP NOTES

Homework Help Students can find help with writing proportions to solve problems such as Exs. 41–47 at **www.mcdougallittell.com.** The information can be printed out for students who don't have access to the Internet.

CAREER NOTE
EXERCISE 43 Additional information about model builders is available at **www.mcdougallittell.com.**

MATHEMATICAL REASONING
Sampling techniques for polls use the reasoning of Exs. 44–46. Pollsters assume that the responses in a small sample represent those of the whole population studied. If so, the field of statistics can be used to analyze the results and predict their accuracy. The challenge is finding a sample that is truly *representative.* For example, in some inaccurate election polls of past decades, names were chosen at random from phone books. Why might this have led to unrepresentative samples? **See sample answer below.**

45. See Additional Answers beginning on page AA1.

ADDITIONAL PRACTICE AND RETEACHING

For Lesson 11.1:
- Practice Levels A, B, and C (*Chapter 11 Resource Book,* p. 13)
- Reteaching with Practice (*Chapter 11 Resource Book,* p. 16)
- See Lesson 11.1 of the *Personal Student Tutor*

For more Mixed Review:
- Search the *Test and Practice Generator* for key words or specific lessons.

4 ASSESS

DAILY HOMEWORK QUIZ

Transparency Available

Solve the proportion. Check for extraneous solutions.

1. $\frac{3}{x} = \frac{5}{2}$ **$\frac{6}{5}$**
2. $\frac{2}{5} = \frac{4}{x-2}$ **12**
3. $\frac{x+3}{5} = \frac{x-5}{4}$ **37**
4. $\frac{1}{1-x} = \frac{9x}{2}$ **$\frac{1}{3}, \frac{2}{3}$**
5. $\frac{11}{x^2+6} = \frac{2}{x}$ **$\frac{3}{2}$, 4**
6. From a jar of pennies, 100 are drawn, marked, and returned. After mixing, a sample of 175 pennies is drawn. Of these, 11 are marked. Estimate the number of pennies in the jar assuming that the ratio of marked pennies in the sample to total pennies in the sample equals the ratio of total marked pennies to the total number of pennies in the jar. **about 1600**

EXTRA CHALLENGE NOTE

Challenge problems for Lesson 11.1 are available in **blackline** format in the *Chapter 11 Resource Book,* p. 20 and at **www.mcdougallittell.com.**

ADDITIONAL TEST PREPARATION

1. **WRITING** Is the proportion $\frac{a}{b} = \frac{c}{d}$ equivalent to $\frac{d}{b} = \frac{c}{a}$? Explain your answer.
   See sample answer below.
2. **OPEN ENDED** Write a problem involving a scale model that you can solve using proportions. Show how to solve your problem.
   See sample answer below.

DATA UPDATE

Updated data for Ex. 48 are available at **www.mcdougallittell.com.**

48. b. $\frac{23.8}{100} = \frac{x}{170{,}581}$**; about 40,598,000 people**

48. d. ***Sample answer:*** **Use the ratio $\frac{7581}{15{,}860}$ to find the percentage and compare the result (47.8%) to 23.8%, which represents the typical value.**

Test Preparation

48. MULTI-STEP PROBLEM Base your answers on the 1997 population data shown in the table for people 25 years and older in the United States.

Total population (in thousands)	Number of people out of 100 with education level			
	Not a high school graduate	High school graduate	Some college, but less than 4 yr	At least 4 yr of college
170,581	17.9	33.8	24.5	23.8

DATA UPDATE of U.S. Bureau of the Census data at www.mcdougallittell.com

a. Out of 200 people aged 25 years or older, about how many would you expect to have just a high school education? to have at least 4 years of college? **about 68 people; about 48 people**

b. Write and solve a proportion to estimate the number of people in the United States aged 25 years or older with at least 4 years of college. **See margin.**

c. Suppose a town has 20,000 residents aged 25 years or older. Write and solve a proportion to estimate the number of town residents 25 years or older who have completed at least 4 years of college. **about 4760 people**

d. *Writing* Suppose another town has 15,860 people aged 25 years or older and that 7581 of these people have completed at least 4 years of college. Explain how you can find out whether the number of college graduates in that town is typical for a town of that size. **See margin.**

★ Challenge

49. LOGICAL REASONING One way to prove that two proportions are equivalent is to apply the properties of equality to transform one of the proportions into the other proportion. Give a sequence of steps that transforms the proportion $\frac{a}{b} = \frac{c}{d}$ into $\frac{a}{c} = \frac{b}{d}$.

Sample answer**: First multiply each side by b to obtain $a = \frac{cb}{d}$, then divide each side by c to obtain $\frac{a}{c} = \frac{b}{d}$.**

EXTRA CHALLENGE

www.mcdougallittell.com

MIXED REVIEW

50. DECIMAL AND PERCENT FORM Copy and complete the table. If necessary, round to the nearest tenth of a percent. (Skills Review, pages 784–785)
0.78; 0.03; 1.76; 1.1; 20%; 7.3%; 66.7%; 200%

Decimal	?	0.2	?	0.073	0.666...	?	?	2
Percent	78%	?	3%	?	?	176%	110%	?

FINDING SQUARE ROOTS Find all square roots of the number or write *no square roots*. Check the results by squaring each root. (Review 9.1)

51. 64 **−8, 8**
52. −9 **no square roots**
53. 12 **$-2\sqrt{3}$, $2\sqrt{3}$**
54. 169 **−13, 13**
55. −20 **no square roots**
56. 50 **$-5\sqrt{2}$, $5\sqrt{2}$**
57. $\frac{9}{25}$ **$-\frac{3}{5}, \frac{3}{5}$**
58. 0.04 **−0.2, 0.2**

SIMPLIFYING RADICAL EXPRESSIONS Simplify the radical expression. (Review 9.2)

59. $\sqrt{18}$ **$3\sqrt{2}$**
60. $\sqrt{20}$ **$2\sqrt{5}$**
61. $\sqrt{80}$ **$4\sqrt{5}$**
62. $\sqrt{162}$ **$9\sqrt{2}$**
63. $9\sqrt{36}$ **54**
64. $\sqrt{\frac{11}{9}}$ **$\frac{\sqrt{11}}{3}$**
65. $\frac{1}{2}\sqrt{28}$ **$\sqrt{7}$**
66. $4\sqrt{\frac{5}{4}}$ **$2\sqrt{5}$**

Additional Test Preparation *Sample answers:*

1. Yes. *Sample answer:* The cross products $ad = bc$ and $da = bc$ are equivalent by the commutative property of multiplication.

2. *Sample answer:* To build a model of a plant cell that is 0.2 mm in length and 0.08 mm in width and height so that the model is 12 cm in width, how long must the model be? (30 cm)

11.2 Percents

What you should learn

GOAL 1 Use equations to solve percent problems.

GOAL 2 Use percents in **real-life** problems, such as finding the number of known insect species in **Example 4**.

Why you should learn it

▼ To model and solve **real-life** problems, such as comparing two discounted prices in **Ex. 45**.

REAL LIFE

GOAL 1 SOLVING PERCENT PROBLEMS

In the percent equation "8 is 40% of 20," you compare 8 to the *base number* 20 by writing $\frac{8}{20} = \frac{40}{100}$. The percent for $\frac{8}{20}$ can be written as 40%, as the fraction $\frac{40}{100}$, or as the decimal 0.4. When you work with percents in equations, you usually write the percent as a fraction or as a decimal. In any percent equation the **base number** is the number you are comparing to. You can write a verbal model to help you solve the equation.

VERBAL MODEL — **Number being compared to base** is a **percent** of **base number**

LABELS

Number compared to base	$=$ a	(same units as b)
Percent	$=$ p	(no units)
Base number	$=$ b	(assigned units)

ALGEBRAIC MODEL — $a = p \cdot b$

In Examples 1–3 you will see how the basic verbal model above can be applied to the three different cases of percent problems—finding a, finding p, or finding b. As shown in these examples, percents are usually converted to decimal form before performing arithmetic operations.

EXAMPLE 1 *Number Compared to Base is Unknown*

What is 30% of 70 feet?

SOLUTION

VERBAL MODEL — a is p percent of b

LABELS

Number compared to base	$=$ a	(feet)
Percent	$= 30\% = \mathbf{0.3}$	(no units)
Base number	$= \mathbf{70}$	(feet)

ALGEBRAIC MODEL

$a = (0.3)(70)$

$a = 21$

▶ 21 feet is 30% of 70 feet.

1 PLAN

PACING

Basic: 1 day
Average: 1 day
Advanced: 1 day
Block Schedule: 0.5 block with 11.1

LESSON OPENER CALCULATOR ACTIVITY

An alternative way to approach Lesson 11.2 is to use the Calculator Activity Lesson Opener:

- Blackline Master (*Chapter 11 Resource Book*, p. 24)
- Transparency (p. 75)

MEETING INDIVIDUAL NEEDS

- ***Chapter 11 Resource Book***
  Prerequisite Skills Review (p. 5)
  Practice Level A (p. 27)
  Practice Level B (p. 28)
  Practice Level C (p. 29)
  Reteaching with Practice (p. 30)
  Absent Student Catch-Up (p. 32)
  Challenge (p. 35)
- ***Resources in Spanish***
- ***Personal Student Tutor***

NEW-TEACHER SUPPORT

See the Tips for New Teachers on pp. 1–2 of the *Chapter 11 Resource Book* for additional notes about Lesson 11.2.

WARM-UP EXERCISES

Transparency Available

Write each number as a percent. Round to the nearest tenth of a percent if necessary.

1. 0.24 **24%**
2. 3.16 **316%**
3. $\frac{4}{5}$ **80%**
4. $\frac{1}{3}$ **33.3%**

2 TEACH

MOTIVATING THE LESSON

Have each student imagine that he or she and a friend have invested money in the stock market. After one month, the value of the student's stocks has increased $58, while the value of the friend's has increased $96. Ask students whose investment was better. Then point out that they cannot answer this question by comparing the increases because they don't know if the initial investments were equal. Instead, they must be able to determine by what percent each investment has increased.

EXTRA EXAMPLE 1
What is 150% of $200? **$300**

EXTRA EXAMPLE 2
Twenty-six is 40% of what number? **65**

EXTRA EXAMPLE 3
Forty-eight is what percent of 160? **30%**

CHECKPOINT EXERCISES

For use after Example 1:

1. What is 75% of 140? **105**

For use after Example 2:

2. 9.6 is 12% of what number? **80**

For use after Example 3:

3. Two hundred forty is what percent of 150? **160%**

STUDENT HELP NOTES

Homework Help Students can find extra examples at **www.mcdougallittell.com** that parallel the examples in the student edition.

EXAMPLE 2 *Base Number is Unknown*

Fourteen dollars is 25% of what amount of money?

SOLUTION

VERBAL MODEL a is p percent of b

LABELS

Number compared to base = **14** (dollars)

Percent = 25% = **0.25** (no units)

Base number = b (dollars)

ALGEBRAIC MODEL

$$14 = (0.25)\,b$$

$$\frac{14}{0.25} = b$$

$$56 = b$$

▶ $14 is 25% of $56.

STUDENT HELP

Study Tip
You can check your work by using a different method. In Example 2, you are comparing $14 to an unknown number of dollars *b*. You can use the proportion $\frac{14}{b} = \frac{25}{100}$.

EXAMPLE 3 *Percent is Unknown*

One hundred thirty-five is what percent of 27?

SOLUTION

VERBAL MODEL a is p percent of b

LABELS

Number compared to base = **135** (no units)

Percent = p (no units)

Base number = **27** (no units)

ALGEBRAIC MODEL

$$135 = p\,(27)$$

$$\frac{135}{27} = p$$

$5 = p$ **Decimal form**

$500\% = p$ **Percent form** $\left(5 = \frac{500}{100}\right)$

CONCEPT SUMMARY THREE FORMS OF PERCENT PROBLEMS $a = pb$

QUESTION	GIVEN	NEED TO FIND	EXAMPLE
What is *p* percent of *b*?	*b* and *p*	***a***	Example 1
a is *p* percent of **what**?	*a* and *p*	***b***	Example 2
a is **what percent** of *b*?	*a* and *b*	***p***	Example 3

STUDENT HELP

INTERNET HOMEWORK HELP
Visit our Web site www.mcdougallittell.com for extra examples.

Study Tip
You can model a percent relationship using a proportion or a decimal equation. This is possible because a percent can be written as either a fraction or a decimal.

GOAL 2 USING PERCENTS IN REAL LIFE

EXAMPLE 4 *Modeling and Using Percents*

SCIENCE CONNECTION There are about 170,000 species of butterflies and moths world-wide. Butterflies and moths make up about 17% of all classified insect species. Estimate how many insect species have been classified.

Insect Species in North America

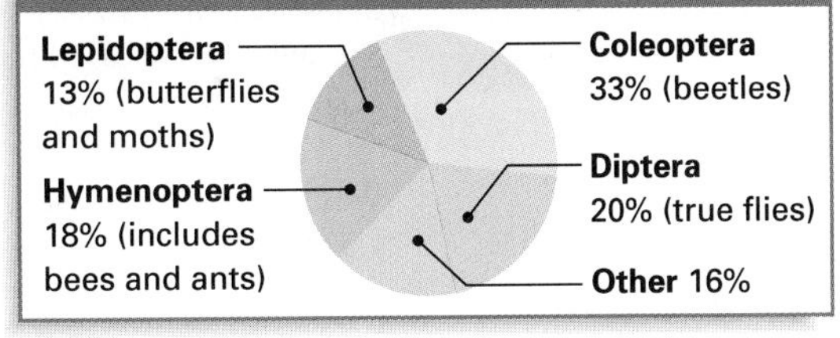

Note: the percents for the world-wide species are slightly different.

SOLUTION

Method 1 Use the percent equation $a = pb$.

PROBLEM SOLVING STRATEGY

VERBAL MODEL **Number of butterfly and moth species** is **p percent** of **Total number of insect species**

LABELS

Number of butterfly and moth species = **170,000** (species)

Percent = 17% = **0.17** (no units)

Total number of insect species = $\mathbf{b}$ (species)

ALGEBRAIC MODEL

$$170{,}000 = 0.17\,b$$

$$\frac{170{,}000}{0.17} = b$$

$$1{,}000{,}000 = b$$

▶ About 1,000,000 species of insects have been classified.

Method 2 Use a proportion.

Write ratios that compare the part to the whole. Let b represent the total number of insect species that have been classified.

$$\frac{\text{Number of butterfly and moth species}}{\text{Total number of insect species}} = \frac{17}{100}$$ Write proportion.

$$\frac{170{,}000}{b} = \frac{17}{100}$$ Substitute.

$$17b = 170{,}000 \cdot 100$$ Use cross products.

$$b = \frac{17{,}000{,}000}{17}$$ Divide each side by 17.

$$b = 1{,}000{,}000$$ Simplify.

▶ About 1,000,000 species of insects have been classified.

FOCUS ON APPLICATIONS

CLASSIFYING INSECTS
Butterflies and moths together make up the insect order *Lepidoptera*. About 17,000 species of butterflies and about 153,000 species of moths have been classified.

APPLICATION LINK
www.mcdougallittell.com

EXTRA EXAMPLE 4
You read in a magazine that an affordable rent is one that is no more than 30% of your salary. The apartment you want rents for $625 per month. At least what must your monthly salary be for you to afford this apartment? **$2083.33**

CHECKPOINT EXERCISES

For use after Example 4:

1. International students make up 12% of the students at a college. If there are 423 international students, how many students are in the entire college? **3525 students**

APPLICATION NOTE
EXAMPLE 4 Additional information about classifying insects is available at **www.mcdougallittell.com.**

EXTRA EXAMPLE 5

The graph shows the results of a poll of students taken to find out the average amount of time spent on various activities in a 24 hour period. If you spent 1.5 hours on homework yesterday, was this percent of time about the same as the poll average? Explain.

How 24 Hours Are Spent

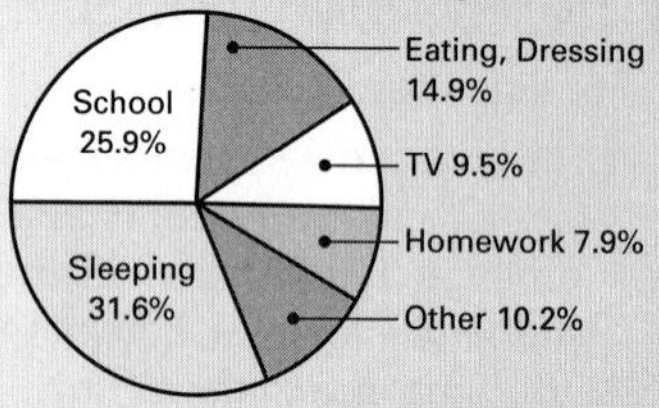

Sample answer: **No. The 6.25% of time you spent on homework is lower than the average of 7.9%. (The difference amounts to about 24 min of study.)**

CHECKPOINT EXERCISES

For use after Example 5:

1. Refer to the graph in Extra Example 5. If you slept 7 hours and 45 minutes last night, was this percent of time about the same as the poll average? Explain. ***Sample answer:*** **Yes. The 32.3% of time you spent sleeping is about the same as the average of 31.6%.**

FOCUS ON VOCABULARY

In the verbal model "*x* is *y* percent of *z*," which variable represents the *base number*? *z*

CLOSURE QUESTION

How do you solve a percent problem using an equation of the form $a = pb$? Explain what each variable represents.

See sample answer below.

DAILY PUZZLER

The \$156.25 six month increase in your stock value reflects a return 1.25 percentage points higher than does the \$200 six month increase in your friend's stock value. Your friend invested \$4000. How much did you invest? **\$2500**

REAL LIFE *Cars*

EXAMPLE 5 *Using Percents to Compare*

The circle graph shows the average costs (in percents) of owning an automobile in 1996.

Suppose that in 1996 a car owner spent \$650 on gasoline for a car whose total costs were \$3750. Was the percent spent on gasoline about the same as the national average?

Automobile Ownership Costs

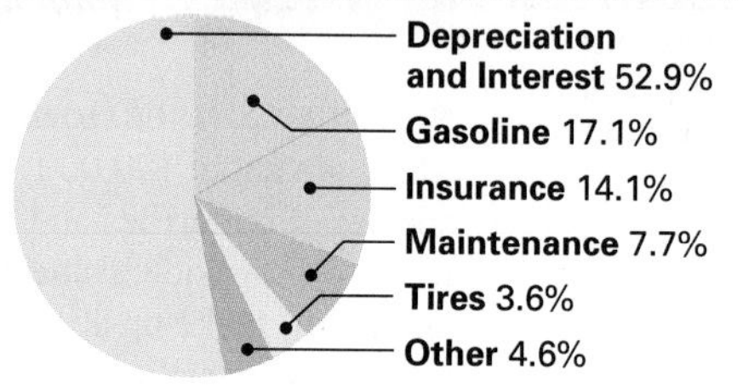

▶ Source: Runzheimer International

SOLUTION Use the percent equation $a = pb$.

PROBLEM SOLVING STRATEGY

VERBAL MODEL [**Amount spent on gasoline**] is **p percent** of [**Total costs of owning a car**]

LABELS

Amount spent on gasoline $= \mathbf{650}$		(dollars)
Percent $= \boldsymbol{p}$		(no units)
Total costs of owning a car $= \mathbf{3750}$		(dollars)

ALGEBRAIC MODEL $\mathbf{650} = \boldsymbol{p}\,\mathbf{(3750)}$

▶ $p \approx 0.173$ or 17.3%, which is about the national average of 17.1%.

GUIDED PRACTICE

Vocabulary Check ✓

1. Write an equation that represents the statement "10% of 160 is 16." What is the base number? **0.10(160) = 16; 160**

Concept Check ✓

DISCOUNTS **In Exercises 2–4, the sale price of a shirt is \$17.25 after a 25% discount is taken.**

2. The sale price is what percent of the regular price? **75%**
3. You can model the situation with an equation of the form *a* is *p* percent of *b*. Is the base *b* the sale price or the regular price? **regular price**
4. Write and solve an equation to find the regular price of the shirt. **$0.75b = 17.25$; \$23**

Skill Check ✓

In Exercises 5–8, solve the percent problem.

5. 35 is what percent of 20? **175%**
6. 12% of 5 is what number? **0.6**
7. 18 is 37.5% of what number? **48**
8. 13.2 is 120% of what number? **11**
9. **SALES TAX** The price of a book without tax is \$5.99 and the sales tax rate is 6%. Find the amount of the tax by using an equation of the form $a = pb$ and by using a proportion. How are the two methods similar? **See margin.**

9. \$.36; *Sample answer:* The equations for the two methods are $a = 0.06(5.99)$ and $\frac{a}{5.99} = \frac{6}{100}$. Notice that the right side of the second equation is the rate expressed as a fraction instead of a decimal. Both methods involve multiplying the base by the rate.

Closure question *Sample answer:*
Substitute for the two variables you know and then solve for the unknown variable. The variable *a* is the number compared to the base, *p* is the percent, and *b* is the base number.

PRACTICE AND APPLICATIONS

STUDENT HELP
Extra Practice to help you master skills is on p. 807.

UNDERSTANDING PERCENT EQUATIONS **Match the percent problem with the equation that represents it.**

A. $a = (0.39)(50)$ **B.** $39 = p(50)$ **C.** $39 = 0.50b$

10. 39 is 50% of what number? C

11. 39% of 50 is what number? A

12. $39 is what percent of $50? B

SOLVING PERCENT PROBLEMS **Solve the percent problem.**

13. What number is 25% of 80? 20

14. 85% of 300 is what number? 255

15. 18 is what percent of 60? 30%

16. 52 is 12.5% of what number? 416

17. 14% of 220 feet is what distance? 30.8 ft

18. How much money is 35% of $750? $262.50

19. 42 feet is 50% of what length? 84 ft

20. What distance is 24% of 710 miles? 170.4 mi

21. 16% of what number is 8? 50

22. $4 is 2.5% of what amount? $160

23. 33 grams is 22% of what weight? 150 g

24. 55 years is what percent of 20 years? 275%

25. How much is 8.2% of 800 tons? 65.6 tons

26. 9 people is what percent of 60 people? 15%

27. 62 hours is what percent of 3 days?

27. $86\frac{1}{9}$%, or about 86.1%

28. 30 inches is what percent of 40 feet? 6.25%

29. $240 is what percent of $50? 480%

30. 2 percent of what amount is $200? $10,000

GEOMETRY CONNECTION **In Exercises 31 and 32, what percent of the region is shaded blue? What percent is shaded yellow? All figures are rectangles.**

31.

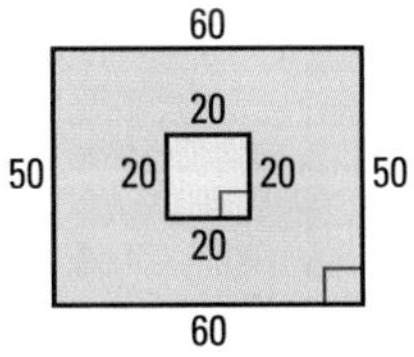

$86\frac{2}{3}$%, or about 86.7%; $13\frac{1}{3}$%, or about 13.3%

32.

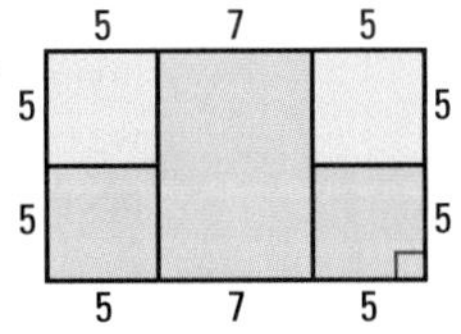

$70\frac{10}{17}$%, or about 70.6%; $29\frac{7}{17}$%, or about 29.4%

OIL CHANGES **The histogram shows how 861 people answered a survey question about when they usually change the oil in their cars.**

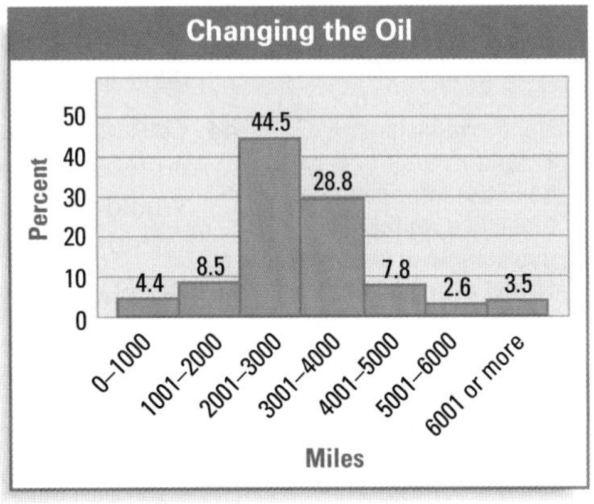

▶ Source: Maritz Marketing Research Inc.

33. How many of the people change their oil between 3001 and 4000 miles? about 248 people

34. How many of the people change their oil between 4001 and 6000 miles? about 90 people

35. If you surveyed 2500 people, about how many people do you expect to answer "2001 to 3000 miles?" about 1113 people

STUDENT HELP
HOMEWORK HELP
Example 1: Exs. 10–30
Example 2: Exs. 10–30
Example 3: Exs. 10–30
Example 4: Exs. 33–39
Example 5: Exs. 42, 43

3 APPLY

ASSIGNMENT GUIDE

BASIC
Day 1: pp. 653–655 Exs. 10–12, 14–30 even, 31–35, 48–50, 55, 60, 65, 70, 72

AVERAGE
Day 1: pp. 653–655 Exs. 10–12, 14–30 even, 31–35, 48–50, 55, 60, 65, 70, 72

ADVANCED
Day 1: pp. 653–655 Exs. 10–12, 20–30 even, 31–39, 45–52, 55, 60, 65, 70, 72

BLOCK SCHEDULE WITH 11.1
pp. 653–655 Exs. 10–12, 14–30 even, 31–35, 48–50, 55, 60, 65, 70, 72

EXERCISE LEVELS

Level A: *Easier*
10–12

Level B: *More Difficult*
13–39, 48–50, 53–72

Level C: *Most Difficult*
40–47, 51, 52

HOMEWORK CHECK

To quickly check student understanding of key concepts, go over the following exercises: Exs. 12, 20, 22, 26, 32, 34. See also the Daily Homework Quiz:

- Blackline Master (*Chapter 11 Resource Book*, p. 38)
- Transparency (p. 87)

COMMON ERROR

EXERCISE 9 In real-life problems such as this, students often have difficulty identifying the number compared to the base, the percent, and the base number. Encourage them to write a verbal model as shown in Examples 4 and 5 prior to assigning labels or substituting any values into $a = pb$.

CAREER NOTE

EXERCISES 36–39 Additional information about careers in College Admissions is available at **www.mcdougallittell.com.**

ADDITIONAL PRACTICE AND RETEACHING

For Lesson 11.2:

- Practice Levels A, B, and C (*Chapter 11 Resource Book,* p. 27)
- Reteaching with Practice (*Chapter 11 Resource Book,* p. 30)
- See Lesson 11.2 of the *Personal Student Tutor*

For more Mixed Review:

- Search the *Test and Practice Generator* for key words or specific lessons.

FOCUS ON CAREERS

REAL LIFE COLLEGE RECRUITER

A college recruiter's duties range from giving tours of the campus, to arranging orientation seminars, to visiting high schools in search of promising students.

INTERNET CAREER LINK
www.mcdougallittell.com

CHOOSING A COLLEGE In Exercises 36–39, use the graph. It shows the responses of 3500 seniors from high schools around the United States.

36. What percent of the seniors said location was the reason for their choice? **21%**

37. What percent of the seniors said academic reputation was the reason for their choice? **18%**

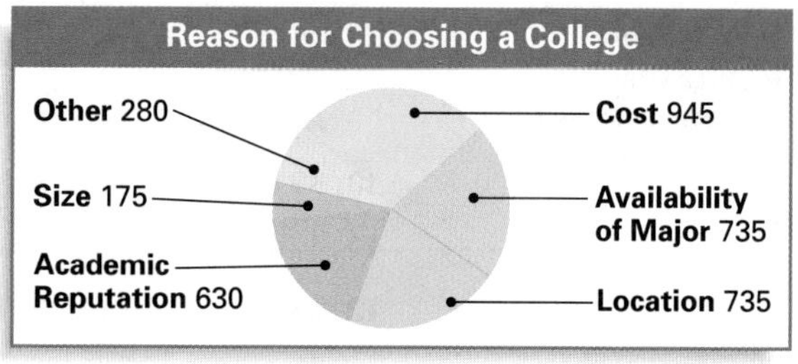

▶ Source: Careers and Colleges

38. What percent of the seniors said size or cost most influences their choice? **32%**

39. Use the survey results to predict the number of seniors in a class of 2000 students who would say that availability of major most influences their choice. **about 420 students**

40. ERROR ANALYSIS Find and correct the mistake in the restaurant bill. The tax rate is 8%. **The correct tax is $2.41, so the total is $32.59.**

Food	$24.93
Beverages	$5.25
Subtotal	$30.18
Tax	$4.22
Total	$34.40

41. TIPPING Use the corrected bill from Exercise 40. In the United States, the standard tip for a waiter or waitress is 15%–20%. You leave a $4.75 tip. Is your tip within the standard range if the tip is figured before tax is added? if the tip is figured on the total including tax? **yes; no**

OZONE LAYER SURVEY In Exercises 42–44, use the graph.

42. In the survey 572 people said they think the ozone layer is getting worse. What was the total number of people surveyed? **1040 people**

43. Use the result from Exercise 42. About how many of the people surveyed think the ozone layer is staying the same? **about 395 people**

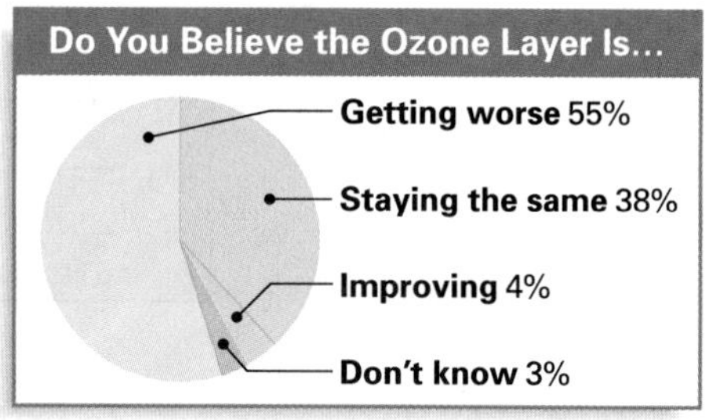

▶ Source: Worthlin Worldwide

44. CRITICAL THINKING In this survey the researchers tried to use a representative sample of people 18 years old and over in the United States. Would this sample be reasonable to use in predicting the responses of scientists? Explain.

44. No; *Sample answer*: **Scientists are likely to be better informed than those in a representative sample of people 18 and over, and their answers are more likely to be based on verified information rather than on personal opinion.**

45. THE BETTER BUY You are shopping and find a coat that is on sale for 30% off. It is regularly priced at $80. Your friend tells you that she saw the same coat for $80 in another store, but it was 20% off plus an additional 10% off. Will you save money by going to the other store? Explain why or why not. **No; the price is $56 at the first store and $57.60 at the second.**

ALTERNATIVE MODELS **In Exercises 46 and 47, the charge for a cab ride is $11.50, and you give a 20% tip. Using the model, find the total cost of the cab ride. Describe what the variable *a* represents.**

46. Model 1: a is 20% of $11.50. **$13.80; the amount of the tip**

47. Model 2: a is 120% of $11.50. **$13.80; the total cost of the ride**

Test Preparation

QUANTITATIVE COMPARISON In Exercises 48–50, choose the statement below that is true about the given numbers.

Ⓐ The number in column A is greater.

Ⓑ The number in column B is greater.

Ⓒ The two numbers are equal.

Ⓓ The relationship cannot be determined from the given information.

	Column A	Column B	
48.	104% of 150	100% of 150 + 4% of 150	**C**
49.	The solution of the equation 24% of $x = 450$	The solution of the equation 12% of $x = 225$	**C**
50.	The solution of the equation 16% of $x = 28$	The solution of the equation $\frac{16}{100} = \frac{x}{28}$	**A**

51. No; let x be the amount your sister earns. Then the amount you earn is $1.1x$ and $\frac{x}{1.1x} \approx 91\%$. So your sister earns about 9% less.

★ Challenge

51. CRITICAL THINKING You earn 10% more money at your summer job than your sister earns at her summer job. Does this mean that your sister earns 10% less money than you? Explain your answer. **See margin.**

52. CRITICAL THINKING A student claims that if a price is now 220% more than it was before, then it is 320% of what it was before, and what it was before is 31.25% of what it is now. Do you agree? Explain your answer.
Yes; let x be the original price; a price that is 220% greater is $x + 2.2x = 3.2x$ or 320% of x; also, 31.25% of 320% of $x = 0.3125(3.2x) = x$.

EXTRA CHALLENGE
→ www.mcdougallittell.com

MIXED REVIEW

FINDING EQUATIONS **The variables x and y vary directly. Use the given values of the variables to write an equation that relates x and y.** **(Review 4.5 for 11.3)**

53. $x = 4, y = 8$ **$y = 2x$**

54. $x = 33, y = 9$ **$y = \frac{3}{11}x$**

55. $x = -2, y = -1$ **$y = \frac{1}{2}x$**

56. $x = 6.3, y = 1.5$ **$y = \frac{5}{21}x$, or $y \approx 0.24x$**

57. $x = 5\frac{1}{3}, y = 8$ **$y = \frac{3}{2}x$**

58. $x = 9.8, y = 3.6$ **$y = \frac{18}{49}x$, or $y \approx 0.37x$**

CHECKING SOLUTIONS **Decide whether the ordered pair is a solution of the inequality.** **(Review 9.7)**

59. $y < x^2 + 6x + 12; (-1, 4)$ **solution**

60. $y \le x^2 - 7x + 9; (-1, 2)$ **solution**

61. $y > 2x^2 - 7x - 15; (2, 5)$ **solution**

62. $y \ge x^2 + 6x + 12; (1, -4)$ **not a solution**

FACTORING EXPRESSIONS **Completely factor the expression.** **(Review 10.8)** **63–71. See margin.**

63. $x^2 + 5x - 14$

64. $7x^2 + 8x + 1$

65. $5x^2 - 51x + 54$

66. $4x^2 - 28x + 49$

67. $6x^2 + 16x$

68. $36x^5 - 90x^3$

69. $3x^3 + 21x^2 + 30x$

70. $36x^3 - 9x$

71. $15x^4 - 50x^3 - 40x^2$

72. TEXAS POPULATION The population P of Texas (in thousands) for 1995 projected through 2025 can be modeled by $P = 18{,}870(1.0124)^t$, where $t = 0$ represents 1995. Find the ratio of the population in 2025 to the population in 2000. **(Review 8.3)** ▶ Source: U.S. Bureau of the Census **about 1.36**

63. $(x + 7)(x - 2)$
64. $(7x + 1)(x + 1)$
65. $(5x - 6)(x - 9)$
66. $(2x - 7)(2x - 7)$
67. $2x(3x + 8)$
68. $18x^3(2x^2 - 5)$
69. $3x(x + 2)(x + 5)$
70. $9x(2x - 1)(2x + 1)$
71. $5x^2(3x + 2)(x - 4)$

4 ASSESS

DAILY HOMEWORK QUIZ

Transparency Available

1. What length is 37.5% of 44 mm? **16.5 mm**

2. 36 ft is what percent of 90 ft? **40%**

3. 18 days is 9% of how many days? **200 days**

4. Given that all figures are rectangles, what percent of the region is shaded? **30%**

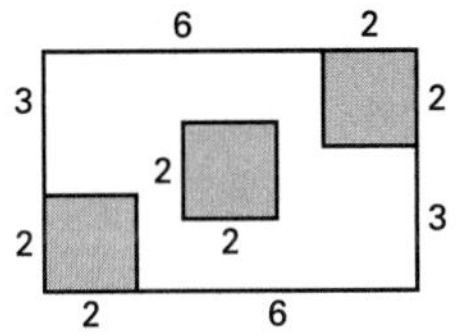

5. In a city council election between Ms. Elizando and Ms. Salazar, 53.5% of the people voted for Ms. Elizando. About how many of the 15,430 voters voted for Ms. Salazar? **7175**

EXTRA CHALLENGE NOTE
→ Challenge problems for Lesson 11.2 are available in **blackline** format in the *Chapter 11 Resource Book*, p. 35 and at **www.mcdougallittell.com.**

ADDITIONAL TEST PREPARATION

1. OPEN ENDED Give one example of a percent problem where the base number is unknown and one example of a percent problem where the number compared to the base is unknown.
Sample answer: **Base unknown: 64 is 160% of what? (40); Number compared unknown: What is 45% of 360? (162)**

1 PLAN

PACING
Basic: 2 days
Average: 2 days
Advanced: 2 days
Block Schedule: 1 block

LESSON OPENER
APPLICATION
An alternative way to approach Lesson 11.3 is to use the Application Lesson Opener:
- Blackline Master (*Chapter 11 Resource Book*, p. 39)
- Transparency (p. 76)

MEETING INDIVIDUAL NEEDS
- ***Chapter 11 Resource Book***
  Prerequisite Skills Review (p. 5)
  Practice Level A (p. 43)
  Practice Level B (p. 44)
  Practice Level C (p. 45)
  Reteaching with Practice (p. 46)
  Absent Student Catch-Up (p. 48)
  Challenge (p. 51)
- ***Resources in Spanish***
- ***Personal Student Tutor***

NEW-TEACHER SUPPORT
See the Tips for New Teachers on pp. 1–2 of the *Chapter 11 Resource Book* for additional notes about Lesson 11.3.

WARM-UP EXERCISES
Transparency Available
The variables *x* and *y* vary directly. Use the given values to write an equation that relates *x* and *y*.

1. $x = 3, y = 12$ $y = 4x$
2. $x = \frac{1}{5}, y = 2$ $y = 10x$
3. $x = 8, y = 2$ $y = \frac{1}{4}x$
4. $x = 3.2, y = 64$ $y = 20x$

EXPLORING DATA AND STATISTICS

11.3 Direct and Inverse Variation

What you should learn

GOAL 1 Use direct and inverse variation.

GOAL 2 Use direct and inverse variation to model **real-life** situations, such as the snowshoe problem in **Exs. 43 and 44.**

Why you should learn it

▼To solve **real-life** situations such as relating the banking angle of a bicycle to its turning radius in **Example 3.**

REAL LIFE

GOAL 1 USING DIRECT AND INVERSE VARIATION

In Lesson 4.5 you studied *direct variation.* Now you will review variation problems and learn about a different type of variation, called **inverse variation.**

MODELS FOR DIRECT AND INVERSE VARIATION

DIRECT VARIATION The variables *x* and *y vary directly* if for a constant *k*

$\frac{y}{x} = k$, or $y = kx$, $k \neq 0$.

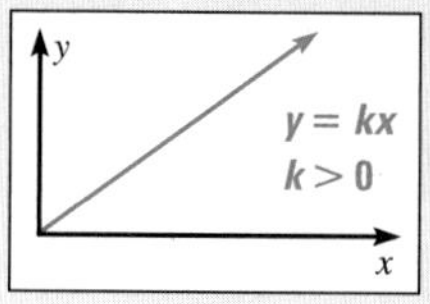

INVERSE VARIATION The variables *x* and *y vary inversely* if for a constant *k*

$xy = k$, or $y = \frac{k}{x}$, $k \neq 0$.

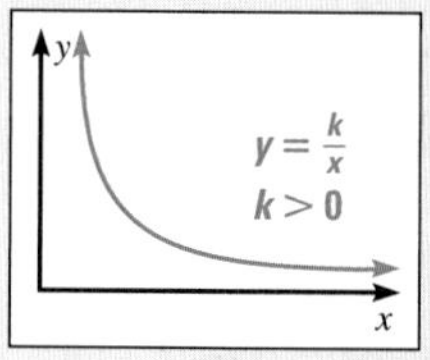

The number *k* is the **constant of variation.**

EXAMPLE 1 *Using Direct and Inverse Variation*

When *x* is 2, *y* is 4. Find an equation that relates *x* and *y* in each case.

a. *x* and *y* vary directly **b.** *x* and *y* vary inversely

SOLUTION

a. $\frac{y}{x} = k$ **Write direct variation model.**

$\frac{4}{2} = k$ **Substitute 2 for *x* and 4 for *y*.**

$2 = k$ **Simplify.**

▶ The direct variation that relates *x* and *y* is $\frac{y}{x} = 2$, or $y = 2x$.

b. $xy = k$ **Write inverse variation model.**

$(2)(4) = k$ **Substitute 2 for *x* and 4 for *y*.**

$8 = k$ **Simplify.**

▶ The inverse variation that relates *x* and *y* is $xy = 8$, or $y = \frac{8}{x}$.

EXAMPLE 2 *Comparing Direct and Inverse Variation*

Compare the direct variation model and the inverse variation model you found in Example 1 using $x = 1, 2, 3,$ and 4.

a. numerically **b.** graphically

SOLUTION

a. Use the models $y = 2x$ and $y = \frac{8}{x}$ to make a table.

x-value	1	2	3	4
Direct, $y = 2x$	2	4	6	8
Inverse, $y = \frac{8}{x}$	8	4	$\frac{8}{3}$	2

Direct Variation: Because k is positive, y increases as x increases. As x increases by 1, y increases by 2.

Inverse Variation: Because k is positive, y decreases as x increases. As x doubles (from 1 to 2), y is halved (from 8 to 4).

b. Use the table of values. Plot the points and then connect the points with a smooth curve.

Direct Variation: The graph for this model is a line passing through the origin.

Inverse Variation: The graph for this model is a curve that approaches the x-axis as x increases and approaches the y-axis as x gets close to 0.

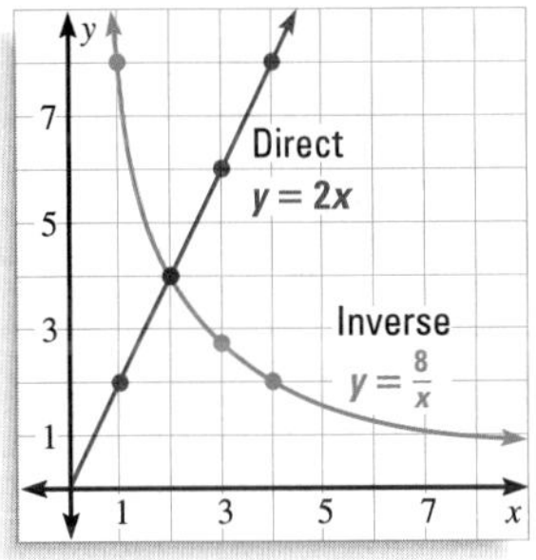

Direct and inverse variation models represent functions because for each value of x there is exactly one value of y. For inverse variation, the domain excludes 0.

ACTIVITY

Developing Concepts

Investigating Direct and Inverse Variation

Make a table and draw a graph for each situation. Then decide whether the quantities vary *directly*, *inversely*, or *neither*.

1. You are walking at a rate of 5 miles per hour. How are your distance d and time t related? **directly**
2. You take a 5-mile walk. How are your rate r and time t related? **inversely**
3. A ball is dropped from a height of 32 feet. How are the height h and time t related? **neither**
4. A rectangle has an area of 12 square inches. How are the length l and width w related? **inversely**

Distance d (in miles)	1	2	3	4	5
Time t (in hours)	$\frac{1}{5}$	$\frac{2}{5}$	$\frac{3}{5}$	$\frac{4}{5}$	1

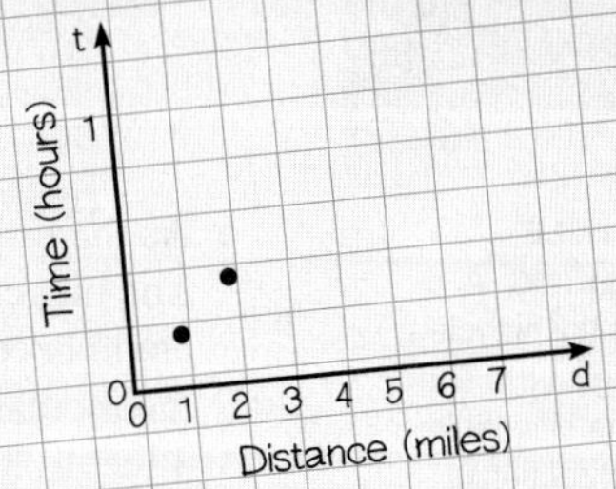

2 TEACH

MOTIVATING THE LESSON

Ask students for real-life examples where as one variable increases, the other increases proportionately. Remind them that this is *direct variation.* Then ask students for examples where as one variable increases, the other *decreases* proportionately. Tell them that this relationship is called *inverse variation.*

ACTIVITY NOTE

In Step 3, the height decreases as time increases, but an inverse variation model is inappropriate. This situation is modeled by the quadratic function $h = 32 - 16t^2$. Students should compare the shape of this graph with the graph in Example 3. Have them multiply corresponding values of h and t to see that the product is not constant.

EXTRA EXAMPLE 1

When x is 3, y is 12. Find an equation that relates x and y in each case.

a. x and y vary directly. **$\frac{y}{x} = 4$, or $y = 4x$**

b. x and y vary inversely. **$xy = 36$, or $y = \frac{36}{x}$**

EXTRA EXAMPLE 2

Compare the direct variation model and the inverse variation model you found in Extra Example 1 using $x = 1, 2, 3, 4$.

a. numerically **See sample answer below.**

b. graphically **See sample answer below.**

CHECKPOINT EXERCISES

For use after Examples 1 and 2:

1. Decide if the data in the table show direct or inverse variation. Write an equation that relates the variables.

x	1	2	3	4
y	12	6	4	3

inverse; $xy = 12$, or $y = \frac{12}{x}$

Extra Example 2 *Sample answers:*

a. Direct: y increases as x increases. As x increases by 1, y increases by 4. Inverse: y decreases as x increases. As x doubles, y is halved.

b. Direct: The graph is a line passing through the origin. Inverse: The graph is a curve that approaches the x-axis as x increases and approaches the y-axis as x gets closer to 0.

EXTRA EXAMPLE 3

The graph models the relationship between the time needed to paint a house and the number of painters. For the values shown, the time needed *T* and the number of painters *n* vary inversely.

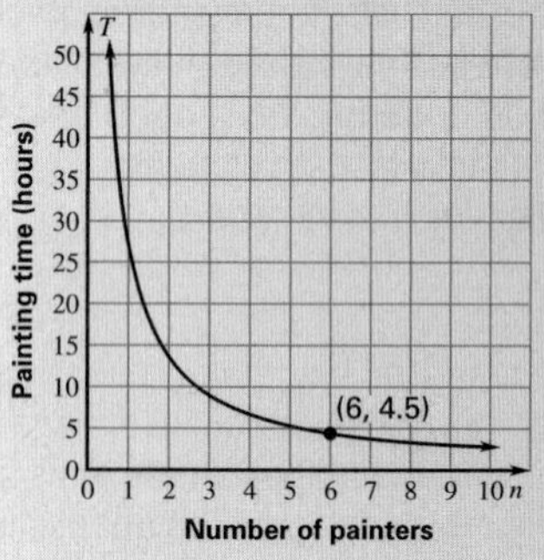

a. Find an inverse variation model relating *T* and *n*. $T = \frac{27}{n}$

b. Use the model to find the time needed by four painters. **6.75 hours**

c. Use the graph to describe how the time needed changes as the number of painters decreases. **It increases.**

CHECKPOINT EXERCISES

For use after Example 3:

1. The area covered by a truckload of mulch and the depth of the mulch are inversely related. A load covers an area of 324 ft^2 six inches deep. Write a model relating the area *A* and the mulch depth *d*. Use the model to find the area covered by mulch 4 inches deep.
$A = 1944d$; 486 ft^2

CLOSURE QUESTION

You are given a table of values of *x* and *y*. How can you tell if the relationship can be modeled by direct variation, inverse variation, or neither?
See sample answer below.

DAILY PUZZLER

If *x* varies directly with *y* and *y* varies inversely with *z*, what is the relationship between *x* and *z*?
***x* varies inversely with *z*.**

GOAL 2 USING DIRECT AND INVERSE VARIATION IN REAL LIFE

EXAMPLE 3 *Writing and Using a Model*

BICYCLING The graph below shows a model for the relationship between the banking angle and the turning radius for a bicycle traveling at a particular speed. For the values shown, the banking angle *B* and the turning radius *r* can be approximated by an inverse variation.

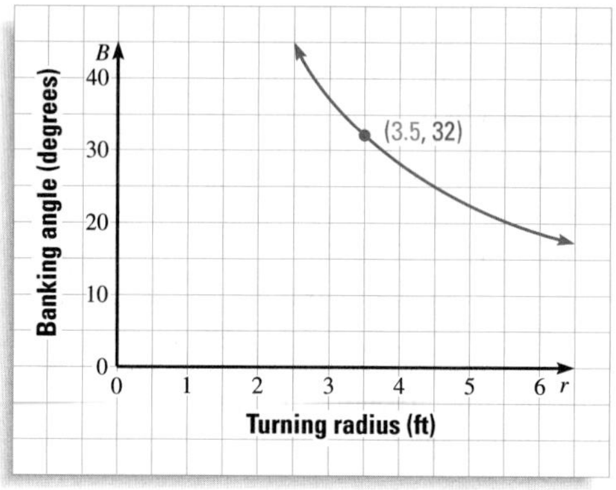

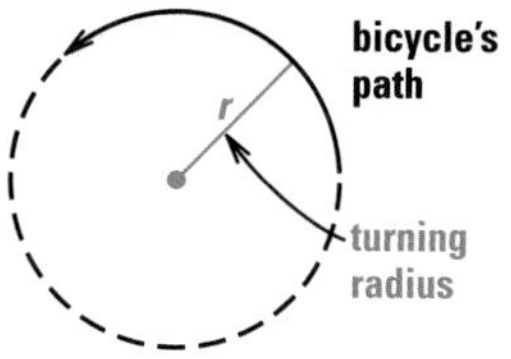

a. Find an inverse variation model that approximately relates *B* and *r*.

b. Use the model to approximate the angle for a radius of 5 feet.

c. Use the graph to describe how the banking angle changes as the turning radius gets smaller.

SOLUTION

a. From the graph, you can see that $B = 32°$ when $r = 3.5$ feet.

$B = \frac{k}{r}$ Write inverse variation model.

$32 = \frac{k}{3.5}$ Substitute 32 for *B* and 3.5 for *r*.

$112 = k$ Solve for *k*.

▶ The model is $B = \frac{112}{r}$, where *B* is in degrees and *r* is in feet.

b. Substitute 5 for *r* in the model found in part (a).

$$B = \frac{112}{5} = 22.4$$

▶ When the turning radius is 5 feet, the banking angle is about 22°.

c. As the turning radius gets smaller, the banking angle becomes greater. The bicyclist leans at greater angles. Notice that the increase in the banking angle becomes more rapid when the turning radius is small. For instance, the increase in the banking angle is about 10° for a 1-foot decrease from 4 feet to 3 feet. The increase in the banking angle is only about 4° for a 1-foot decrease from 6 feet to 5 feet.

FOCUS ON APPLICATIONS

REAL LIFE **BICYCLE BANKING** A bicyclist tips the bicycle when making a turn. The angle *B* of the bicycle from the vertical direction is called the *banking angle.*

Closure Question ***Sample answer:***
If the product of *x* and *y* is constant, it is inverse variation. If the quotient of *x* and *y* is constant, it is direct variation. Otherwise, it is neither.

Guided Practice

Vocabulary Check ✓

Concept Check ✓

1. Two numbers x and y vary directly if there is a constant k, $k \neq 0$, such that $y = kx$. Two numbers x and y vary inversely if there is a constant k, $k \neq 0$, such that $y = \frac{k}{x}$.
2. Direct variation; the equation of the graph is $y = \frac{3}{2}x$.

Skill Check ✓

3. Neither; the equation of the graph is $y = -x^2 + 2$.
4. Inverse variation; the equation of the graph is $y = \frac{4}{x}$.

1. What does it mean for two quantities to vary directly? to vary inversely?
See margin.

Does the graph show *direct variation, inverse variation,* or *neither*? Explain.
2–4. See margin.

2.

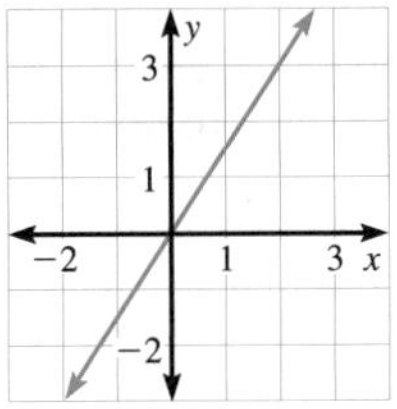

3.

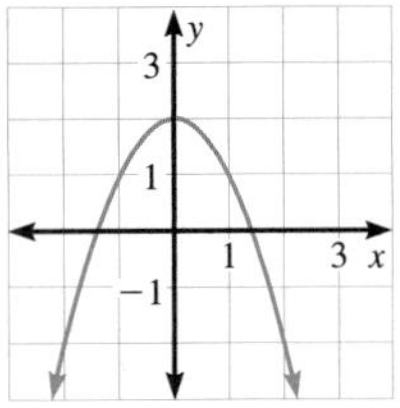

4.

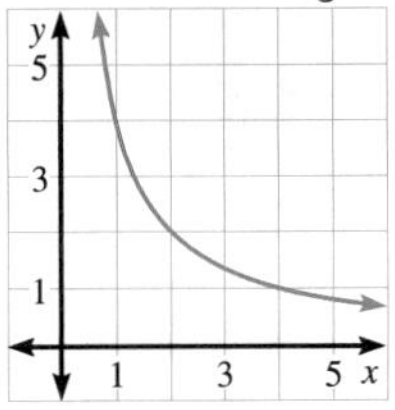

5. One student says that the equation $y = -2x$ is an example of direct variation. Another student says it is inverse variation. Which is correct? Explain.
Direct variation; the equation is of the form $y = kx$, where $k = -2$.

Does the equation model *direct variation, inverse variation,* or *neither*?

6. $x = \frac{4}{y}$ inverse variation
7. $y = 7x - 2$ neither
8. $a = 12b$ direct variation
9. $ab = 9$ inverse variation

When $x = 4$, $y = 6$. For the given type of variation, find an equation that relates x and y. Then find the value of y when $x = 8$.

10. x and y vary directly $y = \frac{3}{2}x$; 12
11. x and y vary inversely $y = \frac{24}{x}$; 3

Practice and Applications

STUDENT HELP

Extra Practice
to help you master skills is on p. 807.

STUDENT HELP

HOMEWORK HELP
Example 1: Exs. 12–32
Example 2: Exs. 33–38
Example 3: Exs. 43–46

DIRECT VARIATION EQUATIONS **The variables x and y vary directly. Use the given values to write an equation that relates x and y.**

12. $x = 3, y = 9$ $y = 3x$
13. $x = 2, y = 8$ $y = 4x$
14. $x = 18, y = 6$ $y = \frac{1}{3}x$
15. $x = 8, y = 24$ $y = 3x$
16. $x = 36, y = 12$ $y = \frac{1}{3}x$
17. $x = 27, y = 3$ $y = \frac{1}{9}x$
18. $x = 24, y = 16$ $y = \frac{2}{3}x$
19. $x = 45, y = 81$ $y = \frac{9}{5}x$
20. $x = 54, y = 27$ $y = \frac{1}{2}x$

INVERSE VARIATION EQUATIONS **The variables x and y vary inversely. Use the given values to write an equation that relates x and y.**

21. $x = 2, y = 5$ $y = \frac{10}{x}$
22. $x = 3, y = 7$ $y = \frac{21}{x}$
23. $x = 16, y = 1$ $y = \frac{16}{x}$
24. $x = 11, y = 2$ $y = \frac{22}{x}$
25. $x = \frac{1}{2}, y = 8$ $y = \frac{4}{x}$
26. $x = \frac{13}{5}, y = 5$ $y = \frac{13}{x}$
27. $x = 12, y = \frac{3}{4}$ $y = \frac{9}{x}$
28. $x = 5, y = \frac{1}{3}$ $y = \frac{5}{3x}$
29. $x = 30, y = 7.5$ $y = \frac{225}{x}$
30. $x = 1.5, y = 50$ $y = \frac{75}{x}$
31. $x = 45, y = \frac{3}{5}$ $y = \frac{27}{x}$
32. $x = 10.5, y = 7$ $y = \frac{73.5}{x}$

DIRECT OR INVERSE VARIATION **Make a table of values for $x = 1, 2, 3,$ and 4. Use the table to sketch a graph. Decide whether x and y vary *directly* or *inversely*.** See margin for graphs.

33. $y = \frac{4}{x}$ 4, 2, $\frac{4}{3}$, 1; inversely
34. $y = \frac{3}{2x}$ $\frac{3}{2}, \frac{3}{4}, \frac{1}{2}, \frac{3}{8}$; inversely
35. $y = 3x$ 3, 6, 9, 12; directly
36. $y = \frac{6}{x}$ 6, 3, 2, $\frac{3}{2}$; inversely

3 APPLY

Assignment Guide

Basic
Day 1: pp. 659–660 Exs. 12–36 even, 37, 38
Day 2: pp. 660–662 Exs. 39–46, 50, 55, 60, 64, 68, Quiz 1 Exs. 1–9

Average
Day 1: pp. 659–660 Exs. 12–36 even, 37, 38
Day 2: pp. 660–662 Exs. 39–50, 55, 60, 64, 68, Quiz 1 Exs. 1–9

Advanced
Day 1: pp. 659–660 Exs. 12–36 even, 37, 38
Day 2: pp. 660–662 Exs. 39–55, 60, 64, 68, Quiz 1 Exs. 1–9

Block Schedule
pp. 659–662 Exs. 12–36 even, 37–50, 55, 60, 64, 68, Quiz 1 Exs. 1–9

Exercise Levels

Level A: *Easier*
12–24, 55–59

Level B: *More Difficult*
25–50, 60–68

Level C: *Most Difficult*
51–54

✓ Homework Check

To quickly check student understanding of key concepts, go over the following exercises: Exs. 14, 22, 32, 34, 38, 42. See also the Daily Homework Quiz:

- Blackline Master (*Chapter 11 Resource Book,* p. 55)
- Transparency (p. 88)

! Common Error

Exercise 5 This lesson has focused on variation problems with k positive. Students should realize that k can be negative. You may want to graph an example of each so that students can see how the graphs differ.

33–36. See Additional Answers beginning on page AA1.

! COMMON ERROR

EXERCISE 41 Students may assume that every linear equation represents direct variation. Or, they may assume here that because b decreases as d increases this situation indicates inverse variation. Point out that direct or inverse variation must have the exact relationships given on p. 656.

STUDENT HELP NOTES

Homework Help Students can find help for Exs. 47–49 at **www.mcdougallittell.com.** The information can be printed out for students who don't have access to the Internet.

ENGLISH LEARNERS

EXERCISES 39–42 Some English language learners may be unfamiliar with passive voice. For Exercises 39–42, explain that *are related by* means the same thing as "you can relate them by (or with)." Consider giving other examples of passive voice (such as *Temperatures are shown in the graph*) to help students gain familiarity with its use; then invite them to suggest another way of saying the same thing (*The graph shows temperatures*).

ADDITIONAL PRACTICE AND RETEACHING

For Lesson 11.3:

- Practice Levels A, B, and C (*Chapter 11 Resource Book,* p. 43)
- Reteaching with Practice (*Chapter 11 Resource Book,* p. 46)
- See Lesson 11.3 of the *Personal Student Tutor*

For more Mixed Review:

- Search the *Test and Practice Generator* for key words or specific lessons.

FOCUS ON APPLICATIONS

REAL LIFE **SNOWSHOES** Snowshoes are made of lightweight materials that distribute a person's weight over a large area. This makes it possible to walk over deep snow without sinking.

VARIATION MODELS FROM DATA **Decide if the data in the table show *direct* or *inverse* variation. Write an equation that relates the variables.**

37.

x	1	3	5	10	0.5
y	5	15	25	50	2.5

direct variation; $y = 5x$

38.

x	1	3	4	10	0.5
y	30	10	7.5	3	60

inverse variation; $y = \frac{30}{x}$

VARIATION MODELS IN CONTEXT **State whether the variables have *direct variation, inverse variation,* or *neither.***

39. **BASE AND HEIGHT** The area B of the base and the height h of a prism with a volume of 10 cubic units are related by the equation $Bh = 10$. **inverse variation**

40. **MASS AND VOLUME** The mass m and the volume V of a substance are related by the equation $2V = m$, where 2 is the density of the substance. **direct variation**

41. **DINNER AND BREAKFAST** Alicia cut a pizza into 8 pieces. The number of pieces d that Alicia ate for dinner and the number of pieces b that she can eat for breakfast are related by the equation $b = 8 - d$. **neither**

42. **HOURS AND PAY RATE** The number of hours h that you must work to earn \$480 and your hourly rate of pay p are related by the equation $ph = 480$. **inverse variation**

SNOWSHOES **In Exercises 43 and 44, use the following information.** When a person walks, the pressure P on each boot sole varies inversely with the area A of the sole. Denise is walking through deep snow, wearing boots that have a sole area of 29 square inches each. The boot-sole pressure is 4 pounds per square inch when she stands on one foot.

43. The constant of variation is Denise's weight in pounds. What is her weight? **116 lb**

44. If Denise wears snowshoes, each with an area 11 times that of her boot soles, what is the snowshoe pressure when she stands on one foot? $\frac{4}{11}$ **lb/in.²**

OCEAN TEMPERATURES **The graph for Exercises 45 and 46 shows water temperatures for part of the Pacific Ocean. Assume temperature varies inversely with depth at depths greater than 900 meters.**

45. Find a model that relates the temperature T and the depth d. Round k to the nearest hundred. $T = \frac{4400}{d}$

46. Find the temperature at a depth of 5000 meters. Round to the nearest tenth of a degree. **0.9°C**

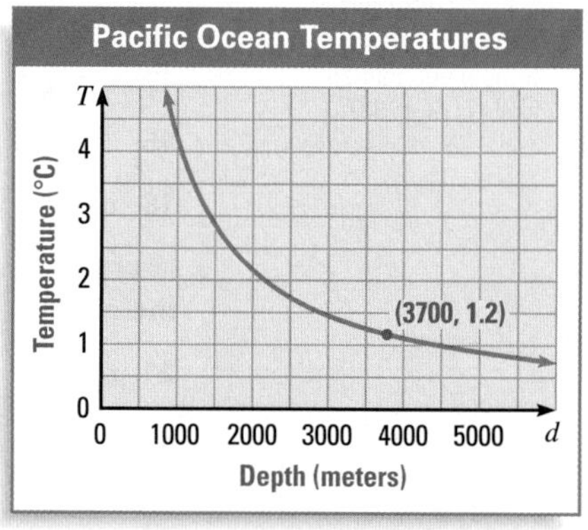

49. Neither; the equation cannot be written either in the form $g = km$ or $g = \frac{k}{m}$ for any constant k.

FUEL MILEAGE **In Exercises 47–49, you are taking a trip on the highway in a car that gets a gas mileage of about 26 miles per gallon for highway driving. You start with a full tank of 12 gallons of gasoline.** $\frac{1}{26}$ **gal/mi ≈ 0.04 gal/mi**

47. Find your rate of gas consumption (the gallons of gas used to drive 1 mile).

48. Use Exercise 47 to write an equation relating the number of gallons of gas in your tank g and the number of miles m that you have driven on your trip. $g = 12 - \frac{m}{26}$

49. Do the variables g and m vary *directly, inversely,* or *neither*? Explain. **See margin.**

STUDENT HELP

INTERNET **HOMEWORK HELP** Visit our Web site www.mcdougallittell.com for help with Exs. 47–49.

Test Preparation

50\. **MULTI-STEP PROBLEM** Use x to represent one dimension of a rectangle, and use y to represent the other dimension.

a. Make a table of possible values of x and y if the area of the rectangle is 12 square inches. Then use your table to sketch a graph. **See margin.**

b. Do x and y vary *directly*, *inversely*, or *neither*? Explain your reasoning. **Inversely; an equation of the graph is $y = \frac{12}{x}$.**

c. Make a table of possible values of x and y if the area of the rectangle is 24 square inches. Then use your table to sketch a graph in the same coordinate plane you used for your graph in part (a). **See margin.**

d. **CRITICAL THINKING** How is the area of the first rectangle related to the area of the second rectangle? For a given value of x, how is the value of y for the first rectangle related to the value of y for the second rectangle? For a given value of y, how are the values of x related?

50d. The area of the first rectangle is half the area of the second; the value of y for the first rectangle is half that for the second; the value of x for the first rectangle is half that for the second.

★ Challenge

CONSTANT RATIOS In Exercises 51 and 52, use the following information.

In a direct variation, the ratio $\frac{y}{x}$ is constant. If (x_1, y_1) and (x_2, y_2) are solutions of the equation $\frac{y}{x} = k$, then $\frac{y_1}{x_1} = k$ and $\frac{y_2}{x_2} = k$. Use the proportion $\frac{y_1}{x_1} = \frac{y_2}{x_2}$ to find the missing value.

51\. Find x_2 when $x_1 = 2, y_1 = 3$, and $y_2 = 6$. **4**

52\. Find y_2 when $x_1 = -4, y_1 = 8$, and $x_2 = -1$. **2**

CONSTANT PRODUCTS In an inverse variation, the product xy is constant. If (x_1, y_1) and (x_2, y_2) are solutions of $xy = k$, then $x_1y_1 = x_2y_2$. Use this equation to find the missing value.

53\. Find y_2 when $x_1 = 4, y_1 = 5$, and $x_2 = 8$. **$2\frac{1}{2}$**

54\. Find x_2 when $x_1 = 9, y_1 = -3$, and $y_2 = 12$. **$-2\frac{1}{4}$**

EXTRA CHALLENGE
www.mcdougallittell.com

MIXED REVIEW

SIMPLIFYING FRACTIONS Simplify the fraction. (Skills Review, p. 781–783)

55\. $\frac{36}{48}$ **$\frac{3}{4}$** 56. $\frac{27}{108}$ **$\frac{1}{4}$** 57. $\frac{96}{180}$ **$\frac{8}{15}$** 58. $\frac{-15}{125}$ **$-\frac{3}{25}$**

PROBABILITY Find the probability. (Review 2.8 for 11.4)

59\. You roll a die. What is the probability that you will roll a four? **$\frac{1}{6}$**

60\. You roll a die. What is the probability that you will roll an odd number? **$\frac{1}{2}$**

SIMPLIFYING FRACTIONS Simplify the fraction. (Review 8.3 for 11.4)

61\. $\frac{y^4 \cdot y^7}{y^5}$ **y^6** 62. $\frac{5xy}{5x^2}$ **$\frac{y}{x}$** 63. $\frac{-3xy^3}{3x^3y}$ **$-\frac{y^2}{x^2}$** 64. $\frac{56x^2y^5}{64x^2y}$ **$\frac{7y^4}{8}$**

CLASSIFYING POLYNOMIALS Identify the leading coefficient, and classify the polynomial by degree and by number of terms. (Review 10.1)

65\. $-5x - 4$ **−5, linear, binomial**
66\. $8x^4 + 625$ **8, quartic, binomial**
67\. $x - x^3$ **−1, cubic, binomial**
68\. $16 - 4x + 3x^2$ **3, quadratic, trinomial**

4 ASSESS

DAILY HOMEWORK QUIZ

Transparency Available

Using the given values of x and y and the type of variation listed, write an equation that relates x and y.

1\. $x = 24, y = 78$; direct **$y = \frac{13}{4}x$**

2\. $x = 4, y = 6$; inverse **$y = \frac{24}{x}$**

3\. Given that $y = -\frac{1}{2}x$, make a table of values for $x = 1, 2, 3,$ and 4. Use the table to sketch a graph. Decide whether x and y vary *directly* or *inversely*.
$-\frac{1}{2}, -1, -\frac{3}{2}, -2$; directly

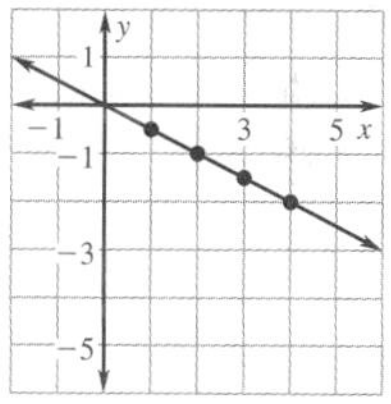

4\. The velocity v of a falling object and the time it has fallen are related by the equation $32t = v$. Do these variables show direct variation, inverse variation, or neither? **direct**

EXTRA CHALLENGE NOTE
Challenge problems for Lesson 11.3 are available in **blackline** format in the *Chapter 11 Resource Book*, p. 51 and at **www.mcdougallittell.com.**

ADDITIONAL TEST PREPARATION

1\. **OPEN ENDED** Make a table of values that can be modeled by direct variation and a table of values that can be modeled by inverse variation. Write the equation for each table.
Check tables and equations.

50a and 50c. See the next page.

ADDITIONAL RESOURCES
An alternative Quiz for Lessons 11.1–11.3 is available in the *Chapter 11 Resource Book,* p. 52.
A **blackline** master with additional Math & History exercises is available in the *Chapter 11 Resource Book,* p. 50.

50a.
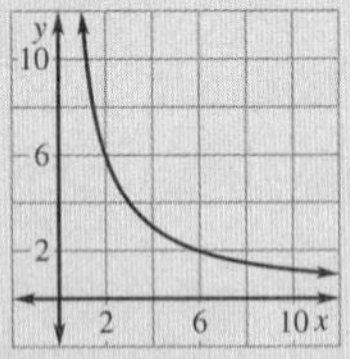

c.
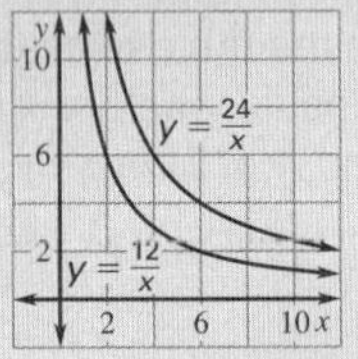

QUIZ 1

Self-Test for Lessons 11.1–11.3

Solve the proportion. (Lesson 11.1)

1. $\frac{x}{10} = \frac{4}{5}$ **8**

2. $\frac{3}{x} = \frac{7}{9}$ $\frac{27}{7}$

3. $\frac{x}{4x - 8} = \frac{2}{x}$ **4**

4. $\frac{6x + 4}{5} = \frac{2}{x}$ $-\frac{5}{3}$, **1**

The variables *x* and *y* vary inversely. Use the given values to write an equation that relates *x* and *y*. (Lesson 11.3)

5. $x = 3, y = 4$ $y = \frac{12}{x}$

6. $x = 12, y = 1$ $y = \frac{12}{x}$

7. $x = 24, y = \frac{3}{4}$ $y = \frac{18}{x}$

CAR WASH In Exercises 8 and 9, use the survey results. (Lesson 11.2)

8. In the survey, 413 people said they hand-wash their cars at home. How many people were surveyed? Round to the nearest whole number.
about 904 people

9. Use the result of Exercise 8 to find the number of people in the survey who wash their cars at an automatic car wash. **about 204 people**

Method	Percent of people
Hand-wash at home	45.7
Automatic car wash	22.6
Manually wash it myself at a car wash	16.3
Others manually wash it at a car wash	14.8
Other	0.7

▶ Source: Maritz Marketing Research

MATH & History — History of Polling

APPLICATION LINK
www.mcdougallittell.com

THEN **IN THE 1930S,** opinion polls were conducted by interviewers knocking on people's doors. Later telephone polling became popular and a sampling method using random digit-dialing was developed.

NOW **DURING THE 1990S,** television stations used computers to evaluate exit poll data and present real-time election results.

After the polls closed in one of the 1998 governor's races, exit poll data reported that the top three candidates had 37%, 34%, and 28% of the vote.

1. A total of 2,075,280 votes had been counted so far. If the exit polls accurately described the vote count, about how many votes did each candidate have?
about 767,854 votes, 705,595 votes, and 581,078 votes
2. Use your results from Exercise 1. The final vote count included 16,486 more votes. If these votes were all for one candidate instead of being distributed as exit polls predicted, would the outcome of the election have been changed? **no**

1948
Opinion polls incorrectly predict Dewey's win.

1970s on

Telephone polling becomes common

Now
Internet sites enable voters to learn early results

Modeling Inverse Variation

You can use a graphing calculator or a computer to decide whether quantities vary directly, inversely, or neither.

▶ EXAMPLE

During a chemistry experiment, the volume of a fixed mass of air was decreased and the pressure at different volumes was recorded. The data are at the left, where x is the volume (in cubic centimeters) and y is the pressure (in atmospheres). Use a graphing calculator to determine if the data vary directly, inversely, or neither. Then make a scatter plot to check your model.

x	y
20	1.06771
19	1.11276
18	1.17583
17	1.24341
16	1.32450
15	1.40559
14	1.51371
13	1.62183
12	1.74347
11	1.90566

▶ SOLUTION

1 Let L_1 represent the volume x and L_2 represent the pressure y. Enter the ordered pairs into the graphing calculator. Then create lists L_3 and L_4 as shown to find whether the data vary directly or inversely.

```
L2/L1→L3
(.0533855 .0585...
L2*L1→L4
(21.3542 21.142...
```

2 Notice that the values in L_3 are all different while the values in L_4 are all about 21.1. This means that x and y vary inversely.

```
L2     |L3     |L4
1.0677 |.05339 |21.354
1.1128 |.05857 |21.143
1.1758 |.06532 |21.164
1.2434 |.07314 |21.138
1.3245 |.08278 |21.192
L4=(21.3542,21.  ...
```

3 To find the constant of variation k, find the mean of the values in L_4. Use this value to write an inverse variation model in the form $y = \frac{k}{x}$.

```
mean(L4)
          21.123503
```

4 Set the viewing rectangle so that $11 \le x \le 20$ and $1 \le y \le 2$. Make a scatter plot of L_1 and L_2. Then graph $y = \frac{21.12}{x}$ on the same screen.

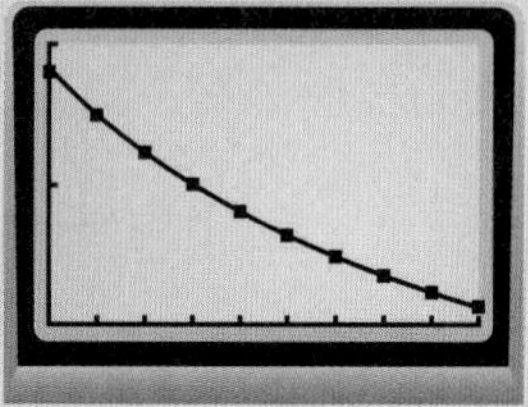

▶ EXERCISES

Use a graphing calculator to decide if the data vary directly or inversely and to find the constant of variation. Then write a model for the data.
1–2. See margin.

1. (10, 8.25), (9, 7.425), (8, 6.6), (7, 5.775), (6, 4.95), (5, 4.125), (4, 3.3)
2. (18, 1.389), (17, 1.471), (16, 1.563), (15, 1.667), (14, 1.786), (13, 1.923)

1. directly; 0.825; $y = 0.825x$
2. inversely; 25; $y = \frac{25}{x}$

STUDENT HELP

INTERNET KEYSTROKE HELP

See keystrokes for several models of calculators at www.mcdougallittell.com

1 Planning the Activity

PURPOSE
To use a calculator to determine if data vary directly or inversely.

MATERIALS
- graphing calculator for each student or pair of students
- Keystroke blackline (*Chapter 11 Resource Book*, p. 40)

PACING
- Example — 15 min
- Exercises — 10 min

▶ *LINK TO LESSON*
Steps 1 and 2 test to see if $\frac{y}{x}$ or xy are constant, which would indicate direct or inverse variation. The graph in Step 4 *appears* to intersect the y-axis, unlike that in Example 2, Lesson 11.3, because the viewing window does not include the y-axis, but starts at $x = 11$.

2 Managing the Activity

COOPERATIVE LEARNING
Working in pairs, one student handles the calculator while the other reads the data and directions. Students then switch roles.

ALTERNATIVE APPROACH
This activity can be done as a class using a graphing calculator display.

3 Closing the Activity

★ ***KEY DISCOVERY***
Tables of data pairs and their products and quotients can be used to identify direct and inverse variation relationships using a graphing calculator.

ACTIVITY ASSESSMENT
Explain how to find an inverse variation model for a set of real-life data. **See sample answer at left.**

Activity Assessment *Sample answer:*
Find the product xy for each data pair. If the products are nearly constant, an inverse variation model is appropriate. Find the mean of these products to estimate k. The model is $xy = k$.

1 PLAN

PACING
Basic: 1 day
Average: 1 day
Advanced: 1 day
Block Schedule: 0.5 block with 11.5

LESSON OPENER ACTIVITY
An alternative way to approach Lesson 11.4 is to use the Activity Lesson Opener:
- Blackline Master (*Chapter 11 Resource Book,* p. 56)
- Transparency (p. 77)

MEETING INDIVIDUAL NEEDS
- ***Chapter 11 Resource Book***
  Prerequisite Skills Review (p. 5)
  Practice Level A (p. 57)
  Practice Level B (p. 58)
  Practice Level C (p. 59)
  Reteaching with Practice (p. 60)
  Absent Student Catch-Up (p. 62)
  Challenge (p. 64)
- ***Resources in Spanish***
- ***Personal Student Tutor***

NEW-TEACHER SUPPORT
See the Tips for New Teachers on pp. 1–2 of the *Chapter 11 Resource Book* for additional notes about Lesson 11.4.

WARM-UP EXERCISES

Transparency Available

Simplify the fraction.

1. $\frac{3x^2}{6xy}$ $\frac{x}{2y}$ **2.** $\frac{-ab}{ab^2}$ $-\frac{1}{b}$

Solve.

3. $x^2 - 16 = 0$ $-4, 4$

4. $x^2 - x - 12 = 0$ $4, -3$

11.4 Simplifying Rational Expressions

What you should learn

GOAL 1 Simplify a rational expression.

GOAL 2 Use rational expressions to find geometric probability.

Why you should learn it

▼ To model **real-life** situations, such as finding the probability of a meteor strike in **Exs. 36–38**.

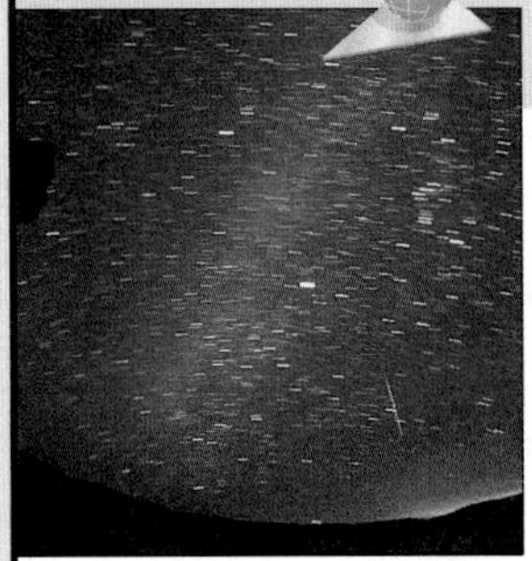

GOAL 1 SIMPLIFYING A RATIONAL EXPRESSION

A **rational number** is a number that can be written as the quotient of two integers, such as $\frac{1}{2}$, $\frac{4}{3}$, and $\frac{7}{1}$. A fraction whose numerator, denominator, or both numerator and denominator are nonzero polynomials is a **rational expression**. Here are some examples.

$$\frac{3}{x+4} \qquad \frac{2x}{x^2-9} \qquad \frac{3x+1}{x^2+1}$$

A rational expression is *undefined* when the denominator is equal to zero. For instance, in the first expression x can be any real number except -4.

To simplify a fraction, you factor the numerator and the denominator and then divide out any common factors. A rational expression is **simplified** if its numerator and denominator have no factors in common (other than ± 1).

SIMPLIFYING FRACTIONS

Let a, b, and c be nonzero numbers.

$$\frac{ac}{bc} = \frac{a \cdot \cancel{c}}{b \cdot \cancel{c}} = \frac{a}{b} \qquad \textbf{Example: } \frac{28}{35} = \frac{4 \cdot \cancel{7}}{5 \cdot \cancel{7}} = \frac{4}{5}$$

EXAMPLE 1 *When and When Not to Divide Out*

Simplify the expression.

a. $\frac{2x}{2(x+5)}$ **b.** $\frac{x(x^2+6)}{x^2}$ **c.** $\frac{x+4}{x}$

SOLUTION

a. $\frac{2x}{2(x+5)} = \frac{\cancel{2} \cdot x}{\cancel{2}(x+5)}$ You can divide out the common factor 2.

$= \frac{x}{x+5}$ Simplified form

b. $\frac{x(x^2+6)}{x^2} = \frac{\cancel{x}(x^2+6)}{\cancel{x} \cdot x}$ You can divide out the common factor x.

$= \frac{x^2+6}{x}$ Simplified form

c. $\frac{x+4}{x}$ You cannot divide out the common term x.

STUDENT HELP

Study Tip
When you simplify rational expressions, you can divide out only factors, not terms.

STUDENT HELP

HOMEWORK HELP Visit our Web site www.mcdougallittell.com for extra examples.

EXAMPLE 2 *Factoring Numerator and Denominator*

Simplify $\frac{2x^2 - 6x}{6x^2}$.

SOLUTION

$$\frac{2x^2 - 6x}{6x^2} = \frac{2x(x-3)}{2 \cdot 3 \cdot x \cdot x} \qquad \text{Factor numerator and denominator.}$$

$$= \frac{\cancel{2x}(x-3)}{\cancel{2x}(3x)} \qquad \text{Divide out common factor } 2x.$$

$$= \frac{x-3}{3x} \qquad \text{Simplified form}$$

EXAMPLE 3 *Recognizing Opposite Factors*

Simplify $\frac{4 - x^2}{x^2 - x - 2}$.

SOLUTION

$$\frac{4 - x^2}{x^2 - x - 2} = \frac{(2-x)(2+x)}{(x-2)(x+1)} \qquad \text{Factor numerator and denominator.}$$

$$= \frac{-(x-2)(2+x)}{(x-2)(x+1)} \qquad \text{Factor } -1 \text{ from } (2 - x).$$

$$= \frac{-\cancel{(x-2)}(x+2)}{\cancel{(x-2)}(x+1)} \qquad \text{Divide out common factor } x - 2.$$

$$= -\frac{x+2}{x+1} \qquad \text{Simplified form}$$

EXAMPLE 4 *Recognizing when an Expression is Undefined*

In Examples 2 and 3, are the original expression and the simplified expression defined for the same values of the variable?

STUDENT HELP

Look Back
For help with using the zero-product property, see p. 597.

SOLUTION A rational expression is undefined when the denominator is 0. Think of setting the denominator equal to 0 and then solving that equation.

Example 2: Yes, $6x^2 = 0$ and $3x = 0$ have the same solution: $x = 0$. Both expressions are undefined when $x = 0$.

Example 3: No, $x^2 - x - 2 = 0$ and $x + 1 = 0$ do not have the same solutions. To see this set the denominator of the original expression equal to zero.

$$x^2 - x - 2 = 0 \qquad \text{Set denominator equal to 0.}$$

$$(x-2)(x+1) = 0 \qquad \text{Factor denominator.}$$

$$x - 2 = 0 \quad \text{or} \quad x + 1 = 0 \qquad \text{Use zero-product property.}$$

$$x = 2 \qquad x = -1 \qquad \text{Solve for } x.$$

▶ The original expression is undefined when $x = 2$ and $x = -1$. The simplified expression is undefined only when $x = -1$.

2 TEACH

EXTRA EXAMPLE 1
Simplify the expression.

a. $\frac{5(3-x)}{5x}$ $\frac{3-x}{x}$

b. $\frac{2x^2}{x(x+5)}$ $\frac{2x}{x+5}$

c. $\frac{5x+3}{x+3}$ $\frac{5x+3}{x+3}$

EXTRA EXAMPLE 2
Simplify $\frac{15x}{5-10x}$. $\frac{3x}{1-2x}$

EXTRA EXAMPLE 3
Simplify $\frac{x^2+6x+9}{9-x^2}$. $\frac{x+3}{3-x}$

EXTRA EXAMPLE 4
In Extra Examples 2 and 3, are the original expression and the simplified expression defined for the same values of the variable?
Example 2: yes; Example 3: no

CHECKPOINT EXERCISES
For use after Examples 1–4:

1. Simplify $\frac{x^2+4x+3}{x^2+x}$. Are the original and simplified expressions defined for the same values of the variable? $\frac{x+3}{x}$; no

STUDENT HELP NOTES

Homework Help Students can find extra examples at **www.mcdougallittell.com** that parallel the examples in the student edition.

Look Back As students look back to p. 597, remind them that the expression must be factored before using the zero product property.

Extra Example 5

A coin is tossed onto the large square region shown below. It is equally likely to land on any point in the region.

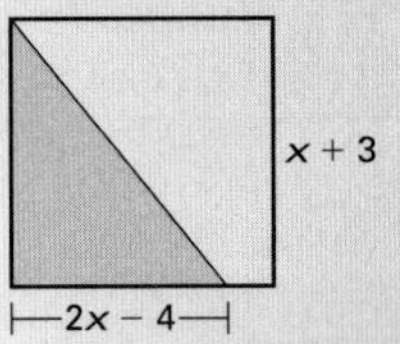

a. Write a model that gives the probability that the coin lands in the triangle. $\frac{x-2}{x+3}$

b. Evaluate the model when $x = 5$. $\frac{3}{8}$

Checkpoint Exercises

For use after Example 5:

1. Write a model for the probability that the coin in Extra Example 5 lands in the square but outside the triangle. (Hint: First subtract the area of the triangle from the area of the square.) Evaluate the model when $x = 5$. $\frac{5}{x+3}$; $\frac{5}{8}$

Focus on Vocabulary

What is *geometric probability*?
Sample answer: **Geometric probability is the ratio of the area of the target region to the area of the possible region.**

Closure Question

Explain the mistake in simplifying $\frac{3x+5}{5}$ as $3x + 1$.
Sample answer: **You can divide out only factors, not terms.**

Daily Puzzler

For how many of the first 100 whole numbers is the rational expression $\frac{x^2 - x - 42}{x(x-1)(x-2)\ldots(x-99)}$ defined? For how many of the first 100 whole numbers is its simplified form defined? **0; 1**

666

GOAL 2 Applying Rational Expressions

Rational expressions can be useful in modeling situations such as finding averages, ratios, and probabilities. In Chapter 2 you learned to find theoretical probability by comparing the number of favorable outcomes to the total number of outcomes. In some cases, you can give theoretical probabilities a geometric interpretation in terms of areas.

GEOMETRIC PROBABILITY

Region B is contained in Region A. An object is tossed onto Region A and is equally likely to land on any point in the region.

The **geometric probability** that it lands in Region B is

$$P = \frac{\text{Area of Region } B}{\text{Area of Region } A}.$$

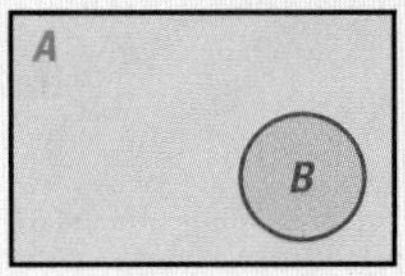

EXAMPLE 5 *Writing and Using a Rational Model*

A coin is tossed onto the large rectangular region shown at the right. It is equally likely to land on any point in the region.

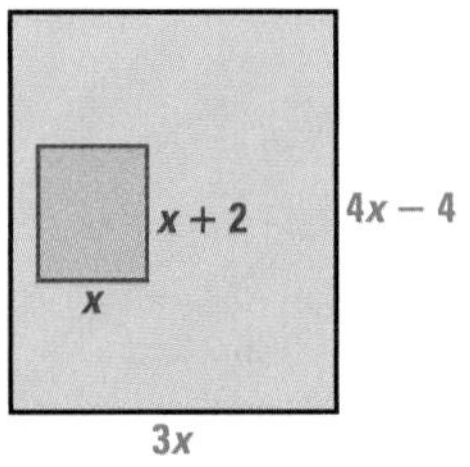

a. Write a model that gives the probability that the coin will land in the small rectangle.

b. Evaluate the model when $x = 10$.

SOLUTION

a. $P = \frac{\text{Area of small rectangle}}{\text{Area of large rectangle}}$ — **Formula for geometric probability**

$= \frac{x(x+2)}{3x(4x-4)}$ — **Find areas.**

$= \frac{x \cdot (x+2)}{3x \cdot 4(x-1)}$ — **Divide out common factors.**

$= \frac{x+2}{12(x-1)}$ — **Simplified form**

b. To find the probability when $x = 10$, substitute 10 for x in the model.

$$P = \frac{x+2}{12(x-1)} = \frac{10+2}{12(10-1)} = \frac{12}{108} = \frac{1}{9}$$

▶ The probability of landing in the small rectangle is $\frac{1}{9}$.

GUIDED PRACTICE

Vocabulary Check ✓

Concept Check ✓

Skill Check ✓

1. Define a rational expression. Then give an example of a rational expression. ***Sample answer:*** **a fraction whose numerator and denominator are nonzero polynomials; examples:** $\frac{x}{x-7}, \frac{2x+1}{x^2+x-3}$

ERROR ANALYSIS Describe the error.

2. 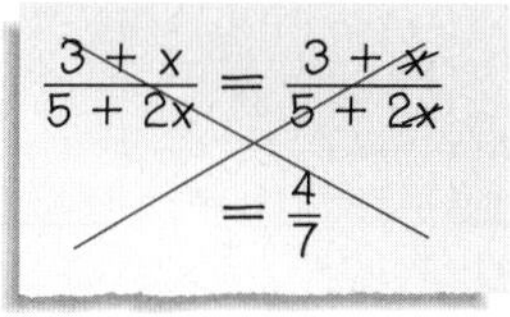

3. 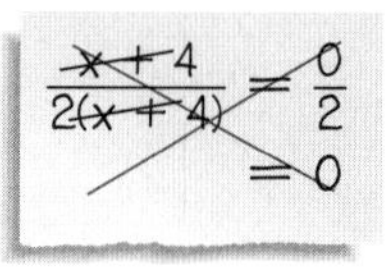

See margin.

For what values of the variable is the rational expression undefined?

4. $\frac{6}{8x}$ **0**

5. $\frac{x-1}{x-5}$ **5**

6. $\frac{2}{x^2-x-2}$ **−1, 2**

7. Which of the following is the simplified form of $\frac{6+2x}{x^2+5x+6}$? **C**

   A. $\frac{2x}{x^2+5x}$ **B.** $\frac{2}{x+5}$ **C.** $\frac{2}{x+2}$

8. Which ratio represents the ratio of the area of the smaller rectangle to the area of the larger rectangle? **B**

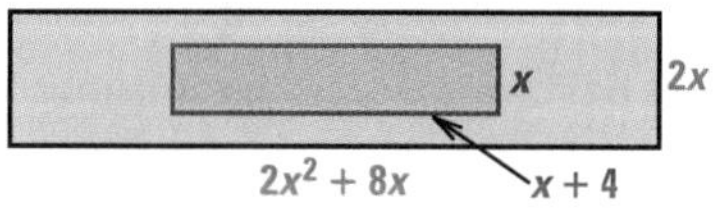

   A. $\frac{1}{2x}$ **B.** $\frac{1}{4x}$

2. *Sample answer:* There are no factors in the numerator and denominator that can be canceled. The rational expression is in simplest form.

3. *Sample answer:* The numerator is equal to $1(x+4)$, so when $x+4$ is canceled in the numerator and denominator, the resulting numerator is 1, not 0; $\frac{x+4}{2(x+4)} = \frac{1}{2}$.

PRACTICE AND APPLICATIONS

STUDENT HELP

Extra Practice to help you master skills is on p. 807.

SIMPLIFYING EXPRESSIONS Simplify the expression if possible.

9. $\frac{4x}{20}$ $\frac{x}{5}$

10. $\frac{15x}{45}$ $\frac{x}{3}$

11. $\frac{-18x^2}{12x}$ $-\frac{3x}{2}$

12. $\frac{14x^2}{50x^4}$ $\frac{7}{25x^2}$

13. $\frac{3x^2-18x}{-9x^2}$ $\frac{6-x}{3x}$

14. $\frac{42x-6x^3}{36x}$ $\frac{7-x^2}{6}$

15. $\frac{7x}{12x+x^2}$ $\frac{7}{x+12}$

16. $\frac{x+2x^2}{x+2}$ **See margin.**

17. $\frac{12-5x}{10x^2-24x}$ $-\frac{1}{2x}$

18. $\frac{x^2+25}{2x+10}$

19. $\frac{5-x}{x^2-8x+15}$ $\frac{1}{3-x}$

20. $\frac{2x^2+11x-6}{x+6}$ **2x − 1**

21. $\frac{x^2+x-20}{x^2+2x-15}$ $\frac{x-4}{x-3}$

22. $\frac{x^3+9x^2+14x}{x^2-4}$ $\frac{x(x+7)}{x-2}$

23. $\frac{x^3-x}{x^3+5x^2-6x}$ $\frac{x+1}{x+6}$

16. cannot be simplified

18. cannot be simplified

UNDEFINED VALUES For what values of the variable is the rational expression undefined?

24. $\frac{7}{x-3}$ **3**

25. $\frac{11}{x-8}$ **8**

26. $\frac{4}{x^2-1}$ **−1, 1**

27. $\frac{x+3}{x^2-9}$ **−3, 3**

28. $\frac{x+9}{x^2+x-12}$ **−4, 3**

29. $\frac{x-3}{x^2+5x-6}$ **−6, 1**

STUDENT HELP

HOMEWORK HELP
Example 1: Exs. 9–23
Example 2: Exs. 9–23
Example 3: Exs. 9–23
Example 4: Exs. 24–29
Example 5: Exs. 30–33

3 APPLY

ASSIGNMENT GUIDE

BASIC
Day 1: pp. 667–669 Exs. 13–33, 39, 46–50, 54, 56

AVERAGE
Day 1: pp. 667–669 Exs. 13–33, 36–39, 46–50, 54, 56

ADVANCED
Day 1: pp. 667–669 Exs. 13–33, 36–41, 46–50, 54, 56

BLOCK SCHEDULE WITH 11.5
pp. 667–669 Exs. 13–33, 36–39, 46–50, 54, 56

EXERCISE LEVELS

Level A: ***Easier***
9–12, 42–45

Level B: ***More Difficult***
13–33, 36–39, 46–57

Level C: ***Most Difficult***
34, 35, 40, 41

✓ HOMEWORK CHECK

To quickly check student understanding of key concepts, go over the following exercises: Exs. 14, 18, 22, 27, 28, 30, 32. See also the Daily Homework Quiz:

- Blackline Master (*Chapter 11 Resource Book*, p. 67)
- Transparency (p. 89)

! COMMON ERROR

EXERCISES 9–23 To help students catch computational and factoring errors, have them check that the original and simplified expressions are equal for $x = 1$ or some other value for which they are defined. This will not guarantee that the simplified version is correct, but it will catch many errors.

APPLICATION NOTE

EXERCISES 36–38 Encourage students to leave their expressions for the area of the Torrid Zone and the total surface area of the earth in terms of π. They will see that π cancels out in Ex. 38.

ADDITIONAL PRACTICE AND RETEACHING

For Lesson 11.4:

- Practice Levels A, B, and C (*Chapter 11 Resource Book,* p. 57)
- Reteaching with Practice (*Chapter 11 Resource Book,* p. 60)
- See Lesson 11.4 of the *Personal Student Tutor*

For more Mixed Review:

- Search the *Test and Practice Generator* for key words or specific lessons.

GEOMETRIC PROBABILITY A coin is tossed onto the large rectangular region shown. It is equally likely to land on any point in the region. Write a model that gives the probability that the coin will land in the red region. Then evaluate the model when $x = 3$.

30.

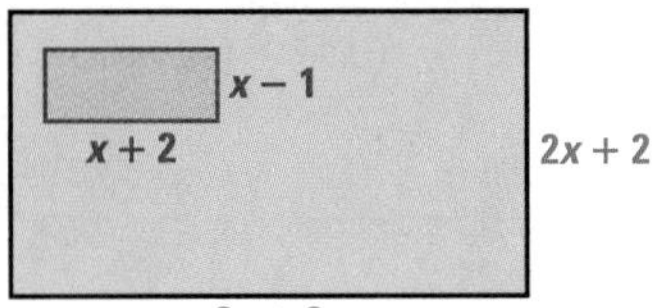

$\frac{x-1}{6(x+1)}; \frac{1}{12}$

31.

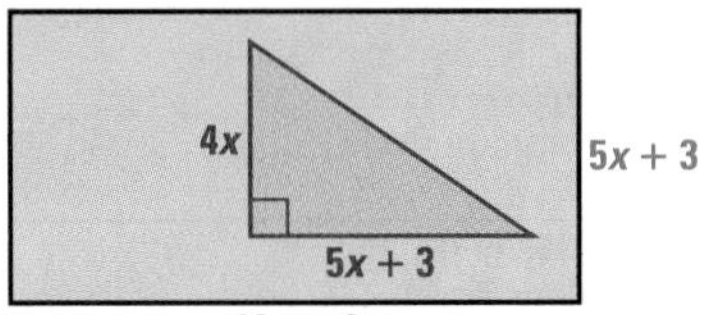

$\frac{x}{5x+3}; \frac{1}{6}$

CARNIVAL GAMES In Exercises 32–34, use the following information.

You are designing a game for a school carnival. Players will drop a coin into a basin of water, trying to hit a target on the bottom. The water is kept moving randomly, so the coin is equally likely to land anywhere. You use a rectangular basin twice as long as it is wide. You place the blue rectangular target an equal distance from each end.

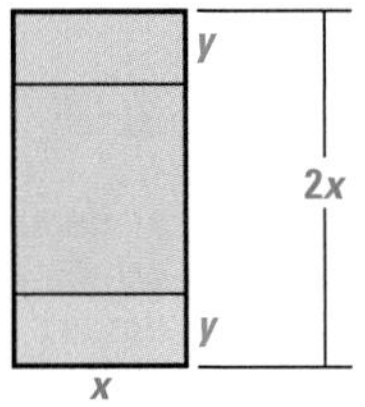

32. Express the two dimensions of the target in terms of the variables x and y. **x and $2(x - y)$**

33. Write a model that gives the probability that the coin will land on the target. $\frac{x-y}{x}$

34. CRITICAL THINKING You want players to win about half the time. Give a set of values you could use for x and y if the basin's area is between 72 and 120 square inches. **See margin.**

35. *Writing* Create three problems of the form $\frac{ax^2 + bx + c}{dx^2 + ex + f}$ in which the numerator and the denominator have a common factor. Describe the process you used to create your problems.

34. Correct values will have $6 \le x \le 2\sqrt{15}$ (≈ 7.75), and choose y so that $y = \frac{1}{2}x$.

35. *Sample answer:* To create such problems, write the numerator and denominator in factored form, then multiply. For example, $\frac{(x+1)(x+2)}{(x+1)(x-2)} = \frac{x^2+3x+2}{x^2-x-2}$.

METEOR STRIKES In Exercises 36–38, use this information. A meteorite is equally likely to hit anywhere on Earth. The probability that a meteorite lands in the Torrid Zone is $\frac{\text{Area of Torrid Zone}}{\text{Total surface area of Earth}}$.

Let R represent Earth's radius.

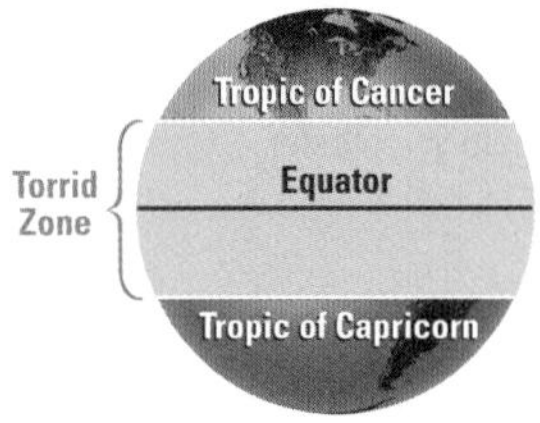

36. Write an expression to estimate the area of the Torrid Zone. You can think of the distance between the tropics (about 3250 miles) as the height of a cylindrical belt around Earth at the equator. The length of the belt is Earth's circumference $2\pi R$. **$6500\pi R$**

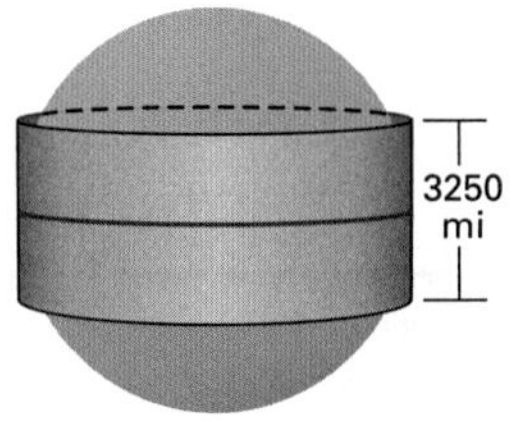

37. The surface area of a sphere with radius R is $4\pi R^2$. Write and simplify an expression for the probability that a meteorite lands in the Torrid Zone. $\frac{1625}{R}$

38. Find the probability in Exercise 37. Use 3963 miles for Earth's radius. **about 0.41**

FOCUS ON APPLICATIONS

REAL LIFE **IMPACT CRATER** at Wolfe Creek, Australia. When a rocky object moving in space enters Earth's atmosphere, it starts to burn causing the flash of light called a *meteor.* Most such objects burn up completely, but about 500 strike Earth each year as *meteorites.*

Test Preparation

39. **MULTI-STEP PROBLEM** In this exercise you will look for a pattern.

a. Copy and complete the table by evaluating the expression for each x-value. **See margin.**

x-values	−2	−1	0	1	2	3	4
$\frac{x^2 - x - 6}{x - 3}$	?	?	?	?	?	?	?
$x + 2$	?	?	?	?	?	?	?

b. Describe the relationship between $\frac{x^2 - x - 6}{x - 3}$ and $x + 2$. What does the table tell you about this relationship?

c. **CRITICAL THINKING** Write another pair of expressions that have this relationship. Describe the procedure you used to find your example. **See margin.**

b. The second expression is the simplified form of the first; the two expressions have the same value unless $x = 3$, in which case the first expression is not defined.

★ Challenge

40. **RATIOS** Write the ratio in simplest form comparing the area of the smaller rectangle to the area of the larger rectangle. $\frac{x+2}{5(x+5)}$

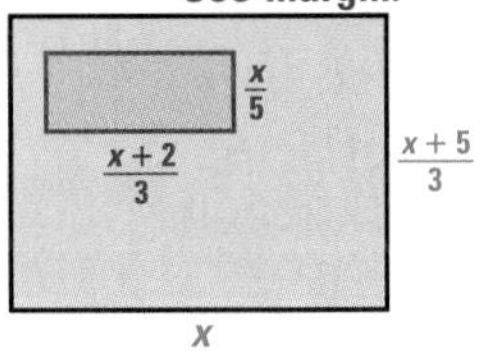

Ex. 40

41. **PROPORTIONS** Solve the proportion.

$$\frac{x^2 + 5x + 6}{x^2 - 2x - 8} = \frac{x^2 - 4x - 5}{x^2 - 8x + 15} \quad 1\frac{2}{3}$$

EXTRA CHALLENGE
www.mcdougallittell.com

MIXED REVIEW

PRODUCTS AND QUOTIENTS Simplify. (Review 2.5 and 2.7 for 11.5)

42. $\left(-\frac{1}{2}\right)\left(\frac{2}{3}\right)$ $-\frac{1}{3}$

43. $(-15)\left(-\frac{5}{6}\right)$ $\frac{25}{2}$

44. $\frac{2}{7} \div \frac{14}{24}$ $\frac{24}{49}$

45. $\frac{4}{9} \div (-36)$ $-\frac{1}{81}$

46. $\left(-\frac{3}{4}\right)\left(\frac{3y}{-5}\right)$ $\frac{9y}{20}$

47. $\frac{2m}{3} \cdot 6m^2$ $4m^3$

48. $\frac{36}{45a} \div \frac{-9a}{5}$ **See margin.**

49. $-18c^3 \div \frac{-27c}{-4}$ **See margin.**

50. **GEOMETRY** The area of the triangle is 192 square meters. What is the value of x? What is the perimeter? (Review 9.1) **4 m; 64 m**

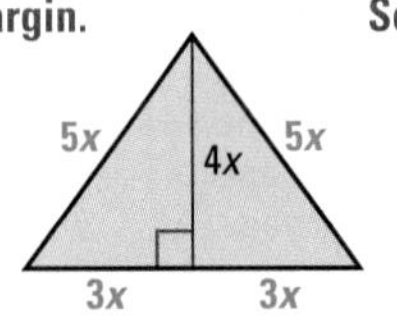

c. *Sample answer:* Any pair of expressions of the form $\frac{(ax+b)(cx+d)}{ax+b}$ and $cx + d$; first I wrote an expression that had a common term in the numerator and denominator, then let the other expression be the other factor of the numerator of the first expression. The two expressions will be equal except when $x = -\frac{b}{a}$.

48. $-\frac{4}{9a^2}$

49. $-\frac{8c^2}{3}$

SKETCHING GRAPHS Sketch the graph of the function. (Review 9.3) **51–56. See margin.**

51. $y = x^2$

52. $y = 4 - x^2$

53. $y = \frac{1}{2}x^2$

54. $y = 5x^2 + 4x - 5$

55. $y = 4x^2 - x + 6$

56. $y = -3x^2 - x + 7$

57. **POPULATION DENSITY** Population density is the number of people per square mile. Suppose the population density of a city decreases by 8% for every mile you travel from the center of the city. In the center of the city, the population density is 2500 people per square mile. Find an exponential decay model for the population density. Copy and complete the table. (Review 8.6) $y = 2500(0.92)^x$; **2116; 1947; 1791; 1648; 1516**

Distance from center of city (miles)	2	3	4	5	6
Population density (people per square mile)	?	?	?	?	?

Additional Test Preparation *Sample answer:*

1. The simplified form of $\frac{x^2+6x+8}{x^2+x-12}$ is $\frac{x+2}{x-3}$. The domain of the original expression is the real numbers excluding −4 and 3. In the simplified form, only 3 is excluded.

4 ASSESS

DAILY HOMEWORK QUIZ

Transparency Available

Simplify the expression if possible.

1. $\frac{9x - 6x^2}{-15x}$ $\frac{2x-3}{5}$

2. $\frac{x^2 - 16}{3x - 12}$ $\frac{x+4}{3}$

3. $\frac{x^3 + 2x^2 - 15x}{x^2 - 5x + 6}$ $\frac{x(x+5)}{x-2}$

For what values of the variable is the rational expression undefined?

4. $\frac{x^2 + 25}{x^2 - 25}$ −5, 5

5. $\frac{7 - 3x}{x^2 + 16x + 63}$ −9, −7

6. A seed is equally likely to fall anywhere on a rectangular patch of garden. Write a model that gives the probability that the seed lands in the shaded region. Evaluate the model for $x = 4$.

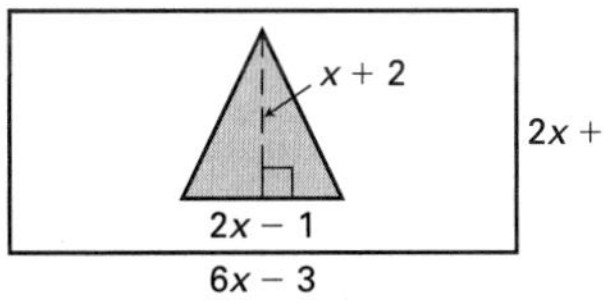

$\frac{x+2}{6(2x+1)}$; $\frac{1}{9}$

EXTRA CHALLENGE NOTE

Challenge problems for Lesson 11.4 are available in **blackline** format in the *Chapter 11 Resource Book*, p. 64 and at **www.mcdougallittell.com.**

ADDITIONAL TEST PREPARATION

1. **OPEN ENDED** Give a rational expression whose simplified form has a different domain than that of the original expression, and tell how the domains differ. **See sample answer at left.**

39a. second row: 0, 1, 2, 3, 4, undef., 6; third row: 0, 1, 2, 3, 4, 5, 6

51–56. See Additional Answers beginning on page AA1.

1 PLAN

PACING
Basic: 1 day
Average: 1 day
Advanced: 1 day
Block Schedule: 0.5 block with 11.4

LESSON OPENER ACTIVITY
An alternative way to approach Lesson 11.5 is to use the Activity Lesson Opener:
- Blackline Master (*Chapter 11 Resource Book,* p. 68)
- Transparency (p. 78)

MEETING INDIVIDUAL NEEDS
- ***Chapter 11 Resource Book***
  Prerequisite Skills Review (p. 5)
  Practice Level A (p. 69)
  Practice Level B (p. 70)
  Practice Level C (p. 71)
  Reteaching with Practice (p. 72)
  Absent Student Catch-Up (p. 74)
  Challenge (p. 77)
- ***Resources in Spanish***
- ***Personal Student Tutor***

NEW-TEACHER SUPPORT
See the Tips for New Teachers on pp. 1–2 of the *Chapter 11 Resource Book* for additional notes about Lesson 11.5.

WARM-UP EXERCISES
Transparency Available
Multiply each expression.
1. $5x^2 \cdot 4x$ $20x^3$
2. $7x(x-1)$ $7x^2-7x$
Factor each expression.
3. $x^2+4x-21$ $(x+7)(x-3)$
4. $5x^2-20x$ $5x(x-4)$

11.5 Multiplying and Dividing Rational Expressions

What you should learn

GOAL 1 Multiply and divide rational expressions.

GOAL 2 Use rational expressions as **real-life** models, as when comparing parts of the service industry to the total in **Exs. 38–41**.

Why you should learn it

▼ To model **real-life** situations, such as describing the average car sales per dealership in **Example 6**.

GOAL 1 FINDING PRODUCTS AND QUOTIENTS

Because the variables in a rational expression represent real numbers, the rules for multiplying and dividing rational expressions are the same as the rules for multiplying and dividing numerical fractions.

MULTIPLYING AND DIVIDING RATIONAL EXPRESSIONS

Let *a, b, c,* and *d* be nonzero polynomials.

TO MULTIPLY, multiply numerators and denominators. $\frac{a}{b} \cdot \frac{c}{d} = \frac{ac}{bd}$

TO DIVIDE, multiply by the reciprocal of the divisor. $\frac{a}{b} \div \frac{c}{d} = \frac{a}{b} \cdot \frac{d}{c}$

EXAMPLE 1 ***Multiplying Rational Expressions Involving Monomials***

Simplify $\frac{3x^3}{4x} \cdot \frac{8x^2}{15x^4}$.

SOLUTION

$\frac{3x^3}{4x} \cdot \frac{8x^2}{15x^4} = \frac{24x^5}{60x^5}$ **Multiply numerators and denominators.**

$= \frac{2 \cdot \cancel{12} \cdot \cancel{x^5}}{5 \cdot \cancel{12} \cdot \cancel{x^5}}$ **Factor, and divide out common factors.**

$= \frac{2}{5}$ **Simplified form**

STUDENT HELP

Study Tip
When multiplying, you usually factor as far as possible to identify all common factors. Note, however, that you do not need to write the prime factorizations of 24 and 60 in Example 1, if you recognize 12 as their greatest common factor.

EXAMPLE 2 ***Multiplying Rational Expressions Involving Polynomials***

Simplify $\frac{x}{3x^2-9x} \cdot \frac{x-3}{2x^2+x-3}$.

SOLUTION

$\frac{x}{3x^2-9x} \cdot \frac{x-3}{2x^2+x-3} = \frac{x(x-3)}{(3x^2-9x)(2x^2+x-3)}$ **Multiply numerators and denominators.**

$= \frac{\cancel{x}\cancel{(x-3)}}{3\cancel{x}\cancel{(x-3)}(x-1)(2x+3)}$ **Factor, and divide out common factors.**

$= \frac{1}{3(x-1)(2x+3)}$ **Simplified form**

STUDENT HELP

Study Tip
When you multiply the numerators and the denominators, leave the products in factored form. At the very end, you may multiply the remaining factors or you may leave your answer in factored form, as in Example 2.

EXAMPLE 3 *Multiplying by a Polynomial*

Simplify $\frac{7x}{x^2 + 5x + 4} \cdot (x + 4)$.

SOLUTION

$$\frac{7x}{x^2 + 5x + 4} \cdot (x + 4) = \frac{7x}{x^2 + 5x + 4} \cdot \frac{x + 4}{1}$$ Write $x + 4$ as $\frac{x+4}{1}$.

$$= \frac{7x(x + 4)}{x^2 + 5x + 4}$$ Multiply numerators and denominators.

$$= \frac{7x\cancel{(x + 4)}}{(x + 1)\cancel{(x + 4)}}$$ Factor, and divide out common factors.

$$= \frac{7x}{x + 1}$$ Simplified form

EXAMPLE 4 *Dividing Rational Expressions*

Simplify $\frac{4n}{n + 5} \div \frac{n - 9}{n + 5}$.

SOLUTION

$$\frac{4n}{n + 5} \div \frac{n - 9}{n + 5} = \frac{4n}{n + 5} \cdot \frac{n + 5}{n - 9}$$ Multiply by reciprocal.

$$= \frac{4n(n + 5)}{(n + 5)(n - 9)}$$ Multiply numerators and denominators.

$$= \frac{4n\cancel{(n + 5)}}{\cancel{(n + 5)}(n - 9)}$$ Divide out common factors.

$$= \frac{4n}{n - 9}$$ Simplified form

EXAMPLE 5 *Dividing by a Polynomial*

Simplify $\frac{x^2 - 9}{4x^2} \div (x - 3)$.

SOLUTION

$$\frac{x^2 - 9}{4x^2} \div (x - 3) = \frac{x^2 - 9}{4x^2} \cdot \frac{1}{x - 3}$$ Multiply by reciprocal.

$$= \frac{x^2 - 9}{4x^2(x - 3)}$$ Multiply numerators and denominators.

$$= \frac{(x + 3)\cancel{(x - 3)}}{4x^2\cancel{(x - 3)}}$$ Factor, and divide out common factors.

$$= \frac{x + 3}{4x^2}$$ Simplified form

STUDENT HELP

INTERNET **HOMEWORK HELP**
Visit our Web site www.mcdougallittell.com for extra examples.

2 TEACH

MOTIVATING THE LESSON
Ask students how many of them plan to have a career in business or someday own their own business. Point out that certain models that will guide their business decisions may involve operations with rational expressions.

EXTRA EXAMPLE 1
Simplify $\frac{7n^5}{5n^2} \cdot \frac{10n^3}{14n}$. n^5

EXTRA EXAMPLE 2
Simplify $\frac{y-5}{3y-3y^2} \cdot \frac{2y^2}{y^2-6y+5}$.
$-\frac{2y}{3(y-1)^2}$

EXTRA EXAMPLE 3
Simplify $\frac{2x+1}{2x^2-x-3} \cdot (2x-3)$.
$\frac{2x+1}{x+1}$

EXTRA EXAMPLE 4
Simplify $\frac{n-2}{2n} \div \frac{n-2}{n+5}$. $\frac{n+5}{2n}$

EXTRA EXAMPLE 5
Simplify $\frac{5x^2-20x}{x+5} \div (x-4)$. $\frac{5x}{x+5}$

CHECKPOINT EXERCISES

For use after Examples 1–3:

1. Simplify $\frac{8x}{x^2+4x+4} \cdot \frac{x+2}{2x^3}$.
$\frac{4}{x^2(x+2)}$

For use after Examples 4 and 5:

2. Simplify $\frac{x^2-x-2}{3x-3} \div (x^2-1)$.
$\frac{x-2}{3(x-1)^2}$

STUDENT HELP NOTES

Homework Help Students can find extra examples at **www.mcdougallittell.com** that parallel the examples in the student edition.

EXTRA EXAMPLE 6

The models below are based on data from a regional crafts fair held annually since 1990. The variable t represents the number of years since 1990.

items sold: $n = \dfrac{20{,}000 + 400t}{1 - 0.03t}$

average cost per item:

$c = \dfrac{850 + 60t}{50 + t}$

a. Find a model for the total yearly sales of the fair.
$\dfrac{4000(85 + 6t)}{1 - 0.03t}$

b. Use the model to predict the total sales in 2003.
about $1,070,000

CHECKPOINT EXERCISES

For use after Example 6:

1. The attendance at the fair in Extra Example 6 is modeled by $a = \dfrac{6100 + 300t}{(1 + 0.01t)}$. Find a model for the average number of items bought per person. Use the model to predict the average number of items bought per person in 1999.
$\dfrac{(200 + 4t)(1 + 0.01t)}{(1 - 0.03t)(61 + 3t)}$; 4 items

FOCUS ON VOCABULARY

What is the difference between a rational expression and a fraction?
***Sample answer:* A fraction is a ratio of numbers, while a rational expression is a ratio of polynomials.**

CLOSURE QUESTION

Compare the procedures used to multiply and divide two rational expressions.
See sample answer below.

DAILY PUZZLER

Multiplying an expression of the form $\dfrac{ax + b}{cx + d}$ by $x^2 - 25$ gives $(x + 5)^2$, and dividing it by $x^2 - 25$ gives $\dfrac{1}{(x - 5)^2}$. What is the expression?
$\dfrac{x + 5}{x - 5}$

GOAL 2 USING RATIONAL MODELS IN REAL LIFE

REAL LIFE
Car Sales

EXAMPLE 6 *Writing and Using a Rational Model*

The models below can be created using data collected by the National Automobile Dealers Association in the United States. Five-year intervals from 1975–1995 were used. Let t represent the number of years since 1975.

Number of new-car dealerships: $D = \dfrac{30{,}000 + 300t}{1 + 0.03t}$

Total sales (in billions of dollars) of new-car dealerships: $S = \dfrac{80 + 10t}{1 - 0.02t}$

a. Find a model for the average sales per new-car dealership.

b. Use the model to predict the average sales in 2005.

SOLUTION

PROBLEM SOLVING STRATEGY

a. VERBAL MODEL

$$\boxed{\text{Average sales per dealership}} = \text{Total sales of dealerships} \div \text{Number of dealerships}$$

LABELS

Average sales per dealership $= \boxed{A}$

Total sales of dealerships $= \dfrac{80 + 10t}{1 - 0.02t}$

Number of dealerships $= \dfrac{30{,}000 + 300t}{1 + 0.03t}$

ALGEBRAIC MODEL

$$\boxed{A} = \frac{80 + 10t}{1 - 0.02t} \div \frac{30{,}000 + 300t}{1 + 0.03t}$$ Write algebraic model.

$$= \frac{80 + 10t}{1 - 0.02t} \cdot \frac{1 + 0.03t}{30{,}000 + 300t}$$ Multiply by reciprocal.

$$= \frac{(80 + 10t)(1 + 0.03t)}{(1 - 0.02t)(30{,}000 + 300t)}$$ Multiply numerators and denominators.

$$= \frac{(\cancel{10})(8 + t)(1 + 0.03t)}{(1 - 0.02t)(\cancel{10})(3000 + 30t)}$$ Divide out common factor 10.

▶ The equation $A = \dfrac{(8 + t)(1 + 0.03t)}{(1 - 0.02t)(3000 + 30t)}$ is a model for the average sales (in billions of dollars) per new-car dealership.

b. In the year 2005, $t = 30$, so substitute 30 for t in the model for A.

$$\frac{(8 + 30)(1 + 0.03 \cdot 30)}{(1 - 0.02 \cdot 30)(3000 + 30 \cdot 30)} = \frac{38 \cdot 1.9}{0.4 \cdot 3900} = \frac{72.2}{1560} \approx 0.04628$$

▶ The model predicts that the average sales in 2005 will be about \$0.0463 billion. Because 1 billion is 1000 million, you can express \$0.0463 billion as 0.0463 • 1000 million, or \$46.3 million.

Closure Question *Sample answer:*
The two procedures differ only in the first step. To divide rational expressions, first rewrite the division as a multiplication by the reciprocal of the divisor. The remaining steps are the same as when multiplying two rational expressions.

GUIDED PRACTICE

Concept Check ✓

1. Describe the steps used to multiply two rational expressions. ***Sample answer:* Multiply the numerators and denominators, then factor and divide out common factors.**
2. Describe the steps used to divide two rational expressions.
3. **ERROR ANALYSIS** Describe the error in the problem at the right. Then do the division correctly. **See margin.**

$$\frac{x+3}{x-3} \div \frac{4x}{x^2-9} = \frac{x+3}{x-3} \cdot \frac{4x}{(x+3)(x-3)} = \frac{4x}{(x-3)^2}$$

Ex. 3

2. *Sample answer:* Multiply the first expression by the reciprocal of the divisor, multiplying the numerators and denominators, then divide out common factors.

3. *Sample answer:* In the first step, the expression on the left should have been multiplied by the reciprocal of the expression on the right; $\frac{(x+3)^2}{4x}$.

Skill Check ✓

Simplify the expression.

4. $\frac{3x}{8x^2} \cdot \frac{4x^3}{3x^4}$ **$\frac{1}{2x^2}$**
5. $\frac{x^2-1}{x} \cdot \frac{2x}{3x-3}$ **$\frac{2(x+1)}{3}$**
6. $\frac{x}{x^2-25} \cdot \frac{x-5}{x+5}$ **$\frac{x}{(x+5)^2}$**
7. $\frac{3x}{x^2-2x-15} \cdot (x+3)$ **$\frac{3x}{x-5}$**
8. $\frac{x}{8-2x} \div \frac{2x}{4-x}$ **$\frac{1}{4}$**
9. $\frac{4x^2-25}{4x} \div (2x-5)$ **$\frac{2x+5}{4x}$**
10. $\frac{x^2-4x+3}{2x} \div \frac{x-1}{2}$ **$\frac{x-3}{x}$**
11. $\frac{9x^2+6x+1}{x+5} \div \frac{3x+1}{x^2+5x}$ **$x(3x+1)$**

PRACTICE AND APPLICATIONS

STUDENT HELP

Extra Practice to help you master skills is on p. 807.

19. $\frac{1}{(x+3)(2x+3)}$

24. $\frac{x+2}{2(x-2)}$

25. $-\frac{x+6}{5x^2}$

SIMPLIFYING EXPRESSIONS **Simplify the expression.**

12. $\frac{4x}{3} \cdot \frac{1}{x}$ **$\frac{4}{3}$**
13. $\frac{9x^2}{4} \cdot \frac{8}{18x}$ **x**
14. $\frac{7x^2}{6x} \cdot \frac{12x^2}{2x}$ **$7x^2$**
15. $\frac{16x^2}{8x} \div \frac{4x^2}{16x}$ **8**
16. $\frac{25x^2}{10x} \div \frac{5x}{10x}$ **$5x$**
17. $\frac{13x^4}{7x} \div \frac{x^3}{7x}$ **$13x$**
18. $\frac{5-2x}{-2} \cdot \frac{24}{10-4x}$ **-6**
19. $\frac{4x}{x^2-9} \cdot \frac{x-3}{8x^2+12x}$ **See margin.**
20. $\frac{-3}{x-4} \cdot \frac{x-4}{12(x-7)}$ **$-\frac{1}{4(x-7)}$**
21. $\frac{3x^2}{10} \div \frac{9x^3}{25}$ **$\frac{5}{6x}$**
22. $\frac{x}{x+2} \div \frac{x+5}{x+2}$ **$\frac{x}{x+5}$**
23. $\frac{5x+15}{3x} \div \frac{x+3}{9x}$ **15**
24. $\frac{2(x+2)}{5(x-3)} \div \frac{4(x-2)}{5x-15}$ **See margin.**
25. $\frac{x^2-36}{-5x^2} \div (x-6)$ **See margin.**
26. $\frac{8}{2+3x} \cdot (8+12x)$ **32**
27. $\frac{3x}{x^2-2x-24} \cdot \frac{x-6}{6x^2+9x}$ **$\frac{1}{(x+4)(2x+3)}$**
28. $\frac{x}{3x^2+2x-8} \cdot (3x-4)$ **$\frac{x}{x+2}$**
29. $\frac{x+1}{x^3(3-x)} \div \frac{5}{x(x-3)}$ **$-\frac{x+1}{5x^2}$**
30. $(4x^2+x-3) \cdot \frac{1}{(4x+3)(x-1)}$ **$\frac{(4x-3)(x+1)}{(4x+3)(x-1)}$**
31. $\frac{x^2-8x+15}{x^2-3x} \div (3x-15)$ **$\frac{1}{3x}$**
32. $\frac{6x^2+7x-33}{x+4} \div (6x-11)$ **$\frac{x+3}{x+4}$**
33. $\left(\frac{x^2}{5} \cdot \frac{x+2}{2}\right) \div \frac{x}{30}$ **$3x(x+2)$**
34. $\left(\frac{2x^2}{3} \cdot \frac{5}{x}\right) \div \frac{6x^2}{25}$ **$\frac{125}{9x}$**

STUDENT HELP

HOMEWORK HELP
Examples 1–5: Exs. 12–34
Example 6: Exs. 35–40

3 APPLY

ASSIGNMENT GUIDE

BASIC
Day 1: pp. 673–675 Exs. 12–34 even, 44, 45, 48–58 even

AVERAGE
Day 1: pp. 673–675 Exs. 12–34 even, 35–37, 44, 45, 48–58 even

ADVANCED
Day 1: pp. 673–675 Exs. 12–34 even, 35–37, 44–47, 48–58 even

BLOCK SCHEDULE WITH 11.4
pp. 673–675 Exs. 12–34 even, 35–37, 44, 45, 48–58 even

EXERCISE LEVELS

Level A: ***Easier***
48–50

Level B: ***More Difficult***
12–37, 44, 45, 51–59

Level C: ***Most Difficult***
38–43, 46, 47

✓ HOMEWORK CHECK

To quickly check student understanding of key concepts, go over the following exercises: Exs. 14, 16, 20, 24, 30, 32. See also the Daily Homework Quiz:

- Blackline Master (*Chapter 11 Resource Book*, p. 80)
- Transparency (p. 90)

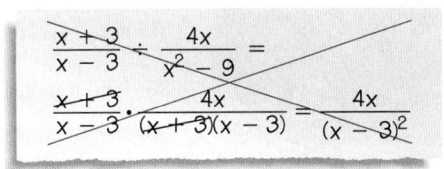

CAREER NOTE

EXERCISES 38–41 Additional information about careers in the service industry is available at **www.mcdougallittell.com.**

MATHEMATICAL REASONING

Point out to students that a proof, like the one they are completing in Ex. 42, is a series of logical steps that moves from a hypothesis to a conclusion. Each step along the way must be justified by a previously proven statement, a definition, or a postulate.

ADDITIONAL PRACTICE AND RETEACHING

For Lesson 11.5:

- Practice Levels A, B, and C (*Chapter 11 Resource Book,* p. 69)
- Reteaching with Practice (*Chapter 11 Resource Book,* p. 72)
- See Lesson 11.5 of the *Personal Student Tutor*

For more Mixed Review:

- Search the *Test and Practice Generator* for key words or specific lessons.

FOCUS ON CAREERS

REAL LIFE **SERVICE INDUSTRY CAREERS** The service industry includes a wide range of careers. Fields of service include health care, automobile and other repair services, legal assistance, education, and recreation.

INTERNET CAREER LINK
www.mcdougallittell.com

35. $A = \dfrac{(3150 - 400t)(10 - t)}{(1 - 0.12t)(111 - 12t)}$

39. $R = \dfrac{46 + 0.7t}{1055 + 23t}$; decreasing; *Sample answer:* The ratios in order (rounded to the nearest hundred-thousandth) are 0.04360, 0.04332, 0.04305, 0.04279, 0.04255, 0.04231, 0.04208, and 0.04186.

40. $R = \dfrac{(48 - t)(1 - 0.04t)}{(1 - 0.06t)(1055 + 23t)}$; the ratio was decreasing from 1990 to 1996, then increasing; *Sample answer:* The ratios in order (rounded to the nearest hundred-thousandth) are 0.04550, 0.04453, 0.04368, 0.04297, 0.04240, 0.04200, 0.04181, and 0.04186.

43. $\dfrac{2x - 4}{x^2 - 4} = \dfrac{2(x - 2)}{(x + 2)(x - 2)} = \dfrac{2}{x + 2} \cdot \dfrac{x - 2}{x - 2} = \dfrac{2}{x + 2} \cdot 1 = \dfrac{2}{x + 2}$

RAILROAD TRAVEL In Exercises 35–37, the models are based on data about train travel from 1990 to 1996 in the United States. Let *t* represent the number of years since 1990. ▶ Source: *Statistical Abstract of the United States*

Miles (in millions) traveled by passengers: $M = \dfrac{6300 - 800t}{1 - 0.12t}$

Passengers (in millions) who traveled by train: $P = \dfrac{222 - 24t}{10 - t}$

35. Find a model for the average number of miles traveled per passenger. **See margin.**

36. Use the model found in Exercise 35 to estimate the average number of miles traveled per passenger in 1995. **282 mi**

37. Use the model to predict the average number of miles traveled per passenger in 2005. **258 mi**

SERVICE INDUSTRY In Exercises 38–41, the models below are based on data collected by the Bureau of Economic Analysis from 1990 to 1997 in the United States. Let *t* represent the number of years since 1990.

Total sales (in billions of dollars) of services: $S = \dfrac{1055 + 23t}{1 - 0.04t}$

Total sales (in billions of dollars) of hotel services: $H = \dfrac{46 + 0.7t}{1 - 0.04t}$

Total sales (in billions of dollars) of auto repair services: $A = \dfrac{48 - t}{1 - 0.06t}$

38. Find the total sales given by each model in 1990.
total sales in billions of dollars: service: \$1055; hotels: \$46; auto repair: \$48

39. Find a model for the ratio of hotel service sales to total service industry sales. Was this ratio increasing or decreasing from 1990 to 1997? Explain. **See margin.**

40. Find a model for the ratio of auto service sales to total service industry sales. Was this ratio increasing or decreasing from 1990 to 1997? Explain. **See margin.**

41. *Writing* What do your answers in Exercises 38 and 39 tell you about how the sales of the service industry were changing in the period from 1990 to 1997? ***Sample answer:*** **In general, sales of hotel services and auto repair services were not increasing as rapidly as the total sales of services.**

PROOF In Exercises 42 and 43, use the *proof* shown below.

Statement	Explanation
1. $\dfrac{ac}{bc} = \dfrac{a}{b} \cdot \dfrac{?}{?}$	1. Apply the rule for multiplying rational expressions.
2. $\underline{?} = \dfrac{a}{b} \cdot \underline{?}$	2. Any nonzero number divided by itself is 1.
3. $\underline{?} = \dfrac{a}{b}$	3. Any nonzero number multiplied by 1 is itself.

42. **LOGICAL REASONING** Copy and complete the *proof* to show why you can divide out common factors. $\dfrac{ac}{bc} = \dfrac{a}{b} \cdot \dfrac{c}{c}$; $\dfrac{ac}{bc} = \dfrac{a}{b} \cdot 1$; $\dfrac{ac}{bc} = \dfrac{a}{b}$

43. Use the method from Exercise 42 to show that $\dfrac{2x - 4}{x^2 - 4} = \dfrac{2}{x + 2}$. **See margin.**

Test Preparation

44. MULTIPLE CHOICE Which of the following represents the expression $\frac{x^2 - 3x}{x^2 - 5x + 6} \cdot \frac{(x-2)^2}{2x}$ in simplified form? **C**

(A) $\frac{x(x-3)}{2}$ (B) $\frac{x}{2}$ (C) $\frac{x-2}{2}$

(D) $\frac{x(x-3)}{x-2}$ (E) $\frac{x^2 - 4x + 4}{x - 2}$

45. MULTIPLE CHOICE Which product equals the quotient $(2x + 2) \div \frac{x^2 + x}{4}$? **D**

(A) $\frac{1}{2x+2} \cdot \frac{x^2+x}{4}$ (B) $\frac{2x+2}{1} \cdot \frac{x^2+x}{4}$ (C) $\frac{1}{2x+2} \cdot \frac{4}{x^2+x}$

(D) $\frac{2x+2}{1} \cdot \frac{4}{x^2+x}$ (E) $\frac{2x+2}{2x+2} \cdot \frac{4}{x^2+x}$

★ Challenge

INDEPENDENT EVENTS **In Exercises 46–47, use the following information.**

Two events are *independent* if the probability that one event will occur is not affected by whether or not the other event occurs. For independent events A and B, the probability that A *and* B will occur equals the probability of A times the probability of B.

For example, if you draw a marble from the jar at the right, put it back, and then draw another one, the probability that both marbles are red is $\frac{3}{5} \cdot \frac{3}{5} = \frac{9}{25}$.

46. A bag contains n marbles. There are r blue marbles and the rest of the marbles are yellow. Find the probability of drawing a yellow marble followed by a blue marble if the first one is put back before drawing again. $\frac{r(n-r)}{n^2}$

EXTRA CHALLENGE
www.mcdougallittell.com

47. Look back at the carnival game in Exercises 32–34 on page 668. Find the probability of hitting the target two times in a row. $\frac{(x-y)^2}{x^2}$

MIXED REVIEW

FINDING THE LCD **Find the least common denominator.** (Skills Review, pp. 781–783)

48. $\frac{3}{4}, \frac{2}{5}$ 20
49. $\frac{2}{9}, \frac{3}{18}$ 18
50. $\frac{1}{16}, \frac{9}{20}$ 80
51. $\frac{14}{54}, \frac{31}{81}$ 162

QUADRATIC FORMULA **Solve the equation.** (Review 9.5)

52. $2x^2 + 12x - 6 = 0$ $-3 - 2\sqrt{3}, -3 + 2\sqrt{3}$
53. $x^2 - 6x + 7 = 0$ $3 - \sqrt{2}, 3 + \sqrt{2}$
54. $3x^2 + 11x + 10 = 0$ $-2, -\frac{5}{3}$

POLYNOMIALS **Add or subtract.** (Review 10.1 for 11.6)

56. $16p^3 + 11p^2 - 8p + 8$

55. $(4t^2 + 5t + 2) - (t^2 - 3t - 8)$ $3t^2 + 8t + 10$
56. $(16p^3 - p^2 + 24) + (12p^2 - 8p - 16)$ See margin.
57. $(a^4 - 12a) + (4a^3 + 11a - 1)$ $a^4 + 4a^3 - a - 1$
58. $(-5x^2 + 2x - 12) - (6 - 9x - 7x^2)$ $2x^2 + 11x - 18$

59. COMPOUND INTEREST After two years, an investment of \$1000 compounded annually at an interest rate r will grow to the amount $1000(1 + r)^2$ in dollars. Write this product as a trinomial. (Review 10.3) $1000 + 2000r + 1000r^2$

4 ASSESS

DAILY HOMEWORK QUIZ

Transparency Available

Simplify the expression.

1. $\frac{4x}{3x^2} \cdot \frac{6x^3}{2x}$ $4x$
2. $\frac{15x^3}{6x} \div \frac{5x}{6x}$ $3x^2$
3. $\frac{x-1}{2x+2} \cdot \frac{4}{3x-3}$ $\frac{2}{3(x+1)}$
4. $\frac{3(2x+1)}{2x-8} \div \frac{8x+4}{3(x-4)}$ $\frac{9}{8}$
5. $(3x-2)(2x+3) \cdot \frac{2}{6x^2 - 13x + 6}$ $\frac{2(2x+3)}{2x-3}$
6. $\frac{5x^2 + 14x - 3}{5x^2 - x} \div (4x + 12)$ $\frac{1}{4x}$

EXTRA CHALLENGE NOTE

Challenge problems for Lesson 11.5 are available in **blackline** format in the *Chapter 11 Resource Book,* p. 77 and at **www.mcdougallittell.com.**

ADDITIONAL TEST PREPARATION

1. **WRITING** Describe the error in the problem below. Then do the division correctly.

$\frac{x^2-4}{x+5} \div \frac{x+5}{x-2} = \frac{x+5}{(x-2)(x+2)} \cdot \frac{x-2}{x+5} = \frac{1}{x+2}$

Sample answer: **You should not take the reciprocal of the dividend when rewriting the problem as multiplication. The correct answer is** $\frac{(x-2)^2(x+2)}{(x+5)^2}$.

1 PLAN

PACING

Basic: 2 days
Average: 2 days
Advanced: 2 days
Block Schedule: 1 block

LESSON OPENER ACTIVITY

An alternative way to approach Lesson 11.6 is to use the Activity Lesson Opener:

- Blackline Master (*Chapter 11 Resource Book*, p. 81)
- Transparency (p. 79)

MEETING INDIVIDUAL NEEDS

- ***Chapter 11 Resource Book***
  Prerequisite Skills Review (p. 5)
  Practice Level A (p. 83)
  Practice Level B (p. 84)
  Practice Level C (p. 85)
  Reteaching with Practice (p. 86)
  Absent Student Catch-Up (p. 88)
  Challenge (p. 90)
- ***Resources in Spanish***
- ***Personal Student Tutor***

NEW-TEACHER SUPPORT

See the Tips for New Teachers on pp. 1–2 of the *Chapter 11 Resource Book* for additional notes about Lesson 11.6.

WARM-UP EXERCISES

Transparency Available

Simplify.

1. $\frac{1}{5}+\frac{2}{15}$ $\frac{1}{3}$

2. $\frac{5}{12}-\frac{3}{8}$ $\frac{1}{24}$

Simplify.

3. $(x^2+4x-3)+(3x^2-1)$
$4x^2+4x-4$

4. $(3x^2-2x+7)-(x^2+4x-2)$
$2x^2-6x+9$

11.6 Adding and Subtracting Rational Expressions

What you should learn

GOAL 1 Add and subtract rational expressions that have like denominators.

GOAL 2 Add and subtract rational expressions that have unlike denominators.

Why you should learn it

▼ To model **real-life** problems, such as planning your route for a car trip in **Example 5**.

REAL LIFE

GOAL 1 FRACTIONS WITH LIKE DENOMINATORS

To add or subtract rational expressions with *like* denominators, combine their numerators, and write the result over the common denominator.

ADDING OR SUBTRACTING WITH LIKE DENOMINATORS

Let *a*, *b*, and *c* be polynomials, with $c \neq 0$.

TO ADD, add the numerators. $\frac{a}{c}+\frac{b}{c}=\frac{a+b}{c}$

TO SUBTRACT, subtract the numerators. $\frac{a}{c}-\frac{b}{c}=\frac{a-b}{c}$

EXAMPLE 1 *Adding and Subtracting Expressions*

a. $\frac{5}{x}+\frac{x-5}{x}=\frac{5+x-5}{x}$ **Add numerators.**

$=\frac{x}{x}$ **Simplify.**

$=1$

b. $\frac{7}{2m-3}-\frac{3m}{2m-3}=\frac{7-3m}{2m-3}$ **Subtract numerators.**

EXAMPLE 2 *Simplifying After Subtracting*

Simplify $\frac{4x}{3x^2-x-2}-\frac{x-2}{3x^2-x-2}$.

SOLUTION

$\frac{4x}{3x^2-x-2}-\frac{x-2}{3x^2-x-2}=\frac{4x-(x-2)}{3x^2-x-2}$ **Subtract.**

$=\frac{3x+2}{3x^2-x-2}$ **Simplify.**

$=\frac{3x+2}{(3x+2)(x-1)}$ **Factor.**

$=\frac{\cancel{3x+2}}{\cancel{(3x+2)}(x-1)}$ **Divide out common factor.**

$=\frac{1}{x-1}$ **Simplified form**

STUDENT HELP

Skills Review
For help with adding fractions with unlike denominators, see pp. 781–783.

GOAL 2 FRACTIONS WITH UNLIKE DENOMINATORS

To add or subtract rational expressions with *unlike* denominators, you must first rewrite the expressions so that they have *like* denominators. The like denominator that you usually use is the least common multiple of the original denominators. It is called the **least common denominator** or LCD.

EXAMPLE 3 *Adding with Unlike Denominators*

Simplify $\frac{7}{6x} + \frac{5}{8x^2}$.

SOLUTION

To find the least common denominator, first completely factor the denominators. You get $6x = 2 \cdot 3 \cdot x$ and $8x^2 = 2^3 \cdot x^2$. The LCD contains the highest power of each factor that appears in either denominator, so the LCD is $2^3 \cdot 3 \cdot x^2$, or $24x^2$.

$\frac{7}{6x} + \frac{5}{8x^2} = \frac{7 \cdot 4x}{6x \cdot 4x} + \frac{5 \cdot 3}{8x^2 \cdot 3}$ — Rewrite fractions using LCD.

$= \frac{28x}{24x^2} + \frac{15}{24x^2}$ — Simplify numerators and denominators.

$= \frac{28x + 15}{24x^2}$ — Add fractions.

STUDENT HELP

Study Tip
You can always find a common denominator by multiplying the two denominators, but it is often easier to use the LCD. For instance, using $24x^2$ rather than $48x^3$ in Example 3 makes the expressions in the numerators simpler.

EXAMPLE 4 *Subtracting with Unlike Denominators*

Simplify $\frac{x+2}{x-1} - \frac{12}{x+6}$.

SOLUTION

Neither denominator can be factored. The least common denominator is the product $(x - 1)(x + 6)$ because it must contain both of these factors.

$\frac{x+2}{x-1} - \frac{12}{x+6} = \frac{(x+2)(x+6)}{(x-1)(x+6)} - \frac{12(x-1)}{(x-1)(x+6)}$ — Rewrite fractions using LCD.

$= \frac{x^2 + 8x + 12}{(x-1)(x+6)} - \frac{12x - 12}{(x-1)(x+6)}$ — Simplify numerators. Leave denominators in factored form.

$= \frac{x^2 + 8x + 12 - (12x - 12)}{(x-1)(x+6)}$ — Subtract fractions.

$= \frac{x^2 + 8x + 12 - 12x + 12}{(x-1)(x+6)}$ — Use distributive property.

$= \frac{x^2 - 4x + 24}{(x-1)(x+6)}$ — Combine like terms.

2 TEACH

EXTRA EXAMPLE 1

a. Simplify $\frac{3y}{y+5} + \frac{4}{y+5}$. $\frac{3y+4}{y+5}$

b. Simplify $\frac{n-3}{2n} - \frac{3}{2n}$. $\frac{n-6}{2n}$

EXTRA EXAMPLE 2

Simplify $\frac{1-x}{2x^2-7x+5} + \frac{3x-3}{2x^2-7x+5}$. $\frac{2}{2x-5}$

EXTRA EXAMPLE 3

Simplify $\frac{5}{3x^3} + \frac{7}{6x}$. $\frac{7x^2+10}{6x^3}$

EXTRA EXAMPLE 4

Simplify $\frac{2x^2}{x+1} - \frac{x-1}{x}$.

$\frac{2x^3 - x^2 + 1}{x(x+1)}$

CHECKPOINT EXERCISES

For use after Examples 1 and 2:

1. Simplify $\frac{2n+1}{3n^2+3n} + \frac{n-1}{3n^2+3n}$.

$\frac{1}{n+1}$

For use after Examples 3 and 4:

2. Simplify $\frac{3y}{2y+1} - \frac{y+3}{y+1}$.

$\frac{y^2-4y-3}{(2y+1)(y+1)}$

STUDENT HELP NOTES

Skills Review As students review adding fractions with unlike denominators on pp. 781–783, remind them that the LCD is the product with the fewest factors that contains all the factors found in either denominator. This is true whether the factors are numbers or expressions. Students may need to do several numerical examples prior to moving on to Example 3.

EXTRA EXAMPLE 5

You and a friend plan to camp on an island 20 miles offshore. Your motor-equipped rowboat travels 12 miles per hour using the motor; by rowing alone, you travel 4 miles per hour. You prefer rowing, but you plan to use the motor partway to save time. Let x represent the number of miles traveled by rowing.

a. Write an expression for the total time. $T = \frac{x+10}{6}$

b. Evaluate the expression at 4 mile intervals.

x	0	4	8	12	16	20
T	1.7	2.3	3	3.7	4.3	5

CHECKPOINT EXERCISES

For use after Example 5:

1. An athlete can steadily run 9 miles per hour or bike 21 miles per hour. A 50 mile event combines running and biking. If x is the running distance, write an expression for the total time. Evaluate the expression at 10 mile intervals. $T = \frac{4x+150}{63}$;

x	0	10	20	30	40	50
T	2.4	3.0	3.7	4.3	4.9	5.6

FOCUS ON VOCABULARY

What is the connection between the *least common multiple* (LCM) and the *least common denominator* (LCD)? ***Sample answer:*** **The LCD of fractions or rational expressions is the LCM of the denominators.**

CLOSURE QUESTION

You want to add two rational expressions. You've found the least common denominator. How do you rewrite the expressions with the same denominator?
See sample answer at right.

DAILY PUZZLER

Which sum equals $\frac{5x+3}{x(x+1)}$?

A. $\frac{5x}{x} + \frac{3}{x+1}$ **B.** $\frac{3}{x} + \frac{2}{x+1}$ B

REAL LIFE *Travel Planning*

EXAMPLE 5 *Writing and Using a Model*

You are planning a 300-mile car trip. You can make the trip using a combination of two roads: a toll road on which you can average 60 miles per hour and a scenic country road on which you can average 40 miles per hour. To avoid tolls and to enjoy the scenery, you want to spend as much time as possible on the country road. However, you do not want your total trip time to be too long.

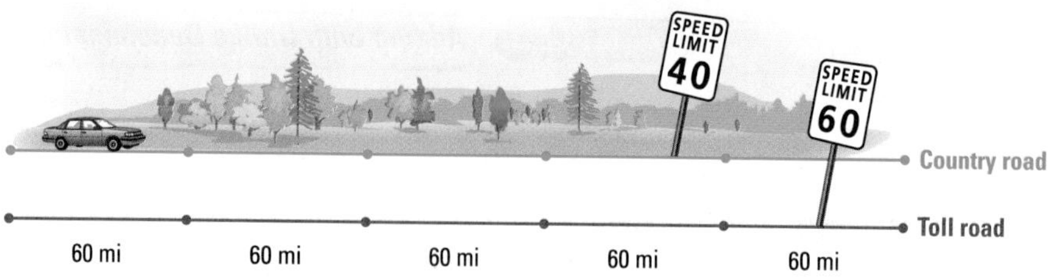

Let x represent the number of miles you travel on the country road.

a. Write an expression for the total time.

b. Evaluate the expression at 60-mile intervals.

SOLUTION

PROBLEM SOLVING STRATEGY

a. VERBAL MODEL

$$\boxed{\text{Total time}} = \frac{\text{Distance on country road}}{\text{Speed on country road}} + \frac{\text{Distance on toll road}}{\text{Speed on toll road}}$$

LABELS

Total time = T		(hours)
Distance on country road = x		(miles)
Speed on country road = **40**		(miles per hour)
Distance on toll road = $300 - x$		(miles)
Speed on toll road = **60**		(miles per hour)

ALGEBRAIC MODEL

$$T = \frac{x}{40} + \frac{300 - x}{60}$$ **Write algebraic model.**

$$= \frac{3x}{120} + \frac{2(300 - x)}{120}$$ **Rewrite fractions using LCD.**

$$= \frac{3x + 600 - 2x}{120}$$ **Add fractions.**

$$= \frac{x + 600}{120}$$ **Combine like terms.**

b. Substitute x-values of 0, 60, 120, 180, 240, and 300 into the expression.

Distance (country) x	0	60	120	180	240	300
Total time T	5.0	5.5	6.0	6.5	7.0	7.5

The table can help you decide how many miles to drive on the country road.

Closure Question *Sample answer:*
For each rational expression, compare the denominator to the LCD to find out which factors are missing. Multiply both the numerator and denominator of the expression by these factors.

GUIDED PRACTICE

Vocabulary Check ✓

1. Explain what is meant by the *least common denominator* of two rational expressions.

Concept Check ✓

2. ERROR ANALYSIS **Explain the error in the following problem.**

$$\frac{3}{x-1}-\frac{2}{x}=\frac{3x}{x(x-1)}-\frac{2(x-1)}{x(x-1)}$$
$$=\frac{3x}{x(x-1)}-\frac{2x-2}{x(x-1)}$$
$$=\frac{3x-2x-2}{x(x-1)}=\frac{x-2}{x(x-1)}$$

When $2x - 2$ is subtracted from $3x$, the result is $3x - 2x + 2$; the correct answer is $\frac{x+2}{x(x-1)}$.

3. You can use $x - 3$ as the LCD when finding the sum $\frac{5}{x-3}+\frac{2}{3-x}$. What number can you multiply the numerator and the denominator of the second fraction by to get an equivalent fraction with $x - 3$ as the new denominator? **-1**

Skill Check ✓

Simplify the expression.

4. $\frac{1}{3x}+\frac{5}{3x}$ **$\frac{2}{x}$** **5.** $\frac{5x}{x+4}+\frac{20}{4+x}$ **5** **6.** $\frac{x}{x^2-9}-\frac{3x+1}{x^2-9}$ **See margin.**

7. $\frac{3}{10x}-\frac{1}{4x^2}$ **$\frac{6x-5}{20x^2}$** **8.** $\frac{x+6}{x+1}-\frac{4}{2x+3}$ **See margin.** **9.** $\frac{x-2}{2x-10}+\frac{x+3}{x-5}$ **See margin.**

1. ***Sample answer*: the least common multiple of the denominators of the expressions, that is, the least number that is a multiple of both denominators**

6. $\frac{-2x-1}{x^2-9}$

8. $\frac{(2x+7)(x+2)}{(x+1)(2x+3)}$

9. $\frac{3x+4}{2(x-5)}$

19. $-\frac{3(x+12)}{(x+4)(x-2)}$

22. $-\frac{5x(x-3)}{(x-1)(x+4)}$

23. $\frac{2(x^2-20)}{(x-10)(x+6)}$

24. $\frac{2x(5x+4)}{(5x-2)(5x+1)}$

PRACTICE AND APPLICATIONS

STUDENT HELP

Extra Practice to help you master skills is on p. 807.

STUDENT HELP

HOMEWORK HELP
Example 1: Exs. 10–31
Example 2: Exs. 10–31
Example 3: Exs. 16–31
Example 4: Exs. 18–31
Example 5: Exs. 36–38

ADDING AND SUBTRACTING **Simplify the expression.**

10. $\frac{7}{2x}+\frac{x+2}{2x}$ **$\frac{x+9}{2x}$** **11.** $\frac{4}{x+1}+\frac{2x-2}{x+1}$ **2** **12.** $\frac{7x}{x^3}-\frac{6x}{x^3}$ **$\frac{1}{x^2}$**

13. $\frac{2}{3x-1}-\frac{5x}{3x-1}$ **$\frac{2-5x}{3x-1}$** **14.** $\frac{3x}{4x+1}+\frac{5x}{4x+1}$ **$\frac{8x}{4x+1}$** **15.** $\frac{-8}{3x^2}+\frac{11}{3x^2}$ **$\frac{1}{x^2}$**

16. $\frac{9}{4x}+\frac{7}{-5x}$ **$\frac{17}{20x}$** **17.** $\frac{11}{6x}+\frac{2}{13x}$ **$\frac{155}{78x}$** **18.** $\frac{9}{5x}-\frac{2}{x^2}$ **$\frac{9x-10}{5x^2}$**

19. $\frac{4}{x+4}-\frac{7}{x-2}$ **See margin.** **20.** $\frac{3}{x+3}+\frac{4x}{2x+6}$ **$\frac{2x+3}{x+3}$** **21.** $\frac{9}{x^2-3x}+\frac{3}{x-3}$ **$\frac{3(x+3)}{x(x-3)}$**

22. $\frac{2x}{x-1}-\frac{7x}{x+4}$ **See margin.** **23.** $\frac{x}{x-10}+\frac{x+4}{x+6}$ **See margin.** **24.** $\frac{4x}{5x-2}-\frac{2x}{5x+1}$ **See margin.**

25. $\frac{x+8}{3x-1}+\frac{x+3}{x+1}$ **See margin.** **26.** $\frac{3x+10}{7x-4}-\frac{x}{4x+3}$ **See margin.** **27.** $\frac{2x+1}{3x-1}-\frac{x+4}{x-2}$ **See margin.**

28. $\frac{x}{x^2+5x-24}+\frac{8}{x^2+5x-24}$ **$\frac{1}{x-3}$** **29.** $\frac{x^2+1}{x^2-4}+\frac{5x}{x^2-4}-\frac{2x+11}{x^2-4}$ **$\frac{x+5}{x+2}$**

30. $\frac{x^2-9}{x+3}+\frac{x^2+9}{x-3}$ **$\frac{2(x^2-3x+9)}{x-3}$** **31.** $\frac{2}{x+1}+\frac{3}{x-2}+\frac{3}{x+4}$ **See margin.**

25. $\frac{4x^2+17x+5}{(3x-1)(x+1)}$

26. $\frac{5x^2+53x+30}{(7x-4)(4x+3)}$

27. $-\frac{x^2+14x-2}{(3x-1)(x-2)}$

31. $\frac{2(4x^2+8x-5)}{(x+1)(x-2)(x+4)}$

3 APPLY

ASSIGNMENT GUIDE

BASIC
Day 1: p. 679 Exs. 10–27
Day 2: pp. 679–682 Exs. 28–33, 42–45, 50–52, 56–64 even, 65, Quiz 2 Exs. 1–13

AVERAGE
Day 1: p. 679 Exs. 10–27
Day 2: pp. 679–682 Exs. 28–33, 36–38, 42–52, 56–64 even, 65, Quiz 2 Exs. 1–13

ADVANCED
Day 1: p. 679 Exs. 10–27
Day 2: pp. 679–682 Exs. 28–33, 36–38, 42–55, 56–64 even, 65, Quiz 2 Exs. 1–13

BLOCK SCHEDULE
pp. 679–682 Exs. 10–33, 36–38, 42–52, 56–64 even, 65, Quiz 2 Exs. 1–13

EXERCISE LEVELS

Level A: ***Easier***
10–12

Level B: ***More Difficult***
13–33, 36, 37, 42–52, 56–65

Level C: ***Most Difficult***
34, 35, 38–41, 53–55

✓ HOMEWORK CHECK

To quickly check student understanding of key concepts, go over the following exercises: Exs. 14, 16, 20, 24, 30, 44. See also the Daily Homework Quiz:

- Blackline Master (*Chapter 11 Resource Book*, p. 94)
- Transparency (p. 91)

! COMMON ERROR

EXERCISE 2 This exercise points out a frequent mistake. Stress that when subtracting a rational expression, students must subtract the entire numerator, not just the first term. Encourage students to use parentheses when combining expressions to avoid this error.

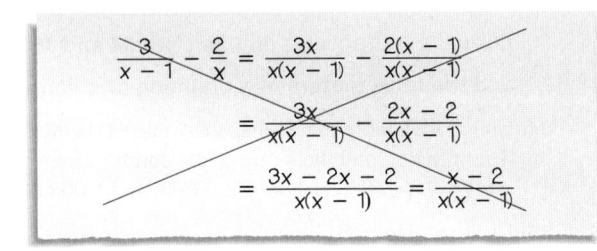

STUDENT HELP NOTES

Homework Help Students can find help for Exs. 39–41 at **www.mcdougallittell.com.** The information can be printed out for students who don't have access to the Internet.

34. Displays may vary depending on use of scientific notation and rounding. Note that the only value that is exact for the entire expression is when $x = 10$. Final zeros obtained by rounding have been dropped.
$x = 10$: 15; 8; 1.875
$x = 100$: 195; 98; 1.9898
$x = 1000$: 1995; 998; 1.999
$x = 10{,}000$: 19,995; 9998; 1.9999
$x = 100{,}000$: 199,995; 99,998; 2
$x = 1{,}000{,}000$: 2E6; 999,998; 2

FOCUS ON APPLICATIONS

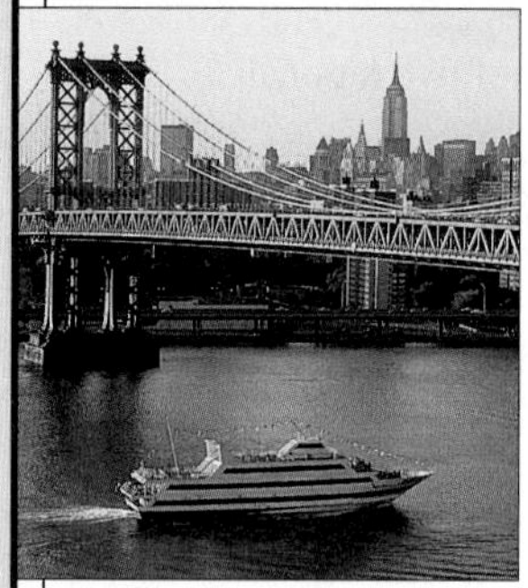

REAL LIFE **TRAVEL BY BOAT** Many cities with harbors offer a variety of types of water transportation. These include regular commuter service, sight-seeing tours, and recreational trips.

35. ***Sample answer*: They get very large; they get very large; they get very close to 2; as x gets very large, the numbers subtracted in the numerator and the denominator become insignificantly small compared to $2x$ and x, so the rational expression gets closer and closer to $\frac{2x}{x} = 2$.**

GEOMETRY CONNECTION **In Exercises 32 and 33, find an expression for the perimeter of the rectangle.**

32. Rectangle with length $\frac{x^2 - 3}{x - 2}$ and width $\frac{2x - 5}{x - 2}$. **$2(x + 4)$**

33. Rectangle with length $\frac{2x - 1}{x - 3}$ and width $\frac{3x + 5}{2x - 1}$. **$\frac{2(7x^2 - 8x - 14)}{(2x - 1)(x - 3)}$**

LOGICAL REASONING **In Exercises 34 and 35, use the expression $\frac{2x - 5}{x - 2}$ and the table feature of a graphing calculator or spreadsheet software.**

34. Construct a table that shows the value of the numerator, the value of the denominator, and the value of the entire rational expression when the value of x is 10, 100, 1000, 10,000, 100,000, and 1,000,000. **See margin.**

35. Use the table from Exercise 34. As x gets large, what happens to the values of the numerator? of the denominator? of the entire rational expression? Why do you think these results occur? **See margin.**

TRAVEL BY BOAT **In Exercises 36–38, use the following information.**
A boat moves through still water at x kilometers (km) per hour. It travels 24 km upstream against a current of 2 km per hour and then returns with the current. The rate upstream is $x - 2$ because the boat moves against the current, and the rate downstream is $x + 2$ because the boat moves with the current.

36. Write an expression for the total time for the round trip. **$\frac{24}{x - 2} + \frac{24}{x + 2}$**

37. Write your answer to Exercise 36 as a single rational expression. **$\frac{48x}{(x - 2)(x + 2)}$**

38. Use your answer to Exercise 37 to find how long the round trip will take if the boat travels 10 kilometers per hour through still water. **5 h**

MOWING LAWNS **In Exercises 39–41, use the following information.**
You are choosing a business partner for a student lawn-care business you are starting. It takes you an average of 35 minutes to mow a lawn, so your rate is 1 lawn in 35 minutes or $\frac{1}{35}$ of a lawn per minute. Let x represent the average time (in minutes) it takes a possible partner to mow a lawn.

39. Write an expression for the partner's rate (that is, the part of a lawn the partner can mow in 1 minute). Then write an expression for the combined rate of you and your partner (the part of a lawn that you both can mow in 1 minute if you work together). **$\frac{1}{x}$; $\frac{1}{35} + \frac{1}{x}$**

40. Write your answer to Exercise 39 as a single rational expression. **$\frac{x + 35}{35x}$**

41. The table shows the mowing times of possible partners.

Possible mowing partner	A	B	C
Mowing time (in minutes)	40	45	55

a. Use your expression from Exercise 40 to find the combined rate of you with each possible partner. **A: $\frac{3}{56}$; B: $\frac{16}{315}$; C: $\frac{18}{385}$**

b. Which partner(s) can you choose if you want a combined rate of $\frac{1}{20}$ of a lawn per minute or faster? **A or B**

STUDENT HELP

INTERNET **HOMEWORK HELP** Visit our Web site www.mcdougallittell.com for help with Exs. 39–41.

COMBINING OPERATIONS **Simplify the expression.**

42. $\frac{2x}{x+5} - \frac{3x+2}{x+5} - \frac{4}{x+5}$ $-\frac{x+6}{x+5}$

43. $\left(\frac{3x^2}{56}\right)\left(\frac{3}{x} + \frac{5}{x}\right)$ $\frac{3x}{7}$

44. $\left(\frac{3x-5}{x} + \frac{1}{x}\right) \div \left(\frac{x}{6x-8}\right)$ $\frac{2(3x-4)^2}{x^2}$

45. $\frac{x-2}{x+6} \div \frac{x+8}{4x-24} \cdot \frac{x-8}{x-2}$ $\frac{4(x-6)(x-8)}{(x+6)(x+8)}$

EXTENSION: TRINOMIAL DENOMINATORS **When you add rational expressions, you may need to factor a trinomial to find the LCD. Study the sample below. Then simplify the expressions in Exercises 46–49.**

Sample: $\frac{2x}{x^2-1} + \frac{3}{x^2+x-2} = \frac{2x}{(x+1)(x-1)} + \frac{3}{(x-1)(x+2)}$

The LCD is $(x+1)(x-1)(x+2)$.

Note: If you just used $(x^2-1)(x^2+x-2)$ as the common denominator, the factor $(x-1)$ would be included twice.

46. $\frac{2}{x-3} + \frac{x}{x^2+3x-18}$ $\frac{3(x+4)}{(x+6)(x-3)}$

47. $\frac{2}{x^2-4} + \frac{3}{x^2+x-6}$ $\frac{5x+12}{(x-2)(x+2)(x+3)}$

48. $\frac{7x+2}{16-x^2} + \frac{7}{x-4}$ $\frac{26}{(x-4)(x+4)}$

49. $\frac{5x-1}{2x^2-7x-15} - \frac{-3x+4}{2x^2+5x+3}$ **See margin.**

49. $\frac{8x^2-15x+19}{(2x+3)(x-5)(x+1)}$

Test Preparation

50. **MULTIPLE CHOICE** Find the LCD of $\frac{-2}{x+9}$ and $\frac{5x}{x^2+9x}$. B

(A) x^2+9 (B) $x(x+9)$ (C) $x(x-9)$ (D) $(x+9)(x^2+9x)$

51. **MULTIPLE CHOICE** Simplify the expression $\frac{x}{x-1} - \frac{1}{2x+1}$. C

(A) $\frac{x-1}{(x-1)(2x+1)}$ (B) $-\frac{x}{x-1}$

(C) $\frac{2x^2+1}{(x-1)(2x+1)}$ (D) $\frac{2x^2-1}{(x-1)(2x+1)}$

52. **MULTIPLE CHOICE** You are making a 350-mile car trip. You decide to drive a little faster to save time. Choose an expression for the time saved if the car's average speed s is increased by 5 miles per hour. C

(A) $\frac{350}{s+5}$ (B) $\frac{s+5}{350} - \frac{s}{350}$ (C) $\frac{350}{s} - \frac{350}{s+5}$ (D) $350(s+5) - 350s$

★ Challenge

AVERAGE SPEED **In Exercises 53–55, you will write and simplify a general expression for the average speed traveled when making a round trip. Let *d* represent the one-way distance. Let *x* represent the speed while traveling there and let *y* represent the speed while traveling back.**

53. Write an expression for the total time for the round trip. Use addition to write your answer as a single rational expression. $\frac{d}{x} + \frac{d}{y}$; $\frac{d(y+x)}{xy}$

54. To find the average speed on the trip, you need to divide the total distance by the total time. Use your answer from Exercise 53 to write an expression for the average speed. Then simplify that expression as far as possible.

54. $2d \div \frac{d(y+x)}{xy}$; $\frac{2xy}{x+y}$

55. What do you notice about the variables in the final answer? If your distance is doubled what happens to the average speed? **See margin.**

EXTRA CHALLENGE

www.mcdougallittell.com

! COMMON ERROR

EXERCISE 45 Some students may multiply the last two terms before dividing. Remind them that the same rules for the order of operations hold for rational expressions as for numbers. So, unless there are other grouping symbols, multiplication and division proceed in order from left to right.

55. The variable *d* that represents the distance is not involved in the expression, which indicates that the average speed is not dependent on the distance traveled; the average speed is not changed.

ADDITIONAL PRACTICE AND RETEACHING

For Lesson 11.6:

- Practice Levels A, B, and C (*Chapter 11 Resource Book,* p. 83)
- Reteaching with Practice (*Chapter 11 Resource Book,* p. 86)
- See Lesson 11.6 of the *Personal Student Tutor*

For more Mixed Review:

- Search the *Test and Practice Generator* for key words or specific lessons.

4 ASSESS

DAILY HOMEWORK QUIZ

Transparency Available

Simplify the expression.

1. $\frac{4x}{7x+2} - \frac{5x}{7x+2}$ $-\frac{x}{7x+2}$
2. $\frac{5}{2x} + \frac{1}{3x}$ $\frac{17}{6x}$
3. $\frac{4}{5x+10} - \frac{2x}{x+2}$ $\frac{4-10x}{5x+10}$
4. $\frac{3x}{3x+1} - \frac{x}{x-3}$ $-\frac{10x}{(3x+1)(x-3)}$
5. $\frac{x^2-4}{x+4} - \frac{x^2+16}{x-4}$ $\frac{-4(2x^2+5x+12)}{(x+4)(x-4)}$
6. $\left(\frac{x-4}{2x} + \frac{3}{2x}\right) \div \left(\frac{x-1}{4x}\right)$ 2

EXTRA CHALLENGE NOTE

Challenge problems for Lesson 11.6 are available in **blackline** format in the *Chapter 11 Resource Book,* p. 90 and at **www.mcdougallittell.com.**

ADDITIONAL TEST PREPARATION

1. **WRITING** Find the sum. Show your work and explain your steps. $\frac{x+1}{2x^2} + \frac{x-1}{3x}$ $\frac{2x^2+x+3}{6x^2}$; Check work and explanations.

ADDITIONAL RESOURCES

An alternative Quiz for Lessons 11.4–11.6 is available in the *Chapter 11 Resource Book,* p. 91.

63.

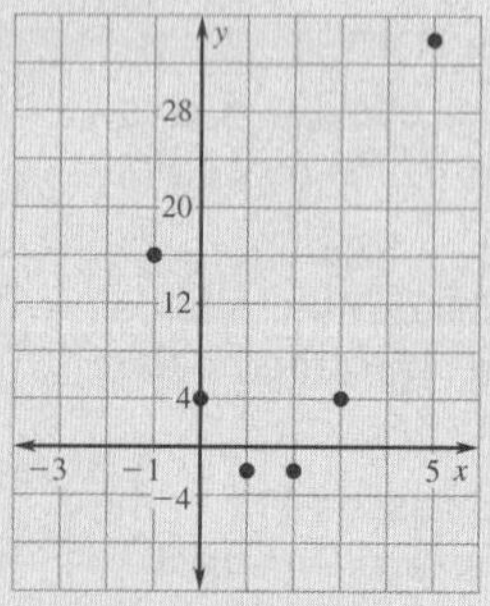

64. See Additional Answers beginning on page AA1.

MIXED REVIEW

SIMPLIFYING EXPRESSIONS **Simplify the expression.** (Review 8.3 for 11.7)

56. $\frac{5}{10x}$ $\frac{1}{2x}$
57. $\frac{4m^2}{6m}$ $\frac{2m}{3}$
58. $\frac{16x^4}{32x^8}$ $\frac{1}{2x^4}$
59. $\frac{42x^4y^3}{6x^3y^9}$ $\frac{7x}{y^6}$

STANDARD FORM **Write the equation in standard form.** (Lesson 9.5 for 11.7)

60. $6x^2 = 5x - 7$ $6x^2 - 5x + 7 = 0$
61. $9 - 6x = 2x^2$ $2x^2 + 6x - 9 = 0$
62. $-4 + 3y^2 = y$ $3y^2 - y - 4 = 0$

CHOOSING MODELS FROM DATA **Make a scatter plot of the data. Then tell whether a *linear, exponential,* or *quadratic* model fits the data.** (Review 9.8) 63–64. See margin for plots.

63. (−1, 16), (0, 4), (1, −2), (2, −2), (3, 4), (5, 34) quadratic
64. (−5, 6), (−4, 3), (−2, −3), (−1, −6), (0, −9), (1, −12) linear

65. **GAME SHOW** A contestant on a television game show must guess the price of a trip within \$1000 of the actual price in order to win. The actual price of the trip is \$8500. Write an absolute-value inequality that shows the range of possible guesses that will win the trip. (Review 6.4) $|x - 8500| \leq 1000$

QUIZ 2

Self-Test for Lessons 11.4–11.6

Simplify the expression if possible. (Lesson 11.4)

1. $\frac{15x^2}{10x}$ $\frac{3x}{2}$
2. $\frac{5x}{11x + x^2}$ $\frac{5}{11+x}$
3. $\frac{3-x}{x^2 - 5x + 6}$ $-\frac{1}{x-2}$
4. $\frac{x^2 - 7x + 12}{x^2 + 3x + 18}$ cannot be simplified

Simplify the expression. (Lessons 11.5 and 11.6)

5. $\frac{5x^2}{2x} \cdot \frac{14x^2}{10x}$ $\frac{7x^2}{2}$
6. $\frac{5}{10+4x} \cdot (20 + 8x)$ 10
7. $\frac{3x+12}{4x} \div \frac{x+4}{2x}$ $\frac{3}{2}$
8. $\frac{5x^2 - 30x + 45}{x+2} \div (5x - 15)$ $\frac{x-3}{x+2}$
9. $\frac{x}{x^2 - 2x - 35} + \frac{5}{x^2 - 2x - 35}$ $\frac{1}{x-7}$
10. $\frac{4x-1}{3x^2 + 8x + 5} - \frac{x-6}{3x^2+8x+5}$ $\frac{1}{x+1}$

CANOEING **In Exercises 11–13, use the following information.**
You are on a canoe trip. You can paddle your canoe at a rate of x miles per hour in still water. The stream is flowing at a rate of 2 miles per hour, so your rate of travel downstream (with the current) is $x + 2$ miles per hour. You travel 15 miles downstream and 15 miles back upstream. (Lesson 11.6)

11. Write an expression for the travel time downstream and an expression for the travel time upstream. $\frac{15}{x+2}$; $\frac{15}{x-2}$
12. Write and simplify an expression for the total travel time. $\frac{15}{x+2} + \frac{15}{x-2} = \frac{30x}{(x+2)(x-2)}$
13. Find the total travel time if your rate of paddling in still water is about 4 miles per hour. 10 h

ACTIVITY 11.7

Developing Concepts

GROUP ACTIVITY
Work with a partner.

MATERIALS
algebra tiles

Group Activity for use with Lesson 11.7

Modeling Polynomial Division

▶ QUESTION **How can you use algebra tiles to model division of polynomials?**

▶ EXPLORING THE CONCEPT

You can use algebra tiles to divide $x^2 + 4x + 4$ by $x + 3$ as follows.

1 Use algebra tiles to model $x^2 + 4x + 4$.

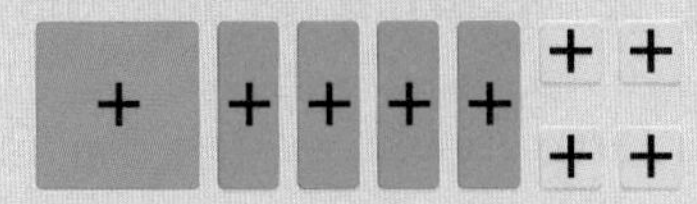

2 Use the tiles to create a length of $x + 3$.

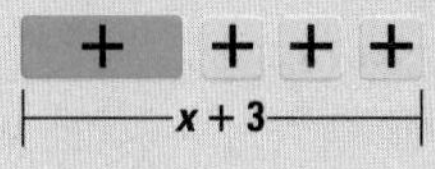

STUDENT HELP

Look Back
For help with algebra tiles, see pp. 575 and 603.

3 Keeping $x + 3$ as the length, try to create a rectangle that uses all the tiles from **Step 1**. Explain why some tiles cannot be used. *Sample answer:* **$x + 3$ does not divide evenly into $x^2 + 4x + 4$.**

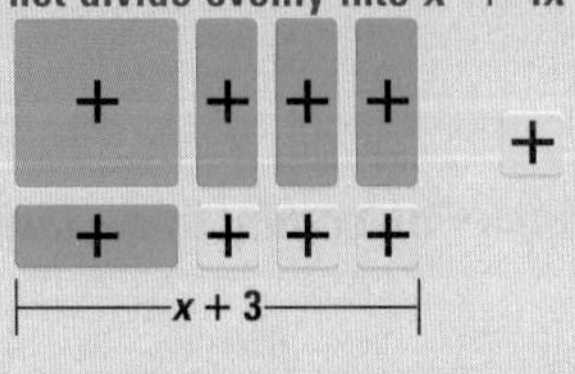

4 The width of the rectangle is the quotient and the leftover tiles are the remainder. Give the quotient and the remainder when you divide $x^2 + 4x + 4$ by $x + 3$. **quotient = $x + 1$, remainder = 1**

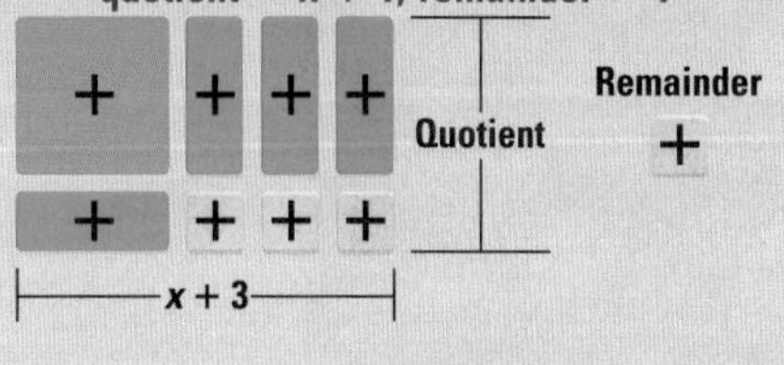

▶ DRAWING CONCLUSIONS

Use algebra tiles to decide whether the polynomial can be divided evenly. Make a sketch of your explanation. Compare your result with that of your partner. Then decide together on the quotient and the remainder if any. **1–4. See margin.**

1. $(x^2 + 4x + 8) \div (x + 2)$

2. $(x^2 + 7x + 8) \div (x + 1)$

3. $(x^2 + 6x + 12) \div (x + 4)$

4. $(x^2 + 9x + 25) \div (x + 5)$

1. **cannot be divided evenly; quotient = $x + 2$, remainder = 4**
2. **cannot be divided evenly; quotient = $x + 6$, remainder = 2**
3. **cannot be divided evenly; quotient = $x + 2$, remainder = 4**
4. **cannot be divided evenly; quotient = $x + 4$, remainder = 5**

5. Use the model at the right to find the missing values in the division.

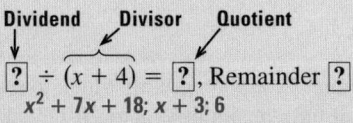

Dividend **Divisor** **Quotient**

$\boxed{?} \div (x + 4) = \boxed{?}$, Remainder $\boxed{?}$

$x^2 + 7x + 18$; $x + 3$; 6

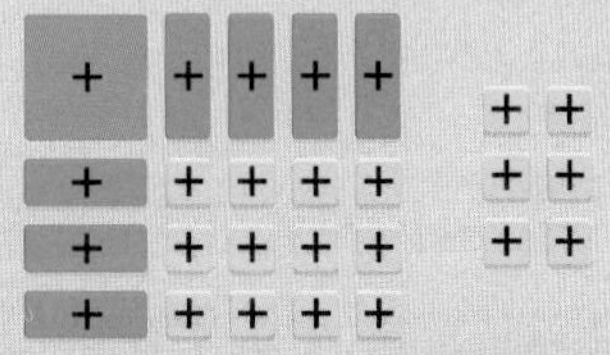

6. With polynomial division, as with whole number division, you can check your work by multiplying the divisor by the quotient and then adding the remainder. Use the model in Exercise 5 to explain why this method works. Then use polynomial multiplication to check the division in Exercise 5. ***Sample answer*: The model without the remainder represents the product $(x + 4)(x + 3)$, so the whole model represents the product of the quotient and the divisor, plus the remainder; $(x + 4)(x + 3) = x^2 + 7x + 12$; $x^2 + 7x + 12 + 6 = x^2 + 7x + 18$.**

1 Planning the Activity

PURPOSE
To explore polynomial division using algebra tiles.

MATERIALS
- algebra tiles for each pair of students
- Activity Support Master (*Chapter 11 Resource Book,* p. 95)

PACING
- Exploring the Concept — 10 min
- Drawing Conclusions — 15 min

▶ *LINK TO LESSON*
As students divide in Example 1 on page 684, they should realize that they could use algebra tiles, though this would be cumbersome with these coefficients.

2 Managing the Activity

COOPERATIVE LEARNING
Each student in a pair can make an algebra tile model of each problem. Students can then compare models.

CLASSROOM MANAGEMENT
You may need to review the definitions of dividend, divisor, quotient, and remainder using real numbers prior to this activity.

3 Closing the Activity

★ *KEY DISCOVERY*
Division of polynomials with positive terms can be modeled using area. In effect, you are looking for the width of a rectangle knowing its area and length.

ACTIVITY ASSESSMENT
JOURNAL Explain how to model dividing polynomials with algebra tiles using $(x^2 + 5x + 8) \div (x + 2)$ as an example. Use the vocabulary *dividend, divisor, quotient,* and *remainder.*
See sample answer at left.

Activity Assessment *Sample answer:*
A good entry should include a description of finding the proper length and the proper width, an account of the remainder tiles, all vocabulary terms, and the correct quotient: $x + 3$, remainder 2.

1 PLAN

PACING
Basic: 2 days
Average: 2 days
Advanced: 2 days
Block Schedule: 1 block

LESSON OPENER
VISUAL APPROACH
An alternative way to approach Lesson 11.7 is to use the Visual Approach Lesson Opener:

- Blackline Master (*Chapter 11 Resource Book,* p. 96)
- Transparency (p. 80)

MEETING INDIVIDUAL NEEDS

- ***Chapter 11 Resource Book***
  Prerequisite Skills Review (p. 5)
  Practice Level A (p. 97)
  Practice Level B (p. 98)
  Practice Level C (p. 99)
  Reteaching with Practice (p. 100)
  Absent Student Catch-Up (p. 102)
  Challenge (p. 104)
- ***Resources in Spanish***
- ***Personal Student Tutor***

NEW-TEACHER SUPPORT
See the Tips for New Teachers on pp. 1–2 of the *Chapter 11 Resource Book* for additional notes about Lesson 11.7.

WARM-UP EXERCISES

Transparency Available

Use long division to find each quotient and remainder.

1. 217 ÷ 31 **7 remainder 0**
2. 257 ÷ 12 **21 remainder 5**
3. 2295 ÷ 51 **45 remainder 0**
4. 448 ÷ 19 **23 remainder 11**

11.7 Dividing Polynomials

What you should learn

GOAL 1 Divide a polynomial by a monomial or by a binomial factor.

GOAL 2 Use polynomial long division.

Why you should learn it

▼ To provide alternative forms of rational expressions to model **real-life** situations, as in the sports equipment problem in **Exs. 50–54.**

1. ***Sample answer*: Write the division as a rational expression; divide by dividing each term by the denominator:** $\frac{2x^2-3}{6x} = \frac{2x^2}{6x} - \frac{3}{6x} = \frac{x}{3} - \frac{1}{2x}$.

2. ***Sample answer*: Multiply** $\frac{x}{3} - \frac{1}{2x}$ **by** $6x$: $6x\left(\frac{x}{3} - \frac{1}{2x}\right) = 2x^2 - 3$.

GOAL 1 DIVIDING A POLYNOMIAL

To divide a polynomial by a *monomial,* divide each term by the monomial, keeping the same signs between terms. Simplify each fraction.

ACTIVITY
Developing Concepts

Investigating Polynomial Division **1–2. See margin.**

1 With a partner, discuss how to get from the original division problem to the simplified expression.

$$(2x^2 - 3) \div (6x) = \frac{?}{6x} = \frac{2x^2}{?} - \frac{?}{?} = \frac{x}{3} - \frac{1}{2x}$$

2 Explain how you could use multiplication to check that $(2x^2 - 3) \div (6x) = \frac{x}{3} - \frac{1}{2x}$. Carry out the check.

EXAMPLE 1 ***Dividing a Polynomial by a Monomial***

Divide $12x^2 - 20x + 8$ by $4x$.

SOLUTION

$$\frac{12x^2 - 20x + 8}{4x} = \frac{12x^2}{4x} - \frac{20x}{4x} + \frac{8}{4x} \qquad \text{Divide each term of numerator by } 4x.$$

$$= \frac{3x(4x)}{4x} - \frac{5(4x)}{4x} + \frac{2(4)}{4x} \qquad \text{Find common factors.}$$

$$= \frac{3x\cancel{(4x)}}{\cancel{4x}} - \frac{5\cancel{(4x)}}{\cancel{4x}} + \frac{2\cancel{(4)}}{\cancel{4}x} \qquad \text{Divide out common factors.}$$

$$= 3x - 5 + \frac{2}{x} \qquad \text{Simplified form}$$

To divide a polynomial by a *binomial,* first look to see whether the numerator and the denominator have a common factor. If they do, you can divide out the common factor.

$$\frac{x^2 - 2x - 3}{x - 3} = \frac{(x + 1)\cancel{(x - 3)}}{\cancel{x - 3}}$$

$$= x + 1$$

GOAL 2 USING LONG DIVISION

You can use **polynomial long division** to divide polynomials that do not have common factors. First review the process for long division in arithmetic.

EXAMPLE 2 *The Long Division Algorithm*

Use long long division to divide 658 by 28.

SOLUTION

$$\begin{array}{r} 23 \\ 28\overline{)658} \\ \underline{56} \\ 98 \\ \underline{84} \\ 14 \end{array}$$

1. Think: $\frac{65}{28} \approx 2$
2. Subtract 2×28 from 65.
3. Bring down the 8. Think: $\frac{98}{28} \approx 3$
4. Subtract 3×28 from 98.
5. Remainder is 14.

Dividend, Quotient, Remainder, Divisor:

$$\frac{658}{28} = 23 + \frac{14}{28} = 23 + \frac{1}{2} = 23\frac{1}{2}$$

STUDENT HELP

Study Tip
To check division, you can multiply the answer by the divisor OR you can multiply the quotient by the divisor and add the remainder. You should get the dividend in either way. In Example 2, $28\left(23\frac{1}{2}\right) = 658$ and $(28)(23) + 14 = 658$.

EXAMPLE 3 *Polynomial Long Division*

Divide $x^2 + 2x + 4$ by $x - 1$.

SOLUTION

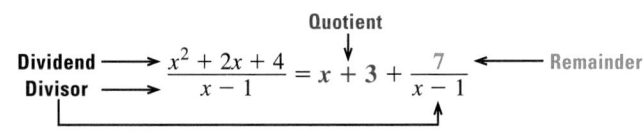

$$\begin{array}{r} x + 3 \\ x - 1\overline{)x^2 + 2x + 4} \\ \underline{x^2 - x} \\ 3x + 4 \\ \underline{3x - 3} \\ 7 \end{array}$$

1. Think: $\frac{x^2}{x} = x$
2. Subtract $x(x - 1)$.
3. Bring down $+ 4$. Think: $\frac{3x}{x} = 3$
4. Subtract $3(x - 1)$.
5. Remainder is 7.

To subtract $(-x)$ from $2x$, you add x.

Dividend, Divisor, Quotient, Remainder:

$$\frac{x^2 + 2x + 4}{x - 1} = x + 3 + \frac{7}{x - 1}$$

▶ The answer is $x + 3 + \frac{7}{x - 1}$.

2 TEACH

MOTIVATING THE LESSON
Dividing polynomials seems very arcane to students—and many have dreaded long division from years back. Inform them that, besides helping to simplify rational models, polynomial division provides an important tool for finding solutions to all sorts of polynomial equations that turn up in science, economics, and other fields.

ACTIVITY NOTE
Inform students that the method developed in the activity on p. 684 applies only when dividing a polynomial by a monomial. It is not helpful to separate the terms of a polynomial when dividing it by a polynomial that is not a monomial.

EXTRA EXAMPLE 1
Divide $9x^2 + 12x - 18$ by $3x$. $3x + 4 - \frac{6}{x}$

EXTRA EXAMPLE 2
Use long division to divide 453 by 18. $25\frac{1}{6}$

EXTRA EXAMPLE 3
Divide $3x^2 + 7x - 12$ by $x + 3$. $3x - 2 - \frac{6}{x + 3}$

CHECKPOINT EXERCISES

For use after Example 1:

1. Divide $4p^2 - 10p$ by $-2p$. $-2p + 5$

For use after Examples 2 and 3:

2. Divide $x^2 - 3x + 5$ by $x + 2$. $x - 5 + \frac{15}{x + 2}$

CONCEPT QUESTION
How can you check the answer to Extra Example 3? *Sample answer:* **Multiply the quotient $3x - 2$ by the divisor $x + 3$. Add the remainder -6. This should equal the dividend $3x^2 + 7x - 12$.**

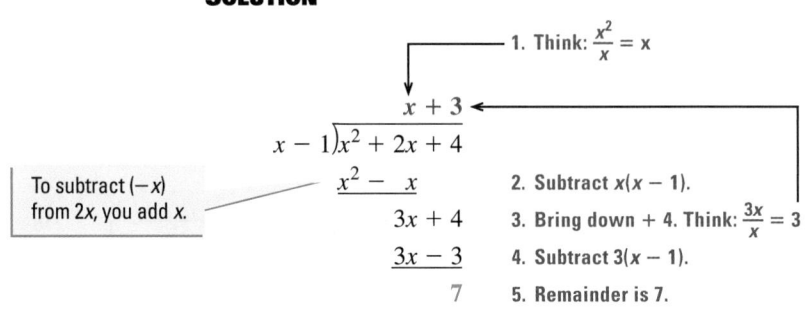

EXTRA EXAMPLE 4
Divide $5n^2 + 2$ by $n + 1$.
$5n - 5 + \frac{7}{n+1}$

EXTRA EXAMPLE 5
Divide $x - 12 + 3x^2$ by $3x + 1$.
$x - \frac{12}{3x+1}$

CHECKPOINT EXERCISES
For use after Examples 4 and 5:
1. Divide $1 - x^2$ by $x - 2$.
$-x - 2 - \frac{3}{x-2}$

STUDENT HELP NOTES
Homework Help Students can find extra examples at **www.mcdougallittell.com** that parallel the examples in the student edition.

FOCUS ON VOCABULARY
What are the *dividend, divisor,* and *quotient* in a division problem?
Sample answer: **The dividend is the expression you are dividing; the divisor is the expression you are dividing the dividend by; and the quotient is the resulting expression.**

CLOSURE QUESTION
When dividing a polynomial by a binomial, what should you do before you begin the long division process?
Sample answer: **Check to see if the dividend and divisor share a common factor. If they do, factor it out.**

DAILY PUZZLER
Find values of a and b so that the following is a true statement.
$(4x^2 + ax + 4) \div (bx + 3) =$
$2x - 1 + \frac{7}{2x+3}$
$a = 4, b = 2$

When dividing polynomials, remember that the polynomials need to be written in standard form. That is, the terms should be written in descending order according to their degrees.

If a term is missing in the standard form of the dividend, insert a term with a coefficient of 0 to hold the place.

EXAMPLE 4 *Adding a Place-holding Zero*

Divide $6x^2 - 11$ by $x + 2$.

SOLUTION

First insert a place-holding term $0x$ in the dividend: $6x^2 - 11 = 6x^2 + 0x - 11$.

$$\begin{array}{r|l} 6x - 12 & \text{1. Think: } \frac{6x^2}{x} = 6x \\ x + 2\overline{)6x^2 + 0x - 11} & \\ \underline{6x^2 + 12x} & \text{2. Subtract } (6x)(x+2). \\ -12x - 11 & \text{3. Bring down } -11\text{. Think: } \frac{-12x}{x} = -12 \\ \underline{-12x - 24} & \text{4. Subtract } -12(x+2). \\ 13 & \text{5. Remainder is 13.} \end{array}$$

▶ The answer is $6x - 12 + \frac{13}{x+2}$.

EXAMPLE 5 *Rewriting in Standard Form*

Divide $m^2 - 16m - 21$ by $4 + 2m$.

SOLUTION

First write the divisor $4 + 2m$ in standard form as $2m + 4$.

$$\begin{array}{r|l} \frac{1}{2}m - 9 & \text{1. Think: } \frac{m^2}{2m} = \frac{1}{2}m \\ 2m + 4\overline{)m^2 - 16m - 21} & \\ \underline{m^2 + 2m} & \text{2. Subtract } \frac{1}{2}m(2m+4). \\ -18m - 21 & \text{3. Bring down } -21\text{. Think: } \frac{-18m}{2m} = -9 \\ \underline{-18m - 36} & \text{4. Subtract } -9(2m+4). \\ 15 & \text{5. Remainder is 15.} \end{array}$$

▶ The answer is $\frac{1}{2}m - 9 + \frac{15}{2m+4}$.

STUDENT HELP
INTERNET
HOMEWORK HELP
Visit our Web site www.mcdougallittell.com for extra examples.

GUIDED PRACTICE

Vocabulary Check ✓

1. How is polynomial long division like long division with whole numbers? How is it different? ***Sample answer:*** **The method is the same; the quantities used are polynomials instead of digits.**

Concept Check ✓

2. For which of the divisions in Exercises 9–14 do you not need to use long division? Explain why not.

2. Exs. 10, 11, 13; *Sample answer:* In Exs. 10 and 13, you are dividing by a monomial. In Ex. 11, you can divide by factoring the dividend and then canceling.

3. For the student's long division work shown at the right, complete the answer. $\boldsymbol{x + 1 - \frac{4}{x-1}}$

$$\frac{x^2 - 5}{x - 1} = \text{Quotient} + \frac{\text{Remainder}}{\text{Divisor}}$$

$$= \boxed{?} + \frac{\boxed{?}}{\boxed{?}}$$

$$\begin{array}{r} x + 1 \\ x - 1 \overline{)x^2 + 0x - 5} \\ \underline{x^2 - x} \\ x - 5 \\ \underline{x - 1} \\ -4 \end{array}$$

4. Explain how to check the answer to the division in Exercise 3. ***Sample answer:*** **Multiply the quotient by the divisor and add the remainder; the result should be the dividend.**

Skill Check ✓

Set up the long division problem, but do not perform the division.

5. Divide $y^2 + 8$ by $y + 2$. $\boldsymbol{y + 2\overline{)y^2 + 0y + 8}}$

6. Divide $-x^2 - 4x + 21$ by $-x + 3$. $\boldsymbol{-x + 3\overline{)-x^2 - 4x + 21}}$

7. Divide $8y^2 - 2y$ by $3y + 5$. $\boldsymbol{3y + 5\overline{)8y^2 - 2y + 0}}$

8. Divide $72 - 18x + x^2$ by $x - 6$. $\boldsymbol{x - 6\overline{)x^2 - 18x + 72}}$

Divide.

9. Divide 856 by 29. $\boldsymbol{29\frac{15}{29}}$

10. Divide $18x^2 + 45x - 36$ by $9x$. $\boldsymbol{2x + 5 - \frac{4}{x}}$

11. Divide $x^2 - 8x + 15$ by $x - 3$. $\boldsymbol{x - 5}$

12. Divide $y^2 + 6y + 2$ by $y + 3$. $\boldsymbol{y + 3 - \frac{7}{y+3}}$

13. Divide $10b^3 - 8b^2 - 5b$ by $-2b$. $\boldsymbol{-5b^2 + 4b + \frac{5}{2}}$

14. Divide $2x^2 - x + 4$ by $3x - 6$. $\boldsymbol{\frac{2}{3}x + 1 + \frac{10}{3x - 6}}$

PRACTICE AND APPLICATIONS

STUDENT HELP

Extra Practice to help you master skills is on p. 807.

DIVIDING BY A MONOMIAL **Divide.**

15. Divide $8x + 13$ by 2. $\boldsymbol{4x + \frac{13}{2}}$

16. Divide $16y - 9$ by 4. $\boldsymbol{4y - \frac{9}{4}}$

17. Divide $9c^2 + 3c$ by c. $\boldsymbol{9c + 3}$

18. Divide $9m^3 + 4m^2 - 8m$ by m. $\boldsymbol{9m^2 + 4m - 8}$

19. Divide $-2x^2 - 12x$ by $-2x$. $\boldsymbol{x + 6}$

20. Divide $7p^3 + 18p^2$ by p^2. $\boldsymbol{7p + 18}$

21. Divide $9a^2 - 54a - 36$ by $3a$. $\boldsymbol{3a - 18 - \frac{12}{a}}$

22. Divide $16y^3 - 36y^2 - 64$ by $-4y^2$. $\boldsymbol{-4y + 9 + \frac{16}{y^2}}$

CHECKING QUOTIENTS AND REMAINDERS **Match the polynomial division problem with the correct answer.**

STUDENT HELP

HOMEWORK HELP
Example 1: Exs. 15–22
Example 2: Exs. 23–26
Example 3: Exs. 23–46
Example 4: Exs. 27–46
Example 5: Exs. 27–46

23. $(5x^2 + 2x + 3) \div (x + 2)$ **D**

24. $(12x - 5) \div (2x - 3)$ **A**

25. $(10x^2 - 7x - 12) \div (2x - 3)$ **B**

26. $(2x^2 + 5x + 2) \div (x + 2)$ **C**

A. $6 + \frac{13}{2x - 3}$

B. $5x + 4$

C. $2x + 1$

D. $5x - 8 + \frac{19}{x + 2}$

3 APPLY

ASSIGNMENT GUIDE

BASIC
Day 1: pp. 687–688 Exs. 16–22 even, 23–26, 28–50 even, 51
Day 2: pp. 687–689 Exs. 15–21 odd, 27–49 odd, 55, 61–65, 69–72

AVERAGE
Day 1: pp. 687–688 Exs. 16–22 even, 23–26, 28–50 even, 51
Day 2: pp. 687–689 Exs. 15–21 odd, 27–49 odd, 52, 53, 55, 61–65, 69–72

ADVANCED
Day 1: pp. 687–688 Exs. 16–22 even, 23–26, 28–50 even, 51
Day 2: pp. 687–689 Exs. 15–21 odd, 27–49 odd, 52–65, 69–72

BLOCK SCHEDULE
pp. 687–689 Exs. 15–53, 55, 61–65, 69–72

EXERCISE LEVELS
Level A: ***Easier***
15–22
Level B: ***More Difficult***
23–53, 55, 61–72
Level C: ***Most Difficult***
54, 56–60

✓ HOMEWORK CHECK
To quickly check student understanding of key concepts, go over the following exercises: Exs. 16, 20, 25, 30, 36, 44, 48. See also the Daily Homework Quiz:
- Blackline Master (*Chapter 11 Resource Book,* p. 107)
- Transparency (p. 92)

Computational errors are common in these problems. Remind students to check their solutions by multiplying the divisor and quotient and then adding the remainder. Caution them first to write the polynomials in standard form and add any necessary place-holding zeros prior to using the long division algorithm.

DATA UPDATE
Updated data for Exs. 50–54 is available at www.mcdougallittell.com.

DIVIDING POLYNOMIALS **Divide.**

27. Divide $a^2 - 3a + 2$ by $a - 1$. $\ a - 2$

28. Divide $y^2 + 5y + 7$ by $y + 2$. $\ y + 3 + \dfrac{1}{y + 2}$

29. Divide $2b^2 - 3b - 4$ by $b - 2$. $\ 2b + 1 - \dfrac{2}{b - 2}$

30. Divide $3p^2 + 10p + 3$ by $p + 3$. $\ 3p + 1$

31. Divide $5g^2 + 14g - 2$ by $g + 3$. $\ 5g - 1 + \dfrac{1}{g + 3}$

32. Divide $c^2 - 25$ by $c - 5$. $\ c + 5$

33. Divide $x^2 - 3x - 59$ by $x + 6$. $\ x - 9 - \dfrac{5}{x - 9}$

34. Divide $d^2 + 15d + 45$ by $d + 5$. $\ d + 10 - \dfrac{5}{d + 5}$

35. Divide $-x^2 - 6x - 16$ by $x + 2$. $\ -x - 4 - \dfrac{8}{x + 2}$

36. Divide $-x^2 + 9x - 12$ by $-x - 2$. $\ x - 11 + \dfrac{34}{x + 2}$

37. Divide $b^2 - 7b + 4$ by $b + 3$. $\ b - 10 + \dfrac{34}{b + 3}$

38. Divide $5 - 7m + 3m^2$ by $m - 3$. $\ 3m + 2 + \dfrac{11}{m - 3}$

39. Divide $x^2 + 9$ by $-x - 4$. $\ -x + 4 - \dfrac{25}{x + 4}$

40. Divide $4x^2 + 12x - 10$ by $x - 2$. $\ 4x + 20 + \dfrac{30}{x - 2}$

41. Divide $-5m^2 + 2$ by $m - 1$. $\ -5m - 5 - \dfrac{3}{m - 1}$

42. Divide $4 + 11q + 6q^2$ by $2q + 1$. $\ 3q + 4$

43. Divide $4 - s^2$ by $s + 5$. $\ -s + 5 - \dfrac{21}{s + 5}$

44. Divide $16a^2 - 25$ by $3 + 4a$. $\ 4a - 3 - \dfrac{16}{4a + 3}$

45. Divide $c^2 - 7c + 21$ by $2c - 6$.
See margin.

46. Divide $b^2 - 7b - 12$ by $4b + 4$.
See margin.

45. $\dfrac{1}{2}c - 2 + \dfrac{9}{2c - 6}$

46. $\dfrac{1}{4}b - 2 - \dfrac{1}{b + 1}$

51. $\dfrac{2}{9} - \dfrac{8}{9t + 99}$

54. The ratio is increasing because as t increases, $9t + 99$ increases, so $\dfrac{8}{9t + 99}$ decreases. That is, the amount subtracted from $\dfrac{2}{9}$ decreases, so the value of the rational expression increases.

GEOMETRY **CONNECTION** The area and one dimension of the rectangle are shown. **Find the missing dimension.**

47.

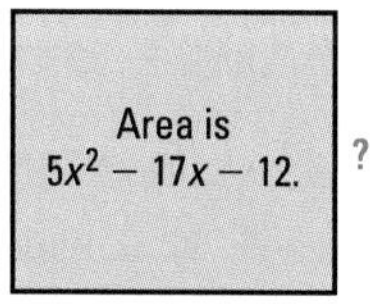

Area is $5x^2 - 17x - 12$. ?

$x - 4$ $5x + 3$

48.

Area is $7x^2 + 17x + 6$. ?

$x + 2$ $7x + 3$

49.

Area is $12x^2 + 5x - 2$. ?

$4x - 1$ $3x + 2$

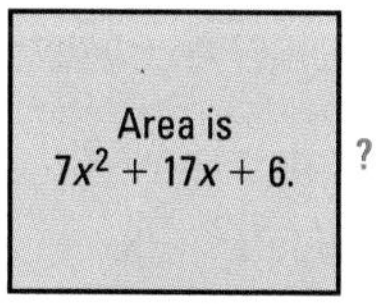

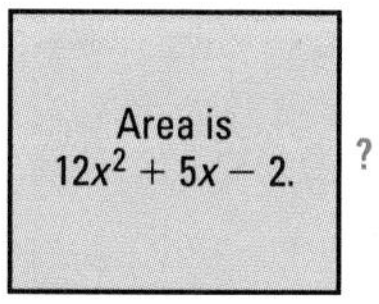

EXERCISE EQUIPMENT
In 1997 Americans spent about 3 billion dollars on exercise equipment.
DATA UPDATE of National Sporting Goods Association data at mcdougallittell.com

EXERCISE AND OTHER SPORTS EQUIPMENT In Exercises 50–54, use the models below that approximate spending in the United States from 1988 to 1997. Let t represent the number of years since 1988.

Dollars spent on exercise equipment (in millions): $E = 200t + 1400$

Total dollars spent on sports equipment (in millions): $S = 900t + 9900$

50. Write a rational model for the ratio of the money spent on exercise equipment to the total money spent on sports equipment. Simplify the model by dividing out the greatest common factor. $\dfrac{2t + 14}{9t + 99}$

51. Use long division to write the model from Exercise 50 in another form. See margin.

52. Copy and complete the table. Use the rewritten model from Exercise 51. Round to the nearest thousandth.

Year t	0	1	2	3	4	5	6	7	8	9
Ratio of E to S	?	?	?	?	?	?	?	?	?	?

0.141 0.148 0.154 0.159 0.163 0.167 0.170 0.173 0.175 0.178

53. Use the table. Was the ratio increasing or decreasing from 1988 to 1997? increasing

54. **CRITICAL THINKING** Look back at the model you found in Exercise 51. How can you tell from the model if the ratio is increasing or if it is decreasing? See margin.

ADDITIONAL PRACTICE AND RETEACHING

For Lesson 11.7:
• Practice Levels A, B, and C (*Chapter 11 Resource Book*, p. 97)
• Reteaching with Practice (*Chapter 11 Resource Book*, p. 100)
• See Lesson 11.7 of the *Personal Student Tutor*

For more Mixed Review:
• Search the *Test and Practice Generator* for key words or specific lessons.

55. c. $\frac{7}{2} = 3.5; \frac{19}{9} \approx 2.111;$

$\frac{12}{7} \approx 1.714; \frac{39}{29} \approx 1.345;$

$\frac{32}{27} \approx 1.185; \frac{47}{42} \approx 1.119$

★ **Challenge**

59. *Sample answer*: The quotient is one more than the number being squared, and the remainder is 1;

$x^2 \div (x - 1) =$

$x + 1 + \frac{1}{x-1}.$

EXTRA CHALLENGE

↳ www.mcdougallittell.com

55. MULTI-STEP PROBLEM Suppose you are 14 years old and your brother is 4 years old.

a. In t years, your age will be $14 + t$. What will your brother's age be? **$4 + t$**

b. Write the ratio of your age in t years to your brother's age in t years. Then use long division to rewrite this ratio. $\frac{14 + t}{4 + t}; 1 + \frac{10}{t + 4}$

c. Use the rewritten ratio to find the ratio of your ages now, in 5 years, in 10 years, in 25 years, in 50 years, and in 80 years. **See margin.**

d. Use your answers to part (c). Is the ratio of your ages getting smaller or larger as time goes by? What value do the ratios approach? **smaller; 1**

e. *Writing* Look at the original form of the ratio and the rewritten form of the ratio. Which form of the ratio makes it easier for you to recognize the trend that you described in part (d)? Explain your choice. **See margin.**

AREA MODELS In Exercises 56–60, use the diagrams.

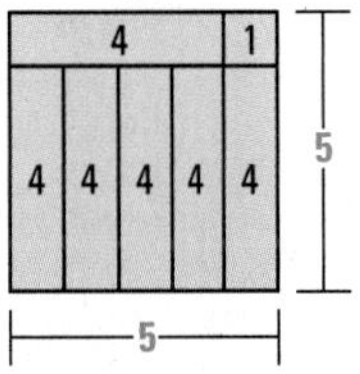

56. Diagram 1 models $5^2 \div 4$. Write the quotient and the remainder. **quotient = 6, remainder = 1**

57. What division problem does Diagram 2 model? Write the quotient and the remainder. **See margin.**

58. Draw diagrams to model $4^2 \div 3$ and $7^2 \div 6$. **See margin.**

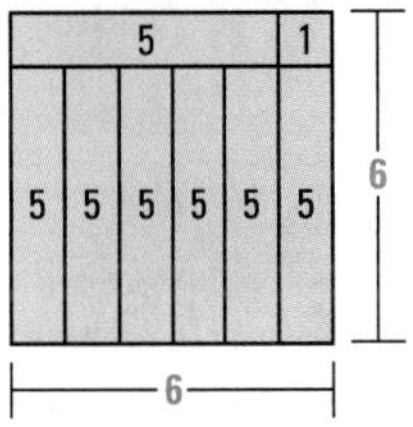

59. LOGICAL REASONING Describe any pattern you observe in Exercises 56–58. Use the pattern to predict the quotient and remainder when x^2 is divided by $x - 1$.

60. Use long division to find $x^2 \div (x - 1)$. Did you get the result you predicted in Exercise 59?

$x + 1 + \frac{1}{x-1}$; yes

Diagram 1

Diagram 2

MIXED REVIEW

55. e. *Sample answer*: Given the form $\frac{14 + t}{4 + t}$, you can see that as t gets larger and larger, the numbers added in the numerator and denominator become insignificant in comparison and that the ratios approach $\frac{t}{t}$ or 1. Given the form $1 + \frac{10}{t + 4}$, you can see that as t gets larger and larger, $\frac{10}{t + 4}$ approaches 0, so that $1 + \frac{10}{t + 4}$ approaches 1.

57. $6^2 \div 5$; quotient = 7, remainder = 1

SOLVING EQUATIONS Solve the equation. (Review 3.2 for 11.8)

61. $\frac{x}{5} = 3$ **15** **62.** $\frac{a}{-3} = 7$ **−21** **63.** $\frac{c}{4} = \frac{6}{8}$ **3** **64.** $\frac{y}{-2} = \frac{5}{4}$ **$-2\frac{1}{2}$**

SOLVING PROPORTIONS In Exercises 65–70, solve the proportion. Check for extraneous solutions. (Review 11.1 for 11.8)

65. $\frac{7}{5} = \frac{2}{x}$ **$1\frac{3}{7}$** **66.** $\frac{2}{x} = \frac{x-1}{6}$ **−3, 4** **67.** $\frac{6x - 7}{4} = \frac{5}{x}$ **$-1\frac{1}{3}, 2\frac{1}{2}$**

68. $\frac{8b^2 + 4b}{4b} = \frac{2b - 5}{3}$ **−2** **69.** $\frac{5p^2 - 9}{5} = \frac{2p^2 + 3p}{2}$ **$-1\frac{1}{5}$** **70.** $\frac{a^2 - 4}{a - 2} = \frac{a + 2}{10}$ **−2**

71. GRAPHING INVERSE VARIATION Use the model $y = \frac{12}{x}$. Make a table of values for $x = 1, 2, 3, 4,$ and 5. Sketch the graph. (Review 11.3 for 11.8) **12; 6; 4; 3; 2.4 See margin.**

72. [BIOLOGY CONNECTION] The largest mammal, a blue whale, has a weight of 1.3×10^5 kilograms. The smallest mammal, a pygmy shrew, has a weight of 2.0×10^{-3} kilogram. What is the ratio of the weight of a blue whale to the weight of a pygmy shrew? (Review 8.4) **6.5×10^7**

DAILY HOMEWORK QUIZ

📑 *Transparency Available*

1. Divide $5 - 3x$ by 5. $1 - \frac{3x}{5}$

2. Divide $4a^3 - 10a$ by $2a$. $2a^2 - 5$

3. Divide $4m^2 + 7m - 15$ by $m + 3$.
$4m - 5$

4. Divide $2x^2 - 3x + 2$ by $x + 2$.
$2x - 7 + \frac{16}{x + 2}$

5. Divide $4y^2 + 12y + 9$ by $2y - 3$.
$2y + 9 + \frac{36}{2y - 3}$

6. The area and one dimension of the rectangle are shown. Find the missing dimension.

Area is $24x^2 - 2x - 15$. ?

$4x + 3$

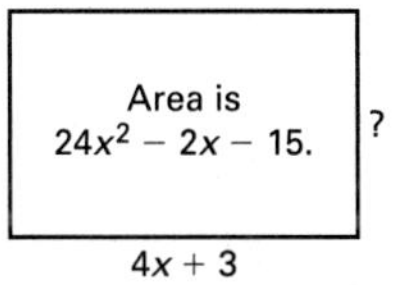

$6x - 5$

EXTRA CHALLENGE NOTE

↳ Challenge problems for Lesson 11.7 are available in **blackline** format in the *Chapter 11 Resource Book*, p. 104 and at **www.mcdougallittell.com**.

ADDITIONAL TEST PREPARATION

1. OPEN ENDED Give an example of polynomial division using the long division algorithm. Explain each step in the process. Check work.

58a.

| 3 | 1 |

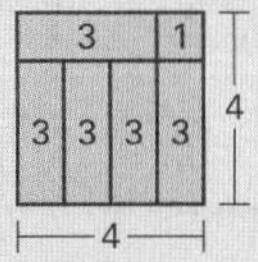

3 3 3 3 4

b.

| 6 | 1 |

6 6 6 6 6 6 7

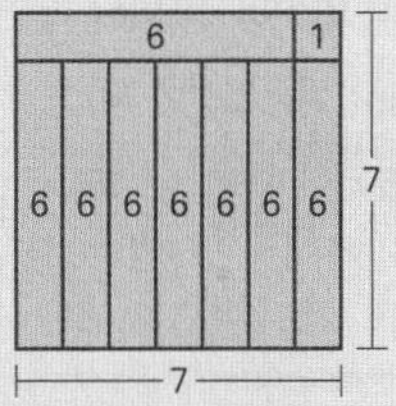

71. See Additional Answers beginning on page AA1.

➤ **LESSON OPENER**
APPLICATION
An alternative way to approach
Lesson 11.8 is to use the
Application Lesson Opener:

• Blackline Master (*Chapter 11 Resource Book,* p. 108)
• Transparency (p. 81)

MEETING INDIVIDUAL NEEDS
• *Chapter 11 Resource Book*
 Prerequisite Skills Review (p. 5)
 Practice Level A (p. 111)
 Practice Level B (p. 112)
 Practice Level C (p. 113)
 Reteaching with Practice (p. 114)
 Absent Student Catch-Up (p. 116)
 Challenge (p. 118)
• *Resources in Spanish*
• *Personal Student Tutor*

NEW-TEACHER SUPPORT
See the Tips for New Teachers on
pp. 1–2 of the *Chapter 11 Resource
Book* for additional notes about
Lesson 11.8.

WARM-UP EXERCISES

 Transparency Available

Solve the proportion.

1. $\frac{15}{2} = \frac{x}{5}$ $\frac{75}{2}$

2. $\frac{12}{x} = \frac{4}{21}$ 63

Find the LCD of the terms.

3. $\frac{16}{3}, \frac{15}{7},$ and $\frac{1}{28}$ 84

4. $\frac{3}{x^2}, \frac{5}{x(x-1)^2},$ and $\frac{7}{x^3(x-1)}$
$x^3(x-1)^2$

11.8 Rational Equations and Functions

What you should learn

GOAL 1 Solve rational equations.

GOAL 2 Graph rational functions.

Why you should learn it

▼ To model **real-life** problems as in the fundraising problem in **Exs. 53–54.**

GOAL 1 SOLVING RATIONAL EQUATIONS

A **rational equation** is an equation that contains rational expressions.
Examples 1 and 2 show the two basic strategies for solving a rational equation.

EXAMPLE 1 *Cross Multiplying*

Solve $\dfrac{5}{y+2} = \dfrac{y}{3}$.

SOLUTION

$\dfrac{5}{y+2} = \dfrac{y}{3}$	Write original equation.
$5(3) = y(y+2)$	Cross multiply.
$15 = y^2 + 2y$	Simplify.
$0 = y^2 + 2y - 15$	Write in standard form.
$0 = (y+5)(y-3)$	Factor right side.

▶ If you set each factor equal to 0, you see that the solutions are -5 and 3.
 Check both solutions in the original equation.

· · · · · · · · · ·

Cross multiplying can be used only for equations in which each side is a single
fraction. The second method, multiplying by the LCD, works for any rational
equation. Multiplying by the LCD always leads to an equation with no fractions.

EXAMPLE 2 *Multiplying by the LCD*

Solve $\dfrac{2}{x} + \dfrac{1}{3} = \dfrac{4}{x}$.

SOLUTION

$\dfrac{2}{x} + \dfrac{1}{3} = \dfrac{4}{x}$	The LCD is $3x$.
$3x \cdot \dfrac{2}{x} + 3x \cdot \dfrac{1}{3} = 3x \cdot \dfrac{4}{x}$	Multiply each side by $3x$.
$6 + x = 12$	Simplify.
$x = 6$	Subtract 6 from each side.
✓ CHECK $\dfrac{2}{6} + \dfrac{1}{3} \stackrel{?}{=} \dfrac{4}{6}$	Substitute 6 for x.
$\dfrac{4}{6} = \dfrac{4}{6}$	Simplify.

EXAMPLE 3 Factoring to Find the LCD

STUDENT HELP

Study Tip
When you solve rational equations, be sure to check for extraneous solutions. Remember, values of the variable that make any denominator equal to 0 are excluded.

Solve $\frac{-4}{y-3} + 1 = \frac{-10}{y^2 + y - 12}$.

SOLUTION The denominator $y^2 + y - 12$ factors as $(y + 4)(y - 3)$, so the LCD is $(y + 4)(y - 3)$. Multiply each side of the equation by $(y + 4)(y - 3)$.

$$\frac{-4}{y-3} \cdot (y+4)(y-3) + 1 \cdot (y+4)(y-3) = \frac{-10}{y^2+y-12} \cdot (y+4)(y-3)$$

$$\frac{-4(y+4)\cancel{(y-3)}}{\cancel{y-3}} + (y+4)(y-3) = \frac{-10\cancel{(y+4)(y-3)}}{\cancel{(y+4)(y-3)}}$$

$$-4(y+4) + (y^2 + y - 12) = -10$$

$$-4y - 16 + y^2 + y - 12 = -10$$

$$y^2 - 3y - 28 = -10$$

$$y^2 - 3y - 18 = 0$$

$$(y-6)(y+3) = 0$$

▶ The solutions are 6 and −3. Check both solutions in the original equation.

EXAMPLE 4 Writing and Using a Rational Equation

BATTING AVERAGE You have 35 hits in 140 times at bat. Your batting average is $\frac{35}{140} = 0.250$. How many consecutive hits must you get to increase your batting average to 0.300?

SOLUTION If your hits are consecutive, then you must get a hit each time you are at bat, so "future hits" is equal to "future times at bat."

VERBAL MODEL

$$\text{Batting average} = \frac{\text{Past hits} + \text{Future hits}}{\text{Past times at bat} + \text{Future times at bat}}$$

LABELS

Batting average = 0.300	(no units)
Past hits = 35	(no units)
Future hits = x	(no units)
Past times at bat = 140	(no units)
Future times at bat = x	(no units)

ALGEBRAIC MODEL

$0.300 = \frac{35 + x}{140 + x}$	Write algebraic model.
$0.300(140 + x) = 35 + x$	Multiply by LCD.
$42 + 0.3x = 35 + x$	Use distributive property.
$7 = 0.7x$	Subtract $0.3x$ and 35 from each side.
$10 = x$	Divide each side by 0.7.

▶ You need to get a hit in each of your next 10 times at bat.

FOCUS ON CAREERS

REAL LIFE **SPORTS REPORTER**
Sports reporters gather statistics and prepare stories that cover all aspects of sports from local sporting events to international sporting events.

INTERNET CAREER LINK
www.mcdougallittell.com

2 TEACH

MOTIVATING THE LESSON
Students are now used to rational expressions being undefined for certain values. Ask them if they know what this means for the graph of a rational expression. Some may respond that there must be a "hole" in the graph. Tell students that this is something that they will be exploring in this lesson.

EXTRA EXAMPLE 1
Solve $\frac{x}{3} = \frac{4}{x-4}$. −2, 6

EXTRA EXAMPLE 2
Solve $\frac{2}{5x} = -\frac{3}{15} + \frac{1}{x}$. 3

EXTRA EXAMPLE 3
Solve $\frac{3}{x-7} + 1 = \frac{8}{x^2 - 9x + 14}$. 0, 6

EXTRA EXAMPLE 4
A college alumni association solicits donations by phone. In 495 calls, 290 alumni have pledged donations. How many consecutive alumni will have to pledge to make the percentage of alumni donating pass 60%? 18

CHECKPOINT EXERCISES

For use after Example 1:

1. Solve $\frac{1}{x} = \frac{x}{x+6}$. −2, 3

For use after Examples 2 and 3:

2. Solve $\frac{x}{x-1} - \frac{3x}{x+5} = \frac{-24}{x^2 + 4x - 5}$. −2, 6

For use after Example 4:

3. What number subtracted from the numerator and denominator of $\frac{151}{181}$ will make it equivalent to $\frac{3}{4}$? 61

CAREER NOTE
EXAMPLE 4 Additional information about a career as a sports reporter is available at **www.mcdougallittell.com.**

EXTRA EXAMPLE 5

Sketch the graph of $y = \frac{2}{x-1} + 3$.

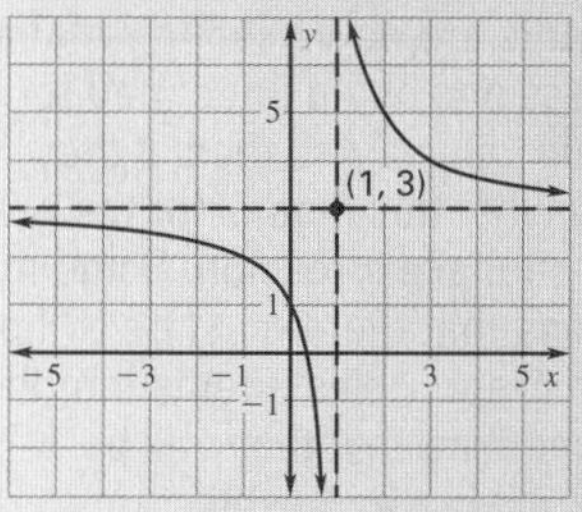

CHECKPOINT EXERCISES

For use after Example 5:

1. Sketch the graph of $y = \frac{-1}{x-3} - 2$.

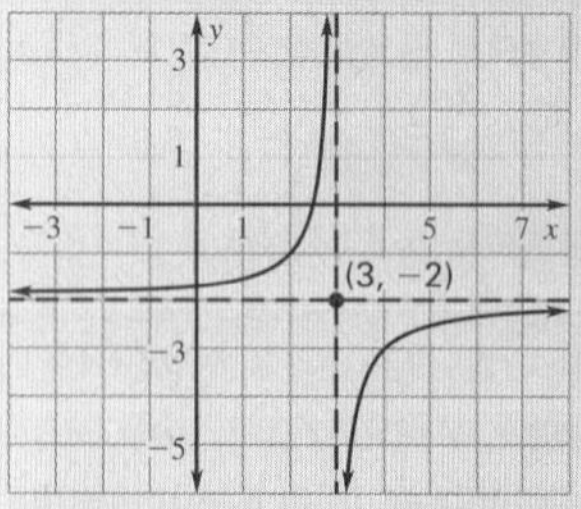

GOAL 2 GRAPHING RATIONAL FUNCTIONS

The inverse variation models you graphed in Lesson 11.3 are a type of *rational function*. A **rational function** is a function of the form

$$f(x) = \frac{\text{polynomial}}{\text{polynomial}}.$$

In this lesson you will learn to graph rational functions whose numerators and denominators are first-degree polynomials. Using long division, such a function can be written in the form

$$y = \frac{a}{x-h} + k.$$

RATIONAL FUNCTIONS WHOSE GRAPHS ARE HYPERBOLAS

The graph of the rational function $y = \frac{a}{x-h} + k$ is a **hyperbola** whose **center** is (h, k).

The vertical and horizontal lines through the center are the *asymptotes* of the hyperbola. An **asymptote** is a line that the graph approaches. While the distance between the graph and the line approaches zero, the asymptote is not part of the graph.

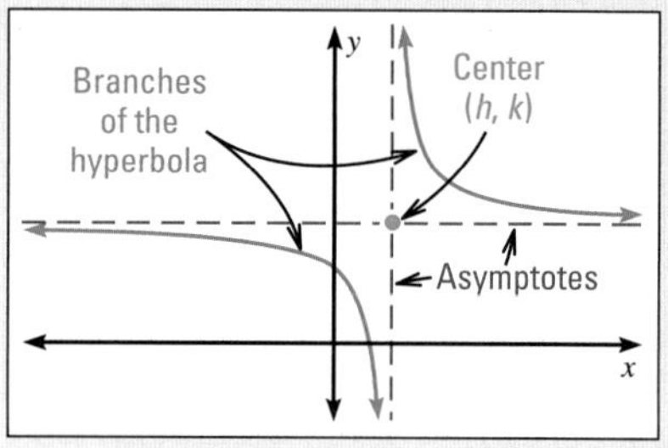

EXAMPLE 5 *Graphing a Rational Function*

Sketch the graph of $y = \frac{1}{x} + 2$.

SOLUTION Think of the function as $y = \frac{1}{x-0} + 2$. You can see that the center is $(0, 2)$. The asymptotes can be drawn as dashed lines through the center. Make a table of values. Then plot the points and connect them with two smooth branches.

x	y
−4	1.75
−2	1.5
−1	1
−0.5	0
0	undefined
0.5	4
1	3
2	2.5
4	2.25

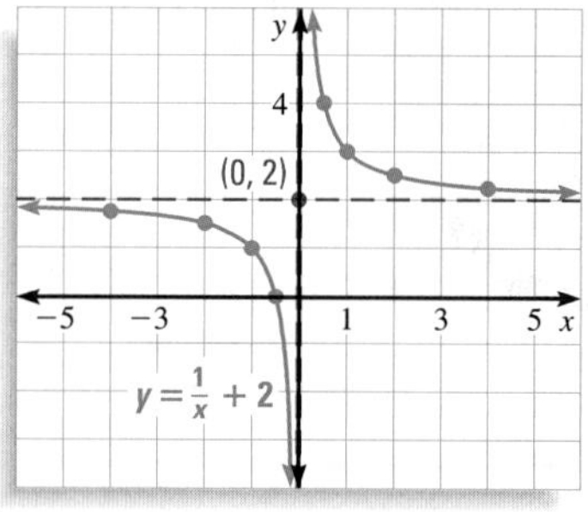

If you have drawn your graph correctly, it should be symmetric about the center (h, k). For example, the points $(0.5, 4)$ and $(-0.5, 0)$ are the same distance from the center $(0, 2)$, but in opposite directions.

EXAMPLE 6 *Graphing a Rational Function When h is Negative*

Sketch the graph of $y = \frac{-1}{x+2} + 1$. Describe the domain.

SOLUTION

For this equation, $x - h = x + 2$, so $h = -2$. The graph is a hyperbola with center at $(-2, 1)$. Plot this point and draw a horizontal and a vertical asymptote through it. Then make a table of values, plot the points, and connect them with two smooth branches.

x	y
-6	1.25
-4	1.5
-3	2
-2.5	3
-2	undefined
-1.5	-1
-1	0
0	0.5
2	0.75

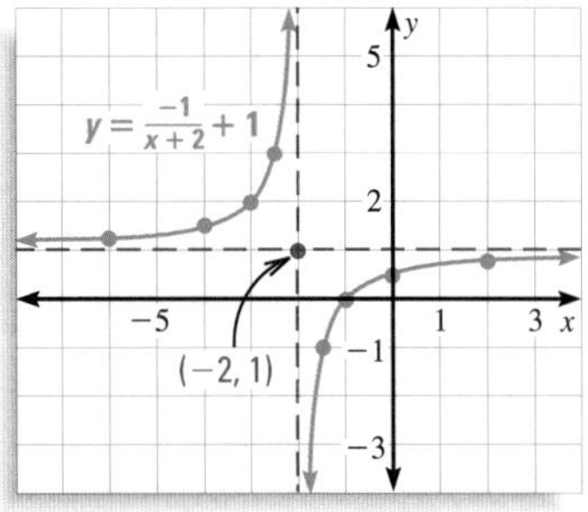

The domain is all real numbers except $x = -2$.

EXAMPLE 7 *Rewriting before Graphing*

Sketch the graph of $y = \frac{2x+1}{x+2}$.

SOLUTION Begin by using long division to rewrite the rational function.

$$\begin{array}{r} 2 \\ x+2\overline{)2x+1} \\ \underline{2x+4} \\ -3 \end{array}$$

Quotient **Remainder**

$$y = 2 + \frac{-3}{x+2}, \text{ or } y = \frac{-3}{x+2} + 2$$

The graph is a hyperbola with center at $(-2, 2)$. Plot this point and draw a horizontal and a vertical asymptote through it. Make a table of values, plot the points, and connect them with two smooth branches.

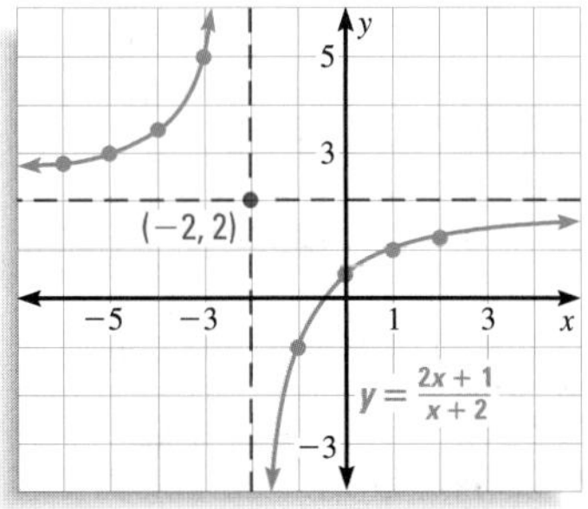

....

When making a table of values, include values of x that are less than h and values of x that are greater than h.

EXTRA EXAMPLE 6

Sketch the graph of $y = \frac{2}{x+1} - 3$.

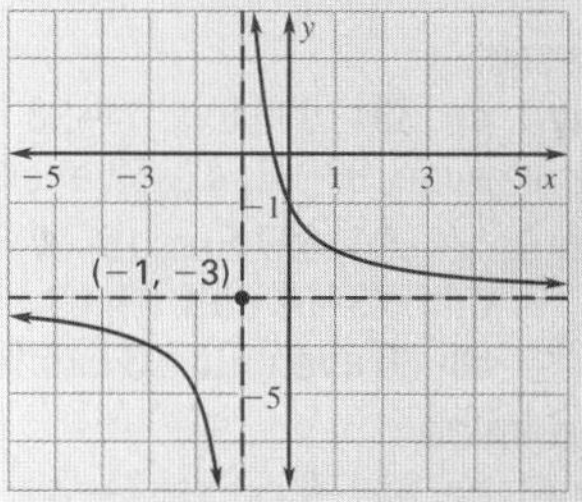

EXTRA EXAMPLE 7

Sketch the graph of $y = \frac{2x-1}{x-2}$.

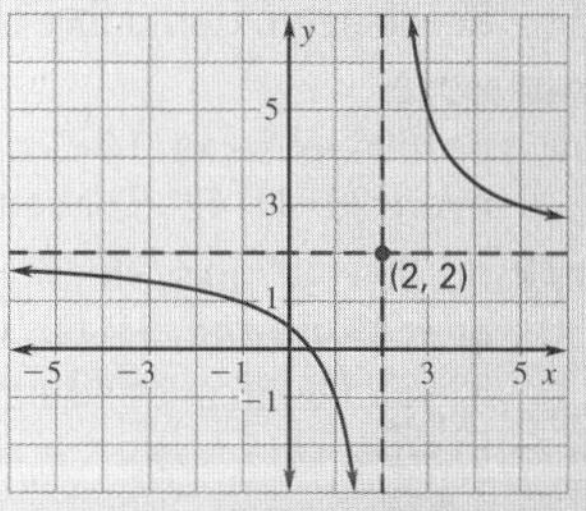

CHECKPOINT EXERCISES

For use after Examples 6 and 7:

1. What are the asymptotes and center of the graph of $y = \frac{3x+7}{x+3}$?
asymptotes: $x = -3$, $y = 3$; center: $(-3, 3)$

FOCUS ON VOCABULARY

What is the *center* of a hyperbola? ***Sample answer:*** **The center is the point where the two asymptotes intersect. It is not a part of the graph of the hyperbola.**

CLOSURE QUESTION

How can you graph an equation of the form $y = \frac{ax+b}{cx+d}$?
See sample answer below.

DAILY PUZZLER

What are the values of a, h, and k in the equation $y = \frac{a}{x-h} + k$ if its graph is the hyperbola with asymptotes $x = 1$ and $y = -2$ that goes through the point $(3, 0)$? **$a = 4$, $h = 1$, $k = -2$**

Closure Question *Sample answer:*
Use long division to write the function in the form $y = \frac{a}{x-h} + k$. Plot the center, (h, k), and draw a horizontal and a vertical asymptote through it. Make a table of values and plot a few points on either side of the vertical asymptote. Then connect the points with two smooth curves.

3 APPLY

ASSIGNMENT GUIDE

BASIC

Day 1: pp. 694–695 Exs. 14–38 even, 39–41, 42–52 even

Day 2: pp. 694–697 Exs. 15–37 odd, 43–51 odd, 62, 63, 68–86 even, Quiz 3 Exs.1–14

AVERAGE

Day 1: pp. 694–695 Exs. 14–38 even, 39–41, 42–52 even

Day 2: pp. 694–697 Exs. 15–37 odd, 43–53 odd, 54, 62, 63, 68–86 even, Quiz 3 Exs.1–14

ADVANCED

Day 1: pp. 694–695 Exs. 14–38 even, 39–41, 42–52 even

Day 2: pp. 694–697 Exs. 15–37 odd, 43–53 odd, 54–60, 62–67, 68–86 even, Quiz 3 Exs.1–14

BLOCK SCHEDULE

pp. 694–697 Exs. 14–54, 62, 63, 68–86 even, Quiz 3 Exs.1–14

EXERCISE LEVELS

Level A: *Easier*
14, 15

Level B: *More Difficult*
16–54, 62, 63, 68–86

Level C: *Most Difficult*
55–61, 64–67

✔ HOMEWORK CHECK

To quickly check student understanding of key concepts, go over the following exercises: Exs. 14, 18, 22, 28, 32, 36, 38, 41, 46, 52. See also the Daily Homework Quiz:

- Blackline Master (*Chapter 12 Resource Book,* p. 11)
- Transparency (p. 94)

12–13. See Additional Answers beginning on page AA1.

GUIDED PRACTICE

Vocabulary Check ✔

1. Describe the shape of a hyperbola. What is an asymptote of a hyperbola? See margin.

Concept Check ✔

In Exercises 2–4, find the least common denominator.

2. $\frac{1}{x}, \frac{x}{3}, \frac{2}{3x}$ $3x$

3. $\frac{3}{4x}, \frac{1}{6x^2}, \frac{1}{8x^2}$ $24x^2$

4. $\frac{1}{x^2+6x+9}, \frac{1}{x+3}$ $(x+3)(x+3)$

5. Which of the equations in Exercises 6–11 can you solve using cross multiplying? Explain your answer. See margin.

Skill Check ✔

Solve the equation. Remember to check for extraneous solutions.

6. $\frac{3}{x} = \frac{x}{12}$ $-6, 6$

7. $\frac{-4x}{x+1} = \frac{2}{x-2}$ $\frac{1}{2}, 1$

8. $\frac{x}{6} + \frac{15}{x} = \frac{5}{6}$ no solution

9. $\frac{4}{x^2-2x} = \frac{4}{3x-6}$ 3

10. $\frac{4}{x(x+1)} = \frac{3}{x}$ $\frac{1}{3}$

11. $\frac{3}{x+4} + \frac{4}{x} = \frac{-5}{x^2+4x}$ -3

Find the center of the hyperbola. Draw the asymptotes and sketch the graph.

12. $y = \frac{4}{x+5} - 3$ $(-5, -3)$; See margin.

13. $y = \frac{2}{x-2} + 3$ $(2, 3)$; See margin.

1. *Sample answer:* A hyperbola consists of two branches, each of which approaches, but does not intersect, two lines called the asymptotes of the hyperbola. The asymptotes intersect at the center of the hyperbola.

PRACTICE AND APPLICATIONS

STUDENT HELP

Extra Practice
to help you master skills is on p. 807.

5. The equations in Exs. 6, 7, 9, and 10; each side of each equation is a single fraction. (Note, however, that solving Exs. 9 and 10 using cross multiplying will introduce extraneous solutions.)

CROSS MULTIPLYING **Solve the equation by cross multiplying.**

14. $\frac{x}{5} = \frac{7}{3}$ $\frac{35}{3}$

15. $\frac{x}{10} = \frac{14}{5}$ 28

16. $\frac{4}{x} = \frac{12}{5(x+2)}$ -5

17. $\frac{5}{x+4} = \frac{5}{3(x+1)}$ $\frac{1}{2}$

18. $\frac{6}{x+2} = \frac{x}{4}$ $-6, 4$

19. $\frac{7}{x+1} = \frac{5}{x-3}$ 13

MULTIPLYING BY THE LCD **Solve the equation by multiplying each side by the least common denominator.**

20. $\frac{56}{x} = \frac{9-x}{2}$ no solution

21. $\frac{x}{x+9} = \frac{9}{x+9} + 2$ -27

22. $\frac{7}{3x-12} - \frac{1}{x-4} = \frac{2}{3}$ 6

23. $\frac{1}{x-4} + \frac{1}{x+4} = \frac{22}{x^2-16}$ 11

24. $\frac{1}{x-4} + 1 = -\frac{7}{x^2+x-20}$ $2, -4$

25. $2 + \frac{8}{x-5} = \frac{x+5}{x^2-25}$ $1\frac{1}{2}$

CHOOSING A METHOD **Solve the equation.**

26. $\frac{1}{4} + \frac{4}{x} = \frac{1}{x}$ -12

27. $\frac{-3x}{x+1} = \frac{-2}{x-1}$ $-\frac{1}{3}, 2$

28. $\frac{1}{5} - \frac{2}{5x} = \frac{1}{x}$ 7

29. $\frac{x}{9} - \frac{8}{x} = \frac{1}{9}$ $-8, 9$

30. $\frac{x+42}{x} = x$ $-6, 7$

31. $\frac{2}{x} - \frac{x}{8} = \frac{3}{4}$ $-8, 2$

32. $\frac{-3}{x+7} = \frac{2}{x+2}$ -4

33. $\frac{2}{x+3} + \frac{1}{x} = \frac{4}{3x}$ $\frac{3}{5}$

34. $\frac{10}{x+3} - \frac{3}{5} = \frac{10x+1}{3x+9}$ 2

35. $\frac{x+3}{x-5} = \frac{56-3x}{x^2-13x+40}$ $-8, 10$

36. $\frac{8}{x+4} + 1 = \frac{5x}{x^2-2x-24}$ $-9, 8$

37. $\frac{x}{x-11} - 1 = \frac{22}{x^2-5x-66}$ -4

38. $\frac{2x}{x+3} - \frac{x}{x+7} = \frac{x^2-1}{x^2+10x+21}$ $-\frac{1}{11}$

STUDENT HELP

HOMEWORK HELP

Example 1: Exs. 14–19, 26–38
Example 2: Exs. 20–38
Example 3: Exs. 20–38
Example 4: Exs. 51, 52
Example 5: Exs. 39–50
Example 6: Exs. 39–50
Example 7: Exs. 39–50

42. domain: all real numbers except 0
43. domain: all real numbers except 3
44. domain: all real numbers except 4
45. domain: all real numbers except –1
46. domain: all real numbers except –9
47. domain: all real numbers except –3
48. domain: all real numbers except 5
49. domain: all real numbers except 1
50. domain: all real numbers except 3
54. *Sample answer:* The smaller the number of sponsors, the more the average contribution is inflated by the $5000 contribution. The fewer the sponsors, the less representative the average is of a typical contribution.

STUDENT HELP

HOMEWORK HELP Visit our Web site www.mcdougallittell.com for help with Exs. 52–58.

57. $\frac{3}{100} + \frac{3}{50x} = \frac{3x+6}{100x}$

FOCUS ON APPLICATIONS

ORIGAMI CRANES The art of paper folding has been practiced in Japan since the seventh century A.D. There is an ancient Japanese belief that folding 1000 cranes will make your wish come true.

MATCHING GRAPHS **Match the function with its graph.**

A.

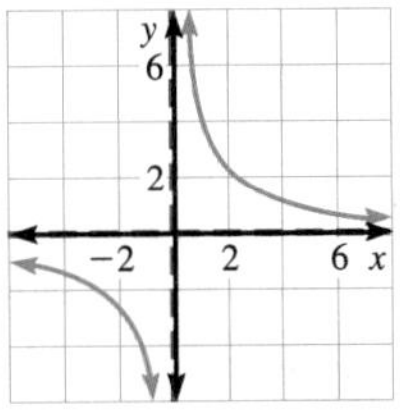

B.

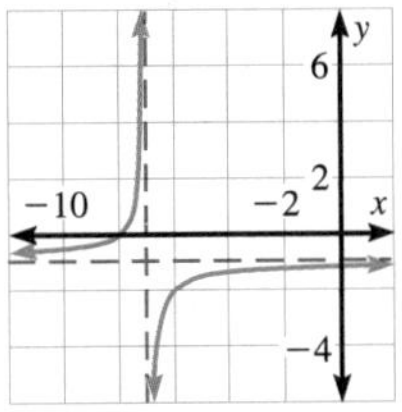

C.

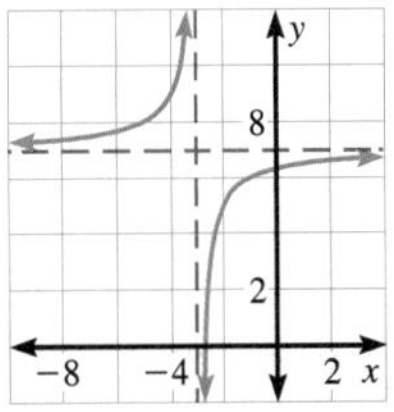

39. $y = -\frac{1}{x+7} - 1$ B

40. $y = \frac{-2}{x+3} + 7$ C

41. $y = \frac{4}{x}$ A

GRAPHING **In Exercises 42–50, graph the function. Describe the domain.** 42–50. See margin for graphs.

42. $y = \frac{1}{x} + 4$

43. $y = \frac{1}{x-3} - 8$

44. $y = \frac{2}{x-4} + 6$

45. $y = -\frac{3}{x+1} + 8$

46. $y = \frac{6}{x+9} - 7$

47. $y = \frac{3x+11}{x+3}$

48. $y = \frac{-2x+11}{x-5}$

49. $y = \frac{9x-6}{x-1}$

50. $y = \frac{-5x+19}{x-3}$

51. **BATTING AVERAGE** After 50 times at bat, a major league baseball player has a batting average of 0.160. How many consecutive hits must the player get to raise his batting average to 0.250? 6 hits

52. **TEST AVERAGES** You have taken 3 tests and have an average of 72 points. If you score 100 points on each of the rest of your tests, how many more tests do you need to take to raise your average to 88 points? 4 tests

FUNDRAISING **In Exercises 53 and 54, a library has received a single large contribution of $5000. A walkathon is also being held in which each sponsor will contribute $10. Let *x* represent the number of sponsors.**

53. Write a function that represents the average contribution per person, including the single large contributor. Sketch the graph of the function. $y = \frac{5000 + 10x}{x+1}$; See margin for graph.

54. Explain how such averages could be used to misrepresent actual contributions. See margin.

ORIGAMI CRANES **In Exercises 55–58, you investigate how long it would take you and a friend to fold 1000 origami cranes. You take 2 minutes to fold 1 crane. Let *x* represent the number of minutes it might take a friend to fold 1 crane.**

55. Working alone, you would take 2000 minutes to fold 1000 cranes, so your work rate is $\frac{1}{2000}$ of the job per minute. Find your work rate per hour. $\frac{60}{2000} = \frac{3}{100}$

56. The expression $\frac{60}{1000x}$ is your friend's work rate per hour. Explain why. See margin.

57. Write an expression for the combined work rate per hour of you and your friend (the part of the job completed in 1 hour if you both work together). See margin.

58. Suppose that together you could complete 1000 cranes in 20 hours. Your combined work rate per hour would be $\frac{1}{20}$. To find the time your friend takes to fold 1 crane, set the expression in Exercise 57 equal to $\frac{1}{20}$ and solve for *x*. 3

STUDENT HELP NOTES

Homework Help Students can find help for Exs. 52–58 at **www.mcdougallittell.com.** The information can be printed out for students who don't have access to the Internet.

42.

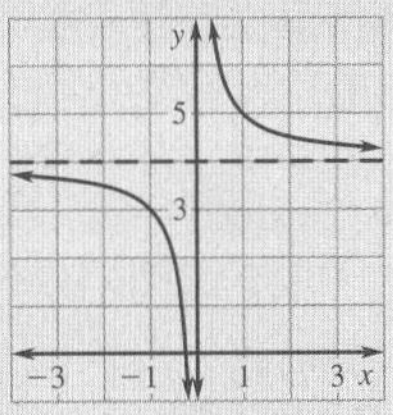

43.

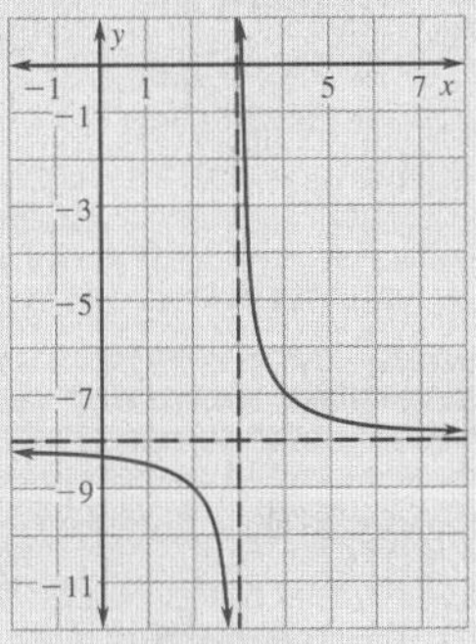

44.

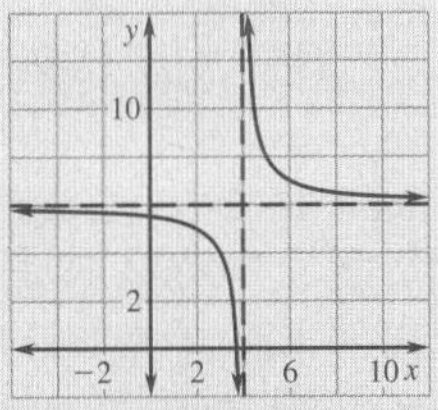

45.

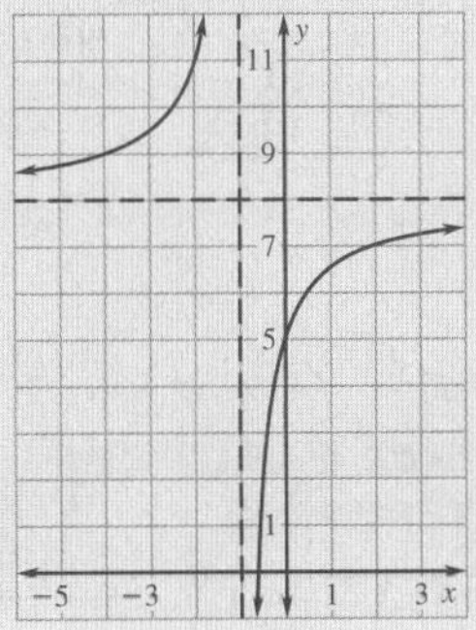

46.

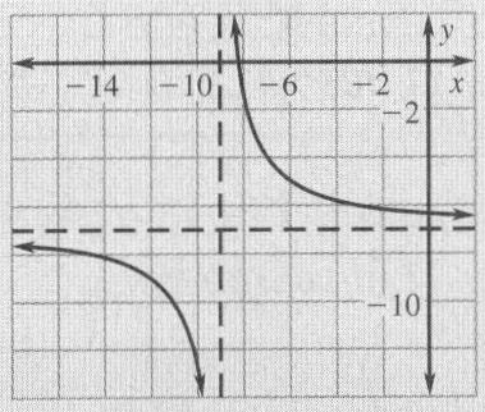

47–49. See the next page.
50, 53, and 56. See Additional Answers beginning on page AA1.

47.
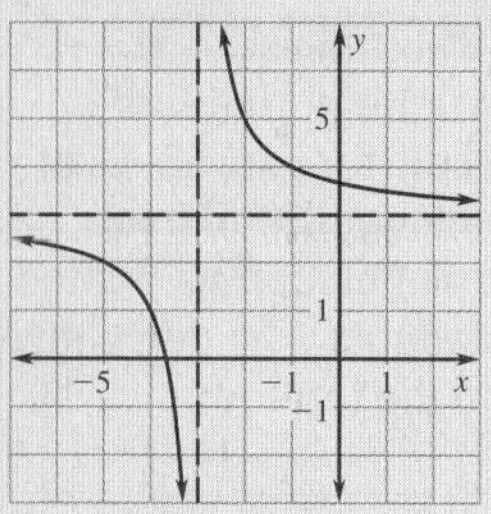

48.
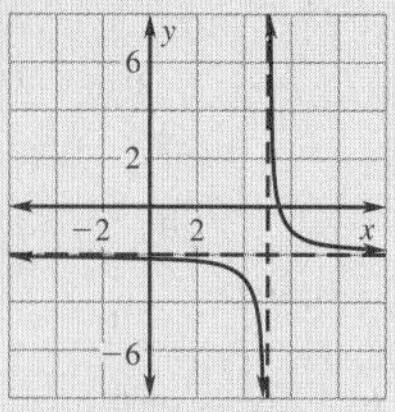

49.
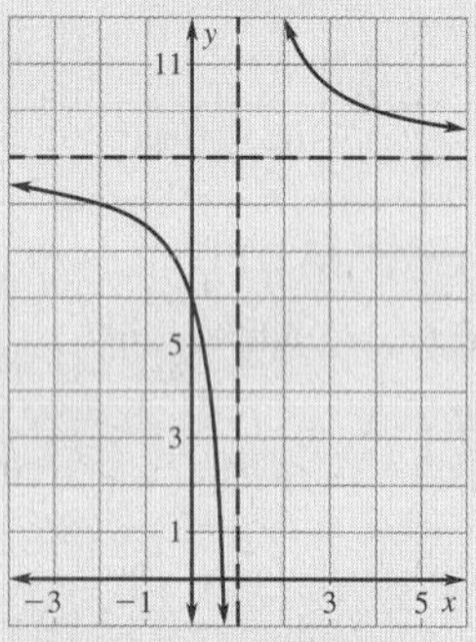

59. See Additional Answers beginning on page AA1.

64. second row: 3, 18.45, 38.24, 58.16, 78.12, 98.10; third row: –2, 18, 38, 58, 78, 98; fourth row: 5, 0.45, 0.24, 0.16, 0.12, 0.10

ADDITIONAL PRACTICE AND RETEACHING

For Lesson 11.8:

- Practice Levels A, B, and C (*Chapter 11 Resource Book*, p. 111)
- Reteaching with Practice (*Chapter 11 Resource Book*, p. 114)
- See Lesson 11.8 of the *Personal Student Tutor*

For more Mixed Review:

- Search the *Test and Practice Generator* for key words or specific lessons.

60. *Sample answer:* The branches of the hyperbolas seem to have the same shape, but the graphs have different asymptotes; yes; in the first equation, the asymptotes are $x = h$ and $y = k$, in the second they are $x = 0$ and $y = 0$. The effect of the h and k in the equation is to shift the position of the branches, not change their shapes.

61. *Sample answer:* For $a > 0$, as a increases, the branches of the hyperbola (which are in the first and third quadrants) become wider and more open. For $a < 0$, as $|a|$ increases, the branches of the hyperbola (which are in the second and fourth quadrants) become wider and more open.

65. *Sample answer:* The values of the first two expressions increase and the value of the third decreases.

67. *Sample answer:* The graph of $y = x - 2$ is an asymptote of the graph of $y = \frac{x^2 + 6}{x + 2}$. Notice that for each column in the table, the value of the first expression is equal to the sum of the other two. $\frac{x^2 + 6}{x + 2} = x - 2 + \frac{10}{x + 2}$. As x gets very large, $\frac{10}{x + 2}$ approaches, but never equals, 0, and $\frac{x^2 + 6}{x + 2}$ approaches, but never equals, $x - 2$.

EXTRA CHALLENGE

www.mcdougallittell.com

INVERSE VARIATION AND RATIONAL FUNCTIONS In Exercises 59 and 60, you will compare the types of graphs in 11.3 with those in this lesson.

59. Graph $f(x) = \frac{6}{x}$ and $f(x) = \frac{6}{x - 2} + 1$ in the same coordinate plane. **See margin.**

60. *Writing* How are the two graphs in Exercise 59 related? Do you think this relationship will hold for any set of functions $f(x) = \frac{a}{x - h} + k$ and $f(x) = \frac{a}{x}$? Explain. **See margin.**

61. CHANGING CONSTANTS Use a graphing calculator to graph $f(x) = \frac{a}{x}$ for at least ten values of a. Include positive, negative, and non-integer values. How does changing the value of a seem to affect the appearance of the graph? **See margin.**

Test Preparation

62. MULTIPLE CHOICE What is the solution of the equation $\frac{9}{x + 5} = \frac{7}{x - 5}$? **D**

(A) 5 (B) 8 (C) 20 (D) 40 (E) 80

63. MULTIPLE CHOICE Choose the graph of the function $y = \frac{4}{x - 2} + 1$. **B**

(A)
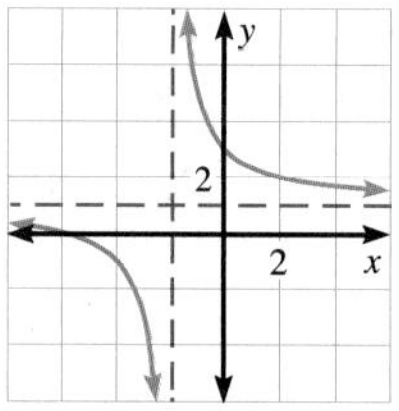

(B)
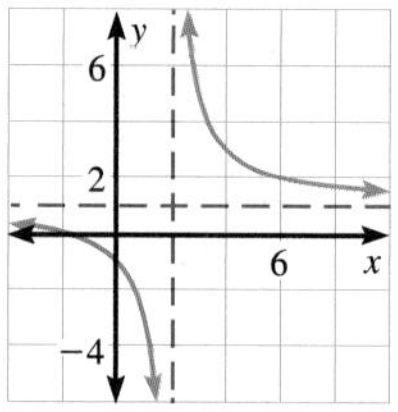

(C)
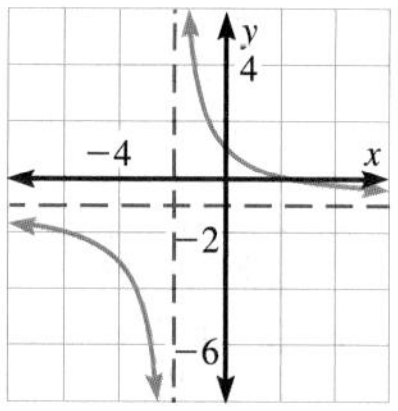

(D) None of the above

★ Challenge

RATIONAL FUNCTIONS In Exercises 64–67, you will look for a pattern.

64. Copy and complete the table. Round your answers to the nearest hundredth. **See margin.**

x	0	20	40	60	80	100
$\frac{x^2 + 6}{x + 2}$	?	?	?	?	?	?
$x - 2$	?	?	?	?	?	?
$\frac{10}{x + 2}$	?	?	?	?	?	?

65. What happens to the values of $\frac{x^2 + 6}{x + 2}$, $(x - 2)$, and $\frac{10}{x + 2}$ as x increases? **See margin.**

66. Which is the graph of $y = x - 2$? of $y = \frac{x^2 + 6}{x + 2}$? **B; A**

67. CRITICAL THINKING Explain the relationship between the two graphs in Exercise 66. Use the table in Exercise 64 to help you. **See margin.**

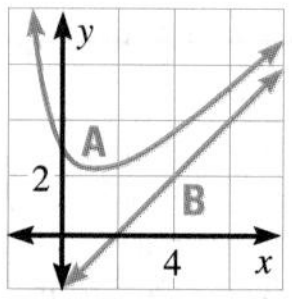

Exs. 66 and 67

MIXED REVIEW

FUNCTION VALUES Evaluate the function for $x = 0, 1, 2, 3,$ and 4. (Review 4.8 for 12.1)

68. $f(x) = 4x$ **0, 4, 8, 12, 16**
69. $f(x) = -x + 9$ **9, 8, 7, 6, 5**
70. $f(x) = 3x + 1$ **1, 4, 7, 10, 13**
71. $f(x) = -x^2$ **0, −1, −4, −9, −16**
72. $f(x) = x^2 - 1$ **−1, 0, 3, 8, 15**
73. $f(x) = \frac{x^2}{2}$ **$0, \frac{1}{2}, 2, 4\frac{1}{2}, 8$**

EVALUATING EXPRESSIONS Evaluate the expression. (Review 8.1, 8.2)

74. $2^4 \cdot 2^3$ **$2^7 = 128$**
75. $6^3 \cdot 6^{-1}$ **$6^2 = 36$**
76. $(3^3)^2$ **$3^6 = 729$**
77. $(-4^{-2})^{-1}$ **$-4^2 = -16$**

RADICAL EXPRESSIONS Simplify the radical expression. (Review 9.2 for 12.1)

78. $\sqrt{50}$ **$5\sqrt{2}$**
79. $\sqrt{72}$ **$6\sqrt{2}$**
80. $\frac{1}{4}\sqrt{112}$ **$\sqrt{7}$**
81. $\frac{1}{2}\sqrt{52}$ **$\sqrt{13}$**
82. $\frac{1}{4}\sqrt{64}$ **2**
83. $\sqrt{256}$ **16**
84. $\frac{1}{5}\sqrt{625}$ **5**
85. $\sqrt{396}$ **$6\sqrt{11}$**

86. ACCOUNT BALANCE A principal of \$500 is deposited in an account that pays 4% interest compounded yearly. Find the balance after 6 years. (Review 8.5) **\$632.66**

QUIZ 3

Self-Test for Lessons 11.7 and 11.8

Divide. (Lesson 11.7)

1. $(x^2 - 8) \div (6x)$ **$\frac{x}{6} - \frac{4}{3x}$**
2. $(6a^3 + 5a^2) \div (10a^2)$ **$\frac{3}{5}a + \frac{1}{2}$**
3. $(x^2 + 16) \div (x + 4)$ **$x - 4 + \frac{32}{x + 4}$**
4. $(2y^2 + 8y - 5) \div (2y - 5)$ **$y + 6.5 + \frac{27.5}{2y - 5}$**
5. $(8 + 4z + z^2) \div (3z - 6)$ **$\frac{1}{3}z + 2 + \frac{20}{3z - 6}$**
6. $(12x^2 + 17x - 5) \div (3x + 2)$ **$4x + 3 - \frac{11}{3x + 2}$**

Solve the equation. (Lesson 11.8)

7. $\frac{1}{2} + \frac{2}{t} = \frac{1}{t}$ **−2**
8. $\frac{3}{x} = \frac{9}{2(x + 2)}$ **4**
9. $\frac{1}{x - 5} + \frac{1}{x + 5} = \frac{x + 3}{x^2 - 25}$ **3**
10. $\frac{7}{8} - \frac{16}{s - 2} = \frac{3}{4}$ **130**

In Exercises 11–13, sketch a graph of the function. (Lesson 11.8) **11–13. See margin.**

11. $f(x) = \frac{3}{x}$
12. $f(x) = \frac{6}{x - 4}$
13. $f(x) = \frac{3}{x + 1} - 2$

14. TELEVISION STATIONS The models are based on data collected by the Television Bureau of Advertising from 1988 to 1998 in the United States. Let t represent the number of years since 1988.

Number of UHF television stations: $U = 14t + 505$

Total number of commercial television stations: $T = 16t + 1048$

Use long division to find a model for the ratio of the number of UHF television stations to the total number of commercial television stations. (Lesson 11.7) **$\frac{7}{8} - \frac{103}{4t + 262}$**

Additional Test Preparation ***Sample answer:***

1. Cross multiplication is fairly easy but works only on proportions. Multiplying by the LCD has the advantage that it works for all rational equations.

11–13. See Additional Answers beginning on page AA1.

4 ASSESS

DAILY HOMEWORK QUIZ

Transparency Available

1. Solve $\frac{4}{2x - 1} = \frac{3}{x - 6}$ by cross multiplying. **$-\frac{21}{2}$**

Solve the equation.

2. $\frac{3}{x} + \frac{2}{3} = \frac{1}{3x}$ **−4**

3. $1 + \frac{x - 1}{x + 8} = -\frac{15}{x^2 + 3x - 40}$ **$-\frac{5}{2}, 4$**

4. $\frac{x}{x - 4} + \frac{3x}{x - 6} = \frac{x^2 - 5x - 4}{x^2 - 10x + 24}$ **$\frac{1}{3}$**

5. Sketch a graph of $y = \frac{3}{x - 2} + 1$. Describe the domain.

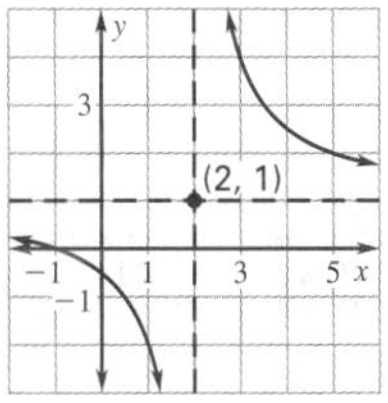

domain: all real numbers except 2

6. After 8 games of bowling in your league, your average score is 145. If you could score 160 on your next few games, how many games would it take to raise your average to 150? **4**

EXTRA CHALLENGE NOTE

Challenge problems for Lesson 11.8 are available in **blackline** format in the *Chapter 11 Resource Book*, p. 118 and at **www.mcdougallittell.com.**

ADDITIONAL TEST PREPARATION

1. WRITING Discuss the two methods of solving rational equations. What are some of the advantages and disadvantages of each? **See sample answer at left.**

2. OPEN ENDED Write an equation for a hyperbola with asymptotes $x = 1$ and $y = -2$. Graph the hyperbola. **The equation will have the form $y = \frac{a}{x - 1} - 2$.**

1 Planning the Activity

PURPOSE
To identify undefined values in a rational function.

MATERIALS
- graphing calculator for each student or pair of students
- Keystroke blackline (*Chapter 11 Resource Book,* p. 109)

PACING
- Exploring the Concept — 10 min
- Drawing Conclusions — 15 min

LINK TO LESSON
Graphing calculators can be used to check the graphs in Examples 5–7 of Lesson 11.8.

2 Managing the Activity

COOPERATIVE LEARNING
Students can work in pairs with a single graphing calculator. Have one student simplify the function while the other enters it in its original form. Students should switch roles for each question.

ALTERNATIVE APPROACH
This activity can be done as a demonstration with a graphing calculator display.

3 Closing the Activity

KEY DISCOVERY
Canceling a common term in a rational function may change the domain of the function.

ACTIVITY ASSESSMENT

Simplify $f(x) = \dfrac{2x^2 - 7x - 4}{x^2 - 7x + 12}$. At what values of x is the original function undefined? At what values of x is the simplified function undefined?

$y = \dfrac{2x+1}{x-3}$; 3 and 4; 3

ACTIVITY 11.8
Using Technology

Graphing Rational Functions

QUESTION **How can you use a graphing calculator or computer to recognize input values that are not in the domain of a rational function?**

EXPLORING THE CONCEPT

1. Simplify the right-hand side of the equation by factoring the numerator and denominator, and dividing out common factors.

$$y = \frac{4x^2 + 9x + 2}{x^2 - 4} \quad \frac{4x + 1}{x - 2}$$

2. Graph the simplified form of the function. Your calculator may show a vertical line at $x = 2$. This is *not* part of the graph. Some calculators draw this line in an attempt to connect the two branches of the graph.

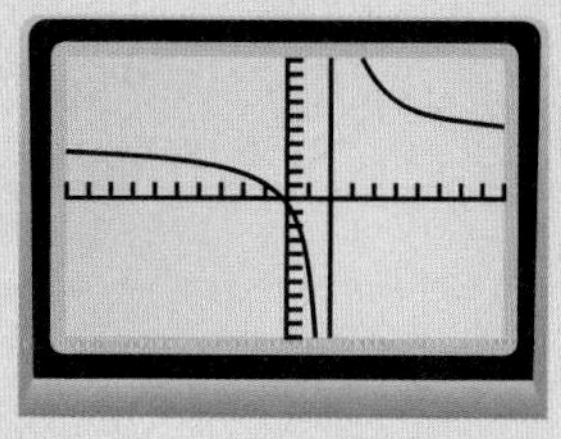

3. Graph the original function in the same viewing rectangle as the simplified form. If you have simplified correctly, the graphs will look identical.

4. To check for undefined values, press ZOOM and choose *Decimal.* Then use the trace feature to find x-values for which no y-values appear in the viewing rectangle. These are the undefined values of the expression. **original: $x = 2$, $x = -2$; simplified: $x = 2$**

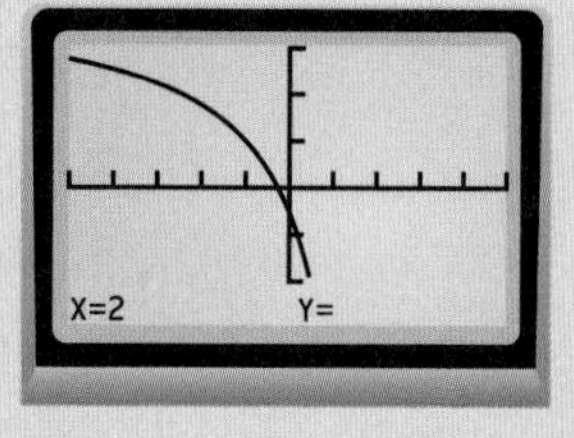

5. Do the original function and the simplified function have the same domain? Explain. **No. *Sample answer:* The original function has the additional restriction that $x \neq -2$, since $x^2 - 4$ cannot equal zero.**

STUDENT HELP
KEYSTROKE HELP
See keystrokes for several models of calculators at www.mcdougallittell.com

DRAWING CONCLUSIONS

In Exercises 1–6, simplify the right-hand side of the rational function. Then use a graphing calculator or computer to check your answer. Give the domain of the original function and of the simplified function. 1–7. See margin.

1. $y = \dfrac{3x^2 - 13x - 10}{x^2 - 25}$ **2.** $y = \dfrac{x^2 - 6x + 8}{x^2 - 2x - 8}$ **3.** $y = \dfrac{2x^2 + 9x + 4}{x^2 + x - 12}$

4. $y = \dfrac{x^2 - x - 20}{x^2 - 16}$ **5.** $y = \dfrac{2x^2 - 3x - 9}{x^2 - 2x - 3}$ **6.** $y = \dfrac{2x^2 - 5x - 3}{x^2 - 8x + 15}$

7. CRITICAL THINKING Are the functions $y = \dfrac{x - 1}{(x + 3)(x - 1)}$ and $y = \dfrac{1}{x + 3}$ equivalent? Explain your reasoning.

Margin answers:

1. $y = \dfrac{3x + 2}{x + 5}$; all real numbers except −5 and 5; all real numbers except −5
2. $y = \dfrac{x - 2}{x + 2}$; all real numbers except −2 and 4; all real numbers except −2
3. $y = \dfrac{2x + 1}{x - 3}$; all real numbers except −4 and 3; all real numbers except 3
4. $y = \dfrac{x - 5}{x - 4}$; all real numbers except −4 and 4; all real numbers except 4
5. $y = \dfrac{2x + 3}{x + 1}$; all real numbers except −1 and 3; all real numbers except −1
6. $y = \dfrac{2x + 1}{x - 5}$; all real numbers except 3 and 5; all real numbers except 5
7. *Sample answer:* No; they do not have the same domain. The domain of the first is all real numbers except −3 and 1; the domain of the second is all real numbers except −3.

CHAPTER 11

Chapter Summary

WHAT did you learn?	*WHY did you learn it?*
Solve proportions. (11.1)	Estimate the number of clay figures in an archaeological dig. (p. 645)
Use equations to solve percent problems. (11.2)	Compare responses to a survey. (p. 654)
Use direct and inverse variation. (11.3)	Relate the banking angle of a bicycle to its turning radius. (p. 658)
Simplify rational expressions. (11.4)	Use rational expressions to model real-life situations. (p. 668)
Find geometric probability. (11.4)	Determine the probability of a meteor strike. (p. 668)
Multiply and divide rational expressions. (11.5)	Analyze service industry sales. (p. 674)
Add and subtract rational expressions. (11.6)	Calculate the time it takes to make a trip. (p. 678)
Divide polynomials. (11.7)	Provide alternate forms of rational expression models. (p. 688)
Solve rational equations. (11.8)	Study changes in batting averages. (p. 691)
Graph rational functions. (11.8)	Provide a visual representation of a fundraising problem. (p. 695)

How does Chapter 11 fit into the BIGGER PICTURE of algebra?

In this chapter you studied rational expressions—fractions whose numerators and denominators are polynomials. Rational expressions occur frequently in real life as proportions, percents, probabilities, and direct and inverse variations. Understanding these will enable you to model and solve a variety of real-life problems.

Techniques used in these problems include simplifying, multiplying, dividing, adding, and subtracting rational expressions. Real-life problems can also be modeled with rational equations and represented by graphing rational functions.

STUDY STRATEGY

How did you use your notes to see what you learned?

The notes you made, using the **Study Strategy** on page 642, may include the ideas shown.

Chapter 11 Notes

Preview—Ratio and Proportion:

- ratio of a to b is $\frac{a}{b}$
- Quantities are measured in the same units.

Review—Ratio and Proportion:

- an equation that states that two ratios are equal
- If $\frac{a}{b} = \frac{c}{d}$, then $\frac{b}{a} = \frac{d}{c}$.
- If $\frac{a}{b} = \frac{c}{d}$, then $ad = bc$.

699

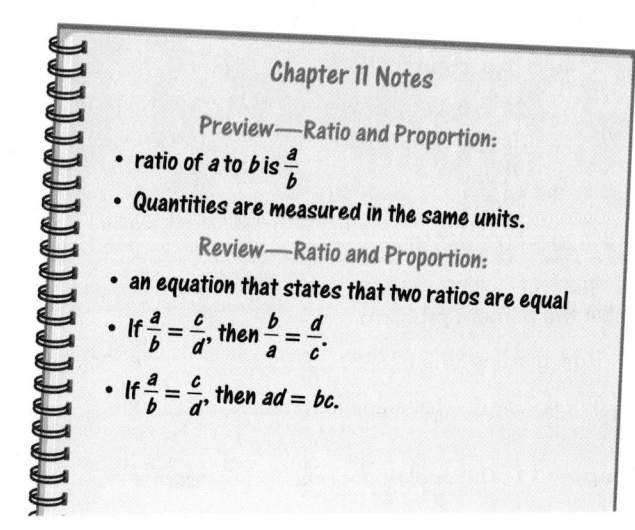

ADDITIONAL RESOURCES

The following resources are available to help review the material in this chapter.

- Chapter Review Games and Activities (*Chapter 11 Resource Book,* p. 119)
- *Instant Replay: Video Review Games*
- *Personal Student Tutor*
- Cumulative Review, Chs. 1–11 (*Chapter 11 Resource Book,* p. 131)

CHAPTER 11 Chapter Review

VOCABULARY

- proportion, p. 643
- extremes of a proportion, p. 643
- means of a proportion, p. 643
- solving a proportion, p. 643
- extraneous solution, p. 644
- base number of a percent equation, p. 649
- inverse variation, p. 656
- constant of variation, p. 656
- rational number, p. 664
- rational expression, p. 664
- simplified rational expression, p. 664
- geometric probability, p. 666
- least common denominator, p. 677
- polynomial long division, p. 685
- rational equation, p. 690
- rational function, p. 692
- hyperbola, p. 692
- center of a hyperbola, p. 692
- asymptote, p. 692

11.1 RATIO AND PROPORTION

Examples on pp. 643–645

EXAMPLE Solve the proportion $\frac{12}{7} = \frac{5}{x}$.

$\frac{12}{7} = \frac{5}{x}$ Write original proportion.

$12 \cdot x = 7 \cdot 5$ Use cross product property.

$x = \frac{35}{12}$ Divide each side by 12.

Solve the proportion. Check for extraneous solutions.

1. $\frac{x}{2} = \frac{4}{7}$ **$1\frac{1}{7}$**
2. $\frac{7}{10} = \frac{9 + x}{x}$ **–30**
3. $\frac{x^2 - 16}{x + 4} = \frac{x - 4}{3}$ **4**
4. $\frac{5}{x + 6} = \frac{x - 6}{x}$ **–4, 9**

11.2 PERCENTS

Examples on pp. 649–652

EXAMPLES

\$20 is 40% of what amount of money?

a is *p* percent of *b*

$20 = 0.4b$

$\frac{20}{0.4} = b$

$50 = b$

▶ \$20 is 40% of \$50.

\$75 is what percent of \$60?

a is *p* percent of *b*

$75 = p(60)$

$\frac{75}{60} = p$

$1.25 = p$, or $p = 125\%$

▶ \$75 is 125% of \$60.

Solve the percent problem.

5. How much is 80% of \$95? **\$76**
6. 24 inches is 250% of what length? **$9\frac{3}{5}$ in.**
7. \$90 is 75% of what amount of money? **\$120**
8. 35 feet is what percent of 175 feet? **20%**

11.3 DIRECT AND INVERSE VARIATION

Examples on pp. 656–658

EXAMPLES When x is 5, y is 25. Write an equation that relates x and y in each case.

x and y vary directly:

$$\frac{y}{x} = k$$
$$\frac{25}{5} = k$$
$$5 = k$$

An equation that relates x and y is $\frac{y}{x} = 5$, or $y = 5x$.

x and y vary inversely:

$$xy = k$$
$$(5)(25) = k$$
$$125 = k$$

An equation that relates x and y is $xy = 125$, or $y = \frac{125}{x}$.

When x is 17, y is 51. Find an equation that relates x and y in each case.

9. x and y vary directly $y = 3x$

10. x and y vary inversely $y = \frac{867}{x}$

11.4 SIMPLIFYING RATIONAL EXPRESSIONS

Examples on pp. 664–666

EXAMPLE To simplify a rational expression, look for common factors.

$$\frac{2x^2 + 3x - 2}{2x^2 + 5x + 2} = \frac{(2x - 1)\cancel{(x + 2)}}{(2x + 1)\cancel{(x + 2)}}$$ **Factor numerator and denominator. Divide out common factor $(x + 2)$.**

$$= \frac{2x - 1}{2x + 1}$$ **Simplified form**

Simplify the expression.

11. $\frac{3x}{9x^2 + 3}$ $\frac{x}{3x^2 + 1}$

12. $\frac{6x^2}{12x^4 + 18x^2}$ $\frac{1}{2x^2 + 3}$

13. $\frac{7x^3 - 21x}{-14x^2}$ $\frac{-x^2 + 3}{2x}$

14. $\frac{5x^2 + 21x + 4}{25x + 100}$ $\frac{5x + 1}{25}$

11.5 MULTIPLYING AND DIVIDING RATIONAL EXPRESSIONS

Examples on pp. 670–672

EXAMPLE To divide rational expressions, multiply by the reciprocal.

$$\frac{6x^2 + x - 1}{2x + 1} \div (9x - 3) = \frac{6x^2 + x - 1}{2x + 1} \cdot \frac{1}{9x - 3}$$ **Multiply by reciprocal.**

$$= \frac{\cancel{(2x + 1)}\cancel{(3x - 1)}}{\cancel{(2x + 1)} \cdot 3\cancel{(3x - 1)}}$$ **Multiply numerators and denominators. Factor and divide out common factors.**

$$= \frac{1}{3}$$ **Simplified form**

Simplify the expression.

15. $\frac{12x^2}{5x^3} \cdot \frac{25x^4}{3x}$ $20x^2$

16. $\frac{9x^3}{x^3 - x^2} \div \frac{x - 8}{x^2 - 9x + 8}$ $9x$

17. $\frac{x^2 + 3x + 2}{x^2 + 7x + 12} \div \frac{x^2 + 5x + 4}{x^2 + 5x + 6}$ $\frac{(x + 2)^2}{(x + 4)^2}$

11.6 ADDING AND SUBTRACTING RATIONAL EXPRESSIONS

Examples on pp. 676–678

EXAMPLE Simplify $\frac{x}{x-5} - \frac{2}{x+2}$. The LCD is $(x-5)(x+2)$.

$\frac{x(x+2)}{(x-5)(x+2)} - \frac{2(x-5)}{(x-5)(x+2)}$ **Rewrite fractions using LCD.**

$= \frac{x^2+2x}{(x-5)(x+2)} - \frac{2x-10}{(x-5)(x+2)}$ **Simplify numerators.**

$= \frac{(x^2+2x)-(2x-10)}{(x-5)(x+2)} = \frac{x^2+10}{(x-5)(x+2)}$ **Subtract fractions and simplify.**

Simplify the expression.

18. $\frac{6x}{x+4} - \frac{5x-4}{x+4}$ **1**

19. $\frac{2x+1}{8x} - \frac{x}{12x}$ $\mathbf{\frac{4x+3}{24x}}$

20. $\frac{x+3}{3x-1} + \frac{4}{x-3}$ $\mathbf{\frac{(x+13)(x-1)}{(3x-1)(x-3)}}$

21. $\frac{-5x-10}{x^2-4} + \frac{4x}{x-2}$ $\mathbf{\frac{4x-5}{x-2}}$

11.7 DIVIDING POLYNOMIALS

Examples on pp. 684–686

EXAMPLES There are two cases to look for when you divide polynomials.

CASE 1: Monomial divisor

To divide a polynomial by a monomial, divide each term by the monomial.

CASE 2: Binomial divisor

To divide a polynomial by a binomial, factor out common factors if possible. If not, use long division.

Divide.

22. Divide $3x^2 - x - 1$ by $x - 2$. $\mathbf{3x+5+\frac{9}{x-2}}$

23. Divide $6x^2 - 36x + 5$ by $6x$. $\mathbf{x-6+\frac{5}{6x}}$

24. Divide $4x^2 + 6x - 5$ by $2x - 1$. $\mathbf{2x+4-\frac{1}{2x-1}}$

25. Divide $5x^2 + 13x - 6$ by $5x - 2$. $\mathbf{x+3}$

11.8 RATIONAL EQUATIONS AND FUNCTIONS

Examples on pp. 690–693

EXAMPLE Solve the equation $\frac{2x}{9} - \frac{1}{x} = \frac{1}{3}$. The LCD is $9x$.

$9x \cdot \frac{2x}{9} - 9x \cdot \frac{1}{x} = 9x \cdot \frac{1}{3}$ **Multiply each side by 9*x*.**

$2x^2 - 9 = 3x$ **Simplify.**

$2x^2 - 3x - 9 = 0$ **Write in standard form.**

$(2x+3)(x-3) = 0$ **Factor left side.**

When you set each factor equal to 0, you find that the solutions are $-\frac{3}{2}$ and 3.

Solve the equation.

26. $\frac{1}{4} - \frac{6}{x} = \frac{3}{x}$ **36**

27. $\frac{x+2}{2} = \frac{4}{x}$ **−4, 2**

28. $\frac{6}{x+4} + \frac{3}{4} = \frac{2x+1}{3x+12}$ **−104**

CHAPTER 11

Chapter Test

Solve the proportion. Check for extraneous solutions.

1. $\frac{6}{x} = \frac{17}{5}$ $1\frac{13}{17}$ **2.** $\frac{x}{4} = \frac{x+8}{x}$ $-4, 8$ **3.** $\frac{x}{-3} = \frac{7}{x-10}$ $3, 7$ **4.** $\frac{x^2-64}{x+8} = \frac{x-8}{2}$ 8

Solve the percent problem.

5. What is 34% of 100 liters? **34 L**

6. What is 86% of \$350? **\$301**

7. 24 yards is 12% of what distance? **200 yd**

8. 36 T-shirts is what percent of 900 T-shirts? **4%**

Make a table of values for $x = 1, 2, 3,$ and 4. Use the table to sketch a graph. Decide whether x and y vary *directly* or *inversely*.

9. $y = 4x$ — 4, 8, 12, 16; directly

10. $y = \frac{50}{x}$ — $50, 25, 16\frac{2}{3}, 12\frac{1}{2}$; inversely

11. $y = \frac{9}{2}x$ — $\frac{9}{2}, 9, \frac{27}{2}, 18$; directly

12. $y = \frac{15}{2x}$ — $\frac{15}{2}, \frac{15}{4}, \frac{5}{2}, \frac{15}{8}$; inversely

Simplify the expression.

13. $\frac{56x^6}{4x^4}$ $14x^2$

14. $\frac{5x^2-15x}{15x^4}$ $\frac{x-3}{3x^3}$

15. $\frac{x^2-x-6}{x^2-4}$ $\frac{x-3}{x-2}$

16. $\frac{6x^2}{8x} \cdot \frac{-4x^3}{2x^2}$ $-\frac{3x^2}{2}$

17. $\frac{x+3}{x^3-x^2-6x} \div \frac{x^2-9}{x^2+x-12}$ See margin.

18. $\frac{x^3+x^2}{x^2-16} \cdot \frac{x+4}{3x^4+x^3-2x^2}$ $\frac{1}{(x-4)(3x-2)}$

19. $\frac{3x^2+6x}{4x} \div \frac{15}{8x^2}$ $\frac{2x^2(x+2)}{5}$

20. $\frac{12x-4}{x-1} + \frac{4x}{x-1}$ $\frac{4(4x-1)}{x-1}$

21. $\frac{5}{2x^2} + \frac{4}{3x}$ $\frac{15+8x}{6x^2}$

22. $\frac{8}{5x} - \frac{4}{x^2}$ $\frac{8x-20}{5x^2}$

23. $\frac{4}{x+3} + \frac{3x}{x-2}$ $\frac{3x^2+13x-8}{(x+3)(x-2)}$

24. $\frac{5x+1}{x-3} - \frac{2x}{x-1}$ $\frac{(3x-1)(x+1)}{(x-3)(x-1)}$

Divide.

25. Divide $4x^3 - 15x^2 - 6x$ by $3x$. $\frac{4}{3}x^2 - 5x - 2$

26. Divide $81x^2 - 25$ by $9x - 5$. $9x + 5$

27. Divide $2x^2 + 11x + 12$ by $x + 4$. $2x + 3$

28. Divide $5x^2 + 4x - 7$ by $x + 2$. $5x - 6 + \frac{5}{x+2}$

Solve the equation.

29. $\frac{3}{4x-9} = \frac{x}{3}$ $-\frac{3}{4}, 3$

30. $\frac{5}{9} + \frac{2}{9x} = \frac{3}{x}$ 5

31. $\frac{5}{x+3} - \frac{3}{x-2} = \frac{5}{3x-6}$ 72

Sketch a graph of the function. 32–34. See margin.

32. $y = \frac{1}{x-4} + 3$ **33.** $y = 5 - \frac{2}{x}$ **34.** $y = \frac{x-5}{x+2}$

35. BAGEL SHOP You invest \$30,000 to start a bagel shop. You can produce bagels for \$1.20 per dozen. How many dozen must you produce before your average cost per dozen (including your initial investment of \$30,000) drops to \$1.80? **50,000 dozen**

36. CARNIVAL GAME At a carnival game, a dart is thrown at the board shown at the right. Assume it is equally likely to land anywhere on the board. Write a model that gives the probability that the dart will land in the small rectangle. Evaluate the model when $x = 5$. $\frac{x+1}{6(x+3)}$; 0.125

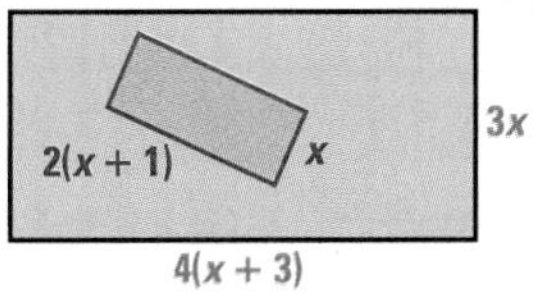

ADDITIONAL RESOURCES

- ***Chapter 11 Resource Book***
  Chapter Test (3 levels) (p. 120)
  SAT/ACT Chapter Test (p. 126)
  Alternative Assessment (p. 127)
- ***Test and Practice Generator***

17. $\frac{x+4}{x(x+2)(x-3)}$

32.

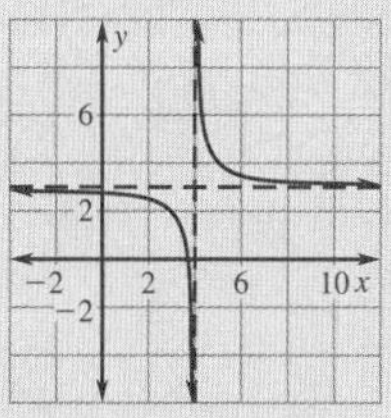

33.

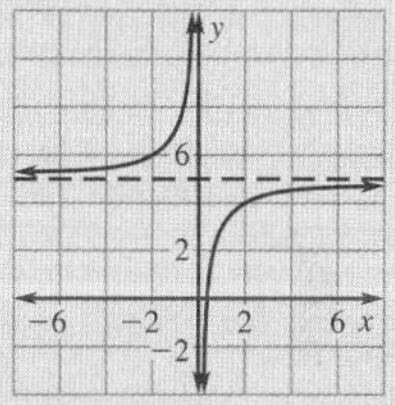

34.

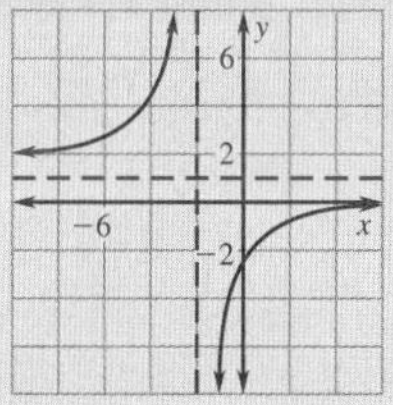

ADDITIONAL RESOURCES

- ***Chapter 11 Resource Book***
  Chapter Test (3 levels) (p. 120)
  SAT/ACT Chapter Test (p. 126)
  Alternative Assessment (p. 127)
- ***Test and Practice Generator***

CHAPTER 11

Chapter Standardized Test

TEST-TAKING STRATEGY Think positively during a test. This will help keep up your confidence and enable you to focus on each question.

1. **MULTIPLE CHOICE** Which of the following is the solution of the proportion $\frac{4}{y+9} = \frac{6}{y-7}$? **B**

   (A) −82 (B) −41 (C) −13
   (D) 7 (E) 41

2. **QUANTITATIVE COMPARISON** Solve each proportion and choose the statement below that is true about the solutions. **B**

COLUMN A	COLUMN B
$\frac{3}{x-2} = \frac{2x+2}{x^2-4}$	$\frac{5}{y+12} = \frac{2}{y+6}$

   (A) The solution in column A is greater.
   (B) The solution in column B is greater.
   (C) The two solutions are equal.
   (D) The relationship cannot be determined from the given information.

3. **MULTIPLE CHOICE** What is 380% of 52? **E**

   (A) 0.07 (B) 13.68 (C) 19.76
   (D) 136.84 (E) 197.6

4. **MULTIPLE CHOICE** 18.6 miles is 15% of what distance? **D**

   (A) 1.24 miles (B) 2.79 miles (C) 105.4 miles
   (D) 124 miles (E) 279 miles

5. **MULTIPLE CHOICE** There are 840 students in your school. At a pep rally for a football game, 614 students attend. What percent of the students in your school attended the pep rally? **C**

   (A) about 136.8% (B) about 78.8%
   (C) about 73.1% (D) about 36.8%
   (E) about 26.9%

6. **MULTIPLE CHOICE** Which of the following equations models inverse variation? **D**

   (A) $y = \frac{x}{3}$ (B) $y = 3x$ (C) $y = x + 3$
   (D) $xy = 3$ (E) $y = 3 - x$

7. **MULTIPLE CHOICE** The variables x and y vary inversely. When x is 9, y is 36. If x is 3, what is y? **E**

   (A) 3 (B) 12 (C) 30
   (D) 36 (E) 108

8. **MULTIPLE CHOICE** What is the simplified form of the expression $\frac{x^3 - 10x^2 + 9x}{x^2 + 5x - 6}$? **D**

   (A) $\frac{x-9}{x+6}$ (B) $\frac{x}{x+6}$ (C) $\frac{1}{x+6}$
   (D) $\frac{x(x-9)}{x+6}$ (E) $\frac{x}{(x-1)(x+6)}$

9. **MULTIPLE CHOICE** A coin is tossed onto the large rectangular region shown below. Assume it is equally likely to land on any point in the region. What is the probability the coin will land in the small rectangle when $x = 3$? **B**

   (A) $\frac{1}{15}$
   (B) $\frac{3}{25}$
   (C) $\frac{1}{5}$
   (D) $\frac{2}{5}$
   (E) $\frac{25}{3}$

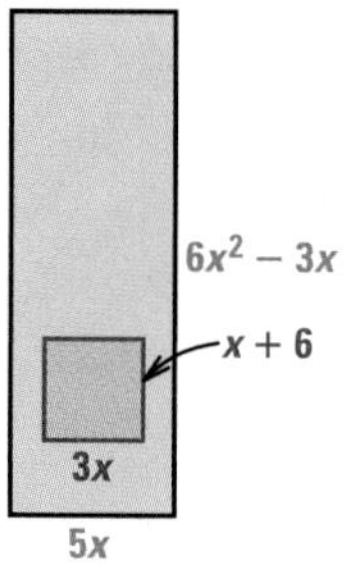

10. **MULTIPLE CHOICE** What is the simplified form of the expression $\frac{9x^2}{4x} \cdot \frac{16x^3}{x^5}$? **C**

   (A) $36x$ (B) $\frac{9}{x}$ (C) $\frac{36}{x}$
   (D) $36x^3$ (E) $\frac{64}{9x^3}$

11. **MULTIPLE CHOICE** What is the simplified form of the expression $\frac{x^2 - 64}{3x^2} \div (x - 8)$? **A**

   (A) $\frac{x+8}{3x^2}$ (B) $\frac{x-8}{3x^2}$ (C) $\frac{1}{3x^2}$
   (D) $\frac{x^3 - 512}{3x}$ (E) $\frac{x+8}{3x^2(x-8)}$

12. MULTIPLE CHOICE Find the LCD of $\frac{7x}{x^2 - 9}$ and $\frac{3}{x^2 + x - 6}$. **D**

Ⓐ $(x + 3)(x - 3)$ Ⓑ $(x^2 - 9)(x^2 + x + 6)$ Ⓒ $(x + 3)(x - 2)$

Ⓓ $(x + 3)(x - 3)(x - 2)$ Ⓔ $(x - 3)(x - 2)$

13. MULTIPLE CHOICE What is the simplified form of the following expression? **D**

$$\frac{2x + 9}{x + 5} - \frac{x - 4}{x - 2}$$

Ⓐ $\frac{x^2 + 6x + 2}{(x + 5)(x - 2)}$ Ⓑ $\frac{x^2 + 6x - 38}{(x + 5)(x - 2)}$ Ⓒ $\frac{x^2 + 6x + 38}{x^2 - 10}$

Ⓓ $\frac{x^2 + 4x + 2}{(x + 5)(x - 2)}$ Ⓔ $\frac{x^2 + 4x - 38}{(x + 5)(x - 2)}$

14. MULTIPLE CHOICE What is $x^2 + 24x - 3$ divided by $x - 4$? **A**

Ⓐ $x + 28 + \frac{109}{x - 4}$ Ⓑ $\frac{x + 28}{x - 4} + 109$ Ⓒ $\frac{x + 28}{x - 4}$

Ⓓ $\frac{x - 4}{x + 28}$ Ⓔ $\frac{x - 4}{x + 28} + 109$

15. MULTIPLE CHOICE Which function represents the graph? **A**

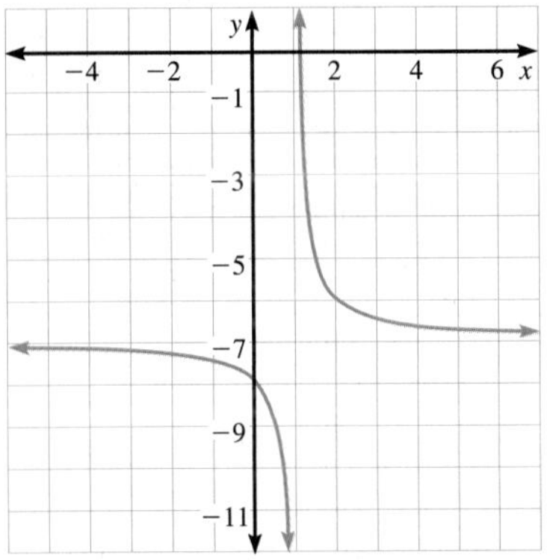

Ⓐ $y = \frac{1}{x - 1} - 7$

Ⓑ $y = \frac{1}{x + 1} - 7$

Ⓒ $y = \frac{1}{x - 7} + 1$

Ⓓ $y = \frac{1}{x - 1} + 7$

Ⓔ $y = \frac{1}{x + 1} + 7$

16. MULTI-STEP PROBLEM You are considering buying a new car and want to compare the yearly fuel costs for your car and for the new car. To find the yearly fuel cost you multiply the distance you travel in a year by the price per gallon and then divide by the number of miles per gallon your car averages.

You determine that you drive approximately 20,000 miles in one year and that you pay an average of $1.20 per gallon of gasoline.

Let x equal the number of miles per gallon your car averages.

a. Write an expression to represent the yearly fuel cost for your car. $\frac{24{,}000}{x}$

b. The new car you want to buy gets five more miles to the gallon than your car. Write an expression to represent the yearly fuel cost for the new car. $\frac{24{,}000}{x + 5}$

c. Suppose your yearly fuel cost is $400 less if you buy the new car. Write and solve an equation to find the number of miles per gallon your old car and the new car will get. $\frac{24{,}000}{x} - 400 = \frac{24{,}000}{x + 5}$; **old car: 15 mi/gal; new car: 20 mi/gal**

d. What is the estimated yearly fuel cost for your old car? for your new car? **$1600; $1200**

CHAPTER 12

PLANNING THE CHAPTER

Radicals and Connections to Geometry

GOALS

Lesson	Goals	NCTM	ITED	SAT9	Terra-Nova	Local
12.1 *pp. 709–715*	GOAL 1 Evaluate and graph a square-root function. GOAL 2 Use square-root functions to model real-life problems. TECHNOLOGY ACTIVITY: 12.1 *Graph a square-root function on a graphing calculator.*	1, 2, 3, 6, 8, 9, 10	MCE, APE	3, 5	11, 14, 16, 18, 52	
12.2 *pp. 716–721*	GOAL 1 Add, subtract, multiply, and divide radical expressions. GOAL 2 Use radical expressions in real-life situations.	1, 2, 6, 8, 9, 10	MCE, APE	3, 5	11, 52	
12.3 *pp. 722–728*	GOAL 1 Solve a radical equation. GOAL 2 Use radical equations to solve real-life problems.	1, 2, 6, 8, 9, 10	MCE, APE	3, 5	17, 52	
12.4 *pp. 729–736*	CONCEPT ACTIVITY: 12.4 *Investigate completing the square using algebra tiles.* GOAL 1 Solve a quadratic equation by completing the square. GOAL 2 Choose a method for solving a quadratic equation.	1, 2	MCE, APE		17, 52	
12.5 *pp. 737–744*	CONCEPT ACTIVITY: 12.5 *Investigate the Pythagorean theorem.* GOAL 1 Use the Pythagorean theorem and its converse. GOAL 2 Use the Pythagorean theorem and its converse in real-life problems.	1, 2, 4, 6, 8, 9, 10	MIGE, RQGE	3	13, 14, 17, 52	
12.6 *pp. 745–750*	GOAL 1 Find the distance between two points in a coordinate plane. GOAL 2 Find the midpoint between two points in a coordinate plane.	1, 3, 4	MIG, RQM, RQGE	29, 30	13, 14	
12.7 *pp. 751–757*	CONCEPT ACTIVITY: 12.7 *Investigate similar triangles.* GOAL 1 Use the sine, cosine, and tangent of an angle. GOAL 2 Use trigonometric ratios in real-life problems.	1, 3, 4, 6, 8, 9, 10		32	13, 14, 17	
12.8 *pp. 758–764*	GOAL 1 Use logical reasoning and proof to prove a statement is true. GOAL 2 Prove that a statement is false.	3, 7, 8			14, 17	

RESOURCES

Chapter Resource Booklets

Chapter Support

Tips for New Teachers	p. 1	Prerequisite Skills Review	p. 5
Parent Guide for Student Success	p. 3	Strategies for Reading Mathematics	p. 7

Lesson Support

Lesson Support	12.1	12.2	12.3	12.4	12.5	12.6	12.7	12.8
Lesson Plans (regular and block)	p. 9	p. 22	p. 35	p. 50	p. 63	p. 80	p. 92	p. 108
Warm-Up Exercises and Daily Quiz	p. 11	p. 24	p. 37	p. 52	p. 65	p. 82	p. 94	p. 110
Activity Support Masters				p. 53	p. 66		p. 95	
Lesson Openers	p. 12	p. 25	p. 38	p. 54	p. 67	p. 83	p. 96	p. 111
Graphing Calculator Activities & Keystrokes	p. 13	p. 26	p. 39		p. 68		p. 97	
Practice (3 levels)	p. 14	p. 27	p. 41	p. 55	p. 70	p. 84	p. 99	p. 112
Reteaching with Practice	p. 17	p. 30	p. 44	p. 58	p. 73	p. 87	p. 102	p. 115
Quick Catch-Up for Absent Students	p. 19	p. 32	p. 46	p. 60	p. 75	p. 89	p. 104	p. 117
Cooperative Learning Activities							p. 105	
Interdisciplinary Applications	p. 20		p. 47			p. 90		p. 118
Real-Life Applications		p. 33		p. 61	p. 76		p. 106	
Math & History Applications					p. 77			
Challenge: Skills and Applications	p. 21	p. 34	p. 48	p. 62	p. 78	p. 91	p. 107	p. 119

Review and Assessment

Quizzes	pp. 49, 79	Alternative Assessment with Math Journal	p. 128
Chapter Review Games and Activities	p. 120	Project with Rubric	p. 130
Chapter Test (3 levels)	pp. 121–126	Cumulative Review	p. 132
SAT/ACT Chapter Test	p. 127	Resource Book Answers	p. A1

Transparencies

	12.1	12.2	12.3	12.4	12.5	12.6	12.7	12.8
Warm-Up Exercises and Daily Quiz	p. 94	p. 95	p. 96	p. 97	p. 98	p. 99	p. 100	p. 101
Alternative Lesson Opener Transparencies	p. 82	p. 83	p. 84	p. 85	p. 86	p. 87	p. 88	p. 89
Examples/Standardized Test Practice	✓	✓	✓	✓	✓	✓	✓	✓
Answer Transparencies	✓	✓	✓	✓	✓	✓	✓	✓

Technology

- EasyPlanner CD
- EasyPlanner Plus Online
- Classzone.com
- Personal Student Tutor CD
- Test and Practice Generator CD
- Instant Replay: Video Review Games
- Electronic Lesson Presentations
- Chapter Audio Summaries CD

Additional Resources

- Basic Skills Workbook
- Explorations and Projects
- Worked-Out Solution Key
- Data Analysis Sourcebook
- Spanish Resources
- Standardized Test Practice Workbook
- Practice Workbook with Examples
- Notetaking Guide

CHAPTER 12

PACING THE CHAPTER

Resource Key
- Student Edition
- Chapter 12 Resource Book
- Teacher's Edition

REGULAR SCHEDULE

Day 1

12.1

Starting Options
- Prereq. Skills Review
- Strategies for Reading
- Warm-Up or Daily Quiz

Teaching Options
- Motivating the Lesson
- Les. Opener (Appl.)
- Graphing Calc. Activity
- Examples 1–4
- Technology Activity
- Closure Question
- Guided Practice Exs.

Apply/Homework
- See Assignment Guide.
- See the CRB: Practice, Reteach, Apply, Extend

Assessment Options
- Checkpoint Exercises
- Daily Quiz (12.1)
- Stand. Test Practice

Day 2

12.2

Starting Options
- Homework Check
- Warm-Up or Daily Quiz

Teaching Options
- Motivating the Lesson
- Les. Opener (Activity)
- Graphing Calc. Activity
- Examples 1–4
- Guided Practice Exs. 1–16

Apply/Homework
- See Assignment Guide.
- See the CRB: Practice, Reteach, Apply, Extend

Assessment Options
- Checkpoint Exercises, p. 717

Day 3

12.2 *(cont.)*

Starting Options
- Homework Check

Teaching Options
- Example 5
- Closure Question
- Guided Practice Ex. 17

Apply/Homework
- See Assignment Guide.
- See the CRB: Practice, Reteach, Apply, Extend

Assessment Options
- Checkpoint Exercises, p. 718
- Daily Quiz (12.2)
- Stand. Test Practice

Day 4

12.3

Starting Options
- Homework Check
- Warm-Up or Daily Quiz

Teaching Options
- Motivating the Lesson
- Les. Opener (Activity)
- Graphing Calc. Activity
- Examples 1–4
- Guided Practice Exs. 1–13

Apply/Homework
- See Assignment Guide.
- See the CRB: Practice, Reteach, Apply, Extend

Assessment Options
- Checkpoint Exercises, p. 723

Day 5

12.3 *(cont.)*

Starting Options
- Homework Check

Teaching Options
- Example 5
- Closure Question
- Guided Practice Ex. 14

Apply/Homework
- See Assignment Guide.
- See the CRB: Practice, Reteach, Apply, Extend

Assessment Options
- Checkpoint Exercises, p. 724
- Daily Quiz (12.3)
- Stand. Test Practice
- Quiz (12.1–12.3)

Day 6

12.4

Starting Options
- Homework Check
- Warm-Up or Daily Quiz

Teaching Options
- Motivating the Lesson
- Concept Act. & Wksht.
- Les. Opener (Application)
- Examples 1–4
- Guided Practice Exs. 1–13

Apply/Homework
- See Assignment Guide.
- See the CRB: Practice, Reteach, Apply, Extend

Assessment Options
- Checkpoint Exercises, p. 731

Day 9

12.5 *(cont.)*

Starting Options
- Homework Check

Teaching Options
- Examples 4–5
- Closure Question

Apply/Homework
- See Assignment Guide.
- See the CRB: Practice, Reteach, Apply, Extend

Assessment Options
- Checkpoint Exercises, p. 740
- Daily Quiz (12.5)
- Stand. Test Practice
- Quiz (12.4–12.5)

Day 10

12.6

Starting Options
- Homework Check
- Warm-Up or Daily Quiz

Teaching Options
- Motivating the Lesson
- Les. Opener (Application)
- Examples 1–5
- Closure Question
- Guided Practice Exs.

Apply/Homework
- See Assignment Guide.
- See the CRB: Practice, Reteach, Apply, Extend

Assessment Options
- Checkpoint Exercises
- Daily Quiz (12.6)
- Stand. Test Practice

Day 11

12.7

Starting Options
- Homework Check
- Warm-Up or Daily Quiz

Teaching Options
- Concept Act. & Wksht.
- Les. Opener (Calc.)
- Graphing Calc. Activity
- Examples 1–3
- Closure Question
- Guided Practice Exs.

Apply/Homework
- See Assignment Guide.
- See the CRB: Practice, Reteach, Apply, Extend

Assessment Options
- Checkpoint Exercises
- Daily Quiz (12.7)
- Stand. Test Practice

Day 12

12.8

Starting Options
- Homework Check
- Warm-Up or Daily Quiz

Teaching Options
- Motivating the Lesson
- Les. Opener (Activity)
- Examples 1–5
- Closure Question
- Guided Practice Exs.

Apply/Homework
- See Assignment Guide.
- See the CRB: Practice, Reteach, Apply, Extend

Assessment Options
- Checkpoint Exercises
- Daily Quiz (12.8)
- Stand. Test Practice
- Quiz (12.6–12.8)

Day 13

Review

Day 13 Start Options
- Homework Check

Reviewing Options
- Chapter 12 Summary
- Chapter 12 Review
- Chapter Review Games and Activities

Apply/Homework
- Chapter 12 Test (practice)
- Ch. Standardized Test (practice)

Day 14

Assess

Day 14 Start Options
- Homework Check

Assessment Options
- Chapter 12 Test
- SAT/ACT Ch. 12 Test
- Alternative Assessment

Day 7

12.4 *(cont.)*

Starting Options
- Homework Check

Teaching Options
- Examples 5–6
- Closure Question
- Guided Practice Exs. 14–19

Apply/Homework
- See Assignment Guide.
- See the CRB: Practice, Reteach, Apply, Extend

Assessment Options
- Checkpoint Exercises, p. 732
- Daily Quiz (12.4)
- Stand. Test Practice

Day 8

12.5

Starting Options
- Homework Check
- Warm-Up or Daily Quiz

Teaching Options
- Motivating the Lesson
- Concept Act. & Wksht.
- Les. Opener (Visual)
- Graphing Calc. Activity
- Examples 1–3
- Guided Practice Exs. 1–12

Apply/Homework
- See Assignment Guide.
- See the CRB: Practice, Reteach, Apply, Extend

Assessment Options
- Checkpoint Exercises, p. 739

BLOCK SCHEDULE

Day 1

12.1 & 12.2

Day 1 Start Options
- Prereq. Skills Review
- Strategies for Reading
- W-Up 12.1 or D. Quiz

Teaching 12.1 Options
- Motivating the Lesson
- Les. Opener (Appl.)
- Graphing Calc. Activity
- Examples 1–4
- Technology Activity
- Closure Question
- Guided Practice Exs.

Beginning 12.2 Options
- Warm-Up (Les. 12.2)
- Motivating the Lesson
- Les. Opener (Activity)
- Graphing Calc. Activity
- Examples 1–4
- Guided Prac. Exs. 1–16

Apply/Homework
- See Assignment Guide.
- See the CRB: Practice, Reteach, Apply, Extend
- Checkpoint Exercises
- Daily Quiz (Les. 12.1)

Day 2

12.2 & 12.3

Day 2 Start Options
- Homework Check
- Daily Quiz (Les. 12.1)

Finishing 12.2 Options
- Example 5
- Closure Question
- Guided Practice Ex. 17

Beginning 12.3 Options
- Warm-Up (Les. 12.3)
- Motivating the Lesson
- Les. Opener (Activity)
- Graphing Calc. Activity
- Examples 1–4
- Guided Practice Exs. 1–13

Apply/Homework
- See Assignment Guide.
- See the CRB: Practice, Reteach, Apply, Extend

Assessment Options
- Checkpoint Exercises
- Daily Quiz (Les. 12.2)
- Stand. Test Prac. (12.2)

Day 3

12.3 & 12.4

Day 3 Start Options
- Homework Check
- Daily Quiz (Les. 12.2)

Finishing 12.3 Options
- Example 5
- Closure Question
- Guided Practice Ex. 14

Beginning 12.4 Options
- Warm-Up (Les. 12.4)
- Motivating the Lesson
- Concept Act. & Wksht.
- Les. Opener (Appl.)
- Examples 1–4
- Guided Practice Exs. 1–13

Apply/Homework
- See Assignment Guide.
- See the CRB: Practice, Reteach, Apply, Extend

Assessment Options
- Checkpoint Exercises
- Daily Quiz (Les. 12.3)
- Stand. Test Prac. (12.3)
- Quiz (12.1–12.3)

Day 4

12.4 & 12.5

Day 4 Start Options
- Homework Check
- Daily Quiz (Les. 12.3)

Finishing 12.4 Options
- Examples 5–6
- Closure Question
- Guided Practice Exs. 14–19

Beginning 12.5 Options
- Warm-Up (Les. 12.5)
- Motivating the Lesson
- Concept Act. & Wksht.
- Les. Opener (Visual)
- Graphing Calc. Activity
- Examples 1–3
- Guided Practice Exs. 1–12

Apply/Homework
- See Assignment Guide.
- See the CRB: Practice, Reteach, Apply, Extend

Assessment Options
- Checkpoint Exercises
- Daily Quiz (Les. 12.4)
- Stand. Test Prac. (12.4)

Day 5

12.5 & 12.6

Day 5 Start Options
- Homework Check
- Daily Quiz (Les. 12.4)

Finishing 12.5 Options
- Examples 4–5
- Closure Question

Teaching 12.6 Options
- Warm-Up (Les. 12.6)
- Motivating the Lesson
- Les. Opener (Appl.)
- Examples 1–5
- Closure Question
- Guided Practice Exs.

Apply/Homework
- See Assignment Guide.
- See the CRB: Practice, Reteach, Apply, Extend

Assessment Options
- Checkpoint Exercises
- D. Quiz (Les. 12.5, 12.6)
- Stand. Test Practice
- Quiz (12.4–12.5)

Day 6

12.7 & 12.8

Day 6 Start Options
- Homework Check
- W-Up 12.7 or D. Quiz

Teaching 12.7 Options
- Concept Act. & Wksht.
- Les. Opener (Calc.)
- Graphing Calc. Activity
- Examples 1–3
- Closure Question
- Guided Practice Exs.

Teaching 12.8 Options
- Warm-Up (Les. 12.8)
- Motivating the Lesson
- Les. Opener (Activity)
- Examples 1–5
- Closure Question
- Guided Practice Exs.

Apply/Homework
- See Assignment Guide.
- See the CRB: Practice, Reteach, Apply, Extend

Assessment Options
- Checkpoint Exercises
- D. Quiz (Les. 12.7, 12.8)
- Quiz (12.6–12.8)

Day 7

Review/Assess

Day 7 Start Options
- Homework Check

Reviewing Options
- Chapter 12 Summary
- Chapter 12 Review
- Chapter Review Games and Activities
- Chapter 12 Test (practice)
- Ch. Standardized Test (practice)

Assessment Options
- Chapter 12 Test
- SAT/ACT Ch. 12 Test
- Alternative Assessment

CHAPTER 12

MEETING INDIVIDUAL NEEDS

BEFORE THE CHAPTER

The *Chapter 12 Resource Book* has the following materials to distribute and use before the chapter:

- **Parent Guide for Student Success (pictured below)**
- **Prerequisite Skills Review**
- **Strategies for Reading Mathematics**

PARENT GUIDE *Pages 3–4*

CHAPTER 12 NAME ______ DATE ______

Parent Guide for Student Success

For use with Chapter 12: Radicals and Connections to Geometry

Chapter Support

Chapter Overview One way that you can help your student succeed in Chapter 12 is by discussing the lesson goals in the chart below. When a lesson is completed, ask your student to interpret the lesson goals for you and to explain how the mathematics of the lesson relates to one of the key applications listed in the chart.

Lesson Title	Lesson Goals	Key Applications
12.1: Square-Root Functions	Evaluate and graph a square-root function. Use square-root functions to model real-life problems.	• Dinosaurs • Medicine • Investigating Accidents
12.2: Operations with Radical Expressions	Add, subtract, multiply, and divide radical expressions. Use radical expressions in real-life situations.	• Sailing • Pole-Vaulting • Falling Objects
12.3: Solving Radical Equations	Solve a radical equation. Use radical equations to solve real-life problems.	• Plane De-icing • Amusement Park Ride • Blotting Paper
12.4: Completing the Square	Solve a quadratic equation by completing the square. Choose a method for solving a quadratic equation.	• Vettisfoss Falls • Diving • Penguins
12.5: The Pythagorean Theorem and Its Converse	Use the Pythagorean Theorem and its converse in real-life problems.	• Baseball Diamond • Designing a Staircase • Planting a New Tree
12.6: The Distance and Midpoint Formulas	Find the distance between two points and the midpoint between two points in a coordinate plane.	• Soccer • Computer Games • Distances on Maps
12.7: Trigonometric Ratios	Use the sine, cosine, and tangent of an angle. Use trigonometric ratios in real-life problems.	• Parasailing • Operatoring a Crane • Cloud Height
12.8: Logical Reasoning: Proof	Use logical reasoning and proof to prove a statement is true. Prove that a statement is false.	• Goldbach's Conjecture • Lawyers • Four-Color Problem

Test-Taking Strategy

On multiple choice tests, try to **make an educated guess** when you are unsure of the answer. First **eliminate answers** that you know cannot be correct. Have your student look at the multiple choice test on pages 770-771. Ask your student to explain how to eliminate some answers to one of the questions without actually finding the correct answer.

PARENT GUIDE FOR STUDENT SUCCESS The first page summarizes the content of Chapter 12. Parents are encouraged to have their students explain how the material relates to key applications in the chapter, such as computer games. The second page (not shown) provides exercises and an activity that parents can do with their students. In the activity, parents and students use a protractor and trigonometric ratios to find the height of their home.

DURING EACH LESSON

The *Chapter 12 Resource Book* has the following alternatives for introducing the lesson:

- **Lesson Openers (pictured below)**
- **Graphing Calculator Activities with Keystrokes**

LESSON OPENER *Page 96*

LESSON 12.7 NAME ______ DATE ______ Available as a transparency

Calculator Activity Lesson Opener

For use with pages 752–757

Lesson 12.7

In Questions 1–4, record the side lengths of the triangle in the table below. Then use your calculator to find the ratios in the last three columns. Round to the nearest thousandth.

	a	b	c	$\frac{a}{b}$	$\frac{a}{c}$	$\frac{b}{c}$
1.						
2.						
3.						
4.						

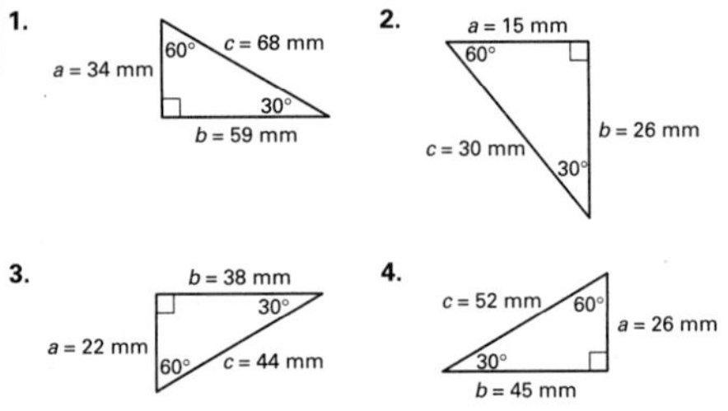

5. Look at the right triangles in Questions 1–4. What is the same about the triangles? What is different?

6. What do you notice about the values in the last three columns of the table?

GRAPHING CALCULATOR LESSON OPENER This Lesson Opener provides an alternative way to start Lesson 12.7 through the use of a graphing calculator. Students use a calculator to find ratios of the lengths of sides of right triangles to develop an understanding of trigonometric ratios.

TECHNOLOGY RESOURCE

Teachers can use the Electronic Lesson Presentations CD-ROM to present Lesson 12.4 or Activity 12.4 using computer animation to step through the Examples.

FOLLOWING EACH LESSON

The *Chapter 12 Resource Book* has a variety of materials to follow-up each lesson. They include the following:

- **Practice (3 levels)**
- **Reteaching with Practice**
- **Quick Catch-Up for Absent Students**
- **Interdisciplinary Applications (pictured below)**
- **Real-Life Applications**

APPLICATION *Page 90*

LESSON 12.6

NAME ______________________ DATE ________

Interdisciplinary Application

For use with pages 745–750

Model Planes (Science)

Your science class is building model planes. The planes can be made of any combination of wood, paper, glue, paper clips, or styrofoam™. Any gliding design is acceptable, but no propelling devices are allowed. Your grade will be determined based on the design of the plane and the distance traveled by the plane.

Every plane will be thrown from a window by your teacher. The window is 33 feet high. You build two designs at home. The first is made of styrofoam™, while the second is made from balsa wood. You test out your planes by throwing them from a window which is 22 feet high and measure the distance they travel. The styrofoam™ design travels 26 feet from the house. The wood design travels 32 feet from the house.

1. Refer to the diagram at the right. Determine the distance traveled from the roof to the ground by the styrofoam™ plane.

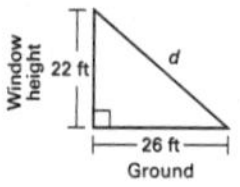

2. Sketch a diagram similar to the one in Exercise 1 that represents the balsa wood plane.

3. Determine the distance traveled by the balsa wood plane.

In Exercises 4 and 5, use the following information

After testing your planes, you decide to make some modifications to the wing design of the balsa wood plane. You redesign your plane with curved wings to increase the lift of the plane.

4. Your teacher throws the redesigned plane from the window and it lands 64 feet from the building. What is the air distance traveled?

5. If all other conditions were the same at school and your house, did the new wing design cause the plane to travel further, or was it due to the greater height of the building?

Lesson 12.6

INTERDISCIPLINARY APPLICATION This application makes a connection between model planes and the distance formula, the topic of Lesson 12.6.

ASSESSING THE CHAPTER

The *Chapter 12 Resource Book* has the following review and assessment materials:

- **Quizzes**
- **Chapter Review Games and Activities**
- **Chapter Test (3 levels)**
- **SAT/ACT Chapter Test**
- **Alternative Assessment and Math Journal (below)**
- **Project with Rubric**
- **Cumulative Review**

ALTERNATIVE ASSESSMENT *Pages 128–129*

CHAPTER 12

NAME ______________________ DATE ________

Alternative Assessment and Math Journal

For use after Chapter 12

JOURNAL **1.** Consider the graphs of $y = |x|$, $y = x^2$, and $y = \sqrt{x}$ to be the "parent" functions.

a. Explain what the graphs of $y = |x| + 2$, $y = x^2 + 2$, and $y = \sqrt{x} + 2$ have in common. Describe how the graphs of $y = |x| - 1$, $y = x^2 - 1$, and $y = \sqrt{x} - 1$ differ from their parent function. What effect would $y = |x + 3|$, $y = (x + 3)^2$, and $y = \sqrt{x + 3}$ have on the graphs compared to the parent function?

b. Given a generic equation $y = |x - h| + k$, $y = (x - h)^2 + k$, or $y = \sqrt{x - h} + k$, what effect do h and k have on the graphs? Write a paragraph to explain different possible scenarios. Be sure to consider values that are positive, negative, or zero.

c. How would a negative sign in form of a function (such as $y = -\sqrt{x - 2}$) change the graph? How would multiplying the function by a whole number (such as $y = 2\sqrt{x}$) make a difference from the graph of the original parent function?

MULTI-STEP PROBLEM **2.** Consider the triangle formed by the coordinates $A(2, -3)$, $B(8, 3)$, and $C(8, -9)$ to answer the following. Graph and label this triangle.

a. Find the midpoints of each side of the triangle.

b. Graph the new triangle formed by connecting the midpoints found in part **a.** Find the lengths of each side of this new triangle. Give exact answers.

c. Determine if the triangle formed in part **b** is a right triangle. Explain how you know.

d. Find the perimeter of the triangle from part **b.**

e. Find the area of the triangle from part **b.**

f. Find and simplify the trigonometric ratios for sin B, cos B, and tan B. Give exact answers.

g. Given point $D(-4, y)$ and the distance from point A to D is 10 units, find the possible value(s) for y. Show all work and give reasons for each step.

3. ***Critical Thinking*** Find the lengths of all three sides of the original triangle ABC. Determine if the original triangle ABC is a right triangle. Make a conjecture about the relationship between a triangle and the new triangle formed by joining the midpoints.

Review and Assess

ALTERNATIVE ASSESSMENT WITH RUBRIC AND MATH JOURNAL The journal exercise asks students to demonstrate their understanding of graphs of square-root functions in writing. The Multi-Step Problem has students pull together a variety of concepts from Chapter 12 to solve a problem about forming a new triangle from the midpoints of an original triangle. Answers and a scoring rubric are provided on a separate sheet.

CHAPTER OVERVIEW

CHAPTER GOALS

Students will investigate and graph square-root functions. They will simplify expressions involving radicals and will solve radical equations by factoring, by using the quadratic formula, or by completing the square. Students will study and use the Pythagorean theorem and its converse. They will be introduced to the midpoint and distance formulas. They will also use similar triangles to explore trigonometric ratios. They will conclude the chapter by using deductive reasoning to complete and construct direct and indirect proofs of theorems in algebra.

APPLICATION NOTE

If a ball tied to a string were swung in a circle at a constant velocity, centripetal force is the force that would keep it moving in a circular path. If the string was cut, the force of the string would no longer be acting on the ball. Then Sir Isaac Newton's first law of motion tells us that, except for the force of gravity, the ball would travel in a straight line at a right angle to the radius of the circle.

Additional information about centripetal force is available at **www.mcdougallittell.com.**

RADICALS AND CONNECTIONS TO GEOMETRY

▶ *How are passengers kept in place on an amusement park ride that spins?*

CHAPTER

12

APPLICATION: Spinning Rides

Some amusement park rides spin so fast that the riders "stick" to the walls of the ride. The force exerted by the wall on the rider is called *centripetal force.*

To design an amusement park ride, engineers must figure out the dimensions of the ride and how many times per minute it will spin. You'll learn to calculate the force generated by such a ride in Chapter 12.

Think & Discuss

1. Based on the numbers in the table, is "revolutions per minute" a function of "height"? **No**

Ride name	Height (feet)	Revolutions per minute
Football Ride	34.4	15
Chaos™	36	12
Centrox	44.3	17.5
Galactica	44.3	17

2. You are designing a spinning ride that is 40 feet high. Use the information in the table to decide on a reasonable range for how many revolutions per minute the ride would make. **between 12 and 17**

Learn More About It

You will calculate the centripetal force exerted on a rider in Exercises 66 and 67 on p. 726.

INTERNET APPLICATION LINK Visit www.mcdougallittell.com for more information about amusement park rides.

ADDITIONAL RESOURCES

Another way to begin the chapter is to show the video clip of a real-life motivator for Chapter 12 on the *Instant Replay: Video Review Games.*

PROJECTS

A project covering Chapters 10–12 appears on pages 774–775 of the Student Edition. An additional project for Chapter 12 is available in the *Chapter 12 Resource Book,* p. 130.

TECHNOLOGY

Software

- *Electronic Teaching Tools*
- *Online Lesson Planner*
- *Personal Student Tutor*
- *Test and Practice Generator*
- *Electronic Lesson Presentations* (Lessons 12.4 and Lab Activity 12.4)

Video

- *Instant Replay: Video Review Games*

INTERNET **Internet Connections**
www.mcdougallittell.com

- **Application Links**
  707, 726, 735, 744
- **Student Help**
  710, 715, 717, 731
- **Career Links**
  711, 713, 749, 760
- **Extra Challenge**
  721, 736, 743, 750, 757, 763

PREPARE

Diagnostic Tools

The **Skill Review** exercises can help you diagnose whether students have the following skills needed in Chapter 12:

- Find the lengths of sides of similar triangles.
- Simplify radical expressions.
- Factor trinomials.

The following resources are available for students who need additional help with these skills:

- Prerequisite Skills Review (*Ch. 12 Resource Book,* p. 5; *Warm-Up Transparencies,* p. 93)
- Reteaching with Practice (Chapter Resource Books for Lessons 3.2, 9.2 and 10.5.)
- *Personal Student Tutor*

Additional Resources

The following resources are provided to help you prepare for the upcoming chapter and customize review materials:

- ***Chapter 12 Resource Book***
  Tips for New Teachers (p. 1)
  Parent Guide (p. 3)
  Lesson Plans (every lesson)
  Lesson Plans for Block Scheduling (every lesson)
- ***Electronic Teaching Tools***
- ***Online Lesson Planner***
- ***Test and Practice Generator***

CHAPTER 12

Study Guide

PREVIEW

What's the chapter about?

Chapter 12 connects **radicals and geometry**. In Chapter 12 you'll learn

- how to solve radical equations and graph radical functions.
- how to apply the Pythagorean theorem.
- how to use trigonometric ratios.
- how to prove theorems by using algebraic properties.

KEY VOCABULARY

▶ Review
- **counterexample, p. 66**
- **similar triangles, p. 140**
- **perfect square trinomial, p. 619**
- **extraneous solutions, p. 644**

▶ New
- **square-root function, p. 709**
- **completing the square, p. 730**
- **Pythagorean theorem, p. 738**
- **hypothesis, p. 739**
- **conclusion, p. 739**
- **converse, p. 739**
- **trigonometric ratios, p. 752**
- **postulate, p. 758**
- **theorem, p. 759**
- **indirect proof, p. 760**

PREPARE

Are you ready for the chapter?

SKILL REVIEW Do these exercises to review skills that you'll apply in this chapter. See the given reference page if there is something you don't understand.

In Exercises 1 and 2, the two triangles are similar. Find the length of the side marked *x*. (Review Example 5, p. 141)

1. 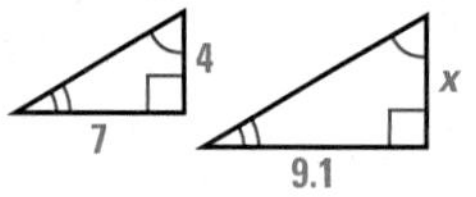

5.2

2. 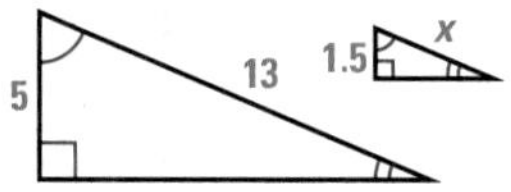

3.9

Simplify the expression. (Review Examples 1 and 2, p. 512)

3. $\sqrt{98}$ $7\sqrt{2}$ **4.** $\sqrt{140}$ $2\sqrt{35}$ **5.** $\sqrt{\frac{7}{4}}$ $\frac{\sqrt{7}}{2}$ **6.** $\frac{\sqrt{144}}{\sqrt{16}}$ 3

Factor the trinomial. (Review Examples 1–4, pp. 604 and 605)

7. $x^2 - 3x - 18$ $(x - 6)(x + 3)$ **8.** $x^2 + 2x - 8$ $(x + 4)(x - 2)$ **9.** $4x^2 + 20x + 25$ $(2x + 5)^2$

STUDENT HELP

Study Tip
"Student Help" boxes throughout the chapter give you study tips and tell you where to look for extra help in this book and on the Internet.

STUDY STRATEGY

Here's a study strategy!

Drawing Diagrams

Sometimes it is helpful to include a diagram or another visual when you take notes. Diagrams and tables can also be used to organize related ideas and terms.

12.1 Functions Involving Square Roots

What you should learn

GOAL 1 Evaluate and graph a function involving square roots.

GOAL 2 Use functions involving square roots to model **real-life** problems, such as an investigation of skid marks in **Ex. 60**.

Why you should learn it

▼ To solve **real-life** problems such as estimating a dinosaur's walking speed in **Example 4**.

REAL LIFE

GOAL 1 GRAPHING FUNCTIONS INVOLVING SQUARE ROOTS

ACTIVITY Developing Concepts

Investigating the Square Root Function

1. Copy the table of values for the function $y = \sqrt{x}$. Use a calculator or the Table of Squares and Square Roots on page 811 to complete your table. Round to the nearest tenth. Plot the points. Graph the function.
0; 1; 1.4; 1.7; 2; 2.2; See margin for graphs.

x	0	1	2	3	4	5
y	?	?	?	?	?	?

2. What are the domain and the range of the function?
Both the domain and range are all the nonnegative numbers.
3. Why are negative values of x not included in the domain?
Square roots are not defined for negative values of x.

The **square root function** is defined by the equation $y = \sqrt{x}$. The domain is all nonnegative numbers and the range is all nonnegative numbers. Understanding this function will help you work with other functions involving square roots.

EXAMPLE 1 *Graphing $y = a\sqrt{x}$*

Sketch the graph of $y = 2\sqrt{x}$. Give the domain and the range.

SOLUTION

The radicand of a square root is always nonnegative, so the domain is the set of all nonnegative numbers. Make a table of values. Then plot the points and connect them with a smooth curve.

x	y
0	$y = 2\sqrt{0} = 0$
1	$y = 2\sqrt{1} = 2$
2	$y = 2\sqrt{2} \approx 2.8$
3	$y = 2\sqrt{3} \approx 3.5$
4	$y = 2\sqrt{4} = 4$
5	$y = 2\sqrt{5} \approx 4.5$

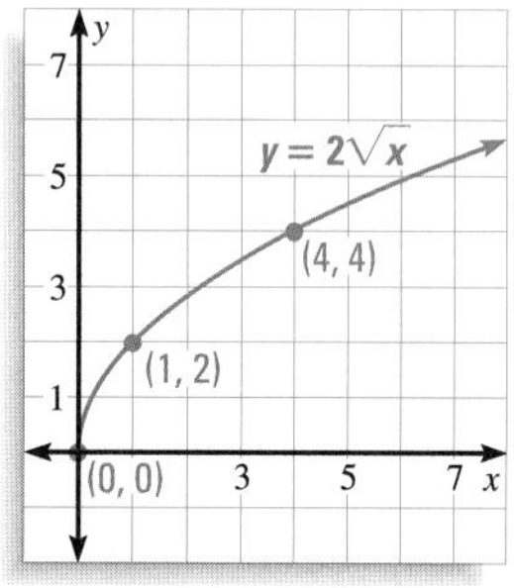

Both the domain and the range are all the nonnegative numbers.

1 PLAN

PACING
Basic: 1 day
Average: 1 day
Advanced: 1 day
Block Schedule: 0.5 block with 12.2

LESSON OPENER
APPLICATION
An alternative way to approach Lesson 12.1 is to use the Application Lesson Opener:
- Blackline Master (*Chapter 12 Resource Book*, p. 12)
- Transparency (p. 82)

MEETING INDIVIDUAL NEEDS
- ***Chapter 12 Resource Book***
  Prerequisite Skills Review (p. 5)
  Practice Level A (p. 14)
  Practice Level B (p. 15)
  Practice Level C (p. 16
  Reteaching with Practice (p. 17)
  Absent Student Catch-Up (p. 19)
  Challenge (p. 21)
- ***Resources in Spanish***
- ***Personal Student Tutor***

NEW-TEACHER SUPPORT
See the Tips for New Teachers on pp. 1–2 of the *Chapter 12 Resource Book* for additional notes about Lesson 12.1.

WARM-UP EXERCISES

Transparency Available

Find the value of each expression to the nearest tenth when $x = 3$.

1. $\sqrt{x}$ 1.7
2. $\sqrt{x+1}$ 2
3. $\sqrt{x^3-2}$ 5
4. Which of the following is not a real number?
a. $\sqrt{-9}$ **b.** $\sqrt{144}$ **c.** $\sqrt{18.5}$ *a*

1. See Additional Answers beginning on page AA1.

2 TEACH

MOTIVATING THE LESSON
You are designing a rectangular garden that will be fenced in and will have an area of 36 ft^2. What could be some possible dimensions of the garden? A square would require the least amount of fencing. Square root functions can be used to find the side length of a square, given its area.

EXTRA EXAMPLE 1
Sketch the graph of $y = -2\sqrt{x}$. Give the domain and range.

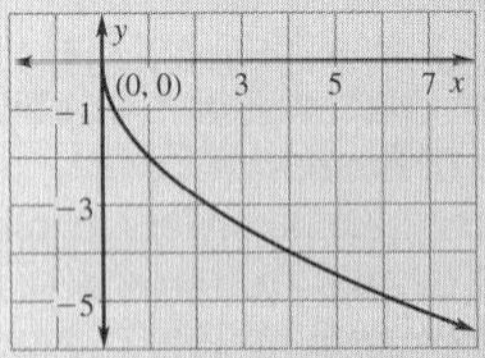

domain: all nonnegative numbers; range: all nonpositive numbers

EXTRA EXAMPLE 2
Find the domain and the range of $y = \sqrt{x} - 1$. Then sketch its graph.

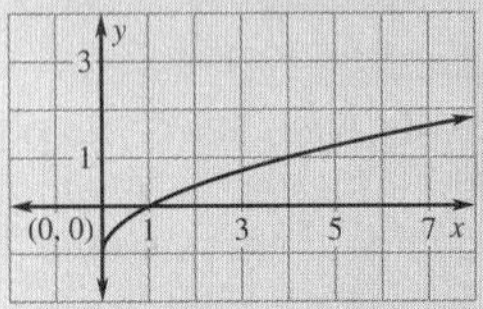

domain: all nonnegative numbers; range: all numbers greater than or equal to −1

CHECKPOINT EXERCISES
For use after Examples 1 and 2:
1. Find the domain and the range of $y = \sqrt{x} + 3$. Then sketch its graph.

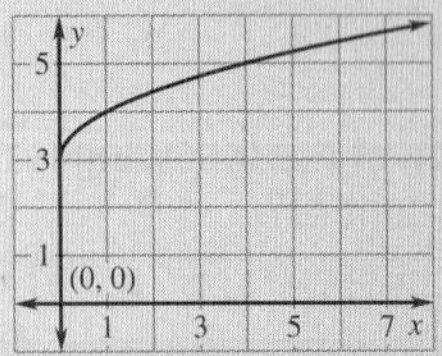

domain: $x \geq 0$; range: $y \geq 3$

STUDENT HELP
HOMEWORK HELP
Visit our Web site www.mcdougallittell.com for extra examples.

It is a good idea to find the domain of a function *before* you make a table of values. This will help you choose values of x for the table.

EXAMPLE 2 *Graphing $y = \sqrt{x} + k$*

Find the domain and the range of $y = \sqrt{x} + 1$. Then sketch its graph.

SOLUTION

The domain is the set of all nonnegative numbers. The range is the set of all numbers that are greater than or equal to 1. Make a table of values, plot the points, and connect them with a smooth curve.

x	y
0	$y = \sqrt{0} + 1 = 1$
1	$y = \sqrt{1} + 1 = 2$
2	$y = \sqrt{2} + 1 \approx 2.4$
3	$y = \sqrt{3} + 1 \approx 2.7$
4	$y = \sqrt{4} + 1 = 3$
5	$y = \sqrt{5} + 1 \approx 3.2$

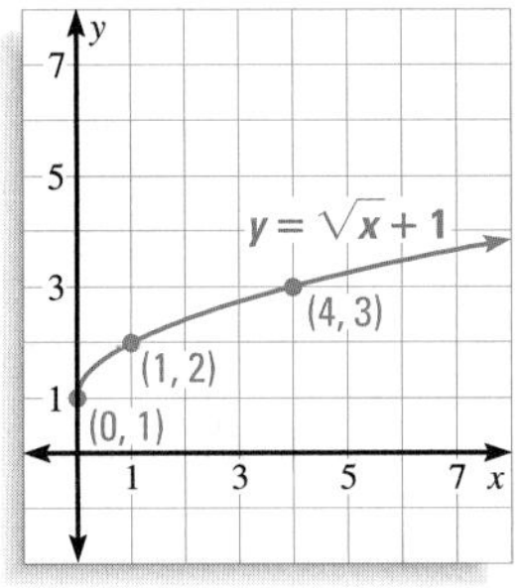

EXAMPLE 3 *Graphing $y = \sqrt{x - h}$*

Find the domain and the range of $y = \sqrt{x - 3}$. Then sketch its graph.

SOLUTION

To find the domain, find the values of x for which the radicand is nonnegative.

$x - 3 \geq 0$ **Write an inequality for the domain.**

$x \geq 3$ **Add 3 to each side.**

The domain is the set of all numbers that are greater than or equal to 3. The range is the set of all nonnegative numbers. Make a table of values, plot the points, and connect them with a smooth curve.

STUDENT HELP
Study Tip
When you make a table of values to sketch the graph of a function involving square roots, choose several values to see the pattern of the curve.

x	y
3	$y = \sqrt{3 - 3} = 0$
4	$y = \sqrt{4 - 3} = 1$
5	$y = \sqrt{5 - 3} \approx 1.4$
6	$y = \sqrt{6 - 3} \approx 1.7$
7	$y = \sqrt{7 - 3} = 2$
8	$y = \sqrt{8 - 3} \approx 2.2$

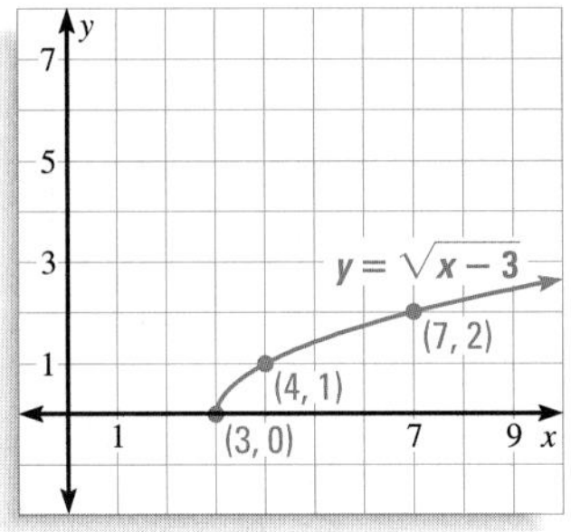

FOCUS ON CAREERS

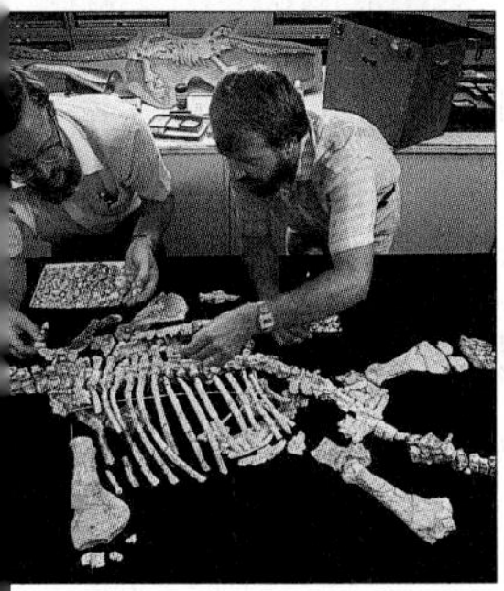

PALEONTOLOGIST A paleontologist studies fossils of animals and plants to better understand the history of life on Earth.

CAREER LINK
www.mcdougallittell.com

GOAL 2 USING SQUARE ROOTS IN REAL LIFE FUNCTIONS

EXAMPLE 4 *Using a Model Involving a Square Root*

DINOSAURS In a natural history museum you see leg bones for two species of dinosaurs and want to know how fast they walked. The maximum walking speed S (in feet per second) of an animal can be modeled by the equation below where $g = 32$ ft/sec^2 and L is the length (in feet) of the animal's leg. ▶ Source: *Discover*

Walking speed model: $S = \sqrt{gL}$

a. Use unit analysis to check the units of the model.

b. Sketch the graph of the model.

c. For one dinosaur the length of the leg is 1 foot. For the other dinosaur the length of the leg is 4 feet. Could the taller dinosaur walk four times as fast as the shorter dinosaur?

SOLUTION

a. **UNIT ANALYSIS** Substitute the units in the model.

$$\sqrt{\frac{\text{ft}}{\text{sec}^2} \cdot \text{ft}} = \sqrt{\frac{\text{ft}^2}{\text{sec}^2}} = \frac{\text{ft}}{\text{sec}}$$

b. To sketch the graph of $S = \sqrt{32L}$, make a table of values, plot the points, and draw a smooth curve through the points.

L	S
0	$S = \sqrt{32 \cdot 0} = 0$
1	$S = \sqrt{32 \cdot 1} \approx 5.66$
2	$S = \sqrt{32 \cdot 2} = 8$
3	$S = \sqrt{32 \cdot 3} \approx 9.80$
4	$S = \sqrt{32 \cdot 4} \approx 11.31$
5	$S = \sqrt{32 \cdot 5} \approx 12.65$
6	$S = \sqrt{32 \cdot 6} \approx 13.86$
7	$S = \sqrt{32 \cdot 7} \approx 14.97$
8	$S = \sqrt{32 \cdot 8} = 16$

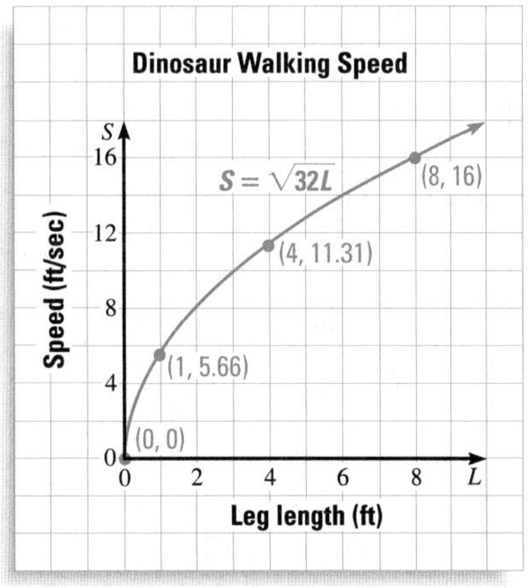

c. Substitute 1 foot and 4 feet into the model.

When L = 1 foot: $S = \sqrt{32 \cdot 1} \approx 5.66$ feet/second

When L = 4 feet: $S = \sqrt{32 \cdot 4} \approx 11.31$ feet/second

▶ The taller dinosaur could walk about twice as fast as the shorter dinosaur, not four times as fast.

EXTRA EXAMPLE 3

Find the domain and the range of $y = \sqrt{x+2}$. Then sketch its graph.

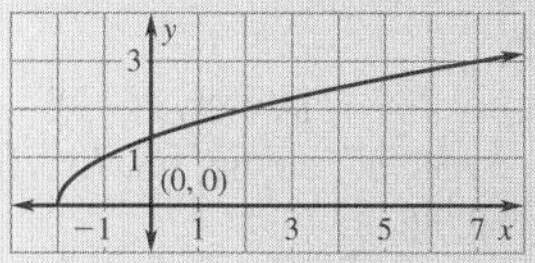

domain: $x \geq -2$; range: $y \geq 0$

EXTRA EXAMPLE 4

A beam's resistance to bending is determined by its moment of inertia, radius of gyration, and area. The length of the radius of gyration, r, can be modeled by $r = \sqrt{\frac{I}{A}}$, where I is the moment of inertia (in cm^4) and A is the area (in cm^2).

a. Use unit analysis to check the units on the model.
$\sqrt{\frac{\text{cm}^4}{\text{cm}^2}} = \sqrt{\text{cm}^2} = \text{cm}$

b. Sketch a graph of the model if $A = 2$ cm^2.

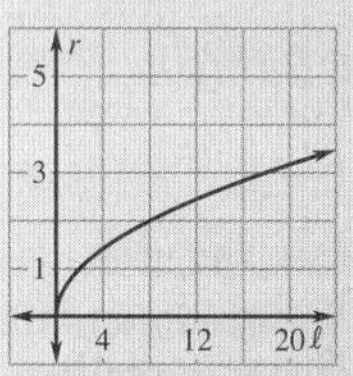

c. Find the radius of gyration to the nearest tenth if $I = 80$ cm^4 and $A = 2$ cm^2. **6.3 cm**

CHECKPOINT EXERCISES

For use after Examples 3 and 4:

1. Find the domain and range of $y = \sqrt{x-1}$ **domain: $x \geq 1$; range: $y \geq 0$**

FOCUS ON VOCABULARY

Is $y = \sqrt{3}x$ a square root function? Explain. **No; a square root function has a variable in the radicand.**

CLOSURE QUESTION

In what quadrant(s) does the graph of $y = \sqrt{x} + 5$ lie? **Quadrant 1**

3 APPLY

ASSIGNMENT GUIDE

BASIC

Day 1: pp. 712–714 Exs. 20–56 even, 57, 60, 63, 65, 70, 75, 80, 86, 87

AVERAGE

Day 1: pp. 712–714 Exs. 20–56 even, 57, 60, 63, 65, 70, 75, 80, 86, 87

ADVANCED

Day 1: pp. 712–714 Exs. 20–56 even, 57, 60, 63–65, 70, 75, 80, 86, 87

BLOCK SCHEDULE WITH 12.2

pp. 712–714 Exs. 20–56 even, 57, 60, 63, 65, 70, 75, 80, 86, 87

EXERCISE LEVELS

Level A: *Easier*
20–29

Level B: *More Difficult*
30–63, 65–88

Level C: *Most Difficult*
64

✓ HOMEWORK CHECK

To quickly check student understanding of key concepts, go over the following exercises: Exs. 22, 24, 32, 34, 44, 48, 52, 54, 60. See also the Daily Homework Quiz:

- Blackline Master (*Chapter 12 Resource Book*, p. 24)
- Transparency (p. 95)

15.

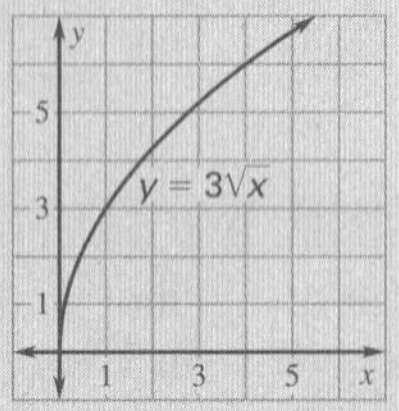

16.

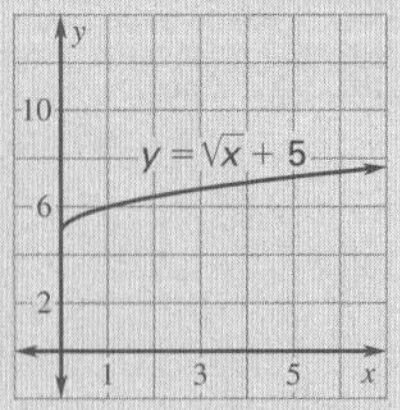

17–18. See Additional Answers beginning on page AA1.

GUIDED PRACTICE

Vocabulary Check ✓

1. Describe the square root function. **1–2. See margin.**

Concept Check ✓

2. Does the domain of $y = \sqrt{x+3}$ include negative values of x? Explain.

Skill Check ✓

Evaluate the function for $x = 0, 1, 2, 3$, and 4. Round your answer to the nearest tenth.

3. $y = 4\sqrt{x}$ **0, 4, 5.7, 6.9, 8**
4. $y = \frac{1}{2}\sqrt{x}$ **0, 0.5, 0.7, 0.9, 1**
5. $y = 3\sqrt{x} + 4$ **4, 7, 8.2, 9.2, 10**
6. $y = 6\sqrt{x} - 3$ **−3, 3, 5.5, 7.4, 9**
7. $y = \sqrt{x+2}$ **1.4, 1.7, 2, 2.2, 2.4**
8. $y = \sqrt{4x-1}$ **not defined, 1.7, 2.6, 3.3, 3.9**

Find the domain and the range of the function. **9–14. See margin.**

9. $y = 3\sqrt{x}$
10. $y = \sqrt{x}$
11. $y = \sqrt{x} - 10$
12. $y = \sqrt{x} + 6$
13. $y = \sqrt{x+5}$
14. $y = \sqrt{x-10}$

Sketch the graph of the function. **15–17. See margin.**

15. $y = 3\sqrt{x}$
16. $y = \sqrt{x} + 5$
17. $y = 3\sqrt{x+1}$

FIRE HOSES For a fire hose with a nozzle that has a diameter of 2 inches, the flow rate f (in gallons per minute) can be modeled by $f = 120\sqrt{p}$ where p is the nozzle pressure in pounds per square inch.

18. Sketch a graph of the model. **See margin.**
19. If the flow rate is 1200 gallons per minute, what is the nozzle pressure? **100 lb/in.²**

1. The square root function is $y = \sqrt{x}$, where the domain and range are all nonnegative numbers.
2. Yes; the domain includes all numbers greater than or equal to −3; for those values, $\sqrt{x+3}$ is defined.
9. Both the domain and range are all the nonnegative numbers.
10. Both the domain and range are all the nonnegative numbers.
11. domain: all nonnegative numbers; range: all numbers greater than or equal to −10
12. domain: all nonnegative numbers; range: all numbers greater than or equal to 6
13. domain: all numbers greater than or equal to −5; range: all nonnegative numbers
14. domain: all numbers greater than or equal to 10; range: all nonnegative numbers

PRACTICE AND APPLICATIONS

STUDENT HELP

Extra Practice to help you master skills is on p. 808.

EVALUATING FUNCTIONS Evaluate the function for the given value of x. Round your answer to the nearest tenth.

20. $y = 3\sqrt{x}$; 9 **9**
21. $y = \frac{1}{2}\sqrt{x} - 1$; 16 **1**
22. $y = \sqrt{x-7}$; 15 **about 2.8**
23. $y = \sqrt{3x-5}$; 7 **4**
24. $y = 6\sqrt{15-x}$; −1 **24**
25. $y = \sqrt{21-2x}$; −2 **5**
26. $y = \sqrt{\frac{x}{2} - 2}$; 22 **3**
27. $y = \sqrt{8x^2 + \frac{3}{2}}$; $\frac{1}{4}$ **about 1.4**
28. $y = \sqrt{\frac{2x}{3} + 5}$; 6 **3**
29. $y = \sqrt{36x - 2}$; $\frac{1}{2}$ **4**

STUDENT HELP

HOMEWORK HELP
Example 1: Exs. 20–56
Example 2: Exs. 20–56
Example 3: Exs. 20–56
Example 4: Exs. 57, 58, 61, 62

FINDING THE DOMAIN Find the domain of the function.

30. $y = 6\sqrt{x}$ **all nonnegative numbers**
31. $y = \sqrt{x-17}$ $x \geq 17$
32. $y = \sqrt{3x-10}$ $x \geq 3\frac{1}{3}$
33. $y = \sqrt{x+5}$ $x \geq -5$
34. $y = 4 + \sqrt{x}$ **all nonnegative numbers**
35. $y = \sqrt{x} - 3$ **all nonnegative numbers**
36. $y = 5 - \sqrt{x}$ **all nonnegative numbers**
37. $y = 4\sqrt{x}$ **all nonnegative numbers**
38. $y = 2\sqrt{4x}$ **all nonnegative numbers**
39. $y = 0.2\sqrt{x}$ **all nonnegative numbers**
40. $y = x\sqrt{x}$ **all nonnegative numbers**
41. $y = \sqrt{x+9}$ $x \geq -9$
42. $y = 8\sqrt{\frac{5}{2}x}$ **all nonnegative numbers**
43. $y = \frac{\sqrt{x}}{5}$ **all nonnegative numbers**
44. $y = \frac{\sqrt{4-x}}{x}$ $x \leq 4, x \neq 0$

45–47. Both the domain and range are all the nonnegative numbers.

48. domain: all nonnegative numbers; range: all numbers greater than or equal to -2

49. domain: all nonnegative numbers; range: all numbers greater than or equal to 4

50. domain: all nonnegative numbers; range: all numbers greater than or equal to -3

51. domain: all numbers greater than or equal to 4; range: all nonnegative numbers

52. domain: all numbers greater than or equal to -1; range: all nonnegative numbers

53. domain: all numbers greater than or equal to 6; range: all nonnegative numbers

54–55. domain: all numbers greater than or equal to -2.5; range: all nonnegative numbers

56. Both the domain and range are all the nonnegative numbers.

FOCUS ON CAREERS

ACCIDENT RECONSTRUCTIONIST

An accident reconstructionist is a police officer who examines physical evidence to find the cause of an accident.

CAREER LINK www.mcdougallittell.com

GRAPHING FUNCTIONS Find the domain and the range of the function. Then sketch the graph of the function. 45–56. See margin.

45. $y = 7\sqrt{x}$ **46.** $y = 4\sqrt{x}$ **47.** $y = 5\sqrt{x}$

48. $y = \sqrt{x} - 2$ **49.** $y = \sqrt{x} + 4$ **50.** $y = \sqrt{x} - 3$

51. $y = \sqrt{x-4}$ **52.** $y = \sqrt{x+1}$ **53.** $y = \sqrt{x-6}$

54. $y = 2\sqrt{4x+10}$ **55.** $y = \sqrt{2x+5}$ **56.** $y = x\sqrt{8x}$

57. MEDICINE A doctor may use a person's body surface area (*BSA*) as an index to prescribe the correct amount of medicine. You can use the model below to approximate a person's *BSA* (in square meters) where *h* represents height (in inches) and *w* represents weight (in pounds.)

$$BSA = \sqrt{\frac{h \cdot w}{3131}}$$

Find the *BSA* of a person who is 5 feet 2 inches tall and weighs 100 pounds. Round your answer to the nearest hundredth. 1.41 m^2

58. PENDULUMS The period *T* (in seconds) of a pendulum is the time it takes for the pendulum to swing back and forth. The period is related to the length *L* (in inches) of the pendulum by the model $T = 2\pi\sqrt{\frac{L}{384}}$. Find the length of a pendulum with a period of eight seconds. Give your answer to the nearest tenth. 622.5 in.

59. GEOMETRY CONNECTION The lateral surface area *S* of a cone with a radius of 14 centimeters can be found using the formula

$$S = \pi \cdot 14\sqrt{14^2 + h^2}$$

where *h* is the height (in centimeters) of the cone.

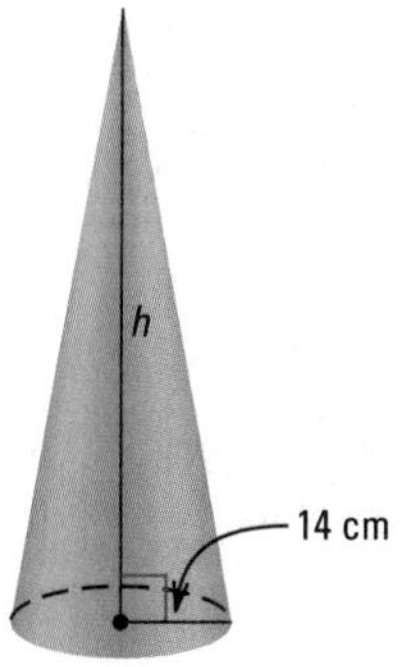

a. Use unit analysis to check the units of the formula. $\text{cm}^2 = \text{cm} \cdot \sqrt{\text{cm}^2 + \text{cm}^2} = \text{cm} \cdot \text{cm}$

b. Sketch the graph of the formula. See margin

c. Find the lateral surface area of a cone with a height of 30 centimeters. 1456 cm^2 (to the nearest cm^2)

60. INVESTIGATING ACCIDENTS An accident reconstructionist is responsible for finding how fast cars were going before an accident. To do this, a reconstructionist uses the model below where *S* is the speed of the car in miles per hour, *d* is the length of the tires' skid marks in feet, and *f* is the coefficient of friction for the road.

Car speed model: $S = \sqrt{30df}$

a. In an accident, a car makes skid marks 74 feet long. The coefficient of friction is 0.5. A witness says that the driver was traveling faster than the speed limit of 45 miles per hour. Can the witness's statement be correct? Explain your reasoning. No; the speed of the car was $\sqrt{30(74)(0.5)} \approx 33.3$ mi/h.

b. How long would the skid marks have to be in order to know that the car was traveling faster than 45 miles per hour? longer than 135 ft

45–56, and 59b. See additional Answers beginning on page AA1.

! COMMON ERROR

EXERCISE 2 Students may easily confuse the domain of this function with the domain of $y = \sqrt{x} + 3$. Ask students to graph the function in the exercise to verify their answer.

APPLICATION NOTE

EXERCISE 18 Two standard fire hose sizes are $1\frac{1}{2}$ in. and $2\frac{1}{2}$ in. A $2\frac{1}{2}$ in. diameter hose puts out about 250 gal/min at 100 lb of pressure per square inch. This is about one ton of water per minute.

EXERCISE 57 It makes sense that body surface area be expressed in a metric unit since intravenous medications are commonly prescribed in cubic centimeters.

EXERCISE 58 The swing of a pendulum is influenced by the acceleration due to gravity, which has been factored into this formula.

EXERCISE 60 The coefficient of friction is determined by the road surface and the road conditions. As the coefficient of friction increases, the speed increases.

FOCUS ON CAREERS

EXERCISE 60 Additional information about accident reconstructionists is available at **www.mcdougallittell.com.**

ADDITIONAL PRACTICE AND RETEACHING

For Lesson 12.1:

- Practice Levels A, B, and C (*Chapter 12 Resource Book*, p. 14)
- Reteaching with Practice (*Chapter 12 Resource Book*, p. 17)
- See Lesson 12.1 of the *Personal Student Tutor*

For more Mixed Review:

- Search the *Test and Practice Generator* for key words or specific lessons.

4 ASSESS

DAILY HOMEWORK QUIZ

Transparency Available

1. Evaluate the square-root function for the given value of x. Round your answer to the nearest tenth.

a. $\sqrt{x-6}$; 20 **3.7**

b. $9\sqrt{16-x}$; -2 **38.2**

2. Find the domain and range of the function.

a. $y=\sqrt{2x-7}$

domain: $x \geq \frac{7}{2}$, range: $y \geq 0$

b. $y=7+\sqrt{x}$

domain: $x \geq 0$, range: $y \geq 7$

c. $y=\frac{\sqrt{3-x}}{x}$ **domain: $x \leq 3$, $x \neq 0$, range: all reals**

EXTRA CHALLENGE NOTE

Challenge problems for Lesson 12.1 are available in **blackline** format in the *Chapter 12 Resource Book,* p. 21 and at **www.mcdougallittell.com.**

ADDITIONAL TEST PREPARATION

1. WRITING How do the domains of the functions $y=\sqrt{x}+4$ and $y=\sqrt{x+4}$ differ? **The domain of $y=\sqrt{x}+4$ cannot include any negative numbers. The domain $y=\sqrt{x+4}$ includes all negative numbers that are greater than or equal to -4.**

2. WRITING Explain why the range of $y=\sqrt{x}$ includes only nonnegative numbers. **The range includes only the principal square root.**

63b.

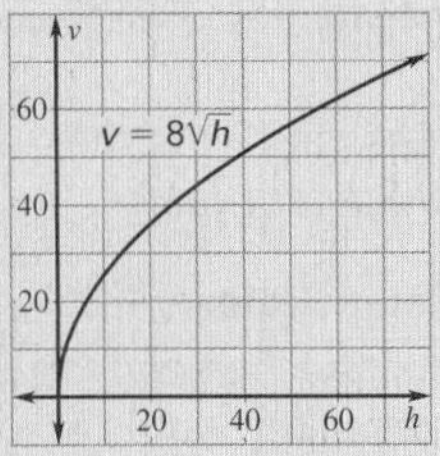

COLLECTING DATA In Exercises 61 and 62, use the information from Example 4 on page 711.

61. Measure the length of your leg and calculate your maximum walking speed. **Answers will vary.**

62. Mark a distance of 80 feet. Time yourself to see how long it takes you to walk the distance. Compare your speed from Exercise 61 with your actual speed. What did you find? **Answers will vary.**

Test Preparation

63. MULTI-STEP PROBLEM The relationship between a roller coaster's velocity v (in feet per second) at the bottom of a drop and the height of the drop h (in feet) can be modeled by the formula $v=\sqrt{2gh}$ where g represents acceleration due to gravity.

a. Use the fact that $g = 32$ ft/sec^2 to show that $v=\sqrt{2gh}$ can be simplified to $v=8\sqrt{h}$. **$v=\sqrt{2(32)h}=\sqrt{64h}=8\sqrt{h}$**

b. Sketch the graph of $v=8\sqrt{h}$. **See margin.**

c. *Writing* Use the formula or the graph to explain why doubling the height of a drop does not double the velocity of a roller coaster. **See margin.**

★ Challenge

64. CRITICAL THINKING Find the domain of $y=\frac{3}{\sqrt{x}-2}$.
all nonnegative numbers except 4

MIXED REVIEW

63. c. *Sample answer:* **When h is doubled, $v=8\sqrt{2h}$. The velocity is multiplied by a factor of $\sqrt{2} \approx 1.4$, not 2. The graph is not linear. If doubling the height doubled the drop, the graph would be linear.**

SIMPLIFYING RADICAL EXPRESSIONS **Simplify the radical expression.** (Review 9.2 for 12.2)

65. $\sqrt{24}$ **$2\sqrt{6}$** **66.** $\sqrt{60}$ **$2\sqrt{15}$** **67.** $\sqrt{175}$ **$5\sqrt{7}$** **68.** $\sqrt{9900}$ **$30\sqrt{11}$**

69. $\sqrt{\frac{20}{25}}$ **$\frac{2\sqrt{5}}{5}$** **70.** $\frac{1}{2}\sqrt{80}$ **$2\sqrt{5}$** **71.** $\frac{3\sqrt{7}}{\sqrt{9}}$ **$\sqrt{7}$** **72.** $4\sqrt{\frac{11}{16}}$ **$\sqrt{11}$**

SOLVING QUADRATIC EQUATIONS **Solve the equation.** (Review 9.5)

73. $x^2+4x-8=0$ **$-2-2\sqrt{3}, -2+2\sqrt{3}$** **74.** $x^2-2x-4=0$ **$1-\sqrt{5}, 1+\sqrt{5}$** **75.** $x^2-6x+1=0$ **$3-2\sqrt{2}, 3+2\sqrt{2}$**

76. $x^2+3x-10=0$ **$-5, 2$** **77.** $2x^2+x=3$ **$-1\frac{1}{2}, 1$** **78.** $4x^2-6x+1=0$ **$\frac{3-\sqrt{5}}{4}, \frac{3+\sqrt{5}}{4}$**

MULTIPLYING EXPRESSIONS **Find the product.** (Review 10.2 for 12.2)

79. $(x-2)(x+11)$ **$x^2+9x-22$** **80.** $(x+4)(3x-7)$ **$3x^2+5x-28$** **81.** $(2x-3)(5x-9)$ **$10x^2-33x+27$**

82. $(x-5)(x-4)$ **$x^2-9x+20$** **83.** $(6x+2)(x^2-x-1)$ **$6x^3-4x^2-8x-2$** **84.** $(2x-1)(x^2+x+1)$ **$2x^3+x^2+x-1$**

SIMPLIFYING EXPRESSIONS **Simplify the expression.** (Review 11.5)

85. $\frac{8x}{3}\cdot\frac{1}{x}$ **$\frac{8}{3}$** **86.** $\frac{8x^2}{3}\cdot\frac{9}{16x}$ **$\frac{3x}{2}$** **87.** $\frac{x}{x+6}\div\frac{x+1}{x+6}$ **$\frac{x}{x+1}$**

88. MOUNT RUSHMORE Four American Presidents' faces are carved on Mount Rushmore: Washington, Jefferson, Roosevelt, and Lincoln. Before the faces were carved on the cliff, scale models were made. The ratio of the faces on the cliff to the models was 12 to 1. If the scale model of President Washington's face was 5 feet tall, how tall is his face on Mount Rushmore? (Review 11.1) **60 ft**

ACTIVITY 12.1

Using Technology

Functions Involving Square Roots

A graphing calculator or a computer can be used to graph functions involving square roots. To graph such a function use the $\sqrt{\ }$ key.

EXPLORING THE CONCEPT

1 Use a graphing calculator to graph $y = a\sqrt{x}$ for $a = 1$. **Steps 1–2 Check graphs.**

2 Repeat **Step 1** for $a = 2, 3,$ and 4. Show all four graphs on the same screen.

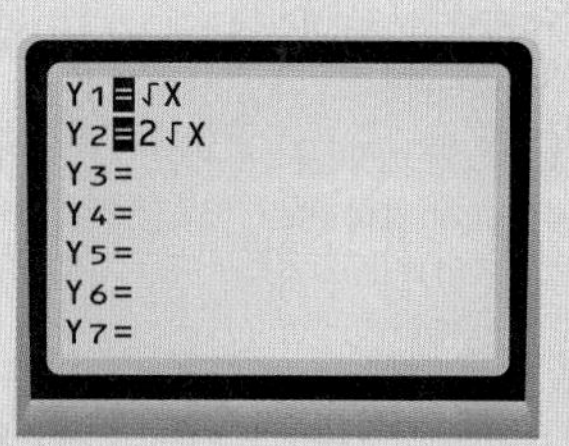

3 Describe the effect that a has on the graph of $y = a\sqrt{x}$.
For $a > 0$, as a increases, the graph of $y = a\sqrt{x}$ gets steeper.

Step 7. For $a < 0$, the graph of $y = a\sqrt{x}$ is in the third quadrant and gets steeper as $|a|$ increases.

EXPLORING THE CONCEPT

4 Use a graphing calculator to graph $y = a\sqrt{x}$ for $a = -1$ and $a = 1$. **Steps 4–6. Check graphs.**

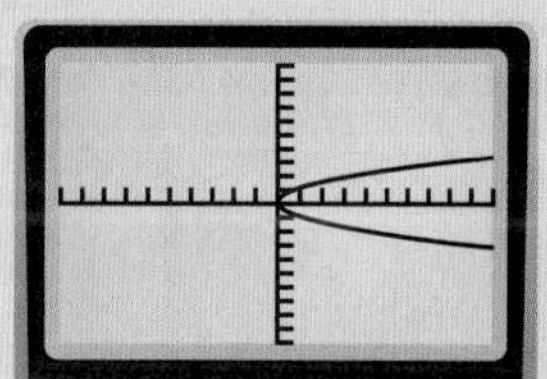

5 Graph $y = a\sqrt{x}$ for $a = -4$ and $a = 4$.

6 Graph $y = a\sqrt{x}$ for $a = -9$ and $a = 9$.

7 Describe the effect that a has on the graph of $y = a\sqrt{x}$ when a is negative. **See margin.**

STUDENT HELP

KEYSTROKE HELP
See keystrokes for several models of calculators at www.mcdougallittell.com

DRAWING CONCLUSIONS

1. Use a graphing calculator to graph the function $y = \sqrt{x} + k$ for $k = -3, -1, 2,$ and 5. Show all four graphs on the same screen. Then describe the effect that k has on the graph of $y = \sqrt{x} + k$.
Check graphs; k moves the graph of $y = \sqrt{x}$ up k units if k is positive and down $|k|$ units if k is negative.

2. Use a graphing calculator to graph the function $y = \sqrt{x + k}$ for $k = -3, -1, 2,$ and 5. You may need to use parentheses around the radicand. Show all four graphs on the same screen. Then describe the effect that k has on the graph of $y = \sqrt{x + k}$.
Check graphs; k moves the graph of $y = \sqrt{x}$ to the left k units if k is positive and $|k|$ units to the right if k is negative.

EXTENDING THE CONCEPT **Find the value of a or k so that the graph of the function matches the given graph.**

3. $y = a\sqrt{x}$ **−2**

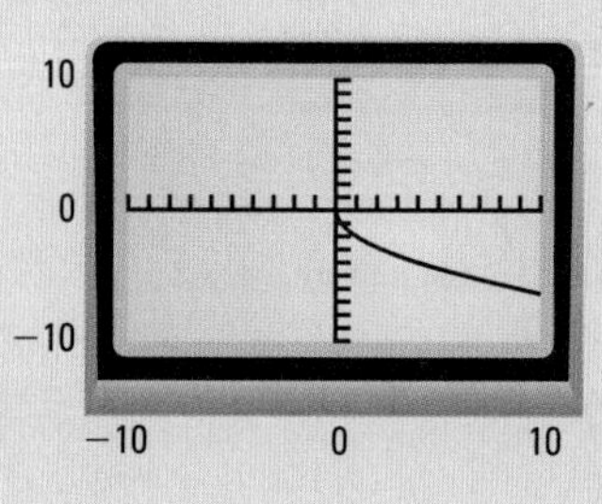

4. $y = \sqrt{x + k}$ **4**

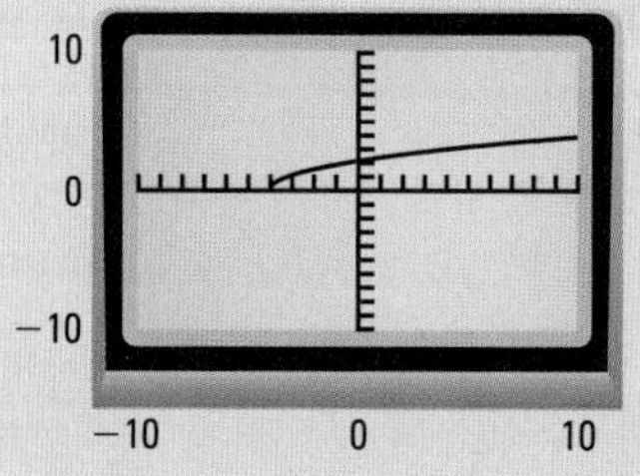

5. $y = \sqrt{x} + k$ **3**

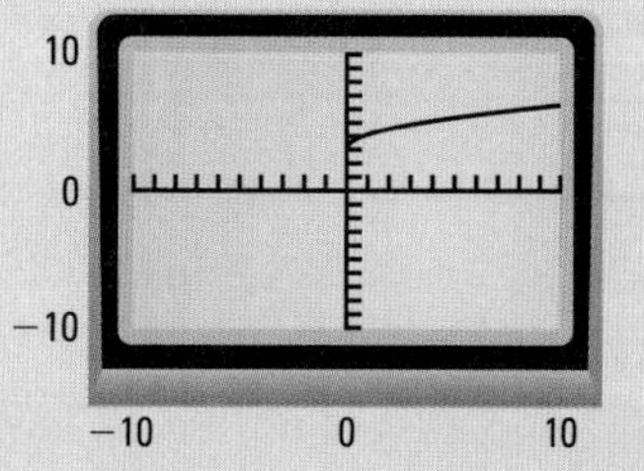

1 Planning the Activity

PURPOSE
To explore the graphs of families of square root functions using a graphing calculator or computer.

MATERIALS
- Keystroke blackline (*Chapter 12 Resource Book*, p. 13)

PACING
- Exploring the Concept — 10 min
- Drawing Conclusions — 10 min

LINK TO LESSON
The graph of $y = a\sqrt{x}$ in Step 4 is the graph of a principal square root of a number. The graph of $y = -a\sqrt{x}$ is the graph of the negative square root of a number. Its range should include the numbers less than or equal to 0.

2 Managing the Activity

ALTERNATIVE APPROACH
Use an overhead projector. For the Exploring the Concept section, first ask students to draw a sketch of what they think the graph will look like. Then ask different students to graph each equation on the overhead.

3 Closing the Activity

KEY DISCOVERY
The graph of $y = \sqrt{x} + k$ is k units above the graph of $y = \sqrt{x}$ when $k > 0$ and $|k|$ units below that graph when $k < 0$. The graph of $y = \sqrt{x + k}$ is k units *left* of the graph of $y = \sqrt{x}$ when $k > 0$ and $|k|$ units *right* of that graph when $k < 0$.

ACTIVITY ASSESSMENT
JOURNAL Where does the graph of $y = a\sqrt{x}$ lie in relation to the graph of $y = -a\sqrt{x}$? **The graphs are symmetric with respect to the x-axis.**

1 PLAN

PACING
Basic: 2 days
Average: 2 days
Advanced: 2 days
Block Schedule: 0.5 block with 12.1
0.5 block with 12.3

LESSON OPENER ACTIVITY
An alternative way to approach Lesson 12.2 is to use the Activity Lesson Opener:
- Blackline Master (*Chapter 12 Resource Book*, p. 25)
- Transparency (p. 83)

MEETING INDIVIDUAL NEEDS
- ***Chapter 12 Resource Book***
  Prerequisite Skills Review (p. 5)
  Practice Level A (p. 27)
  Practice Level B (p. 28)
  Practice Level C (p. 29)
  Reteaching with Practice (p. 30)
  Absent Student Catch-Up (p. 32)
  Challenge (p. 34)
- ***Resources in Spanish***
- ***Personal Student Tutor***

NEW-TEACHER SUPPORT
See the Tips for New Teachers on pp. 1–2 of the *Chapter 12 Resource Book* for additional notes about Lesson 12.2.

WARM-UP EXERCISES
Transparency Available
1. Simplify $2a + a - 6a$. $-3a$
2. Multiply $(x+3)(x-3)$. $x^2 - 9$
3. Multiply $(2x-9)(2x+9)$. $4x^2 - 81$
4. Solve $x^2 - 5x + 3 = 0$. $\frac{5 \pm \sqrt{13}}{2}$

12.2 Operations with Radical Expressions

What you should learn

GOAL 1 Add, subtract, multiply, and divide radical expressions.

GOAL 2 Use radical expressions in **real-life** situations, as in finding the speed of a pole-vaulter in **Ex. 58**.

Why you should learn it

▼ To solve **real-life** problems such as finding distances to the horizon from a schooner in **Example 5**.

GOAL 1 USING RADICAL OPERATIONS

You can use the distributive property to simplify sums and differences of radical expressions when the expressions have the same radicand.

Sum: $\sqrt{2} + 3\sqrt{2} = (1 + 3)\sqrt{2} = 4\sqrt{2}$

Difference: $\sqrt{2} - 3\sqrt{2} = (1 - 3)\sqrt{2} = -2\sqrt{2}$

In part (b) of Example 1, the first step is to remove a perfect square factor from under the radical sign, as you learned on page 512.

EXAMPLE 1 *Adding and Subtracting Radicals*

a. $2\sqrt{2} + \sqrt{5} - 6\sqrt{2} = -4\sqrt{2} + \sqrt{5}$ — **Subtract like radicals.**

b. $4\sqrt{3} - \sqrt{27} = 4\sqrt{3} - \sqrt{9 \cdot 3}$ — **Perfect square factor**

$= 4\sqrt{3} - \sqrt{9} \cdot \sqrt{3}$ — **Use product property.**

$= 4\sqrt{3} - 3\sqrt{3}$ — **Simplify.**

$= \sqrt{3}$ — **Subtract like radicals.**

EXAMPLE 2 *Multiplying Radicals*

a. $\sqrt{2} \cdot \sqrt{8} = \sqrt{16}$ — **Use product property.**

$= 4$ — **Simplify.**

b. $\sqrt{2}(5 - \sqrt{3}) = 5\sqrt{2} - \sqrt{2} \cdot \sqrt{3}$ — **Use distributive property.**

$= 5\sqrt{2} - \sqrt{6}$ — **Use product property.**

c. $(1 + \sqrt{5})^2 = 1^2 + 2\sqrt{5} + (\sqrt{5})^2$ — **Use square of a binomial pattern.**

$= 1 + 2\sqrt{5} + 5$ — **Evaluate powers.**

$= 6 + 2\sqrt{5}$ — **Simplify.**

d. $(a - \sqrt{b})(a + \sqrt{b}) = a^2 - (\sqrt{b})^2$ — **Use sum and difference pattern.**

$= a^2 - b$ — **Simplify.**

The expressions $(a + \sqrt{b})$ and $(a - \sqrt{b})$ are **conjugates**. As you can see in part (d) of Example 2, when a and b are integers, the product $(a + \sqrt{b})(a - \sqrt{b})$ does not involve radicals. Here are some other examples of conjugates.

EXPRESSION	CONJUGATE	PRODUCT
$4 + \sqrt{7}$	$4 - \sqrt{7}$	$4^2 - (\sqrt{7})^2 = 16 - 7 = 9$
$\sqrt{3} - c$	$\sqrt{3} + c$	$(\sqrt{3})^2 - c^2 = 3 - c^2$
$p + \sqrt{q}$	$p - \sqrt{q}$	$p^2 - (\sqrt{q})^2 = p^2 - q$

A simplified fraction does not have a radical in the denominator. To simplify some radical expressions with radicals in the denominator, you can use conjugates. For others, you may be able to write an equivalent expression with a perfect square under the radical sign in the denominator.

STUDENT HELP

HOMEWORK HELP Visit our Web site www.mcdougallittell.com for extra examples.

EXAMPLE 3 Simplifying Radicals

a. $\frac{3}{\sqrt{5}} = \frac{3}{\sqrt{5}} \cdot \frac{\sqrt{5}}{\sqrt{5}}$ Multiply numerator and denominator by $\sqrt{5}$.

$= \frac{3\sqrt{5}}{\sqrt{5} \cdot \sqrt{5}}$ Multiply fractions.

$= \frac{3\sqrt{5}}{5}$ Simplify.

b. $\frac{1}{c - \sqrt{d}} = \frac{1}{c - \sqrt{d}} \cdot \frac{c + \sqrt{d}}{c + \sqrt{d}}$ Multiply numerator and denominator by the conjugate.

$= \frac{c + \sqrt{d}}{(c - \sqrt{d})(c + \sqrt{d})}$ Multiply fractions.

$= \frac{c + \sqrt{d}}{c^2 - d}$ Simplify.

EXAMPLE 4 Checking Quadratic Formula Solutions

Check that $2 + \sqrt{3}$ is a solution of $x^2 - 4x + 1 = 0$.

SOLUTION

You can check the solution by substituting into the equation.

$x^2 - 4x + 1 = 0$ Write original equation.

$(2 + \sqrt{3})^2 - 4(2 + \sqrt{3}) + 1 \stackrel{?}{=} 0$ Substitute $2 + \sqrt{3}$ for x.

$4 + 4\sqrt{3} + 3 - 8 - 4\sqrt{3} + 1 \stackrel{?}{=} 0$ Multiply.

$0 = 0$ Solution is correct.

2 TEACH

MOTIVATING THE LESSON
The masts of sailing ships were equipped with lookout posts, which made it possible to site land from a greater distance than was possible from the deck of the ships. In this lesson you'll use an equation to model the distance you can see to the horizon from a particular height.

EXTRA EXAMPLE 1
Simplify.
a. $3\sqrt{7} - 5\sqrt{7} + 2\sqrt{7}$ **0**
b. $8\sqrt{5} + \sqrt{125}$ **$13\sqrt{5}$**

EXTRA EXAMPLE 2
Multiply.
a. $\sqrt{12} \cdot \sqrt{3}$ **6**
b. $\sqrt{3}(7 + \sqrt{5})$ **$7\sqrt{3} + \sqrt{15}$**
c. $(1 - \sqrt{3})^2$ **$4 - 2\sqrt{3}$**
d. $(6 - \sqrt{2})(6 + \sqrt{2})$ **34**

EXTRA EXAMPLE 3
Simplify.
a. $\frac{1}{\sqrt{3}}$ **$\frac{\sqrt{3}}{3}$**
b. $\frac{4}{3 + \sqrt{2}}$ **$\frac{12 - 4\sqrt{2}}{7}$**

EXTRA EXAMPLE 4
Is $\frac{-5 - \sqrt{33}}{2}$ a solution of $x^2 + 5x - 2$? **Yes**

CHECKPOINT EXERCISES
For use after Examples 1 and 2:
1. Simplify $(2 - \sqrt{7})^2$ **$11 - 4\sqrt{7}$**
For use after Example 3:
2. Simplify $\frac{10}{\sqrt{5}}$. **$2\sqrt{5}$**
For use after Example 4:
3. Is $1 + \sqrt{3}$ a solution of $x^2 - 2x - 3$? **No**

STUDENT HELP NOTES
Homework Help Students can find extra examples at **www.mcdougallittell.com** that parallel the examples in the student edition.

EXTRA EXAMPLE 5

a. Suppose your eye-level height is 20 ft and your friend's is 28 ft. Write an expression that shows how much farther your friend can see than you. Simplify the expression.

$D = \sqrt{\frac{3(28)}{2}} - \sqrt{\frac{3(20)}{2}}$; $D = \sqrt{42} - \sqrt{30}$

b. How much farther is this to the nearest tenth of a mile? **about 1.0 mile**

CHECKPOINT EXERCISES

For use after Example 5:

The maximum walking speed S (in feet per second) of an animal can be modeled by $S = \sqrt{gL}$, where $g = 32$ ft/sec^2 and L is the length (in feet) of the animal's leg. To the nearest tenth, how much faster does a dinosaur with leg length 3 feet walk than a dinosaur with leg length 2 feet? **about 1.8 ft/sec**

FOCUS ON VOCABULARY

Why are $(a + \sqrt{3})$ and $(a - \sqrt{3})$ conjugate pairs? **The binomials differ from each other only in the sign between the two terms and their product is a rational number.**

CLOSURE QUESTION

When is multiplying by a conjugate necessary to simplify a fraction? **when the denominator contains a binomial with a radical term**

DAILY PUZZLER

The sum of three consecutive perfect squares is 110. What is the product of the square roots of these numbers? **210**

GOAL 2 USING RADICALS AS REAL-LIFE MODELS

REAL LIFE Sailing

EXAMPLE 5 *Using a Radical Model*

You and a friend are spending the summer working on a schooner. The distance d (in miles) you can see to the horizon can be modeled by the following equation where h is your eye-level height (in feet) above the water.

Distance to horizon: $d = \sqrt{\frac{3h}{2}}$

a. Your eye-level height is 32 feet and your friend's eye-level height is 18 feet. Write an expression that shows how much farther you can see than your friend. Simplify the expression.

b. To the nearest tenth of a mile, how much farther can you see?

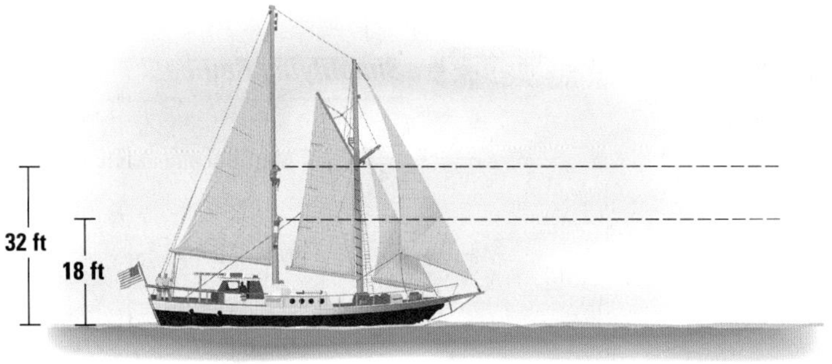

SOLUTION

a. Subtract the distance your friend can see from the distance you can see.

PROBLEM SOLVING STRATEGY

VERBAL MODEL: **Difference in distances** = **Your distance** − **Your friend's distance**

LABELS:

Difference in distances = D (miles)

Your distance = $\sqrt{\frac{3(32)}{2}}$ (miles)

Your friend's distance = $\sqrt{\frac{3(18)}{2}}$ (miles)

ALGEBRAIC MODEL:

$$D = \sqrt{\frac{3(32)}{2}} - \sqrt{\frac{3(18)}{2}}$$
$$= \sqrt{48} - \sqrt{27}$$
$$= \sqrt{16 \cdot 3} - \sqrt{9 \cdot 3}$$
$$= 4\sqrt{3} - 3\sqrt{3}$$
$$= \sqrt{3}$$

▶ You can see $\sqrt{3}$ miles farther than your friend.

b. Use a calculator or the Square Root Table on p. 811 to find that $\sqrt{3} \approx 1.7$.

▶ You can see about 1.7 miles farther than your friend.

GUIDED PRACTICE

Vocabulary Check ✓

1. Complete the following sentence: Two radical expressions are _?_ if they have the same radicand. **like radicals**

2. Write a radical expression and its conjugate. **Accept any expressions of the form $a + \sqrt{b}$ and $a - \sqrt{b}$ where $b > 0$.**

Concept Check ✓

3. Explain how to simplify $\frac{\sqrt{3}}{\sqrt{3} - 1}$. **See margin.**

Skill Check ✓

Simplify the expression.

4. $4\sqrt{5} + 5\sqrt{5}$ **$9\sqrt{5}$**
5. $3\sqrt{7} - 2\sqrt{7}$ **$\sqrt{7}$**
6. $3\sqrt{6} + \sqrt{24}$ **$5\sqrt{6}$**
7. $\sqrt{3} \cdot \sqrt{8}$ **$2\sqrt{6}$**
8. $(2 + \sqrt{3})^2$ **$7 + 4\sqrt{3}$**
9. $\sqrt{3}(5\sqrt{3} - 2\sqrt{6})$ **$15 - 6\sqrt{2}$**
10. $\frac{4}{\sqrt{13}}$ **$\frac{4\sqrt{13}}{13}$**
11. $\frac{3}{8 - \sqrt{10}}$ **$\frac{8 + \sqrt{10}}{18}$**
12. $\frac{6}{\sqrt{10}}$ **$\frac{3\sqrt{10}}{5}$**

Show whether the expression is a solution of the equation.

13. $6g^2 - 156 = 0; \sqrt{26}$
14. $x^2 - 48 = 0; -4\sqrt{3}$
15. $x^2 - 12x + 5 = 0; 6 + \sqrt{31}$
16. $x^2 - 8x + 8 = 0; 4 + 2\sqrt{2}$

17. **SAILING** In Example 5, suppose your eye-level height is 24 feet and your friend's is 12 feet. Write an expression that shows how much farther you can see than your friend. Simplify the expression. **$6 - 3\sqrt{2} \approx 1.8$ mi**

3. Multiply the numerator and denominator of the expression by the conjugate of the denominator, $\sqrt{3} + 1$.

13. $6(\sqrt{26})^2 - 156 = 6(26) - 156 = 0$; solution

14. $(-4\sqrt{3})^2 - 48 = 16 \cdot 3 - 48 = 0$; solution

15. $(6 + \sqrt{31})^2 - 12(6 + \sqrt{31}) + 5 = 67 + 12\sqrt{31} - 72 - 12\sqrt{31} + 5 = 0$; solution

16. $(4 + 2\sqrt{2})^2 - 8(4 + 2\sqrt{2}) + 8 = 24 + 16\sqrt{2} - 32 - 16\sqrt{2} + 8 = 0$; solution

PRACTICE AND APPLICATIONS

STUDENT HELP

Extra Practice to help you master skills is on p. 808.

ADDING AND SUBTRACTING RADICALS **Simplify the expression.**

18. $5\sqrt{7} + 2\sqrt{7}$ **$7\sqrt{7}$**
19. $\sqrt{3} + 5\sqrt{3}$ **$6\sqrt{3}$**
20. $11\sqrt{3} - 12\sqrt{3}$ **$-\sqrt{3}$**
21. $2\sqrt{6} - \sqrt{6}$ **$\sqrt{6}$**
22. $\sqrt{32} + \sqrt{2}$ **$5\sqrt{2}$**
23. $\sqrt{75} + \sqrt{3}$ **$6\sqrt{3}$**
24. $\sqrt{80} - \sqrt{45}$ **$\sqrt{5}$**
25. $\sqrt{72} - \sqrt{18}$ **$3\sqrt{2}$**
26. $\sqrt{147} - 7\sqrt{3}$ **0**
27. $4\sqrt{5} + \sqrt{125} + \sqrt{45}$ **$12\sqrt{5}$**
28. $3\sqrt{11} + \sqrt{176} + \sqrt{11}$ **$8\sqrt{11}$**
29. $\sqrt{24} - \sqrt{96} + \sqrt{6}$ **$-\sqrt{6}$**
30. $\sqrt{243} - \sqrt{75} + \sqrt{300}$ **$14\sqrt{3}$**

MULTIPLYING AND DIVIDING RADICALS **Simplify the expression.**

31. $\sqrt{3} \cdot \sqrt{12}$ **6**
32. $\sqrt{5} \cdot \sqrt{8}$ **$2\sqrt{10}$**
33. $(1 + \sqrt{13})(1 - \sqrt{13})$ **-12**
34. $\sqrt{3}(5\sqrt{2} + \sqrt{3})$ **$5\sqrt{6} + 3$**
35. $\sqrt{6}(7\sqrt{3} + 6)$ **$21\sqrt{2} + 6\sqrt{6}$**
36. $(\sqrt{6} + 5)^2$ **$31 + 10\sqrt{6}$**
37. $(\sqrt{a} - b)^2$ **$a - 2b\sqrt{a} + b^2$**
38. $(\sqrt{c} + d)(3 + \sqrt{5})$ **$3\sqrt{c} + \sqrt{5c} + 3d + d\sqrt{5}$**
39. $(2\sqrt{3} - 5)^2$ **$37 - 20\sqrt{3}$**
40. $\frac{5}{\sqrt{7}}$ **$\frac{5\sqrt{7}}{7}$**
41. $\frac{2}{\sqrt{2}}$ **$\sqrt{2}$**
42. $\frac{9}{5 - \sqrt{7}}$ **$\frac{5 + \sqrt{7}}{2}$**
43. $\frac{3}{\sqrt{48}}$ **$\frac{\sqrt{3}}{4}$**
44. $\frac{6}{10 + \sqrt{2}}$ **$\frac{30 - 3\sqrt{2}}{49}$**
45. $\frac{\sqrt{3}}{\sqrt{3} - 1}$ **$\frac{3 + \sqrt{3}}{2}$**
46. $\frac{14}{60 - \sqrt{578}}$ **$\frac{420 + 119\sqrt{2}}{1511}$**
47. $\frac{12}{7 - \sqrt{3}}$ **$\frac{42 + 6\sqrt{3}}{23}$**
48. $\frac{4 + \sqrt{3}}{a - \sqrt{b}}$ **See margin.**

48. $\frac{4a + a\sqrt{3} + 4\sqrt{b} + \sqrt{3b}}{a^2 - b}$

STUDENT HELP

HOMEWORK HELP
Example 1: Exs. 18–30
Example 2: Exs. 31–39
Example 3: Exs. 40–48
Example 4: Exs. 52–57
Example 5: Exs. 58, 59

3 APPLY

ASSIGNMENT GUIDE

BASIC
Day 1: pp. 719–720 Exs. 18–58 even
Day 2: pp. 719–721 Exs. 19–57 odd, 63–65, 70, 73, 76, 79

AVERAGE
Day 1: pp. 719–720 Exs. 18–58 even
Day 2: pp. 719–721 Exs. 19–57 odd, 60–65, 70, 73, 76, 79

ADVANCED
Day 1: pp. 719–720 Exs. 18–58 even
Day 2: pp. 719–721 Exs. 19–57 odd, 60–68, 70, 73, 76, 79

BLOCK SCHEDULE
pp. 719–720 Exs. 18–58, even (with 12.1)
pp. 719–721 Exs. 18–58, 60–65, 70, 73, 76, 79 (with 12.3)

EXERCISE LEVELS

Level A: ***Easier***
18–21, 69, 70

Level B: ***More Difficult***
22–65, 71–79

Level C: ***Most Difficult***
66–68

✓ HOMEWORK CHECK

To quickly check student understanding of key concepts, go over the following exercises: Exs. 20, 28, 30, 34, 36, 40, 42, 44, 48, 50, 56. See also the Daily Homework Quiz:

- Blackline Master (*Chapter 12 Resource Book*, p. 37)
- Transparency (p. 96)

TEACHING TIP

EXERCISES 59–62 You may wish to connect the formulas in these exercises to the formula for distance an object has fallen after it is dropped, $d = 16t^2$. Solving this formula for t yields $t = \frac{\sqrt{d}}{4}$, the formula in Ex. 59. Substituting $\frac{\sqrt{d}}{4}$ for t in the formula $d = st$ (distance = speed x time) and solving for s yields $s = \frac{d}{\frac{\sqrt{d}}{4}}$, the formula in Exs. 60–62.

71.

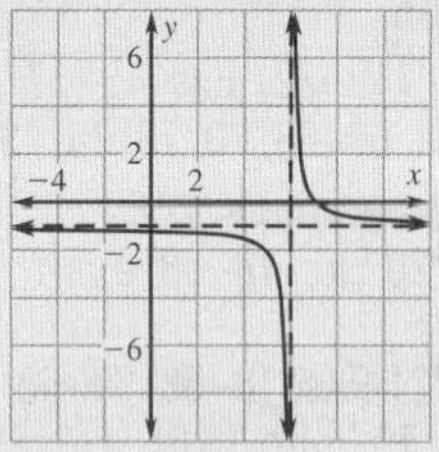

72.

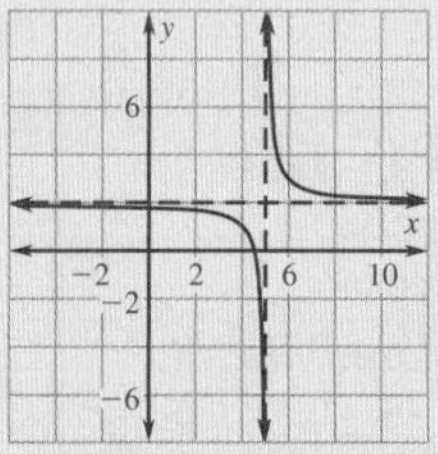

73.

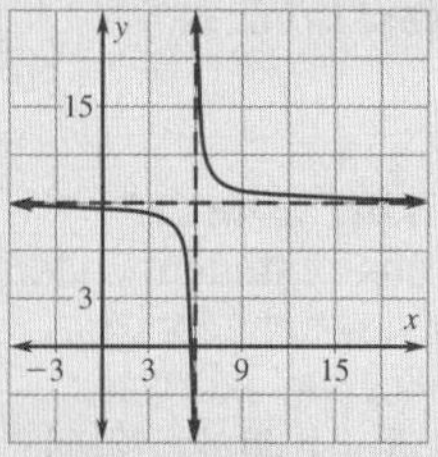

ADDITIONAL PRACTICE AND RETEACHING

For Lesson 12.2:

- Practice Levels A, B, and C (*Chapter 12 Resource Book,* p. 27)
- Reteaching with Practice (*Chapter 12 Resource Book,* p. 30)
- See Lesson 12.2 of the *Personal Student Tutor*

For more Mixed Review:

- Search the *Test and Practice Generator* for key words or specific lessons.

GEOMETRY CONNECTION **Find the area. (See Table of Formulas on page 813.)**

49.

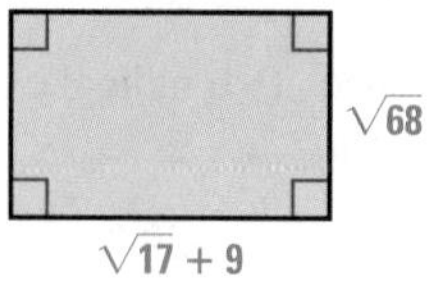

34 + 18√17

50.

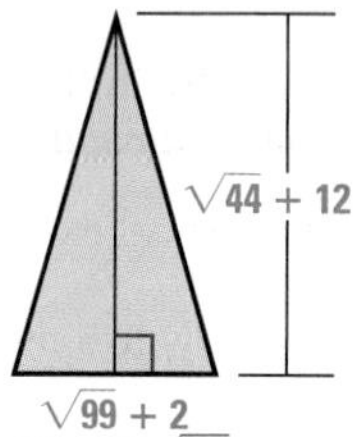

45 + 20√11

51.

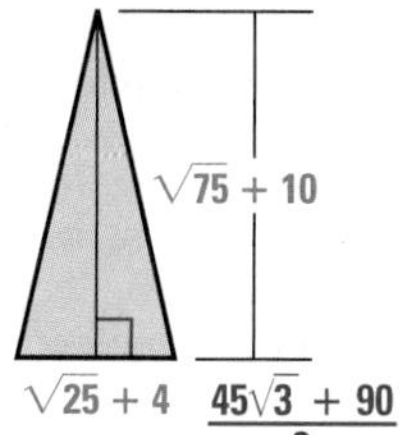

$\mathbf{\frac{45\sqrt{3} + 90}{2}}$

CHECKING SOLUTIONS **In Exercises 52–57, solve the quadratic equation. Check the solutions.**

52. $x^2 + 10x + 13 = 0$ **$-5 - 2\sqrt{3}, -5 + 2\sqrt{3}$**

53. $x^2 - 4x - 6 = 0$ **$2 - \sqrt{10}, 2 + \sqrt{10}$**

54. $x^2 - 4x - 15 = 0$ **$2 - \sqrt{19}, 2 + \sqrt{19}$**

55. $4x^2 - 2x - 1 = 0$ **$\frac{1 - \sqrt{5}}{4}, \frac{1 + \sqrt{5}}{4}$**

56. $x^2 - 6x - 1 = 0$ **$3 - \sqrt{10}, 3 + \sqrt{10}$**

57. $a^2 - 6a - 13 = 0$ **$3 - \sqrt{22}, 3 + \sqrt{22}$**

58. **POLE-VAULTING** A pole-vaulter's approach velocity v (in feet per second) and height reached h (in feet) are related by the following equation.

Pole-vaulter model: $v = 8\sqrt{h}$

If you are a pole-vaulter and reach a height of 20 feet and your opponent reaches a height of 16 feet, approximately how much faster were you running than your opponent? Round your answer to the nearest hundredth. **3.78 ft/sec**

59. **BIOLOGY** Many birds drop clams or other shellfish in order to break the shell and get the food inside. The time t (in seconds) it takes for an object such as a clam to fall a certain distance d (in feet) is given by the equation

$$t = \frac{\sqrt{d}}{4}.$$

A gull drops a clam from a height of 50 feet. A second gull drops a clam from a height of 32 feet. Find the difference in the times that it takes for the clams to reach the ground. Round your answer to the nearest hundredth. **0.35 sec**

FALLING OBJECTS **In Exercises 60–62, use the following information.** The average speed of an object S (in feet per second) that is dropped a certain distance d (in feet) is given by the following equation.

Falling object model: $S = \frac{d}{\frac{\sqrt{d}}{4}}$

60. Rewrite the equation with the right-hand side in simplest form. **$S = 4\sqrt{d}$**

61. Use either equation to find the average speed of an object that is dropped from a height of 400 feet. **80 ft/sec**

62. **CHECKING GRAPHICALLY** Graph both the falling object model and your equation from Exercise 60 on the same screen to check that you simplified correctly. **Check graphs.**

Test Preparation

63. **MULTIPLE CHOICE** Simplify $\sqrt{5}(6 + \sqrt{5})^2$. **C**

(A) $41 + 2\sqrt{5}$ (B) $53\sqrt{5}$ (C) $41\sqrt{5} + 60$ (D) $101\sqrt{5}$

64. **MULTIPLE CHOICE** Which of the following is the difference $\sqrt{3} - 5\sqrt{9}$? **B**

(A) $\sqrt{3} - 3$ (B) $\sqrt{3} - 15$ (C) $-4\sqrt{3}$ (D) $\sqrt{3} - 45$

65. **MULTIPLE CHOICE** Simplify $\frac{3}{5 - \sqrt{5}}$. **A**

(A) $\frac{15 + 3\sqrt{5}}{20}$ (B) $\frac{15 + \sqrt{5}}{20}$ (C) $\frac{15 + \sqrt{15}}{20}$ (D) $\frac{15 - 3\sqrt{5}}{20}$

★ Challenge

SCIENCE CONNECTION **In Exercises 66–68, use the following information.**

Find a partner to help you with the following experiment. Take a ruler and drop it from a fixed height. Once the ruler is released, your partner should catch it as quickly as possible. Measure how far the ruler falls before your partner catches it. The reaction time can be calculated by the equation

$$t = \sqrt{\frac{d}{192}}$$

where t is the reaction time (in seconds) and d is the distance (in inches) the ruler falls.

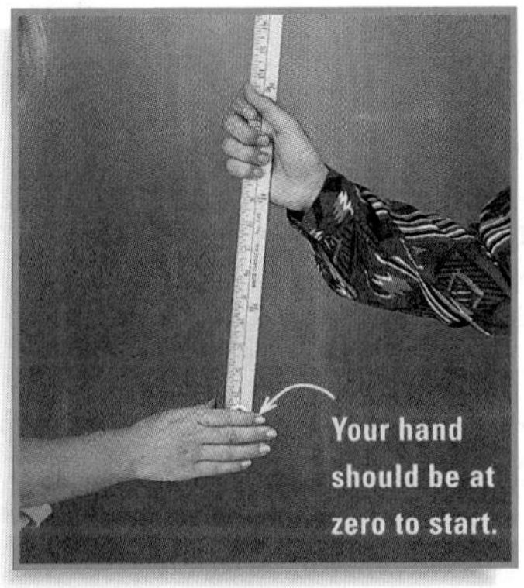

66. Show how to arrive at this equation from the equation $t = \frac{\sqrt{d}}{4}$ in Exercise 59, where d is in feet. **See margin.**

67. Suppose for three trials, the ruler falls 4 inches, 1 inch, and 3 inches. Calculate the average reaction time for the three trials. $\frac{\sqrt{3} + 1}{24} \approx 0.11$ **sec**

68. Perform the experiment several times, recording the distance the ruler falls each time. Organize the results in a table. Calculate the average reaction time for your first five trials. Compare results with your classmates. **Check results.**

66. $t = \frac{\sqrt{d}}{4} = \frac{\sqrt{d}}{\sqrt{16}} = \sqrt{\frac{d}{16}}$ where d is in feet. Changing d from feet to inches gives $\sqrt{\frac{d}{16 \cdot 12}} = \sqrt{\frac{d}{192}}$. So $t = \sqrt{\frac{d}{192}}$.

74. domain: all nonnegative numbers; range: all numbers greater than or equal to –3

75. domain: all numbers greater than or equal to 8; range: all nonnegative numbers

EXTRA CHALLENGE

www.mcdougallittell.com

MIXED REVIEW

PERCENT PROBLEMS **Solve the percent problem.** (Review 11.2)

69. What is 30% of 160? **48**

70. 105 is what percent of 240? **43.75%**

SKETCHING GRAPHS **Sketch the graph of the function.** (Review 11.8)
71–73. See margin for graphs.

71. $y = \frac{1}{x - 6} - 1$

72. $y = \frac{1}{x - 5} + 2$

73. $y = \frac{2}{x - 6} + 9$

FINDING DOMAINS AND RANGES **Find the domain and the range of the function.** (Review 12.1 for 12.3) **74–79. See margin.**

74. $f(x) = \sqrt{x} - 3$

75. $f(x) = \sqrt{x - 8}$

76. $f(x) = \sqrt{\frac{1}{2}x^2}$

77. $f(x) = \sqrt{x} + 4$

78. $f(x) = 6x$

79. $f(x) = \sqrt{x + 3}$

76. domain: all real numbers; range: all nonnegative numbers

77. domain: all nonnegative numbers; range: all numbers greater than or equal to 4

78. Both the domain and range are all the real numbers.

79. domain: all numbers greater than or equal to –3; range: all nonnegative numbers

4 ASSESS

DAILY HOMEWORK QUIZ

Transparency Available

1. Simplify the expression.
   - **a.** $5\sqrt{6} - 9\sqrt{6}$ $-4\sqrt{6}$
   - **b.** $3\sqrt{17} + 9\sqrt{11} + \sqrt{17}$ $4\sqrt{17} + 9\sqrt{11}$
   - **c.** $\sqrt{7} \cdot \sqrt{3}$ $\sqrt{21}$
   - **d.** $\sqrt{2}(7\sqrt{3} + \sqrt{2})$ $7\sqrt{6} + 4$
   - **e.** $(\sqrt{3} + 4)^2$ $19 + 8\sqrt{3}$
   - **f.** $\frac{5}{8 - \sqrt{6}}$ $\frac{40 + 5\sqrt{6}}{58}$
2. Solve $x^2 - 8x - 1 = 0$. $4 \pm \sqrt{17}$

EXTRA CHALLENGE NOTE

Challenge problems for Lesson 12.2 are available in **blackline** format in the *Chapter 12 Resource Book*, p. 34 and at **www.mcdougallittell.com.**

ADDITIONAL TEST PREPARATION

1. **WRITING** How do the procedures for simplifying $\frac{x}{\sqrt{7}}$ and $\frac{x}{3 + \sqrt{7}}$ differ? **In $\frac{x}{\sqrt{7}}$ both numerator and denominator are multiplied by the denominator of the fraction. In $\frac{x}{3 + \sqrt{7}}$, the numerator and denominator are multiplied by the conjugate of the denominator.**

1 PLAN

PACING

Basic: 2 days
Average: 2 days
Advanced: 2 days
Block Schedule: 0.5 block with 12.2
0.5 block with 12.4

LESSON OPENER ACTIVITY

An alternative way to approach Lesson 12.3 is to use the Activity Lesson Opener:

- Blackline Master (*Chapter 12 Resource Book*, p. 38)
- Transparency (p. 84)

MEETING INDIVIDUAL NEEDS

- ***Chapter 12 Resource Book***
  Prerequisite Skills Review (p. 5)
  Practice Level A (p. 41)
  Practice Level B (p. 42)
  Practice Level C (p. 43)
  Reteaching with Practice (p. 44)
  Absent Student Catch-Up (p. 46)
  Challenge (p. 48)
- ***Resources in Spanish***
- ***Personal Student Tutor***

NEW-TEACHER SUPPORT

See the Tips for New Teachers on pp. 1–2 of the *Chapter 12 Resource Book* for additional notes about Lesson 12.3.

WARM-UP EXERCISES

Transparency Available

1. Simplify $(\sqrt{6})^2$. **6**
2. Solve $\sqrt{x} = 4$. **16**
3. Solve $\sqrt{x} = 17$. **289**
4. Solve $x^2 - 5x + 6 = 0$. **$x = 2$ or $x = 3$**
5. Solve $\frac{x}{3} = \frac{3}{5}$. **$x = 1.8$**

12.3 Solving Radical Equations

What you should learn

GOAL 1 Solve a radical equation.

GOAL 2 Use radical equations to solve **real-life** problems, such as finding the centripetal force a person experiences on a ride in **Exs. 66 and 67**.

Why you should learn it

▼ To solve **real-life** problems such as finding the nozzle pressure of antifreeze used to de-ice a plane in **Example 5**.

GOAL 1 SOLVING A RADICAL EQUATION

In solving an equation involving radicals, the following property can be extremely useful.

SQUARING BOTH SIDES OF AN EQUATION

Suppose a and b are algebraic expressions.

If $a = b$, then $a^2 = b^2$. **Example:** $\sqrt{4} = 2$, so $(\sqrt{4})^2 = 2^2$.

EXAMPLE 1 *Solving a Radical Equation*

Solve $\sqrt{x} - 7 = 0$.

SOLUTION Isolate the radical expression on one side of the equation.

$\sqrt{x} - 7 = 0$	Write original equation.
$\sqrt{x} = 7$	Add 7 to each side.
$(\sqrt{x})^2 = 7^2$	Square each side.
$x = 49$	Simplify.

▶ The solution is 49. Check the solution in the original equation.

EXAMPLE 2 *Solving a Radical Equation*

To solve the equation $\sqrt{2x - 3} + 3 = 4$, you need to isolate the radical expression first.

$\sqrt{2x - 3} + 3 = 4$	Write original equation.
$\sqrt{2x - 3} = 1$	Subtract 3 from each side.
$(\sqrt{2x - 3})^2 = 1^2$	Square each side.
$2x - 3 = 1$	Simplify.
$2x = 4$	Add 3 to each side.
$x = 2$	Divide each side by 2.

▶ The solution is 2. Check the solution in the original equation.

EXTRANEOUS SOLUTIONS Squaring both sides of an equation often introduces an extraneous solution of $a^2 = b^2$ that is *not* a solution of $a = b$. When you square both sides of an equation, check each solution in the *original* equation.

EXAMPLE 3 *Checking for Extraneous Solutions*

Solve the equation.

a. $\sqrt{x+2} = x$ **b.** $\sqrt{x} + 13 = 0$

SOLUTION

a.

$\sqrt{x+2} = x$	Write original equation.
$(\sqrt{x+2})^2 = x^2$	Square each side.
$x + 2 = x^2$	Simplify.
$0 = x^2 - x - 2$	Write in standard form.
$0 = (x-2)(x+1)$	Factor.
$x = 2$ or $x = -1$	Zero-product property

✓**CHECK** Substitute 2 and −1 in the original equation.

$\sqrt{2+2} \stackrel{?}{=} 2$ $\sqrt{-1+2} \stackrel{?}{=} -1$

$2 = 2$ ✓ $1 \neq -1$

▶ The only solution is 2, because $x = -1$ is not a solution.

b.

$\sqrt{x} + 13 = 0$	Write original equation.
$\sqrt{x} = -13$	Subtract 13 from each side.
$(\sqrt{x})^2 = (-13)^2$	Square each side.
$x = 169$	Simplify.

▶ $\sqrt{169} + 13 \neq 0$, so the equation has no solution.

STUDENT HELP

Study Tip

In part (b) of Example 3 you can also use critical thinking to find that there is no solution. Because $\sqrt{x}$ represents a nonnegative number, there is *no* value for $\sqrt{x}$ that is the opposite of 13.

EXAMPLE 4 *Using a Geometric Mean*

The *geometric mean* of a and b is $\sqrt{ab}$. If the geometric mean of a and 6 is 12, what is the value of a?

SOLUTION Geometric mean $= \sqrt{ab}$

$12 = \sqrt{6a}$	Substitute.
$12^2 = (\sqrt{6a})^2$	Square each side.
$144 = 6a$	Simplify.
$24 = a$	Solve for a.

2 TEACH

MOTIVATING THE LESSON

When you ride on a Ferris wheel centripetal force acts on your body to keep it moving in a circular direction with the wheel. If you were not holding on and there were no bar across your seat and gravity were absent, at some point you would begin to move in a straight line right out of the Ferris wheel chair. In this lesson, you will solve radical equations to find the centripetal force acting on a body in other amusement park rides.

EXTRA EXAMPLE 1

Solve $\sqrt{x} - 10 = 0$. **$x = 100$**

EXTRA EXAMPLE 2

Solve $\sqrt{3x+1} - 2 = 6$. **$x = 21$**

EXTRA EXAMPLE 3

a. Solve $\sqrt{4x-3} = x$. **$x = 1$ or $x = 3$**

b. Solve $\sqrt{x} + 9 = 0$. **no solution**

EXTRA EXAMPLE 4

The geometric mean of x and 4 is 10. What is the value of x? **$x = 25$**

CHECKPOINT EXERCISES

For use after Example 1:

1. Solve $\sqrt{2x} - 6 = 0$. **$x = 18$**

For use after Example 2:

2. Solve $\sqrt{5x} + 7 = 17$. **$x = 20$**

For use after Example 3:

3. Solve $\sqrt{3x-2} = x$. **$x = 1$ or $x = 2$**

4. Solve $\sqrt{x} + 144 = 0$. **no solution**

For use after Example 4:

5. The geometric mean of x and 6 is 30. What is the value of x? **$x = 150$**

CONCEPT QUESTION

EXAMPLE 3 How do you know from the equivalent equation $\sqrt{x} = -13$ that the original equation has no solutions? **The value of $\sqrt{x}$ must be a nonnegative number.**

EXTRA EXAMPLE 5

Use the relationship described in Example 5.

a. Find the nozzle pressure for a nozzle whose diameter is 1 inch. **about 69.4 pounds per square inch**

b. When you decrease the diameter of the nozzle by 20% from 1.25 inches to 1 inch, do you also change the nozzle pressure by 20%? Explain. **No, when you decrease the nozzle pressure by 20%, the pressure increases by about 144%.**

CHECKPOINT EXERCISES

For use after Example 5:

1. The horizontal distance S (in meters) traveled by a projectile is related to its initial velocity v (in m/s) by the formula $\sqrt{\frac{S}{.03}} = v$. If $v = 224$ m/s, find S. **about 1500 m**

ENGLISH LEARNERS

Keep in mind that many word problems describe situations with which English language learners may be unfamiliar. For Example 5, briefly describe the terms *commercial airline, gallons, antifreeze, de-icing, nozzle, nozzle diameter,* and *nozzle pressure* so that students can imagine the situation described in the word problem.

FOCUS ON VOCABULARY

In your own words describe an extraneous solution. **See margin.**

CLOSURE QUESTION

Explain the procedure to solve a radical equation. **See margin.**

DAILY PUZZLER

A pyramid is formed with a series of consecutive integers. The first four rows of the pyramid are shown below. What is the last number of the seventh row? The 20th row? **28; 210**

1
2 3
4 5 6
7 8 9 10

GOAL 2 SOLVING RADICAL EQUATIONS IN REAL LIFE

REAL LIFE *Plane De-icing*

EXAMPLE 5 *Using a Radical Model*

You work for a commercial airline and remove ice from planes. The relationship among the flow rate r (in gallons per minute) of the antifreeze for de-icing, the nozzle diameter d (in inches), and the nozzle pressure P (in pounds per square inch) is shown in the diagram. You want a flow rate of 250 gallons per minute.

a. Find the nozzle pressure for a nozzle whose diameter is 1.25 inches.

b. Find the nozzle pressure for a nozzle whose diameter is 1.75 inches.

c. When you increase the diameter of the nozzle by 40%, do you also change the nozzle pressure by 40%? Explain.

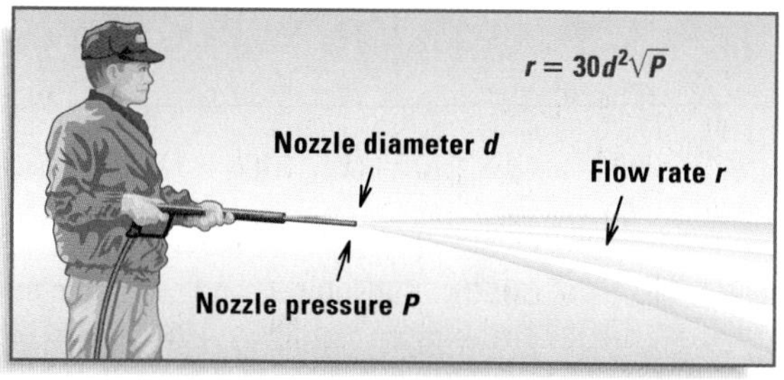

SOLUTION

a.

$r = 30d^2\sqrt{P}$	Write model for water flow.
$250 = 30(1.25)^2\sqrt{P}$	Substitute 1.25 for d and 250 for r.
$250 = 46.875\sqrt{P}$	Simplify.
$\frac{250}{46.875} = \sqrt{P}$	Divide each side by 46.875.
$\left(\frac{250}{46.875}\right)^2 = P$	Square each side.
$28.4 \approx P$	Use a calculator.

▶ The nozzle pressure will be about 28.4 pounds per square inch.

b.

$r = 30d^2\sqrt{P}$	Write model for water flow.
$250 = 30(1.75)^2\sqrt{P}$	Substitute 1.75 for d and 250 for r.
$\frac{250}{91.875} = \sqrt{P}$	Simplify and divide each side by 91.875.
$\left(\frac{250}{91.875}\right)^2 = P$	Square each side.
$7.4 \approx P$	Use a calculator.

▶ The nozzle pressure will be about 7.4 pounds per square inch.

c. No, when you increase the nozzle diameter by 40%, the pressure decreases by about 74%.

Focus on Vocabulary *Sample answer:*
a solution of a derived equation that is not a solution of the original equation

Closure Question *Sample answer:*
First isolate the radical expression. Then square both sides. Solve the resulting equation. Check for extraneous solutions.

GUIDED PRACTICE

Vocabulary Check ✓

1. Explain what an *extraneous solution* is. **a solution of the equation produced by the solution method that is not a solution of the original equation**

Concept Check ✓

2. One reason for checking a solution in the original equation is to look for an error in one of the steps of the solution. Give another reason. **to check for extraneous solutions**
3. Is 36 a solution of $\sqrt{x} = -6$? Why or why not? **No; $\sqrt{x}$ is the positive square root of x, so $\sqrt{36} = 6$, not -6. ($\sqrt{x} = -6$ has no solution.)**

Skill Check ✓

Solve the equation. Check for extraneous solutions.

4. $\sqrt{x} - 20 = 0$ **400**
5. $\sqrt{5x + 1} + 8 = 12$ **3**
6. $\sqrt{4x} - 1 = 3$ **4**
7. $\sqrt{x} + 6 = 0$ **no solution**
8. $\sqrt{4x + 5} = x$ **5**
9. $\sqrt{x + 6} - x = 0$ **3**
10. $x = \sqrt{x + 12}$ **4**
11. $-5 + \sqrt{x} = 0$ **25**
12. $x = \sqrt{5x + 24}$ **8**
13. Find a if the geometric mean of 12 and a is 6. **3**
14. PLANE DE-ICING Use Example 5 to answer the question. What will the nozzle pressure be if the nozzle diameter is 2 inches? **about 4.3 lb/in.^{2}**

PRACTICE AND APPLICATIONS

STUDENT HELP

Extra Practice to help you master skills is on p. 808.

SOLVING RADICAL EQUATIONS **Solve the equation. Check for extraneous solutions.**

15. $\sqrt{x} - 9 = 0$ **81**
16. $\sqrt{x} - 1 = 0$ **1**
17. $\sqrt{x} + 5 = 0$ **no solution**
18. $\sqrt{x} - 10 = 0$ **100**
19. $\sqrt{x} - 15 = 0$ **225**
20. $\sqrt{x} - 0 = 0$ **0**
21. $\sqrt{6x} - 13 = 23$ **216**
22. $\sqrt{4x + 1} + 5 = 10$ **6**
23. $\sqrt{9 - x} - 10 = 14$ **−567**
24. $\sqrt{5x + 1} + 2 = 6$ **3**
25. $\sqrt{6x - 2} - 3 = 7$ **17**
26. $4 = 7 - \sqrt{33x - 2}$ **$\frac{1}{3}$**
27. $10 = 4 + \sqrt{5x + 11}$ **5**
28. $-5 - \sqrt{10x - 2} = 5$ **no solution**
29. $\sqrt{-x} - \frac{3}{2} = \frac{3}{2}$ **−9**
30. $\sqrt{x} + \frac{1}{3} = \frac{13}{3}$ **16**
31. $\sqrt{\frac{1}{5}x - 2} - \frac{1}{10} = \frac{7}{10}$ **$13\frac{1}{5}$**
32. $x = \sqrt{\frac{3}{2}x + \frac{5}{2}}$ **$\frac{5}{2}$**
33. $\sqrt{\frac{1}{4}x - 4} - 3 = 5$ **272**
34. $6 - \sqrt{7x - 9} = 3$ **$2\frac{4}{7}$**
35. $\sqrt{\frac{1}{9}x + 1} - \frac{2}{3} = \frac{5}{3}$ **40**
36. $x = \sqrt{35 + 2x}$ **7**
37. $x = \sqrt{-4x - 4}$ **no solution**
38. $x = \sqrt{6x - 9}$ **3**
39. $x = \sqrt{30 - x}$ **5**
40. $x = \sqrt{11x - 28}$ **4, 7**
41. $\sqrt{-10x - 4} = 2x$ **no solution**
42. $x = \sqrt{200 - 35x}$ **5**
43. $\sqrt{110 - x} = x$ **10**
44. $2x = \sqrt{-13x - 10}$ **no solution**
45. $x = \sqrt{4x + 45}$ **9**
46. $\frac{1}{5}x = \sqrt{x - 6}$ **10, 15**
47. $\frac{2}{3}x = \sqrt{24x - 128}$ **6, 48**

STUDENT HELP

HOMEWORK HELP
Example 1: Exs. 15–21
Example 2: Exs. 22–47
Example 3: Exs. 15–47
Example 4: Exs. 48–53
Example 5: Exs. 69, 70

GEOMETRIC MEAN **Two numbers and their geometric mean are given. Find the value of *a*.**

48. 12 and a; 27 **60.75**
49. 4 and a; 16 **64**
50. 6 and a; 27 **121.5**
51. 4 and a; 14 **49**
52. 6 and a; 72 **864**
53. 8 and a; 104 **1352**

3 APPLY

ASSIGNMENT GUIDE

BASIC
Day 1: pp. 725–727 Exs. 16–60 even, 66, 74, 76
Day 2: pp. 725–728 Exs. 15–61 odd, 67, 69, 75, 77–80, 85, 90, 95, 99, 100, Quiz 1 Exs. 1–16

AVERAGE
Day 1: pp. 725–727 Exs. 16–60 even, 66, 74, 76
Day 2: pp. 725–728 Exs. 15–61 odd, 67, 69, 75, 77–80, 85, 90, 95, 99, 100, Quiz 1 Exs. 1–16

ADVANCED
Day 1: pp. 725–727 Exs. 16–60 even, 66, 74, 76
Day 2: pp. 725–728 Exs. 15–61 odd, 67, 69, 75, 77–81, 85, 90, 95, 99, 100, Quiz 1 Exs. 1–16

BLOCK SCHEDULE
pp. 725–727 Exs. 16–60 even, 66, 74, 76 (with 12.2)
pp. 725–728 Exs. 15–61, 66, 67, 69, 74–80, 85, 90, 95, 99, 100, Quiz 1 Exs. 1–16 (with 12.4)

EXERCISE LEVELS

Level A: *Easier*
15–27
Level B: *More Difficult*
28–67, 69, 71–86, 88–100
Level C: *Most Difficult*
68, 70, 87

✓ HOMEWORK CHECK

To quickly check student understanding of key concepts, go over the following exercises: Exs. 18, 26, 40, 44, 50, 60, 66, 74. See also the Daily Homework Quiz:

- Blackline Master (*Chapter 12 Resource Book*, p. 52)
- Transparency (p. 97)

STUDENT HELP NOTES

Look Back As students look back to page 359 ask them what a nonnegative value of $-x$ in Exercise 65 means about the value of $11x-30$? **It must be negative.**

APPLICATION NOTE

EXERCISES 69 AND 70
Additional information about centripetal force is available at **www.mcdougallittell.com.**

MATHEMATICAL REASONING

You can also find a geometric mean by solving a proportion. To say that 12 is the geometric mean of a and 6 means that a is related to 12 as 12 is related to 6. Thus $6a = 144$, and $a = 24$. If x is the geometric mean of 0.5 and 32, what proportion can be used to find x? $\frac{0.5}{x} = \frac{x}{32}$

ADDITIONAL PRACTICE AND RETEACHING

For Lesson 12.3:

- Practice Levels A, B, and C (*Chapter 12 Resource Book,* p. 41)
- Reteaching with Practice (*Chapter 12 Resource Book,* p. 44)
- See Lesson 12.3 of the *Personal Student Tutor*

For more Mixed Review:

- Search the *Test and Practice Generator* for key words or specific lessons.

ERROR ANALYSIS **Find and correct the error.**

54.

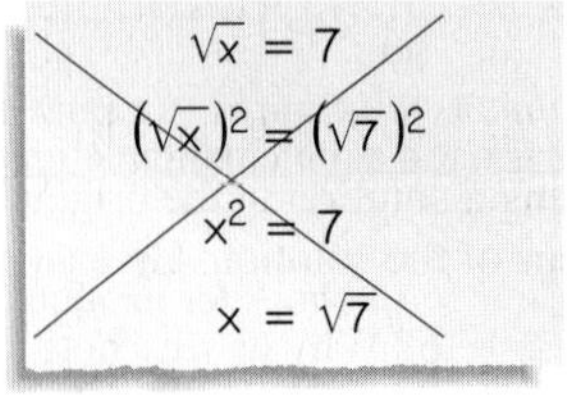

54. $(\sqrt{x})^2 = x$, not x^2; Also, right side of second equation should be 7^2, not $(\sqrt{7})^2$; $x = 49$

55.

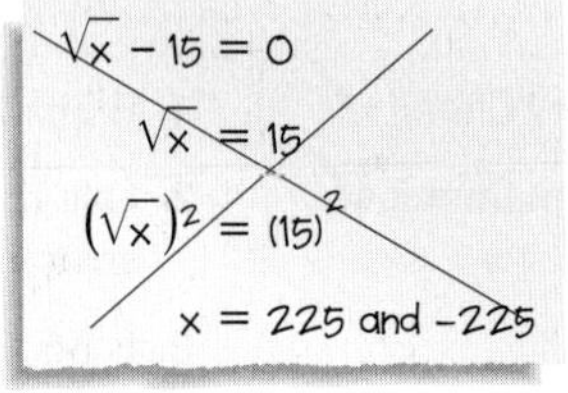

$15^2 = 225$, not -225; $x = 225$

SOLVING GRAPHICALLY **In Exercises 56–64, use a graphing calculator to graphically solve the radical equation. Check the solution algebraically.**

STUDENT HELP

Look Back
For help with solving an equation using a graphing calculator, see p. 359.

56. $\sqrt{x+4} = 3$ **5**   **57.** $\sqrt{x-5.6} = 2.5$ **11.85**   **58.** $\sqrt{9.2-x} = 1.8$ **5.96**

59. $\sqrt{6x-2} - 3 = 7$ **17**   **60.** $4 + \sqrt{x} = 9$ **25**   **61.** $\sqrt{7x-12} = x$ **3, 4**

62. $\sqrt{2x+7} = x+2$ **1**   **63.** $\sqrt{3x-2} = 4-x$ **2**   **64.** $\sqrt{15-4x} = 2x$ **1.5**

65. **VISUAL THINKING** Use a graphing calculator or a computer to show that the radical equation $\sqrt{11x-30} = -x$ has no solution. Explain.

65. *Sample answer:* The graphs do not intersect; the graph of $y = \sqrt{11x-30}$ is entirely in the first quadrant, while the graph of $y = -x$ is entirely in the second and fourth quadrants.

GEOMETRY CONNECTION **Find the value of *x*.**

66. Perimeter = 30 **75**   **67.** Area = $82\frac{1}{2}$ **113**

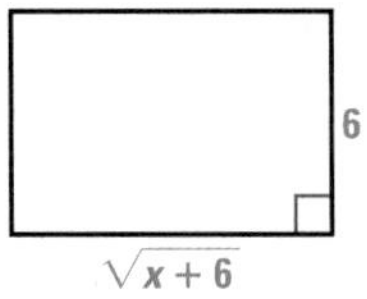

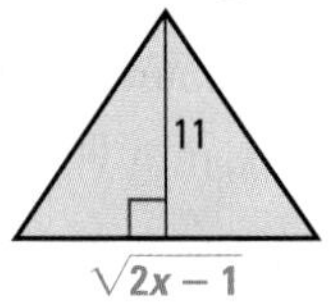

68. **LOGICAL REASONING** Write a radical equation that has a solution of 18.
Sample answer: $\sqrt{x+7} = 5$

AMUSEMENT PARK RIDE **In Exercises 69 and 70, use the following information.**

A ride at an amusement park spins in a circle of radius r (in meters). The centripetal force F experienced by a passenger on the ride is modeled by the equation below, where t is the number of seconds the ride takes to complete one revolution and m is the mass (in kilograms) of the passenger.

$$t = \sqrt{\frac{4\pi^2 mr}{F}}$$

70. Yes; if you solve the equation for F, $F = \frac{4\pi^2 mr}{t^2}$, and if r and t do not vary, F is directly proportional to m.

69. A person whose mass is 67.5 kilograms is on a ride that is spinning in a circle at a rate of 10 seconds per revolution. The radius of the circle is 6 meters. How much centripetal force does the person experience? **about 160**

70. **CRITICAL THINKING** A person whose mass is 54 kilograms is on the ride. A mass of 54 kilograms is 80% of 67.5 kilograms. Is the centripetal force experienced by the person 80% of the force in Exercise 69? Explain. **See margin.**

STUDENT HELP

APPLICATION LINK
Visit our Web site www.mcdougallittell.com for more information about centripetal force.

BLOTTING PAPER In Exercises 71 and 72, use the following information. Blotting paper is a thick, soft paper used for absorbing fluids such as water or ink. The distance d (in centimeters) that tap water is absorbed up a strip of blotting paper at a temperature of 28.4°C is given by the equation

$d = 0.444\sqrt{t}$ where t is the time (in seconds).

71. Approximately how many minutes would it take for the water to travel a distance of 28 centimeters up the strip of blotting paper? **about 66.3 min**

72. How far up the blotting paper would the water be after $33\frac{1}{3}$ seconds? **about 2.6 cm**

GEOMETRIC MEAN In Exercises 73–76, use the proportion $\frac{a}{b} = \frac{b}{d}$, where a, b, and d are positive numbers.

73. In the proportion $\frac{a}{b} = \frac{b}{d}$, b is called the geometric mean of a and d. Use the cross product property to show that $b = \sqrt{ad}$.
By the cross product property, $ad = b^2$ so, $b = \sqrt{ad}$.

74. Two numbers have a geometric mean of 4. One number is 6 more than the other.

a. Use the proportion in Exercise 73. Rewrite the proportion substituting the given value for the geometric mean. $\frac{a}{4} = \frac{4}{d}$

b. Let x represent one of the numbers. How can you represent "one number is 6 more than the other" in the proportion? Rewrite the proportion using x. $x + 6$; $\frac{x}{4} = \frac{4}{x + 6}$

c. Solve the proportion in part (b) to find the numbers. **2 and 8**

75. Two numbers have a geometric mean of 10. One number is 21 less than the other. Find the numbers. **4 and 25**

76. Two numbers have a geometric mean of 12. One number is 32 more than the other. Find the numbers. **4 and 36**

Test Preparation

QUANTITATIVE COMPARISON In Exercises 77–80, choose the statement below that is true about the given quantities.

Ⓐ The quantity in column A is greater.

Ⓑ The quantity in column B is greater.

Ⓒ The two quantities are equal.

Ⓓ The relationship cannot be determined from the given information.

	Column A	Column B
77.	The geometric mean of 3 and 12	The geometric mean of 1 and 36
78.	The solution of $\sqrt{x} + 4 = 5$	The solution of $\sqrt{x} - 4 = 5$
79.	The solution of $\sqrt{x} - 3 = 5$	The solution of $\sqrt{x - 3} = 5$
80.	The geometric mean of -1 and -64	The geometric mean of -2 and -32

★ Challenge

81. **LOGICAL REASONING** Using $4\sqrt{x} = 2x + k$, find three different expressions that can be substituted for k so that the equation has two solutions, one solution, and no solution. Describe how you found the equations.

77. C

78. B

79. A

80. C

81. The equation has two solutions when $k < 2$, one solution when $k = 2$, and no solution when $k > 2$; squaring both sides of the equation yields the quadratic equation $4x^2 + (4k - 16)x + k^2 = 0$, which has determinant $-128k + 256$. The equation has two solutions if the determinant is positive, one solution if the determinant is zero, and no solution if the determinant is negative.

4 ASSESS

DAILY HOMEWORK QUIZ

Transparency Available

1. Solve the equation.

a. $\sqrt{x} - 5 = 0$ **25**

b. $\sqrt{2x - 7} - 5 = 5$ $\frac{107}{2}$

c. $x = \sqrt{15x - 14}$ **14, 1**

2. The geometric mean of 3 and a is 9. What is a? **27**

3. The perimeter of a rectangle is 20. Its dimensions are 4 and $\sqrt{x - 1}$. Find the value of x. **37**

EXTRA CHALLENGE NOTE

Challenge problems for Lesson 12.3 are available in **blackline** format in the *Chapter 12 Resource Book*, p. 48 and at **www.mcdougallittell.com.**

ADDITIONAL TEST PREPARATION

1. OPEN ENDED Give an example of a radical equation that can be transformed into a quadratic equation. Give an example of a radical equation that can be transformed into a linear equation. ***Sample answer:*** $\sqrt{x + 1} = x$; $\sqrt{x} = 2$

ADDITIONAL RESOURCES
An alternative Quiz for Lessons 12.1–12.3 is available in the *Chapter 12 Resource Book,* p. 49.)

1. 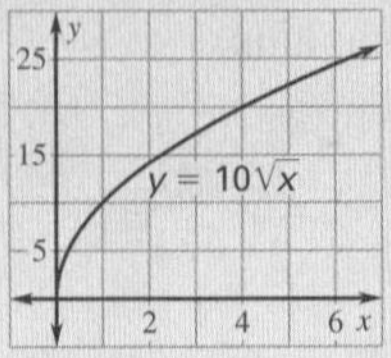

2. 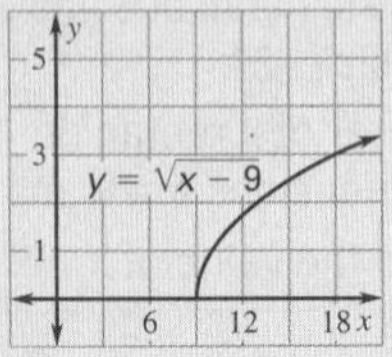

3. 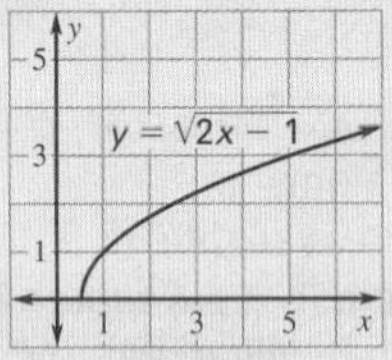

4. 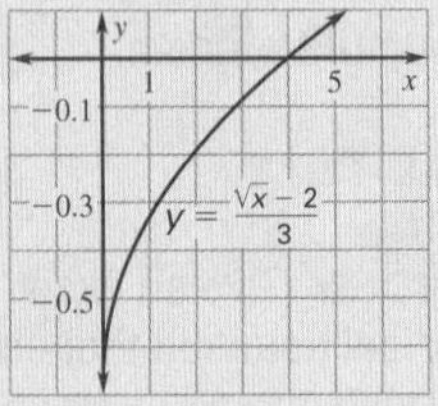

MIXED REVIEW

SOLVING EQUATIONS Solve the equation. (Review 9.1, 9.2 for 12.4)

82. $x^2 = 36$ **−6, 6**
83. $x^2 = 11$ $-\sqrt{11}, \sqrt{11}$
84. $7x^2 = 700$ **−10, 10**
85. $25x^2 - 9 = -5$ $-\frac{2}{5}, \frac{2}{5}$
86. $\frac{1}{7}x^2 - 7 = -7$ **0**
87. $-16t^2 + 48 = 0$ $-\sqrt{3}, \sqrt{3}$

FINDING PRODUCTS Multiply. (Review 10.3 for 12.4)

88. $(x + 5)^2$ $x^2 + 10x + 25$
89. $(2x - 3)^2$ $4x^2 - 12x + 9$
90. $(3x + 5y)(3x - 5y)$ $9x^2 - 25y^2$
91. $(6y - 4)(6y + 4)$ $36y^2 - 16$
92. $(x + 7y)^2$ $x^2 + 14xy + 49y^2$
93. $(2a - 9b)^2$ $4a^2 - 36ab + 81b^2$

FACTORING TRINOMIALS Factor the trinomial. (Review 10.7 for 12.4)

94. $x^2 + 18x + 81$ $(x + 9)^2$
95. $x^2 - 12x + 36$ $(x - 6)^2$
96. $4x^2 + 28x + 49$ $(2x + 7)^2$

UNDEFINED VALUES For what values of the variable is the rational expression undefined? (Review 11.4)

97. $\frac{5}{x - 4}$ **4**
98. $\frac{x + 2}{x^2 - 4}$ **−2 and 2**
99. $\frac{x + 4}{x^2 + x - 6}$ **−3 and 2**

100. **COMPARING COSTS** A discount grocery store offers two types of memberships with annual membership fees of \$25 and \$100. If you pay the \$25 membership fee, you get regular discounted prices. If you pay the \$100 membership fee, you get an additional 10% discount. How much money would you have to spend on groceries for the two memberships to cost the same? (Review 7.2) **\$750**

QUIZ 1

Self-Test for Lessons 12.1–12.3

Identify the domain and the range of the function. Then sketch the graph of the function. (Lesson 12.1) **1–4. See margin.**

1. $y = 10\sqrt{x}$
2. $y = \sqrt{x - 9}$
3. $y = \sqrt{2x - 1}$
4. $y = \frac{\sqrt{x} - 2}{3}$

1. Both the domain and range are all nonnegative numbers.
2. domain: all numbers greater than or equal to 9; range: all nonnegative numbers
3. domain: all numbers greater than or equal to $\frac{1}{2}$; range: all nonnegative numbers
4. domain: all nonnegative numbers; range: all numbers greater than or equal to $-\frac{2}{3}$

Simplify the expression. (Lesson 12.2)

5. $7\sqrt{10} + 11\sqrt{10}$ $18\sqrt{10}$
6. $2\sqrt{7} - 5\sqrt{28}$ $-8\sqrt{7}$
7. $4\sqrt{5} + \sqrt{125} - \sqrt{80}$ $5\sqrt{5}$
8. $\sqrt{3}(3\sqrt{2} + \sqrt{3})$ $3\sqrt{6} + 3$
9. $(\sqrt{7} + 5)^2$ $32 + 10\sqrt{7}$
10. $\frac{15}{8 + \sqrt{7}}$ $\frac{40 - 5\sqrt{7}}{19}$

Solve the equation. (Lesson 12.3)

11. $\sqrt{x} - 12 = 0$ **144**
12. $\sqrt{x} - 8 = 0$ **64**
13. $\sqrt{3x + 2} + 2 = 3$ $-\frac{1}{3}$
14. $\sqrt{3x - 2} + 3 = 7$ **6**
15. $\sqrt{77 - 4x} = x$ **7**
16. $x = \sqrt{2x + 3}$ **3**

17. **NOZZLE PRESSURE** The nozzle pressure P (in pounds per square inch) of a hose with diameter d (in inches) and water-flow rate r (in gallons per minute) is given by the equation $r = 30d^2\sqrt{P}$. Find the nozzle pressure in a hose that has a water-flow rate of 250 gallons per minute and a diameter of 2.5 inches. (Lesson 12.3) $1\frac{7}{9}$ lb/in.^{2}

ACTIVITY 12.4

Developing Concepts

Modeling Completing the Square

GROUP ACTIVITY
Work with a partner.

MATERIALS
algebra tiles

STUDENT HELP

Look Back
For help with algebra tiles, see p. 575.

QUESTION How can you use algebra tiles to represent perfect square trinomials?

EXPLORING THE CONCEPT

1. Use algebra tiles to model the expression $x^2 + 6x$.

2. Arrange the x^2-tile and the x-tiles to form part of a square.

3. To complete the square, you need to add nine 1-tiles.

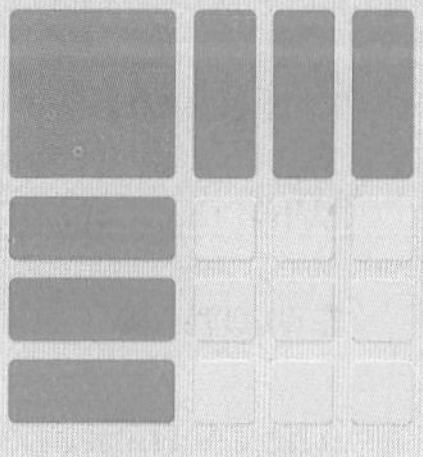

You have completed the square: $x^2 + 6x + 9 = (x + 3)^2$.

DRAWING CONCLUSIONS

1. Use algebra tiles to complete the square for each expression in the table. Then copy and complete the table to show your results.

Expression	Number of tiles to complete the square	Number of tiles as a perfect square
$x^2 + 6x$	9	3^2
$x^2 + 8x$	? **16**	? 4^2
$x^2 + 4x$	? **4**	? 2^2
$x^2 + 2x$	? **1**	? 1^2

2. How is the number in the third column related to the coefficient of x in the first column? **The base of the expression is half the coefficient of x.**

3. Use the pattern you found in Exercise 2 to predict how many tiles you would need to add to complete the square for the expression $x^2 + 14x$. **49 tiles**

1 Planning the Activity

PURPOSE
To demonstrate completing a perfect square trinomial using algebra tiles.

MATERIALS
- algebra tiles
- Activity Support Master (*Chapter 12 Resource Book,* p. 53)

PACING
- Exploring the Concept — 5 min
- Drawing Conclusions — 10 min

LINK TO LESSON
In Example 2 of Lesson 12.4 students can make an algebra tile model to complete the square of $x^2 + 10x$.

2 Managing the Activity

CLASSROOM MANAGEMENT
Do the first exercise together with the students, demonstrating the algebra tile model at the chalkboard. Then assign the students to work together with their partners to complete the remaining exercises. Students may be confused about how to place the x-tiles. Remind them that the model is a square, so equal numbers of x-tiles must be placed along each edge.

3 Closing the Activity

KEY DISCOVERY
To complete a perfect square trinomial, add the square of one-half the coefficient of the middle term.

ACTIVITY ASSESSMENT
If $x^2 + nx + z = \left(x + \frac{n}{2}\right)^2$, what is the relationship between $\frac{n}{2}$ and z?
$\left(\frac{n}{2}\right)^2 = \frac{n^2}{4} = z$

1 PLAN

PACING
Basic: 2 days
Average: 2 days
Advanced: 2 days
Block Schedule: 0.5 block with 12.3
0.5 block with 12.5

LESSON OPENER APPLICATION
An alternative way to approach Lesson 12.4 is to use the Application Lesson Opener:

- Blackline Master (*Chapter 12 Resource Book,* p. 54)
- Transparency (p. 85)

MEETING INDIVIDUAL NEEDS

- ***Chapter 12 Resource Book***
  Prerequisite Skills Review (p. 5)
  Practice Level A (p. 55)
  Practice Level B (p. 56)
  Practice Level C (p. 57)
  Reteaching with Practice (p. 58)
  Absent Student Catch-Up (p. 60)
  Challenge (p. 62)
- ***Resources in Spanish***
- ***Personal Student Tutor***

NEW-TEACHER SUPPORT
See the Tips for New Teachers on pp. 1–2 of the *Chapter 12 Resource Book* for additional notes about Lesson 12.4.

WARM-UP EXERCISES

Transparency Available

Simplify.

1. $(x + 9)^2$ $x^2 + 18x + 81$

2. $(2x - 3)^2$ $4x^2 - 12x + 9$

Factor.

3. $x^2 - 10x + 25$ $(x - 5)^2$

4. $9x^2 + 24x + 16$ $(3x + 4)^2$

12.4 Completing the Square

What you should learn

GOAL 1 Solve a quadratic equation by completing the square.

GOAL 2 Choose a method for solving a quadratic equation.

Why you should learn it

▼ To solve **real-life** problems such as finding the path of the diver in **Ex. 71.**

GOAL 1 SOLVING BY COMPLETING THE SQUARE

In the activity on page 729, you completed the square for expressions of the form $x^2 + bx$ where $b = 2, 4, 6,$ and 8. In each case, $x^2 + bx + \left(\frac{b}{2} \cdot \frac{b}{2}\right)$ was modeled by a square with sides of length $x + \frac{b}{2}$. By using FOIL to expand $\left(x + \frac{b}{2}\right)\left(x + \frac{b}{2}\right)$, you can show that this pattern holds for any real number b.

COMPLETING THE SQUARE

To complete the square of the expression $x^2 + bx$, add the square of half the coefficient of x.

$$x^2 + bx + \left(\frac{b}{2}\right)^2 = \left(x + \frac{b}{2}\right)^2$$

EXAMPLE 1 Completing the Square

What term should you add to $x^2 - 8x$ so that the result is a perfect square?

SOLUTION

The coefficient of x is -8, so you should add $\left(\frac{-8}{2}\right)^2$, or 16, to the expression.

$$x^2 - 8x + \left(\frac{-8}{2}\right)^2 = x^2 - 8x + 16 = (x - 4)^2$$

EXAMPLE 2 Solving a Quadratic Equation

Solve $x^2 + 10x = 24$ by completing the square.

SOLUTION

$x^2 + 10x = 24$	**Write original equation.**
$x^2 + 10x + 5^2 = 24 + 5^2$	**Add $\left(\frac{10}{2}\right)^2$, or 5^2, to each side.**
$(x + 5)^2 = 49$	**Write left side as perfect square.**
$x + 5 = \pm 7$	**Find square root of each side.**
$x = -5 \pm 7$	**Subtract 5 from each side.**
$x = 2$ or $x = -12$	**Simplify.**

▶ The solutions are 2 and -12. Check these in the original equation to see that both are solutions.

STUDENT HELP

Study Tip
When completing the square to solve an equation, remember that you must always add the term $\left(\frac{b}{2}\right)^2$ to *both* sides of the equation.

EXAMPLE 3 *Solving a Quadratic Equation*

Solve $x^2 - x - 3 = 0$ by completing the square.

SOLUTION

$x^2 - x - 3 = 0$ — Write original equation.

$x^2 - x = 3$ — Add 3 to each side.

$x^2 - x + \left(-\frac{1}{2}\right)^2 = 3 + \frac{1}{4}$ — Add $\left(-\frac{1}{2}\right)^2$, or $\frac{1}{4}$, to each side.

$\left(x - \frac{1}{2}\right)^2 = \frac{13}{4}$ — Write left side as perfect square.

$x - \frac{1}{2} = \pm\frac{\sqrt{13}}{2}$ — Find square root of each side.

$x = \frac{1}{2} \pm \frac{\sqrt{13}}{2}$ — Add $\frac{1}{2}$ to each side.

▶ The solutions are $\frac{1}{2} + \frac{\sqrt{13}}{2}$ and $\frac{1}{2} - \frac{\sqrt{13}}{2}$. Check these in the original equation.

STUDENT HELP

Study Tip
To solve an equation by completing the square, you first need to isolate the x^2-term and the x-term on one side of the equation.

If the leading coefficient of the quadratic is not 1, you should divide each side of the equation by this coefficient *before* completing the square.

EXAMPLE 4 *The Leading Coefficient is Not 1*

Solve $2x^2 - x - 2 = 0$ by completing the square.

SOLUTION

$2x^2 - x - 2 = 0$ — Write original equation.

$2x^2 - x = 2$ — Add 2 to each side.

$x^2 - \frac{1}{2}x = 1$ — Divide each side by 2.

$x^2 - \frac{1}{2}x + \left(-\frac{1}{4}\right)^2 = 1 + \frac{1}{16}$ — Add $\left(-\frac{1}{4}\right)^2$, or $\frac{1}{16}$, to each side.

$\left(x - \frac{1}{4}\right)^2 = \frac{17}{16}$ — Write left side as perfect square.

$x - \frac{1}{4} = \pm\frac{\sqrt{17}}{4}$ — Find square root of each side.

$x = \frac{1}{4} \pm \frac{\sqrt{17}}{4}$ — Add $\frac{1}{4}$ to each side.

▶ The solutions are $\frac{1}{4} + \frac{\sqrt{17}}{4} \approx 1.28$ and $\frac{1}{4} - \frac{\sqrt{17}}{4} \approx -0.78$.

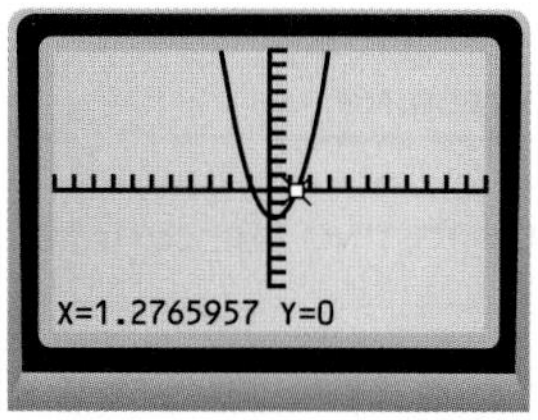

✓**CHECK** The graph generated by a graphing calculator appears to confirm the solutions.

STUDENT HELP

INTERNET **HOMEWORK HELP**
Visit our Web site www.mcdougallittell.com for extra examples.

2 TEACH

MOTIVATING THE LESSON
A penguin's wings serve as flippers for underwater swimming. Some species move along in a series of dives that resemble the diving arcs made by a porpoise. In this lesson you will learn a method of solving a quadratic equation called completing the square. Then you will be able to determine the horizontal distance covered by a porpoising penguin.

EXTRA EXAMPLE 1
What term should you add to $x^2 + 20x$ so that the result is a perfect square? **100**

EXTRA EXAMPLE 2
Solve $x^2 + 2x = 35$ by completing the square. **−7 and 5**

EXTRA EXAMPLE 3
Solve $x^2 - 8x + 2 = 0$ by completing the square. **$4 \pm \sqrt{14}$**

EXTRA EXAMPLE 4
Solve $3x^2 + 5x - 7 = 0$ by completing the square.
$-\frac{5}{6} + \frac{\sqrt{109}}{6} \approx 0.91$ and $-\frac{5}{6} - \frac{\sqrt{109}}{6} \approx -2.57$

✓ CHECKPOINT EXERCISES

For use after Example 1:
1. What term added to $x^2 - 12x$ results in a perfect square? **36**

For use after Example 2:
2. Solve $x^2 - 3x = 18$. **6 and −3**

For use after Example 3:
3. Solve $x^2 - 4x - 11 = 0$ by completing the square. **$2 \pm \sqrt{15}$**

For use after Example 4:
4. Solve $3x^2 + 6x - 2 = 0$ by completing the square. **$-1 + \frac{\sqrt{15}}{3}$ and $-1 - \frac{\sqrt{15}}{3}$**

EXTRA EXAMPLE 5

Choose a method to solve the quadratic equation. Explain your choice.

a. $x^2 - 3x - 10 = 0$ **the quadratic can be easily factored; −2 and 5**

b. $2x^2 - 6x - 3 = 0$ $\frac{3}{2} \pm \frac{\sqrt{15}}{2}$; **because the quadratic trinomial cannot be easily factored, use the quadratic formula**

EXTRA EXAMPLE 6

A company that makes cell phones has an average weekly profit of $20,000. The company models its profit p using $p = x^2 + 15x - 48$, where x is the number of cell phones it sells. How many cell phones must it sell to make a weekly profit of $20,000? **135**

CHECKPOINT EXERCISES

For use after Example 5:

1. Solve $16x^2 + 20x - 1 = 0$ using any method. $-\frac{5}{8} \pm \frac{\sqrt{29}}{8}$

For use after Example 6:

2. Marlene jogs from her house to school. Her route takes her south, then east. The most direct route, however, can be modeled by $-3x^2 + 2x + 4 = y$, where x is the number of miles east and y is the number of miles south that Marlene jogs. How many miles east does she jog? **about 1.5 mi**

GOAL 2 CHOOSING A SOLUTION METHOD

There are many connections among the five methods for solving a quadratic equation. For instance, in Example 4 you used completing the square to solve the equation, and you used a graphing approach to confirm that the solutions are reasonable.

In the following activity, you will use completing the square to develop the quadratic formula.

ACTIVITY Developing Concepts

Investigating the Quadratic Formula

Consider a general quadratic equation

$ax^2 + bx + c = 0$ where $a \neq 0$.

Perform the following steps. Then describe how your result is related to the quadratic formula.

1. Subtract c from each side of the equation $ax^2 + bx + c = 0$.
2. Divide each side by a.
3. Add the square of half the coefficient of x to each side.
4. Write the left side as a perfect square.
5. Use a common denominator to express the right side as a single faction.
6. Find the square root of each side. Include $\pm$ on the right side.
7. Solve for x by subtracting the same term from each side.
8. Use a common denominator to express the right side as a single fraction.

Step 1. $ax^2 + bx = -c$

Step 2. $x^2 + \frac{bx}{a} = -\frac{c}{a}$

Step 3. $x^2 + \frac{bx}{a} + \frac{b^2}{4a^2} = -\frac{c}{a} + \frac{b^2}{4a^2}$

Step 4. $\left(x + \frac{b}{2a^2}\right)^2 = -\frac{c}{a} + \frac{b^2}{4a^2}$

Step 5. $\left(x + \frac{b}{2a}\right)^2 = \frac{-4ac + b^2}{4a^2}$

Step 6. $x + \frac{b}{2a} = \pm\frac{\sqrt{b^2 - 4ac}}{2a}$

Step 7. $x = -\frac{b}{2a} \pm \frac{\sqrt{b^2 - 4ac}}{2a}$

Step 8. $x = \frac{-b \pm \sqrt{b^2 - 4ac}}{2a}$

You have learned the following five methods for solving quadratic equations. The table will help you choose which method to use to solve an equation.

CONCEPT SUMMARY METHODS FOR SOLVING $ax^2 + bx + c = 0$

Method	Lesson	Comments
FINDING SQUARE ROOTS	9.1	Efficient way to solve $ax^2 + c = 0$.
GRAPHING	9.4	Can be used for *any* quadratic equation. May give only approximate solutions.
USING THE QUADRATIC FORMULA	9.5	Can be used for *any* quadratic equation. Always gives exact solutions to the equation.
FACTORING	10.5–10.8	Efficient way to solve a quadratic equation if the quadratic expression can be factored easily.
COMPLETING THE SQUARE	12.4	Can be used for *any* quadratic equation but is simplest to apply when $a = 1$ and b is an even number.

EXAMPLE 5 *Choosing a Solution Method*

Choose a method to solve the quadratic equation. Explain your choice.

a. $3x^2 - 15 = 0$ **b.** $2x^2 + 3x - 4 = 0$

SOLUTION

a. Because this quadratic equation has the form $ax^2 + c = 0$, it is most efficiently solved by finding square roots.

$3x^2 - 15 = 0$ **Write original equation.**

$3x^2 = 15$ **Add 15 to each side.**

$x^2 = 5$ **Divide each side by 3.**

$x = \pm\sqrt{5}$ **Find square roots.**

▶ The solutions are $\sqrt{5}$ and $-\sqrt{5}$.

b. ▶ Because this quadratic equation is not easily factored, you can use a graphing calculator to approximate the solutions. TRACE to estimate the x-intercepts. The approximate solutions are -2.3 and 0.9.

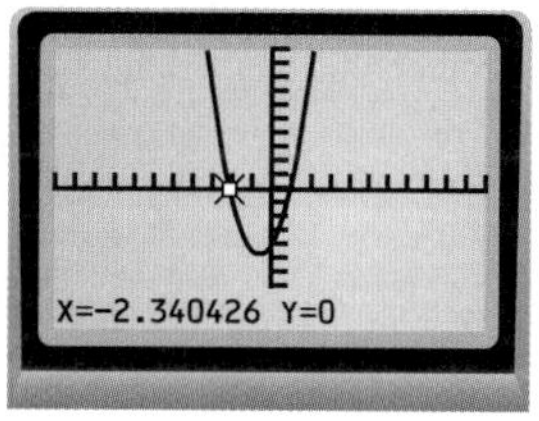

✓**CHECK** Using the quadratic formula, the exact solutions are

$$\frac{-3 + \sqrt{41}}{4} \text{ and } \frac{-3 - \sqrt{41}}{4}.$$

EXAMPLE 6 *Choosing the Quadratic Formula*

VETTISFOSS FALLS The Vettisfoss waterfall falls over a vertical cliff. The path of the water can be modeled by $h = -6.03x^2 + 901$ where h is the height (in feet) above the lower river and x is the horizontal distance (in feet) from the base of the cliff. How far from the base of the cliff does the water hit the lower river?

SOLUTION

When the falling water hits the lower river, $h = 0$. The quadratic equation $0 = -6.03x^2 + 901$ cannot be factored easily and cannot be solved easily by completing the square. The quadratic formula is a good choice.

$x = \dfrac{-b \pm \sqrt{b^2 - 4ac}}{2a}$ **Write quadratic formula.**

$x = \dfrac{0 \pm \sqrt{0 - 4(6.03)901}}{2(-6.03)}$ **Substitute values for a, b, and c.**

≈ 12.3 or -12.3 **Use a calculator.**

▶ Distance must be positive, so the negative solution is extraneous. The water hits the lower river about 12.3 feet from the base of the cliff. Check this solution in the original equation.

FOCUS ON APPLICATIONS

REAL LIFE **VETTISFOSS FALLS** is Norway's second-highest waterfall. There are many waterfalls in Norway. In fact, Norway generates more hydroelectric power per person than any other country.

FOCUS ON VOCABULARY
Why is adding 9 to both sides of $x^2 + 6x = 2$ called completing the square? **Adding 9 to the left side makes the binomial a perfect trinomial square.**

CLOSURE QUESTION
What kinds of quadratic equations are more efficiently solved by completing the square? **those equations with $a = 1$ and b an even number.**

DAILY PUZZLER
A dartboard consists of a square inscribed in a circle and another circle inscribed in that square. How many times more likely are you to hit within the larger circle than you are to hit within the smaller circle, assuming your dart lands on the dartboard? **two times**

3 APPLY

ASSIGNMENT GUIDE

BASIC

Day 1: pp. 734–735 Exs. 20–66 even

Day 2: pp. 734–736 Exs. 33–67 odd, 68–70, 74–76, 82, 86, 90, 94, 98, 102, 106

AVERAGE

Day 1: pp. 734–735 Exs. 20–66 even

Day 2: pp. 734–736 Exs. 33–67 odd, 68–71, 74–76, 82, 86, 90, 94, 98, 102, 106

ADVANCED

Day 1: pp. 734–735 Exs. 20–66 even

Day 2: pp. 734–736 Exs. 33–67 odd, 68–80, 82, 86, 90, 94, 98, 102, 106

BLOCK SCHEDULE

pp. 734–735 Exs. 20–66 even (with 12.3)
pp. 734–736 Exs. 20–32 even, 33–71, 74–76, 82, 86, 90, 94, 98, 102, 106 (with 12.5)

EXERCISE LEVELS

Level A: *Easier*
20–37

Level B: *More Difficult*
38–71, 74–76, 81, 108

Level C: *Most Difficult*
72, 73, 77–80

✓ HOMEWORK CHECK

To quickly check student understanding of key concepts, go over the following exercises: Exs. 20, 22, 30, 34, 38, 48, 68, 70. See also the Daily Homework Quiz:

- Blackline Master (*Chapter 12 Resource Book*, p. 65)
- Transparency (p. 98)

GUIDED PRACTICE

Vocabulary Check ✓

1. The leading coefficient of the polynomial $3x^2 - 8x + 4$ is _?_. 3

Concept Check ✓

2. Explain why completing the square of the expression $x^2 + bx$ is easier to do when b is an even number.
To complete the square you need to square $\frac{b}{2}$, which is an integer if b is even.

Skill Check ✓

Find the term that should be added to the expression to create a perfect square trinomial.

3. $x^2 + 20x$ 100
4. $x^2 + 50x$ 625
5. $x^2 - 10x$ 25
6. $x^2 - 14x$ 49
7. $x^2 - 22x$ 121
8. $x^2 + 100x$ 2500
9. Solve $x^2 - 3x = 8$ by completing the square. Solve the equation by using the quadratic formula. Which method did you find easier? $\frac{3-\sqrt{41}}{2}, \frac{3+\sqrt{41}}{2}$; choices may vary.

Solve by completing the square.

10. $x^2 - 2x - 18 = 0$ $1-\sqrt{19}, 1+\sqrt{19}$
11. $x^2 + 14x + 13 = 0$ $-13, -1$
12. $3x^2 + 4x - 1 = 0$ $\frac{-2-\sqrt{7}}{3}, \frac{-2+\sqrt{7}}{3}$
13. $3x^2 - 7x + 6 = 0$ no solution

Choose a method to solve the quadratic equation. What method did you use? Explain your choice. 14–19. Sample explanations are given.

14. $x^2 - x - 2 = 0$
15. $3x^2 + 17x + 10 = 0$
16. $x^2 - 9 = 0$
17. $-3x^2 + 5x + 5 = 0$ $\frac{5+\sqrt{85}}{6}, \frac{5-\sqrt{85}}{6}$; Quadratic formula; The expression is not factorable, and the coefficient of x is odd.
18. $x^2 + 2x - 14 = 0$ $1+\sqrt{15}, 1-\sqrt{15}$; Completing the square; the coefficient of x^2 is 1 and the coefficient of x is even.
19. $3x^2 - 2 = 0$ $\frac{\sqrt{6}}{3}, \frac{-\sqrt{6}}{3}$; Finding square roots; the expression is equivalent to $x^2 = \frac{2}{3}$.

14. 2, −1; Factoring; the expression $x^2 - x - 2$ can be factored as $(x-2)(x+1)$.
15. $\frac{-2}{3}$, −5; Factoring; the expression $3x^2 + 17x + 10$ can be factored $(3x+2)(x+5)$.
16. 3, −3; Factoring or finding square roots; both methods are easily applied.
38. $\frac{1-2\sqrt{7}}{3}, \frac{1+2\sqrt{7}}{3}$
39. $\frac{-3-\sqrt{109}}{10}, \frac{-3+\sqrt{109}}{10}$
40. $\frac{-1-\sqrt{5}}{2}, \frac{-1+\sqrt{5}}{2}$
41. $\frac{-1-2\sqrt{3}}{2}, \frac{-1+2\sqrt{3}}{2}$
42. $\frac{12-7\sqrt{3}}{3}, \frac{12+7\sqrt{3}}{3}$
43. $\frac{10-\sqrt{107}}{2}, \frac{10+\sqrt{107}}{2}$
47. $\frac{-3-\sqrt{41}}{4}, \frac{-3+\sqrt{41}}{4}$
48. $\frac{-12-\sqrt{390}}{6}, \frac{-12+\sqrt{390}}{6}$
49. $\frac{30-17\sqrt{5}}{10}, \frac{30+17\sqrt{5}}{10}$

PRACTICE AND APPLICATIONS

STUDENT HELP

Extra Practice to help you master skills is on p. 808.

PERFECT SQUARE TRINOMIALS **Find the term that should be added to the expression to create a perfect square trinomial.**

20. $x^2 - 12x$ 36
21. $x^2 + 8x$ 16
22. $x^2 + 21x$ $\frac{441}{4}$
23. $x^2 - 22x$ 121
24. $x^2 + 11x$ $\frac{121}{4}$
25. $x^2 - 40x$ 400
26. $x^2 + 0.4x$ 0.04
27. $x^2 + \frac{3}{4}x$ $\frac{9}{64}$
28. $x^2 + \frac{4}{5}x$ $\frac{4}{25}$
29. $x^2 - 5.2x$ 6.76
30. $x^2 - 0.3x$ 0.0225
31. $x^2 + \frac{2}{3}x$ $\frac{1}{9}$

COMPLETING THE SQUARE **Solve the equation by completing the square.**

32. $x^2 + 10x = 39$ −13, 3
33. $x^2 + 16x = 17$ −17, 1
34. $x^2 - 24x = -44$ 2, 22
35. $x^2 - 8x + 12 = 0$ 2, 6
36. $x^2 + 5x - \frac{11}{4} = 0$ $-\frac{11}{2}, \frac{1}{2}$
37. $x^2 + 11x + \frac{21}{4} = 0$ $-\frac{21}{2}, -\frac{1}{2}$
38. $x^2 - \frac{2}{3}x - 3 = 0$ See margin.
39. $x^2 + \frac{3}{5}x - 1 = 0$ See margin.
40. $x^2 + x - 1 = 0$ See margin.
41. $4x^2 + 4x - 11 = 0$ See margin.
42. $3x^2 - 24x - 1 = 0$ See margin.
43. $4x^2 - 40x - 7 = 0$ See margin.
44. $2x^2 - 8x - 13 = 7$ $2+\sqrt{14}, 2-\sqrt{14}$
45. $5x^2 - 20x - 20 = 5$ −1, 5
46. $3x^2 + 4x + 4 = 3$ $-1, -\frac{1}{3}$
47. $4x^2 + 6x - 6 = 2$ See margin.
48. $6x^2 + 24x - 41 = 0$ See margin.
49. $20x^2 - 120x - 109 = 0$ See margin.

STUDENT HELP

HOMEWORK HELP

Example 1: Exs. 20–31
Example 2: Exs. 32–40
Example 3: Exs. 35–40
Example 4: Exs. 41–49
Example 5: Exs. 50–67
Example 6: Ex. 71

50. $\frac{5-\sqrt{29}}{2}, \frac{5+\sqrt{29}}{2}$; Quadratic formula; the expression can't be factored and the coefficient of x is not even.

51. $-\sqrt{3}, \sqrt{3}$; Factoring or finding square roots; both methods are easily applied.

52. -8; 3; Factoring; the expression $n^2 + 5n - 24$ can be factored as $(n + 8)(n - 3)$.

53. $\frac{5}{3}, -\frac{5}{3}$; Factoring or finding square roots; both methods are easily applied.

54. 5, -4; Factoring; the expression $x^2 - x - 20$ can be factored as $(x - 5)(x + 4)$.

55. 5, -11; completing the square or factoring; $a = 1$ and b is an even number.

61. $\frac{12-\sqrt{174}}{6}, \frac{12+\sqrt{174}}{6}$

70. $\frac{b}{2}(2b + 4) = 60$; base $= -1 + \sqrt{61} \approx 6.8$ cm, height $= 2 + 2\sqrt{61} \approx 17.6$ cm

FOCUS ON APPLICATIONS

PENGUINS dive to capture food. Adelie penguins can stay underwater for seven minutes.

APPLICATION LINK
www.mcdougallittell.com

CHOOSING A METHOD **Choose a method to solve the quadratic equation. Explain your choice.** 50–55. Sample explanations are given. See margin.

50. $x^2 - 5x - 1 = 0$ **51.** $4x^2 - 12 = 0$ **52.** $n^2 + 5n - 24 = 0$

53. $9a^2 - 25 = 0$ **54.** $x^2 - x - 20 = 0$ **55.** $x^2 + 6x - 55 = 0$

SOLVING EQUATIONS **Solve the quadratic equation.**

56. $x^2 - 10x = 0$ **0, 10**

57. $c^2 + 2c - 26 = 0$ $-1 - 3\sqrt{3}, -1 + 3\sqrt{3}$

58. $8x^2 + 14x = -5$ $-1\frac{1}{4}, -\frac{1}{2}$

59. $x^2 - 16 = 0$ **-4, 4**

60. $x^2 + 12x + 20 = 0$ $-10, -2$

61. $x^2 - 4x = \frac{5}{6}$ See margin.

62. $4x^2 + 4x + 1 = 0$ $-\frac{1}{2}$

63. $13x^2 - 26x = 0$ 0, 2

64. $4p^2 - 12p + 5 = 0$ $\frac{1}{2}, 2\frac{1}{2}$

65. $7z^2 - 46z = 21$ $-\frac{3}{7}, 7$

66. $11x^2 - 22 = 0$ $-\sqrt{2}, \sqrt{2}$

67. $x^2 + 20x + 10 = 0$ $-10 - 3\sqrt{10}, -10 + 3\sqrt{10}$

GEOMETRY CONNECTION **In Exercises 68–70, make a sketch and write a quadratic equation to model the situation. Then solve the equation.**

68. In art class you are designing the floor plan of a house. The kitchen is supposed to have 150 square feet of space. What should the dimensions of the kitchen be if you want it to be square?
$x^2 = 150$; $5\sqrt{6}$ ft by $5\sqrt{6}$ ft ≈ 12.25 ft by 12.25 ft

69. A rectangle is $2x$ feet long and $x + 5$ feet wide. The area is 600 square feet. What are the dimensions of the rectangle? $2x(x + 5) = 600$; 30 ft by 20 ft

70. The height of a triangle is 4 more than twice its base. The area of the triangle is 60 square centimeters. What are the dimensions of the triangle? See margin.

71. **DIVING** The path of a diver diving from a 10-foot high diving board is

$$h = -0.44x^2 + 2.61x + 10$$

where h is the height of the diver above water (in feet) and x is the horizontal distance (in feet) from the end of the board. How far from the end of the board will the diver enter the water? **about 8.58 ft**

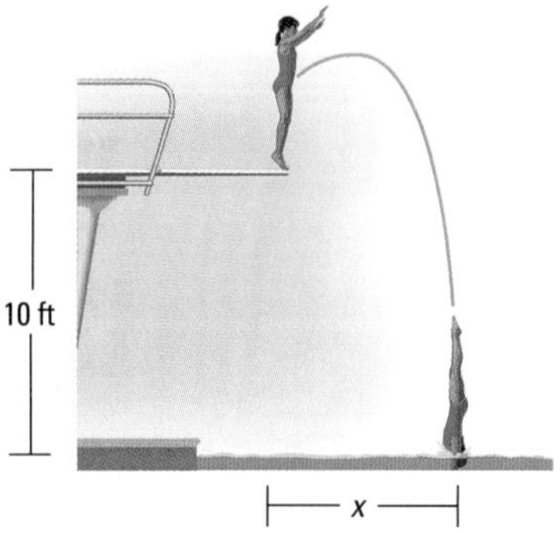

PENGUINS **In Exercises 72 and 73, use the following information.**
You are on a research boat in the ocean. You see a penguin jump out of the water. The path followed by the penguin is given by

$$h = -0.05x^2 + 1.178x$$

where h is the height (in feet) the penguin jumps out of the water and x is the horizontal distance (in feet) traveled by the penguin over the water.

72. Sketch a graph of the equation. See margin.

73. How many horizontal feet did the penguin travel over the water before reaching its maximum height? **about 11.8 ft**

! COMMON ERROR

EXERCISES 5–7 If students add a negative number to try to make a perfect square trinomial, ask them to try to factor their trinomial.

APPLICATION NOTE

EXERCISE 71 The solutions to $0 = -0.44x^2 + 2.61x + 10$ are the points on the graph of the diving path equation that represent the two positions when the diver's height is 0 feet above the water. The negative value, however, is not an answer to this exercise as it makes no sense in a real world context.

EXERCISES 72 AND 73
Additional information about penguins is available at **www.mcdougallittell.com.**

72.

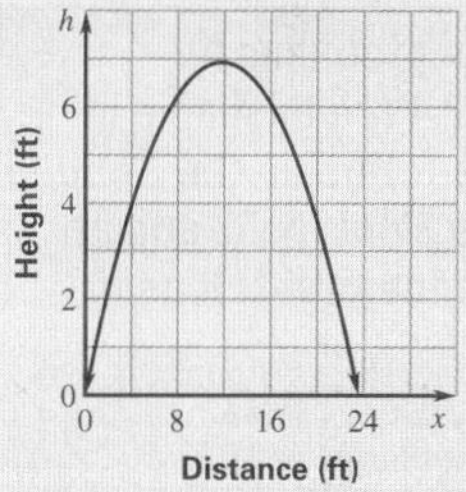

ADDITIONAL PRACTICE AND RETEACHING

For Lesson 12.4:

- Practice Levels A, B, and C (*Chapter 12 Resource Book*, p. 55)
- Reteaching with Practice (*Chapter 12 Resource Book*, p. 58)
- See Lesson 12.4 of the *Personal Student Tutor*

For more Mixed Review:

- Search the *Test and Practice Generator* for key words or specific lessons.

4 ASSESS

DAILY HOMEWORK QUIZ

Transparency Available

1. What term should be added to $x^2 - 16x$ to make it a perfect square trinomial? **64**
2. Solve $x^2 - 20x - 2 = 0$.
   $10 \pm \sqrt{102}$
3. Solve $4x^2 - 2x - 1 = 0$.
   $\frac{1 \pm \sqrt{5}}{4}$
4. The height of a triangle is 2 cm more than its base. The area of the triangle is 30 cm^2. What are the base and height of the triangle? **about 6.8 cm and 8.8 cm**

EXTRA CHALLENGE NOTE

Challenge problems for Lesson 12.4 are available in **blackline** format in the *Chapter 12 Resource Book*, p. 62 and at **www.mcdougallittell.com.**

ADDITIONAL TEST PREPARATION

1. **WRITING** Explain why $\frac{9}{4}$ should be added to $x^2 + 3x = 1$ so that the result is a perfect square trinomial. **Since $\frac{9}{4}$ is the square of one-half the linear term, 3, the resulting trinomial $x^2 + 3x + \frac{9}{4}$ is a perfect square trinomial. It can be factored into $\left(x + \frac{3}{2}\right)^2$.**
2. **WRITING** Explain why the quadratic formula is not the most efficient means of solving $16x^2 - 40x = -25$. **When the equation is written in standard form, it can be solved using factoring since $16x^2 - 40x + 25$ is a perfect square trinomial.**

94–96. See Additional Answers beginning on page AA1.

Test Preparation

74. MULTIPLE CHOICE Which of the following is a solution of the equation $2x^2 + 8x - 25 = 5$? **B**

Ⓐ $\sqrt{17} + 1$ Ⓑ $-\sqrt{19} - 2$ Ⓒ $\sqrt{17} - 2$ Ⓓ $\sqrt{21} - 2$

75. MULTIPLE CHOICE What term should you add to $x^2 - \frac{1}{2}x$ so that the result is a perfect square trinomial? **C**

Ⓐ $\frac{1}{2}$ Ⓑ $\frac{1}{4}$ Ⓒ $\frac{1}{16}$ Ⓓ $\frac{1}{32}$

76. MULTIPLE CHOICE Solve $x^2 + 8x - 2 = 0$. **A**

Ⓐ $-4 \pm 3\sqrt{2}$ Ⓑ $4 \pm 3\sqrt{2}$ Ⓒ $-4 \pm 2\sqrt{2}$ Ⓓ $4 \pm \sqrt{16}$

★ Challenge

79. $y = -\left(x + \frac{5}{2}\right)^2 + \frac{49}{4}$; $\left(-\frac{5}{2}, \frac{49}{4}\right)$

EXTRA CHALLENGE

www.mcdougallittell.com

VERTEX FORM **The *vertex form* of a quadratic function is $y = a(x - h)^2 + k$. Its graph is a parabola with vertex at (h, k). In Exercises 77–79, use completing the square to write the quadratic function in vertex form. Then give the coordinates of the vertex of the graph of the function.**

77. $y = x^2 + 10x + 25$ **$y = (x + 5)^2$; $(-5, 0)$**

78. $y = 2x^2 + 12x + 13$ **$y = 2(x + 3)^2 - 5$; $(-3, -5)$**

79. $y = -x^2 - 5x + 6$ **See margin.**

80. QUADRATIC FORMULA Explain why the quadratic formula gives solutions only if $a \neq 0$ and $b^2 - 4ac \geq 0$.
If $a = 0$, the rational expression $\frac{-b \pm \sqrt{b^2 - 4ac}}{2a}$ is not defined because the denominator is zero. If $b^2 - 4ac < 0$, the rational expression is not defined because the numerator is the square root of a negative number.

MIXED REVIEW

MEASURES OF CENTRAL TENDENCY **Find the mean, the median, and the mode of the collection of numbers.** (Review 6.6)

81. 1, 5, 2, 4, 3, 6, 1 **$3\frac{1}{7}$, 3, 1**

82. 9, 6, 10, 14, 10, 3 **$8\frac{2}{3}$, $9\frac{1}{2}$, 10**

83. −6, 20, −8, −18, 10 **$-\frac{2}{5}$, −6, none**

84. 17, 9, 11, 15, 4, 15, 8, 3, 11 **$10\frac{1}{3}$, 11, 11 and 15**

SOLVING LINEAR SYSTEMS **Solve the linear system.** (Review 7.2, 7.3)

85. $y = 4x$, $x + y = 10$ **(2, 8)**

86. $3x + y = 12$, $9x - y = 36$ **(4, 0)**

87. $2x - y = 8$, $2x + 2y = 2$ **(3, −2)**

SOLVING EQUATIONS **Solve the equation.** (Review 9.1, 9.2 for 12.5)

88. $16 + x^2 = 64$ **$-4\sqrt{3}$, $4\sqrt{3}$**

89. $x^2 + 81 = 144$ **$-3\sqrt{7}$, $3\sqrt{7}$**

90. $x^2 + 25 = 81$ **$-2\sqrt{14}$, $2\sqrt{14}$**

91. $4x^2 - 144 = 0$ **−6, 6**

92. $x^2 - 30 = -3$ **$-3\sqrt{3}$, $3\sqrt{3}$**

93. $x^2 = \frac{9}{25}$ **$-\frac{3}{5}$, $\frac{3}{5}$**

SKETCHING GRAPHS **Sketch the graph of the function.** (Review 9.3) **94–96. See margin.**

94. $y = x^2 + x + 2$

95. $y = -3x^2 - x - 4$

96. $y = 2x^2 - 3x + 4$

SOLVING EQUATIONS **Solve the equation.** (Review 10.4)

97. $(x + 4)(x - 8) = 0$ **−4, 8**

98. $(x - 3)(x - 2) = 0$ **2, 3**

99. $(x + 5)(x + 6) = 0$ **−6, −5**

100. $(x + 4)^2 = 0$ **−4**

101. $(x - 3)^2 = 0$ **3**

102. $6(x - 14)^2 = 0$ **14**

FACTORING TRINOMIALS **Factor the trinomial if possible.** (Review 10.5, 10.6)

103. $x^2 + x - 20$ **$(x + 5)(x - 4)$**

104. $x^2 - 10x + 24$ **$(x - 6)(x - 4)$**

105. $x^2 + 2x + 4$ **cannot be factored**

106. $3x^2 - 15x + 18$ **$3(x - 2)(x - 3)$**

107. $2x^2 - x - 3$ **$(2x - 3)(x + 1)$**

108. $14x^2 - 19x - 3$ **$(7x + 1)(2x - 3)$**

ACTIVITY 12.5

Developing Concepts

Group Activity for use with Lesson 12.5

Investigating the Pythagorean Theorem

GROUP ACTIVITY
Work in a small group.

MATERIALS
- ruler
- scissors
- glue or tape

QUESTION **Is the Pythagorean theorem true only for right triangles?**

EXPLORING THE CONCEPT

For the right triangle shown, a, b, and c are the lengths of the sides of the triangle. The Pythagorean theorem states that $a^2 + b^2 = c^2$.

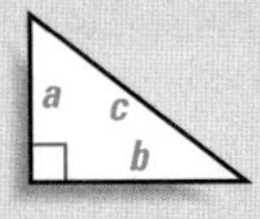

1 Copy the triangle on a piece of paper. Draw a square on each of the triangle's three sides. Label the areas a^2, b^2, and c^2. Then cut out the squares.

2 Cut and rearrange the areas that represent the squares of a and b to see if they fit exactly into the area that represents the square of c.

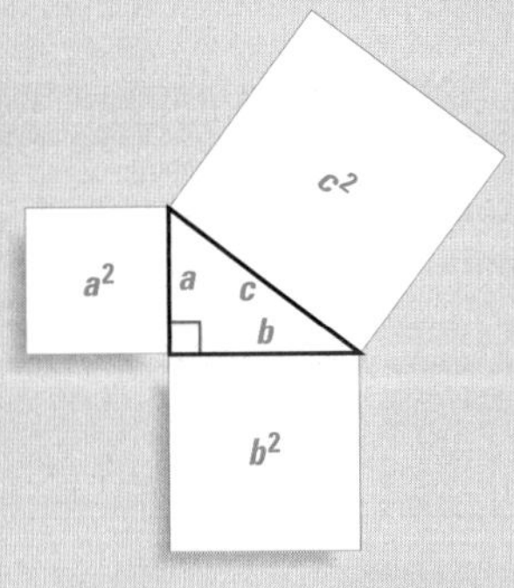

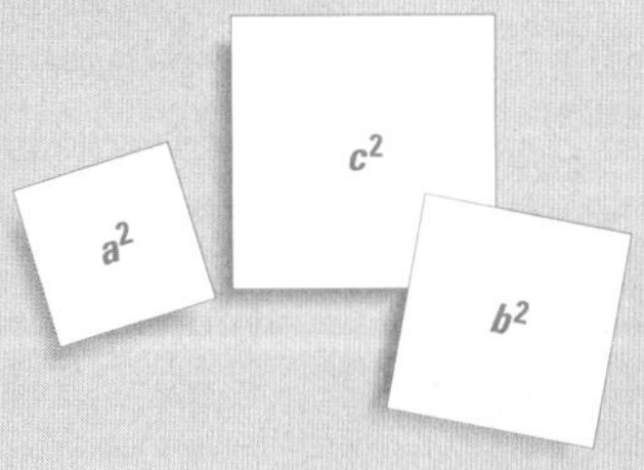

DRAWING CONCLUSIONS

1. Each group member should copy a different one of the triangles below. Then draw the areas that represent the squares of the sides. Cut and rearrange them to see how the larger area relates to the sum of the two smaller areas. **Check results.**

Obtuse triangle

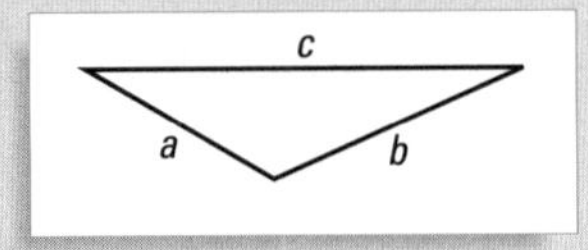

Acute triangle

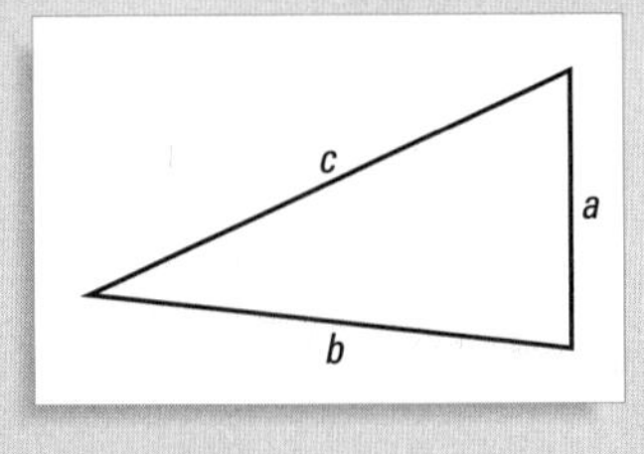

2. Draw any triangle and follow the directions of Exercise 1. **Check results.**

In Exercises 3–5, a, b, and c are the lengths of the sides of a triangle, and c is the greatest length. Use the results of your investigations with triangles to complete the statement with >, <, or =.

3. In an obtuse triangle, $a^2 + b^2$ _?_ c^2. $<$

4. In a right triangle, $a^2 + b^2$ _?_ c^2. $=$

5. In an acute triangle, $a^2 + b^2$ _?_ c^2. $>$

1 Planning the Activity

PURPOSE
To investigate the relationships between the squares of the lengths of the sides in right, obtuse, and acute triangles.

MATERIALS
- ruler
- scissors
- glue or tape
- Activity Support Master (*Chapter 12 Resource Book,* p. 66)

PACING
- Exploring the Concept — 10 min
- Drawing Conclusions — 15 min

LINK TO LESSON
Students may wish to use the geometric procedure presented here to verify part *a* of Example 1 in Lesson 12.5.

2 Managing the Activity

CLASSROOM MANAGEMENT
As a time-saver you may wish to have right, acute, and obtuse triangles already drawn on paper for each group. You may wish to have one type of triangle available for each student.

COOPERATIVE LEARNING
Have each student in the group complete Steps 1 and 2 individually, then compare his or her results with the other members of the group.

3 Closing the Activity

KEY DISCOVERY
If a triangle is not a right triangle, then $a^2 + b^2 \neq c^2$.

ACTIVITY ASSESSMENT
The lengths of the sides of a triangle are 5 cm, 12 cm and 13 cm. Is the triangle right, obtuse, or acute?
right

1 PLAN

PACING
Basic: 2 days
Average: 2 days
Advanced: 2 days
Block Schedule: 0.5 block with 12.4
0.5 block with 12.6

LESSON OPENER
VISUAL APPROACH
An alternative way to approach Lesson 12.5 is to use the Visual Approach Lesson Opener:

- Blackline Master (*Chapter 12 Resource Book,* p. 67)
- Transparency (p. 86)

MEETING INDIVIDUAL NEEDS

- ***Chapter 12 Resource Book***
  Prerequisite Skills Review (p. 5)
  Practice Level A (p. 70)
  Practice Level B (p. 71)
  Practice Level C (p. 72)
  Reteaching with Practice (p. 73)
  Absent Student Catch-Up (p. 75)
  Challenge (p. 78)
- ***Resources in Spanish***
- ***Personal Student Tutor***

NEW-TEACHER SUPPORT
See the Tips for New Teachers on pp. 1–2 of the *Chapter 12 Resource Book* for additional notes about Lesson 12.5.

WARM-UP EXERCISES
Transparency Available
Give the positive solution of each equation. Round to the nearest tenth where necessary.

1. $x^2 = 225$ 15
2. $x^2 = 1700$ 41.2
3. $\sqrt{400} = x$ 20
4. $\sqrt{185} = x$ 13.6

12.5 The Pythagorean Theorem and Its Converse

What *you should learn*

GOAL 1 Use the Pythagorean theorem and its converse.

GOAL 2 Use the Pythagorean theorem and its converse in **real-life** problems such as finding distance on a baseball field in **Example 4**.

Why *you should learn it*

▼ To solve **real-life** problems such as installing guy wires in **Ex. 44**.

GOAL 1 USING THE THEOREM AND ITS CONVERSE

A *theorem* is a statement that can be proven to be true. The activity on page 737 uses area to explore the *Pythagorean theorem*, but does not prove the theorem.

The **Pythagorean theorem** states a relationship among the sides of a right triangle. The **hypotenuse** is the side opposite the right angle. The other two sides are the **legs**. The theorem is named after the Greek mathematician Pythagoras. However, there are records that suggest earlier use of the basic principle in northern Africa, in Babylonia, and in India.

THE PYTHAGOREAN THEOREM

If a triangle is a right triangle, then the sum of the squares of the lengths of the legs a and b equals the square of the length of the hypotenuse c.

$$a^2 + b^2 = c^2$$

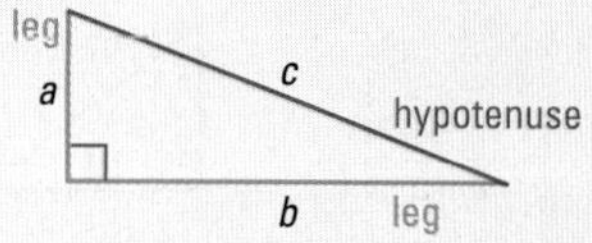

EXAMPLE 1 ***Using the Pythagorean Theorem***

a. Given $a = 6$ and $b = 8$, find c.

$a^2 + b^2 = c^2$
$6^2 + 8^2 = c^2$
$100 = c^2$
$\sqrt{100} = c$
$10 = c$

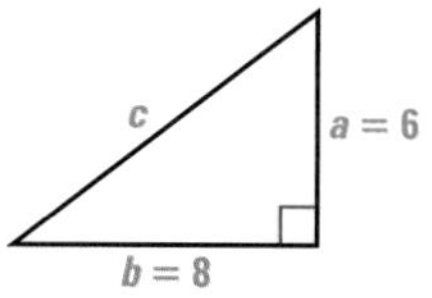

b. Given $a = 5$ and $c = 6$, find b.

$a^2 + b^2 = c^2$
$5^2 + b^2 = 6^2$
$b^2 = 6^2 - 5^2$
$b^2 = 11$
$b = \sqrt{11}$, or about 3.32

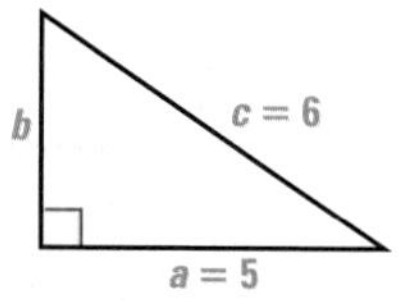

EXAMPLE 2 *Using the Pythagorean Theorem*

A right triangle has one leg that is 3 inches longer than the other leg. The hypotenuse is 15 inches. Find the missing lengths.

PROBLEM SOLVING STRATEGY

SOLUTION

DRAW A DIAGRAM Sketch a right triangle and label the sides. Let x be the length of the shorter leg. Use the Pythagorean theorem to solve for x.

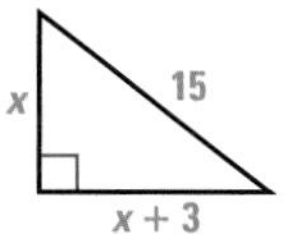

$a^2 + b^2 = c^2$	Write Pythagorean theorem.
$x^2 + (x + 3)^2 = 15^2$	Substitute for *a*, *b*, and *c*.
$x^2 + x^2 + 6x + 9 = 225$	Simplify.
$2x^2 + 6x - 216 = 0$	Write in standard form.
$2(x - 9)(x + 12) = 0$	Factor.
$x = 9$ or $x = -12$	Zero-product property

▶ Length is positive. The sides have lengths 9 inches and $9 + 3 = 12$ inches.

....

STUDENT HELP

Look Back
For help with if-then statements, see p. 189.

LOGICAL REASONING In mathematics an if-then statement is a statement of the form "If *p*, then *q*," where *p* is the **hypothesis** and *q* is the **conclusion**.

The **converse** of the statement "If *p*, then *q*" is the related statement "If *q*, then *p*," in which the hypothesis and conclusion are interchanged. The converse of a true statement may be false. However, the converse of the Pythagorean theorem is true.

CONVERSE OF THE PYTHAGOREAN THEOREM

If a triangle has side lengths a, b, and c such that $a^2 + b^2 = c^2$, then the triangle is a right triangle.

EXAMPLE 3 *Determining Right Triangles*

Determine whether the given lengths are sides of a right triangle.

a. 11.9, 12.0, 16.9 **b.** 5, 11, 12

STUDENT HELP

Study Tip
In a right triangle the hypotenuse is always the longest side.

SOLUTION Use the converse of the Pythagorean theorem.

a. The lengths are sides of a right triangle because

$$11.9^2 + 12.0^2 = 141.61 + 144 = 285.61 = 16.9^2.$$

b. The lengths are not sides of a right triangle because

$$5^2 + 11^2 = 25 + 121 = 146 \neq 12^2.$$

2 TEACH

MOTIVATING THE LESSON
When building a frame for a wall, a carpenter sometimes uses the "3-4-5" rule to determine if two boards are joined at a 90° angle. This is done by placing a mark 3 in. from the corner on one board and 4 in. from the corner on the other board. If the straight-line distance between these two points is 5 in., then the carpenter knows that the angle is 90°. You will study a method for determining right triangles in this lesson.

EXTRA EXAMPLE 1

a. Given $a = 8$ and $b = 15$, find c. **17**

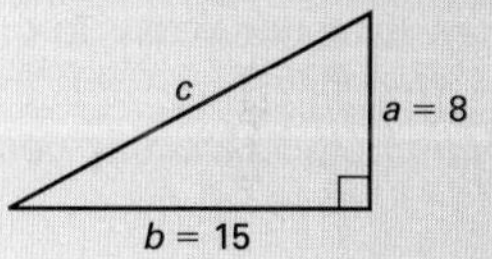

b. Given $a = 2$ and $c = 7$, find b. $3\sqrt{5} \approx 6.7$

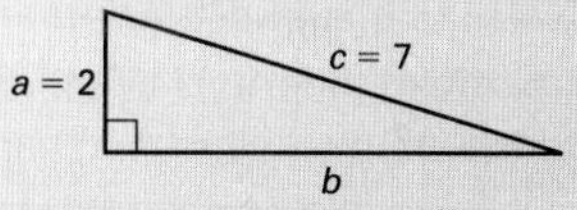

EXTRA EXAMPLE 2
A right triangle has a hypotenuse that is 1 ft longer than the longer leg. The shorter leg is 9 ft. Find the missing lengths. **The longer leg is 40 ft and the hypotenuse is 41 ft.**

EXTRA EXAMPLE 3
Determine whether the lengths are sides of a right triangle.
a. 0.5, 0.8, 1.0 **No**
b. 20, 21, 29 **Yes**

CHECKPOINT EXERCISES

For use after Examples 1 and 2:

1. A triangle has a hypotenuse with length 34 in. One leg is 14 in. shorter than the other leg. Find the lengths of the legs. **16 in. and 30 in.**

For use after Example 3:

2. Determine whether 1.5, 2.0, and 2.5 are the lengths of the sides of a right triangle. **Yes**

EXTRA EXAMPLE 4
A ladder 8 m long is resting against a building. The bottom of the ladder is 3 m from the wall. To the nearest tenth, how far up the wall does the ladder reach?
7.4 m

EXTRA EXAMPLE 5
Erica is setting out the boundary of a tool shed that she is going to build. The building will be 8 ft long and 6 ft wide. The diagonal of the boundary measures 11 ft. How does Erica know that the corners of her boundaries are not square? $6^2 + 8^2 \neq 11^2$

CHECKPOINT EXERCISES
For use after Examples 4 and 5:
1. A plane flies 150 mi due south, then 75 mi due west. To the nearest mile, how far is the plane from its starting point?
168 mi

FOCUS ON VOCABULARY
What is the hypothesis of the Pythagorean theorem? What is the conclusion? **hypothesis: A triangle is a right triangle; conclusion: The sum of the squares of the lengths of the legs equals the square of the length of the hypotenuse.**

CLOSURE QUESTION
The sides of a triangle are 5, 8, and 10. How do you know that the triangle is not a right triangle?
$5^2 + 8^2 \neq 10^2$

DAILY PUZZLER
A right triangle has sides with integer lengths. The product of the legs is 120. The sum of the three sides is 40. What are the lengths of the three sides of the triangle? **8, 15, 17**

GOAL 2 USING THE PYTHAGOREAN THEOREM IN REAL LIFE

REAL LIFE
Baseball

EXAMPLE 4 *Using the Pythagorean Theorem*

The length of each side of a baseball diamond is 90 feet. What is the distance from home plate to second base?

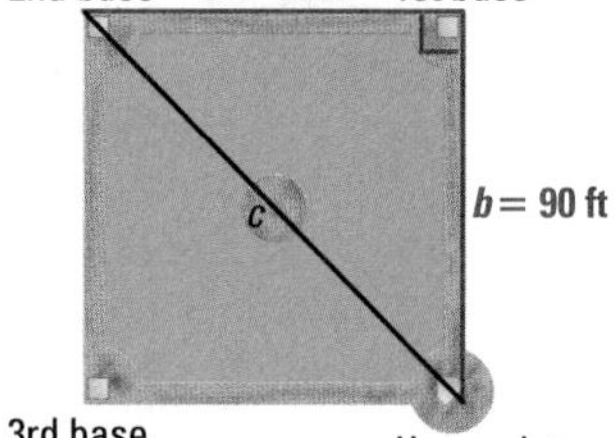

SOLUTION

The diagonal from home plate to second base is the hypotenuse c of a right triangle. The length of each leg is 90 feet.

$c^2 = a^2 + b^2$	Write Pythagorean theorem.
$c^2 = 90^2 + 90^2$	Substitute 90 for a and for b.
$c^2 = 8100 + 8100$	Multiply.
$c^2 = 16{,}200$	Add.
$c = \sqrt{16{,}200}$	Find positive square root of each side.
$c = 90\sqrt{2}$	Simplify.
$c \approx 127.3$	Use a calculator.

▶ The distance from home plate to second base is about 127.3 feet.

EXAMPLE 5 *Using the Pythagorean Converse*

You can take a rope and tie 12 equally spaced knots in it. You can then use the rope to check that a corner is a right angle. Why does this method work?

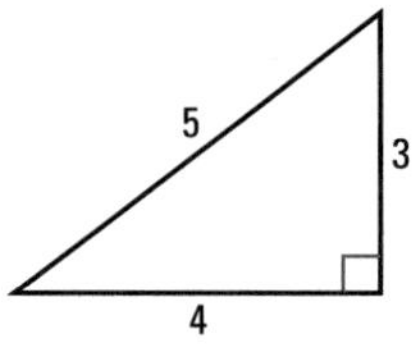

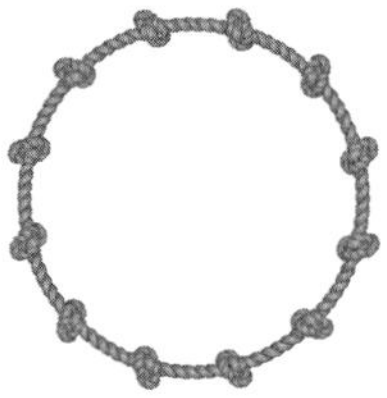

SOLUTION

You can use the rope to form a triangle with a hypotenuse length of 5 and leg lengths of 3 and 4. Use the converse of the Pythagorean theorem to check.

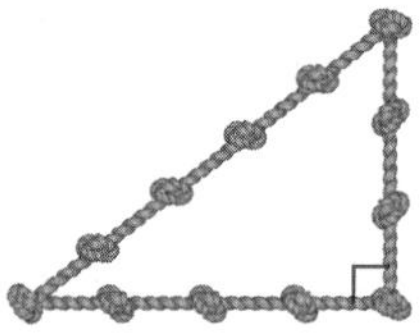

$5^2 \stackrel{?}{=} 3^2 + 4^2$

$25 \stackrel{?}{=} 9 + 16$

$25 = 25$ ✓

▶ Because you can use the knots to form the sides of a right triangle, one angle of the triangle must measure 90°. So you can check with the rope that a corner is a right angle.

GUIDED PRACTICE

Vocabulary Check ✓

1. The two sides of a right triangle that are not the hypotenuse are the __?__. legs

2. State the hypothesis and the conclusion of the statement "If x is an even number, then x^2 is an even number." **hypothesis: x is an even number; conclusion: x^2 is an even number**

Concept Check ✓

3. Explain how you can use the converse of the Pythagorean theorem to tell whether three given lengths can be sides of a right triangle. See margin.

Skill Check ✓

3. Check to see that the sum of the squares of the two shortest lengths is equal to the square of the longest length.

Find the missing length of the right triangle if *a* and *b* are the lengths of the legs and *c* is the length of the hypotenuse.

4. $a = 7, b = 24$ 25
5. $a = 5, c = 13$ 12
6. $b = 15, c = 17$ 8
7. $a = 9, c = 41$ 40
8. $b = 11, c = 61$ 60
9. $a = 12, b = 35$ 37

Find each missing length.

10. 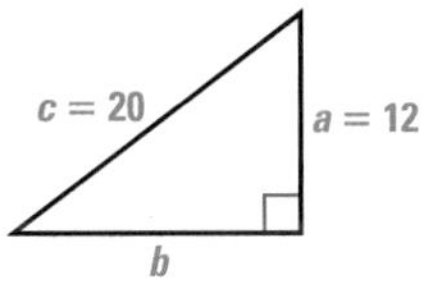

16

11. 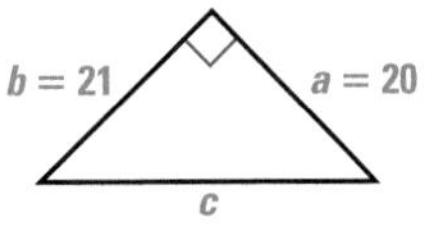

29

12. 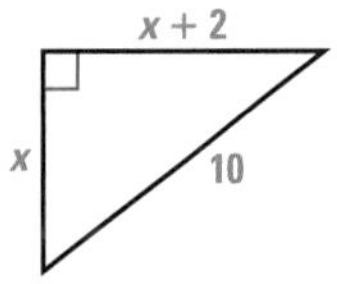

6, 8

PRACTICE AND APPLICATIONS

STUDENT HELP

Extra Practice to help you master skills is on p. 808.

USING THE PYTHAGOREAN THEOREM **Find the missing length of the right triangle if *a* and *b* are the lengths of the legs and *c* is the length of the hypotenuse.**

13. $a = 3, b = 4$ 5
14. $a = 5, c = 10$ $5\sqrt{3}$
15. $b = 3, c = 7$ $2\sqrt{10}$
16. $a = 10, b = 24$ 26
17. $b = 9, c = 16$ $5\sqrt{7}$
18. $a = 14, c = 21$ $7\sqrt{5}$

MISSING LENGTH **Find each missing length.**

19. 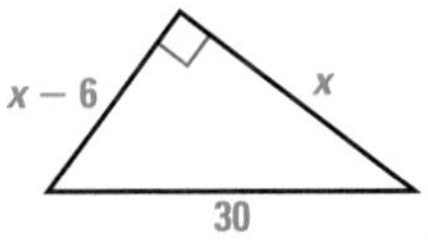

24, 18

20. 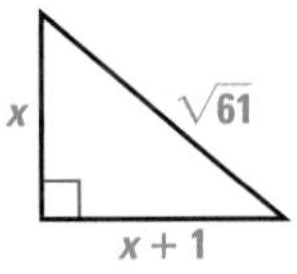

5, 6

21. 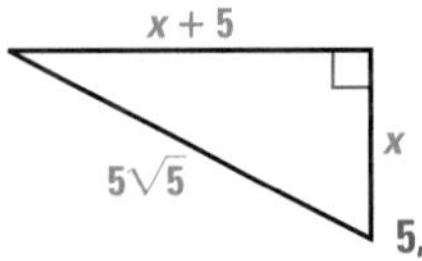

5, 10

22. 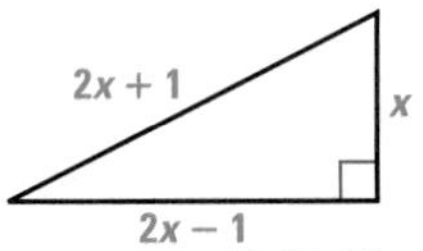

8, 15, 17

23. 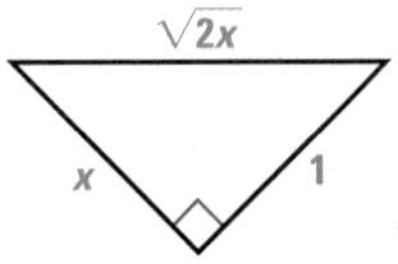

1, $\sqrt{2}$

24. 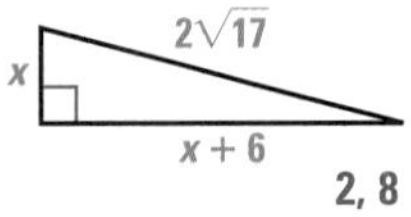

2, 8

DETERMINING RIGHT TRIANGLES **Determine whether the given lengths are sides of a right triangle. Explain your reasoning.** 25–27. See margin.

25. 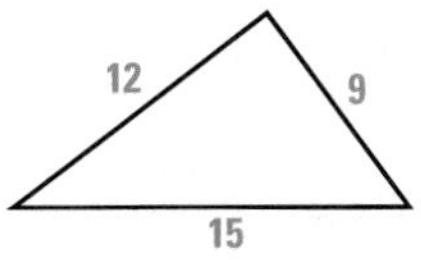

26. 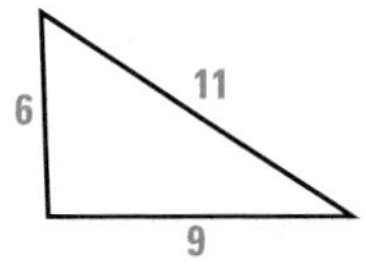

27. 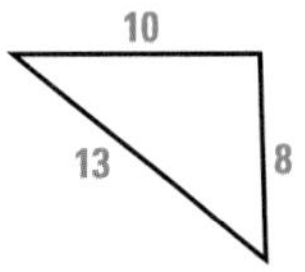

25. Yes; $9^2 + 12^2 = 225 = 15^2$.

26. No; $6^2 + 9^2 = 117$ and $11^2 = 121$.

27. No; $8^2 + 10^2 = 164$ and $13^2 = 169$.

STUDENT HELP

HOMEWORK HELP
Example 1: Exs. 13–18
Example 2: Exs. 19–24
Example 3: Exs. 25–33
Example 4: Exs. 39, 41–44
Example 5: Ex. 40

3 APPLY

ASSIGNMENT GUIDE

BASIC
Day 1: p. 741 Exs. 13–27
Day 2: pp. 742–744 Exs. 33–39, 41–45, 50, 55, 60, 65, 70, 71, Quiz 2 Exs. 1–10

AVERAGE
Day 1: p. 741 Exs. 13–27
Day 2: pp. 742–744 Exs. 33–39, 41–45, 50, 55, 60, 65, 70, 71, Quiz 2 Exs. 1–10

ADVANCED
Day 1: p. 741 Exs. 13–27
Day 2: pp. 742–744 Exs. 33–48, 50, 55, 60, 65, 70, 71, Quiz 2 Exs. 1–10

BLOCK SCHEDULE
p. 741 Exs. 13–27 (with 12.4)
pp. 741–744 Exs. 13–27, 33–39, 41–45, 50, 55, 60, 65, 70, 71, Quiz 2 Exs. 1–10 (with 12.6)

EXERCISE LEVELS

Level A: *Easier*
13–19, 25–27, 33
Level B: *More Difficult*
20–24, 28–32, 34–39, 41–45, 49–71
Level C: *Most Difficult*
40, 46–48

✓ HOMEWORK CHECK

To quickly check student understanding of key concepts, go over the following exercises: Exs. 18, 20, 24, 26, 28, 30, 31 38. See also the Daily Homework Quiz:

- Blackline Master (*Chapter 12 Resource Book*, p. 82)
- Transparency (p. 99)

APPLICATION NOTE

EXERCISE 40 Some of the earliest surveyors were hired to replace landmarks that were carried off when the Nile River flooded its banks. The surveyors made measurements with ropes that were marked at certain intervals. For this reason they were called rope stretchers.

EXERCISE 41 Frontage along a street or shoreline is often used to assess property value. All towns assess the value of the properties in the town for tax purposes using their tax maps. These maps include the property boundaries of all the properties in the town.

EXERCISES 42 AND 43 A staircase must be built with attention to the rise to run ratio. If the tread is too short or the riser too high, the stairs will be dangerous.

45b.

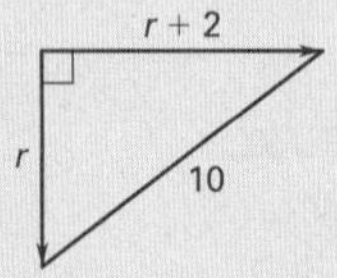

49.

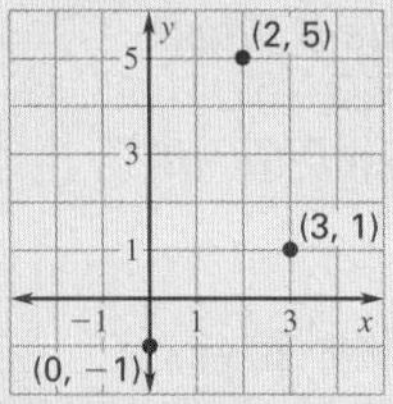

50–54. See Additional Answers beginning on page AA1.

ADDITIONAL PRACTICE AND RETEACHING

For Lesson 12.5:

- Practice Levels A, B, and C (*Chapter 12 Resource Book,* p. 70)
- Reteaching with Practice (*Chapter 12 Resource Book,* p. 73)
- See Lesson 12.5 of the *Personal Student Tutor*

For more Mixed Review:

- Search the *Test and Practice Generator* for key words or specific lessons.

28. No; $2^2 + 10^2 = 104$ and $11^2 = 121$.

29. Yes; $15^2 + 20^2 = 625 = 25^2$.

30. Yes; $5^2 + 12^2 = 169 = 13^2$.

31. Yes; $11^2 + 60^2 = 3721 = 61^2$.

32. No; $7^2 + 24^2 = 625$ and $26^2 = 676$.

33. Yes; $9.9^2 + 2^2 = 102.01 = 10.1^2$.

38. hypothesis: a triangle has sides of 8 in. and 9 in.; conclusion: the third side is greater than 1 in. and less than 17 in.

40. Use the rope to form a triangle with a hypotenuse of length 13 and leg lengths 5 and 12. The rope forms a right triangle since $5^2 + 12^2 = 169 = 13^2$. Then one of the angles of the triangle is a right angle.

DETERMINING RIGHT TRIANGLES Determine whether the given lengths are sides of a right triangle. Explain your reasoning. See margin.

28. 2, 10, 11 **29.** 15, 20, 25 **30.** 5, 12, 13

31. 11, 60, 61 **32.** 7, 24, 26 **33.** 9.9, 2, 10.1

IF-THEN STATEMENTS In Exercises 34–38, state the hypothesis and the conclusion of the statement.

34. If today is Tuesday, then yesterday was Monday. **hypothesis: today is Tuesday; conclusion: yesterday was Monday**

35. If a polygon is a square, then it is a parallelogram. **hypothesis: a polygon is a square; conclusion: it is a parallelogram**

36. If $\frac{x}{3} = -15$, then $x = -45$. **hypothesis: $\frac{x}{3} = -15$; conclusion: $x = -45$**

37. If the area of a square is 25 square feet, then the length of a side is 5 feet. **hypothesis: the area of a square is 25 sq. ft; conclusion: the length of a side is 5 ft**

38. If a triangle has sides that are 8 inches and 9 inches long, then the length of the third side is greater than 1 inch and less than 17 inches. **See margin.**

39. DIAGONAL OF A FIELD A field hockey field is a rectangle 60 yards by 100 yards. What is the length of the diagonal from one corner of the field to the opposite corner? **about 116.6 ft**

40. *Writing* You have a rope with 30 equally spaced knots in it. How can you use the rope to check that a corner is a right angle? **See margin.**

41. SURVEYING LAND You are surveying a triangular-shaped piece of land. You have measured and recorded two lengths on a plot plan. What is the length of the property along the street? Round your answer to the nearest hundredth. **113.58 ft**

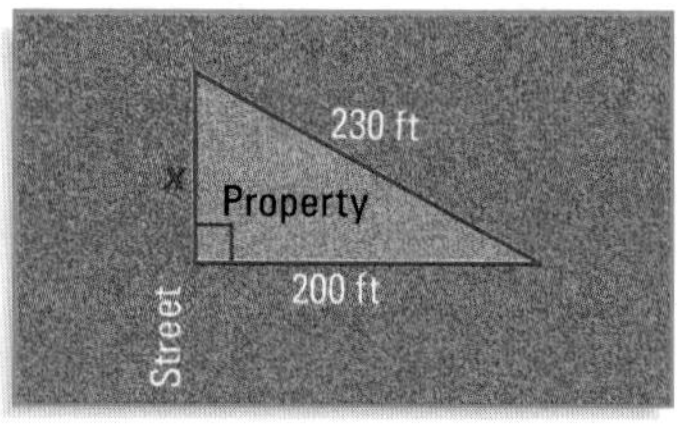

DESIGNING A STAIRCASE In Exercises 42 and 43, you are building the staircase shown at the right.

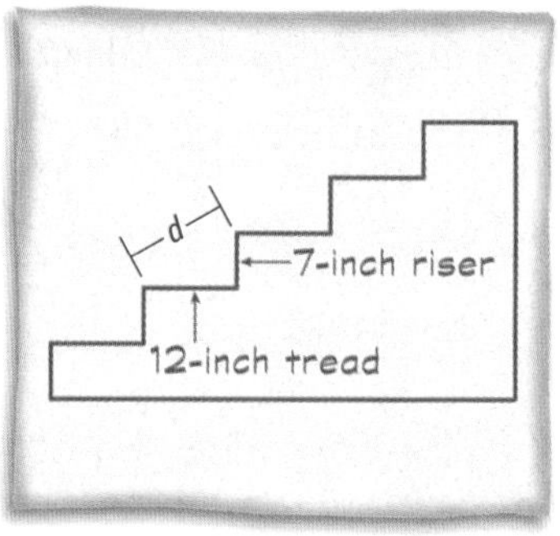

42. Find the distance d between the edges of each step. **about 13.9 in.**

43. The staircase will also have a handrail that is as long as the distance between the edge of the first step and the edge of the top step. How long is the handrail? **about 55.6 in.**

44. PLANTING A NEW TREE You have just planted a new tree. To support the tree in bad weather, you attach guy wires from the trunk of the tree to stakes in the ground. You cut 30 feet of wire into four equal lengths to make the guy wires. You attach four guy wires, evenly spaced around the tree. You put the stakes in the ground five feet from the base of the trunk. Approximately how far up the trunk should you attach the guy wires? **about 5.6 ft**

Test Preparation

45. d. *Sample answer:* I used factoring because the simplified quadratic equation that results is $r^2 + 2r - 48 = 0$, and $r^2 + 2r - 48 = (r + 8)(r - 6)$.

45. MULTI-STEP PROBLEM Amalia and Cindy leave from the same point at the same time. Cindy bicycles east at a rate that is 2 miles per hour faster than Amalia, who bicycles south. After one hour they are 10 miles apart.

a. Let r represent Amalia's rate in miles per hour. Write an expression for the distance each girl has traveled in one hour. **Amalia: r mi, Cindy $r + 2$ mi**

b. Draw and label a diagram of the situation. **See margin.**

c. Use the Pythagorean theorem to find how fast each person is traveling. **Amalia: 6 mi/h Cindy: 8 mi/h**

d. *Writing* Which method did you use to solve the quadratic equation? Give a reason for your choice. **See margin.**

★ Challenge

In Exercises 46–48, use the fact that a Pythagorean triple is a group of three integers, such as 3, 4, and 5, that could be the lengths of the sides of a right triangle.

46. Find two other Pythagorean triples that are not multiples of 3, 4, 5 or of each other. ***Sample answers*: 5, 12, 13; 7, 24, 25; 8, 15, 17; 9, 40, 41; 11, 60, 61**

47. Notice that $3 \cdot 4 \cdot 5 = 60$. Is the product of the three numbers in each Pythagorean triple evenly divisible by 3? by 4? by 5? **Yes; yes; yes**

EXTRA CHALLENGE
www.mcdougallittell.com

48. Do your observations in Exercise 47 suggest a statement about Pythagorean triples that might be true? Explain. **Conjecture: The product of the numbers in any given Pythagorean triple is a multiple of 60. Each of the numbers 3, 4, and 5 is a factor of at least one of the three numbers in a Pythagorean triple.**

MIXED REVIEW

PLOTTING ORDERED PAIRS Plot the ordered pairs in a coordinate plane. (Review 4.1 for 12.6) **49–54. See margin.**

49. (2, 5), (0, −1), (3, 1)
50. (2, −5), (2, 4), (−3, 0)
51. (−1, −2), (−4, 5), (0, 2)
52. (1, 4), (−2, −1), (3, −1)
53. (2, 3), (−2, −3), (4, −2)
54. (1, 3), (−3, 1), (3, −4)

FINDING INTERCEPTS Find the x-intercepts of the graph of the equation. (Review 9.5)

55. $y = x^2 + 2x + 15$ **no x-intercepts**
56. $y = x^2 + 8x + 12$ **$-6, -2$**
57. $y = x^2 + x - 10$
58. $y = x^2 + 8x + 16$ **-4**
59. $y = x^2 + 3x + 1$
60. $y = x^2 - 8x - 11$ **$4 - 3\sqrt{3}, 4 + 3\sqrt{3}$**
61. $y = x^2 + 8x - 10$ **$-4 - \sqrt{26}, -4 + \sqrt{26}$**
62. $y = 3x^2 + 20x + 1$
63. $y = -x^2 + 4x + 1$ **$2 - \sqrt{5}, 2 + \sqrt{5}$**

57. $\frac{-1 - \sqrt{41}}{2}, \frac{-1 + \sqrt{41}}{2}$

59. $\frac{-3 - \sqrt{5}}{2}, \frac{-3 + \sqrt{5}}{2}$

62. $\frac{-10 - \sqrt{97}}{3}, \frac{-10 + \sqrt{97}}{3}$

FACTORING SPECIAL PRODUCTS Factor the expression. (Review 10.7)

64. $x^2 - 64$ **$(x + 8)(x - 8)$**
65. $16x^2 - 25$ **$(4x - 5)(4x + 5)$**
66. $x^2 + 18x + 81$ **$(x + 9)^2$**
67. $7x^2 - 28x + 28$ **$7(x - 2)^2$**
68. $45x^2 - 60x + 20$ **$5(3x - 2)^2$**
69. $-48x^2 + 216x - 243$ **$-3(4x - 9)^2$**

70. GEOMETRIC MEAN The geometric mean of 16 and a is 32. What is the value of a? (Review 12.3) **64**

71. POSTAGE STAMPS In 1960, a first-class United States postage stamp cost \$.04. In 1999, a first-class United States postage stamp cost \$.33. Write a compound inequality that represents the different prices that a postage stamp could have cost between 1960 and 1999. (Review 6.3) **$0.04 \le x \le 0.33$**

4 ASSESS

DAILY HOMEWORK QUIZ

Transparency Available

1. Find the missing lengths of the right triangle if a and b are the lengths of the legs and c is the length of the hypotenuse.
a. $a = 12, b = x, c = 25$ **$b = \sqrt{481}$**
b. $a = x + 2, b = x, c = \sqrt{13}$ **$a = 3.35, b = 1.35$**

2. Are 8, 12, and 17 lengths of a right triangle? Explain your reasoning. **No, $8^2 + 12^2 \neq 17^2$**

3. State the hypothesis and the conclusion of the following statement. "If a right triangle has legs that are 3 inches and 4 inches long, then the hypotenuse is 5 inches long. **hypothesis: a right triangle has legs that are 3 inches and 4 inches long; conclusion: the hypotenuse is 5 inches long**

EXTRA CHALLENGE NOTE
Challenge problems for Lesson 12.5 are available in **blackline** format in the *Chapter 12 Resource Book*, p. 78 and at **www.mcdougallittell.com.**

ADDITIONAL TEST PREPARATION

1. WRITING Each side of the cube below is 8 inches long. Explain how you could apply the Pythagorean theorem twice to find the length of diagonal $\overline{AE}$.

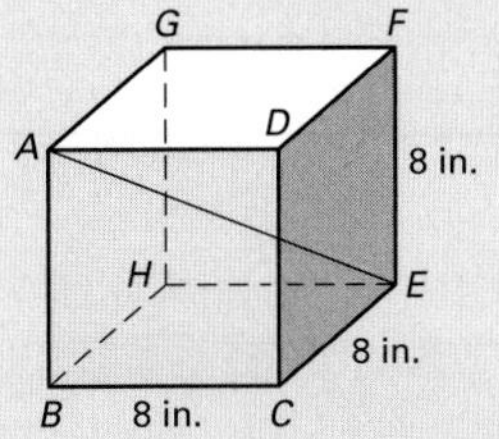

First use the theorem to find BE in $\triangle BCE$. Then use $\triangle ABE$ to find AE.

ADDITIONAL RESOURCES
An alternative Quiz for Lessons 12.4 and 12.5 is available in the *Chapter 12 Resource Book,* p. 79.

A **blackline** master with additional Math & History exercises is available in the *Chapter 12 Resource Book,* p. 77.

QUIZ 2

Self-Test for Lessons 12.4 and 12.5

Solve the equation by completing the square. (Lesson 12.4)

2. $\frac{-1-\sqrt{10}}{2}, \frac{-1+\sqrt{10}}{2}$

1. $2x^2 - 6x - 15 = 5$ $-2, 5$ **2.** $4x^2 + 4x - 9 = 0$ **3.** $x^2 + 2x = 2$ $-1 - \sqrt{3}, -1 + \sqrt{3}$

Tell whether the given lengths are sides of a right triangle. (Lesson 12.5)

4. 6, 7, 8 no **5.** 9, 40, 41 yes **6.** 12, 35, 37 yes

Find the missing length of the right triangle if *a* and *b* are the lengths of the legs and *c* is the length of the hypotenuse. (Lesson 12.5)

7. $a = 5, b = 12$ 13 **8.** $a = 11, c = 61$ 60 **9.** $b = 63, c = 65$ 16

10. **DEPTH OF A SUBMARINE** The sonar of a Navy cruiser detects a submarine that is 2500 feet away. The point on the water directly above the submarine is 1500 feet away from the front of the cruiser. What is the depth of the submarine? (Lesson 12.5) 2000 ft

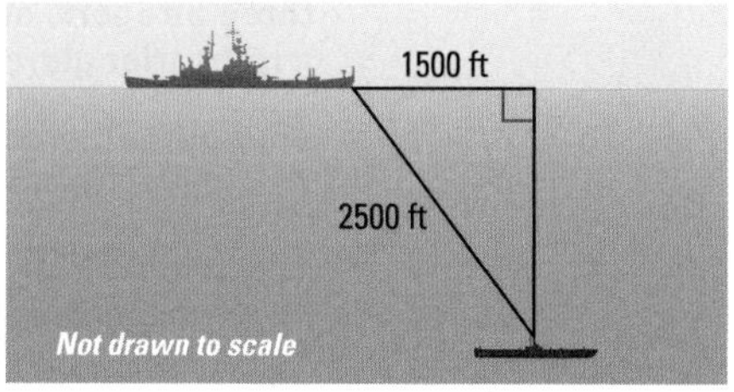
1500 ft
2500 ft
Not drawn to scale

MATH & History: History of Pythagorean Theorem

APPLICATION LINK
www.mcdougallittell.com

THEN

THE ANCIENT BABYLONIAN TABLET below suggests that the Babylonians were aware of the theorem now known as the Pythagorean theorem as early as 1650 B.C.

1. The table gives two of three numbers found on the tablet that could be sides of a right triangle. Find the length of the other leg.

NOW

TODAY architects often use right triangles, whose sides show the relationships stated in the Pythagorean theorem.

	Leg	Leg	Hypotenuse
120	119	?	169
90	56	?	106
240	161	?	289
72	65	?	97
3600	2700	?	4500
2700	1771	?	3229

c.1650 B.C.
Babylonian tablet lists Pythagorean triples.
Pythagoras discovers his theorem.
c. 540 B.C.
Now
The National Aquarium in Baltimore, Maryland, was designed using right triangles.

12.6 The Distance and Midpoint Formulas

What you should learn

GOAL 1 Find the distance between two points in a coordinate plane.

GOAL 2 Find the midpoint between two points in a coordinate plane.

Why you should learn it

▼ To find distances in **real-life** situations, such as the length of a soccer kick in **Example 3**.

REAL LIFE

GOAL 1 FINDING THE DISTANCE BETWEEN TWO POINTS

ACTIVITY Developing Concepts

Investigating Distance

1. Plot $A(2, 1)$ and $B(6, 4)$ on graph paper. Then draw a right triangle that has $\overline{AB}$ as its hypotenuse. **There are two possible triangles.**
2. Label the coordinates of the vertex C.
3. Find the lengths of the legs of $\triangle ABC$. **3 units and 4 units**
4. Use the Pythagorean theorem to find AB. **5 units**
5. Check the distance by actual measurement. **Check results.**

The steps used in the investigation can be used to develop a general formula for the distance between two points $A(x_1, y_1)$ and $B(x_2, y_2)$. Using the Pythagorean theorem, you can write the equation

$$d^2 = (x_2 - x_1)^2 + (y_2 - y_1)^2.$$

Solving this equation for d produces the **distance formula**.

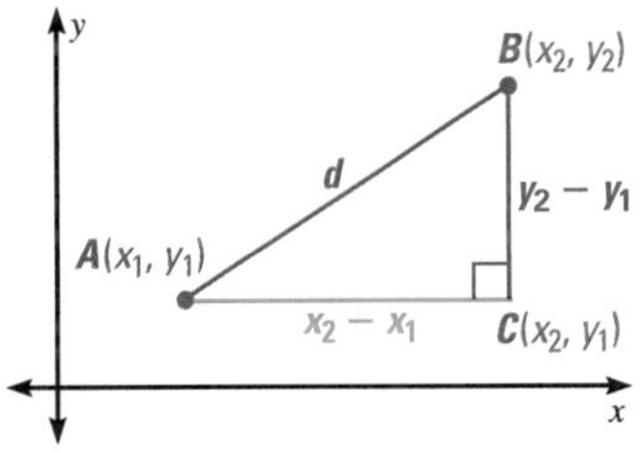

THE DISTANCE FORMULA

The distance d between the points (x_1, y_1) and (x_2, y_2) is

$$d = \sqrt{(x_2 - x_1)^2 + (y_2 - y_1)^2}.$$

EXAMPLE 1 *Finding the Distance Between Two Points*

Find the distance between $(1, 4)$ and $(-2, 3)$.

SOLUTION To find the distance, use the distance formula.

$d = \sqrt{(x_2 - x_1)^2 + (y_2 - y_1)^2}$ **Write distance formula.**

$= \sqrt{(-2 - 1)^2 + (3 - 4)^2}$ **Substitute.**

$= \sqrt{10}$ **Simplify.**

≈ 3.16 **Use a calculator.**

1 PLAN

PACING
Basic: 1 day
Average: 1 day
Advanced: 1 day
Block Schedule: 0.5 block with 12.5

LESSON OPENER APPLICATION
An alternative way to approach Lesson 12.6 is to use the Application Lesson Opener:
- Blackline Master (*Chapter 12 Resource Book*, p. 83)
- Transparency (p. 87)

MEETING INDIVIDUAL NEEDS
- ***Chapter 12 Resource Book***
  Prerequisite Skills Review (p. 5)
  Practice Level A (p. 84)
  Practice Level B (p. 85)
  Practice Level C (p. 86)
  Reteaching with Practice (p. 87)
  Absent Student Catch-Up (p. 89)
  Challenge (p. 91)
- ***Resources in Spanish***
- ***Personal Student Tutor***

NEW-TEACHER SUPPORT
See the Tips for New Teachers on pp. 1–2 of the *Chapter 12 Resource Book* for additional notes about Lesson 12.6.

WARM-UP EXERCISES

Transparency Available

1. To get home from school Jan walks 1.5 miles south, then 0.75 miles west. To get home from school Celia walks 0.5 miles east from school, then 1.85 miles north. Whose home is farther from school? **Celia's**

Step 1–2.

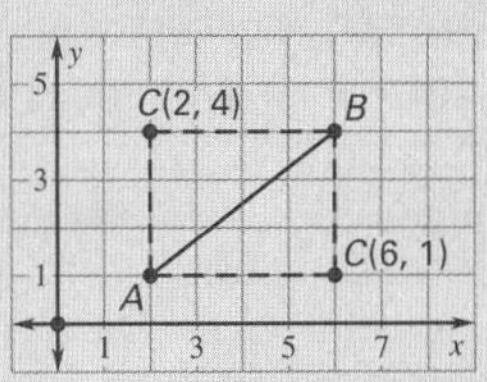

2 TEACH

MOTIVATING THE LESSON
A resort hotel has two golf courses. One is located 2 mi north and 3 mi east of the hotel. The other is located 3 mi south and 2 mi west of the hotel. The straight-line distance between the two courses can be found using the distance formula, the topic of this lesson.

EXTRA EXAMPLE 1
Find the distance between (3, –1) and (4, 0). **$\sqrt{2} \approx 1.41$**

EXTRA EXAMPLE 2
Are (–2, 3), (–1, 1), and (2, 3) the vertices of a right triangle? **No; the lengths of the sides are $\sqrt{5}$, $\sqrt{13}$, and 4 and $5 + 13 \neq 16$.**

EXTRA EXAMPLE 3
A player kicks a soccer ball from a position that is 8 yd from a sideline and 15 yd from a goal line. The ball lands at a position that is 30 yd from the same goal line and 55 yd from the same sideline. To the nearest yard, how far was the ball kicked? **49 yd**

CHECKPOINT EXERCISES

For use after Examples 1 and 2:

1. Which points are the vertices of a right triangle? **a**
   a. (6, 4), (–3, 1), (9, –5)
   b. (0, 0), (4, 3), (1, –1)
   c. (2, –5), (3, –3), (–3, 2)

For use after Example 3:

2. A garage is 5 ft from a driveway and 15 ft from a walkway. A house is 30 ft from the driveway and 4 ft from the walkway. How far is the garage from the house, to the nearest foot? **about 27 ft**

CONCEPT QUESTION
EXAMPLE 1 What does the expression $(x_2 - x_1)$ represent? What does the expression $(y_2 - y_1)$ represent? **the horizontal distance between the points; the vertical distance between the points**

EXAMPLE 2 *Checking a Right Triangle*

Decide whether the points (3, 2), (2, 0), and (–1, 4) are vertices of a right triangle.

SOLUTION

Use the distance formula to find the lengths of the three sides.

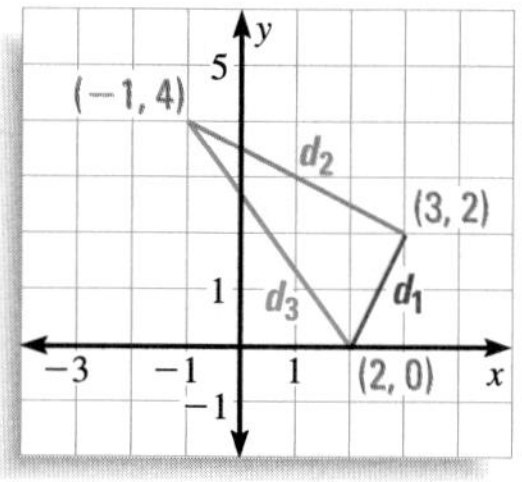

$$d_1 = \sqrt{(3-2)^2 + (2-0)^2} = \sqrt{1+4} = \sqrt{5}$$

$$d_2 = \sqrt{[3-(-1)]^2 + (2-4)^2} = \sqrt{16+4} = \sqrt{20}$$

$$d_3 = \sqrt{[2-(-1)]^2 + (0-4)^2} = \sqrt{9+16} = \sqrt{25}$$

Next find the sum of the squares of the lengths of the two shorter sides.

$d_1^2 + d_2^2 = (\sqrt{5})^2 + (\sqrt{20})^2$ **Substitute for d_1 and d_2.**

$= 5 + 20$ **Simplify.**

$= 25$ **Add.**

The sum of the squares of the lengths of the shorter sides is 25, which is equal to the square of the length of the longest side, $(\sqrt{25})^2$.

▶ The given points are vertices of a right triangle.

When you want to use the distance formula to find a distance in a real-life problem, the first step is to draw a diagram and assign coordinates to the points. This process is called *superimposing* a coordinate system on the diagram.

REAL LIFE
Soccer

EXAMPLE 3 *Applying the Distance Formula*

A player kicks a soccer ball from a position that is 10 yards from a sideline and 5 yards from a goal line. The ball lands at a position that is 45 yards from the same goal line and 40 yards from the same sideline. How far was the ball kicked?

SOLUTION

You can begin by superimposing a coordinate system on the soccer field. The ball is kicked from the point (10, 5). It lands at the point (40, 45). Use the distance formula.

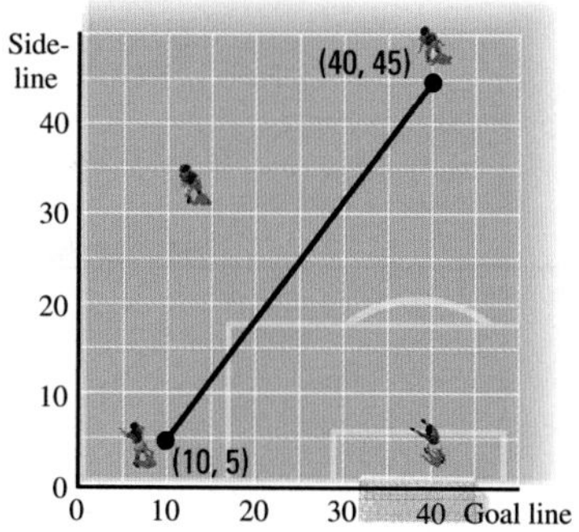

$$d = \sqrt{(40-10)^2 + (45-5)^2}$$

$$= \sqrt{900 + 1600}$$

$$= \sqrt{2500}$$

$$= 50$$

▶ The ball was kicked 50 yards.

GOAL 2 FINDING THE MIDPOINT BETWEEN TWO POINTS

The **midpoint** of a line segment is the point on the segment that is equidistant from its end-points. The *midpoint between two points* is the midpoint of the line segment connecting them.

THE MIDPOINT FORMULA

The midpoint between (x_1, y_1) and (x_2, y_2) is $\left(\frac{x_1 + x_2}{2}, \frac{y_1 + y_2}{2}\right)$.

EXAMPLE 4 *Finding the Midpoint Between Two Points*

Find the midpoint between $(-2, 3)$ and $(4, 2)$. Use a graph to check the result.

SOLUTION

$$\left(\frac{-2+4}{2}, \frac{3+2}{2}\right) = \left(\frac{2}{2}, \frac{5}{2}\right) = \left(1, \frac{5}{2}\right)$$

▶ The midpoint is $\left(1, \frac{5}{2}\right)$.

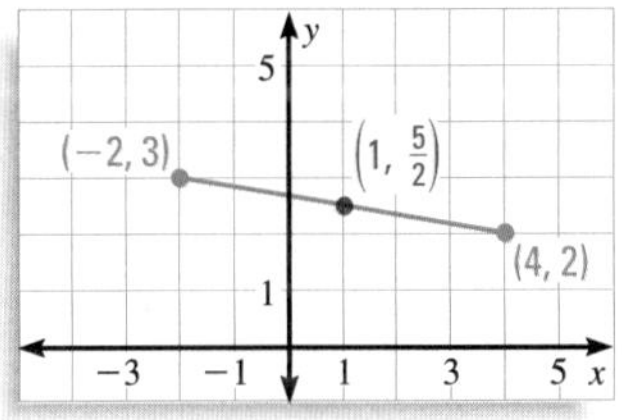

✓**CHECK** From the graph, you can see that the point $\left(1, \frac{5}{2}\right)$ appears halfway between $(-2, 3)$ and $(4, 2)$. You can also use the distance formula to check that the distances from the midpoint to each given point are equal.

REAL LIFE
Computers

EXAMPLE 5 *Applying the Midpoint Formula*

You are using computer software to design a video game. You want to place a buried treasure chest halfway between the center of the base of a palm tree and the corner of a large boulder. Find where you should place the treasure chest.

SOLUTION

Begin by assigning coordinates to the locations of the two landmarks. The center of the base of the palm tree is at $(200, 75)$. The corner of the boulder is at $(25, 175)$. Then use the midpoint formula to find the point that is halfway between the two landmarks.

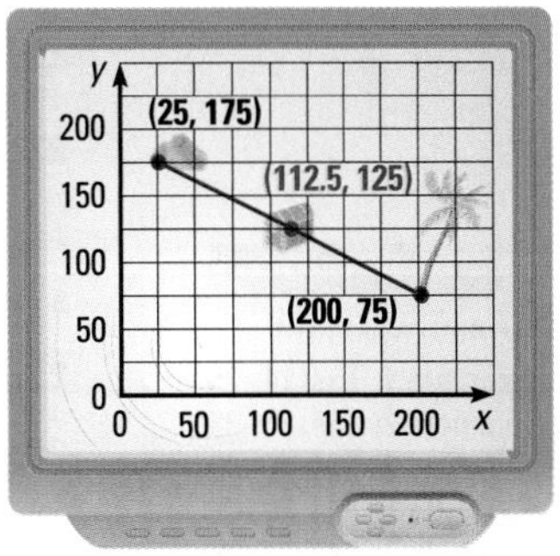

$$\left(\frac{25+200}{2}, \frac{175+75}{2}\right) = \left(\frac{225}{2}, \frac{250}{2}\right)$$

$$= (112.5, 125)$$

▶ You should place the treasure chest at $(112.5, 125)$.

EXTRA EXAMPLE 4

Find the midpoint between $(5, -1)$ and $(-3, 7)$. **(1, 3)**

EXTRA EXAMPLE 5

There are 90 ft between consecutive bases on a baseball diamond. Suppose third base is located at $(10, 0)$ and first base is located at $(100, -90)$. A ball is hit and lands halfway between first base and third base. Where does the ball land? **(55, −45)**

CHECKPOINT EXERCISES

For use after Example 4:

1. Find the midpoint between $(-10, 10)$ and $(15, 18)$. $\left(\frac{5}{2}, 14\right)$

For use after Example 5:

2. A grid shows the position of a mall and your house. The mall is located at $(1, 14)$ and your house is located at $(9, 2)$. Your friend's house is at the midpoint of the line between the mall and your house. Where is your friend's house on the grid? **(5, 8)**

FOCUS ON VOCABULARY

What are the distance and midpoint formulas used for? ***Sample answer:*** **The distance formula gives the distance between two points. The midpoint formula gives the midpoint between two points.**

CLOSURE QUESTION

Would you use the distance formula or the midpoint formula to find how far two traffic lights were from each other? **the distance formula**

DAILY PUZZLER

The map of a treasure hunt is laid out in a square on a coordinate grid. The treasure is at the center of the square. The coordinates of one pair of the square's adjacent vertices are $(4, 1)$ and $(4, 13)$. What are the possible coordinates of the treasure? **(−2, 7) and (10, 7)**

3 APPLY

ASSIGNMENT GUIDE

BASIC
Day 1: pp. 748–750 Exs. 14–48 even, 49–53, 59–61, 66–72 even

AVERAGE
Day 1: pp. 748–750 Exs. 14–48 even, 49–53, 56–61, 66–72 even

ADVANCED
Day 1: pp. 748–750 Exs. 14–48 even, 49–53, 56–64, 66–72 even

BLOCK SCHEDULE WITH 12.5
pp. 748–750 Exs. 14–48 even, 49–53, 56–61, 66–72 even

EXERCISE LEVELS

Level A: *Easier*
14–22, 34–37

Level B: *More Difficult*
23–33, 38–61

Level C: *Most Difficult*
62–64

✔ HOMEWORK CHECK

To quickly check student understanding of key concepts, go over the following exercises: Exs. 14, 22, 28, 32, 48. See also the Daily Homework Quiz:

- Blackline Master (*Chapter 12 Resource Book,* p. 94)
- Transparency (p. 100)

GUIDED PRACTICE

Vocabulary Check ✓

1. What is meant by the *midpoint* between two points?
   the point that lies on the line connecting two points and is halfway between them

Concept Check ✓

2. Explain how you can use the Pythagorean theorem to find the distance between any two points in a coordinate plane. See margin.

2. Draw a right triangle that has a horizontal leg and a vertical leg and whose hypotenuse has the two points as endpoints. Then use the lengths of the legs and the Pythagorean theorem to determine the length of the hypotenuse.

Skill Check ✓

Use the coordinate plane to estimate the distance between the two points. Then use the distance formula to find the distance between the points. Round the result to the nearest hundredth. 3–5. Estimates may vary.

3. $(1, 5), (-3, 1)$ 5.66

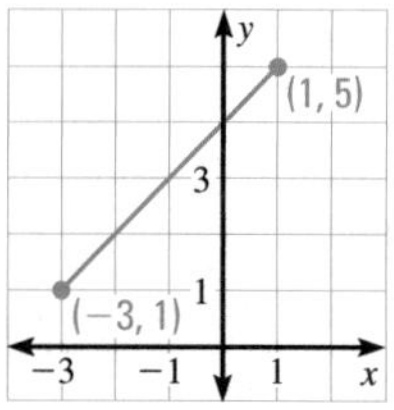

4. $(-3, -2), (4, 1)$ 7.62

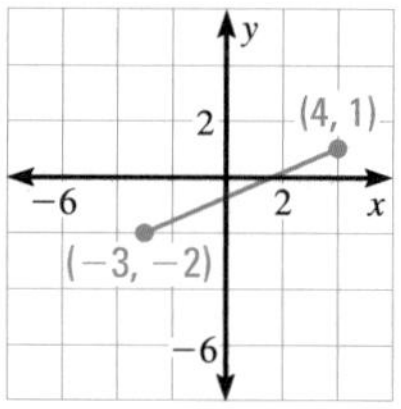

5. $(5, -2), (-1, 1)$ 6.71

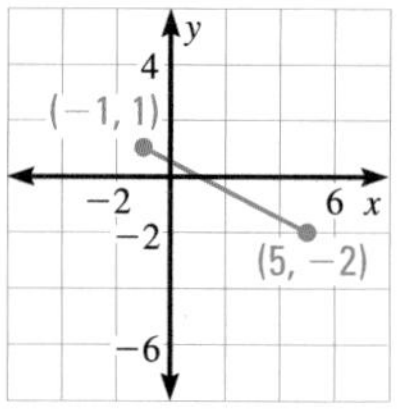

Decide whether the points are vertices of a right triangle.

6. $(0, 0), (20, 0), (20, 21)$ yes

7. $(4, 0), (4, -4), (10, -4)$ yes

8. $(-2, 0), (-1, 0), (1, 7)$ no

9. $(2, 0), (-2, 2), (-3, -5)$ no

Find the midpoint between the two points.

10. $(4, 4), (-1, 2)$ $\left(\frac{3}{2}, 3\right)$

11. $(6, 2), (2, -3)$ $\left(4, -\frac{1}{2}\right)$

12. $(-5, 3), (-3, -3)$ $(-4, 0)$

13. SOCCER Suppose the soccer ball in Example 3 lands in a position that is 25 yards from the same goal line and 25 yards from the same sideline. How far was the ball kicked? 25 yd

PRACTICE AND APPLICATIONS

STUDENT HELP
Extra Practice to help you master skills is on p. 808

FINDING DISTANCE **Find the distance between the two points. Round the result to the nearest hundredth if necessary.**

14. $(2, 0), (8, -3)$ 6.71

15. $(2, -8), (-3, 3)$ 12.08

16. $(3, -1), (0, 3)$ 5

17. $(5, 8), (-2, 3)$ 8.60

18. $(-3, 1), (2, 6)$ 7.07

19. $(-6, -2), (-3, -5)$ 4.24

20. $(4, 5), (-1, 3)$ 5.39

21. $(-6, 1), (3, 1)$ 9

22. $(-2, -1), (3, -3)$ 5.39

23. $(3.5, 6), (-3.5, -2)$ 10.63

24. $\left(\frac{1}{2}, \frac{1}{4}\right), (2, 1)$ 1.68

25. $\left(\frac{1}{3}, \frac{1}{6}\right), \left(-\frac{2}{3}, \frac{8}{3}\right)$ 2.69

STUDENT HELP
HOMEWORK HELP
Example 1: Exs. 14–25
Example 2: Exs. 26–33
Example 3: Exs. 46–53
Example 4: Exs. 34–45
Example 5: Exs. 54, 55

RIGHT TRIANGLES **Graph the points. Decide whether they are vertices of a right triangle.**

26. $(4, 0), (2, 1), (-1, -5)$ yes

27. $(5, 4), (2, 1), (-3, 2)$ no

28. $(1, -5), (2, 3), (-3, 4)$ no

29. $(-1, 1), (-3, 3), (-7, -1)$ yes

30. $(-3, 2), (-3, 5), (0, 2)$ yes

31. $(3, -1), (2, 4), (-3, 0)$ no

32. $(-2, 2), (3, 4), (4, 2)$ no

33. $(0, -4), (4, -1), (4, -4)$ yes

40. $\left(-\frac{5}{2}, -\frac{3}{2}\right)$

41. $\left(-\frac{5}{2}, -\frac{5}{2}\right)$

43. $\left(-2, -\frac{1}{2}\right)$

45. $\left(-\frac{5}{2}, -4\right)$

FOCUS ON CAREERS

CARTOGRAPHER Cartographers prepare maps using information from surveys, aerial photographs, and satellite data.

CAREER LINK www.mcdougallittell.com

51. $(-1, 0)$, $\left(1\frac{1}{2}, 0\right)$, $\left(-\frac{1}{2}, -2\right)$

52. $2\sqrt{2}$, $\frac{\sqrt{17}}{2}$, $2\frac{1}{2}$

53. The perimeter of the original triangle, 14.78, is twice that of the perimeter of the new triangle, 7.39.

55. You could meet at a point 1 mi east and 1 mi north of the starting point. Both you and your friend would each have to walk $\sqrt{13} \approx 3.6$ mi.

FINDING THE MIDPOINT Find the midpoint between the two points.

34. (3, 0), (−5, 4) **(−1, 2)** **35.** (0, 0), (0, 8) **(0, 4)** **36.** (1, 2), (5, 4) **(3, 3)**

37. (−1, 2), (7, 4) **(3, 3)** **38.** (−3, 3), (2, −2) $\left(-\frac{1}{2}, \frac{1}{2}\right)$ **39.** (2, 7), (4, 3) **(3, 5)**

40. (−1, 1), (−4, −4) **41.** (−4, 0), (−1, −5) **42.** (5, 1), (1, −5) **(3, −2)**

43. (0, −3), (−4, 2) **44.** (5, −5), (−5, 1) **(0, −2)** **45.** (−4, −3), (−1, −5)

DISTANCES ON MAPS In Exercises 46–48, use the map. Each side of a square in the coordinate plane that is superimposed on the map represents 95 miles. The points represent city locations.
46-48. Estimates may vary.

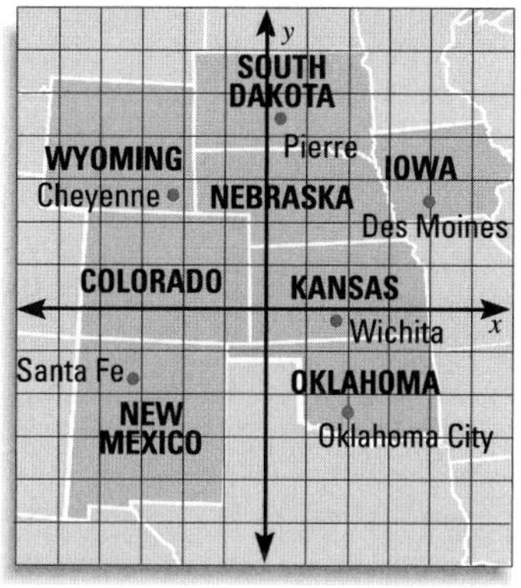

46. Use the distance formula to estimate the distance between Pierre, South Dakota, and Santa Fe, New Mexico. **about 660 mi**

47. Use the distance formula to estimate the distance between Wichita, Kansas, and Cheyenne, Wyoming. **about 460 mi**

48. Use the distance formula to estimate the distance between Oklahoma City, Oklahoma, and Des Moines, Iowa. **about 500 mi**

GEOMETRY CONNECTION In Exercises 49–53, use the diagram below.

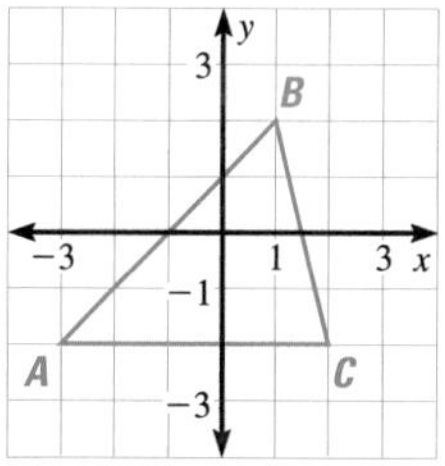

49. Copy the diagram of triangle *ABC* on graph paper. **Check drawings.**

50. Find the length of each side of the triangle. $4\sqrt{2}, \sqrt{17}, 5$

51. Find the midpoint of each side of the triangle.

52. Join the midpoints to form a new triangle. Find the length of each of its sides. **See margin.**

53. Compare the perimeters of the two triangles.

HIKING TRIP In Exercises 54 and 55, use the following information. You and a friend go hiking. You hike 3 miles north and 2 miles west. Starting from the same point, your friend hikes 4 miles east and 1 mile south.

54. How far apart are you and your friend? (*Hint*: Draw a diagram on a grid.) $2\sqrt{13} \approx 7.2$ **mi**

55. If you and your friend want to meet for lunch, where could you meet so that both of you hike the same distance? How far do you have to hike?

GEOMETRY CONNECTION In Exercises 56–58, use the following information. A trapezoid is *isosceles* if its two opposite nonparallel sides have the same length.

56. Draw the polygon whose vertices are $A(1, 1)$, $B(5, 9)$, $C(2, 8)$, and $D(0, 4)$. **See margin.**

57. Show that the polygon is a trapezoid by showing that only two of the sides are parallel. **See margin.**

58. Use the distance formula to show that the trapezoid is isosceles. **Sides** AD **and** BC **both have length** $\sqrt{10}$.

! COMMON ERROR

EXERCISES 10–12 Students may confuse the midpoint formula with the distance formula and divide the differences of the *x*-coordinates and the *y*-coordinates by two. Remind students that they are finding an average *x*-coordinate and an average *y*-coordinate, which requires adding the coordinates.

52.

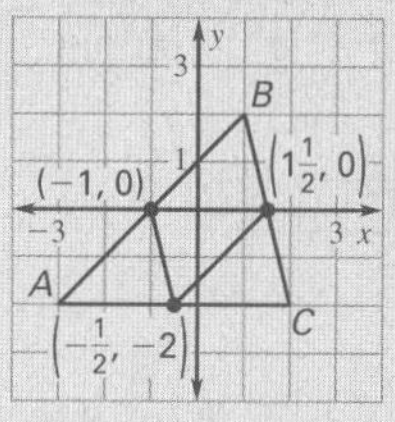

56.

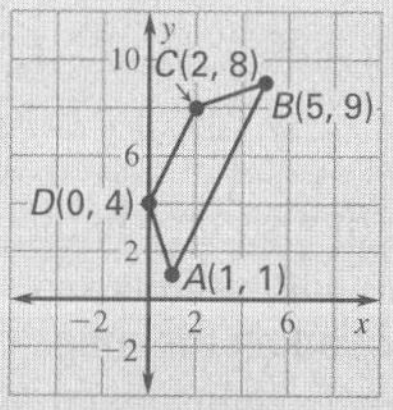

57. Side $\overline{DC}$ is parallel to side $\overline{AB}$ because both have slope 2. Sides $\overline{CB}$ and $\overline{DA}$ are not parallel because $\overline{DA}$ has slope −3 and $\overline{CB}$ has slope $\frac{1}{3}$.

ADDITIONAL PRACTICE AND RETEACHING

For Lesson 12.6:

- Practice Levels A, B, and C (*Chapter 12 Resource Book*, p. 84)
- Reteaching with Practice (*Chapter 12 Resource Book*, p. 87)
- See Lesson 12.6 of the *Personal Student Tutor*

For more Mixed Review:

- Search the *Test and Practice Generator* for key words or specific lessons.

4 ASSESS

DAILY HOMEWORK QUIZ

Transparency Available

1. Find the distance to the nearest tenth between the points (4, 8) and (–3, 5). **7.6**
2. Graph (1, 3), (–2, 3) and (1, 1). Decide whether they are vertices of a right triangle. **Yes,**

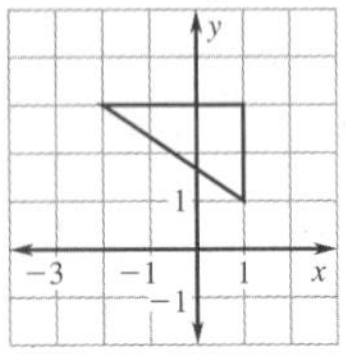

3. Find the midpoint of (–9, –1) and (4, 11). $\left(-\frac{5}{2}, 5\right)$
4. Each side of a square in the coordinate plane that is superimposed on the map represents 0.5 miles. Estimate the distance between school and home. **2.5 miles**

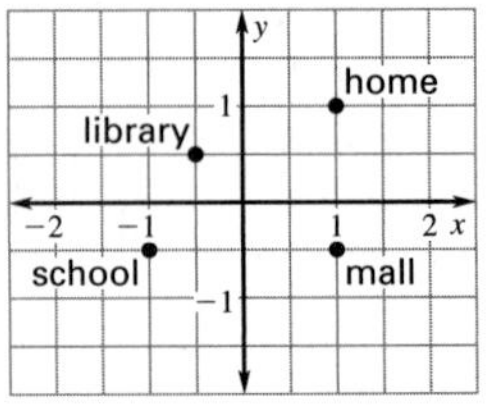

EXTRA CHALLENGE NOTE

Challenge problems for Lesson 12.6 are available in **blackline** format in the *Chapter 12 Resource Book,* p. 91 and at **www.mcdougallittell.com.**

ADDITIONAL TEST PREPARATION

1. **OPEN ENDED** Describe a real-life situation in which the midpoint formula would be used. ***Sample answer:* If the designers of a mall wanted to locate an information booth halfway between two entrances to the mall, they could superimpose a diagram of the mall on a grid and then use the midpoint formula to locate the halfway point.**

Test Preparation

59. MULTIPLE CHOICE What is the distance between (–6, –2) and (2, 4)? **C**

Ⓐ $2\sqrt{5}$ Ⓑ $2\sqrt{7}$ Ⓒ 10 Ⓓ 28

60. MULTIPLE CHOICE What is the midpoint between (–2, –3) and $\left(1, \frac{1}{2}\right)$? **D**

Ⓐ $\left(-1, -2\frac{1}{2}\right)$ Ⓑ $\left(-\frac{1}{2}, -2\frac{1}{2}\right)$ Ⓒ $\left(-1, -1\frac{1}{4}\right)$ Ⓓ $\left(-\frac{1}{2}, -1\frac{1}{4}\right)$

61. MULTIPLE CHOICE The vertices of a right triangle are (0, 0), (0, 6), and (6, 0). What is the length of the hypotenuse? **B**

Ⓐ 6 Ⓑ $6\sqrt{2}$ Ⓒ 36 Ⓓ 72

★ Challenge

TRIP PLANNING **In Exercises 62–64, use the following information. You are planning a family vacation. Each side of a square in the coordinate plane that is superimposed on the map represents 50 miles.**

62. How far is it from your home to the amusement park? **about 269 mi**
63. You leave your home and go to the amusement park. After visiting the amusement park, you go to the beach. You return home. How far did you travel? **about 1309 mi**
64. During your vacation, you want to visit all of the sites on the map. There are two orders in which to visit the sites so that you travel the shortest distance. What are the two orders? **Visit the sites in one of these orders: (1) amusement park, beach, campground, zoo or (2) zoo, campground, beach, amusement park.**

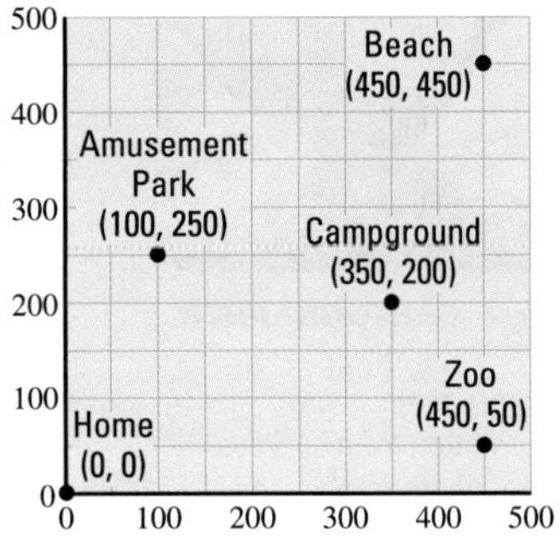

EXTRA CHALLENGE
www.mcdougallittell.com

MIXED REVIEW

FACTORING **Factor the expression completely.** (Review 10.8)

65. $3x^3 + 12x^2 - 15x$ **$3x(x + 5)(x - 1)$**

66. $x^4 - 3x^2 - 25x^2 + 75$ **$(x + 5)(x - 5)(x^2 - 3)$**

CLASSIFYING EQUATIONS **Does the equation model *direct variation, inverse variation,* or *neither*?** (Review 11.3)

67. $x = \frac{7}{y}$ **inverse variation**

68. $y = 8x$ **direct variation**

69. $y = 9x + 1$ **neither**

DIVIDING POLYNOMIALS **Divide.** (Review 11.7)

70. $(6z + 10) \div 2$ **$3z + 5$**

71. $(7x^3 - 2x^2) \div 14x$ **$\frac{7x^2 - 2x}{14}$**

GEOMETRY CONNECTION **In Exercises 72 and 73, the two triangles are similar. Write an equation and solve it to find the length of the side marked *x*.** (Review 3.2 for 12.7)

72.

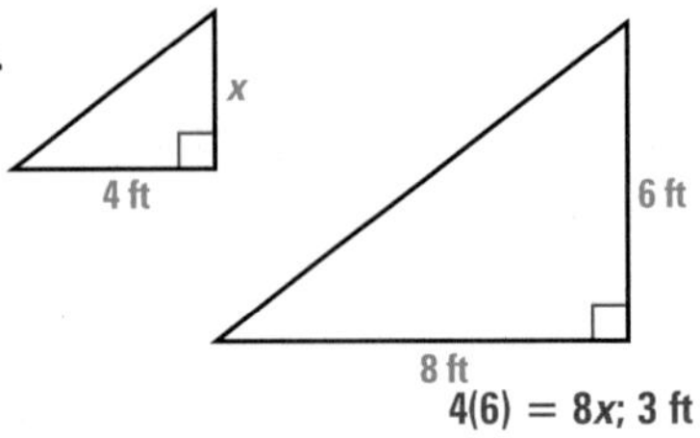

4(6) = 8x; 3 ft

73.

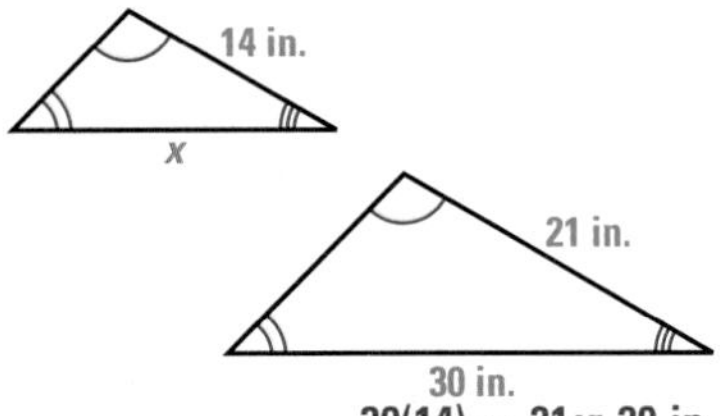

30(14) = 21x; 20 in.

ACTIVITY 12.7

Developing Concepts

Investigating Similar Triangles

GROUP ACTIVITY
Work in a small group.

MATERIALS
- graph paper
- ruler
- protractor

QUESTION What relationships are there between the sides of similar right triangles?

EXPLORING THE CONCEPT

1. Copy the three triangles onto a sheet of paper. They are similar because their corresponding angles are the same.

   Check drawings.

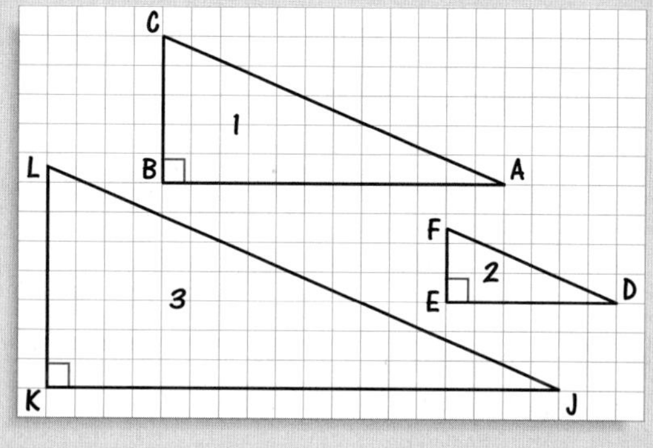

2. Label the lengths of the sides of each triangle to the nearest half unit. Use a strip of graph paper to measure the length of the hypotenuse.

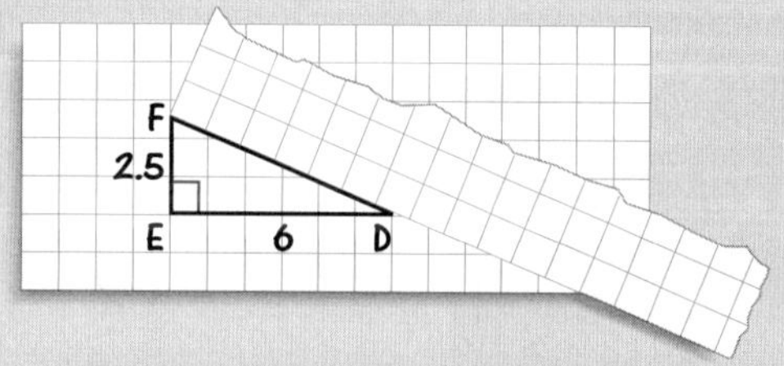

AB = 12; *BC* = 5; *AC* = 13; *DE* = 6, *EF* = 2.5, *FD* = 6.5; *KL* = 7.5, *JL* = 18, *KJ* = 19.5

3. Copy and complete the table. Find the ratios to the nearest hundredth.

	Shorter leg	Longer leg	Hypotenuse	$\frac{\text{Shorter leg}}{\text{Hypotenuse}}$	$\frac{\text{Longer leg}}{\text{Hypotenuse}}$	$\frac{\text{Shorter leg}}{\text{Longer leg}}$
△1	?	?	?	?	?	?
△2	?	?	?	?	?	?
△3	?	?	?	?	?	?

△1: 5, 12, 13, 0.38, 0.92, 0.42; △2: 2.5, 6, 6.5, 0.38, 0.92, 0.42; △3: 7.5, 18, 19.5, 0.38, 0.92, 0.42

DRAWING CONCLUSIONS

1. What do you notice about the ratios you found for all three similar triangles?
   Each ratio is the same for all three triangles.
2. Each person in the group should draw a different triangle that is similar to the triangles at the top of the page. Exchange triangles and use a protractor to check that the triangles are similar.
   Check results.
3. For your triangle, predict what the following ratios will be. Then measure the sides and find each ratio to test your prediction.

   a. $\frac{\text{shorter leg}}{\text{hypotenuse}}$ 0.38 **b.** $\frac{\text{longer leg}}{\text{hypotenuse}}$ 0.92 **c.** $\frac{\text{shorter leg}}{\text{longer leg}}$ 0.42

1 Planning the Activity

PURPOSE
To investigate the ratios of corresponding sides of similar triangles by measuring.

MATERIALS
- graph paper
- ruler
- protractor
- Activity Support Master (*Chapter 12 Resource Book,* p. 95)

PACING
- Exploring the Concept — 10 min
- Drawing Conclusions — 15 min

LINK TO LESSON
As students study Lesson 12.7, they should recognize that all similar right triangles will have the same sine, cosine, and tangent ratios.

2 Managing the Activity

ALTERNATIVE APPROACH
Find the class averages of the table entries and record them in a table drawn on an overhead. Record the ratios in this table. Lead a class discussion of Exercise 1.

CLASSROOM MANAGEMENT
You may want to assign responsibilities to each person in the group to ensure participation by all students.

3 Closing the Activity

KEY DISCOVERY
The ratio between any pair of corresponding sides in a right triangle is constant for a given angle measure.

ACTIVITY ASSESSMENT
What is the ratio $\frac{\text{shorter leg}}{\text{hypotenuse}}$ in a right triangle with one angle measure equal to 23°? about 0.3907

1 PLAN

PACING
Basic: 1 day
Average: 1 day
Advanced: 1 day
Block Schedule: 0.5 block with 12.8

LESSON OPENER
CALCULATOR ACTIVITY
An alternative way to approach Lesson 12.7 is to use the Calculator Activity Lesson Opener:
- Blackline Master (*Chapter 12 Resource Book,* p. 96)
- Transparency (p. 88)

MEETING INDIVIDUAL NEEDS
- ***Chapter 12 Resource Book***
  Prerequisite Skills Review (p. 5)
  Practice Level A (p. 99)
  Practice Level B (p. 100)
  Practice Level C (p. 101)
  Reteaching with Practice (p. 102)
  Absent Student Catch-Up (p. 104)
  Challenge (p. 107)
- ***Resources in Spanish***
- ***Personal Student Tutor***

NEW-TEACHER SUPPORT
See the Tips for New Teachers on pp. 1–2 of the *Chapter 12 Resource Book* for additional notes about Lesson 12.7.

WARM-UP EXERCISES
Transparency Available

1. Which equation is equivalent to $0.2458 = \frac{20}{s}$? b

a. $s = \frac{0.2458}{20}$

b. $s = \frac{20}{0.2458}$

c. $0.2458 = \frac{s}{20}$

Solve to the nearest hundredth.

2. $0.3558 = \frac{12}{q}$ 33.73

3. $0.7288 = \frac{b}{252}$ 183.66

EXPLORING DATA AND STATISTICS

12.7 Trigonometric Ratios

What you should learn

GOAL 1 Use the sine, cosine, and tangent of an angle.

GOAL 2 Use trigonometric ratios in **real-life** problems, such as finding cloud height in **Ex. 20.**

Why you should learn it

▼ To solve **real-life** problems, such as finding the height of a parasailer in **Example 3.**

REAL LIFE

GOAL 1 USING TRIGONOMETRIC RATIOS

In the ancient Greek language, the word *trigonometry* means *measurement of triangles.* A **trigonometric ratio** is a ratio of the lengths of two sides of a right triangle. The three basic trigonometric ratios are **sine**, **cosine**, and **tangent**. You can abbreviate them as *sin*, *cos*, and *tan*.

TRIGONOMETRIC RATIOS

$$\sin A = \frac{\text{side opposite } \angle A}{\text{hypotenuse}} = \frac{a}{c}$$

$$\cos A = \frac{\text{side adjacent to } \angle A}{\text{hypotenuse}} = \frac{b}{c}$$

$$\tan A = \frac{\text{side opposite } \angle A}{\text{side adjacent to } \angle A} = \frac{a}{b}$$

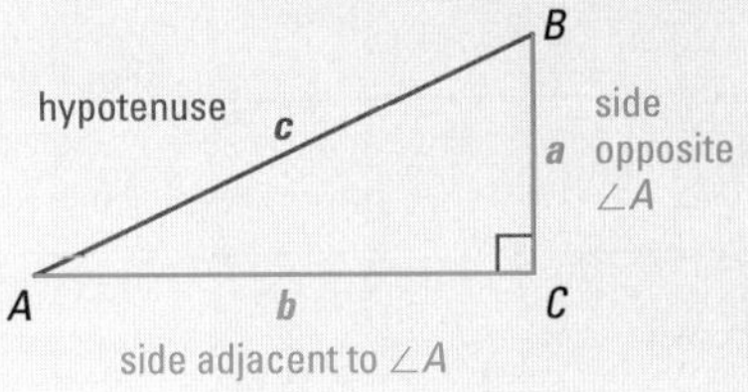

Because all right triangles with a given measure for $\angle A$ are similar, the value of a trigonometric ratio depends only on the measure of $\angle A$. It does not depend on the triangle's size.

EXAMPLE 1 *Finding Trigonometric Ratios*

For $\triangle DEF$, find the sine, the cosine, and the tangent of the angle.

a. $\angle D$

b. $\angle E$

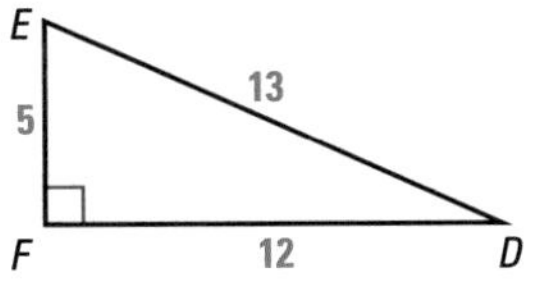

SOLUTION

a. For $\angle D$, the opposite side is 5, and the adjacent side is 12. The hypotenuse is 13.

$$\sin D = \frac{\text{opposite}}{\text{hypotenuse}} = \frac{5}{13}$$

$$\cos D = \frac{\text{adjacent}}{\text{hypotenuse}} = \frac{12}{13}$$

$$\tan D = \frac{\text{opposite}}{\text{adjacent}} = \frac{5}{12}$$

b. For $\angle E$, the opposite side is 12, and the adjacent side is 5. The hypotenuse is 13.

$$\sin E = \frac{\text{opposite}}{\text{hypotenuse}} = \frac{12}{13}$$

$$\cos E = \frac{\text{adjacent}}{\text{hypotenuse}} = \frac{5}{13}$$

$$\tan E = \frac{\text{opposite}}{\text{adjacent}} = \frac{12}{5}$$

STUDENT HELP
Trig Table
For a table of trigonometric ratios, see p. 812.

SOLVING RIGHT TRIANGLES If you know the measure of one angle and the length of one side of a right triangle, then you can use trigonometric ratios and a calculator or a table to find the lengths of the other two sides. This is called *solving a right triangle.*

EXAMPLE 2 *Solving a Right Triangle*

For $\triangle PQR$, $p = 5$ and the measure of $\angle P$ is 30°.

a. Find the length q.

b. Find the length r.

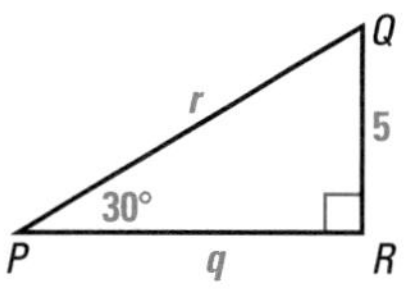

SOLUTION

a. You are given the side opposite $\angle P$, and you need to find the length of the adjacent side.

$\tan P = \frac{\text{opposite}}{\text{adjacent}}$	Definition of tangent
$\tan 30° = \frac{5}{q}$	Substitute 5 for p and 30° for $\angle P$.
$q = \frac{5}{\tan 30°}$	Solve for q.
$q \approx \frac{5}{0.5774}$	Use a calculator or a table.
$q \approx 8.66$	Simplify.

▶ The length q is about 8.66 units.

b. You are given the side opposite $\angle P$ and you need to find the length of the hypotenuse.

$\sin P = \frac{\text{opposite}}{\text{hypotenuse}}$	Definition of sine
$\sin 30° = \frac{5}{r}$	Substitute 5 for p and 30° for $\angle P$.
$r = \frac{5}{\sin 30°}$	Solve for r.
$r = \frac{5}{0.5}$	Use a calculator or a table.
$r = 10$	Simplify.

▶ The length r is 10 units.

✓**CHECK** You can use the Pythagorean theorem to check that the results are reasonable. Because the value of q was rounded, the check will not be exact.

$p^2 + q^2 = r^2$	Pythagorean theorem
$5^2 + 8.66^2 \stackrel{?}{=} 10^2$	Substitute for p, q, and r.
$99.996 \approx 100$	Side lengths are approximately correct.

STUDENT HELP

Study Tip
When you use a calculator to find trigonometric ratios, be sure that the calculator is set in *degree* mode.

2 TEACH

EXTRA EXAMPLE 1
For $\triangle ABC$, find the sine, the cosine, and the tangent of the angle.

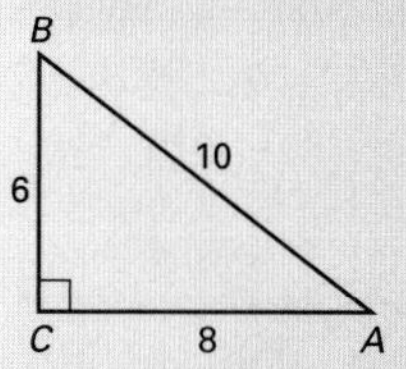

a. $\angle A$ $\frac{3}{5}; \frac{4}{5}; \frac{3}{4}$

b. $\angle B$ $\frac{4}{5}; \frac{3}{5}; \frac{4}{3}$

EXTRA EXAMPLE 2
For $\triangle MNQ$, $n = 10$ and the measure of $\angle N$ is 55°.

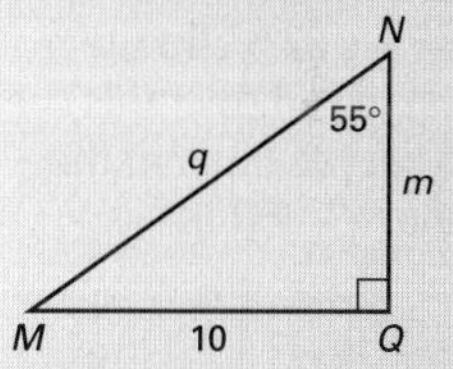

a. Find the length m. **about 7.00**
b. Find the length q. **about 12.21**

CHECKPOINT EXERCISES

For use after Example 1 :

1. Find the sine, cosine, and tangent of $\angle J$ in $\triangle JKL$.

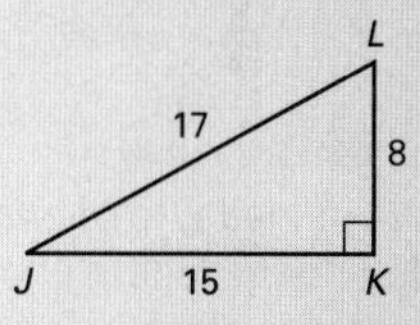

$\sin \angle J = \frac{8}{17}$; $\cos \angle J = \frac{15}{17}$; $\tan \angle J = \frac{8}{15}$

For use after Example 2 :

2. Find the length of b in $\triangle ABC$. **4**

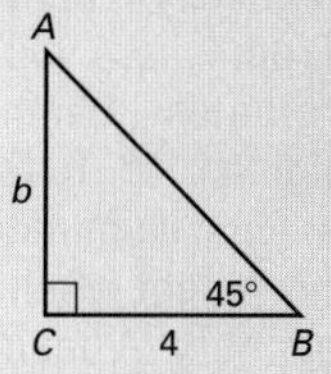

GOAL 2 USING TRIGONOMETRIC RATIOS IN REAL LIFE

REAL LIFE
Parasailing

EXAMPLE 3 *Solving a Right Triangle*

A boat is pulling a parasailer. The line to the parasailer is 800 feet long. The angle between the line and the water is about 25°.

a. How high is the parasailer?

b. How far back is the parasailer from the boat?

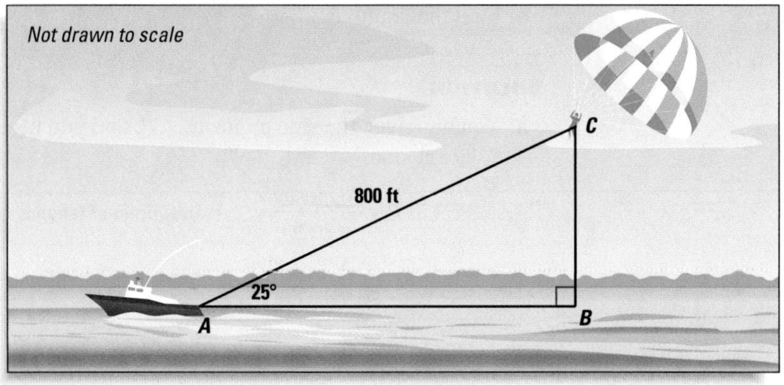

SOLUTION

a. Find the length of the side opposite $\angle A$.

$$\sin A = \frac{\text{opposite}}{\text{hypotenuse}}$$ **Definition of sine**

$$\sin 25° = \frac{a}{800}$$ **Substitute 800 for *c* and 25° for ∠*A*.**

$$800 \cdot \sin 25° \approx a$$ **Multiply each side by 800.**

$$0.4226(800) \approx a$$ **Use a calculator or a table.**

$$338.08 \approx a$$ **Simplify.**

▶ The parasailer is about 338 feet above the water.

b. Find the length of the side adjacent to $\angle A$.

$$\cos A = \frac{\text{adjacent}}{\text{hypotenuse}}$$ **Definition of cosine**

$$\cos 25° = \frac{b}{800}$$ **Substitute 800 for *c* and 25° for ∠*A*.**

$$800 \cdot \cos 25° = b$$ **Multiply each side by 800.**

$$0.9063(800) \approx b$$ **Use a calculator or a table.**

$$725.04 \approx b$$ **Simplify.**

▶ The parasailer is about 725 feet back from the boat.

EXTRA EXAMPLE 3

A nature preserve is building a bridge from point C to point A. The surveying crew lays off a distance of 59 m at a right angle to $\overline{CA}$. A 50° angle is then sited at B. What will be the length of the bridge? **about 70.3 m**

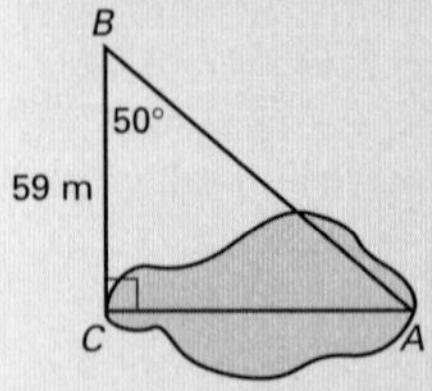

CHECKPOINT EXERCISES

For use after Example 3:

1. The captain of a boat, located at R, sights the top of a 160-ft lighthouse at S. The measure of $\angle SRT$ is 32.5°. To the nearest mile, how far is the boat from the base of the lighthouse? **251 ft**

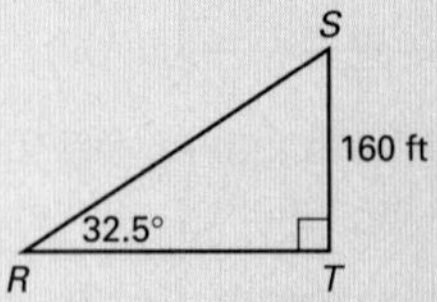

FOCUS ON VOCABULARY

Define the three basic trigonometric ratios. **See box on page 752.**

CLOSURE QUESTION

If you know the measure of one acute angle and the length of one side of a right triangle, how can you *solve the right triangle?* **Use trigonometric ratios and a calculator or a table to find the lengths of the other two sides.**

DAILY PUZZLER

You are playing flashlight tag with some friends. From where you are, your flashlight creates a beam of light with an angle of 15°. To tag your friend, you must illuminate at least 2 feet of his height when the flashlight is stationary. What is the closest distance you can be to your friend to make a tag? **7.6 ft**

GUIDED PRACTICE

Vocabulary Check ✓ **1.** Define *trigonometric ratio.*
a ratio of the lengths of two sides of a right triangle

Concept Check ✓ **2.** Is it *true* or *false* that for any right triangle with a 30° angle, $\sin 30° = 0.5$? Explain. **True; the value of a trigonometric ratio depends only on the measure of the angle, not on the size of the triangle.**

Skill Check ✓ **In Exercises 3–5, use $\triangle ABC$ at the right.**

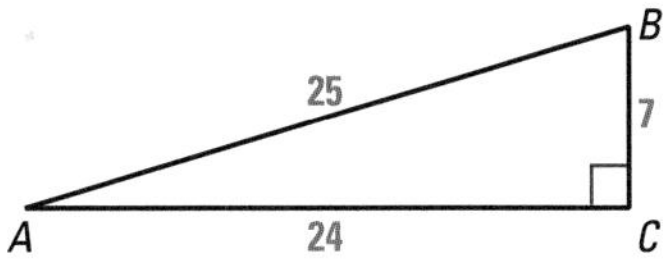

3. Find the sine of $\angle A$. $\frac{7}{25}$

4. Find the cosine of $\angle A$. $\frac{24}{25}$

5. Find the tangent of $\angle A$. $\frac{7}{24}$

Find the missing lengths of the sides of the triangle. Round to the nearest hundredth. Use the Pythagorean theorem to check the result.

6.
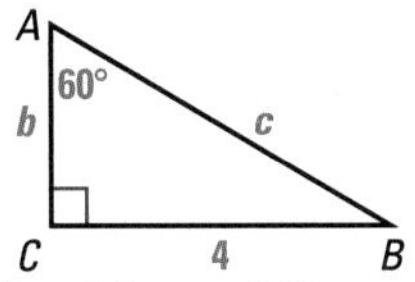

$b \approx 2.3$, $c \approx 4.62$

7.
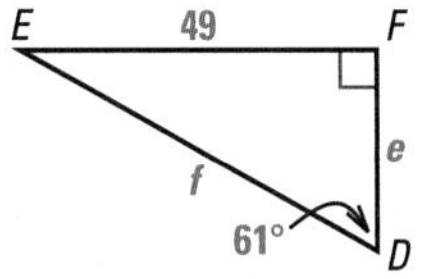

$e \approx 27.16$, $f \approx 56.02$

8.
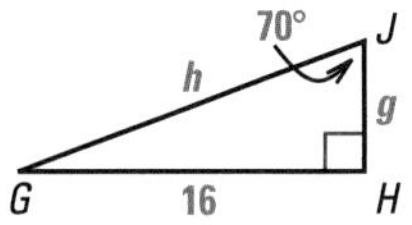

$g \approx 5.82$, $h \approx 17.03$

9. PARASAILING In Example 3, suppose the angle is 35°. How high is the parasailer? How far back is the parasailer? **about 459 ft; about 655 ft**

10. $\sin R = \frac{4}{5}$, $\cos R = \frac{3}{5}$, $\tan R = \frac{4}{3}$; $\sin S = \frac{3}{5}$, $\cos S = \frac{4}{5}$, $\tan S = \frac{3}{4}$

PRACTICE AND APPLICATIONS

STUDENT HELP
Extra Practice to help you master skills is on p. 808.

11. $\sin R = \frac{5}{13}$, $\cos R = \frac{12}{13}$, $\tan R = \frac{5}{12}$; $\sin S = \frac{12}{13}$, $\cos S = \frac{5}{13}$, $\tan S = \frac{12}{5}$

12. $\sin R = \frac{11}{61}$, $\cos R = \frac{60}{61}$, $\tan R = \frac{11}{60}$; $\sin S = \frac{60}{61}$, $\cos S = \frac{11}{61}$; $\tan S = \frac{60}{11}$

STUDENT HELP
HOMEWORK HELP
Example 1: Exs. 10–12
Example 2: Exs. 13–18
Example 3: Exs. 19, 20

FINDING TRIGONOMETRIC RATIOS **Find the sine, the cosine, and the tangent of $\angle R$ and of $\angle S$.** **See margin.**

10.
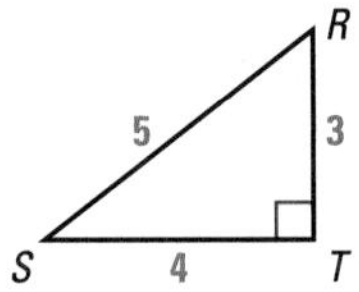

11.
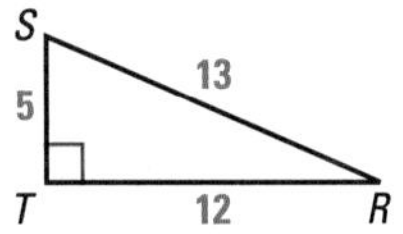

12.
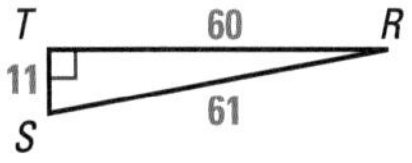

SOLVING RIGHT TRIANGLES **Find the missing lengths of the sides of the triangle. Round your answers to the nearest hundredth. Use the Pythagorean theorem to check.**

13.
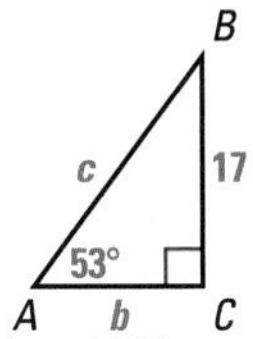

$b \approx 12.81$, $c \approx 21.29$

14.
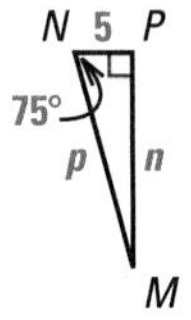

$n \approx 18.66$, $p \approx 19.32$

15.
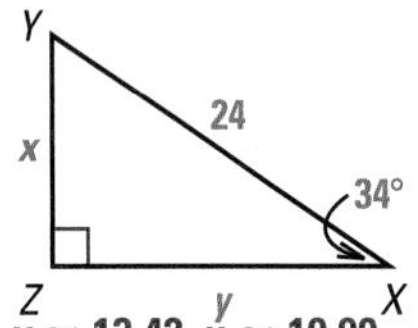

$x \approx 13.42$, $y \approx 19.90$

16.
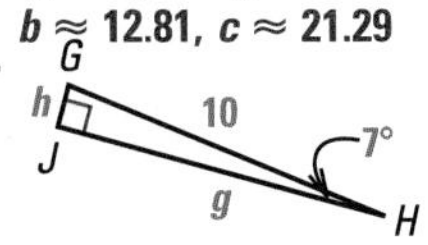

$g \approx 9.93$, $h \approx 1.22$

17.
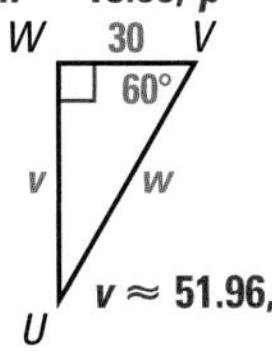

$v \approx 51.96$, $w = 60$

18.
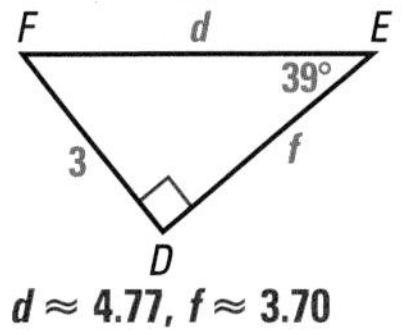

$d \approx 4.77$, $f \approx 3.70$

3 APPLY

ASSIGNMENT GUIDE

BASIC
Day 1: pp. 755–757 Exs. 10–19, 22, 24, 28–44 even

AVERAGE
Day 1: pp. 755–757 Exs. 10–19, 22, 24, 28–44 even

ADVANCED
Day 1: pp. 755–757 Exs. 10–19, 22–27, 28–44 even

BLOCK SCHEDULE WITH 12.8
pp. 755–757 Exs. 10–19, 22, 24, 28–44 even

EXERCISE LEVELS

Level A: *Easier*
10–18, 22
Level B: *More Difficult*
19–21, 24, 28–45
Level C: *Most Difficult*
23, 25–27

✓ HOMEWORK CHECK

To quickly check student understanding of key concepts, go over the following exercises: Exs. 10, 12, 14, 18, 22. See also the Daily Homework Quiz:

- Blackline Master (*Chapter 12 Resource Book,* p. 110)
- Transparency (p. 101)

Application Note

Exercise 21 The model assumes that the full 5 feet will be added to the height from eye level to the top of the bottle. In a more realistic model, however, only about 57 inches, or eye level, should be added to the height.

Exercise 22 Students will need to know that the altitude to the base of an isosceles triangle bisects the vertex angle.

Mathematical Reasoning

If the tangent ratio of one acute angle of a right triangle is $\frac{a}{b}$, what is the tangent ratio of the other acute angle of the triangle? $\frac{b}{a}$

20.

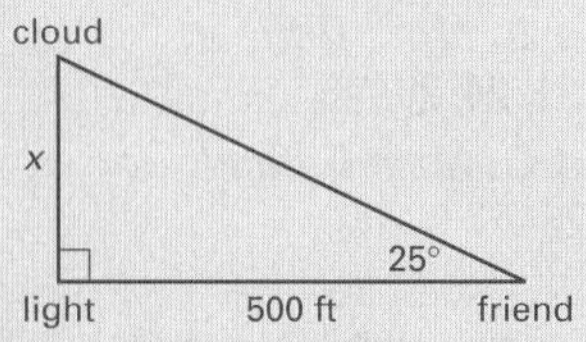

Additional Practice and Reteaching

For Lesson 12.7:

- Practice Levels A, B, and C (*Chapter 12 Resource Book,* p. 99)
- Reteaching with Practice (*Chapter 12 Resource Book,* p. 102)
- See Lesson 12.7 of the *Personal Student Tutor*

For more Mixed Review:

- Search the *Test and Practice Generator* for key words or specific lessons.

19. **Crane Operator** You are a crane operator using a crane with a 210-foot boom. You lift a load so that the load is directly in front of you. You estimate that the angle between the boom and the line between you and the load is 35°. About how far are you from the load? **about 172 ft**

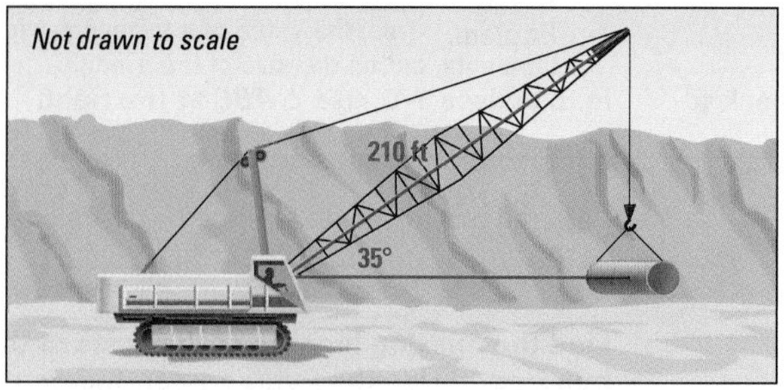

20. **Cloud Height** You are doing an experiment for your science class. You shine a spotlight from the ground straight up onto a cloud to measure the height of the cloud. Your friend stands 500 feet from the spotlight. She estimates that the angle formed between the ground and the line from her feet to the cloud is 25°. Draw a sketch to model the situation. Then find the height of the cloud. **about 233 ft; see margin for graph.**

21. **Collinsville Catsup Bottle** You are standing 197 feet from the base of the world's largest catsup bottle located in Collinsville, Illinois. You estimate that the angle between your eye level and the line from your eyes to the top of the bottle is 40°. If you are 5 feet tall, about how high is the top of the bottle? **about 170 ft**

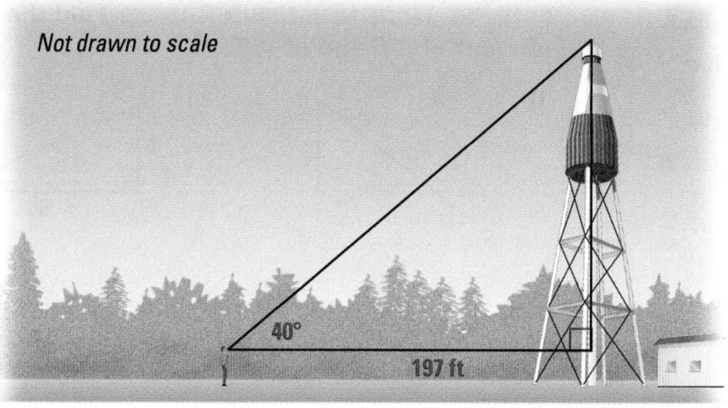

22. **Dinosaur Tracks** You are a paleontologist. You are estimating the length of a dinosaur's hind legs based on its track marks. You measure the pace length as three feet. The pace length represents the base of an isosceles triangle, where the equal sides are the dinosaur's hind legs that enclose the walking angle. Assuming the walking angle is 40°, what is the length of the dinosaur's hind legs? **4.39 ft**

Test Preparation

24b. Use $\sin 30° = \frac{3}{r}$ or since you know the value of q, use the Pythagorean theorem.

23. **VIEWING ANGLE** A car is traveling on a level road toward a mountain two kilometers high. The angle of elevation from the car to the top of the mountain changes from 6° to 15°. How far has the car traveled? **11.56 km**

24. **MULTI-STEP PROBLEM** Use the triangle below.

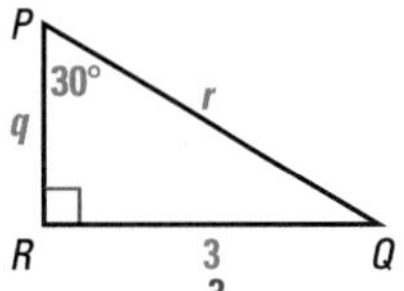

a. Find the length q. **5.20**

b. Describe two methods for finding the length r.

c. Find the length r. **6**

d. *Writing* Which method did you use in part (c)? Tell why you chose that method. ***Sample answer:*** **I used $\sin 30° = \frac{3}{r}$ since the calculation is simpler than that involved in using the Pythagorean theorem.**

★ Challenge

EXTENSION: SLOPE AND ANGLE In Exercises 25 and 26, find the slope of the line that passes through the points. Explain how the slope of the line is related to $\angle A$.

25.

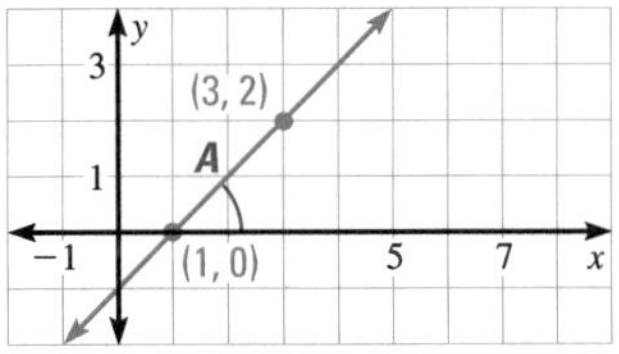

1; slope = tan A

26.

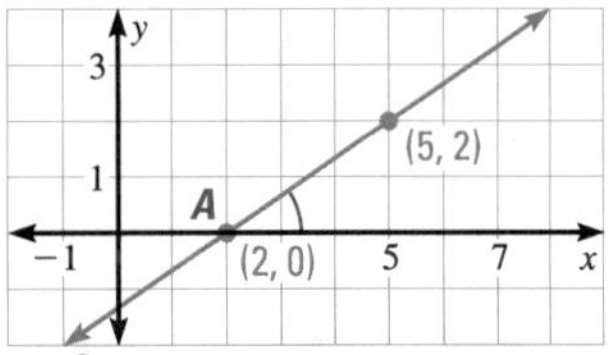

$\frac{2}{3}$; slope = tan A

EXTRA CHALLENGE

www.mcdougallittell.com

27. **LOGICAL REASONING** A line with a positive slope passes through the origin, making a 60° angle with the positive x-axis. Write an equation of the line. **$y = x \tan 60° \approx 1.73x$**

MIXED REVIEW

COUNTEREXAMPLES Decide whether the statement is *true* or *false*. If it is false, give a counterexample. (Review 2.1 for 12.8)

28. The absolute value of a number is always positive. **False; the absolute value of zero is not positive.**

29. The opposite of a number is always positive. **False; the opposite of a positive number is negative.**

USING THE DISTRIBUTIVE PROPERTY Use the distributive property to simplify the expression. (Review 2.6 for 12.8)

30. $6(w - 3)$ **$6w - 18$**

31. $-p(p + 1)$ **$-p^2 - p$**

32. $-(x - 8)$ **$-x + 8$**

33. $(x + 3)x$ **$x^2 + 3x$**

34. $(x - 2)5x$ **$5x^2 - 10x$**

35. $(4 + x)(-6x)$ **$-24x - 6x^2$**

SUBTRACTING VERTICALLY Use a vertical format to subtract the second polynomial from the first polynomial. (Review 10.1)

36. $6x^2 - 3x + 2,\ 2x^2 + x + 7$ **$4x^2 - 4x - 5$**

37. $4x^3 + 3x^2 + 8x + 6,\ 2x^3 - 3x^2 - 7x$ **$2x^3 + 6x^2 + 15x + 6$**

38. $10x^3 + 15,\ 17x^3 - 4x + 5$ **$-7x^3 + 4x + 10$**

39. $-2x^3 + 5x^2 - x + 8,\ -2x^3 + 3x - 4$ **$5x^2 - 4x + 12$**

SIMPLIFYING RATIONAL EXPRESSIONS Simplify the expression. (Review 11.6)

40. $\frac{3}{x} + \frac{x + 9}{x}$ **$\frac{x + 12}{x}$**

41. $\frac{8}{4a + 1} - \frac{5}{4a + 1}$ **$\frac{3}{4a + 1}$**

42. $\frac{2}{2x} + \frac{12}{x}$ **$\frac{13}{x}$**

43. $\frac{5}{4x} - \frac{7}{3x}$ **$-\frac{13}{12x}$**

44. $\frac{2x}{x + 1} + \frac{5}{x + 3}$ **See margin.**

45. $\frac{6x}{x + 1} - \frac{2x + 4}{x + 1}$ **$\frac{4(x - 1)}{x + 1}$**

44. $\frac{(2x + 1)(x + 5)}{(x + 1)(x + 3)}$

4 ASSESS

DAILY HOMEWORK QUIZ

Transparency Available

1. Find the sine, the cosine, and the tangent of $\angle A$ and $\angle B$.

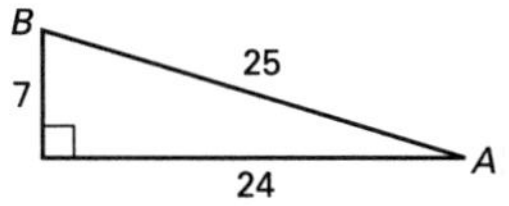

sine $\angle A = \frac{7}{25}$, sine $\angle B = \frac{24}{25}$;
cos $\angle A = \frac{24}{25}$, cos $\angle B = \frac{7}{25}$;
tan $\angle A = \frac{7}{24}$, tan $\angle B = \frac{24}{7}$

2. Find the missing lengths of the sides of the triangle. Round your answer to the nearest hundredth.

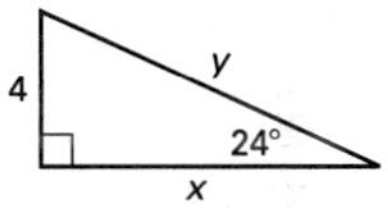

$y = 9.83$, $x = 8.98$

3. A string tied to a stake is attached to a kite. The string is 30 ft long. The angle between the ground and the string is 65°. How high is the kite? **27.2 ft**

EXTRA CHALLENGE NOTE

Challenge problems for Lesson 12.7 are available in **blackline** format in the *Chapter 12 Resource Book*, p. 107 and at **www.mcdougallittell.com.**

ADDITIONAL TEST PREPARATION

1. **WRITING** How is the sine ratio of one acute angle of a right triangle related to the cosine ratio of the other acute angle of the triangle? **They are equal.**

1 PLAN

PACING
Basic: 1 day
Average: 1 day
Advanced: 1 day
Block Schedule: 0.5 block with 12.7

LESSON OPENER ACTIVITY
An alternative way to approach Lesson 12.8 is to use the Activity Lesson Opener:
- Blackline Master (*Chapter 12 Resource Book,* p. 111)
- Transparency (p. 89)

MEETING INDIVIDUAL NEEDS
- ***Chapter 12 Resource Book***
  Prerequisite Skills Review (p. 5)
  Practice Level A (p. 112)
  Practice Level B (p. 113)
  Practice Level C (p. 114)
  Reteaching with Practice (p. 115)
  Absent Student Catch-Up (p. 117)
  Challenge (p. 119)
- ***Resources in Spanish***
- ***Personal Student Tutor***

NEW-TEACHER SUPPORT
See the Tips for New Teachers on pp. 1–2 of the *Chapter 12 Resource Book* for additional notes about Lesson 12.8.

WARM-UP EXERCISES
Transparency Available

Which property of real numbers justifies each statement?

1. $3(a + b) = 3a + 3b$
   distributive property
2. If $x = 5$, then $x + 2 = 5 + 2$
   addition property of equality
3. $4 + z = z + 4$
   commutative property of addition
4. $(p + q) + 0 = p + q$
   addition identity
5. $5\left(\frac{1}{5}\right) = 1$ **multiplication inverse**

12.8 Logical Reasoning: Proof

What you should learn

GOAL 1 Use logical reasoning and proof to prove a statement is true.

GOAL 2 Prove that a statement is false.

Why you should learn it

▼ To help solve **real-life** problems, such as defending a client in court in **Example 4.**

GOAL 1 PROVING A STATEMENT IS TRUE

LOGICAL REASONING Mathematics is believed to have begun with practical "rules of thumb" that were developed to deal with real-life problems, such as tax records, the surveying of fields, inventories of storehouses, business transactions between merchants, and astronomical calculations. Then, about 2500 years ago, Greek geometers developed a different approach toward mathematics. Starting with a handful of properties that they believed to be true, these geometers insisted on logical reasoning as the basis for developing more elaborate mathematical tools. These more elaborate tools, which can be proved to be true, are called *theorems*.

The basic properties of mathematics that mathematicians accept without proof are called **postulates** or **axioms**. Many of the rules that were discussed in Chapter 2, such as the properties of addition, the properties of multiplication, and the distributive property, fall into this category. The following chart provides a summary of the rules that underlie algebra.

CONCEPT SUMMARY — THE BASIC AXIOMS OF ALGEBRA

Let a, b, and c be real numbers.

Axioms of Addition and Multiplication

CLOSURE:	$a + b$ is a real number.	ab is a real number.
COMMUTATIVE:	$a + b = b + a$	$ab = ba$
ASSOCIATIVE:	$(a + b) + c = a + (b + c)$	$(ab)c = a(bc)$
IDENTITY:	$a + 0 = a, 0 + a = a$	$a(1) = a, 1(a) = a$
INVERSE:	$a + (-a) = 0$	$a\left(\frac{1}{a}\right) = 1, a \neq 0$

Axiom Relating Addition and Multiplication

DISTRIBUTIVE:	$a(b + c) = ab + ac$	$(a + b)c = ac + bc$

Axioms of Equality

ADDITION:	If $a = b$, then $a + c = b + c$.
MULTIPLICATION:	If $a = b$, then $ac = bc$.
SUBSTITUTION:	If $a = b$, then a can be substituted for b.

Once a list of axioms has been accepted, you can add more rules, formulas, and properties to the list. Some of the new concepts are definitions. You don't have to prove a definition, but it must be consistent with previous definitions and axioms. For instance, the definition of *reciprocal* does not have to be proved.

Other new statements, called **theorems**, do have to be proved. For instance, basic axioms have been used to prove the theorem that for all real numbers b and c, $c(-b) = -cb$. Once a theorem is proved, it can be used as a reason in proofs of other theorems.

STUDENT HELP

Study Tip
When you are proving a new theorem, every step must be justified by an axiom, a definition, given information, or a previously proved theorem.

EXAMPLE 1 *Proving a Theorem*

Use the definition of subtraction, $a - b = a + (-b)$, to prove the following theorem: $c(a - b) = ca - cb$.

SOLUTION

$c(a - b) = c[a + (-b)]$	Definition of subtraction
$= ca + c(-b)$	Distributive property
$= ca + (-cb)$	Theorem stated above
$= ca - cb$	Definition of subtraction

A **conjecture** is a statement that is thought to be true but is not yet proved. Often it is a statement based on observation.

EXAMPLE 2 *Goldbach's Conjecture*

Christian Goldbach (1690–1764) thought the following statement might be true. It is now referred to as *Goldbach's Conjecture.*

Every even integer, except 2, is equal to the sum of two prime numbers.

The following list shows that every even number between 4 and 30 is equal to the sum of two prime numbers. Does this list prove that *every* even number greater than 2 is equal to the sum of two prime numbers?

$4 = 2 + 2$	$6 = 3 + 3$	$8 = 3 + 5$	$10 = 3 + 7$
$12 = 5 + 7$	$14 = 3 + 11$	$16 = 3 + 13$	$18 = 5 + 13$
$20 = 3 + 17$	$22 = 3 + 19$	$24 = 5 + 19$	$26 = 3 + 23$
$28 = 5 + 23$	$30 = 7 + 23$		

SOLUTION

This list of examples *does not* prove the conjecture. No number of examples can prove that the rule is true for *every* even integer greater than 2. (At the time this book is being written, no one has been able to prove or disprove Goldbach's Conjecture.)

2 TEACH

MOTIVATING THE LESSON

A container of low-fat ice cream states that it has 50% less fat than regular ice cream. The regular brand has 9 g of fat per serving. The low-fat brand has 6 g of fat per serving. Do you think the low-fat brand is advertising truthfully?

EXTRA EXAMPLE 1

Use the associative property of addition to prove $[b + (-c)] + c = b$.

$[b + (-c)] + c = b + [(-c) + c]$ **associative**
$= b + 0$ **addition inverse**
$= b$ **addition identity**

EXTRA EXAMPLE 2

The sides of some right triangles are listed. For each triangle $a^2 + b^2 = c^2$. Does this prove the Pythagorean theorem?

3 4 5	6 8 10	5 12 13
8 15 17	1.5 2 2.5	9 40 41

No; these examples only demonstrate the theorem. A number of examples cannot prove that the rule is true for all numbers.

CHECKPOINT EXERCISES

For use after Example 1:

1. Use the definition of multiplication of fractions, $\frac{a}{b}\left(\frac{c}{d}\right) = \frac{ac}{bd}$, and the axioms of algebra to supply the reasons for the proof that $\frac{a}{a} = 1$.

$\frac{a}{a} = \frac{a \cdot 1}{a}$ **multiplication identity**
$= a \cdot \frac{1}{a}$ **multiplying of fractions**
$= 1$ **multiplication inverse**

For use after Example 2:

2. Give 3 examples to show that when $x < 0$ and n is an odd integer, then x^n is negative. Do your examples prove that x^n is negative for these conditions? ***Sample answer:*** **$(-3)^3 = -27$, $(-2)^1 = -2$, $(-4)^3 = -64$; No**

GOAL 2 PROVING A STATEMENT IS FALSE

Sometimes people propose new rules but later find that they *are not* consistent with the axioms, definitions, and theorems in the accepted list. To show that a conjecture is false, you need only one counterexample.

EXAMPLE 3 *Finding a Counterexample*

STUDENT HELP

Look Back
For help with counterexamples, see p. 66.

Assign values to a and b to show that the following conjecture is false.

$a + (-b) = (-a) + b.$

SOLUTION

You can choose any values of a and b. For instance, let $a = 1$ and let $b = 2$.

Evaluate the left side of the equation.

$a + (-b) = 1 + (-2)$ Substitute 1 for a and 2 for b.

$= -1$ Simplify.

Evaluate the right side of the equation.

$(-a) + b = (-1) + 2$ Substitute 1 for a and 2 for b.

$= 1$ Simplify.

Because $-1 \neq 1$, you have shown one case in which $a + (-b) = (-a) + b$ is false. The counterexample of $a = 1$ and $b = 2$ is sufficient to prove that $a + (-b) = (-a) + b$ is false. If the conjecture were true, it would have to work for *any* choices of a and b.

....

INDIRECT PROOF In this lesson you have used direct proofs to prove that statements are true, and you have used counterexamples to prove that statements are false.

Another type of proof is **indirect proof**. To prove a statement indirectly, assume that the statement is false. If this assumption leads to an impossibility, then you have proved that the original statement is true. An indirect proof is also called a *proof by contradiction*.

EXAMPLE 4 *Use of Contradiction in Real Life*

FOCUS ON CAREERS

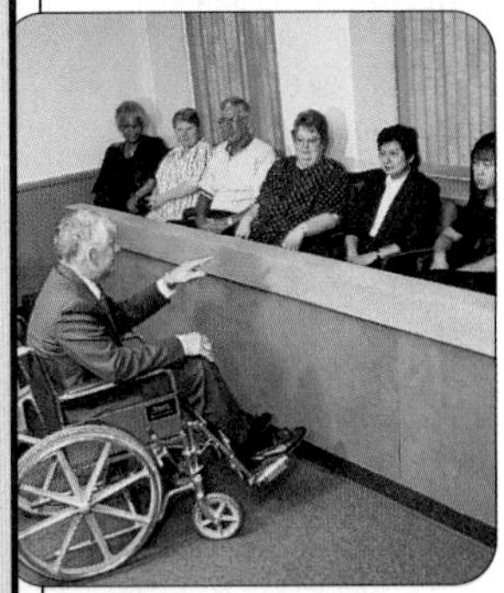

REAL LIFE **LAWYERS** Lawyers represent people in criminal and civil trials by presenting evidence supporting their client's case. They also give advice on legal matters.

INTERNET CAREER LINK www.mcdougallittell.com

LAWYERS You are a lawyer. You have been assigned to defend someone accused of violating a law uptown at 10:00 A.M. on March 22. You argue that if guilty, your client must have been there at that time. You have a video of your client being interviewed by a TV reporter at a political rally on the other side of town at the same time.

You argue that it would be impossible for your client to be uptown at 10:00 A.M. and on the other side of town at 10:00 A.M. Therefore your client cannot be guilty.

EXTRA EXAMPLE 3

Assign values to a and b to show that $a^2 + b^2 = (a + b)^2$ is false. *Sample answer:* Let $a = 2$ and $b = -1$. Then $2^2 + (-1)^2 = 5$, but $(2 + (-1))^2 = 1$.

EXTRA EXAMPLE 4

A car dealership advertises that nobody can beat their prices. You find another dealership that offers the same car for $500 less than the other dealership. What can you conclude? The ad does not tell the truth.

CHECKPOINT EXERCISES

For use after Example 3:

1. Assign values to x and y to show that $-(x + y) = -x + y$ is false. *Sample answer:* If $x = -1$ and $y = 10$, $-(-1 + 10) = -9$, but $-(-1) + 10 = 11$.

For use after Example 4:

2. You and a friend are solving a problem using algebra. Your friend says the answer is $\sqrt{63}$. Your teacher tells you that the answer must be greater than 8. What can you conclude? If $\sqrt{63}$ is the correct answer, then $\sqrt{63} > 8$. Since $\sqrt{63}$ is between $\sqrt{49}$ and $\sqrt{64}$, or between 7 and 8, you reason that $\sqrt{63}$ is not the correct answer.

APPLICATION NOTE

EXAMPLE 4 This kind of reasoning involves an argument form called *modus tollens or the Law of the Contrapositive.* The statement "If my client is not guilty, then he was not uptown at 10 A.M." is the contrapositive of the if-then statement "If my client was uptown at 10 A.M., then he is guilty." and is equivalent to the if-then statement. The lawyer uses the contrapositive to prove his point.

EXAMPLE 5 *Using an Indirect Proof*

Use an indirect proof to prove that $\sqrt{2}$ is an irrational number.

SOLUTION

Assume that $\sqrt{2}$ is *not* irrational. Then $\sqrt{2}$ is rational and can be written as the quotient of two integers a and b that have no common factors other than 1.

$\sqrt{2} = \frac{a}{b}$	Assume $\sqrt{2}$ is a rational number.
$2 = \frac{a^2}{b^2}$	Square each side.
$2b^2 = a^2$	Multiply each side by b^2.

This implies that 2 is a factor of a^2. Therefore 2 is also a factor of a. (You will prove this in Exercise 25 on page 763.) So a can be written as $2c$.

$2b^2 = (2c)^2$	Substitute $2c$ for a.
$2b^2 = 4c^2$	Simplify.
$b^2 = 2c^2$	Divide each side by 2.

This implies that 2 is a factor of b^2 and also a factor of b. So 2 is a factor of both a and b. But this is impossible because a and b have no common factors. Therefore it must be impossible that $\sqrt{2}$ is a rational number. So you can conclude that $\sqrt{2}$ must be an irrational number.

EXTRA EXAMPLE 5

Use an indirect proof to prove that if $a < 0$, then $\frac{1}{a} < 0$.

Assume $\frac{1}{a} > 0$. (Note that $\frac{1}{a}$ can never equal 0.) Since $a < 0$, then $\frac{1}{a} \cdot a < 0 \cdot \frac{1}{a}$. (multiplication axiom of inequality) $\frac{1}{a} \cdot a < 0$ (multiplication property of zero) $1 < 0$ (axiom of multiplicative inverses). Since this is impossible, it must be impossible that $\frac{1}{a} > 0$ if $a < 0$. So you can conclude that $\frac{1}{a} < 0$.

CHECKPOINT EXERCISES

For use after Example 5:

1. State which assumption you would make to begin an indirect proof of each statement.
   a. If $a > 1$, then $\frac{1}{a} < 1$. $\frac{1}{a} \geq 1$
   b. If $ac = bc$ and $c \neq 0$, then $a = b$. $a \neq b$

GUIDED PRACTICE

Vocabulary Check ✓

1. Explain the difference between an axiom and a theorem.
   An axiom is a rule accepted without proof; a theorem must be proved.
2. What is the first step of an indirect proof?
   Assume that the opposite of what you want to prove is true.

Concept Check ✓

In Exercises 3–11, state the basic axiom of algebra that is represented. See margin.

3. $y(1) = y$
4. $2x + 3 = 3 + 2x$
5. $5(x + y) = 5x + 5y$
6. $(4x)y = 4(xy)$
7. $y + 0 = y$
8. $x + (-x) = 0$
9. $x\left(\frac{1}{x}\right) = 1$
10. $cd = dc$
11. $(x + y) + z = x + (y + z)$

3. Identity property of multiplication
4. Commutative property of addition
5. Distributive property
6. Associative property of multiplication
7. Identity property of addition
8. Inverse property of addition
9. Inverse property of multiplication
10. Commutative property of multiplication
11. Associative property of addition

12. Is subtraction closed for the positive real numbers? That is, if a and b are positive real numbers, must $(a - b)$ be a positive real number? Explain your thinking.
    No; not necessarily; if $b \geq a$, then $a - b$ is not positive.

Skill Check ✓

13. Copy and complete the proof of the statement: If $a = b$ and $c = d$, then $ac = bd$. Each variable represents any real number.
    Multiplication axiom of equality; Substitution property of equality

$a = b$	Given
$ac = bc$	Multiplication axiom of equality
$c = d$	Given
$bc = bd$	?
$ac = bd$	?

FOCUS ON VOCABULARY

What is the difference between a postulate and a theorem? Postulates are mathematical rules that are accepted as true without proof. Theorems are statements that have been proved.

CLOSURE QUESTION

What techniques can you use to prove a statement is true. How can you prove that a statement is false? direct or indirect proof; show a counterexample

DAILY PUZZLER

The student council officers, Yolanda, Juan, Carl, and Danielle, sat around a circular table. The vice-president sat to the right of the secretary. The president sat to the right of the vice-president. Juan sat across from Carl. Yolanda sat to the left of the treasurer. The treasurer sat to the left of the secretary. Juan sat to the right of Yolanda. Who holds which office? Yolanda is president, Carl is vice-president, Juan is treasurer, and Danielle is secretary.

3 APPLY

ASSIGNMENT GUIDE

BASIC
Day 1: pp. 762–764 Exs. 14–21, 29, 35, 40, 45, 50, Quiz 3 Exs. 1–14

AVERAGE
Day 1: pp. 762–764 Exs. 14–21, 24, 25, 29, 35, 40, 45, 50, Quiz 3 Exs. 1–14

ADVANCED
Day 1: pp. 762–764 Exs. 14–30, 35, 40, 45, 50, Quiz 3 Exs. 1–14

BLOCK SCHEDULE WITH 12.7
pp. 762–764 Exs. 14–21, 24, 25, 29, 35, 40, 45, 50, Quiz 3 Exs. 1–14

EXERCISE LEVELS

Level A: ***Easier***
48–53

Level B: ***More Difficult***
14, 19–21, 29, 31–47

Level C: ***Most Difficult***
15–18, 22–28, 30

✔ HOMEWORK CHECK

To quickly check student understanding of key concepts, go over the following exercises: Exs. 14, 16, 18, 20, 22, 24, 26, 28. See also the Daily Homework Quiz:

- Transparency (p. 102)

28. By the distance formula, $BC = \sqrt{(0-x)^2 + (y-0)^2} = \sqrt{x^2 + y^2}$ so $BD = CD = \frac{1}{2}\sqrt{x^2 + y^2}$. The midpoint D has coordinates $\left(\frac{0+x}{2}, \frac{y+0}{2}\right) = \left(\frac{x}{2}, \frac{y}{2}\right)$. Using the distance formula, $AD = \sqrt{\left(\frac{x}{2} - 0\right)^2 + \left(\frac{y}{2} - 0\right)^2} = \sqrt{\frac{x^2}{4} + \frac{y^2}{4}} = \frac{1}{2}\sqrt{x^2 + y^2}$. Therefore $AD = BD = CD$ and the midpoint of the hypotenuse of the right triangle is equidistant from the vertices.

PRACTICE AND APPLICATIONS

STUDENT HELP
Extra Practice to help you master skills is on p. 808.

15. $a - b = a + (-b)$
Definition of subtraction
$a - b = -b + a$
Commutative property of addition

16. $(a - b) - c = [a + (-b)] + (-c)$
Definition of subtraction
$= a + [(-b) + (-c)]$
Associative property of addition
$= a + (-1)(b + c)$
Distributive property
$= a - (b + c)$
Definition of subtraction

17. $a + (-1)(a) = 1(a) + (-1)(a)$
Identity property of multiplication
$= [1 + (-1)](a)$
Distributive property
$= (0)(a)$
Inverse property of addition
$= 0$
Multiplication property of 0
Since $a + (-1)(a) = 0$, $(-1)(a) = -a$ by definition.

18. No; no number of examples can prove a conjecture is true for all real numbers; the conjecture is true.

21. 2 and 3 are integers, but $\frac{2}{3}$ is not an integer.

STUDENT HELP
HOMEWORK HELP
Example 1: Exs. 14–17
Example 2: Ex. 18
Example 3: Exs. 19–21
Example 4: Ex. 24
Example 5: Exs. 25–27

14. SUPPLYING REASONS Copy and complete the proof of the statement: For all real numbers a and b, $(a + b) - b = a$.

$(a + b) - b = (a + b) + (-b)$	Definition of subtraction
$(a + b) - b = a + [b + (-b)]$	Associative property of addition
$(a + b) - b = a + 0$	? Inverse property of addition
$(a + b) - b = a$	? Identity property of addition

PROVING THEOREMS **In Exercises 15–17, prove the theorem. (Use the basic axioms of algebra and the definition of subtraction given in Example 1.)**

15. If a and b are real numbers, then $a - b = -b + a$. See margin.

16. If a, b, and c are real numbers, then $(a - b) - c = a - (b + c)$. See margin.

17. If a is any real number, then $-1(a) = -a$. See margin.

18. PROVING A CONJECTURE A student proposes the following conjecture. *The sum of the first n odd integers is* n^2. She gives four examples: $1 = 1^2$, $1 + 3 = 4 = 2^2$, $1 + 3 + 5 = 9 = 3^2$, and $1 + 3 + 5 + 7 = 16 = 4^2$. Do the examples prove her conjecture? Explain. Do you think the conjecture is true? See margin.

FINDING A COUNTEREXAMPLE **In Exercises 19–21, find a counterexample to show that the statement is *not* true.** 19–21. Sample answers are given.

19. If a and b are real numbers, then $(a + b)^2 = a^2 + b^2$. $(1 + 2)^2 \neq 1^2 + 2^2$

20. If a, b, and c are nonzero real numbers, then $(a \div b) \div c = a \div (b \div c)$. $(12 \div 6) \div 2 \neq 12 \div (6 \div 2)$

21. If a and b are integers, then $a \div b$ is an integer. See margin.

22. GEOMETRY CONNECTION Explain how the diagrams below can be used to give a geometrical argument to support the conjecture in Exercise 18.

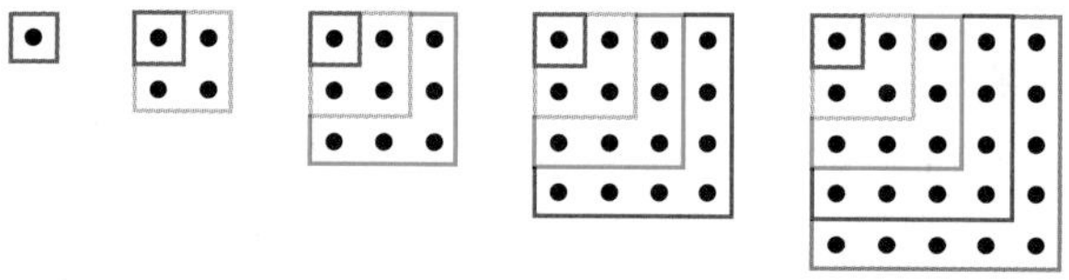

Each block represents a sum of odd integers; $1 = 1^2$, $1 + 3 = 2^2$, $1 + 3 + 5 = 3^2$, $1 + 3 + 5 + 7 = 4^2$, $1 + 3 + 5 + 7 + 9 = 5^2$.

23. THE FOUR-COLOR PROBLEM A famous theorem states that any map can be colored with four different colors so that no two countries that share a border have the same color. No matter how the map shown at the right is colored with three different colors, at least two countries having a common border will have the same color. Does this map serve as a counterexample to the following proposal? Explain. Yes; this map shows that the proposal is false.

Any map can be colored with three different colors so that no two countries that share a border have the same color.

INDIRECT PROOF In Exercises 24–27, use an indirect proof to prove that the conclusion is true.

24. Your bus leaves a track meet at 4:30 P.M. and does not travel faster than 60 miles per hour. The meet is 45 miles from home. Your bus will not get you home in time for dinner at 5 P.M.

25. If p is an integer and p^2 is divisible by 2, then p is divisible by 2. (*Hint*: An odd number can be written as $2n + 1$, where n is an integer. An even number can be written as $2n$.)

26. If $a < b$, then $a + c < b + c$.

27. If $ac > bc$ and $c > 0$, then $a > b$.

28. PROOF USING THE MIDPOINT Let D represent the midpoint between B and C, as shown at the right. Prove that for any right triangle, the midpoint of its hypotenuse is equidistant from the three vertices of the triangle. In order to prove this, you must first find the distance, BC, between B and C. Using the distance formula, you get $BC = \sqrt{x^2 + y^2}$, so BD and CD must be $\frac{1}{2}\sqrt{x^2 + y^2}$. **See margin.**

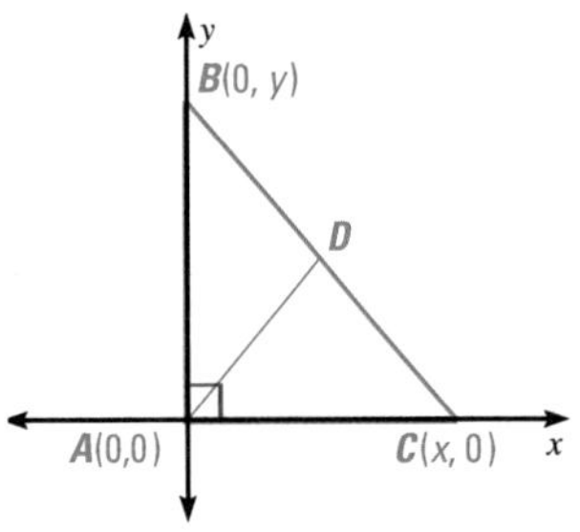

To help you with your proof, use the distance formula to find AD.

29. MULTI-STEP PROBLEM In graph theory, a *complete graph* is one in which every pair of vertices is connected by part of the graph called an *edge*. The following graphs are complete.

Two vertices **Three vertices** **Four vertices** **Five vertices**

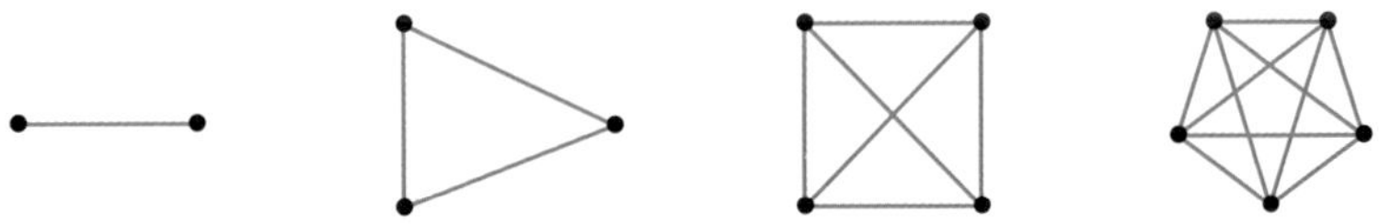

a. Copy and complete the table. **6; 10; 15**

Vertices	2	3	4	5	6
Edges	1	3	?	?	?

b. Make a conjecture about the relationship between the number of vertices and the number of edges in a complete graph.

c. Use your conjecture to predict how many edges a complete graph with 10 vertices would have. **45 edges**

30. PYTHAGOREAN THEOREM Explain how the following diagrams could be used to give a geometrical proof of the Pythagorean theorem. **See margin.**

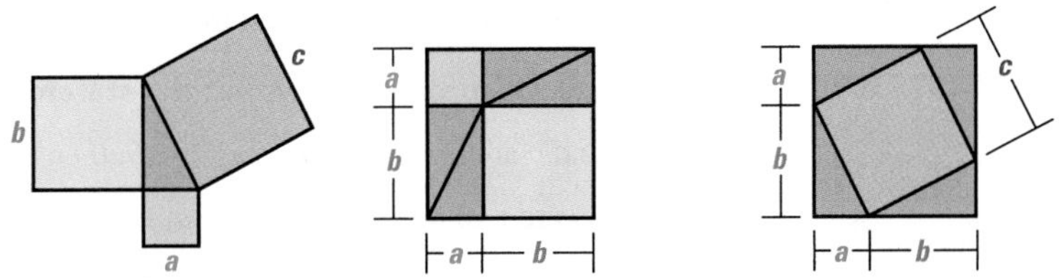

24. Assume that the bus will get you home in time for dinner at 5 P.M. The bus made the 45-mile trip home in half an hour. That means the bus was traveling 90 mi/h, which is more than its maximum speed.

25. Assume that p is an integer, p^2 is odd, and p is even. Let $p = 2n$. Then $p^2 = (2n)^2 = 4n^2 = 2(2n^2)$, an even number, which is impossible, since p^2 is odd. Then p must be odd.

26. Assume that $a < b$ and $a + c \geq b + c$. Then $(a + c) - (b + c) \geq 0$. But $(a + c) - (b + c) = a - b$, so $a - b \geq 0$. However, since $a < b$, it is impossible for $a - b$ to be greater than or equal to 0. Then $a + c < b + c$.

Test Preparation

27. Assume that $ac > bc$, $c > 0$, and $a \leq b$. By the Multiplication property of inequality, $ac \leq bc$, which is impossible since it was given that $ac > bc$. Then $a > b$.

29b. A complete graph with n vertices has $\frac{n(n-1)}{2}$ edges.

★ Challenge

EXTRA CHALLENGE

www.mcdougallittell.com

! COMMON ERROR

EXERCISE 16 Students may begin the proof with the incorrect statement $(a - b) - c = a - (b - c)$. Remind them that the real numbers are not associative for subtraction. Simplified, $a - b - c \neq a - b + c$.

APPLICATION NOTE

EXERCISE 28 Students are actually proving the theorem for the particular case where A is at the origin and B and C lie on the coordinate axes. A proof of the general case would involve $A(0, 0)$, $B(x_1, y_1)$ and $C(x_2, y_2)$ where $\overline{AC} \perp \overline{AB}$. Then it can be shown that $AD = CD = BD = \frac{1}{2}\sqrt{(x_2 - x_1)^2 + (y_2 - y_1)^2}$.

EXERCISE 29 Airline companies use graph theory to determine the possible flight patterns between major cities. Utilities can also use graph theory to determine the flow of electric current between power stations and their customers.

30. *Sample answer:* In the figure at the left, c is the length of the hypotenuse of a right triangle and a and b are the lengths of the legs. The next two figures are squares having the same area. The area of the square on the left is $(a + b)^2$ or $a^2 + 2ab + b^2$. The area of the square on the right can be written as the sum of its parts $4\left(\frac{1}{2}ab\right) + c^2$ or $2ab + c^2$. So, $a^2 + 2ab + b^2 = 2ab + c^2$, and $a^2 + b^2 = c^2$.

ADDITIONAL PRACTICE AND RETEACHING

For Lesson 12.8:

- Practice Levels A, B, and C (*Chapter 12 Resource Book*, p. 112)
- Reteaching with Practice (*Chapter 12 Resource Book*, p. 115)
- See Lesson 12.8 of the *Personal Student Tutor*

For more Mixed Review:

- Search the *Test and Practice Generator* for key words or specific lessons.

4 ASSESS

DAILY HOMEWORK QUIZ

Transparency Available

1. A student wrote the following rule: *The sum of the digits of the cube of a number equals 1, 8, or 9.* He gives the following three examples:
   $3^3 = 27, 2 + 7 = 9$;
   $5^3 = 125, 1 + 2 + 5 = 8$;
   $10^3 = 1000, 1 + 0 + 0 + 0 = 1$
   Do the examples prove his conjecture? Explain. Do you think the conjecture is true?
   No; no number of examples can prove that a rule is true; it appears to be true
2. Use an indirect proof to prove that if $a > b$, then $a + c > b + c$.
   Assume that $a > b$ and $a + c \le b + c$. Then $(a + c) - (b + c) \le 0$. But $(a + c) - (b + c) = a - b$, and, since $a > b$, it is impossible for $a - b$ to be less than or equal to 0. Then $a + c > b + c$.

EXTRA CHALLENGE NOTE

Challenge problems for Lesson 12.8 are available in **blackline** format in the *Chapter 12 Resource Book,* p. 119 and at **www.mcdougallittell.com.**

ADDITIONAL TEST PREPARATION

1. **WRITING** Explain the difference between proving a statement true and proving a statement false.
   To prove a statement is true, you need to prove the statement for all cases. To prove a statement is false, you need show only one counterexample.

MIXED REVIEW

FINDING SOLUTIONS **Decide how many solutions the equation has.** (Review 9.6)

31. $x^2 - 2x + 4 = 0$ **no real solutions**
32. $-2x^2 + 4x - 2 = 0$ **one solution**
33. $8x^2 - 8x + 2 = 0$ **one solution**
34. $x^2 - 14x + 49 = 0$ **one solution**
35. $-3x^2 - 5x + 1 = 0$ **two solutions**
36. $6x^2 - x + 5 = 0$ **no real solutions**
37. $x^2 - 2x - 15 = 0$ **two solutions**
38. $x^2 + 16x + 64 = 0$ **one solution**
39. $x^2 + 11x + 30 = 0$ **two solutions**

FINDING SOLUTIONS **Decide whether the ordered pair is a solution of the inequality.** (Review 9.7)

40. $y < x^2 - 2x - 5$; $(1, 1)$ **not a solution**
41. $y \ge 2x^2 - 8x + 8$; $(3, -2)$ **not a solution**
42. $y \le 2x^2 - 3x + 10$; $(-2, 20)$ **solution**
43. $y > 4x^2 - 48x + 61$; $(1, 17)$ **not a solution**
44. $y \ge x^2 + 4x$; $(-2, -4)$ **solution**
45. $y < 3x^2 - 2x$; $(5, 10)$ **solution**
46. $y > 3x^2 + 50x + 500$; $(-6, 100)$ **not a solution**
47. $y \ge -x^2 + 3x - \frac{15}{4}$; $(2, -3)$ **not a solution**

PERCENTS **Solve the percent problem.** (Review 11.2)

48. How much is 15% of \$15? **\$2.25**
49. 41 inches is what percent of 50 inches? **82%**
50. 100 is 1% of what number? **10,000**
51. 6 inches is what percent of 3 inches? **200%**
52. 1240 is 80% of what number? **1550**
53. \$5 is 33% of what amount of money? **\$15.15**

QUIZ 3

Self-Test for Lessons 12.6–12.8

Find the distance between the two points. Round the result to the nearest hundredth if necessary. Then find the midpoint between the two points. (Lesson 12.6)

1. $(1, 3), (7, -9)$ **13.42; (4, −3)**
2. $(-2, -5), (6, -11)$ **10; (2, −8)**
3. $(0, 0), (8, -14)$ **16.12; (4, −7)**
4. $(-8, -8), (-8, 8)$ **16; (−8, 0)**
5. $(3, 4), (-3, 4)$ **6; (0, 4)**
6. $(1, 7), (-4, -2)$ **10.30; $\left(-\frac{3}{2}, \frac{5}{2}\right)$**
7. $(2, 0), (-2, -3)$ **5; $\left(0, -\frac{3}{2}\right)$**
8. $(-3, 3), (4, 1)$ **7.28; $\left(\frac{1}{2}, 2\right)$**
9. $(3, 4), (2, -4)$ **8.06; $\left(\frac{5}{2}, 0\right)$**

Find the sine, the cosine, and the tangent of angles *A* and *B*. (Lesson 12.7) **10–12. See margin.**

10.

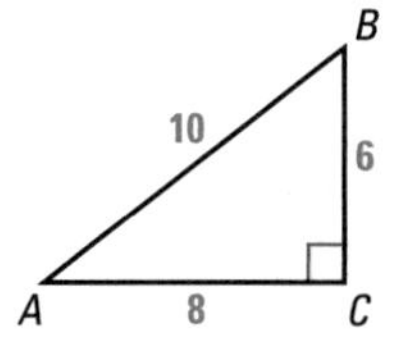

11.

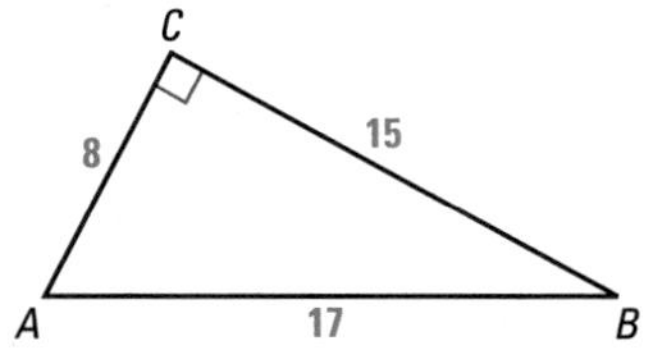

12. Prove the theorem $(a - b)c = ac - bc$. Use only the basic axioms of algebra, the definition of subtraction, and the theorem $(-b)c = -bc$. (Lesson 12.8)

Find a counterexample to show that the statement is *not* true. (Lesson 12.8) **13, 14. See margin.**

13. If a, b, and c are real numbers and $a < b$, then $ac < bc$.
14. If a and b are real numbers, then $-(a + b) = (-a) - (-b)$.

10. $\sin A = \frac{3}{5}$, $\cos A = \frac{4}{5}$, $\tan A = \frac{3}{4}$; $\sin B = \frac{4}{5}$, $\cos B = \frac{3}{5}$, $\tan B = \frac{4}{3}$

11. $\sin A = \frac{15}{17}$, $\cos A = \frac{8}{17}$, $\tan A = \frac{15}{8}$; $\sin B = \frac{8}{17}$, $\cos B = \frac{15}{17}$, $\tan B = \frac{8}{15}$

12. $(a - b)c = [a + (-b)]c$ Definition of subtraction
$= a(c) + (-b)c$ Distributive property
$= ac + (-bc)$ Property of opposites
$= ac - bc$ Definition of subtraction

13. Accept any real numbers a, b, and c with $a < b$ and $c \le 0$. *Sample answer:* $a = 1$, $b = 2$, and $c = -1$: $1 < 2$, and $1(-1) > 2(-1)$.

14. Accept any real numbers a and b with $b \ne 0$. *Sample answer:* $a = 1$ and $b = 2$; $-(1 + 2) \ne (-1) - (-2)$.

CHAPTER 12

Chapter Summary

WHAT did you learn?	*WHY did you learn it?*
Evaluate and graph a square-root function. **(12.1)**	Investigate walking speeds of dinosaurs. **(p. 711)**
Add, subtract, multiply, and divide radical expressions. **(12.2)**	Compare speeds of pole-vaulters. **(p. 720)**
Solve a radical equation. **(12.3)**	Find how much centripetal force a person experiences on an amusement park ride. **(p. 726)**
Solve a quadratic equation by completing the square. **(12.4)**	Investigate a proof of the quadratic formula. **(p. 732)**
Choose a method for solving a quadratic equation. **(12.4)**	Develop efficient problem solving skills. **(p. 732)**
Use the Pythagorean theorem and its converse. **(12.5)**	Calculate the distance from home plate to second base. **(p. 740)**
Find the distance and midpoint between two points in a coordinate plane. **(12.6)**	Apply the distance and midpoint formulas to model real-life situations. **(pp. 746, 747)**
Find the trigonometric ratios *sine*, *cosine*, and *tangent*. **(12.7)**	Determine cloud height above the ground. **(p. 756)**
Use logical reasoning and proof. **(12.8)**	Prove that statements are true or false. **(pp. 759, 760)**

How does Chapter 12 fit into the BIGGER PICTURE of algebra?

Radicals have many applications in geometry and in other fields. In this chapter you learned to add, subtract, multiply, and divide radical expressions and to solve radical equations.

The chapter also contains material that helps prepare you for other courses in mathematics. There are many applications involving geometry; an introduction to the trigonometric ratios sine, cosine, and tangent; and an introduction to some of the formal properties of algebra explored in Algebra 2.

STUDY STRATEGY

How did you use the notes in your notebook?

A map of ideas you made, using the **Study Strategy** on page 708, may look like this one.

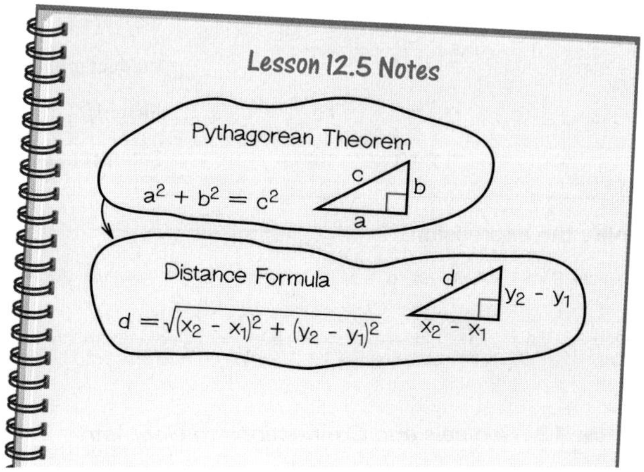

765

ADDITIONAL RESOURCES

The following resources are available to help review the material in this chapter.

- Chapter Review Games and Activities (*Chapter 12 Resource Book,* p. 120)
- *Instant Replay: Video Review Games*
- *Personal Student Tutor*
- Cumulative Review, Chs. 1–12 (*Chapter 12 Resource Book,* p. 132)

1.

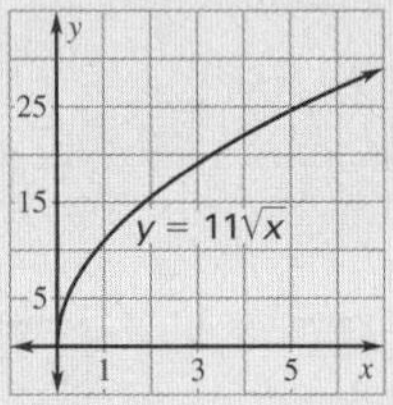

2.

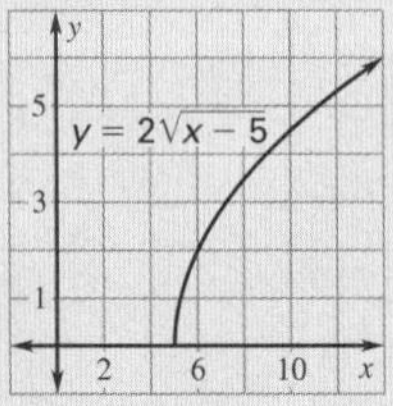

3.

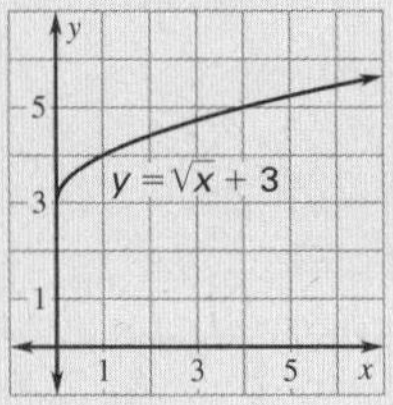

CHAPTER 12

Chapter Review

VOCABULARY

- square-root function, p. 709
- conjugates, p. 717
- complete the square, p. 730
- Pythagorean theorem, p. 738
- hypotenuse, p. 738
- legs of a right triangle, p. 738
- hypothesis, p. 739
- conclusion, p. 739
- converse, p. 739
- distance formula, p. 745
- midpoint between two points, p. 747
- midpoint formula, p. 747
- trigonometric ratio, p. 752
- sine, cosine, tangent, p. 752
- postulates, or axioms, p. 758
- theorems, p. 759
- conjecture, p. 759
- indirect proof, p. 760

12.1 FUNCTIONS INVOLVING SQUARE ROOTS

Examples on pp. 709–711

EXAMPLE To sketch the graph of $y = \sqrt{x} - 2$, note that the domain is the set of all nonnegative numbers. Then make a table of values, plot the points, and connect them with a smooth curve. The range is all numbers greater than or equal to -2.

x	y
0	$y = \sqrt{0} - 2 = -2$
1	$y = \sqrt{1} - 2 = -1$
2	$y = \sqrt{2} - 2 \approx -0.6$
⋮	⋮

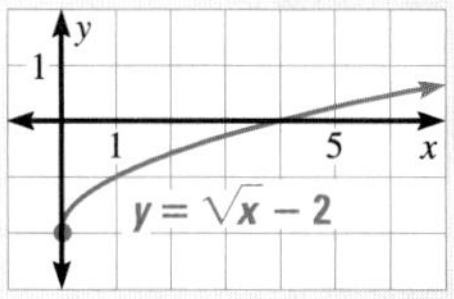

Identify the domain and the range of the function. Then graph the function.

1. $y = 11\sqrt{x}$ 1. Both the domain and range are all nonnegative numbers. See margin.

2. $y = 2\sqrt{x-5}$ 2. domain: all numbers greater than or equal to 5; range: all nonnegative numbers. See margin.

3. $y = \sqrt{x} + 3$ 3. domain: all nonnegative numbers; range: all numbers greater than or equal to 3. See margin.

12.2 OPERATIONS WITH RADICAL EXPRESSIONS

Examples on pp. 716–718

EXAMPLE You can use radical operations to simplify radical expressions.

$$4\sqrt{20} - 3\sqrt{5} = 4\sqrt{4 \cdot 5} - 3\sqrt{5}$$ **Perfect square factor**

$$= 4\sqrt{2 \cdot 2} \cdot \sqrt{5} - 3\sqrt{5}$$ **Product property**

$$= 8\sqrt{5} - 3\sqrt{5}$$ **Simplify.**

$$= 5\sqrt{5}$$ **Subtract like radicals.**

Simplify the expression.

4. $\sqrt{5} + 2\sqrt{5} - \sqrt{3}$ $3\sqrt{5} - \sqrt{3}$

5. $\sqrt{6}\,(2\sqrt{3} - 4\sqrt{2})$ $6\sqrt{2} - 8\sqrt{3}$

6. $(3 - \sqrt{10})^2$ $19 - 6\sqrt{10}$

7. $(\sqrt{8} + \sqrt{3})^2$ $11 + 4\sqrt{6}$

8. $\dfrac{21}{\sqrt{3}}$ $7\sqrt{3}$

9. $\dfrac{8}{6 - \sqrt{4}}$ 2

12.3 SOLVING RADICAL EQUATIONS

Examples on pp. 722–724

EXAMPLE Solve $\sqrt{3x - 2} = x$.

$(\sqrt{3x - 2})^2 = x^2$	**Square both sides.**
$3x - 2 = x^2$	**Simplify.**
$0 = x^2 - 3x + 2$	**Write in standard form.**
$0 = (x - 2)(x - 1)$	**Factor.**
$x = 2$ or $x = 1$	**Zero-product property**

Solve the equation.

10. $2\sqrt{x} - 4 = 0$ **4**

11. $x = \sqrt{-4x - 4}$ **no solutions**

12. $\sqrt{x - 3} + 2 = 8$ **39**

12.4 COMPLETING THE SQUARE

Examples on pp. 730–733

EXAMPLE Solve $x^2 - 6x - 1 = 6$ by completing the square.

$x^2 - 6x = 7$	**Isolate the x^2-term and the x-term.**
$x^2 - 6x + 9 = 7 + 9$	**Add $\left(-\frac{6}{2}\right)^2$ to each side.**
$(x - 3)^2 = 16$	**Write left side as perfect square.**
$x - 3 = \pm 4$	**Find square root of each side.**
$x = 7$ or $x = -1$	**Solve for x.**

Solve the equation by completing the square.

13. $x^2 - 4x - 1 = 7$ **$2 - 2\sqrt{3}, 2 + 2\sqrt{3}$**

14. $x^2 + 20x + 19 = 0$ **−19, −1**

15. $2x^2 - x - 4 = 10$ **$\frac{1 - \sqrt{113}}{4}, \frac{1 + \sqrt{113}}{4}$**

12.5 THE PYTHAGOREAN THEOREM AND ITS CONVERSE

Examples on pp. 738–740

EXAMPLE Given $a = 6$ and $c = 12$, find b.

$a^2 + b^2 = c^2$	**Write Pythagorean theorem.**
$6^2 + b^2 = 12^2$	**Substitute for a and c.**
$b = 6\sqrt{3}$	**Solve for b.**

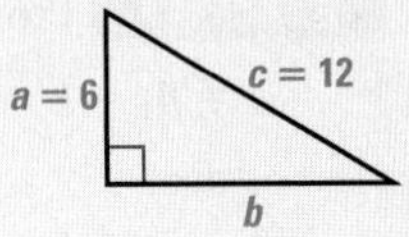

Find each missing length.

16.

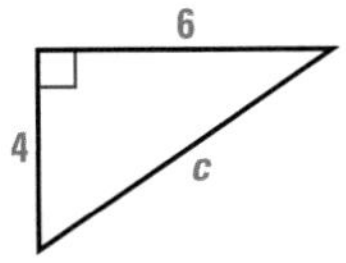

$2\sqrt{13}$

17.

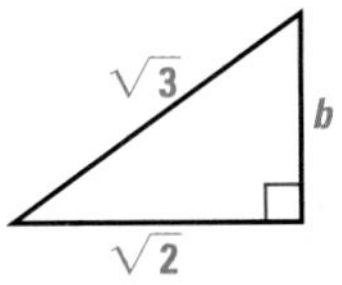

1

18.

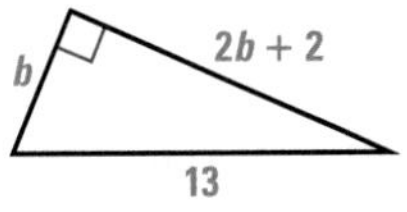

5, 12

24. $c(-b) = c((-1) \cdot b)$ Property of opposites
$= (c(-1))b$ Associative property of multiplication
$= -1(c)b$ Commutative property of multiplication
$= -cb$ Property of opposites

12.6 THE DISTANCE AND MIDPOINT FORMULAS

Examples on pp. 745–747

EXAMPLE Find the distance d and the midpoint m between $(-6, -2)$ and $(4, 3)$.

$$d = \sqrt{(x_2 - x_1)^2 + (y_2 - y_1)^2}$$
$$= \sqrt{[4 - (-6)]^2 + [3 - (-2)]^2}$$
$$= \sqrt{125}$$
$$= 5\sqrt{5}$$
$$m = \left(\frac{x_1 + x_2}{2}, \frac{y_1 + y_2}{2}\right)$$
$$= \left(\frac{-6 + 4}{2}, \frac{-2 + 3}{2}\right)$$
$$= \left(-1, \frac{1}{2}\right)$$

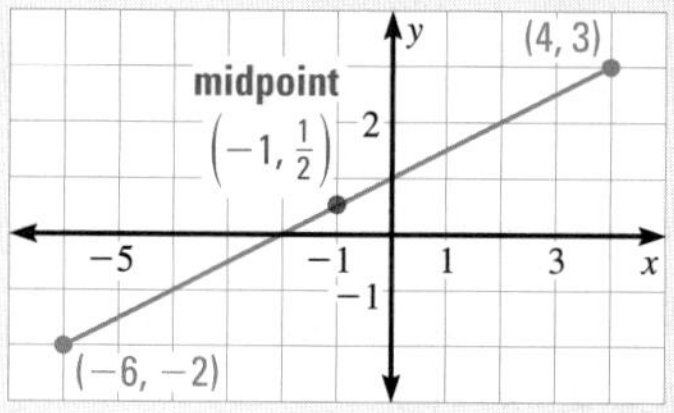

Find the distance and the midpoint between the two points.

19. $(8, 5)$ and $(11, -4)$ **$3\sqrt{10}, \left(\frac{19}{2}, \frac{1}{2}\right)$**

20. $(-3, 6)$ and $(1, 7)$ **$\sqrt{17}, \left(-1, \frac{13}{2}\right)$**

21. $(-2, -2)$ and $(2, 8)$ **$2\sqrt{29}, (0, 3)$**

12.7 TRIGONOMETRIC RATIOS

Examples on pp. 752–754

EXAMPLES

$$\sin A = \frac{\text{side opposite } \angle A}{\text{hypotenuse}} = \frac{a}{c} = \frac{6}{10} = 0.6$$

$$\cos A = \frac{\text{side adjacent to } \angle A}{\text{hypotenuse}} = \frac{b}{c} = \frac{8}{10} = 0.8$$

$$\tan A = \frac{\text{side opposite } \angle A}{\text{side adjacent to } \angle A} = \frac{a}{b} = \frac{6}{8} = 0.75$$

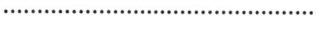

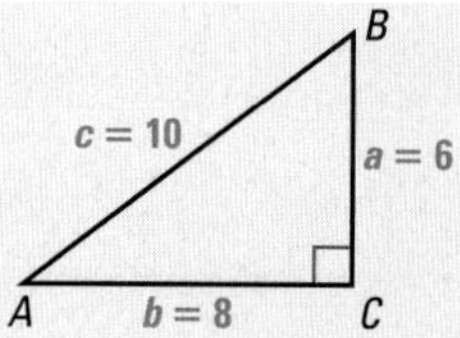

22. Find the sine, the cosine, and the tangent of $\angle B$. **0.8, 0.6, $1.\overline{3}$**

12.8 LOGICAL REASONING: PROOF

Examples on pp. 758–761

EXAMPLE Prove that for all numbers a and b, $(a + b) - b = a$.

$(a + b) - b = (a + b) + (-b)$	**Definition of subtraction**
$= a + [b + (-b)]$	**Associative property of addition**
$= a + 0$	**Inverse property of addition**
$= a$	**Identity property of addition**

23. Which basic axiom of algebra is represented by $\left(\frac{2}{3}\right)\left(\frac{4}{5}\right) = \left(\frac{4}{5}\right)\left(\frac{2}{3}\right)$? **Commutative property of multiplication**

24. Prove that $(c)(-b) = -cb$ for all real numbers c and b. **See margin.**

CHAPTER 12

Chapter Test

Identify the domain and the range of the function. Then sketch its graph. 1–4. See margin.

1. $y = 12\sqrt{x}$ **2.** $y = \sqrt{2x + 7}$ **3.** $y = \sqrt{3x} - 3$ **4.** $y = \sqrt{x - 5}$

Simplify the expression.

5. $3\sqrt{2} - \sqrt{2}$ $2\sqrt{2}$ **6.** $(4 + \sqrt{7})(4 - \sqrt{7})$ 9 **7.** $(4\sqrt{5} + 1)^2$ $81 + 8\sqrt{5}$ **8.** $\frac{8}{3 - \sqrt{5}}$ $6 + 2\sqrt{5}$

Solve the equation.

9. $\sqrt{y} + 6 = 10$ 16 **10.** $\sqrt{2m + 3} - 6 = 4$ $48\frac{1}{2}$ **11.** $n = \sqrt{9n - 18}$ 3, 6 **12.** $p = \sqrt{-3p + 18}$ 3

Solve the equation by completing the square.

13. $x^2 - 6x = -5$ 1, 5 **14.** $x^2 - 2x = 2$ $1 - \sqrt{3}, 1 + \sqrt{3}$ **15.** $x^2 + \frac{4}{5}x - 1 = 0$ $\frac{-2 - \sqrt{29}}{5}, \frac{-2 + \sqrt{29}}{5}$

Determine whether the given lengths can be the sides of a right triangle.

16. 6, 18, 36 no **17.** 9, 40, 41 yes **18.** 1.5, 3.6, 3.9 yes

In Exercises 19 and 20, use the diagram.

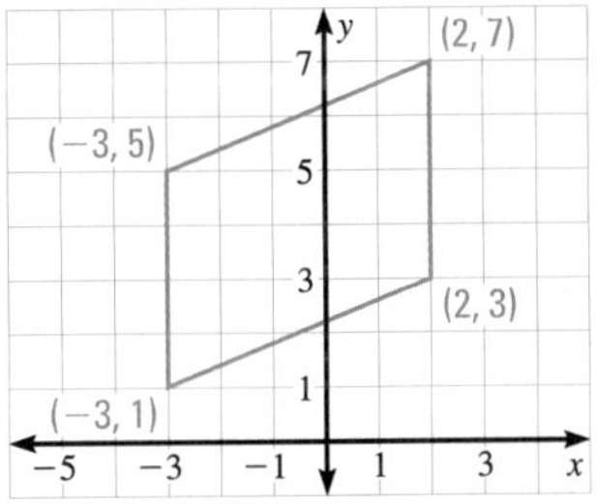

19. Find the perimeter of the parallelogram. $8 + 2\sqrt{29} \approx 18.77$

20. Find the coordinates of the midpoint of each side of the parallelogram. 20. $(-3, 3), \left(-\frac{1}{2}, 6\right), (2, 5), \left(-\frac{1}{2}, 2\right)$

21. Find the perimeter of the parallelogram whose vertices are the midpoints you found in Exercise 20. $\sqrt{61} + \sqrt{29} \approx 13.20$

22. Compare the perimeters you found in Exercises 19 and 21. The perimeter of the original parallelogram is greater than the perimeter of the new parallelogram.

Find the missing lengths of the sides of the triangles. Round your answer to the nearest hundredth.

23. 10, 14.14

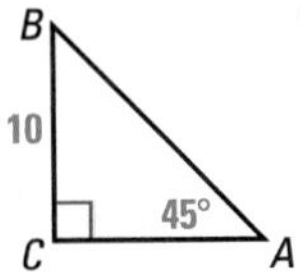

24. 8.71, 13.42

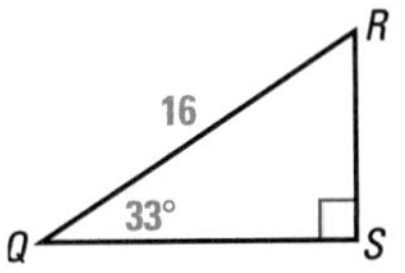

25. 15.45, 14.93

26. Prove that if a, b, and c are real numbers, and $a + c = b + c$, then $a = b$.

26. $(a + c) + (-c) = (b + c) + (-c)$
Addition property of equality
$(a + [c + (-c)] = b + [c + (-c)]$
Associative property of addition
$a + 0 = b + 0$
Inverse property of addition
$a = b$
Identity property of addition

27. PONY EXPRESS Pony express stations were about 10 miles apart. The latitude-longitude coordinates of two former stations in western Nevada are (39.2 N, 119.0 W) and (39.2 N, 118.5 W). Find the coordinates of the station that was halfway between them. about (39.2 N, 118.75 W)

28. GATEWAY ARCH You are standing 134 feet from the Gateway Arch in St. Louis, Missouri. You estimate that the angle between the ground and the line from where you stand to the top of the arch is 78°. Estimate the height of the Gateway Arch. about 630 ft

ADDITIONAL RESOURCES

- ***Chapter 12 Resource Book***
  Chapter Test (3 levels) (p. 121)
  SAT/ACT Chapter Test (p. 127)
  Alternative Assessment (p. 128)
- ***Test and Practice Generator***

1. Both the domain and range are all nonnegative numbers.

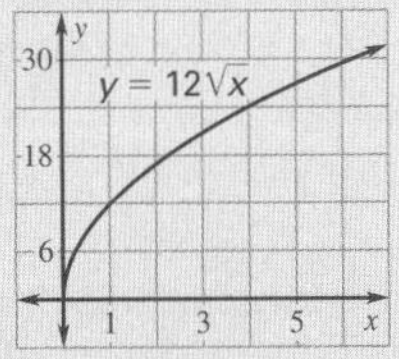

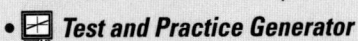

2. domain: all numbers greater than or equal to $-3\frac{1}{2}$; range: all nonnegative numbers

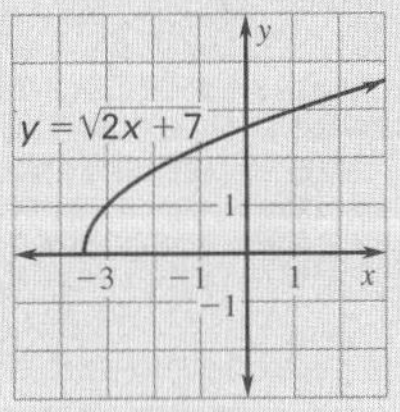

3. domain: all nonnegative numbers; range: all numbers greater than or equal to −3

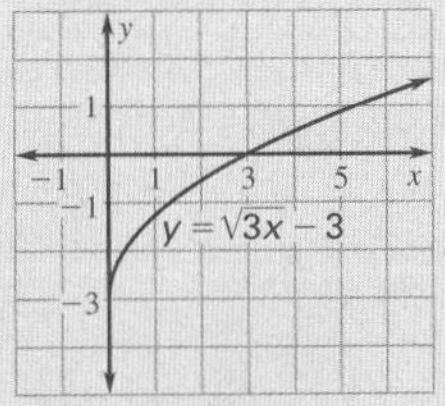

4. domain: all numbers greater than or equal to 5; range: all nonnegative numbers

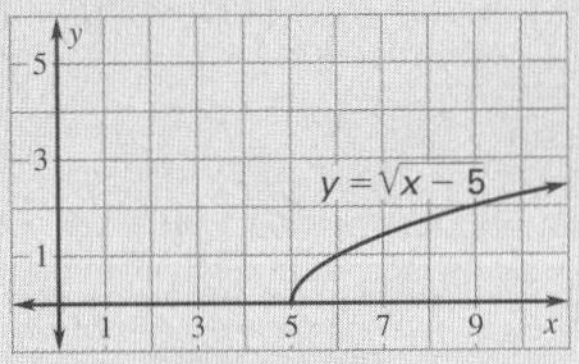

ADDITIONAL RESOURCES

- ***Chapter 12 Resource Book***
  Chapter Test (3 levels) (p. 121)
  SAT/ACT Chapter Test (p. 127)
  Alternative Assessment (p. 128)
- ***Test and Practice Generator***

CHAPTER 12

Chapter Standardized Test

TEST-TAKING STRATEGY Learn as much as you can about a test ahead of time, such as the types of questions and the topics that the test will cover.

1. **MULTIPLE CHOICE** What is the value of $f(x) = \frac{x\sqrt{x^2 - 1}}{x^2 + 8}$ when $x = 8$? **C**

   Ⓐ $\frac{3\sqrt{7}}{16}$ Ⓑ $\frac{7}{8}$ Ⓒ $\frac{\sqrt{7}}{3}$
   Ⓓ $\frac{8}{9}$ Ⓔ $3\sqrt{7}$

2. **MULTIPLE CHOICE** Which quadrants of the coordinate plane will contain the graph of $y = \sqrt{3x - 1} - 2$? **B**

   Ⓐ Quadrant I
   Ⓑ Quadrants I and IV
   Ⓒ Quadrant IV
   Ⓓ Quadrants II and III
   Ⓔ Quadrants I and II

3. **MULTIPLE CHOICE** Evaluate the expression $5\sqrt{7} + \sqrt{448} + \sqrt{175} - \sqrt{63}$. **A**

   Ⓐ $15\sqrt{7}$ Ⓑ $16\sqrt{7}$ Ⓒ $18\sqrt{7}$
   Ⓓ $20\sqrt{7}$ Ⓔ $21\sqrt{7}$

4. **MULTIPLE CHOICE** Which one of the following is the simplified expression of $(3 - \sqrt{6})^2$? **D**

   Ⓐ $9 - 5\sqrt{6}$ Ⓑ $3 - 6\sqrt{6}$
   Ⓒ 15 Ⓓ $15 - 6\sqrt{6}$
   Ⓔ None of these

5. **MULTIPLE CHOICE** The geometric mean of 144 and a is 6. What is a? **B**

   Ⓐ $\frac{1}{24}$ Ⓑ $\frac{1}{4}$ Ⓒ $\frac{1}{2}$
   Ⓓ 12 Ⓔ 36

6. **MULTIPLE CHOICE** Which one of the following is a solution of the equation $x = \sqrt{880 - 18x}$? **C**

   Ⓐ −22 Ⓑ 0 Ⓒ 22
   Ⓓ 40 Ⓔ 49

7. **MULTIPLE CHOICE** What term should be added to $x^2 - \frac{4}{3}x$ so that the result is a perfect square trinomial? **C**

   Ⓐ $\frac{-2}{3}$ Ⓑ $\frac{64}{9}$ Ⓒ $\frac{4}{9}$
   Ⓓ $\frac{-8}{3}$ Ⓔ $\frac{2}{3}$

8. **MULTIPLE CHOICE** What is the length of the missing side of the triangle? **D**

   Ⓐ 10 Ⓑ 11
   Ⓒ 12 Ⓓ 13
   Ⓔ 15

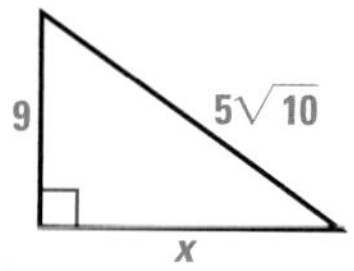

9. **MULTIPLE CHOICE** What is the distance between points P and Q? **E**

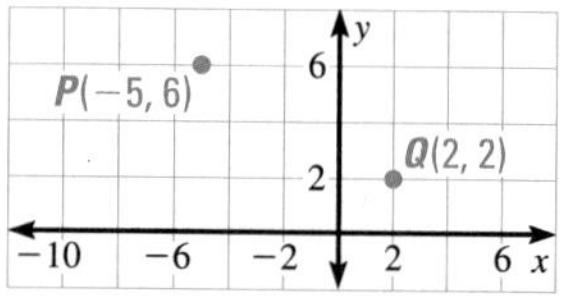

   Ⓐ 5 Ⓑ $\sqrt{33}$ Ⓒ $\sqrt{113}$
   Ⓓ $\sqrt{73}$ Ⓔ $\sqrt{65}$

10. **MULTIPLE CHOICE** Use the graph in Exercise 9. Find the midpoint between points P and Q. **A**

   Ⓐ $\left(-\frac{3}{2}, 4\right)$ Ⓑ $\left(-\frac{7}{2}, 2\right)$ Ⓒ $\left(-\frac{3}{2}, 2\right)$
   Ⓓ $\left(-\frac{7}{2}, 4\right)$ Ⓔ $\left(\frac{7}{2}, 2\right)$

11. **MULTIPLE CHOICE** The length of side a is 18. The tangent of angle A is $\frac{9}{5}$. What is the length of the hypotenuse? **C**

   Ⓐ 10
   Ⓑ 20
   Ⓒ $2\sqrt{106}$
   Ⓓ $4\sqrt{106}$
   Ⓔ $8\sqrt{53}$

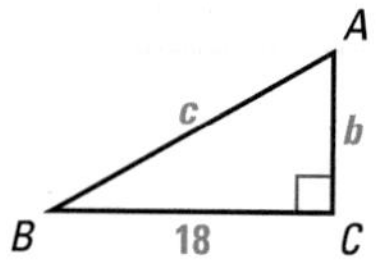

QUANTITATIVE COMPARISON In Exercises 12–15, choose the statement below that is true about the numbers.

Ⓐ The number in column A is greater.

Ⓑ The number in column B is greater.

Ⓒ The two numbers are equal.

Ⓓ The relationship cannot be determined from the given information.

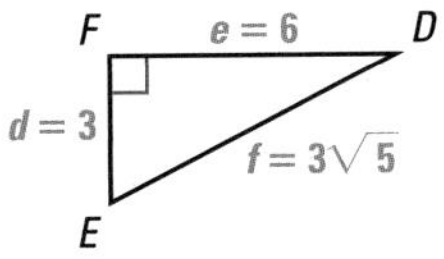

	Column A	Column B	
12.	$\sin D$	$\sin E$	B
13.	$\cos D$	$\sin D$	A
14.	$\cos D$	$\cos E$	A
15.	$\sin E$	$\cos E$	A

16. MULTIPLE CHOICE Choose the missing reason in the following proof that for all real numbers a and b, $-(a + b) = (-a) + (-b)$. D

STATEMENTS	REASONS
1. a and b are real numbers	**1.** Given
2. $-(a + b) = (-1)(a + b)$	**2.** Multiplicative property of -1
3. $-(a + b) = (-1)a + (-1)b$	**3.** ______________________
4. $-(a + b) = (-a) + (-b)$	**4.** Multiplicative property of -1

Ⓐ Definition of subtraction

Ⓑ Associative property of addition

Ⓒ Inverse property of addition

Ⓓ Distributive property

Ⓔ Multiplication axiom of equality

17. MULTI-STEP PROBLEM You have a very small television set in your home now. The length of your little TV screen is 8 inches, but the TV is called a 10-inch TV because the length of the screen's diagonal is 10 inches. After shopping for a larger television, you find two televisions that interest you. One is a 25-inch TV with a screen length of 20 inches. The other TV is advertised as a big screen because it is "double the dimensions of the 25-inch TV."

a. Write an equation and solve it to find the width of the 25-inch screen. $w^2 + 20^2 = 25^2$; 15 in.

b. Find the length and width of the big screen television. 30 in. by 40 in.

c. What is the area of the 25-inch screen? of the big screen? 300 in.^{2}; 1200 in.^{2}

d. How many times larger is the area of the big screen than the area of the 25-inch screen? 4 times

e. Suppose that the television manufacturer produced a 75-inch screen TV. How many times larger is the area of the 75-inch screen than the area of the 25-inch screen? 9 times

ADDITIONAL RESOURCES

A Cumulative Review covering Chapters 1–12 is available in the *Chapter 12 Resource Book,* p. 132.

19. Accept any equation of the form $y = \frac{2}{3}x + b$ for any number b except -1.

20. Accept any equation of the form $y = -\frac{3}{2}x + b$ for any number b; check graphs.

27. $-\frac{5}{4}$

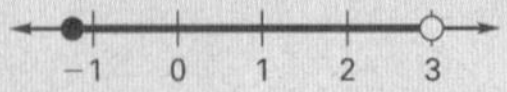

28.

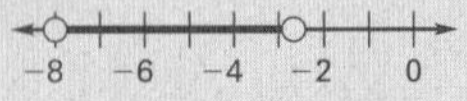

29.

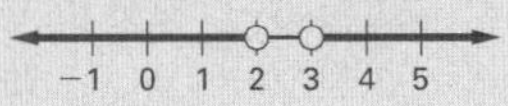

CHAPTER 12

Cumulative Practice

Translate the verbal sentence into an equation or an inequality. Then use the information to solve for the variable(s). (1.4–1.5)

1. The quotient of m and 7 is greater than or equal to 16. $\frac{m}{7} \geq 16$; $m \geq 112$
2. The sum of 4 and the second power of b is equal to 104. $4 + b^2 = 104$; $-10, 10$
3. The distance t you travel by train is $3\frac{2}{3}$ times the distance d you live from the train station. You drive 3 miles to get from your house to the train station. $t = 3\frac{2}{3}(3)$; 11 mi

Evaluate the expression for the given value of *x*. (2.1–2.3, 2.5–2.7)

4. $3 + x + (-4), x = 5$ 4
5. $-x + 12 - 5, x = 9$ -2
6. $3.5 - (-x), x = 1.5$ 5
7. $-(-3)^2(x), x = 7$ -63
8. $6x(x + 2), x = 2$ 48
9. $(8x + 1)(-3), x = \frac{1}{2}$ -15
10. $\frac{1}{4}\left|(-x)(-x)(-x)\right|, x = 4$ 16
11. $\frac{x^2 + 4}{6}, x = 8$ $11\frac{1}{3}$
12. $(-5)\left(-\frac{3}{4}x\right), x = 6$ $22\frac{1}{2}$

Solve the equation. Round the result to the nearest hundredth. (3.1–3.4, 3.6)

13. $-\frac{2}{9}(x - 5) = 12$ -49
14. $7x - (3x - 2) = 38$ 9
15. $\frac{1}{3}x + 7 = -7x - 5$ $-1\frac{7}{11}$
16. $8(x + 3) - 2x = 4(x - 8)$ -28
17. $11.47 + 6.23x = 7.62 + 5.51x$ -5.35
18. $-3(2.98 - 4.1x) = 9.2x + 6.25$ 4.9

In Exercises 19 and 20, use the graph. (4.3–4.4, 4.6, 5.3)
19–20. See margin.

19. Write an equation of a line that is parallel to the line shown.
20. Write an equation of a line that is perpendicular to the line shown. Graph the equation in the same coordinate plane to check your answer.

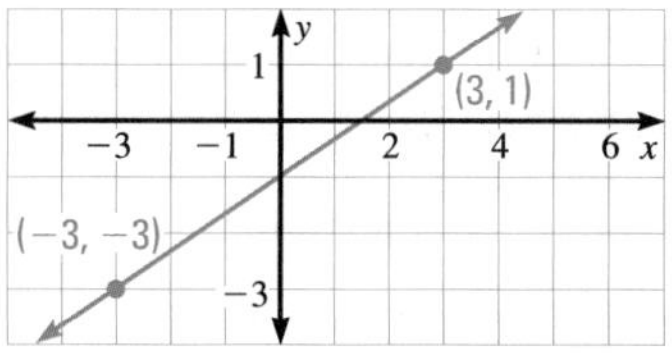

Decide whether the relation is a function. If it is a function, give the domain and the range. (4.8)

21. function; domain: −1, 1, 3, 5, 7; range: −1, 1, 3, 5

Input	Output
−1	−1
1	−1
3	1
5	3
7	5

22.

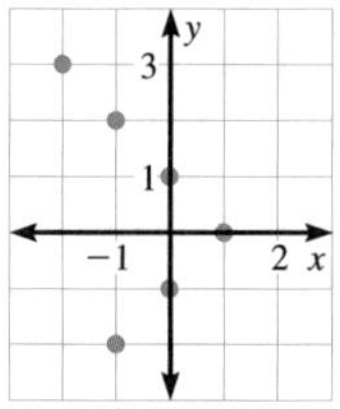

not a function

23. function; domain: −2, −1, 0, 1, 2; range: −2, −1, 0, 1

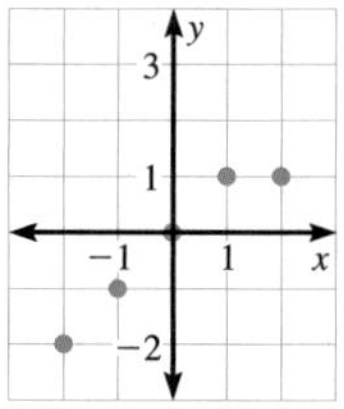

24.

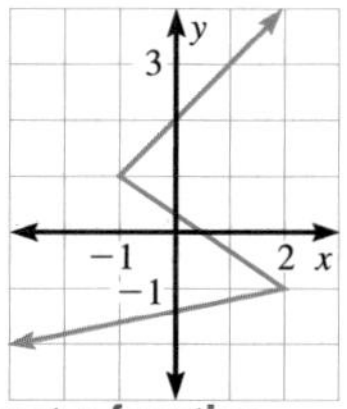

not a function

Use the information provided to write an equation of the line in standard form. (5.1–5.2, 5.5–5.6)

25. The slope is $\frac{4}{5}$; the y-intercept is -3. $4x - 5y = 15$
26. The line passes through $(-1, 2)$; the slope is $\frac{1}{3}$. $x - 3y = -7$

Solve the inequality and graph the solution. (6.3–6.4)

27. $-3 < -4x + 9 \leq 14$
$-\frac{5}{4} \leq x < 3$. See margin.
28. $|3x + 16| + 2 < 10$
$-8 < x < -2\frac{2}{3}$. See margin.
29. $3x - 4 > 5$ or $5x + 1 < 11$
$x > 3$ or $x < 2$. See margin.

Solve the system of linear equations. (7.2–7.3)

30. $4y = -8x + 16$ (1, 2)
$2y = 11x - 7$

31. $-2x + 3y = 15$ (24, 21)
$10x - 11y = 9$

32. $y = 5x - 2$ $\left(\frac{1}{2}, \frac{1}{2}\right)$
$3x + 7y = 5$

Simplify. Then evaluate the expression when $a = 1$ and $b = 2$. (8.1–8.3)

33. $\frac{b^8}{b^2}$ b^6; 64

34. $3a^4 \cdot a^{-3}$ $3a$; 3

35. $(-a^3)(2b^2)^3$ $-8a^3b^6$; -512

36. $4b^3 \cdot (2 + b)^2$
$16b^3 + 16b^4 + 4b^5$; 512

37. $\frac{4a^{-3}b^3}{ab^{-2}}$ $\frac{4b^5}{a^4}$; 128

38. $\frac{(5ab^2)^{-2}}{a^{-3}b}$ $\frac{a}{25b^5}$; $\frac{1}{800}$

Decide how many solutions the equation has. Then solve the equation. (9.1, 9.5–9.6, 10.5)

39. $6x^2 + 8 = 34$
two solutions; $\frac{\sqrt{39}}{3}$, $-\frac{\sqrt{39}}{3}$

40. $4x^2 - 9x + 5 = 0$
two solutions; 1, $1\frac{1}{4}$

41. $3x^2 + 6x + 3 = 0$
one solution; -1

Completely factor the expression. (10.5–10.7)

42. $x^2 + 6x + 8$ $(x + 4)(x + 2)$

43. $x^2 - 24x - 112$ $(x - 28)(x + 4)$

44. $3x^2 + 17x - 6$ $(3x - 1)(x + 6)$

45. $4x^2 + 12x + 9$ $(2x + 3)^2$

46. $x^2 + 10x + 25$ $(x + 5)^2$

47. $x^2 - 14x + 49$ $(x - 7)^2$

Solve the equation. (10.4–10.8)

48. $(3x + 1)(2x + 7) = 0$ $-3\frac{1}{2}$, $-\frac{1}{3}$

49. $6x^2 - x - 7 = 8$ $-1\frac{1}{2}$, $1\frac{2}{3}$

50. $4x^2 + 16x + 16 = 0$ -2

51. $x^3 + 5x^2 - 4x - 20 = 0$
-5, -2, 2

52. $x^4 + 9x^3 + 18x^2 = 0$ -6, -3, 0

53. $x^2 - \frac{4}{3}x + \frac{4}{9} = 0$ $\frac{2}{3}$

Simplify the expression. (11.4–11.6)

54. $\frac{4x}{12x^2}$ $\frac{1}{3x}$

55. $\frac{2x + 6}{x^2 - 9}$ $\frac{2}{x - 3}$

56. $\frac{3x}{x^2 - 2x - 24} \cdot \frac{x - 6}{6x^2 + 9x}$ $\frac{1}{(x + 4)(2x + 3)}$

57. $\frac{x^2 - 6x + 8}{x^2 - 2x} \div (3x - 12)$ $\frac{1}{3x}$

58. $\frac{4}{x + 2} + \frac{15x}{3x + 6}$ $\frac{5x + 4}{x + 2}$

59. $\frac{3x}{x + 4} - \frac{x}{x - 1}$ $\frac{x(2x - 7)}{(x + 4)(x - 1)}$

Simplify the expression. (12.2)

60. $4\sqrt{7} + 3\sqrt{7}$ $7\sqrt{7}$

61. $9\sqrt{2} - 12\sqrt{8}$ $-15\sqrt{2}$

62. $\sqrt{6}(5\sqrt{3} + 6)$
$15\sqrt{2} + 6\sqrt{6}$

63. $\frac{11}{7 - \sqrt{3}}$ $\frac{77 + 11\sqrt{3}}{46}$

In Exercises 64 and 65, use the triangles at the right. (12.7)

64. Find the sine, the cosine, and the tangent of $\angle Q$ and $\angle R$. Round your answer to the nearest hundredth. See margin.

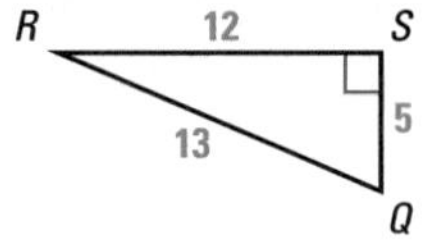

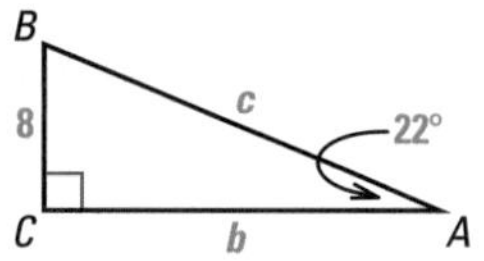

65. Find the lengths of the sides of $\triangle ABC$. Tan 22° = 0.4040. Round your answer to the nearest hundredth. $b = 19.80$, $c = 21.36$

64. $\sin Q = \frac{12}{13}$, $\cos Q = \frac{5}{13}$,
$\tan Q = \frac{12}{5}$; $\sin R = \frac{5}{13}$,
$\cos R = \frac{12}{13}$, $\tan R = \frac{5}{12}$

PVR SUBSCRIPTIONS In Exercises 66 and 67, Company A sells a Personal Video Recorder (PVR) for \$700 and charges a \$5 monthly subscription fee. Company B sells a PVR for \$500 and charges a \$20 monthly fee. (6.2, 7.3–7.4)

66. What is the total cost of each PVR after one year? after three years?
Company A PVR: \$760, \$880; Company B PVR: \$740, \$1220

67. How many months must the PVRs be used in order for the total costs of the two models to be the same? $13\frac{1}{3}$ months

PROJECT OVERVIEW

MATHEMATICAL GOALS

- Complete a table of values.
- Determine a pattern in a table.
- Solve a quadratic equation.

MANAGING THE PROJECT

CLASSROOM MANAGEMENT
The Chapter 12 project may be completed by individual students or in groups of two. If groups are used, one student should draw the figures. The second student should record the results. Both members should discuss the results. They should complete Step 4 together.

If the members choose to write a report, they should create an outline for the report together. Both members can then write a draft of the report. They can collaborate to create the finished report. Each group member should include a copy of the report in his or her portfolio.

If the members choose to make a poster, they should discuss the elements to be put on the poster and then decide which elements each member will create.

ALTERNATIVE APPROACH
Ask for three volunteers to draw the figures at the chalkboard by approximating a square. Lead a discussion of the results and ask a fourth student to record the results in a table at the chalkboard. Have each student complete Step 4. Then compare results. Point out errors in calculations, where necessary.

CONCLUDING THE PROJECT

Have one group member present the report or project to the class. The other member can present and explain the sketch and table.

GRADING THE PROJECT

4 Students complete the drawing and record the results correctly. They record the correct ratios and clearly and correctly explain how the ratios are related to the golden ratio. They correctly show the value of b when $a = 1$. They predict the

Investigating the Golden Ratio

OBJECTIVE **Explore what the golden ratio is and how it is used.**

Materials: graph paper, metric ruler, graphing calculator (optional)

Over the centuries, the golden rectangle has fascinated artists, architects, and mathematicians. A golden rectangle has the special shape such that when a square is cut from one end, the ratio of length to width of the remaining rectangle is equal to the ratio of length to width of the original rectangle. This ratio, the *golden ratio*, is $\frac{1 + \sqrt{5}}{2}$, or about 1.618034.

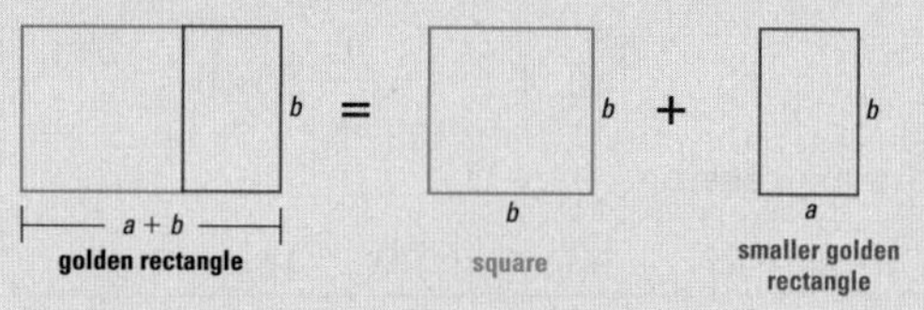

It is *not* possible to constuct a golden rectangle with integer sidelengths. However, it *is* possible to construct rectangles with integer sidelengths whose ratios of length to width approximate the golden ratio.

APPROXIMATING A GOLDEN RECTANGLE

1. On graph paper, draw a 1-by-1 square.

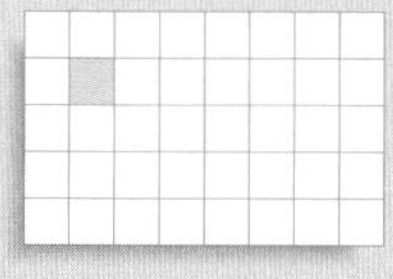

2. On one side of the square, add another 1-by-1 square.

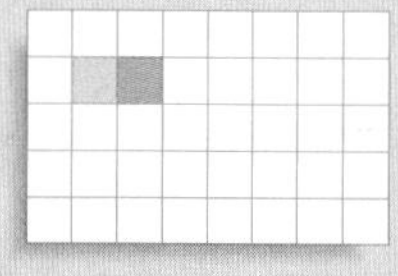

3. Build a 2-by-2 square on the longest side of your 1-by-2 rectangle.

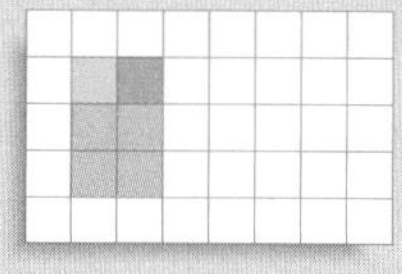

4. Build a 3-by-3 square on the longest side of your 2-by-3 rectangle.

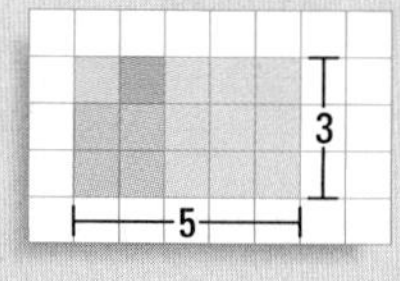

Steps 1–4. Check drawings.

INVESTIGATING THE GOLDEN RATIO

One property of golden rectangles is that their lengths and widths are proportional. For example, in the above diagram, $\frac{a + b}{b} = \frac{b}{a}$.

1. Copy and complete the table by using this property to predict the lengths and widths of the next four rectangles you can build on the drawing from **Step 4**. **length, b: 8, 13, 21, 34; width; a: 5, 8, 13, 21**

length, b	3	5	?	?	?	?
width, a	2	3	?	?	?	?

2. Draw the next four rectangles to check your predictions. **Check drawings.**
3. Add a row to your table and record the ratio of the length to the width of the rectangles. Explain how the ratios are related to the golden ratio. **See margin.**
4. Let $r = \frac{b}{a}$ represent the golden ratio. Use $\frac{a + b}{b} = \frac{b}{a}$ to show that $r = \frac{1 + \sqrt{5}}{2}$ if $a = 1$. **See margin.**

3. $\frac{\text{length}}{\text{width}}$**: 1.5, $1.\overline{6}$, 1.625, about 1.62, about 1.62; *Sample answer*: The ratios get closer and closer to the golden ratio.**

4. $b = \frac{b + 1}{b}$**, or** $b^2 = b + 1$**; using the quadratic formula and noting that b (which is a length) must be positive,** $b = \frac{1 + \sqrt{5}}{2}$**.**

PRESENTING YOUR RESULTS

Write a report or make a poster to present your results.

- Include a sketch of a golden rectangle.
- Include the table you made in Exercise 1.
- Include your answers to Exercises 2–4.
- Describe what you learned about the golden ratio.

EXTENSIONS

- Before the pediment on top of the Parthenon in Athens was destroyed, the front of the building fit almost exactly into a golden rectangle. Do some research on the Parthenon to find out if the golden ratio was used in other aspects of its design.

- The golden rectangle can be found in the works of many artists and in art objects around the world, such as the proportions of a jar from the Ch'ing dynasty of China, shown at the right. Find pictures of works of art and see if you can find any golden rectangles in them.

- The golden ratio can also be found in nature. For example, the average chicken egg fits nicely inside a golden rectangle. Measure at least six chicken eggs and find the average of their ratios.

- Do some research and find other examples of the golden ratio in art, nature, and architecture.
- Find some rectangular objects around you that you think look nice. Measure them to see if they are golden rectangles. You might try a picture frame, a $1 bill, a business card, a 3-by-5 index card, or a TV screen.

dimensions of the next few rectangles and extend the sketch to check their predictions. They calculate the ratio of the lengths to the widths for the next few rectangles. They extended the project by completing any of the research activities or by calculating the average length of six eggs. Their report or poster contains a clear sketch and a table of values that is easy to read. Both posters and reports are clearly presented.

3 Students complete the drawing and table, calculate the ratios, and solve the quadratic equation, but they do not calculate all the ratios correctly or they do not solve the equation correctly. They do not relate all the mathematical ideas to the data. Students include all the information required for the report or poster and show they extended the project by completing any of the research activities or by calculating the average length of six eggs. The report and the poster are clear and show effort.

2 Students complete the drawing but may not accurately record, calculate, or describe the data, or solve the quadratic equation. They do not complete any of the extensions. A report is written or a poster is created, but either may be missing a sketch of the golden rectangle or the table. The report or poster does not include all the ideas necessary to convey meaning.

1 Students do not show that they understand the mathematical ideas and their work is incomplete. The explanations of the relationships between the rectangle ratios and the golden ratio are incorrect and the entries in the table are incomplete. The equation is not solved and the report or poster does not convey understanding or show a significant effort to complete the activity. Students should speak with the teacher to review their work and make a new start.

Contents of Student Resources

Skills Review Handbook

FACTORS AND MULTIPLES

The **natural numbers** are all the numbers in the sequence 1, 2, 3, 4, 5, ... Natural numbers that are multiplied are **factors**. For example, 3 and 7 are factors of 21, because $3 \cdot 7 = 21$. A **prime number** is a natural number that has exactly two factors, itself and 1.

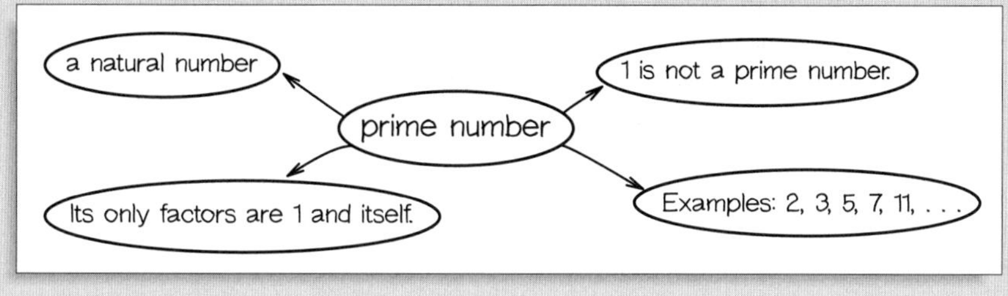

To write the **prime factorization** of a number, write the number as a product of prime numbers.

EXAMPLE Write the prime factorization of 315.

SOLUTION Use a tree diagram to factor the number until all factors are prime numbers. To determine the factors, test the prime numbers in order.

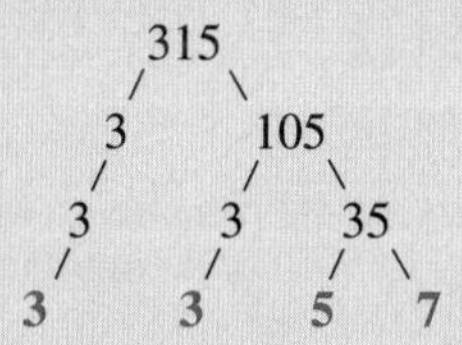

▶ The prime factorization of 315 is $3 \cdot 3 \cdot 5 \cdot 7$. This may also be written as $3^2 \cdot 5 \cdot 7$.

A **common factor** of two natural numbers is a number that is a factor of both numbers. For example, 7 is a common factor of 35 and 56, because $35 = 5 \cdot 7$ and $56 = 8 \cdot 7$. The **greatest common factor** (GCF) of two natural numbers is the largest number that is a factor of both.

EXAMPLE Find the greatest common factor of 180 and 84.

SOLUTION First write the prime factorization of each number. Multiply the common prime factors to find the greatest common factor.

$180 = 2 \cdot 2 \cdot 3 \cdot 3 \cdot 5$

$84 = 2 \cdot 2 \cdot 3 \cdot 7$

▶ The greatest common factor is $2 \cdot 2 \cdot 3 = 12$.

A **common multiple** of two natural numbers is a number that is a multiple of both numbers. For example, **42** is a common multiple of 6 and 14, because **42** = 6 • 7 and **42** = 14 • 3. The **least common multiple** (LCM) of two natural numbers is the smallest number that is a multiple of both.

EXAMPLE Find the least common multiple of 24 and 30.

SOLUTION First write the prime factorization of each number.

24 = 2 • 2 • 2 • 3

30 = 2 • 3 • 5

▶ The least common multiple is the product of the common prime factors and all the prime factors that are not common. The least common multiple of 24 and 30 is 2 • 3 • 2 • 2 • 5 = 120.

PRACTICE

List all the factors of the number.

1. 18 **1, 2, 3, 6, 9, 18**
2. 10 **1, 2, 5, 10**
3. 77 **1, 7, 11, 77**
4. 35 **1, 5, 7, 35**
5. 27 **1, 3, 9, 27**
6. 100 **1, 2, 4, 5, 10, 20, 25, 50, 100**
7. 42 **1, 2, 3, 6, 7, 14, 21, 42**
8. 49 **1, 7, 49**
9. 52 **1, 2, 4, 13, 26, 52**
10. 81 **1, 3, 9, 27, 81**
11. 121 **1, 11, 121**
12. 150 **1, 2, 3, 5, 6, 10, 15, 25, 30, 50, 75, 150**

Write the prime factorization of the number if it is not a prime number. If a number is prime, write *prime*.

13. 27 3^3
14. 24 $2^3 \cdot 3$
15. 32 2^5
16. 61 **prime**
17. 55 $5 \cdot 11$
18. 68 $2^2 \cdot 17$
19. 48 $2^4 \cdot 3$
20. 225 $3^2 \cdot 5^2$
21. 90 $2 \cdot 3^2 \cdot 5$
22. 75 $3 \cdot 5^2$
23. 39 $3 \cdot 13$
24. 1000 $2^3 \cdot 5^3$
25. 728 $2^3 \cdot 7 \cdot 13$
26. 101 **prime**
27. 512 2^9
28. 210 $2 \cdot 3 \cdot 5 \cdot 7$

List all the common factors of the pair of numbers.

29. 15, 30 **1, 3, 5, 15**
30. 36, 54 **1, 2, 3, 6, 9, 18**
31. 5, 20 **1, 5**
32. 14, 21 **1, 7**
33. 9, 36 **1, 3, 9**
34. 24, 28 **1, 2, 4**
35. 20, 55 **1, 5**
36. 12, 30 **1, 2, 3, 6**

Find the greatest common factor of the pair of numbers.

37. 25, 30 **5**
38. 32, 40 **8**
39. 17, 24 **1**
40. 35, 150 **5**
41. 14, 28 **14**
42. 65, 39 **13**
43. 102, 51 **51**
44. 128, 104 **8**
45. 36, 50 **2**
46. 45, 135 **45**
47. 29, 87 **29**
48. 35, 48 **1**
49. 56, 70 **14**
50. 88, 231 **11**
51. 47, 48 **1**
52. 93, 124 **31**

Find the least common multiple of the pair of numbers.

53. 5, 7 **35**
54. 3, 4 **12**
55. 3, 16 **48**
56. 7, 12 **84**
57. 4, 12 **12**
58. 9, 15 **45**
59. 12, 35 **420**
60. 6, 14 **42**
61. 20, 25 **100**
62. 10, 24 **120**
63. 3, 17 **51**
64. 15, 40 **120**
65. 70, 14 **70**
66. 36, 50 **900**
67. 22, 30 **330**

COMPARING AND ORDERING NUMBERS

When you compare two numbers a and b, there are exactly three possibilities. The possibilities are described at the right in words and in symbols. To compare two whole numbers or decimals, compare the digits of the two numbers from left to right. Find the first place in which the digits are different.

a is *less than b.*	$a < b$
a is *equal to b.*	$a = b$
a is *greater than b.*	$a > b$

EXAMPLE Compare the two numbers. Write the answer using <, =, or >.

a. 4723 and 4732 **b.** 27.52 and 27.39

SOLUTION

a. 4 7 2 3

4 7 3 2

▶ 2 < 3, so 4723 < 4732.

You can picture this on a number line. The numbers on a number line increase from left to right.

4723 is *less* than 4732.

4723 is to the *left* of 4732.

b. 2 7 . 5 2

2 7 . 3 9

▶ 5 > 3, so 27.52 > 27.39.

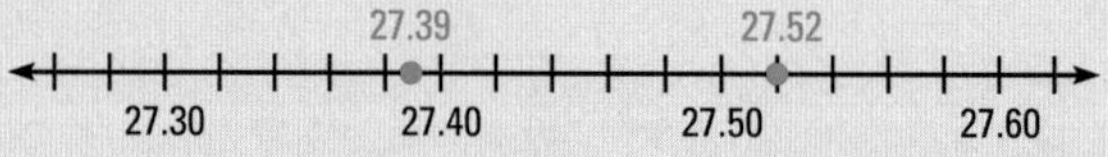

27.52 is *greater* than 27.39.

27.52 is to the *right* of 27.39.

To compare two fractions that have the same denominator, compare the numerators. If the fractions have different denominators, first rewrite one or both fractions to produce equivalent fractions with a common denominator. The **least common denominator** (LCD) is the least common multiple of the denominators.

EXAMPLE Write the numbers $\frac{3}{4}$, $\frac{7}{8}$, and $\frac{5}{12}$ in order from least to greatest.

SOLUTION The LCD of the fractions is 24.

$\frac{3}{4} = \frac{3 \cdot 6}{4 \cdot 6} = \frac{18}{24}$ $\frac{7}{8} = \frac{7 \cdot 3}{8 \cdot 3} = \frac{21}{24}$ $\frac{5}{12} = \frac{5 \cdot 2}{12 \cdot 2} = \frac{10}{24}$

Compare the numerators: $10 < 18 < 21$, so $\frac{5}{12} < \frac{3}{4} < \frac{7}{8}$.

▶ In order from least to greatest, the fractions are $\frac{5}{12}$, $\frac{3}{4}$, and $\frac{7}{8}$.

779

EXAMPLE Compare $4\frac{3}{4}$ and $4\frac{2}{3}$. Write the answer using $<$, $=$, or $>$.

SOLUTION The whole number parts of the mixed numbers are the same, so compare the fraction parts.

The LCD of $\frac{3}{4}$ and $\frac{2}{3}$ is 12.

$$\frac{3}{4} = \frac{3 \cdot 3}{4 \cdot 3} = \frac{9}{12} \qquad \frac{2}{3} = \frac{2 \cdot 4}{3 \cdot 4} = \frac{8}{12}$$

Compare the numerators: $9 > 8$, so $\frac{9}{12} > \frac{8}{12}$.

▶ Since $\frac{3}{4} > \frac{2}{3}$, it follows that $4\frac{3}{4} > 4\frac{2}{3}$.

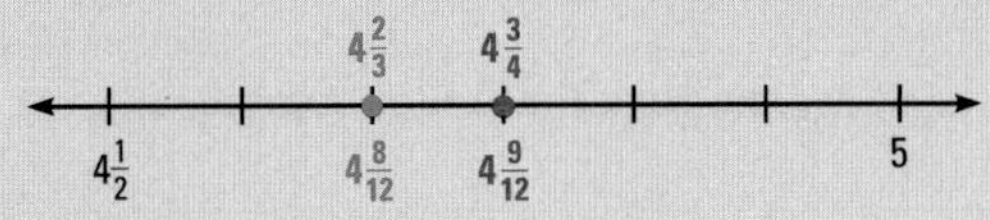

$4\frac{3}{4}$ is to the right of $4\frac{2}{3}$.

PRACTICE

Compare the two numbers. Write the answer using <, =, or >.

1. 12,428 and 15,116 $<$
2. 905 and 961 $<$
3. 142,109 and 140,999 $>$
4. 16.82 and 14.09 $>$
5. 0.00456 and 0.40506 $<$
6. 23.03 and 23.3 $<$
7. 1005.2 and 1050.7 $<$
8. 932,778 and 934,112 $<$
9. 0.058 and 0.102 $<$
10. $\frac{7}{13}$ and $\frac{3}{13}$ $>$
11. $17\frac{1}{4}$ and $15\frac{11}{12}$ $>$
12. $\frac{7}{10}$ and $\frac{3}{4}$ $<$
13. $\frac{5}{9}$ and $\frac{15}{27}$ $=$
14. $\frac{1}{2}$ and $\frac{3}{8}$ $>$
15. $\frac{1}{8}$ and $\frac{1}{9}$ $>$
16. $\frac{4}{5}$ and $\frac{2}{3}$ $>$
17. $42\frac{1}{5}$ and $41\frac{7}{8}$ $>$
18. 508.881 and 508.793 $>$
19. 32,227 and 32,226.5 $>$
20. $\frac{5}{8}$ and $\frac{2}{3}$ $<$
21. $17\frac{5}{6}$ and $17\frac{5}{7}$ $>$

Write the numbers in order from least to greatest.

22. 1207, 1702, 1220, 1772
    1207, 1220, 1702, 1772
23. 45,617, 45,242, 40,099, 40,071
    40,071; 40,099; 45,242; 45,617
24. 23.12, 23.5, 24.0, 23.08, 24.01
    23.08, 23.12, 23.5, 24.0, 24.01
25. 9.027, 9.10, 9.003, 9.3, 9.27
    9.003, 9.027, 9.10, 9.27, 9.3
26. 4.07, 4.5, 4.01, 4.22
    4.01, 4.07, 4.22, 4.5
27. $\frac{1}{3}, \frac{5}{6}, \frac{3}{8}, \frac{5}{4}$ **$\frac{1}{3}, \frac{3}{8}, \frac{5}{6}, \frac{5}{4}$**
28. $\frac{3}{5}, \frac{3}{2}, \frac{3}{4}, \frac{3}{10}, \frac{3}{7}$ **$\frac{3}{10}, \frac{3}{7}, \frac{3}{5}, \frac{3}{4}, \frac{3}{2}$**
29. $1\frac{2}{5}, \frac{7}{4}, \frac{5}{3}, 1\frac{1}{8}, \frac{15}{16}$ **$\frac{15}{16}, 1\frac{1}{8}, 1\frac{2}{5}, \frac{5}{3}, \frac{7}{4}$**
30. $14\frac{7}{9}, 15\frac{1}{3}, 14\frac{5}{6}, 15\frac{1}{4}$
    $14\frac{7}{9}, 14\frac{5}{6}, 15\frac{1}{4}, 15\frac{1}{3}$
31. $\frac{7}{8}, \frac{5}{4}, 1\frac{1}{3}, \frac{5}{12}$ **$\frac{5}{12}, \frac{7}{8}, \frac{5}{4}, 1\frac{1}{3}$**
32. You need a piece of trim $6\frac{5}{8}$ yards long to complete a craft project. You have a piece $6\frac{3}{4}$ yards long left over from another project. Is the trim long enough? **Yes.**
33. Goran finished a ski race in 53.56 seconds. Markus finished in 53.78 seconds. Which skier finished first? **Goran**

FRACTION OPERATIONS

To add or subtract two fractions with the same denominator, add or subtract the numerators.

EXAMPLE

$\frac{3}{5} + \frac{4}{5} = \frac{3+4}{5}$ **Add numerators.**

$= \frac{7}{5}$, or $1\frac{2}{5}$ **Simplify.**

EXAMPLE

$\frac{7}{10} - \frac{2}{10} = \frac{7-2}{10}$ **Subtract numerators.**

$= \frac{5}{10}$ **Simplify.**

$= \frac{\cancel{5}}{2 \cdot \cancel{5}}$ **Factor numerator and denominator.**

$= \frac{1}{2}$ **Simplify fraction to lowest terms.**

To add or subtract two fractions with different denominators, write equivalent fractions with a common denominator.

EXAMPLE $\frac{3}{5} + \frac{5}{6} = \frac{18}{30} + \frac{25}{30}$ **Use the LCD, 30.**

$= \frac{18+25}{30}$ **Add numerators.**

$= \frac{43}{30}$, or $1\frac{13}{30}$ **Simplify.**

To add or subtract mixed numbers, you can first rewrite them as fractions.

EXAMPLE $3\frac{2}{3} - 2\frac{1}{4} = \frac{11}{3} - \frac{9}{4}$ **Rewrite mixed numbers as fractions.**

$= \frac{44}{12} - \frac{27}{12}$ **The LCD is 12.**

$= \frac{44-27}{12}$ **Subtract numerators.**

$= \frac{17}{12}$, or $1\frac{5}{12}$ **Simplify.**

To multiply two fractions, multiply the numerators and multiply the denominators.

EXAMPLE $\frac{3}{4} \times \frac{5}{6} = \frac{3 \times 5}{4 \times 6}$ **Multiply numerators and multiply denominators.**

$= \frac{15}{24}$ **Simplify.**

$= \frac{\cancel{3} \cdot 5}{\cancel{3} \cdot 8}$ **Factor numerator and denominator.**

$= \frac{5}{8}$ **Simplify fraction to lowest terms.**

Two numbers are **reciprocals** of each other if their product is 1. Every number except 0 has a reciprocal.

$5 \times \frac{1}{5} = 1$, so 5 and $\frac{1}{5}$ are reciprocals.

$\frac{2}{3} \times \frac{3}{2} = 1$, so $\frac{2}{3}$ and $\frac{3}{2}$ are reciprocals.

$1\frac{1}{4} \times \frac{4}{5} = \frac{5}{4} \cdot \frac{4}{5} = 1$, so $1\frac{1}{4}$ and $\frac{4}{5}$ are reciprocals.

To find the reciprocal of a number, write the number as a fraction. Then interchange the numerator and the denominator.

EXAMPLE Find the reciprocal of $3\frac{1}{4}$.

SOLUTION $3\frac{1}{4} = \frac{13}{4}$ Write $3\frac{1}{4}$ as a fraction.

$\frac{13}{4} \rightarrow \frac{4}{13}$ Interchange numerator and denominator.

▶ The reciprocal of $3\frac{1}{4}$ is $\frac{4}{13}$.

✓**CHECK** $3\frac{1}{4} \times \frac{4}{13} = \frac{13}{4} \times \frac{4}{13} = \frac{13 \times 4}{4 \times 13} = 1$

To divide by a fraction, multiply by its reciprocal.

EXAMPLE

$\frac{3}{4} \div \frac{5}{6} = \frac{3}{4} \times \frac{6}{5}$	The reciprocal of $\frac{5}{6}$ is $\frac{6}{5}$.
$= \frac{3 \times 6}{4 \times 5}$	Multiply numerators and denominators.
$= \frac{18}{20}$	Simplify.
$= \frac{\cancel{2} \cdot 9}{\cancel{2} \cdot 10}$	Factor numerator and denominator.
$= \frac{9}{10}$	Simplify fraction to lowest terms.

EXAMPLE

$2\frac{1}{2} \div 4\frac{1}{6} = \frac{5}{2} \div \frac{25}{6}$	Write mixed numbers as fractions.
$= \frac{5}{2} \times \frac{6}{25}$	The reciprocal of $\frac{25}{6}$ is $\frac{6}{25}$.
$= \frac{5 \times 6}{2 \times 25}$	Multiply numerators and denominators.
$= \frac{30}{50}$	Simplify.
$= \frac{3 \times \cancel{10}}{5 \times \cancel{10}}$	Factor numerators and denominators.
$= \frac{3}{5}$	Simplify fraction to lowest terms.

PRACTICE

Add or subtract. Write the answer as a fraction or a mixed number in lowest terms.

1. $\frac{1}{6} + \frac{4}{6}$ **$\frac{5}{6}$**
2. $\frac{5}{8} - \frac{3}{8}$ **$\frac{1}{4}$**
3. $\frac{4}{9} - \frac{1}{9}$ **$\frac{1}{3}$**
4. $\frac{5}{12} + \frac{3}{12}$ **$\frac{2}{3}$**

5. $\frac{1}{2} + \frac{1}{8}$ **$\frac{5}{8}$**
6. $\frac{3}{5} - \frac{1}{10}$ **$\frac{1}{2}$**
7. $\frac{7}{10} + \frac{1}{3}$ **$1\frac{1}{30}$**
8. $\frac{15}{24} - \frac{7}{12}$ **$\frac{1}{24}$**

9. $5\frac{1}{8} - 2\frac{3}{4}$ **$2\frac{3}{8}$**
10. $1\frac{3}{7} + \frac{1}{2}$ **$1\frac{13}{14}$**
11. $4\frac{3}{8} - 2\frac{5}{6}$ **$1\frac{13}{24}$**
12. $\frac{3}{7} + \frac{3}{4}$ **$1\frac{5}{28}$**

13. $7\frac{1}{2} + \frac{7}{10}$ **$8\frac{1}{5}$**
14. $5\frac{5}{9} - 2\frac{1}{3}$ **$3\frac{2}{9}$**
15. $4\frac{5}{8} - 1\frac{3}{16}$ **$3\frac{7}{16}$**
16. $9\frac{2}{5} + 3\frac{1}{3}$ **$12\frac{11}{15}$**

Find the reciprocal of each number.

17. 7 **$\frac{1}{7}$**
18. $\frac{1}{14}$ **14**
19. $\frac{7}{12}$ **$1\frac{5}{7}$**
20. $\frac{5}{8}$ **$1\frac{3}{5}$**

21. $\frac{1}{20}$ **20**
22. 100 **$\frac{1}{100}$**
23. $\frac{5}{13}$ **$2\frac{3}{5}$**
24. $\frac{6}{7}$ **$1\frac{1}{6}$**

25. $1\frac{1}{5}$ **$\frac{5}{6}$**
26. $2\frac{3}{5}$ **$\frac{5}{13}$**
27. $\frac{3}{9}$ **3**
28. $\frac{12}{17}$ **$1\frac{5}{12}$**

29. $6\frac{2}{5}$ **$\frac{5}{32}$**
30. $10\frac{1}{3}$ **$\frac{3}{31}$**
31. $\frac{2}{7}$ **$3\frac{1}{2}$**
32. $4\frac{3}{4}$ **$\frac{4}{19}$**

Multiply or divide. Write the answer as a fraction or a mixed number in lowest terms.

33. $\frac{1}{2} \times \frac{1}{2}$ **$\frac{1}{4}$**
34. $\frac{2}{3} \times \frac{4}{5}$ **$\frac{8}{15}$**
35. $\frac{5}{8} \times \frac{4}{15}$ **$\frac{1}{6}$**
36. $\frac{3}{7} \times \frac{7}{9}$ **$\frac{1}{3}$**

37. $\frac{3}{4} \times \frac{8}{9}$ **$\frac{2}{3}$**
38. $1\frac{2}{3} \times \frac{3}{5}$ **1**
39. $3 \times 2\frac{5}{9}$ **$7\frac{2}{3}$**
40. $5\frac{1}{4} \times 1\frac{1}{7}$ **6**

41. $\frac{7}{8} \div \frac{3}{4}$ **$1\frac{1}{6}$**
42. $\frac{5}{12} \div \frac{1}{2}$ **$\frac{5}{6}$**
43. $\frac{4}{5} \div \frac{2}{3}$ **$1\frac{1}{5}$**
44. $\frac{11}{16} \div 1\frac{1}{2}$ **$\frac{11}{24}$**

45. $4\frac{1}{2} \div \frac{3}{4}$ **6**
46. $2\frac{1}{4} \div 1\frac{1}{3}$ **$1\frac{11}{16}$**
47. $3\frac{2}{5} \div 4$ **$\frac{17}{20}$**
48. $7\frac{1}{5} \div 2\frac{1}{4}$ **$3\frac{1}{5}$**

Add, subtract, multiply, or divide. Write the answer as a fraction or a mixed number in lowest terms.

49. $\frac{15}{16} - \frac{1}{8}$ **$\frac{13}{16}$**
50. $\frac{5}{9} \times 1\frac{1}{2}$ **$\frac{5}{6}$**
51. $\frac{12}{13} \div \frac{12}{13}$ **1**
52. $\frac{24}{25} + \frac{1}{5}$ **$1\frac{4}{25}$**

53. $5\frac{1}{2} - \frac{1}{8}$ **$5\frac{3}{8}$**
54. $\frac{3}{10} \div \frac{1}{5}$ **$1\frac{1}{2}$**
55. $\frac{7}{8} \times \frac{4}{9}$ **$\frac{7}{18}$**
56. $\frac{1}{3} + \frac{1}{6}$ **$\frac{1}{2}$**

57. $4\frac{1}{4} \times \frac{2}{3}$ **$2\frac{5}{6}$**
58. $9\frac{2}{5} + 3\frac{1}{2}$ **$12\frac{9}{10}$**
59. $\frac{4}{5} \div \frac{1}{2}$ **$1\frac{3}{5}$**
60. $6\frac{5}{7} - 2\frac{1}{5}$ **$4\frac{18}{35}$**

61. $\frac{9}{10} + \frac{3}{8}$ **$1\frac{11}{40}$**
62. $8\frac{1}{2} \times \frac{1}{4}$ **$2\frac{1}{8}$**
63. $\frac{11}{15} \times \frac{3}{8}$ **$\frac{11}{40}$**
64. $\frac{4}{7} \div \frac{4}{5}$ **$\frac{5}{7}$**

65. You went trout fishing. You caught trout with the lengths (in inches) of $8\frac{5}{8}$, $10\frac{1}{2}$, $9\frac{1}{4}$, $8\frac{3}{4}$, $10\frac{2}{3}$, and $9\frac{5}{6}$. Find the average length of the trout you caught. **$9\frac{29}{48}$ in.**

Fractions, Decimals, and Percents

Percent (%) means "divided by 100."

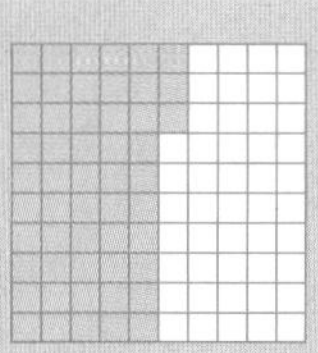

53% = 53 divided by 100 = $\frac{53}{100}$

The grid contains 100 squares.

53 of the 100 squares are shaded.

53% of the squares are shaded.

To write a percent as a decimal, move the decimal point two places to the *left* and remove the percent symbol.

EXAMPLES

a. 85% = 85% = 0.85

b. 3% = 03% = 0.03

c. 427% = 427% = 4.27

d. 12.5% = 12.5% = 0.125

To write a percent as a fraction in lowest terms, first write the percent as a fraction with a denominator of 100. Then simplify if possible.

EXAMPLES

a. $71\% = \frac{71}{100}$

b. $10\% = \frac{10}{100} = \frac{1}{10}$

c. $4\% = \frac{4}{100} = \frac{1}{25}$

d. $350\% = \frac{350}{100} = \frac{7}{2} = 3\frac{1}{2}$

To write a decimal as a percent, move the decimal point two places to the *right* and add a percent symbol.

EXAMPLES

a. 0.93 = 0.93 = 93%

b. 1.47 = 1.47 = 147%

c. 0.025 = 0.025 = 2.5%

d. 0.005 = 0.005 = 0.5%

To write a decimal as a fraction in lowest terms, first write the decimal as a fraction with a denominator of 100. Then simplify if possible.

EXAMPLES

a. $0.73 = \frac{73}{100}$

b. $0.25 = \frac{25}{100} = \frac{1}{4}$

c. $0.05 = \frac{5}{100} = \frac{1}{20}$

d. $2.75 = 2\frac{75}{100} = 2\frac{3}{4}$

It is simple to write a fraction as a percent if the denominator of the fraction is a factor of 100. If not, divide the numerator by the denominator.

EXAMPLES

a. $\frac{17}{25}$ → 25 is a factor of 100, so write $\frac{17}{25} = \frac{17 \cdot 4}{25 \cdot 4} = \frac{68}{100} = 68\%$.

b. $\frac{1}{8}$ → 8 is not a factor of 100, so divide: $1 \div 8 = 0.125 = 12.5\%$.

c. $\frac{1}{6}$ → 6 is not a factor of 100, so divide: $1 \div 6 = 0.1666\ldots \approx 0.167 = 16.7\%$.

You should memorize the percent, decimal, and fraction relationships in this chart.

Equivalent percents, decimals, and fractions		
$1\% = 0.01 = \frac{1}{100}$	$33\frac{1}{3}\% = 0.\overline{3} = \frac{1}{3}$	$66\frac{2}{3}\% = 0.\overline{6} = \frac{2}{3}$
$10\% = 0.1 = \frac{1}{10}$	$40\% = 0.4 = \frac{2}{5}$	$75\% = 0.75 = \frac{3}{4}$
$20\% = 0.2 = \frac{1}{5}$	$50\% = 0.5 = \frac{1}{2}$	$80\% = 0.8 = \frac{4}{5}$
$25\% = 0.25 = \frac{1}{4}$	$60\% = 0.6 = \frac{3}{5}$	$100\% = 1$

PRACTICE

Write the percent as a decimal and as a fraction or a mixed number in lowest terms.

1. 63% 0.63, $\frac{63}{100}$ **2.** 7% 0.07, $\frac{7}{100}$ **3.** 24% 0.24, $\frac{6}{25}$ **4.** 35% 0.35, $\frac{7}{20}$ **5.** 17% 0.17, $\frac{17}{100}$

6. 125% 1.25, $1\frac{1}{4}$ **7.** 45% 0.45, $\frac{9}{20}$ **8.** 250% 2.5, $2\frac{1}{2}$ **9.** $33\frac{1}{3}\%$ $0.\overline{3}$, $\frac{1}{3}$ **10.** 96% 0.96, $\frac{24}{25}$

11. 62.5% 0.625, $\frac{5}{8}$ **12.** 725% 7.25, $7\frac{1}{4}$ **13.** 5.2% 0.052, $\frac{13}{250}$ **14.** 0.8% 0.008, $\frac{1}{125}$ **15.** 0.12% 0.0012, $\frac{3}{2500}$

Write the decimal as a percent and as a fraction or a mixed number in lowest terms.

16. 0.39 39%, $\frac{39}{100}$ **17.** 0.08 8%, $\frac{2}{25}$ **18.** 0.12 12%, $\frac{3}{25}$ **19.** 1.5 150%, $1\frac{1}{2}$ **20.** 0.72 72%, $\frac{18}{25}$

21. 0.05 5%, $\frac{1}{20}$ **22.** 2.08 208%, $2\frac{2}{25}$ **23.** 4.8 480%, $4\frac{4}{5}$ **24.** 0.02 2%, $\frac{1}{50}$ **25.** 3.75 375%, $3\frac{3}{4}$

26. 0.85 85%, $\frac{17}{20}$ **27.** 0.52 52%, $\frac{13}{25}$ **28.** 0.9 90%, $\frac{9}{10}$ **29.** 0.005 0.5%, $\frac{1}{200}$ **30.** 2.01 201%, $2\frac{1}{100}$

Write the fraction or mixed number as a decimal and as a percent. Round decimals to the nearest thousandth. Round percents to the nearest tenth of a percent.

31. $\frac{7}{10}$ 0.7, 70% **32.** $\frac{13}{20}$ 0.65, 65% **33.** $\frac{11}{25}$ 0.44, 44% **34.** $\frac{3}{10}$ 0.3, 30% **35.** $\frac{3}{8}$ 0.375, 37.5%

36. $2\frac{3}{4}$ 2.75, 275% **37.** $5\frac{1}{8}$ 5.125, 512.5% **38.** $\frac{19}{20}$ 0.95, 95% **39.** $\frac{7}{8}$ 0.875, 87.5% **40.** $3\frac{7}{25}$ 3.28, 328%

41. $\frac{5}{6}$ 0.833, 83.3% **42.** $3\frac{3}{5}$ 3.6, 360% **43.** $\frac{8}{15}$ 0.533, 53.3% **44.** $8\frac{1}{5}$ 8.2, 820% **45.** $1\frac{5}{12}$ 1.417, 141.7%

USING PERCENT

A fraction shows what part one number is of another. You can first write a fraction to determine what percent one number is of another.

EXAMPLE What percent of 120 is 48?

SOLUTION First write a fraction that compares 48 to 120: $\frac{48}{120}$.

$\frac{48}{120} = 0.4$ **Write the fraction as a decimal.**

$= 40\%$ **Write the decimal as a percent.**

▶ 48 is 40% of 120.

To find a percent of a given number, first write the percent as a decimal or as a fraction. Then multiply.

EXAMPLE What is 75% of 160?

SOLUTION $75\% = \frac{3}{4}$ **Write 75% as a fraction.**

$\frac{3}{4} \times 160 = 120$ **Multiply.**

▶ 75% of 160 is 120.

EXAMPLE What is 6% of $29.95?

SOLUTION $6\% = 0.06$ **Write 6% as a decimal.**

$0.06 \times \$29.95 = \$1.797$ **Multiply.**

$= \$1.80$ **Round to the nearest cent.**

▶ To the nearest cent, 6% of $29.95 is $1.80.

PRACTICE

Find the answer.

1. What percent of 90 is 15? **$16\frac{2}{3}\%$** **2.** 12 is what percent of 60? **20%** **3.** What percent of 80 is 30? **37.5%**

4. What percent of 60 is 90? **150%** **5.** What percent of 90 is 60? **$66\frac{2}{3}\%$** **6.** 6 is what percent of 120? **5%**

7. 15 is what percent of 90? **$16\frac{2}{3}\%$** **8.** What percent of 18 is 4.5? **25%** **9.** 18 is what percent of 96? **18.75%**

Find the number. Round sums of money to the nearest cent.

10. 38% of 250 **95** **11.** 12% of 75 **9** **12.** 25% of 84 **21** **13.** 50% of 96 **48**

14. 42% of 115 **48.3** **15.** $33\frac{1}{3}\%$ of 114 **38** **16.** 5.5% of $102.95 **$5.66** **17.** 70% of 60 **42**

18. $66\frac{2}{3}\%$ of 48 **32** **19.** 8% of $12.99 **$1.04** **20.** 150% of 90 **135** **21.** 4.5% of $75 **$3.38**

22. You are charged 6.5% tax on a $42 purchase. Find the amount of the tax. **$2.73**

RATIO AND RATE

A ratio compares two numbers using division. If a and b are two quantities measured in the same units, then the **ratio of a to b** is $\frac{a}{b}$. The ratio of a to b can be written in three ways: a to b, $a : b$, and $\frac{a}{b}$.

EXAMPLE Write the ratio 7 to 15 in three ways.

SOLUTION 7 to 15 $\quad 7 : 15 \quad \frac{7}{15}$

EXAMPLE Write the ratio 12 to 60 in lowest terms.

SOLUTION 12 to 60 $= \frac{12}{60}$ First write the ratio as a fraction.

$= \frac{1}{5}$ Then write the fraction in lowest terms.

If a and b are two quantities measured in different units, then the **rate of a per b** is $\frac{a}{b}$. A **unit rate** is a rate per one unit of a given quantity. To determine a unit rate, write the rate with a denominator of 1.

EXAMPLE A car traveled 648 miles using 18 gallons of gas. Find the unit rate in miles per gallon.

SOLUTION $\frac{\text{miles}}{\text{gallons}} \rightarrow \frac{648}{18} = \frac{36}{1}$

▶ The unit rate is 36 miles per gallon.

PRACTICE

Write each ratio in lowest terms.

1. 75 to 20 **15 to 4**
2. 162 : 36 **9 : 2**
3. $\frac{36}{60}$ $\frac{3}{5}$
4. 21 : 15 **7 : 5**
5. $\frac{60}{85}$ $\frac{12}{17}$
6. 42 to 72 **7 to 12**
7. 136 : 24 **17 : 3**
8. $\frac{216}{120}$ $\frac{9}{5}$
9. 24 cars to 36 cars **2 to 3**
10. 39 students : 104 students **3 : 8**
11. 75 hours to 35 hours **15 to 7**
12. 8 hours : 2 hours **4 : 1**
13. 9 inches to 36 inches **1 to 4**
14. 48 hours to 10 hours **24 to 5**

Find each unit rate.

15. \$90 for 4 tickets **\$22.50 per ticket**
16. \$51 for 6 hours **\$8.50 per hour**
17. 208 miles in 4 hours **52 miles per hour**
18. \$1050 in 6 weeks **\$175 per week**
19. 476 miles for 17 gallons **28 miles per gallon**
20. 128 ounces for 16 people **8 ounces per person**
21. \$2.70 for 27 minutes **\$.10 per minute**
22. \$8.67 for 3 notebooks **\$2.89 per notebook**
23. 65 meters in 3 seconds **21.6 meters per second**
24. 18 hours in 4 days **4.5 hours per day**
25. \$476 for 425 miles **\$1.12 per mile**
26. \$50 for 5 hours **\$10 per hour**

COUNTING METHODS

Suppose your family's favorite pizza toppings are hamburger, sausage, pepperoni, mushroom, onion, and pepper. You have to choose one meat topping and one vegetable topping. How many choices are possible?

There are many ways to answer that question, including **making a list**, drawing a **tree diagram**, and using the **counting principle**.

EXAMPLE Make a list to determine the number of topping choices.

SOLUTION Pair each meat topping with each vegetable topping.

hamburger, mushroom	**hamburger**, onion	**hamburger**, pepper
sausage, mushroom	sausage, onion	sausage, pepper
pepperoni, mushroom	pepperoni, onion	pepperoni, pepper

▶ There are nine possible choices.

EXAMPLE Use a tree diagram to determine the number of topping choices.

SOLUTION Draw a tree diagram and count the number of branches.

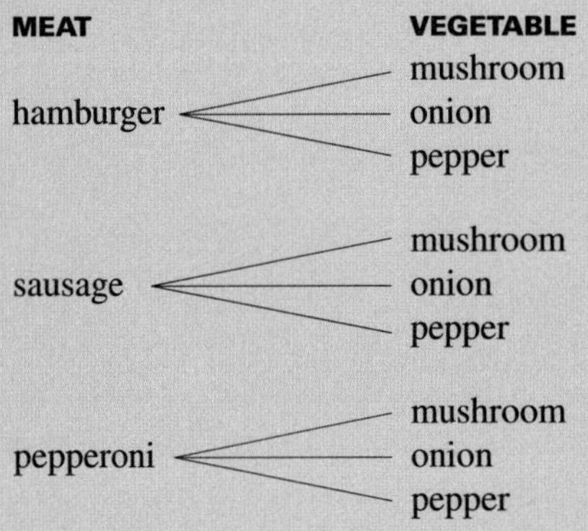

▶ The tree has nine branches. There are nine possible choices.

THE COUNTING PRINCIPLE When one item is selected from each of two or more lists, the total number of possible combinations is the product of the number of items in each list.

EXAMPLE Use the counting principle to determine the number of topping choices.

SOLUTION Multiply the number of meat choices by the number of vegetable choices.

3 meat choices $\times$ 3 vegetable choices $=$ 9 possible topping choices

▶ There are nine possible choices.

788

In situations involving a greater number of items, it may not be practical to make a list or to draw a tree diagram. Use the counting principle instead.

EXAMPLE One state plans to issue a series of license plates using codes consisting of three letters followed by two digits. Letters and digits may both be repeated. How many different plates are possible?

SOLUTION Use the Counting Principle.

26 choices for first letter	$\times$	26 choices for second letter	$\times$	26 choices for third letter	$\times$	10 choices for first digit	$\times$	10 choices for second digit

$26 \times 26 \times 26 \times 10 \times 10 = 1{,}757{,}600$

▶ There are 1,757,600 different license plates possible.

PRACTICE

Use the indicated method to answer each question.

1. Jacob is choosing an outfit from three pairs of jeans and three shirts. The jeans are blue, black, and gray. The shirts are white, blue, and yellow. How many such outfits are possible? (Make a list.) **9 outfits**

2. The student council is selling sweatshirts with the school name and emblem. The shirts come in the colors, styles, and sizes shown at the right. How many different types of sweatshirt are available? (Make a tree diagram.) **18 types**

3. There are 270 girls and 210 boys enrolled in the Jefferson School. The cover of the yearbook will show a photograph of a pair of students, one girl and one boy. How many different pairs of students are possible? (Use the counting principle.) **56,700 pairs**

Sweatshirt Choices

Colors: white, gray, black

Styles: hooded, plain

Sizes: small, medium, large

Ex. 2

Answer each question using any method you choose.

4. A student can choose one sandwich, one salad, and one piece of fruit in the cafeteria. The choices include the following: grilled cheese sandwich, hamburger, tuna sandwich, garden salad, fruit salad, apple, banana, and orange. How many different lunches are possible? **18 lunches**

5. In a given calling area, telephone numbers have the same three first digits. The four-digit number that follows is different for each telephone number. How many numbers are possible? **10,000 numbers**

6. A radio host plans to play one song from each of three CDs by a single artist. The first CD includes 12 songs. The second includes 14 songs, and the third includes 13 songs. How many different combinations are possible? **2184 combinations**

7. Members of a video club choose a personal identification number (PIN) consisting of two letters followed by two digits. How many PINs are possible if letters and digits may be repeated? How many are possible if neither letters nor digits may be repeated? **67,600 PINs; 58,500 PINs**

789

PERIMETER, AREA, AND VOLUME

The **perimeter** P of a figure is the distance around it.

EXAMPLES

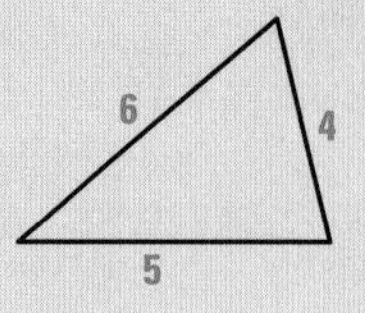

$P = 6 + 5 + 4$
$= 15$

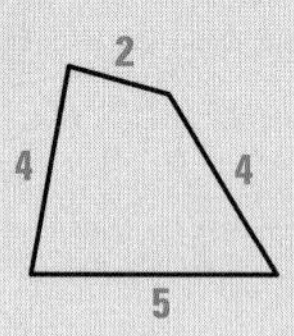

$P = 4 + 2 + 4 + 5$
$= 15$

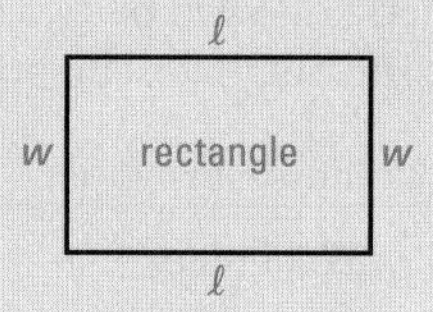

$P = l + w + l + w$
$= 2l + 2w$

EXAMPLE Find the perimeter of a rectangle with length 14 centimeters and width 6 centimeters.

SOLUTION $P = 2l + 2w = (2 \times 14) + (2 \times 6) = 28 + 12 = 40$

▶ The perimeter is 40 centimeters.

A **regular polygon** is a polygon in which all the angles have the same measure and all the sides have the same length. The perimeter of a regular polygon can be found by multiplying the length of a side by the number of sides.

EXAMPLES

regular (equilateral) triangle

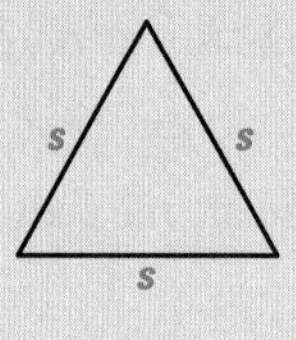

$P = 3s$

square

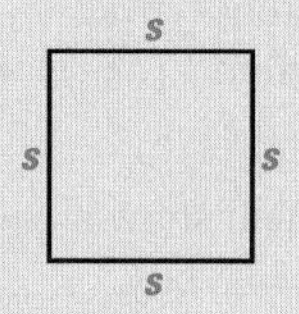

$P = 4s$

regular pentagon

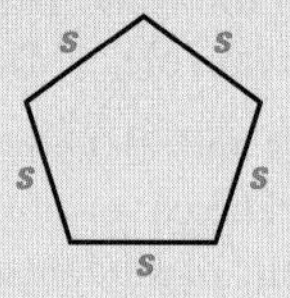

$P = 5s$

The **area** A of a figure is the number of square units enclosed by the figure.

EXAMPLES

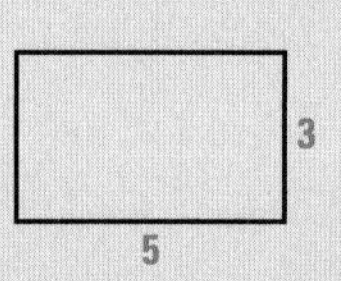

$A = 5 \times 3$
$= 15$ square units

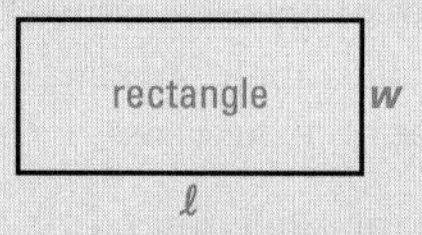

$A =$ length $\times$ width
$= l \times w$
$= lw$

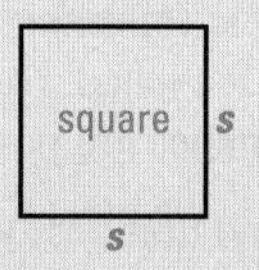

$A =$ side $\times$ side
$= s \times s$
$= s^2$

EXAMPLE Find the area of a square with sides 15 inches long.

SOLUTION $A = s^2 = 15^2 = 225$

▶ The area is 225 square inches.

The **volume** of a space figure is the amount of space enclosed by the figure. Volume is measured in cubic units.

One such unit is the cubic centimeter (cm^3). It is the amount of space enclosed by a cube whose length, width, and height are each 1 cm.

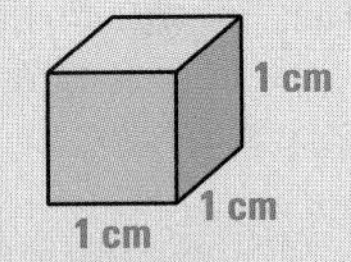

EXAMPLES

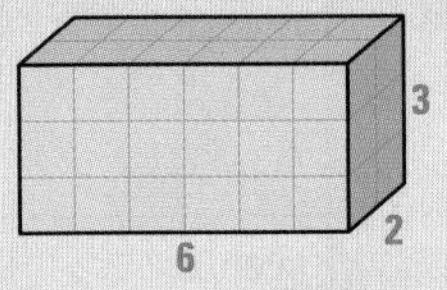

$V = 6 \times 2 \times 3$

$= 36$ cubic units

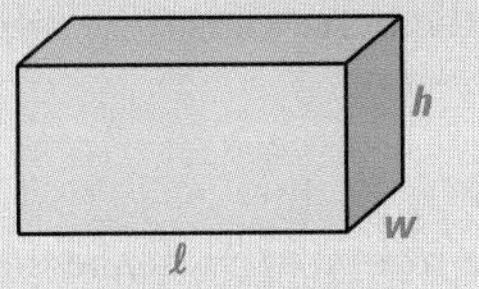

$V = l \times w \times h$

EXAMPLE Find the volume of a box with length 8 feet, width 5 feet, and height 9 feet.

SOLUTION $V = l \times w \times h = 8 \times 5 \times 9 = 360$

▶ The volume is 360 cubic feet (ft^3).

PRACTICE

Find the perimeter of each figure.

1.

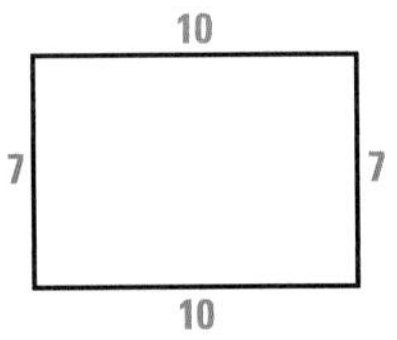

34 units

2.

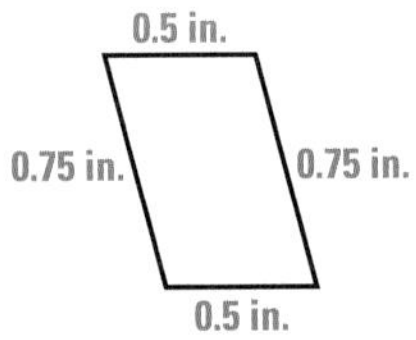

2.5 in.

3.

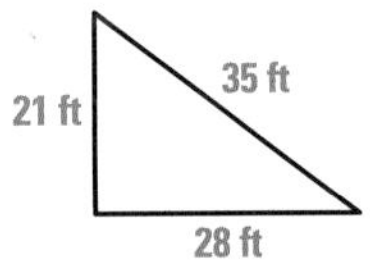

84 ft

4.

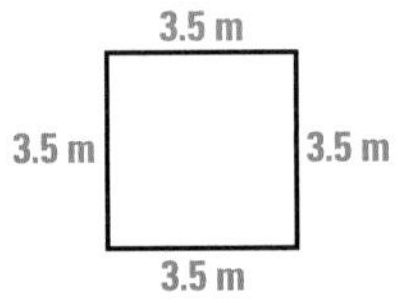

14 m

5. a square with sides of length 18 ft **72 ft**

6. a rectangle with length 6 m and width 7 m **26 m**

Find the area of each figure.

7. a square with sides of length 29 yd **841 yd^2**

8. a rectangle with length 7 km and width 4 km **28 km^2**

9. a square with sides of length 3.5 in. **12.25 $in.^2$**

10. a rectangle with length 24 ft and width 6 ft **144 ft^2**

11. a square with sides of length 7.2 cm **51.84 cm^2**

12. a rectangle with length 7.5 in. and width 8 in. **60 $in.^2$**

13. a square with sides of length 45 km **2025 km^2**

14. a rectangle with length 5.3 m and width 4 m **21.2 m^2**

Find the volume of each box.

15. a cube with sides of length 25 ft **15,625 ft^3**

16. a cube with sides of length 4.2 cm **74.088 cm^3**

17. a box with length 15 yd, width 7 yd, and height 4 yd **420 yd^3**

18. a box with length 7.3 cm, width 5 cm, and height 3.2 cm **116.8 cm^3**

19. a box with length 5.3 in., width 4 in., and height 10 in. **212 $in.^3$**

DATA DISPLAYS

A **bar graph** can be used to display data that fall into distinct categories. The bars in a bar graph are the same width. The height or length of each bar is determined by the data it represents and by the scale you choose.

Field location	Number of home runs
left	31
left-center	21
center	15
right-center	3
right	0

EXAMPLE In 1998, baseball player Mark McGwire hit a record 70 home runs. The table shows the locations to which the home runs were hit. Draw a bar graph to display the data. ▶ Source: *The Boston Globe*

SOLUTION

1. First, choose a scale. Since the data range from 0 to 31, make the scale increase from 0 to 35 by fives.
2. Draw and label the axes. Mark intervals on the vertical axis according to the scale you chose.
3. Draw a bar for each category.
4. Give the bar graph a title.

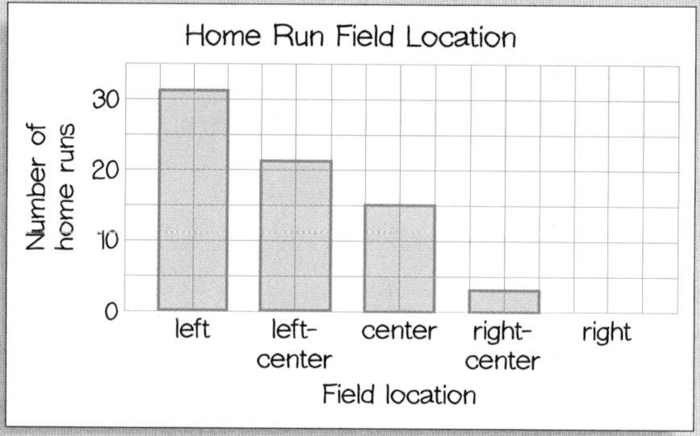

A **histogram** is a bar graph that shows how many data items occur within given intervals. The number of data items in an interval is the **frequency** for that interval.

Distance (ft)	Frequency
301–350	4
351–400	24
401–450	27
451–500	11
501–550	4

EXAMPLE The table shows the distances of McGwire's home runs. Draw a histogram to display them.

SOLUTION Use the same method you used for drawing the bar graph above. However, do not leave spaces between the bars.

Since the frequencies range from 4 to 27, make the scale increase from 0 to 30 by fives.

Draw and label the axes. Mark intervals on the vertical axis and draw a bar for each category. Do not leave spaces between the bars.

Give the histogram a title.

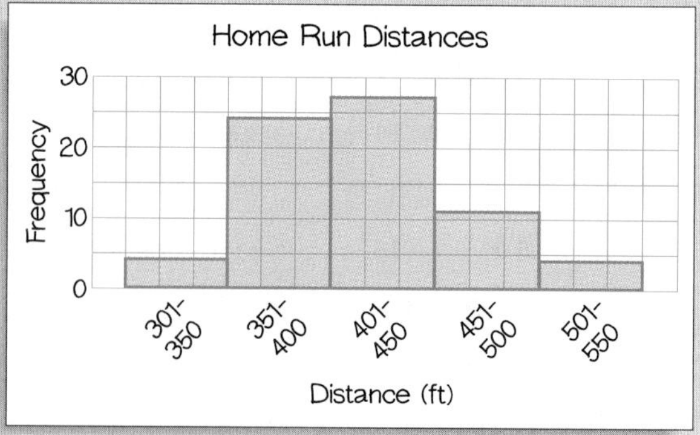

A **line graph** can be used to show how data change over time.

EXAMPLE A science class recorded the highest temperature each day from December 1 to December 14. The temperatures are given in the table. Draw a line graph to display the data.

Date	1	2	3	4	5	6	7
Temperature (°F)	40	48	49	61	24	35	34

Date	8	9	10	11	12	13	14
Temperature (°F)	42	41	40	22	20	28	30

SOLUTION

1. Choose a scale.
2. Draw and label the axes. Mark evenly spaced intervals on both axes.
3. Graph each data item as a point. Connect the points.
4. Give the line graph a title.

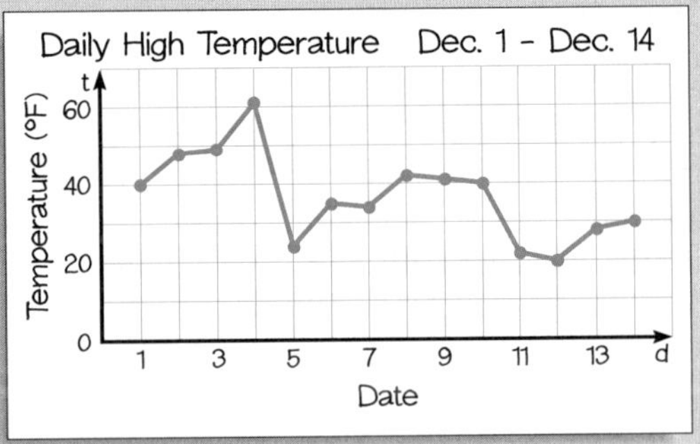

A **circle graph** can be used to show how parts relate to a whole and to each other.

EXAMPLE The table shows the number of sports-related injuries treated in a hospital emergency room in one year. Draw a circle graph to display the data.

Related sport	Number of injuries
basketball	56
football	34
skating/hockey	22
track and field	10
other	28

SOLUTION

1. To begin, find the total number of injuries.

   $56 + 34 + 22 + 10 + 28 = \mathbf{150}$

   To find the degree measure for each sector of the circle, write a fraction comparing the number of injuries to the total. Then multiply the fraction by 360°. For example:

   Football: $\frac{34}{150} \cdot 360° = 81.6°$

2. Next, draw a circle. Use a protractor to draw the angle for each sector.
3. Label each sector.
4. Give the circle graph a title.

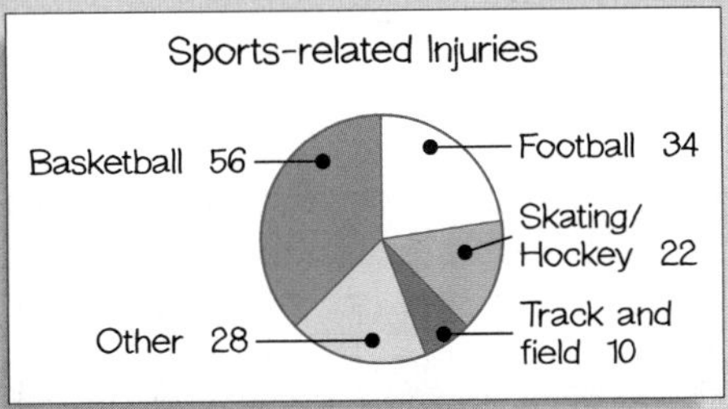

1–10. Sample answers are given.

2. Home Run Field Locations

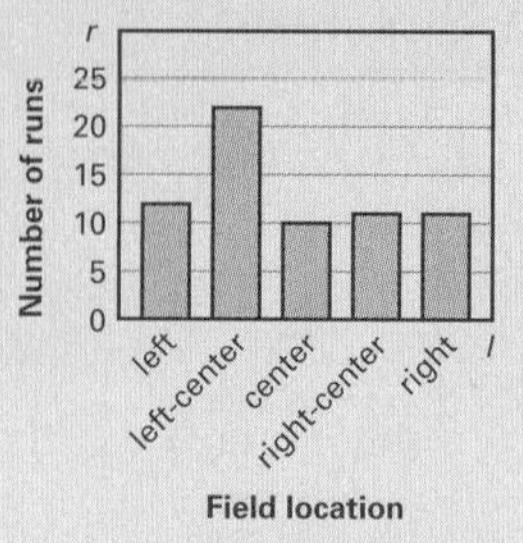

4. Home Run Distances

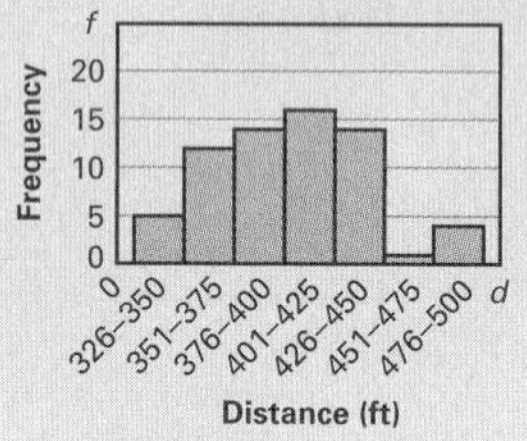

5. Ski Runs

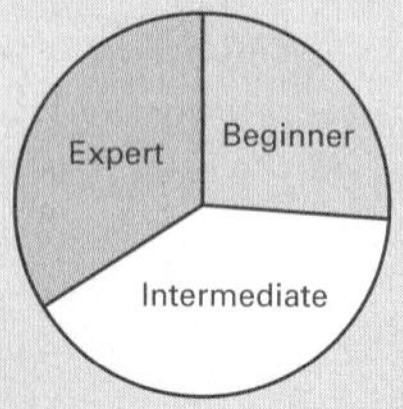

6. Patient's Temperature

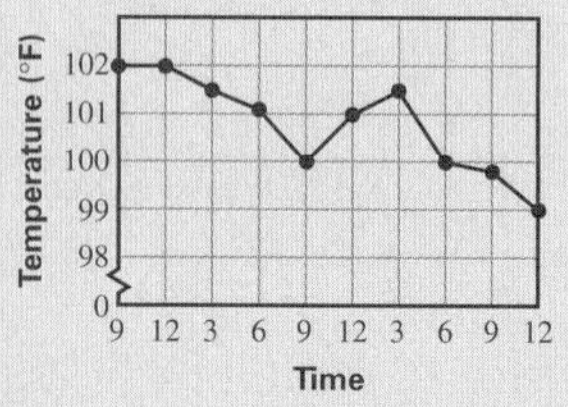

7. Stock Prices

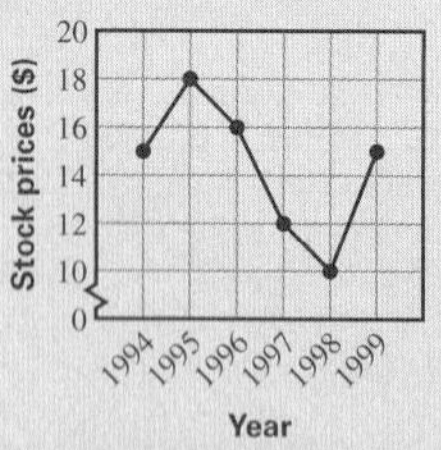

8. Passenger Car Stopping Distance (dry road)

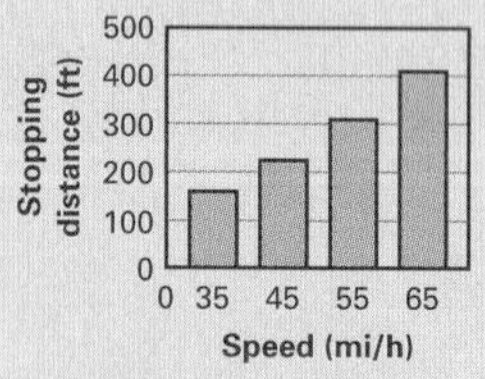

PRACTICE

In 1998, baseball player Sammy Sosa hit 66 home runs. The tables show the field locations and distances of his home runs. ▶ Source: *The Boston Globe* **1–10. Sample responses are given.**

1. The location data range from 10 to 22. The scale must start at 0. Choose a reasonable scale for a bar graph. **0–25 with increases of 5**
2. Draw a bar graph to display the field locations of Sosa's home runs. **See margin.**
3. The distance data range from 1 to 16. The scale must start at 0. Choose a reasonable scale for a histogram. **0–20 with increases of 5**
4. Draw a histogram to display the distances of Sosa's home runs. **See margin.**

Field location	Number of home runs
left	12
left-center	22
center	10
right-center	11
right	11

Exs. 1 and 2

Distance (ft)	Frequency
326–350	5
351–375	12
376–400	14
401–425	16
426–450	14
451–475	1
476–500	4

Exs. 3 and 4

5. There are 150 runs at the Mountain Mania ski resort, including 51 expert runs, 60 intermediate runs, and 39 beginner runs. Draw a circle graph to display the data. **See margin.**
6. A nurse recorded a patient's temperature (°F) every 3 hours from 9 A.M. until noon of the following day. The temperatures were 102°, 102°, 101.5°, 101.1°, 100°, 101°, 101.5°, 100°, 99.8°, and 99°. Draw a line graph to display the data. **See margin.**

Choose an appropriate graph to display the data. Draw the graph.

7. **See margin.**

Value of one share of Company Stock						
Year	1994	1995	1996	1997	1998	1999
Value ($)	15	18	16	12	10	15

8. **See margin.**

Passenger Car Stopping Distance (dry road)				
Speed (mi/h)	35	45	55	65
Distance (ft)	160	225	310	410

9. **See margin.**

Fat in One Tablespoon of Canola Oil	
Type of fat	Number of Grams
saturated	1
polyunsaturated	4
monounsaturated	8

▶ Source: U.S. Department of Agriculture

10. **See margin.**

Population of Meridian City by Age					
Age	Under 5	5–19	20–44	45–64	65 and older
Population	912	2556	4812	2232	1502

PROBLEM SOLVING

One of your primary goals in mathematics should be to become a good problem solver. It will help to approach every problem with an organized plan.

STEP 1 UNDERSTAND THE PROBLEM.
Read the problem carefully. Organize the information you are given and decide what you need to find. Determine whether some of the information given is unnecessary, or whether enough information is given. Supply missing facts, if possible.

STEP 2 MAKE A PLAN TO SOLVE THE PROBLEM.
Choose a strategy. (Get ideas from the list given on page 796.) Choose the correct operations. Decide if you will use a tool such as a calculator, a graph, or a spreadsheet.

STEP 3 CARRY OUT THE PLAN TO SOLVE THE PROBLEM.
Use the strategy and any tools you have chosen. Estimate before you calculate, if possible. Do any calculations that are needed. Answer the question that the problem asks.

STEP 4 CHECK TO SEE IF YOUR ANSWER IS REASONABLE.
Reread the problem and see if your answer agrees with the given information.

EXAMPLE How many segments can be drawn between 7 points, no three of which lie on the same line?

1. You are given a number of points, along with the information that no three points lie on the same line. You need to determine how many segments can be drawn between the points.
2. Some strategies to consider are: draw a diagram, solve a simpler problem and look for a pattern.
3. Consider the problem for fewer points.

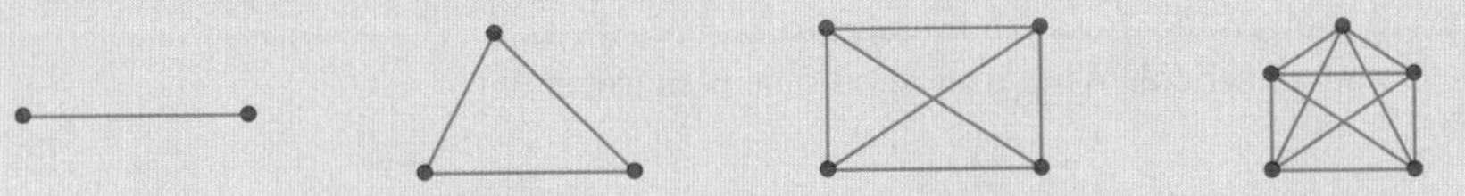

2 points 1 segment | **3 points 3 segments** | **4 points 6 segments** | **5 points 10 segments**

Look for a pattern. Then continue the pattern to find the number of segments for 7 points.

Number of points	2	3	4	5	6	7
Number of segments	1	3	6	10	15	21

+2 +3 +4 +5 +6

▶ Given 7 points, no three of which lie on the same line, 21 segments can be drawn between the points.

4. You can check your solution by making a sketch.

9.

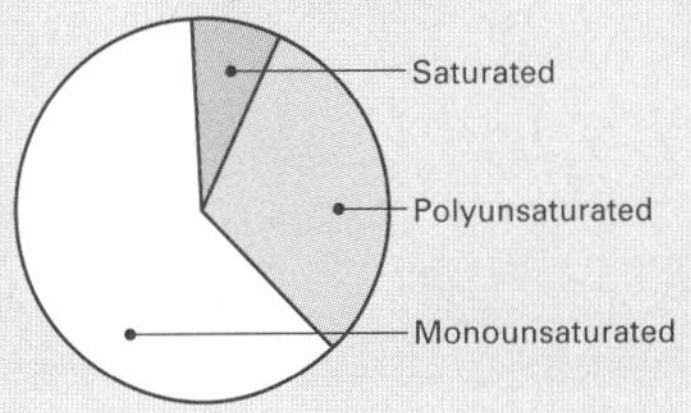

10.

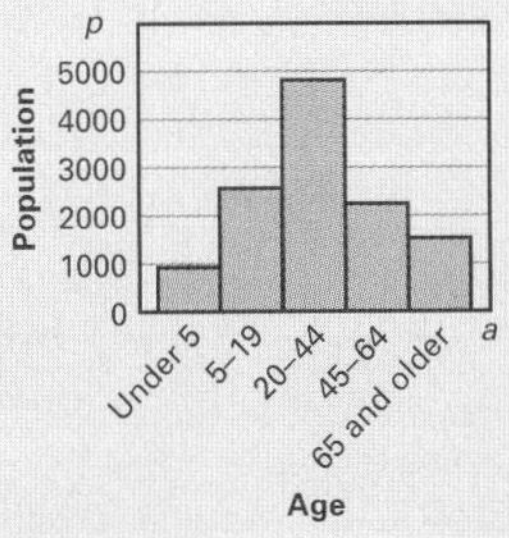

In **Step 2** of the problem solving plan, you may want to consider the following strategies.

PROBLEM SOLVING STRATEGIES

• **Guess, check, and revise.**	When you do not seem to have enough information.
• **Draw a diagram or a graph.**	When words describe a picture.
• **Make a table or an organized list.**	When you have data or choices to organize.
• **Use an equation or a formula.**	When you know a relationship between quantities.
• **Use a proportion.**	When you know that two ratios are equal.
• **Look for a pattern.**	When you can examine several cases.
• **Break the problem into simpler parts.**	When you have a multi-step problem.
• **Solve a simpler problem.**	When easier numbers help you understand a problem.
• **Work backward.**	When you are looking for a fact leading to a known result.

PRACTICE

1. Tasha bought salads at \$2.75 each and cartons of milk at \$.80 each. The total cost was \$16.15. How many of each did Tasha buy? **5 sandwiches, 3 cartons of milk**
2. A rectangular garden is 45 feet long and has perimeter 150 feet. Rows of plants are planted 3 feet apart. Find the area of the garden. **1350 ft^2**
3. If five turkey club sandwiches cost \$18.75, how much would seven sandwiches cost? **\$26.25**
4. How many diagonals can be drawn from one vertex of a 12-sided polygon? **9 diagonals**
5. Nguyen wants to arrive at school no later than 7:25 A.M. for his first class. It takes him 25 minutes to shower and dress, 15 minutes to eat breakfast, and at least 20 minutes to get to school. What time should he plan to get out of bed? **no later than 6:25 A.M.**
6. There are 32 players in a single-elimination chess tournament. That is, a player who loses once is eliminated. Assuming that no ties are allowed, how many games must be played to determine a champion? **31 games**
7. Andrea, Betty, Joyce, Karen, and Paula are starters on their school basketball team. How many different groups of three can be chosen for a newspaper photo? **10 groups**
8. Carl has \$135 in the bank and plans to save \$5 per week. Jean has \$90 in the bank and plans to save \$10 per week. How many weeks will it be before Jean has at least as much in the bank as Carl? **9 weeks**
9. The Peznolas are planning to use square tiles to tile a kitchen floor that is 18 feet long and 15 feet wide. Each tile covers one square foot. A carton of tiles costs \$18. How much will it cost to cover the entire kitchen floor? **The problem cannot be solved; not enough information is given.**

Extra Practice

CHAPTER 1

Evaluate the expression for the given value of the variable. (Lesson 1.1)

1. $15a$ when $a = 7$ **105** **2.** $0.75 + x$ when $x = 2.25$ **3** **3.** $y - 14$ when $y = 32$ **18**

4. $\frac{c}{23}$ when $c = 391$ **17** **5.** $\frac{1578}{d}$ when $d = 3$ **526** **6.** $\frac{3}{4} \cdot z$ when $z = \frac{2}{3}$ $\frac{1}{2}$

Evaluate the expression for the given value(s) of the variable(s). (Lesson 1.2)

7. $3y^2$ when $y = 5$ **75** **8.** $(4x)^3$ when $x = 2$ **512** **9.** $6x^4$ when $x = 4$ **1536**

10. $a^4 - 5$ when $a = 3$ **76** **11.** $(x + 2)^3$ when $x = 4$ **216** **12.** $(c - d)^2$ when $c = 10$ and $d = 3$ **49**

Evaluate the expression for the given value of the variable. (Lesson 1.3)

13. $3a^2 + 8$ when $a = 5$ **83** **14.** $44 - 4c$ when $c = 7$ **16** **15.** $(d^4 - 6) \div 5$ when $d = 3$ **15**

Evaluate the expression. (Lesson 1.3)

16. $3^3 - 12 \div 4$ **24** **17.** $10^2 \div 4 + 6$ **31** **18.** $10^2 \div (4 + 6)$ **10** **19.** $\frac{9 \cdot 7^2}{5 + 8^2 - 6}$ **7**

20. $3 + 7(3.5 \div 5)$ **7.9** **21.** $2 + 21 \div 3 - 6$ **3** **22.** $50 \div (6^2 - 11) - 2$ **0** **23.** $[(5 \cdot 2^3) + 8] \div 16$ **3**

Use mental math to solve the equation. (Lesson 1.4)

24. $x + 7 = 13$ **6** **25.** $3y = 21$ **7** **26.** $8t - 1 = 23$ **3** **27.** $\frac{m}{4} = 6$ **24**

Check whether the given number is a solution of the inequality. (Lesson 1.4)

28. $y + 10 < 22; 12$ **not a solution** **29.** $2n + 5 > 15; 8$ **solution** **30.** $7d > 15 - d; 2$ **solution** **31.** $\frac{c + 1}{2} \le 3; 5$ **solution**

32. $b(2b - 5) < 40; 6$ **not a solution** **33.** $12x > x + 27; 3$ **solution** **34.** $17 - 3t \ge 3; 4$ **solution** **35.** $r(r - 7) \le 16; 9$ **not a solution**

36. Your friend's score in a game is 225. This is 25 more than twice your score. Write an equation to model the situation. What is your score? (Lesson 1.5) **$2x + 25 = 225$; 100**

Write the verbal sentence as an equation or an inequality. (Lesson 1.5)

37. Thirteen less than the product of 25 and a number n is 37. **$25n - 13 = 37$**

38. The sum of five and a number y is less than twelve. **$y + 5 < 12$**

39. WIRELESS INDUSTRY The table shows the estimated number of cellular telephone subscribers (in millions) in the United States. Make a bar graph and a line graph of the data. (Lesson 1.6) **See margin.**

Year	1993	1994	1995	1996	1997	1998
Subscribers (millions)	16	24	34	44	55	69

▶ Source: Cellular Telecommunications Industry Association

Make an input-output table for the function. Use 0, 1, 2, and 3 as the domain. Then draw a line graph to represent the data in the table. (Lesson 1.7) **40–43. See margin.**

40. $y = 8 - 2x$ **41.** $y = 4.5 + x$ **42.** $y = 7x + 1$ **43.** $y = x$

39.

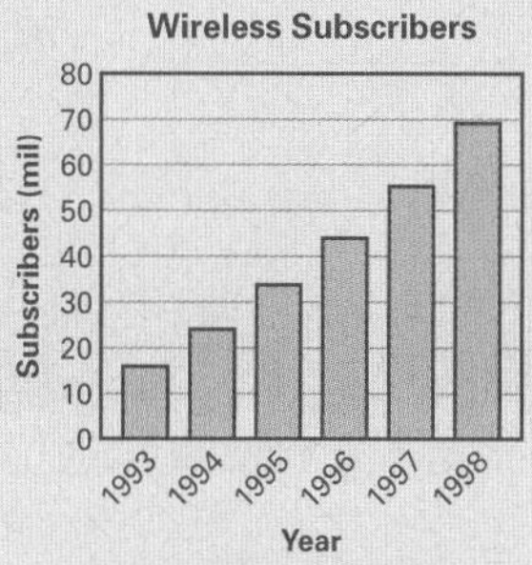

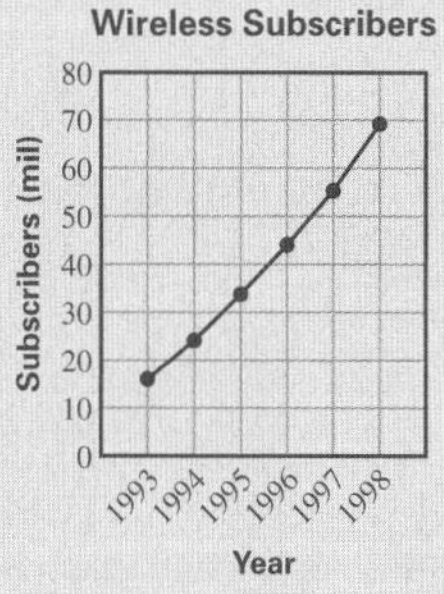

40. Input: 0, 1, 2, 3; Output: 8, 6, 4, 2

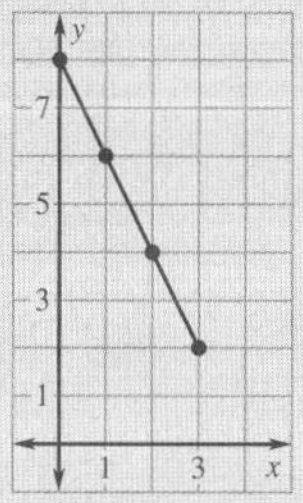

41. Input: 0, 1, 2, 3; Output: 4.5, 5.5, 6.5, 7.5

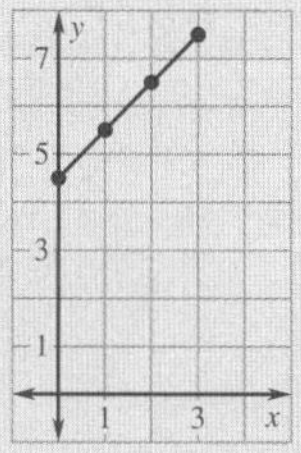

42. Input: 0, 1, 2, 3; Output: 1, 8, 15, 22

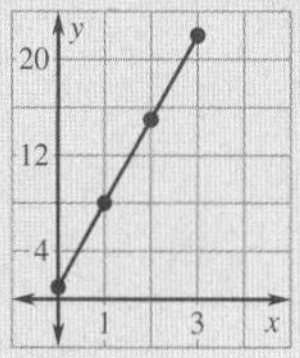

43. Input: 0, 1, 2, 3; Output: 0, 1, 2, 3

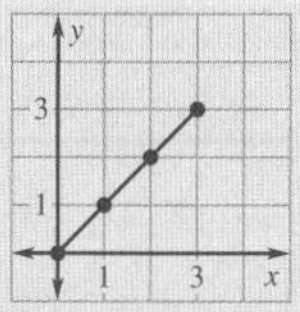

1.

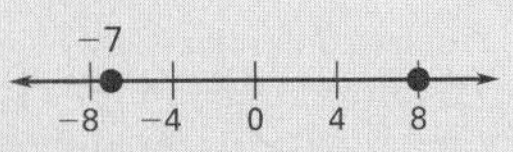

2.

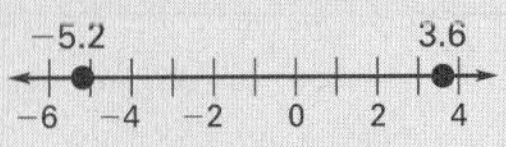

3.

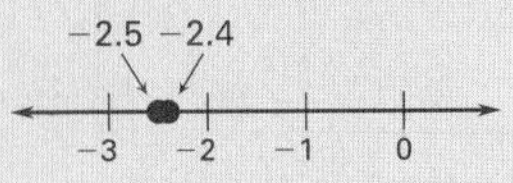

4.

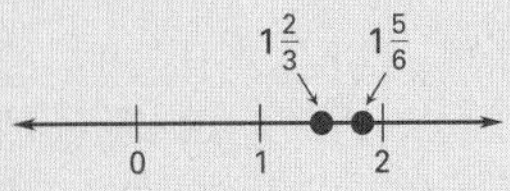

CHAPTER 2

Graph the numbers on a number line. Then write two inequalities that compare the two numbers. (Lesson 2.1) **1–4. See margin for graphs.**

1. -7 and 8 **$-7 < 8, 8 > -7$**

2. -5.2 and 3.6 **$-5.2 < 3.6, 3.6 > -5.2$**

3. -2.4 and -2.5 **$-2.5 < -2.4, -2.4 > -2.5$**

4. $1\frac{2}{3}$ and $1\frac{5}{6}$ **$1\frac{2}{3} < 1\frac{5}{6}$, $1\frac{5}{6} > 1\frac{2}{3}$**

Evaluate the expression. (Lesson 2.1)

5. $|8.5|$ **8.5** **6.** $|-3|$ **3** **7.** $|-4| + 3$ **7** **8.** $7 - |-5|$ **2**

Find the sum. (Lesson 2.2)

9. $-3 + 8$ **5** **10.** $18 + 27$ **45** **11.** $5 + (-7)$ **−2** **12.** $-4 + (-11)$ **−15**

13. $-4 + 13 + (-6)$ **3** **14.** $15 + (-12) + (-4)$ **−1** **15.** $-2 + (-9) + 8$ **−3** **16.** $17 + (-5) + 15$ **27**

Evaluate the expression. (Lesson 2.3)

17. $-8 - 5$ **−13** **18.** $4.1 - 6.3$ **−2.2** **19.** $-3 - (-7)$ **4** **20.** $-6 + (-3) - 4$ **−13**

21. $3.6 - 2.4 - (-6.1)$ **7.3** **22.** $-15 + 4 - 12$ **−23** **23.** $-11 - (-6) - 7$ **−12** **24.** $\frac{9}{10} - \frac{1}{2} + \left(-\frac{1}{5}\right)$ **$\frac{1}{5}$**

Find the sum or the difference of the matrices. (Lesson 2.4)

25. $\begin{bmatrix} 8 & -4 \\ 9 & 3 \end{bmatrix} + \begin{bmatrix} -2 & 6 \\ -1 & 5 \end{bmatrix}$ **$\begin{bmatrix} 6 & 2 \\ 8 & 8 \end{bmatrix}$**

26. $\begin{bmatrix} 3 & -6 \\ 1 & 4 \end{bmatrix} - \begin{bmatrix} 5 & 9 \\ 2 & -2 \end{bmatrix}$ **$\begin{bmatrix} -2 & -15 \\ -1 & 6 \end{bmatrix}$**

27. $\begin{bmatrix} -6 & 8 & 3 \\ 4 & 2 & 6 \end{bmatrix} + \begin{bmatrix} 3 & -6 & 7 \\ -4 & -5 & 8 \end{bmatrix}$ **$\begin{bmatrix} -3 & 2 & 10 \\ 0 & -3 & 14 \end{bmatrix}$**

Simplify the expression. (Lesson 2.5)

28. $(-6)(-7)$ **42** **29.** $(-5)(9)$ **−45** **30.** $(3)(-8)(-2)$ **48** **31.** $(-8)(-4x)$ **$32x$**

32. $-3(-y)(-y)$ **$-3y^2$** **33.** $(-c)^3(c)$ **$-c^4$** **34.** $(-7)^2(b)(-b)$ **$-49b^2$** **35.** $-4(-a^4)$ **$4a^4$**

Use the distributive property to rewrite the expression without parentheses. (Lesson 2.6)

36. $6(y + 5)$ **$6y + 30$** **37.** $4(a - 6)$ **$4a - 24$** **38.** $(x + 3)(-5)$ **$-5x - 15$** **39.** $-r(r - 5)$ **$-r^2 + 5r$**

40. $-k(7 + k)$ **$-7k - k^2$** **41.** $(x + 4)6x$ **$6x^2 + 24x$** **42.** $s(s - s^2)$ **$s^2 - s^3$** **43.** $(0.5z - 1.4)6$ **$3z - 8.4$**

Simplify the expression. (Lesson 2.6)

44. $3x + 7x$ **$10x$** **45.** $5.4m - 2.3m$ **$3.1m$** **46.** $82p - (-29p)$ **$111p$** **47.** $6 - 4t - 4$ **$2 - 4t$**

48. $5 + 4(x - 2)$ **$-3 + 4x$** **49.** $8x^2 + 5 - 2x^2$ **$6x^2 + 5$** **50.** $2x(7 - x) + 3x^2$ **$14x + x^2$** **51.** $\frac{2}{3}x + \left(-\frac{1}{6}\right)x$ **$\frac{1}{2}x$**

Simplify the expression. (Lesson 2.7)

52. $18 \div (-2)$ **−9** **53.** $-48 \div 12$ **−4** **54.** $16 \div \left(-\frac{4}{5}\right)$ **−20** **55.** $\frac{3x}{8} \div \frac{1}{2}$ **$\frac{3x}{4}$**

56. $21x \div 7$ **$3x$** **57.** $8x \div \left(-\frac{1}{4}\right)$ **$-32x$** **58.** $-24x \div \left(-\frac{2}{3}\right)$ **$36x$** **59.** $\frac{-22}{-\frac{1}{3}}$ **66**

60. PROBABILITY You randomly choose a button from a bag containing 9 yellow, 7 green, 11 blue, and 9 red buttons. What is the probability that you choose a red button? (Lesson 2.8) **$\frac{1}{4}$ or 25%**

Given the probability, find the odds. (Lesson 2.8)

61. The probability of scoring the basket is 0.20. **1 to 4** **62.** The probability of winning the election is $\frac{1}{3}$. **1 to 2**

CHAPTER 3

Solve the equation. (Lesson 3.1)

1. $y - 6 = 8$ **14** **2.** $n + 5 = -10$ **−15** **3.** $a - (-6) = 22$ **16** **4.** $14 - r = 3$ **11**

5. $|-7| + k = 4$ **−3** **6.** $\frac{1}{5} + m = -\frac{2}{5}$ $-\frac{3}{5}$ **7.** $-t - (-3) = 0$ **3** **8.** $-b + 3 - 1 = 3 \cdot 4$ **−10**

Solve the equation. (Lesson 3.2)

9. $7x = 35$ **5** **10.** $6a = 3$ $\frac{1}{2}$ **11.** $\frac{x}{10} = -2$ **−20** **12.** $-\frac{3}{8}t = 0$ **0**

13. $|-5| = 15b$ $\frac{1}{3}$ **14.** $-\frac{7}{8}r = \frac{3}{4}$ $-\frac{6}{7}$ **15.** $\frac{y}{10} = \frac{-2}{5}$ **−4** **16.** $-\frac{2}{3} = \frac{1}{9}k$ **−6**

Solve the equation. (Lesson 3.3)

17. $6x + 8 = 32$ **4** **18.** $2x - 1 = 11$ **6** **19.** $-x - 5 + 3x = 1$ **3** **20.** $4(x - 9) = 8$ **11**

21. $\frac{3}{5}x - 7 = 17$ **40** **22.** $x = 2(x - 1) + 6$ **−4** **23.** $\frac{x}{4} = -\frac{x}{2} - 1$ $-\frac{4}{3}$ **24.** $-5 = \frac{3}{8}(x - 1)$ $-12\frac{1}{3}$

Solve the equation if possible. (Lesson 3.4)

25. $-6 + 5x = 8x - 9$ **1** **26.** $8x + 6 = 3(4 - x)$ $\frac{6}{11}$ **27.** $4\left(\frac{1}{2}x + \frac{1}{2}\right) = 2x + 2$ **all real numbers**

28. $-3(-x - 4) = 2x + 1$ **−11** **29.** $4x = -2(-2x + 3)$ **no solution** **30.** $-4(x - 3) = 6(x + 5)$ **−1.8**

31. PHOTO FRAMING A photo measuring 5 inches by 7 inches is placed in a frame that adds a 1-inch border around the photo. Sketch a diagram of the photo and frame. What is the total width and length of the frame? (Lesson 3.5) **See margin for diagram. 7 in., 9 in.**

Solve the equation. Round the result to the nearest hundredth. (Lesson 3.6)

32. $4x = 82.50$ **20.63** **33.** $-26x - 59 = 135$ **−7.46** **34.** $2.3 - 4.8x = 8.2x + 5.6$ **−0.25**

35. $18.25x - 4.15 = 2.75x$ **0.27** **36.** $3(3.1x - 4.2) = 6.2x + 3.1$ **5.06** **37.** $8.4x - 3.2 = 4.1(3.4 - 2.1x)$ **1.01**

38. BICYCLING If you ride your bike for 2.5 hours and travel 37.5 miles, what is your average speed? (*Hint*: Solve $d = rt$ for r.) (Lesson 3.7) **15 mi/h**

Rewrite the equation so that *y* is a function of *x*. Then use the result to find *y* when *x* = −2, −1, 0, and 1. (Lesson 3.7)

39. $3 - y = x$ **$y = -x + 3$; 5, 4, 3, 2** **40.** $6x - 2y = 10$ **$y = 3x - 5$; −11, −8, −5, −2** **41.** $-6x - 1 = -4y + 2$ **$y = \frac{3}{2}x + \frac{3}{4}$; $-2\frac{1}{4}$, $-\frac{3}{4}$, $\frac{3}{4}$, $2\frac{1}{4}$**

42. $-2y - 3x - 16 = 0$ **$y = -\frac{3}{2}x - 8$; −5, $-6\frac{1}{2}$, −8, $-9\frac{1}{2}$** **43.** $\frac{1}{2}x - \frac{1}{2}y = 6$ **$y = x - 12$; −14, −13, −12, −11** **44.** $-2(x - 4) = 2(y + 5)$ **$y = -x - 1$; 1, 0, −1, −2**

Find the unit rate. (Lesson 3.8)

45. \$3 for 5 yogurt snacks **\$.60 per yogurt snack** **46.** \$50.75 for working 7 hours **\$7.25 per hour** **47.** Drive 122 miles in 2.5 hours **48.8 mi/h**

Convert the measure. Round your answer to the nearest hundredth. (Lesson 3.8)

48. 24 miles to kilometers (1 mile = 1.609 km) **38.62 km** **49.** 6 fluid ounces to cups (1 cup = 8 fluid ounces) **0.75 cup**

Find the percent. Round to the nearest whole percent. (Lesson 3.8)

50. Tax of \$1.00 on an item priced at \$19.99 **5%** **51.** 9 field goals made out of 14 attempted **64%**

52. \$4.50 tip on a restaurant bill of \$22 **20%** **53.** 127 surveys returned out of 350 **36%**

31.

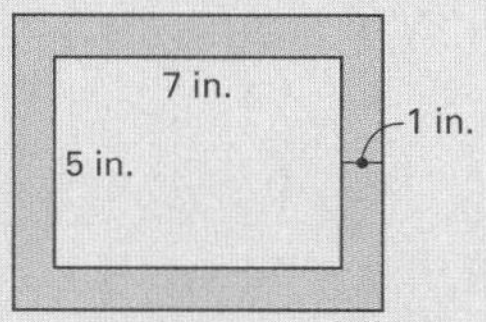

1.
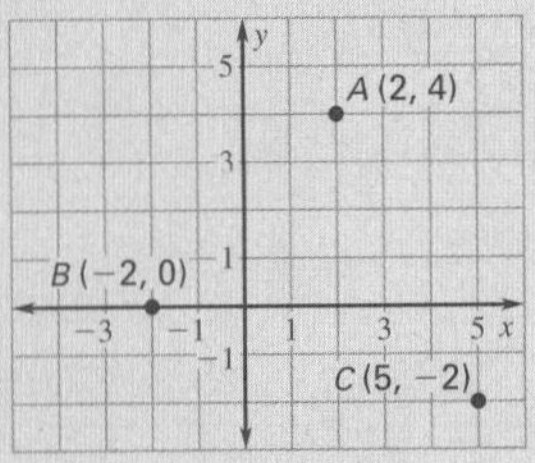

2.
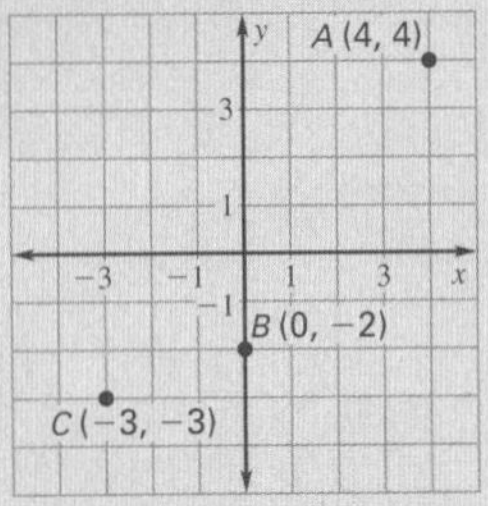

3.
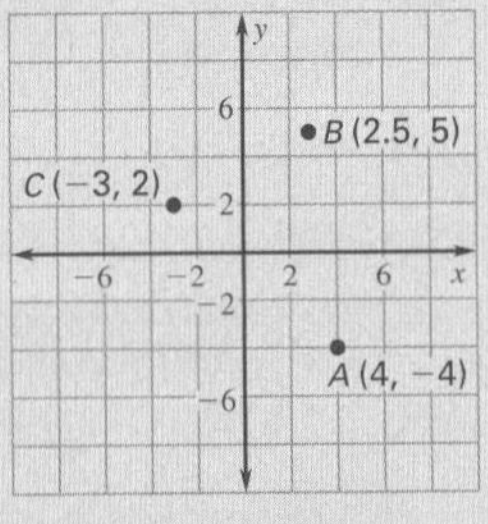

4.
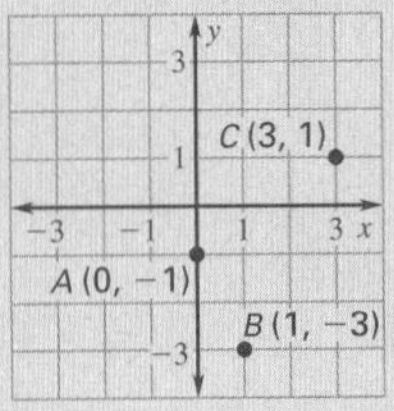

5.
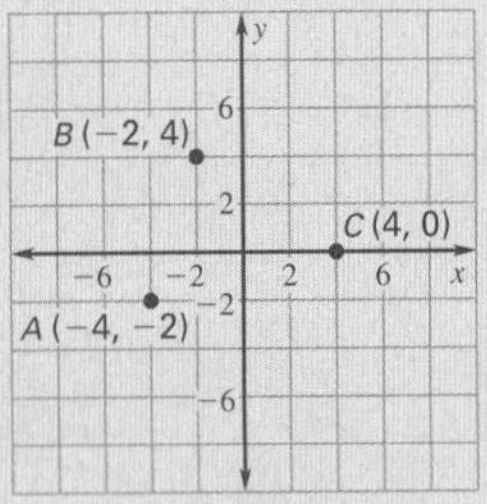

6.
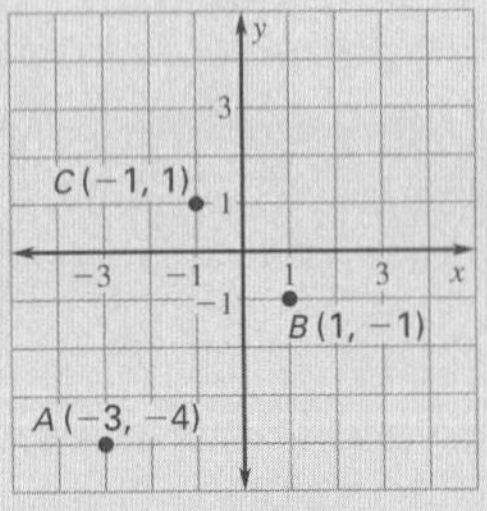

7–22, 40–47. See Additional Answers beginning on page AA1.

CHAPTER 4

Plot and label the ordered pairs in a coordinate plane. (Lesson 4.1) **1–6. See margin.**

1. $A(2, 4), B(-2, 0), C(5, -2)$ **2.** $A(4, 4), B(0, -2), C(-3, -3)$ **3.** $A(4, -4), B(2.5, 5), C(-3, 2)$

4. $A(0, -1), B(1, -3), C(3, 1)$ **5.** $A(-4, -2), B(-2, 4), C(4, 0)$ **6.** $A(-3, -4), B(1, -1), C(-1, 1)$

Use a table of values to graph the equation. (Lesson 4.2) **7–14. See margin.**

7. $y = 5x + 1$ **8.** $y = -2x + 4$ **9.** $y = -2$ **10.** $4x + y = -8$

11. $x = 3$ **12.** $y = -(6 + x)$ **13.** $y = \frac{1}{2}x - 1$ **14.** $y = -\frac{1}{3}x - 4$

Find the *x*-intercept and the *y*-intercept of the line. Graph the equation. Label the points where the line crosses the axes. (Lesson 4.3) **15–22. See margin for graphs.**

15. $5x + y = -5$ **−1, −5** **16.** $2x - y = 6$ **3, −6** **17.** $y = 2x - 5$ $2\frac{1}{2}, -5$ **18.** $6y + 2x = 12$ **6, 2**

19. $14 = y - 2x$ **−7, 14** **20.** $8x + 2y = -16$ **−2, −8** **21.** $y = 6 - 3x$ **2, 6** **22.** $1.5y = x - 3$ **3, −2**

Find the slope of the line passing through the points. (Lesson 4.4)

23. $(6, 1), (-4, 1)$ **0** **24.** $(2, 2), (-1, 4)$ $-\frac{2}{3}$ **25.** $(0, 6), (-3, 0)$ **2** **26.** $(4, 5), (2, 2)$ $\frac{3}{2}$

27. $(-4, 2), (-3, -5)$ **−7** **28.** $(3, 6), (3, -1)$ **no slope** **29.** $(-1, 0), (0, -1)$ **−1** **30.** $(-2, -3), (1, -2)$ $\frac{1}{3}$

The variables *x* and *y* vary directly. Use the given values to write an equation that relates *x* and *y*. (Lesson 4.5)

31. $x = 6, y = 18$ $y = 3x$ **32.** $x = 4, y = 1$ $y = \frac{1}{4}x$ **33.** $x = 8, y = -7$ $y = -\frac{7}{8}x$ **34.** $x = -\frac{1}{2}, y = -10$ $y = 20x$

35. $x = -2, y = -2$ $y = x$ **36.** $x = 8, y = -4$ $y = -\frac{1}{2}x$ **37.** $x = 2.5, y = 2$ $y = \frac{4}{5}x$ **38.** $x = 2.1, y = -6.3$ $y = -3x$

39. **EARNINGS** Your earnings vary directly with the number of hours you work. If you earn \$60 for 4 hours of work, how much will you earn for working 35 hours? (Lesson 4.5) **\$525**

Write the equation in slope-intercept form. Then graph the equation. (Lesson 4.6) **40–47. See margin for graphs.**

40. $x - y = 1$ $y = x - 1$ **41.** $y = 3$ $y = 3$ **42.** $-3x + 2y = 6$ $y = \frac{3}{2}x + 3$ **43.** $y + 4 = 0$ $y = -4$

44. $x = 2y + 1$ $y = \frac{1}{2}x - \frac{1}{2}$ **45.** $2(x + y + 1) = 4y$ $y = x + 1$ **46.** $2x - 4y + 6 = 0$ $y = \frac{1}{2}x + \frac{3}{2}$ **47.** $5x - 3y + 2 = 14 - 4x$ $y = 3x - 4$

Solve the equation graphically. Check your solution algebraically. (Lesson 4.7)

48. $6 - 5x = 1$ **1** **49.** $2x = -12$ **−6** **50.** $6x + 8 = 2x$ **−2** **51.** $8x + 3 = 19$ **2**

52. $\frac{4}{3}x - 2 = -6$ **−3** **53.** $15 - 2x = 3x$ **3** **54.** $\frac{2}{3}x + 5 = -2$ $-\frac{21}{2}$ **55.** $7x - 4 = -3 + 4x$ $\frac{1}{3}$

Decide whether the relation is a function. (Lesson 4.8)

56.
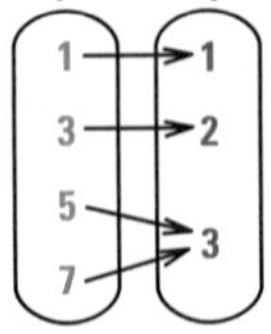

function

57.

Input	Output
4	−2
0	0
4	2
2	4

not a function

58.
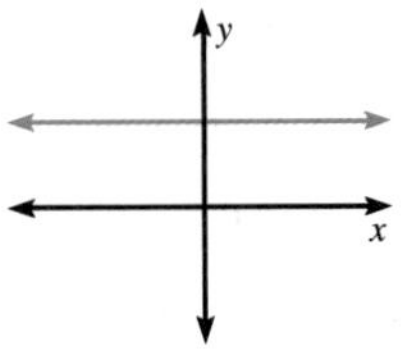

function

59.
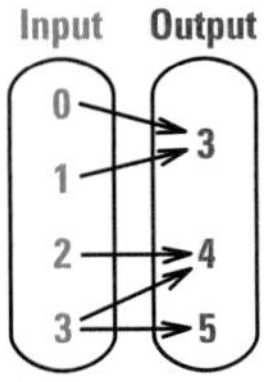

not a function

CHAPTER 5

Write an equation of the line in slope-intercept form. (Lesson 5.1)

1. The slope is 2; the y-intercept is 1. $y = 2x + 1$
2. The slope is -3; the y-intercept is -2. $y = -3x - 2$
3. The slope is 1; the y-intercept is -3. $y = x - 3$
4. The slope is -4; the y-intercept is 0. $y = -4x$

Write an equation of the line that passes through the point and has the given slope. Write the equation in slope-intercept form. (Lesson 5.2)

5. $(-1, 0), m = 3$ $y = 3x + 3$
6. $(5, 2), m = -2$ $y = -2x + 12$
7. $(3, 6), m = 0$ $y = 6$
8. $(-2, 1), m = -5$ $y = -5x - 9$
9. $(-3, -1), m = 4$ $y = 4x + 11$
10. $(1, 5), m = 8$ $y = 8x - 3$
11. $(2, -1), m = \frac{1}{2}$ $y = \frac{1}{2}x - 2$
12. $(-4, 3), m = -\frac{1}{3}$ $y = -\frac{1}{3}x + \frac{5}{3}$

Write an equation in slope-intercept form of the line that passes through the points. (Lesson 5.3)

13. $(3, -2), (5, 4)$ $y = 3x - 11$
14. $(2, 17), (-1, -1)$ $y = 6x + 5$
15. $(5, 1), (0, -6)$ $y = \frac{7}{5}x - 6$
16. $(-2, -1), (4, -4)$ $y = -\frac{1}{2}x - 2$
17. $(-1, 7), (5, 7)$ $y = 7$
18. $(0, 0), (2, 3)$ $y = \frac{3}{2}x$
19. $(-3, 5), (-6, 8)$ $y = -x + 2$
20. $(5, 2), (1, 14)$ $y = -3x + 17$

Draw a scatter plot for the data. Draw a best-fitting line and write an equation of the line. (Lesson 5.4) 21-22. Sample equations are given.

21.

x	-1	0	1	3	4	5
y	0	-1	1	3	2	4

See margin. $y = \frac{3}{4}x - \frac{1}{2}$

22.

x	2	2.5	3	3	3.5	4
y	2	-1	1	-3	-2	-4

See margin. $y = -5x + 14$

Write the point-slope form of the equation of the line that passes through the point and has the given slope. Then rewrite the equation in slope-intercept form. (Lesson 5.5)

23. $(5, 3), m = -2$ $y - 3 = -2(x - 5)$; $y = -2x + 13$
24. $(3, 7), m = -3$ $y - 7 = -3(x - 3)$; $y = -3x + 16$
25. $(5, 4), m = 5$ $y - 4 = 5(x - 5)$; $y = 5x - 21$
26. $(-2, -4), m = \frac{3}{4}$ $y + 4 = \frac{3}{4}(x + 2)$; $y = \frac{3}{4}x - \frac{5}{2}$
27. $(-3, -5), m = -\frac{5}{3}$
28. $(0, 8), m = \frac{1}{4}$
29. $(2, 4), m = -\frac{1}{2}$
30. $(-1, -7), m = 4$

27. $y + 5 = -\frac{5}{3}(x + 3)$; $y = -\frac{5}{3}x - 10$ 28. $y - 8 = \frac{1}{4}x$; $y = \frac{1}{4}x + 8$ 29. $y - 4 = -\frac{1}{2}(x - 2)$; $y = -\frac{1}{2}x + 5$ 30. $y + 7 = 4(x + 1)$; $y = 4x - 3$

Write an equation in standard form of the line that passes through the given point and has the given slope, or that passes through the two points. (Lesson 5.6)

31. $(5, -2), m = 3$ $3x - y = 17$
32. $(1, 2), (3, 4)$ $x - y = -1$
33. $(-1, 0), (-5, -3)$ $3x - 4y = -3$
34. $(-4, 3), m = -\frac{5}{6}$ $5x + 6y = -2$
35. $(-3, 7), (1, 7)$ $y = 7$
36. $(5, 7), m = -\frac{3}{4}$ $3x + 4y = 43$
37. $(-2, -5), m = 5$ $5x - y = -5$
38. $(4, -2), (4, 5)$ $x = 4$

WIND CHILL **Wind chill is based on the combined effects of temperature and wind speed. The table shows the wind chill for various temperatures (in degrees Fahrenheit) when the wind speed is 5 miles per hour.** (Lesson 5.7)

Temperature	35	30	25	20	15	10
Wind chill	32	27	22	16	11	6

39. Make a scatter plot of the data. See margin.
40. Write a linear model for the data. $w = t - 4$
41. Use the linear model to estimate the wind chill at 5° Fahrenheit. 1°F

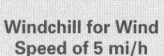

40–41. Equation and solution may vary depending on the method used to derive the linear model.

21.

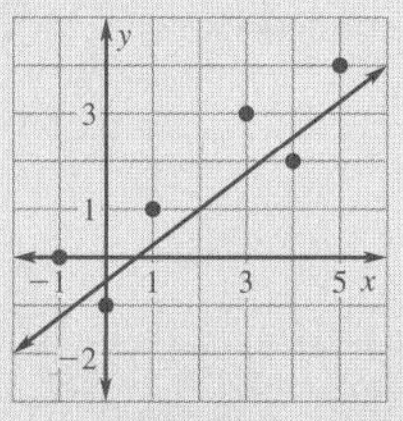

22.

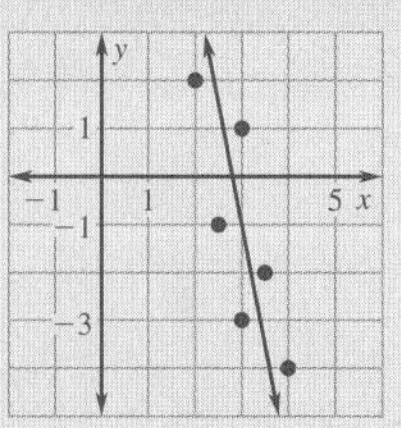

39.

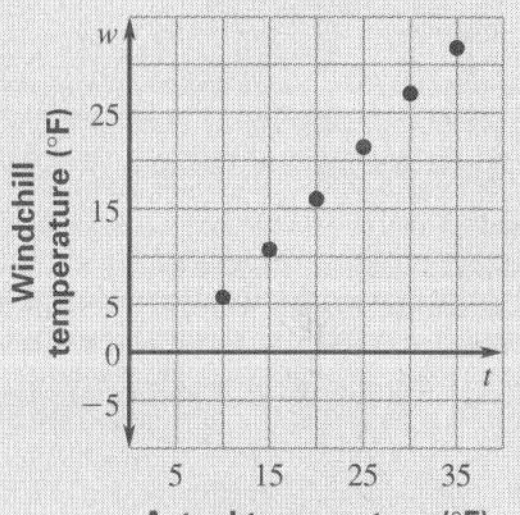

1. 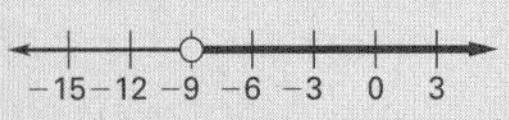

2. 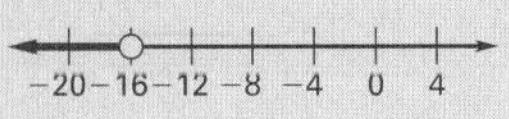

3. 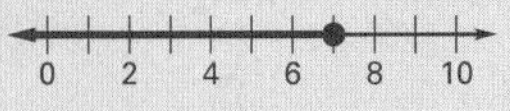

4. 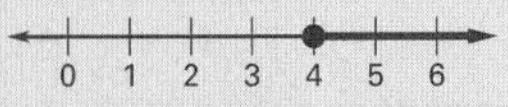

5. 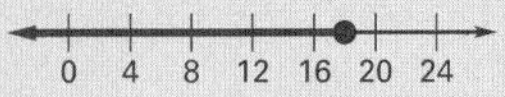

6. 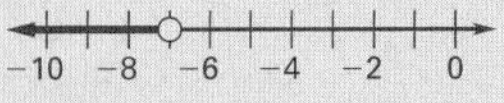

7. 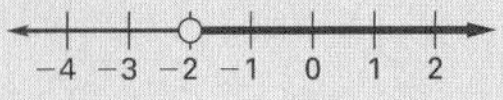

8. 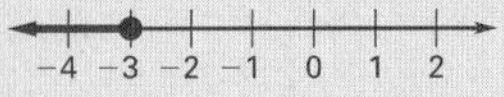

15. 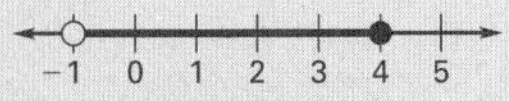

16. 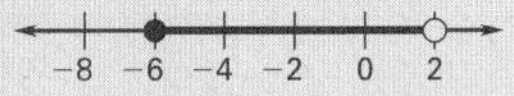

17. 

18. 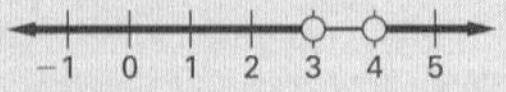

19. 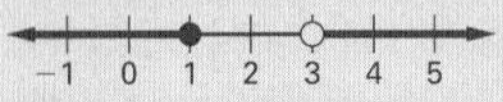

20. 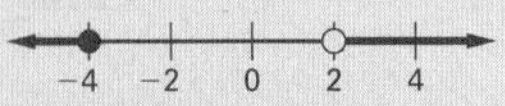

27. 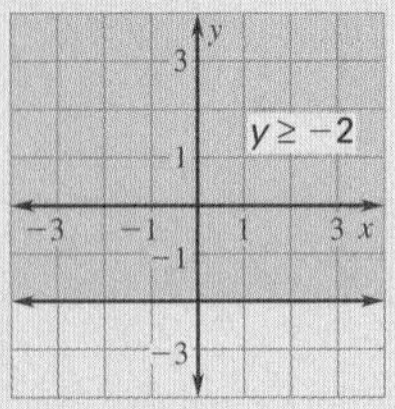

28. 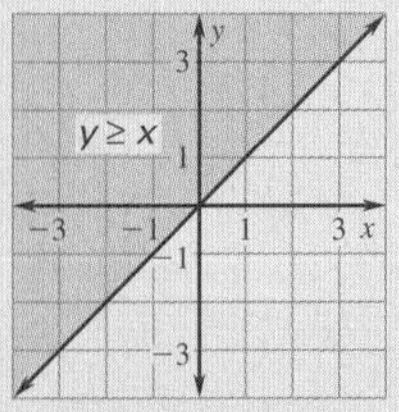

29. 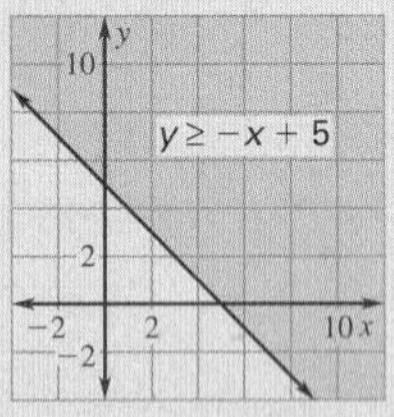

30. 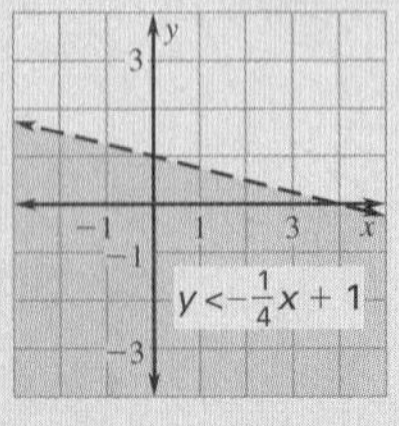

31–39, 48–51. See Additional Answers beginning on page AA1.

CHAPTER 6

Solve the inequality and graph its solution. (Lesson 6.1) **1–8. See margin for graphs.**

1. $x + 5 > -4$ **$x > -9$** 2. $m - 4 < -20$ **$m < -16$** 3. $3 \geq y - 4$ **$y \leq 7$** 4. $9x \geq 36$ **$x \geq 4$**

5. $\frac{k}{9} \leq 2$ **$k \leq 18$** 6. $-5a > 35$ **$a < -7$** 7. $-\frac{x}{10} < \frac{1}{5}$ **$x > -2$** 8. $0.5 \leq -\frac{b}{6}$ **$b \leq -3$**

Solve the inequality. (Lesson 6.2)

9. $0 \leq \frac{1}{2}x + 6$ **$x \geq -12$** 10. $\frac{3}{4}x + 5 \leq 8$ **$x \leq 4$** 11. $3x + 8 \geq -2x + 3$ **$x \geq -1$**

12. $-3x - 7 < 2$ **$x > -3$** 13. $-(x + 5) < -4x - 11$ **$x < -2$** 14. $-(4 + x) > 2(x - 5)$ **$x < 2$**

Solve the inequality and graph the solution. (Lesson 6.3) **15–20. See margin for graphs.**

15. $8 \geq x + 4 > 3$ **$4 \geq x > -1$** 16. $-36 \leq 6x < 12$ **$-6 \leq x < 2$** 17. $-15 < -3x < 18$ **$-6 < x < 5$**

18. $2x + 1 > 9$ or $3x - 5 < 4$ **$x > 4$ or $x < 3$** 19. $x + 1 > 4$ or $2x + 3 \leq 5$ **$x > 3$ or $x \leq 1$** 20. $-4x + 1 \geq 17$ or $5x - 4 > 6$ **$x \leq -4$ or $x > 2$**

Solve the equation or the inequality. (Lesson 6.4)

21. $|10 + x| = 4$ **$-14, -6$** 22. $|2x + 3| = 9$ **$-6, 3$** 23. $|x - 4| + 4 = 7$ **$1, 7$**

24. $|6x - 5| + 1 < 8$ **$-\frac{1}{3} < x < 2$** 25. $|3x + 4| - 6 \geq 14$ **$x \leq -8$ or $x \geq 5\frac{1}{3}$** 26. $|10 - 4x| \leq 2$ **$2 \leq x \leq 3$**

Sketch the graph of the inequality. (Lesson 6.5) **27–35. See margin.**

27. $y \geq -2$ 28. $x - y \leq 0$ 29. $x + y \geq 5$ 30. $4y + x < 4$

31. $x - 3y \leq 0$ 32. $3y - 2x < 6$ 33. $5x - 3y > 9$ 34. $2y - x > 10$

35. PURCHASING SNACKS You have \$20 to spend on snacks for your friends. Each juice drink costs \$2 and each bag of popcorn costs \$1.25. Let x represent the number of drinks. Let y represent the number of bags of popcorn. Write and graph an inequality to model the numbers of juice drinks and popcorn bags you can buy. (Lesson 6.5) **$2x + 1.25y \leq 20$; See margin for graph.**

Make a stem-and-leaf plot for the data. (Lesson 6.6) **36–39. See margin.**

36. 18, 31, 98, 71, 22, 14, 20, 66, 18, 76, 33, 13

37. 13, 22, 40, 21, 45, 65, 66, 16, 31, 33, 41, 35, 51

38. 3, 22, 41, 5, 63, 8, 17, 24, 60, 64, 6, 55, 78, 91, 13, 32, 54, 72

39. 44, 19, 51, 2, 46, 8, 39, 41, 38, 39, 56, 27, 54, 43, 60, 62, 21, 49, 22, 63

Find the mean, the median, and the mode of the collection of numbers. (Lesson 6.6)

40. 8, 3, 2, 8, 1, 4, 8, 5, 6 **5, 5, 8**

41. 13, 15, 17, 13, 11, 13, 9, 17, 4, 8 **12, 13, 13**

42. 3.2, 5.4, 2.3, 1.2, 3.2, 5.1, 4.2 **about 3.51, 3.2, 3.2**

43. 123, 151, 121, 112, 146, 112, 138 **129, 123, 112**

Find the first, second, and third quartiles of the data. (Lesson 6.7)

44. 8, 6, 7, 4, 9, 7, 5 **5, 7, 8**

45. 16, 20, 76, 31, 14, 88, 54 **16, 31, 76**

46. 41, 45, 10, 11, 31, 15, 18, 25, 33, 32 **15, 28, 33**

47. 19, 72, 93, 82, 22, 21, 14, 31 **20, 26.5, 77**

Draw a box-and-whisker plot of the data. (Lesson 6.7) **48–51. See margin.**

48. 6, 14, 8, 13, 5, 11, 9, 18, 3

49. 2, 7, 8, 3, 5, 7, 9, 3

50. 18, 22, 13, 9, 4, 31, 52, 40

51. 1, 8, 21, 12, 5, 2, 9, 10, 4, 15, 3, 23, 31

CHAPTER 7

Graph and check to solve the linear system. **(Lesson 7.1)** **1–8. See margin.**

1. $y = 5$
$x = -2$

2. $x = 0$
$y = 3x + 7$

3. $y = \frac{1}{2}x + 5$
$y = -3x + 5$

4. $5x + 3y = 15$
$4x - 3y = 12$

5. $y = -2x - 6$
$y = -4$

6. $x + y = 10$
$x - y = -2$

7. $-2x + 4y = 12$
$5x - 2y = 10$

8. $\frac{1}{8}(x + y) = 1$
$x - y = 4$

Use the substitution method to solve the linear system. **(Lesson 7.2)**

9. $x = 5y$ **(−5, −1)**
$2x + 3y = -13$

10. $y = -2x$ **(−7, 14)**
$x + y = 7$

11. $x + y = 9$ **(6, 3)**
$x - y = 3$

12. $2a + 3b = 3$ **(0, 1)**
$a - 6b = -6$

13. $s - t = 5$ $\left(5\frac{1}{7}, \frac{1}{7}\right)$
$3s + 4t = 16$

14. $5x - 8y = -17$ **(3, 4)**
$3x = 5 + y$

15. $2m + n = 7$ $\left(15\frac{1}{2}, -24\right)$
$4m + 3n = -10$

16. $5a + b = 4$ $\left(\frac{1}{2}, 1\frac{1}{2}\right)$
$7a + 5b = 11$

Use linear combinations to solve the system of linear equations. **(Lesson 7.3)**

17. $3x + 3y = 6$ **(2, 0)**
$2x - 3y = 4$

18. $3x + 7y = -1$ $\left(\frac{1}{3}, -\frac{2}{7}\right)$
$7y = -6x$

19. $\frac{2}{5}x - \frac{1}{2}y = 1$ **(0, −2)**
$\frac{1}{5}x + \frac{1}{2}y = -1$

20. $x + 4y = \frac{9}{2}$ $\left(\frac{9}{10}, \frac{9}{10}\right)$
$\frac{1}{2}(x - y) = 0$

21. $2x + y = 3(x - 5)$
$x + 5 = 4y + 2x$
(13, −2)

22. $2x + 3y = 15$ $\left(-1, 5\frac{2}{3}\right)$
$3y + 5x = 12$

23. $y = 2x - 36$
$3x - 0.5y = 26$
(4, −28)

24. $-4x - 15 = 5y$
$2y = 11 - 5x$
(5, −7)

25. CONCERT TICKETS You pay \$105 for 8 tickets to attend a folk festival. Tickets for students cost \$10 each and tickets for adults cost \$15 each. How many of each type of ticket did you buy? **(Lesson 7.4)** **3 student, 5 adult**

PRINTING T-SHIRTS **The tennis team wants to purchase T-shirts for its members. Company A charges a \$20 setup fee and \$8 per shirt. Company B charges a \$10 setup fee and \$10 per shirt.** **(Lesson 7.4)**

26. How many shirts would have to be purchased for the costs to be equal? **5 shirts**

27. Which company offers a better price for 12 shirts? **Company A**

Use any method to solve the linear system and tell how many solutions the system has. **(Lesson 7.5)**

28. $x + y = 4$
$2x + 3y = 9$
one solution, (3, 1)

29. $x + y = 6$
$3x + 3y = 3$
no solution

30. $x + 2y = 5$
$3x - 15 = -6y$
infinitely many solutions

31. $12x - y = 5$
$-8x + y = -5$
one solution, (0, −5)

32. $y = -3x$
$6y - x = 38$
one solution, (−2, 6)

33. $2x - 3y = 3$
$6x - 9y = 9$
infinitely many solutions

34. $3x + 6 = 7y$
$x + 2y = 11$
one solution, (5, 3)

35. $3x - 8y = 4$
$6x - 42 = 16y$
no solution

Graph the system of linear inequalities. **(Lesson 7.6)** **36–43. See margin.**

36. $y \ge 0$
$x \le 0$

37. $y > x + 1$
$y < x + 3$

38. $x \ge 1$
$y + x \le 5$

39. $y + 2 < -x$
$2y - 4 > 3x$

40. $x < 5$
$x \ge 1$
$y \ge -2$
$y < 7$

41. $y > x - 4$
$y \ge -x - 1$
$y \le 0$

42. $y > x - 3$
$x - y > -2$
$x \le 3$

43. $3x - 1 < 5$
$-x + y \le 10$
$-5x + 2 < 12$

1.
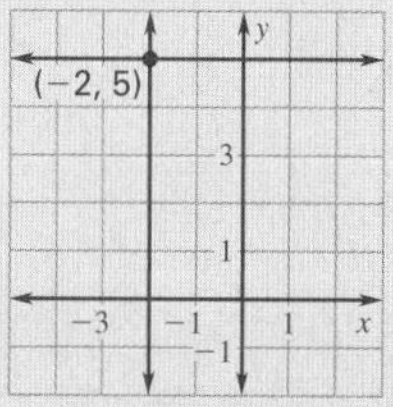

2.
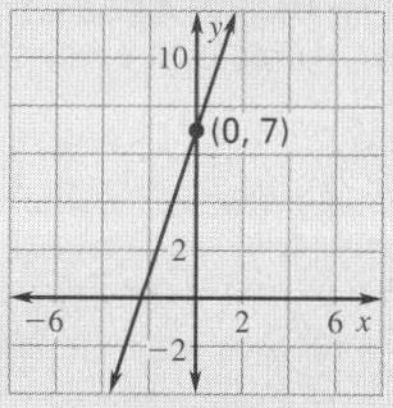

3.
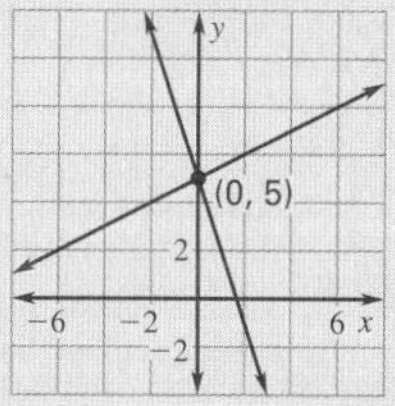

4.
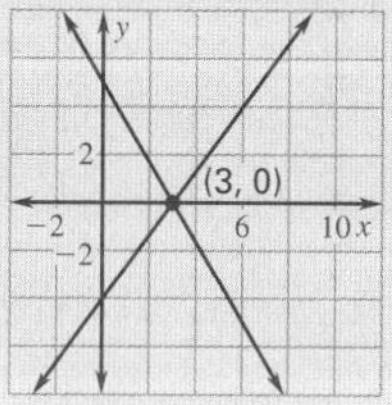

5.
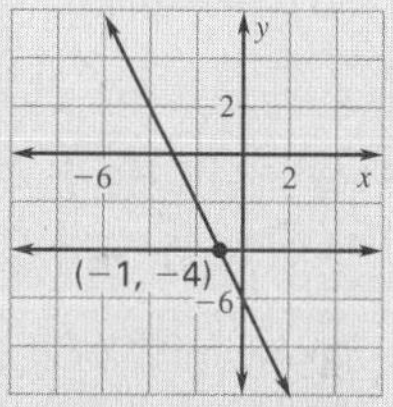

6.
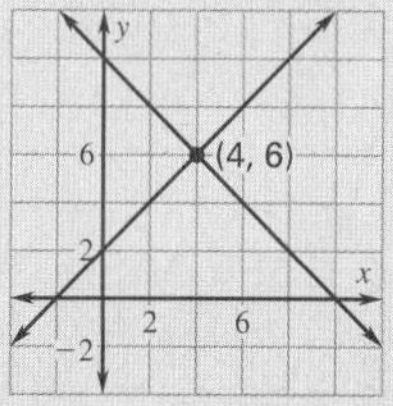

7.
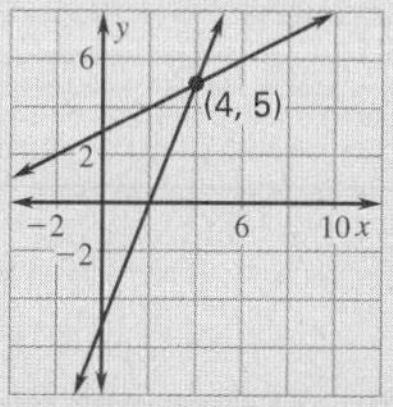

8.
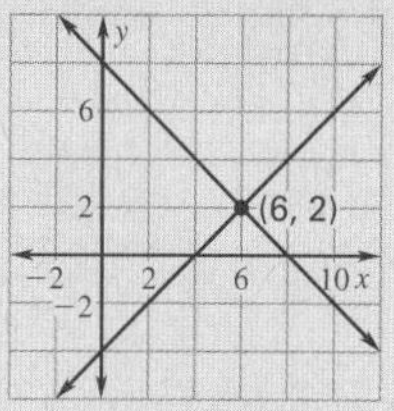

36–43. See Additional Answers beginning on page AA1.

17.
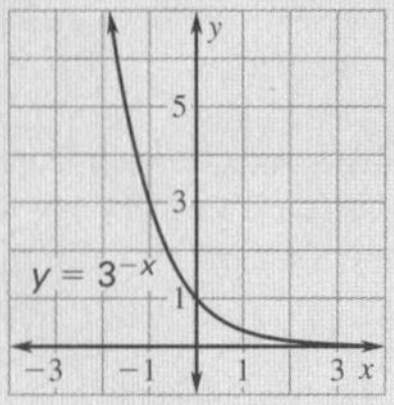

18.
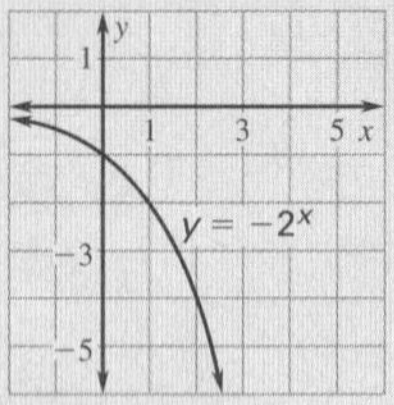

19.
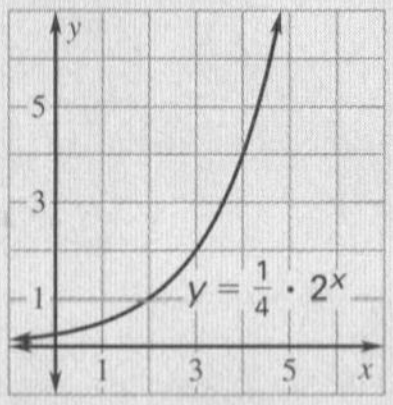

20.
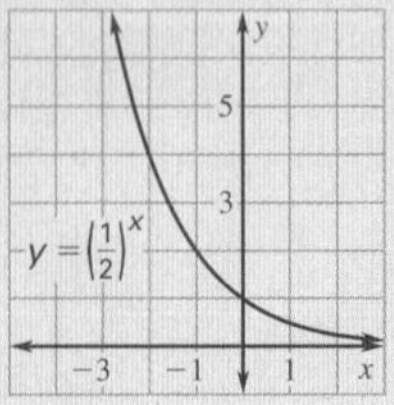

CHAPTER 8

Simplify, if possible. Write your answer as a power or as a product of powers. (Lesson 8.1)

1. $(7^2)(7^3)$ **7^5, or 16,807** **2.** $2^3 \cdot 2^4$ **2^7, or 128** **3.** $(12x)^3$ **$1728x^3$** **4.** $-(4x)^2 \cdot (5x^3)$ **$-80x^5$**

5. $(4r^2s)^2(-2s^2)^3$ **$-128r^4s^8$** **6.** $(7x^3y) \cdot (-2x^4)$ **$-14x^7y$** **7.** $(3x)^3(-5y)^2$ **$675x^3y^2$** **8.** $(-x^3)^2(x)^2(-x^4)^3$ **$-x^{20}$**

Rewrite the expression with positive exponents. (Lesson 8.2)

9. m^{-4} **$\frac{1}{m^4}$** **10.** $\left(\frac{x}{2}\right)^{-2}$ **$\frac{4}{x^2}$** **11.** $\frac{y}{x^{-2}}$ **x^2y** **12.** $\frac{6}{(2a)^{-3}}$ **$48a^3$**

13. $\frac{3}{3x^{-4}y^3}$ **$\frac{x^4}{y^3}$** **14.** $(-3t)^0 \cdot \frac{2}{s^{-2}}$ **$2s^2$** **15.** $6x \cdot \frac{1}{x^{-3}}$ **$6x^4$** **16.** $\left(\frac{4b^{-1}}{2a^4}\right)^{-2}$ **$\frac{a^8b^2}{4}$**

Graph the exponential function. (Lesson 8.2) **17–20. See margin.**

17. $y = 3^{-x}$ **18.** $y = -2^x$ **19.** $y = \frac{1}{4} \cdot 2^x$ **20.** $y = \left(\frac{1}{2}\right)^x$

Simplify the expression. The simplified expression should have no negative exponents. (Lesson 8.3)

21. $\frac{2^{11}}{2^8}$ **8** **22.** $\frac{6^5}{6^7}$ **$\frac{1}{36}$** **23.** $\left(\frac{3x^2z^4}{2xz}\right)^3$ **$\frac{27x^3z^9}{8}$** **24.** $\frac{(a^3)^5}{(a^4)^5}$ **$\frac{1}{a^5}$**

25. $\frac{18b^2c}{4bc^3} \cdot \frac{(3ab)^{-2}}{5a^2c^3}$ **$\frac{1}{10a^4bc^5}$** **26.** $\frac{(rst)^{-2}}{rs} \cdot \frac{(t^2)^3}{(s^{-3})^4}$ **$\frac{s^9t^4}{r^3}$** **27.** $5^0 \cdot \frac{(5xy)^2}{(x^3y^{-5})^3}$ **$\frac{25y^{17}}{x^7}$** **28.** $\left(\frac{3}{8}\right)^{-1} \cdot \frac{(2a^3x^5)^2}{(8a^{-3}x^{-1})^{-3}}$ **$\frac{16{,}384x^7}{3a^3}$**

29. You roll a die three times. What is the probability that you would roll a two each time? (Lesson 8.3) **$\frac{1}{216} \approx 0.005$**

Rewrite in decimal form. (Lesson 8.4)

30. 4.813×10^{-6} **0.000004813** **31.** 3.11×10^4 **31,100** **32.** 8.4162×10^{-2} **0.084162** **33.** 9.43×10^0 **9.43**

34. 5.0645×10^{10} **50,645,000,000** **35.** 1.2468×10^{-3} **0.0012468** **36.** 2.34×10^{-8} **0.0000000234** **37.** 6.09013×10^{10} **60,901,300,000**

Rewrite in scientific notation. (Lesson 8.4)

38. 5280 **5.28×10^3** **39.** 0.0378 **3.78×10^{-2}** **40.** 11.38 **1.138×10^1** **41.** 33,000,000 **3.3×10^7**

42. 827.66 **8.2766×10^2** **43.** 0.208054 **2.08054×10^{-1}** **44.** 16.354 **1.6354×10^1** **45.** 0.000891 **8.91×10^{-4}**

INTEREST **A principal of $1100 is deposited in an account that pays 5% interest compounded yearly. Find the balance after the given number of years.** (Lesson 8.5)

46. 1 year **$1155** **47.** 10 years **$1791.78** **48.** 15 years **$2286.82** **49.** 25 years **$3724.99**

Classify the model as *exponential growth* or *exponential decay*. Identify the growth or decay factor and the percent of increase or decrease per time period. (Lesson 8.6)

50. $y = 16(0.99)^t$ **exponential decay; 0.99; 1%** **51.** $y = 22(0.5)^t$ **exponential decay; 0.5; 50%** **52.** $y = 8(1.12)^t$ **exponential growth; 1.12; 12%** **53.** $y = 35(1.04)^t$ **exponential growth; 1.04; 4%**

54. $y = 13\left(\frac{7}{6}\right)^t$ **exponential growth; $\frac{7}{6}$; $16\frac{2}{3}$%** **55.** $y = 110\left(\frac{2}{3}\right)^t$ **exponential decay; $\frac{2}{3}$; $33\frac{1}{3}$%** **56.** $y = 4(1.5)^t$ **exponential growth; 1.5; 50%** **57.** $y = 9(0.68)^t$ **exponential decay; 0.68; 32%**

58. DEPRECIATION A piece of equipment originally costs $120,000. Its value decreases at a rate of 10% per year. Estimate its value after 6 years. (Lesson 8.6) **about $63,773**

CHAPTER 9

Evaluate the expression. Give the exact value if possible. Otherwise, approximate to the nearest hundredth. (Lesson 9.1)

1. $-\sqrt{100}$ **−10**
2. $\pm\sqrt{676}$ **± 26**
3. $-\sqrt{0.25}$ **−0.5**
4. $\pm\sqrt{169}$ **± 13**
5. $\sqrt{379}$ **19.47**
6. $-\sqrt{36}$ **−6**
7. $\sqrt{220}$ **14.83**
8. $\sqrt{0.01}$ **0.1**

Solve the equation or write *no solution*. Write the solutions as integers if possible. Otherwise, write them as radical expressions. (Lesson 9.1)

9. $x^2 = 25$ **−5, 5**
10. $4x^2 - 8 = 0$ $-\sqrt{2}, \sqrt{2}$
11. $x^2 = -16$ **no solution**
12. $x^2 + 1 = 1$ **0**
13. $3x^2 - 48 = 0$ **−4, 4**
14. $6x^2 + 6 = 4$ **no solution**
15. $2x^2 - 6 = 0$ $-\sqrt{3}, \sqrt{3}$
16. $x^2 - 4 = -3$ **−1, 1**
17. FALLING OBJECT A ball is dropped from a bridge 150 feet above a river. How long will it take for the ball to hit the surface of the water? (Lesson 9.1) **about 3.06 sec**

Simplify the expression. (Lesson 9.2)

18. $\sqrt{250}$ $5\sqrt{10}$
19. $6\sqrt{8} \cdot 7\sqrt{2}$ **168**
20. $\sqrt{4} \cdot 3\sqrt{2}$ $6\sqrt{2}$
21. $-2\sqrt{6} \cdot 7\sqrt{30}$ $-84\sqrt{5}$
22. $\sqrt{\frac{11}{16}}$ $\frac{\sqrt{11}}{4}$
23. $\frac{\sqrt{20}}{\sqrt{5}}$ **2**
24. $\frac{1}{2}\sqrt{\frac{8}{50}}$ $\frac{1}{5}$
25. $\sqrt{\frac{2}{3}} \cdot \sqrt{\frac{5}{3}}$ $\frac{\sqrt{10}}{3}$

Sketch the graph of the function. Label the vertex. (Lesson 9.3) **26–33. See margin.**

26. $y = 3x^2$
27. $y = x^2 - 4$
28. $y = -x^2 - 2x$
29. $y = x^2 - 6x + 8$
30. $y = 4x^2 + 4x - 5$
31. $y = x^2 - 2x + 3$
32. $y = -x^2 + 3x + 2$
33. $y = -3x^2 + 12x - 1$

Represent the solution graphically. Check the solutions algebraically. (Lesson 9.4)

34. $x^2 - 6x = -5$ **1, 5**
35. $x^2 + 5x = -6$ **−3, −2**
36. $x^2 - 3x = 4$ **−1, 4**
37. $\frac{1}{4}x^2 = 9$ **−6, 6**
38. $x^2 + 3x = 10$ **−5, 2**
39. $x^2 - 9 = 0$ **−3, 3**
40. $\frac{1}{2}x^2 + 2x = 6$ **−6, 2**
41. $-2x^2 + 4x + 6 = 0$ **−1, 3**

Write the quadratic equation in standard form. Solve using the quadratic formula. (Lesson 9.5)

42. $x^2 + x = 12$ $x^2 + x - 12 = 0$; **−4, 3**
43. $3r^2 + 2 = -8r$ $3r^2 + 8r + 2 = 0$; $\frac{-4 - \sqrt{10}}{3}, \frac{-4 + \sqrt{10}}{3}$
44. $3k^2 + 11k = 4$ $3k^2 + 11k - 4 = 0$; $-4, \frac{1}{3}$
45. $-x^2 + 5x = 4$ $-x^2 + 5x - 4 = 0$; **1, 4**
46. $m^2 - 2m - 4 = -3$ $m^2 - 2m - 1 = 0$; $1 - \sqrt{2}, 1 + \sqrt{2}$
47. $2x^2 - 6x = 5$ $2x^2 - 6x - 5 = 0$; $\frac{3 - \sqrt{19}}{2}, \frac{3 + \sqrt{19}}{2}$
48. $b^2 - 8 = 7b$ $b^2 - 7b - 8 = 0$; **−1, 8**
49. $10 - 2x^2 = -x$ $-2x^2 + x + 10 = 0$; $-2, \frac{5}{2}$

Find the discriminant for the equation. Then tell if the equation has *two solutions, one solution,* or *no real solution*. (Lesson 9.6)

50. $3x^2 + 14x - 5 = 0$ **256; two solutions**
51. $4x^2 + 12x + 9 = 0$ **0; one solution**
52. $x^2 + 10x + 9 = 0$ **64; two solutions**
53. $2x^2 + 8x + 8 = 0$ **0; one solution**
54. $5x^2 + 125 = 0$ **−2500; no real solution**
55. $x^2 - 2x + 35 = 0$ **−136; no real solution**
56. $2x^2 - x - 3 = 0$ **25; two solutions**
57. $-3x^2 + 5x - 6 = 0$ **−47; no real solution**

Sketch the graph of the inequality. (Lesson 9.7) **58–65. See margin.**

58. $y > -x^2 + 4$
59. $y \le 4x^2$
60. $y > 5x^2 + 10x$
61. $y < x^2 + 5x$
62. $y \ge -3x^2 - 6x + 4$
63. $y < x^2 - 4x + 3$
64. $y \le -x^2 + 4x + 5$
65. $y \ge -x^2 + 8x - 1$

Make a scatter plot of the data. Then name the type of model that best fits the data. (Lesson 9.8) **66–67. See margin for plots.**

66. (−1, −1.5), (2.5, 2), (0.5, 0), (−0.5, −1), (1.5, 1), (−2, −2.5), (0, −0.5) **linear**
67. (4, 6), (−2, 12), (3, 2), (2, 0), (1, 0), (0, 2), (5, 12), (−1, 6) **quadratic**

26.
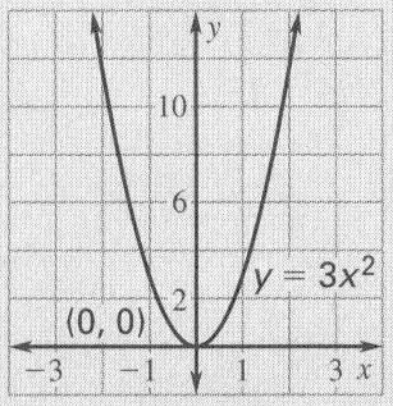

27.
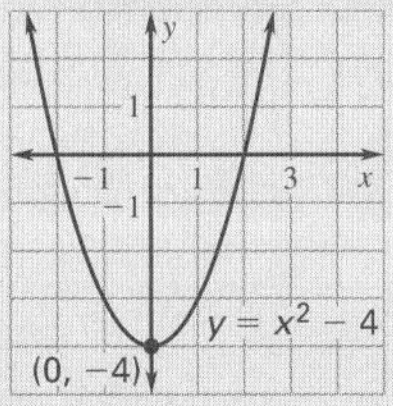

28.
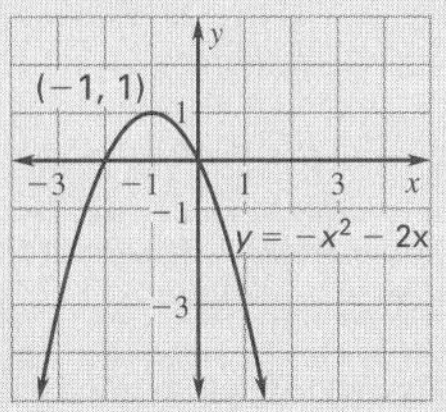

29.
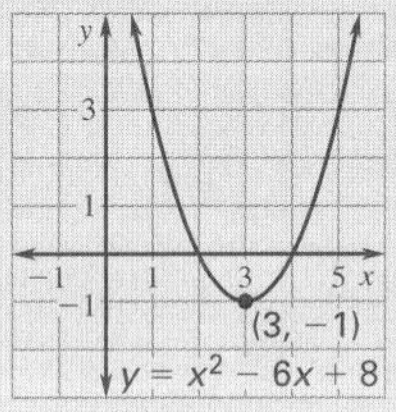

30.
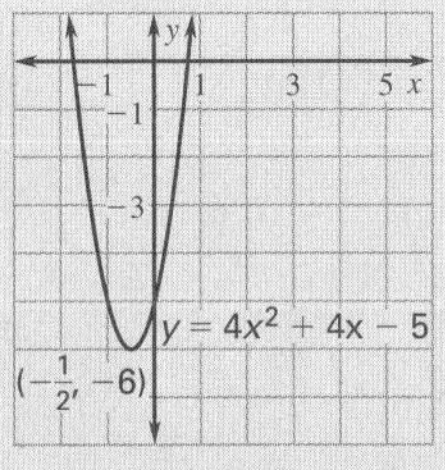

31.
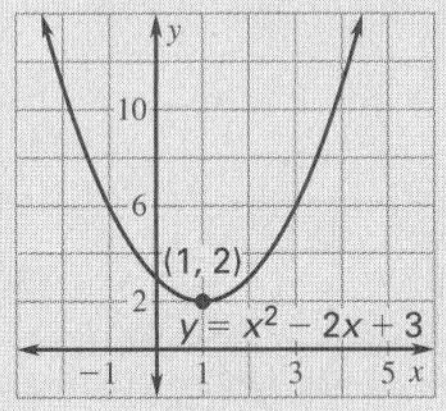

32.
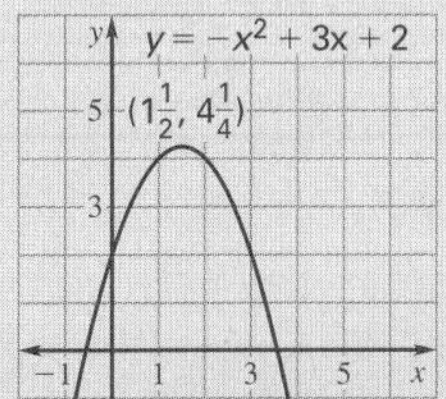

33, 58–67. See Additional Answers beginning on page AA1.

31.
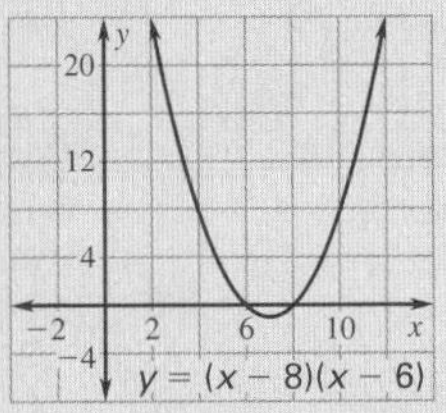

32.
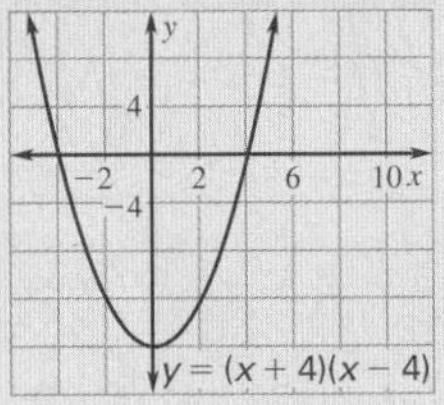

33.
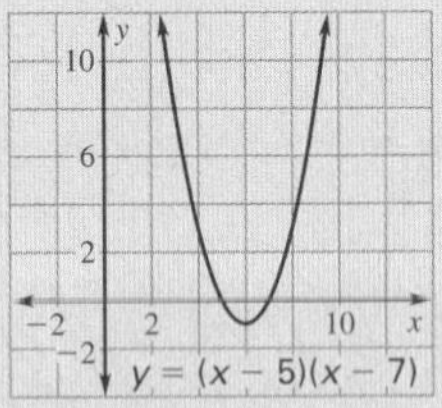

34.
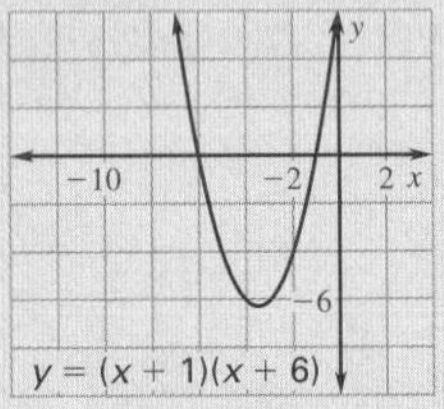

35.
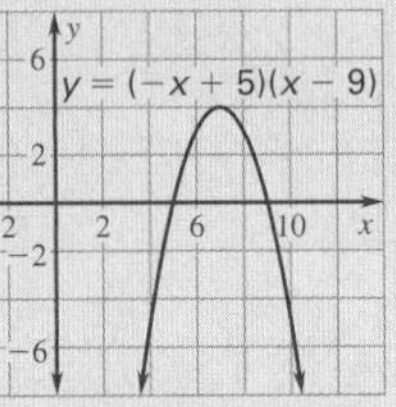

36.
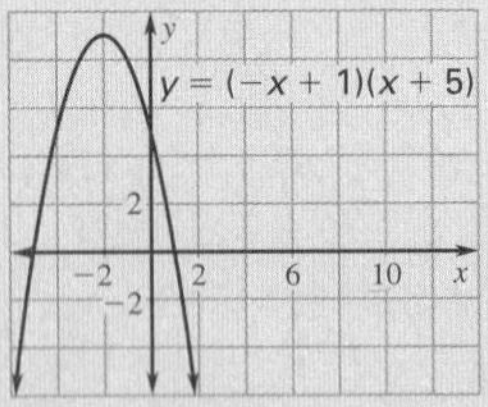

37.
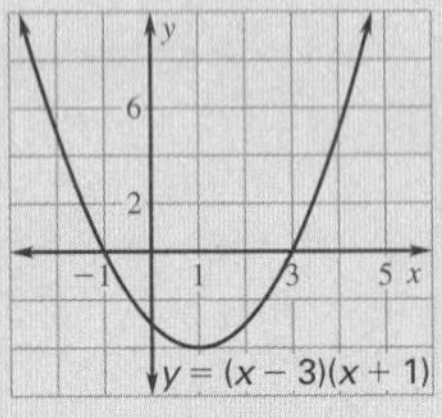

38.
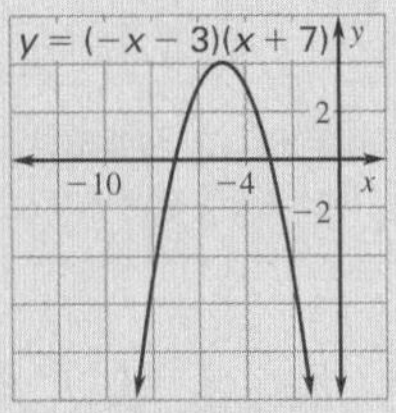

CHAPTER 10

Use a vertical format or a horizontal format to find the sum or the difference. (Lesson 10.1)

1. $(7x^2 - 4) + (x^2 + 5)$ **$8x^2 + 1$**
2. $(3x^2 - 2) - (2x - 6x^2)$ **$9x^2 - 2x - 2$**
3. $(8x^2 - 3x + 7) + (6x^2 - 4x + 1)$ **$14x^2 - 7x + 8$**
4. $(-z^3 + 3z) + (-z^2 - 4z - 6)$ **$-z^3 - z^2 - z - 6$**
5. $(5x^2 + 7x - 4) - (4x^2 - 2x)$ **$x^2 + 9x - 4$**
6. $(3a + 2a^4 - 5) - (a^3 + 2a^4 + 5a)$ **$-a^3 - 2a - 5$**

Find the product. (Lesson 10.2)

7. $x(4x^2 - 8x + 7)$ **$4x^3 - 8x^2 + 7x$**
8. $-3x(x^2 + 5x - 5)$ **$-3x^3 - 15x^2 + 15x$**
9. $5b^2(3b^3 - 2b^2 + 1)$ **$15b^5 - 10b^4 + 5b^2$**
10. $(t + 9)(2t + 1)$ **$2t^2 + 19t + 9$**
11. $(d - 1)(d + 5)$ **$d^2 + 4d - 5$**
12. $(3z + 4)(5z - 8)$ **$15z^2 - 4z - 32$**
13. $(x + 3)(x^2 - 2x + 6)$ **$x^3 + x^2 + 18$**
14. $(3 + 2s - s^2)(s - 1)$ **$-3 + s + 3s^2 - s^3$**

Find the product. (Lesson 10.3)

15. $(x + 9)^2$ **$x^2 + 18x + 81$**
16. $(-c - d)^2$ **$c^2 + 2cd + d^2$**
17. $(a - 2)(a + 2)$ **$a^2 - 4$**
18. $(-7 + m)(-7 - m)$ **$49 - m^2$**
19. $(4x + 5)^2$ **$16x^2 + 40x + 25$**
20. $(5p - 6q)^2$ **$25p^2 - 60pq + 36q^2$**
21. $(2a + 3b)(2a - 3b)$ **$4a^2 - 9b^2$**
22. $(10x - 5y)(10x + 5y)$ **$100x^2 - 25y^2$**

Solve the factored equation. (Lesson 10.4)

23. $(x + 3)(x + 6) = 0$ **$-3, -6$**
24. $(x - 11)^2 = 0$ **11**
25. $(z - 1)(z + 5) = 0$ **$1, -5$**
26. $-8(5w - 2)(w + 4) = 0$ **$\frac{2}{5}, -4$**
27. $(6n - 9)(n - 7) = 0$ **$\frac{3}{2}, 7$**
28. $3(x + 2)^2 = 0$ **-2**
29. $(2d - 2)(4d - 8) = 0$ **$1, 2$**
30. $2(3x - 1)(2x + 5) = 0$ **$\frac{1}{3}, -\frac{5}{2}$**

Find the *x*-intercepts and the vertex of the graph of the function. Then sketch the graph of the function. (Lesson 10.4) **31–38. See margin for graphs.**

31. $y = (x - 8)(x - 6)$ **6, 8; (7, −1)**
32. $y = (x + 4)(x - 4)$ **−4, 4; (0, −16)**
33. $y = (x - 5)(x - 7)$ **5, 7; (6, −1)**
34. $y = (x + 1)(x + 6)$ **$-1, -6; \left(-3\frac{1}{2}, -6\frac{1}{4}\right)$**
35. $y = (-x + 5)(x - 9)$ **5, 9; (7, 4)**
36. $y = (-x + 1)(x + 5)$ **−5, 1; (−2, 9)**
37. $y = (x - 3)(x + 1)$ **−1, 3; (1, −4)**
38. $y = (-x - 3)(x + 7)$ **−7, −3; (−5, 4)**

Solve the equation by factoring. (Lesson 10.5)

39. $x^2 + 6x + 9 = 0$ **-3**
40. $x^2 + 2x - 35 = 0$ **$-7, 5$**
41. $x^2 - 12x = -36$ **6**
42. $-x^2 - 4x = 3$ **$-3, -1$**
43. $x^2 - 15x = -54$ **$6, 9$**
44. $-x^2 + 14x = 48$ **$6, 8$**
45. $x^2 - 2x = 24$ **$-4, 6$**
46. $x^2 - 5x + 4 = 0$ **$1, 4$**

Solve the equation by factoring. (Lesson 10.6)

47. $2x^2 + x - 6 = 0$ **$-2, 1\frac{1}{2}$**
48. $2x^2 + 7x = -3$ **$-3, -\frac{1}{2}$**
49. $9x^2 + 24x = -16$ **$-\frac{4}{3}$**
50. $20x^2 + 23x + 6 = 0$ **$-\frac{2}{5}, -\frac{3}{4}$**
51. $4x^2 - 5x = 6$ **$-\frac{3}{4}, 2$**
52. $3x^2 - 5 = -14x$ **$-5, \frac{1}{3}$**
53. $3x^2 - 17x = 56$ **$-2\frac{1}{3}, 8$**
54. $12x^2 + 46x - 36 = 0$ **$-4\frac{1}{2}, \frac{2}{3}$**

Factor the expression. Tell which special product factoring pattern you used. (Lesson 10.7)

55. $x^2 - 1$ **$(x + 1)(x - 1)$; difference of two squares**
56. $9b^2 - 81$ **$(3b + 9)(3b - 9)$; difference of two squares**
57. $121 - x^2$ **$(11 - x)(11 + x)$; difference of two squares**
58. $12 - 27x^2$ **$3(2 + 3x)(2 - 3x)$; difference of two squares**
59. $t^2 + 2t + 1$ **$(t + 1)^2$; perfect square trinomial**
60. $x^2 + 20x + 100$ **$(x + 10)^2$; perfect square trinomial**
61. $64y^2 + 48y + 9$ **$(8y + 3)^2$; perfect square trinomial**
62. $20x^2 - 100x + 125$ **$5(2x - 5)^2$; perfect square trinomial**

Factor the expression completely. (Lesson 10.8)

63. $x^4 - 9x^2$ **$x^2(x + 3)(x - 3)$**
64. $m^3 + 11m^2 + 28m$ **$m(m + 4)(m + 7)$**
65. $x^4 + 4x^3 - 45x^2$ **$x^2(x + 9)(x - 5)$**
66. $x^3 + 2x^2 - 4x - 8$ **$(x - 2)(x + 2)^2$**
67. $-3y^3 - 15y^2 - 12y$ **$-3y(y + 4)(y + 1)$**
68. $x^3 - x^2 + 4x - 4$ **$(x^2 + 4)(x - 1)$**
69. $7x^6 - 21x^4$ **$7x^4(x^2 - 3)$**
70. $8t^3 - 3t^2 + 16t - 6$ **$(t^2 + 2)(8t - 3)$**

71. GEOMETRY The width of a box is 4 inches less than the length. The height is 9 inches greater than the length. The box has a volume of 180 cubic inches. What are the dimensions of the box? (Lesson 10.8) **2 in. by 6 in. by 15 in.**

CHAPTER 11

Solve the proportion. Check for extraneous solutions. (Lesson 11.1)

1. $\frac{x}{2} = \frac{8}{x}$ **−4, 4** **2.** $\frac{9}{m} = \frac{15}{10}$ **6** **3.** $\frac{c^2 - 16}{c + 4} = \frac{c - 4}{3}$ **4** **4.** $\frac{3}{5} = \frac{x + 2}{6}$ $\mathbf{\frac{8}{5}}$

5. $\frac{14}{2} = \frac{7}{n - 5}$ **6** **6.** $\frac{12}{8} = \frac{5 + t}{t - 3}$ **19** **7.** $\frac{x + 15}{16} = \frac{-9}{x - 10}$ **−6, 1** **8.** $\frac{x + 30}{-4} = \frac{143}{x - 18}$ **−8, −4**

Solve the percent problem. (Lesson 11.2)

9. What number is 40% of 60? **24** **10.** 75 is what percent of 250? **30%** **11.** How much is 4% of 525 pounds? **21 lb**

12. 8 is 20% of what number? **40** **13.** 35% of 720 is what number? **252** **14.** 12% of what amount is \$1482? **\$12,350**

Decide if the data in the table show *direct* or *inverse* variation. Write an equation that relates the variables. (Lesson 11.3)

15.

x	0.5	1	2	4
y	4	2	1	0.5

inverse variation; $xy = 2$

16.

x	0.5	1	2	2.5
y	1.5	3	6	7.5

direct variation; $y = 3x$

Simplify the expression if possible. (Lesson 11.4)

17. $\frac{12x^4}{42x}$ $\mathbf{\frac{2x^3}{7}}$ **18.** $\frac{5x^2 - 15x^3}{10x}$ $\mathbf{\frac{x - 3x^2}{2}}$ **19.** $\frac{x + 6}{x^2 + 7x + 6}$ $\mathbf{\frac{1}{x + 1}}$ **20.** $\frac{x^2 - 8x + 15}{x - 3}$ **$x - 5$**

GEOMETRIC PROBABILITY **A coin is tossed onto the large square region shown. It is equally likely to land on any point in the region.** (Lesson 11.4)

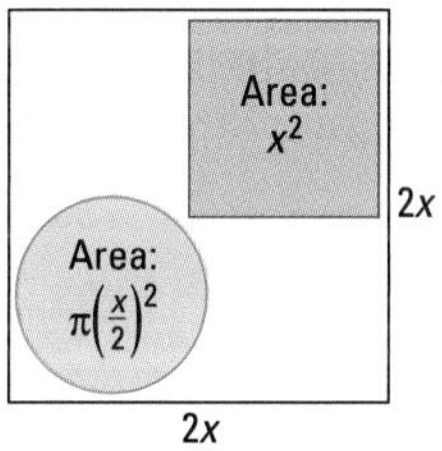

21. Find the probability that the coin lands in the small square. $\mathbf{\frac{1}{4}}$

22. Find the probability that the coin lands in the circle. $\mathbf{\frac{\pi}{16}}$

Simplify the expression. (Lesson 11.5)

23. $\frac{3x}{5} \cdot \frac{15}{18x}$ $\mathbf{\frac{1}{2}}$ **24.** $\frac{4x^2}{7} \cdot \frac{14}{8x}$ **x** **25.** $\frac{2x^2}{3x} \cdot \frac{6x^3}{20x^2}$ $\mathbf{\frac{x^2}{5}}$ **26.** $\frac{x - 3}{5(x + 4)} \cdot \frac{4(x + 4)}{x - 3}$ $\mathbf{\frac{4}{5}}$

27. $\frac{1}{4x} \div \frac{6x}{15}$ $\mathbf{\frac{5}{8x^2}}$ **28.** $\frac{10x^2}{x^2 - 25} \cdot (x - 5)$ $\mathbf{\frac{10x^2}{x + 5}}$ **29.** $\frac{5x}{x - 3} \div \frac{x - 8}{x - 3}$ $\mathbf{\frac{5x}{x - 8}}$ **30.** $\frac{x^2 + 5x - 36}{x^2 - 81} \div (x^2 - 16)$ $\mathbf{\frac{1}{(x - 9)(x + 4)}}$

Simplify the expression. (Lesson 11.6)

31. $\frac{20}{5x} - \frac{15}{5x}$ $\mathbf{\frac{1}{x}}$ **32.** $\frac{4}{x + 2} + \frac{8}{x + 2}$ $\mathbf{\frac{12}{x + 2}}$ **33.** $\frac{3}{x + 7} - \frac{7}{x - 4}$ $\mathbf{\frac{-4x - 61}{(x + 7)(x - 4)}}$

34. $\frac{5x + 3}{x^2 - 25} + \frac{5}{x - 5}$ $\mathbf{\frac{10x + 28}{(x + 5)(x - 5)}}$ **35.** $\frac{7}{2x} - \frac{1}{9x^2}$ $\mathbf{\frac{63x - 2}{18x^2}}$ **36.** $\frac{3x}{6x^2 + 13x + 2} + \frac{x + 1}{x^2 + 5x + 6}$ $\mathbf{\frac{9x^2 + 16x + 1}{(6x + 1)(x + 2)(x + 3)}}$

Divide. (Lesson 11.7)

37. Divide $14x^3 + 7x^2 + x$ by $7x$. $\mathbf{2x^2 + x + \frac{1}{7}}$ **38.** Divide $x^2 + 6x - 16$ by $x + 8$. **$x - 2$**

39. Divide $12x^2 - 2x - 2$ by $3x + 1$. **$2(2x - 1)$** **40.** Divide $2x^3 + 9x^2 - 6x + 2$ by $2x + 1$. $\mathbf{x^2 + 4x - 5 + \frac{7}{2x + 1}}$

Solve the equation. (Lesson 11.8)

41. $\frac{4}{x - 6} = \frac{x}{10}$ **−4, 10** **42.** $\frac{-2}{3x} = \frac{4 + x}{6}$ **−2** **43.** $\frac{4}{x} + \frac{2}{3} = \frac{6}{x}$ **3**

44. $\frac{x}{x - 5} - \frac{11}{x - 5} = 7$ **4** **45.** $\frac{1}{x - 3} = \frac{5}{x + 9}$ **6** **46.** $\frac{5}{x - 1} + 1 = \frac{4}{x^2 + 3x - 4}$ **−6, −2**

1. all nonnegative numbers; all nonnegative numbers

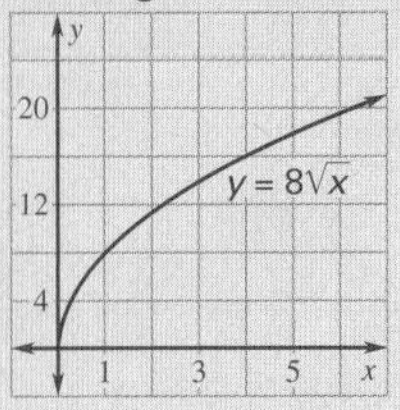

2. all nonnegative numbers; all numbers greater than or equal to −5

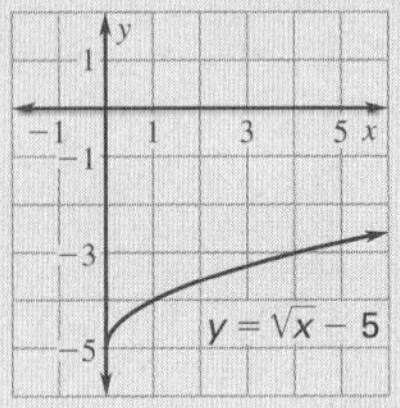

3. all numbers greater than or equal to −3; all nonnegative numbers

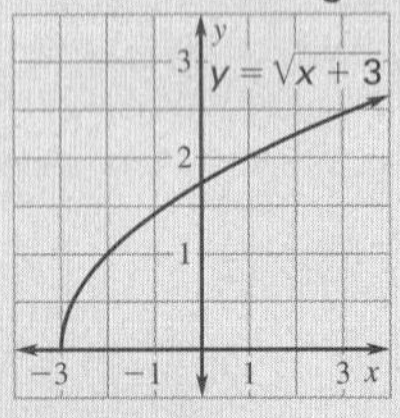

4. all nonnegative numbers; all nonnegative numbers

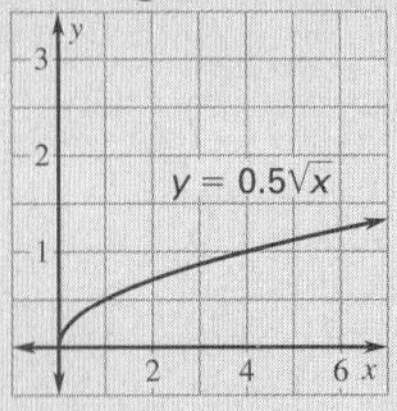

5. all numbers greater than or equal to 2; all nonnegative numbers

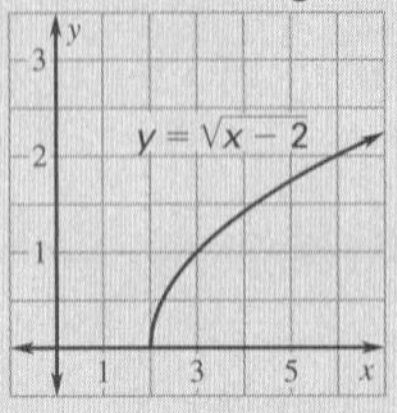

6. all nonnegative numbers; all numbers greater than or equal to 1

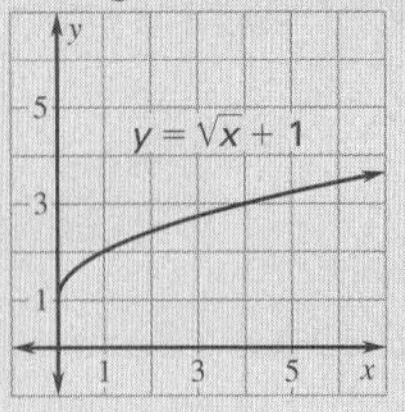

7, 8, 26, 27, 43–45. See Additional Answers beginning on page AA1.

CHAPTER 12

Find the domain and the range of the function. Then sketch its graph. (Lesson 12.1) 1–8. See margin.

1. $y = 8\sqrt{x}$ **2.** $y = \sqrt{x} - 5$ **3.** $y = \sqrt{x + 3}$ **4.** $y = 0.5\sqrt{x}$

5. $y = \sqrt{x - 2}$ **6.** $y = \sqrt{x} + 1$ **7.** $y = \sqrt{3x + 2}$ **8.** $y = \sqrt{4x - 3}$

Simplify the expression. (Lesson 12.2)

9. $3\sqrt{5} + 2\sqrt{5}$ $5\sqrt{5}$ **10.** $8\sqrt{7} - 15\sqrt{7}$ $-7\sqrt{7}$ **11.** $2\sqrt{8} + 3\sqrt{32}$ $16\sqrt{2}$ **12.** $\sqrt{20} - \sqrt{45} + \sqrt{80}$ $3\sqrt{5}$

13. $\sqrt{3}(7 - \sqrt{6})$ $7\sqrt{3} - 3\sqrt{2}$ **14.** $(4 + \sqrt{10})^2$ $26 + 8\sqrt{10}$ **15.** $\frac{4}{\sqrt{24}}$ $\frac{\sqrt{6}}{3}$ **16.** $\frac{3}{5 - \sqrt{2}}$ $\frac{15 + 3\sqrt{2}}{23}$

Solve the equation. Check for extraneous solutions. (Lesson 12.3)

17. $\sqrt{x} - 11 = 0$ 121 **18.** $\sqrt{2x - 1} + 4 = 7$ 5 **19.** $\sqrt{x + 20} = x$ 5

20. $12 = \sqrt{3x + 1} + 7$ 8 **21.** $\frac{1}{2}x = \sqrt{2x - 3}$ 2, 6 **22.** $\sqrt{18 - 2x} + 5 = x$ 7

Solve the equation by completing the square. (Lesson 12.4)

23. $x^2 + 10x = 56$ −14, 4 **24.** $x^2 - 6x + 4 = 20$ −2, 8 **25.** $x^2 - 7x = -10$ 2, 5

26. $x^2 + x = 3$ See margin. **27.** $5x^2 - 12x - 15 = 0$ See margin. **28.** $3x^2 - 10x - 29 = 0$ $\frac{5 - 4\sqrt{7}}{3}, \frac{5 + 4\sqrt{7}}{3}$

Find the missing length of the right triangle if *a* and *b* are the lengths of the legs and *c* is the length of the hypotenuse. (Lesson 12.5)

29. $a = 1, b = 1$ $c = \sqrt{2}$ **30.** $a = 1, c = 2$ $b = \sqrt{3}$ **31.** $b = 6, c = 10$ $a = 8$

32. $a = 7, b = 10$ $c = \sqrt{149}$ **33.** $b = 15, c = 25$ $a = 20$ **34.** $a = 30, c = 50$ $b = 40$

Find the distance and the midpoint between the two points. Round the distance to the nearest hundredth if necessary. (Lesson 12.6)

35. (0, 4), (4, 5) 4.12; $\left(2, 4\frac{1}{2}\right)$ **36.** (−3, 3), (6, −1) 9.85; $\left(1\frac{1}{2}, 1\right)$ **37.** (1, 0), (4, −4) 5; (2.5, −2) **38.** (0, 0), (3, −2) 3.61; $\left(1\frac{1}{2}, -1\right)$

39. (7, −6), (−1, −6) 8; (3, −6) **40.** (5, 2), (5, −4) 6; (5, −1) **41.** (12, −7), (−4, 2) 18.36; $\left(4, -2\frac{1}{2}\right)$ **42.** (−4, −5), (−8, −9) 5.66; (−6, −7)

Find the sine, the cosine, and the tangent of ∠*D* and ∠*E*. (Lesson 12.7) 43–45. See margin.

43.

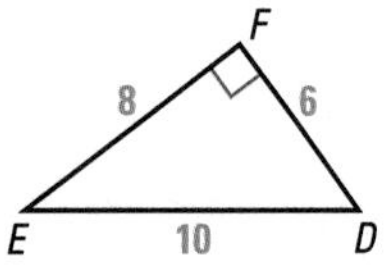

44.

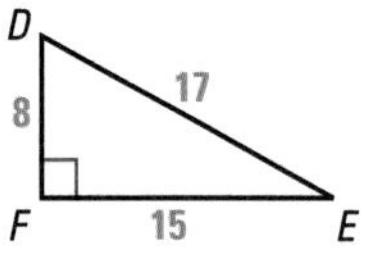

45.

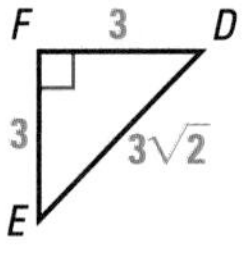

Find the lengths of the sides of the triangle. Round your answer to the nearest hundredth. (Lesson 12.7)

46.

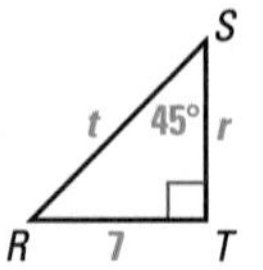

r = 7, t = 9.90

47.

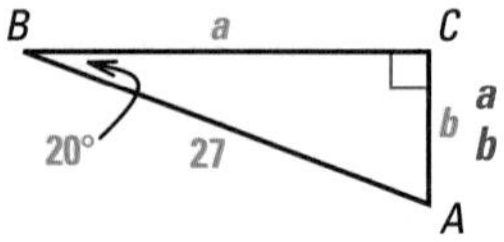

a = 25.37, b = 9.23

48.

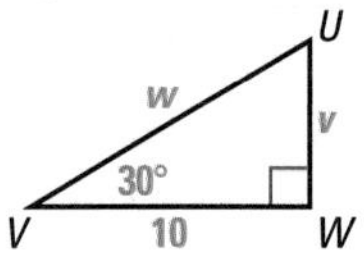

v = 5.77, w = 11.55

49. Find a counterexample to show that the statement is not true. *If a, b, and c are real numbers, then a • (b • c) = (a • b) • (a • c).* (Lesson 12.8) any nonzero numbers *b* and *c* and any nonzero number *a* except 1

Table of Symbols

Symbol		Page
$\cdot$	multiplication, times ($\times$)	3
a^n	*n*th power of *a*	9
$\ldots$	and so on	9
( )	parentheses	10
[ ]	brackets	10
$=$	equal sign, is equal to	24
$\stackrel{?}{=}$	is this statement true?	24
$\neq$	is not equal to	24
$<$	is less than	26
$\leq$	is less than or equal to	26
$>$	is greater than	26
$\geq$	is greater than or equal to	26
$^\circ$	degree(s)	64
$-a$	the opposite of *a*	65
$\lvert a \rvert$	absolute value of *a*	65
$\begin{bmatrix} 1 & 0 \\ 0 & 1 \end{bmatrix}$	matrix	86
$\frac{1}{a}$	reciprocal of *a*, $a \neq 0$	108
P	probability of an event	114
$\frac{a}{b}$	ratio of *a* to *b*	140

Symbol		Page
$\overline{AB}$	segment *AB*	140
AB	the length of $\overline{AB}$	140
$\approx$	is approximately equal to	166
$\frac{a}{b}$	rate of *a* per *b*	180
(x, y)	ordered pair	203
m	slope	226
k	constant of variation	234
b	*y*-intercept	241
π	pi, an irrational number approximately equal to 3.14	452
a^{-n}	$\frac{1}{a^n}$, $a \neq 0$	456
$c \times 10^n$	scientific notation, where $1 \leq c < 10$ and n is an integer	470
$\sqrt{a}$	the nonnegative square root of *a*	503
$\pm$	plus or minus	503
$\triangle ABC$	triangle *ABC*	745
$\angle A$	angle *A*	752
$\sin A$	sine of $\angle A$	752
$\cos A$	cosine of $\angle A$	752
$\tan A$	tangent of $\angle A$	752

809

Table of Measures

Time

60 seconds (sec) = 1 minute (min)
60 minutes = 1 hour (h)
24 hours = 1 day
7 days = 1 week
4 weeks (approx.) = 1 month

365 days, 52 weeks (approx.), 12 months = 1 year
10 years = 1 decade
100 years = 1 century

Metric

Length

10 millimeters (mm) = 1 centimeter (cm)
100 cm, 1000 mm = 1 meter (m)
1000 m = 1 kilometer (km)

Area

100 square millimeters (mm^2) = 1 square centimeter (cm^2)
10,000 cm^2 = 1 square meter (m^2)
10,000 m^2 = 1 hectare (ha)

Volume

1000 cubic millimeters (mm^3) = 1 cubic centimeter (cm^3)
1,000,000 cm^3 = 1 cubic meter (m^3)

Liquid Capacity

1000 milliliters (mL) = 1 liter (L)
1000 L = 1 kiloliter (kL)

Mass

1000 milligrams (mg) = 1 gram (g)
1000 g = 1 kilogram (kg)
1000 kg = 1 metric ton (t)

Temperature **Degrees Celsius (°C)**

0°C = freezing point of water
37°C = normal body temperature
100°C = boiling point of water

United States Customary

Length

12 inches (in.) = 1 foot (ft)
36 in., 3 ft = 1 yard (yd)
5280 ft, 1760 yd = 1 mile (mi)

Area

144 square inches ($in.^2$) = 1 square foot (ft^2)
9 ft^2 = 1 square yard (yd^2)
43,560 ft^2, 4840 yd^2 = 1 acre (A)

Volume

1728 cubic inches ($in.^3$) = 1 cubic foot (ft^3)
27 ft^3 = 1 cubic yard (yd^3)

Liquid Capacity

8 fluid ounces (fl oz) = 1 cup (c)
2 c = 1 pint (pt)
2 pt = 1 quart (qt)
4 qt = 1 gallon (gal)

Weight

16 ounces (oz) = 1 pound (lb)
2000 lb = 1 ton (t)

Temperature **Degrees Fahrenheit (°F)**

32°F = freezing point of water
98.6°F = normal body temperature
212°F = boiling point of water

Table of Squares and Square Roots

No.	Square	Sq. Root
1	1	1.000
2	4	1.414
3	9	1.732
4	16	2.000
5	25	2.236
6	36	2.449
7	49	2.646
8	64	2.828
9	81	3.000
10	100	3.162
11	121	3.317
12	144	3.464
13	169	3.606
14	196	3.742
15	225	3.873
16	256	4.000
17	289	4.123
18	324	4.243
19	361	4.359
20	400	4.472
21	441	4.583
22	484	4.690
23	529	4.796
24	576	4.899
25	625	5.000
26	676	5.099
27	729	5.196
28	784	5.292
29	841	5.385
30	900	5.477
31	961	5.568
32	1,024	5.657
33	1,089	5.745
34	1,156	5.831
35	1,225	5.916
36	1,296	6.000
37	1,369	6.083
38	1,444	6.164
39	1,521	6.245
40	1,600	6.325
41	1,681	6.403
42	1,764	6.481
43	1,849	6.557
44	1,936	6.633
45	2,025	6.708
46	2,116	6.782
47	2,209	6.856
48	2,304	6.928
49	2,401	7.000
50	2,500	7.071

No.	Square	Sq. Root
51	2,601	7.141
52	2,704	7.211
53	2,809	7.280
54	2,916	7.348
55	3,025	7.416
56	3,136	7.483
57	3,249	7.550
58	3,364	7.616
59	3,481	7.681
60	3,600	7.746
61	3,721	7.810
62	3,844	7.874
63	3,969	7.937
64	4,096	8.000
65	4,225	8.062
66	4,356	8.124
67	4,489	8.185
68	4,624	8.246
69	4,761	8.307
70	4,900	8.367
71	5,041	8.426
72	5,184	8.485
73	5,329	8.544
74	5,476	8.602
75	5,625	8.660
76	5,776	8.718
77	5,929	8.775
78	6,084	8.832
79	6,241	8.888
80	6,400	8.944
81	6,561	9.000
82	6,724	9.055
83	6,889	9.110
84	7,056	9.165
85	7,225	9.220
86	7,396	9.274
87	7,569	9.327
88	7,744	9.381
89	7,921	9.434
90	8,100	9.487
91	8,281	9.539
92	8,464	9.592
93	8,649	9.644
94	8,836	9.695
95	9,025	9.747
96	9,216	9.798
97	9,409	9.849
98	9,604	9.899
99	9,801	9.950
100	10,000	10.000

No.	Square	Sq. Root
101	10,201	10.050
102	10,404	10.100
103	10,609	10.149
104	10,816	10.198
105	11,025	10.247
106	11,236	10.296
107	11,449	10.344
108	11,664	10.392
109	11,881	10.440
110	12,100	10.488
111	12,321	10.536
112	12,544	10.583
113	12,769	10.630
114	12,996	10.677
115	13,225	10.724
116	13,456	10.770
117	13,689	10.817
118	13,924	10.863
119	14,161	10.909
120	14,400	10.954
121	14,641	11.000
122	14,884	11.045
123	15,129	11.091
124	15,376	11.136
125	15,625	11.180
126	15,876	11.225
127	16,129	11.269
128	16,384	11.314
129	16,641	11.358
130	16,900	11.402
131	17,161	11.446
132	17,424	11.489
133	17,689	11.533
134	17,956	11.576
135	18,225	11.619
136	18,496	11.662
137	18,769	11.705
138	19,044	11.747
139	19,321	11.790
140	19,600	11.832
141	19,881	11.874
142	20,164	11.916
143	20,449	11.958
144	20,736	12.000
145	21,025	12.042
146	21,316	12.083
147	21,609	12.124
148	21,904	12.166
149	22,201	12.207
150	22,500	12.247

811

Table of Trigonometric Ratios

Angle	Sine	Cosine	Tangent
1°	.0175	.9998	.0175
2°	.0349	.9994	.0349
3°	.0523	.9986	.0524
4°	.0698	.9976	.0699
5°	.0872	.9962	.0875
6°	.1045	.9945	.1051
7°	.1219	.9925	.1228
8°	.1392	.9903	.1405
9°	.1564	.9877	.1584
10°	.1736	.9848	.1763
11°	.1908	.9816	.1944
12°	.2079	.9781	.2126
13°	.2250	.9744	.2309
14°	.2419	.9703	.2493
15°	.2588	.9659	.2679
16°	.2756	.9613	.2867
17°	.2924	.9563	.3057
18°	.3090	.9511	.3249
19°	.3256	.9455	.3443
20°	.3420	.9397	.3640
21°	.3584	.9336	.3839
22°	.3746	.9272	.4040
23°	.3907	.9205	.4245
24°	.4067	.9135	.4452
25°	.4226	.9063	.4663
26°	.4384	.8988	.4877
27°	.4540	.8910	.5095
28°	.4695	.8829	.5317
29°	.4848	.8746	.5543
30°	.5000	.8660	.5774
31°	.5150	.8572	.6009
32°	.5299	.8480	.6249
33°	.5446	.8387	.6494
34°	.5592	.8290	.6745
35°	.5736	.8192	.7002
36°	.5878	.8090	.7265
37°	.6018	.7986	.7536
38°	.6157	.7880	.7813
39°	.6293	.7771	.8098
40°	.6428	.7660	.8391
41°	.6561	.7547	.8693
42°	.6691	.7431	.9004
43°	.6820	.7314	.9325
44°	.6947	.7193	.9657
45°	.7071	.7071	1.0000

Angle	Sine	Cosine	Tangent
46°	.7193	.6947	1.0355
47°	.7314	.6820	1.0724
48°	.7431	.6691	1.1106
49°	.7547	.6561	1.1504
50°	.7660	.6428	1.1918
51°	.7771	.6293	1.2349
52°	.7880	.6157	1.2799
53°	.7986	.6018	1.3270
54°	.8090	.5878	1.3764
55°	.8192	.5736	1.4281
56°	.8290	.5592	1.4826
57°	.8387	.5446	1.5399
58°	.8480	.5299	1.6003
59°	.8572	.5150	1.6643
60°	.8660	.5000	1.7321
61°	.8746	.4848	1.8040
62°	.8829	.4695	1.8807
63°	.8910	.4540	1.9626
64°	.8988	.4384	2.0503
65°	.9063	.4226	2.1445
66°	.9135	.4067	2.2460
67°	.9205	.3907	2.3559
68°	.9272	.3746	2.4751
69°	.9336	.3584	2.6051
70°	.9397	.3420	2.7475
71°	.9455	.3256	2.9042
72°	.9511	.3090	3.0777
73°	.9563	.2924	3.2709
74°	.9613	.2756	3.4874
75°	.9659	.2588	3.7321
76°	.9703	.2419	4.0108
77°	.9744	.2250	4.3315
78°	.9781	.2079	4.7046
79°	.9816	.1908	5.1446
80°	.9848	.1736	5.6713
81°	.9877	.1564	6.3138
82°	.9903	.1392	7.1154
83°	.9925	.1219	8.1443
84°	.9945	.1045	9.5144
85°	.9962	.0872	11.4301
86°	.9976	.0698	14.3007
87°	.9986	.0523	19.0811
88°	.9994	.0349	28.6363
89°	.9998	.0175	57.2900

Table of Formulas

Geometric Formulas	
Perimeter of a polygon	$P = a + b + \ldots + z$ where $a, b, \ldots, z$ = side lengths
Area of a triangle	$A = \frac{1}{2}bh$ where b = base and h = height
Area of a square	$A = s^2$ where s = side length
Area of a rectangle	$A = lw$ where l = length and w = width
Area of a trapezoid	$A = \frac{1}{2}h(b_1 + b_2)$ where h = height and b_1, b_2 = bases
Volume of a cube	$V = s^3$ where s = edge length
Volume of a rectangular prism	$V = lwh$ where l = length, w = width, and h = height
Circumference of a circle	$C = \pi d$ where $\pi \approx 3.14$ and d = diameter $C = 2\pi r$ where $\pi \approx 3.14$ and r = radius
Area of a circle	$A = \pi r^2$ where $\pi \approx 3.14$ and r = radius
Surface area of a sphere	$S = 4\pi r^2$ where $\pi \approx 3.14$ and r = radius
Volume of a sphere	$V = \frac{4}{3}\pi r^3$ where $\pi \approx 3.14$ and r = radius
Other Formulas	
Average speed	$r = \frac{d}{t}$ or $d = rt$ where r = average rate or speed, d = distance, and t = time
Probability of an event	$P = \frac{\text{Number of favorable outcomes}}{\text{Total number of outcomes}}$ where $0 \le P \le 1$
Algebraic Formulas	
Slope formula	$m = \frac{y_2 - y_1}{x_2 - x_1}$ where m = slope and (x_1, y_1) and (x_2, y_2) are two points
Quadratic formula	The solutions of $ax^2 + bx + c = 0$ are $x = \frac{-b \pm \sqrt{b^2 - 4ac}}{2a}$ when $a \neq 0$ and $b^2 - 4ac \ge 0$.
Pythagorean theorem	$a^2 + b^2 = c^2$ where a, b = length of the legs and c = length of the hypotenuse of a right triangle
Distance formula	$d = \sqrt{(x_2 - x_1)^2 + (y_2 - y_1)^2}$ where d = distance and (x_1, y_1) and (x_2, y_2) are coordinates of two points
Midpoint formula	The midpoint between (x_1, y_1) and (x_2, y_2) is $\left(\frac{x_1 + x_2}{2}, \frac{y_1 + y_2}{2}\right)$.

Table of Properties

Basic Properties

	Addition	Multiplication
Closure	$a + b$ is a real number.	ab is a real number.
Commutative	$a + b = b + a$	$ab = ba$
Associative	$(a + b) + c = a + (b + c)$	$(ab)c = a(bc)$
Identity	$a + 0 = a, 0 + a = a$	$a(1) = a, 1(a) = a$
Inverse Property	$a + (-a) = 0$	$a \cdot \frac{1}{a} = 1, a \neq 0$
Prop. of zero in mult.		$a(0) = 0$
Property of opposites		$(-1)a = -a$ or $a(-1) = -a$
Distributive	$a(b + c) = ab + ac$ or $(b + c)a = ba + ca$	

Properties of Equality

Addition	If $a = b$, then $a + c = b + c$.
Subtraction	If $a = b$, then $a - c = b - c$.
Multiplication	If $a = b$, then $ca = cb$.
Division	If $a = b$, and $c \neq 0$, then $\frac{a}{c} = \frac{b}{c}$.

Properties of Exponents

Product of Powers	$a^m \cdot a^n = a^{m+n}$
Power of a Power	$(a^m)^n = a^{m \cdot n}$
Power of a Product	$(a \cdot b)^m = a^m \cdot b^m$
Quotient of Powers	$\frac{a^m}{a^n} = a^{m-n}, a \neq 0$
Power of a Quotient	$\left(\frac{a}{b}\right)^m = \frac{a^m}{b^m}, b \neq 0$
Negative Exponent	$a^{-n} = \frac{1}{a^n}, a \neq 0$
Zero Exponent	$a^0 = 1, a \neq 0$

Properties of Radicals

Product Property	$\sqrt{ab} = \sqrt{a} \cdot \sqrt{b}$
Quotient Property	$\sqrt{\frac{a}{b}} = \frac{\sqrt{a}}{\sqrt{b}}$

Properties of Proportions

Reciprocal	If $\frac{a}{b} = \frac{c}{d}$, then $\frac{b}{a} = \frac{d}{c}$.
Cross-multiplying	If $\frac{a}{b} = \frac{c}{d}$, then $ad = bc$.

Special Products and Their Factors

Sum and Difference Pattern	$(a + b)(a - b) = a^2 - b^2$
Square of a Binomial Pattern	$(a + b)^2 = a^2 + 2ab + b^2$ $(a - b)^2 = a^2 - 2ab + b^2$
FOIL	$(a + b)(c + d) = a \cdot c + a \cdot d + b \cdot c + b \cdot d$ First Outer Inner Last

Properties of Rational Expressions

Multiplication	$\frac{a}{b} \cdot \frac{c}{d} = \frac{ac}{bd}$
Division	$\frac{a}{b} \div \frac{c}{d} = \frac{a}{b} \cdot \frac{d}{c}$
Addition	$\frac{a}{c} + \frac{b}{c} = \frac{a + b}{c}$
Subtraction	$\frac{a}{c} - \frac{b}{c} = \frac{a - b}{c}$

Mixed Problem Solving

CHAPTER 1

1. **SOFTBALL** A softball player's batting average is $\frac{h}{a}$, where h is the number of hits and a is the number of at bats. So far this season, Jessica has 8 hits in 17 at bats. Find Jessica's batting average. Round your answer to the nearest thousandth. **(Lesson 1.1)** **0.471**

2. **CANDLES** A candle company makes candles shaped like pyramids. One of the molds for these candles has a height of 8 centimeters and a square base that measures 9 centimeters on each side. Find the amount of wax needed to fill the mold by finding the volume of the pyramid. (The formula for the volume of a pyramid is $\frac{1}{3}$ times the height times the area of the base.) **(Lesson 1.2)** **216 cm^3**

AMUSEMENT PARKS In Exercises 3–7, use the following information. (Lesson 1.3)
A family of six has coupons for discounted admission to an amusement park. The members of the family are the following ages: 67, 40, 39, 14, 8, and 4.

Amusement Park Admission Prices

Age	Regular price	Discounted price
13–59	\$10.00	\$7.50
5–12 and 60+	\$8.00	\$6.00
under 5	free	free

3. Write an expression that represents the total cost of admission for the family at the regular price. **3(10) + 2(8)**

4. Evaluate the expression from Exercise 3. **46**

5. Write an expression that represents the total cost of admission for the family at the discounted price. **3(7.50) + 2(6)**

6. Evaluate the expression from Exercise 5. **34.50**

7. How much money did the family save by using the coupons? **\$11.50**

8. **TRAVEL** While planning a vacation, your family budgets \$1200 to spend on lodging. The hotel costs \$129 per night. Will you have enough money in the budget to stay for 7 nights? You write the inequality $129h \leq 1200$ to model the situation. What do 129, h, and 1200 represent? **(Lesson 1.4)** **See margin.**

BULK FOODS In Exercises 9–11, use the following information. (Lesson 1.5)
The trail mix in the bulk foods section of a grocery store costs \$1.90 per pound. Martin paid \$5.70 for the bag that he filled with trail mix. Martin wants to know how many pounds of trail mix he bought.

9. Write a verbal model that relates the cost per pound of the trail mix, the pounds of trail mix, and the total cost of the trail mix. **See margin.**

10. Assign labels and write an algebraic model based on your verbal model in Exercise 9. **Let x = the number of pounds; $1.90x = 5.70$.**

11. Use mental math to solve the equation in Exercise 10. How many pounds of trail mix did Martin buy? **3 pounds**

12. **CONCERT BAND** The table below shows the number of students in the concert band at a school for the given years. Make a line graph of the data. Discuss what the line graph shows. **(Lesson 1.6)** **See margin.**

Year	1997	1999	2001	2003
Students	62	51	54	57

TELEPHONE CALLS In Exercises 13 and 14, use the following information. (Lesson 1.7)
The total cost C (in dollars) of a telephone call from the United States to Moscow, Russia is given by $C = 0.4t$, where t represents the length of the call in minutes.

13. Evaluate the equation for t = 5, 10, 15, 20, 25. Organize your results in an input-output table. **See margin.**

14. Describe the domain and range of the function whose values are shown in the table. **domain: 5, 10, 15, 20, 25; range: 2, 4, 6, 8, 10.**

8. **Yes; 129 represents the cost per night, h represents the number of nights, and 1200 represents the amount of money budgeted for lodging.**

9. **The cost per pound times the number of pounds equals the total cost.**

12.

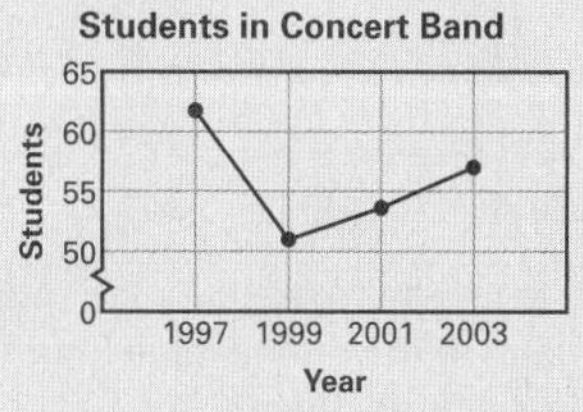

Sample answer: **the line graph shows that the number of students sharply decreased from 1997 to 1999, then steadily increased from 1999 to 2003.**

13.

Input	5	10	15	20	25
Output	2	4	6	8	10

8.

	S	L
Popcorn	24	13
Nachos	18	11
Soda	31	28

9.

	S	L
Popcorn	62	55
Nachos	29	20
Soda	53	66

10.

	S	L
Popcorn	86	68
Nachos	47	31
Soda	84	94

The matrix represents the total sales at the concession stand during both events.

CHAPTER 2

TEMPERATURES In Exercises 1–3, use the table below, which shows the daily low temperatures in Petoskey, Michigan during a week in January. (Lesson 2.1)

Day	Temperature
Monday	7°F
Tuesday	−3°F
Wednesday	2°F
Thursday	−7°F
Friday	−5°F
Saturday	−2°F
Sunday	9°F

1. Which low temperature was the coldest? **Thursday**
2. Which days had low temperatures above −2°F? **Monday, Wednesday, Sunday**
3. What is the absolute value of the low temperature on Tuesday? **3**
4. MONEY On a Saturday morning, your parents give you $15 for your allowance. You spend $18.50 at the mall. Then you make $20 baby-sitting that night. Do you end the day with more money than you started with? If so, how much more? (Lesson 2.2) **yes; $16.50**
5. STOCK You purchase some shares of stock. The price of each share fell by $3.20 in the first month, rose by $5.45 in the second month, and fell by $1.12 in the third month. Did the price of a share increase or decrease over the three months? (Lesson 2.2) **increase**

WEATHER In Exercises 6 and 7, use the following information. (Lesson 2.3)
On Sunday, there were 14 inches of snow on the ground. About 1.5 inches of the snow melted on Monday, 2 inches melted on Tuesday, and 1.5 inches melted on Wednesday.

6. How much snow remained at the end of the day on Wednesday? **9 in.**
7. At the end of the day on Thursday, there were 8.5 inches of snow on the ground. How much snow melted on Thursday? **0.5 in.**

CONCESSION STAND In Exercises 8–10, use the table below, which shows the sales at a concession stand during two different events. (Lesson 2.4)

	Soccer Game		Football Game	
	Small	Large	Small	Large
Popcorn	24	13	62	55
Nachos	18	11	29	20
Soda	31	28	53	66

8. Write a matrix for the concession stand sales at the soccer game. **See margin.**
9. Write a matrix for the concession stand sales at the football game. **See margin.**
10. Find the sum of the matrices from Exercises 8 and 9. Explain what the resulting matrix represents. **See margin.**
11. HOT AIR BALLOON A hot air balloon descends with a velocity of −678 feet per minute. Find the vertical distance traveled in 3.5 minutes. (Lesson 2.5) **−2373 ft**
12. GEOMETRY Write and simplify an expression for the perimeter of the figure. (Lesson 2.6) **$8x - 10$**

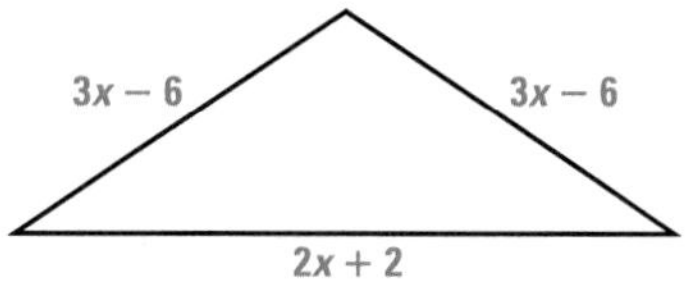

13. FOOD There are $19\frac{1}{2}$ ounces of pretzels in a bag. If a serving size is $\frac{3}{4}$ ounce, then how many servings are in the bag of pretzels? (Lesson 2.7) **26**
14. MUSIC Your compact disc collection contains 27 rock discs, 15 country discs, 28 pop discs and 10 hip-hop discs. You randomly choose a compact disc from your collection. What is the probability that the disc is a pop disc? (Lesson 2.8) **0.35**

CHAPTER 3

1. **CATS** A veterinary assistant steps on a scale while holding a cat. The scale reads 161 pounds. Alone the assistant weighs 147 pounds. Write and solve an equation to find the weight of the cat. **(Lesson 3.1)**
Sample equation: $x + 147 = 161$; 14 lb

2. **SNOWFALL** The average snowfall in Alpena during the month of February is 3 times the average snowfall in Madison during the month of February. The average snowfall in Alpena during February is 15 inches. Write and solve an equation to find the average snowfall in Madison during the month of February. **(Lesson 3.2)**
Sample equation: $3x = 15$; 5 in.

3. **GEOMETRY** The triangles below are similar. Write and solve an equation to find the length of the side marked x.
(Lesson 3.2) _Sample equation:_ $\frac{15}{18} = \frac{x}{21}$; 17.5 cm

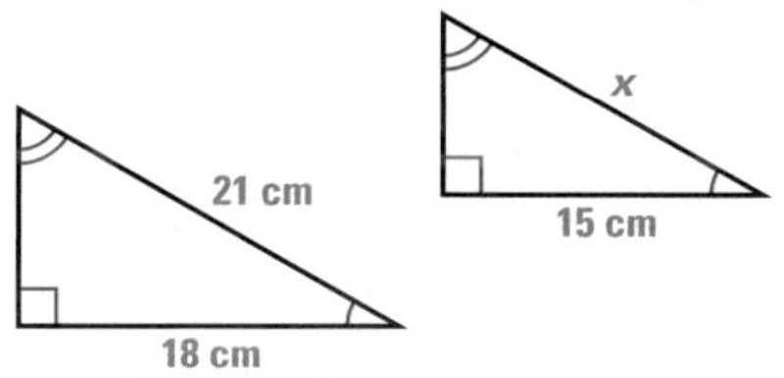

4. **GARDENING** An online garden supply company offers daffodil bulbs for $.59 each plus $2.75 for shipping and handling for the entire order. You want to spend $15 on an order of daffodil bulbs. Write and solve an equation to find the number of daffodil bulbs you can purchase. **(Lesson 3.3)**
Sample equation: $0.59x + 2.75 = 15$; 20 bulbs

5. **BOOKS** You spend $13.25 to buy 6 books and a bookmark at a used book sale. You pay the same amount for each book you buy, and the bookmark costs $1.25. Write and solve an equation to find the cost of one book. **(Lesson 3.3)**
Sample equation: $6x + 1.25 = 13.25$; $2

6. **EXERCISE** Each yoga class at a gym costs $10 for nonmembers. If you are a member of the gym, you pay a one-time joining fee of $45, and each yoga class costs $5. Write and solve an equation to find the number of yoga classes you must take to justify becoming a member. **(Lesson 3.4)**
Sample equation: $5x + 45 = 10x$; 9 classes

TRANSPORTATION In Exercises 7 and 8, use the following information. (Lesson 3.5)
Two planes leave an airport at 1:00 P.M. One plane heads south and the second plane heads north. The plane heading north is traveling 150 miles per hour. At 3:00 P.M. the two planes are 650 miles apart.

7. At 3:00 P.M., how far has the plane heading north traveled? **300 mi**

8. At 3:00 P.M., how far has the plane heading south traveled? **350 mi**

9. **TELEPHONES** Your long-distance telephone company charges a monthly fee of $4.75 plus $.10 for each minute of long-distance telephone calls. Your bill for last month was $15.25. Write and solve an equation to find the number of minutes of long-distance telephone calls for last month.
(Lesson 3.6) _Sample equation:_ $0.10x + 4.75 = 15.25$; 105 minutes

ELECTRICAL POWER In Exercises 10 and 11, use the following information. (Lesson 3.7)
The electrical power rating P, in watts, of an electrical appliance is given by the formula $P = VI$, where V is the voltage and I is the current in amps.

10. Solve the formula for I. $I = \frac{P}{V}$

11. Use the new formula to find the current running through the appliance given its power rating and voltage. **Stereo, 0.27, Coffeemaker, 8.18**

Appliance	Stereo	Coffeemaker
Power rating	30 watts	900 watts
Voltage	110 volts	110 volts

12. **FOOD** You spend $2.34 on 6 pounds of bananas. What is the cost per pound?
(Lesson 3.8) $.39 per pound

13. **JEWELRY** A craft store sells assorted collections of beads. One assortment contains 100 beads and costs $8.00. A larger assortment contains 250 beads and costs $12.50. Which assortment is the better buy? Explain how you decided. **(Lesson 3.8)**
The larger assortment is the better buy; the price per bead in the smaller assortment is $.08 and the price per bead in the larger assortment is $.05, and $.08 > $.05.

1. Shots vs. Goals

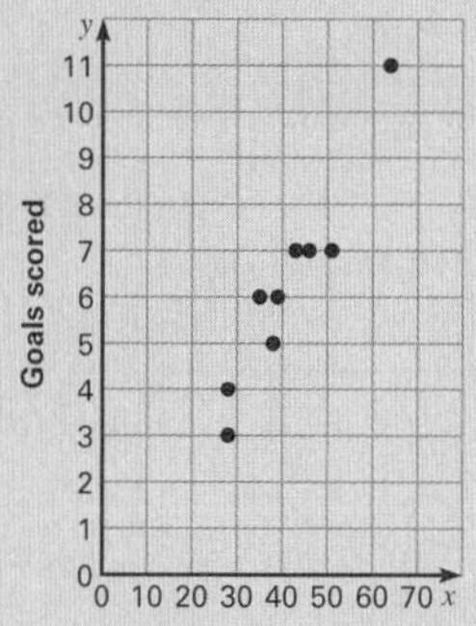

2. *Sample answer:* **As the number of shots on goal increases, the number of goals scored tends to increase.**

3.

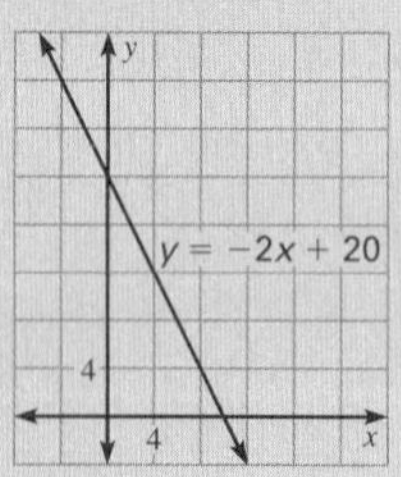

5. Museum Attendance

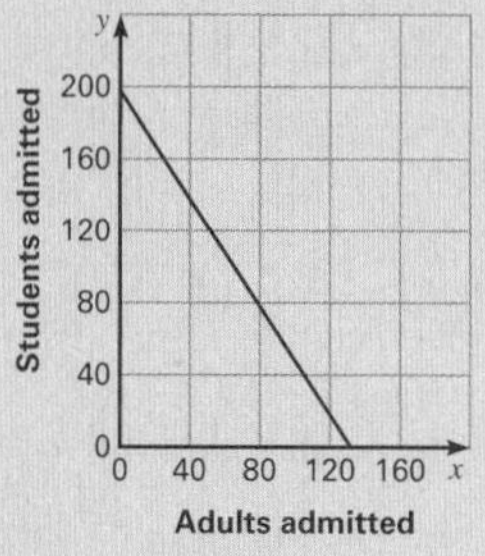

9. **2.5; 15; The slope is the cost per letter and the *y*-intercept is the fee.**

CHAPTER 4

ICE HOCKEY In Exercises 1 and 2, use the table below, which shows the number of shots on goal and the number of goals scored by players on an ice hockey team during one season. (Lesson 4.1)

Shots	Goals
64	11
39	6
38	5
43	7
46	7
28	4
51	7
35	6
28	3

1. Make a scatter plot that shows the number of shots on goal and the number of goals scored for the players. Use the horizontal axis to represent the number of shots on goal. **See margin.**

2. What conclusions (if any) can you make from the scatter plot? **See margin.**

SUMMER JOBS In Exercises 3 and 4, use the following information. (Lesson 4.2)
You charge \$50 for cleaning a large office and \$25 for cleaning a small office. You want to make \$500 in one week. An algebraic model for your earnings is $50x + 25y = 500$, where x is the number of large offices you clean per week, and y is the number of small offices you clean.

3. Write the algebraic model in function form. Then use a table of values to graph the equation. **$y = -2x + 20$; See margin for art.**

4. If you do not clean any large offices during the week, how many small offices do you have to clean to earn \$500? **20 small offices**

5. **MUSEUMS** The cost of admission to a science museum is \$18 for adults and \$12 for students. The museum collected \$2376 from admissions for one day. Make a graph that shows the possible numbers of adults and students admitted to the museum. (Lesson 4.3) **See margin.**

6. **BOWLING** In 1995, the cost of bowling one string at a bowling alley was \$1.75. In 2000, the cost of one string was \$2.25. If the cost of bowling one string increased the same amount each year, find the rate of change of the cost in dollars per year. (Lesson 4.4) **\$.10 per year**

HEALTH CARE In Exercises 7 and 8, use the information below. (Lesson 4.5)
The dosage of a medication for a child is based on the child's weight. Examples of such recommendations are shown in the table below. The recommended dosage (in milligrams) varies directly with the weight (in pounds) of the child.

Weight (lb)	Dosage (mg)
20	125
28	175
32	200
40	250

7. Write a direct variation model that relates the weight w to the dosage d. **$d = 6.25w$**

8. Estimate the dosage for a child who weighs 45 pounds. **about 280 mg**

9. **CLOTHING** A tailor charges a \$15 fee plus \$2.50 per letter to sew a name onto a jacket. The equation $y = 2.5x + 15$ models the total cost y of sewing a name with x letters. Find the slope and y-intercept of the equation. Explain what the slope and y-intercept mean in the cost model. (Lesson 4.6) **See margin.**

10. **CLUB MEMBERSHIP** The total cost y of a family membership to a health club for x months can be modeled by the equation $y = 30x + 75$. How many months of family membership do you have if your family pays \$225? (Lesson 4.7) **5 months**

11. **BICYCLES** You ride your bicycle from your house to school. Your average speed is 15 miles per hour. Write a linear function that models the distance you travel $d(t)$ as a function of t, the time spent traveling. Then use your model to estimate the distance traveled after you ride for 0.2 hours. (Lesson 4.8) **$d(t) = 15t$; 3 miles**

CHAPTER 5

DOGS In Exercises 1–3, use the following information. (Lesson 5.1)
A group of dogsledders set off on an expedition with 91 pounds of dog food for their dogs. The dogs eat 7 pounds of dog food each day.

1. Write an equation in slope-intercept form to model the total remaining dog food y (in pounds) in terms of x, the days on the expedition. $y = -7x + 91$

2. Copy and complete the table using your equation from Exercise 1.

Days (x)	3	5	7	9
Remaining dog food (y)	?	?	?	?

70 56 42 28

3. Find the number of days that the supply of dog food will last. **13 days**

VINTAGE CARS In Exercises 4 and 5, use the following information. (Lesson 5.2)
You bought a vintage car for \$4500 last year. You are told that the car is worth \$600 more this year, and you should expect the value of the car to increase by \$600 every year.

4. Write an equation that models the total value v of the car in terms of n, the number of years you have owned the car. $v = 600n + 4500$

5. Calculate the value of the car after you have owned it for 7 years. **\$8700**

MOVING VANS In Exercises 6–8, use the following information. (Lesson 5.3)
The cost of renting a moving van for a 26 mile trip is \$62.50. The cost of renting the same moving van for a 38 mile trip is \$65.50.

6. Write a linear equation that models the cost c of renting the van in terms of m, the length (in miles) of the trip. $c = 0.25m + 56$

7. Use your equation from Exercise 6 to find the cost of renting the van for a 50 mile trip. **\$68.50**

8. You have only \$75 to spend on the moving van. What is the longest trip you can take? **76 miles**

BASKETBALL In Exercises 9 and 10, use the following information. (Lesson 5.4)
You stood in 6 locations on a basketball court and made 20 attempts at a basket from each location. The table shows the distance each location was from the basket and the number of shots made.

Distance from basket (feet)	Shots made
5	18
10	15
15	10
20	10
25	7
30	4

9. Make a scatter plot of the data. Let x equal the distance from the basket and let y equal the shots made. **See margin.**

10. Find an equation of the line that you think best fits the data. ***Sample answer:*** $y = -0.6x + 22$

11. CANOEING You are on an 18 mile long canoe trip and you are paddling at a constant rate. You started the trip at 9:00 A.M., and you traveled 6 miles by 10:30 A.M. Write an equation that gives the distance d (in miles) remaining in terms of the time t (in hours). What is the slope of the line? (Lesson 5.5) $d = -4t + 18$; slope $= -4$

12. FUNDRAISING Your scout troop holds a breakfast to raise money for a local charity. Your troop charges \$8 for an adult and \$4 for a child, and you raise a total of \$1200. Write an equation that represents the different amounts of adults x and children y that could have attended the breakfast. (Lesson 5.6) $8x + 4y = 1200$

13. WEB SITES The table shows the hits received by a Web site each week after its launch. Write a linear model for the data. Then predict the number of hits the Web site will receive in week 7. (Lesson 5.7) ***Sample answer:*** $y = 97x - 118$; 561

Week	Hits
1	57
2	76
3	107
4	236
5	367
6	499

9.

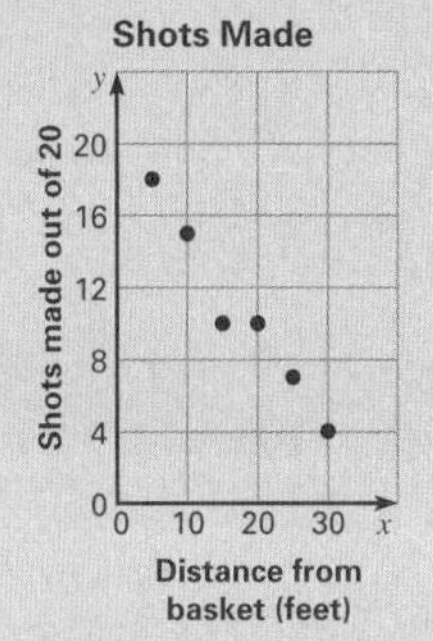

7.

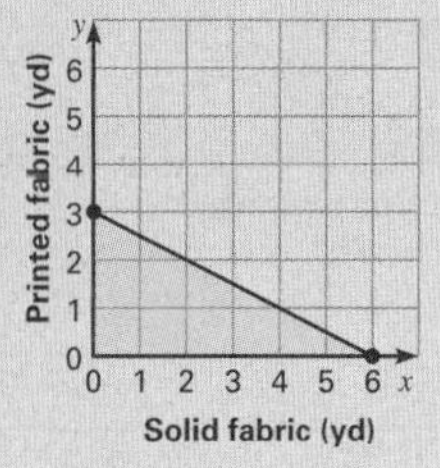

8.

0	5 7
1	0 3 3 5 5 7 8
2	0 0 1 2 2 3 5
3	0 0 0 0

Key: 1|3 = 13

10. ***Sample answer:*** **The mean (19.3) and the median (20) are more representative than the mode (30) because the mode is the greatest travel time.**

11.

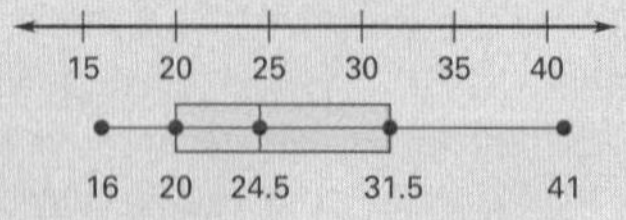

CHAPTER 6

1. **MONEY** You want to purchase a bicycle that costs \$165. You have already saved \$118 for the bicycle. Let x represent the additional amount of money you need to purchase the bicycle. Write an inequality for x. What is the least additional amount of money that you need to purchase the bicycle? **(Lesson 6.1)** **$118 + x \le 165$; \$47**

2. **CATERING COSTS** Dave is planning the awards banquet for his bowling league. He has \$1500 in the budget to spend on the catering for the 108 people that will be attending the banquet. The catering company charges a \$150 for a setup fee plus a set amount per person. Dave needs to decide how much he can afford to spend per person. Write and solve an inequality to determine the maximum amount of money that he can spend per person. **(Lesson 6.2)** **$150 + 108x \le 1500$; \$12.50**

3. **GEOMETRY** Write an inequality for the value of x. **(Lesson 6.2)** **$x > 10$ ft**

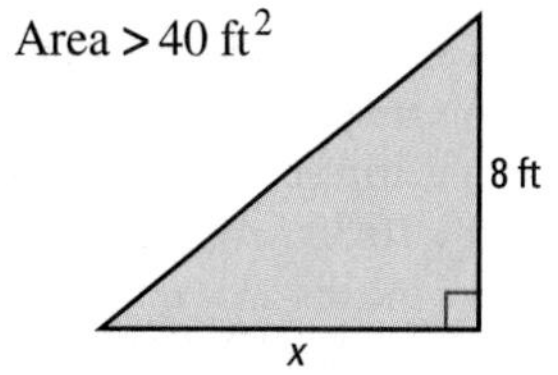

4. **TEMPERATURE** The table shows the normal daily high temperatures for the month of January in four cities in California. Write a compound inequality that represents the different temperatures. **(Lesson 6.3)** **$52.7 \le x \le 65.9$**

City	Temperature (°F)
Los Angeles	65.7
Sacramento	52.7
San Diego	65.9
San Francisco	55.6

5. **CALCIUM** The recommended daily intake of calcium for teenagers is estimated between 1200 milligrams and 1500 milligrams. Write an absolute-value inequality to represent the recommended daily intake of calcium for teenagers. **(Lesson 6.4)** **$|x - 1350| \le 150$**

6. **QUALITY CONTROL** Dara works in the quality control department at a company that packages baking mixes. Each box of vanilla cake mix that she inspects must weigh between 509 grams and 525 grams. Write an absolute-value inequality to represent this situation. **(Lesson 6.4)** **$|x - 517| \le 8$**

7. **SEWING** You have \$15 to spend on two different fabrics for a sewing project. The solid fabric that you want costs \$2.50 per yard and the printed fabric that you want costs \$4.99 per yard. Let x represent the number of yards of the solid fabric you can buy. Let y represent the number of yards of the printed fabric you can buy. Write and graph an inequality to model the amounts of each fabric you can buy. **(Lesson 6.5)** **$2.50x + 4.99y \le 15$; See margin for art.**

TRAVEL TIME In Exercises 8–10, use the following information. (Lesson 6.6)

A survey asks 20 students the average time (in minutes) it takes them to get to school in the morning. The results are listed below.

18, 23, 17, 30, 13, 22, 20, 20, 7, 22,
25, 15, 13, 30, 15, 5, 30, 21, 30, 10

8. Use a stem-and-leaf plot to order the times from least to greatest. **See margin.**

9. Find the mean, median, and mode of the data. **19.3 min; 20 min; 30 min**

10. Is one measure of central tendency most representative of a typical travel time? Explain your reasoning. **See margin.**

11. **AGES** The ages of the employees at a roller rink are listed below. Draw a box-and-whisker plot of the data. **(Lesson 6.7)** **See margin.**
24, 22, 30, 18, 29, 38, 33, 17, 22, 25, 16, 41

12. **TEST SCORES** Create a collection of 16 test scores that could be represented by the box-and-whisker plot shown below. **(Lesson 6.7)** ***Sample answer:*** **64, 70, 70, 71, 71, 73, 80, 82, 86, 86, 87, 87, 88, 90, 92, 99**

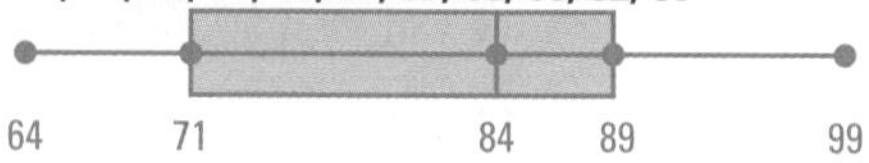

CHAPTER 7

1. **TENNIS** A tennis club charges \$55 for a summer membership plus \$12 for each hour of court use. A second club charges \$20 for a summer membership plus \$17 for each hour of court use. Use the graph below to determine the number of hours of court use for which the total cost of playing tennis at each club is the same. **(Lesson 7.1) 7 hours**

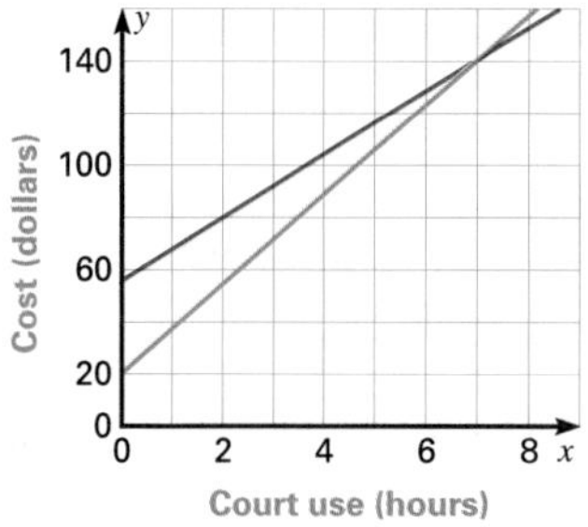

2. **CARPETS** Store A charges \$3 per square foot for a wall-to-wall carpet and charges \$85 for installation. Store B charges \$2 per square foot for the same carpet and charges \$100 for installation. Use a graph to determine the square footage for which the total cost of the carpet and installation at each store is the same. **(Lesson 7.1) 15 ft^2**

3. **MONEY** A bag contains quarters and dimes. There are 15 coins in the bag. The value of the coins is \$2.70. Use substitution to determine the number of quarters and dimes in the bag. **(Lesson 7.2) 8 quarters, 7 dimes**

4. **MOVIES** A movie rental store sells previously viewed movies. The movies in DVD format cost \$10 each and the movies in VHS format cost \$7 each. Jana bought 6 movies for \$51, excluding tax. Use substitution to determine the number of DVDs that Jana bought. **(Lesson 7.2) 3 DVDs**

5. **TELEPHONE RATES** A telephone company offers two long distance plans. One plan charges \$4.95 per month and \$.07 per minute. The other plan does not have a monthly fee and charges \$.10 per minute. Use linear combinations to determine the number of minutes for which the cost of each plan is the same. **(Lesson 7.3) 165 minutes**

6. **AGES** Twice Mark's age minus Tom's age is equal to 9. Twice Tom's age minus 8 is equal to twice Mark's age. Use linear combinations to determine Mark's age and Tom's age. **(Lesson 7.3) Mark is 13 years old, Tom is 17 years old.**

7. **PLANETARIUM SHOWS** The prices for admission tickets to the planetarium shows at a science museum are listed in the table below. On Monday, 540 admission tickets for the planetarium shows are purchased. The museum collected \$3257.50 in admission ticket sales for the planetarium shows. How many student admission tickets were sold? **(Lesson 7.4) 425 student tickets**

Admission Ticket Prices	
Adult	\$8
Student	\$5.50

ART SUPPLIES In Exercises 8 and 9, use the following information. (Lesson 7.5)
At an art supply store, all of the tubes of oil paint cost the same, and all of the tubes of acrylic paint cost the same. An art teacher paid \$43.71 for 3 tubes of oil paint and 6 tubes of acrylic paint. The next week he paid \$14.57 for one tube of oil paint and 2 tubes of acrylic paint.

8. Can you determine how much the art teacher paid for one tube of oil paint? Explain. **See margin.**

9. The following week, the art teacher paid \$39.89 for 3 tubes of oil paint and 5 tubes of acrylic paint. Can you determine how much he paid for one tube of oil paint? Explain. **See margin.**

JOBS In Exercises 10 and 11, use the following information. (Lesson 7.6)
During a summer, you can work a total of no more than 30 hours a week at your two jobs. Your camp counseling job pays \$8 an hour, and your job at a retail store pays \$10 an hour. You want to earn at least \$250 a week.

10. Write a system of linear inequalities that shows the various number of hours you can work at each job.
$x + y \leq 30$, $8x + 10y \geq 250$, $x \geq 0$, $y \geq 0$

11. Graph the system of linear inequalities. **See margin.**

8. No; the system of equations that describe the situation is $3x + 6y = 43.71$ and $x + 2y = 14.57$, which has infinitely many solutions.

9. Yes; you can combine the new equation, $3x + 5y = 39.89$, with either of the original equtions to produce a system that has the solution (6.93, 3.82). A tube of oil paint costs \$6.93.

11.

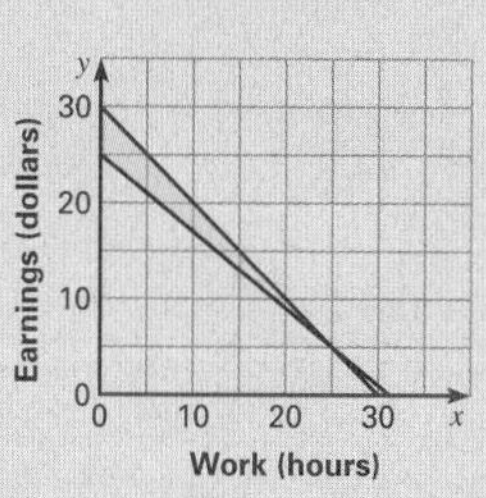

CHAPTER 8

1. **GEOMETRY** The volume V of a cylinder is given by $V = \pi r^2 h$, where r is the radius of a base, h is the height, and $\pi \approx 3.14$. What is the volume of the cylinder in terms of t? **(Lesson 8.1)** $1017.36t^2$ cm^3

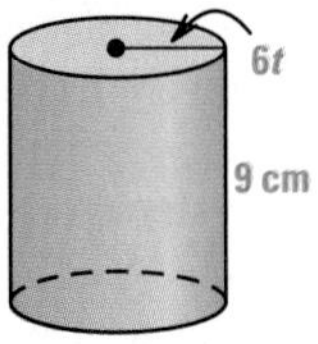

TESTS In Exercises 2 and 3, use the following information. (Lesson 8.1)
Part A of a test has 15 true-false questions. Part B has 5 multiple choice questions, each of which has 4 choices. There are 2^{15} ways to answer the questions in Part A. There are 4^5 ways to answer the questions in Part B.

2. How many ways are there to answer all 20 questions? 2^{25} or 33,554,432 ways

3. If you guess the answer to each question, what is the probability that you will get them all right? $\left(\frac{1}{2}\right)^{25}$ or about 0.00000003

4. **SAVINGS ACCOUNTS** You opened a savings account in 1994. The balance A is modeled by $A = 320(1.03)^t$, where $t = 0$ represents the year 2004. What is the account balance in 1994? in 2004? in 2014? **(Lesson 8.2)** $238.11; $320; $430.05

5. **SPORTS** The volume V of a ball is given by $V = \frac{4}{3}\pi r^3$, where r is the radius of the ball. The radius of a baseball is about $\frac{1}{3}$ the radius of a soccer ball. Find the ratio of the volume of the baseball to the volume of the soccer ball. Let r represent the radius of the soccer ball. **(Lesson 8.3)** $\frac{1}{27}$

PROBABILITY In Exercises 6 and 7, you roll a number cube five times. (Lesson 8.3)

6. What is the probability that you roll a three each time? $\left(\frac{1}{6}\right)^5 \approx 0.0001$

7. What is the probability that you roll five odd numbers in a row? $\left(\frac{1}{2}\right)^5 \approx 0.03$

ASTRONOMY In Exercises 8–11, write the mass (in kilograms) of the planet in scientific notation. (Lesson 8.4)

8. Mercury: 330,000,000,000,000,000,000,000 3.3×10^{23} kg
9. Venus: 4,900,000,000,000,000,000,000,000 4.9×10^{24} kg
10. Mars: 640,000,000,000,000,000,000,000 6.4×10^{23} kg
11. Pluto: 13,000,000,000,000,000,000,000 1.3×10^{22} kg
12. **HEARTBEATS** Consider a person whose heart beats 65 times per minute and who lives to be 90 years old. Estimate the number of times the person's heart beats during his or her life. Do not acknowledge leap years. Write your answer in decimal form and in scientific notation. **(Lesson 8.4)** 3,074,760,000 or 3.07476×10^9 times
13. **INVESTING** Which description best represents the growth curve in the graph? **(Lesson 8.5)** C

A. Deposit = $150, annual rate = 5%

B. Deposit = $350, annual rate = 11%

C. Deposit = $150, annual rate = 18%

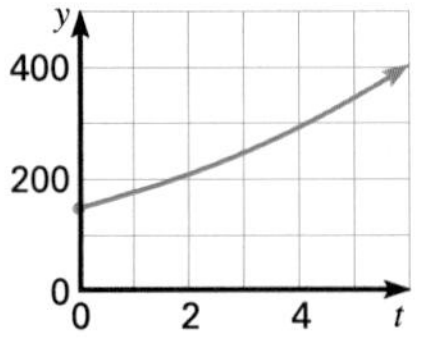

STEREOS In Exercises 14 and 15, use the following information. (Lesson 8.6)
You want to sell your old stereo system. You place an advertisement about your stereo system in the classified section of your local newspaper. The selling price listed in the advertisement is $80.50. You decide to decrease the price by 10% each week that you do not sell the stereo system.

14. Write an exponential decay model to represent the situation. $P = 80.5(0.9)^t$

15. You sell your stereo system for the price listed in the advertisement for the third week. How much money did you get for your stereo system? $58.68

CHAPTER 9

1. **DIVING** You dive from a cliff that is 20 feet above the water. Your height h after t seconds can be modeled by $h = -16t^2 + 20$. How long will it take you to reach the water? **(Lesson 9.1) about 1.1 sec**

2. **GEOMETRY** Find the area of the rectangle below. Give both the exact answer in simplified form and the decimal approximation rounded to the nearest hundredth. **(Lesson 9.2) $10\sqrt{15}$ cm^2, 38.73 cm^2**

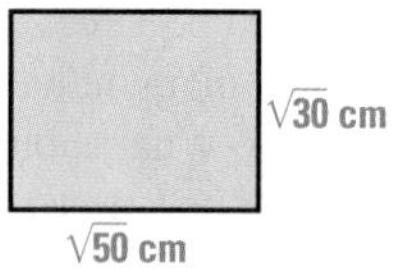

LAWN SPRINKLERS In Exercises 3–5, use the following information. (Lesson 9.3)
The arc of the water sprayed from a lawn sprinkler is given by the model

$$y = -0.04x^2 + 0.93x$$

where x is the distance (in feet) across the lawn and y is the height (in feet) of the arc.

3. Sketch the graph of the function. Label the vertex. **See margin.**

4. What is the maximum height of the water sprayed from the sprinkler? **about 5.4 ft**

5. The distance across the lawn from the sprinkler to where the arc lands is called the *distance of throw.* What is the distance of throw for the sprinkler? **23.25 ft**

SKI JUMPS In Exercises 6 and 7, use the following information. (Lesson 9.4)
The path of a ski jumper after takeoff can be modeled by the equation

$$y = -0.036x^2 + 1.38x + 20$$

where x is the horizontal distance (in feet) and y is the vertical distance (in feet).

6. Sketch a graph of the model for positive values of x and y. **See margin.**

7. Use the graph to estimate a positive root of the equation below. **about 49.5**

$$0 = -0.036x^2 + 1.38x + 20$$

8. **SOFTBALL** You and a partner are practicing catching fly balls. When you throw the softball upward for your partner to catch, it leaves your hand when it is 5.5 feet off the ground. The softball has an initial velocity of 35 feet per second. Suppose that your partner misses the ball and it hits the ground. How long does it take for the softball to hit the ground? **(Lesson 9.5) about 2.3 sec**

9. **CHEERLEADING** A cheerleading squad does a stunt that has a flyer, who is thrown in the air and caught. The stunt works best when the flyer's feet reach a height of at least 30 feet off the ground. The flyer starts with her feet 4.5 feet off the ground and her initial velocity is 31 feet per second. Will the flyer's feet reach a height of at least 30 feet off the ground? Explain. **(Lesson 9.6) See margin.**

10. **BRIDGES** The parabolic opening in the bridge below can be modeled by $y = -0.04x^2 + 25$, where x is the distance (in feet) from the center of the opening and y is the height (in feet) above the road. Write a system of inequalities that describes the shaded region. **(Lesson 9.7)**
$y \geq -0.04x^2 + 25$, $-35 \leq x \leq 35$, $0 \leq y \leq 35$

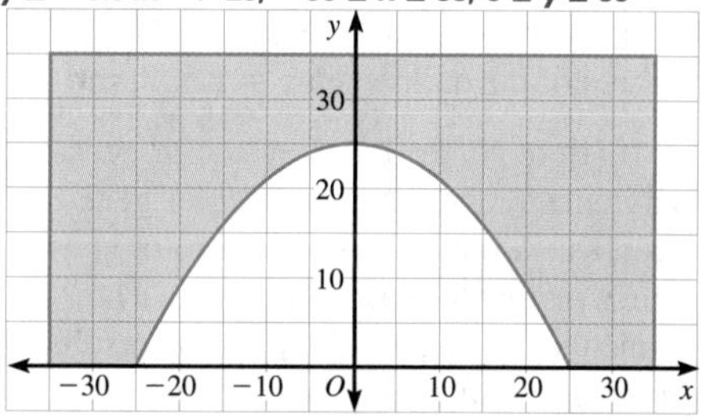

11. **AUTOMOBILES** The table below lists the value of an automobile over time. Which type of model best represents the data? Write a model. **(Lesson 9.8)**

Year	Value (dollars)
2002	12,500.00
2003	11,250.00
2004	10,125.00
2005	9112.50

exponential; $v = 12{,}500(0.9)^n$, where n represents the number of years since 2002

3.

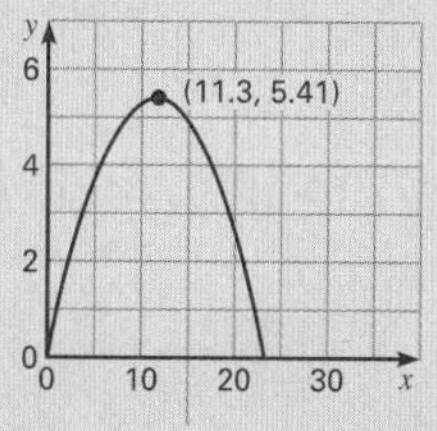

6.

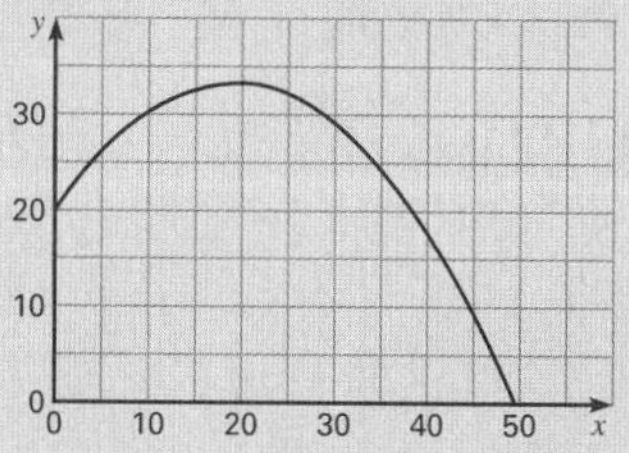

9. **No; since the discriminant of the equation $0 = -16t^2 + 31t - 25.5$ is negative, there is no real solution. That is, there is no time when the flyer's feet will be 30 feet off the ground.**

CHAPTER 10

1. GEOMETRY Write an expression in standard form for the perimeter of the triangle shown below. **(Lesson 10.1)**
$P = 5x^2 + 4x + 4$

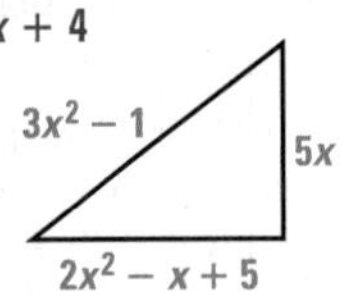

QUILTS In Exercises 2 and 3, use the information below. (Lesson 10.2)
You are making the quilt shown below, which has 6 squares in the center. The border around the squares is 0.5 foot wide, and the border around the outside of the quilt is 1.5 feet wide.

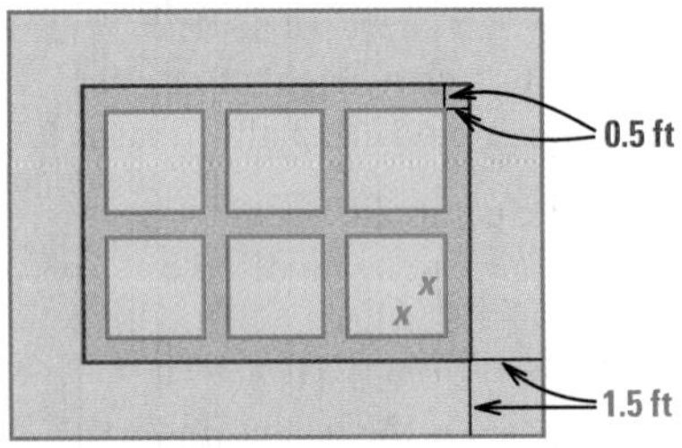

2. Write a polynomial expression that represents the total area of the quilt. Give your answer as a quadratic trinomial.
$A = 6x^2 + 23.5x + 22.5$

3. Use the expression from Exercise 2 to find the area of the quilt when $x = 1.5, 2,$ and 2.5.
$71.25\ ft^2$, $93.5\ ft^2$, $118.75\ ft^2$

4. MICE In mice, the black eye color gene E is dominant, and the red eye color gene e is recessive. A mouse whose eye color genes are EE or Ee will have black eyes. The Punnett square shows the possible results of crossing two mice that have recessive eye color genes. Write a polynomial model that can be used to represent the Punnett square. **(Lesson 10.3)** **$0.25E^2 + 0.5Ee + 0.25e^2$**

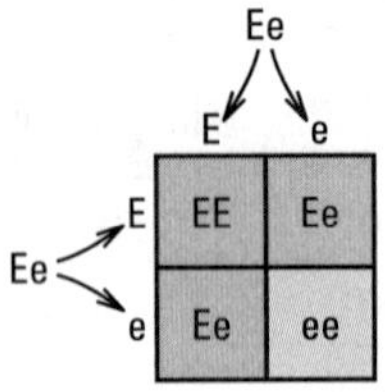

ARCHITECTURE In Exercises 5 and 6, use the following information. (Lesson 10.4)
An architect designs a large arched doorway for the main entrance to an office building. The shape of the arch can be modeled by $y = -0.04(x + 22)(x - 22)$, where x and y are measured in feet.

5. How wide is the arch at the base? **44 ft**

6. How high is the arch? **19.36 ft**

7. GARDENS You decide to plant a bulb garden. You want the length of the rectangular garden to be 3 feet longer than its width. The bulbs that you have will cover 88 square feet. What should the length and the width be? **(Lesson 10.5)**
length: 11 ft, width: 8 ft

SPORTS In Exercises 8 and 9, use the following information. (Lessons 10.6, 10.7)
In the vertical motion model $h = -16t^2 + vt + s$, h is the height (in feet), t is the time in motion (in seconds), v is the initial velocity (in feet per second), and s is the initial height (in feet).

8. A quarterback releases a football when it is 6 feet off the ground, and the football has an initial upward velocity of 46 feet per second. The receiver misses the pass and the football hits the ground. How long does it take for the football to hit the ground? **3 sec**

9. A volleyball is hit when it is 4 feet off the ground. The volleyball has an initial upward velocity of 32 feet per second. Does the volleyball reach a height of 20 feet? If so, how long does it take to reach that height?
yes; 1 sec

GEOMETRY In Exercises 10 and 11, the rectangle below has an area of 32 square meters. (Lesson 10.8)

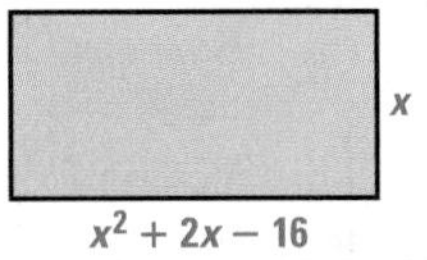

10. Write a polynomial equation that you can solve to find the length and the width of the rectangle. **$x^3 + 2x^2 - 16x = 32$ or $x^3 + 2x^2 - 16x - 32 = 0$**

11. What are the dimensions of the rectangle?
4 m, 8 m

CHAPTER 11

1. **SCALE DRAWINGS** You want to make a scale drawing of your classroom. The actual length of your classroom is 36 feet and the actual width is 32 feet. In your drawing, the length of your classroom is 9 inches. What should the width of your classroom in your drawing be? **(Lesson 11.1)** **8 in.**

MUSIC In Exercises 2 and 3, use the following information. **(Lesson 11.2)**
The graph shows the responses of 200 students when asked to name their favorite type of music.

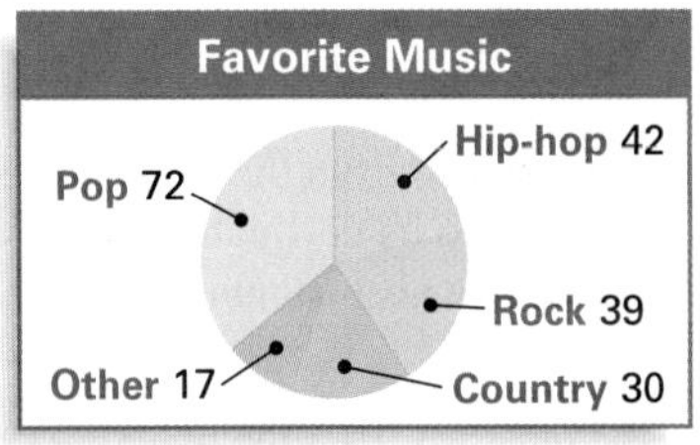

2. What percent of the students said that rock is their favorite type of music? **19.5%**

3. What percent of the students said that pop is their favorite type of music? **36%**

4. **BABY-SITTING** The table below lists the money you receive for baby-sitting. Decide if the data show *direct* or *inverse* variation. Then write an equation that relates the hours worked h and the total pay p. **(Lesson 11.3)** **direct variation; $p = 8.5h$**

Hours, h	2	3	5
Pay (dollars), p	17	25.50	42.50

5. **PROBABILITY** A coin is tossed onto the large rectangular region below. It is equally likely to land on any point in the region. Write a model that gives the probability that the coin will land in the small red rectangle. **(Lesson 11.4)** $P = \frac{x}{5x - 1}$

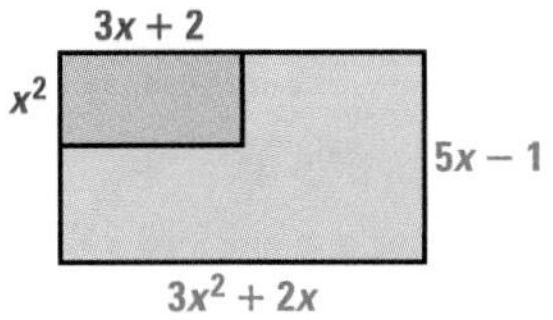

REAL ESTATE In Exercises 6 and 7, use the following information. **(Lesson 11.5)**
The models show the total value (in millions of dollars) of homes sold each year and the number of real estate agents selling homes each year at a real estate agency. Let t represent the number of years since 2000.

Total value of homes sold: $V = \frac{1000 + 140t}{2 - 0.1t}$

Number of agents: $A = \frac{210 + 10t}{1 - 0.05t}$

6. Find a model for the average value of homes sold each year per agent. $\frac{V}{A} = \frac{50 + 7t}{21 + t}$

7. Use your model from Exercise 6 to find the average value of homes sold per agent for 2005. **about \$3.27 million**

HIKING In Exercises 8–10, use the following information. **(Lesson 11.6)**
You are planning a 20-mile hiking trip. Some of the trails that you plan to hike are steep and some are relatively flat. You average 2 miles per hour on the steep trails and 4 miles per hour on the flat trails.

8. Write an expression for the total time of the hike. Let x represent the miles of steep trails. $\frac{x}{2} + \frac{20 - x}{4}$

9. Write your answer to Exercise 8 as a single rational expression. $\frac{x + 20}{4}$

10. Use your expression from Exercise 9 to determine the total time of the hike if you hike 5 miles of steep trails. **6.25 h**

11. **GEOMETRY** The area of the rectangle below is $3x^2 + 3x - 6$. Find the length. **(Lesson 11.7)** $3x - 3$

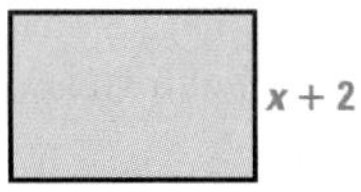

12. **BASKETBALL** You have made 27 free-throws in 45 attempts. Your free-throw average is $\frac{27}{45} = 0.6$. How many consecutive free-throws must you make to increase your free-throw average to 0.75? **(Lesson 11.8)** **27**

12. Let $a = 4$ and $b = 2$. $\frac{1}{4} < \frac{1}{2}$.

CHAPTER 12

1. **BASEBALL** The radius of a sphere can be found using the formula $r = \sqrt{\frac{S}{4\pi}}$, where S is the surface area and r is the radius of the sphere. Find the radius of a baseball that has a surface area of 168 square centimeters. Round your answer to the nearest hundredth. **(Lesson 12.1)** **3.66 cm**

PENDULUMS In Exercises 2 and 3, use the following information. **(Lesson 12.2)**
The *period* of a pendulum is the time it takes for the pendulum to swing back and forth. The model $T = 2\pi\sqrt{\frac{L}{981}}$ relates the period T (in seconds) of a pendulum to the length L (in centimeters) of the pendulum.

2. A pendulum has a length of 50 centimeters. A second pendulum has a length of 32 centimeters. Write an expression that can be used to find the difference of the periods of the pendulums. $2\pi\sqrt{\frac{50}{981}} - 2\pi\sqrt{\frac{32}{981}}$
3. Simplify your expression from Exercise 2. Round your answer to the nearest hundredth. $\frac{2\pi\sqrt{218}}{327}$; **0.28 sec**

ROLLER COASTERS In Exercises 4 and 5, use the following information. **(Lesson 12.3)**
The model $v = 8\sqrt{h}$ relates the velocity v (in feet per second) of a roller coaster at the bottom of a drop to the height h (in feet) of the drop.

4. The velocity of a roller coaster at the bottom of a drop is 44 feet per second. Find the height of the drop. **30.25 ft**
5. The velocity of a roller coaster at the bottom of a drop is 68 feet per second. Find the height of the drop. **72.25 ft**
6. **FOOTBALL** The path of a kicked football can be modeled by

$$y = -0.012x^2 + 1.11x$$

where x is the horizontal distance (in feet) and y is the vertical distance (in feet). Suppose that the football is not caught and it hits the ground. Find the horizontal distance covered by the football. **(Lesson 12.4)** **92.5 ft**

7. **PARKS** The dimensions of a rectangular park are shown below. Find the length of the path that connects opposite corners of the park. **(Lesson 12.5)** **125 ft**

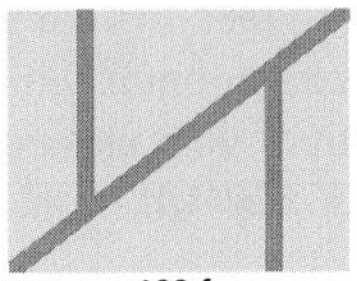

75 ft

100 ft

HELICOPTERS In Exercises 8 and 9, use the following information. **(Lesson 12.6)**
A helicopter is located 4 miles east and 2 miles north of a hospital. It travels to a destination 8 miles east and 12 miles north of the hospital.

8. What is the distance from the helicopter's original location to its destination? Round your answer to the nearest hundredth of a mile. **10.77 mi**
9. What is the distance from the helicopter's destination to the hospital? Round your answer to the nearest hundredth of a mile. **14.42 mi**
10. **SKATEBOARDING** You are building a skateboarding ramp. You want the angle of incline of the ramp to be 15°, and you want the height to be 3 feet, as shown below. What is the horizontal distance x? Round your answer to the nearest tenth of a foot. **(Lesson 12.7)** **11.2 ft**

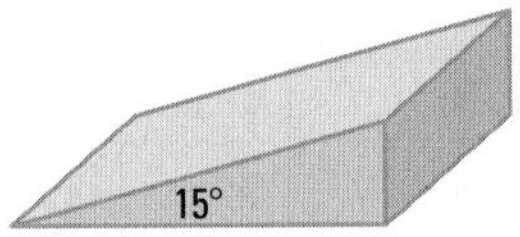

3 ft

x

COUNTEREXAMPLES In Exercises 11–13, find a counterexample to show that the statement is not true. **(Lesson 12.8)**

11. If a and b are odd integers, then $a + b$ is an odd integer. ***Sample answer:*** **3 + 5 = 8, which is not an odd integer.**
12. If a and b are nonzero real numbers and $a > b$, then $\frac{1}{a} > \frac{1}{b}$. **See margin.**
13. For any integer a, $a^3 > a^2$.
Sample answer: **Let $a = -2$. $-8 < 4$.**

Appendix 1 Samples and Surveys

GOAL ***Recognize bias in samples and survey questions; analyze the results of surveys.***

Surveys are often conducted to gather data about a *population*. A **population** is the entire group of people or objects that you want information about. When it is difficult to survey an entire population, a **sample**, or a part of the group, is surveyed.

TYPES OF SAMPLES

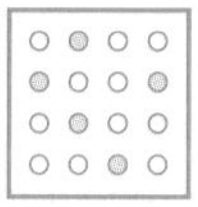

SIMPLE RANDOM SAMPLE

Members are chosen using a method that gives everyone an equally likely chance of being selected.

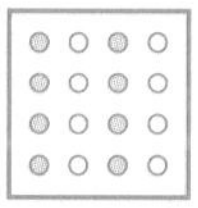

SYSTEMATIC SAMPLE

Members are chosen using a pattern, such as selecting every other person.

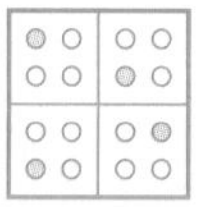

STRATIFIED SAMPLE

The population is first divided into groups. Then members are randomly chosen from each group.

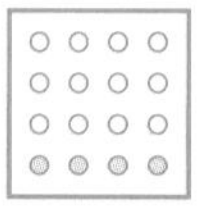

CONVENIENCE SAMPLE

Members are chosen because they are easily accessible.

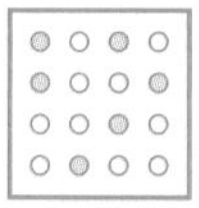

SELF-SELECTED SAMPLE

Members volunteer to participate.

EXAMPLE 1 *Classifying a Sample*

A teacher wants to select a group of 5 students. Classify the sample.

a. The teacher selects the first 5 students who raise their hands.

b. The teacher draws the names of 5 students from a hat.

SOLUTION

a. This is a self-selected sample because the students volunteer.

b. This is a simple random sample because each student has an equally likely chance of being selected.

PLANNING
You can use Appendix 1 after completing Lesson 1.6.

In a **random sample**, each person or object has an equally likely chance of being selected. A random sample is most likely to produce a representative sample of a population. A sample that is not representative of a population is a **biased sample**. In a biased sample, one or more of the parts of a population are overrepresented or underrepresented. A non-random sample can result in a biased sample.

Simple random, systematic, and stratified samples are preferred because the resulting samples are usually random. The results of convenience and self-selected samples are more likely to be biased.

EXAMPLE 2 *Identifying Potentially Biased Samples*

Administrators at your school are considering adding lacrosse as a school-sponsored sport. They decide to survey students to see if there is enough interest in the sport. Decide whether the sampling method could result in a biased sample. Explain your reasoning.

a. Survey students that currently participate in sports at your school.

b. Survey every tenth student who enters the school in the morning.

SOLUTION

a. This method could result in a biased sample because students who already participate in sports are more likely to favor the addition of a new sport than other students in the school.

b. This method is not likely to result in a biased sample because a wide range of students will be surveyed.

. . . .

When conducting a survey, it is important that the survey questions are carefully written. If a question is poorly written, then the responses of the people surveyed may not accurately reflect their opinions or actions. These types of questions are called **biased questions**.

EXAMPLE 3 *Identifying Biased Questions*

Decide whether the survey question may be biased. Explain.

a. Do you agree with your school's attendance policy?

b. Due to the large number of accidents among drivers that are under the age of 18, do you support raising the minimum driving age to 18?

SOLUTION

a. This question may be biased. It assumes that the student being surveyed is familiar with the attendance policy.

b. This question may be biased. Respondents may think a "no" response means that they are in favor of drivers under 18 getting into accidents.

828

ANALYZING DATA To help you make conclusions about data collected from a survey, the data needs to be organized. Data displays are often used to organize and present data collected from a survey. However, the data should be accurately displayed so that it can be correctly interpreted.

EXAMPLE 4 *Making Conclusions About Data from a Survey*

You conduct a survey that asks 30 students the average number of minutes they spend talking on the phone each day. The information collected from the survey is shown at the right.

Minutes	Tally
0–14	𝍸 III
15–29	𝍸 𝍸 II
30–44	𝍸 I
45–59	IIII

a. Make a histogram of the data.

b. Make a conclusion about the data.

STUDENT HELP

Look Back
For help with making histograms, see p. 792.

SOLUTION

a. Draw and label the horizontal and vertical axes.

List each interval from the table on the horizontal axis.

The greatest frequency is 12. So, start the vertical axis at 0 and end at 14, using increments of 2.

Draw a bar for each interval. The bars should have the same width.

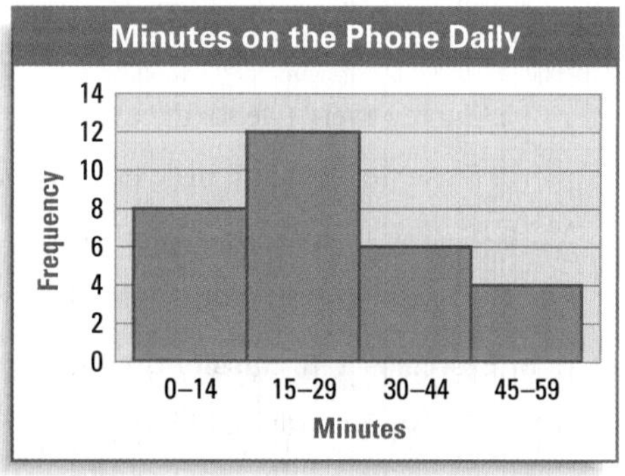

b. The histogram shows that the interval with greatest frequency is the 15–29 minute interval.

In general, more than half of the students spend less than 30 minutes talking on the phone each day.

....

ANALYZING SURVEYS The results of public opinion surveys often appear in newspapers and magazines. You should be critical of these surveys, especially if the results are meant to persuade you in some way. If you notice non-random sampling, biased questions, or poorly displayed data, then it is likely that the survey results are biased.

EXAMPLE 5 *Analyzing the Results of a Survey*

A survey asks shoppers whether they prefer to shop for clothes on-line or at a store. The results are shown at the right.

a. Tell why the results are biased.

b. Suggest a way to eliminate the bias.

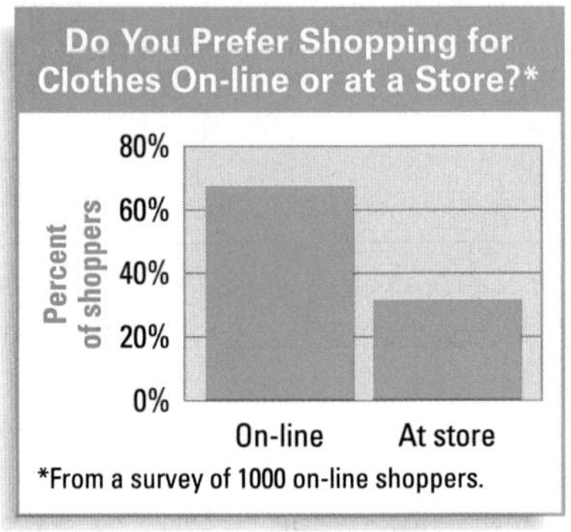

SOLUTION

a. The results of the survey are biased because the sampling method is biased. The sample contains only on-line shoppers, so the group of people who prefer shopping for clothes at a store are underrepresented.

b. The survey should be conducted using a random sample. For example, conduct a telephone survey.

EXERCISES

In Exercises 1–4, classify the sample involving the following situation. A fast-food restaurant decides to conduct a survey to find out if their customers are satisfied with the service at the restaurant.

1. Survey every twentieth customer. **systematic sample**
2. Survey every tenth male customer and every tenth female customer. **stratified sample**
3. Survey customers randomly chosen by the computer system. **simple random sample**
4. Survey customers only when the restaurant is not busy. **convenience sample**

In Exercises 5–8, classify the sample.

5. The staff at a town library want to know what new programs residents would be interested in. A computer generates a list of 100 randomly chosen town residents to be telephoned and surveyed. **simple random sample**
6. The manager of a grocery store wants to know if shoppers are satisfied with the selection of items in the deli case. A deli customer is surveyed every 30 minutes. **systematic sample**
7. The editor of a school newspaper wants to know how students feel about the extracurricular activities offered by the school. Twenty students from each class (freshman, sophomore, junior, and senior) are randomly chosen from the school directory and surveyed. **stratified sample**
8. The manager of a hotel wants to know how guests rate the cleanliness of the rooms. Response cards are left in each room for guests to fill out during their stay. **self-selected sample**

11. No; this survey uses a systematic sample and is not likely to result in a biased sample; a wide range of students will be surveyed.

12. Yes; teenagers in one city are not likely to be representative of average teenagers.

13. Yes; self-selected samples are generally not representative of the entire population.

14. No; this survey uses a stratified sample and is not likely to result in a biased sample; a wide range of menu items will be tasted.

17. Yes; the word "exciting" associated with the laser show and "boring" associated with the art museum encourages respondents to choose the laser show.

18. Yes; the first sentence is an opinion that is not part of the question, yet influences respondents to agree that more money should be budgeted for trash.

21. 50 students randomly chosen from a class of 300; this results in one of every six students in the population chosen, whereas the other results in one of every fifteen students in the population chosen. The greater the fraction of students surveyed, the more accurate the results will be.

Decide whether the sampling method could result in a random sample. Explain your reasoning.

9. A dentist wants to expand her office hours to better accommodate her patients. She decides to list her patients alphabetically and survey every tenth patient on the list about his or her time preferences. **Yes; patients are equally likely to be surveyed.**

10. The manager of a radio station wants to know the listeners' favorite songs. He decides to have listeners call the radio station and request their favorite songs. **No; this method is a self-selected sample. People do not have the same chance of responding.**

Decide whether the sampling method could result in a biased sample. Explain your reasoning. **11–14. See margin.**

11. Administrators at your school want to know if more vegetarian items should be added to the lunch menu. They decide to survey every fifth student waiting in line to purchase lunch.

12. A Boston-based market research company wants to learn more about the television viewing habits of average teenagers. They decide to survey randomly chosen teenagers who live in Boston.

13. The editor of a snowboarding magazine wants to know the readers' favorite places to ride. The latest issue of the magazine includes a survey for readers to complete and return.

14. A restaurant critic wants to taste a variety of items at a restaurant. She randomly chooses two items from each of the following groups on the menu: appetizers, entrées, and desserts.

Decide whether the survey question may be biased. Explain.

15. Do you agree with the rules established for traveling on public transportation with a dog? **Yes; the question assumes that respondents are familiar with the rules.**

16. How often do you read the local newspaper? **No; the question is objectively worded.**

17. Would you rather see an exciting laser show or visit a boring art museum? **See margin.**

18. The fewer trash cans that a community has, the more litter the community has. Should our community include more money in its budget for more trash cans and trash pick up? **See margin.**

In Exercises 19–21, use the following information.
Your class officers are planning a dance. They need to decide whether to have a disc jockey or a live band at the dance. They decide to survey a sample of students to determine their preference.

19. Describe a sampling method that can be used so that the sample will most likely be random. ***Sample answer:* Get a list of all the students in your class. Survey every tenth person on the list.**

20. Write a question to ask the members of the sample. The question should be carefully written so that it is not biased. ***Sample answer:* Should the music at the dance be supplied by a disc jockey or a live band?**

21. Which is more likely to give accurate results, 50 students randomly chosen from a class of 300, or 20 students randomly chosen from a class of 300? Explain your reasoning. **See margin.**

22.

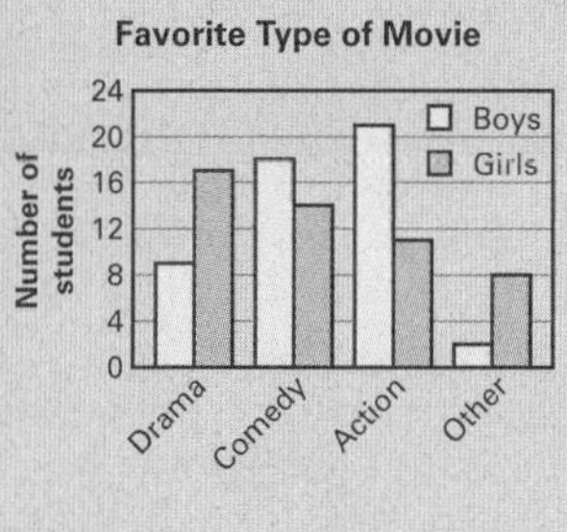

24.

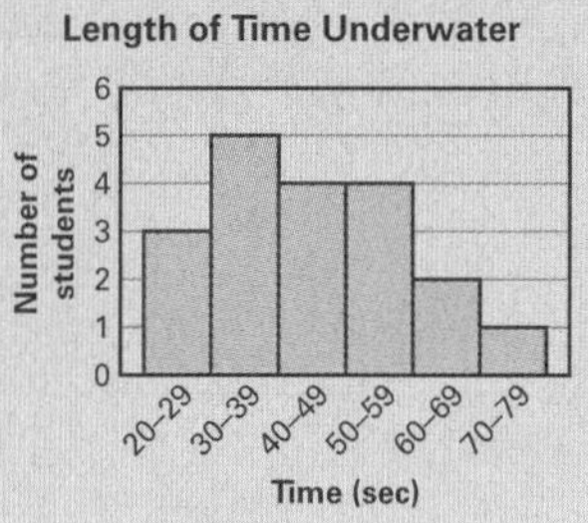

23. *Sample answer:* The boys tended to prefer comedies and action movies, while the girls tended to prefer dramas.

26. *Sample answer:* The only people surveyed were customers of Sal's, so people who prefer Pizza Primo are underrepresented. Also, there may be restaurants other than the two mentioned that sell pizza. The bias can be eliminated by asking people at the post office which restaurant's pizza is their favorite.

27. *Sample answer:* The survey question is biased by the phrase "allowed to endanger the safety of pedestrians." The bias can be eliminated by asking an unbiased question of randomly selected residents, such as "Should skateboarders be allowed to skateboard on sidewalks, or should skateboarders be required to skateboard only in skateboard parks?"

30. *Sample answer:* What is your favorite type of music? How many hours a day do you listen to that kind of music on a typical day? Do you usually listen to music on a CD player, an MP3 player, the Internet, the radio, or some other source?

In Exercises 22 and 23, use the following information.
A sample of 100 students, 50 boys and 50 girls, are asked to name their favorite type of movie. The results are shown in the table below.

Group	Drama	Comedy	Action	Other
Boys	9	18	21	2
Girls	17	14	11	8

22. Make a double bar graph of the data. **See margin.**

23. Use your display from Exercise 22 to make a conclusion about the data. **See margin.**

In Exercises 24 and 25, use the following information.
A sample of 19 students enrolled in a swimming class are asked to hold their breath underwater for as long as possible. The results (in seconds) are listed below.

65, 29, 38, 50, 60, 43, 27, 48, 29, 79, 37, 45, 48, 32, 57, 35, 54, 53, 37

24. Make a histogram of the data using the following intervals: 20–29, 30–39, 40–49, 50–59, 60–69, 70–79. **See margin.**

25. Use your display from Exercise 24 to make a conclusion about the data. ***Sample answer:*** **Over two thirds of the times were between 30 sec and 59 sec.**

In Exercises 26 and 27, tell why the results of the survey are biased. Then tell how to eliminate the bias.

26.

Sal's Pizza
is your favorite!

*Which pizza do you prefer?**

Sal's Pizza	72%
Pizza Primo	28%

* Based on a survey of 250 Sal's Pizza customers.

27.

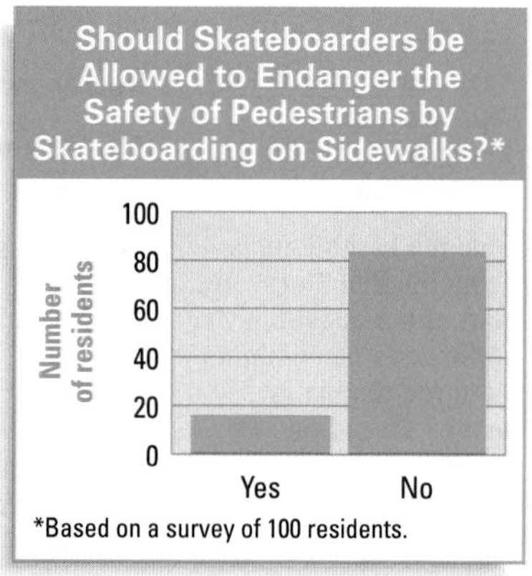

28. Find the results of a survey published in a magazine or newspaper. Do you think that the results of the survey are biased? Explain your reasoning. **Answers will vary.**

In Exercises 29–32, you want to learn more about the music that students in your school listen to.

29. Describe how you would choose a sample of students to survey. ***Sample answer:*** **Ask every tenth student who enters the cafeteria.**

30. Write a survey that consists of 3 or 4 questions.

31. Give the survey to the sample and present your results. **Answers will vary.**

32. Review your results. Would you change anything about the way the survey was designed or conducted? Explain. **Answers will vary.**

Appendix 2

Probability of Compound Events

GOAL ***Find the probability of independent, dependent, mutually exclusive, and overlapping events.***

When you consider the probability of two events occurring, the events are called **compound events**. *Independent, dependent, mutually exclusive,* and *overlapping events* are examples of compound events.

Two events are **independent events** if the occurrence of one event does not affect the occurrence of the other. Two events are **dependent events** if the occurrence of one event *does* affect the occurrence of the other.

EXAMPLE 1 *Deciding Whether Events Are Independent or Dependent*

Decide whether the events are *independent* or *dependent.*

a. Each whole number from 1 through 10 is written on a piece of paper and placed in a hat. You randomly choose a piece of paper, do not put it back, then randomly choose another piece of paper.
Event *A*: Choose the 5 first.
Event *B*: Choose an odd number second.

b. You flip a coin and roll a number cube.
Event *A*: Get tails when flipping the coin.
Event *B*: Get a 2 when rolling the number cube.

SOLUTION

a. The events are dependent. Choosing the 5 and not putting it back affects the outcomes of choosing an odd number second.

b. The events are independent. Getting tails when flipping a coin does not affect the outcomes of rolling a number cube.

PROBABILITIES OF INDEPENDENT AND DEPENDENT EVENTS

For two independent events A and B, the probability that both events occur is the product of the probabilities of the events.

$P(A \text{ and } B) = P(A) \bullet P(B)$ Events A and B are independent.

For two dependent events A and B, the probability that both occur is the product of the probability of the first event and the conditional probability of the second event given the first event.

$P(A \text{ and } B) = P(A) \bullet P(B \mid A)$ Events A and B are dependent.

STUDENT HELP

Study Tip
In the formula for the probability of dependent events, $P(B \mid A)$ represents the probability that B will occur given that A has occurred.

PLANNING
You can use Appendix 2 after completing Chapter 2.

EXAMPLE 2 Finding the Probability of Independent Events

Of the 35 CDs in your collection, 21 are rock and 14 are hip-hop. You randomly choose a CD and put it back. Then you randomly choose another CD. Find the probability of choosing a rock CD first and choosing a hip-hop CD second.

SOLUTION

Let event A be "choose a rock CD first" and let event B be "choose a hip-hop CD second." Since the first CD chosen is replaced before the second CD is chosen, events A and B are independent. The probability of A and B is shown below.

$$P(A \text{ and } B) = \frac{21}{35} \cdot \frac{14}{35} = \frac{6}{25} = 0.24$$

▶ The probability is 0.24.

EXAMPLE 3 Finding the Probability of Dependent Events

A drawer contains 9 white socks, 7 black socks, and 5 blue socks. You randomly choose one sock from the drawer. Then you randomly choose another sock without replacing the first. Find the probability of events A and B.

a. Event *A*: Choose a white sock first.
Event *B*: Choose a black sock second.

b. Event *A*: Choose a blue sock first.
Event *B*: Choose a blue sock second.

SOLUTION

a. First, find the probability of each event.

$$P(A) = \frac{9}{21}$$ **Of the 21 socks, 9 are white.**

$$P(B \mid A) = \frac{7}{20}$$ **Of the remaining 20 socks, 7 are black.**

Then multiply the probabilities.

$$P(A \text{ and } B) = \frac{9}{21} \cdot \frac{7}{20} = \frac{3}{20} = 0.15$$

▶ The probability is 0.15.

b. First, find the probability of each event.

$$P(A) = \frac{5}{21}$$ **Of the 21 socks, 5 are blue.**

$$P(B \mid A) = \frac{4}{20}$$ **Of the remaining 20 socks, 4 are blue.**

Then multiply the probabilities.

$$P(A \text{ and } B) = \frac{5}{21} \cdot \frac{4}{20} = \frac{1}{21} \approx 0.048$$

▶ The probability is about 0.048.

Mutually exclusive events are events that have no outcomes in common. Use the rule below to find the probability of mutually exclusive events.

PROBABILITY OF MUTUALLY EXCLUSIVE EVENTS

For two mutually exclusive events A and B, the probability that either of the events occurs is the sum of the probabilities of the events.

$P(A \text{ or } B) = P(A) + P(B)$ Events A and B are mutually exclusive.

EXAMPLE 4 *Finding the Probability of Mutually Exclusive Events*

You are playing a game using the spinner at the right, which is divided into equal parts. On your spin, what is the probability that the spinner lands on an even number or a prime number?

SOLUTION

First, find the probabilities of the events. Let event A be getting an even number and let event B be getting a prime number.

$P(A) = \frac{4}{8}$ Of the 8 numbers, 4 are even.

$P(B) = \frac{3}{8}$ Of the 8 numbers, 3 are prime.

Then find the sum of the probabilities.

$$P(A \text{ or } B) = \frac{4}{8} + \frac{3}{8} = \frac{7}{8} = 0.875$$

▶ The probability that the spinner lands on an even number or a prime number is 0.875, or 87.5%.

....

Overlapping events are events that have one or more outcomes in common. Use the rule below to find the probability of overlapping events.

PROBABILITY OF OVERLAPPING EVENTS

For two overlapping events A and B, the probability that either of the events occurs is the sum of the probabilities of the events, minus the probability that both events occur.

$P(A \text{ or } B) = P(A) + P(B) - P(A \text{ and } B)$ Events A and B are overlapping.

EXAMPLE 5 *Finding the Probability of Overlapping Events*

You roll a number cube. What is the probability that you get an odd number or a number greater than 3?

SOLUTION

Find the probabilities of the events. Let event A be getting an odd number and let event B be getting a number greater than 3.

$P(A) = \frac{3}{6}$ — Of the 6 numbers, 3 are odd (1, 3, 5).

$P(B) = \frac{3}{6}$ — Of the 6 numbers, 3 are greater than 3 (4, 5, 6).

$P(A \text{ and } B) = \frac{1}{6}$ — Of the 6 numbers, 1 is odd and greater than 3 (5).

Find the probability of the events.

$$P(A \text{ or } B) = \frac{3}{6} + \frac{3}{6} - \frac{1}{6} = \frac{5}{6} \approx 0.83$$

▸ The probability that the number is odd or greater than 3 is about 0.83.

EXERCISES

Decide whether the events are *independent* or *dependent*.

1. You toss a coin two times.
   Event *A*: Get tails on the first toss.
   Event *B*: Get tails on the second toss. **independent**

2. A box contains blue and yellow paper clips. You randomly choose a paper clip, then choose another without replacing the first paper clip.
   Event *A*: The first paper clip is blue.
   Event *B*: The second paper clip is yellow. **dependent**

3. A bag contains tulip bulbs and daffodil bulbs. You randomly choose a bulb, plant it, and randomly choose another bulb.
   Event *A*: The first bulb is a daffodil bulb.
   Event *B*: The second bulb is a daffodil bulb. **dependent**

4. You randomly choose a card from a standard deck of 52 cards. You put the card back in the deck and then your friend randomly chooses a card.
   Event *A*: You choose the queen of diamonds.
   Event *B*: Your friend chooses the six of clubs. **independent**

5. Each week you and your sister assign chores by writing the chores on pieces of paper and randomly choose them from a hat.
   Event *A*: You choose "rake the leaves" the first week of October.
   Event *B*: You choose "rake the leaves" the second week of October. **independent**

Events A and B are independent. Find the indicated probability.

6. $P(A) = 0.5$
$P(B) = 0.8$
$P(A \text{ and } B) = \underline{?}$ 0.4

7. $P(A) = 0.6$
$P(B) = \underline{?}$ 0.4
$P(A \text{ and } B) = 0.24$

8. $P(A) = \underline{?}$ 0.11
$P(B) = 0.5$
$P(A \text{ and } B) = 0.055$

Events A and B are dependent. Find the indicated probability.

9. $P(A) = 0.9$
$P(B \mid A) = 0.7$
$P(A \text{ and } B) = \underline{?}$ 0.63

10. $P(A) = 0.2$
$P(B \mid A) = \underline{?}$ 0.21
$P(A \text{ and } B) = 0.042$

11. $P(A) = \underline{?}$ 0.85
$P(B \mid A) = 0.4$
$P(A \text{ and } B) = 0.34$

Decide whether the situation describes *independent* or *dependent* events. Then answer the question.

12. You plant a fruit tree. The probability that the tree will survive its first winter is 0.8. If it survives the first winter, the probability that the tree will bear fruit by its fourth year is 0.6. What is the probability that the tree will bear fruit by its fourth year? **dependent; 0.48**

13. You toss a coin and roll a number cube. What is the probability that you get tails and a 3? **independent; $\frac{1}{12} \approx 0.083$**

14. A bucket of tennis balls contains 12 yellow, 5 orange, and 3 pink tennis balls. You randomly choose a ball, put it back, and randomly choose another ball. What is the probability that the first ball is yellow and the second ball is orange? **independent; $\frac{3}{20} = 0.15$**

15. A bowl contains 22 green grapes and 28 purple grapes. You randomly choose a grape, eat it, and randomly choose another grape. What is the probability that both grapes are green? **dependent; $\frac{33}{175} \approx 0.189$**

In Exercises 16 and 17, use the following information.
The formula for finding the probability of independent events can be extended to three or more events. This is also true for dependent events.

16. $\frac{1}{10,000} = 0.0001$

16. You are assigned a 4 digit personal identification number (PIN) for use with your bank card. Each of the digits can be a whole number from 0 to 9, and the digits can repeat. What is the probability that your PIN is 1234?

17. In Exercise 16, suppose that the digits cannot repeat. What is the probability that your PIN is 1234? **$\frac{1}{5040} \approx 0.0002$**

In Exercises 18 and 19, read the first 100 words of a book. Count the number of words that satisfy each of the events below, which are independent.

Event *A*: The word starts with a vowel.
Event *B*: The word has more than four letters.
Event *C*: The word has one letter.

18. You randomly choose a word from the first 100 words. What is the probability that the word starts with a vowel and has more than four letters? **Answers will vary.**

19. You randomly choose a word from the first 100 words. What is the probability that the word starts with a vowel and has one letter? **Answers will vary.**

Events *A* and *B* are mutually exclusive events. Find *P*(*A* or *B*).

20. $P(A) = 0.2$
$P(B) = 0.12$ **0.32**

21. $P(A) = 0.07$
$P(B) = 0.33$ **0.4**

22. $P(A) = 62\%$
$P(B) = 15\%$ **77%**

Events *A* and *B* are mutually exclusive events that involve rolling a number cube. Find *P*(*A* or *B*).

23. Event *A*: Get a number greater than 4.
Event *B*: Get a number less than 4. $\frac{5}{6} \approx 0.83$

24. Event *A*: Get a 2.
Event *B*: Get an odd number. $\frac{2}{3} \approx 0.67$

In Exercises 25–27, you randomly choose a letter from the word INDIANA. Find the probability of choosing the given letters.

25. I or N $\frac{4}{7} \approx 0.57$

26. N or D $\frac{3}{7} \approx 0.43$

27. A or N $\frac{4}{7} \approx 0.57$

28. A wallet contains six \$1 bills, one \$5 bill, and three \$10 bills. A bill is randomly chosen from the wallet. What is the probability that the bill is a \$1 bill or a \$5 bill? $\frac{7}{10} = 0.7$

Events *A* and *B* are overlapping events. Find *P*(*A* or *B*).

29. $P(A) = 0.4$
$P(B) = 0.5$
$P(A \text{ and } B) = 0.2$ **0.7**

30. $P(A) = 0.5$
$P(B) = 0.5$
$P(A \text{ and } B) = 0.375$ **0.625**

31. $P(A) = 11\%$
$P(B) = 58\%$
$P(A \text{ and } B) = 23\%$ **46%**

In Exercises 32–35, events *A* and *B* are overlapping events that involve spinning the spinner below. Find *P*(*A* or *B*).

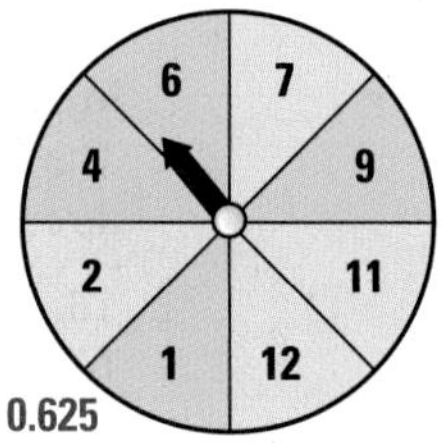

32. Event *A*: Lands on an odd number.
Event *B*: Lands on green. $\frac{3}{4} = 0.75$

33. Event *A*: Lands on an even number.
Event *B*: Lands on yellow. $\frac{3}{4} = 0.75$

34. Event *A*: Lands on green.
Event *B*: Lands on a number divisible by 3. $\frac{5}{8} = 0.625$

35. Event *A*: Lands on a number greater than 4.
Event *B*: Lands on a number less than 11. $\frac{3}{8} = 0.375$

36. A standard deck of 52 playing cards contains 13 cards in each of the following four suits: hearts, diamonds, clubs, and spades. You randomly choose a card from the deck. What is the probability that the card is a 4 or a club? $\frac{4}{13} \approx 0.31$

Events *A* and *B* are *complementary*. Use the following information to find the indicated probability.

Two mutually exclusive events in which one or the other event must occur are called *complementary events*. The sum of the probabilities of two complementary events is 1. So, if *A* and *B* are complementary events, then $P(A) + P(B) = 1$.

37. $P(A) = \underline{?}$ **0.4**
$P(B) = 0.6$

38. $P(A) = 0.14$
$P(B) = \underline{?}$ **0.86**

39. $P(A) = \underline{?}$ **72%**
$P(B) = 28\%$

Appendix 3

Estimation and Reasonableness

GOAL ***Use estimation to solve problems and to determine whether answers are reasonable.***

Some real-world problems require only an estimate for an answer instead of an exact answer. Before using estimation, you need to decide whether estimation is appropriate.

EXAMPLE 1 *Deciding Whether Estimation is Appropriate*

A music store is having a sale on CDs. All of the new releases have a sale price of $10.88, which includes tax. Erik wants to buy 3 CDs that are new releases. Decide whether an *estimate* or an *exact answer* is appropriate. Then solve.

a. How much money should Erik take with him to pay for the CDs?

b. What is the total cost of the CDs?

SOLUTION

a. An estimate is appropriate. To be sure that Erik has enough money, he should round up: $11 • 3 = $33. So, Erik should bring $33.

b. An exact answer is appropriate.

$10.88 • 3 = $32.64

....

Estimation can often be used to check the reasonableness of an answer.

EXAMPLE 2 *Using Estimation to Check Reasonableness*

Determine whether the answers to the problem are reasonable.

Problem Use the formula $S = 2\pi r^2 + 2\pi rh$ to find the surface area S (in square meters) of the cylinder at the right.

Susan's answer: about 324 m^2

Len's answer: about 525 m^2

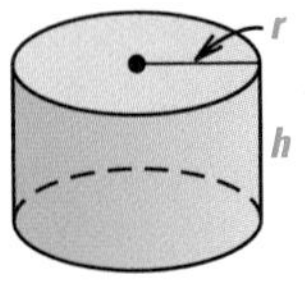

$r = 6.1$ m
$h = 7.6$ m

STUDENT HELP

Study Tip
The Greek letter π (pi) is often approximated by the number 3.14.

SOLUTION

To find an estimate of the surface area, round the radius, height, and pi to the nearest whole number: $S \approx 2(3)(6^2) + 2(3)(6)(8) = 504$ m^2.

The estimate of 504 square meters is not close to Susan's answer, but it is close to Len's answer. So, Susan's answer is not reasonable and Len's answer is reasonable.

PLANNING
You can use Appendix 3 after completing Lesson 9.2.

ESTIMATION METHODS *Front-end estimation* can be used to estimate a sum when all the numbers have the same number of digits. When numbers being added have about the same value, *clustering* can be used to estimate their sum. Different types of estimation may yield different levels of accuracy.

EXAMPLE 3 *Comparing Estimation Methods*

You recently started working at a grocery store. The table below shows the amounts of your first 6 paychecks.

Week	1	2	3	4	5	6
Pay	\$30.44	\$31.62	\$38.12	\$40.01	\$43.19	\$46.88

a. Use front-end estimation to estimate your total pay for the 6 weeks.

b. Use clustering to estimate your total pay for the 6 weeks.

c. Find the actual total. Which method gives a closer estimate? Why?

SOLUTION

a. To use front-end estimation, add the values of the greatest digits to get a low estimate. Then use the remaining digits to adjust the sum and get a closer estimate.

1 Add the values of the tens' digits.

```
  30.44
  31.62
  38.12
  40.01
  43.19
+ 46.88
-------
 210
```

2 Approximate the sum of the remaining digits.

```
  30.44 ↘
  31.62 → about 10 more
  38.12 ↗
  40.01 ↘
  43.19 → about 10 more
+ 46.88 ↗
--------------------------
          about 20 more
```

3 Add the two sums.

```
  210
+  20
-----
  230
```

▶ Your total pay for the first 6 weeks is about \$230.

b. When a group of data are clustered, you can choose a convenient value near the center of the cluster to represent each data value in the group.

In the table, the amounts cluster around \$40. An estimate of the sum is given by $6 \cdot 40 = 240$.

▶ Your total pay for the first 6 weeks is about \$240.

c. The actual total is \$230.26. The estimate obtained by front-end estimation is closer than the estimate obtained by clustering. The estimate obtained by clustering is greater than the estimate obtained by front-end estimation because the value chosen to represent the data in the cluster group is greater than the average data value in the cluster.

EXERCISES

Decide whether an *estimate* or an *exact answer* is appropriate.

1. the balance of a bank account on a bank statement **exact answer**
2. the number of hours spent doing homework during a week **estimate**
3. the amount of money left as a tip for a waiter **estimate**
4. a baseball player's batting average on a baseball card **exact answer**

Decide whether an *estimate* or an *exact answer* is appropriate. Then solve the problem.

5. You are helping prepare fundraising materials. It takes you 28 seconds to stuff an envelope and put an address label on the envelope. How many minutes should it take you to prepare 50 envelopes? **estimate; about 25 min**
6. You have received the following scores on your science tests: 76, 82, 91, 87, 89. What is your average? **exact answer; 85**

Determine whether the answer to the problem is reasonable. Explain your reasoning.

7. **Problem** Evaluate $5x - 5$ when $x = 6.1$. **Answer** 25.5
   Reasonable; $5x - 5 \approx 5(6) - 5 = 25$, and 25 is close to 25.5.
8. **Problem** Evaluate $3z^2 + z$ when $z = -1.9$. **Answer** -12.73
   Not reasonable; $3(-2)^2 + (-2) = 10$, and 10 is not close to -12.73.
9. **Problem** Evaluate $2\pi r$ when $r = 4.8$ ft. **Answer** 9.6 ft
   Not reasonable; $2\pi r \approx 2(3)(5) = 30$, and 30 is not close to 9.6.
10. **Problem** Solve $9.85x = 66.98$. **Answer** $x = 6.8$
    Reasonable; $10x \approx 70$, $x = 7$, and 7 is close to 6.8.
11. An airplane travels 1448 miles in 3 hours 20 minutes. A passenger calculates the average speed of the aircraft to be 580 miles per hour. Use an estimate to decide if the calculation is reasonable. **The calculation is not reasonable.**

In Exercises 12–15, use the following information.
You are waiting in line to pay for several items at a grocery store. You want to be sure that you will have enough money for the items. The prices of the items are \$5.89, \$3.23, \$1.79, \$2.97, \$6.78, \$3.44, and \$4.88.

12. Estimate the total price by adding the values of the ones' digits, ignoring the decimal parts. **\$24**
13. Is your estimate in Exercise 12 high or low? Is your estimate appropriate for the situation? Explain.
14. Estimate the total price by rounding each dollar amount up to the next whole number and adding the rounded amounts. **\$31**
15. Is your estimate in Exercise 14 high or low? Is your estimate appropriate for the situation? Explain.
16. The weights (in pounds) of the starting defensive players for a football team are listed below. Use clustering to estimate the total weight of the starting defensive players. ***Sample answer:* about 2640 lb**

    216, 218, 220, 224, 230, 245, 251, 265, 268, 272, 278

13. Low; no; the actual total price is several dollars higher than the estimate in Exercise 12.

15. High; yes; you can be sure that you have enough money for all the items using the estimate in Exercise 14.

In Exercises 17–19, use the following information.
You are mailing a box of items to a relative. The weights (in ounces) of the items are listed below.

12.5, 13.2, 13.6, 20.8, 21.9, 22.7, 22.9

17. Use front-end estimation to estimate the total weight of the items in the box. ***Sample answer:*** **about 125 oz**

18. Use clustering to estimate the total weight of the items in the box. ***Sample answer:*** **about 126 oz**

19. Find the actual weight of the items in the box. Which method gives a closer estimate? Explain. **127.6 oz; clustering; the estimate using clustering yielded an answer closer to the actual weight of 127.6 ounces.**

In Exercises 20 and 21, use the cube at the right.

20. Estimate the volume of the cube. (The volume V of a cube is given by the formula $V = s^3$, where s is the length of a side of the cube.) **about 1728 in.^{3}**

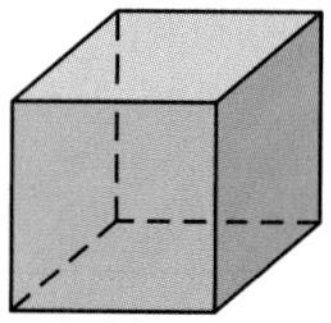

21. Use a calculator to compare the actual volume of the cube to your estimate from Exercise 20. By what percent does your estimate differ from the actual volume? **The actual volume is 1643.032 in.^{3}; the estimate is about 5.2% greater than the actual volume.**

In Exercises 22–25, use the following information.
Recall that if $a = b^2$, then $\sqrt{a} = b$. For example, because $9 = 3^2$, $\sqrt{9} = 3$.

22. 5; $\sqrt{21}$ is closer to 5 than 4 because 21 is closer to 25 than 16.

22. Determine whether $\sqrt{21}$ is closer to 4 or 5. Explain your reasoning.

23. On a number line, between what two consecutive integers does $\sqrt{154}$ lie? **12 and 13**

24. Find the value of $\sqrt{227}$ to the nearest whole number. **15**

25. Estimate the value of $7.15 + 3.9\sqrt{89}$ by using whole number estimates. **43**

In Exercises 26–28, use estimation to solve the problem.

26. The time it takes a pendulum to swing back and forth is called its *period*. The formula $T = 2\pi\sqrt{\frac{L}{384}}$ can be used to find the period T (in seconds) of a pendulum with a length of L inches. Estimate the period of a pendulum that is 90 inches long. **about 3 sec**

27. A sphere with radius r has a surface area S of 220 square centimeters. Use the formula $S = 4\pi r^2$ to estimate the length of the radius. **about 4 cm**

28. As Earth orbits the sun, the distance from Earth to the sun varies from about 9.14×10^7 to 9.45×10^7 miles. The speed of light is 186,282 miles per second. Estimate the time, in seconds, it takes for light from the sun to reach Earth when Earth is 9.1421555×10^7 miles from the sun. **about 450 sec**

Glossary

A

absolute value (p. 65) The distance between the origin and the point representing the real number. The symbol $|a|$ represents the absolute value of a number a.

algorithm (p. 39) A step-by-step process used to solve a problem.

asymptote (p. 692) A line that a graph approaches. The distance between the graph and the line approaches zero.

axioms (p. 758) A The basic properties of mathematics that mathematicians accept without proof.

axis of symmetry of a parabola (p. 518) The line passing through the vertex that divides the parabola into two symmetric parts. The two symmetric parts are mirror images of each other. *See also* parabola.

B

bar graph (p. 41) A graph that organizes a collection of data by using horizontal or vertical bars to display how many times each event or number occurs in the collection.

base number of a percent equation (p. 649) The number that is being compared *to* in any percent equation. The number b in the verbal model a is p percent of b.

base of a power (p. 9) The number or variable that is used as a factor in repeated multiplication. For example, in the expression 4^6, 4 is the base.

best-fitting line (p. 292) A line that best fits the data points on a scatter plot.

binomial (p. 576) A polynomial with two terms.

box-and-whisker plot (p. 375) A data display that divides a set of data into four parts. The box represents half of the data. The segments, or whiskers, extend to the least and greatest data items.

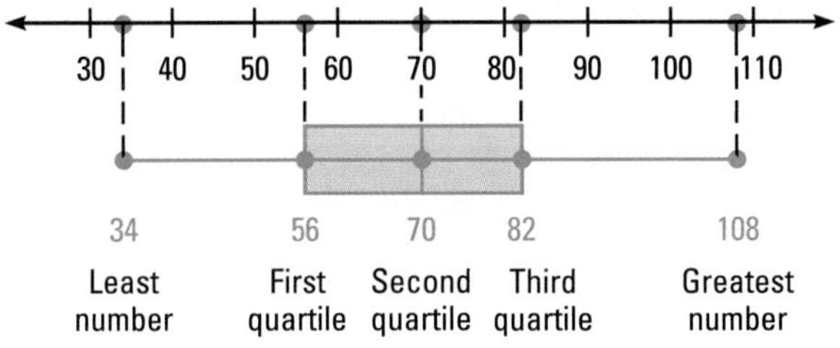

C

center of a hyperbola (p. 692) The point (h, k) in the graph of the rational function $f(x) = \frac{a}{x-h} + k$.

closed set (p. 113) A set of numbers is closed under an operation if applying the operation to any two numbers in the set results in another number in that set. This set is said to have closure.

coefficient (p. 102) A number multiplied by a variable in a term. The number is the coefficient of the variable.

completing the square (p. 730) The process of rewriting a quadratic equation so that one side is a perfect square trinomial.

compound inequality (p. 346) Two inequalities connected by *and* or *or*.

conclusion (pp. 187, 739) The "then" part of an if-then statement. In the statement "If p, then q," q is the conclusion.

conjecture (p. 759) A statement that is thought to be true but is not yet proved. Often it is a statement based on observation.

conjugates (p. 717) The expressions $(a + \sqrt{b})$ and $(a - \sqrt{b})$ are conjugates of each other.

consecutive integers (p. 149) Integers that follow each other in order. For example, 4, 5, 6.

constant terms (p. 102) Terms with no variable factors. For example, in $x + 2 - 5x - 4$, the constant terms are 2 and -4.

constant of variation (p. 234) The constant in a variation model. Represented by the variable k. *See also* direct variation *and* inverse variation.

converse of a statement (p. 739) A related statement in which the hypothesis and conclusion are interchanged. The converse of the statement "If p, then q" is "If q, then p."

coordinate plane (p. 203) A plane formed by two real number lines that intersect at a right angle.

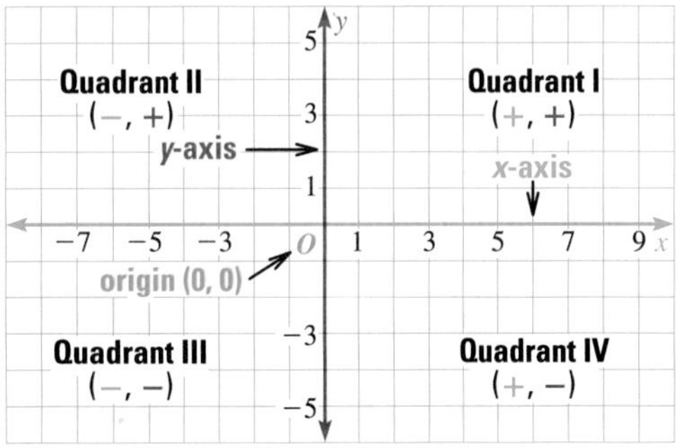

correlation (p. 295) The relationship between two data sets.

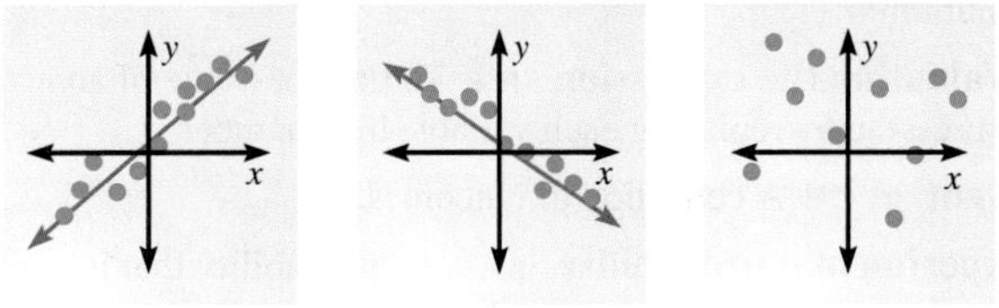

Positive correlation | Negative correlation | Relatively no correlation

cosine (p. 752) *See* trigonometric ratio.

counterexample (p. 66) An example used to show that a given statement is false.

D

data (p. 40) Information, facts, or numbers used to describe something.

decay factor (p. 484) The expression $1 - r$ in the exponential decay model where r is the decay rate. *See also* exponential decay.

decay rate (p. 484) The variable r in the exponential decay model. *See also* exponential decay.

deductive reasoning (p. 187) Reasoning where a conclusion is reached based on facts, definitions, rules, or properties.

degree of a polynomial (p. 576) The largest degree of the terms of the polynomial.

degree of a term (p. 576) The exponent of the variable of the term.

direct variation (pp. 234, 656) The relationship of two variables x and y if there is a nonzero number k such that $y = kx$, or $\frac{y}{x} = k$. The variables x and y *vary directly*.

discriminant (p. 541) The expression $b^2 - 4ac$ where a, b, and c are coefficients of the quadratic equation $ax^2 + bx + c = 0$.

distance formula (p. 745) The distance d between the points (x_1, y_1) and (x_2, y_2) is

$$d = \sqrt{(x_2 - x_1)^2 + (y_2 - y_1)^2}.$$

distributive property (p. 100) The product of a and $(b + c)$: $a(b + c) = ab + ac$ or $(b + c)a = ba + ca$. The product of a and $(b - c)$: $a(b - c) = ab - ac$ or $(b - c)a = ba - ca$.

domain of a function (p. 47) The collection of all input values.

E

entry or element of a matrix (p. 86) Each number in the matrix. *See also* matrix.

equation (p. 24) A statement formed when an equal sign is placed between two expressions.

equivalent equations (p. 132) Equations with the same solutions as the original equation.

equivalent inequalities (p. 335) Inequalities with the same solution(s).

evaluating the expression (p. 3) To find the value of an expression by replacing each variable by a number.

event (p. 114) A collection of outcomes.

experimental probability (p. 115) A probability that is based on repetitions of an actual experiment.

exponent (p. 9) The number or variable that represents the number of times the base is used as a factor. For example, in the expression 4^6, 6 is the exponent.

exponential decay (p. 484) A quantity that is decreasing by the same percent in each unit of time t where C is the initial amount.

Exponential decay model: $y = C(1 - r)^t$

exponential function (p. 458) A function of the form $y = ab^x$, where $b > 0$ and $b \neq 1$.

exponential growth (p. 477) A quantity that is increasing by the same percent in each unit of time t where C is the initial amount.

Exponential growth model: $y = C(1 + r)^t$

extraneous solution (pp. 644, 723) A trial solution that does not satisfy the original equation.

extremes of a proportion (p. 643) In the proportion $\frac{a}{b} = \frac{c}{d}$, a and d are the extremes.

F

factor (p. 777) The numbers and variables that are multiplied in an expression. For example, 4 and 9 are factors of 36 and 6 and x are factors of $6x$.

factor a polynomial completely (p. 625) To write a polynomial as the product of:

- monomial factors
- prime factors with at least two terms.

factor a quadratic expression (p. 604) To write a quadratic expression as the product of two linear expressions.

factored form of a polynomial (pp. 597, 625) A polynomial that is written as the product of two or more prime factors.

favorable outcomes (p. 114) The outcomes for a particular event that are being considered. *See also* outcomes.

FOIL pattern (p. 585) A pattern used to multiply two binomials. Multiply the First, Outer, Inner, and Last terms.

For example, $(x + 4)(2x + 3) = 2x^2 + 3x + 8x + 12$

$= 2x^2 + 11x + 12$

formula (p. 174) An algebraic equation that relates two or more real-life quantities.

function (p. 46) A rule that establishes a relationship between two quantities, called the input and the output. For each input, there is exactly one output.

function form (p. 176) A two-variable equation is written in function form if one of its variables is isolated on one side of the equation. The isolated variable is the output and is a function of the input.

function notation (p. 257) A way to name a function that is defined by an equation. For an equation in x and y, the symbol $f(x)$ replaces y and is read as "the value of f at x" or simply as "f of x."

G

generalization (p. 187) A conclusion based on several observations.

geometric mean (p. 723) The geometric mean of a and b is $\sqrt{ab}$.

geometric probability (p. 666) The probability P that an object is tossed onto Region A and lands in Region B where Region B is contained in Region A is

$$P = \frac{\text{Area of Region } B}{\text{Area of Region } A}.$$

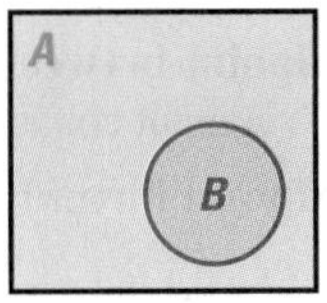

graph of an equation in two variables (p. 210) The set of *all* points (x, y) that are solutions of the equation.

graph of a function (p. 257) The set of all points $(x, f(x))$, where x is in the domain of the function.

graph of a linear inequality in one variable (p. 334) The set of points on a number line that represents all solutions of the inequality.

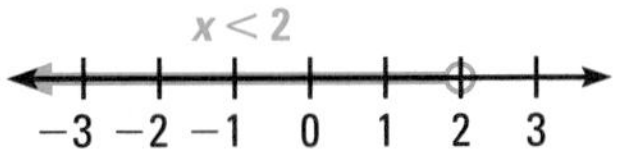

graph of a linear inequality in two variables (p. 360) The graph of all ordered pairs (x, y) that are solutions of the inequality.

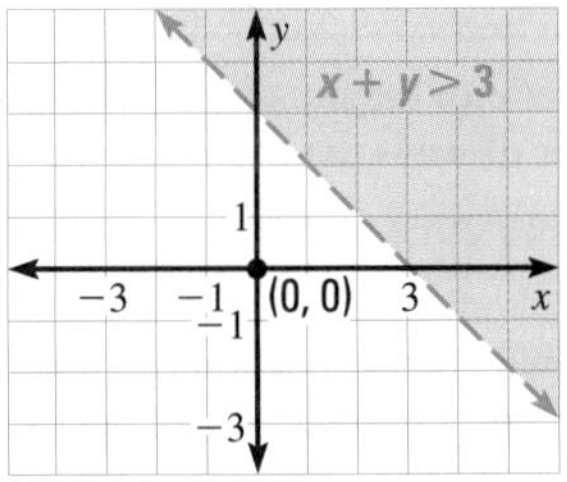

graph of a number (p. 63) The point that corresponds to a number.

graph of an ordered pair (p. 203) The point in the plane that corresponds to an ordered pair (x, y).

graph of a quadratic inequality (p. 548) The graph of all ordered pairs (x, y) that are solutions of the inequality.

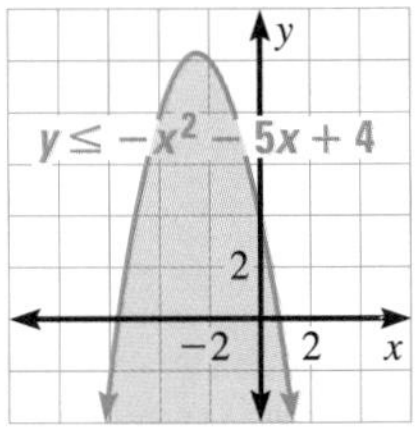

graph of a system of linear inequalities (p. 432) The graph of all solutions of the system.

grouping symbols (p. 10) Symbols such as parentheses ( ) or brackets [ ] that indicate the order in which operations should be performed. Operations within the innermost set of grouping symbols are done first.

growth factor (p. 477) The expression $1 + r$ in the exponential growth model where r is the growth rate.

growth rate (p. 477) The variable r in the exponential growth model. *See also* exponential growth.

H

half-plane (p. 360) In a coordinate plane, the region on either side of a boundary line.

hyperbola (p. 692) The graph of a rational function $f(x) = \frac{a}{x - h} + k$, whose center is (h, k).

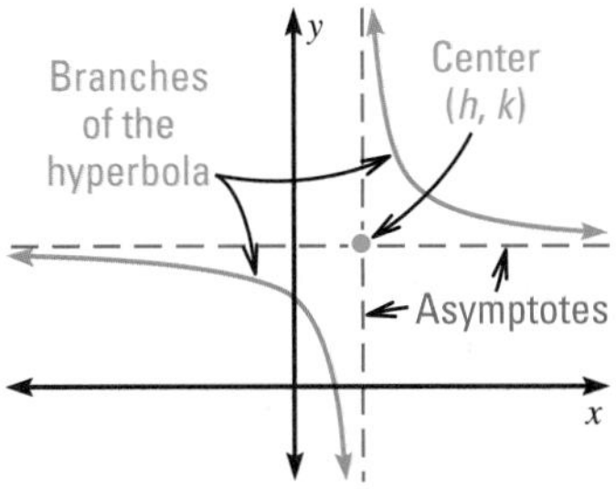

hypotenuse (p. 738) The side opposite the right angle in a right triangle.

hypothesis (pp. 187, 739) The "if" part of an if-then statement. In the statement "If p, then q," p is the hypothesis.

I

identity (p. 155) An equation that is true for all values of the variables.

if-then statement (p. 187) A statement of the form "If p, then q," where p is the *hypothesis* and q is the *conclusion*.

indirect proof (p. 760) A type of proof where it is first assumed that the statement is false. If this assumption leads to an impossibility, then the original statement has been proved to be true.

inductive reasoning (p. 187) A form of reasoning in which a conclusion is reached based on several observations.

inequality (p. 26) A sentence formed when an inequality symbol is placed between two expressions.

initial amount (pp. 477, 484) The variable C in the exponential growth or decay model. *See also* exponential growth *and* exponential decay.

input (p. 46) *See* function.

input-output table (p. 46) A table used to describe a function by listing the inputs and outputs.

integers (p. 63) Any of the numbers ... −3, −2, −1, 0, 1, 2, 3, ...

intersecting lines (p. 426) Two lines that share exactly one point.

inverse operations (p. 132) Operations that undo each other, such as addition and subtraction.

inverse variation (p. 656) The relationship of two variables x and y if there is a nonzero number k such that $xy = k$, or $y = \frac{k}{x}$. The variables x and y *vary inversely*.

irrational number (p. 504) A number that cannot be written as the quotient of two integers.

L

leading coefficient (pp. 505, 576) The coefficient of the first term in a polynomial written in standard form.

least common denominator, LCD (p. 677) The least common multiple of the denominators of two or more fractions.

legs of a right triangle (p. 738) The two sides of a right triangle that are not opposite the right angle.

like terms (p. 102) Terms that have the same variable raised to the same power.

line graph (p. 42) A graph that uses line segments to connect data points. It is especially useful for showing changes in data over time.

linear combination (p. 411) An equation obtained by adding one of two equations (or a multiple of one of the equations) to the other equation in a linear system.

linear equation in one variable (p. 133) An equation in which the variable is raised to the first power and does not occur in a denominator, inside a square root symbol, or inside an absolute value symbol.

linear extrapolation (p. 318) A method of estimating the coordinates of a point that lies to the right or left of all of the given data points.

linear inequality in *x* and *y* (p. 360) An inequality that can be written as follows:

$ax + by < c \qquad ax + by \le c$

$ax + by > c \qquad ax + by \ge c$

linear interpolation (p. 318) A method of estimating the coordinates of a point that lies between two given data points.

linear model (p. 274) A linear function that is used to model a real-life situation. In the linear model $y = mx + b$, m is the rate of change and b is the initial amount.

M

mathematical model (p. 33) An expression, equation, or inequality that represents a real-life situation.

matrix (p. 86) A rectangular arrangement of numbers into horizontal rows and vertical columns. The plural of *matrix* is *matrices.*

column

row

$$\begin{bmatrix} 3 & 1 & 0 & 8 \\ -1 & 2 & 4 & -2 \\ 5 & -7 & 3 & 6 \end{bmatrix}$$

mean or average (p. 369) The sum of n numbers divided by n.

means of a proportion (p. 643) In the proportion $\frac{a}{b} = \frac{c}{d}$, b and c are the means.

measure of central tendency (p. 369) A number that is used to represent a typical number in a data set. *See also* mean, median, *and* mode.

median (p. 369) The middle number of a collection of n numbers when the numbers are written in numerical order. If n is even, the median is taken to be the average of the two middle numbers.

midpoint between two points (p. 747) The midpoint of the line segment connecting them.

midpoint formula (p. 747) The midpoint between (x_1, y_1) and (x_2, y_2) is $\left(\frac{x_1 + x_2}{2}, \frac{y_1 + y_2}{2}\right)$.

mode (p. 369) The number that occurs most frequently in a collection of n numbers. A set of data can have more than one mode or no mode.

modeling (p. 33) Writing algebraic expressions, equations, or inequalities that represent real-life situations.

monomial (p. 576) A polynomial with only one term.

N

negative numbers (p. 63) Any of the numbers less than zero. *See also* real number line.

negative square root (p. 503) One of two square roots of a positive real number.

O

odds (p. 116) The ratio of the number of ways an event can occur to the number of ways the event cannot occur.

open sentence (p. 24) An equation that contains one or more variables.

opposites (p. 65) Two points on a number line that are the same distance from the origin but are on opposite sides of it.

order of operations (p. 16) The rules established to evaluate an expression involving more than one operation.

ordered pair (p. 203) A pair of numbers used to identify a point in a plane. *See also* coordinate plane.

origin of a coordinate plane (p. 203) The point (0, 0) in a coordinate plane at which the horizontal axis intersects the vertical axis. *See also* coordinate plane.

origin of a number line (p. 63) The point labeled zero on a number line.

outcomes (p. 114) The different possible results of a probability experiment.

output (p. 46) *See* function.

P

parabola (p. 518) The U-shaped graph of a quadratic function.

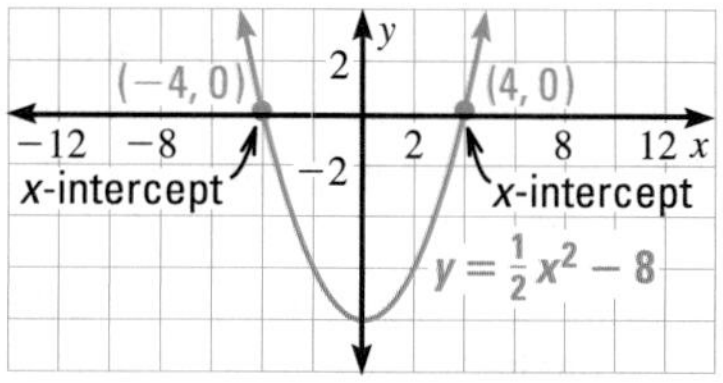

parallel lines (p. 242) Two different lines in the same plane that do not intersect.

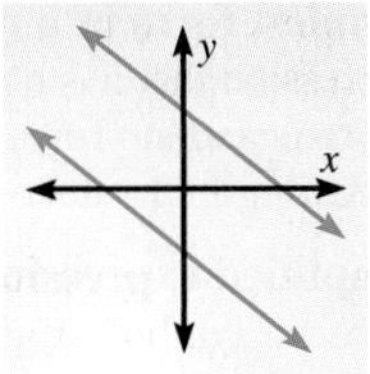

percent of decrease (p. 484) The expression $100r$ is the percent of decrease where r is the decay rate in the exponential decay model.

percent of increase (p. 477) The expression $100r$ is the percent of increase where r is the growth rate in the exponential growth model.

perfect square (p. 504) A number whose square roots are integers or quotients of integers.

perpendicular lines (p. 246) Two nonvertical lines in the same plane such that the slope of one line is the negative reciprocal of the slope of the other.

A vertical line and a horizontal line in the same plane are also perpendicular.

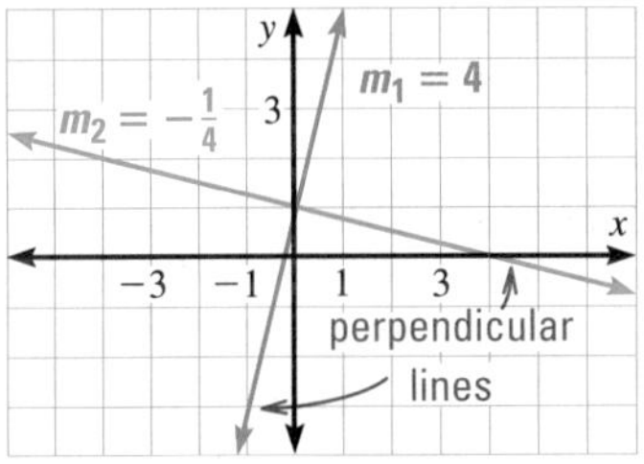

plotting a point (pp. 63, 203) Drawing the point on a number line that corresponds to a number. Drawing the point in a coordinate plane that corresponds to an ordered pair of numbers.

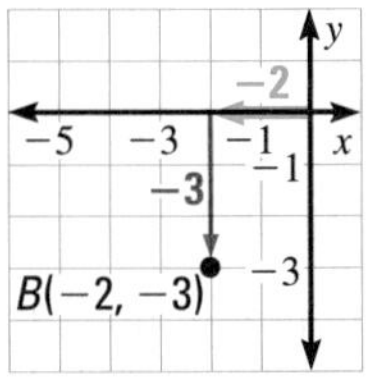

point-slope form (p. 300) The equation of a nonvertical line $y - y_1 = m(x - x_1)$ that passes through a given point (x_1, y_1) with a slope of m.

polynomial (p. 576) An expression which is the sum of terms of the form ax^k where k is a nonnegative integer.

positive numbers (p. 63) Any of the numbers greater than zero. *See also* real number line.

positive square root, or principal square root (p. 503) One of two square roots of a positive real number.

postulates (p. 758) The basic properties of mathematics that mathematicians accept without proof.

power (p. 9) The result of repeated multiplication. For example, in the expression $4^2 = 16$, 16 is the second power of 4.

prime factor (p. 625) A factor that is not the product of polynomials having integer coefficients.

probability of an event (p. 114) A measure of the likelihood that the event will occur due to chance. It is a number between 0 and 1, inclusive.

properties of equality (p. 139) The rules of algebra used to transform equations into equivalent equations.

proportion (p. 643) An equation that states that two ratios are equal. For example, $\frac{a}{b} = \frac{c}{d}$, where a, b, c, and d are nonzero real numbers.

Pythagorean theorem (p. 738) If a triangle is a right triangle, then the sum of the squares of the lengths of the legs a and b equals the square of the length of the hypotenuse c.

$a^2 + b^2 = c^2$

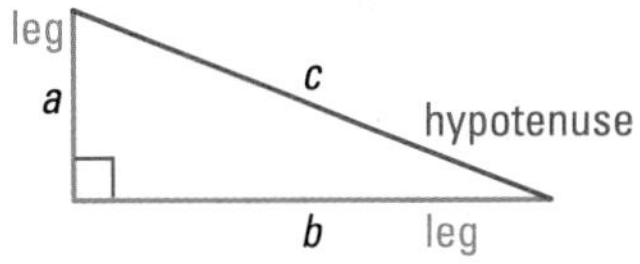

Q

quadrant (p. 203) One of four parts into which the axes divide a coordinate plane. *See also* coordinate plane.

quadratic equation in standard form (p. 505) An equation written in the form $ax^2 + bx + c = 0$, where $a \neq 0$.

quadratic formula (p. 533) The formula used to find the solutions of the quadratic equation $ax^2 + bx + c = 0$, when $a \neq 0$ and $b^2 - 4ac \geq 0$.

$$x = \frac{-b \pm \sqrt{b^2 - 4ac}}{2a}$$

quadratic function in standard form (p. 518) A function written in the form $y = ax^2 + bx + c$, where $a \neq 0$.

quadratic inequality (p. 548) An inequality that can be written as follows:

$y < ax^2 + bx + c$ $\quad$ $y \leq ax^2 + bx + c$
$y > ax^2 + bx + c$ $\quad$ $y \geq ax^2 + bx + c$

quadratic model (p. 554) A function used to model a collection of data or a real-life situation.

Quadratic model: $y = ax^2 + bx + c$

quartiles (p. 375) Three numbers that separate a set of data into four parts.

- The *first quartile* is the median of the lower half of the data.
- The *second quartile* (or median) separates the data into two halves: the numbers that are below the median and the numbers that are above the median.
- The *third quartile* is the median of the upper half of the data.

R

radicand (p. 503) The number or expression inside a radical symbol.

range of a function (p. 47) The collection of all output values.

rate of *a* per *b* (p. 180) The relationship $\frac{a}{b}$ of two quantities a and b that are measured in different units.

847

rate of change (p. 229) A comparison of two different quantities that are changing. Slope provides an important way of visualizing a rate of change.

ratio of *a* to *b* (p. 140) The relationship $\frac{a}{b}$ of two quantities a and b that are measured in the same unit.

rational equation (p. 690) An equation that contains rational expressions.

rational expression (p. 664) A fraction whose numerator, denominator, or both numerator and denominator are nonzero polynomials.

rational function (p. 692) A function of the form $f(x) = \frac{\text{polynomial}}{\text{polynomial}}$.

rational number (p. 664) A number that can be written as the quotient of two integers.

real number line (p. 63) A line that pictures real numbers as points.

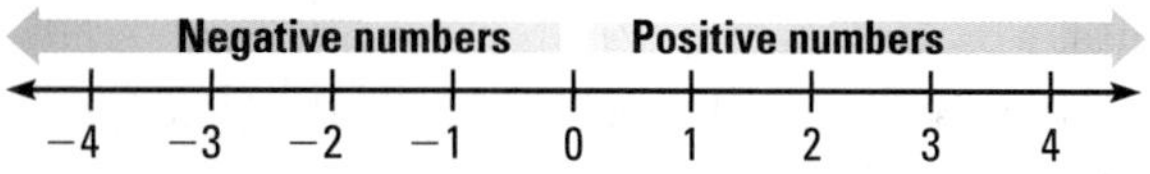

real numbers (p. 63) The set of numbers consisting of the positive numbers, the negative numbers, and zero. *See also* real number line.

reciprocal (p. 108) If $\frac{a}{b}$ is a nonzero number, then its reciprocal is $\frac{b}{a}$. The product of a number and its reciprocal is 1.

relation (p. 256) Any set of ordered pairs (x, y).

roots of a quadratic equation (p. 526) The solutions of $ax^2 + bx + c = 0$.

round-off error (p. 166) The error produced when a decimal result is rounded in order to provide a meaningful answer.

S

scalar multiplication (p. 97) Multiplication of a matrix by a real number.

scatter plot (p. 204) A graph of pairs of numbers that represent real-life situations. It is a way to analyze the relationship between two quantities.

scientific notation (p. 470) A number expressed in the form $c \times 10^n$, where $1 \le c < 10$ and n is an integer.

similar triangles (p. 140) Two triangles are similar if they have equal corresponding angles. It can be shown that this is equivalent to the ratios of the lengths of the corresponding sides being equal.

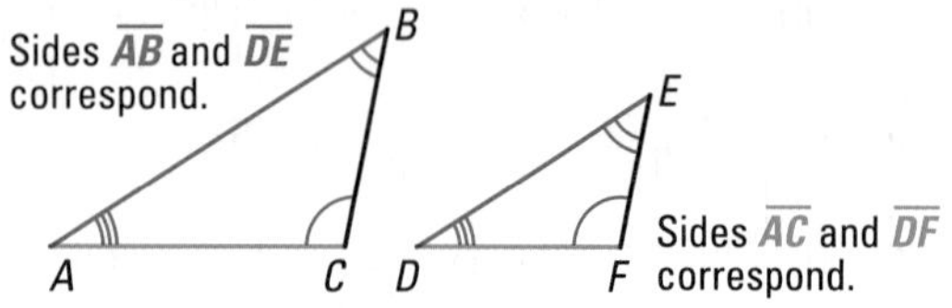

simplest form of a radical expression (p. 512) An expression that has no perfect square factors other than 1 in the radicand, no fractions in the radicand, and no radicals appearing in the denominator of a fraction.

simplified expression (p. 102) An expression is simplified if it has no symbols of grouping and if all the like terms have been combined.

simplified rational expression (p. 664) A rational expression is simplified if its numerator and denominator have no factors in common (other than ± 1).

sine (p. 752) *See* trigonometric ratio.

slope (p. 226) The number of units a nonvertical line rises or falls for each unit of horizontal change from left to right. The slope m is $m = \frac{y_2 - y_1}{x_2 - x_1}$.

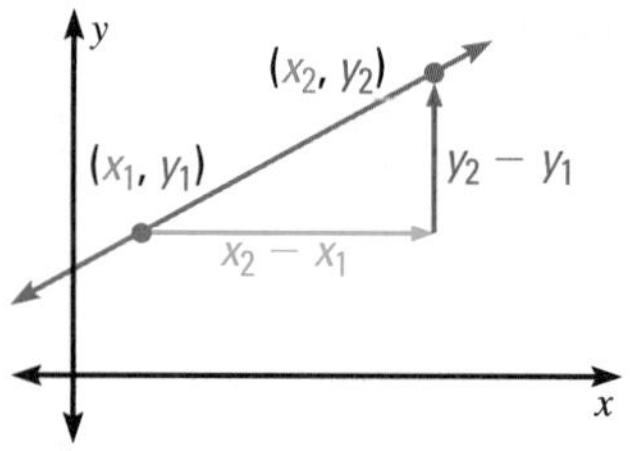

slope-intercept form (pp. 241, 273) A linear equation written in the form $y = mx + b$. The slope of the line is m. The y-intercept is b. *See also* slope *and* y-intercept.

$y = 2x + 3$
Slope is 2.
y-intercept is 3.

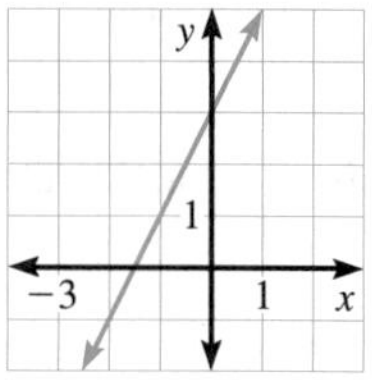

solution of an equation (p. 24) A number that, when substituted for the variable in an equation, results in a true statement.

solution of an inequality (p. 26) A number that, when substituted for the variable in an inequality, results in a true statement.

solution of a linear equation (p. 210) An ordered pair (x, y) is a solution of a linear equation if the equation is true when the values of x and y are substituted into the equation.

solution of a linear inequality (p. 360) An ordered pair (x, y) is a solution of a linear inequality if the inequality is true when the values of x and y are substituted into the inequality.

solution step (p. 133) The result of applying a transformation to an equation when solving the equation.

solution of a system of linear equations in two variables (p. 398) An ordered pair (x, y) that satisfies each equation in the system.

solution of a system of linear inequalities (p. 432) An ordered pair (x, y) that is a solution of each inequality in the system.

solving an equation (p. 25) Finding all the solutions of an equation.

solving a right triangle (p. 753) Finding the lengths of the other two sides of a right triangle, given the measure of one acute angle and the length of one side of the right triangle.

square root (p. 503) If $b^2 = a$, then b is a square root of a. Square roots are written with a radical symbol $\sqrt{\ }$.

square root function (p. 709) The function defined by $y = \sqrt{x}$.

standard form of an equation of a line (p. 308) A linear equation of the form $Ax + By = C$ where A, B, and C are real numbers and A and B are not both zero.

standard form of a polynomial (p. 576) A polynomial whose terms are placed in descending order, from largest degree to smallest degree.

stem-and-leaf plot (p. 368) An arrangement of digits that is used to display and order numerical data.

system of linear equations (p. 398) Two or more linear equations in the same variables. This is also called a linear system.

system of linear inequalities (p. 432) Two or more linear inequalities in the same variables. This is also called a system of inequalities.

T

tangent (p. 752) *See* trigonometric ratio.

terms of an expression (p. 80) The parts that are added in an expression. For example, in the expression $5 - x$, the terms are 5 and $-x$.

theorem (pp. 738, 759) A statement that can be proven to be true.

theoretical probability (p. 114) A type of probability that is based on the number of favorable outcomes divided by the total number of outcomes.

time period (p. 477) The variable t in the exponential growth and decay models. *See also* exponential growth *and* exponential decay.

transform an equation (p. 132) To change an equation into an equivalent equation.

trigonometric ratio (p. 752) The ratio of the lengths of two sides of a right triangle. For example:

$$\sin A = \frac{\text{side opposite } \angle A}{\text{hypotenuse}}$$

$$\cos A = \frac{\text{side adjacent to } \angle A}{\text{hypotenuse}}$$

$$\tan A = \frac{\text{side opposite } \angle A}{\text{side adjacent to } \angle A}$$

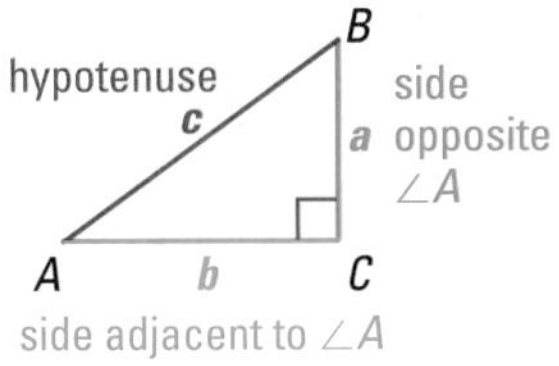

trinomial (p. 576) A polynomial with three terms.

U

unit analysis (p. 5) Writing the units of each variable in a real-life problem to help determine the units for the answer.

unit rate (p. 180) A rate per one given unit.

V

values (p. 3) The numbers represented by variables.

variable (p. 3) A letter that is used to represent one or more numbers.

variable expression (p. 3) A collection of numbers, variables, and operations.

velocity (p. 66) The speed and direction in which an object is traveling (up is positive and down is negative). The speed of an object is the absolute value of its velocity.

verbal model (p. 5) An expression that uses words to represent a real-life situation.

vertex of a parabola (p. 518) The lowest point of a parabola that opens up or the highest point of a parabola that opens down. *See also* parabola.

vertical motion models (p. 535) A model for the height of a falling object that is dropped and the model for an object thrown down or up.

X

***x*-axis** (p. 203) The horizontal axis in a coordinate plane. *See also* coordinate plane.

***x*-coordinate** (p. 203) The first number in an ordered pair. *See also* ordered pair *and* plotting a point.

***x*-intercept** (p. 218) The x-coordinate of a point where a graph crosses the x-axis.

Y

***y*-axis** (p. 203) The vertical axis in a coordinate plane. *See also* coordinate plane.

***y*-coordinate** (p. 203) The second number in an ordered pair. *See also* ordered pair *and* plotting a point.

***y*-intercept** (p. 218) The y-coordinate of a point where a graph crosses the y-axis.

Z

zero-product property (p. 597) If a and b are real numbers and $ab = 0$, then $a = 0$ or $b = 0$.

English-to-Spanish Glossary

A

absolute value (p. 65) **valor absoluto** La distancia entre el origen y el punto que representa el número real. El símbolo $|a|$ representa el valor absoluto de un número a.

algorithm (p. 39) **algoritmo** Un proceso para resolver un problema que se lleva a cabo paso a paso.

asymptote (p. 692) **asíntota** Una recta a la que se acerca una gráfica. La distancia entre la gráfica y la recta va acercándose a cero.

axioms (p. 758) **axiomas** Las propiedades fundamentales de las matemáticas que los matemáticos aceptan sin pruebas.

axis of symmetry of a parabola (p. 518) **eje de simetría de una parábola** La recta que al pasar por el vértice divide una parábola en dos partes simétricas. Las dos partes simétricas son imágenes especulares entre sí. *Ver también* parabola / parábola.

B

bar graph (p. 41) **gráfica de barras** Una gráfica que organiza un conjunto de datos mediante el uso de barras horizontales o verticales para mostrar cuántas veces cada suceso o número ocurre en el conjunto.

base number of a percent equation (p. 649) **número básico de una ecuación de porcentaje** El número *al que* se compara en una ecuación de porcentaje. El número b del modelo verbal: a es el porcentaje p de b.

base of a power (p. 9) **base de una potencia** El número o variable que se usa como factor en la multiplicación repetida. Por ejemplo, en la expresión 4^6, 4 es la base.

best-fitting line (p. 292) **recta que mejor se ajusta** La recta que se ajusta mejor a los puntos de datos de un diagrama de dispersión.

binomial (p. 576) **binomio** Un polinomio con dos términos.

box-and-whisker plot (p. 375) **gráfica de frecuencias acumuladas** Una manera de mostrar datos que divide un conjunto de datos en cuatro partes. La caja representa la mitad de los datos. Los segmentos se extienden hacia los datos mayor y menor.

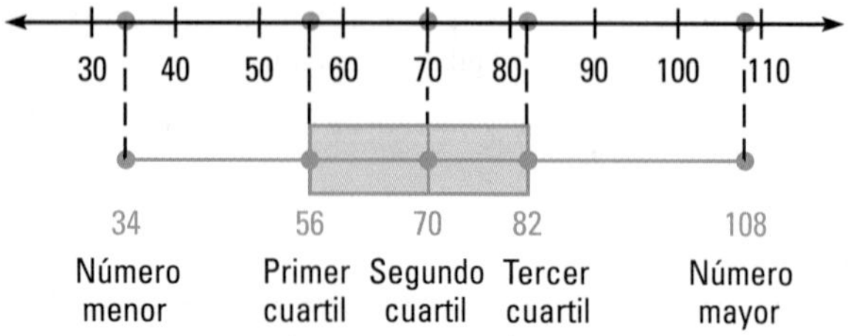

C

center of a hyperbola (p. 692) **centro de una hipérbola** El punto (h, k) de la gráfica de la función racional $f(x) = \frac{a}{x - h} + k$.

closed set (p. 113) **conjunto cerrado** Un conjunto de números se cierra bajo una operación si al aplicar la operación a dos números cualesquiera en el conjunto resulta otro número contenido en el conjunto. Se dice que tal conjunto está cerrado.

coefficient (p. 102) **coeficiente** Un número multiplicado por una variable en un término. El número es el coeficiente de la variable.

completing the square (p. 730) **completar el cuadrado** El proceso de volver a escribir una ecuación cuadrática de manera que un lado sea un trinomio cuadrado perfecto.

compound inequality (p. 346) **desigualdad compuesta** Dos desigualdades conectadas por *y* u *o*.

conclusion (pp. 187, 739) **conclusión** La parte "entonces" de un enunciado "si…, entonces…" En el enunciado "si p, entonces q", q es la conclusión.

conjecture (p. 759) **conjetura** Un enunciado que se piensa es verdadero pero que todavía no ha sido probado. A menudo se trata de un enunciado basado en la observación.

conjugates (p. 717) **conjugados** Las expresiones $(a + \sqrt{b})$ y $(a - \sqrt{b})$ son conjugados entre sí.

consecutive integers (p. 149) **números enteros consecutivos** Números enteros que van seguidos. Por ejemplo, 4, 5, 6.

constant terms (p. 102) **términos constantes** Términos sin factores variables. Por ejemplo, en $x + 2 - 5x - 4$, los términos constantes son 2 y -4.

constant of variation (p. 234) **constante de variación** La constante de un modelo de variación. Se representa con la variable k. *Ver también* direct variation / variación directa *e* inverse variation / variación inversa.

converse of a statement (p. 739) **recíproco de un enunciado** Un enunciado relacionado en el que se intercambian la hipótesis y la conclusión. El recíproco del enunciado "Si p, entonces q", es "Si q, entonces p".

coordinate plane (p. 203) **plano de coordenadas** Un plano formado por dos rectas numéricas reales que se cortan en un ángulo recto.

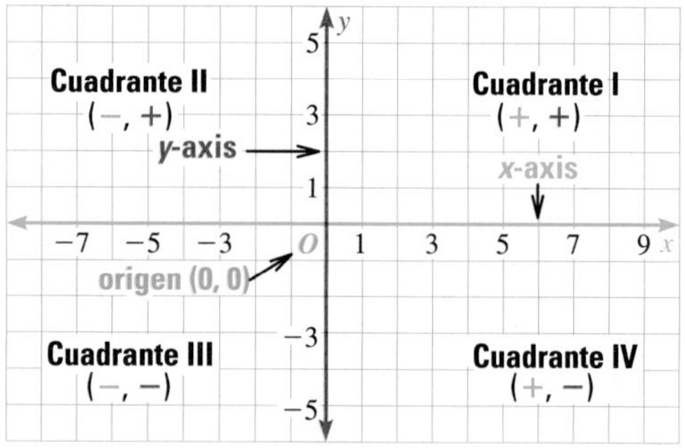

correlation (p. 295) **correlación** La relación entre dos conjuntos de datos.

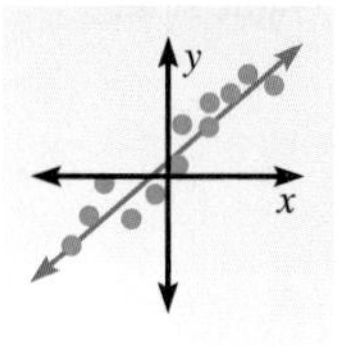

Correlación positiva

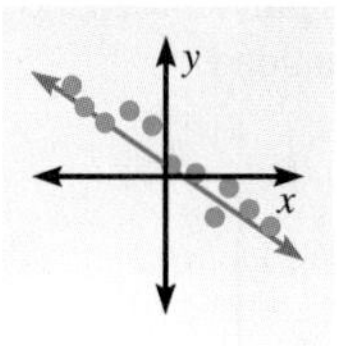

Correlación negativa

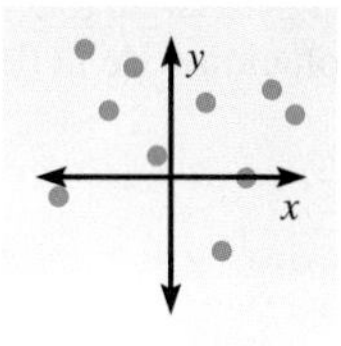

Ninguna correlación

cosine (p. 752) **coseno** *Ver* trigonometric ratio / razón trigonométrica

counterexample (p. 66) **contraejemplo** Un ejemplo que se usa para mostrar que un enunciado es falso.

D

data (p. 40) **datos** Información, hechos o números que se usan para describir algo.

decay factor (p. 484) **factor de decrecimiento** La expresión $1 - r$ del modelo de decrecimiento donde r es la tasa de decrecimiento. *Ver también* decay rate / tasa de decrecimiento.

decay rate (p. 484) **tasa de decrecimiento** La variable r del modelo de decrecimiento exponencial. *Ver también* exponential decay / decrecimiento exponencial.

deductive reasoning (p. 187) **razonamiento deductivo** Razonamiento mediante el que se llega a una conclusión basándose en datos, definiciones, reglas o propiedades.

degree of a polynomial (p. 576) **grado de un polinomio** El mayor grado de los términos del polinomio.

degree of a term (p. 576) **grado de un término** El exponente de la variable del término.

direct variation (pp. 234, 656) **variación directa** La relación entre dos variables x e y cuando hay un número k distinto a cero donde $y = kx$, o $\frac{y}{x} = k$. Las variables x e y varían directamente.

discriminant (p. 541) **discriminante** La expresión $b^2 - 4ac$ donde a, b y c son coeficientes de la ecuación cuadrática $ax^2 + bx + c = 0$.

distance formula (p. 745) **fórmula de la distancia** La distancia d entre los puntos (x_1, y_1) y (x_2, y_2) es

$$d = \sqrt{(x_2 - x_1)^2 + (y_2 - y_1)^2}.$$

distributive property (p. 100) **propiedad distributiva** El producto de a y $(b + c)$: $a(b + c) = ab + ac$ o $(b + c)a = ba + ca$. El producto de a y $(b - c)$: $a(b - c) = ab - ac$ o $(b - c)a = ba - ca$.

domain of a function (p. 47) **dominio de una función** El conjunto de todos los valores de entrada.

E

entry or element of a matrix (p. 86) **entrada o elemento de una matriz** Cada uno de los números de la matriz. *Ver también* matrix / matriz.

equation (p. 24) **ecuación** Un enunciado formado cuando un signo de igualdad se pone entre dos expresiones.

equivalent equations (p. 132) **ecuaciones equivalentes** Ecuaciones con la misma solución que la ecuación original.

equivalent inequalities (p. 335) **desigualdades equivalentes** Desigualdades con la misma solución o las mismas soluciones.

evaluating the expression (p. 3) **evaluar la expresión** Hallar el valor de una expresión mediante la sustitución de una variable por un número.

event (p. 114) **suceso** Un conjunto de casos.

experimental probability (p. 115) **probabilidad experimental** Una probabilidad que se basa en repeticiones de un experimento real.

exponent (p. 9) **exponente** El número o variable que representa el número de veces que la base se usa como factor. Por ejemplo, en la expresión 4^6, 6 es el exponente.

exponential decay (p. 484) **decrecimiento exponencial** Una cantidad que disminuye en el mismo porcentaje durante cada unidad de tiempo t donde C es la cantidad inicial.

Modelo de decrecimiento exponencial: $y = C(1 - r)^t$

exponential function (p. 458) **función exponencial** Una función que tiene la forma $y = ab^x$, donde $b > 0$ y $b \neq 1$.

exponential growth (p. 477) **crecimiento exponencial** Una cantidad que aumenta en el mismo porcentaje durante cada unidad de tiempo t donde C es la cantidad inicial.

Modelo de crecimiento exponencial: $y = C(1 + r)^t$

extraneous solution (pp. 644, 723) **solución extraña** Una solución tentativa que no satisface la ecuación original.

extremes of a proportion (p. 643) **extremos de una proporción** En la proporción $\frac{a}{b} = \frac{c}{d}$, a y d son los extremos.

F

factor (p. 777) **factor** Los números y las variables que se multiplican en una expresión. Por ejemplo, 4 y 9 son factores de 36 y 6 y x son factores de $6x$.

factor a polynomial completely (p. 625) **factorizar un polinomio completamente** Escribir un polinomio como el producto de:

- factores monomiales
- factores primos con por lo menos dos términos

factor a quadratic expression (p. 604) **factorizar una expresión cuadrática** Escribir una expresión cuadrática como el producto de dos expresiones lineales.

factored form of a polynomial (pp. 597, 625) **forma factorizada de un polinomio** Un polinomio que se escribe como el producto de dos o más factores primos.

favorable outcomes (p. 114) **casos favorables** Los casos de un suceso determinado que se están considerando. *Ver también* outcomes / casos.

FOIL pattern (p. 585) **patrón FOIL** Un patron que se usa para multiplicar dos binomios. Se multiplican los primeros términos (*First*), los términos exteriores (*Outer*), los interiores (*Inner*) y los últimos (*Last*).

Por ejemplo, $(x + 4)(2x + 3) = 2x^2 + 3x + 8x + 12$
$= 2x^2 + 11x + 12$

formula (p. 174) **fórmula** Una ecuación algebraica que relaciona dos o más cantidades de la vida real.

function (p. 46) **función** Una regla que establece la relación entre dos cantidades, llamadas entrada y salida. Para cada entrada, hay exactamente una salida.

function form (p. 176) **forma de función** Una ecuación de dos variables se escribe en forma de función si una de sus variables se aísla en un lado de la ecuación. La variable aislada es la salida y es una función de la entrada.

function notation (p. 257) **notación de función** Una manera de nombrar una función definida por una ecuación. Para una ecuación en x e y, el símbolo $f(x)$ reemplaza a y y se lee como "el valor de f en x", o simplemente como "f de x".

G

generalization (p. 187) **generalización** Una conclusión basada en varias observaciones.

geometric mean (p. 723) **media geométrica** La media geométrica de a y b es $\sqrt{ab}$.

geometric probability (p. 666) **probabilidad geométrica** La probabilidad P de que un objeto lanzado en la región A caiga en la región B cuando la región B se sitúa dentro de la región A es:

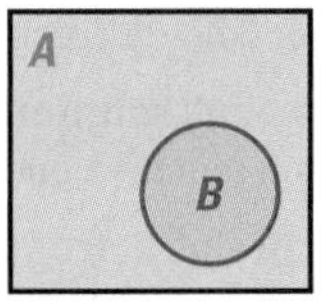

$$P = \frac{\text{Área de la región } B}{\text{Área de la región } A}$$

graph of an equation in two variables (p. 210) **gráfica de una ecuación de dos variables** El conjunto de todos los puntos (x, y) que sean soluciones de la ecuación.

graph of a function (p. 257) **gráfica de una función** El conjunto de todos los puntos $(x, f(x))$, donde x está en el dominio de la función.

graph of a linear inequality in one variable (p. 334) **gráfica de una desigualdad lineal de una variable** El conjunto de todos los puntos de una recta numérica que representen todas las soluciones de la desigualdad.

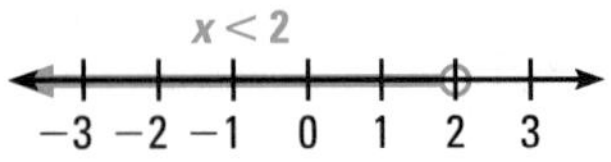

graph of a linear inequality in two variables (p. 360) **gráfica de una desigualdad lineal de dos variables** La gráfica de todos los pares ordenados (x, y) que sean soluciones de la desigualdad.

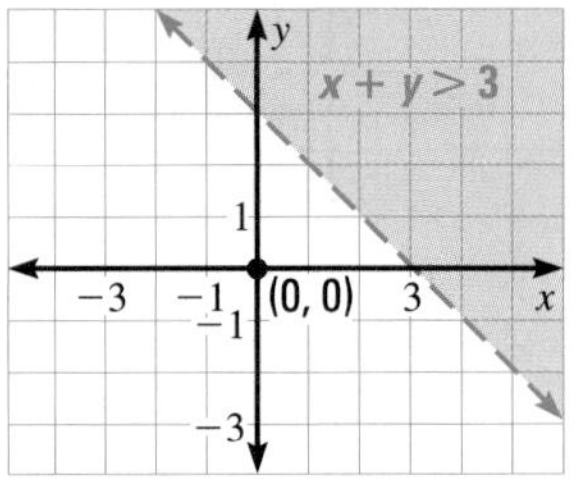

graph of a number (p. 63) **gráfica de un número** El punto que corresponde a un número.

graph of an ordered pair (p. 203) **gráfica de un par ordenado** El punto del plano que corresponde a un par ordenado (x, y).

graph of a quadratic inequality (p. 548) **gráfica de una desigualdad cuadrática** La gráfica de todos los pares ordenados (x, y) que sean soluciones de la desigualdad.

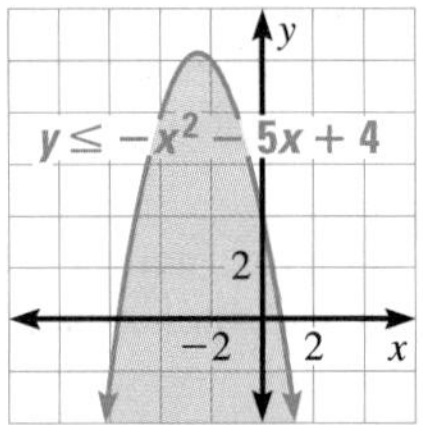

graph of a system of linear inequalities (p. 432) **gráfica de un sistema de desigualdades lineales** La gráfica de todas las soluciones del sistema.

grouping symbols (p. 10) **signos de agrupación** Signos tales como paréntesis ( ) o corchetes [ ] que indican el orden en que las operaciones deben realizarse. Se realizan primero las operaciones dentro de los signos más interiores.

growth factor (p. 477) **factor de crecimiento** La expresión $1 + r$ del modelo de crecimiento exponencial donde r es la tasa de crecimiento.

growth rate (p. 477) **tasa de crecimiento** La variable r del modelo de crecimiento exponencial. *Ver también* exponential growth / crecimiento exponencial.

H

half-plane (p. 360) **semiplano** En un plano de coordenadas, la región de un lado o del otro lado de una recta límite.

hyperbola (p. 692) **hipérbola** La gráfica de una función racional $f(x) = \frac{a}{x - h} + k$, cuyo centro es (h, k).

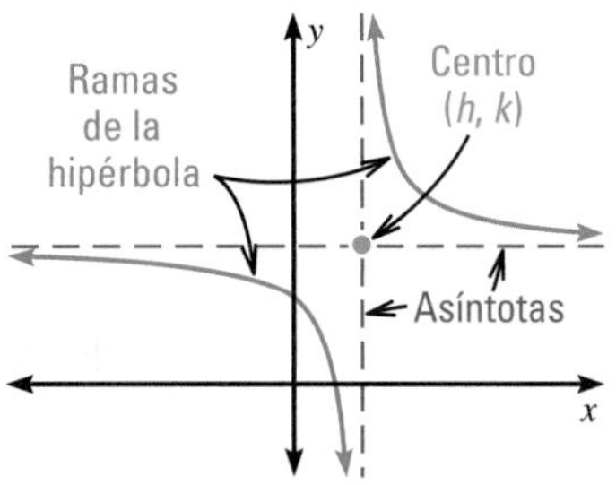

hypotenuse (p. 738) **hipotenusa** El lado opuesto al ángulo recto de un triángulo rectángulo.

hypothesis (pp. 187, 739) **hipótesis** La parte "si…," de un enunciado "si…, entonces…". En el enunciado "Si *p*, entonces *q*", *p* es la hipótesis.

I

identity (p. 155) **identidad** Una ecuación que es verdadera para todos los valores de las variables.

if-then statement (p. 187) **enunciado de "si…, entonces**…" Un enunciado de la forma "Si *p*, entonces *q*" donde *p* es la hipótesis y *q* es la conclusión.

indirect proof (p. 760) **prueba indirecta** Un tipo de prueba donde primero se asume que el enunciado es falso. Si esto lleva a una imposibilidad, entonces se ha probado que el enunciado original es verdadero.

inductive reasoning (p. 187) **razonamiento inductivo** Una forma de razonamiento en el que se llega a una conclusión basándose en varias observaciones.

inequality (p. 26) **desigualdad** Una oración formada al colocar un signo de desigualdad entre dos expresiones.

initial amount (pp. 477, 484) **cantidad inicial** La variable *C* del modelo de crecimiento o decrecimiento exponencial. *Ver también* exponential growth / crecimiento exponencial y exponential decay / decrecimiento exponencial.

input (p. 46) **entrada** *Ver* function / función.

input-output table (p. 46) **tabla de entradas y salidas** Una tabla usada para describir una función mediante una lista de las entradas y salidas.

integers (p. 63) **números enteros** Cualquiera de los números ... $-3, -2, -1, 0, 1, 2, 3$, ...

intersecting lines (p. 426) **rectas que se cortan, rectas secantes** Dos rectas que comparten exactamente un punto.

inverse operations (p. 132) **operaciones inversas** Operaciones que se anulan entre sí, como la suma y la resta.

inverse variation (p. 656) **variación inversa** La relación entre dos variables *x* e *y* si hay un número *k* distinto a cero tal que $xy = k$, o $y = \frac{k}{x}$. Las variables *x* e *y* varían inversamente.

irrational number (p. 504) **número irracional** Un número que no puede escribirse como el cociente de dos números enteros.

L

leading coefficient (pp. 505, 576) **coeficiente inicial** El coeficiente del primer término de un polinomio escrito en forma general.

least common denominator, LCD (p. 677) **mínimo común denominador, m.c.d**. El mínimo común múltiplo de los denominadores de dos o más fracciones.

legs of a right triangle (p. 738) **catetos de un triángulo rectángulo** Los dos lados de un triángulo rectángulo que no están opuestos al ángulo recto.

like terms (p. 102) **términos semejantes** Términos que tienen la misma variable elevada a la misma potencia.

line graph (p. 42) **gráfica lineal** Una gráfica que usa segmentos de recta para conectar puntos de datos. Sobre todo es útil para mostrar cambios en los datos a través del tiempo.

linear combination (p. 411) **combinación lineal** Una ecuación que se obtiene al sumar una de dos ecuaciones (o un múltiplo de una de dos ecuaciones) a la otra ecuación de un sistema lineal.

linear equation in one variable (p. 133) **ecuación lineal de una variable** Una ecuación en la que la variable se eleva a la primera potencia y que no se encuentra en un denominador, ni dentro del signo de raíz cuadrada, ni dentro de un símbolo de valor absoluto.

linear extrapolation (p. 318) **extrapolación lineal** Un método de estimar las coordenadas de un punto que está situado a la derecha o a la izquierda de todos los puntos de los datos mostrados.

linear inequality in *x* and *y* (p. 360) **desigualdad lineal con *x* e *y*** Una desigualdad que puede escribirse como sigue:

$ax + by < c \quad ax + by \le c$

$ax + by > c \quad ax + by \ge c$

linear interpolation (p. 318) **interpolación lineal** Un método de estimar las coordenadas de un punto que está situado entre dos puntos de los datos dados.

linear model (p. 274) **modelo lineal** Una función lineal que se usa para representar una situación de la vida real. En el modelo lineal $y = mx + b$, *m* es la razón de cambio y *b* es la cantidad inicial.

M

mathematical model (p. 33) **modelo matemático** Una expresión, ecuación o desigualdad que representa una situación de la vida real.

matrix (p. 86) **matriz** Un arreglo rectangular de números en filas horizontales y columnas verticales.

columna (arrow to third column)

fila (arrow to second row)

$$\begin{bmatrix} 3 & 1 & 0 & 8 \\ -1 & 2 & 4 & -2 \\ 5 & -7 & 3 & 6 \end{bmatrix}$$

mean or average (p. 369) **media o promedio** La suma de *n* números dividida por *n*.

means of a proportion (p. 643) **medios de una proporción** En la proporción $\frac{a}{b} = \frac{c}{d}$, *b* y *c* son los medios.

measure of central tendency (p. 369) **medida de tendencia central** Un número que se usa para representar un número típico de un conjunto de datos. *Ver también* mean / media, median / mediana *y* mode / moda.

853

median (p. 369) **mediana** El número en el medio de un conjunto de n números cuando los números se escriben en orden numérico. Si n es par, es el promedio de los dos números en el medio del grupo.

midpoint between two points (p. 747) **punto medio entre dos puntos** El punto medio del segmento de recta que los conecta.

midpoint formula (p. 747) **fórmula del punto medio** El punto medio entre (x_1, y_1) y (x_2, y_2) es $\left(\frac{x_1 + x_2}{2}, \frac{y_1 + y_2}{2}\right)$.

mode (p. 369) **moda** El número que ocurre con más frecuencia en un conjunto de n números. Un conjunto de datos puede tener más de una moda o ninguna.

modeling (p. 33) **hacer un modelo, representar** Escribir expresiones algebraicas o desigualdades que representan situaciones de la vida real.

monomial (p. 576) **monomio** Un polinomio con un solo término.

N

negative numbers (p. 63) **números negativos** Cualquiera de los números menos de cero. *Ver también* real number line / recta numérica real.

negative square root (p. 503) **raíz cuadrada negativa** Una de las dos raíces cuadradas de un número positivo real.

O

odds (p. 116) **probabilidad de ocurrencia** La razón entre el número de maneras de que puede ocurrir un suceso y el número de maneras de que no puede ocurrir.

open sentence (p. 24) **expresión con variables** Una ecuación que contiene una o más variables.

opposites (p. 65) **opuestos** Dos puntos de una recta numérica que están a la misma distancia del origen pero situados en lados opuestos.

order of operations (p. 16) **orden de las operaciones** Las reglas establecidas para evaluar una expresión cuando hay que realizar más de una operación.

ordered pair (p. 203) **par ordenado** Un par de números que se usan para identificar un punto de un plano. *Ver también* coordinate plane / plano de coordenadas.

origin of a coordinate plane (p. 203) **origen de un plano de coordenadas** El punto (0, 0) de un plano de coordenadas donde el eje horizontal corta el eje vertical. *Ver también* coordinate plane / plano de coordenadas.

origin of a number line (p. 63) **origen de una recta numérica** El punto marcado cero de una recta numérica.

outcomes (p. 114) **casos** Los posibles resultados de un experimento de probabilidad.

output (p. 46) **salida** *Ver* function/función.

P

parabola (p. 518) **parábola** La gráfica con forma de U de una función cuadrática.

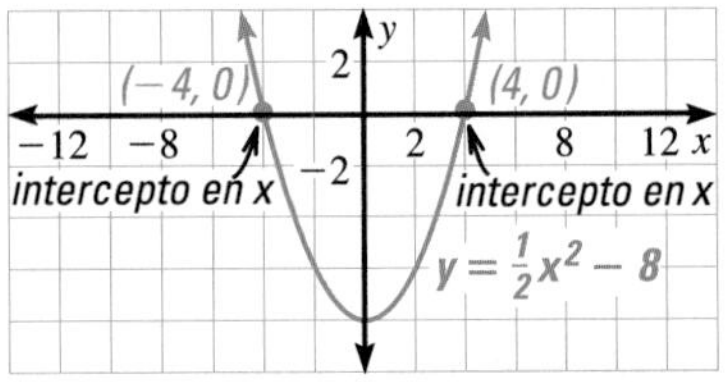

parallel lines (p. 242) **rectas paralelas** Dos rectas diferentes del mismo plano que no se cortan.

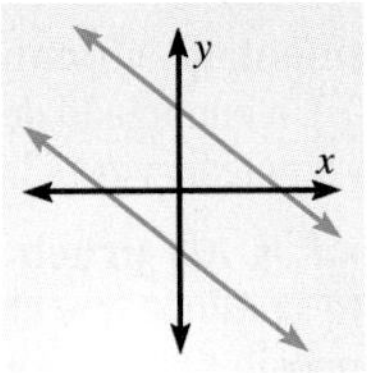

percent of decrease (p. 484) **porcentaje de disminución** La expresión $100r$ es el porcentaje de disminución donde r es la tasa de decrecimiento del modelo de decrecimiento exponencial.

percent of increase (p. 477) **porcentaje de aumento** La expresión $100r$ es el porcentaje de aumento donde r es la tasa de crecimiento del modelo de crecimiento exponencial.

perfect square (p. 504) **cuadrado perfecto** Un número cuyas raíces cuadradas son números enteros o cocientes de números enteros.

perpendicular lines (p. 246) **rectas perpendiculares** Dos rectas no verticales de un plano situadas de tal manera que la pendiente de una es el recíproco negativo de la otra.

Una recta vertical y una recta horizontal situadas en el mismo plano también son perpendiculares.

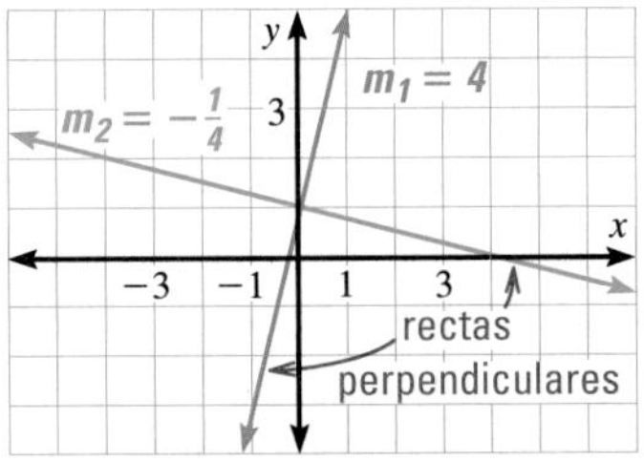

plotting a point (pp. 63, 203) **marcar un punto** Dibujar el punto en una recta numérica que corresponde a un número. Dibujar el punto en un plano de coordenadas que corresponde a un par ordenado de números.

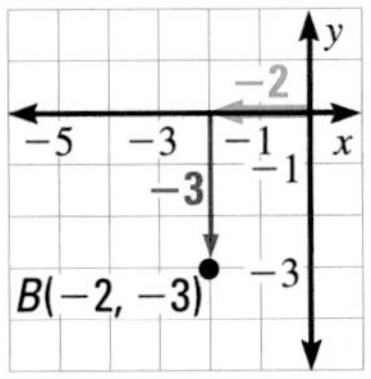

point-slope form (p. 300) **ecuación punto-pendiente** La ecuación de una recta no vertical $y - y_1 = m(x - x_1)$ que pasa por el punto dado (x_1, y_1) con una pendiente de m.

polynomial (p. 576) **polinomio** Una expresión que es la suma de los términos de la forma ax^k donde k es un número entero no negativo.

positive numbers (p. 63) **números positivos** Cualquiera de los números mayores que cero. *Ver también* real number line / recta numérica real.

positive square root, or principal square root (p. 503) **raíz cuadrada positiva o principal** Una de dos raíces cuadradas de un número real positivo.

postulates (p. 758) **postulados** Las propiedades básicas de las matemáticas que los matemáticos aceptan sin pruebas.

power (p. 9) **potencia** El resultado de la multiplicación repetida. Por ejemplo, en la expresión $4^2 = 16$, 16 es la segunda potencia de 4.

prime factor (p. 625) **factor primo** Un factor que no es el producto de polinomios con coeficientes que son enteros.

probability of an event (p. 114) **probabilidad de un suceso** Una medida de las probabilidades de que el suceso ocurra al azar. Es un número entre 0 y 1, inclusive.

properties of equality (p. 139) **propiedades de igualdad** Reglas de álgebra usadas para transformar ecuaciones en ecuaciones equivalentes.

proportion (p. 643) **proporción** Una ecuación que enuncia que dos razones son iguales. Por ejemplo, $\frac{a}{b} = \frac{c}{d}$, donde *a, b, c* y *d* son números reales distintos a cero.

Pythagorean theorem (p. 738) **teorema de Pitágoras** Si un triángulo es rectángulo, entonces la suma de los cuadrados de las longitudes de los catetos *a* y *b* es igual al cuadrado de la longitud de la hipotenusa *c*.

$a^2 + b^2 = c^2$

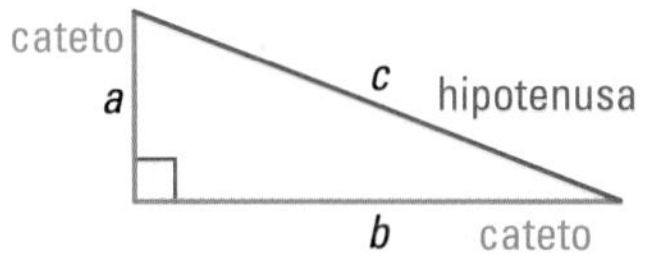

Q

quadrant (p. 203) **cuadrante** Una de las cuatro partes en las que los ejes dividen un plano de coordenadas. *Ver también* coordinate plane / plano de coordenadas.

quadratic equation in standard form (p. 505) **ecuación cuadrática en forma general** Una ecuación escrita en la forma $ax^2 + bx + c = 0$, donde $a \neq 0$.

quadratic formula (p. 533) **fórmula cuadrática** La fórmula que se usa para hallar las soluciones de la ecuación cuadrática $ax^2 + bx + c = 0$, donde $a \neq 0$ y $b^2 - 4ac \geq 0$.

$$x = \frac{-b \pm \sqrt{b^2 - 4ac}}{2a}$$

quadratic function in standard form (p. 518) **función cuadrática en forma general** Una función escrita en la forma $y = ax^2 + bx + c$, donde $a \neq 0$.

quadratic inequality (p. 548) **desigualdad cuadrática** Una desigualdad que puede escribirse así:

$y < ax^2 + bx + c \qquad y \leq ax^2 + bx + c$

$y > ax^2 + bx + c \qquad y \geq ax^2 + bx + c$

quadratic model (p. 554) **modelo cuadrático** Una función que se usa para representar un conjunto de datos o una situación de la vida real.

Modelo cuadrático: $y = ax^2 + bx + c$

quartiles (p. 375) **cuartiles** Tres números que separan un conjunto de datos en cuatro partes.

- El primer cuartil es la mediana de la mitad inferior de los datos.
- El segundo cuartil (o mediana) separa los datos en dos mitades: los números por debajo de la mediana y los números por encima de la mediana.
- El tercero es la mediana de la mitad superior de los datos.

R

radicand (p. 503) **radicando** El número o expresión dentro del símbolo radical.

range of a function (p. 47) **rango de una función** El conjunto de todos los valores de salida.

rate of *a* per *b* (p. 180) **relación de *a* por *b*** La relación $\frac{a}{b}$ entre dos cantidades *a* y *b* que se miden con unidades distintas.

rate of change (p. 229) **razón de cambio** Una comparación de dos cantidades diferentes que están cambiando. La pendiente sirve de modelo importante para visualizar la razón de cambio.

ratio of *a* to *b* (p. 140) **razón de *a* a *b*** La relación $\frac{a}{b}$ entre dos cantidades *a* y *b* que se miden con las mismas unidades.

rational equation (p. 690) **ecuación racional** Una ecuación que contiene expresiones racionales.

rational expression (p. 664) **expresión racional** Una fracción cuyo numerador, denominador o ambos son polinomios distintos a cero.

rational function (p. 692) **función racional** Una función de la forma $f(x) = \frac{\text{polinomio}}{\text{polinomio}}$.

rational number (p. 664) **número racional** Un número que puede escribirse como el cociente de dos números enteros.

real number line (p. 63) **recta numérica real** Una recta que representa números reales como puntos.

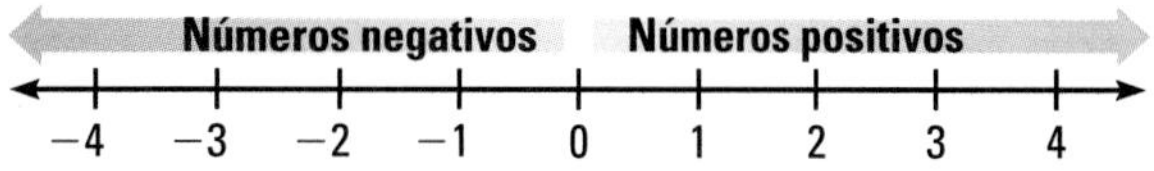

real numbers (p. 63) **números reales** El conjunto de números que consiste en los números positivos, los números negativos y cero. *Ver también* real number line / recta numérica real.

reciprocal (p. 108) **recíproco** Si $\frac{a}{b}$ es un número distinto a cero, entonces su recíproco es $\frac{b}{a}$. El producto de un número y su recíproco es 1.

relation (p. 256) **relación** Cualquier conjunto de pares ordenados (x, y).

roots of a quadratic equation (p. 526) **raíces de una ecuación cuadrática** Las soluciones de $ax^2 + bx + c = 0$.

round-off error (p. 166) **error de redondeo** El error que se produce cuando se redondea un resultado decimal para llegar a una respuesta significante.

S

scalar multiplication (p. 97) **multiplicación escalar** La multiplicación de una matriz por un número real.

scatter plot (p. 204) **diagrama de dispersión** Una gráfica de pares de números que representan situaciones de la vida real. Es una manera de analizar la relación entre dos cantidades.

scientific notation (p. 470) **notación científica** Un número expresado en la forma $c \times 10^n$, donde $1 \le c < 10$ y n es un número entero.

similar triangles (p. 140) **triángulos semejantes** Dos triángulos son semejantes si tienen ángulos correspondientes iguales. Puede mostrarse también que son semejantes si son iguales las razones de las longitudes de los lados correspondientes.

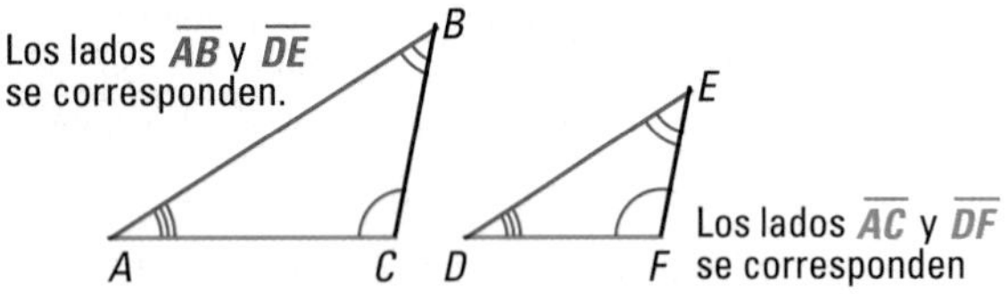

simplest form of a radical expression (p. 512) **forma más simple de una expresión radical** Una expresión que no tiene factores cuadrados perfectos distintos de 1 en el radicando, ninguna fracción en el radicando y ningún radical en el denominador de una fracción.

simplified expression (p. 102) **expresión simplificada** Una expresión está simplificada si no tiene signos de agrupación y si todos los términos semejantes se han combinado.

simplified rational expression (p. 664) **expresión racional simplificada** Una expresión racional está simplificada si el numerador y denominador no tienen factores en común (distinos de ± 1).

sine (p. 752) **seno** *Ver* trigonometric ratio / razón trigonométrica.

slope (p. 226) **pendiente** El número de unidades que sube o baja una recta no vertical por cada unidad de cambio horizontal de izquierda a derecha. La pendiente de m es $m = \dfrac{y_2 - y_1}{x_2 - x_1}$.

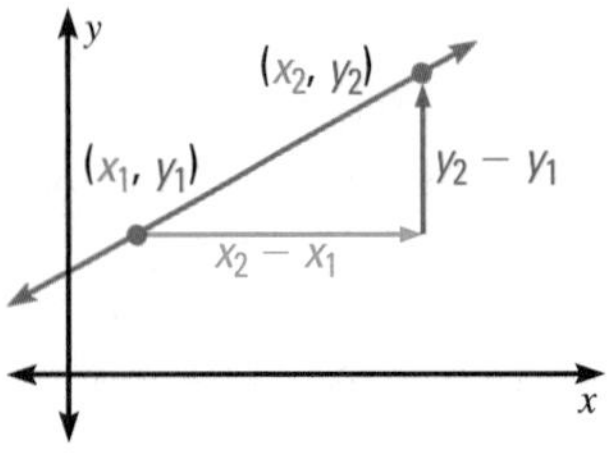

slope-intercept form (pp. 241, 273) **ecuación de pendiente-intercepto** Una ecuación lineal escrita en forma de $y = mx + b$. La pendiente de la recta es m. El intercepto en y es b. *Ver también* slope / pendiente y y-intercept / intercepto en y.

$y = 2x + 3$
La pendiente es 2.
El intercepto en y es 3.

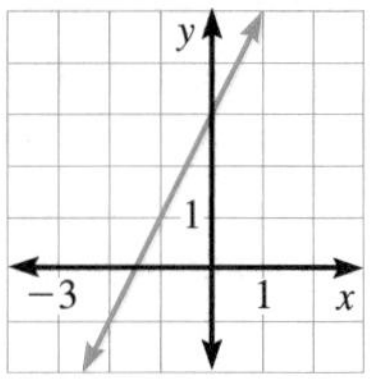

solution of an equation (p. 24) **solución de una ecuación** Un número que cuando se sustituye por la variable de una ecuación produce un enunciado verdadero.

solution of an inequality (p. 26) **solución de una desigualdad** Un número que cuando se sustituye por la variable de una desigualdad produce un enunciado verdadero.

solution of a linear equation (p. 210) **solución de una ecuación lineal** Un par ordenado (x, y) es una solución de una ecuación lineal si la ecuación es verdadera cuando se sustituyen los valores de x e y en la ecuación.

solution of a linear inequality (p. 360) **solución de una desigualdad lineal** Un par ordenado (x, y) es una solución de una desigualdad lineal si la desigualdad es verdadera cuando se sustityen los valores de x e y en la desigualdad.

solution step (p. 133) **paso de una solución** El resultado de aplicar una transformación a una ecuación al resolverla.

solution of a system of linear equations in two variables (p. 398) **solución de un sistema de ecuaciones lineales de dos variables** Un par ordenado (x, y) que satisface cada ecuación del sistema.

solution of a system of linear inequalities (p. 432) **solución de un sistema de desigualdades lineales** Un par ordenado (x, y) que es una solución de todas las desigualdades del sistema.

solving an equation (p. 25) **resolver una ecuación** Hallar todas las soluciones de una ecuación.

solving a right triangle (p. 753) **resolver un triángulo rectángulo** Hallar las longitudes de los otros dos lados de un tríangulo rectángulo, dadas la medida de uno de los ángulos agudos y la longitud de un lado del triángulo.

square root (p. 503) **raíz cuadrada** Si $b^2 = a$, entonces b es una raíz cuadrada de a. Las raíces cuadradas se escriben con un signo de radical $\sqrt{\ }$.

square root function (p. 709) **función con raíces cuadradas** La función definida por $y = \sqrt{x}$.

standard form of an equation of a line (p. 308) **forma general de la ecuación de una recta** Una ecuación lineal de la forma $Ax + By = C$ donde A, B y C son números reales y A y B no son ambos cero.

standard form of a polynomial (p. 576) **forma general de un polinomio** Un polinomio cuyos términos se sitúan en orden descendiente, de grado mayor a grado menor.

stem-and-leaf plot (p. 368) **tabla arborescente** Un arreglo de dígitos que se usa para mostrar y ordenar datos numéricos.

system of linear equations (p. 398) **sistema de ecuaciones lineales** Dos o más ecuaciones lineales con las mismas variables. También se llama sistema lineal.

system of linear inequalities (p. 432) **sistema de desigualdades lineales** Dos o más desigualdades lineales con las mismas variables. También se llama sistema de desigualdades.

T

tangent (p. 752) **tangente** *Ver* trigonometric ratio / razón trigonométrica.

terms of an expression (p. 80) **términos de una expresión** Las partes de una expresión que se suman. Por ejemplo, en la expresión $5 - x$, los términos son 5 y $-x$.

theorem (pp. 738, 759) **teorema** Un enunciado que puede probarse verdadero.

theoretical probability (p. 114) **probabilidad teórica** Un tipo de probabilidad que se basa en el número de casos favorables dividido por el número total de casos.

time period (p. 477) **período de tiempo** La variable *t* en los modelos de crecimiento y decrecimiento exponencial. *Ver también* exponential growth / crecimiento exponencial y exponential decay / decrecimiento exponencial.

transform an equation (p. 132) **transformar una ecuación** Cambiar una ecuación en una ecuación equivalente.

trigonometric ratio (p. 752) **razón trigonométrica** La razón de las longitudes de dos lados de un triángulo rectángulo. Por ejemplo:

$$\text{sen } A = \frac{\text{lado opuesto a } \angle A}{\text{hipotenusa}}$$

$$\cos A = \frac{\text{lado adyacente a } \angle A}{\text{hipotenusa}}$$

$$\tan A = \frac{\text{lado opuesto a } \angle A}{\text{lado adyacente a } \angle A}$$

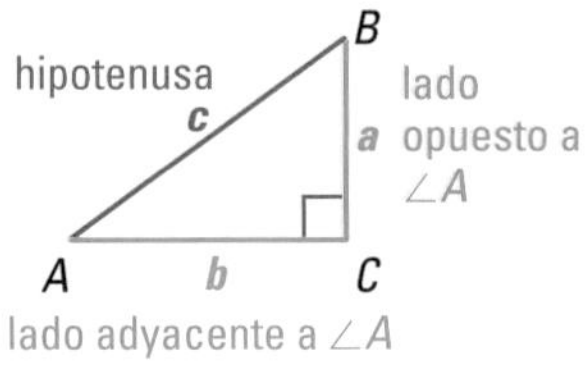

trinomial (p. 576) **trinomio** Un polinomio con tres términos.

U

unit analysis (p. 5) **análisis dimensional** Escribir las unidades de cada variable en un problema de la vida real para ayudar a determinar las unidades de la respuesta.

unit rate (p. 180) **relación unitaria** Una relación por una unidad dada.

V

values (p. 3) **valores** Los números representados por variables.

variable (p. 3) **variable** Una letra que se usa para representar uno o más números.

variable expression (p. 3) **expresión algebraica** Un conjunto de números, variables y operaciones.

velocity (p. 66) **velocidad** La rapidez y dirección con las que se mueve un objeto (hacia arriba es positiva y hacia abajo es negativa.) La rapidez de un objeto es el valor absoluto de su velocidad.

verbal model (p. 5) **modelo verbal** Una expresión en la que se usan palabras para representar una situación de la vida real.

vertex of a parabola (p. 518) **vértice de una parábola** El punto más bajo de una parábola que abre hacia arriba o el punto más alto de una parábola que se abre hacia abajo. *Ver también* parabola / parábola.

vertical motion models (p. 535) **modelos del movimiento vertical** Un modelo para representar la altura de un objeto en movimiento que se ha dejado caer y para representar un objeto lanzado hacia abajo o hacia arrriba.

X

x*-axis** (p. 203) **eje de *x El eje horizontal en un plano de coordenadas. *Ver también* plano de coordenadas / coordinate plane.

x*-coordinate** (p. 203) **coordenada *x El primer número en un par ordenado. *Ver también* ordered pair / par ordenado y plotting a point / marcar un punto.

x*-intercept** (p. 218) **intercepto en *x La coordenada *x* de un punto donde una gráfica cruza el eje de *x*.

Y

y*-axis** (p. 203) **eje de *y El eje vertical en un plano de coordenadas. *Ver también* plano de coordenadas / coordinate plane.

y*-coordinate** (p. 203) **coordenada *y El segundo número de un par ordenado. *Ver también* ordered pair / par ordenado y plotting a point / marcar un punto.

y*-intercept** (p. 218) **intercepto en *y La coordenada *y* de un punto donde una gráfica cruza el eje de *y*.

Z

zero-product property (p. 597) **propiedad del producto nulo** Si a y b son números reales y $ab = 0$, entonces $a = 0$ o $b = 0$.

857

Answers and Index

The Teacher's Edition supplements the answers and index material in the Student Edition. The credits are also included.

Credits

Cover Credits

Earth Imaging/Tony Stone Images (background); Dan Bosler/Tony Stone Images (t); T. Kevin Smyth/The Stock Market (r); courtesy, NASA (b); Don Mason/The Stock Market (l).

Text Credits

143 Adapted from *The Pleasures of Japanese Cooking* by Heihachi Tanaka with Betty Nicholas. Copyright © 1963 by Prentice-Hall, Inc., renewed in 1991 by Heihachi Tanaka and Betty Nicholas. Reprinted with the permission of Simon & Schuster, Inc.

Photography

i, ii Earth Imaging/Tony Stone Images (background); Dan Bosler/Tony Stone Images (t); T. Kevin Smyth/The Stock Market (r); courtesy, NASA (b); Don Mason/The Stock Market (l); **iv** Sovfoto/Eastfoto/PNI/PictureQuest (tl); Dean Siracusa/FPG International (tc); **v** RMIP/Richard Haynes (tc); CORBIS/Tim Mosenfelder (tr); **vi** Stuart Westmorland/Photo Researchers, Inc.; **vii** Baron Wolman/Tony Stone Images; **viii** Darrell Gulin/Tony Stone Images; **ix** Dennis Hallinan/FPG International; **x** Melissa Farlow/National Geographic Image Collection; **xi** Bob Daemmrich/The Image Works; **xii** Billy Hustace/Tony Stone Images; **xiii** Dean Abramson/Stock Boston/PNI/PictureQuest; **xiv** Vincent Laforet/Allsport; **xv** David Parker/Science Photo Library/Photo Researchers, Inc.; **xvi** CORBIS/Phillip Gould; **xvii** Rex A. Butcher/Tony Stone Images; **xix** Robert Rathe/Stock Boston (t); Keith Wood/Tony Stone Images (c); Rachel Epstein/PhotoEdit (cl); Chuck Savage/The Stock Market (cr); Roger Tully/Tony Stone Images (b); **xxii, 1** Stuart Westmorland/Photo Researchers, Inc.; **3** T. Kevin Smyth/The Stock Market; **5** Ken Biggs/Tony Stone Images; **7** courtesy, Heidi Zimmer; **9** Wolfgang Kaehler; **13** John Kuhn (l); Howell/Liaison Agency, Inc. (br); **16** Mark Gibson; **20** RMIP/Richard Haynes; **24** Don Smetzer/Tony Stone Images; **26** Zigy Kaluzny/Tony Stone Images; **29** Hurlin-Saola/Liaison Agency; **32** Ted Streshinsky/Photo 20-20; **34** David H. Frazier/Tony Stone Images; **37** David Young-Wolff/PhotoEdit; **38** Robert Ginn/PhotoEdit; **39** Bibliothèque Nationale, Paris (bl); X512.Sch2. Rare Book and Manuscript Library, Columbia University (bcl); Stock Montage (bcr); courtesy, IBM Archives (br); **40** Mark Stouffer/Bruce Coleman Inc.; **42** Tom McHugh/Photo Researchers, Inc.; **46** Mark Wagner/Tony Stone Images; **48** Stephen Frink/The Stock Market; **60** Baron Wolman/Tony Stone Images; **61** George Hall/Woodfin Camp and Associates; **63** Everett Johnson/Tony Stone Images; **66** courtesy, NASA; **68** Jay M. Pasachoff/Visuals Unlimited; **69** AP Photo/Osamu Honda; **72** Eric Berndt/Unicorn Stock Photo; **74** Stephen Frisch/Stock Boston; **76** CC **79** Marilyn "Angel" Wynn/Sun Valley Video and Photography; **81** Alan Schein/The Stock Market; **83** James Hazelwood; Lockwood/DRK Photo; **86** AP/Wide World Photos; **87** UPI/CORBIS/Bettmann; **90** Jim Shippee/Unicorn Stock Photo; **95** Nick Bergkessel/Photo Researchers, Inc.; **97** MONKMEYER/Grantpix; **100** Carl J. Single/The Image Works; **102** MONKMEYER/Wyman; **104** Phil Degginger/Bruce Coleman Inc; **107** The Bodleian Library, University of Oxford, MS Sansk. d.14 folio 47r (cr); *The Emperor SiuenLi,* 17th century, Bibliothèque Nationale, Paris, France/Bridgeman Art Library (bl); Science & Society Picture Library (bc); Gail McCawn/Photo Researchers, Inc. (br); **108** Tony Freeman/PhotoEdit; **110** James Handkley/Tony Stone Images; **112** Stephen Dunn/Allsport; **114** Doug Perrine/DRK Photo; **115** Bob Daemmrich Photos, Inc.; **118** David Parker/Photo Researchers, Inc.; **128** Darrell Gulin/Tony Stone Images; **129** Tom & Pat Leeson/DRK Photo; **132** Peter Pearson/Tony Stone Images; **136** Robert Brenner/PhotoEdit; **138** Rob Matheson/The Stock Market; **140** Mel Traxel/Motion Picture & Television Photo Archive; **142** Antman/The Image Works; **143** A & J Verkaik/The Stock Market; **145** Bachmann/PhotoEdit; **147** Jackson Smith/Uniphoto; **149** Reproduced with permission of the Thornton Wilder Collection, and the Barbara Hogenson Agency/Yale Collection of American Literature, Beinecke Rare Book and Manuscript Library.; **150** J. F. Towers/The Stock Market; **152** CORBIS/Macduff Everton (bl); CORBIS/Bettmann (bc); CORBIS/Gail Mooney (br); **154** Don Mason/The Stock Market; **158** courtesy, Gary Scott, Callison Architecture (bl); courtesy, Callison Architecture (br); **160** Steve Woit/Stock Boston; **162** Ahup & Manoj Sham/Animals Animals; **164** Rick Edwards/Animals Animals; **166** Peter Menzel/Stock Boston; **170** Aaron Haupt/Photo Researchers, Inc.; **171** Flip Nicklin/Minden Pictures; **174** Mike Brown/Florida Today/Liaison Agency; **175** National Geographic Image Collection; **180** Chuck Savage/The Stock Market; **181** Wolfgang Kaehler; **182** Paolo Koch/Photo Researchers, Inc.; **185** David Young-Wolff/Tony Stone Images; **198** David Young-Wolff/PhotoEdit; **199** Bob Daemmrich/Tony Stone Images (bl); David Sutherland/Tony Stone Images (br); **200, 201** Dennis Hallinan/FPG International; **203** Duomo Photography, Inc.; **205** Eric Berndt/Unicorn Stock Photo; **207** CC Lockwood/Animals Animals; **210** Bongarts Photography/Sportschrome; **216** Thomas Zimmerman/Tony Stone Images; **218** Aaron Haupt/Stock Boston; **220** John Warden/Tony Stone Images; **222** Mark Harris/Tony Stone Images; **224** René Descartes/from a painting by Frans Hals. © North Wind Picture Archives (bl); Publication of Descartes text, *La Géométrie* (1637). Stock Montage, Inc. (bcl); courtesy, Casio, Inc. (bcr); Barbara Alper/Stock Boston/PNI/PictureQuest & Sovfoto/Eastfoto/PNI/PictureQuest (montage, br); **225** School Division, Houghton Mifflin Company (background, cl); RMIP/Richard Haynes (inset, cl); Alan Kearney/Viesti Associates, Inc. (tr); **226** Dean Siracusa/FPG International; **231** Robert Rathe/Stock Boston; **232** TriStar Pictures/Shooting Star; **234** Bob Daemmrich/Stock Boston; **235** courtesy, Zeppelin Library Archive; **236** Keith Wood/Tony Stone Images; **238** Sovfoto/Eastfoto (tl); Artville, LLC.(br); **240** RMIP/Richard Haynes (all); **241** Nathan Benn/Stock Boston; **243** Charles Gupton/Stock Boston; **245** Matthew J. Atanian/Tony Stone Images; **250** Tony Freeman/PhotoEdit; **252** Roger Tully/Tony Stone Images; **254** Charlie Westerman/Liaison Agency; **256** David Young-Wolff/PhotoEdit; **258** Ron Sanford/Tony Stone Images; **261** CORBIS/Tim Mosenfelder; **270** Melissa Farlow/National Geographic Image Collection; **271** Mastrorillo/The Stock Market (r); Werner Forman Archive, Maxwell Museum of Anthropology, Albuquerque, NM, USA/Art Resource (bc); **273** Eastcott/The Image Works; **274** Bill Aron/PhotoEdit; **279** David Sailors/The Stock Market; **281** Mark Burnett/Stock Boston; **285** Uniphoto; **287** Kenneth Garrett/National Geographic Image Collection; **292** RMIP/Richard Haynes; **294** Steven E. Sutton/Duomo; **295** Luis Bongarts/Allsport; **300** Tom Stewart/The Stock Market; **302** Matthew Stockman/Allsport; **306** Grant/Mesa Verde National Park, 09116.4 (bl); Special Collections, The University of Arizona Library (bc); CORBIS/Richard A. Cooke (br); **308** Len Rue Jr./Photo Researchers, Inc.; **310** courtesy, Texas Gas Transmission; **312** Wayne Lankinen/DRK Photo; **316** Kevin Horan/Stock Boston; **318** Dan Bosler/Tony Stone Images; **320** Matthew Borkoski/Stock Boston; **330, 331** Bob Daemmrich/The Image Works; **334** Tony Freeman/PhotoEdit; **337** Gunther/Motion Picture & Television Photo Archive; **338** James Sugar/Black Star/PNI (l); Frederica Georgia/Photo Researchers, Inc. (br); **340** Donald C. Johnson/The Stock Market; **342** David J. Sams/Stock Boston; **344** Tony

Teacher's Edition Credits

T6 Stuart Westmorland/Photo Researchers, Inc.; T7 Baron Wolman/Tony Stone Images; T8 Darrell Gulin/Tony Stone Images; T9 Dennis Hallinan/FPG International; T10 Melissa Farlow/National Geographic Image Collection; T11 Bob Daemmrich/The Image Works; T12 Billy Hustace/Tony Stone Images; T13 Dean Abramson/Stock Boston/PNI/PictureQuest; T14 Vincent Laforet/Allsport; T15 David Parker/Science Photo Library/Photo Researchers, Inc.; T16 CORBIS/Phillip Gould; T17 Rex A. Butcher/Tony Stone Images; T19 Robert Rathe/Stock Boston (t); Keith Wood/Tony Stone Images (c); Rachel Epstein/PhotoEdit (cl); Chuck Savage/The Stock Market (cr); Roger Tully/Tony Stone Images (b); T22 Aaron Haupt/Stock Boston; T23 RMIP/Richard Haynes; T28 Michael Goss; T35 Steve Woit/Stock Boston; T41 Meridian Creative Group; T44 CORBIS/LWA-Dann Tardif; T48 CORBIS/Charles Gupton; T50 CORBIS/LWA-Dann Tardif; T56 CORBIS/LWA-Dann Tardif.

Freeman/PhotoEdit; **346** Chuck Pefley/Tony Stone Images (cl); Rich Iwaski/Tony Stone Images (br); **350** Erich Lessing/Art Resource; **352** The Granger Collection (cr); Stock Montage (bl, bcl, bcr); Superstock (br); **353** Ken Clark; **355** Kaluzny/Thatcher/Tony Stone Images; **360** Al Giddings Images, Inc.; **362** Al Giddings Images, Inc.; **365** Yoav Levy/Phototake; **368** Harry M. Walker; **375** Porter Gifford/Liaison Agency; **379** David Shopper/Stock Boston; **392** RMIP/Richard Haynes (all); **393** Jon Riley/Tony Stone Images; **394** Billy Hustace/Tony Stone Images; **395** Bob Daemmrich/Uniphoto; **398** Mark E. Gibson; **400** Chuck Savage/The Stock Market; **403** Monkmeyer/Goodwin; **405** David Young-Wolff/PhotoEdit; **407** AP Photo; **409** RMIP/Richard Haynes; **411** Uniphoto; **413** North Wind Picture Archives; **415** Tom Evans/Photo Researchers, Inc.; **417** Reproduced by Permission of the Commercial Press (Hong Kong) Limited from the publication of *Chinese Mathematics: A Concise History*. (cl); Michael W. Thomas/Stock South/PNI (bl); Artville, LLC. (bc); Chris Sorensen/The Stock Market & Artville, LLC. (montage, br); **422** CORBIS/Kevin R. Morris; **426** James Amos/National Geographic Image Collection; **428** Richard Harrington/FPG International (cl, mid-cr); James Amos/National Geographic Image Collection (mid-cl, cr) **430** Mark E. Gibson; **432** Lee Greenwood/Stock Boston; **434** MONKMEYER/Anderson; **436** M. Granitsas/The Image Works; **446** Dean Abramson/Stock Boston/PNI; **447** David Madison; **450** Peter Menzel/Stock Boston; **452** Mark Wagner/Tony Stone Images; **454** Keith Wood/Tony Stone Images; **456** Brian Drake/Sportschrome; **460** courtesy, Texas Historical Commission; **463** Duomo Photography, Inc.; **465** Tim Flach/Tony Stone Images; **467** John Eastcott/DRK Photo; **468** Randy Wells/Tony Stone Images; **470** Jacques Jangoux/Tony Stone Images; **474** North Wind Picture Archives; **477** Tom McHugh/Photo Researchers, Inc.; **481** Marcia Griffen/Animals Animals; **483** RMIP/Richard Haynes; **484** Sears Roebuck & Co./Brown Brothers; **489** Patrick Ramsey/International Stock Photo; **491** CORBIS/Bettmann (cr); Andrew Syred/Science Photo Library/Photo Researchers, Inc. & CORBIS/Bettmann (montage, bl); CORBIS (bc); Secchi-Lecaque/CNRI/Science Photo/Photo Researchers, Inc. (br); **492** Dr. David Jones/Science Photo Library/Photo Researchers, Inc.; **500, 501** Vincent Laforet/Allsport; **503** Richard B. Levine; **509** Ken M. Johns/Photo Researchers, Inc. (tl); Photo Researchers, Inc. (cl); Charles D. Winters/Photo Researchers, Inc.(tc); E.R. Degginger/Photo Researchers, Inc. (cr, bl); Tom McHugh/Photo Researchers, Inc. (bc); Biophoto Associates/Photo Researchers, Inc. (br); **511** Stephen Munday/Allsport; **513** Mike Hewitt/Allsport; **518** Tim Davis/Photo Researchers, Inc.; **520** Tony Duffy/Allsport; **522** Gordon R. Gainer/The Stock Market; **523** Michel Hans/Vandystadt/Allsport; **526** David Madison/Duomo; **530** Kevin Vandivier/Viesti Associates, Inc.; **533** Stephen Dunn/Allsport; **535** Ferne Saltzman/Albuquerque International Balloon Fiesta; **537** Gordon & Cathy Illg/Animals Animals; **541** Daniel D. Lamoreux/Visuals Unlimited; **545** Gary A. Conner/PhotoEdit; **546** Chad Slattery/Tony Stone Images; **547** Yale University Library, Babylonian Collection (bl); Stock Montage (bc); courtesy, NASA (br); **548** Susan Van Etten/PhotoEdit; **550** Fugifotos/The Image Works; **552** Dave Bartruff/Stock Boston; **554** Roger Klocek/Visuals Unlimited; **555** Kevin Schafer/Peter Arnold, Inc.; **556** Jonathan Nouros/PhotoEdit; **570** Christopher M. Schnurr; **571** RMIP/Richard Haynes (background); David Pollack/The Stock Market (inset); **572, 573** David Parker/Science Photo Library/Photo Researchers, Inc.; **576** Spencer Grant/PhotoEdit; **578** Elizabeth Tustian; **581** Rathe/Stock Boston; **584** David Young-Wolff/Tony Stone Images; **590** CORBIS/Tom Brakefield; **592** Mark E. Gibson/The Stock Market; **594** Michael Schimpf; **596** Amy Reichman/Envision (cr); Fred Hultstrand History in Picture Collection, NDSU, Fargo, N.D. (bl); CORBIS/Bettmann (bc); reproduced with permission from the Sweden Post (br); **597** Randy Wells/Tony Stone Images; **601** Allen E. Morton/Visuals Unlimited; **604** Orion Press/Tony Stone Images; **606** Rachel Epstein/PhotoEdit; **608** Richard Vogel/Liaison Agency; **611** CORBIS/Charles & Josette Lenars; **615** Vandystadt/Allsport; **619** Rob Tringali Jr./Sportschrome; **621** Tony Freeman/PhotoEdit; **623** Darren Carroll/Duomo (cr); Mike Powell/Allsport (bl); **625** courtesy, NASA; **630** Account Phototake/Phototake; **640, 641** CORBIS/Phillip Gould; **643** Cordelia Williams; **645** Louis Mazzatenta/National Geographic Image Collection; **647** Cordelia Williams (tr); Don Hamerman/New England Stock Photo (bl); **649** David Young-Wolff/PhotoEdit; **651** Burke/Triolo/Artville; **654** David Young-Wolff/PhotoEdit; **656** Karl Weatherly/Allstock/PNI; **660** Superstock; **662** CORBIS/Bettmann (bl); Artville, LLC. (bcl); CORBIS/Joseph Sohm/ChromoSohm Inc. (bcr); School Division/Houghton Mifflin Company (br); **664** Jonathan Blair/Woodfin Camp and Associates; **668** Bill Bachman/Photo Researchers, Inc.; **674** Tony Page/Tony Stone Images; **676** Ron Dorsey/Stock Boston; **680** Joseph Pobereskin/Tony Stone Images; **684** David Young-Wolff/PhotoEdit; **688** Frank Siteman/Stock Boston; **690** C.J. Allen/Stock Boston; **691** CORBIS/Neal Preston; **695** CORBIS/Keren Su (l); RMIP/Richard Haynes (cr); **706, 707** Rex A. Butcher/Tony Stone Images; **709** Louis Mazzatenta/National Geographic Image Collection; **711** Peter Menzel/Tony Stone Images; **713** Tony Freeman/PhotoEdit; **716** Chuck Fishman/Woodfin Camp and Associates; **720** John Bova/Photo Researchers, Inc.; **721** RMIP/Richard Haynes; **722** Art Montes De Oca/FPG International; **726** Jeff Persons/Stock Boston; **730** William Sallaz/Duomo; **733** Paolo Koch/Photo Researchers, Inc.; **735** Norbert Wu (all); **738** Elizabeth Tustian; **744** Rare Book and Manuscript Library, Columbia University. Cunieform, Plimpton 322 (cr, bl); SEF/Art Resource (bc); CORBIS/Joseph Sohm/Chromosohm Inc. (br); **745** Doug Pensinger/Allsport; **749** Bob Daemmrich/PNI/PictureQuest; **752** John Neubauer/PhotoEdit; **758** John Neubauer/PhotoEdit; **760** Dennis MacDonald/PhotoEdit; **775** RMIP/Richard Haynes (tl, br); CORBIS/Jim Winkley/Ecoscene (tr); CORBIS/Peter Harholdt (cr).

Illustration

Steve Cowden **658, 718, 735, 746**
Laurie O'Keefe **162, 236, 415, 537, 567**
School Division, Houghton Mifflin Company **397, 413, 501, 513, 515, 522, 550, 552, 556, 571, 581, 586, 592, 594, 606, 608, 616, 630, 641, 662, 668, 675, 678, 747, 754, 756**
Doug Stevens **357, 419, 543, 599, 573, 601, 613, 724**

Selected Answers

CHAPTER 1

SKILL REVIEW (p. 2) **1.** 0.6, $\frac{3}{5}$ **2.** 0.33, $\frac{33}{100}$ **3.** 0.0008, $\frac{1}{1250}$ **4.** 1.5, $1\frac{1}{2}$ **5.** < **6.** > **7.** > **8.** > **9.** 12.25 cm^2, 14 cm **10.** 6 ft^2, 12 ft **11.** 8.32 mi^2, 13.6 mi

1.1 PRACTICE (pp. 6–8) **5.** 30 **7.** 25 **9.** 2 **11.** 15 **13.** 3 h; $\frac{\text{mi}}{\text{mi/h}} = \text{mi} \cdot \frac{\text{h}}{\text{mi}} = \text{h}$ **15.** 138 mi; $\frac{\text{mi}}{\text{h}} \cdot \text{h} = \text{mi}$ **17.** time = $\frac{\text{distance}}{\text{rate}}$; time = t h, distance = 10 mi, rate = 1.25 mi/h; $t = \frac{10}{1.25}$; 8 h **19.** 60 **21.** 0.25 **23.** 82 **25.** 2.1 **27.** 5 **29.** $\frac{7}{16}$ **31.** \$2.40 **33.** She did not use the correct decimal equivalent of 2.5%, 0.025. **35.** $1\frac{4}{11}$ mi/min ≈ 1.36 mi/min **37.** $\frac{2}{15}$ km/min ≈ 0.13 km/min **39.** 3 h **41.** 4.3 cal/min **43.** Distances are given to the nearest foot. Mt. McKinley: 20,322 ft and 8707 ft; Mt. Elbrus: 18,481 ft and 10,548 ft; Mt. Kilimanjaro: 19,564 and 9465 ft; Mt. Aconcagua: 22,831 ft and 6198 ft

1.1 MIXED REVIEW (p. 8) **51.** $\frac{3}{10}$ **53.** $2\frac{3}{4}$ **55.** 4 **57.** $10\frac{1}{10}$ **59.** 70 in.^{2} **61.** 4 **63.** 6 **65.** 8 **67.** 10

1.2 PRACTICE (pp. 12–14) **5.** D **7.** A **9.** 64 **11.** 216 **13.** 125 **15.** 1000 **17.** 2^3 **19.** 9^y **21.** 3^4y **23.** c^6 **25.** $(4x)^3$ **27.** 25 **29.** 1296 **31.** 2401 **33.** 729 **35.** 1024 **37.** 1,000,000 **39.** 2197 **41.** 64 **43.** 16 **45.** 100 **47.** 59,049 **49.** 1,953,125 **51.** 262,144 **53.** 1,679,616 **55.** 15,625 **57.** 343 **59.** 4096 **61.** 1, 4, 9, 16, 25 **63.** 148.84 ft^2 **65.** 1728 in.^{3} **67.** about 1,333,333 ft^3

1.2 MIXED REVIEW (p. 14) **73.** 10.2 **75.** 0.625, 62.5% **77.** 0.55, 55% **79.** 21 **81.** 4 **83.** $2\frac{1}{4}$

TECHNOLOGY ACTIVITY 1.2 (p. 15) **1.** 33,554,432; 1.13×10^{15}; 3.78×10^{22}; 1.27×10^{30} **3.** 16; 64; 256; 1024; 4096 **5.** 36; 1296; 46,656; 1,679,616; 60,466,176

1.3 PRACTICE (pp. 19–21) **5.** 150 **7.** 2 **9.** 0 **11.** 3 **13.** 19 **15.** 300 **17.** 34 **19.** 18 **21.** 17 **23.** 12 **25.** 16 **27.** 8 **29.** 11 **31.** 27 **33.** 27 **35.** 27 **37.** 0.125 **39.** 3 **43.** Calculator B; $15 - [(6 \div 3) \times 4]$ **45.** Calculator A; $15 + (10 \div 5) + 4$ **47.** $(2 \times \$49.99) + (3 \times \$44.10)$ **49.** \$28,475 **51.** \$30 **53.** $2\pi rh + 2\pi r^2$ or $2\pi r(h + r)$

1.3 MIXED REVIEW (p. 22) **59.** 8 **61.** $\frac{2}{9}$ **63.** $-\frac{1}{16}$ **65.** 108 **67.** 1280 **69.** 6144 **71.** $2\frac{1}{2}$ h

QUIZ 1 (p. 22) **1.** 18 **2.** 12 **3.** 32 **4.** 1 **5.** 1 **6.** 18 **7.** 13.5 **8.** 16.7 **9.** 40 mi/h **10.** 60 mi/h **11.** 60 mi/h **12.** 6^3 **13.** 4^5 **14.** $(5y)^3$ **15.** 5^2 **16.** $3^6 \cdot t$ **17.** $(7x)^2$ **18.** 7776 **19.** 432 **20.** 1728 **21.** 33 **22.** $1\frac{2}{7}$ **23.** $\frac{1}{2}$ **24.** $\frac{1}{3}$ **25.** 2 **26.** 1.728 ft^3

1.4 PRACTICE (pp. 27–29) **9.** B **11.** C **13.** B **15.** not a solution **17.** solution **19.** not a solution **21.** not a solution **23.** solution **25.** not a solution **27.** What number can be added to 6 to get 11? 5 **29.** What number can be multiplied by 3 to get 12? 4 **31.** What number can be multiplied by 4 to get 36? 9 **33.** One less than two times what number is 9? 5 **35.** What number when divided by 7 gives 3? 21 **37.** What number cubed gives 125? 5 **39.** solution **41.** not a solution **43.** not a solution **45.** not a solution **47.** solution **49.** solution **51.** solution **53.** solution **55.** not a solution **57.** G **59.** H **61.** B **63.** A **65.** the number of ft^2 per station, the number of stations, the total number of ft^2 available; 20 stations **67.** cost each time you fill the tank, number of times you fill the tank, total amount to be paid for gas; yes **69.** 400 m/min **71.** 5 h **73.** 0°C **75.** 8 ft **77. a.** The length of the jet is 1.212 times the wingspan. **b.** It is longer than it is wide; since $y > 0$, $1.212y > y$.

1.4 MIXED REVIEW (p. 30) **85.** $\frac{1}{4}$ **87.** $\frac{4}{25}$ **89.** y^2 **91.** c^4 **93.** $(5y)^4$ **95.** $(9a)^6$ **97.** 1^3 **99.** t^6

1.5 PRACTICE (pp. 35–38) **5.** C **7.** B **9.** $x + 10 = 24$ **11.** $\frac{20}{n} \le 2$ **13.** $x + 9$ **15.** $\frac{1}{2}x + 3$ **17.** $\frac{x}{0.2}$ **19.** $\frac{2^3}{x}$ **21.** $5^2 - x$ **23.** $9 > 3s$ **25.** $14x = 1$ **27.** $3(x - 2) = 10$ **29.** $(38 - n) - 23 < 8$ **31.** $(20 - x) - 5 \ge 10$ **33.** $9 + \frac{b}{10} \ge 11$ **35.** $q \ge 100$ **37.** $\frac{35}{t} \le 7$ **39.** $b = e - 1.50$ **41.** $c = 3r + 229$ **43.** $V \le 30 - 3$ or $s^3 \le 30 - 3$ **45.** $P = 4(s - 2)$ **47.** $A = \frac{1}{2}(7 + 9)(h + 7)$ **49.** The number of subscriptions times the cost of each subscription equals the amount the club needs to raise. **51.** 21 **53.** $21 \cdot 15 = 315$; the answer is reasonable. **55.** $20m = 260$ **57.** $\frac{\text{dollars}}{\text{mi/h}} \cdot \text{mi/h} = \text{dollars}$ **59.** The number of weeks times the total of \$20 plus the additional amount saved per week equals the total saved; let x = the additional amount saved per week; $12(x + 20) = 480$. **61.** Let x = the total number of points the team must equal or exceed; $x \ge 18\left(\frac{544}{17} + 5\right)$.

SELECTED ANSWERS

1.5 MIXED REVIEW (p. 38) **67.** = **69.** > **71.** <

73.

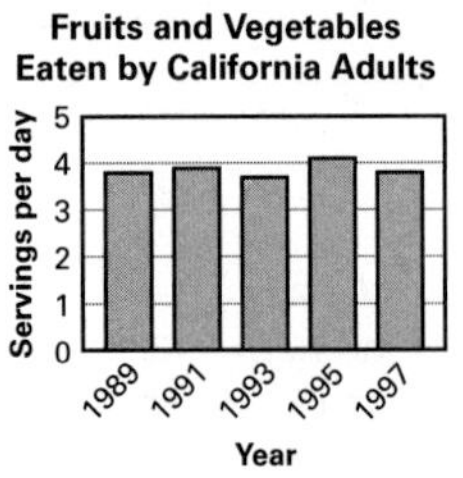

75. 10 **77.** 360

QUIZ 2 (p. 39) **1.** $2\frac{3}{4}$ **2.** 21 **3.** 34 **4.** 12 **5.** 42 **6.** 54 **7.** not a solution **8.** solution **9.** solution **10.** not a solution **11.** solution **12.** not a solution **13.** Let x = the total cost of the pizza; $\frac{1}{4}x = 2.65$; $10.60.

1.6 PRACTICE (pp. 43–45) **5.** true **7.** false **9.** false

11.

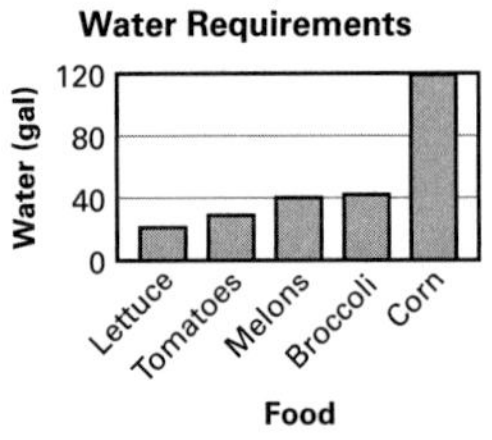

13. paper; about 33 million tons **15.** 6 years **17.** 1991

19.

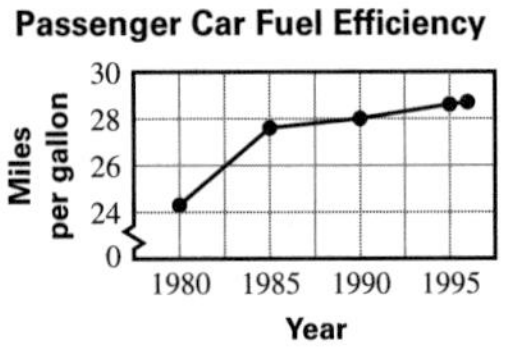

1.6 MIXED REVIEW (p. 45) **23.** not a solution **25.** solution **27.** $12 = \frac{n}{3}$

1.7 PRACTICE (pp. 49–51)

7. 44.5; 89; 133.5; 178; 222.5; 267

9.

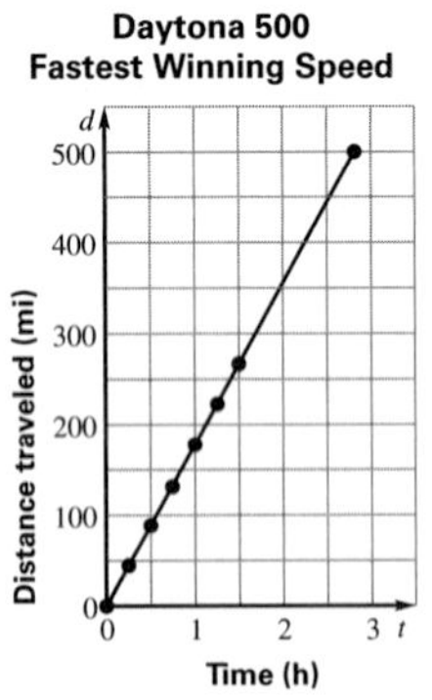

13. Output: 2, 5, 8, 11 **15.** Output: 0, 5, 10, 15 **17.** Output: 1, 3, 5, 7 **19.** Output: 6.5, 8.5, 14.5, 20.5, 26.5 **21.** Output: 19, 16, 13, 12, 11.5 **23.** Output: 0.5, 1.75, 8.5, 19.75, 35.5 **25.** 360 · 2, 720; 360 · 3, 1080; 360 · 4, 1440; 360 · 5; 1800

27.

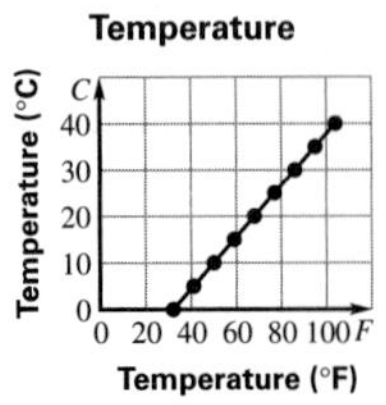

29. domain: 0, 5, 10, 15, 20, 25, 30, 35, 40; range: 32, 41, 50, 59, 68, 77, 86, 95, 104 **31.** domain: 1, 2, 3, 4, 5, 6, 7; range: 6.25, 7.50, 8.75, 10.00, 11.25, 12.50, 13.75

33. $C = 0.375n$ **35.** Input values represent the number of posters; corresponding output numbers represent the cost of the posters. **37.** Let n = the number of days and a = the amount in dollars that you earn; $a = 2n + 5$.

1.7 MIXED REVIEW (p. 52) **45.** x^6 **47.** y^7 **49.** 6 **51.** 219 **53.** solution **55.** solution **57.** not a solution **59.** $250 million

QUIZ 3 (p. 52)

1.

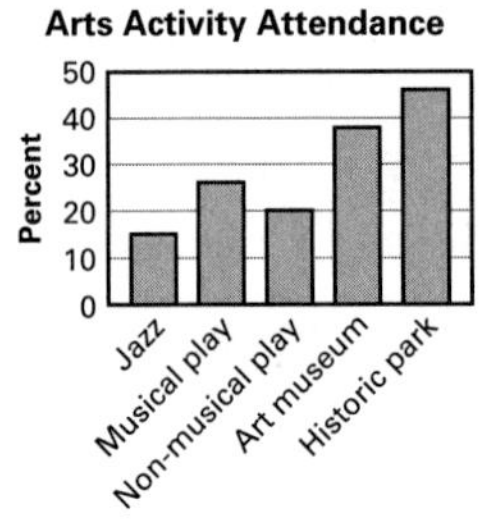

2. Conclusions may include the following:

- Nearly half of all 18-to-24-year-olds attended a historic park.
- More of the 18-to-24-year-olds attended a musical play than attended a non-musical play.
- The events in increasing order of popularity are jazz, non-musical play, musical play, art museum, historic park.
- More than three times as many 18-to-24-year-olds attended a historic park as attended a jazz event.
- About twice as many 18-to-24-year-olds attended an art museum as attended a non-musical play.

3. B **4.** Let t = the number of minutes after you turn on the heat and a = your altitude; $a = 25t + 200$. **5.** Input: 0, 1, 2, 3, 4, 5, 6; Output: 200, 225, 250, 275, 300, 325, 350

CHAPTER 1 REVIEW (pp. 54–56) **1.** 37 **3.** 10 **5.** 10 **7.** 2 h **9.** 3125 **11.** 33 **13.** 15,625 **15.** 9 **17.** 1.5 **19.** solution **21.** solution **23.** *Sample answer:* The women have won 41 Wimbledon titles, while the men have won 30, so the women have won 11 more Wimbledon titles than the men have won.

25.

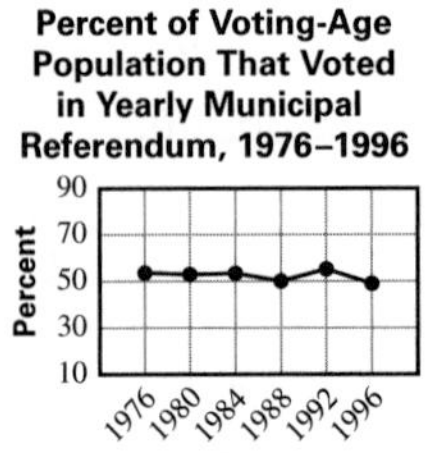

27. Perimeter = Length + Width + Length + Width; $P = 10w$

29. domain = 1, 2, 3, 4, 5; range = 10, 20, 30, 40, 50

CHAPTER 2

SKILL REVIEW (p. 62) **1.** 4.07, 4.6, 5.06, 5.46, 5.5, 6.1 **2.** $\frac{25}{11}, 2\frac{3}{5}, 2\frac{5}{8}, 2\frac{2}{3}, \frac{11}{4}, 2\frac{5}{6}$ **3.** $\frac{3}{14}$ **4.** $1\frac{1}{2}$ **5.** $3\frac{7}{8}$ **6.** $5\frac{2}{3}$

7. $2\frac{1}{2}$ **8.** $\frac{1}{6}$ **9.** 28 **10.** $6\frac{2}{3}$ **11.** 5 **12.** 2 **13.** 19 **14.** 1 **15.** 0 **16.** 11

2.1 PRACTICE (pp. 67–70) **5.** < **7.** <

9. 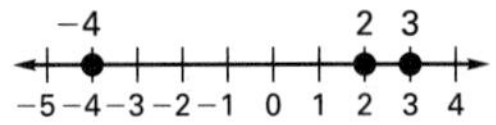

−4, 2, 3

11. 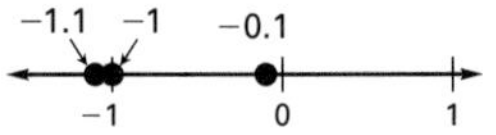

−1.1, −1, −0.1

13. 4.1 **15.** 103 **17.** negative

19. 

$-6.4 < -6.3, -6.3 > -6.4$

23. 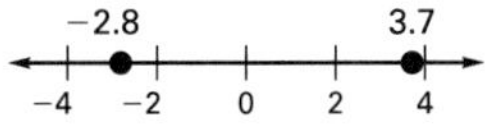

$-2.8 < 3.7, 3.7 > -2.8$

25. 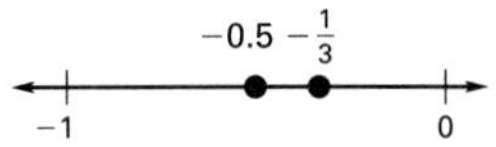

$-0.5 < -\frac{1}{3}, -\frac{1}{3} > -0.5$

27. −1.8, −0.66, 0.7, 3, 4.6, 4.66 **29.** $-3, -2.6, -\frac{1}{2}, 0, \frac{1}{2}, 4.8$ **31.** $-5.8, -\frac{3}{4}, -\frac{1}{2}, \frac{1}{3}, 2.4, 7$ **33.** −8 **35.** −3.8 **37.** $\frac{5}{6}$ **39.** 2.01 **41.** 7 **43.** 4.5 **45.** $\frac{4}{5}$ **47.** 4.3 **49.** 0.09 **51.** 0 **53.** −1, 1 **55.** False; 0 and all negative numbers are counterexamples. **57.** False; all real numbers except 0 are counterexamples. **59.** ≥ **61.** ≤ **63.** −8 **65.** 0 **67.** Regulus **69.** Canopus, Vega, Sirius, and Arcturus **71.** Kobayashi **73.** Bemvenuti, Early, Kobayashi, Scranton **75.** Sörenstam **77.** 429 ft/min, 429 ft/min **79.** 10°F, 25 mi/h **81.** 10°F, 20 mi/h

2.1 MIXED REVIEW (p. 70) **89.** $\frac{5}{6}$ **91.** $\frac{23}{24}$ **93.** $6\frac{13}{63}$ **95.** 5 **97.** 53 **99.** $\frac{4}{5}$ **101.** $r + 8 = 17$

2.2 PRACTICE (pp. 75–77) **5.** −2 **7.** −5 **9.** 5 **11.** 118°F **13.** −3 **15.** −10 **17.** −6 **19.** −4 **21.** 2 **23.** 19 **25.** −19 **27.** 2 **29.** −8.2 **31.** 6 **33.** identity property of addition **35.** property of zero **37.** −85.79 **39.** 126.04 **41.** 5304.128 **43.** −1 **45.** −10 **47.** 3 **49.** −24 **51.** no; 9 + (−5) + 4 + 5 = 13 **53.** F atom 1: 0, not an ion; F atom 2: −1, ion; F atom 2 **55.** No; the sum of the losses is greater than the sum of the profits by $1011.01. **57.** 10:00 A.M.

2.2 MIXED REVIEW (p. 77) **63.** $\frac{2}{9}$ **65.** $\frac{5}{8}$ **67.** $\frac{13}{18}$ **69.** $\frac{11}{48}$ **71.** 15 **73.** 37 **75.** no **77.** yes **79.** yes **81.** domain = 1, 2, 3, 4, 5; range = 6.85, 7.70, 8.55, 9.40, 10.25

2.3 PRACTICE (pp. 82–84) **5.** 3 + 8; 11 **7.** 2 + 3 + (−6); −1 **9.** −3.6 + 6; 2.4 **11.** $\frac{2}{3} + \frac{1}{6} + \left(-\frac{1}{3}\right); \frac{1}{2}$ **13.** $-7y^2$, $12y$, and −6

15.

Input	Function	Output
−2	$y = -(-2) - 3$	−1
−1	$y = -(-1) - 3$	−2
0	$y = -0 - 3$	−3
1	$y = -1 - 3$	−4
2	$y = -2 - 3$	−5

Each time x increases by 1, y decreases by 1.

17. −7 **19.** 5 **21.** −15 **23.** −6 **25.** −4.9 **27.** 9.5 **29.** 3 **31.** $-1\frac{3}{20}$ **33.** 9 **35.** −23.1 **37.** 3 **39.** 2 **41.** −28 **43.** 23 **45.** −8 **47.** −28.6 **49.** $-1\frac{17}{18}$ **51.** −4 and $-y$ **53.** $-3x$ and 6 **55.** −9 and $4b$ **57.** x, $-y$, and −7

59.

Input	Function	Output
−2	$y = -2 - 8$	−10
−1	$y = -1 - 8$	−9
0	$y = 0 - 8$	−8
1	$y = 1 - 8$	−7

61.

Input	Function	Output
−2	$y = -(-2) + 12.1$	14.1
−1	$y = -(-1) + 12.1$	13.1
0	$y = -0 + 12.1$	12.1
1	$y = -1 + 12.1$	11.1

63.

Input	Function	Output
−2	$y = 27 + (-2)$	25
−1	$y = 27 + (-1)$	26
0	$y = 27 + 0$	27
1	$y = 27 + 1$	28

65. 3.1 **67.** 1.54 **69.** 5.8 **71.** 275 ft; up **73.** −$.66/oz **75.** −$.19/oz **77.** from start of trail to cave: 410 ft; from cave to Oak Ridge: 1040 ft; from Oak Ridge to trail junction: −235 ft; from trail junction to Mt. Parker: 1110 ft **79.** −7301 ft, −662 ft, −1883 ft, 77 ft, −1311 ft, 8021 ft

2.3 MIXED REVIEW (p. 85) **85.** $5\frac{2}{3}$ **87.** 16 **89.** 88 **91.** $1\frac{1}{2}$ **93.** false **95.** Yes; the sum of the profits, $7932.18, exceeds the sum of the losses, $6934.36, by $997.82.

QUIZ 1 (p. 85) **1.** −6.2, −5.32, 5.04, 5.3, 5.31, 6.3 **2.** −1.6, −1.07, −0.28, 0.18, 1.06, 1.16 **3.** $-7.5, -7\frac{1}{3}, 7.3, 7\frac{1}{3}, 7.5, 7\frac{2}{3}$ **4.** $-\frac{33}{5}, -6\frac{2}{5}, -6.3, 6.05, 6.42, \frac{33}{5}$ **5.** 3.76 **6.** 75 **7.** −345 **8.** 27.5 **9.** 7 **10.** 67.4 **11.** 24 **12.** −472 **13.** −3 **14.** −1 **15.** −14 **16.** 54 **17.** 3.8 **18.** −91 **19.** −0.4 **20.** 47.3 + (−2.1) + (−11.3) + 12.9; 46.8 in.

2.4 PRACTICE (pp. 89–90)

5. $\begin{bmatrix} -1 & -4 \\ -5 & 1 \\ 0 & 5 \end{bmatrix}; \begin{bmatrix} -5 & 4 \\ -7 & 7 \\ 2 & -13 \end{bmatrix}$ **7.** yes **9.** no **11.** $\begin{bmatrix} 7 & -5 \\ -3 & -1 \end{bmatrix}$

13. $\begin{bmatrix} 4 & -6 & 7 \\ -8 & -2 & 10 \end{bmatrix}$ **15.** $\begin{bmatrix} 7.7 & 8 \\ 4.1 & -6.6 \\ 9.1 & -12.9 \end{bmatrix}$ **17.** $\begin{bmatrix} 1 & -10 \\ 6 & 4 \end{bmatrix}$

19. $\begin{bmatrix} 2 & -2 \\ 5 & -21 \\ 0 & -1 \end{bmatrix}$ **21.** −4, −1, 7, −3 **23.**

	Sale	Reg.
CD	52	3300
Tape	28	1600

25. $\begin{bmatrix} 76 & 59 \\ 50 & 46 \\ 97 & 66 \end{bmatrix}$ **27.** $\begin{bmatrix} 0.20 & 0.20 & 0.20 \\ 0.35 & 0.35 & 0.35 \\ 0.45 & 0.45 & 0.45 \end{bmatrix}$

2.4 Mixed Review (p. 91) **33.** 5^2 **35.** x^2y^3 **37.** $3t^4$ **39.** –43.7 **41.** 36 **43.** 15 **45.** –21 **47.** –3 **49.** –14 **51.** –3

Technology Activity 2.4 (p. 92)

1. $\begin{bmatrix} 7 & -6 \\ 6 & -3 \\ -10 & -14 \end{bmatrix}$; $\begin{bmatrix} 7 & -4 \\ 4 & -3 \\ -8 & -2 \end{bmatrix}$ **3.** $\begin{bmatrix} 10.7 & -1.9 \\ -3.8 & -1.9 \\ 1.7 & -11.1 \\ 7.2 & -8.4 \\ -4.4 & 13.2 \end{bmatrix}$; $\begin{bmatrix} 4.1 & -8.3 \\ 2.6 & -7.3 \\ 18.1 & -7.7 \\ -3 & 1.2 \\ 3.4 & 1 \end{bmatrix}$

2.5 Practice (pp. 96–98) **5.** –8 **7.** 35 **9.** 72 **11.** 70 **13.** 48 **15.** –47 ft **17.** –16 **19.** 5 **21.** $-\frac{11}{8}$ **23.** 24 **25.** 78 **27.** 12 **29.** 8 **31.** $-7x$ **33.** $-4x^3$ **35.** $-16y$ **37.** $-y^5$ **39.** x **41.** $3w^3$ **43.** –12 **45.** 21 **47.** –20 **49.** 76 **51.** –70.70 **53.** 1809.86 **55.** –89.04 **57.** 14.8 **59.** False; any nonpositive number a is a counterexample. **61.** The total amount lost is the loss per pound times the number of pounds of bananas sold. **63.** \$248.56 **65.** –\$125; \$50 **67.** 940,000 visitors **69.** $\begin{bmatrix} -42 & 28 & -21 \\ 7 & -28 & 63 \end{bmatrix}$

71. Sample answers are given. **a.** $\frac{1}{3}(x)(-24)$ **b.** $-\frac{1}{6}(x) + 3.5x$

2.5 Mixed Review (p. 98) **77.** 10 **79.** 32 **81.** 11, –11

83.

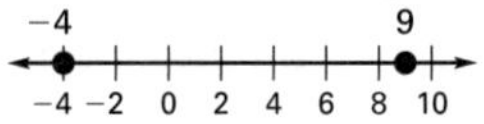

$-4 < 9, 9 > -4$

85.

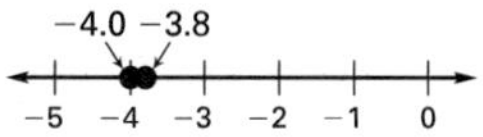

$-3.8 > -4.0, -4.0 < -3.8$

89. $-t$ and 5 **91.** 31 and $-15n$ **93.** m, $-2n$, and $-t^2$ **95.** y, 6, and $-8x$ **97.** –\$11.9 billion

2.6 Practice (pp. 103–105) **5.** $5w - 40$ **7.** $12y - xy$ **9.** $-\frac{5}{6}b + 5$ **11.** 6.9 **13.** simplified **15.** simplified **17.** $8t^2 + 3t - 4$ **19.** false; 3(2) + 3(7) **21.** true **23.** true **25.** $3x + 12$ **27.** $5y - 10$ **29.** $-y + 9$ **31.** $-4t + 32$ **33.** $x^2 + x$ **35.** $-r^2 + 9r$ **37.** $6x - 2$ **39.** $-6x + 12$ **41.** $4x^2 + 32x$ **43.** $15y^2 - 10y$ **45.** $9t + 27$ **47.** $\frac{5}{2}x - \frac{10}{3}$ **49.** $11x$ **51.** $-17b$ **53.** $y + 4$ **55.** $-0.8t$ **57.** $-101a$ **59.** $5x^3 - 2$ **61.** $-5y - 2$ **63.** $10s + 14$ **65.** $4x - 3x^2$ **67.** $9t^2 - 15t$ **69.** $2w^2 + 3w$ **71.** no **73.** 16,098 tons **75.** $T = 5000 + 0.02s + 0.06(5000 - s)$; $T = 5300 - 0.04s$ **77.** $w = 45b + 240$ **79.** $4x - 14$ **81.** $(x + 5)11 = 11x + 55$; $11x + 55$ **83.** $T = 0.10C + 0.5(0.10C)$ **85.** Yes; the simplified version of your friend's equation is identical to your equation. **87.** \$3150 **89.** no

2.6 Mixed Review (p. 106) **95.** $\frac{11}{6}$ **97.** 1 **99.** $\frac{1}{435}$ **101.** $\frac{8}{27}$ **103.** 1 **105.** 3 **107.** 5 **109.** –7 **111.** –6 **113.** 8.8

Quiz 2 (p. 107)

1. $\begin{bmatrix} 5 & -13 \\ -4 & 9 \\ -9 & -8 \end{bmatrix}$; $\begin{bmatrix} -1 & 1 \\ -6 & 1 \\ -5 & 8 \end{bmatrix}$ **2.** –63 **3.** –21 **4.** 14 **5.** –2800 **6.** –3 **7.** 10.8 **8.** 110 **9.** 270 **10.** $7.1t$ **11.** $-13x$ **12.** $-10b^3$ **13.** $-\frac{1}{2}x^4$ **14.** $-43x$ **15.** $2y$ **16.** $-8t$ **17.** $-24 + 3b$ **18.** $20y + 10$

19.

	ft	m
Sears Tower	1454	443
Empire State Building	1250	381
Aon Center	1136	346

2.7 Practice (pp. 111–112) **7.** $-\frac{1}{8}$ **9.** $-\frac{5}{11}$ **11.** –50 **13.** –72 **15.** –\$.17 **17.** –5 **19.** –7 **21.** –11 **23.** $-\frac{1}{3}$ **25.** $-21\frac{7}{9}$ **27.** $-43\frac{1}{5}$ **29.** –7 **31.** 145 **33.** $-12t$ **35.** $-\frac{x}{36}$ **37.** $\frac{3y}{2}$ **39.** $121x$ **41.** $10x^2$ **43.** $30x^2$ **45.** x **47.** $-2w$ **49.** $-11x - 5$ **51.** $9 - x$ **53.** 4 **55.** $217\frac{1}{2}$ **57.** 24 **59.** all real numbers except 2 **61.** all real numbers except 0 **63.** –2.5 ft/sec **65.** $100x$; $1000x$; $10{,}000x$; $100{,}000x$ **67.** $10{,}000{,}000x$ **69.** –11 ft/sec

2.7 Mixed Review (p. 113) **75.** 0.75 **77.** 0.1 **79.** 0.17 **81.** 0.75 **83.** 147 **85.** –250 **87.** 20,736 **89.** 0.45 **91.** –27 **93.** 15 **95.** 12 **97.** –1 **99.** –8

2.8 Practice (pp. 117–119)

5. Unlikely; the probability is 0.25. **7.** 0.2 **9.** 0.25 **11.** 0.65 **13.** 4 to 7 **15.** 1 to 3 **17.** 1 to 3 **19.** 0.5 **21. a.** 0.6 **b.** 3 to 2 **23.** 0.45 **25.** 690 to 298, or about 7 to 3 **27.** 1 to 19 **29.** $\frac{253}{1310} \approx 0.19$ **31.** 1 to 3

2.8 Mixed Review (p. 120) **39.** 8 **41.** $5\frac{1}{2}$ **43.** 120 **45.** $z - 17 = 9$ **47.** $-3 = y + (-6)$ **49.** –13 **51.** –5 **53.** –11.7 **55.** –17.4 **57.** $-4\frac{1}{8}$

Quiz 3 (p. 120) **1.** –49 **2.** 54 **3.** 128 **4.** –4 **5.** –5 **6.** –16 **7.** –320 **8.** $\frac{1}{3}$ **9.** $-7x$ **10.** $\frac{7}{x}$ **11.** $18x$ **12.** $5x$ **13.** $35x$ **14.** $-20t$ **15.** $-55y$ **16.** $\frac{x}{36}$ **17.** 0.32 **18.** Unlikely; the probability is less than 0.25. **19.** Certain; the definition of a certain event is an event with probability 1. **20.** Unlikely; the probability is $\frac{1}{30{,}001}$, which is very small but not equal to 0. **21.** Quite likely; the probability is about 0.7.

Chapter 2 Review (pp. 122–124)

1.

$7 > -6, -6 < 7$

3.

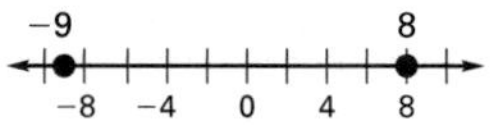

$-9 < 8, 8 > -9$

5. 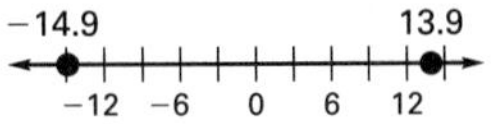

13.9 > –14.9, –14.9 < 13.9

7. 5 **9.** –0.7 **11.** –12 **13.** –1 **15.** –17 **17.** $-4\frac{1}{16}$

19. $\begin{bmatrix} 1 & -4 \\ 1 & 9 \end{bmatrix}; \begin{bmatrix} -7 & 0 \\ 15 & -1 \end{bmatrix}$ **21.** –36 **23.** 600 **25.** $-3\frac{7}{9}$

27. 49.5 **29.** –84 **31.** 14 **33.** $5x + 60$ **35.** $5.5b - 55$ **37.** $-3t - 33$ **39.** $-2.5z + 12.5$ **41.** –4 **43.** –3 **45.** –50 **47.** –64 **49.** 0.25, 1 to 3 **51.** 0.1, 1 to 9

CHAPTER 3

SKILL REVIEW (p. 130) **1.** 71.5 **2.** 29.315 **3.** 15 **4.** 2.52 **5.** no **6.** yes **7.** yes **8.** $12r + 18$ **9.** $-18 + 30z$ **10.** $3x - 21$ **11.** $-9 + 3x$ **12.** $3a + 2$ **13.** $1 + 4y$

3.1 PRACTICE (pp. 135–137) **5.** 13 **7.** 3 **9.** 3 **11.** Add 4.65 to both sides. **13.** Subtract 28. **15.** Add 15. **17.** Subtract –3 (or add 3). **19.** Add –45 (or subtract 45). **21.** –3 **23.** 8 **25.** –11 **27.** 10 **29.** $-\frac{3}{4}$ **31.** 5 **33.** 6 **35.** 29 **37.** –6 **39.** –16 **41.** –9 **43.** B; 8 members **45–49 odd:** Sample equations are given. **45.** $x + 7 = 24$; \$17 **47.** $43{,}368 + x = 49{,}831$; 6463 seats **49.** $x - 732 = 645$; 1377 min **51.** 20 cm **53.** $c + 248 = 4218 + 3800 + 2764$; 10,534 acres

3.1 MIXED REVIEW (p. 137) **57.** $\frac{n}{6} = 32$ **59.** $\frac{2}{3}n = 8$ **61.** y **63.** x **65.** y **67.** $-a$ **69.** $21 - 14y$ **71.** $-6x - 16$ **73.** $-5xy - 15x$ **75.** $18xy + 54y^2$

3.2 PRACTICE (pp. 141–144) **5.** 32 **7.** –12 **9.** $-\frac{5}{3}$ **11.** 5 **13.** *Sample equation:* $\frac{x}{3} = \frac{4}{6}$; 2 **15.** Divide by –2. **17.** Divide by $\frac{2}{3}$ (or multiply by $\frac{3}{2}$). **19.** Multiply by $-\frac{4}{3}$. **21.** no **23.** no **25.** –2 **27.** $\frac{1}{6}$ **29.** 18 **31.** $\frac{3}{8}$ **33.** 84 **35.** –40 **37.** 3 **39.** –90 **41.** –1 **43.** $-7\frac{1}{3}$ **45.** 20 **47.** $4\frac{1}{2}$ **49–53 odd:** Sample equations are given. **49.** $52x = 676$; 13 pieces **51.** $\frac{3}{8}x = 3.3$; \$8.80 **53.** $30{,}000x = 10{,}000$; $\frac{1}{3}$ ft^2 **57. a.** 12 min **b.** 36 sec **59.** $\frac{x}{12} = \frac{4}{8}$; 6 in. **61.** $\frac{5}{4} = \frac{x}{1}$; $1\frac{1}{4}$ c

3.2 MIXED REVIEW (p. 144) **67.** $9n - 12 = 60$ **69.** $11 = \frac{2}{5}(n - 13)$ **71.** $7y - 9$ **73.** $12y + 15$ **75.** $-19y + 54$ **77.** 3 **79.** 30 **81.** 7

3.3 PRACTICE (pp. 148–151) **3.** 2 **5.** 13 **7.** 10 **9.** 6 errands **11.** yes **13.** no **15.** no **17.** 3 **19.** 14 **21.** 4 **23.** 12 **25.** 2 **27.** 1 **29.** 8 **31.** 3 **33.** 3 **35.** 8 **37.** –13 **39.** 2 **41.** The distributive property was applied incorrectly; $2(x - 3) = 2x - 6$; $5\frac{1}{2}$. **43.** The left side was simplified incorrectly; $4\left(\frac{1}{4}x - 2\right) = x - 8$; 36.

47, 49. Preferences may vary. **47.** –24 **49.** –1 **51.** 21, 33, and 69 **53.** *Sample equation:* $34x + 339 = 458$; 3.5 h **55.** 125 students **57.** 48 pounds per square inch **59.** Let t = time to complete; $1800t + 2400t = 8400$; 2 h **61.** *Sample equation:* $15x + 20x = 910$; 26 h **63.** *Sample equation:* $25q + 10(q - 4) + 5(2q) = 500$; 12 quarters, 24 nickels, and 8 dimes

3.3 MIXED REVIEW (p. 151) **67.** 4^2 **69.** a^6 **71.** 2^4 **73.** 10 **75.** 13.9 **77.** –22 **79.** $\frac{1}{2}$ **81.** –3 **83.** hamburgers: \$134 income, \$24 expense; hot dogs: \$137 income, \$29 expense; tacos: \$118 income, \$45 expense **85.** tacos

QUIZ 1 (p. 152) **1.** 17 **2.** $4\frac{1}{2}$ **3.** –2 **4.** $\frac{5}{8}$ **5.** 32 **6.** 3 **7.** –46 **8.** 12 **9.** –5 **10.** *Sample equation:* $x + 4 = 91$; 87 **11.** $22\frac{2}{9}$°C

3.4 PRACTICE (pp. 157–158) **5.** –7; one solution **7.** no solution **9.** 5; one solution **11.** 10 pies **13.** 1 **15.** $\frac{1}{4}$ **17.** 1 **19.** –7 **21.** $-\frac{3}{32}$ **23.** $\frac{1}{4}$ **25.** 2 **27.** $\frac{1}{2}$ **29.** 1 **31.** all real numbers **33.** 4 **35.** $-\frac{1}{2}$ **37.** no solution **39.** 1 **41.** 1 **43.** Since –6 = –6, you can conclude that the given equation is true for all real numbers, not that $b = -6$. **45.** *Sample equation:* $3x + 25 = 5x$; 12.5 sessions; it is worthwhile to become a member if you expect to attend 13 or more sessions. **47.** The possible number is $3\frac{1}{2}$ more than the actual number, 50. **49.** about 12 ft

3.4 MIXED REVIEW (p. 159) **55.** 0.28 **57.** 0.03 **59.** 37.8 **61.** 410.4 **63.** \$74.28 **65.** dollars **67.** hours **69.** 6 cm

3.5 PRACTICE (pp. 163–165) **3.** rate your friend is driving = 52 (mi/h); time after you start driving = t (h); friend's distance when you leave = 32 (mi); rate you are driving = 60 (mi/h); $52t + 32 = 60t$.

5. *Sample answer:*

Hours	Your distance (mi)	Friend's distance (mi)
1	60	84
2	120	136
3	180	188
4	240	240

7. C; 18 in. **9.** hours **11.** ft **13.** No; according to the graph the solution is 5 weeks, not 25 weeks. **15.** Let x = the width in inches of the photos; $2x + 2 = 6\frac{1}{2}$. **17.** *Sample equation:* $45 + 3x = 108 - 4x$; in 9 years **19.** $\frac{1}{2}$ mi; $x + \frac{1}{2}$

3.5 MIXED REVIEW (p. 165) **27.** 0.27; 27% **29.** 0.81; 81% **31.** –25 **33.** –1.5 **35.** –27 **37.** $1\frac{3}{4}$ **39.** $-\frac{1}{2}$

3.6 PRACTICE (pp. 169–171) **5.** 23.4 **7.** –13.9 **9.** 6.82 **11.** 3.17 **13.** \$33.38 **15.** 36.3; 36.26 **17.** 7.3; 7.32 **19.** 34.7; 34.70 **21.** 4.33 **23.** 5.78 **25.** 1.71 **27.** –2.88 **29.** 0.11 **31.** 0.14 **33.** 1.60 **35.** 12.17 **37.** 2.75 **39.** 0.31 **41.** $-625y - 184 = 2506y$ **43.** $4500n - 375 = 750n + 2000$ **49.** \$7.29 **51.** 0.24, or 24% **53.** *Sample equation:* $500 + 440x = 2975$; 6 min **55.** 9.75 h; the total number of hours that you used your Internet service last month **57.** $(t - 10)$°C, $0.37(t - 10)$ mm, $[16.8 - 0.37(t - 10)]$ mm

3.6 MIXED REVIEW (p. 172)

61.

Input	Output
2	13
3	15.5
4	18
5	20.5
6	23

domain: 2, 3, 4, 5, 6; range: 13, 15.5, 18, 20.5, 23

63. 3 **65.** –7.9 **67.** $\frac{8}{15} \approx 0.533$ **69.** It increased by \$85.25. **71.** –\$234.87

QUIZ 2 (p. 172) **1.** $\frac{1}{2}$ **2.** $3\frac{1}{4}$ **3.** no solution **4.** $1\frac{2}{5}$ **5.** no solution **6.** $\frac{1}{4}$ **7.** 0.45 **8.** 1.29 **9.** –0.10 **10.** 0.41 **11.** C **12.** 10 dogs

TECHNOLOGY ACTIVITY 3.6 (p. 173) **1.** 12.3 **3.** 5.3

3.7 PRACTICE (pp. 177–179) **9.** $y = 5 - x$ **11.** $b = \frac{2A}{h}$ **13.** $h = \frac{V}{\pi r^2}$ **15.** $y = 5 - 2x$ **17.** $y = 6x - \frac{13}{2}$ **19.** $y = -1.5x + 9$ **21.** $y = -10x + 35$ **23.** $y = \frac{3}{4}x - \frac{9}{4}$ **25.** $y = 12x - 2$ **27.** $y = x - \frac{9}{4}$ **29.** $y = \frac{1}{2}x + 1$ **31.** $x = 3y - 12$; –18, –15, –12, –9 **33.** $x = \frac{5}{2}y - 3$; –8, $-5\frac{1}{2}$, –3, $-\frac{1}{2}$ **35. a.** 0.2; 20% **b.** 0.4; 40% **c.** 0.5; 50% **37.** $d = \frac{P - 2112}{64}$ **39.** *P* is a function of *d*; *P* is the isolated variable. **41.** $2\pi X$ **43.** The first and third columns are identical.

3.7 MIXED REVIEW (p. 179) **51.** no **53.** $-8b$ **55.** $-3x^3$ **57.** $16t^2$ **59.** pro sports **61.** 31

3.8 PRACTICE (pp. 183–185) **5.** \$213 **7.** $0.45(280) = x$; 126 **9.** $0.2x = 15$; 75 **11.** \$.40/can **13.** about \$.93/qt **15.** 300 mi/h **17.** 0.05 mi/min **21.** 21,000,000 mi/wk **23.** $1\frac{1}{2}$ c, or 1.5 c **25.** 90.1 km **27.** about 300 mi **29.** *Sample equation:* $50 \cdot 47 = x$; about 2350 more books **31.** \$5 **33.** 16% **35.** 65% **37.** about 28.9% **39.** about 1013 people **41.** *Sample estimate:* about 350 students

3.8 MIXED REVIEW (p. 186)

49.

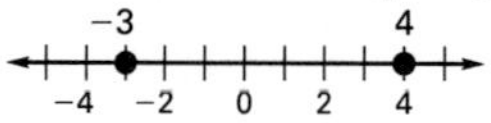

$4 > -3$; $-3 < 4$

51.

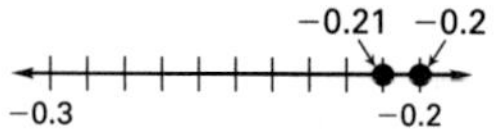

$-0.2 > -0.21$; $-0.21 < -0.2$

53.

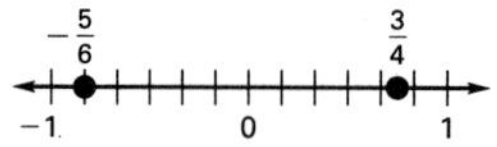

$\frac{3}{4} > -\frac{5}{6}$; $-\frac{5}{6} < \frac{3}{4}$

55. –1.91 **57.** 1.47 **59.**

Projected Number of People 85 Years and Older

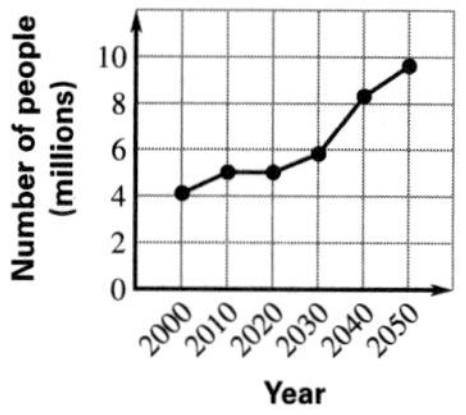

QUIZ 3 (p. 186) **1.** $y = -\frac{3}{2}x + 6$; 9, 6, 3 **2.** $y = 3x + 12$; 6, 12, 18 **3.** $y = -\frac{1}{5}x + 1\frac{3}{5}$; 2, $1\frac{3}{5}$, $1\frac{1}{5}$ **4.** $w = \frac{V}{\ell h}$ **5.** $h = \frac{2A}{b}$ **6.** $r = \frac{P}{C} - 1$ **7.** \$93 **8.** 20 in. **9–12.** Equations may vary. **9.** $33 \cdot 12 = x$; 396 mi **10.** $0.03x = 42$; 1400 students **11.** $900 \cdot \frac{210}{525} = x$; about 360 people **12.** 40%

CHAPTER 3 EXTENSION (p. 188) **1.** Inductive reasoning; a conclusion is reached based on observation of a pattern. **3.** Deductive reasoning; a conclusion is reached because the hypothesis is true. **5.** *Sample counterexample:* The square of 1 is equal to itself. **7.** 21, 24, 27 **9.** hypothesis: "you are in Minnesota in January"; conclusion: "you will be cold"

CHAPTER 3 REVIEW (pp. 190–192) **1.** 11 **3.** –60 **5.** –432 **7.** –12 **9.** 3 **11.** –8 **13.** –1 **15.** $\frac{15}{2}$ **17.** –2 **19.** all real numbers **21.** –15 **23.** $12\left(\frac{5}{12}\right) = x\left(\frac{1}{2}\right)$; 10 km/h **25.** 2.6 **27.** 0.4 **29.** $r = \frac{S}{\pi s} - R$ **31.** $y = -\frac{3}{4}x + \frac{7}{2}$; 5, $\frac{7}{2}$, $2\frac{3}{4}$, $-\frac{1}{4}$ **33.** \$200

CUMULATIVE PRACTICE (pp. 196–197) **1.** 14 **3.** 81 **5.** 1.01 **7.** –3 **9.** 2 **11.** –20 **13.** 4 **15.** $\frac{5}{6}$ **17.** yes **19.** yes **21.** yes **23.** $4n + 17$ **25.** $-3n > 12$ **27.** $\begin{bmatrix} 5 & 4 \\ -6 & -5 \end{bmatrix}$ **29.** $\begin{bmatrix} 26 & 14 & -6 \\ -54 & -3 & 35 \end{bmatrix}$ **31.** $\frac{2}{11} \approx 0.18$ **33.** $\frac{1}{6} \approx 0.17$ **35.** 8 **37.** 15 **39.** 16 **41.** –19 **43.** $-1\frac{4}{11}$ **45.** –2.30 **47.** 1.51 **49.** $x = \frac{3}{2}y - \frac{1}{2}$; $-3\frac{1}{2}$, $-\frac{1}{2}$, $1\frac{3}{4}$, 4 **51.** $27\frac{7}{9}$°C **53.** 300 rolls **55.** 3639 kronor **57.** 4852 kronor

SELECTED ANSWERS

SA6

CHAPTER 4

SKILL REVIEW (p. 202) **1.** $0.5, \frac{1}{2}$ **2.** $0.75, \frac{3}{4}$ **3.** $0.01, \frac{1}{100}$ **4.** $0.2, \frac{1}{5}$ **5–7.** Sample answers are given.

5.

x	y
0	70
1	75
2	80
5	95
10	120

6.

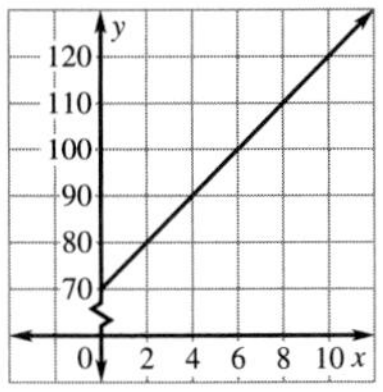

7. domain: $x \geq 0$; range: $y \geq 70$ **8.** -1 **9.** 2

4.1 PRACTICE (pp. 206–208) **3.** $A(1, 1)$, $B(-2, 1)$, $C(0, -2)$

5.

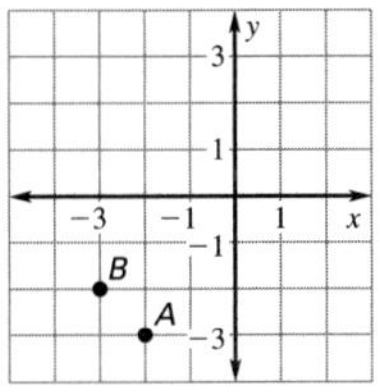

7.

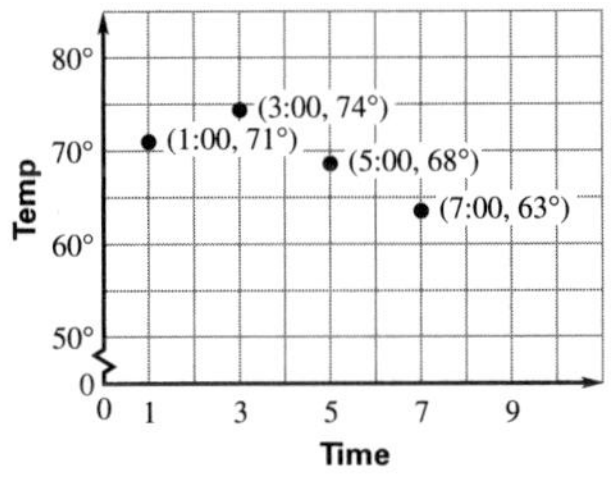

9. *Sample answer:* Yes; The amount spent on snowmobiles increased but not rapidly. **11.** $A(2, 4)$, $B(0, -1)$, $C(-1, 0)$, $D(-2, -1)$

13.

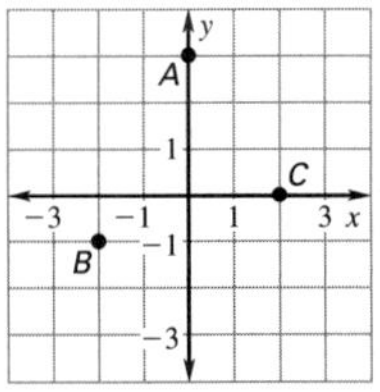

17.

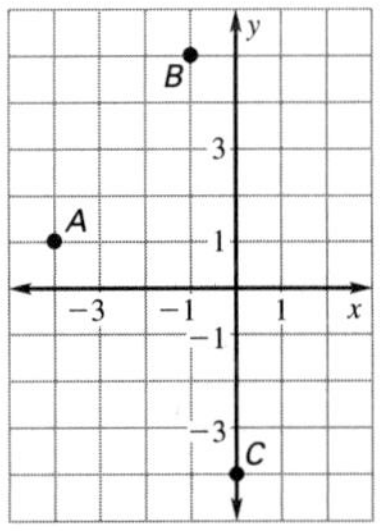

19. Quadrant IV **21.** Quadrant I **23.** Quadrant III **25.** Quadrant III **27.** pounds; inches **29.** C **31.** It decreases. **35.**

Wing Length vs. Wing-beat Rate

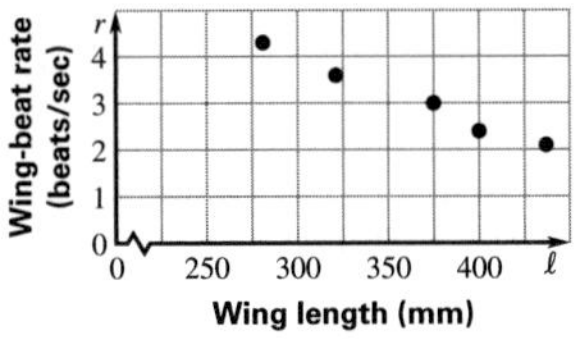

37. As wing length increases, wing-beat rate decreases.

4.1 MIXED REVIEW (p. 208) **43.** 7 **45.** 15 **47.** 39 **49.** 1.07 **51.** $\frac{2}{3}$

TECHNOLOGY ACTIVITY 4.1 (p. 209) **1.** As age increases, time decreases. **3.** From ages 10–14, as age increases, time decreases. From ages 14–16, time increases, then decreases at age 17.

4.2 PRACTICE (pp. 214–217)

5.

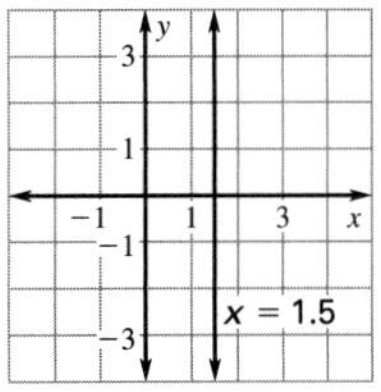

7. no **9.** no **11.** about 35,180,000 households **13. a.** no; $y \neq 5$ **b.** yes; $y = 5$ **15.** no **17.** yes **19.** no **21–29 odd:** Sample answers are given. **21.** $(-1, -8)$, $(0, -5)$, $(1, -2)$ **23.** $(-1, -4)$, $(0, -6)$, $(1, -8)$ **25.** $\left(\frac{1}{2}, -1\right)$, $\left(\frac{1}{2}, 0\right)$, $\left(\frac{1}{2}, 1\right)$ **27.** $(-1, 3)$, $(0, 2)$, $(1, 1)$ **29.** $(-1, -6)$, $(0, -4)$, $(1, -2)$ **31.** $y = -\frac{2}{3}x + 2$ **33.** $y = -x + \frac{19}{5}$ **35.** $y = -x - 5$

37.

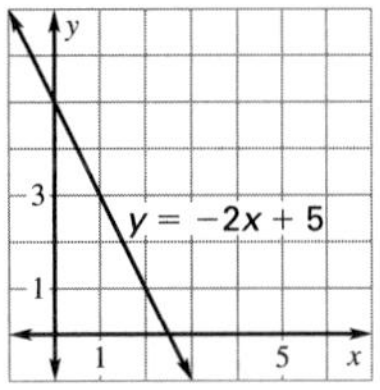

39.

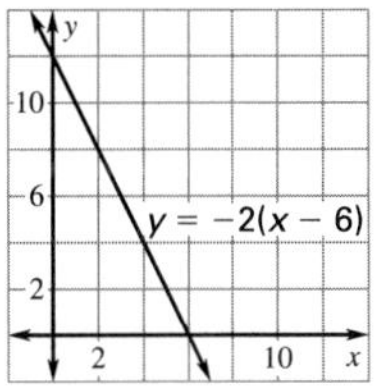

41.

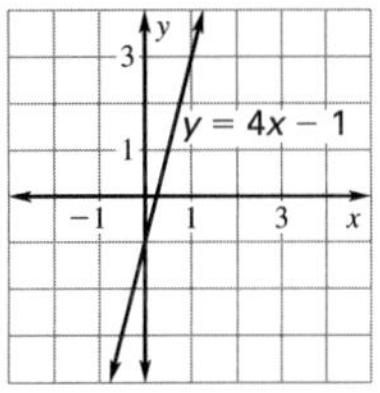

45.

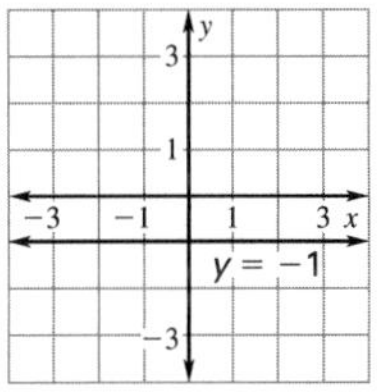

49.

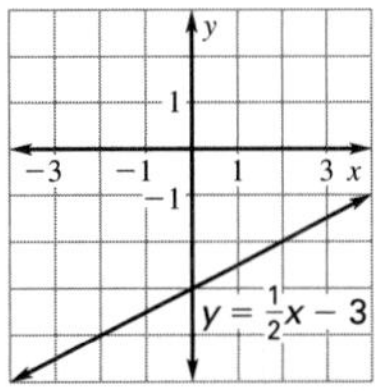

51.

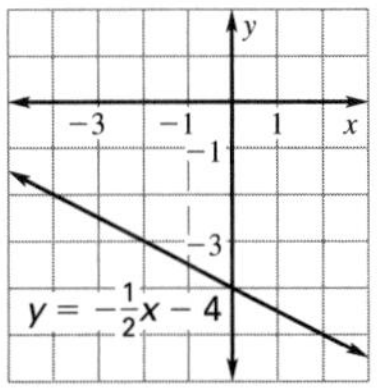

53. D **55.** C **57.**

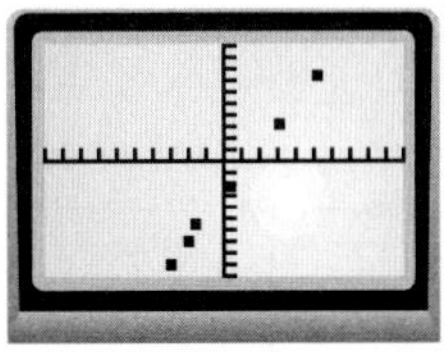

59. $x = -1.6$; $2(-1.6) - 3 = -6.2$, but $y = -5.5$. **61.** $(-8, 11)$ **63.** $(4, -4)$ **65.** $(3, 0)$ **67.** $y = -2x + 30$ **69.** yes **71.** Running time; Calories burned while swimming; 7.1; Calories burned while swimming; Swimming time; $7.1x + 10.1y = 800$ **73.** about 48 min **75.** Horizontal; the number of senators does not vary.

4.2 MIXED REVIEW (p. 217) **81.** 4 **83.** -29 **85.** 0 **87.** $\begin{bmatrix} 16 & 3 \\ -4 & 18 \end{bmatrix}$ **89.** $\begin{bmatrix} -16 & 50 \\ -9 & 19 \end{bmatrix}$ **91.** 13 **93.** -16 **95.** -15 **97.** 9

SA7

4.3 PRACTICE (pp. 221–223)

5. –3 **7.** –2; 2

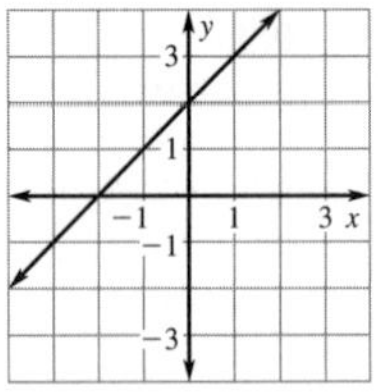

9. 2; –4

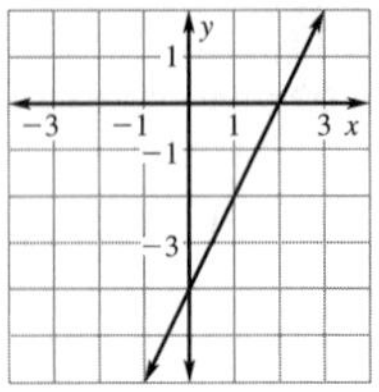

11. –3; 3

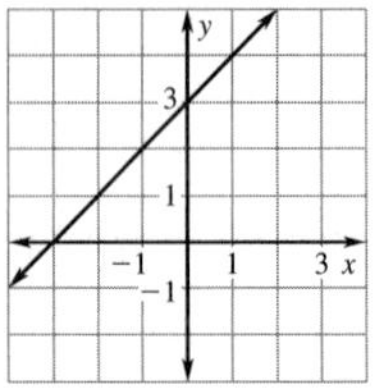

13. *Sample answer:* \$8 for adult tickets and \$4 for student tickets or \$9 for adult tickets and \$2 for student tickets
15. –2; 4 **17.** 5 **19.** –5 **21.** 9 **23.** –6 **25.** –5 **27.** –4
29. –15 **31.** –8 **33.** 3 **34.** 3

35.

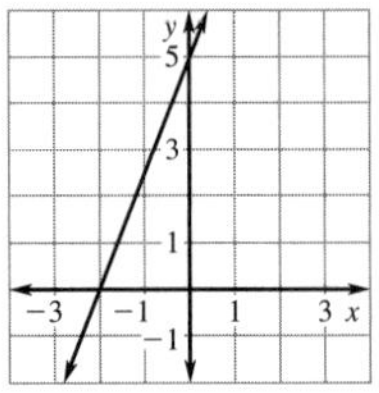

37.

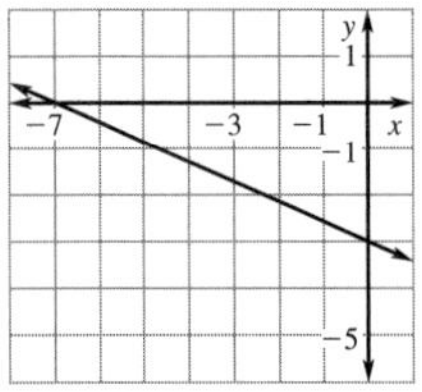

41. A **43.** B **45.** 3; –3

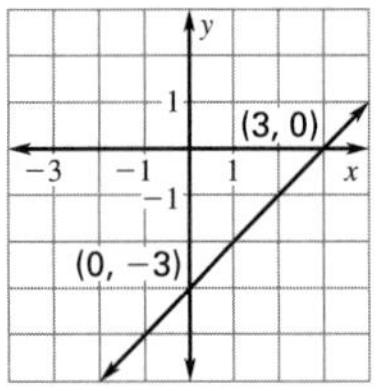

47. 2; –6

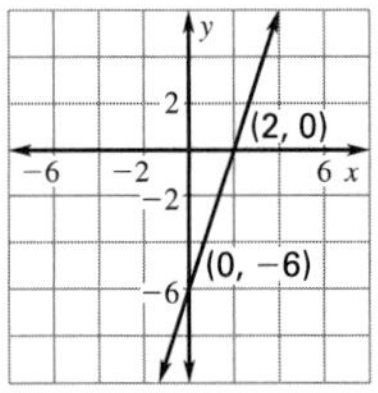

49. 8; 4

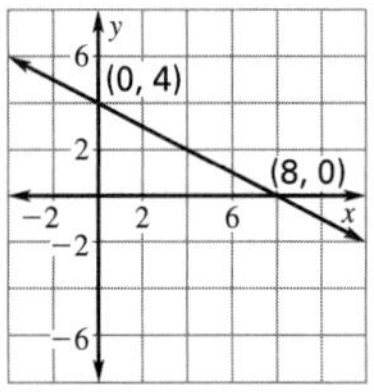

53. –18; –4

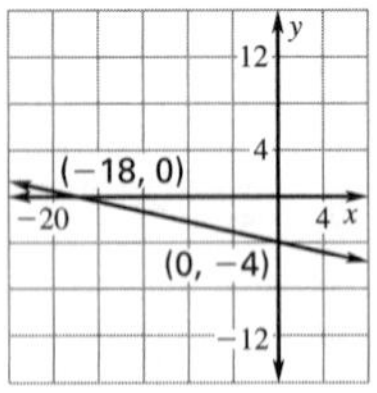

55. 4; 16

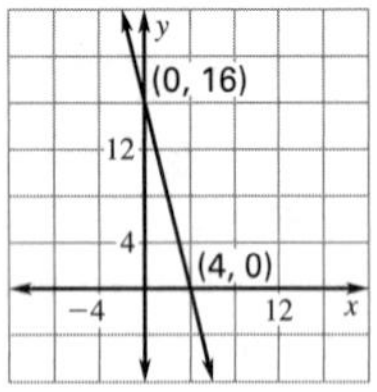

57. True; when $y = 0$, $3x = 30$ and $x = 10$. **59.** False; the graph is a vertical line. **61.** 75; the number of adult tickets that must be sold if no student tickets are sold
63. The possible pairs (x, y) are (0, 120), (5, 112), (10, 104), (15, 96), (20, 88), (25, 80), (30, 72), (35, 64), (40, 56), (45, 48), (50, 40), (55, 32), (60, 24), (65, 16), (70, 8), and (75, 0).

65. Running Time vs. Walking Time

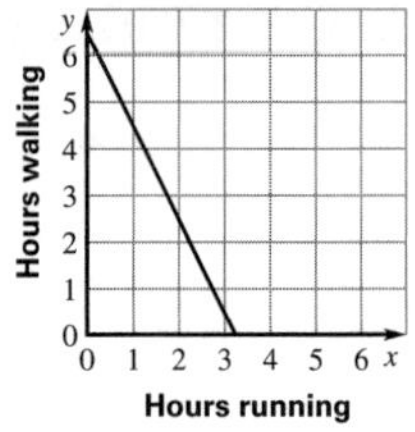

Sample answers:
(1.5, 3.55), (2.25, 2.05), (2.5, 1.55), (2.75, 1.05), (3, 0.55)

67. Theater Attendance

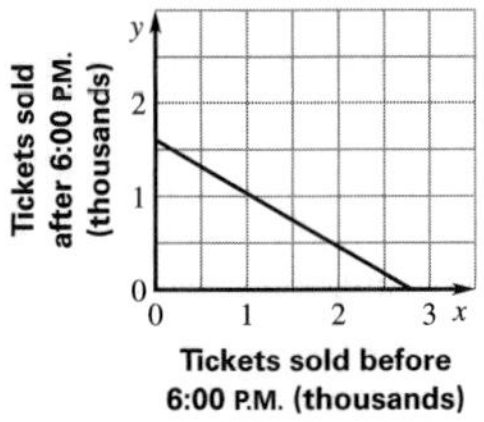

4.3 MIXED REVIEW (p. 223)

71. –4 **73.** –17 **75.** $\frac{5}{3}$ **77.** 6
79. 6 **81.** –60 **83.** –2 **85.** 8 **87.** B

QUIZ 1 (p. 224)

1.

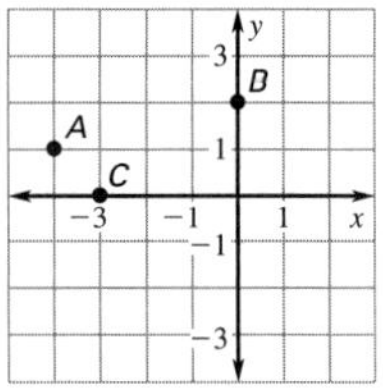

2.

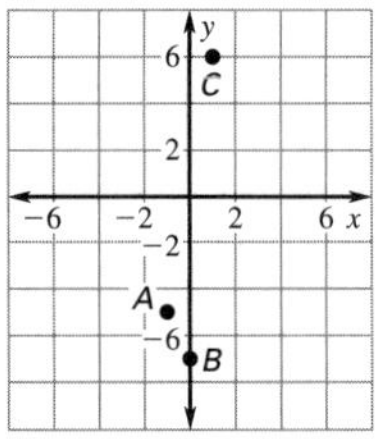

3.

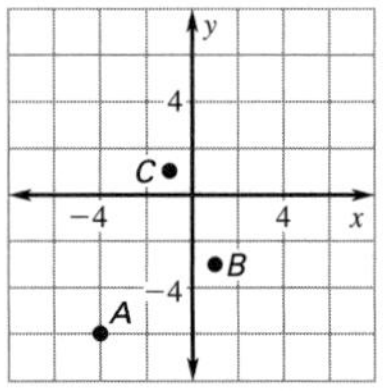

4.

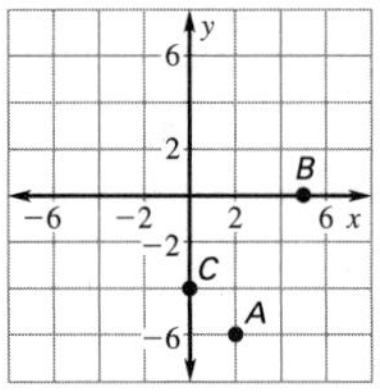

5–10. Sample answers are given.

5. (0, –6), (3, 0), (2, –2)

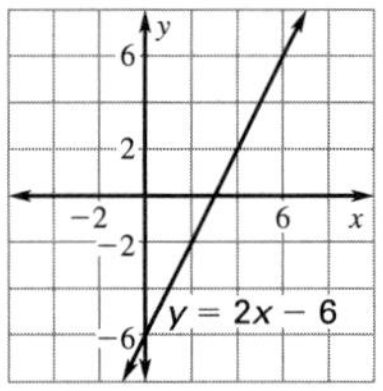

6. (0, –1), $\left(\frac{1}{2}, 1\right)$, $\left(-\frac{1}{2}, -3\right)$

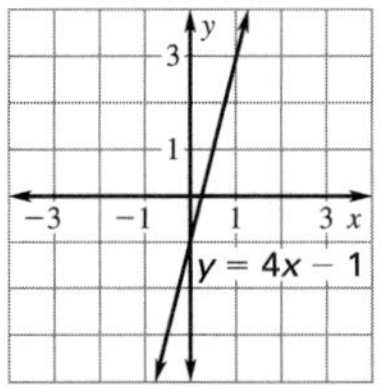

7. (0, 2), (1, –4), (–1, 8)

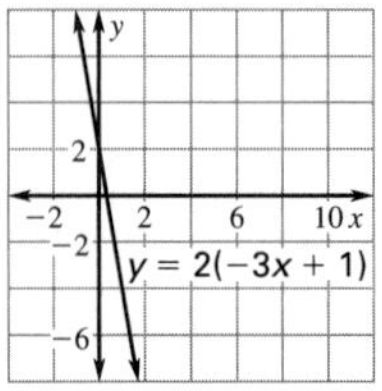

8. (3, –1), (3, 0), (3, 1)

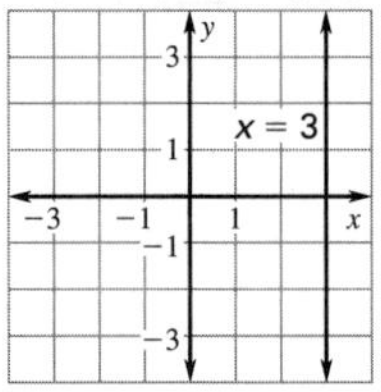

9. (0, 12), (1, 9), (4, 0)

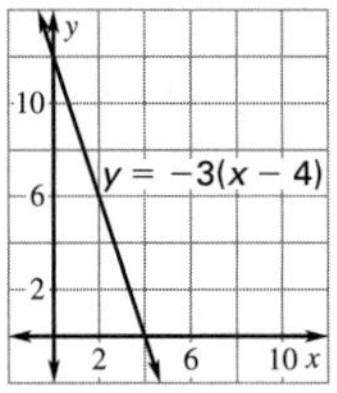

10. (–3, –5), (9, –5), (2, –5)

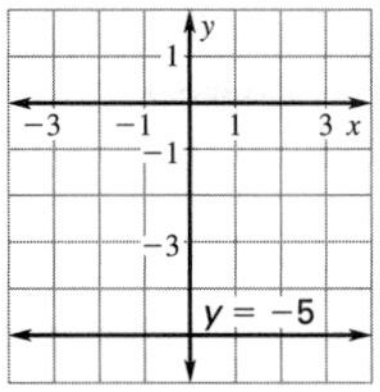

11. 4; 4

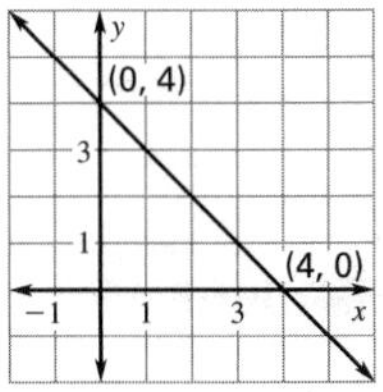

12. $2\frac{1}{2}$; –5

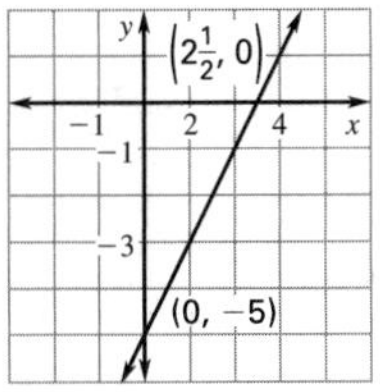

13. –4; 12

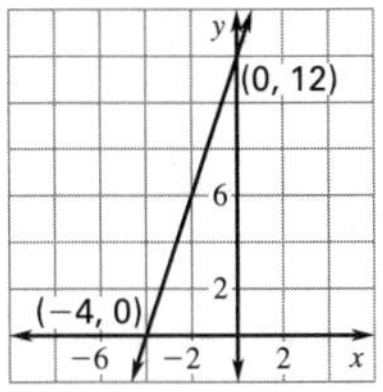

14. 9; 9

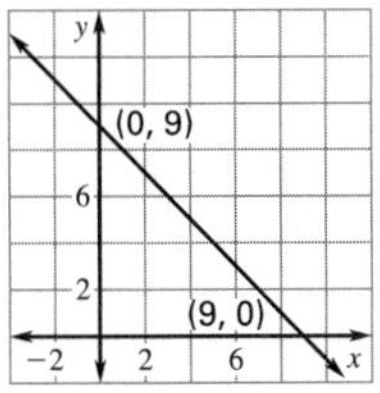

15. $\frac{1}{2}$; –3

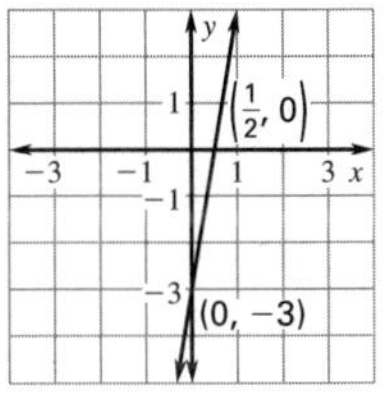

16. –5; 50

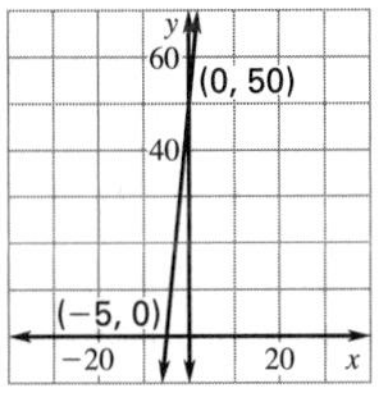

4.4 PRACTICE (pp. 230–233) **5.** 2 **7.** –1 **9.** 3 **11.** –9 **13.** Negative; the line falls from left to right. **15.** Negative; the line falls from left to right. **17.** Negative; the line falls from left to right. **19.** Undefined; the line is vertical. **21.** $-\frac{3}{4}$ **23.** $\frac{3}{4}$ **25.** –4 **27.** $-\frac{5}{2}$ **29.** undefined **31.** $\frac{5}{2}$ **33.** $-\frac{3}{5}$ **35.** *a:* –1; *b:* 1 **37.** *a:* –2; *b:* undefined **39.** 4 **41.** 1 **43.** –3 **45.** 3 **47.** No matter which pairs of points are chosen, the slope is the same, $\frac{1}{2}$. **49.** No; the slope of the line that passes through (–2, 4) and (2, –2) is $-\frac{3}{2}$. The slope of the line that passes through (2, –2) and (6, 0) is $\frac{1}{2}$. The slope of the line that passes through (–2, 4) and (6, 0) is $-\frac{1}{2}$. Three distinct lines, each with a different slope, are determined by the three points. **51.** $\frac{3}{50}$; The road grade is the slope of the road expressed as a percent. **53.** \$8/year **55.** section 1: $-\frac{1}{20}$; section 2: $\frac{3}{20}$; section 3: $\frac{1}{6}$ **57.** \$5,500,000/year **59.** about \$.15/year **63.** 1960–1970; about 0.5 million people/year

4.4 MIXED REVIEW (p. 233) **73.** 12 **75.** 11 **77.** 8 **79.** x^2 **81.** $-2x^2$ **83.** $-\frac{1}{2}x^2$ **85.** $y = -\frac{5}{9}x + 2$ **87.** $y = \frac{1}{4}x + 9$ **89.** $y = -\frac{3}{5}x + \frac{17}{5}$ **91.** –6; 12 **93.** 3; –6 **95.** $-\frac{1}{2}$; 1

4.5 PRACTICE (pp. 237–238) **3.** yes; 1; 1 **5.** yes; $\frac{1}{2}$; $\frac{1}{2}$ **7.** no **9.** 6; 6; 6; 6 **11.** \$36 **13.** $-\frac{2}{5}$; $-\frac{2}{5}$ **15.** –3; –3 **17.** 0.4; 0.4 **19.** $\frac{5}{4}$; $\frac{5}{4}$ **21.** yes **23.** $y = 3x$ **25.** $y = \frac{2}{9}x$ **27.** $y = 0.2x$ **29.** $y = x$ **31.** $y = -\frac{1}{3}x$ **33.** 324 lb **35.** yes **37.** 45 in.; lower

4.5 MIXED REVIEW (p. 239) **45.** 2 **47.** –5 **49.** –3 **51.** $x = -\frac{5}{2}y + 6$ **53.**

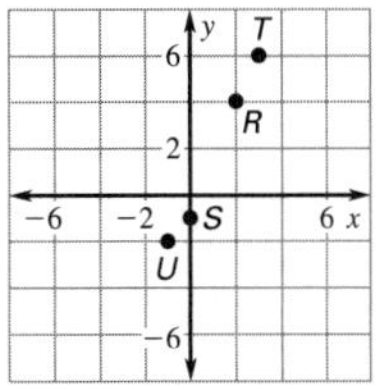

55. No; 3(–4) – 6(–2) = 0 and 0 ≠ –2

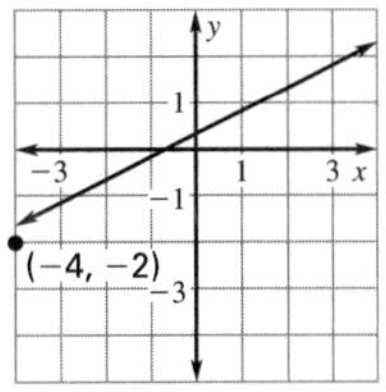

57. No; –4(–4) – 3(2) = 10 and 10 ≠ –8

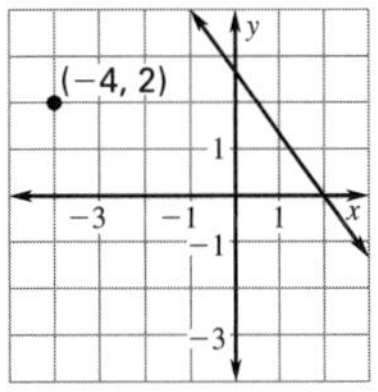

59. 10; $-9\frac{1}{11}$ **61.** 10; 2 **63.** 7; $-\frac{7}{3}$

4.6 PRACTICE (pp. 244–246) **5.** 2; 1 **7.** 1; 3 **9.** 5; –3 **11.** 30; 50 **13.** 6; 4 **15.** 2; –3 **17.** 0; –2 **19.** $\frac{1}{4}$; $\frac{1}{2}$ **21.** –3; $\frac{1}{2}$ **23.**

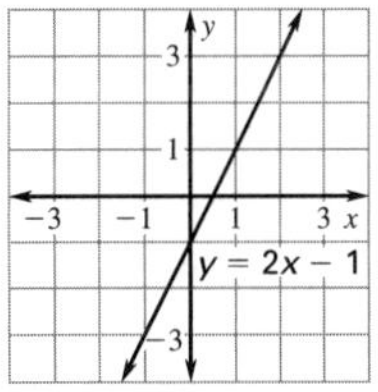

25.

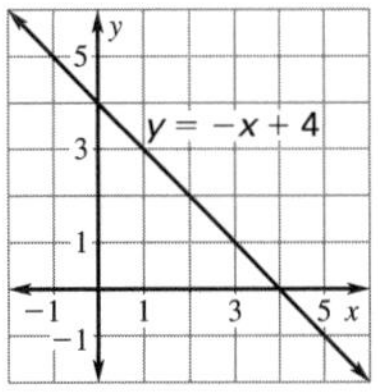

29.

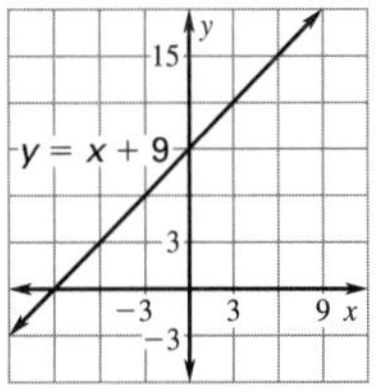

31.

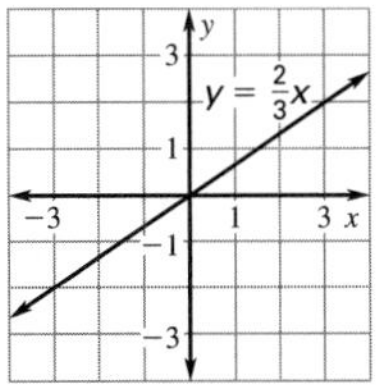

33.

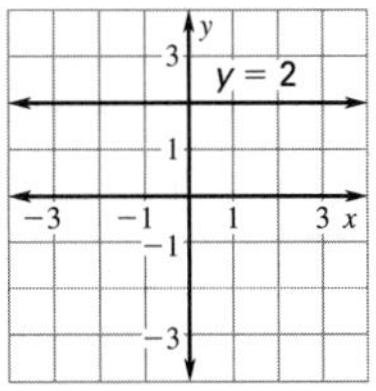

35. $y = \frac{1}{2}x - \frac{3}{2}$

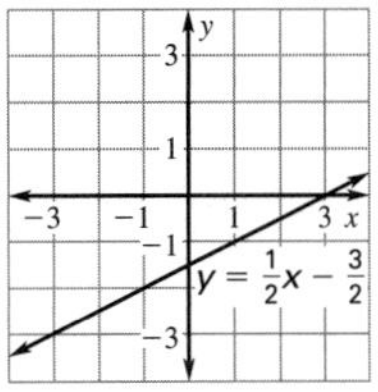

37. $y = 4x - 3$

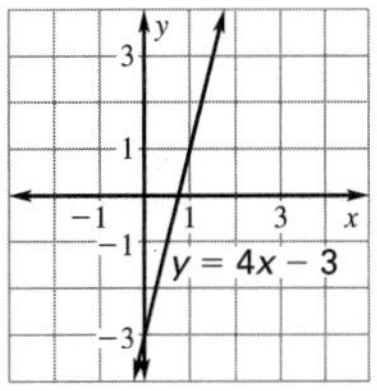

SELECTED ANSWERS

39. $y = -x$

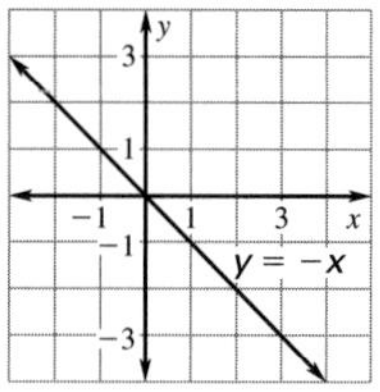

41. $y = -\frac{1}{3}x + 1$

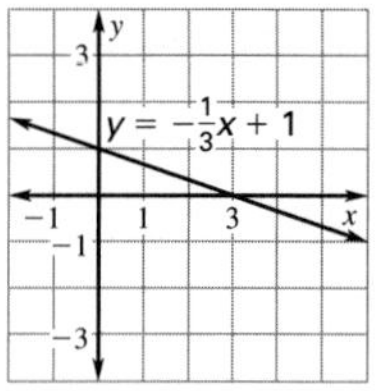

47. Yes; both have slope 2. **49.** No; one has slope $\frac{4}{3}$, the other has slope $-\frac{4}{3}$. **51.** Yes; both have slope 3. **53.** C **55.** B **63.** A triangle; 9 square units; the base is 6 units long and the height is 3 units.

65.

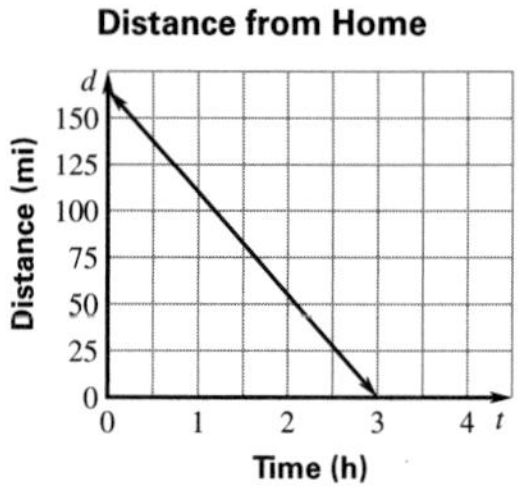

67. 4 ft

4.6 MIXED REVIEW (p. 247) **79.** 5 **81.** 12 **83.** 6 **85.** –1 **87.** A **89.** from 2001 to 2002

QUIZ 2 (p. 247) **1.** $\frac{2}{5}$ **2.** $\frac{2}{5}$ **3.** $\frac{7}{9}$ **4.** 2 **5.** 0 **6.** –1 **7.** $y = 5x$ **8.** $y = 8x$ **9.** $y = 6x$ **10.** $y = \frac{3}{4}x$ **11.** $y = 0.7x$ **12.** $y = \frac{4}{9}x$

13. $y = 2x - 4$

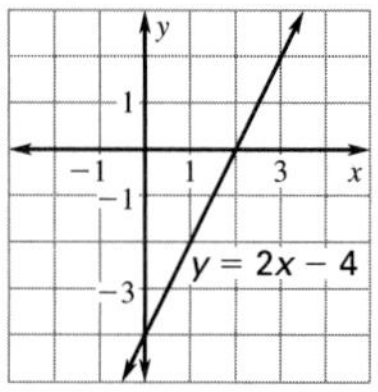

14. $y = -\frac{1}{2}x - \frac{1}{3}$

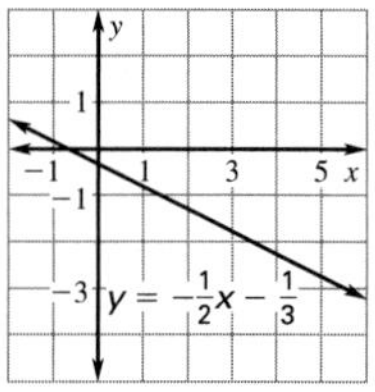

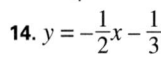

15. $y = \frac{1}{2}x - 6$

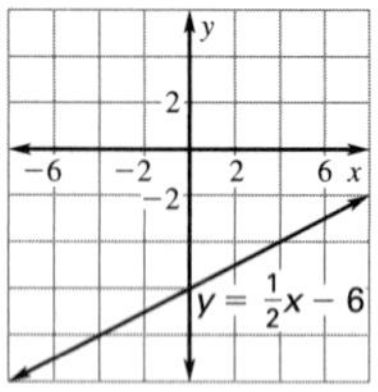

16. $y = x + 2$

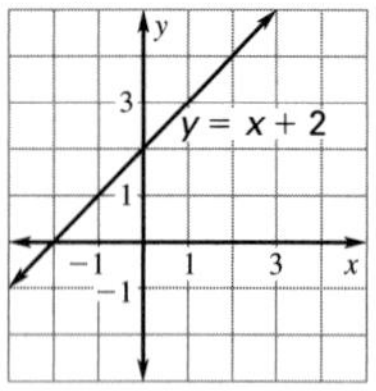

17. $y = -\frac{1}{2}x + 1$

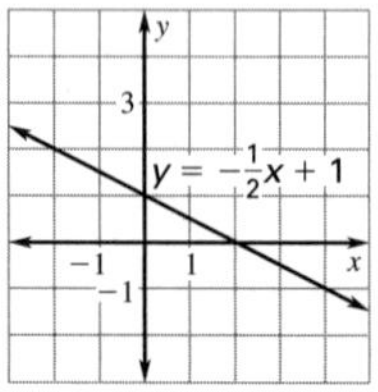

18. $y = -x + \frac{1}{2}$

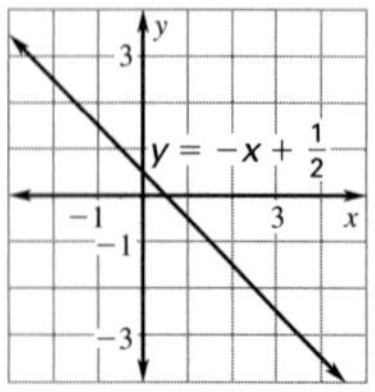

19. –\$30,080/month

TECHNOLOGY ACTIVITY 4.6 (p. 248–249)

1.

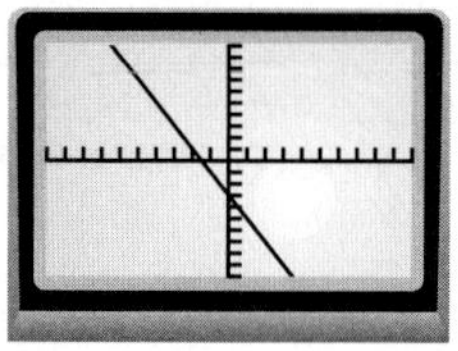

3.

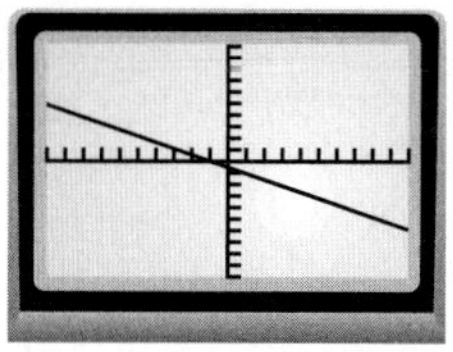

5.

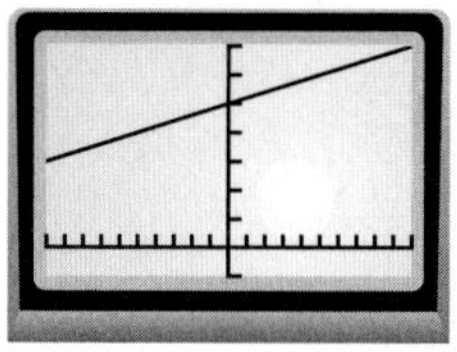

7.

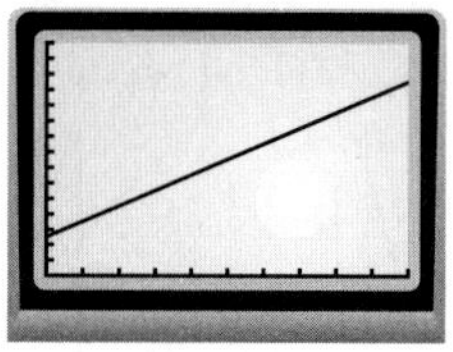

9. *Sample answer:* standard viewing window
11. *Sample answer:* Xmin = –10, Xmax = 30, Xscl = 5, Ymin = –10000, Ymax = 50000, Yscl = 5000
13. about 662.5 **15.** about –1911

4.7 PRACTICE (pp. 253–254) **7.** A **9.** B **11.** C **13.** B **15.** $-4x - 7 = 0$; $y = -4x - 7$ **17.** $4x + 5 = 0$; $y = 4x + 5$ **19.** $x + 7 = 0$; $y = x + 7$ **21.** 3 **23.** $1\frac{1}{2}$ **25.** –4 **27.** 1 **29.** 1 **31.** 3 **33.** 2 **35.** –4 **37.** –4 **39.** –5 **41.** 6 **43.** –2 **45.** 1999 **47.** 2001

4.7 MIXED REVIEW (p. 255) **53.** no **55.** no **57.** no **59.** $-64 + 8y$ **61.** $-3q^2 - 12q$ **63.** $-2d + 10$ **65.** $15w - 10w^2$

4.8 PRACTICE (pp. 259–261) **3.** –10 **5.** –16 **7.** yes; 10, 20, 30, 40, 50; 100, 200, 300, 400, 500 **9.** no **11.** Yes; no vertical line passes through two points on the graph. **13.** No; there are vertical lines that pass through two or more points on the graph. **15.** yes; 1, 2, 3, 4; 0 **17.** yes; 0, 1, 2, 3; 2, 4, 6, 8 **19.** yes; 1, 3, 5, 7; 1, 2, 3 **21.** 14; –2; –26 **23.** 2.5; 0; –3.75 **25.** –1.34; –2; –2.99 **27.** 7.8; 7; 5.8 **29.** C **31.** B

33.

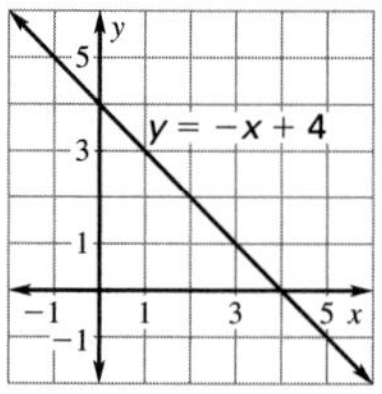

35.

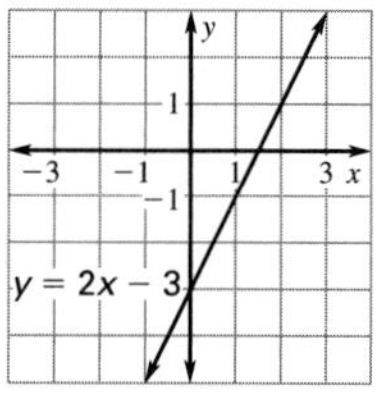

37.

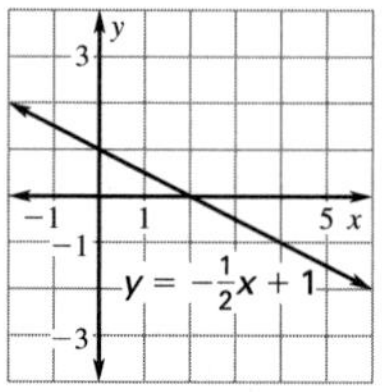

39.

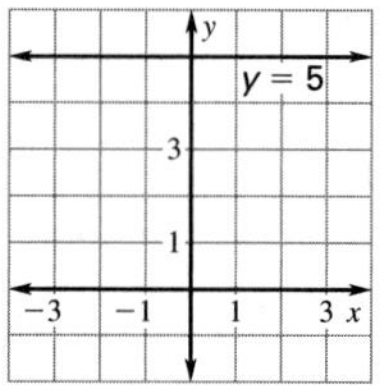

43. –1 **45.** –3 **47.** 1.75 **49.** 0 **51.** about 16 million **53.** Yes; each input is paired with exactly one output. **55.** $f(t) = \frac{100}{17}t$, or $f(t) \approx 5.88t$

4.8 MIXED REVIEW (p. 262)

63. $\begin{bmatrix} 9 & -11 \\ 12 & 7 \end{bmatrix}$ **65.** $\begin{bmatrix} 9.8 & -17.9 \\ -12.3 & -0.1 \\ 12.4 & -13.2 \end{bmatrix}$ **67.** 4 **69.** -3 **71.** $2\frac{1}{2}$

73. $y = 2x + 3$

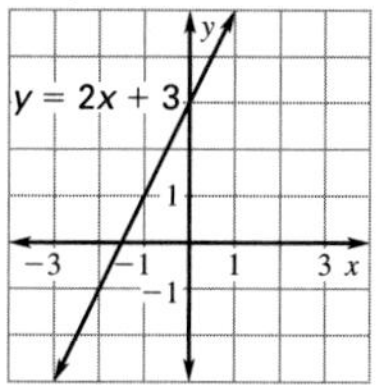

75. $y = 2x - 7$

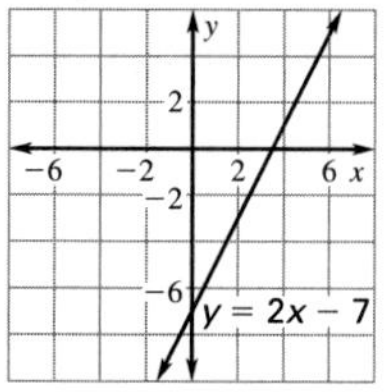

77. $y = \frac{1}{2}x + 1$

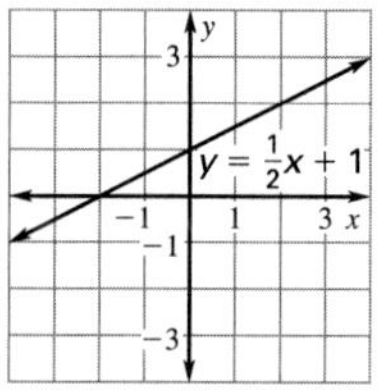

79.

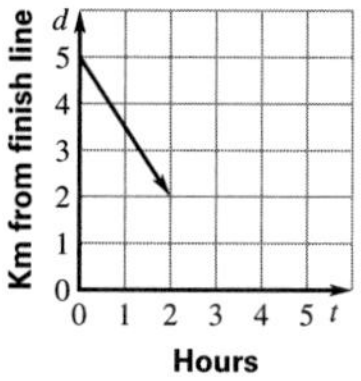

QUIZ 3 (p. 262) **1.** -2 **2.** $\frac{1}{2}$ **3.** 1 **4.** $-\frac{1}{2}$ **5.** -13 **6.** $-5\frac{1}{3}$
7. 6; -9; -29 **8.** -9; 3; 19 **9.** 3.25; -2; -9 **10.** -4.2; 0; 5.6
11. $9\frac{3}{4}$; 9; 8 **12.** 2; $\frac{2}{7}$; -2 **13.**

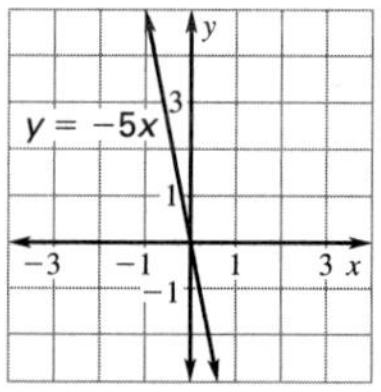

14.

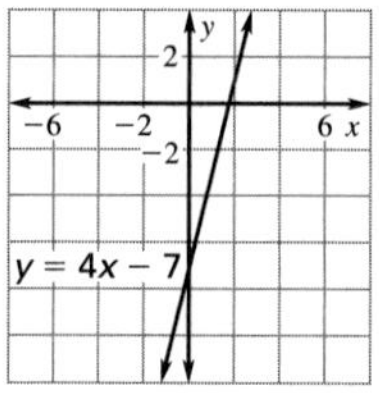

15.

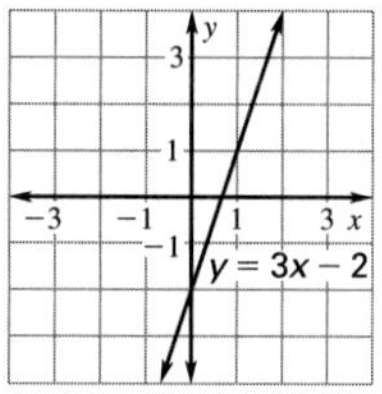

16.

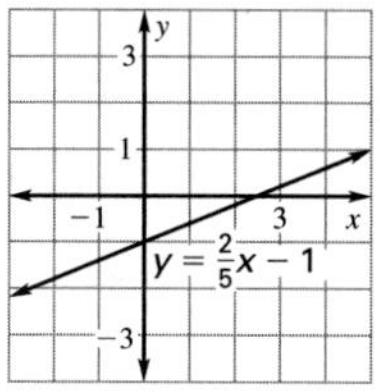

17.

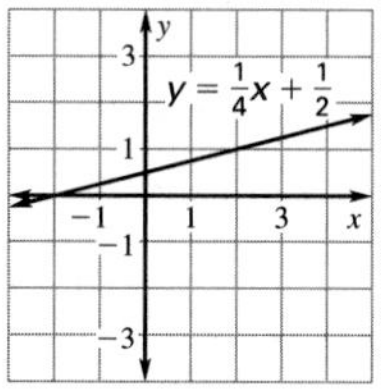

18.

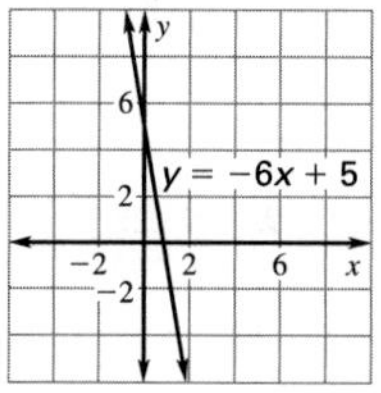

19. 250 lunches

CHAPTER 4 REVIEW (pp. 264–266)

1.

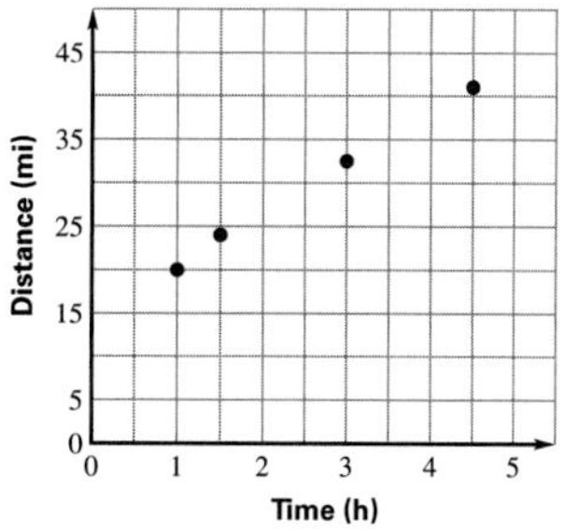

3, 5.

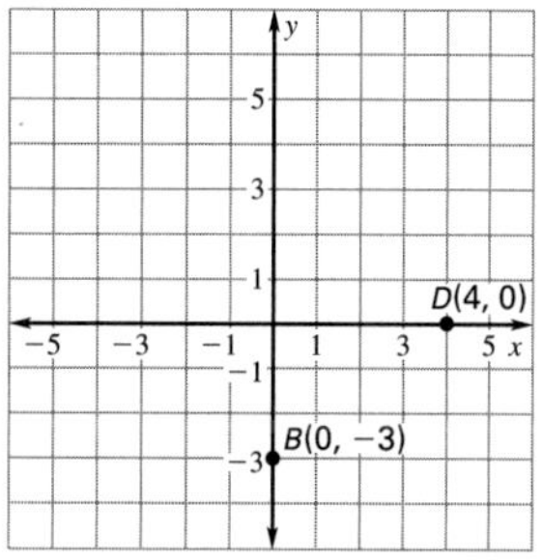

7.

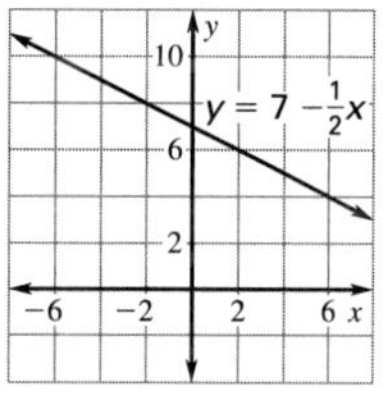

9.

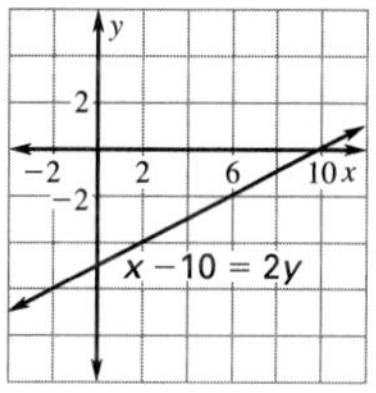

11.

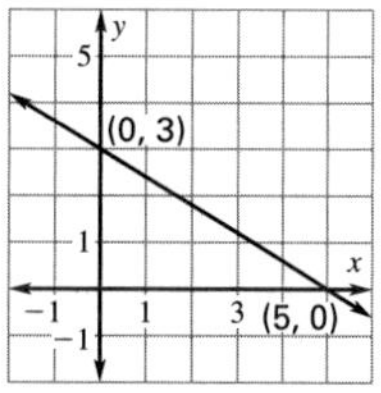

15. 6 **17.** 0 **19.** $y = -\frac{1}{3}x$

21. $y = 3.5x$ **23.**

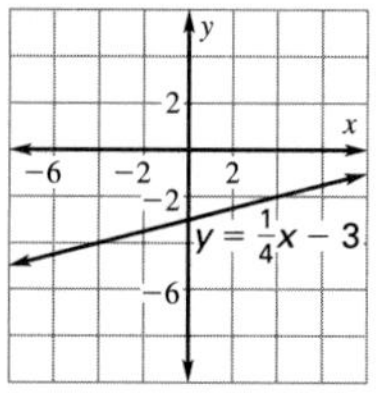

25. 2 **27.** 8

29. -9

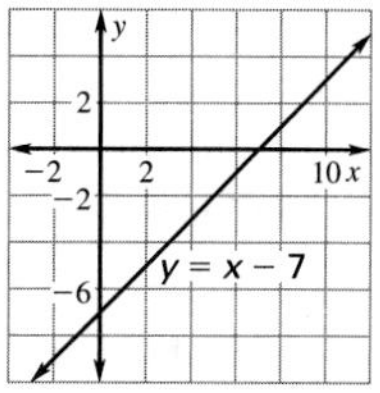

31. 0

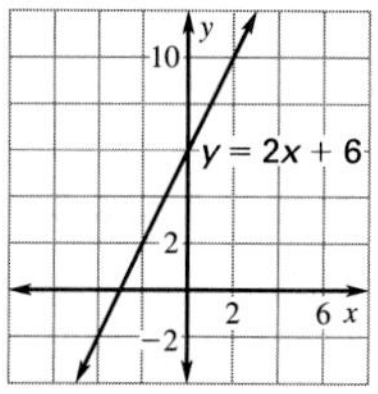

SELECTED ANSWERS

CHAPTER 5

SKILL REVIEW (p. 272) **1.** 2 **2.** 10 **3.** –6

4.

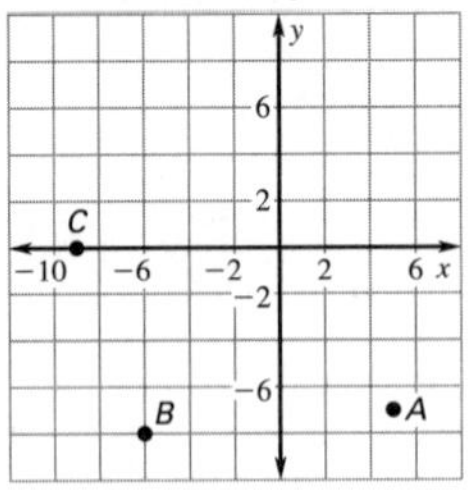

5.

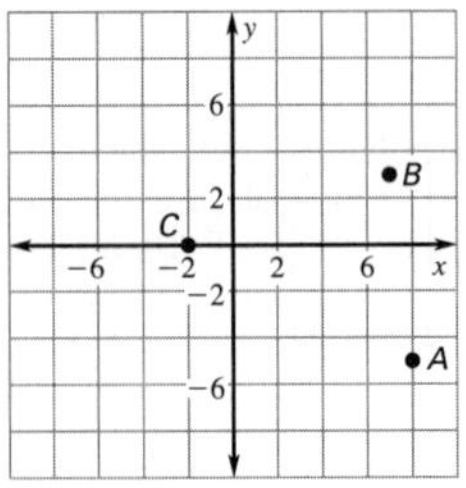

6. 4.5, 2

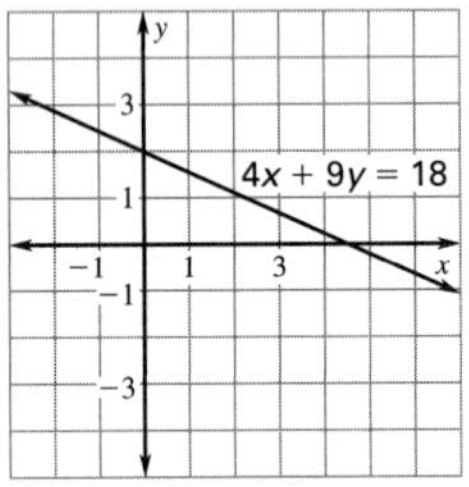

7. –5, –3

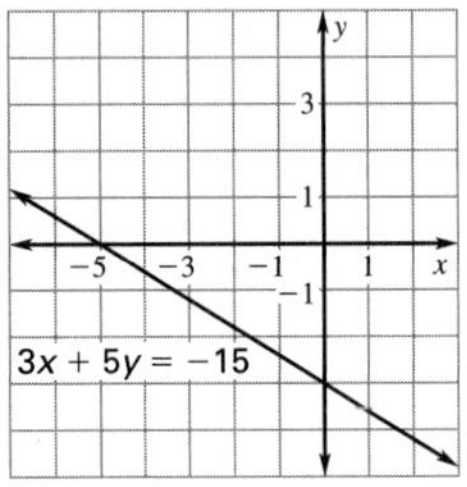

8. $-\frac{4}{3}$, –4

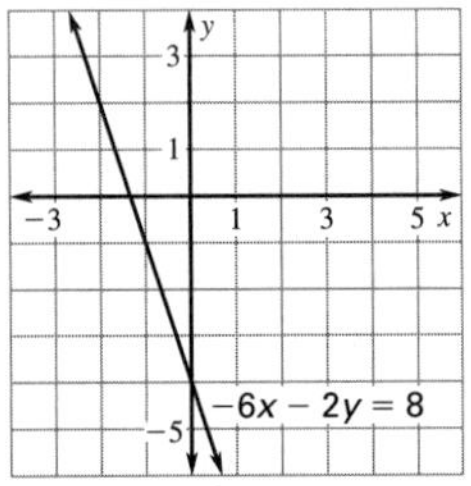

5.1 PRACTICE (pp. 276–277) **3.** $y = x$ **5.** $y = -x + 3$ **7.** $y = -3x + \frac{1}{2}$ **9.** $y = 1.5x + 4$ **11.** \$22 **13.** $y = x + 2$ **15.** $y = 2x - 1$ **17.** $y = -\frac{1}{4}x + 1$ **19.** $y = -3x - \frac{1}{2}$ **21.** $y = x - 3$ **23.** *a:* $y = 2x + 3$; *b:* $y = 2x$; *c:* $y = 2x - 4$ **25.** *a:* $y = 4$; *b:* $y = 1$; *c:* $y = -2$ **27.** about 3,710,400 **29.** \$36.25; \$42.50; \$48.75; \$55 **31.** Let y = your uncle's weight in lb and x = the number of months from the start of his weight loss; $y = 180 - 2x$; lb = lb – $\frac{\text{lb}}{\text{month}}$ · months. **33.** $y = 200 - 50t$; the distance from home decreases as time increases. **35.** $T = 5x$ **37.** *Sample answer:* If the club members can sell only 200 calendars, the fundraising project will not raise any money at all because the cost for producing 200 calendars will equal the money raised by selling 200 calendars. Therefore, it is not a very effective project.

5.1 MIXED REVIEW (p. 278) **41.** –3 **43.** –11 **45.** Let x = the width of each photograph; $3x + 3\frac{1}{2} = 13\frac{1}{4}$.

47. $y = 10x$

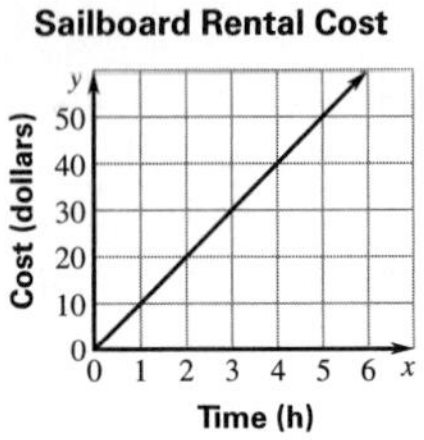

49. –2, 2

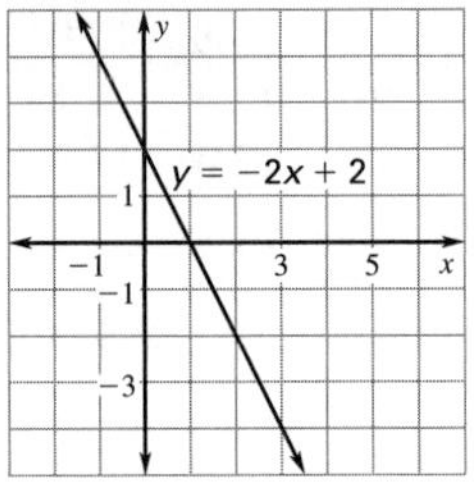

51. $\frac{3}{2}$, 3

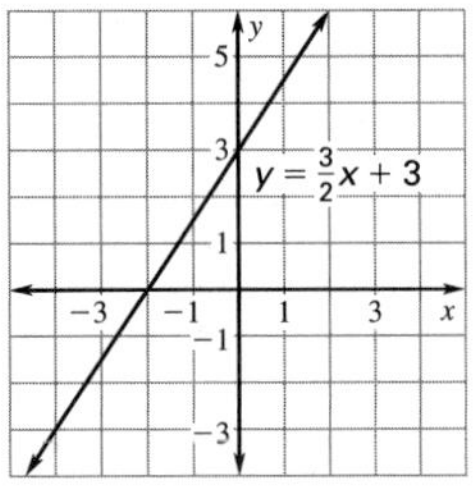

53. –3, 4

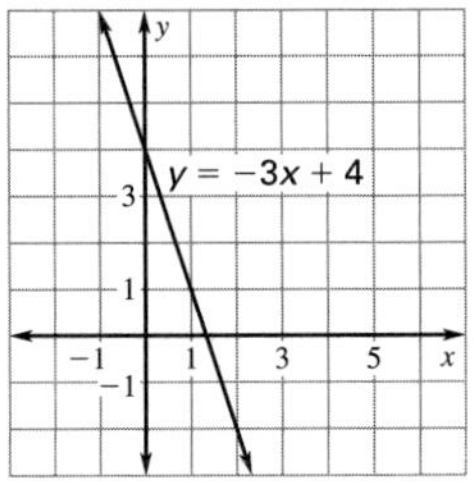

55. $-\frac{1}{3}$, 5

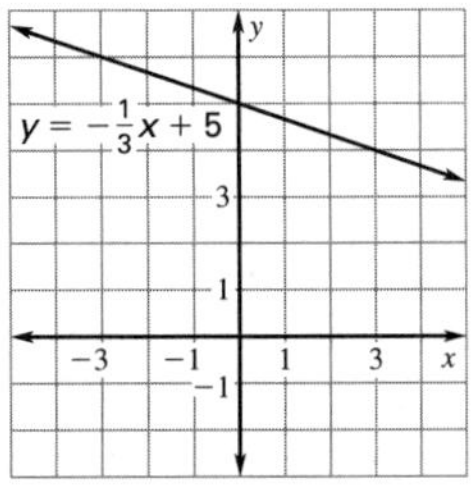

57. 1, –10

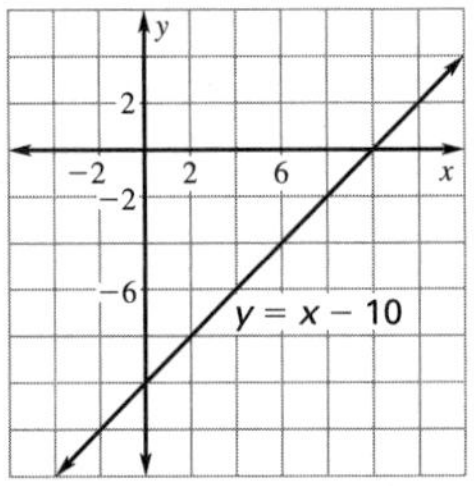

5.2 PRACTICE (pp. 282–283) **3.** $y = \frac{1}{2}x + \frac{5}{2}$ **5.** $y = \frac{2}{3}x - \frac{50}{3}$ **7.** $y = 5x - 12$ **9.** $y = \frac{1}{2}x + 7$ **11.** $y = -8.25x + 616.5$ **13.** $y = 2x - 7$ **15.** $y = -4x + 1$ **17.** $y = 4x - 1$ **19.** $y = \frac{1}{3}x + 3$ **21.** $y = 4x - 2$ **23.** $y = 4$ **25.** $y = \frac{1}{2}x + \frac{3}{2}$ **27.** $y = \frac{3}{5}x$ **29.** $x = 2$ **31.** $y = 3x - 12$ **33.** $y = x + 2$ **35.** $y = -3x + 14$ **37.** $y = -\frac{1}{3}x + \frac{7}{3}$ **39.** $y = 6x - 33$ **41.** $y = \frac{1}{2}x - 2$ **43.** about 162,000 **45.** \$21,750 **47.** \$39.80 **49.** \$2

5.2 MIXED REVIEW (p. 284) **53.** 32,768 **55.** 531,441 **57.** 64 **59.** 10,000 **61.** 10,077,696 **63.** about 2,541,865,828,000 **65.** $-\frac{1}{2}$ **67.** –11 **69.** –20 **71.** 14 **73.** $\frac{3}{4}$ **75.** $-\frac{2}{3}$

5.3 PRACTICE (pp. 288–290) **3.** $\frac{1}{4}$ **5.** –1 **7.** $y = \frac{4}{5}x - \frac{1}{5}$ **9.** $y = \frac{4}{5}x + \frac{9}{5}$ **11.** $y = -\frac{10}{3}x - \frac{31}{3}$ **13.** $y = \frac{1}{2}x$ **15.** $y = 6x - 27$ **17.** $y = 2$ **19.** $y = -3x - 3$ **21.** $y = \frac{8}{3}x + \frac{2}{3}$ **23.** $y = -2$ **25.** $y = -2x + 1$ **27.** $y = -\frac{6}{7}x + \frac{8}{7}$

SA12

29. $y = \frac{1}{2}x + \frac{1}{2}$ **31.** $x = 5$ **33.** $y = -\frac{1}{11}x + \frac{17}{11}$ **35.** $y = -\frac{14}{13}x + \frac{116}{13}$ **37.** $y = 3$ **39.** $y = -\frac{3}{5}x + \frac{27}{5}$ **41.** $y = -\frac{2}{3}x + \frac{11}{3}$ **43.** $y = -\frac{15}{7}x + \frac{4}{7}$ **45.** Lines p and r; their slopes are negative reciprocals. **47.** $y = -2x + 13$ **49.** $\overline{WX}$: $y = -\frac{4}{3}x + 4$; $\overline{XY}$: $y = \frac{3}{4}x - \frac{9}{4}$; $\overline{ZY}$: $y = -\frac{4}{3}x - \frac{13}{3}$ **51.** $y = -\frac{26}{3}x + 60$; $-\frac{26}{3}$ **53.** $s = \frac{3}{5}T + 331$ **55.** 25°C **57.** 686 m

5.3 MIXED REVIEW (p. 291) **67.** $\frac{6a - 7b}{2}$ **69.** -5 **71.** -35 **73.** $\frac{1}{3}$ **75.** positive

QUIZ 1 (p. 291) **1.** $y = \frac{4}{3}x + 4$ **2.** $y = -2x + 2$ **3.** $y = \frac{4}{3}x - 4$ **4.** $y = 2x + 14$ **5.** $y = -3x + 7$ **6.** $y = x - 3$ **7.** $y = x - 2$ **8.** $y = -\frac{1}{3}x - \frac{1}{3}$ **9.** $y = \frac{13}{3}x + \frac{44}{3}$ **10.** $y = 4$ **11.** $y = 28x + 10$

12. Canoe Rental Cost

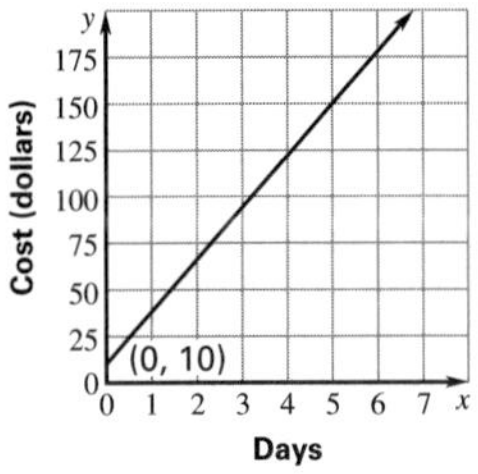

13. \$94

5.4 PRACTICE (pp. 296–297) **7.** negative correlation; *Sample answer:* $y = -1.54x + 3.23$ **9.** positive correlation; *Sample answer:* $y = 1.16x - 6.22$ **11.** *Sample answer:* $y = 1.19x + 1.19$ **13.** *Sample answer:* $y = 1.64x + 2.57$ **15.** *Sample answer:* $y = -1.79x + 15.45$ **17.** positive correlation **19.** negative correlation **21.** positive correlation **23.** *Sample answer:* $y = 25x + 50$

25. Sample Pupil Diameters (Daylight)

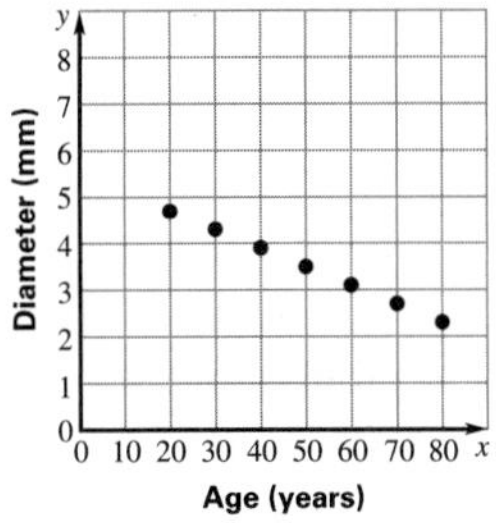

Sample Pupil Diameters (Night)

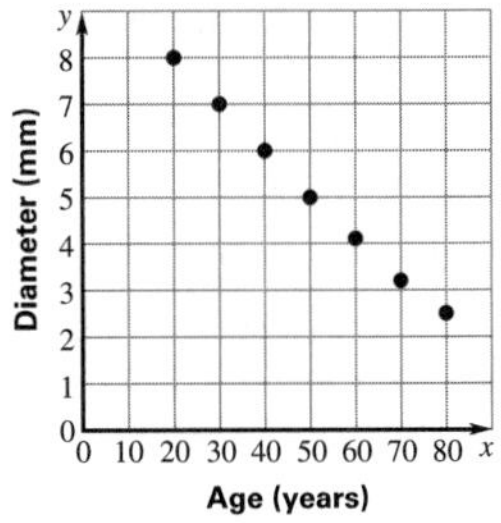

5.4 MIXED REVIEW (p. 298) **35.** horizontal **37.** vertical **39.** positive **41.** zero **43.** $y = -\frac{1}{4}x + \frac{11}{2}$ **45.** $y = -x + 10$ **47.** $y = -\frac{3}{7}x - \frac{2}{7}$ **49.** $y = \frac{7}{2}x - \frac{33}{2}$

TECHNOLOGY ACTIVITY 5.4 (p. 299) **1.** $y = 0.47x + 2.01$ **3.** $y = 0.95x + 1.41$

5.5 PRACTICE (pp. 303–304) **3.** $y - 4 = \frac{1}{2}(x - 3)$ **5.** $y + 2 = 0$ **7.** $y - 4 = \frac{1}{2}(x - 3)$ **9.** $y - 10 = 4(x + 3)$ **11.** $y + 7 = -8(x - 7)$ **13.** $y + 1 = -\frac{3}{5}(x + 4)$ or $y - 2 = -\frac{3}{5}(x + 9)$ **15.** $y - 12 = -2(x - 2)$ or $y - 2 = -2(x - 7)$ **17.** $y + 7 = \frac{3}{2}(x + 4)$ or $y - 2 = \frac{3}{2}(x - 2)$ **19.** $y + 3 = -\frac{1}{2}(x + 1)$ **21.** $y - 5 = 5(x - 1)$ or $y + 5 = 5(x + 1)$ **23.** $y - 10 = -\frac{13}{5}(x + 9)$ or $y + 3 = -\frac{13}{5}(x + 4)$ **25.** $y - 10 = -12(x + 5)$ or $y + 2 = -12(x + 4)$ **27.** $y + 9 = -\frac{1}{3}(x + 3)$ or $y + 8 = -\frac{1}{3}(x + 6)$ **29.** $y + 7 = -(x - 1)$ or $y + 5 = -(x + 1)$ **31.** $y - 2 = -5(x + 6)$ **33.** $y + 2 = 2(x + 8)$ **35.** $y - 4 = 6(x + 3)$ **37.** $y + 1 = 0$ **39.** $y - 4 = 2(x - 1)$; $y = 2x + 2$ **41.** $y - 2 = \frac{1}{2}(x - 6)$; $y = \frac{1}{2}x - 1$ **43.** $y + 1 = -1(x - 5)$; $y = -x + 4$ **45.** $y - 1 = -\frac{1}{8}(x + 1)$; $y = -\frac{1}{8}x + \frac{7}{8}$ **47.** $y + 6 = -\frac{2}{3}(x + 9)$; $y = -\frac{2}{3}x - 12$ **49.** $y = -\frac{1}{3}x - \frac{4}{3}$ **51.** $y - 3 = 2(x - 1)$ or $y - 5 = 2(x - 2)$ **53.** $y + 10 = -\frac{3}{2}(x - 7)$ or $y + 22 = -\frac{3}{2}(x - 15)$ **55.** $y + 4 = \frac{3}{4}(x + 3)$ or $y + 1 = \frac{3}{4}(x - 1)$ **57.** $d = -0.76t + 120$ **59.** at about 11:38 A.M. **61.** at about 11:49 A.M. **63.** at about 2:25 P.M.

5.5 MIXED REVIEW (p. 305) **71.** solution **73.** solution **75.** solution **77.**

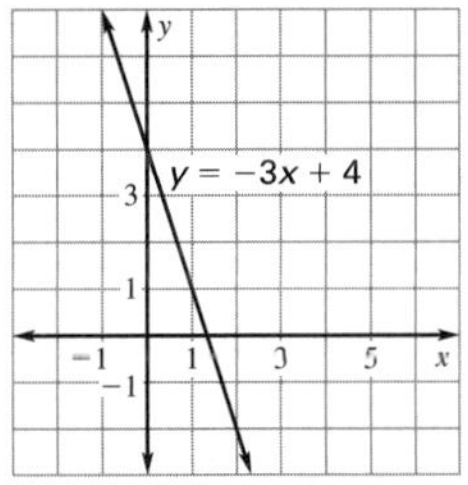

79.

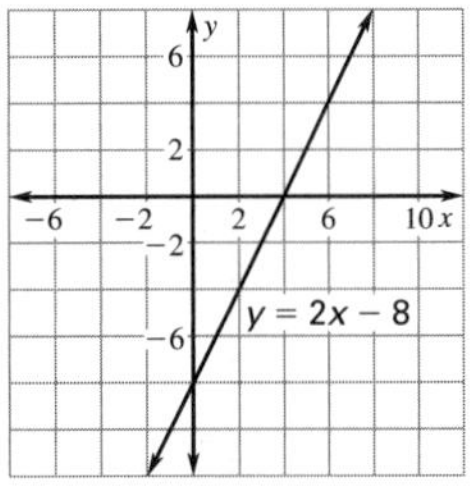

81.

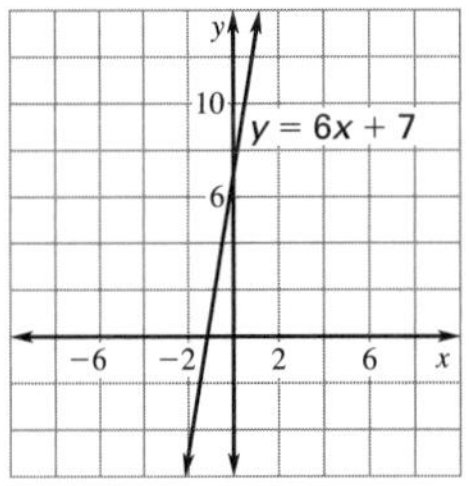

83.

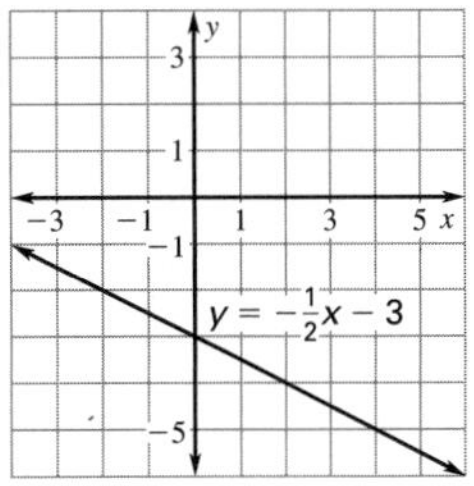

85.

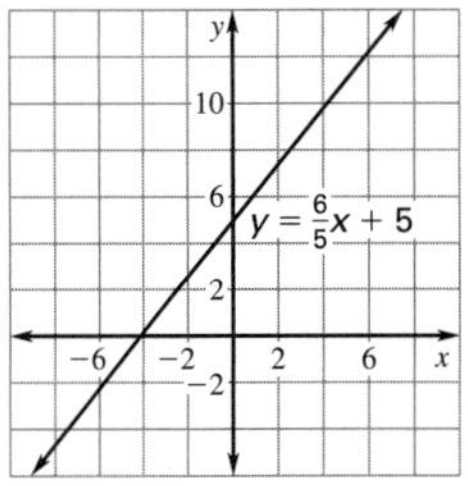

87. $33\frac{1}{3}\%$

QUIZ 2 (p. 306) **1.** *Sample equation:* $y = 0.82x - 2.78$; positive **2.** *Sample equation:* $y = -0.53x - 0.73$; negative **3.** *Sample equation:* $y = 1.38x + 0.99$; positive **4.** $y + 4 = -1(x - 1)$ **5.** $y - 5 = 2(x - 2)$ **6.** $y + 2 = -2(x + 3)$ **7.** $y + 3 = 0$ **8.** $y - 6 = \frac{1}{2}(x + 6)$

9. $y - 8 = 7(x - 3)$ **10.** $y = \frac{1}{3}x$ or $y + 1 = \frac{1}{3}(x + 3)$
11. $y + 2 = 0$ **12.** $y + 3 = \frac{3}{2}(x + 5)$ or $y - 6 = \frac{3}{2}(x - 1)$

5.6 PRACTICE (pp. 311–313) **5–15 odd:** Sample answers are given. **5.** $5x + y = 6$ **7.** $x - 2y = -16$ **9.** $3x - 2y = 4$ **11.** $5x - y = 7$ **13.** $x + y = -3$ **15.** $2x + y = 5$ **17.** 8; 6; 4; 2; 0 **19–53 odd:** Sample answers are given. **19.** $x + 3y = 4$ **21.** $2x - 3y = 14$ **23.** $y = -3$ **25.** $3x - y = 8$ **27.** $6x - 7y = -18$ **29.** $5x - 2y = -18$ **31.** $x + 7y = 6$ **33.** $2x - y = -19$ **35.** $3x + y = 1$ **37.** $5x - y = 17$ **39.** $7x + y = 23$ **41.** $2x + y = 17$ **43.** $x - y = -3$ **45.** $3x - 4y = -13$ **47.** $x + y = -3$ **49.** $y = 0$ **51.** $y = -4$, $x = 2$ **53.** $y = -3$, $x = 0$ **55.** $y = 4$, $x = -4$ **57.** $y = -1$, $x = 6$ **59.** $y = 0$, $x = -9$ **61.** $y = -3$, $x = 10$ **63.** The expression inside the brackets was simplified incorrectly; $x - (-6) = x + 6$. **65.**

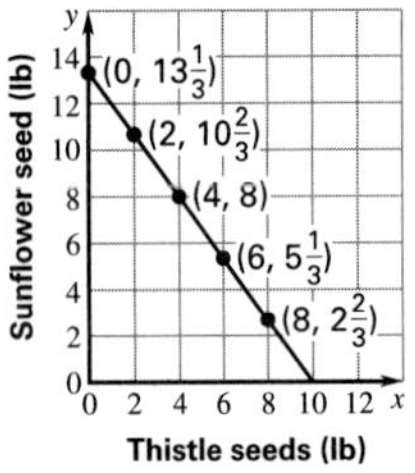

67. *Sample answer:* $2x + 3y = 10$ **69.** 3 students **71.** 0% blue, 60% red; 10% blue, 50% red; 20% blue, 40% red; 30% blue, 30% red; 40% blue, 20% red; 50% blue, 10% red; 60% blue, 0% red **73.** 0% blue, 80% red; 10% blue, 70% red; 20% blue, 60% red; 30% blue, 50% red; 40% blue, 40% red; 50% blue, 30% red; 60% blue, 20% red; 70% blue, 10% red; 80% blue, 0% red

5.6 MIXED REVIEW (p. 314) **85.** 11 **87.** 4 **89.** $5\frac{1}{4}$

91.

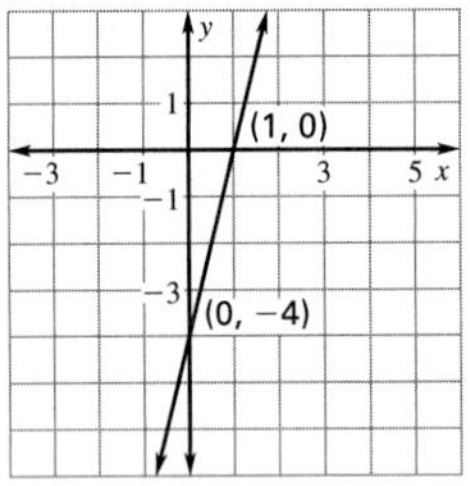

93.

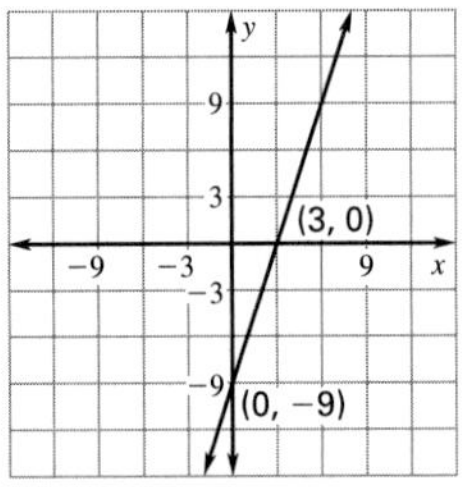

95.

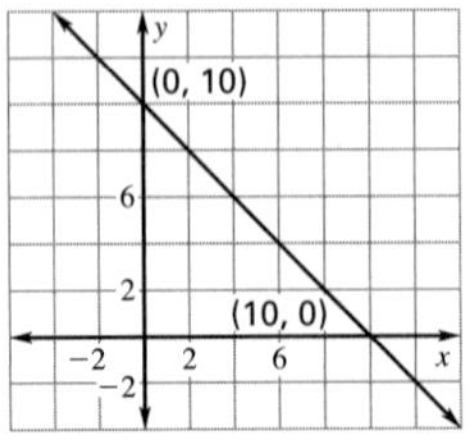

97. $y = \frac{15}{2}x - \frac{93}{2}$

99. $y = \frac{6}{11}x - \frac{58}{11}$ **101.** $y = -\frac{1}{15}x - \frac{11}{15}$ **103.** $y = -\frac{6}{19}x - \frac{87}{19}$ **105.** \$118.67 **107.** 25%

5.7 PRACTICE (pp. 319–321) **7.** *Sample answer:* $y = 0.62x + 1.96$ **9.** about 20.56 gal; extrapolation **11.** no **13.** yes **15.** yes **17.**

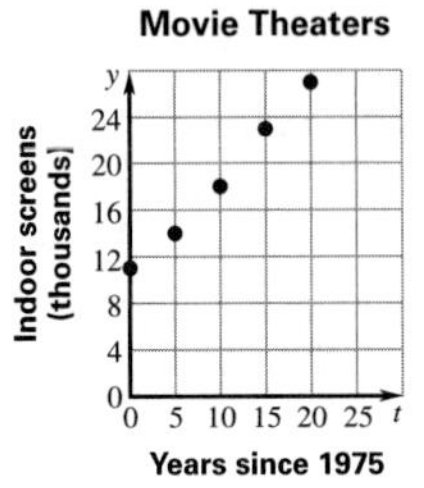

19. indoor screens **21.** about 35,000 theaters **23.** $y = 0.94x + 16.17$ **25.** $y = 2.24x + 30.91$ **27.** about \$64.51 billion; linear extrapolation

5.7 MIXED REVIEW (p. 322)

39.

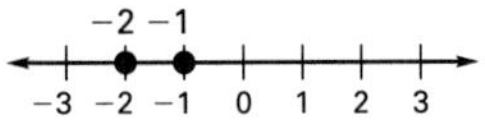

$-2 < -1$, $-1 > -2$

41.

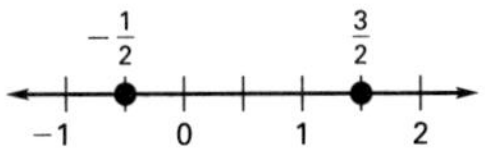

$-\frac{1}{2} < \frac{3}{2}$, $\frac{3}{2} > -\frac{1}{2}$

43.

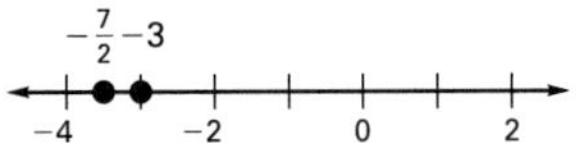

$-3 > -\frac{7}{2}$, $-\frac{7}{2} < -3$

45. $y = \frac{1}{2}x - 8$ **47.** $y = -2x - 6$ **49.** $y = 7x + 28$ **51.** $y = -5x + 44$ **53.** $y = 2x + 23$

QUIZ 3 (p. 322) **1–6.** Sample answers are given. **1.** $x - 2y = -6$ **2.** $9x + 3y = 18$ **3.** $-2x + 6y = 8$ **4.** $9x - y = -12$ **5.** $2x - 3y = -18$ **6.** $14x - 16y = 1$ **7.** $2x + y = 4$ **8.** $x + 2y = 6$ **9.** $3x + y = 8$ **10.** $2x - y = -5$ **11.** $3x - y = -40$ **12.** $6x - y = -94$ **13.** Let x = the number of years since 1985; $y = 0.13x + 74.80$. **14.** about 75.32 years **15.** about 78.7 years

CHAPTER 5 REVIEW (pp. 324–326) **1.** $y = 2x - 2$ **3.** $y = -8x - 3$ **5.** $y = 2x + 22$ **7.** $y = -\frac{11}{7}x - \frac{19}{7}$ **9.** $y = \frac{3}{10}x + \frac{22}{5}$ **11.** *Sample answer:* $y = 1.98x + 1.11$ **13.** $y - 4 = \frac{1}{6}(x + 4)$ or $y - 5 = \frac{1}{6}(x - 2)$; $y = \frac{1}{6}x + \frac{14}{3}$ **15.** $y + 2 = -5(x - 1)$ or $y - 8 = -5(x + 1)$; $y = -5x + 3$ **17–21 odd:** Sample answers are given. **17.** $8x + 3y = 2$ **19.** $x + 3y = 2$ **21.** $4x - 3y = 12$ **23.** about 242 million; linear extrapolation **25.** about 178 million; linear extrapolation

CHAPTER 6

SKILL REVIEW (p. 332) **1.** solution **2.** not a solution **3.** solution **4.** not a solution **5.** not a solution **6.** solution **7.** 10 **8.** 24 **9.** 9 **10.** 6 **11.** 7 **12.** 2

13.

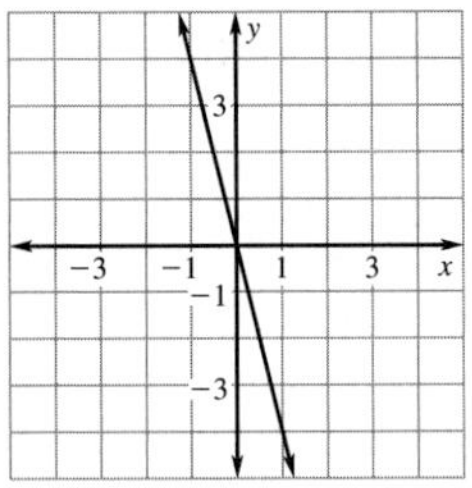

14.

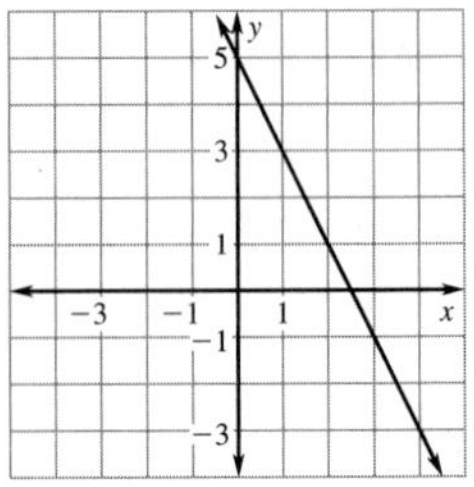

SELECTED ANSWERS

15. 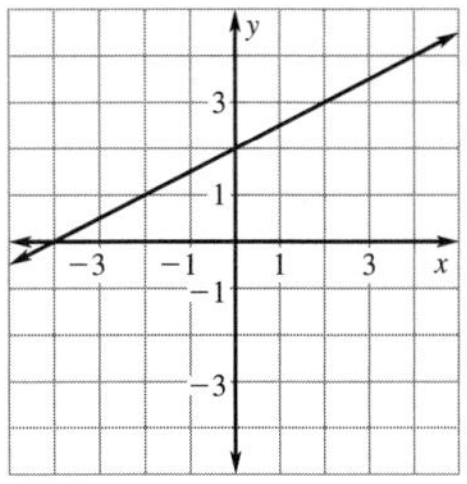

16. 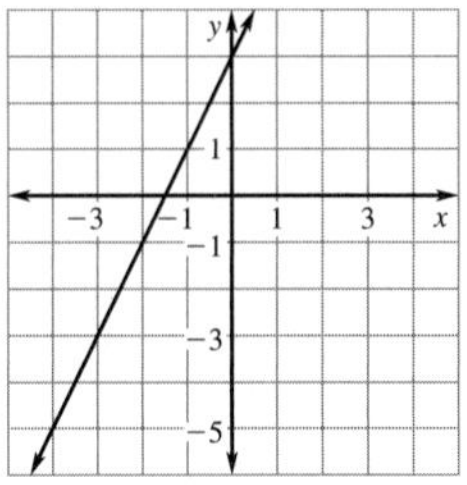

6.1 Practice (pp. 337–338)

9. open dot **11.** open dot **13.** closed dot **15.** $y \geq -5$ 

17. $x > 14$ 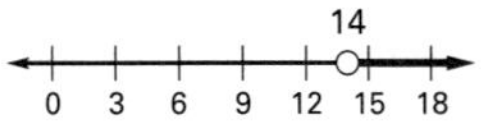

19. $y \geq -6$ 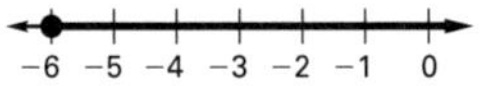

21. 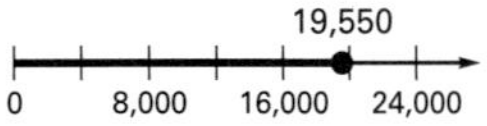

23. 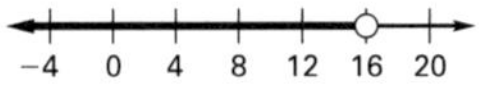

25. 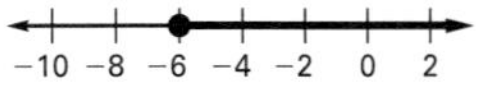

27. 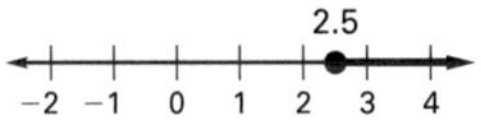

29. 

31. $x < 2$ 

33. $x < 24$ **35.** $p \geq 11$ **37.** $x > 5$ 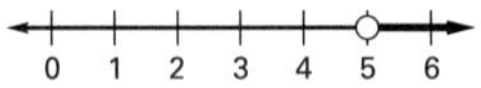

39. $c \geq 4$ 

41. $m \geq -17$ **43.** $x < 2$

45. $x \geq -10$ 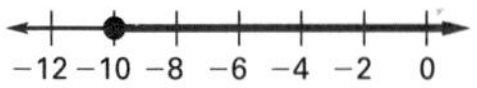

47. $a < -10$

49. $m \geq -5$ **51.** $x \leq -5$ **53.** $a \geq 20$ **55.** $x < 5$; E **57.** $x > 5$; C **59.** $x \geq 0$; A **61.** $x < \frac{6}{55}$

63. $M < 1376$ 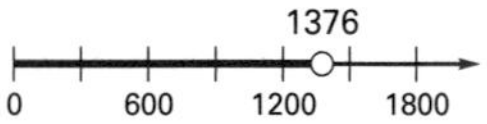

65. $\ell < 1700$ 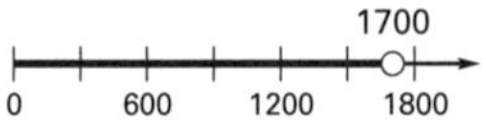

67. $131 \leq f \leq 587$

6.1 Mixed Review (p. 339)

71. $\frac{1}{9}$ **73.** 14 **75.** 56 **77.** 3 **79.** −4

81. $y = \frac{1}{2}x - 1$; (0, −1) and (2, 0) 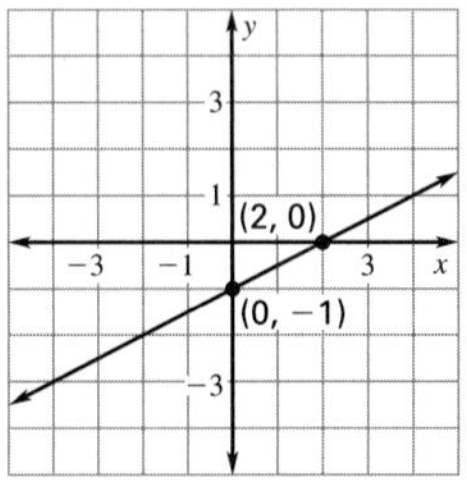

83. $y = 4$; the line does not cross the x-axis and crosses the y-axis at (0, 4). 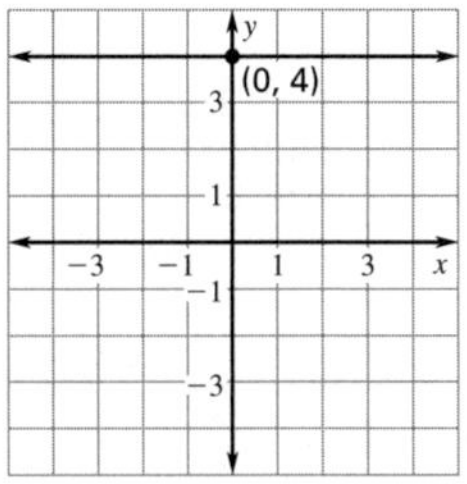

85. $x = 5$; the line crosses the x-axis at (5, 0) and does not cross the y-axis. 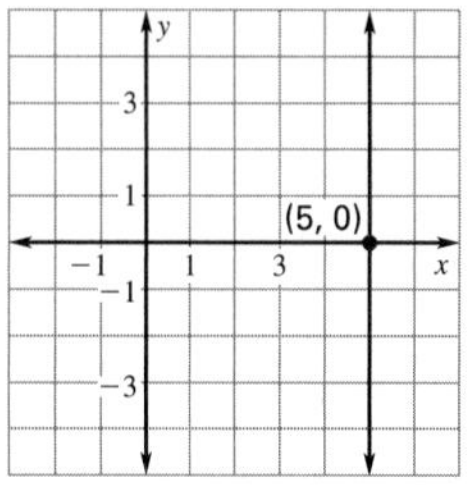

87. $d = 0.12t$

6.2 Practice (pp. 343–344)

5. $y > -3$ **7.** $x \leq 3$ **9.** $x \leq 10$ **11.** $x < -4$ **13.** $x < -\frac{3}{4}$ **15.** $x > -18$ **17.** $x \geq \frac{7}{6}$ **19.** $x \geq -3$ **21.** $x < 3$ **23.** $x \leq -5$ **25.** $x > -9$ **27.** $x \geq 12$ **29.** $x > -8$ **31.** $x \geq 11$ **33.** $x > 6\frac{2}{3}$ **35.** $x > \frac{1}{6}$ **37.** $1.25r + 5 \leq 25$; 16 rides **39.** $0.75t + 14 \leq 18.50$; 6 toppings **41.** 2 years **43.** $x > 4$ m **45.** $x < 10$ ft **47.** $x \leq 5$ in.

6.2 Mixed Review (p. 345)

55. −10 **57.** 12 **59.** $\frac{16}{17}$ **61.** $t > 2\ell$ **63.** 3 **65.** 0.75

6.3 Practice (pp. 349–350)

3. $2 < x < 5$ 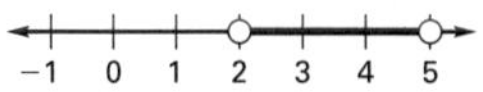

5. $3 < x < 4$

7. $2 < x \leq 4$ **9.** $x < -2$ or $x > 1$ **11.** $1 \leq d \leq 3$

13. $x > 7$ or $x < 5$ 

15. $-9 \leq x < 4$ 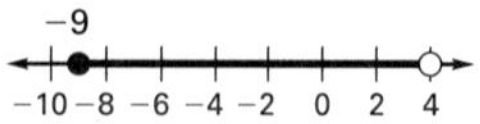

17. $-6 < x < -1$ 

19. x is greater than −2 and less than 9. **21.** x is greater than or equal to 1 and less than or equal to 6. **23.** x is less than or equal to −5 or greater than or equal to 10. **25.** x is greater than −2 and less than or equal to 9.

27. $-5 < x < -4$ 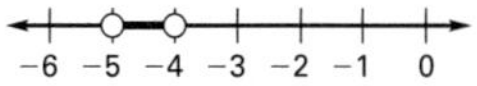

29. $-\frac{1}{3} \leq x < \frac{2}{3}$ 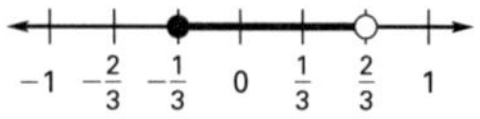

31. $x > 3$ or $x \leq 1$ 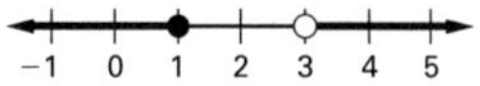

33. 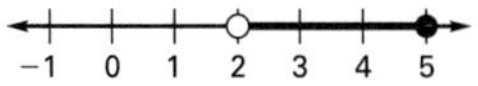

$2 < x \leq 5$; solution

35. 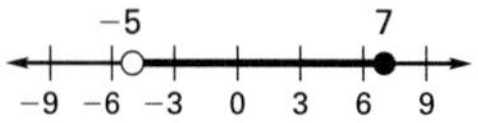

$-5 < x \leq 7$; solution

37. $85{,}000 \leq x \leq 2{,}600{,}000$ **39.** *Sample answer:* A solution would be a number on the graph of both inequalities. Since the graphs do not overlap, there are no such numbers. **41.** A **43.** $10^8 < d < 10^9$

6.3 Mixed Review (p. 351)

49. 7 **51.** 10 **53.** 16.5 **55.** −8 **57.** 16 **59.** −12 **61.** −7 **63.** 7.8 **65.** 3 **67.** at least 30 times **69.** 4 mi

Quiz 1 (p. 352)

1. $x < 5$ 

2. $x \leq -8$ 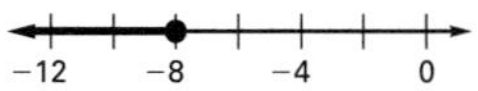

3. $x \leq 4$ 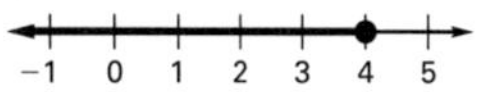

SA15

4. $x \le -10$

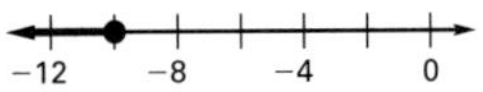

5. $x \le -4$

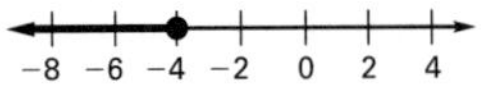

6. $3 < x < 12$

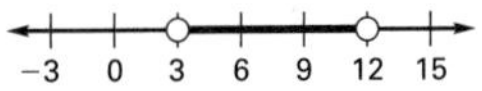

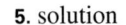

7. $x < 2$

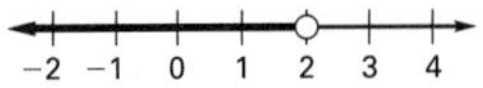

8. $x \ge 17$

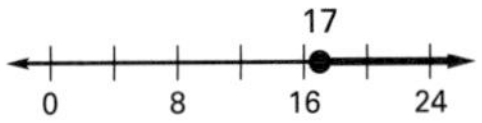

9. $-12 \le x \le -2$

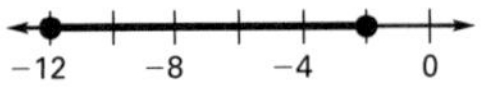

10. $x > -9$ or $x < -12$

11. $x > 5$ or $x < -1$

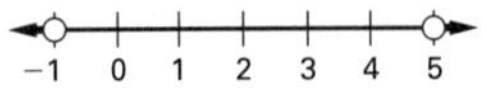

12. $x < -1$ or $x > 5\frac{4}{5}$

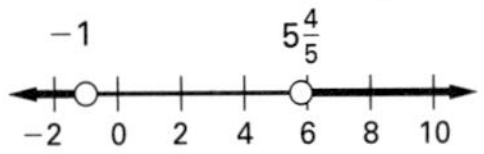

13. $x > -\frac{5}{6}$ or $x < -4$

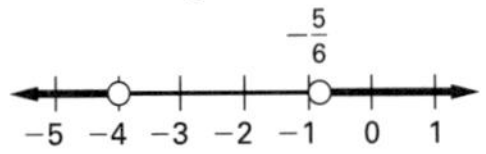

14. $x \ge 52$ **15.** $-128.6 < T < 136$

6.4 Practice (pp. 356–357)

7. 0 **9.** $-6, 14$ **11.** $\frac{1}{3}, -1\frac{2}{3}$ **13.** C **15.** $-10 < x < -2$ **17.** $-2 \le x \le 1\frac{1}{3}$ **19.** $-7, 7$ **21.** $-25, 25$ **23.** $-16, 6$ **25.** 3, 7 **27.** $-12, 6$ **29.** $-3.8, 10.2$ **31.** $-5, 6$ **33.** $-9\frac{1}{3}, 6$ **35.** $-7, 11$ **37.** $-8.2, 1$ **39.** $-2, 3$ **41.** $-1 \le x \le 10$ **43.** $x < -0.8$ or $x > 5.2$ **45.** $x \le -2$ or $x \ge 8$ **47.** $-1 < x < 9$

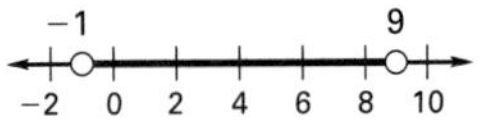

49. $-14 \le x \le 20$

51. $x \le -11$ or $x \ge 5$

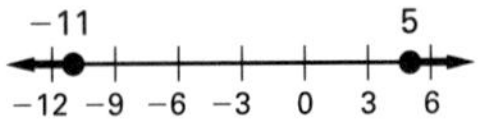

53. $x < -3\frac{1}{2}$ or $x > \frac{1}{2}$

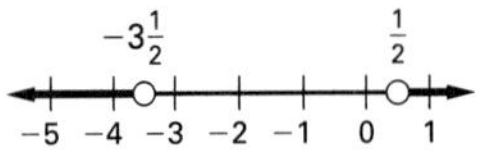

55. $-6 < x < 0$ **57.** $-5 \le x \le 4\frac{3}{5}$

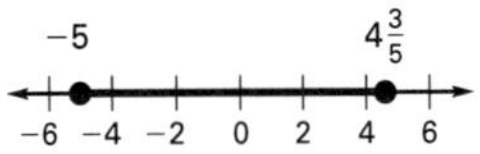

59. $-6 < x < 1$

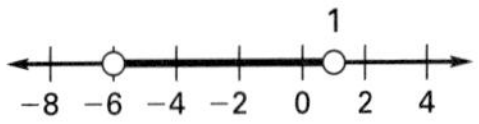

61. $|x - 15| \le 7$ **63.** $|x - 28| \le 4$ **65.** orange or red **67.** green

6.4 Mixed Review (p. 358)

75. $\begin{bmatrix} 8 & -17 \\ 9 & 5 \end{bmatrix}$ **77.** $y = -\frac{2}{3}x + 4$

79.

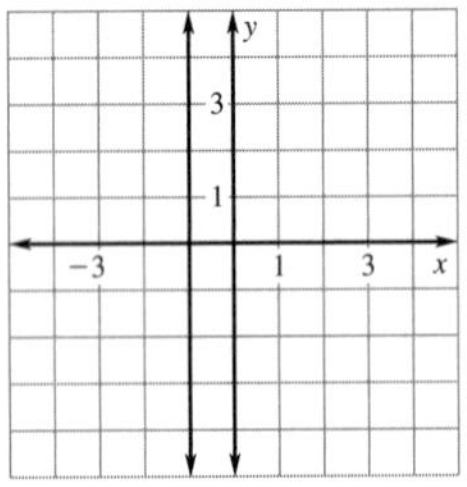

81.

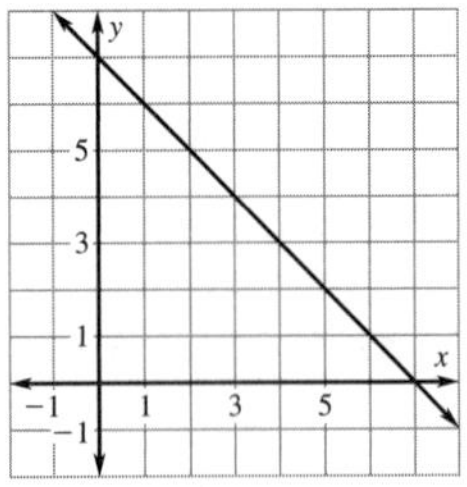

83. $y + 5 = -2(x - 2)$; $y = -2x - 1$ **85.** 57 mi/h

Technology Activity 6.4 (p. 359)

1. $-7, 1$ **3.** $-10, 18$ **5.** $-4, 64$ **7.** 1.8, 6.8 **9.** $-50, -22$

6.5 Practice (pp. 363–365)

5. solution

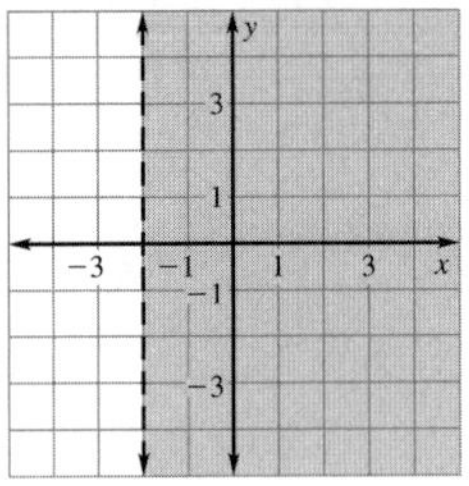

7. not a solution

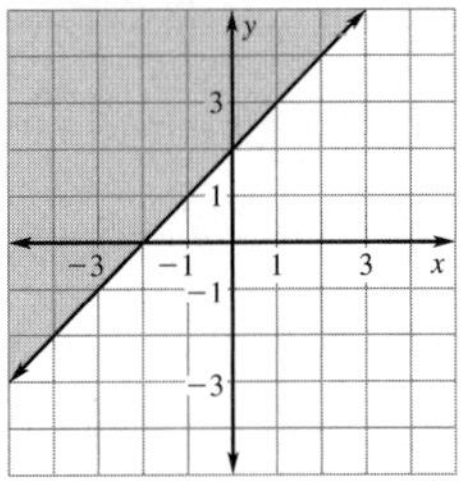

9. solution **11.** solution

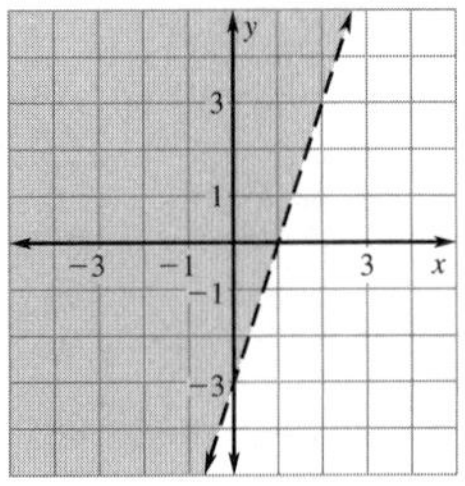

13. Let x = the number of 2-point shots and y = the number of 3-point shots; $2x + 3y \ge 12$. **15.** solution, not a solution **17.** solution, solution **19.** not a solution, not a solution **21.** solution, not a solution

31.

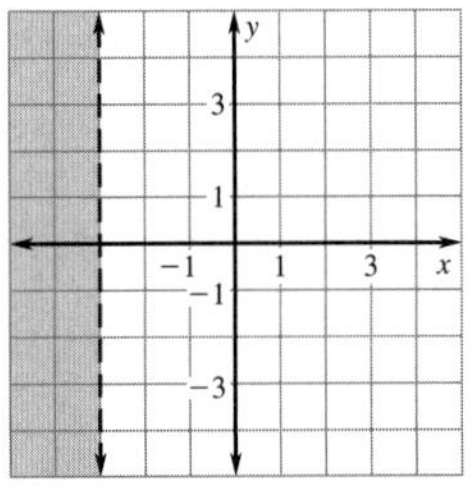

33.

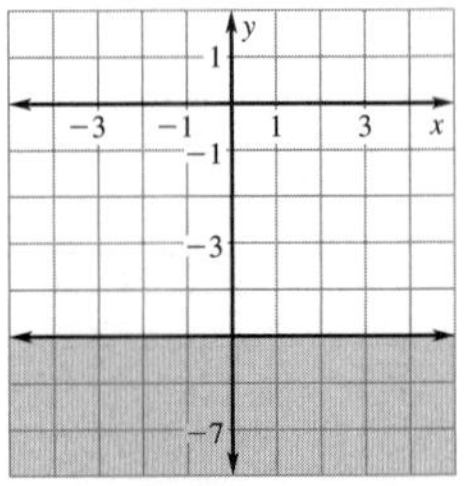

35.

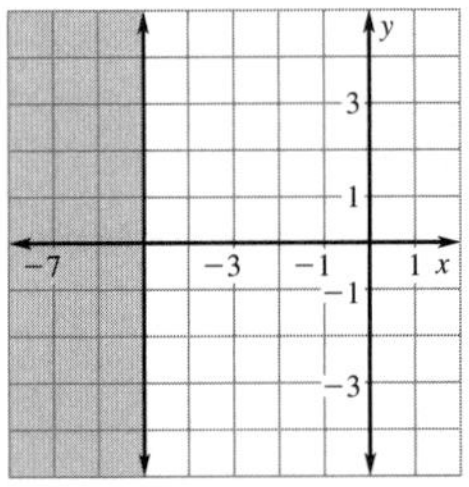

37.

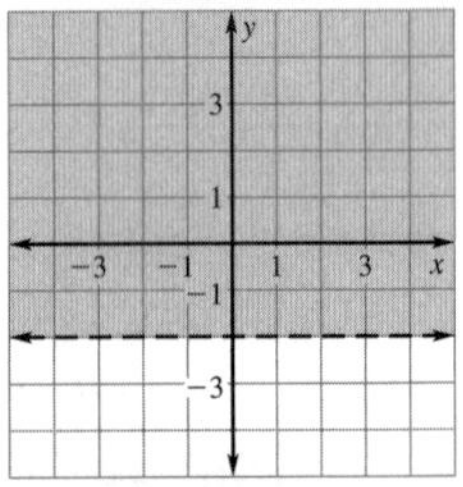

43. D **45.** E **47.** C **49.**

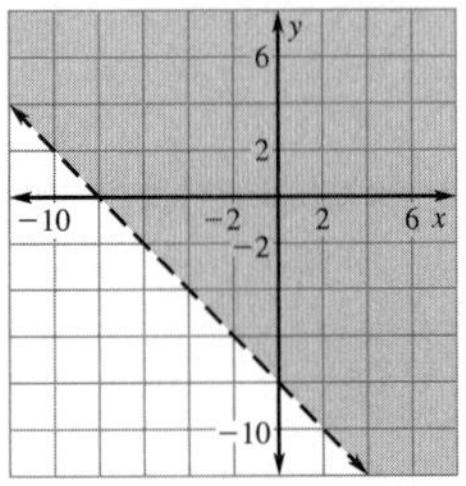

53.

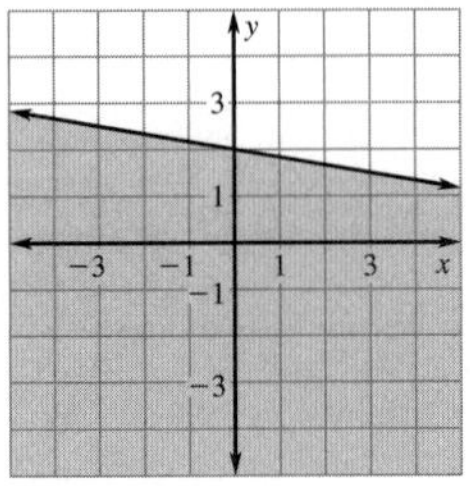

59.

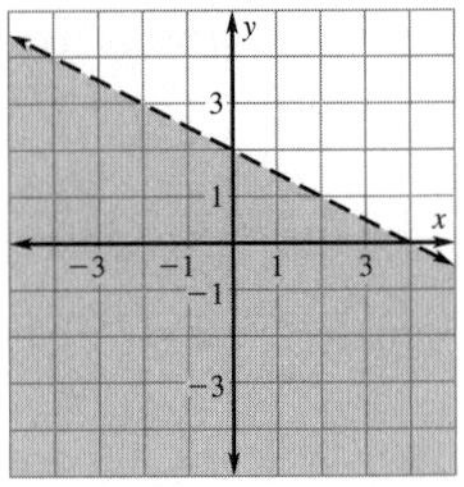

SA16

61. $y \geq -2x + 2$ or $2x + y \geq 2$ **63.** $y < 2x$ or $2x - y > 0$

65. a. $3x + 7y < 21$

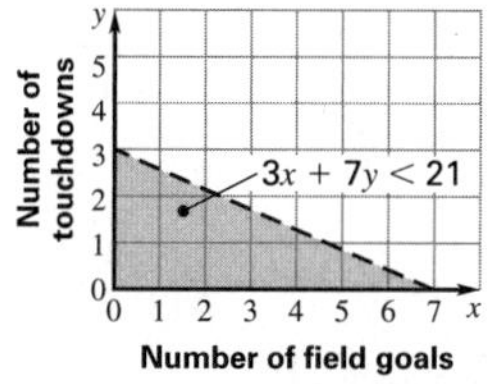

b. No; only points for which both coordinates are nonnegative integers represent solutions, since only a whole number of touchdowns or field goals can be scored. The real-life solutions are (0, 0), (0, 1), (0, 2), (1, 0), (1, 1), (1, 2), (2, 0), (2, 1), (2, 2), (3, 0), (3, 1), (4, 0), (4, 1), (5, 0), and (6, 0). **67. a.** Let (x, y) represent x cups of cereal and y cups of skim milk; the possible solutions are (1, 1), (1, 2), (1, 3), (1, 4), (2, 1), (2, 2), (2, 3), (3, 1), (3, 2), (4, 1). **b.** *Sample answer:* 1 glass of tomato juice, 2 cups of cereal, and 2 or 3 cups of milk **c.** The possible solutions are all those in part (a) as well as (1, 5), (2, 4), (3, 3), (4, 2), and (5, 1). *Sample answer:* one glass of tomato juice, 3 cups of cereal, and 2 cups of milk

6.5 MIXED REVIEW (p. 366) **75.** 19 **77.** $r = \frac{d}{t}$

79.

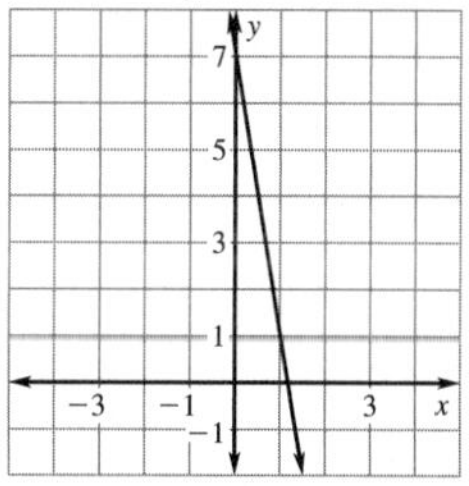

81.

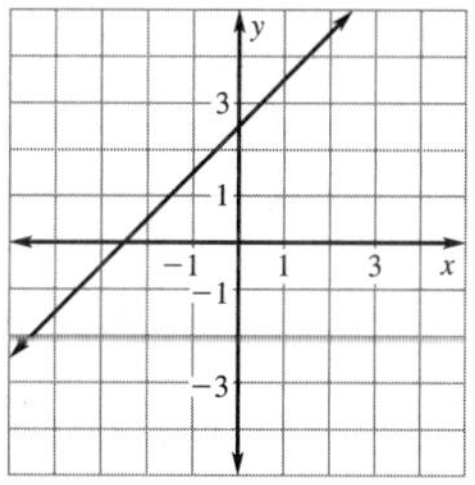

83. −5, 2 **85.** $1, -\frac{1}{5}$ **87.** 0, −2 **89.** 3, −6.5 **91.** $-\frac{3}{2}, 5$

QUIZ 2 (p. 366) **1.** −15, 15 **2.** −22, 22 **3.** −9, 3 **4.** −8, 12 **5.** −7, 0 **6.** $-\frac{5}{3}, 3$

7. $x < 3$ or $x > 5$

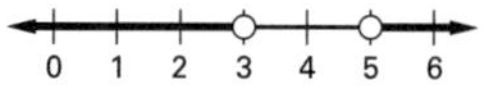

8. $-9 < x < -5$

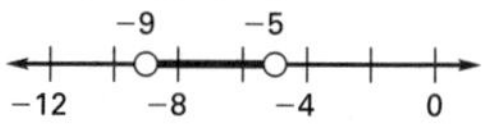

9. $3 \leq x \leq 21$

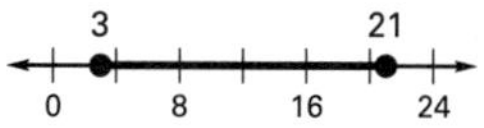

10. $x < -2$ or $x > \frac{5}{2}$

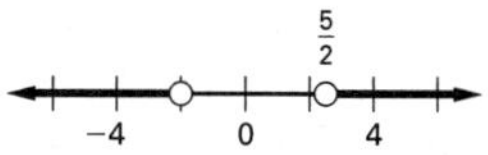

11. $-16 \leq x \leq 9$

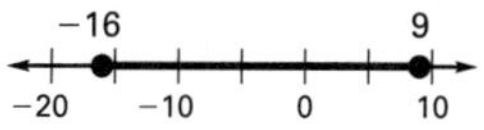

12. $x < -6$ or $x > 5$

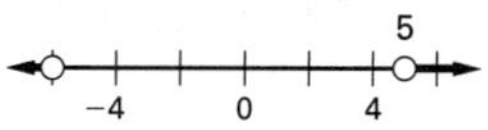

13. solution, solution **14.** not a solution, solution **15.** not a solution, solution **16.** solution, not a solution

17.

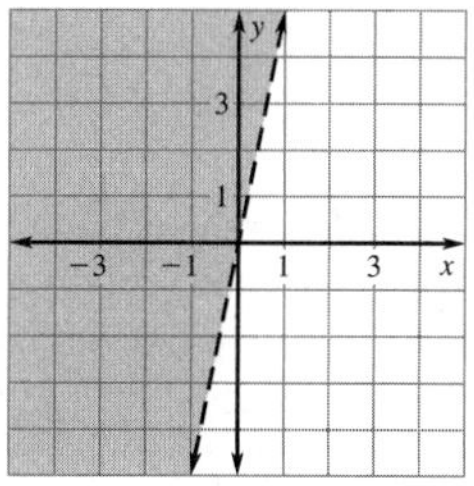

18.

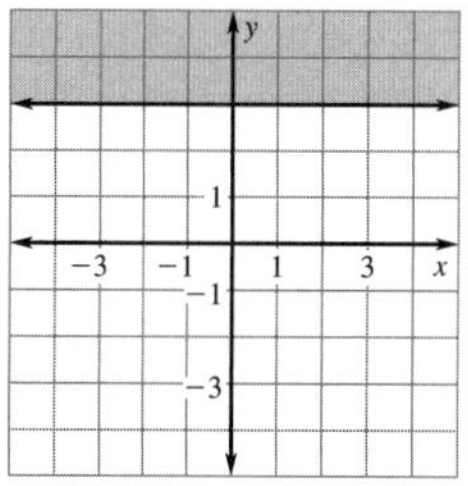

19.

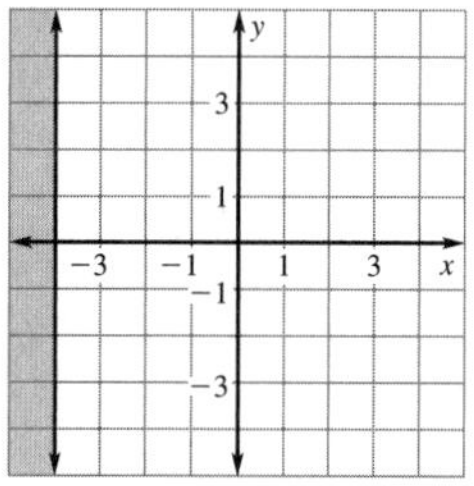

20.

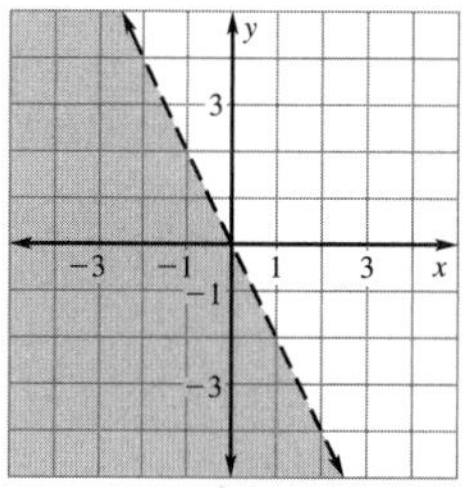

21.

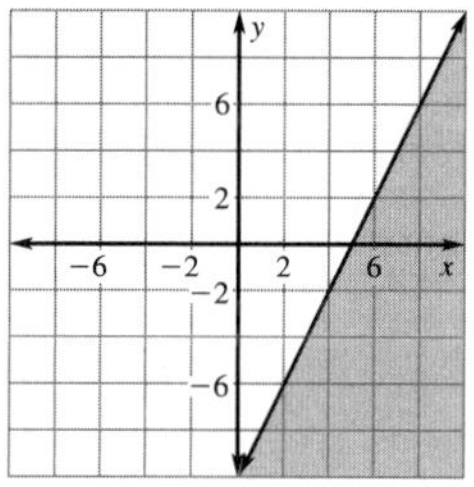

22.

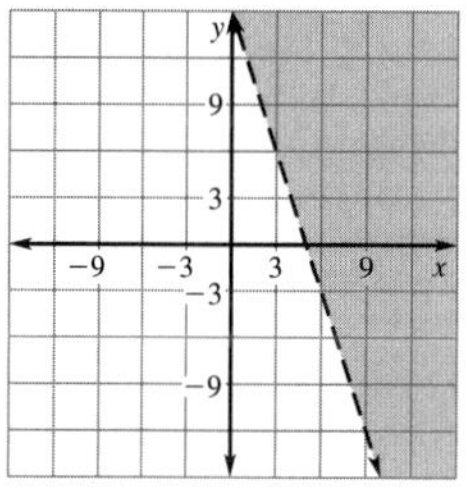

23. $|x - 104| \leq 39$

TECHNOLOGY ACTIVITY 6.5 (p. 367)

1, 5. The lines are not part of the solution.

1.

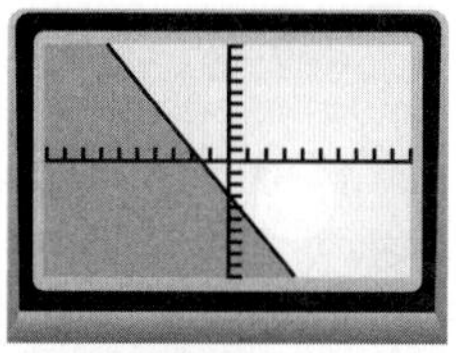

3.

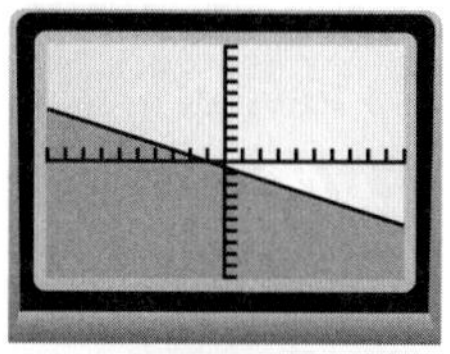

5.

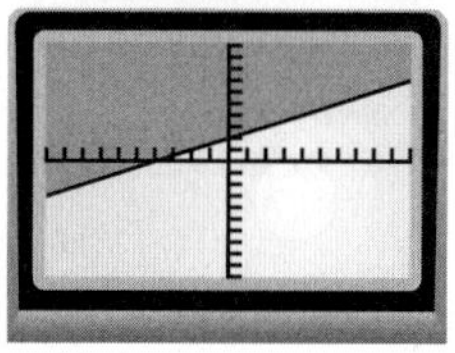

7.

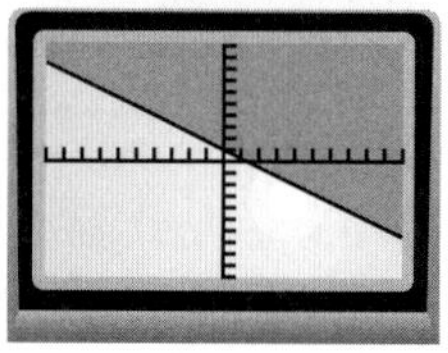

13. $y > x$

6.6 PRACTICE (pp. 371–373) **5.** 3, 2, 2 **7.** $1, -\frac{1}{2}$, no mode **9.** 11, 13, 13, 15, 16, 17, 18, 20, 20, 22, 25, 27, 27, 28, 31, 33, 34, 37, 38 **11.** 50, 52, 54, 60, 63, 65, 70, 74, 74, 74, 75, 78, 78

13.

0	1 2 3 4 5
1	
2	0 0 2 4 6 8 8 8
3	0 1 7 9

Key: 2 | 0 = 20

1, 2, 3, 4, 5, 20, 20, 22, 24, 26, 28, 28, 28, 30, 31, 37, 39

15. 30, 33, 37, 39, 44, 48, 52, 54, 61, 61, 62, 68, 68, 76, 76, 76, 77, 79, 80, 81, 82, 84, 86, 87, 87 **17.** 7, 7, 10 **19.** 6, 6, 6 **21.** 8, 7.5, 5 and 10 **23.** 4.19, 3.98, 1.2 **25.** 72.8, 66 **27.** yes; 61

SA17

29. No; the mean salary of the other 99 adults is $25,000. The median and mode (both $25,000) are much more representative of the "average" salary. **31.** *Sample answer:* The median (619) and the mode (619) are both slightly more representative than the mean (753) because they are closer to more scores in the lower three fourths of the data.

33.

1	3 24 25
2	3 20 22
3	17
4	1 8 14 17 30 30
5	10
6	3 5 13 24
7	31
8	21 26
9	12
10	11 17
11	4 11 28
12	9 15 28

Key: 12 | 9 = 12–9 (Dec. 9)

1–3, 1–24, 1–25, 2–3, 2–20, 2–22, 3–17, 4–1, 4–8, 4–14, 4–17, 4–30, 4–30, 5–10, 6–3, 6–5, 6–13, 6–24, 7–31, 8–21, 8–26, 9–12, 10–11, 10–17, 11–4, 11–11, 11–28, 12–9, 12–15, 12–28

35.

20	44
21	93
22	14 17 26 52
23	93
24	78
25	70 74

Key: 25 | 70 = 257.0

204.4, 219.3, 221.4, 221.7, 222.6, 225.2, 239.3, 247.8, 257.0, 257.4

37. 300

39. *Sample answers:*

- About half of the 17 top male and female golfers had fewer than 40 tournament wins.
- For both males and females, by far the most common stem was 3, meaning between 30 and 39 tournament wins.
- There were 4 male and 2 female golfers with 60 or more tournament wins.

6.6 Mixed Review (p. 374)

45. 358 **47.** not a solution **49.** A **51.** –1 **53.** $-\frac{1}{2}$

55. $x \geq 9$

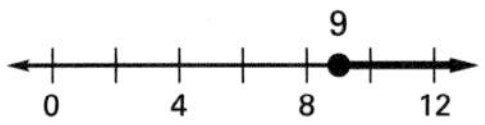

57. $-8 \leq x < -2$

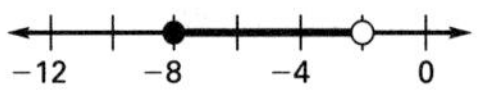

59. $x \leq -25$ or $x \geq -7$

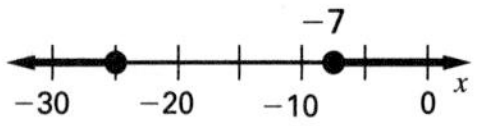

6.7 Practice (pp. 378–380)

5. 2, 4, 6 **7.** 6, 11, 16 **9.** C **11.** 5, 6.5, 9 **13.** 4, 5, 7

15.

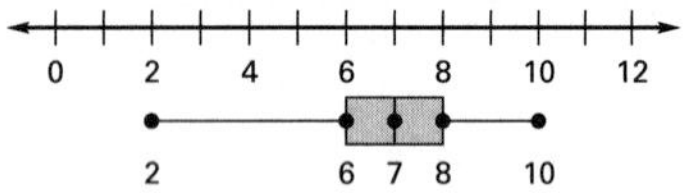

17.

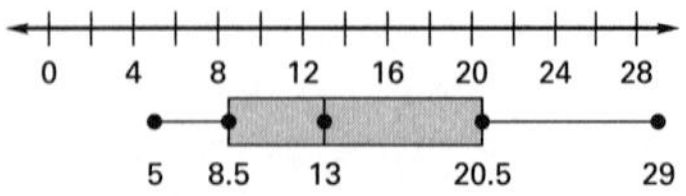

19. Days of Rainfall

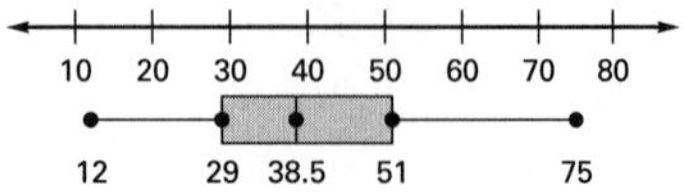

21. Movie Attendance

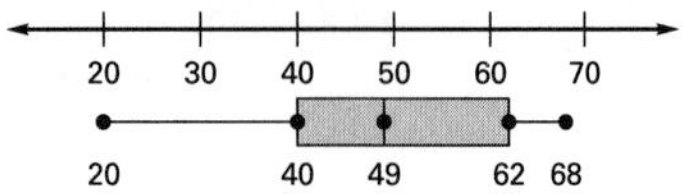

23. *Sample answer:* 10, 11, 11, 12, 14, 15, 16, 18, 20, 21, 25, 27, 27, 29, 30, 38

25. 3.6, 3.8, 4.8, 6.2, 6.7, 7.0, 7.1, 7.7, 8.4, 8.8

3	6 8
4	8
5	
6	2 7
7	0 1 7
8	4 8

Key: 8 | 4 = 8.4 oz

27. 60, 70, 75, 80, 90, 90, 90, 135, 140, 260

6	0
7	0 5
8	0
9	0 0 0
10	
11	
12	
13	5
14	0
26	0

Key: 14 | 0 = 140 min

31. 0.75 h **33.** No. *Sample answer:* More people travel 0.5–1 hour; the box represents about 50% of the data, each whisker about 25%.

35.

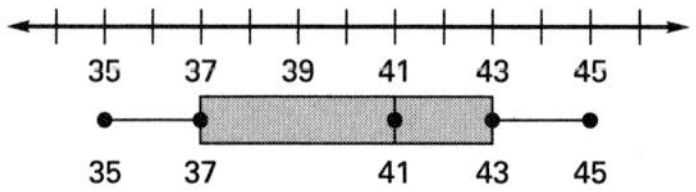

6.7 Mixed Review (p. 381)

39. $x = \frac{1}{4}y + 1$; $\frac{1}{2}, \frac{3}{4}, 1, 1\frac{1}{4}$

41. $x = \frac{3}{2}y - \frac{5}{2}$; $-\frac{11}{2}, -4, -2\frac{1}{2}, -1$ **43.** $10.50/h

45. solution **47.** not a solution

49. *Sample answer:* $y = 1430x + 64{,}500$ (in thousands)

Quiz 3 (p. 381)

1.

0	5 7
1	0 6
2	0 3 4 5 9
3	1 2 7 8

Key: 3 | 1 = 31

5, 7, 10, 16, 20, 23, 24, 25, 29, 31, 32, 37, 38

2.

1	2 8
2	6 7
3	3 3
4	2 4 6 7
5	9
6	1

Key: 5 | 9 = 59

12, 18, 26, 27, 33, 33, 42, 44, 46, 47, 59, 61

3. 8, 9, 9 and 10 **4.** 37.5, 41, no mode

5. 8.5, 13, 18

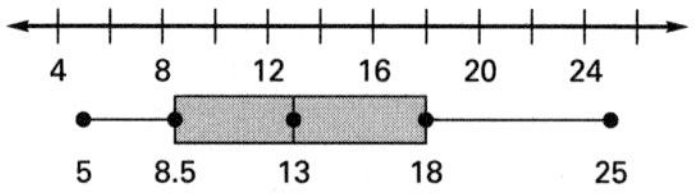

6. 20, 45, 70

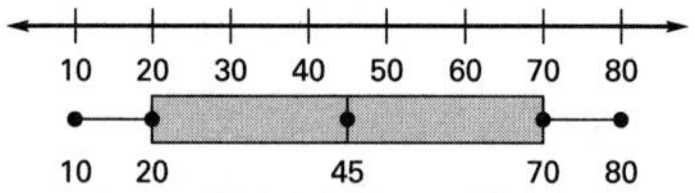

7. 9, 40.5, 56

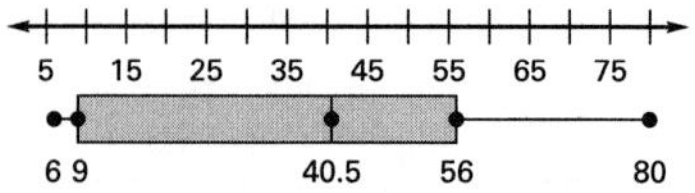

8. 4, 8.5, 13

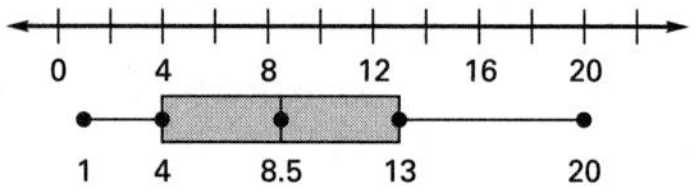

9. Mean: 39
Median: 40
Mode: 57

TECHNOLOGY ACTIVITY 6.7 (p. 382)

1. (least: 4; greatest: 19); quartiles: 10, 13.5, 18 **3.** (least: 20.1; greatest: 43.4); quartiles: 23.2, 25.7, 38.65

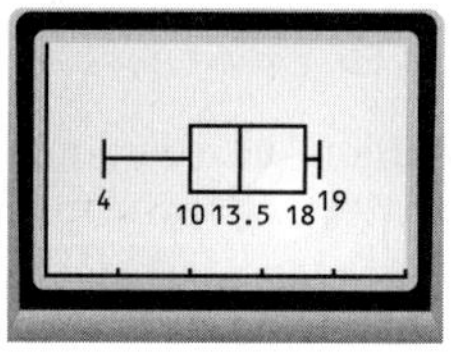

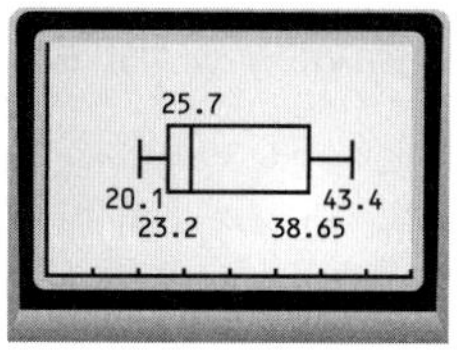

CHAPTER 6 REVIEW (pp. 384–386)

1. $x \le 2$

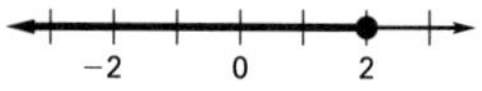

3. $x < -15$

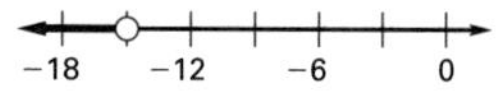

5. $x > -\frac{2}{3}$ **7.** $x \le 5$ **9.** *Sample answer:* x is greater than 2 and less than 5. **11.** x is less than or equal to $-1\frac{1}{4}$ or x is greater than 7. **13.** –6, 16 **15.** $-11 \le x \le 1$

17. $-4\frac{2}{3} \le x \le 3\frac{1}{3}$

19.

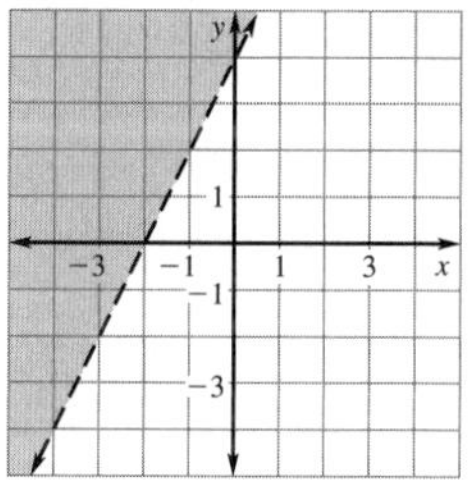

21.

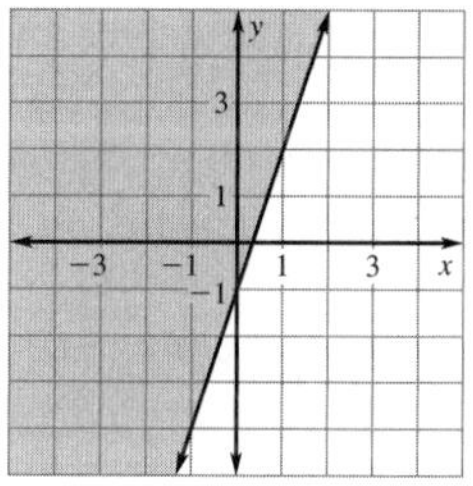

23. 72, 74.5, 58 and 78 **25.** *Sample answer:* The temperatures in Tokyo are much higher. The highest Paris temperature is only slightly higher than the median Tokyo temperature. Only 25% of the Paris temperatures are at least 15.3°, while half of the Tokyo temperatures are at least 15.85°.

CUMULATIVE PRACTICE (pp. 390–391) **1.** 7 **3.** 45 **5.** 6

7. –6 **9.** 14.5 **11.** $\begin{bmatrix} 6 & 7 & -6 \\ -5 & -1 & 5 \end{bmatrix}$ **13.** $18 + 3x$

15. $-15t + 5t^2$ **17.** $11b + 7$ **19.** $5.3y$ **21.** 12 **23.** 24

25. 12 **27.** $1\frac{11}{14}$ **29.** 4 **31.** 1 **33.** $y = \frac{4}{5}x - \frac{13}{5}$ **35.** \$.50/can

37.

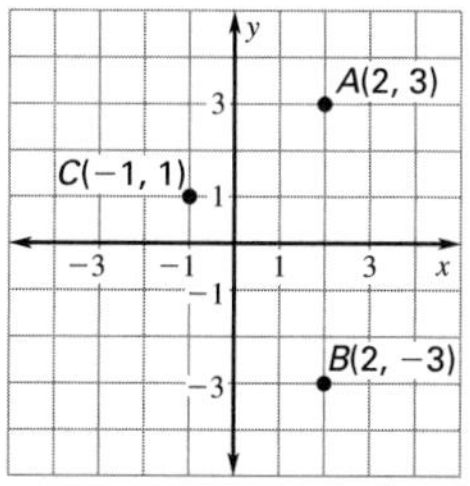

39.

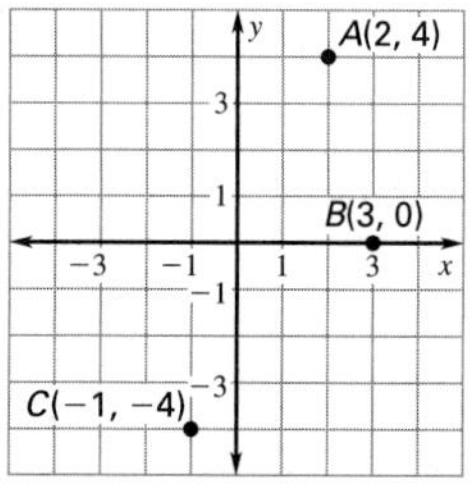

41. $\frac{1}{3}$

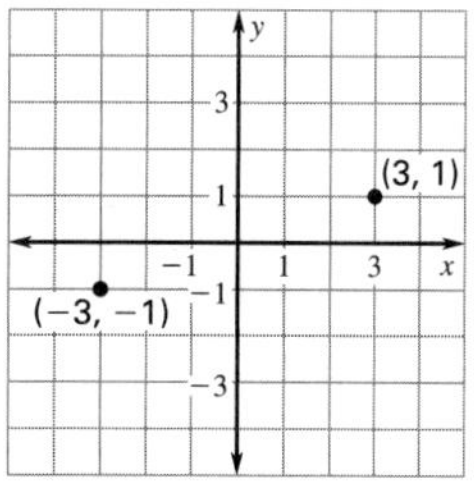

43. undefined

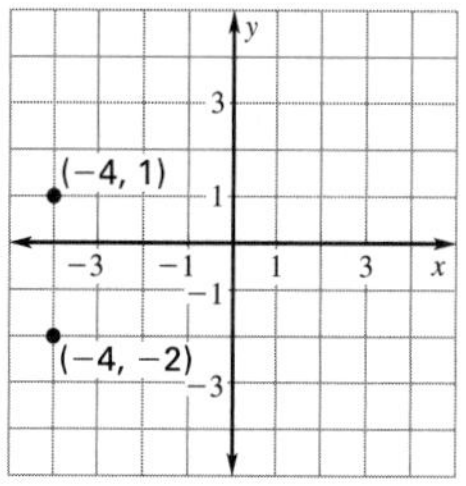

45.

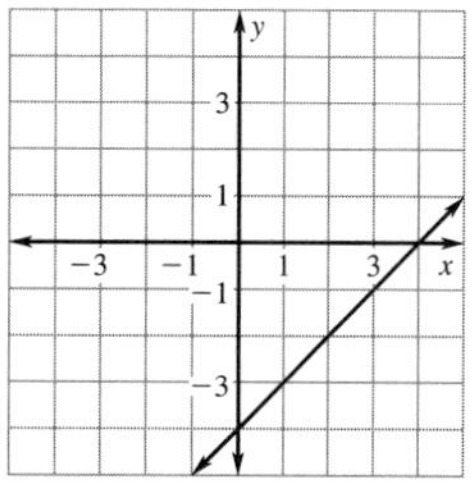

47.

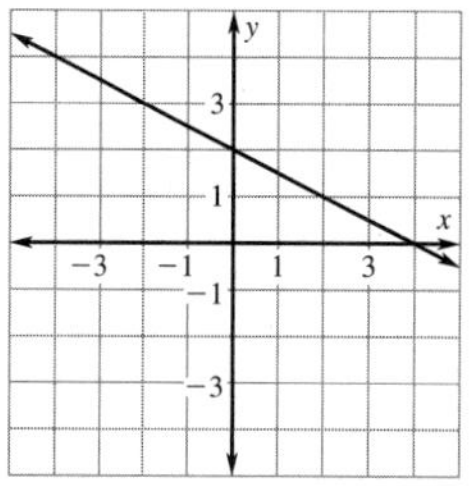

49.

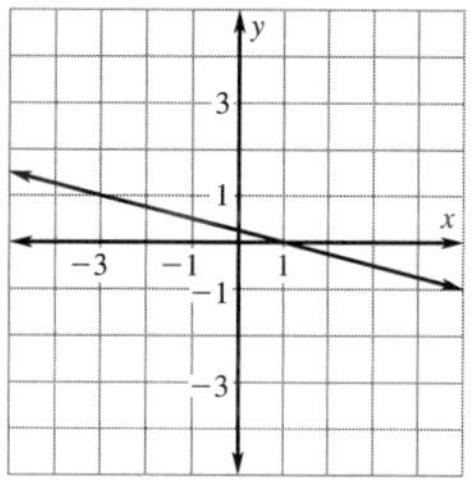

51. $y = x - 3$ **53.** $y = 2x + 3$

55. $y = -5x - 9$ **57.** $y = \frac{1}{2}x + 3$ **59.** $x < 2$ **61.** $x \le -48$

63. $x < 6$ **65.** $x \ge 10.1$ **67.** $4 < x < 6$ **69.** $x < -2$ or $x > 18$

71. 16, 12.5, 5 **73.** No. *Sample answer:* By estimation, you have about \$25 – 2(\$10) = \$5 left. The actual amount is \$25 – 2(\$9.99) = \$5.02, which is less than \$5.95.

75. –100 ft/min, 100 ft/min **77.** \$9.90 **79.** *Sample answer:* $y = 365x + 4896$

CHAPTER 7

SKILL REVIEW (p. 396) **1.** $21x$ **2.** $-36r$ **3.** $-d$ **4.** $\frac{5}{4}w$

5. $-\frac{16}{21}g$ **6.** $\frac{23}{40}y$ **7.** –1 **8.** no solution **9.** all real numbers

10. $\frac{1}{2}$ **11.** solution **12.** not a solution **13.** not a solution

14. not a solution

7.1 PRACTICE (pp. 401–403) **3.** not a solution

5. not a solution **7.** $-(-4) + (-2) = 2$; $2(-4) + (-2) = -10$; (–4, –2) is not a solution of either equation. **9.** (2, –1)

11. solution **13.** not a solution **15.** not a solution **17.** (4, 5)

19. (3, 0) **21.** (6, –6) **23.** (–3, –5) **25.** (–4, –5)

27. (1, 4) **29.** (–8, –4) **31.** (8, 6) **33.** $\left(-1\frac{1}{2}, -2\frac{1}{2}\right)$

35. 125,000 mi **37.** *Sample answer:* coastal: $y = 0.12t + 46$; inland: $y = -0.12t + 54$ **39.** *Sample answer:* 1973

7.1 MIXED REVIEW (p. 403) **47.** –3 **49.** 5 **51.** 9 **53.** $y = -4x + 12$ **55.** $y = 4x - 9$ **57.** $y = 2x - 1$

59. $x < 4260$

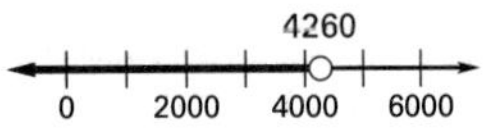

TECHNOLOGY ACTIVITY 7.1 (p. 404) **1.** (–3.5, 2.5) **3.** (–1, –2) **5.** The lines are parallel. Since the lines do not intersect, the system has no solution.

7.2 PRACTICE (pp. 408–410) **7.** (–2, –3) **9.** (1, 2) **11.** (1, 0) **13.** $\left(1\frac{1}{2}, -\frac{1}{2}\right)$ **15.** *Sample answer:* I would isolate *m* in the second equation because it is easy to do so and the value is easy to substitute in the other equation. **17.** (6, 2) **19.** (3, 8) **21.** (2, 9) **23.** (5, 15) **25.** (–3, –3) **27.** (12, 17) **29.** (–168, –108) **31.** (4, 1) **33.** $\left(\frac{1}{3}, -4\frac{1}{3}\right)$

35. The variables are eliminated, leaving a statement that is always true. You solved for one variable in one equation and substituted for that variable in the same equation; Substitute $y = -3x + 9$ into Equation 2 to get (1, 6) for the solution. **37.** (100, 50) **39.** (7, –6) **41.** (24, –4) **43.** 30 11-in. softballs, 50 12-in. softballs
45. \$3375 in company EFG, \$1125 in company PQR
47. 882 m uphill, 675 m downhill

7.2 MIXED REVIEW (p. 410) **55.** $-2x$ **57.** $26y$

59. $y = 2x + 4$

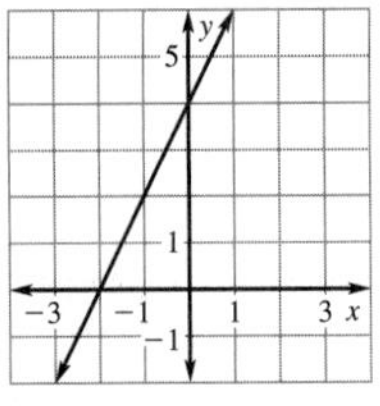

61. $y = -\frac{5}{3}x + 1$

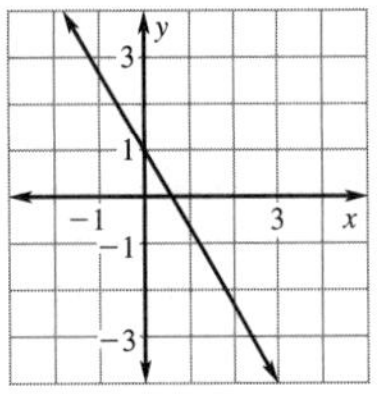

63. $y = -2$

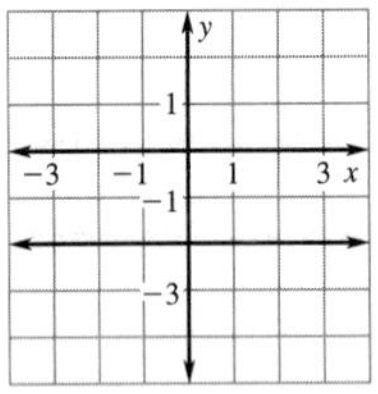

65. $-19 \le x \le 9$

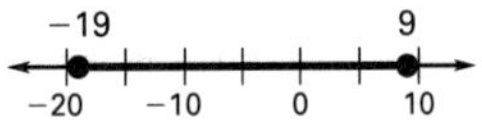

67. $x < -\frac{1}{2}$ or $x > 5\frac{1}{2}$

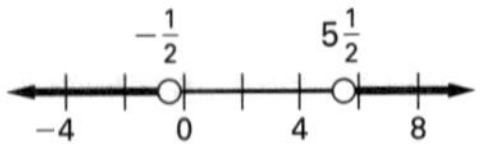

7.3 PRACTICE (pp. 414–416) **5.** *Sample answer:* Multiply either equation by –1, solve for *y*, then substitute in either original equation to solve for *x;* (3, 1). **7.** *Sample answer:* Subtract 2*x* from each side of the first equation, solve for *y*, then substitute in either original equation to solve for *x*; (3, 2). **9.** (14, 6) **11.** (–1, 1) **13.** (1, 1) **15.** (1, 1) **17.** (–2, 3) **19.** (3, 2) **21.** (2, 0) **23.** $\left(13, \frac{5}{8}\right)$ **25.** (21, –3) **27.** (8, –1) **29.** (50, 50) **31.** (1, 2) **33.** (4, 3) **35.** (2, 1) **37.** (2, 0) **39.** (3, 2) **41.** (0, 4) **43.** 5 g **45.** Let *s* = speed in still air and *w* = wind speed; $s - w = 300$, $s + w = 450$. **47.** 375 mi/h; 75 mi/h

7.3 MIXED REVIEW (p. 416) **57.** $y = \frac{1}{2}x$ **59.** $y = x - 1$ **61.** $y = -\frac{1}{3}x + 6$ **63.** solution; not a solution **65.** solution; not a solution **67.** (–3, –2) **69.** (10, –2) **71.** (3, 5)

QUIZ 1 (p. 417) **1.** (3, –4) **2.** (6, 8) **3.** (0, 0) **4.** (1, 9) **5.** (–1, 3) **6.** (10, –6) **7.** (5, 1) **8.** $\left(-\frac{1}{2}, \frac{1}{2}\right)$ **9.** $\left(2\frac{2}{3}, -6\frac{2}{3}\right)$ **10.** 4 CDs; 6 CDs

7.4 PRACTICE (pp. 421–424) **3.** *Sample answer:* Substitution or linear combinations; it would be simple to write either variable in terms of the other or to eliminate *x* by multiplying either equation by –1.
5. *Sample answer:* Any of the three methods would be reasonable; it would be simple to write either variable in terms of the other, both would be simple to graph, and *y* could be eliminated by multiplying either equation by –1.
7. Let *x* = price per gallon of regular and *y* = price per gallon of premium. **9.** regular: \$1.19; premium: \$1.39
11. (3, 2) **13.** *Sample answer:* Substitution or linear combinations; it would be simple to write *y* in terms of *x*, and *y* could be eliminated by adding the equations.
15. *Sample answer:* Substitution; either variable may be easily eliminated using one of the equations.
17. *Sample answer:* Linear combinations; no variable has a coefficient of 1 or –1. **19.** (2, 1) **21.** (2, –1) **23.** (1, –1) **25.** (2, –1) **27.** $\left(5\frac{5}{9}, 10\right)$ **29.** (–4, 4) **31.** (3, 0) **33.** (4, 2) **35.** $\left(-\frac{3}{4}, 3\frac{1}{2}\right)$ **37.** (6, 6) **39.** (6, –5) **41.** (–4, 5) **43.** $\left(\frac{1}{3}, 0\right)$ **45.** (2, 6) **47.** 20 mL of the 5% solution, 40 mL of the 2% solution **49.** 4 children **51.** $y = 172x + 6200$ and $y = -16x + 6500$; $\left(1\frac{28}{47}, 6474\frac{22}{47}\right)$; the point of intersection represents the number of years after 1970 (about 1.6) when the demand for low-income housing and the availability were equal (about 6,474,500). **53.** $y = 7x + 50$ **55.** 1 h at 4 mi/h and $\frac{1}{2}$ h at 6 mi/h **57.** substitution, graphing, linear combinations **59.** No; solving the system involving any two of the equations will produce the desired solution.
61. $r + b = 12$; $r = 3b$ **63.** 0.75

7.4 MIXED REVIEW (p. 424) **67.** parallel **69.** not parallel

71.

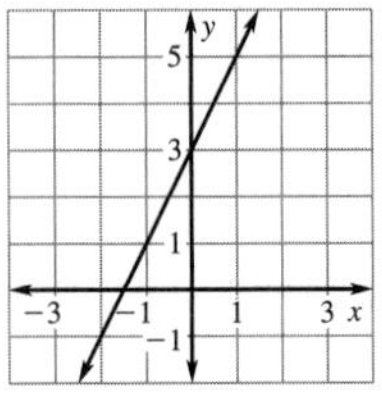

73.

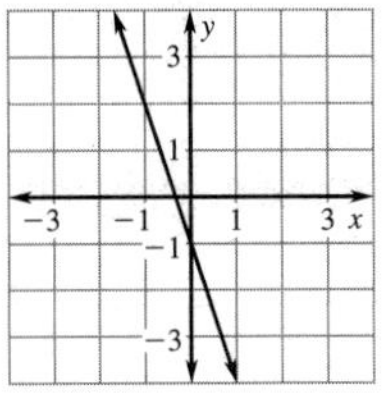

75. 70 out of 100 households

SA20

7.5 Practice (pp. 429–431)

7. 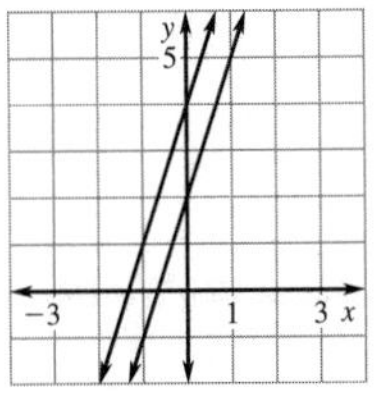

no solution **9.** no solution

11. exactly one solution, (5, 12) **13.** D; no solution **15.** B; infinitely many solutions **17.** C; infinitely many solutions **19.** infinitely many solutions **21.** infinitely many solutions **23.** no solution **25.** infinitely many solutions **27.** no solution **29.** no solution **31.** No; the system of equations that describes the situation is $4x + 2y = 99.62$, $12x + 6y = 298.86$, which has infinitely many solutions.

7.5 Mixed Review (p. 431)

43. A **45.** D **47.** 55 ft/h

7.6 Practice (pp. 435–437)

5. 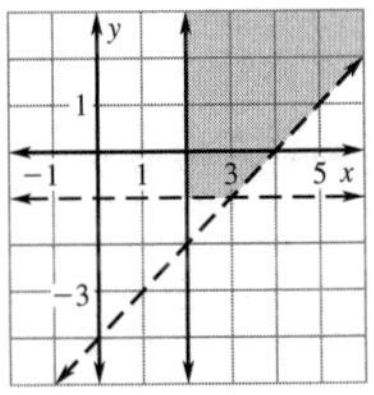

7. 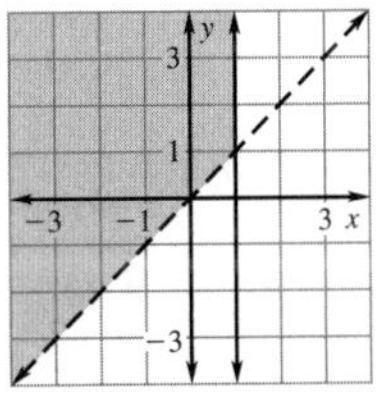

9. C **11.** B **13.** $y \le \frac{1}{2}x + 2,\ y \ge \frac{1}{2}x - 2$

15. 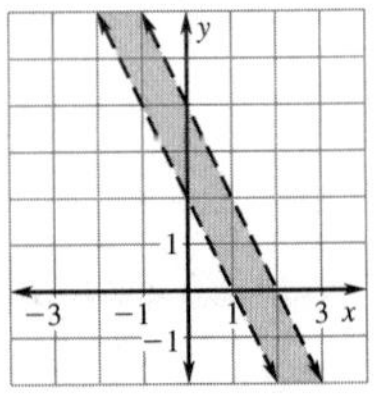

17. 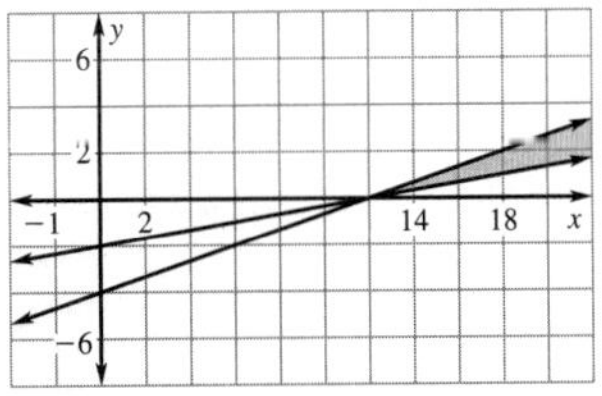

19. 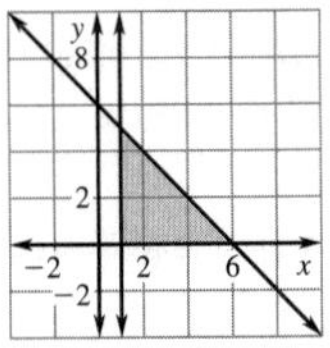

21. 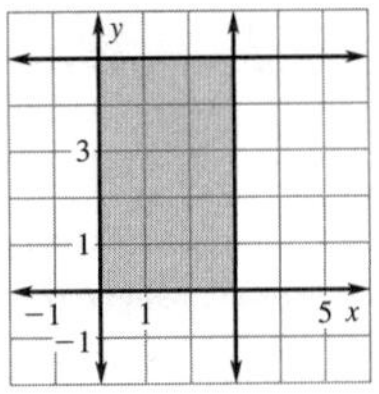

23. 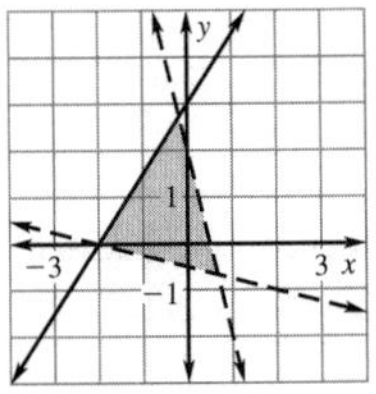

25. 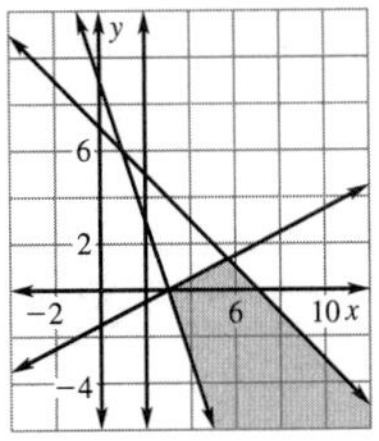

27. (0, −5), (0, 11), (6, 5) **29.** (0, 0), (−2, 1), (2, 5)

31. $1 \le x \le 7,\ 1 \le y \le 6$

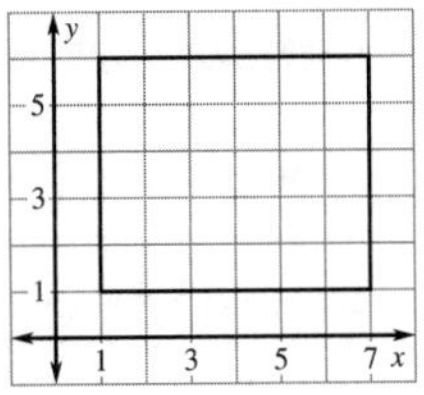

33. $y \le x + 2,\ y \le -x + 7,\ 1 \le y \le 3$

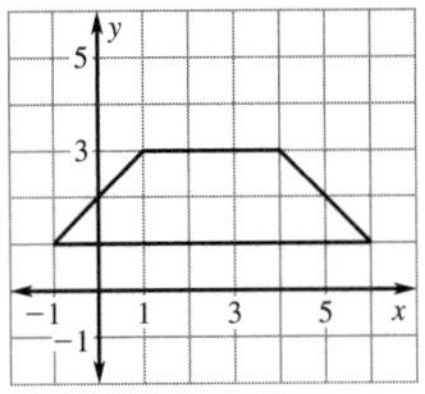

35. 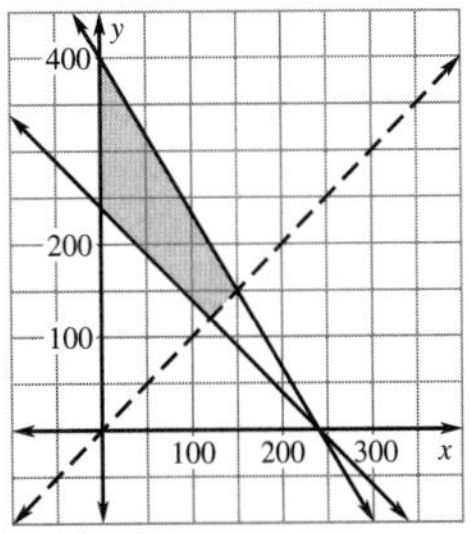

37. 8 square units

39. *Sample answers:* 4 h baby-sitting and 14 h as cashier, 12 h baby-sitting and 6 h as cashier **41. a.** (0, 0), (0, 2400), (900, 600), (1050, 0) **b.** 0; 72,000; 27,000; 10,500 **c.** $72,000

7.6 Mixed Review (p. 438)

45. 243 **47.** 137 **49.** 33 **51.** 49 **53.** −60 **55.** 9 **57.** 47, 56, 57

Quiz 2 (p. 438)

1. length: 8 ft, width: 3 ft **2.** no solution **3.** no solution **4.** infinitely many solutions **5.** exactly one solution, (5, −6) **6.** exactly one solution, (0, 1) **7.** infinitely many solutions **8.** 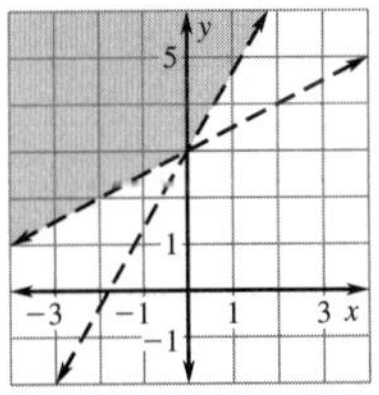

9. 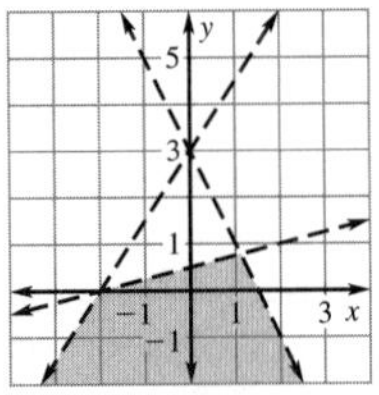

10. 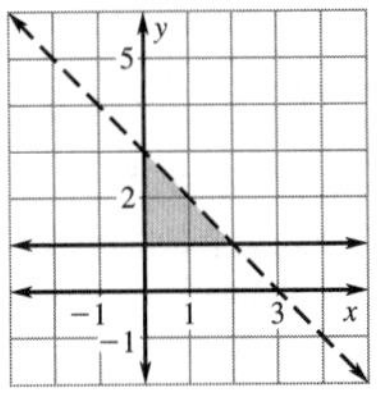

11. 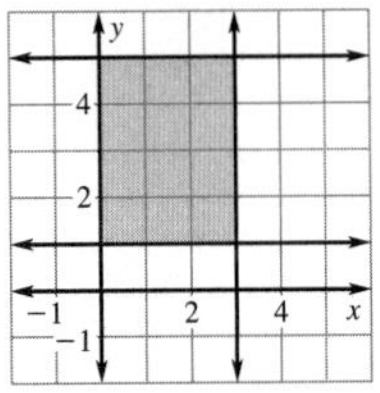

12. 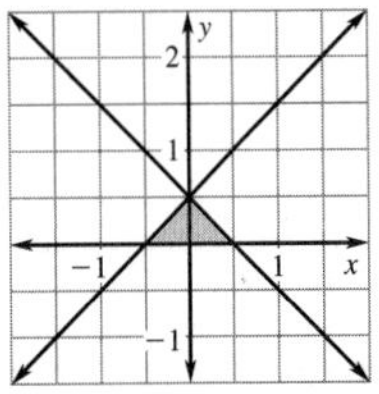

13. 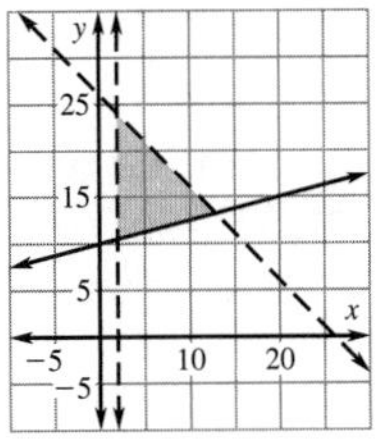

14. $10,000 at 5%, $6000 at 6%

Chapter 7 Review (pp. 440–442)

1.
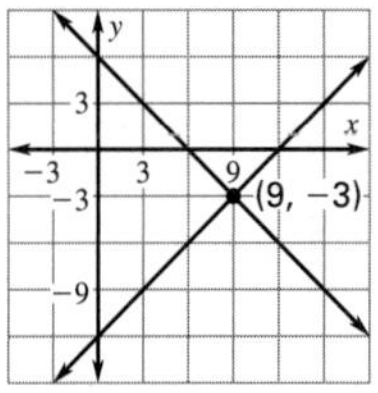

3.
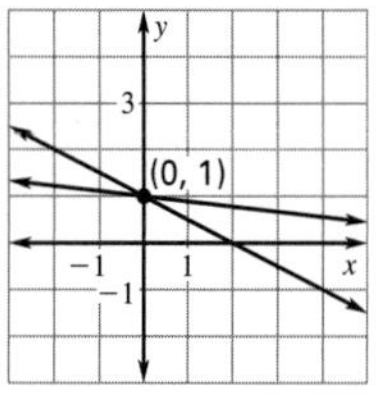

5.
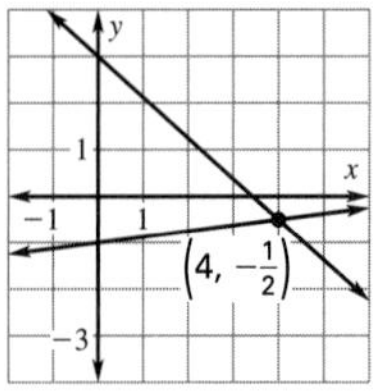

7. (0, 3) **9.** $\left(\frac{5}{8}, 1\frac{1}{2}\right)$ **11.** $\left(\frac{1}{2}, 0\right)$ **13.** $\left(-5, 2\frac{3}{5}\right)$ **15.** (3, –5) **17.** (–1, 1) **19.** Ferris wheel: 5 times, roller coaster: 7 times

21. infinitely many solutions **23.** exactly one solution, (2, 6) **25.** exactly one solution, $\left(-3\frac{1}{4}, 6\frac{1}{2}\right)$

27.
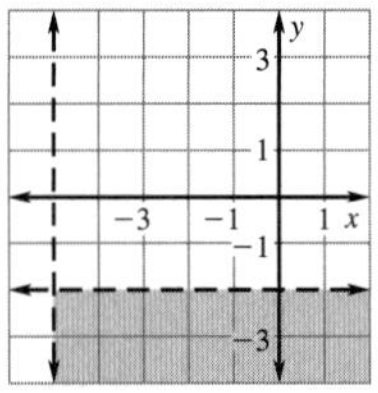

29.
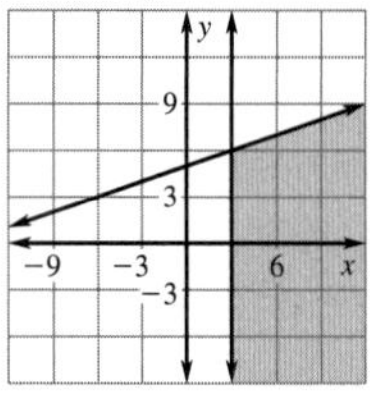

31.
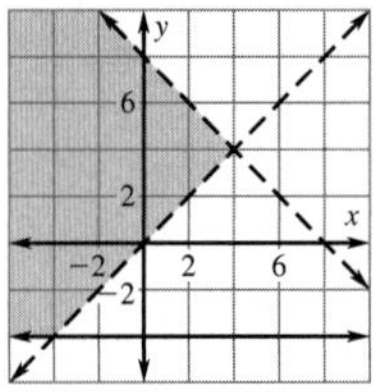

CHAPTER 8

Skill Review (p. 448) **1.** 6^2 **2.** 4^3 **3.** $(2y)^5$ **4.** $\frac{1}{4}$ **5.** $\frac{1}{9}$ **6.** $\frac{3}{4}$ **7.** $\frac{1}{2}$ **8.** \$8.25/h **9.** \$.79/cantaloupe **10.** \$.50/can **11.** 45 mi/gal

8.1 Practice (pp. 453–455) **5.** m^3 **7.** 3^7, or 2187 **9.** x^9 **11.** 4 **13.** 4^9, or 262,144 **15.** m^{32} **17.** a^2b^4 **19.** $x^{12}y^{20}$ **21.** $(-2)^3x^9$, or $-8x^9$ **23.** 5^{11}, or 48,828,125 **25.** 7^8, or 5,764,801 **27.** $3^2 \cdot 7^2$, or 441 **29.** $(-5)^3a^3$, or $-125a^3$ **31.** $(-4)^6$, or 4096 **33.** $(5 + x)^{18}$ **35.** 3^3b^4, or $27b^4$ **37.** $4x^9$ **39.** $-3^2 \cdot 7^2 \cdot x^{10}$, or $-441x^{10}$ **41.** $3 \cdot 2^3 \cdot y^5$, or $24y^5$ **43.** $-r^3s^7$ **45.** $-(-3)^3c^3d^5$, or $27c^3d^5$ **47.** $6^2 \cdot \left(\frac{1}{4}\right)^2 \cdot a^{14}$, or $\frac{9}{4}a^{14}$ **49.** -3^2w^7, or $-9w^7$ **51.** $(-x)^9$, or $-x^9$ **53.** $-r^4s^{14}t^9$ **55.** a^5; 1 **57.** a^6; 1 **59.** a^2b^4; 16 **61.** $-(a^2b^6)$; –64 **63.** > **65.** < **67.** > **69.** 788.9 **71.** 308.9 **73.** 333,446,267.9 **75.** approximately $113.04a^3$ **77.** 8 to 1 **79. a.** 2^{30} or 1,073,741,824 ways **b.** $\left(\frac{1}{2}\right)^{30}$, or about 0.0000000009 **81.** 3^{14}; $\frac{1}{3^{14}}$, or about 0.0000002

8.1 Mixed Review (p. 455) **87.** 10,000 **89.** $\frac{1}{25}$ **91.** $\frac{45}{4}$

93.
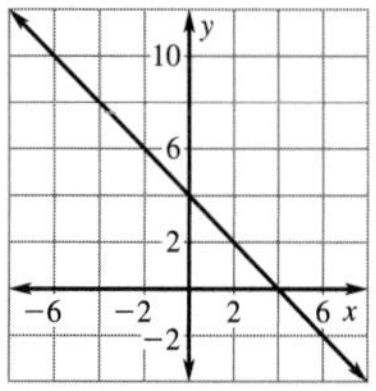

95.
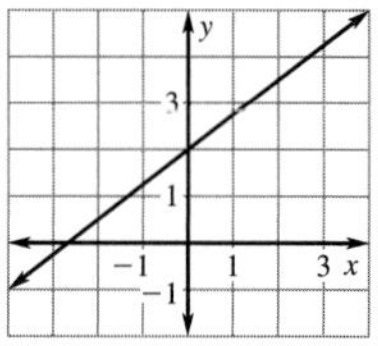

97.
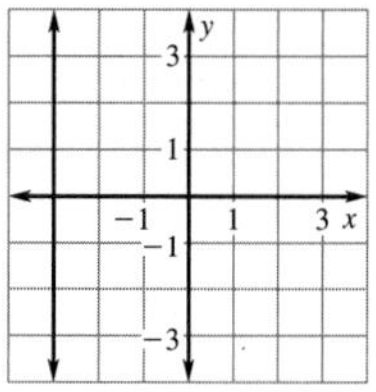

99.

101.

103. $x \geq -\frac{10}{3}$

105. $\$.07 \leq p \leq \$.71$, where p is the price of a quart of milk

8.2 Practice (pp. 459–461) **3.** $\frac{1}{3}$ **5.** not defined **7.** $\frac{1}{m^2}$ **9.** $\frac{a^5}{b^8}$ **11.** yes **13.** about 26.9 points per game **15.** $\frac{1}{81}$ **17.** 32 **19.** 100 **21.** $\frac{1}{2}$ **23.** 1 **25.** 1 **27.** –9 **29.** $\frac{1}{125}$ **31.** $\frac{3}{x^4}$ **33.** $\frac{y^4}{x^2}$ **35.** $\frac{8}{x^2y^6}$ **37.** $\frac{x^{10}}{4y^{14}}$ **39.** $-\frac{1}{64x^3}$ **41.** $\frac{1}{9x^2y^2}$ **43.** $8m^2$ **45.** $-\frac{1}{2x^3}$ **47.** C **49.** 0.03125 **51.** 0.00352 **53.** no **55.** no

57.
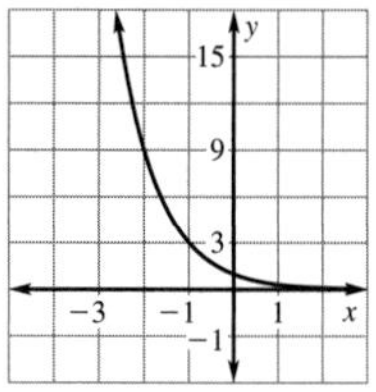

59.
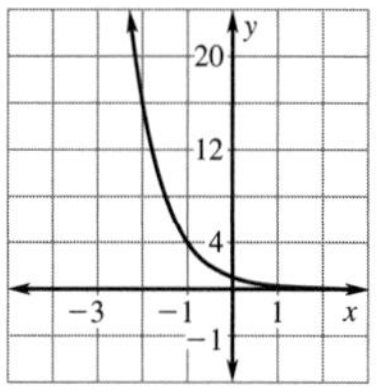

61.
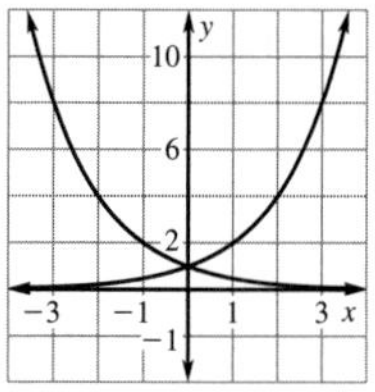

Sample answer: Both graphs are curves that pass through (0, 1) and both increase without limit in one direction and get very close to the x-axis in the other.

63. (0, 1); yes; (0, 2)

65. a. 117; 141; 203; 243; 292 **b.**
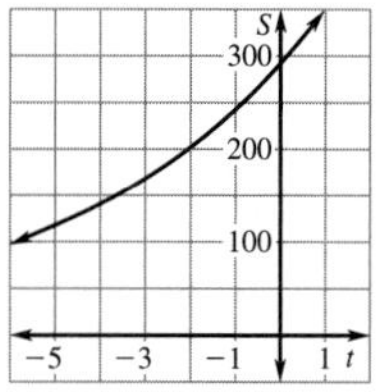

8.2 Mixed Review (p. 461) **71.** $\frac{4}{25}$ **73.** $-\frac{729}{1000}$

75. $-12 \leq x \leq 2$

77. $-11 \leq x \leq 7$
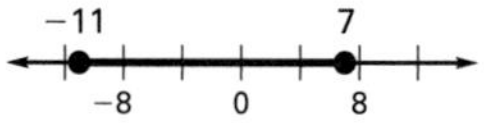

79. $4 < x < 5$
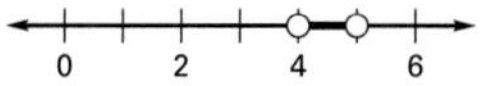

81. 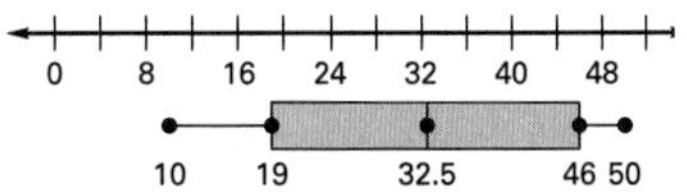

83. $\left(\frac{1}{2}, 3\right)$

85. (100, 50) **87.** (–129, 9)

TECHNOLOGY ACTIVITY 8.2 (p. 462) **7.** *Sample answer:* The graph is close to the negative x-axis and curves upward to the right. **9.** *Sample answer:* The graph is close to the negative x-axis and curves downward to the right.

8.3 PRACTICE (pp. 466–468) **3.** 125 **5.** a^3 **7.** a^3 **9.** $\frac{1}{5}$ **11.** $\frac{1}{32}$ **13.** $\frac{25}{m^2}$ **15.** $\frac{64}{125}$ **17.** $\frac{x^{18}}{y^{30}}$ **19.** 125 **21.** –1 **23.** 2187 **25.** $\frac{1}{125}$ **27.** 27 **29.** $\frac{9}{25}$ **31.** $\frac{81}{x^4}$ **33.** $\frac{1}{x^5}$ **35.** $\frac{1}{x^3}$ **37.** y^2 **39.** $\frac{1}{r^{12}}$ **41.** $\frac{8x^6y^9}{27}$ **43.** $5x^3y^3$ **45.** $\frac{x^6}{256y^8}$ **47.** $\frac{5}{4x^{11}y^4}$ **49.** $\frac{6^3}{6} = 6^{3-1} = 6^2 = 36$ **51.** The volume of Earth is about 64 times the volume of the Moon. **53.** $\frac{1}{1.2^6} \approx \frac{1}{3}$ **55.** about 3 times **57.** $\frac{1}{1.13^2}$ or about 0.8 **59. a.** 200; 160; 128; 102; 82; 66; 52 **61.** $\left(\frac{1}{2}\right)^6 \approx 0.02$

8.3 MIXED REVIEW (p. 469) **67.** 100,000 **69.** $\frac{1}{10,000}$

71. 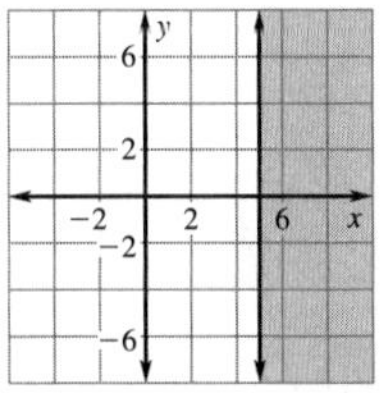

73. 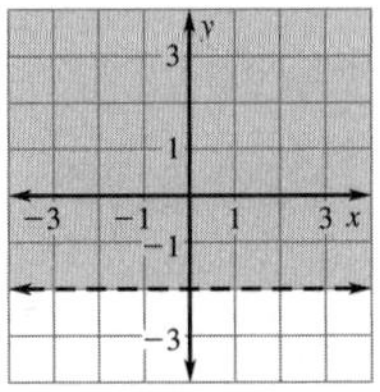

75. 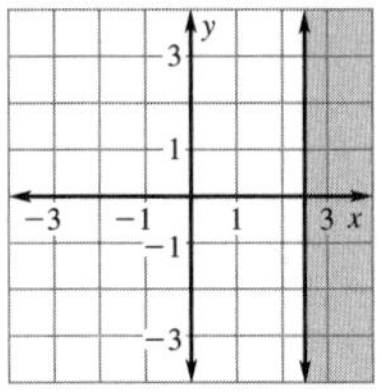

77. 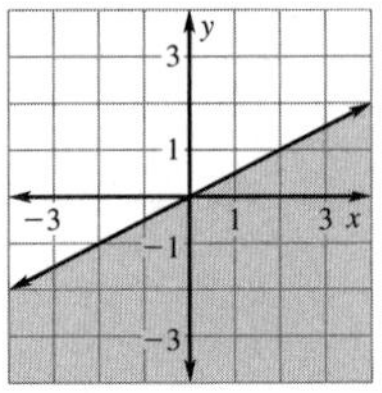

79. solution **81.** solution **83.** (8, 4) **85.** (4, 3)

QUIZ 1 (p. 469) **1.** 2187 **2.** 256 **3.** 10,000 **4.** $\frac{1}{2401}$ **5.** $\frac{1}{16,384}$ **6.** $\frac{7}{6}$ **7.** $\frac{1}{3125}$ **8.** 2187 **9.** $\frac{64}{125}$ **10.** –128 **11.** 64 **12.** 4 **13.** x^9 **14.** $-32x^5$ **15.** $-3a^5$ **16.** c^5 **17.** x^2 **18.** x **19.** $\frac{16m^4}{81n^4}$ **20.** x **21.** $-1728a^6$ **22.** $16m^{10}$ **23.** $30x^2$ **24.** \$145.87; \$214.33; \$250

8.4 PRACTICE (pp. 473–475) **3.** 430 **5.** 0.245 **7.** 3.96×10 **9.** 1.2×10^3 **11.** 6.9×10^6 **13.** 7.2×10^7 **15.** approximately 1.97×10^4 sec, or about 5.5 h **17.** 0.98 **19.** 8652.1 **21.** 0.000006002 **23.** 100,012,000 **25.** 5×10^{-2} **27.** 4.22×10^{-2} **29.** 7×10^8 **31.** 8.551×10^{-3} **33.** 4.59×10^{-4} **35.** 8.8×10^7 **37.** 1.2×10^5; 120,000 **39.** 1.5×10^5; 150,000 **41.** 1.6×10^2; 160 **43.** 6.0×10^0; 6 **45.** 8.1×10^7; 81,000,000 **47.** 2.4×10^{10}; 24,000,000,000 **49.** 1.09926×10^6; 1,099,260 **51.** 5.76×10^{-8}; 0.0000000576 **53.** 1.2×10^8 **55.** 5.0819×10^{13} **57.** about 3.57×10^{14} mi^3 **59.** Louisiana Purchase: \$.03/acre; Gadsden Purchase: \$.53/acre **61.** 4.4064×10^{10} m^3 **63.** Answers are rounded to the nearest whole number; TX: 2167; MN: 1522; PA; 1583; VT: 810; CA: 1806

8.4 MIXED REVIEW (p. 475) **67.** 0.22 **69.** 0.0007 **71.** 0.005 **73.** 2.55 **75.** (1, 4)

77. 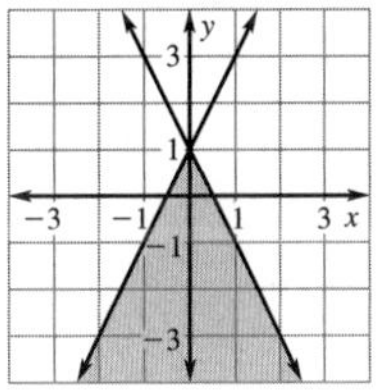

79. 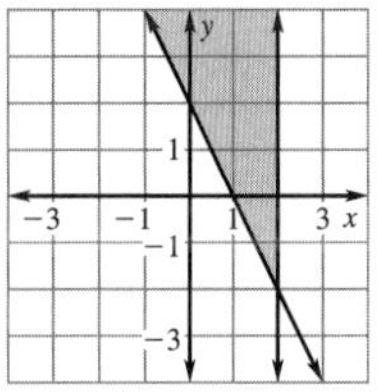

81. 1000 **83.** $\frac{1}{49}$

8.5 PRACTICE (pp. 480–482) **5.** B **7.** \$2231.39 **9.** \$4489.99 **11.** \$379.25 **13.** \$505.67 **15.** \$325.40 **17.** about 46.3 L/min; about 86.5 L/min **19.** A **21.** $P = 10,000(1.25)^t$

23. 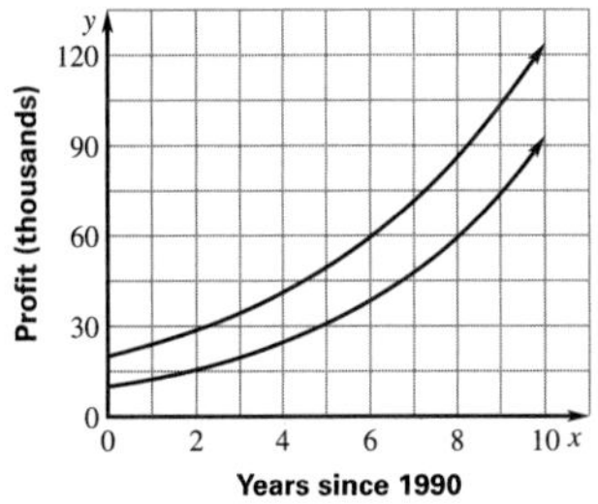

Sample answer: I would base my choice on how long I thought I would own the business (and how realistic I thought the continued annual increase was). The profits for the business in Ex. 22 started higher than those for the other business and would remain higher for 16 years. After that, the profits of the other business would increase more rapidly and be much higher. For example, after 20 years the profits would be more than \$100,000 greater. **25.** B

8.5 MIXED REVIEW (p. 482) **33.** 176.25 **35.** 260 **37.** 0.96 **39.** $-\frac{16}{15}$ **41.** –0.29999952 **43.** 0.41405 **45.** 2.4 **47.** 3

8.6 PRACTICE (pp. 488–490) **5.** \$5137 **7.** \$3770 **9.** C **11.** \$5450 **13.** \$2845 **15.** B **17.** exponential growth; 1.18; 18% **19.** exponential growth; $\frac{5}{4}$; 25% **21.** exponential decay; $\frac{2}{5}$; 60% **23.** 160 mg **25.** 246 mg

27. $N = 64\left(\frac{1}{2}\right)^t$ **31.** $M = 302(0.9)^t$

33.

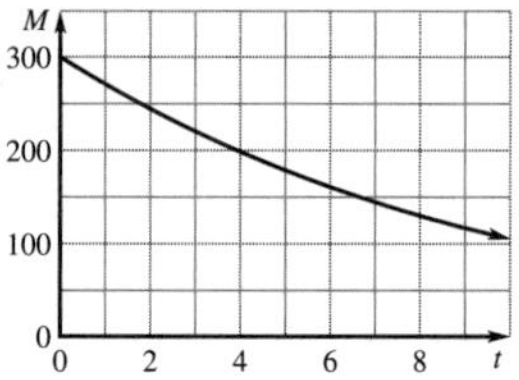

8.6 MIXED REVIEW (p. 490) **41.** 20 **43.** 7 **45.** −1.3 **47.** 2.4

QUIZ 2 (p. 491) **1.** 1.1205×10^{-2} **2.** 1.4×10^8 **3.** 6.7×10^{-8} **4.** 3.072×10^{10} **5.** 4820 **6.** 5,000,000,000 **7.** 0.00000704 **8.** 0.01112 **9.** $y = 50(1.05)^t$; about \$70 **10.** $y = 20{,}000(0.85)^t$; about \$8874

TECHNOLOGY ACTIVITY 8.6 (p. 492) **1.** $y = 1.0339(1.6293)^x$

CHAPTER 8 REVIEW (pp. 494–496) **1.** 2^9, or 512 **3.** $3^3 \cdot 2^2 \cdot a^5$, or $108a^5$ **5.** s^7; 128 **7.** s^6t^2; 576 **9.** $\frac{1}{125}$ **11.** 32 **13.** $\frac{x^6}{y^6}$ **15.** 1

17.

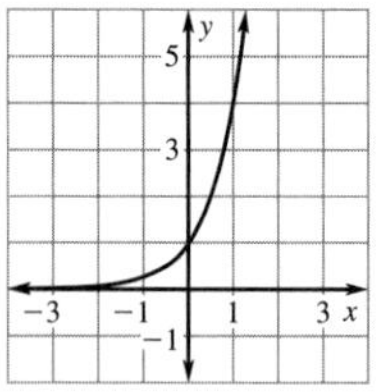

19.

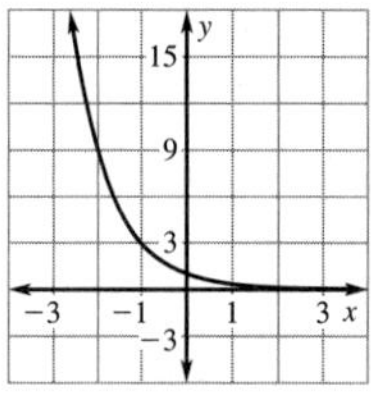

21. $\frac{1}{27}$ **23.** $\frac{16}{81}$ **25.** $\frac{531{,}441}{b^6}$ **27.** m^6 **29.** $-27a^3b^3$ **31.** $-24a^2b^2$ **33.** $\frac{1}{1.17^5} \approx 0.46$ **35.** 768,000 **37.** 0.0002 **39.** 6.79×10^{-4} **41.** 7.52×10^7 **43.** $w = 2(1.05)^x$ **45.** $e = 125(0.97)^x$

CHAPTER 9

SKILL REVIEW (p. 502) **1.** −60 **2.** 12 **3.** 44 **4.** 2

5.

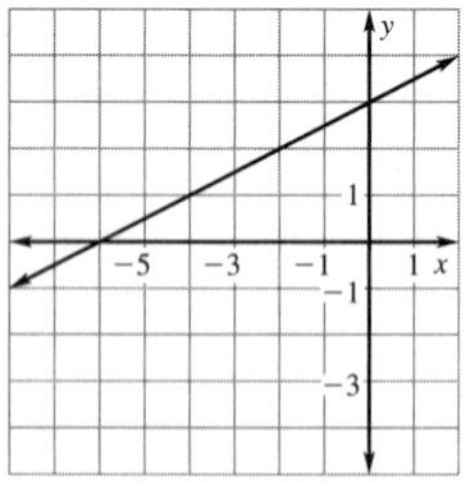

6.

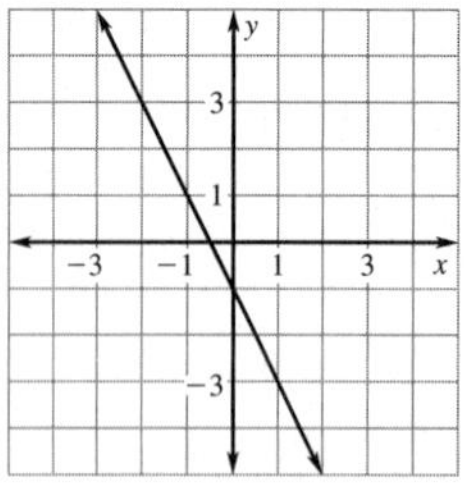

7.

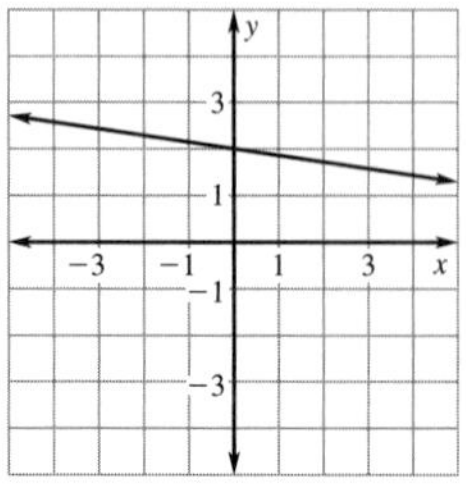

8. not a solution **9.** solution **10.** solution

9.1 PRACTICE (pp. 507–509) **9.** 0.9 **11.** ±3 **13.** 3, 7 **15.** −2, 2 **17.** −1.2, 1.2 **19.** about 1.7 sec **21.** about 3.5 sec **23.** −7, 7 **25.** no square roots **27.** 0 **29.** −0.3, 0.3 **31.** −13 **33.** 3.61 **35.** 0.2 **37.** −0.32 **39.** 3 **41.** undefined **43.** $\sqrt{109}$ **45.** −2.27, 4.27 **47.** −1.33, 2.13 **49.** −11.24, −2.76 **51.** −1.80, 5.13 **53.** −1.34, −0.99 **55.** −8, 8 **57.** −4, 4 **59.** no solution **61.** −3, 3 **63.** −4, 4 **65.** no solution **67.** −3, 3 **69.** −3.87, 3.87 **71.** −2.65, 2.65 **73.** no solution **75.** −2.83, 2.83 **77.** −1.41, 1.41 **79.** $h = -16t^2 + 60$ **81.** $\sqrt{3.75} \approx 1.94$ sec **83.** 3 sec **85.** 5 sec **87.** 2003 **89.** 0.40 mm **91.** 0.15 mm **93.** 0.12 mm

9.1 MIXED REVIEW (p. 510) **101.** 11 **103.** $2^3 \times 3^2$ **105.** (−1, −2) **107.** (−48, 14) **109.** (−4, −19) **111.** 1225 student tickets, 1117 general admission tickets

9.2 PRACTICE (pp. 514–515) **5.** A **7.** C **9.** *Sample answer:* The factor 5 cannot be taken outside the radicand. The first step should be $\sqrt{50} = \sqrt{25 \cdot 2}$. Then, by the product property, this equals $\sqrt{25} \cdot \sqrt{2}$, or $5\sqrt{2}$. **11.** $3\sqrt{3}$ **13.** $5\sqrt{3}$ **15.** $5\sqrt{5}$ **17.** $4\sqrt{5}$ **19.** $\sqrt{6}$ **21.** $4\sqrt{3}$ **23.** $\frac{\sqrt{11}}{4}$ **25.** $2\sqrt{5}$ **27.** $3\sqrt{3}$ **29.** $\frac{3\sqrt{2}}{4}$ **31.** $\frac{3\sqrt{3}}{4}$ **33.** $8\sqrt{5}$ **35.** $\frac{\sqrt{3}}{2}$ **37.** 2 **39.** $\frac{4\sqrt{3}}{9}$ **41.** $\sqrt{30}$ **43.** $18\sqrt{7}$ **45.** −18 **47.** −3 **49.** $\frac{-2\sqrt{5}}{5}$ **51.** $\frac{1}{4}$; 0.25 **53.** $\frac{\sqrt{15}}{2} \approx 1.94\ \text{m}^2$ **55.** $70\sqrt{2}$ m/sec

9.2 MIXED REVIEW (p. 516)

65.

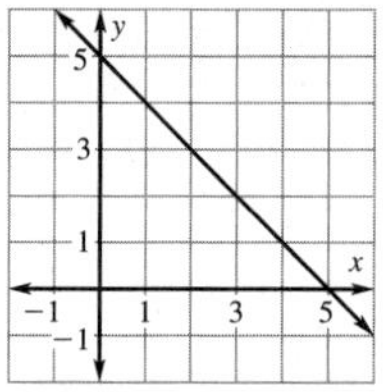

67.

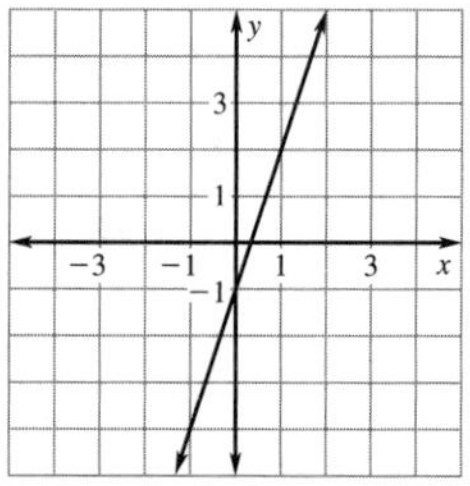

69. b^8; 256 **71.** $\frac{x^5}{2}$ **73.** $\frac{y^3}{x^4}$ **75.** $(1.3)^4 \approx 2.86$

9.3 PRACTICE (pp. 521–523) **5.** up; $x = -2$ **7.** up; $x = -\frac{7}{2}$ **9.** up; $x = \frac{1}{5}$ **11.** (0, 0)

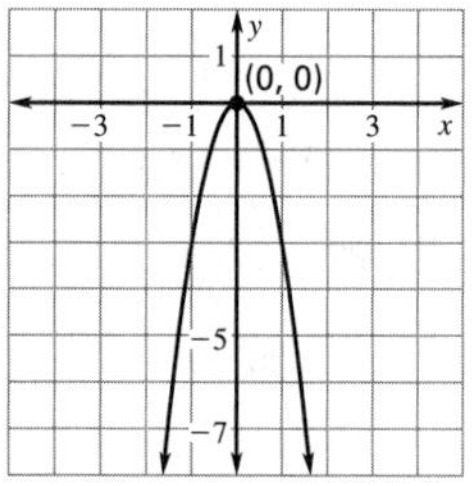

13. (0, 10)

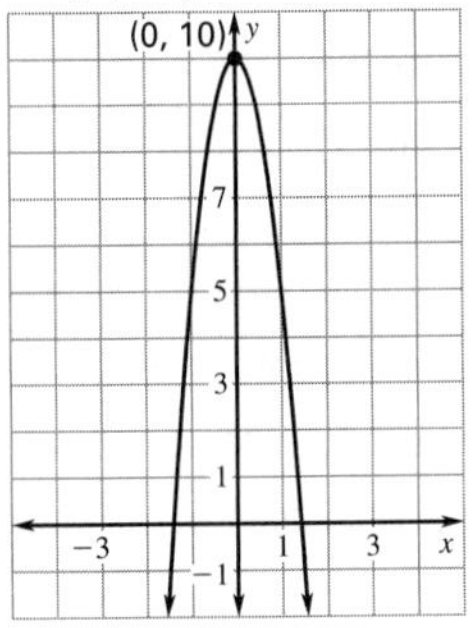

15. (3, –1)

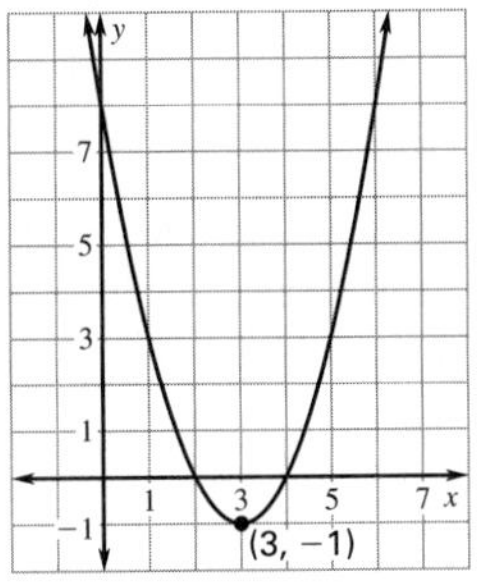

17. $\left(-\frac{1}{2}, 13\right)$ **19.** (2, –5)

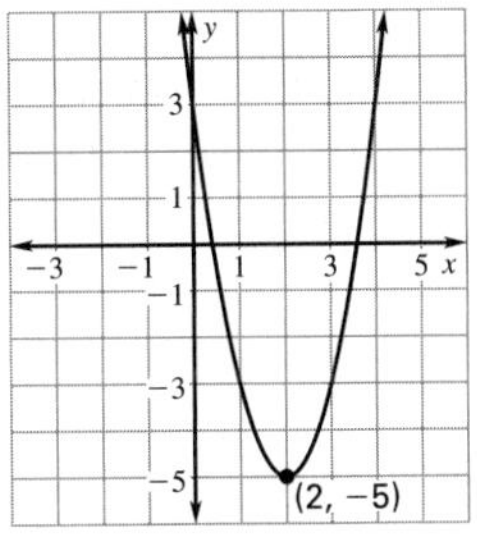

21. a. up **b.** (0, 0) **c.** $x = 0$ **23. a.** up **b.** (0, 0) **c.** $x = 0$ **25. a.** down **b.** (0, 0) **c.** $x = 0$ **27. a.** down **b.** (0, 0) **c.** $x = 0$ **29. a.** up **b.** $\left(2\frac{1}{2}, -12\frac{1}{2}\right)$ **c.** $x = 2\frac{1}{2}$ **31. a.** down **b.** $\left(\frac{3}{5}, 3\frac{3}{5}\right)$ **c.** $x = \frac{3}{5}$ **33. a.** up **b.** (–1, 2) **c.** $x = -1$ **35. a.** up **b.** $\left(1\frac{3}{4}, -14\frac{1}{8}\right)$ **c.** $x = 1\frac{3}{4}$ **37. a.** down **b.** (4, 48) **c.** $x = 4$ **39. a.** up **b.** $\left(-\frac{1}{32}, -8\frac{1}{256}\right)$ **c.** $x = -\frac{1}{23}$ **41. a.** up **b.** (2.56, –13.1) **c.** $x = 2.56$ **43. a.** down **b.** (–0.35, 3.885) **c.** $x = -0.35$

45. (0, 0)

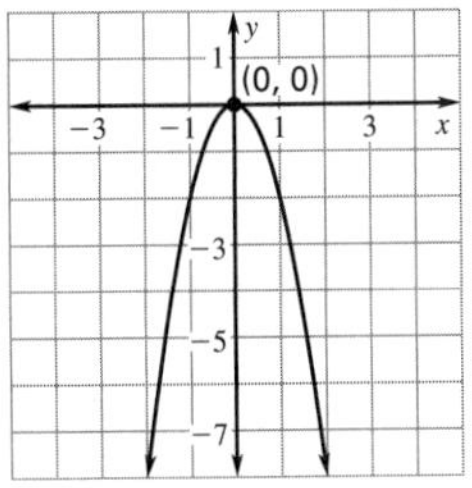

47. (–2, –5)

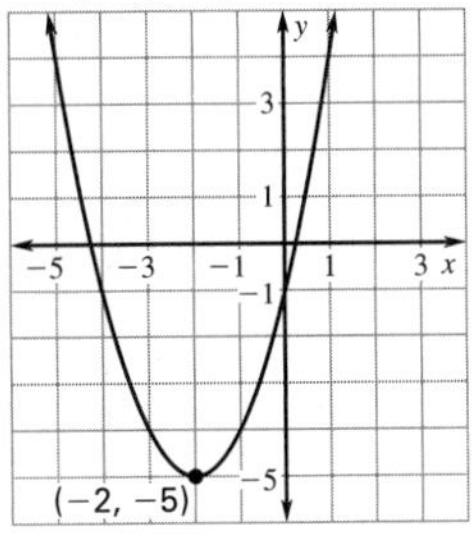

49. (–1, –7)

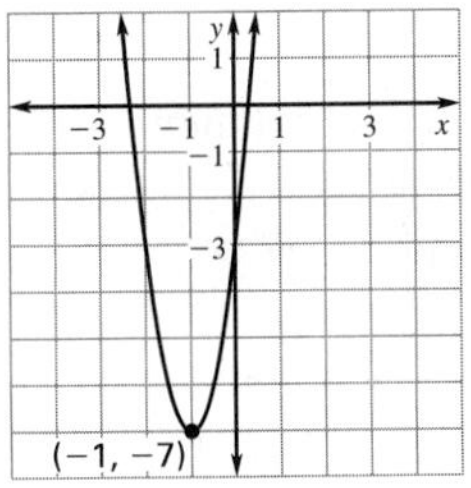

51. $\left(\frac{1}{3}, -\frac{2}{3}\right)$

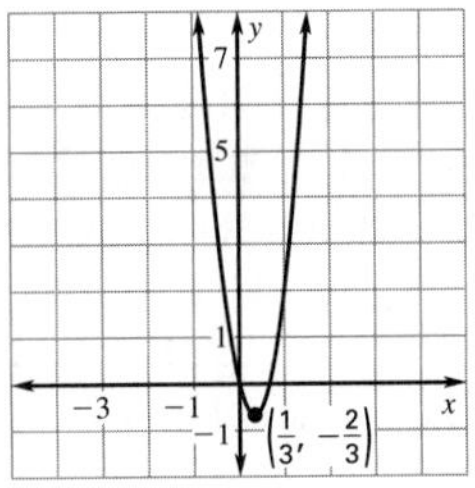

53. $\left(-\frac{1}{2}, \frac{15}{4}\right)$ **55.** $\left(\frac{1}{3}, -\frac{4}{3}\right)$ **57.** $\left(-\frac{1}{3}, -\frac{2}{3}\right)$ **59.** $\left(\frac{1}{2}, 8\right)$

61. $\left(\frac{3}{2}, -\frac{1}{2}\right)$

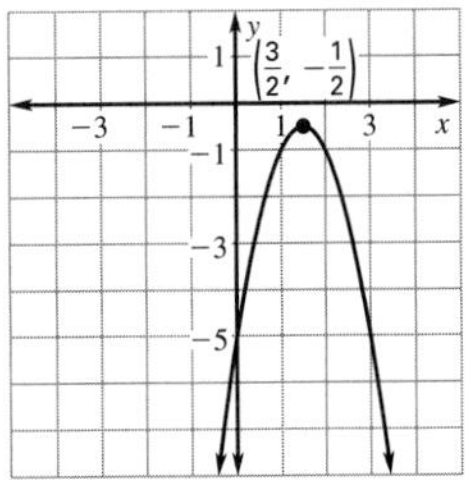

63. (–4, 14)

65. 5 ft **67.** 70 ft **69.** 1980 to 1984 **71.** The vertex is the point at which the graph changes direction, that is, when the values of G stop decreasing and begin increasing. **73.** about 0.21 sec

9.3 Mixed Review (p. 524)

81. $y = 3x - 6$

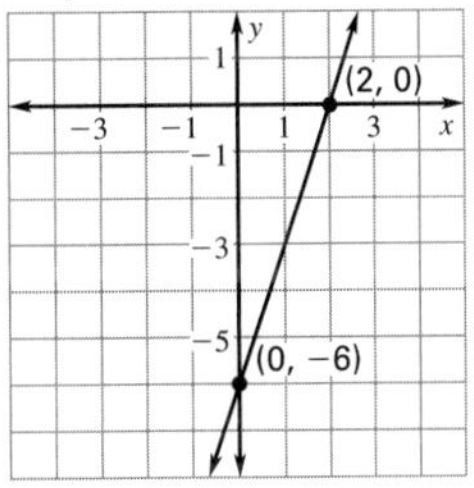

83. $y = -2x + 6$

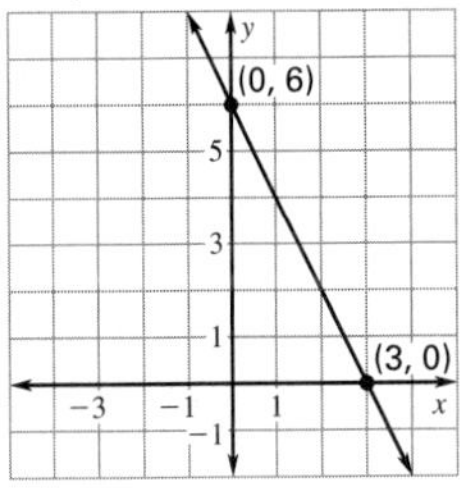

85.

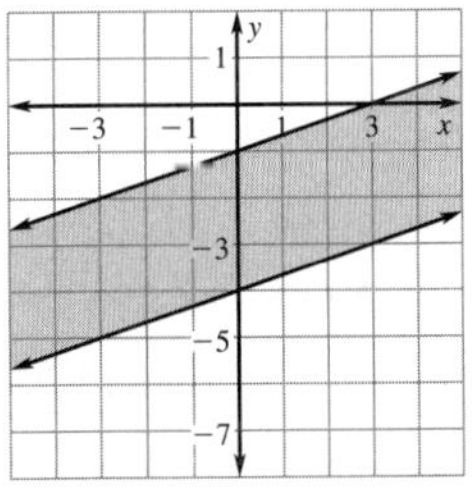

87.

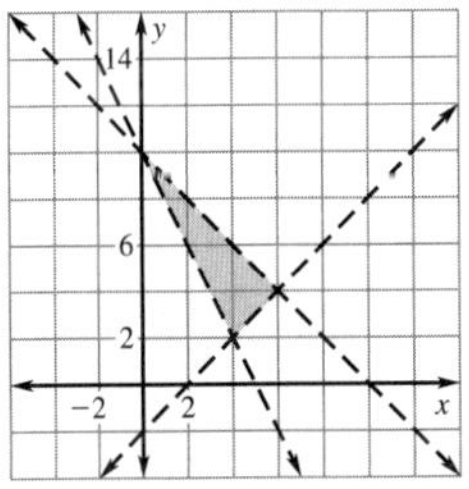

89. 3^6, or 729 **91.** a^6 **93.** $216x^3$ **95.** $-27a^6b^6$ **97.** 9.87×10^5 **99.** 1.229×10^9 **101.** 9.99×10^{-3}

Quiz 1 (p. 524) **1.** 12 **2.** –14 **3.** –26 **4.** –5.20 **5.** 2.45 **6.** 1.22 **7.** 0.4 **8.** 1.5 **9.** –13, 13 **10.** –4, 4 **11.** $-\sqrt{10}, \sqrt{10}$ **12.** $-2\sqrt{2}, 2\sqrt{2}$ **13.** $3\sqrt{2}$ **14.** 10 **15.** 11 **16.** $\frac{\sqrt{5}}{2}$ **17.** up; (–1, –12); $x = -1$ **18.** up; (2, –14); $x = 2$ **19.** up; (–1, –13); $x = -1$ **20.** up; $\left(-5, -15\frac{1}{2}\right)$; $x = -5$ **21.** up; $\left(\frac{1}{2}, 5\frac{1}{4}\right)$; $x = \frac{1}{2}$ **22.** up; $\left(-4\frac{1}{2}, -20\frac{1}{4}\right)$; $x = -4\frac{1}{2}$

23.

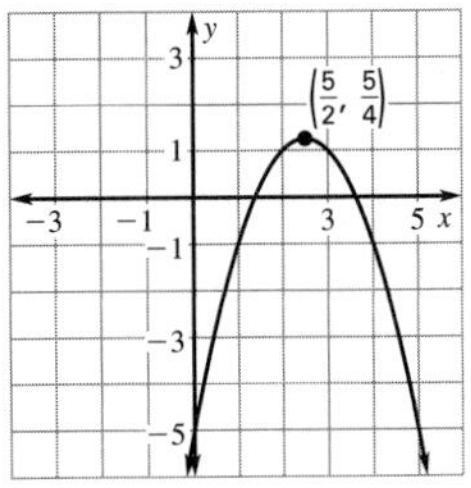

24.

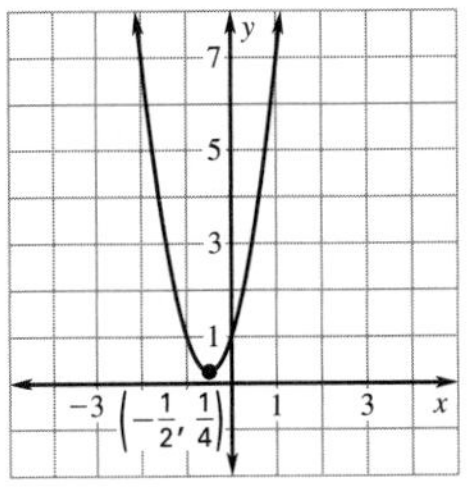

SELECTED ANSWERS

25. 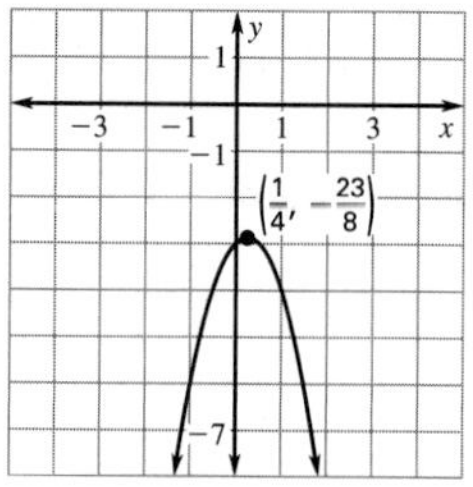

TECHNOLOGY ACTIVITY 9.3 (p. 525)

1. $y = 5.25x^2 + 67.21x - 81.46$

9.4 PRACTICE (pp. 529–530)

7. −2, 2 **9.** −3, 3 **11.** −3, 3 **13.** −3, 3 **15.** −2, −1 **17.** −4, 5 **19.** −4.2, 0.2 **21.** −4, 4 **23.** −5, 5 **25.** −12, 12 **27.** −5, 5 **29.** −4, 4 **31.** −9, 9 **33.** −2, 3 **35.** −1, 2 **37.** −3, 1 **39.** −1, 5 **41.** −7, 3 **43.** −2, −1 **45.** −1, 4 **47.** −1.17, 0.17 **49.** −8, 4
51. 69.19 ft; the distance was greater than that for an initial angle of 65°, but less than that for an initial angle of 35°.
53. 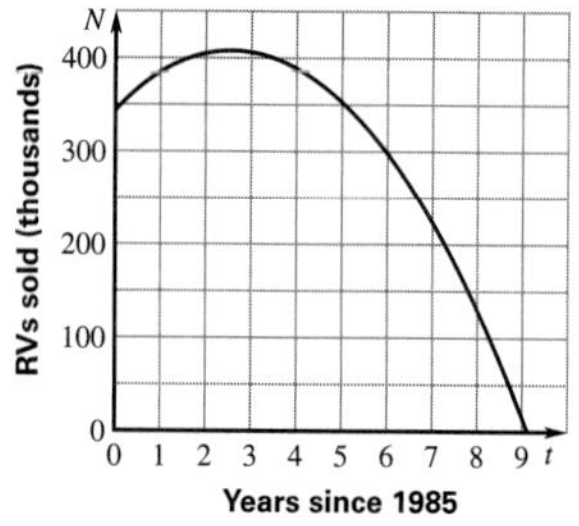

55. 1994

9.4 MIXED REVIEW (p. 531)

65. $\left(-\frac{5}{2}, \frac{3}{4}\right)$; one solution **67.** infinitely many solutions **69.** no solution **71.** −1.17, 6.17 **73.** −1.47, 7.47 **75.** $2\sqrt{10}$ **77.** $2\sqrt{15}$ **79.** $2\sqrt{5}$ **81.** $\frac{1}{2}\sqrt{2}$ **83.** pasta dishes: \$5.95, salads: \$1.95

TECHNOLOGY ACTIVITY 9.4 (p. 532)

1. 0.26, 6.41 **3.** 0.21, 4.79 **5.** 1.71, 2 **7.** 3.04, 6.40

9.5 PRACTICE (pp. 536–537)

5. 5, −3 **7.** 0.5, 1.5 **9.** $-3 + 2\sqrt{3}, -3 - 2\sqrt{3}$ **11.** $8x^2 + 6x - 2 = 0; -1, \frac{1}{4}$ **13.** $2x^2 - 14x - 36 = 0; -2, 9$ **15.** $4x^2 + 4x + 1 = 0; -0.5$ **17.** −2, −8 **19.** $\frac{-3 + 3\sqrt{3}}{2} \approx 1.10, \frac{-3 - 3\sqrt{3}}{2} \approx -4.10$ **21.** $\frac{-2 + \sqrt{5}}{2} \approx 0.12, \frac{-2 - \sqrt{5}}{2} \approx -2.12$ **23.** 25 **25.** 1 **27.** 1 **29.** 21 **31.** 39 **33.** −10, −1 **35.** $-\frac{4}{3}, 2$ **37.** $\frac{-7 + \sqrt{22}}{9} \approx -0.26, \frac{-7 - \sqrt{22}}{9} \approx -1.30$ **39.** $\frac{-1 + 2\sqrt{2}}{7} \approx 0.26, \frac{-1 - 2\sqrt{2}}{7} \approx -0.55$ **41.** $6 + \sqrt{62} \approx 13.87, 6 - \sqrt{62} \approx -1.87$ **43.** $\frac{5 + \sqrt{177}}{-4} \approx -4.58, \frac{5 - \sqrt{177}}{-4} \approx 2.08$ **45.** $x^2 - 3x + 2 = 0; 1, 2$ **47.** $6x^2 + 5x - 1 = 0; -1, \frac{1}{6}$ **49.** $2z^2 - 5z - 7 = 0; -1, \frac{7}{2}$ **51.** $5c^2 - 9c + 4 = 0; \frac{4}{5}, 1$ **53.** 4, −2 **55.** −2, 3 **57.** −2.56, 1.56 **59.** $-1, \frac{1}{3}$ **61.** −0.62, 1.62 **63–69 odd:** Sample answers are given. All the equations can be solved using the quadratic formula. **63.** $4\sqrt{2} \approx 5.66, -4\sqrt{2} \approx -5.66$; finding square roots; simplest method **65.** $-\frac{7}{2}, \frac{7}{2}$; finding square roots; simplest method **67.** −9; finding square roots; the expression on the left can be written $(h + 9)^2$. **69.** $\frac{1 + \sqrt{5}}{6} \approx 0.54, \frac{1 - \sqrt{5}}{6} \approx -0.21$; quadratic formula; when written in standard form, the expression cannot be written as a perfect square. **71.** 2.30 sec **73.** 2.21 sec **75.** 1.31 sec **77.** 1.37 sec **79.** 5.70 sec **81.** 0.44 sec **83.** 0.79 sec

9.5 MIXED REVIEW (p. 538)

87. 25 **89.** −48 **91.** 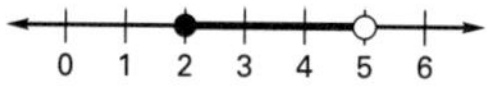

93. $-3 < x < 5$
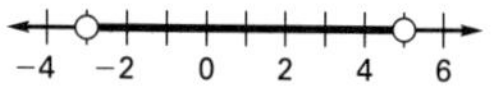

95. $x \le -15$ or $x \ge 5$
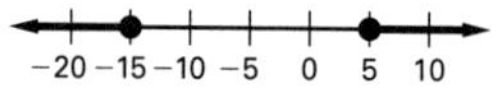

97. $\left(\frac{1}{3}, -\frac{5}{3}\right)$
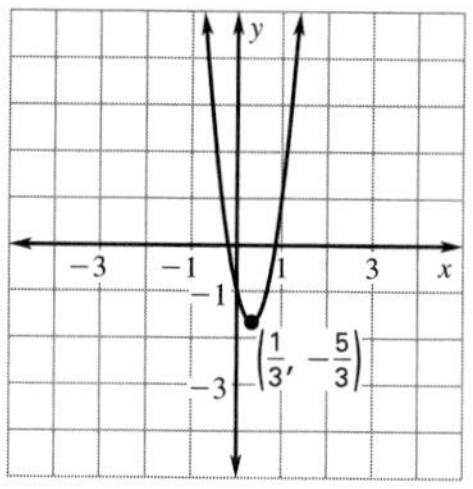

99. $\left(-\frac{3}{4}, \frac{25}{8}\right)$
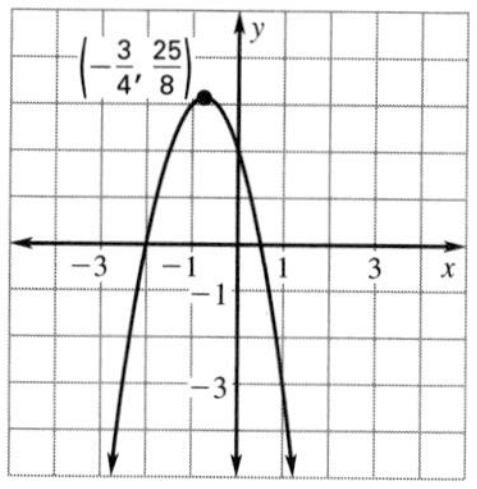

101. $\left(\frac{1}{32}, 3\frac{255}{256}\right)$
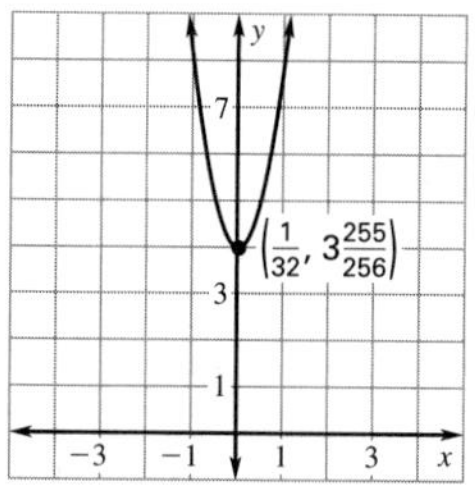

103. about 3.84×10^4 people; about 1.17×10^6 people

TECHNOLOGY ACTIVITY 9.5 (p. 539)

1. −6, 5 **3.** −2 **5.** no solution **7.** −2.4, 1 **9.** −6.35, −2.16

9.6 PRACTICE (pp. 544–545)

3. −56; no real solution **5.** 65; two solutions **7.** A **9.** two solutions **11.** two solutions **13.** no real solution **15.** one x-intercept; B **17.** no x-intercept; A **19.** two solutions **21.** The equation has two solutions for all values of c less than 4, one solution for $c = 4$, and no real solution for all values of c greater than 4.

23. The equation has two solutions for all values of c less than $1\frac{1}{8}$, one solution for $c = 1\frac{1}{8}$, and no real solution for all values of c greater than $1\frac{1}{8}$.
25. Yes; no; *Sample answer:* Because the discriminant for you is positive, your equation has two solutions; because the discriminant for your friend is negative, your friend's equation has no real solution. That is, there is no time when your friend's jump height will be 3.4 ft.
27. *Sample answer:* The payroll will reach \$80 billion if the discriminant of the equation $26t^2 + 1629t - 60{,}042 = 0$ is nonnegative; since the discriminant is 8,898,009, the payroll will reach that amount. **29.** yes

9.6 Mixed Review (p. 546)

35.
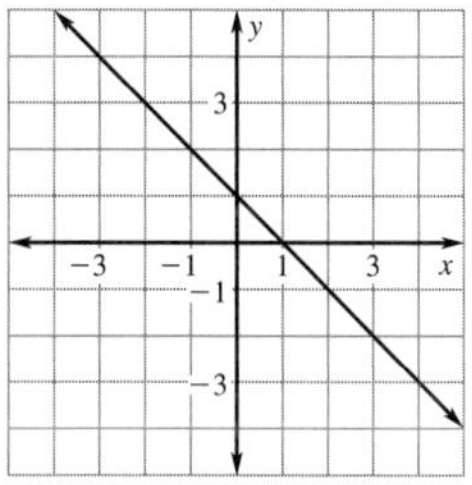

37.
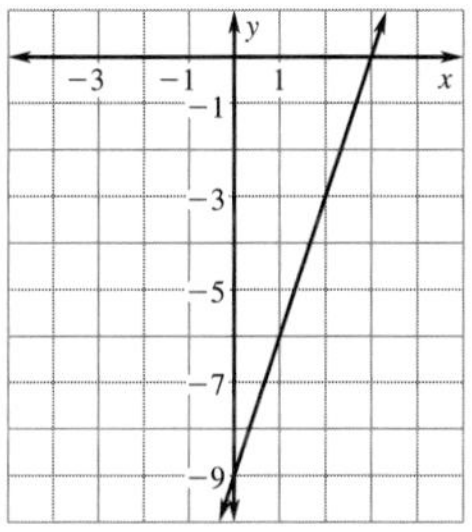

39. $x \le -45$ **41.** $x \le 12$

43.
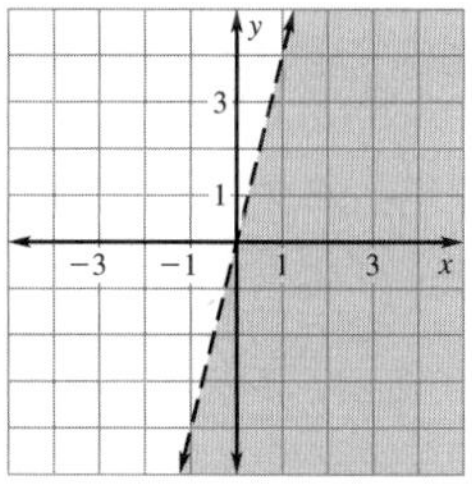

45.
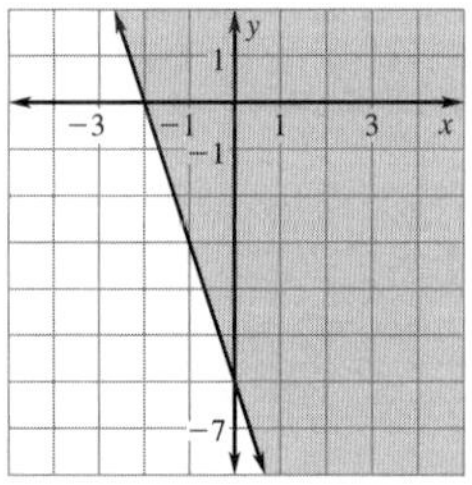

47. Modem Prices
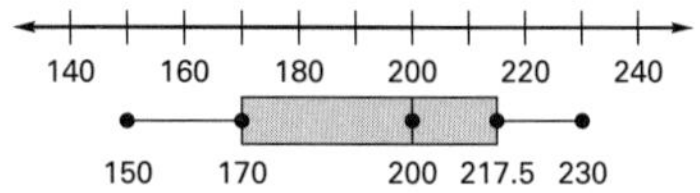

Quiz 2 (p. 547)
1. −2, 5 **2.** 6 **3.** −3, −1 **4.** −3
5. −6, −0.5 **6.** −2, 8 **7.** 1.5, 2 **8.** $-3, 1\frac{1}{3}$ **9.** −1, −0.6
10. two solutions **11.** one solution **12.** no real solution
13. Yes; *Sample answer:* The discriminant of the equation $-16t^2 + 50t + 6 = 45$ is positive, so the equation has two real solutions, meaning that the ball would reach a height of 45 ft.

9.7 Practice (pp. 551–552)
5. (1, −2) is not a solution; (0, 0) is a solution. **7.**
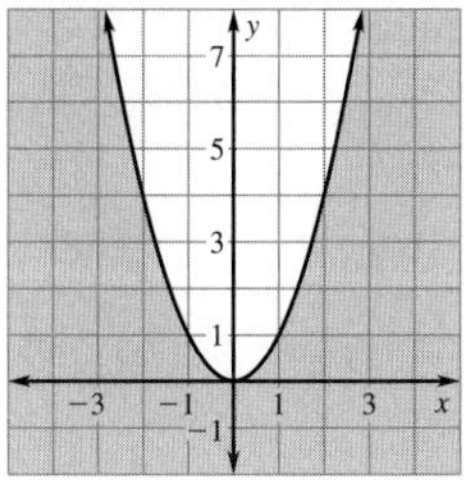

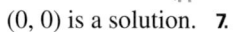

9.
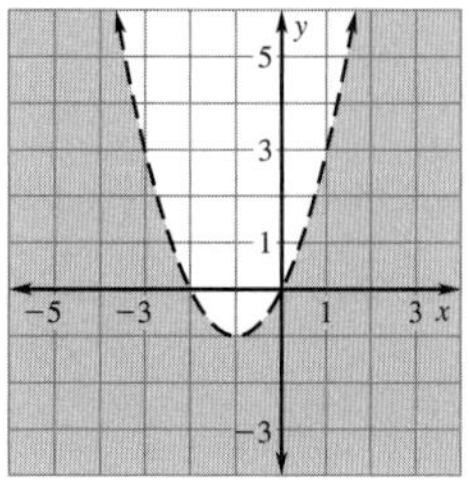

11.
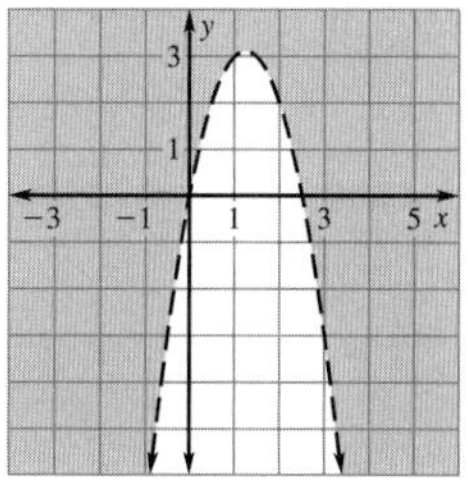

13. solution **15.** not a solution **17.** C **19.** E **21.** D

23.
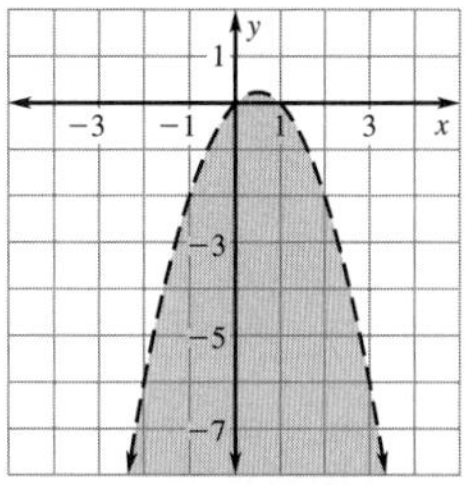

25.
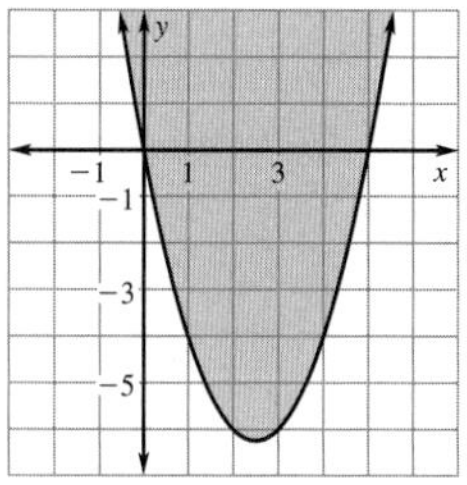

29.
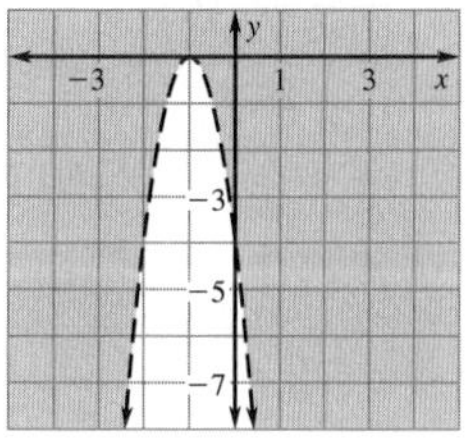

33.
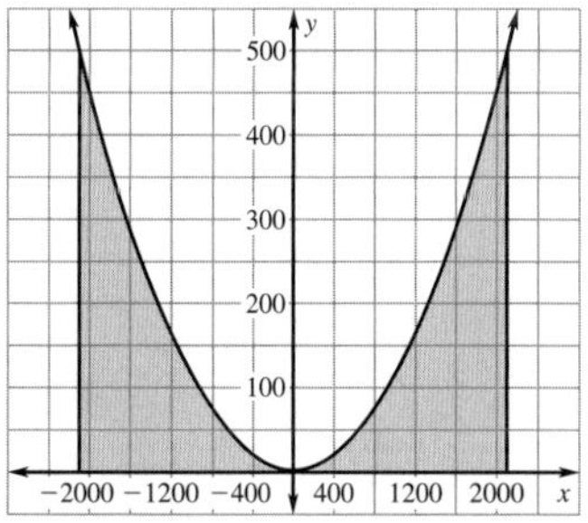

35. I3; for every value of x, the value of y (that is, the cost of the diamond) is lower for that grade than for either of the others.

37. A; VS2

9.7 Mixed Review (p. 553)

45.
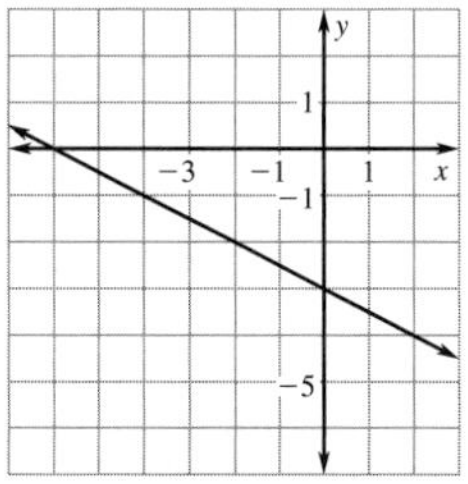

47.
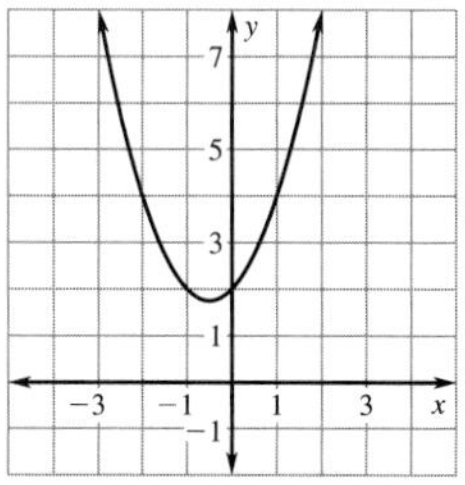

49.
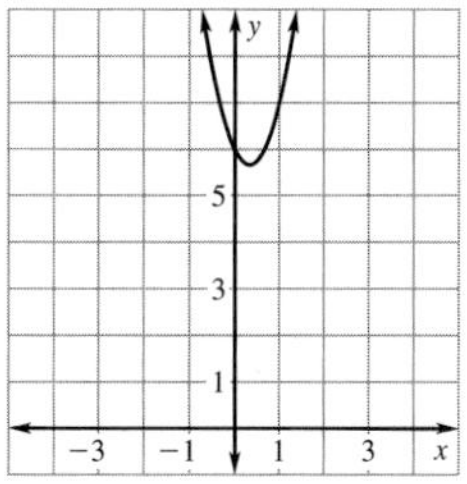

51. $y = -\frac{1}{6}x$ **53.** $y = 4x$

SELECTED ANSWERS

55. $y = \frac{6}{23}x$ **57.**

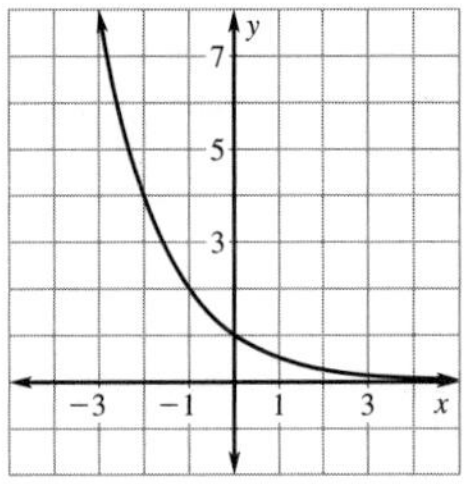

59.

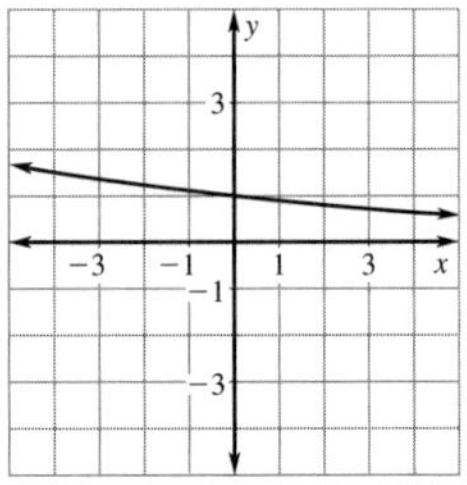

61.

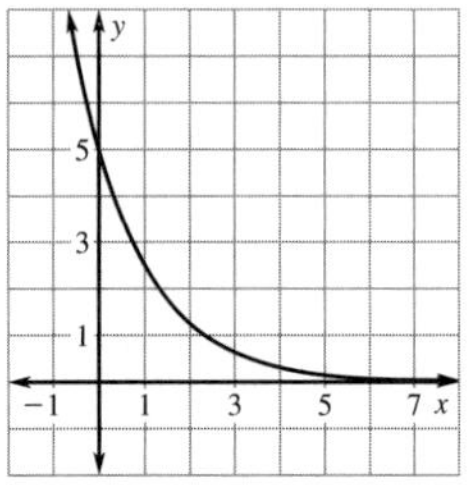

63. 1, 2 **65.** $\frac{9+\sqrt{105}}{-4} \approx -4.81, \frac{9-\sqrt{105}}{-4} \approx 0.31$

67. $\frac{-2+\sqrt{10}}{2} \approx 0.58, \frac{-2-\sqrt{10}}{2} \approx -2.58$

9.8 PRACTICE (pp. 557–559) **3.** quadratic **5.** linear
7. Quadratic; the graph is a curve but the ratios of consecutive y-coordinates are not the same.

9.

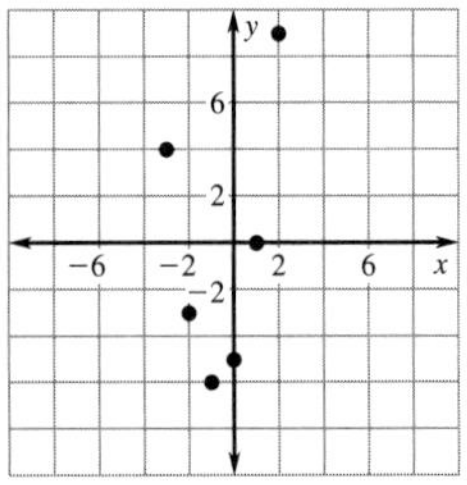

quadratic

11.

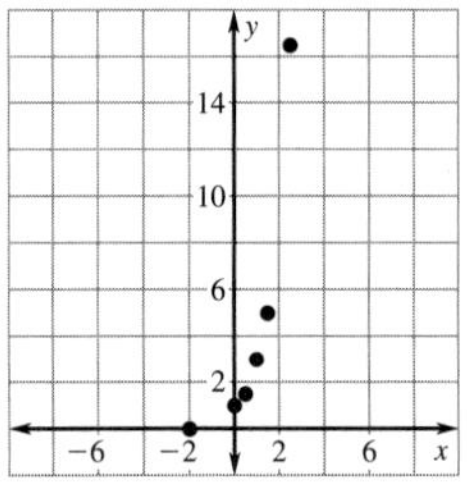

exponential

13. exponential **15.** quadratic

17.

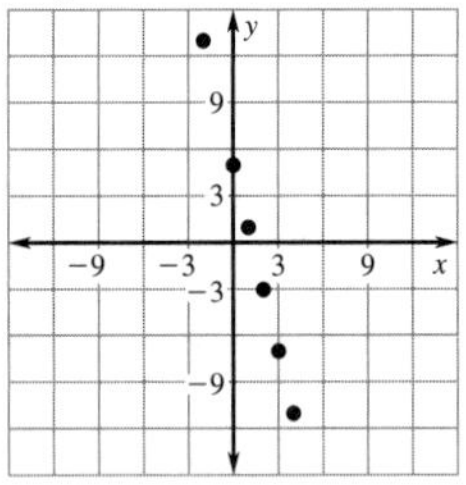

linear **19.** linear; $B = 0.43w$

21.

Year t	Attendance A	Change in A from previous year
0	12,580,660	—
1	12,338,980	–241,680
2	12,781,160	442,180
3	13,907,200	1,126,040
4	15,717,100	1,809,900

23. $y = -2x + 9$; –2 **25.** (4, 1)

9.8 MIXED REVIEW (p. 560) **33.** $-3, x$ **35.** $-5, -8x$
37. $-5y^2$ **39.** $2r^3$ **41.** $-y^3$ **43.** 0.73 **45.** –0.41
47. $y = 3x - 11$ **49.** $y = -\frac{1}{2}x + 4$

QUIZ 3 (p. 560) **1.** not a solution **2.** solution **3.** solution
4. solution **5.** B **6.** C **7.** A

8.

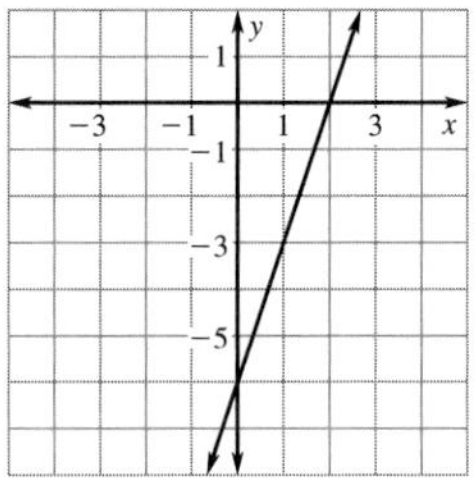

linear

9.

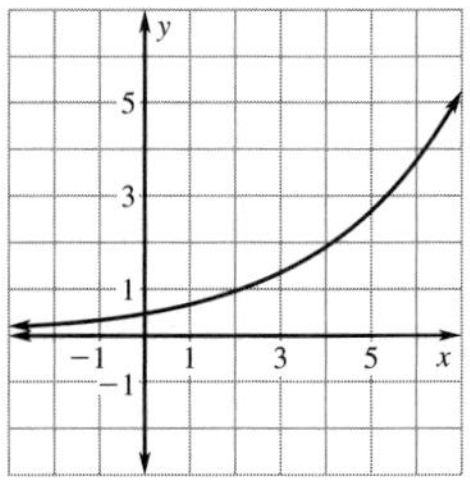

exponential

10.

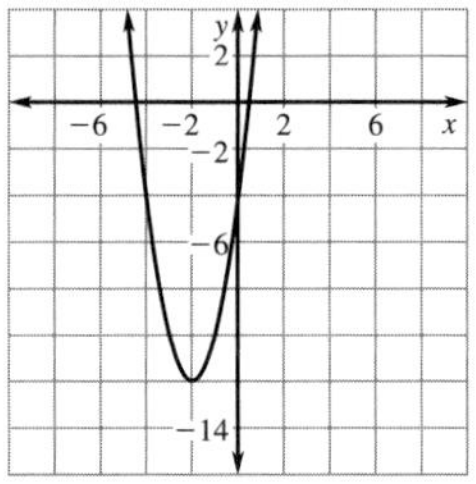

quadratic

11.

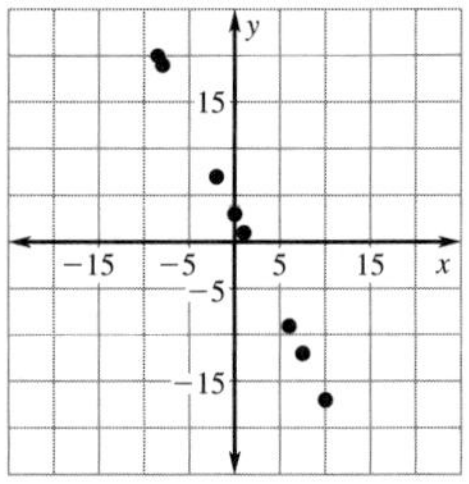

linear; $y = -2x + 3$

12.

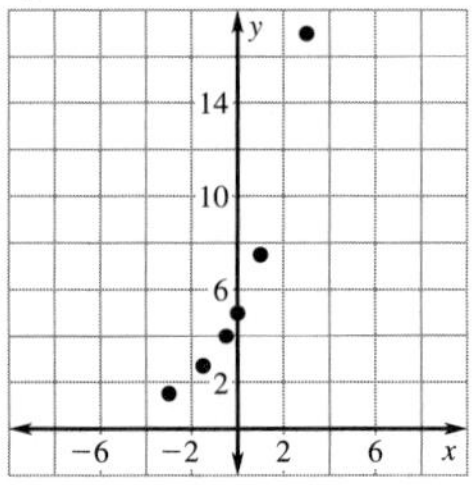

exponential; $y = 5(1.5)^x$

13.

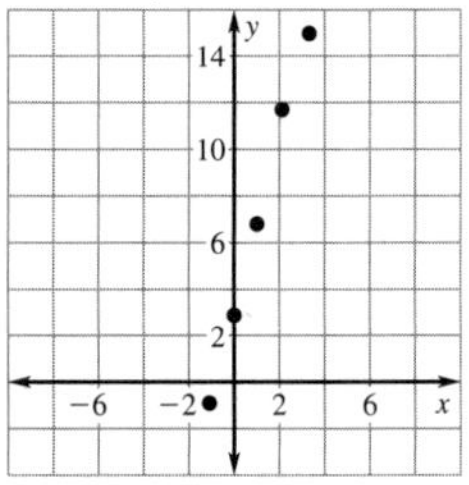

linear; $y = 3.72x + 3.16$

CHAPTER 9 REVIEW (pp. 562–564)
1. –12, 12 **3.** 0 **5.** –5, 5 **7.** $3\sqrt{5}$ **9.** $\frac{3}{4}$

11.

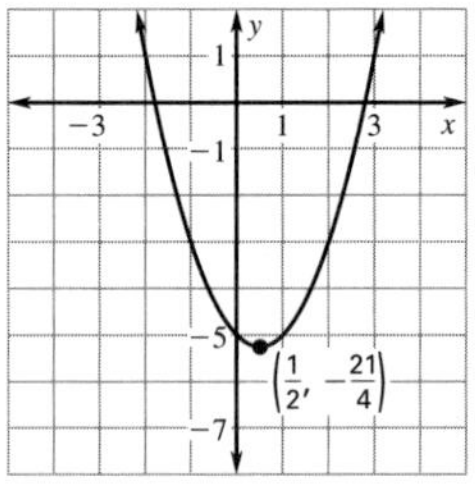

13.

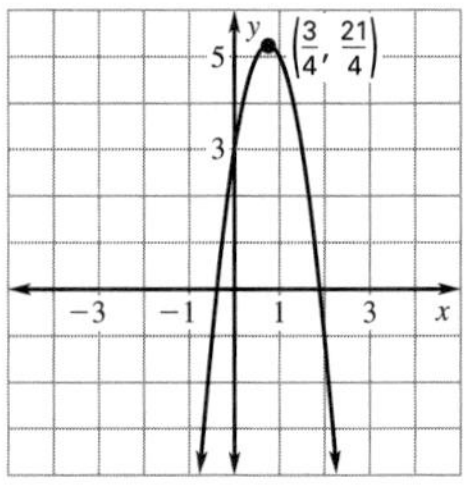

15. –3, 5 **17.** –6, 4 **19.** $-\frac{3}{2}$, 2 **21.** one solution
23. no real solution **25.**

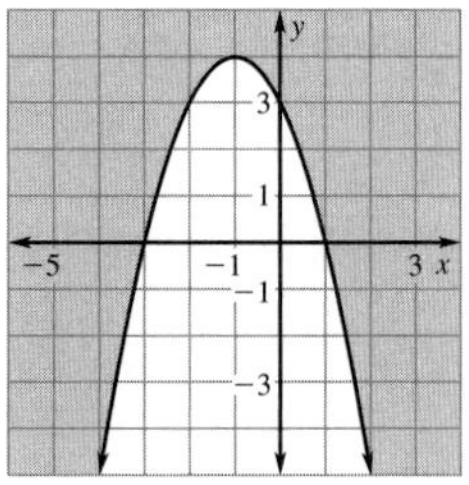

SELECTED ANSWERS

27. 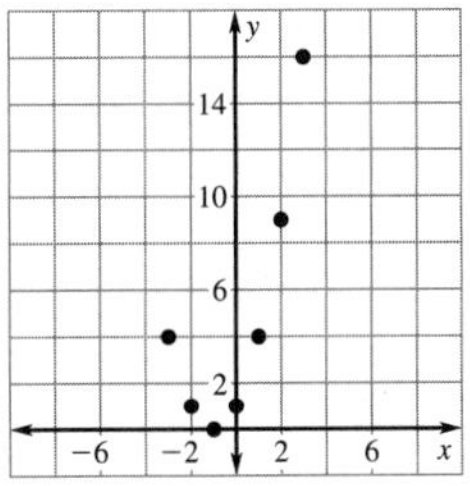

quadratic

29. 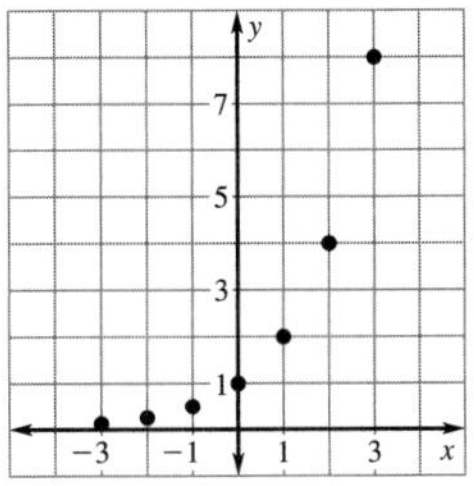

exponential

CUMULATIVE PRACTICE (pp. 568–569) **1.** Yes; no input value has two different output values. **3.** Yes; no input value has two different output values. **5.** $\frac{2}{9} \approx 22\%$; 2 to 7 **7.** *Sample equation:* $\frac{x}{35} = \frac{6}{15}$; 14 in. **9.** 4 **11.** 2 **13.** 6 **15.** $2x - 6y = -4$ **17.** 5, 7, 8 **19.** 4, 4, 8 **21.** $\left(14, -3\frac{1}{2}\right)$ **23.** (–5, –3) **25.** (6.8, –3.7)

27. 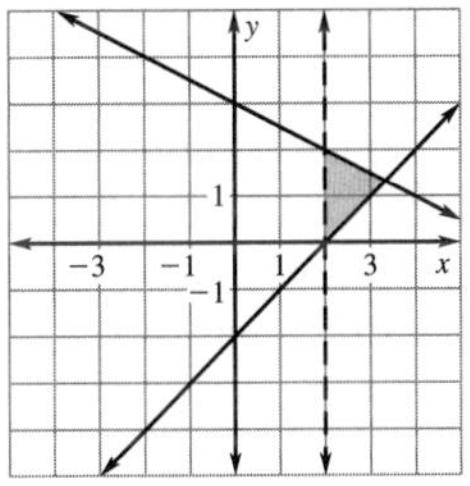

29. 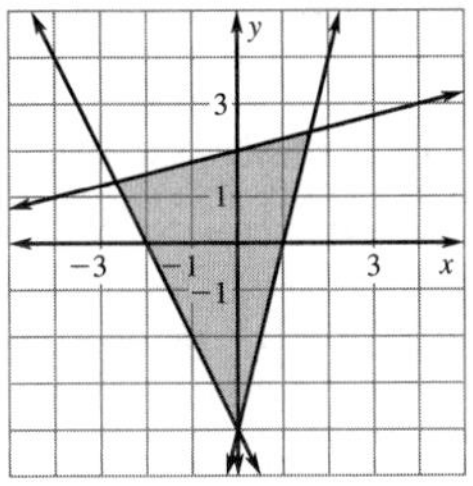

31. $\frac{1}{x^2}$ **33.** $\frac{4x^4}{y^{10}}$ **35.** $9a^{12}$ **37.** $5x^5y^4$ **39.** $4.2 \times 10^{-8} = 0.000000042$ **41.** $2.92 \times 10^{-12} = 0.00000000000292$ **43.** $2.1952 \times 10^{-5} = 0.000021952$ **45.** B **47.** A **49.** $\frac{\sqrt{7}}{3}$ **51.** $\frac{4\sqrt{7}}{7}$ **53.** $\sqrt{3}$ **55.** 15 **59.** –9, –1 **61.** $\frac{-4 - \sqrt{31}}{3} \approx -3.19$, $\frac{-4 + \sqrt{31}}{3} \approx 0.52$ **63.** 1, 5 **65.** \$10.31; \$26.28 **67.** No; the discriminant of the equation $-16t^2 + 100t = 180$ is negative, so the equation has no real solution. That is, there is no time t at which the flare will reach a height of 180 ft.

CHAPTER 10

SKILL REVIEW (p. 574) **1.** $3x^2 + 18x$ **2.** $14 - 28x$ **3.** $-4x - 20$ **4.** $-64 + 16x$ **5.** $3x - 6$ **6.** $2x + 3$ **7.** $-2x - 4$ **8.** $x^2 - x$ **9.** x^{10} **10.** x^{12} **11.** $-8a^3b^3$ **12.** $9x^4y^8$ **13.** –56; no real solution **14.** 196; two solutions **15.** 0; one solution

10.1 PRACTICE (pp. 579–581) **7.** linear, binomial **9.** cubic, trinomial **11.** constant, monomial **13.** $4x^2 - 7x - 2$ **15.** $-4x^2 + 9x + 4$ **17.** $-x^2 - x - 16$ **19.** –3; linear, binomial **21.** 1; cubic, polynomial **23.** –6; constant, monomial **25.** 5; quartic, trinomial **27.** –4; quartic, polynomial **29.** –16; cubic, monomial **31.** $-6x^3 + 3x^2 + 4$ **33.** $-7m^2 + 7m - 3$ **35.** –6 **37.** $8y^3 + 6y^2 - 3y - 3$ **39.** $3x^2 - 5$ **41.** $x^3 + 1$ **43.** $2n^3 + 3n - 5$ **45.** $10b^4 - 3b^3 - 13b^2 + 20b - 4$ **47.** $25x^3 - 4x + 14$ **49.** $x^2 + 2x + 2$ **51.** $-3x^2 + 6$ **53.** $-9z^3 - 8z^2 + 10z$ **55.** $-19t + 2$ **57.** $x^4 + x^3 + \frac{1}{12}x^2 - 9$ **59.** $9x^4 - 3x^3 - 7x^2 + 6x - 17$ **61.** $-0.7y^3 + 6.9y^2 - 3.9y + 10.4$ **63.** $1.5x^2 + 60x$ **65.** $F = 1223.58t + 79{,}589.03$ **67.** $A = 1.381t^2 + 3.494t + 235.325$ **69.** From top to bottom the models are *A*, *E*, and *N*.

10.1 MIXED REVIEW (p. 582) **73.** $-3x - 5$ **75.** $8x + 24$ **77.** $3 + 4x$ **79.** *Sample answer:* $y = -1.42x + 7.35$ **81.** 2^{16}; 65,536 **83.** $(1.1^3)(3.3^3)$; 47.83 **85.** 2.9^{15}; 8,629,188.75

TECHNOLOGY ACTIVITY 10.1 (p. 583) **1.** correct **3.** $3x^2 - 9x$ **5.** $-2x^2 + 14x + 4$ **7.** $6x^2 + 12x - 1$

10.2 PRACTICE (pp. 587–588) **5.** $8x^2 + 14x + 3$

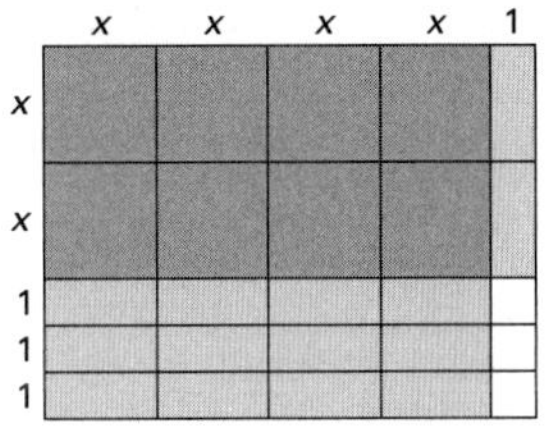

7. $2x(4x) + 2x(1) + 3(4x) + 3(1) = 8x^2 + 2x + 12x + 3 = 8x^2 + 14x + 3$ **9.** $-8x^2 - 14x$ **11.** $-y^2 - 10y - 16$ **13.** $x^2 + 15x + 54$ **15.** $-2b^2 - 10b + 48$ **17.** $15w^3 - 3w^2 - 45w + 9$ **19.** $21t^3 - 3t^5 - 9t^2$ **21.** $-6y^3 - 5y^2$ **23.** $-6b^5 + 16b^3 - 11b^2$ **25.** $6d^2 + 11d + 3$ **27.** $3s^3 + 5s^2 - 3s - 2$ **29.** $x^3 - 38x - 12$ **31.** $6z^2 + 25z + 14$ **33.** $2w^2 + 5w - 25$ **35.** $10t^2 + 9t - 9$ **37.** $4x^2 - 31x - 8$ **39.** $99w^2 - 2w - 80$ **41.** $68.62y^2 - 151.91y + 62.22$ **43.** $4n^2 - \frac{26}{5}n - 12$ **45.** $2.5z^2 + 4.65z - 26.23$ **47.** $-4s^3 - 15s^2 + 3s - 4$ **49.** $2x^2 - 11x - 6$ **51.** $-x^3 - 13x^2 - 40x + 14$ **53.** $3x^2 + \frac{19}{2}x - 10$ **55.** 300 **57.** The revenue decreases from \$989.12 million to \$678.08 million. **59.** *Sample answer:* As time goes on, the revenue will increase because the value of the model $1.5125t^2 + 156.1846t + 3066.344$ will increase as t increases.

10.2 MIXED REVIEW (p. 589) **63.** $\frac{1}{9}m^2$ **65.** $0.25w^2$ **67.** 4^8, or 65,536 **69.** $256c^8$ **71.** $-w^{12}$ **73.** $512x^6y^{24}$ **75.** one solution **77.** two solutions **79.** two solutions

81. 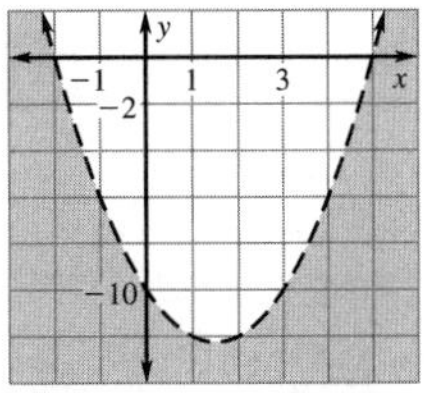

83. 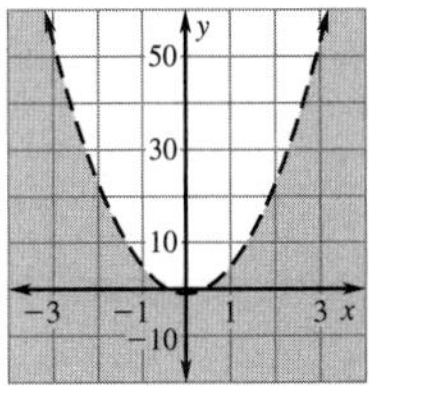

85. 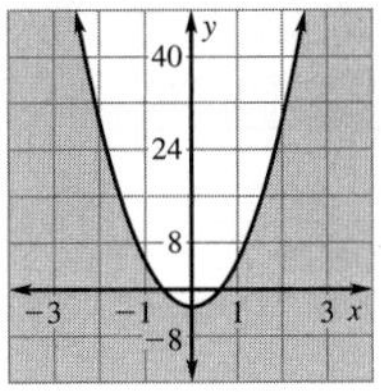

10.3 Practice (pp. 593–594)

5. 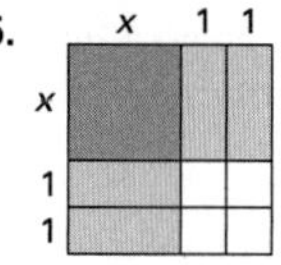

$x^2 + 4x + 4$ **7.** $w^2 - 121$ **9.** $4y^2 - 12y + 9$

11. $t^2 - 36$ **13.** $4y^2 - 25$ **15.** $x^2 - 9$ **17.** $4m^2 - 4$ **19.** $9 - 4x^2$ **21.** $x^2 + 10x + 25$ **23.** $9t^2 + 6t + 1$ **25.** $16b^2 - 24b + 9$ **27.** $x^2 - 16$ **29.** $9x^2 - 1$ **31.** $a^2 - 4b^2$ **33.** $9y^2 + 48y + 64$ **35.** $4x^2 - \frac{1}{4}$ **37.** $9s^2 - 16t^2$ **39.** true **41.** true **43.** 884 **45.** 256 **47.** $(x + 3)^2$, $x^2 + 6x + 9$; square of a binomial **49.** $(0.5C + 0.5c)^2 = 0.25C^2 + 0.5Cc + 0.25c^2$ **51.** $(0.5F + 0.5f)^2 = 0.25F^2 + 0.5Ff + 0.25f^2$ **53.** $P(1 - r)(1 + r) = P(1 - r^2)$ **55.** a decrease of Pr^2 dollars

10.3 Mixed Review (p. 594)

63. $\frac{x^3}{64}$ **65.** $\frac{64x^3}{y^9}$ **67.** $9x^2$ **69.** $\left(-\frac{3}{4}, 4\frac{7}{8}\right)$; $x = -\frac{3}{4}$ **71.** (2, 20); $x = 2$ **73.** (6, 14); $x = 6$ **75.** $-4, 4$ **77.** $-5, 5$ **79.** $-3, 3$ **81.** Yes; *Sample answer:* Let x be the length of the rectangle and y be the width. Then $2x + 2y = 52$ and $xy = 148.75$. Substituting, $x(26 - x) = 148.75$, which is equivalent to $x^2 - 26x + 148.75 = 0$. Since the discriminant of the equation is 81, the equation has two solutions. The rectangle is 17.5 by 8.5.

Quiz 1 (p. 596)

1. $3x^2 + 5x + 9$ **2.** $-6x^3 - 14x^2 + 2x - 2$ **3.** $3t^2 - 13t + 14$ **4.** $6x^3 - 5x^2 + 4x + 3$ **5.** $x^2 + 7x - 8$ **6.** $16n^2 - 49$ **7.** $-2x^3 + 5x^2 - 6x + 8$ **8.** $27m^2 - 3m - 4$ **9.** $\frac{1}{4}x^2 + x - 15$ **10.** $-12x^5 + 11x^4 - 3x^2$ **11.** $x^2 - 36$ **12.** $16x^2 - 9$ **13.** $25 - 9b^2$ **14.** $4x^2 - 49y^2$ **15.** $9x^2 + 36x + 36$ **16.** $36 + 96x + 64x^2$

17. 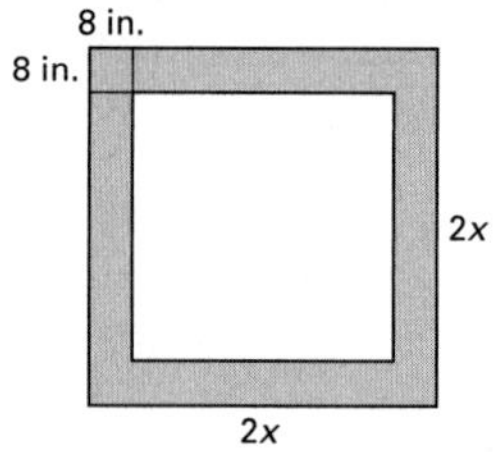

$(2x - 16)^2 = (4x^2 - 64x + 256)$ in.^{2}

10.4 Practice (pp. 600–601)

9. no **11.** yes **13.** 3, 5 **15.** $-9, 2$ **17.** $-\frac{1}{4}$ **19.** $-4, -1$ **21.** $-8, 6$ **23.** $9, -8$ **25.** $2, -1$ **27.** 7, 5 **29.** -9 **31.** 5.6 **33.** 5 **35.** -4 **37.** $-3, \frac{3}{2}$ **39.** $\frac{5}{4}, 3, \frac{4}{3}$ **41.** $1.9, -2.1$ **43.** $\frac{1}{8}, -\frac{3}{50}, \frac{2}{9}$ **45.** B

47. 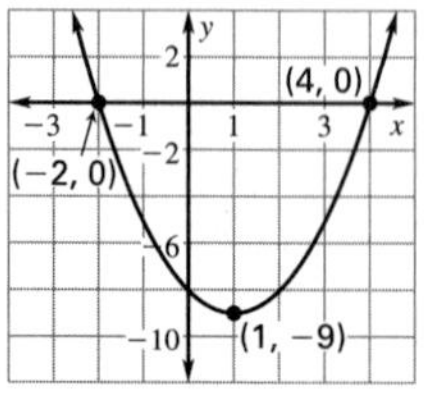

49. 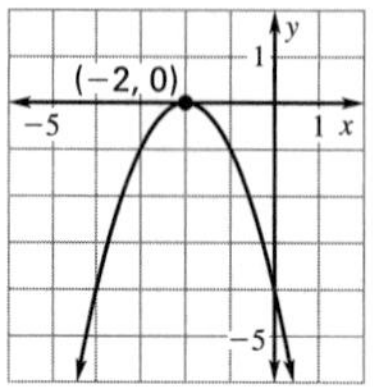

51. 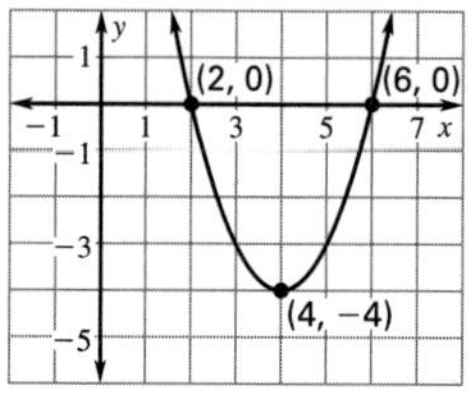

53. Use the distance between the x-intercepts at $(-500, 0)$ and $(500, 0)$. **55.** 600 ft **57.** 1200 m

10.4 Mixed Review (p. 602)

61. 210,000 **63.** 857,000,000 **65.** 0.00371 **67.** 0.000722 **69.** $x^2 - 9x + 14$ **71.** $x^2 + x - 20$ **73.** $x^2 + \frac{1}{3}x - \frac{2}{9}$ **75.** $6x^2 + 19x - 7$ **77.** $24x^2 - x - 3$ **79.** $x^2 + 20x + 100$ **81.** Exponential decay; let P be the price in 1996, t the number of years since 1996, and y the price after t years; $y = P(0.84)^t$. **83.** Exponential decay; let n be the number of members in 1996, t the number of years since 1996, and y the number of members after t years; $y = n(0.97)^t$.

10.5 Practice (pp. 607–608)

5. D **7.** B **9.** The equation can be solved by factoring because the discriminant, 0, is a perfect square. **11.** The equation cannot be solved by factoring because the discriminant, 40, is not a perfect square. **13.** B **15.** $(x + 9)(x - 1)$ **17.** $(b + 8)(b - 3)$ **19.** $(y - 6)(y + 3)$ **21.** $(m - 10)(m + 3)$ **23.** $(11 - s)(4 - s)$ **25.** $(x - 15)(x - 30)$ **27.** $-2, -5$ **29.** 2, 7 **31.** $-1, -15$ **33.** $-13, 5$ **35.** 3, 17 **37.** $-7, 4$ **39.** $-9, 10$ **41.** 2, 9 **43.** yes; $(y + 15)(y + 4)$ **45.** no **47.** no **49.** $-15, 7$ **51.** $-32, -20$ **53.** *Sample answer:* $x^2 + 11x + 30 = 0$ **55.** *Sample answer:* $x^2 - 427x = 0$ **57.** 18 or -1 **59.** $x - 9$ **61.** $x^2 - x \le 20$ **63.** platform: 95 m by 95 m; building: 57 m by 57 m

10.5 Mixed Review (p. 609)

73. 1 **75.** 7 **77.** 18 **79.** $y^2 + 5y - 36$ **81.** $60 + 3w - 3w^2$ **83.** $2b^4 + 4b^3 - 6b^2 - 20b$ **85.** $36z^2 + 4z + \frac{1}{9}$ **87.** $-12, -7$ **89.** 19 **91.** 9, 6 **93.** $1, -\frac{1}{2}$ **95.** $\frac{3}{2}$

10.6 Practice (pp. 614–616)

5. B **7.** A **9.** $(2x + 1)(x - 2)$ **11.** cannot be factored **13.** $-8, \frac{1}{2}$ **15.** A **17.** A **19.** $(6b + 1)(b - 2)$ **21.** $(5w + 1)(w - 2)$ **23.** $(2y - 5)(3y + 2)$ **25.** $(3c - 4)(c - 11)$ **27.** $(7y - 4)(2y - 1)$ **29.** $(2t + 7)(3t - 10)$ **31.** $(z + 10)(2z - 1)$ **33.** $-2\frac{1}{2}, 7$ **35.** $-11, -\frac{1}{3}$ **37.** $-1, 9\frac{1}{2}$ **39.** $-8, -1\frac{1}{2}$ **41.** $-\frac{2}{3}, 4\frac{1}{2}$ **43.** $-1\frac{1}{2}, 1\frac{1}{5}$ **45.** $-1, -\frac{5}{8}$ **47.** $-1\frac{1}{9}, 6\frac{1}{2}$ **49.** 5 **51.** $-\frac{1}{2}, 0$ **53.** $\frac{3}{5}, 1\frac{1}{7}$ **55.** $-2, 2\frac{1}{2}$ **57.** $-\frac{2}{7}, 1\frac{3}{8}$ **59.** $\frac{-23 - \sqrt{2089}}{24}, \frac{-23 + \sqrt{2089}}{24}$ **61.** $\frac{8}{23}, 2$ **63.** $\frac{11}{17}, \frac{5}{7}$ **65.** 3 sec; $t = 0.125$ is not a reasonable solution because at that time the acrobat is still rising. **67.** $2.5x^2 = 9000$ **69.** $x^2 - x - 12 = 0$; explanations will vary. **71.** $6x^2 + x - 1 = 0$; explanations will vary.

10.6 MIXED REVIEW (p. 617) **79.** $(-2, 3)$ **81.** $(40, 40)$ **83.** not a solution **85.** $b^2 - 81$ **87.** $4a^2 - 49$ **89.** $10{,}000 + 5400x + 729x^2$

QUIZ 2 (p. 617) **1.** -5 **2.** $\frac{3}{2}, 4$ **3.** $-\frac{7}{2}, 4$ **4.** $-\frac{1}{12}$ **5.** -4 **6.** $4, -7, -1$ **7.** $-2, \frac{1}{2}, \left(-\frac{3}{4}, -\frac{25}{16}\right)$

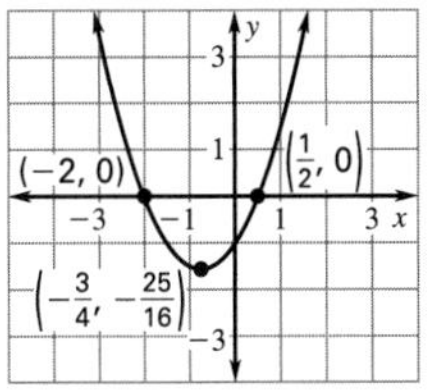

8. $-4, 4, (0, -16)$

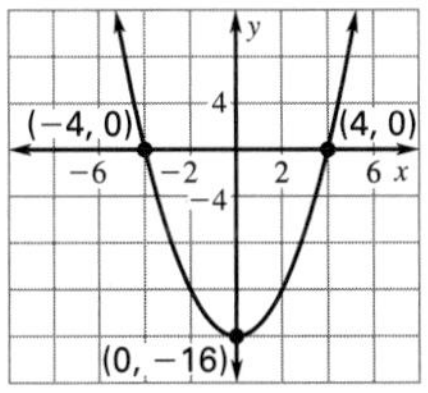

9. $\frac{1}{3}, -2, \left(-\frac{5}{6}, -\frac{49}{12}\right)$

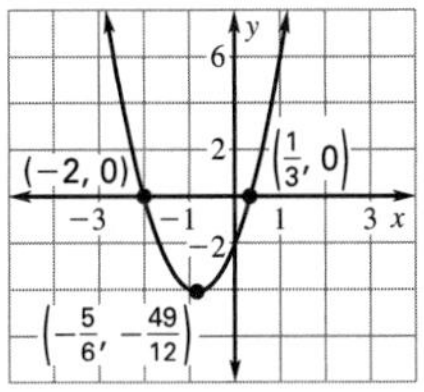

10. $(y + 4)(y - 1)$ **11.** $(w + 2)(w + 11)$ **12.** $(n + 19)(n - 3)$ **13.** $(x + 6)(x + 11)$ **14.** $(t - 43)(t + 2)$ **15.** $(9 - z)(-5 + z)$ **16.** $(4b + 9)(3b - 11)$ **17.** $(t + 29)(2t + 5)$ **18.** $2(3d - 2)(3d - 7)$ **19.** $-6, 1$ **20.** $-25, -1$ **21.** $2, 9$ **22.** $-5, 34$ **23.** $-37, 2$ **24.** $-\frac{17}{2}, -8$ **25.** $-\frac{7}{3}, -\frac{2}{5}$ **26.** $\frac{4}{9}, \frac{9}{2}$ **27.** $-\frac{11}{2}, \frac{16}{3}$

CONCEPT ACTIVITY 10.7 (p. 618)
1. $(x + 2)^2$ **3.** $(2x + 1)^2$ **5.** $a + b$; $a + b$; $(a + b)^2$ **7.** $(4x + 7)^2$ **9.** $(a - b)$; $(a - b)$; $(a - b)^2$

10.7 PRACTICE (pp. 622–623) **7.** $(t + 5)^2$ **9.** $(4 + t)(4 - t)$ **11.** $2(3 + z)(3 - z)$ **13.** $-12, 12$ **15.** $-5, 5$ **17.** 3 **19.** $(10x + 11)(10x - 11)$ **21.** $60(y + 3)(y - 3)$ **23.** $2(7 + t)(7 - t)$ **25.** $(3t - 2q)(3t + 2q)$ **27.** $(x + 4)^2$ **29.** $(y + 15)^2$ **31.** $(3x + 1)^2$ **33.** $(5n - 2)^2$ **35.** $2(3x + 1)^2$ **37.** $-4(2w + 5)^2$ **39.** $(z + 5)(z - 5)$; difference of two squares **41.** $4(n + 3)(n - 3)$; difference of two squares **43.** $4(b - 5)^2$; perfect square trinomial **45.** $-2(x - 13)^2$; perfect square trinomial **47.** $(x + 100w)(x - 100w)$; difference of two squares **49.** $\left(x + \frac{1}{3}\right)^2$; perfect square trinomial **51.** $-6, 6$ **53.** $-\frac{2}{5}, \frac{2}{5}$ **55.** $-\frac{3}{2}, \frac{3}{2}$ **57.** 9 **59.** $\frac{5}{6}$ **61.** $\frac{7}{4}$ **63.** 1.5 in. **65.** $\frac{1}{2}$ sec **67.** yes; 7 sec **69.** 32 ft/sec

10.7 MIXED REVIEW (p. 624) **75.** 15 **77.** 2 **79.** not a solution **81.** solution **83.** $(-2, -6)$ **85.** $(-4, 1)$ **87.** $5\sqrt{3}$ **89.** 18 **91.** $4\sqrt{2}$ **93.** $-6\sqrt{3}$ **95.** $\frac{4}{9}, 6$

10.8 PRACTICE (pp. 629–631) **7.** $3x^2(2 + x^2)$ **9.** no; $x(7x^2 - 11)$ **11.** no; $3w(3w + 4)(3w - 4)$ **13.** $-5, -6$; *Sample answer:* factoring **15.** $6v(v^2 - 3)$ **17.** $3x(1 - 3x)$ **19.** $2a^2(2a^3 + 4a - 1)$ **21.** $6x^2(4x + 3)$ **23.** $2y(y - 6)(y + 1)$ **25.** $-7m(m - 3)(m - 1)$ **27.** $4t(t + 6)(t - 6)$ **29.** $(c^3 - 12)(c + 1)$ **31.** $(b^3 - 4)(6b + 5)$ **33.** $(a + 6)(a + 2)(a - 2)$ **35.** $3(m^2 - 2)(m - 5)$ **37.** $-3, -4$; *Sample answer:* factoring **39.** $-13, 9$; *Sample answer:* factoring **41.** $-3, 9$; *Sample answer:* factoring **43.** $0, -4, 4$; *Sample answer:* factoring **45.** $0, \frac{29 - \sqrt{593}}{2}, \frac{29 + \sqrt{593}}{2}$; factoring and quadratic formula **47.** $\frac{-9 - \sqrt{305}}{16}, \frac{-9 + \sqrt{305}}{16}$; quadratic formula **49.** $0, \frac{-3 - \sqrt{457}}{8}, \frac{-3 + \sqrt{457}}{8}$; factoring and quadratic formula **51.** 3 sec **55.** $L = h - 3$, $w = h - 9$ **57.** $L = 9$ in., $w = 3$ in., $h = 12$ in. **59.** $(3x + 7)(x + 2)$

10.8 MIXED REVIEW (p. 632)
69. $x \le -16$

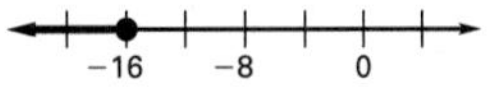

71. $x \ge -7$

73. $-2, 12$ **75.** no solution **77.** $-3, \frac{3}{2}$

79.

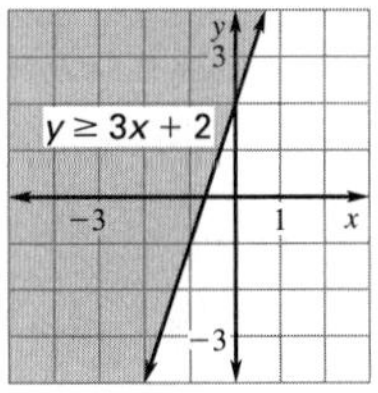

81. $\frac{31}{23} \approx 134.78\%$

QUIZ 3 (p. 632) **1.** $(7x + 8)(7x - 8)$; difference of two squares **2.** $(11 + 3x)(11 - 3x)$; difference of two squares **3.** $(2t + 5)^2$; perfect square trinomial **4.** $2(6 + 5y)(6 - 5y)$; difference of two squares **5.** $(3y + 7)^2$; perfect square trinomial **6.** $3(n - 6)^2$; perfect square trinomial **7.** $-8, 8$ **8.** -4 **9.** $-\frac{16}{3}$ **10.** 4 **11.** $-\frac{2}{3}, \frac{2}{3}$ **12.** $-\frac{1}{3}$ **13.** $3x^2(x + 4)$ **14.** $3x(2x + 1)$ **15.** $9x^3(2x - 1)$ **16.** $2x(4x^4 + 2x - 1)$ **17.** $2x(x - 1)(x - 2)$ **18.** $3x(4x - 5)(4x + 5)$ **19.** $(x^2 + 4)(x + 3)$ **20.** $0, -\frac{3}{2}, \frac{3}{2}$; factoring special products **21.** $-3, \frac{5}{3}$; factoring $ax^2 + bx + c$ **22.** 2; graphing **23.** 0, about 0.77, about 2.26; quadratic formula **24.** $0, -3, 4$; factoring **25.** 0, about -7.15, about -0.35; quadratic formula

CHAPTER 10 REVIEW (pp. 634–636) **1.** $2x^2 + 5x + 7$ **3.** $x^3 + 2x^2 + 2x - 2$ **5.** $-8x^4 + 12x^3$ **7.** $x^2 + 14x + 33$ **9.** $2x^3 + 11x^2 - 5x - 50$ **11.** $x^2 - 225$ **13.** $x^2 + 4x + 4$ **15.** $-1, -10$ **17.** -9 **19.** $9, 12$ **21.** -13 **23.** $(2x - 3)(x + 4)$ **25.** $(2x - 3)(2x - 3)$ **27.** $-\frac{3}{4}$ **29.** $-2x^3(x + 2)(x - 1)$ **31.** $(x^2 + 5)(3x + 1)$

SA31

CHAPTER 11

SKILL REVIEW (p. 642) **1.** $\frac{1}{2}$ **2.** $1\frac{3}{10}$ **3.** $-\frac{3}{55}$ **4.** $1\frac{5}{18}$ **5.** $\frac{4x}{3}$ **6.** $-21x$ **7.** $3x + 5$ **8.** $\frac{4-x}{2}$ **9.** $-9, 9$ **10.** $1, 1\frac{1}{2}$ **11.** $-1, 4$

11.1 PRACTICE (pp. 646–647) **9.** $\frac{6}{7}$ **11.** $4\frac{1}{2}$ **13.** $-2, 3$

15. Other proportions include:

$\frac{\text{height of actual statue}}{\text{height of model}} = \frac{\text{length of actual statue}}{\text{length of model}}$,

$\frac{\text{height of model}}{\text{height of actual statue}} = \frac{\text{length of model}}{\text{length of actual statue}}$,

$\frac{\text{length of actual statue}}{\text{height of actual statue}} = \frac{\text{length of model}}{\text{height of model}}$.

17. 3 **19.** $5\frac{5}{8}$ **21.** $2\frac{1}{2}$ **23.** -10 **25.** $-1\frac{1}{2}$ **27.** 27 **29.** $2\frac{4}{5}$ **31.** 2, 5 **33.** 4 **35.** 6 **37.** $1\frac{2}{3}$ **39.** $-\frac{1}{3}$ **41.** 22 in.

43. *Sample answer:* By rewriting 1 ft as 12 in., you can set up the proportion $\frac{\frac{1}{16}\text{ in.}}{12\text{ in.}} = \frac{1\text{ in.}}{192\text{ in.}}$. Now that all the units are the same, you can use the cross product property. This gives 12 in. = 12 in., which shows that the proportion is correct. **45.** food: about 16; clothes, accessories: about 4; books, magazines, comics: about 3; toys, stickers, games: about 3; movie tickets: about 3; arcade games: about 3; gifts: about 3; movie rentals: about 3; music: about 2; footwear: about 2; grooming products: about 2
47. about 4771

11.1 MIXED REVIEW (p. 648) **51.** $-8, 8$ **53.** $-2\sqrt{3}, 2\sqrt{3}$ **55.** no square roots **57.** $-\frac{3}{5}, \frac{3}{5}$ **59.** $3\sqrt{2}$ **61.** $4\sqrt{5}$ **63.** 54 **65.** $\sqrt{7}$

11.2 PRACTICE (pp. 652–654) **5.** 175% **7.** 48 **9.** \$.36; *Sample answer:* The equations for the two methods are $a = 0.06(5.99)$ and $\frac{a}{5.99} = \frac{6}{100}$. Notice that the right side of the second equation is the rate expressed as a fraction instead of a decimal. Both methods involve multiplying the base by the rate. **11.** A **13.** 20 **15.** 30% **17.** 30.8 ft **19.** 84 ft **21.** 50 **23.** 150 g **25.** 65.6 tons **27.** $86\frac{1}{9}$%, or about 86.1% **29.** 480% **31.** $86\frac{2}{3}$%, or about 86.7%; $13\frac{1}{3}$%, or about 13.3% **33.** about 248 people **35.** about 1113 people **37.** 18% **39.** about 420 students **41.** yes; no **43.** about 395 people **45.** No; the price is \$56 at the first store and \$57.60 at the second. **47.** \$13.80; the total cost of the ride

11.2 MIXED REVIEW (p. 655) **53.** $y = 2x$ **55.** $y = \frac{1}{2}x$ **57.** $y = \frac{3}{2}x$ **59.** solution **61.** solution **63.** $(x + 7)(x - 2)$ **65.** $(5x - 6)(x - 9)$ **67.** $2x(3x + 8)$ **69.** $3x(x + 2)(x + 5)$ **71.** $5x^2(3x + 2)(x - 4)$

11.3 PRACTICE (pp. 659–660) **7.** neither **9.** inverse variation **11.** $y = \frac{24}{x}$; 3 **13.** $y = 4x$ **15.** $y = 3x$ **17.** $y = \frac{1}{9}x$ **19.** $y = \frac{9}{5}x$ **21.** $y = \frac{10}{x}$ **23.** $y = \frac{16}{x}$ **25.** $y = \frac{4}{x}$ **27.** $y = \frac{9}{x}$ **29.** $y = \frac{225}{x}$ **31.** $y = \frac{27}{x}$ **33.** 4, 2, $\frac{4}{3}$, 1

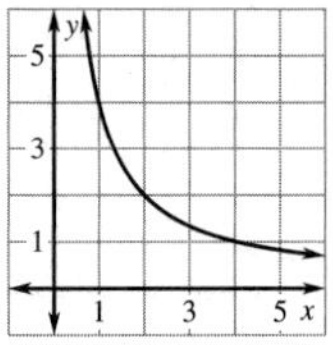

inversely

35. 3, 6, 9, 12

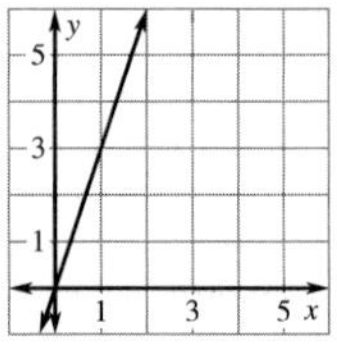

directly

37. direct variation; $y = 5x$ **39.** inverse variation **41.** neither **43.** 116 lb **45.** $T = \frac{4440}{d}$ **47.** $\frac{1}{26}$ gal/mi ≈ 0.04 gal/mi **49.** Neither; the equation cannot be written either in the form $g = km$ or $g = \frac{k}{m}$ for any constant k.

11.3 MIXED REVIEW (p. 661) **55.** $\frac{3}{4}$ **57.** $\frac{8}{15}$ **59.** $\frac{1}{6}$ **61.** y^6 **63.** $-\frac{y^2}{x^2}$ **65.** -5, linear, binomial **67.** -1, cubic, binomial

QUIZ 1 (p. 662) **1.** 8 **2.** $\frac{27}{7}$ **3.** 4 **4.** $-\frac{5}{3}$, 1 **5.** $y = \frac{12}{x}$ **6.** $y = \frac{12}{x}$ **7.** $y = \frac{18}{x}$ **8.** about 904 people **9.** about 204 people

TECHNOLOGY ACTIVITY 11.3 (p. 663)
1. directly; 0.825; $y = 0.825x$

11.4 PRACTICE (pp. 667–668) **5.** 5 **7.** C **9.** $\frac{x}{5}$ **11.** $-\frac{3x}{2}$ **13.** $\frac{6-x}{3x}$ **15.** $\frac{7}{x+12}$ **17.** $-\frac{1}{2x}$ **19.** $\frac{1}{3-x}$ **21.** $\frac{x-4}{x-3}$ **23.** $\frac{x+1}{x+6}$ **25.** 8 **27.** $-3, 3$ **29.** $-6, 1$ **31.** $\frac{x}{5x+3}$; $\frac{1}{6}$ **33.** $\frac{x-y}{x}$ **37.** $\frac{1625}{R}$

11.4 MIXED REVIEW (p. 669) **43.** $\frac{25}{2}$ **45.** $-\frac{1}{81}$ **47.** $4m^3$ **49.** $-\frac{8c^2}{3}$ **51.**

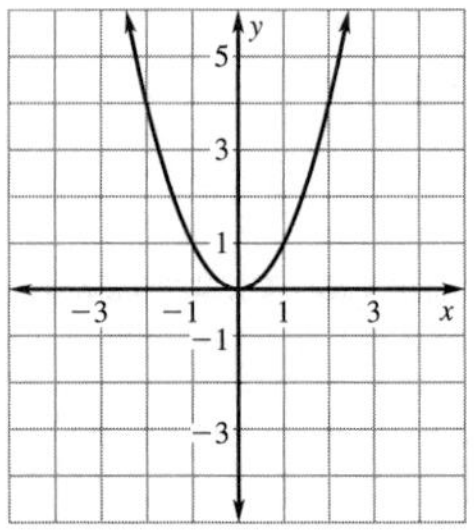

53. 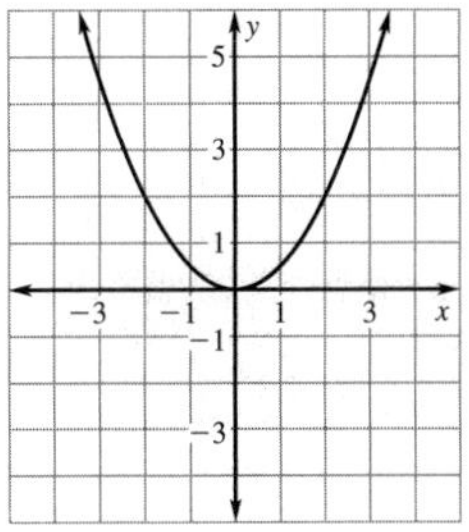

55. 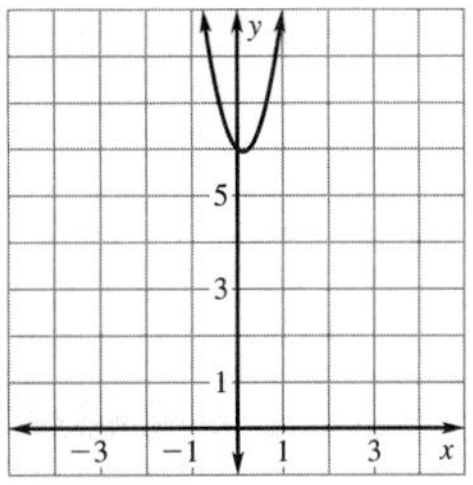

57. $y = 2500(0.92)^x$; 2116; 1947; 1791; 1648; 1516

11.5 PRACTICE (pp. 673–674) 5. $\frac{2(x+1)}{3}$ 7. $\frac{3x}{x-5}$ 9. $\frac{2x+5}{4x}$ 11. $x(3x+1)$ 13. x 15. 8 17. $13x$ 19. $\frac{1}{(x+3)(2x+3)}$ 21. $\frac{5}{6x}$ 23. 15 25. $-\frac{x+6}{5x^2}$ 27. $\frac{1}{(x+4)(2x+3)}$ 29. $-\frac{x+1}{5x^2}$ 31. $\frac{1}{3x}$ 33. $3x(x+2)$ 35. $A = \frac{(3150-400t)(10-t)}{(1-0.12t)(111-12t)}$

37. 258 mi 39. $R = \frac{46+0.7t}{1055+23t}$; decreasing; *Sample answer:* The ratios in order (rounded to the nearest hundred-thousandth) are 0.04360, 0.04332, 0.04305, 0.04279, 0.04255, 0.04231, 0.04208, and 0.04186.

43. $\frac{2x-4}{x^2-4} = \frac{2(x-2)}{(x+2)(x-2)} = \frac{2}{x+2} \cdot \frac{x-2}{x-2} = \frac{2}{x+2} \cdot 1 = \frac{2}{x+2}$

11.5 MIXED REVIEW (p. 675) 49. 18 51. 162 53. $3-\sqrt{2}$, $3+\sqrt{2}$ 55. $3t^2+8t+10$ 57. a^4+4a^3-a-1 59. $1000+2000r+1000r^2$

11.6 PRACTICE (pp. 679–681) 5. 5 7. $\frac{6x-5}{20x^2}$ 9. $\frac{3x+4}{2(x-5)}$

11. 2 13. $\frac{2-5x}{3x-1}$ 15. $\frac{1}{x^2}$ 17. $\frac{155}{78x}$ 19. $-\frac{3(x+12)}{(x+4)(x-2)}$

21. $\frac{3(x+3)}{x(x-3)}$ 23. $\frac{2(x^2-20)}{(x-10)(x+6)}$ 25. $\frac{4x^2+17x+5}{(3x-1)(x+1)}$

27. $-\frac{x^2+14x-2}{(3x-1)(x-2)}$ 29. $\frac{x+5}{x+2}$ 31. $\frac{2(4x^2+8x-5)}{(x+1)(x-2)(x+4)}$

33. $\frac{2(7x^2-8x-14)}{(2x-1)(x-3)}$ 35. *Sample answer:* They get very large; they get very large; they get very close to 2; as x gets very large, the numbers subtracted in the numerator and the denominator become insignificantly small compared to $2x$ and x, so the rational expression gets closer and closer to $\frac{2x}{x} = 2$. 37. $\frac{48x}{(x-2)(x+2)}$ 39. $\frac{1}{x}$; $\frac{1}{35}+\frac{1}{x}$ 41. a. A: $\frac{3}{56}$; B: $\frac{16}{315}$; C: $\frac{18}{385}$ b. A or B 43. $\frac{3x}{7}$ 45. $\frac{4(x-6)(x-8)}{(x+6)(x+8)}$

47. $\frac{5x+12}{(x-2)(x+2)(x+3)}$ 49. $\frac{8x^2-15x+19}{(2x+3)(x-5)(x+1)}$

11.6 MIXED REVIEW (p. 682) 57. $\frac{2m}{3}$ 59. $\frac{7x}{y^6}$

61. $2x^2+6x-9=0$ 63. 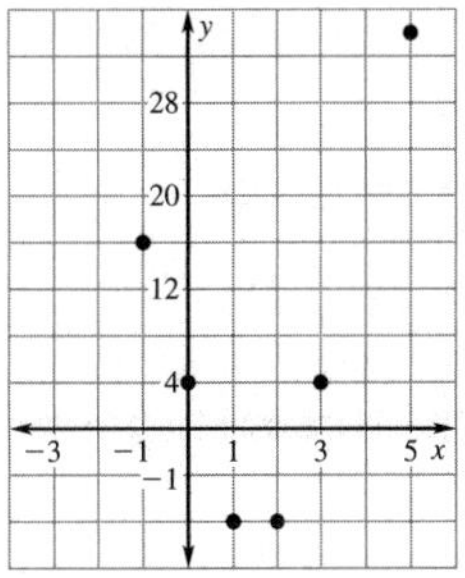

quadratic

65. $|x-8500| \le 1000$

QUIZ 2 (p. 682) 1. $\frac{3x}{2}$ 2. $\frac{5}{11+x}$ 3. $-\frac{1}{x-2}$ 4. cannot be simplified 5. $\frac{7x^2}{2}$ 6. 10 7. $\frac{3}{2}$ 8. $\frac{x-3}{x+2}$ 9. $\frac{1}{x-7}$

10. $\frac{1}{x+1}$ 11. $\frac{15}{x+2}$; $\frac{15}{x-2}$ 12. $\frac{15}{x+2}+\frac{15}{x-2}=\frac{30x}{(x+2)(x-2)}$ 13. 10 h

11.7 PRACTICE (pp. 687–688) 5. $y+2\overline{)y^2+0y+8}$ 7. $3y+5\overline{)8y^2-2y+0}$ 9. $29\frac{15}{29}$ 11. $x-5$ 13. $-5b^2+4b+\frac{5}{2}$ 15. $4x+\frac{13}{2}$ 17. $9c+3$ 19. $x+6$

21. $3a-18-\frac{12}{a}$ 23. D 25. B 27. $a-2$

29. $2b+1-\frac{2}{b-2}$ 31. $5g-1+\frac{1}{g+3}$ 33. $x+6-\frac{5}{x-9}$

35. $-x-4-\frac{8}{x+2}$ 37. $b-10+\frac{34}{b+3}$ 39. $-x+4-\frac{25}{x+4}$

41. $-5m-5-\frac{3}{m-1}$ 43. $-s+5-\frac{21}{s+5}$ 45. $\frac{1}{2}c-2+\frac{9}{2c-6}$

47. $5x+3$ 49. $3x+2$ 51. $\frac{2}{9}-\frac{8}{9t+99}$ 53. increasing

11.7 MIXED REVIEW (p. 689) 61. 15 63. 3 65. $1\frac{3}{7}$ 67. $-1\frac{1}{3}, 2\frac{1}{2}$ 69. $-1\frac{1}{5}$ 71. 12; 6; 4; 3; 2.4

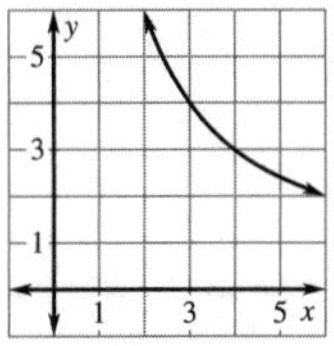

11.8 PRACTICE (pp. 694–696) 7. $\frac{1}{2}$, 1 9. 3 11. −3

13. (2, 3) 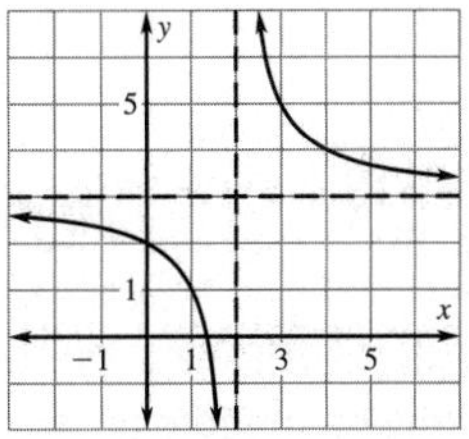

15. 28 17. $\frac{1}{2}$ 19. 13

21. −27 23. 11 25. $1\frac{1}{2}$ 27. $-\frac{1}{3}$, 2 29. −8, 9 31. −8, 2

33. $\frac{3}{5}$ 35. −8, 10 37. −4 39. B 41. A

SELECTED ANSWERS

43. domain: all real numbers except 3

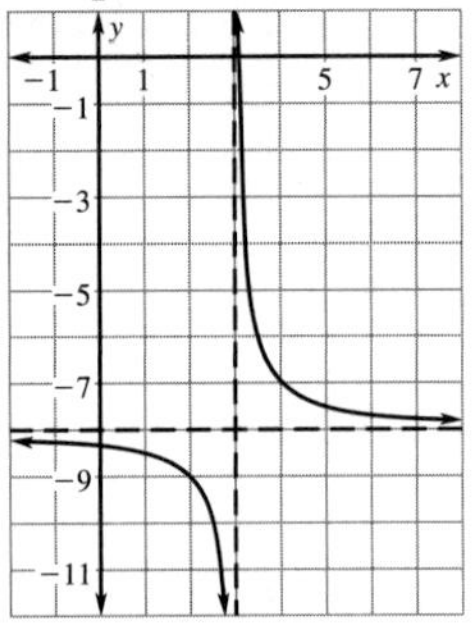

45. domain: all real numbers except –1

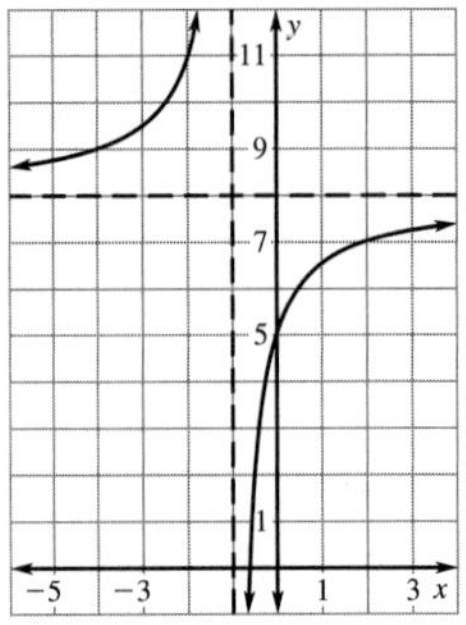

47. domain: all real numbers except –3

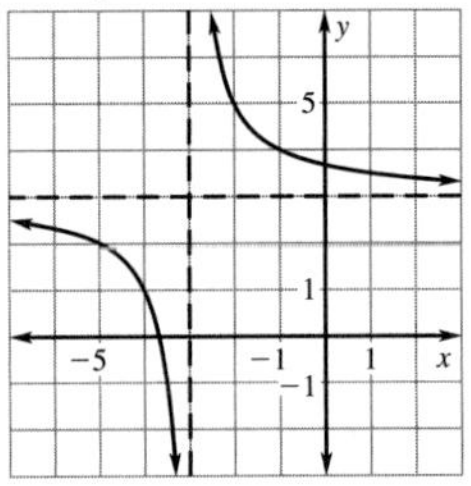

49. domain: all real numbers except 1

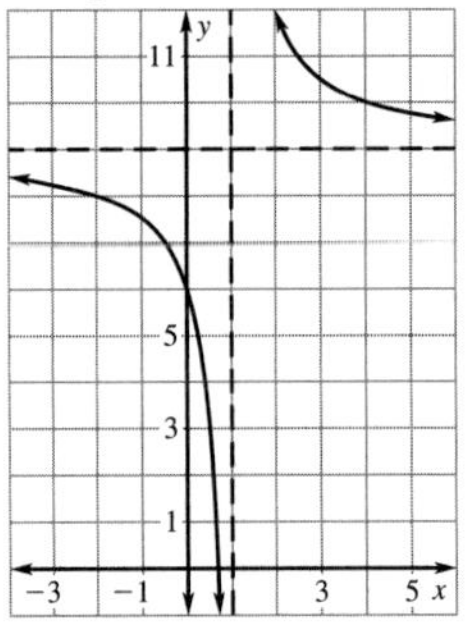

51. 6 hits **53.** $y = \frac{5000 + 10x}{x + 1}$

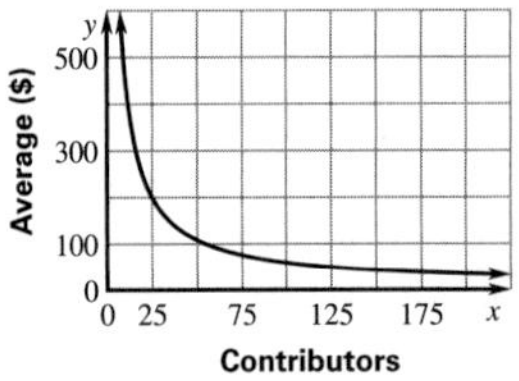

55. $\frac{60}{2000} = \frac{3}{100}$ **57.** $\frac{3}{100} + \frac{3}{50x} = \frac{3x + 6}{100x}$

59.

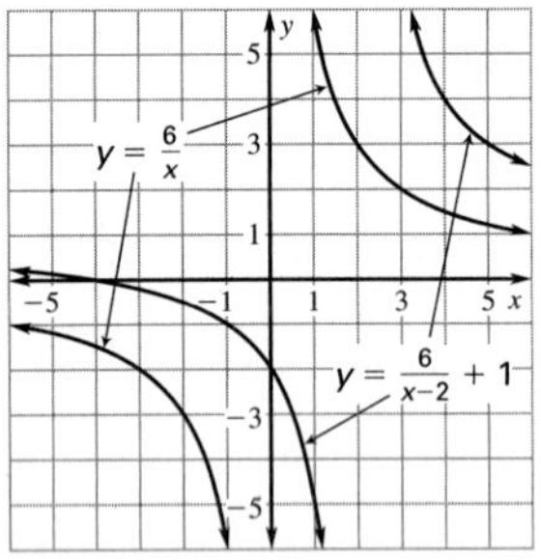

61. *Sample answer:* For $a > 0$, as a increases, the branches of the hyperbola (which are in the first and third quadrants) become wider and more open. For $a < 0$, as $|a|$ increases, the branches of the hyperbola (which are in the second and fourth quadrants) become wider and more open.

11.8 MIXED REVIEW (p. 697) **69.** 9, 8, 7, 6, 5 **71.** 0, –1, –4, –9, –16 **73.** $0, \frac{1}{2}, 2, 4\frac{1}{2}, 8$ **75.** $6^2 = 36$ **77.** $-4^2 = -16$ **79.** $6\sqrt{2}$ **81.** $\sqrt{13}$ **83.** 16 **85.** $6\sqrt{11}$

QUIZ 3 (p. 697) **1.** $\frac{x}{6} - \frac{4}{3x}$ **2.** $\frac{3}{5}a + \frac{1}{2}$ **3.** $x - 4 + \frac{32}{x + 4}$ **4.** $y + 6.5 + \frac{27.5}{2y - 5}$ **5.** $\frac{1}{3}z + 2 + \frac{20}{3z - 6}$ **6.** $4x + 3 - \frac{11}{3x + 2}$ **7.** –2 **8.** 4 **9.** 3 **10.** 130 **11.**

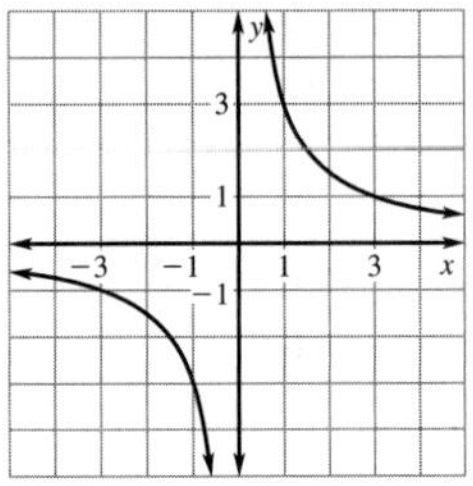

12.

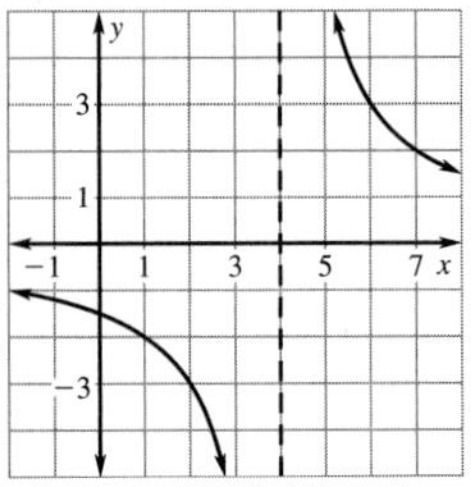

13.

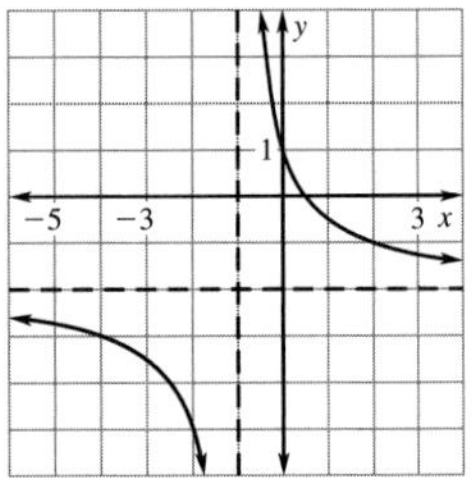

14. $\frac{7}{8} - \frac{103}{4t + 262}$

TECHNOLOGY ACTIVITY 11.8 (p. 698) **1.** $y = \frac{3x + 2}{x + 5}$; all real numbers except –5 and 5; all real numbers except –5 **3.** $y = \frac{2x + 1}{x - 3}$; all real numbers except –4 and 3; all real numbers except 3 **5.** $y = \frac{2x + 3}{x + 1}$; all real numbers except –1 and 3; all real numbers except –1 **7.** *Sample answer:* No; they do not have the same domain. The domain of the first is all real numbers except –3 and 1; the domain of the second is all real numbers except –3.

CHAPTER 11 REVIEW (pp. 700–702) **1.** $1\frac{1}{7}$ **3.** 4 **5.** \$76 **7.** \$120 **9.** $y = 3x$ **11.** $\frac{x}{3x^2 + 1}$ **13.** $\frac{-x^2 + 3}{2x}$ **15.** $20x^2$ **17.** $\frac{(x + 2)^2}{(x + 4)^2}$ **19.** $\frac{4x + 3}{24x}$ **21.** $\frac{4x - 5}{x - 2}$ **23.** $x - 6 + \frac{5}{6x}$ **25.** $x + 3$ **27.** –4, 2

CHAPTER 12

SKILL REVIEW (p. 708) **1.** 5.2 **2.** 3.9 **3.** $7\sqrt{2}$ **4.** $2\sqrt{35}$ **5.** $\frac{\sqrt{7}}{2}$ **6.** 3 **7.** $(x - 6)(x + 3)$ **8.** $(x + 4)(x - 2)$ **9.** $(2x + 5)^2$

12.1 PRACTICE (pp. 712–714) **3.** 0, 4, 5.7, 6.9, 8 **5.** 4, 7, 8.2, 9.2, 10 **7.** 1.4, 1.7, 2, 2.2, 2.4 **9.** Both the domain and range are all the nonnegative numbers. **11.** domain: all nonnegative numbers; range: all numbers greater than or equal to –10 **13.** domain: all numbers greater than or equal to –5; range: all nonnegative numbers

15.

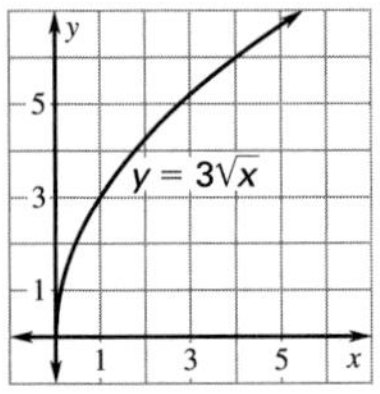

17.

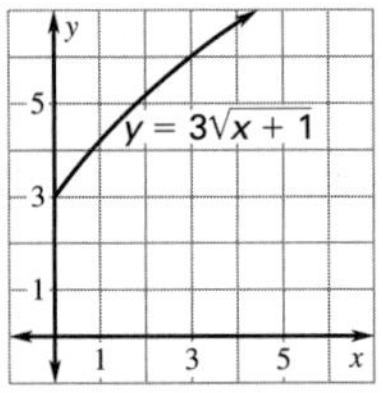

19. 100 lb/in.^{2} **21.** 1 **23.** 4 **25.** 5 **27.** about 1.4 **29.** 4 **31.** $x \geq 17$ **33.** $x \geq -5$ **35.** all nonnegative numbers **37.** all nonnegative numbers **39.** all nonnegative numbers **41.** $x \geq -9$ **43.** all nonnegative numbers

45. Both the domain and range are all the nonnegative numbers.

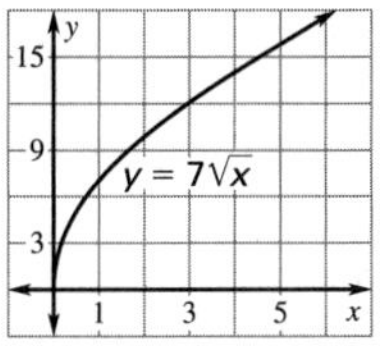

49. domain: all nonnegative numbers; range: all numbers greater than or equal to 4

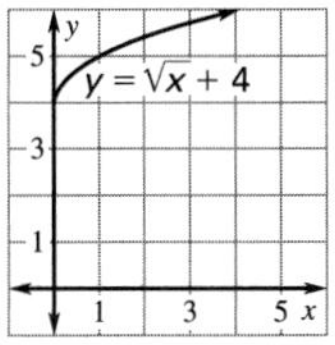

51. domain: all numbers greater than or equal to 4; range: all nonnegative numbers

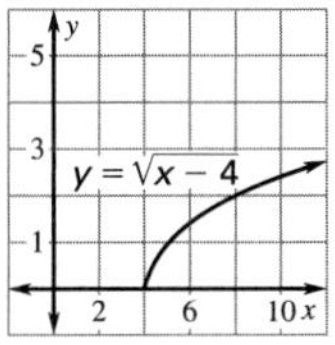

55. domain: all numbers greater than or or equal to –2.5; range: all nonnegative numbers

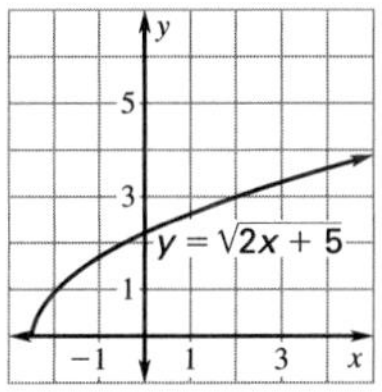

57. 1.41 m^2 **59. a.** $\text{cm}^2 = \text{cm} \cdot \sqrt{\text{cm}^2 + \text{cm}^2} = \text{cm} \cdot \text{cm}$

b.

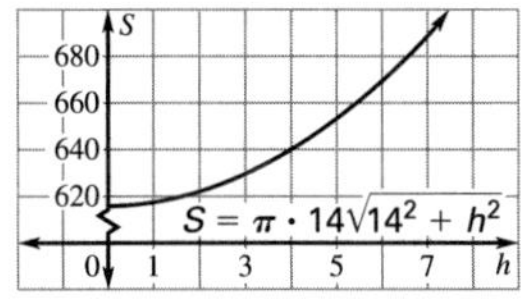

c. 1456 cm^2 (to the nearest cm^2)

12.1 Mixed Review (p. 714)

65. $2\sqrt{6}$ **67.** $5\sqrt{7}$ **69.** $\frac{2\sqrt{5}}{5}$ **71.** $\sqrt{7}$ **73.** $-2 - 2\sqrt{3}, -2 + 2\sqrt{3}$ **75.** $3 - 2\sqrt{2}, 3 + 2\sqrt{2}$ **77.** $-1\frac{1}{2}, 1$ **79.** $x^2 + 9x - 22$ **81.** $10x^2 - 33x + 27$ **83.** $6x^3 - 4x^2 - 8x - 2$ **85.** $\frac{8}{3}$ **87.** $\frac{x}{x + 1}$

Technology Activity 12.1 (p. 715)

1. k moves the graph of $y = \sqrt{x}$ up k units if k is positive and down $|k|$ units if k is negative. **3.** –2 **5.** 3

12.2 Practice (pp. 719–720)

5. $\sqrt{7}$ **7.** $2\sqrt{6}$ **9.** $15 - 6\sqrt{2}$ **11.** $\frac{8 + \sqrt{10}}{18}$ **13.** $6(\sqrt{26})^2 - 156 = 6(26) - 156 = 0$; solution **15.** $(6 + \sqrt{31})^2 - 12(6 + \sqrt{31}) + 5 = 67 + 12\sqrt{31} - 72 - 12\sqrt{31} + 5 = 0$; solution **17.** $6 - 3\sqrt{2} \approx 1.8$ mi **19.** $6\sqrt{3}$ **21.** $\sqrt{6}$ **23.** $6\sqrt{3}$ **25.** $3\sqrt{2}$ **27.** $12\sqrt{5}$ **29.** $-\sqrt{6}$ **31.** 6 **33.** –12 **35.** $21\sqrt{2} + 6\sqrt{6}$ **37.** $a - 2b\sqrt{a} + b^2$ **39.** $37 - 20\sqrt{3}$ **41.** $\sqrt{2}$ **43.** $\frac{\sqrt{3}}{4}$ **45.** $\frac{3 + \sqrt{3}}{2}$ **47.** $\frac{42 + 6\sqrt{3}}{23}$ **49.** $34 + 18\sqrt{17}$ **51.** $\frac{45\sqrt{3} + 90}{2}$ **53.** $2 - \sqrt{10}, 2 + \sqrt{10}$ **55.** $\frac{1 - \sqrt{5}}{4}, \frac{1 + \sqrt{5}}{4}$ **57.** $3 - \sqrt{22}, 3 + \sqrt{22}$ **59.** 0.35 sec **61.** 80 ft/sec

12.2 Mixed Review (p. 721)

69. 48

71.

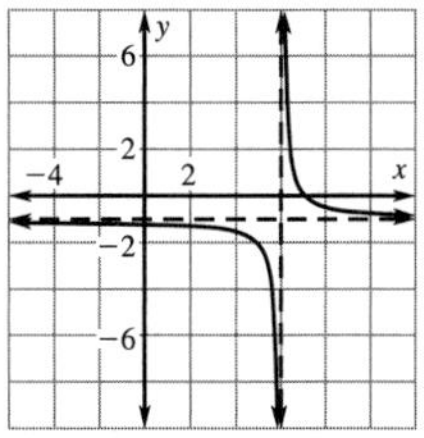

73.

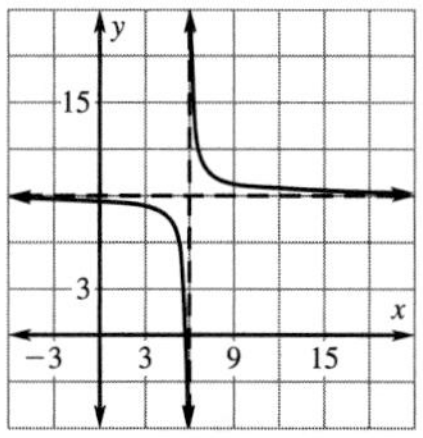

75. domain: all numbers greater than or equal to 8; range: all nonnegative numbers **77.** domain: all nonnegative numbers; range: all numbers greater than or equal to 4 **79.** domain: all numbers greater than or equal to –3; range: all nonnegative numbers

12.3 Practice (pp. 725–727)

5. 3 **7.** no solution **9.** 3 **11.** 25 **13.** 3 **15.** 81 **17.** no solution **19.** 225 **21.** 216 **23.** –567 **25.** 17 **27.** 5 **29.** –9 **31.** $13\frac{1}{5}$ **33.** 272 **35.** 40 **37.** no solution **39.** 5 **41.** no solution **43.** 10 **45.** 9 **47.** 6, 48 **49.** 64 **51.** 49 **53.** 1352 **55.** $15^2 = 225$, not –225; $x = 225$ **57.** 11.85 **59.** 17 **61.** 3, 4 **63.** 2 **65.** *Sample answer:* The graphs do not intersect; the graph of $y = \sqrt{11x - 30}$ is entirely in the first quadrant, while the graph of $y = -x$ is entirely in the second and fourth quadrants. **67.** 113 **69.** 160 **71.** about 66.3 min **73.** By the cross product property, $ad = b^2$, so $b = \sqrt{ad}$. **75.** 4 and 25

12.3 Mixed Review (p. 728)

83. $-\sqrt{11}, \sqrt{11}$ **85.** $-\frac{2}{5}, \frac{2}{5}$ **87.** $-\sqrt{3}, \sqrt{3}$ **89.** $4x^2 - 12x + 9$ **91.** $36y^2 - 16$ **93.** $4a^2 - 36ab + 81b^2$ **95.** $(x - 6)^2$ **97.** 4 **99.** –3 and 2

Quiz 1 (p. 728)

1. Both the domain and range are all nonnegative numbers.

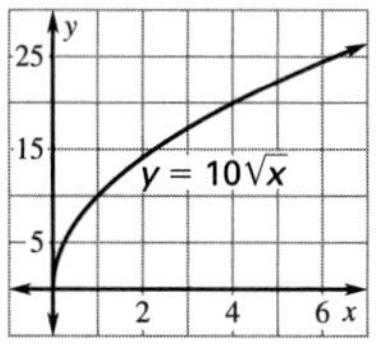

2. domain: all numbers greater than or equal to 9; range: all nonnegative numbers

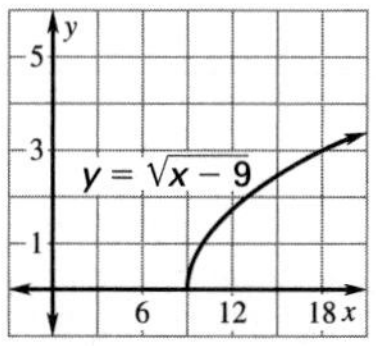

3. domain: all numbers greater than or equal to $\frac{1}{2}$; range: all nonnegative numbers

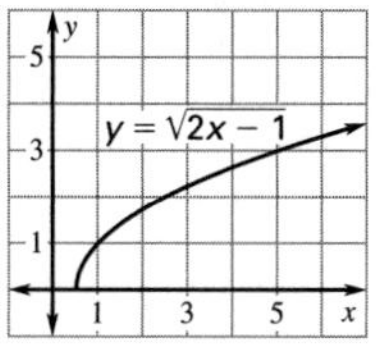

SA35

4. all nonnegative numbers, all numbers greater than or equal to $-\frac{2}{3}$

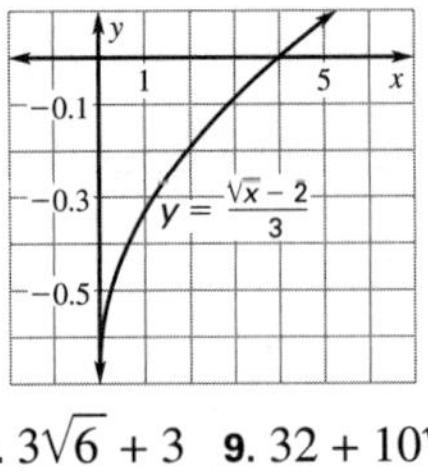

5. $18\sqrt{10}$ 6. $-8\sqrt{7}$ 7. $5\sqrt{5}$ 8. $3\sqrt{6} + 3$ 9. $32 + 10\sqrt{7}$ 10. $\frac{40 - 5\sqrt{7}}{19}$ 11. 144 12. 64 13. $-\frac{1}{3}$ 14. 6 15. 7 16. 3 17. $1\frac{7}{9}$ lb/in.^{2}

12.4 PRACTICE (pp. 734–735) 3. 100 5. 25 7. 121 9. $\frac{3 - \sqrt{41}}{2}, \frac{3 + \sqrt{41}}{2}$; choices may vary 11. −13, −1 13. no solution 15. $-\frac{2}{3}$, −5; factoring; the expression $3x^2 + 17x + 10$ can be factored $(3x + 2)(x + 5)$. 17. $\frac{5 + \sqrt{85}}{6}$, $\frac{5 - \sqrt{85}}{6}$; quadratic formula; the expression is not factorable, and the coefficient of x is odd. 19. $\frac{\sqrt{6}}{3}, \frac{-\sqrt{6}}{3}$; finding square roots; the expression is equivalent to $x^2 = \frac{2}{3}$. 21. 16 23. 121 25. 400 27. $\frac{9}{64}$ 29. 6.76 31. $\frac{1}{9}$ 33. −17, 1 35. 2, 6 37. $-\frac{21}{2}, -\frac{1}{2}$ 39. $\frac{-3 - \sqrt{109}}{10}, \frac{-3 + \sqrt{109}}{10}$ 41. $\frac{-1 - 2\sqrt{3}}{2}, \frac{-1 + 2\sqrt{3}}{2}$ 43. $\frac{10 - \sqrt{107}}{2}, \frac{10 + \sqrt{107}}{2}$ 45. −1, 5 47. $\frac{-3 - \sqrt{41}}{4}$, $\frac{-3 + \sqrt{41}}{4}$ 49. $\frac{30 - 17\sqrt{5}}{10}, \frac{30 + 17\sqrt{5}}{10}$ 51. $-\sqrt{3}, \sqrt{3}$; factoring or finding square roots; both methods are easily applied. 53. $\frac{5}{3}, -\frac{5}{3}$; factoring or finding square roots; both methods are easily applied. 55. −11, 5; completing the square or factoring; $a = 1$ and b is an even number. 57. $-1 - 3\sqrt{3}, -1 + 3\sqrt{3}$ 59. −4, 4 61. $\frac{12 - \sqrt{174}}{6}, \frac{12 + \sqrt{174}}{6}$ 63. 0, 2 65. $-\frac{3}{7}$, 7 67. $-10 - 3\sqrt{10}, -10 + 3\sqrt{10}$ 69. $2x(x + 5) = 600$; 30 ft by 20 ft 71. about 8.58 ft 73. about 11.8 ft

12.4 MIXED REVIEW (p. 736) 81. $3\frac{1}{7}$, 3, 1 83. $-\frac{2}{5}$, −6, none 85. (2, 8) 87. (3, −2) 89. $-3\sqrt{7}, 3\sqrt{7}$ 91. −6, 6 93. $-\frac{3}{5}, \frac{3}{5}$ 95.

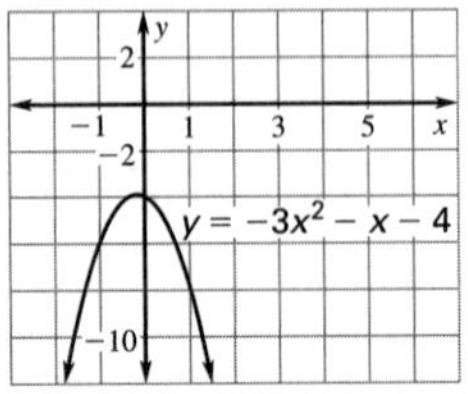

97. −4, 8 99. −6, −5 101. 3 103. $(x + 5)(x - 4)$ 105. cannot be factored 107. $(2x - 3)(x + 1)$

12.5 PRACTICE (pp. 741–742) 5. 12 7. 40 9. 37 11. 29 13. 5 15. $2\sqrt{10}$, or about 6.32 17. $5\sqrt{7}$, or about 13.23 19. 24, 18 21. 5, 10 23. 1, $\sqrt{2}$, or 1, about 1.41 25. Yes; $9^2 + 12^2 = 225 = 15^2$. 27. No; $8^2 + 10^2 = 164$ and $13^2 = 169$. 29. Yes; $15^2 + 20^2 = 625 = 25^2$. 31. Yes; $11^2 + 60^2 = 3721 = 61^2$. 33. Yes; $9.9^2 + 2^2 = 102.01 = 10.1^2$. 35. hypothesis: a polygon is a square; conclusion: it is a parallelogram 37. hypothesis: the area of a square is 25 sq. ft; conclusion: the length of a side is 5 ft 39. about 116.6 ft 41. 113.58 ft 43. about 55.6 in.

12.5 MIXED REVIEW (p. 743)

51.

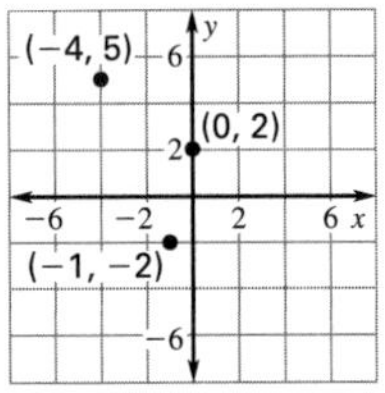

53.

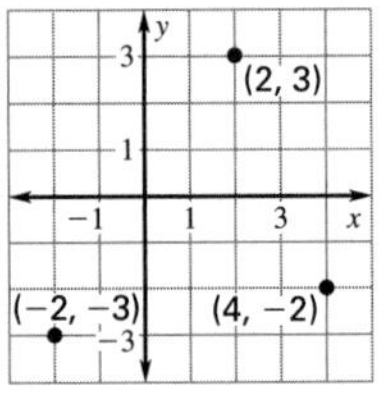

55. no x-intercepts 57. $\frac{-1 - \sqrt{41}}{2}, \frac{-1 + \sqrt{41}}{2}$ 59. $\frac{-3 - \sqrt{5}}{2}, \frac{-3 + \sqrt{5}}{2}$ 61. $-4 - \sqrt{26}, -4 + \sqrt{26}$ 63. $2 - \sqrt{5}, 2 + \sqrt{5}$ 65. $(4x - 5)(4x + 5)$ 67. $7(x - 2)^2$ 69. $-3(4x - 9)^2$ 71. $0.04 \le x \le 0.33$

QUIZ 2 (p. 744) 1. −2, 5 2. $\frac{-1 - \sqrt{10}}{2}, \frac{-1 + \sqrt{10}}{2}$ 3. $-1 - \sqrt{3}, -1 + \sqrt{3}$ 4. no 5. yes 6. yes 7. 13 8. 60 9. 16 10. 2000 ft

12.6 PRACTICE (pp. 748–749) 3. Estimates may vary; 5.66 5. Estimates may vary; 6.71 7. yes 9. no 11. $\left(4, -\frac{1}{2}\right)$ 13. 25 yd 15. 12.08 17. 8.60 19. 4.24 21. 9 23. 10.63 25. 2.69 27. no 29. yes 31. no 33. yes 35. (0, 4) 37. (3, 3) 39. (3, 5) 41. $\left(-\frac{5}{2}, -\frac{5}{2}\right)$ 43. $\left(-2, -\frac{1}{2}\right)$ 45. $\left(-\frac{5}{2}, -4\right)$ 47. Estimates may vary; about 460 mi 51. (−1, 0), $\left(1\frac{1}{2}, 0\right)$, $\left(-\frac{1}{2}, -2\right)$ 53. The perimeter of the original triangle, 14.78, is twice that of the perimeter of the new triangle, 7.39. 55. You could meet at a point 1 mi east and 1 mi north of the starting point. Both you and your friend would each have to walk $\sqrt{13} \approx 3.6$ mi. 57. Side $\overline{DC}$ is parallel to side $\overline{AB}$ because both have slope 2. Sides $\overline{CB}$ and $\overline{DA}$ are not parallel because $\overline{DA}$ has slope −3 and $\overline{CB}$ has slope $\frac{1}{3}$.

12.6 MIXED REVIEW (p. 750) 65. $3x(x + 5)(x - 1)$ 67. inverse variation 69. neither 71. $\frac{7x^2 - 2x}{14}$ 73. $30(14) = 21x$; 20 in.

12.7 PRACTICE (pp. 755–757) **3.** $\frac{7}{25}$ **5.** $\frac{7}{24}$ **7.** $e \approx 27.16$, $f \approx 56.02$ **9.** about 459 ft; about 655 ft **11.** $\sin R = \frac{5}{13}$, $\cos R = \frac{12}{13}$, $\tan R = \frac{5}{12}$; $\sin S = \frac{12}{13}$, $\cos S = \frac{5}{13}$, $\tan S = \frac{12}{5}$ **13.** $b \approx 12.81$, $c \approx 21.29$ **15.** $x \approx 13.42$, $y \approx 19.90$ **17.** $v \approx 51.96$, $w = 60$ **19.** about 172 ft **21.** about 170 ft **23.** 11.56 km

12.7 MIXED REVIEW (p. 757) **29.** False; the opposite of a positive number is negative. **31.** $-p^2 - p$ **33.** $x^2 + 3x$ **35.** $-24x - 6x^2$ **37.** $2x^3 + 6x^2 + 15x + 6$ **39.** $5x^2 - 4x + 12$ **41.** $\frac{3}{4a+1}$ **43.** $-\frac{13}{12x}$ **45.** $\frac{4(x-1)}{(x+1)}$

12.8 PRACTICE (pp. 761–763) **13.** Multiplication axiom of equality; Substitution property of equality

15. $a - b = a + (-b)$ Definition of subtraction
$a - b = -b + a$ Commutative property of addition

17. $a + (-1)(a) = 1(a) + (-1)(a)$ Identity property of multiplication
$= [1 + (-1)](a)$ Distributive property
$= (0)(a)$ Inverse property of addition
$= 0$ Multiplication property of 0

Since $a + (-1)(a) = 0$, $(-1)(a) = -a$ by definition.

19. *Sample answer*: $(1 + 2)^2 \neq 1^2 + 2^2$ **21.** *Sample answer*: 2 and 3 are integers, but $\frac{2}{3}$ is not an integer. **23.** Yes; this map shows that the proposal is false. **25.** Assume that p is an integer, p^2 is odd, and p is even. Let $p = 2n$. Then $p^2 = (2n)^2 = 4n^2 = 2(2n^2)$, an even number, which is impossible since p^2 is odd. Then p must be odd. **27.** Assume that $ac > bc$, $c > 0$, and $a \leq b$. By the multiplication property of inequality, $ac \leq bc$, which is impossible since it was given that $ac > bc$. Then $a > b$.

12.8 MIXED REVIEW (p. 764) **31.** no real solutions **33.** one solution **35.** two solutions **37.** two solutions **39.** two solutions **41.** not a solution **43.** not a solution **45.** solution **47.** not a solution **49.** 82% **51.** 200% **53.** \$15.15

QUIZ 3 (p. 764) **1.** 13.42; $(4, -3)$ **2.** 10; $(2, -8)$ **3.** 16.12; $(4, -7)$ **4.** 16; $(-8, 0)$ **5.** 6; $(0, 4)$ **6.** 10.30; $\left(-\frac{3}{2}, \frac{5}{2}\right)$ **7.** 5; $\left(0, -\frac{3}{2}\right)$ **8.** 7.28; $\left(\frac{1}{2}, 2\right)$ **9.** 8.06; $\left(\frac{5}{2}, 0\right)$ **10.** $\sin A = \frac{3}{5}$, $\cos A = \frac{4}{5}$, $\tan A = \frac{3}{4}$; $\sin B = \frac{4}{5}$, $\cos B = \frac{3}{5}$, $\tan B = \frac{4}{3}$ **11.** $\sin A = \frac{15}{17}$, $\cos A = \frac{8}{17}$, $\tan A = \frac{15}{8}$; $\sin B = \frac{8}{17}$, $\cos B = \frac{15}{17}$, $\tan B = \frac{8}{15}$

12. $(a - b)c = [a + (-b)]c$ Definition of subtraction
$= a(c) + (-b)c$ Distributive property
$= ac + (-bc)$ $(-b)c = -bc$
$= ac - bc$ Definition of subtraction

13. Any real numbers a, b, and c with $a < b$ and $c \leq 0$. *Sample answer:* $a = 1$, $b = 2$, and $c = -1$: $1 < 2$, and $1(-1) > 2(-1)$. **14.** Accept any real numbers a and b with $b \neq 0$. *Sample answer:* $a = 1$ and $b = 2$; $-(1 + 2) \neq (-1) - (-2)$.

CHAPTER 12 REVIEW (pp. 766–768)
1. Both the domain and range are all nonnegative numbers.

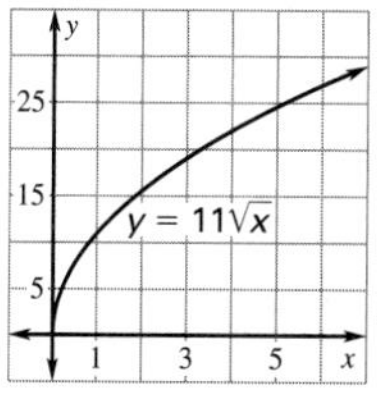

3. domain: all nonnegative numbers; range: all numbers greater than or equal to 3

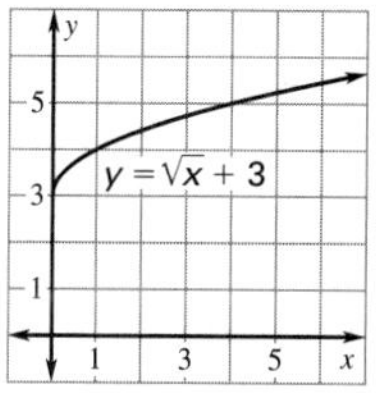

5. $6\sqrt{2} - 8\sqrt{3}$ **7.** $11 + 4\sqrt{6}$ **9.** 2 **11.** no solutions **13.** $2 - 2\sqrt{3}, 2 + 2\sqrt{3}$ **15.** $\frac{1-\sqrt{113}}{4}, \frac{1+\sqrt{113}}{4}$ **17.** 1 **19.** $3\sqrt{10}, \left(\frac{19}{2}, \frac{1}{2}\right)$ **21.** $2\sqrt{29}$, $(0, 3)$ **23.** commutative property of multiplication

CUMULATIVE PRACTICE (pp. 772–773) **1.** $\frac{m}{7} \geq 16$; $m \geq 112$ **3.** $t = 3\frac{2}{3}(3)$; 11 mi **5.** -2 **7.** -63 **9.** -15 **11.** $11\frac{1}{3}$ **13.** -49 **15.** $-1\frac{7}{11}$ **17.** -5.35 **19.** Accept any equation of the form $y = \frac{2}{3}x + b$ for any number b except -1. **21.** function; domain: $-1, 1, 3, 5, 7$; range: $-1, 1, 3, 5$ **23.** function; domain: $-2, -1, 0, 1, 2$; range: $-2, -1, 0, 1$ **25.** $4x - 5y = 15$ **27.** $-\frac{5}{4} \leq x < 3$

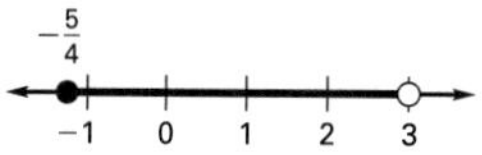

29. $x > 3$ or $x < 2$

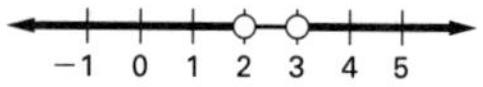

31. $(24, 21)$ **33.** b^6; 64 **35.** $-8a^3b^6$; -512 **37.** $\frac{4b^5}{a^4}$; 128 **39.** two solutions; $\frac{\sqrt{39}}{3}, -\frac{\sqrt{39}}{3}$ **41.** one solution; -1 **43.** $(x - 28)(x + 4)$ **45.** $(2x + 3)^2$ **47.** $(x - 7)^2$ **49.** $-1\frac{1}{2}, 1\frac{2}{3}$ **51.** $-5, -2, 2$ **53.** $\frac{2}{3}$ **55.** $\frac{2}{x-3}$ **57.** $\frac{1}{3x}$ **59.** $\frac{x(2x-7)}{(x+4)(x-1)}$ **61.** $-15\sqrt{2}$ **63.** $\frac{77 + 11\sqrt{3}}{46}$ **65.** $b = 19.80$, $c = 21.36$ **67.** $13\frac{1}{3}$ months

SELECTED ANSWERS

SKILLS REVIEW HANDBOOK

FACTORS AND MULTIPLES (p. 778) **1.** 1, 2, 3, 6, 9, 18 **3.** 1, 7, 11, 77 **5.** 1, 3, 9, 27 **7.** 1, 2, 3, 6, 7, 14, 21, 42 **9.** 1, 2, 4, 13, 26, 52 **11.** 1, 11, 121 **13.** 3^3 **15.** 2^5 **17.** $5 \cdot 11$ **19.** $2^4 \cdot 3$ **21.** $2 \cdot 3^2 \cdot 5$ **23.** $3 \cdot 13$ **25.** $2^3 \cdot 7 \cdot 13$ **27.** 2^9 **29.** 1, 3, 5, 15 **31.** 1, 5 **33.** 1, 3, 9 **35.** 1, 5 **37.** 5 **39.** 1 **41.** 14 **43.** 51 **45.** 2 **47.** 29 **49.** 14 **51.** 1 **53.** 35 **55.** 48 **57.** 12 **59.** 420 **61.** 100 **63.** 51 **65.** 70 **67.** 330

COMPARING AND ORDERING NUMBERS (p. 780) **1.** $<$ **3.** $>$ **5.** $<$ **7.** $<$ **9.** $<$ **11.** $>$ **13.** $=$ **15.** $>$ **17.** $>$ **19.** $>$ **21.** $>$ **23.** 40,071; 40,099; 45,242; 45,617 **25.** 9.003, 9.027, 9.10, 9.27, 9.3 **27.** $\frac{1}{3}, \frac{3}{8}, \frac{5}{6}, \frac{5}{4}$ **29.** $\frac{15}{16}, 1\frac{1}{8}, 1\frac{2}{5}, \frac{5}{3}, \frac{7}{4}$ **31.** $\frac{5}{12}, \frac{7}{8}, \frac{5}{4}, 1\frac{1}{3}$ **33.** Goran

FRACTION OPERATIONS (p. 783) **1.** $\frac{5}{6}$ **3.** $\frac{1}{3}$ **5.** $\frac{5}{8}$ **7.** $1\frac{1}{30}$ **9.** $2\frac{3}{8}$ **11.** $1\frac{13}{24}$ **13.** $8\frac{1}{5}$ **15.** $3\frac{7}{16}$ **17.** $\frac{1}{7}$ **19.** $1\frac{5}{7}$ **21.** 20 **23.** $2\frac{3}{5}$ **25.** $\frac{5}{6}$ **27.** 3 **29.** $\frac{5}{32}$ **31.** $3\frac{1}{2}$ **33.** $\frac{1}{4}$ **35.** $\frac{1}{6}$ **37.** $\frac{2}{3}$ **39.** $7\frac{2}{3}$ **41.** $1\frac{1}{6}$ **43.** $1\frac{1}{5}$ **45.** 6 **47.** $\frac{17}{20}$ **49.** $\frac{13}{16}$ **51.** 1 **53.** $5\frac{3}{8}$ **55.** $\frac{7}{18}$ **57.** $2\frac{5}{6}$ **59.** $1\frac{3}{5}$ **61.** $1\frac{11}{40}$ **63.** $\frac{11}{40}$ **65.** $9\frac{29}{48}$ in.

FRACTIONS, DECIMALS, AND PERCENTS (p. 785) **1.** 0.63, $\frac{63}{100}$ **3.** 0.24, $\frac{6}{25}$ **5.** 0.17, $\frac{17}{100}$ **7.** 0.45, $\frac{9}{20}$ **9.** $0.\overline{3}$, $\frac{1}{3}$ **11.** 0.625, $\frac{5}{8}$ **13.** 0.052, $\frac{13}{250}$ **15.** 0.0012, $\frac{3}{2500}$ **17.** 8%, $\frac{2}{25}$ **19.** 150%, $1\frac{1}{2}$ **21.** 5%, $\frac{1}{20}$ **23.** 480%, $4\frac{4}{5}$ **25.** 375%, $3\frac{3}{4}$ **27.** 52%, $\frac{13}{25}$ **29.** 0.5%, $\frac{1}{200}$ **31.** 0.7, 70% **33.** 0.44, 44% **35.** 0.375, 37.5% **37.** 5.125, 512.5% **39.** 0.875, 87.5% **41.** 0.833, 83.3% **43.** 0.533, 53.3% **45.** 1.417, 141.7%

USING PERCENT (p. 786) **1.** $16\frac{2}{3}\%$ **3.** 37.5% **5.** $66\frac{2}{3}\%$ **7.** $16\frac{2}{3}\%$ **9.** 18.75% **11.** 9 **13.** 48 **15.** 38 **17.** 42 **19.** \$1.04 **21.** \$3.38

RATIO AND RATE (p. 787) **1.** 15 to 4 **3.** $\frac{3}{5}$ **5.** $\frac{12}{17}$ **7.** 17:3 **9.** 2 to 3 **11.** 15 to 7 **13.** 1 to 4 **15.** \$22.50 per ticket **17.** 52 miles per hour **19.** 28 miles per gallon **21.** \$.10 per minute **23.** $21.\overline{6}$ meters per second **25.** \$1.12 per mile

COUNTING METHODS (p. 789) **1.** 9 outfits **3.** 56,700 pairs **5.** 10,000 numbers **7.** 67,600 PINs; 58,500 PINs

PERIMETER, AREA, AND VOLUME (p. 791) **1.** 34 units **3.** 84 ft **5.** 72 ft **7.** 841 yd^2 **9.** 12.25 $in.^2$ **11.** 51.84 cm^2 **13.** 2025 km^2 **15.** 15,625 ft^3 **17.** 420 yd^3 **19.** 212 $in.^3$

DATA DISPLAYS (p. 794) **1–9 odd:** Sample answers are given. **1.** 0–25 with increases of 5 **3.** 0–20 with increases of 5

5.

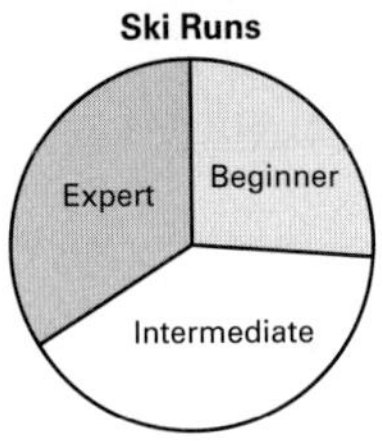

7.

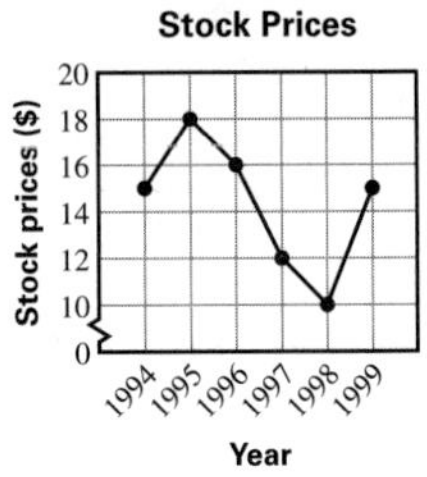

9.

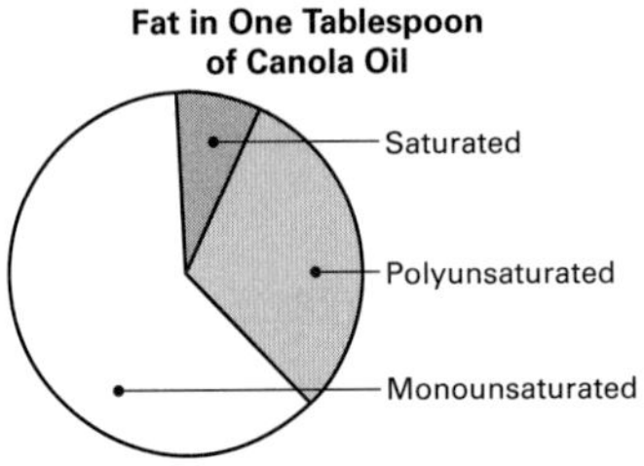

PROBLEM SOLVING (p. 796) **1.** 5 sandwiches, 3 cartons of milk **3.** \$26.25 **5.** no later than 6:25 A.M. **7.** 10 groups **9.** The problem cannot be solved; not enough information is given.

EXTRA PRACTICE

CHAPTER 1 (p. 797) **1.** 105 **3.** 18 **5.** 526 **7.** 75 **9.** 1536 **11.** 216 **13.** 83 **15.** 15 **17.** 31 **19.** 7 **21.** 3 **23.** 3 **25.** 7 **27.** 24 **29.** solution **31.** solution **33.** solution **35.** not a solution **37.** $25n - 13 = 37$

39.

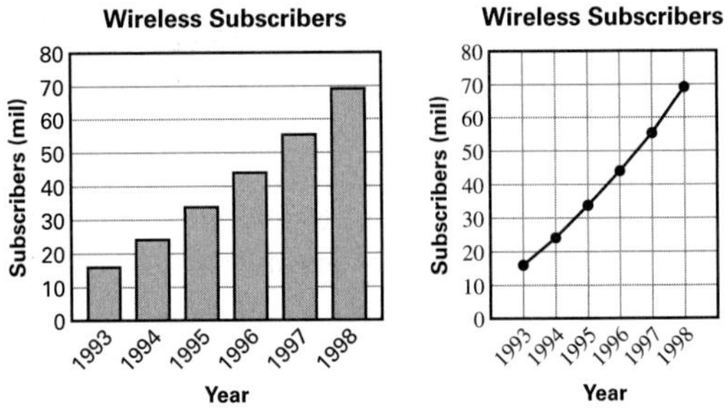

41. Input: 0, 1, 2, 3; Output: 4.5, 5.5, 6.5, 7.5

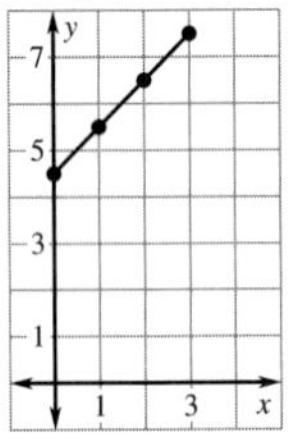

43. Input: 0, 1, 2, 3; Output: 0, 1, 2, 3

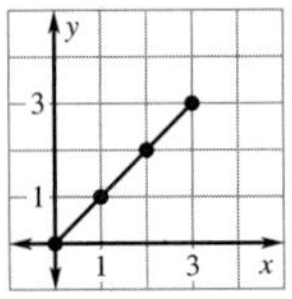

CHAPTER 2 (p. 798)

1.

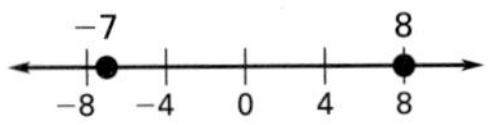

$-7 < 8, 8 > -7$

3.

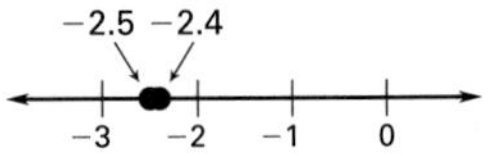

$-2.5 < -2.4, -2.4 > -2.5$

5. 8.5 **7.** 7 **9.** 5 **11.** −2 **13.** 3 **15.** −3 **17.** −13 **19.** 4 **21.** 7.3 **23.** −12 **25.** $\begin{bmatrix} 6 & 2 \\ 8 & 8 \end{bmatrix}$ **27.** $\begin{bmatrix} -3 & 2 & 10 \\ 0 & -3 & 14 \end{bmatrix}$ **29.** −45 **31.** $32x$ **33.** $-c^4$ **35.** $4a^4$ **37.** $4a - 24$ **39.** $-r^2 + 5r$

41. $6x^2 + 24x$ **43.** $3z - 8.4$ **45.** $3.1m$ **47.** $2 - 4t$ **49.** $6x^2 + 5$
51. $\frac{1}{2}x$ **53.** -4 **55.** $\frac{3x}{4}$ **57.** $-32x$ **59.** 66 **61.** 1 to 4

CHAPTER 3 (p. 799)

1. 14 **3.** 16 **5.** -3 **7.** 3 **9.** 5 **11.** -20 **13.** $\frac{1}{3}$ **15.** -4 **17.** 4 **19.** 3 **21.** 40 **23.** $-\frac{4}{3}$
25. 1 **27.** all real numbers **29.** no solution
31.

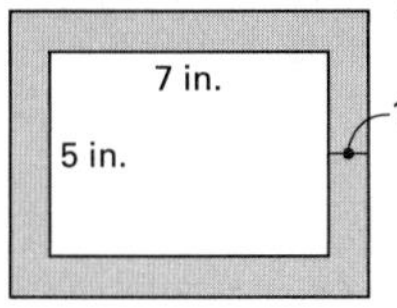

7 in., 9 in. **33.** -7.46

35. 0.27 **37.** 1.01 **39.** $y = -x + 3$; 5, 4, 3, 2 **41.** $y = \frac{3}{2}x + \frac{3}{4}$; $-2\frac{1}{4}, -\frac{3}{4}, \frac{3}{4}, 2\frac{1}{4}$ **43.** $y = x - 12$; $-14, -13, -12, -11$
45. \$.60 per yogurt snack **47.** 48.8 mi/h **49.** 0.75 cup
51. 64% **53.** 36%

CHAPTER 4 (p. 800)

1.

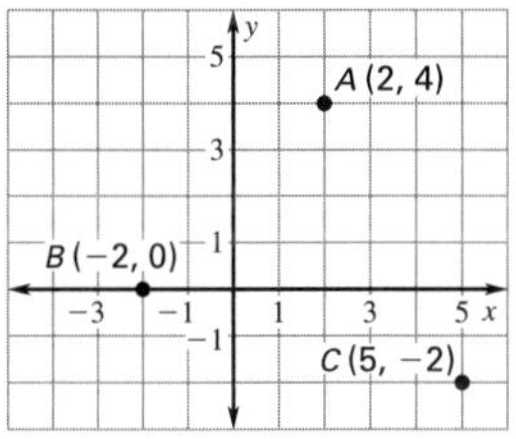

7.

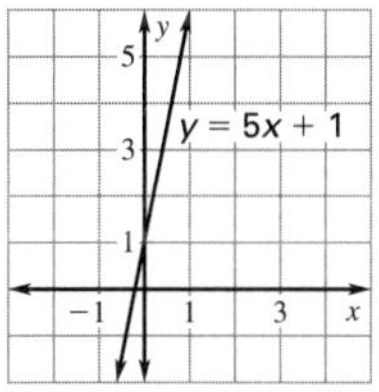

15. $-1, -5$

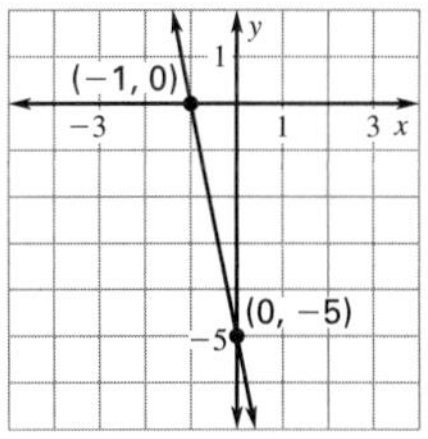

17. $2\frac{1}{2}, -5$ **19.** $-7, 14$ **21.** 2, 6 **23.** 0 **25.** 2 **27.** -7 **29.** -1 **31.** $y = 3x$

33. $y = -\frac{7}{8}x$ **35.** $y = x$ **37.** $y = \frac{4}{5}x$ **39.** \$525 **41.** $y = 3$
43. $y = -4$ **45.** $y = x + 1$ **47.** $y = 3x - 4$

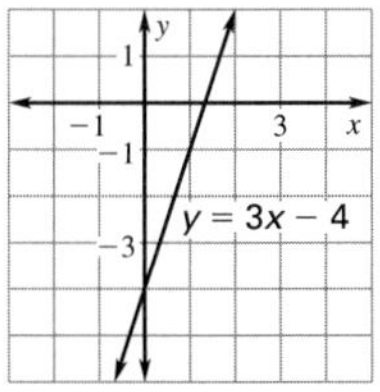

49. -6 **51.** 2 **53.** 3 **55.** $\frac{1}{3}$
57. not a function **59.** not a function

CHAPTER 5 (p. 801)

1. $y = 2x + 1$ **3.** $y = x - 3$ **5.** $y = 3x + 3$
7. $y = 6$ **9.** $y = 4x + 11$ **11.** $y = \frac{1}{2}x - 2$ **13.** $y = 3x - 11$
15. $y = \frac{7}{5}x - 6$ **17.** $y = 7$ **19.** $y = -x + 2$
21.

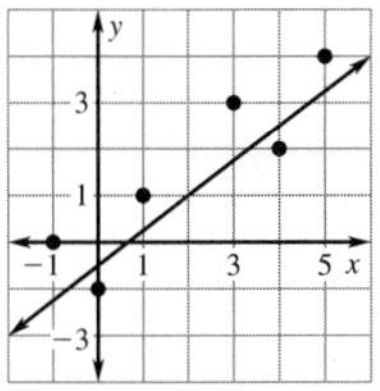

Sample answer: $y = \frac{3}{4}x - \frac{1}{2}$

23. $y - 3 = -2(x - 5)$; $y = -2x + 13$ **25.** $y - 4 = 5(x - 5)$; $y = 5x - 21$ **27.** $y + 5 = -\frac{5}{3}(x + 3)$; $y = -\frac{5}{3}x - 10$

29. $y - 4 = -\frac{1}{2}(x - 2)$; $y = -\frac{1}{2}x + 5$ **31.** $3x - y = 17$
33. $3x - 4y = -3$ **35.** $y = 7$ **37.** $5x - y = -5$
39.

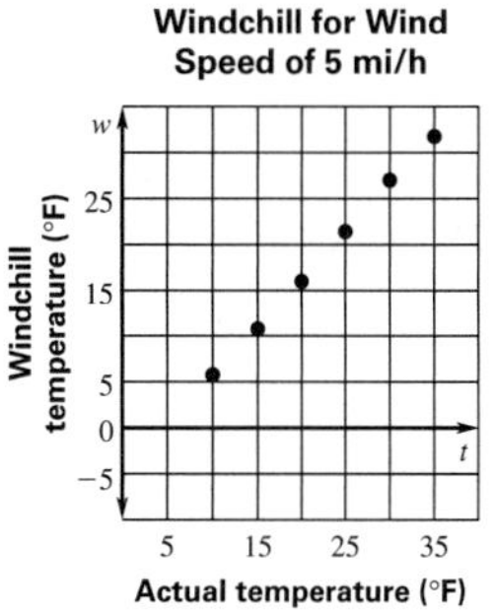

41. Solution may vary depending upon method used to derive the linear model; -1°F.

CHAPTER 6 (p. 802)

1. $x > -9$

3. $y \le 7$

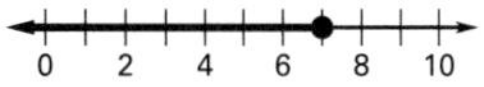

5. $k \le 18$ **7.** $x > -2$
9. $x \ge -12$ **11.** $x \ge -1$ **13.** $x < -2$
15. $4 \ge x > -1$

17. $-6 < x < 5$

19. $x > 3$ or $x \le 1$

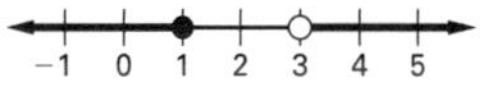

21. $-14, -6$
23. 1, 7 **25.** $x \le -8$ or $x \ge 5\frac{1}{3}$ **27.**

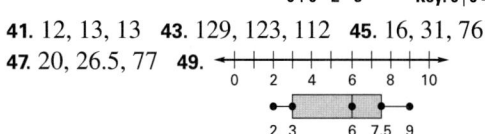

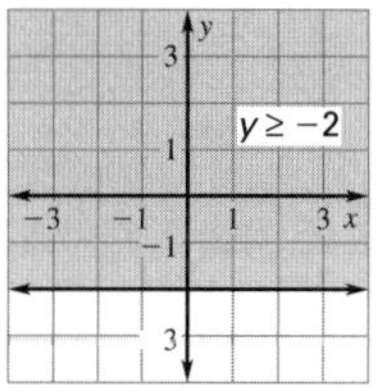

35. $2x + 1.25y \le 20$

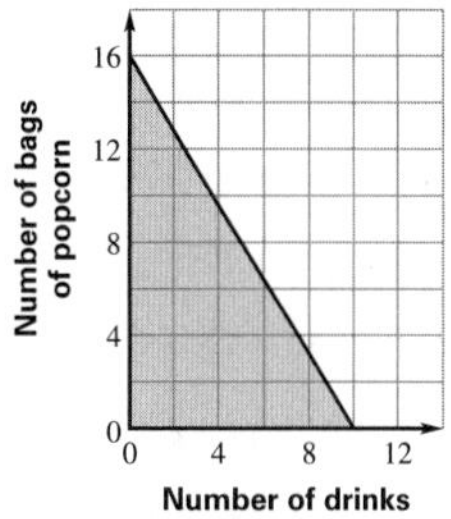

37.

1	3 6
2	1 2
3	1 3 5
4	0 1 5
5	1
6	5 6

Key: 6 | 5 = 65

39.

0	2 8
1	9
2	1 2 7
3	8 9 9
4	1 3 4 6 9
5	1 4 6
6	0 2 3

Key: 6 | 0 = 60

41. 12, 13, 13 **43.** 129, 123, 112 **45.** 16, 31, 76
47. 20, 26.5, 77 **49.**

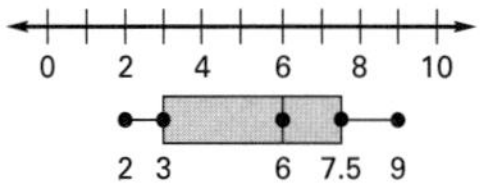

CHAPTER 7 (p. 803)

1. $(-2, 5)$ **3.**

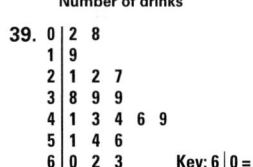

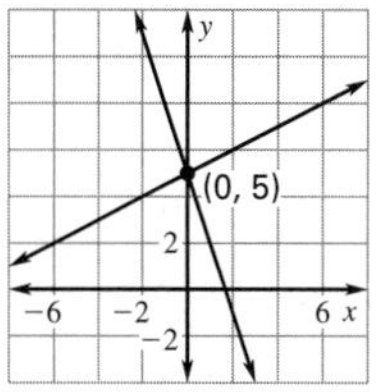

SELECTED ANSWERS

5. (–1, –4) **7.** (4, 5) **9.** (–5, –1) **11.** (6, 3) **13.** $\left(5\frac{1}{7}, \frac{1}{7}\right)$
15. $\left(15\frac{1}{2}, -24\right)$ **17.** (2, 0) **19.** (0, –2) **21.** (13, –2)
23. (4, –28) **25.** 3 student, 5 adult **27.** Company A
29. no solution **31.** one solution, (0, –5)
33. infinitely many solutions **35.** no solution

37.

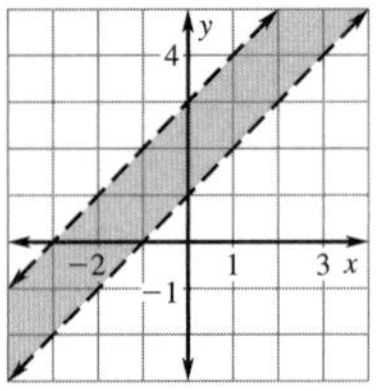

41.

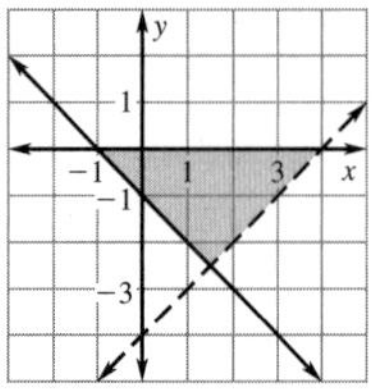

CHAPTER 8 (p. 804) **1.** 7^5, or 16,807 **3.** $1728x^3$
5. $-128r^4s^8$ **7.** $675x^3y^2$ **9.** $\frac{1}{m^4}$ **11.** x^2y **13.** $\frac{x^4}{y^3}$ **15.** $6x^4$

17.

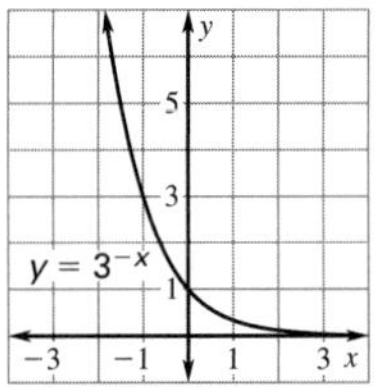

19.

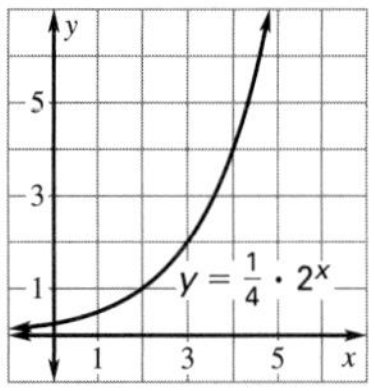

21. 8

23. $\frac{27x^3z^9}{8}$ **25.** $\frac{1}{10a^4bc^5}$ **27.** $\frac{25y^{17}}{x^7}$ **29.** $\frac{1}{216} \approx 0.005$
31. 31,100 **33.** 9.43 **35.** 0.0012468 **37.** 60,901,300,000
39. 3.78×10^{-2} **41.** 3.3×10^7 **43.** 2.08054×10^{-1}
45. 8.91×10^{-4} **47.** \$1791.78 **49.** \$3724.99
51. exponential decay; 0.5; 50% **53.** exponential growth; 1.04; 4% **55.** exponential decay; $\frac{2}{3}$; $33\frac{1}{3}$%
57. exponential decay; 0.68; 32%

CHAPTER 9 (p. 805) **1.** –10 **3.** –0.5 **5.** 19.47 **7.** 14.83
9. –5, 5 **11.** no solution **13.** –4, 4 **15.** $-\sqrt{3}, \sqrt{3}$
17. about 3.06 sec **19.** 168 **21.** $-84\sqrt{5}$ **23.** 2 **25.** $\frac{\sqrt{10}}{3}$

27.

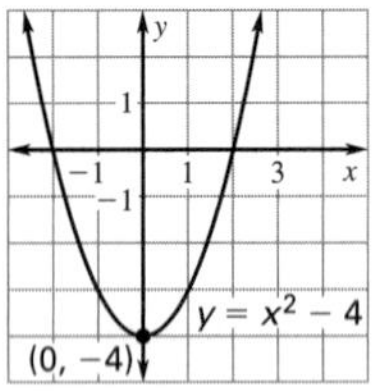

33.

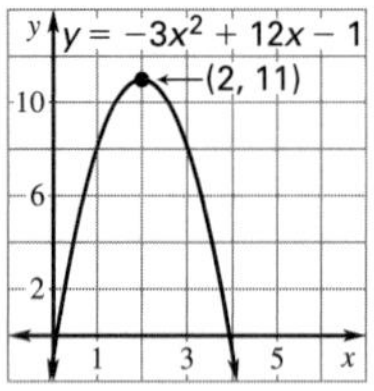

35. –3, –2 **37.** –6, 6 **39.** –3, 3 **41.** –1, 3
43. $3r^2 + 8r + 2 = 0$; $\frac{-4 - \sqrt{10}}{3}, \frac{-4 + \sqrt{10}}{3}$
45. $-x^2 + 5x - 4 = 0$; 1, 4
47. $2x^2 - 6x - 5 = 0$; $\frac{3 - \sqrt{19}}{2}, \frac{3 + \sqrt{19}}{2}$
49. $-2x^2 + x + 10 = 0$; $-2, \frac{5}{2}$ **51.** 0; one solution
53. 0; one solution **55.** –136; no real solution
57. –47; no real solution **67.** quadratic

CHAPTER 10 (p. 806) **1.** $8x^2 + 1$ **3.** $14x^2 - 7x + 8$
5. $x^2 + 9x - 4$ **7.** $4x^3 - 8x^2 + 7x$ **9.** $15b^5 - 10b^4 + 5b^2$
11. $d^2 + 4d - 5$ **13.** $x^3 + x^2 + 18$ **15.** $x^2 + 18x + 81$
17. $a^2 - 4$ **19.** $16x^2 + 40x + 25$ **21.** $4a^2 - 9b^2$
23. –3, –6 **25.** 1, –5 **27.** $\frac{3}{2}$, 7 **29.** 1, 2
31. 6, 8; (7, –1)

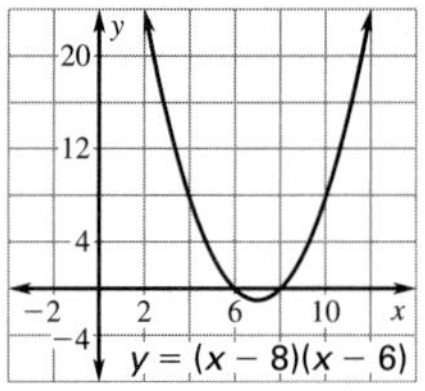

33. 5, 7; (6, –1)
35. 5, 9; (7, 4)
37. –1, 3; (1, –4)

39. –3 **41.** 6 **43.** 6, 9 **45.** –4, 6 **47.** $-2, 1\frac{1}{2}$ **49.** $-\frac{4}{3}$
51. $-\frac{3}{4}$, 2 **53.** $-2\frac{1}{3}$, 8 **55.** $(x + 1)(x - 1)$; difference of two squares **57.** $(11 - x)(11 + x)$; difference of two squares **59.** $(t + 1)^2$; perfect square trinomial
61. $(8y + 3)^2$; perfect square trinomial **63.** $x^2(x + 3)(x - 3)$
65. $x^2(x + 9)(x - 5)$ **67.** $-3y(y + 4)(y + 1)$ **69.** $7x^4(x^2 - 3)$
71. 2 in. by 6 in. by 15 in.

CHAPTER 11 (p. 807) **1.** –4, 4 **3.** 4 **5.** 6 **7.** –6, 1 **9.** 24
11. 21 lb **13.** 252 **15.** inverse variation; $xy = 2$
17. $\frac{2x^3}{7}$ **19.** $\frac{1}{x + 1}$ **21.** $\frac{1}{4}$ **23.** $\frac{1}{2}$ **25.** $\frac{x^2}{5}$ **27.** $\frac{5}{8x^2}$
29. $\frac{5x}{x - 8}$ **31.** $\frac{1}{x}$ **33.** $\frac{-4x - 61}{(x + 7)(x - 4)}$ **35.** $\frac{63x - 2}{18x^2}$
37. $2x^2 + x + \frac{1}{7}$ **39.** $2(2x - 1)$ **41.** –4, 10 **43.** 3 **45.** 6

CHAPTER 12 (p. 808)
1. all nonnegative numbers; all nonnegative numbers
3. all numbers greater than than or equal to –3; all nonnegative numbers

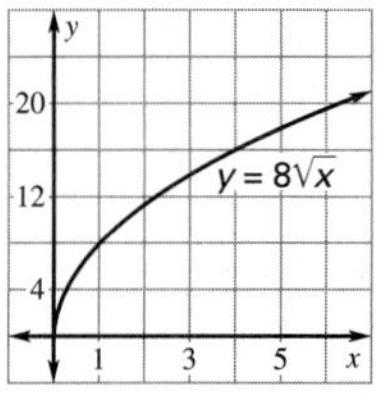

5. all numbers greater than or equal to 2; all nonnegative numbers **7.** all numbers greater than or equal to $-\frac{2}{3}$; all nonnegative numbers **9.** $5\sqrt{5}$ **11.** $16\sqrt{2}$
13. $7\sqrt{3} - 3\sqrt{2}$ **15.** $\frac{\sqrt{6}}{3}$ **17.** 121 **19.** 5 **21.** 2, 6
23. –14, 4 **25.** 2, 5 **27.** $\frac{6 - \sqrt{111}}{5}, \frac{6 + \sqrt{111}}{5}$ **29.** $c = \sqrt{2}$
31. $a = 8$ **33.** $a = 20$ **35.** 4.12; $\left(2, 4\frac{1}{2}\right)$ **37.** 5; (2.5, –2)
39. 8; (3, –6) **41.** 18.36; $\left(4, -2\frac{1}{2}\right)$ **43.** $\sin D = \frac{4}{5}$, $\cos D = \frac{3}{5}$, $\tan D = \frac{4}{3}$; $\sin E = \frac{3}{5}$, $\cos E = \frac{4}{5}$, $\tan E = \frac{3}{4}$ **45.** $\sin D = \frac{\sqrt{2}}{2}$, $\cos D = \frac{\sqrt{2}}{2}$, $\tan D = 1$; $\sin E = \frac{\sqrt{2}}{2}$, $\cos E = \frac{\sqrt{2}}{2}$, $\tan E = 1$
47. $a = 25.37$, $b = 9.23$ **49.** any nonzero numbers b and c and any nonzero number a except 1

SA40

TEACHER'S EDITION INDEX

Student Edition entries in black.
Teacher's Edition entries in blue.

A

M

T

U

V

X

Y

Z

ADDITIONAL ANSWERS

CHAPTER 1

1.6 PRACTICE AND APPLICATIONS (pp. 43–45)

11. Water Requirements

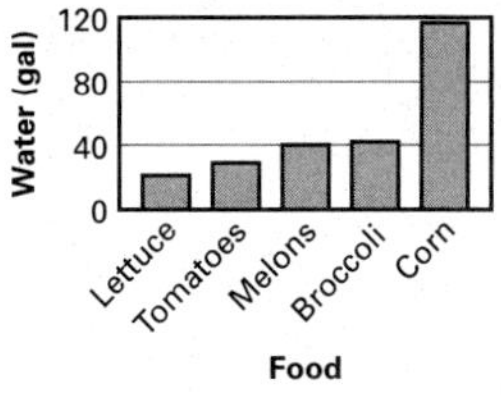

CHAPTER 2

2.3 PRACTICE AND APPLICATIONS (pp. 82–84)

59.

Input	Function	Output
–2	$y = -2 - 8$	–10
–1	$y = -1 - 8$	–9
0	$y = 0 - 8$	–8
1	$y = 1 - 8$	–7

60.

Input	Function	Output
–2	$y = 12 - (-2)$	14
–1	$y = 12 - (-1)$	13
0	$y = 12 - 0$	12
1	$y = 12 - 1$	11

61.

Input	Function	Output
–2	$y = -(-2) + 12.1$	14.1
–1	$y = -(-1) + 12.1$	13.1
0	$y = -0 + 12.1$	12.1
1	$y = -1 + 12.1$	11.1

62.

Input	Function	Output
–2	$y = -8.5 - (-(-2))$	–10.5
–1	$y = -8.5 - (-(-1))$	–9.5
0	$y = -8.5 - (-0)$	–8.5
1	$y = -8.5 - (-1)$	–7.5

63.

Input	Function	Output
–2	$y = 27 + (-2)$	25
–1	$y = 27 + (-1)$	26
0	$y = 27 + 0$	27
1	$y = 27 + 1$	28

64.

Input	Function	Output
–2	$y = -(-2) + 13 - (-2)$	17
–1	$y = -(-1) + 13 - (-1)$	15
0	$y = -0 + 13 - 0$	13
1	$y = -1 + 13 - 1$	11

77. from start of trail to Cave: 410 ft; from Cave to Oak Ridge: 1040 ft; from Oak Ridge to Trail Junction: –235 ft; from Trail Junction to Mt. Parker: 1110 ft

2.4 PRACTICE AND APPLICATIONS (pp. 89–91)

29. a.

	May	June	July
Atlanta	79.6	85.8	88.0
San Francisco	66.5	70.3	71.6
Anchorage	54.4	61.6	65.2

b.

	May	June	July
Atlanta	58.7	66.2	69.5
San Francisco	49.7	52.6	53.9
Anchorage	38.8	47.2	51.7

c. Subtract the second matrix from the first.

	May	June	July
Atlanta	20.9	19.6	18.5
San Francisco	16.8	17.7	17.7
Anchorage	15.6	14.4	13.5

2.6 CONCEPT ACTIVITY (p. 99)

2.

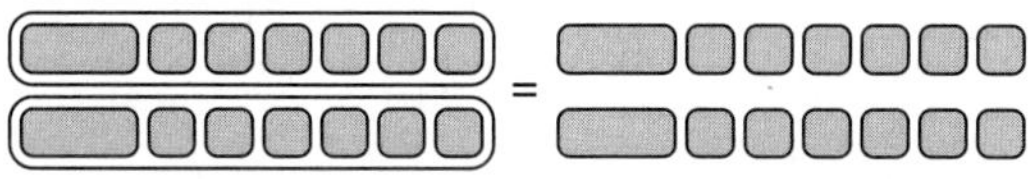

3.

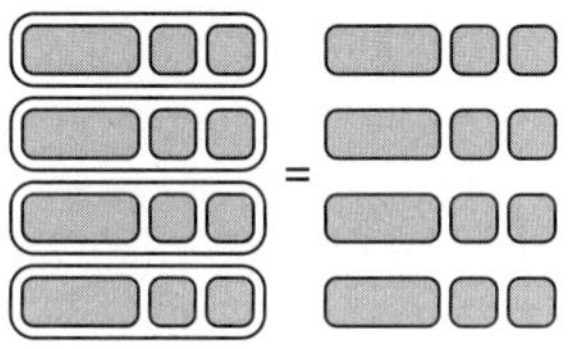

4.

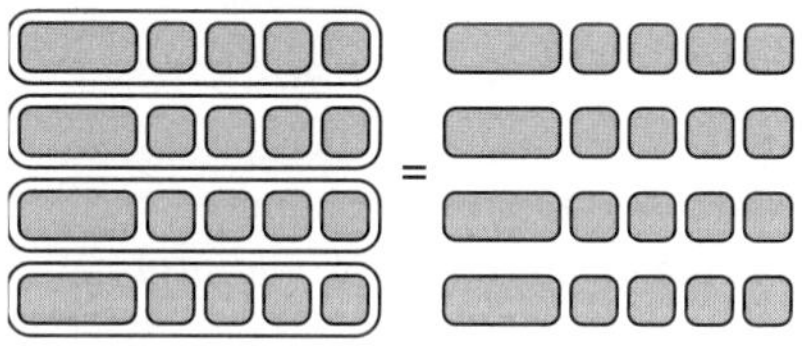

5.

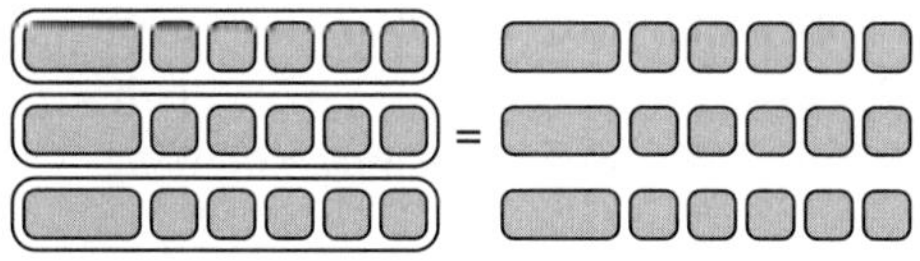

CHAPTER 3

3.5 GUIDED PRACTICE (p. 163)

1.

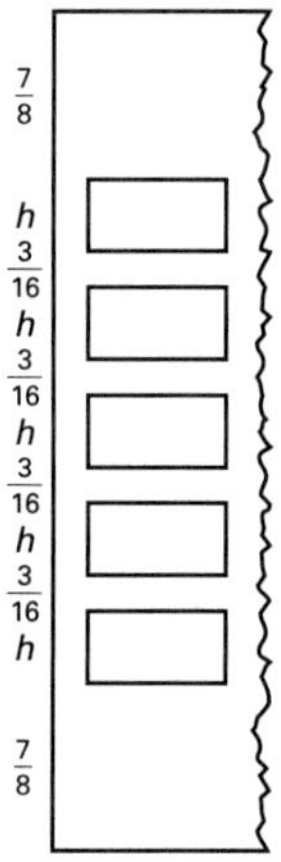

; Let h = picture height; *sample equation:* $2 \cdot \frac{7}{8} + 4 \cdot \frac{3}{16} + 5 \cdot h = 11$; 1.7 in.

3. rate your friend is driving = 52 (mi/h); time after you start driving = t (h); friend's distance when you leave = 32 (mi); rate you are driving = 60 (mi/h); $52t + 32 = 60t$.

5. *Sample answer:*

Hours	Your distance (mi)	Friend's distance (mi)
1	60	84
2	120	136
3	180	188
4	240	240

3.5 Practice and Applications (pp. 163–165)

8.

Width and height (in.)	Girth (in.)	Combined length and girth (in.)
5	20	56
8	32	68
12	48	84
16	64	100
18	72	108
20	80	116

; the package that is 18 in. wide and 18 in. high

14.

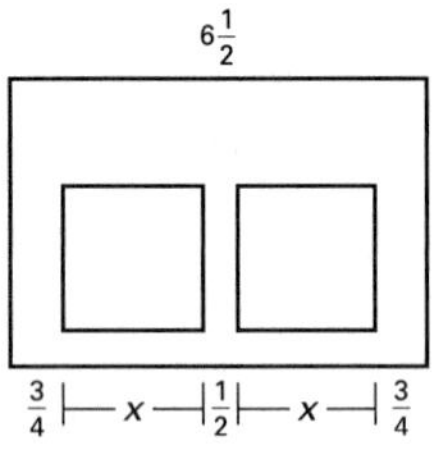

3.6 Practice and Applications (pp. 169–171)

44. *Sample answer:* A is the more accurate answer, but since masses are not usually described to the nearest hundred-millionth of a kilogram, B is a more reasonable answer and would be sufficient for most situations. **48.** No, the two methods do not give the same final result; Method 1: **a.** 13.1 m/s **b.** 786 m/min **c.** 47,160 m/h **d.** 47.2 km/h Method 2: **a.** 13.0701869 m/s **b.** 784.2112142 m/min **c.** 47,052.67285 m/h **d.** 47.05267285 km/h ≈ 47.1 km/h **60.** *Sample answer:* Rounding early produces a significant difference for an equation such as $5.24(7.85 + 0.88x) = 4.62x + 72.37$, where the final solution step is of the form $ax = b$, with $|a|$ very small and $|b|$ comparatively much larger than $|a|$.

3.8 Mixed Review (p. 186)

49.

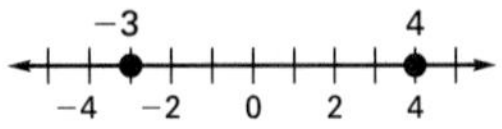

50.

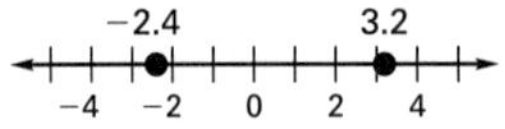

51.

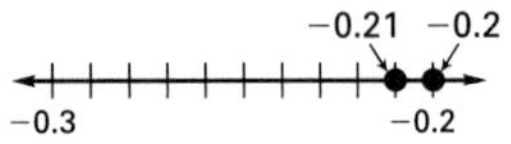

52.

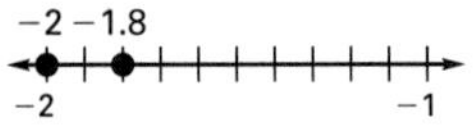

53.

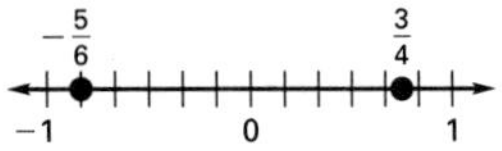

54.

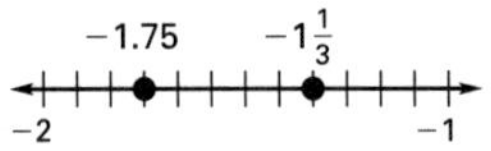

59. **Projected Number of People 85 Years and Older**

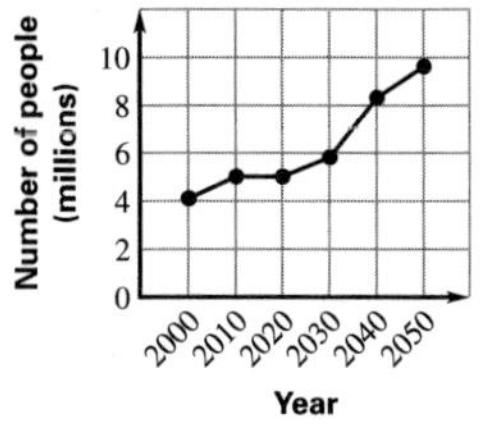

Cumulative Practice, Chapters 1–3 (pp. 196–197)

52. **Silver Produced in the U.S.**

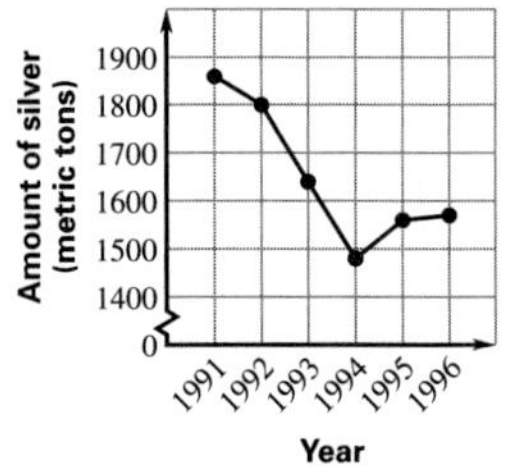

CHAPTER 4

4.1 Practice and Applications (pp. 206–208)

13.

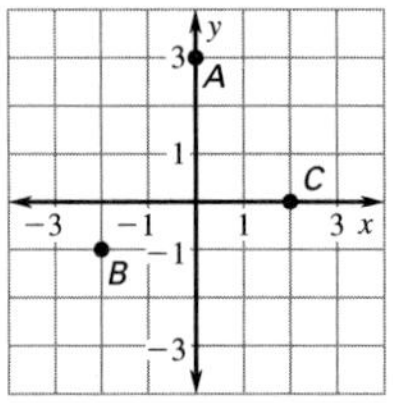

14.

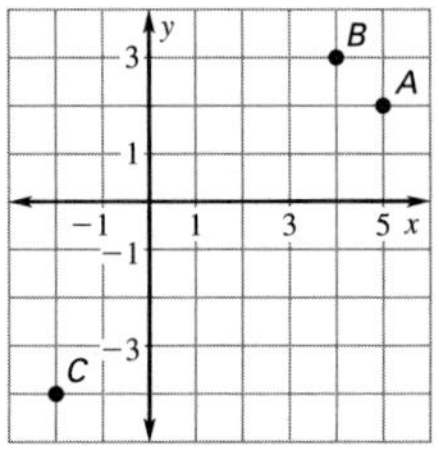

15.

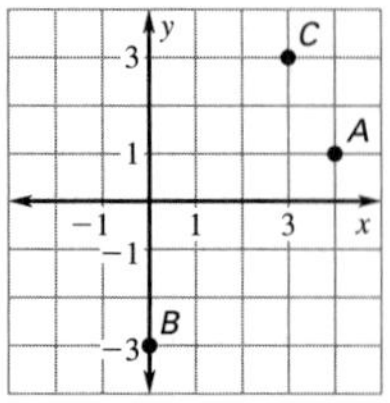

16.

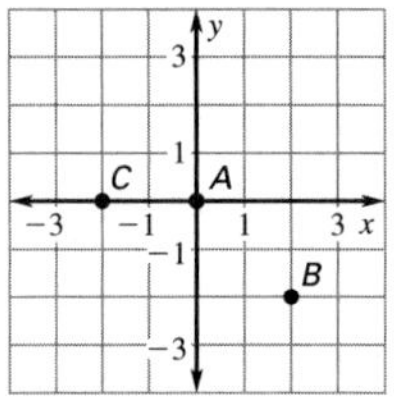

17.

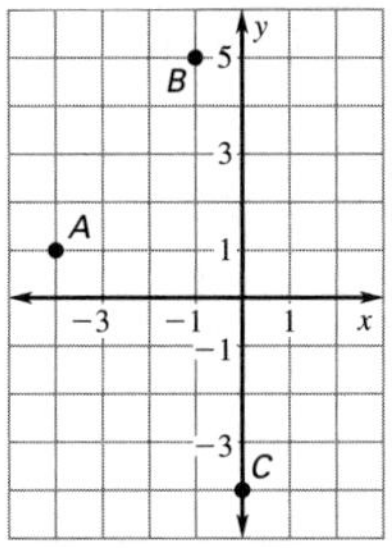

18.

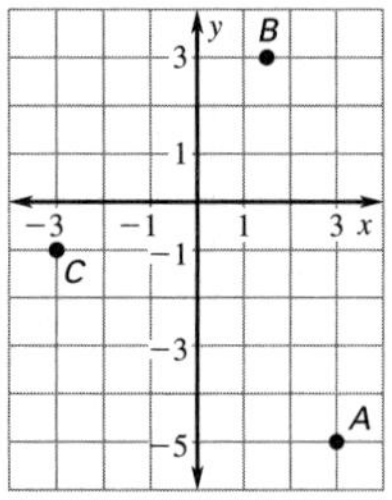

38. a. **Film Developed for Winter Olympics**

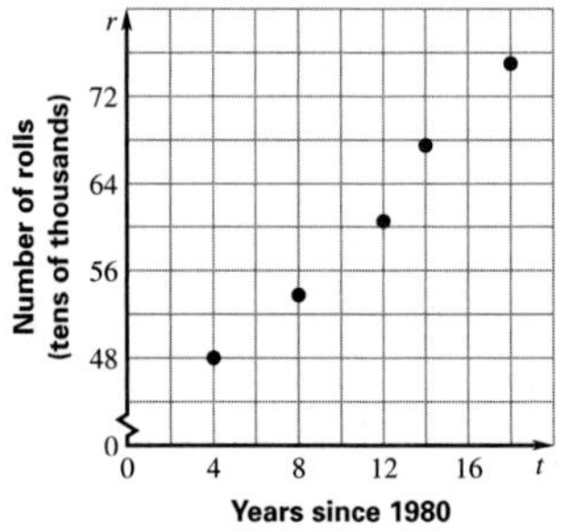

39. **CD Songs vs. Minutes**

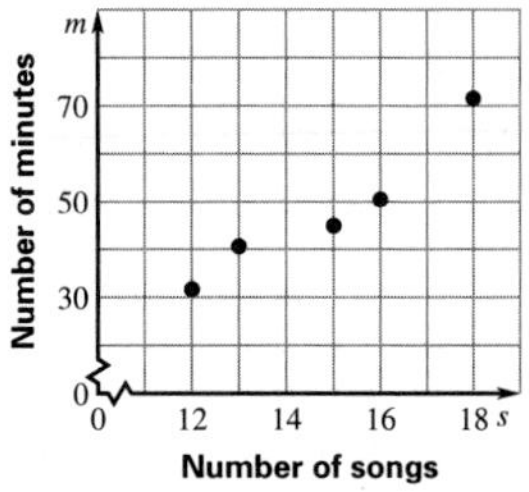

40.

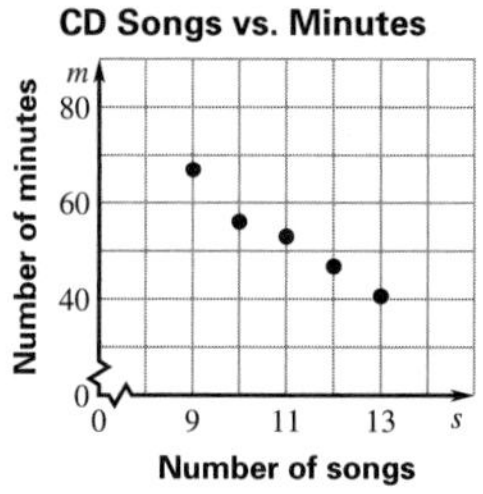

4.1 Technology Activity (p. 209)

2.

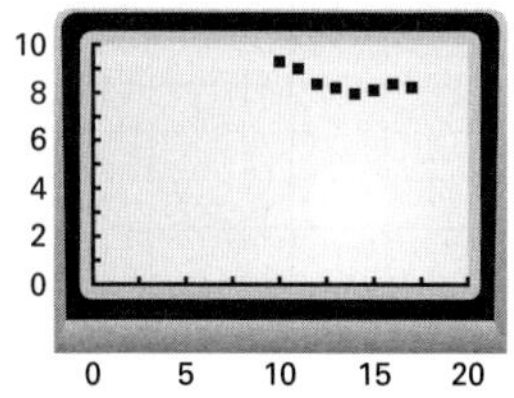

3. From ages 10–14, as age increases, time decreases. From ages 14–16, time increases, then decreases at age 17.

4.2 Practice and Applications (pp. 214–217)

37.

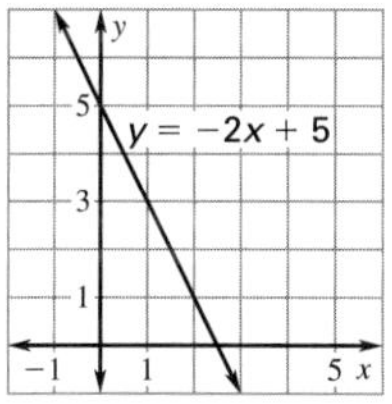

38.

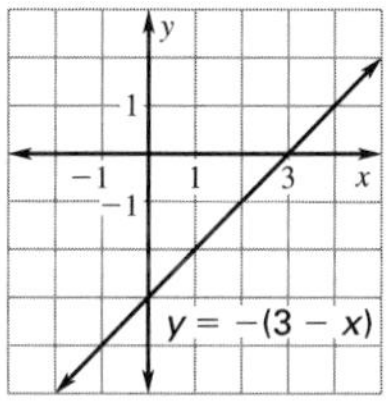

39.

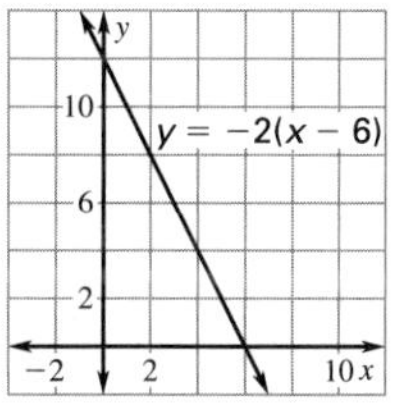

40.

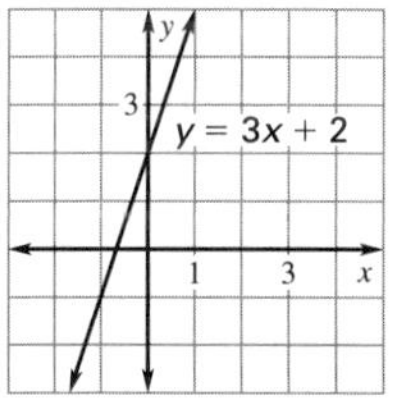

41.

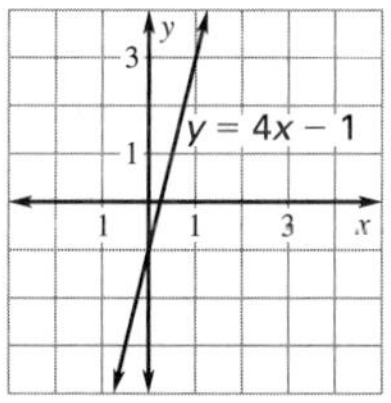

42.

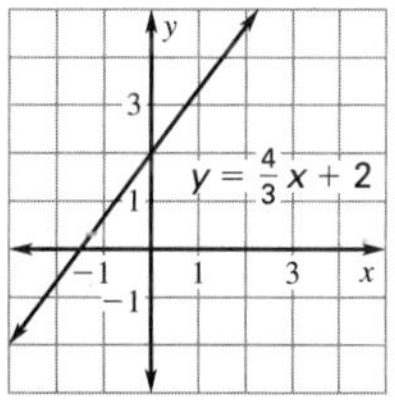

43.

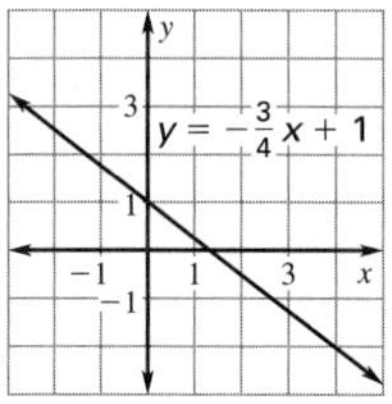

44.

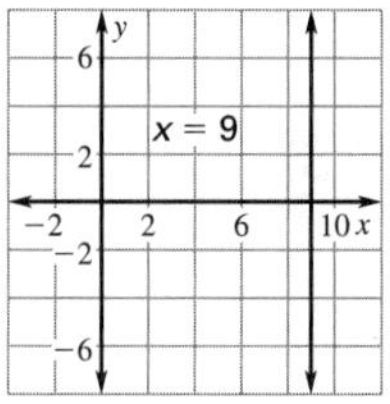

45.

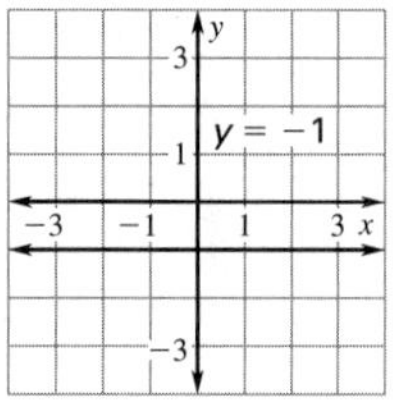

46.

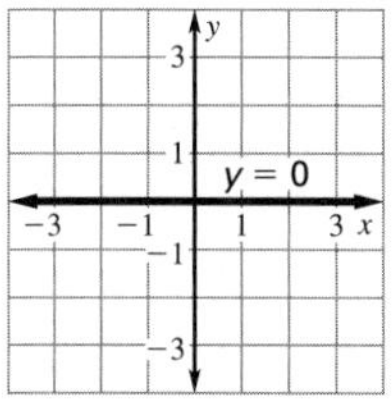

47.

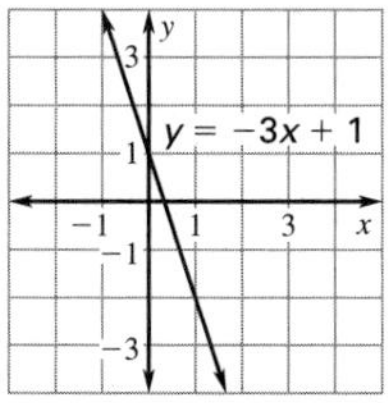

48.

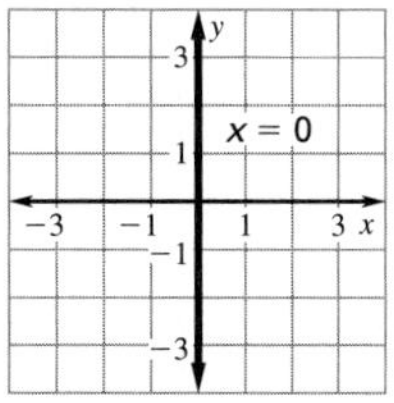

49.

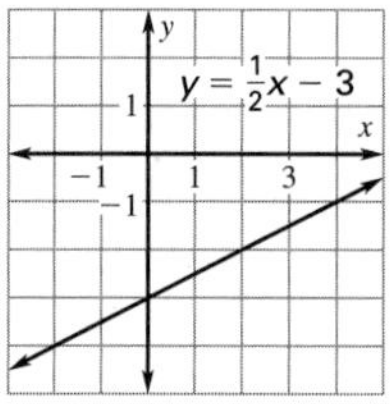

50.

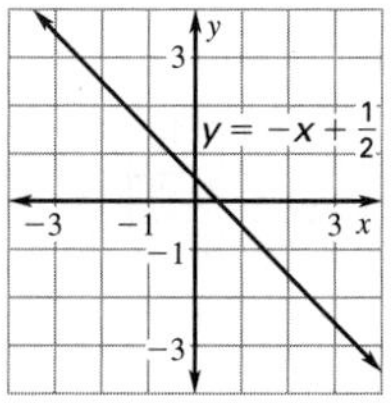

51.

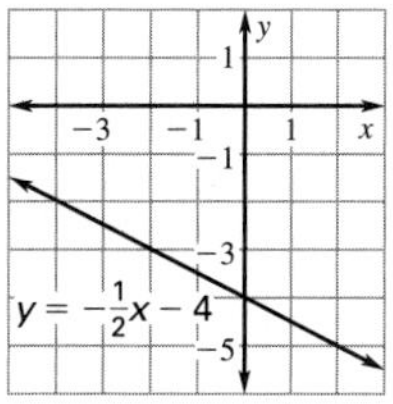

57.

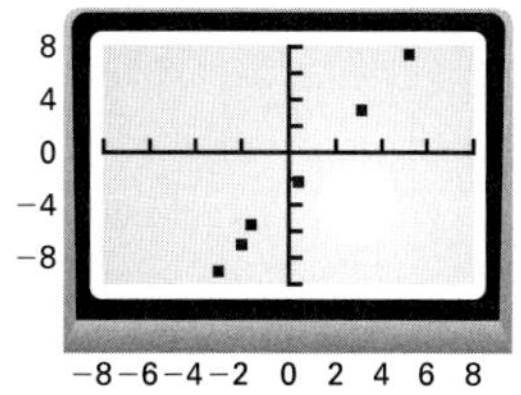

79.

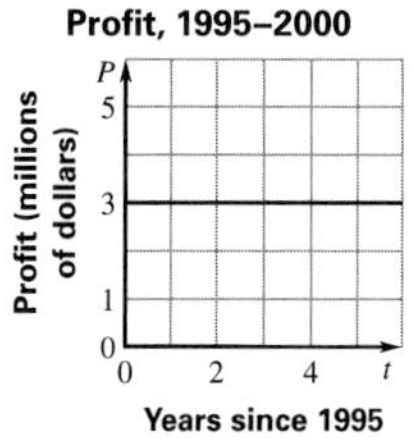

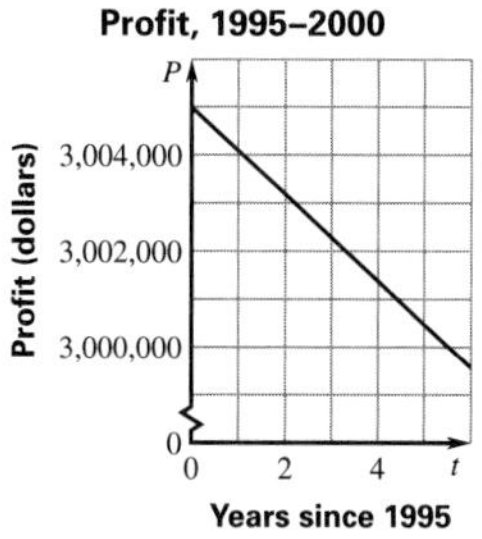

4.3 Guided Practice (p. 221)

8.

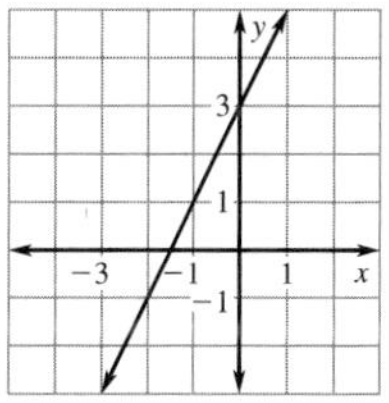

9.

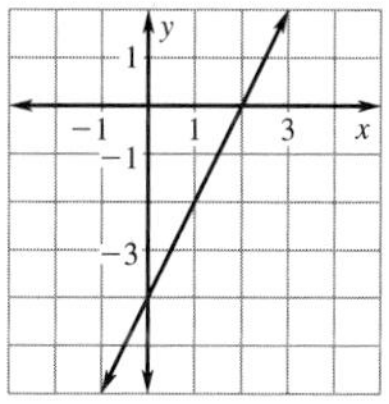

10.

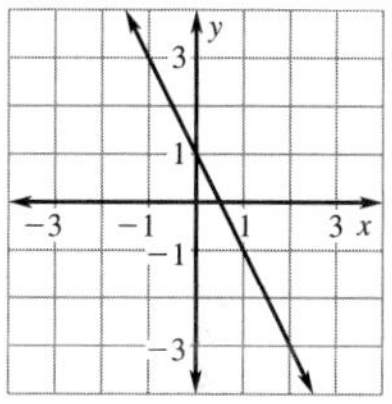

11.

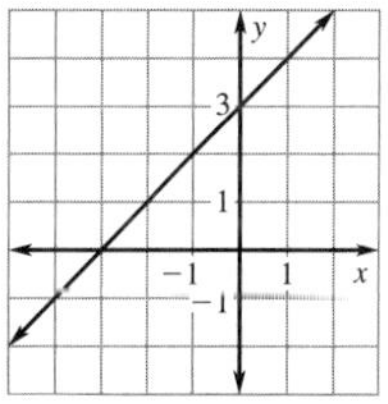

12.

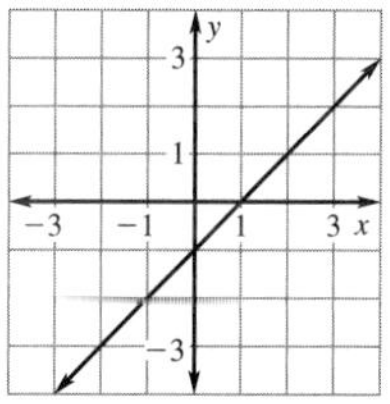

4.3 Practice and Applications (pp. 221–223)

35.

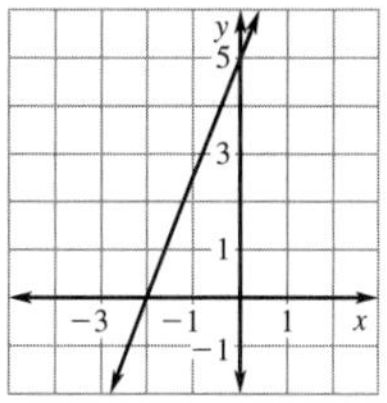

36.

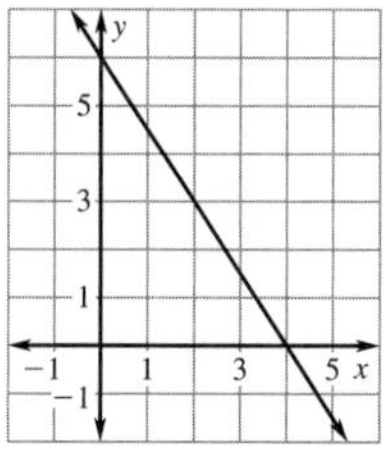

37.

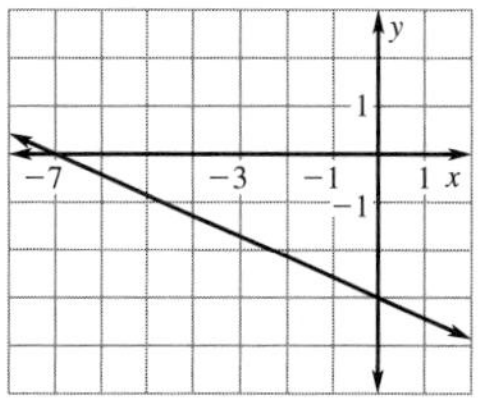

38.

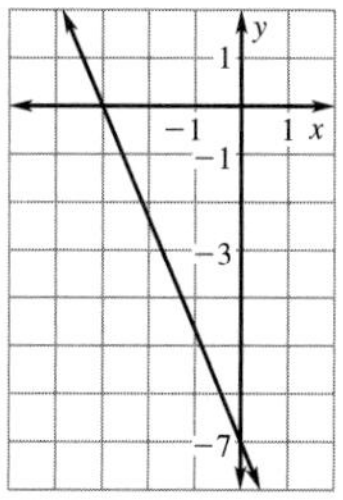

39.
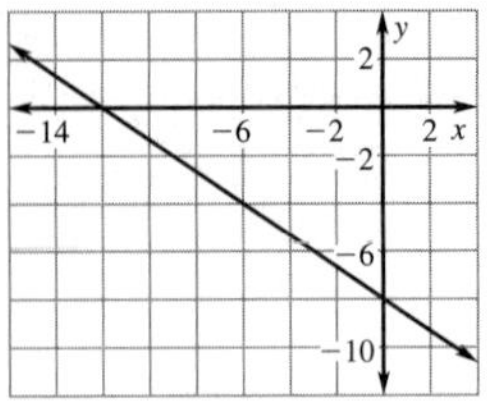

40.
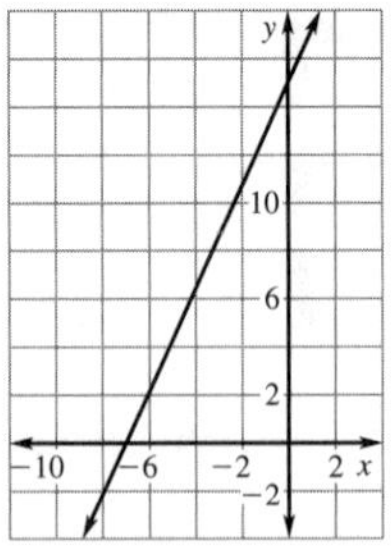

46.
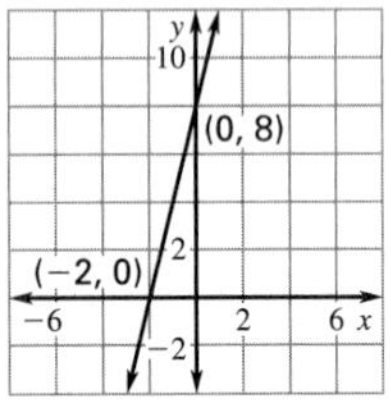

47.
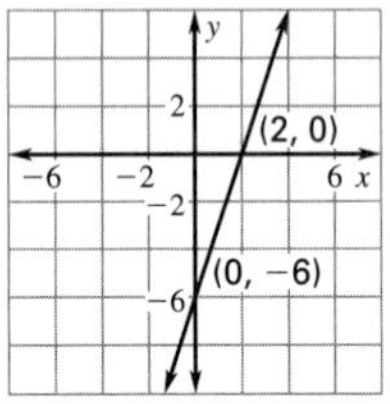

48.
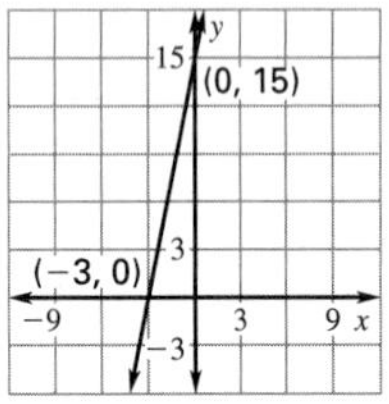

49.
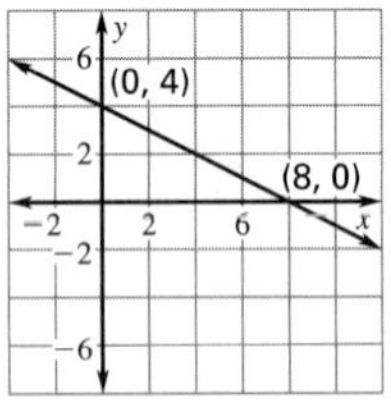

50.
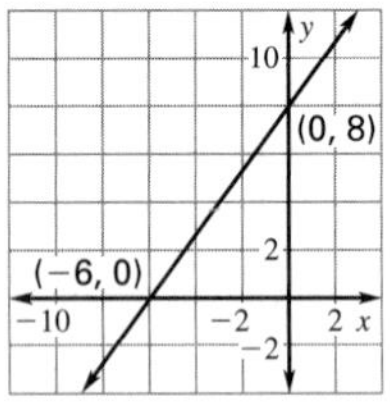

51.
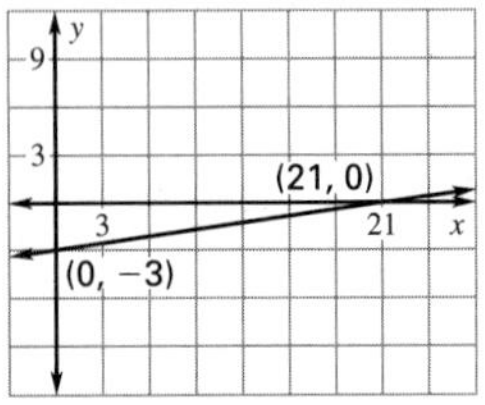

52.
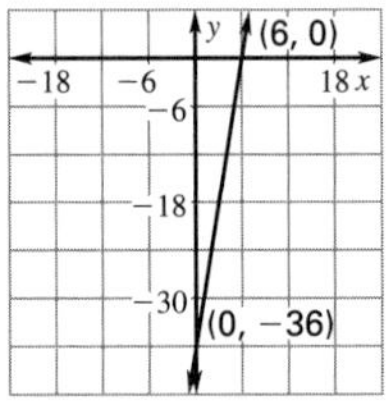

53.
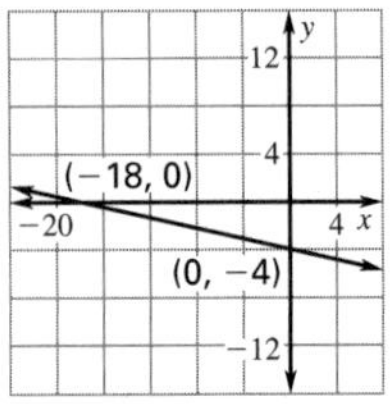

54.
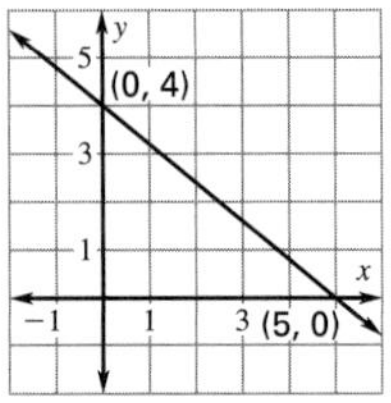

55.
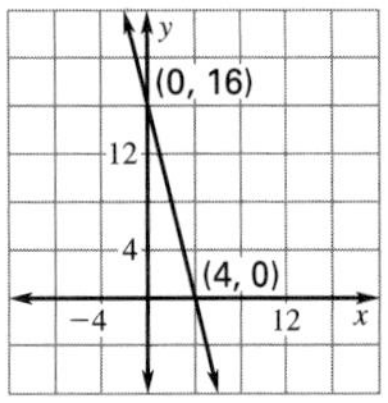

60.
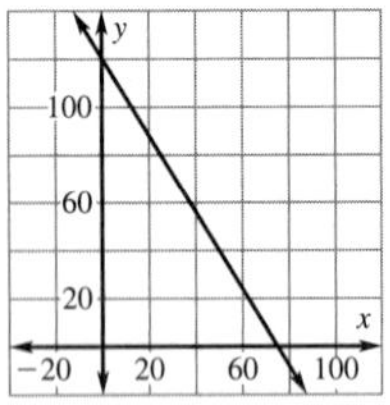

65. **Running Time vs. Walking Time**

67. **Theater Attendance**
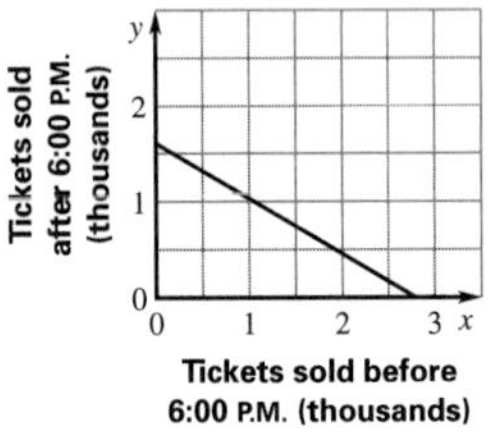

4.3 Quiz 1 (p. 224)

5.
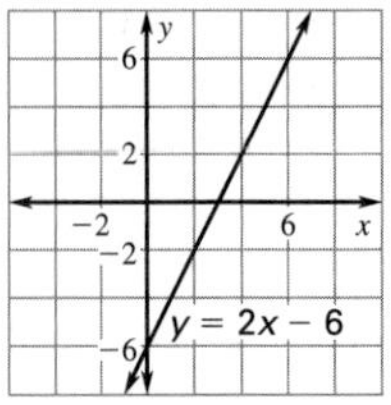

6.
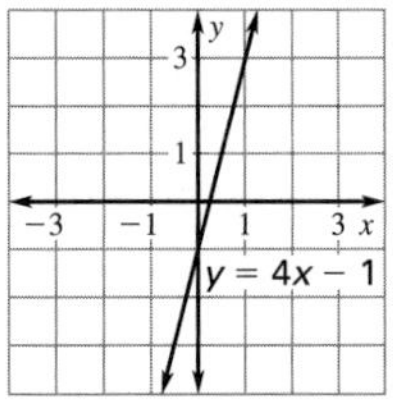

7.
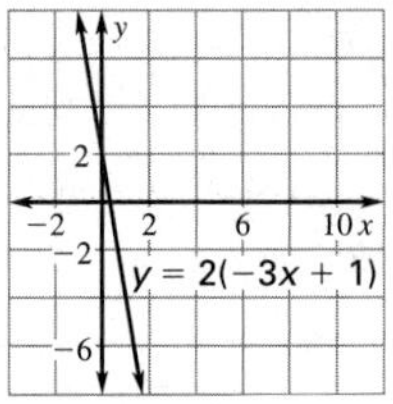

8.
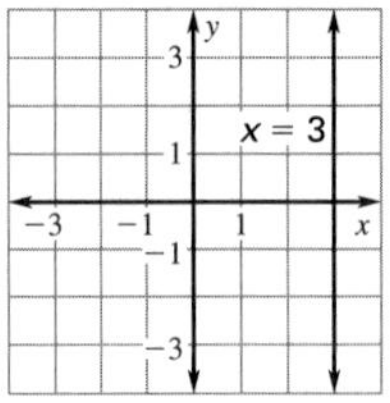

9.
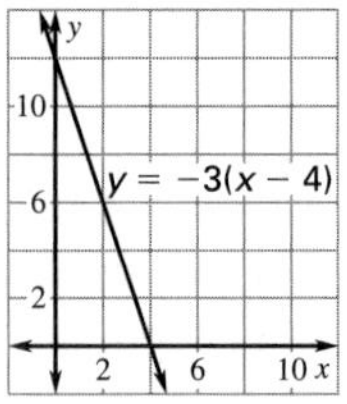

10.
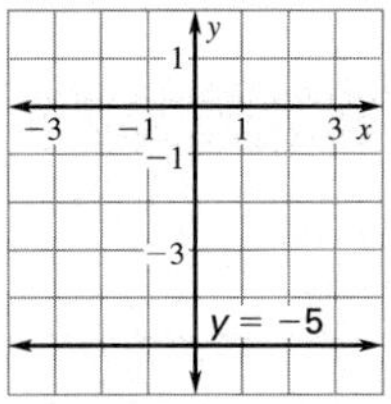

11.
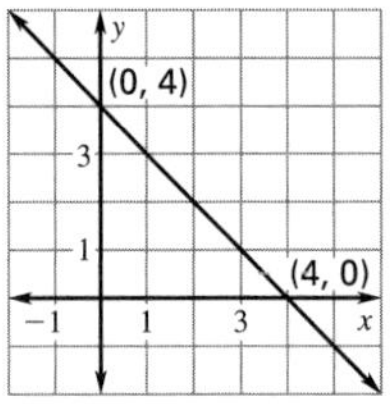

12.
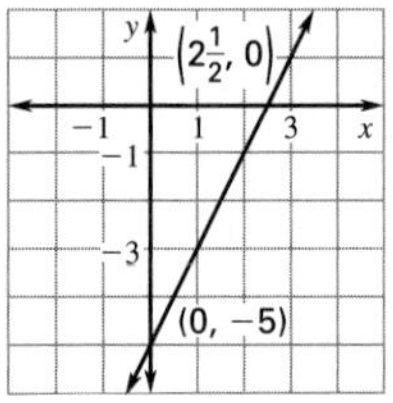

13.
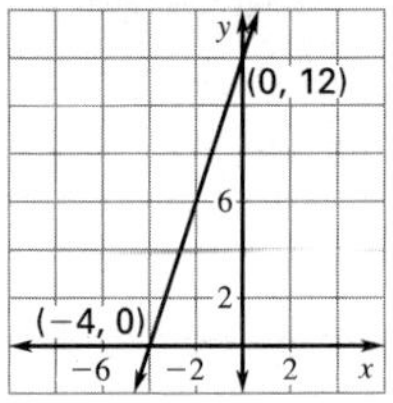

14.
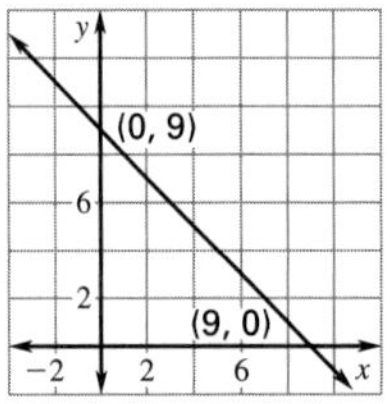

15.
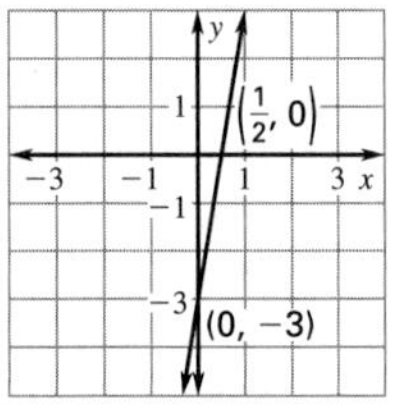

16.
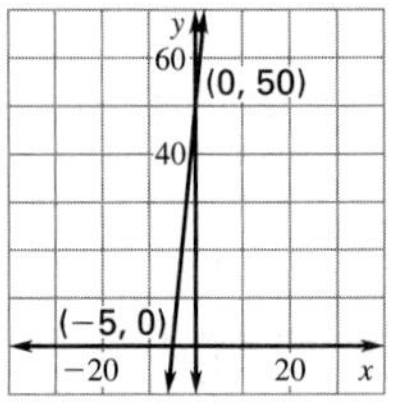

4.5 Practice and Applications (pp. 237–239)

34. **Chlorine Needed vs. Amount of Water in Pool**
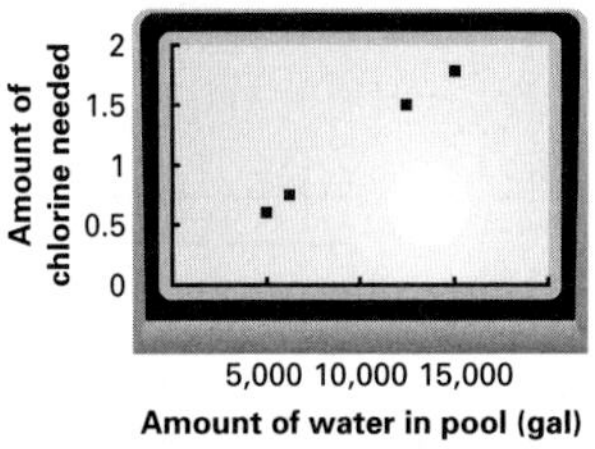

4.5 Mixed Review (p. 239)

53.
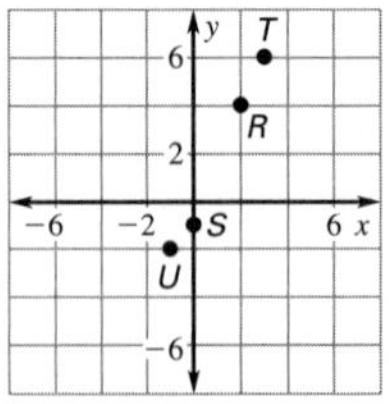

54.
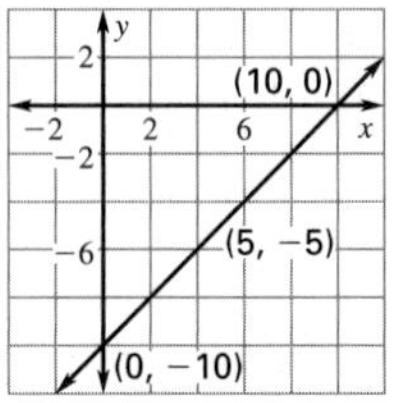

55.
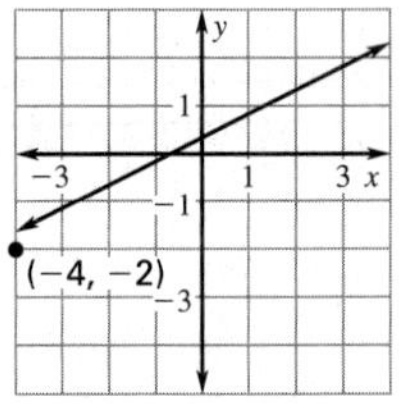

56.
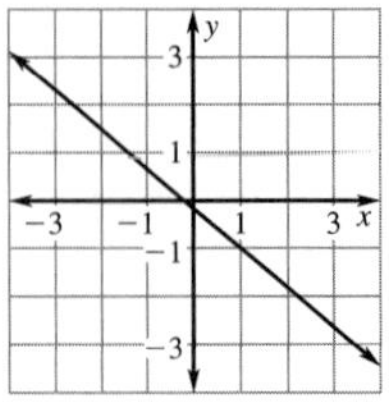

57.
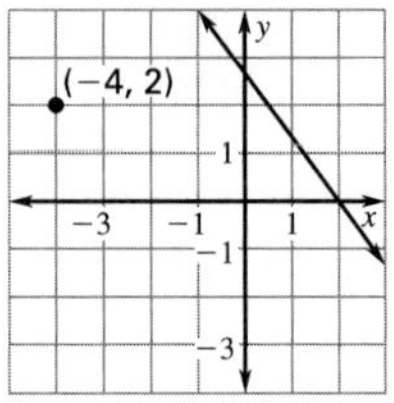

4.6 GUIDED PRACTICE (p. 244)

12. Amount of Money in Savings Account

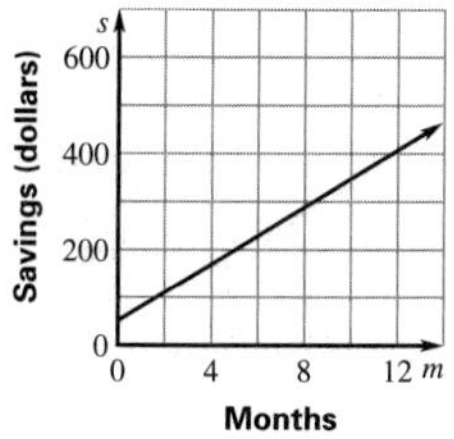

4.6 PRACTICE AND APPLICATIONS (pp. 244–246)

22.

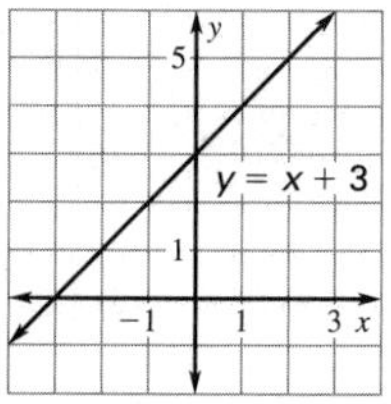

23.

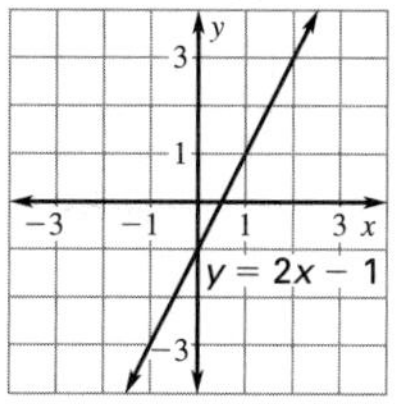

24.

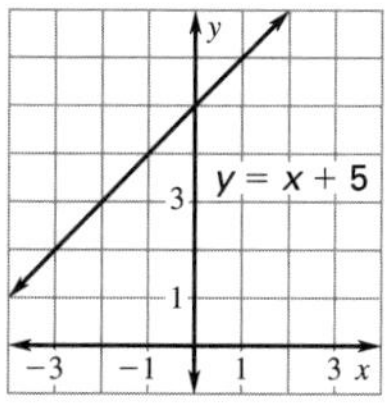

25.

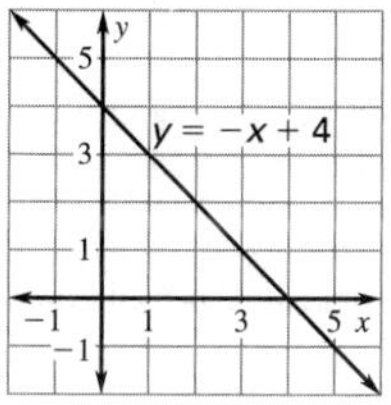

26.

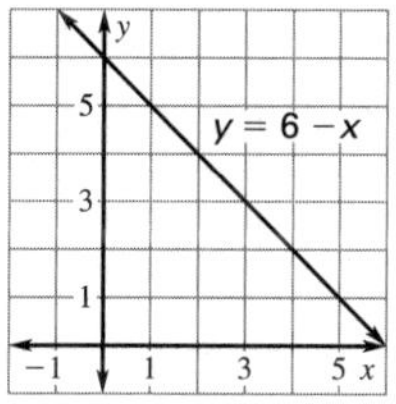

27.

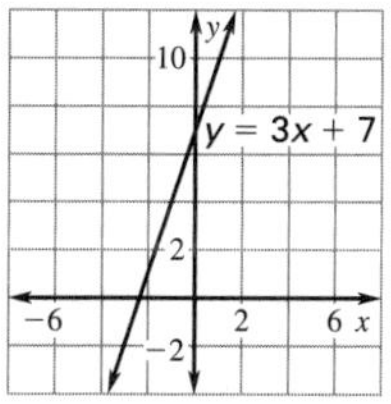

28.

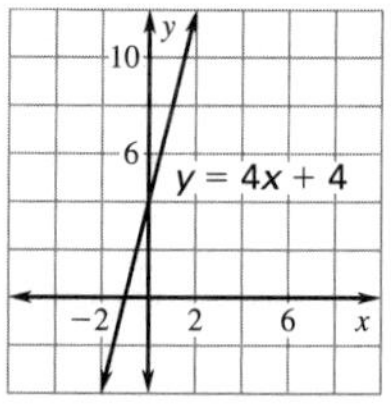

29.

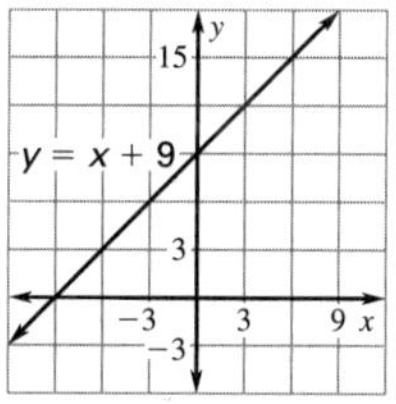

30.

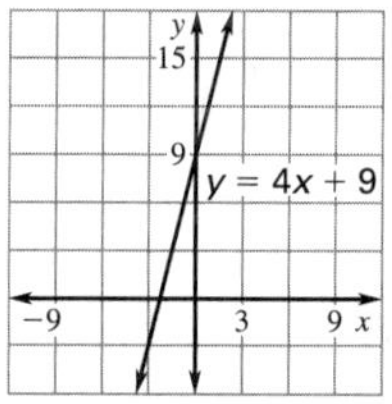

31.

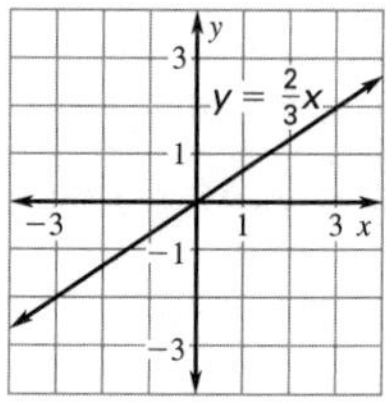

32.

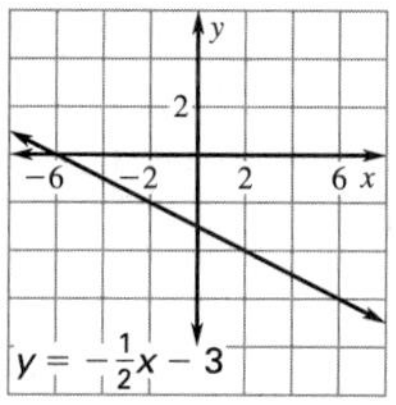

33.

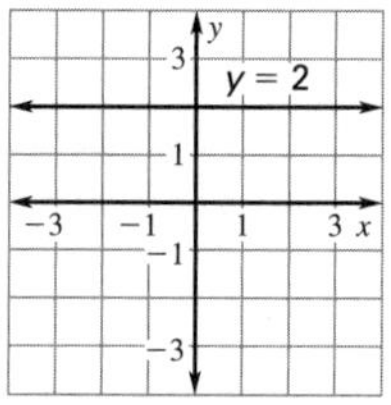

34.

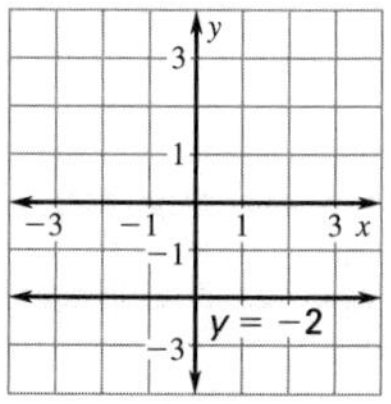

35.

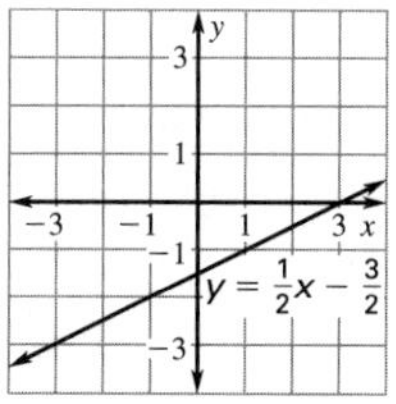

36.

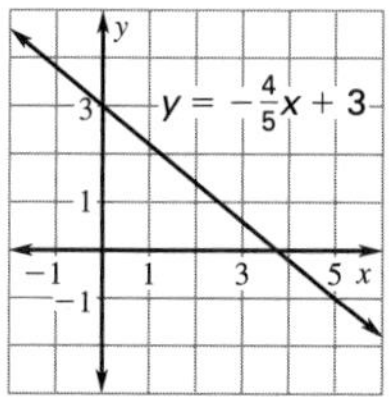

37.

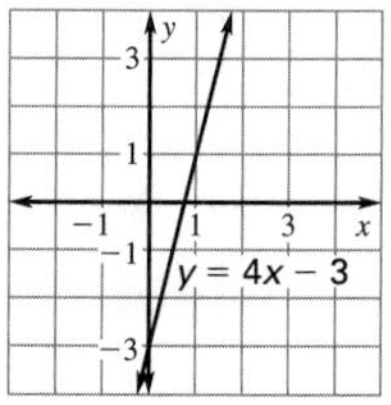

38.

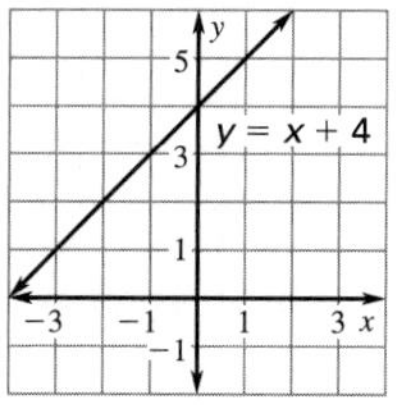

39.

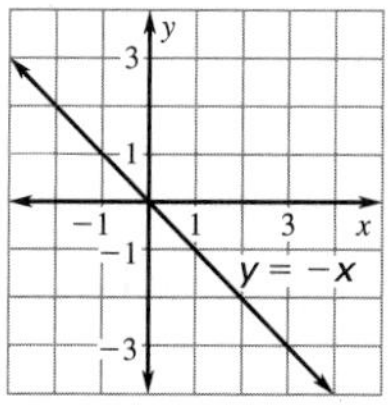

40.

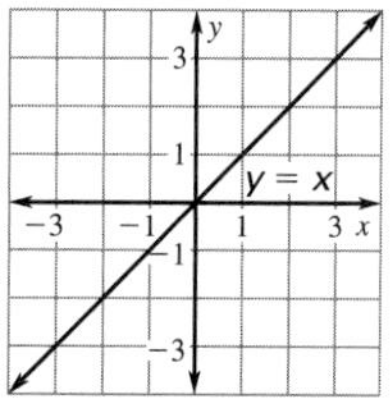

41.

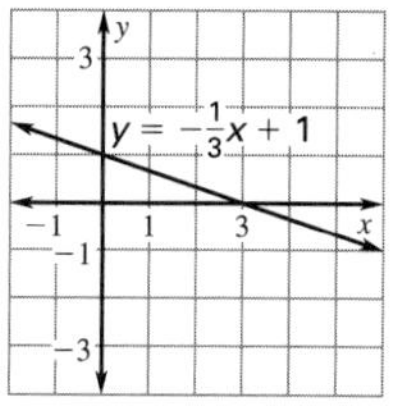

42.

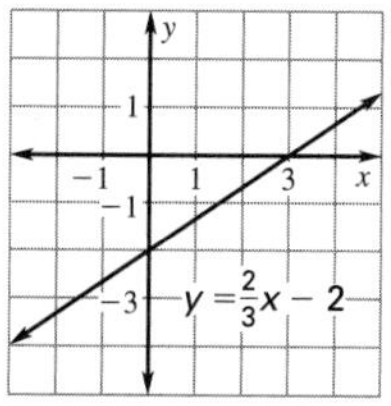

43.

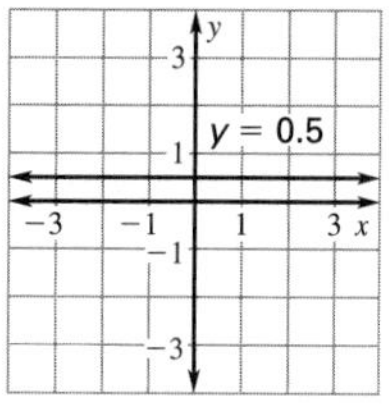

44.

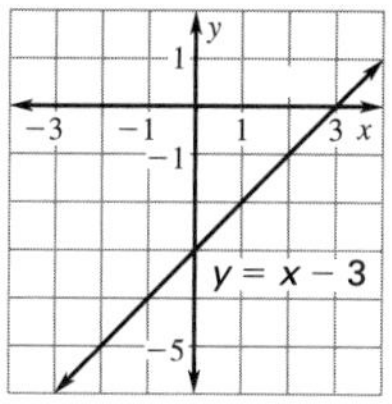

45.

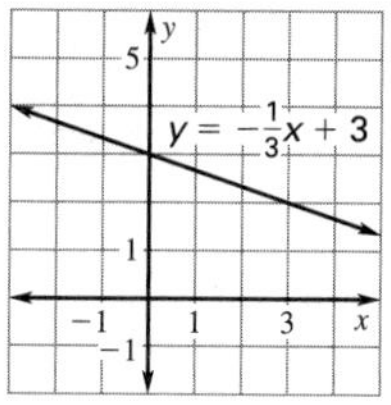

64. Distance from Home

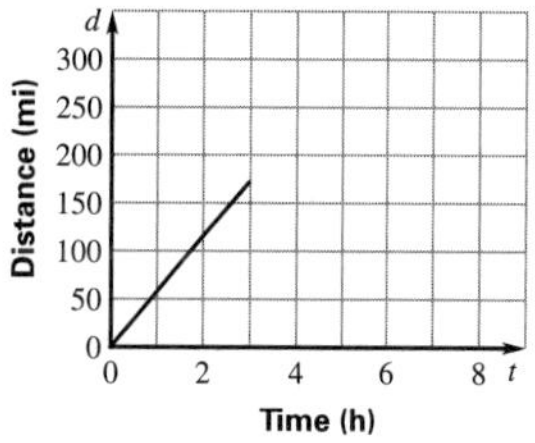

65. Distance from Home

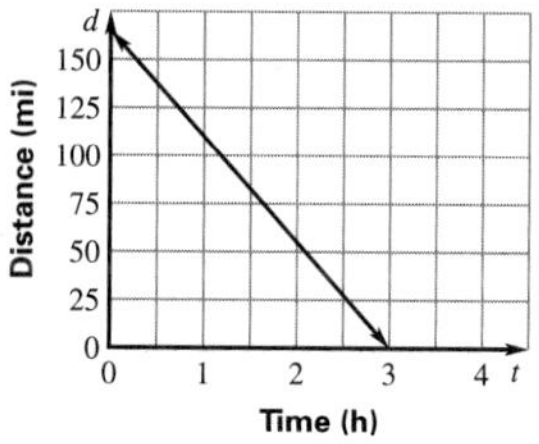

QUIZ 2 (p. 247)

14.

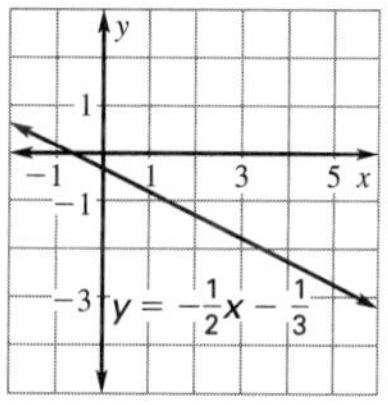

15.

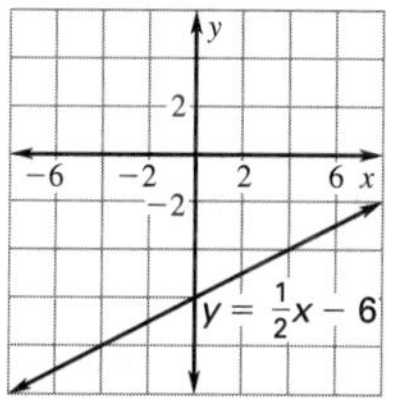

16.

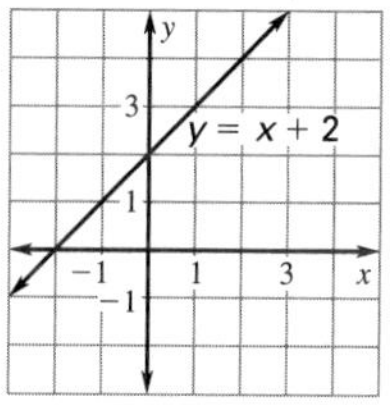

17.

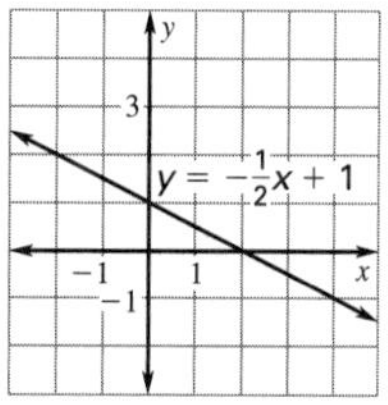

18.

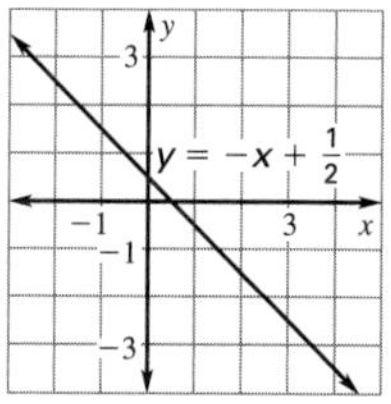

4.6 TECHNOLOGY ACTIVITY (p. 248–249)

5.

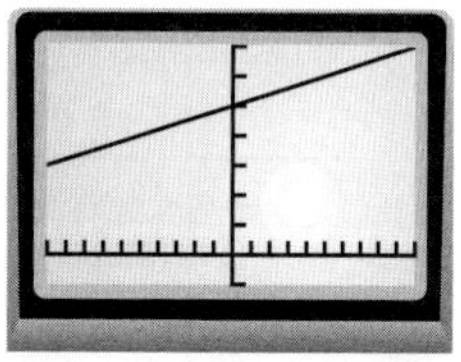

6.

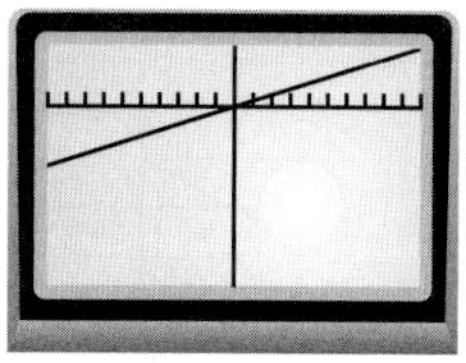

7.

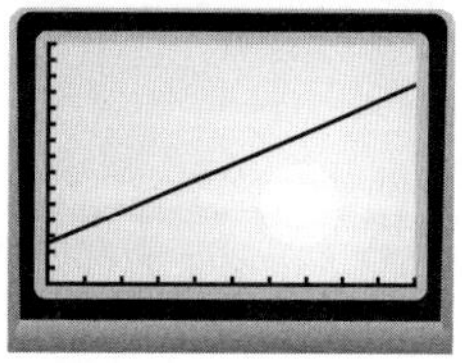

4.7 Mixed Review (p. 255)

66.

Input	Output
2	48
4	96
6	144
8	192
10	240

4.8 Practice and Applications (pp. 259–261)

37.
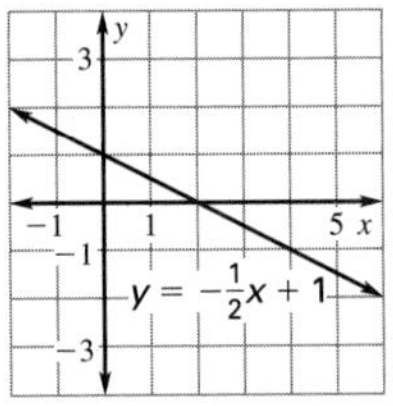

38.
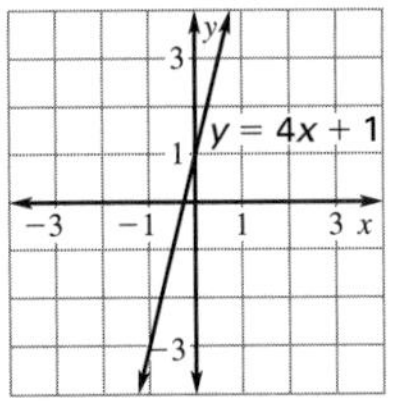

39.
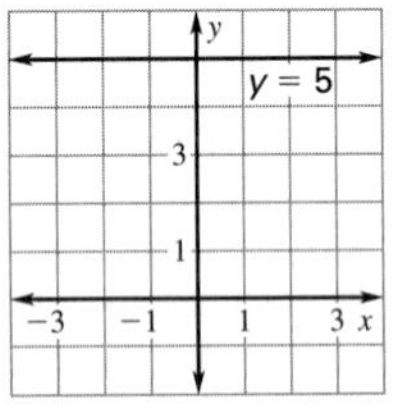

40.
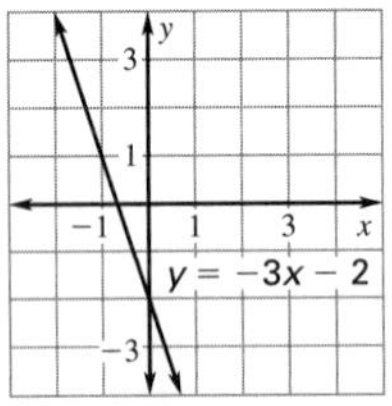

59.
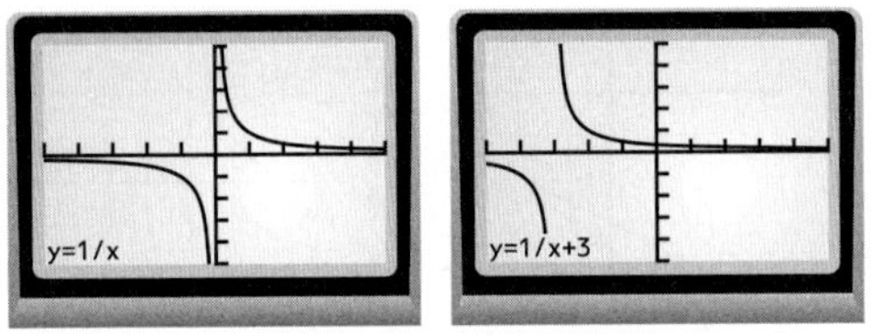

4.8 Mixed Review (p. 262)

73.
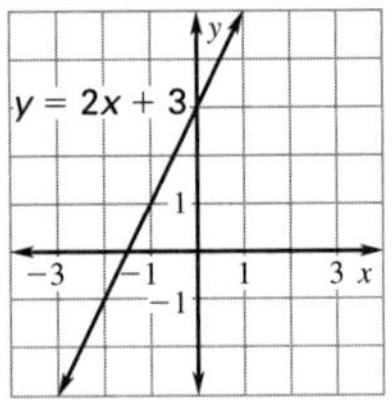

74.
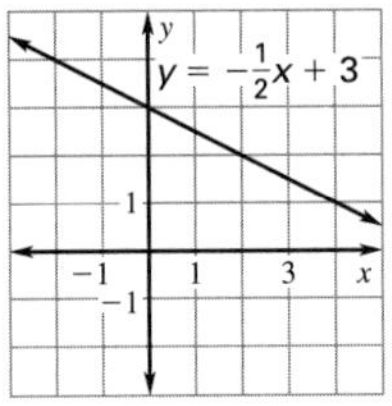

75.
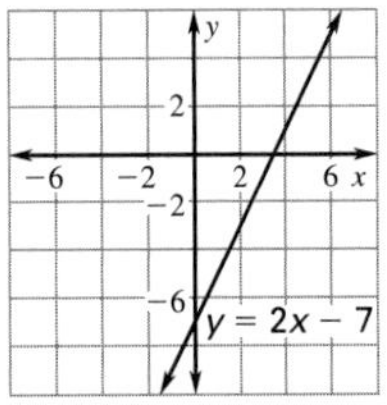

76.
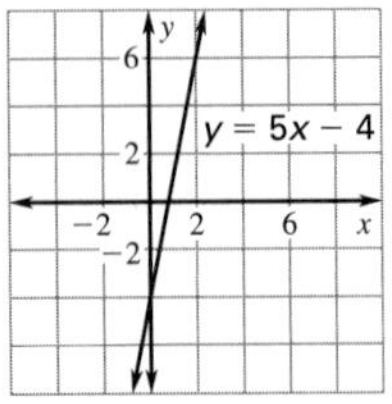

77.
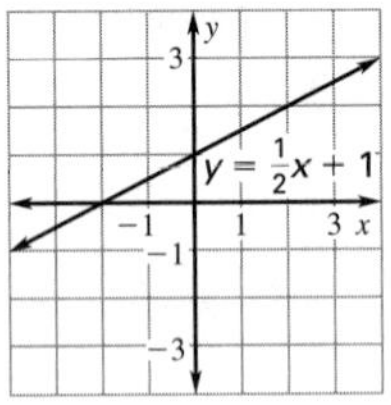

78.
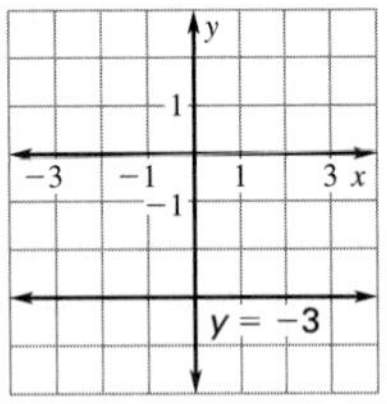

79.
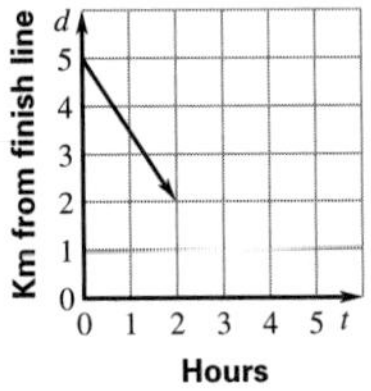

Quiz 3 (p. 262)

17.
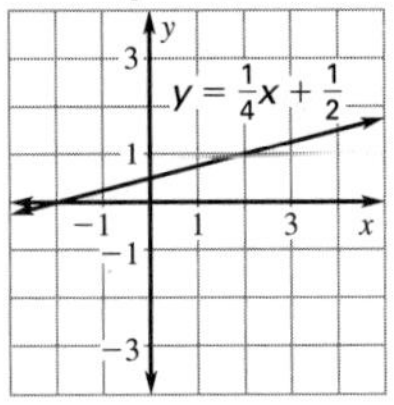

18.
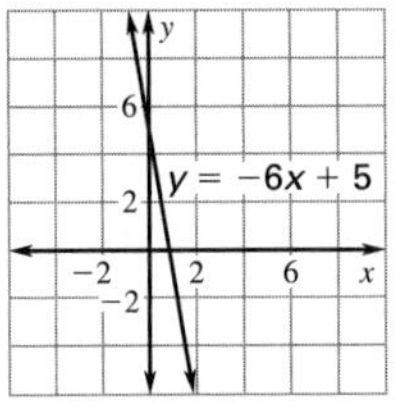

Chapter 4 Review (pp. 264–266)

10.
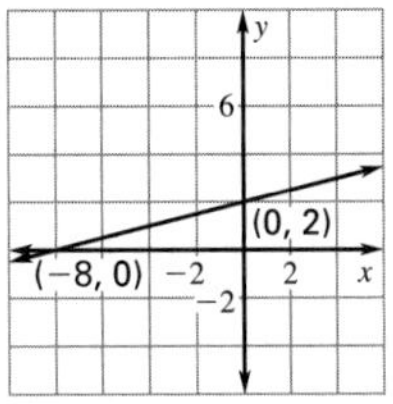

11.
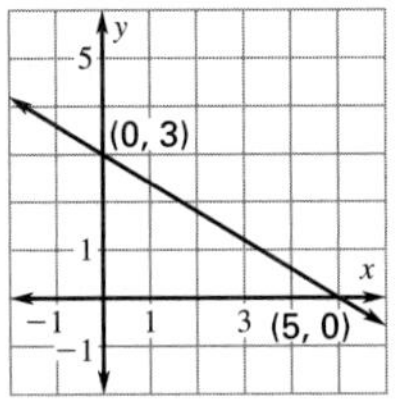

12.
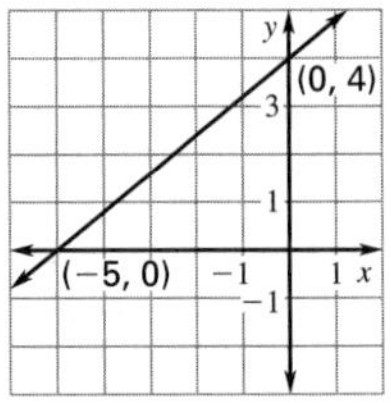

13.
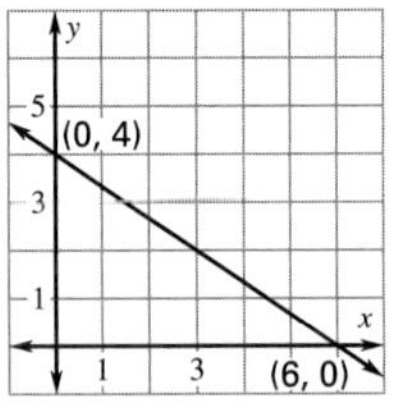

30.
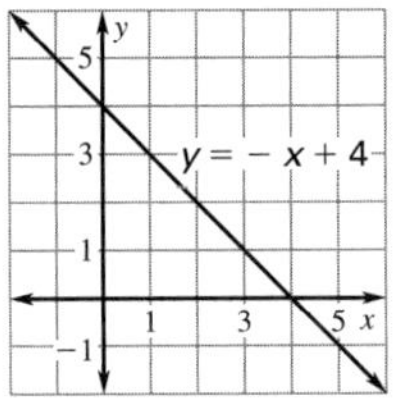

31.
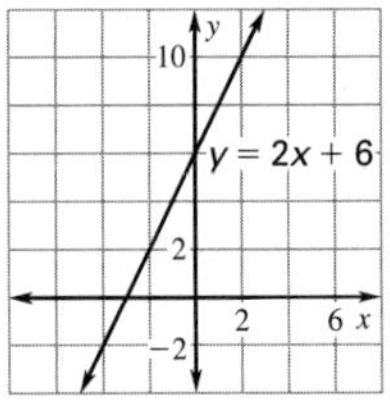

32.
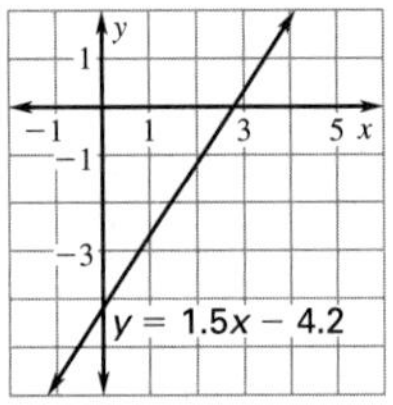

Chapter 4 Test (p. 267)

5.
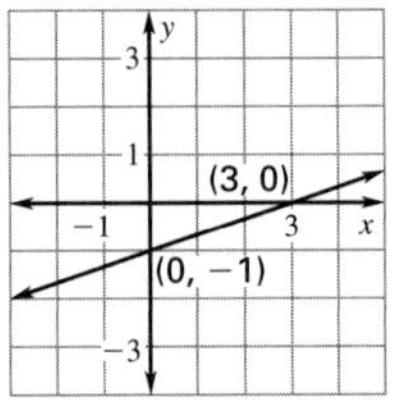

6.
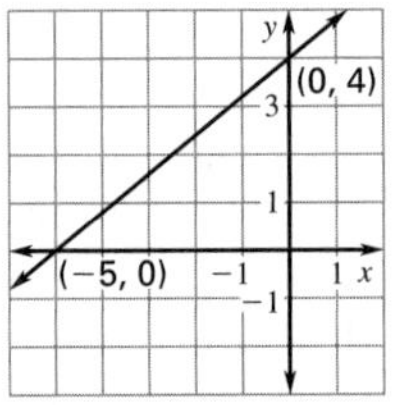

7.
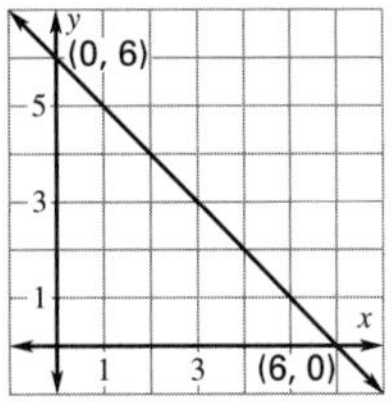

8.
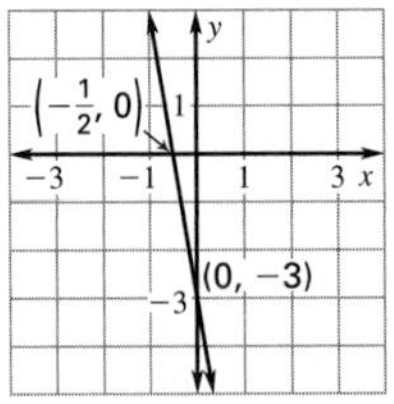

9.
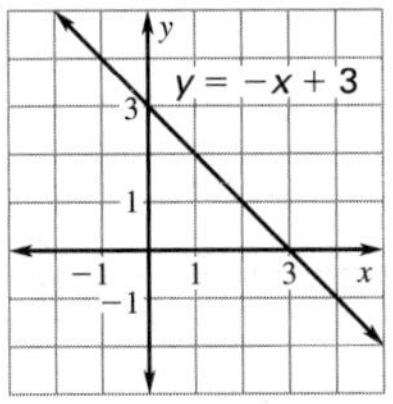

10.
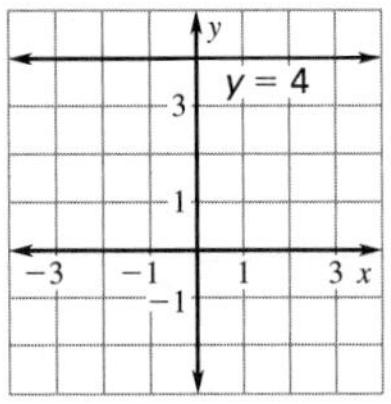

11.
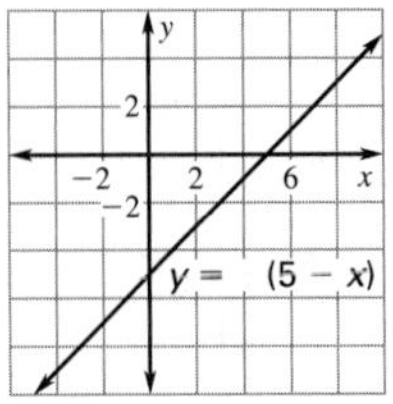

12.
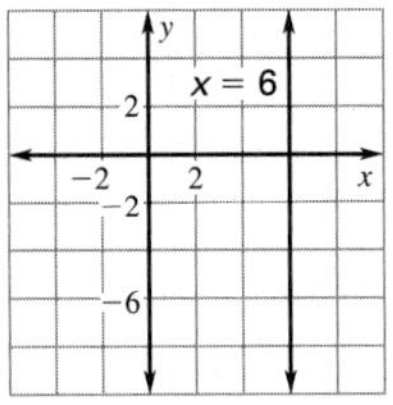

13.
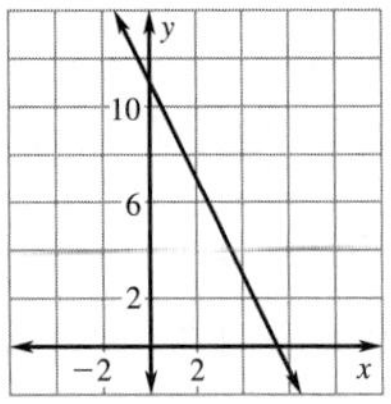

14.
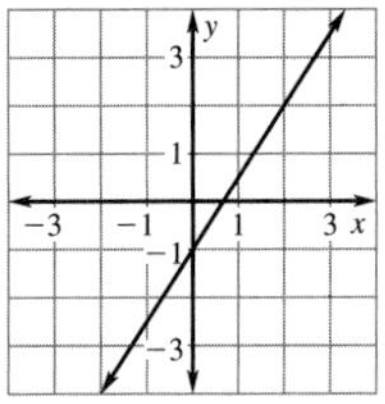

15.
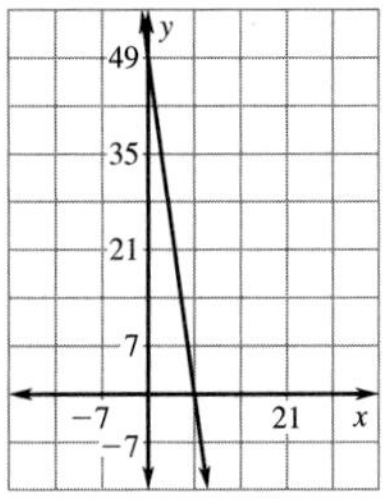

16.
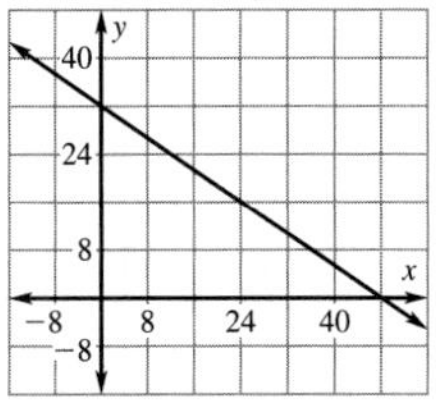

33.
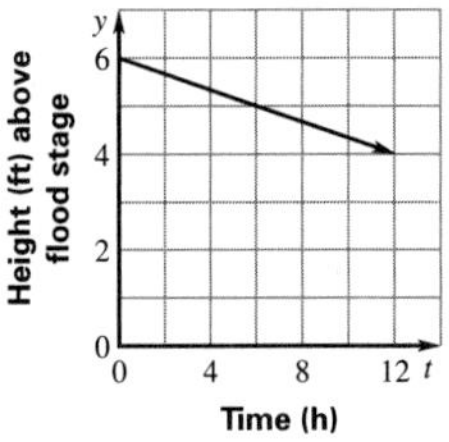

CHAPTER 5

SKILL REVIEW (p. 272)

4.
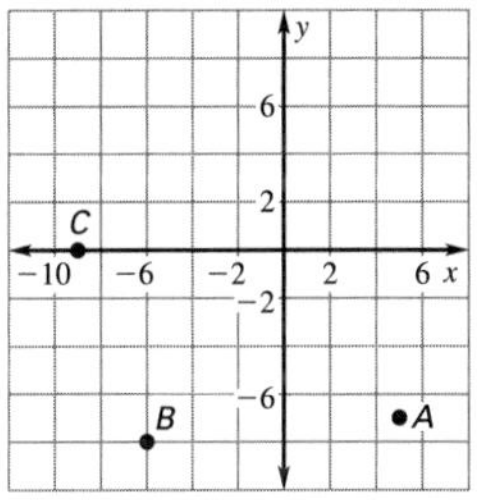

5.
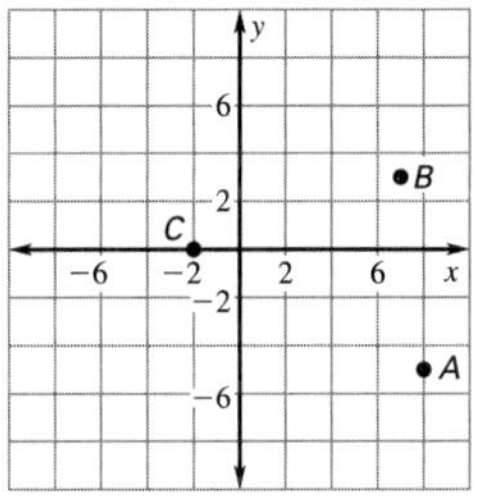

6.
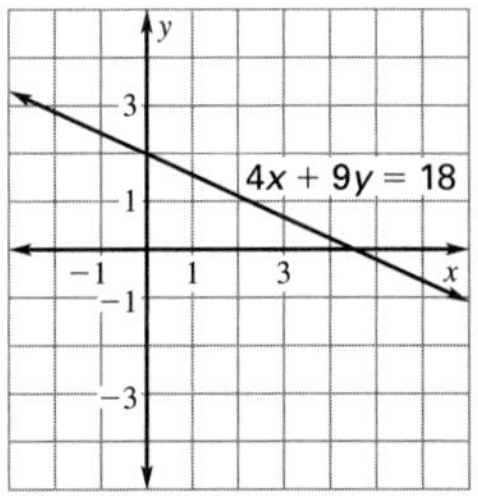

7.
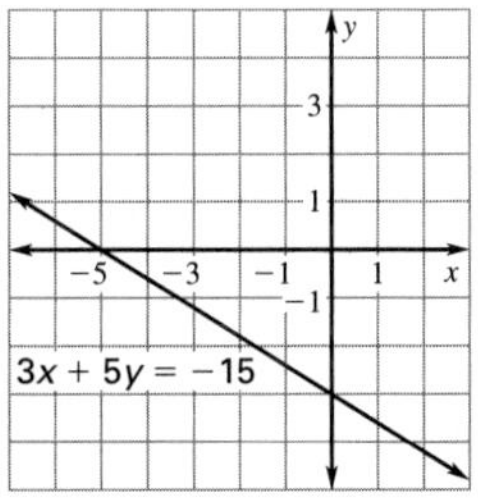

8.
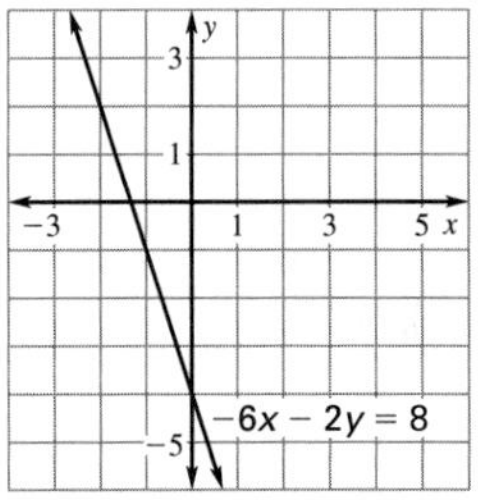

5.1 PRACTICE AND APPLICATIONS (pp. 276–278)

40. **Ceramics Class Students**

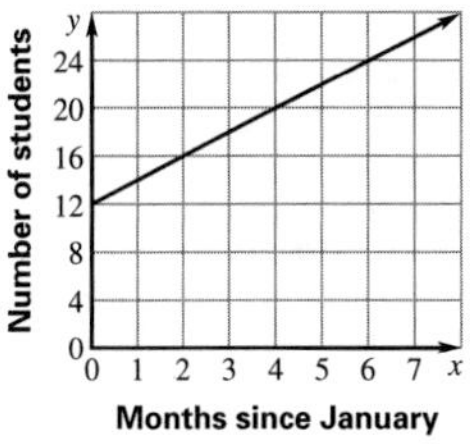

5.1 MIXED REVIEW (p. 278)

47. **Sailboard Rental Cost**

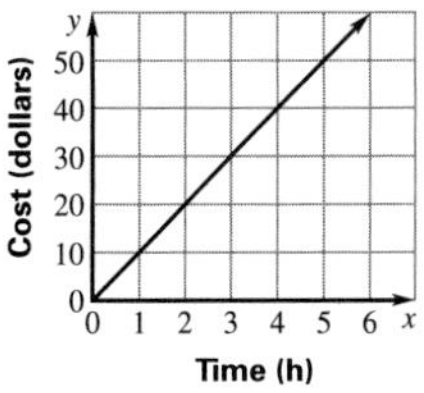

48. **Life Jacket Rental Cost**

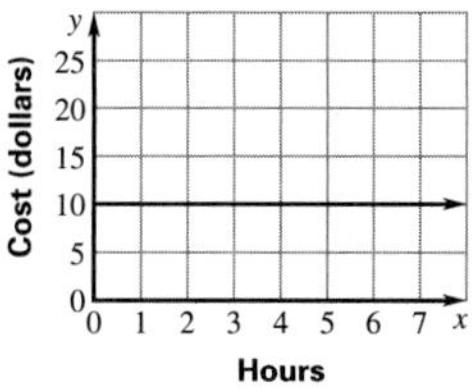

49.
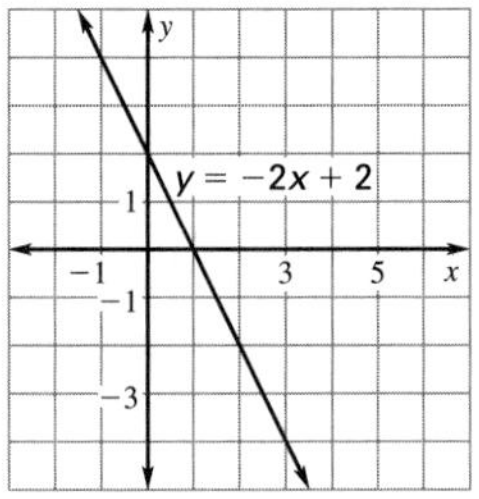

50.
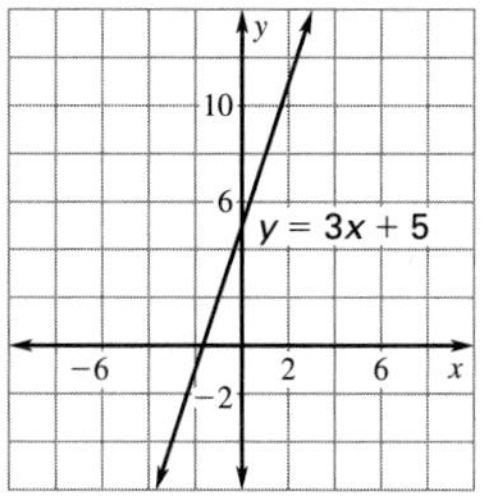

51.
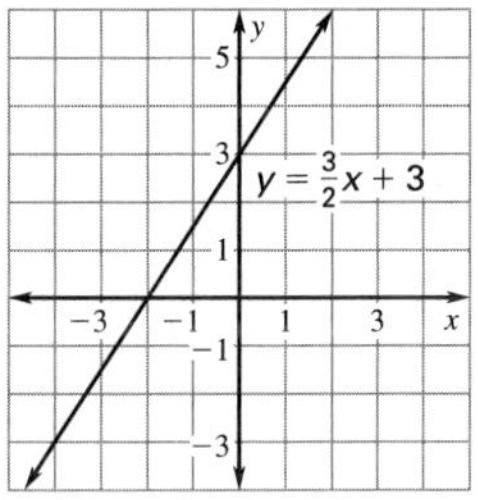

52.
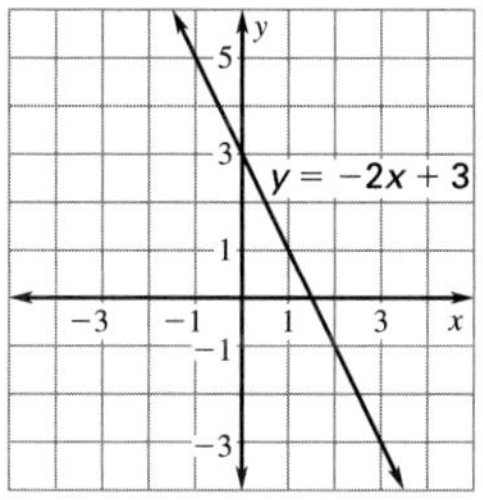

53.
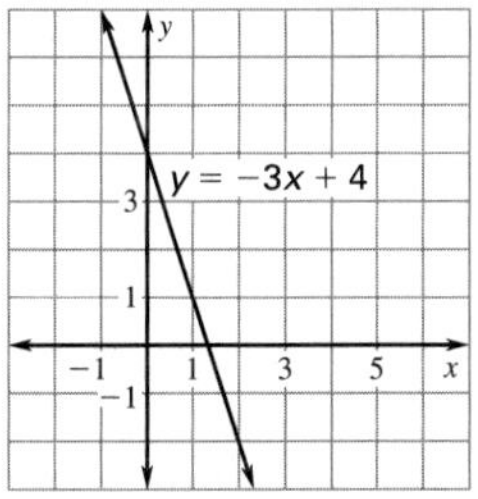

54.
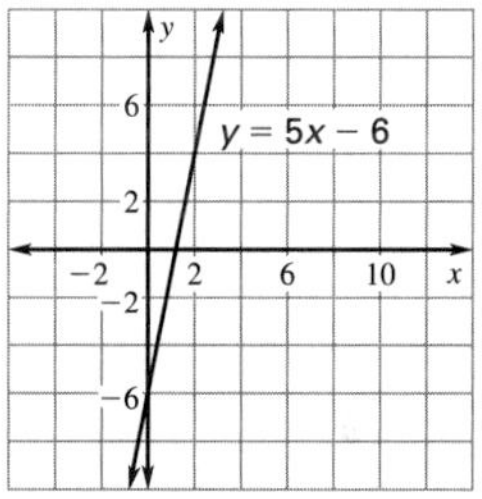

55.
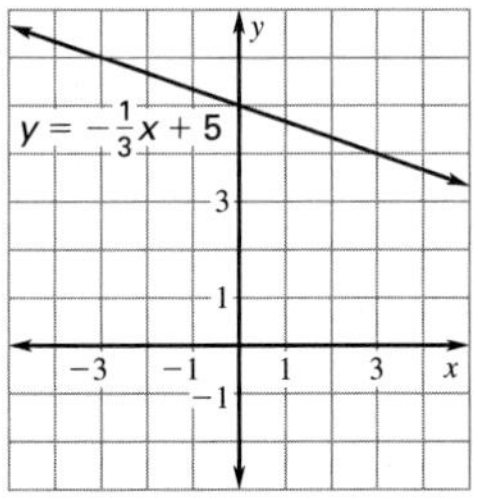

56.
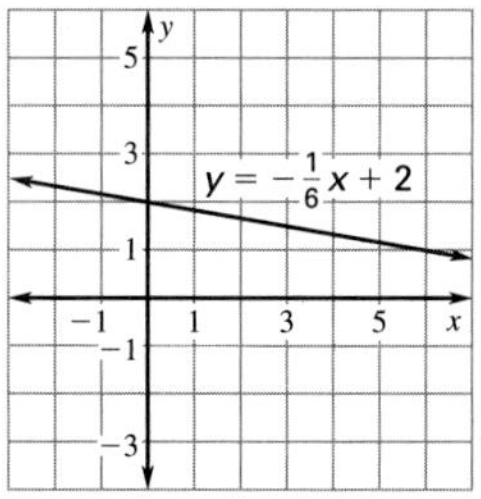

57.
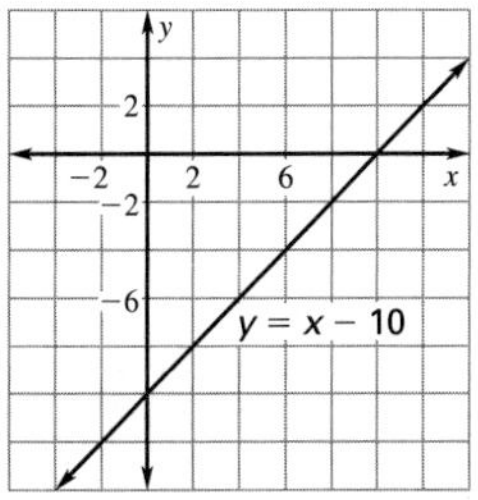

5.2 Practice and Applications (pp. 282–284)

51. b.

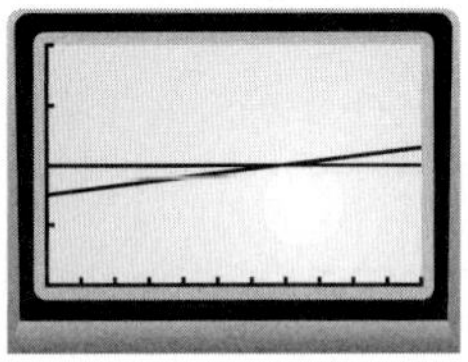

5.4 Activity (p. 292)

1, 2.

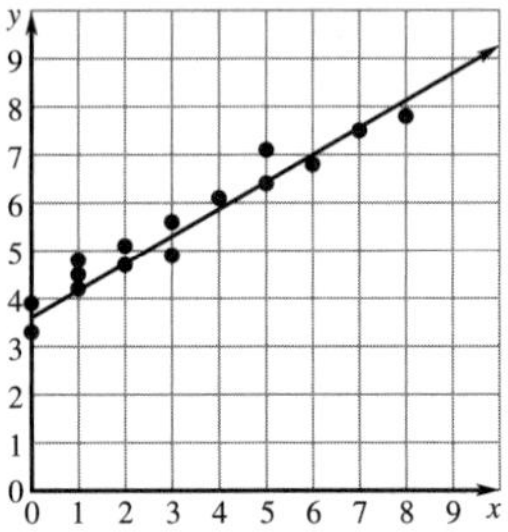

2. See graph for Step 1. *Sample answer:* I drew a line so that there were about the same number of points above and below the line.
3. *Sample answer:* (0, 3.6) and (6, 7)
4. *Sample answer:* $y = 0.57x + 3.6$

5.4 Practice and Applications (pp. 296–298)

25.

Sample Pupil Diameters (Daylight)

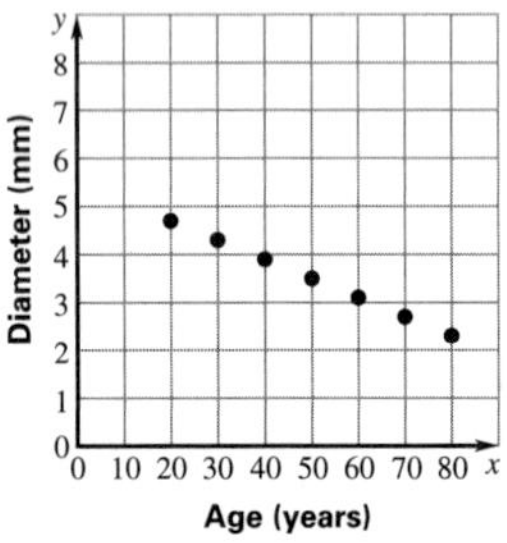

Sample Pupil Diameters (Night)

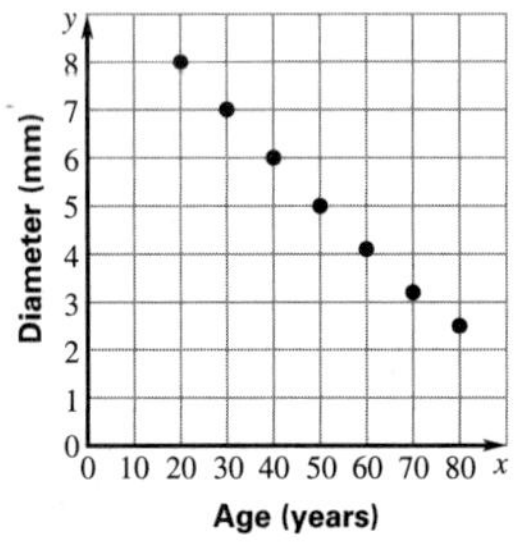

32. The line through those points does not approximate the data points very well. The data points for 2000–2003, for example, are not at all close to the line. **33.** No; the best-fitting line depends on all the data points and not on any two in particular. In fact, there may be no best-fitting line. Check sketches.

5.5 Mixed Review (p. 305)

80.

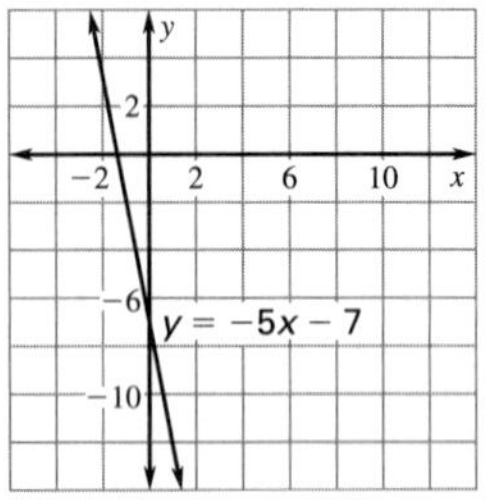

81.

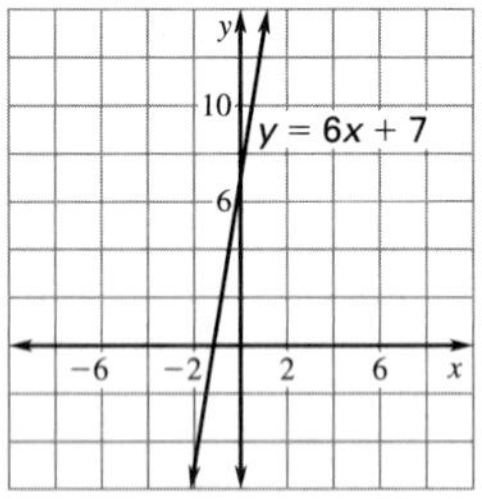

82.

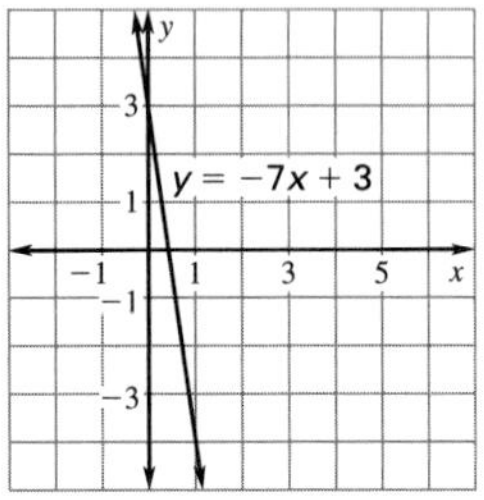

83.

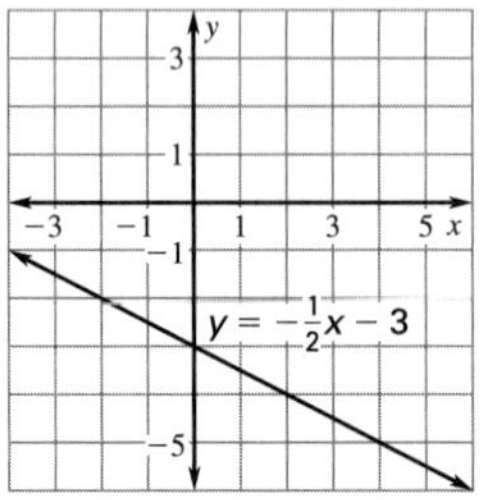

84.

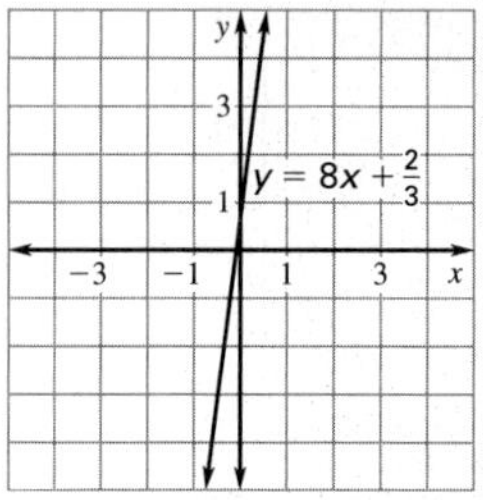

85.

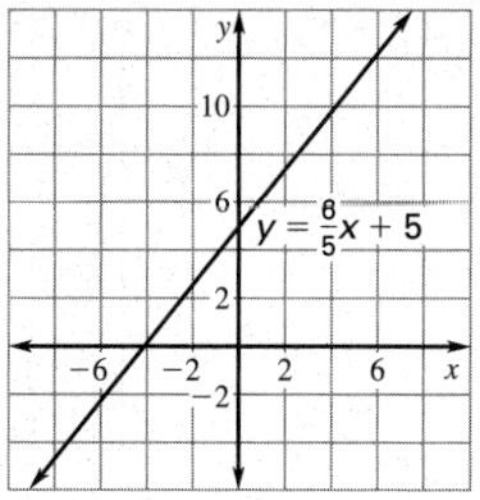

5.6 Concept Activity (p. 307)

2.

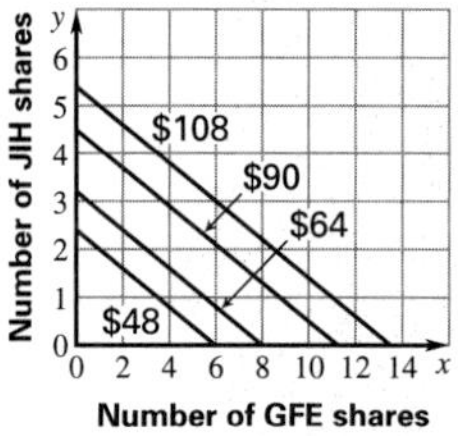

5.6 Activity (p. 308)

1. a–c.

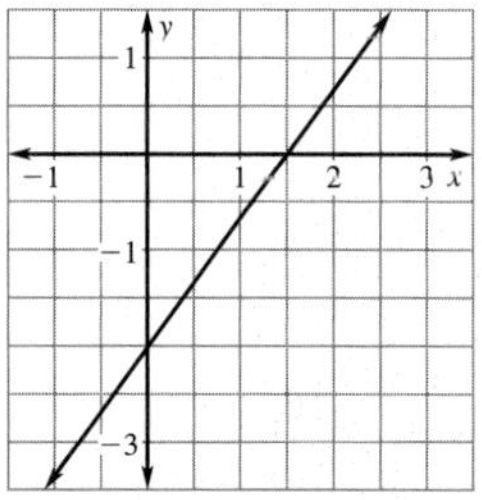

2. The three equations are equivalent; because they all have the same graph.

5.6 Mixed Review (p. 314)

90.

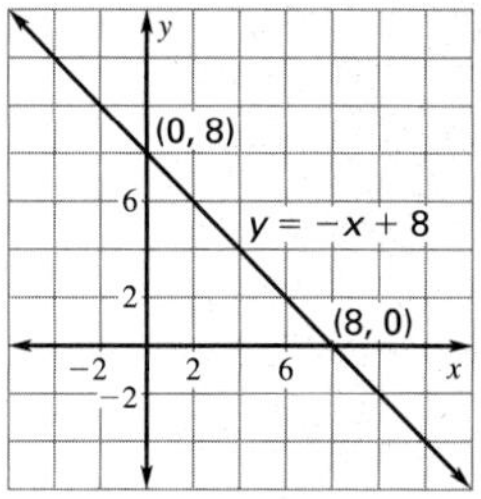

91.

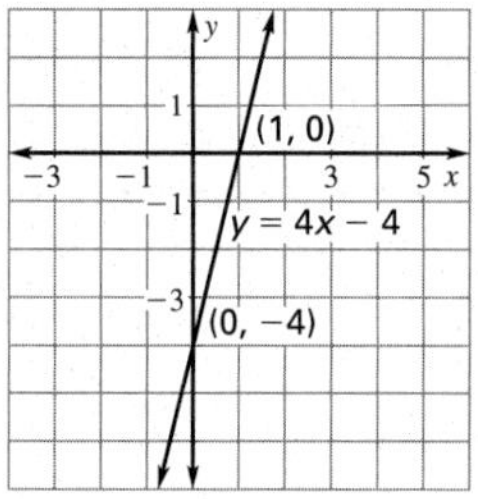

92.

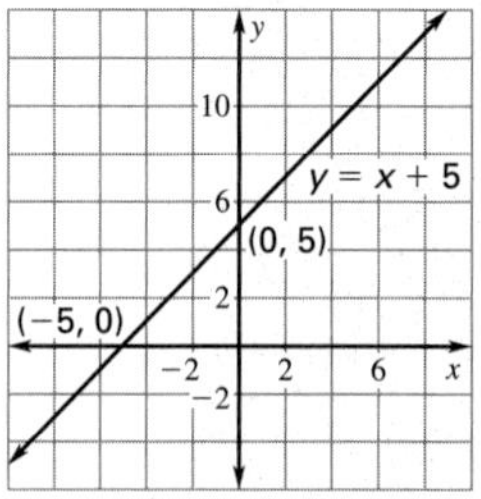

93.

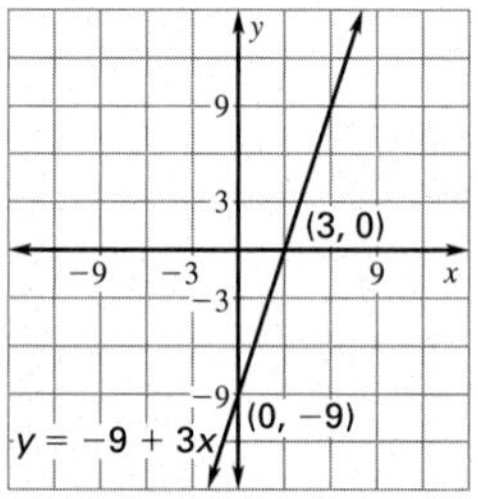

94.

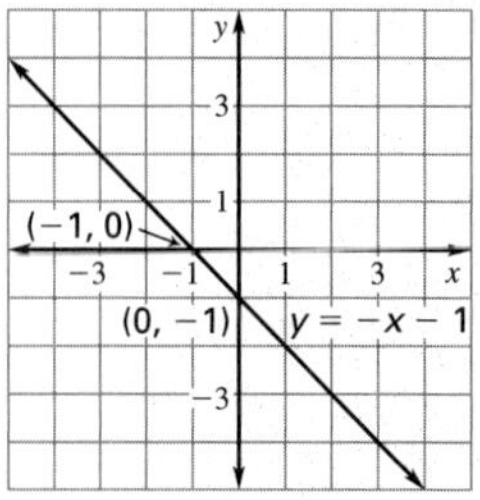

95.

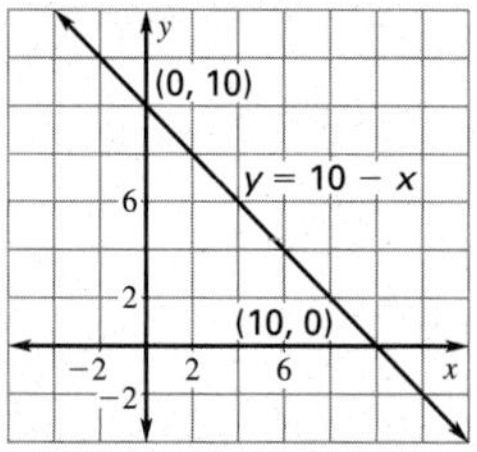

5.7 CONCEPT ACTIVITY (p. 315)

Step 2.

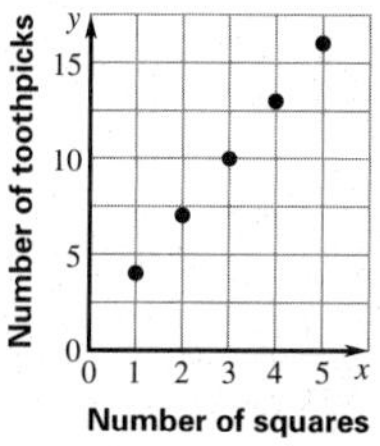

1. A: B: C:

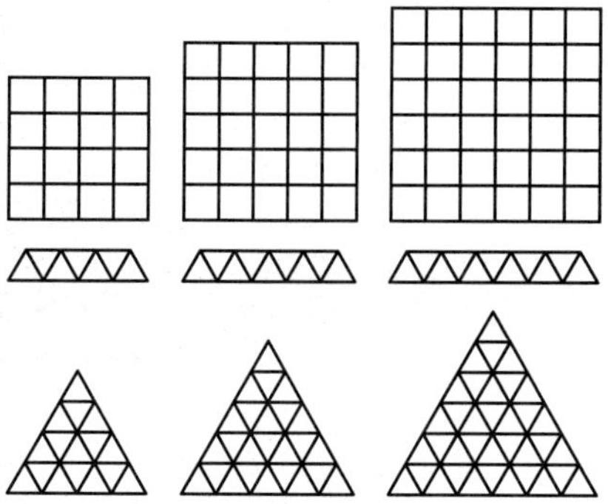

2. A:

Stage number	1	2	3	4	5	6
Number of squares	1	4	9	16	25	36
Number of toothpicks	4	12	24	40	60	84

B:

Stage number	1	2	3	4	5	6
Number of triangles	1	3	5	7	9	11
Number of toothpicks	3	7	11	15	19	23

C:

Stage number	1	2	3	4	5	6
Number of triangles	1	4	9	16	25	36
Number of toothpicks	3	9	18	30	45	63

3.

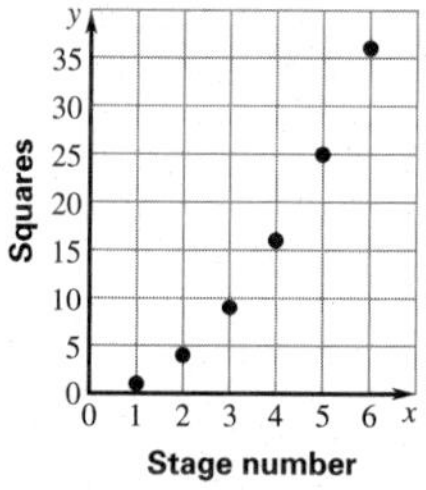

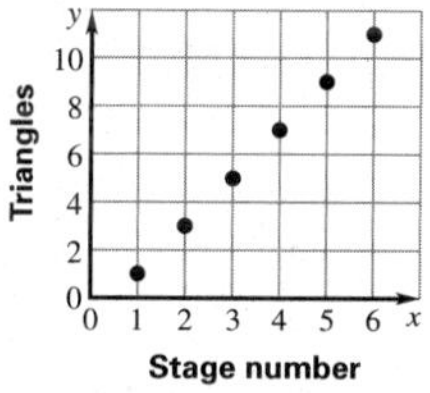

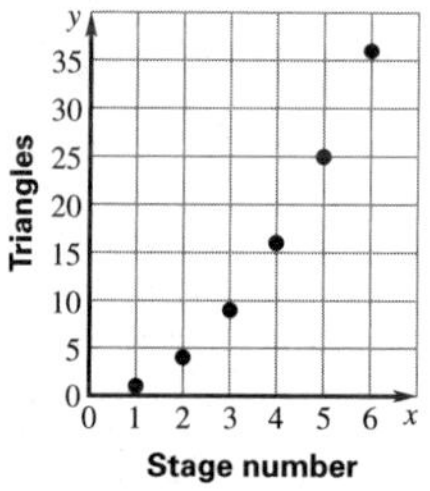

4.

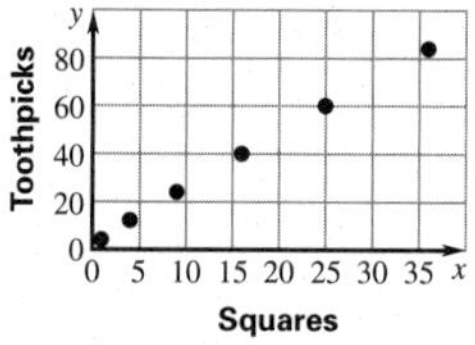

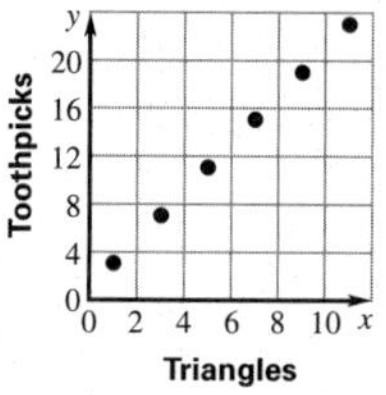

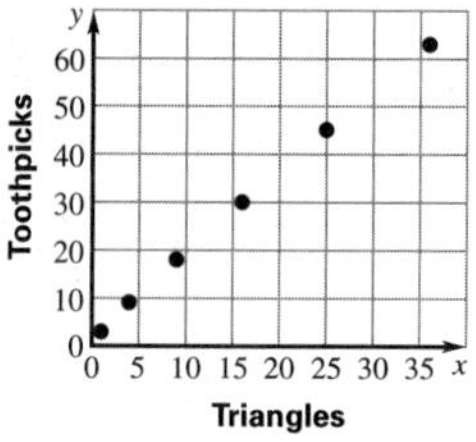

CHAPTER 6

SKILL REVIEW (p. 332)

15.

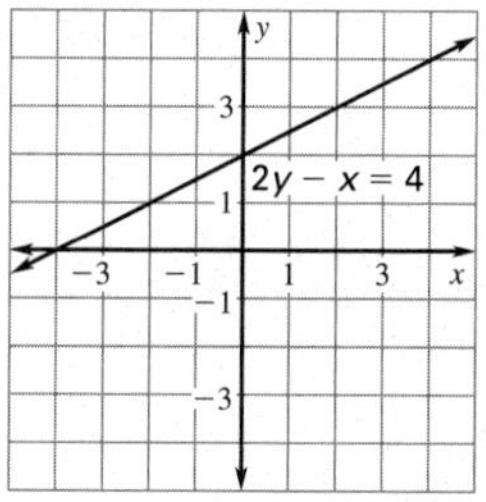

16.

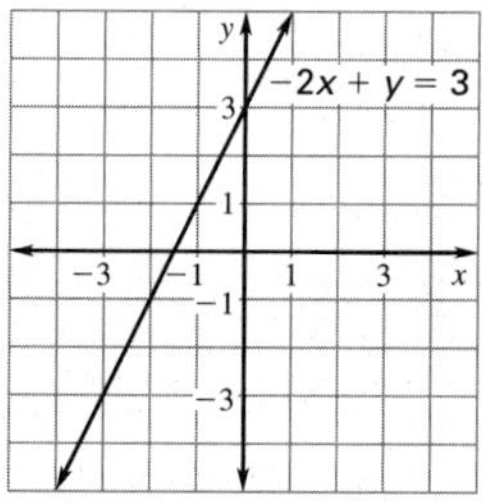

6.1 PRACTICE AND APPLICATIONS (pp. 337–339)

22.

23.

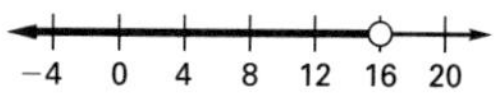

24.

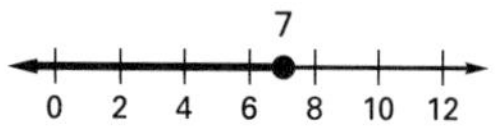

25.

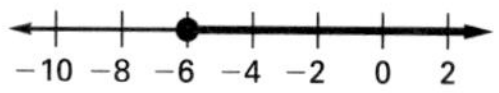

26.

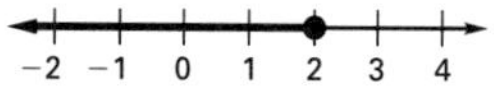

27.

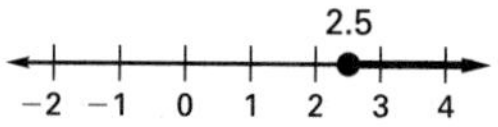

28.

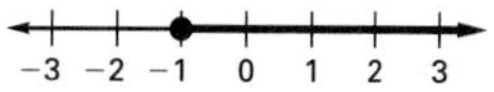

29.

30.

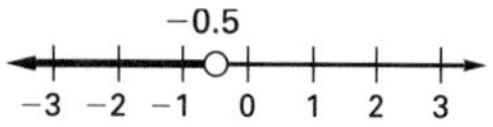

31.

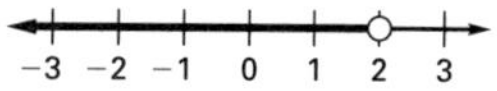

32.

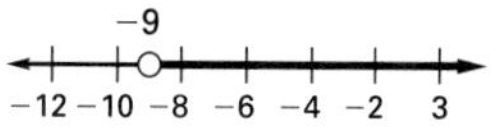

33.

34.

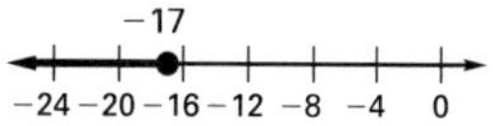

35.

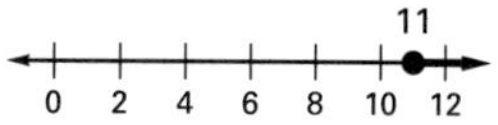

36.

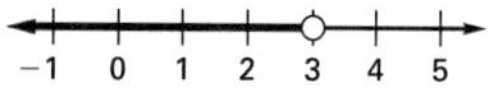

37.

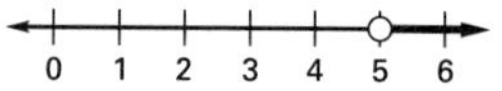

38.

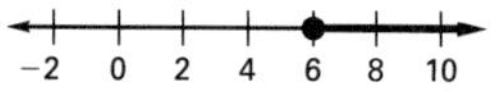

39.

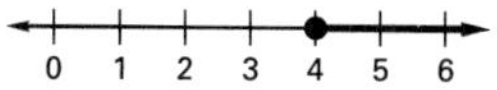

40.

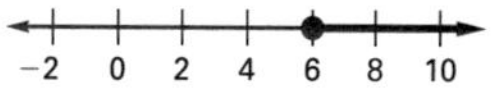

41.

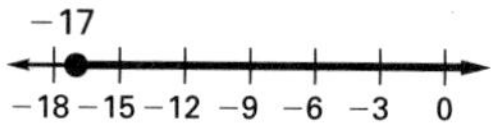

42.

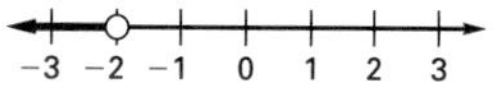

43.

44.

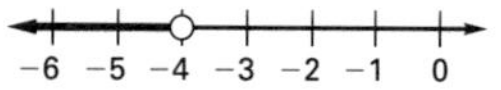

45.

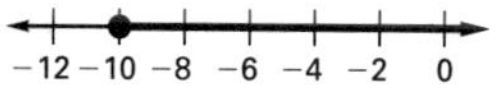

6.1 MIXED REVIEW (p. 339)

80.

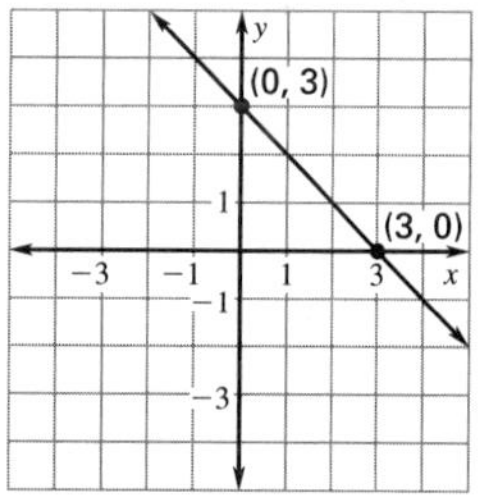

81.

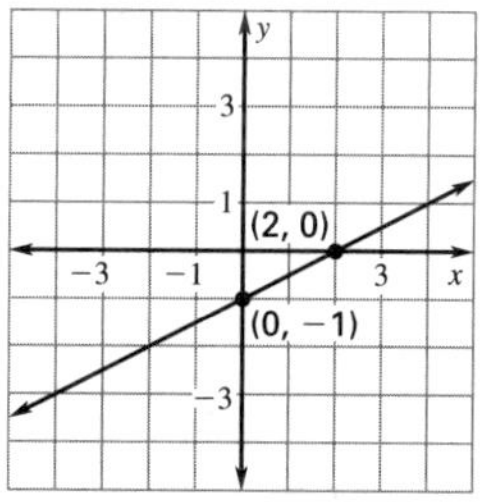

82.

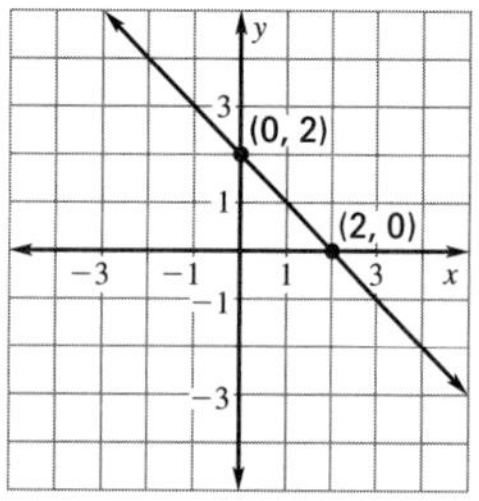

83.

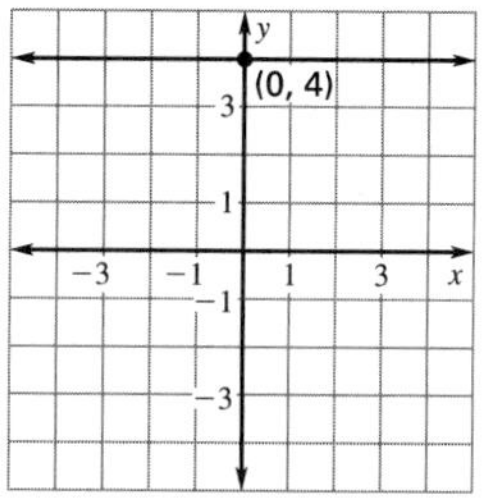

84.

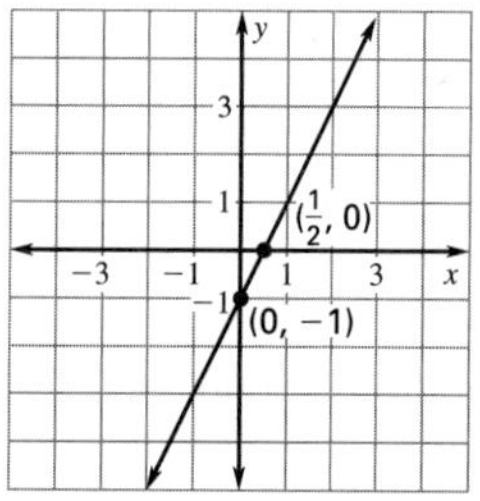

85.

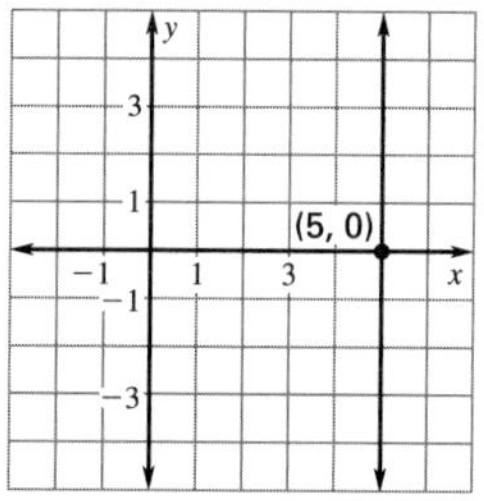

6.3 Practice and Applications (pp. 349–351)

26.

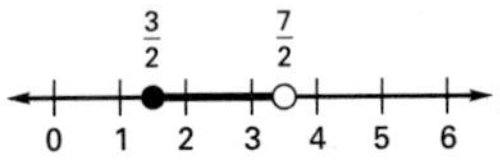

27.

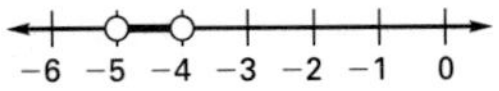

28.

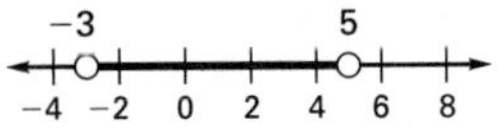

29.

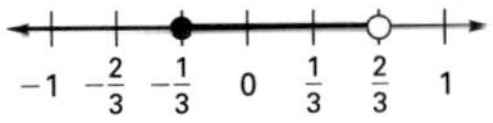

30.

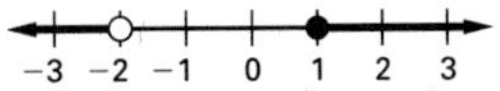

31.

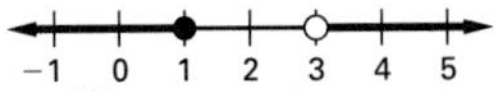

32.

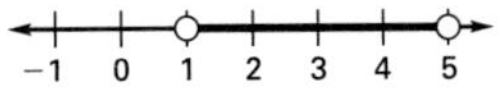

33.

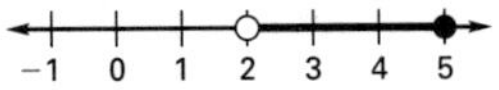

34.

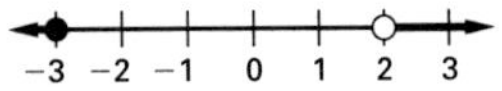

35.

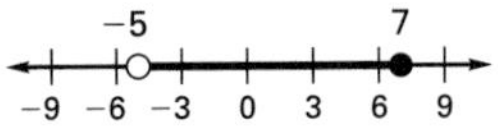

6.4 Activity (p. 354)

1. **2.** **3.**

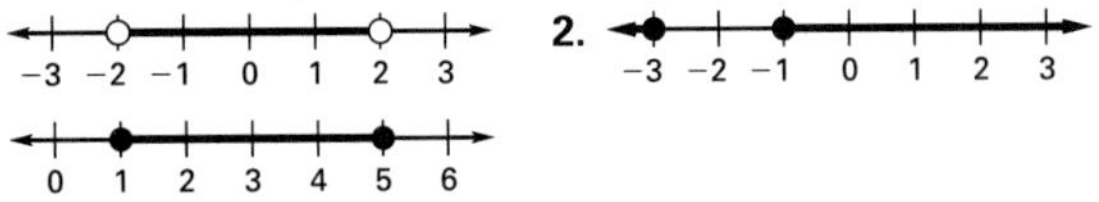

6.4 Practice and Applications (pp. 356–358)

46.

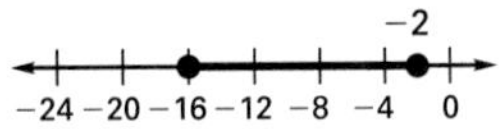

47.

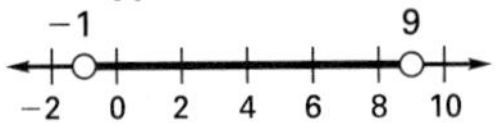

48.

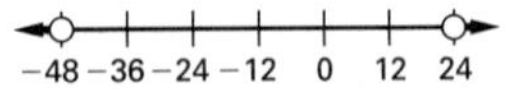

49.

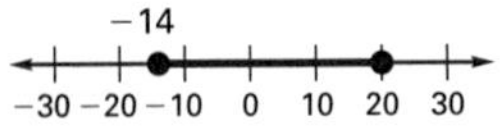

50.

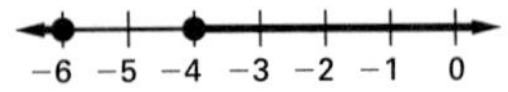

51.

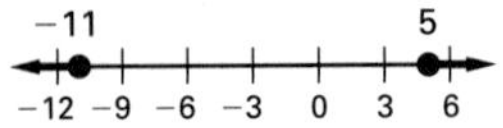

52.

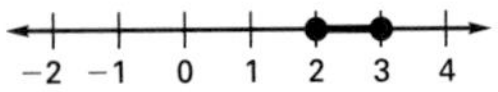

53.

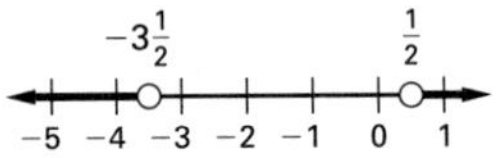

54.

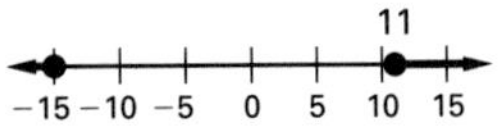

55.

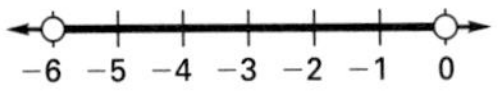

56.

57.

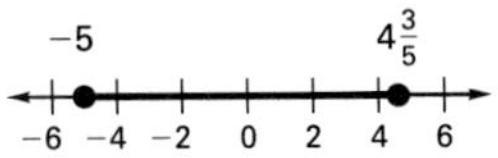

58.

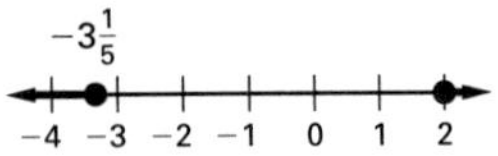

59.

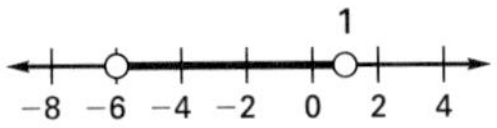

60.

6.4 Mixed Review (p. 358)

80.

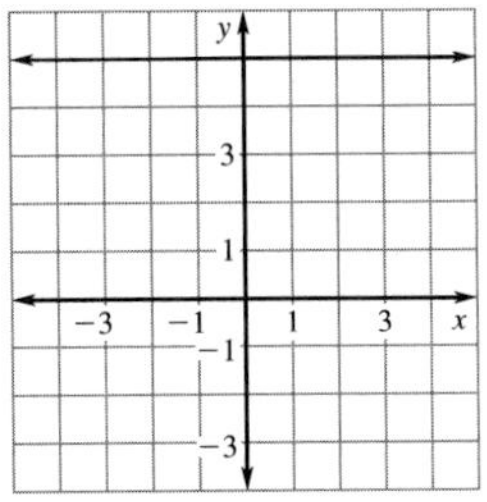

81.

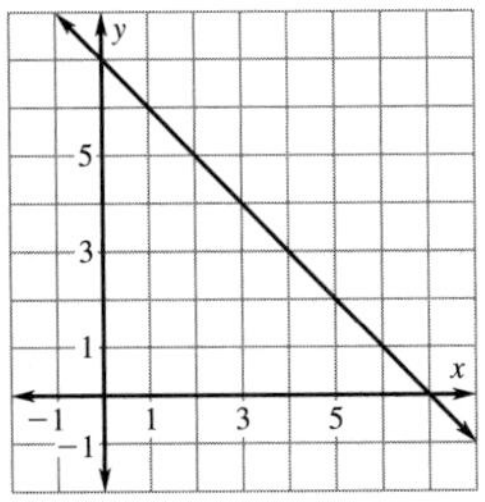

6.5 Guided Practice (p. 363)

4.

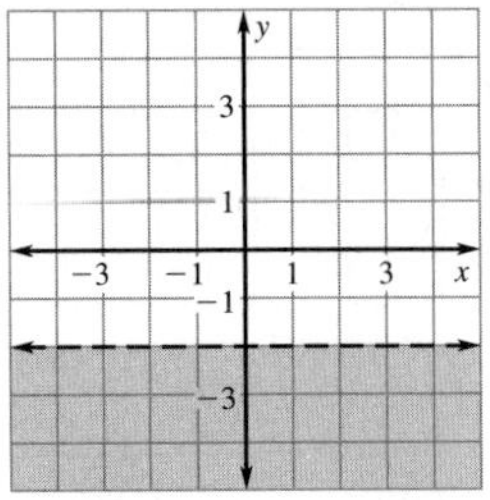

5.

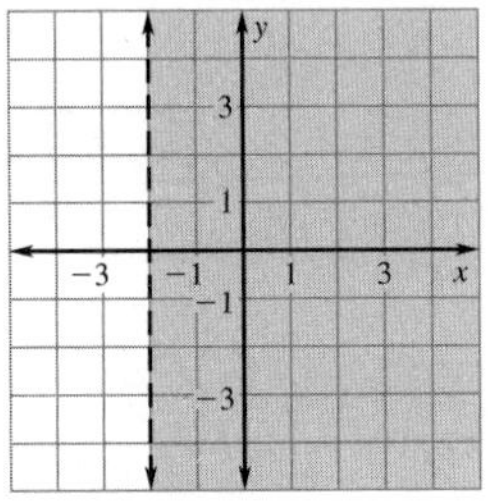

6.

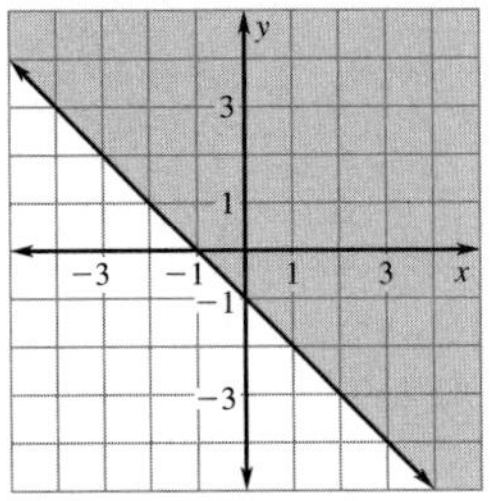

7.

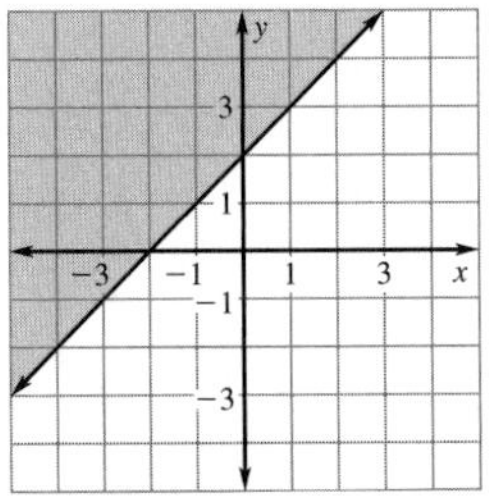

8.

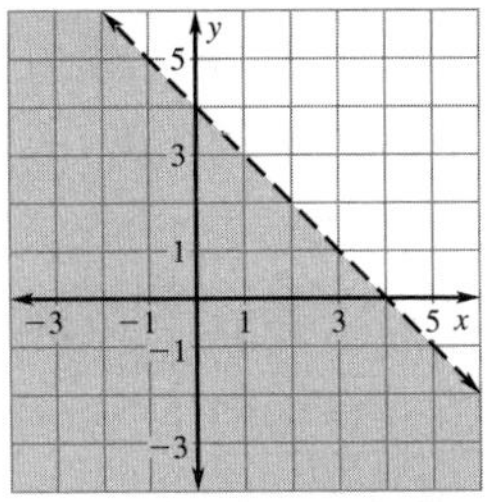

9.

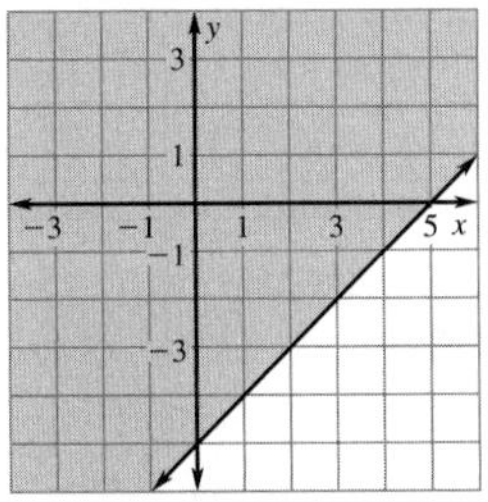

10.

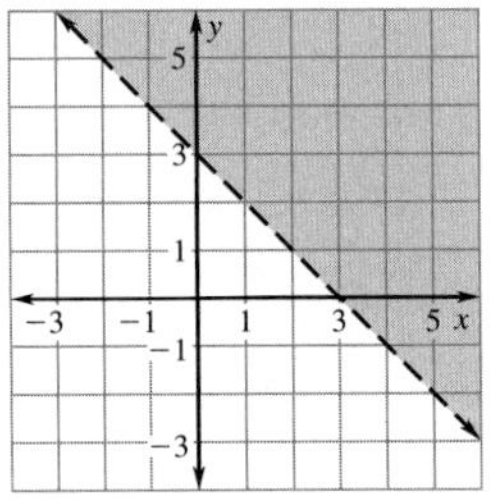

11.

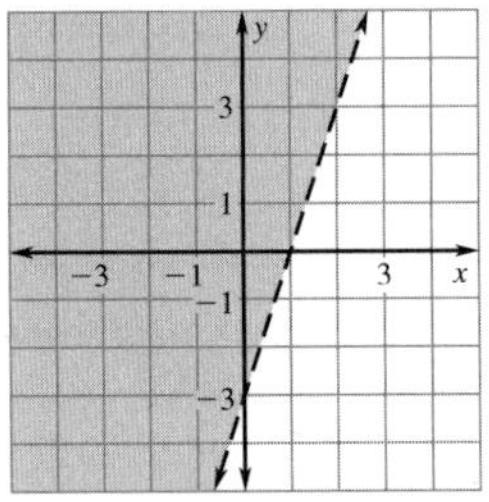

12.

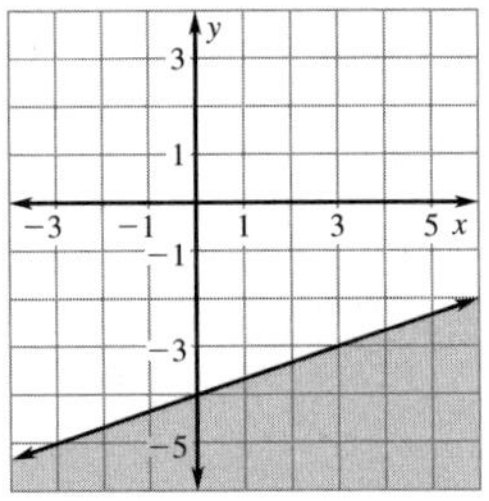

14.

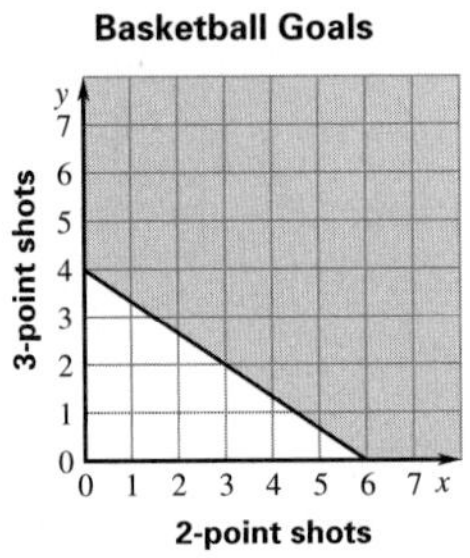

6.5 Practice and Applications (pp. 363–365)

23.

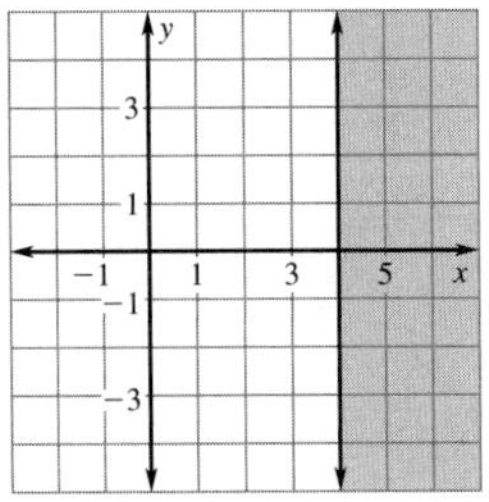

24.

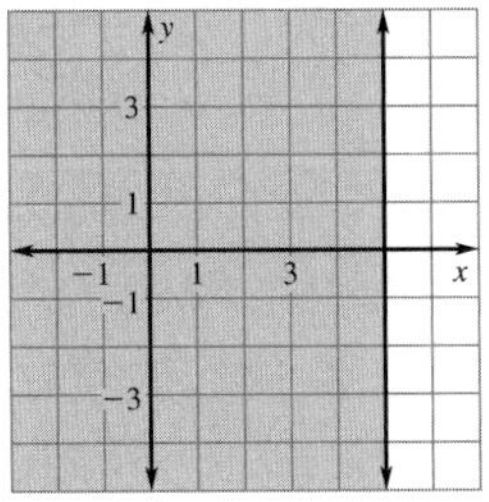

25.

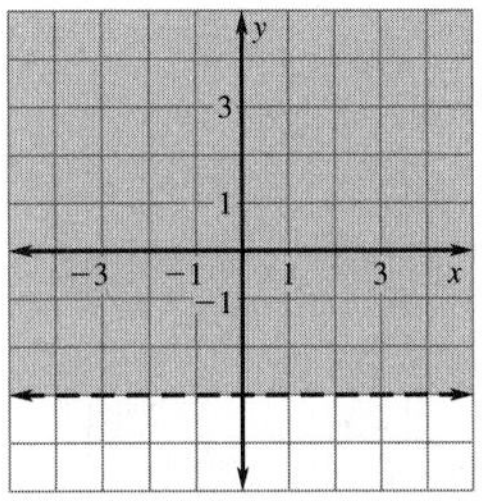

26.

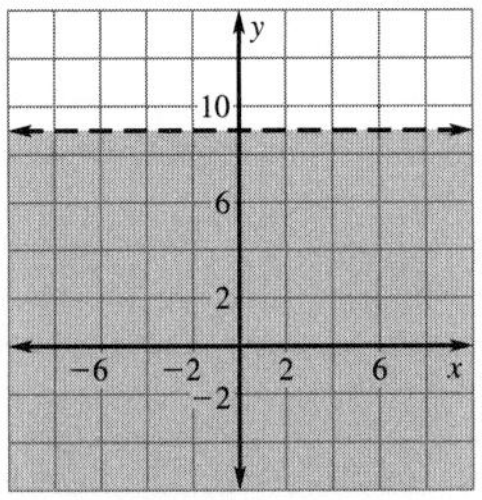

27.

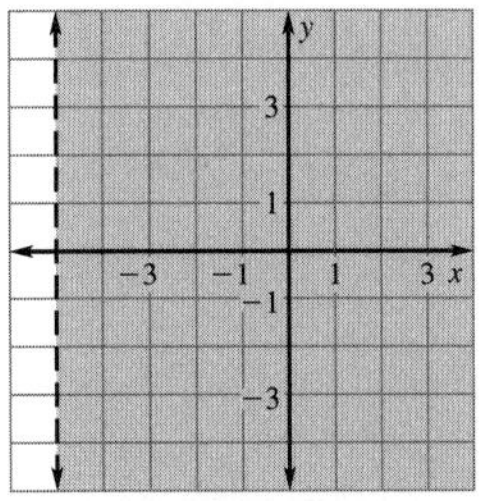

28.

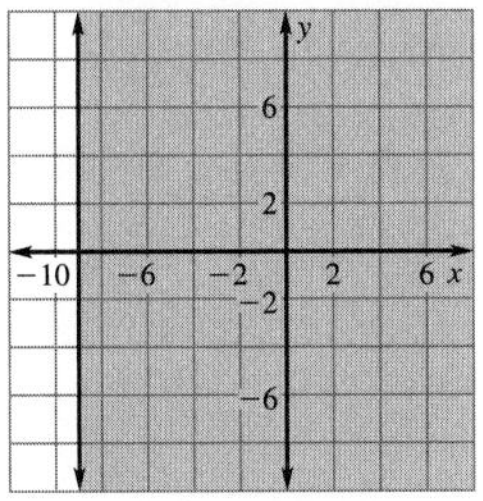

29.

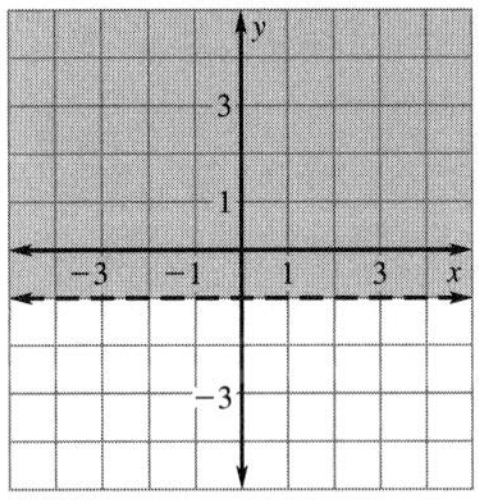

30.

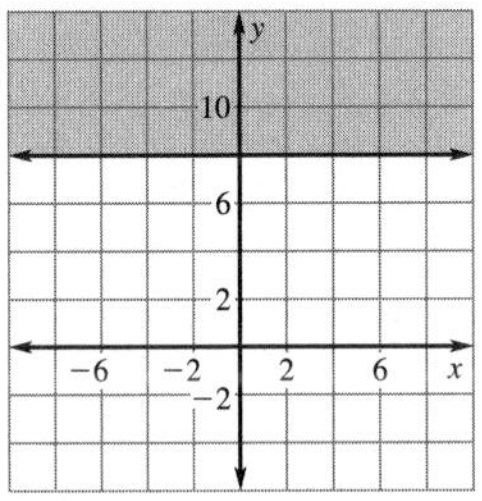

31.

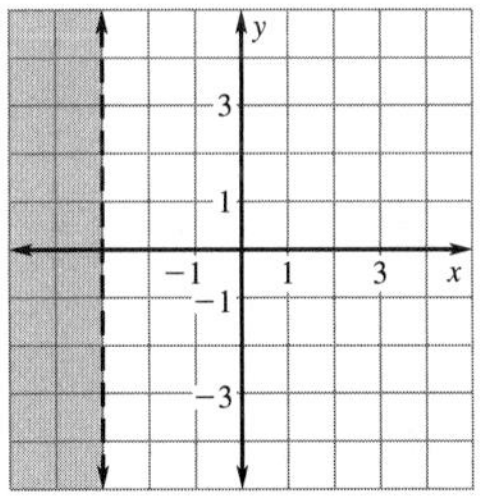

32.

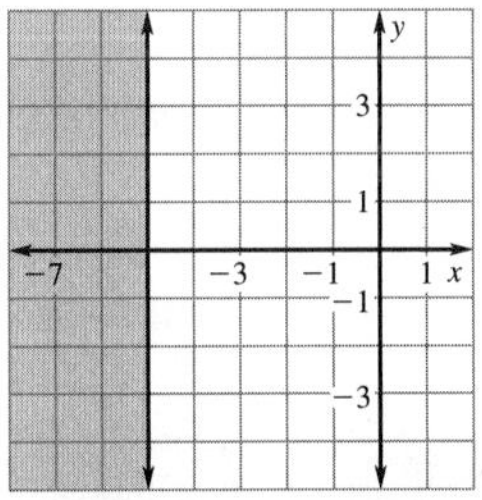

33.

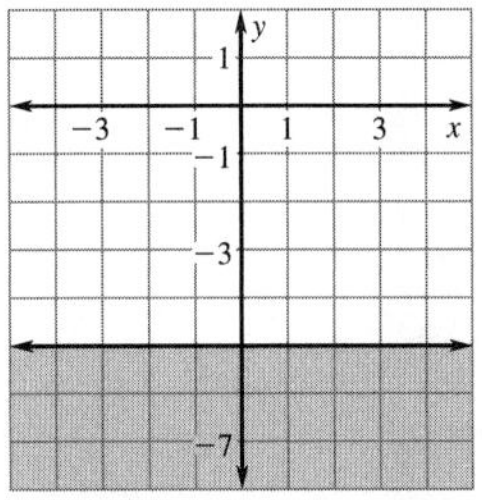

34.

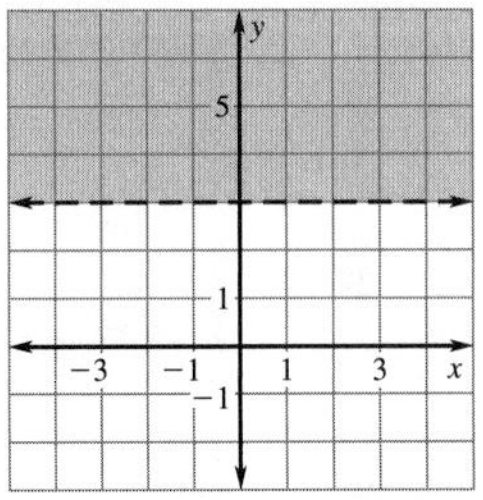

35.

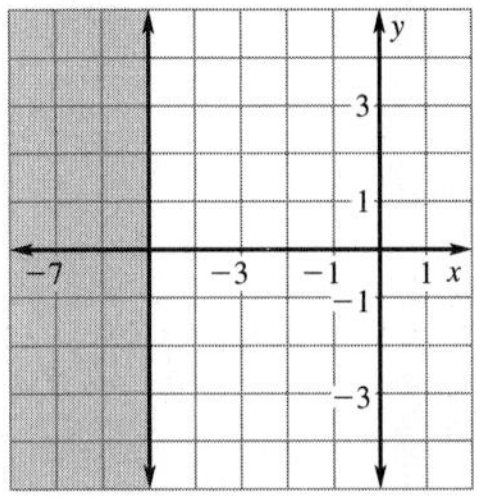

36.

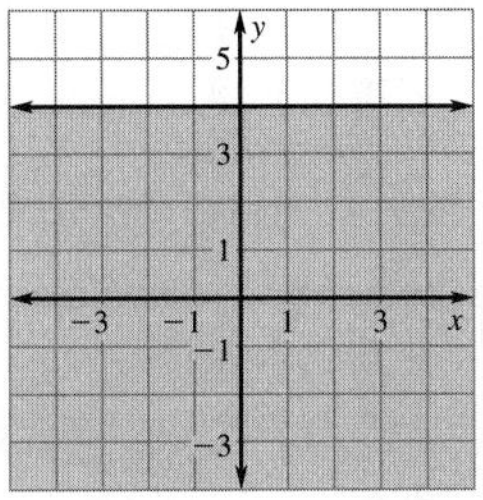

37.

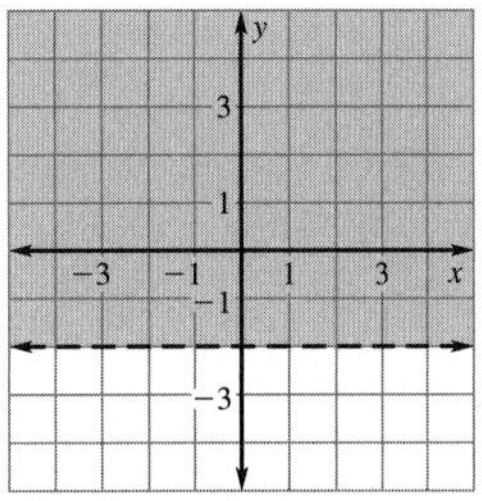

38.

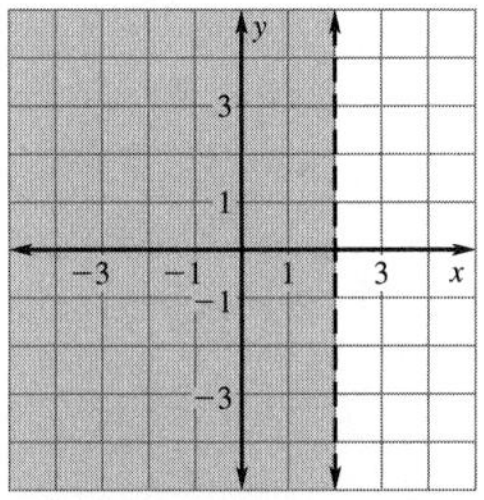

39.

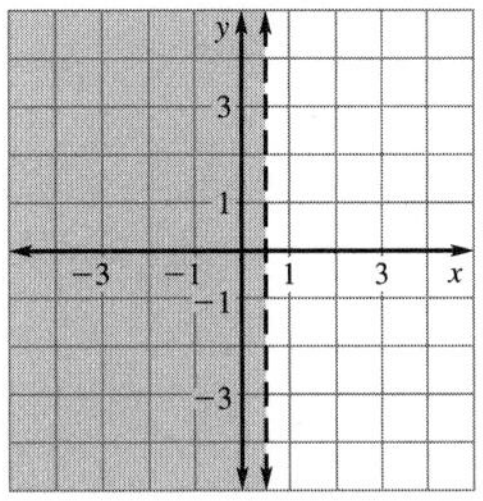

40.

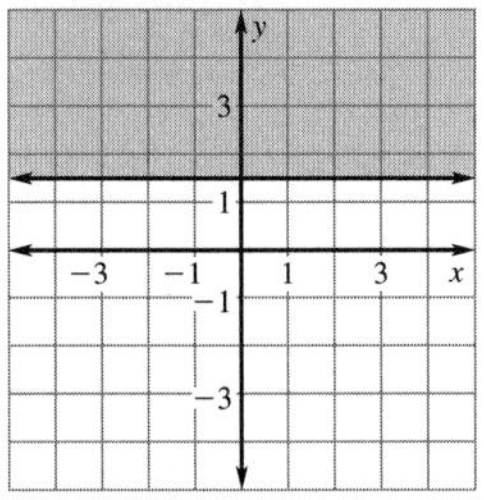

41.

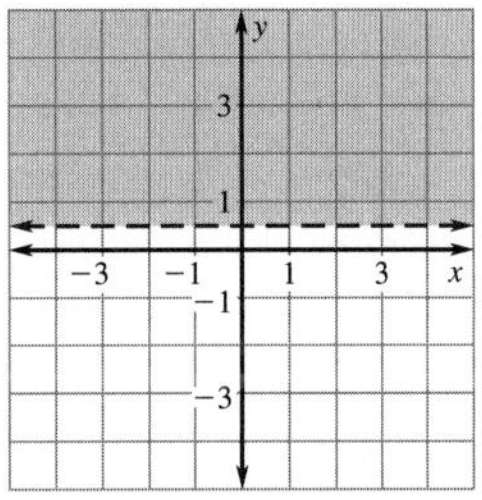

42.

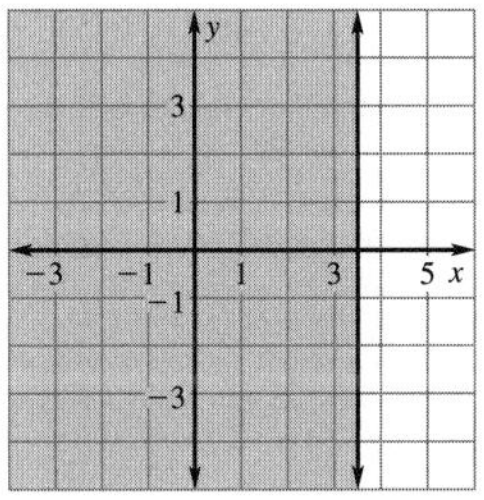

49.

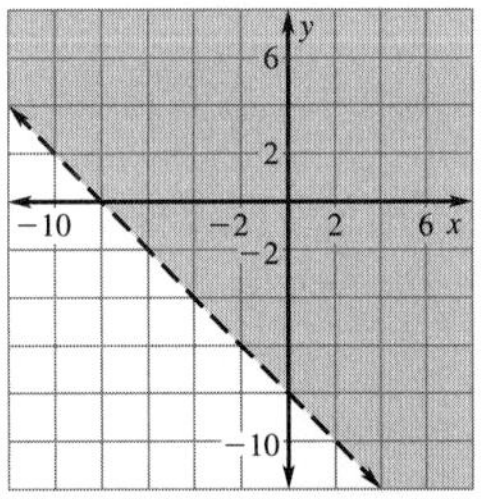

50.

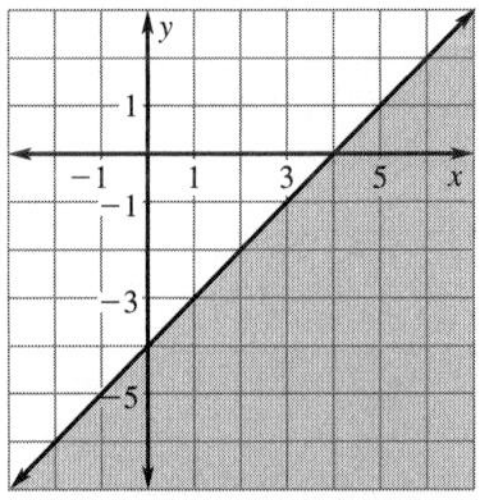

51.

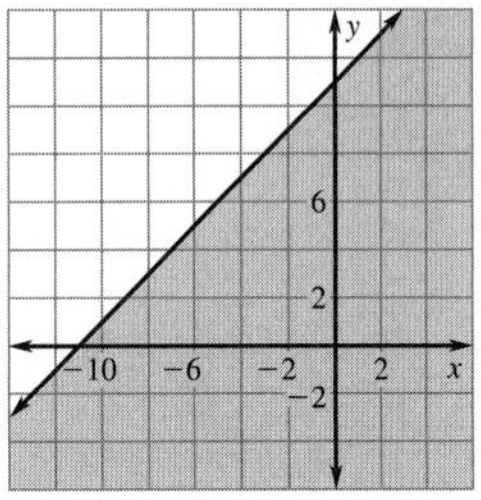

52.

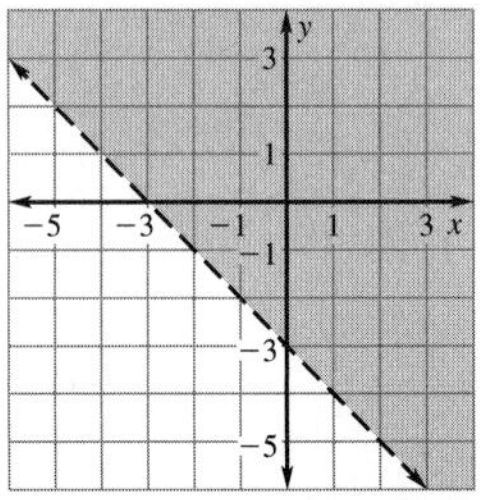

53.

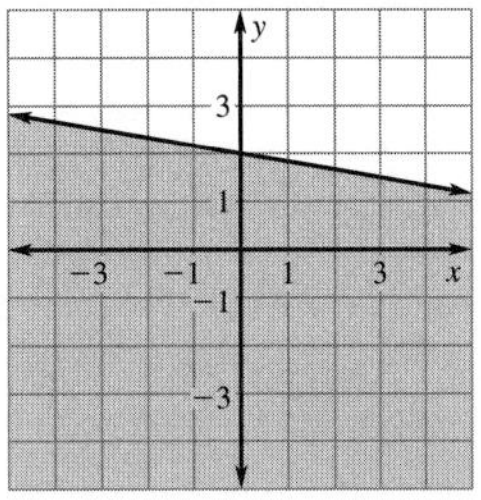

54.

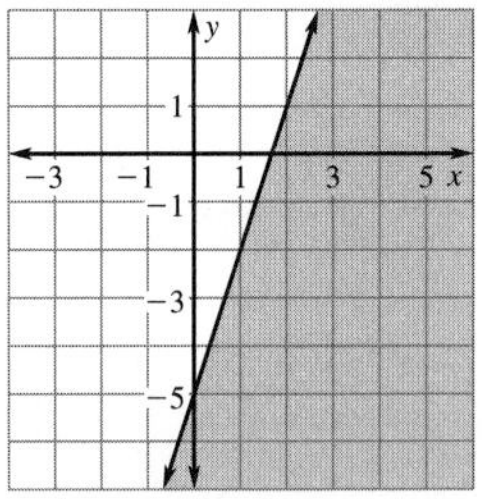

55.
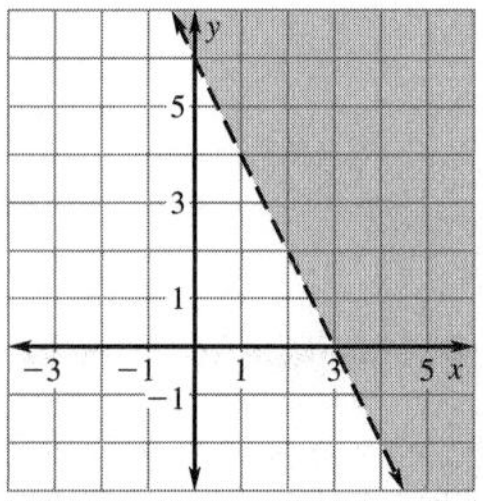

56.
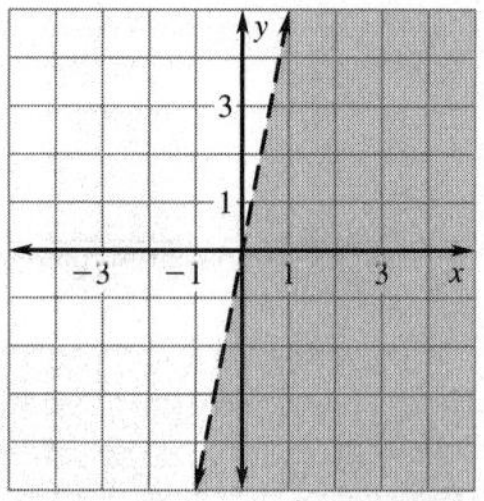

57.
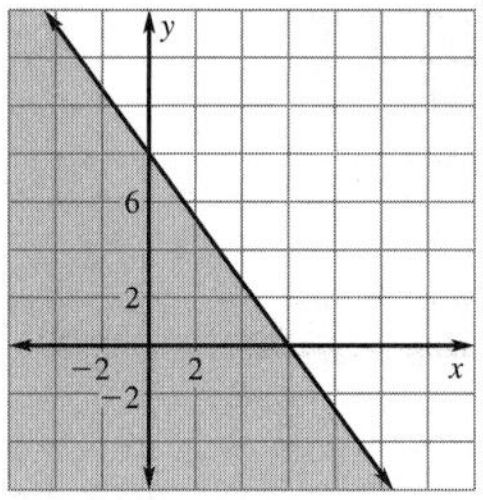

58.
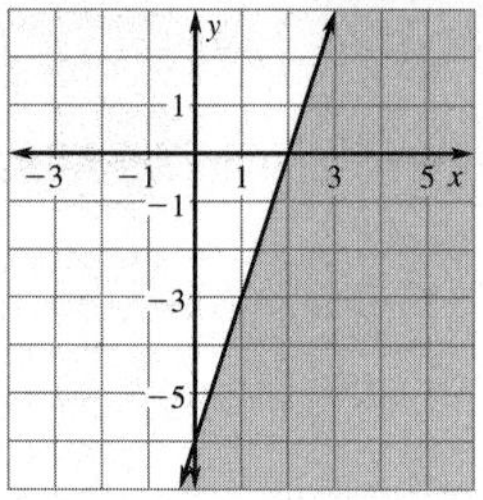

59.
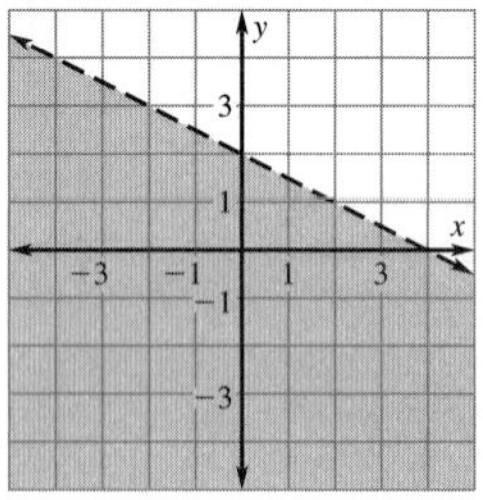

60.
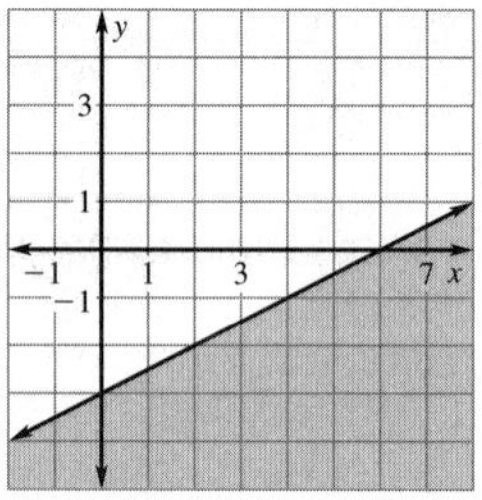

17.
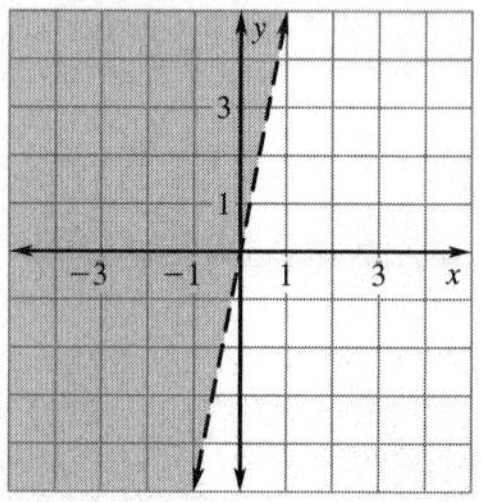

18.
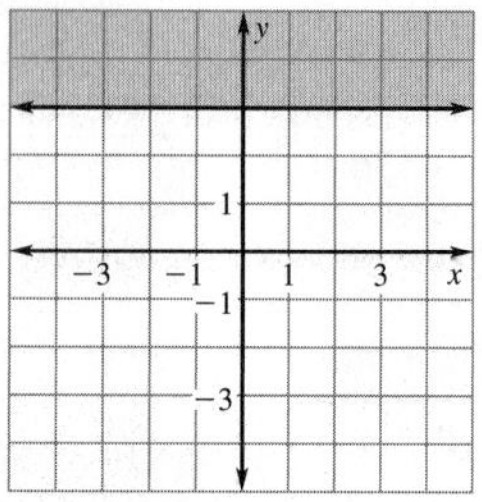

19.
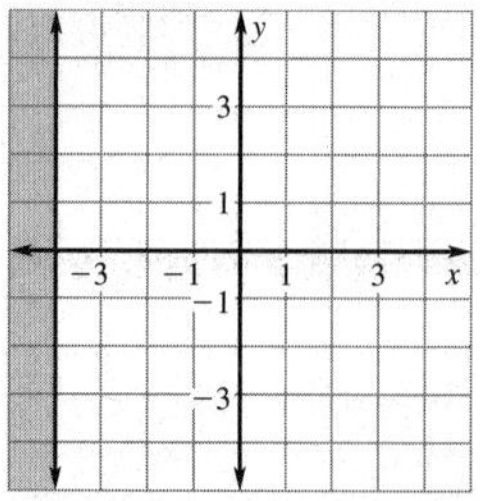

20.
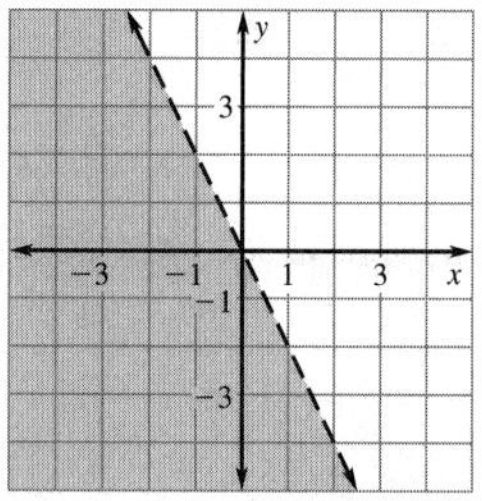

21.
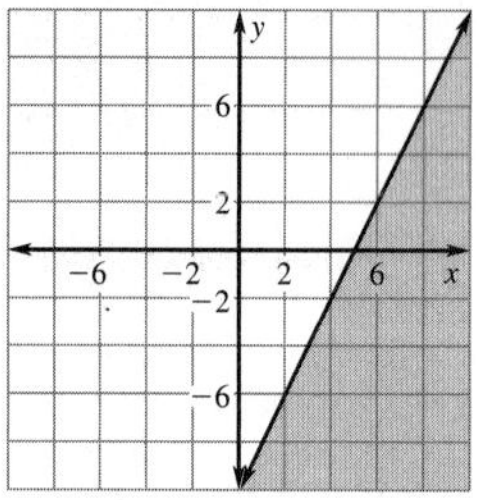

22.
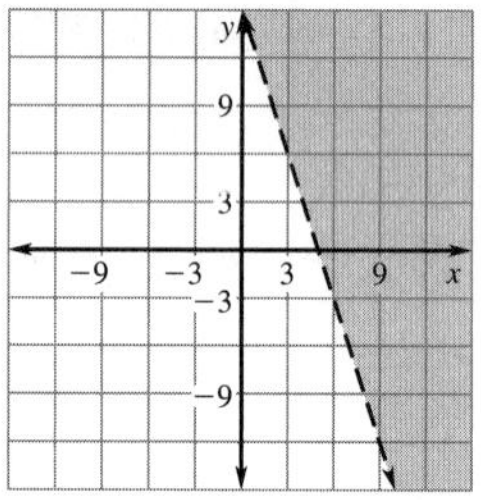

6.5 Mixed Review (p. 366)

79.
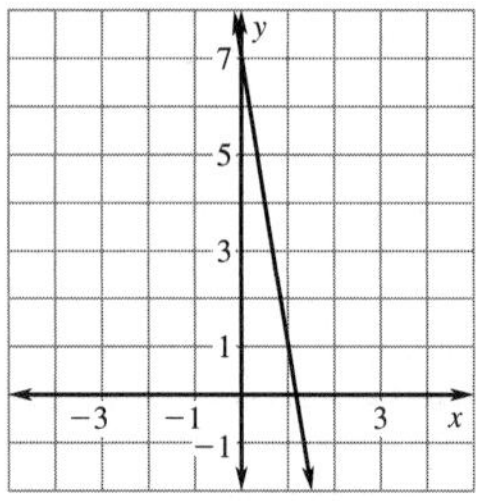

80.
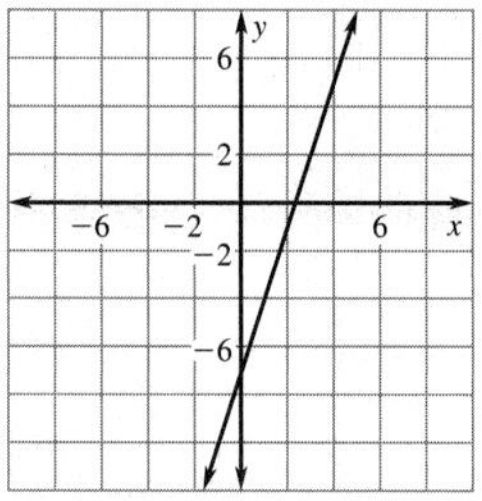

81.
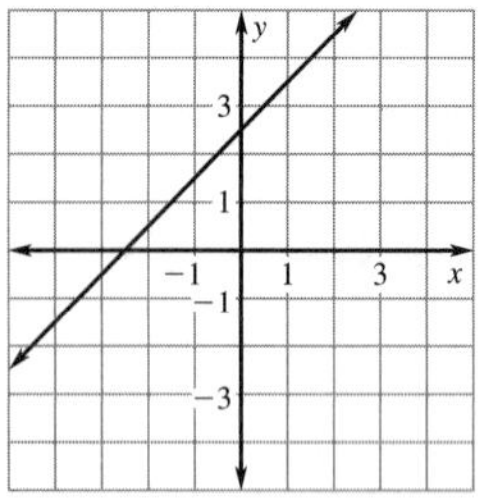

82.
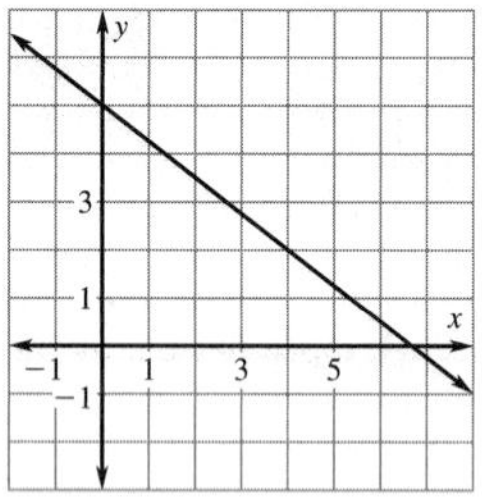

Quiz 2 (p. 366)

7.
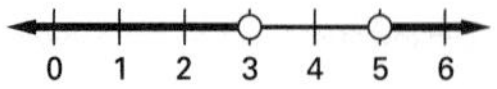

8.
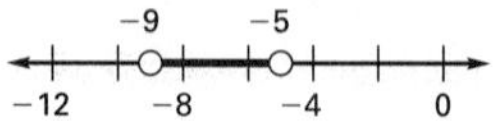

9.
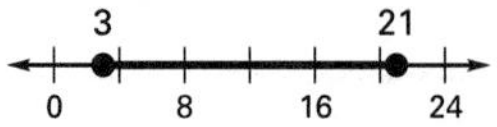

10.
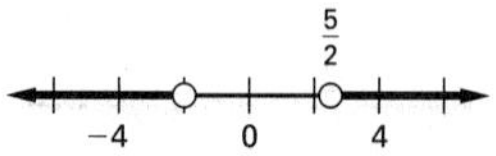

11.
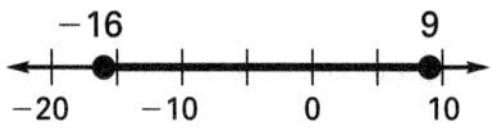

12.
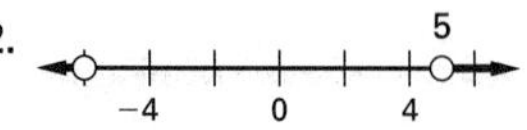

6.5 Technology Activity (p. 367)

1, 2, 5, 6, 9, 10. The lines are not part of the solution.

1.
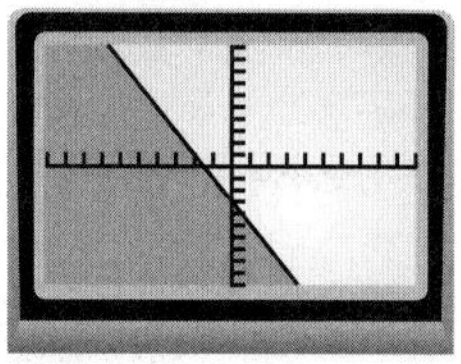

2.
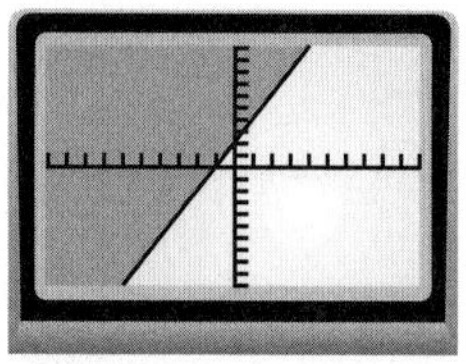

3.
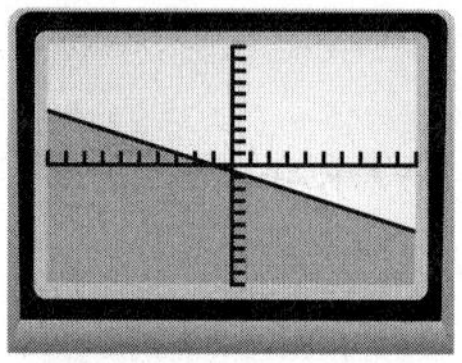

4.
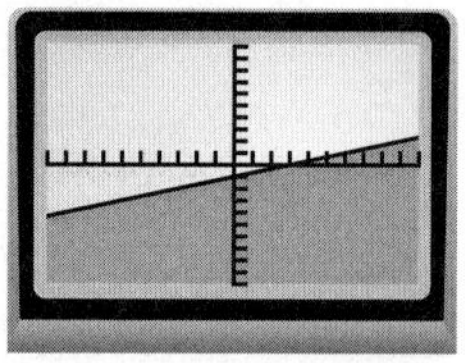

5.
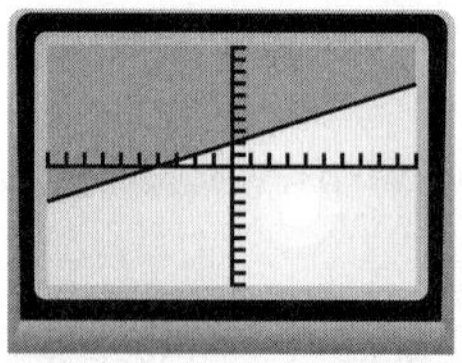

6.
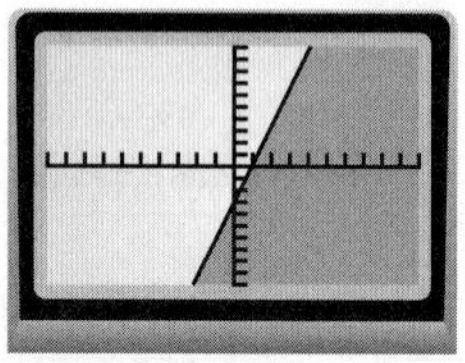

7.
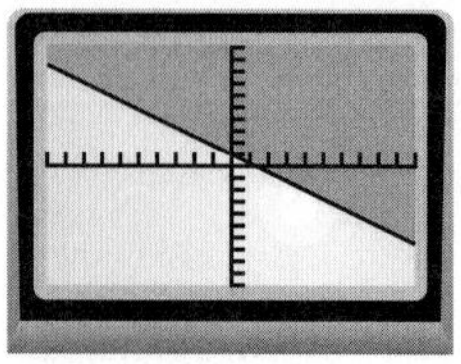

8.
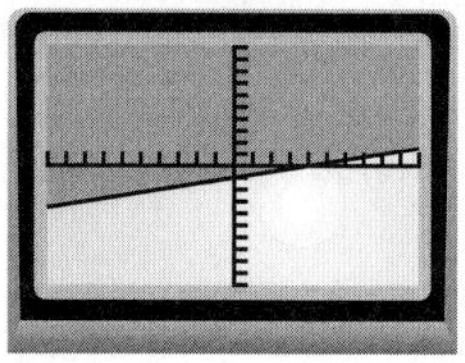

9.

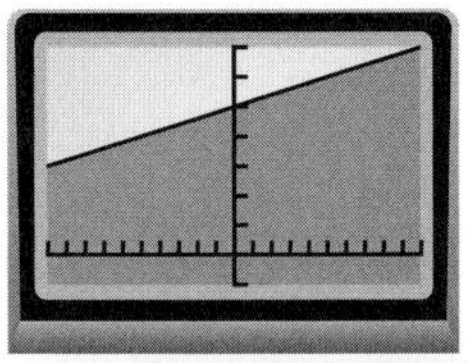

10.

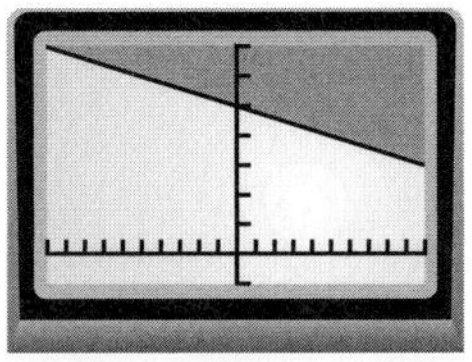

11.

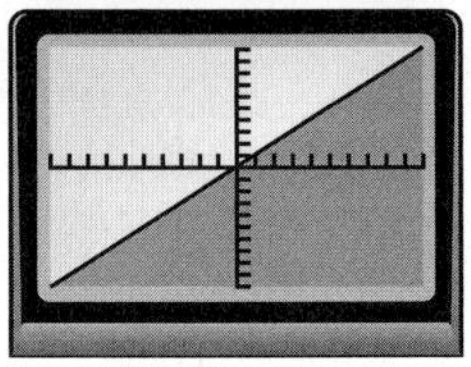

12.

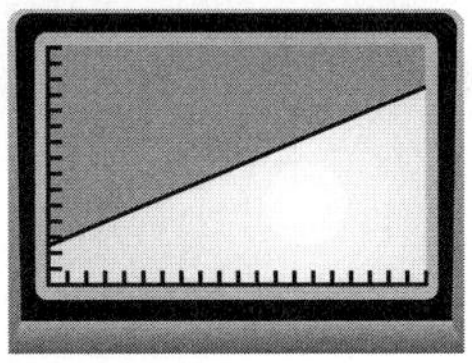

6.6 Guided Practice (p. 371)

2. Answers may include the following. A stem-and-leaf plot makes it simple to: order data; group data into subdivisions; determine the median of the data; determine the range of the data; compare all data values to a particular value.

4.

1	1 3
2	1 2 2
3	0 4 8 9
4	6 8
5	5

Key: 5 | 5 = 55

6.6 Practice and Applications (pp. 371–374)

33.

1	3 24 25
2	3 20 22
3	17
4	1 8 14 17 30 30
5	10
6	3 5 13 24
7	31
8	21 26
9	12
10	11 17
11	4 11 28
12	9 15 28

Key: 12 | 9 = 12–9 (Dec. 9)

1–3, 1–24, 1–25, 2–3, 2–20, 2–22, 3–17, 4–1, 4–8, 4–14, 4–17, 4–30, 4–30, 5–10, 6–3, 6–5, 6–13, 6–24, 7–31, 8–21, 8–26, 9–12, 10–11, 10–17, 11–4, 11–11, 11–28, 12–9, 12–15, 12–28

40. Male:

2	9 9
3	0 1 1 2 3 6 8
4	0 0
5	1 2
6	0 3
7	0
8	1

Key: 8 | 1 = 81

Female:

2	6 9 9
3	1 1 1 2 5 8
4	2 2 4
5	0 5 7
6	
7	
8	2 8

Key: 8 | 2 = 82

6.6 Mixed Review (p. 374)

54.

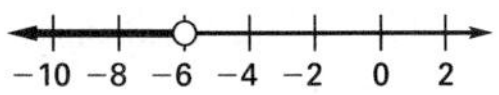

55.

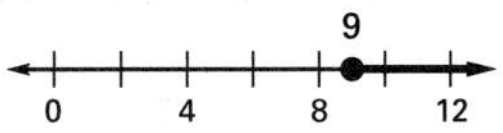

56.

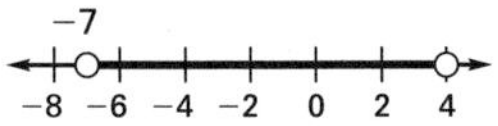

57.

58.

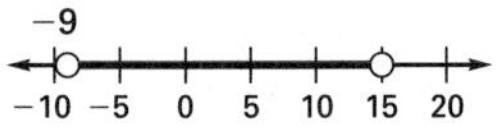

59.

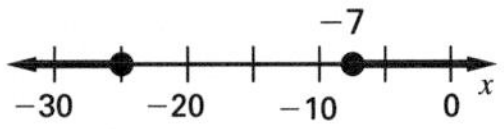

6.7 Guided Practice (p. 378)

8. **Ages of Horses**

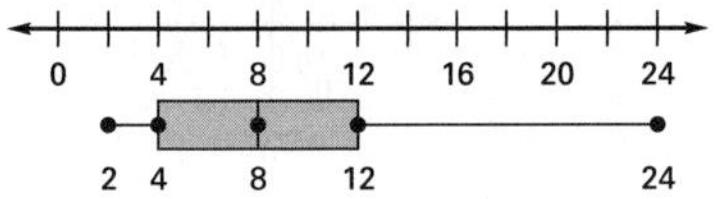

6.7 Practice and Applications (pp. 378–380)

15.

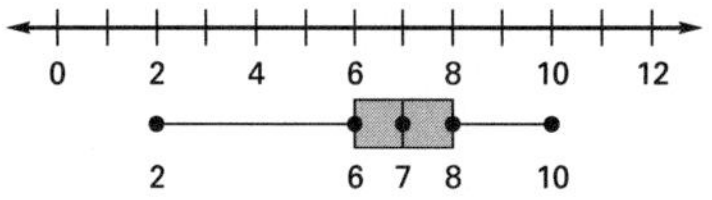

16.

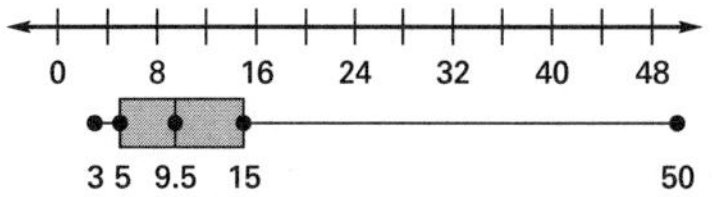

17.

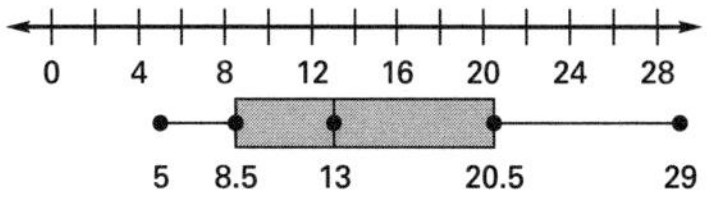

18.

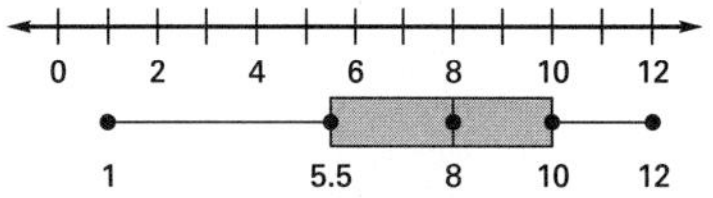

19. **Days of Rainfall**

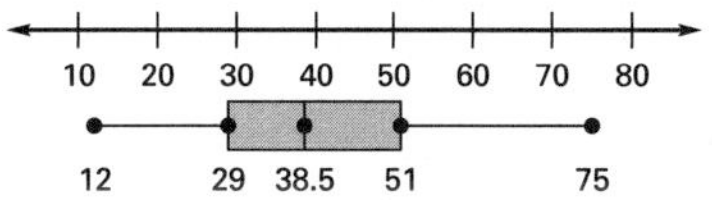

20. **Weights of Dogs**

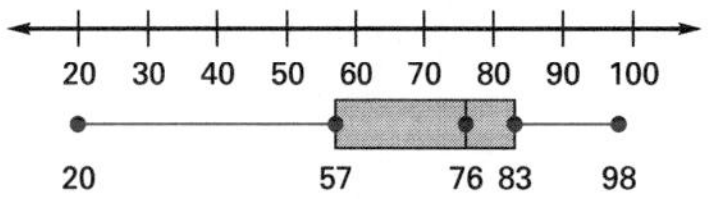

21. **Movie Attendance**

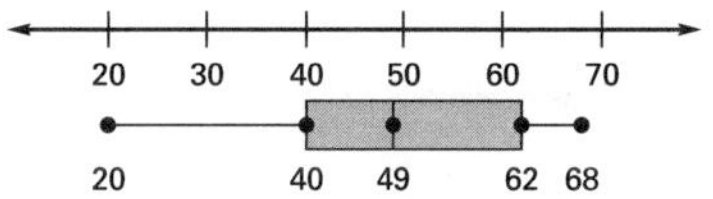

22. **Weekly Tips**

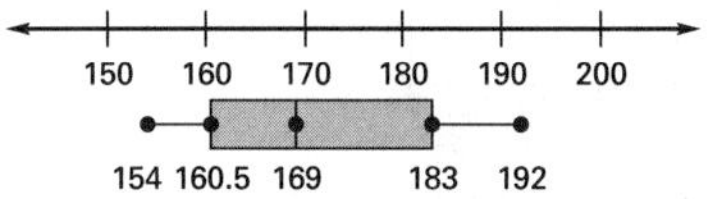

25.

3	6 8
4	8
5	
6	2 7
7	0 1 7
8	4 8

Key: 8 | 4 = 8.4 oz

26. **Phone Weights (oz)**

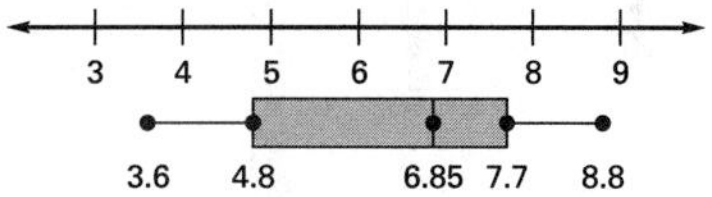

27.

6	0
7	0 5
8	0
9	0 0 0
10	
11	
12	
13	5
14	0
⋮	
26	0

Key: 14 | 0 = 140 min

28. **Operating Times (min)**

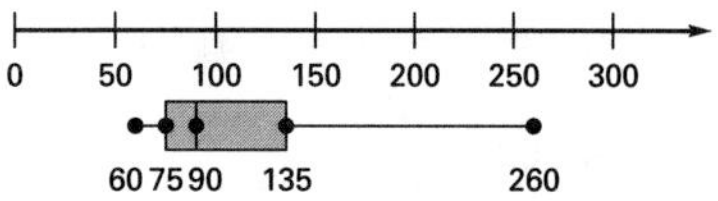

29. *Sample answer*: The lowest 25% of the data are clumped closely together while the upper 25% of the data extend far to the right due to an outlier. The range of the lower three fourths of the data is less than half the range of all the data. The median operating time is less than the mean. At least one phone has a high operating time, but most do not.

ADDITIONAL ANSWERS

30.

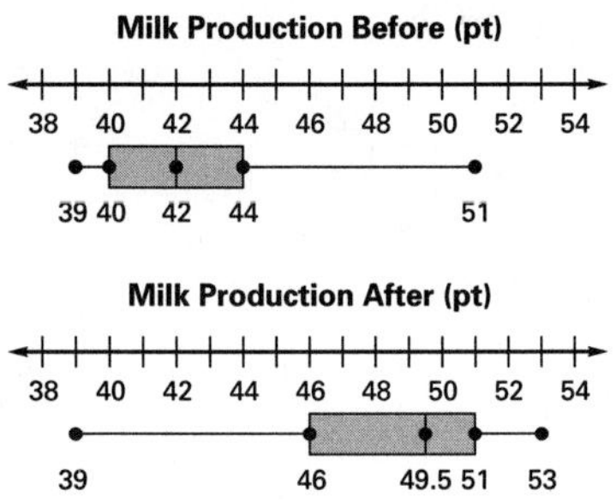

QUIZ 3 (p. 381)

5.

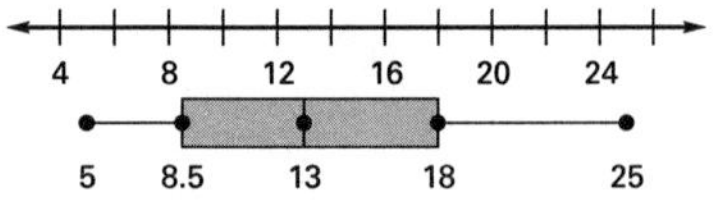

6.

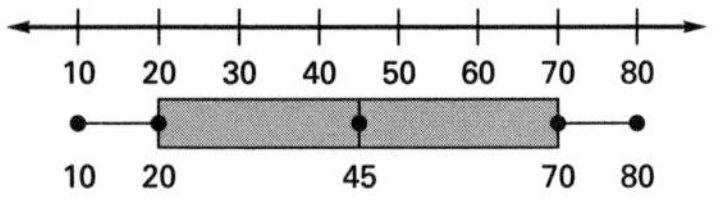

7.

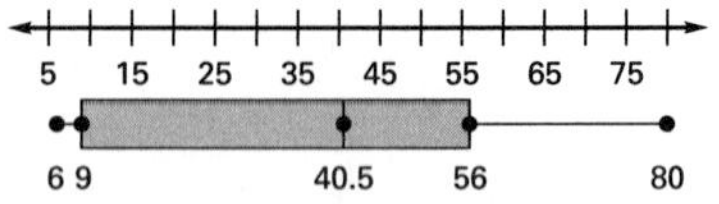

8.

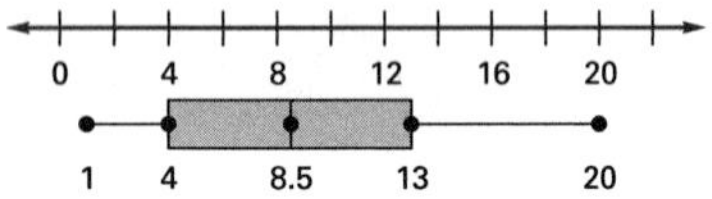

6.7 TECHNOLOGY ACTIVITY (p. 382)

1.

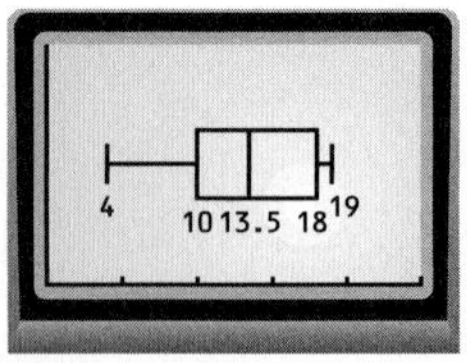

2.

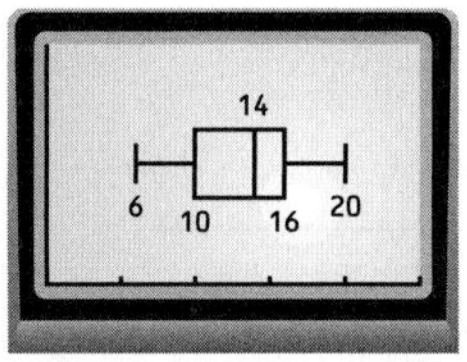

3.

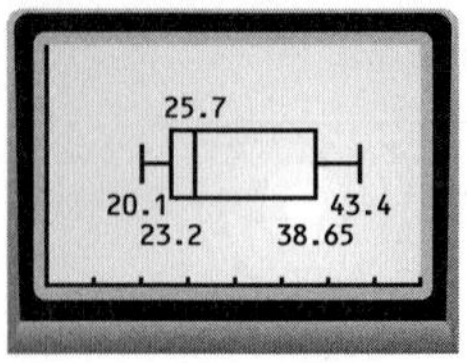

4.

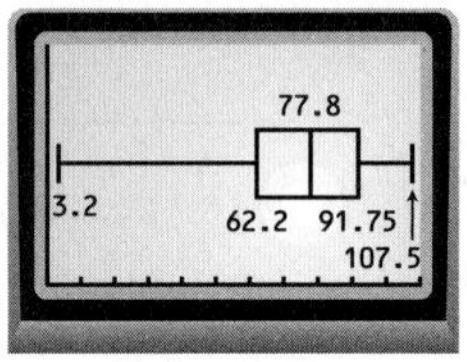

CUMULATIVE PRACTICE, CHAPTERS 1–6 (pp. 390–391)

78.

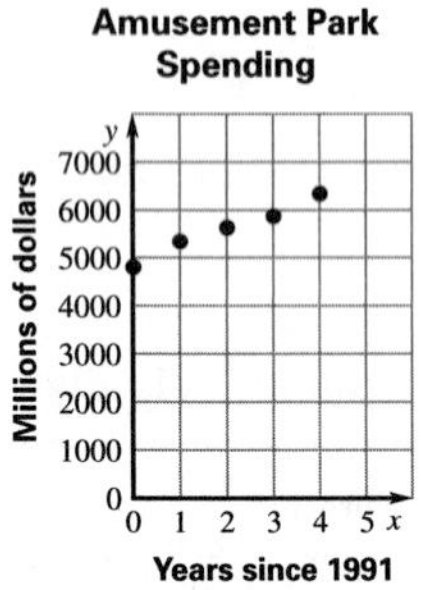

CHAPTER 7

7.2 PRACTICE AND APPLICATIONS (pp. 408–410) **16.** *Sample answer:* I would isolate d in the second equation because it is easier to do so than to isolate d in the first equation or c in either equation.

48.

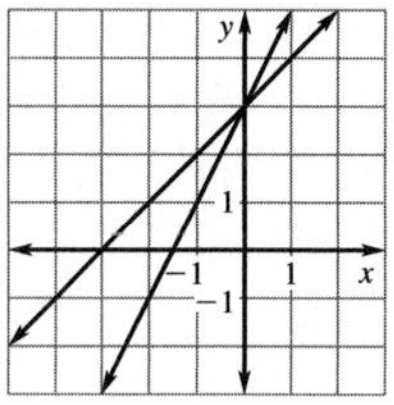

7.2 MIXED REVIEW (p. 410)

58.

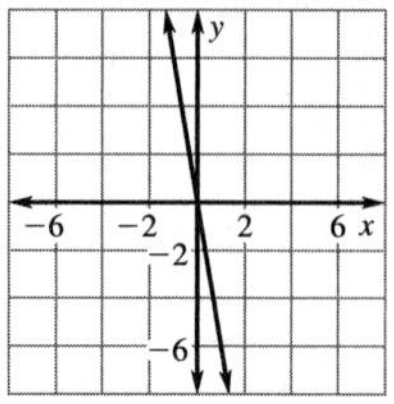

59.

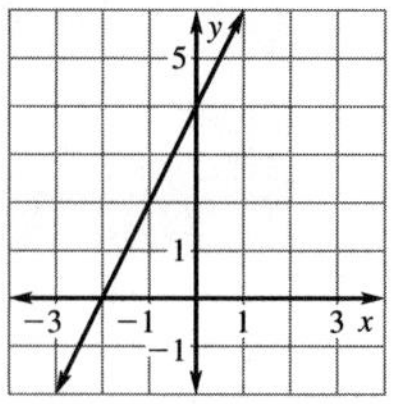

60.

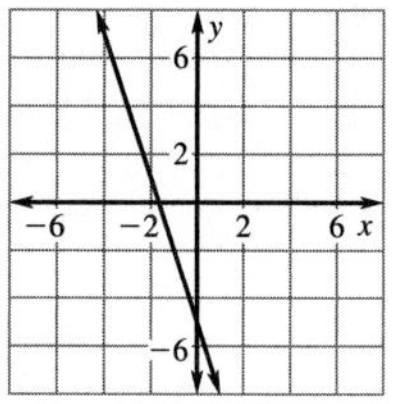

61.

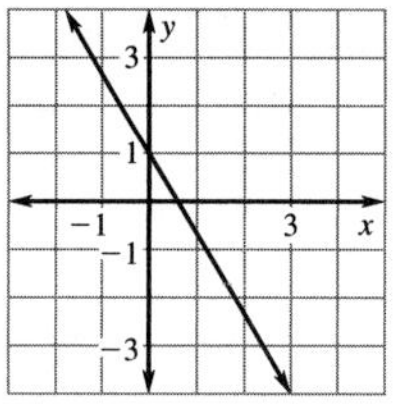

62.

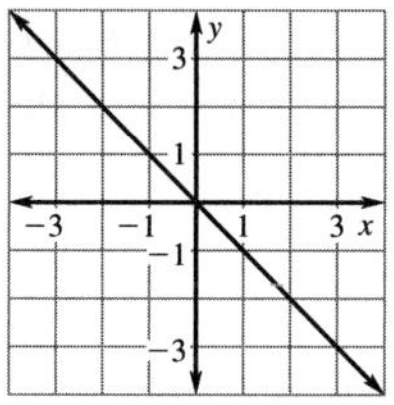

63.

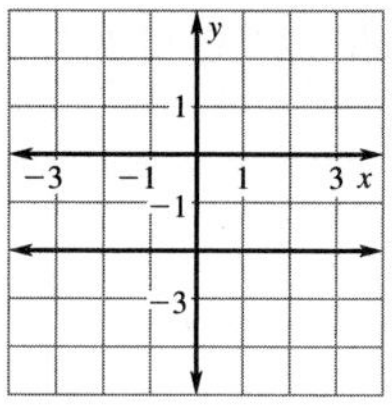

64.

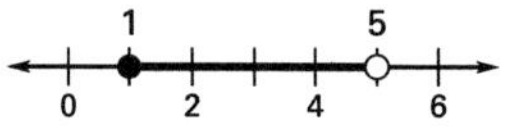

65.

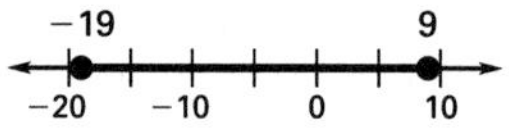

66.

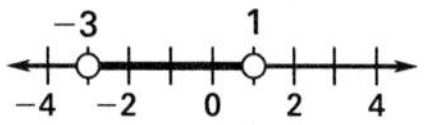

67.

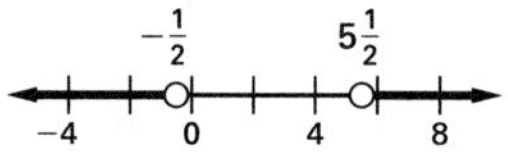

68.

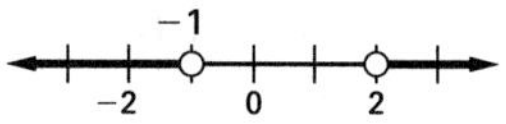

7.5 CONCEPT ACTIVITY (p. 425)

Step 1. a.

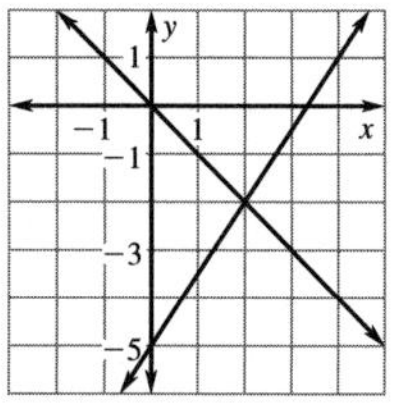

b.

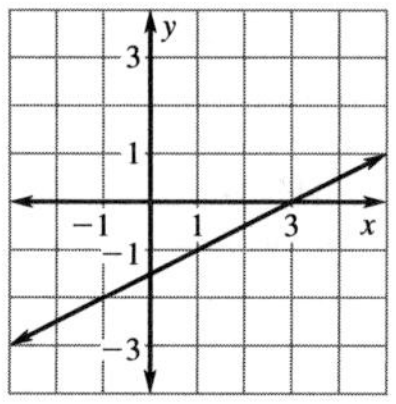

c.

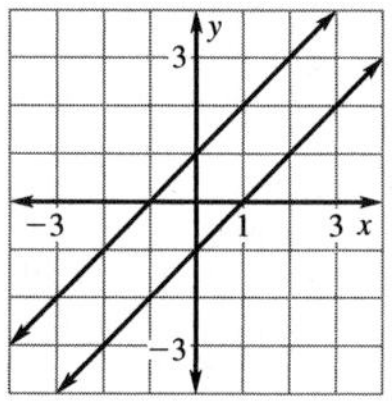

1. a.

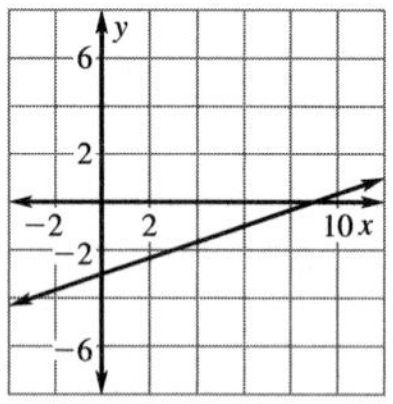

b.

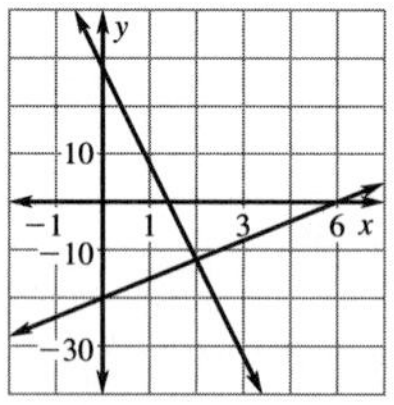

c.

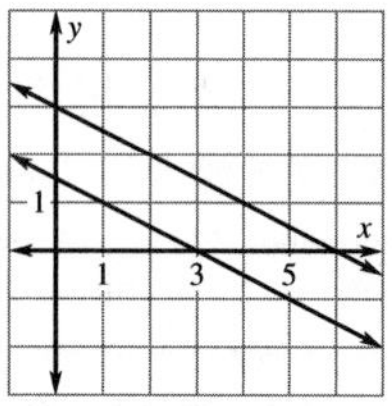

7.5 GUIDED PRACTICE (p. 429)

6.

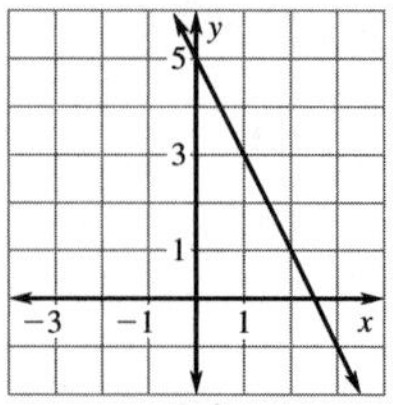

7.

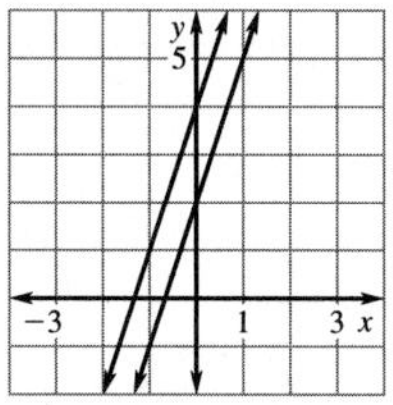

8.

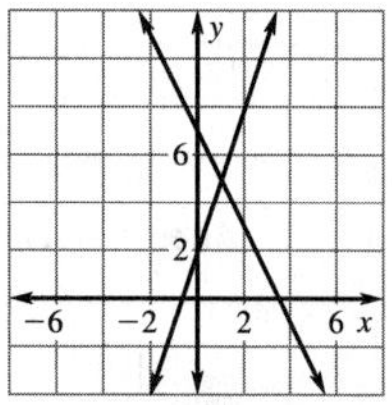

7.6 GUIDED PRACTICE (p. 435)

3. Choose a point in the region you think is correct and substitute its coordinates in each of the inequalities in the system.

ADDITIONAL ANSWERS

4. The student graphed $y > 1$ instead of $y > -1$; the boundary line for $y > -1$ should be a dashed line rather than a solid line; the student graphed $x \le 2$, instead of $x \ge 2$; the student shaded the wrong half-plane for $y > x - 4$; the student drew the boundary line for $y > x - 4$ as a solid line rather than a dashed line.

5. 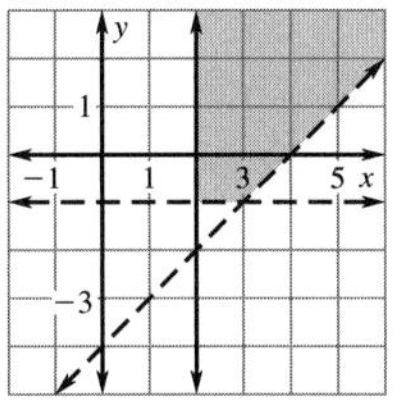

6. 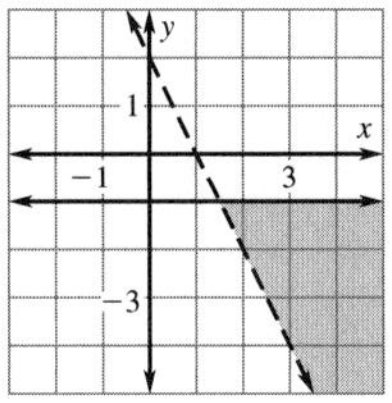

7. 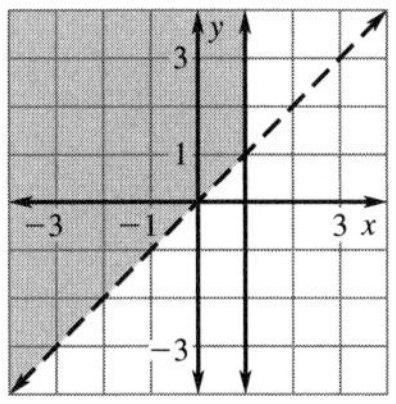

8. 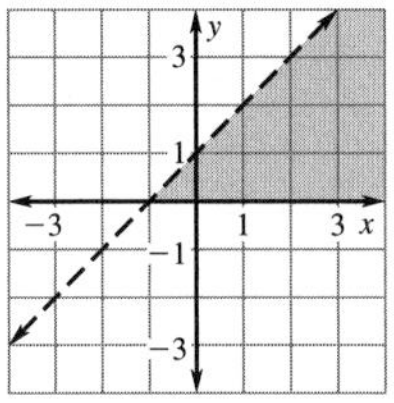

7.6 Practice and Applications (pp. 435–437)

15. 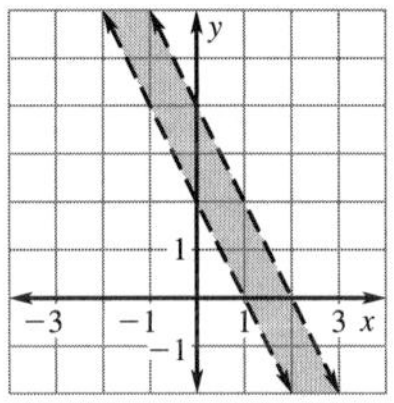

16. 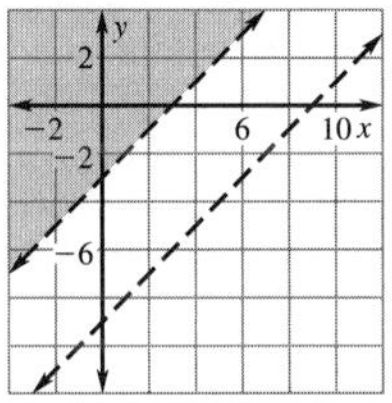

17. 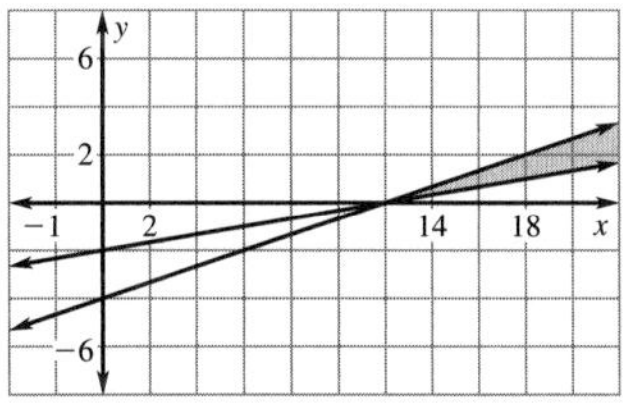

18. 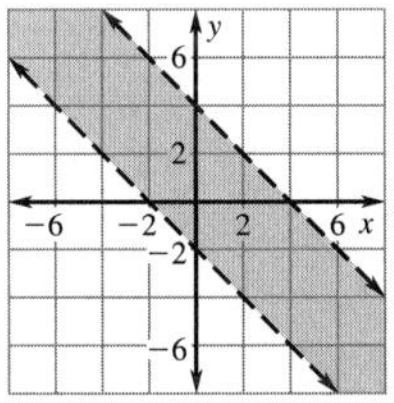

19. 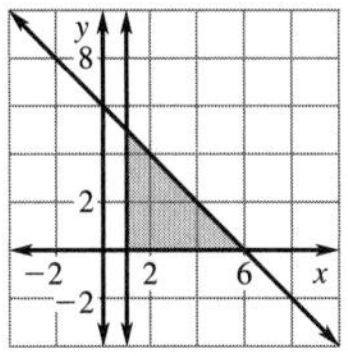

20. 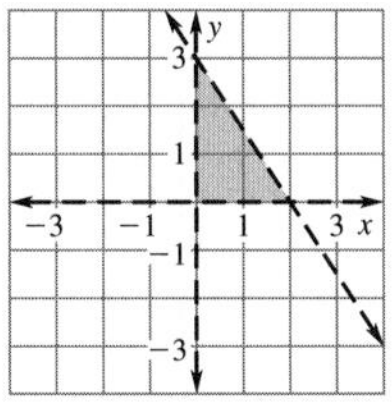

21. 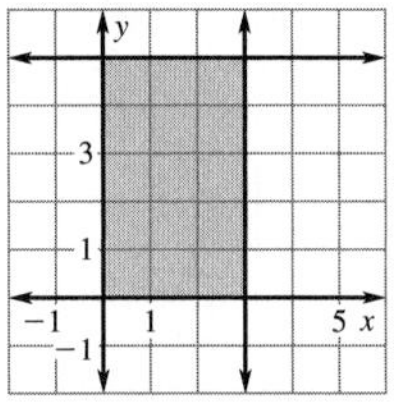

22. 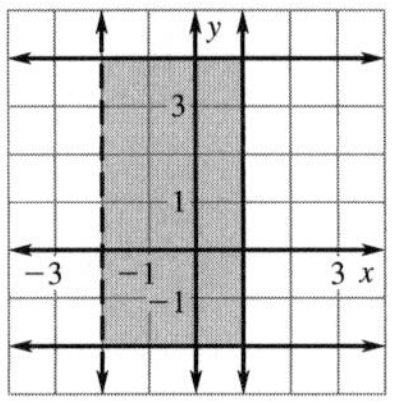

23. 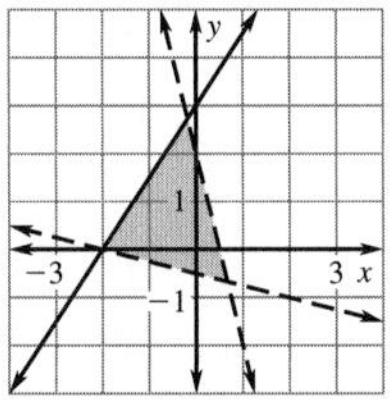

24. 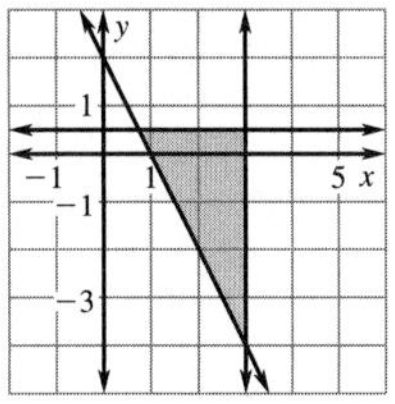

25. 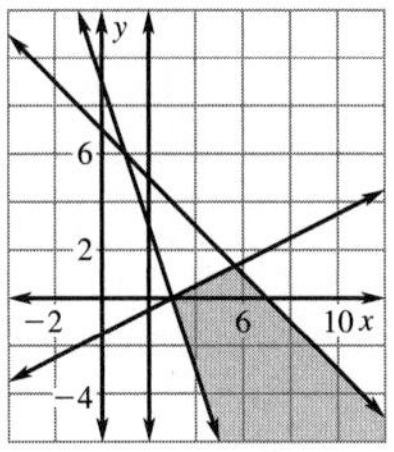

26. 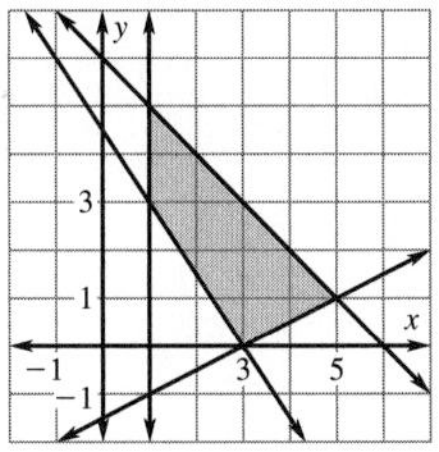

35. 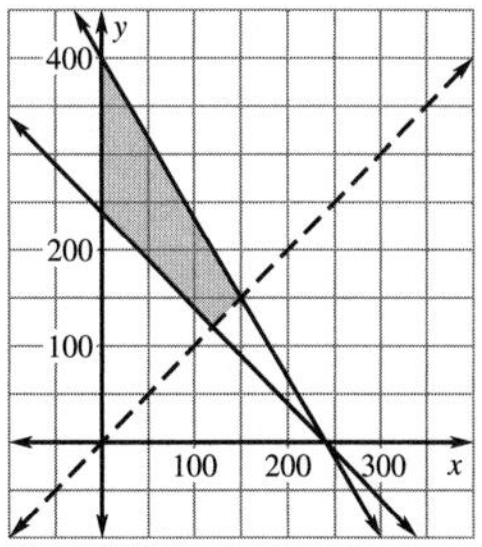

36. 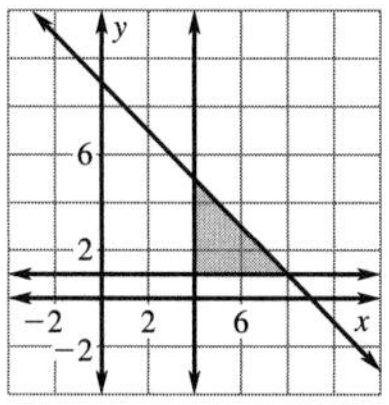

38. 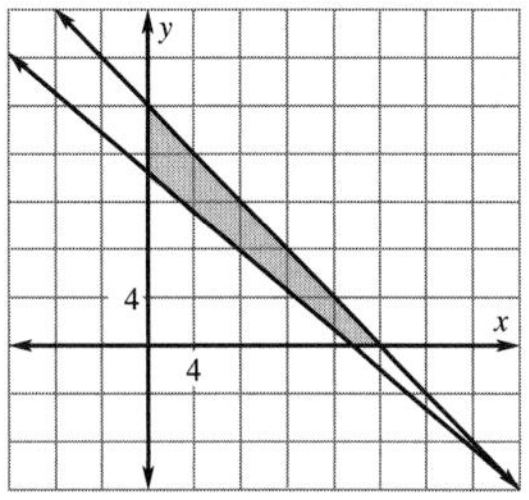

40. 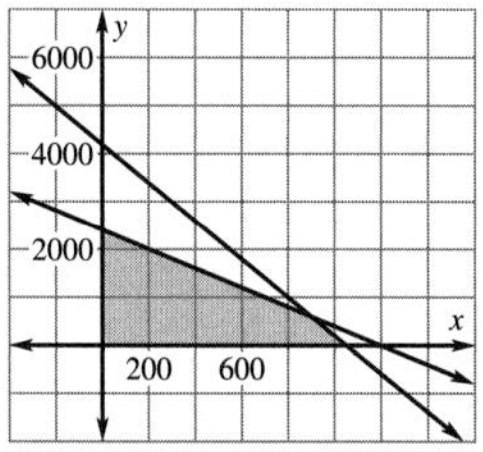

CHAPTER 8

8.1 Mixed Review (p. 455)

92. 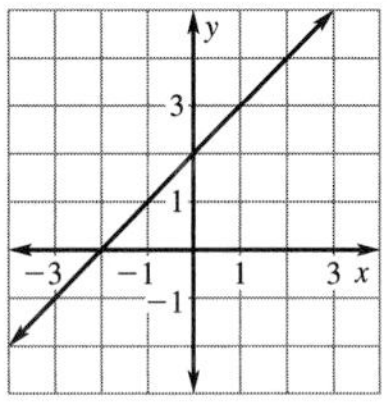

93. 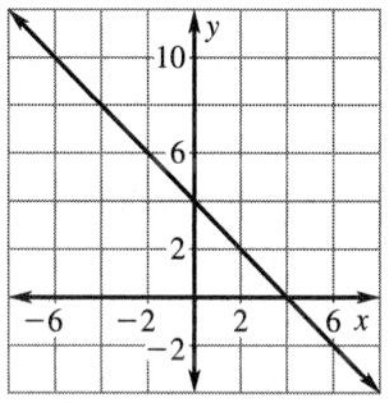

94. 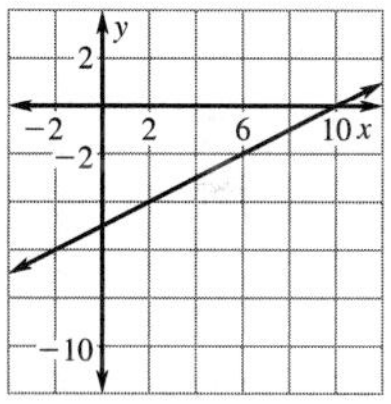

95. 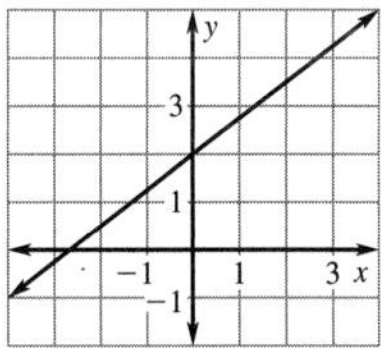

96. 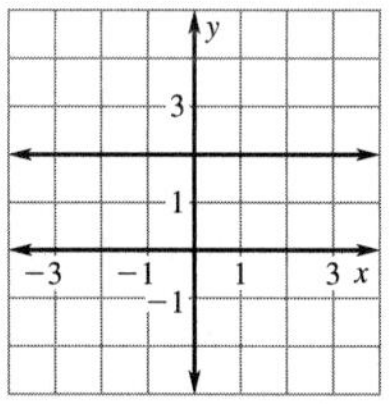

97. 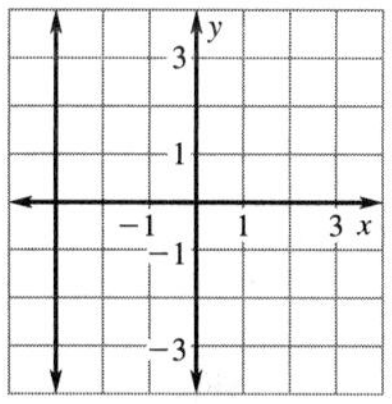

98. 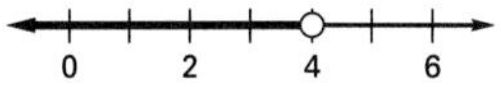

99. 

100. 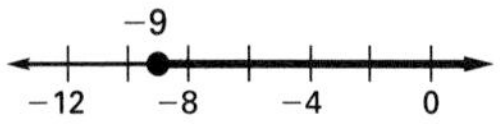

101. 

8.2 Practice and Applications (pp. 459–461)

57. 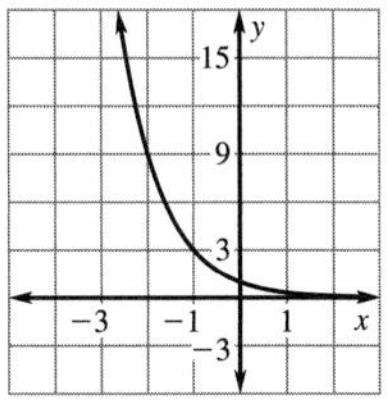

58. 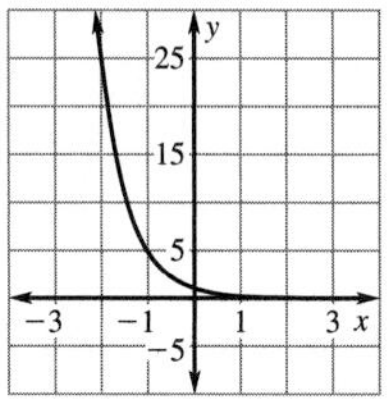

59. 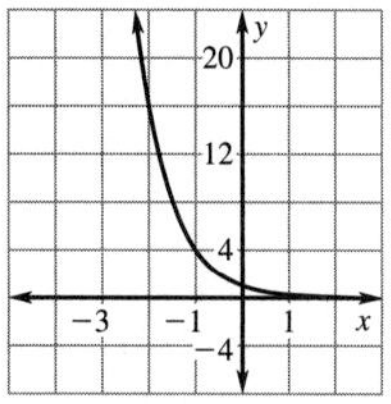

60. 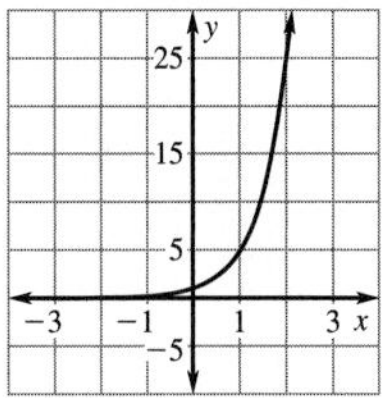

61. 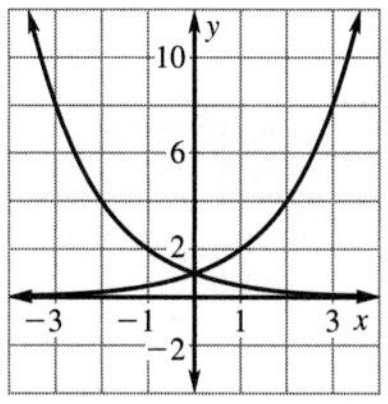

62. 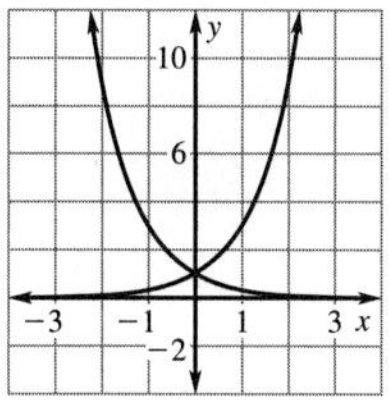

65. b. 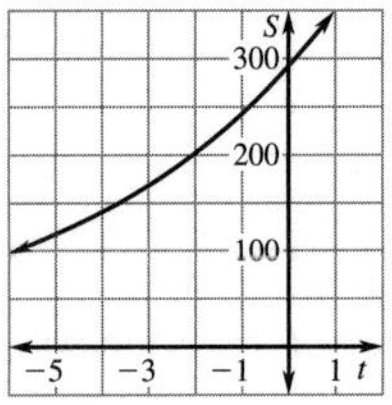

8.2 Mixed Review (p. 461)

75. 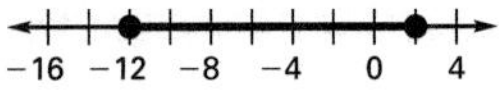

76. 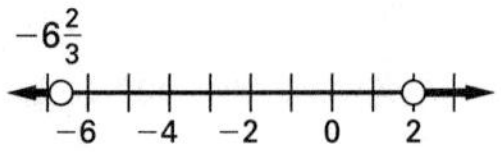

77. 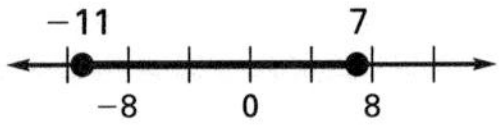

78. 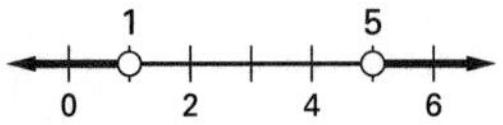

79. 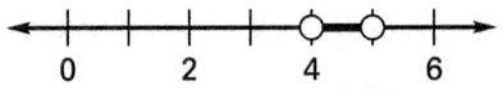

80. 

81. 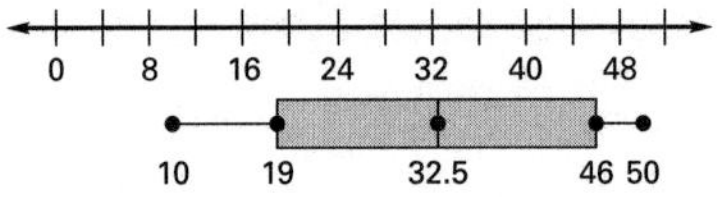

82. 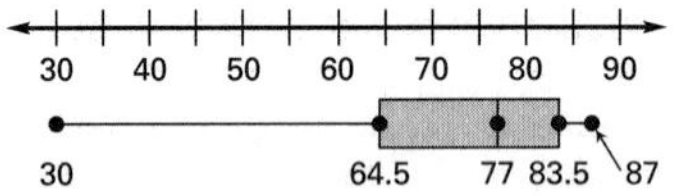

8.3 Mixed Review (p. 469)

71. 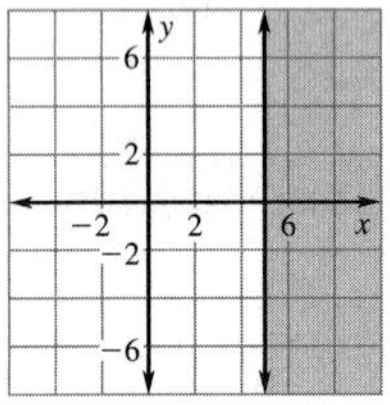

72. 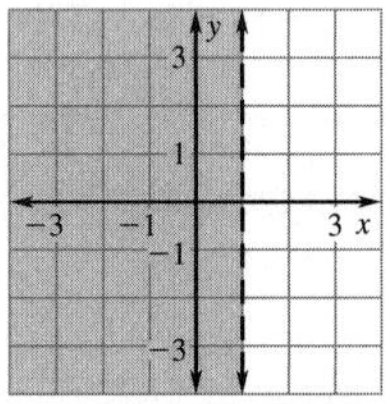

73. 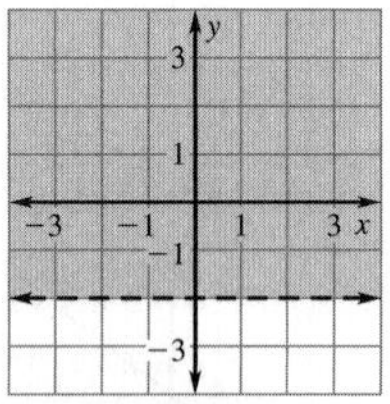

74. 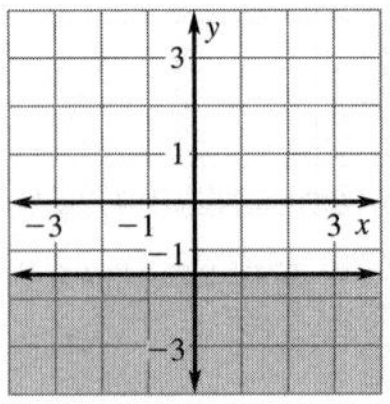

75. 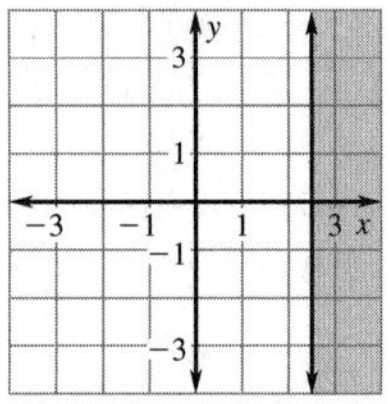

76. 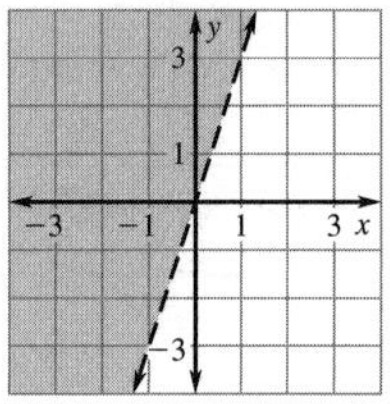

77. 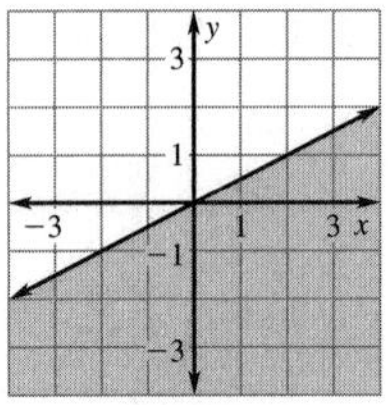

78. 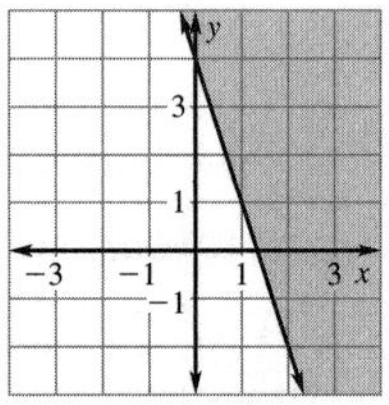

8.4 Mixed Review (p. 475)

77. 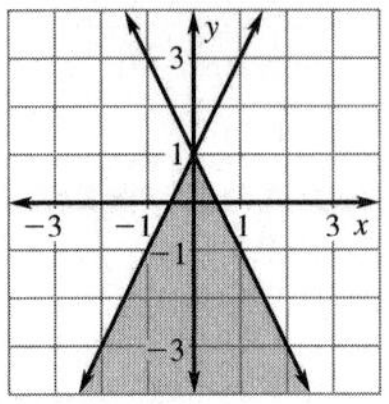

78. 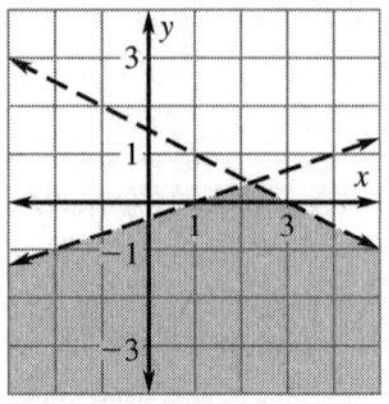

79. 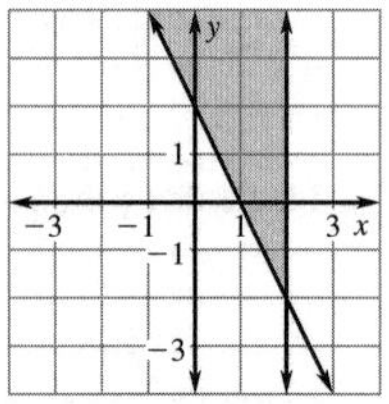

8.6 Concept Activity (p. 483)

4. *Sample answer:* All the graphs go through (0, 100). All the points lie along more or less smooth curves that at first descend rapidly and then level off rapidly. The graphs show that by the time there were 7 tosses, most of the cups had no more pennies in them.

8.6 Practice and Applications (pp. 488–490)

33. 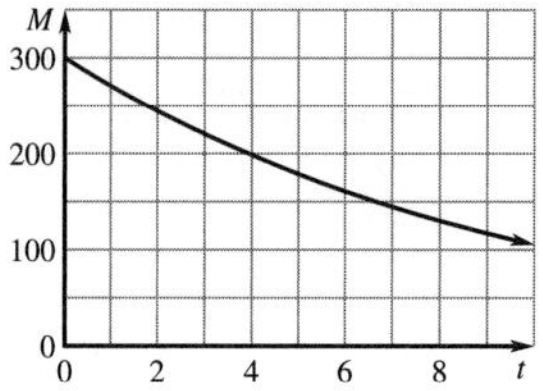

CHAPTER 9

9.2 Mixed Review (p. 516)

66. 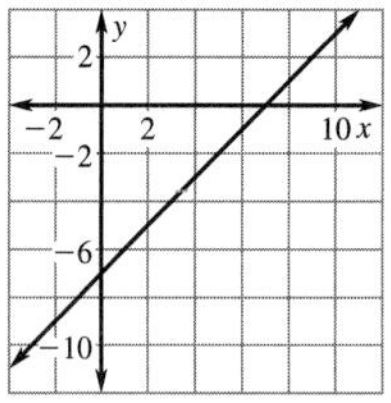

67. 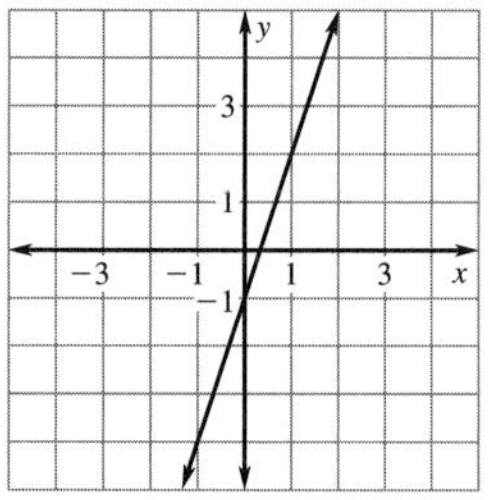

9.3 Concept Activity (p. 517)

4. 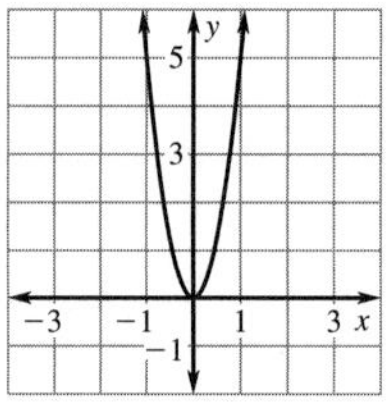

5. 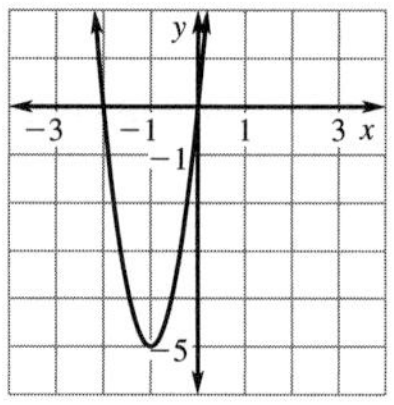

6. 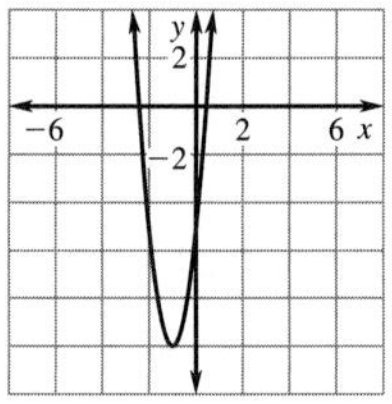

7. 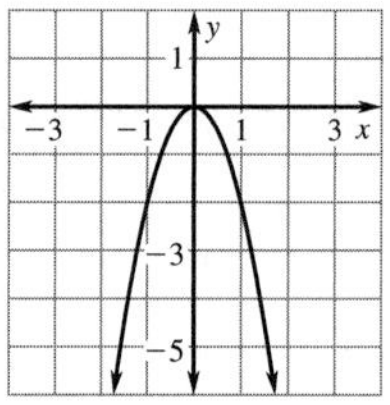

8. 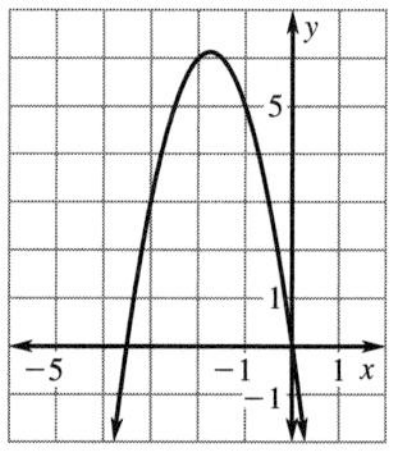

9. 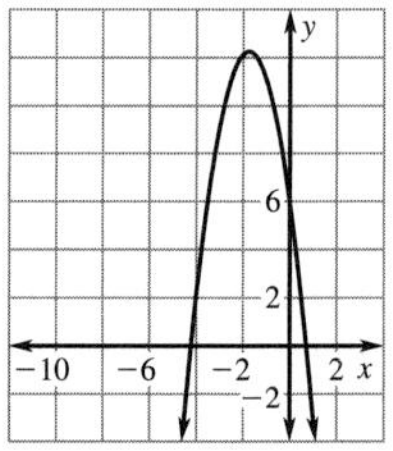

9.3 Guided Practice (p. 521)

11. 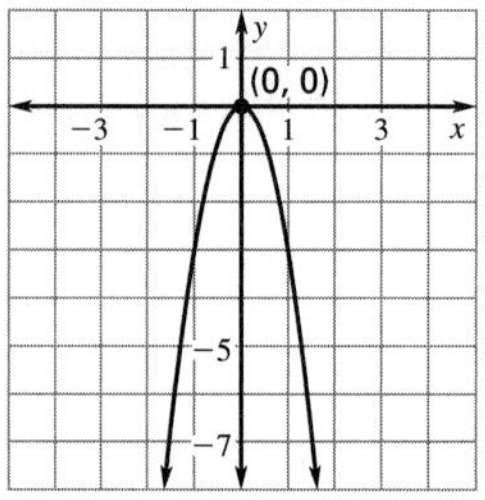

12. 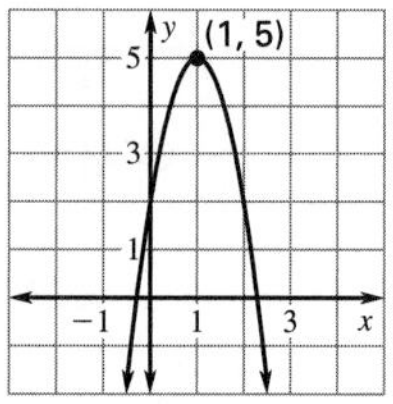

13.
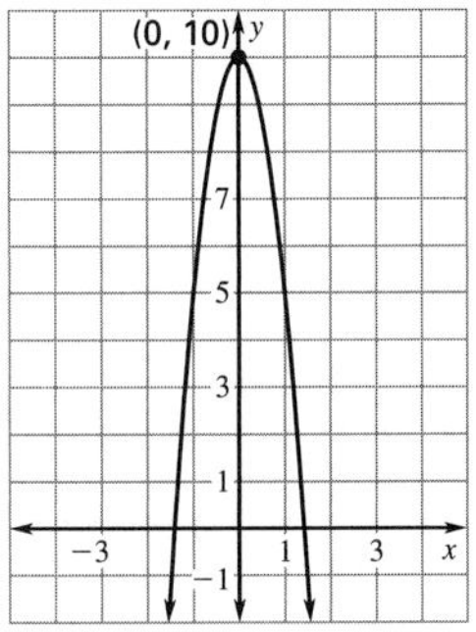

14.
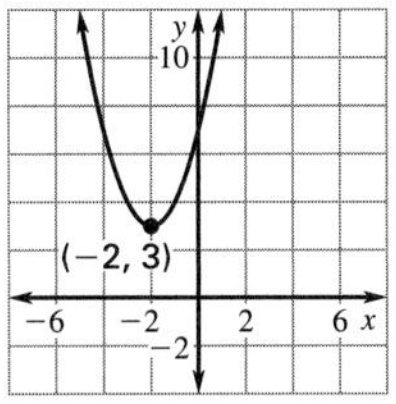

15.
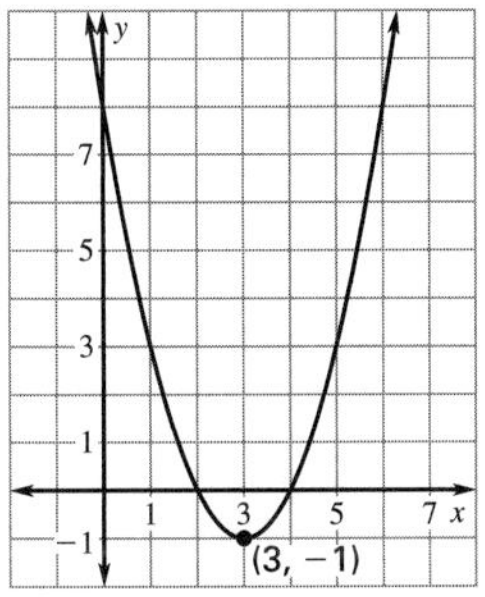

16.
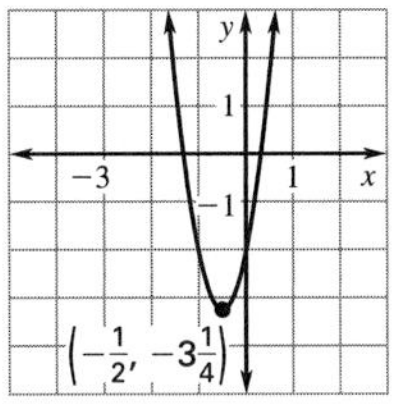

17.
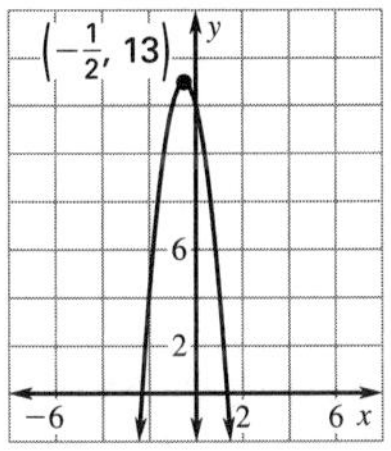

18.
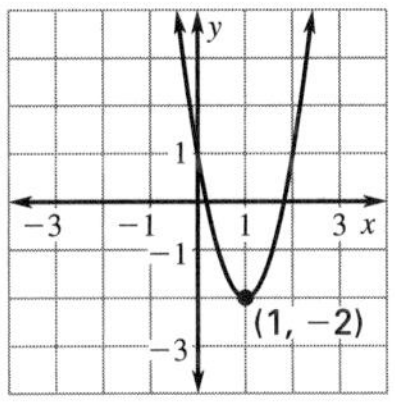

19.
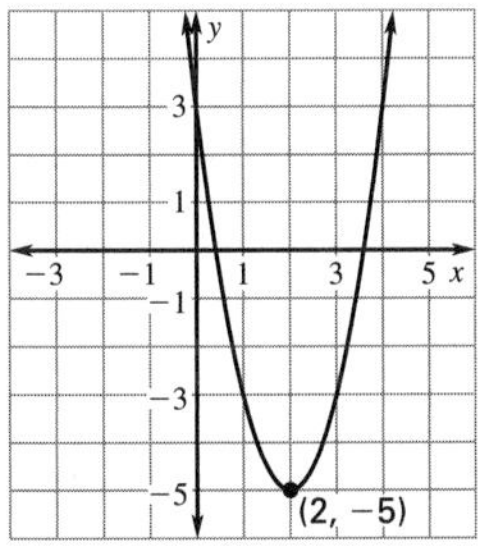

9.3 PRACTICE AND APPLICATIONS (pp. 521–523) **21. a.** up **b.** (0, 0) **c.** $x = 0$ **22. a.** down **b.** (0, 0) **c.** $x = 0$ **23. a.** up **b.** (0, 0) **c.** $x = 0$ **24. a.** up **b.** (0, 0) **c.** $x = 0$ **25. a.** down **b.** (0, 0) **c.** $x = 0$ **26. a.** down **b.** (0, 0) **c.** $x = 0$ **27. a.** down **b.** (0, 0) **c.** $x = 0$ **28. a.** up **b.** $\left(\frac{1}{10}, -\frac{1}{20}\right)$ **c.** $x = \frac{1}{10}$ **29. a.** up **b.** $\left(2\frac{1}{2}, -12\frac{1}{2}\right)$ **c.** $x = 2\frac{1}{2}$ **30. a.** down **b.** $\left(\frac{1}{7}, \frac{1}{7}\right)$ **c.** $x = \frac{1}{7}$ **31. a.** down **b.** $\left(\frac{3}{5}, 3\frac{3}{5}\right)$ **c.** $x = \frac{3}{5}$ **32. a.** up **b.** $\left(-\frac{1}{6}, 3\frac{5}{6}\right)$ **c.** $x = -\frac{1}{6}$ **33. a.** up **b.** (−1, 2) **c.** $x = -1$ **34. a.** down **b.** $\left(-\frac{1}{2}, 9\right)$ **c.** $x = -\frac{1}{2}$ **35. a.** up **b.** $\left(1\frac{3}{4}, -14\frac{1}{8}\right)$ **c.** $x = 1\frac{3}{4}$ **36. a.** up **b.** $\left(-1\frac{3}{4}, -27\frac{1}{8}\right)$ **c.** $x = -1\frac{3}{4}$ **37. a.** down **b.** (4, 48) **c.** $x = 4$ **38. a.** up **b.** $\left(-3, -11\frac{1}{2}\right)$ **c.** $x = -3$ **39. a.** up **b.** $\left(-\frac{1}{32}, -8\frac{1}{256}\right)$ **c.** $x = -\frac{1}{32}$ **40. a.** down **b.** $\left(\frac{1}{4}, -2\frac{3}{8}\right)$ **c.** $x = \frac{1}{4}$ **41. a.** up **b.** (2.56, −13.1) **c.** $x = 2.56$ **42. a.** up **b.** $\left(-\frac{2}{7}, -8\frac{2}{7}\right)$ **c.** $x = -\frac{2}{7}$ **43. a.** down **b.** (−0.35, 3.885) **c.** $x = -0.35$

46.
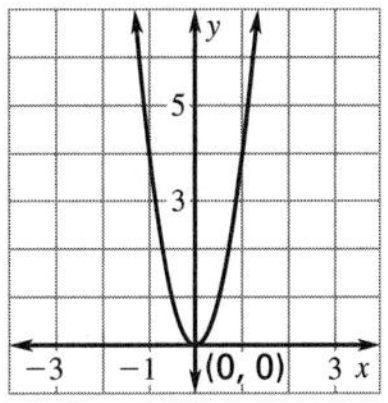

47.
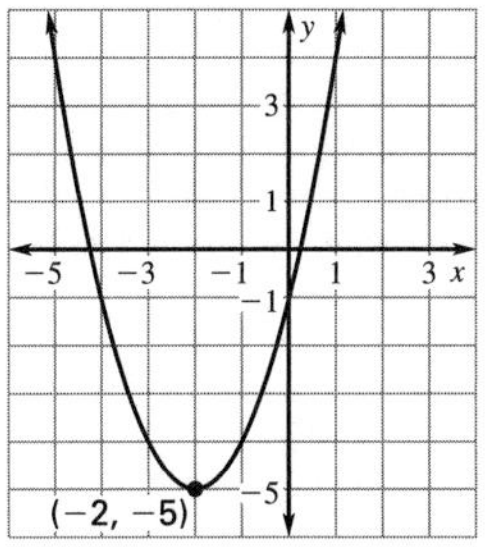

48.
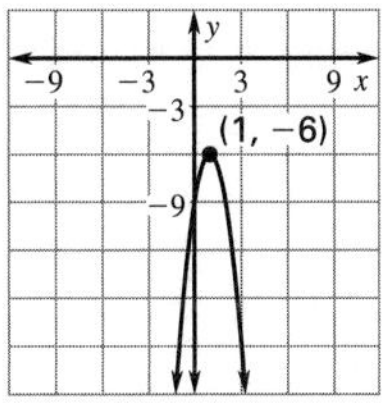

49.
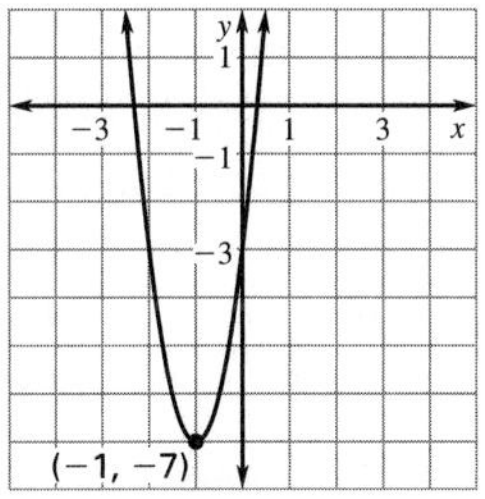

50.
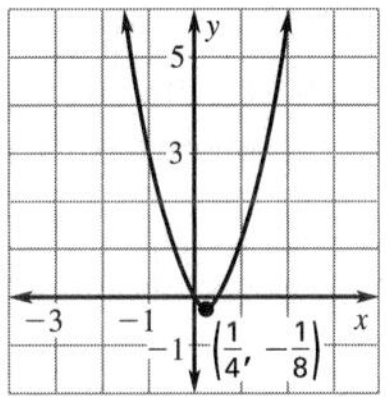

51.
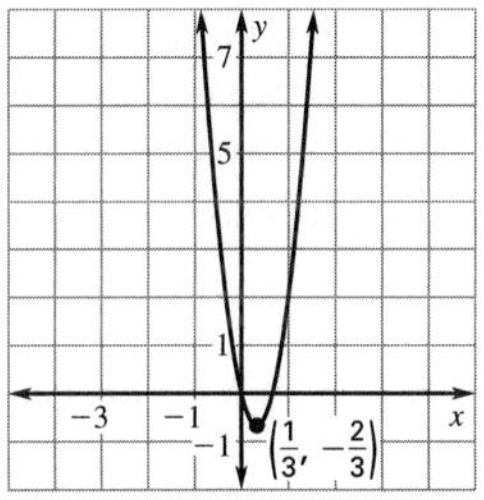

52.
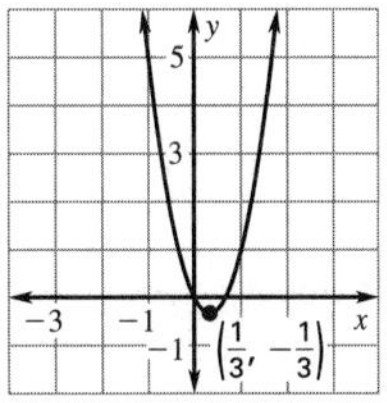

53.
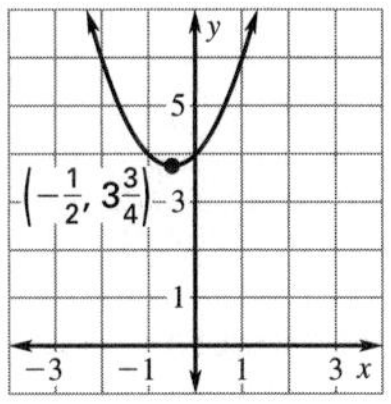

54.
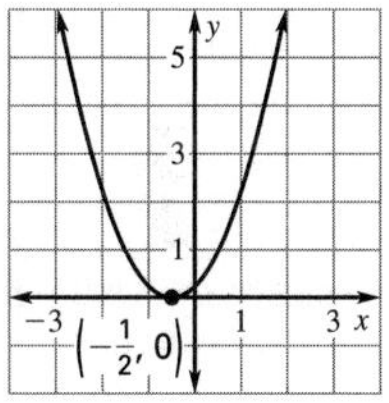

55.
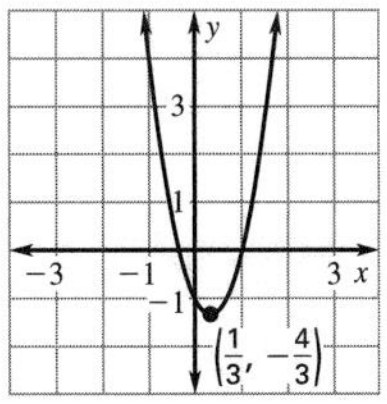

56.
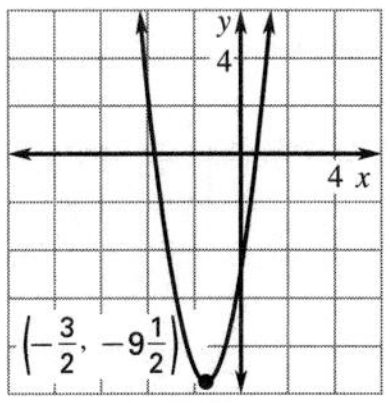

57.
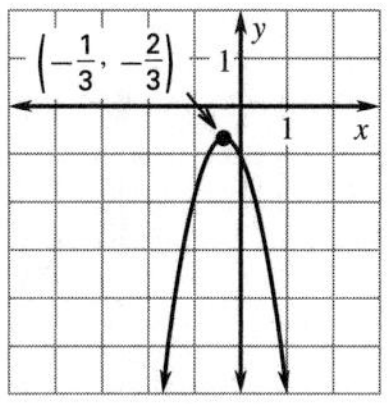

58.
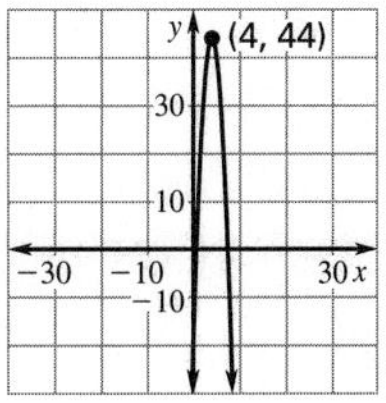

59.
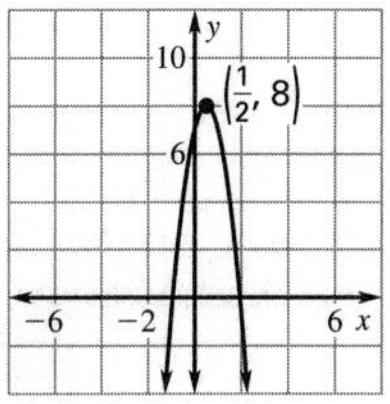

60.
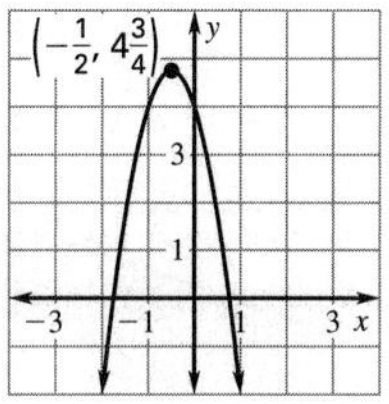

61. 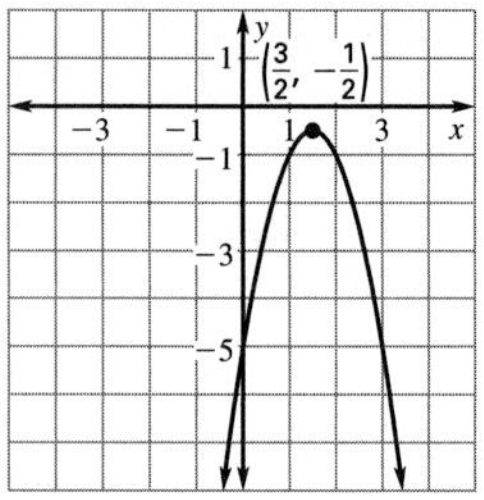

62. 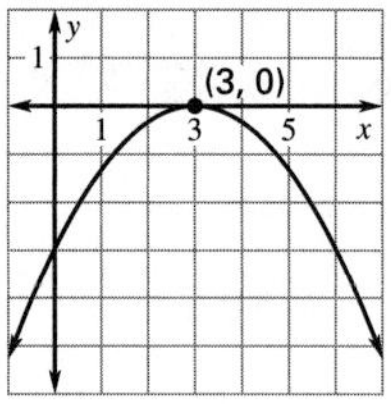

63. 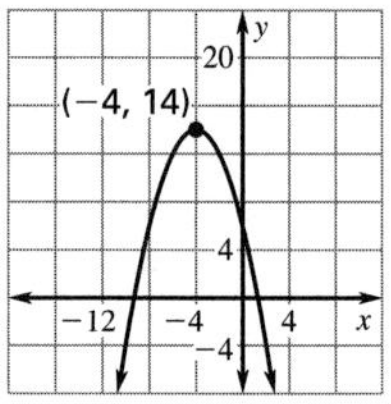

64. 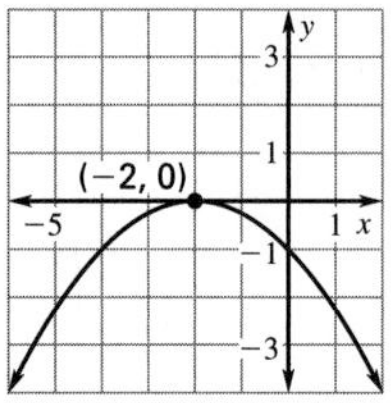

77. b. 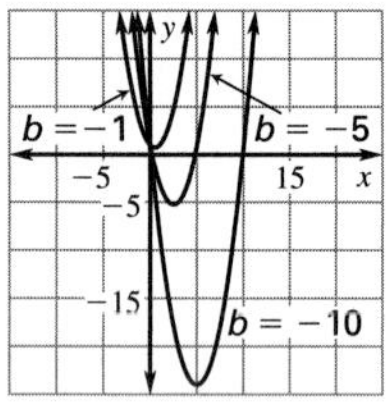

; The functions differ in the value of *b,* and vary in the position of the parabola. When *b* changes from –1 to –5, the vertex of the graph moves to the right and down. When *b* changes from –5 to –10, the vertex of the graph moves farther to the right and down.

c. 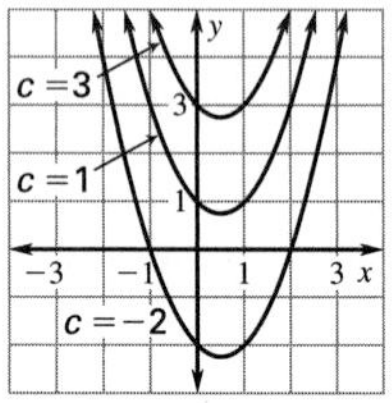

; The functions differ in the value of *c,* and vary in the position of the parabola. When *c* changes from 1 to 3, the vertex of the graph moves up 2 units. When *c* changes from 1 to –2, the vertex of the graph moves down 3 units.

9.3 Mixed Review (p. 524)

81. 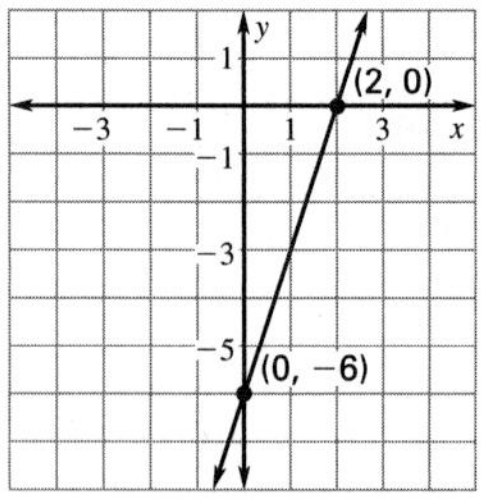

82. 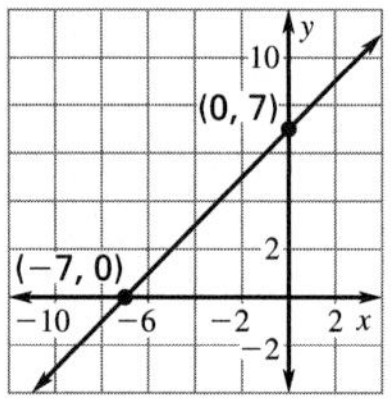

83. 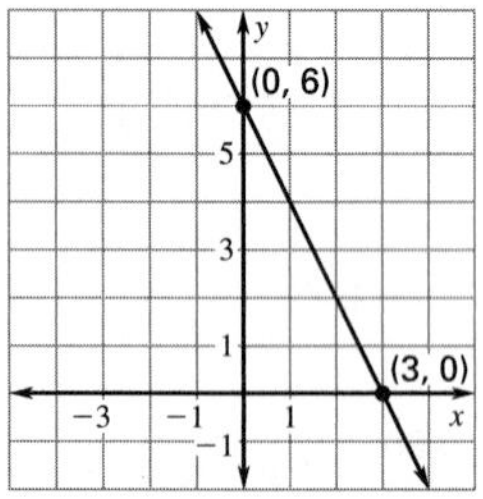

84. 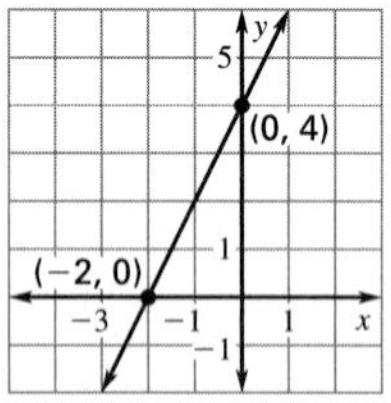

85. 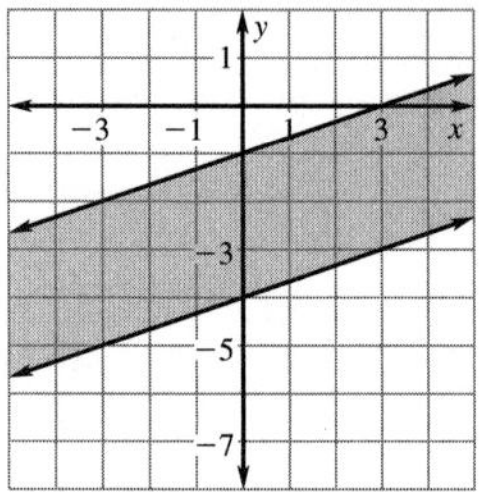

86. 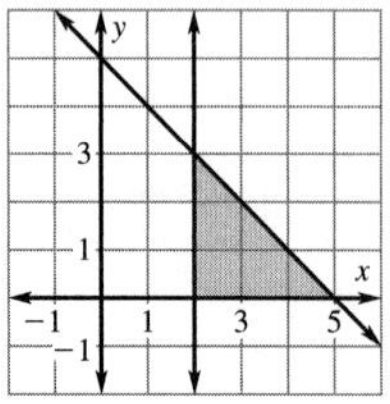

87. 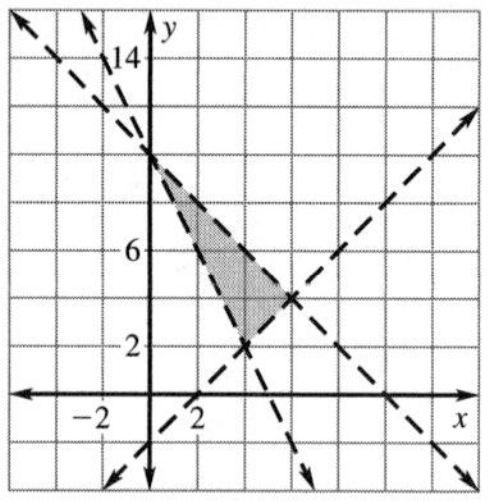

Quiz 1 (p. 524)

23. 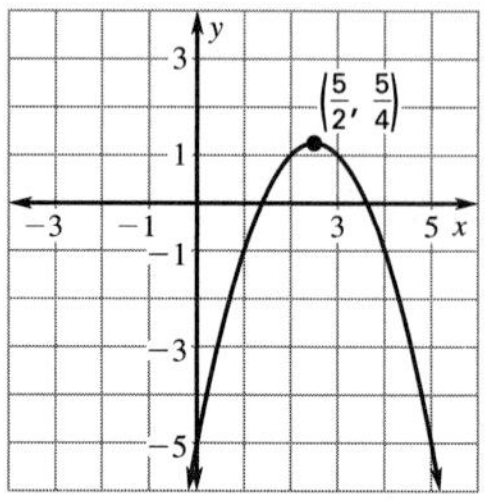

24. 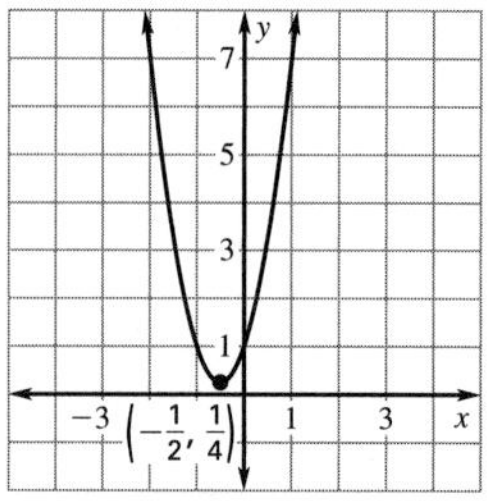

25. 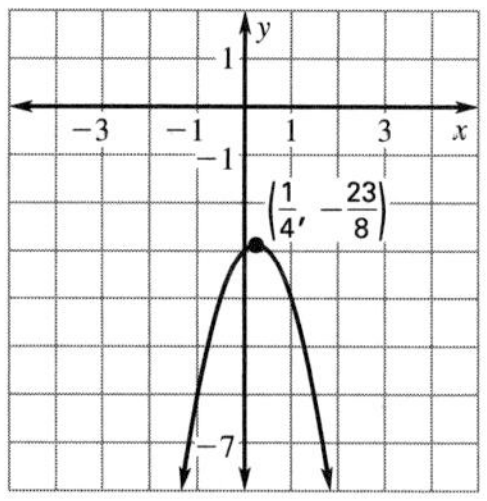

9.5 Practice and Applications (pp. 536–538)

36. $\frac{5-\sqrt{7}}{6} \approx 0.39, \frac{5+\sqrt{7}}{6} \approx 1.27$ **37.** $\frac{-7-\sqrt{22}}{9} \approx -1.30,$ $\frac{-7+\sqrt{22}}{9} \approx -0.26$ **38.** $\frac{-3-\sqrt{17}}{8} \approx -0.89, \frac{-3+\sqrt{17}}{8} \approx 0.14$

39. $\frac{-1-2\sqrt{2}}{7} \approx -0.55, \frac{-1+2\sqrt{2}}{7} \approx 0.26$

40. $\frac{3+\sqrt{57}}{-12} \approx -0.88, \frac{3-\sqrt{57}}{-12} \approx 0.38$

41. $6-\sqrt{62} \approx -1.87, 6+\sqrt{62} \approx 13.87$

42. $\frac{-3+\sqrt{89}}{-20} \approx -0.32, \frac{-3-\sqrt{89}}{-20} \approx 0.62$

43. $\frac{5+\sqrt{177}}{-4} \approx -4.58, \frac{5-\sqrt{177}}{-4} \approx 2.08$

69. $\frac{1+\sqrt{5}}{6} \approx 0.54, \frac{1-\sqrt{5}}{6} \approx -0.21$; quadratic formula; when written in standard form, the expression cannot be written as a perfect square. **70.** $4+3\sqrt{2} \approx 8.24, 4-3\sqrt{2} \approx -0.24$; quadratic formula; when written in standard form, the expression cannot be written as a perfect square.

9.5 Mixed Review (p. 538)

94. 

95. 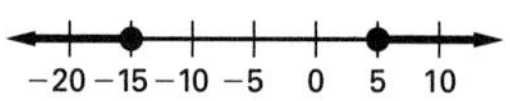

96. 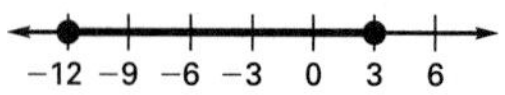

97. 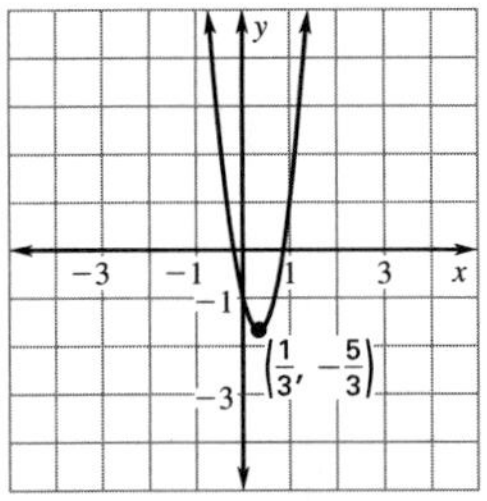

98. 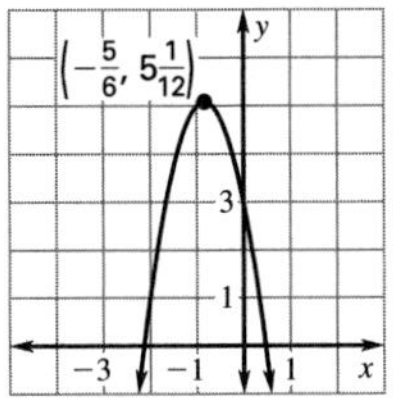

99. 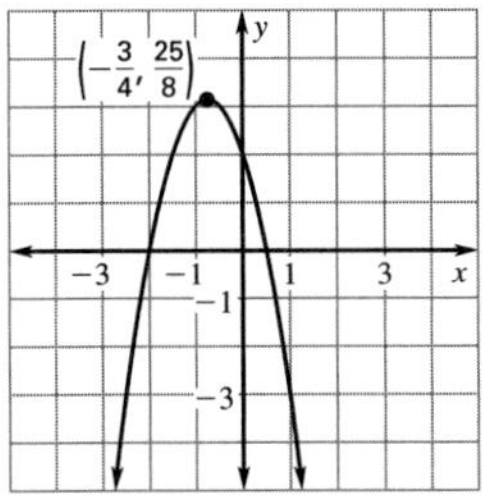

100. 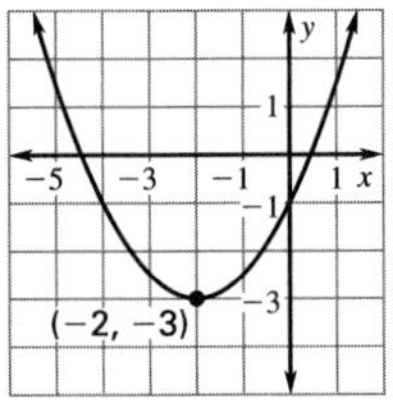

101. 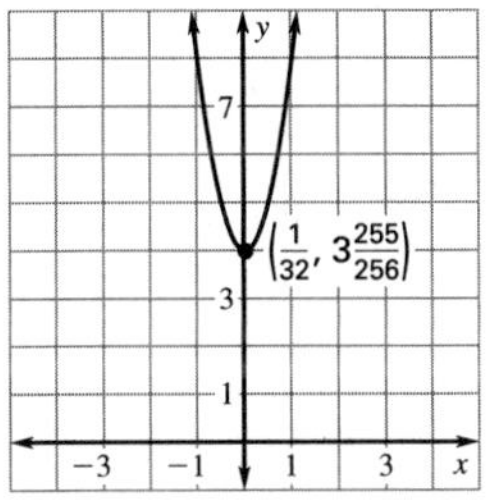

102. 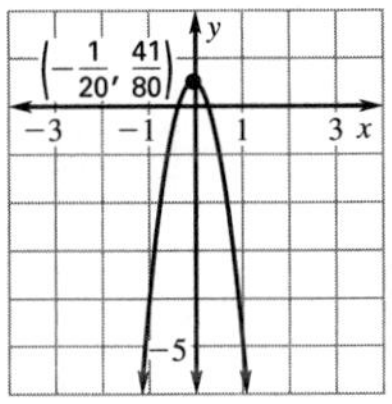

9.6 CONCEPT ACTIVITY (p. 540)

1. (1) 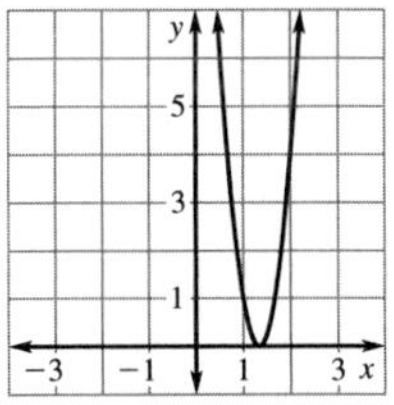

2. (1) 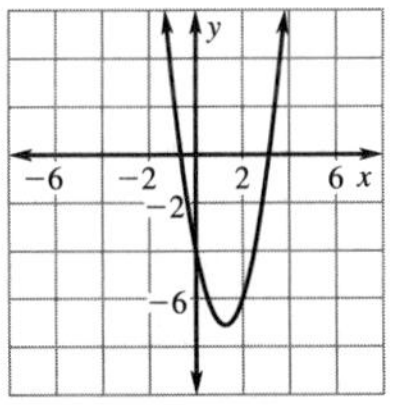

3. (1) 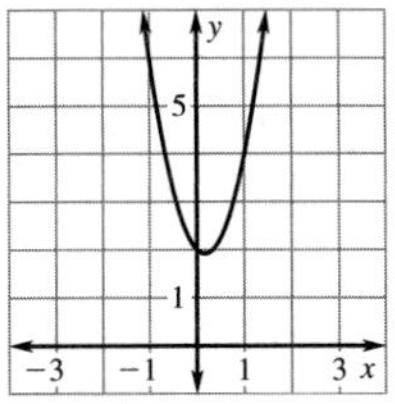

4. (1) 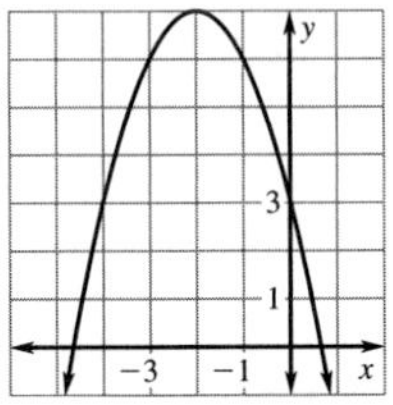

5. (1) 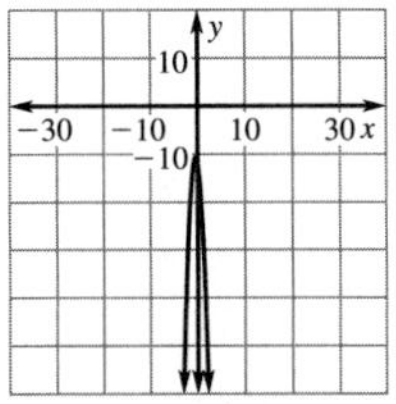

6. (1) 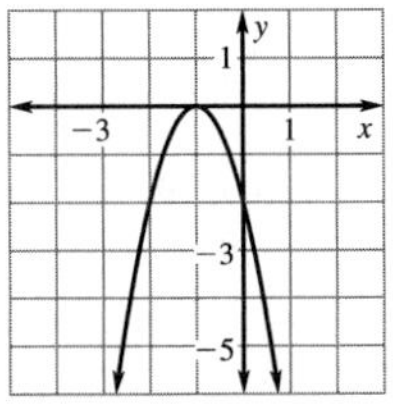

7. (1) 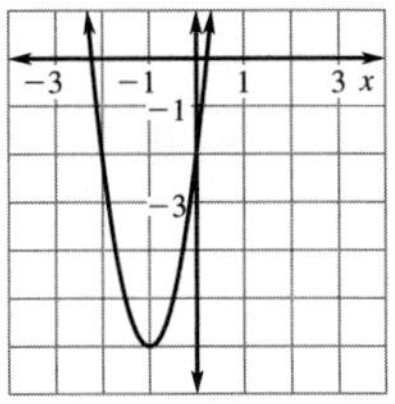

8. (1) 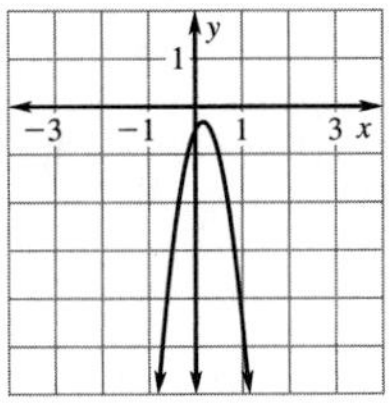

9. (1) 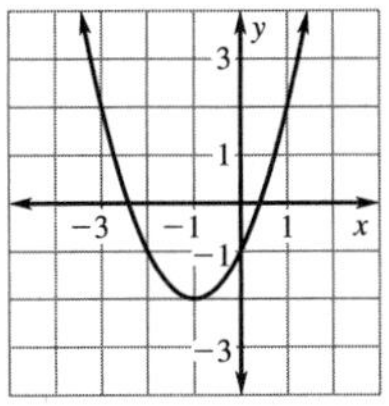

10. *Sample answer:* All of the graphs are parabolas, but some are wide and some are narrow, some open up and some open down. Some intersect the x-axis in one point, some in two points, and some do not intersect the x-axis at all.

11. (1) 0 (2) 57 (3) −23 (4) 28 (5) −215 (6) 0 (7) 96 (8) −8 (9) 8

9.6 MIXED REVIEW (p. 546)

45. 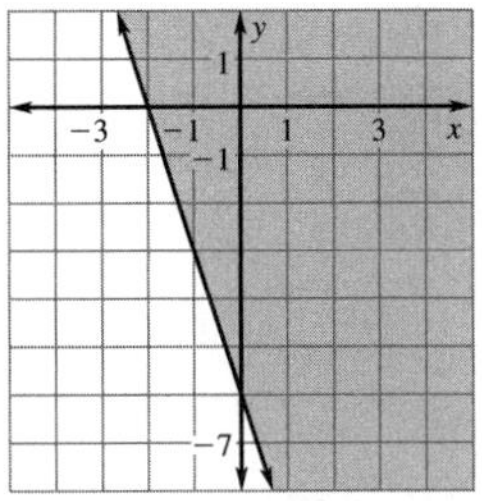

46.

15	0
16	
17	0 0
18	
19	0
20	0 0
21	5
22	0
23	0

Key: 23 | 0 = 230

47. Modem Prices

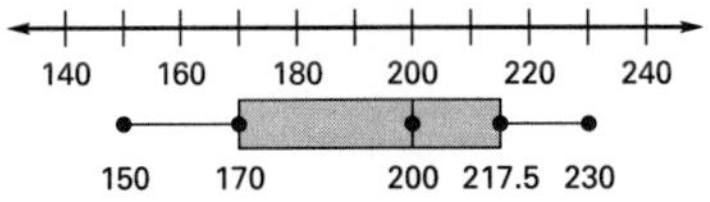

9.7 GUIDED PRACTICE (p. 551)

9. 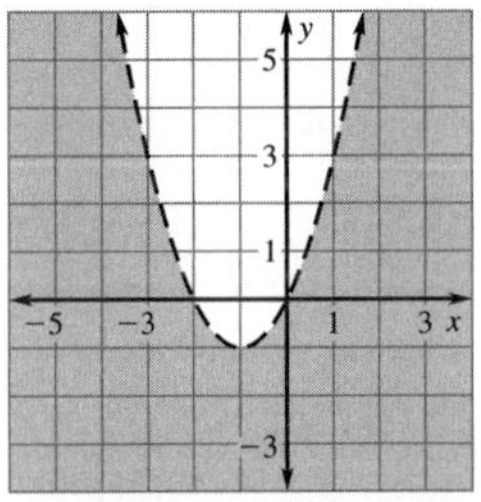

10. 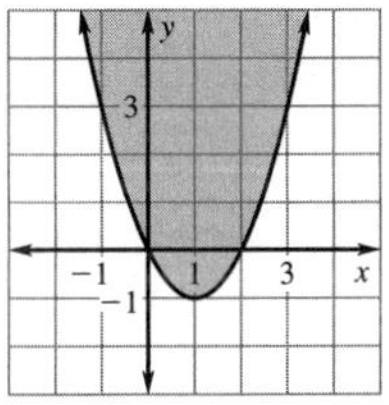

11. 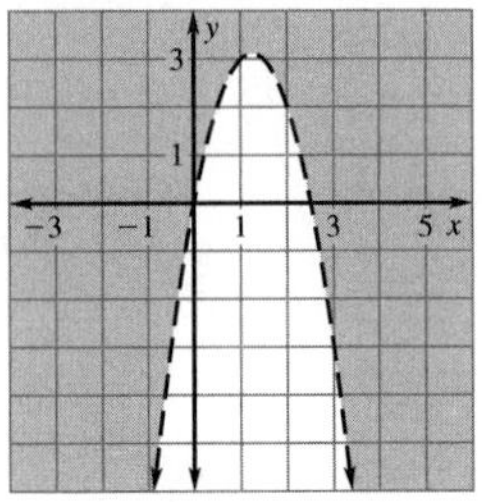

12. 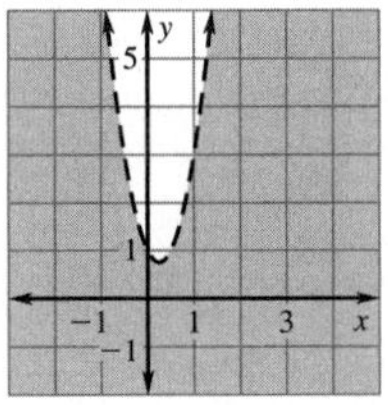

9.7 PRACTICE AND APPLICATIONS (pp. 551–553)

25. 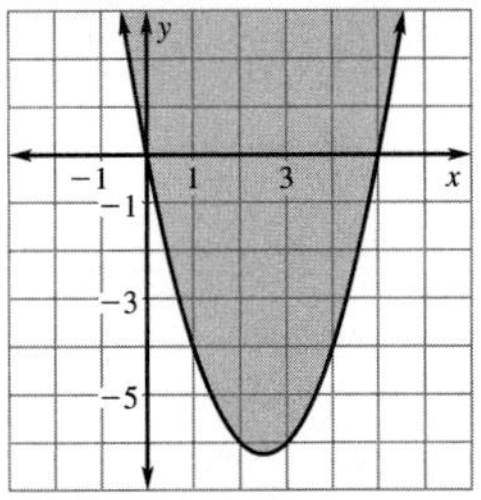

26. 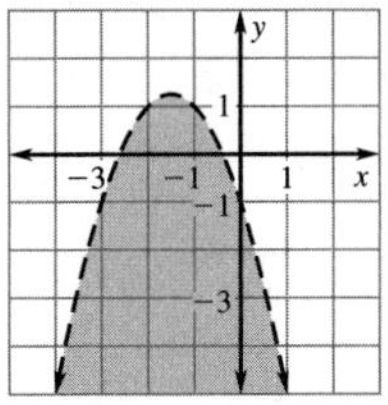

27. 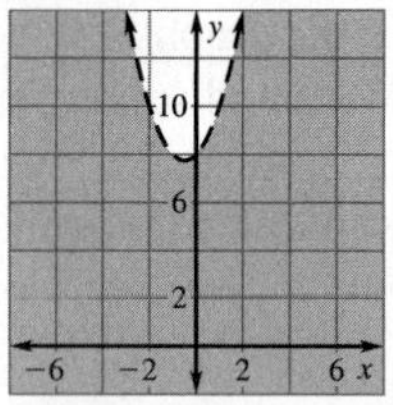 **28.** 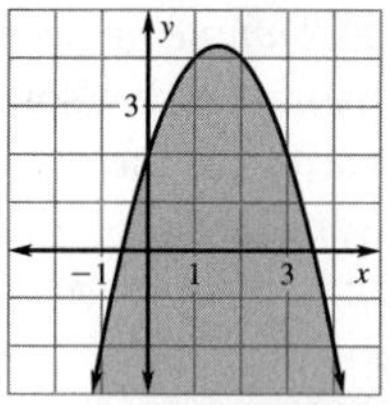

29. 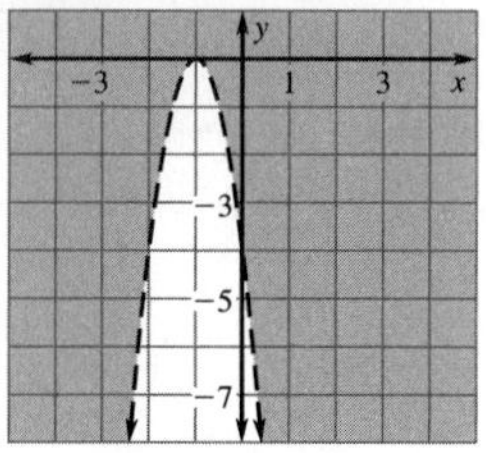 **30.** 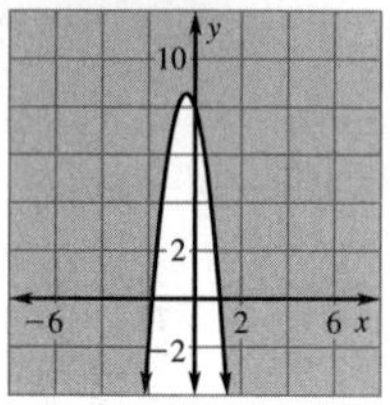

31. 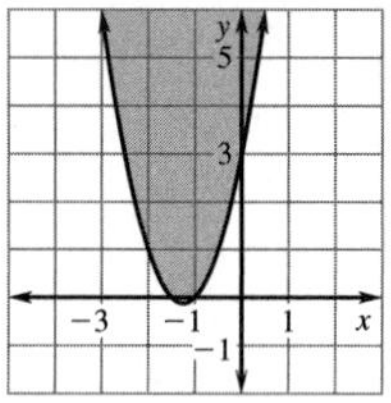 **33.** 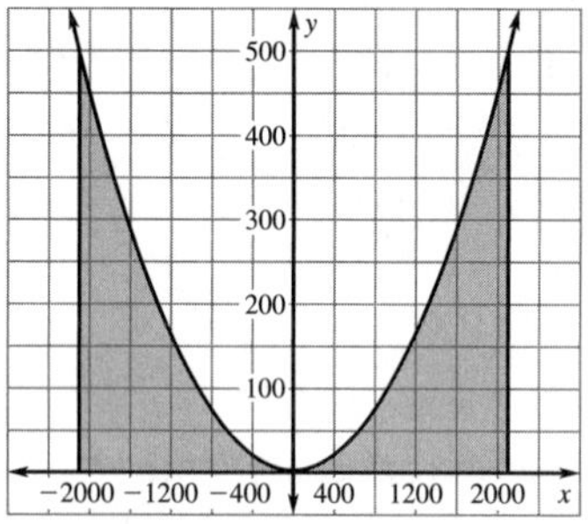

41–42. **Per Capita Milk Consumption**

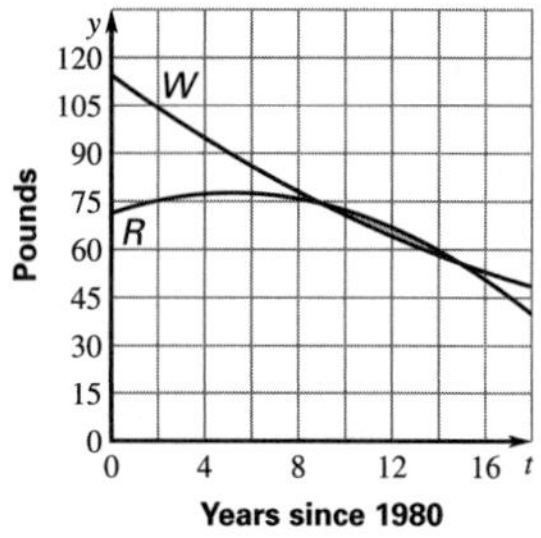

9.7 Mixed Review (p. 553)

44. 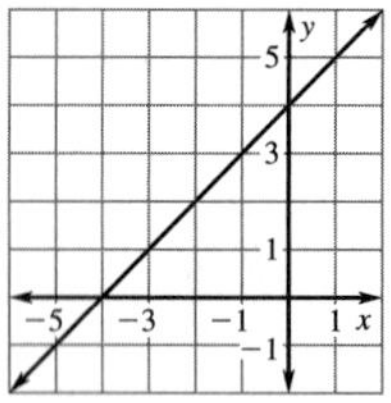 **45.** 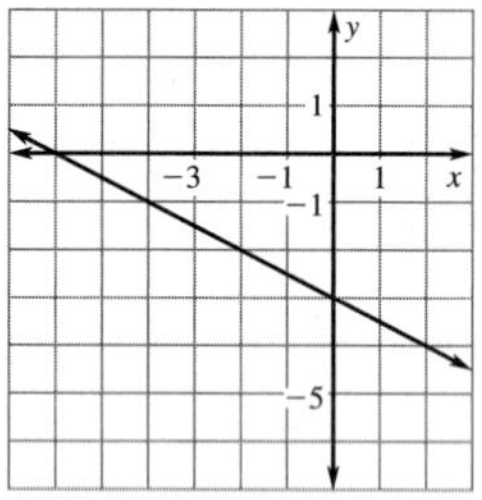

46. 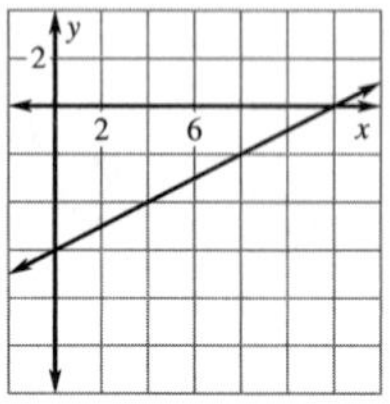 **47.** 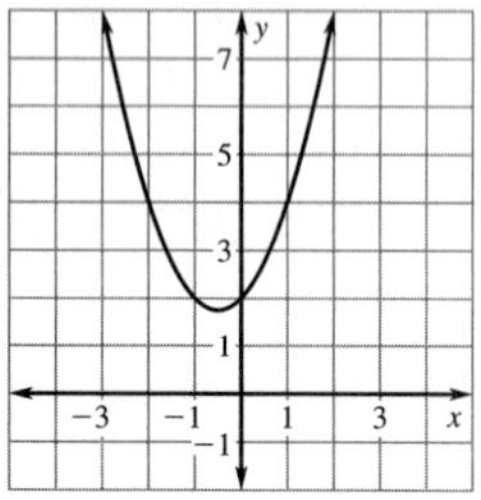

48. 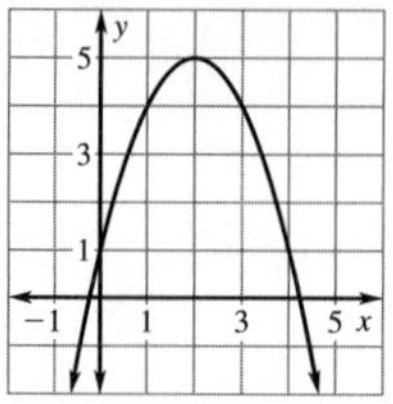 **49.** 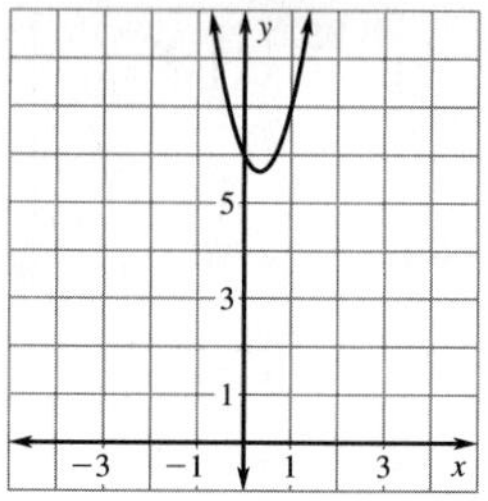

56. 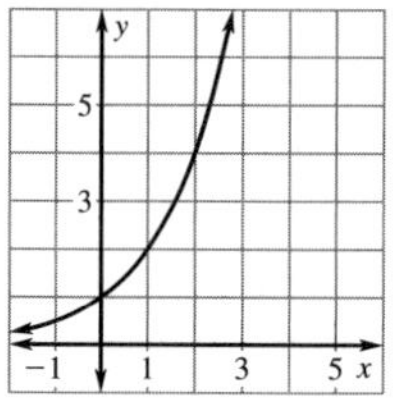 **57.** 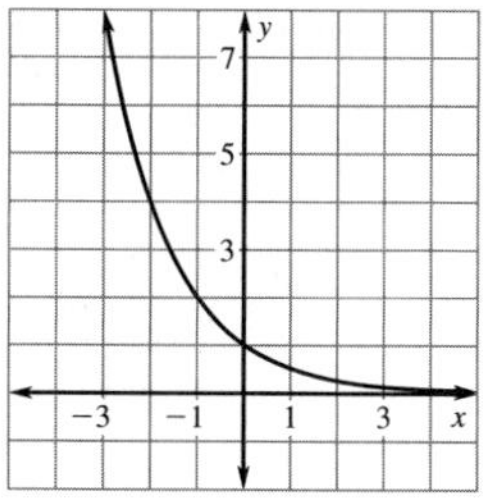

58. 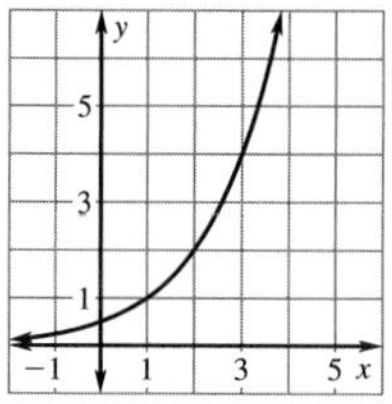 **59.** 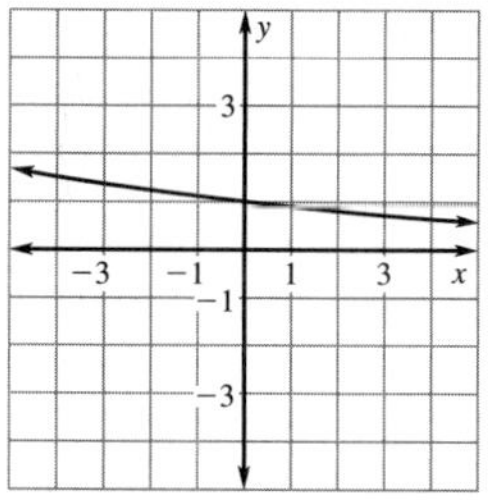

60. 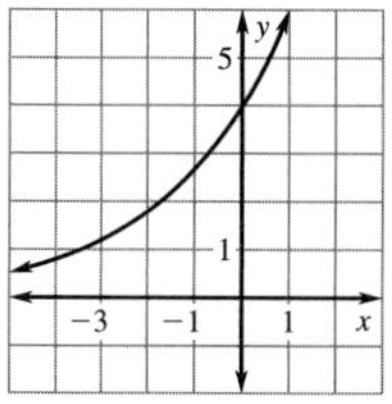 **61.** 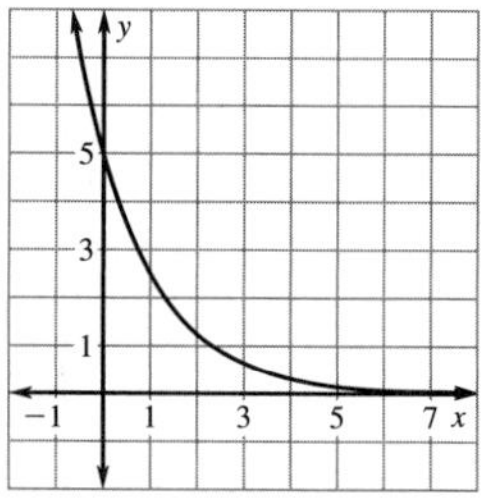

9.8 Guided Practice (p. 557)

6. 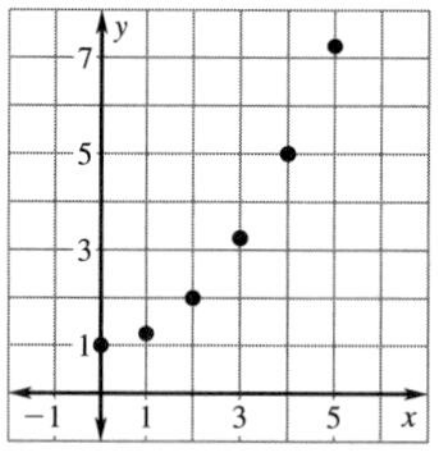

9.8 Practice and Applications (pp. 557–559)

9. 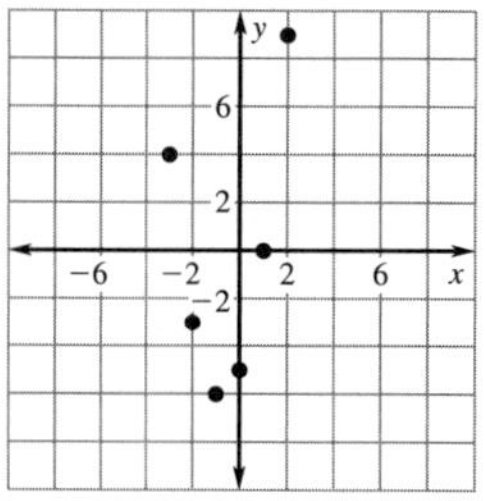 **10.** 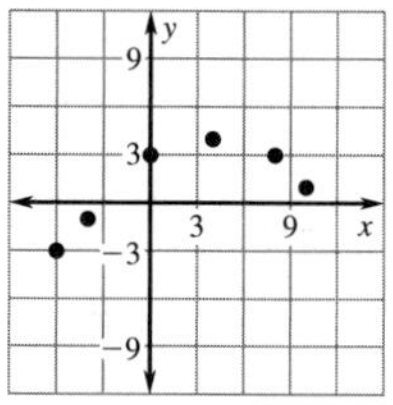

11. 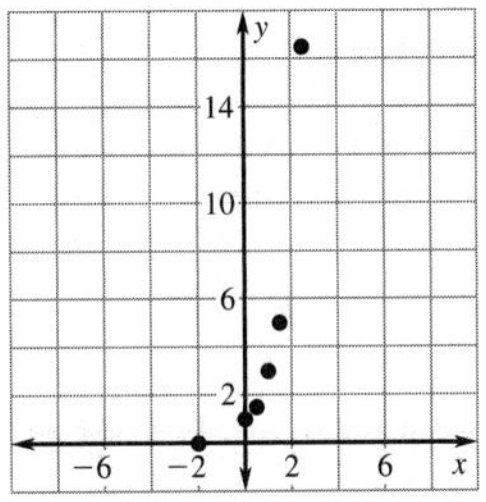

12. 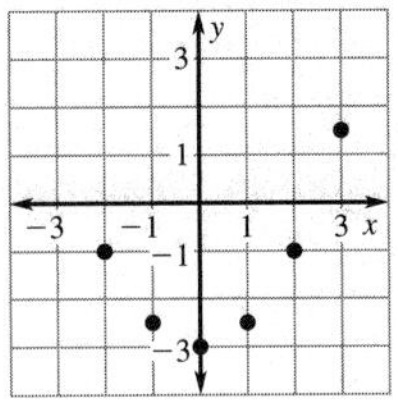

13. 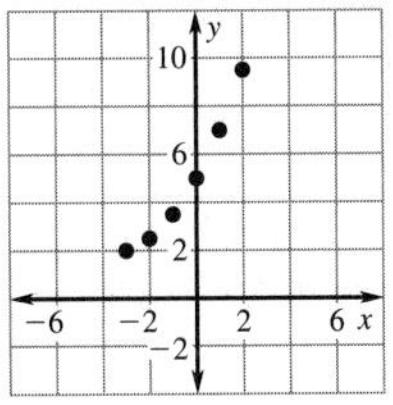

14. 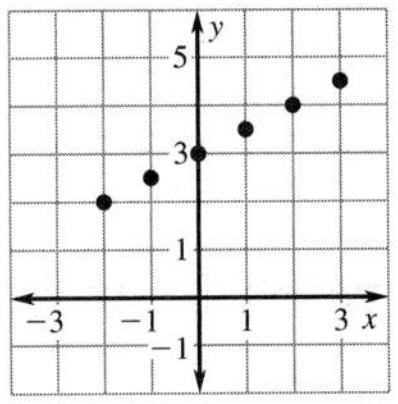

15. 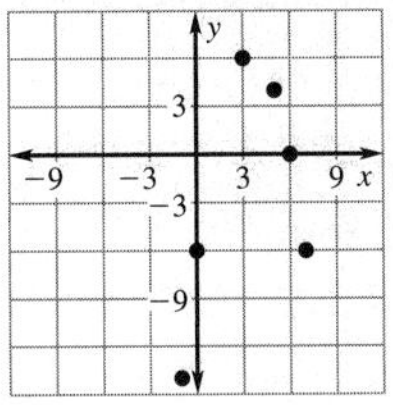

16. 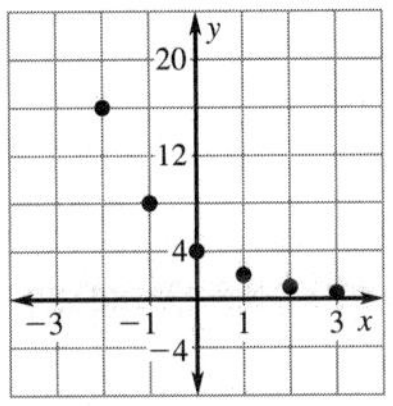

17. 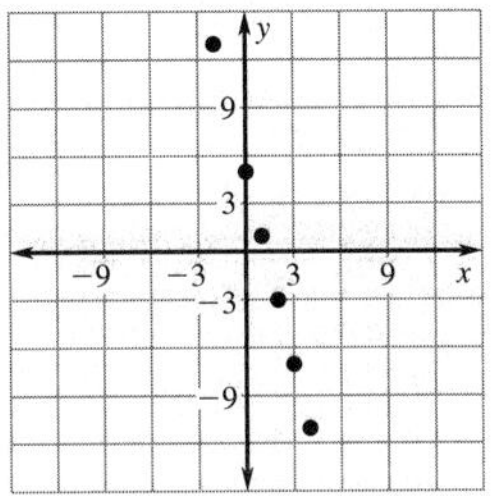

21.

Year t	Attendance A	Change in A from previous year
0	12,580,660	—
1	12,338,980	–241,680
2	12,781,160	442,180
3	13,907,200	1,126,040
4	15,717,100	1,809,900

QUIZ 3 (p. 560)

8. 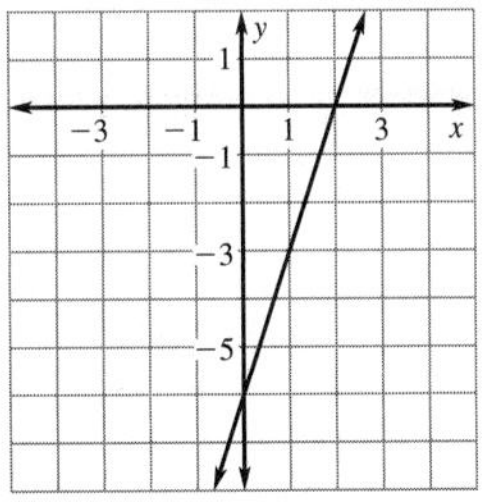

9. 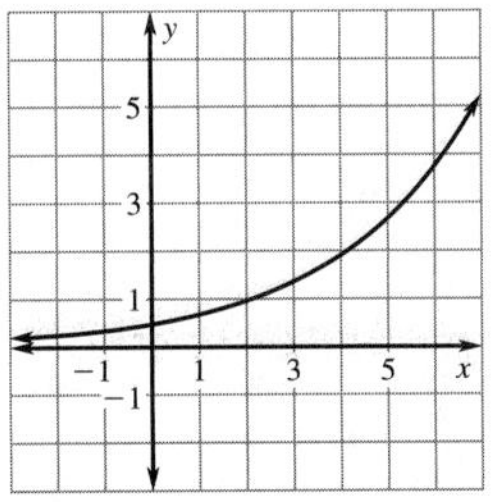

10. 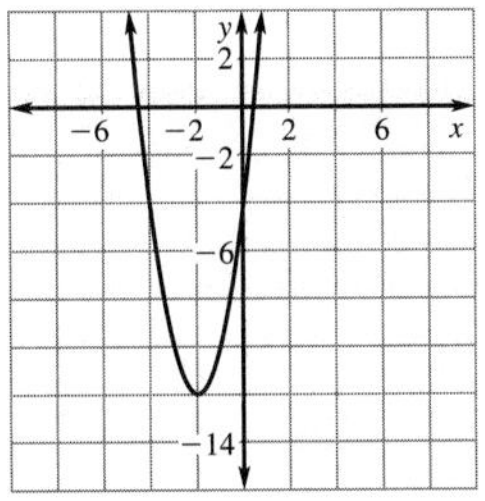

11. 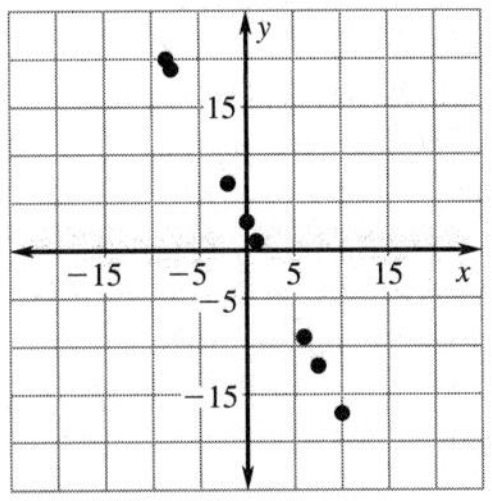

12. 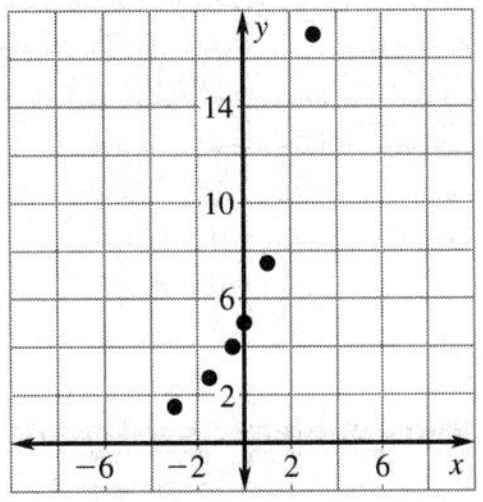

13. 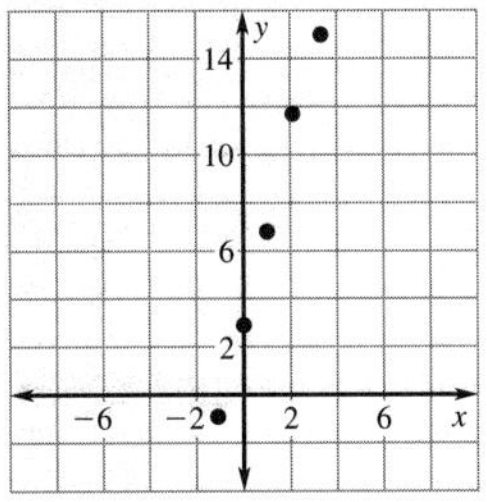

PROJECT, CHAPTERS 7–9 (pp. 570–571)

2. Pizza Sales

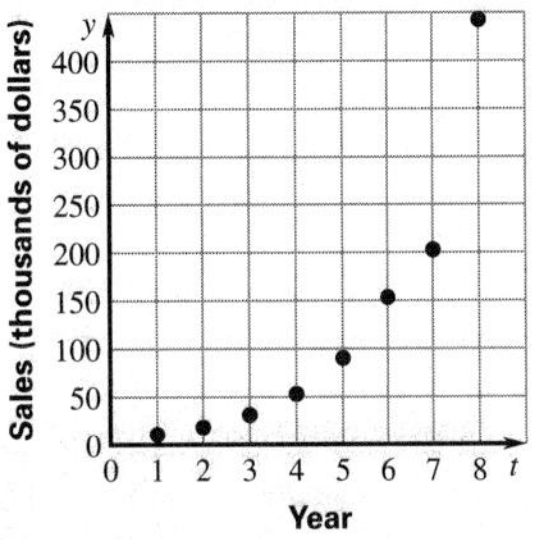

CHAPTER 10

10.1 CONCEPT ACTIVITY (p. 575)

1. 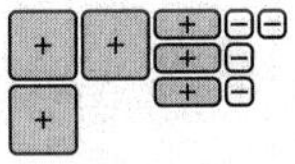

2. 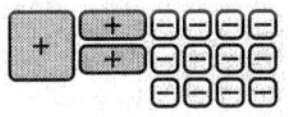

3. 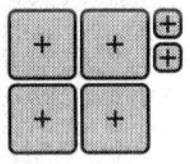

5. 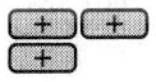

6. 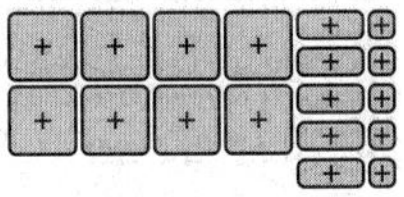

10.2 GUIDED PRACTICE (p. 587)

5. 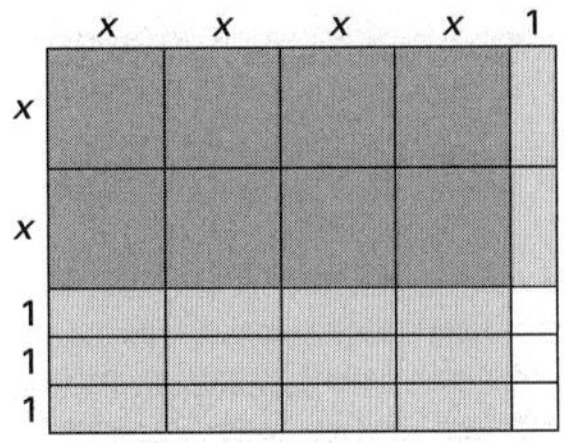

10.2 MIXED REVIEW (p. 589)

80. 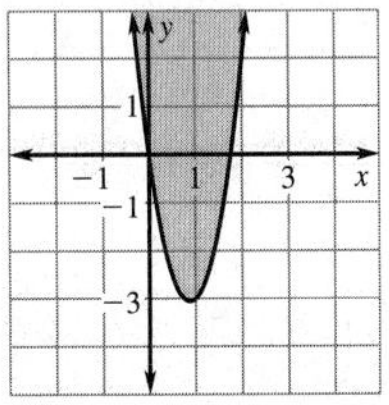

81. 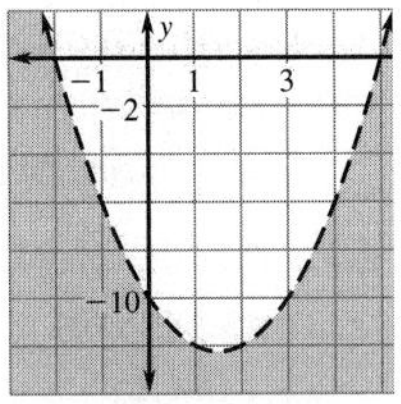

82. 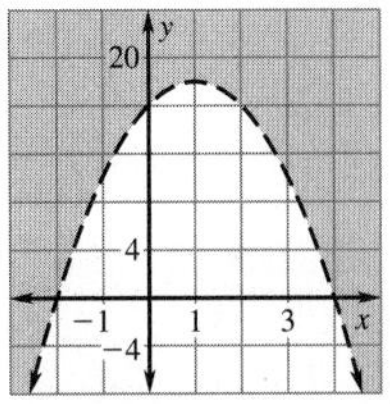

83. 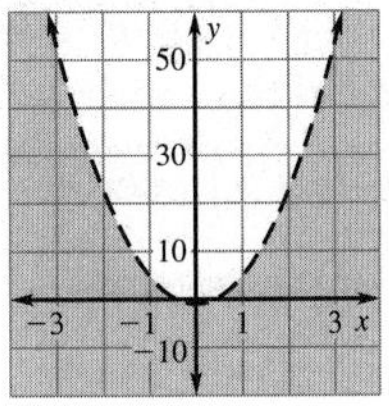

84. 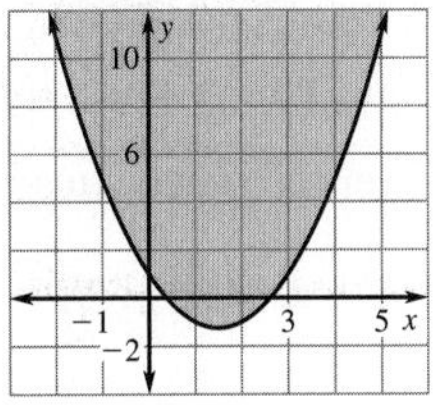

85. 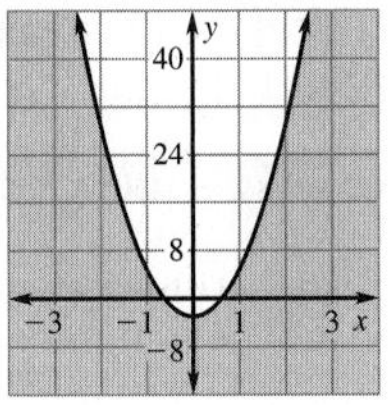

10.4 Activity (p. 597)

1.

Expression	–3	–2	–1	0	1	2	3
$(x-3)(x+2)$	6	0	–4	–6	–6	–4	0
$(x+1)(x+3)$	0	–1	0	3	8	15	24
$(x-2)(x-2)$	25	16	9	4	1	0	1
$(x-1)(x+2)$	4	0	–2	–2	0	4	10

3. *Sample answer:* To solve an equation in factored form, set each factor equal to zero and solve the resulting linear equations. The solution to each of the linear equations is a solution of the original equation.

10.4 Practice and Applications (pp. 600–602)

47.
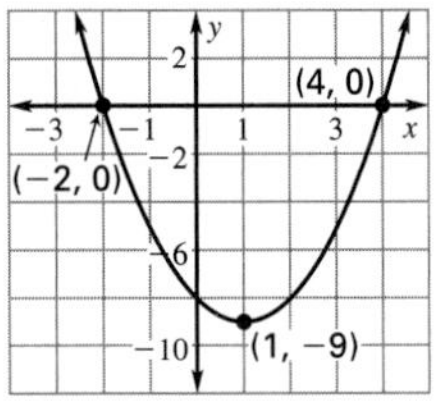

48.
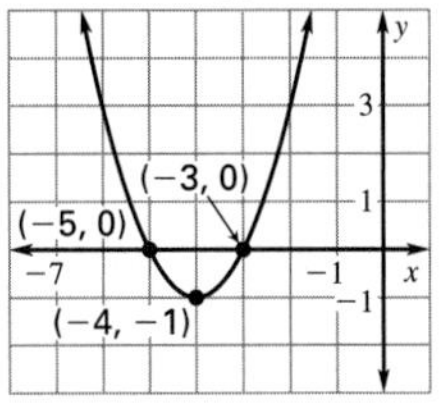

49.
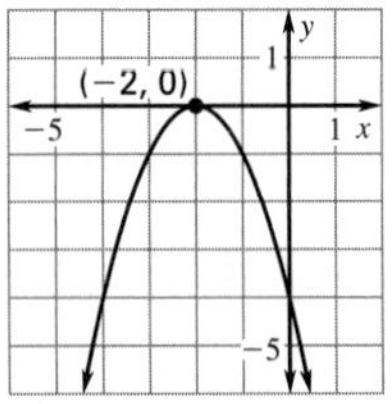

50.
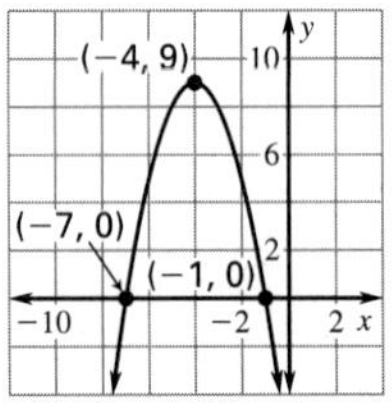

51.
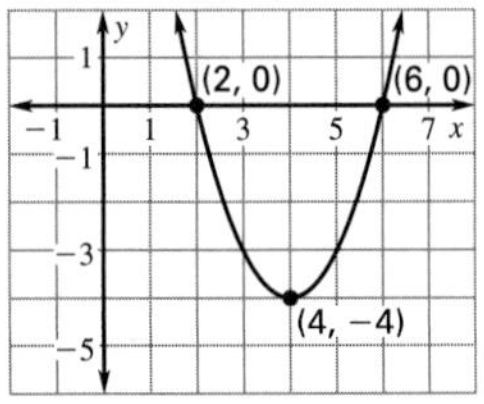

52.
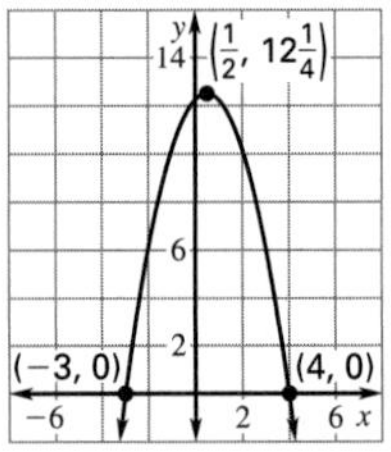

10.6 Concept Activity (p. 610)

7.
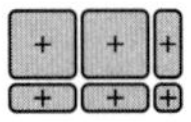

8.
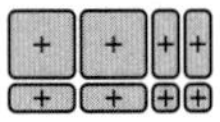

9.
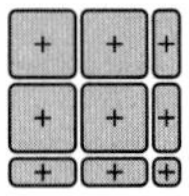

10.7 Guided Practice (p. 622)

4. (1) $a^2 - b^2 = (a+b)(a-b)$; *sample answer:* $x^2 - 100 = (x+10)(x-10)$; (2) $a^2 + 2ab + b^2 = (a+b)^2$; *sample answer:* $x^2 + 18x + 81 = (x+9)^2$; (3) $a^2 - 2ab + b^2 = (a-b)^2$; *sample answer:* $4x^2 - 12x + 9 = (2x-3)^2$;

10.7 Practice and Applications (pp. 622–624)

42. $2(4+3x)(4-3x)$; difference of two squares **43.** $4(b-5)^2$; perfect square trinomial **44.** $-3(3t+1)^2$; perfect square trinomial **45.** $-2(x-13)^2$; perfect square trinomial **46.** $(13+x)(13-x)$; difference of two squares **47.** $(x+100w)(x-100w)$; difference of two squares **48.** $3(7x-6)(7x+6)$; difference of two squares **49.** $\left(x+\frac{1}{3}\right)^2$; perfect square trinomial **50.** $\frac{3}{4}(1+4x)(1-4x)$; difference of two squares

Chapter 10 Test (p. 637)

44.
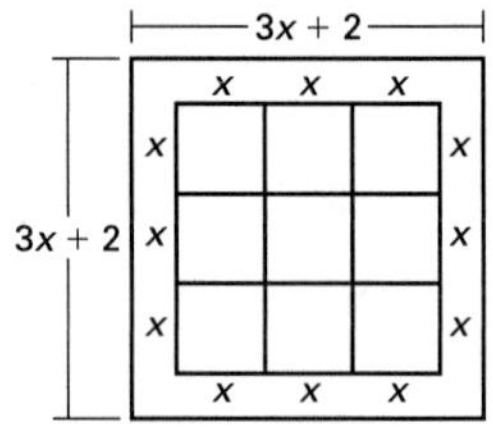

CHAPTER 11

11.1 Practice and Applications (pp. 646–648)

45. food: about 16; clothes, accessories: about 4; books, magazines, comics: about 3; toys, stickers, games: about 3; movie tickets: about 3; arcade games: about 3; gifts: about 3; movie rentals: about 3; music: about 2; footwear: about 2; grooming products: about 2

11.3 Practice and Applications (pp. 659–661)

33.
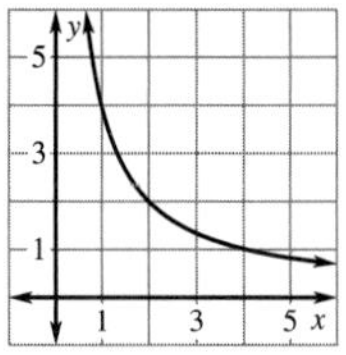

34.
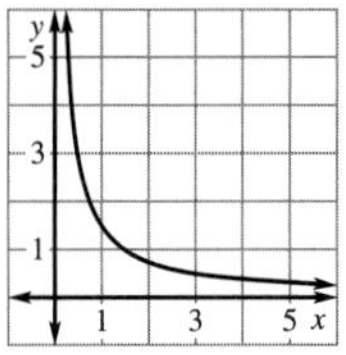

35.
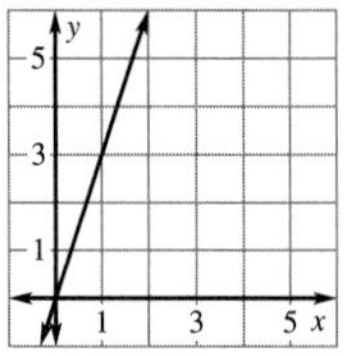

36.
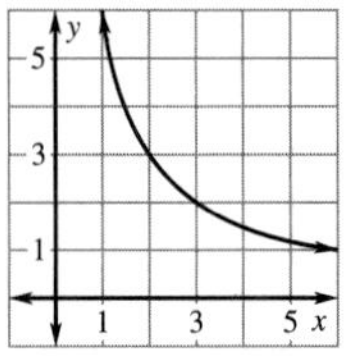

11.4 Mixed Review (p. 669)

51.
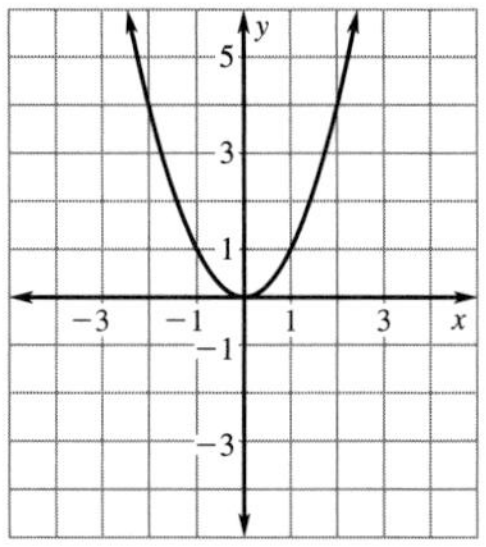

52.
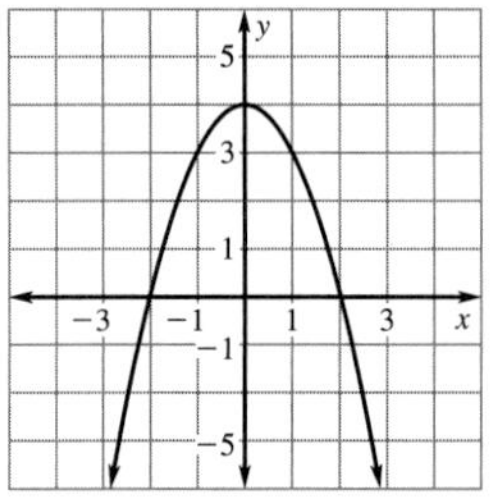

53.
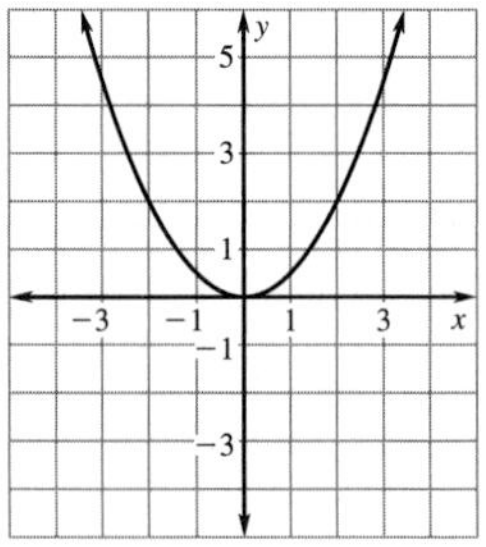

54.
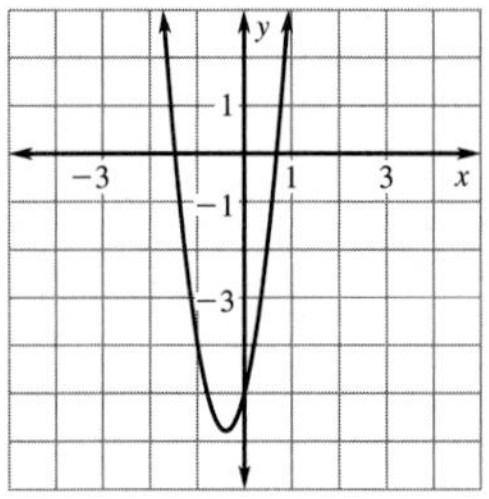

55.
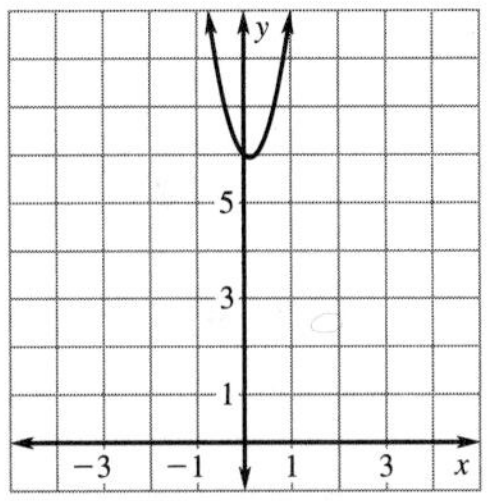

56.
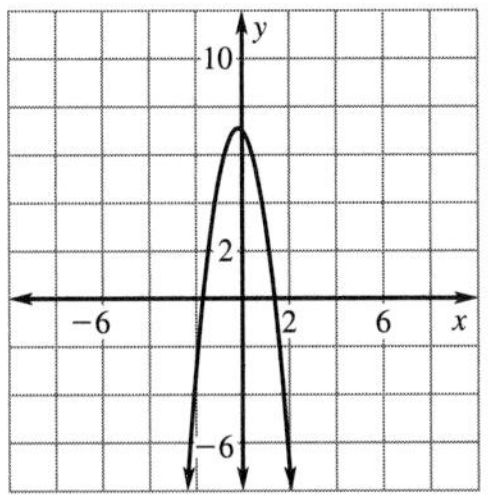

11.6 Mixed Review (p. 682)

64.
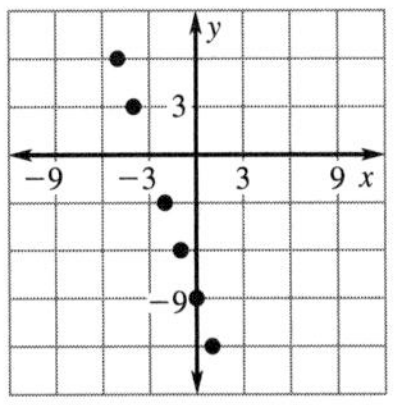

11.7 Mixed Review (p. 689)

71.
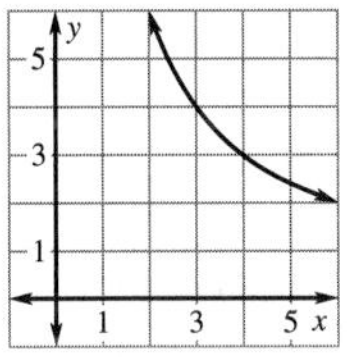

11.8 Guided Practice (p. 694)

12.
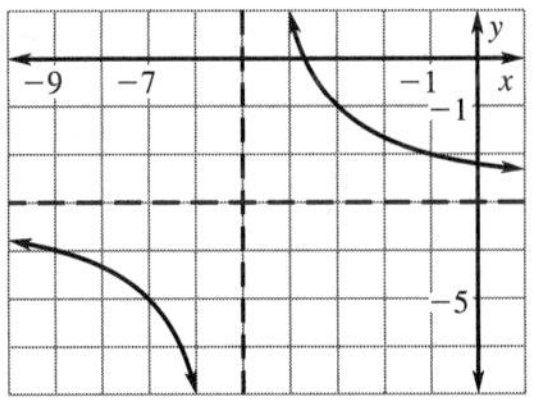

13.
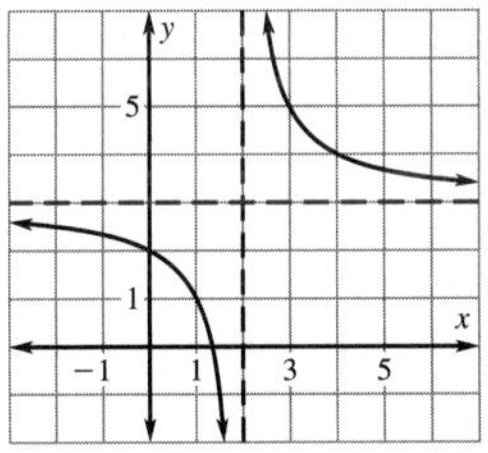

11.8 Practice and Applications (pp. 694–696)

50.
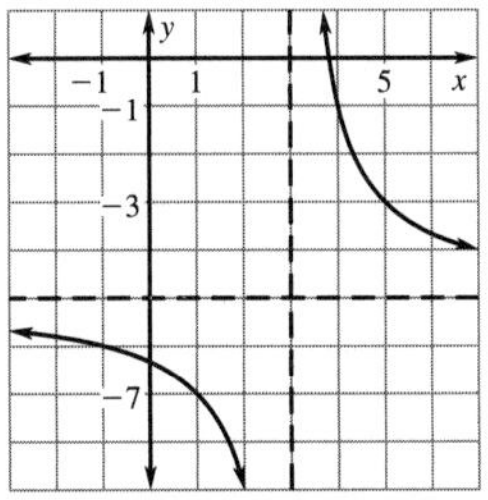

53.
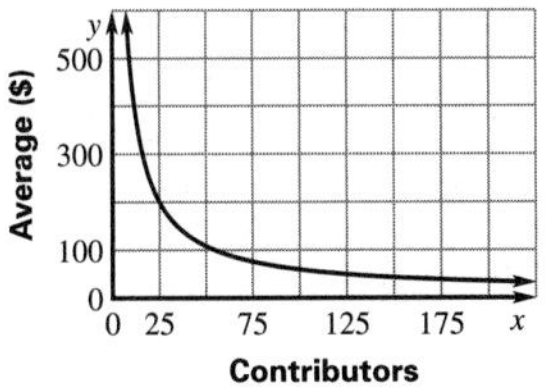

56. *Sample answer:* If the friend's work rate is x minutes per crane, then the work rate per minute for 1000 cranes is $\frac{1}{1000x}$ and the work rate per hour is $\frac{60}{1000x}$.

59.
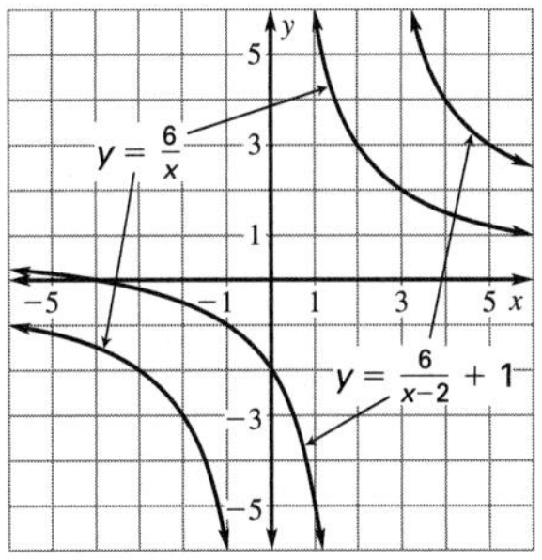

Quiz 3 (p. 697)

11.
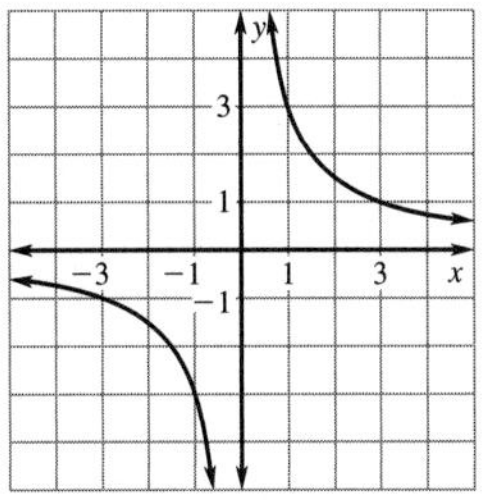

12.
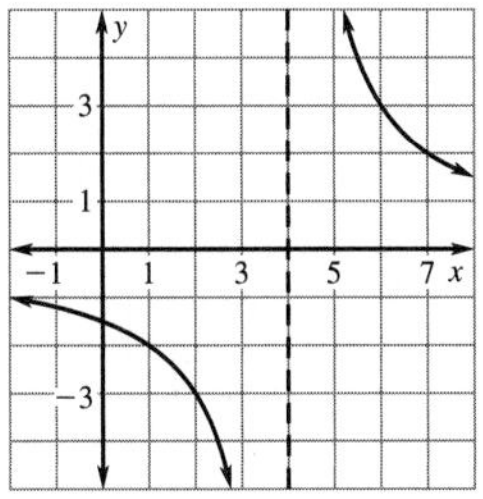

13.
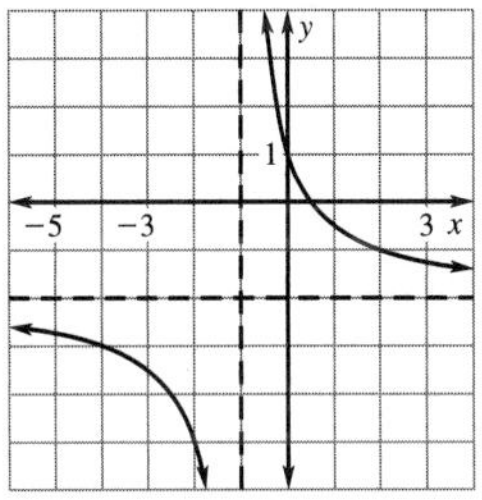

CHAPTER 12

12.1 Activity (p. 709)

1.
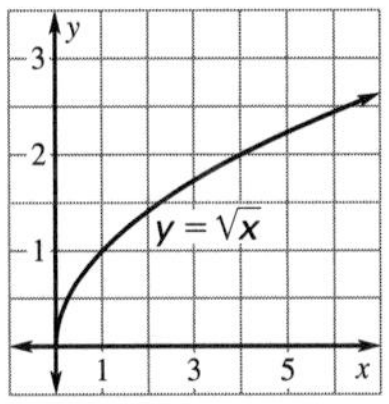

12.1 Guided Practice (p. 712)

17.
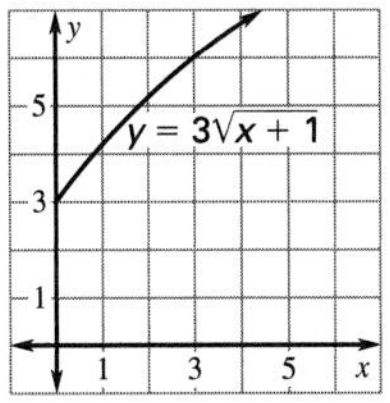

18.
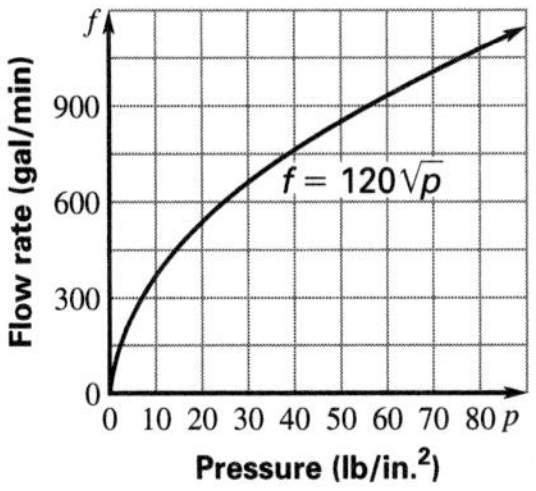

12.1 PRACTICE AND APPLICATIONS (pp. 712–714)

45. 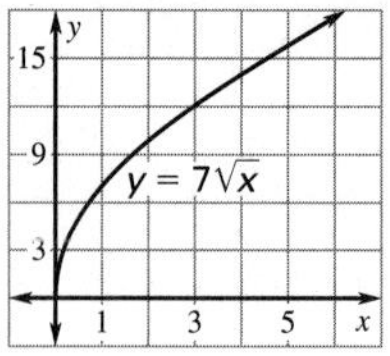

46. 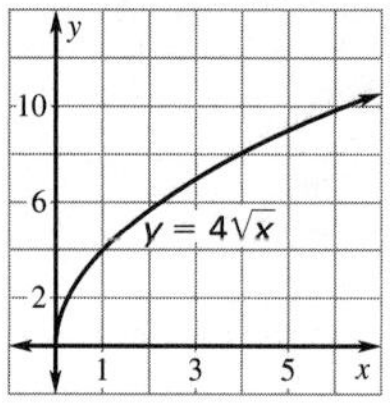

47. 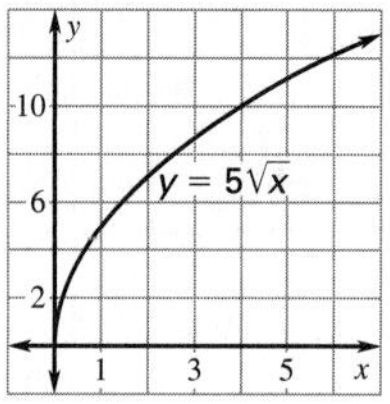

48. 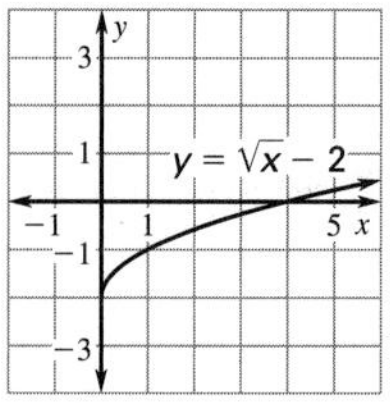

49. 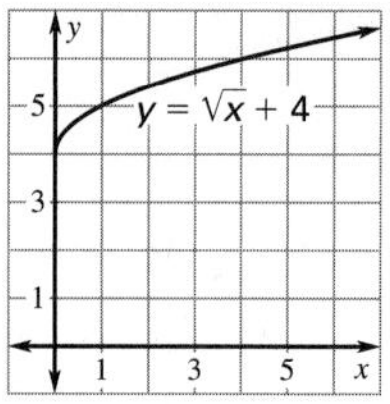

50. 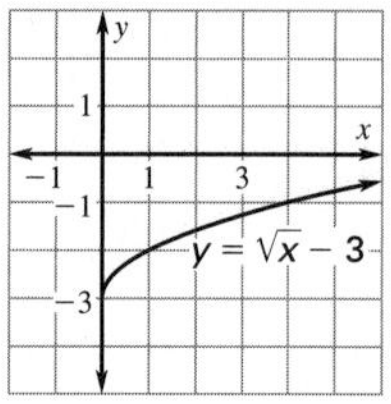

51. 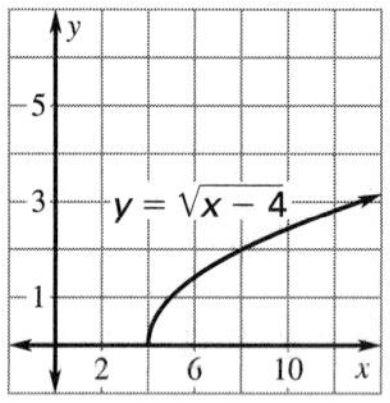

52. 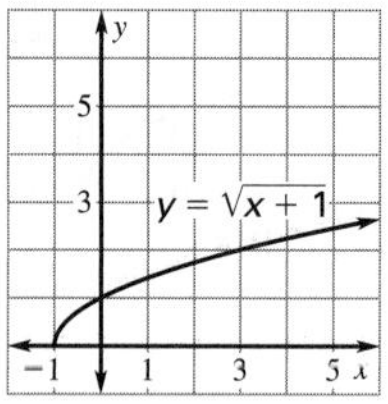

53. 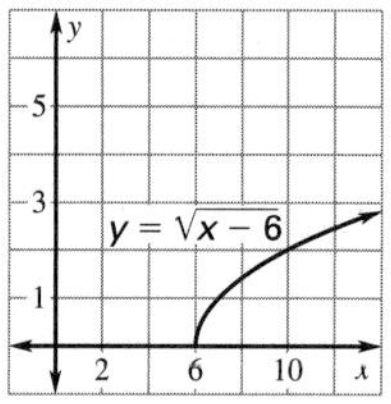

54. 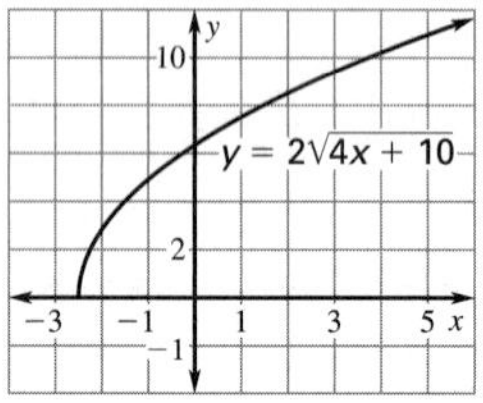

55. 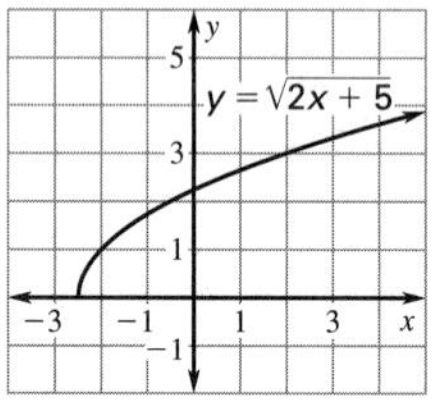

56. 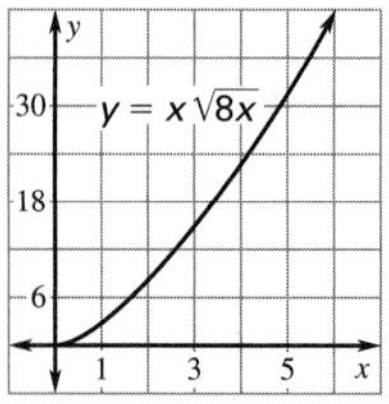

59. b. 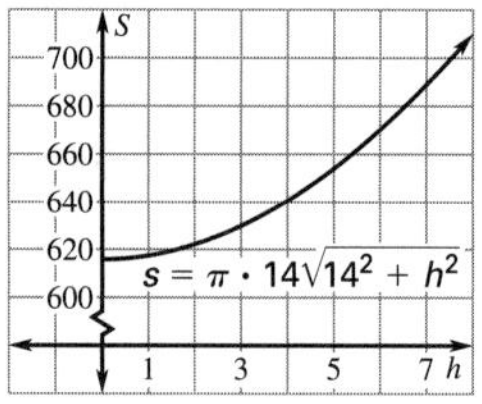

12.4 MIXED REVIEW (p. 736)

94. 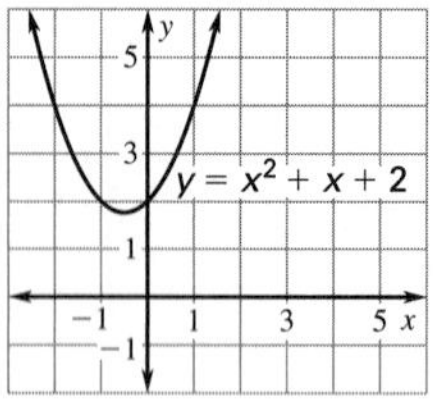

95. 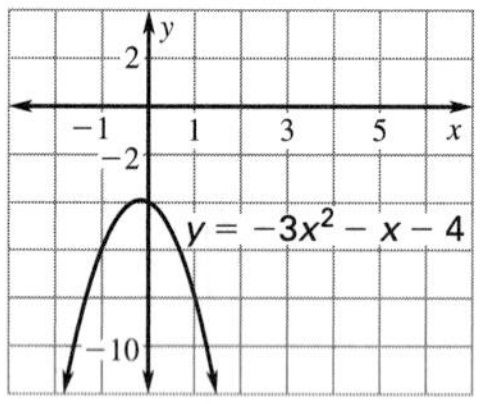

96. 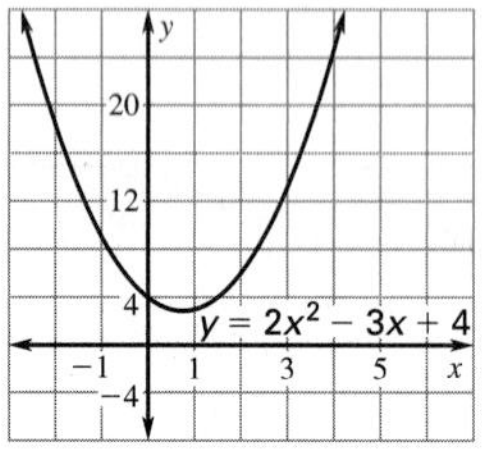

12.5 MIXED REVIEW (p. 743)

50. 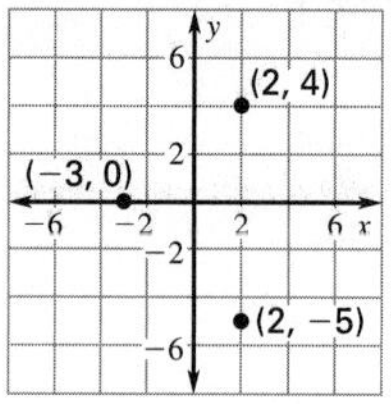

51. 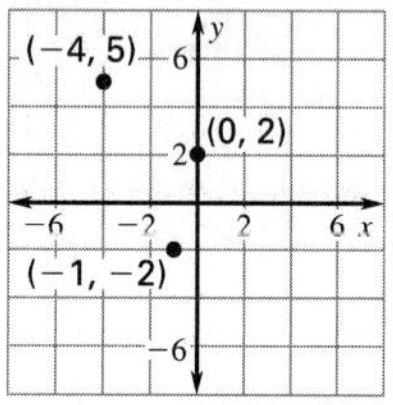

52. 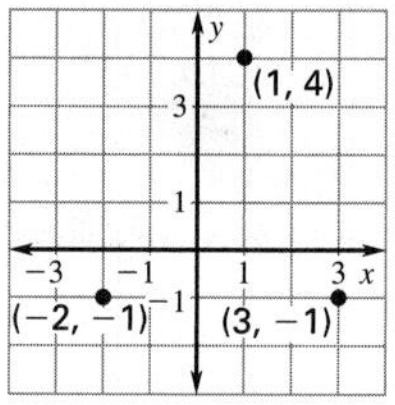

53. 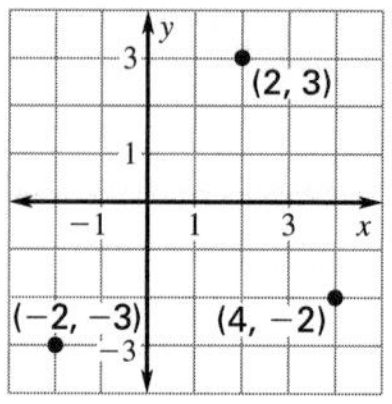

54. 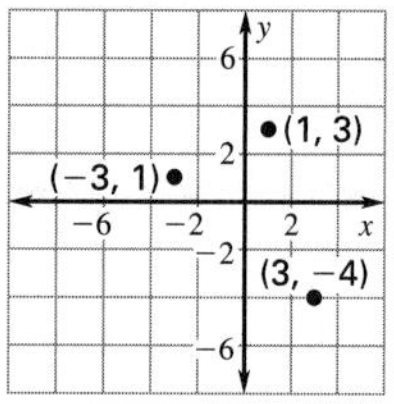

12.8 PRACTICE AND APPLICATIONS (pp. 762–763)

28. By the distance formula, $BC = \sqrt{(0-x)^2 + (y-0)^2} = \sqrt{x^2 + y^2}$, so $BD = CD = \frac{1}{2}\sqrt{x^2 + y^2}$. The midpoint D has coordinates $\left(\frac{0+x}{2}, \frac{y+0}{2}\right) = \left(\frac{x}{2}, \frac{y}{2}\right)$. Using the distance formula, $AD = \sqrt{\left(\frac{x}{2} - 0\right)^2 + \left(\frac{y}{2} - 0\right)^2} = \sqrt{\frac{x^2}{4} + \frac{y^2}{4}} = \frac{1}{2}\sqrt{x^2 + y^2}$. Therefore, $AD = BD = CD$ and the midpoint of the hypotenuse of the right triangle is equidistant from the vertices. **30.** *Sample answer:* In the figure at the left, c is the length of the hypotenuse of a right triangle and a and b are the lengths of the legs. The next two figures are squares having the same area. The area of the square on the left is $(a+b)^2$ or $a^2 + 2ab + b^2$. The area of the square on the right can be written as the sum of its parts, $4\left(\frac{1}{2}ab\right) + c^2$ or $2ab + c^2$. So, $a^2 + 2ab + b^2 = 2ab + c^2$, and $a^2 + b^2 = c^2$.

EXTRA PRACTICE

CHAPTER 4 (p. 800)

7. 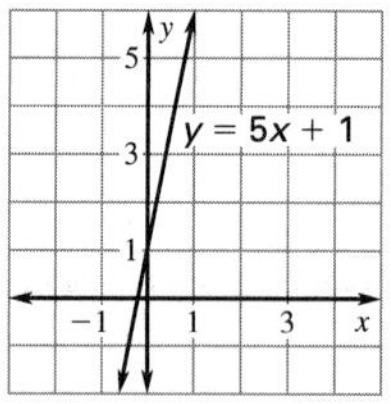

8. 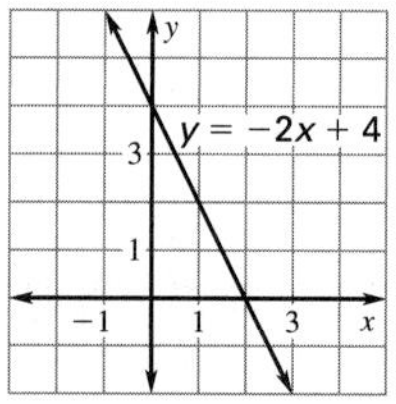

9. 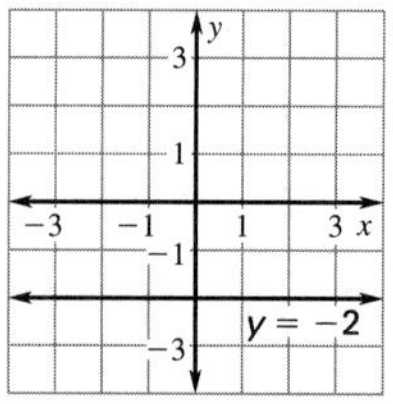

10. 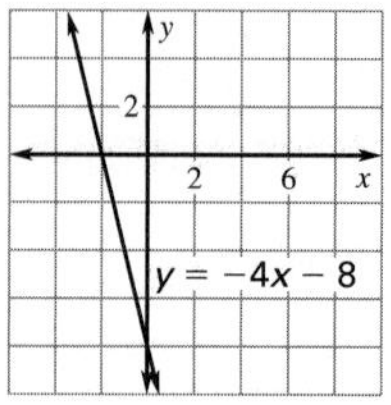

11. 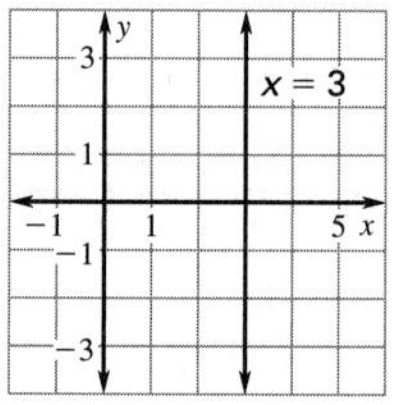

12. 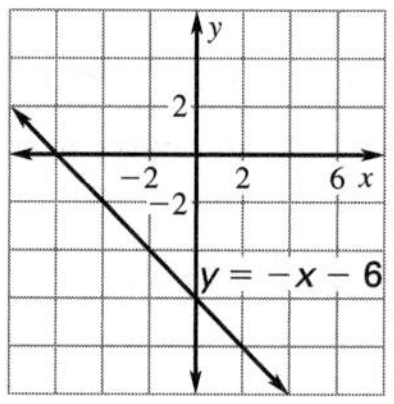

13. 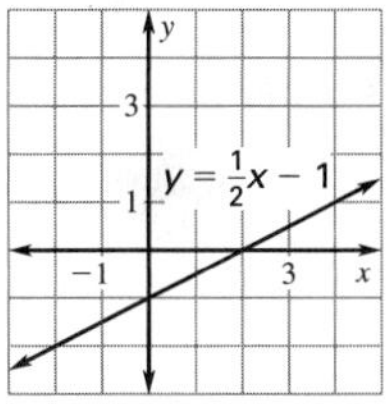

14. 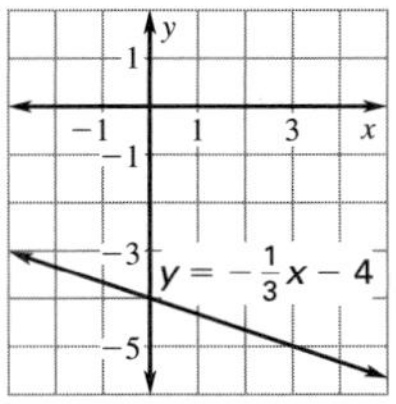

15. 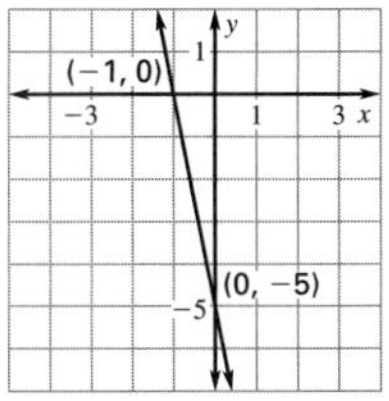

16. 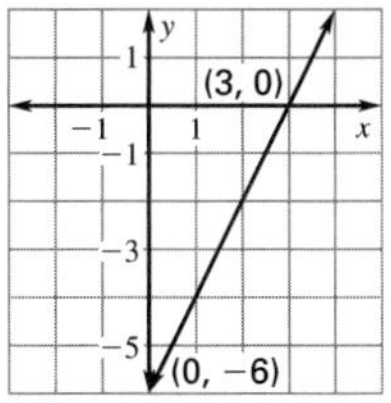

17. 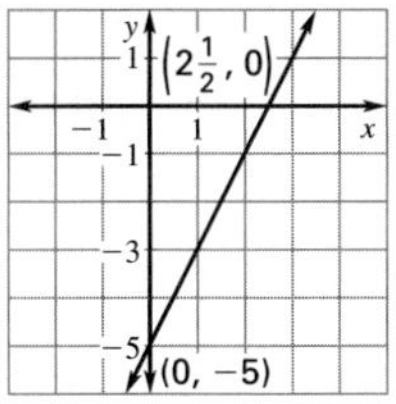

18. 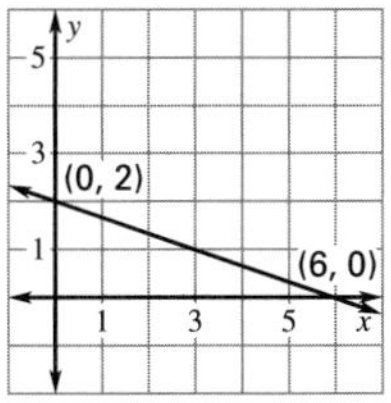

19. 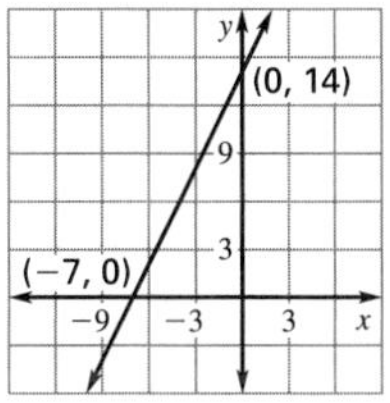

20. 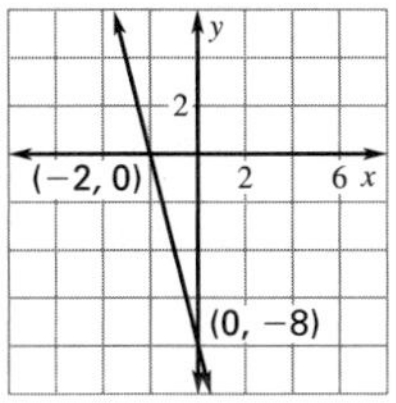

21. 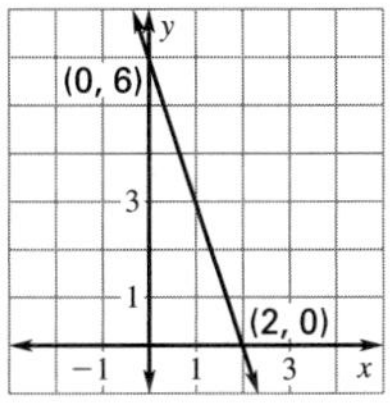

22. 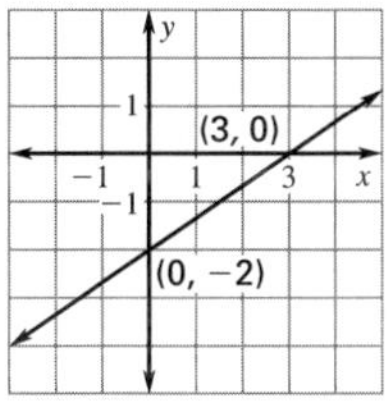

40. 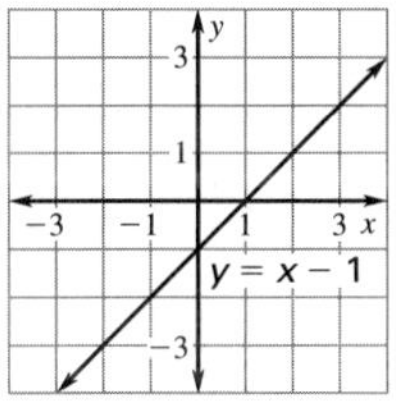

41. 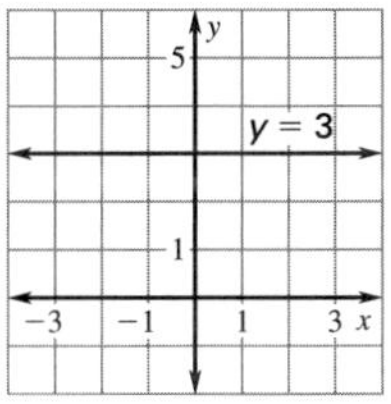

42. 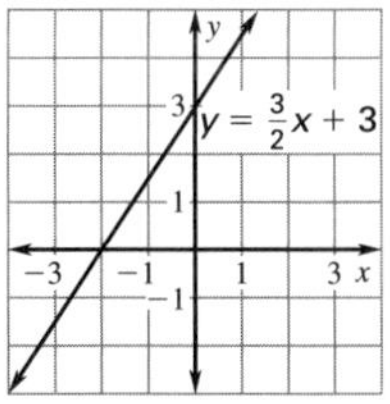

43. 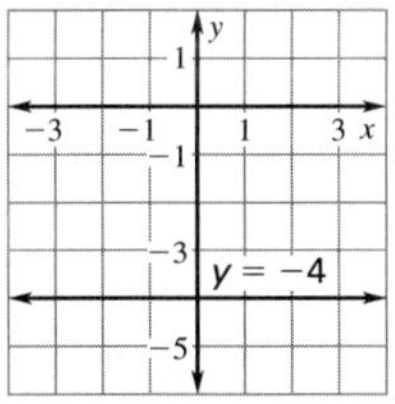

44. 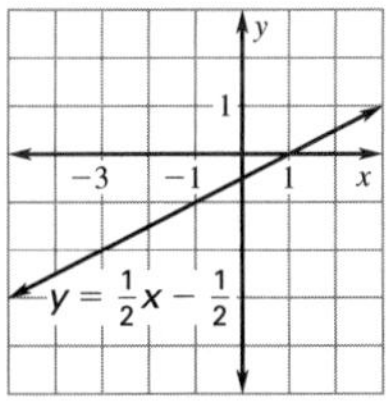

45. 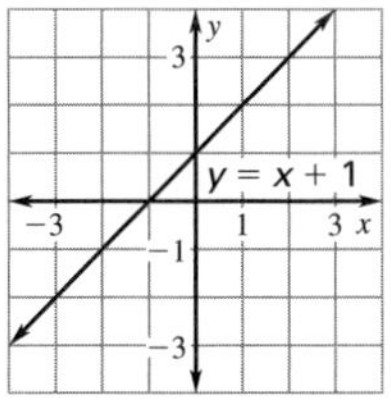

46. 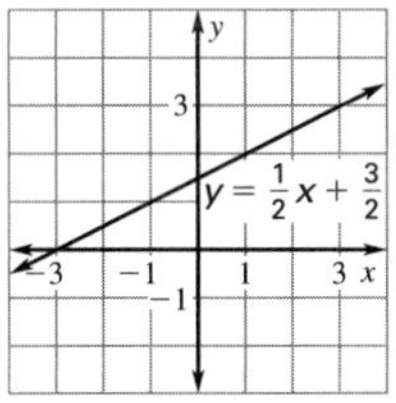

47. 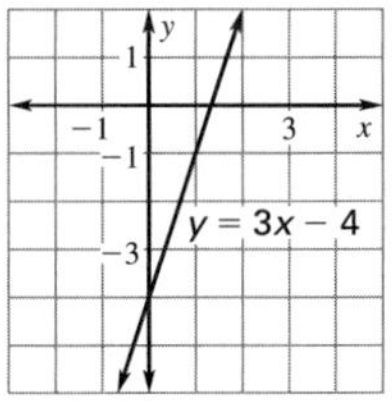

CHAPTER 6 (p. 802)

31. 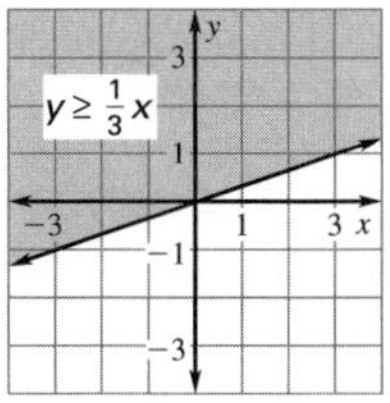

32. 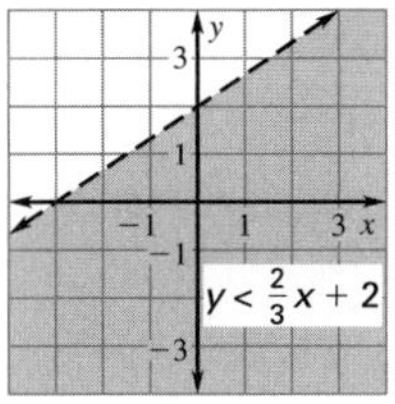

33. 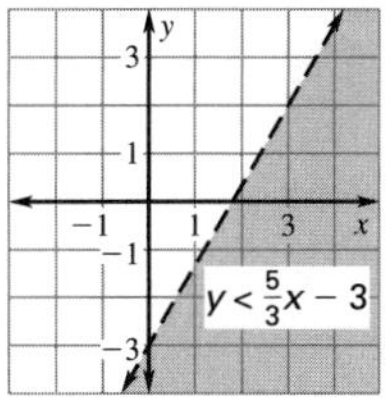

34. 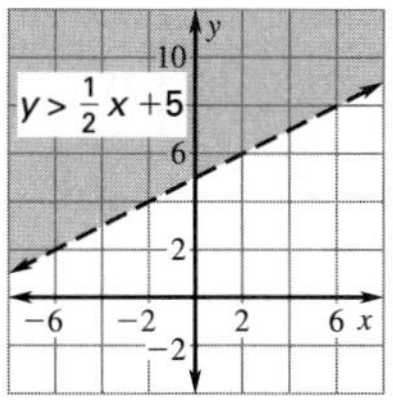

35. 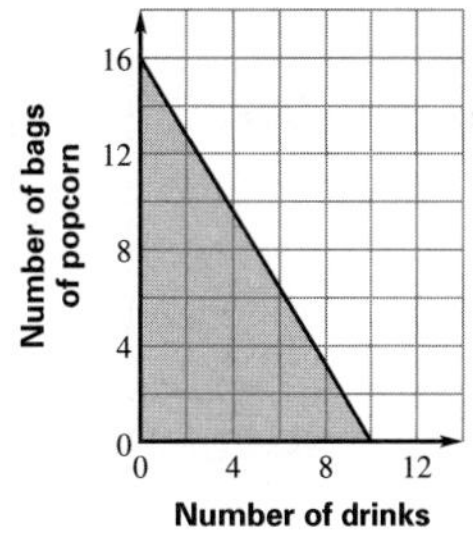

36.

1	3 4 8 8
2	0 2
3	1 3
4	
5	
6	6
7	1 6
8	
9	8

Key: 9 | 8 = 98

37.

1	3 6
2	1 2
3	1 3 5
4	0 1 5
5	1
6	5 6

Key: 6 | 5 = 65

38.

0	3 5 6 8
1	3 7
2	2 4
3	2
4	1
5	4 5
6	0 3 4
7	2 8
8	
9	1

Key: 9 | 1 = 91

39.

0	2 8
1	9
2	1 2 7
3	8 9 9
4	1 3 4 6 9
5	1 4 6
6	0 2 3

Key: 6 | 3 = 63

48. 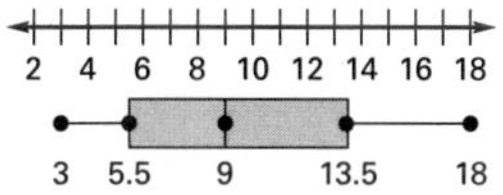

49. 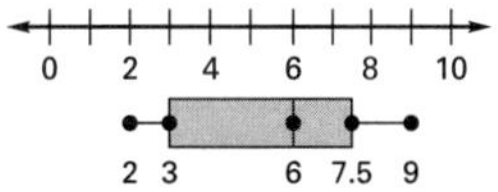

50. 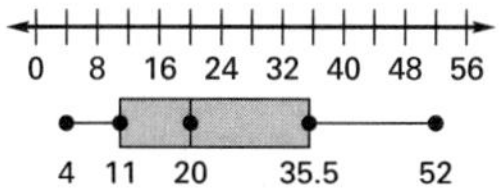

51. 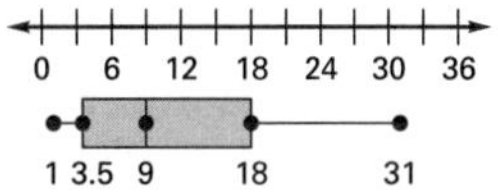

CHAPTER 7 (p. 803)

36. 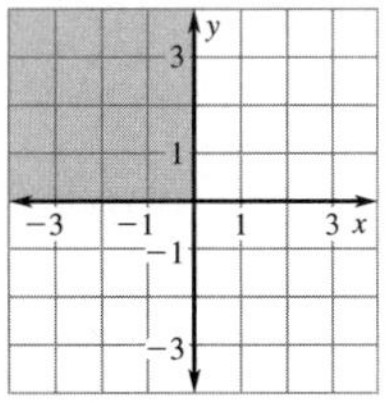

37. 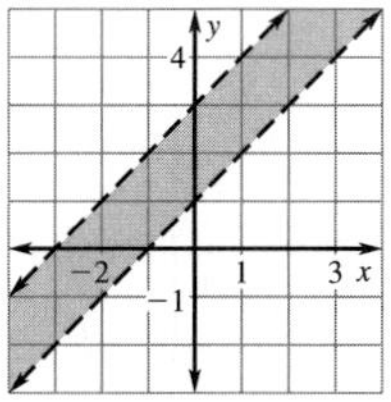

38. 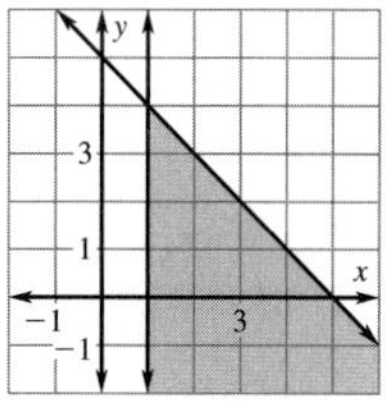

39. 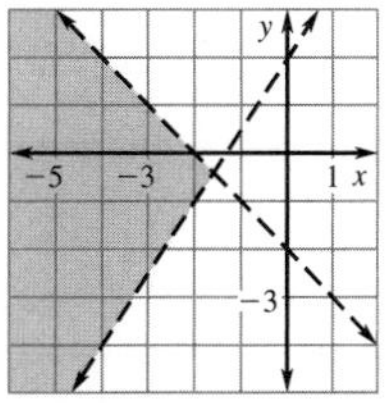

40. 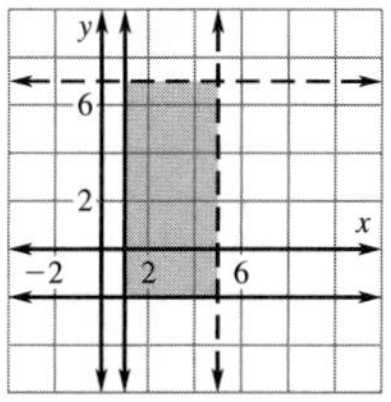

41. 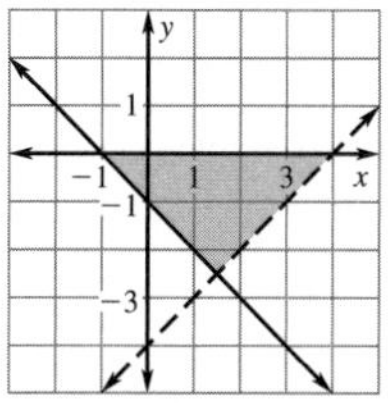

42. 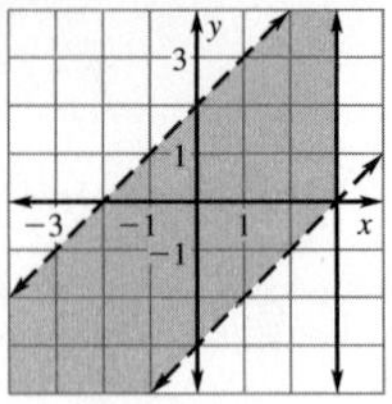

43. 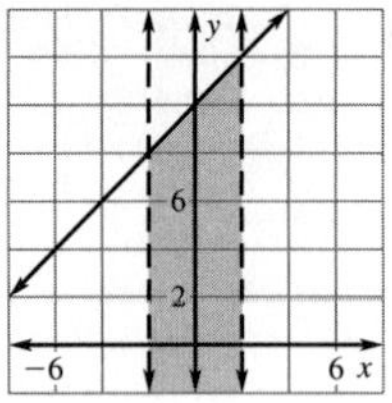

CHAPTER 9 (p. 805)

33. 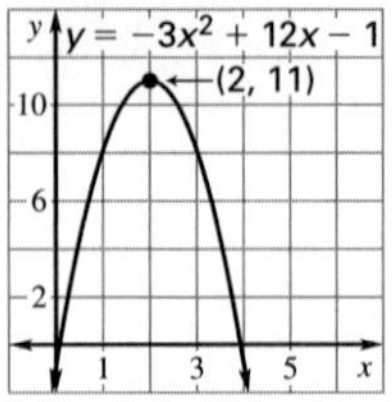

58. 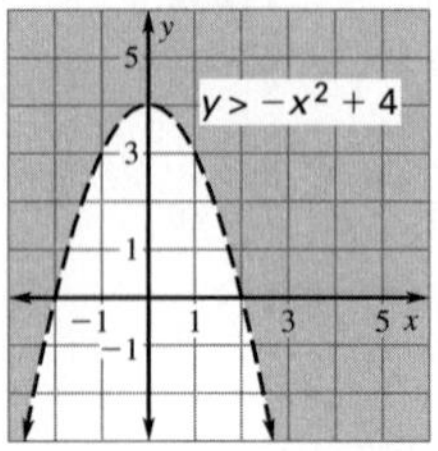

59. 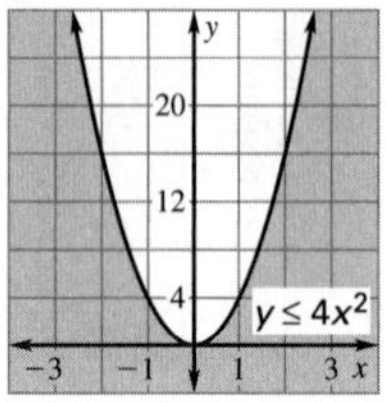

60. 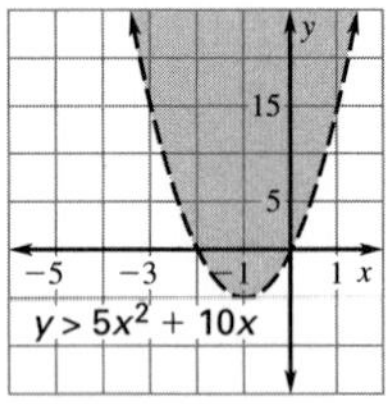

61. 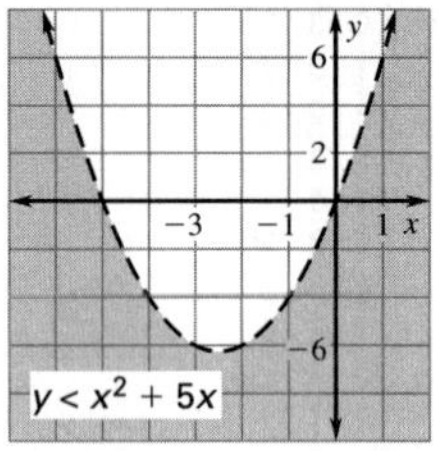

62. 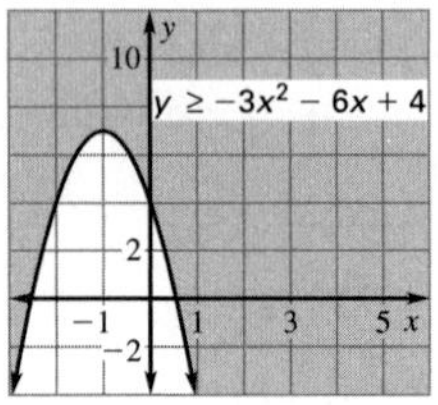

63. 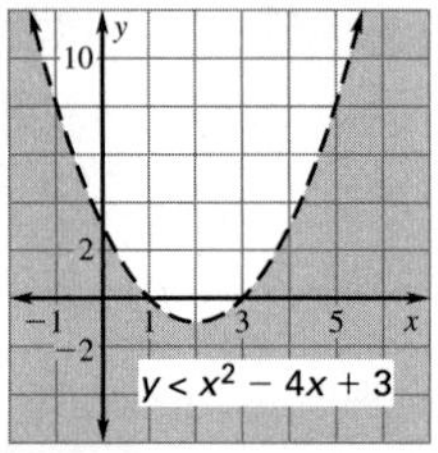

64. 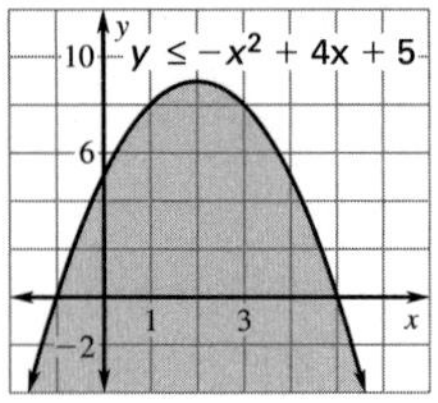

65. 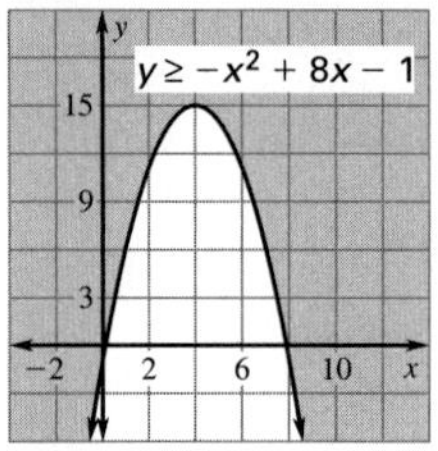

66. 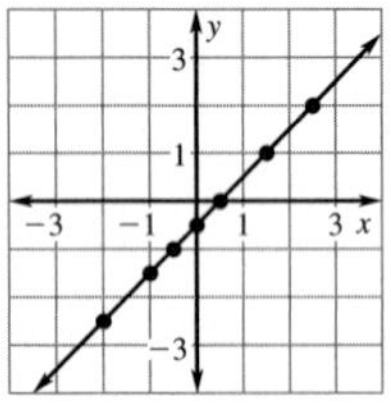

67. 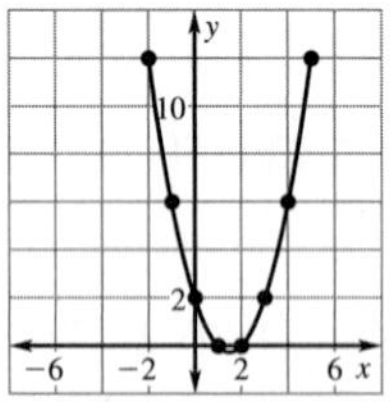

CHAPTER 12 (p. 808)

7. all numbers greater than or equal to $-\frac{2}{3}$; all nonnegative numbers;

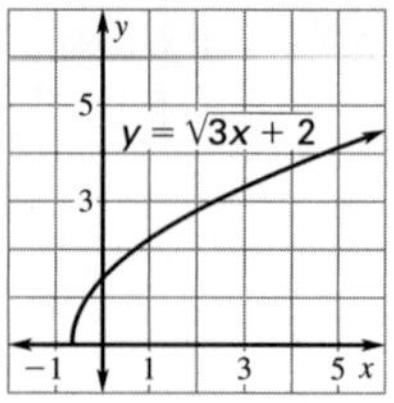

8. all numbers greater than or equal to $\frac{3}{4}$; all nonnegative numbers;

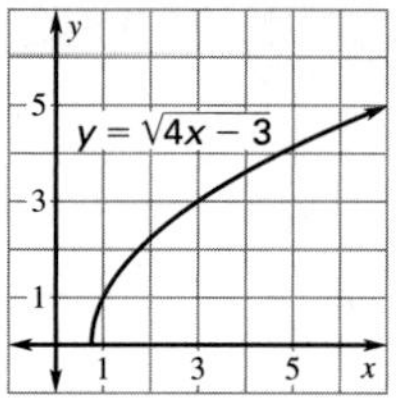

26. $\frac{-1 - \sqrt{13}}{2}, \frac{-1 + \sqrt{13}}{2}$ **27.** $\frac{6 - \sqrt{111}}{5}, \frac{6 + \sqrt{111}}{5}$

43. $\sin D = \frac{4}{5}$, $\cos D = \frac{3}{5}$, $\tan D = \frac{4}{3}$, $\sin E = \frac{3}{5}$, $\cos E = \frac{4}{5}$, $\tan E = \frac{3}{4}$

44. $\sin D = \frac{15}{17}$, $\cos D = \frac{8}{17}$, $\tan D = \frac{15}{8}$, $\sin E = \frac{8}{17}$, $\cos E = \frac{15}{17}$, $\tan E = \frac{8}{15}$

45. $\sin D = \frac{\sqrt{2}}{2}$, $\cos D = \frac{\sqrt{2}}{2}$, $\tan D = 1$, $\sin E = \frac{\sqrt{2}}{2}$, $\cos E = \frac{\sqrt{2}}{2}$, $\tan E = 1$

MATHEMATICS REFERENCE SHEET

Area

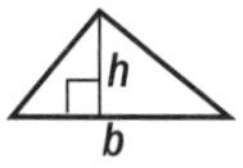

Triangle $A = \frac{1}{2}bh$

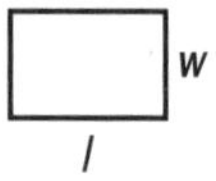

Rectangle $A = lw$

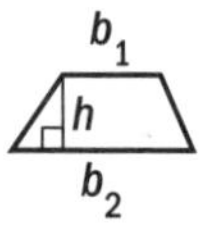

Trapezoid $A = \frac{1}{2}h(b_1 + b_2)$

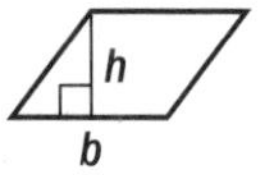

Parallelogram $A = bh$

Circle $A = \pi r^2$

Key

b = base
h = height
l = length
w = width
ℓ = slant height
$S.A.$ = surface area
d = diameter
r = radius
A = area
C = circumference
V = volume

Use 3.14 or $\frac{22}{7}$ for π.

Circumference

$C = \pi d = 2\pi r$

		Volume	Total Surface Area
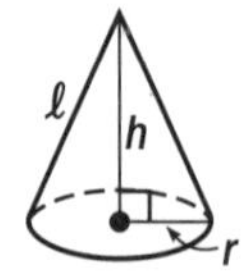	**Right Circular Cone**	$V = \frac{1}{3}\pi r^2 h$	$S.A. = \frac{1}{2}(2\pi r)\ell + \pi r^2 = \pi r\ell + \pi r^2$
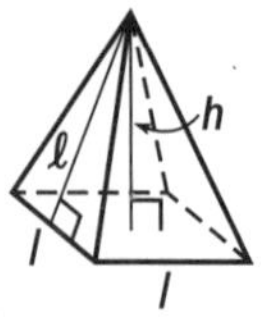	**Square Pyramid**	$V = \frac{1}{3}l^2 h$	$S.A. = 4\left(\frac{1}{2}l\ell\right) + l^2 = 2l\ell + l^2$
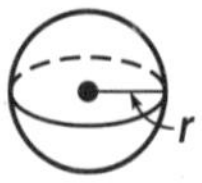	**Sphere**	$V = \frac{4}{3}\pi r^3$	$S.A. = 4\pi r^2$
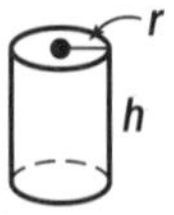	**Right Circular Cylinder**	$V = \pi r^2 h$	$S.A. = 2\pi rh + 2\pi r^2$
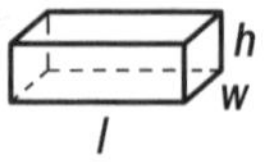	**Rectangular Solid**	$V = lwh$	$S.A. = 2(lw) + 2(hw) + 2(lh)$

In the following formulas, n represents the number of sides:

In a polygon, the sum of the measures of the interior angles is equal to $180(n - 2)$.

In a regular polygon, the measure of an interior angle is equal to $\frac{180(n-2)}{n}$.